AF861243

MECHANICS OF MATERIALS

An Integrated Learning System, 4th Edition

TIMOTHY A. PHILPOT

Author
Timothy A. Philpot

ISBN 9781119429975

Copyright © 2017 John Wiley & Sons, Inc. All rights reserved. No part of this publication may be reproduced, stored in a retrieval system or transmitted in any form or by any means, electronic, mechanical, photocopying, recording, scanning or otherwise, except as permitted under Sections 107 or 108 of the 1976 United States Copyright Act, without either the prior written permission of the Publisher, or authorization through payment of the appropriate per-copy fee to the Copyright Clearance Center, Inc. 222 Rosewood Drive, Danvers, MA 01923, Web site: www.copyright.com. Requests to the Publisher for permission should be addressed to the Permissions Department, John Wiley & Sons, Inc., 111 River Street, Hoboken, NJ 07030-5774, (201)748-6011, fax (201)748-6008, Web site: http://www.wiley.com/go/permissions.

Printed in the United States of America 10 9 8 7 6 5 4 3 2 1

List of Titles

Mechanics of Materials: An Integrated Learning System, 4th edition
by Timothy A. Philpot
Copyright © 2016, ISBN: 978-1-119-32961-9

Table of Contents

WileyPLUS

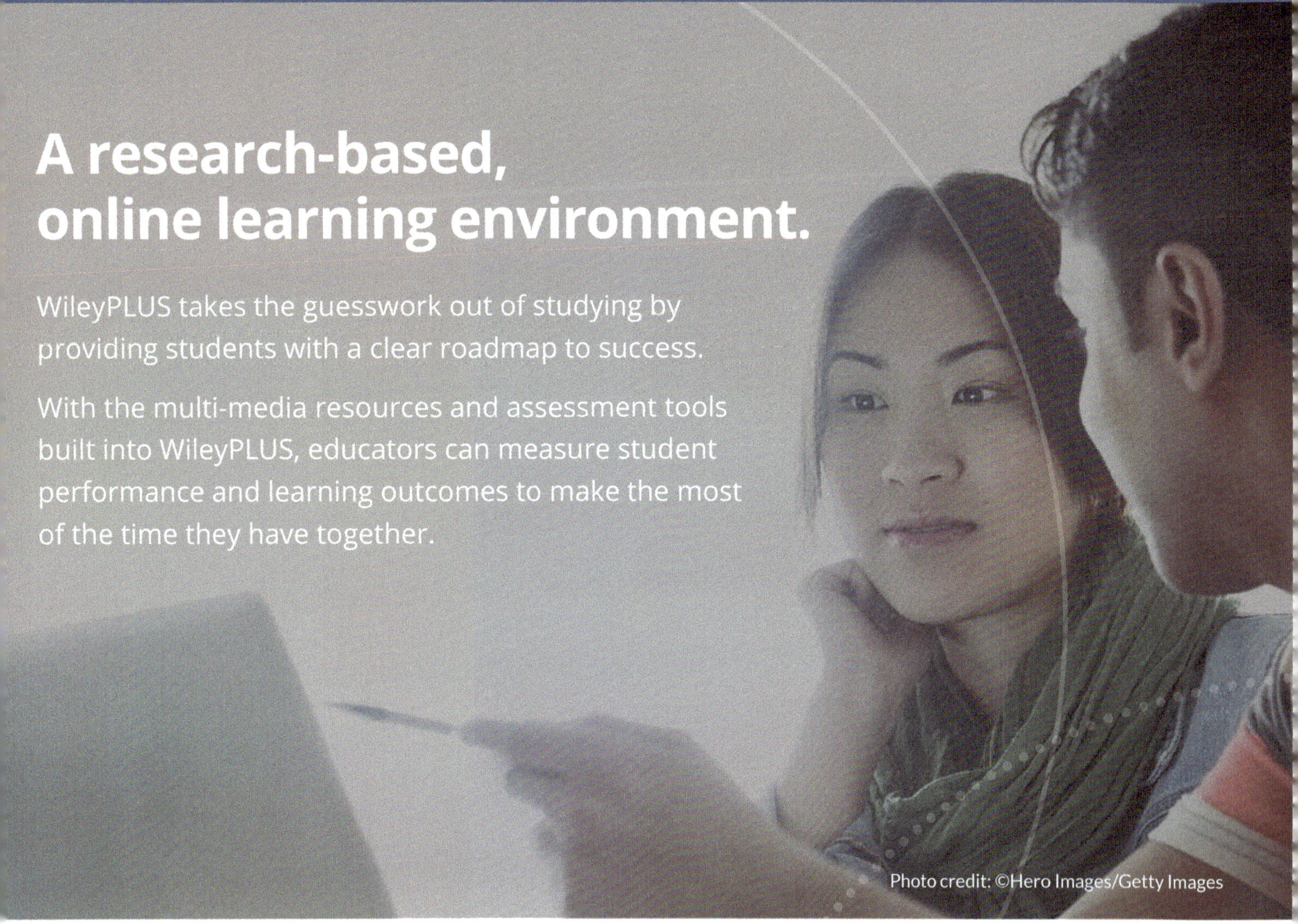

A research-based, online learning environment.

WileyPLUS takes the guesswork out of studying by providing students with a clear roadmap to success.

With the multi-media resources and assessment tools built into WileyPLUS, educators can measure student performance and learning outcomes to make the most of the time they have together.

Photo credit: ©Hero Images/Getty Images

Identify which students are struggling early in the semester.

Educators can track the students' progress through required or practice assignments, and know which students are keeping up with the pace of the course.

Help students organize their learning and get the practice they need.

The Read, Study & Practice area, organized by learning objective, provides students with the complete e-textbook and a wealth of resources for learning the material.

Measure outcomes to promote continuous improvement.

With Gradebook and learning objective-based results, both instructors and students can gauge problem areas and act on what's most important.

www.wileyplus.com

WILEY

MECHANICS OF MATERIALS:
An Integrated Learning System

MECHANICS OF MATERIALS: An Integrated Learning System

Timothy A. Philpot
Missouri University of Science and Technology
Rolla, Missouri

WILEY

VICE PRESIDENT AND DIRECTOR	Don Fowley
EXECUTIVE EDITOR	Linda Ratts
ASSOCIATE DEVELOPMENTAL EDITOR	Adria Giattino
EDITORIAL ASSISTANT	Adriana Alecci
SENIOR CONTENT MANAGER	Valerie Zaborski
SENIOR PRODUCTION EDITOR	Ken Santor
EXECUTIVE MARKETING MANAGER	Dan Sayre
PRODUCT DESIGNER	Brad Franklin
SENIOR DESIGNER	Thomas Nery
SENIOR PHOTO EDITOR	Billy Ray
ELECTRONIC ILLUSTRATIONS	Timothy A. Philpot
COVER DESIGNER	Thomas Nery
PRODUCTION MANAGEMENT SERVICES	Aptara
COVER PHOTO	©Charles Schug/iStockphoto

This book was set in 10/12 TimesLTStd by Aptara® Inc., and printed and bound by Quad Graphics. The cover was printed by Quad Graphics.

This book is printed on acid free paper. ∞

Founded in 1807, John Wiley & Sons, Inc. has been a valued source of knowledge and understanding for more than 200 years, helping people around the world meet their needs and fulfill their aspirations. Our company is built on a foundation of principles that include responsibility to the communities we serve and where we live and work. In 2008, we launched a Corporate Citizenship Initiative, a global effort to address the environmental, social, economic, and ethical challenges we face in our business. Among the issues we are addressing are carbon impact, paper specifications and procurement, ethical conduct within our business and among our vendors, and community and charitable support. For more information, please visit our website: www.wiley.com/go/citizenship.

Copyright © 2017, 2013, 2011, 2008 John Wiley & Sons, Inc. All rights reserved. No part of this publication may be reproduced, stored in a retrieval system or transmitted in any form or by any means, electronic, mechanical, photocopying, recording, scanning or otherwise, except as permitted under Sections 107 or 108 of the 1976 United States Copyright Act, without either the prior written permission of the Publisher, or authorization through payment of the appropriate per-copy fee to the Copyright Clearance Center, Inc. 222 Rosewood Drive, Danvers, MA 01923, website www.copyright.com. Requests to the Publisher for permission should be addressed to the Permissions Department, John Wiley & Sons, Inc., 111 River Street, Hoboken, NJ 07030-5774, (201)748-6011, fax (201)748-6008, website http://www.wiley.com/go/permissions.

Evaluation copies are provided to qualified academics and professionals for review purposes only, for use in their courses during the next academic year. These copies are licensed and may not be sold or transferred to a third party. Upon completion of the review period, please return the evaluation copy to Wiley. Return instructions and a free of charge return shipping label are available at www.wiley.com/go/returnlabel. Outside of the United States, please contact your local representative.

Library of Congress Cataloging-in-Publication Data

Philpot, Timothy A.
Mechanics of materials : an integrated learning system / Timothy A. Philpot. — 4th ed.
p. cm.
ISBN: 978-1-119-32088-3 (ePub)
1. Materials—Mechanical properties. 2. Strength of materials. I. Title.

TA405.P4884 2012
620.1'123—dc23

2012017380

The inside back cover will contain printing identification and country of origin if omitted from this page. In addition, if the ISBN on the back cover differs from the ISBN on this page, the one on the back cover is correct.

ePub: 978-1-119-32088-3
EVAL: 978-1-118-98912-8
BRV: 978-1-119-22748-9

Printed in the United States of America
10 9 8 7 6 5 4 3 2 1

About the Author

Timothy A. Philpot is an Associate Professor in the Department of Civil, Architectural, and Environmental Engineering at the Missouri University of Science and Technology (formerly known as the University of Missouri–Rolla). He received his B.S. degree from the University of Kentucky in 1979, his M.Engr. degree from Cornell University in 1980, and his Ph.D. degree from Purdue University in 1992. In the 1980s, he worked as a structural engineer in the offshore construction industry in New Orleans, London, Houston, and Singapore. He joined the faculty at Murray State University in 1986, and since 1999, he has been on the faculty at Missouri S & T.

Dr. Philpot's primary areas of teaching and research are in engineering mechanics and the development of interactive, multimedia educational software for the introductory engineering mechanics courses. He is the developer of *MDSolids* and *MecMovies*, two award-winning instructional software packages. *MDSolids–Educational Software for Mechanics of Materials* won a 1998 Premier Award for Excellence in Engineering Education Courseware by NEEDS, the National Engineering Education Delivery System. *MecMovies* was a winner of the 2004 NEEDS Premier Award competition as well as a winner of the 2006 MERLOT Classics and MERLOT Editors' Choice Awards for Exemplary Online Learning Resources. Dr. Philpot is also a certified *Project Lead the Way* affiliate professor for the Principles of Engineering course, which features *MDSolids* in the curriculum.

He is a licensed professional engineer and a member of the American Society of Civil Engineers and the American Society for Engineering Education. He has been active in leadership of the ASEE Mechanics Division.

Preface

At the beginning of each semester, I always tell my students the story of my undergraduate Mechanics of Materials experience. While I somehow managed to make an A in the course, Mechanics of Materials was one of the most confusing courses in my undergraduate curriculum. As I continued my studies, I found that I really didn't understand the course concepts well, and this weakness hindered my understanding of subsequent design courses. It wasn't until I began my career as an engineer that I began to relate the Mechanics of Materials concepts to specific design situations. Once I made that real-world connection, I understood the design procedures associated with my discipline more completely and I developed confidence as a designer. My educational and work-related experiences convinced me of the central importance of the Mechanics of Materials course as the foundation for advanced design courses and engineering practice.

The Education of the Mind's Eye

As I gained experience during my early teaching career, it occurred to me that I was able to understand and explain the Mechanics of Materials concepts because I relied upon a set of mental images that facilitated my understanding of the subject. Years later, during a formative assessment of the MecMovies software, Dr. Andrew Dillon, Dean of the School of Information at the University of Texas at Austin, succinctly expressed the role of mental imagery in the following way: "A defining characteristic of an expert is that an expert has a strong mental image of his or her area of expertise while a novice does not." Based on this insight, it seemed logical that one of the instructor's primary objectives should be to teach to the mind's eye—conveying and cultivating relevant mental images that inform and guide students in the study of Mechanics of Materials. The illustrations as well as the MecMovies software integrated in this book have been developed with this objective in mind.

MecMovies Instructional Software

Computer-based instruction often enhances the student's understanding of Mechanics of Materials. With three-dimensional modeling and rendering software, it is possible to create photo-realistic images of various components and to show these components from various viewpoints. In addition, animation software allows objects or processes to be shown in motion. By combining these two capabilities, a fuller description of a physical object can be presented, which can facilitate the mental visualization so integral to understanding and solving engineering problems.

Animation also offers a new generation of computer-based learning tools. The traditional instructional means used to teach Mechanics of Materials—example problems—can be greatly enhanced through animation by emphasizing and illustrating desired problem-solving processes in a more memorable and engaging way. Animation can be used to create interactive tools that focus on specific skills students need to become proficient problem-solvers. These computer-based tools can provide not only the correct solution, but also a detailed visual and verbal explanation of the process needed to arrive at the solution. The feedback provided by the software can lessen some of the anxiety typically associated with traditional homework assignments, while also enabling learners to build their competence and confidence at a pace that is right for them.

This book integrates computer-based instruction into the traditional textbook format with the addition of the MecMovies instructional software. At present, MecMovies consists of over 160 animated "movies" on topics spanning the breadth of the Mechanics of Materials course. Most of these animations present detailed example problems, and about 80 movies are interactive, providing learners with the opportunity to apply concepts and receive immediate feedback that includes key considerations, calculation details, and intermediate results. MecMovies was a winner of the 2004 Premier Award for Excellence in Engineering Education Courseware presented by NEEDS (the National Engineering Education Delivery System, a digital library of learning resources for engineering education).

Hallmarks of the Textbook

In 30 years of teaching the fundamental topics of strength, deformation, and stability, I have encountered successes and frustrations, and I have learned from both. This book has grown out of a passion for clear communication between instructor and student and a drive for documented effectiveness in conveying this foundational material to the differing learners in my classes. With this book and the MecMovies instructional software that is integrated throughout, my desire is to present and develop the theory and practice of Mechanics of Materials in a straightforward plain-speaking manner that addresses the needs of varied learners. The text and software strive to be "student-friendly" without sacrificing rigor or depth in the presentation of topics.

Communicating visually: I invite you to thumb through this book. My hope is that you will find a refreshing clarity in both the text and the illustrations. As both the author and the illustrator, I've tried to produce visual content that will help illuminate the subject matter for the mind's eye of the reader. The illustrations use color, shading, perspective, texture, and dimension to convey concepts clearly, while aiming to place these concepts in the context of real-world components and objects. These illustrations have been prepared by an engineer to be used by engineers to train future engineers.

Problem-solving schema: Educational research suggests that transfer of learning is more effective when students are able to develop *problem-solving schema*, which Webster's Dictionary defines as "a mental codification that includes an organized way of responding to a complex situation." In other words, understanding and proficiency are enhanced if students are encouraged to build a structured framework for mentally organizing concepts and their method of application. This book and software include a number of features aimed at helping students to organize and categorize the Mechanics of Materials concepts and problem-solving procedures. For instance, experience has shown that statically indeterminate axial and torsion structures are among the most difficult topics for students. To help organize the solution process for these topics, a five-step method is utilized. This approach provides students with a problem-solving method that systematically transforms a potentially confusing situation into an easily understandable calculation procedure. Summary tables are also presented in these topics to help students place common statically indeterminate structures into categories based on the specific geometry of the structure. Another topic that students typically find confusing is the use of the superposition method to determine beam deflections. This topic is introduced in the text through enumeration of eight simple skills commonly used in solving problems of this type. This organizational scheme allows students to develop proficiency incrementally before considering more complex configurations.

Style and clarity of examples: To a great extent, the Mechanics of Materials course is taught through examples, and consequently, this book places great emphasis on the presentation and quality of example problems. The commentary and the illustrations associated with example problems are particularly important to the learner. The commentary explains why various steps are taken and describes the rationale for each step in the solution process, while the illustrations help build the mental imagery needed to transfer the concepts to differing situations. Students have found the step-by-step approach used in MecMovies to be particularly helpful, and a similar style is used in the text. Altogether, this book and the MecMovies software present more than 270 fully illustrated example problems that provide both the breadth and the depth required to develop competency and confidence in problem-solving skills.

Homework philosophy: Since Mechanics of Materials is a problem-solving course, much deliberation has gone into the development of homework problems that elucidate and reinforce the course concepts. This book includes 1200 homework problems in a range of difficulty suitable for learners at various stages of development. These problems have been designed with the intent of building the technical foundation and skills that will be necessary in subsequent engineering design courses. The problems are intended to be challenging, and at the same time, practical and pertinent to traditional engineering practice.

New in the Fourth Edition

- Several new topics have been added to the fourth edition:
    - **8.10 Bending of Curved Bars**
    - **13.9 Generalized Hooke's Law for Orthotropic Materials**
    - **14.5 Stresses in Thick-Walled Cylinders**
    - **14.6 Deformations in Thick-Walled Cylinders**
    - **14.7 Interference Fits**
- Additional examples concerning shear stress in thin-walled members have been added to Chapter 9.
- A straightforward procedure for determining three principal stresses and their associated direction cosines has been added to Section 12.11 General State of Stress at a Point.
- In Section 13.6, the procedure for constructing Mohr's circle for plane strain has been simplified.
- Additional examples related to three-dimensional stress and strain relations have been added to Chapter 13. Further, a discussion of the inclusion of temperature effects in the generalized Hooke's law relationships has been added.
- Design equations in Chapter 16 for the critical buckling stress of wood columns have been updated to conform to the latest provisions of the *National Design Specification for Wood Construction.*

- Appendix E Fundamental Mechanics of Materials Equations has been added.
- An extensive number of changes have been made to the textbook problems. More than 430 new problems have been developed. In ten of the seventeen chapters, more than 60% of the textbook problems are new for this edition.

Incorporating MecMovies into Course Assignments

Some instructors may have had unsatisfying experiences with instructional software in the past. Often, the results have not matched the expectations, and it is understandable that instructors may be reluctant to incorporate computer-based instructional content into their course. For those instructors, this book can stand completely on its own merits without the need for the MecMovies software. Instructors will find that this book can be used to successfully teach the time-honored Mechanics of Materials course without making use of the MecMovies software in any way. However, the MecMovies software integrated into this book is a new and valuable instructional medium that has proven to be both popular and effective with Mechanics of Materials students. Naysayers may argue that for many years instructional software has been included as supplemental material in textbooks, and it has not produced significant changes in student performance. While I cannot disagree with this assessment, let me try to persuade you to view MecMovies differently.

Experience has shown that the *manner* in which instructional software is integrated into a course is just as important as the quality of the software itself. Students have many demands on their study time, and in general, they will not invest their time and effort in software that they perceive to be peripheral to the course requirements. In other words, *supplementary* software is doomed to failure, regardless of its quality or merit. To be effective, instructional software must be *integrated into the course assignments* on a regular and frequent basis. Why would you as an instructor alter your traditional teaching routine to integrate computer-based assignments into your course? The answer is because the unique capabilities offered by MecMovies can (a) provide individualized instruction to your students, (b) enable you to spend more time discussing advanced rather than introductory aspects of many topics, and (c) make your teaching efforts more effective.

The computer as an instructional medium is well suited for individualized interactive learning exercises, particularly for those skills that require repetition to master. MecMovies has many interactive exercises, and at a minimum, these features can be utilized by instructors to (a) ensure that students have the appropriate skills in prerequisite topics such as centroids and moments of inertia, (b) develop necessary proficiency in specific problem-solving skills, and (c) encourage students to stay up to date with lecture topics. Three types of interactive features are included in MecMovies:

1. **Concept Checkpoints** – This feature is used for rudimentary problems requiring only one or two calculations. It is also used to build proficiency and confidence in more complicated problems by subdividing the solution process into a sequence of steps that can be mastered sequentially.
2. **Try One problems** – This feature is appended to specific example problems. In a Try One problem, the student is presented with a problem similar to the example so that he or she has the opportunity to immediately apply the concepts and problem-solving procedures illustrated in the example.
3. **Games** – Games are used to develop proficiency in specific skills that require repetition to master. For example, games are used to teach centroids, moments of inertia, shear-force and bending-moment diagrams, and Mohr's circle.

With each of these software features, numeric values in the problem statement are dynamically generated for each student, the student's answers are evaluated, and a summary report suitable for printing is generated. *This enables daily assignments to be collected without imposing a grading burden on the instructor.*

Many of the interactive MecMovies exercises assume no prior knowledge of the topic. Consequently, an instructor can require a *MecMovies* feature to be completed *before giving a lecture on the topic*. For example, Coach Mohr's Circle of Stress guides students step by step through the details of constructing Mohr's circle for plane stress. If students complete this exercise before attending the first Mohr's circle lecture, then the instructor can be confident that students will have at least a basic understanding of how to use Mohr's circle to determine principal stresses. The instructor is then free to build upon this basic level of understanding to explain additional aspects of Mohr's circle calculations.

Student response to MecMovies has been excellent. Many students report that they prefer studying from MecMovies rather than from the text. Students quickly find that MecMovies does indeed help them understand the course material better and thus score better on exams. Furthermore, less quantifiable benefits have been observed when MecMovies is integrated into the course. Students are able to ask better, more specific questions in class concerning aspects of theory that they don't yet fully understand, and students' attitudes about the course overall seem to improve.

WileyPLUS

WileyPLUS, Wiley's digital learning platform, provides instructor and student resources. WileyPLUS builds student confidence by taking the guesswork out of studying and providing them with a clear roadmap: what is assigned, what is required for each assignment, and whether assignments are done correctly. Independent research shows that students using WileyPLUS take more initiative, so the instructor has a greater impact on achievement in the classroom and beyond. WileyPLUS also helps students study at a pace that's right for them. Our integrated resources—available 24/7—function like a personal tutor, directly addressing each student's needs by providing specific problem-solving techniques.

Sign up for access at www.wileyplus.com.

What Do Students Receive with *WileyPLUS*?

- E-book version of the print text, and also, features hyperlinks to questions, definitions, and supplements for quick and easy support.
- Immediate feedback and question assistance, including links to relevant sections in the online digital textbook.
- Integrated, multimedia resources—including MecMovies, Demonstration Videos, and Problem-solving Videos—that provide multiple study paths and encourage active learning.
- Reading & Concept Check Questions which students can use to test their understanding of topics, or instructors can assign as a Pre-Lecture Quiz.
- NEW Practice Problems which students can use to test themselves and hone problem-solving skills.

What Do Instructors Receive with *WileyPLUS*?

- Homework management tools, which enable the instructor to easily assign and automatically grade problems.
- QuickStart reading and homework assignments that can be used as-is or customized to fit the needs of your course.
- Media-rich course materials including Instructor Solutions Manual, Art PPT presentations, and Image Gallery.
- Auto-gradable Guided Online (GO) Tutorials and Multistep Problems, which enable students to learn problem-solving strategies step-by-step and pinpoint exactly where they are making mistakes.
- NEW Practice Problem PPTs that show worked examples for use in lecture by instructors, or provided to students for review.

Selected instructor and student resources are available on the book's companion site: www.wiley.com/college/philpot.

Acknowledgements

- Thanks to Linda Ratts, Adria Giattino, Brad Franklin, Agie Sznajdrowicz, Ken Santor, and the hard working, efficient, and focused staff at Wiley for keeping this project on track.
- Thanks to Dr. Jeffrey S. Thomas at Missouri State University for his excellent work on the WileyPLUS content for this textbook. Thanks for your innovation, dedication, and commitment to this effort.
- Thanks to all those who contributed to the WileyPLUS course. This includes John Baker, who developed stellar GO Multistep and Tutorial Problems. Thanks also to Mark Hanson, Patrick Kwon, and Michael Watson for their talent in accuracy checking problems and solutions.
- Thanks to Jackie Henry of Aptara, Inc., and Brian Baker of Write With, Inc., for their great work in editing and preparing the manuscript for production.
- And finally to Pooch, my wife, the mother of my children, the love of my life, and my constant companion for the past 44 years. Words are inadequate to convey the depth of love, support, strength, encouragement, optimism, wisdom, enthusiasm, good humor, and sustenance that you give so freely to me.

The following colleagues in the engineering teaching profession reviewed parts or all of the manuscript, and I am deeply indebted to them for their constructive criticism and words of encouragement.

Third Edition. Leticia Anaya, *University of North Texas*; Stanton Apple, *Arkansas Tech University*; Andy Bazar, *California State University*; Gary Butson, *Southern Illinois University, Carbon*; David Jack, *Baylor University*; Lin-Mih Hsia, *California State University, Los Angeles*; Eric Kasper, *California Polytechnic, San Luis Obispo*; Jeff Kuo, *California State University, Fullerton*; Yabin Lao, *Arizona State*; Gabriel Portirniche, *University of Idaho*; Sangram Redkar, *Arizona State University East*; Candy Sulzbach, *Colorado School of Mines*; and Rafi Tarefder, *University of New Mexico*.

Second Edition. John Baker, *University of Kentucky*; George R. Buchanan, *Tennessee Technological University*; Debra Dudick, *Corning Community College*; Yalcin Ertekin, *Trine University*; Nicholas Xuanlai Fang, *University of Illinois Urbana-Champaign*; Noe Vargas Hernandez, *University of Texas at El Paso*; Ernst W. Kiesling, *Texas Tech University*; Lee L. Lowery, Jr., *Texas A&M University*; Kenneth S. Manning, *Adirondack Community College*; Prasad S. Mokashi, *Ohio State University*; Ardavan Motahari, *University of Texas at Arlington*; Dustyn Roberts, *New York University*; Zhong-Sheng Wang, *Embry-Riddle Aeronautical University*.

First Edition. Stanton Apple, *Arkansas Tech University*; John Baker, *University of Kentucky*; Kenneth Belanus, *Oklahoma State University*; Xiaomin Deng, *University of South Carolina*; Udaya Halahe, *West Virginia University*; Scott Hendricks, *Virginia Polytechnic Institute and State University*; Tribikram Kundu, *University of Arizona*; Patrick Kwon, *Michigan State University*; Shaofan Li, *University of California, Berkeley*; Cliff Lissenden, *Pennsylvania State University*; Vlado Lubarda, *University of California, San Diego*; Gregory Olsen, *Mississippi State University*; Ramamurthy Prabhakaran, *Old Dominion University*; Oussama Safadi, *University of Southern California*; Hani Salim, *University of Missouri–Columbia*; Scott Schiff, *Clemson University*; Justin Schwartz, *Florida State University*; Lisa Spainhour, *Florida State University*; and Leonard Spunt, *California State University, Northridge*.

Contact Information

I would greatly appreciate your comments and suggestions concerning this book and the MecMovies software. Please feel free to send me an e-mail message at *philpott@mst.edu* or *philpott@mdsolids.com*.

CHAPTER 1

Stress

1.1 Introduction

The three fundamental areas of engineering mechanics are statics, dynamics, and mechanics of materials. Statics and dynamics are devoted primarily to the study of *external* forces and motions associated with particles and rigid bodies (i.e., idealized objects in which any change of size or shape due to forces can be neglected). Mechanics of materials is the study of the *internal* effects caused by external loads acting on real bodies that deform (meaning objects that can stretch, bend, or twist). Why are the internal effects in an object important? The reason is that engineers are called upon to design and produce a variety of objects and structures, such as automobiles, airplanes, ships, pipelines, bridges, buildings, tunnels, retaining walls, motors, and machines—and these objects and structures are all subject to internal forces, moments, and torques that affect their properties and operation. Regardless of the application, a safe and successful design must address the following three mechanical concerns:

1. **Strength:** Is the object strong enough to withstand the loads that will be applied to it? Will it break or fracture? Will it continue to perform properly under repeated loadings?
2. **Stiffness:** Will the object deflect or deform so much that it cannot perform its intended function?
3. **Stability:** Will the object suddenly bend or buckle out of shape at some elevated load so that it can no longer continue to perform its function?

Addressing these concerns requires both an assessment of the intensity of the internal forces and deformations acting within the body and an understanding of the mechanical characteristics of the material used to make the object.

Mechanics of materials is a basic subject in many engineering fields. The course focuses on several types of components: bars subjected to axial loads, shafts in torsion, beams in bending, and columns in compression. Numerous formulas and rules for design found in engineering codes and specifications are based on mechanics-of-materials fundamentals associated with these types of components. With a strong foundation in mechanics-of-materials concepts and problem-solving skills, the student is well equipped to continue into more advanced engineering design courses.

1.2 Normal Stress Under Axial Loading

In every subject area, there are certain fundamental concepts that assume paramount importance for a satisfactory comprehension of the subject matter. In mechanics of materials, such a concept is that of **stress**. In the simplest qualitative terms, *stress is the intensity of internal force*. Force is a vector quantity and, as such, has both magnitude and direction. Intensity implies an area over which the force is distributed. Therefore, stress can be defined as

$$\text{Stress} = \frac{\text{Force}}{\text{Area}} \tag{1.1}$$

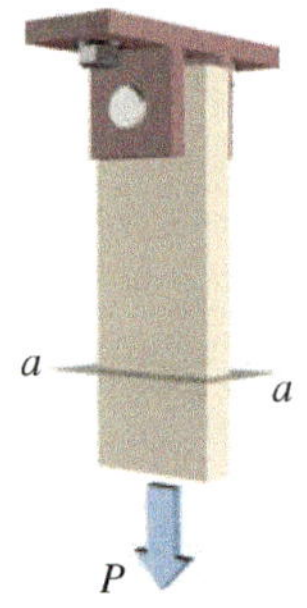

FIGURE 1.1*a* Bar with axial load P.

To introduce the concept of a **normal stress**, consider a rectangular bar subjected to an axial force (Figure 1.1*a*). An **axial force** is a load that is directed along the longitudinal axis of the member. Axial forces that tend to elongate a member are termed **tension forces**, and forces that tend to shorten a member are termed **compression forces**. The axial force P in Figure 1.1*a* is a tension force. To investigate internal effects, the bar is cut by a transverse plane, such as plane *a–a* of Figure 1.1*a*, to expose a free-body diagram of the bottom half of the bar (Figure 1.1*b*). Since this cutting plane is perpendicular to the longitudinal axis of the bar, the exposed surface is called a **cross section**.

The technique of cutting an object to expose the internal forces acting on a plane surface is often referred to as the **method of sections**. The cutting plane is called the **section plane**. To investigate internal effects, one might simply say something like "Cut a section through the bar" to imply the use of the method of sections. This technique will be used throughout the study of mechanics of materials to investigate the internal effects caused by external forces acting on a solid body.

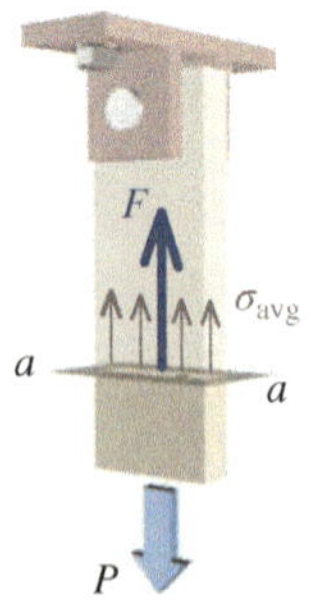

FIGURE 1.1*b* Average stress.

Equilibrium of the lower portion of the bar is attained by a distribution of internal forces that develops on the exposed cross section. This distribution has a resultant internal force F that is normal to the exposed surface, is equal in magnitude to P, and has a line of action that is collinear with the line of action of P. The intensity of F acting in the material is referred to as stress.

In this instance, the stress acts on a surface that is *perpendicular* to the direction of the internal force F. A stress of this type is called a **normal stress**, and it is denoted by the

Greek letter σ (sigma). To determine the magnitude of the normal stress in the bar, the average intensity of the internal force on the cross section can be computed as

$$\sigma_{avg} = \frac{F}{A} \tag{1.2}$$

where A is the cross-sectional area of the bar.

The **sign convention** for normal stresses is defined as follows:

- A positive sign indicates a *tensile normal stress*, and
- a negative sign denotes a *compressive normal stress*.

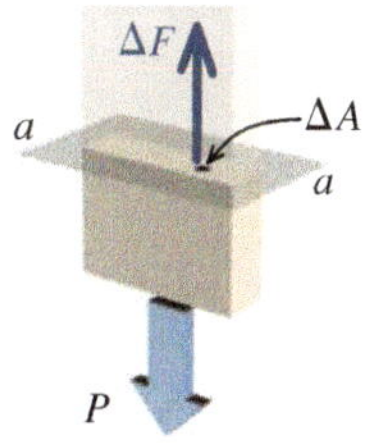

FIGURE 1.1c Stress at a point.

Consider now a small area ΔA on the exposed cross section of the bar, as shown in Figure 1.1*c*, and let ΔF represent the resultant of the internal forces transmitted in this small area. Then the average intensity of the internal force being transmitted in area ΔA is obtained by dividing ΔF by ΔA. If the internal forces transmitted across the section are assumed to be uniformly distributed, the area ΔA can be made smaller and smaller, until, in the limit, it will approach a point on the exposed surface. The corresponding force ΔF also becomes smaller and smaller. The stress at the point on the cross section to which ΔA converges is defined as

$$\sigma = \lim_{\Delta A \to 0} \frac{\Delta F}{\Delta A} \tag{1.3}$$

If the distribution of stress is to be uniform, as in Equation (1.2), the resultant force must act through the centroid of the cross-sectional area. For long, slender, axially loaded members, such as those found in trusses and similar structures, it is generally assumed that the normal stress is uniformly distributed except near the points where the external load is applied. Stress distributions in axially loaded members are not uniform near holes, grooves, fillets, and other features. These situations will be discussed in later sections on stress concentrations. *In this book, it is understood that axial forces are applied at the centroids of the cross sections unless specifically stated otherwise.*

Stress Units

Since the normal stress is computed by dividing the internal force by the cross-sectional area, stress has the dimensions of force per unit area. When U.S. customary units are used, stress is commonly expressed in pounds per square inch (psi) or kips per square inch (ksi) where 1 kip = 1,000 lb. When the International System of Units, universally abbreviated SI (from the French *Système International d'Unités*), is used, stress is expressed in pascals (Pa) and computed as force in newtons (N) divided by area in square meters (m^2). For typical engineering applications, the pascal is a very small unit and, therefore, stress is more commonly expressed in megapascals (MPa) where 1 MPa = 1,000,000 Pa. A convenient alternative when calculating stress in MPa is to express force in newtons and area in square millimeters (mm^2). Therefore,

$$1 \text{ MPa} = 1{,}000{,}000 \text{ N/m}^2 = 1 \text{ N/mm}^2 \tag{1.4}$$

Significant Digits

In this book, final numerical answers are usually presented with three significant digits when a number begins with the digits 2 through 9 and with four significant digits when the number begins with the digit 1. Intermediate values are generally recorded with additional digits to minimize the loss of numerical accuracy due to rounding.

In developing stress concepts through example problems and exercises, it is convenient to use the notion of a **rigid element**. Depending on how it is supported, a rigid element may move vertically or horizontally, or it may rotate about a support location. The rigid element is assumed to be infinitely strong.

EXAMPLE 1.1

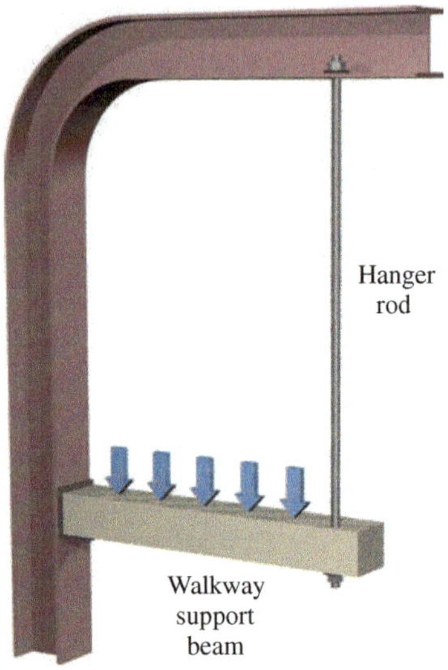

A solid 0.5 in. diameter steel hanger rod is used to hold up one end of a walkway support beam. The force carried by the rod is 5,000 lb. Determine the normal stress in the rod. (Disregard the weight of the rod.)

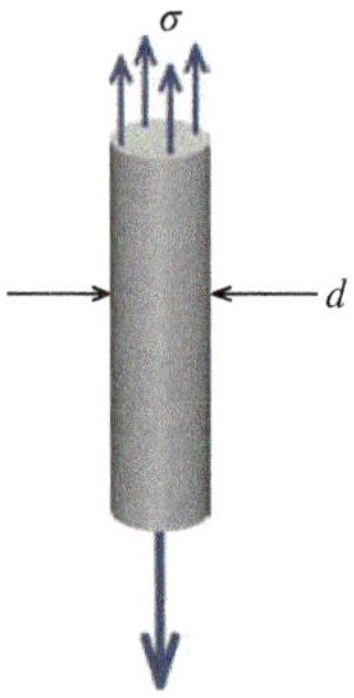

Free-body diagram of hanger rod.

SOLUTION

A free-body diagram of the rod is shown. The solid rod has a circular cross section, and its area is computed as

$$A = \frac{\pi}{4}d^2 = \frac{\pi}{4}(0.5\text{ in.})^2 = 0.19635\text{ in.}^2$$

where d = rod diameter.

Since the force in the rod is 5,000 lb, the normal stress in the rod can be computed as

$$\sigma = \frac{F}{A} = \frac{5{,}000\text{ lb}}{0.19635\text{ in.}^2} = 25{,}464.73135\text{ psi}$$

Although this answer is numerically correct, it would not be proper to report a stress of 25,464.73135 psi as the final answer. A number with this many digits implies an accuracy that we have no right to claim. In this instance, both the rod diameter and the force are given with only one significant digit of accuracy; however, the stress value we have computed here has 10 significant digits.

In engineering, it is customary to round final answers to three significant digits (if the first digit is not 1) or four significant digits (if the first digit is 1). Using this guideline, the normal stress in the rod would be reported as

$$\sigma = 25{,}500\text{ psi}$$ **Ans.**

In many instances, the illustrations in this book attempt to show objects in realistic three-dimensional perspective. Wherever possible, an effort has been made to show free-body diagrams within the actual context of the object or structure. In these illustrations, the free-body diagram is shown in full color while other portions of the object or structure are faded out.

EXAMPLE 1.2

Rigid bar *ABC* is supported by a pin at *A* and axial member (1), which has a cross-sectional area of 540 mm². The weight of rigid bar *ABC* can be neglected. (**Note:** 1 kN = 1,000 N.)

(a) Determine the normal stress in member (1) if a load of $P = 8$ kN is applied at *C*.

(b) If the maximum normal stress in member (1) must be limited to 50 MPa, what is the maximum load magnitude *P* that may be applied to the rigid bar at *C*?

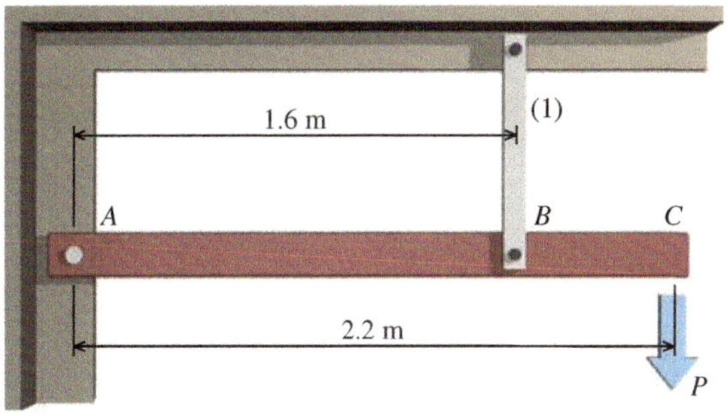

Plan the Solution

(Part a)

Before the normal stress in member (1) can be computed, its axial force must be determined. To compute this force, consider a free-body diagram of rigid bar *ABC* and write a moment equilibrium equation about pin *A*.

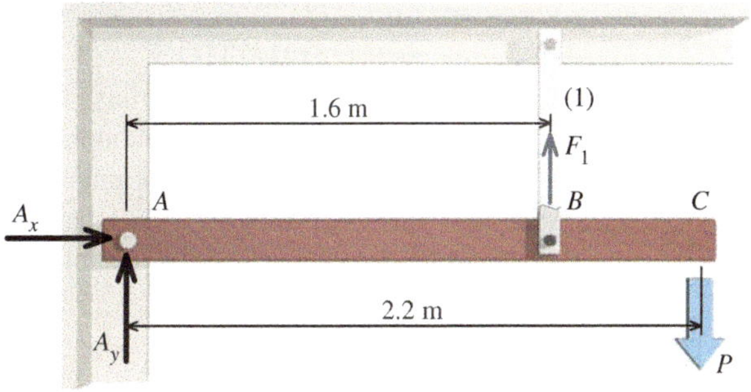

Free-body diagram of rigid bar *ABC*.

SOLUTION

(Part a)

For rigid bar *ABC*, write the equilibrium equation for the sum of moments about pin *A*. Let F_1 = internal force in member (1) and assume that F_1 is a tension force. Positive moments in the equilibrium equation are defined by the right-hand rule. Then

$$\Sigma M_A = -(8\text{ kN})(2.2\text{ m}) + (1.6\text{ m})F_1 = 0$$

$$\therefore F_1 = 11\text{ kN}$$

The normal stress in member (1) can be computed as

$$\sigma_1 = \frac{F_1}{A_1} = \frac{(11\text{ kN})(1{,}000\text{ N/kN})}{540\text{ mm}^2} = 20.370\text{ N/mm}^2 = 20.4\text{ MPa} \qquad \textbf{Ans.}$$

(Note the use of the conversion factor 1 MPa = 1 N/mm².)

Plan the Solution

(Part b)

Using the stress given, compute the maximum force that member (1) may safely carry. Once this force is computed, use the moment equilibrium equation to determine the load *P*.

SOLUTION

(Part b)

Determine the maximum force allowed for member (1):

$$\sigma = \frac{F}{A}$$

$$\therefore F_1 = \sigma_1 A_1 = (50\text{ MPa})(540\text{ mm}^2) = (50\text{ N/mm}^2)(540\text{ mm}^2) = 27{,}000\text{ N} = 27\text{ kN}$$

Compute the maximum allowable load *P* from the moment equilibrium equation:

$$\Sigma M_A = -(2.2\text{ m})P + (1.6\text{ m})(27\text{ kN}) = 0$$

$$\therefore P = 19.64\text{ kN} \qquad \textbf{Ans.}$$

5

EXAMPLE 1.3

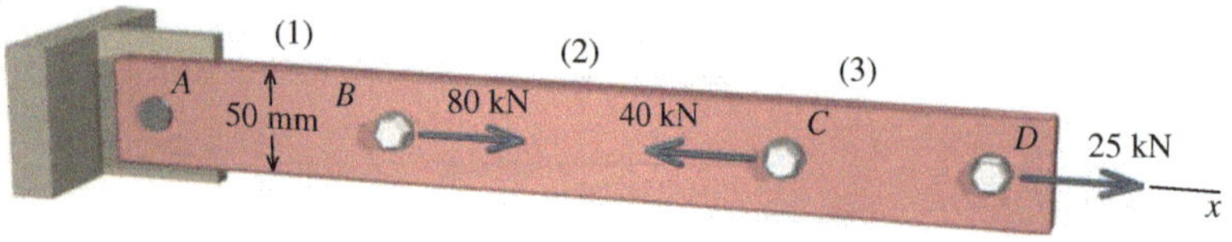

A 50 mm wide steel bar has axial loads applied at points B, C, and D. If the normal stress magnitude in the bar must not exceed 60 MPa, determine the minimum thickness that can be used for the bar.

Plan the Solution

Draw free-body diagrams that expose the internal force in each of the three segments. In each segment, determine the magnitude and direction of the internal axial force required to satisfy equilibrium. Use the largest-magnitude internal axial force and the allowable normal stress to compute the minimum cross-sectional area required for the bar. Divide the cross-sectional area by the 50 mm bar width to compute the minimum bar thickness.

SOLUTION

Begin by drawing a free-body diagram (FBD) that exposes the internal force in segment (3). Since the reaction force at A has not been calculated, it will be easier to cut through the bar in segment (3) and consider the portion of the bar starting at the cut surface and extending to the free end of the bar at D. An unknown internal axial force F_3 exists in segment (3), and it is helpful to establish a consistent convention for problems of this type.

> **Problem-Solving Tip:** When cutting an FBD through an axial member, assume that the internal force is tension and draw the force arrow directed *away from the cut surface*. If the computed value of the internal force turns out to be a positive number, then the assumption of tension is confirmed. If the computed value turns out to be a negative number, then the internal force is actually compressive.

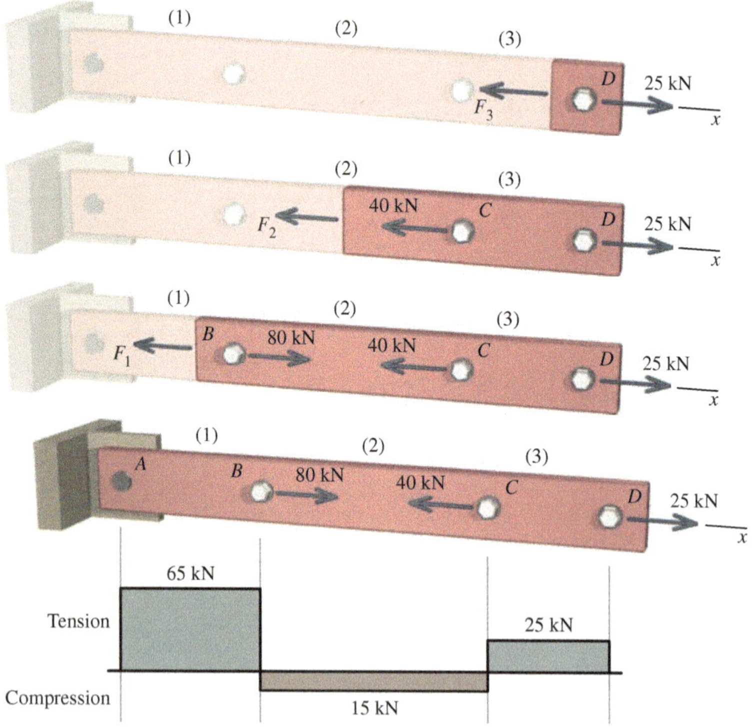

Axial-force diagram showing internal forces in each bar segment.

On the basis of an FBD cut through axial segment (3), the equilibrium equation is

$$\Sigma F_x = -F_3 + 25\text{ kN} = 0$$

$$\therefore F_3 = 25\text{ kN} = 25\text{ kN (T)}$$

Repeat this procedure for an FBD exposing the internal force in segment (2):

$$\Sigma F_x = -F_2 - 40\text{ kN} + 25\text{ kN} = 0$$

$$\therefore F_2 = -15\text{ kN} = 15\text{ kN (C)}$$

Then repeat for an FBD exposing the internal force in segment (1):

$$\Sigma F_x = -F_1 + 80\text{ kN} - 40\text{ kN} + 25\text{ kN} = 0$$

$$\therefore F_1 = 65\text{ kN (T)}$$

It is always a good practice to construct a simple plot that graphically summarizes the internal axial forces along the bar. The axial-force diagram on the left shows internal tension forces above the axis and internal compression forces below the axis.

The required cross-sectional area will be computed on the basis of (the absolute value of) the largest-magnitude internal force. The normal stress in the bar must be limited to 60 MPa. *To facilitate the calculation, the conversion 1 MPa = 1 N/mm² is used; therefore, 60 MPa = 60 N/mm²*, and we have

$$\sigma = \frac{F}{A} \qquad \therefore A \geq \frac{F}{\sigma} = \frac{(65\text{ kN})(1{,}000\text{ N/kN})}{60\text{ N/mm}^2} = 1{,}083.333\text{ mm}^2$$

Since the flat steel bar is 50 mm wide, the minimum thickness that can be used for the bar is

$$t_{\text{min}} \geq \frac{1{,}083{,}333\text{ mm}^2}{50\text{ mm}} = 21.667\text{ mm} = 21.7\text{ mm} \qquad \textbf{Ans.}$$

In practice, the bar thickness would be rounded up to the next-larger standard size.

Review

Recheck your calculations, paying particular attention to the units. Always show the units in your calculations because doing so is an easy and fast way to discover mistakes. Are the answers reasonable? If the bar thickness had been 0.0217 mm instead of 21.7 mm, would your solution have been reasonable, based on your common sense and intuition?

MecMovies

EXAMPLE

M1.4 Two axial members are used to support a load P applied at joint B.

- Member (1) has a cross-sectional area of A_1 = 3,080 mm² and an allowable normal stress of 180 MPa.
- Member (2) has a cross-sectional area of A_2 = 4,650 mm² and an allowable normal stress of 75 MPa.

Determine the maximum load P that may be supported without exceeding either allowable normal stress.

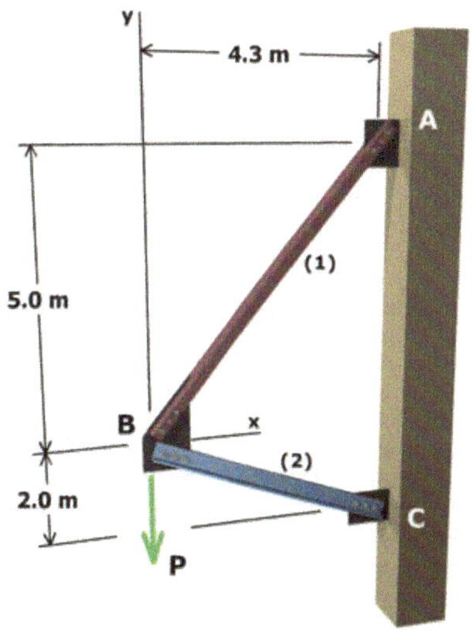

1.3 Direct Shear Stress

Loads applied to a structure or a machine are generally transmitted to individual members through connections that use rivets, bolts, pins, nails, or welds. In all of these connections, one of the most significant stresses induced is a *shear stress*. In the previous section, normal stress was defined as the intensity of an internal force acting on a surface *perpendicular* to the direction of the internal force. Shear stress is also the intensity of an internal force, but shear stress acts on a surface that is *parallel* to the internal force.

To investigate shear stress, consider a simple connection in which the force carried by an axial member is transmitted to a support by means of a solid circular pin (Figure 1.2*a*). The load is transmitted from the axial member to the support by a **shear force** (i.e., a force

FIGURE 1.2*a* Single-shear pin connection.

7

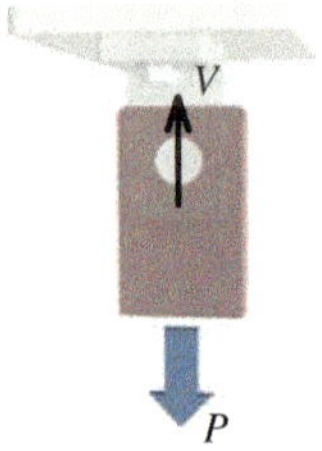

FIGURE 1.2*b* Free-body diagram showing shear force transmitted by pin.

MecMovies 1.7 and 1.8 present animated illustrations of single- and double-shear bolted connections.

MecMovies 1.9 presents an animated illustration of a shear key connection between a gear and a shaft.

that tends to cut) distributed on a transverse cross section of the pin. A free-body diagram of the axial member with the pin is shown in Figure 1.2*b*. In this diagram, a resultant shear force V has replaced the distribution of shear forces on the transverse cross section of the pin. Equilibrium requires that the resultant shear force V equal the applied load P. Since only one cross section of the pin transmits load between the axial member and the support, the pin is said to be in **single shear**.

From the definition of stress given by Equation (1.1), an average shear stress on the transverse cross section of the pin can be computed as

$$\tau_{avg} = \frac{V}{A_V} \tag{1.5}$$

where A_V = area transmitting shear stress. The Greek letter τ (tau) is commonly used to denote shear stress. A sign convention for shear stress will be presented in a later section of the book.

The stress at a point on the transverse cross section of the pin can be obtained by using the same type of limit process that was used to obtain Equation (1.3) for the normal stress at a point. Thus,

$$\tau = \lim_{\Delta A_V \to 0} \frac{\Delta V}{\Delta A_V} \tag{1.6}$$

It will be shown later in this text that the shear stresses cannot be uniformly distributed over the transverse cross section of a pin or bolt and that the *maximum shear stress* on the transverse cross section may be much larger than the average shear stress obtained by using Equation (1.5). The design of simple connections, however, is usually based on average-stress considerations, and this procedure will be followed in this book.

The key to determining shear stress in connections is to visualize the failure surface or surfaces that will be created if the connectors (i.e., pins, bolts, nails, or welds) actually break (i.e., fracture). The shear area A_V that transmits shear force is the area exposed when the connector fractures. Two common types of shear failure surfaces for pinned connections are shown in Figures 1.3 and 1.4. Laboratory specimens that have failed on a single shear plane

Jeffery S. Thomas

FIGURE 1.3 Single-shear failure in pin specimens.

Jeffery S. Thomas

FIGURE 1.4 Double-shear failure in a pin specimen.

are shown in Figure 1.3. Similarly, a pin that has failed on two parallel shear planes is shown in Figure 1.4.

EXAMPLE 1.4

Chain members (1) and (2) are connected by a shackle and pin. If the axial force in the chains is $P = 28$ kN and the allowable shear stress in the pin is $\tau_{\text{allow}} = 90$ MPa, determine the minimum acceptable diameter d for the pin.

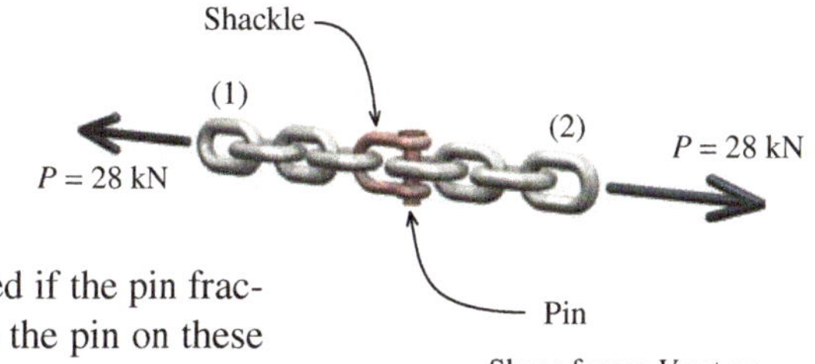

Shear forces V act on two surfaces of the pin.

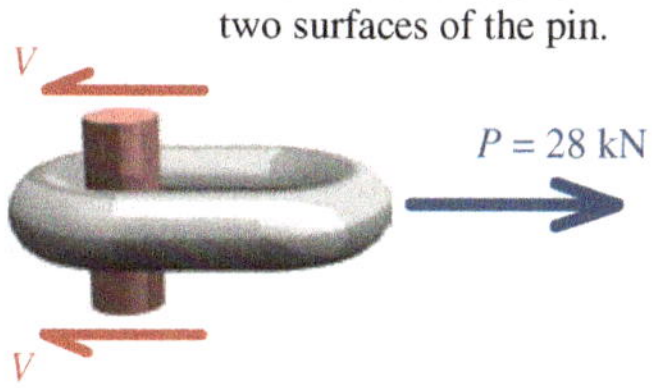

Free-body diagram of pin.

Plan the Solution

To solve the problem, first visualize the surfaces that would be revealed if the pin fractured because of the applied load P. Shear stress will be developed in the pin on these surfaces, at the two interfaces (i.e., common boundaries) between the pin and the shackle. The shear area needed to resist the shear force acting on each of the surfaces must be found, and from this area the minimum pin diameter can be calculated.

SOLUTION

Draw a free-body diagram (FBD) of the pin, which connects chain (2) to the shackle. Two shear forces V will resist the applied load $P = 28$ kN. The shear force V acting on each surface must equal one-half of the applied load P; therefore, $V = 14$ kN.

Next, the area of each surface is simply the cross-sectional area of the pin. The average shear stress acting on each of the pin failure surfaces is, therefore, the shear force V divided by the cross-sectional area of the pin. Since the average shear stress must be limited to 90 MPa, the minimum cross-sectional area required to satisfy the allowable shear stress requirement can be computed as

$$\tau = \frac{V}{A_{\text{pin}}} \qquad \therefore A_{\text{pin}} \geq \frac{V}{\tau_{\text{allow}}} = \frac{(14\text{ kN})(1{,}000\text{ N/kN})}{90\text{ N/mm}^2} = 155.556\text{ mm}^2$$

The minimum pin diameter required for use in the shackle can be determined from the required cross-sectional area:

$$A_{\text{pin}} \geq \frac{\pi}{4} d_{\text{pin}}^2 = 155.556\text{ mm}^2 \qquad \therefore d_{\text{pin}} \geq 14.07\text{ mm} \qquad \text{say, } d_{\text{pin}} = 15\text{ mm} \qquad \textbf{Ans.}$$

In this connection, two cross sections of the pin are subjected to shear forces V; consequently, the pin is said to be in **double shear**.

MecMovies

EXAMPLE

M1.5 A pin at C and a round aluminum rod at B support the rigid bar BCD. If the allowable pin shear stress is 50 MPa, what is the minimum diameter required for the pin at C?

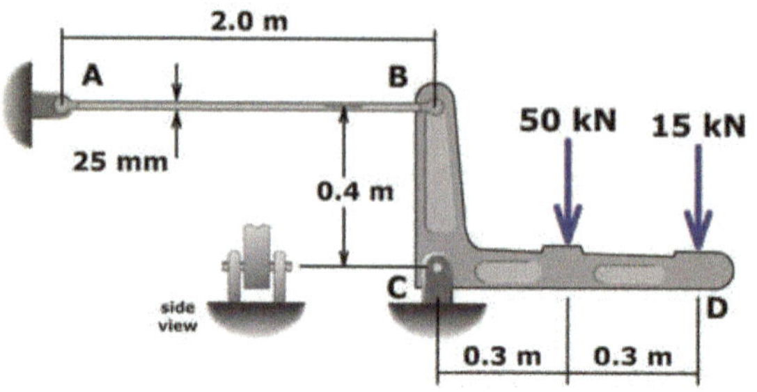

EXAMPLE 1.5

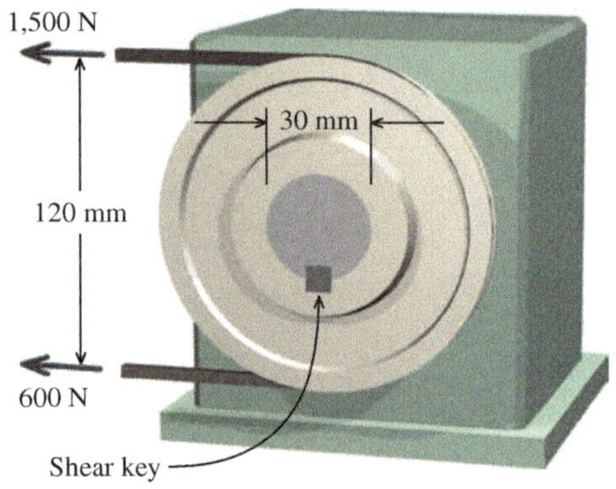

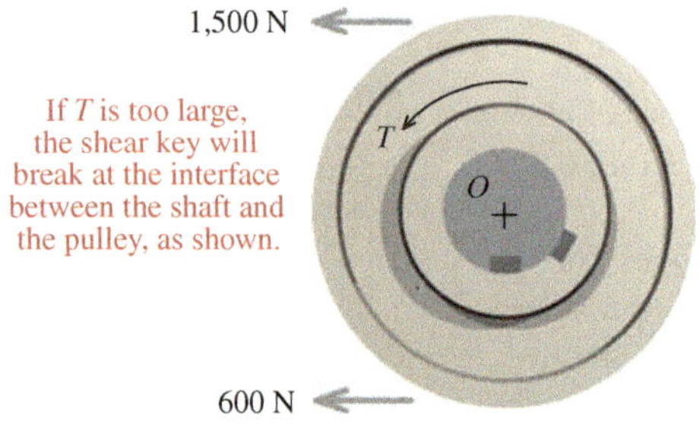

Visualize failure surface in shear key.

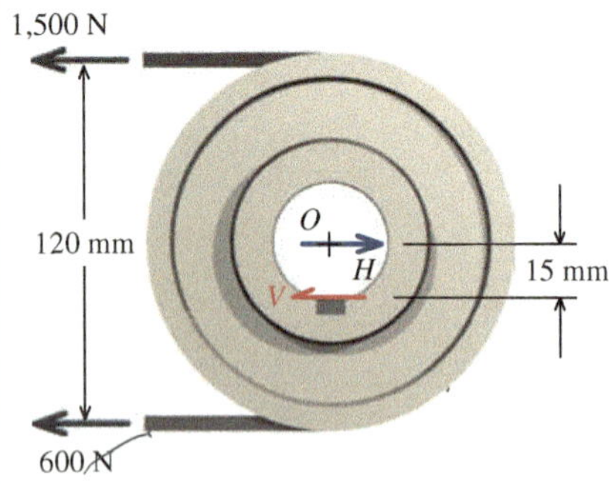

Free-body diagram of pulley.

A belt pulley used to drive a device is attached to a 30 mm diameter shaft with a square shear key. The belt tensions are 1,500 N and 600 N, as shown. The shear key dimensions are 6 mm by 6 mm by 25 mm long. Determine the shear stress produced in the shear key.

Plan the Solution

A shear key is a common component used to connect pulleys, chain sprockets, and gears to solid circular shafts. A rectangular slot is cut in the shaft, and a matching notch of the same width is cut in the pulley. After the slot and the notch are aligned, a square metal piece is inserted in the opening. This metal piece is called a shear key; it forces the shaft and the pulley to rotate together.

Before beginning the calculations, try to visualize the failure surface in the shear key. Since the belt tensions are unequal, a moment is created about the center of the shaft. This type of moment, called a **torque**, causes the shaft and pulley to rotate. If the torque T created by the unequal belt tensions is too large, the shear key will break at the interface between the shaft and the pulley, allowing the pulley to spin freely on the shaft. This failure surface is the plane at which shear stress is created in the shear key.

From the belt tensions and the pulley diameter, determine the torque T exerted on the shaft by the pulley. From a free-body diagram (FBD) of the pulley, determine the force that must be supplied by the shear key to satisfy equilibrium. Once the force in the shear key is known, the shear stress in the key can be computed by using the shear key dimensions.

SOLUTION

Consider an FBD of the pulley. This FBD includes the belt tensions, but it specifically excludes the shaft. The FBD cuts through the shear key at the interface between the pulley and the shaft. We will assume that there could be an internal force acting on the exposed surface of the shear key. This force will be denoted as shear force V. The distance from V to the center O of the shaft is equal to the radius of the shaft. Since the shaft diameter is 30 mm, the distance from O to shear force V is 15 mm. The magnitude of shear force V can be found from a moment equilibrium equation about

10

point O, which is the center of rotation for both the pulley and the shaft. In this equation, positive moments are defined by the right-hand rule:

$$\Sigma M_O = (1{,}500 \text{ N})(60 \text{ mm}) - (600 \text{ N})(60 \text{ mm}) - (15 \text{ mm})V = 0$$

$$\therefore V = 3{,}600 \text{ N}$$

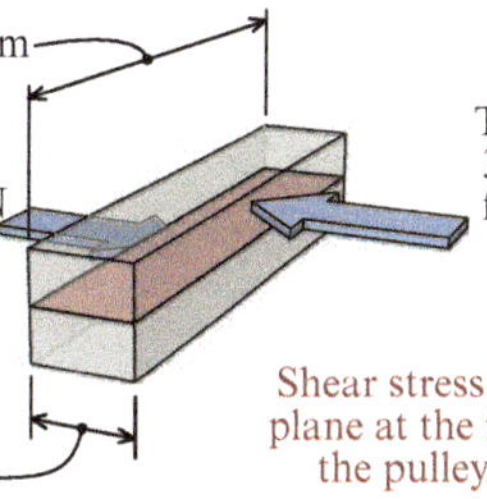

Enlarged view of shear key.

For the pulley to be in equilibrium, a shear force of $V = 3{,}600$ N must be supplied by the shear key.

An enlarged view of the shear key is shown on the right. The torque created by the belt tensions exerts a force of 3,600 N on the shear key. For equilibrium, a force equal in magnitude, but opposite in direction, must be exerted on the key by the shaft. This pair of forces tends to cut the key, producing a shear stress. The shear stress acts on the plane highlighted in red.

An internal force of $V = 3{,}600$ N must exist on an internal plane of the shear key if the pulley is to be in equilibrium. The area of this plane surface is the product of the shear key width and length:

$$A_V = (6 \text{ mm})(25 \text{ mm}) = 150 \text{ mm}^2$$

The shear stress produced in the shear key can now be computed:

$$\tau = \frac{V}{A_V} = \frac{3{,}600 \text{ N}}{150 \text{ mm}^2} = 24.0 \text{ N/mm}^2 = 24.0 \text{ MPa} \qquad \textbf{Ans.}$$

MecMovies

EXAMPLE

M1.6 A torque of $T = 10$ kN·m is transmitted between two flanged shafts by means of four 22 mm diameter bolts. Determine the average shear stress in each bolt if the diameter of the bolt circle is 250 mm. (Disregard friction between the flanges.)

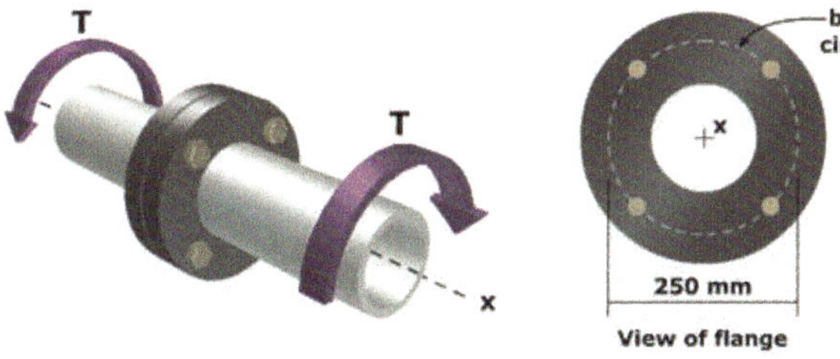

Another common type of shear loading is termed **punching shear**. Examples of this type of loading include the action of a punch in forming rivet holes in a metal plate, the tendency of building columns to punch through footings, and the tendency of a tensile axial load on a bolt to pull the shank of the bolt through the head. Under a punching shear load, the significant stress is the average shear stress on the surface defined by the *perimeter* of the punching member and the *thickness* of the punched member. Punching shear is illustrated by the three composite wood specimens shown in Figure 1.5. The central hole in each specimen is a pilot hole used to guide the punch. The specimen on the left shows the surface initiated at the outset of the shear failure. The center specimen reveals the failure surface after the punch is driven partially through the block. The specimen on the right shows the block after the punch has been driven completely through the block.

MecMovies 1.10 presents an animated illustration of punching shear.

11

Jeffery S. Thomas

FIGURE 1.5 Punching shear failure in composite wood block specimens.

EXAMPLE 1.6

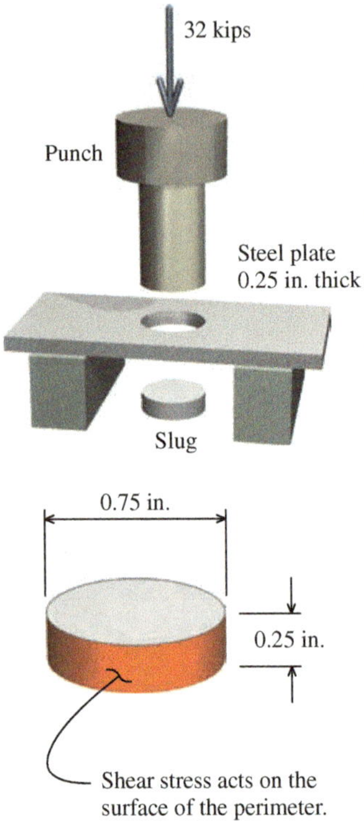

A punch for making holes in steel plates is shown. A downward punching force of 32 kips is required to punch a 0.75 in. diameter hole in a steel plate that is 0.25 in. thick. Determine the average shear stress in the steel plate at the instant the circular slug (the portion of the steel plate removed to create the hole) is torn away from the plate.

Plan the Solution

Visualize the surface that is revealed when the slug is removed from the plate. Compute the shear stress from the applied punching force and the area of the exposed surface.

SOLUTION

The area subjected to shear stress occurs around the perimeter of the slug. Use the slug diameter d and the plate thickness t to compute the shear area A_V:

$$A_V = \pi\, dt = \pi(0.75 \text{ in.})(0.25 \text{ in.}) = 0.58905 \text{ in.}^2$$

The average shear stress τ is computed from the punching force $P = 32$ kips and the shear area:

$$\tau = \frac{P}{A_V} = \frac{32 \text{ kips}}{0.58905 \text{ in.}^2} = 54.3 \text{ ksi}$$ **Ans.**

1.4 Bearing Stress

A third type of stress, **bearing stress**, is actually a special category of normal stress. Bearing stresses are compressive normal stresses that occur on the surface of contact *between two separate interacting members*. This type of normal stress is defined in the same manner as normal and shear stresses (i.e., force per unit area); therefore, the average bearing stress σ_b is expressed as

$$\sigma_b = \frac{F}{A_b} \tag{1.7}$$

where A_b = area of contact between the two components.

EXAMPLE 1.7

A steel pipe column (6.5 in. outside diameter; 0.25 in. wall thickness) supports a load of 11 kips. The steel pipe rests on a square steel base plate, which in turn rests on a concrete slab.

(a) Determine the bearing stress between the steel pipe and the steel plate.
(b) If the bearing stress of the steel plate on the concrete slab must be limited to 90 psi, what is the minimum allowable plate dimension a?

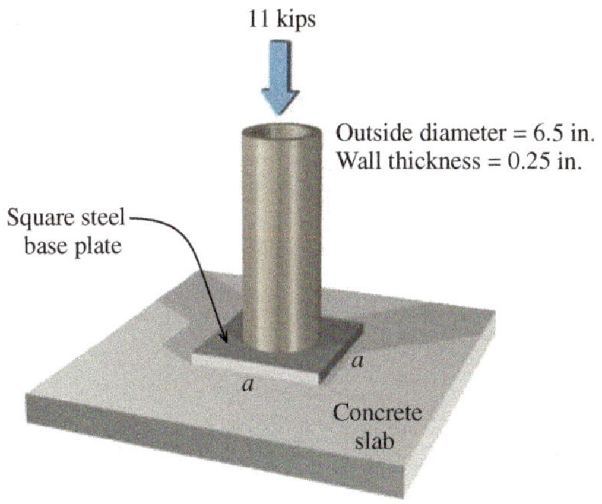

Plan the Solution

To compute bearing stress, the area of contact between two objects must be determined.

SOLUTION

(a) The cross-sectional area of the pipe is required in order to compute the compressive bearing stress between the column post and the base plate. The cross-sectional area of a pipe is given by

$$A_{\text{pipe}} = \frac{\pi}{4}(D^2 - d^2)$$

where D = outside diameter and d = inside diameter. The inside diameter d is related to the outside diameter D by

$$d = D - 2t$$

where t = wall thickness. Therefore, with D = 6.5 in. and d = 6.0 in., the area of the pipe is

$$A_{\text{pipe}} = \frac{\pi}{4}(D^2 - d^2) = \frac{\pi}{4}[(6.5\text{ in.})^2 - (6.0\text{ in.})^2] = 4.9087\text{ in.}^2$$

The bearing stress between the pipe and the base plate is

$$\sigma_b = \frac{F}{A_b} = \frac{11\text{ kips}}{4.9087\text{ in.}^2} = 2.24\text{ ksi}$$

(b) The minimum area required for the steel plate in order to limit the bearing stress to 90 psi is

$$\sigma_b \geq \frac{F}{A_b} \qquad \therefore A_b \geq \frac{F}{\sigma_b} = \frac{(11\text{ kips})(1{,}000\text{ lb/kip})}{90\text{ psi}} = 122.222\text{ in.}^2$$

Since the steel plate is square, its area of contact with the concrete slab is

$$A_b = a \times a \geq 122.222\text{ in.}^2 \qquad \therefore a \geq \sqrt{122.222\text{ in.}^2} = 11.06\text{ in.} \qquad \text{say, 12 in.}$$ **Ans.**

Bearing stresses also develop on the contact surface between a plate and the body of a bolt or a pin. A bearing failure at a bolted connection in a thin steel component is shown in Figure 1.6. A tension load was applied upward to the steel component, and a bearing failure occurred below the bolt hole.

FIGURE 1.6 Bearing stress failure at a bolted connection.

The distribution of bearing stresses on a semicircular contact surface is quite complicated, and an average bearing stress is often used for design purposes. This average bearing stress σ_b is computed by dividing the transmitted force by the **projected area** of contact between a plate and the bolt or pin, instead of the actual contact area. This approach is illustrated in the next example.

EXAMPLE 1.8

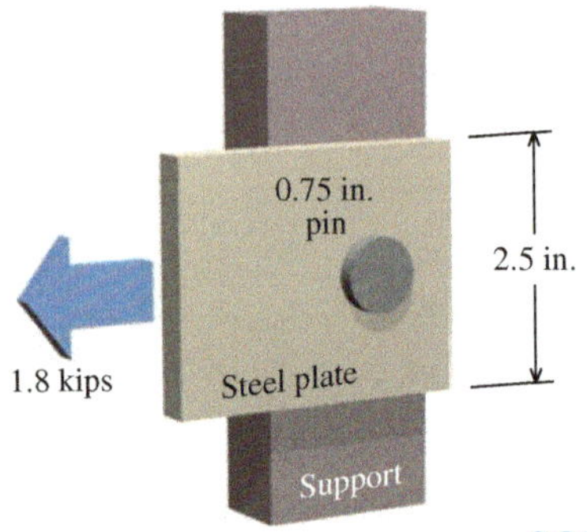

A 2.5 in. wide by 0.125 in. thick steel plate is connected to a support with a 0.75 in. diameter pin. The steel plate carries an axial load of 1.8 kips. Determine the bearing stress in the steel plate.

Plan the Solution

Bearing stresses will develop on the surface where the steel plate contacts the pin. This surface is the right side of the hole in the illustration. To determine the average bearing stress, the projected area of contact between the plate and the pin must be calculated.

SOLUTION

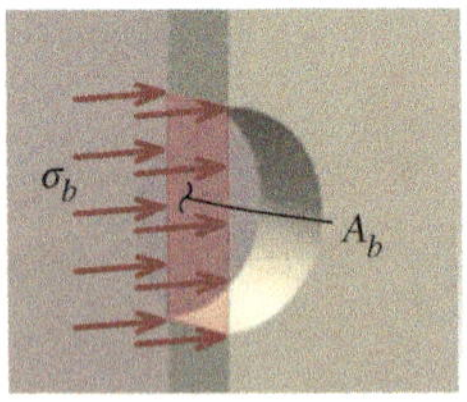

Enlarged view of projected contact area.

The 1.8 kip load pulls the steel plate to the left, bringing the right side of the hole into contact with the pin. Bearing stresses will occur on the right side of the hole (in the steel plate) and on the right half of the pin.

Since the actual distribution of bearing stress on a semicircular surface is complicated, an average bearing stress is typically used for design purposes. Instead of the actual contact area, the projected area of contact is used in the calculation.

The figure at the left shows an enlarged view of the projected contact area between the steel plate and the pin. An average bearing stress σ_b is exerted on the steel plate by the pin. Not shown is the equal-magnitude bearing stress exerted on the pin by the steel plate.

The projected area A_b is equal to the product of the pin (or bolt) diameter d and the plate thickness t. For the pinned connection shown, the projected area A_b between the 0.125 in. thick steel plate and the 0.75 in. diameter pin is calculated as

$$A_b = dt = (0.75 \text{ in.})(0.125 \text{ in.}) = 0.09375 \text{ in.}^2$$

The average bearing stress between the plate and the pin is therefore

$$\sigma_b = \frac{F}{A_b} = \frac{1.8 \text{ kips}}{0.09375 \text{ in.}^2} = 19.20 \text{ ksi} \qquad \textbf{Ans.}$$

MecMovies

EXAMPLE

M1.1 A 60 mm wide by 8 mm thick steel plate is connected to a gusset plate by a 20 mm diameter pin. If a load of P = 70 kN is applied, determine the normal, shear, and bearing stresses in this connection.

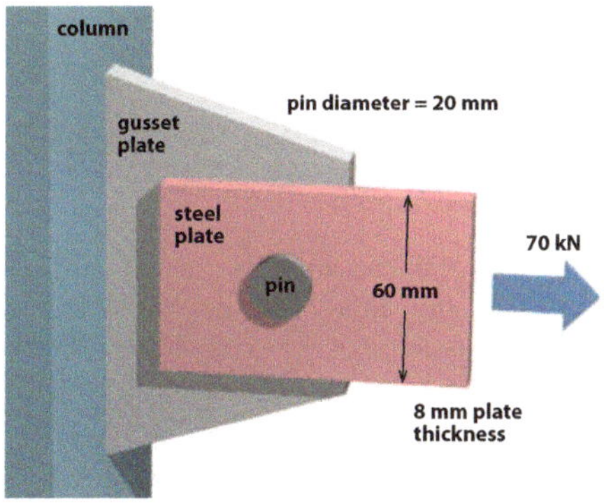

EXERCISES

M1.1 For the pin connection shown, determine the normal stress acting on the gross area, the normal stress acting on the net area, the shear stress in the pin, and the bearing stress in the steel plate at the pin.

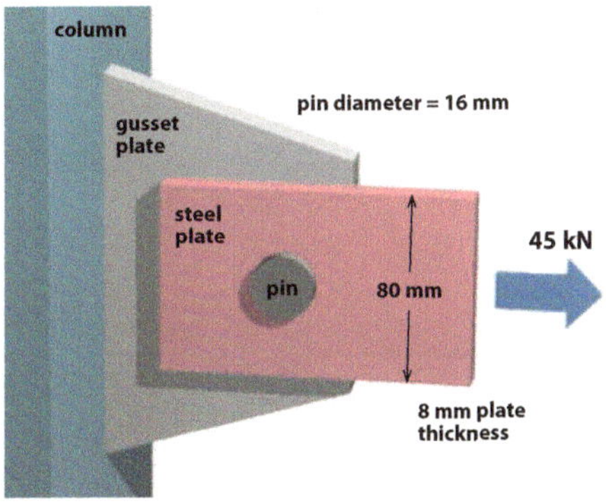

FIGURE M1.1

M1.2 Use normal stress concepts for four introductory problems.

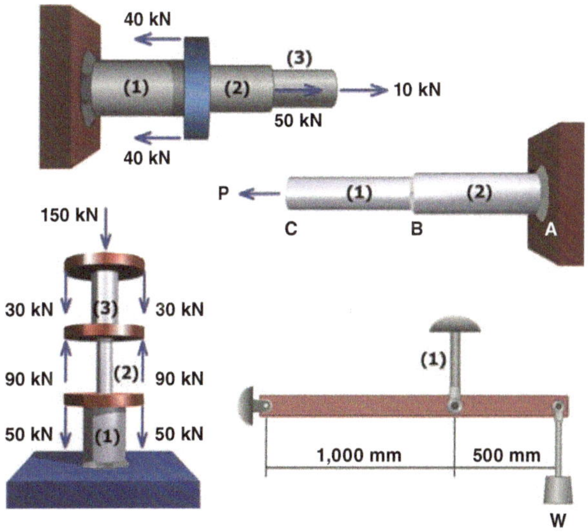

FIGURE M1.2

M1.3 Use shear stress concepts for four introductory problems.

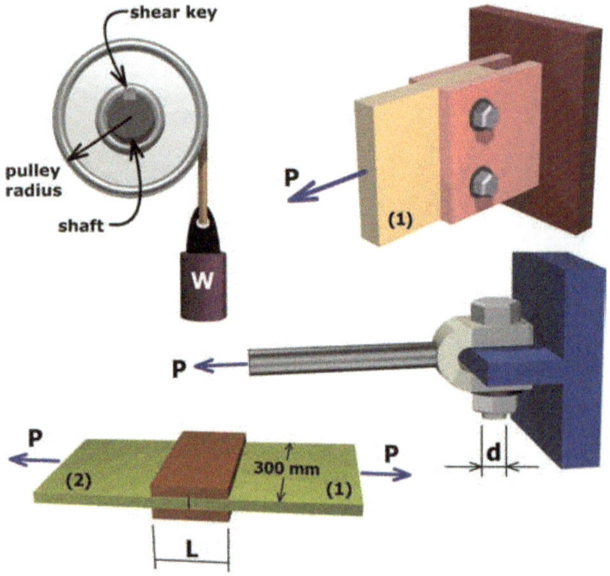

FIGURE M1.3

M1.4 Given the areas and allowable normal stresses for members (1) and (2), determine the maximum load P that may be supported by the structure without exceeding either allowable stress.

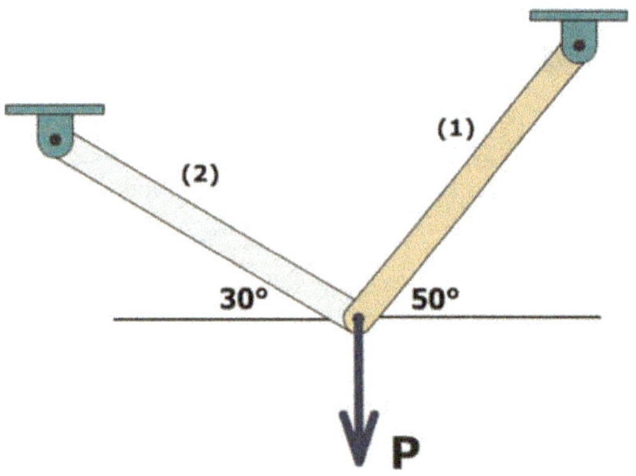

FIGURE M1.4

M1.5 For the pin at C, determine the resultant force, the shear stress, and the minimum required pin diameter for six configuration variations.

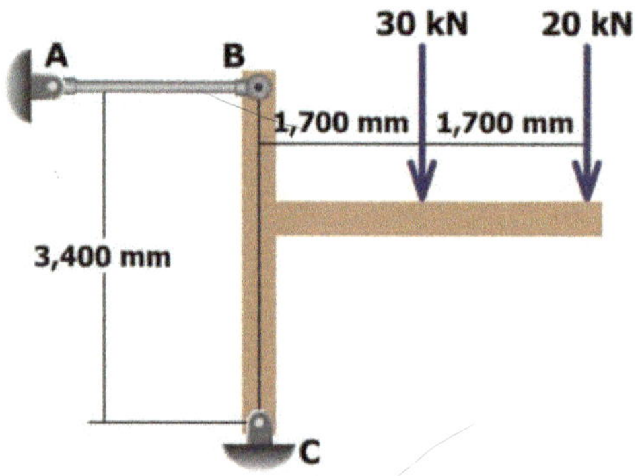

FIGURE M1.5

M1.6 A torque T is transmitted between two flanged shafts by means of six bolts. If the shear stress in the bolts must be limited to a specified value, determine the minimum bolt diameter required for the connection.

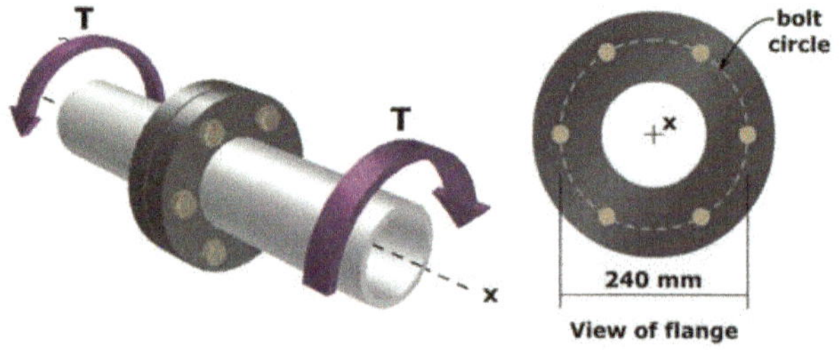

FIGURE M1.6

PROBLEMS

P1.1 A steel bar of rectangular cross section 15 mm by 60 mm is loaded by a compressive force of 110 kN that acts in the longitudinal direction of the bar. Compute the average normal stress in the bar.

P1.2 A circular pipe with outside diameter 4.5 in. and wall thickness 0.375 in. is subjected to an axial tensile force of 42,000 lb. Compute the average normal stress in the pipe.

P1.3 A circular pipe with an outside diameter of 80 mm is subjected to an axial compressive force of 420 kN. The average normal stress must not exceed 130 MPa. Compute the minimum wall thickness required for the pipe.

P1.4 Three solid bars, each with square cross sections, make up the axial assembly shown in Figure P1.4/5. Two loads of $P = 30$ kN are applied to the assembly at flange B, two loads of $Q = 18$ kN are applied at C, and one load of $R = 42$ kN is applied at end D. The bar dimensions are $b_1 = 60$ mm, $b_2 = 20$ mm, and $b_3 = 40$ mm. Determine the normal stress in each bar.

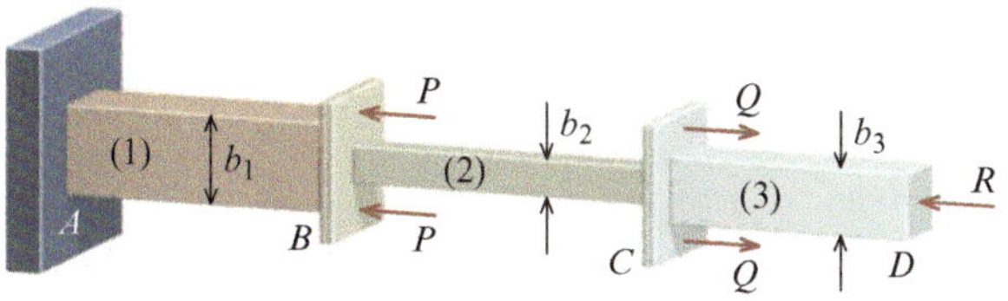

FIGURE P1.4/5

P1.5 Three solid bars, each with square cross sections, make up the axial assembly shown in Figure P1.4/5. Two loads of $P = 25$ kN are applied to the assembly at flange B, two loads of $Q = 15$ kN are applied at C, and one load of $R = 35$ kN is applied at end D. Bar (1) has a width of $b_1 = 90$ mm. Calculate the width b_2 required for bar (2) if the normal stress magnitude in bar (2) must equal the normal stress magnitude in bar (1).

P1.6 Axial loads are applied with rigid bearing plates to the solid cylindrical rods shown in Figure P1.6/7. One load of $P =$ 1,500 lb is applied to the assembly at A, two loads of $Q = 900$ lb are applied at B, and two loads of $R = 1{,}300$ lb are applied at C. The diameters of rods (1), (2), and (3) are, respectively, $d_1 = 0.625$ in., $d_2 = 0.500$ in., and $d_3 = 0.875$ in. Determine the axial normal stress in each of the three rods.

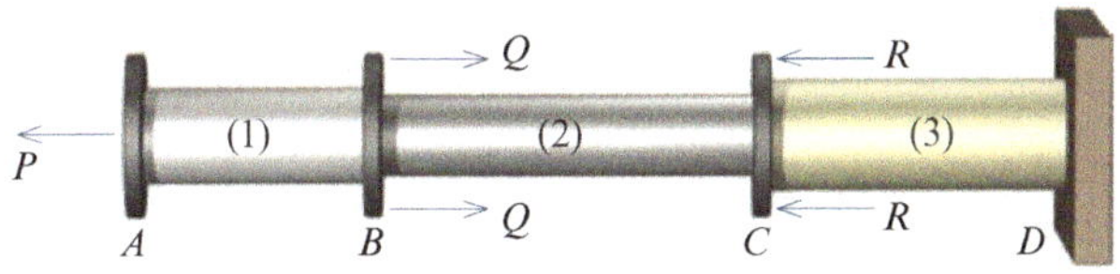

FIGURE P1.6/7

P1.7 Axial loads are applied with rigid bearing plates to the solid cylindrical rods shown in Figure P1.6/7. One load of $P = 30$ kips is applied to the assembly at A, two loads of $Q = 25$ kips are applied at B, and two loads of $R = 35$ kips are applied at C. The normal stress magnitude in aluminum rod (1) must be limited to 20 ksi. The normal stress magnitude in steel rod (2) must be limited to 35 ksi. The normal stress magnitude in brass rod (3) must be limited to 25 ksi. Determine the minimum diameter required for each of the three rods.

P1.8 Determine the normal stress in rod (1) for the mechanism shown in Figure P1.8. The diameter of rod (1) is 8 mm, and load $P = 2{,}300$ N. Use the following dimensions: $a = 120$ mm, $b = 200$ mm, $c = 170$ mm, and $d = 90$ mm.

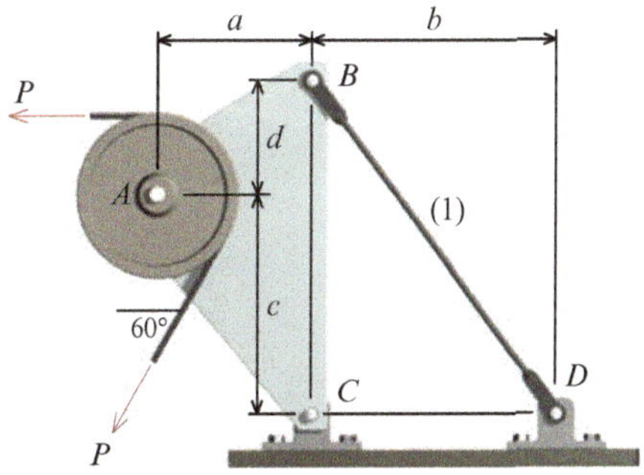

FIGURE P1.8

P1.9 Determine the normal stress in bar (1) for the mechanism shown in Figure P1.9. The area of bar (1) is 2,600 mm². The distributed load intensities are $w_C = 12$ kN/m and $w_D = 30$ kN/m. Use the following dimensions: $a = 7.5$ m and $b = 3.0$ m.

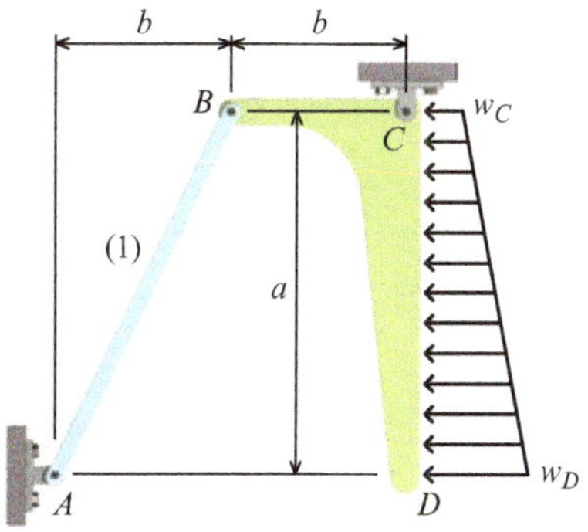

FIGURE P1.9

P1.10 The rigid beam BC shown in Figure P1.10 is supported by rods (1) and (2), which have diameters of 0.875 in. and 1.125 in., respectively. For a uniformly distributed load of $w = 4{,}200$ lb/ft, determine the normal stress in each rod. Assume that $L = 14$ ft and $a = 9$ ft.

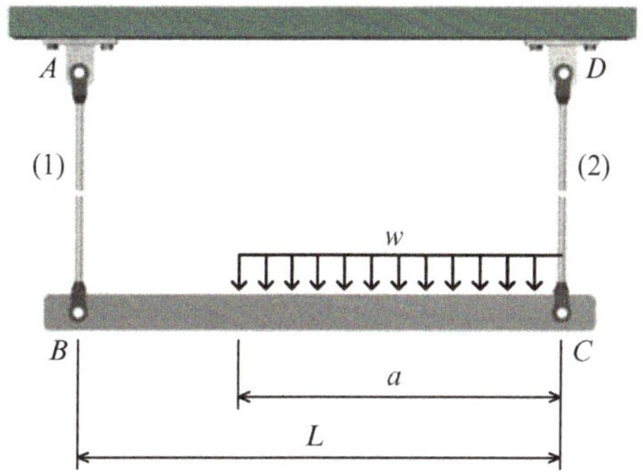

FIGURE P1.10

P1.11 The rigid beam ABC shown in Figure P1.11 is supported by a pin connection at C and by steel rod (1), which has a diameter of 10 mm. If the normal stress in rod (1) must not exceed 225 MPa, what is the maximum uniformly distributed load w that may be

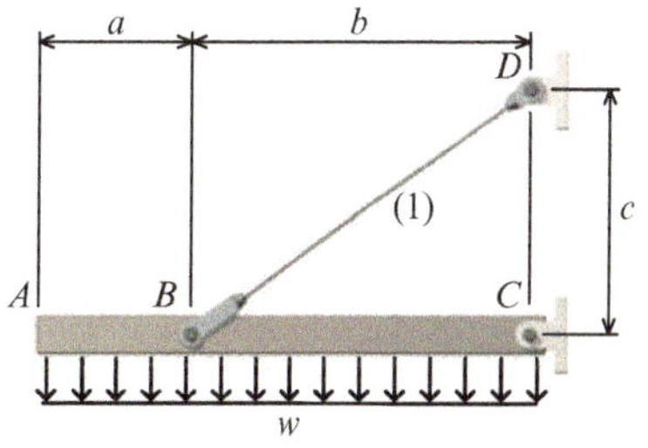

FIGURE P1.11

applied to beam ABC? Use dimensions of $a = 340$ mm, $b = 760$ mm, and $c = 550$ mm.

P1.12 A simple pin-connected truss is loaded and supported as shown in Figure P1.12. The load P is 200 kN. All members of the truss are aluminum pipes that have an outside diameter of 115 mm and a wall thickness of 6 mm. Determine the normal stress in each truss member. Assume truss dimensions of $a = 12.0$ m, $b = 7.5$ m, and $c = 6.0$ m.

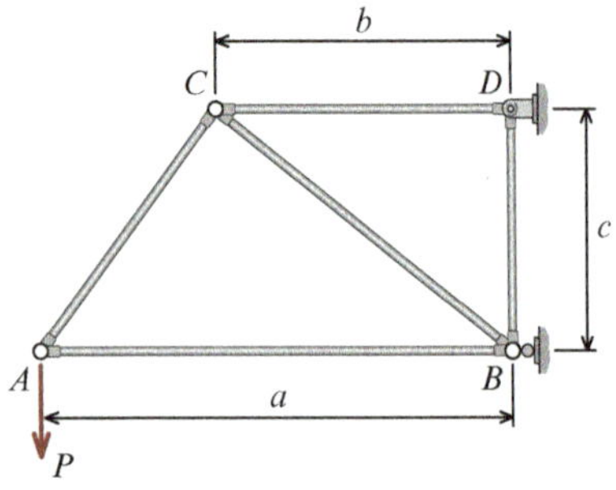

FIGURE P1.12

P1.13 A horizontal load P is applied to an assembly consisting of two inclined bars, as shown in Figure 1.13. The cross-sectional area of bar (1) is 1.5 in.^{2}, and the cross-sectional area of bar (2) is 1.8 in.^{2}. The normal stress in either bar must not exceed 24 ksi. Determine the maximum load P that may be applied to this assembly. Assume dimensions of $a = 16$ ft, $b = 8$ ft, and $c = 13$ ft.

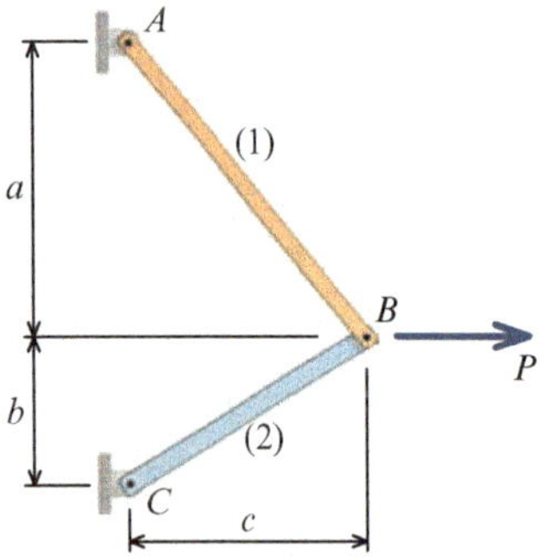

FIGURE P1.13

P1.14 The rectangular bar shown in Figure P1.14 is subjected to a uniformly distributed axial loading of $w = 13$ kN/m and a concentrated force of $P = 9$ kN at B. Determine (1) the magnitude of the maximum normal stress in the bar and (2) its location x. Assume $a = 0.5$ m, $b = 0.7$ m, $c = 15$ mm, and $d = 40$ mm.

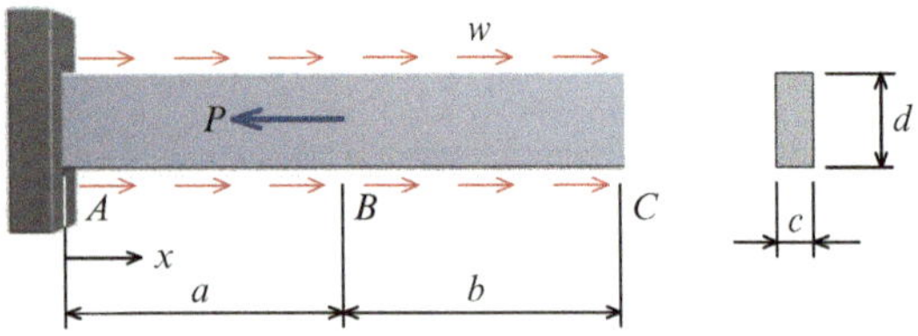

FIGURE P1.14

P1.15 The solid 1.25 in. diameter rod shown in Figure P1.15 is subjected to a uniform axial distributed loading of $w = 750$ lb/ft along its length. Two concentrated loads also act on the rod: $P = 2{,}000$ lb and $Q = 1{,}000$ lb. Assume that $a = 16$ in. and $b = 32$ in. Determine the normal stress in the rod at the following locations:

(a) $x = 10$ in.
(b) $x = 30$ in.

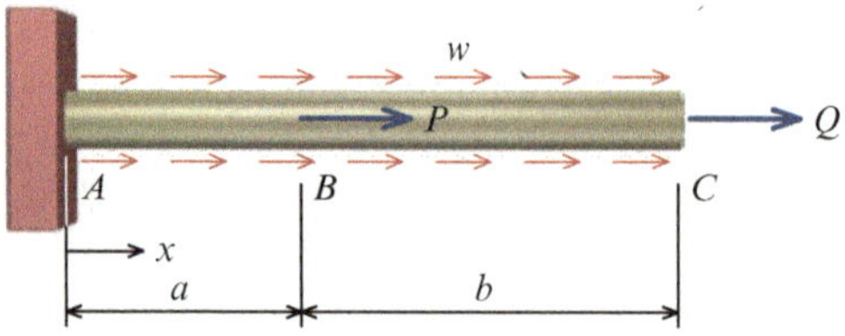

FIGURE P1.15

P1.16 A block of wood is tested in direct shear with the use of the test fixture shown in Figure P1.16. The dimensions of the test specimen are $a = 3.75$ in., $b = 1.25$ in., $c = 2.50$ in., and $d = 6.0$ in. During the test, a load of $P = 1{,}590$ lb produces a shear failure in the wood specimen. What is the magnitude of the average shear stress in the specimen at failure?

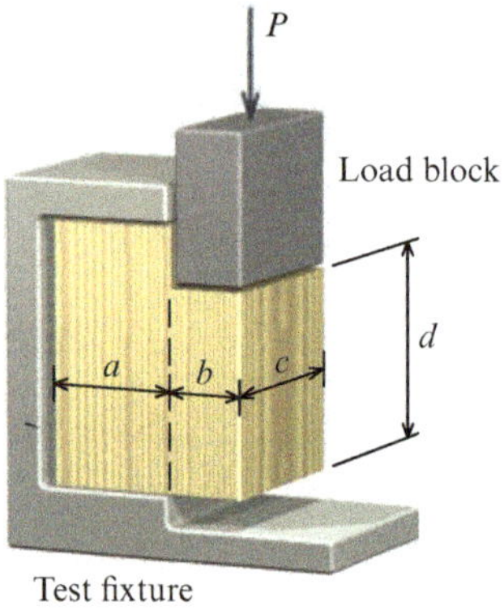

FIGURE P1.16

P1.17 A cylindrical rod of diameter $d = 0.625$ in. is attached to a plate by a cylindrical rubber grommet, as shown in Figure P1.17. The plate has a thickness $t = 0.875$ in. If the axial load on the rod is $P = 175$ lb, what is the average shear stress on the cylindrical surface of contact between the rod and the grommet?

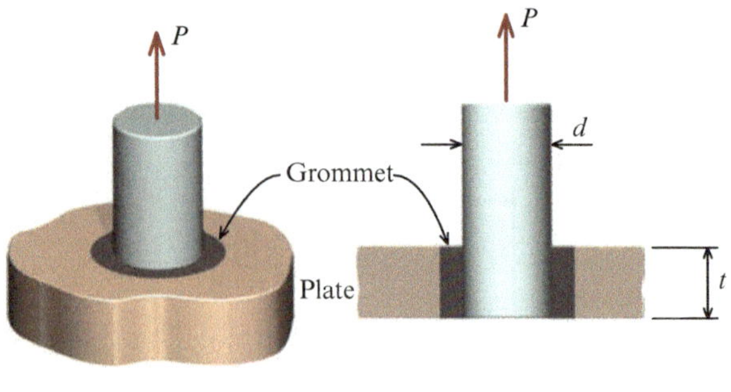

FIGURE P1.17

P1.18 Two wood boards, each 19 mm thick, are joined by the glued finger joint shown in Figure P1.18. The finger joint will fail when the average shear stress in the glue reaches 940 kPa.

Determine the shortest allowable length d of the cuts if the joint is to withstand an axial load of $P = 5.5$ kN. Use $a = 23$ mm and $b = 184$ mm.

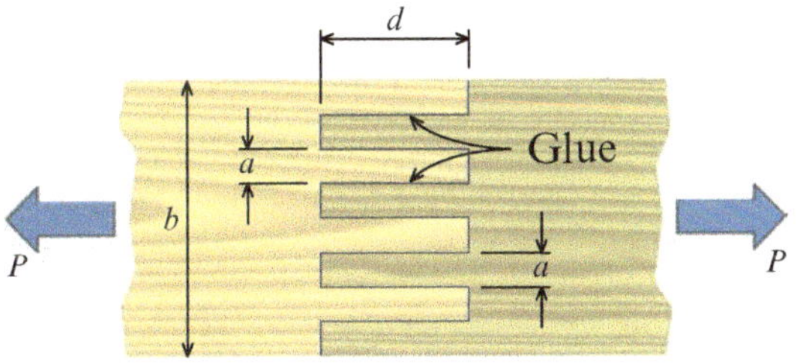

FIGURE P1.18

P1.19 For the connection shown in Figure P1.19, determine the average shear stress produced in the 7/8 in. diameter bolts if the applied load is $P = 32{,}000$ lb.

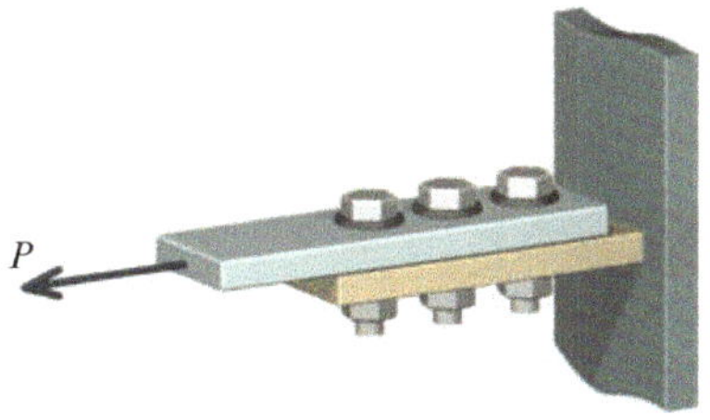

FIGURE P1.19

P1.20 For the clevis connection shown in Figure P1.20, determine the maximum applied load P that can be supported by the 15 mm diameter pin if the average shear stress in the pin must not exceed 130 MPa.

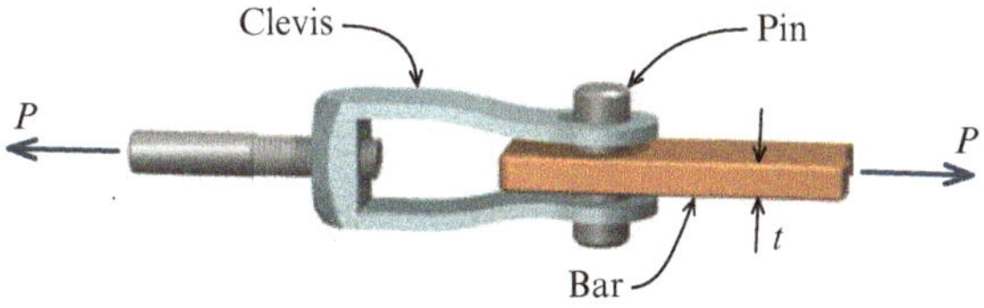

FIGURE P1.20

P1.21 The five-bolt connection shown in Figure P1.21 must support an applied load of $P = 160$ kips. If the average shear stress in the bolts must be limited to 30 ksi, what is the minimum bolt diameter that may be used for this connection?

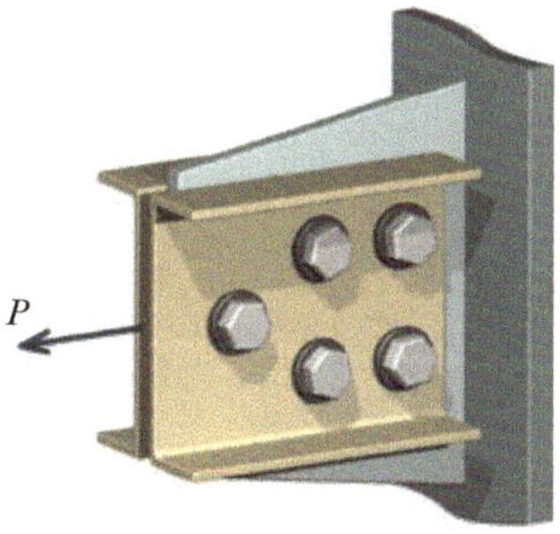

FIGURE P1.21

P1.22 The handle shown in Figure P1.22 is attached to a 40 mm diameter shaft with a square shear key. The forces applied to the lever are each of magnitude $P = 1{,}300$ N. If the average shear stress in the key must not exceed 150 MPa, determine the minimum dimension a that must be used if the key is 25 mm long. The overall length of the handle is $L = 0.70$ m.

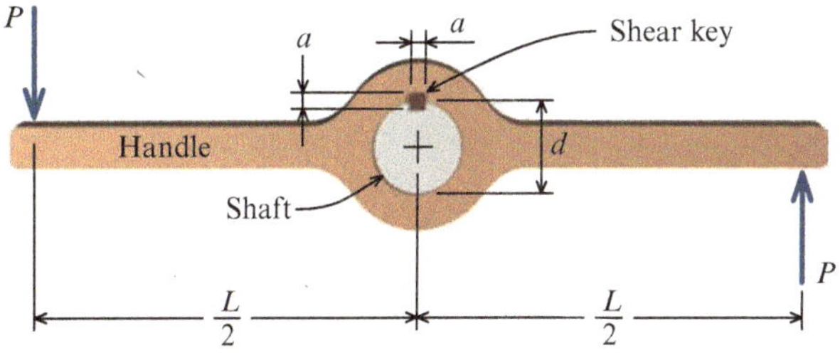

FIGURE P1.22

P1.23 An axial load P is supported by the short steel column shown in Figure P1.23. The column has a cross-sectional area of 14,500 mm². If the average normal stress in the steel column must not exceed 75 MPa, determine the minimum required dimension a so that the bearing stress between the base plate and the concrete slab does not exceed 8 MPa. Assume that $b = 420$ mm.

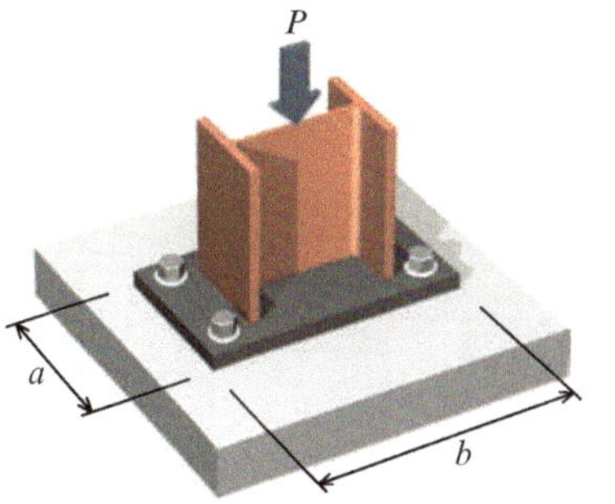

FIGURE P1.23

19

P1.24 The two wooden boards shown in Figure P1.24 are connected by a 0.5 in. diameter bolt. Washers are installed under the head of the bolt and under the nut. The washer dimensions are D = 2 in. and d = 5/8 in. The nut is tightened to cause a tensile stress of 9,000 psi in the bolt. Determine the bearing stress between the washer and the wood.

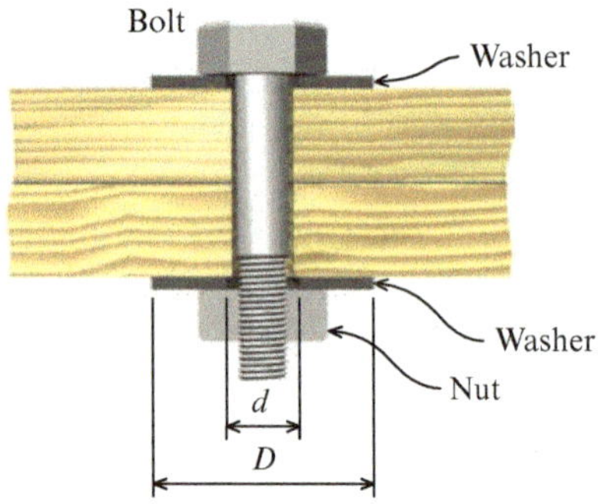

FIGURE P1.24

P1.25 For the beam shown in Figure P1.25, the allowable bearing stress for the material under the supports at A and B is $\sigma_b = 800$ psi. Assume that w = 2,100 lb/ft, P = 4,600 lb, a = 20 ft, and b = 8 ft. Determine the size of *square* bearing plates required to support the loading shown. Dimension the plates to the nearest ½ in.

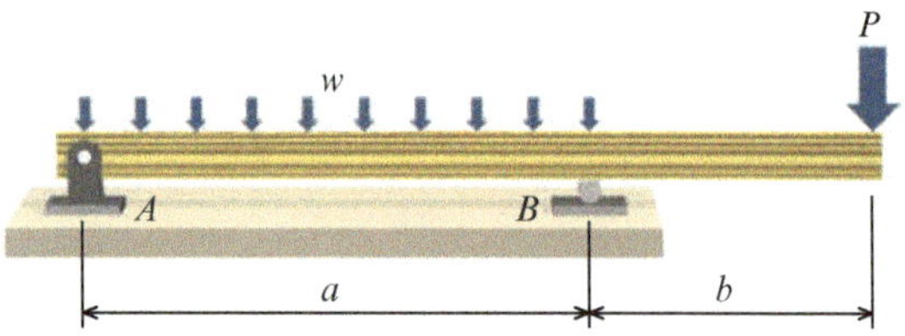

FIGURE P1.25

P1.26 A wooden beam rests on a square post. The vertical reaction force of the beam at the post is P = 1,300 lb. The post has cross-sectional dimensions of a = 6.25 in. The beam has a width b = 1.50 in. and a depth d = 7.50 in. What is the average bearing stress in the beam?

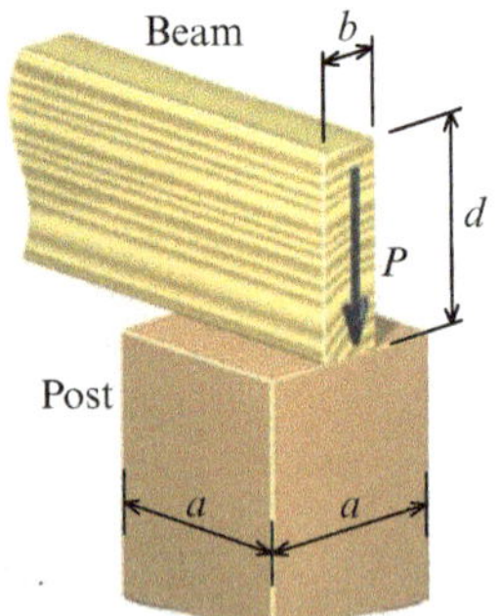

FIGURE P1.26

P1.27 The pulley shown in Figure P1.27 is connected to a bracket with a circular pin of diameter d = 6 mm. Each vertical side of the bracket has a width b = 25 mm and a thickness t = 4 mm. If the pulley belt tension is P = 570 N, what is the average bearing stress produced in the bracket by the pin?

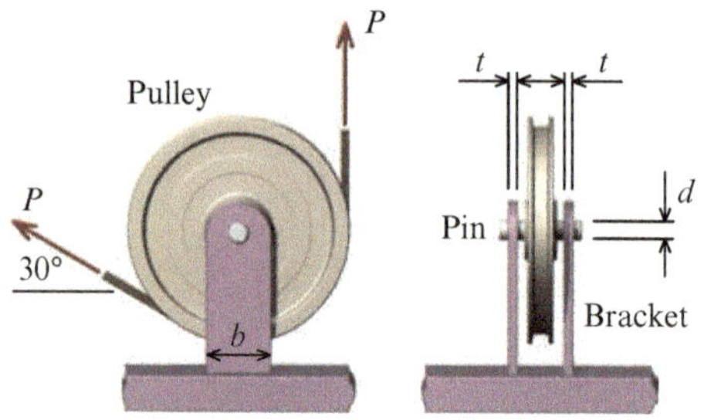

FIGURE P1.27

P1.28 The solid rod shown in Figure P1.28 is of diameter d = 15 mm and passes through a hole of diameter D = 20 mm in the support plate. When a load P is applied to the rod, the rod head rests on the plate, which has a thickness b = 12 mm. The rod head has a diameter a = 30 mm and a thickness t = 10 mm. If the normal stress produced in the rod by load P is 225 MPa, determine

(a) the average bearing stress acting between the support plate and the rod head.
(b) the average shear stress produced in the rod head.
(c) the punching shear stress produced in the support plate by the rod head.

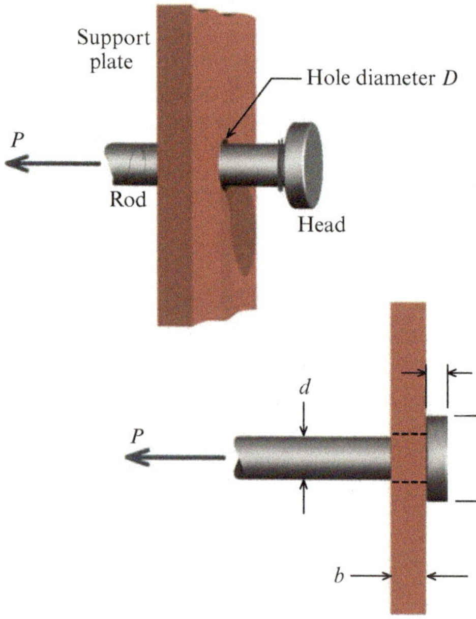

FIGURE P1.28

20

P1.29 A hollow box beam $ABCD$ is supported at A by a pin that passes through the beam as shown in Figure P1.29. The beam is also supported by a roller located at B. The beam dimensions are $a = 2.5$ ft, $b = 5.5$ ft, and $c = 3.5$ ft. Two equal concentrated loads of $P = 2{,}750$ lb are placed on the beam at points C and D. The beam has a wall thickness $t = 0.375$ in., and the pin at A has a diameter of 0.750 in. Determine

(a) the average shear stress in the pin at A.
(b) the average bearing stress in the box beam at A.

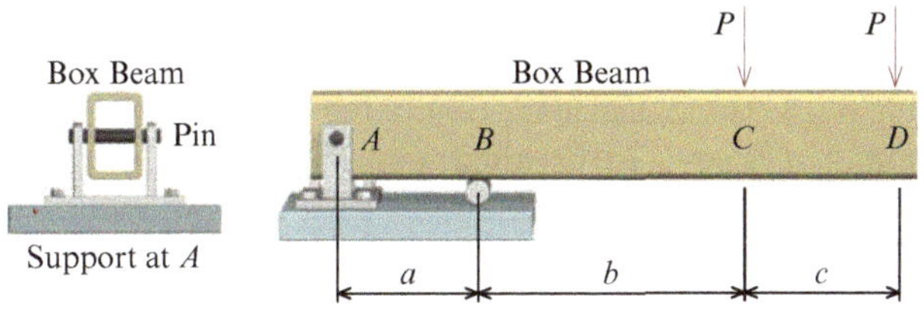

FIGURE P1.29

P1.30 Rigid bar ABC shown in Figure P1.30 is supported by a pin at bracket A and by tie rod (1). Tie rod (1) has a diameter of 5 mm and is supported by double-shear pin connections at B and D. The pin at bracket A is a single-shear connection. All pins are 7 mm in diameter. Assume that $a = 600$ mm, $b = 300$ mm, $h = 450$ mm, $P = 900$ N, and $\theta = 55°$. Determine

(a) the normal stress in rod (1).
(b) the average shear stress in pin B.
(c) the average shear stress in pin A.

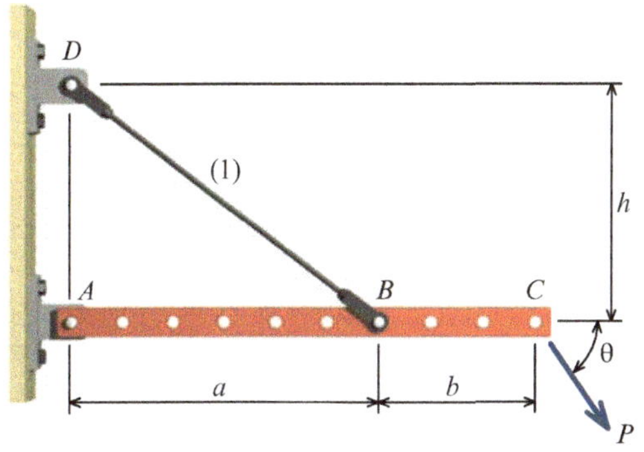

FIGURE P1.30

P1.31 The bell crank shown in Figure P1.31 is in equilibrium for the forces acting in rods (1) and (2). The crank is supported by a 10 mm diameter pin at B that acts in single shear. The thickness of the crank is 5 mm. Assume that $a = 65$ mm, $b = 150$ mm, $F_1 = 1{,}100$ N, and $\theta = 50°$. Determine

(a) the average shear stress in pin B.
(b) the average bearing stress in the bell crank at B.

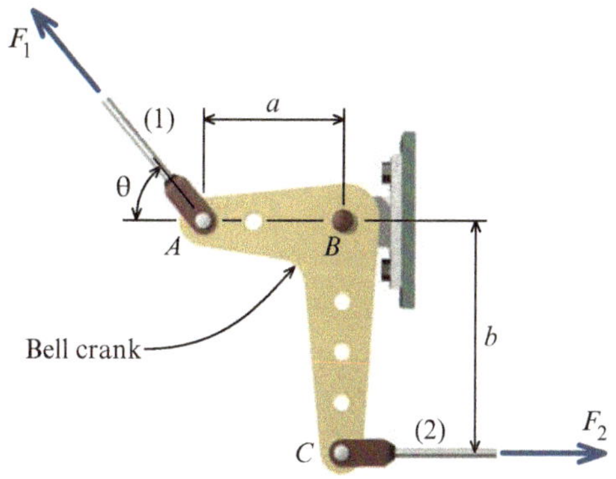

FIGURE P1.31

P1.32 The beam shown in Figure P1.32 is supported by a pin at C and by a short link AB. If $w = 30$ kN/m, determine the average shear stress in the pins at A and C. Each pin has a diameter of 25 mm. Assume that $L = 1.8$ m and $\theta = 35°$.

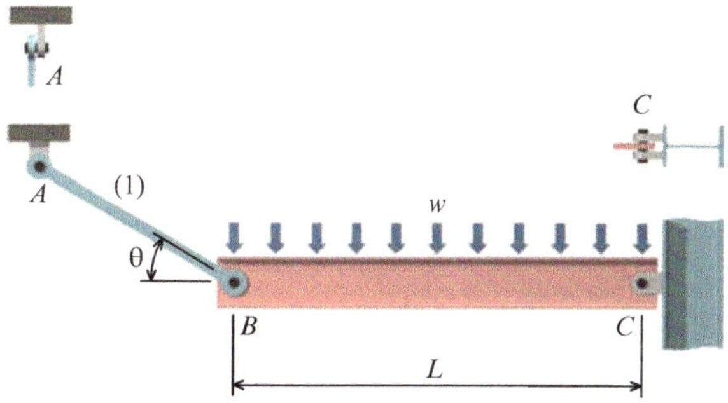

FIGURE P1.32

P1.33 The bell-crank mechanism shown in Figure P1.33 is in equilibrium for an applied load of $P = 7$ kN at A. Assume that $a = 200$ mm, $b = 150$ mm, and $\theta = 65°$. Determine the minimum diameter d required for pin B for each of the following conditions:

(a) The shear stress in the pin must not exceed 40 MPa.
(b) The bearing stress in the bell crank must not exceed 100 MPa.
(c) The bearing stress in the support bracket must not exceed 165 MPa.

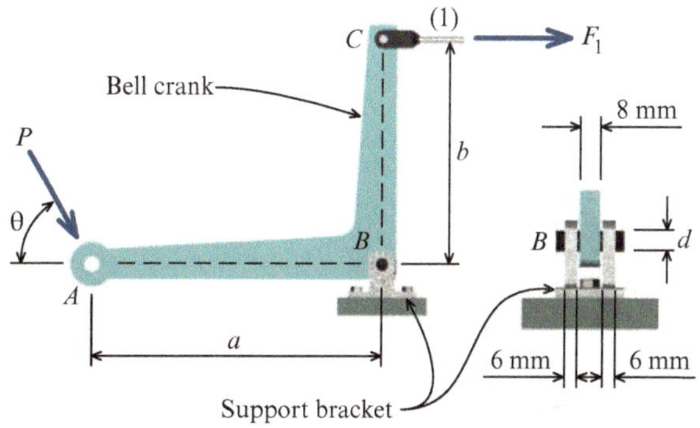

FIGURE P1.33

1.5 Stresses on Inclined Sections

In previous sections, normal, shear, and bearing stresses on planes parallel and perpendicular to the axes of centrically loaded members were introduced. Stresses on planes inclined to the axes of axially loaded bars will now be examined.

MecMovies 1.11 is an animated presentation of the theory of stresses on an inclined plane.

Consider a prismatic bar subjected to an axial force P applied to the centroid of the bar (Figure 1.7*a*). Loading of this type is termed **uniaxial**, since the force applied to the bar acts in one direction (i.e., either tension or compression). The cross-sectional area of the bar is A. To investigate the stresses that are acting internally in the material, we will cut through the bar at section *a–a*. The free-body diagram (Figure 1.7*b*) exposes the normal stress σ that is distributed over the cut section of the bar. The normal stress magnitude may be calculated from $\sigma = P/A$, provided that the stress is uniformly distributed. In this case, the stress will be uniform because the bar is prismatic and the force P is applied at the centroid of the cross section. The resultant of this normal stress distribution is equal in magnitude to the applied load P and has a line of action that is coincident with the axes of the bar, as shown. Note that there will be no shear stress τ, since the cut surface is perpendicular to the direction of the resultant force.

Section *a–a* is unique, however, because it is the only surface that is perpendicular to the direction of force P. A more general case would take into account a section cut through the bar at an arbitrary angle. In that regard, consider a free-body diagram along section *b–b* (Figure 1.7*c*). Because the stresses are the same throughout the entire bar, the stresses on the inclined surface must be uniformly distributed. Since the bar is in equilibrium, the resultant of the uniformly distributed stress must equal P even though the stress acts on a surface that is inclined.

In referencing planes, the orientation of the plane is specified by the normal to the plane. The inclined plane shown in Figure 1.7*d* is termed the n face because the n axis is the normal to that plane.

The orientation of the inclined surface can be defined by the angle θ between the x axis and an axis *normal* to the plane, which is the n axis, as shown in Figure 1.7*d*. A positive angle θ is defined as a counterclockwise rotation from the x axis to the n axis. The t axis is *tangential* to the cut surface, and the n–t axes form a right-handed coordinate system.

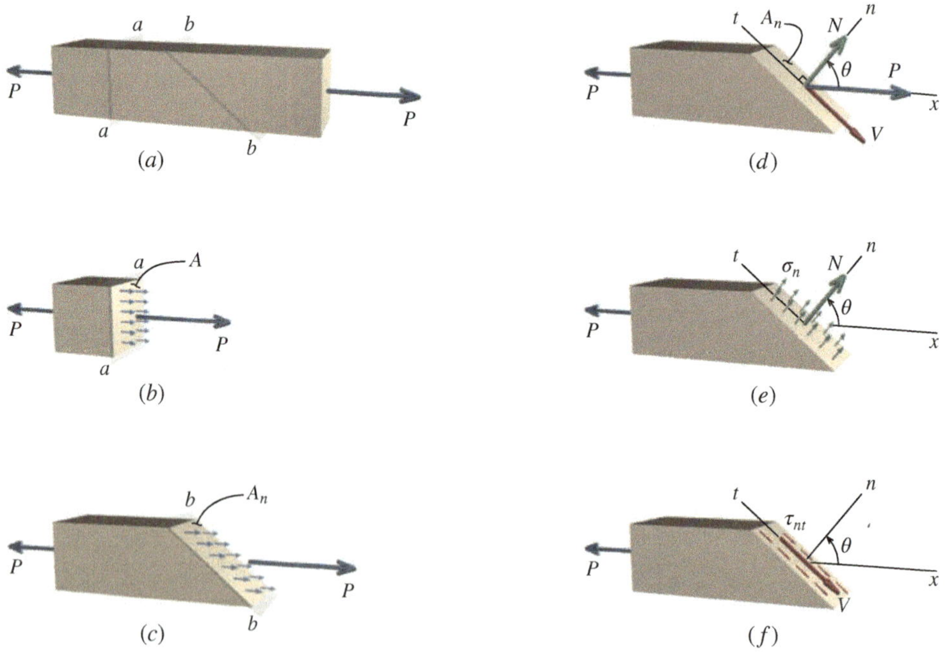

FIGURE 1.7 (*a*) Prismatic bar subjected to axial force P. (*b*) Normal stresses on section *a–a*. (*c*) Stresses on inclined section *b–b*. (*d*) Force components acting perpendicular and parallel to inclined plane. (*e*) Normal stresses acting on inclined plane. (*f*) Shear stresses acting on inclined plane.

To investigate the stresses acting on the inclined plane (Figure 1.7*d*), the components of resultant force P acting perpendicular and parallel to the plane must be computed. Using θ as defined previously, we find that the perpendicular force component (i.e., normal force) is $N = P \cos\theta$ and the parallel force component (i.e., shear force) is $V = -P \sin\theta$. (The negative sign indicates that the shear force acts in the $-t$ direction, as shown in Figure 1.7*d*.) The area of the inclined plane $A_n = A/\cos\theta$, where A is the cross-sectional area of the axially loaded member. The normal and shear stresses acting on the inclined plane (Figures 1.7*e* and 1.7*f*) can now be determined by dividing the force component by the area of the inclined plane:

$$\sigma_n = \frac{N}{A_n} = \frac{P\cos\theta}{A/\cos\theta} = \frac{P}{A}\cos^2\theta = \frac{P}{2A}(1 + \cos 2\theta) \tag{1.8}$$

$$\tau_{nt} = \frac{V}{A_n} = \frac{-P\sin\theta}{A/\cos\theta} = -\frac{P}{A}\sin\theta\cos\theta = -\frac{P}{2A}\sin 2\theta \tag{1.9}$$

Since both the area A_n of the inclined surface and the values for the normal and shear forces, N and V, respectively, on the surface depend on the angle of inclination θ, the normal and shear stresses σ_n and τ_{nt} also depend on the angle of inclination θ of the plane. *This dependence of stress on both force and area means that stress is not a vector quantity*; therefore, the laws of the vector addition do not apply to stresses.

A graph showing the values of σ_n and τ_{nt} as a function of θ is given in Figure 1.8. These plots indicate that σ_n is largest when θ is 0° or 180°, that τ_{nt} is largest when θ is 45° or 135°, and also that $\tau_{max} = \sigma_{max}/2$. Therefore, the maximum normal and shear stresses in an axial member that is subjected to an uniaxial tension or compression force applied through the centroid of the member (termed a **centric loading**) are

$$\sigma_{max} = \frac{P}{A} \quad \text{and} \quad \tau_{max} = \frac{P}{2A} \tag{1.10}$$

Note that the normal stress is either maximum or minimum on planes for which the shear stress is zero. It can be shown that the shear stress is always zero on the planes of maximum or minimum normal stress. The concepts of maximum and minimum normal stress and maximum shear stress for more general cases will be treated in later sections of this book.

The plot of normal and shear stresses for axial loading, shown in Figure 1.8, indicates that the sign of the shear stress changes when θ is greater than 90°. The magnitude of the shear stress for any angle θ, however, is the same as that for 90° + θ. The sign change merely indicates that the shear force V changes direction.

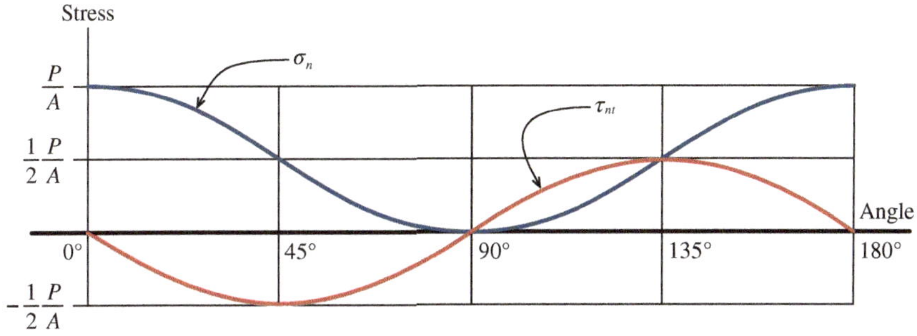

FIGURE 1.8 Variation of normal and shear stress as a function of the orientation θ of the inclined plane.

Significance

Although one might think that there is only a single stress in a material (particularly in a simple axial member), the preceding discussion has demonstrated that there are many different combinations of normal and shear stress in a solid object. The magnitude and direction of the normal and shear stresses at any point depend on the orientation of the plane being considered.

Why Is This Important? In designing a component, an engineer must be mindful of all possible combinations of normal stress σ_n and shear stress τ_{nt} that exist on internal surfaces of the object, not just the most obvious ones. Further, different materials are sensitive to different types of stress. For example, laboratory tests on specimens loaded in uniaxial tension reveal that brittle materials tend to fail in response to the magnitude of normal stress. These materials fracture on a transverse plane (i.e., a plane such as section *a–a* in Figure 1.7*a*). Ductile materials, by contrast, are sensitive to the magnitude of shear stress. A ductile material loaded in uniaxial tension will fracture on a 45° plane, since the maximum shear stress occurs on this surface.

1.6 Equality of Shear Stresses on Perpendicular Planes

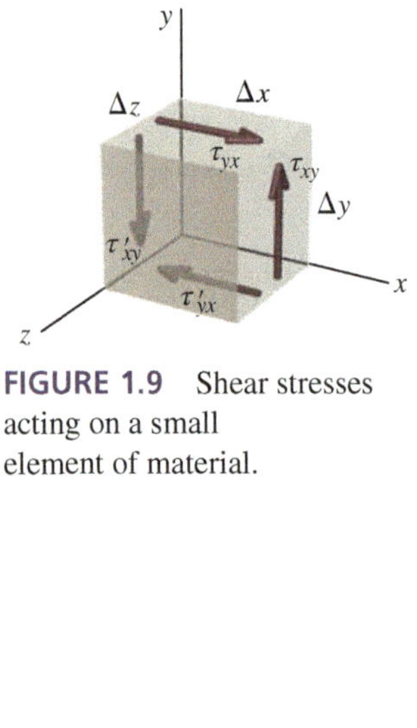

FIGURE 1.9 Shear stresses acting on a small element of material.

If an object is in equilibrium, then any portion of the object that one chooses to examine must also be in equilibrium, no matter how small that portion may be. Therefore, let us consider a small element of material that is subjected to shear stress, as shown in Figure 1.9. The front and rear faces of this small element are free of stress.

Equilibrium involves forces, not stresses. For us to consider the equilibrium of this element, we must find the forces produced by the stresses that act on each face, by multiplying the stress acting on each face by the area of the face. For example, the horizontal force acting on the top face of this element is given by $\tau_{yx}\Delta x\Delta z$, and the vertical force acting on the right face of the element is given by $\tau_{xy}\Delta y\Delta z$. Equilibrium in the horizontal direction gives

$$\Sigma F_x = \tau_{yx}\Delta x\Delta z - \tau'_{yx}\Delta x\Delta z = 0 \qquad \therefore \tau_{yx} = \tau'_{yx}$$

Equilibrium in the vertical direction gives

$$\Sigma F_y = \tau_{xy}\Delta y\Delta z - \tau'_{xy}\Delta y\Delta z = 0 \qquad \therefore \tau_{xy} = \tau'_{xy}$$

Finally, taking moments about the z axis gives

$$\Sigma M_z = (\tau_{xy}\Delta y\Delta z)\Delta x - (\tau_{yx}\Delta x\Delta z)\Delta y = 0 \qquad \therefore \tau_{xy} = \tau_{yx}$$

Shear stress arrows on adjacent faces act either toward each other or away from each other. In other words, the arrows are arranged head-to-head or tail-to-tail—never head-to-tail—on intersecting perpendicular planes.

Consequently, equilibrium requires that

$$\tau_{xy} = \tau_{yx} = \tau'_{xy} = \tau'_{yx} = \tau$$

In other words, if a shear stress acts on one plane in the object, then shear stresses of equal magnitude act on three other planes. The shear stresses must be oriented either as shown in Figure 1.9 or in the opposite directions on each face.

EXAMPLE 1.9

A 120 mm wide steel bar with a butt-welded joint, as shown, will be used to carry an axial tension load of $P = 180$ kN. If the normal and shear stresses on the plane of the butt weld must be limited to 80 MPa and 45 MPa, respectively, determine the minimum thickness required for the bar.

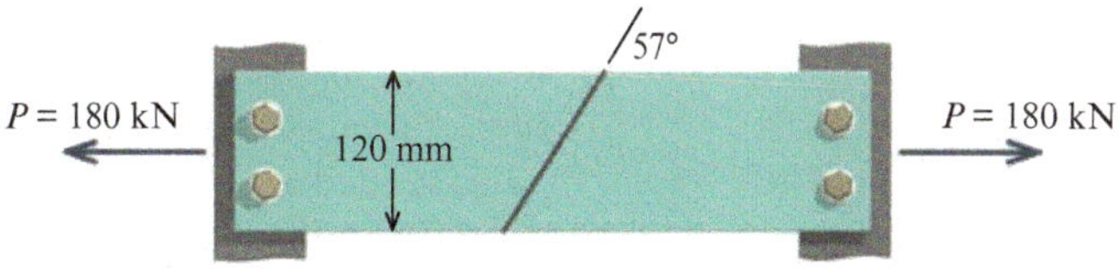

Plan the Solution

Either the normal stress limit or the shear stress limit will dictate the area required for the bar. There is no way to know beforehand which stress will control; therefore, both possibilities must be checked. The minimum cross-sectional area required for each limit must be determined. Using the larger of these two results, we will determine the minimum bar thickness. For illustration, this example will be worked in two ways:

(a) by directly using the normal and shear components of force P,
(b) by using Equations (1.8) and (1.9).

SOLUTION

(a) Solution Using Normal and Shear Force Components

Consider a free-body diagram (FBD) of the left half of the member. Resolve the axial force $P = 180$ kN into a force component N perpendicular to the weld and a force component V parallel to the weld.

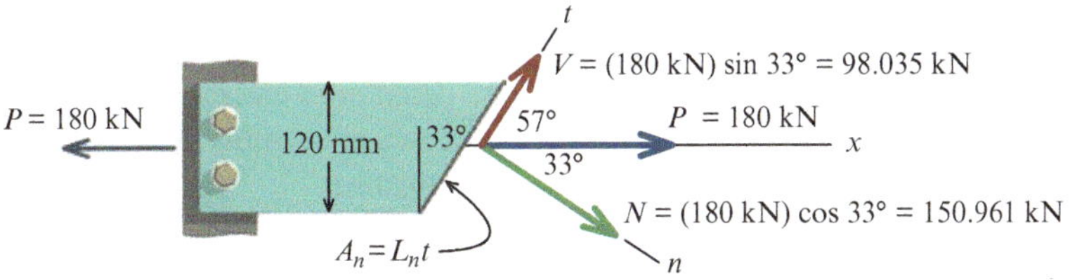

The minimum cross-sectional area A_n of the weld needed to limit the normal stress on the weld to 80 MPa can be computed from

$$\sigma_n \geq \frac{N}{A_n} \qquad \therefore A_n \geq \frac{(150.961 \text{ kN})(1{,}000 \text{ N/kN})}{80 \text{ N/mm}^2} = 1{,}887.013 \text{ mm}^2$$

Similarly, the minimum cross-sectional area A_n of the weld needed to limit the shear stress on the weld to 45 MPa can be computed from

$$\tau_{nt} \geq \frac{V}{A_n} \qquad \therefore A_n \geq \frac{(98.035 \text{ kN})(1{,}000 \text{ N/kN})}{45 \text{ N/mm}^2} = 2{,}178.556 \text{ mm}^2$$

To satisfy both normal and shear stress limits, the minimum cross-sectional area A_n needed for the weld is $A_n \geq 2{,}178.556 \text{ mm}^2$. Next, we can determine the length L_n of the weld along the inclined surface. From the geometry of the surface,

$$\cos 33° = \frac{120 \text{ mm}}{L_n} \qquad \therefore L_n = \frac{120 \text{ mm}}{\cos 33°} = 143.084 \text{ mm}$$

Thus, to provide the necessary weld area, the minimum thickness is computed as

$$t_{\min} \geq \frac{2{,}178.556 \text{ mm}^2}{143.084 \text{ mm}} = 15.23 \text{ mm} \qquad \textbf{Ans.}$$

(b) Solution Using Equations (1.8) and (1.9)

Determine the angle θ needed for Equations (1.8) and (1.9). The angle θ is defined as the angle between the transverse cross section (i.e., the section perpendicular to the applied load) and the inclined surface, with positive angles defined in a counterclockwise direction. Although the butt-weld angle is labeled 57° in the problem sketch, that is not the value needed for θ. For use in the equations, $\theta = -33°$.

The normal and shear stresses on the inclined plane can be computed from

$$\sigma_n = \frac{P}{A}\cos^2\theta \quad \text{and} \quad \tau_{nt} = -\frac{P}{A}\sin\theta\cos\theta$$

According to the 80 MPa normal stress limit, the minimum cross-sectional area required for the bar is

$$A_{\min} \geq \frac{P}{\sigma_n}\cos^2\theta = \frac{(180 \text{ kN})(1{,}000 \text{ N/kN})}{80 \text{ N/mm}^2}\cos^2(-33°) = 1{,}582.58 \text{ mm}^2$$

Similarly, the minimum area required for the bar, on the basis of the 45 MPa shear stress limit, is

$$A_{\min} \geq -\frac{P}{\tau_{nt}}\sin\theta\cos\theta = -\frac{(180 \text{ kN})(1{,}000 \text{ N/kN})}{45 \text{ N/mm}^2}\sin(-33°)\cos(-33°) = 1{,}827.09 \text{ mm}^2$$

Note: Here we are concerned with force and area *magnitudes*. If the area calculations had produced a negative value, we would have considered only the absolute value.

To satisfy both stress limits, the larger of the two areas must be used. Since the steel bar is 120 mm wide, the minimum bar thickness must be

$$t_{\min} \geq \frac{1{,}827.09 \text{ mm}^2}{120 \text{ mm}} = 15.23 \text{ mm} \qquad \textbf{Ans.}$$

MecMovies

EXAMPLES

M1.12 The steel bar shown has a 100 mm by 25 mm rectangular cross section. If an axial force of P = 40 kN is applied to the bar, determine the normal and shear stresses acting on the inclined surface *a–a*.

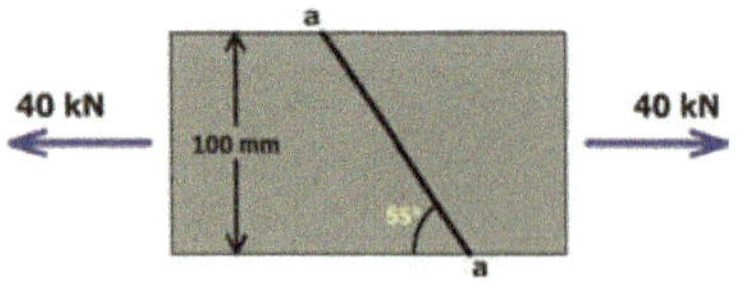

M1.13 The steel bar shown has a 50 mm by 10 mm rectangular cross section. The allowable normal and shear stresses on the inclined surface must be limited to 40 MPa and 25 MPa, respectively. Determine the magnitude of the maximum axial force of P that can be applied to the bar.

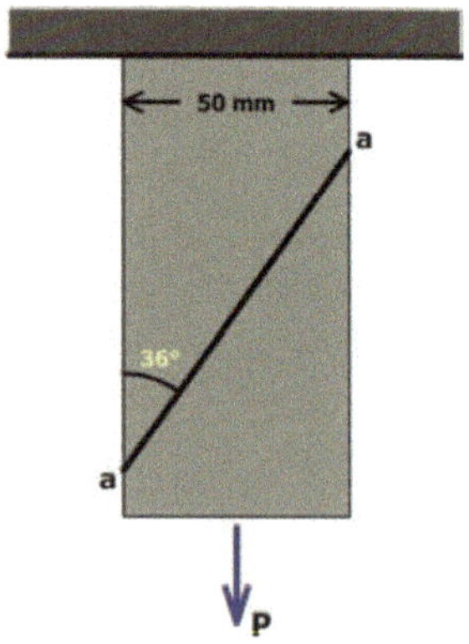

EXERCISES

M1.12 The bar shown has a rectangular cross section. For a given load *P*, determine (1) the force components perpendicular and parallel to section *a–a*, (2) the inclined surface area, and (3) the normal and shear stress magnitudes acting on surface *a–a*.

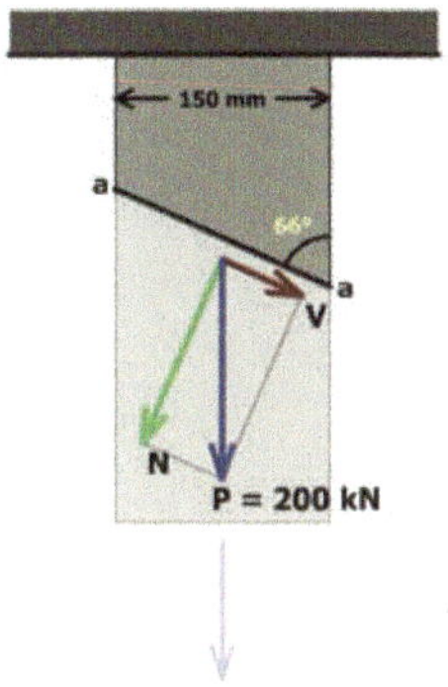

FIGURE M1.12

M1.13 The bar shown has a rectangular cross section. The allowable normal and shear stresses on inclined surface *a–a* are given. Determine (1) the magnitude of the maximum axial force *P* that can be applied to the bar and (2) the actual normal and shear stresses acting on inclined plane *a–a*.

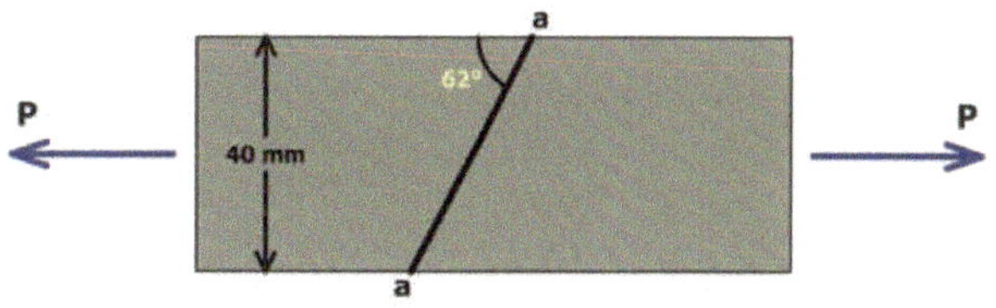

FIGURE M1.13

PROBLEMS

P1.34 A structural steel bar with a 4.0 in. × 0.875 in. rectangular cross section is subjected to an axial load of 45 kips. Determine the maximum normal and shear stresses in the bar.

P1.35 A stainless steel rod of circular cross section will be used to carry an axial load of 30 kN. The maximum stresses in the rod must be limited to 100 MPa in tension and 60 MPa in shear. Determine the required minimum diameter for the rod.

P1.36 Two wooden members, each having a width b = 1.50 in. and a depth d = 0.5 in., are joined by the simple glued scarf joint shown in Figure P1.36/37. Assume that $\beta = 40°$. If the allowable shear stress for the glue used in the joint is 90 psi, what is the largest axial load *P* that may be applied?

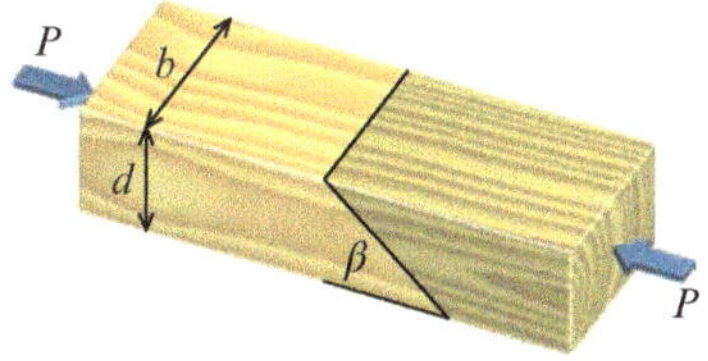

FIGURE P1.36/37

P1.37 Two wooden members, each having a width b = 4.50 in. and a depth d = 1.75 in., are joined by the simple glued scarf joint shown in Figure P1.36/37. Assume that $\beta = 35°$. Given that the compressive axial load is P = 900 lb, what are the normal stress and shear stress magnitudes in the glued joint?

P1.38 Two aluminum plates, each having a width b = 7.0 in. and a thickness t = 0.625 in., are welded together as shown in Figure P1.38/39. Assume that a = 4.0 in. Given that the tensile axial load is P = 115 kips, determine (a) the normal stress that acts perpendicular to the weld and (b) the shear stress that acts parallel to the weld.

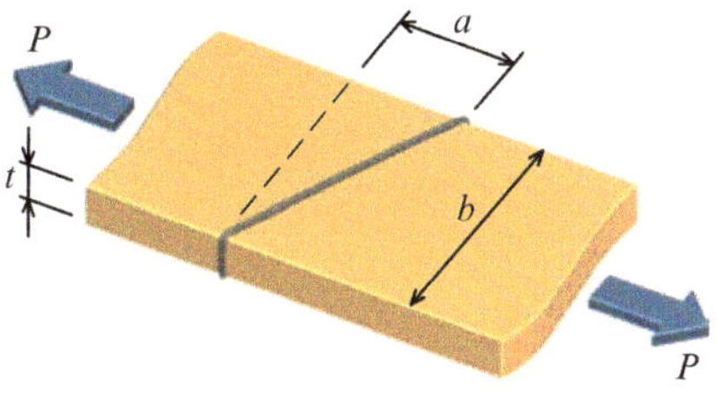

FIGURE P1.38/39

P1.39 Two aluminum plates, each having a width b = 5.0 in. and a thickness t = 0.75 in., are welded together as shown in Figure P1.38/39. Assume that a = 2.0 in. Specifications require that the normal and shear stress magnitudes acting in the weld material may not exceed 35 ksi and 24 ksi, respectively. Determine the largest axial load *P* that can be applied to the aluminum plates.

P1.40 Two wooden members are glued together as shown in Figure P1.40. Each member has a width b = 1.50 in. and a depth d = 3.50 in. Use β = 75°. Determine the average shear stress magnitude in the glue joint.

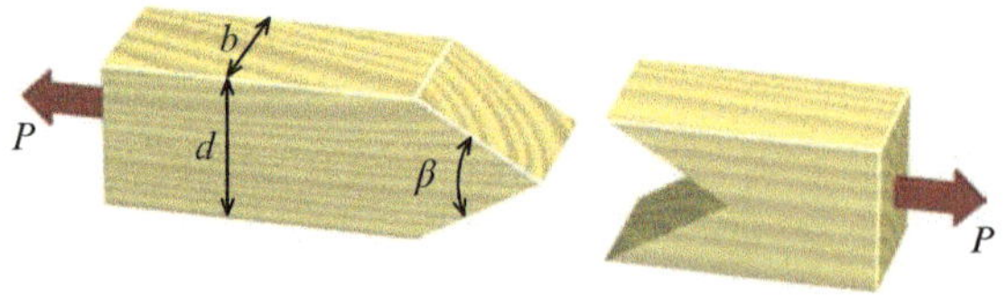

FIGURE P1.40

P1.41 Two bars are connected with a welded butt joint as shown in Figure P1.41. The bar dimensions are b = 200 mm and t = 50 mm, and the angle of the weld is α = 35°. The bars transmit a force of P = 250 kN. What is the magnitude of the average shear stress that acts on plane AB?

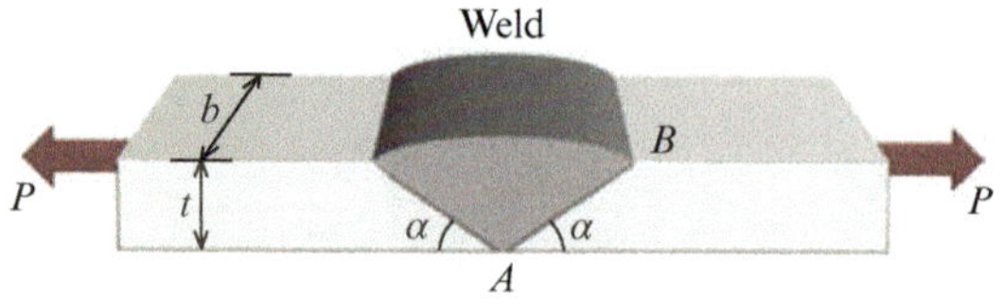

FIGURE P1.41

28

CHAPTER 2

Strain

2.1 Displacement, Deformation, and the Concept of Strain

In the design of structural elements or machine components, the deformations sustained by the body because of applied loads often represent a design consideration equally as important as stress. For this reason, the nature of the deformations sustained by a real deformable body as a result of internal stress will be studied, and methods for calculating deformations will be established.

Displacement

When a system of loads is applied to a machine component or structural element, individual points of the body generally move. This movement of a point with respect to some convenient reference system of axes is a vector quantity known as a **displacement**. In some instances, displacements are associated with a translation and/or rotation of the body as a whole. The size and shape of the body are not changed by this type of displacement, which is termed a **rigid-body displacement**. In Figure 2.1*a*, consider points H and K on a solid body. If the body is displaced (both translated and rotated), points H and K will move to new locations H' and K'. The position vector between H' and K', however, has the same length as the

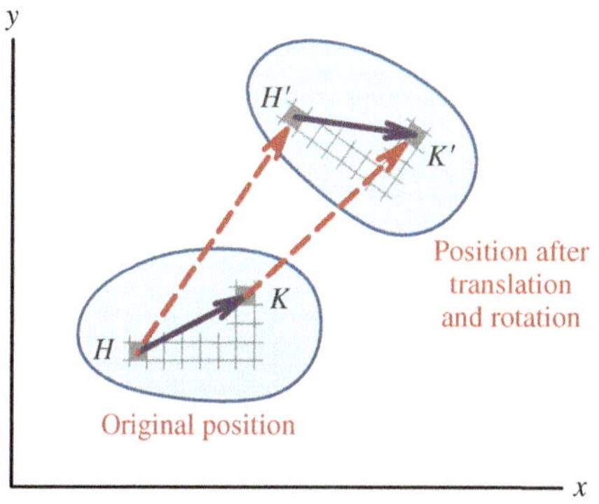

FIGURE 2.1a Rigid-body displacement.

position vector between H and K. In other words, the orientation of H and K relative to each other does not change when a body undergoes a displacement.

Deformation

When displacements are caused by an applied load or a change in temperature, individual points of the body move relative to each other. The change in any dimension associated with these load- or temperature-induced displacements is known as **deformation**. Figure 2.1*b* shows a body both before and after a deformation. For simplicity, the deformation shown in the figure is such that point H does not change location; however, point K on the undeformed body moves to location K' after the deformation. Because of the deformation, the position vector between H and K' is much longer than the HK vector in the undeformed body. Also, notice that the grid squares shown on the body before deformation (Figure 2.1*a*) are no longer squares after the deformation: Both the size and the shape of the body have been altered by the deformation.

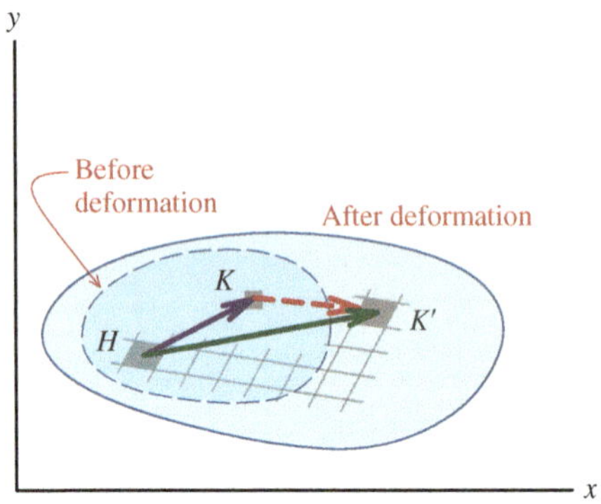

FIGURE 2.1*b* Deformation of a body.

Under general conditions of loading, deformations will not be uniform throughout the body. Some line segments will experience extensions, while others will experience contractions. Different segments (of the same length) along the same line may experience different amounts of extension or contraction. Similarly, changes in the angles between line segments may vary with position and orientation in the body. This nonuniform nature of load-induced deformations will be investigated in more detail in Chapter 13.

Strain

Strain is a quantity used to provide a measure of the intensity of a deformation (deformation per unit length), just as stress is used to provide a measure of the intensity of an internal force (force per unit area). In Sections 1.2 and 1.3, two types of stresses were defined: normal stresses and shear stresses. The same classification is used for strains. **Normal strain**, designated by the Greek letter ε (epsilon), is used to provide a measure of the elongation or contraction of an arbitrary line segment in a body after deformation. **Shear strain**, designated by the Greek letter γ (gamma), is used to provide a measure of angular distortion (change in the angle between two lines that are orthogonal in the undeformed state). The deformation, or strain, may be the result of a change in temperature, of a stress, or of some other physical phenomenon, such as grain growth or shrinkage. In this book, only strains resulting from changes in stress or temperature are considered.

2.2 Normal Strain

Average Normal Strain

The deformation (change in length and width) of a simple bar under an axial load (see Figure 2.2) can be used to illustrate the idea of a normal strain. The average normal strain ε_{avg} over the length of the bar is obtained by dividing the axial deformation δ of the bar by its initial length L; thus,

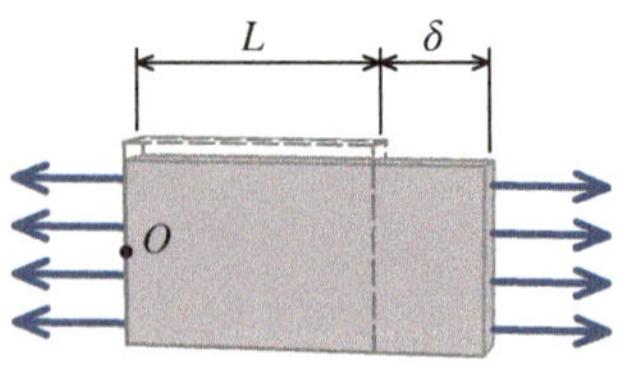

FIGURE 2.2 Normal strain.

$$\varepsilon_{avg} = \frac{\delta}{L} \tag{2.1}$$

Accordingly, a positive value of δ indicates that the axial member gets longer, and a negative value of δ indicates that the axial member gets shorter (termed *contraction*).

A normal strain in an axial member is also termed an **axial strain**.

Normal Strain at a Point

In those cases in which the deformation is nonuniform along the length of the bar (e.g., a long bar hanging under its own weight), the average normal strain given by Equation (2.1) may be significantly different from the normal strain at an arbitrary point O along the bar. The normal strain at a point can be determined by decreasing the length over which the actual deformation is measured. In the limit, a quantity defined as the normal strain at the point $\varepsilon(O)$ is obtained. This limit process is indicated by the expression

$$\varepsilon(O) = \lim_{\Delta L \to 0} \frac{\Delta\delta}{\Delta L} = \frac{d\delta}{dL} \tag{2.2}$$

Strain Units

Equations (2.1) and (2.2) indicate that normal strain is a dimensionless quantity; however, normal strains are frequently expressed in units of in./in., mm/mm, m/m, μin./in., μm/m, or με. The symbol μ in the context of strain is spoken as "micro," and it denotes a factor of 10^{-6}. The conversion from dimensionless quantities such as in./in. or m/m to units of "microstrain" (such as μin./in., μm/m, or με) is

$$\mathbf{1\,\mu\varepsilon = 1 \times 10^{-6}\ in./in. = 1 \times 10^{-6}\ m/m}$$

Since normal strains are small, dimensionless numbers, it is also convenient to express strains in terms of *percent*. For most engineered objects made from metals and alloys, normal strains seldom exceed values of 0.2%, which is equivalent to 0.002 m/m.

Measuring Normal Strains Experimentally

Normal strains can be measured with a simple loop of wire called a **strain gage**. The common strain gage (Figure 2.3) consists of a thin metal-foil grid that is bonded to the surface of a machine part or a structural element. When loads or temperature changes are imposed, the object being tested elongates or contracts, creating normal strains. Since the strain gage is bonded to the object, it undergoes the same strain as the object. As the strain gage elongates or contracts, the electrical resistance of the metal-foil grid changes proportionately. The relationship between strain in the gage and its corresponding change in resistance is predetermined by the strain gage manufacturer through a calibration procedure for each type of gage. Consequently, the precise measurement of changes in resistance in the gage serves as an indirect measure of strain. Strain gages are accurate and extremely sensitive, enabling normal strains as small as 1 με to be measured. Applications involving strain gages will be discussed in more detail in Chapter 13.

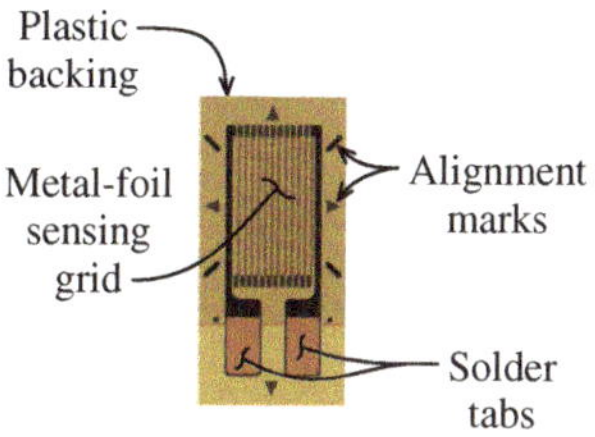

FIGURE 2.3

Sign Conventions for Normal Strains

From the definitions given by Equation (2.1) and Equation (2.2), normal strain is positive when the object elongates and negative when the object contracts. In general, elongation will occur if the axial stress in the object is tension. Therefore, positive normal strains are referred to as *tensile strains*. The opposite will be true for compressive axial stresses; therefore, negative normal strains are referred to as *compressive strains*.

In developing the concept of normal strain through example problems and exercises, it is convenient to use the notion of a **rigid bar**. A rigid bar is meant to represent an object that undergoes no deformation of any kind. Depending on how it is supported, the rigid bar may translate (i.e., move up/down or left/right) or rotate about a support location (see Example 2.1), but it does not bend or deform in any way regardless of the loads acting on it. If a rigid bar is straight before loads are applied, then it will be straight after loads are applied. The bar may translate or rotate, but it will remain straight.

EXAMPLE 2.1

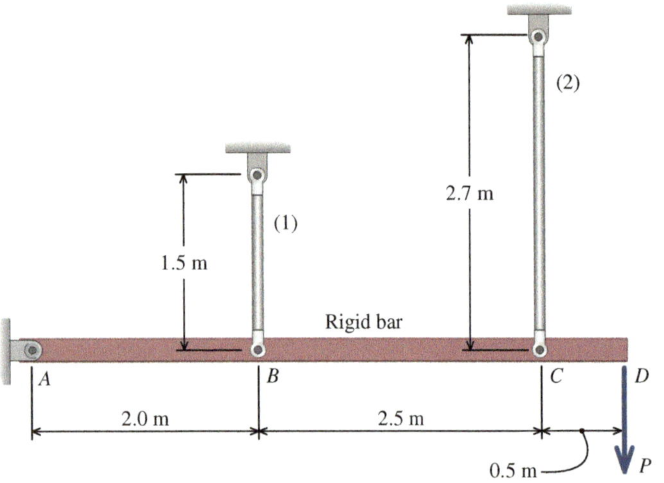

A rigid bar $ABCD$ is pinned at A and supported by two steel rods connected at B and C, as shown. There is no strain in the vertical rods before load P is applied. After load P is applied, the normal strain in rod (2) is 800 με. Determine

(a) the axial normal strain in rod (1).
(b) the axial normal strain in rod (1) if there is a 1 mm gap in the connection between the rigid bar and rod (2) before the load is applied.

Plan the Solution

For this problem, the definition of normal strain will be used to relate strain and elongation for each rod. Since the rigid bar is pinned at A, it will rotate about the support; however, it will remain straight. The deflections at points B, C, and D along the rigid bar can be determined by similar triangles. In part (b), the 1 mm gap will cause an increased deflection in the rigid bar at point C, and this deflection will in turn lead to increased strain in rod (1).

SOLUTION

(a) The normal strain is given for rod (2); therefore, the deformation in that rod can be computed as follows:

$$\varepsilon_2 = \frac{\delta_2}{L_2} \qquad \therefore\ \delta_2 = \varepsilon_2 L_2 = (800\ \mu\varepsilon)\left[\frac{1\text{ mm/mm}}{1{,}000{,}000\ \mu\varepsilon}\right](2{,}700\text{ mm}) = 2.16\text{ mm}$$

Note that the given strain value ε_2 must be converted from units of με into dimensionless units (i.e., mm/mm). Since the strain is positive, rod (2) elongates.

Because rod (2) is connected to the rigid bar and because rod (2) elongates, the rigid bar must deflect 2.16 mm downward at joint C. However, rigid bar $ABCD$ is supported by a pin at joint A, so deflection is prevented at its left end. Therefore, rigid bar $ABCD$ rotates about pin A. Sketch the configuration of the rotated rigid bar, showing the deflection that takes place at C. Sketches of this type are known as **deformation diagrams**.

Although the deflections are very small, they have been greatly exaggerated here for clarity in the sketch. For problems of this type, the small-deflection approximation

$$\sin\theta \approx \tan\theta \approx \theta$$

is used, where θ is the rotation angle of the rigid bar in radians.

To distinguish clearly between elongations that occur in the rods and deflections at locations along the rigid bar, rigid-bar *transverse deflections* (i.e., deflections up or down in this case) will be denoted by the symbol v. Therefore, the rigid-bar deflection at joint C is designated v_C.

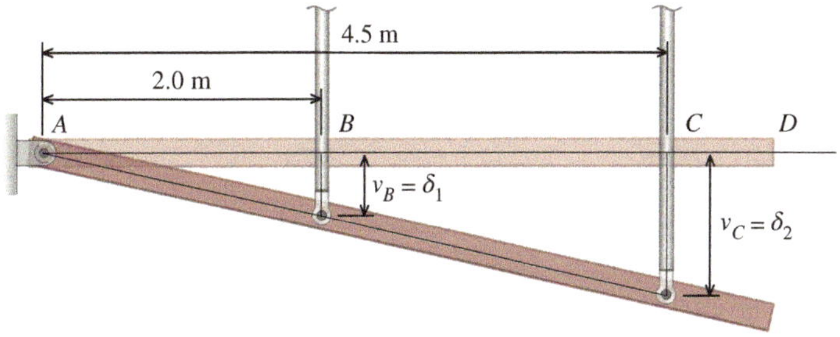

We will assume that there is a perfect fit in the pin connection at joint C; therefore, the rigid-bar deflection at C is equal to the elongation that occurs in rod (2) ($v_C = \delta_2$).

From the deformation diagram of the rigid-bar geometry, the rigid-bar deflection v_B at joint B can be determined from **similar triangles**:

$$\frac{v_B}{2.0\text{ m}} = \frac{v_C}{4.5\text{ m}} \qquad \therefore\ v_B = \frac{2.0\text{ m}}{4.5\text{ m}}(2.16\text{ mm}) = 0.96\text{ mm}$$

If there is a perfect fit in the connection between rod (1) and the rigid bar at joint B, then rod (1) elongates by an amount equal to the rigid-bar deflection at B; hence, $\delta_1 = v_B$. Knowing the deformation produced in rod (1), we can now compute its strain:

$$\varepsilon_1 = \frac{\delta_1}{L_1} = \frac{0.96\text{ mm}}{1{,}500\text{ mm}} = 0.000640\text{ mm/mm} = 640\ \mu\varepsilon \qquad \textbf{Ans.}$$

(b) As in part (a), the deformation in the rod can be computed as

$$\varepsilon_2 = \frac{\delta_2}{L_2} \qquad \therefore\ \delta_2 = \varepsilon_2 L_2 = (800\,\mu\varepsilon)\left[\frac{1\text{ mm/mm}}{1{,}000{,}000\ \mu\varepsilon}\right](2{,}700\text{ mm}) = 2.16\text{ mm}$$

Sketch the configuration of the rotated rigid bar for case (b). In this case, there is a 1 mm gap between rod (2) and the rigid bar at C. Because of this gap, the rigid bar will deflect 1 mm downward at C before it begins to stretch rod (2). The total deflection of C is made up of the 1 mm gap plus the elongation that occurs in rod (2); hence, $v_C = 2.16\text{ mm} + 1\text{ mm} = 3.16\text{ mm}$.

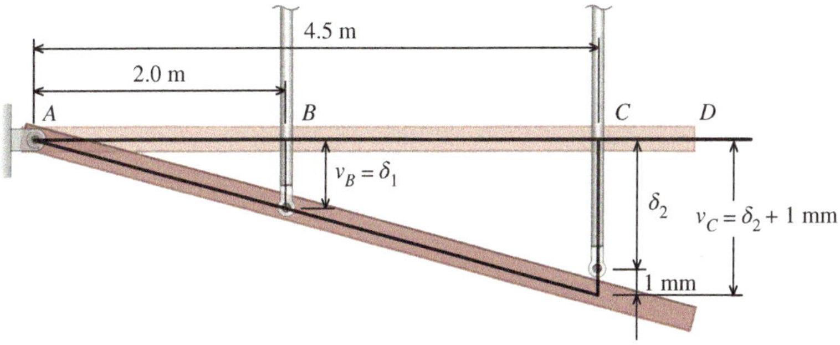

As before, the rigid-bar deflection v_B at joint B can be determined from similar triangles:

$$\frac{v_B}{2.0\text{ m}} = \frac{v_C}{4.5\text{ m}} \qquad \therefore\ v_B = \frac{2.0\text{ m}}{4.5\text{ m}}(3.16\text{ mm}) = 1.404\text{ mm}$$

Since there is a perfect fit in the connection between rod (1) and the rigid bar at joint B, it follows that $\delta_1 = v_B$, and the strain in rod (1) can be computed:

$$\varepsilon_1 = \frac{\delta_1}{L_1} = \frac{1.404\text{ mm}}{1{,}500\text{ mm}} = 0.000936\text{ mm/mm} = 936\ \mu\varepsilon \qquad \textbf{Ans.}$$

Compare the strains in rod (1) for cases (a) and (b). Notice that a very small gap at C caused the strain in rod (1) to increase markedly.

EXAMPLES

M2.1 A rigid steel bar *ABC* is supported by three rods. There is no strain in the rods before load *P* is applied. After load *P* is applied, the axial strain in rod (1) is 1,200 με.

(a) Determine the axial strain in rods (2).

(b) Determine the axial strain in rods (2) if there is a 0.5 mm gap in the connections between rods (2) and the rigid bar before the load is applied.

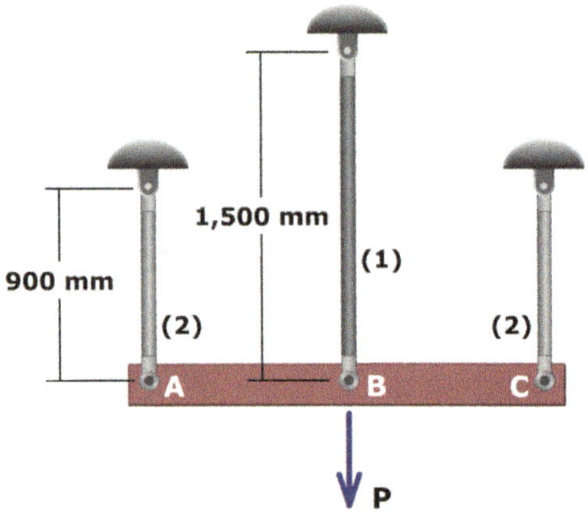

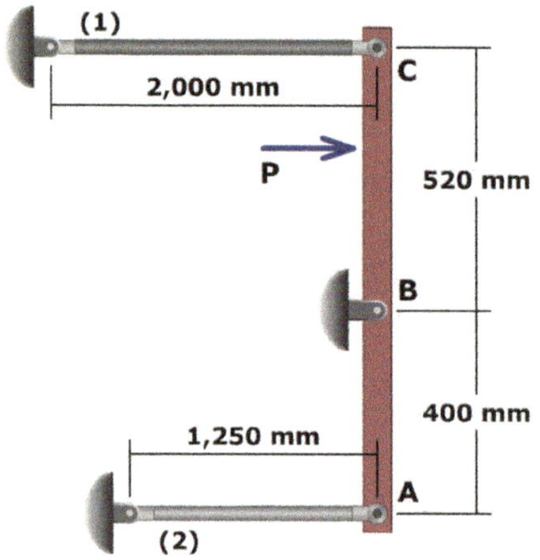

M2.2 A rigid steel bar *ABC* is pinned at *B* and supported by two rods at *A* and *C*. There is no strain in the rods before load *P* is applied. After load *P* is applied, the axial strain in rod (1) is +910 με. Determine the axial strain in rod (2).

M2.4 The load *P* produces an axial strain of −1,800 με in post (2). Determine the axial strain in rod (1).

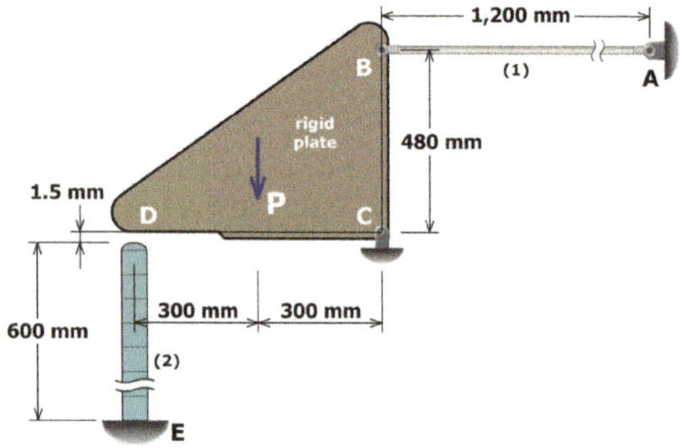

EXERCISES

M2.1 A rigid horizontal bar *ABC* is supported by three vertical rods. There is no strain in the rods before load *P* is applied. After load *P* is applied, the axial strain is a specified value. Determine the deflection of the rigid bar at *B* and the normal strain in rods (2) if there is a specified gap between rod (1) and the rigid bar before the load is applied.

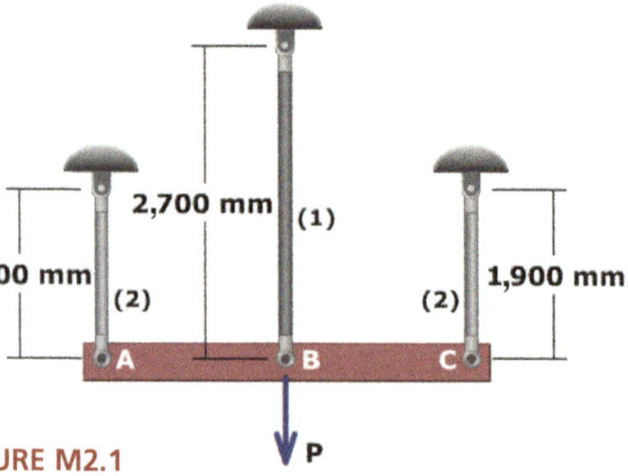

FIGURE M2.1

M2.2 A rigid steel bar AB is pinned at A and supported by two rods. There is no strain in the rods before load P is applied. After load P is applied, the axial strain in rod (1) is a specified value. Determine the axial strain in rod (2) and the downward deflection of the rigid bar at B.

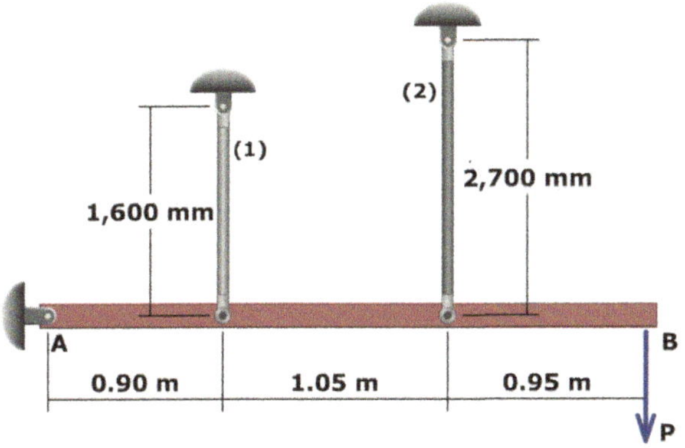

FIGURE M2.2

M2.3 Use normal-strain concepts for four introductory problems using these two structural configurations.

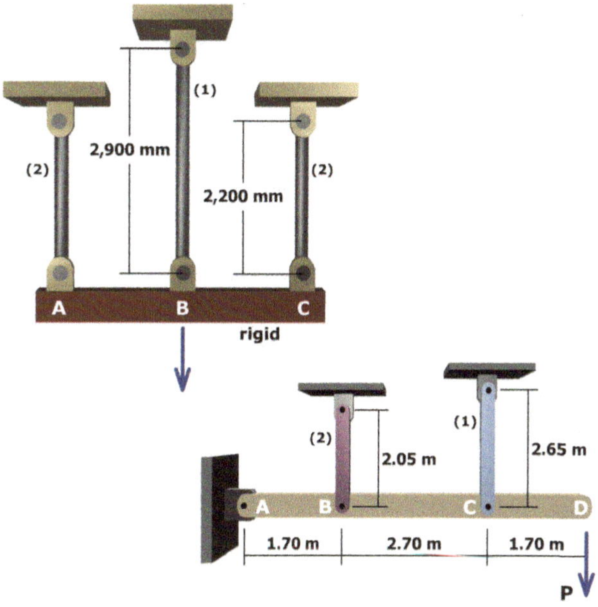

FIGURE M2.3

PROBLEMS

P2.1 When an axial load is applied to the ends of the two-segment rod shown in Figure P2.1, the total elongation between joints A and C is 7.5 mm. The segment lengths are a = 1.2 m and b = 2.8 m. In segment (2), the normal strain is measured as 2,075 μm/m. Determine

(a) the elongation of segment (2).
(b) the normal strain in segment (1) of the rod.

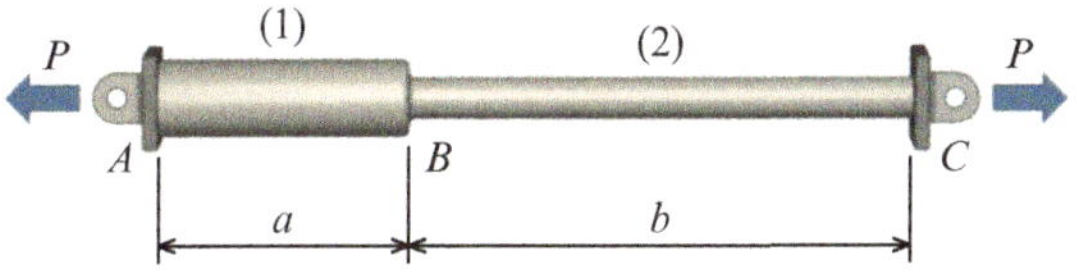

FIGURE P2.1

P2.2 The two bars shown in Figure P2.2 are used to support load P. When unloaded, joint B has coordinates (0, 0). After load P is applied, joint B moves to the coordinate position (–0.55 in., –0.15 in.). Assume that a = 15 ft, b = 27 ft, c = 11 ft, and d = 21 ft. Determine the normal strain in each bar.

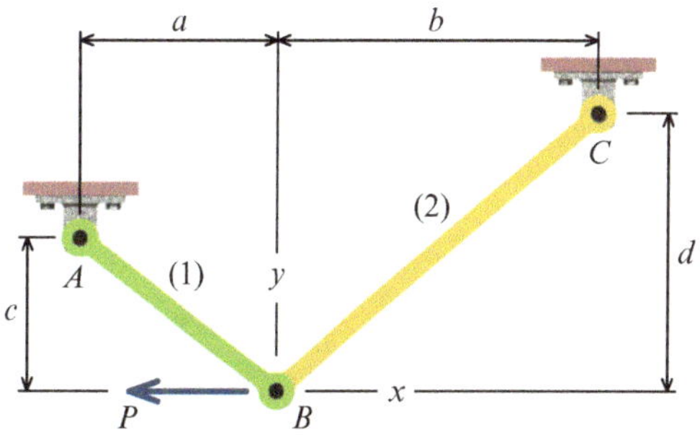

FIGURE P2.2

P2.3 Pin-connected rigid bars AB, BC, and CD are initially held in the positions shown in Figure P2.3 by taut wires (1) and (2). The bar lengths are a = 24 ft and b = 18 ft. Joint C is given a horizontal displacement of 5 in. to the right. (Note that this displacement causes both joints B and C to move to the right and slightly downward.) What is the change in the average normal strain in wire (1) after the displacement?

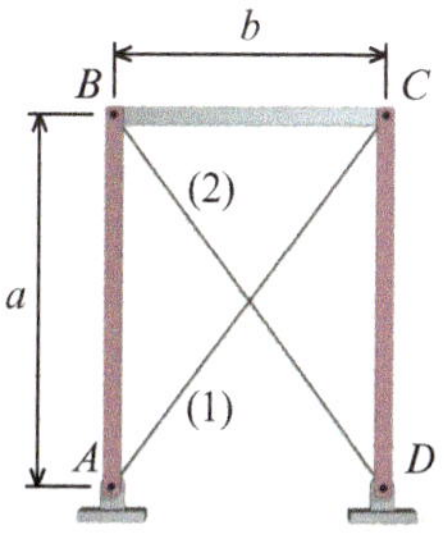

FIGURE P2.3

P2.4 Bar (1) has a length of $L_1 = 2.50$ m, and bar (2) has a length of $L_2 = 0.65$ m. Initially, there is a gap of $\Delta = 3.5$ mm between the rigid plate at B and bar (2). After application of the loads P to the rigid plate at B, the rigid plate moved to the right, stretching bar (1) and compressing bar (2). The normal strain in bar (1) was measured as 2,740 μm/m after the loads P were applied. Determine the normal strain produced in bar (2).

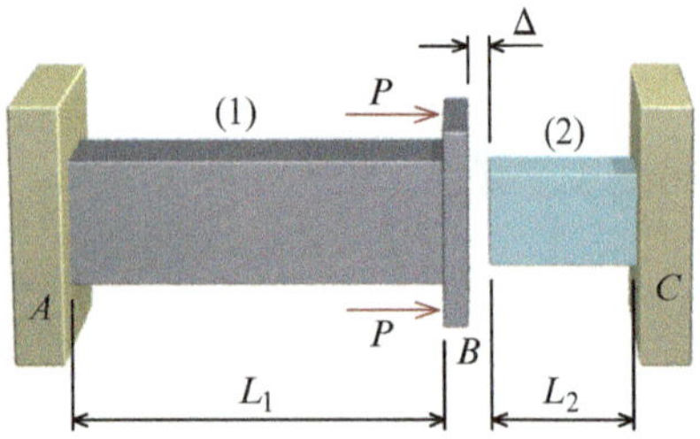

FIGURE P2.4

P2.5 In Figure P2.5, rigid bar ABC is supported by a pin at B and by post (1) at A. However, there is a gap of $\Delta = 10$ mm between the rigid bar at A and post (1). After load P is applied to the rigid bar, point C moves to the left by 8 mm. If the length of post (1) is $L_1 =$ 1.6 m, what is the average normal strain that is produced in post (1)? Use dimensions of $a = 1.25$ m and $b = 0.85$ m.

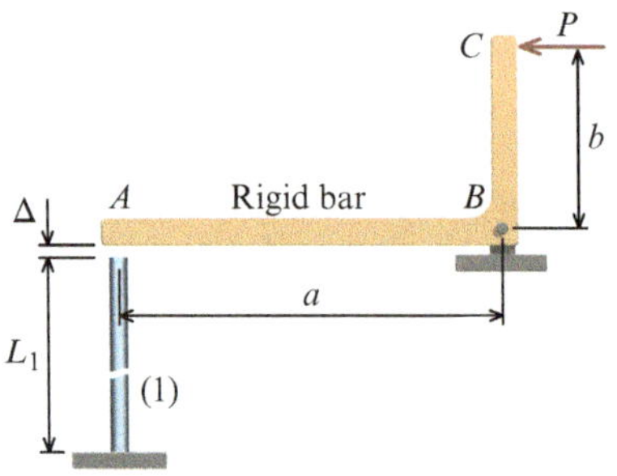

FIGURE P2.5

P2.6 The rigid bar ABC is supported by three bars as shown in Figure P2.6. Bars (1) attached at A and C are identical, each having a length $L_1 = 160$ in. Bar (2) has a length $L_2 = 110$ in.; however, there is a clearance $c = 0.25$ in. between bar (2) and the pin in the rigid bar at B. There is no strain in the bars before load P is applied, and $a = 50$ in. After application of load P, the tensile normal strain in bar (2) is measured as 960 με. What is the normal strain in bars (1)?

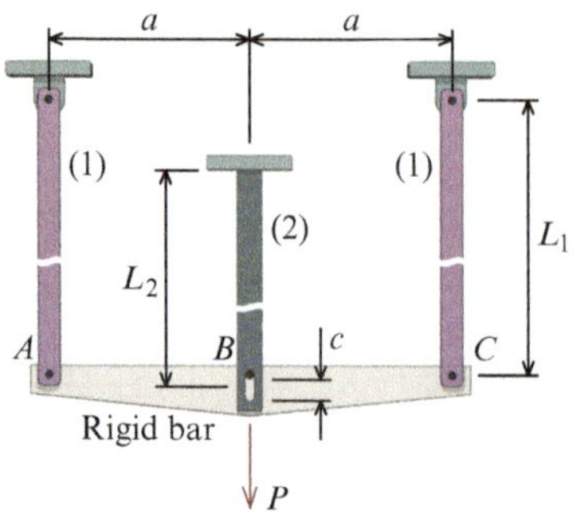

FIGURE P2.6

P2.7 Rigid bar $ABCD$ is supported by two bars as shown in Figure P2.7. There is no strain in the vertical bars before load P is applied. After load P is applied, the normal strain in bar (2) is measured as −3,300 μm/m. Use the dimensions $L_1 = 1{,}600$ mm, $L_2 =$ 1,200 mm, $a = 240$ mm, $b = 420$ mm, and $c = 180$ mm. Determine

(a) the normal strain in bar (1).
(b) the normal strain in bar (1) if there is a 1 mm gap in the connection at pin C before the load is applied.
(c) the normal strain in bar (1) if there is a 1 mm gap in the connection at pin B before the load is applied.

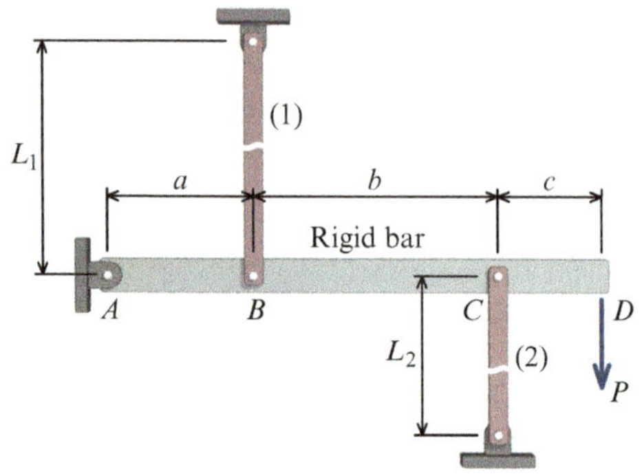

FIGURE P2.7

P2.8 The sanding-drum mandrel shown in Figure P2.8 is made for use with a hand drill. The mandrel is made from a rubberlike material that expands when the nut is tightened to secure the sanding sleeve placed over the outside surface. If the diameter D of the mandrel increases from 2.00 in. to 2.15 in. as the nut is tightened, determine

(a) the average normal strain along a diameter of the mandrel.
(b) the circumferential strain at the outside surface of the mandrel.

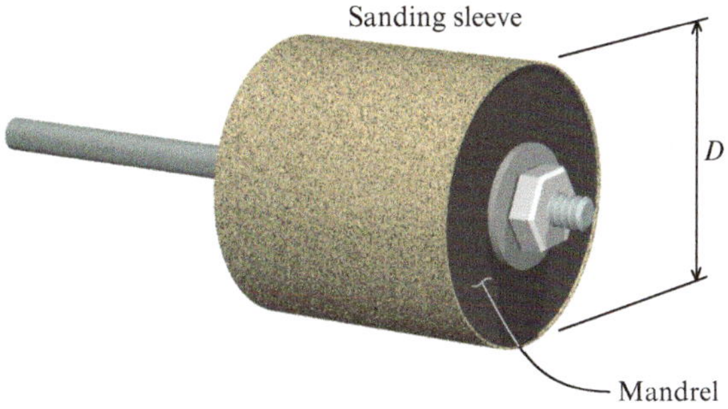

FIGURE P2.8

P2.9 The normal strain in a suspended bar of material of varying cross section due to its own weight is given by the expression $\gamma y/3E$, where γ is the specific weight of the material, y is the distance from the free (i.e., bottom) end of the bar, and E is a material constant. Determine, in terms of γ, L, and E,

(a) the change in length of the bar due to its own weight.
(b) the average normal strain over the length L of the bar.
(c) the maximum normal strain in the bar.

P2.10 A steel cable is used to support an elevator cage at the bottom of a 2,000 ft deep mine shaft. A uniform normal strain of 250 μin./in. is produced in the cable by the weight of the cage. At each point, the weight of the cable produces an additional normal strain that is proportional to the length of the cable below the point. If the total normal strain in the cable at the cable drum (upper end of the cable) is 700 μin./in., determine

(a) the strain in the cable at a depth of 500 ft.
(b) the total elongation of the cable.

2.3 Shear Strain

A deformation involving a change in shape (a distortion) can be used to illustrate a shear strain. An average shear strain γ_{avg} associated with two reference lines that are orthogonal in the undeformed state (two edges of the element shown in Figure 2.4) can be obtained by dividing the shear deformation δ_x (the displacement of the top edge of the element with respect to the bottom edge) by the perpendicular distance L between these two edges. If the deformation is small, meaning that $\sin\gamma \approx \tan\gamma \approx \gamma$ and $\cos\gamma \approx 1$, then shear strain can be defined as

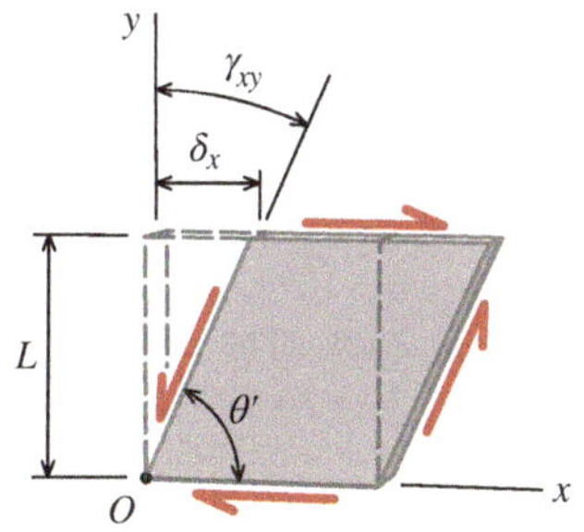

FIGURE 2.4 Shear strain.

$$\gamma_{avg} = \frac{\delta_x}{L} \tag{2.3}$$

For those cases in which the deformation is nonuniform, the shear strain at a point, $\gamma_{xy}(O)$, associated with two orthogonal reference lines x and y is obtained by measuring the shear deformation as the size of the element is made smaller and smaller. In the limit,

$$\gamma_{xy}(O) = \lim_{\Delta L \to 0} \frac{\Delta\delta_x}{\Delta L} = \frac{d\delta_x}{dL} \tag{2.4}$$

Since shear strain is defined as the tangent of the angle of distortion, and since the tangent of that angle is equal to the angle in radians for small angles, an equivalent expression for shear strain that is sometimes useful for calculations is

$$\gamma_{xy}(O) = \frac{\pi}{2} - \theta' \tag{2.5}$$

In this expression, θ' is the angle in the deformed state between two initially orthogonal reference lines.

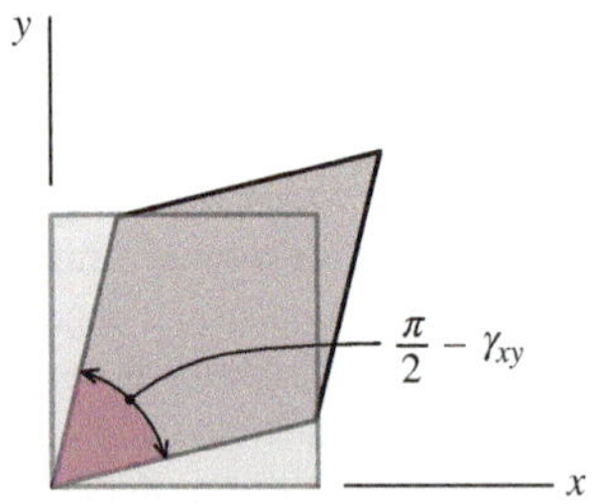

FIGURE 2.5*a* A positive value for the shear strain γ_{xy} means that the angle θ' between the *x* and *y* axes decreases in the deformed object.

Units of Strain

Equations (2.3) through (2.5) indicate that shear strains are dimensionless angular quantities, expressed in radians (rad) or microradians (μrad). The conversion from radians, a dimensionless quantity, to microradians is **1 μrad = 1 × 10^{-6} rad**.

Measuring Shear Strains Experimentally

Shear strain is an angular measure, and it is not possible to directly measure the extremely small angular changes typical of engineered structures. However, shear strain can be determined experimentally by using an array of three strain gages called a **strain rosette**. Strain rosettes will be discussed in more detail in Chapter 13.

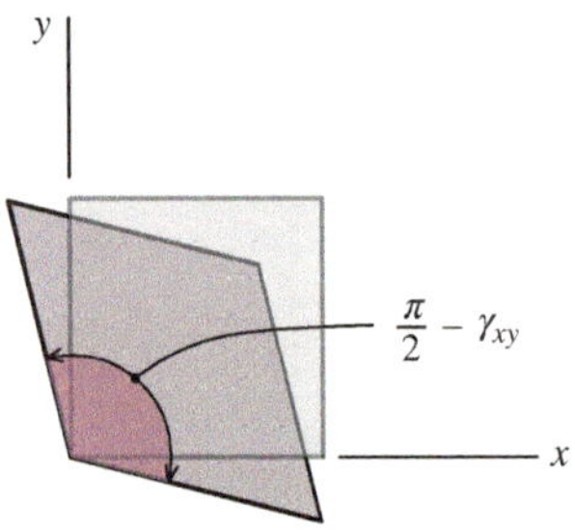

FIGURE 2.5*b* The angle between the *x* and *y* axes increases when the shear strain γ_{xy} has a negative value.

Sign Conventions for Shear Strains

Equation (2.5) shows that shear strains will be positive if the angle θ' between the *x* and *y* axes decreases. If the angle θ' increases, the shear strain is negative. To state this relationship another way, Equation (2.5) can be rearranged to give the angle θ' in the deformed state between two reference lines that are initially 90° apart:

$$\theta' = \frac{\pi}{2} - \gamma_{xy}$$

If the value of γ_{xy} is positive, then the angle θ' in the deformed state will be less than 90° (i.e., less than $\pi/2$ rad) (Figure 2.5*a*). If the value of γ_{xy} is negative, then the angle θ' in the deformed state will be greater than 90° (Figure 2.5*b*). Positive and negative shear strains are not given special or distinctive names.

EXAMPLE 2.2

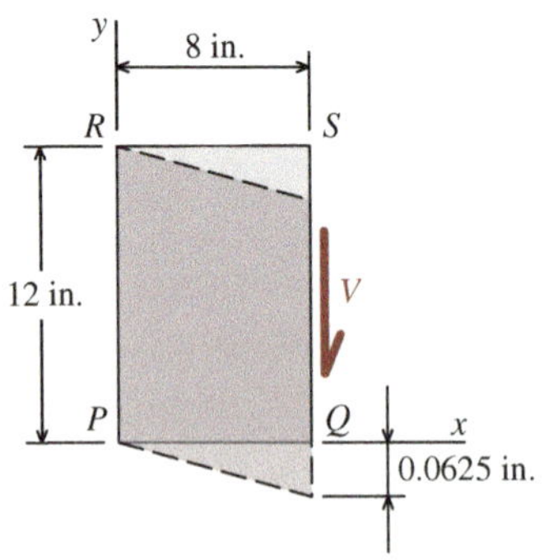

The shear force *V* shown causes side *QS* of the thin rectangular plate to displace downward 0.0625 in. Determine the shear strain γ_{xy} at *P*.

Plan the Solution

Shear strain is an angular measure. Determine the angle between the *x* axis and side *PQ* of the deformed plate.

SOLUTION

Determine the angles created by the 0.0625 in. deformation. **Note:** The small-angle approximation will be used here; therefore, $\sin\gamma \approx \tan\gamma \approx \gamma$, and we have

$$\gamma = \frac{0.0625 \text{ in.}}{8 \text{ in.}} = 0.0078125 \text{ rad}$$

In the undeformed plate, the angle at P is $\pi/2$ rad. After the plate is deformed, the angle at P increases. Since the angle after deformation is equal to $(\pi/2) - \gamma$, the shear strain at P must be a negative value. A simple calculation shows that the shear strain at P is the shear strain at P is

$$\gamma = -0.00781 \text{ rad} \qquad \textbf{Ans.}$$

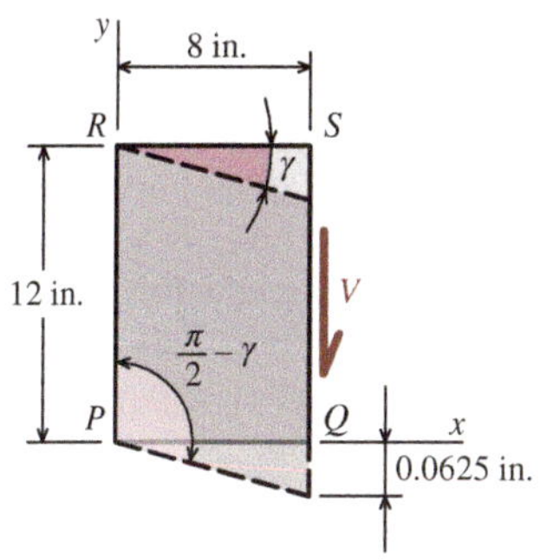

EXAMPLE 2.3

A thin rectangular plate is uniformly deformed as shown. Determine the shear strain γ_{xy} at P.

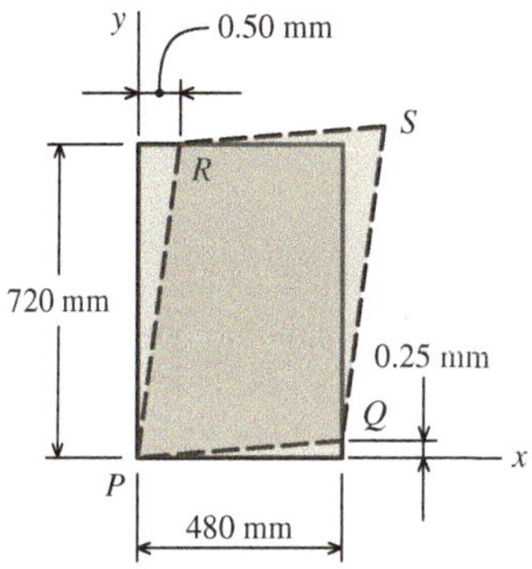

Plan the Solution

Shear strain is an angular measure. Determine the two angles created by the 0.25 mm deflection and the 0.50 mm deflection. Add these two angles together to determine the shear strain at P.

SOLUTION

Determine the angles created by each deformation. **Note:** The small-angle approximation will be used here; therefore, $\sin\gamma \approx \tan\gamma \approx \gamma$. From the given data, we obtain

$$\gamma_1 = \frac{0.50 \text{ mm}}{720 \text{ mm}} = 0.000694 \text{ rad}$$

$$\gamma_2 = \frac{0.25 \text{ mm}}{480 \text{ mm}} = 0.000521 \text{ rad}$$

The shear strain at P is simply the sum of these two angles:

$$\begin{aligned}\gamma &= \gamma_1 + \gamma_2 = 0.000694 \text{ rad} + 0.000521 \text{ rad} = 0.001215 \text{ rad} \\ &= 1{,}215 \text{ μrad} \qquad \textbf{Ans.}\end{aligned}$$

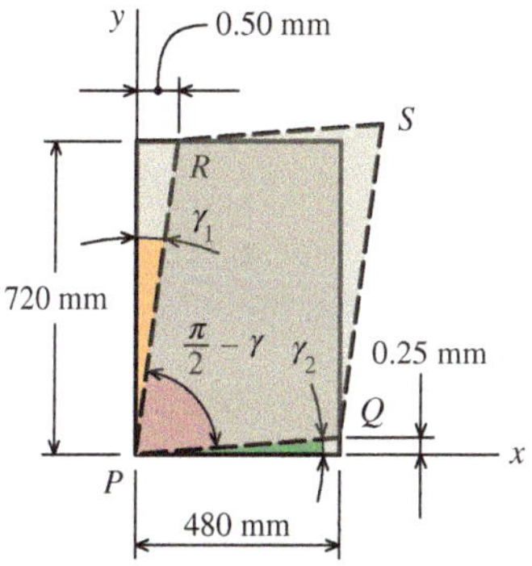

Note: The angle at P in the deformed plate is less than $\pi/2$, as it should be for a positive shear strain. Although not requested in the problem, the shear strain at corners Q and R will be negative and will have the same magnitude as the shear strain at corner P.

MecMovies

EXAMPLE

M2.5 A thin triangular plate is uniformly deformed. Determine the shear strain at P after point P has been displaced 1 mm downward.

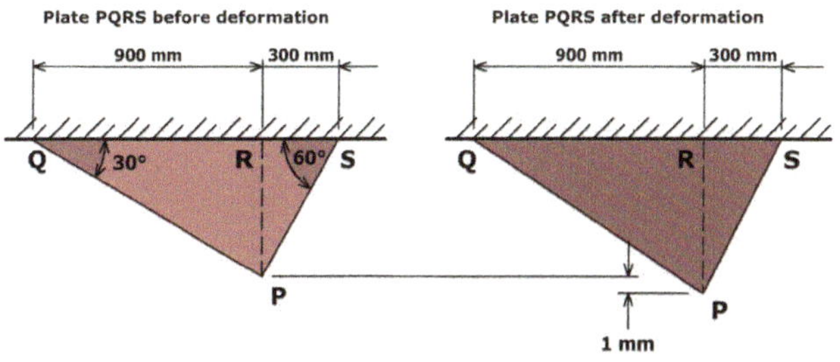

PROBLEMS

P2.11 A thin rectangular polymer plate *PQRS* of width $b = 400$ mm and height $a = 180$ mm is shown in Figure P2.11. The plate is deformed so that corner *Q* is displaced upward by $c = 3.0$ mm and corner *S* is displaced leftward by the same amount. Determine the shear strain at corner *P* after deformation.

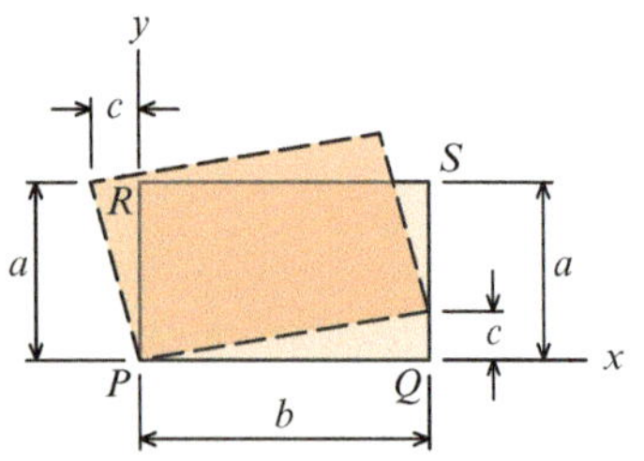

FIGURE P2.11

P2.12 A thin triangular plate *PQR* forms a right angle at point *Q*. During deformation, point *Q* moves to the right by $u = 0.8$ mm and upward by $v = 1.3$ mm to new position *Q*′, as shown in Figure P2.12. Determine the shear strain γ at corner *Q*′ after deformation. Use $a = 225$ mm, $b = 455$ mm, and $d = 319.96$ mm.

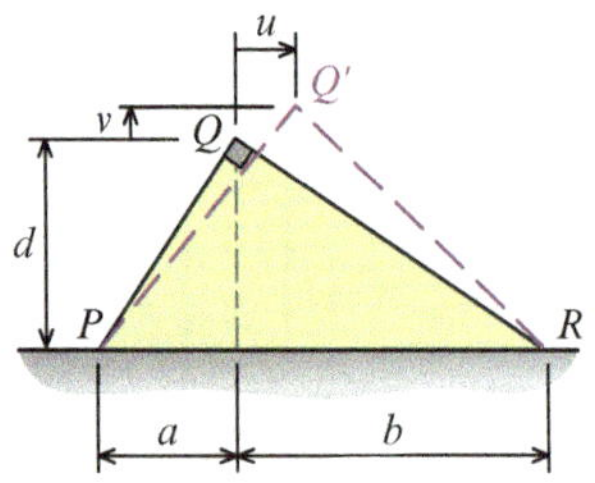

FIGURE P2.12

P2.13 A thin triangular plate *PQR* forms a right angle at point *Q*. During deformation, point *Q* moves to the left by $u = 2.0$ mm and upward by $v = 5.0$ mm to new position *Q*′, as shown in Figure P2.13. Determine the shear strain γ at corner *Q*′ after deformation. Use $c = 700$ mm, $\alpha = 28°$, and $\beta = 62°$.

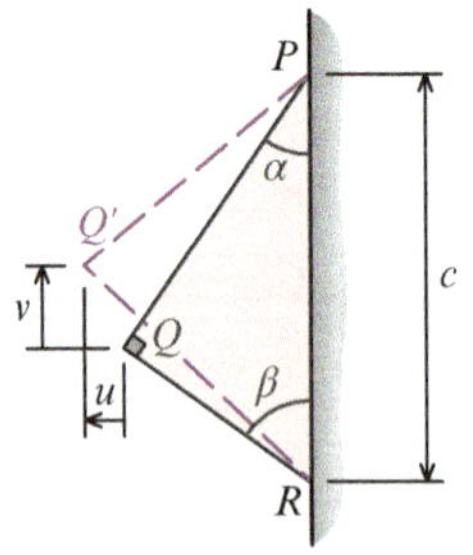

FIGURE P2.13

P2.14 A thin square polymer plate is deformed into the position shown by the dashed lines in Figure P2.14. Assume that $a = 800$ mm, $b = 85$ mm, and $c = 960$ mm. Determine the shear strain γ_{xy} (a) at corner *P* and (b) at corner *Q*, after deformation.

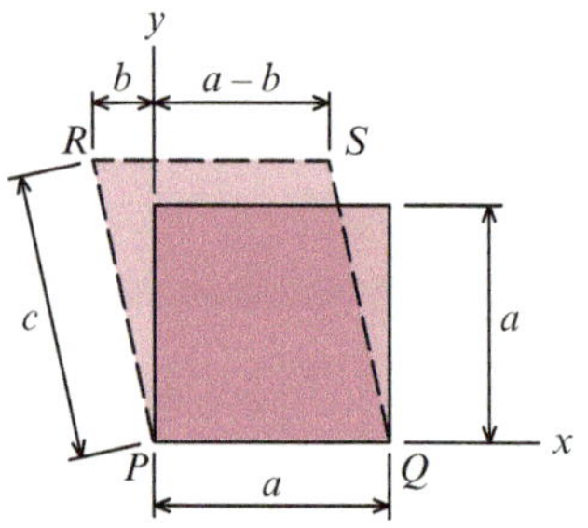

FIGURE P2.14

P2.15 A thin square plate *PQRS* is symmetrically deformed into the shape shown by the dashed lines in Figure P2.15. The initial length of diagonals *PR* and *QS* is $d = 295$ mm. After deformation, diagonal *PR* has a length of $d_{PR} = 295.3$ mm and diagonal *QS* has a length of $d_{QS} = 293.7$ mm. For the deformed plate, determine

(a) the normal strain of diagonal *QS*.
(b) the shear strain γ_{xy} at corner *P*.

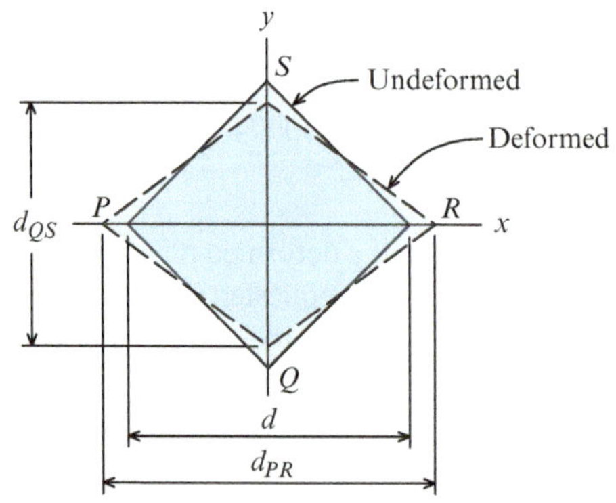

FIGURE P2.15

2.4 Thermal Strain

When unrestrained, most engineering materials expand when heated and contract when cooled. The thermal strain caused by a one-degree (1°) change in temperature is designated by the Greek letter α (alpha) and is known as the **coefficient of thermal expansion**. The strain due to a temperature change of ΔT is

$$\boxed{\varepsilon_T = \alpha \Delta T} \tag{2.6}$$

The coefficient of thermal expansion is approximately constant over a considerable range of temperatures. (In general, the coefficient increases with an increase in temperature.) For a uniform material (termed a **homogeneous material**) that has the same mechanical properties in every direction (termed an **isotropic material**), the coefficient applies to all dimensions (i.e., all directions). Values of the coefficient of expansion for common materials are included in Appendix D.

A material of uniform composition is called a **homogeneous material**. In materials of this type, local variations in composition can be considered negligible for engineering purposes. Furthermore, homogeneous materials cannot be mechanically separated into different materials (the way carbon fibers in a polymer matrix can). Common homogeneous materials are metals, alloys, ceramics, glass, and some types of plastics.

Total Strains

Strains caused by temperature changes and strains caused by applied loads are essentially independent. The total normal strain in a body acted on by both temperature changes and an applied load is given by

$$\boxed{\varepsilon_{\text{total}} = \varepsilon_\sigma + \varepsilon_T} \tag{2.7}$$

An **isotropic material** has the same mechanical properties in all directions.

Since homogeneous, isotropic materials, when unrestrained, expand uniformly in all directions when heated (and contract uniformly when cooled), neither the shape of the body nor the shear stresses and shear strains are affected by temperature changes.

EXAMPLE 2.4

A steel bridge beam has a total length of 150 m. Over the course of a year, the bridge is subjected to temperatures ranging from –40°C to +40°C, and the associated temperature changes cause the beam to expand and contract. Expansion joints between the bridge beam and the supports at the ends of the bridge (called abutments) are installed to allow this change in length to take place without restraint. Determine the change in length that must be accommodated by the expansion joints. Assume that the coefficient of thermal expansion for steel is $11.9 \times 10^{-6}/°\text{C}$.

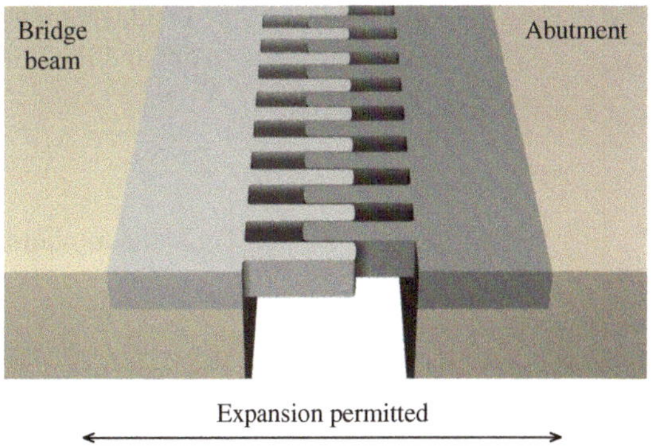

Typical "finger-type" expansion joint for bridges.

Plan the Solution

Determine the thermal strain from Equation (2.6) for the total temperature variation. The change in length is the product of the thermal strain and the beam length.

SOLUTION

The thermal strain for a temperature variation of 80°C (40°C – (–40°C) = 80°C) is

$$\varepsilon_T = \alpha \Delta T = (11.9 \times 10^{-6}/°\text{C})(80°\text{C}) = 0.000952 \text{ m/m}$$

The total change in the beam length is, therefore,

$$\delta_T = \varepsilon L = (0.000952 \text{ m/m})(150 \text{ m}) = 0.1428 \text{ m} = 142.8 \text{ mm} \qquad \textbf{Ans.}$$

Thus, the expansion joint must accommodate at least 142.8 mm of horizontal movement.

EXAMPLE 2.5

Cutting tools such as mills and drills are connected to machining equipment by means of toolholders. The cutting tool must be firmly clamped by the toolholder to achieve precise machining, and shrink-fit toolholders take advantage of thermal expansion properties to achieve this strong, concentric clamping force. To insert a cutting tool, the shrink-fit holder is rapidly heated while the cutting tool remains at room temperature. When the holder has expanded sufficiently, the cutting tool drops into the holder. The holder is then cooled, clamping the cutting tool with a very large force exerted directly on the tool shank.

At 20°C, the cutting tool shank has an outside diameter of 18.000 ± 0.005 mm and the toolholder has an inside diameter of 17.950 ± 0.005 mm. If the tool shank is held at 20°C, what is the minimum temperature to which the toolholder must be heated in order to insert the cutting tool shank? Assume that the coefficient of thermal expansion of the toolholder is $11.9 \times 10^{-6}/°\text{C}$.

Plan the Solution

Use the diameters and tolerances to compute the maximum outside diameter of the shank and the minimum inside diameter of the holder. The difference between these two diameters is the amount of expansion that must occur in the holder. For the tool shank to drop into the holder, the inside diameter of the holder must equal or exceed the shank diameter.

SOLUTION

The maximum outside diameter of the shank is 18.000 + 0.005 mm = 18.005 mm. The minimum inside diameter of the holder is 17.950 – 0.005 mm = 17.945 mm. Therefore, the inside diameter of the holder must be increased by 18.005 – 17.945 mm = 0.060 mm. To expand the holder by this amount requires a temperature increase

$$\delta_T = \alpha \Delta T d = 0.060 \text{ mm} \qquad \therefore \Delta T = \frac{0.060 \text{ mm}}{(11.9 \times 10^{-6}/°\text{C})(17.945 \text{ mm})} = 281°\text{C}$$

Thus, the toolholder must attain a minimum temperature of

$$20°\text{C} + 281°\text{C} = 301°\text{C} \qquad \textbf{Ans.}$$

42

PROBLEMS

P2.16 An airplane has a half-wingspan of 96 ft. Determine the change in length of the aluminum alloy [$\alpha = 13.1 \times 10^{-6}/°F$] wing spar if the plane leaves the ground at a temperature of 59°F and climbs to an altitude where the temperature is −70°F.

P2.17 A square high-density polyethylene [$\alpha = 158 \times 10^{-6}/°C$] plate has a width of 300 mm. A 180 mm diameter circular hole is located at the center of the plate. If the temperature of the plate increases by 40°C, determine

(a) the change in width of the plate.
(b) the change in diameter of the hole.

P2.18 A circular steel [$\alpha = 6.5 \times 10^{-6}/°F$] band is to be mounted on a circular steel drum. The outside diameter of the drum is 50 in. The inside diameter of the circular band is 49.95 in. The band will be heated and then slipped over the drum. After the band cools, it will grip the drum tightly. This process is called *shrink fitting*. If the temperature of the band is 72°F before heating, compute the minimum temperature to which the band must be heated so that it can be slipped over the drum. Assume that an extra 0.05 in. in diameter is needed for clearance so that the band can be easily slipped over the drum. Assume that the drum diameter remains constant.

P2.19 At a temperature of 60°F, a gap of $a = 0.125$ in. exists between the two polymer bars shown in Figure P2.19. Bar (1) has a length $L_1 = 40$ in. and a coefficient of thermal expansion of $\alpha_1 = 47 \times 10^{-6}/°F$. Bar (2) has a length $L_2 = 24$ in. and a coefficient of thermal expansion of $\alpha_2 = 66 \times 10^{-6}/°F$. The supports at A and D are rigid. What is the lowest temperature at which the gap is closed?

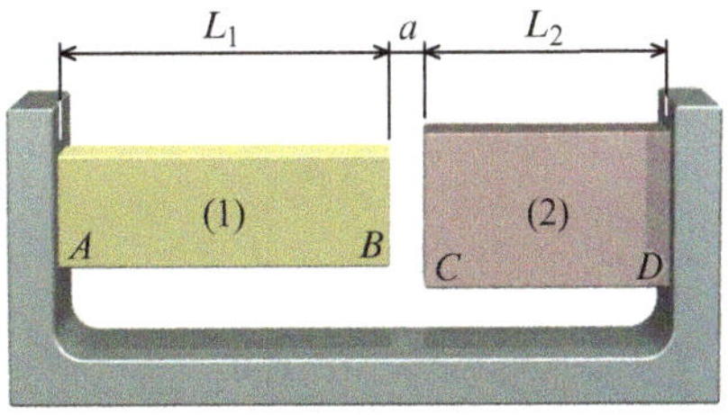

FIGURE P2.19

P2.20 An aluminum pipe has a length of 60 m at a temperature of 10°C. An adjacent steel pipe at the same temperature is 5 mm longer. At what temperature will the aluminum pipe be 15 mm longer than the steel pipe? Assume that the coefficient of thermal expansion of the aluminum is $22.5 \times 10^{-6}/°C$ and that the coefficient of thermal expansion of the steel is $12.5 \times 10^{-6}/°C$.

P2.21 The simple mechanism shown in Figure P2.21 can be calibrated to measure temperature change. Use dimensions of a = 25 mm, b = 90 mm, and L_1 = 180 mm. The coefficient of thermal expansion of member (1) is $23.0 \times 10^{-6}/°C$. Determine the horizontal displacement of pointer tip D for the mechanism shown in response to a temperature increase of 35°C. Assume that pointer BCD is not affected significantly by temperature change.

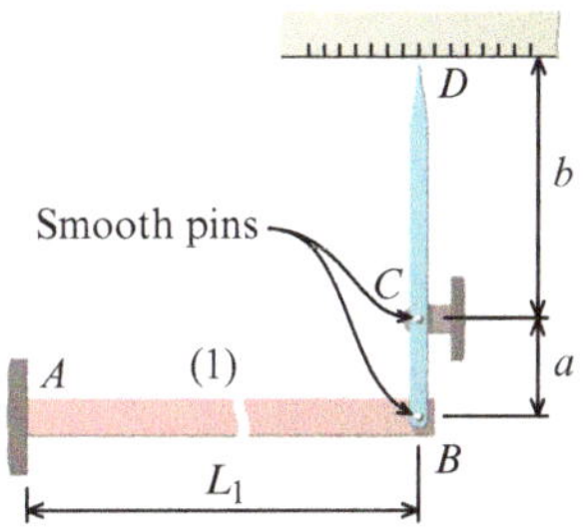

FIGURE P2.21

P2.22 For the assembly shown in Figure P2.22, high-density polyethylene bars (1) and (2) each have coefficients of thermal expansion of $\alpha = 88 \times 10^{-6}/°F$. If the temperature of the assembly is decreased by 50°F from its initial temperature, determine the resulting displacement of pin B. Assume that b = 32 in. and $\theta = 55°$.

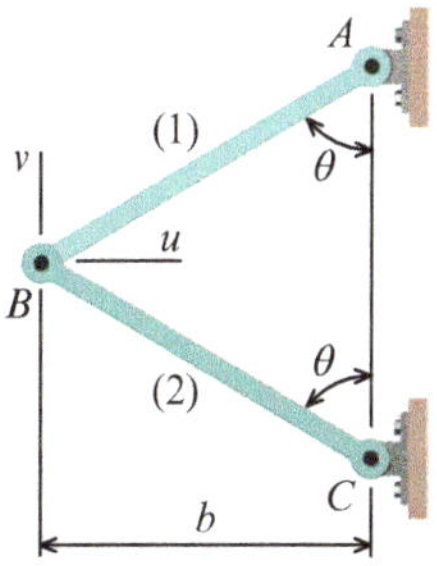

FIGURE P2.22

CHAPTER 3

Mechanical Properties of Materials

3.1 The Tension Test

To design a structural or mechanical component properly, the engineer must understand the characteristics of the component and work within the limitations of the material used in it. Materials such as steel, aluminum, plastic, and wood each respond uniquely to applied loads and stresses. To determine the strength and characteristics of materials such as these requires laboratory testing. One of the simplest and most effective laboratory tests for obtaining engineering design information about a material is called the **tension test**.

The tension test is very simple. A specimen of the material, usually a round rod or a flat bar, is pulled with a controlled tension force. As the force is increased, the elongation of the specimen is measured and recorded. The relationship between the applied load and the resulting deformation can be observed from a plot of the data. This load–deformation plot has limited direct usefulness, however, because it applies only to the specific specimen (meaning the specific diameter or cross-sectional dimensions) used in the test procedure.

A more useful diagram than the load–deformation plot is a plot showing the relationship between stress and strain, called the **stress–strain diagram**. The stress–strain diagram is more useful because it applies to the material in general rather than to the particular specimen used in the test. The information obtained from the stress–strain diagram can be applied to all components, regardless of their dimensions. The load and elongation data obtained in the tension test can be readily converted to stress and strain data.

Tension Test Setup

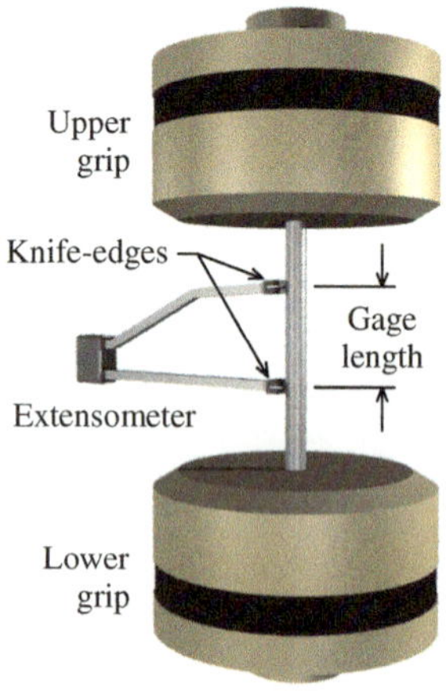

FIGURE 3.1 Tension test setup.

To conduct the tension test, the test specimen is inserted into grips that hold the specimen securely while a tension force is applied by the testing machine (Figure 3.1). Generally, the lower grip remains stationary while the upper grip moves upward, thus creating tension in the specimen.

Several types of grips are commonly used, depending on the specimen being tested. For plain round or flat specimens, wedge-type grips are often used. The wedges are used in pairs that ride in a V-shaped holder. The wedges have teeth that bite into the specimen. The tension force applied to the specimen drives the wedges closer together, increasing the clamping force on the specimen. More sophisticated grips use fluid pressure to actuate the wedges and increase their holding power.

Some tension specimens are machined by cutting threads on the rod ends and reducing the diameter between the threaded ends (Figure 3.2). Threads of this sort are called *upset threads*. Since the rod diameter at the ends is larger than the diameter of the specimen, the presence of the threads does not reduce the strength of the specimen. Tension specimens with upset threads are attached to the testing machine by means of threaded specimen holders, which eliminate any possibility that the specimen will slip or pull out of the grips during the test.

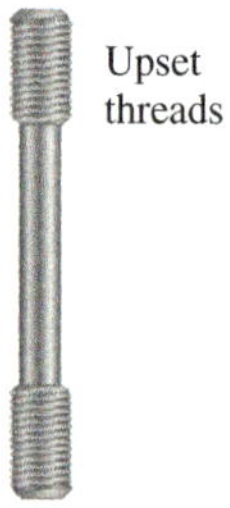

FIGURE 3.2 Tension test specimen with upset threads.

An instrument called an *extensometer* is used to measure the elongation in the tension test specimen. The extensometer has two knife-edges, which are clipped to the test specimen (clips not shown in Figure 3.1). The initial distance between the knife-edges is called the *gage length*. As tension is applied, the extensometer measures the elongation that occurs in the specimen within the gage length. Extensometers are capable of very precise measurements—elongations as small as 0.0001 in. or 0.002 mm. They are available in a range of gage lengths, with the most common models ranging from 0.3 in. to 2 in. (in U.S. units) and from 8 mm to 100 mm (in SI units).

Tension Test Measurements

Several measurements are made before, during, and after the test. Before the test, the cross-sectional area of the specimen must be determined. The area of the specimen will be used with the force data to compute the normal stress. The gage length of the extensometer should also be noted. Normal strain will be computed from the deformation of the specimen (i.e., its axial elongation) and the gage length. During the test, the force applied to the specimen is recorded and the elongation in the specimen between the extensometer knife-edges is measured. After the specimen has broken, the two halves of the specimen are fitted together so that the final gage length and the diameter of the cross section at the fracture location can be measured. The average engineering strain determined from the final and initial gage lengths provides one measure of ductility. The reduction in area (between the area of the fracture surface and the original cross-sectional area) divided by the original cross-sectional area provides a second measure of the ductility of the material. The term **ductility** describes the amount of strain that the material can withstand before fracturing.

MecMovies 3.1 shows an animated tension test.

Tension Test Results. The typical results from a tension test of a ductile metal are shown in Figure 3.3. Several characteristic features are commonly found on the load–deformation plot. As the load is applied, there is a range in which the deformation is linearly related to the load (1). At some load, the load–deformation plot will begin to curve and there will be noticeably larger deformations in response to relatively small increases in load (2). As the

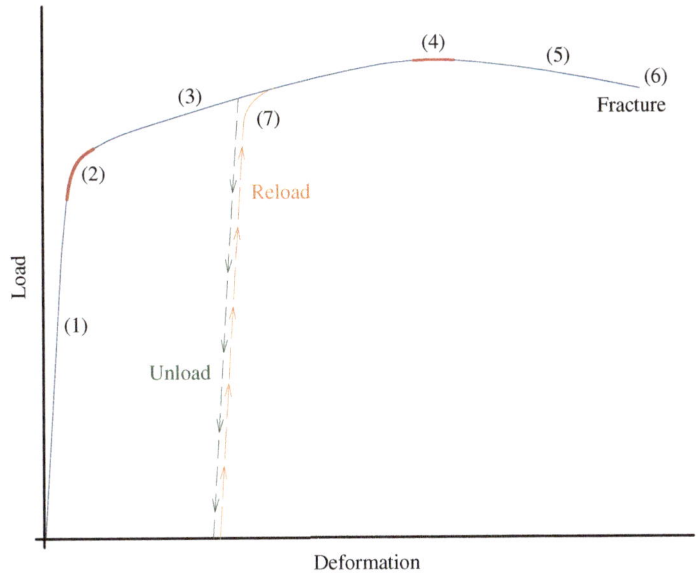

FIGURE 3.3 Load–deformation plot from tension test.

load is continually increased, stretching in the specimen will be obvious (3). At some point, a maximum load intensity will be reached (4). Immediately following this peak, the specimen will begin to narrow and elongate markedly at one specific location, causing the load acting in the specimen to decrease (5). Shortly thereafter, the specimen will fracture (6), breaking into two pieces at the narrowest cross section.

Another interesting characteristic of materials, particularly metals, can be observed if the test is interrupted at a point beyond the linear region. For the test depicted in Figure 3.3, the specimen was loaded into region (3) and then the load was removed. In that case, the specimen does not unload along the original loading curve. Rather, it unloads along a path that is parallel to the initial linear plot (1). Then, when the load is completely removed, the deformation of the specimen is not zero, as it was at the outset of the test. In other words, the specimen has been permanently and irreversibly deformed. When the test resumes and the load is increased, the reloading path follows the unloading path exactly. As it approaches the original load–deformation plot, the reloading plot begins to curve (7) in a fashion similar to region (2) on the original plot. However, the load at which the reloading plot markedly turns (7) is larger than it was in the original loading (2). The process of unloading and reloading has strengthened the material so that it can withstand a larger load before it becomes distinctly nonlinear. The unload–reload behavior seen here is a very useful characteristic, particularly for metals. One technique for increasing the strength of a material is a process of stretching and relaxing called **work hardening**.

Preparing the Stress–Strain Diagram. The load–deformation data that are obtained in the tension test provide information about only one specific size of specimen. The test results are more useful if they are generalized into a stress–strain diagram. To construct a stress–strain diagram from tension test results,

(a) divide the specimen elongation data by the extensometer gage length to obtain the normal strain,

(b) divide the load data by the initial specimen cross-sectional area to obtain the normal stress, and

(c) plot strain on the horizontal axis and stress on the vertical axis.

3.2 The Stress–Strain Diagram

Typical stress–strain diagrams for an aluminum alloy and a low-carbon steel are shown in Figure 3.4. Material properties essential for engineering design are obtained from the stress–strain diagram. These stress–strain diagrams will be examined to determine several important properties, including the proportional limit, the elastic modulus, the yield strength, and the ultimate strength. The difference between engineering stress and true stress will be discussed, and the concept of ductility in metals will be introduced.

MecMovies 3.1 shows an animated discussion of stress–strain diagrams.

Proportional Limit

The **proportional limit** is the stress at which the stress–strain plot is no longer linear. Strains in the linear portion of the stress–strain diagram typically represent only a small fraction of the total strain at fracture. Consequently, it is necessary to enlarge the scale to observe the linear portion of the curve clearly. The linear region of the aluminum alloy stress–strain diagram is enlarged in Figure 3.5. A best-fit line is plotted through the stress–strain data points. The stress at which the stress–strain data begin to curve away from this line is called the proportional limit. The proportional limit for this material is approximately 43.5 ksi.

Recall the unload–reload behavior shown in Figure 3.3. As long as the stress in the material remains below the proportional limit, no permanent damage will be caused during loading and unloading. In an engineering context, this property means that a component can be loaded and unloaded many, many times and it will still behave "just like new." The property is called **elasticity**, and it means that a material returns to its original dimensions during unloading. The material itself is said to be **elastic** in the linear region.

Most engineered components are designed to function elastically to avoid permanent deformations that occur after the proportional limit is exceeded. In addition, the size and shape of an object are not significantly changed if strains and deformations are kept small. This property can be a particularly important consideration for mechanisms and machines, which consist of many parts that must fit together to operate properly.

Elastic Modulus

Most components are designed to function elastically. Consequently, the relationship between stress and strain in the initial, linear region of the stress–strain diagram is of particular interest regarding engineering materials. In 1807, Thomas Young proposed characterizing the material's behavior in the elastic region by the ratio between normal stress

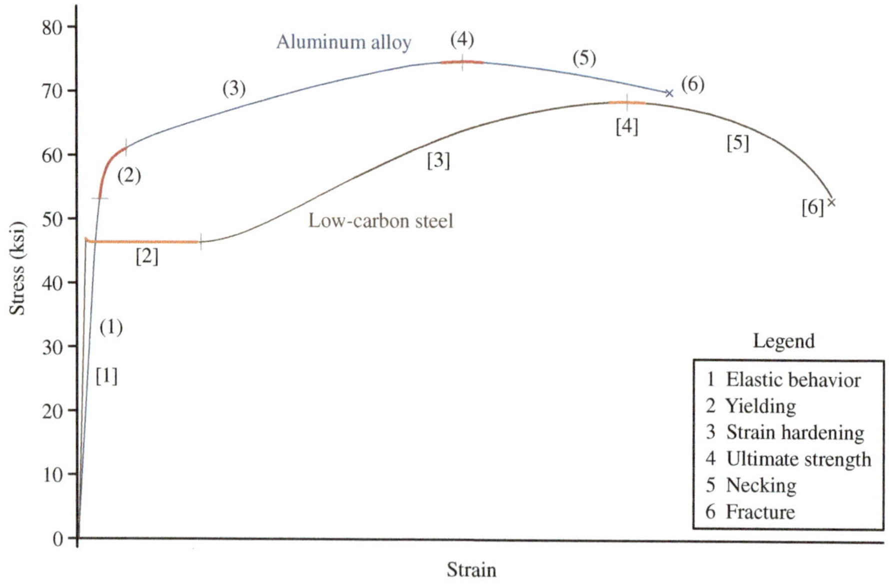

FIGURE 3.4 Typical stress–strain diagrams for two common metals.

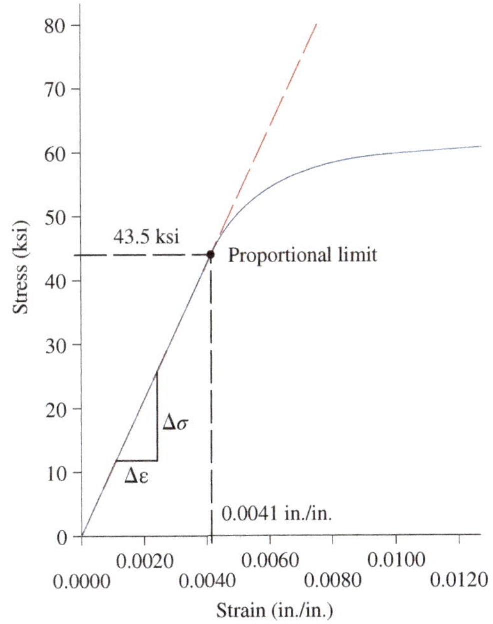

FIGURE 3.5 Proportional limit.

and normal strain. This ratio is the slope of the initial straight-line portion of the stress–strain diagram. It is called **Young's modulus**, the **elastic modulus**, or the **modulus of elasticity**, and it is denoted by the symbol E:

$$E = \frac{\Delta\sigma}{\Delta\varepsilon} \tag{3.1}$$

The elastic modulus E is a measure of the material's *stiffness*. In contrast to strength measures, which predict how much load a component can withstand, a stiffness measure such as the elastic modulus E is important because it defines how much stretching, compressing, bending, or deflecting will occur in a component in response to the loads acting on it.

In any experimental procedure, there is some amount of error associated with making a measurement. To minimize the effect of this measurement error on the computed elastic modulus value, it is better to use widely separated data points to calculate E. In the linear portion of the stress–strain diagram, the two most widely spaced data points are the proportional limit point and the origin. Using the proportional limit and the origin, we would compute the elastic modulus E as

$$E = \frac{43.5 \text{ ksi}}{0.0041 \text{ in./in.}} = 10{,}610 \text{ ksi} \tag{3.2}$$

In practice, the best value for the elastic modulus E is obtained from a least-squares fit of a line to the data between the origin and the proportional limit. Using a least-squares analysis, we find that the elastic modulus for this material is $E = 10{,}750$ ksi.

Work Hardening

The effect of unloading and reloading on the load–deformation plot was shown in Figure 3.3. The effect of unloading and reloading on the stress–strain diagram is shown

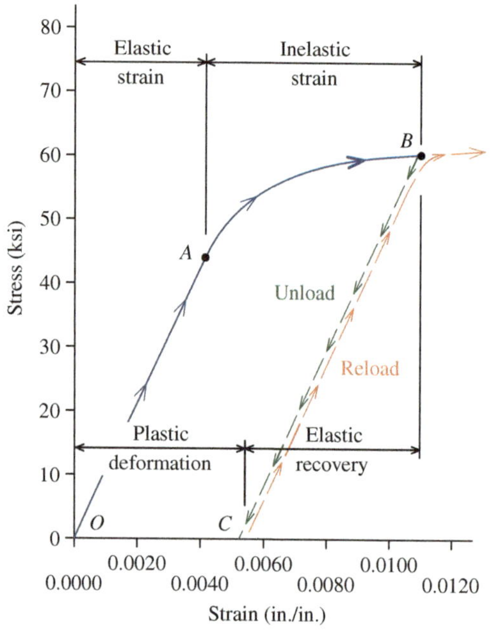

FIGURE 3.6 Work hardening.

in Figure 3.6. Suppose that the stress acting on a material is increased above the proportional limit stress to point *B*. The strain between the origin *O* and the proportional limit *A* is termed **elastic strain**. This strain will be fully recovered after the stress is removed from the material. The strain between the points *A* and *B* is termed **inelastic strain**. When the stress is removed (i.e., unloaded), only a portion of this strain will be recovered. As stress is removed from the material, it unloads on a path parallel to the elastic modulus line—that is, parallel to path *OA*. A portion of the strain at *B* is recovered elastically. However, a portion of the strain remains in the material permanently. This strain is referred to as **residual strain** or **permanent set** or **plastic deformation**. As stress is reapplied, the material reloads along path *CB*. Upon reaching point *B*, the material will resume following the original stress–strain curve. The proportional limit after reloading becomes the stress at point *B*, which is greater than the proportional limit for the original loading (i.e., point *A*). This phenomenon is called **work hardening**, because it has the effect of increasing the proportional limit for the material.

In general, a material acting in the linear portion of the stress–strain curve is said to exhibit **elastic behavior**. Strains in the material are temporary, meaning that all strain is recovered when the stress on the material is removed. Beyond the elastic region, a material is said to exhibit **plastic behavior**. Although some strain in the plastic region is temporary and can be recovered upon removal of the stress, a portion of the strain in the material is permanent. The permanent strain is termed **plastic deformation**.

Elastic Limit

Most engineered components are designed to act elastically, meaning that when loads are released, the component will return to its original, undeformed configuration. For proper design, therefore, it is important to define the stress at which the material will no longer behave elastically. With most materials, there is a gradual transition from elastic to plastic behavior, and the point at which plastic deformation begins is difficult to define with precision. One measure that has been used to establish this threshold is termed the elastic limit.

The **elastic limit** is the largest stress that a material can withstand without any measurable permanent strain remaining after complete release of the stress. The procedure required to determine the elastic limit involves cycles of loading and unloading, each time incrementally increasing the applied stress (Figure 3.7). For instance, stress is increased to point A and then removed, with the strain returning to the origin O. This process is repeated for points B, C, D, and E. In each instance, the strain returns to the origin O upon unloading. Eventually, a stress will be reached (point F) such that not all of the strain will be recovered during unloading (point G). The elastic limit is the stress at point F.

How does the elastic limit differ from the proportional limit? Although such materials are not common in engineered applications, a material can be elastic even though its stress–strain relationship is nonlinear. For a nonlinear elastic material, the elastic limit could be substantially greater than the proportional limit stress. Nevertheless, the proportional limit is generally favored in practice since the procedure required to establish the elastic limit is tedious.

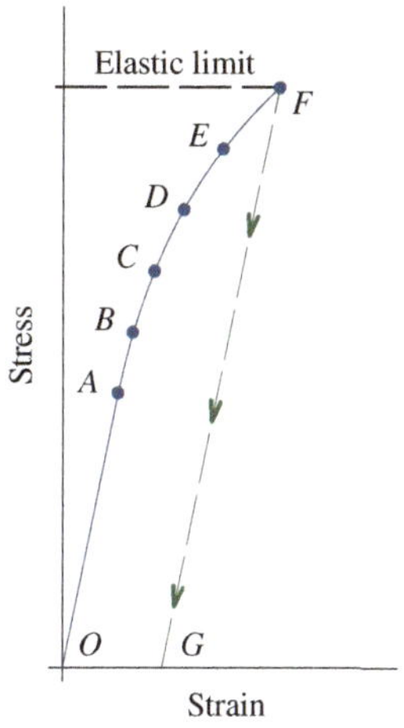

FIGURE 3.7 Elastic limit.

Yielding

For many common materials (such as the low-carbon steel shown in Figure 3.4 and enlarged in Figure 3.8), the elastic limit is indistinguishable from the proportional limit. Past the elastic limit, relatively large deformations will occur for small or almost negligible increases in stress. This behavior is termed **yielding**.

A material that behaves in the manner depicted in Figure 3.8 is said to have a **yield point**. The yield point is the stress at which there is an appreciable increase in strain with no increase in stress. Low-carbon steel, in fact, has two yield points. Upon reaching the upper yield point, the stress drops abruptly to a sustained lower yield point. When a material yields without an increase in stress, the material is often referred to as being **perfectly plastic**. Materials having a stress–strain diagram similar to Figure 3.8 are termed **elastoplastic**.

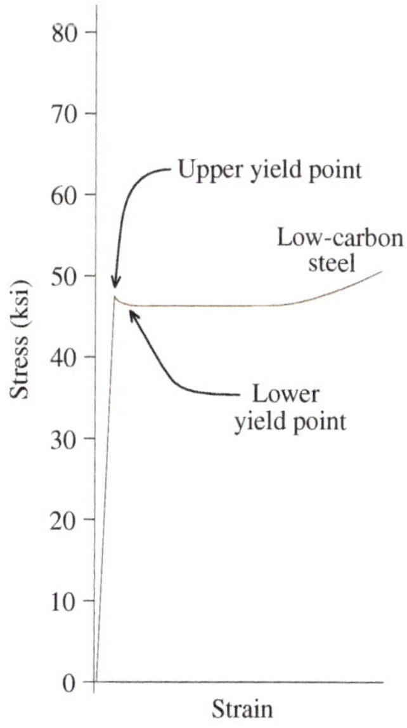

FIGURE 3.8 Yield point for low-carbon steel.

Not every material has a yield point. Materials such as the aluminum alloy shown in Figure 3.4 do not have a clearly defined yield point. While the proportional limit marks the uppermost end of the linear portion of the stress–strain curve, it is sometimes difficult in practice to determine the proportional limit stress, particularly for materials with a gradual transition from a straight line to a curve. For such materials, a yield strength is defined. The **yield strength** is the stress that will induce a specified permanent set (i.e., plastic deformation) in the material, usually 0.05% or 0.2%. (**Note:** A permanent set of 0.2% is another way of expressing a strain value of 0.002 in./in., or 0.002 mm/mm.) To determine the yield strength from the stress–strain diagram, mark a point on the strain axis at the specified permanent set (Figure 3.9). Through this point, draw a line that is parallel to the initial elastic modulus line. The stress at which the offset line intersects the stress–strain diagram is termed the yield strength.

Strain Hardening and Ultimate Strength

After yielding has taken place, most materials can withstand additional stress before fracturing. The stress–strain curve rises continuously toward a peak stress value, which is termed the **ultimate strength**. The ultimate strength may also be called the tensile strength or the ultimate tensile strength (UTS). The rise in the curve is called **strain hardening**. The strain-hardening regions and the ultimate strength points for a low-carbon steel and an aluminum alloy are indicated on the stress–strain diagrams in Figure 3.4.

Necking

In the yield and strain-hardening regions, the cross-sectional area of the specimen decreases uniformly and permanently. Once the specimen reaches the ultimate strength, however, the change in the specimen cross-sectional area is no longer uniform throughout the gage length. The cross-sectional area begins to decrease in a localized

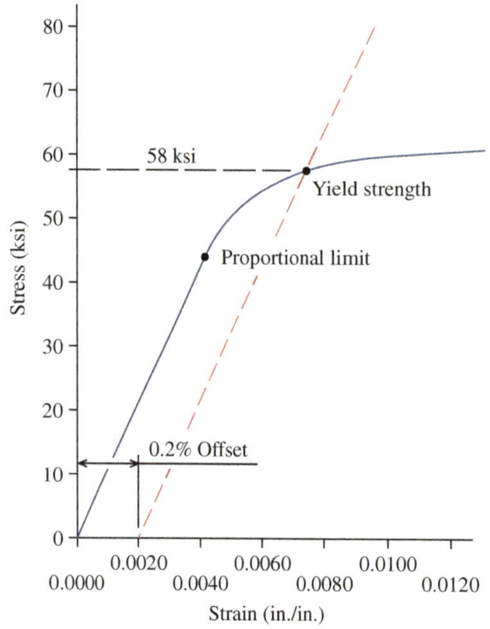

FIGURE 3.9 Yield strength determined by offset method.

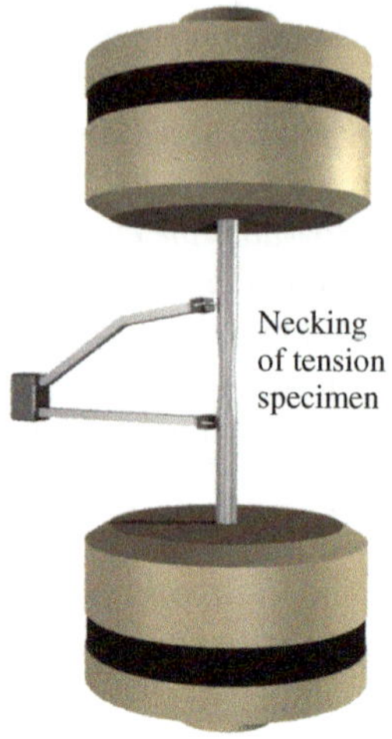

FIGURE 3.10 Necking in a tension specimen.

region of the specimen, forming a contraction, or "neck." This behavior is referred to as **necking** (Figure 3.10 and Figure 3.11). Necking occurs in ductile materials, but not in brittle materials. (See discussion of ductility, to follow.)

Fracture

Many ductile materials break in what is termed a cup-and-cone fracture (Figure 3.12). In the region of maximum necking, a circular fracture surface forms at an angle of roughly 45° with respect to the tensile axis. This failure surface appears as a cup on one portion of the broken specimen and as a cone on the other portion. In contrast, brittle materials often fracture on a flat surface that is oriented perpendicular to the tensile axis. The stress at which the specimen breaks into two pieces is called the **fracture stress**. Examine the relationship between the ultimate strength and the fracture stress in Figure 3.4. *Does it seem odd that the fracture stress is less than the ultimate strength?* If the specimen did not

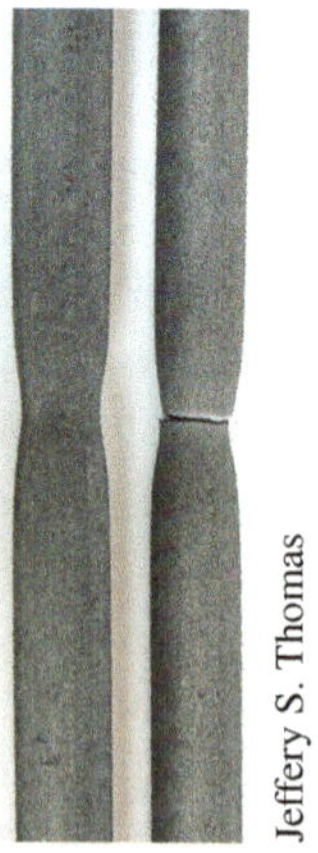

FIGURE 3.11 Necking in a ductile metal specimen.

FIGURE 3.12 Cup-and-cone failure surfaces.

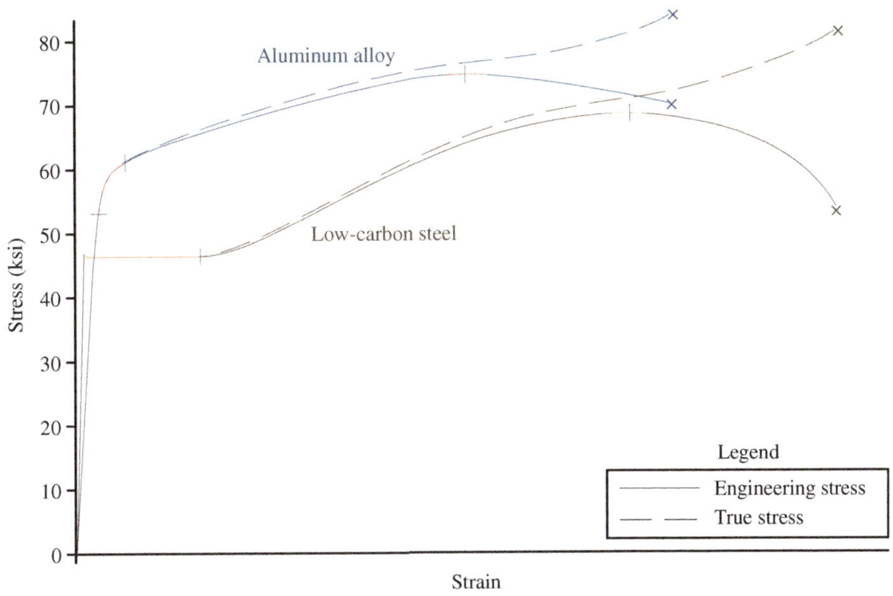

FIGURE 3.13 True stress versus engineering stress.

break at the ultimate strength, why would it break at a lower stress? Recall that the normal stress in the specimen was computed by dividing the specimen load by the original cross-sectional area. This method of calculating stresses is known as **engineering stress**. Engineering stress does not take into account any changes in the specimen's cross-sectional area during application of the load. After the ultimate strength is reached, the specimen starts to neck. As contraction within the localized neck region grows more pronounced, the cross-sectional area decreases continually. The engineering stress calculations, however, are based on the original specimen cross-sectional area. Consequently, the engineering stress computed at fracture and shown on the stress–strain diagram is not an accurate reflection of the **true stress** in the material. If one were to measure the diameter of the specimen during the tension test and compute the true stress according to the reduced diameter, one would find that the true stress continues to increase above the ultimate strength (Figure 3.13).

Ductility

Strength and stiffness are not the only properties of interest to a design engineer. Another important property is ductility. **Ductility** describes the material's capacity for plastic deformation.

A material that can withstand large strains before fracture is called a **ductile material**. Materials that exhibit little or no yielding before fracture are called **brittle materials**. Ductility is not necessarily related to strength. Two materials could have exactly the same strength, but very different strains at fracture (Figure 3.14).

Often, increased material strength is achieved at the cost of reduced ductility. In Figure 3.15, stress–strain curves for four different types of steel are compared. All four curves branch from the same elastic modulus line; therefore, each of the steels has the same stiffness. The steels range from a brittle steel (1) to a ductile steel (4). Steel (1) represents a hard tool steel, which exhibits no plastic deformation before fracture. Steel (4) is typical of low-carbon steel, which exhibits extensive plastic deformation before fracture. Of these steels, steel (1) is the strongest, but also the least ductile. Steel (4) is the weakest, but also the most ductile.

For the engineer, ductility is important in that it indicates the extent to which a metal can be deformed without fracture in metalworking operations such as bending, rolling, forming, drawing, and extruding. In fabricated structures and machine components, ductility also gives an indication of the material's ability to deform at holes, notches, fillets, grooves, and other

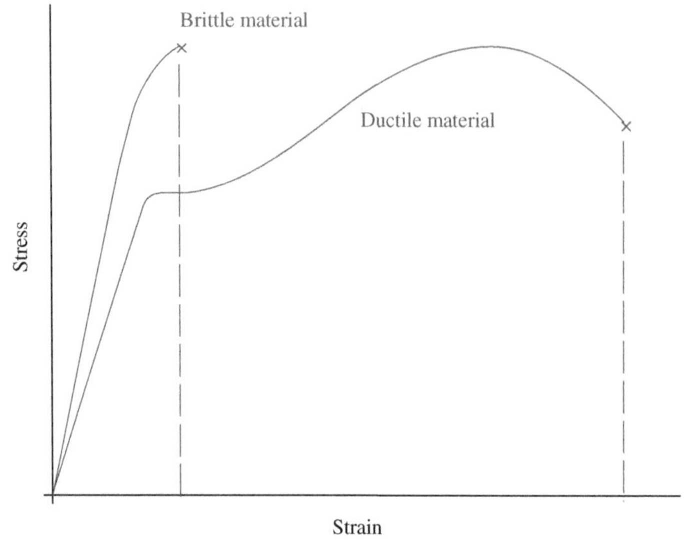

FIGURE 3.14 Ductile versus brittle materials.

discontinuities that cause stresses to intensify locally. Plastic deformation in a ductile material allows stress to flow to a larger region around discontinuities. This redistribution of stress minimizes peak stress magnitudes and helps to prevent fracture in the component. Since ductile materials stretch greatly before fracturing, excessive component deformations in buildings, bridges, and other structures can warn of impending failure, providing opportunities for safe exit from the structure and allowing for repairs. Brittle materials exhibit sudden failure with little or no warning. Ductile materials also give the structure some capacity to absorb and redistribute the effects of extreme load events such as earthquakes.

Ductility Measures. Two measures of ductility are obtained from the tension test. The first is the engineering strain at fracture. To determine this measure, the two halves of the broken specimen are fitted together, the final gage length is measured, and then the average strain is calculated from the initial and final gage lengths. This value is usually expressed as a percentage, and it is referred to as the **percent elongation**.

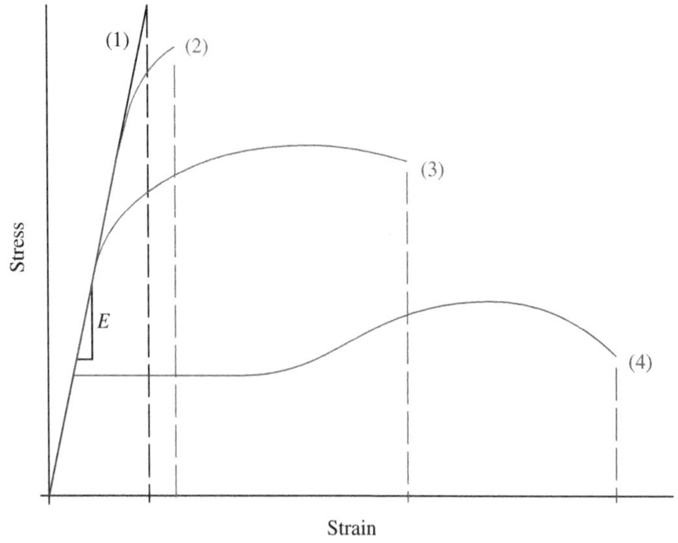

FIGURE 3.15 Trade-off between strength and ductility for steels.

The second measure is the reduction in area at the fracture surface. This value is also expressed as a percentage and is referred to as the **percent reduction of area**. It is calculated as

$$\text{Percent reduction of area} = \frac{A_0 - A_f}{A_0}(100\%) \tag{3.3}$$

where A_0 = original cross-sectional area of the specimen and A_f = cross-sectional area on the fracture surface of the specimen.

Review of Significant Features

The stress–strain diagram provides essential engineering design information that is applicable to components of any shape or size. While each material has its particular characteristics, several important features are found on stress–strain diagrams for materials commonly used in engineering applications. These features are summarized in Figure 3.16.

Strain hardening

- As the material stretches, it can withstand increasing amounts of stress.

Ultimate strength

- According to the engineering definition of stress, the ultimate strength is the largest stress that the material can withstand.

Yield

- A slight increase in stress causes a marked increase in strain.
- Beginning at yield, the material is permanently altered. Only a portion of the strain will be recovered after the stress has been removed.
- Strains are termed inelastic since only a portion of the strain will be recovered upon removal of the stress.
- The yield strength is an important design parameter for the material.

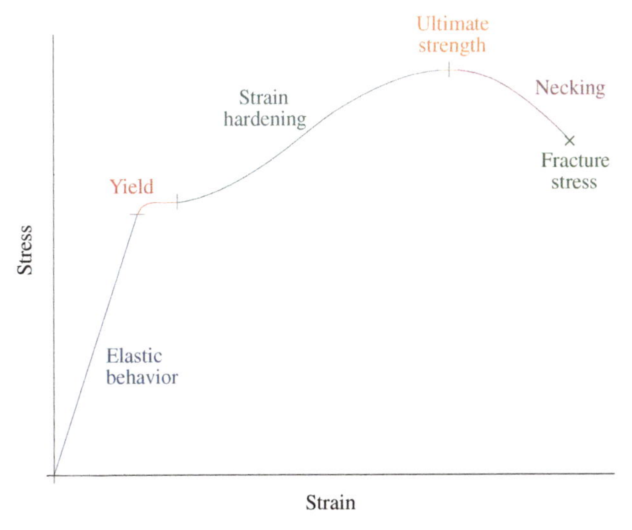

Necking

- The cross-sectional area begins to decrease markedly in a localized region of the specimen.
- The tension force required to produce additional stretch in the specimen decreases as the area is reduced.
- Necking occurs in ductile materials, but not in brittle materials.

Elastic behavior

- In general, the initial relationship between stress and strain is linear.
- Elastic strain is temporary, meaning that all strain is fully recovered upon removal of the stress.
- The slope of this line is called the elastic modulus or the modulus of elasticity.

Fracture stress

- The fracture stress is the engineering stress at which the specimen breaks into two pieces.

FIGURE 3.16 Review of significant features on the stress–strain diagram.

3.3 Hooke's Law

As discussed previously, the initial portion of the stress–strain diagram for most materials used in engineering structures is a straight line. The stress–strain diagrams for some materials, such as gray cast iron and concrete, show a slight curve even at very small stresses, but it is common practice to neglect the curvature and draw a straight line in order to average the data for the first part of the diagram. The proportionality of load to deflection was first recorded by Robert Hooke, who observed in 1678, *Ut tensio sic vis* ("As the stretch, so the force"). This relationship is referred to as **Hooke's law**. For normal stress σ and normal strain ε acting in one direction (termed **uniaxial** stress and strain), Hooke's law is written as

$$\boxed{\sigma = E\varepsilon} \tag{3.4}$$

where E is the elastic modulus.

Hooke's law also applies to shear stress τ and shear strain γ,

$$\boxed{\tau = G\gamma} \tag{3.5}$$

where G is called the **shear modulus** or the **modulus of rigidity**.

3.4 Poisson's Ratio

A material loaded in one direction will undergo strains perpendicular to the direction of the load as well as parallel to it. In other words,

- If a solid body is subjected to an axial tension, it contracts in the lateral directions.
- If a solid body is compressed, it expands in the lateral directions.

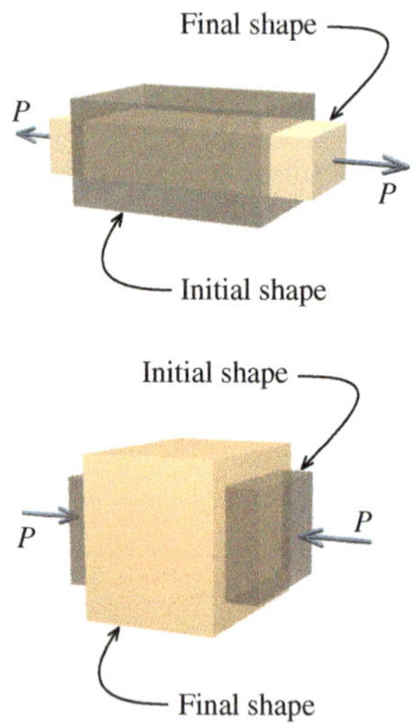

FIGURE 3.17 Lateral contraction and lateral expansion of a solid body subjected to axial forces.

This phenomenon is illustrated in Figure 3.17, where the deformations are *greatly exaggerated*. Experiments have shown that the relationship between lateral and longitudinal strains caused by an axial force remains constant, provided that the material remains *elastic* and is *homogeneous* and *isotropic* (as defined in Section 2.4). This constant is a property of the material, just like other properties, such as the elastic modulus E. The ratio of the lateral or transverse strain (ε_{lat} or ε_t) to the longitudinal or axial strain ($\varepsilon_{\text{long}}$ or ε_a) for a uniaxial state of stress is called **Poisson's ratio**, after Siméon D. Poisson, who identified the constant in 1811. Poisson's ratio is denoted by the Greek symbol ν (nu) and is defined as follows:

$$\boxed{\nu = -\frac{\varepsilon_{\text{lat}}}{\varepsilon_{\text{long}}} = -\frac{\varepsilon_t}{\varepsilon_a}} \tag{3.6}$$

The ratio $\nu = -\varepsilon_t/\varepsilon_a$ is valid only for a uniaxial state of stress (i.e., simple tension or compression). The negative sign appears in Equation (3.6) because the lateral and longitudinal strains are always of opposite signs for uniaxial stress (i.e., if one strain is elongation, the other strain is contraction).

Values vary for different materials, but for most metals, Poisson's ratio has a value between 1/4 and 1/3. Because the volume of material must remain constant, the largest possible value for Poisson's ratio is 0.5. Values approaching this upper limit are found only for materials such as rubber.

Relationship Between *E*, *G*, and *ν*

Poisson's ratio is related to the elastic modulus *E* and the shear modulus *G* by the formula

$$G = \frac{E}{2(1 + \nu)} \tag{3.7}$$

The Poisson effect exhibited by materials causes no additional stresses in the lateral direction unless the transverse deformation is inhibited or prevented in some manner.

EXAMPLE 3.1

A tension test was conducted on a 1.975 in. wide by 0.375 in. thick specimen of a nylon plastic. A 4.000 in. gage length was marked on the specimen before application of the load. In the elastic portion of the stress–strain curve at an applied load of $P = 6{,}000$ lb, the elongation in the gage length was measured as 0.023 in. and the contraction in the width of the bar was measured as 0.004 in. Determine

(a) the elastic modulus *E*.
(b) Poisson's ratio *ν*.
(c) the shear modulus *G*.

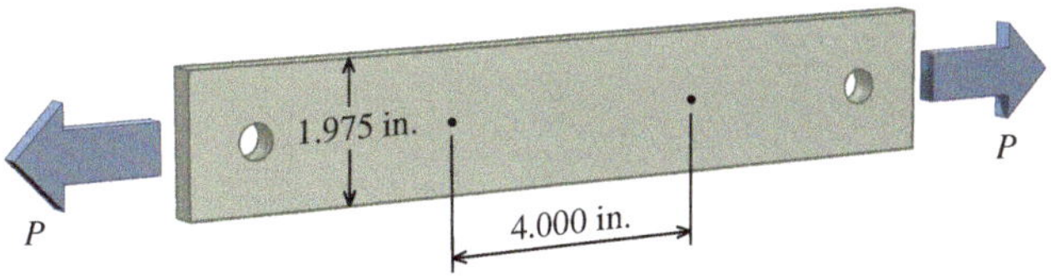

Plan the Solution

(a) From the load and the initial measured dimensions of the bar, the normal stress can be computed. The normal strain in the longitudinal (i.e., axial) direction, $\varepsilon_{\text{long}}$, can be computed from the elongation in the gage length and the initial gage length. With these two quantities, the elastic modulus *E* can be calculated from Equation (3.4).
(b) From the contraction in the width and the initial width of the bar, the strain in the lateral (i.e., transverse) direction, ε_{lat}, can be computed. Poisson's ratio can then be found from Equation (3.6).
(c) The shear modulus can be calculated from Equation (3.7).

SOLUTION

(a) The normal stress in the plastic specimen is

$$\sigma = \frac{6{,}000 \text{ lb}}{(1.975 \text{ in.})(0.375 \text{ in.})} = 8{,}101.27 \text{ psi}$$

The longitudinal strain is

$$\varepsilon_{\text{long}} = \frac{0.023 \text{ in.}}{4.000 \text{ in.}} = 0.005750 \text{ in./in.}$$

Therefore, the elastic modulus is

$$E = \frac{\sigma}{\varepsilon} = \frac{8{,}101.27 \text{ psi}}{0.005750 \text{ in./in.}} = 1{,}408{,}916 \text{ psi} = 1{,}409{,}000 \text{ psi} \qquad \textbf{Ans.}$$

(b) The lateral strain is

$$\varepsilon_{lat} = \frac{-0.004 \text{ in.}}{1.975 \text{ in.}} = -0.002025 \text{ in./in.}$$

From Equation (3.6), Poisson's ratio can be computed as

$$\nu = -\frac{\varepsilon_{lat}}{\varepsilon_{long}} = -\frac{-0.002025 \text{ in./in.}}{0.005750 \text{ in./in.}} = 0.352 \qquad \textbf{Ans.}$$

(c) The shear modulus is then computed from Equation (3.7) as

$$G = \frac{E}{2(1+\nu)} = \frac{1{,}408{,}916 \text{ psi}}{2(1+0.352)} = 521{,}049 \text{ psi} = 521{,}000 \text{ psi} \qquad \textbf{Ans.}$$

EXAMPLE 3.2

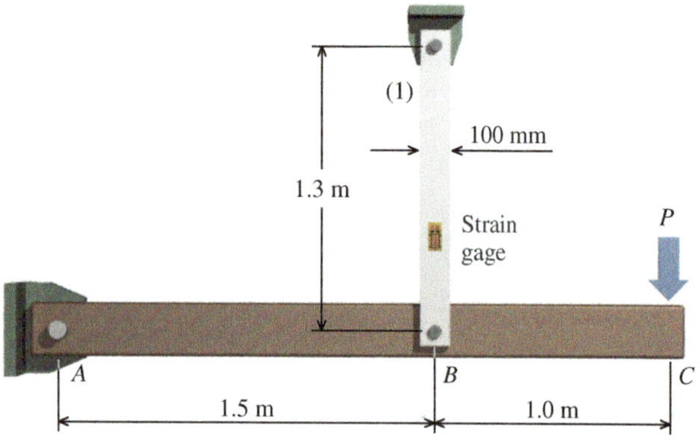

Rigid bar *ABC* is supported by a pin at *A* and a 100 mm wide by 6 mm thick aluminum [E = 70 GPa; $\alpha = 22.5 \times 10^{-6}$/°C; ν = 0.33] alloy bar at *B*. A strain gage affixed to the surface of the aluminum bar is used to measure its longitudinal strain. Before load *P* is applied to the rigid bar at *C*, the strain gage measures zero longitudinal strain at an ambient temperature of 20°C. After load *P* is applied to the rigid bar at *C* and the temperature drops to −10°C, a longitudinal strain of 2,400 με is measured in the aluminum bar. Determine

(a) the stress in member (1).
(b) the magnitude of load *P*.
(c) the change in the width of the aluminum bar (i.e., the 100 mm dimension).

Plan the Solution

This problem illustrates some misconceptions that are common in applying Hooke's law and Poisson's ratio, particularly when temperature change is a factor in the analysis.

SOLUTION

(a) Since the elastic modulus E and the longitudinal strain ε are given in the problem, one might be tempted to compute the normal stress in aluminum bar (1) from Hooke's law [Equation (3.4)]:

$$\sigma_1 = E_1\varepsilon_1 = (70 \text{ GPa})(2{,}400 \ \mu\varepsilon)\left[\frac{1{,}000 \text{ MPa}}{1 \text{ GPa}}\right]\left[\frac{1 \text{ mm/mm}}{1{,}000{,}000 \ \mu\varepsilon}\right] = 168 \text{ MPa}$$

This calculation is not correct for the normal stress in member (1). Why is it incorrect?

From Equation (2.7), the total strain ε_{total} in an object includes a portion ε_σ due to stress and a portion ε_T due to temperature change. The strain gage affixed to member (1) has measured the total strain in the aluminum bar as ε_{total} = 2,400 με = 0.002400 mm/mm. In this problem, however, the temperature of member (1) has dropped 30°C

58

before the strain measurement. From Equation (2.6), the strain caused by the temperature change in the aluminum bar is

$$\varepsilon_T = \alpha\,\Delta T = (22.5 \times 10^{-6}/°\text{C})(-30°\text{C}) = -0.000675 \text{ mm/mm}$$

Hence, the strain caused by normal stress in member (1) is

$$\varepsilon_{\text{total}} = \varepsilon_\sigma + \varepsilon_T$$

$$\therefore \varepsilon_\sigma = \varepsilon_{\text{total}} - \varepsilon_T = 0.002400 \text{ mm/mm} - (-0.000675 \text{ mm/mm})$$
$$= 0.003075 \text{ mm/mm}$$

Using this strain value, we can now compute the normal stress in member (1) from Hooke's law:

$$\sigma_1 = E\varepsilon = (70 \text{ GPa})(0.003075 \text{ mm/mm}) = 215.25 \text{ MPa} = 215 \text{ MPa}$$ **Ans.**

(b) The axial force in member (1) is computed from the normal stress and the bar area:

$$F_1 = \sigma_1 A_1 = (215.25 \text{ N/mm}^2)(100 \text{ mm})(6 \text{ mm}) = 129{,}150 \text{ N}$$

Now write an equilibrium equation for the sum of moments about joint A, and solve for load P:

$$\Sigma M_A = (1.5 \text{ m})(129{,}150 \text{ N}) - (2.5 \text{ m})P = 0$$

$$\therefore P = 77{,}490 \text{ N} = 77.5 \text{ kN}$$ **Ans.**

(c) The change in the width of the bar is computed by multiplying the lateral (i.e., transverse) strain ε_{lat} by the 100 mm initial width. To determine ε_{lat}, the definition of Poisson's ratio [Equation (3.6)] is used:

$$\nu = -\frac{\varepsilon_{\text{lat}}}{\varepsilon_{\text{long}}} \qquad \therefore \varepsilon_{\text{lat}} = -\nu\varepsilon_{\text{long}}$$

Using the given value of Poisson's ratio and the measured strain, we could calculate ε_{lat} as

$$\varepsilon_{\text{lat}} = -\nu\varepsilon_{\text{long}} = -(0.33)(2{,}400\ \mu\varepsilon) = -792\ \mu\varepsilon$$

This calculation is not correct for the lateral strain in member (1). Why is it incorrect?

The Poisson effect applies only to strains caused by stresses (i.e., mechanical effects). When they are unrestrained, homogeneous, isotropic materials expand uniformly in all directions as they are heated (and contract uniformly as they cool). Consequently, thermal strains should not be included in the calculation of Poisson's ratio. For this problem, the lateral strain should be calculated as

$$\varepsilon_{\text{lat}} = -(0.33)(0.003075 \text{ mm/mm}) + (-0.000675 \text{ mm/mm}) = -0.0016898 \text{ mm/mm}$$

The change in the width of the aluminum bar is, therefore,

$$\delta_{\text{width}} = (-0.0016898 \text{ mm/mm})(100 \text{ mm}) = -0.1690 \text{ mm}$$ **Ans.**

EXAMPLE 3.3

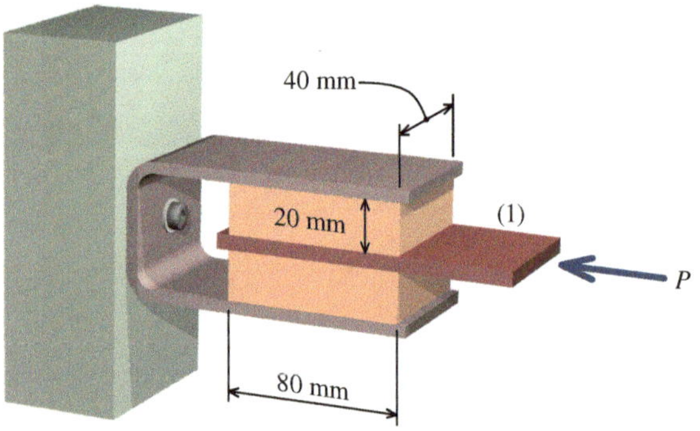

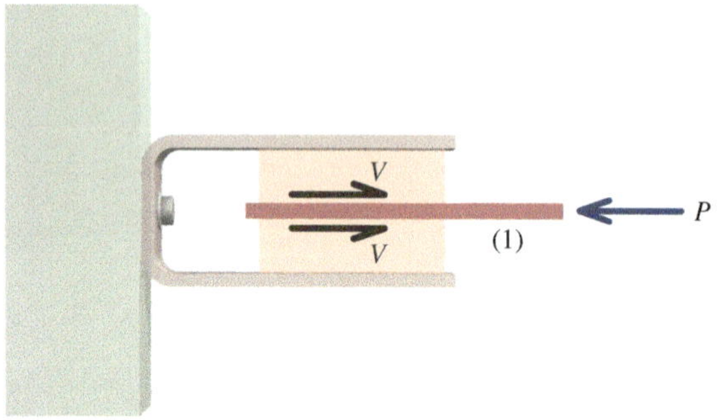

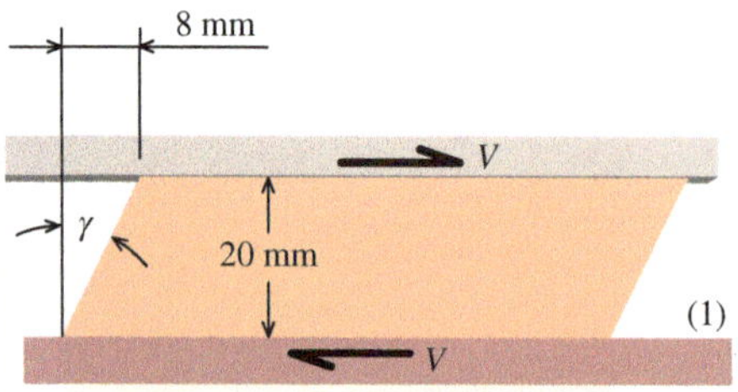

Two blocks of rubber, each 80 mm long by 40 mm wide by 20 mm thick, are bonded to a rigid support mount and to a movable plate (1). When a force $P = 2{,}800$ N is applied to the assembly, plate (1) deflects 8 mm horizontally. Determine the shear modulus G of the rubber used for the blocks.

Plan the Solution

Hooke's law [Equation (3.5)] expresses the relationship between shear stress and shear strain. The shear stress can be determined from the applied load P and the area of the rubber blocks that contact the movable plate (1). Shear strain is an angular measure that can be determined from the horizontal deflection of plate (1) and the thickness of the rubber blocks. The shear modulus G is computed by dividing the shear stress by the shear strain.

SOLUTION

Consider a free-body diagram of movable plate (1). Each rubber block provides a shear force that opposes the applied load P. From a consideration of equilibrium, the sum of forces in the horizontal direction is

$$\Sigma F_x = 2V - P = 0$$

$$\therefore V = P/2 = (2{,}800 \text{ N})/2 = 1{,}400 \text{ N}$$

Next, consider a free-body diagram of the upper rubber block in its deflected position. The shear force V acts on a surface that is 80 mm long and 40 mm wide. Therefore, the average shear stress in the rubber block is

$$\tau = \frac{1{,}400 \text{ N}}{(80 \text{ mm})(40 \text{ mm})} = 0.4375 \text{ MPa}$$

The 8 mm horizontal deflection causes the block to skew as shown. The angle γ (measured in radians) is the shear strain:

$$\tan\gamma = \frac{8 \text{ mm}}{20 \text{ mm}} \qquad \therefore \gamma = 0.3805 \text{ rad}$$

The shear stress τ, the shear modulus G, and the shear strain γ are related by Hooke's law:

$$\tau = G\gamma$$

Therefore, the shear modulus of the rubber used for the blocks is

$$G = \frac{\tau}{\gamma} = \frac{0.4375 \text{ MPa}}{0.3805 \text{ rad}} = 1.150 \text{ MPa}$$ **Ans.**

MecMovies

EXERCISE

M3.1 Figure M3.1 depicts basic problems requiring the use of Hooke's law.

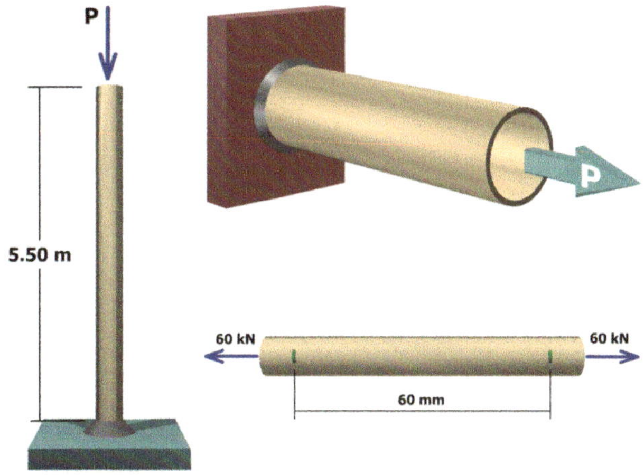

FIGURE M3.1

PROBLEMS

P3.1 At the proportional limit, a 2 in. gage length of a 0.500 in. diameter alloy rod has elongated 0.0035 in. and the diameter has been reduced by 0.0003 in. The total tension force on the rod was 5.45 kips. Determine the following properties of the material:

(a) the proportional limit.
(b) the modulus of elasticity.
(c) Poisson's ratio.

P3.2 A solid circular rod with a diameter $d = 16$ mm is shown in Figure P3.2. The rod is made of an aluminum alloy that has an elastic modulus $E = 72$ GPa and a Poisson's ratio $\nu = 0.33$. When subjected to the axial load P, the diameter of the rod decreases by 0.024 mm. Determine the magnitude of load P.

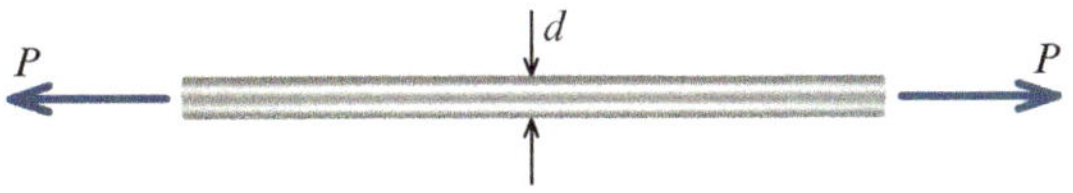

FIGURE P3.2

P3.3 The polymer bar shown in Figure P3.3 has a width $b = 50$ mm, a depth $d = 100$ mm, and a height $h = 270$ mm. At a compressive load $P = 135$ kN, the bar height contracts by $\Delta h = -2.50$ mm and the bar depth elongates by $\Delta d = 0.38$ mm. At this load, the stress in the polymer bar is less than its proportional limit. Determine

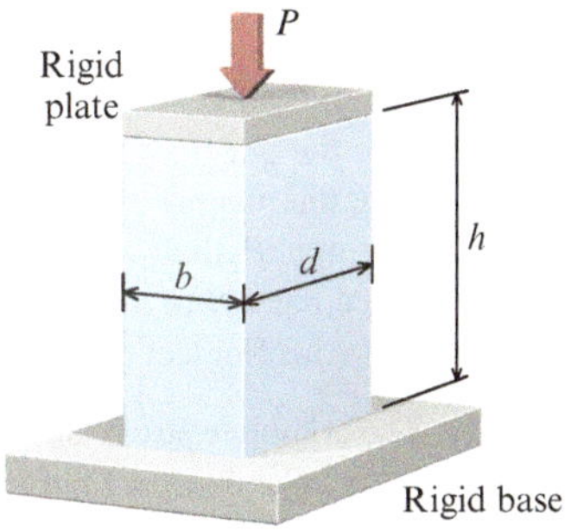

FIGURE P3.3

(a) the modulus of elasticity.
(b) Poisson's ratio.
(c) the change in the bar width b.

P3.4 A 0.625 in. thick rectangular alloy bar is subjected to a tensile load P by pins at A and B as shown in Figure P3.4. The width of the bar is $w = 2.00$ in. Strain gages bonded to the specimen measure the following strains in the longitudinal (x) and transverse (y) directions: $\varepsilon_x = 1{,}140\ \mu\varepsilon$ and $\varepsilon_y = -315\ \mu\varepsilon$.

(a) Determine Poisson's ratio for this specimen.
(b) If the measured strains were produced by an axial load $P =$ 17.4 kips, what is the modulus of elasticity for this specimen?

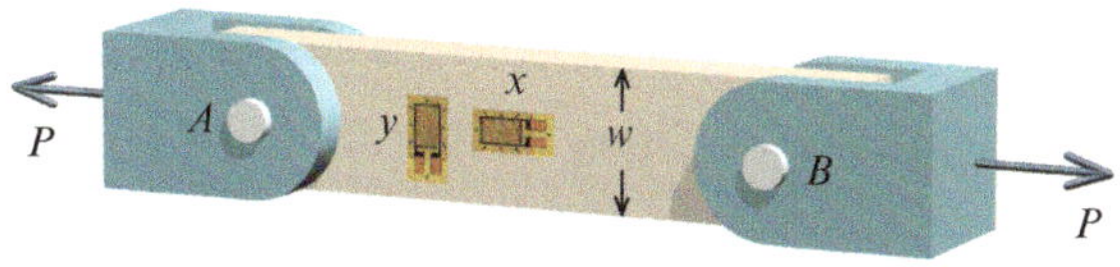

FIGURE P3.4

P3.5 A 40 mm by 40 mm square $ABCD$ (i.e., $a = 40$ mm) is drawn on a rectangular bar prior to loading. (See Figure P3.5a). A uniform normal stress $\sigma = 54$ MPa is then applied to the ends of the rectangular bar, and square $ABCD$ is deformed into the shape of a rhombus, as shown in the Figure P3.5b. The dimensions of the rhombus after loading are $b = 56.88$ mm and $c = 55.61$ mm. Determine the modulus of elasticity for the material. Assume that the material behaves elastically for the applied stress.

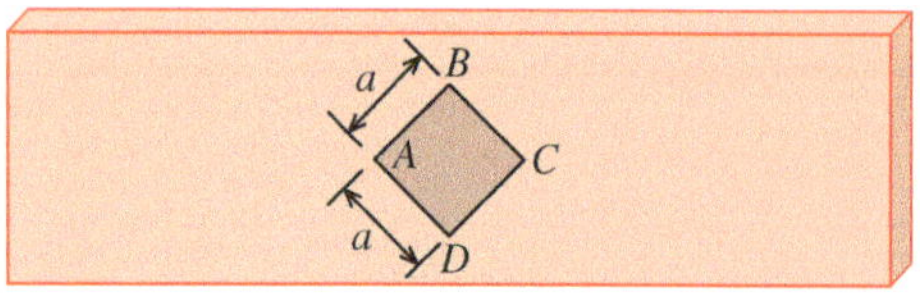

Initial square drawn on bar before loading.

FIGURE P3.5*a*

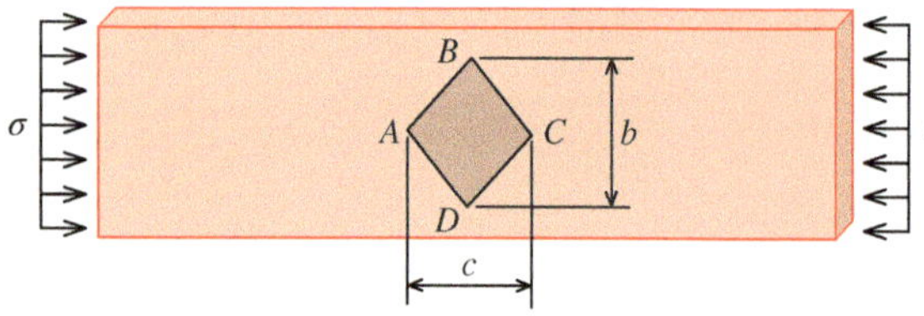

Rhombus after bar is loaded by stress σ.

FIGURE P3.5*b*

P3.6 A nylon [E = 2,500 MPa; ν = 0.4] bar is subjected to an axial load that produces a normal stress σ. Before the load is applied, a line having a slope of 3:2 (i.e., 1.5) is marked on the bar as shown in Figure P3.6. Determine the slope of the line when σ = 105 MPa.

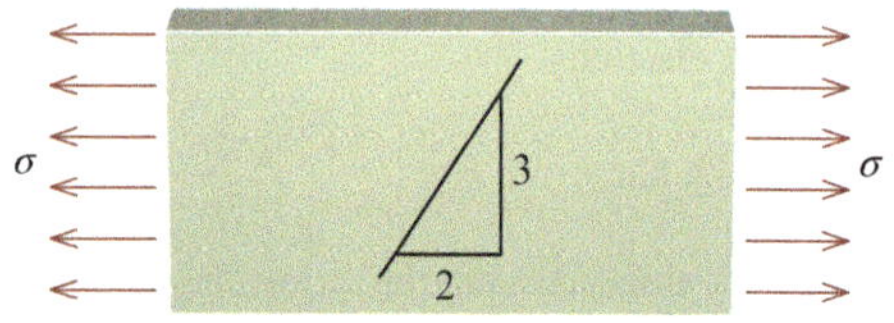

FIGURE P3.6

P3.7 A nylon [E = 360 ksi; ν = 0.4] rod (1) having a diameter d_1 = 2.00 in. is placed inside a steel [E = 29,000 ksi; ν = 0.29] tube (2) as shown in Figure P3.7. The inside diameter of the tube is d_2 = 2.02 in. An external load P is applied to the rod, compressing it. At what value of P will the space between the rod and the tube be closed?

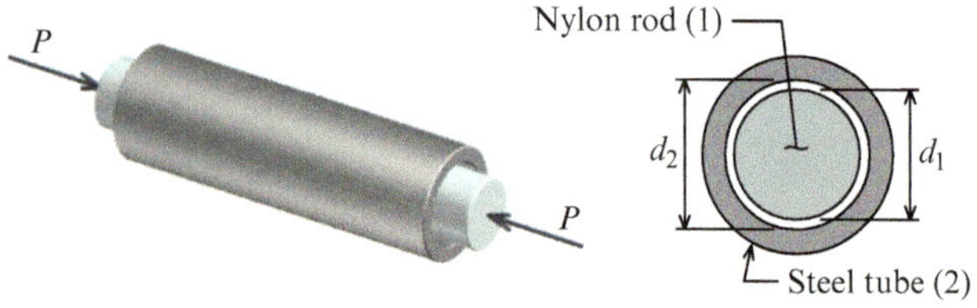

FIGURE P3.7

P3.8 A metal specimen with an original diameter of 0.500 in. and a gage length of 2.000 in. is tested in tension until fracture occurs. At the point of fracture, the diameter of the specimen is 0.260 in. and the fractured gage length is 3.08 in. Calculate the ductility in terms of percent elongation and percent reduction in area.

P3.9 A portion of the stress–strain curve for a stainless steel alloy is shown in Figure P3.9. A 350 mm long bar is loaded in tension until it elongates 2.0 mm, and then the load is removed.

(a) What is the permanent set in the bar?

(b) What is the length of the unloaded bar?

(c) If the bar is reloaded, what will be the proportional limit?

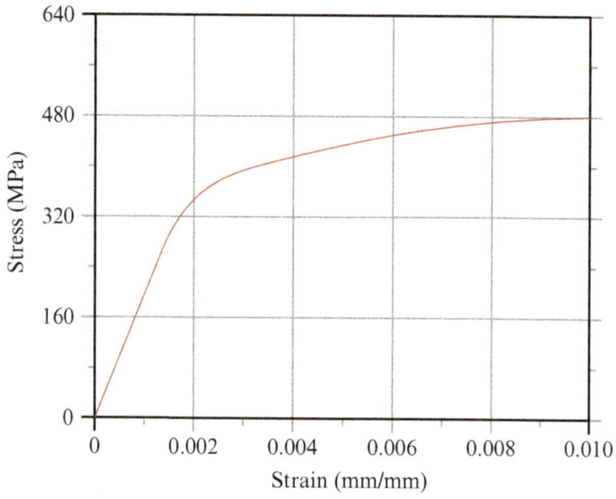

FIGURE P3.9

P3.10 A plastic block is bonded to a fixed base and to a horizontal rigid plate as shown in Figure P3.10. The shear modulus of the plastic is G = 45,000 psi, and the block dimensions are a = 4.0 in., b = 2.0 in., and c = 1.50 in. A horizontal force P = 8,500 lb is applied to the plate. Determine the horizontal deflection of the plate.

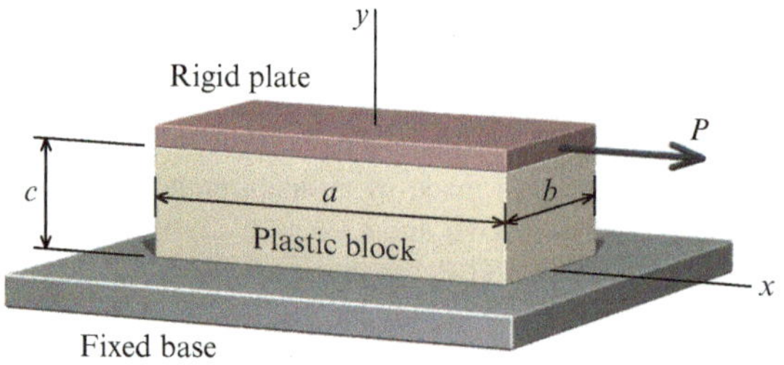

FIGURE P3.10

P3.11 A 0.5 in. thick plastic panel is bonded to the pin-jointed steel frame shown in Figure P3.11. Assume that a = 4.0 ft, b = 6.0 ft, and G = 70,000 psi for the plastic, and determine the magnitude of the force P that would displace bar AB to the right by 0.8 in. Neglect the deformation of the steel frame.

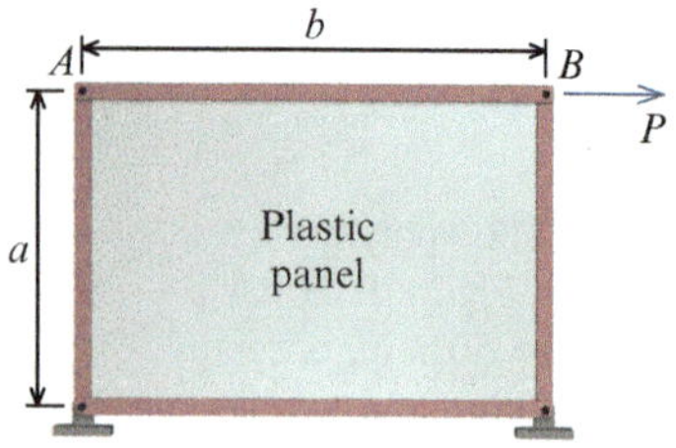

FIGURE P3.11

P3.12 The complete stress–strain diagram for a particular stainless steel alloy is shown in Figure P3.12*a*/13*a*. This diagram has been enlarged in Figure P3.12*b*/13*b* to show in more detail the linear portion of the stress-strain diagram. A rod made from this material is initially 800 mm long at a temperature of 20°C. After a tension force is applied to the rod and the temperature is increased by 200°C, the length of the rod is 804 mm. Determine the stress in the rod, and state whether the elongation in the rod is elastic or inelastic. Assume the coefficient of thermal expansion for this material is $18 \times 10^{-6}/°C$.

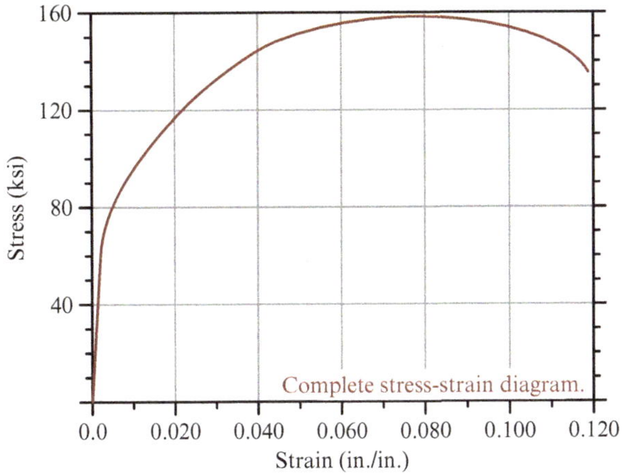

FIGURE P3.12*a*/13*a*

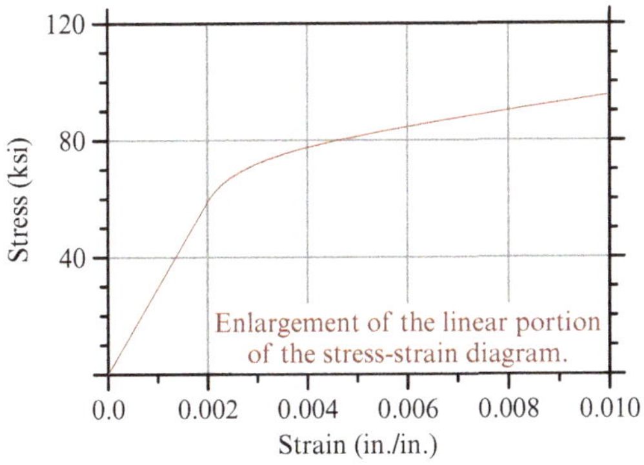

FIGURE P3.12*b*/13*b*

P3.13 A tensile test specimen of stainless steel alloy having a diameter of 12.6 mm and a gage length of 50 mm was tested to fracture. The complete stress–strain diagram for this specimen is shown in Figure P3.12*a*/13*a*. This diagram has been enlarged in Figure P3.12*b*/13*b* to show in more detail the linear portion of the stress-strain diagram. Determine

(a) the modulus of elasticity.
(b) the proportional limit.
(c) the ultimate strength.
(d) the yield strength (0.20% offset).
(e) the fracture stress.
(f) the true fracture stress if the final diameter of the specimen at the location of the fracture was 8.89 mm.

P3.14 A 7075-T651 aluminum alloy specimen with a diameter of 0.500 in. and a 2.0 in. gage length was tested to fracture. Load and deformation data obtained during the test are given in the accompanying table. Determine

(a) the modulus of elasticity.
(b) the proportional limit.
(c) the yield strength (0.20% offset).
(d) the ultimate strength.
(e) the fracture stress.
(f) the true fracture stress if the final diameter of the specimen at the location of the fracture was 0.387 in.

Load (lb)	Change in Length (in.)	Load (lb)	Change in Length (in.)
0	0	14,690	0.0149
1,221	0.0012	14,744	0.0150
2,479	0.0024	15,119	0.0159
3,667	0.0035	15,490	0.0202
4,903	0.0048	15,710	0.0288
6,138	0.0060	16,032	0.0581
7,356	0.0072	16,295	0.0895
8,596	0.0085	16,456	0.1214
9,783	0.0096	16,585	0.1496
11,050	0.0110	16,601	0.1817
12,247	0.0122	16,601	0.2278
13,434	0.0134	16,489	0.2605
		16,480	fracture

P3.15 A Grade 2 Titanium tension test specimen has a diameter of 12.60 mm and a gage length of 50 mm. In a test to fracture, the stress and strain data shown in the accompanying table were obtained. Determine

(a) the modulus of elasticity.
(b) the proportional limit.
(c) the yield strength (0.20% offset).
(d) the ultimate strength.
(e) the fracture stress.
(f) the true fracture stress if the final diameter of the specimen at the location of the fracture was 9.77 mm.

Load (kN)	Change in Length (mm)	Load (kN)	Change in Length (mm)
0.00	0.000	52.74	0.314
4.49	0.017	56.95	0.480
8.84	0.032	60.76	0.840
13.29	0.050	63.96	1.334
17.57	0.064	66.61	1.908
22.10	0.085	68.26	2.562
26.46	0.103	69.08	3.217
30.84	0.123	69.41	3.938
35.18	0.144	69.39	4.666
39.70	0.171	69.25	5.292
43.95	0.201	68.82	6.023
48.44	0.241	68.35	6.731
		68.17	fracture

P3.16 Compound axial member ABC shown in Figure P3.16 has a uniform diameter $d = 1.50$ in. Segment (1) is an aluminum [$E_1 = 10{,}000$ ksi] alloy rod with length $L_1 = 90$ in. Segment (2) is a copper [$E_2 = 17{,}000$ ksi] alloy rod with length $L_2 = 130$ in. When axial force P is applied, a strain gage attached to copper segment (2) measures a normal strain of $\varepsilon_2 = 2{,}100$ μin./in. in the longitudinal direction. What is the total elongation of member ABC?

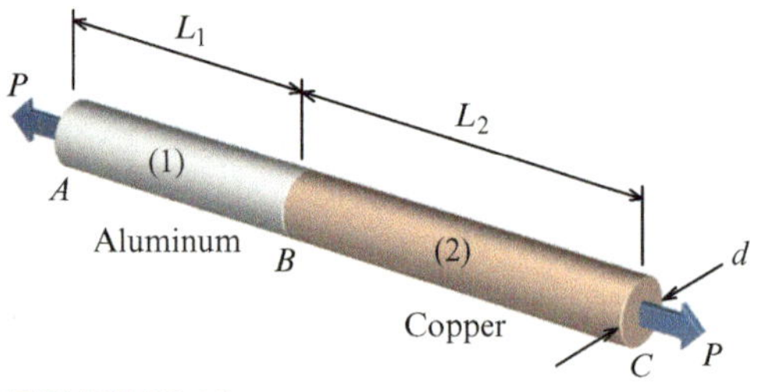

FIGURE P3.16

P3.17 An aluminum alloy [$E = 70$ GPa; $\nu = 0.33$; $\alpha = 23.0 \times 10^{-6}/°C$] plate is subjected to a tensile load P as shown in Figure P3.17. The plate has a depth $d = 260$ mm, a cross-sectional area $A = 6{,}500$ mm^2, and a length $L = 4.5$ m. The initial longitudinal normal strain in the plate is zero. After load P is applied and the temperature of the plate has been increased by $\Delta T = 56°C$, the longitudinal normal strain in the plate is found to be 2,950 με. Determine

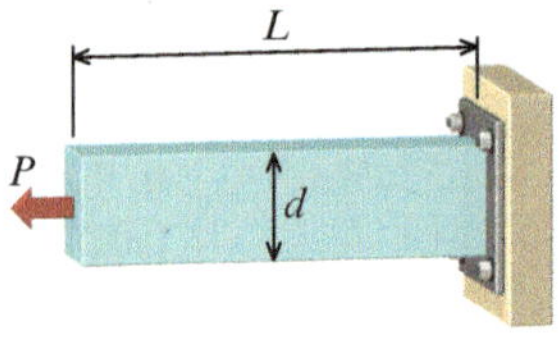

FIGURE P3.17

(a) the magnitude of load P.
(b) the change Δd in plate depth.

P3.18 The rigid plate in Figure P3.18 is supported by bar (1) and by a double-shear pin connection at B. Bar (1) has a length $L_1 = 60$ in., a cross-sectional area $A_1 = 0.47$ in.^{2}, an elastic modulus $E = 10{,}000$ ksi, and a coefficient of thermal expansion of $\alpha = 13 \times 10^{-6}/°F$. The pin at B has a diameter of 0.438 in. After load P has been applied and the temperature of the entire assembly has been *decreased* by 30°F, the total strain in bar (1) is measured as 570 με (elongation). Assume dimensions of $a = 12$ in. and $b = 20$ in. Determine

(a) the magnitude of load P.
(b) the average shear stress in pin B.

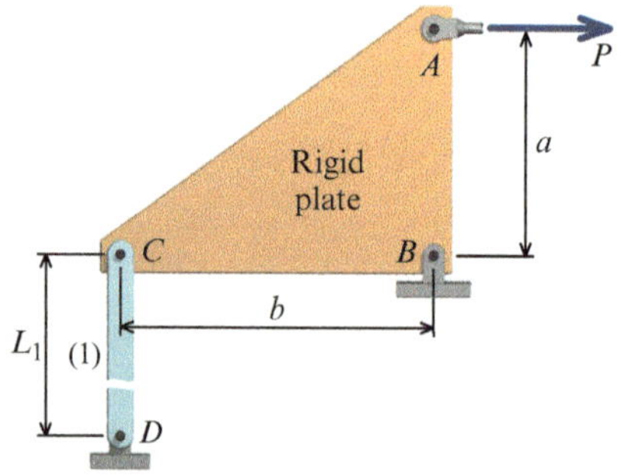

FIGURE P3.18

P3.19 In Figure P3.19, member (1) is an aluminum bar that has a cross-sectional area $A = 1.05$ in.^{2}, an elastic modulus $E = 10{,}000$ ksi, and a coefficient of thermal expansion of $\alpha = 12.5 \times 10^{-6}/°F$. After a load P of unknown magnitude is applied to the structure and the temperature is *increased* by 65°F, the normal strain in bar (1) is measured as −540 με. Use dimensions of $a = 24.6$ ft, $b = 11.7$ ft, and $c = 14.0$ ft. Determine the magnitude of load P.

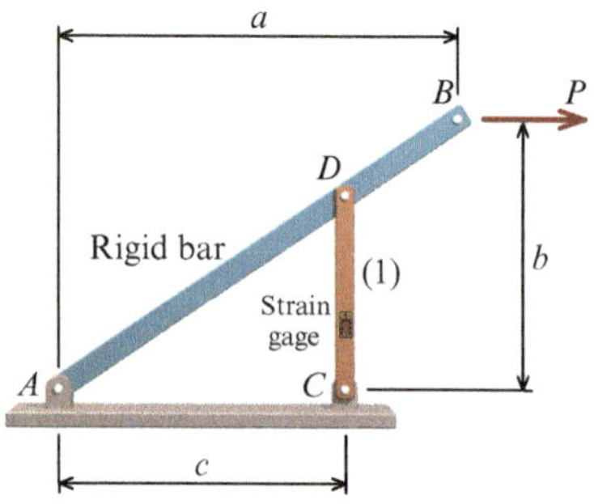

FIGURE P3.19

CHAPTER 4

Design Concepts

4.1 Introduction

The design problems faced by engineers involve many considerations, such as function, safety, initial cost, life-cycle cost, environmental impacts, efficiency, and aesthetics. In mechanics of materials, however, our interest focuses on three mechanical considerations: strength, stiffness, and stability. In addressing these concerns, a number of uncertainties must be considered and accounted for in a successful design.

The loads that act on structures or machines are generally estimated, and there may be substantial variation in these loads, such as the following:

- The rate of loading may differ from design assumptions.
- There is uncertainty associated with the material used in a structure or machine. Since testing usually damages the material, the mechanical properties of the material cannot be evaluated directly, but rather are determined by testing specimens of a similar material. For a material such as wood, there may be substantial variation in the strength and stiffness of individual boards and timbers.
- Material strengths may change over time, through corrosion and other effects.
- Environmental conditions, such as temperature, humidity, and exposure to rain and snow, may differ from design assumptions.

- Although their chemical composition may be the same, the materials used in prototypes or test components may differ from those used in production components because of such factors as microstructure, size, rolling or forming effects, and surface finish.
- Stresses may be created in a component during the fabrication process, and poor workmanship could diminish the strength of a design.
- Models and methods used in analysis may oversimplify or incorrectly idealize a structure and thereby inadequately represent its true behavior.

Textbook problems may convey the impression that analysis and design are processes of applying rigorous calculation procedures to perfectly defined structures and machines in order to obtain definitive results. In practice, however, design procedures must make allowances for many factors that cannot be quantified with great certainty.

4.2 Types of Loads

The forces that act on a structure or machine are called **loads**. The specific types of load that act on a structure or machine depend on the particular application. Several types of load that act on building structures are discussed next.

Dead Loads

Dead loads consist of the weight of various structural members and the weights of objects that are permanently attached to a structure. For a building, the self-weight of the structure includes items such as beams, columns, floor slabs, walls, plumbing, electrical fixtures, permanent mechanical equipment, and the roof. The magnitudes and locations of these loads are unchanging throughout the lifetime of the structure.

In designing a structure, the size of each individual beam, floor, column, and other component is unknown at the outset. An analysis of the structure must be performed before final member sizes can be determined; however, the analysis must include the weight of the members. Consequently, it is often necessary to perform design calculations iteratively: estimating the weight of various components; performing an analysis; selecting appropriate member sizes; and, if significant differences are present, repeating the analysis with improved estimates for the member weights.

Although the self-weight of a structure is generally well defined, the dead load may be underestimated because of the uncertainty of other dead-load components, such as the weight of permanent equipment, room partitions, roofing materials, floor coverings, fixed service equipment, and other immovable fixtures. Future modifications to the structure may also need to be considered. For instance, additional highway paving materials may be added to the deck of a bridge structure at a future time.

Live Loads

Live loads are loads in which the magnitude, duration, and location of the loading vary throughout the lifetime of the structure. They may be caused by the weight of objects temporarily placed on the structure, moving vehicles or people, or natural forces. The live load on floors and decks is typically modeled as a uniformly distributed area loading that accounts for items normally associated with the intended use of the space. For typical office and residential structures, these items include occupants, furnishings, and storage.

For structures such as bridges and parking garages, a concentrated live load (or loads) representing the weight of vehicles or other heavy items must be considered in addition to the distributed uniform area loading. In the analysis, the effects of such concentrated loads at various potentially critical locations must be investigated.

A load suddenly applied to a structure is termed **impact**. A crate dropped on the floor of a warehouse or a truck bouncing on uneven pavement creates a greater force in a structure than would normally occur if the load were applied slowly and gradually. Specified live loads generally include an appropriate allowance for impact effects of normal use and traffic. Special impact consideration may be necessary for structures supporting elevator machinery, large reciprocating or rotating machinery, and cranes.

By their nature, live loads are known with much less certainty than dead loads. Live loads vary in intensity and location throughout the lifetime of the structure. In a building, for example, unanticipated crowding of people in a space may occur on occasion or perhaps a space may be subjected to unusually large loads during renovation as furnishings or other materials are temporarily relocated.

Snow Load

In colder climates, snow load may be a significant design consideration for roof elements. The magnitude and duration of snow loads cannot be known with great certainty. Further, the distribution of snow generally will not be uniform on a roof structure because wind will cause snow to drift. Large accumulations of snow often will occur near locations where a roof changes height, creating additional loading effects.

Wind Loads

Wind exerts pressure on a building in proportion to the square of its velocity. At any given moment, wind velocities consist of an average velocity plus a superimposed turbulence known as a wind gust. Wind pressures are distributed over a building's exterior surfaces, both as positive pressures that push on walls or roof surfaces and as negative pressures (or suction) that uplift roofs and pull walls outward. Wind load magnitudes acting on structures vary with geographic location, heights above ground, surrounding terrain characteristics, building shape and features, and other factors. Wind is capable of striking a structure from any direction. Altogether, these characteristics make it very difficult to predict the magnitude and distribution of wind loading accurately.

4.3 Safety

Engineers seek to produce objects that are sufficiently strong to perform their intended function safely. To achieve safety in design with respect to strength, structures and machines are always designed to withstand loads above what would be expected under ordinary conditions. (Such loads are termed **overload**). While this reserve capacity is needed to ensure safety in response to an extreme load event, it also allows the structure or machine to be used in ways not originally anticipated during design.

The crucial question, however, is "How safe is safe enough?" On the one hand, if a structure or machine does not have enough extra capacity, there is a significant probability that an overload could cause failure, where failure is defined as breakage, rupture, or collapse. On the other hand, if too much reserve capacity is incorporated into the design of a component, the potential for failure may be slight but the object may be unnecessarily bulky, heavy, or expensive to build. The best designs strike a balance between economy and a conservative, but reasonable, margin of safety against failure.

Two philosophies for addressing safety are commonly used in current engineering design practice for structures and machines. These two approaches are called *allowable stress design* and *load and resistance factor design*.

4.4 Allowable Stress Design

The **allowable stress design (ASD)** method focuses on loads that exist at normal or typical conditions. These loads are termed **service loads**, and they consist of dead, live, wind, and other loads that are expected to occur while the structure is in service. In the ASD method, a structural element is designed so that elastic stresses produced by service loads do not exceed some fraction of the specified minimum yield stress of the material—a stress limit that is termed the **allowable stress** (Figure 4.1). If stresses under ordinary conditions are maintained at or below the allowable stress, a reserve capacity of strength will be available should an unanticipated overload occur, thus providing a margin of safety for the design.

The allowable stress used in design calculations is found by dividing the failure stress by a **factor of safety (FS)**:

$$\sigma_{\text{allow}} = \frac{\sigma_{\text{failure}}}{\text{FS}} \quad \text{or} \quad \tau_{\text{allow}} = \frac{\tau_{\text{failure}}}{\text{FS}} \tag{4.1}$$

Failure may be defined in several ways. It may be that "failure" refers to an actual fracture of the component, in which case the ultimate strength of the material (as determined from the stress–strain curve) is used as the failure stress in Equation (4.1). Alternatively, failure may refer to an excessive deformation in the material associated with yielding that renders the component unsuitable for its intended function. In this situation, the failure stress in Equation (4.1) is the yield stress.

Factors of safety are established by groups of experienced engineers who write the codes and specifications used by other designers. The provisions of codes and specifications are intended to provide reasonable levels of safety without unreasonable cost. The type of failure anticipated, as well as the history of similar components, the consequences of failure, and other uncertainties, are considered in deciding on appropriate factors of safety for various situations. Typical factors of safety range from 1.5 to 3, although larger values may be found in specific applications.

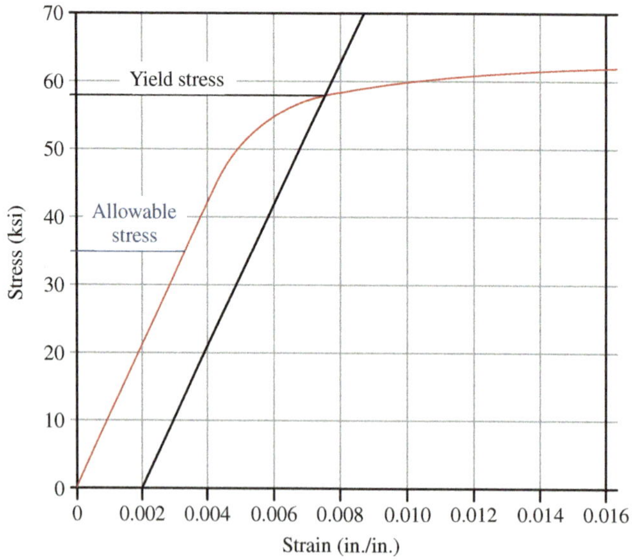

FIGURE 4.1 Allowable stress on the stress–strain curve.

68

In some instances, engineers may need to assess the level of safety in an existing or a proposed design. For this purpose, the factor of safety may be computed as the ratio of the anticipated failure stress to the estimated actual stress:

$$\text{FS} = \frac{\sigma_{\text{failure}}}{\sigma_{\text{actual}}} \quad \text{or} \quad \text{FS} = \frac{\tau_{\text{failure}}}{\tau_{\text{actual}}} \tag{4.2}$$

Factor-of-safety calculations need not be limited to stresses. The factor of safety may also be defined as the ratio between a failure-producing force and the estimated actual force—for instance,

$$\text{FS} = \frac{P_{\text{failure}}}{P_{\text{actual}}} \quad \text{or} \quad \text{FS} = \frac{V_{\text{failure}}}{V_{\text{actual}}} \tag{4.3}$$

EXAMPLE 4.1

A load of 8.9 kN is applied to a 6 mm thick steel plate, as shown. The steel plate is supported by a 10 mm diameter steel pin in a single-shear connection at A and a 10 mm diameter steel pin in a double-shear connection at B. The ultimate shear strength of the steel pins is 280 MPa, and the ultimate bearing strength of the steel plate is 530 MPa. Determine

(a) the factors of safety for pins A and B with respect to the ultimate shear strength.

(b) the factor of safety with respect to the ultimate bearing strength for the steel plate at pin B.

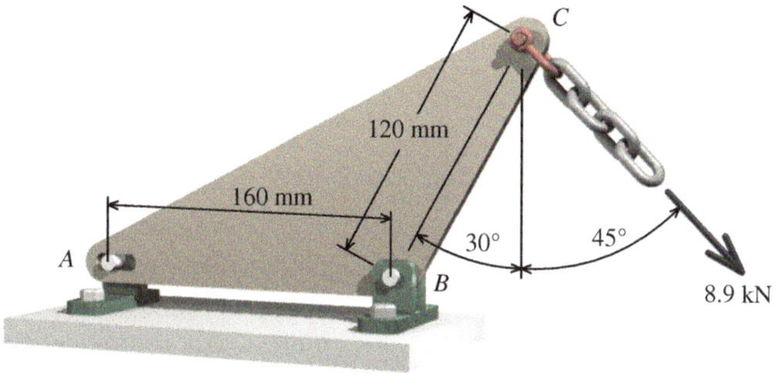

Plan the Solution

From equilibrium considerations, the reaction forces at pins A and B will be computed. In particular, the resultant force at B must be computed from the horizontal and vertical reactions at B. Once the pin forces have been determined, we will calculate the average shear stresses in pins A and B, taking into account whether the pin is used in a single-shear connection, as is the case for pin A, or a double-shear connection, as is the case for pin B. The factor of safety for each pin is found by dividing the ultimate shear strength by the average shear stress in the pin.

The average bearing stress in the plate at B is based on the projected area of contact between the plate and the pin, an area that is simply the product of the pin diameter and the plate thickness. We will divide the resultant force at B by the projected area of contact to obtain the average bearing stress in the plate. The factor of safety for the bearing at B is calculated by dividing the ultimate bearing strength of the plate by the average bearing stress.

SOLUTION

From equilibrium considerations, the reaction forces at pins A and B can be determined. **Note:** The pin at A rides in a slotted hole; therefore, it exerts only a vertical force on the steel plate.

The reaction forces are shown on the sketch along with pertinent dimensions.

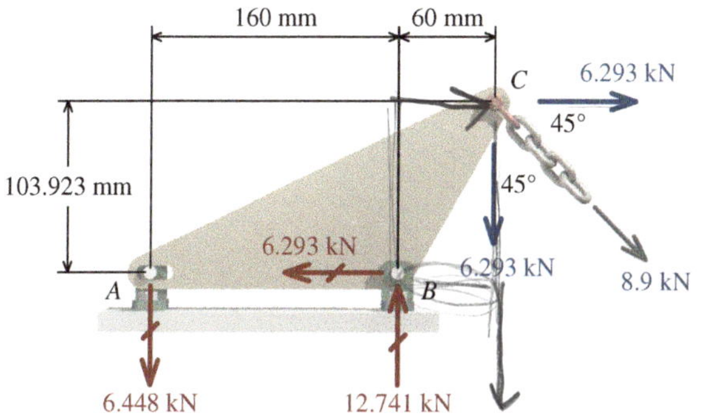

The resultant force exerted by pin B on the plate is

$$R_B = \sqrt{(6.293 \text{ kN})^2 + (12.741 \text{ kN})^2} = 14.210 \text{ kN}$$

Note: The resultant force should always be used in computing the shear stress in a pin or bolt.

(a) The cross-sectional area of a 10 mm diameter pin is $A_{pin} = 78.540\ \text{mm}^2$. Since pin A is a single-shear connection, its shear area A_V is equal to the pin cross-sectional area A_{pin}. The average shear stress in pin A is found from the shear force V_A that acts on the pin (i.e., the 6.448 kN reaction force) and A_V:

$$\tau_A = \frac{V_A}{A_V} = \frac{(6.448\ \text{kN})(1{,}000\ \text{N/kN})}{78.540\ \text{mm}^2} = 82.1\ \text{MPa}$$

Pin B is a double-shear connection; therefore, the pin area subjected to shear stress A_V is equal to twice the pin cross-sectional area A_{pin}. The shear force V_B that acts on the pin equals the resultant force at B. Thus, the average shear stress in pin B is

$$\tau_B = \frac{V_B}{A_V} = \frac{(14.210\ \text{kN})(1{,}000\ \text{N/kN})}{2(78.540\ \text{mm}^2)} = 90.5\ \text{MPa}$$

By Equation (4.2), the pin factors of safety with respect to the 280 MPa ultimate shear strength are

$$\text{FS}_A = \frac{\tau_{\text{failure}}}{\tau_{\text{actual}}} = \frac{280\ \text{MPa}}{82.1\ \text{MPa}} = 3.41 \quad \text{and} \quad \text{FS}_B = \frac{\tau_{\text{failure}}}{\tau_{\text{actual}}} = \frac{280\ \text{MPa}}{90.5\ \text{MPa}} = 3.09 \qquad \textbf{Ans.}$$

(b) The bearing stress at B occurs on the contact surface between the 10 mm diameter pin and the 6 mm thick steel plate. Although the actual stress distribution in the steel plate at this contact point is quite complex, the average bearing stress is customarily computed from the contact force and a projected area equal to the product of the pin diameter and the plate thickness. Therefore, the average bearing stress in the steel plate at pin B is computed as

$$\sigma_b = \frac{R_B}{d_B t} = \frac{(14.210\ \text{kN})(1{,}000\ \text{N/kN})}{(10\ \text{mm})(6\ \text{mm})} = 236.8\ \text{MPa}$$

The factor of safety of the plate with respect to the 530 MPa ultimate bearing strength is

$$\text{FS}_{\text{bearing}} = \frac{530\ \text{MPa}}{236.8\ \text{MPa}} = 2.24 \qquad \textbf{Ans.}$$

EXAMPLE 4.2

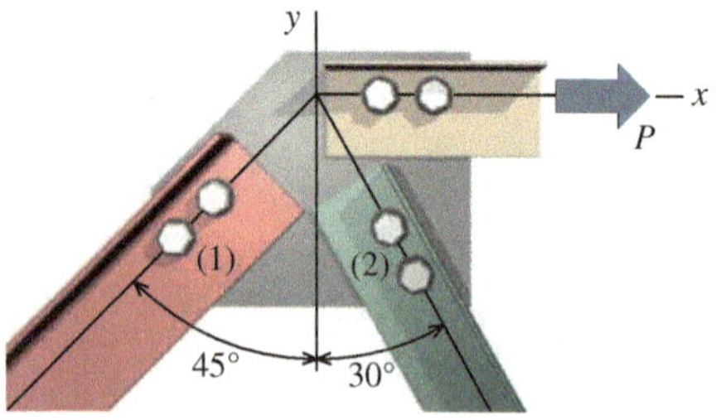

A truss joint is shown in the sketch. Member (1) has a cross-sectional area of 7.22 in.², and member (2) has a cross-sectional area of 3.88 in.². Both members are A36 steel with a yield strength of 36 ksi. If a factor of safety of 1.5 is required, determine the maximum load P that may be applied to the joint.

Plan the Solution

Since truss members are two-force members, two equilibrium equations can be written for the concurrent force system. From these equations, the unknown load P can be expressed in terms of member forces F_1 and F_2. An allowable stress can then be determined from the yield strength of the steel and the specified factor of safety. With the allowable stress and the cross-sectional area, the maximum allowable member force can be determined. However, it is not likely that both members will be stressed to their allowable limit. It is more probable that one member will *control* the design. Using the equilibrium results and the allowable member forces, we can determine the controlling member and, in turn, compute the maximum load P.

SOLUTION

Equilibrium

The free-body diagram (FBD) for the truss joint is shown. From the FBD, two equilibrium equations in terms of three unknowns—F_1, F_2, and P—can be written. **Note:** We will assume that internal member forces F_1 and F_2 are tension forces (even though we may expect member (2) to be in compression). We have, then,

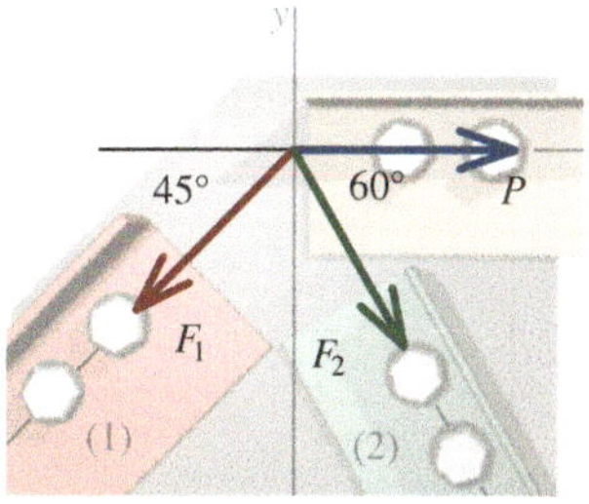

$$\Sigma F_x = -F_1 \cos 45° + F_2 \cos 60° + P = 0 \tag{a}$$

$$\Sigma F_y = -F_1 \sin 45° - F_2 \sin 60° = 0 \tag{b}$$

From these two equations, expressions for the unknown load P can be derived in terms of member forces F_1 and F_2:

$$P = \left[\cos 45° + \frac{\sin 45°}{\sin 60°}\cos 60°\right] F_1 \tag{c}$$

$$P = -\left[\frac{\sin 60°}{\sin 45°}\cos 45° + \cos 60°\right] F_2 \tag{d}$$

Allowable stress: The allowable normal stress in the steel members can be computed from Equation (4.1):

$$\sigma_{\text{allow}} = \frac{\sigma_Y}{\text{FS}} = \frac{36\text{ ksi}}{1.5} = 24\text{ ksi} \tag{e}$$

Allowable member force: The allowable stress can be used to calculate the allowable force in each member:

$$F_{1,\text{allow}} = \sigma_{\text{allow}} A_1 = (24\text{ ksi})(7.22\text{ in.}^2) = 173.28\text{ kips} \tag{f}$$

$$F_{2,\text{allow}} = \sigma_{\text{allow}} A_2 = (24\text{ ksi})(3.88\text{ in.}^2) = 93.12\text{ kips} \tag{g}$$

Problem-Solving Tip: A common mistake at this point in the solution would be to compute P by substituting the two allowable forces into Equation (a). This approach, however, does not work, because equilibrium will not be satisfied in Equation (b). *Equilibrium must always be satisfied.*

Compute maximum P: Next, two possibilities must be investigated: Either member (1) controls, or member (2) controls. First, assume that the allowable force in member (1) controls the design. Substitute the allowable force for member (1) into Equation (c) to compute the maximum load P that would be permitted:

$$P = \left[\cos 45^\circ + \frac{\sin 45^\circ}{\sin 60^\circ}\cos 60^\circ\right]F_1 = 1.11536F_{1,\text{allow}} \tag{h}$$
$$= (1.11536)(173.28 \text{ kips})$$
$$\therefore P \leq 193.27 \text{ kips}$$

Next, use Equation (d) to compute the maximum load P that would be permitted if member (2) controls:

$$P = -\left[\frac{\sin 60^\circ}{\sin 45^\circ}\cos 45^\circ + \cos 60^\circ\right]F_2 = -1.36603F_{2,\text{allow}} \tag{i}$$
$$= -(1.36603)(93.12 \text{ kips})$$
$$\therefore P \leq -127.20 \text{ kips}$$

Why is P negative in Equation (i), and, more important, how do we interpret this negative value? The allowable stress computed in Equation (e) made no distinction between tension and compression stress. Accordingly, the allowable member forces computed in Equations (f) and (g) were *magnitudes* only. These member forces could be tension (i.e., positive values) or compression (i.e., negative values) forces. In Equation (i), a maximum load was computed as $P = -127.20$ kips. This result implies that the load P acts in the $-x$ direction, and, clearly, that is not what the problem intends. Therefore, we must conclude that the allowable force in member (2) is actually a compression force:

$$P \leq -(1.36603)(-93.12 \text{ kips}) = 127.20 \text{ kips} \tag{j}$$

Compare the results from Equations (h) and (j) to conclude that the maximum load that may be applied to this truss joint is

$$P = 127.20 \text{ kips}$$ **Ans.**

Member forces at maximum load P: Member (2) has been shown to *control* the design; in other words, the strength of member (2) is the limiting factor or the most critical consideration. At the maximum load P, use Equations (c) and (d) to compute the actual member forces:

$$F_1 = 114.05 \text{ kips (T)}$$

and

$$F_2 = -93.12 \text{ kips} = 93.12 \text{ kips (C)}$$

The actual normal stresses in the members are

$$\sigma_1 = \frac{F_1}{A_1} = \frac{114.05 \text{ kips}}{7.22 \text{ in.}^2} = 15.80 \text{ ksi (T)}$$

and

$$\sigma_2 = \frac{F_2}{A_2} = \frac{-93.12 \text{ kips}}{3.88 \text{ in.}^2} = 24.0 \text{ ksi (C)}$$

Note: The normal stress *magnitudes* in both members are less than or equal to the 24-ksi allowable stress.

MecMovies

EXAMPLES

M4.1 The structure shown is used to support a distributed load of w = 15 kN/m. Each bolt at A, B, and C has a diameter of 16 mm, and each bolt is used in a double-shear connection. The cross-sectional area of axial member (1) is 3,080 mm^2.

The limiting stress in axial member (1) is 50 MPa, and the limiting stress in the bolts is 280 MPa. Determine the factors of safety with respect to the specified limiting stresses for axial member (1) and bolt C.

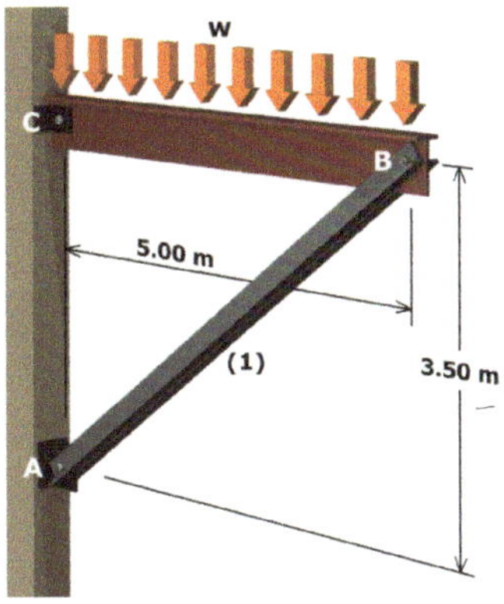

M4.2 Two steel plates are connected by a pair of splice plates with eight bolts, as shown. The ultimate strength of the bolts is 270 MPa. An axial tension load of P = 480 kN is transmitted by the steel plates.

If a factor of safety of 1.6 with respect to failure by fracture is specified, determine the minimum acceptable diameter of the bolts.

M4.3 The structure shown supports a distributed load of w kN/m. The 16 mm diameter bolts at A, B, and C are each used in double-shear connections. The cross-sectional area of axial member (1) is 3,080 mm^2.

The limiting normal stress in axial member (1) is 50 MPa, and the limiting stress in the bolts is 280 MPa. If a minimum factor of safety of 2.0 is required for all components, determine the maximum allowable distributed load w that may be supported by the structure.

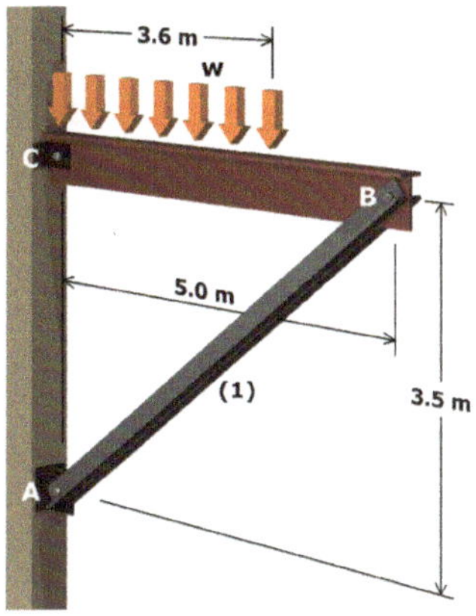

EXERCISES

M4.1 The structure shown supports a specified distributed load. Suppose the limiting stresses for rod (1) and pins A, B, and C are given. Determine the axial force in rod (1), the resultant force in pin C, and the factors of safety with respect to the specified limiting stresses for rod (1) and pins B and C.

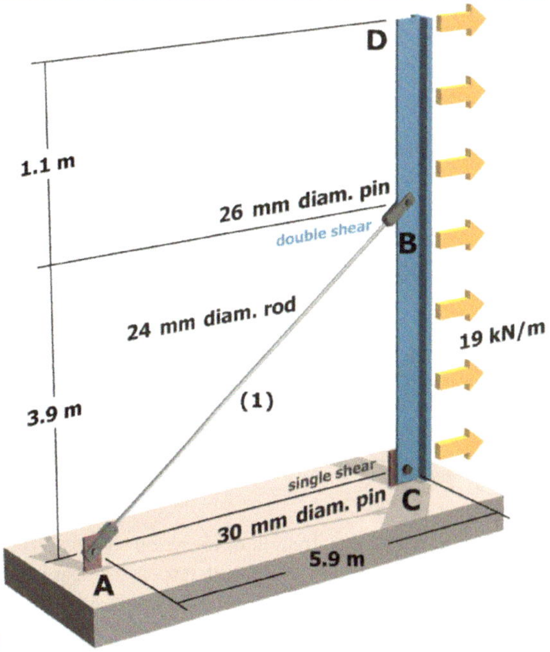

FIGURE M4.1

M4.2 The single-shear connection consists of a number of bolts, as shown. Given the bolt diameter and the ultimate strength of the bolts, determine the factor of safety for the connection for a specified tension load *P*.

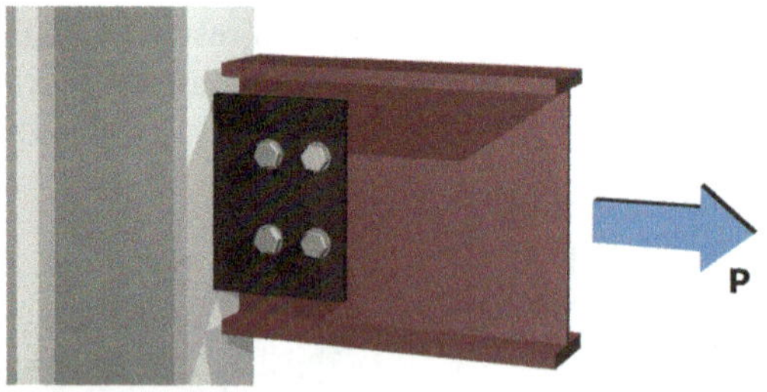

FIGURE M4.2

M4.3 The structure shown supports an unspecified load *w*. Limiting stresses are given for rod (1) and the pins. For a specified minimum factor of safety, determine (a) the maximum load magnitude *w* that may be applied to the structure and (b) the stresses in the rod and pins at the maximum load *w*.

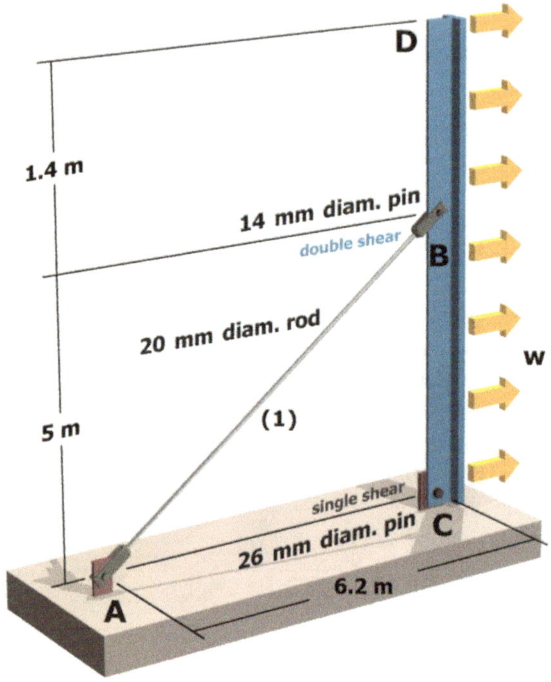

FIGURE M4.3

PROBLEMS

P4.1 Seven bolts are used in the connection between the bar and the support shown in Figure P4.1. The ultimate shear strength of the bolts is 320 MPa, and a factor of safety of 2.5 is required with respect to fracture. Determine the minimum allowable bolt diameter required to support an applied load of $P = 1{,}225$ kN.

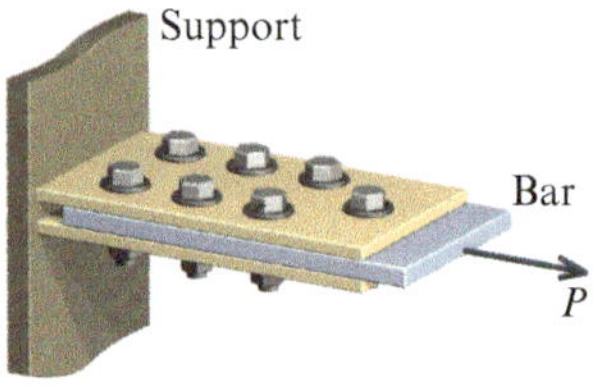

FIGURE P4.1

P4.2 A thrust collar that rests on a 0.75 in. thick polymer plate supports the vertical nylon shaft shown in Figure P4.2. The shaft has a diameter $d = 0.50$ in. The load acting on the shaft is $P = 600$ lb.

(a) The ultimate shear strength of the nylon shaft is 2,700 psi, and a factor of safety of 2.5 with respect to shear is required. Determine the minimum thickness *t* required for the thrust collar.

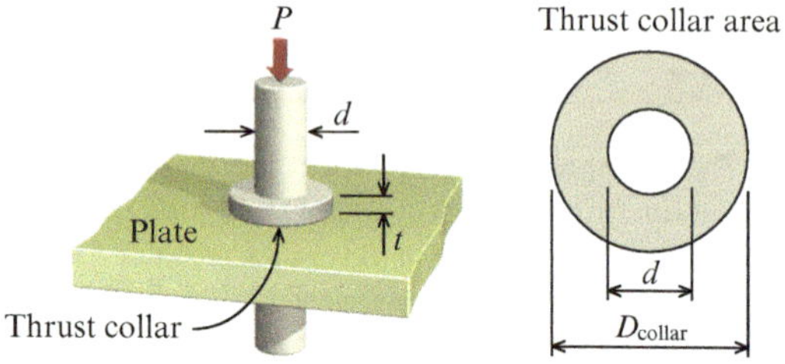

FIGURE P4.2

(b) The ultimate bearing strength of the polymer plate is 2,000 psi, and a factor of safety of 3.0 with respect to bearing is required. Determine the minimum outer diameter D_{collar} required for the thrust collar.

P4.3 Two flat bars loaded in tension by force *P* are spliced with the use of two rectangular splice plates and two 0.375 in. diameter bolts as shown in Figure P4.3. Away from the connection, the bars have a width $b = 0.625$ in. and a thickness $t = 0.25$ in. The bars are made of polyethylene having an ultimate tensile strength of 3,100 psi and an ultimate bearing strength of 2,000 psi. The bolts are made of nylon that has an ultimate shear strength of 4,000 psi. Determine the allowable load *P* for this connection if a safety factor of 3.5 is required. Consider tension and bearing in the bars and shear in the bolts. Disregard friction between the plates.

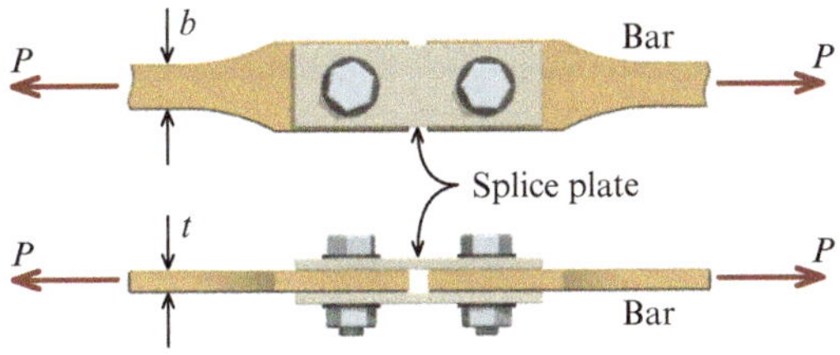

FIGURE P4.3

P4.4 A bar with a rectangular cross section is connected to a support bracket with a circular pin, as shown in Figure P4.4. The bar has a width $b = 40$ mm and a thickness of 10 mm. The pin has a diameter of 12 mm. The yield strength of the bar is 415 MPa, and a factor of safety of 1.67 with respect to yield is required. The bearing strength of the bar is 380 MPa, and a factor of safety of 2.0 with respect to bearing is required. The shear strength of the pin is 550 MPa, and a factor of safety of 2.5 with respect to shear failure is required. Determine the magnitude of the maximum load P that may be applied to the bar.

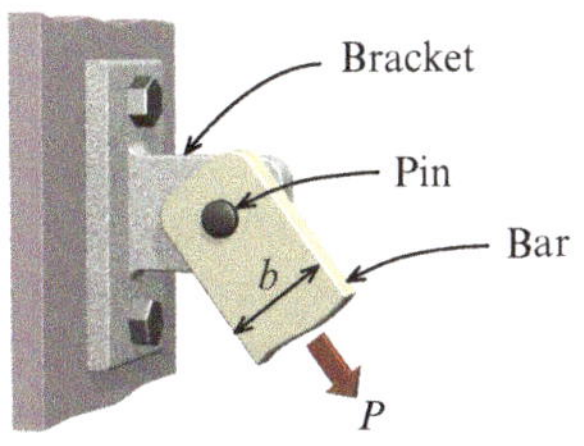

FIGURE P4.4

P4.5 In Figure P4.5, member (1) has a cross-sectional area of 0.7 in.^{2} and a yield strength of 50 ksi. Member (2) has a cross-sectional area of 1.8 in.^{2} and a yield strength of 36 ksi. A factor of safety of 1.67 with respect to yield is required for both members. Use dimensions of $a = 60$ in., $b = 28$ in., $c = 54$ in., and $d = 72$ in. Determine the maximum allowable load P that may be applied to the assembly. Report the factors of safety for both members at the allowable load.

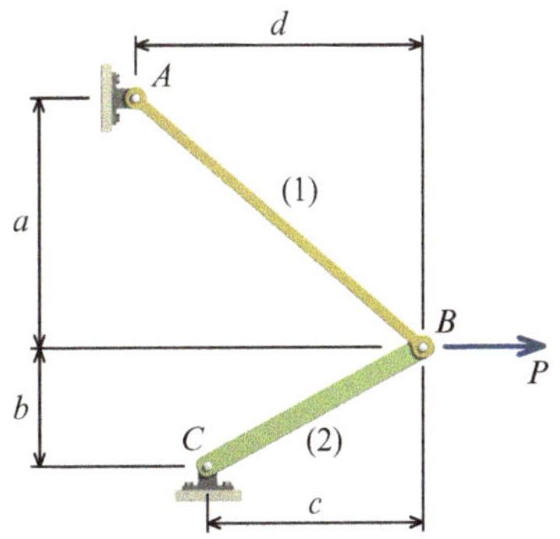

FIGURE P4.5

P4.6 The bell-crank mechanism shown in Figure P4.6 is supported by a single-shear pin connection at B and a roller support at C. A load P acts at an angle of $\theta = 40°$ at joint A. The pin at B has a diameter of 0.375 in., and the bell crank has a thickness of 0.188 in. The ultimate shear strength of the pin material is 28 ksi, and the ultimate bearing strength of the crank material is 52 ksi. A minimum factor of safety of 3.0 with respect to shear and bearing is required. Assume that $a = 9.0$ in. and $b = 5.0$ in. What is the maximum load P that may be applied at joint A of the bell crank?

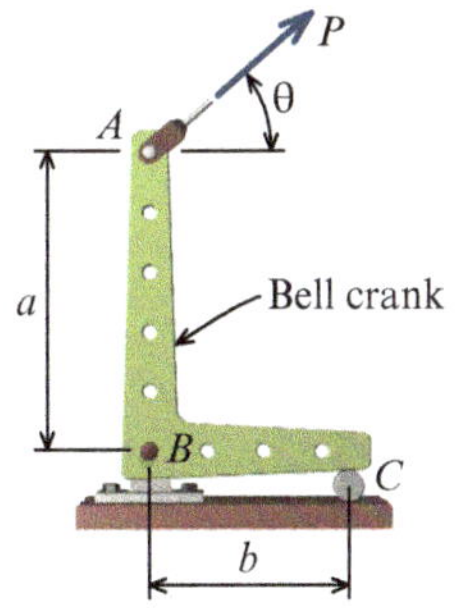

FIGURE P4.6

P4.7 The bell-crank mechanism shown in Figure P4.7 is in equilibrium for a load $F_1 = 7$ kN applied at A. Assume that $a = 150$ mm, $b = 100$ mm, $c = 80$ mm, and $\theta = 55°$. The bell crank has a thickness $t = 10$ mm, and the support bracket has a thickness $t_b = 8$ mm. The pin at B has a diameter $d = 12$ mm and an ultimate shear strength of 290 MPa. The bell crank and the support bracket each have an ultimate bearing strength of 380 MPa. Determine

(a) the factor of safety in pin B with respect to the ultimate shear strength.
(b) the factor of safety of the bell crank at pin B with respect to the ultimate bearing strength.
(c) the factor of safety in the support bracket with respect to the ultimate bearing strength.

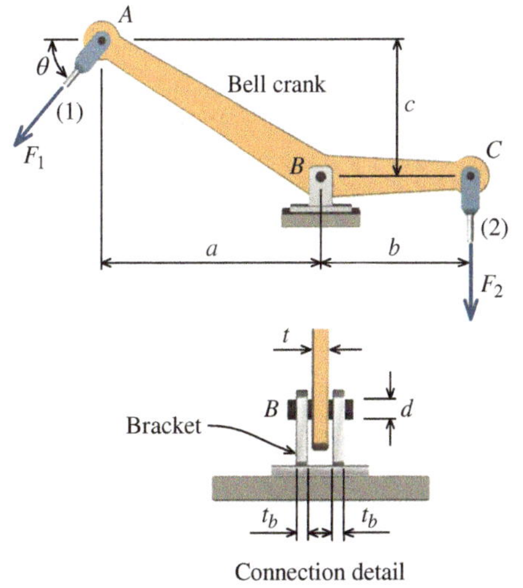

FIGURE P4.7

P4.8 Rigid bar *ABC* is supported by member (1) at *B* and by a pin connection at *C*, as shown in Figure P4.8. Member (1) has a cross-sectional area of 1.25 in.^{2} and a yield strength of 40 ksi. The pins at *B*, *C*, and *D* each have a diameter of 0.5 in. and an ultimate shear strength of 72 ksi. Each pin connection is a double-shear connection. Specifications call for a minimum factor of safety in member (1) of 1.67 with respect to its yield strength. The minimum factor of safety required for pins *B*, *C*, and *D* is 3.0 with respect to the pin ultimate shear strength. Determine the allowable load *P* that may be applied to the rigid bar at *A*. Use overall dimensions of a = 54 in. and b = 110 in.

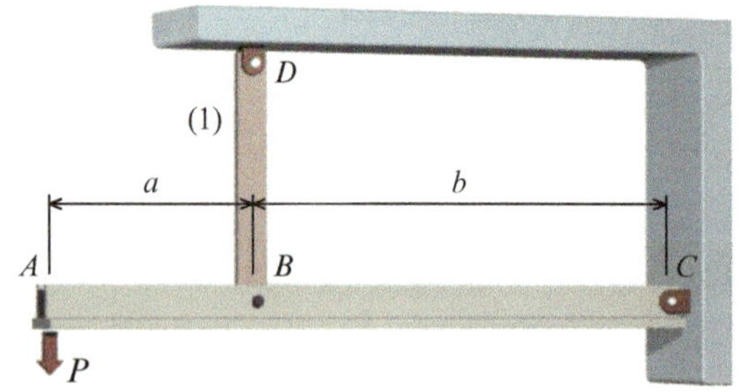

FIGURE P4.8

P4.9 In Figure P4.9, rigid beam *AB* is subjected to a distributed load that increases linearly from zero to a maximum intensity of w_B = 17.5 kN/m. Beam *AB* is supported by a single-shear pin connection at joint *A* and by a double-shear connection to member (1) at joint *B*. Member (1) is connected to the support at *C* with a double-shear pin connection. Use dimensions of a = 2.8 m, b = 1.0 m, and c = 1.2 m.

(a) The yield strength of member (1) is 340 MPa. A factor of safety of 1.67 with respect to the yield strength is required for the normal stress of member (1). Determine the minimum cross-sectional area required for member (1).
(b) The ultimate shear strength of the material used for pins *A*, *B*, and *C* is 270 MPa. A factor of safety of 2.50 with respect to the ultimate shear strength is required for the pins, and all three pins are to have the same diameter. Determine the minimum pin diameter that may be used.

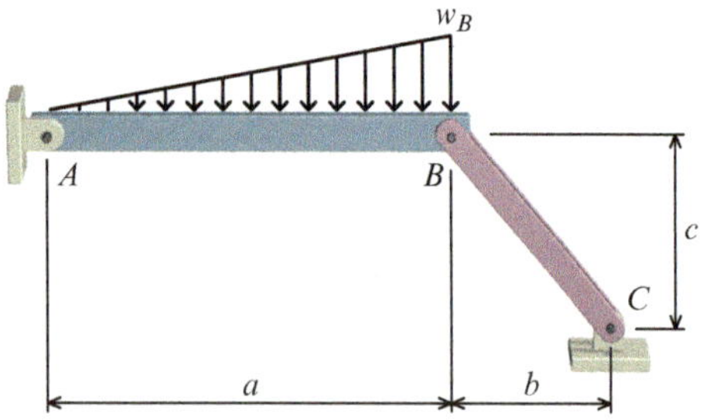

FIGURE P4.9

P4.10 Beam *AB* is supported as shown in Figure P4.10/11. Tie rod (1) is attached at *B* and *C* with double-shear pin connections, while the pin at *A* is attached with a single-shear connection. The pins at *A*, *B*, and *C* each have an ultimate shear strength of 54 ksi, and tie rod (1) has a yield strength of 36 ksi. A uniformly distributed load w = 1,800 lb/ft is applied to the beam as shown. Dimensions are a = 15 ft, b = 30 in., and c = 7 ft. A factor of safety of 2.5 is required for all components. Determine

(a) the minimum required diameter for tie rod (1).
(b) the minimum required diameter for the double-shear pins at *B* and *C*.
(c) the minimum required diameter for the single-shear pin at *A*.

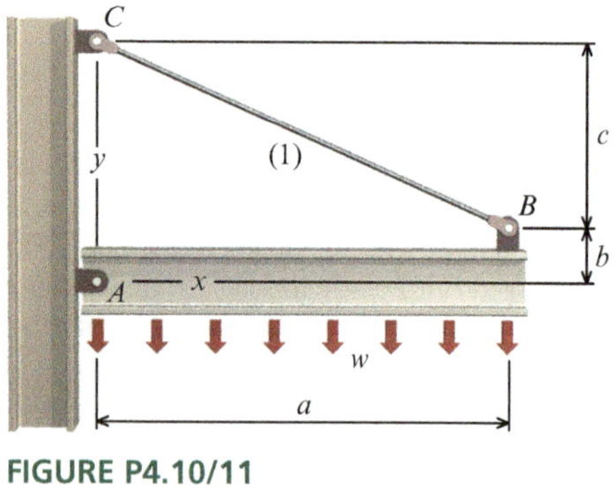

FIGURE P4.10/11

P4.11 Beam *AB* is supported as shown in Figure P4.10/11. Tie rod (1) has a diameter of 50 mm, and it is attached at *B* and *C* with 24 mm diameter double-shear pin connections. The pin connection at *A* consists of a 40 mm diameter single-shear pin. The pins at *A*, *B*, and *C* each have an ultimate shear strength of 520 MPa, and tie rod (1) has a yield strength of 280 MPa. A uniformly distributed load *w* is applied to the beam as shown. Dimensions are a = 4.0 m, b = 0.5 m, and c = 1.5 m. A minimum factor of safety of 2.5 is required for all components. What is the maximum loading *w* that may be applied to the structure?

P4.12 The idler pulley mechanism shown in Figure P4.12 must support belt tensions of P = 130 lb. Rigid bar *ABC* is supported by rod (1), which has a diameter of 0.188 in. and a yield strength of 35,000 psi. Rod (1) is connected at *A* and *B* with 0.25 in. diameter pins in double-shear connections. A 0.25 in. diameter pin in a single-shear connection holds rigid bar *ABC* at support *C*. Each pin has an ultimate shear strength of 42,000 psi. Overall dimensions of the mechanism are a = 12 in., b = 9 in., c = 7 in., and d = 3 in. Determine

(a) the factor of safety for rod (1) with respect to its yield strength.
(b) the factor of safety for pin *B* with respect to its ultimate shear strength.
(c) the factor of safety for pin *C* with respect to its ultimate shear strength.

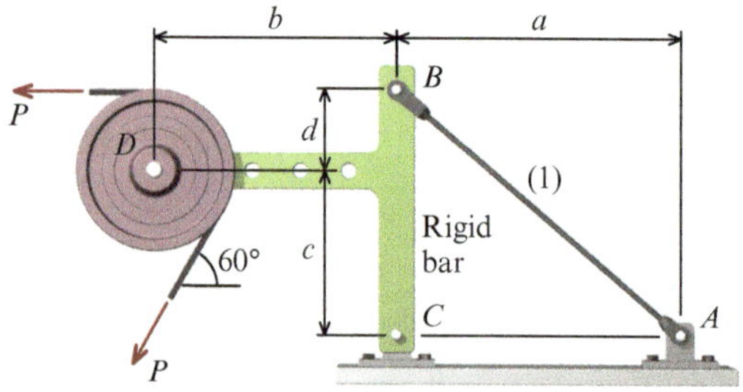

FIGURE P4.12

4.5 Load and Resistance Factor Design

A second common design philosophy is termed **load and resistance factor design (LRFD)**. This approach is most widely used in the design of reinforced concrete, steel, and wood structures.

To illustrate the differences between the ASD and LRFD philosophies, consider the following example: Suppose that an engineer using ASD calculates that a certain member of a steel bridge truss will be subjected to a load of 100 kN. Using an appropriate factor of safety for this type of member—say, 1.6—the engineer properly designs the truss member so that it can support a load of 160 kN. Since the member strength is greater than the load acting on it, the truss member performs its intended function. However, we know that the load on the truss member will change throughout the lifetime of the structure. There will be many times when no vehicles are crossing the bridge, and consequently, the member load will be much less than 100 kN. There may also be instances in which the bridge is completely filled with vehicles and the member load will be greater than 100 kN. The engineer has properly designed the truss member to support a load of 160 kN, but suppose that the steel material was not quite as strong as expected or that stresses were created in the member during the construction process. Then the actual strength of the member could be, say, 150 kN, rather than the expected strength of 160 kN. If the actual load on our hypothetical truss member exceeds 150 kN, the member will fail. Thus, the question is, "How likely is it that this situation will occur?" The ASD approach cannot answer that question in any quantitative manner.

Design provisions in LRFD are based on probability concepts. Strength design procedures in LRFD recognize that the actual loads acting on structures and the true strength of structural components (termed **resistance** in LRFD) are in fact random variables that cannot be known with complete certainty. With the use of statistics to characterize both the load and resistance variables, design procedures are developed so that properly designed components have an acceptably small, but quantifiable, probability of failure, and this probability of failure is consistent among structural elements (e.g., beams, columns, connections, etc.) of different materials (e.g., steel vs. wood vs. concrete) used for similar purposes.

Probability Concepts

To illustrate the concepts inherent in LRFD (without delving too deeply into probability theory), consider the aforementioned truss member example. Suppose that 1,000 truss bridges were investigated and that, in each of those bridges, a typical tension member was singled out. For that tension member, two load magnitudes were recorded. First, the service load effect used in the design calculations (i.e., the design tension force in this case) for a truss member was noted. For purposes of illustration, this service load effect will be denoted Q^*. Second, the maximum tension load effect that acted on the truss member at any time throughout the entire lifetime of the structure was identified. For each case, the maximum tension load effect is compared with the service load effect Q^* and the results are displayed on a histogram showing the frequency of occurrence of differing load levels (Figure 4.2). For example, in 128 out of 1,000 cases, the maximum tension load in the truss member was 20 percent larger than the tension used in the design calculations.

For the same tension members, suppose that two strength magnitudes were recorded. First, the calculated strength of the member was noted. For purposes of illustration, this design strength will be denoted as resistance R^*. Second, the maximum tension strength actually available in the member was determined. This value represents the tension load that would cause the member to fail if it were tested to destruction. The maximum tension strength can be compared with the design resistance R^*, and the results can be displayed on

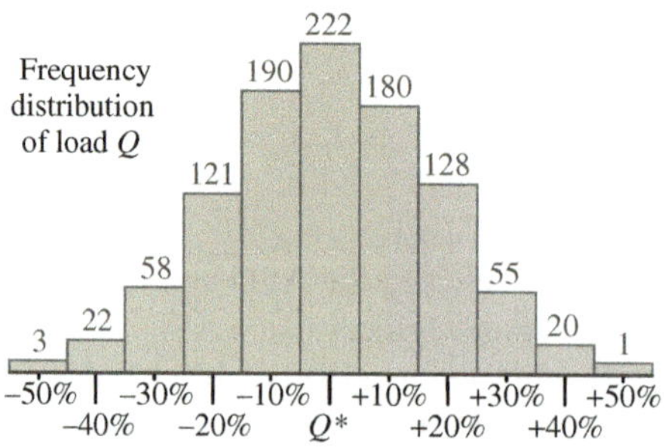

FIGURE 4.2 Histogram of load effects.

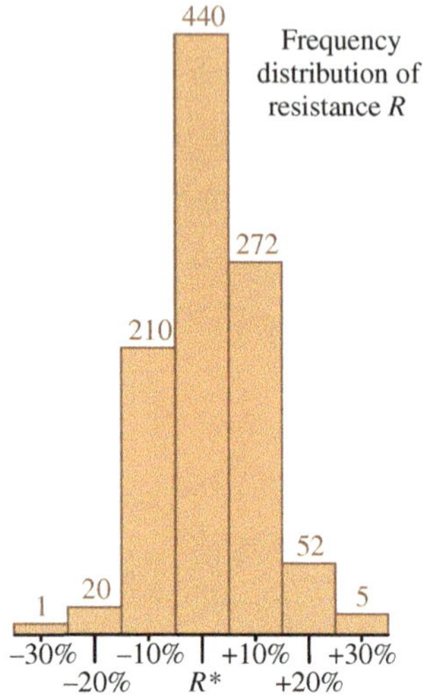

FIGURE 4.3 Resistance histogram.

a histogram showing the frequency of occurrence of differing resistance levels (Figure 4.3). For example, in 210 out of 1,000 cases, the maximum tension strength in the truss member was 10 percent less than the nominal strength predicted by the design calculations.

A structural component will not fail as long as the strength provided by the component is greater than the effect caused by the loads. In LRFD, the general format for a strength design provision is expressed as

$$\phi R_n \geq \Sigma \gamma_i Q_{ni} \tag{4.4}$$

where ϕ = resistance factor corresponding to the type of component (i.e., beam, column, connection, etc.), R_n = nominal component resistance (i.e., strength), γ_i = load factors corresponding to each type of load (i.e., dead load, live load, etc.), and Q_{ni} = nominal service load effects (such as axial force, shear force, and bending moments) for each type of load. In general, the resistance factors ϕ are less than 1 and the load factors γ_i are greater than 1. In nontechnical language, the resistance of the structural component is *underrated* (to account for the possibility that the actual member strength may be less than predicted) whereas the load effect on the member is *overrated* (to account for extreme load events made possible by the variability inherent in the loads).

Regardless of the design philosophy, a properly designed component must be stronger than the load effects acting on it. In LRFD, however, the process of establishing appropriate design factors considers the member resistance R and load effect Q as random variables rather than quantities that are known exactly. Suitable factors for use in LRFD design equations, as typified by Equation (4.4), are determined through a process that takes into account the relative positions of the member resistance distribution R (Figure 4.3) and the load effects distribution Q (Figure 4.2). Appropriate values of the ϕ and γ_i factors are determined through a procedure known as **code calibration** that uses a **reliability analysis** in which the ϕ and γ_i factors are chosen so that a specific target probability of failure is achieved. The design strength of members is based on the load effects; therefore, the design factors "shift" the resistance distribution to the right of the load distribution so that the strength is greater than the load effect (Figure 4.4).

To illustrate this concept, consider the data obtained from the 1,000-bridge example. The use of very small ϕ factors and very large γ_i factors would ensure that all truss members are strong enough to withstand all load effects (Figure 4.4). This situation, however, would be overly conservative and might produce structures that are unnecessarily expensive.

The use of relatively large ϕ factors and relatively small γ_i factors would create a region in which the resistance distribution R and the load distribution Q overlap (Figure 4.5); in other words, the member strength will be less than or equal to the load effect. From Figure 4.5, one would predict that 22 out of 1,000 truss members will fail. (**Note:** The truss members are properly designed. The failure discussed here is due to random variation rather

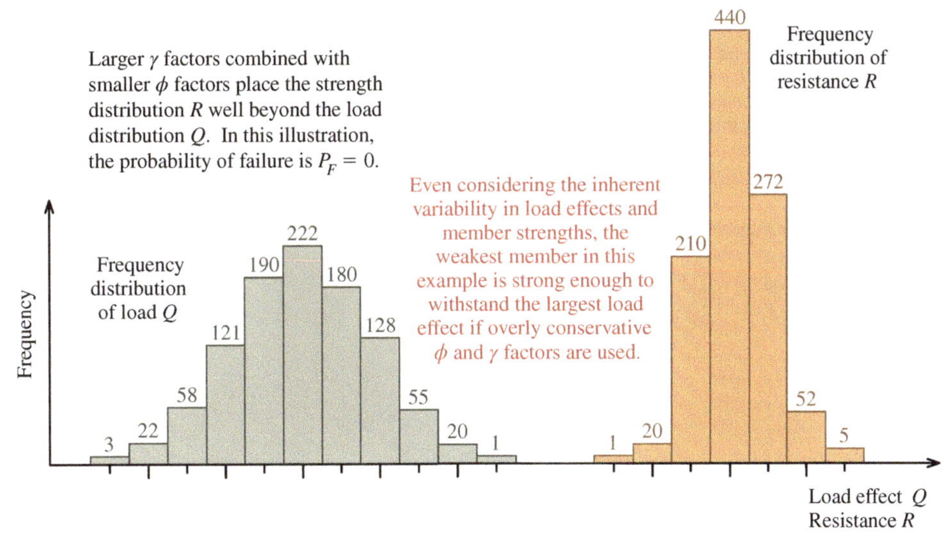

FIGURE 4.4 Overly conservative load and resistance factors produce designs with a near-zero probability of failure.

than error or incompetence.) A probability of failure $P_F = 0.022$ represents too much risk to be acceptable, particularly where public safety is directly concerned.

An appropriate combination of ϕ and γ_i factors creates a small region of overlap between R and Q (Figure 4.6). From the figure, the probability of failure is 1 out of 1,000 truss members, or $P_F = 0.001$. This rate might represent an acceptable trade-off between risk and cost. (The value $P_F = 0.001$ is known as a **notional failure rate**. The true failure rate is always much less, as engineering experience has shown over years of successful practice. In reliability analyses, often only the means and standard deviations of many variables can be estimated and the true shape of the random variable distributions is generally not known. These and other considerations lead to higher predicted failure rates than actually occur in practice.)

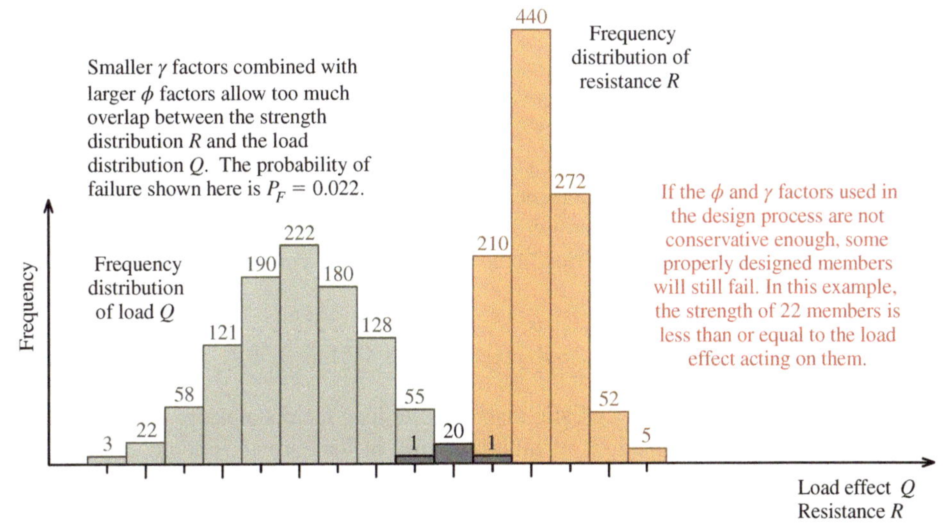

FIGURE 4.5 Unconservative load and resistance factors produce an unacceptable probability of failure.

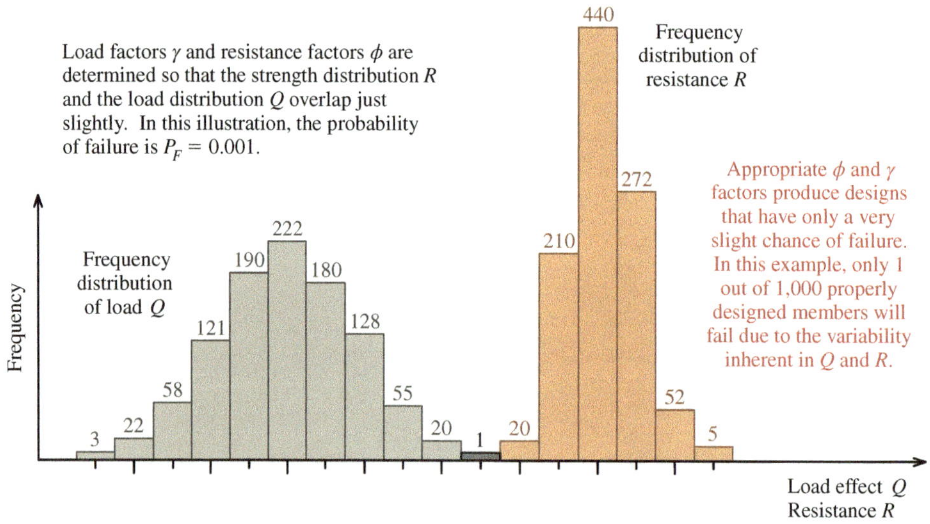

FIGURE 4.6 Appropriate load and resistance factors produce a satisfactory probability of failure.

Load Combinations

Loads that act on structures are inherently variable. Although the designer may make a reasonable estimate of the service loads that are expected to act on a structure, it is likely that the actual loads will differ from the service loads. Further, the range of variation expected for each type of load is different. For example, live loads could be expected to vary more widely than dead loads. To account for load variability, LRFD multiplies each type of load by specific load factors γ_i and sums the load components to obtain an ultimate load at which failure (i.e., rupture or collapse) is considered imminent. The structure or structural component is then proportioned so that the nominal strength ϕR_n of the component is equal to or greater than the ultimate load U.

For example, the ultimate load U due to a combination of dead load D and live load L acting simultaneously on a structural steel component would be computed with the load factors shown in the following equation

$$U = \Sigma \gamma_i Q_{ni} = 1.2D + 1.6L \tag{4.5}$$

The larger load factor $\gamma_L = 1.6$ associated with the live load L reflects the greater uncertainty inherent in this type of load compared with the dead load D, which is known with much greater certainty and, accordingly, has a smaller load factor $\gamma_D = 1.2$.

Various possible load combinations must be checked, and each combination has a unique set of load factors. For example, the ultimate load U acting on a structural steel member through a combination of dead load D, live load L, wind load W, and snow load S would be calculated as

$$U = \Sigma \gamma_i Q_{ni} = 1.2D + 1.3\,W + 0.5L + 0.5S \tag{4.6}$$

While load factors are generally greater than 1, lesser load factors are appropriate for some types of loads when combinations of multiple types of load are considered. These lesser load factors reflect the low probability that extreme events in multiple types of loads will occur simultaneously. For example, it is not likely that the largest snow load will occur at the same moment as the extreme wind load and the extreme live load. (It is not even likely that the extreme wind load and the extreme live load will occur simultaneously.)

Limit States

LRFD is based on a **limit states** philosophy. In this context, the term *limit state* is used to describe a condition under which a structure or some portion of the structure ceases to perform its intended function. Two general kinds of limit states apply to structures: **strength limit states** and **serviceability limit states**. Strength limit states define safety with regard to extreme load events during which the overriding concern is the protection of human life from sudden or catastrophic structural failure. Serviceability limit states pertain to the satisfactory performance of structures under ordinary load conditions. These limit states include considerations such as excessive deflections, vibrations, cracking, and other concerns that may have functional or economic consequences but do not threaten public safety.

EXAMPLE 4.3

A rectangular steel plate is subjected to an axial dead load of 30 kips and a live load of 48 kips. The yield strength of the steel is 36 ksi.

(a) *ASD Method*: If a factor of safety of 1.5 with respect to yielding is required, determine the required cross-sectional area of the plate according to the ASD method.

(b) *LRFD Method*: Use the LRFD method to determine the required cross-sectional area of the plate on the basis of yielding of the gross section. Use a resistance factor $\phi_t =$ 0.9 and load factors of 1.2 and 1.6 for the dead and live loads, respectively.

Plan the Solution

A simple design problem illustrates how the two methods are used.

SOLUTION

(a) ASD Method

Determine the allowable normal stress from the specified yield stress and the factor of safety:

$$\sigma_{\text{allow}} = \frac{\sigma_Y}{\text{FS}} = \frac{36 \text{ ksi}}{1.5} = 24 \text{ ksi}$$

The service load acting on the tension member is the sum of the dead and live components:

$$P = D + L = 30 \text{ kips} + 48 \text{ kips} = 78 \text{ kips}$$

The cross-sectional area required to support the service load is computed as

$$A \geq \frac{P}{\sigma_{\text{allow}}} = \frac{78 \text{ kips}}{24 \text{ ksi}} = 3.25 \text{ in.}^2 \qquad \textbf{Ans.}$$

(b) LRFD Method

The factored load acting on the tension member is computed as

$$P_u = 1.2D + 1.6L = 1.2(30 \text{ kips}) + 1.6(48 \text{ kips}) = 112.8 \text{ kips}$$

The nominal strength of the tension member is the product of the yield stress and the cross-sectional area:

$$P_n = \sigma_Y A$$

The design strength is the product of the nominal strength and the resistance factor for this type of component (i.e., a tension member). The design strength must equal or exceed the factored load acting on the member:

$$\phi_t P_n \geq P_u$$

Thus, the cross-sectional area required to support the given loading is

$$\phi_t P_n = \phi_t \sigma_Y A \geq P_u$$

$$\therefore A \geq \frac{P_u}{\phi_t \sigma_Y} = \frac{112.8 \text{ kips}}{0.9(36 \text{ ksi})} = 3.48 \text{ in.}^2$$ **Ans.**

PROBLEMS

P4.13 A 25 mm thick steel plate will be used as an axial member to support a dead load of 280 kN and a live load of 410 kN. The yield strength of the steel is 250 MPa.

(a) Use the ASD method to determine the minimum plate width b required for the axial member if a factor of safety of 1.67 with respect to yielding is required.
(b) Use the LRFD method to determine the minimum plate width b required for the axial member on the basis of yielding of the gross section. Use a resistance factor $\phi_t = 0.9$ and load factors of 1.2 and 1.6 for the dead and live loads, respectively.

P4.14 A round steel tie rod is used as a tension member to support a dead load of 90 kN and a live load of 120 kN. The yield strength of the steel is 320 MPa.

(a) Use the ASD method to determine the minimum diameter required for the tie rod if a factor of safety of 2.0 with respect to yielding is required.
(b) Use the LRFD method to determine the minimum diameter required for the tie rod on the basis of yielding of the gross section. Use a resistance factor $\phi_t = 0.9$ and load factors of 1.2 and 1.6 for the dead and live loads, respectively.

CHAPTER 5

Axial Deformation

5.1 Introduction

In Chapter 1, the concept of stress was developed as a means of measuring the force distribution within a body. In Chapter 2, the concept of strain was introduced to describe the deformation produced in a body. Chapter 3 discussed the behavior of typical engineering materials and how this behavior can be idealized by equations that relate stress and strain. Of particular interest are materials that behave in a linear-elastic manner. For these materials, there is a proportional relationship between stress and strain—a relationship that can be idealized by Hooke's law. Chapter 4 discussed two general approaches to designing components and structures that perform their intended function while maintaining an appropriate margin of safety. In the remaining chapters of the book, these concepts will be employed to investigate a wide variety of structural members subjected to axial, torsional, and flexural loadings.

The problem of determining forces and deformations at all points within a body subjected to external forces is extremely difficult when the loading or geometry of the body is complicated. Therefore, practical solutions to most design problems employ what has become known as the *mechanics of materials approach.* With this approach, real structural elements are analyzed as idealized models subjected to simplified loadings and restraints. The resulting solutions are approximate, since they consider only the effects that significantly affect the magnitudes of stresses, strains, and deformations.

More powerful computational methods derived from the *theory of elasticity* are available to analyze objects that involve complicated loading and geometry. Of these methods, the *finite element method* is the most widely used. Although the *mechanics of materials approach* presented here is somewhat less rigorous than the *theory of elasticity approach*, experience indicates that the results obtained from the mechanics of materials approach are quite satisfactory for a wide variety of important engineering problems. One of the primary reasons for this is **Saint-Venant's principle**.

5.2 Saint-Venant's Principle

Consider a rectangular bar subjected to a compressive axial force P (Figure 5.1). The bar is fixed at its base, and the total force P is applied to the top of the bar in three equal portions distributed as shown over a narrow region equal to one-fourth of the bar's width. The magnitude of force P is such that the material behaves elastically; therefore, Hooke's law applies. The deformations of the bar are indicated by the grid lines shown. In particular, notice that the grid lines are distorted in the regions near force P and near the fixed base. Away from these two regions, however, the grid lines are not distorted, remaining orthogonal and uniformly compressed in the direction of the applied force P.

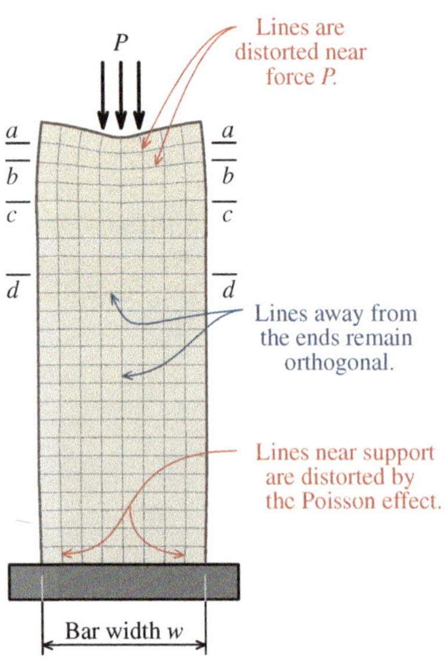

FIGURE 5.1 Rectangular bar subjected to compressive force.

Since Hooke's law applies, stress is proportional to strain (and, in turn, deformation). Therefore, stress will become more uniformly distributed throughout the bar as the distance from the load P increases. To illustrate the variation of stress with distance from P, the normal stresses acting in the vertical direction on Sections *a–a*, *b–b*, *c–c*, and *d–d* (see Figure 5.1) are shown in Figure 5.2. On Section *a–a* (Figure 5.2*a*), normal stresses directly under P are quite large while stresses on the remainder of the cross section are very small. On Section *b–b* (Figure 5.2*b*), stresses in the middle of the bar are still pronounced but stresses away from the middle are significantly larger than those on Section *a–a*. Stresses are more uniform on Section *c–c* (Figure 5.2*c*). On Section *d–d* (Figure 5.2*d*), which is located below P at a distance equal to the bar width w, stresses are essentially constant across the width of the rectangular bar. This comparison shows that localized effects caused by a load tend to vanish as the distance from the load increases. In general, the stress distribution becomes nearly uniform at a distance equal to the bar width w from the end of the bar, where w is the largest lateral dimension of the axial member (such as the bar width or the rod diameter). The maximum stress at this distance is only a few percent larger than the average stress.

In Figure 5.1, the grid lines are also distorted near the base of the axial bar because of the Poisson effect. The bar ordinarily would expand in width in response to the compressive normal strain caused by P. The fixity of the base prevents this expansion, and consequently, additional stresses are created. Using an argument similar to that just given, we could show that this increase in stress becomes negligible at a distance w above the base.

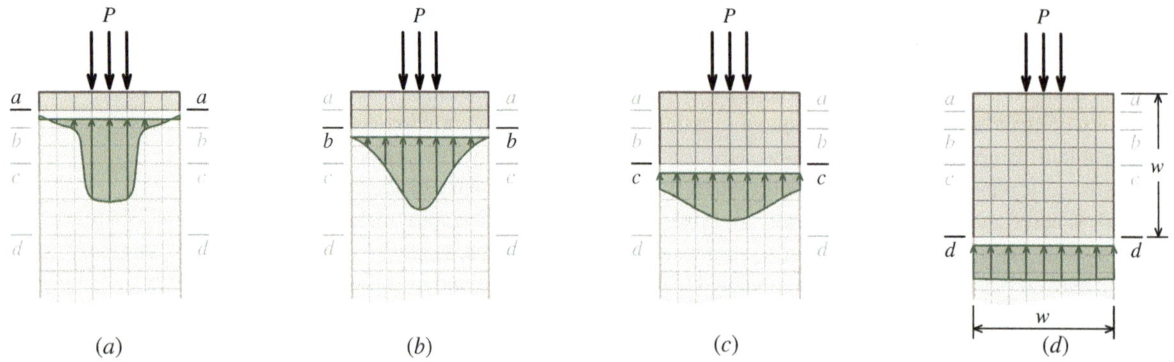

FIGURE 5.2 Normal stress distributions on sections.

poisson ratio: tension ⇒ contracts laterally
compressed ⇒ expands laterally

$\nu = -\dfrac{\varepsilon_{lat}}{\varepsilon_{long}}$

The increased normal stress magnitudes near P and near the fixed base are examples of **stress concentrations**. Stress concentrations occur where loads are applied, and they also occur in the vicinity of holes, grooves, notches, fillets, and other changes in shape that interrupt the smooth flow of stress through a solid body. Stress concentrations associated with axial loads will be discussed in more detail in Section 5.7, and stress concentrations associated with other types of loading will be discussed in subsequent chapters.

The behavior of strain near points where loads are applied was discussed in 1855 by Barré de Saint-Venant (1797–1886), a French mathematician. Saint-Venant observed that localized effects disappeared at some distance from such points. Furthermore, he observed that the phenomenon was independent of the distribution of the applied load as long as the resultant forces were "equipollent" (i.e., statically equivalent). This idea is known as **Saint-Venant's principle** and is widely used in engineering design.

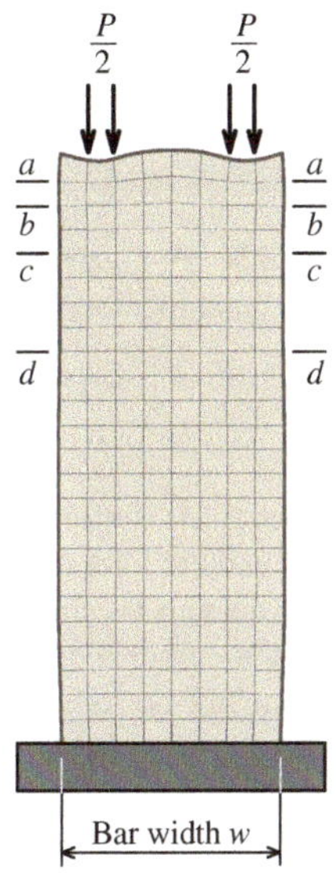

FIGURE 5.3 Rectangular bar with a different, but equivalent, applied load distribution.

Saint-Venant's principle is independent of the distribution of the applied load, provided that the resultant forces are equivalent. To illustrate this independence, consider the same axial bar as discussed before; however, in this instance, the force P is split into four equal portions and applied to the upper end of the bar, as shown in Figure 5.3. As in the previous case, the grid lines are distorted near the applied loads, but they become uniform at a moderate distance away from the point where the load is applied. Normal stress distributions on Sections a–a, b–b, c–c, and d–d are shown in Figure 5.4. On Section a–a (Figure 5.4a), normal stresses directly under the applied loads are quite large while stresses in the middle of the cross section are very small. As the distance from the load increases, the peak stresses diminish (Figure 5.4b; Figure 5.4c) until the stresses become essentially uniform at Section d–d (Figure 5.4d), which is located below P at a distance equal to the bar width w.

To summarize, peak stresses (Figure 5.2a; Figure 5.4a) may be several times the average stress (Figure 5.2d; Figure 5.4d); however, the maximum stress diminishes rapidly as the distance from the point where the load is applied increases. This observation is also generally true for most stress concentrations (such as holes, grooves, and fillets). Thus, the complex localized stress distribution that occurs near loads, supports, or other stress concentrations will not significantly affect stresses in a body at sections *sufficiently distant* from them. In other words, localized stresses and deformations have little effect on the overall behavior of a body.

Expressions will be developed throughout the study of mechanics of materials for stresses and deformations in various members under various types of loadings. According to Saint-Venant's principle, these expressions are valid for *entire members*, except for those regions very near load application points, very near supports, or very near abrupt changes in member cross section.

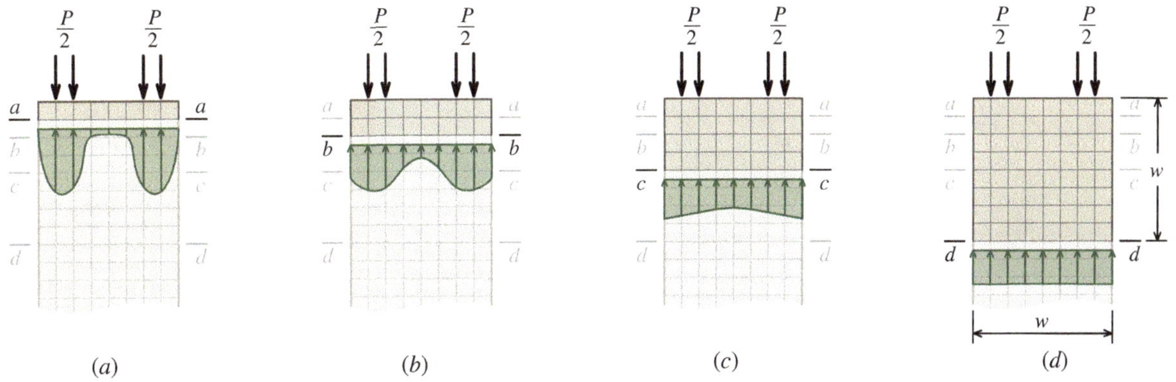

FIGURE 5.4 Normal stress distributions on sections.

5.3 Deformations in Axially Loaded Bars

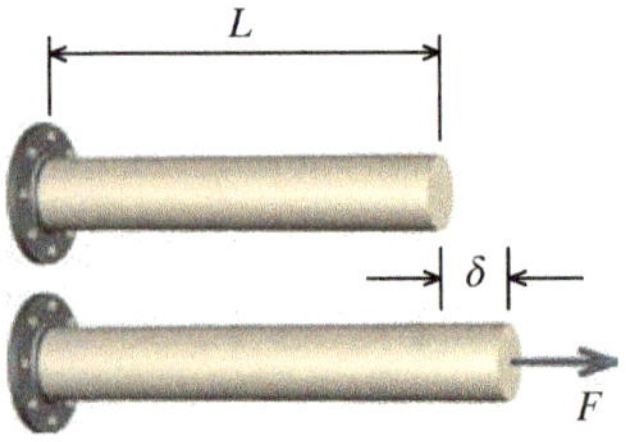

FIGURE 5.5 Elongation of a prismatic axial member.

When a bar of uniform cross section is axially loaded by forces applied at the ends (such a bar is called a two-force member), the normal strain along the length of the bar is assumed to have a constant value. By definition, the deformation (Figure 5.5) of the bar resulting from the axial force F may be expressed as $\delta = \varepsilon L$. The axial stress in the bar is given by $\sigma = F/A$, where A is the cross-sectional area. If the axial stress does not exceed the proportional limit of the material, Hooke's law may be applied to relate stress and strain: $\sigma = E\varepsilon$. Thus, the axial deformation may be expressed in terms of stress or load as:

$$\delta = \varepsilon L = \frac{\sigma L}{E} \tag{5.1}$$

or

$$\delta = \frac{FL}{AE} \tag{5.2}$$

A member that is subjected to no moments and has forces applied at only two points is called a **two-force member**. For equilibrium, the line of action of both forces must pass through the two points where the forces are applied.

A material of uniform composition is called a **homogeneous material.** The term **prismatic** describes a structural member that has a straight longitudinal axis and a constant cross section.

Equation (5.2) is termed the *force–deformation relationship* for axial members and is the more general of these two equations. Equation (5.1) frequently proves convenient when a situation involves a limiting axial stress, such as the design of a bar so that both its deformation and its axial stress are within prescribed limits. In addition, Equation (5.1) often proves to be convenient when a problem involves the comparison of stresses in assemblies consisting of multiple axial members.

Equations (5.1) and (5.2) may be used only if the axial member

- is homogeneous (i.e., E is constant),
- is prismatic (i.e., has a uniform cross-sectional area A), and
- has a constant internal force (i.e., is loaded only by forces at its ends).

If the member is subjected to axial loads at intermediate points (i.e., points other than the ends) or if it consists of various cross-sectional areas or materials, then the axial member must be divided into segments that satisfy the three requirements just listed. For compound axial members consisting of two or more segments, the overall deformation of the axial member can be determined by algebraically adding the deformations of all the segments:

$$\delta = \sum_i \frac{F_i L_i}{A_i E_i} \tag{5.3}$$

Here, F_i, L_i, A_i, and E_i are the internal force, length, cross-sectional area, and elastic modulus, respectively, for individual segments i of the compound axial member.

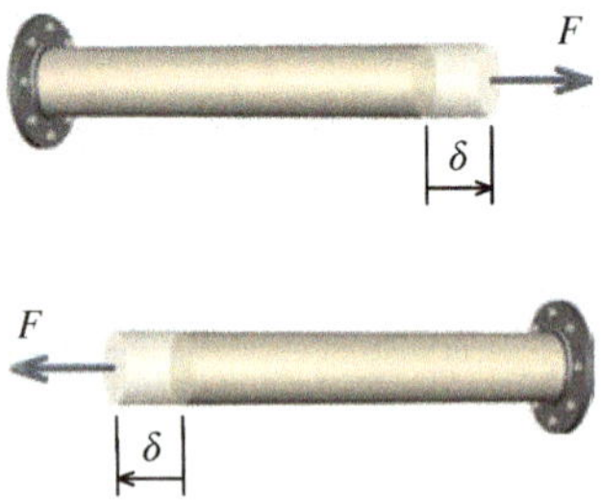

FIGURE 5.6 Positive sign convention for internal force F and deformation δ.

In Equation (5.3), a consistent sign convention is necessary to calculate the deformation δ produced by an internal force F. The **sign convention** (Figure 5.6) for deformation is defined as follows:

- A positive value of δ indicates that the axial member gets longer; accordingly, a positive internal force F produces tensile normal stress.
- A negative value of δ indicates that the axial member gets shorter (termed *contraction*). A negative internal force F produces compressive normal stress.

A three-segment compound axial member is shown in Figure 5.7*a*. To determine the overall deformation of this axial member, the deformations for each of the three segments

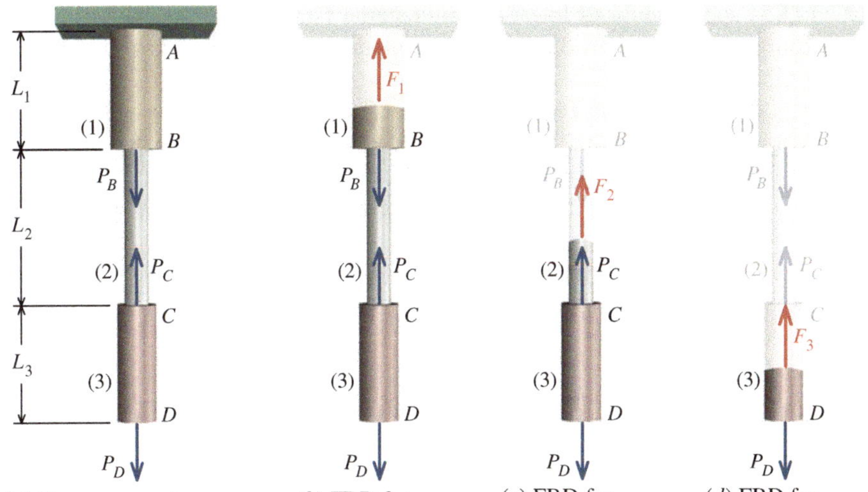

(*a*) Three-segment axial member

(*b*) FBD for segment (1)

(*c*) FBD for segment (2)

(*d*) FBD for segment (3)

FIGURE 5.7 Compound axial member and associated free-body diagrams (FBDs).

are first calculated individually. Then, the three deformation values are added together to give the overall deformation. The internal force F_i in each segment is determined from the free-body diagrams shown in Figure 5.7*b*–*d*.

For those cases in which the axial force or the cross-sectional area varies continuously along the length of the bar (Figure 5.8*a*), Equations (5.1), (5.2), and (5.3) are not valid. In Section 2.2, the normal strain at a point for the case of nonuniform deformation was defined as $\varepsilon = d\delta/dL$. Thus, the increment of deformation associated with a differential element of length $dL = dx$ may be expressed as $d\delta = \varepsilon\,dx$. If Hooke's law applies, the strain may again be expressed as $\varepsilon = \sigma/E$, where $\sigma = F(x)/A(x)$ and both the internal force F and the cross-sectional area A may be functions of position x along the bar (Figure 5.8*b*). Thus,

$$d\delta = \frac{F(x)}{A(x)E}dx \tag{5.4}$$

Integrating Equation (5.4) yields the following expression for the total deformation of the bar:

$$\delta = \int_0^L d\delta = \int_0^L \frac{F(x)}{A(x)E}dx \tag{5.5}$$

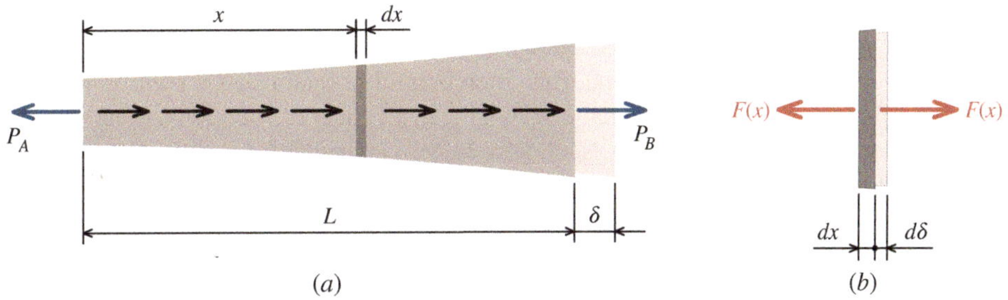

FIGURE 5.8 Axial member with varying internal force and cross-sectional area.

Equation (5.5) applies only to linear-elastic material (since Hooke's law was assumed). Also, Equation (5.5) was derived under the assumption that the stress distribution was uniformly distributed over every cross section [i.e., $\sigma = F(x)/A(x)$]. While this is true for prismatic bars, it is not true for tapered bars. However, Equation (5.5) gives acceptable results if the angle between the sides of the bar is small. For example, if the angle between the sides of the bar does not exceed 20°, there is less than a 3 percent difference between the results obtained from Equation (5.5) and the results obtained from more advanced elasticity methods.

MecMovies

EXAMPLE

M5.3 A load of $P = 50$ kN is applied to a compound axial member. Segment (1) is a solid brass [E = 100 GPa] rod 20 mm in diameter. Segment (2) is a solid aluminum [E = 70 GPa] rod. Determine the minimum diameter of the aluminum segment if the axial displacement of C relative to support A must not exceed 5 mm.

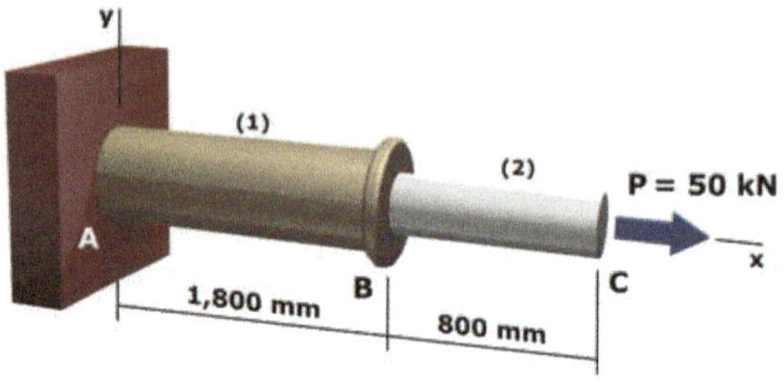

EXAMPLE 5.1

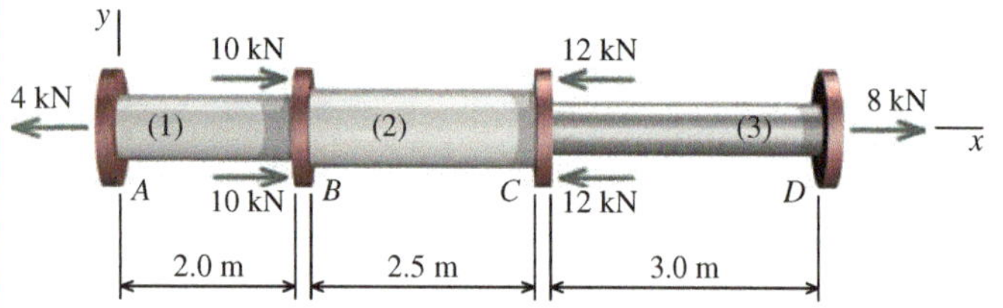

The compound axial member shown consists of a solid aluminum [E = 70 GPa] segment (1) 20 mm in diameter, a solid aluminum segment (2) 24 mm in diameter, and a solid steel [E = 200 GPa] segment (3) 16 mm in diameter. Determine the displacements of points B, C, and D relative to end A.

Plan the Solution

Free-body diagrams (FBDs) will be drawn to expose the internal axial forces in each segment. With the use of the internal force and the cross-sectional area, the normal stress can be computed. The deformation of each segment can be computed from Equation (5.2), and Equation (5.3) will be used to compute the displacements of points B, C, and D relative to end A.

Nomenclature

Before we begin the solution, we will define the terms used to discuss problems of this type. Segments (1), (2), and (3) will be referred to as *axial members* or simply *members*. Members are deformable: They either elongate or contract in response to their internal axial force. As a rule, the internal axial force in a member will be assumed to be *tension*. While this convention is not essential, it is often helpful to establish a repetitive solution procedure that can be applied as a matter of course in a variety of situations. Members are labeled by a number in parentheses, such as member (1), and deformations in a member are denoted δ_1, δ_2, etc.

Points A, B, C, and D refer to *joints*. A joint is the connection point between components (adjacent members in this example), or a joint may simply denote a specific location (such as joints A and D). Joints do not elongate or contract—they *move*, either in translation or in rotation. Therefore, a joint may be said to undergo *displacement*. (In other contexts, a joint might also *rotate* or *deflect*.) Joints are denoted by a capital letter. A joint displacement in the horizontal direction is denoted by u together with a subscript identifying the joint (e.g., u_A). A positive displacement u means that the joint moves to the right.

SOLUTION

Equilibrium

Draw an FBD that exposes the internal axial force in member (1). Assume tension in member (1).

The equilibrium equation for this FBD is

$$\Sigma F_x = F_1 - 4\text{ kN} = 0$$

$$\therefore F_1 = 4\text{ kN} = 4\text{ kN (T)}$$

Draw an FBD for member (2) and assume tension in member (2).

The equilibrium equation for this FBD is

$$\Sigma F_x = F_2 + 2(10\text{ kN}) - 4\text{ kN} = 0$$

$$\therefore F_2 = -16\text{ kN} = 16\text{ kN (C)}$$

Similarly, draw an FBD for member (3) and assume tension in the member. Although two different FBDs are possible, the simpler one is shown.

The equilibrium equation for this FBD is

$$\Sigma F_x = -F_3 + 8\text{ kN} = 0$$

$$\therefore F_3 = 8\text{ kN} = 8\text{ kN (T)}$$

Before proceeding, plot the *internal forces* F_1, F_2, and F_3 acting in the compound member. It is the *internal forces* that create deformations in the axial members, not the external forces applied at joints A, B, C, and D.

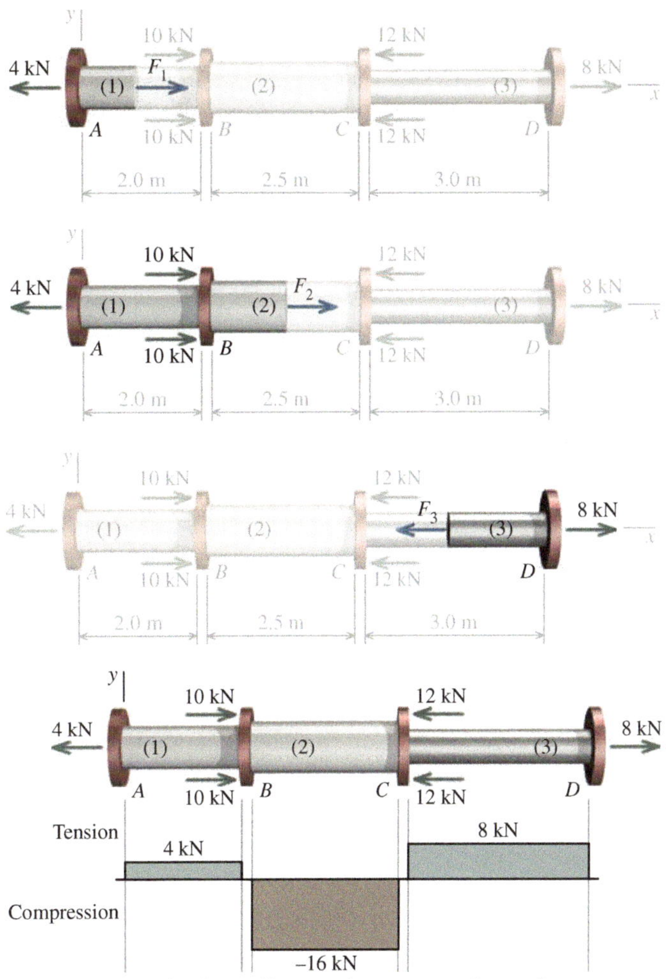

Axial force diagram for compound member.

Problem-Solving Tip: When drawing an FBD that cuts through an axial member, assume that the internal force is tension and draw the force arrow so that it is directed *away from the cut surface*. If the computed internal force value turns out to be a positive number, then the assumption of tension is confirmed. If the computed value turns out to be a negative number, then the internal force is actually compression.

Force–Deformation Relationships

The relationship between the deformation of an axial member and its internal force is expressed by Equation (5.2):

$$\delta = \frac{FL}{AE}$$

Since the internal force is assumed to be a tensile force, the axial deformation is assumed to be an elongation. If the internal force is compressive, then the use of a negative value for the internal force F in the preceding equation will produce a *negative deformation*—in other words, a *contraction*.

Now compute the deformations in each of the three members. Member (1) is a solid aluminum rod 20 mm in diameter; therefore, its cross-sectional area is $A_1 = 314.159\ \text{mm}^2$, and

$$\delta_1 = \frac{F_1 L_1}{A_1 E_1} = \frac{(4\ \text{kN})(1{,}000\ \text{N/kN})(2.0\ \text{m})(1{,}000\ \text{mm/m})}{(314.159\ \text{mm}^2)(70\ \text{GPa})(1{,}000\ \text{MPa/GPa})} = 0.364\ \text{mm}$$

Member (2) has a diameter of 24 mm; therefore, its cross-sectional area is $A_2 = 452.389\ \text{mm}^2$, and we have

$$\delta_2 = \frac{F_2 L_2}{A_2 E_2} = \frac{(-16\ \text{kN})(1{,}000\ \text{N/kN})(2.5\ \text{m})(1{,}000\ \text{mm/m})}{(452.389\ \text{mm}^2)(70\ \text{GPa})(1{,}000\ \text{MPa/GPa})} = -1.263\ \text{mm}$$

The negative value of δ_2 indicates that member (2) contracts.

Member (3) is a solid steel rod 16 mm in diameter. Its cross-sectional area is $A_3 = 201.062\ \text{mm}^2$, and

$$\delta_3 = \frac{F_3 L_3}{A_3 E_3} = \frac{(8\ \text{kN})(1{,}000\ \text{N/kN})(3.0\ \text{m})(1{,}000\ \text{mm/m})}{(201.062\ \text{mm}^2)(200\ \text{GPa})(1{,}000\ \text{MPa/GPa})} = 0.597\ \text{mm}$$

Geometry of Deformations

Since the joint displacements of B, C, and D relative to joint A are desired, joint A will be taken as the origin of the coordinate system. *How are the joint displacements related to the member deformations in the compound axial member?* The deformation of an axial member can be expressed as the difference between the displacements of the end joints of the member. For example, the deformation of member (1) can be expressed as the difference between the displacement of joint A (i.e., the $-x$ end of the member) and the displacement of joint B (i.e., the $+x$ end of the member):

$$\delta_1 = u_B - u_A$$

Similarly, for members (2) and (3),

$$\delta_2 = u_C - u_B \qquad \text{and} \qquad \delta_3 = u_D - u_C$$

Since the displacements are measured relative to joint A, define the displacement of joint A as $u_A = 0$. The preceding equations can then be solved for the joint displacements in terms of the member elongations:

$$u_B = \delta_1, \qquad u_C = u_B + \delta_2 = \delta_1 + \delta_2, \qquad \text{and} \qquad u_D = u_C + \delta_3 = \delta_1 + \delta_2 + \delta_3$$

Using these expressions, we can now compute the joint displacements:

$$u_B = \delta_1 = 0.364\ \text{mm} = 0.364\ \text{mm} \rightarrow$$

$$u_C = \delta_2 + \delta_2 = 0.364\ \text{mm} + (-1.263\ \text{mm}) = -0.899\ \text{mm} = 0.899\ \text{mm} \leftarrow$$

$$u_D = \delta_1 + \delta_2 + \delta_3 = 0.364\ \text{mm} + (-1.263\ \text{mm}) + 0.597\ \text{mm} = -0.302\ \text{mm}$$
$$= 0.302\ \text{mm} \leftarrow$$

Ans.

A positive value for u indicates a displacement in the $+x$ direction, and a negative u indicates a displacement in the $-x$ direction. Thus, joint D moves to the left even though tension exists in member (3).

The nomenclature and sign conventions introduced in this example may seem unnecessary for such a simple problem. However, the calculation procedure established here will prove quite powerful as problems that are more complex are introduced, particularly problems that cannot be solved with statics alone.

MecMovies

EXAMPLE

M5.2 The roof and second floor of a building are supported by the column shown. The structural steel [E = 200 GPa] column has a constant cross-sectional area of 7,500 mm^2. Determine the deflection of joint C relative to foundation A.

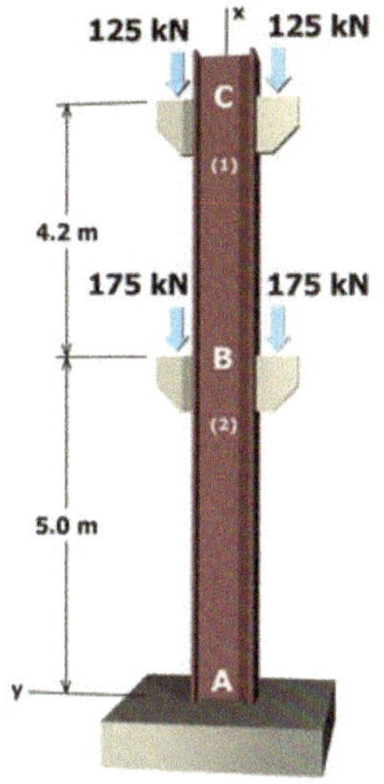

EXAMPLE 5.2

A steel [E = 30,000 ksi] bar of rectangular cross section consists of a uniform-width segment (1) and a tapered segment (2), as shown. The width of the tapered segment varies linearly from 2 in. at the bottom to 5 in. at the top. The bar has a constant thickness of 0.50 in. Determine the elongation of the bar resulting from application of the 30 kip load. Neglect the weight of the bar.

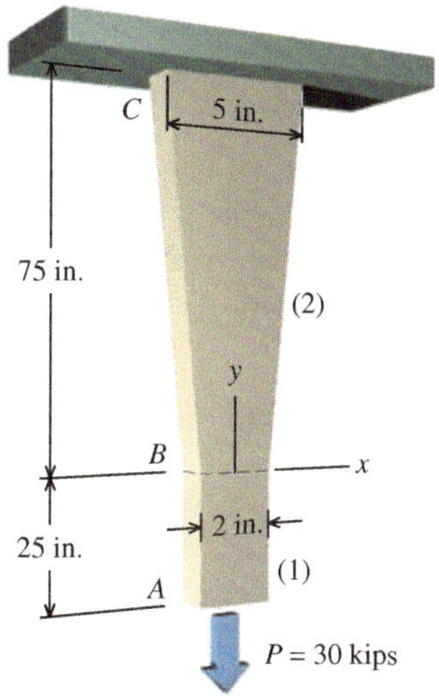

Plan the Solution

The elongation of uniform-width segment (1) may be determined from Equation (5.2). The tapered segment (2) requires the use of Equation (5.5). An expression for the varying cross-sectional area of segment (2) must be derived and used in the integral for the 75 in. length of the tapered segment.

SOLUTION

For the uniform-width segment (1), the deformation from Equation (5.2) is

$$\delta_1 = \frac{F_1 L_1}{A_1 E_1} = \frac{(30 \text{ kips})(25 \text{ in.})}{(2 \text{ in.})(0.5 \text{ in.})(30{,}000 \text{ ksi})} = 0.0250 \text{ in.}$$

For tapered section (2), the width w of the bar varies linearly with position y. The cross-sectional area in the tapered section can be expressed as

$$A_2(y) = wt = \left[2 \text{ in.} + \frac{3 \text{ in.}}{75 \text{ in.}}(y \text{ in.})\right](0.5 \text{ in.}) = 1 + 0.02y \text{ in.}^2$$

Since the weight of the bar is neglected, the force in the tapered segment is constant and simply equal to the 30 kip applied load. Integrate Equation (5.5) to obtain

$$\delta_2 = \int_0^{75} \frac{F_2}{A_2(y)E_2}\,dy = \frac{F_2}{E_2}\int_0^{75} \frac{1}{A_2(y)}\,dy = \frac{30 \text{ kips}}{30{,}000 \text{ ksi}}\int_0^{75} \frac{1}{(1+0.02y)}\,dy$$

$$= (0.001 \text{ in.}^2)\left(\frac{1}{0.02 \text{ in.}}\right)[\ln(1+0.02y)]_0^{75} = 0.0458 \text{ in.}$$

91

The total elongation of the bar is the sum of the elongations of the two segments:

$$u_A = \delta_1 + \delta_2 = 0.0250 \text{ in.} + 0.0458 \text{ in.} = 0.0708 \text{ in.} \downarrow \qquad \textbf{Ans.}$$

Note: If the weight of the bar had not been neglected, the internal force F in both uniform-width segment (1) and tapered segment (2) would not have been constant and Equation (5.5) would be required for both segments. To include the weight of the bar in the analysis, a function expressing the change in internal force as a function of the vertical position y should be derived for each segment. The internal force F at any position y is the sum of a constant force equal to P and a varying force equal to the self-weight of the axial member below position y. The force due to self-weight will be a function that expresses the volume of the bar below any position y, multiplied by the specific weight of the material that the bar is made of. Since the internal force F varies with y, it must be included inside the integral in Equation (5.5).

MecMovies

EXERCISES

M5.1 Use the axial deformation equation for three introductory problems.

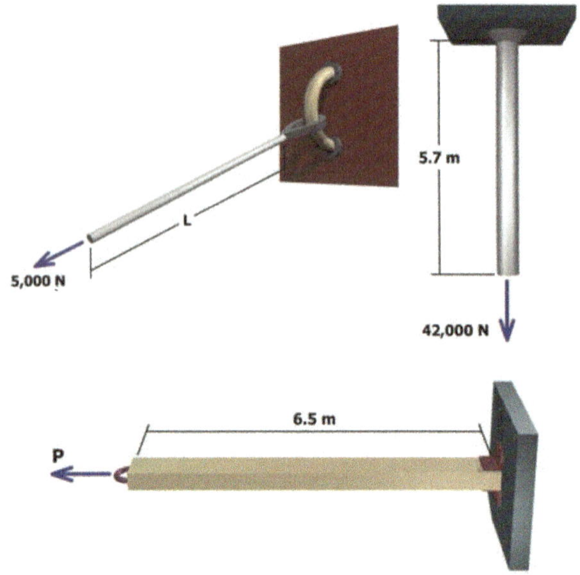

FIGURE M5.1

M5.2 Apply the axial deformation concept to compound axial members.

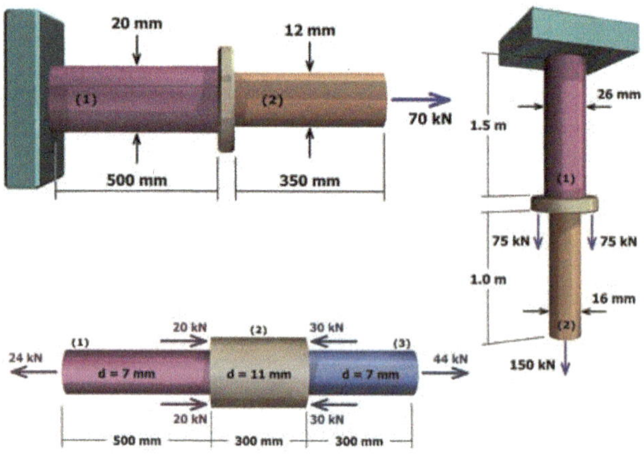

FIGURE M5.2

PROBLEMS

P5.1 A steel [E = 200 GPa] rod with a circular cross section is 15 m long. Determine the minimum diameter required if the rod must transmit a tensile force of 300 kN without exceeding an allowable stress of 250 MPa or stretching more than 10 mm.

P5.2 A rectangular bar of length L has a slot in the central half of its length, as shown in Figure P5.2. The bar has width b, thickness t, and elastic modulus E. The slot has width $b/3$. If L = 400 mm, b = 45 mm,

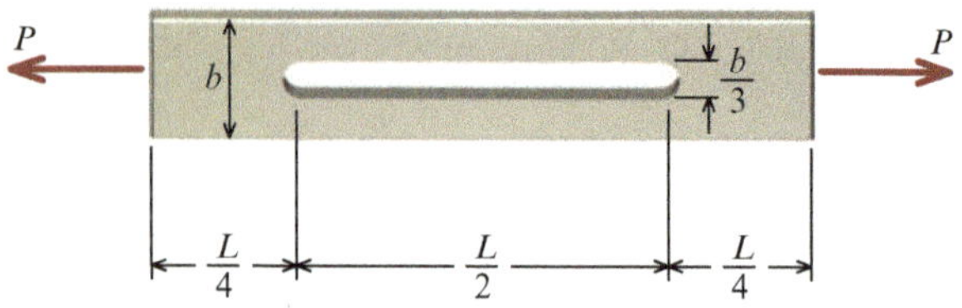

FIGURE P5.2

$t = 8$ mm, and $E = 72$ GPa, determine the overall elongation of the bar for an axial force of $P = 18$ kN.

P5.3 Compound axial member ABC shown in Figure P5.3 has a uniform diameter $d = 1.25$ in. Segment (1) is an aluminum [$E_1 =$ 10,000 ksi] alloy and segment (2) is a copper [$E_2 =$ 17,000 ksi] alloy. The lengths of segments (1) and (2) are $L_1 = 84$ in. and $L_2 =$ 130 in., respectively. Determine the force P required to stretch compound member ABC by a total of 0.25 in.

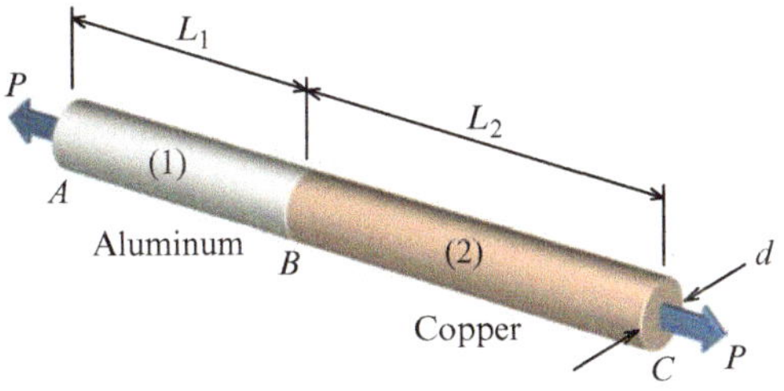

FIGURE P5.3

P5.4 Three solid cylindrical rods are welded together to form the compound axial member shown in Figure P5.4. The member is attached to a fixed support at A. Each rod has an elastic modulus $E =$ 40 GPa. Use the following values for the rod lengths and areas: $L_1 = 1{,}440$ mm, $L_2 = 1{,}680$ mm, $L_3 = 1{,}200$ mm, $A_1 = 260$ mm^2, $A_2 = 130$ mm^2, and $A_3 = 65$ mm^2. What magnitude of external load P is needed to displace end D a distance $u_D = 60$ mm to the right?

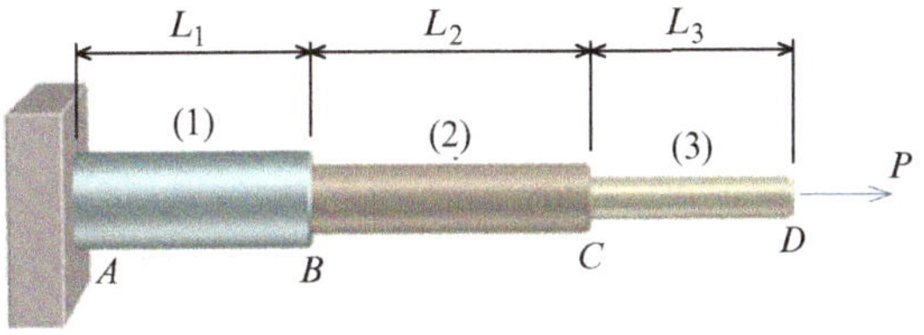

FIGURE P5.4

P5.5 The assembly shown in Figure P5.5 consists of a bronze rod (1) and an aluminum rod (2), each having a cross-sectional area of 0.15 in.^{2}. The assembly is subjected to loads $P = 1{,}900$ lb and $Q = 3{,}200$ lb, acting in the directions shown. Determine the horizontal displacement u_B of coupling B with respect to rigid support A. Use $L_1 = 160$ in., $L_2 = 40$ in., $E_1 = 16 \times 10^6$ psi, and $E_2 = 10 \times 10^6$ psi.

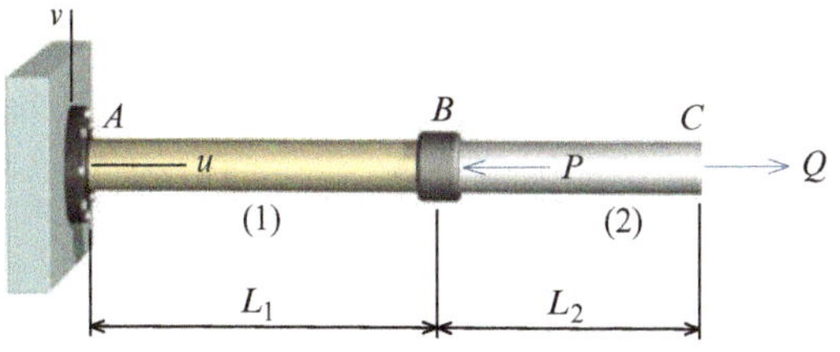

FIGURE P5.5

P5.6 A compound steel ($E =$ 30,000 ksi) bar is subjected to the loads shown in Figure P5.6. Bar (1) has a cross-sectional area $A_1 =$ 0.35 in.^{2} and a length $L_1 = 105$ in. Bar (2) has an area $A_2 = 0.35$ in.^{2} and a length $L_2 = 85$ in. Bar (3) has an area $A_3 = 0.09$ in.^{2} and a length $L_3 = 40$ in. The loads and angles are $P = 40$ kips, $\alpha = 36°$, $Q = 13$ kips, $\beta = 58°$, and $R =$ 9 kips. Determine the vertical deflection of end A.

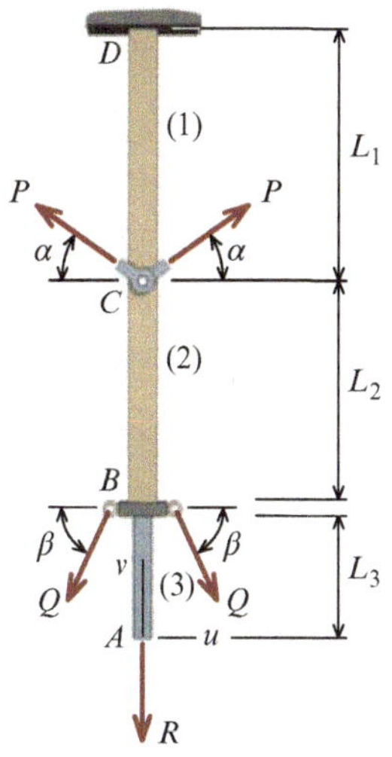

FIGURE P5.6

P5.7 Rigid bar DB in Figure P5.7 is supported at pin B by axial member ABC, which has a cross-sectional area $A_1 = A_2 = 75$ mm^2, an elastic modulus $E_1 = E_2 = 900$ MPa, and lengths $L_1 = 400$ mm and $L_2 = 600$ mm. A load $Q = 500$ N is applied at end C of member (2), and a load $P = 1{,}500$ N is applied to the rigid bar at a distance $a = 80$ mm from pin support D and $b = 50$ mm from pin B. Determine the downward deflections of

(a) pin B.
(b) pin C.

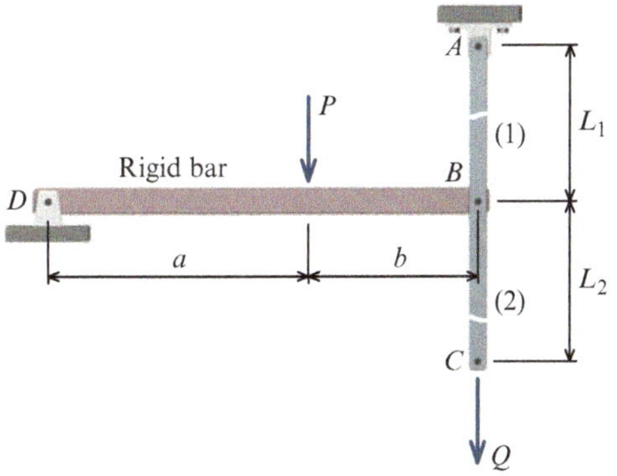

FIGURE P5.7

P5.8 A steel [$E = 200$ GPa] bar (1) supports beam AB, as shown in Figure P5.8. The cross-sectional area of the bar is 600 mm^2. If the stress in the bar must not exceed 330 MPa and the maximum deformation in the bar must not exceed 15 mm, determine the maximum load P that may be supported by this structure. Use dimensions $a = 5.0$ m, $b = 3.0$ m, and $c = 6.5$ m.

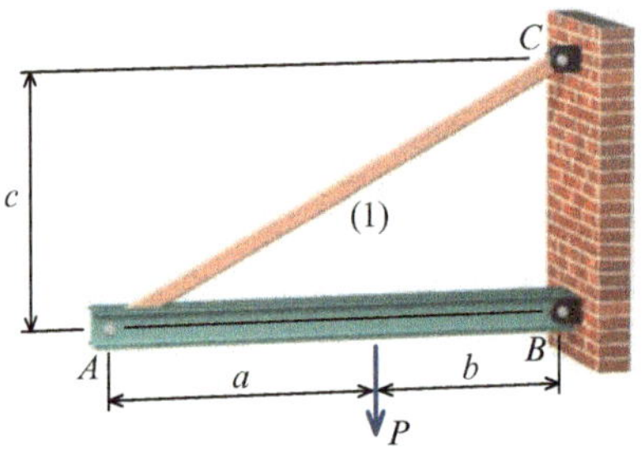

FIGURE P5.8

P5.9 The assembly shown in Figure P5.9 consists of rod (1) and tube (2), both made of steel (E = 30,000 ksi). The tube is attached to a fixed support at C. A rigid end plate is welded to the tube at B, and the rod is welded to the end plate. The rod is solid, with a diameter d_1 = 0.375 in. The tube has an outside diameter D_2 = 2.25 in. and a wall thickness t_2 = 0.125 in. The dimensions of the assembly are a = 60 in. and b = 32 in. Determine the horizontal deflection at rod end A for an applied load P = 4 kips.

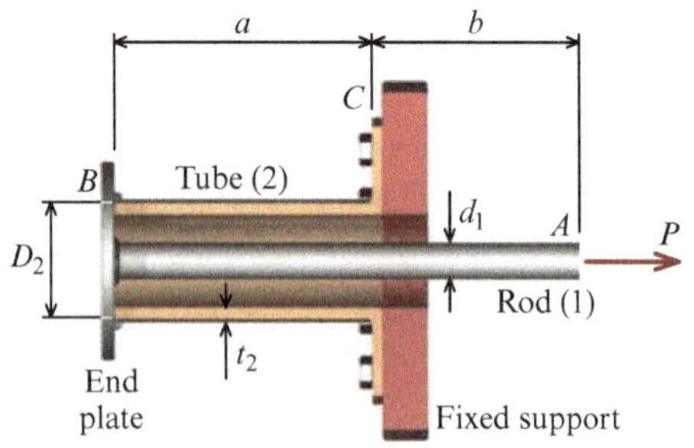

FIGURE P5.9

P5.10 The system shown in Figure P5.10 consists of a rod, a spring, a tube, and a fixed base. Polymer rod (1) is attached to rod cap A at its upper end and passes freely through the spring, tube cap B, tube (2), and the fixed base to rod end D, where a load P is applied. The load applied at rod end D is transmitted by the rod to rod cap A, causing the spring to compress. The force in the spring is then transmitted to the tube, in which it creates compressive deformation. The polymer rod (1) has a diameter d_1 = 12 mm, an elastic modulus E_1 = 1,400 MPa, and a length L_1 = 640 mm. The linear compression spring has a spring constant k = 850 N/mm. Tube (2) has an outside diameter D_2 = 50 mm, an inside diameter d_2 = 46 mm, an elastic modulus E_2 = 900 MPa, and a length L_2 = 350 mm. For an applied load P = 750 N, determine the vertical displacement of (a) rod cap A, (b) tube cap B, and (c) rod end D relative to C.

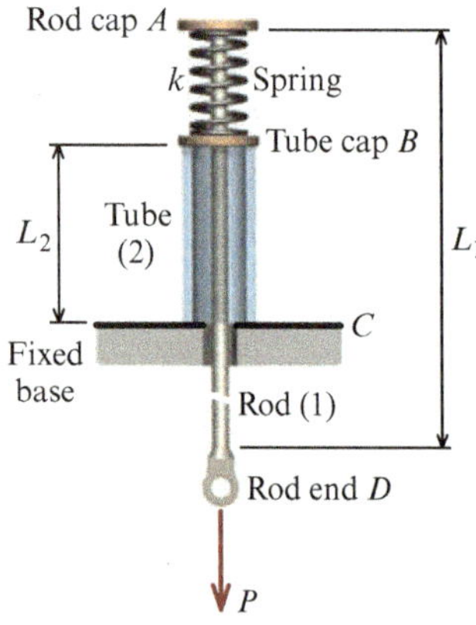

FIGURE P5.10

P5.11 The mechanism shown in Figure P5.11 consists of rigid bar ABC pinned at C to nylon [E = 200 ksi] rod (1), which has a diameter 0.25 in. and a length L_1 = 14 in. The linear compression spring, which is placed between the base of the mechanism at E and a rigid plate at D, has a spring constant k = 940 lb/in. The base is rigid and fixed in position, and rigid bar ABC is horizontal before the load P is applied. Rod (1) is attached to plate D at the lower end of the spring. The nylon rod passes freely through the compression spring and through a hole in the base at E. Determine the value of load P that is needed to deflect end A downward by 2.0 in. Use dimensions a = 9.0 in., b = 3.0 in., and c = 5.0 in.

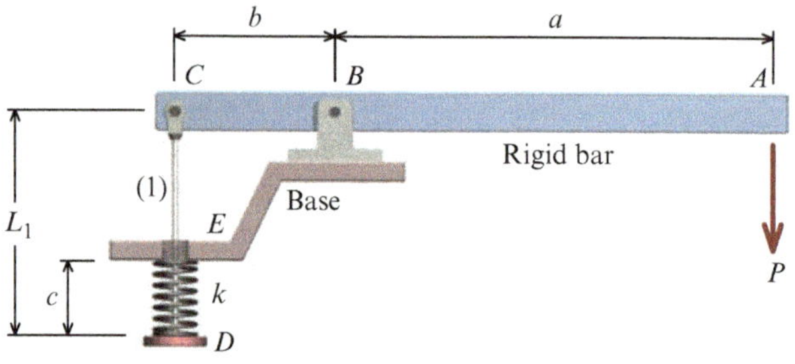

FIGURE P5.11

P5.12 The wooden pile shown in Figure P5.12 has a diameter of 100 mm and is subjected to a load P = 75 kN. Along the length of the pile and around its perimeter, soil supplies a constant frictional resistance w = 3.70 kN/m. The length of the pile is L = 5.0 m and its elastic modulus is E = 8.3 GPa. Calculate

(a) the force F_B needed at base of the pile for equilibrium.
(b) the magnitude of the downward displacement at A relative to B.

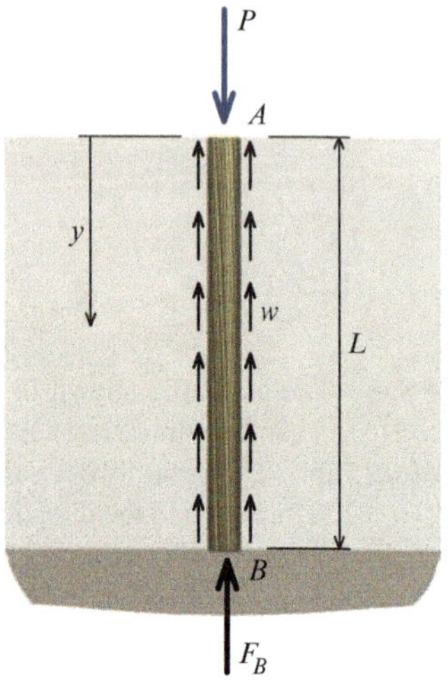

FIGURE P5.12

P5.13 A 1 in. diameter steel [E = 29,000 ksi and γ = 490 lb/ft^3] rod hangs vertically while suspended from one end. The length of the rod is 320 ft. Determine the change in length of the bar due to its own weight.

P5.14 A homogenous rod of length L and elastic modulus E is a truncated cone with diameter that varies linearly from d_0 at one end to $2d_0$ at the other end. A concentrated axial load P is applied to the ends of the rod, as shown in Figure P5.14. Assume that the taper of the cone is slight enough for the assumption of a uniform axial stress distribution over a cross section to be valid.

(a) Determine an expression for the stress distribution over an arbitrary cross section at x.
(b) Determine an expression for the elongation of the rod.

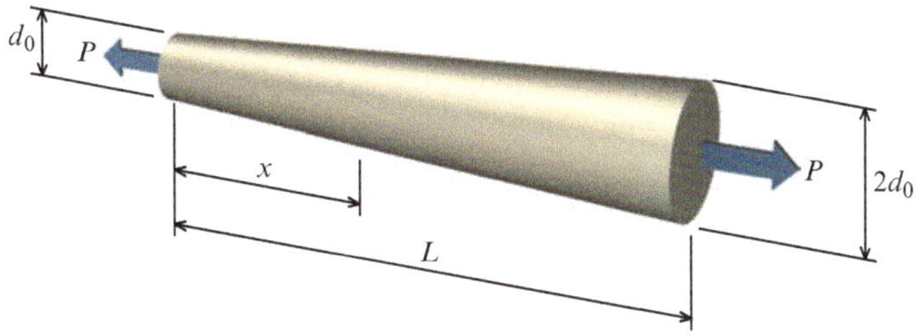

FIGURE P5.14

P5.15 The conical bar shown in Figure P5.15 is made of aluminum alloy [E = 10,600 ksi and γ = 0.100 lb/in.[3]]. The bar has a 2 in. radius at its upper end and a length L = 20 ft. Assume that the taper of the bar is slight enough for the assumption of a uniform axial stress distribution over a cross section to be valid. Determine the extension of the bar due to its own weight.

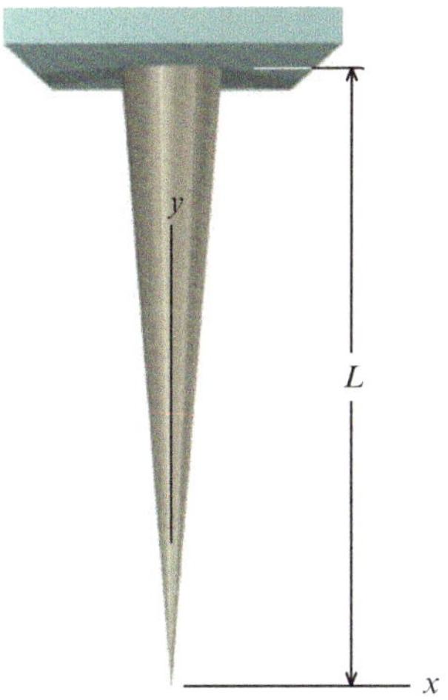

FIGURE P5.15

5.4 Deformations in a System of Axially Loaded Bars

Many structures consist of more than one axially loaded member, and for these structures, axial deformations and stresses for a *system* of pin-connected deformable bars must be determined. The problem is approached through a study of the geometry of the deformed system, from which the axial deformations of the various bars in the system are obtained.

In this section, the analysis of statically determinate structures consisting of homogeneous, prismatic axial members will be considered. In analyzing these types of structures, we begin with a free-body diagram showing all forces acting on the key elements of the structure. Then, we investigate how the structure as a whole deflects in response to the deformations that occur in the axial members.

A homogeneous, prismatic member (a) is straight, (b) has a constant cross-sectional area, and (c) consists of a single material (so it has exactly one value of E).

EXAMPLE 5.3

The assembly shown consists of rigid bar ABC, two fiber-reinforced plastic (FRP) rods (1) and (3), and FRP post (2). The modulus of elasticity for the FRP is E = 18 GPa. Determine the vertical deflection of joint D relative to its initial position after the 30 kN load is applied.

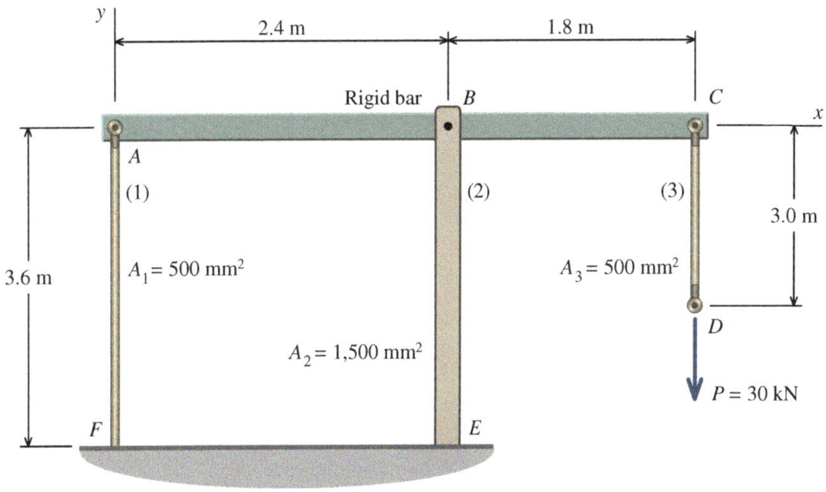

The three axial members are connected to the rigid beam by pins. Assume that member (1) is pinned to the foundation at F and member (2) is fixed in the foundation at E.

Plan the Solution

The deflection of joint D relative to its initial position must be computed. The deflection of D relative to joint C is simply the elongation of member (3). The challenge in this problem, however, lies in computing the deflection at C. The rigid bar will deflect and rotate because of the elongation and contraction in members (1) and (2). To determine the final position of the rigid bar, we must first compute the forces in the three axial members, using equilibrium equations. Then, Equation (5.2) can be used to compute the deformation in each member. A *deformation diagram* can be drawn to define the relationships between the rigid-bar deflections at A, B, and C. Then, the member deformations will be related to those deflections. Finally, the deflection of joint D can be computed from the sum of the rigid-bar deflection at C and the elongation in member (3).

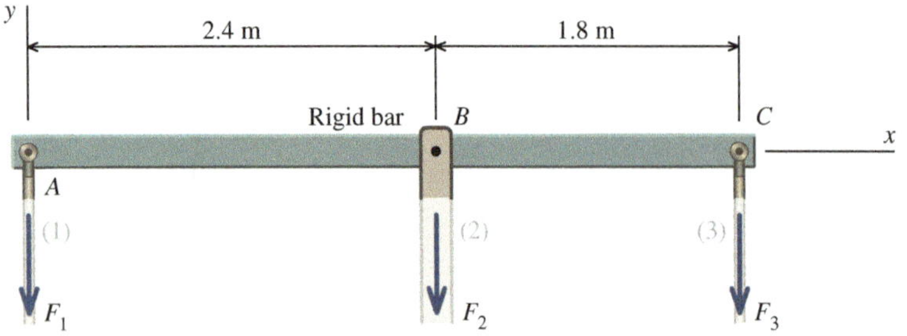

SOLUTION

Equilibrium

Draw a free-body diagram (FBD) of the rigid bar and write two equilibrium equations:

$$\Sigma F_y = -F_1 - F_2 - F_3 = 0$$
$$\Sigma M_B = (2.4\text{ m})F_1 - (1.8\text{ m})F_3 = 0$$

By inspection, $F_3 = P = 30$ kN. Using this result, we can simultaneously solve the two equations to give $F_1 = 22.5$ kN and $F_2 = -52.5$ kN.

Force–Deformation Relationships

Compute the deformations in each of the three members:

$$\delta_1 = \frac{F_1 L_1}{A_1 E_1} = \frac{(22.5\text{ kN})(1{,}000\text{ N/kN})(3.6\text{ m})(1{,}000\text{ mm/m})}{(500\text{ mm}^2)(18\text{ GPa})(1{,}000\text{ MPa/GPa})} = 9.00\text{ mm}$$

$$\delta_2 = \frac{F_2 L_2}{A_2 E_2} = \frac{(-52.5\text{ kN})(1{,}000\text{ N/kN})(3.6\text{ m})(1{,}000\text{ mm/m})}{(1{,}500\text{ mm}^2)(18\text{ GPa})(1{,}000\text{ MPa/GPa})} = -7.00\text{ mm}$$

$$\delta_3 = \frac{F_3 L_3}{A_3 E_3} = \frac{(30\text{ kN})(1{,}000\text{ N/kN})(3.0\text{ m})(1{,}000\text{ mm/m})}{(500\text{ mm}^2)(18\text{ GPa})(1{,}000\text{ MPa/GPa})} = 10.00\text{ mm}$$

The negative value of δ_2 indicates that member (2) contracts.

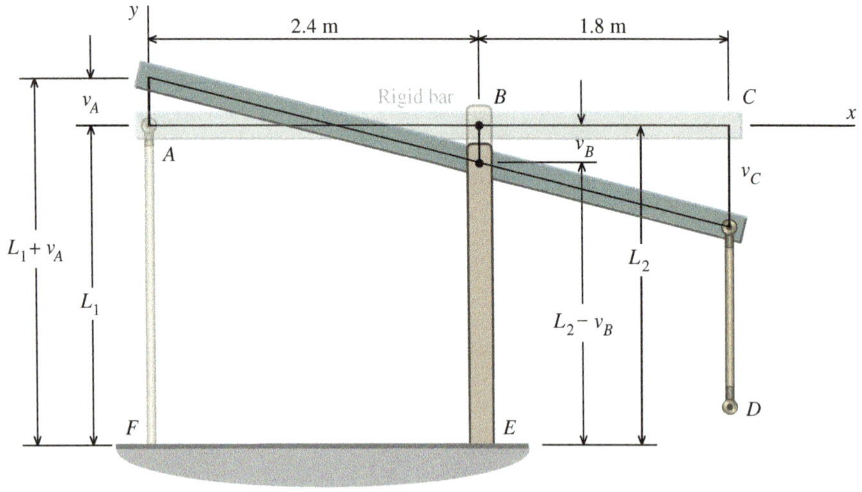

Geometry of Deformations

Sketch the final deflected shape of the rigid bar. Member (1) elongates, so A will deflect upward. Member (2) contracts, so B will deflect downward. The deflection of C must be determined.

(**Note:** Vertical deflections of the rigid-bar joints are denoted by v. In this example, the displacements v_A, v_B, and v_C are treated as absolute values and the sketch is used to establish the relationship between the joint displacements.)

The rigid-bar deflections at joints A, B, and C can be related by similar triangles:

$$\frac{v_A + v_B}{2.4\text{ m}} = \frac{v_C - v_B}{1.8\text{ m}} \qquad \therefore v_C = \frac{1.8\text{ m}}{2.4\text{ m}}(v_A + v_B) + v_B = 0.75(v_A + v_B) + v_B$$

How are the rigid-bar deflections v_A and v_B that are shown on the sketch related to the member deformations δ_1 and δ_2? By definition, deformation is the difference between the initial and final lengths of an axial member. Using the sketch of the deflected rigid bar, we can define the deformation in member (1) in terms of its initial and final lengths:

$$\delta_1 = L_{\text{final}} - L_{\text{initial}} = (L_1 + v_A) - L_1 = v_A \qquad \therefore v_A = \delta_1 = 9.00 \text{ mm}$$

Similarly, for member (2),

$$\delta_2 = L_{\text{final}} - L_{\text{initial}} = (L_2 - v_B) - L_2 = -v_B \qquad \therefore v_B = -\delta_2 = -(-7.00 \text{ mm}) = 7.00 \text{ mm}$$

With these results, the magnitude of the rigid-bar deflection at C can now be computed:

$$v_C = 0.75(v_A + v_B) + v_B = 0.75(9.00 \text{ mm} + 7.00 \text{ mm}) + 7.00 \text{ mm} = 19.00 \text{ mm}$$

The direction of the deflection is shown on the deformation diagram; that is, joint C deflects 19.00 mm downward.

Deflection of *D*

The downward deflection of joint D is the sum of the rigid-bar deflection at C and the elongation in member (3):

$$v_D = v_C + \delta_3 = 19.00 \text{ mm} + 10.00 \text{ mm} = 29.0 \text{ mm} \qquad \textbf{Ans.}$$

MecMovies

EXAMPLE

M5.4 An assembly consists of three rods attached to rigid bar AB. Rod (1) is steel, and rods (2) and (3) are aluminum. The area and elastic modulus of each rod is noted on the sketch. A force of 80 kN is applied at D. Determine the vertical deflections of points A, B, C, and D.

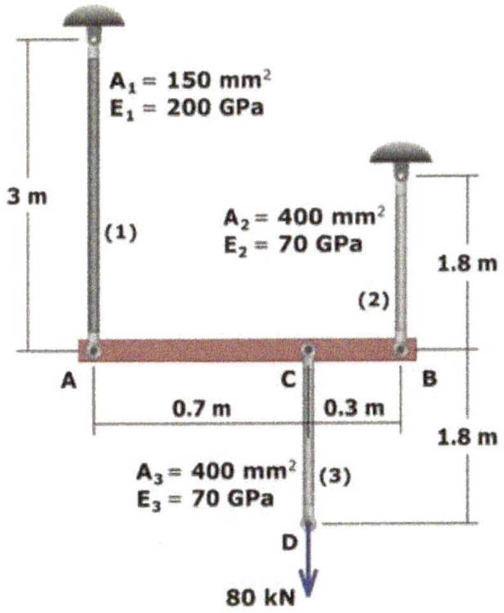

The preceding examples considered structures consisting of parallel axial bars, making the geometry of deformation for the structure relatively straightforward to analyze. Suppose, however, that one is interested in a structure in which the axial members are *not* parallel. The structure shown in Figure 5.9 consists of three axial members (AB, BC, and BD) connected to a common joint at B. In the figure, the solid lines represent the unstrained (i.e., unloaded) configuration of the system and the dashed lines represent the

97

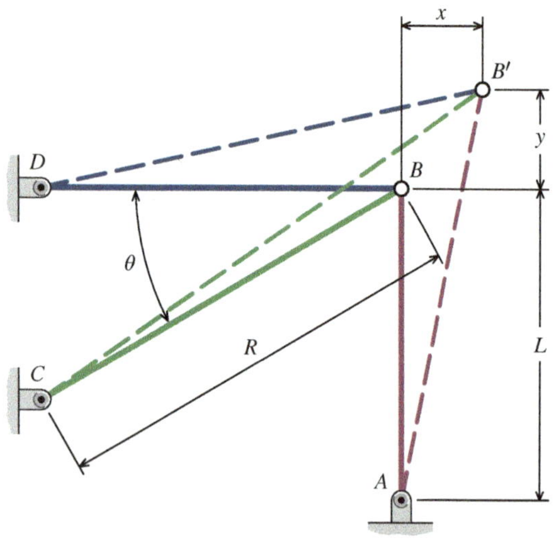

FIGURE 5.9 Axial structure with intersecting members.

configurations due to a force applied at joint B. From the Pythagorean theorem, the actual deformation in bar AB is

$$\delta_{AB} = \sqrt{(L + y)^2 + x^2} - L$$

Transposing the last term and squaring both sides gives

$$\delta_{AB}^2 + 2L\delta_{AB} + L^2 = L^2 + 2Ly + y^2 + x^2$$

If the displacements are small (the usual case for stiff materials and elastic action), the terms involving the squares of the displacements may be neglected; hence, the deformation in bar AB is

$$\delta_{AB} \approx y$$

In a similar manner, the deformation in bar BD is

$$\delta_{BD} \approx x$$

The axial deformation of bar BC is

$$\delta_{BC} = \sqrt{(R\cos\theta + x)^2 + (R\sin\theta + y)^2} - R$$

Transposing the last term and squaring both sides gives

$$\delta_{BC}^2 + 2R\delta_{BC} + R^2 = R^2\cos^2\theta + 2Rx\cos\theta + x^2 + R^2\sin^2\theta + 2Ry\sin\theta + y^2$$

The second-degree displacement terms can be neglected since the displacements are small. Using the trigonometric identity $\sin^2\theta + \cos^2\theta = 1$, we can express the deformation in member BC as

$$\delta_{BC} \approx x\cos\theta + y\sin\theta$$

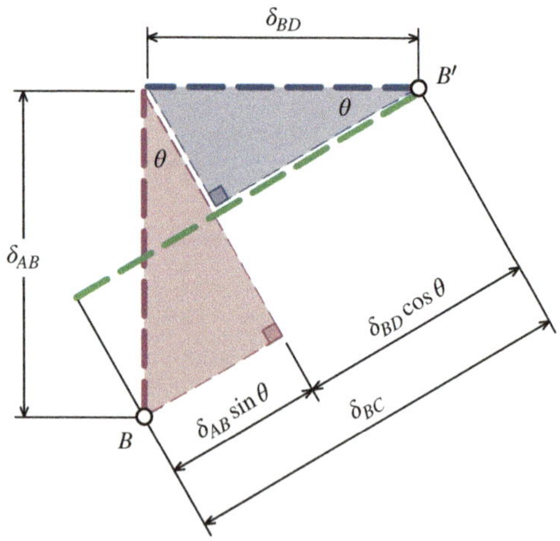

FIGURE 5.10 Geometric interpretation of member deformations.

or, in terms of the deformations of the other two bars,

$$\delta_{BC} \approx \delta_{BD}\cos\theta + \delta_{AB}\sin\theta$$

The geometric interpretation of this equation is indicated by the shaded right triangles in Figure 5.10.

The general conclusion that may be drawn from the preceding discussion is that, *for small displacements*, the axial deformation in any bar may be assumed to be equal to the component of the displacement of one end of the bar (relative to the other end), taken in the direction of the unstrained orientation of the bar. Rigid members of the system may change orientation or position, but they will not be deformed in any manner. For example, if bar BD of Figure 5.9 were rigid and subjected to a small upward rotation, then point B could be assumed to be displaced vertically through a distance y, and δ_{BC} would be equal to $y\sin\theta$.

EXAMPLE 5.4

A tie rod (1) and a pipe strut (2) are used to support a 50 kN load, as shown. The cross-sectional areas are $A_1 = 650\ \text{mm}^2$ for tie rod (1) and $A_2 = 925\ \text{mm}^2$ for pipe strut (2). Both members are made of structural steel that has an elastic modulus $E = 200$ GPa.

(a) Determine the normal stresses in tie rod (1) and pipe strut (2).
(b) Determine the elongation or contraction of each member.
(c) Sketch a deformation diagram that shows the displaced position of joint B.
(d) Compute the horizontal and vertical displacements of joint B.

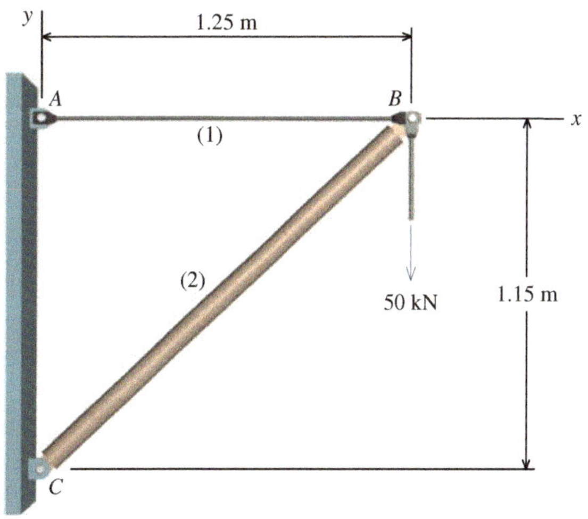

Plan the Solution

From a free-body diagram of joint B, the internal axial forces in members (1) and (2) can be calculated. The elongation (or contraction) of each member can then be computed from Equation (5.2). To determine the displaced position of joint B, the following approach will be used: We will imagine that the pin at joint B is temporarily removed, allowing members (1) and (2) to deform either in elongation or contraction. Then, member (1) will be rotated about joint A, member (2) will be rotated about joint C, and the intersection point of these two members will be located. We will imagine that the pin at B is now reinserted in the joint at this location. The deformation diagram describing the preceding movements will be used to compute the horizontal and vertical displacements of joint B.

SOLUTION

(a) Member Stresses

The internal axial forces in members (1) and (2) can be determined from equilibrium equations based on a free-body diagram of joint B. The sum of forces in the horizontal (x) direction can be written as

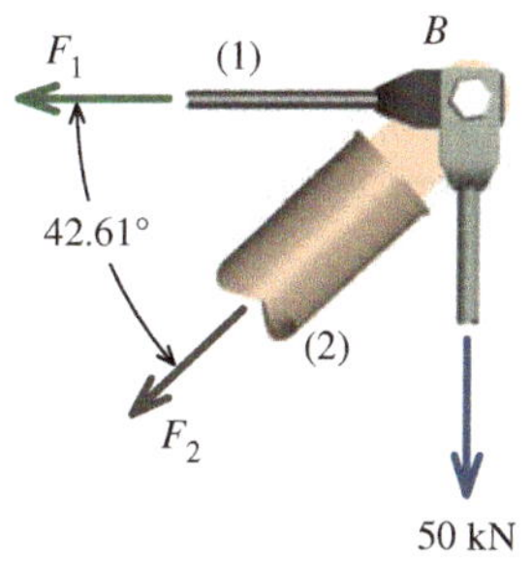

$$\Sigma F_x = -F_1 - F_2 \cos 42.61° = 0$$

and the sum of forces in the vertical (y) direction can be expressed as

$$\Sigma F_y = -F_1 \sin 42.61° - 50\ \text{kN} = 0$$

$$\therefore\ F_2 = -73.85\ \text{kN}$$

Substituting this result into the previous equation gives

$$F_1 = 54.36\ \text{kN}$$

The normal stress in tie rod (1) is

$$\sigma_1 = \frac{F_1}{A_1} = \frac{(54.36\ \text{kN})(1{,}000\ \text{N/kN})}{650\ \text{mm}^2} = 83.63\ \text{N/mm}^2\ \text{(T)} = 83.6\ \text{MPa (T)} \qquad \textbf{Ans.}$$

and the normal stress in pipe strut (2) is

$$\sigma_2 = \frac{F_2}{A_2} = \frac{(73.85\ \text{kN})(1{,}000\ \text{N/kN})}{925\ \text{mm}^2} = 79.84\ \text{N/mm}^2\ \text{(C)} = 79.8\ \text{MPa (C)} \qquad \textbf{Ans.}$$

(b) Member Deformations

The deformations in the members are determined from either Equation (5.1) or Equation (5.2). The elongation in tie rod (1) is

$$\delta_1 = \frac{\sigma_1 L_1}{E_1} = \frac{(83.63 \text{ N/mm}^2)(1.25 \text{ m})(1{,}000 \text{ mm/m})}{200{,}000 \text{ N/mm}^2} = 0.5227 \text{ mm} \qquad \textbf{Ans.}$$

The length of inclined pipe strut (2) is

$$L_2 = \sqrt{(1.25 \text{ m})^2 + (1.15 \text{ m})^2} = 1.70 \text{ m}$$

and its deformation is

$$\delta_2 = \frac{\sigma_2 L_2}{E_2} = \frac{(-79.84 \text{ N/mm}^2)(1.70 \text{ m})(1{,}000 \text{ mm/m})}{200{,}000 \text{ N/mm}^2} = -0.6786 \text{ mm} \qquad \textbf{Ans.}$$

The negative sign indicates that member (2) contracts (i.e., gets shorter).

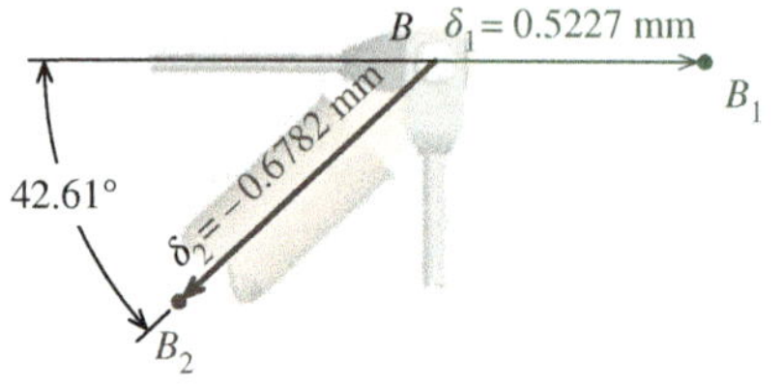

(c) Deformation Diagram

Step 1: To determine the displaced position of joint B, let us first imagine that the pin at joint B is temporarily removed, allowing members (1) and (2) to deform freely by the amounts computed in part (b). Since joint A of the tie rod is fixed to a support, it remains stationary. Thus, when tie rod (1) elongates by 0.5227 mm, joint B moves to the right, *away* from joint A to the displaced position B_1.

Similarly, joint C of the pipe strut remains stationary. When member (2) contracts by 0.6782 mm, joint B of the pipe strut moves *toward* joint C, ending up in displaced position B_2. These deformations are shown in the accompanying figure.

Step 2: In the previous step, we imagined removing the pin at B and allowing each member to deform freely, either elongating or contracting, as dictated by the internal forces acting in each member. In actuality, however, the two members are connected by pin B. The second step of this process requires finding the displaced position B' of the pin connecting tie rod (1) and pipe strut (2) such that B' is consistent with member elongations δ_1 and δ_2.

Because of the axial deformations, both tie rod (1) and pipe strut (2) must rotate slightly if they are to remain connected at pin B. Tie rod (1) will pivot about stationary end A, and pipe strut (2) will pivot about stationary end C. If the rotation angles are small, the circular arcs that describe possible displaced positions of joint B can be replaced by straight lines that are perpendicular to the unloaded orientations of the members.

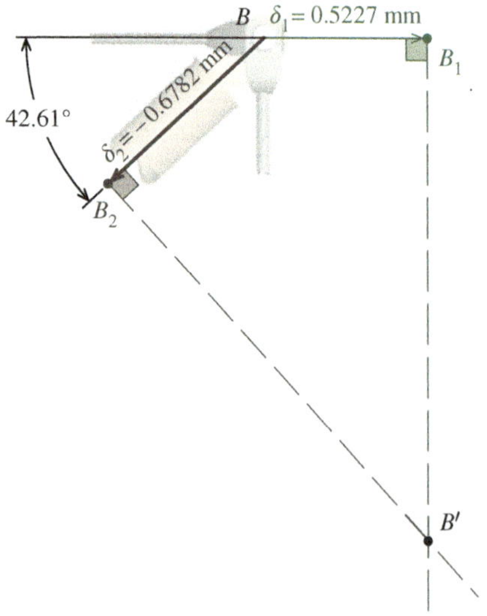

Consider the figure shown. As tie rod (1) rotates clockwise about stationary end A, joint B_1 moves downward. If the rotation angle is small, the circular arc describing the possible displaced positions of joint B_1 can be approximated by a line that is perpendicular to the original orientation of tie rod (1).

Similarly, as pipe strut (2) rotates clockwise about stationary end C, the circular arc describing the possible displaced positions of joint B_2 can be approximated by a line that is perpendicular to the original orientation of member (2).

The intersection of these two perpendiculars at B' marks the final displaced position of joint B.

Step 3: For the two-member structure considered here, the deformation diagram forms a quadrilateral shape. The angle between member (2) and the x axis is 42.61°; therefore, the obtuse angle at B must equal 180° − 42.61° = 137.39°.

Since the sum of the four interior angles of a quadrilateral must equal 360° and since the angles at B_1 and B_2 are each 90°, the acute angle at B' must equal $360° - 90° - 90° - 137.39° = 42.61°$.

Using this deformation diagram, we can determine the horizontal and vertical distances between initial joint position B and displaced joint position B'.

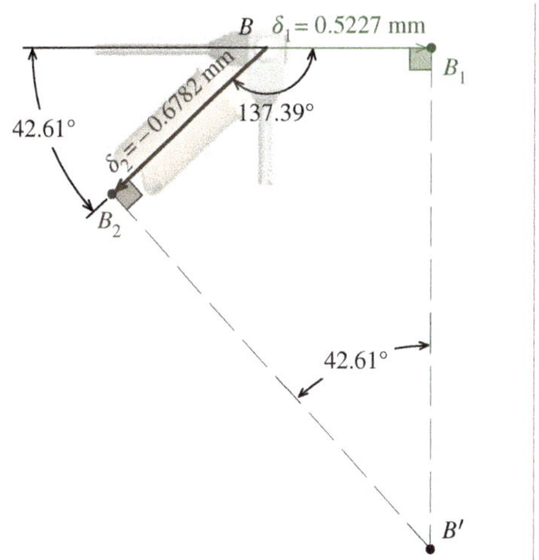

(d) Joint Displacement

The deformation diagram can now be analyzed to determine the location of B', which is the final position of joint B. By inspection, the horizontal translation Δx of joint B is

$$\Delta x = \delta_1 = 0.5227 \text{ mm} = 0.523 \text{ mm} \qquad \textbf{Ans.}$$

Computation of the vertical translation Δy requires several intermediate steps. From the deformation diagram, the distance labeled b is simply equal to the magnitude of deformation δ_2; therefore, $b = |\delta_2| = 0.6782$ mm. The distance a is found from

$$\cos 42.61° = \frac{a}{0.5227 \text{ mm}}$$

$$\therefore a = (0.5227 \text{ mm})\cos 42.61° = 0.3847 \text{ mm}$$

The vertical translation Δy can now be computed as

$$\sin 42.61° = \frac{a+b}{\Delta y}$$

$$\therefore \Delta y = \frac{a+b}{\sin 42.61°} = \frac{0.3847 \text{ mm} + 0.6782 \text{ mm}}{\sin 42.61°} = 1.570 \text{ mm} \qquad \textbf{Ans.}$$

By inspection, joint B is displaced downward and to the right.

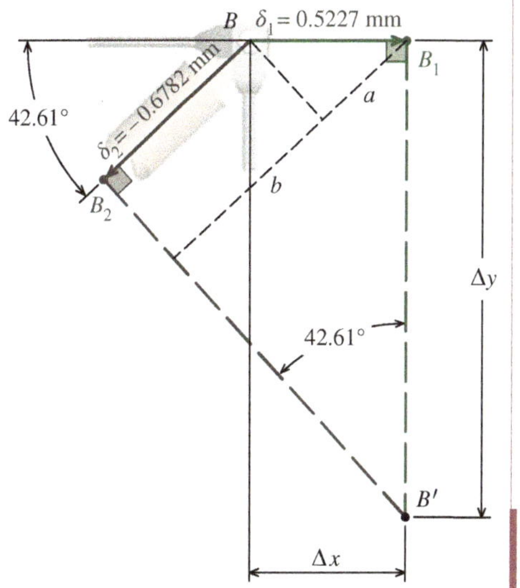

PROBLEMS

P5.16 Rigid beam ABC shown in Figure P5.16 is supported by rods (1) and (2) that have identical lengths $L = 7.0$ m. Rod (1) is made of steel [E = 200 GPa] and has a cross-sectional area of 175 mm². Rod (2) is made of aluminum [E = 70 GPa] and has a cross-sectional area of 300 mm². Assume dimensions of $a = 1.5$ m and $b = 3.0$ m. For a uniformly distributed load $w = 15$ kN/m, determine the deflection of the rigid beam at point B.

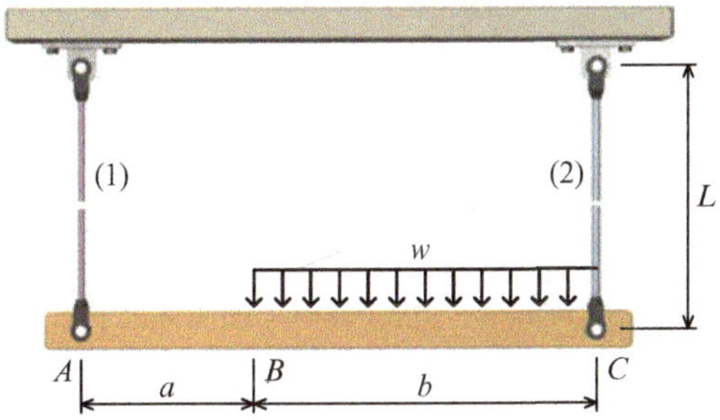

FIGURE P5.16

P5.17 In Figure P5.17, the horizontal rigid beam $ABCD$ is supported by vertical bars (1) and (2) and is loaded at points A and D by vertical forces $P = 30$ kN and $Q = 40$ kN, respectively. Bar (1) is made of aluminum [E = 70 GPa] and has a cross-sectional area of 225 mm² and a length $L_1 = 8.0$ m. Bar (2) is made of steel [E = 200 GPa] and has a cross-sectional area of 375 mm² and a length $L_2 = 5.0$ m. Assume dimensions $a = 2.5$ m, $b = 2.0$ m, and $c = 1.50$ m. Determine the deflection of the rigid beam (a) at point A and (b) at point D.

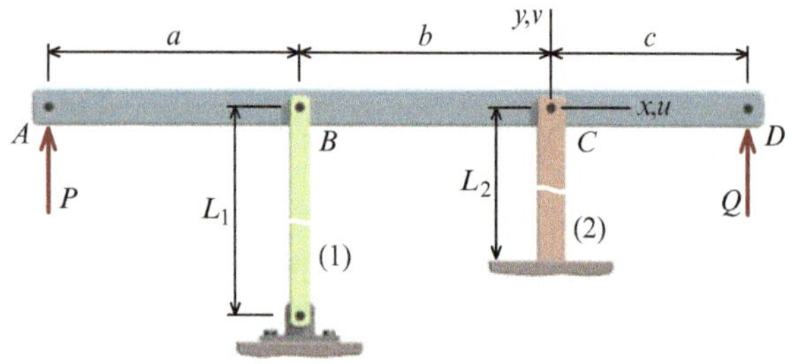

FIGURE P5.17

P5.18 The truss shown in Figure P5.18 is constructed from three aluminum alloy members, each having a cross-sectional area A = 850 mm^2 and an elastic modulus E = 70 GPa. Assume that a = 7.0 m, b = 4.5 m, and c = 5.0 m. Calculate the horizontal displacement of roller A when the truss supports loads of P = 12 kN and Q = 30 kN acting in the directions shown.

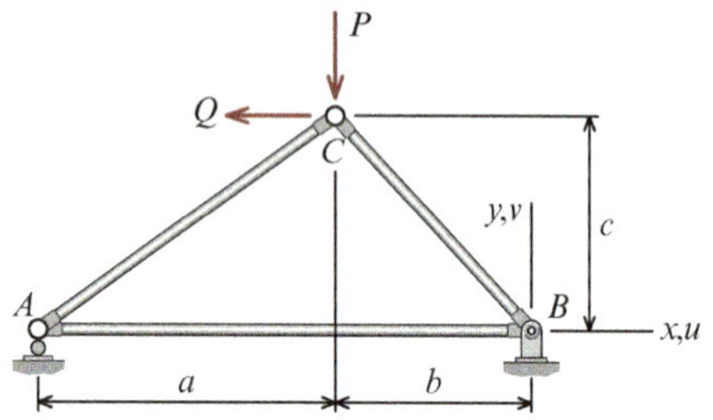

FIGURE P5.18

P5.19 Three bars, each with a cross-sectional area of 750 mm^2 and a length of 5.0 m, are connected and loaded as shown in Figure P5.19. Bars (1) and (3) are made of steel [E = 200 GPa] and bar (2) is made of an aluminum alloy [E = 70 GPa]. At D, all three bars are connected to a slider block that is constrained to travel in a smooth, horizontal slot. When the load P is applied, the strain in bar (2) is found to be 0.0012 mm/mm. Assume that a = 2.4 m and b = 1.6 m. Determine

(a) the horizontal deflection of D.
(b) the deformations in each of the three bars.
(c) the magnitude of load P.

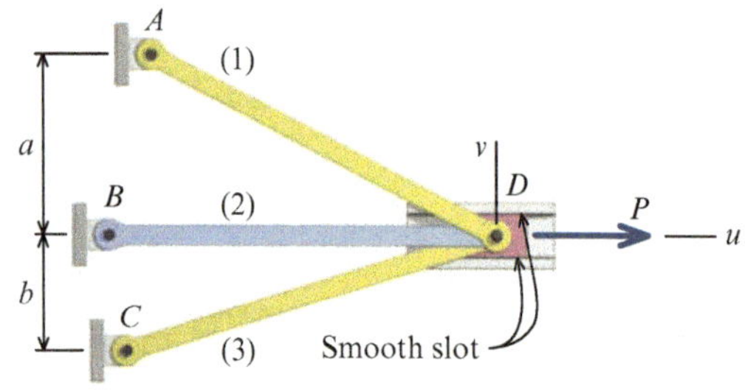

FIGURE P5.19

P5.20 The pin-connected assembly shown in Figure P5.20 consists of solid aluminum [E = 70 GPa] rods (1) and (2) and solid steel [E = 200 GPa] rod (3). Each rod has a diameter of 16 mm. Assume that a = 2.5 m, b = 1.6 m, and c = 0.8 m. If the normal stress in any rod must not exceed 150 MPa, determine

(a) the maximum load P that may be applied at A.
(b) the magnitude of the resulting deflection at A.

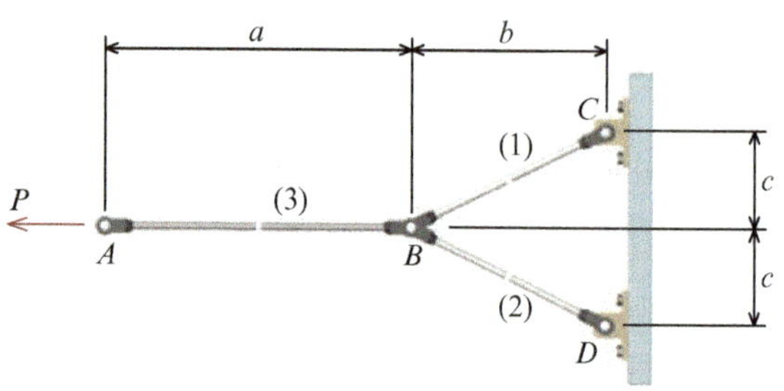

FIGURE P5.20

P5.21 A tie rod (1) and a pipe strut (2) are used to support a load P = 25 kips as shown in Figure P5.21. Pipe strut (2) has an outside diameter of 6.625 in. and a wall thickness of 0.280 in. Both the tie rod and the pipe strut are made of structural steel with a modulus of elasticity E = 29,000 ksi and a yield strength σ_Y = 36 ksi. For the tie rod, the minimum factor of safety with respect to yield is 1.5 and the maximum allowable axial elongation is 0.30 in.

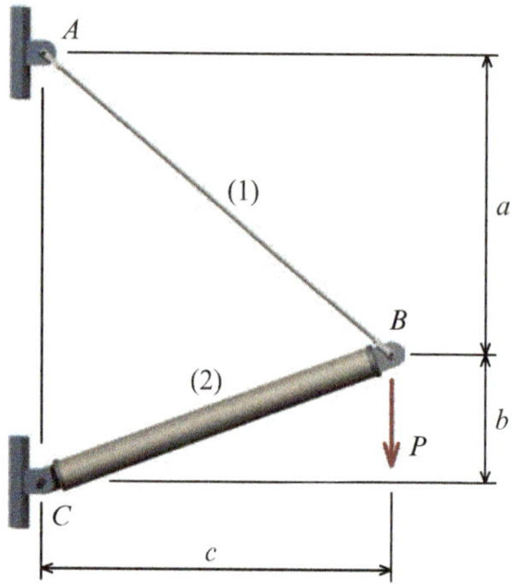

FIGURE P5.21

(a) Determine the minimum diameter required to satisfy both constraints for tie rod (1).
(b) Draw a deformation diagram showing the final position of joint B.

P5.22 Two axial members are used to support a load P = 72 kips as shown in Figure P5.22. Member (1) is 12 ft long, has a cross-sectional area A_1 = 1.75 in.^{2}, and is made of structural steel [E = 29,000 ksi]. Member (2) is 16 ft long, has a cross-sectional area of A_2 = 4.50 in.^{2}, and is made of an aluminum alloy [E = 10,000 ksi].

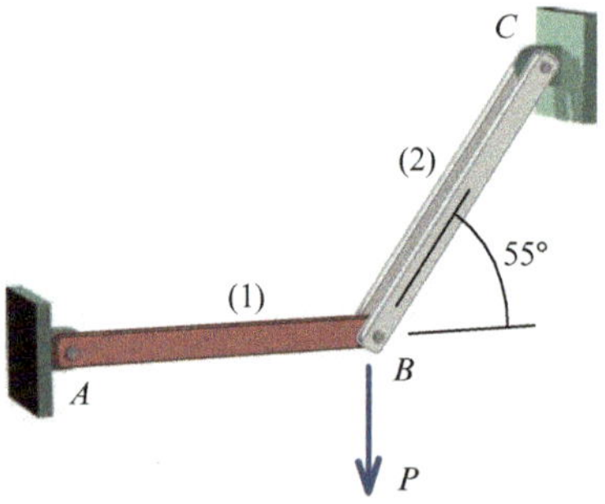

FIGURE P5.22

(a) Compute the normal stress in each axial member.
(b) Compute the deformation of each axial member.
(c) Draw a deformation diagram showing the final position of joint B.
(d) Compute the horizontal and vertical displacements of joint B.

5.5 Statically Indeterminate Axially Loaded Members

In many simple structures and mechanical systems constructed with axially loaded members, it is possible to determine the reactions at supports and the internal forces in the individual members by drawing free-body diagrams and solving equilibrium equations. Such structures and systems are classified as **statically determinate**.

For other structures and mechanical systems, the equations of equilibrium alone are not sufficient for the determination of axial forces in the members and reactions at supports. In other words, there are not enough equilibrium equations to solve for all of the unknowns in the system. These structures and systems are termed **statically indeterminate**. Structures of this type can be analyzed by supplementing the equilibrium equations with additional equations involving the geometry of the deformations in the members of the structure or system. The general solution process can be organized into a five-step procedure:

Step 1 — Equilibrium Equations: Equations expressed in terms of the unknown axial forces are derived for the structure on the basis of equilibrium considerations.

Step 2 — Geometry of Deformation: The geometry of the specific structure is evaluated to determine how the deformations of the axial members are related.

Step 3 — Force–Deformation Relationships: The relationship between the internal force in an axial member and its corresponding elongation is expressed by Equation (5.2).

Step 4 — Compatibility Equation: The force–deformation relationships are substituted into the geometry-of-deformation equation to obtain an equation that is based on the structure's geometry, but expressed in terms of the unknown axial forces.

Step 5 — Solve the Equations: The equilibrium equations and the compatibility equation are solved simultaneously to compute the unknown axial forces.

In engineering literature, **force–deformation relationships** are also called **constitutive relationships** since these relationships idealize the physical properties of the material—in other words, the *constitution* of the material.

The use of this procedure to analyze a statically indeterminate axial structure is illustrated in the next example.

As discussed in Chapters 1 and 2, it is convenient to use the notion of a **rigid element** to develop axial deformation concepts. A rigid element (such as a bar, a beam, or a plate) represents an object that is infinitely stiff, meaning that it does not deform in any manner. While it may translate or rotate, a rigid element does not stretch, compress, skew, or bend.

EXAMPLE 5.5

A 1.5 m long rigid beam *ABC* is supported by three axial members, as shown in the figure that follows. A concentrated load of 220 kN is applied to the rigid beam directly under *B*.

The axial members (1) connected at *A* and at *C* are identical aluminum alloy [E = 70 GPa] bars, each having a cross-sectional area A_1 = 550 mm^2 and a length L_1 = 2 m. Member (2) is a steel [E = 200 GPa] bar with a cross-sectional area A_2 = 900 mm^2 and a length L_2 = 2 m. All members are connected with simple pins.

If all three bars are initially unstressed, determine

(a) the normal stresses in the aluminum and steel bars, and
(b) the deflection of the rigid beam after application of the 220 kN load.

Plan the Solution

A free-body diagram (FBD) of rigid beam *ABC* will be drawn, and from this sketch, equilibrium equations will be derived in terms of the unknown member forces F_1 and F_2.

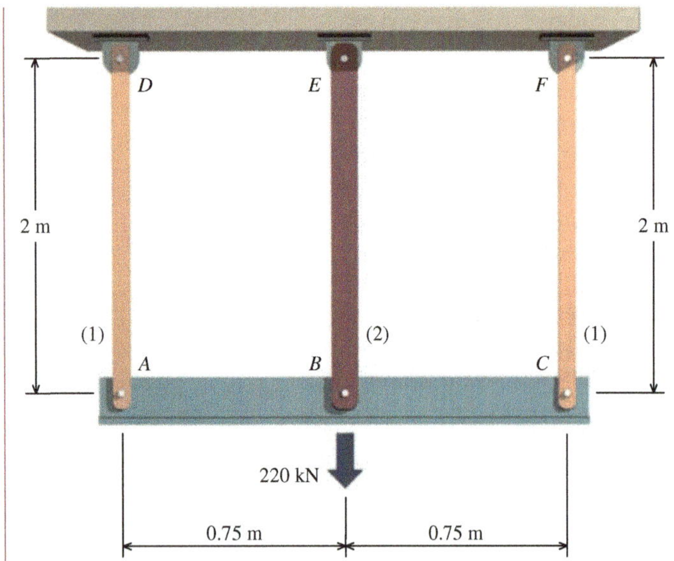

Since the axial members and the 220 kN load are arranged symmetrically relative to midpoint B of the rigid beam, the forces in the two aluminum bars (1) must be identical. The internal forces in the axial members are related to their deformations by Equation (5.2). Because members (1) and (2) are connected to rigid beam ABC, they are not free to deform independently of each other. On the basis of this observation and considering the symmetry of the structure, we can assert that the deformations in members (1) and (2) must be equal. This fact can be combined with the relationship between the internal force in a member and the member's deformation [Equation (5.2)] to derive another equation, which is expressed in terms of the unknown member forces F_1 and F_2. This equation is called a *compatibility equation*. The equilibrium and compatibility equations can be solved simultaneously to calculate the member forces. Then, after F_1 and F_2 have been determined, the normal stresses in each bar and the deflection of rigid beam ABC can be calculated.

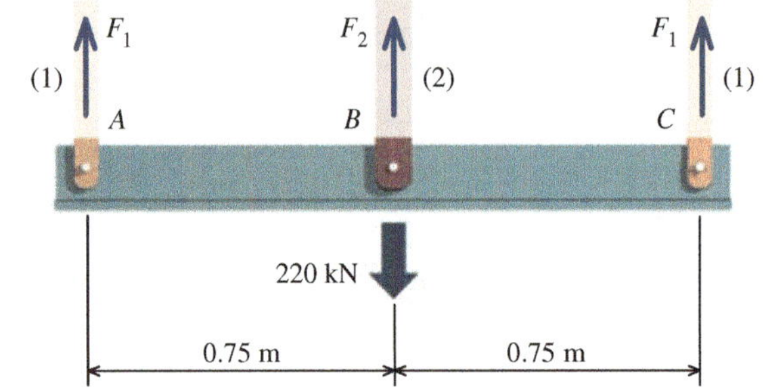

SOLUTION

Step 1 — Equilibrium Equations: An FBD of rigid beam ABC is shown. From the overall symmetry of the structure and the loads, we know that the forces in members AD and CF must be identical; therefore, we will denote the internal forces in each of these members as F_1. The internal force in member BE will be denoted F_2.

From this FBD, equilibrium equations can be written for (a) the sum of forces in the vertical direction (i.e., the y direction) and (b) the sum of moments about joint A:

$$\Sigma F_y = 2F_1 + F_2 - 220 \text{ kN} = 0 \tag{a}$$

$$\Sigma M_A = (1.5 \text{ m})F_1 + (0.75 \text{ m})F_2 - (0.75 \text{ m})(220 \text{ kN}) = 0 \tag{b}$$

Two unknowns appear in these equations (F_1 and F_2), and at first glance it seems as though we should be able to solve them simultaneously for F_1 and F_2. However, if Equation (b) is divided by 0.75 m, then Equations (a) and (b) are identical. Consequently, a second equation that is independent of the equilibrium equation must be derived in order to solve for F_1 and F_2.

Step 2 — Geometry of Deformation: By symmetry, we know that rigid beam ABC must remain horizontal after the 220 kN load is applied. Consequently, joints A, B, and C must all displace downward by the same amount: $v_A = v_B = v_C$. *How are these rigid-beam joint displacements related to member deformations δ_1 and δ_2?* Since the members are connected directly to the rigid beam (and there are no other considerations, such as gaps or clearances in the pin connections),

$$v_A = v_C = \delta_1 \quad \text{and} \quad v_B = \delta_2 \tag{c}$$

Step 3 — Force–Deformation Relationships: We know that the elongation in an axial member can be expressed by Equation (5.2). Therefore, the relationship between the internal axial force in a member and the member's deformation can be expressed for each member as

$$\delta_1 = \frac{F_1 L_1}{A_1 E_1} \quad \text{and} \quad \delta_2 = \frac{F_2 L_2}{A_2 E_2} \tag{d}$$

Step 4 — Compatibility Equation: The force–deformation relationships [Equation (d)] can be substituted into the geometry-of-deformation equation [Equation (c)] to obtain a new equation based on deformations but expressed in terms of the unknown member forces F_1 and F_2:

$$v_A = v_B = v_C \qquad \therefore \frac{F_1 L_1}{A_1 E_1} = \frac{F_2 L_2}{A_2 E_2} \tag{e}$$

Step 5 — Solve the Equations: From compatibility equation (e), derive an expression for F_1:

$$F_1 = F_2 \frac{L_2}{L_1}\frac{A_1}{A_2}\frac{E_1}{E_2} = F_2 \frac{(2\text{ m})}{(2\text{ m})}\frac{(550\text{ mm}^2)}{(900\text{ mm}^2)}\frac{(70\text{ GPa})}{(200\text{ GPa})} = 0.2139\, F_2 \tag{f}$$

Substitute Equation (f) into Equation (a) and solve for F_1 and F_2:

$$\Sigma F_y = 2F_1 + F_2 = 2(0.2139\, F_2) + F_2 = 220\text{ kN}$$

$$\therefore F_2 = 154.083\text{ kN} \qquad \text{and} \qquad F_1 = 32.958\text{ kN}$$

The normal stress in aluminum bars (1) is

$$\sigma_1 = \frac{F_1}{A_1} = \frac{32{,}958\text{ N}}{550\text{ mm}^2} = 59.9\text{ MPa (T)}$$ **Ans.**

and the normal stress in steel bar (2) is

$$\sigma_2 = \frac{F_2}{A_2} = \frac{154{,}083\text{ N}}{900\text{ mm}^2} = 171.2\text{ MPa (T)}$$ **Ans.**

From Equation (c), the deflection of the rigid beam is equal to the deformation of the axial members. Since both members (1) and (2) elongate by the same amount, either term in Equation (d) can be used. We thus have

$$\delta_1 = \frac{F_1 L_1}{A_1 E_1} = \frac{(32{,}958\text{ N})(2{,}000\text{ mm})}{(550\text{ mm}^2)(70{,}000\text{ N/mm}^2)} = 1.712\text{ mm}$$

Therefore, the rigid-beam deflection is $v_A = v_B = v_C = \delta_1 = 1.712$ mm. **Ans.**

By inspection, the rigid beam deflects downward.

The five-step procedure demonstrated in the previous example provides a versatile method for the analysis of statically indeterminate structures. Additional problem-solving considerations and suggestions for each step of the process are discussed in the table that follows.

Solution Method for Statically Indeterminate Axial Structures

Step 1	Equilibrium Equations	Draw one or more free-body diagrams (FBDs) for the structure, focusing on the joints that connect the members. Joints are located wherever (a) an external force is applied, (b) the cross-sectional properties (such as area or diameter) change, (c) the properties of the material (i.e., E) change, or (d) a member connects to a rigid element (such as a rigid bar, beam, plate, or flange). Generally, FBDs of reaction joints are not useful. Write equilibrium equations for the FBDs. Note the number of unknowns involved and the number of independent equilibrium equations. If the number of unknowns exceeds the number of equilibrium equations, a deformation equation must be written for each extra unknown. Comments: • Label the joints with capital letters and label the members with numbers. This simple scheme can help you to clearly recognize effects that occur in members (such as deformations) and effects that pertain to joints (such as deflections of rigid elements). • As a rule, *assume that the internal force in an axial member is tensile*, an assumption that is consistent with a positive deformation (i.e., an elongation) of the axial member. This practice will make it easier for us to incorporate the effects of temperature change into our analyses of the deformations in structures made up of axial members. Temperature change effects will be discussed in Section 5.6.
Step 2	Geometry of Deformation	This step is unique to statically indeterminate problems. The structure or system should be scrutinized to assess how the deformations in the members are related to each other. An equilibrium equation and the corresponding geometry-of-deformation equation must be *consistent*. This condition means that, when a tensile force is assumed for a member in an FBD, a tensile deformation must be indicated for the same member in the deformation diagram. Most of the statically indeterminate axial structures fall into one of three general configurations: 1. Coaxial or parallel axial members. 2. Axial members connected end to end in series. 3. Axial members connected to a rotating rigid element. Characteristics of these three categories are discussed in more detail shortly.
Step 3	Force–Deformation Relationships	The relationship between the internal force in, and the deformation of, axial member i is expressed by $$\delta_i = \frac{F_i L_i}{A_i E_i}$$ As a practical matter, writing down force–deformation relationships for the axial members at this stage of the solution is a helpful routine. These relationships will be used to construct the compatibility equation(s) in Step 4.
Step 4	Compatibility Equation	The force–deformation relationships (from Step 3) are incorporated into the geometric relationship of member deformations (from Step 2) to derive a new equation, which is expressed in terms of the unknown member forces. Together, the compatibility and equilibrium equations provide sufficient information to solve for the unknown variables.
Step 5	Solve the Equations	The compatibility equation and the equilibrium equation(s) are solved simultaneously. While conceptually straightforward, this step requires careful attention to calculation details such as sign conventions and consistency of units.

Successful application of the five-step solution method depends in no small part on the ability to understand how axial deformations are related in a structure. The table which follows highlights three common categories of statically indeterminate structures that are made up of axial members. For each general category, possible geometry-of-deformation equations are discussed.

Geometry of Deformations for Typical Statically Indeterminate Axial Structures

Equation Form	Comments	Typical Problems
1. Coaxial or parallel axial members.		
	Problems in this category include side-by-side plates, a tube with a filled core, a concrete column with embedded reinforcing steel, and three parallel rods symmetrically connected to a rigid bar.	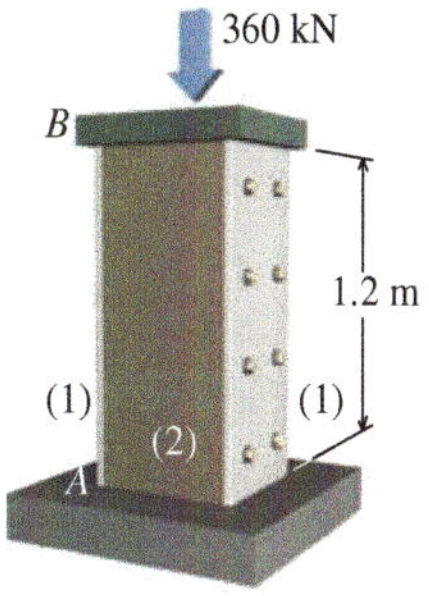 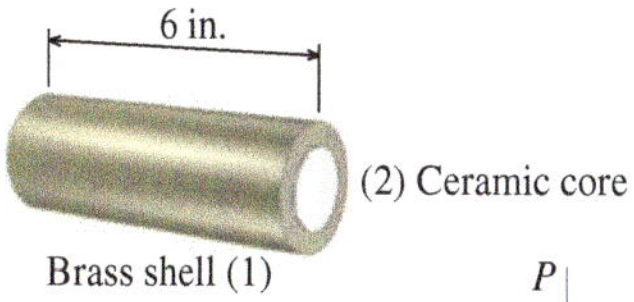 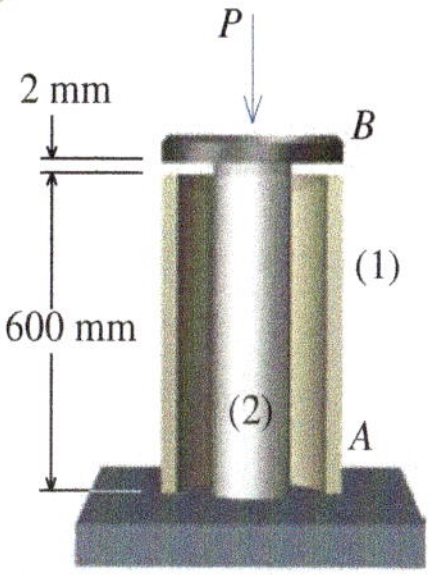 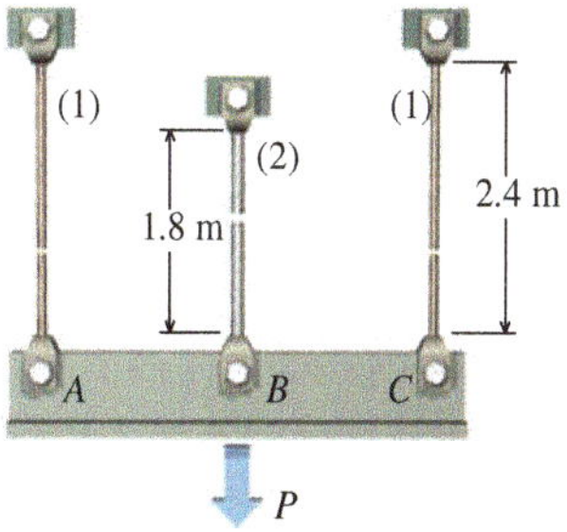
$\delta_1 = \delta_2$	The deformation of each axial member must be the same, unless there is a gap or clearance in the connections.	
$\delta_1 + \text{gap} = \delta_2$ $\delta_1 = \delta_2 + \text{gap}$	If there is a gap, then the deformation of one member equals the deformation of the other member plus the length of the gap.	
2. Axial members connected end to end in series.		
	Problems in this category include two or more members connected end to end.	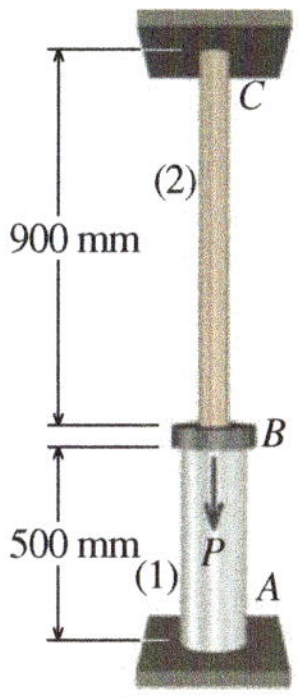 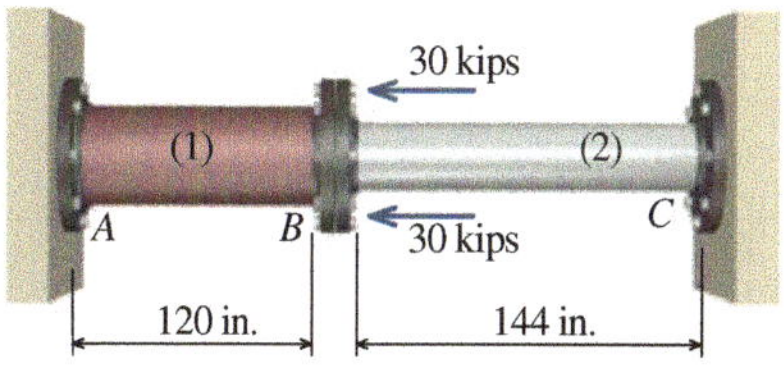 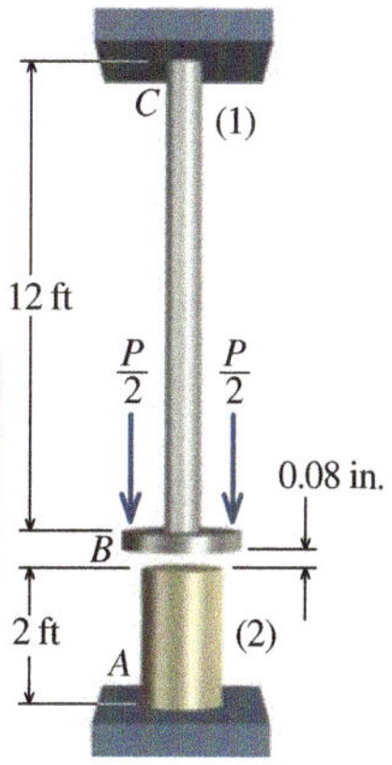 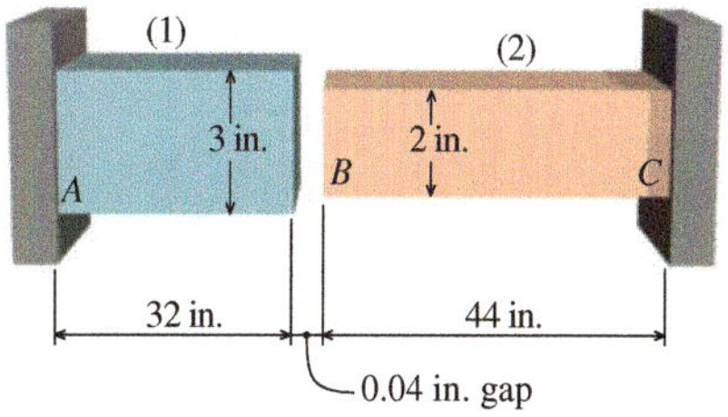
$\delta_1 + \delta_2 = 0$	If there are no gaps or clearances in the configuration, then the member deformations must sum to zero; in other words, an elongation in member (1) is accompanied by an equal contraction in member (2).	
$\delta_1 + \delta_2 = \text{constant}$	If there is a gap or clearance between the two members or if the supports move as the load is applied, then the sum of the deformations of the members equals the specified distance.	

Equation Form	Comments	Typical Problems
	3. Axial members connected to a rotating rigid element.	
	Problems in this category feature a rigid bar or a rigid plate to which the axial members are attached. The rigid element is pinned so that it rotates about a fixed point. Since the axial members are attached to the rotating element, their deformations are constrained by the position of the deflected rigid bar. The relationship between member deformations can be found from the principle of similar triangles.	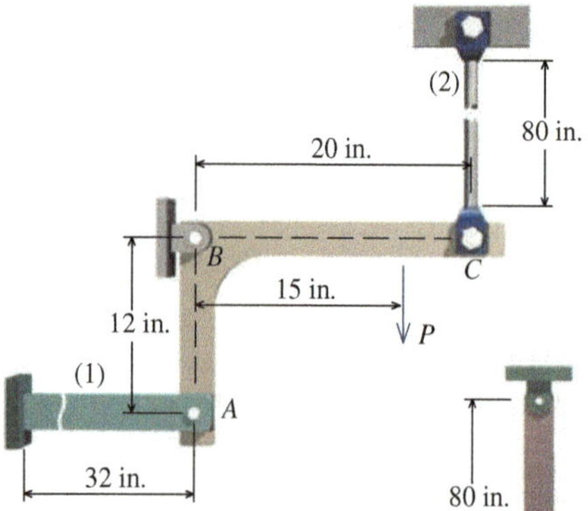 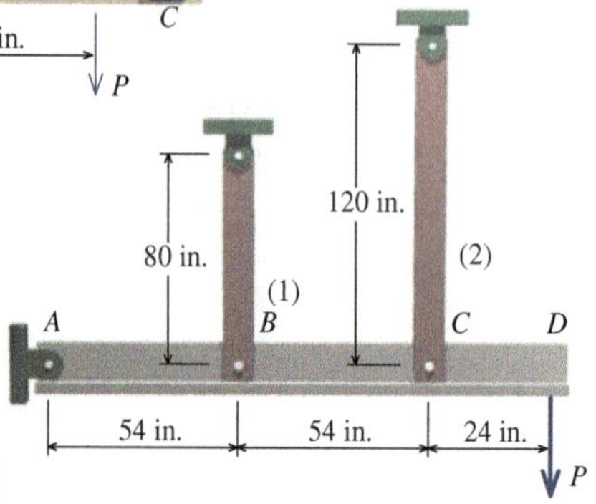
$\frac{\delta_1}{a} = \frac{\delta_2}{b}$	If both members elongate or both members contract as the rigid bar rotates, then the first equation form is obtained.	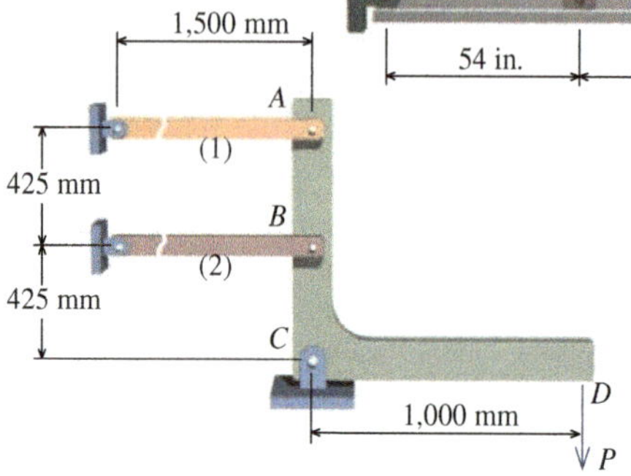
$\frac{\delta_1}{a} = -\frac{\delta_2}{b}$	If one member elongates while the other member contracts as the rigid bar rotates, then the geometry-of-deformation equation takes the second form.	
$\frac{\delta_1 + \text{gap}}{a} = \frac{\delta_2}{b}$	If there is a gap or clearance in a joint, then the geometry-of-deformation equation takes the third form.	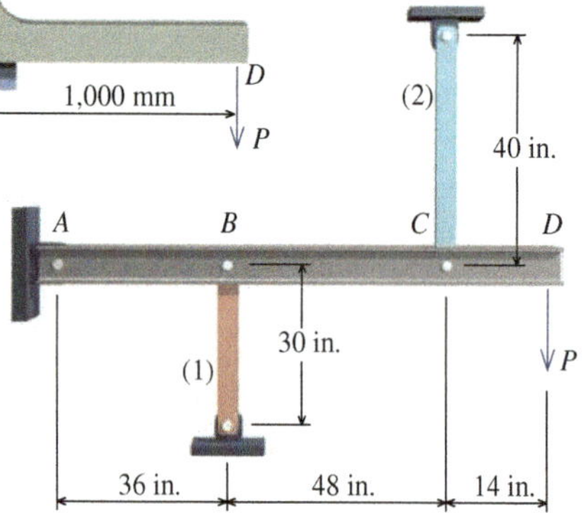

EXAMPLE 5.6

A steel pipe (1) is attached to an aluminum pipe (2) at flange B. The pipes are attached to rigid supports at A and C, respectively.

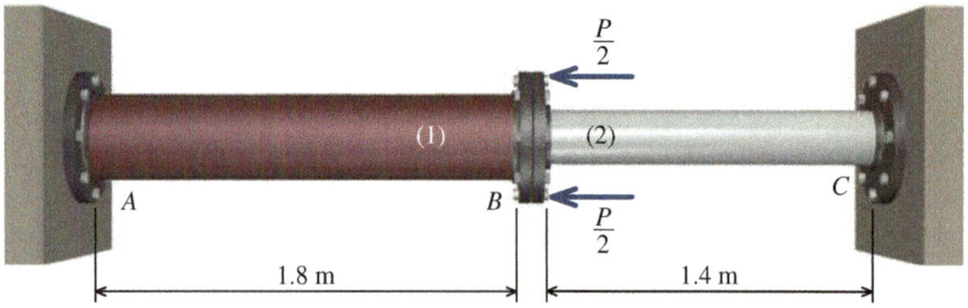

Member (1) has a cross-sectional area $A_1 = 3{,}600 \text{ mm}^2$, an elastic modulus $E_1 = 200$ GPa, and an allowable normal stress of 160 MPa. Member (2) has a cross-sectional area $A_2 = 2{,}000 \text{ mm}^2$, an elastic modulus $E_2 = 70$ GPa, and an allowable normal stress of 120 MPa. Determine the maximum load P that can be applied to flange B without exceeding either allowable stress.

Plan the Solution

Consider a free-body diagram (FBD) of flange B, and write the equilibrium equation for the sum of forces in the x direction. This equation will have three unknowns: F_1, F_2, and P.

Determine the geometry-of-deformation equation and write the force–deformation relationships for members (1) and (2). Substitute the force–deformation relationships into the geometry-of-deformation equation to obtain the compatibility equation. Then, use the allowable stress and area of member (1) to compute a value for P. Repeat this procedure, using the allowable stress and area of member (2), to compute a second value for P. Choose the smaller of these two values as the maximum force P that can be applied to flange B.

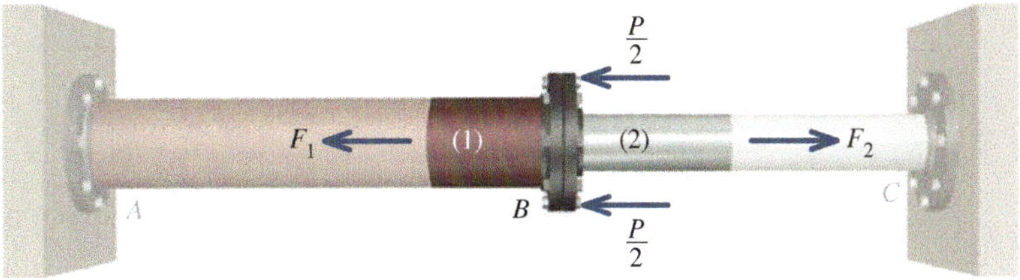

SOLUTION

Step 1 — Equilibrium Equations: The free-body diagram for joint B is shown. Notice that internal tensile forces are assumed in both member (1) and member (2) [even though one would expect to find that member (1) would actually be in compression].

The equilibrium equation for joint B is simply

$$\Sigma F_x = F_2 - F_1 - P = 0 \qquad \text{(a)}$$

Step 2 — Geometry of Deformation: Since the compound axial member is attached to rigid supports at A and C, the overall deformation of the structure must be zero. In other words,

$$\delta_1 + \delta_2 = 0 \qquad \text{(b)}$$

Step 3 — Force–Deformation Relationships: Write generic force–deformation relationships for the members:

$$\delta_1 = \frac{F_1 L_1}{A_1 E_1} \quad \text{and} \quad \delta_2 = \frac{F_2 L_2}{A_2 E_2} \qquad \text{(c)}$$

Step 4 — Compatibility Equation: Substitute Equations (c) into Equation (b) to obtain the compatibility equation:

$$\frac{F_1 L_1}{A_1 E_1} + \frac{F_2 L_2}{A_2 E_2} = 0 \qquad \text{(d)}$$

Step 5 — Solve the Equations: First, we will substitute for F_2 in Equation (a). To accomplish this, solve Equation (d) for F_2:

$$F_2 = -F_1 \frac{L_1}{L_2} \frac{A_2}{A_1} \frac{E_2}{E_1} \qquad \text{(e)}$$

Now substitute Equation (e) into Equation (a) to obtain

$$-F_1 \frac{L_1}{L_2} \frac{A_2}{A_1} \frac{E_2}{E_1} - F_1 = -F_1\left[\frac{L_1}{L_2} \frac{A_2}{A_1} \frac{E_2}{E_1} + 1\right] = P$$

There are still two unknowns in this equation; consequently, another equation is necessary to obtain a solution. Let F_1 equal the force corresponding to the allowable stress $\sigma_{allow,1}$ in member (1), and solve for the applied load P. (**Note:** The negative sign attached to F_1 can be omitted here since we are interested only in the magnitude of load P.) First, we have

$$\sigma_{allow,1}A_1\left[\frac{L_1}{L_2}\frac{A_2}{A_1}\frac{E_2}{E_1}+1\right] = (160\text{ N/mm}^2)(3{,}600\text{ mm}^2)\left[\left(\frac{1.8}{1.4}\right)\left(\frac{2{,}000}{3{,}600}\right)\left(\frac{70}{200}\right)+1\right]$$

$$= (576{,}000\text{ N})[1.25] = 720{,}000\text{ N} = 720\text{ kN} \geq P$$

Next, repeat this process for member (2). Then solve Equation (e) for F_1:

$$F_1 = -F_2\frac{L_2}{L_1}\frac{A_1}{A_2}\frac{E_1}{E_2} \tag{f}$$

Now substitute Equation (f) into Equation (a) to obtain

$$F_2 + F_2\frac{L_2}{L_1}\frac{A_1}{A_2}\frac{E_1}{E_2} = F_2\left[1+\frac{L_2}{L_1}\frac{A_1}{A_2}\frac{E_1}{E_2}\right] = P$$

Finally, let F_2 equal the allowable force and solve for the corresponding applied force P:

$$\sigma_{allow,2}A_2\left[1+\frac{L_2}{L_1}\frac{A_1}{A_2}\frac{E_1}{E_2}\right] = (120\text{ N/mm}^2)(2{,}000\text{ mm}^2)\left[1+\left(\frac{1.4}{1.8}\right)\left(\frac{3{,}600}{2{,}000}\right)\left(\frac{200}{70}\right)\right]$$

$$= (240{,}000\text{ N})[5.0] = 1{,}200{,}000\text{ N} = 1{,}200\text{ kN} \geq P$$

Therefore, the maximum load P that can be applied to the flange at B is $P = 720$ kN. **Ans.**

MecMovies

EXAMPLES

M5.5 A steel rod (1) is attached to a steel post (2) at flange B. A downward load of 110 kN is applied to flange B. Rod (1) and post (2) are attached to rigid supports at A and C, respectively. Rod (1) has a cross-sectional area of 800 mm² and an elastic modulus of 200 GPa. Post (2) has a cross-sectional area of 1,600 mm² and an elastic modulus of 200 GPa.

(a) Compute the normal stress in rod (1) and post (2).
(b) Compute the deflection of flange B.

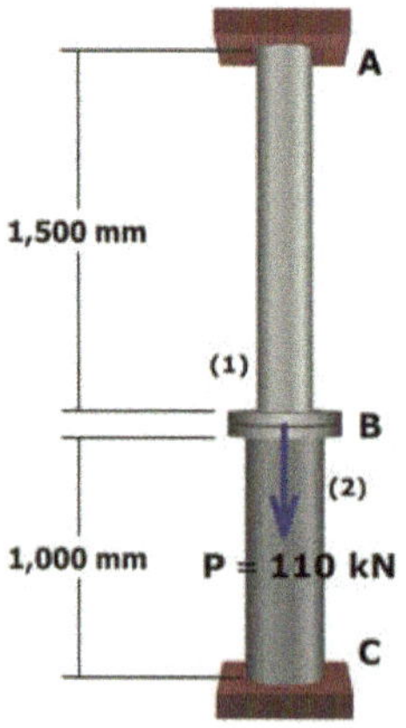

M5.6 An aluminum tube (1) encases a brass core (2). The two components are bonded together to form an axial member that is subjected to a downward force of 30 kN. Tube (1) has an outer diameter $D = 30$ mm and an inner diameter $d = 22$ mm. The elastic modulus of the aluminum is 70 GPa. The brass core (2) has a diameter d = 22 mm and an elastic modulus of 105 GPa. Compute the normal stresses in tube (1) and core (2).

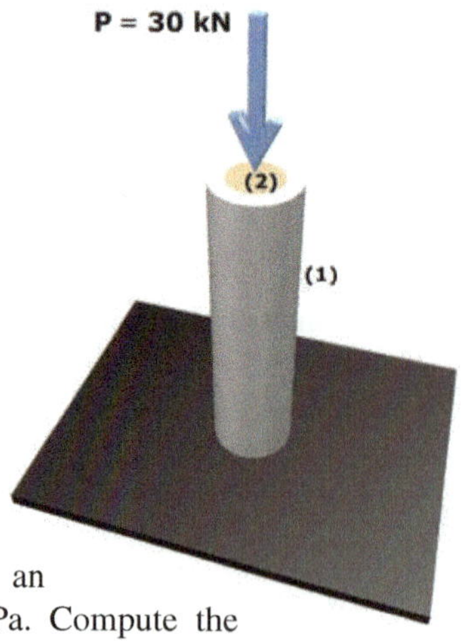

110

Structures with a Rotating Rigid Bar

Problems involving a rotating rigid element can be particularly difficult to solve. For these structures, a deformation diagram should be drawn at the outset. This diagram is essential to obtaining the correct geometry-of-deformation equation. The assumed internal force and the assumed deformation must be consistent. Thus, if elongation is assumed for a member in the deformation diagram, then the positive internal force for the member is tensile. MecMovies Example M5.7 illustrates problems of this type.

MecMovies

EXAMPLE

M5.7 Rigid bar AD is pinned at A and supported by bars (1) and (2) at B and C, respectively. Bar (1) is aluminum and bar (2) is brass. A concentrated load $P = 36$ kN is applied to the rigid bar at D. Compute the normal stress in each bar and the downward deflection of the rigid bar at D.

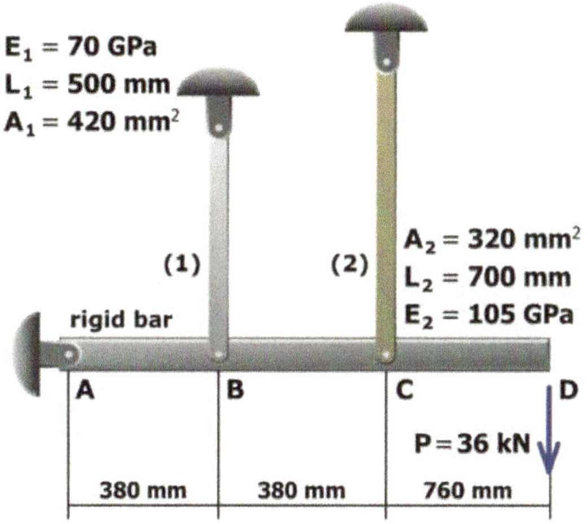

Some structures with rotating rigid bars have opposing members; that is, one member is elongated, while the other member is compressed. Figure 5.11 illustrates the subtle difference between these two types of configuration.

For the structure with two tension members (Figure 5.11*a*), the geometry of deformations in terms of vertical joint deflections v_B and v_C is found by similar triangles (Figure 5.11*b*):

$$\frac{v_B}{x_B} = \frac{v_C}{x_C}$$

From Figure 5.11*c*, the member deformations δ_1 and δ_2 are related to the joint deflections v_B and v_C by

$$\delta_1 = L_{\text{final}} - L_{\text{initial}} = (L_1 + v_B) - L_1 = v_B \qquad \therefore v_B = \delta_1$$

and

$$\delta_2 = L_{\text{final}} - L_{\text{initial}} = (L_2 + v_C) - L_2 = v_C \qquad \therefore v_C = \delta_2 \tag{5.6}$$

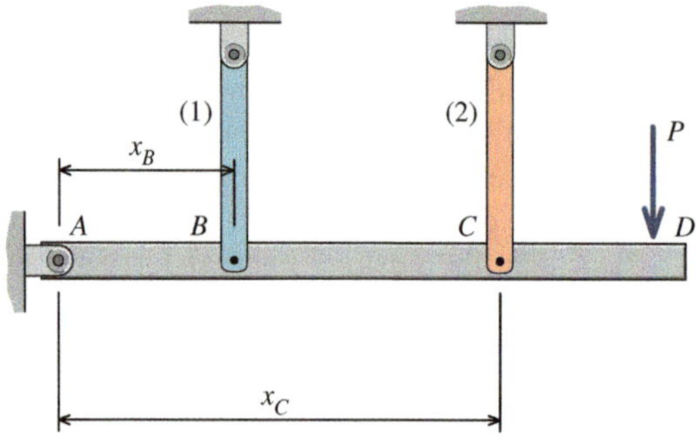

FIGURE 5.11*a* Configuration with two tension members.

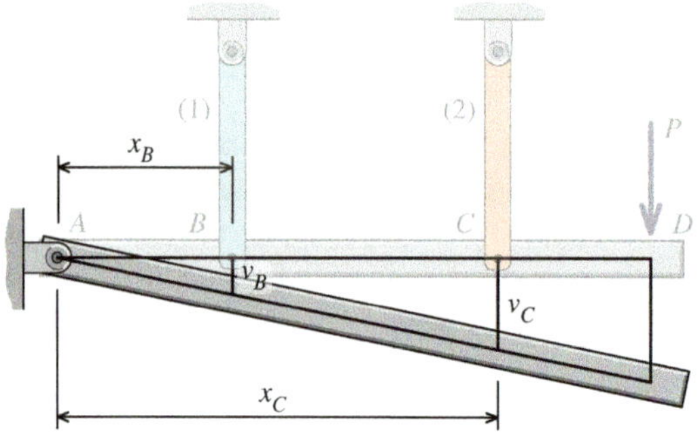

FIGURE 5.11*b* Deformation diagram.

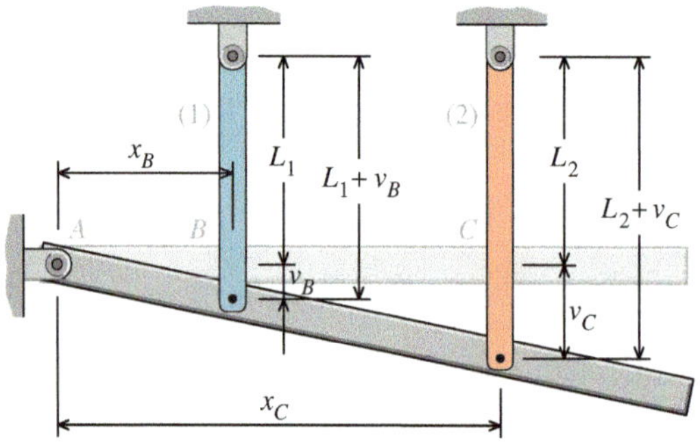

FIGURE 5.11*c* Showing member deformations.

Therefore, the geometry-of-deformation equation can be written in terms of member deformations as

$$\frac{\delta_1}{x_B} = \frac{\delta_2}{x_C} \tag{5.7}$$

For the structure with two opposing axial members (Figure 5.11*d*), the geometry-of-deformation equation in terms of joint deflections v_B and v_C is the same as before (Figure 5.11*e*):

$$\frac{v_B}{x_B} = \frac{v_C}{x_C}$$

From Figure 5.11*f*, the member deformations δ_1 and δ_2 are related to the joint deflections v_B and v_C by

$$\delta_1 = L_{\text{final}} - L_{\text{initial}} = (L_1 + v_B) - L_1 = v_B \qquad \therefore v_B = \delta_1$$

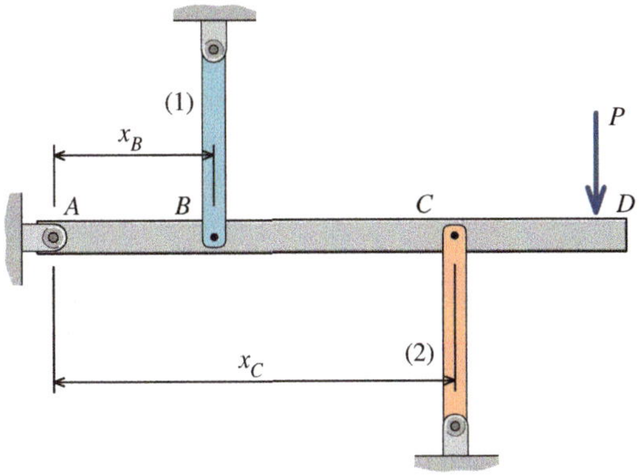

FIGURE 5.11*d* Configuration with opposing members.

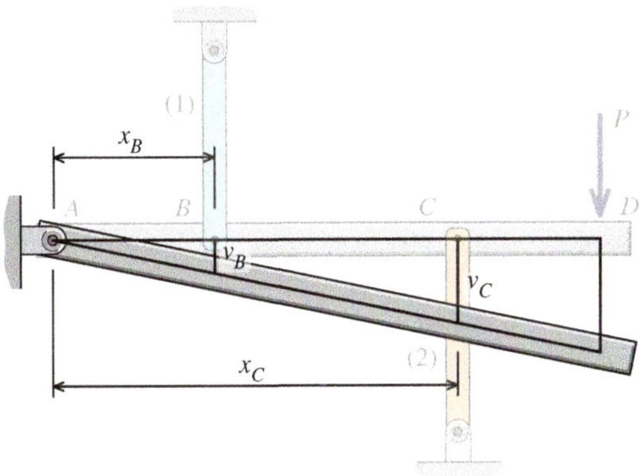

FIGURE 5.11*e* Deformation diagram.

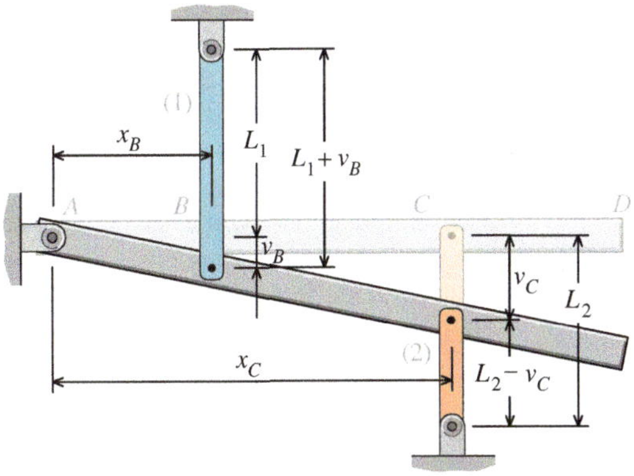

FIGURE 5.11*f* Showing member deformations.

and

$$\delta_2 = L_{\text{final}} - L_{\text{initial}} = (L_2 - v_C) - L_2 = -v_C \qquad \therefore\ v_C = -\delta_2 \tag{5.8}$$

Note the subtle difference between Equations (5.6) and Equations (5.8). The geometry-of-deformation equation for the configuration with opposing members, in terms of member deformations, is, therefore,

$$\frac{\delta_1}{x_B} = -\frac{\delta_2}{x_C} \tag{5.9}$$

An equilibrium equation and the corresponding deformation equation must be compatible; that is, when a tensile force is assumed for a member in a free-body diagram, a tensile deformation must be indicated for the same member in the deformation diagram. In the configurations shown here, internal tensile forces have been assumed for all the axial members. For the structure shown in Figure 5.11*d*, the displacement of the rigid bar at *C* (Figure 5.11*e*) causes a contraction in member (2). As shown in Equations (5.8), this condition produces a negative sign for δ_2, and as a result, the geometry-of-deformation equation in Equation (5.9) is slightly different from the geometry-of-deformation equation found for the structure with two tension members [Equation (5.7)].

Rigid-bar structures with opposing axial members are analyzed in MecMovies Examples M5.8 and M5.9.

MecMovies

EXAMPLES

M5.8 A pin-connected structure is loaded and supported as shown. Member *ABCD* is a rigid bar that is horizontal before the load *P* is applied. Members (1) and (2) are aluminum [E = 70 GPa], with cross-sectional areas $A_1 = A_2 =$ 160 mm². Member (1) is 900 mm in length, and member (2) is 1,250 mm. A load P = 35 kN is applied to the structure at *D*.

(a) Calculate the axial forces in members (1) and (2).
(b) Compute the normal stress in members (1) and (2).
(c) Compute the downward deflection of the rigid bar at *D*.

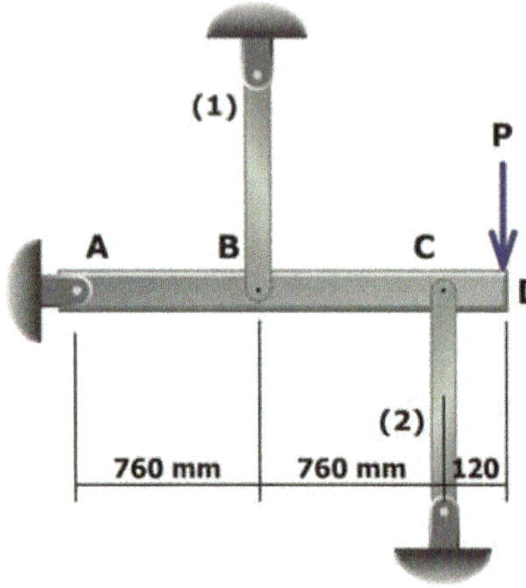

M5.9 Rigid bar *ABCD* is pinned at *C* and supported by bars (1) and (2) at *A* and *D*, respectively. Bar (1) is aluminum and bar (2) is bronze. A concentrated load P = 80 kN is applied to the rigid bar at *B*. Compute the normal stress in each bar and the downward deflection at *A* of the rigid bar.

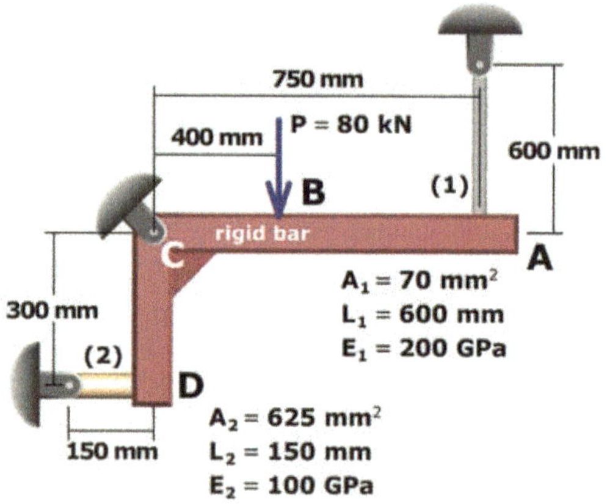

M5.10 An aluminum bar (2) is to be connected to a brass post (1). When the two axial members were installed, however, it was discovered that there was a 1/16 in. gap between flange *B* and the post. The brass post (1) has a cross-sectional area A_1 = 0.60 in.² and an elastic modulus E_1 = 16,000 ksi. The aluminum bar (2) has properties of A_2 = 0.20 in.² and E_2 = 10,000 ksi.

If bolts are inserted through the flange at *B* and tightened until the gap is closed, how much stress will be induced in each of the axial members?

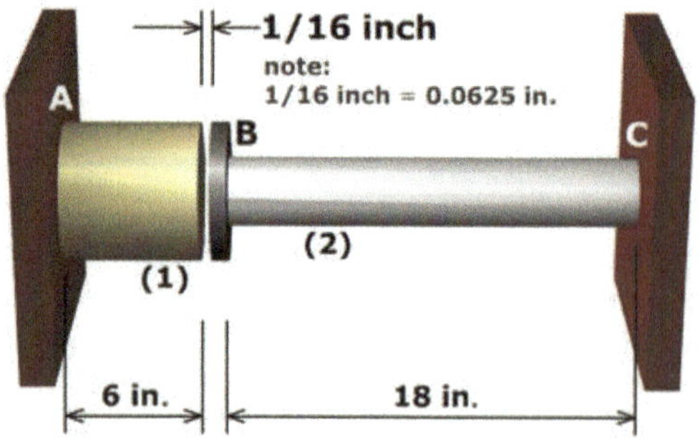

EXERCISES

M5.5 A composite axial structure consists of two rods joined at flange *B*. Rods (1) and (2) are attached to rigid supports at *A* and *C*, respectively. A concentrated load *P* is applied to flange *B* in the direction shown. Determine the internal forces and normal stresses in each rod. Also, determine the deflection of flange *B* in the *x* direction.

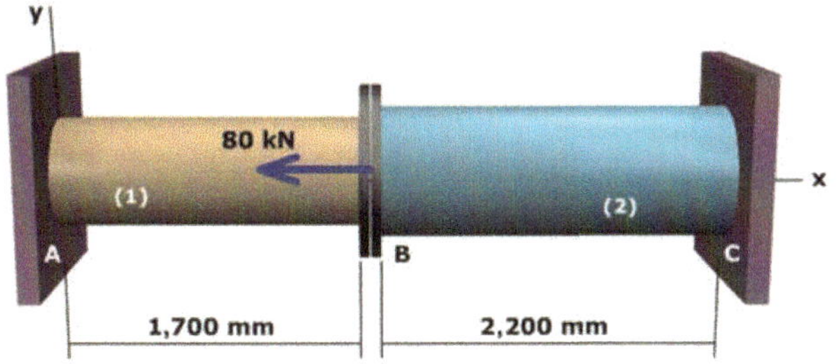

FIGURE M5.5

M5.6 A composite axial structure consists of a tubular shell (1) bonded to length *AB* of a continuous solid rod that extends from *A* to *C*. The rod is labeled (2) and (3). A concentrated load *P* is applied to the free end *C* of the rod in the direction shown. Determine the internal forces and normal stresses in shell (1) and core (2) (i.e., between *A* and *B*). Also, determine the deflection in the *x* direction of end *C* relative to support *A*.

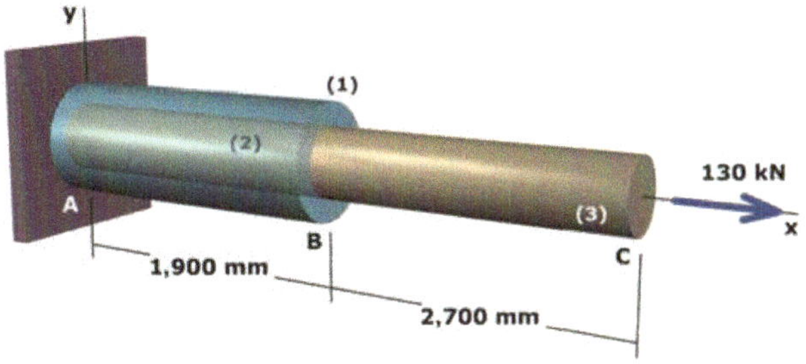

FIGURE M5.6

M5.7 Determine the internal forces and normal stresses in bars (1) and (2). Also, determine the deflection of the rigid bar in the *x* direction at *C*.

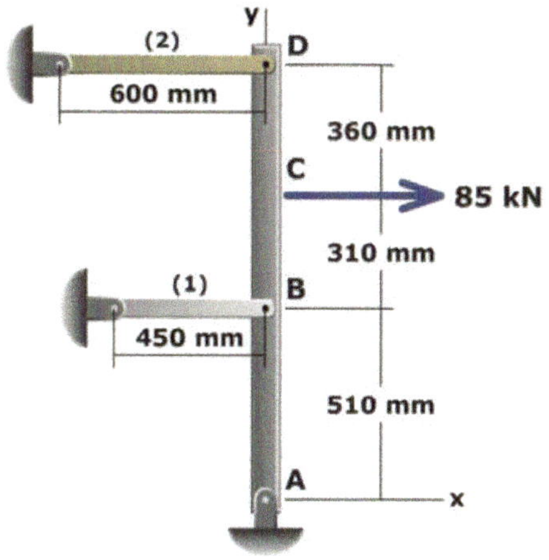

FIGURE M5.7

M5.8 Determine the internal forces and normal stresses in bars (1) and (2). Also, determine the deflection of the rigid bar in the *x* direction at *C*.

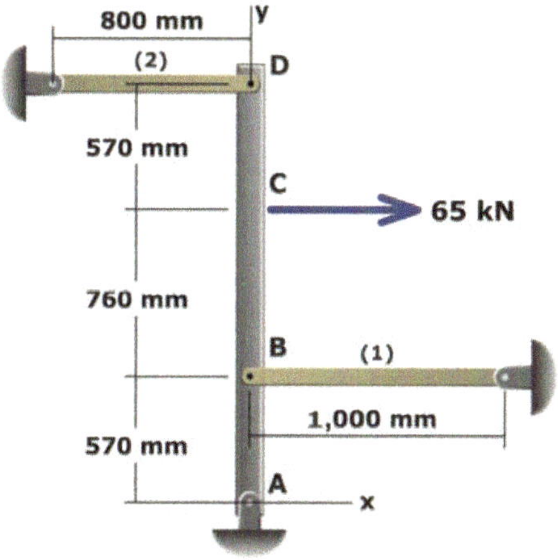

FIGURE M5.8

PROBLEMS

P5.23 A load *P* is supported by a structure consisting of rigid bar *ABC*, two identical solid aluminum alloy [$E = 73$ GPa] rods (1), and a solid bronze [$E = 105$ GPa] rod (2), as shown in Figure P5.23. The aluminum rods (1) each have a diameter of 12 mm and a length $L_1 = 2.3$ m. They are symmetrically positioned relative to middle rod (2) and the applied load *P*. Bronze rod (2) has a diameter of 18 mm and a length $L_2 = 1.6$ m. If all bars are unstressed before the load *P* is applied, determine the normal stresses in the aluminum and bronze rods after a load $P = 50$ kN is applied to the rigid bar.

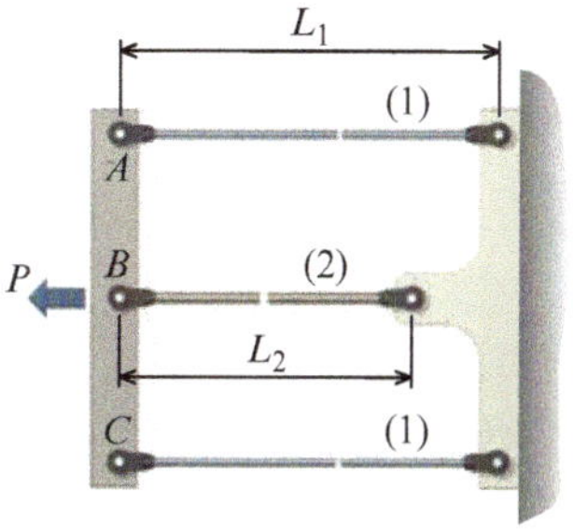

FIGURE P5.23

P5.24 A composite bar is fabricated by brazing aluminum alloy [$E = 10{,}000$ ksi] bars (1) to a center brass [$E = 17{,}000$ ksi] bar (2) as shown in Figure P5.24. Assume that $d = 0.75$ in., $a = 0.1875$ in., and $L = 20$ in. If the total axial force carried by the two aluminum bars must equal the axial force carried by the brass bar, calculate the thickness b required for brass bar (2).

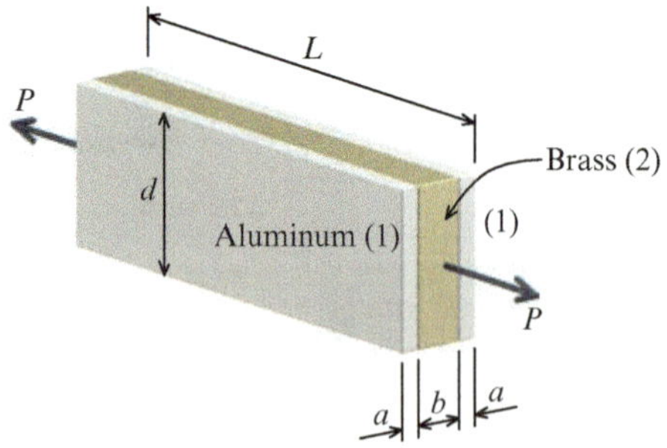

FIGURE P5.24

P5.25 An aluminum alloy [$E = 10{,}000$ ksi; $\sigma_Y = 40$ ksi] pipe (1) is connected to a bronze [$E = 16{,}000$ ksi; $\sigma_Y = 45$ ksi] pipe at flange B, as shown in Figure P5.25. The pipes are attached to rigid supports at A and C, respectively. Pipe (1) has an outside diameter of 2.375 in., a wall thickness of 0.203 in., and a length $L_1 = 6$ ft. Pipe (2) has an outside diameter of 4.50 in., a wall thickness of 0.226 in., and a length $L_2 = 10$ ft. If a minimum factor of safety of 1.67 is required for each pipe, determine

(a) the maximum load P that may be applied at flange B.
(b) the deflection of flange B at the load determined in part (a).

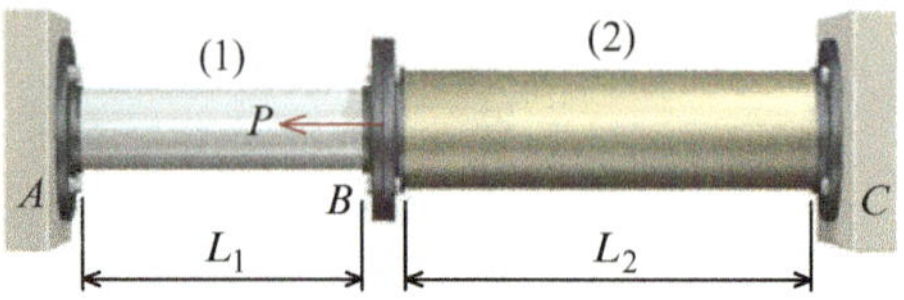

FIGURE P5.25

P5.26 A rectangular polypropylene [$E = 6{,}200$ MPa] bar (1) is connected to a rectangular nylon [$E = 1{,}400$ MPa] bar (2) at flange B. The assembly (shown in Figure P5.26) is connected to rigid supports at A and C. Bar (1) has a cross-sectional area $A_1 = 1{,}100$ mm^2 and a length $L_1 = 1{,}450$ mm. Bar (2) has a cross-sectional area $A_2 = 2{,}800$ mm^2 and a length $L_2 = 550$ mm. After two loads of $P = 4$ kN are applied to flange B, determine

(a) the forces in bars (1) and (2).
(b) the deflection of flange B.

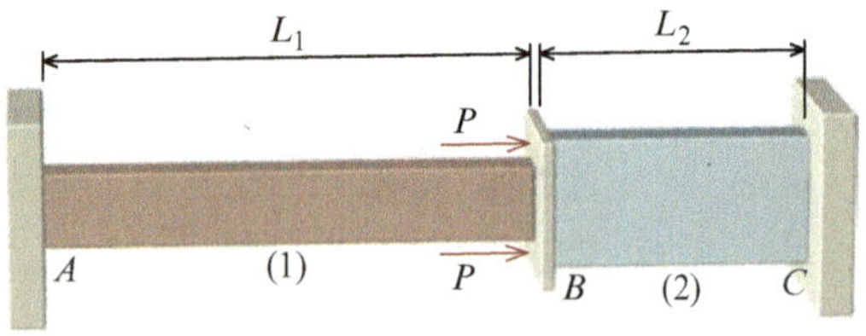

FIGURE P5.26

P5.27 In Figure P5.27, the two vertical steel [$E = 200$ GPa] rods that support rigid bar $ABCD$ are initially free of stress. Rod (1) has an area $A_1 = 450$ mm^2 and a length $L_1 = 2.2$ m. Rod (2) has an area $A_2 = 325$ mm^2 and a length $L_2 = 1.6$ m. Assume dimensions of $a = 3.0$ m, $b = 1.0$ m, and $c = 1.25$ m. After a load $P = 80$ kN is applied to the rigid bar at D, determine

(a) the normal stresses in rods (1) and (2).
(b) the deflection of the rigid bar at D.

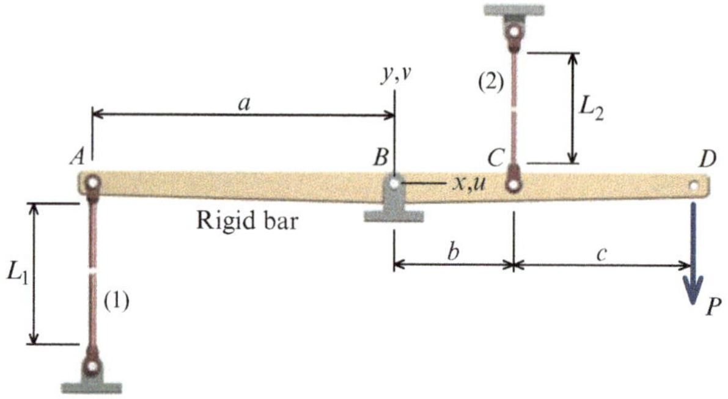

FIGURE P5.27

P5.28 The pin-connected structure shown in Figure P5.28 consists of a rigid bar $ABCD$ and two aluminum alloy [$E = 100$ GPa] bars, each with a cross-sectional area of 340 mm^2. Bar (1) has a length $L_1 = 810$ mm and bar (2) has a length $L_2 = 1{,}080$ mm. Assume dimensions of $a = 480$ mm, $b = 380$ mm, and $c = 600$ mm. If the allowable normal stress in bars (1) and (2) must be limited to 165 MPa each, determine

(a) the maximum load P that may be applied at D to the rigid bar.
(b) the deflection at D for the load determined in part (a).

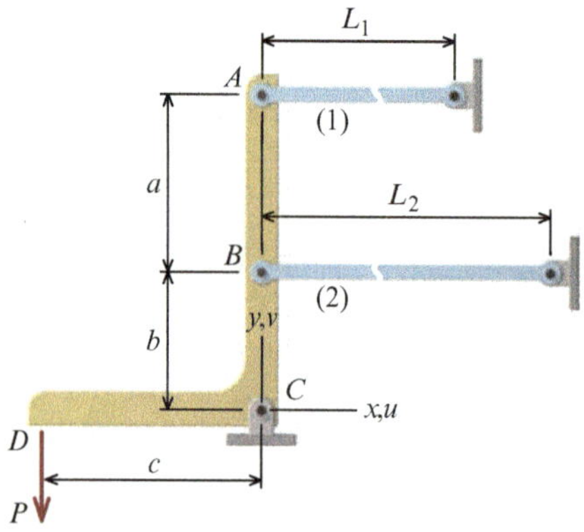

FIGURE P5.28

P5.29 In Figure P5.29, a load P is supported by a structure consisting of rigid bar BDF and three identical aluminum [E = 10,000 ksi] rods, each having a diameter of 0.625 in. Use dimensions of a = 72 in., b = 60 in., and L = 90 in. For a load P = 16 kips, determine

(a) the tensile force produced in each rod.
(b) the vertical deflection at B of the rigid bar.

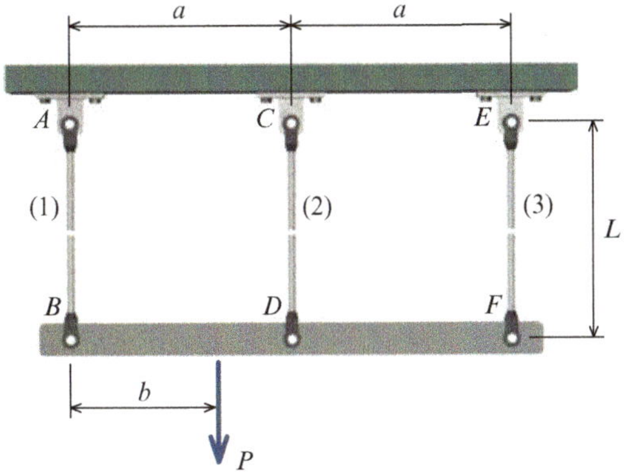

FIGURE P5.29

P5.30 A uniformly distributed load w is supported by a structure consisting of rigid bar BDF and three rods, as shown in Figure P5.30. Rods (1) and (2) are 0.75 in. diameter stainless steel rods that have an elastic modulus E = 28,000 ksi and a yield strength σ_Y = 36 ksi. Rod (3) is a 1.25 in. diameter bronze rod that has an elastic modulus E = 16,000 ksi and a yield strength σ_Y = 48 ksi. Use a = 4 ft and L = 9 ft. If a minimum factor of safety of 1.8 is specified for the normal stress in each rod, calculate the maximum magnitude of the distributed load w that can be supported.

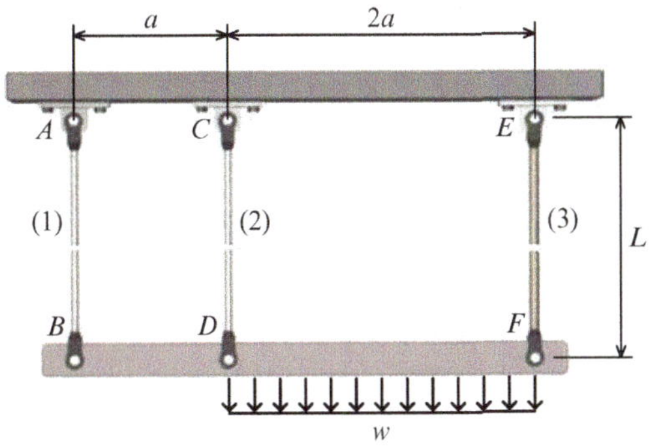

FIGURE P5.30

P5.31 Links (1) and (2) support rigid bar $ABCD$ shown in Figure P5.31. Link (1) is bronze [E = 15,200 ksi] with a cross-sectional area A_1 = 0.50 in.² and a length L_1 = 24 in. Link (2) is cold-rolled steel [E = 30,000 ksi] with a cross-sectional area A_2 = 0.375 in.² and a length L_2 = 32 in. Use dimensions of a = 14 in., b = 16 in., and c = 18 in. For an applied load of P = 9 kips, determine

(a) the normal stresses in links (1) and (2).
(b) the deflection of end D of the rigid bar.

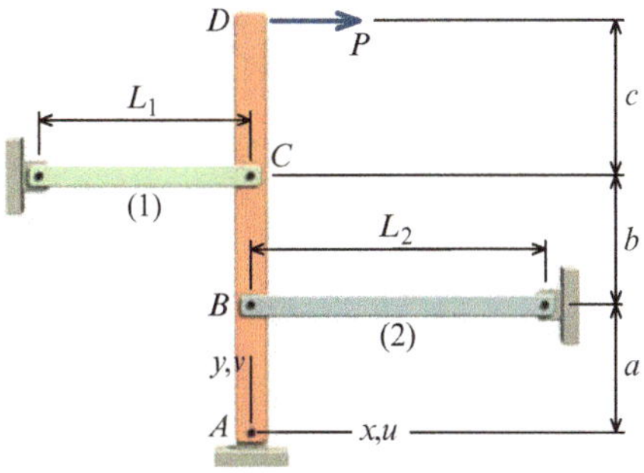

FIGURE P5.31

P5.32 In Figure P5.32, polycarbonate bars (1) and (2) are attached at their lower ends to a slider block at B that travels vertically inside of a fixed smooth slot. Load P is applied to the slider block. Each bar has a cross-sectional area of 85 mm², a length of 1.25 m, and an elastic modulus E = 8 GPa. Use dimensions of a = 1.15 m and b = 0.90 m. For an applied load of P = 1,700 N, determine

(a) the forces in bars (1) and (2).
(b) the deflection at B of the slider block.

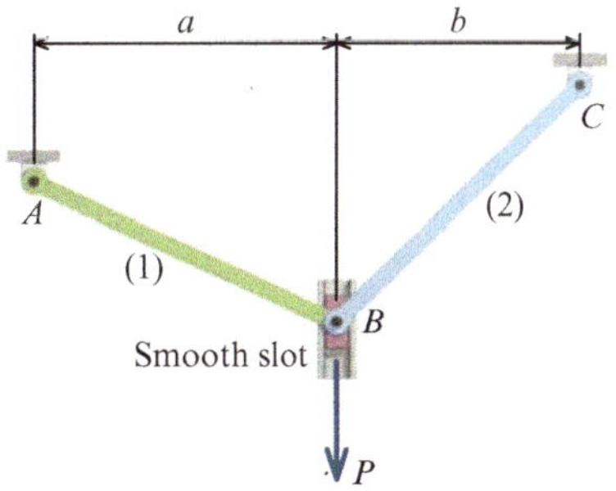

FIGURE P5.32

P5.33 The pin-connected structure shown in Figure P5.33 consists of two cold-rolled steel [E = 207 GPa] bars (1) and a bronze [E = 105 GPa] bar (2) that are connected at pin D. All three bars have cross-sectional areas of 650 mm². A load P = 275 kN is applied to the structure at D. Using a = 2.4 m and b = 3.2 m, calculate

(a) the normal stresses in bars (1) and (2).
(b) the displacement of pin D.

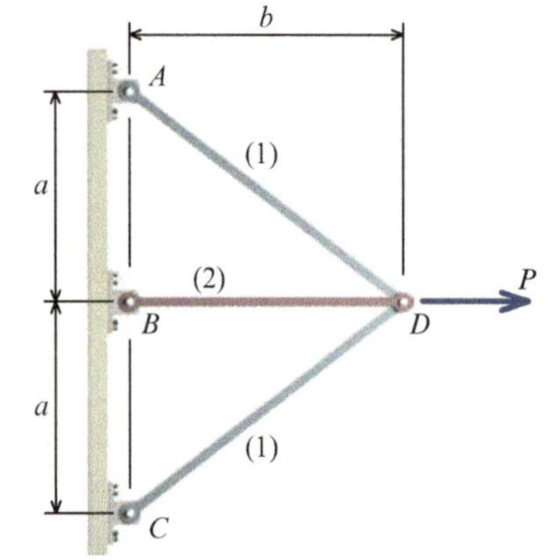

FIGURE P5.33

P5.34 A pin-connected structure is loaded and supported as shown in Figure P5.34. Member *ABCD* is a rigid bar that is horizontal before a load $P = 6$ kips is applied at end *D*. Bars (1) and (2) are made of steel [E = 30,000 ksi], and each bar has a cross-sectional area of 0.45 in.^{2}. Dimensions of the structure are $a = 220$ in., $b = 140$ in., $c = 100$ in., and $d = 220$ in. Determine

(a) the normal stress in each bar.
(b) the downward deflection of end *D* of the rigid bar.

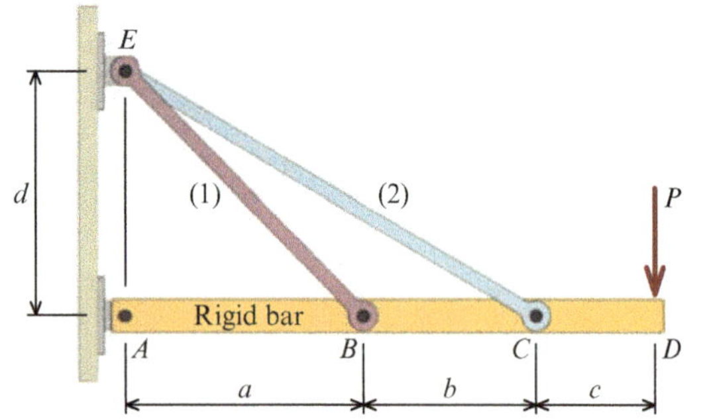

FIGURE P5.34

P5.35 A load $P = 100$ kN is supported by an assembly consisting of rigid bar *ABC*, two identical solid bronze [E = 105 GPa] bars, and a solid aluminum alloy [E = 70 GPa] bar, as shown in Figure P5.35. The bronze bars (1) each have a cross-sectional area of 125 mm^2 and

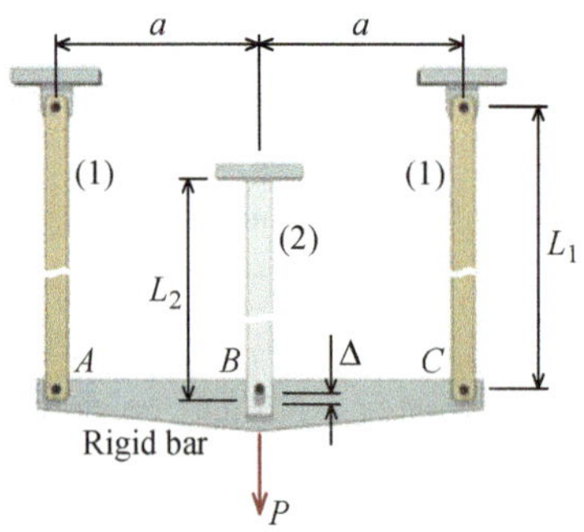

FIGURE P5.35

a length $L_1 = 2{,}400$ mm. Aluminum bar (2) has a cross-sectional area of 375 mm^2 and a length $L_2 = 1{,}500$ mm. Assume that $a = 500$ mm. All bars are unstressed before the load *P* is applied; however, there is a gap of $\Delta = 2$ mm in the connection at *B*. Determine

(a) the axial forces in the bronze and aluminum bars.
(b) the downward deflection of rigid bar *ABC*.

P5.36 A bronze pipe (1) is to be connected to an aluminum alloy pipe (2) at flange *B*, as shown in Figure P5.36. When the assembly is put in place, however, a gap of $\Delta = 0.25$ in. exists between the two pipes. Bronze pipe (1) has an elastic modulus $E_1 = 16{,}000$ ksi, a cross-sectional area $A_1 = 2.23$ in.^{2}, and a length $L_1 = 5.0$ ft. Aluminum alloy pipe (2) has an elastic modulus $E_2 = 10{,}000$ ksi, a cross-sectional area $A_2 = 1.07$ in.^{2}, and a length $L_2 = 9.0$ ft. If bolts are inserted into the flanges and tightened so that the gap at *B* is closed, determine

(a) the normal stresses produced in each of the members.
(b) the final position of flange *B* with respect to support *A*.

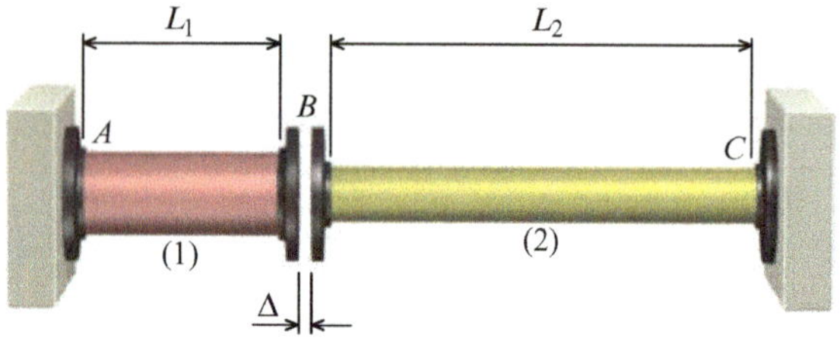

FIGURE P5.36

P5.37 Rigid bar *ABCD* shown in Figure P5.37 is supported by a smooth pin at *D* and by vertical aluminum alloy [$E = 10{,}000$ ksi] bars attached at joints *A* and *C*. Bar (1) was fabricated to its intended length $L_1 = 40$ in. Bar (2) was intended to have a length of 60 in.; however, its actual length after fabrication was found to be $L_2 = 59.88$ in. To connect it to the pin at *C* on the rigid bar, bar (2) will need to be manually stretched by $\Delta = 0.12$ in. After bar (2) has been stretched and connected to the pin at *C*, a load $P = 20$ kips is applied to the rigid bar at *B*. Use the following additional properties and dimensions: $A_1 = A_2 = 0.75$ in.^{2}, $a = 36$ in., $b = 54$ in., and $c = 48$ in. Determine

(a) the normal stresses in bars (1) and (2).
(b) the deflection at *A* of the rigid bar.

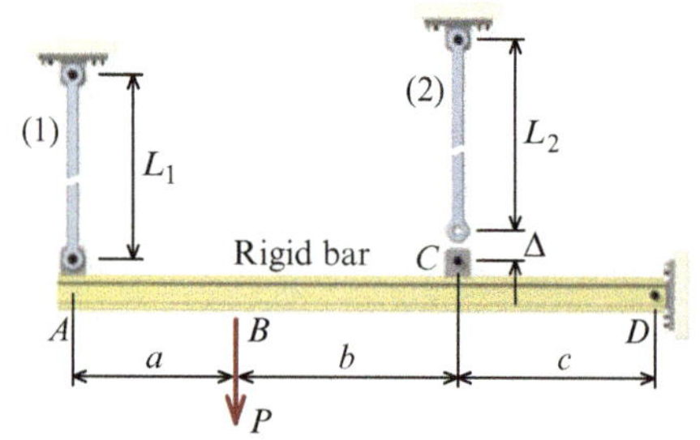

FIGURE P5.37

P5.38 A 0.625 in. diameter steel [$E = 30{,}000$ ksi] bolt (1) is placed in a copper [$E = 16{,}000$ ksi] tube (2), as shown in Figure P5.38. The tube has an outside diameter of 1.50 in., a wall thickness of 0.1875 in., and a length $L = 9.0$ in. Rigid washers, each with a thickness $t = 0.125$ in., cap the ends of the copper tube. The bolt has 20 threads per inch. That is, each time the nut is turned one complete revolution, the nut advances 0.05 in. (i.e., 1/20 in.). The nut is hand tightened on the bolt until the bolt, nut, washers, and tube are just snug, meaning that all slack has been removed from the assembly but no stress has yet been induced. What stresses are produced in the bolt and in the tube if the nut is tightened an additional quarter turn past the snug-tight condition?

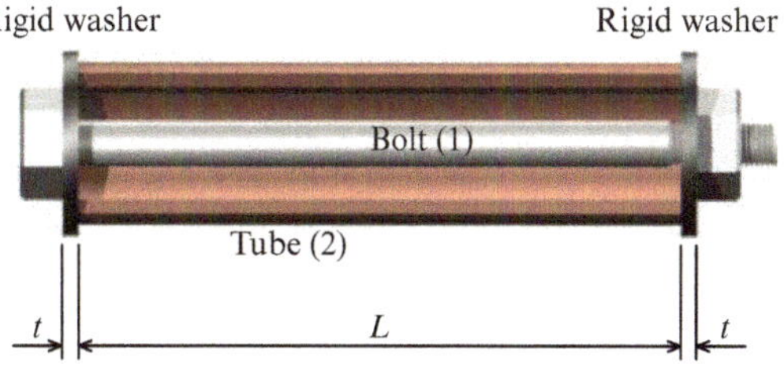

FIGURE P5.38

5.6 Thermal Effects on Axial Deformation

As discussed in Section 2.4, a temperature change ΔT creates normal strains

$$\varepsilon_T = \alpha \Delta T \tag{5.10}$$

in a solid material. Over the length L of an axial member, the deformation resulting from a temperature change is

$$\delta_T = \varepsilon_T L = \alpha \Delta T L \tag{5.11}$$

If an axial member is allowed to elongate or contract freely, temperature change creates no stress in a material. However, substantial stresses can result in an axial member if elongation or contraction is inhibited.

Force–Temperature–Deformation Relationship

The relationship between internal force and axial deformation developed in Equation (5.2) can be enhanced to include the effects of temperature change:

$$\delta = \frac{FL}{AE} + \alpha \Delta T L \tag{5.12}$$

The deformation of a statically determinate axial member can be computed from Equation (5.12) since the member is free to elongate or contract in response to a change in temperature. In a statically indeterminate axial structure, however, the deformation due to temperature changes may be constrained by supports or other components in the structure. Restrictions of this sort inhibit the elongation or contraction of a member, causing normal stresses to develop. These stresses are often referred to as *thermal stresses*, even though temperature change by itself causes no stress.

MecMovies

EXAMPLES

M5.11 A 20 mm diameter steel [$E = 200$ GPa; $\alpha = 12.0 \times 10^{-6}$/°C] rod is held snugly between rigid walls, as shown. Calculate the temperature drop ΔT at which the shear stress in the 15 mm diameter bolt becomes 70 MPa.

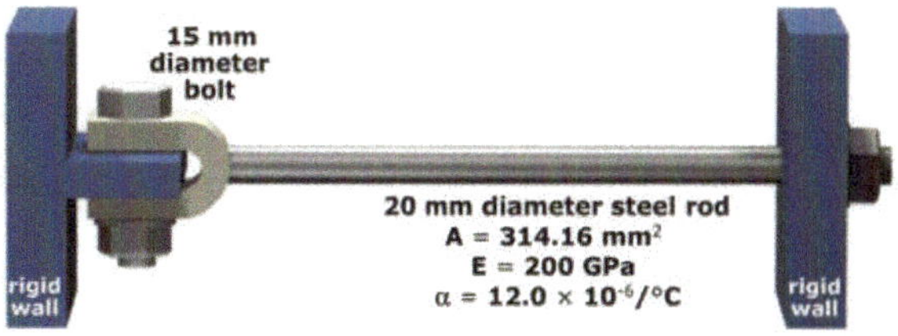

119

M5.12 A rigid bar *ABC* is pinned at *A* and supported by a steel wire at *B*. Before weight *W* is attached to the rigid bar at *C*, the rigid bar is horizontal. After weight *W* is attached and the temperature of the assembly has been increased by 50°C, careful measurements reveal that the rigid bar has deflected downward 2.52 mm at point *C*. Determine

(a) the normal strain in wire (1).
(b) the normal stress in wire (1).
(c) the magnitude of weight *W*.

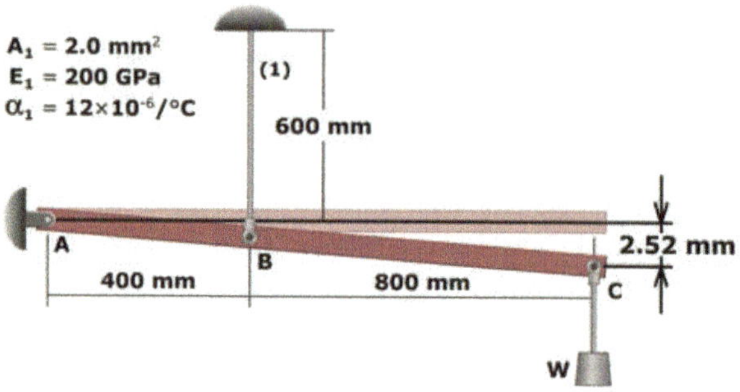

Incorporating Temperature Effects in Statically Indeterminate Structures

In Section 5.5, a five-step procedure for analyzing statically indeterminate axial structures was outlined. Temperature effects can be easily incorporated into this procedure by using Equation (5.12) to define the force–temperature–deformation relationships for the axial members, instead of Equation (5.2). With the five-step procedure, analyzing indeterminate structures involving temperature change is no more difficult conceptually than analyzing those same structures without thermal effects. The addition of the $\alpha \Delta T L$ term in Equation (5.12) does increase the computational difficulty, but the overall procedure is the same. In fact, it is the more challenging problems, such as those involving temperature change, in which the advantages and potential of the five-step procedure are most evident.

It is essential that Equation (5.12) be consistent, meaning that a positive internal force F (i.e., a tensile force) and a positive ΔT should produce a positive member deformation (i.e., an elongation). The need for consistency explains the emphasis on assuming an internal tensile force in all axial members, even if, intuitively, one might anticipate that an axial member should act in compression.

MecMovies

EXAMPLE

M5.13 An aluminum bar (1) is attached to a steel post (2) at rigid flange *B*. Bar (1) and post (2) are initially stress free when they are connected to the flange at a temperature of 20°C. The aluminum bar (1) has a cross-sectional area $A_1 = 200$ mm², a modulus of elasticity $E_1 = 70$ GPa, and a coefficient of thermal expansion $\alpha_1 = 23.6 \times 10^{-6}$/°C. The steel post (2) has properties $A_2 = 450$ mm², $E_2 = 200$ GPa, and $\alpha_2 = 12.0 \times 10^{-6}$/°C. Determine the normal stresses in members (1) and (2) and the deflection at flange *B* after the temperature increases to 75°C.

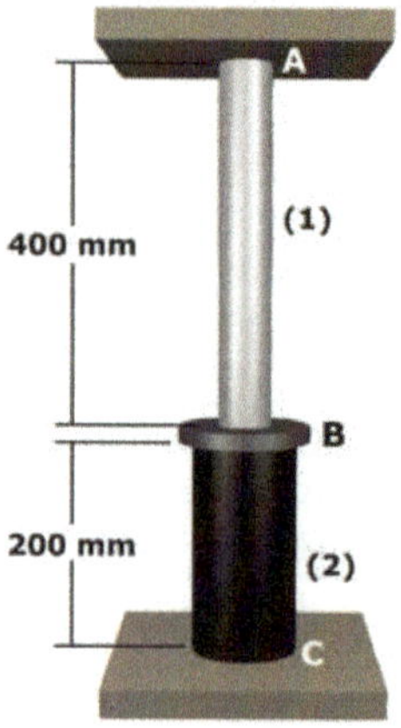

120

EXAMPLE 5.7

An aluminum rod (1) [E = 70 GPa; α = $22.5 \times 10^{-6}/°C$] and a brass rod (2) [E = 105 GPa; $\alpha = 18.0 \times 10^{-6}/°C$] are connected to rigid supports, as shown. The cross-sectional areas of rods (1) and (2) are 2,000 mm^2 and 3,000 mm^2, respectively. The temperature of the structure will increase.

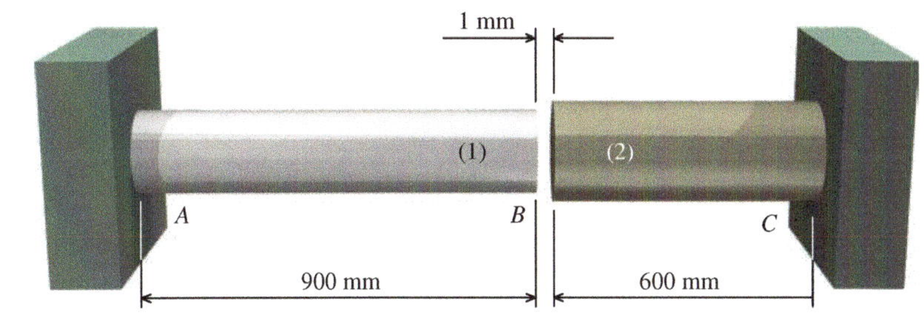

(a) Determine the temperature increase that will close the initial 1 mm gap between the two axial members.
(b) Compute the normal stress in each rod if the total temperature increase is 60°C.

Plan the Solution

First, we must determine whether the temperature increase will cause sufficient elongation to close the 1 mm gap. If the two axial members come into contact, the problem becomes statically indeterminate and the solution will proceed with the five-step procedure outlined in Section 5.5. To maintain consistency in the force–temperature–deformation relationships, tension will be assumed in both members (1) and (2), even though it is apparent that both members will be compressed because of the temperature increase. Accordingly, the values obtained for the internal axial forces F_1 and F_2 should be negative.

SOLUTION

(a) The axial elongation in the two rods due solely to a temperature increase can be expressed as

$$\delta_{1,T} = \alpha_1 \Delta T L_1 \qquad \text{and} \qquad \delta_{2,T} = \alpha_2 \Delta T L_2$$

If the two rods are to touch at B, the sum of the elongations in the rods must equal 1 mm:

$$\delta_{1,T} + \delta_{2,T} = \alpha_1 \Delta T L_1 + \alpha_2 \Delta T L_2 = 1 \text{ mm}$$

We solve this equation for ΔT:

$$(22.5 \times 10^{-6}/°C)\Delta T(900 \text{ mm}) + (18.0 \times 10^{-6}/°C)\Delta T(600 \text{ mm}) = 1 \text{ mm}$$

$$\therefore \Delta T = 32.2°C \qquad \textbf{Ans.}$$

(b) Given that a temperature increase of 32.2°C closes the 1 mm gap, a larger temperature increase (i.e., 60°C in this instance) will cause the aluminum and brass rods to compress each other, since the rods are prevented from expanding freely by the supports at A and C.

Step 1 — Equilibrium Equations: Consider a free-body diagram (FBD) of joint B after the aluminum and brass rods have come into contact. The sum of forces in the horizontal direction consists exclusively of the internal member forces.

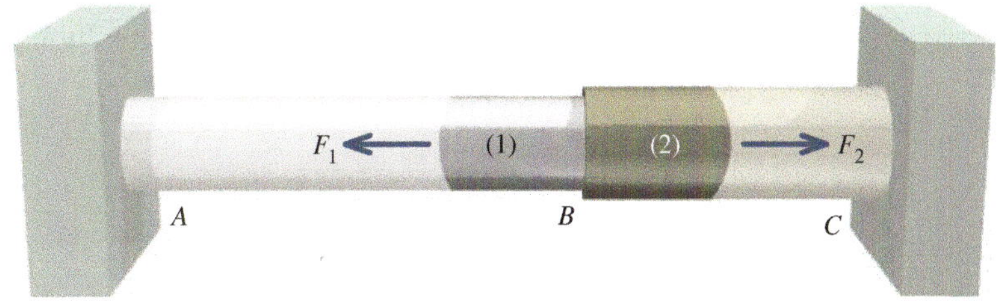

$$\Sigma F_x = F_2 - F_1 = 0 \qquad \therefore F_1 = F_2$$

Step 2 — Geometry of Deformation: Since the compound axial member is attached to rigid supports at A and C, the overall elongation of the structure can be no more than 1 mm. In other words,

$$\delta_1 + \delta_2 = 1 \text{ mm} \tag{a}$$

Step 3 — Force–Temperature–Deformation Relationships: Write the force–temperature–deformation relationships for the two members:

$$\delta_1 = \frac{F_1 L_1}{A_1 E_1} + \alpha_1 \Delta T L_1 \quad \text{and} \quad \delta_2 = \frac{F_2 L_2}{A_2 E_2} + \alpha_2 \Delta T L_2 \tag{b}$$

Step 4 — Compatibility Equation: Substitute Equations (b) into Equation (a) to obtain the compatibility equation:

$$\frac{F_1 L_1}{A_1 E_1} + \alpha_1 \Delta T L_1 + \frac{F_2 L_2}{A_2 E_2} + \alpha_2 \Delta T L_2 = 1 \text{ mm} \tag{c}$$

Step 5 — Solve the Equations: Substitute $F_2 = F_1$ (from the equilibrium equation) into Equation (c), and solve for the internal force F_1:

$$F_1\left[\frac{L_1}{A_1 E_1} + \frac{L_2}{A_2 E_2}\right] = 1 \text{ mm} - \alpha_1 \Delta T L_1 - \alpha_2 \Delta T L_2 \tag{d}$$

In computing the value for F_1, pay close attention to the units, making sure that they are consistent:

$$F_1\left[\frac{900 \text{ mm}}{(2{,}000 \text{ mm}^2)(70{,}000 \text{ N/mm}^2)} + \frac{600 \text{ mm}}{(3{,}000 \text{ mm}^2)(105{,}000 \text{ N/mm}^2)}\right]$$
$$= 1 \text{ mm} - (22.5 \times 10^{-6}/°\text{C})(60°\text{C})(900 \text{ mm}) - (18.0 \times 10^{-6}/°\text{C})(60°\text{C})(600 \text{ mm}) \tag{e}$$

Therefore,

$$F_1 = -103{,}560 \text{ N} = -103.6 \text{ kN}$$

The normal stress in rod (1) is

$$\sigma_1 = \frac{F_1}{A_1} = \frac{-103{,}560 \text{ N}}{2{,}000 \text{ mm}^2} = -51.8 \text{ MPa} = 51.8 \text{ MPa (C)}$$ **Ans.**

and the normal stress in rod (2) is

$$\sigma_2 = \frac{F_2}{A_2} = \frac{-103{,}560 \text{ N}}{3{,}000 \text{ mm}^2} = -34.5 \text{ MPa} = 34.5 \text{ MPa (C)}$$ **Ans.**

MecMovies

EXAMPLE

M5.14 A rectangular bar 30 mm wide and 24 mm thick made of aluminum [$E = 70$ GPa; $\alpha = 23.0 \times 10^{-6}$/°C] and two rectangular copper [$E = 120$ GPa; $\alpha = 16.0 \times 10^{-6}$/°C] bars 30 mm wide and 12 mm thick are connected by two smooth 11 mm diameter pins. When the pins are initially inserted into the bars, both the copper and aluminum bars are stress free. After the temperature of the assembly has increased by 65°C, determine

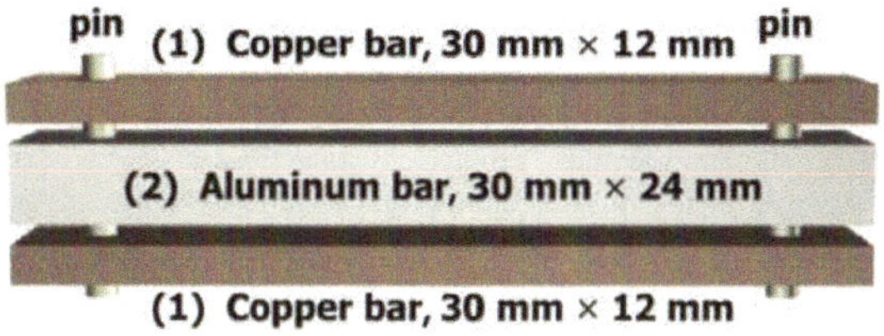

(a) the internal axial force in the aluminum bar.
(b) the normal strain in the copper bars.
(c) the shear stress in the 11 mm diameter pins.

EXAMPLE 5.8

A pin-connected structure is loaded and supported as shown. Member *BCDF* is a rigid plate. Member (1) is a steel [$E = 200$ GPa; $A_1 = 310$ mm^2; $\alpha = 11.9 \times 10^{-6}$/°C] bar, and member (2) is an aluminum [$E = 70$ GPa; $A_2 = 620$ mm^2; $\alpha = 22.5 \times 10^{-6}$/°C] bar. A load of 6 kN is applied to the plate at *F*. If the temperature increases by 20°C, compute the normal stresses in members (1) and (2).

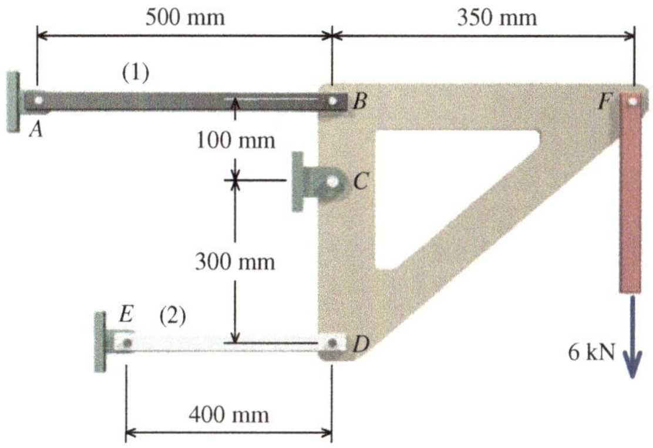

Plan the Solution

The five-step procedure for solving indeterminate problems will be used. Since the rigid plate is pinned at *C*, it will rotate about *C*. A deformation diagram will be sketched to show the relationship between the rigid-plate deflections at joints *B* and *D*, based on the assumption that the plate rotates clockwise about *C*. The joint deflections will be related to the deformations δ_1 and δ_2, which will lead to a compatibility equation expressed in terms of the member forces F_1 and F_2.

SOLUTION

Step 1 — Equilibrium Equations:

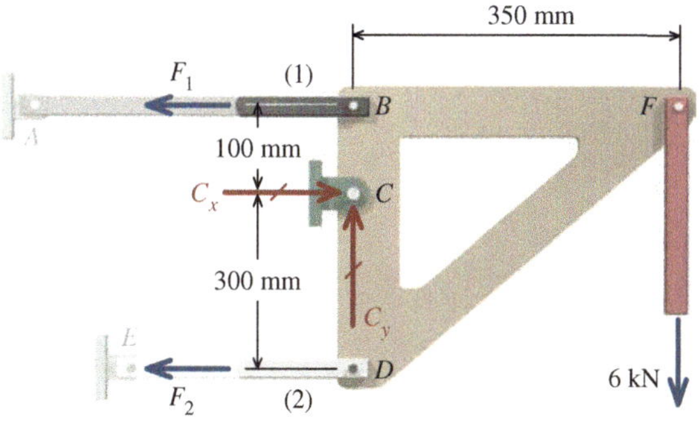

$$\Sigma M_C = F_1(100 \text{ mm}) - F_2(300 \text{ mm}) - (6 \text{ kN})(350 \text{ mm}) = 0 \quad \text{(a)}$$

Step 2 — Geometry of Deformation: Sketch the deflected position of the rigid plate. Since the plate is pinned at *C*, the plate will rotate about *C*. The relationship between the deflections of joints *B* and *D* can be expressed by similar triangles:

$$\frac{v_B}{100 \text{ mm}} = \frac{v_D}{300 \text{ mm}} \quad \text{(b)}$$

How are the deformations in members (1) and (2) related to the joint deflections at B and D?

123

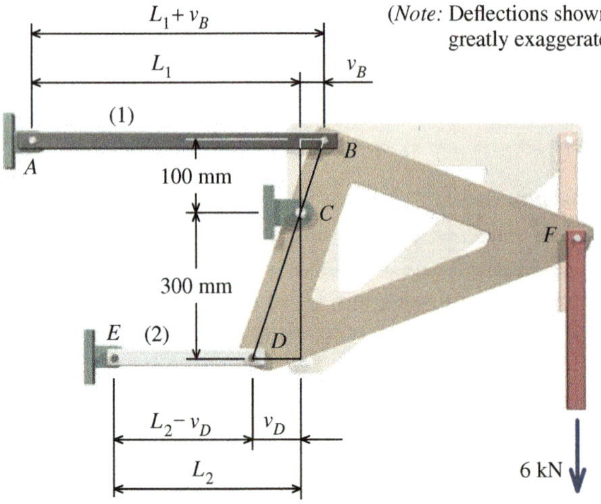

(*Note:* Deflections shown greatly exaggerated.)

By definition, the deformation in a member is the difference between its final length (i.e., after the load is applied and the temperature is increased) and its initial length. For member (1), therefore,

$$\delta_1 = L_{\text{final}} - L_{\text{initial}} = (L_1 + v_B) - L_1 = v_B \qquad \therefore\ v_B = \delta_1 \tag{c}$$

Similarly, for member (2),

$$\delta_2 = L_{\text{final}} - L_{\text{initial}} = (L_2 - v_D) - L_2 = -v_D \qquad \therefore\ v_D = -\delta_2 \tag{d}$$

Substitute the results from Equations (c) and (d) into Equation (b) to obtain

$$\frac{\delta_1}{100\text{ mm}} = -\frac{\delta_2}{300\text{ mm}} \tag{e}$$

Step 3 — Force–Temperature–Deformation Relationships: Write the general force–temperature–deformation relationships for the two axial members:

$$\delta_1 = \frac{F_1 L_1}{A_1 E_1} + \alpha_1 \Delta T L_1 \quad \text{and} \quad \delta_2 = \frac{F_2 L_2}{A_2 E_2} + \alpha_2 \Delta T L_2 \tag{f}$$

Step 4 — Compatibility Equation: Substitute the force–temperature–deformation relationships from Equation (f) into Equation (e) to obtain the compatibility equation:

$$\frac{1}{100\text{ mm}}\left[\frac{F_1 L_1}{A_1 E_1} + \alpha_1 \Delta T L_1\right] = -\frac{1}{300\text{ mm}}\left[\frac{F_2 L_2}{A_2 E_2} + \alpha_2 \Delta T L_2\right] \tag{g}$$

This equation is derived from information about the deflected position of the structure and is expressed in terms of the two unknown member forces F_1 and F_2.

Step 5 — Solve the Equations: In the compatibility equation [Equation (g)], group the terms that include F_1 and F_2 on the left-hand side of the equation:

$$\frac{F_1 L_1}{(100\text{ mm})A_1 E_1} + \frac{F_2 L_2}{(300\text{ mm})A_2 E_2} = -\frac{1}{100\text{ mm}}\alpha_1 \Delta T L_1 - \frac{1}{300\text{ mm}}\alpha_2 \Delta T L_2 \tag{h}$$

Equilibrium equation (a) can be treated in the same manner:

$$F_1(100\text{ mm}) - F_2(300\text{ mm}) = (6\text{ kN})(350\text{ mm}) \tag{i}$$

124

Equations (h) and (i) can be solved simultaneously in several ways. The hand solution here will use the substitution method. First, solve Equation (i) for F_2:

$$F_2 = \frac{F_1(100\text{ mm}) - (6\text{ kN})(350\text{ mm})}{300\text{ mm}} \tag{j}$$

Next, substitute this expression into Equation (h) and collect terms with F_1 on the left-hand side of the equation:

$$\frac{F_1 L_1}{(100\text{ mm})A_1 E_1} + \frac{[(100\text{ mm}/300\text{ mm})F_1]L_2}{(300\text{ mm})A_2 E_2}$$
$$= -\frac{1}{100\text{ mm}}\alpha_1 \Delta T L_1 - \frac{1}{300\text{ mm}}\alpha_2 \Delta T L_2 + (6\text{ kN})\left[\frac{350\text{ mm}}{300\text{ mm}}\right]\frac{L_2}{(300\text{ mm})A_2 E_2}$$

Simplifying and solving for F_1 then gives

$$F_1\left[\frac{500\text{ mm}}{(100\text{ mm})(310\text{ mm}^2)(200{,}000\text{ N/mm}^2)} + \frac{(1/3)(400\text{ mm})}{(300\text{ mm})(620\text{ mm}^2)(70{,}000\text{ N/mm}^2)}\right]$$
$$= -\frac{1}{100\text{ mm}}(11.9 \times 10^{-6}/^\circ\text{C})(20^\circ\text{C})(500\text{ mm})$$
$$- \frac{1}{300\text{ mm}}(22.5 \times 10^{-6}/^\circ\text{C})(20^\circ\text{C})(400\text{ mm})$$
$$+ (6{,}000\text{ N})\left[\frac{350\text{ mm}}{300\text{ mm}}\right]\frac{400\text{ mm}}{(300\text{ mm})(620\text{ mm}^2)(70{,}000\text{ N/mm}^2)}$$

Therefore,

$$F_1 = -17{,}328.8\text{ N} = -17.33\text{ kN} = 17.33\text{ kN (C)}$$

Substituting back into Equation (j) gives

$$F_2 = -12{,}776.3\text{ N} = -12.78\text{ kN} = 12.78\text{ kN (C)}$$

The normal stresses in members (1) and (2) can now be determined:

$$\sigma_1 = \frac{F_1}{A_1} = \frac{-17{,}328.8\text{ N}}{310\text{ mm}^2} = -55.9\text{ MPa} = 55.9\text{ MPa (C)}$$
$$\sigma_2 = \frac{F_2}{A_2} = \frac{-12{,}776.3\text{ N}}{620\text{ mm}^2} = -20.6\text{ MPa} = 20.6\text{ MPa (C)}$$

Ans.

Note: The deformation of member (1) can be computed as

$$\delta_1 = \frac{F_1 L_1}{A_1 E_1} + \alpha_1 \Delta T L_1 = \frac{(-17{,}328.8\text{ N})(500\text{ mm})}{(310\text{ mm}^2)(200{,}000\text{ N/mm}^2)}$$
$$+ (11.9 \times 10^{-6}/^\circ\text{C})(20^\circ\text{C})(500\text{ mm})$$
$$= -0.1397\text{ mm} + 0.1190\text{ mm} = -0.0207\text{ mm}$$

and the deformation of member (2) is

$$\delta_2 = \frac{F_2 L_2}{A_2 E_2} + \alpha_2 \Delta T L_2 = \frac{(-12{,}776.3\ \text{N})(400\ \text{mm})}{(620\ \text{mm}^2)(70{,}000\ \text{N/mm}^2)}$$

$$+ (22.5 \times 10^{-6}/^\circ\text{C})(20^\circ\text{C})(400\ \text{mm})$$

$$= -0.1178\ \text{mm} + 0.1800\ \text{mm} = 0.0622\ \text{mm}$$

Contrary to our initial assumption in the deformation diagram, member (1) actually contracts and member (2) elongates. This outcome is explained by the elongation caused by the temperature increase. The rigid plate actually rotates counterclockwise about C.

MecMovies

EXAMPLE

M5.15 A brass link and a steel rod have the dimensions shown at a temperature of 20°C. The steel rod is cooled until it fits freely into the link. The temperature of the entire link-and-rod assembly is then warmed to 40°C. Determine

(a) the final normal stress in the steel rod.

(b) the deformation of the steel rod.

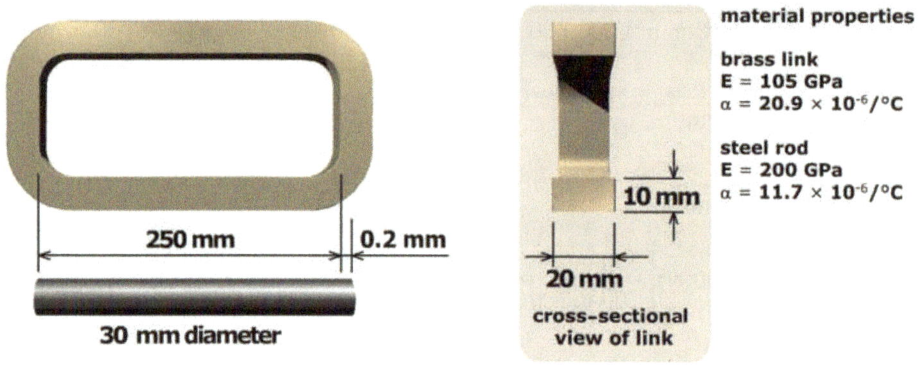

EXERCISES

M5.13 A composite axial structure consists of two rods joined at flange B. Rods (1) and (2) are attached to rigid supports at A and C, respectively. A concentrated load P is applied to flange B in the direction shown. Determine the internal forces and normal stresses in each rod after the temperature changes by the indicated ΔT. Also, determine the deflection of flange B in the x direction.

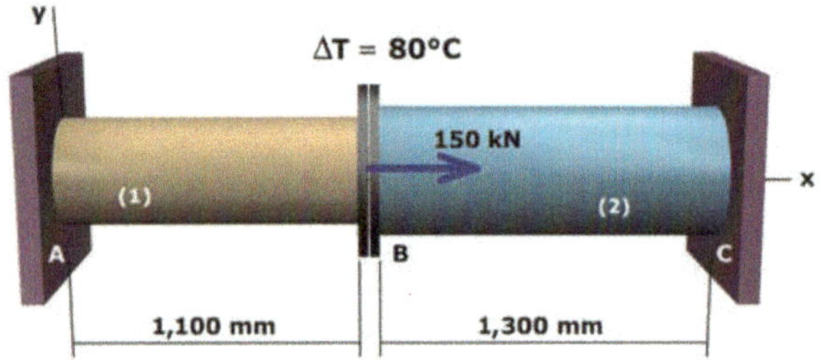

FIGURE M5.13

M5.14 A rigid horizontal bar ABC is supported by three vertical rods as shown. The system is stress free before the load is applied. After the load P is applied, the temperature of all three rods is raised by the indicated ΔT. Determine

(a) the internal force in rod (1).

(b) the normal stress in rod (2).

(c) the normal strain in rod (1).

(d) the downward deflection of the rigid bar at B.

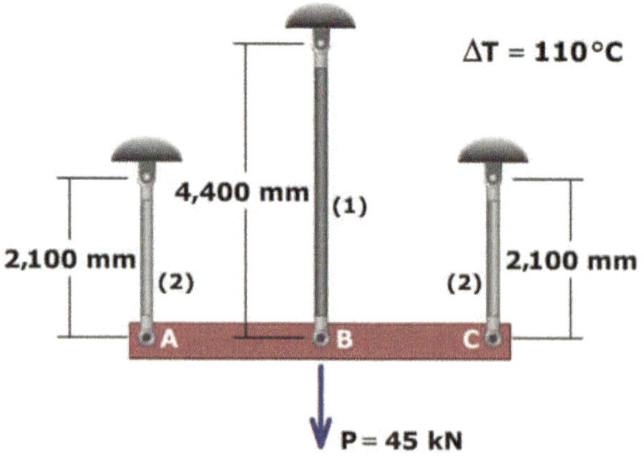

FIGURE M5.14

PROBLEMS

P5.39 A circular aluminum alloy [$E = 70$ GPa; $\alpha = 22.5 \times 10^{-6}$/°C; $\nu = 0.33$] pipe has an outside diameter of 220 mm, a wall thickness of 8 mm, and a length of 5 m. The pipe supports a compressive load of 650 kN. After the temperature of the pipe drops by 45°C, determine

(a) the axial deformation of the pipe.
(b) the change in diameter of the pipe.

P5.40 A solid 0.125 in. diameter steel [$E = 30{,}000$ ksi; $\alpha = 6.5 \times 10^{-6}$/°F] wire is stretched between fixed supports so that it is under an initial tensile force of 40 lb. If the temperature of the wire drops by 75°F, what is the tensile stress in the wire?

P5.41 An American Iron and Steel Institute (AISI) 1040 hot-rolled steel [$E = 207$ GPa; $\alpha = 11.3 \times 10^{-6}$/°C] bar is held between two rigid supports. The bar is stress free at a temperature of 25°C. The bar is then heated uniformly. If the yield strength of the steel is 414 MPa, determine the temperature at which yield first occurs.

P5.42 A high-density polyethylene [$E = 128$ ksi; $\alpha = 88 \times 10^{-6}$/°F; $\nu = 0.4$] bar is positioned between two rigid supports, as shown in Figure P5.42. The bar is square with cross-sectional dimensions of 2.50 in. by 2.50 in. and length $L = 48$ in. At room temperature, a gap of $\Delta = 0.25$ in. exists between the block and the rigid support at B. After a temperature increase of 120°F, determine

(a) the normal stress in the bar.
(b) the longitudinal strain in the bar.
(c) the transverse (i.e., lateral) strain in the bar.

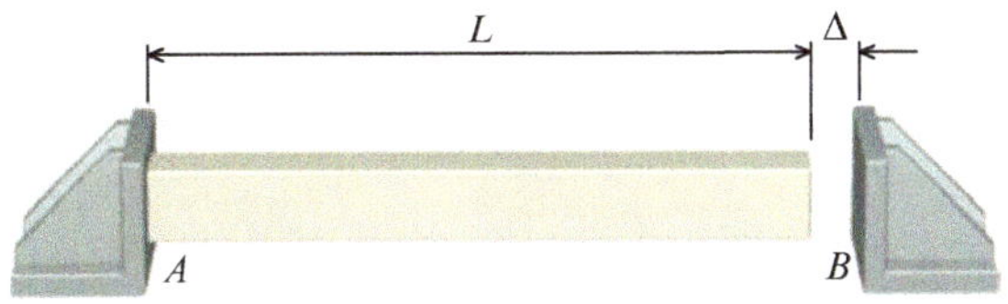

FIGURE P5.42

P5.43 Rigid bar ABC is supported by bronze rod (1) and stainless steel rod (2) as shown in Figure P5.43. A concentrated load $P = 24$ kips is applied to the free end of bronze rod (3). Determine the deflection of rod end D after the temperature of all rods has increased by 130°F. Use the following dimensions and properties:

$a = 3.5$ ft, $b = 6.5$ ft.

Rod (1): $d_1 = 1.00$ in., $L_1 = 12$ ft, $E_1 = 15{,}200$ ksi, $\alpha_1 = 12.20 \times 10^{-6}$/°F.

Rod (2): $d_2 = 0.75$ in., $L_2 = 9$ ft, $E_2 = 28{,}000$ ksi, $\alpha_2 = 9.60 \times 10^{-6}$/°F.

Rod (3): $d_3 = 1.25$ in., $L_3 = 6$ ft, $E_3 = 15{,}200$ ksi, $\alpha_3 = 12.20 \times 10^{-6}$/°F.

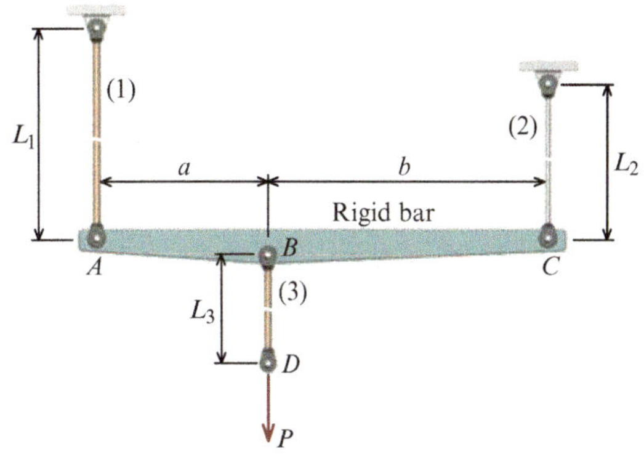

FIGURE P5.43

P5.44 Two ductile cast iron [$E = 24{,}400$ ksi; $\alpha = 6.0 \times 10^{-6}$/°F] bars are connected with a pin at B, as shown in Figure P5.44. Each bar has a cross-sectional area of 1.50 in.^{2}. If the temperature of the bars is decreased by 75°F from their initial temperature, what force P would need to be applied at B so that the total displacement (caused both by the temperature change and by the applied load) of joint B is zero? Use $a = 10$ ft and $\beta = 55°$.

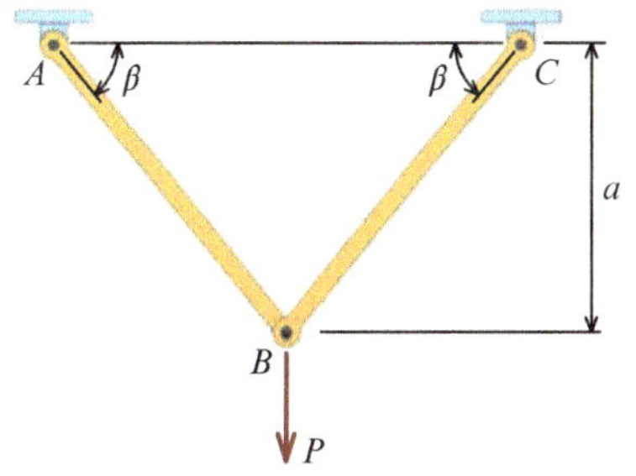

FIGURE P5.44

P5.45 A solid aluminum alloy [$E = 69$ GPa; $\alpha = 23.6 \times 10^{-6}$/°C] rod (1) is attached rigidly to a solid brass [$E = 115$ GPa; $\alpha = 18.7 \times 10^{-6}$/°C] rod (2), as shown in Figure P5.45. The compound rod is subjected to a tensile load $P = 6$ kN. The diameter of each rod is 10 mm. The rods lengths are $L_1 = 525$ mm and $L_2 = 675$ mm. Compute the change in temperature required to produce zero horizontal deflection at end C of the compound rod.

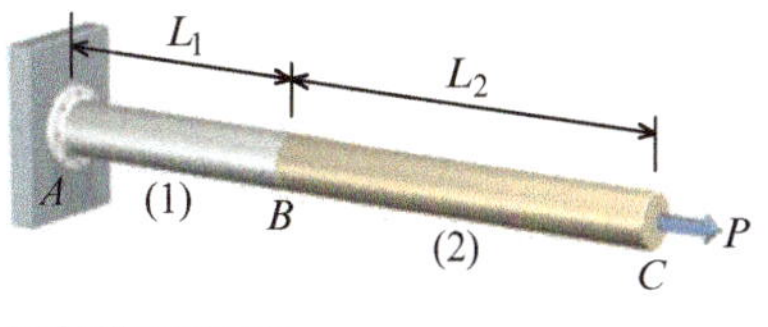

FIGURE P5.45

P5.46 At a temperature of 15°C, a gap of $\Delta = 5$ mm exists between the two polymer bars shown in the accompanying figure. Bar (1) has a length $L_1 = 700$ mm, a cross-sectional area $A_1 = 1{,}250$ mm^2, a coefficient of thermal expansion $\alpha_1 = 130 \times 10^{-6}$/°C, and an elastic modulus $E_1 = 1{,}400$ MPa. Bar (2) has a length $L_2 = 500$ mm, a cross-sectional area $A_2 = 3{,}750$ mm^2, a coefficient of thermal expansion $\alpha_2 = 80 \times 10^{-6}$/°C, and an elastic modulus $E_2 = 3{,}700$ MPa. The supports at A and D are rigid. Determine

(a) the lowest temperature at which the gap is closed.
(b) the normal stress in the two bars at a temperature of 100°C.
(c) the longitudinal normal strain in the two bars at 100°C.

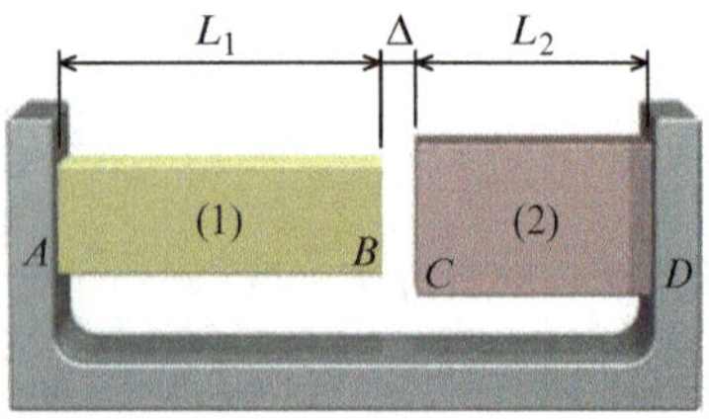

FIGURE P5.46

P5.47 A bronze [$E = 15{,}200$ ksi; $\alpha = 12.2 \times 10^{-6}$/°F] rod (2) is placed within a steel [$E = 30{,}000$ ksi; $\alpha = 6.3 \times 10^{-6}$/°F] tube (1). The two components are connected at each end by 0.5 in. diameter pins that pass through both the tube and the rod, as shown in Figure P5.47. The outside diameter of tube (1) is 3.00 in. and its wall thickness is 0.125 in. The diameter of rod (2) is 2.25 in. and the distance between pins is $L = 21$ in. What is the average shear stress in the pins if the temperature of the entire assembly is increased by 90°F?

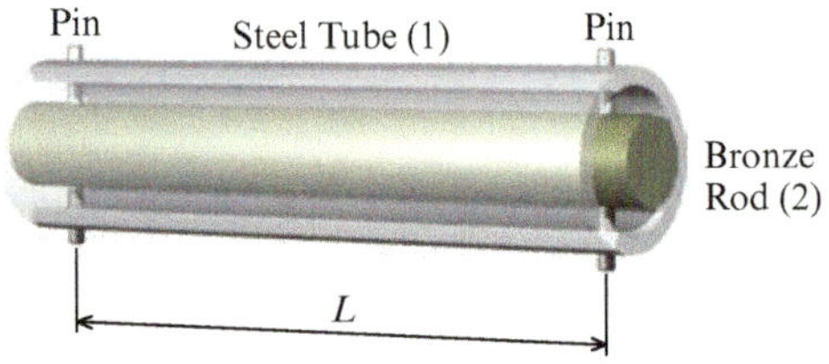

FIGURE P5.47

P5.48 A titanium [$E = 16{,}500$ ksi; $\alpha = 5.3 \times 10^{-6}$/°F] bar (1) and a bronze [$E = 15{,}200$ ksi; $\alpha = 12.2 \times 10^{-6}$/°F] bar (2), each restrained at one end, are fastened at their free ends by a pin of diameter $d = 0.4375$ in., as shown in Figure P5.48. The length of bar (1) is $L_1 = 19$ in. and its cross-sectional area is $A_1 = 0.50$ in.^{2}. The length of bar (2) is $L_2 = 27$ in. and its cross-sectional area is $A_2 = 0.65$ in.^{2}. What is the average shear stress in the pin at B if the temperature changes by 40°F?

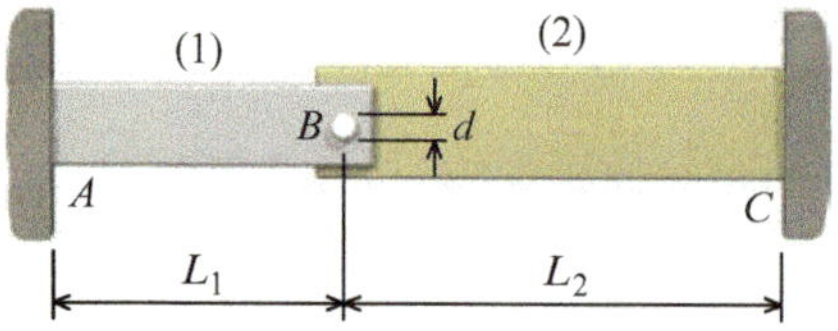

FIGURE P5.48

P5.49 Rigid bar ABC is supported by two identical solid bronze [$\sigma_Y = 331$ MPa; $E = 105$ GPa; $\alpha = 22.0 \times 10^{-6}$/°C] rods and a single solid steel [$\sigma_Y = 250$ MPa; $E = 200$ GPa; $\alpha = 11.7 \times 10^{-6}$/°C] rod, as shown in Figure P5.49. Bronze rods (1) each have diameter $d_1 = 25$ mm and length $L_1 = 3.0$ m. Steel rod (2) has length $L_2 = 4.0$ m.

(a) Calculate the diameter d_2 required for steel rod (2) so that the deflection of joint B is zero for any change in temperature.
(b) Using the value of d_2 determined in part (a), determine the maximum temperature decrease that is allowable for this assembly if a factor of safety of 2.0 is specified for the normal stress in each rod.

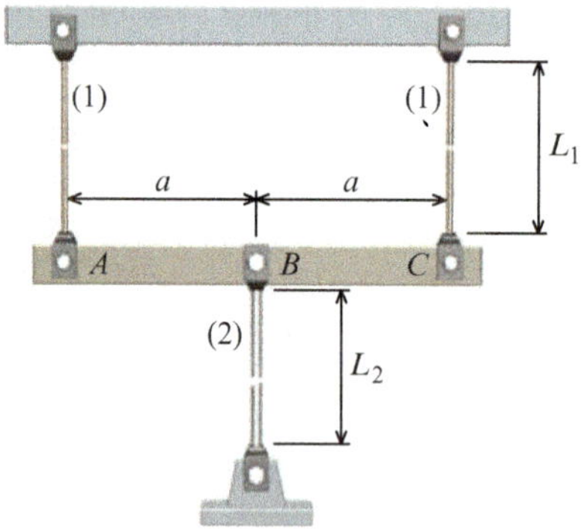

FIGURE P5.49

P5.50 A polymer cylinder (2) is clamped between rigid heads by two steel bolts (1), as shown in Figure P5.50. The steel [$E = 29{,}000$ ksi; $\alpha = 6.5 \times 10^{-6}$/°F] bolts have a diameter of 0.50 in. The

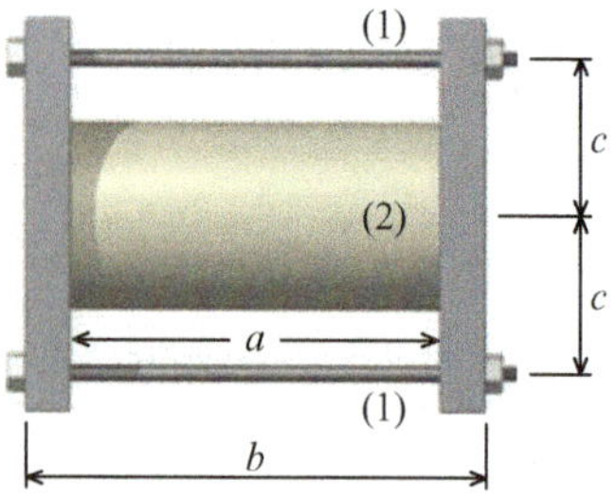

FIGURE P5.50

polymer [$E = 370$ ksi; $\alpha = 39.0 \times 10^{-6}/°F$] cylinder has an outside diameter of 6.625 in. and a wall thickness of 0.432 in. Assume that $a = 24$ in. and $b = 28$ in. If the temperature of this assembly changes by $\Delta T = 120°F$, determine

(a) the normal stress in the polymer cylinder.
(b) the normal strain in the polymer cylinder.
(c) the normal strain in the steel bolts.

P5.51 A load P will be supported by a structure consisting of a rigid bar $ABCD$, a steel [$E = 29{,}000$ ksi; $\alpha = 6.5 \times 10^{-6}/°F$] bar (1), and an aluminum alloy [$E = 10{,}000$ ksi; $\alpha = 12.5 \times 10^{-6}/°F$] bar (2), as shown in Figure P5.51. The bars have cross-sectional areas $A_1 = 0.80$ in.^{2} and $A_2 = 1.30$ in.^{2}, respectively. Dimensions of the structure are $a = 3$ ft, $b = 3.75$ ft, $c = 5.0$ ft, $L_1 = 12$ ft, and $L_2 = 20$ ft. The bars are unstressed when the structure is assembled. After a concentrated load $P = 35$ kips is applied and the temperature has been increased by 60°F, determine

(a) the normal stresses in bars (1) and (2).
(b) the vertical deflection of joint D.

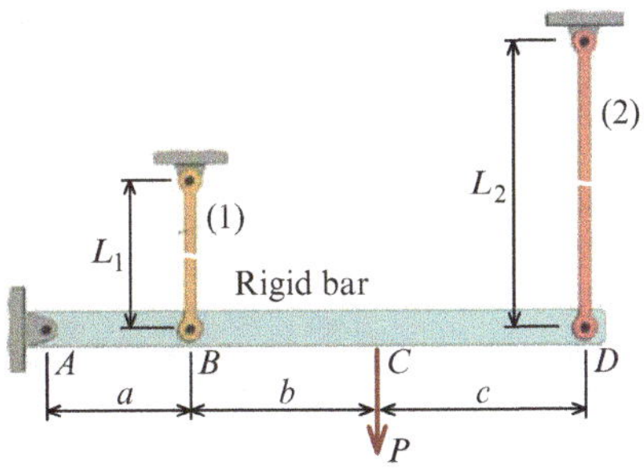

FIGURE P5.51

P5.52 The pin-connected assembly shown in Figure P5.52 consists of two aluminum alloy [$E = 69$ GPa; $\alpha = 23.6 \times 10^{-6}/°C$] bars (1) and a titanium alloy [$E = 114$ GPa; $\alpha = 9.5 \times 10^{-6}/°C$] bar (2) that are connected at pin D. Each of the aluminum bars has cross-sectional area $A_1 = 300$ mm^2, and the titanium bar has cross-sectional area $A_2 = 500$ mm^2. Dimensions for the assembly are $a = 1.80$ m and $b = 2.50$ m. All bars are unstressed at a temperature of 25°C. Calculate (a) the normal stresses in the aluminum and titanium bars and (b) the vertical displacement of pin D when the temperature of the assembly reaches 85°C.

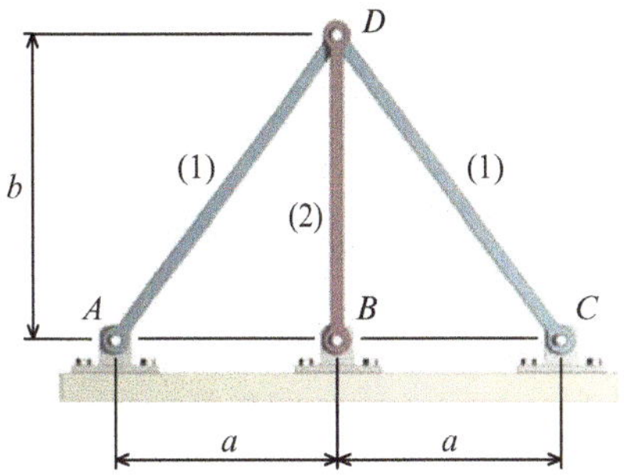

FIGURE P5.52

P5.53 The pin-connected assembly shown in Figure P5.53 consists of a rigid bar ABC, a cast iron [$E = 24{,}400$ ksi; $\alpha = 6.0 \times 10^{-6}/°F$] bar (1), and an aluminum alloy [$E = 10{,}000$ ksi; $\alpha = 13.1 \times 10^{-6}/°F$] bar (2). Bar (1) has cross-sectional area $A_1 = 0.50$ in.^{2} and length $L_1 = 18$ in. Bar (2) has cross-sectional area $A_2 = 1.60$ in.^{2} and length $L_2 = 45$ in. Dimensions of the assembly are $a = 25$ in. and $b = 80$ in. The bars are unstressed when the structure is assembled at 75°F. After assembly, the temperature of bar (2) is decreased by 90°F while the temperature of bar (1) remains constant at 75°F. Determine the normal strains in both bars for this condition.

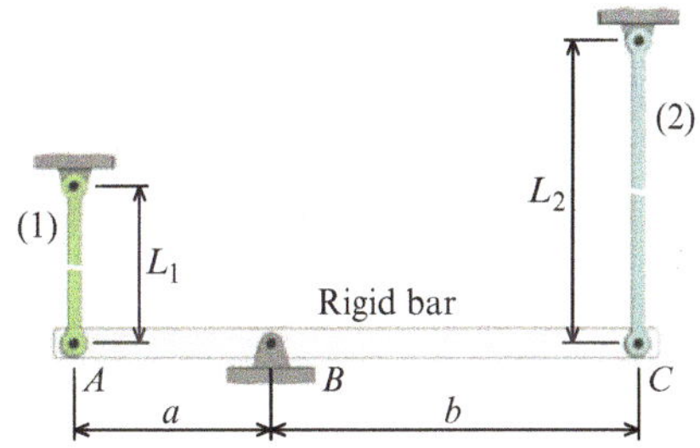

FIGURE P5.53

5.7 Stress Concentrations

In the preceding sections, it was assumed that the average normal stress, defined as $\sigma = P/A$, is the significant or critical stress in an axial member. While this is true for many problems, the maximum normal stress on a given section may be substantially greater than the average normal stress, and for certain combinations of loading and material, the maximum, rather than the average, normal stress is the more important consideration. If there exists in the structure or machine element a discontinuity that interrupts the stress path (called a *stress trajectory*), the stress at the discontinuity may be considerably greater than the

A stress trajectory is a line that is parallel to the maximum normal stress everywhere.

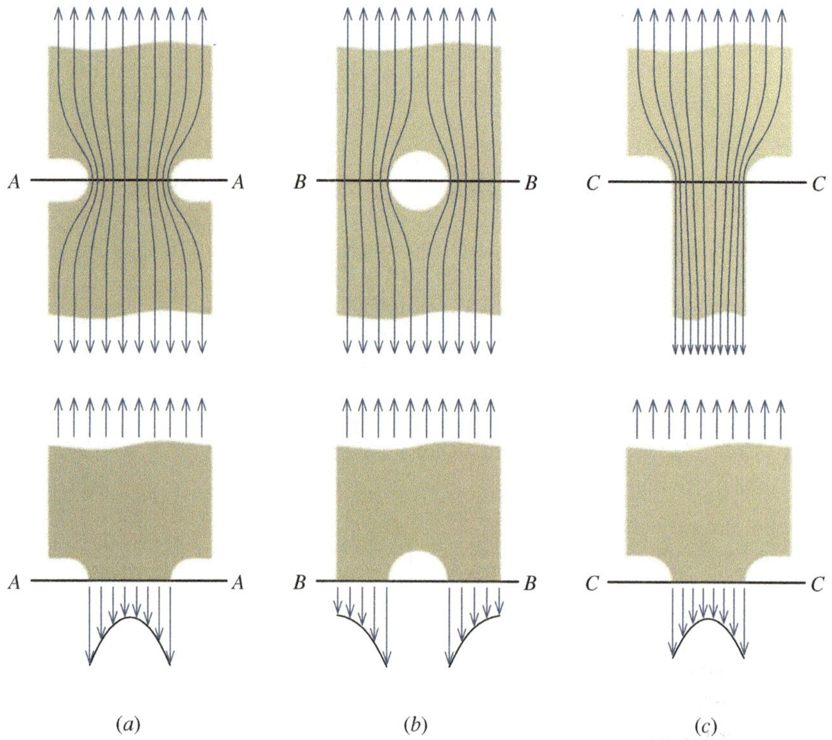

FIGURE 5.12 Typical stress trajectories and normal stress distributions for flat bars with (*a*) notches, (*b*) a centrally located hole, and (*c*) shoulder fillets.

average stress on the section (termed the *nominal* stress). We then say that a *stress concentration* exists at the discontinuity. The effect of a stress concentration is illustrated in Figure 5.12, in which a type of discontinuity is shown in the upper figure and the approximate distribution of normal stress on a transverse plane is shown in the accompanying lower figure. The ratio of the maximum stress to the nominal stress on the section is known as the *stress-concentration factor K*. Thus, the maximum normal stress in an axially loaded member can be calculated as the product of K and the nominal stress:

$$\boxed{\sigma_{\text{max}} = K\sigma_{\text{nom}}} \tag{5.13}$$

Curves similar to those shown in Figures 5.13, 5.14, and 5.15[1] can be found in numerous design handbooks. It is important that the user of such curves (or tables of factors) ascertain whether the factors are based on the gross or net section. In this book, the stress-concentration factors K are to be used in conjunction with the nominal stresses produced at the minimum or net cross-sectional area, as shown in Figure 5.12. The K factors shown in Figures 5.13, 5.14, and 5.15 are based on the stresses at the net section.

The case of a small circular hole in a wide plate under uniform unidirectional tension (Figure 5.16) offers an excellent illustration of localized stress redistribution. The theory of

[1] Adapted from Walter D. Pilkey, *Peterson's Stress Concentration Factors*, 2nd ed. (New York: John Wiley & Sons, Inc., 1997).

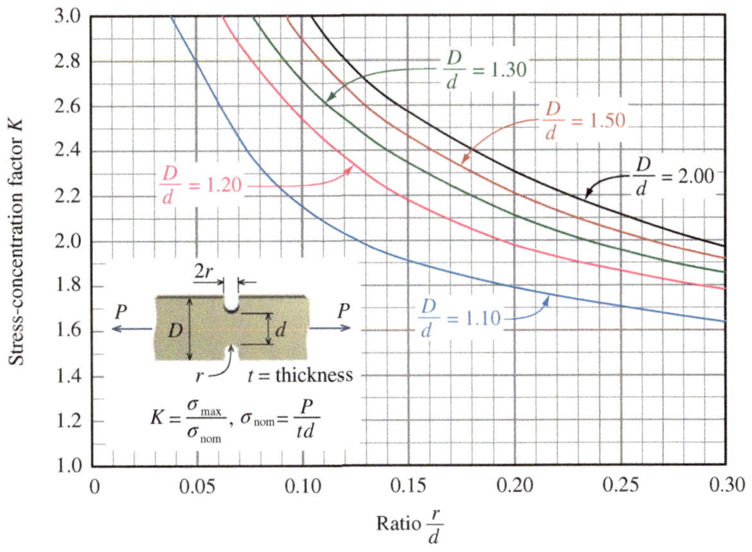

FIGURE 5.13 Stress-concentration factors K for a flat bar with opposite U-shaped notches.

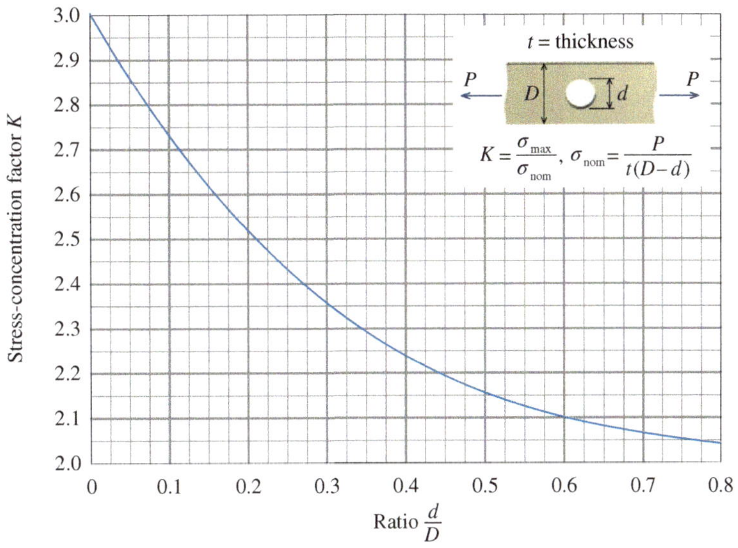

FIGURE 5.14 Stress-concentration factors K for a flat bar with a centrally located circular hole.

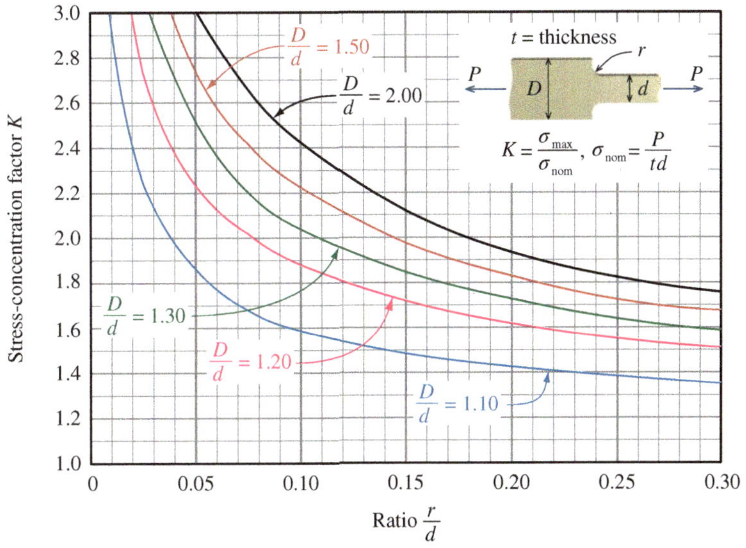

FIGURE 5.15 Stress-concentration factors K for a flat bar with shoulder fillets.

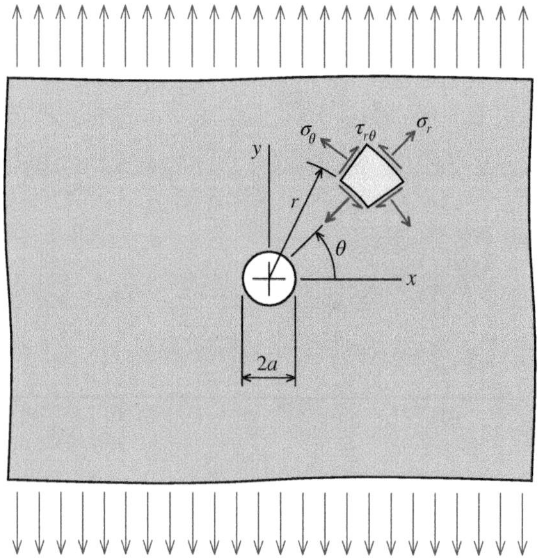

FIGURE 5.16 Circular hole in a wide plate subjected to uniform unidirectional tension.

elasticity solution is expressed in terms of a radial stress σ_r, a tangential stress σ_θ, and a shearing stress $\tau_{r\theta}$, as shown in Figure 5.16. The equations are as follows:

$$\sigma_r = \frac{\sigma}{2}\left(1-\frac{a^2}{r^2}\right)-\frac{\sigma}{2}\left(1-\frac{4a^2}{r^2}+\frac{3a^4}{r^4}\right)\cos 2\theta$$

$$\sigma_\theta = \frac{\sigma}{2}\left(1+\frac{a^2}{r^2}\right)+\frac{\sigma}{2}\left(1+\frac{3a^4}{r^4}\right)\cos 2\theta$$

$$\tau_{r\theta} = \frac{\sigma}{2}\left(1+\frac{2a^2}{r^2}-\frac{3a^4}{r^4}\right)\sin 2\theta$$

On the boundary of the hole (at $r = a$) these equations reduce to the following:

$$\sigma_r = 0$$
$$\sigma_\theta = \sigma(1 + 2\cos 2\theta)$$
$$\tau_{r\theta} = 0$$

At $\theta = 0°$, the tangential stress $\sigma_\theta = 3\sigma$, where σ is the uniform tensile stress in the plate in regions far removed from the hole. Thus, the stress-concentration factor associated with this type of discontinuity is 3.

The localized nature of a stress concentration can be evaluated by considering the distribution of the tangential stress σ_θ along the x axis ($\theta = 0°$). Here,

$$\sigma_\theta = \frac{\sigma}{2}\left(2+\frac{a^2}{r^2}+\frac{3a^4}{r^4}\right)$$

At a distance $r = 3a$ (i.e., one hole diameter from the hole boundary), this equation yields $\sigma_\theta = 1.074\sigma$. Thus, the stress that began as three times the nominal stress at the boundary of the hole has decayed to a value only 7 percent greater than the nominal at a distance of one diameter from the hole. This rapid decay is typical of the redistribution of stress in the neighborhood of a discontinuity.

For a ductile material, the stress concentration associated with static loading does not cause concern, because the material will yield in the region of high stress. With the redistribution of stress that accompanies this local yielding, equilibrium will be attained and no fracture will occur. However, if the load is an impact load or a repeated load, instead of a static load, the material may fracture. Also, if the material is brittle, even a static load may cause fracture. Therefore, in the case of an impact load or a repeated load on any material, or static loading on a brittle material, the presence of a stress concentration must not be ignored.

Specific stress-concentration factors depend on both geometric considerations and the type of loading. In this section, stress-concentration factors pertaining to axial loading have been discussed. Stress-concentration factors for torsion and bending will be discussed in subsequent chapters.

EXAMPLE 5.9

The machine part shown is 20 mm thick and is made of C86100 bronze. (See Appendix D for properties.) Determine the maximum safe load P if a factor of safety of 2.5 with respect to failure by yield is specified.

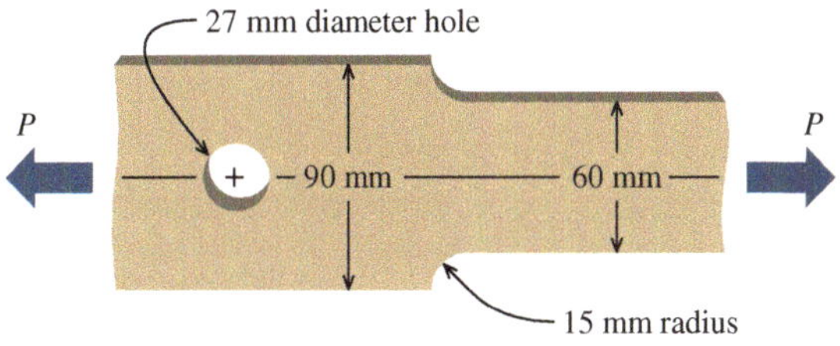

SOLUTION

The yield strength of C86100 bronze is 331 MPa. (See Appendix D for properties.) The allowable stress, based on a factor of safety of 2.5, is 331/2.5 = 132.4 MPa. The maximum stress in the machine part will occur either in the fillet between the two sections or on the boundary of the circular hole.

At the Fillet

$$\frac{D}{d} = \frac{90 \text{ mm}}{60 \text{ mm}} = 1.5 \quad \text{and} \quad \frac{r}{d} = \frac{15 \text{ mm}}{60 \text{ mm}} = 0.25$$

From Figure 5.15, $K \cong 1.73$. Thus,

$$P = \frac{\sigma_{\text{allow}} A_{\text{min}}}{K} = \frac{(132.4 \text{ N/mm}^2)(60 \text{ mm})(20 \text{ mm})}{1.73} = 91{,}838 \text{ N} = 91.8 \text{ kN}$$

At the Hole

$$\frac{d}{D} = \frac{27 \text{ mm}}{90 \text{ mm}} = 0.3$$

From Figure 5.14, $K \cong 2.36$. Thus,

$$P = \frac{\sigma_{\text{allow}} A_{\text{net}}}{K} = \frac{(132.4 \text{ N/mm}^2)(90 \text{ mm} - 27 \text{ mm})(20 \text{ mm})}{2.36} = 70{,}688 \text{ N} = 70.7 \text{ kN}$$

Therefore,

$$P_{\text{max}} = 70.7 \text{ kN} \qquad \textbf{Ans.}$$

PROBLEMS

P5.54 A 100 mm wide by 8 mm thick steel bar is transmitting an axial tensile load of 3,000 N. After the load is applied, a 4 mm diameter hole is drilled through the bar, as shown in Figure P5.54. The hole is centered in the bar.

(a) Determine the stress at point *A* (on the edge of the hole) in the bar before and after the hole is drilled.
(b) Does the axial stress at point *B* on the edge of the bar increase or decrease as the hole is drilled? Explain.

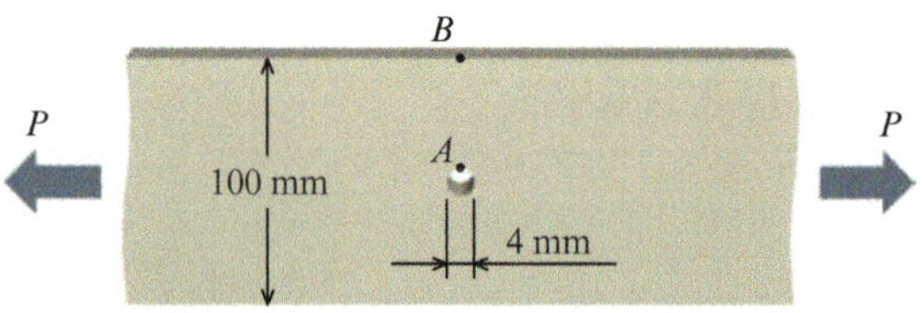

FIGURE P5.54

P5.55 The machine part shown in Figure P5.55 is 8 mm thick and is made of AISI 1020 cold-rolled steel. (See Appendix D for properties.) Determine the maximum safe load *P* if a factor of safety of 3 with respect to failure by yield is specified.

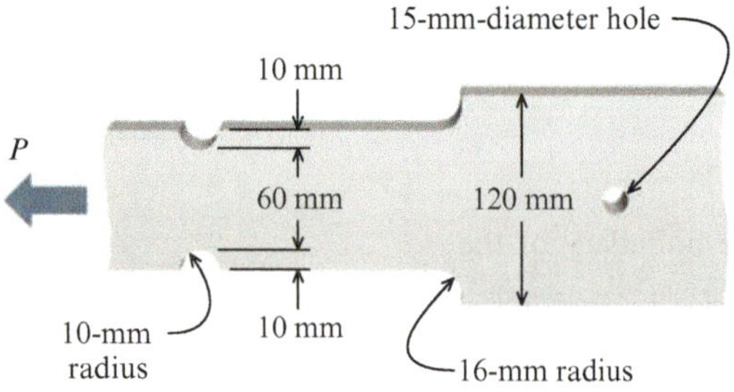

FIGURE P5.55

P5.56 The 0.25 in thick bar shown in Figure P5.56 is made of 2014-T4 aluminum (see Appendix D for properties) and will be subjected to an axial tensile load $P = 1{,}500$ lb. A 0.5625 in. diameter hole is located on the centerline of the bar. Determine the minimum safe width *D* for the bar if a factor of safety of 2.5 with respect to failure by yield must be maintained.

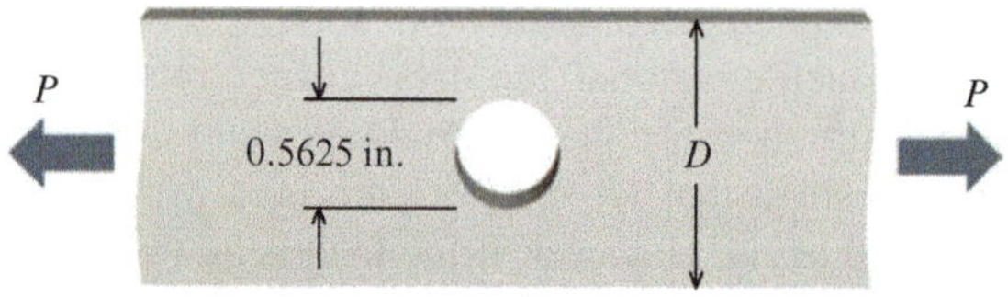

FIGURE P5.56

P5.57 The machine part shown in Figure P5.57 is 10 mm thick, is made of AISI 1020 cold-rolled steel (see Appendix D for properties), and is subjected to a tensile load $P = 45$ kN. Determine the minimum radius *r* that can be used between the two sections if a factor of safety of 2 with respect to failure by yield is specified. Round the minimum fillet radius up to the nearest 1 mm multiple.

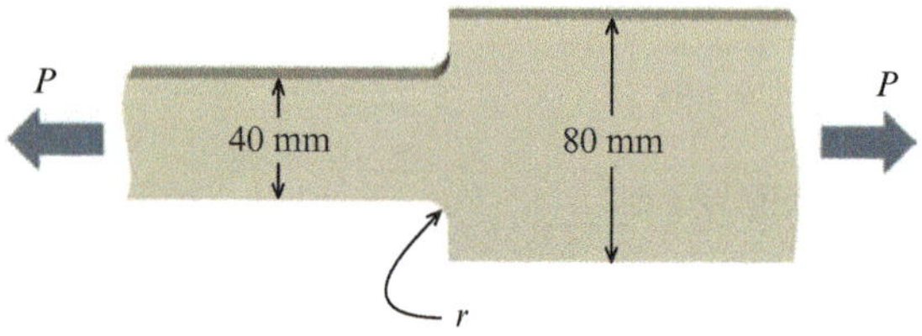

FIGURE P5.57

P5.58 The stepped bar with a circular hole shown in Figure P5.58 is made of annealed 18-8 stainless steel. The bar is 12 mm thick and will be subjected to an axial tensile load $P = 70$ kN. The normal stress in the bar is not to exceed 150 MPa. To the nearest millimeter, determine

(a) the maximum allowable hole diameter *d*.
(b) the minimum allowable fillet radius *r*.

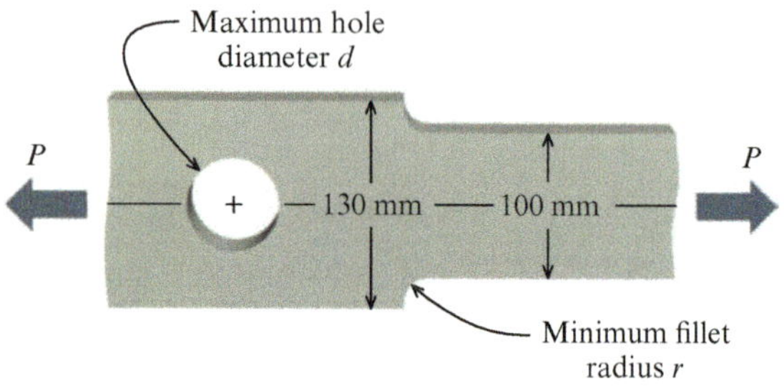

FIGURE P5.58

CHAPTER 6

Torsion

6.1 Introduction

Torque is a moment that tends to twist a member *about its longitudinal axis.* In the design of machinery (and some structures), the problem of transmitting a torque from one plane to a parallel plane is frequently encountered. The simplest device for accomplishing this function is called a *shaft*. Shafts are commonly used to connect an engine or a motor to a pump, a compressor, an axle, or a similar device. Shafts connecting gears and pulleys are a common application involving torsion members. Most shafts have circular cross sections, either solid or tubular. A modified free-body diagram of a typical device is shown in Figure 6.1. The weight and bearing reactions are not shown on the diagram, since they do not contribute useful information to the torsion problem. The resultant of the electromagnetic forces applied to the armature A of the motor is a moment that is resisted by the resultant of the bolt forces (another moment) acting on the flange coupling B. The circular shaft (1) transmits the torque from the armature to the coupling. The torsion problem is concerned with the determination of stresses in shaft (1) and the deformation of the shaft. For the elementary analysis developed in this book, shaft segments such as the segment between transverse planes a–a and b–b in Figure 6.1 will be considered. By limiting the analysis to shaft segments such as this, the complicated states of stress that occur at the locations of the torque-applying components (i.e., the armature and flange coupling) can be avoided. Recall that Saint-Venant's principle states that the effects introduced by attaching the armature and coupling to the shaft will cease to be evident in the shaft at a distance of approximately one shaft diameter from these components.

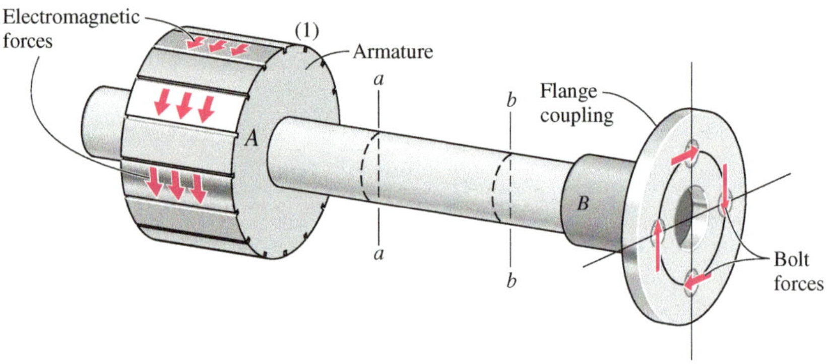

FIGURE 6.1 Modified free-body diagram of a typical electric motor shaft.

Torsion of noncircular shapes produces *warping*, in which planar cross sections before application of the loading become nonplanar, or *warped*, after a torque is applied.

In 1784, C. A. Coulomb, a French engineer, experimentally developed the relationship between the applied torque and the angle of twist for circular bars.[1] A. Duleau, another French engineer, in a paper published in 1820, analytically derived the same relationship by making the assumptions *that a plane section before twisting remains plane after twisting* and *that a radial line on the cross section remains plane after twisting*. Visual examination of twisted models indicates that these assumptions are apparently correct for either solid or hollow circular sections (provided that the hollow section is circular and symmetrical with respect to the axis of the shaft), but incorrect for any other shape. For example, compare the distortions evident in the two prismatic rubber shaft models shown in Figure 6.2. Figures 6.2*a* and 6.2*b* show a circular rubber shaft before and after an external torque T is applied to its ends. When torque T is applied to the end of the round shaft, the circular cross sections and longitudinal grid lines marked on the shaft deform into the pattern shown in Figure 6.2*b*. Each longitudinal grid line is twisted into a helix that intersects the circular cross sections at equal angles. The length of the shaft and its radius remain unchanged. Each cross section remains plane and undistorted as it rotates

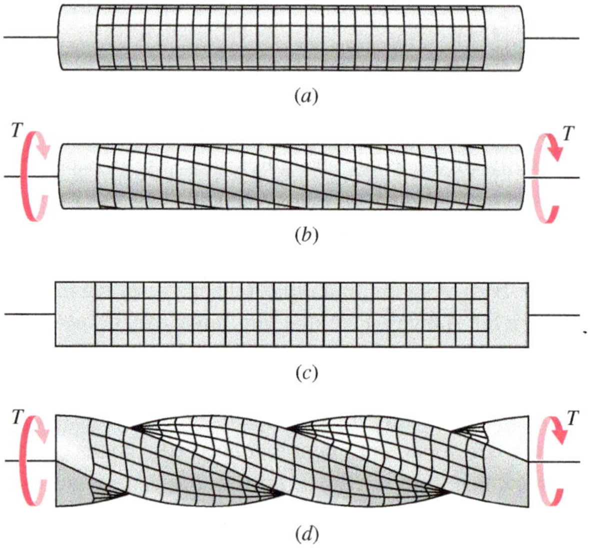

FIGURE 6.2 Torsional deformations illustrated by rubber models with circular (*a, b*) and square (*c, d*) cross sections.

[1] From S. P. Timoshenko, *History of Strength of Materials* (New York: McGraw-Hill, 1953).

with respect to an adjacent cross section. Figures 6.2*c* and 6.2*d* show a square rubber shaft before and after an external torque T is applied to its ends. Plane cross sections in Figure 6.2*c* before the torque is applied do not remain plane after T is applied (Figure 6.2*d*). The behavior exhibited by the square shaft is characteristic of all but circular sections; therefore, the analysis that follows is valid *only for solid or hollow circular shafts*.

6.2 Torsional Shear Strain

Consider a long, slender shaft of length L and radius c that is fixed at one end, as shown in Figure 6.3*a*. When an external torque T is applied to the free end of the shaft at B, the shaft deforms as shown in Figure 6.3*b*. All cross sections of the shaft are subjected to the same internal torque T; therefore, the shaft is said to be in *pure torsion*. Longitudinal lines in Figure 6.3*a* are twisted into helixes as the free end of the shaft rotates through an angle ϕ. This angle of rotation is known as the *angle of twist*. The angle of twist changes along the length L of the shaft. For a prismatic shaft, the angle of twist will vary linearly between the ends of the shaft. The twisting deformation does not distort cross sections of the shaft in any way, and the overall shaft length remains constant. As discussed in Section 6.1, the following assumptions can be applied to torsion of shafts that have circular—either solid or hollow—cross sections:

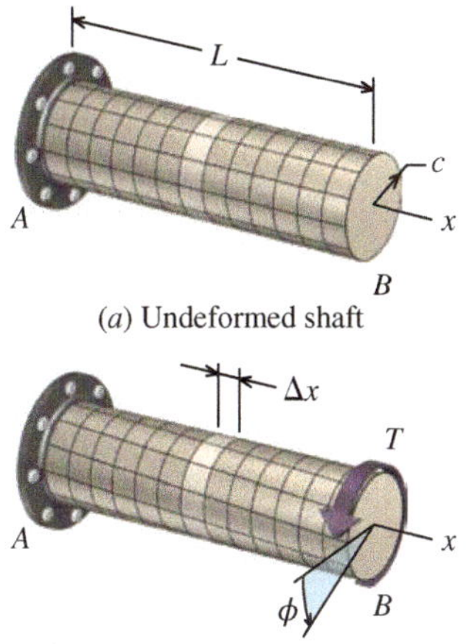

(*a*) Undeformed shaft

(*b*) Deformed shaft in response to torque T

FIGURE 6.3 Prismatic shaft subjected to pure torsion.

- A plane section before twisting remains plane after twisting. In other words, circular cross sections do not *warp* as they twist.
- Cross sections rotate about, and remain perpendicular to, the longitudinal axis of the shaft.
- Each cross section remains undistorted as it rotates relative to neighboring cross sections. In other words, the cross section remains circular and there is no strain in the plane of the cross section. Radial lines remain straight and radial as the cross section rotates.
- The distances between cross sections remain constant during the twisting deformation. In other words, no axial strain occurs in a round shaft as it twists.

To help us investigate the deformations that occur during twisting, a short segment Δx of the shaft shown in Figure 6.3 is isolated in Figure 6.4*a*. The shaft radius is c; however, for more generality, an interior cylindrical portion at the core of the shaft will be examined (Figure 6.4*b*). The radius of this core portion is denoted by ρ, where $0 < \rho \leq c$. As the shaft twists, the two cross sections of the segment rotate about the x axis and line element CD on the undeformed shaft is twisted into helix $C'D'$. The angular difference between the rotations of the two cross sections is equal to $\Delta\phi$. This angular difference creates a shear strain γ in the shaft. The shear strain γ is equal to the angle between line elements $C'D'$ and $C'D''$, as shown in Figure 6.4*b*. The value of the angle γ is given by

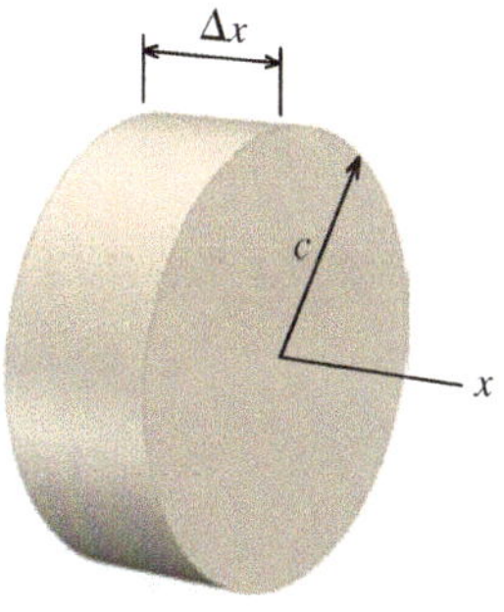

FIGURE 6.4*a* Shaft segment of length Δx.

$$\tan\gamma = \frac{D'D''}{\Delta x}$$

The distance $D'D''$ can also be expressed by the arc length $\rho\Delta\phi$, giving

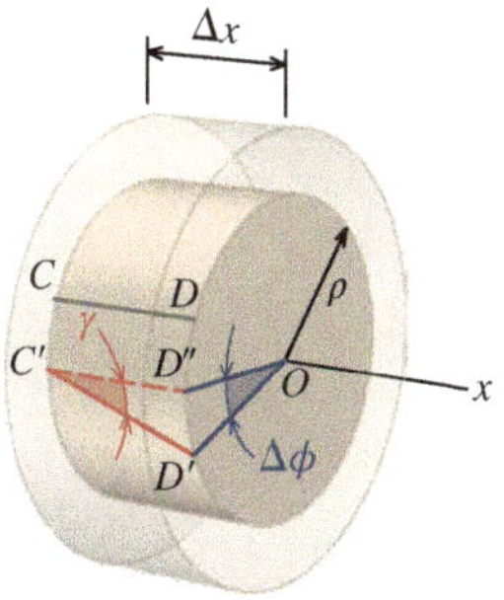

FIGURE 6.4*b* Torsional deformation of shaft segment.

$$\tan\gamma = \frac{\rho\Delta\phi}{\Delta x}$$

If the strain is small, tan $\gamma \approx \gamma$; therefore,

$$\gamma = \rho\frac{\Delta\phi}{\Delta x}$$

As the length Δx of the shaft segment decreases to zero, the shear strain becomes

$$\gamma = \rho\frac{d\phi}{dx} \tag{6.1}$$

The quantity $d\phi/dx$ is the *angle of twist per unit length*. Note that Equation (6.1) is linear with respect to the radial coordinate ρ; therefore, the shear strain at the shaft centerline (i.e., $\rho = 0$) is zero, while the largest shear strain occurs for the largest value of ρ (i.e., $\rho = c$), which occurs on the outermost surface of the shaft.

$$\gamma_{\text{max}} = c\frac{d\phi}{dx} \tag{6.2}$$

Equations (6.1) and (6.2) can be combined to express the shear strain at any radial coordinate ρ in terms of the maximum shear strain:

$$\gamma_\rho = \frac{\rho}{c}\gamma_{\text{max}} \tag{6.3}$$

Note that these equations are valid for elastic or inelastic action and for homogeneous or heterogeneous materials, provided that the strains are not too large (i.e., tan $\gamma \approx \gamma$). Problems and examples in this book will be assumed to satisfy that requirement.

6.3 Torsional Shear Stress

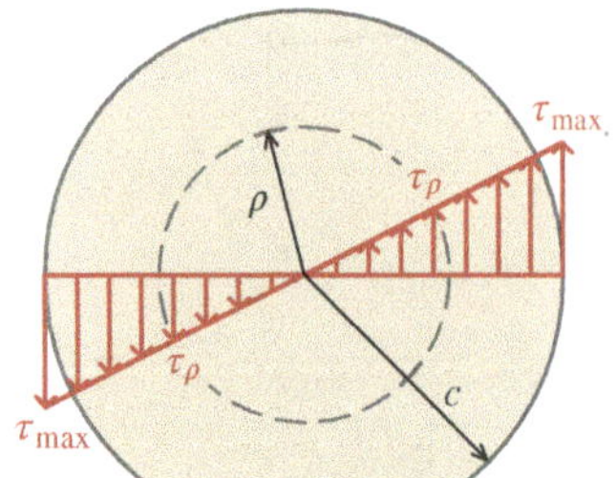

FIGURE 6.5 Linear variation of shear stress intensity as a function of radial coordinate ρ.

If the assumption is now made that Hooke's law applies, then the shear strain γ can be related to the shear stress τ by the relationship $\tau = G\gamma$ [Equation (3.5)], where G is the shear modulus (also called the modulus of rigidity). This assumption is valid if the shear stresses remain below the proportional limit for the shaft material. Using Hooke's law, we can express Equation (6.3) in terms of τ to give the relationship between the shear stress τ_ρ at any radial coordinate ρ and the maximum shear stress τ_{max}, which occurs on the outermost surface of the shaft (i.e., $\rho = c$):[2]

$$\tau_\rho = \frac{\rho}{c}\tau_{\text{max}} \tag{6.4}$$

As with shear strain, shear stress in a circular shaft increases linearly in intensity as the radial distance ρ from the centerline of the shaft increases. The maximum shear stress intensity occurs on the outermost surface of the shaft. The variation in the magnitude (intensity) of the shear stress is illustrated in Figure 6.5. **Furthermore, shear stress never acts solely on a single surface. Shear stress on a cross-sectional surface is always accompanied by shear stress of equal magnitude acting on a longitudinal surface, as depicted in Figure 6.6.**

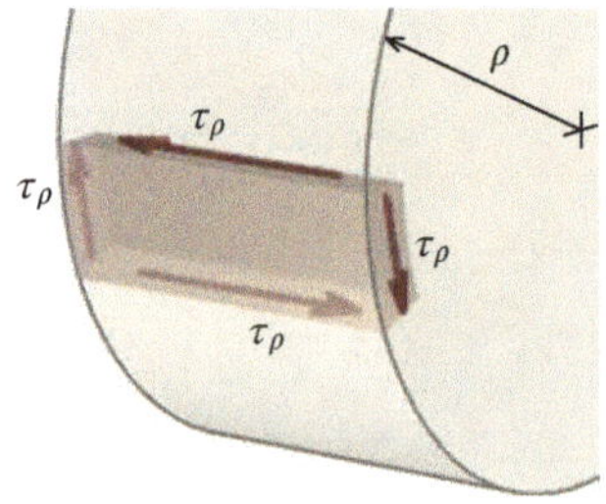

FIGURE 6.6 Shear stresses act on both cross-sectional and longitudinal planes.

The relationship between the torque T transmitted by a shaft and the shear stress τ_ρ developed internally in the shaft must be developed. Consider a very small portion dA of a cross-sectional surface (Figure 6.7). In response to torque T, shear stresses τ_ρ are developed on the surface of the cross section on area dA, which is located at a radial distance of ρ from

[2] In keeping with the notation presented in Section 1.5, the shear stress τ_ρ should actually be designated $\tau_{x\theta}$ to indicate that it acts on the x face in the direction of increasing θ. However, for the elementary theory of torsion of circular sections discussed in this book, the shear stress on any transverse plane *always acts perpendicular to the radial direction* at any point. Consequently, the formal double-subscript notation for shear stress is not needed for accuracy and can be omitted here.

the longitudinal axis of the shaft. The resultant shear force dF acting on the small element is given by the product of the shear stress τ_ρ and the area dA. The force dF produces a moment $dM = \rho\, dF = \rho(\tau_\rho\, dA)$ about the shaft centerline O. The resultant moment produced by the shear stress about the shaft centerline is found by integrating dM over the cross-sectional area:

$$\int dM = \int_A \rho \tau_\rho \, dA$$

If Equation (6.4) is substituted into this equation, the result is

$$\int dM = \int_A \rho \frac{\tau_{\max}}{c} \rho dA = \int_A \frac{\tau_{\max}}{c} \rho^2 \, dA$$

Since $\tau_{\max}$ and c do not vary with dA, these terms can be moved outside of the integral. Furthermore, the sum of all elemental moments dM must equal the torque T to satisfy equilibrium; therefore,

$$T = \int dM = \frac{\tau_{\max}}{c} \int_A \rho^2 \, dA \tag{a}$$

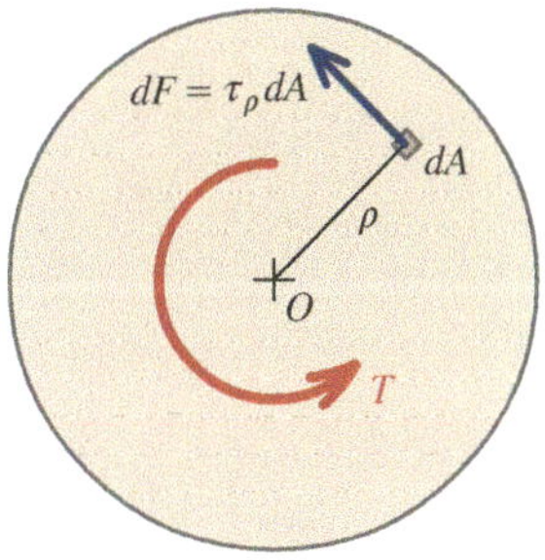

FIGURE 6.7 Calculating the resultant moment produced by torsion shear stress.

The integral in Equation (a) is called the *polar moment of inertia, J*:

$$J = \int_A \rho^2 \, dA \tag{b}$$

Substituting Equation (b) into Equation (a) gives a relationship between the torque T and the maximum shear stress $\tau_{\max}$:

$$T = \frac{\tau_{\max}}{c} J \tag{c}$$

Alternatively, solving equation (c) for the maximum shear stress gives

$$\tau_{\max} = \frac{Tc}{J} \tag{6.5}$$

If Equation (6.4) is substituted into Equation (6.5), a more general relationship can be obtained for the shear stress τ_ρ at any radial distance ρ from the shaft centerline:

$$\tau_\rho = \frac{T\rho}{J} \tag{6.6}$$

Equation (6.6), for which Equation (6.5) is a special case, is known as the *elastic torsion formula*. In general, the internal torque T in a shaft or shaft segment is obtained from a free-body diagram and an equilibrium equation. **Note:** Equations (6.5) and (6.6) apply only to linearly elastic action in homogeneous and isotropic materials.

MecMovies 6.2 presents an animated derivation of the elastic torsion formula.

Polar Moment of Inertia, *J*

The polar moment of inertia for a solid circular shaft is given by

$$J = \frac{\pi}{2} r^4 = \frac{\pi}{32} d^4 \tag{6.7}$$

The polar moment of inertia is also known as the polar second moment of area.

where r is the radius of the shaft and d is the diameter. For a hollow circular shaft, the polar moment of inertia is given by

$$J = \frac{\pi}{2}[R^4 - r^4] = \frac{\pi}{32}[D^4 - d^4] \tag{6.8}$$

where R is the outside radius, r is the inside radius, D is the outside diameter, and d is the inside diameter, of the shaft.

Typically, J has units of in.^{4} in the U.S. Customary System and mm^4 in SI.

6.4 Stresses on Oblique Planes

The elastic torsion formula [Equation (6.6)] can be used to calculate the maximum shear stress produced on a transverse section in a circular shaft by a torque. It is necessary to establish whether the transverse section is a plane of maximum shear stress and whether there are other significant stresses induced by torsion. For this study, the stresses at point A in the shaft of Figure 6.8*a* will be analyzed. Figure 6.8*b* shows a differential element taken from the shaft at A, as well as the shear stresses acting on transverse and longitudinal planes. The stress τ_{xy} may be determined by means of the elastic torsion formula and the equation $\tau_{yx} = \tau_{xy}$ (see Section 1.6). If the equations of equilibrium are applied to the free-body diagram of Figure 6.8*c*, the following result is obtained:

$$\Sigma F_t = \tau_{nt}\, dA - \tau_{xy}(dA\cos\theta)\cos\theta + \tau_{yx}(dA\sin\theta)\sin\theta = 0$$

We then have

$$\tau_{nt} = \tau_{xy}(\cos^2\theta - \sin^2\theta) = \tau_{xy}\cos 2\theta \tag{6.9}$$

and

$$\Sigma F_n = \sigma_n dA - \tau_{xy}(dA\cos\theta)\sin\theta - \tau_{yx}(dA\sin\theta)\cos\theta = 0$$

from which it follows that

$$\sigma_n = 2\tau_{xy}\sin\theta\cos\theta = \tau_{xy}\sin 2\theta \tag{6.10}$$

These results are shown in the graph of Figure 6.9, from which it is apparent that the maximum shear stress occurs on transverse and longitudinal diametral planes (i.e., longitudinal planes that include the centerline of the shaft). The graph also shows that the maximum normal stresses occur on planes oriented at 45° with the axis of the shaft and perpendicular to the surface of the shaft. On one of these planes ($\theta = 45°$ in Figure 6.8*b*), the normal stress is tensile, and on the other ($\theta = 135°$), the normal stress is compressive. Furthermore, *the maximum magnitudes for both σ and τ are equal.* Therefore, the maximum shear stress given by the elastic torsion formula is also numerically equal to the maximum normal stress that occurs at a point in a circular shaft subjected to pure torsion.

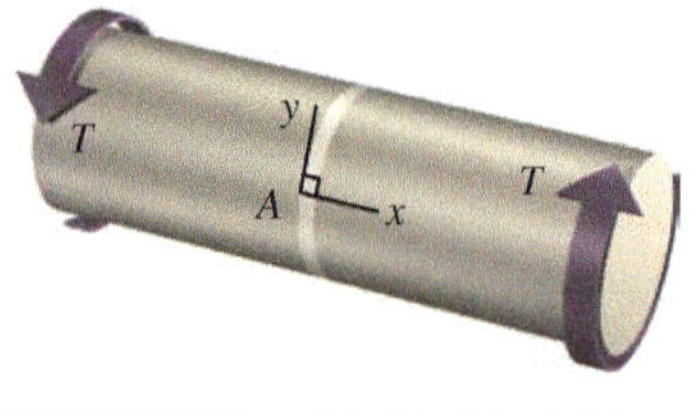

FIGURE 6.8*a* Shaft subjected to pure torsion.

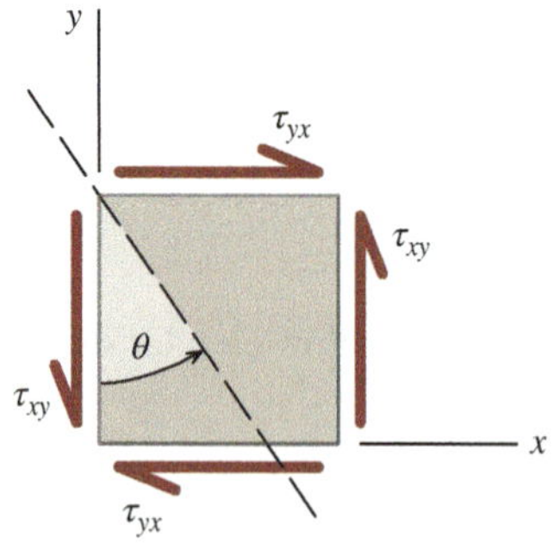

FIGURE 6.8*b* Differential element at point A on the shaft.

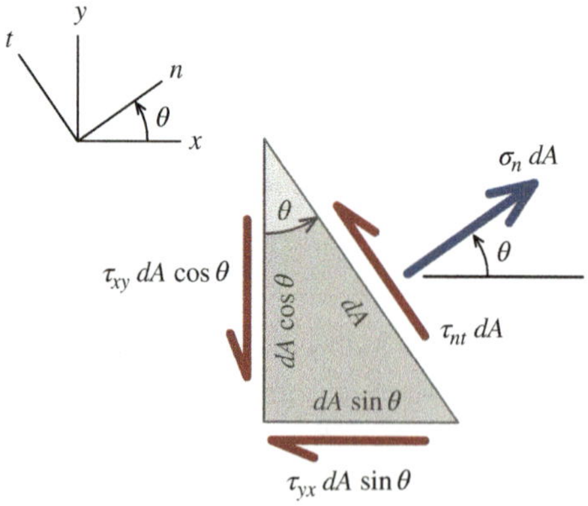

FIGURE 6.8*c* Free-body diagram of a wedge-shaped portion of the differential element.

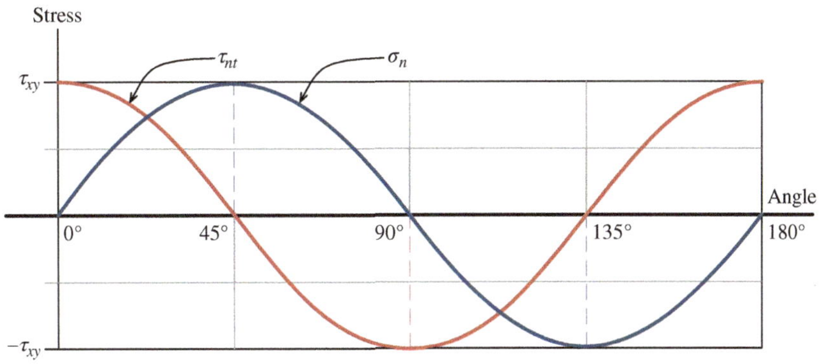

FIGURE 6.9 Variation of normal and shear stresses with angle θ on the surface of a shaft.

Any of the stresses discussed in the preceding paragraph may be significant in a particular problem. Compare, for example, the failures shown in Figure 6.10. In Figure 6.10*a*, the steel axle of a truck split longitudinally. One would expect this type of failure to occur also in a shaft of wood with the grain running longitudinally. In Figure 6.10*b*, compressive stress caused the thin-walled aluminum alloy tube to buckle along one 45° plane while tensile stress caused tearing on the other 45° plane. Buckling of thin-walled tubes (and other shapes) subjected to torsional loading is a matter of great importance to the designer. In Figure 6.10*c*, tensile normal stresses caused the gray cast iron shaft to fail in tension—a failure typical of any brittle material subjected to torsion. In Figure 6.10*d*, the low-carbon steel failed in shear on a plane that is almost transverse—a typical failure for ductile material. The reason the fracture in Figure 6.10*d* did not occur on a transverse plane is that, under the large plastic twisting deformation before rupture (note the spiral lines indicating elements originally parallel to the axis of the bar), longitudinal elements were subjected to axial tensile loading. This axial loading was induced because the testing machine grips would not permit the torsion specimen to shorten as the elements were twisted into spirals. This axial tensile stress (not shown in Figure 6.8) changes the plane of maximum shear stress from a transverse to an oblique plane (resulting in a warped surface of rupture).[3]

Buckling is a *stability failure*. The phenomenon of stability failure is discussed in Chapter 16.

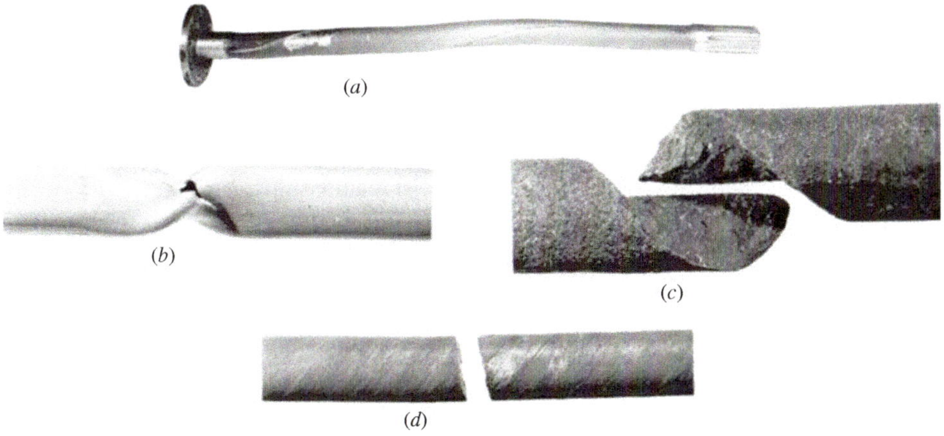

FIGURE 6.10 Photos of actual shaft failures.

[3] The tensile stress is not entirely due to the grips, because the plastic deformation of the outer elements of the bar is considerably greater than that of the inner elements. This difference in deformation results in a spiral tensile stress in the outer elements and a similar compressive stress in the inner elements.

6.5 Torsional Deformations

If the shear stresses in a shaft are below the proportional limit of the shaft material, then Hooke's law, $\tau = G\gamma$, relates shear stress and shear strain in the torsion member. The relationship between the shear stress in a shaft at any radial coordinate ρ and the internal torque T is given by Equation (6.6):

$$\tau_\rho = \frac{T\rho}{J}$$

The shear strain is related to the angle of twist per unit length by Equation (6.1):

$$\gamma = \rho \frac{d\phi}{dx}$$

Equations (6.6) and (6.1) can be substituted into Hooke's law,

$$\tau_\rho = G\gamma \qquad \therefore \frac{T\rho}{J} = G\rho \frac{d\phi}{dx}$$

The resulting equation expresses the angle of twist per unit length in terms of the torque T:

$$\frac{d\phi}{dx} = \frac{T}{JG} \tag{6.11}$$

MecMovies 6.2 presents an animated derivation of the angle-of-twist relationship.

To obtain the angle of twist for a specific shaft segment, Equation (6.11) can be integrated with respect to the longitudinal coordinate x over the length L of the segment:

$$\int d\phi = \int_L \frac{T}{JG}\, dx$$

If the shaft is homogeneous (i.e., G is constant) and prismatic (meaning that the diameter is constant and, in turn, J is constant), and if the shaft has a constant internal torque T, then the *angle of twist* ϕ in the shaft can be expressed as

$$\phi = \frac{TL}{JG} \tag{6.12}$$

The units of ϕ are radians in both SI and the U.S. customary system.

Alternatively, Hooke's law and Equations (6.1), (6.2), (6.5), and (6.6) can be combined to give the following additional angle-of-twist relationships:

$$\phi = \frac{\gamma_\rho L}{\rho} = \frac{\tau_\rho L}{\rho G} = \frac{\tau_{\text{max}} L}{cG} \tag{6.13}$$

These relationships are often useful in dual-specification problems, such as those in which limiting values of ϕ and τ are both specified.

To reiterate, Equations (6.12) and (6.13) may be used to compute the angle of twist ϕ only if the torsional member

- is homogeneous (i.e., G is constant),
- is prismatic (i.e., the diameter is constant and, in turn, J is constant), and
- has a constant internal torque T.

If a torsion member is subjected to external torques at intermediate points (i.e., points other than the ends) or if it consists of segments of various diameters or materials, then the torsion member must be divided into segments that satisfy the three requirements just listed. For compound torsion members comprising two or more segments, the overall angle of twist can be determined by algebraically adding the segment twist angles:

$$\phi = \sum_i \frac{T_i L_i}{J_i G_i} \tag{6.14}$$

Here, T_i, L_i, G_i, and J_i are the internal torque, length, shear modulus, and polar moment of inertia, respectively, for individual segments i of the compound torsion member.

The amount of twist in a shaft (or a structural element) is frequently a key consideration in design. The angle of twist ϕ determined from Equations (6.12) and (6.13) is applicable to a constant-diameter shaft segment that is sufficiently removed from sections to which pulleys, couplings, or other mechanical devices are attached (so that Saint-Venant's principle is applicable). However, for practical purposes, it is customary to neglect local distortion at all connections and compute twist angles as though there were no discontinuities.

Rotation Angles

It is often necessary to determine angular displacements at particular points in a compound torsional member or within a system of several torsional members. For example, the proper operation of a system of shafts and gears may require that the angular displacement at a specific gear not exceed a limiting value. The term *angle of twist* pertains to the torsional deformation in shafts or shaft segments. The term *rotation angle* is used in referring to the angular displacement at a specific point in the torsion system or at rigid components, such as pulleys, gears, couplings, and flanges.

6.6 Torsion Sign Conventions

A consistent sign convention is very helpful to us when we analyze torsion members and assemblies of torsion members. The sign conventions that follow will be used for

- internal torques in shafts or shaft segments,
- angles of twist in shafts or shaft segments, and
- rotation angles of specific points or rigid components.

Internal Torque Sign Convention

Moments in general, and internal torques specifically, are conveniently represented by a double-headed vector arrow. This convention is based on the right-hand rule:

- Curl the fingers of your right hand in the direction that the moment tends to rotate. The direction that your right thumb points indicates the direction of the double-headed vector arrow.
- Conversely, point your right-hand thumb in the direction of the double-headed vector arrow, and the fingers of your right hand curl in the direction that the moment tends to rotate.

A positive internal torque T in a shaft or other torsion member tends to rotate in a right-hand rule sense about the outward normal to an exposed section. In other words, an internal torque is positive if the right-hand thumb points outward away from the sectioned surface when the fingers of the right hand are curled in the direction that the internal torque tends to rotate. This sign convention is illustrated in Figure 6.11.

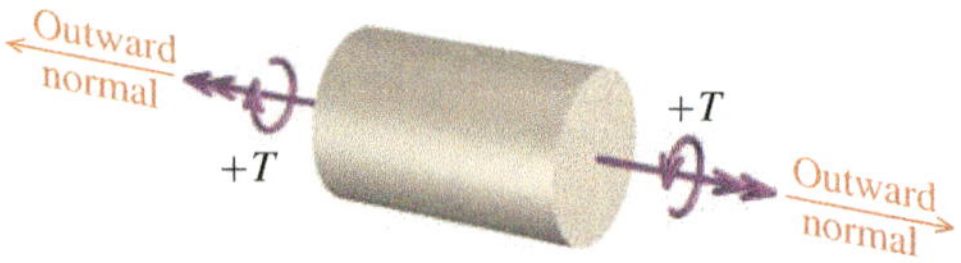

MecMovies 6.3 presents an animation of the sign conventions used for internal torques, shaft element twist angles, and rotation angles.

FIGURE 6.11 Sign convention for internal torque.

Sign Convention for Angles of Twist

The sign convention for angles of twist is consistent with the internal torque sign convention. A positive angle of twist ϕ in a shaft or other torsion member acts in a right-hand rule sense about the outward normal to an exposed section. In other words,

- At an exposed section of the torsion member, curl the fingers of your right hand in the direction of the twisting deformation.
- If your right-hand thumb points outward, away from the sectioned surface, the angle of twist is positive.

This sign convention is illustrated in Figure 6.12.

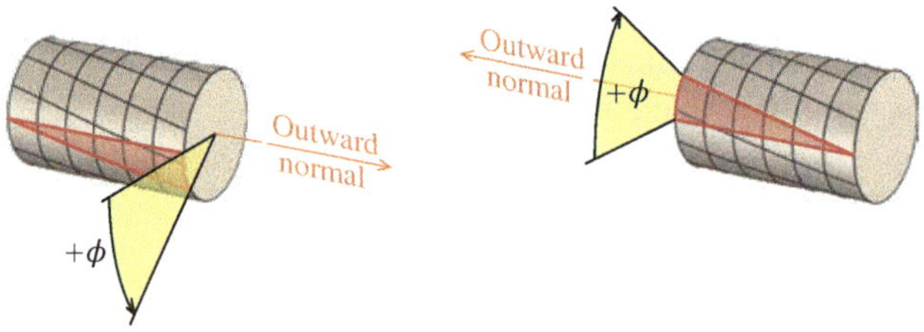

FIGURE 6.12 Sign convention for angles of twist.

Sign Convention for Rotation Angles

Let the longitudinal axis of a shaft be defined as the x axis. Then, a positive rotation angle acts in a right-hand rule sense about the positive x axis. For this sign convention, an origin must be defined for the coordinate system of the torsion member. If two parallel shafts are considered, then the two positive x axes should extend in the same direction. This sign convention is illustrated in Figure 6.13.

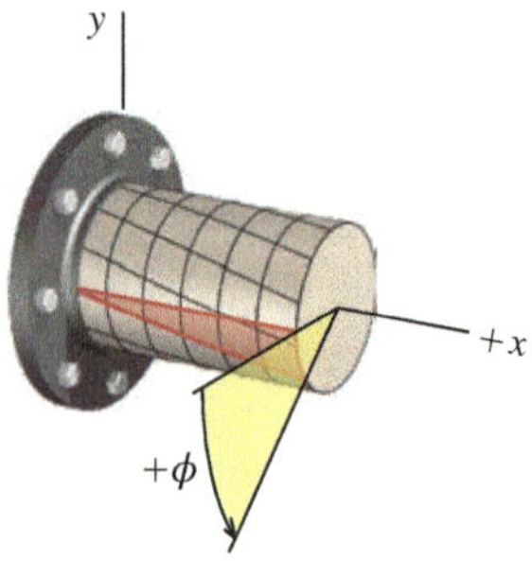

FIGURE 6.13 Sign convention for rotation angles.

EXAMPLE 6.1

A hollow circular steel shaft with an outside diameter of 1.50 in. and a wall thickness of 0.125 in. is subjected to a pure torque of 140 lb·ft. The shaft is 90 in. long. The shear modulus of the steel is $G = 12{,}000$ ksi. Determine

(a) the maximum shear stress in the shaft.
(b) the magnitude of the angle of twist in the shaft.

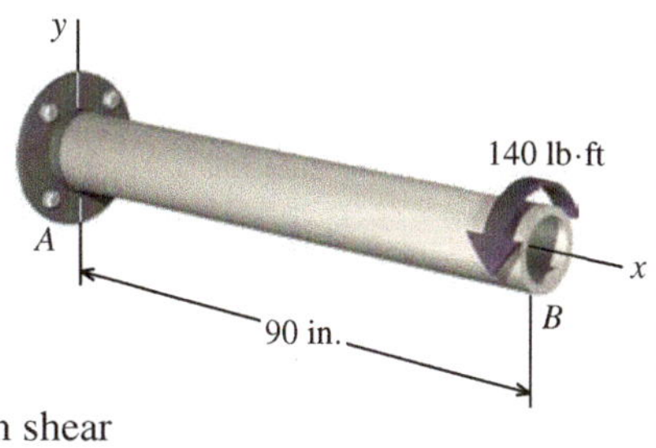

Plan the Solution

The elastic torsion formula [Equation (6.5)] will be used to compute the maximum shear stress, and the angle-of-twist equation [Equation (6.12)] will be used to determine the angle of twist in the hollow shaft.

SOLUTION

The polar moment of inertia J for the hollow shaft will be required for these calculations. The shaft has an outside diameter $D = 1.50$ in. and a wall thickness $t = 0.125$ in. The inside diameter of the shaft is $d = D - 2t = 1.50\text{ in.} - 2(0.125\text{ in.}) = 1.25$ in. The polar moment of inertia for the hollow shaft is

$$J = \frac{\pi}{32}[D^4 - d^4] = \frac{\pi}{32}[(1.50\text{ in.})^4 - (1.25\text{ in.})^4] = 0.257325\text{ in.}^4$$

(a) The maximum shear stress is computed from the elastic torsion formula:

$$\tau = \frac{Tc}{J} = \frac{(140\text{ lb}\cdot\text{ft})(1.50\text{ in.}/2)(12\text{ in./ft})}{0.257325\text{ in.}^4} = 4{,}896.5\text{ psi} = 4{,}900\text{ psi} \qquad \textbf{Ans.}$$

(b) The magnitude of the angle of twist in the 90 in. long shaft is

$$\phi = \frac{TL}{JG} = \frac{(140\text{ lb}\cdot\text{ft})(90\text{ in.})(12\text{ in./ft})}{(0.257325\text{ in.}^4)(12{,}000{,}000\text{ lb/in.}^2)} = 0.0490\text{ rad} \qquad \textbf{Ans.}$$

EXAMPLE 6.2

A 500 mm long solid steel [$G = 80$ GPa] shaft is being designed to transmit a torque $T = 20$ N·m. The maximum shear stress in the shaft must not exceed 70 MPa, and the angle of twist must not exceed 3° in the 500 mm length. Determine the minimum diameter d required for the shaft.

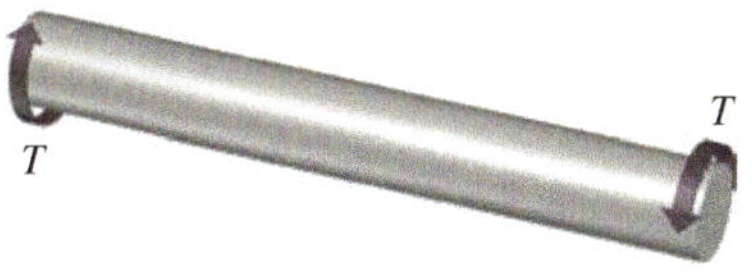

Plan the Solution

The elastic torsion formula [Equation (6.5)] and the angle-of-twist equation [Equation (6.12)] will be algebraically manipulated to solve for the minimum diameter required to satisfy each consideration. The larger of the two diameters will dictate the minimum diameter d that can be used for the shaft.

SOLUTION

The elastic torsion formula relates shear stress and torque:

$$\tau = \frac{Tc}{J}$$

In this instance, the torque and the allowable shear stress are known for the shaft. Putting the known terms on the right-hand side of the equation gives

$$\frac{J}{c} = \frac{T}{\tau}$$

Next, express the left-hand side of this equation in terms of the shaft diameter d:

$$\frac{(\pi/32)d^4}{d/2} = \frac{\pi}{16}d^3 = \frac{T}{\tau}$$

Now solve for the minimum diameter that will satisfy the 80 MPa allowable shear stress limit:

$$d^3 \geq \frac{16}{\pi}\frac{T}{\tau} = \frac{16(20\ \text{N}\cdot\text{m})(1{,}000\ \text{mm/m})}{\pi(70\ \text{N/mm}^2)} = 1{,}455.1309\ \text{mm}^3$$

$$\therefore d \geq 11.33\ \text{mm}$$

The angle of twist in the shaft must not exceed 3° in a 500 mm length. Solving the angle-of-twist equation

$$\phi = \frac{TL}{JG}$$

for the moment of inertia J yields

$$J = \frac{TL}{G\phi}$$

Finally, express the polar moment of inertia in terms of the diameter d, and solve for the minimum diameter that will satisfy the 3° limit:

$$d^4 \geq \frac{32TL}{\pi G\phi} = \frac{32(20\ \text{N}\cdot\text{m})(500\ \text{mm})(1{,}000\ \text{mm/m})}{\pi(80{,}000\ \text{N/mm}^2)(3°)(\pi\ \text{rad}/180°)} = 24{,}317.084\ \text{mm}^4$$

$$\therefore d \geq 12.49\ \text{mm}$$

On the basis of these two calculations, the minimum diameter that is acceptable for the shaft is $d \geq 12.49$ mm. **Ans.**

EXAMPLE 6.3

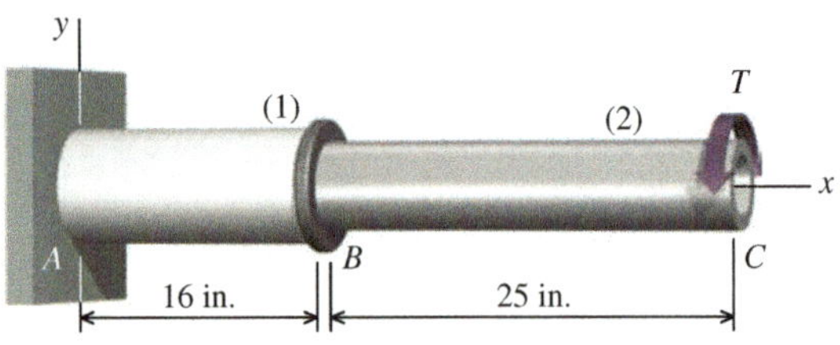

A compound shaft consists of a solid aluminum segment (1) and a hollow steel segment (2). Segment (1) is a solid 1.625 in. diameter aluminum shaft with an allowable shear stress of 6,000 psi and a shear modulus of 4×10^6 psi. Segment (2) is a hollow steel shaft with an outside diameter of 1.25 in., a wall thickness of 0.125 in., an allowable shear stress of 9,000 psi, and a shear modulus of 11×10^6 psi. In addition to the allowable shear stresses, specifications require that the rotation angle at the free end of the shaft must not exceed 2°. Determine the magnitude of the largest torque T that may be applied to the compound shaft at C.

Plan the Solution

To determine the largest torque T that can be applied at C, we must consider the maximum shear stresses and the angles of twist in both shaft segments.

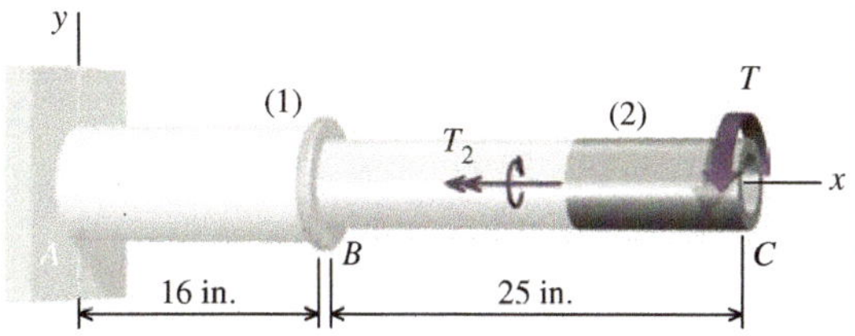

SOLUTION

The internal torques acting in segments (1) and (2) can be easily determined from free-body diagrams cut through each segment.

Cut a free-body diagram through segment (2), and include the free end of the shaft. A positive internal torque T_2 is assumed to act in segment (2). The following equilibrium equation is obtained:

$$\Sigma M_x = T - T_2 = 0 \qquad \therefore T_2 = T$$

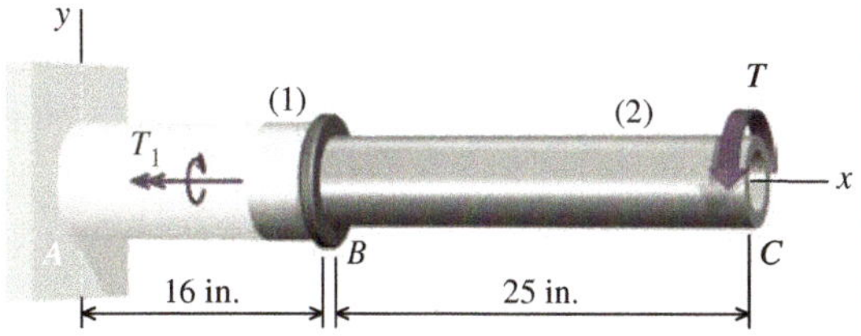

Repeat the process with a free-body diagram cut through segment (1) that includes the free end of the shaft. From this free-body diagram, a similar equilibrium equation is obtained:

$$\Sigma M_x = T - T_1 = 0 \qquad \therefore T_1 = T$$

Therefore, the internal torque in both segments of the shaft is equal to the external torque applied at C.

Shear Stress

In this compound shaft, the diameters and allowable shear stresses in segments (1) and (2) are known. Thus, the elastic torsion formula can be solved for the allowable torque that may be applied to each segment:

$$T_1 = \frac{\tau_1 J_1}{c_1} \qquad T_2 = \frac{\tau_2 J_2}{c_2}$$

Segment (1) is a solid 1.625 in. diameter aluminum shaft. The polar moment of inertia for this segment is

$$J_1 = \frac{\pi}{32}(1.625 \text{ in.})^4 = 0.684563 \text{ in.}^4$$

Use this value along with the 6,000 psi allowable shear stress to determine the allowable torque T_1:

$$T_1 \le \frac{\tau_1 J_1}{c_1} = \frac{(6{,}000 \text{ psi})(0.684563 \text{ in.}^4)}{(1.625 \text{ in.}/2)} = 5{,}055.2 \text{ lb}\cdot\text{in.} \qquad \text{(a)}$$

Segment (2) is a hollow steel shaft with an outside diameter $D = 1.25$ in. and a wall thickness $t = 0.125$ in. The inside diameter of this segment is $d = D - 2t = 1.25$ in. $- 2(0.125$ in.$) =$ 1.00 in. The polar moment of inertia for segment (2) is

$$J_2 = \frac{\pi}{32}[(1.25 \text{ in.})^4 - (1.00 \text{ in.})^4] = 0.141510 \text{ in.}^4$$

Use this value along with the 9,000 psi allowable shear stress to determine the allowable torque T_2:

$$T_2 \le \frac{\tau_2 J_2}{c_2} = \frac{(9{,}000 \text{ psi})(0.141510 \text{ in.}^4)}{(1.25 \text{ in.}/2)} = 2{,}037.7 \text{ lb}\cdot\text{in.} \qquad \text{(b)}$$

Rotation Angle at *C*

The angles of twists in segments (1) and (2) can be expressed, respectively, as

$$\phi_1 = \frac{T_1 L_1}{J_1 G_1} \quad \text{and} \quad \phi_2 = \frac{T_2 L_2}{J_2 G_2}$$

The rotation angle at C is the sum of these two angles of twist:

$$\phi_C = \phi_1 + \phi_2 = \frac{T_1 L_1}{J_1 G_1} + \frac{T_2 L_2}{J_2 G_2}$$

Consequently, since $T_1 = T_2 = T$, it follows that

$$\phi_C = T\left[\frac{L_1}{J_1 G_1} + \frac{L_2}{J_2 G_2}\right]$$

Solving for the external torque T gives

$$T \le \frac{\phi_C}{\dfrac{L_1}{J_1 G_1} + \dfrac{L_2}{J_2 G_2}}$$

$$\le \frac{(2^\circ)(\pi \text{ rad}/180^\circ)}{\dfrac{16 \text{ in.}}{(0.684563 \text{ in.}^4)(4{,}000{,}000 \text{ psi})} + \dfrac{25 \text{ in.}}{(0.141510 \text{ in.}^4)(11{,}000{,}000 \text{ psi})}} \qquad \text{(c)}$$

$$= 1{,}593.6 \text{ lb}\cdot\text{in.}$$

External Torque *T*

Compare the three torque limits obtained in Equations (a), (b), and (c). On the basis of these results, the maximum external torque that can be applied to the shaft at C is

$$T = 1{,}594 \text{ lb}\cdot\text{in.} = 132.8 \text{ lb}\cdot\text{ft} \qquad \textbf{Ans.}$$

EXAMPLE 6.4

A solid steel [$G = 80$ GPa] shaft of variable diameter is subjected to the torques shown. Segment (1) of the shaft has a diameter of 36 mm, segment (2) has a diameter of 30 mm, and segment (3) has a diameter of 25 mm. The bearing shown allows the shaft to turn freely. Additional bearings have been omitted for clarity.

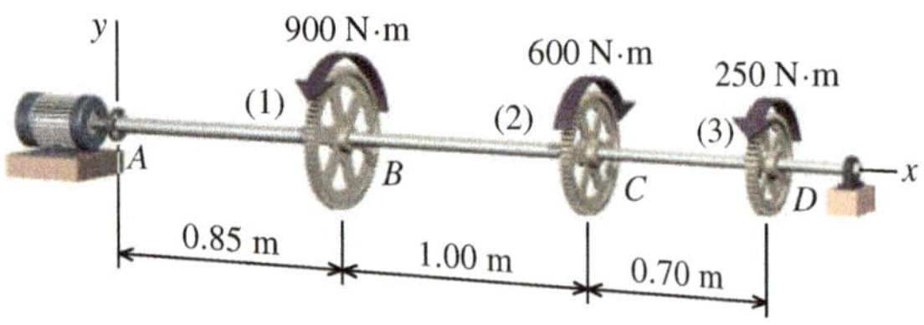

(a) Determine the internal torque in segments (1), (2), and (3) of the shaft. Plot a diagram showing the internal torques in all segments of the shaft. Use the sign convention presented in Section 6.6.

(b) Compute the magnitude of the maximum shear stress in each segment of the shaft.

(c) Determine the rotation angles along the shaft measured at gears B, C, and D relative to flange A. Plot a diagram showing the rotation angles at all points on the shaft.

Plan the Solution

The internal torques in the three shaft segments will be determined from free-body diagrams and equilibrium equations. The elastic torsion formula [Equation (6.5)] will be used to compute the maximum shear stress in each segment once the internal torques are known. The angle-of-twist equations [Equations (6.12) and (6.14)] will be used to determine the twisting in individual shafts as well as the rotation angles at gears *B*, *C*, and *D*.

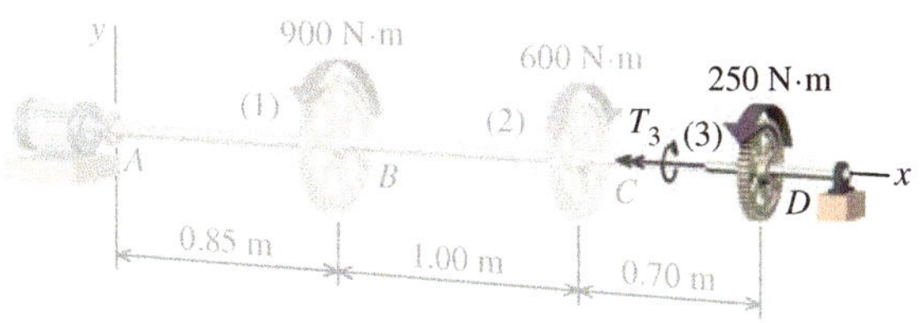

SOLUTION

Equilibrium

Consider a free-body diagram that cuts through shaft segment (3) and includes the free end of the shaft. A positive internal torque T_3 is assumed to act in segment (3). The equilibrium equation obtained from this free-body diagram gives the internal torque in segment (3) of the shaft:

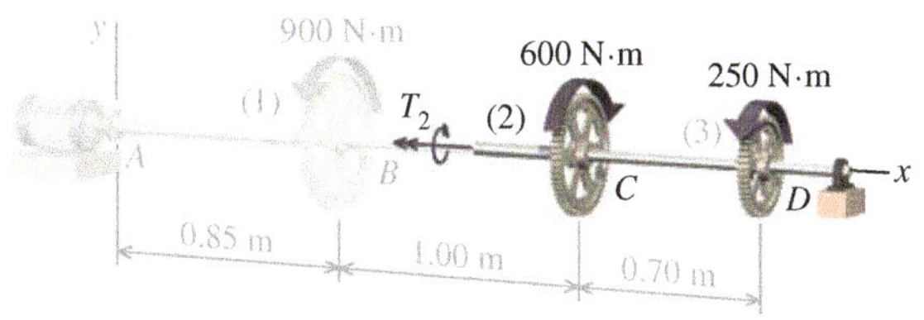

$$\Sigma M_x = 250\text{ N}\cdot\text{m} - T_3 = 0$$

$$\therefore T_3 = 250\text{ N}\cdot\text{m}$$

Similarly, the internal torque in segment (2) is found from an equilibrium equation obtained from a free-body diagram that cuts through segment (2) of the shaft. A positive internal torque T_2 is assumed to act in segment (2). Thus,

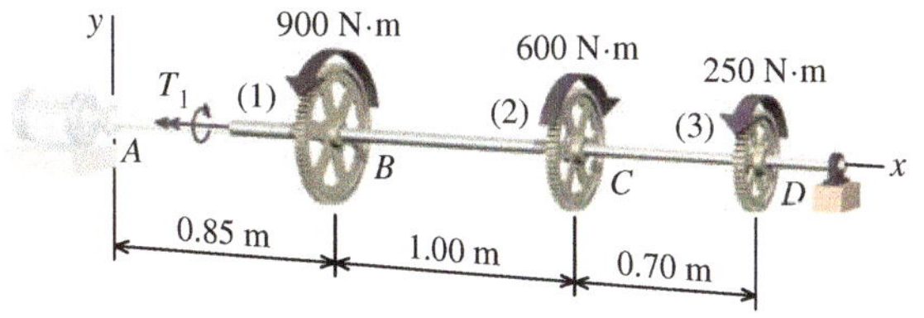

$$\Sigma M_x = 250\text{ N}\cdot\text{m} - 600\text{ N}\cdot\text{m} - T_2 = 0$$

$$\therefore T_2 = -350\text{ N}\cdot\text{m}$$

And for segment (1),

$$\Sigma M_x = 250\text{ N}\cdot\text{m} - 600\text{ N}\cdot\text{m} + 900\text{ N}\cdot\text{m} - T_1 = 0$$

$$\therefore T_1 = 550\text{ N}\cdot\text{m}$$

A torque diagram is produced by plotting these three results.

Polar Moments of Inertia

The elastic torsion formula will be used to compute the maximum shear stress in each shaft segment. For this calculation, the polar moments of inertia must be computed for each segment. Segment (1) is a solid 36 mm diameter shaft. The polar moment of inertia for this shaft segment is

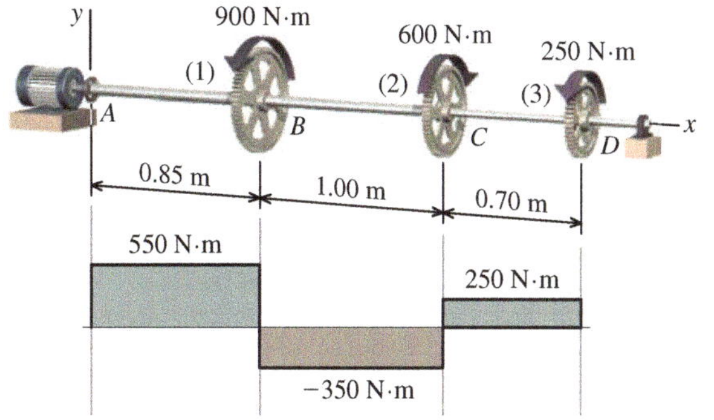

Internal torque diagram for compound shaft.

$$J_1 = \frac{\pi}{32}(36\text{ mm})^4 = 164{,}895.9\text{ mm}^4$$

Shaft segment (2), which is a solid 30 mm diameter shaft, has a polar moment of inertia of

$$J_2 = \frac{\pi}{32}(30\text{ mm})^4 = 79{,}521.6\text{ mm}^4$$

The polar moment of inertia for shaft segment (3), which is a solid 25 mm diameter shaft, has a value of

$$J_3 = \frac{\pi}{32}(25\text{ mm})^4 = 38{,}349.5\text{ mm}^4$$

Shear Stresses

The maximum shear stress magnitude in each segment can be calculated with the use of the elastic torsion formula:

$$\tau_1 = \frac{T_1 c_1}{J_1} = \frac{(550\text{ N}\cdot\text{m})(36\text{ mm}/2)(1{,}000\text{ mm/m})}{164{,}895.9\text{ mm}^4} = 60.0\text{ MPa} \qquad \textbf{Ans.}$$

$$\tau_2 = \frac{T_2 c_2}{J_2} = \frac{(350\text{ N}\cdot\text{m})(30\text{ mm}/2)(1{,}000\text{ mm/m})}{79{,}521.6\text{ mm}^4} = 66.0\text{ MPa} \qquad \textbf{Ans.}$$

$$\tau_3 = \frac{T_3 c_3}{J_3} = \frac{(250\text{ N}\cdot\text{m})(25\text{ mm}/2)(1{,}000\text{ mm/m})}{38{,}349.5\text{ mm}^4} = 81.5\text{ MPa} \qquad \textbf{Ans.}$$

Angles of Twist

Before rotation angles can be determined, the angles of twist in each segment must be determined. In the preceding calculation, the sign of the internal torque was not considered because only the *magnitude* of the shear stress was desired. For the angle-of-twist calculations here, the sign of the internal torque must be included. We obtain

$$\phi_1 = \frac{T_1 L_1}{J_1 G_1} = \frac{(550\text{ N}\cdot\text{m})(850\text{ mm})(1{,}000\text{ mm/m})}{(164{,}895.9\text{ mm}^4)(80{,}000\text{ N/mm}^2)} = 0.035439\text{ rad}$$

$$\phi_2 = \frac{T_2 L_2}{J_2 G_2} = \frac{(-350\text{ N}\cdot\text{m})(1{,}000\text{ mm})(1{,}000\text{ mm/m})}{(79{,}521.6\text{ mm}^4)(80{,}000\text{ N/mm}^2)} = -0.055017\text{ rad}$$

$$\phi_3 = \frac{T_3 L_3}{J_3 G_3} = \frac{(250\text{ N}\cdot\text{m})(700\text{ mm})(1{,}000\text{ mm/m})}{(38{,}349.5\text{ mm}^4)(80{,}000\text{ N/mm}^2)} = 0.057041\text{ rad}$$

Rotation Angles

The angles of twist can be defined in terms of the rotation angles at the ends of each segment:

$$\phi_1 = \phi_B - \phi_A \qquad \phi_2 = \phi_C - \phi_B \qquad \phi_3 = \phi_D - \phi_C$$

The origin of the coordinate system is located at flange *A*. We will arbitrarily define the rotation angle at flange *A* to be zero ($\phi_A = 0$). The rotation angle at gear *B* can be calculated from the angle of twist in segment (1):

$$\phi_1 = \phi_B - \phi_A$$

$$\therefore \phi_B = \phi_A + \phi_1 = 0 + 0.035439\text{ rad}$$
$$= 0.035439\text{ rad} = 0.0354\text{ rad}$$

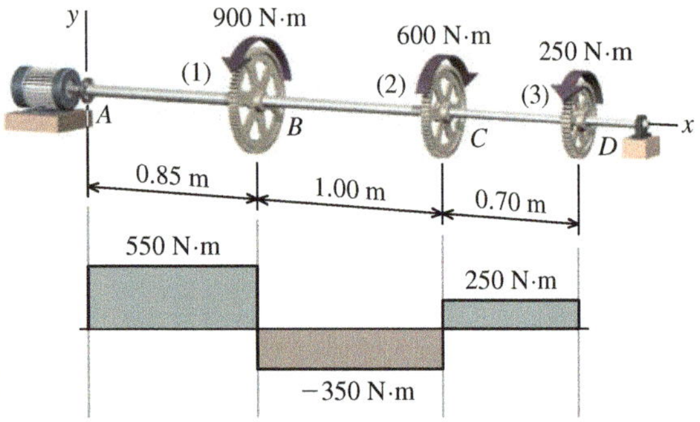

Internal torque diagram for compound shaft.

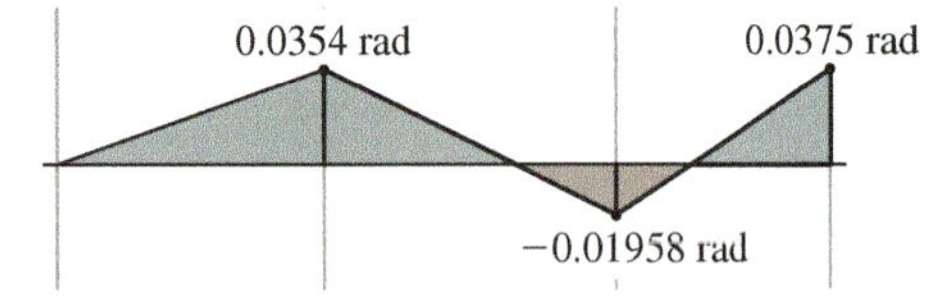

Rotation angle diagram for compound shaft.

Similarly, the rotation angle at *C* is determined from the angle of twist in segment (2) and the rotation angle of gear *B*:

$$\phi_2 = \phi_C - \phi_B$$

$$\therefore \phi_C = \phi_B + \phi_2 = 0.035439\text{ rad} + (-0.055017\text{ rad})$$
$$= -0.019578\text{ rad} = -0.01958\text{ rad}$$

Finally, the rotation angle at gear *D* is

$$\phi_3 = \phi_D - \phi_C$$

$$\therefore \phi_D = \phi_C + \phi_3 = -0.019578\text{ rad} + 0.057041\text{ rad}$$
$$= 0.037464\text{ rad} = 0.0375\text{ rad}$$

A plot of the rotation angle results can be added to the torque diagram to give a complete report for the three-segment shaft.

MecMovies

EXAMPLES

M6.4 Determine the torque T that causes a maximum shearing stress of 50 MPa in the hollow shaft. The outside diameter of the shaft is 40 mm, and the wall thickness is 5 mm.

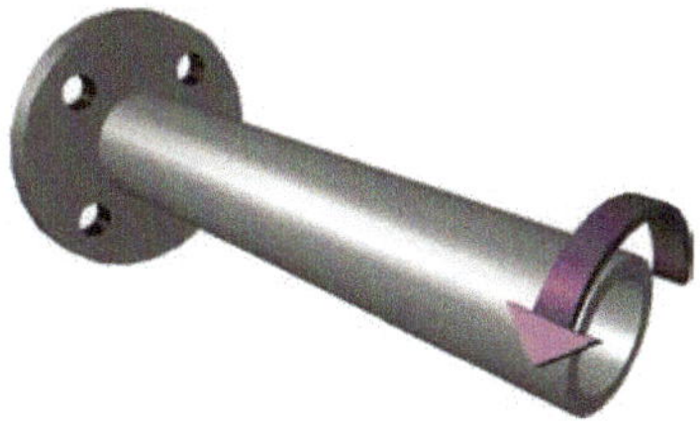

M6.5 Determine the minimum permissible diameter for a solid shaft subjected to a torque of 5 kN · m. The allowable shear stress for the shaft is 65 MPa.

M6.6 A single torque of $T = 50$ N · m is applied to a compound torsion member. Segment (1) is a 32 mm diameter solid brass [$G = 37$ GPa] rod. Segment (2) is a solid aluminum [$G = 26$ GPa] rod. Determine the minimum diameter of the aluminum segment if the rotation angle at C relative to the support A must not exceed 3°.

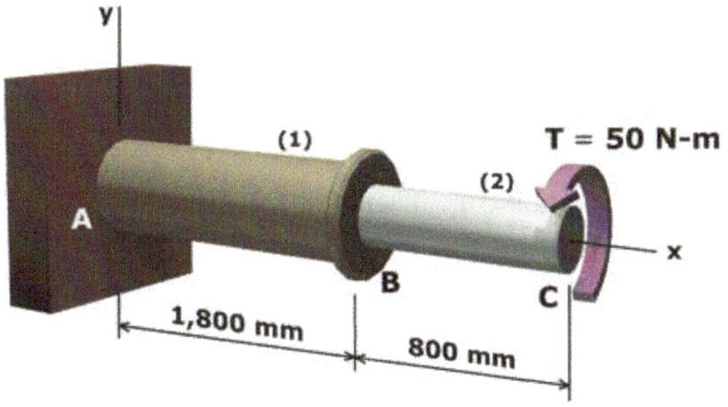

M6.7 A solid circular driveshaft connects a motor A to gears B and C. The torque on gear B is 600 N · m, and the torque on gear C is 200 N · m, acting in the directions shown. The driveshaft is steel [$G = 66$ MPa] with a diameter of 25 mm.

(a) Determine the maximum shear stress in shafts (1) and (2).
(b) Determine the rotation angle of C with respect to A.

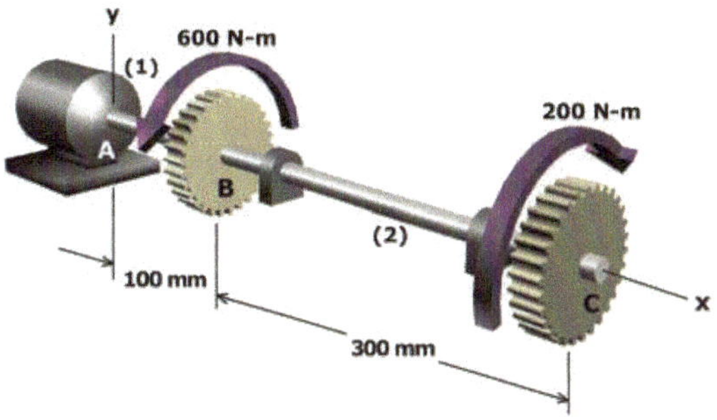

M6.8 The solid steel [$G = 80$ GPa] shaft between coupling A and gear B has a diameter of 35 mm. Between gears B and C, the diameter of the solid shaft is reduced to 25 mm. At gear B, a 20 N · m concentrated torque is applied to the shaft in the direction indicated. A concentrated torque T_C will be applied at gear C. If the total angle of rotation at C is not to exceed 1°, determine the magnitude of torque T_C that can be applied in the direction shown.

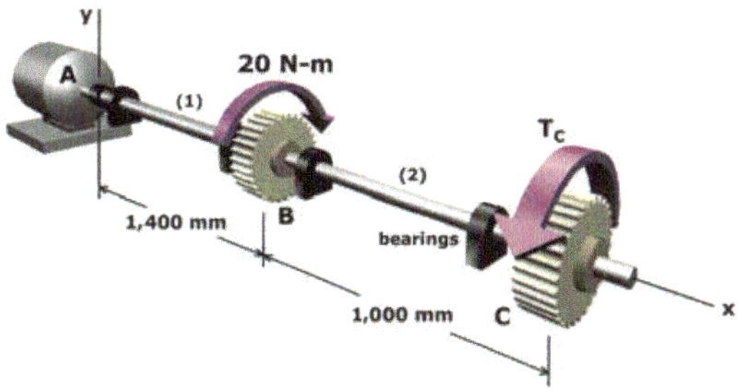

EXERCISE

M6.1 Ten basic torsion problems involving internal torques, shear stress, and angles of twist for a multisegment shaft.

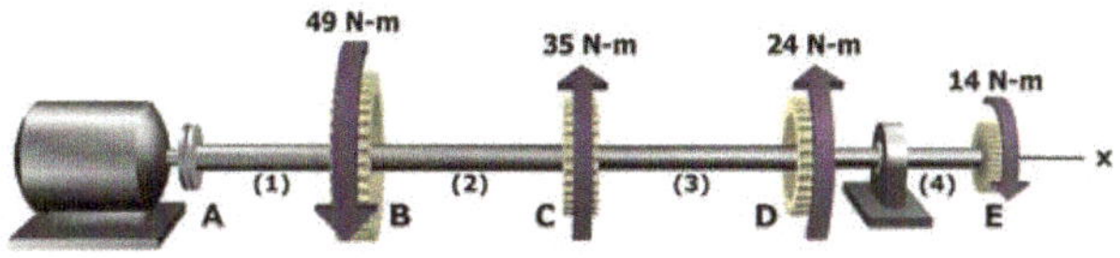

FIGURE M6.1

PROBLEMS

P6.1 A solid circular steel shaft having an outside diameter $d = 60$ mm is subjected to a pure torque $T = 2{,}400$ N·m. What is the maximum shear stress in the shaft?

P6.2 A hollow stainless steel shaft with an outside diameter of 42 mm and a wall thickness of 4 mm has an allowable shear stress of 100 MPa. Determine the maximum torque T that may be applied to the shaft.

P6.3 What is the minimum diameter required for a solid steel shaft to transmit a torque of 20,000 lb·ft if the maximum shear stress in the shaft must not exceed 8,000 psi?

P6.4 The compound shaft shown in Figure P6.4 consists of two steel [$G = 80$ GPa] pipe segments. The shaft is subjected to external torques T_B and T_C that act in the directions shown. Segment (1) has an outside diameter of 206 mm and a wall thickness of 10 mm. Segment (2) has an outside diameter of 138 mm and a wall thickness of 14 mm. The torque is known to have a magnitude of $T_C = 22$ kN·m. The shear strain in segment (1) is measured as 500 μrad. Determine the magnitude of the external torque T_B.

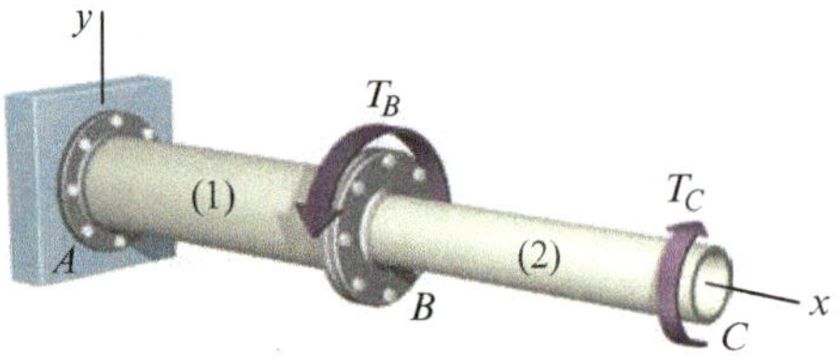

FIGURE P6.4

P6.5 A solid constant-diameter circular shaft is subjected to torques $T_A = 420$ lb·ft, $T_B = 1{,}040$ lb·ft, $T_C = 850$ lb·ft, and $T_D = 230$ lb·ft, acting in the directions shown in Figure 6.5. The bearings shown allow the shaft to turn freely.

(a) Plot a torque diagram showing the internal torque in segments (1), (2), and (3) of the shaft. Use the sign convention presented in Section 6.6.
(b) If the allowable shear stress in the shaft is 6,000 psi, what is the minimum acceptable diameter for the shaft?

FIGURE P6.5

P6.6 The nut on a bolt is tightened by applying a force $P = 16$ lb to the end of a wrench at a distance $a = 6$ in. from the axis of the bolt, as shown in Figure P6.6. The body of the bolt has an outside diameter $d = 0.375$ in. What is the maximum torsional shear stress in the body of the bolt?

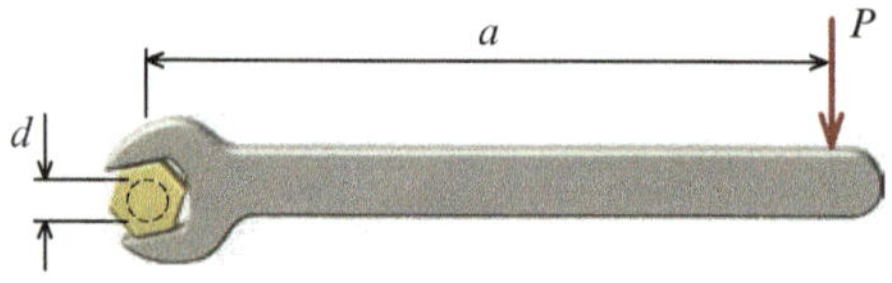

FIGURE P6.6

P6.7 A solid circular bronze [$G = 6{,}500$ ksi] shaft 5 ft long and 1.375 in. in diameter carries a torque of 80 lb·ft. For this shaft, determine

(a) the maximum shear stress.
(b) the magnitude of the angle of twist.

P6.8 A solid titanium alloy [$G = 114$ GPa] shaft that is 900 mm long will be subjected to a pure torque $T = 160$ N·m. Determine the minimum diameter required if the shear stress must not exceed 165 MPa and the angle of twist must not exceed 5°. Report both the maximum shear stress τ and the angle of twist ϕ at this minimum diameter.

P6.9 A compound shaft (Figure P6.9) consists of two segments. Segment (1) is a solid bronze [$G = 45$ GPa] shaft with an outside diameter of 80 mm, a length $L_1 = 780$ mm, and an allowable shear stress of 65 MPa. Segment (2) is a solid aluminum alloy [$G = 26$ GPa ksi] shaft with an outside diameter of 95 mm, a length $L_2 = 620$ mm, and an allowable shear stress of 40 MPa. The maximum rotation angle at the free end of the compound shaft must be limited to $\phi_A \leq 2.5°$. Determine the magnitude of the largest torque T_A that may be applied at A.

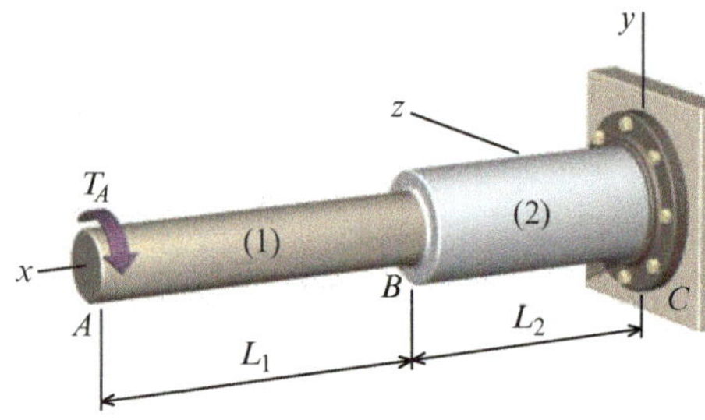

FIGURE P6.9

P6.10 The mechanism shown in Figure P6.10 is in equilibrium for an applied load P = 20 kN. Specifications for the mechanism limit the shear stress in the steel [G = 80 GPa] shaft BC to 70 MPa, the shear stress in bolt A to 100 MPa, and the vertical deflection of joint D to a maximum value of 25 mm. Assume that the bearings allow the shaft to rotate freely. Using L = 1,200 mm, a = 110 mm, and b = 210 mm, calculate

(a) the minimum diameter required for shaft BC.
(b) the minimum diameter required for bolt A.

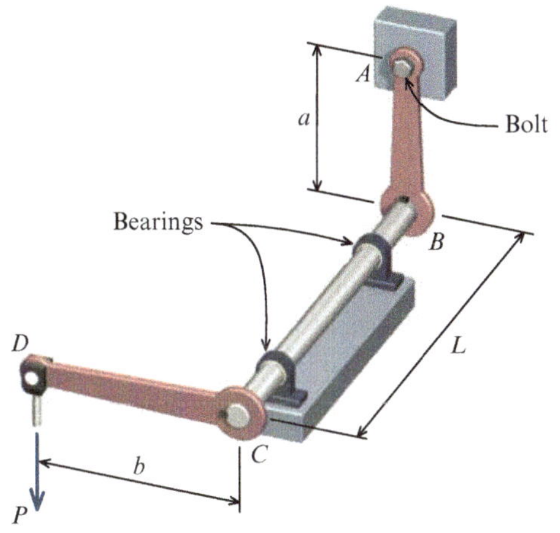

FIGURE P6.10

P6.11 A simple torsion-bar spring is shown in Figure P6.11. The shear stress in the steel [G = 11,500 ksi] shaft is not to exceed 10,000 psi, and the vertical deflection of joint D is not to exceed 0.5 in. when a load P = 3,400 lb is applied. Neglect the bending of the shaft, and assume that the bearing at C allows the shaft to rotate freely. Determine the minimum diameter required for the shaft. Use dimensions of a = 72 in., b = 30 in., and c = 18 in.

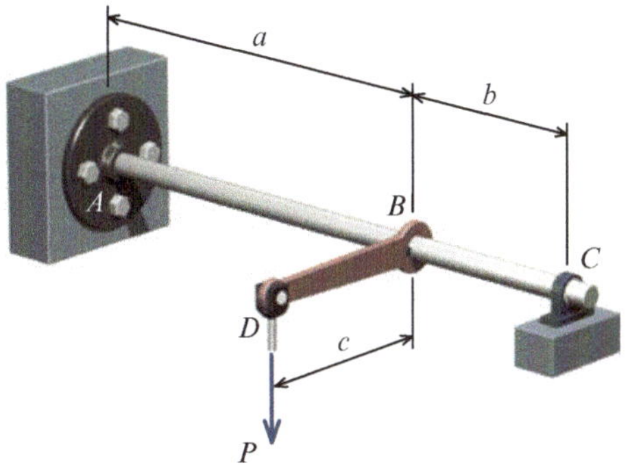

FIGURE P6.11

P6.12 A rod specimen of ductile cast iron was tested in a torsion-testing machine. The rod diameter was 22 mm, and the rod length was 300 mm. When the applied torque reached 271.4 N·m, a shear strain of 2,015 microradians was measured in the specimen. What was the angle of twist in the specimen?

P6.13 A two-segment shaft is used to transmit power at constant speed through the pulleys shown in Figure P6.13. Power is input to the shaft at B through a torque T_B = 260 N·m. A torque T_A = 90 N·m is removed from the shaft at A, and a torque T_C = 170 N·m is removed from the shaft at C. The external torques act in the directions indicated in the figure. Both shaft segments are made of phosphor bronze [G = 42 GPa], and both segments are solid shafts. Segment (1) has a diameter d_1 = 25 mm and a length L_1 = 300 mm. Segment (2) has a diameter d_2 = 30 mm and a length L_2 = 900 mm. Determine the rotation angle of pulley C relative to pulley A.

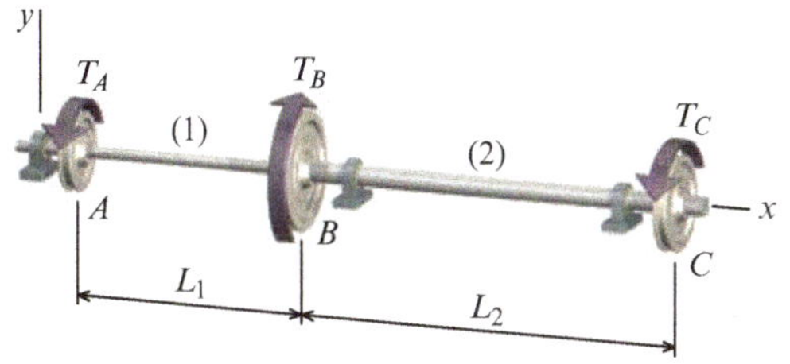

FIGURE P6.13

P6.14 A compound shaft drives three gears, as shown in Figure P6.14. Segments (1) and (2) of the shaft are hollow bronze [G = 6,500 ksi] tubes with an outside diameter of 1.75 in. and a wall thickness of 0.1875 in. Segments (3) and (4) are solid 1.00 in. diameter steel [G = 11,500 ksi] shafts. The shaft lengths are L_1 = 60 in., L_2 = 14 in., L_3 = 20 in., and L_4 = 26 in. The torques applied to the shafts have magnitudes T_B = 960 lb·ft, T_D = 450 lb·ft, and T_E = 130 lb·ft, acting in the directions shown. The bearings shown allow the shaft to turn freely. Calculate

(a) the maximum shear stress in the compound shaft.
(b) the rotation angle of flange C with respect to flange A.
(c) the rotation angle of gear E with respect to flange A.

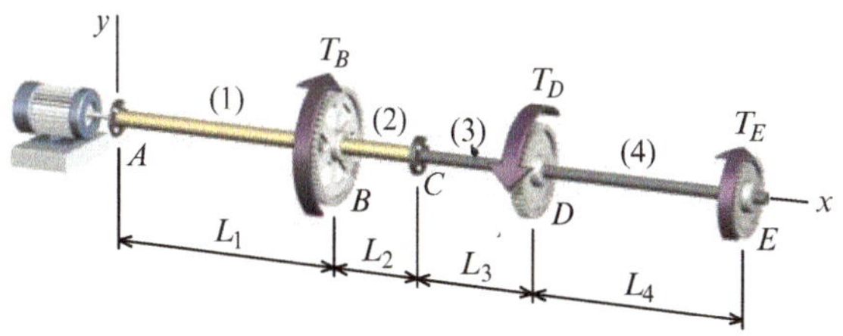

FIGURE P6.14

P6.15 Figure P6.15 shows a cutaway view of an assembly in which a solid bronze [G = 6,500 ksi] rod (1) is fitted inside of an aluminum alloy [G = 4,000 ksi] tube (2). The tube is attached to a fixed plate at C, and both the rod and the tube are welded to a rigid end plate at A. The rod diameter is d_1 = 1.75 in. The outside diameter of the tube is D_2 = 3.00 in., and its wall thickness is t_2 = 0.125 in. Overall dimensions of the assembly are a = 40 in. and b = 8 in. The allowable shear stresses for the bronze and aluminum materials are 10,000 psi and 8,000 psi, respectively. Further, the rotation angle of end B must be limited to a magnitude of 5°. On the basis of these constraints, what is the maximum torque T_B that can be applied at end B?

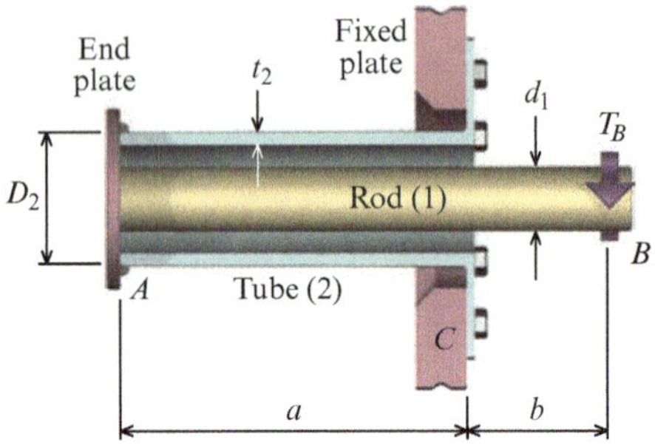

FIGURE P6.15

P6.16 A compound shaft (Figure P6.16) consists of a titanium alloy [G = 6,200 ksi] tube (1) and a solid stainless steel [G = 11,500 ksi] shaft (2). Tube (1) has a length L_1 = 40 in., an outside diameter D_1 = 1.75 in., and a wall thickness t_1 = 0.125 in. Shaft (2) has a length L_2 = 50 in. and a diameter d_2 = 1.25 in. If an external torque T_B = 580 lb·ft acts at pulley B in the direction shown, calculate the torque T_C required at pulley C so that the rotation angle of pulley C relative to A is zero.

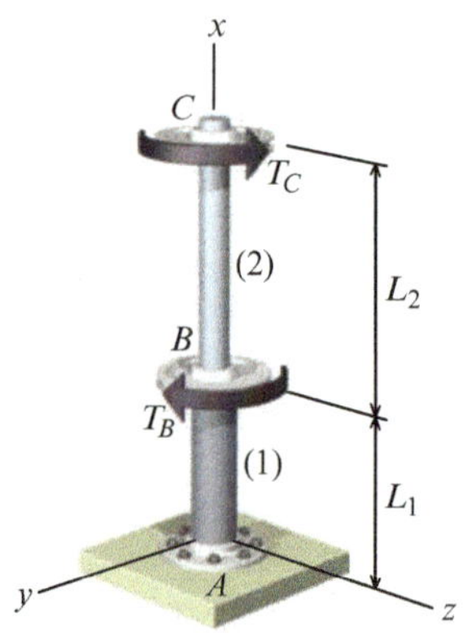

FIGURE P6.16

P6.17 The copper pipe shown in Figure P6.17 has an outside diameter of 70 mm and a wall thickness of 5 mm. The pipe is subjected to a uniformly distributed torque t = 500 N·m/m along its entire length. Using a = 1.0 m, b = 1.6 m, and c = 3.2 m, calculate

(a) the shear stress at A on the outer surface of the pipe.
(b) the shear stress at B on the outer surface of the pipe.

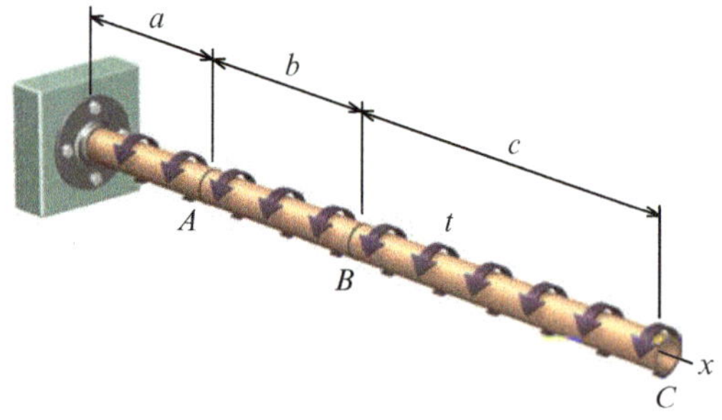

FIGURE P6.17

6.7 Gears in Torsion Assemblies

Gears are a fundamental component found in many types of mechanisms and devices—particularly those devices which are driven by motors or engines. Gears are used for many purposes, such as

- transmitting torque from one shaft to another,
- increasing or decreasing torque in a shaft,
- increasing or decreasing the rate of rotation of a shaft,
- changing the rotation direction of two shafts, and
- changing rotational motion from one orientation to another—for instance, changing rotation about a horizontal axis to rotation about a vertical axis.

Furthermore, since gears have *teeth*, shafts connected by gears are always synchronized exactly with one another.

A basic gear assembly is shown in Figure 6.14. In this assembly, torque is transmitted from shaft (1) to shaft (2) by means of gears A and B, which have radii of R_A and R_B, respectively. The number of teeth on each gear is denoted by N_A and N_B. Positive internal torques T_1 and T_2 are assumed in shafts (1) and (2). For clarity, bearings necessary to support the two shafts have been omitted. This configuration will be used to illustrate basic relationships involving torque, rotation angle, and rotation speed in torsion assemblies with gears.

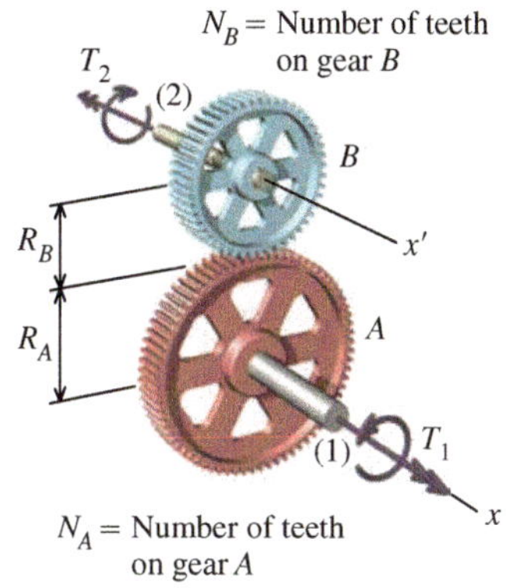

FIGURE 6.14 Basic gear assembly.

Torque

To illustrate the relationship between the internal torques in shafts (1) and (2), free-body diagrams of each gear are shown in Figure 6.15. If the system is to be in equilibrium, then each gear must be in equilibrium. Consider the free-body diagram of gear A. The internal torque T_1 acting in shaft (1) is transmitted directly to gear A. This torque causes gear A to rotate counterclockwise. As gears A and B rotate, the teeth of gear B exert a force on gear A that acts tangential to both gears. This force, which opposes the rotation of gear A, is denoted by F. A moment equilibrium equation about the x axis gives the relationship between T_1 and F for gear A:

$$\Sigma M_x = T_1 - F \times R_A = 0 \qquad \therefore F = \frac{T_1}{R_A} \tag{a}$$

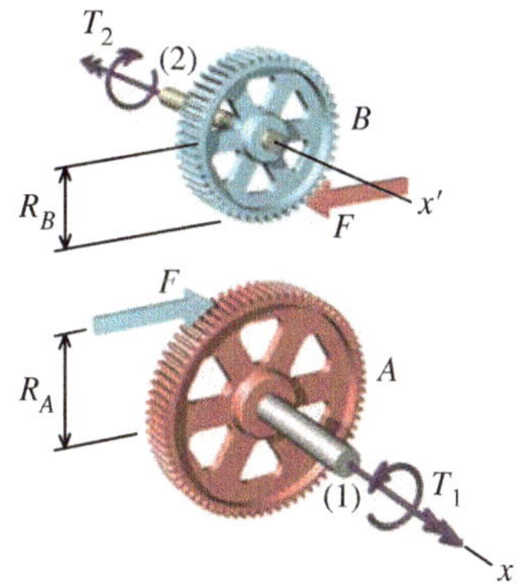

FIGURE 6.15 Free-body diagrams of gears A and B.

Next, consider the free-body diagram of gear B. If the teeth of gear B exert a force F on gear A, then the teeth of gear A must exert a force of equal magnitude on gear B, but that acts in the opposite direction. This force causes gear B to rotate clockwise. A moment equilibrium equation about the x' axis then gives

$$\Sigma M_{x'} = -F \times R_B - T_2 = 0 \tag{b}$$

If the expression for F determined in Equation (a) is substituted into Equation (b), then the torque T_2 required to satisfy equilibrium can be expressed in terms of the torque T_1:

$$-\frac{T_1}{R_A} R_B - T_2 = 0 \qquad \therefore T_2 = -T_1 \frac{R_B}{R_A} \tag{c}$$

The magnitude of T_2 is related to T_1 by the ratio of the gear radii. Since the two gears rotate in opposite directions, however, the sign of T_2 is opposite from the sign of T_1.

MecMovies 6.9 presents an animation that illustrates basic gear relationships for torque, rotation angle, rotation speed, and power transmission.

Gear Ratio. The ratio R_B/R_A in Equation (c) is called the *gear ratio*, and this ratio is the key parameter that dictates relationships between shafts connected by gears. The gear ratio in Equation (c) is expressed in terms of the gear radii; however, it can also be expressed in terms of gear diameters or gear teeth.

The diameter D of a gear is simply two times its radius R. Accordingly, the gear ratio in Equation (c) could also be expressed as D_B/D_A, where D_A and D_B are the diameters of gears A and B, respectively.

For two gears to interlock properly, the teeth on both gears must be the same size. In other words, the arclength of a single tooth—a quantity termed the *pitch p*—must be the same for both gears. The circumference C of gears A and B can be expressed either in terms of the gear radius,

$$C_A = 2\pi R_A \qquad C_B = 2\pi R_B$$

or in terms of the pitch p and the number of teeth N on the gear:

$$C_A = pN_A \qquad C_B = pN_B$$

The expressions for the circumference of each gear can be equated and solved for the pitch p on each gear:

$$p = \frac{2\pi R_A}{N_A} \qquad p = \frac{2\pi R_B}{N_B}$$

Moreover, since the tooth pitch p must be the same for both gears, it follows that

$$\frac{R_B}{R_A} = \frac{N_B}{N_A}$$

In sum, the gear ratio between any two gears A and B can be expressed equivalently by either gear radii, gear diameters, or numbers of gear teeth:

$$\text{Gear ratio} = \frac{R_B}{R_A} = \frac{D_B}{D_A} = \frac{N_B}{N_A} \tag{d}$$

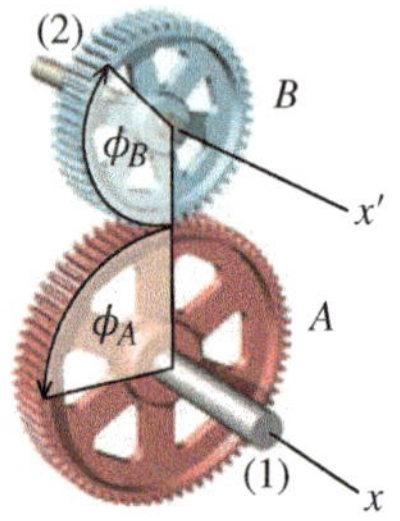

FIGURE 6.16 Rotation angles for gears A and B.

Rotation Angle. When gear A turns through an angle ϕ_A as shown in Figure 6.16, the arclength along the perimeter of gear A is $s_A = R_A\phi_A$. Similarly, the arclength along the perimeter of gear B is $s_B = R_B\phi_B$. Since the teeth on each gear must be the same size, the arclengths that are turned by the two gears must be equal in magnitude. The two gears, however, turn in opposite directions. If s_A and s_B are equated, and rotation in the opposite direction is accounted for, then the rotation angle ϕ_A is given by

$$R_A\phi_A = -R_B\phi_B \qquad \therefore \phi_A = -\frac{R_B}{R_A}\phi_B \tag{e}$$

Note: The term R_B/R_A in Equation (e) is simply the gear ratio; therefore,

$$\phi_A = -(\text{Gear ratio})\phi_B \tag{f}$$

Rotation Speed. The rotation speed ω is the rotation angle ϕ turned by the gear in a unit of time; therefore, the rotation speeds of two interlocked gears are related in the same manner as described for rotation angles—that is,

$$\omega_A = -(\text{Gear ratio})\omega_B \tag{g}$$

EXAMPLE 6.5

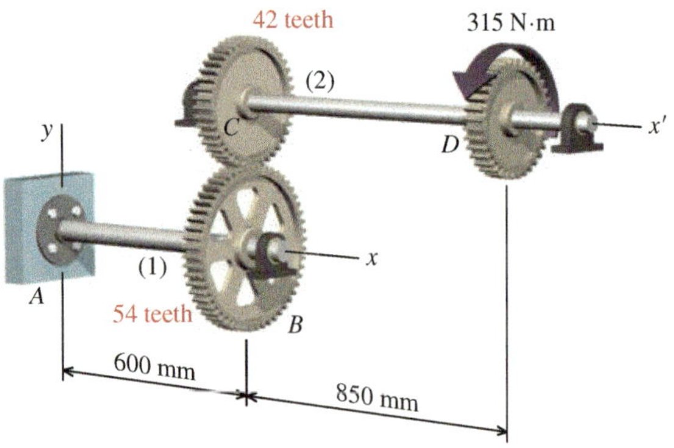

Two solid steel [$G = 80$ GPa] shafts are connected by the gears shown. Shaft (1) has a diameter of 35 mm, and shaft (2) has a diameter of 30 mm. Assume that the bearings shown allow free rotation of the shafts. If a 315 N·m torque is applied at gear D, determine

(a) the maximum shear stress magnitudes in each shaft.
(b) the angles of twist ϕ_1 and ϕ_2.
(c) the rotation angles ϕ_B and ϕ_C of gears B and C, respectively.
(d) the rotation angle of gear D.

Plan the Solution

The internal torque in shaft (2) can easily be determined from a free-body diagram of gear D; however, the internal torque in shaft (1) will be dictated by the ratio of gear sizes. Once you have determined the internal torques in both shafts, calculate the angles of twist in each shaft, paying particular attention to the signs of the twist angles. The twist angle in shaft (1) will dictate how much gear B rotates, which in turn will dictate the rotation angle of gear C. The rotation angle of gear D will depend upon the rotation angle of gear C and the angle of twist in shaft (2).

SOLUTION

Equilibrium

Consider a free-body diagram that cuts through shaft (2) and includes gear D. A positive internal torque will be assumed in shaft (2). From this free-body diagram, a moment equilibrium equation about the x' axis can be written to determine the internal torque T_2 in shaft (2):

$$\Sigma M_{x'} = 315 \text{ N·m} - T_2 = 0 \qquad \therefore T_2 = 315 \text{ N·m} \tag{a}$$

Next, consider a free-body diagram that cuts through shaft (2) and includes gear C. Once again, a positive internal torque will be assumed in shaft (2). The teeth of gear B exert a force F on the teeth of gear C. If the radius of gear C is denoted by R_C, a moment equilibrium equation about the x' axis can be written as

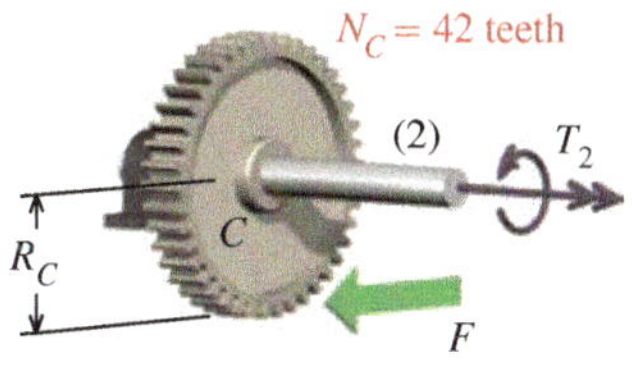

$$\Sigma M_{x'} = T_2 - F \times R_C = 0 \qquad \therefore F = \frac{T_2}{R_C} \tag{b}$$

A free-body diagram of gear B that cuts through shaft (1) is shown. A positive internal torque T_1 is assumed to act in shaft (1). If the teeth of gear B exert a force F on the teeth of gear C, then equilibrium requires that the teeth of gear C exert a force of equal magnitude in the opposite direction on the teeth of gear B. With the radius of gear B denoted by R_B, a moment equilibrium equation about the x axis can be written as

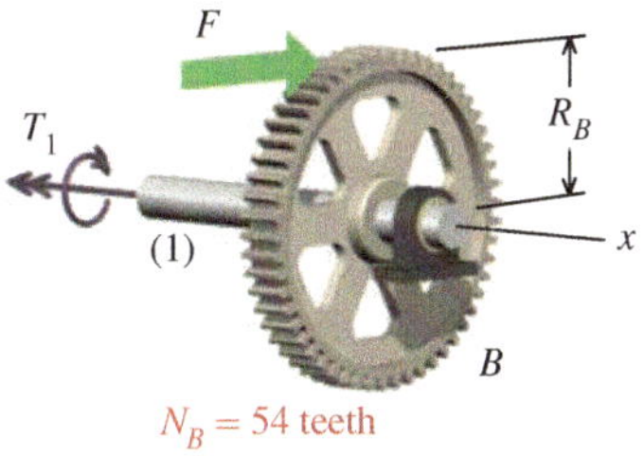

$$\Sigma M_x = -T_1 - F \times R_B = 0 \qquad \therefore T_1 = -F \times R_B \tag{c}$$

The internal torque in shaft (2) is given by Equation (a). The internal torque in shaft (1) can be determined by substituting Equation (b) into Equation (c):

$$T_1 = -F \times R_B = -\frac{T_2}{R_C} R_B = -T_2 \frac{R_B}{R_C}$$

The gear radii R_B and R_C are not known. However, the ratio R_B/R_C is simply the gear ratio between gears B and C. Since the teeth on both gears must be the same size in order for the gears to mesh properly, the ratio of teeth on each gear is equivalent to the ratio of the gear radii. Consequently, the torque in shaft (1) can be expressed in terms of N_B and N_C, the number of teeth on gears B and C, respectively:

$$T_1 = -T_2 \frac{R_B}{R_C} = -T_2 \frac{N_B}{N_C} = -(315 \text{ N·m}) \frac{54 \text{ teeth}}{42 \text{ teeth}} = -405 \text{ N·m}$$

Shear Stresses

The maximum shear stress magnitude in each shaft will be calculated from the elastic torsion formula. The polar moments of inertia for each shaft will be required for this calculation. Shaft (1) is a solid 35 mm diameter shaft that has a polar moment of inertia

$$J_1 = \frac{\pi}{32}(35 \text{ mm})^4 = 147{,}324 \text{ mm}^4$$

Shaft (2) is a solid 30 mm diameter shaft that has a polar moment of inertia

$$J_2 = \frac{\pi}{32}(30 \text{ mm})^4 = 79{,}552 \text{ mm}^4$$

To calculate the maximum shear stress magnitudes, the absolute values of T_1 and T_2 will be used. The maximum shear stress magnitude in the 35 mm diameter shaft (1) is

$$\tau_1 = \frac{T_1 c_1}{J_1} = \frac{(405\ \text{N}\cdot\text{m})(35\ \text{mm}/2)(1{,}000\ \text{mm/m})}{147{,}324\ \text{mm}^4} = 48.1\ \text{MPa} \qquad \textbf{Ans.}$$

and the maximum shear stress magnitude in the 30 mm diameter shaft (2) is

$$\tau_2 = \frac{T_2 c_2}{J_2} = \frac{(315\ \text{N}\cdot\text{m})(30\ \text{mm}/2)(1{,}000\ \text{mm/m})}{79{,}552\ \text{mm}^4} = 59.4\ \text{MPa} \qquad \textbf{Ans.}$$

Angles of Twist

The angles of twist must be calculated with the signed values of T_1 and T_2. Shaft (1) is 600 mm long, and its shear modulus is $G = 80$ GPa = 80,000 MPa. The angle of twist in this shaft is

$$\phi_1 = \frac{T_1 L_1}{J_1 G_1} = \frac{(-405\ \text{N}\cdot\text{m})(600\ \text{mm})(1{,}000\ \text{mm/m})}{(147{,}324\ \text{mm}^4)(80{,}000\ \text{N/mm}^2)} = -0.020618\ \text{rad} = -0.0206\ \text{rad}$$

Ans.

Shaft (2) is 850 mm long; therefore, its angle of twist is

$$\phi_2 = \frac{T_2 L_2}{J_2 G_2} = \frac{(315\ \text{N}\cdot\text{m})(850\ \text{mm})(1{,}000\ \text{mm/m})}{(79{,}522\ \text{mm}^4)(80{,}000\ \text{N/mm}^2)} = 0.042087\ \text{rad} = 0.0421\ \text{rad} \qquad \textbf{Ans.}$$

Rotation Angles of Gears *B* and *C*

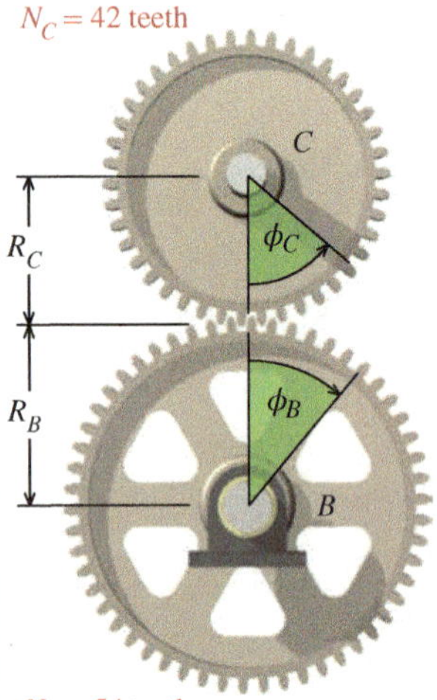

The rotation of gear *B* is equal to the angle of twist in shaft (1):

$$\phi_B = \phi_1 = -0.020618\ \text{rad} = -0.0206\ \text{rad} \qquad \textbf{Ans.}$$

Note: From the sign convention for rotation angles described in Section 6.6 and illustrated in Figure 6.13, a negative rotation angle for gear *B* indicates that gear *B* rotates clockwise, as shown in the accompanying figure.

The rotation angles of gears *B* and *C* are related because the arclengths associated with the respective rotations must be equal. Why? Because the gear teeth are interlocked. The gears turn in opposite directions, however. In this instance, gear *B* turns clockwise, which causes gear *C* to rotate in a counterclockwise direction. This change in the direction of rotation is accounted for in the calculations by a negative sign, so that

$$R_C \phi_C = -R_B \phi_B$$

where R_B and R_C are the radii of gears *B* and *C*, respectively. Using this relationship, we can express the rotation angle of gear *C* as

$$\phi_C = -\frac{R_B}{R_C}\phi_B$$

However, the ratio R_B/R_C is simply the gear ratio between gears *B* and *C*, and this ratio can be equivalently expressed in terms of N_B and N_C, the number of teeth on gears *B* and *C*, respectively. Thus,

$$\phi_C = -\frac{N_B}{N_C}\phi_B$$

Therefore, the rotation angle of gear *C* is

$$\phi_C = -\frac{N_B}{N_C}\phi_B = -\frac{54\ \text{teeth}}{42\ \text{teeth}}(-0.020618\ \text{rad}) = 0.026509\ \text{rad} = 0.0265\ \text{rad} \qquad \textbf{Ans.}$$

Rotation Angle of Gear *D*

The rotation angle of gear *D* is equal to the rotation angle of gear *C* plus the twist that occurs in shaft (2):

$$\phi_D = \phi_C + \phi_2 = 0.026509\ \text{rad} + 0.042087\ \text{rad} = 0.068596\ \text{rad} = 0.0686\ \text{rad} \qquad \textbf{Ans.}$$

MecMovies

EXAMPLE

M6.13 Two solid steel [$G = 80$ GPa] shafts are connected by the gears shown. The diameter of each shaft is 35 mm. A torque $T = 685$ N · m is applied to the system at D. Determine

(a) the maximum shear stress in each shaft.
(b) the angle of rotation at D.

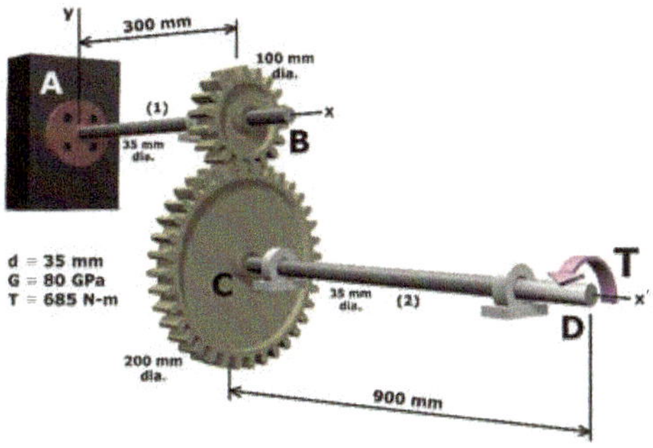

EXERCISES

M6.9 Six multiple-choice questions concerning torque, rotation angle, and rotation speed of gears.

FIGURE M6.9

M6.10 Six basic calculations involving two shafts connected by gears.

FIGURE M6.10

M6.11 Six basic calculations involving three shafts connected by gears.

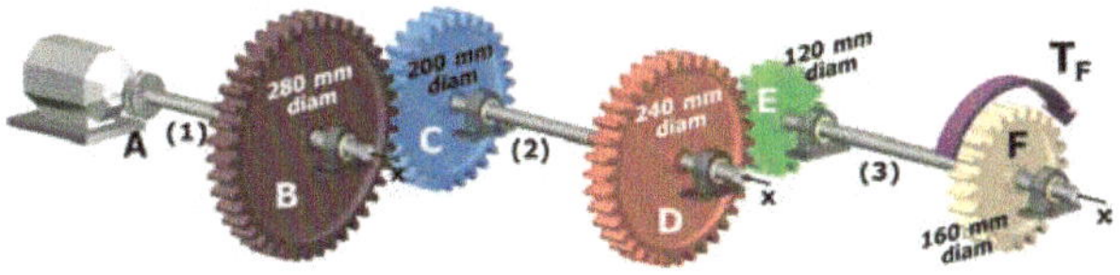

FIGURE M6.11

M6.12 Five basic twist and rotation angle calculations involving two shafts connected by gears.

FIGURE M6.12

PROBLEMS

P6.18 The gear train system shown in Figure P6.18/19 includes shafts (1) and (2), which are solid 15 mm diameter steel shafts. The allowable shear stress of each shaft is 85 MPa. The diameter of gear B is $D_B = 200$ mm, and the diameter of gear C is $D_C = 150$ mm. The bearings shown allow the shafts to rotate freely. Determine the maximum torque T_D that can be applied to the system without exceeding the allowable shear stress in either shaft.

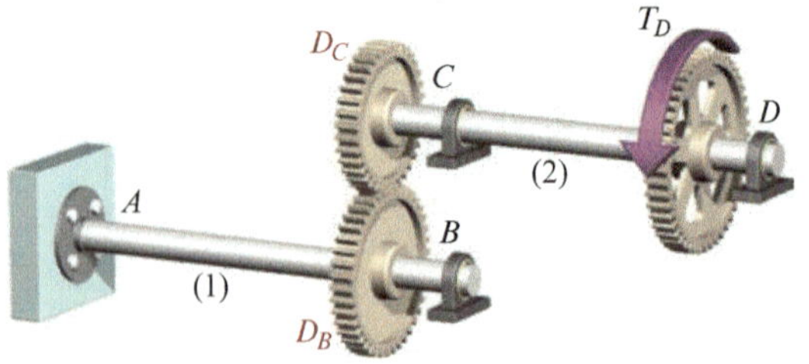

FIGURE P6.18/19

P6.19 The gear train system shown in Figure P6.18/19 includes shafts (1) and (2), which are solid 0.75 in. diameter steel shafts. The diameter of gear B is $D_B = 8$ in., and the diameter of gear C is $D_C =$ 5 in. The bearings shown allow the shafts to rotate freely. An external torque $T_D = 45$ lb · ft is applied at gear D. Determine the maximum shear stress produced in shafts (1) and (2).

P6.20 In the system shown in Figure P6.20, the motor applies a torque $T_A = 40$ N · m to the pulley at A. Through a sequence of pulleys and belts, this torque is amplified to drive a gear at E. The pulleys have diameters $D_A = 50$ mm, $D_B = 150$ mm, $D_C = 50$ mm, and $D_D = 250$ mm.

(a) Calculate the torque T_E that is produced at gear E.
(b) Shaft (2) is to be a solid shaft, and the maximum shear stress must be limited to 60 MPa. What is the minimum diameter that may be used for shaft (2)?

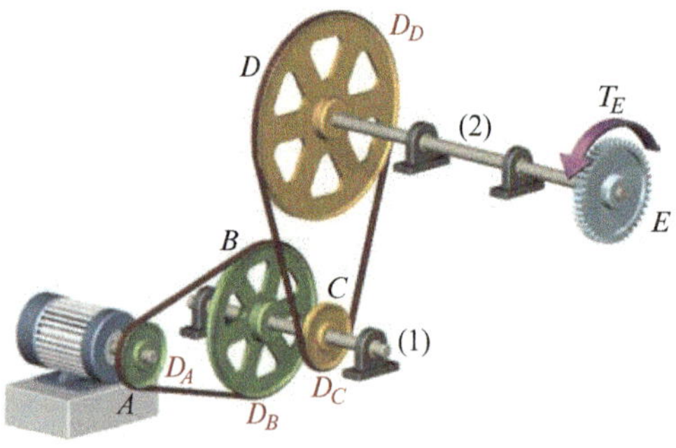

FIGURE P6.20

P6.21 A motor provides a torque of 1,500 lb · ft to gear B of the system shown in Figure P6.21. Gear A takes off 900 lb · ft from shaft (1), and gear C takes off the remaining torque. Both shafts (1) and (2) are solid and made of steel [$G = 11{,}500$ ksi]. The shaft lengths are $L_1 = 15$ ft and $L_2 = 9$ ft, respectively. If the angle of twist in each shaft must not exceed 3.0°, calculate the minimum diameter required for each shaft.

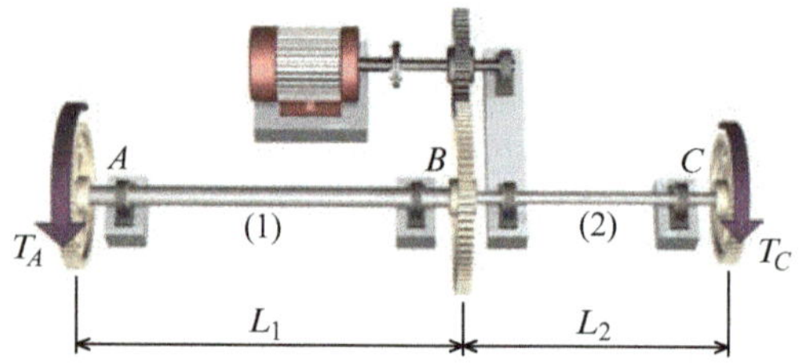

FIGURE P6.21

P6.22 Two solid steel shafts are connected by the gears shown in Figure P6.22/23. The design requirements for the system specify (1) that both shafts must have the same diameter, (2) that the maximum shear stress in each shaft must be less than 10,000 psi, and (3) that the rotation angle of gear D must not exceed 8°. Determine the minimum required diameter of the shafts if the torque applied at gear D is $T_D = 350$ lb · ft. The shaft lengths are $L_1 = 78$ in. and $L_2 = 60$ in. The number of teeth on gears B and C are $N_B = 90$ and $N_C = 52$, respectively. Assume that the shear modulus of both shafts is $G = 11{,}500$ ksi and that the bearings shown allow free rotation of the shafts.

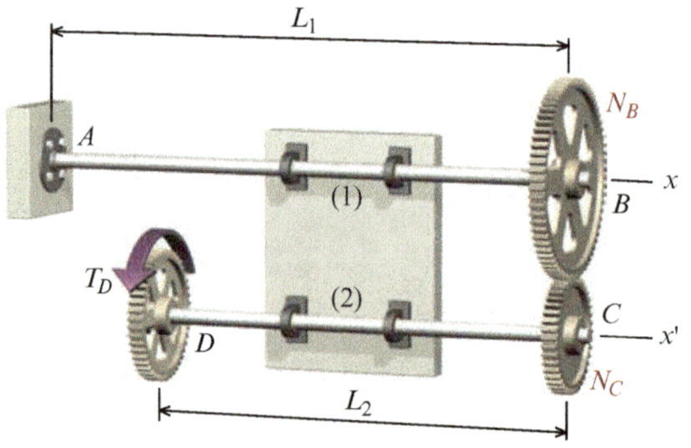

FIGURE P6.22/23

P6.23 Two solid 120 mm diameter steel shafts are connected by the gears shown in Figure P6.22/23. The shaft lengths are $L_1 = 4$ m and $L_2 = 3$ m. The number of teeth on gears B and C are $N_B = 200$ and $N_C = 115$, respectively. Assume that the shear modulus of both shafts is $G = 80$ MPa and that the bearings shown allow free rotation of the shafts. If the torque applied at gear D is $T_D = 6{,}000$ N · m, determine

(a) the internal torques T_1 and T_2 in the two shafts.
(b) the angles of twist ϕ_1 and ϕ_2.
(c) the rotation angles ϕ_B and ϕ_C of gears B and C.
(d) the rotation angle of gear D.

6.8 Power Transmission

One of the most common uses for a circular shaft is the transmission of power from motors or engines to other devices and components. **Power** is defined as the *work* performed in a unit of time. The work W done by a torque T of constant magnitude is equal to the product of the torque T and the angle ϕ through which the torque rotates:

$$W = T\phi \tag{6.15}$$

Power is the *rate* at which the work is done. Therefore, Equation (6.15) can be differentiated with respect to time t to give an expression for the power P transmitted by a shaft subjected to a constant torque T:

$$P = \frac{dW}{dt} = T\frac{d\phi}{dt} \tag{6.16}$$

The rate of change of the angular displacement $d\phi/dt$ is the rotational speed or angular velocity ω. Therefore, the power P transmitted by a shaft is a function of the magnitude of the torque T in the shaft and its rotational speed ω; that is

$$P = T\omega \tag{6.17}$$

where ω is measured in radians per second.

Power Units

In SI, an appropriate unit for torque is the newton-meter (N·m). The corresponding SI unit for power is termed a *watt*:

$$P = T\omega = (\text{N}\cdot\text{m})(\text{rad/s}) = \frac{\text{N}\cdot\text{m}}{\text{s}} = 1 \text{ watt} = 1\text{ W}$$

In U.S. customary units, torque is often measured in lb·ft; thus, the corresponding power unit is

$$P = T\omega = (\text{lb}\cdot\text{ft})(\text{rad/s}) = \frac{\text{lb}\cdot\text{ft}}{\text{s}}$$

In U.S. practice, power is typically expressed in terms of *horsepower* (hp), which has the following conversion factor:

$$1\text{ hp} = 550\frac{\text{lb}\cdot\text{ft}}{\text{s}} \tag{6.18}$$

Rotational Speed Units

The rotational speed ω of a shaft is commonly expressed either as frequency f or as revolutions per minute (rpm). Frequency f is the number of revolutions per unit of time. The standard unit of frequency is the hertz (Hz), which is equal to one revolution per second (s^{-1}). Since a shaft turns through an angle of 2π radians in one revolution (rev), rotational speed ω can be expressed in terms of frequency f measured in Hz:

$$\omega = \left(\frac{f\text{ rev}}{\text{s}}\right)\left(\frac{2\pi\text{ rad}}{\text{rev}}\right) = 2\pi f\text{ rad/s}$$

Accordingly, Equation (6.17) can be written in terms of frequency f (measured in Hz) as

$$P = T\omega = 2\pi f T \tag{6.19}$$

Another common measure of rotational speed is revolutions per minute (rpm). Rotational speed ω can be expressed in terms of revolutions per minute, n, as

$$\omega = \left(\frac{n \text{ rev}}{\text{min}}\right)\left(\frac{2\pi \text{ rad}}{\text{rev}}\right)\left(\frac{1 \text{ min}}{60 \text{ s}}\right) = \frac{2\pi n}{60} \text{ rad/s}$$

Equation (6.17) can be written in terms of rpm n as

$$P = T\omega = \frac{2\pi n T}{60} \tag{6.20}$$

EXAMPLE 6.6

A solid 0.75 in. diameter steel shaft transmits 7 hp at 3,200 rpm. Determine the maximum shear stress magnitude produced in the shaft.

Plan the Solution

The power transmission equation [Equation (6.17)] will be used to calculate the torque in the shaft. The maximum shear stress in the shaft can then be calculated from the elastic torsion formula [Equation (6.5)].

SOLUTION

Power P is related to torque T and rotation speed ω by the relationship $P = T\omega$. Since information about the power and rotation speed is given, this relationship can be solved for the unknown torque T. The conversion factors required in the process, however, can be confusing at first. We have

$$T = \frac{P}{\omega} = \frac{(7 \text{ hp})\left(\dfrac{550 \text{ lb}\cdot\text{ft/s}}{1 \text{ hp}}\right)}{\left(\dfrac{3{,}200 \text{ rev}}{\text{min}}\right)\left(\dfrac{2\pi \text{ rad}}{1 \text{ rev}}\right)\left(\dfrac{1 \text{ min}}{60 \text{ s}}\right)} = \frac{3{,}850 \ \dfrac{\text{lb}\cdot\text{ft}}{\text{s}}}{335.1032 \ \dfrac{\text{rad}}{\text{s}}} = 11.4890 \text{ lb}\cdot\text{ft}$$

The polar moment of inertia for a solid 0.75 in. diameter shaft is

$$J = \frac{\pi}{32}(0.75 \text{ in.})^4 = 0.0310631 \text{ in.}^4$$

Therefore, the maximum shear stress produced in the shaft is

$$\tau = \frac{Tc}{J} = \frac{(11.4890 \text{ lb}\cdot\text{ft})(0.75 \text{ in.}/2)(12 \text{ in./ft})}{0.0310631 \text{ in.}^4} = 1{,}664 \text{ psi}$$ **Ans.**

MecMovies

EXAMPLES

M6.16 A 2 m long hollow steel [G = 75 GPa] shaft has an outside diameter of 75 mm and an inside diameter of 65 mm. If the maximum shear stress in the shaft must be limited to 50 MPa and the angle of twist must be limited to 1°, determine the maximum power that can be transmitted by this shaft when it is rotating at 600 rpm.

M6.17 A motor shaft is being designed to transmit 40 kW of power at 900 rpm. If the shearing stress in the shaft must be limited to 75 MPa, determine

(a) the minimum diameter required for a solid shaft.
(b) the minimum outside diameter required for a hollow shaft if the inside diameter of the shaft is assumed to be 80 percent of its outside diameter.

M6.18 The motor shown supplies 15 hp at 1,800 rpm at flange A. Shaft (1) is a solid 0.75 in. diameter shaft, and shaft (2) is a solid 1.50 in. diameter shaft. Both shafts are made of steel [$G = 12{,}000$ ksi]. The bearings shown permit free rotation of the shafts. Determine

(a) the maximum shear stress produced in each shaft.
(b) the rotation angle of gear D with respect to flange A.

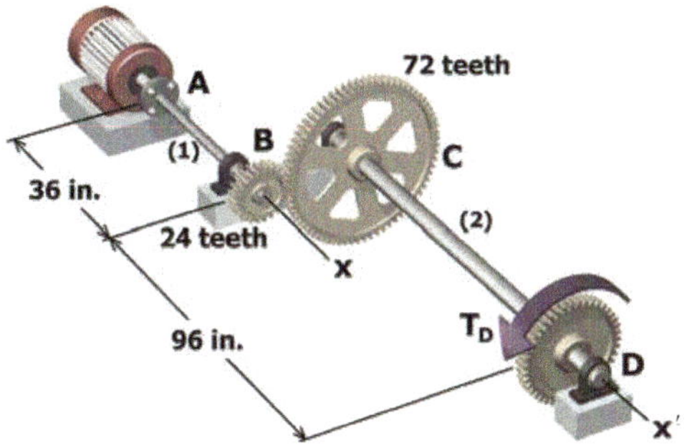

EXAMPLE 6.7

Two solid 25 mm diameter steel shafts are connected by the gears shown. A motor supplies 20 kW at 15 Hz to the system at A. The bearings shown permit free rotation of the shafts. Determine

(a) the torque available at gear D.
(b) the maximum shear stress magnitudes in each shaft.

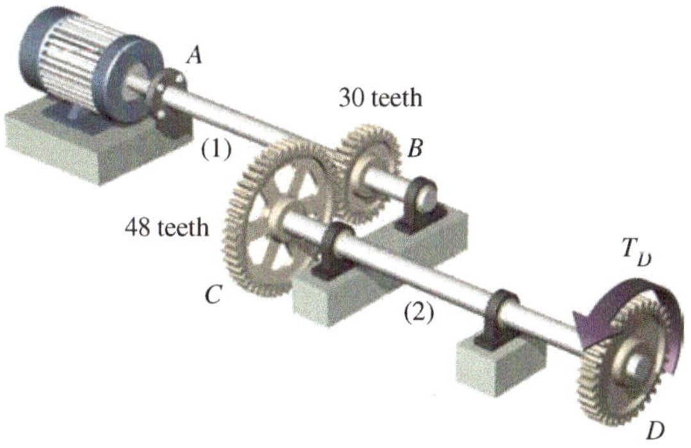

Plan the Solution

The torque in shaft (1) can be calculated from the power transmission equation. The torque in shaft (2) can then be determined from the gear ratio. Once the torques are known, the maximum shear stress magnitudes will be determined from the elastic torsion formula.

SOLUTION

The torque in shaft (1) can be calculated from the power transmission equation. The power supplied by the motor is 20 kW, so

$$P = (20 \text{ kW})\left(\frac{1{,}000 \text{ W}}{1 \text{ kW}}\right) = 20{,}000 \text{ W} = 20{,}000 \frac{\text{N}\cdot\text{m}}{\text{s}}$$

The motor rotates at 15 Hz. This rotation speed must be converted to units of rad/s:

$$\omega = 15 \text{ Hz} = \left(\frac{15 \text{ rev}}{\text{s}}\right)\left(\frac{2\pi \text{ rad}}{1 \text{ rev}}\right) = 94.24778 \frac{\text{rad}}{\text{s}}$$

The torque in shaft (1) is therefore

$$T_1 = \frac{P}{\omega} = \frac{20{,}000 \text{ N}\cdot\text{m/s}}{94.24778 \text{ rad/s}} = 212.2066 \text{ N}\cdot\text{m}$$

The torque in shaft (2) will be increased because gear C is larger than gear B. Use the number of teeth on each gear to establish the gear ratio, and compute the magnitude of the torque in shaft (2) as

$$T_2 = (212.2066 \text{ N}\cdot\text{m})\left(\frac{48 \text{ teeth}}{30 \text{ teeth}}\right) = 339.5306 \text{ N}\cdot\text{m}$$

Note: Only the magnitude of the torque is needed in this instance; consequently, the absolute value of T_2 is computed here.

The torque available at gear D in this system is therefore $T_D = 340\ \text{N}\cdot\text{m}$. **Ans.**

Shear Stresses

The polar moment of inertia for the solid 25 mm diameter shafts is

$$J = \frac{\pi}{32}(25\ \text{mm})^4 = 38{,}349.5\ \text{mm}^4$$

The maximum shear stress magnitudes in each segment can be calculated by the elastic torsion formula:

$$\tau_1 = \frac{T_1 c_1}{J_1} = \frac{(212.2066\ \text{N}\cdot\text{m})(25\ \text{mm}/2)(1{,}000\ \text{mm/m})}{38{,}349.5\ \text{mm}^4} = 69.2\ \text{MPa} \qquad \textbf{Ans.}$$

$$\tau_2 = \frac{T_2 c_2}{J_2} = \frac{(339.5306\ \text{N}\cdot\text{m})(25\ \text{mm}/2)(1{,}000\ \text{mm/m})}{38{,}349.5\ \text{mm}^4} = 110.7\ \text{MPa} \qquad \textbf{Ans.}$$

MecMovies

EXERCISES

M6.14 Six basic calculations involving power transmission in two shafts connected by gears.

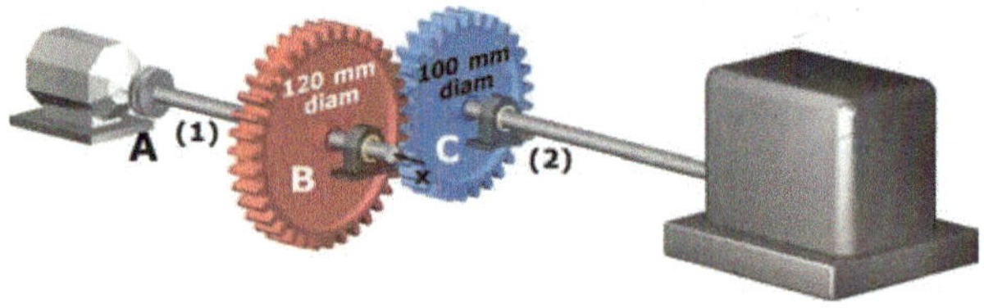

FIGURE M6.14

M6.15 Six basic calculations involving power transmission in three shafts connected by gears.

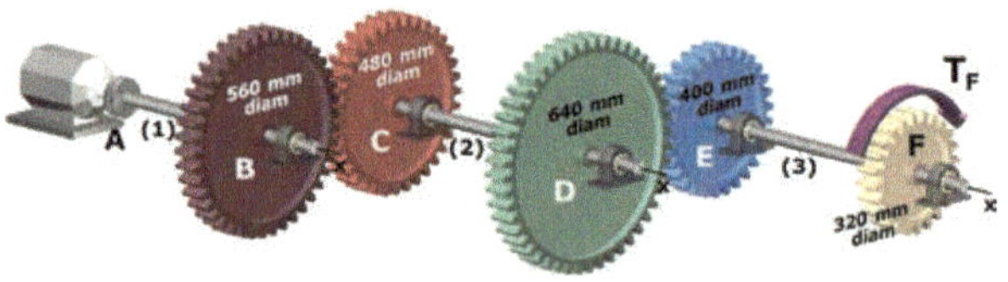

FIGURE M6.15

Problems

P6.24 A solid 40 mm diameter shaft is rotating at a speed of 560 rpm with a maximum shear stress of 85 MPa. Determine the power in kW being delivered by the shaft.

P6.25 A solid 1.50 in. diameter shaft is used to synchronize multiple mechanisms in a manufacturing plant. The torque delivered by the shaft is 3,100 lb · in. What horsepower must be delivered to the shaft to maintain a shaft speed of 600 rpm? What is the maximum shear stress developed in the shaft?

P6.26 A solid steel [$G = 11{,}500$ ksi] shaft that has a diameter of 2.00 in. and a length of 30 ft transmits 14 hp while rotating at 90 rpm. What is the angle of twist in the shaft?

P6.27 A solid stainless steel [$G = 86$ GPa] shaft with a diameter of 30 mm and a length of 900 mm transmits 42 kW from an electric motor to a compressor. If the allowable shear stress is 50 MPa and the allowable angle of twist is 1.5°, what is the minimum allowable speed of rotation?

P6.28 A hollow aluminum alloy [$G = 3{,}800$ ksi] shaft having a length of 12 ft, an outside diameter of 4.50 in., and a wall thickness of 0.50 in. rotates at 3 Hz. The allowable shear stress is 6 ksi, and the allowable angle of twist is 5°. What horsepower may the shaft transmit?

P6.29 The impeller shaft of a fluid agitator transmits 28 kW at 440 rpm. If the allowable shear stress in the shaft must be limited to 80 MPa, determine

(a) the minimum diameter required for a solid impeller shaft.
(b) the maximum inside diameter permitted for a hollow impeller shaft if the outside diameter is 40 mm.
(c) the percent savings in weight realized if the hollow shaft is used instead of the solid shaft. (*Hint*: The weight of a shaft is proportional to its cross-sectional area.)

P6.30 A pulley with a diameter $D = 8$ in. is mounted on a shaft with a diameter $d = 1.25$ in. as shown in Figure P6.30. Around the pulley is a belt having tensions $F_1 = 120$ lb and $F_2 = 480$ lb. If the shaft turns at 180 rpm, calculate

(a) the horsepower being transmitted by the shaft.
(b) the maximum shear stress in the shaft.

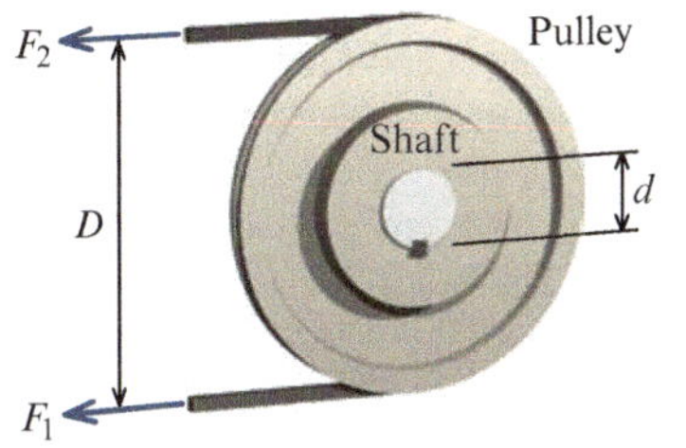

FIGURE P6.30

P6.31 A conveyor belt is driven by a 20 kW motor turning at 9 Hz. Through a series of gears that reduce the speed, the motor drives the belt drum shaft at a speed of 0.5 Hz. If the allowable shear stress is 70 MPa and both shafts are solid, calculate

(a) the required diameter of the motor shaft.
(b) the required diameter of the belt drum shaft.

P6.32 A motor at A supplies 25 hp to the system shown in Figure P6.32/33. Sixty percent of the power supplied by the motor is taken off by gear C, and the remaining 40 percent of the power is taken off by gear D. Power shaft segments (1) and (2) are hollow steel tubes with an outside diameter of 1.50 in. and a wall thickness of 0.1875 in. The diameters of pulleys A and B are $D_A = 4$ in. and $D_B = 14$ in., respectively. If the allowable shear stress for the steel tubes is 8,000 psi, what is the slowest permissible rotation speed for the motor?

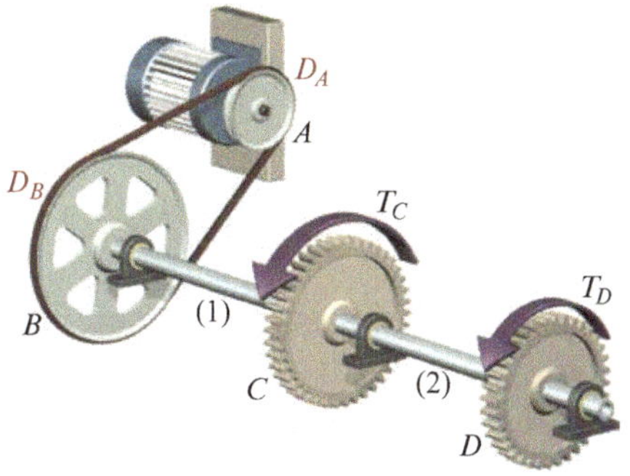

FIGURE P6.32/33

P6.33 A motor supplies sufficient power to the system shown in Figure P6.32/33 so that gears C and D provide torques $T_C = 2.6$ kN·m and $T_D = 1.2$ kN·m, respectively, to machinery in a factory. Power shaft segments (1) and (2) consist of a hollow steel tube with an outside diameter of 90 mm and a wall thickness of 8 mm. The diameters of pulleys A and B are $D_A = 100$ mm and $D_B = 360$ mm, respectively. If the power shaft [i.e., segments (1) and (2)] rotates at 160 rpm, determine

(a) the maximum shear stress in power shaft segments (1) and (2).
(b) the power (in kW) that must be provided by the motor.
(c) the rotation speed (in rpm).
(d) the torque applied to pulley A by the motor.

P6.34 In Figure P6.34/35, the motor at A produces 30 kW of power while turning at 8.0 Hz. The gears at A and B connecting the motor to line shaft $BCDE$ have $N_A = 30$ teeth and $N_B = 96$ teeth, respectively. On the line shaft, gear C takes off $P_C = 50\%$ of the power, gear D removes $P_D = 30\%$, and gear E removes the remainder. The line shaft will be solid, all segments will have the same diameter, and the shaft will be made of steel [$G = 80$ GPa]. The lengths of the segments are $L_1 = 4.6$ m, $L_2 = 2.3$ m, and $L_3 = 2.3$ m, and the bearings shown permit free rotation of the shaft. Determine the minimum diameter required for line shaft $BCDE$ for an allowable shear stress of 65 MPa and an allowable rotation angle of 10° of gear E relative to gear B.

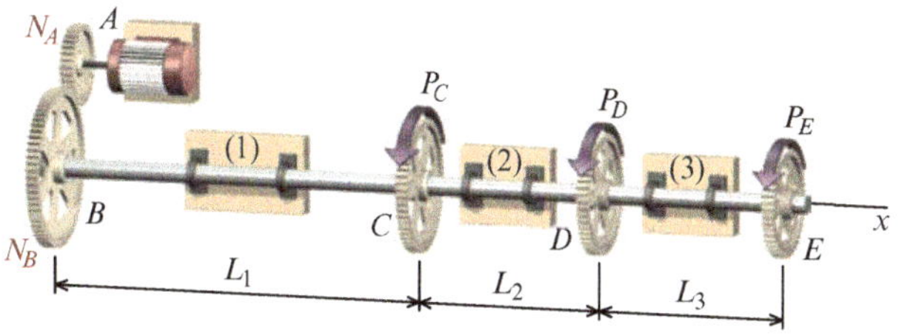

FIGURE P6.34/35

P6.35 In Figure P6.34/35, the motor at A produces 25 hp while turning at 2,600 rpm. The gears at A and B connecting the motor to line shaft $BCDE$ have $N_A = 30$ teeth and $N_B = 140$ teeth, respectively. On the shaft, gear C takes off $P_C = 40\%$ of the power, gear D removes $P_D = 40\%$, and gear E removes the remainder. Line shaft $BCDE$ is a solid 1.25 in. diameter shaft made of an aluminum alloy that has a shear modulus of 3,800 ksi. The lengths of the segments are $L_1 = 80$ in., $L_2 = 42$ in., and $L_3 = 36$ in. The bearings shown permit free rotation of the line shaft. Calculate

(a) the magnitude of the maximum shear stress in each shaft segment.
(b) the rotation angle of gear E with respect to gear B.

P6.36 The motor at A is required to provide 40 hp of power to line shaft $BCDE$ shown in Figure P6.36/37, turning gears B and E at 900 rpm. Gear B removes 70% of the power from the line shaft, and gear E removes 30%. Shafts (1) and (2) are solid aluminum alloy shafts with an outside diameter of 1.25 in. and a shear modulus of 3,800 ksi. Shaft (3) is a solid steel shaft that has an outside diameter of 0.75 in. and a shear modulus of 11,600 ksi.

The shaft lengths are $L_1 = 8.25$ ft, $L_2 = 4.25$ ft, and $L_3 = 7.0$ ft. The diameters of pulleys A and C are $D_A = 2$ in. and $D_C = 10$ in., respectively. The bearings shown allow free rotation of the shafts. Calculate

(a) the magnitude of the maximum shear stress in each shaft.
(b) the rotation angle of gear E with respect to pulley C.
(c) the magnitude of the torque and the rotation speed required for motor A.

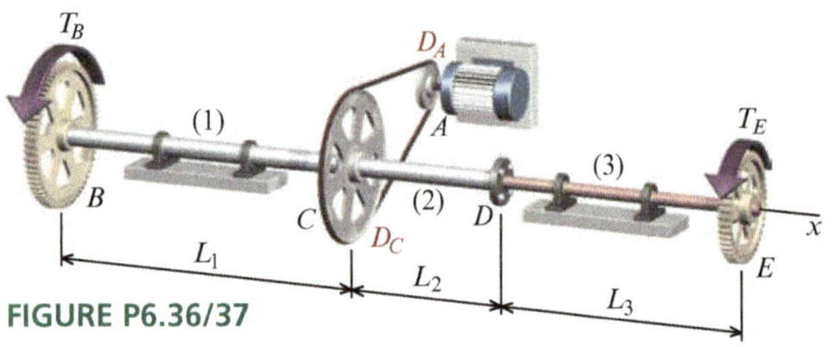

FIGURE P6.36/37

P6.37 The motor at A is required to provide 50 kW of power to line shaft $BCDE$ shown in Figure P6.36/37, turning gears B and E at 8 Hz. Gear B removes 65% of the power from the line shaft, and gear E removes 35%. Shafts (1) and (2) are solid aluminum alloy [G = 26 GPa] shafts having an allowable shear stress of 45 MPa. Shaft (3) is a solid steel [G = 80 MPa] shaft that has an allowable shear stress of 60 MPa. The shaft lengths are L_1 = 2.1 m, L_2 = 1.2 m, and L_3 = 1.8 m. The diameters of pulleys A and C are D_A = 70 mm and D_C = 300 mm, respectively. The bearings shown allow free rotation of the shafts. Calculate

(a) the minimum permissible diameter for aluminum shafts (1) and (2).
(b) the minimum permissible diameter for steel shaft (3).
(c) the rotation angle of gear E with respect to pulley C if the shafts have the minimum permissible diameters as determined in (a) and (b).
(d) the magnitude of the torque and the rotation speed (in Hz) required for motor A.

P6.38 The motor shown in Figure P6.38 supplies 15 kW at 1,700 rpm at A. Shafts (1) and (2) are each solid 30 mm diameter shafts. Shaft (1) is made of an aluminum alloy [G = 26 GPa], and shaft (2) is made of bronze [G = 45 GPa]. The shaft lengths are L_1 = 3.4 m and L_2 = 2.7 m, respectively. Gear B has 54 teeth, and gear C has 96 teeth. The bearings shown permit free rotation of the shafts. Determine

(a) the maximum shear stress produced in shafts (1) and (2).
(b) the rotation angle of gear D with respect to flange A.

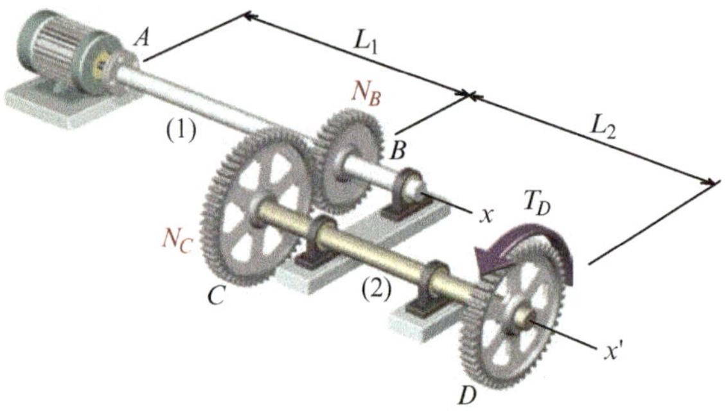

FIGURE P6.38

6.9 Statically Indeterminate Torsion Members

In many simple mechanical and structural systems subjected to torsional loading, it is possible to determine the reactions at supports and the internal torques in the individual members by drawing free-body diagrams and solving equilibrium equations. Such torsional systems are classified as **statically determinate**.

For many mechanical and structural systems, however, the equations of equilibrium alone are not sufficient for the determination of internal torques in the members and reactions at supports. In other words, there are not enough equilibrium equations to solve for all of the unknowns in the system. These structures and systems are termed **statically indeterminate**. We can analyze structures of this type by supplementing the equilibrium equations with additional equations involving the geometry of the deformations in the members of the structure or system. The general solution process can be organized into a five-step procedure analogous to the procedure developed for statically indeterminate axial structures in Section 5.5:

Step 1 — Equilibrium Equations: Equations expressed in terms of the unknown internal torques are derived for the system on the basis of equilibrium considerations.

Step 2 — Geometry of Deformation: The geometry of the specific system is evaluated to determine how the deformations of the torsion members are related.

Step 3 — Torque–Twist Relationships: The relationships between the internal torque in a member and its corresponding angle of twist are expressed by Equation (6.12).

Step 4 — Compatibility Equation: The torque–twist relationships are substituted into the geometry-of-deformation equation to obtain an equation that is based on the structure's geometry, but expressed in terms of the unknown internal torques.

Step 5 — Solve the Equations: The equilibrium equations and the compatibility equation are solved simultaneously to compute the unknown internal torques.

The use of this procedure to analyze a statically indeterminate torsion system is illustrated in the next example.

EXAMPLE 6.8

A compound shaft consists of two solid shafts that are connected at flange B and securely attached to rigid walls at A and C. Shaft (1) is a 3.00 in. diameter solid aluminum [G = 4,000 ksi] shaft that is 60 in. long. Shaft (2) is a 2.00 in. diameter solid bronze [G = 6,500 ksi] shaft that is 40 in. long. If a concentrated torque of 32 kip·in. is applied to flange B, determine

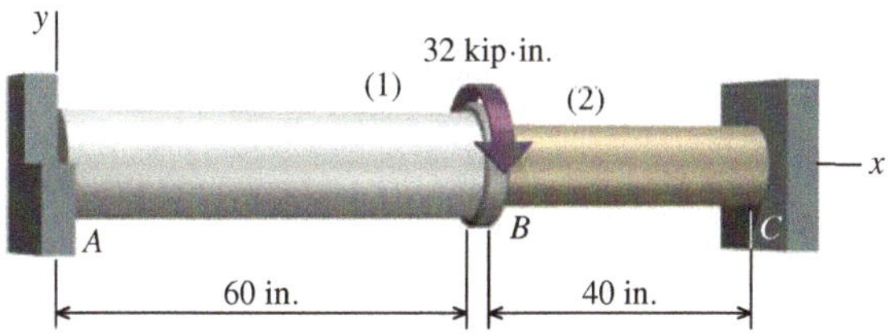

(a) the maximum shear stress magnitudes in shafts (1) and (2).
(b) the rotation angle of flange B relative to support A.

Plan the Solution

The solution begins with a free-body diagram at flange B. The equilibrium equation obtained from this free-body diagram reveals that the compound shaft is statically indeterminate. We can obtain the additional information needed to solve the problem by considering the relationship between the angles of twist in the aluminum and bronze segments of the shaft.

SOLUTION

Step 1 — Equilibrium Equation: Draw a free-body diagram of flange B. Assume *positive internal torques* in shaft segments (1) and (2). [See the sign convention detailed in Section 6.6.] From this free-body diagram, the following moment equilibrium equation can be obtained:

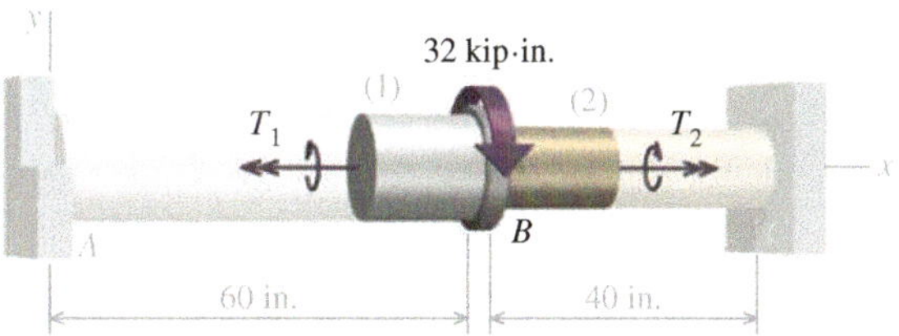

$$\Sigma M_x = -T_1 + T_2 + 32 \text{ kip·in.} = 0 \quad \text{(a)}$$

There are two unknowns in Equation (a): T_1 and T_2. Consequently, statics alone does not provide enough information for this problem to be solved. To obtain another relationship involving the unknown torques T_1 and T_2, we will consider the general relationship between the twist angles in the compound shaft.

Step 2 — Geometry of Deformation: The next question is, "How are the angles of twist in the two shaft segments related?" The compound shaft is attached to rigid walls at A and C; therefore, the twisting that occurs in shaft segment (1) plus the twisting in shaft segment (2) cannot result in any net rotation of the compound shaft. In other words, the sum of these angles of twist must equal zero:

$$\phi_1 + \phi_2 = 0 \tag{b}$$

Step 3 — Torque–Twist Relationships: The angles of twists in shaft segments (1) and (2) can be expressed by the angle-of-twist equation [Equation (6.12)]. Angle-of-twist equations can be written for both segment (1) and segment (2):

$$\phi_1 = \frac{T_1 L_1}{J_1 G_1} \qquad \phi_2 = \frac{T_2 L_2}{J_2 G_2} \tag{c}$$

Step 4 — Compatibility Equation: The torque–twist relationships [Equation (c)] can be substituted into the geometry-of-deformation equation [Equation (b)] to obtain a new relationship between the unknown torques T_1 and T_2:

$$\frac{T_1 L_1}{J_1 G_1} + \frac{T_2 L_2}{J_2 G_2} = 0 \tag{d}$$

Notice that this relationship is based, not on equilibrium, but rather on the relationship between deformations that occur in the compound shaft. This type of equation is termed a *compatibility equation*.

Step 5 — Solve the Equations: Two equations have been developed in terms of the internal torques T_1 and T_2:

$$\Sigma M_x = -T_1 + T_2 + 32 \text{ kip}\cdot\text{in.} = 0 \tag{a}$$

$$\frac{T_1 L_1}{J_1 G_1} + \frac{T_2 L_2}{J_2 G_2} = 0 \tag{d}$$

These two equations must be solved simultaneously for us to determine the torques in each shaft segment. First, the compatibility equation [Equation (d)] can be solved for the internal torque T_2:

$$T_2 = -T_1\left(\frac{L_1}{J_1 G_1}\right)\left(\frac{J_2 G_2}{L_2}\right) = -T_1\left(\frac{L_1}{L_2}\right)\left(\frac{J_2}{J_1}\right)\left(\frac{G_2}{G_1}\right)$$

Next, substitute this result into the equilibrium equation [Equation (a)]:

$$-T_1 - T_1\left(\frac{L_1}{L_2}\right)\left(\frac{J_2}{J_1}\right)\left(\frac{G_2}{G_1}\right) + 32 \text{ kip}\cdot\text{in.} = 0$$

Then solve for the internal torque T_1:

$$T_1 = \frac{32 \text{ kip}\cdot\text{in.}}{\left[1 + \left(\frac{L_1}{L_2}\right)\left(\frac{J_2}{J_1}\right)\left(\frac{G_2}{G_1}\right)\right]} \tag{e}$$

Polar moments of inertia for the aluminum and bronze shaft segments are needed for this calculation. Aluminum segment (1) is a solid 3.00 in. diameter shaft that is 60 in. long and has a shear modulus of 4,000 ksi. The polar moment of inertia for segment (1) is

$$J_1 = \frac{\pi}{32}(3.00 \text{ in.})^4 = 7.952156 \text{ in.}^4$$

Bronze segment (2) is a solid 2.00 in. diameter shaft that is 40 in. long and has a shear modulus of 6,500 ksi. Its polar moment of inertia is

$$J_2 = \frac{\pi}{32}(2.00 \text{ in.})^4 = 1.570796 \text{ in.}^4$$

The internal torque T_1 is computed by substitution of all values into Equation (e):

$$T_1 = \frac{32 \text{ kip}\cdot\text{in.}}{\left[1 + \left(\frac{60 \text{ in.}}{40 \text{ in.}}\right)\left(\frac{1.570796 \text{ in.}^4}{7.952156 \text{ in.}^4}\right)\left(\frac{6{,}500 \text{ ksi}}{4{,}000 \text{ ksi}}\right)\right]} = \frac{32 \text{ kip}\cdot\text{in.}}{1.481481} = 21.600 \text{ kip}\cdot\text{in.}$$

The internal torque T_2 can be found by substitution back into Equation (a):

$$T_2 = T_1 - 32 \text{ kip}\cdot\text{in.} = 21.600 \text{ kip}\cdot\text{in.} - 32 \text{ kip}\cdot\text{in.} = -10.400 \text{ kip}\cdot\text{in.}$$

Shear Stresses

Since the internal torques are now known, the maximum shear stress magnitudes can be calculated for each segment from the elastic torsion formula [Equation (6.5)]. In calculating the maximum shear stress magnitude, only the absolute value of the internal torque is used. In segment (1), the maximum shear stress magnitude in the 3.00 in. diameter aluminum shaft is

$$\tau_1 = \frac{T_1 c_1}{J_1} = \frac{(21.600 \text{ kip}\cdot\text{in.})(3.00 \text{ in.}/2)}{7.952156 \text{ in.}^4} = 4.07 \text{ ksi} \qquad \textbf{Ans.}$$

The maximum shear stress magnitude in the 2.00 in. diameter bronze shaft segment (2) is

$$\tau_2 = \frac{T_2 c_2}{J_2} = \frac{(10.400 \text{ kip}\cdot\text{in.})(2.00 \text{ in.}/2)}{1.570796 \text{ in.}^4} = 6.62 \text{ ksi} \qquad \textbf{Ans.}$$

Rotation Angle of Flange *B*

The angle of twist in shaft segment (1) can be expressed as the difference between the rotation angles at the $+x$ and $-x$ ends of the segment:

$$\phi_1 = \phi_B - \phi_A$$

Since the shaft is rigidly fixed to the wall at A, $\phi_A = 0$. The rotation angle of flange B, therefore, is simply equal to the angle of twist in shaft segment (1). **Note:** The proper sign of the internal torque T_1 must be used in the angle of twist calculation. Thus,

$$\phi_B = \phi_1 = \frac{T_1 L_1}{J_1 G_1} = \frac{(21.600 \text{ kip}\cdot\text{in.})(60 \text{ in.})}{(7.952156 \text{ in.}^4)(4{,}000 \text{ ksi})} = 0.040744 \text{ rad} = 0.0407 \text{ rad} \qquad \textbf{Ans.}$$

The five-step procedure demonstrated in the previous example provides a versatile method for the analysis of statically indeterminate torsion structures. Additional problem-solving considerations and suggestions for each step of the process are discussed in the following table:

Solution Method for Statically Indeterminate Torsion Systems

Step 1	Equilibrium Equations	Draw one or more free-body diagrams (FBDs) for the structure, focusing on the joints, which connect the members. Joints are located wherever (a) an external torque is applied, (b) one or more cross-sectional properties (such as the diameter) change, (c) the material properties (i.e., G) change, or (d) a member connects to a rigid element (such as a gear, pulley, support, or flange). Generally, FBDs of reaction joints are not useful. Write equilibrium equations for the FBDs. Note the number of unknowns involved and the number of independent equilibrium equations. If the number of unknowns exceeds the number of equilibrium equations, a deformation equation must be written for each extra unknown. Comments: • Label the joints with capital letters, and label the members with numbers. This simple scheme can help you clearly recognize effects that occur in members (such as angles of twist) and effects that pertain to joints (such as rotation angles of rigid elements). • As a rule, when cutting an FBD through a torsion member, *assume that the internal torque is positive*, as detailed in Section 6.6. The consistent use of positive internal torques along with positive angles of twist (in Step 3) proves quite effective for many situations.
Step 2	Geometry of Deformation	This step is unique to statically indeterminate problems. The structure or system should be studied to assess how the deformations of the torsion members are related to each other. Most of the statically indeterminate torsion systems can be categorized as either 1. systems with coaxial torsion members or 2. systems with torsion members connected end to end in series.
Step 3	Torque–Twist Relationships	The relationship between the internal torque and the angle of twist in a torsion member is expressed by $$\phi_i = \frac{T_i L_i}{J_i G_i}$$ As a practical matter, writing down torque–twist relationships for the torsion members is a helpful routine at this stage of the calculation procedure. These relationships will be used to construct the compatibility equation(s) in Step 4.
Step 4	Compatibility Equation(s)	The torque–twist relationships (from Step 3) are incorporated into the geometric relationship of member angles of twist (from Step 2) to derive a new equation, which is expressed in terms of the unknown internal torques. Together, the compatibility and equilibrium equations provide sufficient information to solve for the unknown variables.
Step 5	Solve the Equations	The compatibility equation(s) and the equilibrium equation(s) are solved simultaneously. While conceptually straightforward, this step requires careful attention to calculation details, such as sign conventions and consistency of units.

Successful application of the five-step solution method depends on the ability to understand how twisting deformations are related in a system. The table that follows presents considerations for two common categories of statically indeterminate torsion systems. For each general category, possible geometry-of-deformation equations are discussed.

Geometry of Deformations for Typical Statically Indeterminate Torsion Systems

Equation Form	Comments	Typical Problems
	1. Coaxial torsion members.	
$\phi_1 = \phi_2$	Problems in this category include a tube surrounding an inner shaft. The angles of twist for both torsional members must be identical for this type of system.	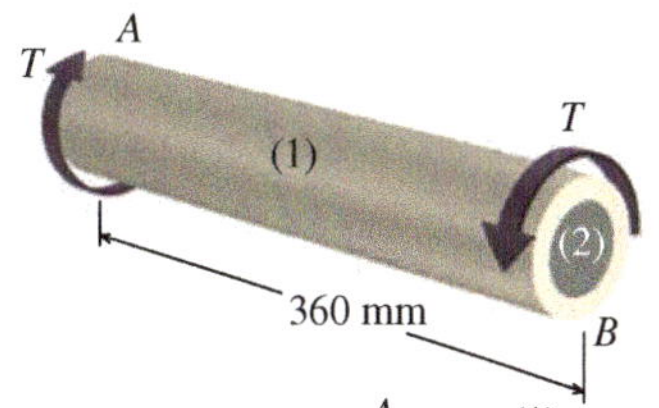 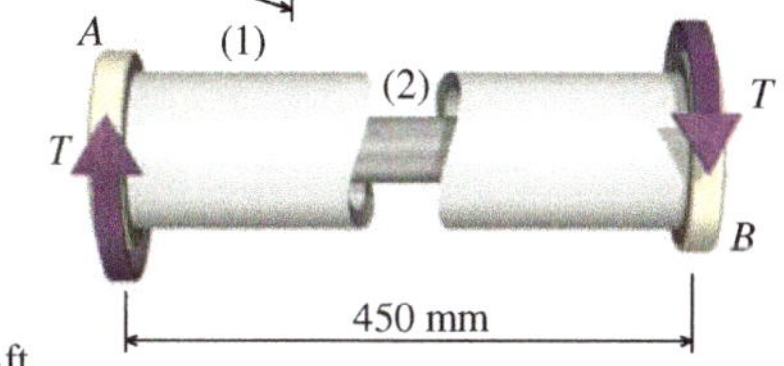 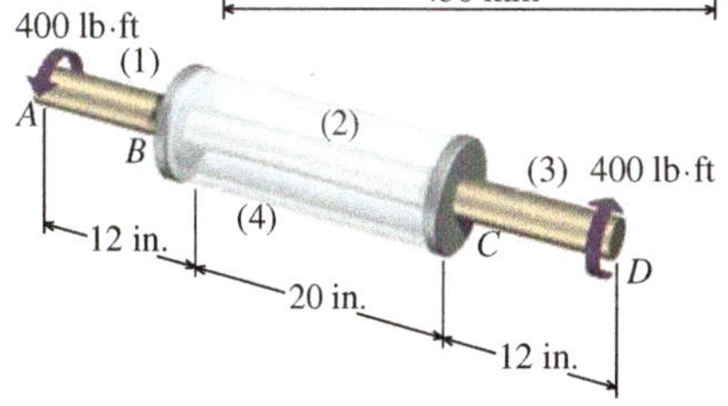
	2. Torsion members connected end to end in series.	
	Problems in this category include two or more members connected end to end.	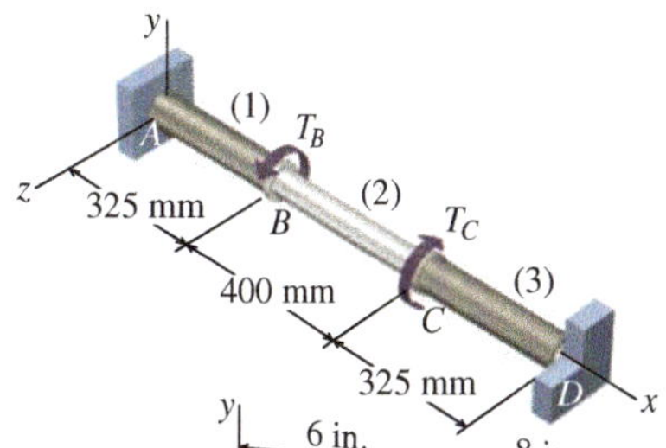
$\phi_1 + \phi_2 = 0$	If there are no gaps or clearances in the configuration, the member angles of twist must sum to zero.	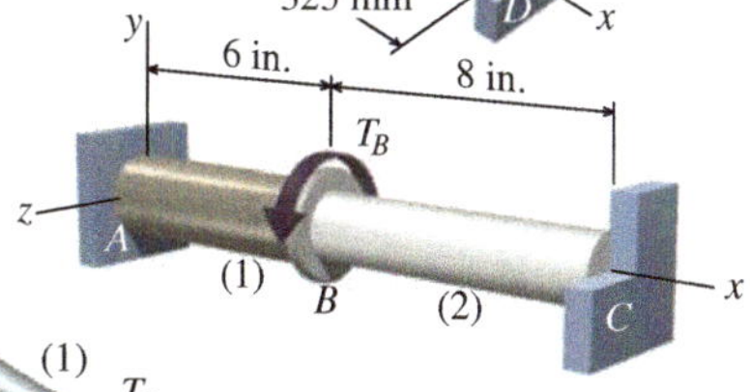
$\phi_1 + \phi_2 = \text{constant}$	If there is a misfit between two members or if the supports move as the torque or torques are applied, then the sum of the angles of twist of the members equals the specified angular rotation.	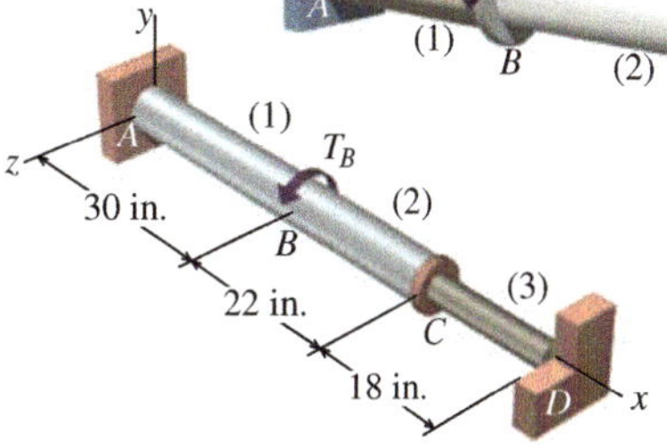

MecMovies

EXAMPLES

M6.19 A composite shaft consists of a hollow aluminum [G = 26 GPa] shaft (1) bonded to a hollow bronze [G = 38 GPa] shaft (2). The outside diameter of shaft (1) is 50 mm, and the inside diameter is 42 mm. The outside diameter of shaft (2) is 42 mm, and the inside diameter is 30 mm. A concentrated torque T = 1,400 N·m is applied to the composite shaft at the free end B. Determine

(a) the torques T_1 and T_2 developed in the aluminum and bronze shafts.
(b) the maximum shear stresses τ_1 and τ_2 in each shaft.
(c) the angle of rotation of end B.

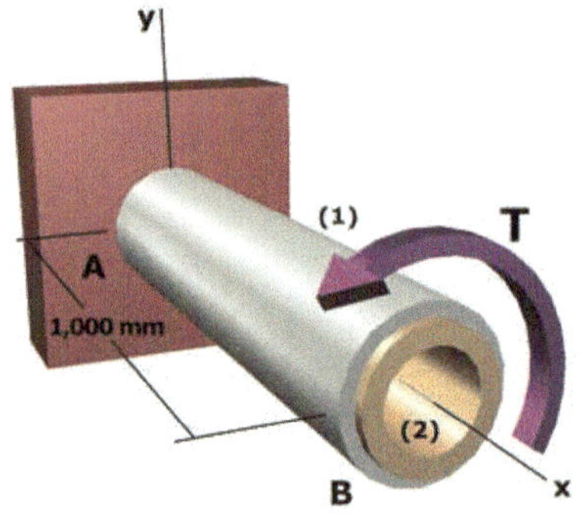

M6.20 A composite shaft consists of a hollow steel [G = 75 GPa] shaft (1) connected to a solid brass [G = 40 GPa] shaft (2) at flange B. The outside diameter of shaft (1) is 50 mm, and the inside diameter is 40 mm. The outside diameter of shaft (2) is 50 mm. A concentrated torque T = 1,000 N·m is applied to the composite shaft at flange B. Determine

(a) the torques T_1 and T_2 developed in the steel and brass shafts.
(b) the maximum shear stresses τ_1 and τ_2 in each shaft.
(c) the angle of rotation of flange B.

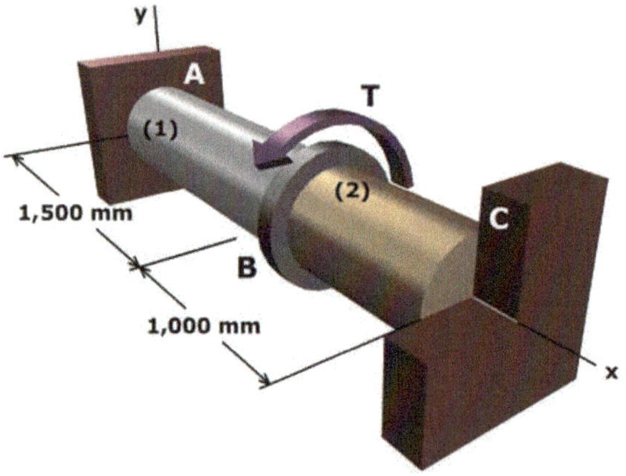

EXAMPLE 6.9

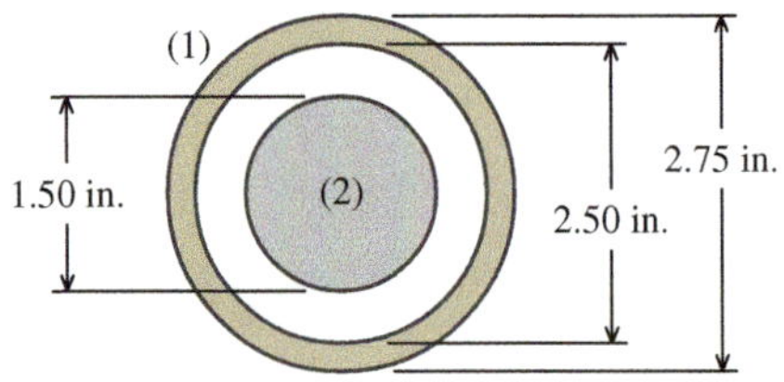

Cross-sectional dimensions.

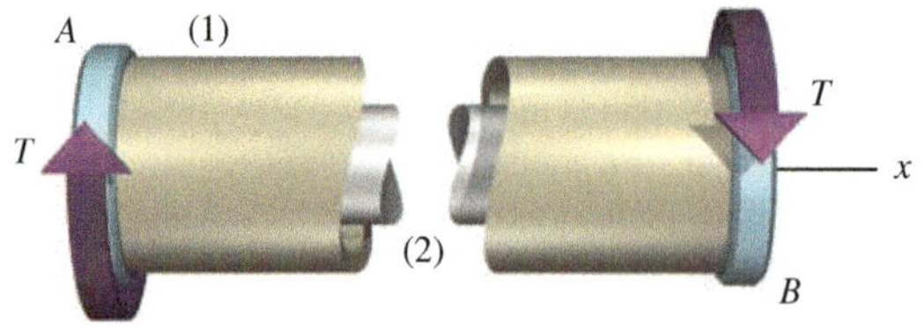

A composite shaft assembly consists of an inner stainless steel [G = 12,500 ksi] core (2) connected by rigid plates at A and B to the ends of a brass [G = 5,600 ksi] tube (1). The cross-sectional dimensions of the assembly are shown. The allowable shear stress of the brass tube (1) is 12 ksi, and the allowable shear stress of the stainless steel core (2) is 18 ksi. Determine the maximum torque T that can be applied to the composite shaft.

Plan the Solution

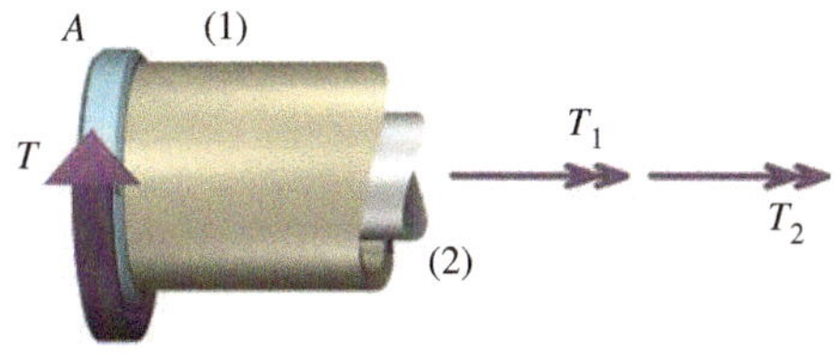

A free-body diagram cut through the assembly will expose the internal torques in the tube and the core. Since there are two internal torques and only one equilibrium equation, the assembly is statically indeterminate. The tube and the core are attached to rigid end plates; therefore, as the assembly twists, both the tube and the core will twist by the same amount. This relationship will be used to derive a compatibility equation in terms of the unknown internal torques. Information about the allowable shear stresses will then be used to determine which of the two components controls the torque capacity of the composite shaft assembly.

SOLUTION

Step 1 — Equilibrium Equation: Cut a free-body diagram through the assembly around rigid end plate A. From this free-body diagram, the following equilibrium equation can be obtained:

$$\Sigma M_x = -T + T_1 + T_2 = 0 \tag{a}$$

Since there are three unknowns—T_1, T_2, and the external torque T—this assembly is statically indeterminate.

Step 2 — Geometry of Deformation: The tube and the core are both attached to rigid end plates. Therefore, when the assembly is twisted, both components must twist the same amount:

$$\phi_1 = \phi_2 \tag{b}$$

Step 3 — Torque–Twist Relationships: The angles of twists in tube (1) and core (2) can be expressed as

$$\phi_1 = \frac{T_1 L_1}{J_1 G_1} \quad \text{and} \quad \phi_2 = \frac{T_2 L_2}{J_2 G_2} \tag{c}$$

Step 4 — Compatibility Equation: Substitute the torque–twist relationships [Equation (c)] into the geometry-of-deformation equation [Equation (b)] to obtain the compatibility equation:

$$\frac{T_1 L_1}{J_1 G_1} = \frac{T_2 L_2}{J_2 G_2} \tag{d}$$

Step 5 — Solve the Equations: Two equations have been derived in terms of the three unknown torques (T_1, T_2, and the external torque T). Additional information is needed to solve for the unknown torques.

Allowable Shear Stresses

The maximum shear stress in the tube and in the core will be determined by the elastic torsion formula. Since allowable shear stresses are specified for both components, the elastic torsion formula can be written for each component and rearranged to solve for the torque. For brass tube (1),

$$\tau_1 = \frac{T_1 c_1}{J_1} \quad \therefore T_1 = \frac{\tau_1 J_1}{c_1} \tag{e}$$

For stainless steel core (2),

$$\tau_1 = \frac{T_2 c_2}{J_2} \qquad \therefore T_2 = \frac{\tau_2 J_2}{c_2} \tag{f}$$

Substitute Equations (e) and (f) into the compatibility equation [Equation (d)], and simplify:

$$T_1 \frac{L_1}{J_1 G_1} = T_2 \frac{L_2}{J_2 G_2}$$

$$\frac{\tau_1 J_1}{c_1} \frac{L_1}{J_1 G_1} = \frac{\tau_2 J_2}{c_2} \frac{L_2}{J_2 G_2}$$

$$\frac{\tau_1 L_1}{c_1 G_1} = \frac{\tau_2 L_2}{c_2 G_2} \tag{g}$$

Note: Equation (g) is simply Equation (6.13) written for tube (1) and core (2). Since the tube and the core are both the same length, Equation (g) can be simplified to

$$\frac{\tau_1}{c_1 G_1} = \frac{\tau_2}{c_2 G_2} \tag{h}$$

We cannot know beforehand which component will control the capacity of the torsional assembly. Let us assume that the maximum shear stress in the stainless steel core (2) will control; that is, $\tau_2 = 18$ ksi. In that case, the corresponding shear stress in brass tube (1) can be calculated from Equation (h):

$$\tau_1 = \tau_2 \left(\frac{c_1}{c_2}\right)\left(\frac{G_1}{G_2}\right) = (18 \text{ ksi})\left(\frac{2.75 \text{ in.}/2}{1.50 \text{ in.}/2}\right)\left(\frac{5{,}600 \text{ ksi}}{12{,}500 \text{ ksi}}\right) = 14.784 \text{ ksi} > 12 \text{ ksi} \quad \text{N.G.}$$

This shear stress exceeds the 12 ksi allowable shear stress for the brass tube. Therefore, our initial assumption is proved incorrect—the maximum shear stress in the brass tube actually controls the torque capacity of the assembly.

Equation (h) is then solved for τ_2, given that the allowable shear stress of the brass tube is $\tau_1 = 12$ ksi:

$$\tau_2 = \tau_1 \left(\frac{c_2}{c_1}\right)\left(\frac{G_2}{G_1}\right) = (12 \text{ ksi})\left(\frac{1.50 \text{ in.}/2}{2.75 \text{ in.}/2}\right)\left(\frac{12{,}500 \text{ ksi}}{5{,}600 \text{ ksi}}\right) = 14.610 \text{ ksi} < 18 \text{ ksi} \quad \text{O.K.}$$

Allowable Torques

On the basis of the compatibility equation, we now know the maximum shear stresses that will be developed in each of the components. From these shear stresses, we can determine the torques in each component by using Equations (e) and (f).

The polar moments of inertia for each component are required. For the brass tube (1),

$$J_1 = \frac{\pi}{32}[(2.75 \text{ in.})^4 - (2.50 \text{ in.})^4] = 1.779801 \text{ in.}^4$$

and for the stainless steel core (2),

$$J_2 = \frac{\pi}{32}(1.50 \text{ in.})^4 = 0.497010 \text{ in.}^4$$

From Equation (e), the allowable internal torque in brass tube (1) can be calculated as

$$T_1 = \frac{\tau_1 J_1}{c_1} = \frac{(12\text{ ksi})(1.779801\text{ in.}^4)}{2.75\text{ in.}/2} = 15.533\text{ kip}\cdot\text{in.}$$

and from Equation (f), the corresponding internal torque in the stainless steel core (2) is

$$T_2 = \frac{\tau_2 J_2}{c_2} = \frac{(14.610\text{ ksi})(0.497010\text{ in.}^4)}{1.50\text{ in.}/2} = 9.682\text{ kip}\cdot\text{in.}$$

Finally, substitute these results into the equilibrium equation [Equation (a)] to determine the magnitude of the external torque T that may be applied to the composite shaft assembly:

$$T = T_1 + T_2 = 15.533\text{ kip}\cdot\text{in.} + 9.682\text{ kip}\cdot\text{in.} = 25.2\text{ kip}\cdot\text{in.} \qquad \textbf{Ans.}$$

MecMovies

EXAMPLES

M6.21 A composite shaft consists of a hollow steel [G = 75 GPa] shaft (1) connected to a solid bronze [G = 38 GPa] shaft (2) at flange B. The outside diameter of shaft (1) is 80 mm, and the inside diameter is 65 mm. The outside diameter of shaft (2) is 80 mm. The allowable shear stresses for the steel and bronze materials are 90 MPa and 50 MPa, respectively. Determine

(a) the maximum torque T that can be applied to flange B.
(b) the stresses τ_1 and τ_2 developed in the steel and bronze shafts.
(c) the angle of rotation of flange B.

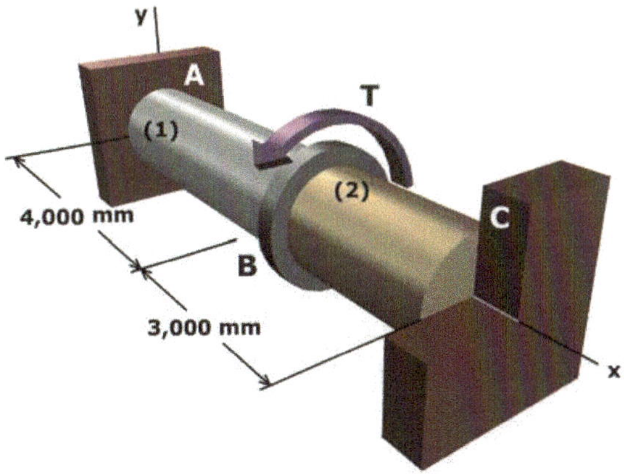

M6.22 A composite shaft consists of a hollow aluminum [G = 26 GPa] shaft (1) bonded to a hollow bronze [G = 38 GPa] shaft (2) at flange B. The outside diameter of shaft (1) is 50 mm, and the inside diameter is 42 mm. The outside diameter of shaft (2) is 42 mm, and the inside diameter is 30 mm. The allowable shear stresses for the aluminum and bronze materials are 85 MPa and 100 MPa, respectively. Determine

(a) the maximum torque T that can be applied to the free end B of the shaft.
(b) the stresses τ_1 and τ_2 developed in the shafts.
(c) the angle of rotation of end B.

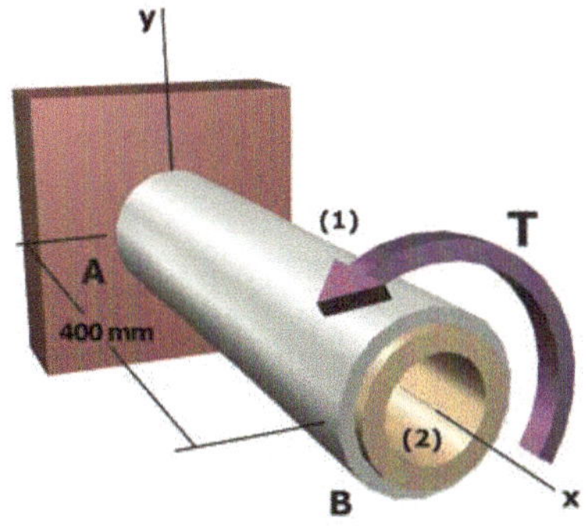

M6.23 A composite shaft consists of a hollow stainless steel [$G = 86$ GPa] shaft (1) connected to a solid bronze [$G = 38$ GPa] shaft (2) at flange B. The outside diameter of shaft (1) is 75 mm, and the inside diameter is 55 mm. The outside diameter of shaft (2) is 75 mm. A concentrated torque T will be applied to the composite shaft at flange B. Determine

(a) the maximum magnitude of the concentrated torque T if the angle of rotation at flange B cannot exceed 3°.
(b) the maximum shear stresses τ_1 and τ_2 in each shaft.

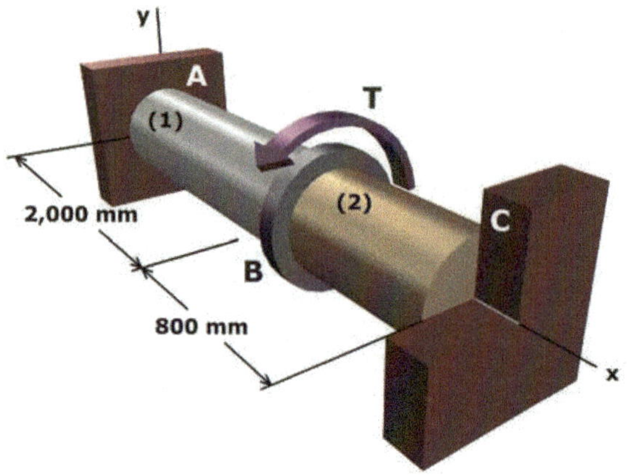

EXAMPLE 6.10

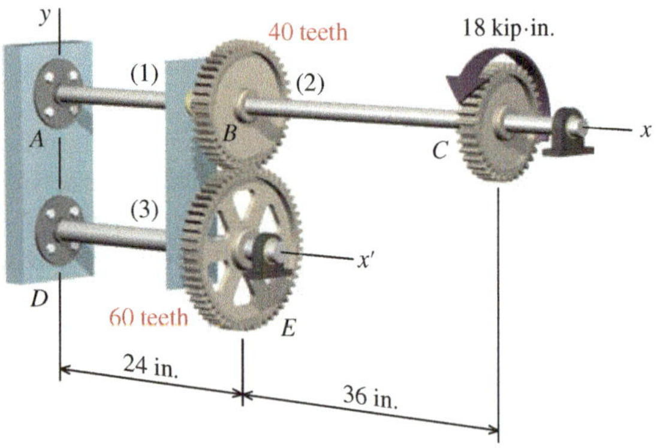

A torque of 18 kip·in. acts on gear C of the assembly shown. Shafts (1) and (2) are solid 2.00 in. diameter steel shafts, and shaft (3) is a solid 2.50 in. diameter steel shaft. Assume that $G = 12{,}000$ ksi for all shafts. The bearings shown allow free rotation of the shafts. Determine

(a) the maximum shear stress magnitudes in shafts (1), (2), and (3).
(b) the rotation angle of gear E.
(c) the rotation angle of gear C.

Plan the Solution

A torque of 18 kip·in. is applied to gear C. This torque is transmitted by shaft (2) to gear B, causing it to rotate and, in turn, twist shaft (1). The rotation of gear B also causes gear E to rotate, which causes shaft (3) to twist. Therefore, the torque of 18 kip·in. on gear C will produce torques in all three shafts. The rotation angle of gear B will be dictated by the angle of twist in shaft (1). Similarly, the rotation angle of gear C will be dictated by the angle of twist in shaft (3). Furthermore, the relative rotation of gears B and E will be a function of the gear ratio. These relationships will be considered in analyzing the internal torques produced in the three shafts. Once the internal torques are known, the maximum shear stresses, twist angles, and rotation angles can be determined.

SOLUTION

Step 1 — Equilibrium Equations: Consider a free-body diagram that cuts through shaft (2) and includes gear C. A positive internal torque will be assumed in shaft (2). From this free-body diagram, a moment equilibrium equation about the x axis can be written to determine the internal torque T_2 in shaft (2).

$$\Sigma M_x = 18 \text{ kip·in.} - T_2 = 0 \qquad \therefore T_2 = 18 \text{ kip·in.} \tag{a}$$

Next, consider a free-body diagram that cuts through shafts (1) and (2) and includes gear B. Once again, a positive internal torque will be assumed in shafts (1) and (2). The teeth of gear E exert a force F on the teeth of gear B. If the radius of gear B is denoted by R_B, a moment equilibrium equation about the x axis can be written as

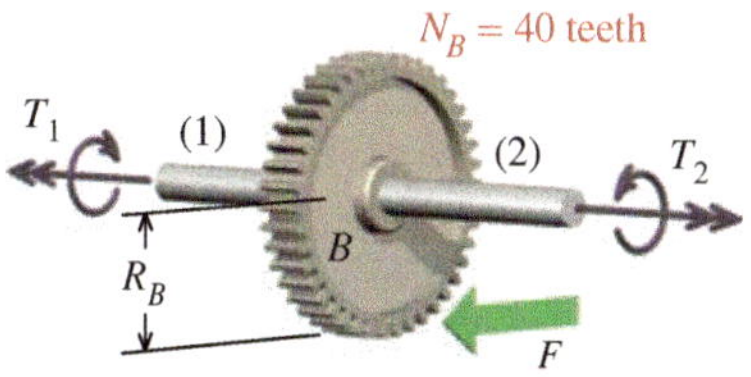

$$\Sigma M_x = T_2 - T_1 - F \times R_B = 0 \qquad \text{(b)}$$

Next, consider a free-body diagram that cuts through shaft (3) and includes gear E as shown. A positive internal torque T_3 is assumed to act in shaft (3). Since the teeth of gear E exert a force F on the teeth of gear B, equilibrium requires that the teeth of gear B exert a force of equal magnitude in the opposite direction on the teeth of gear E. With the radius of gear E denoted by R_E, a moment equilibrium equation about the x' axis can be written as

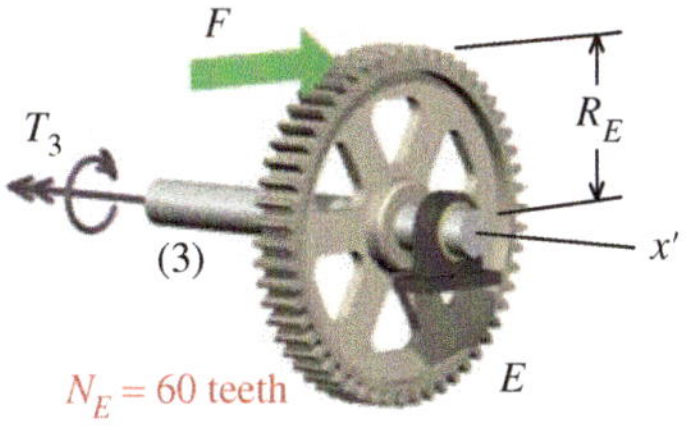

$$\Sigma M_{x'} = -T_3 - F \times R_E = 0 \qquad \therefore F = -\frac{T_3}{R_E} \qquad \text{(c)}$$

The results of Equations (a) and (c) can be substituted into Equation (b) to give

$$T_1 = T_2 - F \times R_B = 18 \text{ kip·in.} - \left(-\frac{T_3}{R_E}\right) R_B = 18 \text{ kip·in.} + T_3 \frac{R_B}{R_E}$$

The gear radii R_B and R_E are not known. However, the ratio R_B/R_E is simply the gear ratio between gears B and E. Since the teeth on both gears must be the same size in order for the gears to mesh properly, the ratio of the teeth on each gear is equivalent to the ratio of the gear radii. Consequently, the torque in shaft (1) can be expressed in terms of N_B and N_E, the number of teeth on gears B and E, respectively:

$$T_1 = 18 \text{ kip·in.} + T_3 \frac{N_B}{N_E} \qquad \text{(d)}$$

Equation (d) summarizes the results of the equilibrium considerations, but there are still two unknowns in it: T_1 and T_3. Consequently, this problem is statically indeterminate. To solve the problem, an additional equation must be developed. This second equation will be derived from the relationship between the angles of twist in shafts (1) and (3).

Step 2 — Geometry of Deformation: The rotation of gear B is equal to the angle of twist in shaft (1):

$$\phi_B = \phi_1$$

Similarly, the rotation of gear E is equal to the angle of twist in shaft (3):

$$\phi_E = \phi_3$$

However, since the gear teeth mesh, the rotation angles for gears B and E are not independent. The arclengths associated with the respective rotations must be equal, but the gears turn in opposite directions. Thus, the relationship between the gear rotations can be stated as

$$R_B \phi_B = -R_E \phi_E$$

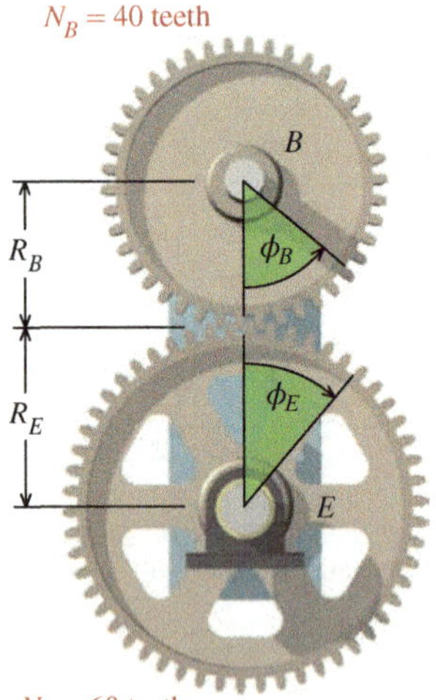

where R_B and R_E are the radii of gears B and E, respectively. Since the gear rotation angles are related to the shaft angles of twist, the preceding equation can be expressed as

$$R_B\phi_1 = -R_E\phi_3 \tag{e}$$

Step 3 — Torque–Twist Relationships: The angles of twist in shafts (1) and (3) can be expressed as

$$\phi_1 = \frac{T_1L_1}{J_1G_1} \quad \text{and} \quad \phi_3 = \frac{T_3L_3}{J_3G_3} \tag{f}$$

Step 4 — Compatibility Equation: Substitute the torque–twist relationships [Equation (f)] into the geometry-of-deformation equation [Equation (e)] to obtain

$$R_B\frac{T_1L_1}{J_1G_1} = -R_E\frac{T_3L_3}{J_3G_3}$$

which can be rearranged and expressed in terms of the gear ratio N_B/N_E:

$$\frac{N_B}{N_E}\frac{T_1L_1}{J_1G_1} = -\frac{T_3L_3}{J_3G_3} \tag{g}$$

Note: The compatibility equation has two unknowns: T_1 and T_3. This equation can be solved simultaneously with the equilibrium equation [Equation (d)] to calculate the internal torques in shafts (1) and (3).

Step 5 — Solve the Equations: Solve for the internal torque T_3 in Equation (g):

$$T_3 = -T_1\frac{N_B}{N_E}\left(\frac{L_1}{L_3}\right)\left(\frac{J_3}{J_1}\right)\left(\frac{G_3}{G_1}\right)$$

Then substitute the result into Equation (d):

$$\begin{aligned} T_1 &= 18 \text{ kip·in.} + T_3\frac{N_B}{N_E} \\ &= 18 \text{ kip·in.} + \left[-T_1\frac{N_B}{N_E}\left(\frac{L_1}{L_3}\right)\left(\frac{J_3}{J_1}\right)\left(\frac{G_3}{G_1}\right)\right]\frac{N_B}{N_E} \\ &= 18 \text{ kip·in.} - T_1\left(\frac{N_B}{N_E}\right)^2\left(\frac{L_1}{L_3}\right)\left(\frac{J_3}{J_1}\right)\left(\frac{G_3}{G_1}\right) \end{aligned}$$

Now group the T_1 terms to obtain

$$T_1\left[1 + \left(\frac{N_B}{N_E}\right)^2\left(\frac{L_1}{L_3}\right)\left(\frac{J_3}{J_1}\right)\left(\frac{G_3}{G_1}\right)\right] = 18 \text{ kip·in.} \tag{h}$$

Polar moments of inertia for the shafts are needed for this calculation. Shaft (1) is a solid 2.00 in. diameter shaft, and shaft (3) is a solid 2.50 in. diameter shaft. The polar moments of inertia for these shafts are as follows:

$$J_1 = \frac{\pi}{32}(2.00 \text{ in.})^4 = 1.570796 \text{ in.}^4$$

$$J_3 = \frac{\pi}{32}(2.50 \text{ in.})^4 = 3.834952 \text{ in.}^4$$

Both shafts have the same length, and both have the same shear modulus. Therefore, Equation (h) reduces to

$$T_1\left[1 + \left(\frac{40 \text{ teeth}}{60 \text{ teeth}}\right)^2 (1)\left(\frac{3.834952 \text{ in.}^4}{1.570796 \text{ in.}^4}\right)(1)\right] = T_1(2.085070) = 18 \text{ kip}\cdot\text{in.}$$

From this equation, the internal torque in shaft (1) is computed as $T_1 = 8.6328$ kip · in. Then, substitute this result back into Equation (d) to find that the internal torque in shaft (3) is $T_3 = -14.0508$ kip · in.

Shear Stresses

The maximum shear stress magnitudes in the three shafts can now be calculated from the elastic torsion formula:

$$\tau_1 = \frac{T_1 c_1}{J_1} = \frac{(8.6328 \text{ kip}\cdot\text{in.})(2.00 \text{ in.}/2)}{1.570796 \text{ in.}^4} = 5.50 \text{ ksi} \qquad \textbf{Ans.}$$

$$\tau_2 = \frac{T_2 c_2}{J_2} = \frac{(18 \text{ kip}\cdot\text{in.})(2.00 \text{ in.}/2)}{1.570796 \text{ in.}^4} = 11.46 \text{ ksi} \qquad \textbf{Ans.}$$

$$\tau_3 = \frac{T_3 c_3}{J_3} = \frac{(14.0508 \text{ kip}\cdot\text{in.})(2.50 \text{ in.}/2)}{3.834952 \text{ in.}^4} = 4.58 \text{ ksi} \qquad \textbf{Ans.}$$

Since only the shear stress magnitudes are required here, the absolute value of T_3 is used.

Rotation Angle of Gear *E*

The rotation angle of gear *E* is equal to the angle of twist in shaft (3):

$$\phi_E = \phi_3 = \frac{T_3 L_3}{J_3 G_3} = \frac{(-14.0508 \text{ kip}\cdot\text{in.})(24 \text{ in.})}{(3.834952 \text{ in.}^4)(12{,}000 \text{ ksi})} = -0.007328 \text{ rad} = -0.00733 \text{ rad}$$

Ans.

Rotation Angle of Gear *C*

The rotation angle of gear *C* is equal to the rotation angle of gear *B* plus the additional twist that occurs in shaft (2):

$$\phi_C = \phi_B + \phi_2$$

The rotation angle of gear *B* is equal to the angle of twist in shaft (1):

$$\phi_B = \phi_1 = \frac{T_1 L_1}{J_1 G_1} = \frac{(8.6328 \text{ kip}\cdot\text{in.})(24 \text{ in.})}{(1.570796 \text{ in.}^4)(12{,}000 \text{ ksi})} = 0.010992 \text{ rad}$$

Note: The rotation angle of gear *B* can also be found from the rotation angle of gear *E*:

$$\phi_B = -\frac{N_E}{N_B}\phi_E = -\frac{60}{40}(-0.007328 \text{ rad}) = 0.010992 \text{ rad}$$

The angle of twist in shaft (2) is

$$\phi_2 = \frac{T_2 L_2}{J_2 G_2} = \frac{(18 \text{ kip}\cdot\text{in.})(36 \text{ in.})}{(1.570796 \text{ in.}^4)(12{,}000 \text{ ksi})} = 0.034377 \text{ rad}$$

Therefore, the rotation angle of gear C is

$$\phi_C = \phi_B + \phi_2 = 0.010992 \text{ rad} + 0.034377 \text{ rad} = 0.045369 \text{ rad} = 0.0454 \text{ rad} \qquad \textbf{Ans.}$$

MecMovies

EXAMPLE

M6.24 An assembly of two solid brass [$G = 44$ GPa] shafts connected by gears is subjected to a concentrated torque of 240 N·m, as shown. Shaft (1) has a diameter of 20 mm, while the diameter of shaft (2) is 16 mm. Rotation at the lower end of each shaft is prevented. Determine the maximum shear stress in shaft (2) and the rotation angle at A.

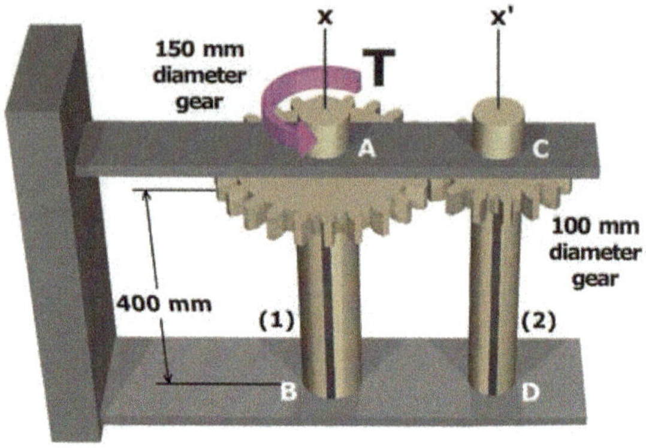

EXERCISES

M6.19 A composite torsion member consists of a tubular shell (1) bonded to length AB of a continuous solid shaft that extends from A to C, which is labeled (2) and (3). A concentrated torque T is applied to the free end C of the shaft in the direction shown. Determine the internal torques and shear stresses in shell (1) and core (2) (i.e., between A and B). Also, determine the rotation angle at end C.

M6.20 A composite torsion member consists of two solid shafts joined at flange B. Shafts (1) and (2) are attached to rigid supports at A and C, respectively. A concentrated torque T is applied to flange B in the direction shown. Determine the internal torques and shear stresses in each shaft. Also, determine the rotation angle of flange B.

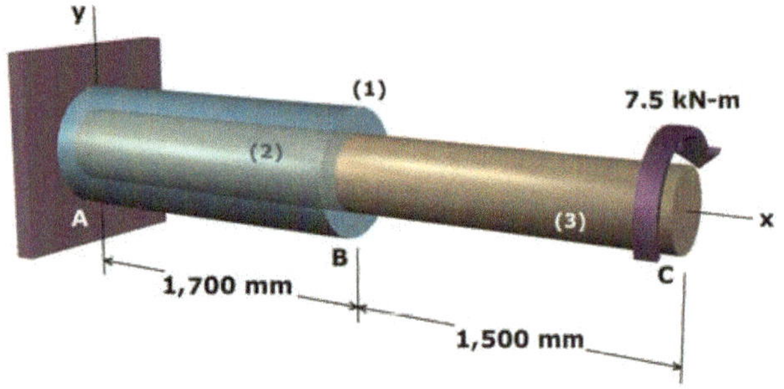

FIGURE M6.19

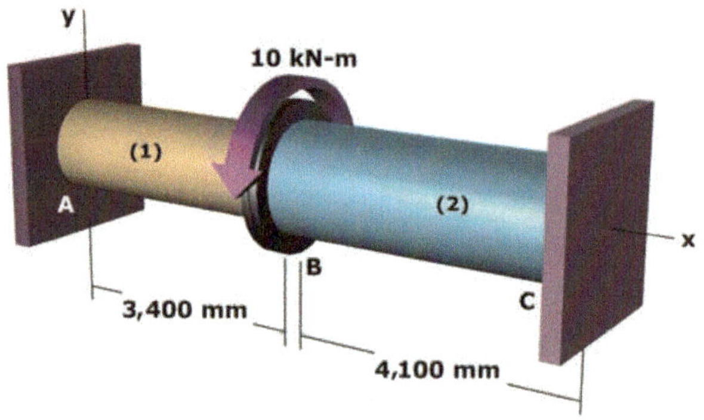

FIGURE M6.20

M6.21 A composite torsion member consists of two solid shafts joined at flange B. Shafts (1) and (2) are attached to rigid supports at A and C, respectively. Using the allowable shear stresses indicated on the sketch, determine the maximum torque T that may be applied to flange B in the direction shown. Determine the maximum shear stress in each shaft and the rotation angle of flange B at the maximum torque.

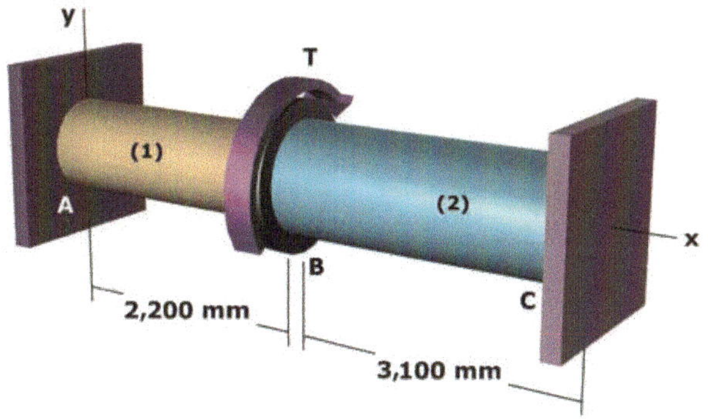

FIGURE M6.21

PROBLEMS

P6.39 A hollow circular cold-rolled stainless steel [G_1 = 12,500 ksi] tube (1) with an outside diameter of 2.25 in. and an inside diameter of 2.00 in. is securely shrink fitted to a solid 2.00 in. diameter cold-rolled bronze [G_2 = 6,500 ksi] core (2) as shown in Figure P6.39/40. The length of the assembly is L = 20 in. The allowable shear stress of tube (1) is 60 ksi, and the allowable shear stress of core (2) is 25 ksi. Determine

(a) the allowable torque T that can be applied to the tube-and-core assembly.
(b) the corresponding torques produced in tube (1) and core (2).
(c) the angle of twist produced in the assembly by the allowable torque T.

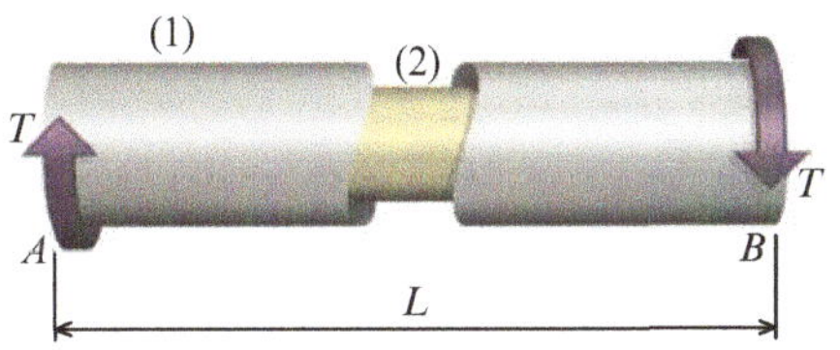

FIGURE P6.39/40

P6.40 The composite shaft shown in Figure P6.39/40 consists of an aluminum alloy tube (1) securely bonded to an inner brass core (2). The aluminum tube has an outside diameter of 60 mm, an inside diameter of 50 mm, and a shear modulus G_1 = 26 GPa. The solid brass core has a diameter of 50 mm and a shear modulus G_2 = 44 GPa. The length of the assembly is L = 600 mm. If the composite shaft is subjected to a torque T = 5.0 kN · m, determine

(a) the maximum shear stresses in the aluminum tube and the brass core.
(b) the rotation angle of end B relative to end A.

P6.41 A solid 2.50 in. diameter cold-rolled brass [G = 6,400 ksi] shaft that is $L_1 + L_2$ = 115 in. long extends through and is *completely bonded* to a hollow aluminum [G = 3,800] tube, as shown in Figure P6.41. The aluminum tube (1) has an outside diameter of 3.50 in., an inside diameter of 2.50 in., and a length L_1 = 70 in. Both the brass shaft and the aluminum tube are securely attached to the wall support at A. External torques in the directions shown are applied at B and C. The torque magnitudes are T_B = 6.0 kip · ft and T_C = 2.5 kip · ft, respectively. After T_B and T_C are applied to the composite shaft, determine

(a) the maximum shear stress magnitude in aluminum tube (1).
(b) the maximum shear stress magnitude in brass shaft segment (2).
(c) the maximum shear stress magnitude in brass shaft segment (3).
(d) the rotation angle of joint B.
(e) the rotation angle of end C.

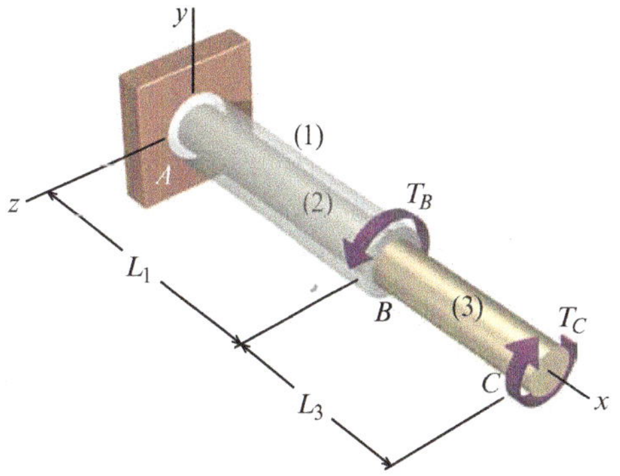

FIGURE P6.41

P6.42 The compound shaft shown in Figure P6.42/43 consists of two pipes that are connected at flange B and securely attached to rigid walls at A and C. Pipe (1) is made of an aluminum alloy [G = 26 GPa]. It has an outside diameter of 140 mm, a wall thickness of 7 mm, and a length L_1 = 9.0 m. Pipe (2) is made of steel [G = 80 GPa]. It has an outside diameter of 100 mm, a wall thickness of 6 mm, and a length L_2 = 7.5 m. If a concentrated torque of 15 kN · m is applied to flange B, determine

(a) the maximum shear stress magnitudes in pipes (1) and (2).
(b) the rotation angle of flange B relative to support A.

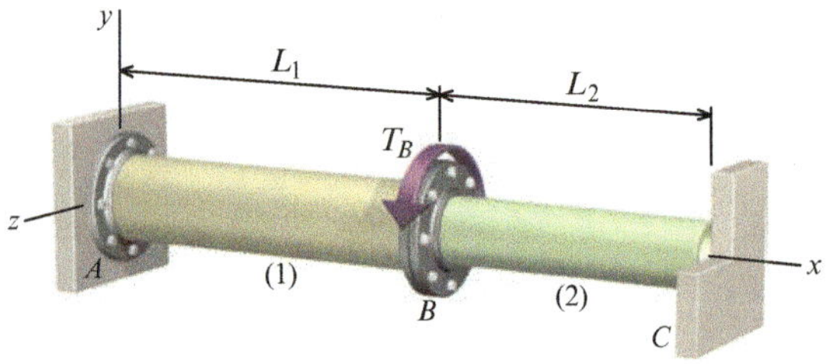

FIGURE P6.42/43

P6.43 The compound shaft shown in Figure P6.42/43 consists of a solid aluminum segment (1) and a hollow brass segment (2) that are connected at flange B and securely attached to rigid supports at A and C. Aluminum segment (1) has a diameter of 0.875 in., a length $L_1 = 45$ in., a shear modulus of 3,800 ksi, and an allowable shear stress of 6 ksi. Brass segment (2) has an outside diameter of 0.68 in., a wall thickness of 0.09 in., a length $L_2 = 30$ in., a shear modulus of 6,400 ksi, and an allowable shear stress of 8 ksi. Determine

(a) the allowable torque T_B that can be applied to the compound shaft at flange B.
(b) the magnitudes of the internal torques in segments (1) and (2).
(c) the rotation angle of flange B that is produced by the allowable torque T_B.

P6.44 In Figure P6.44, shafts (1) and (2) are identical solid 1.25 in. diameter steel [$G = 11{,}500$ ksi] shafts. Shaft (1) is fixed to a rigid support at A and shaft (2) is fixed to a rigid support at C. Shaft (1) has a length $L_1 = 4.25$ ft, and shaft (2) has a length $L_2 = 2.75$ ft. The allowable shear stress of the steel is 8,000 psi. The number of teeth on gears B and D is $N_B = 72$ teeth and $N_D = 54$ teeth, respectively. Determine the allowable magnitude of the torque T_E.

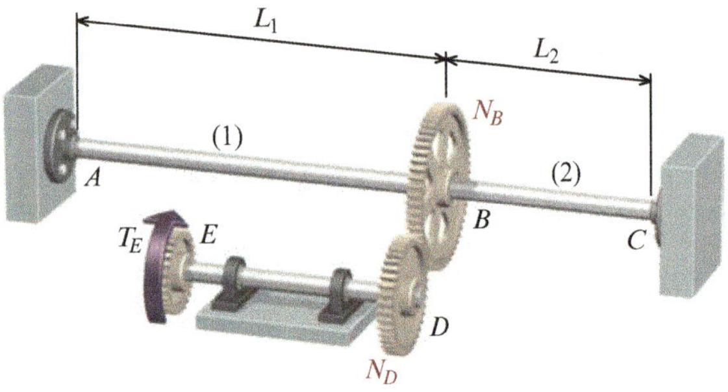

FIGURE P6.44

P6.45 The torsional assembly shown in Figure P6.45 consists of solid 2.50 in. diameter aluminum [$G = 4{,}000$ ksi] segments (1) and (3) and a central solid 3.00 in. diameter bronze [$G = 6{,}500$ ksi] segment (2). The segment lengths are $L_1 = L_3 = 36$ in. and $L_2 = 54$ in. Concentrated torques $T_B = 2T_0$ and $T_C = T_0$ are applied to the assembly at B and C, respectively. If the rotation angle at joint B must not exceed 3°, determine

(a) the maximum magnitude of T_0 that may be applied to the assembly.
(b) the maximum shear stress magnitude in aluminum segments (1) and (3).
(c) the maximum shear stress magnitude in bronze segment (2).

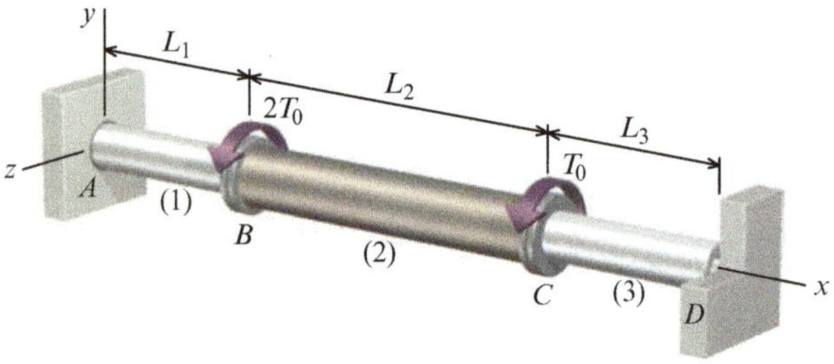

FIGURE P6.45

P6.46 The torsional assembly shown in Figure P6.46 consists of a hollow aluminum alloy [$G = 28$ GPa] segment (2) and two brass [$G = 44$ GPa] tube segments (1) and (3). The brass segments have an outside diameter of 114.3 mm, a wall thickness of 6.1 mm, and lengths $L_1 = L_3 = 2{,}400$ mm. The aluminum segment (2) has an outside diameter of 63.5 mm, a wall thickness of 3.2 mm, and a length $L_2 = 4{,}200$ mm. If concentrated torques $T_B = 4$ kN·m and $T_C = 7$ kN·m are applied in the directions shown, determine

(a) the maximum shear stress magnitude in brass tube segments (1) and (3).
(b) the maximum shear stress magnitude in aluminum segment (2).
(c) the rotation angle of joint C.

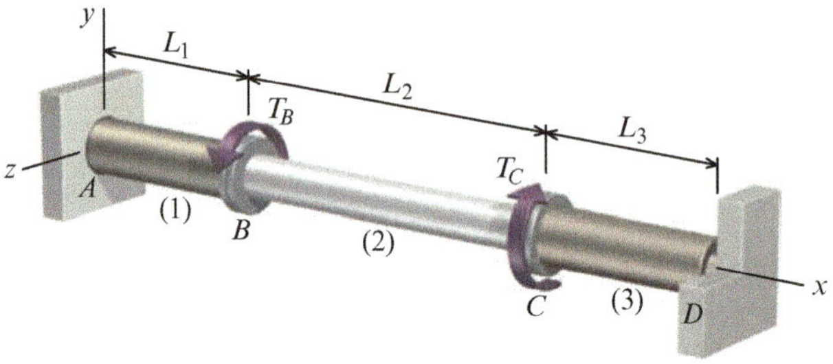

FIGURE P6.46

P6.47 The gear assembly shown in Figure P6.47 is subjected to a torque $T_C = 300$ N·m. Shafts (1) and (2) are solid 30 mm diameter steel [$G = 80$ GPa] shafts, and shaft (3) is a solid 35 mm diameter steel shaft. The shaft lengths are $L_1 = L_3 = 1{,}250$ mm and $L_2 = 850$ mm. The number of teeth on gears B and E is $N_B = 60$ teeth and $N_E = 108$ teeth, respectively. Determine

(a) the maximum shear stress magnitude in shaft (1).
(b) the maximum shear stress magnitude in shaft (3).
(c) the rotation angle of gear E.
(d) the rotation angle of gear C.

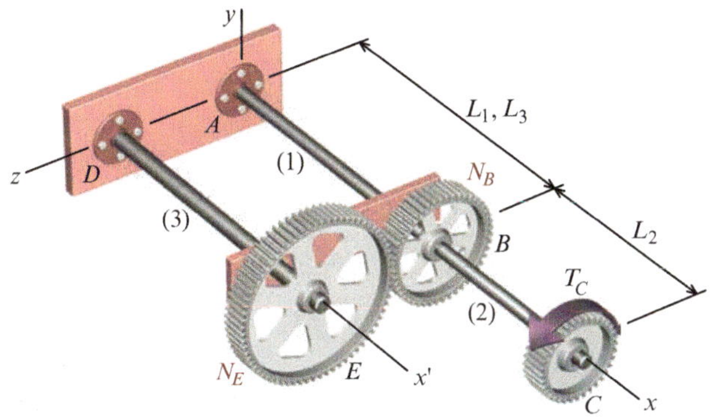

FIGURE P6.47

P6.48 The aluminum alloy [$G = 4{,}000$ ksi] pipe shown in Figure P6.48 is fixed to the wall support at C. The bolt holes in the flange at A were supposed to align with mating holes in the wall support; however, an angular misalignment of 4° was found to exist. To connect the pipe to its supports, a temporary installation torque T'_B must be applied at B to align flange A with the mating holes in the wall support. The outside diameter of the pipe is 5.5625 in. and its wall thickness is 0.258 in. The segment lengths are $L_1 = 16$ ft and $L_2 = 24$ ft.

(a) Determine the temporary installation torque T'_B that must be applied at B to align the bolt holes at A.
(b) Determine the maximum shear stress τ_{initial} in the pipe after the bolts are connected and the temporary installation torque at B is removed.
(c) Determine the magnitude of the maximum shear stress in segments (1) and (2) if an external torque $T_B = 150$ kip · in. is applied at B after the bolts are connected.

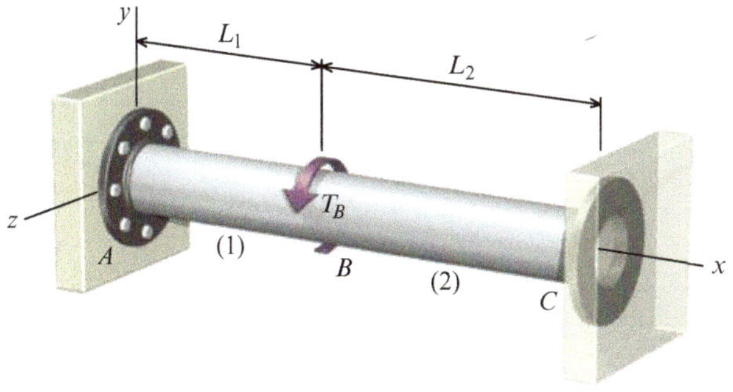

FIGURE P6.48

6.10 Stress Concentrations in Circular Shafts Under Torsional Loadings

In Section 5.7, it was shown that the introduction of a circular hole or other geometric discontinuity into an axially loaded member causes a significant increase in the magnitude of the stress in the immediate vicinity of the discontinuity. This phenomenon, called stress concentration, also occurs for circular shafts under torsional forms of loading.

Previously in this chapter, the maximum shear stress in a circular shaft of uniform cross section and made of a linearly elastic material was given by Equation (6.5):

$$\tau_{\max} = \frac{Tc}{J}$$

In the context of stress concentrations in circular shafts, this stress is considered a **nominal stress**, meaning that it gives the shear stress in regions of the shaft that are sufficiently removed from discontinuities. Shear stresses become much more intense near abrupt changes in shaft diameter, and Equation (6.5) does not predict the maximum stresses near discontinuities such as grooves or fillets. The maximum shear stress at discontinuities is expressed in terms of a stress-concentration factor

$$K = \frac{\tau_{\max}}{\tau_{\text{nom}}} \tag{6.21}$$

In this equation, τ_{nom} is the stress given by Tc/J for the minimum diameter of the shaft (termed the **minor diameter**) at the discontinuity.

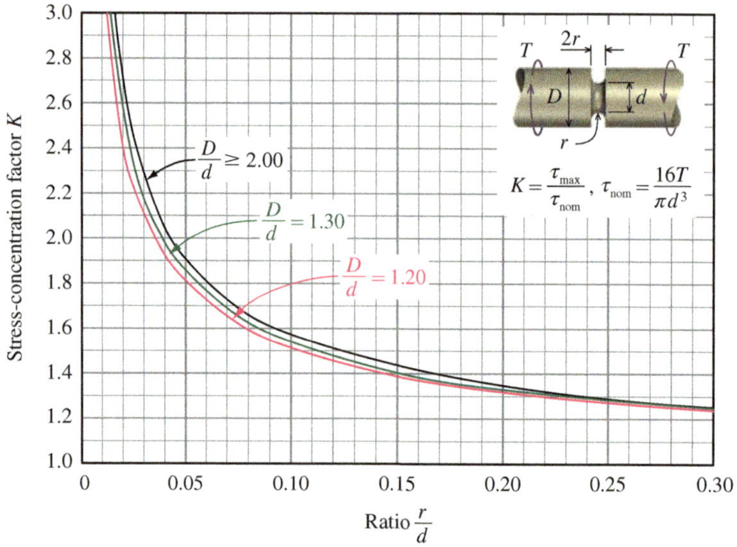

FIGURE 6.17 Stress-concentration factors K for a circular shaft with a U-shaped groove.

The full shaft diameter D at the discontinuity is termed the **major diameter**. The reduced shaft diameter d at the discontinuity is termed the **minor diameter**.

Stress-concentration factors K for circular shafts with U-shaped grooves and for stepped circular shafts are shown in Figures 6.17 and 6.18, respectively.[4] For both types of discontinuity, the stress-concentration factors K depend upon (a) the ratio D/d of the major diameter D to the minor diameter d and (b) the ratio r/d of the groove or fillet radius r to the minor diameter d. An examination of Figures 6.17 and 6.18 suggests that a generous fillet radius r should be used wherever a change in shaft diameter occurs. Equation (6.21) can be used to determine localized maximum shear stresses as long as the value of τ_{max} does not exceed the proportional limit of the material.

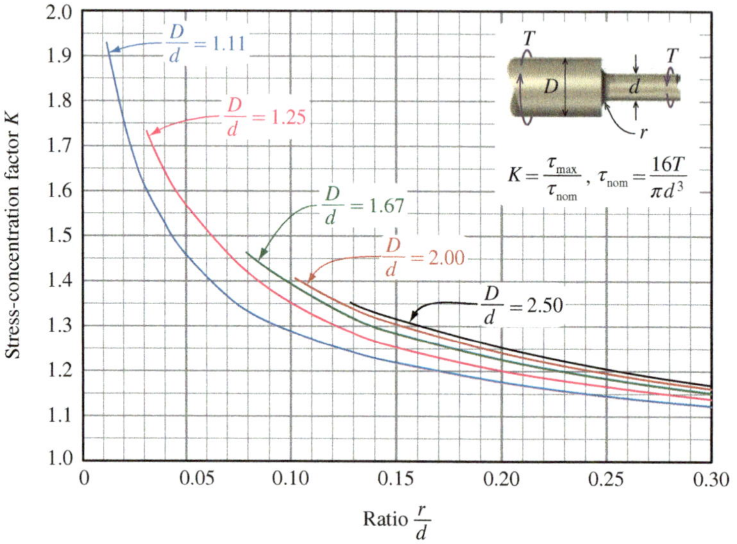

FIGURE 6.18 Stress-concentration factors K for a stepped shaft with shoulder fillets.

[4] Adapted from Walter D. Pilkey, *Peterson's Stress Concentration Factors*, 2nd ed. (New York: John Wiley & Sons, Inc., 1997).

Stress concentrations also occur at other features commonly found in circular shafts, such as oil holes and keyways used for attaching pulleys and gears to the shaft. Each of these discontinuities requires special consideration during the design process.

EXAMPLE 6.11

A stepped shaft has a 3 in. diameter for one-half of its length and a 1.5 in. diameter for the other half. If the maximum shear stress in the shaft must be limited to 8,000 psi when the shaft is transmitting a torque of 4,400 lb·in., determine the minimum fillet radius r needed at the junction between the two portions of the shaft.

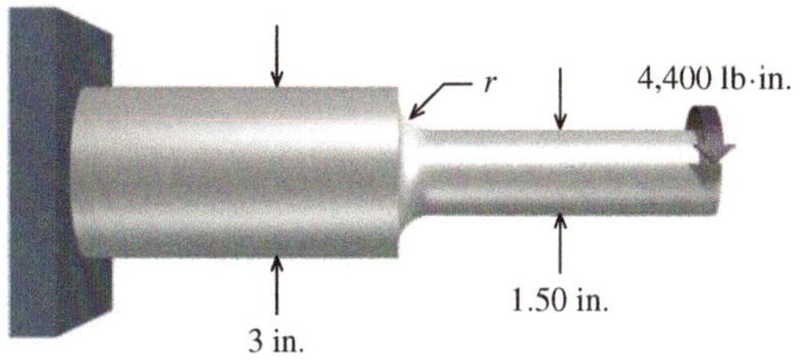

Plan the Solution

The maximum shear stress produced in the smaller diameter (i.e., minor-diameter) segment of the shaft will be determined. From this shear stress and the allowable shear stress, the maximum allowable stress-concentration factor K can be determined. With the allowable K and the other parameters of the shaft, Figure 6.18 can be used to determine the minimum permissible fillet radius.

SOLUTION

The maximum shear stress produced by the 4,400 lb·in. torque in the minor-diameter shaft segment is

$$\tau_{nom} = \frac{Tc}{J} = \frac{(4{,}400 \text{ lb}\cdot\text{in.})(0.75 \text{ in.})}{\dfrac{\pi}{32}(1.5 \text{ in.})^4} = 6{,}639.7 \text{ psi}$$

Since the maximum shear stress in the fillet between the two portions of the shaft must be limited to 8,000 psi, the maximum permissible value for the stress-concentration factor K, on the basis of the nominal shear stress in the minor-diameter section, is

$$K = \frac{\tau_{max}}{\tau_{nom}} \qquad \therefore K \le \frac{8{,}000 \text{ psi}}{6{,}639.7 \text{ psi}} = 1.20$$

The stress-concentration factor K depends on two ratios: D/d and r/d. For the 3 in. diameter shaft with the 1.5 in. diameter reduced section, the ratio $D/d = (3.00 \text{ in.})/(1.50 \text{ in.}) = 2.00$. From the curves in Figure 6.18, a ratio $r/d = 0.238$ together with a ratio $D/d = 2.00$ will produce a stress-concentration factor $K = 1.20$. Thus, the minimum permissible radius for the fillet between the two portions of the shaft is

$$\frac{r}{d} \ge 0.238 \qquad \therefore r \ge 0.238\,(1.50 \text{ in.}) = 0.357 \text{ in.} \qquad \textbf{Ans.}$$

PROBLEMS

P6.49 A stepped shaft with a major diameter $D = 20$ mm and a minor diameter $d = 16$ mm is subjected to a torque of 25 N·m. A full quarter-circular fillet having a radius $r = 2$ mm is used to transition from the major diameter to the minor diameter. Determine the maximum shear stress in the shaft.

P6.50 A fillet with a radius of 1/2 in. is used at the junction of a stepped shaft where the diameter is reduced from 8.00 in. to 6.00 in. Determine the maximum torque that the shaft can transmit if the maximum shear stress in the fillet must be limited to = ksi.

P6.51 A stepped shaft with a major diameter $D = 2.50$ in. and a minor diameter $d = 1.25$ in. is subjected to a torque of 1,200 lb·in. If the maximum shear stress must not exceed 4,000 psi, determine the minimum radius r that may be used for a fillet at the junction of the two shaft segments. The fillet radius must be chosen as a multiple of 0.05 in.

P6.52 A stepped shaft has a major diameter $D = 100$ mm and a minor diameter $d = 75$ mm. A fillet with a 10 mm radius is used to transition between the two shaft segments. The maximum shear stress in the shaft must be limited to 60 MPa. If the shaft rotates at a constant angular speed of 500 rpm, determine the maximum power that may be delivered by the shaft.

P6.53 A semicircular groove with a 6 mm radius is required in a 50 mm diameter shaft. If the maximum allowable shear stress in the shaft must be limited to 40 MPa, determine the maximum torque that can be transmitted by the shaft.

P6.54 A 1.25 in. diameter shaft contains a 0.25 in. deep U-shaped groove that has a 1/8 in. radius at the bottom of the groove. The maximum shear stress in the shaft must be limited to 12,000 psi. If the shaft rotates at a constant angular speed of 6 Hz, determine the maximum power that may be delivered by the shaft.

6.11 Torsion of Noncircular Sections

Prior to 1820, the year that A. Duleau published experimental results to the contrary, it was thought that the shear stresses in any torsionally loaded member were proportional to the distance from the longitudinal axis of the member. Duleau proved experimentally that relationship does not hold for rectangular cross sections. An examination of Figure 6.19 will verify Duleau's conclusion. If the stresses in the rectangular bar were proportional to the distance from its axis, the maximum stress would occur at the corners. However, if there is a stress of any magnitude at a corner, as indicated in Figure 6.19*a*, it could be resolved into the components indicated in Figure 6.19*b*. If these components existed, the two components shown by the blue arrows would also exist. But these last components cannot exist, since the surfaces on which they are shown are free boundaries. Therefore, the shear stresses at the corners of the rectangular bar must be zero.

The first correct analysis of the torsion of a prismatic bar of noncircular cross section was published by Saint-Venant in 1855; however, the scope of this analysis is beyond the elementary discussions of this book.[5] The results of Saint-Venant's analysis indicate that, in general, every section will warp (i.e., not remain plane) when twisted, *except for members with circular cross sections.*

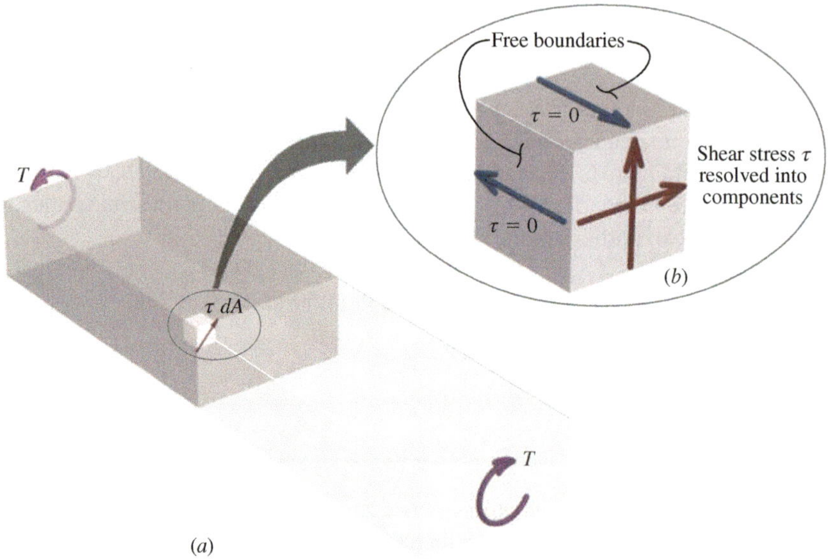

FIGURE 6.19 Torsional shear stresses in a rectangular bar.

[5] A complete discussion of the theory is presented in various books, such as I. S. Sokolnikoff, *Mathematical Theory of Elasticity*, 2nd. ed. (New York: McGraw-Hill, 1956), pp. 109–134.

For the case of the rectangular bar shown in Figure 6.2*d*, the distortion of the small squares is greatest at the midpoint of a side of the cross section and disappears at the corners. Since this distortion is a measure of shear strain, Hooke's law requires that the shear stress be largest at the midpoint of a side of the cross section and zero at the corners. Equations for the maximum shear stress and the angle of twist for a rectangular section obtained from Saint-Venant's theory are

$$\tau_{max} = \frac{T}{\alpha a^2 b} \tag{6.22}$$

and

$$\phi = \frac{TL}{\beta a^3 bG} \tag{6.23}$$

where a and b are the lengths of the short and long sides of the rectangle, respectively. The numerical constants α and β can be obtained from Table 6.1.[6]

Table 6.1 Table of Constants for Torsion of a Rectangular Bar

Ratio b/a	α	β
1.0	0.208	0.1406
1.2	0.219	0.166
1.5	0.231	0.196
2.0	0.246	0.229
2.5	0.258	0.249
3.0	0.267	0.263
4.0	0.282	0.281
5.0	0.291	0.291
10.0	0.312	0.312
∞	0.333	0.333

Narrow Rectangular Cross Sections

In Table 6.1, we observe that values for α and β are equal for $b/a \geq 5$. For aspect ratios $b/a \geq 5$, the coefficients α and β that respectively appear in Equations (6.22) and (6.23) can be calculated from the following equation:

$$\alpha = \beta = \frac{1}{3}\left(1 - 0.630\frac{a}{b}\right) \tag{6.24}$$

As a practical matter, an aspect ratio $b/a \geq 21$ is sufficiently large that values of $\alpha = \beta = 0.333$ can be used to calculate maximum shear stresses and deformations in narrow rectangular bars within an accuracy of 3 percent. Accordingly, equations for the maximum shear stress and angle of twist in narrow rectangular bars can be expressed as

$$\tau_{max} = \frac{3T}{a^2 b} \tag{6.25}$$

[6] See S. P. Timoshenko and J. N. Goodier, *Theory of Elasticity*, 3rd ed. (New York: McGraw-Hill, 1969), Section 109.

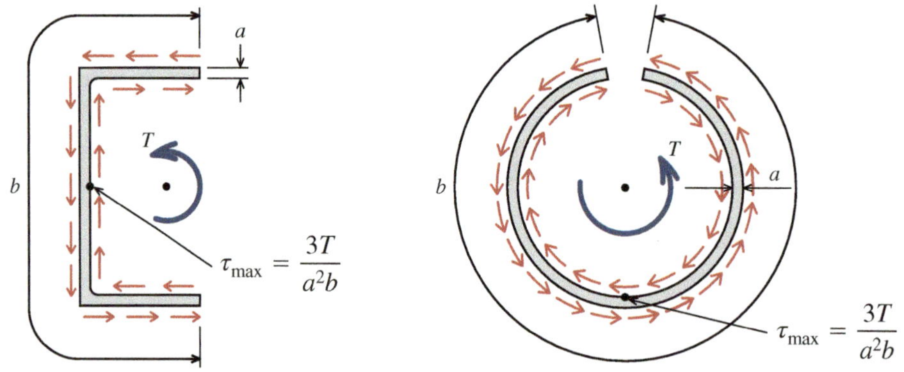

FIGURE 6.20 Equivalent narrow rectangular sections with shear stress distribution.

and

$$\phi = \frac{3TL}{a^3bG} \tag{6.26}$$

The absolute value of the maximum shear stress in a narrow rectangular bar occurs on the edge of the bar in the middle of the long side. For a thin-walled member of uniform thickness and arbitrary shape, the maximum shear stress and the shear stress distribution are equivalent to those quantities in a rectangular bar with a large b/a ratio. Thus, Equations (6.25) and (6.26) can be used to compute the maximum shear stress and the angle of twist for thin-walled shapes such as those shown in Figure 6.20. For use in these equations, the length a is taken as the thickness of the thin-walled shape. The length b is equal to the length of the thin-walled shape as measured along the centerline of the wall.

EXAMPLE 6.12

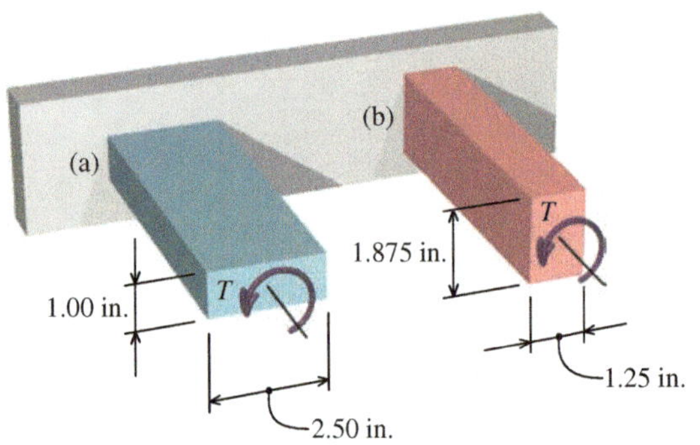

The two rectangular polymer bars shown are each subjected to a torque $T = 2{,}000$ lb·in. For each bar, determine

(a) the maximum shear stress.
(b) the rotation angle at the free end if the bar has a length of 12 in. Assume that $G = 500$ ksi for the polymer material.

Plan the Solution

The aspect ratio b/a for each bar will be computed. On the basis of this ratio, the constants α and β will be determined from Table 6.1. The maximum shear stress and rotation angles will be computed from Equations (6.22) and (6.23), respectively.

SOLUTION

For bar (a), the long side of the bar is $b = 2.50$ in. and the short side is $a = 1.00$ in.; therefore, $b/a = 2.5$. From Table 6.1, $\alpha = 0.258$ and $\beta = 0.249$.

The maximum shear stress produced in bar (a) by a torque $T = 2{,}000$ lb·in. is

$$\tau_{max} = \frac{T}{\alpha a^2 b} = \frac{2{,}000 \text{ lb·in.}}{(0.258)(1.00 \text{ in.})^2(2.50 \text{ in.})} = 3{,}100 \text{ psi} \qquad \textbf{Ans.}$$

and the angle of twist for a 12 in. long bar is

$$\phi = \frac{TL}{\beta a^3 bG} = \frac{(2{,}000 \text{ lb}\cdot\text{in.})(12 \text{ in.})}{(0.249)(1.00 \text{ in.})^3(2.50 \text{ in.})(500{,}000 \text{ psi})} = 0.0771 \text{ rad} \qquad \textbf{Ans.}$$

For bar (b), the long side of the bar is $b = 1.875$ in. and the short side is $a = 1.25$ in.; therefore, $b/a = 1.5$. From Table 6.1, $\alpha = 0.231$ and $\beta = 0.196$.

The maximum shear stress produced in bar (b) by a torque $T = 2{,}000$ lb·in. is

$$\tau_{\text{max}} = \frac{T}{\alpha a^2 b} = \frac{2{,}000 \text{ lb}\cdot\text{in.}}{(0.231)(1.25 \text{ in.})^2(1.875 \text{ in.})} = 2{,}960 \text{ psi} \qquad \textbf{Ans.}$$

and the angle of twist for a 12 in. long bar is

$$\phi = \frac{TL}{\beta a^3 bG} = \frac{(2{,}000 \text{ lb}\cdot\text{in.})(12 \text{ in.})}{(0.196)(1.25 \text{ in.})^3(1.875 \text{ in.})(500{,}000 \text{ psi})} = 0.0669 \text{ rad} \qquad \textbf{Ans.}$$

6.12 Torsion of Thin-Walled Tubes: Shear Flow

The elementary torsion theory presented in Sections 6.1, 6.2, and 6.3 is limited to circular sections; however, one class of noncircular sections can be readily analyzed by elementary methods. These shapes are thin-walled tubes such as the one illustrated in Figure 6.21*a*, which represents a noncircular section with a wall of variable thickness (i.e., t varies).

A useful concept associated with the analysis of thin-walled sections is the **shear flow** q, defined as the internal shearing force per unit of length of the thin section. Typical units for q are pounds per inch or newtons per meter. In terms of stress, q equals $\tau \times t$, where τ is the average shear stress across the thickness t.

First, we will demonstrate that the shear flow on a cross section is constant even though the wall thickness of the section may vary. Figure 6.21*b* shows a block cut from the member of Figure 6.21*a* between A and B. Since the member is subjected to pure torsion,

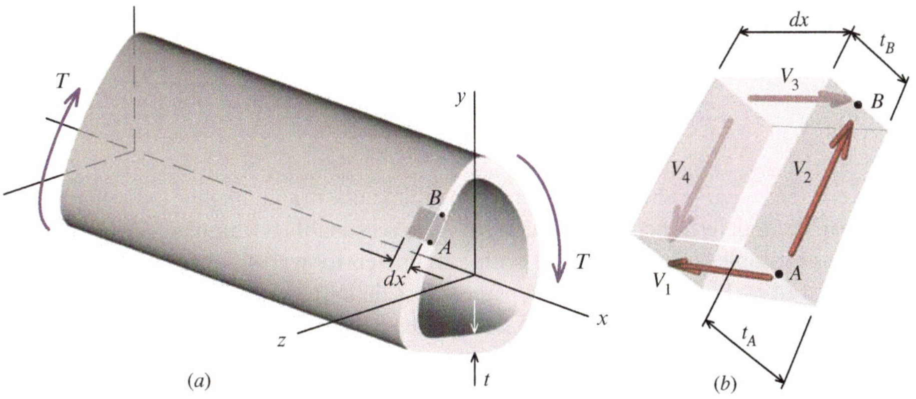

FIGURE 6.21 Shear flow in thin-walled tubes.

the shear forces V_1, V_2, V_3, and V_4 alone are necessary and sufficient for equilibrium (i.e., no normal forces are involved). Summing forces in the x direction gives

$$V_1 = V_3$$

or

$$q_1\,dx = q_3\,dx$$

from which it follows that

$$q_1 = q_3$$

Note that the shear flow and the shear stress always act tangent to the wall of the tube.

and, since $q = \tau \times t$,

$$\boxed{\tau_1 t_A = \tau_3 t_B} \tag{a}$$

The shear stresses at point A on the longitudinal and transverse planes have the same magnitude; likewise, the shear stresses at point B on the longitudinal and transverse planes have the same magnitude. Consequently, Equation (a) may be written as

$$\tau_A t_A = \tau_B t_B$$

or

$$q_A = q_B$$

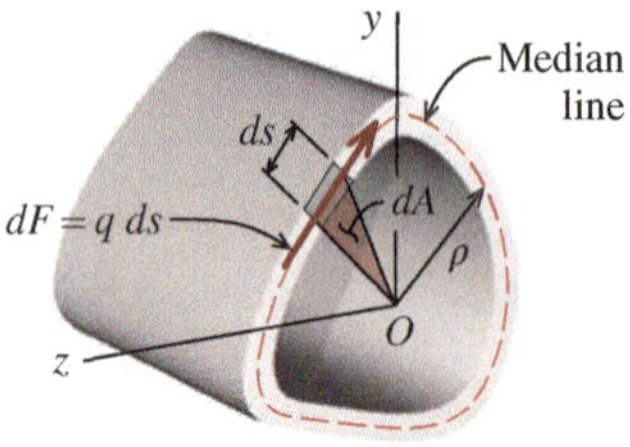

FIGURE 6.22 Deriving a relationship between internal torque and shear stress in a thin-walled section.

which demonstrates that the *shear flow on a cross section is constant* even though the wall thickness of the section varies. Since q is constant over a cross section, the *largest* average shear stress will occur where the wall thickness is the *smallest.*

Next, an expression relating torque and shear stress will be developed. Consider the force dF acting through the center of a differential length of perimeter ds, as shown in Figure 6.22. The differential moment produced by dF about the origin O is simply $\rho \times dF$, where ρ is the mean radial distance from the perimeter element to the origin. The internal torque equals the resultant of all of the differential moments; that is,

$$T = \int (dF)\rho = \int (q\,ds)\,\rho = q\int \rho\,ds$$

This integral may be difficult to integrate by formal calculus; however, the quantity $\rho\,ds$ is twice the area of the triangle shown shaded in Figure 6.22, which makes the integral equal to twice the area A_m *enclosed by the median line.* In other words, A_m is the mean area enclosed within the boundary of the tube wall *centerline.* The resulting expression relates the torque T and shear flow q:

$$\boxed{T = q(2A_m)} \tag{6.27}$$

Or, in terms of stress,

$$\boxed{\tau = \frac{T}{2A_m t}} \tag{6.28}$$

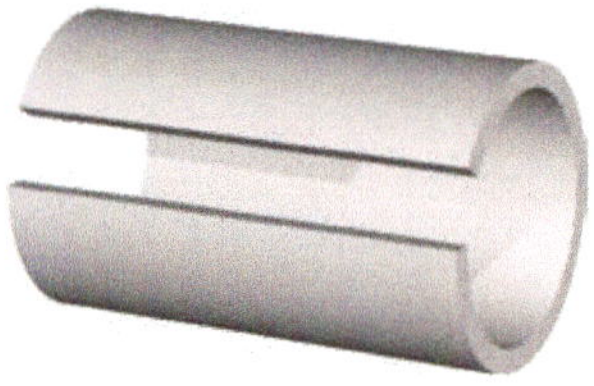

FIGURE 6.23 Thin-walled shape with an "open" cross section.

where τ is the *average* shear stress across the thickness t (and tangent to the perimeter). The shear stress determined by Equation (6.28) is reasonably accurate when t is relatively small. For example, in a round tube with a diameter–to–wall-thickness ratio of 20, the stress as given by Equation (6.28) is 5 percent less than that given by the torsion formula. It must be emphasized that Equation (6.28) applies only to "closed" sections—that is, sections with a continuous periphery. If the member were slotted longitudinally (e.g., see Figure 6.23), the resistance to torsion would be diminished considerably from that of the closed section.

EXAMPLE 6.13

A rectangular box section of aluminum alloy has outside dimensions of 100 mm by 50 mm. The plate thickness is 2 mm for the 50 mm sides and 3 mm for the 100 mm sides. If the maximum shear stress must be limited to 95 MPa, determine the maximum torque T that can be applied to the section.

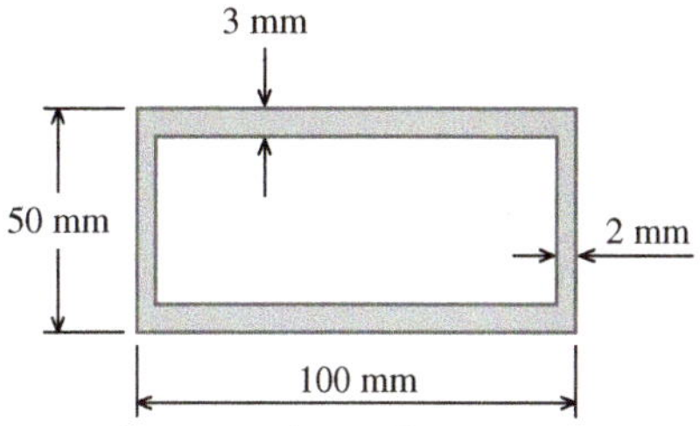

Cross-sectional dimensions.

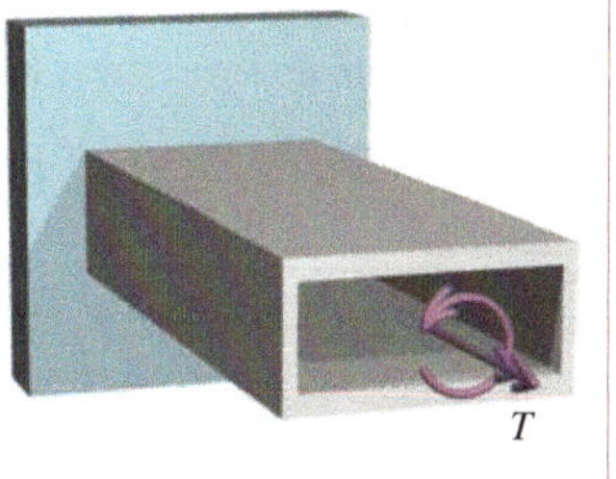

Plan the Solution

The maximum shear stress will occur in the thinnest plate. From the allowable shear stress, the shear flow in the thinnest plate will be calculated. Next, the area A enclosed by the median line (see Figure 6.22) of the section wall will be calculated. Finally, the maximum torque will be computed from Equation (6.27).

SOLUTION

The maximum shear stress will occur in the thinnest plate; therefore, the critical shear flow q is

$$q = \tau t = (95 \text{ N/mm}^2)(2 \text{ mm}) = 190 \text{ N/mm}$$

The area enclosed by the median line is

$$A_m = (100 \text{ mm} - 2 \text{ mm})(50 \text{ mm} - 3 \text{ mm}) = 4{,}606 \text{ mm}^2$$

Finally, the torque that can be transmitted by the section is computed from Equation (6.27):

$$T = q(2A_m) = (190 \text{ N/mm})(2)(4{,}606 \text{ mm}^2) = 1{,}750{,}280 \text{ N}\cdot\text{mm} = 1{,}750 \text{ N}\cdot\text{m} \quad \textbf{Ans.}$$

PROBLEMS

P6.55 A solid 1.50 in. × 1.50 in. square shaft is made of steel [G = 11,500 ksi] that has an allowable shear stress of 16 ksi. The shaft is used to transmit power from a tractor to farm implements, and it has a length of 6 ft. Determine

(a) the largest torque T that may be transmitted by the shaft.
(b) the angle of twist in the shaft at the torque found in part (a).

P6.56 The allowable shear stress for each bar shown in Figure P6.56/57 is 50 ksi. Assume that a = 0.5 in. and determine

(a) the largest torque T that may be applied to each bar.
(b) the corresponding rotation angle at the free end if each bar has a length of 48 in. Assume that G = 12,500 ksi.

P6.57 The bars shown in Figure P6.56/57 have equal cross-sectional areas, and they are each subjected to a torque T = 550 N · m. Using a = 10 mm, determine

(a) the maximum shear stress in each bar.
(b) the rotation angle at the free end if each bar has a length of 900 mm. Assume that G = 28 GPa.

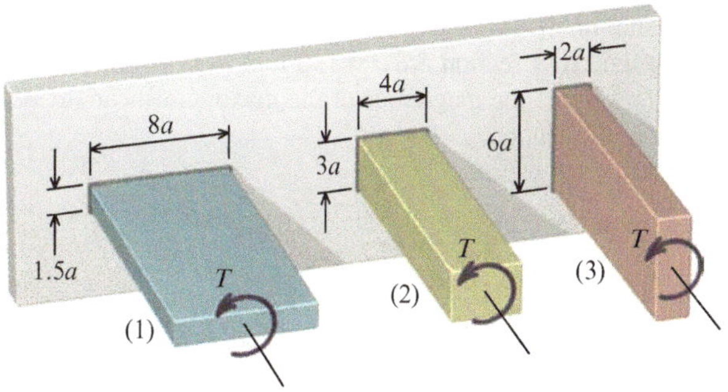

FIGURE P6.56/57

P6.58 The steel [G = 11,200 ksi] angle shape shown in Figure P6.58 is subjected to twisting moments of T = 400 lb · ft about its longitudinal axis (i.e., the x axis) at its ends. The angle shape is

36 in. long, and its cross-sectional dimensions are $b = 3.0$ in., $d = 5.0$ in., and $t = 0.375$ in. Determine

(a) the angle of twist of one end relative to the other end.
(b) the maximum shear stress in the angle.

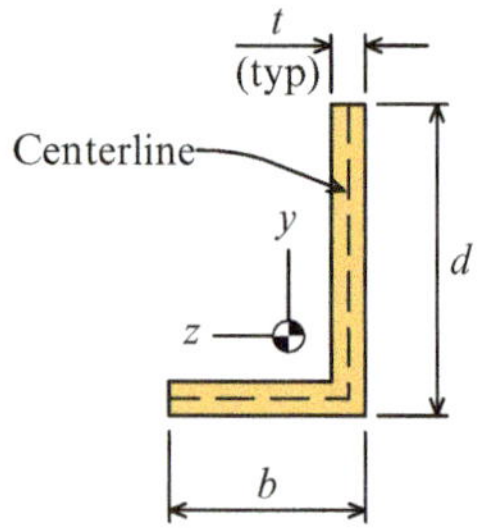

FIGURE P6.58

P6.59 A torque $T = 9{,}000$ lb·ft will be applied to the hollow, thin-walled stainless steel section shown in Figure P6.59. The dimensions of the cross section are $a = 4.0$ in., $b = 6.0$ in., $c = 5.0$ in., and $d = 8.0$ in. If the maximum shear stress must be limited to 12 ksi, determine the minimum thickness t required for the section. (*Note:* The dimensions shown are measured to the wall centerline.)

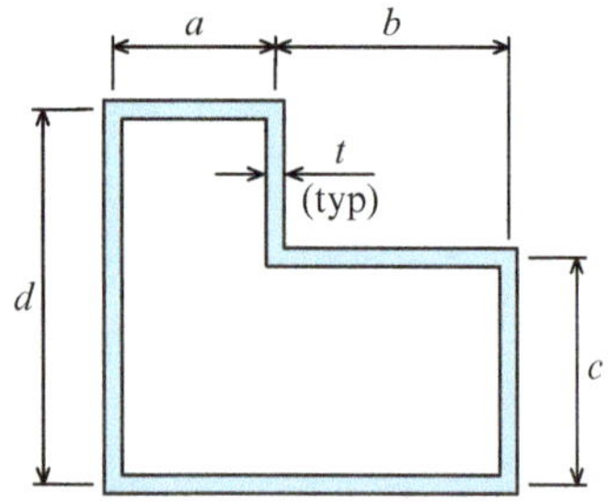

FIGURE P6.59

P6.60 A torque $T = 9.7$ kN·m will be applied to the hollow, thin-walled aluminum alloy section shown in Figure P6.60. The dimensions of the cross section are $a = 150$ mm, $t_1 = 15$ mm, and $t_2 = 6$ mm. Determine the magnitude of the maximum shear stress developed in the section.

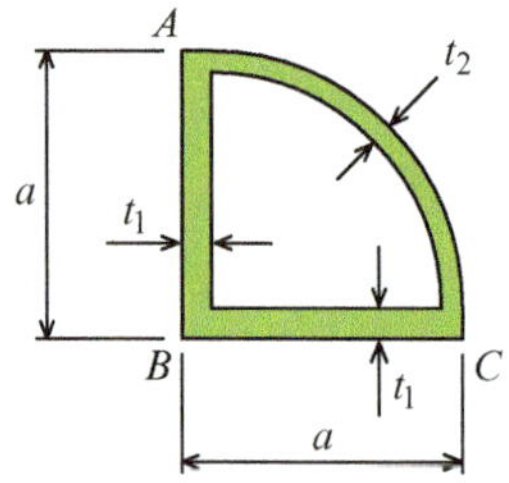

FIGURE P6.60

P6.61 A torque $T = 4{,}200$ kN·m is applied to a torsion member whose cross section is shown in Figure P6.61. The dimensions of the cross section are $a = 600$ mm, $b = 1{,}080$ mm, $d = 1{,}660$ mm, $t_1 = 18$ mm, $t_2 = 9$ mm, and $t_3 = 15$ mm. The dimensions given for a, b, and d are measured to the wall centerline. Determine the average shear stress acting at points A, B, and C.

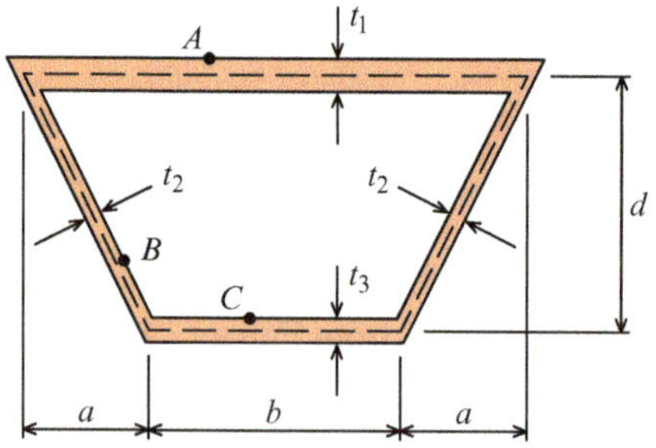

FIGURE P6.61

P6.62 A torque $T = 190$ lb·ft is applied to a torsion member whose cross section is shown in Figure P6.62. The dimensions of the cross section are $a = 0.75$ in., $b = 2.25$ in., $c = 1.00$ in., $d = 2.75$ in., $t_1 = 0.065$ in., and $t_2 = 0.120$ in. Determine the maximum average shear stress developed in the section. (*Note:* The dimensions shown are measured to the wall centerline.)

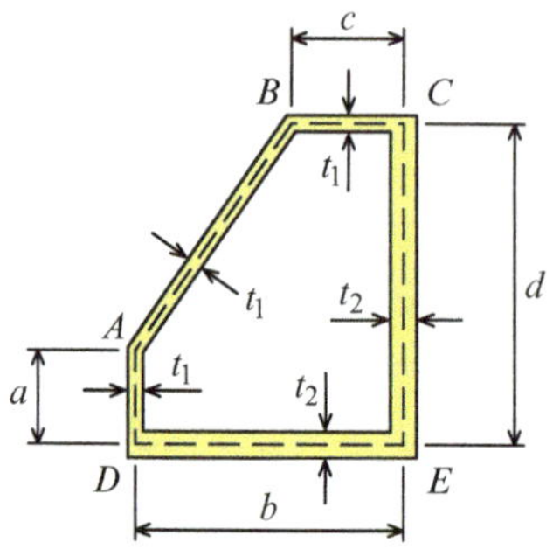

FIGURE P6.62

CHAPTER 7

Equilibrium of Beams

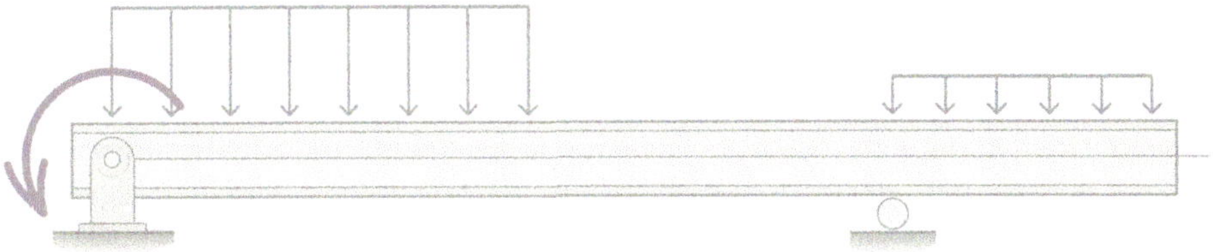

7.1 Introduction

The behavior of slender structural members subjected to axial loads and to torsional loadings was discussed in Chapters 5 and 6, respectively. This chapter begins the consideration of beams, one of the most common and important components used in structural and mechanical applications. **Beams** are usually long (compared with their cross-sectional dimensions), straight, prismatic members which support loads that act perpendicular to the longitudinal axis of the member. They resist transverse applied loads by a combination of internal shear force and bending moment.

The term **transverse** refers to loads and sections that are perpendicular to the longitudinal axis of the member.

Types of Supports

Beams are normally classified by the manner in which they are supported. Figure 7.1 shows graphic symbols used to represent three types of supports:

- Figure 7.1*a* shows a **pin support**. A pin support prevents translation in two orthogonal directions. For beams, this condition means that displacements parallel to the longitudinal axis of the beam (i.e., the x direction in Figure 7.1*a*) and displacements perpendicular to the longitudinal axis (i.e., the y direction in Figure 7.1*a*) are restrained at the supported joint. Note that, while translation is restrained by a pin support, rotation of the joint is permitted. In Figure 7.1*a*, the beam is free to rotate about the z axis and reaction forces act on the beam in the x and y directions.

- Figure 7.1*b* shows a **roller support**. A roller support prevents translation perpendicular to the longitudinal axis of the beam (i.e., the *y* direction in Figure 7.1*b*); however, the joint is free to translate in the *x* direction and to rotate about the *z* axis. Unless specifically stated otherwise, a roller support should be assumed to prevent joint displacement both in the +*y* and −*y* directions. The roller support in Figure 7.1*b* provides a reaction force to the beam in the *y* direction only.
- Figure 7.1*c* shows a **fixed support**. A fixed support prevents both translation and rotation at the supported joint. The fixed support shown in Figure 7.1*c* provides reaction forces to the beam in the *x* and *y* directions, as well as a reaction moment in the *z* direction. This type of support is sometimes called a **moment connection**.

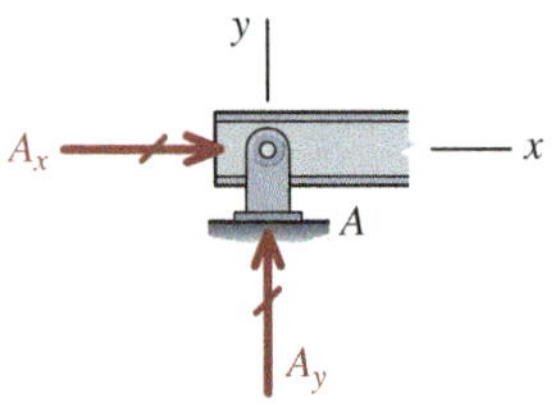

(*a*) Pin support

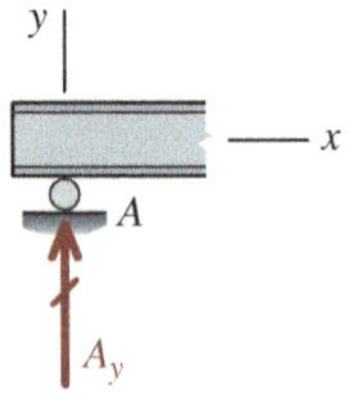

(*b*) Roller support

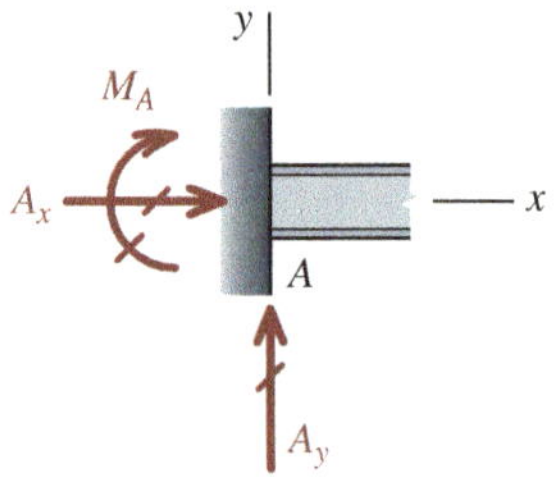

(*c*) Fixed support

FIGURE 7.1 Types of supports.

Figure 7.1 shows *symbols* that represent three types of supports commonly associated with beams. It is important to keep in mind that these symbols are simply graphic shorthand used to communicate the beam support conditions easily. Actual pin, roller, and fixed supports may take many configurations. Figure 7.2 shows one possibility for each type of connection.

One type of pin support is shown in Figure 7.2*a*. In this connection, three bolts are used to attach the beam to a small component called a *clip angle*, which in turn is bolted to the vertical supporting member (called a **column**). The bolts prevent the beam from moving either horizontally or vertically. Strictly speaking, the bolts also provide some resistance against rotation at the joint. Since the bolts are located close to the middle of the beam, however, they are not capable of fully restraining rotation at the connection. This type of connection permits enough rotation so that the joint is classified as a pin connection.

Figure 7.2*b* shows one type of roller connection. The bolts are inserted into slotted holes in a small plate called a *shear tab*. Since the bolts are in slots, the beam is free to deflect in the horizontal direction, but it is restrained from deflecting either upward or downward. Slotted holes are sometimes used to facilitate the construction process, making it easier for heavy beams to be quickly attached to columns.

Figure 7.2*c* shows a welded steel moment connection. Notice that extra plates are welded to the top and bottom surfaces of the beam and that these plates are connected directly to the column. These extra plates prevent the beam from rotating at the joint.

Types of Statically Determinate Beams

Beams are further classified by the manner in which the supports are arranged. Figure 7.3 shows three common statically determinate beams. Figure 7.3*a* shows a **simply supported**

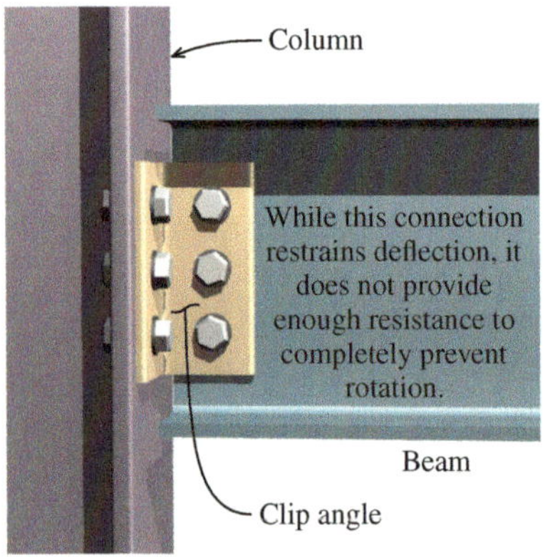

(*a*) Pin support

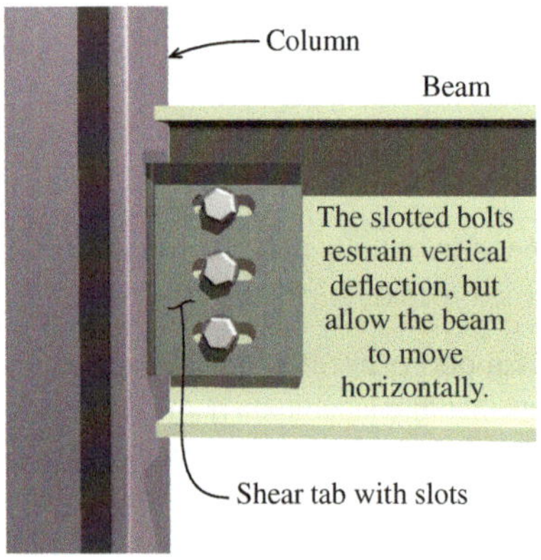

(*b*) Roller support

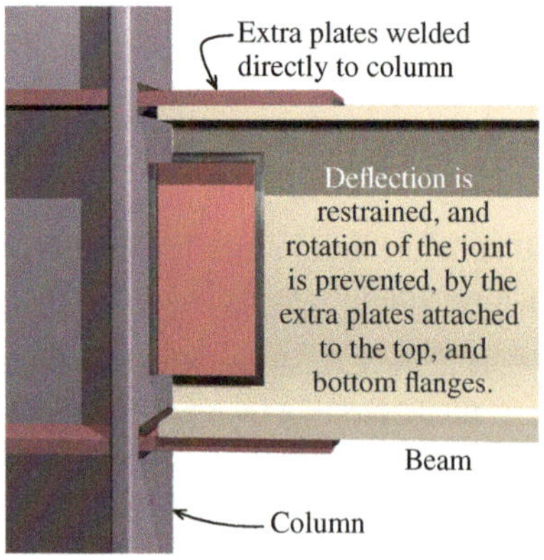

(*c*) Fixed support

FIGURE 7.2 Examples of actual beam supports.

beam (also called a **simple beam**). A simply supported beam has a pin support at one end and a roller support at the opposite end. Figure 7.3*b* shows a variation of the simply supported beam in which the beam continues across the support in what is termed an **overhang**. In both cases, the pin and roller supports provide three reaction forces for the simply supported beam: a horizontal reaction force at the pin and vertical reaction forces at both the pin and the roller. Figure 7.3*c* shows a **cantilever beam**. A cantilever beam has a fixed support at one end only. The fixed support provides three reactions to the beam: horizontal and vertical reaction forces and a reaction moment. These three unknown reaction forces can be determined from the three equilibrium equations (i.e., $\Sigma F_x = 0$, $\Sigma F_y = 0$, and $\Sigma M = 0$) available for a rigid body.

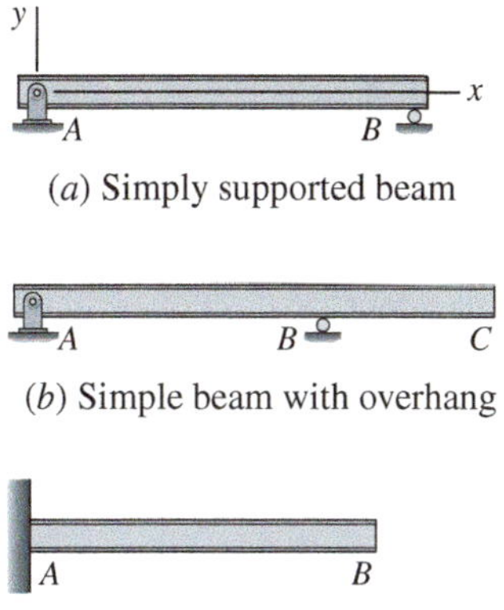

FIGURE 7.3 Types of statically determinate beams.

Types of Loads

Several types of loads are commonly supported by beams (Figure 7.4). Loads focused on a small length of the beam are called **concentrated loads**. Loads from columns or from other members, as well as support reaction forces, are typically represented by concentrated loads. Concentrated loads may also represent wheel loads from vehicles or forces applied by machinery to the structure. Loads that are spread along a portion of the beam are termed **distributed loads**. Distributed loads that are constant in magnitude are termed **uniformly distributed loads**. Examples of uniformly distributed loads include the weight of a concrete floor slab or the forces created by wind. In some instances, the load may be **linearly distributed**, which means that the distributed load, as the term implies, changes linearly in magnitude over the span of the loading. Snow, soil, and fluid pressure, for example, can create linearly distributed loads. A beam may also be subjected to **concentrated moments**, which tend to bend and rotate the beam. Concentrated moments are most often created by other members that connect to the beam.

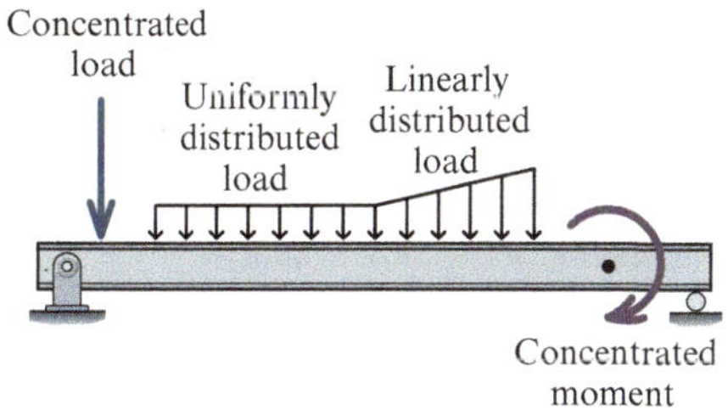

FIGURE 7.4 Symbols used for various types of loads.

7.2 Shear and Moment in Beams

To determine the stresses created by applied loads, it is first necessary to determine the internal shear force V and the internal bending moment M acting in the beam at any point of interest. The general approach for finding V and M is illustrated in Figure 7.5. In this figure, a simply supported beam with an overhang is subjected to two concentrated loads P_1 and P_2 as well as to a uniformly distributed load w. A free-body diagram is obtained by cutting a section at a distance x from pin support A. The cutting plane exposes an internal shear force V and an internal bending moment M. If the beam is in equilibrium, then any portion of the beam that we consider must also be in equilibrium. Consequently, the free body with shear force V and bending moment M must satisfy equilibrium. Thus, equilibrium considerations can be used to determine values for V and M acting at location x.

Because of the applied loads, beams develop internal shear forces V and bending moments M that vary along the length of the beam. For us to properly analyze the stresses produced in a beam, we must determine V and M at all locations along the beam span. These results are typically plotted as a function of x in what is known as a **shear-force and bending-moment diagram**. This diagram summarizes all shear forces and bending moments along the beam, making it straightforward to identify the maximum and minimum values for both V and M. These extreme values are required for calculating the largest stresses.

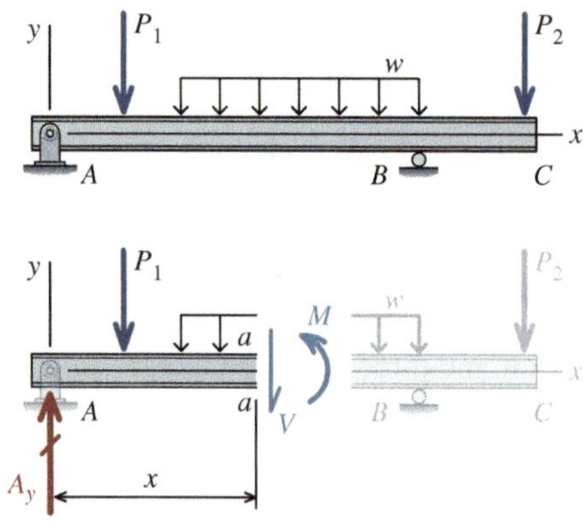

FIGURE 7.5 Method of sections applied to beams.

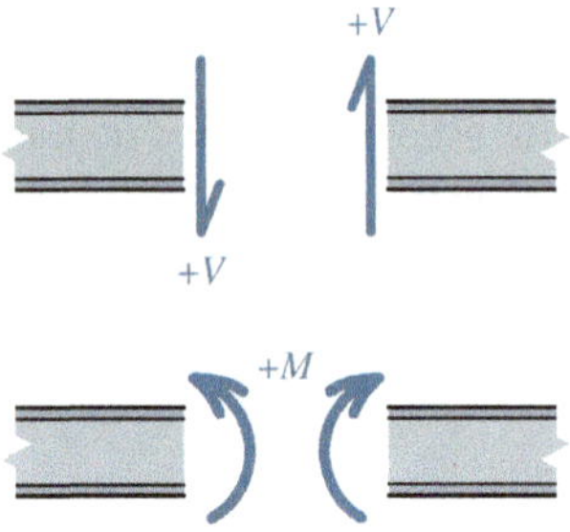

FIGURE 7.6 Sign conventions for internal shear force V and bending moment M.

Since many different loads may act on a beam, functions describing the variation of $V(x)$ and $M(x)$ may not be continuous throughout the entire span of the beam. Because of this consideration, shear-force and bending-moment functions must be determined for a number of intervals along the beam. In general, intervals are delineated by

(a) the locations of concentrated loads, concentrated moments, and support reactions or
(b) the span of distributed loads.

The examples that follow illustrate how shear-force and bending-moment functions can be derived for various intervals by the use of equilibrium considerations.

Sign Conventions for Shear-Force and Bending-Moment Diagrams. Before deriving internal shear-force and bending-moment functions, we must develop consistent sign conventions. These sign conventions are illustrated in Figure 7.6.

A positive internal shear force *V*

- acts downward on the right-hand face of a beam.
- acts upward on the left-hand face of a beam.

A positive internal bending moment *M*

- acts counterclockwise on the right-hand face of a beam.
- acts clockwise on the left-hand face of a beam.

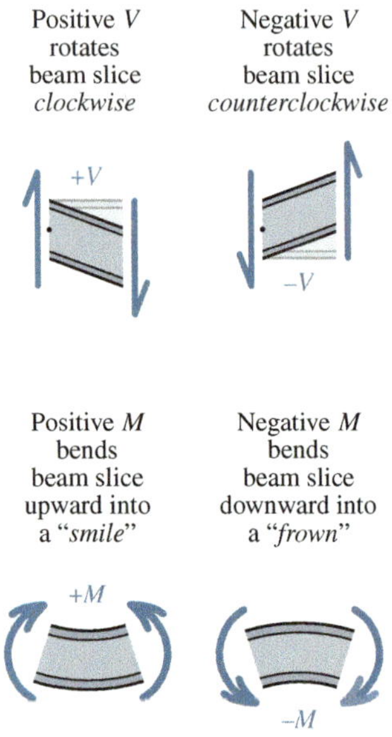

FIGURE 7.7 Sign conventions for V and M shown on beam slice.

These sign conventions can also be expressed by the directions of V and M that act on a small slice of the beam. This alternative statement of the V and M sign conventions is illustrated in Figure 7.7.

A positive internal shear force *V* causes a beam element to rotate clockwise.
A positive internal bending moment *M* bends a beam element concave upward.

A shear-force and bending-moment diagram will be created for each beam by plotting shear-force and bending-moment functions. To ensure consistency among the functions, it is very important that these sign conventions be observed.

EXAMPLE 7.1

Draw the shear-force and bending-moment diagrams for the simply supported beam shown.

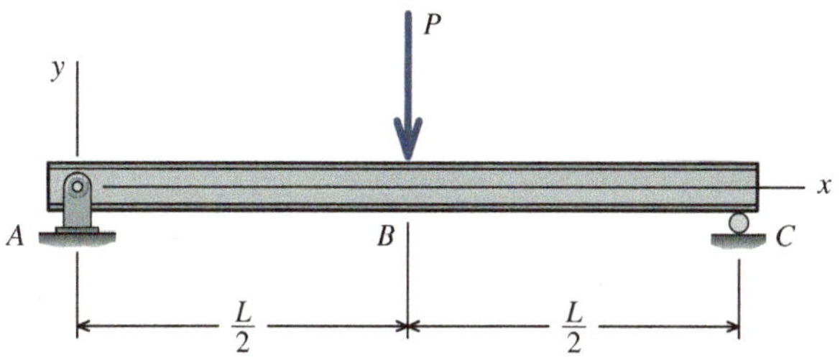

Plan the Solution

First, determine the reaction forces at pin A and roller C. Then, consider two intervals along the beam span: between A and B, and between B and C. Cut a section in each interval and draw the appropriate free-body diagram (FBD), showing the unknown internal shear force V and internal bending moment M acting on the exposed surface. Write the equilibrium equations for each FBD, and solve them for functions describing the variation of V and M with location x along the span. Plot these functions to complete the shear-force and bending-moment diagrams.

SOLUTION

Support Reactions

Since this beam is symmetrically supported and symmetrically loaded, the reaction forces must also be symmetric. Therefore, each support exerts an upward force equal to $P/2$. Because no applied loads act in the x direction, the horizontal reaction force at pin support A is zero.

Shear and Moment Functions

In general, the beam will be sectioned at an arbitrary distance x from pin support A and all forces acting on the free body will be shown, including the unknown internal shear force V and internal bending moment M acting on the exposed surface.

Interval $0 \le x < L/2$: The beam is cut on section a–a, which is located an arbitrary distance x from pin support A. An unknown shear force V and an unknown bending moment M are shown on the exposed surface of the beam. Note that positive directions are assumed for both V and M. (See Figure 7.6 for sign conventions.)

Since no forces act in the x direction, the equilibrium equation $\Sigma F_x = 0$ is trivial. The sum of forces in the vertical direction yields the desired function for V:

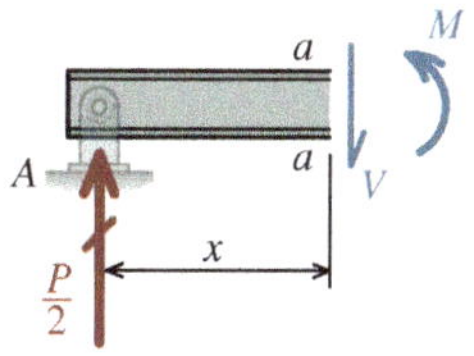

$$\Sigma F_y = \frac{P}{2} - V = 0 \qquad \therefore V = \frac{P}{2} \tag{a}$$

The sum of moments about section a–a gives the desired function for M:

$$\Sigma M_{a-a} = -\frac{P}{2}x + M = 0 \qquad \therefore M = \frac{P}{2}x \tag{b}$$

These results show that the internal shear force V is constant and the internal bending moment M varies linearly in the interval $0 \le x < L/2$.

Interval $L/2 \le x < L$: The beam is cut on section b–b, which is located an arbitrary distance x from pin support A. Section b–b, however, is located beyond B, where the concentrated load P is applied. As before, an unknown shear force V and an unknown bending moment M are shown on the exposed surface of the beam and positive directions are assumed for both V and M.

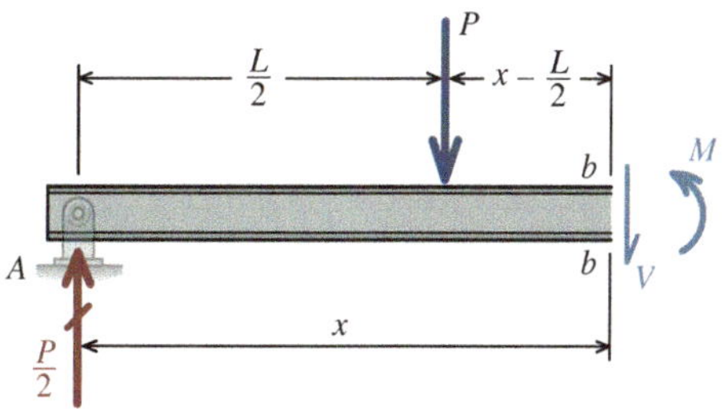

The sum of forces in the vertical direction yields the desired function for V:

$$\Sigma F_y = \frac{P}{2} - P - V = 0 \qquad \therefore V = -\frac{P}{2} \tag{c}$$

The equilibrium equation for the sum of moments about section b–b gives the desired function for M:

$$\Sigma M_{b-b} = P\left(x - \frac{L}{2}\right) - \frac{P}{2}x + M = 0$$
$$\therefore M = -\frac{P}{2}x + \frac{PL}{2} \tag{d}$$

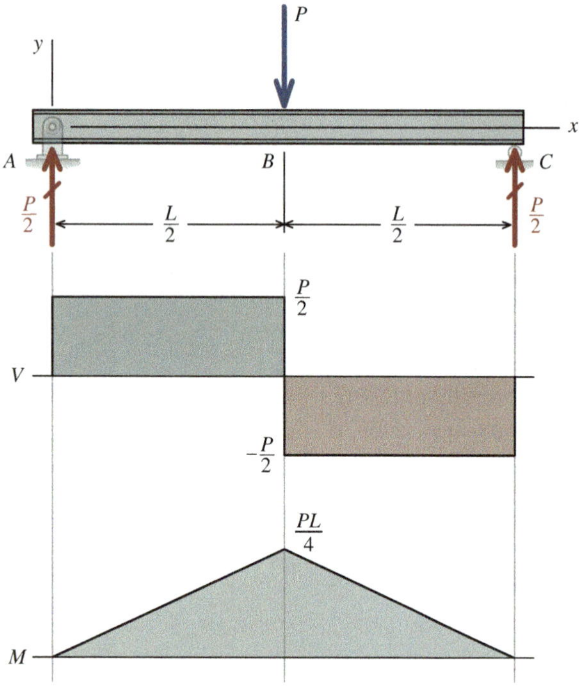

Again, the internal shear force V is constant and the internal bending moment M varies linearly in the interval $L/2 \leq x < L$.

Plot the Functions

Plot the functions given in Equations (a) and (b) for the interval $0 \leq x < L/2$, and the functions defined by Equations (c) and (d) for the interval $L/2 \leq x < L$, to create the shear-force and bending-moment diagram shown.

The maximum internal shear force is $V_{max} = \pm P/2$. The maximum internal bending moment is $M_{max} = PL/4$, and it occurs at $x = L/2$.

Notice that the concentrated load causes a discontinuity at its point of application. In other words, the shear-force diagram "jumps" by an amount equal to the magnitude of the concentrated load. The jump in this case is downward, which is the same direction as that of the concentrated load P.

EXAMPLE 7.2

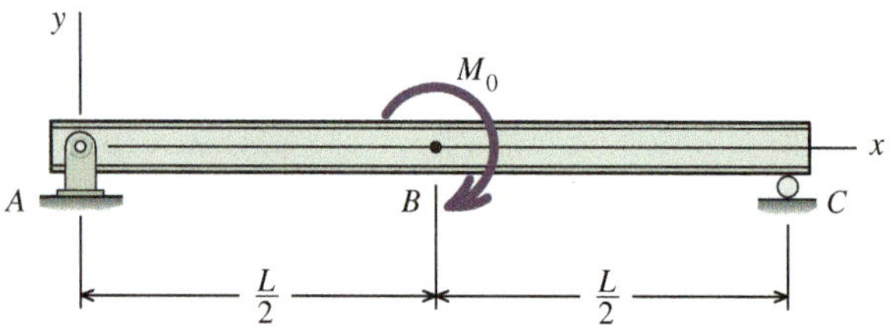

Draw the shear-force and bending-moment diagrams for the simple beam shown.

Plan the Solution

The solution process outlined in Example 7.1 will be used to derive V and M functions for this beam.

SOLUTION

Support Reactions

An FBD of the beam is shown. The equilibrium equations are as follows:

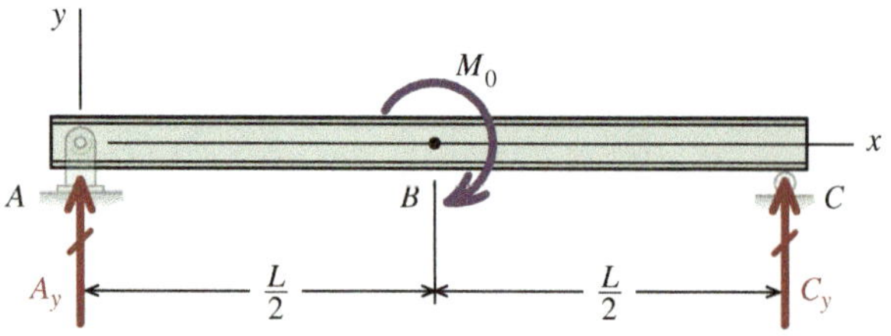

$$\Sigma F_y = A_y + C_y = 0$$
$$\Sigma M_A = -M_0 + C_y L = 0$$

From these equations, the beam reactions are

$$C_y = \frac{M_0}{L} \qquad \text{and} \qquad A_y = -\frac{M_0}{L}$$

The negative value for A_y indicates that this reaction force acts opposite to the direction assumed initially. Subsequent free-body diagrams will be revised to show A_y acting downward.

Interval $0 \le x < L/2$: Section the beam at an arbitrary distance x between A and B. Show the unknown shear force V and the unknown bending moment M on the exposed surface of the beam. Assume positive directions for both V and M, according to the sign convention given in Figure 7.6.

The sum of forces in the vertical direction yields the desired function for V:

$$\Sigma F_y = -\frac{M_0}{L} - V = 0 \qquad \therefore V = -\frac{M_0}{L} \tag{a}$$

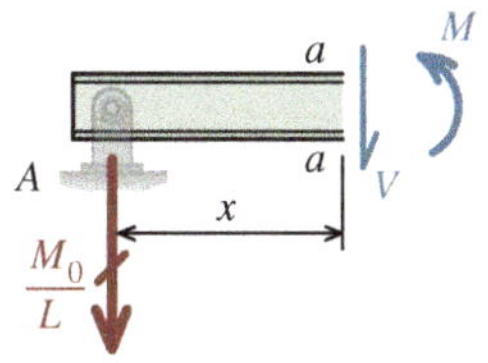

The sum of moments about section a–a gives the desired function for M:

$$\Sigma M_{a-a} = \frac{M_0}{L}x + M = 0 \qquad \therefore M = -\frac{M_0}{L}x \tag{b}$$

These results indicate that the internal shear force V is constant and the internal bending moment M varies linearly in the interval $0 \le x < L/2$.

Interval $L/2 \le x < L$: The beam is cut on section b–b, which is at an arbitrary location between B and C. The sum of forces in the vertical direction yields the desired function for V:

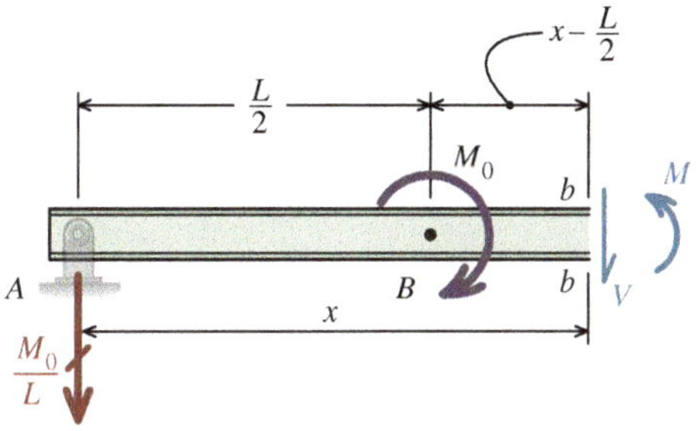

$$\Sigma F_y = -\frac{M_0}{L} - V = 0 \qquad \therefore V = -\frac{M_0}{L} \tag{c}$$

The equilibrium equation for the sum of moments about section b–b gives the desired function for M:

$$\begin{aligned} \Sigma M_{b-b} &= \frac{M_0}{L}x - M_0 + M = 0 \\ \therefore M &= M_0 - \frac{M_0}{L}x \end{aligned} \tag{d}$$

Again, the internal shear force V is constant and the internal bending moment M varies linearly in the interval $L/2 \le x < L$.

Plot the Functions

Plot the functions given in Equations (a) and (b) for the interval $0 \le x < L/2$, and the functions defined by Equations (c) and (d) for the interval $L/2 \le x < L$, to create the shear-force and bending-moment diagram shown.

The maximum internal shear force is $V_{max} = -M_0/L$. The maximum internal bending moment is $M_{max} = \pm M_0/2$, and it occurs at $x = L/2$.

Notice that the concentrated moment does not affect the shear-force diagram at B. It does, however, create a discontinuity in the bending-moment diagram at the point of application of the concentrated moment: The bending-moment diagram "jumps" by an amount equal to the magnitude of the concentrated moment. *The clockwise concentrated external moment M_0 causes the bending-moment diagram to "jump" upward* at B by an amount equal to the magnitude of the concentrated moment.

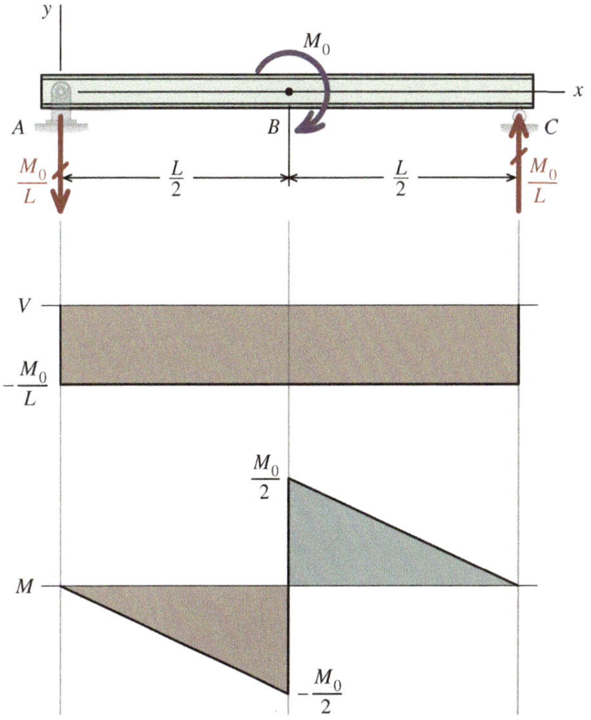

EXAMPLE 7.3

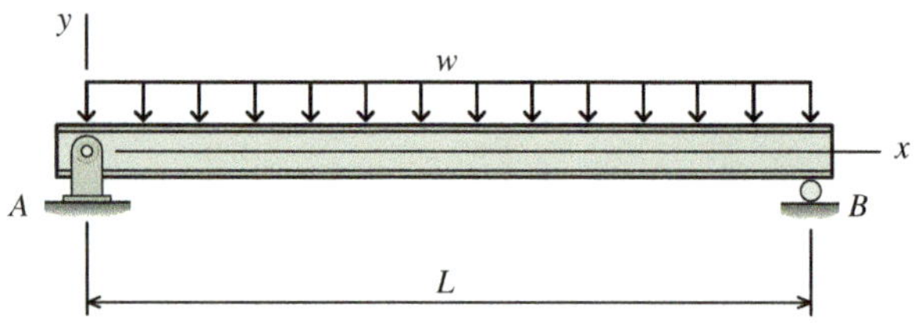

Draw the shear-force and bending-moment diagrams for the simply supported beam shown.

Plan the Solution

After the support reactions at pin A and roller B have been determined, cut a section at an arbitrary location x and draw the corresponding free-body diagram (FBD), showing the unknown internal shear force V and internal bending moment M acting on the cut surface. Develop the equilibrium equations for the FBD, and solve these two equations for functions describing the variation of V and M with location x along the span. Plot these functions to complete the shear-force and bending-moment diagrams.

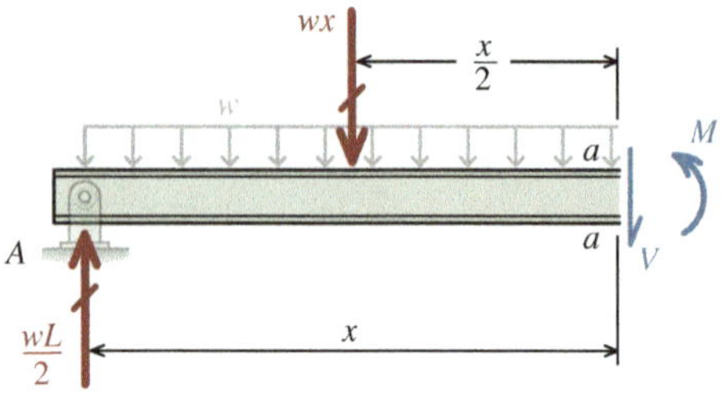

SOLUTION

Support Reactions

Since this beam is symmetrically supported and symmetrically loaded, the reaction forces must also be symmetric. The total load acting on the beam is wL; therefore, each support exerts an upward force equal to half of this load: $wL/2$.

Interval $0 \leq x < L$: Section the beam at an arbitrary distance x between A and B. **Make sure that the original distributed load w is shown on the FBD at the outset.** Show the unknown shear force V and the unknown bending moment M on the exposed surface of the beam. Assume positive directions for both V and M, according to the sign convention given in Figure 7.6. The resultant of the uniformly distributed load w acting on a beam of length x is equal to wx. The resultant force acts at the middle of this loading (i.e., at the centroid of the rectangle that has width x and height w). The sum of forces in the vertical direction yields the desired function for V:

$$\Sigma F_y = \frac{wL}{2} - wx - V = 0$$
$$\therefore V = \frac{wL}{2} - wx = w\left(\frac{L}{2} - x\right) \qquad \text{(a)}$$

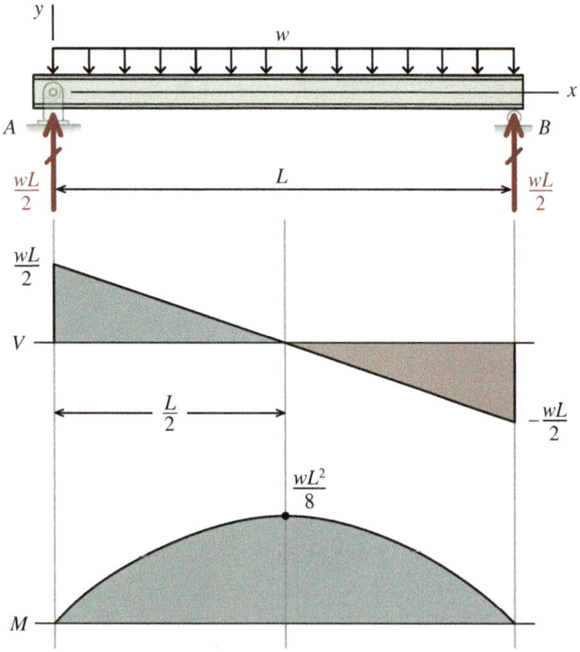

The shear-force function is linear (i.e., it is a first-order function), and the slope of this line is equal to $-w$ (which is the intensity of the distributed load).

The sum of moments about section a–a gives the desired function for M:

$$\Sigma M_{a-a} = -\frac{wL}{2}x + wx\frac{x}{2} + M = 0$$
$$\therefore M = \frac{wL}{2}x - \frac{wx^2}{2} = \frac{wx}{2}(L - x) \qquad \text{(b)}$$

The internal bending moment M is a quadratic function (i.e., a second-order function).

Plot the Functions

Plot the functions given in Equations (a) and (b) to create the shear-force and bending-moment diagram shown.

The maximum internal shear force is $V_{max} = \pm wL/2$, and it is found at A and B. The maximum internal bending moment is $M_{max} = wL^2/8$, and it occurs at $x = L/2$.

Note that the maximum bending moment occurs at a location where the shear force V is equal to zero.

EXAMPLE 7.4

Draw the shear-force and bending-moment diagrams for the simply supported beam shown.

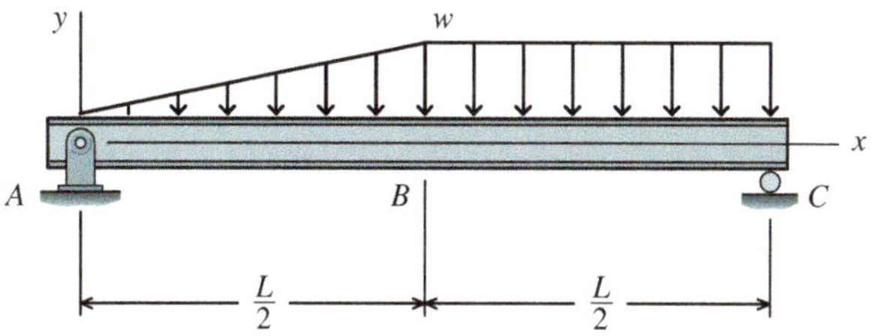

Plan the Solution

After determining the support reactions at pin A and roller C, cut sections between A and B (in the linearly distributed loading) and between B and C (in the uniformly distributed loading). Draw the appropriate free-body diagrams, work out the equilibrium equations for each FBD, and solve the equations for functions describing the variation of V and M with location x along the span. Plot these functions to complete the shear-force and bending-moment diagrams.

SOLUTION

Support Reactions

The FBD for the entire beam is shown. The resultant force of the linearly distributed loading is equal to the area of the triangle that has base $L/2$ and height w:

$$\frac{1}{2}\left(\frac{L}{2}\right)w = \frac{wL}{4}$$

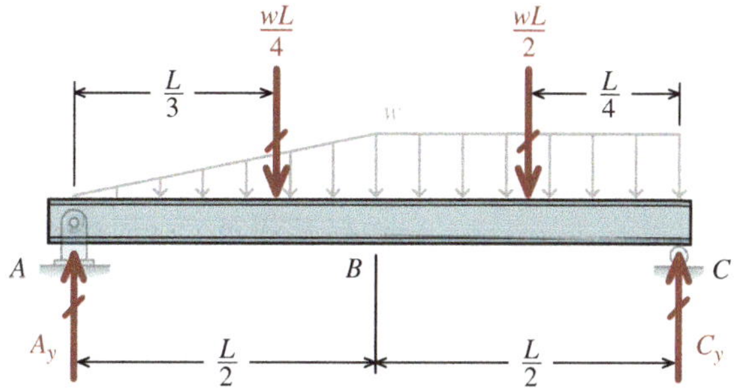

The resultant force acts at the centroid of this triangle, which is located at two-thirds of the base dimension, measured from the point of the triangle:

$$\frac{2}{3}\left(\frac{L}{2}\right) = \frac{L}{3}$$

Equilibrium equations for the beam can be written as

$$\Sigma F_y = A_y + C_y - \frac{wL}{4} - \frac{wL}{2} = 0 \quad \text{and} \quad \Sigma M_A = C_y L - \frac{wL}{4}\left(\frac{L}{3}\right) - \frac{wL}{2}\left(\frac{3L}{4}\right) = 0$$

which can be solved to determine the reaction forces:

$$A_y = \frac{7}{24}wL \quad \text{and} \quad C_y = \frac{11}{24}wL$$

Interval $0 \le x < L/2$: Section the beam at an arbitrary distance x between A and B. **Make sure that you replace the original linearly distributed load on the FBD.** A new resultant force for the linearly distributed load must be derived specifically for this FBD.

The slope of the linearly distributed load is equal to $w/(L/2) = 2w/L$. Accordingly, the height of the triangular loading at section a–a is equal to the product of this slope and the distance x—that is, $(2w/L)x$. Therefore, the resultant of the linearly distributed

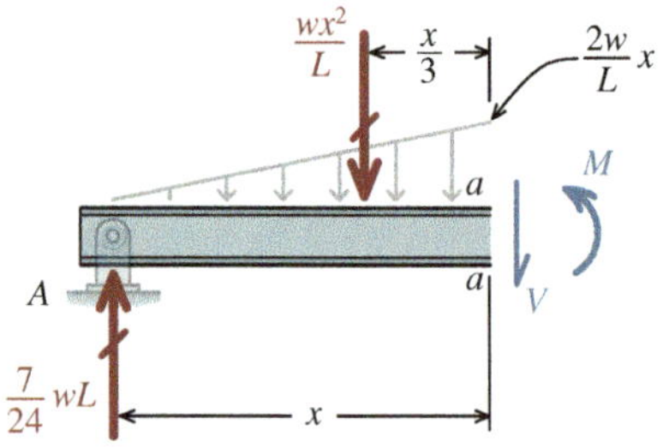

load that is acting on this FBD is $(1/2)x\ [(2w/L)x] = (wx^2/L)$, and it acts at a distance $x/3$ from section a–a.

V and M functions applicable for the interval $0 \le x < L/2$ can be derived from the equilibrium equations for the FBD:

$$\Sigma F_y = \frac{7}{24}wL - \frac{wx^2}{L} - V = 0 \qquad \therefore V = -\frac{wx^2}{L} + \frac{7}{24}wL \tag{a}$$

$$\Sigma M_{a-a} = -\frac{7}{24}wLx + \frac{wx^2}{L}\left(\frac{x}{3}\right) + M = 0 \qquad \therefore M = -\frac{wx^3}{3L} + \frac{7}{24}wLx \tag{b}$$

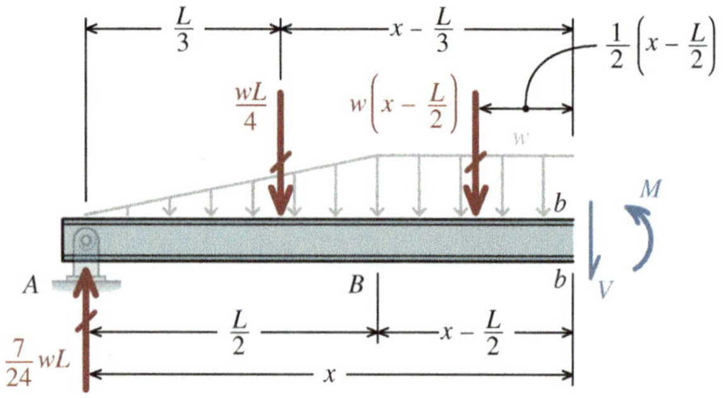

The shear-force function is quadratic (i.e., a second-order function), and the bending-moment function is cubic (i.e., a third-order function).

Interval $L/2 \le x < L$: Section the beam at an arbitrary distance x between B and C. **Make sure that you replace the original distributed loads on the FBD before deriving the V and M functions.**

On the basis of this FBD, the equilibrium equations can be written as follows:

$$\Sigma F_y = \frac{7}{24}wL - \frac{wL}{4} - w\left(x - \frac{L}{2}\right) - V = 0 \qquad \therefore V = \frac{7}{24}wL - \frac{wL}{4} - w\left(x - \frac{L}{2}\right) \tag{c}$$

$$\Sigma M_{a-a} = -\frac{7}{24}wLx + \frac{wL}{4}\left(x - \frac{L}{3}\right) + w\left(x - \frac{L}{2}\right)\frac{1}{2}\left(x - \frac{L}{2}\right) + M = 0$$
$$\therefore M = \frac{7}{24}wLx - \frac{wL}{4}\left(x - \frac{L}{3}\right) - \frac{w}{2}\left(x - \frac{L}{2}\right)^2 \tag{d}$$

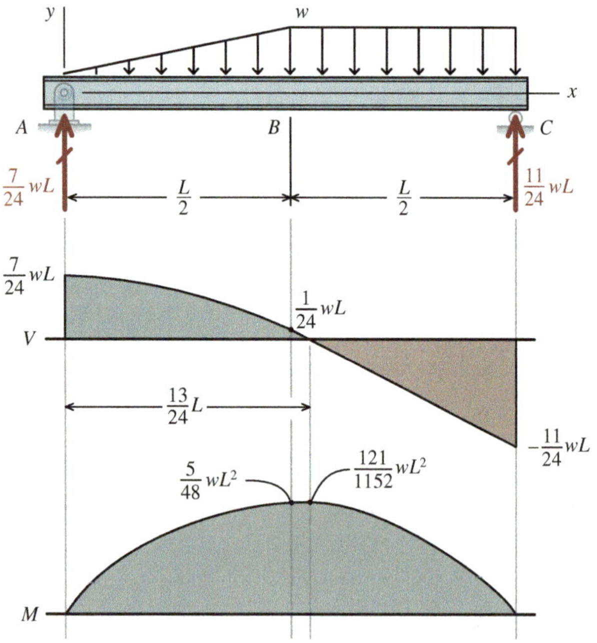

These equations can be simplified to

$$V = w\left(\frac{13}{24}L - x\right) \quad \text{and}$$

$$M = \frac{w}{24}(-12x^2 + 13Lx - L^2)$$

The shear-force function is linear (i.e., a first-order function), and the bending-moment function is quadratic (i.e., a second-order function), between B and C.

Plot the Functions

Plot the V and M functions to create the shear-force and bending-moment diagram shown.

Notice that the maximum bending moment occurs at a location where the shear force V is equal to zero.

EXAMPLE 7.5

Draw the shear-force and bending-moment diagrams for the cantilever beam shown.

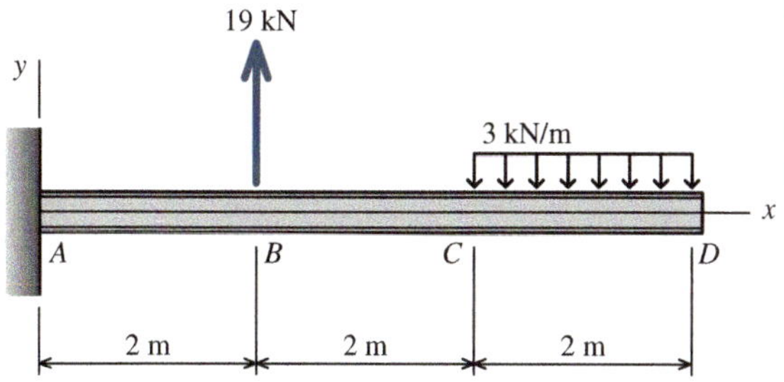

Plan the Solution

Initially, determine the reactions at fixed support A. Three sections will need to be considered, one each for the intervals between AB, BC, and CD. For each section, draw the appropriate free-body diagram, develop the equilibrium equations, and solve the equations for functions describing the variation of V and M with location x along the span. Plot these functions to complete the shear-force and bending-moment diagrams.

SOLUTION

Support Reactions

An FBD of the entire beam is shown. Since no forces act in the x direction, the reaction force $A_x = 0$ will be omitted from the FBD. The nontrivial equilibrium equations are as follows:

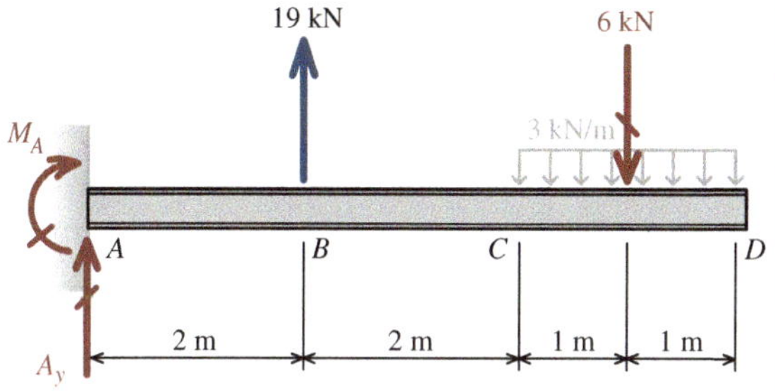

$$\Sigma F_y = A_y + 19 \text{ kN} - 6 \text{ kN} = 0$$
$$\Sigma M_A = -M_A + (19 \text{ kN})(2 \text{ m}) - (6 \text{ kN})(5 \text{ m}) = 0$$

From these equations, the beam reactions are found to be

$$A_y = -13 \text{ kN} \quad \text{and} \quad M_A = 8 \text{ kN}\cdot\text{m}$$

Since A_y is negative, it really acts downward. The correct direction of this reaction force will be shown in subsequent free-body diagrams.

Interval $0 \le x < 2$ m: Section the beam at an arbitrary distance x between A and B. The FBD for this section is shown. From the equilibrium equations for this FBD, determine the desired functions for V and M:

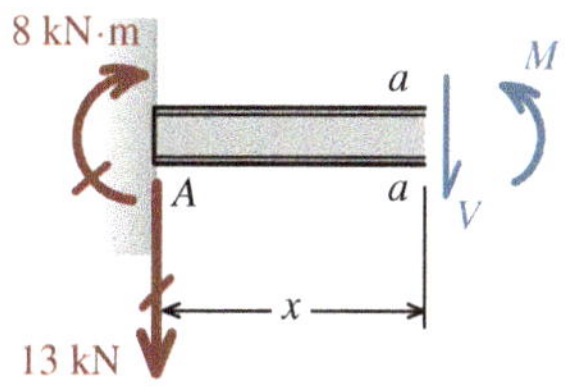

$$\Sigma F_y = -13 \text{ kN} - V = 0 \qquad \therefore V = -13 \text{ kN} \tag{a}$$

$$\Sigma M_{a-a} = (13 \text{ kN})x - 8 \text{ kN}\cdot\text{m} + M = 0 \qquad \therefore M = -(13 \text{ kN})x + 8 \text{ kN}\cdot\text{m} \tag{b}$$

Interval $2 \text{ m} \le x < 4$ m: From a section cut between B and C, determine the desired shear and moment functions:

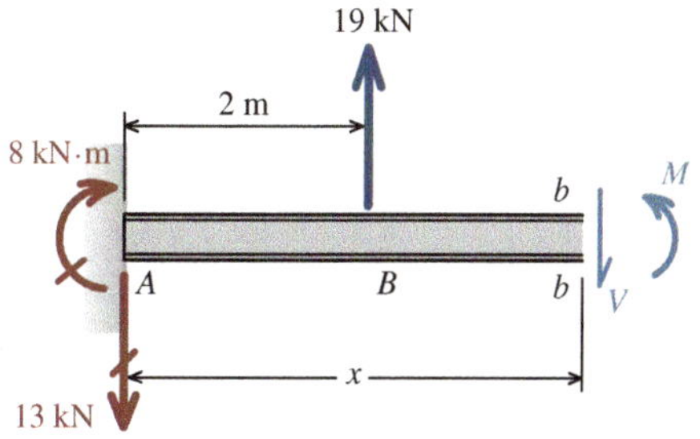

$$\Sigma F_y = -13 \text{ kN} + 19 \text{ kN} - V = 0 \qquad \therefore V = 6 \text{ kN} \tag{c}$$

$$\Sigma M_{b-b} = (13 \text{ kN})x - (19 \text{ kN})(x - 2 \text{ m}) - 8 \text{ kN}\cdot\text{m} + M = 0 \qquad \therefore M = (6 \text{ kN})x - 30 \text{ kN}\cdot\text{m} \tag{d}$$

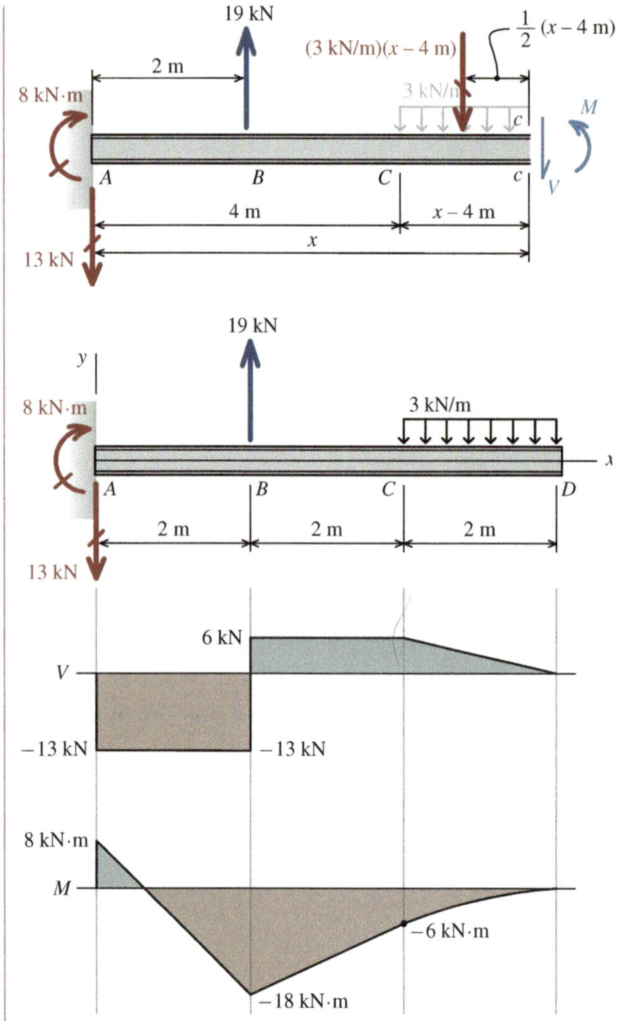

Interval 4 m ≤ x < 6 m: From a section cut between *C* and *D*, determine the desired shear and moment functions:

$$\Sigma F_y = -13\text{ kN} + 19\text{ kN} - (3\text{ kN/m})(x - 4\text{ m}) - V = 0$$
$$\therefore V = (3\text{ kN/m})x + 18\text{ kN} \quad \text{(e)}$$

$$\Sigma M_{c-c} = (13\text{ kN})x - (19\text{ kN})(x - 2\text{ m}) + (3\text{ kN/m})(x - 4\text{ m})\frac{(x - 4\text{ m})}{2} - 8\text{ kN}\cdot\text{m} + M = 0$$
$$\therefore M = -(1.5\text{ kN/m})x^2 + (18\text{ kN})x - 54\text{ kN}\cdot\text{m} \quad \text{(f)}$$

Plot the Functions

Plot the functions given in Equations (a) through (f) to construct the shear-force and bending-moment diagram shown.

Notice that the shear-force diagram is constant in intervals *AB* and *BC* (i.e., it is a zero-order function) and linear in interval *CD* (i.e., it is a first-order function). The bending-moment function is linear in intervals *AB* and *BC* (i.e., it is a first-order function) and quadratic in interval *CD* (i.e., it is a second-order function).

PROBLEMS

P7.1–P7.7 For the beams shown in Figures P7.1–P7.7,

(a) derive equations for the shear force *V* and the bending moment *M* for any location in the beam. (Place the origin at point *A*.)
(b) use the derived functions to plot the shear-force and bending-moment diagrams for the beam.
(c) specify the values for key points on the diagrams.

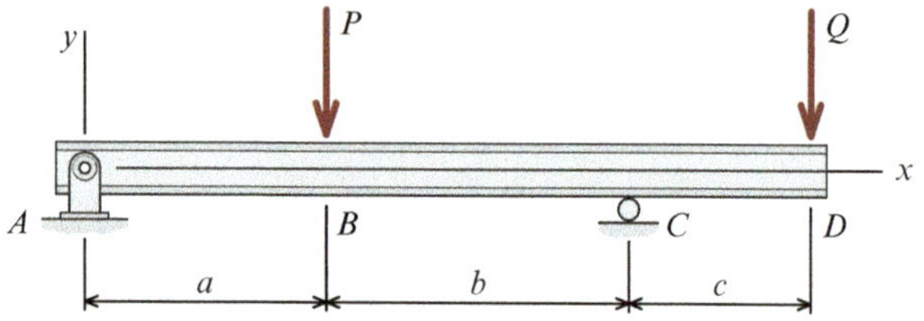

FIGURE P7.2

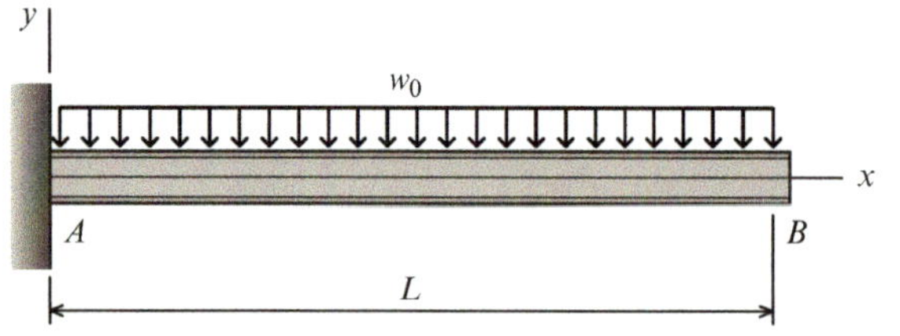

FIGURE P7.1

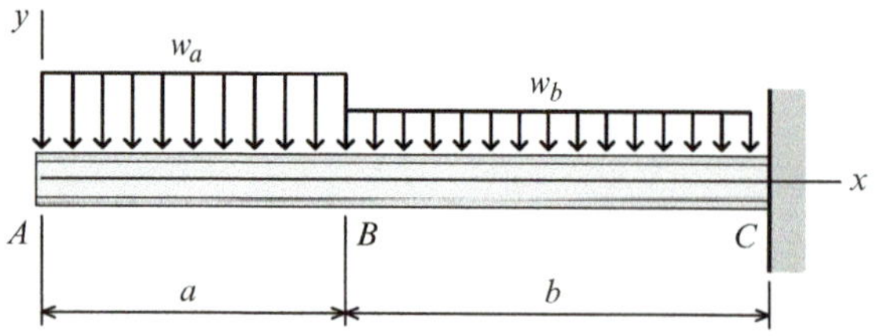

FIGURE P7.3

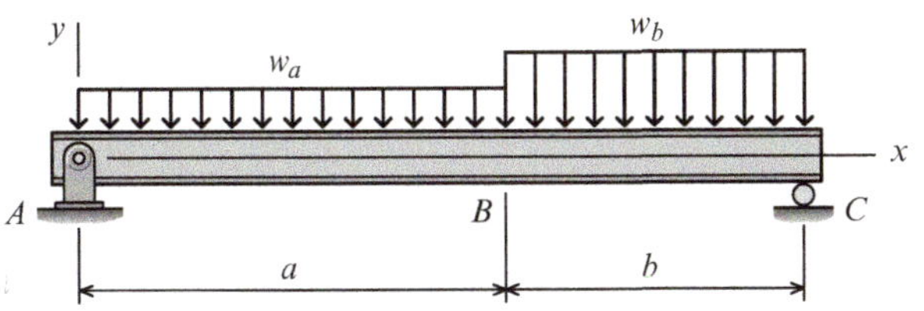

FIGURE P7.4

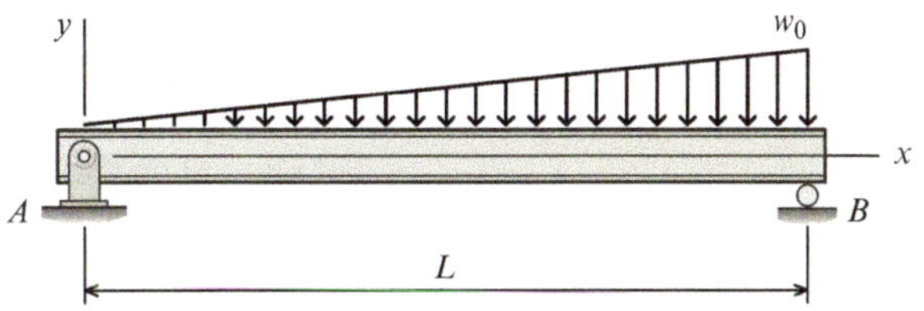

FIGURE P7.6

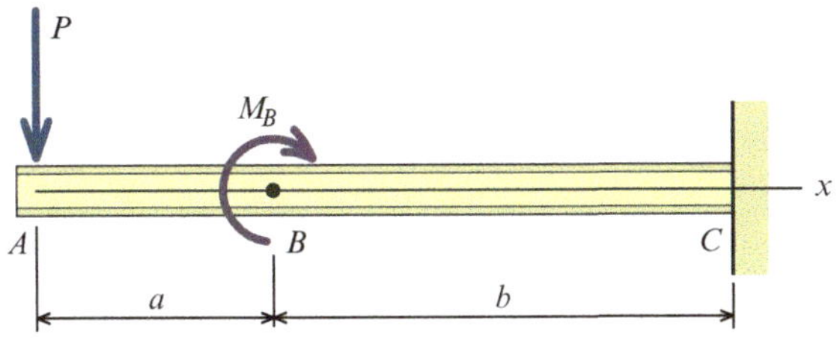

FIGURE P7.5

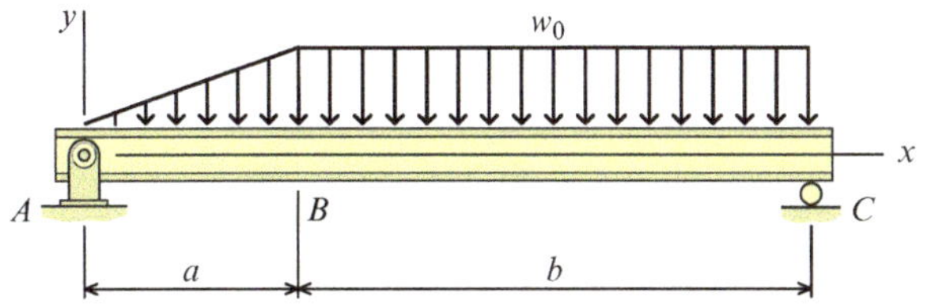

FIGURE P7.7

7.3 Graphical Method for Constructing Shear and Moment Diagrams

As shown in Section 7.2, we can construct shear and moment diagrams by developing functions that express the variation of the internal shear force $V(x)$ and the internal bending moment $M(x)$ along the beam and then plotting these functions. When a beam has several loads, however, this approach can be quite time consuming and a simpler method is desired. The process of constructing shear and moment diagrams is much easier if specific relationships among load, shear, and moment are taken into consideration.

Consider a beam subjected to several loads, as shown in Figure 7.8*a*. **All loads are shown in their respective positive directions.** We will investigate a small portion of the beam where there are no external concentrated loads or concentrated moments. This small beam element has length Δx (Figure 7.8*b*). An internal shear force V and an internal bending moment M act on the left side of the beam element. Because the distributed load is acting on this element, the shear force and bending moment on the right side must be slightly different in order to satisfy equilibrium. Specifically, they must have values of $V + \Delta V$ and $M + \Delta M$, respectively. All shear forces and bending moments are assumed to act in their positive directions, as defined by the sign convention shown in Figure 7.6. The distributed load can be replaced by its resultant force $w(x)\,\Delta x$, which acts at a fractional distance $k\,\Delta x$

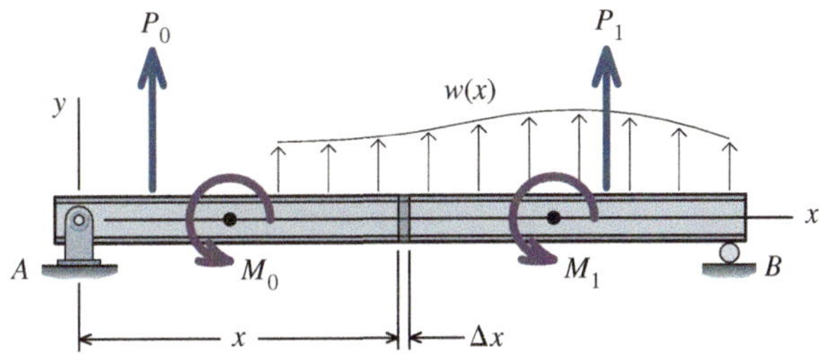

FIGURE 7.8*a* Generalized beam subjected to positive external loads.

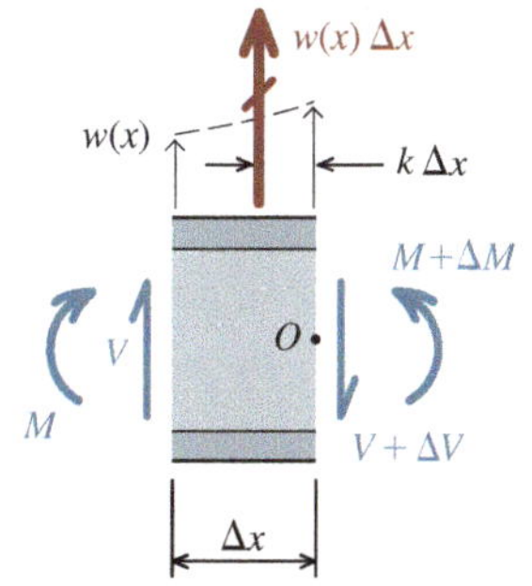

FIGURE 7.8*b* Beam element showing internal shear forces and bending moments.

from the right side, where $0 < k < 1$ (e.g., if the distributed load is uniform, then $k = 0.5$). This small portion of the beam must satisfy equilibrium; therefore, we can consider two equilibrium conditions—the sum of forces in the vertical direction and the sum of moments about point O on the right side of the element:

$$\Sigma F_y = V + w(x)\Delta x - (V + \Delta V) = 0$$

$$\therefore \Delta V = w(x)\Delta x$$

$$\Sigma M_O = -V\Delta x - w(x)\Delta x(k\Delta x) - M + (M + \Delta M) = 0$$

$$\therefore \Delta M = V\Delta x + w(x)\Delta x(k\Delta x)$$

Dividing each of the resulting equations by Δx and taking the limit as $\Delta x \to 0$ gives the following relationships:

$$\frac{dV}{dx} = w(x) \tag{7.1}$$

slope of shear force diagram

$$\frac{dM}{dx} = V \tag{7.2}$$

slope of bending-moment diagram

A **positive slope** inclines upward in moving from left to right or downward in moving from right to left.

Equation (7.1) indicates that the *slope* of the shear-force diagram is equal to the numerical value of the distributed load intensity at any location x. Similarly, Equation (7.2) indicates that the *slope* of the bending-moment diagram is equal to the numerical value of the shear force at any location x.

To illustrate the meaning of Equation (7.1), consider the beam shown in Figure 7.9. This beam is subjected to a distributed load $w(x)$ that increases from $w(x) = w_A = 0$ at A to

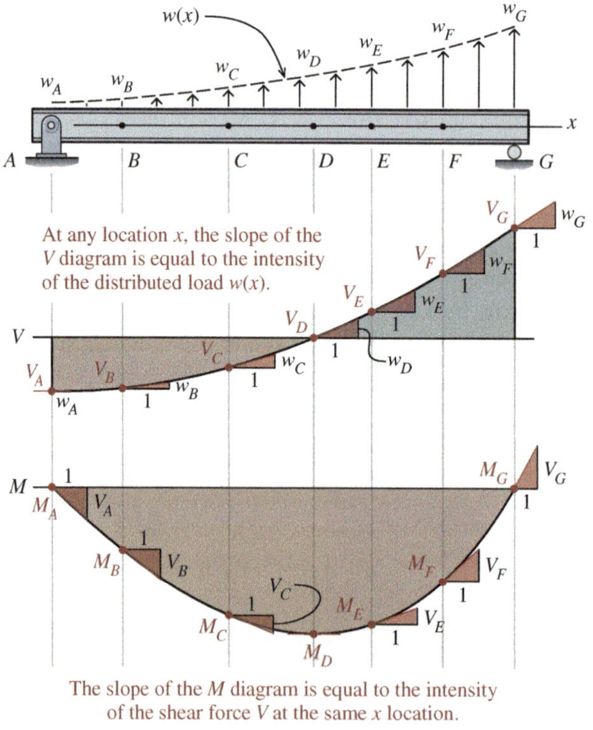

FIGURE 7.9 Relationships between slopes for the load, shear, and moment diagrams.

$w(x) = w_G$ at G. At A, where the distributed load w is zero, the slope of the shear-force diagram is also zero. Moving to the right along the beam span, the distributed load increases to a small positive value at B and, accordingly, the slope of the shear-force diagram at B is a small positive value (i.e., the shear-force diagram slopes slightly upward). At points C through G, the magnitude of the distributed load gets larger and larger (i.e., more and more positive). Similarly, the slope of the shear-force diagram at these points becomes increasingly more positive. In other words, the V curve gets increasingly steeper as the distributed load w gets larger.

In a similar manner, Equation (7.2) states that the slope of the bending-moment diagram at any point is equal to the shear force V at that same point. At point A in Figure 7.9, the shear force V_A is a relatively large negative value; therefore, the slope of the bending-moment diagram is a relatively large negative value. In other words, the M diagram slopes sharply downward. At points B and C, the shear forces V_B and V_C are still negative, but not as negative as V_A. Consequently, the M diagram still slopes downward, but not as steeply as at A. At point D, the shear force V_D is zero, meaning that the slope of the M diagram is zero. (This fact represents an important detail because the maximum and minimum values of a function are at those locations where the slope of the V diagram is zero.) At point E, the shear force V_E becomes a small positive number and, accordingly, the M diagram begins to slope upward slightly. At points F and G, the shear forces V_F and V_G are relatively large positive numbers, meaning that the M diagram slopes sharply upward.

For brevity, the shear-force diagram is also termed the V diagram or the V curve. The bending-moment diagram is also termed the M diagram or the M curve.

Equations (7.1) and (7.2) may be rewritten in the form $dV = w(x)\,dx$ and $dM = V\,dx$. The terms $w(x)\,dx$ and $V\,dx$ represent differential areas under the distributed-load and shear-force diagrams, respectively. Equation (7.1) can be integrated between any two locations x_1 and x_2 on the beam:

$$\int_{V_1}^{V_2} dV = \int_{x_1}^{x_2} w(x)\,dx$$

Integrating gives the following relationship:

$$\Delta V = V_2 - V_1 = \int_{x_1}^{x_2} w(x)\,dx \qquad (7.3)$$

= area under distributed load curve b/w those two points

The terms **load diagram** and **distributed-load curve** are synonyms. For brevity, the distributed-load curve is referred to as the w diagram or the w curve.

Similarly, Equation (7.2) can be expressed in integral form as

$$\int_{M_1}^{M_2} dM = \int_{x_1}^{x_2} V\,dx$$

which gives the relationship

$$\Delta M = M_2 - M_1 = \int_{x_1}^{x_2} V\,dx \qquad (7.4)$$

= area under the shear-force curve

Equation (7.3) reveals that the *change in shear force*, ΔV, between any two points on the beam is equal to the area under the distributed-load curve between those same two points. Similarly, Equation (7.4) states that the *change in bending moment*, ΔM, between any two points is equal to the corresponding area under the shear-force curve.

To illustrate the significance of Equations (7.3) and (7.4), consider the beam shown in Figure 7.10. The change in shear force between points E and F can be found from the area under the distributed-load curve between those same two points. Similarly, the change in bending moment between points B and C is given by the area under the V curve between those same two points.

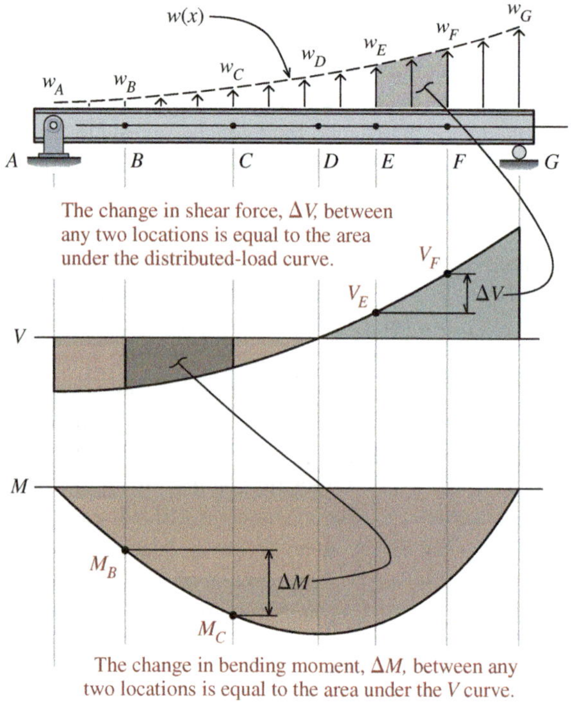

FIGURE 7.10 Relationships between areas for the load, shear, and moment diagrams.

The positive direction for a concentrated load is upward.

An upward load P causes the shear diagram to jump upward. Similarly, a downward load P causes the shear diagram to jump downward.

Regions of Concentrated Loads and Moments

Equations (7.1) through (7.4) were derived for a portion of the beam subjected to a distributed load only. Next, consider the free-body diagram of a very thin portion of the beam (see Figure 7.8*a*) directly beneath one of the concentrated loads (Figure 7.11*a*). Force equilibrium for this free body can be stated as

$$\Sigma F_y = V + P_0 - (V + \Delta V) = 0 \qquad \therefore \Delta V = P_0 \tag{7.5}$$

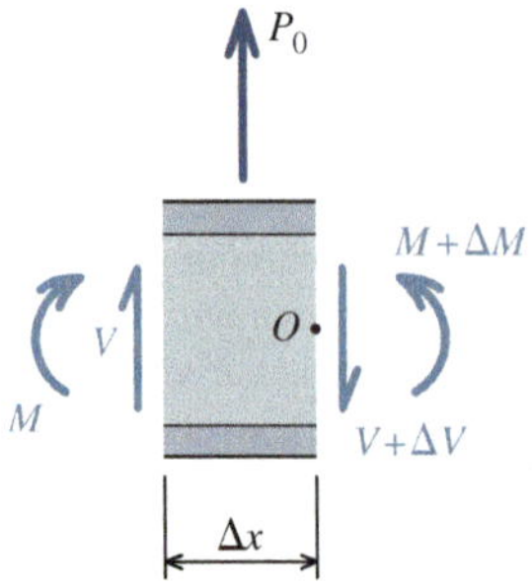

FIGURE 7.11*a* Free-body diagram of beam element subjected to concentrated load P_0.

This equation shows that the change in shear force, ΔV, between the left and right sides of a thin beam element is equal to the intensity of the external concentrated load P_0 acting on the beam element. At the location of a positive external load, the shear-force diagram is discontinuous: It "jumps" upward by an amount equal to the intensity of an upward concentrated load. A downward external concentrated load causes the shear-force diagram to jump downward. (See Example 7.1.)

Next, consider a thin beam element located at a concentrated moment (Figure 7.11*b*). Moment equilibrium for this element can be expressed as

$$\Sigma M_O = -M - V\Delta x + M_0 + (M + \Delta M) = 0$$

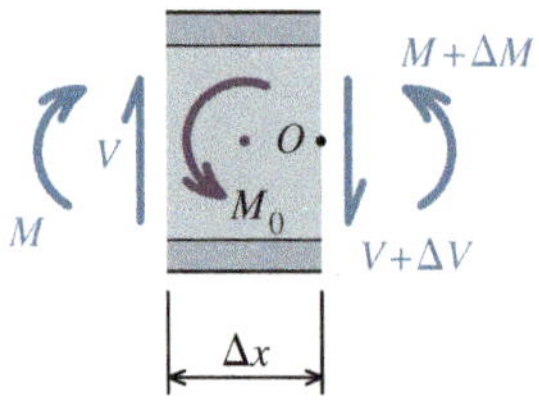

FIGURE 7.11*b* Free-body diagram of beam element subjected to concentrated moment M_0.

As Δx approaches zero,

$$\Delta M = -M_0 \tag{7.6}$$

The moment diagram is discontinuous at locations where external concentrated moments are applied. Equation (7.6) reveals that the change in internal bending moment, ΔM, between the left and right sides of a thin beam element is equal to the negative of the external concentrated moment M_0 acting on the beam element. If a positive external moment is defined as acting counterclockwise, then a positive external moment causes the bending-moment diagram to "jump" downward. Conversely, a negative external moment (i.e., a moment that acts clockwise) causes the internal bending-moment diagram to jump upward (see Example 7.2).

Maximum and Minimum Bending Moments

In mathematics, we find the maximum value of a function $f(x)$ by first taking the derivative of the function, setting the derivative equal to zero, and determining the location x at which the derivative is zero. Then, once this value of x is known, it can be substituted into $f(x)$ and the maximum value ascertained.

In the context of shear and moment diagrams, the function of interest is the bending-moment function $M(x)$. The derivative of this function is dM/dx, and accordingly, the maximum bending moment will occur at locations where $dM/dx = 0$. Notice, however, Equation (7.2), which states that $dM/dx = V$. If these two equations are combined, we can conclude that the maximum or minimum bending moment occurs at locations where $V = 0$. This conclusion will be true unless there is a discontinuity in the M diagram caused by an external concentrated moment. Consequently, maximum and minimum bending moments will occur at points where the V curve crosses the $V = 0$ axis, as well as at points where external concentrated moments are applied to the beam. Bending moments corresponding to the location of discontinuities also should be computed to check for maximum or minimum bending moment values.

Six Rules for Constructing Shear-Force and Bending-Moment Diagrams

Equations (7.1) through (7.6) constitute six rules that can be used to construct shear-force and bending-moment diagrams for any beam. These rules, grouped according to usage, can be stated as follows:

Rules for the Shear-Force Diagram

Rule 1: The shear-force diagram is discontinuous at points subjected to concentrated loads P. An upward P causes the V diagram to jump upward, and a downward P causes the V diagram to jump downward [Equation (7.5)].

Rule 2: The *change* in internal shear force between any two locations x_1 and x_2 is equal to the area under the distributed-load curve [Equation (7.3)].

A negative area results from negative w (i.e., a downward distributed load).

Rule 3: At any location x, the slope of the V diagram is equal to the intensity of the distributed load w [Equation (7.1)].

Rules for the Bending-Moment Diagram

Rule 4: The *change* in internal bending moment between any two locations x_1 and x_2 is equal to the area under the shear-force diagram [Equation (7.4)].

The area computed from negative shear force values is considered negative.

Rule 5: At any location x, the *slope* of the M diagram is equal to the intensity of the internal shear force V [Equation (7.2)].

Rule 6: The bending-moment diagram is discontinuous at points subjected to external concentrated moments. A clockwise external moment causes the M diagram to jump upward, and a counterclockwise external moment causes the M diagram to jump downward [Equation (7.6)].

For convenience, these six rules are presented, along with illustrations, in Table 7.1.

Table 7.1 Construction Rules for Shear-Force and Bending-Moment Diagrams

<table>
<tr><th>Equation</th><th>Load Diagram w</th><th>Shear-Force Diagram V</th><th>Bending-Moment Diagram M</th></tr>
<tr><td colspan="4">Rule 1: Concentrated loads create discontinuities in the shear-force diagram. [Equation (7.5)]</td></tr>
<tr><td>$\Delta V = P_0$</td><td>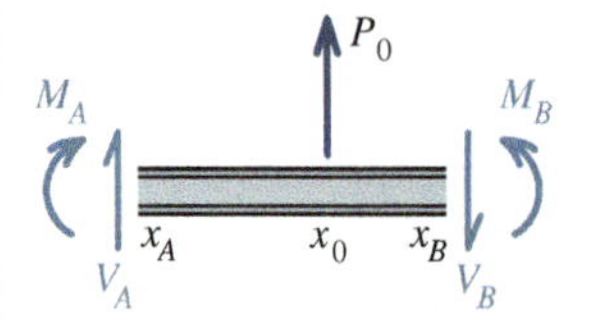
</td><td>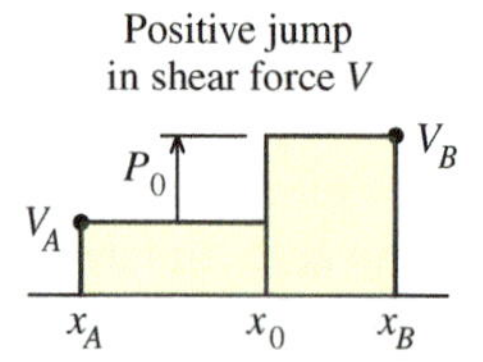
</td><td>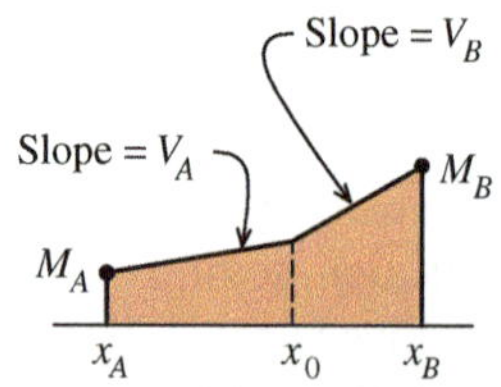
</td></tr>
<tr><td colspan="4">Rule 2: The change in shear force is equal to the area under the distributed-load curve. [Equation (7.3)]</td></tr>
<tr><td>$V_B - V_A = \int_{x_A}^{x_B} w(x)\,dx$</td><td>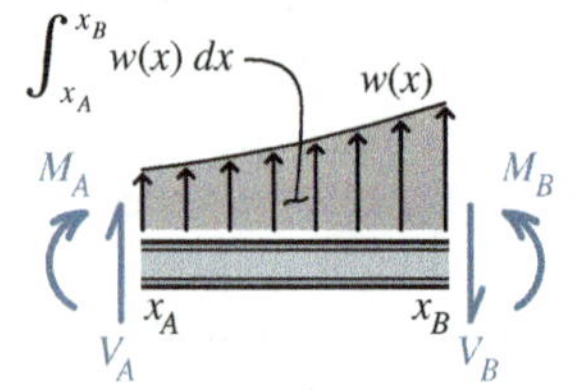
</td><td>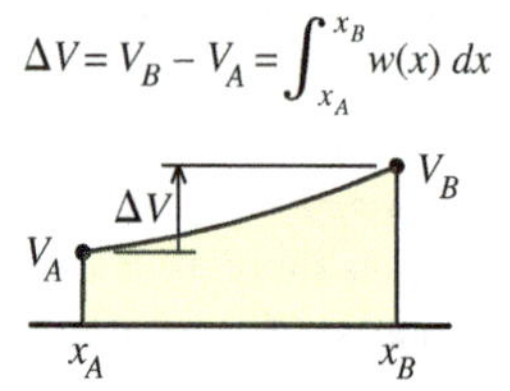
</td><td>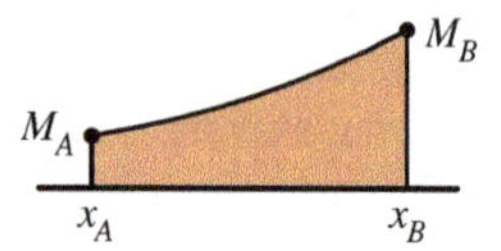
</td></tr>
<tr><td colspan="4">Rule 3: The slope of the V diagram is equal to the intensity of the distributed load w. [Equation (7.1)]</td></tr>
<tr><td>$\frac{dV}{dx} = w(x)$</td><td>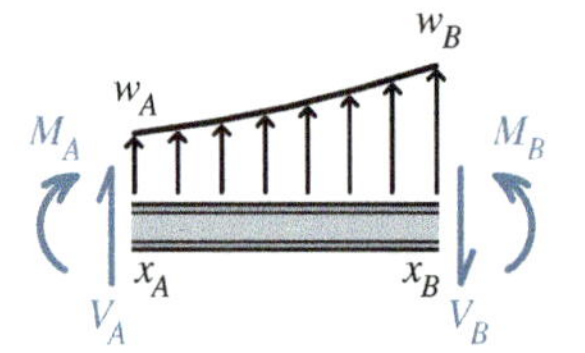
</td><td>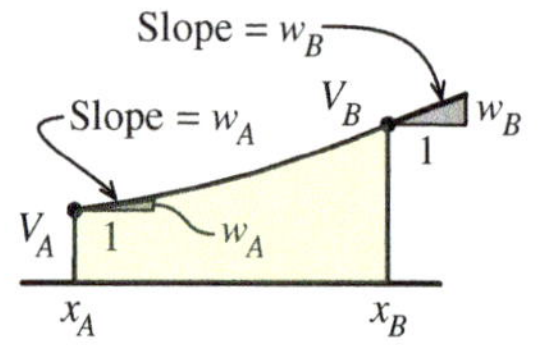
</td><td>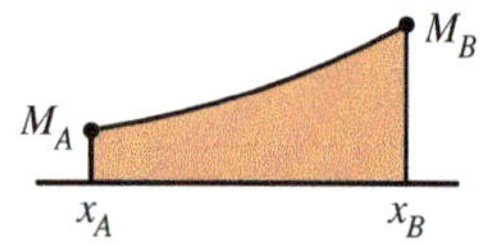
</td></tr>
<tr><td colspan="4">Rule 4: The change in bending moment is equal to the area under the shear-force diagram. [Equation (7.4)]</td></tr>
<tr><td>$M_B - M_A = \int_{x_A}^{x_B} V\,dx$</td><td>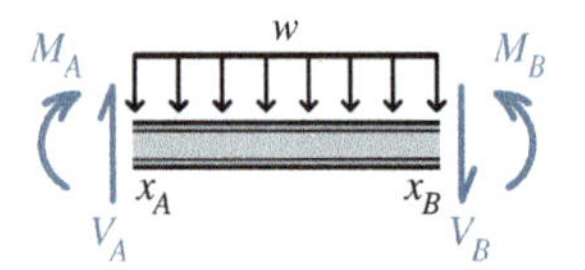
</td><td>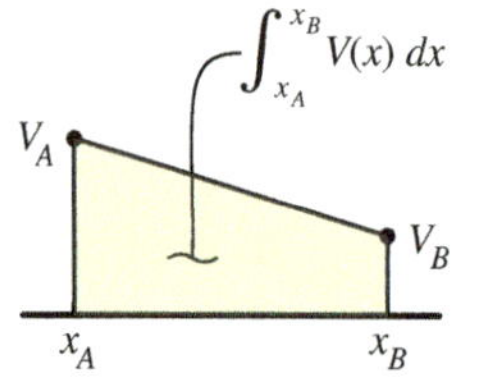
</td><td>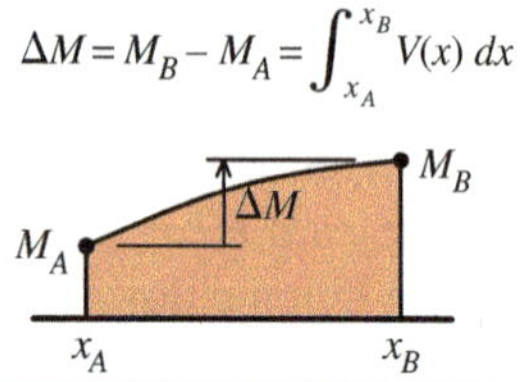
</td></tr>
<tr><td colspan="4">Rule 5: The slope of the M diagram is equal to the intensity of the shear force V. [Equation (7.2)]</td></tr>
<tr><td>$\frac{dM}{dx} = V$</td><td>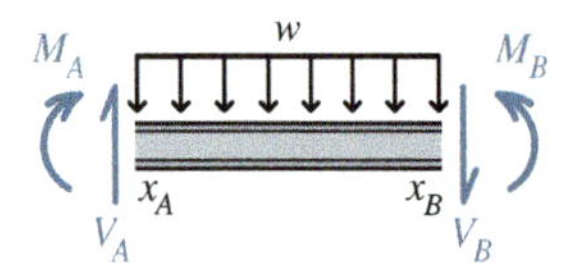
</td><td>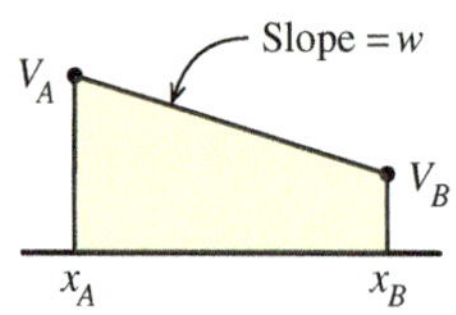
</td><td>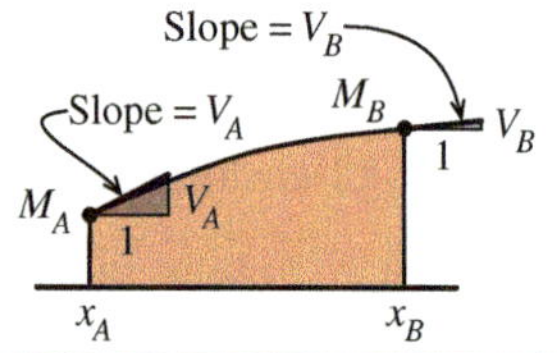
</td></tr>
<tr><td colspan="4">Rule 6: Concentrated moments create discontinuities in the bending-moment diagram. [Equation (7.6)]</td></tr>
<tr><td>$\Delta M = -M_0$</td><td>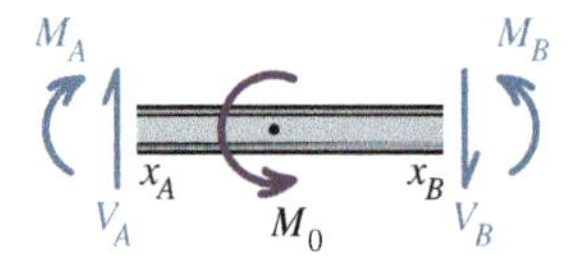
</td><td>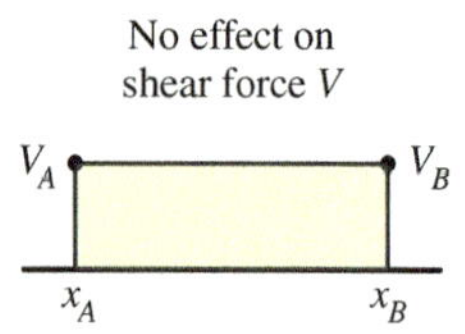
</td><td>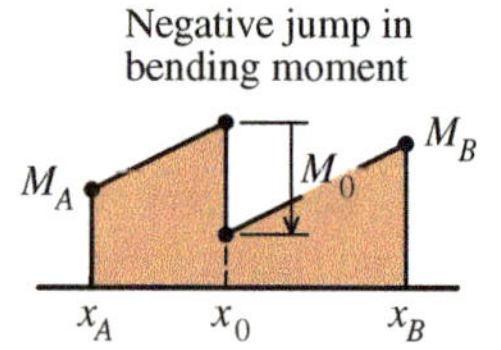
</td></tr>
</table>

General Procedure for Constructing Shear-Force and Bending-Moment Diagrams

The method for constructing V and M diagrams presented here is called the **graphical method** because the load diagram is used to construct the shear-force diagram and then the shear-force diagram is used to construct the bending-moment diagram. The six rules just outlined are used to make these constructions. The graphical method is much less time consuming than the process of deriving $V(x)$ and $M(x)$ functions for the entire beam, and it provides the information necessary to analyze and design beams. The general procedure can be summarized by the following steps:

Step 1 — Complete the Load Diagram: Sketch the beam, including the supports, loads, and key dimensions. Calculate the external reaction forces, and if the beam is a cantilever, find the external reaction moment. Show these reaction forces and moments on the load diagram, using arrows to indicate the direction in which they act.

Step 2 — Construct the Shear-Force Diagram: The shear-force diagram will be constructed directly beneath the load diagram. For that reason, it is convenient to draw a series of vertical lines beneath significant locations on the beam in order to help align the diagrams. Begin the shear-force diagram by drawing a horizontal axis, which will serve as the x axis for the V diagram. The shear-force diagram should always start and end on the value $V = 0$. Construct the V diagram from the leftmost end of the beam toward the rightmost end, using the rules outlined on p. 209. Rules 1 and 2 will be the rules most frequently used to determine shear-force values at important points. Rule 3 is useful for sketching the proper shape of the diagram between these key points. Label all points where the shear force changes abruptly and all locations where maximum or minimum (i.e., maximum negative values) shear forces occur.

Step 3 — Locate Key Points on the Shear-Force Diagram: Special attention should be paid to locating points where the V diagram crosses the $V = 0$ axis, because these points indicate locations where the bending moment will be either a maximum or a minimum value. *For beams with distributed loadings, Rule 3 will be essential for this task.*

Step 4 — Construct the Bending-Moment Diagram: The bending-moment diagram will be constructed directly beneath the shear-force diagram. Begin the bending-moment diagram by drawing a horizontal axis, which will serve as the x axis for the M diagram. The bending-moment diagram should always start and end on the value $M = 0$. Construct the M diagram from the leftmost end of the beam toward the rightmost end, using the rules outlined on p. 209. Rules 4 and 6 will be the rules most frequently used to determine bending-moment values at important points. Rule 5 is useful for sketching the proper diagram shape between these key points. Label all points where the bending moment changes abruptly and all locations where maximum or minimum (i.e., maximum negative values) bending moments occur.

The graphical method is most useful when the areas associated with Equations (7.3) and (7.4) are simple rectangles or triangles. These types of areas exist when beam loadings are concentrated loads or uniformly distributed loads.

The idea of starting and ending at $V = 0$ is related to the beam equilibrium equation $\Sigma F_y = 0$. A shear-force diagram that does not return to $V = 0$ at the rightmost end of the beam indicates that equilibrium has not been satisfied. The most common cause of this error in the V diagram is a mistake in the calculated beam reaction forces.

The idea of starting and ending at $M = 0$ is related to the beam equilibrium equation $\Sigma M = 0$. A bending-moment diagram that does not return to $M = 0$ at the rightmost end of the beam indicates that equilibrium has not been satisfied. The most common cause of this error in the M diagram is a mistake in the calculated beam reaction forces. If the applied loads included concentrated moments, another common error is "jumping" the wrong direction at the discontinuities.

In the example problems that follow, a special notation is used to denote values at discontinuities on the V and M diagrams. To illustrate this notation, suppose that a discontinuity occurs at $x = 15$ on the shear-force diagram. Then the shear value on the $-x$ side of the discontinuity will be denoted $V(15^-)$, and the value on the $+x$ side will be denoted $V(15^+)$. Similarly, if a bending-moment discontinuity occurs at $x = 0$, then the moment values at the discontinuity will be denoted $M(0^-)$ and $M(0^+)$.

EXAMPLE 7.6

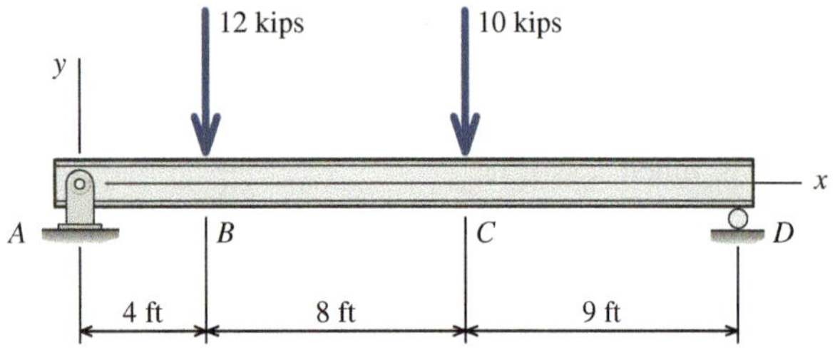

Draw the shear-force and bending-moment diagrams for the simply supported beam shown. Determine the maximum bending moment that occurs in the span.

Plan the Solution

Complete the load diagram by calculating the reaction forces at pin A and roller D. Since only concentrated loads act on this beam, use Rule 1 to construct the shear-force diagram from the load diagram. Construct the bending-moment diagram from the shear-force diagram, using Rule 4 to calculate the change in bending moments between key points.

SOLUTION

Support Reactions

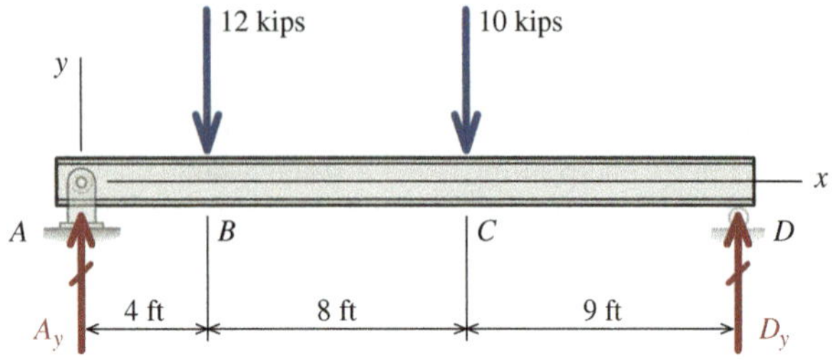

An FBD of the entire beam is shown. Since no loads act in the horizontal direction, the equilibrium equation $\Sigma F_x = 0$ is trivial and will not be considered further. The nontrivial equilibrium equations are as follows:

$$\Sigma F_y = A_y + D_y - 12 \text{ kips} - 10 \text{ kips} = 0$$

$$\Sigma M_A = -(12 \text{ kips})(4 \text{ ft}) - (10 \text{ kips})(12 \text{ ft}) + D_y(21 \text{ ft}) = 0$$

The following beam reactions can be computed from these equations:

$$A_y = 14 \text{ kips} \qquad \text{and} \qquad D_y = 8 \text{ kips}$$

Construct the Shear-Force Diagram

On the load diagram, show the reaction forces acting in their proper directions. Draw a series of vertical lines beneath key points on the beam, and draw a horizontal line that will define the axis for the V diagram. Use the steps outlined next to construct the V diagram. (**Note:** The lowercase letters on the V diagram correspond to the explanations given for each step.)

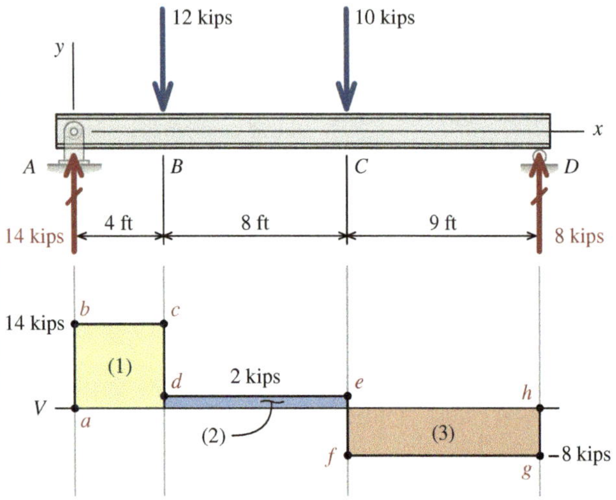

a $V(0^-) = 0$ kips (zero shear at end of beam).

b $V(0^+) = 14$ kips (**Rule 1:** V diagram jumps up by an amount equal to the 14 kip reaction).

c $V(4^-) = 14$ kips (**Rule 2:** Since $w = 0$, the area under the w curve is also zero. Hence, there is no change in the shear-force diagram).

d $V(4^+) = 2$ kips (**Rule 1:** V diagram jumps down by 12 kips).

e $V(12^-) = 2$ kips (**Rule 2:** The area under the w curve is zero; therefore, $\Delta V = 0$).

f $V(12^+) = -8$ kips (**Rule 1:** V diagram jumps down by 10 kips).

g $V(21^-) = -8$ kips (**Rule 2:** The area under the w curve is zero; therefore, $\Delta V = 0$).

h $V(21^+) = 0$ kips (**Rule 1:** V diagram jumps up by an amount equal to the 8 kip reaction force and returns to $V = 0$ kips).

Notice that the V diagram started at $V_a = 0$ and finished at $V_h = 0$.

Construct the Bending-Moment Diagram

Starting with the V diagram, use the steps that follow to construct the M diagram. (**Note:** The lowercase letters on the M diagram correspond to the explanations given for each step.)

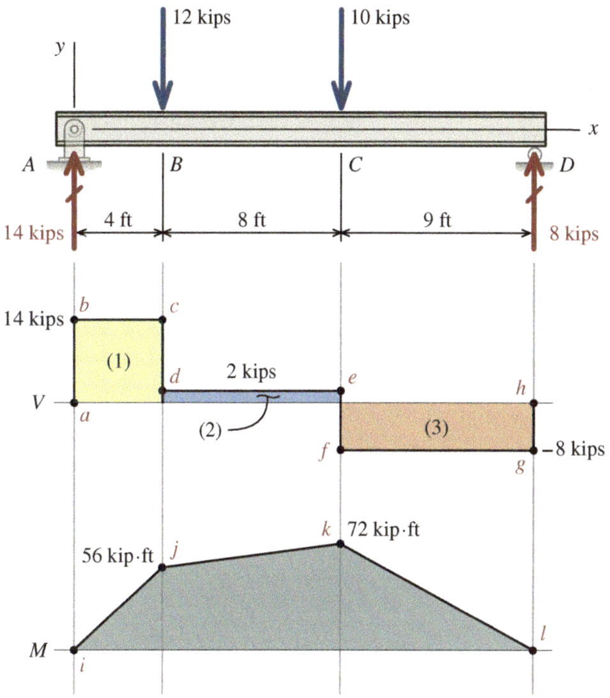

i $M(0) = 0$ (zero moment at the pinned end of a simply supported beam).

j $M(4) = 56$ kip·ft (**Rule 4:** The change in bending moment ΔM between any two points is equal to the area under the V diagram). The area under the V diagram between $x = 0$ ft and $x = 4$ ft is simply the area of rectangle (1), which is 4 ft wide and +14 kips tall. The area of this rectangle is (+14 kips)(4 ft) = +56 kip·ft (a positive value). Since $M = 0$ kip·ft at $x = 0$ ft and the change in bending moment is $\Delta M = +56$ kip·ft, the bending moment at $x = 4$ ft is $M_j = 56$ kip·ft.

k $M(12) = 72$ kip·ft (**Rule 4:** ΔM = area under the V diagram). ΔM is equal to the area under the V diagram between $x = 8$ ft and $x = 12$ ft. The area of rectangle (2) is (+2 kips)(8 ft) = +16 kip·ft. Therefore, $\Delta M = +16$ kip·ft (a positive value). Since $M = +56$ kip·ft at j and $\Delta M = +16$ kip·ft, the bending moment at k is $M_k = +56$ kip·ft + 16 kip·ft = +72 kip·ft. Even though the shear force decreases from +14 kips to +2 kips, notice that the bending moment continues to increase in this region.

l $M(21) = 0$ kip·ft (**Rule 4:** ΔM = area under the V diagram). The area under the V diagram between $x = 12$ ft and $x = 21$ ft is the area of rectangle (3), which is (−8 kips)(9 ft) = −72 kip·ft (a negative value); therefore, $\Delta M = -72$ kip·ft. At point k, $M = +72$ kip·ft. The bending moment changes by $\Delta M = -72$ kip·ft between k and l; consequently, the bending moment at $x = 21$ ft is $M_l = 0$ kip·ft. This result is correct, since we know that the bending moment at roller D must be zero.

Notice that the M diagram started at $M_i = 0$ and finished at $M_l = 0$. Notice also that the M diagram consists of linear segments. From **Rule 5** (the slope of the M diagram is equal to the intensity of the shear force V), we can observe that the slope of the M diagram must be constant between points i, j, k, and l, because the shear force is constant in the corresponding regions. The slope of the M diagram between points i and j is +14 kips, the M slope between points j and k is +2 kips, and the M slope between points k and l is −8 kips. The only type of curve that has a constant slope is a line.

The maximum shear force is $V = 14$ kips. The maximum bending moment is $M = +72$ kip·ft, at $x = 12$ ft. Notice that the maximum bending moment occurs where the shear-force diagram crosses the $V = 0$ axis (between points e and f).

Relationships Among the Diagram Shapes

Equation (7.3) reveals that the V diagram is obtained by integrating the distributed load w, and Equation (7.4) shows that the M diagram is obtained by integrating the shear force V. Consider, for example, a beam segment that has no distributed load ($w = 0$). For this case, integrating w gives a constant shear-force function [i.e., a zero-order function $f(x^0)$], and integrating a constant V gives a linear function for the bending moment [i.e., a first-order function $f(x^1)$]. If a beam segment has constant w [i.e., a zero-order function $f(x^0)$], then the V diagram is a first-order function $f(x^1)$ and the M diagram is a second-order function $f(x^2)$. As can be seen, the order of the function increases successively by 1 in going from the w to the V to the M diagrams.

If the V diagram is constant for a beam segment, then the M diagram will be linear, making the M diagram relatively straightforward to sketch. If the V diagram is linear for a beam segment, then the M diagram will be quadratic (i.e., a parabola). A parabola can take one of two shapes: either concave or convex. The proper shape for the M diagram can be determined from information found on the V diagram, since the slope of the M diagram is equal to the intensity of the shear force V [**Rule 5:** Equation (7.2)]. Various shear-force diagram shapes and their corresponding bending-moment shapes are illustrated in Figure 7.12.

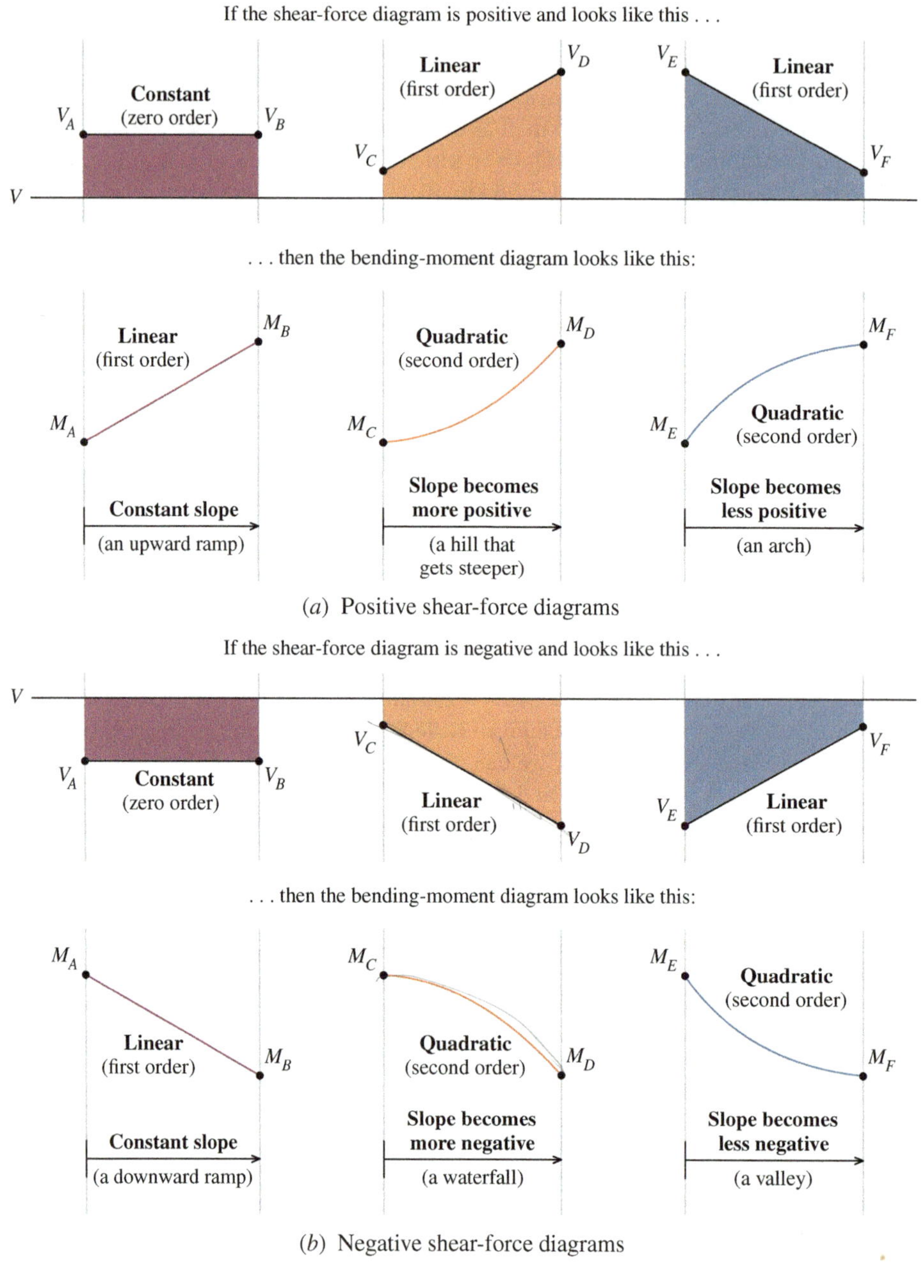

FIGURE 7.12 Relationships between V and M diagram shapes.

EXAMPLE 7.7

Draw the shear-force and bending-moment diagrams for the simply supported beam shown. Determine the maximum bending moment that occurs in the span.

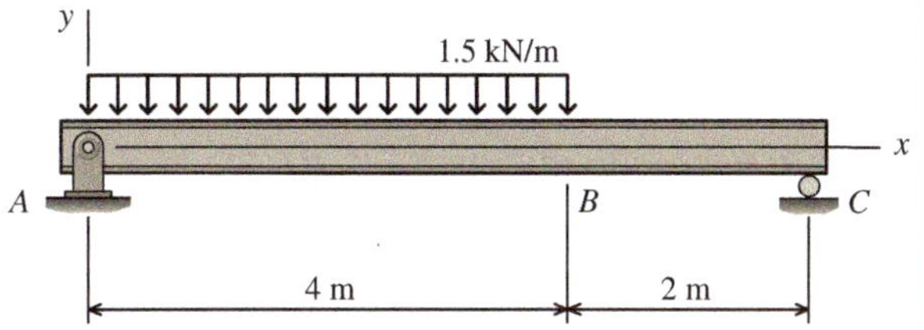

Plan the Solution

This example focuses on finding the maximum moment in a beam that has a uniformly distributed load. To calculate the maximum moment, we must first find the location where $V = 0$. To do this, we will determine the slope of the shear-force diagram from the intensity of the distributed loading, using Rule 3. Once the location where $V = 0$ is established, the maximum bending moment can be calculated from Rule 4.

SOLUTION

Support Reactions

An FBD of the beam is shown. For the purpose of calculating external beam reactions, the -1.5 kN/m distributed load can be replaced by its resultant force of (1.5 kN/m)(4 m) = 6 kN acting downward at the centroid of the loading. The equilibrium equations are as follows:

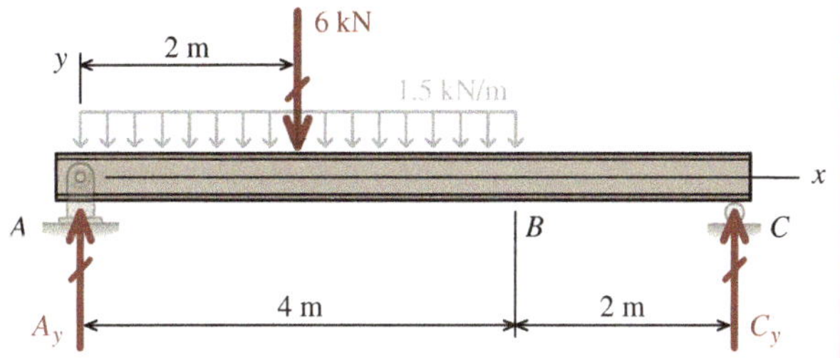

$$\Sigma F_y = A_y + C_y - 6 \text{ kN} = 0$$

$$\Sigma M_A = -(6 \text{ kN})(2 \text{ m}) + C_y(6 \text{ m}) = 0$$

From these equations, the beam reactions are

$$A_y = 4 \text{ kN} \qquad \text{and} \qquad C_y = 2 \text{ kN}$$

Construct the Shear-Force Diagram

On the load diagram, show the reaction forces acting in their proper directions. The original distributed load—*not the 6 kN resultant force*—should be used to construct the V diagram. The resultant force can be used to determine the external beam reactions; however, it cannot be used to determine the variation in shear force in the beam.

The steps that follow are used to construct the V diagram. (**Note:** The lowercase letters on the diagram correspond to the explanations given for each step.)

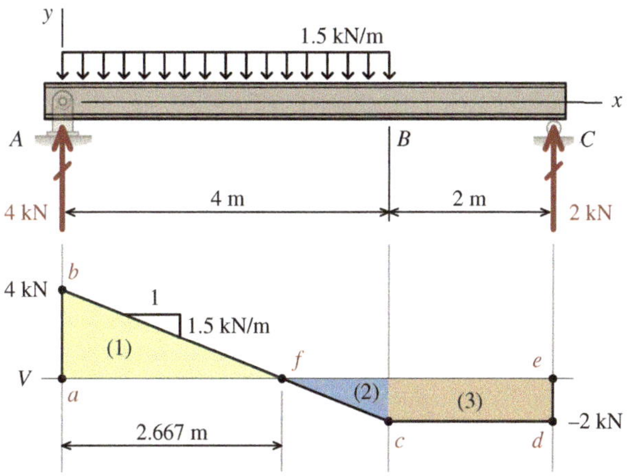

a $V(0^-) = 0$ kN (zero shear at end of beam).

b $V(0^+) = 4$ kN (**Rule 1:** V diagram jumps up by an amount equal to the 4 kN reaction force).

c $V(4) = -2$ kN (**Rule 2:** The change in shear force, ΔV, is equal to the area under the w curve). The area under the w curve between A and B is $(-1.5 \text{ kN/m})(4 \text{ m}) = -6$ kN; therefore, $\Delta V = -6$ kN. Since $V_b = +4$ kN, the shear force at c is $V_b = +4 \text{ kN} - 6 \text{ kN} = -2$ kN.

Because w is constant between A and B, the slope of the V diagram is also constant (**Rule 3**) and equal to -1.5 kN/m between b and c. Consequently, the V diagram is linear in this region.

d $V(6^-) = -2$ kN (**Rule 2:** The area under the w curve is zero between B and C; therefore, $\Delta V = 0$).

e $V(6^+) = 0$ kN (**Rule 1:** V diagram jumps up by an amount equal to the 2 kN reaction force and returns to $V = 0$ kN).

f Before the *V* diagram is complete, we must locate the point between *b* and *c* where $V = 0$. To do this, recall that the slope of the shear-force diagram (dV/dx) is equal to the intensity of the distributed load *w* (**Rule 3**). In this instance, a finite length Δx of the beam is considered rather than an infinitesimal length *dx*. Accordingly, Equation (7.1) can be expressed as

$$\text{Slope of } V \text{ diagram} = \frac{\Delta V}{\Delta x} = w \qquad \text{(a)}$$

Given that the distributed load is $w = -1.5$ kN/m between points *A* and *B*, the slope of the *V* diagram between points *b* and *c* is equal to -1.5 kN/m. Since $V = 4$ kN at point *b*, the shear force must change by $\Delta V = -4$ kN to cross the $V = 0$ axis. Use the known slope and the required ΔV to solve for Δx from Equation (a):

$$\Delta x = \frac{\Delta V}{w} = \frac{-4\text{ kN}}{-1.5\text{ kN/m}} = 2.667\,\text{m}$$

Since $x = 0$ m at *b*, point *f* is located 2.667 m from the left end of the beam.

Construct the Bending-Moment Diagram

Starting with the *V* diagram, the steps that follow are used to construct the *M* diagram. (**Note:** The lowercase letters on the *M* diagram correspond to the explanations given for each step.)

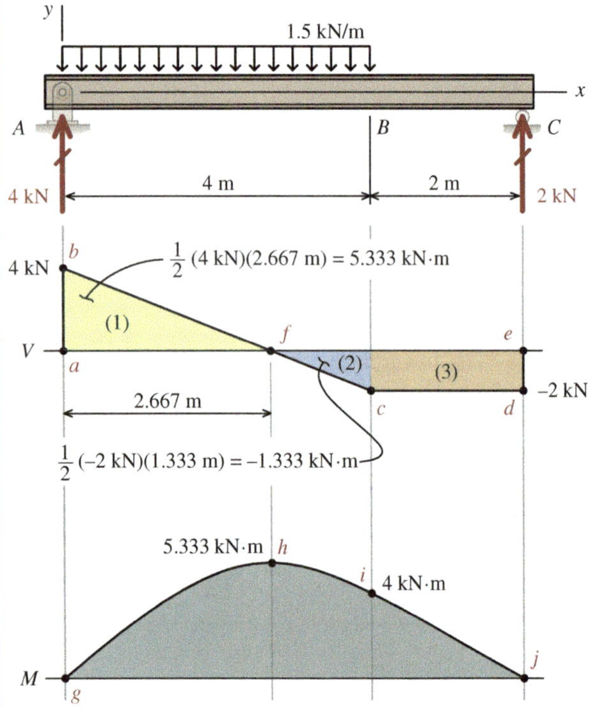

g $M(0) = 0$ (zero moment at the pinned end of the simply supported beam).

h $M(2.667) = +5.333$ kN·m (**Rule 4:** The change in bending moment, ΔM, between any two points is equal to the area under the *V* diagram). The *V* diagram between *b* and *f* is a triangle (1) with a width of 2.667 m and a height of +4 kN. The area of this triangle is +5.333 kN·m; therefore, $\Delta M = +5.333$ kN·m. Since $M = 0$ kN·m at $x = 0$ m and $\Delta M = +5.333$ kN·m, the bending moment at $x = 2.667$ m is $M_h = +5.333$ kN·m.

The shape of the bending-moment diagram between *g* and *h* can be sketched from **Rule 5** (the slope of the *M* diagram is equal to the shear force *V*). The shear force at *b* is +4 kN; therefore, the *M* diagram has a large positive slope at *g*. Between *b* and *f*, the shear force is still positive but decreases in magnitude; consequently, the slope of the *M* diagram is positive but becomes less steep as *x* increases. At *f*, $V = 0$, so the slope of the *M* diagram becomes zero.

i $M(4) = +4$ kN·m (**Rule 4:** ΔM = area under the *V* diagram). The shear-force diagram between *f* and *c* is a triangle (2) with a width of 1.333 m and a height of -2 kN. This triangle has a negative area of -1.333 kN·m; therefore, $\Delta M = -1.333$ kN·m. At *h*, $M = +5.333$ kN·m. Adding $\Delta M = -1.333$ kN·m to this value gives the bending moment at $x = 4$ m: $M_i = +4$ kN·m.

The shape of the bending-moment diagram between *h* and *i* can be sketched from **Rule 5** (the slope of the *M* diagram is equal to the shear force *V*). The slope of the *M* diagram is zero at *h*, corresponding to $V = 0$ at *f*. As *x* increases, *V* becomes increasingly negative; consequently, the slope dM/dx of the *M* diagram becomes more and more negative until it reaches -2 kN at point *i*.

j $M(6) = 0$ kN·m (**Rule 4:** ΔM = area under the V diagram). The area under the V diagram between $x = 4$ m and $x = 6$ m is simply the area of rectangle (3): $(-2 \text{ kN}) \times (2 \text{ m}) = -4$ kN·m. Adding $\Delta M = -4$ kN·m to the bending moment at point i ($M_i = +4$ kN·m) gives the bending moment at point j: $M_j = 0$ kN·m. This result is correct, since we know that the bending moment at roller C must be zero.

The maximum shear force is $V = 4$ kN. The maximum bending moment is $M = +5.333$ kN·m, at $x = 2.667$ m, where the shear-force diagram crosses $V = 0$ (between points b and c).

EXAMPLE 7.8

Draw the shear-force and bending-moment diagrams for the simply supported beam shown. Determine the maximum positive bending moment and the maximum negative bending moment that occur in the beam.

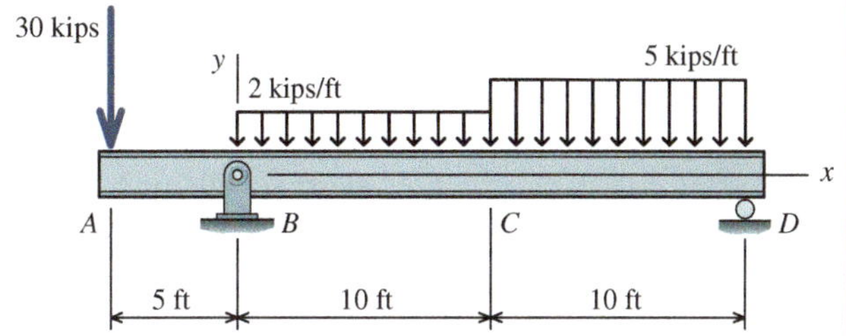

Plan the Solution

The challenges of this problem are

(a) to determine both the largest positive and largest negative moments and

(b) to sketch the proper shape of the M curve as it goes from negative to positive values.

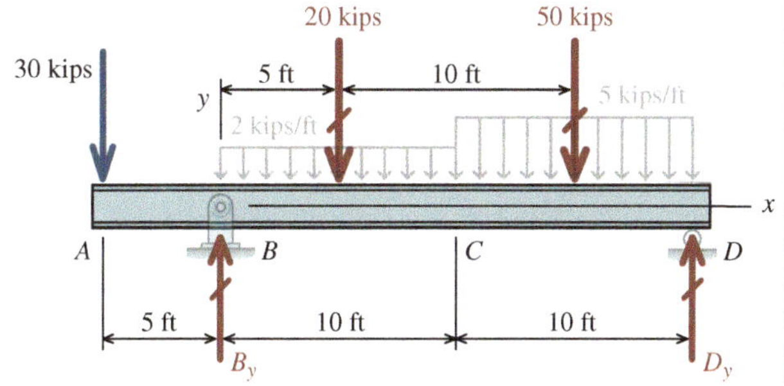

SOLUTION

Support Reactions

An FBD of the beam is shown. For the purpose of calculating external beam reactions, the distributed loads are replaced by their resultant forces. The equilibrium equations are as follows:

$$\Sigma F_y = B_y + D_y - 30 \text{ kips} - 20 \text{ kips} - 50 \text{ kips} = 0$$

$$\Sigma M_B = (30 \text{ kips})(5 \text{ ft}) - (20 \text{ kips})(5 \text{ ft}) - (50 \text{ kips})(15 \text{ ft}) + D_y(20 \text{ ft}) = 0$$

From these equations, the beam reactions are $B_y = 65$ kips and $D_y = 35$ kips.

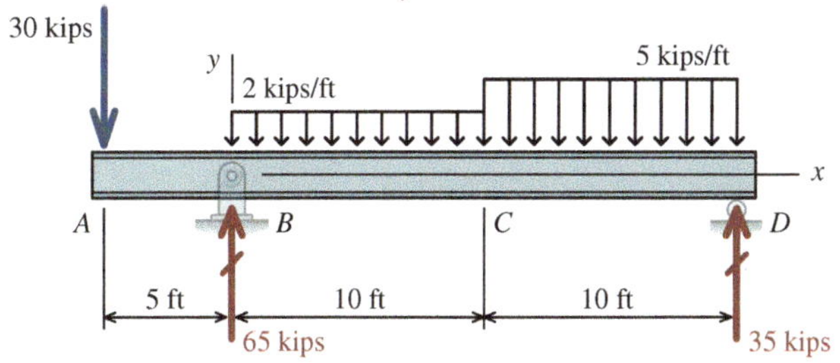

Construct the Shear-Force Diagram

Before beginning, complete the load diagram by noting the reaction forces and using arrows to indicate their proper directions. Use the *original distributed loads—not the resultant forces*—to construct the V diagram.

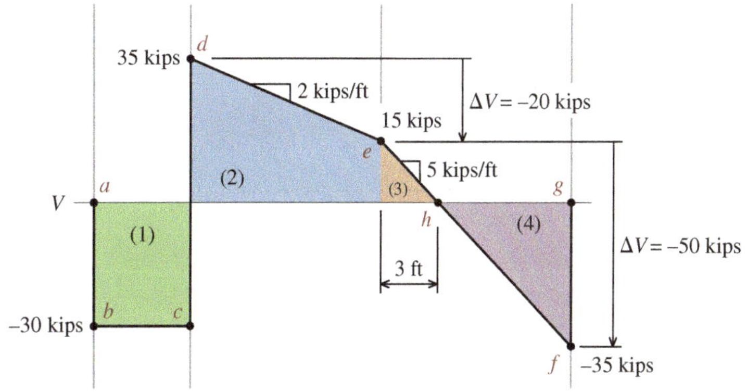

a $V(-5^-) = 0$ kips.

b $V(-5^+) = -30$ kips (**Rule 1**).

c $V(0^-) = -30$ kips (**Rule 2**).
There is zero area under the w curve between A and B; therefore, $\Delta V = 0$ between b and c.

d $V(0^+) = +35$ kips (**Rule 1**).

e $V(10) = +15$ kips (**Rule 2:** ΔV = area under w curve). The area under the w curve between B and C is -20 kips. Since w is constant in this region, the slope of the V diagram is also constant (**Rule 3**) and equal to -2 kips/ft between d and e.

f $V(20^-) = -35$ kips (**Rule 2:** ΔV = area under w curve). The area under the w curve between C and D is -50 kips. The slope of the V diagram is constant (**Rule 3**) and equal to -5 kips/ft between e and f.

g $V(20^+) = 0$ kips (**Rule 1**).

h To complete the V diagram, locate the point between e and f where $V = 0$. The slope of the V diagram in this interval is -5 kips/ft (**Rule 3**).

At point e, $V = +15$ kips; consequently, the shear force must change by $\Delta V = -15$ kips in order for it to intersect the $V = 0$ axis. Use the known slope and the required ΔV to find Δx:

$$\Delta x = \frac{\Delta V}{w} = \frac{-15 \text{ kips}}{-5 \text{ kips/ft}} = 3.0 \text{ ft}$$

Since $x = 10$ ft at point e, point h is located at $x = 13$ ft.

Construct the Bending-Moment Diagram

Starting with the V diagram, the steps that follow are used to construct the M diagram:

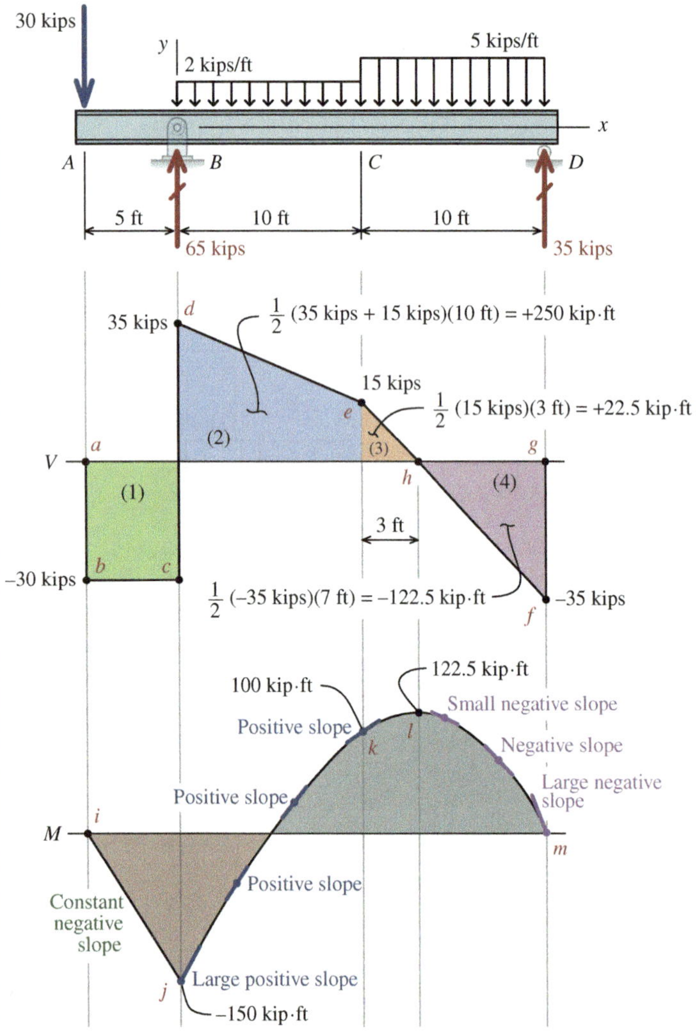

i $M(-5) = 0$ (there is zero moment at the free end of a simply supported beam).

j $M(0) = -150$ kip·ft (**Rule 4:** ΔM = area under V diagram). The area of region (1) is (-30 kips)(5 ft) $= -150$ kip·ft; therefore, $\Delta M = -150$ kip·ft. The M diagram is linear between points i and j, having a constant negative slope of -30 kips.

k $M(10) = +100$ kip·ft (**Rule 4:** ΔM = area under V diagram). The area of trapezoid (2) is $+250$ kip·ft; hence, $\Delta M = +250$ kip·ft. Adding $\Delta M = +250$ kip·ft to the -150 kip·ft moment at j gives $M_k = +100$ kip·ft at $x = 10$ ft.

Use **Rule 5** (slope of M diagram = shear force V) to sketch the M diagram between j and k. Since $V_d = +35$ kips, the M diagram has a large positive slope at j. As x increases, the shear force stays positive but decreases to a value $V_e = +15$ kips at point e. As a result, the slope of the M diagram will be positive between j and k but will flatten as it nears point k.

l $M(13) = +122.5$ kip·ft (**Rule 4:** ΔM = area under V diagram). Area (3) under the V diagram is $+22.5$ kip·ft; thus, $\Delta M = +22.5$ kip·ft. Add $+22.5$ kip·ft to $M_k = +100$ kip·ft to compute $M_l = +122.5$ kip·ft at point l. Since $V = 0$ at this location, the slope of the M diagram is zero at point l.

m $M(20) = 0$ kip·ft (**Rule 4:** ΔM = area under V diagram). The area of triangle (4) is -122.5 kip·ft; therefore, $\Delta M = -122.5$ kip·ft.

The shape of the bending-moment diagram between l and m can be sketched from **Rule 5** (slope of M diagram = shear force V). The slope of the M diagram is zero at l. As x increases, V becomes increasingly negative; consequently, the slope of the M diagram becomes more and more negative until it reaches its most negative slope at $x = 20$ ft.

The maximum positive bending moment is +122.5 kip·ft, and it occurs at $x = 13$ ft. The maximum negative bending moment is –150 kip·ft, and this bending moment occurs at $x = 0$.

EXAMPLE 7.9

Draw the shear-force and bending-moment diagrams for the cantilever beam shown. Determine the maximum bending moment that occurs in the beam.

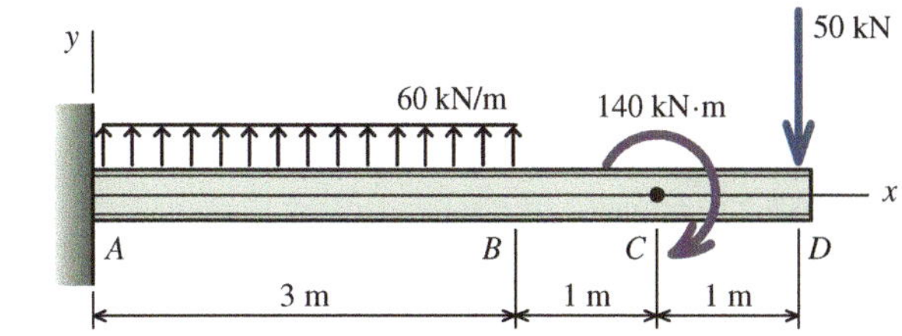

Plan the Solution

The effects of external concentrated moments on the V and M diagrams are sometimes confusing. Two external concentrated moments act on this cantilever beam.

SOLUTION

Support Reactions

An FBD of the beam is shown. For the purpose of calculating external beam reactions, the distributed loads are replaced by their resultant forces. The equilibrium equations are as follows:

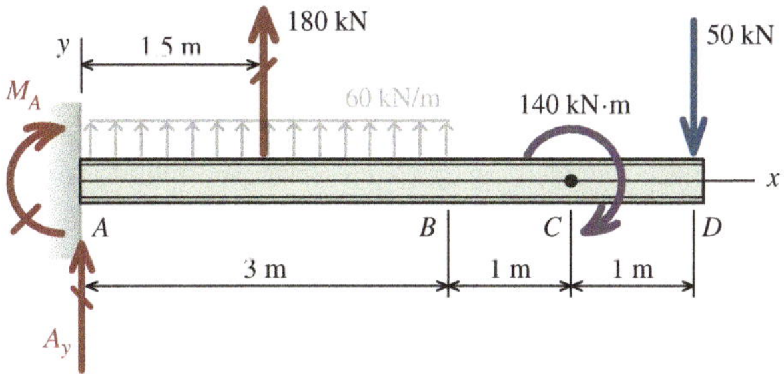

$$\Sigma F_y = A_y + 180 \text{ kN} - 50 \text{ kN} = 0$$

$$\Sigma M_A = (180 \text{ kN})(1.5 \text{ m}) - (50 \text{ kN})(5 \text{ m}) - 140 \text{ kN·m} - M_A = 0$$

From these equations, the beam reactions are $A_y = -130$ kN and $M_A = -120$ kN·m.

Construct the Shear-Force Diagram

Before beginning, complete the load diagram by noting the reaction forces and using arrows to indicate their proper directions. Use the *original distributed loads—not the resultant forces*—to construct the V diagram.

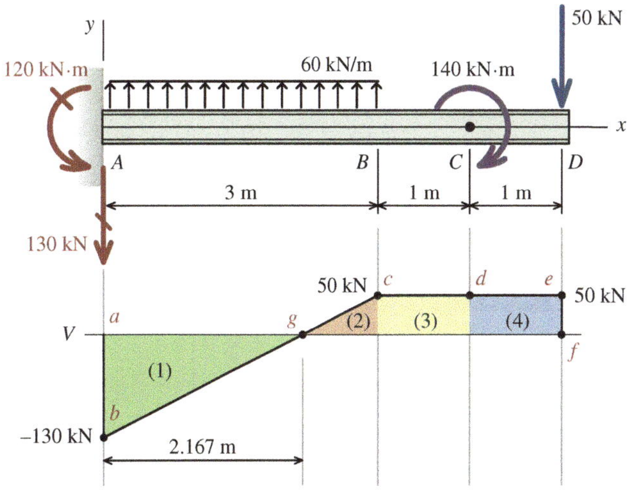

a $V(0^-) = 0$ kN.

b $V(0^+) = -130$ kN (**Rule 1**).

c $V(3) = +50$ kN (**Rule 2**). The area under the w curve between A and B is +180 kN; therefore, $\Delta V = +180$ kN between b and c.

d $V(4) = +50$ kN (**Rule 2:** ΔV = area under w curve). There is zero area under the w curve between B and C; therefore, no change occurs in V.

e $V(5^-) = +50$ kN (**Rule 2:** ΔV = area under w curve). There is zero area under the w curve between C and D; therefore, no change occurs in V.

f $V(5^+) = 0$ kN (**Rule 1**).

g To complete the V diagram, locate the point between b and c at which $V = 0$. The slope of the V diagram in this interval is +60 kN/m (**Rule 3**). At point b, $V = -130$ kN; consequently, the shear force must change by $\Delta V = +130$ kN in order for

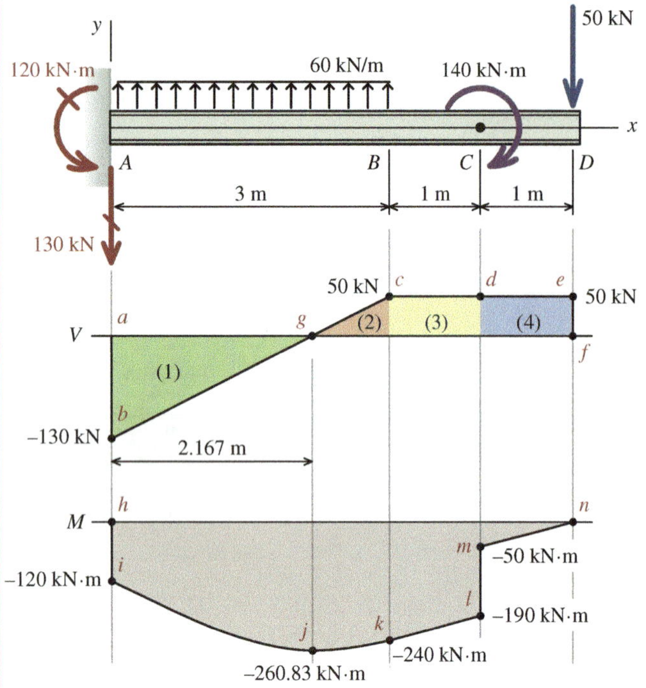

it to intersect the $V = 0$ axis. Use the known slope and the required ΔV to find Δx:

$$\Delta x = \frac{\Delta V}{w} = \frac{+130 \text{ kN}}{+60 \text{ kN/m}} = 2.1667 \text{ m}$$

Construct the Bending-Moment Diagram

Starting with the V diagram, use the following steps to construct the M diagram:

h $M(0^-) = 0$.

i $M(0^+) = -120$ kN·m (**Rule 6:** For a counterclockwise external moment, the M diagram jumps down by an amount equal to the 120 kN·m reaction).

j $M(2.1667) = -260.836$ kN·m (**Rule 4:** ΔM = area under V diagram). Area (1) = −140.836 kN·m; therefore, $\Delta M = -140.836$ kN·m.

Use **Rule 5** (slope of M diagram = shear force V) to sketch the M diagram between i and j. Since $V_b = -130$ kN, the M diagram has a large negative slope at i. As x increases, the shear force becomes less negative until it reaches zero at g. As a result, the slope of the M diagram will be negative between i and j but will flatten as it reaches point j.

k $M(3) = -240$ kN·m (**Rule 4:** ΔM = area under V diagram). Area (2) = +20.833 kN·m; hence, $\Delta M = +20.833$ kN·m. Adding ΔM to the −260.836 kN·m moment at j gives $M_k = -240$ kN·m at $x = 3$ m.

Use **Rule 5** (slope of M diagram = shear force V) to sketch the M diagram between j and k. Since $V_g = 0$, the M diagram has zero slope at j. As x increases, the shear force becomes increasingly positive until it reaches its largest positive value at point c. This situation means that the slope of the M diagram will be positive between j and k, curving upward more and more as x increases.

l $M(4^-) = -190$ kN·m (**Rule 4:** ΔM = area under V diagram). Area (3) = +50 kN·m. Adding $\Delta M = +50$ kN·m to the −240 kN·m moment at k gives $M_l = -190$ kN·m at $x = 4$ m.

m $M(4^+) = -50$ kN·m (**Rule 6:** For a clockwise external moment, the M diagram jumps up by an amount equal to the 140 kN·m external concentrated moment).

n $M(5) = 0$ kN·m (**Rule 4:** ΔM = area under V diagram). Area (4) = +50 kN·m.

The maximum bending moment is −260.8 kN·m, and it occurs at $x = 2.1667$ m.

MecMovies

EXAMPLES

M7.1 Six rules for constructing shear-force and bending-moment diagrams.

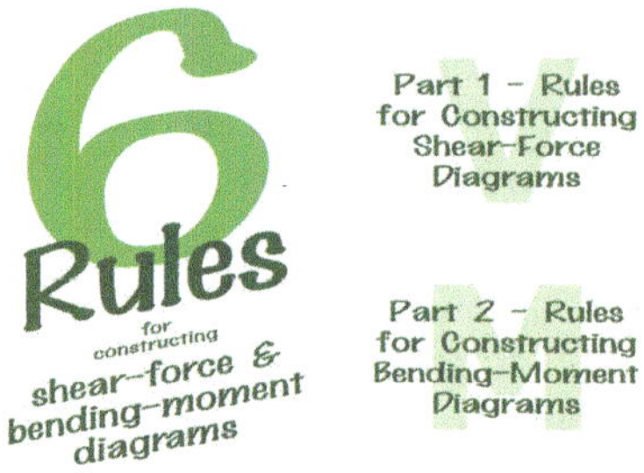

M7.3 Dynamically generated shear-force and bending-moment diagrams for 48 beams with various support and loading configurations. Brief explanations are given for all key points on both the V and M diagrams.

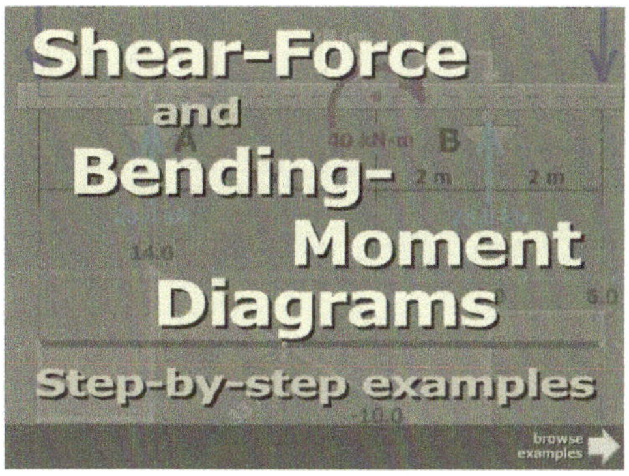

EXERCISES

M7.1 **Six Rules for Constructing Shear-Force and Bending-Moment Diagrams.** Score at least 40 points for each of the six rules. (Minimum total score = 240 points.)

M7.2 Following the rules, score at least 350 points out of 400 possible points.

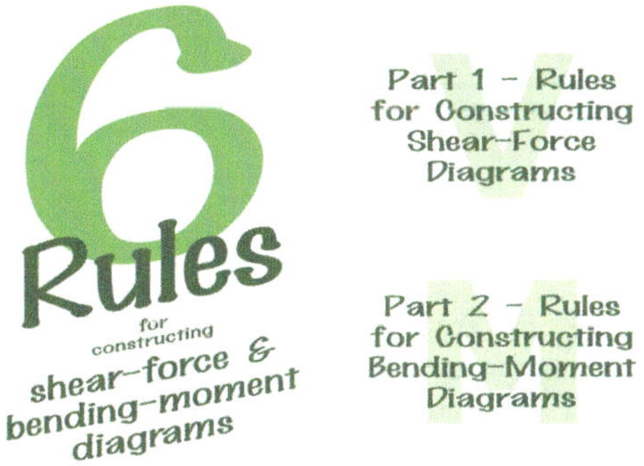

FIGURE M7.1

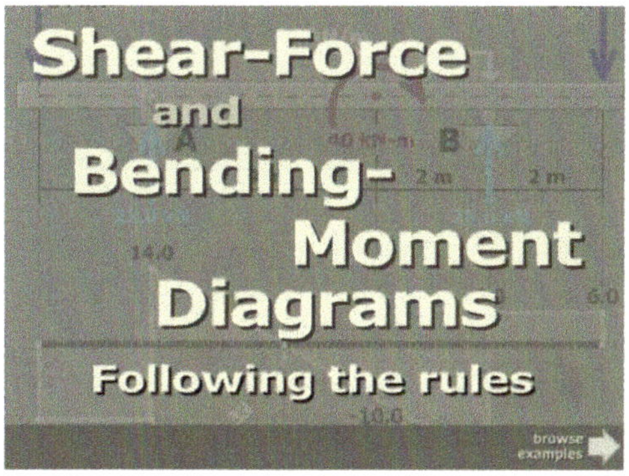

FIGURE M7.2

PROBLEMS

P7.8–P7.16 Use the graphical method to construct the shear-force and bending-moment diagrams for the beams shown in Figures P7.8–P7.16. Label all significant points on each diagram, and identify the maximum moments (both positive and negative) along with their respective locations. Clearly differentiate straight-line and curved portions of the diagrams.

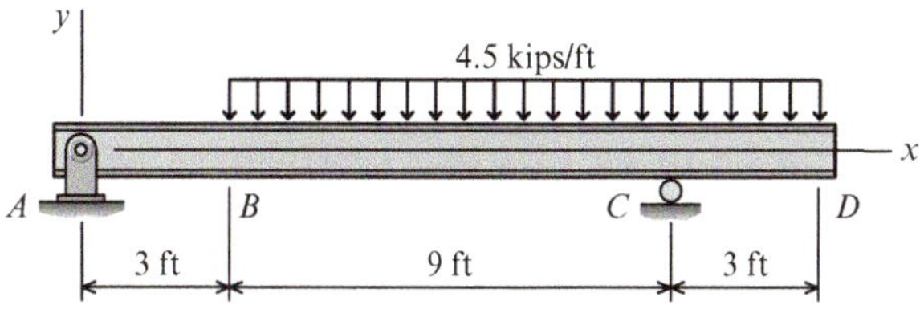

FIGURE P7.9

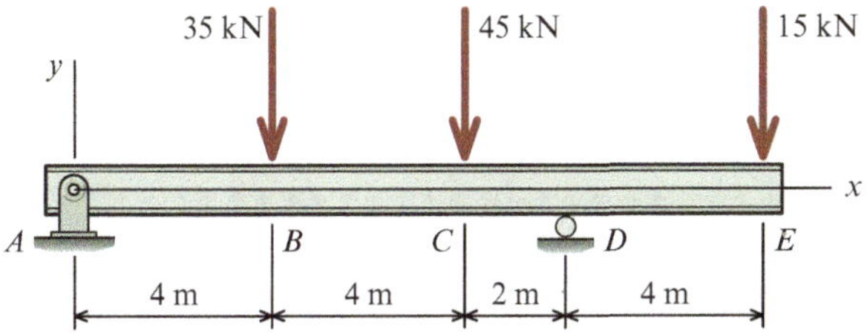

FIGURE P7.8

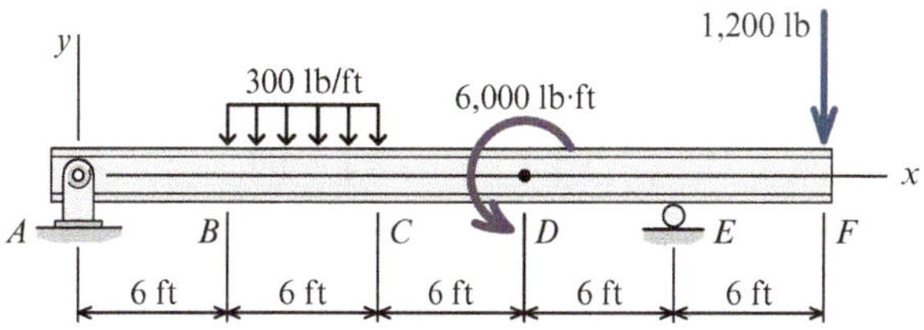

FIGURE P7.10

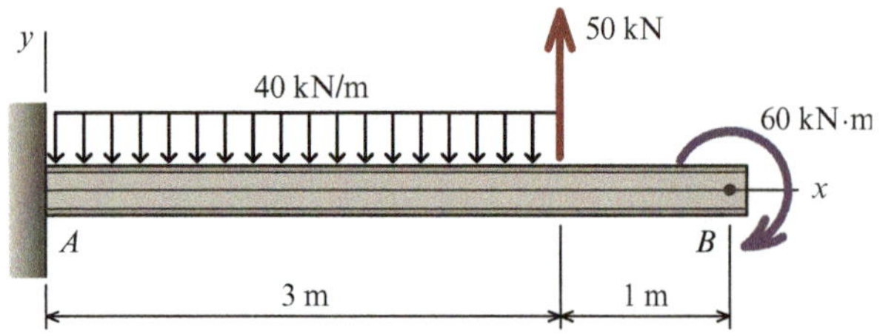

FIGURE P7.11

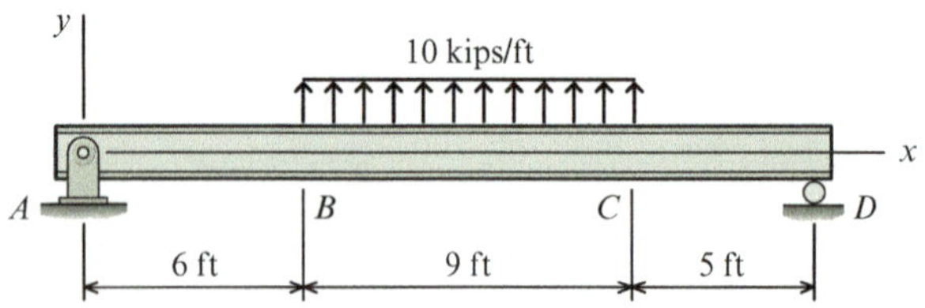

FIGURE P7.12

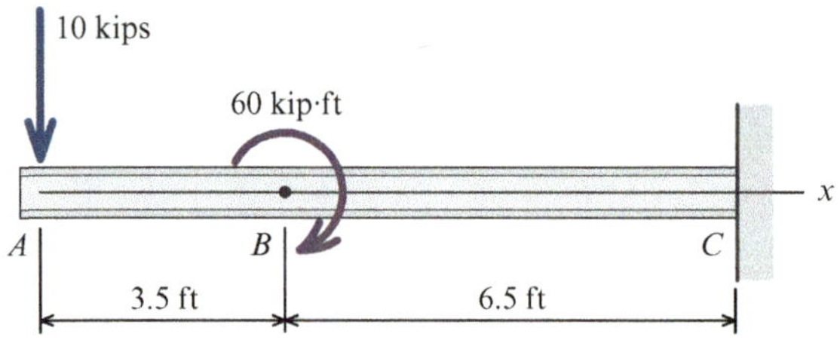

FIGURE P7.13

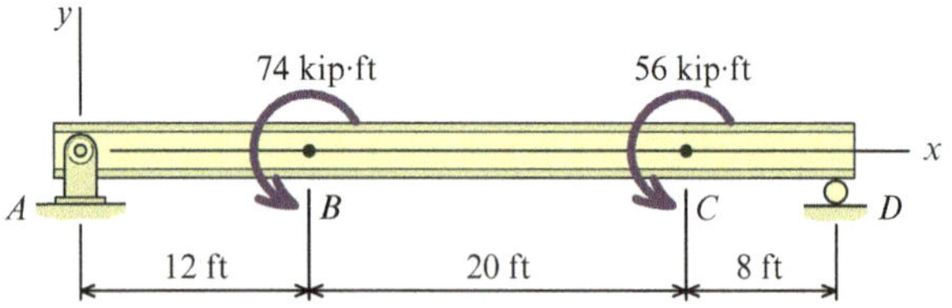

FIGURE P7.14

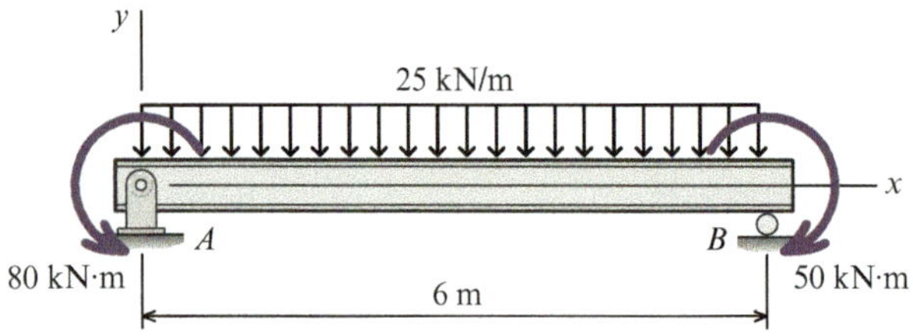

FIGURE P7.15

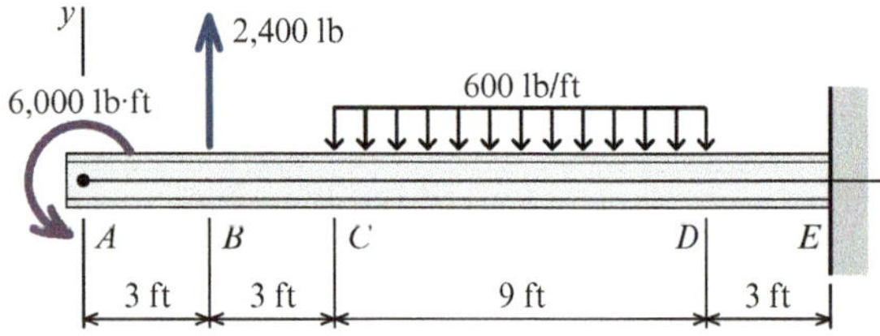

FIGURE P7.16

P7.17–P7.18 Draw the shear-force and bending-moment diagrams for the beams shown in Figures P7.17 and P7.18. Assume the upward reaction provided by the ground to be uniformly distributed. Label all significant points on each diagram. Determine the maximum value of (a) the internal shear force and (b) the internal bending moment.

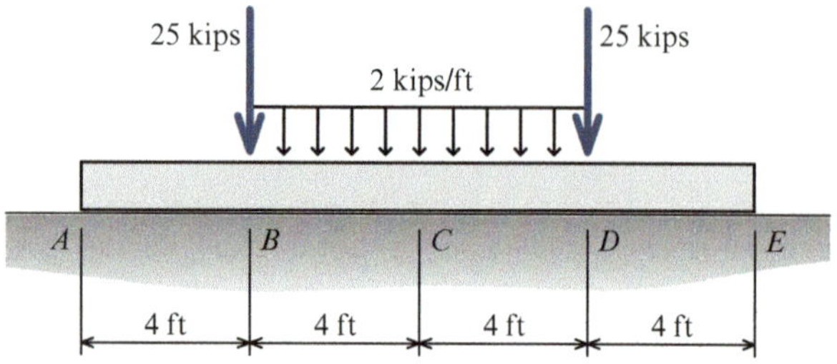

FIGURE P7.17

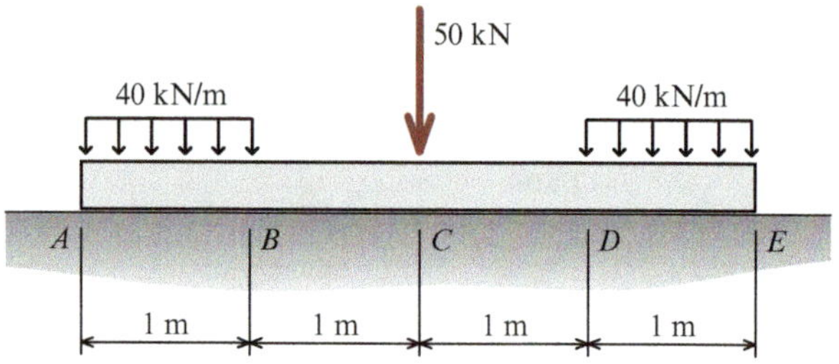

FIGURE P7.18

P7.19–P7.20 Use the graphical method to construct the shear-force and bending-moment diagrams for the beams shown in Figures P7.19 and P7.20. Label all significant points on each diagram, and identify the maximum moments along with their respective locations. In addition,

(a) determine V and M in the beam at a point located 0.75 m to the right of B.
(b) determine V and M in the beam at a point located 1.25 m to the left of C.

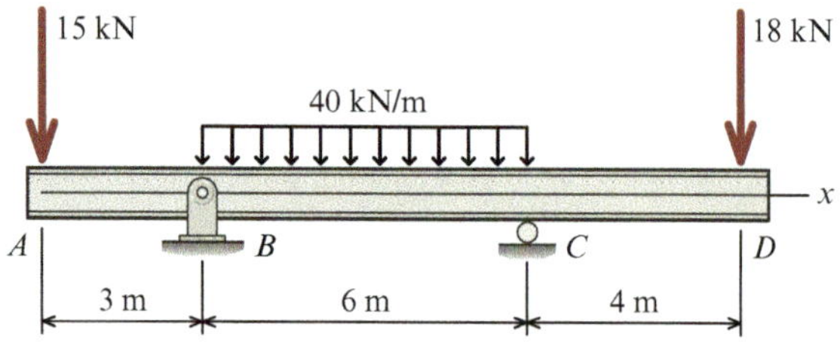

FIGURE P7.19

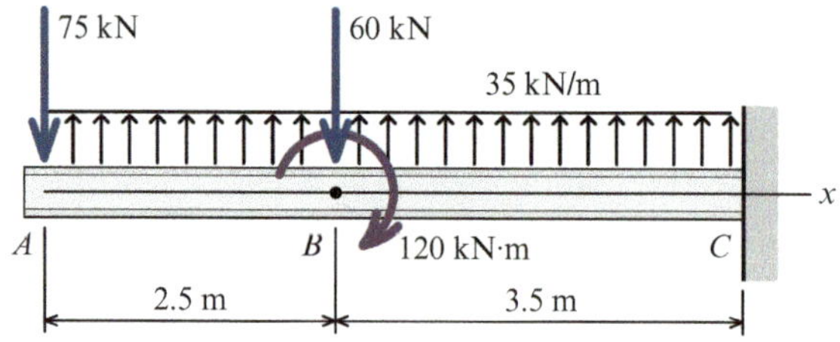

FIGURE P7.20

P7.21–P7.22 Use the graphical method to construct the shear-force and bending-moment diagrams for the beams shown in Figures P7.21 and P7.22. Label all significant points on each diagram, and identify the maximum moments along with their respective locations. In addition,

(a) determine V and M in the beam at a point located 1.50 m to the right of B.
(b) determine V and M in the beam at a point located 1.25 m to the left of D.

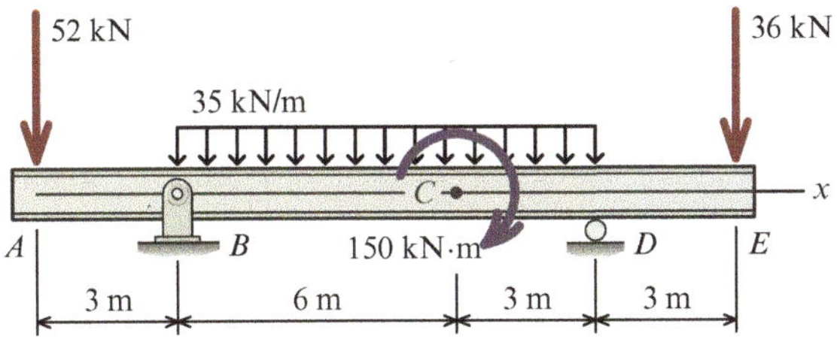

FIGURE P7.21

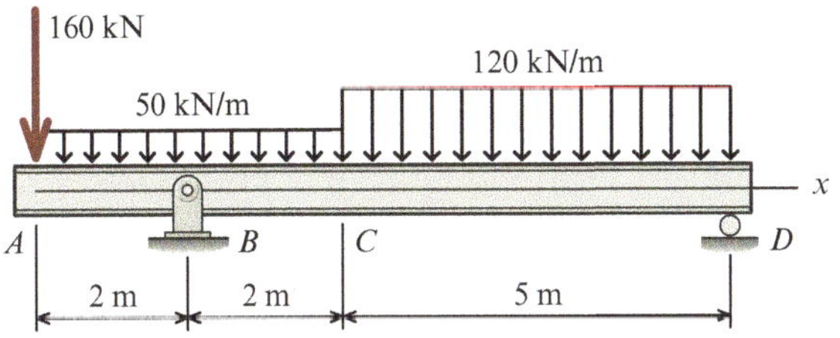

FIGURE P7.22

P7.23–P7.31 Use the graphical method to construct the shear-force and bending-moment diagrams for the beams shown in Figures P7.23–P7.31. Label all significant points on each diagram, and identify the maximum moments along with their respective locations. Clearly differentiate straight-line and curved portions of the diagrams.

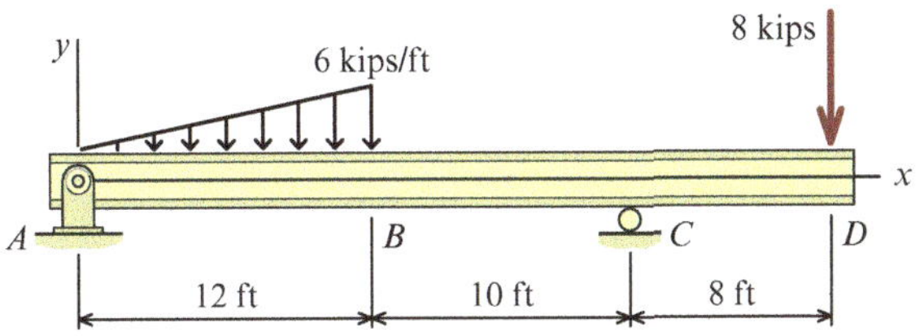

FIGURE P7.23

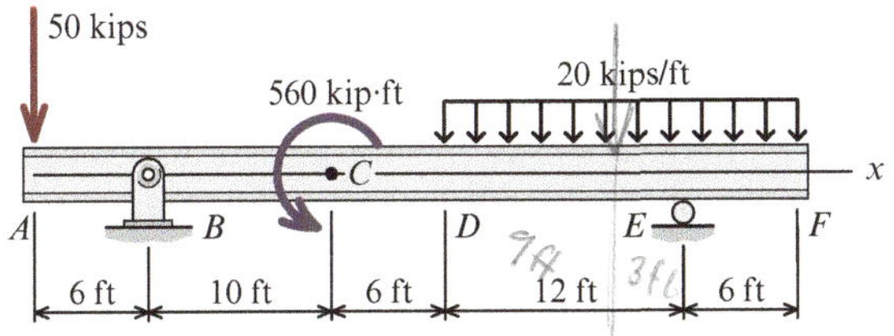

FIGURE P7.24

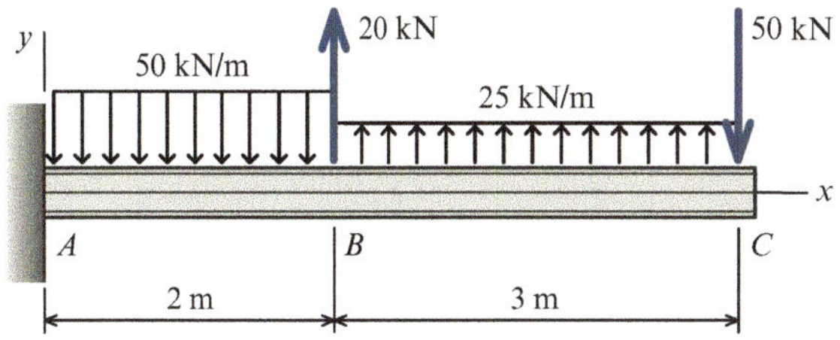

FIGURE P7.25

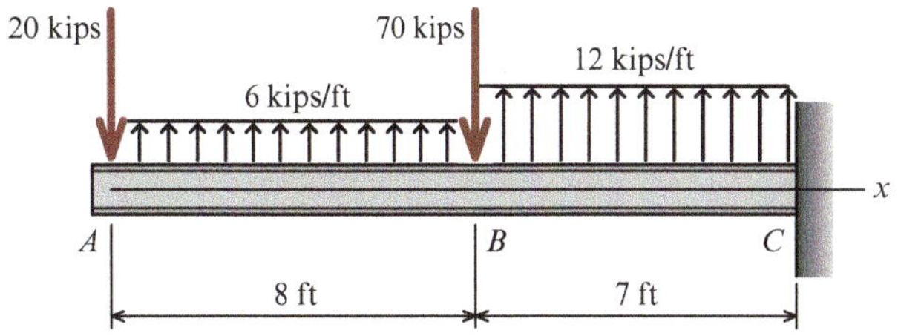

FIGURE P7.26

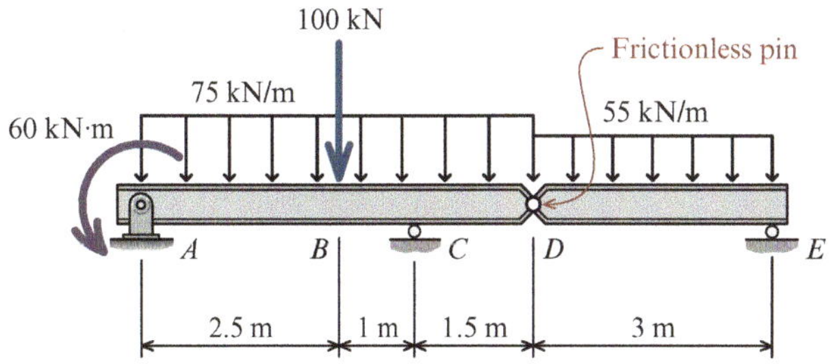

FIGURE P7.27

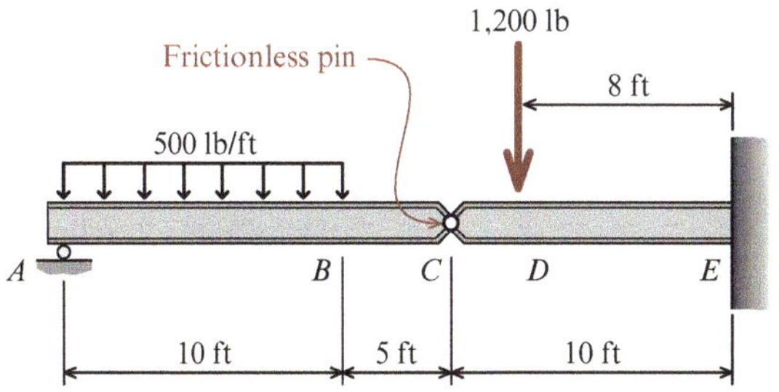

FIGURE P7.28

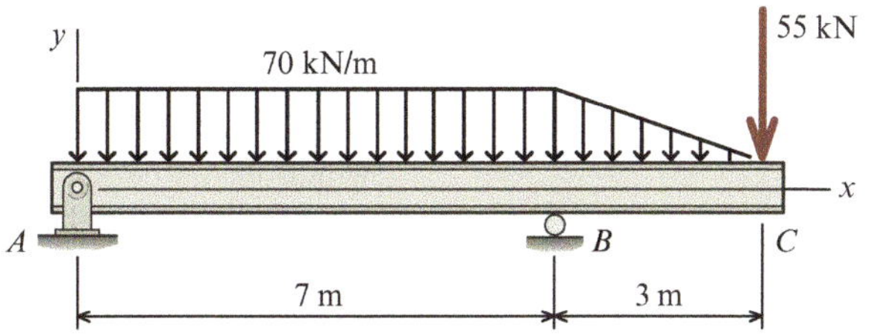

FIGURE P7.29

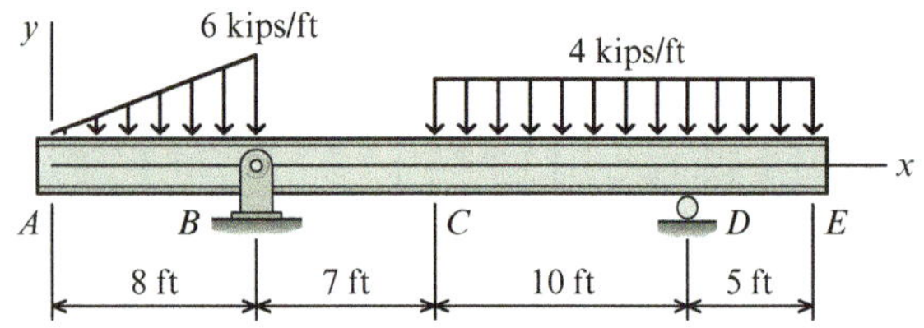

FIGURE P7.30

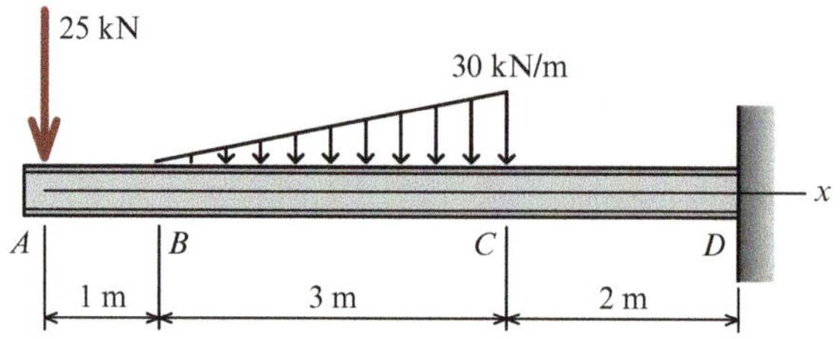

FIGURE P7.31

7.4 Discontinuity Functions to Represent Load, Shear, and Moment

In Section 7.2, we constructed shear and moment diagrams by developing functions that express the variation of internal shear force $V(x)$ and internal bending moment $M(x)$ along the beam and then plotting these functions. The method of integration used in Section 7.2 is convenient if the loads can be expressed as continuous functions acting over the entire length of the beam. However, if several loadings act on the beam, this approach can become extremely tedious and time consuming because a new set of functions must be developed for each interval of the beam.

In this section, a method will be presented in which a single function is formulated that incorporates all loads acting on the beam. This single load function $w(x)$ will be constructed in such a way that it will be continuous for the entire length of the beam even though the loads may not be. The load function $w(x)$ can then be integrated twice—first to derive $V(x)$ and a second time to obtain $M(x)$. To express the load on the beam in a single function, two types of mathematical operators will be employed. **Macaulay functions** will be used to describe distributed loads, and **singularity functions** will be used to represent concentrated forces and concentrated moments. Together, these functions are termed **discontinuity functions**. Their usage has restrictions that distinguish them from ordinary functions. To provide a clear indication of these restrictions, the traditional parentheses used with functions are replaced by angle brackets, called **Macaulay brackets**, that take the form $\langle x - a \rangle^n$.

Macaulay Functions

Distributed loadings can be represented by Macaulay functions, which are defined in general terms as follows:

$$\langle x - a \rangle^n = \begin{cases} 0 & \text{when } x < a \\ (x - a)^n & \text{when } x \geq a \end{cases} \quad \text{for } n \geq 0 \ (n = 0, 1, 2, \ldots) \tag{7.7}$$

Whenever the term inside the brackets is less than zero, the Macaulay function equals zero and it is as if the function does not exist. However, when the term inside the brackets is greater than or equal to zero, the Macaulay function behaves like an ordinary function, which would be written with parentheses. In other words, the Macaulay function acts like a switch in which the function turns on for values of x greater than or equal to a.

Three Macaulay functions corresponding, respectively, to $n = 0$, $n = 1$, and $n = 2$ are plotted in Figure 7.13. In Figure 7.13a, the function $\langle x - a \rangle^0$ is discontinuous at $x = a$, producing a plot in the shape of a step. Accordingly, this function is termed a **step function**.

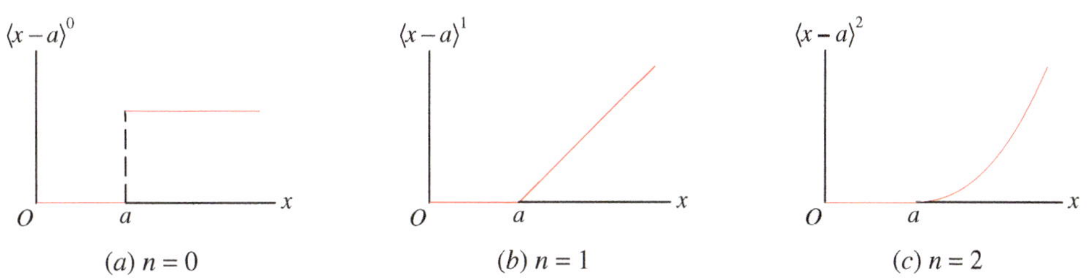

FIGURE 7.13 Graphs of Macaulay functions.

From the definition given in Equation (7.7), and with the recognition that any number raised to the zero power is defined as unity, the step function can be summarized as

$$\langle x - a\rangle^0 = \begin{cases} 0 & \text{when } x < a \\ 1 & \text{when } x \geq a \end{cases} \tag{7.8}$$

When scaled by a constant value equal to the load intensity, the step function $\langle x - a\rangle^0$ can be used to represent uniformly distributed loadings. In Figure 7.13*b*, the function $\langle x - a\rangle^1$ is termed a **ramp function** because it produces a linearly increasing plot beginning at $x = a$. Accordingly, the ramp function $\langle x - a\rangle^1$, combined with the appropriate load intensity, can be used to represent linearly distributed loadings. The function $\langle x - a\rangle^2$ in Figure 7.13*c* produces a parabolic plot beginning at $x = a$.

Observe that the quantity inside of the Macaulay brackets is a measure of length; therefore, it will include a length dimension, such as meters or feet. The Macaulay functions will be scaled by a constant to account for the intensity of the loading and to ensure that all terms included in the load function $w(x)$ have consistent units of force per unit length. Table 7.2 gives discontinuity expressions for various types of loads.

Singularity Functions

Singularity functions are used to represent concentrated forces P_0 and concentrated moments M_0. A concentrated force P_0 can be considered a special case of a distributed load in which an extremely large load P_0 acts over a distance ε that approaches zero (Figure 7.14*a*). Thus, the intensity of the loading is $w = P_0/\varepsilon$, and the area under the loading is equivalent to P. This condition can be expressed by the singularity function

$$w(x) = P_0\langle x - a\rangle^{-1} = \begin{cases} 0 & \text{when } x \neq a \\ P_0 & \text{when } x = a \end{cases} \tag{7.9}$$

in which the function has a value of P_0 only at $x = a$ and is otherwise zero. Observe that $n = -1$. Since the bracketed term has a length unit, the result of the function has units of force per unit length, as is required for dimensional consistency.

Similarly, a concentrated moment M_0 can be considered as a special case involving two distributed loadings, as shown in Figure 7.14*b*. For this type of load, the following singularity function can be employed:

$$w(x) = M_0\langle x - a\rangle^{-2} = \begin{cases} 0 & \text{when } x \neq a \\ M_0 & \text{when } x = a \end{cases} \tag{7.10}$$

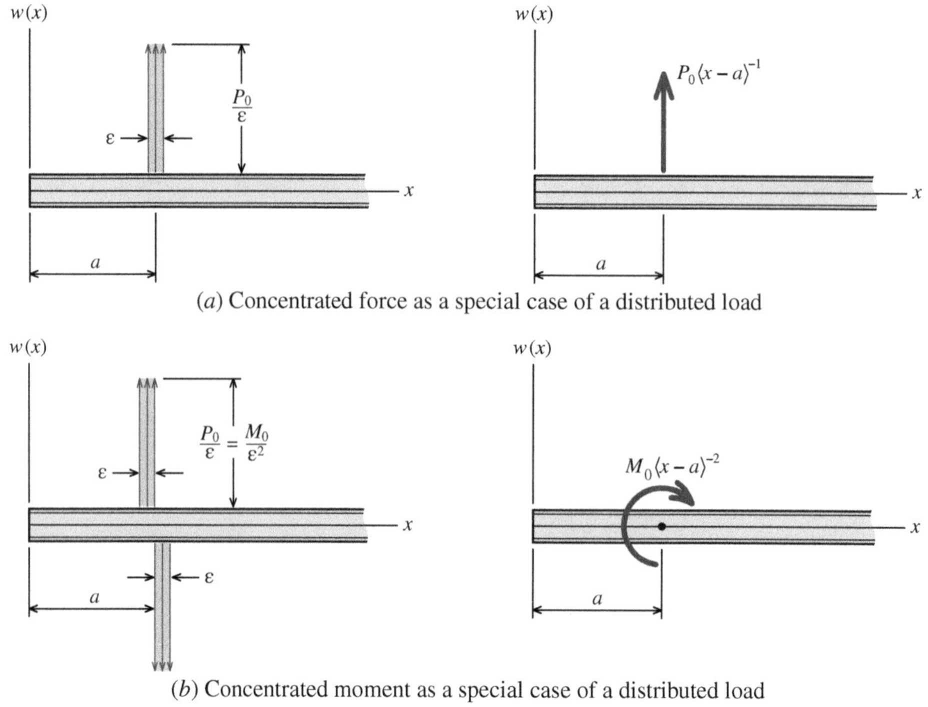

FIGURE 7.14 Singularity functions to represent (*a*) concentrated forces and (*b*) concentrated moments.

As before, the function has a value of M_0 only at $x = a$ and is otherwise zero. In Equation (7.10), notice that $n = -2$, which ensures that the result of the function has consistent units of force per unit length.

Integrals of Discontinuity Functions

Integration of discontinuity functions is defined by the following rules:

$$\int \langle x - a\rangle^n \, dx = \begin{cases} \dfrac{\langle x - a\rangle^{n+1}}{n+1} & \text{for } n \geq 0 \\ \langle x - a\rangle^{n+1} & \text{for } n < 0 \end{cases} \tag{7.11}$$

Notice that, for negative values of the exponent n, the only effect of integration is that n increases by 1.

Constants of Integration. The integration of Macaulay functions produces constants of integration. The constant of integration that results from the integration of $w(x)$ to obtain $V(x)$ is simply the shear force at $x = 0$—that is, $V(0)$. Similarly, the second constant of integration that results when $V(x)$ is integrated to obtain $M(x)$ is the bending moment at $x = 0$—that is, $M(0)$. If the loading function $w(x)$ is written solely in terms of the applied loads, then constants of integration must be included in the integration process and evaluated with the use of boundary conditions. As these constants of integration are introduced into either the $V(x)$ or $M(x)$ functions, they are expressed by singularity functions in the form $C\langle x\rangle^0$. After their introduction into either $V(x)$ or $M(x)$, the constants are integrated in the usual manner in subsequent integrals.

However, the same result for both $V(x)$ and $M(x)$ can be obtained by including the reaction forces and moments in the load function $w(x)$. The inclusion of reaction forces and moments in $w(x)$ has considerable appeal, since the constants of integration for both $V(x)$ and $M(x)$ are automatically determined without the need for explicit reference to boundary conditions. The reactions for statically determinate beams are easily computed in a fashion that is familiar to all engineering students. Accordingly, beam reaction forces and moments will be incorporated into the load function $w(x)$ in the examples presented subsequently in this section.

To summarize, constants of integration arise in the double integration of $w(x)$ to obtain $V(x)$ and $M(x)$. If $w(x)$ is formulated solely in terms of the applied loads, then these constants must be explicitly determined with the use of boundary conditions. However, if beam reaction forces and moments are included in $w(x)$ along with the applied loads, then constants of integration are redundant and thus unnecessary for the $V(x)$ and $M(x)$ functions.

Application of Discontinuity Functions to Determine V and M. Table 7.2 summarizes discontinuity expressions for $w(x)$ that are required for various common loadings. It is important to keep in mind that Macaulay functions continue indefinitely for $x > a$. In other words, once a Macaulay function is switched on, it stays on for all increasing values of x. In accordance with the concept of superposition, a Macaulay function is cancelled by the addition of another Macaulay function to the beam's $w(x)$ function.

Macaulay functions continue indefinitely for $x > a$. Therefore, a new Macaulay function (or, in some cases, several functions) must be introduced to terminate a previous function.

EXAMPLE 7.10

Use discontinuity functions to obtain expressions for the internal shear force $V(x)$ and internal bending moment $M(x)$ in the beam shown. Then, use these expressions to plot the shear-force and bending-moment diagrams for the beam.

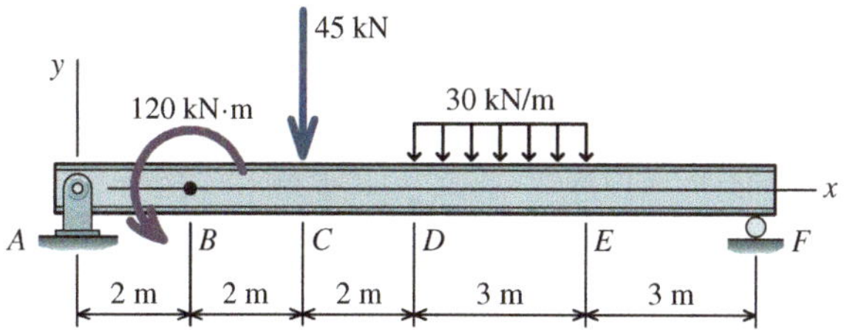

Plan the Solution

Determine the reactions at simple supports A and F. Using Table 7.2, write expressions for $w(x)$ for each of the three loads acting on the beam, as well as for the two support reactions. Integrate $w(x)$ to determine an equation for the shear force $V(x)$, and then integrate $V(x)$ to determine an equation for the bending moment $M(x)$. Plot these functions to complete the shear-force and bending-moment diagrams.

SOLUTION

Support Reactions

An FBD of the beam is shown. The equilibrium equations are as follows:

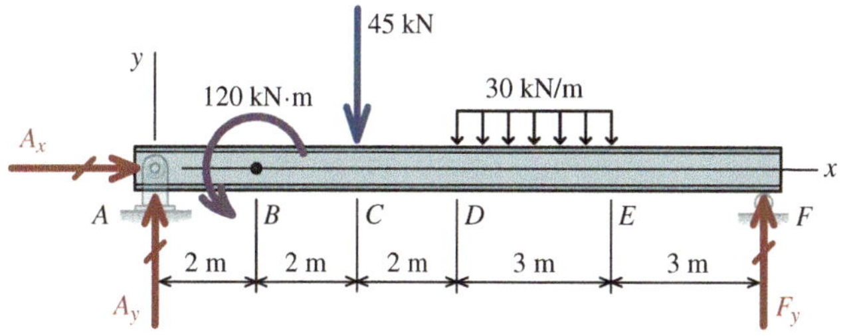

$$\Sigma F_x = A_x = 0 \qquad \text{(trivial)}$$

$$\Sigma F_y = A_y + F_y - 45 \text{ kN} - (30 \text{ kN/m})(3 \text{ m}) = 0$$

$$\Sigma M_A = 120 \text{ kN}\cdot\text{m} - (45 \text{ kN})(4 \text{ m}) - (30 \text{ kN/m})(3 \text{ m})(7.5 \text{ m}) + F_y(12 \text{ m}) = 0$$

From these equations, the beam reactions are

$$A_y = 73.75 \text{ kN} \qquad \text{and} \qquad F_y = 61.25 \text{ kN}$$

Discontinuity Expressions

Reaction force A_y: The upward reaction force at A is expressed by

$$w(x) = A_y\langle x - 0 \text{ m}\rangle^{-1} = 73.75 \text{ kN}\langle x - 0 \text{ m}\rangle^{-1} \qquad \text{(a)}$$

Table 7.2 Basic Loads Represented by Discontinuity Functions

Case	Load on Beam	Discontinuity Expressions
1	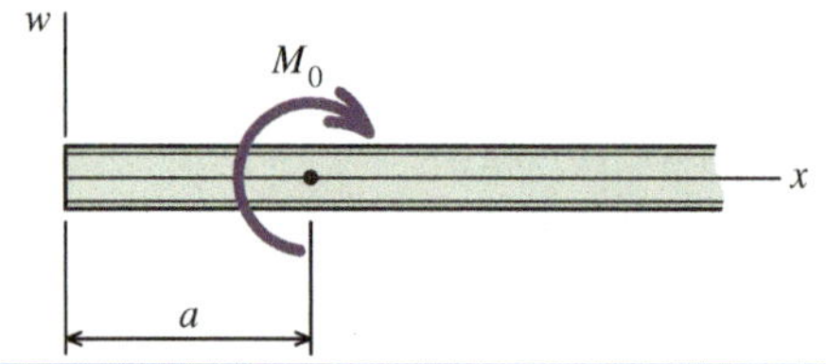	$w(x) = M_0\langle x-a\rangle^{-2}$ $V(x) = M_0\langle x-a\rangle^{-1}$ $M(x) = M_0\langle x-a\rangle^{0}$
2	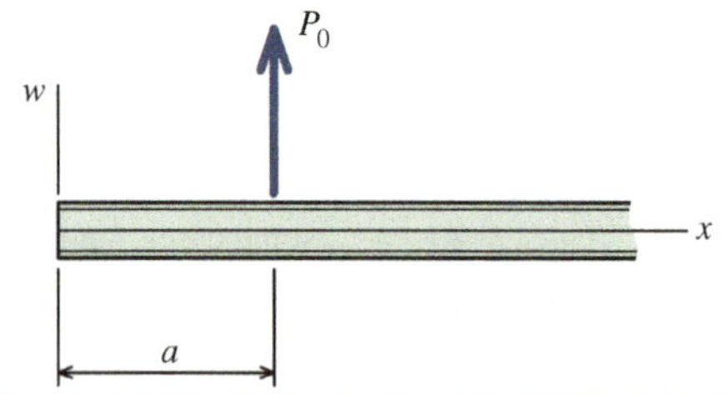	$w(x) = P_0\langle x-a\rangle^{-1}$ $V(x) = P_0\langle x-a\rangle^{0}$ $M(x) = P_0\langle x-a\rangle^{1}$
3	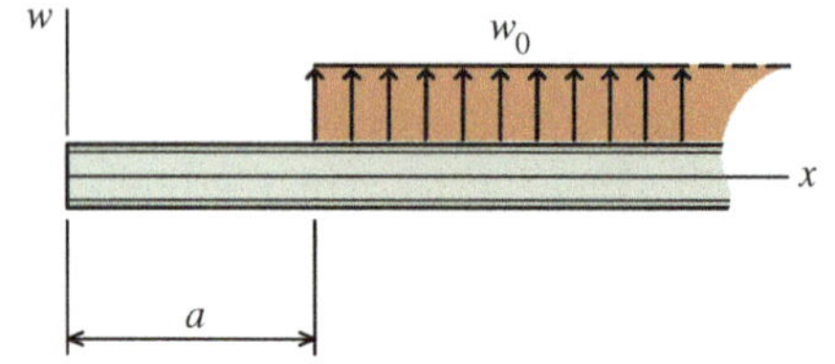	$w(x) = w_0\langle x-a\rangle^{0}$ $V(x) = w_0\langle x-a\rangle^{1}$ $M(x) = \dfrac{w_0}{2}\langle x-a\rangle^{2}$
4	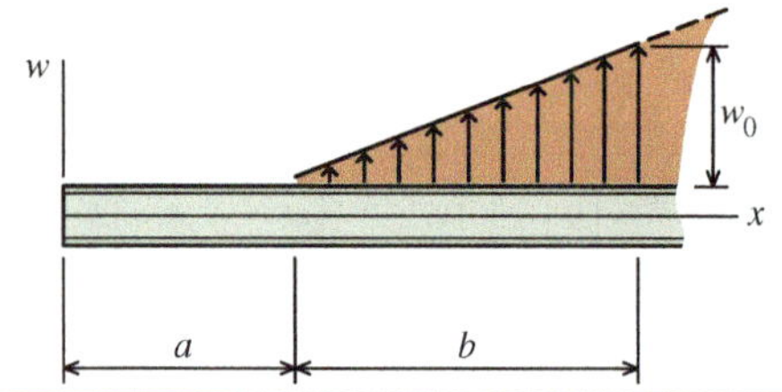	$w(x) = \dfrac{w_0}{b}\langle x-a\rangle^{1}$ $V(x) = \dfrac{w_0}{2b}\langle x-a\rangle^{2}$ $M(x) = \dfrac{w_0}{6b}\langle x-a\rangle^{3}$
5	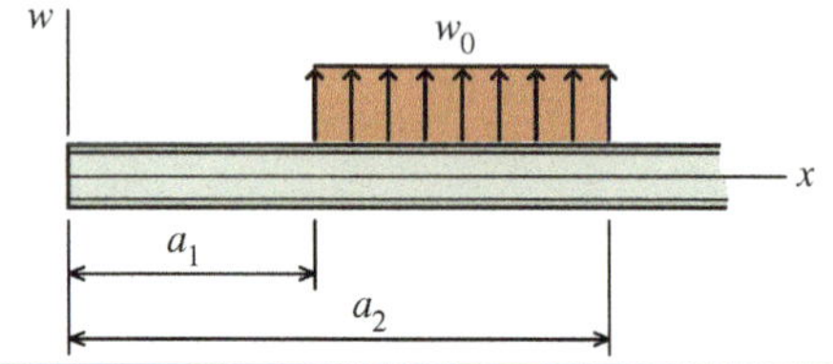	$w(x) = w_0\langle x-a_1\rangle^{0} - w_0\langle x-a_2\rangle^{0}$ $V(x) = w_0\langle x-a_1\rangle^{1} - w_0\langle x-a_2\rangle^{1}$ $M(x) = \dfrac{w_0}{2}\langle x-a_1\rangle^{2} - \dfrac{w_0}{2}\langle x-a_2\rangle^{2}$
6	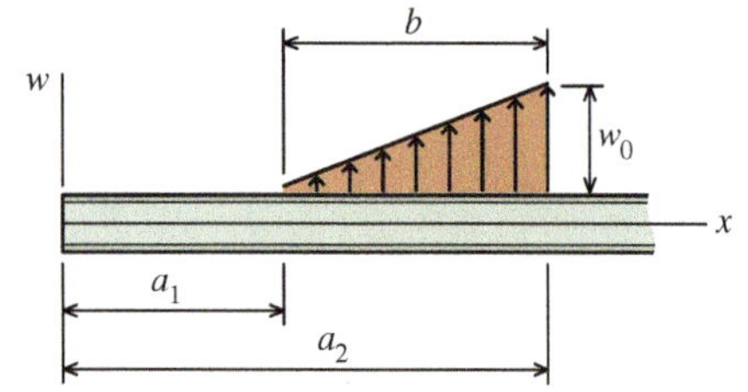	$w(x) = \dfrac{w_0}{b}\langle x-a_1\rangle^{1} - \dfrac{w_0}{b}\langle x-a_2\rangle^{1} - w_0\langle x-a_2\rangle^{0}$ $V(x) = \dfrac{w_0}{2b}\langle x-a_1\rangle^{2} - \dfrac{w_0}{2b}\langle x-a_2\rangle^{2} - w_0\langle x-a_2\rangle^{1}$ $M(x) = \dfrac{w_0}{6b}\langle x-a_1\rangle^{3} - \dfrac{w_0}{6b}\langle x-a_2\rangle^{3} - \dfrac{w_0}{2}\langle x-a_2\rangle^{2}$
7	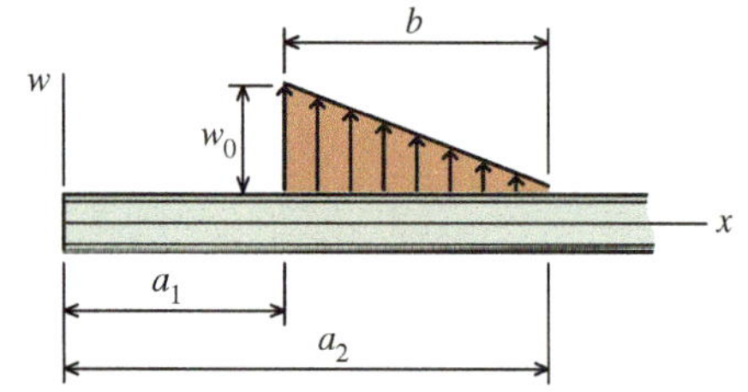	$w(x) = w_0\langle x-a_1\rangle^{0} - \dfrac{w_0}{b}\langle x-a_1\rangle^{1} + \dfrac{w_0}{b}\langle x-a_2\rangle^{1}$ $V(x) = w_0\langle x-a_1\rangle^{1} - \dfrac{w_0}{2b}\langle x-a_1\rangle^{2} + \dfrac{w_0}{2b}\langle x-a_2\rangle^{2}$ $M(x) = \dfrac{w_0}{2}\langle x-a_1\rangle^{2} - \dfrac{w_0}{6b}\langle x-a_1\rangle^{3} + \dfrac{w_0}{6b}\langle x-a_2\rangle^{3}$

120 kN·m concentrated moment: From case 1 of Table 7.2, the 120 kN·m concentrated moment acting at $x = 2$ m is represented by the singularity function

$$w(x) = -120 \text{ kN·m}\langle x - 2 \text{ m}\rangle^{-2} \quad \text{(b)}$$

Note that the negative sign is included to account for the counterclockwise moment rotation shown on this beam.

45 kN concentrated load: From case 2 of Table 7.2, the 45 kN concentrated load acting at $x = 4$ m is represented by the singularity function

$$w(x) = -45 \text{ kN}\langle x - 4 \text{ m}\rangle^{-1} \quad \text{(c)}$$

Note that the negative sign is included to account for the downward direction of the 45 kN concentrated load shown on the beam.

30 kN/m uniformly distributed load: The uniformly distributed load requires the use of two terms. Term 1 applies the 30 kN/m downward load at point D, where $x = 6$ m:

$$w(x) = -30 \text{ kN/m}\langle x - 6 \text{ m}\rangle^{0}$$

The uniformly distributed load represented by this term continues to act on the beam for values of x greater than $x = 6$ m. For the beam and loading considered here, the distributed load should act only within the interval $6 \text{ m} \leq x \leq 9$ m. To terminate the downward distributed load at $x = 9$ m requires the superposition of a second term. This second term applies an equal-magnitude upward uniformly distributed load that begins at E, where $x = 9$ m:

$$w(x) = -30 \text{ kN/m}\langle x - 6 \text{ m}\rangle^{0} + 30 \text{ kN/m}\langle x - 9 \text{ m}\rangle^{0} \quad \text{(d)}$$

The addition of these two terms produces a downward 30 kN/m distributed load that begins at $x = 6$ m and terminates at $x = 9$ m.

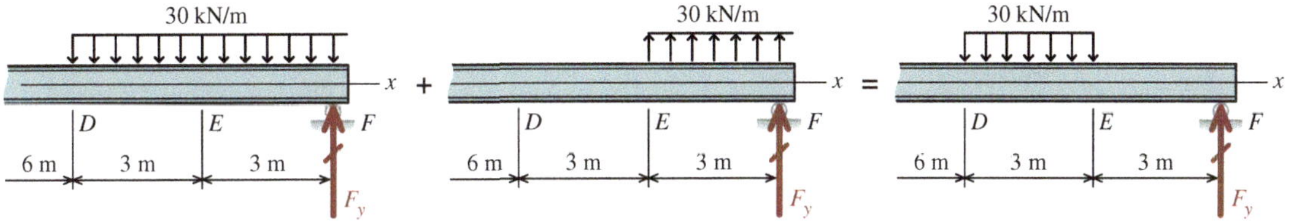

Term 1: Downward uniformly distributed load beginning at D
$-(30 \text{ kN/m})\langle x - 6 \text{ m}\rangle^0$

Term 2: Upward uniformly distributed load beginning at E
$+(30 \text{ kN/m})\langle x - 9 \text{ m}\rangle^0$

30 kN/m distributed load beginning at D and ending at E

Reaction force F_y: The upward reaction force at F is expressed by

$$w(x) = F_y\langle x - 12 \text{ m}\rangle^{-1} = 61.25 \text{ kN}\langle x - 12 \text{ m}\rangle^{-1} \quad \text{(e)}$$

As a practical matter, this term has no effect, since the value of Equation (e) is zero for all values of $x \leq 12$ m. Because the beam is only 12 m long, values of $x > 12$ m make no sense in this situation. However, the term will be retained here for completeness and clarity.

Complete-beam loading expression: The sum of Equations (a) through (e) gives the load expression $w(x)$ for the entire beam:

$$\begin{aligned} w(x) = {} & 73.75 \text{ kN}\langle x - 0 \text{ m}\rangle^{-1} - 120 \text{ kN·m}\langle x - 2 \text{ m}\rangle^{-2} - 45 \text{ kN}\langle x - 4 \text{ m}\rangle^{-1} \\ & - 30 \text{ kN/m}\langle x - 6 \text{ m}\rangle^{0} + 30 \text{ kN/m}\langle x - 9 \text{ m}\rangle^{0} + 61.25 \text{ kN}\langle x - 12 \text{ m}\rangle^{-1} \end{aligned} \quad \text{(f)}$$

Shear-force equation: Using the integration rules given in Equation (7.11), integrate Equation (f) to derive the shear-force equation for the beam:

$$\begin{aligned} V(x) &= \int w(x)\,dx \\ &= 73.75\text{ kN}\langle x - 0\text{ m}\rangle^0 - 120\text{ kN}\cdot\text{m}\langle x - 2\text{ m}\rangle^{-1} - 45\text{ kN}\langle x - 4\text{ m}\rangle^0 \\ &\quad - 30\text{ kN/m}\langle x - 6\text{ m}\rangle^1 + 30\text{ kN/m}\langle x - 9\text{ m}\rangle^1 + 61.25\text{ kN}\langle x - 12\text{ m}\rangle^0 \end{aligned} \tag{g}$$

Bending-moment equation: Similarly, integrate Equation (g) to derive the bending-moment equation for the beam:

$$\begin{aligned} M(x) &= \int V(x)\,dx \\ &= 73.75\text{ kN}\langle x - 0\text{ m}\rangle^1 - 120\text{ kN}\cdot\text{m}\langle x - 2\text{ m}\rangle^0 - 45\text{ kN}\langle x - 4\text{ m}\rangle^1 \\ &\quad - \frac{30\text{ kN/m}}{2}\langle x - 6\text{ m}\rangle^2 + \frac{30\text{ kN/m}}{2}\langle x - 9\text{ m}\rangle^2 + 61.25\text{ kN}\langle x - 12\text{ m}\rangle^1 \end{aligned} \tag{h}$$

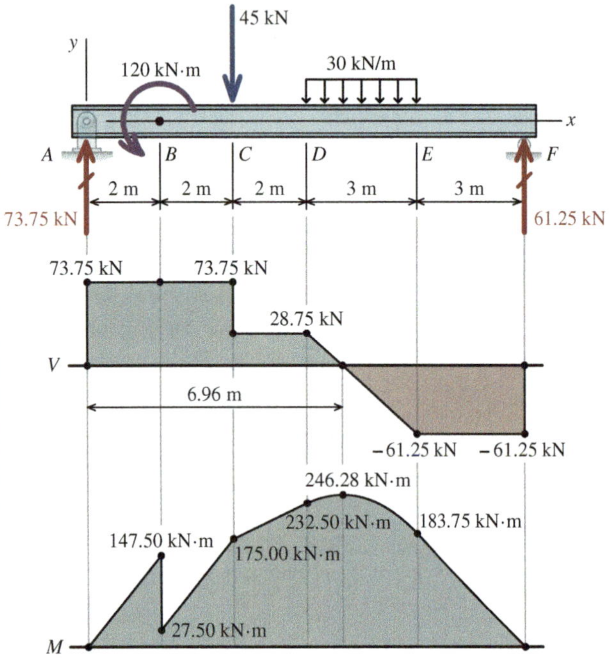

Plot the Functions

Plot the $V(x)$ and $M(x)$ functions given in Equations (g) and (h) for $0\text{ m} \le x \le 12\text{ m}$ to create the shear-force and bending-moment diagram shown.

EXAMPLE 7.11

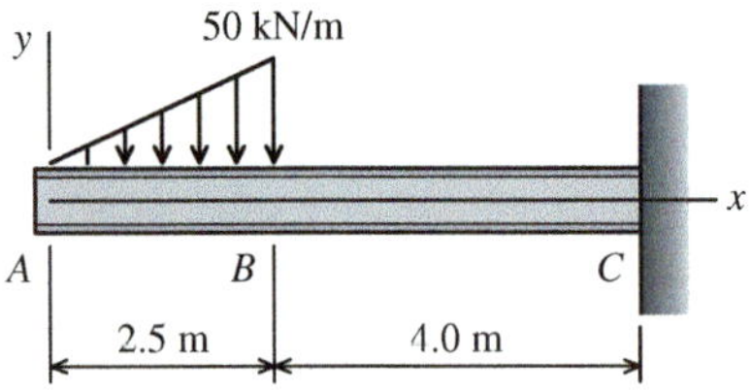

Use discontinuity functions to express the linearly distributed load acting on the beam between A and B.

Plan the Solution

The expressions found in Table 7.2 are explained by means of the example of the loading shown on the beam to the left.

SOLUTION

When we refer to case 4 of Table 7.2, our first instinct might be to represent the linearly distributed load on the beam with just a single term:

$$w(x) = -\frac{50\text{ kN/m}}{2.5\text{ m}}\langle x - 0\text{ m}\rangle^1$$

However, this term by itself produces a load that continues to increase as x increases. But we need to terminate the linear load at B, so we might try adding the algebraic inverse of the linearly distributed load to the $w(x)$ equation:

$$w(x) = -\frac{50\text{ kN/m}}{2.5\text{ m}}\langle x - 0\text{ m}\rangle^1 + \frac{50\text{ kN/m}}{2.5\text{ m}}\langle x - 2.5\text{ m}\rangle^1$$

The sum of these two expressions represents the loading shown next. While the second expression has indeed cancelled out the linearly distributed load from B onward, a uniformly distributed loading remains.

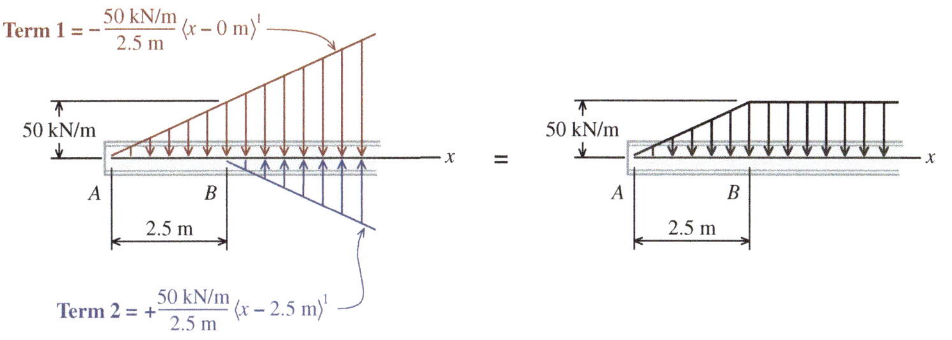

Adding the inverse of the linearly increasing load to the beam at B . . .

. . . eliminates the linear portion of the load; however, a uniformly distributed load remains.

To cancel this uniformly distributed load, a third term that begins at B is required:

$$w(x) = -\frac{50\text{ kN/m}}{2.5\text{ m}}\langle x - 0\text{ m}\rangle^1 + \frac{50\text{ kN/m}}{2.5\text{ m}}\langle x - 2.5\text{ m}\rangle^1 + 50\text{ kN/m}\langle x - 2.5\text{ m}\rangle^0$$

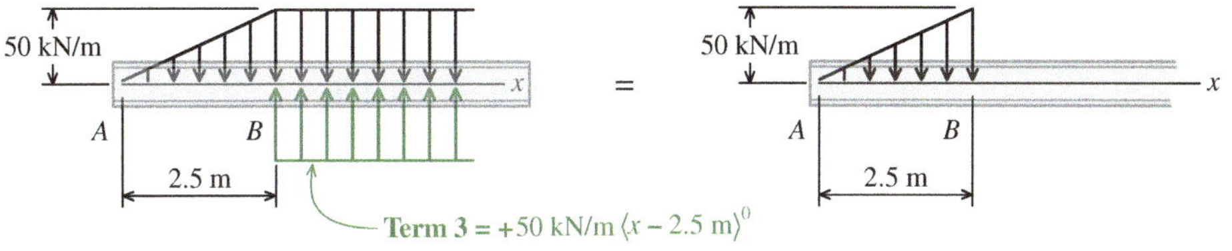

An additional uniform-load term that begins at B is required in order to cancel the remaining uniform load.

Therefore, three terms must be superimposed in order to obtain the desired linearly distributed load between A and B.

As shown in this example, three terms are required to represent the linearly increasing load that acts between A and B. Case 6 of Table 7.2 summarizes the general discontinuity expressions for a linearly increasing load. Similar reasoning is used to develop case 7 of Table 7.2 for a linearly decreasing distributed loading.

EXAMPLE 7.12

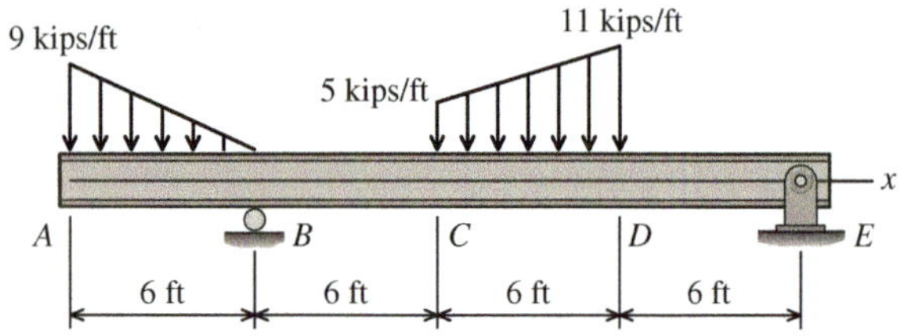

Use discontinuity functions to obtain expressions for the internal shear force $V(x)$ and internal bending moment $M(x)$ in the beam shown. Then, use these expressions to plot the shear-force and bending-moment diagrams for the beam.

Plan the Solution

Determine the reactions at simple supports A and E. Using Table 7.2, write $w(x)$ expressions for the linearly decreasing load between A and B and for the linearly increasing load between C and D, as well as for the two support reactions. Integrate $w(x)$ to determine an equation for the shear force $V(x)$, and then integrate $V(x)$ to determine an equation for the bending moment $M(x)$. Plot these functions to complete the shear-force and bending-moment diagrams.

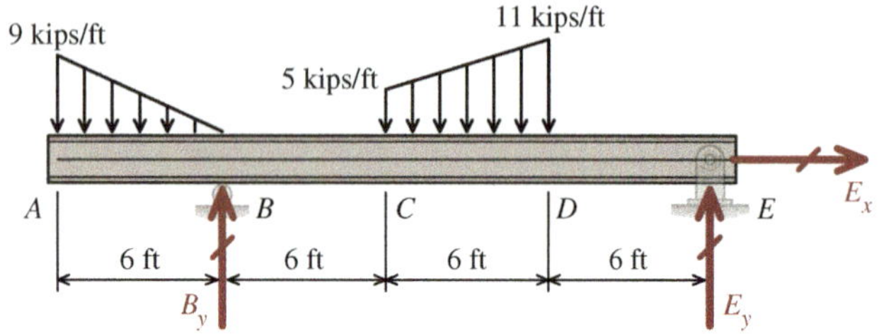

SOLUTION

Support Reactions

An FBD of the beam is shown. Before beginning, it is convenient to subdivide the linearly increasing load between C and D into

(a) a uniformly distributed load that has an intensity of 5 kips/ft and

(b) a linearly distributed load that has a maximum intensity of 6 kips/ft.

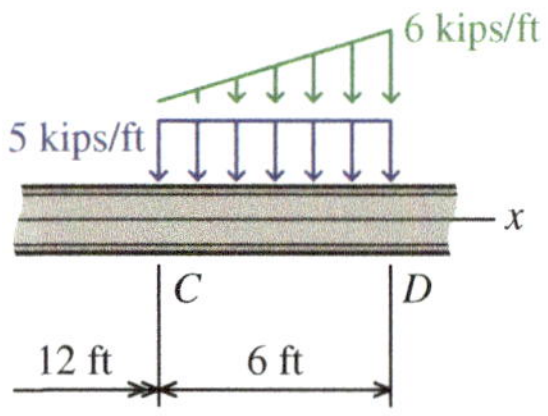

Accordingly, the beam equilibrium equations are as follows:

$$\Sigma F_x = E_x = 0 \qquad \text{(trivial)}$$

$$\Sigma F_y = B_y + E_y - \frac{1}{2}(9\text{ kips/ft})(6\text{ ft}) - (5\text{ kips/ft})(6\text{ ft}) - \frac{1}{2}(6\text{ kips/ft})(6\text{ ft}) = 0$$

$$\Sigma M_B = \frac{1}{2}(9\text{ kips/ft})(6\text{ ft})(4\text{ ft}) - (5\text{ kips/ft})(6\text{ ft})(9\text{ ft})$$
$$- \frac{1}{2}(6\text{ kips/ft})(6\text{ ft})(10\text{ ft}) + E_y(18\text{ ft}) = 0$$

From these equations, the beam reactions are

$$B_y = 56.0\text{ kips} \qquad \text{and} \qquad E_y = 19.0\text{ kips}$$

Discontinuity Expressions

Decreasing linearly distributed load between A and B: Use case 7 of Table 7.2 to write the following expression for the 9 kips/ft linearly distributed loading:

$$w(x) = -9\text{ kips/ft}\langle x - 0\text{ ft}\rangle^0 + \frac{9\text{ kips/ft}}{6\text{ ft}}\langle x - 0\text{ ft}\rangle^1 - \frac{9\text{ kips/ft}}{6\text{ ft}}\langle x - 6\text{ ft}\rangle^1 \qquad \text{(a)}$$

Reaction force B_y: The upward reaction force at B is expressed with the use of case 2 of Table 7.2:

$$w(x) = 56.0\text{ kips}\langle x - 6\text{ ft}\rangle^{-1} \qquad \text{(b)}$$

Uniformly distributed load between C and D: The uniformly distributed load requires the use of two terms. From case 5 of Table 7.2, express this loading as

$$w(x) = -5\text{ kips/ft}\langle x - 12\text{ ft}\rangle^0 + 5\text{ kips/ft}\langle x - 18\text{ ft}\rangle^0 \quad \text{(c)}$$

Increasing linearly distributed load between C and D: Use case 6 of Table 7.2 to write the following expression for the 6 kips/ft linearly distributed loading:

$$w(x) = -\frac{6\text{ kips/ft}}{6\text{ ft}}\langle x - 12\text{ ft}\rangle^1 + \frac{6\text{ kips/ft}}{6\text{ ft}}\langle x - 18\text{ ft}\rangle^1 + 6\text{ kips/ft}\langle x - 18\text{ ft}\rangle^0 \quad \text{(d)}$$

Reaction force E_y: The upward reaction force at E is expressed by

$$w(x) = 19\text{ kips}\langle x - 24\text{ ft}\rangle^{-1} \quad \text{(e)}$$

As a practical matter, this term has no effect, since the value of Equation (e) is zero for all values of $x \leq 24$ ft. However, the term will be retained here for completeness and clarity.

Complete-beam loading expression: The sum of Equations (a) through (e) gives the load expression $w(x)$ for the entire beam:

$$\begin{aligned} w(x) = &-9\text{ kips/ft}\langle x - 0\text{ ft}\rangle^0 + \frac{9\text{ kips/ft}}{6\text{ ft}}\langle x - 0\text{ ft}\rangle^1 - \frac{9\text{ kips/ft}}{6\text{ ft}}\langle x - 6\text{ ft}\rangle^1 \\ &+ 56.0\text{ kips}\langle x - 6\text{ ft}\rangle^{-1} - 5\text{ kips/ft}\langle x - 12\text{ ft}\rangle^0 + 5\text{ kips/ft}\langle x - 18\text{ ft}\rangle^0 \\ &- \frac{6\text{ kips/ft}}{6\text{ ft}}\langle x - 12\text{ ft}\rangle^1 + \frac{6\text{ kips/ft}}{6\text{ ft}}\langle x - 18\text{ ft}\rangle^1 \\ &+ 6\text{ kips/ft}\langle x - 18\text{ ft}\rangle^0 + 19\text{ kips}\langle x - 24\text{ ft}\rangle^{-1} \end{aligned} \quad \text{(f)}$$

Shear-force equation: Integrate Equation (f), using the integration rules given in Equation (7.11), to derive the shear-force equation for the beam:

$$\begin{aligned} V(x) &= \int w(x)dx \\ &= -9\text{ kips/ft}\langle x - 0\text{ ft}\rangle^1 + \frac{9\text{ kips/ft}}{2\,(6\text{ ft})}\langle x - 0\text{ ft}\rangle^2 - \frac{9\text{ kips/ft}}{2\,(6\text{ ft})}\langle x - 6\text{ ft}\rangle^2 \\ &\quad + 56.0\text{ kips}\langle x - 6\text{ ft}\rangle^0 - 5\text{ kips/ft}\langle x - 12\text{ ft}\rangle^1 + 5\text{ kips/ft}\langle x - 18\text{ ft}\rangle^1 \\ &\quad - \frac{6\text{ kips/ft}}{2(6\text{ ft})}\langle x - 12\text{ ft}\rangle^2 + \frac{6\text{ kips/ft}}{2(6\text{ ft})}\langle x - 18\text{ ft}\rangle^2 \\ &\quad + 6\text{ kips/ft}\langle x - 18\text{ ft}\rangle^1 + 19\text{ kips}\langle x - 24\text{ ft}\rangle^0 \end{aligned} \quad \text{(g)}$$

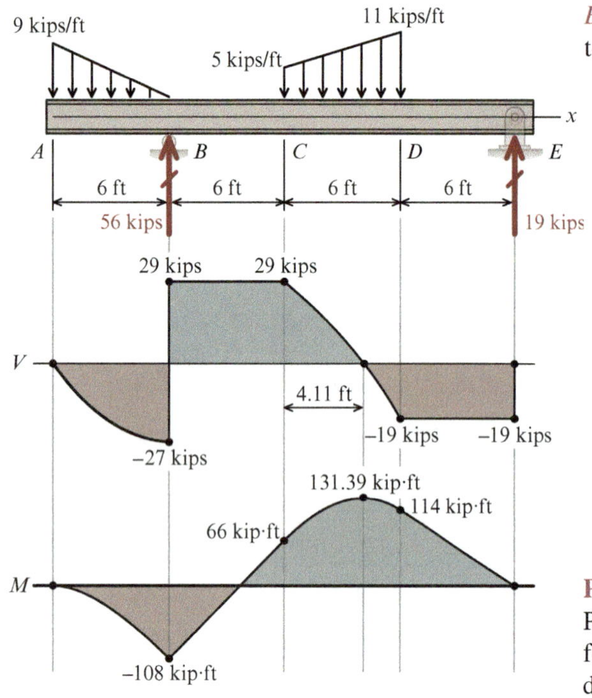

Bending-moment equation: Similarly, integrate Equation (g) to derive the bending-moment equation for the beam:

$$
\begin{aligned}
M(x) &= \int V(x)dx \\
&= -\frac{9\text{ kips/ft}}{2}\langle x - 0\text{ ft}\rangle^2 + \frac{9\text{ kips/ft}}{6(6\text{ ft})}\langle x - 0\text{ ft}\rangle^3 \\
&\quad - \frac{9\text{ kips/ft}}{6(6\text{ ft})}\langle x - 6\text{ ft}\rangle^3 + 56.0\text{ kips}\langle x - 6\text{ ft}\rangle^1 \\
&\quad - \frac{5\text{ kips/ft}}{2}\langle x - 12\text{ ft}\rangle^2 + \frac{5\text{ kips/ft}}{2}\langle x - 18\text{ ft}\rangle^2 \\
&\quad - \frac{6\text{ kips/ft}}{6(6\text{ ft})}\langle x - 12\text{ ft}\rangle^3 + \frac{6\text{ kips/ft}}{6(6\text{ ft})}\langle x - 18\text{ ft}\rangle^3 \\
&\quad + \frac{6\text{ kips/ft}}{2}\langle x - 18\text{ ft}\rangle^2 + 19\text{ kips}\langle x - 24\text{ ft}\rangle^1
\end{aligned} \tag{h}
$$

Plot the Functions

Plot the $V(x)$ and $M(x)$ functions given in Equations (g) and (h) for $0\text{ ft} \le x \le 24\text{ ft}$ to create the shear-force and bending-moment diagram shown.

PROBLEMS

P7.32–P7.42 For the beams and loadings shown in Figures P7.32–P7.42,

(a) use discontinuity functions to write the expression for $w(x)$; include the beam reactions in this expression.
(b) integrate $w(x)$ twice to determine $V(x)$ and $M(x)$.
(c) use $V(x)$ and $M(x)$ to plot the shear-force and bending-moment diagrams.

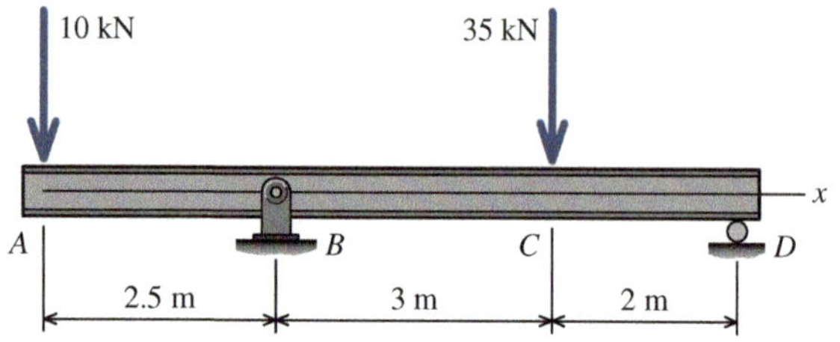

FIGURE P7.33

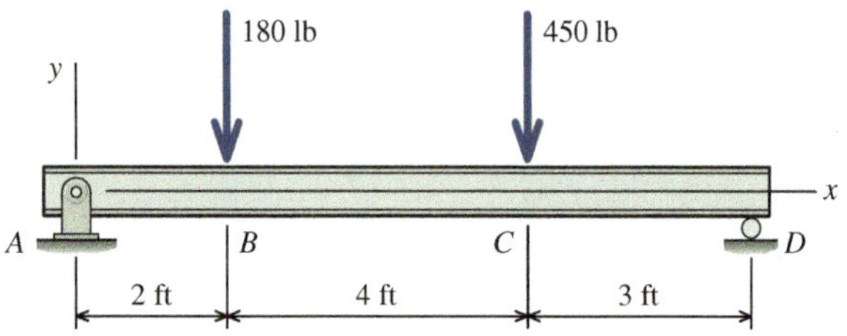

FIGURE P7.32

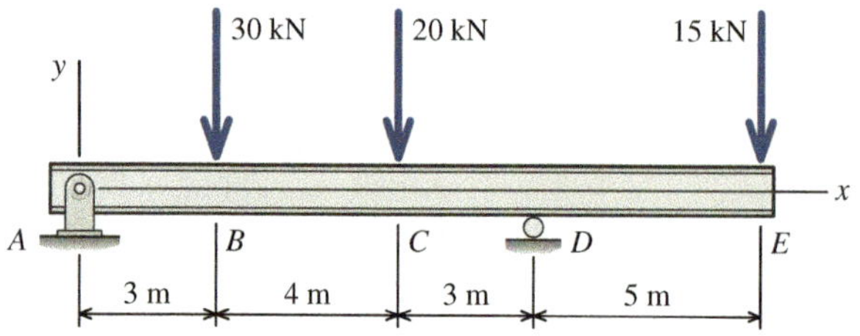

FIGURE P7.34

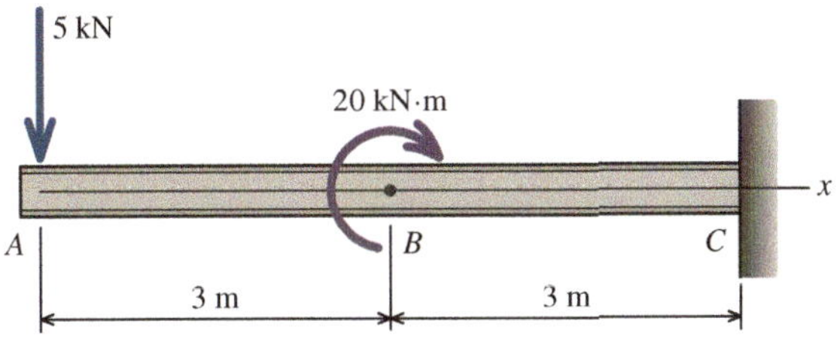

FIGURE P7.35

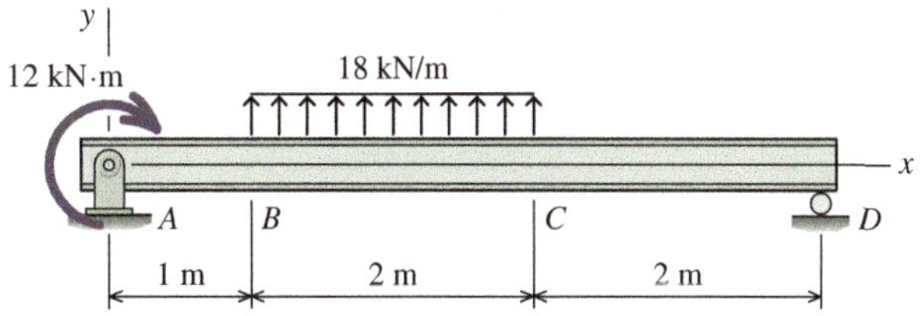

FIGURE P7.40

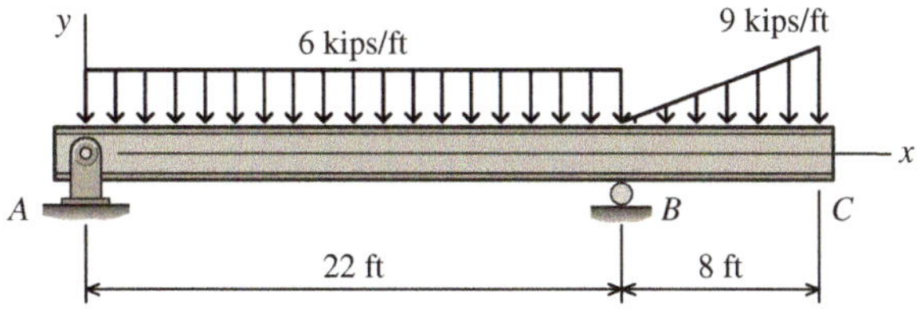

FIGURE P7.41

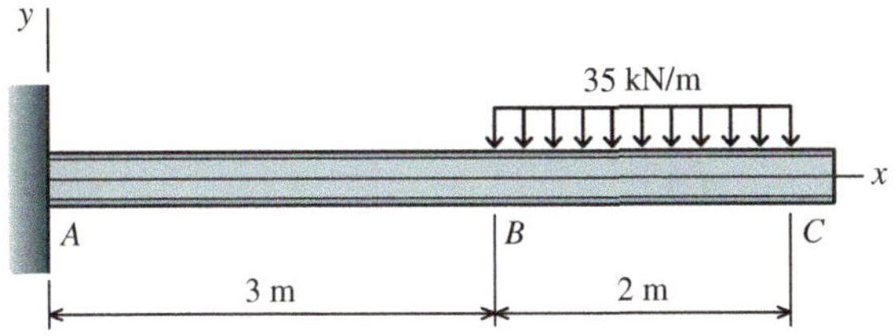

FIGURE P7.36

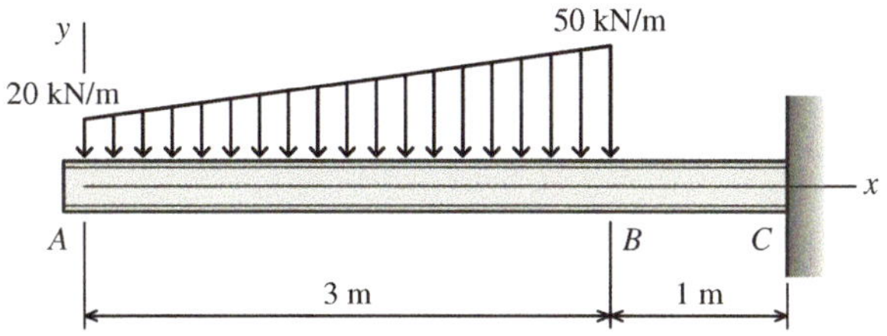

FIGURE P7.42

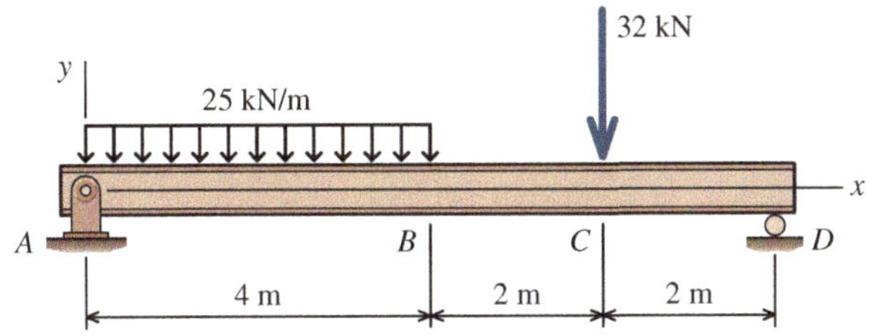

FIGURE P7.37

P7.43–P7.48 For the beams and loadings shown in Figures P7.43–P7.48,

(a) use discontinuity functions to write the expression for $w(x)$; include the beam reactions in this expression.

(b) integrate $w(x)$ twice to determine $V(x)$ and $M(x)$.

(c) determine the maximum bending moment in the beam between the two simple supports.

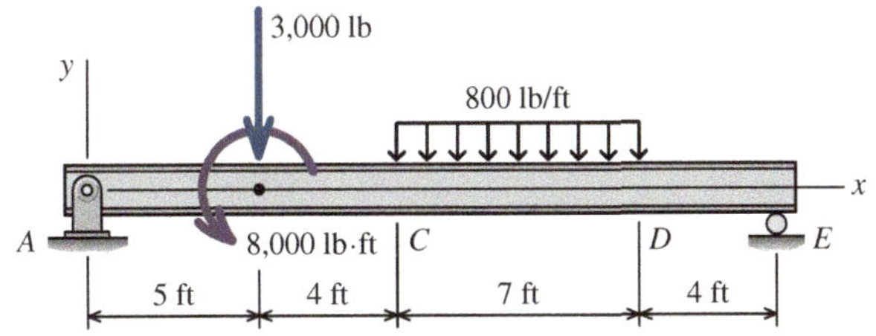

FIGURE P7.38

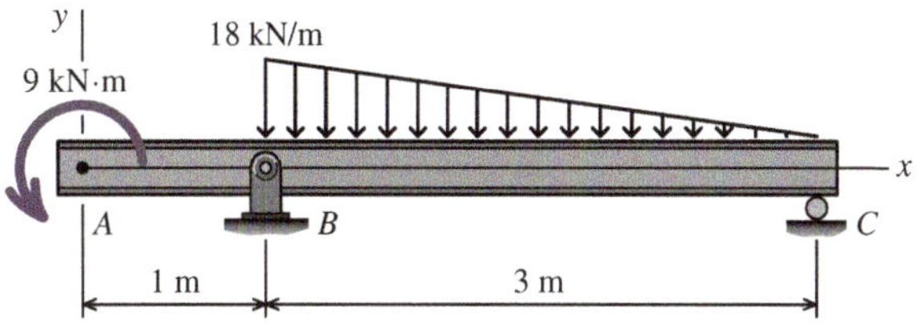

FIGURE P7.43

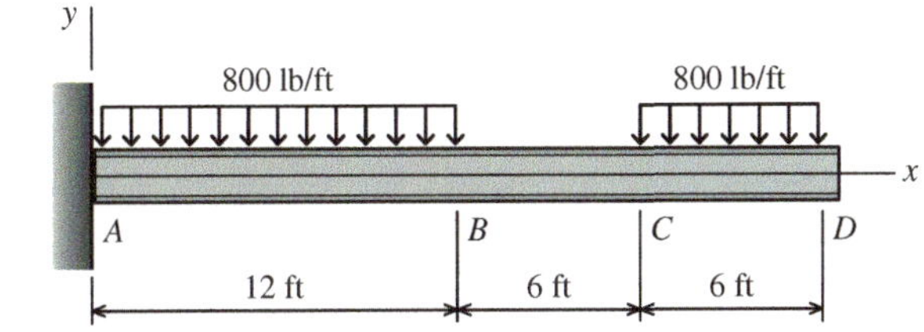

FIGURE P7.39

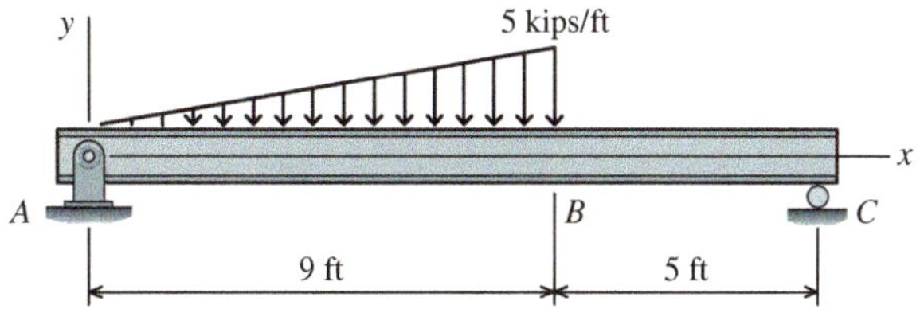

FIGURE P7.44

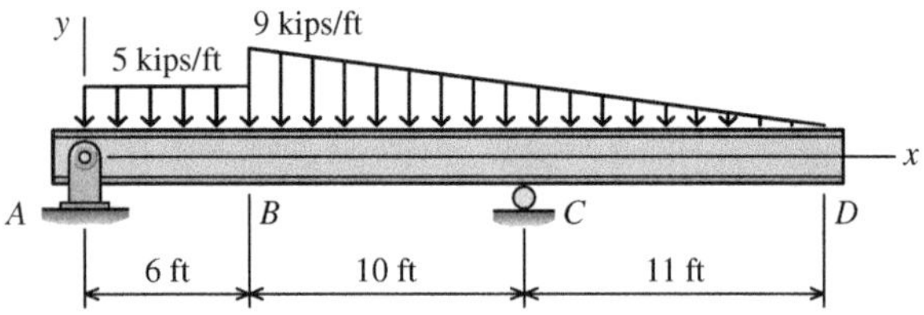

FIGURE P7.45

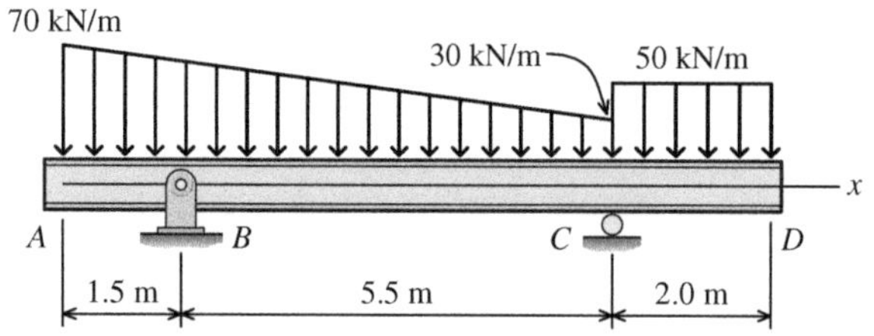

FIGURE P7.47

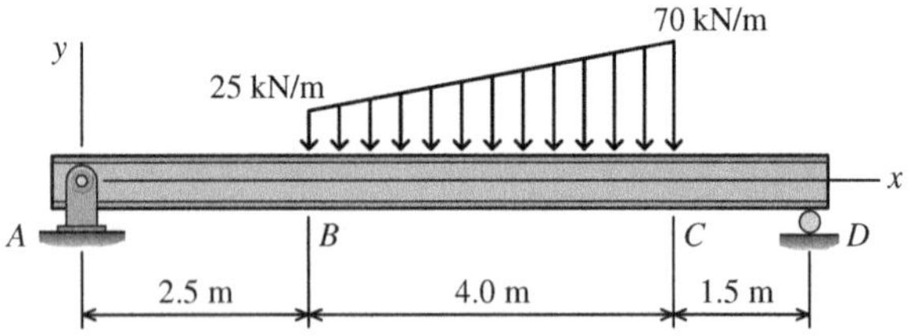

FIGURE P7.46

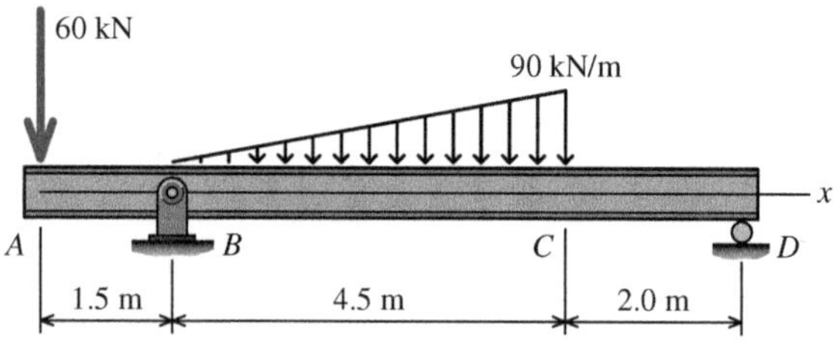

FIGURE P7.48

CHAPTER 8

Bending

8.1 Introduction

Perhaps the most common type of structural member is the beam. In actual structures and machines, beams can be found in a wide variety of sizes, shapes, and orientations. The elementary stress analysis of the beam constitutes one of the more interesting facets of mechanics of materials.

Beams are usually long (compared with their cross-sectional dimensions), straight, prismatic members that support **transverse loads**, which are loads that act perpendicular to the longitudinal axis of the member (Figure 8.1*a*). Loads on a beam cause it to **bend** (or **flex**) as opposed to stretching, compressing, or twisting. The applied loads cause the initially straight member to deform into a curved shape (Figure 8.1*b*), which is called the **deflection curve** or the **elastic curve**.

The term **transverse** refers to loads and sections that are perpendicular to the longitudinal axis of the member.

In this chapter, we will consider beams that are initially straight and that have a **longitudinal plane of symmetry** (Figure 8.2*a*). The member cross section, the support

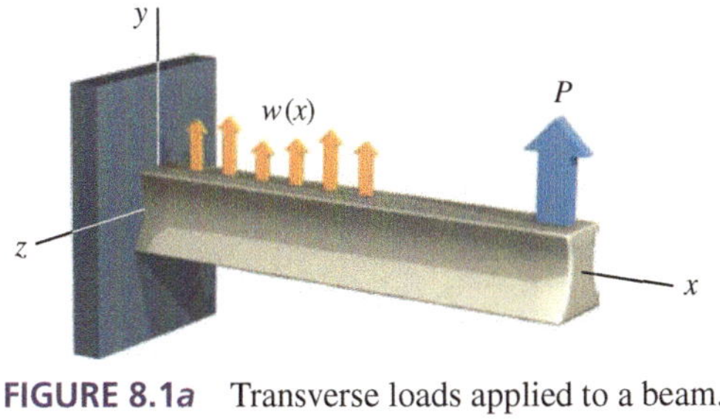

FIGURE 8.1*a* Transverse loads applied to a beam.

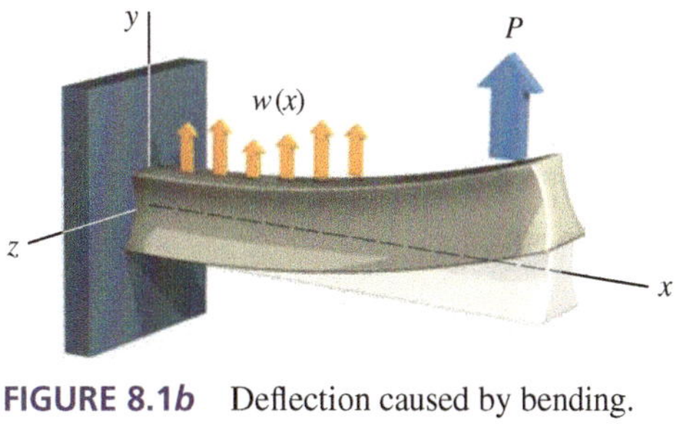

FIGURE 8.1*b* Deflection caused by bending.

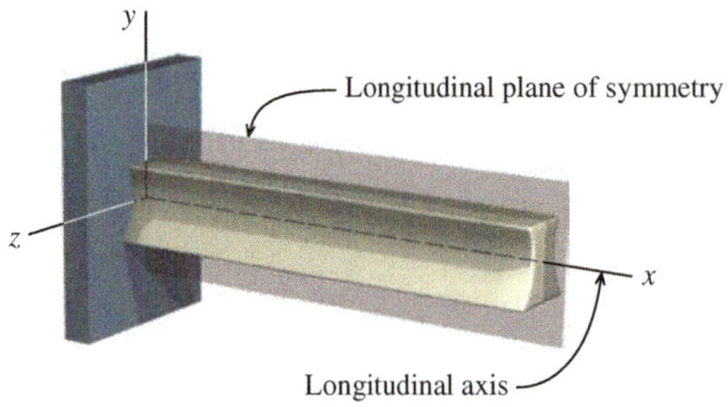

FIGURE 8.2*a* Longitudinal plane of symmetry.

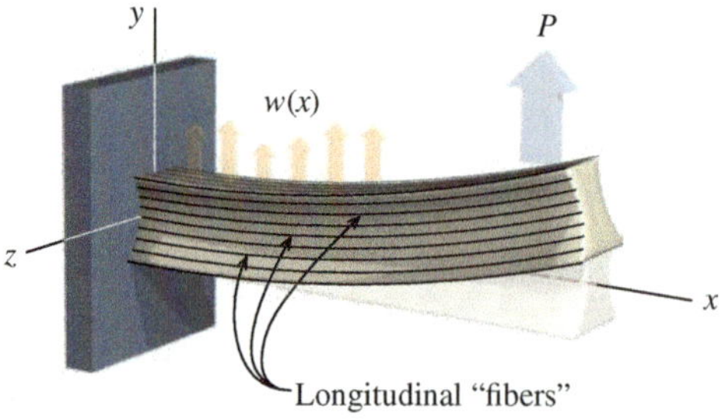

FIGURE 8.2*b* The notion of longitudinal "fibers."

conditions, and the applied loads are symmetric with respect to this plane of symmetry. Coordinate axes used for beams will be defined so that the **longitudinal axis** of the member will be designated the x axis; the y axis will be directed vertically upward, and the z axis will be oriented so that the x–y–z axes form a right-hand coordinate system. In Figure 8.1*b*, the x–y plane is called the **plane of bending**, since the loads and the member deflection occur in this plane. Bending (also termed **flexure**) is said to occur about the z axis.

In discussing and understanding the behavior of beams, it is convenient to imagine the beam to be a bundle of many *longitudinal fibers*, which run parallel to the longitudinal axis (or, simply, the **axis**) of the beam (Figure 8.2*b*). This terminology originated when the most common material used to construct beams was wood, which is a fibrous material. Although metals such as steel and aluminum do not contain fibers, the terminology is nevertheless quite useful for describing and understanding bending behavior. As shown in Figure 8.2*b*, bending causes fibers in the upper portion of the beam to be shortened or compressed while fibers in the lower portion are elongated in tension.

Pure Bending

Pure bending refers to the flexure of a beam in response to constant (i.e., equal) bending moments. For example, the region between points B and C of the beam shown in Figure 8.3

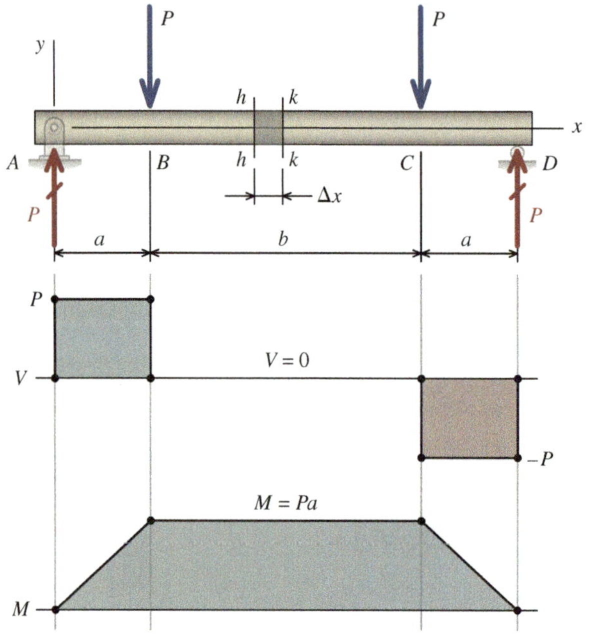

FIGURE 8.3 Example of pure bending in a region of a beam.

has a constant bending moment M, and consequently, this region is said to be in pure bending. Pure bending occurs only in regions where the transverse shear force V is equal to zero. Recall Equation (7.2), $V = dM/dx$. If the bending moment M is constant, then $dM/dx = 0$, and thus, $V = 0$. Pure bending also implies that no axial forces act in the beam.

In contrast, **nonuniform bending** refers to flexure in which the shear force V is not equal to zero. If $V \neq 0$, then $dM/dx \neq 0$, which means that the bending moment changes along the span of the beam.

In the sections that follow, the strains and stresses in beams subjected to pure bending will be investigated. Fortunately, the results obtained for pure bending can be applied to beams with nonuniform bending if the beam is relatively long compared with its cross-sectional dimensions—in other words, if the beam is "slender."

8.2 Flexural Strains

To investigate the strains produced in a beam subjected to pure bending, consider a short segment of the beam shown in Figure 8.3. The segment, located between sections h–h and k–k, is shown in Figure 8.4 with the deformations greatly exaggerated. The beam is assumed to be straight before bending occurs, and the cross section of the beam is constant. (In other words, the beam is a prismatic member.) Sections h–h and k–k, which were plane surfaces before deformation, remain plane surfaces after deformation.

If the beam is initially straight, then all beam fibers between sections h–h and k–k are initially the same length Δx. After bending occurs, the fibers in the upper portions of the cross section become shortened and the fibers in the lower portions become elongated. However, a single surface where the fibers neither shorten nor elongate exists between the upper and lower surfaces of the beam. This surface is called the **neutral surface** of the beam, and its intersection with any cross section is called the **neutral axis** of the section. All fibers on one side of the neutral surface are compressed, and those on the opposite side are elongated.

When subjected to pure bending, the beam deforms into the shape of a circular arc. The center O of this arc is called the **center of curvature**. The radial distance from the center of curvature to the neutral surface of the beam is called the **radius of curvature**, and it is designated by the Greek letter ρ (rho).

Consider a longitudinal fiber located at some distance y above the neutral surface. In other words, the origin of the y coordinate axis will be located on the neutral surface. Before bending, the fiber has a length Δx. After bending, it becomes shorter, and its deformed length will be denoted $\Delta x'$. From the definition of normal strain given in Equation (2.1), the normal strain of this longitudinal fiber can be expressed as

$$\varepsilon_x = \frac{\delta}{L} = \lim_{\Delta x \to 0} \frac{\Delta x' - \Delta x}{\Delta x}$$

The beam segment subjected to pure bending deflects into the shape of a circular arc, and the interior angle of this arc will be denoted $\Delta\theta$. According to the geometry shown in Figure 8.4, the lengths Δx and $\Delta x'$ can be expressed in terms of arc lengths so that the longitudinal strain ε_x can be related to the radius of curvature ρ as

$$\varepsilon_x = \lim_{\Delta x \to 0} \frac{\Delta x' - \Delta x}{\Delta x} = \lim_{\Delta\theta \to 0} \frac{(\rho - y)\Delta\theta - \rho\Delta\theta}{\rho\Delta\theta} = -\frac{1}{\rho}y \tag{8.1}$$

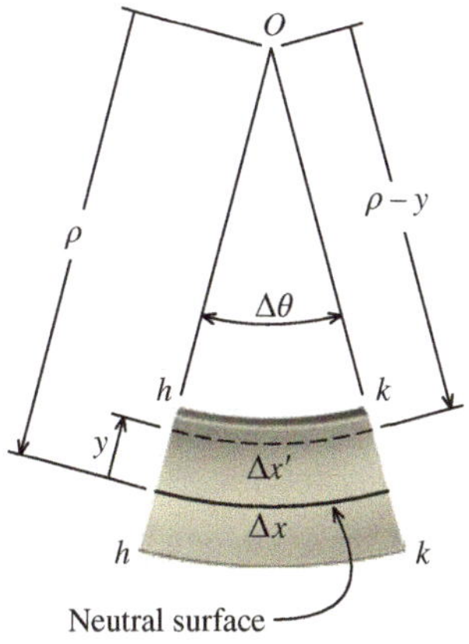

FIGURE 8.4 Flexural deformation.

Equation (8.1) indicates that the normal strain developed in any fiber is directly proportional to the distance of the fiber from the neutral surface. Equation (8.1) is valid for beams of any material, whether the material is elastic or inelastic, linear or nonlinear. Notice that the strain determined here occurs in the x direction, even though the loads applied to the beam act in the y direction and the beam bends about the z axis. For a positive value of ρ (as defined shortly), the negative sign in Equation (8.1) indicates that compressive strain will be developed in the fibers above the neutral surface (i.e., where y values are positive) while tensile strain will occur below the neutral surface (where y values are negative). Note that the sign convention for ε_x is the same as that defined for normal strains in Chapter 2; specifically, elongation is positive and shortening is negative.

Curvature κ (Greek letter kappa) is a measure of how sharply a beam is bent, and it is related to the radius of curvature ρ by

$$\kappa = \frac{1}{\rho} \tag{8.2}$$

If the load on a beam is small, then the beam deflection will be small, the radius of curvature ρ will be very large, and the curvature κ will be very small. Conversely, a beam with a large deflection will have a small radius of curvature ρ and a large curvature κ. For the x–y–z coordinate axes used here, the sign convention for κ is defined such that κ is positive if the center of curvature is located above a beam. The center of curvature O for the beam segment shown in Figure 8.4 is located above the beam; therefore, this beam has a positive curvature κ, and in accordance with Equation (8.2), the radius of curvature ρ must be positive, too. To summarize, κ and ρ always have the same sign. They are both positive if the center of curvature is located above the beam, and they are both negative if the center of curvature is located below the beam.

Transverse Deformations

Longitudinal strains ε_x in the beam are accompanied by deformations in the plane of the cross section (i.e., strains in the y and z directions) because of the Poisson effect. Since most beams are slender, the deformations in the y–z plane due to Poisson effects are very small. If the beam is free to deform laterally (as is usually the case), normal strains in the y and z directions do not cause transverse stresses. This situation is comparable to that of a prismatic bar in tension or compression, and therefore, the longitudinal fibers in a beam subjected to pure bending are in a state of *uniaxial stress.*

8.3 Normal Stresses in Beams

For pure bending, the longitudinal strain ε_x that occurs in the beam varies in proportion to the fiber's distance from the neutral surface of the beam. The variation of normal stress σ_x acting on a transverse cross section can be determined from a stress–strain curve for the specific material used to fabricate the beam. For most engineering materials, the stress–strain diagrams for both tension and compression are identical in the elastic range. Although the diagrams may differ somewhat in the inelastic range, the differences can be neglected in many instances. *For the beam problems considered in this book, the tension and compression stress–strain diagrams will be assumed identical.*

The most common stress–strain relationship encountered in engineering is the equation for a linear elastic material, defined by Hooke's law: $\sigma = E\varepsilon$. If the strain relationship defined in Equation (8.1) is combined with Hooke's law, then the variation of normal stress with distance y from the neutral surface can be expressed as

$$\sigma_x = E\varepsilon_x = -\frac{E}{\rho}y = -E\kappa y \tag{8.3}$$

Equation (8.3) shows that the normal stress σ_x on the transverse section of the beam varies linearly with distance y from the neutral surface. This type of stress distribution is shown in Figure 8.5*a* for the case of a bending moment M that produces compressive stresses above the neutral surface and tensile stresses below the neutral surface.

Since plane cross sections remain plane, the normal stress σ_x caused by bending is also uniformly distributed in the z direction.

While Equation (8.3) describes the variation of normal stress over the depth of a beam, its usefulness depends upon knowing the location of the neutral surface. Moreover, the radius of curvature ρ is generally not known, whereas the internal bending moment M is readily available from shear-force and bending-moment diagrams. A more useful relationship than Equation (8.3) would be one that related the normal stresses produced in the beam to the internal bending moment M. Both of these objectives can be accomplished by determining the resultant of the normal stress σ_x acting over the depth of the cross section.

In general, the resultant of the normal stresses in a beam consists of two components:

(a) a resultant force acting in the x direction (i.e., the longitudinal direction) and
(b) a resultant moment acting about the z axis.

If the beam is subjected to pure bending, the resultant force in the longitudinal direction must be zero and the resultant moment must equal the internal bending moment M in the beam. Then, on the basis of the stress distribution shown in Figure 8.5*a*, two equilibrium equations can be written: $\Sigma F_x = 0$ and $\Sigma M_z = 0$. From these two equations,

(a) the location of the neutral surface can be determined and
(b) the relationship between bending moment and normal stress can be established.

Location of the Neutral Surface

The cross section of the beam is shown in Figure 8.5*b*. We will consider a small element dA of the cross-sectional area A. The beam is assumed to be homogeneous, and the bending stresses are produced at an arbitrary radius of curvature ρ. The distance from the area dA to the neutral axis is measured by the coordinate y. The normal stresses acting on area dA

The intersection of the **neutral surface** (which is a plane) and any cross section of the beam (also a plane surface) is a line, which is termed the **neutral axis**.

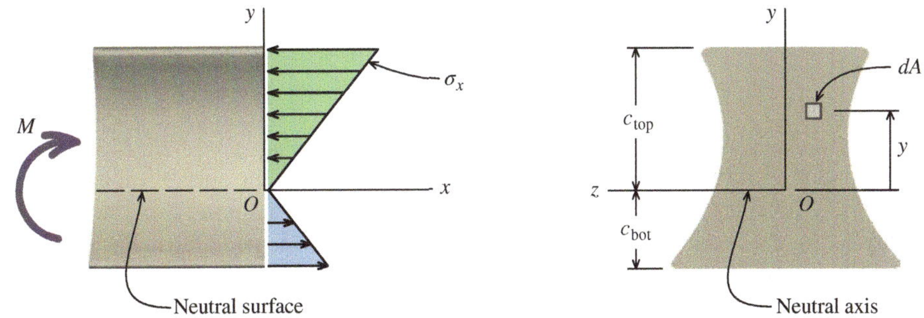

(*a*) Side view of beam, showing normal stress distribution

(*b*) Beam cross section

FIGURE 8.5 Normal stresses in a beam of linearly elastic material.

In Figure 8.5*a*, compressive stresses are indicated by arrows pointing toward the cross section and tensile stresses are indicated by arrows pointing away from the cross section.

produce a resultant force dF given by $\sigma_x\, dA$. (Recall that force can be thought of as the product of stress and area.) In order to satisfy horizontal equilibrium, all forces dF for the beam in Figure 8.5a must sum to zero, or, as expressed in terms of calculus,

$$\Sigma F_x = \int dF = \int_A \sigma_x\, dA = 0$$

Substitution of Equation (8.3) for σ_x yields

$$\Sigma F_x = \int_A \sigma_x\, dA = -\int_A \frac{E}{\rho} y\, dA = -\frac{E}{\rho}\int_A y\, dA = 0 \tag{8.4}$$

In Equation (8.4), the elastic modulus E cannot be zero for a solid material. The radius of curvature ρ could equal infinity; however, this would imply that the beam does not bend at all. Consequently, horizontal equilibrium of the normal stresses can be satisfied only if

$$\int_A y\, dA = 0 \tag{a}$$

This equation states that the first moment of area of the cross section with respect to the z axis must equal zero. From statics, recall that the definition of the centroid of an area with respect to a horizontal axis also includes the first-moment-of-area term:

$$\bar{y} = \frac{\int_A y\, dA}{\int_A dA} \tag{b}$$

Keep in mind that this conclusion assumes pure bending of an elastic material. If an axial force exists in the flexural member or if the material is inelastic, the neutral surface will not pass through the centroid of the cross-sectional area.

Substituting Equation (a) into Equation (b) shows that equilibrium can be satisfied only if $\bar{y} = 0$; in other words, the distance $\bar{y}$ measured from the neutral surface to the centroid of the cross-sectional area must be zero. Thus, for pure bending, **the neutral axis must pass through the centroid of the cross-sectional area.**

As discussed in Section 8.1, the study of bending presented here applies to beams that have a longitudinal plane of symmetry. Consequently, the y axis must pass through the centroid. The origin O of the beam coordinate system (see Figure 8.5b) is located at the centroid of the cross-sectional area. The x axis lies in the plane of the neutral surface and is coincident with the longitudinal axis of the member. The y axis lies in the longitudinal plane of symmetry, originates at the centroid of the cross section, and is directed vertically upward (for a horizontal beam). The z axis also originates at the centroid and acts in the direction that produces a right-hand x–y–z coordinate system.

Moment–Curvature Relationship

A moment is composed of a force term and a distance term. The distance term is often called a *moment arm*. On area dA, the force is $\sigma_x\, dA$. The moment arm is y, which is the distance from the neutral surface to dA.

The second equilibrium equation to be satisfied requires that the sum of moments must equal zero. Consider again the area element dA and the normal stress that acts upon it (Figure 8.5b). Since the resultant force dF acting on dA is located a distance y from the z axis, it produces a moment dM about the z axis. The resultant force can be expressed as $dF = \sigma_x\, dA$. A positive normal stress σ_x (i.e., a tensile normal stress) acting on area dA, which is located a positive distance y above the neutral axis, produces a moment dM that rotates in a negative right-hand rule sense about the z axis; therefore, the incremental moment dM is expressed as $dM = -y\sigma_x\, dA$.

All such moment increments that act on the cross section, along with the internal bending moment M, must sum to zero in order to satisfy equilibrium about the z axis:

$$\Sigma M_z = -\int_A y\sigma_x \, dA - M = 0$$

If Equation (8.3) is substituted for σ_x, then the bending moment M can be related to the radius of curvature ρ:

$$M = -\int_A y\sigma_x \, dA = \frac{E}{\rho}\int_A y^2 \, dA \tag{8.5}$$

Again from statics, recall that the integral term in Equation (8.5) is called the second moment of area or, more commonly, the **area moment of inertia**:

$$I_z = \int_A y^2 \, dA$$

The subscript z indicates an area moment of inertia determined with respect to the z centroidal axis (i.e., the axis about which the bending moment M acts). The integral term in Equation (8.5) can be replaced by the moment of inertia I_z, where

$$M = \frac{EI_z}{\rho}$$

to give an expression relating the beam curvature to its internal bending moment:

$$\kappa = \frac{1}{\rho} = \frac{M}{EI_z} \tag{8.6}$$

This relationship is called the **moment–curvature equation**, and it shows that the beam curvature is directly related to the bending moment and inversely related to the quantity EI_z. In general, the term EI is known as the **flexural rigidity**, and it is a measure of the bending resistance of a beam.

In the context of mechanics of materials, the area moment of inertia is usually referred to as, simply, the **moment of inertia**.

The radius of curvature ρ is measured from the center of curvature to the neutral surface of the beam. (See Figure 8.5*b*.)

Flexure Formula

The relationship between the normal stress σ_x and the curvature was developed in Equation (8.3), and the relationship between the curvature and the bending moment M is given by Equation (8.6). These two relationships can be combined, giving

$$\sigma_x = -E\kappa y = -E\left(\frac{M}{EI_z}\right)y$$

to define the stress produced in a beam by a bending moment:

$$\sigma_x = -\frac{My}{I_z} \tag{8.7}$$

Equation (8.7) is known as the **elastic flexure formula** or, simply, the **flexure formula**. As developed here, a bending moment M that acts about the z axis produces normal stresses that act in the x direction (i.e., the longitudinal direction) of the beam. The stresses vary linearly in intensity over the depth of the cross section. The normal stresses produced in a beam by a bending moment are commonly referred to as **bending stresses** or **flexural stresses**.

Examination of the flexure formula reveals that a positive bending moment causes negative normal stresses (i.e., compression) for portions of the cross section above the neutral axis (i.e., positive y values) and positive normal stresses (i.e., tension) for portions below the neutral axis (i.e., negative y values). The opposite stresses occur for a negative bending moment. The distributions of bending stresses for both positive and negative bending moments are illustrated in Figure 8.6.

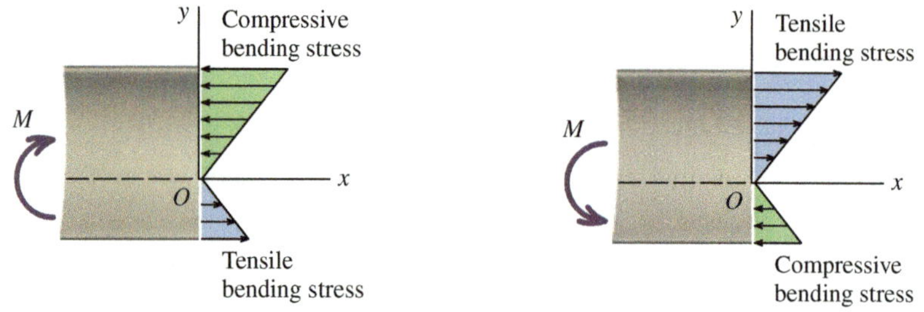

(*a*) Bending stresses caused by positive *M*

(*b*) Bending stresses caused by negative *M*

FIGURE 8.6 Relationship between bending moment *M* and bending stress.

In Chapter 7, a positive internal bending moment was defined as a moment that

- acts counterclockwise on the right-hand face of a beam; or
- acts clockwise on the left-hand face of a beam.

This sign convention can now be enhanced by taking into account the bending stresses produced by the internal moment. The enhanced bending-moment sign convention is illustrated in Figure 8.7.

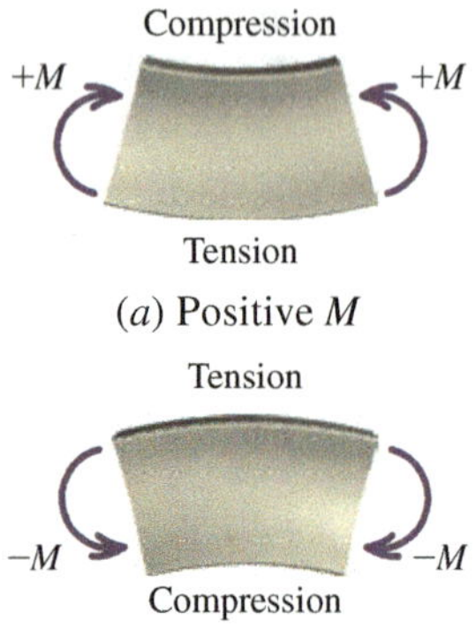

(*a*) Positive *M*

(*b*) Negative *M*

FIGURE 8.7 Enhanced bending-moment sign convention.

A positive internal bending moment *M* causes

- compressive bending stresses above the neutral axis;
- tensile bending stresses below the neutral axis; and
- a positive curvature κ.

A negative internal bending moment *M* causes

- tensile bending stresses above the neutral axis;
- compressive bending stresses below the neutral axis; and
- a negative curvature κ

Maximum Stresses on a Cross Section

Since the intensity of the bending stress σ_x varies linearly with distance y from the neutral surface [see Equation (8.3)], the maximum bending stress σ_{max} occurs on either the top or the bottom surface of the beam, depending on which surface is farther from the neutral surface. In Figure 8.5*b*, the distances from the neutral axis to either the top or the bottom of the cross section are denoted by c_{top} and c_{bot}, respectively. In this context, c_{top} and c_{bot} are taken as absolute values of the y coordinates of the top and bottom surfaces. The corresponding bending stress *magnitudes* are given by the following equations:

$$\sigma_{max} = \frac{Mc_{top}}{I_z} = \frac{M}{S_{top}} \quad \text{for the top surface of the beam}$$
$$\sigma_{max} = \frac{Mc_{bot}}{I_z} = \frac{M}{S_{bot}} \quad \text{for the bottom surface of the beam} \tag{8.8}$$

The sense of σ_x (either tension or compression) is dictated by the sign of the bending moment. The quantities S_{top} and S_{bot} are called the **section moduli** of the cross section, and they are defined as

$$S_{top} = \frac{I_z}{c_{top}} \quad \text{and} \quad S_{bot} = \frac{I_z}{c_{bot}} \tag{8.9}$$

The section modulus is a convenient property for beam design because it combines two important cross-sectional properties into a single quantity.

The beam cross section shown in Figure 8.5 is symmetric about the y axis. If a beam cross section is also symmetric about the z axis, it is called a **doubly symmetric cross section**. For a doubly symmetric shape, $c_{top} = c_{bot} = c$ and the bending stress *magnitudes* at the top and bottom of the cross section are equal and given by

$$\sigma_{max} = \frac{Mc}{I_z} = \frac{M}{S} \quad \text{where} \quad S = \frac{I_z}{c} \tag{8.10}$$

Again, Equation (8.10) gives only the magnitude of the stress. The sense of σ_x (either tension or compression) is dictated by the sense of the bending moment.

Nonuniform Bending

The preceding analysis assumed that a slender, homogeneous, prismatic beam was subjected to pure bending. If the beam is subjected to nonuniform bending, which occurs when a transverse shear force V exists, then the shear force produces out-of-plane distortions of the cross sections. Strictly speaking, these out-of-plane distortions violate the initial assumption that cross-sectional surfaces that are planar before bending remain planar after bending. However, the distortion caused by transverse shear forces is not significant for common beams, and its effect may be neglected. Therefore, the equations developed in this section may be used to calculate flexural stresses for beams subjected to nonuniform bending.

Summary

Bending stresses in a beam are evaluated in a three-step process.

Step 1—Determine the Internal Bending Moment *M*: The bending moment may be specified, but more typically, it is determined by constructing a shear-force and bending-moment diagram.

Step 2 — Calculate Properties for the Cross Section of the Beam: The location of the centroid must be determined first, since the centroid defines the neutral surface for pure bending. Next, the moment of inertia of the cross-sectional area must be calculated about the centroidal axis that corresponds to the bending moment M. For example, if the bending moment M acts about the z axis, then the moment of inertia about the z axis is required. (Appendix A presents a review of area geometric properties, such as centroids and moments of inertia.) Finally, bending stresses within the cross section vary with depth. Therefore, the y coordinate at which stresses are to be calculated must be established.

Step 3 — Use the Flexure Formula to Calculate Bending Stresses: Equations (8.7) and (8.10) for bending stresses were derived:

$$\sigma_x = -\frac{My}{I_z}$$

$$\sigma_x = \frac{Mc}{I_z} = \frac{M}{S}$$

In common practice, each of these equations is often called the *flexure formula*. The first form is more useful for calculating the bending stress at locations other than the top or the bottom of the cross section of the beam. Use of this form requires careful attention to the sign conventions for M and y. The second form is more useful for calculating maximum bending stress *magnitudes*. If it is important to determine whether the bending stress is either tension or compression, then that is done by inspection, using the sense of the internal bending moment M.

EXAMPLE 8.1

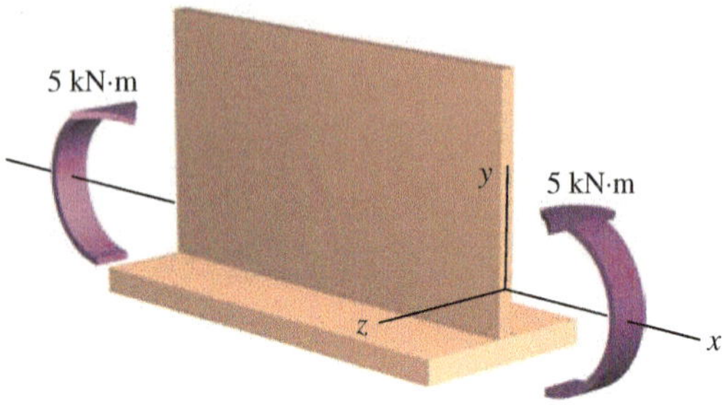

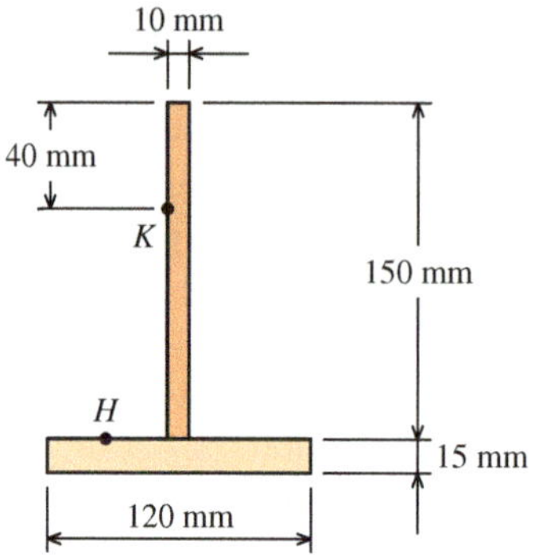

A beam with an inverted tee-shaped cross section is subjected to positive bending moments of $M_z = 5$ kN · m. The cross-sectional dimensions of the beam are shown. Determine

(a) the centroid location, the moment of inertia about the z axis, and the controlling section modulus about the z axis.

(b) the bending stress at points H and K. State whether the normal stress is *tension* or *compression*.

(c) the maximum bending stress produced in the cross section. State whether the stress is *tension* or *compression*.

Plan the Solution

The normal stresses produced by the bending moment will be determined from the flexure formula [Equation (8.7)]. Before the flexure formula is applied, however, the section properties of the beam cross section must be calculated. The bending moment acts about the z centroidal axis; therefore, the location of the centroid in the y direction must be determined. Once the centroid has been located, the moment of inertia of the cross section about the z centroidal axis will be calculated. When the centroid location and the moment of inertia about the centroidal axis are known, the bending stresses can be readily calculated from the flexure formula.

SOLUTION

(a) The centroid location in the horizontal direction can be determined from symmetry alone. The centroid location in the y direction must be determined for the inverted-tee cross section. The tee shape is first subdivided into rectangular shapes (1) and (2), and the area A_i for each of these shapes is computed. For calculation purposes, a reference axis is arbitrarily established. In this example, the reference axis will be placed at the bottom surface of the tee shape. The distance y_i in the vertical direction from the reference axis to the centroid of each rectangular area A_i is determined, and the product $y_i A_i$ (termed the *first moment of area*) is computed. The centroid location $\bar{y}$ measured from the reference axis is computed as the sum of the first moments of area $y_i A_i$ divided by the sum of the areas A_i. The calculation for the inverted-tee cross section is summarized in the accompanying table.

	A_i (mm²)	y_i (mm)	y_iA_i (mm³)
(1)	1,500	90	135,000
(2)	1,800	7.5	13,500
	3,300		148,500

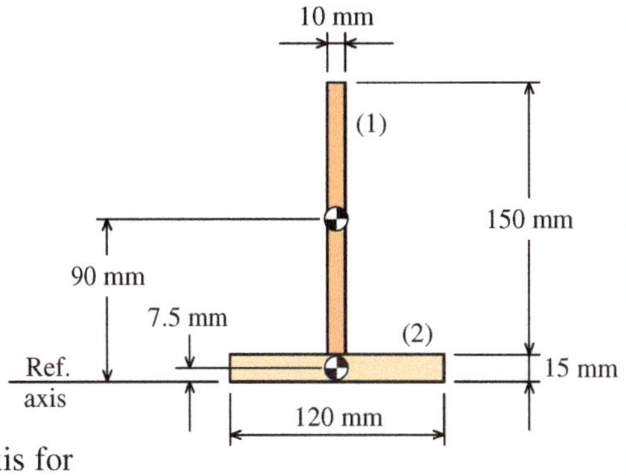

$$\bar{y} = \frac{\Sigma y_i A_i}{\Sigma A_i} = \frac{148{,}500 \text{ mm}^3}{3{,}300 \text{ mm}^2} = 45.0 \text{ mm}$$

Thus, the z centroidal axis is located 45.0 mm above the reference axis for the inverted-tee cross section. **Ans.**

The internal bending moment acts about the z centroidal axis, and consequently, the moment of inertia must be determined about this same axis for the inverted-tee cross section. Since the centroids of areas (1) and (2) do not coincide with the z centroidal axis for the entire cross section, the parallel-axis theorem must be used to calculate the moment of inertia for the inverted tee shape.

See Section A.2 of Appendix A for a review of the parallel-axis theorem.

The moment of inertia I_{ci} of each rectangular shape about its own centroid must be computed for the calculation to begin. For example, the moment of inertia of area (1) about the z centroidal axis for area (1) is calculated as $I_{c1} = bh^3/12 = (10 \text{ mm})(150 \text{ mm})^3/12 = 2{,}812{,}500 \text{ mm}^4$. Next, the perpendicular distance d_i between the z centroidal axis for the inverted-tee shape and the z centroidal axis for area A_i must be determined. The term d_i is squared and multiplied by A_i, and the result is added to I_{ci} to give the moment of inertia for each rectangular shape about the z centroidal axis of the inverted-tee cross section. The results for all areas A_i are then summed to determine the moment of inertia of the cross section about its centroidal axis. The complete calculation procedure is summarized in the following table:

	I_{ci} (mm⁴)	$\lvert d_i \rvert$ (mm)	$d_i^2A_i$ (mm⁴)	I_z (mm⁴)
(1)	2,812,500	45.0	3,037,500	5,850,000
(2)	33,750	37.5	2,531,250	2,565,000
				8,415,000

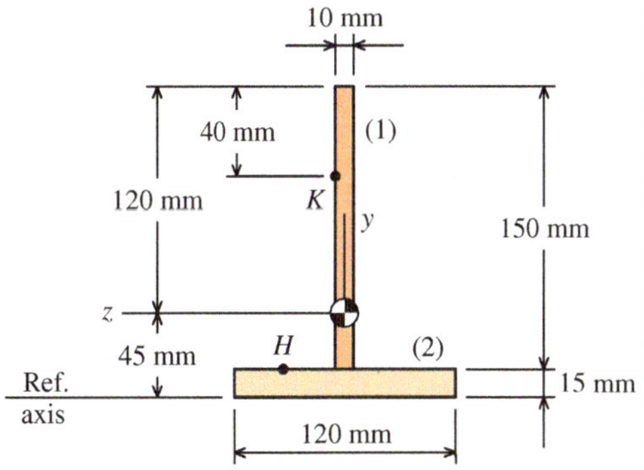

The moment of inertia of the cross section about its z centroidal axis is thus $I_z = 8{,}415{,}000 \text{ mm}^4$. **Ans.**

Since the inverted-tee cross section is not symmetric about its z centroidal axis, two section moduli are possible. [See Equation (8.9).] The distance from the z axis to the upper surface of the cross section will be denoted c_{top}. The section modulus calculated with this value is

$$S_{top} = \frac{I_z}{c_{top}} = \frac{8{,}415{,}000 \text{ mm}^4}{120 \text{ mm}} = 70{,}125 \text{ mm}^3$$

Let the distance from the z axis to the lower surface of the cross section be denoted c_{bot}. Then, the corresponding section modulus is

$$S_{bot} = \frac{I_z}{c_{bot}} = \frac{8{,}415{,}000 \text{ mm}^4}{45 \text{ mm}} = 187{,}000 \text{ mm}^3$$

The controlling section modulus is the smaller of these two values; therefore, the section modulus for the inverted tee cross section is

$$S = 70{,}125 \text{ mm}^3 \qquad \textbf{Ans.}$$

Why is the smaller section modulus said to control in this context? The maximum bending stress is calculated with the use of the section modulus from the following form of the flexure formula [see Equation (8.10)]:

$$\sigma_{max} = \frac{M}{S}$$

The section modulus S appears in the denominator of this formula; consequently, there is an inverse relationship between the section modulus and the bending stress. The smaller value of S corresponds to the larger bending stress.

(b) Since the centroid location and the moment of inertia about the centroidal axis have been determined, the flexure formula [Equation (8.7)] can now be used to determine the bending stress at any coordinate location y. (Recall that the y coordinate axis has its origin at the centroid.) Point H is located at $y = -30$ mm; therefore, the bending stress at H is given by

$$\sigma_x = -\frac{My}{I_z} = -\frac{(5 \text{ kN}\cdot\text{m})(-30 \text{ mm})(1{,}000 \text{ N/kN})(1{,}000 \text{ mm/m})}{8{,}415{,}000 \text{ mm}^4}$$
$$= 17.83 \text{ MPa} = 17.83 \text{ MPa (T)} \qquad \textbf{Ans.}$$

Point K is located at $y = +80$ mm; hence, the bending stress at K is calculated as

$$\sigma_x = -\frac{My}{I_z} = -\frac{(5 \text{ kN}\cdot\text{m})(80 \text{ mm})(1{,}000 \text{ N/kN})(1{,}000 \text{ mm/m})}{8{,}415{,}000 \text{ mm}^4}$$
$$= -47.5 \text{ MPa} = 47.5 \text{ MPa (C)} \qquad \textbf{Ans.}$$

(c) Regardless of the particular cross-sectional geometry, the largest bending stress in any beam will occur at either the top surface or the bottom surface of the beam. If the cross section is not symmetric about the axis of bending, then the largest bending stress (for any given moment M) will occur at the location farthest from the neutral axis—in other words, at the point that has the largest y coordinate. For the inverted-tee cross section, the largest bending stress will occur at the upper surface:

$$\sigma_x = -\frac{My}{I_z} = -\frac{(5 \text{ kN}\cdot\text{m})(120 \text{ mm})(1{,}000 \text{ N/kN})(1{,}000 \text{ mm/m})}{8{,}415{,}000 \text{ mm}^4}$$
$$= -71.3 \text{ MPa} = 71.3 \text{ MPa (C)} \qquad \textbf{Ans.}$$

Alternatively, the section modulus S could be used in Equation (8.10) to determine the magnitude of the maximum bending stress:

$$\sigma_{max} = \frac{M}{S} = \frac{(5 \text{ kN}\cdot\text{m})(1{,}000 \text{ N/kN})(1{,}000 \text{ mm/m})}{70{,}125 \text{ mm}^3}$$
$$= 71.3 \text{ MPa} = 71.3 \text{ MPa (C) by inspection}$$

If Equation (8.10) is used to calculate the maximum bending stress, the sense of the stress (either tension or compression) must be determined by inspection.

EXAMPLE 8.2

The cross-sectional dimensions of a beam are shown in the accompanying diagram. If the maximum allowable bending stress is 230 MPa, determine the magnitude of the maximum internal bending moment M that can be supported by the beam. (**Note:** The rounded corners of the cross section can be neglected in performing the section property calculations.)

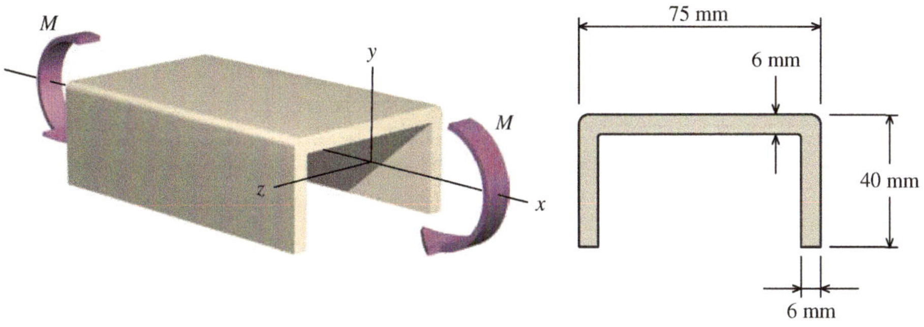

Plan the Solution

The centroid location and the moment of inertia of the beam cross section must be calculated at the outset. Once the section properties have been computed, the flexure formula will be rearranged to determine the maximum bending moment that can be applied without exceeding the 230 MPa allowable bending stress.

SOLUTION

The centroid location in the horizontal direction can be determined from symmetry. The cross section can be subdivided into three rectangular shapes. In accordance with the procedure described in Example 8.1, the centroid calculation for this shape is summarized in the following table:

	A_i (mm^2)	y_i (mm)	y_iA_i (mm^3)
(1)	450	37	16,650
(2)	204	17	3,468
(3)	204	17	3,468
	858		23,586

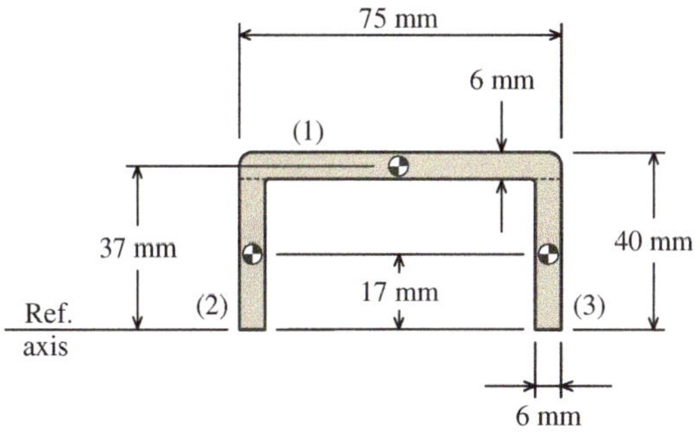

$$\bar{y} = \frac{\Sigma y_i A_i}{\Sigma A_i} = \frac{23{,}586 \text{ mm}^3}{858 \text{ mm}^2} = 27.49 \text{ mm}$$

Therefore, the z centroidal axis is located 27.49 mm above the reference axis for this cross section. **Ans.**

The calculation of the moment of inertia about this axis is summarized in the following table:

	I_c (mm^4)	$\lvert d_i \rvert$ (mm)	$d_i^2A_i$ (mm^4)	I_z (mm^4)
(1)	1,350	9.51	40,698.0	42,048.0
(2)	19,652	10.49	22,448.2	42,100.2
(3)	19,652	10.49	22,448.2	42,100.2
				126,248.4

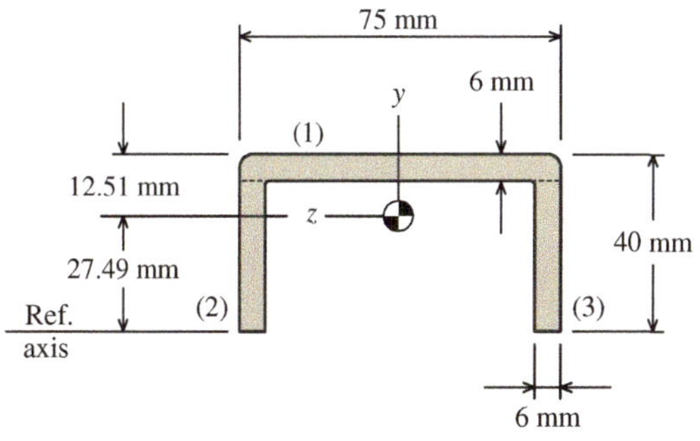

The moment of inertia of the cross section about its z centroidal axis is $I_z = 126{,}248.4\ \text{mm}^4$. **Ans.**

The largest bending stress in any beam will occur at either the top or the bottom surface of the beam. For this cross section, the distance to the bottom of the beam is greater than the distance to the top of the beam. Therefore, the largest bending stress will occur on the bottom surface of the cross section, at $y = -27.49$ mm. In this situation, it is convenient to use the flexure formula in the form of Equation (8.10), setting $c = 27.49$ mm. Equation (8.10) can be rearranged to solve for the bending moment M that will produce a bending stress of 230 MPa on the bottom surface of the beam:

$$M \leq \frac{\sigma_x I_z}{C} = \frac{(230\ \text{N/mm}^2)(126{,}248.4\ \text{mm}^4)}{27.49\ \text{mm}}$$
$$= 1{,}056{,}280\ \text{N}\cdot\text{mm} = 1{,}056\ \text{N}\cdot\text{m} \qquad \textbf{Ans.}$$

For the bending moment direction indicated in the sketch on the previous page, a bending moment of $M = 1{,}056\ \text{N}\cdot\text{m}$ will produce a compressive stress of 230 MPa on the bottom surface of the beam.

MecMovies

EXAMPLES

M8.4 Investigate bending stresses acting on various portions of a cross section, and determine internal bending moments, given bending stresses.

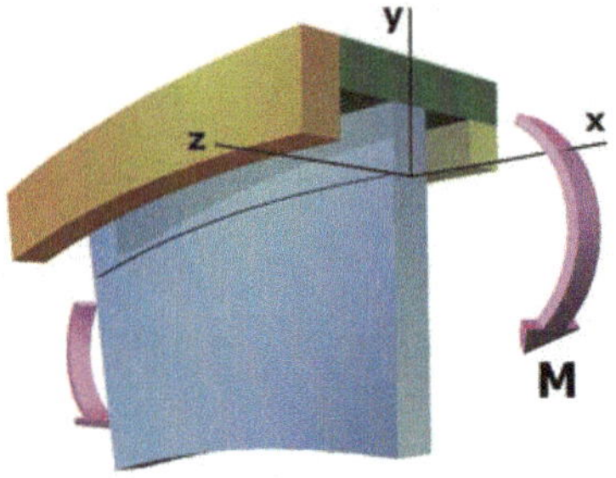

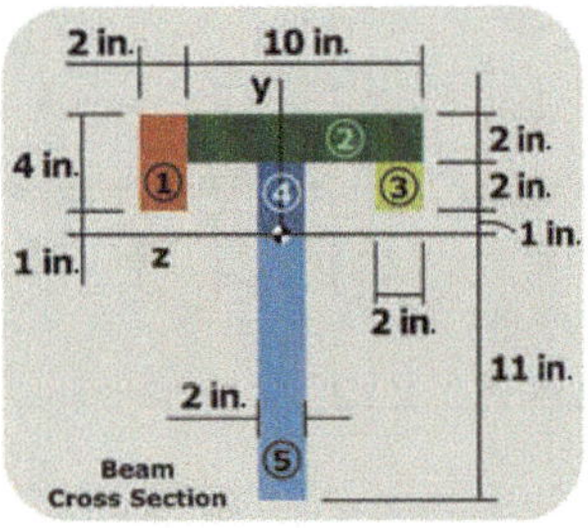

M8.5 Animated example of the procedure for calculating the centroid of a tee shape.

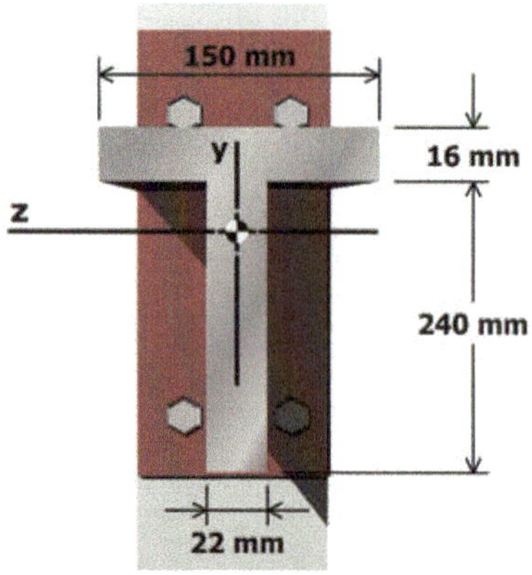

M8.6 Animated example of the procedure for calculating the centroid of a U shape.

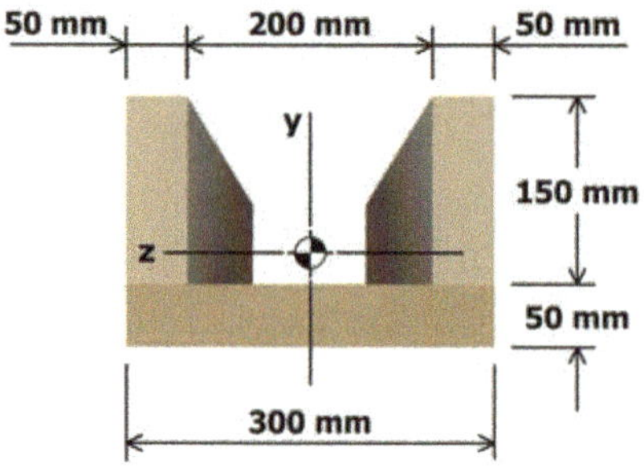

250

M8.7 Determine the centroid location and the moment of inertia about the centroidal axis for a tee shape.

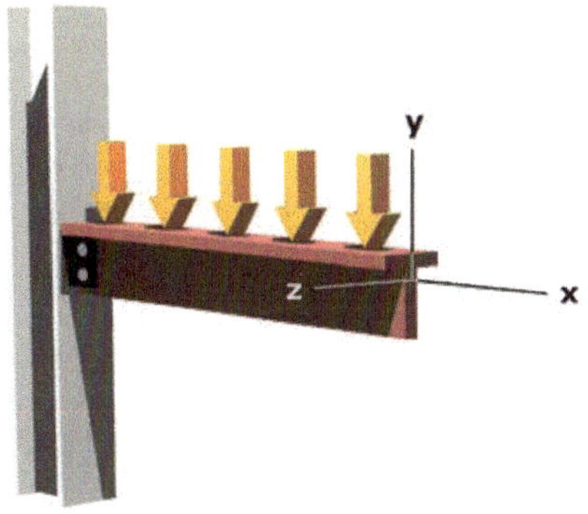

EXERCISES

M8.1 **The Centroids Game: Learning the Ropes.** Score at least 90 percent on the game.

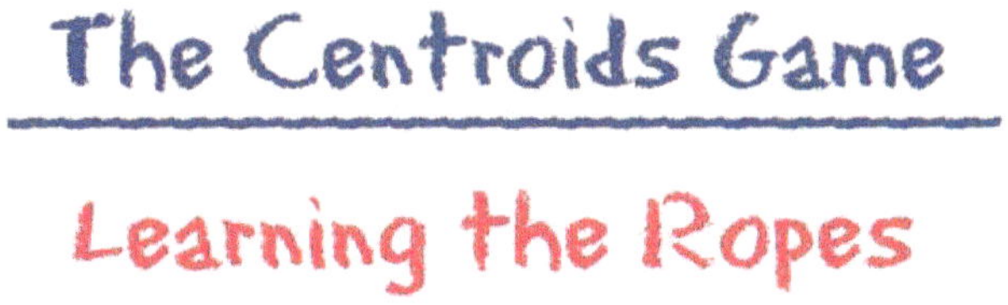

FIGURE M8.1

M8.2 **The Moment of Inertia Game: Starting from Square One.** Score at least 90 percent on the game.

FIGURE M8.2

M8.3 Use the flexure formula to determine bending stresses in a flanged shape.

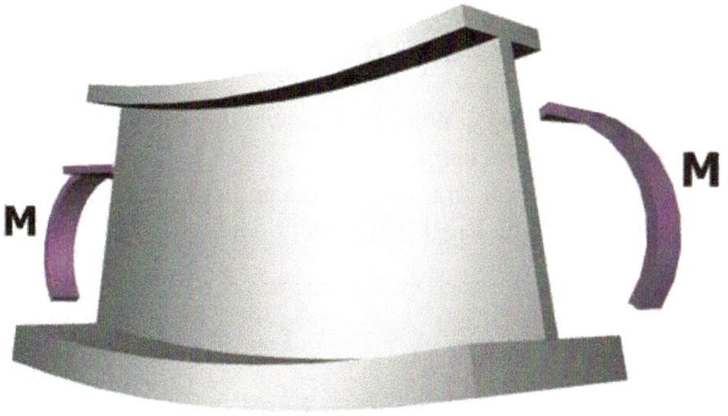

FIGURE M8.3

PROBLEMS

P8.1 During the fabrication of a laminated-timber arch, one of the 10 in. wide by 1 in. thick Douglas fir [E = 1,900 ksi] planks is bent to a radius of curvature of 40 ft. Determine the maximum bending stress developed in the plank.

P8.2 A copper wire of diameter d = 2 mm is coiled around a spool of radius r. The elastic modulus of the copper is E = 117 GPa and its yield strength is 310 MPa. Determine the minimum spool radius r that may be used if the bending stress in the wire is not to exceed the yield strength.

P8.3 The boards for a concrete form are to be bent into a circular shape having an inside radius of 10 m. What maximum thickness can be used for the boards if the normal stress is not to exceed 7 MPa? Assume that the modulus of elasticity of the wood is 12 GPa.

P8.4 A beam is subjected to equal bending moments of M_z = 3,200 N·m, as shown in Figure P8.4a. The cross-sectional dimensions (Figure P8.4b) are b = 150 mm, c = 30 mm, d = 70 mm, and t = 6 mm. Determine

(a) the centroid location, the moment of inertia about the z axis, and the controlling section modulus about the z axis.
(b) the bending stress at point H. State whether the normal stress at H is *tension* or *compression.*
(c) the bending stress at point K. State whether the normal stress at K is *tension* or *compression.*
(d) the maximum bending stress produced in the cross section. State whether the stress is *tension* or *compression.*

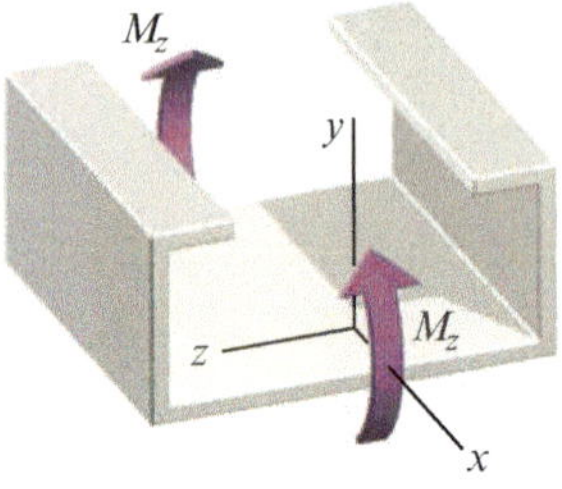

FIGURE P8.4a

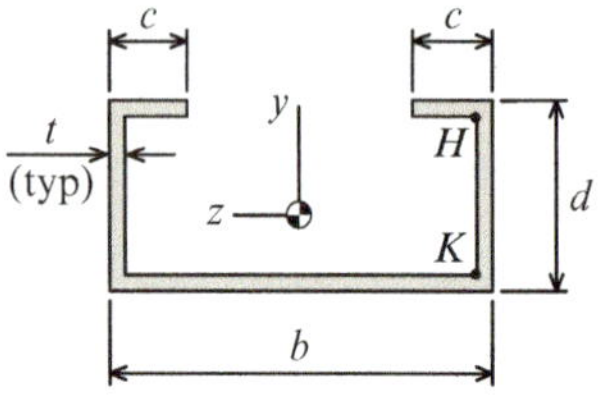

FIGURE P8.4b

P8.5 A beam is subjected to equal bending moments of M_z = 45 kip·ft, as shown in Figure P8.5a. The cross-sectional dimensions (Figure P8.5b) are b_1 = 7.5 in., d_1 = 1.5 in., b_2 = 0.75 in., d_2 = 6.0 in., b_3 = 3.0 in., and d_3 = 2.0 in. Determine

(a) the centroid location, the moment of inertia about the z axis, and the controlling section modulus about the z axis.
(b) the bending stress at point H. State whether the normal stress at H is *tension* or *compression.*
(c) the bending stress at point K. State whether the normal stress at K is *tension* or *compression.*
(d) the maximum bending stress produced in the cross section. State whether the stress is *tension* or *compression.*

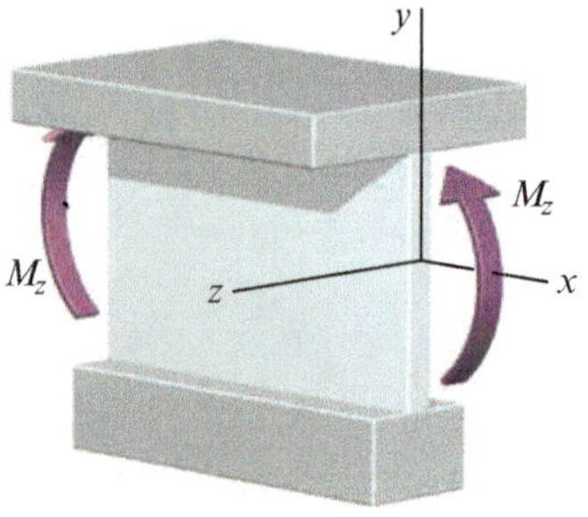

FIGURE P8.5a

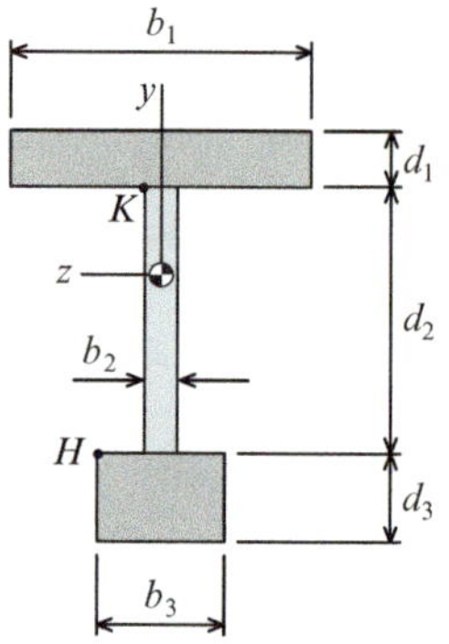

FIGURE P8.5b

P8.6 A beam is subjected to equal bending moments of M_z = 240 N·m, as shown in Figure P8.6a. The cross-sectional dimensions (Figure P8.6b) are a = 20 mm, b = 40 mm, d = 80 mm, and r = 12 mm. Determine

(a) the centroid location, the moment of inertia about the z axis, and the controlling section modulus about the z axis.
(b) the bending stress at point H. State whether the normal stress at H is *tension* or *compression.*
(c) the bending stress at point K. State whether the normal stress at K is *tension* or *compression.*
(d) the maximum bending stress produced in the cross section. State whether the stress is *tension* or *compression.*

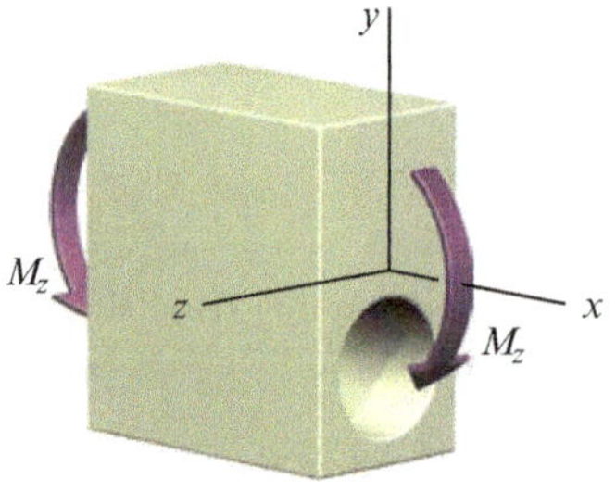

FIGURE P8.6a

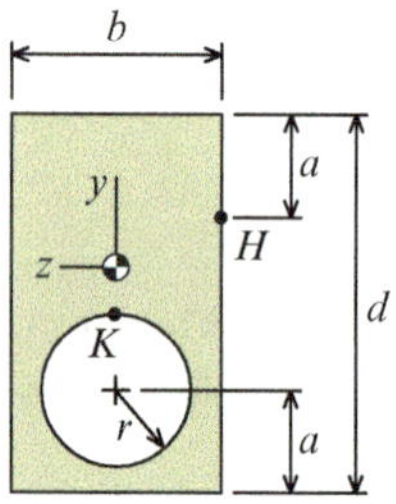

FIGURE P8.6b

P8.7 The dimensions of the beam cross section shown in Figure P8.7 are a = 7 mm and b = 45 mm. The internal bending moment about the z centroidal axis is M_z = 325 N·m. What is the magnitude of the maximum bending stress in the beam?

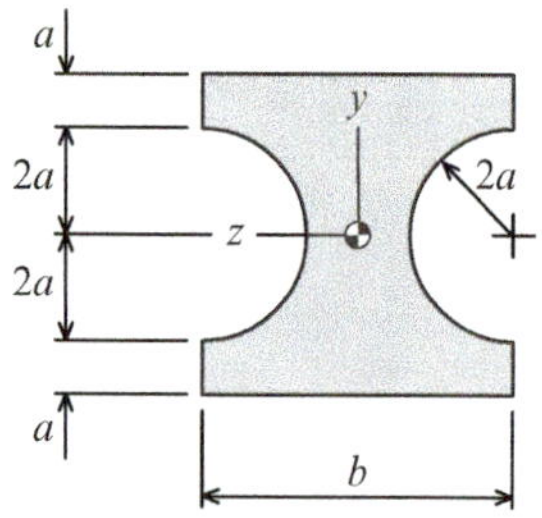

FIGURE P8.7

P8.8 The dimensions of the double-box beam cross section shown in Figure P8.8 are b = 150 mm, d = 50 mm, and t = 4 mm. If the maximum allowable bending stress is σ_b = 17 MPa, determine the magnitude of the maximum internal bending moment M_z that can be applied to the beam.

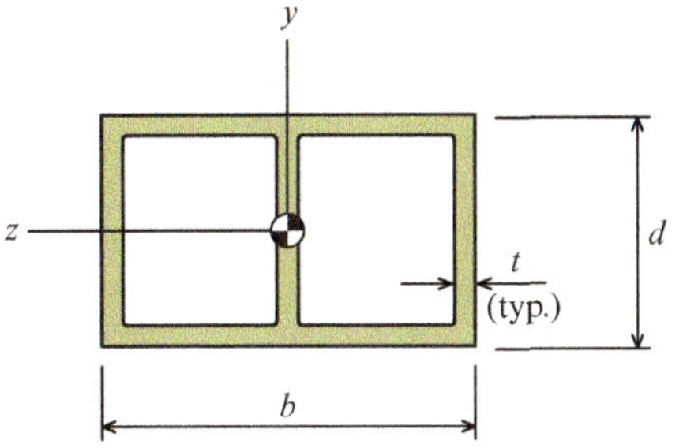

FIGURE P8.8

P8.9 An aluminum alloy [E = 10,300 ksi] tee-shaped bar is used as a beam. The cross section of the tee shape, shown in Figure P8.9, has a total depth of d = 7.5 in. The centroidal z axis of this cross section is located 2.36 in. above point H. The beam spans in the x direction and it bends about the z centroidal axis. A strain gage is affixed to the side of the tee stem at a distance c = 1.25 in. below the top surface of the tee shape. After loads are applied to the beam, a normal strain of $\varepsilon_x = -810\ \mu\varepsilon$ is measured by the strain gage. What is the bending stress at point H?

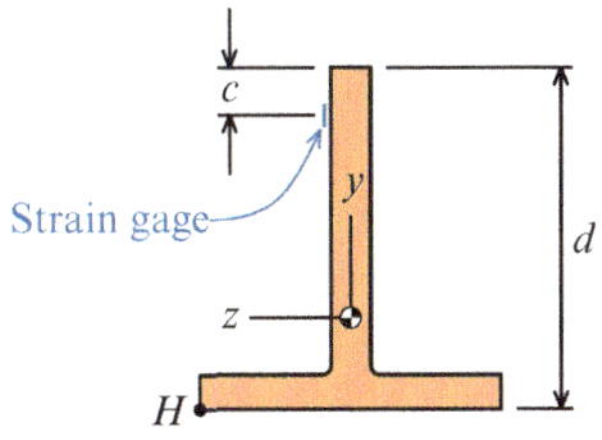

FIGURE P8.9

P8.10 The cross-sectional dimensions of the beam shown in Figure P8.10 are d = 17 in., b_f = 10 in., t_f = 1.0 in., t_w = 0.60 in., and a = 3.5 in.

(a) If the bending stress at point K is 5.4 ksi (C), what is the bending stress at point H? State whether the normal stress at H is *tension* or *compression*.
(b) If the allowable bending stress is 30 ksi, what is the magnitude of the maximum bending moment M_z that can be supported by the beam?

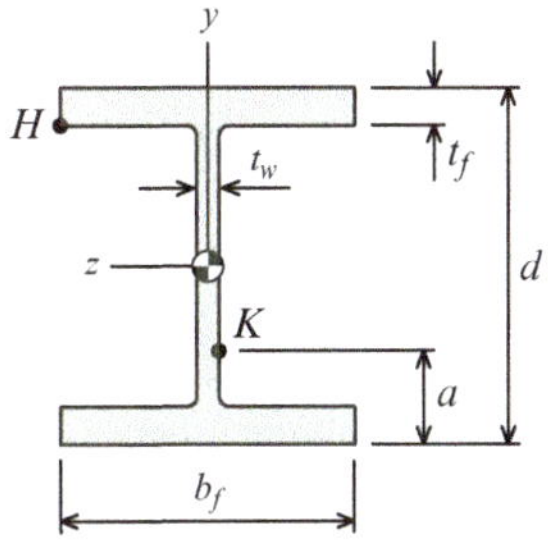

FIGURE P8.10

P8.11 The cross-sectional dimensions of the beam shown in Figure P8.11 are r_o = 115 mm and r_i = 95 mm. Given M_z = 16 kN·m, α = 30°, and β = 55°, what are the bending stresses at points H and K?

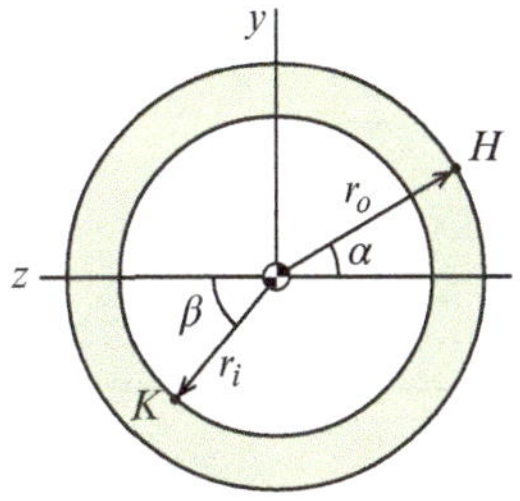

FIGURE P8.11

P8.12 The cross-sectional dimensions of the beam shown in Figure P8.12 are a = 5.0 in., b = 6.0 in., d = 4.0 in., and t = 0.5 in. The internal bending moment about the z centroidal axis is M_z = −4.25 kip·ft. Determine

(a) the maximum tensile bending stress in the beam.
(b) the maximum compressive bending stress in the beam.

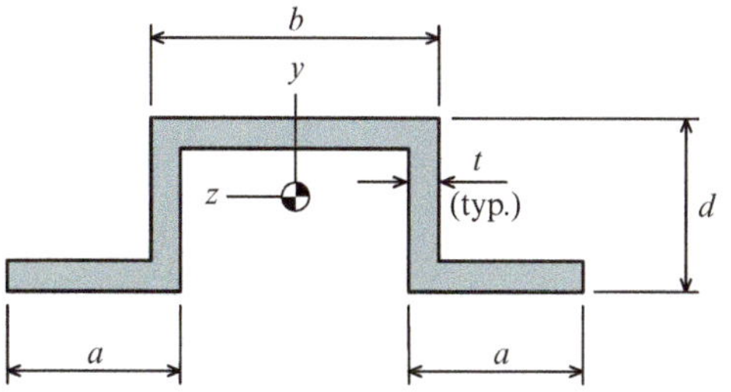

FIGURE P8.12

P8.13 Two uniformly distributed loads w = 3,600 lb/ft act on the simply supported beam shown in Figure P8.13*a*. The beam spans are a = 8 ft and L = 16 ft. The beam cross section shown in Figure P8.13*b* has dimensions of b_1 = 16 in., d_1 = 6 in., b_2 = 10 in., and d_2 = 10 in. Calculate the maximum tensile and compressive bending stresses produced in segment BC of the beam.

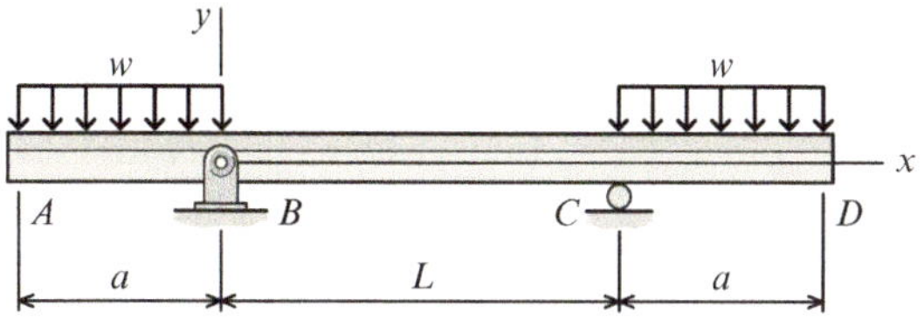

FIGURE P8.13*a*

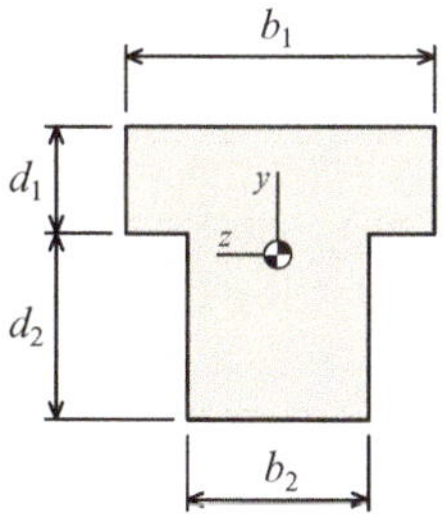

FIGURE P8.13*b*

P8.14 An extruded polymer beam is subjected to a bending moment M as shown in Figure P8.14*a*. The length of the beam is L = 800 mm. The cross-sectional dimensions (Figure P8.14*b*) of the beam are b_1 = 34 mm, d_1 = 100 mm, b_2 = 20 mm, d_2 = 20 mm, and a = 7 mm. For this material, the allowable tensile bending stress is 16 MPa and the allowable compressive bending stress is 12 MPa. Determine the largest moment M that can be applied as shown to the beam.

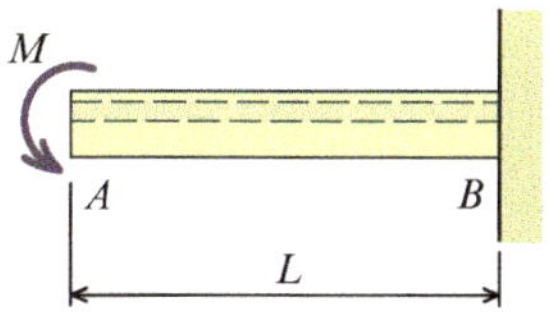

FIGURE P8.14*a*

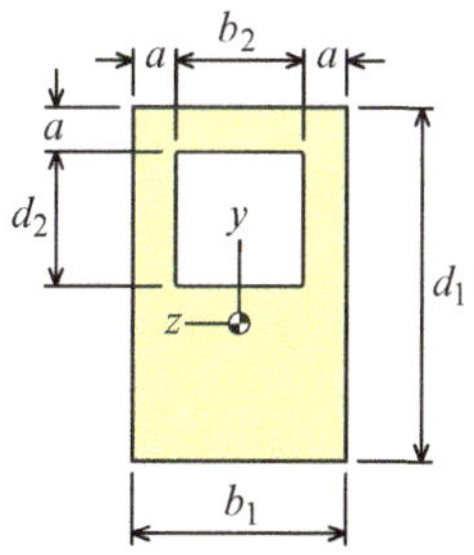

FIGURE P8.14*b*

8.4 Analysis of Bending Stresses in Beams

In this section, the flexure formula will be applied in the analysis of bending stresses for statically determinate beams subjected to various applied loads. The analysis process begins with the construction of shear-force and bending-moment diagrams for the specific span and loading. The cross-sectional properties of the beam will be determined next. Essential properties include

(a) the centroid of the cross section,

(b) the moment of inertia of the cross-sectional area about the centroidal axis of bending, and

(c) the distances from the centroidal axis to both the top and bottom surfaces of the beam.

After these prerequisite calculations have been completed, bending stresses can be calculated from the flexure formula at any location on the beam.

Beams can be supported and loaded in a variety of ways; consequently, the distribution and intensity of positive and negative bending moments are unique for each beam. Understanding the significance of the bending-moment diagram as it relates to flexural stresses is essential for the analysis of beams. For instance, consider a reinforced-concrete beam with an overhang, as shown in Figure 8.8. Concrete is a material with substantial strength in compression, but very low strength in tension. When concrete is used to construct a beam, steel bars must be placed in those regions where tensile stresses occur, in order to reinforce the concrete. In some portions of the overhang beam, tensile stresses will develop below the neutral axis, while tensile stresses will occur above the neutral axis in other portions. The engineer must define these regions of tensile stress so that the reinforcing steel is placed where it is needed. In sum, the engineer must be attentive not only to the magnitude of bending stresses, but also to the sense (either tension or compression) of stresses that occur above and below the neutral axis and that vary with positive and negative bending moments along the span.

Cross-Sectional Shapes for Beams

Beams can be constructed from many different cross-sectional shapes, such as squares, rectangles, solid circular shapes, and round pipe or tube shapes. A number of additional shapes are available for use in structures made of steel, aluminum, and fiber-reinforced

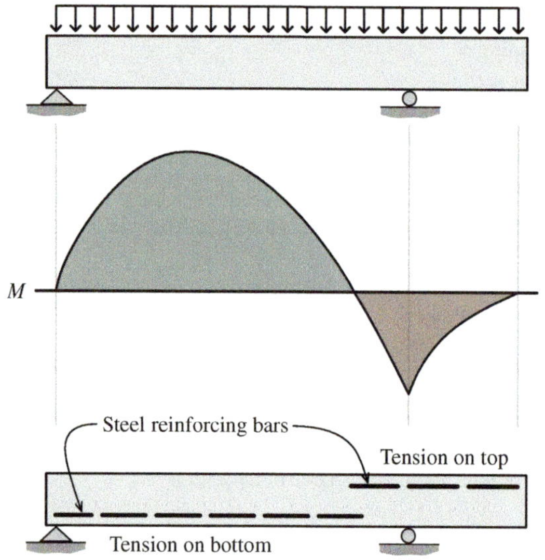

FIGURE 8.8 Reinforced-concrete beam.

254

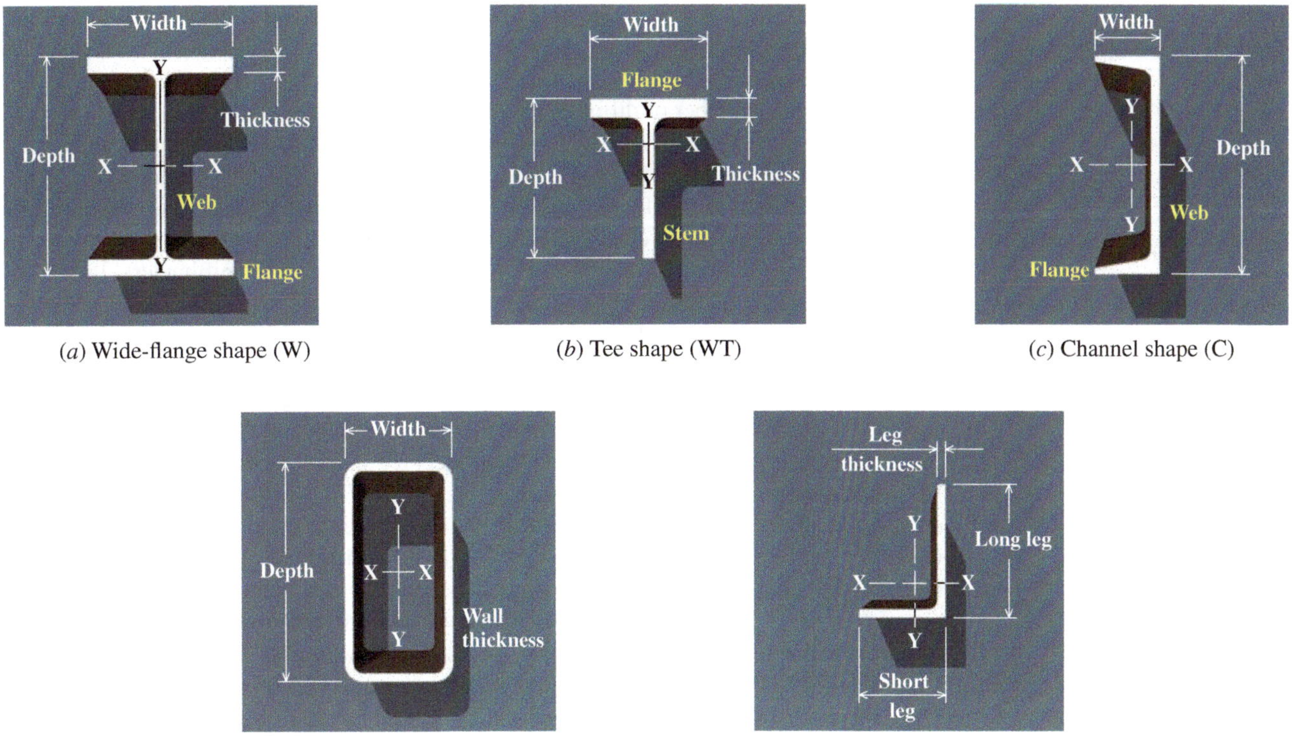

(*a*) Wide-flange shape (W) (*b*) Tee shape (WT) (*c*) Channel shape (C)

(*d*) Hollow structural section (HSS) (*e*) Angle shape (L)

FIGURE 8.9 Standard steel shapes.

plastics, and it is worthwhile to discuss some terminology associated with these standard shapes. Since steel is perhaps the most common material used in structures, this discussion will focus on the five standard rolled structural steel shapes shown in Figure 8.9.

The most commonly used steel shape for beams is called a **wide-flange shape** (Figure 8.9*a*). The wide-flange shape is optimized for economy in bending applications. As shown by Equation (8.10), the bending stress in a beam is inversely related to its section modulus S. If a choice is given between two shapes having the same allowable stress, the shape with the larger S is the better choice because it will be able to withstand more bending moment than the one with the smaller S. The weight of a beam is proportional to its cross-sectional area, and typically, the cost of a beam is directly related to its weight. Therefore, a shape that is optimized for bending is configured so that it provides the largest possible section modulus S for a given cross-sectional area of material. The area of a wide-flange shape is concentrated in its **flanges**. The area of the **web**, which connects the two flanges, is relatively small. By increasing the distance between the centroid and each flange, the shape's moment of inertia (about the X–X axis) can be increased dramatically, roughly in proportion to the square of this distance. Consequently, the section modulus of the shape can be substantially increased with a minimal overall increase in area.

For a wide-flange shape, the moment of inertia I and the section modulus S about the X–X centroidal axis (shown in Figure 8.9*a*) are much larger than I and S about the Y–Y centroidal axis. As a result, a shape that is aligned so that bending occurs about the X–X axis is said to be bending about its **strong axis**. Conversely, bending about the Y–Y axis is termed bending about the **weak axis**.

In U.S. customary units, a wide-flange shape is designated by the letter W followed by the nominal depth of the shape measured in inches and its weight per unit length measured in pounds per foot. A typical U.S. customary designation is W12 × 50, which is spoken as

"W12 by 50." This shape is nominally 12 in. deep, and it weighs 50 lb/ft. W shapes are manufactured by passing a hot billet of steel through several sets of rollers, arrayed in series, that incrementally transform the hot steel into the desired shape. By varying the spacing between rollers, a number of different shapes of the same nominal dimensions can be produced, giving the engineer a finely graduated selection of shapes. In making W shapes, the distance between flanges is kept constant while the flange thickness is increased. Consequently, the **actual depth** of a W shape is generally not equal to its **nominal depth**. For example, the nominal depth of a W12 × 50 shape is 12 in., but its actual depth is 12.2 in.

In SI units, the nominal depth of the W shape is measured in millimeters. Instead of weight per unit length, the shape designation gives mass per unit length, where mass is measured in kilograms and length is measured in meters. A typical SI designation is W310 × 74. This shape is nominally 310 mm deep, and it has a mass of 74 kg/m.

Figure 8.9*b* shows a **tee shape**, which consists of a flange and a **stem**. Figure 8.9*c* shows a **channel shape**, which is similar to a W shape, except that the flanges are truncated so that the shape has one flat vertical surface. Steel tee shapes are designated by the letters WT, and channel shapes are designated by the letter C. WT shapes and C shapes are named in a fashion similar to the way W shapes are named, where the nominal depth and either the weight per unit length or the mass per unit length is specified. Steel WT shapes are manufactured by cutting a W shape at middepth; therefore, the nominal depth of a WT shape is generally not equal to its actual depth. C shapes are rolled so that the actual depth is equal to the nominal depth. Both WT shapes and C shapes have strong and weak axes for bending.

Figure 8.9*d* shows a rectangular tube shape called a **hollow structural section (HSS)**. The designation used for HSS shapes gives the overall depth, followed by the outside width, followed by the wall thickness. For example, an HSS10 × 6 × 0.50 is 10 in. deep and 6 in. wide and has a wall thickness of 0.50 in.

Figure 8.9*e* shows an **angle shape**, which consists of two **legs**. Angle shapes are designated by the letter L followed by the **long leg** dimension, the **short leg** dimension, and the leg thickness (e.g., L6 × 4 × 0.50). Although angle shapes are versatile members that can be used for many purposes, single L shapes are rarely used as beams because they are not very strong and they tend to twist about their longitudinal axis as they bend. However, pairs of angles connected back-to-back are regularly used as flexural members in a configuration that is called a **double-angle shape (2L)**.

Cross-sectional properties of standard shapes are presented in Appendix B. While one could calculate the area and moment of inertia of a W shape or a C shape from the specified flange and web dimensions, the numerical values given in the tables in Appendix B are preferred, since they take into account specific section details, such as fillets.

EXAMPLE 8.3

A flanged cross section is used to support the loads shown on the beam in the accompanying diagrams. The dimensions of the shape are given. Consider the entire 20 ft length of the beam, and determine

(a) the maximum tensile bending stress at any location along the beam, and
(b) the maximum compressive bending stress at any location along the beam.

Plan the Solution

The flexure formula will be used to determine the bending stresses in this beam. However, the internal bending moments that are produced in the beam and the properties of the cross section must be determined before the stress calculations can be performed. With the use of the graphical method presented in Section 7.3, the shear-force and bending-moment diagrams for the beam and loading will be constructed. Then, the centroid

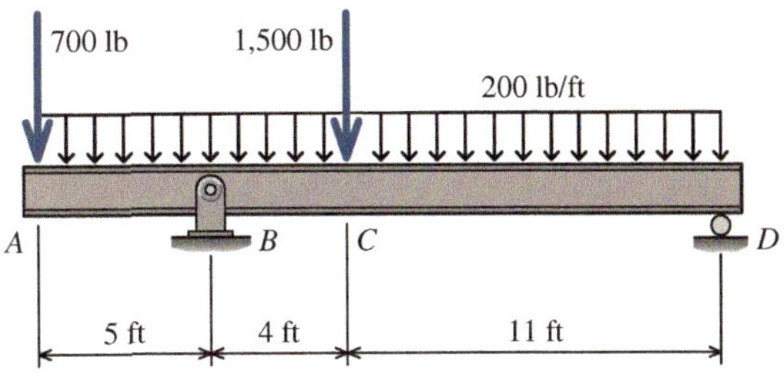

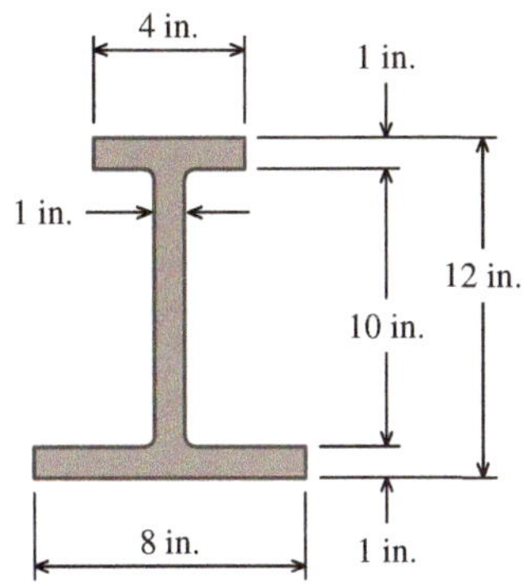

location and the moment of inertia will be calculated for the cross section of the beam. Since the cross section is not symmetric about the axis of bending, bending stresses must be investigated for both the largest positive and largest negative internal bending moments that occur along the entire beam span.

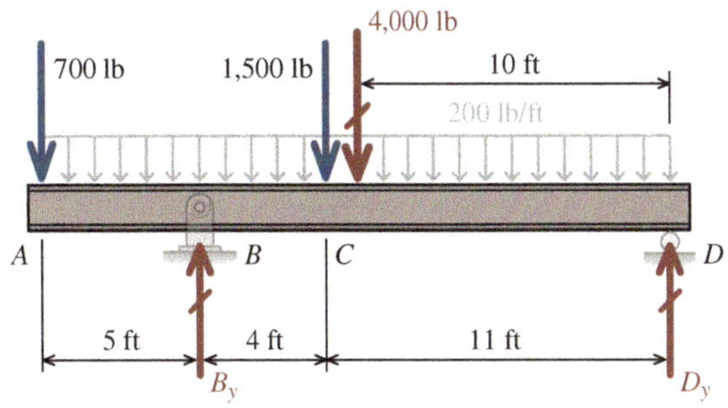

SOLUTION

Support Reactions

A free-body diagram (FBD) of the beam is shown. For the purpose of calculating the external beam reactions, the downward 200 lb/ft distributed load can be replaced by a resultant force of (200 lb/ft)(20 ft) = 4,000 lb acting downward at the centroid of the loading. The equilibrium equations are as follows:

$$\Sigma F_y = B_y + D_y - 700\text{ lb} - 1{,}500\text{ lb} - 4{,}000\text{ lb} = 0$$

$$\Sigma M_D = (700\text{ lb})(20\text{ ft}) + (1{,}500\text{ lb})(11\text{ ft})$$
$$+ (4{,}000\text{ lb})(10\text{ ft}) - B_y(15\text{ ft}) = 0$$

From these equilibrium equations, the beam reactions at pin support B and roller support D are, respectively,

$$B_y = 4{,}700\text{ lb} \qquad \text{and} \qquad D_y = 1{,}500\text{ lb}$$

Construct the Shear-Force and Bending-Moment Diagrams

The shear-force and bending-moment diagrams can be constructed with the six rules outlined in Section 7.3.

The maximum positive internal bending moment occurs 3.5 ft to the right of C and has a value of $M =$ 5,625 lb · ft.

The maximum negative internal bending moment occurs at pin support B and has a value of $M = -6{,}000$ lb · ft.

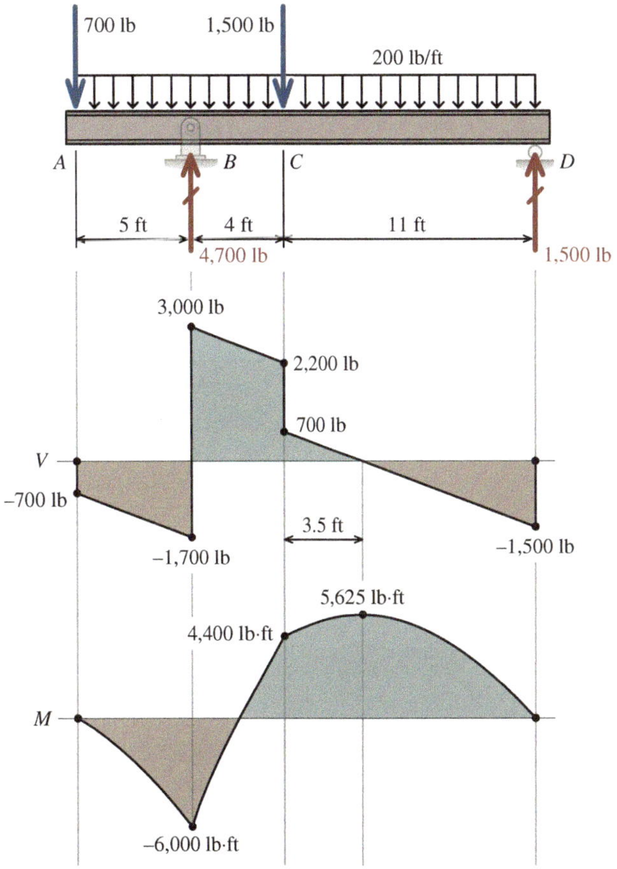

Centroid Location

The centroid location in the horizontal direction can be determined from symmetry alone. To determine the vertical location of the centroid, the flanged cross section is subdivided into three rectangular shapes. A reference axis for the calculation is established at the bottom surface of the lower flange. The centroid calculation for the flanged shape is summarized in the accompanying table.

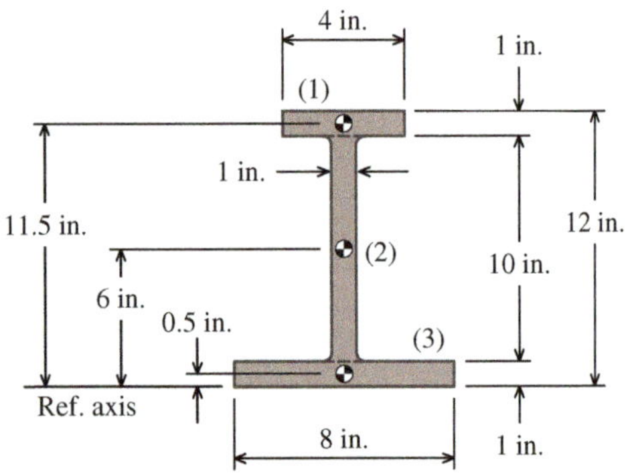

	A_i (in.^{2})	y_i (in.)	y_iA_i (in.^{3})
(1)	4.0	11.5	46.0
(2)	10.0	6.0	60.0
(3)	8.0	0.5	4.0
	22.0		110.0

$$\bar{y} = \frac{\Sigma y_i A_i}{\Sigma A_i} = \frac{110.0\ \text{in.}^3}{22.0\ \text{in.}^2} = 5.0\ \text{in.}$$

Therefore, the z centroidal axis is located 5.0 in. above the reference axis for this cross section.

Moment of Inertia

Since the centroids of areas (1), (2), and (3) do not coincide with the z centroidal axis for the entire cross section, the parallel-axis theorem must be used to calculate the moment of inertia of the cross section about this axis. The complete calculation is summarized in the following table:

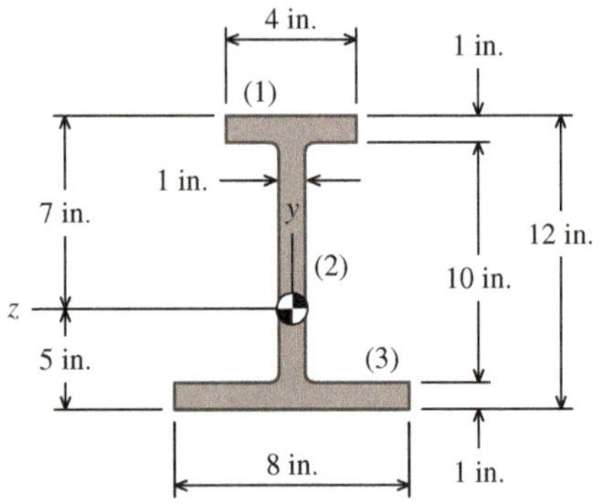

	I_c (in.^{4})	$\lvert d_i \rvert$ (in.)	$d_i^2A_i$ (in.^{4})	I_z (in.^{4})
(1)	0.333	6.5	169.000	169.333
(2)	83.333	1.0	10.000	93.333
(3)	0.667	4.5	162.000	162.667
				425.333

Thus, the moment of inertia of the cross section about its z centroidal axis is $I_z = 425.333\ \text{in.}^4$.

Flexure Formula

A positive bending moment produces compressive normal stress at the top of the beam and tensile normal stress at the bottom. Since the cross section of the beam is not symmetric about the axis of bending (i.e., the z axis), the magnitude of the bending stress at the top of the beam will be greater than that at the bottom of the beam.

The maximum positive internal bending moment is $M = 5{,}625\ \text{lb}\cdot\text{ft}$. For this positive moment, the compressive bending stress produced on the top of the flanged shape (at $y = +7$ in.) is calculated as

$$\sigma_x = -\frac{My}{I_z} = -\frac{(5{,}625\ \text{lb}\cdot\text{ft})(7\ \text{in.})(12\ \text{in./ft})}{425.333\ \text{in.}^4} = -1{,}111\ \text{psi} = 1{,}111\ \text{psi (C)}$$

and the tensile bending stress produced on the bottom of the flanged shape (at $y = -5$ in.) is calculated as

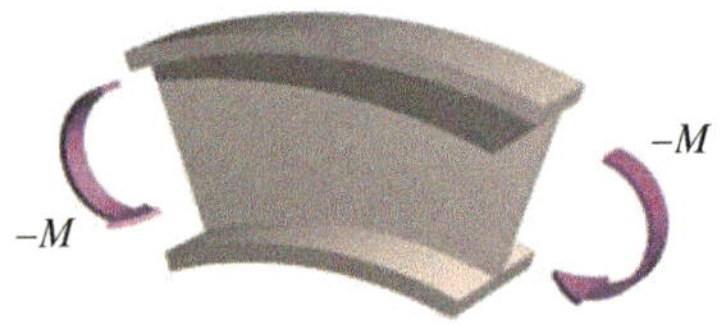

$$\sigma_x = -\frac{My}{I_z} = -\frac{(5{,}625\ \text{lb}\cdot\text{ft})(-5\ \text{in.})(12\ \text{in./ft})}{425.333\ \text{in.}^4} = +793\ \text{psi} = 793\ \text{psi (T)}$$

A negative bending moment produces tensile stress at the top of the beam and compressive stress at the bottom. The maximum negative internal bending moment is

$M = -6{,}000$ lb·ft. For this negative moment, the tensile bending stress produced on the top of the flanged shape (at $y = +7$ in.) is calculated as

$$\sigma_x = -\frac{My}{I_z} = -\frac{(-6{,}000 \text{ lb}\cdot\text{ft})(7 \text{ in.})(12 \text{ in./ft})}{425.333 \text{ in.}^4} = +1{,}185 \text{ psi} = 1{,}185 \text{ psi (T)}$$

and the compressive bending stress produced on bottom of the flanged shape (at $y = -5$ in.) is

$$\sigma_x = -\frac{My}{I_z} = -\frac{(-6{,}000 \text{ lb}\cdot\text{ft})(-5 \text{ in.})(12 \text{ in./ft})}{425.333 \text{ in.}^4} = -846 \text{ psi} = 846 \text{ psi (C)}$$

(a) Maximum tensile bending stress: For this beam, the maximum tensile bending stress occurs on top of the beam, at the location of the maximum negative internal bending moment. The maximum tensile bending stress is $\sigma_x = 1{,}185$ psi (T). **Ans.**

(b) Maximum compressive bending stress: The maximum compressive bending stress also occurs on top of the beam; however, it occurs at the location of the maximum positive internal bending moment. The maximum compressive bending stress is $\sigma_x = 1{,}111$ psi (C). **Ans.**

EXAMPLE 8.4

A solid steel shaft 40 mm in diameter supports the loads shown. Determine the magnitude and location of the maximum bending stress in the shaft.

Note: For the purposes of this analysis, the bearing at *B* can be idealized as a pin support and the bearing at *E* can be idealized as a roller support.

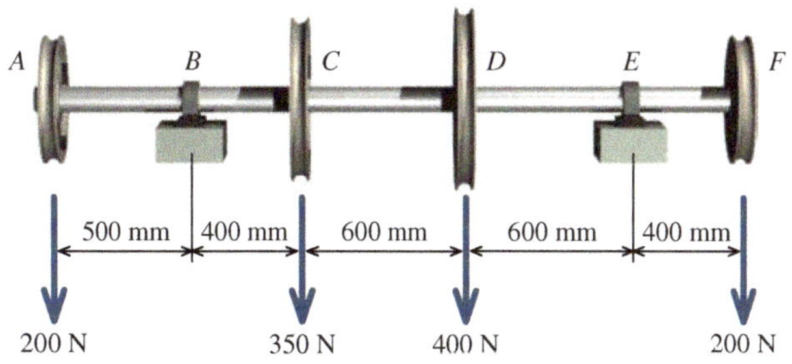

Plan the Solution

The shear-force and bending-moment diagrams for the shaft and loading will be constructed by the graphical method presented in Section 7.3. Since the circular cross section is symmetric about the axis of bending, the maximum bending stress will occur at the location of the maximum internal bending moment.

SOLUTION

Support Reactions

An FBD of the beam is shown. From this FBD, the equilibrium equations can be written as follows:

$$\Sigma F_y = B_y + E_y - 200 \text{ N} - 350 \text{ N} - 400 \text{ N} - 200 \text{ N} = 0$$

$$\begin{aligned}\Sigma M_B &= (200 \text{ N})(500 \text{ mm}) - (350 \text{ N})(400 \text{ mm}) - (400 \text{ N})(1{,}000 \text{ mm}) \\ &\quad - (200 \text{ N})(2{,}000 \text{ mm}) + E_y(1{,}600 \text{ mm}) = 0\end{aligned}$$

From these equilibrium equations, the beam reactions at pin support *B* and roller support *E* are, respectively,

$$B_y = 625 \text{ N} \qquad \text{and} \qquad E_y = 525 \text{ N}$$

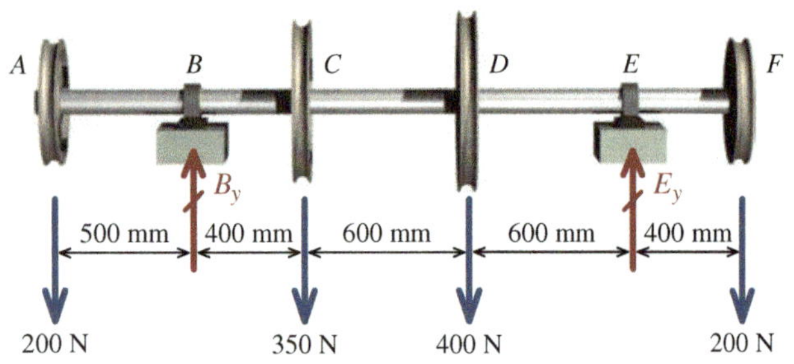

Construct the Shear-Force and Bending-Moment Diagrams

The shear-force and bending-moment diagrams can be constructed in accordance with the six rules outlined in Section 7.3.

The maximum internal bending moment occurs at D and has a magnitude of $M = 115\ \text{N}\cdot\text{m}$.

Moment of Inertia

The moment of inertia for the 40 mm diameter solid steel shaft is

$$I_z = \frac{\pi}{64}d^4 = \frac{\pi}{64}(40\ \text{mm})^4 = 125{,}664\ \text{mm}^4$$

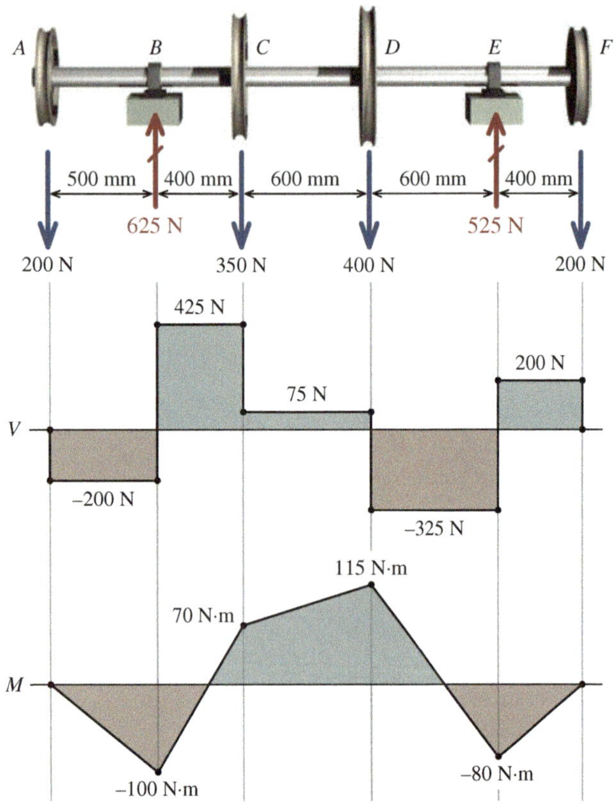

Flexure Formula

The maximum bending stress in the shaft occurs at D. Since the circular cross section is symmetric about the axis of bending, both the tensile and compressive bending stresses have the same magnitude. In this situation, the flexure formula in the form of Equation (8.10) is convenient for calculating bending stresses. The distance c used in Equation (8.10) is simply the shaft radius. From this form of the flexure formula, the magnitude of the maximum bending stress in the shaft is

$$\sigma_{\text{max}} = \frac{Mc}{I_z} = \frac{(115\ \text{N}\cdot\text{m})(20\ \text{mm})(1{,}000\ \text{mm/m})}{125{,}664\ \text{mm}^4}$$
$$= 18.30\ \text{MPa} \qquad \textbf{Ans.}$$

Section Modulus for a Solid Circular Section

Alternatively, the magnitude of the maximum bending stress in the shaft can be computed from the section modulus. For a solid circular section, the following formula can be derived for the section modulus:

$$S = \frac{I_z}{c} = \frac{(\pi/64)d^4}{d/2} = \frac{\pi}{32}d^3$$

For the 40 mm diameter solid steel shaft considered here, the section modulus is, therefore,

$$S = \frac{\pi}{32}d^3 = \frac{\pi}{32}(40\ \text{mm})^3 = 6{,}283\ \text{mm}^3$$

and the magnitude of the maximum bending stress in the shaft can be computed as

$$\sigma_{\text{max}} = \frac{M}{S} = \frac{(115\ \text{N}\cdot\text{m})(1{,}000\ \text{mm/m})}{6{,}283\ \text{mm}^3} = 18.30\ \text{MPa} \qquad \textbf{Ans.}$$

MecMovies

EXAMPLES

M8.9 Determine the bending-moment diagram and the maximum tension and compression bending stresses for a tee shape.

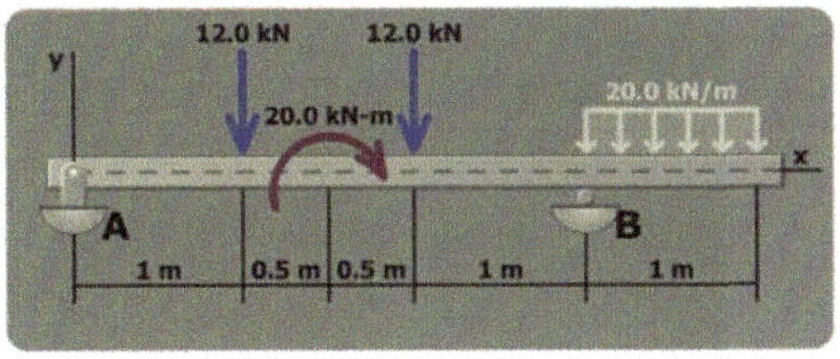

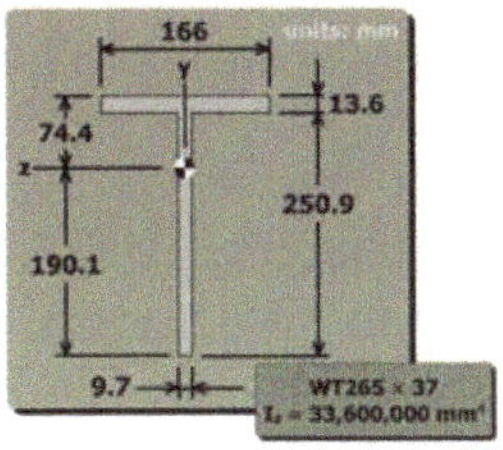

M8.10 Determine maximum bending moments, given allowable tension and compression bending stresses.

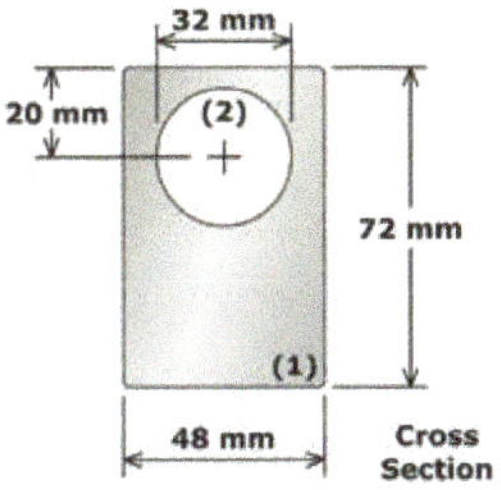

M8.11 Determine the bending-moment diagram, the moment of inertia, and the bending stress produced in a simply supported wide-flange steel beam.

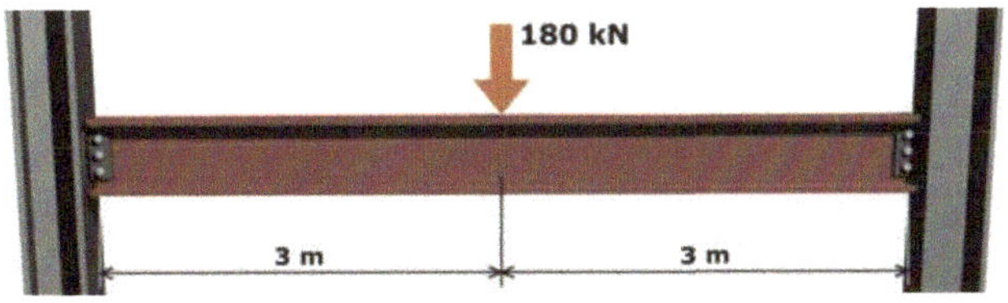

M8.12 Determine the bending-moment diagram, the moment of inertia, and the bending stress produced in a cantilever beam consisting of a tee shape.

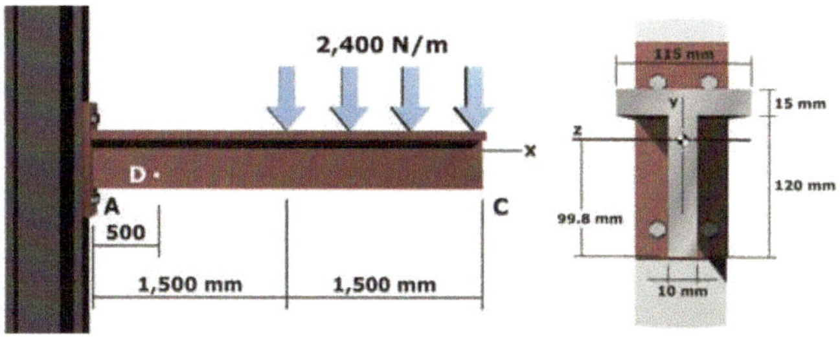

M8.13 Determine the bending-moment diagram, the centroid location, the moment of inertia, and the bending stress for a simply supported beam with a U-shaped cross section.

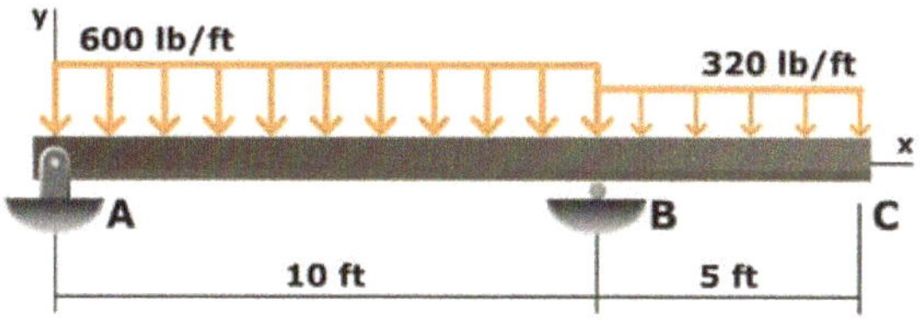

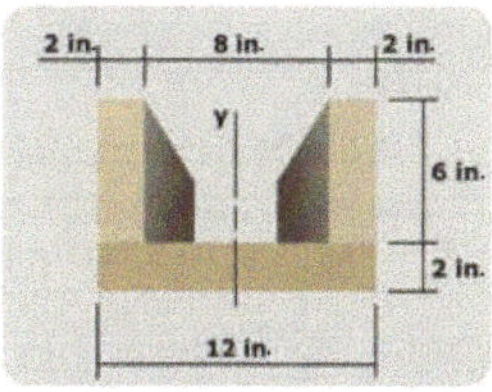

M8.14 Determine the bending-moment diagram and the bending stress for a standard steel shape that is used as a simply supported beam with an overhang.

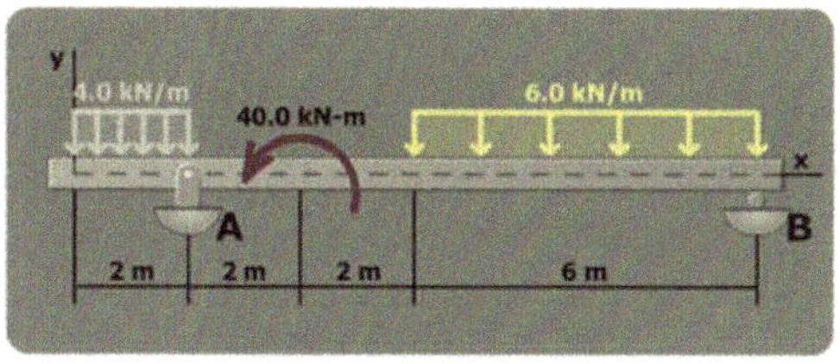

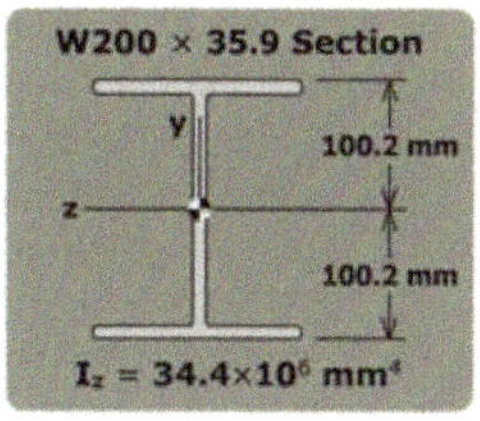

M8.15 Moment-of-inertia calculations involving shapes built up from standard steel shapes.

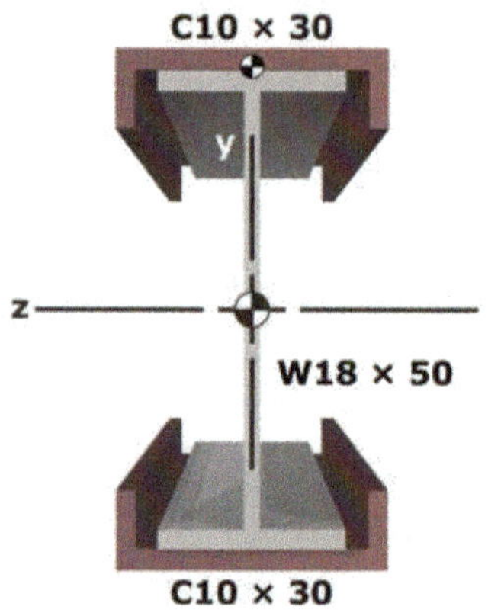

EXERCISES

M8.8 Calculate the tension and compression bending stresses produced in singly symmetric cross sections.

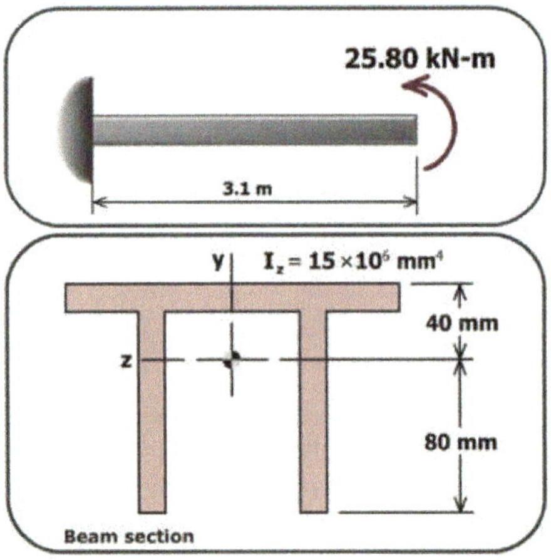

FIGURE M8.8

M8.9 Given a specific bending-moment diagram of a beam, compute the maximum tension and compression bending stresses produced at any location along the beam span.

M8.10 Given an allowable tension bending stress and an allowable compression bending stress, determine the maximum internal bending-moment magnitude that may be applied to a beam.

PROBLEMS

P8.15 A solid steel shaft 1.125 in. in diameter supports loads P_A = 65 lb, P_C = 150 lb, and P_E = 55 lb as shown in Figure P8.15. Assume that a = 10 in., b = 25 in., c = 13 in., and d = 16 in. The bearing at B can be idealized as a roller support, and the bearing at D can be idealized as a pin support. Determine the magnitude and location of the maximum bending stress in the shaft.

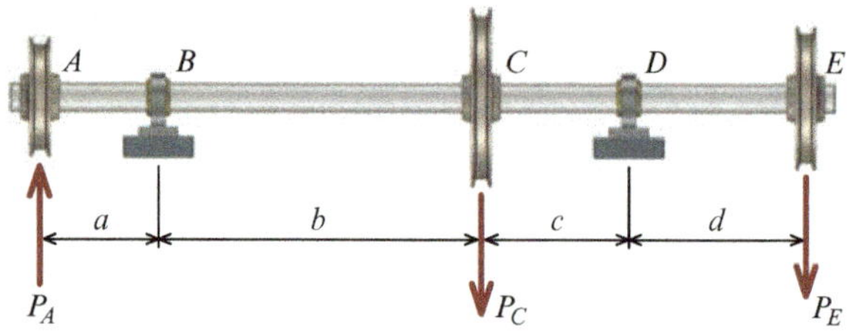

FIGURE P8.15

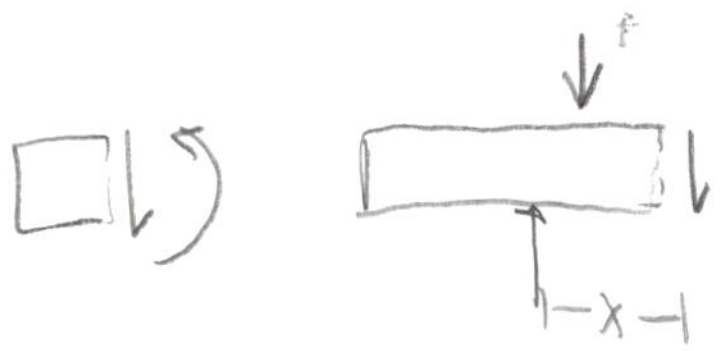

P8.16 A W18 × 40 standard steel shape is used to support the loads shown on the beam in Figure P8.16. Determine the magnitude and location of the maximum bending stress in the beam.

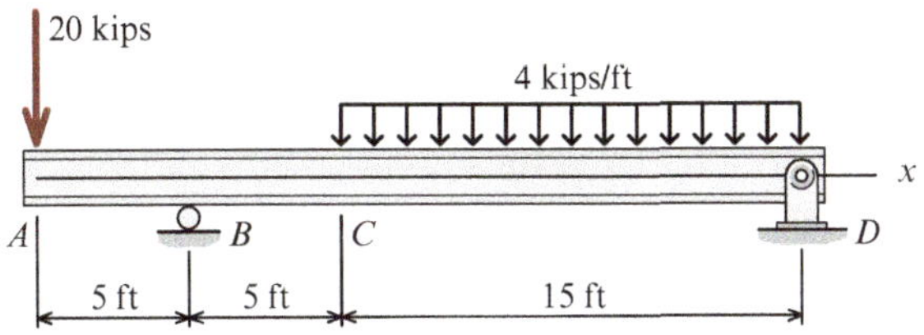

FIGURE P8.16

P8.17 A W410 × 60 standard steel shape is used to support the loads shown on the beam in Figure P8.17. Assume that $P = 125$ kN, $w = 30$ kN/m, $a = 3$ m, and $b = 2$ m. Determine the magnitude and location of the maximum bending stress in the beam.

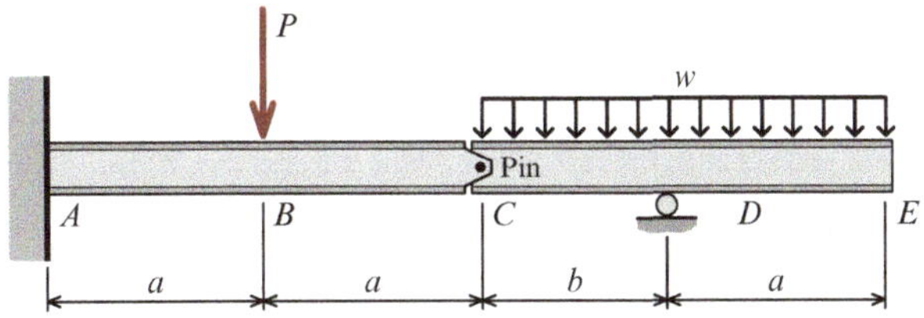

FIGURE P8.17

P8.18 A W16 × 50 standard steel shape is used to support the loads shown on the beam in Figure P8.18. Assume that $P = 12$ kips, $w = 1.0$ kips/ft, $a = 30$ ft, and $b = 10$ ft. Determine the magnitude and location of the maximum bending stress in the beam.

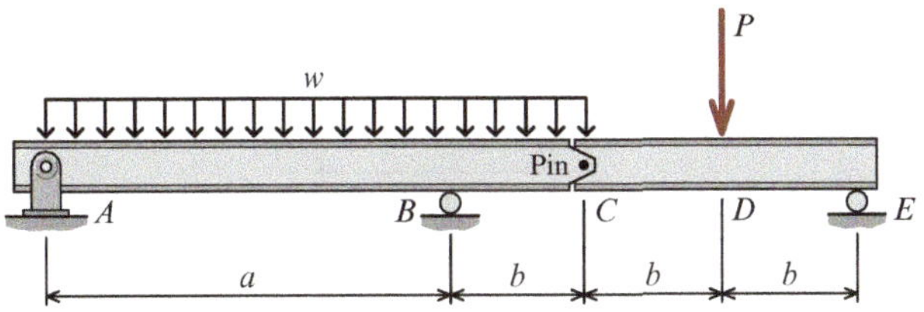

FIGURE P8.18

P8.19 An HSS12 × 8 × 1/2 standard steel shape is used to support the loads shown on the beam in Figure P8.19. The shape is oriented so that bending occurs about the strong axis. Determine the magnitude and location of the maximum bending stress in the beam.

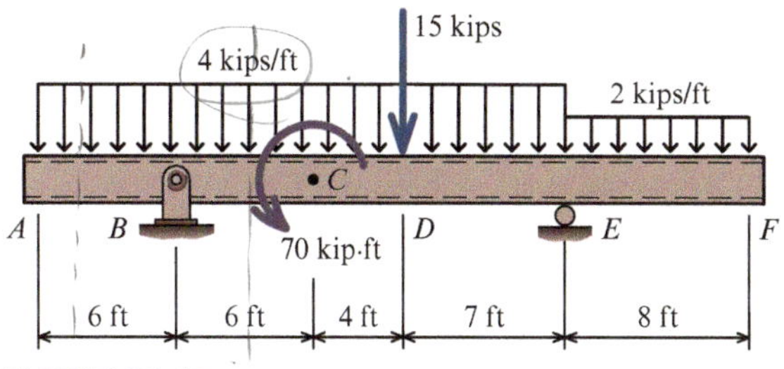

FIGURE P8.19

P8.20 A W360 × 72 standard steel shape is used to support the loads shown on the beam in Figure P8.20*a*. The shape is oriented so that bending occurs about the weak axis, as shown in Figure P8.20*b*. Consider the entire 6 m length of the beam, and determine

(a) the maximum tensile bending stress at any location along the beam and

(b) the maximum compressive bending stress at any location along the beam.

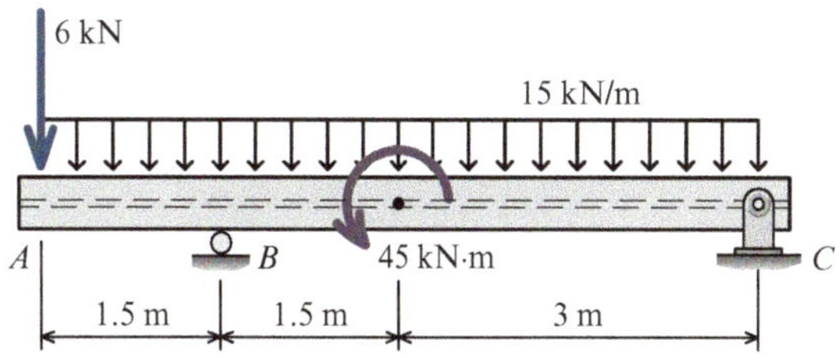

FIGURE P8.20*a*

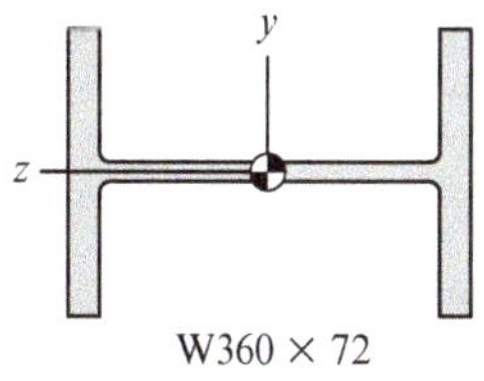

FIGURE P8.20*b*

P8.21 A W310 × 38.7 standard steel shape is used to support the loads shown on the beam in Figure P8.21. What is the magnitude of the maximum bending stress at any location along the beam?

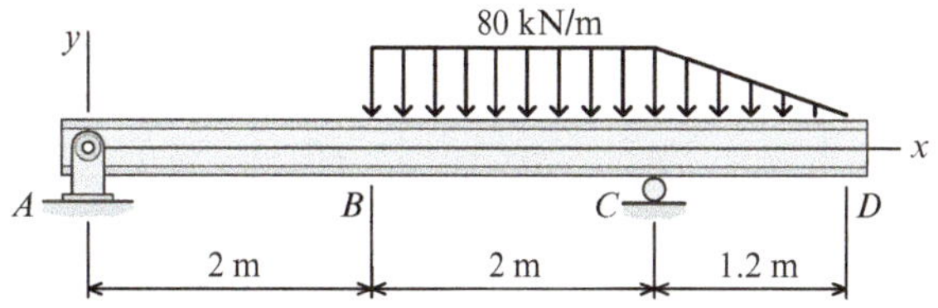

FIGURE P8.21

P8.22 A WT305 × 41 standard steel shape is used to support the loads shown on the beam in Figure P8.22*a*. The dimensions from the top and bottom of the shape to the centroidal axis are shown on the sketch of the cross section (Figure P8.22*b*). Consider the entire 4.8 m length of the beam, and determine

(a) the maximum tensile bending stress at any location along the beam and
(b) the maximum compressive bending stress at any location along the beam.

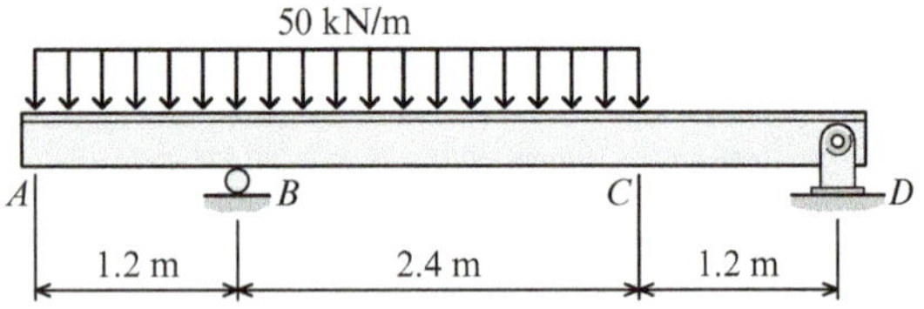

FIGURE P8.22*a*

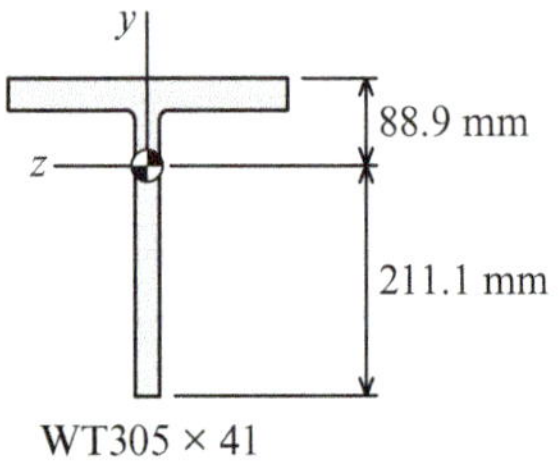

FIGURE P8.22*b*

P8.23 A flanged wooden shape is used to support the loads shown on the beam in Figure P8.23*a*. The dimensions of the shape are shown in Figure P8.23*b*. Consider the entire 18 ft length of the beam, and determine

(a) the maximum tensile bending stress at any location along the beam and
(b) the maximum compressive bending stress at any location along the beam.

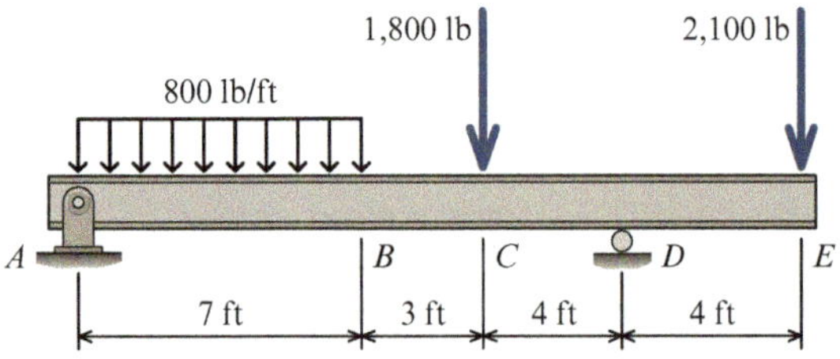

FIGURE P8.23*a*

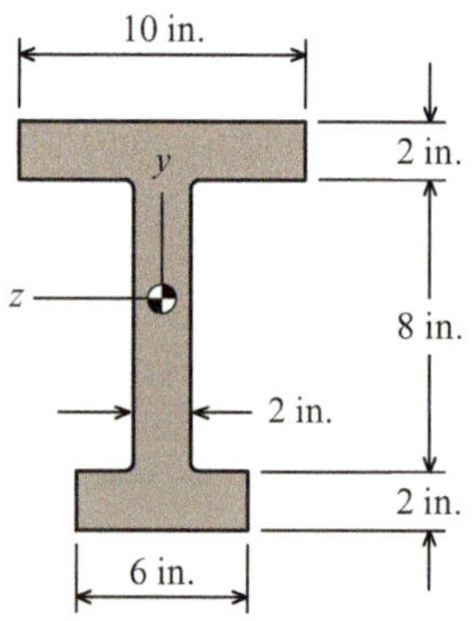

FIGURE P8.23*b*

P8.24 The steel beam in Figure P8.24*a*/25*a* has the cross section shown in Figure P8.24*b*/25*b*. The beam length is $L = 22$ ft, and the cross-sectional dimensions are $d = 16.3$ in., $b_f = 10.0$ in., $t_f = 0.665$ in., and $t_w = 0.395$ in. Calculate the maximum bending stress in the beam if $w_0 = 6$ kips/ft.

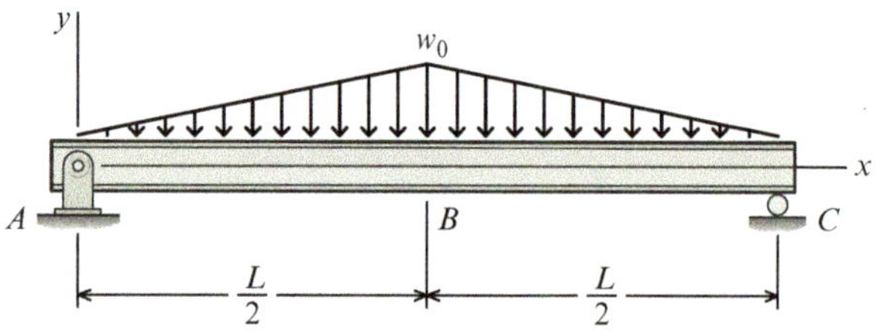

FIGURE P8.24*a*/25*a*

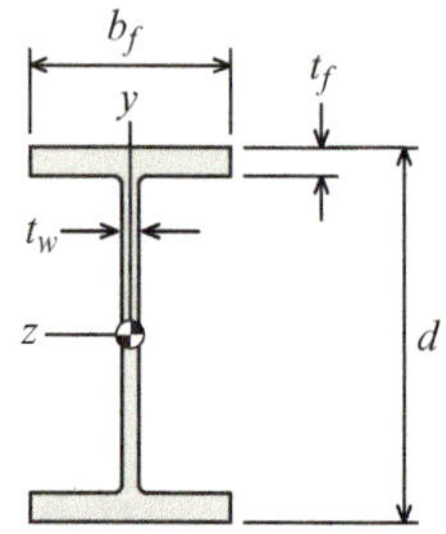

FIGURE P8.24*b*/25*b*

P8.25 The steel beam in Figure P8.24*a*/25*a* has the cross section shown in Figure P8.24*b*/25*b*. The beam length is $L = 6.0$ m, and the cross-sectional dimensions are $d = 350$ mm, $b_f = 205$ mm, $t_f = 14$ mm, and $t_w = 8$ mm. Calculate the largest intensity of distributed load w_0 that can be supported by this beam if the allowable bending stress is 200 MPa.

8.5 Introductory Beam Design for Strength

At a minimum, a beam must be designed so that it is capable of supporting the loads acting on it without exceeding allowable bending stresses. A successful design involves the determination of an *economical* cross section for the beam—one that performs the intended function but does not waste materials. Elementary design generally involves either

(a) the determination of appropriate dimensions for basic shapes, such as rectangular or circular cross sections, or

(b) the selection of satisfactory standard manufactured shapes that are available for the preferred material.

A complete beam design requires attention to many concerns. This discussion, however, will be limited to the task of proportioning cross sections so that allowable bending stresses are satisfied, thus ensuring that a beam has sufficient strength to support the loads that act upon it.

The section modulus S is a particularly convenient property for beam strength design. One form of the flexure formula given by Equation (8.10) for doubly symmetric shapes was

$$\sigma_{max} = \frac{Mc}{I} = \frac{M}{S} \qquad \text{where} \qquad S = \frac{I}{c}$$

If an allowable bending stress is specified for the beam material, then the flexure formula can be solved for the minimum required section modulus S_{min}:

$$S_{min} \geq \left| \frac{M}{\sigma_{allow}} \right| \tag{8.11}$$

Using Equation (8.11), the engineer may either

(a) determine the cross-sectional dimensions necessary to attain the minimum section modulus or

(b) select a standard shape that offers a section modulus equal to or greater than S_{min}.

The maximum bending moment in the beam is found from a bending-moment diagram. If the cross section to be used for the beam is doubly symmetric, then the maximum bending-moment magnitude (i.e., either positive or negative M) should be used in Equation (8.11). In some instances, it may be necessary to investigate both the maximum positive bending moment and the maximum negative bending moment. One such situation arises when different allowable tensile and compressive bending stresses are specified for a cross section that is not doubly symmetric, such as a tee shape.

If a beam has a simple cross-sectional shape, such as a circle, a square, or a rectangle of specified height-to-width proportions, then its dimensions can be determined directly from S_{min}, since, by definition, $S = I/c$. If a more complex shape (e.g., a W shape) is to be used for the beam, then tables of cross-sectional properties such as those included in Appendix B are utilized. The general process for selecting an economical standard steel shape from a table of section properties is outlined in Table 8.1.

The ratio of one dimension to another is called an **aspect ratio**. For a rectangular cross section, the ratio of height h to width b is the aspect ratio of the beam.

Table 8.1 Selecting Standard Steel Shapes for Beams

Step 1: Calculate the minimum section modulus required for the specific span and loading.

Step 2: In the tables of section properties (such as those presented in Appendix B), locate the section modulus values. Typically, the beam will be oriented so that bending occurs about the strong axis; therefore, find the column that gives S for the strong axis (which is typically designated as the X–X axis).

Step 3: Start your search at the bottom of the table. Shapes are typically sorted from heaviest to lightest; therefore, the shapes at the bottom of the table are usually the lightest-weight members. Scan up the column until a section modulus equal to or slightly greater than S_{min} is found. This shape is acceptable, and its designation should be noted.

Step 4: Continue scanning upwards until several acceptable shapes have been determined.

Step 5: After several acceptable shapes have been identified, select one shape for use as the beam cross section. The lightest-weight cross section is usually chosen, because the cost of the beam is directly related to the weight of the beam. However, other considerations could affect the choice. For example, a limited height might be available for the beam, thus necessitating a shorter and heavier cross section instead of a taller, but lighter, shape.

EXAMPLE 8.5

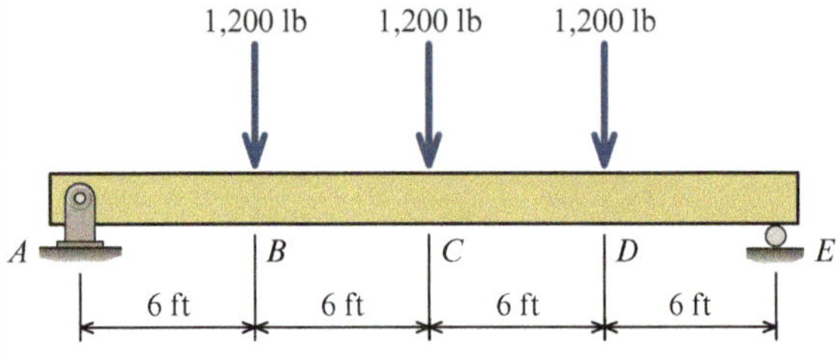

Beam and loading.

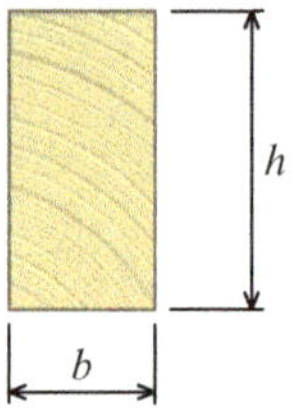

Cross section.

A 24 ft long simply supported wooden beam supports three 1,200 lb concentrated loads that are located at the quarter points of the span. The allowable bending stress of the wood is 1,800 psi. If the aspect ratio of the solid rectangular wooden beam is specified as $h/b = 2.0$, determine the minimum width b that can be used for the beam.

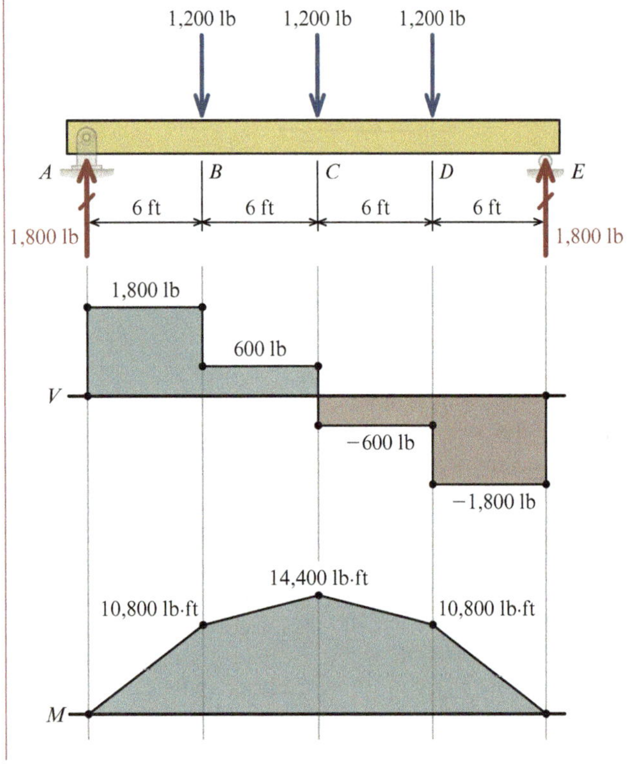

Plan the Solution

By the graphical method presented in Section 7.3, the shear-force and bending-moment diagrams for the beam and loading will be constructed at the outset. With the use of the maximum internal bending moment and the specified allowable bending stress, the required section modulus can be determined from the flexure formula [Equation (8.10)]. The beam cross section can then be proportioned so that its height is twice as large as its width.

SOLUTION

Construct the Shear-Force and Bending-Moment Diagrams

The shear-force and bending-moment diagrams for the beam and loading are shown. The maximum internal bending moment occurs at C.

Required Section Modulus

The flexure formula can be solved for the minimum section modulus required to support a maximum internal bending moment of $M = 14{,}400$ lb·ft without exceeding the 1,800 psi allowable bending stress:

$$\sigma_{max} = \frac{M}{S} \leq \sigma_{allow}$$

$$\therefore S \geq \frac{M}{\sigma_{\text{allow}}} = \frac{(14{,}400 \text{ lb} \cdot \text{ft})(12 \text{ in./ft})}{1{,}800 \text{ psi}}$$

$$= 96.0 \text{ in.}^3$$

Section Modulus for a Rectangular Section

For a solid rectangular section with width b and height h, the following formula can be derived for the section modulus:

$$S = \frac{I_z}{c} = \frac{bh^3/12}{h/2} = \frac{bh^2}{6}$$

The aspect ratio specified for the beam in this problem is $h/b = 2$; therefore, $h = 2b$. Substituting this requirement into the section modulus formula gives

$$S = \frac{bh^2}{6} = \frac{b(2b)^2}{6} = \frac{4}{6}b^3 = \frac{2}{3}b^3$$

The minimum required beam width can now be determined:

$$\frac{2}{3}b^3 \geq 96.0 \text{ in.}^3 \qquad \therefore b \geq 5.24 \text{ in.}$$ **Ans.**

EXAMPLE 8.6

The beam shown will be constructed from a standard steel W shape with an allowable bending stress of 30 ksi.

(a) Develop a list of acceptable shapes that could be used for this beam. Include the most economical W8, W10, W12, W14, W16, and W18 shapes on the list of possibilities.

(b) Select the most economical W shape for the beam.

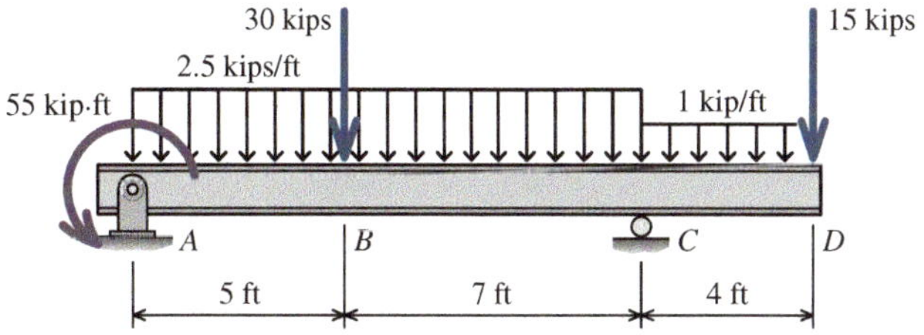

Plan the Solution

By the graphical method presented in Section 7.3, the shear-force and bending-moment diagrams for the beam and loading will be constructed at the outset. With the use of the maximum internal bending moment and the specified allowable bending stress, the required section modulus can be determined from the flexure formula [Equation (8.10)]. Acceptable standard steel W shapes will be selected from Appendix B, and the lightest of those shapes will be chosen as the most economical shape for this application.

SOLUTION

Support Reactions

An FBD of the beam is shown. From this FBD, the equilibrium equations can be written as follows:

$$\Sigma F_y = A_y + C_y - 30 \text{ kips} - 15 \text{ kips} - 30 \text{ kips} - 4 \text{ kips} = 0$$

$$\Sigma M_C = (30 \text{ kips})(7 \text{ ft}) + (30 \text{ kips})(6 \text{ ft}) - (4 \text{ kips})(2 \text{ ft}) - (15 \text{ kips})(4 \text{ ft}) + 55 \text{ kip}\cdot\text{ft} - A_y(12 \text{ ft}) = 0$$

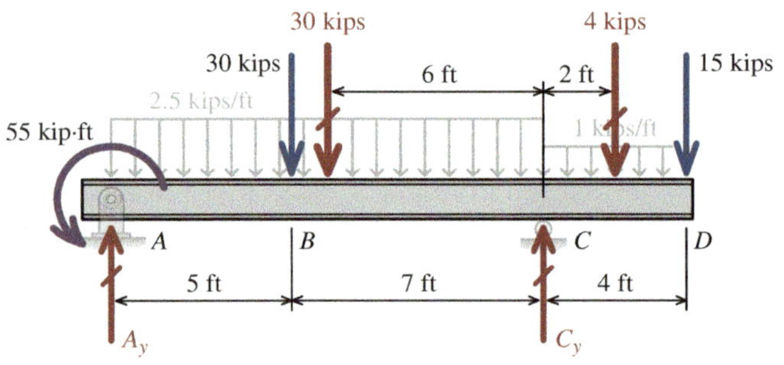

From these equilibrium equations, the beam reactions at pin support A and roller support C are

$$A_y = 31.42 \text{ kips} \qquad \text{and} \qquad C_y = 47.58 \text{ kips}$$

Shear-Force and Bending-Moment Diagrams
The shear-force and bending-moment diagrams for the beam and loading are shown. The maximum internal bending moment in the beam is $M = 70.83$ kip·ft, and it occurs at B.

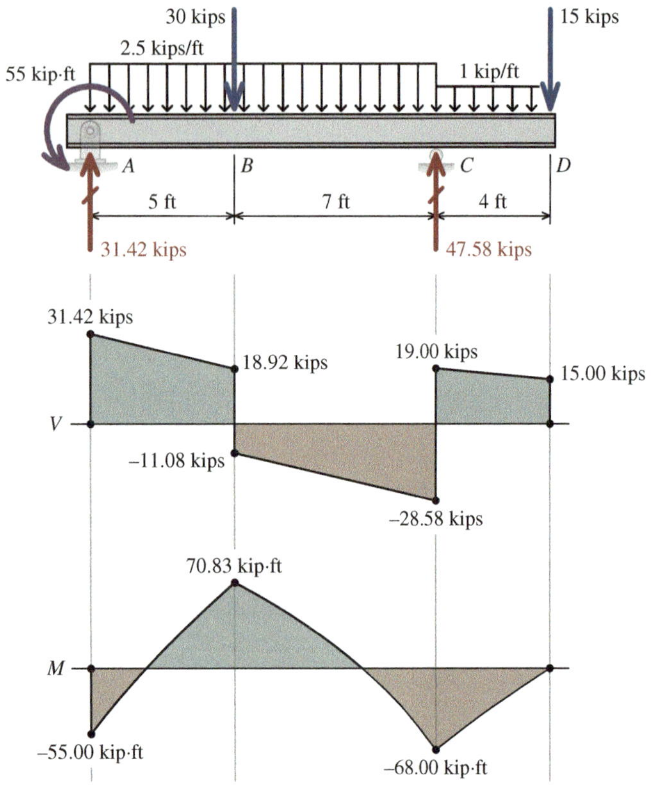

Required Section Modulus
The flexure formula can be solved for the minimum section modulus required to support the maximum internal bending moment without exceeding the 30 ksi allowable bending stress:

$$\sigma_{\text{max}} = \frac{M}{S} \leq \sigma_{\text{allow}}$$

$$\therefore S \geq \frac{M}{\sigma_{\text{allow}}} = \frac{(70.83 \text{ kip·ft})(12 \text{ in./ft})}{30 \text{ ksi}} = 28.33 \text{ in.}^3$$

(a) Select acceptable steel shapes: The properties of selected standard steel wide-flange shapes are presented in Appendix B. W shapes having a section modulus greater than or equal to 28.33 in.^{3} are acceptable for the beam and loading considered here. Since the cost of a steel beam is proportional to its weight, it is generally preferable to select the lightest acceptable shape for use.

Follow the procedure for selecting standard steel shapes presented in Table 8.1. By this process, the following shapes are identified as being acceptable for the beam and loading:

W8 × 40, $S = 35.5$ in.^{3}
W10 × 30, $S = 32.4$ in.^{3}
W12 × 26, $S = 33.4$ in.^{3}
W14 × 22, $S = 29.0$ in.^{3}
W16 × 31, $S = 47.2$ in.^{3}
W18 × 35, $S = 57.6$ in.^{3}

(b) Select the most economical W-shape: The most economical W shape can now be selected from the short list of acceptable shapes. From this list, a W14 × 22 standard steel wide-flange shape is identified as the lightest-weight section for this beam and loading. **Ans.**

PROBLEMS

P8.26 A small aluminum alloy [$E = 70$ GPa] tee shape is used as a simply supported beam as shown in Figure P8.26*a*. For this beam, $a = 160$ mm. The cross-sectional dimensions (Figure P8.26*b*) of the tee shape are $b = 20$ mm, $d = 30$ mm, and $t = 6$ mm. After loads are applied to the beam at *B*, *C*, and *D*, a compressive normal strain of 440 με is measured from a strain gage located at $c = 8$ mm below the topmost edge of the tee stem at section 1–1. What is the applied load *P*?

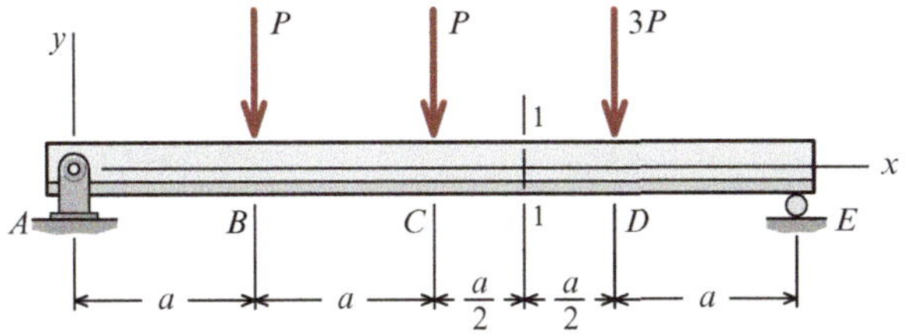

FIGURE P8.26*a*

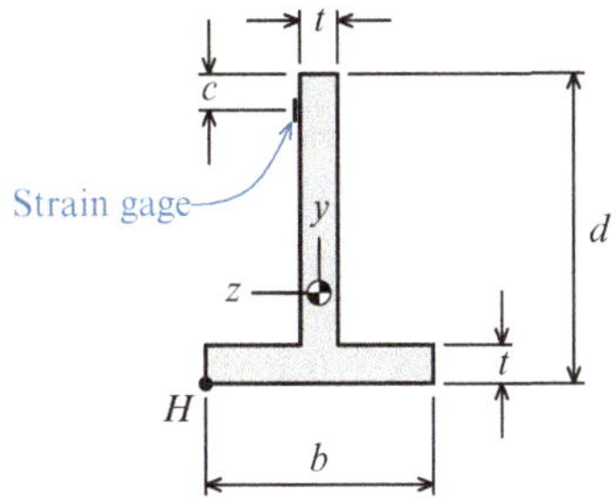

FIGURE P8.26*b*

P8.27 A solid steel shaft supports loads $P_A = 250$ N and $P_C = 620$ N as shown in Figure P8.27. Assume that $a = 500$ mm, $b = 700$ mm, and $c = 600$ mm. The bearing at *B* can be idealized as a roller support, and the bearing at *D* can be idealized as a pin support. If the allowable bending stress is 105 MPa, determine the minimum diameter that can be used for the shaft.

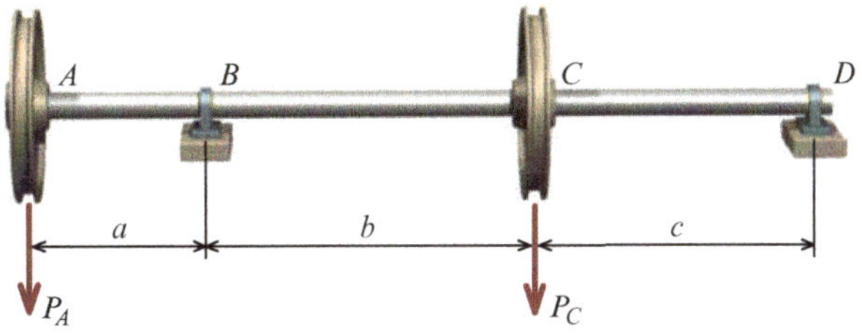

FIGURE P8.27

P8.28 A simply supported wooden beam (Figure P8.28*a*/29*a*) with a span of $L = 6$ m supports a uniformly distributed load w_0. The beam width is $b = 120$ mm and the beam height is $h = 300$ mm (Figure P8.28*b*/29*b*). The allowable bending stress of the wood is 8.0 MPa. Calculate the magnitude of the maximum load w_0 that may be carried by the beam.

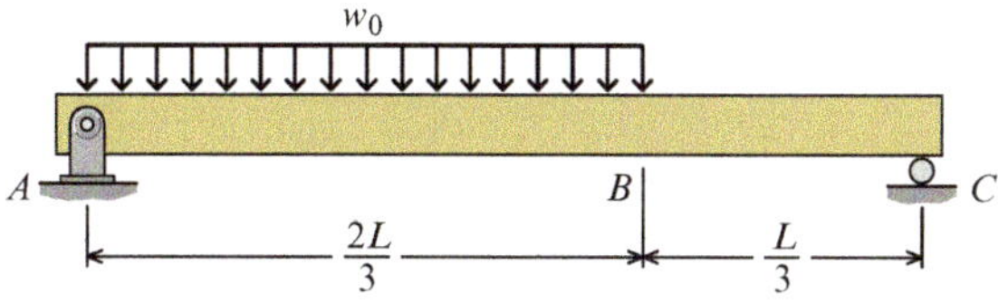

FIGURE P8.28*a*/29*a*

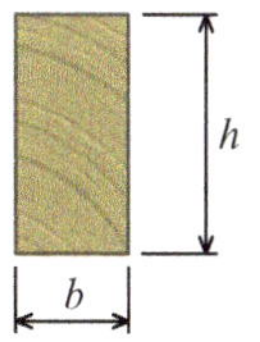

FIGURE P8.28*b*/29*b*

P8.29 A simply supported wooden beam (Figure P8.28*a*/29*a*) with a span of $L = 24$ ft supports a uniformly distributed load $w_0 = 450$ lb/ft. The allowable bending stress of the wood is 1,200 psi. If the aspect ratio of the solid rectangular wood beam is specified as $h/b = 3.0$ (Figure P8.28*b*/29*b*), calculate the minimum width *b* that can be used for the beam.

P8.30 A cantilever timber beam (Figure P8.30*a*/31*a*) with a span of $L = 4.25$ m supports a linearly distributed load with maximum intensity of $w_0 = 5.5$ kN/m. The allowable bending stress of the wood is 7.0 MPa. If the aspect ratio of the solid rectangular timber is specified as $h/b = 0.67$ (Figure P8.30*b*/31*b*), determine the minimum width *b* that can be used for the beam.

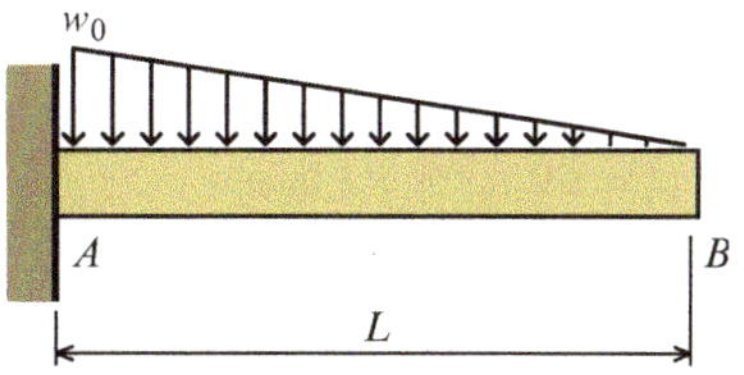

FIGURE P8.30*a*/31*a*

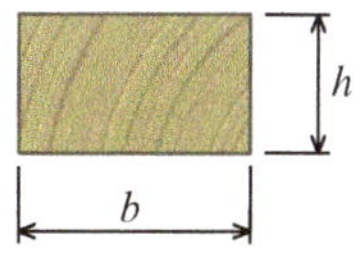

FIGURE P8.30*b*/31*b*

P8.31 A cantilever timber beam (Figure P8.30*a*/31*a*) with a span of $L = 14$ ft supports a linearly distributed load with maximum intensity of w_0. The beam width is $b = 15$ in. and the beam height is $h = 10$ in. (Figure P8.30*b*/31*b*). The allowable bending stress of the wood is 1,000 psi. Calculate the magnitude of the maximum load w_0 that may be carried by the beam.

P8.32 The lever shown in Figure P8.32 must exert a force of $P = 2{,}700$ lb at *B*, as shown. The lever lengths are $a = 10.5$ in. and $b = 46.0$ in. The allowable bending stress for the lever is 12,000 psi. If the height *h* of the lever is to be three times the thickness *t* (i.e., $h/t = 3$), what is the minimum thickness *t* that can be used for the lever?

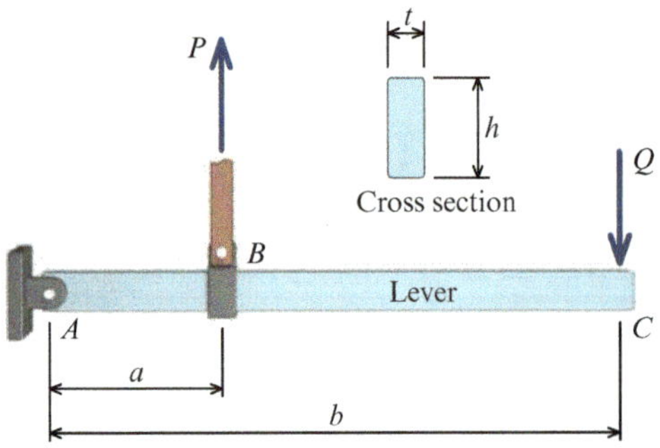

FIGURE P8.32

P8.33 The beam shown in Figure P8.33 will be constructed from a standard steel W shape using an allowable bending stress of 30 ksi. The beam span is $L = 24$ ft, and the beam loads are $w = 1.5$ kips/ft and $P = 20$ kips.

(a) Develop a list of five acceptable shapes that could be used for this beam. Include the most economical W14, W16, W18, W21, and W24 shapes on the list of possibilities.
(b) Select the most economical W shape for this beam.

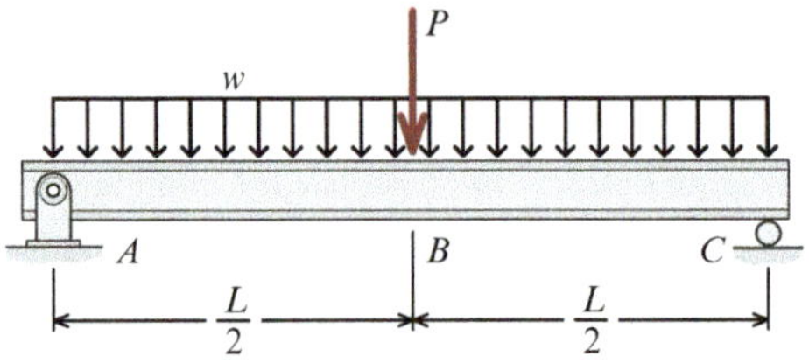

FIGURE P8.33

P8.34 The beam shown in Figure P8.34 will be constructed from a standard steel W shape using an allowable bending stress of 30 ksi. The beam span is $L = 24$ ft, and the beam loads are $w = 1{,}100$ lb/ft and $2w = 2{,}200$ lb/ft.

(a) Develop a list of four acceptable shapes that could be used for this beam. Include the most economical W10, W12, W14, and W16 shapes on the list of possibilities.
(b) Select the most economical W shape for this beam.

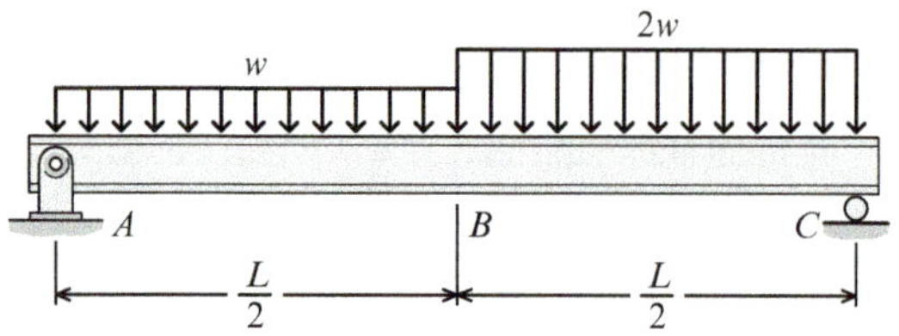

FIGURE P8.34

P8.35 The beam shown in Figure P8.35 will be constructed from a standard steel W shape using an allowable bending stress of 165 MPa.

(a) Develop a list of four acceptable shapes that could be used for this beam. Include the most economical W310, W360, W410, and W460 shapes on the list of possibilities.
(b) Select the most economical W shape for this beam.

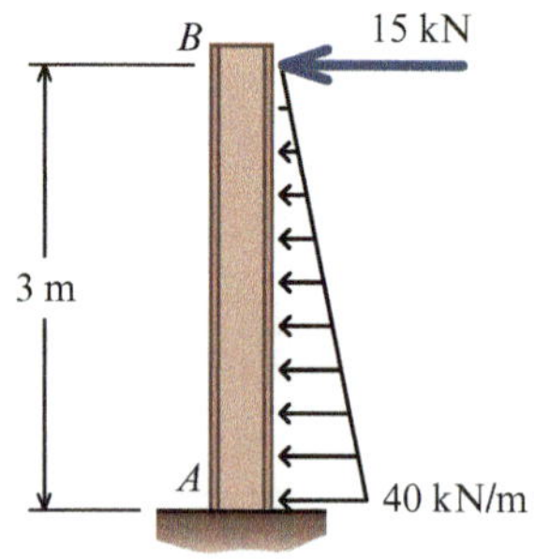

FIGURE P8.35

P8.36 The beam shown in Figure P8.36 will be constructed from a standard steel HSS shape using an allowable bending stress of 30 ksi.

(a) Develop a list of three acceptable shapes that could be used for this beam. On this list, include the most economical HSS8, HSS10, and HSS12 shapes.
(b) Select the most economical HSS shape for this beam.

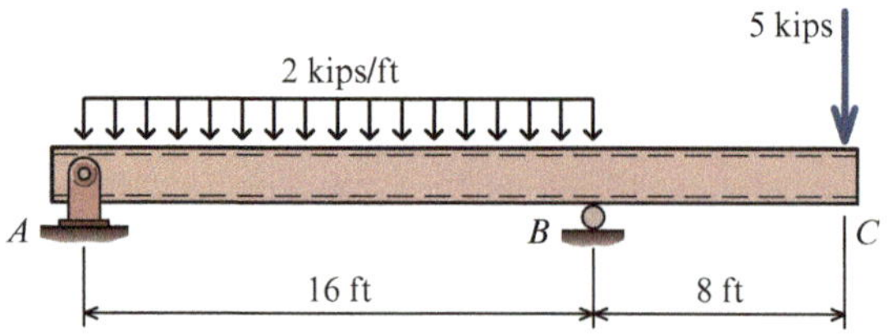

FIGURE P8.36

8.6 Flexural Stresses in Beams of Two Materials

Many structural applications involve beams made of two materials. These types of beams are called **composite beams**. Examples include wooden beams reinforced with steel plates attached to the top and bottom surfaces, and reinforced concrete beams in which steel reinforcing bars are embedded to resist tensile stresses. Engineers purposely design beams in this manner so that advantages offered by each material can be efficiently utilized.

The flexure formula was derived for homogeneous beams—that is, beams consisting of a single, uniform material characterized by an elastic modulus E. As a result, the flexure formula cannot be used to determine the normal stresses in composite beams without some additional modifications. In this section, a computational method will be developed so that a beam cross section that consists of two different materials can be "transformed" into an *equivalent* cross section consisting of a single material. The flexure formula can then be used to evaluate bending stresses in this equivalent homogeneous beam.

Equivalent Beams

Before considering a beam made of two materials, let us first examine what is required so that two beams of different materials can be considered *equivalent*. Suppose that a small rectangular aluminum bar having an elastic modulus $E_{\text{alum}} = 70$ GPa is used as a beam in pure bending (Figure 8.10*a*). The bar is subjected to an internal bending moment $M = 140{,}000$ N·mm, which causes the bar to bend about the z axis. The width of the bar is 15 mm, and the height of the bar is 40 mm (Figure 8.10*b*); therefore, its moment of inertia about the z axis is $I_{\text{alum}} = 80{,}000$ mm^4. The radius of curvature ρ of this beam can be computed from Equation (8.6):

$$\frac{1}{\rho} = \frac{M}{EI_{\text{alum}}} = \frac{140{,}000 \text{ N}\cdot\text{mm}}{(70{,}000 \text{ N/mm}^2)(80{,}000 \text{ mm}^4)}$$

$$\therefore \rho = 40{,}000 \text{ mm}$$

The maximum bending strain caused by the bending moment can be determined from Equation (8.1):

$$\varepsilon_x = -\frac{1}{\rho}y = -\frac{1}{40{,}000 \text{ mm}}(\pm 20 \text{ mm}) = \pm 0.0005 \text{ mm/mm}$$

Next, suppose that we want to replace the aluminum bar with wood, which has an elastic modulus $E_{\text{wood}} = 10$ GPa. In addition, we require that the wooden beam be equivalent to the aluminum bar. The question, then, becomes, "What dimensions are required in order for the wooden beam to be equivalent to the aluminum bar?"

But what is meant by "equivalent" in this context? To be equivalent, the wooden beam must have the same radius of curvature ρ and the same distribution of bending strains ε_x as the aluminum bar for the given internal bending moment M. To produce the same ρ for the 140 N·m bending moment, the moment of inertia of the wooden beam must be increased to

$$I_{\text{wood}} = \frac{M}{E}\rho = \frac{140{,}000 \text{ N}\cdot\text{mm}}{10{,}000 \text{ N/mm}^2}(40{,}000 \text{ mm}) = 560{,}000 \text{ mm}^4$$

Thus, the wooden beam must be larger than the aluminum bar in order to have the same radius of curvature. However, equivalence also requires that the wood beam exhibit the same distribution of strains. So, since strains are directly proportional to y, *the wooden beam must have the same y coordinates as the aluminum bar*; in other words, the height of the wooden beam must also be 40 mm.

The moment of inertia of the wooden beam must be larger than that of the aluminum bar, *but the height of both must be the same.* Therefore, the wooden beam must be wider than the aluminum bar if the two beams are to be equivalent:

$$I_{\text{wood}} = \frac{bh^3}{12} = \frac{b_{\text{wood}}(40 \text{ mm})^3}{12} = 560{,}000 \text{ mm}^4$$

$$\therefore b_{\text{wood}} = 105 \text{ mm}$$

In this example, a wooden beam that is 105 mm wide and 40 mm tall is equivalent to an aluminum bar that is 15 mm wide and 40 mm tall (Figure 8.10*c*). Since the elastic moduli of the two materials are different (by a factor of 7 in this case), the wooden beam (which has the lesser E) must be wider than the aluminum bar (which has the greater E)—wider in this case by a factor of 7.

Now, if the two beams are equivalent, are the bending stresses the same? The bending stress produced in the aluminum bar can be calculated from the flexure formula:

$$\sigma_{\text{alum}} = \frac{(140{,}000 \text{ N}\cdot\text{mm})(20 \text{ mm})}{80{,}000 \text{ mm}^4} = 35 \text{ MPa}$$

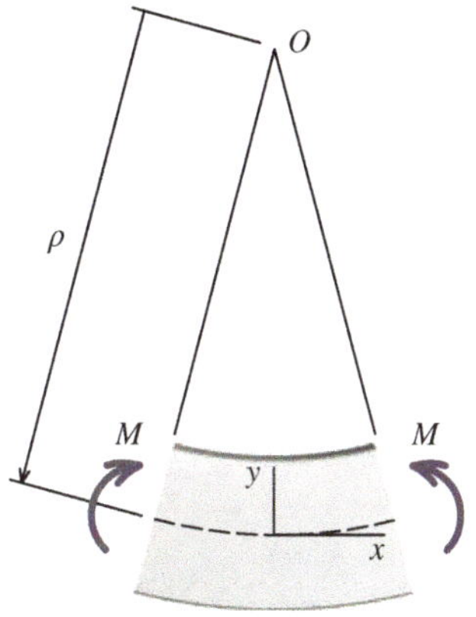

(*a*) Bar subjected to pure bending

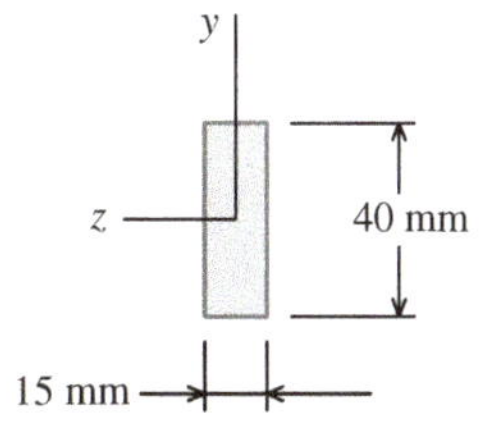

(*b*) Cross-sectional dimensions of aluminum bar

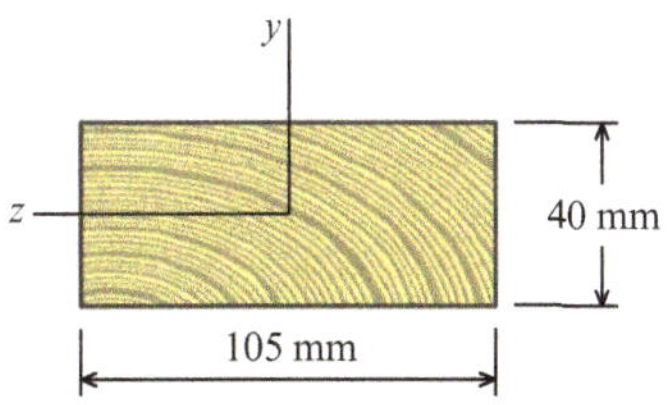

(*c*) Cross-sectional dimensions of equivalent wooden beam

FIGURE 8.10 Equivalent beams of aluminum and wood.

Similarly, the bending stress in the wooden beam is

$$\sigma_{\text{wood}} = \frac{(140{,}000\ \text{N}\cdot\text{mm})(20\ \text{mm})}{560{,}000\ \text{mm}^4} = 5\ \text{MPa}$$

Thus, the bending stress in the wood is one-seventh of the stress in the aluminum; therefore, equivalent beams do not necessarily have the same bending stresses, only the same ρ and ε.

In this example, the elastic moduli, the beam widths, and the bending stresses all differ by a factor of 7. Compare the moment–curvature relationships for the two beams:

$$\frac{1}{\rho} = \frac{M}{E_{\text{alum}} I_{\text{alum}}} = \frac{M}{E_{\text{wood}} I_{\text{wood}}}$$

Expressing the moments of inertia in terms of the respective beam widths b_{alum} and b_{wood} and the common beam height h gives

$$\frac{M}{E_{\text{alum}}\left(\dfrac{b_{\text{alum}} h^3}{12}\right)} = \frac{M}{E_{\text{wood}}\left(\dfrac{b_{\text{wood}} h^3}{12}\right)}$$

which can be simplified to

$$\frac{b_{\text{wood}}}{b_{\text{alum}}} = \frac{E_{\text{alum}}}{E_{\text{wood}}}$$

The ratio of the elastic moduli will be termed the **modular ratio** and denoted by the symbol n. For the two materials considered here, the modular ratio has the value

$$n = \frac{E_{\text{alum}}}{E_{\text{wood}}} = \frac{70\ \text{GPa}}{10\ \text{GPa}} = 7$$

Hence, the factor of 7 that appears throughout this example stems from the modular ratio for the two materials. The required width of the wooden beam can be expressed in terms of the modular ratio n as

$$\frac{b_{\text{wood}}}{b_{\text{alum}}} = \frac{E_{\text{alum}}}{E_{\text{wood}}} = n \qquad \therefore\ b_{\text{wood}} = n b_{\text{alum}} = 7(15\ \text{mm}) = 105\ \text{mm}$$

As mentioned, the bending stresses from the two beams also differ by a factor of 7. Consequently, since the aluminum and wooden beams are equivalent, the bending strains are the same for the two beams:

$$(\varepsilon_x)_{\text{alum}} = (\varepsilon_x)_{\text{wood}}$$

Stress is related to strain by Hooke's law; therefore, the bending strains can be expressed as

$$(\varepsilon_x)_{\text{alum}} = \left(\frac{\sigma}{E}\right)_{\text{alum}} \qquad \text{and} \qquad (\varepsilon_x)_{\text{wood}} = \left(\frac{\sigma}{E}\right)_{\text{wood}}$$

The relationship between the bending stresses of the two materials can now be expressed in terms of the modular ratio n:

$$\frac{\sigma_{\text{alum}}}{E_{\text{alum}}} = \frac{\sigma_{\text{wood}}}{E_{\text{wood}}} \qquad \text{or} \qquad \frac{\sigma_{\text{alum}}}{\sigma_{\text{wood}}} = \frac{E_{\text{alum}}}{E_{\text{wood}}} = n$$

Once again, the ratio of the bending stresses differs by an amount equal to the modular ratio n.

To summarize, a beam made of one material is transformed into an equivalent beam of a different material by modifying the beam width (*and only the beam width*). The ratio between the elastic moduli of the two materials (termed the modular ratio) dictates the change in width required for equivalence. Bending stresses are not equal for equivalent beams; rather, they, too, differ by a factor equal to the *modular ratio*.

Transformed-Section Method

The concepts introduced in the preceding example can be used to develop a method for analyzing beams made up of two materials. The basic idea is to transform a cross section that consists of two different materials into an equivalent cross section of only one material. Once this transformation is completed, techniques developed previously for flexure of homogeneous beams can be used to determine the bending stresses.

Consider a beam cross section that is made up of two linear elastic materials (designated Material 1 and Material 2) perfectly bonded together (Figure 8.11*a*). This composite beam will bend as described in Section 8.2. If a bending moment is applied to such a beam, then, as with a homogeneous beam, the total cross-sectional area will remain plane after bending. It then follows that the normal strains will vary linearly with the y coordinate measured from the neutral surface and that Equation (8.1) is valid:

$$\varepsilon_x = -\frac{1}{\rho}y$$

In this situation, however, the neutral surface cannot be assumed to pass through the centroid of the composite area.

We next wish to transform Material 2 into an equivalent amount of Material 1 and, in so doing, define a new cross section made entirely of Material 1. In order for this transformed cross section to be valid for calculation purposes, it must be equivalent to the actual cross section (which consists of Material 1 and Material 2), meaning that the strains and curvature of the transformed section must be the same as the strains and curvature of the actual cross section.

In this procedure, Material 1 can be thought of as a "common currency" for the transformation. All areas are converted to their equivalents in the common currency.

How much area of Material 1 is equivalent to an area dA of Material 2? Consider a cross section consisting of two materials such that Material 2 is stiffer than Material 1; in other words, $E_2 > E_1$ (Figure 8.11*b*). We will investigate the force transmitted by an area element dA_2 of Material 2. Element dA has width dz and height dy. The force dF transmitted by this element of area is given by $dF = \sigma_x\, dz\, dy$. From Hooke's law, the stress σ_x can be expressed as the product of the elastic modulus and the strain; therefore,

$$dF = (E_2\varepsilon)dz\,dy$$

Since Material 2 is stiffer than Material 1, more area of Material 1 will be required to transmit a force equal to dF. Now, the distribution of strain in the transformed section must be the same as that in the actual cross section. For that reason, the y dimensions (i.e., the dimensions perpendicular to the neutral axis) in the transformed section must be the same

Suppose that Material 2 was a "hard" material like steel and Material 1 was a "soft" material like rubber. If the strains in both the rubber and the steel were the same, then a much greater area of rubber would be required to transmit the same force that could be transmitted by a small area of steel.

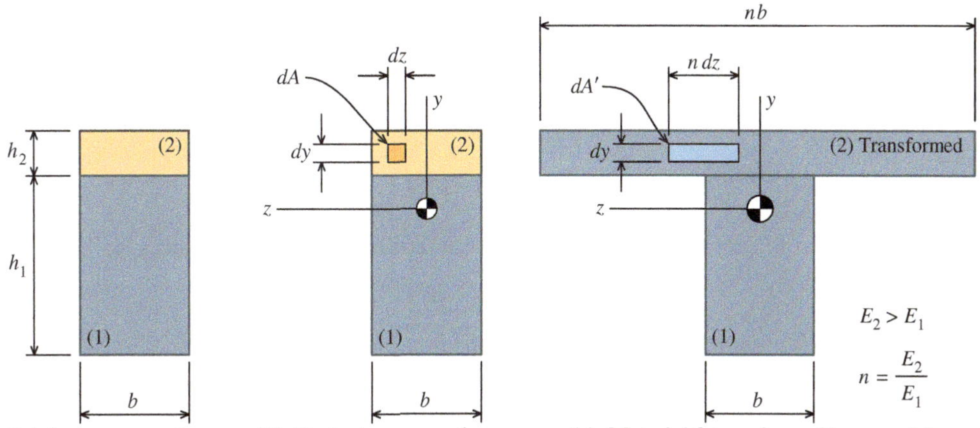

(*a*) Original cross section (*b*) Orginal cross section with $dA = dy\,dz$ (*c*) Material 2 transformed by use of the modular ratio n

FIGURE 8.11 Beam with two materials: basic geometry and transformed geometry of the cross section.

as those in the actual cross section. The width dimension (i.e., the dimension parallel to the neutral axis), however, can be modified. Let the equivalent area dA' of Material 1 be given by the height dy and a modified width $n\ dz$, where n is a factor to be determined (Figure 8.11*c*). Then the force transmitted by this area of Material 1 can be expressed as

$$dF' = (E_1\varepsilon)(n\,dz)dy$$

If the transformed section is to be equivalent to the actual cross section, then the forces dF' and dF must be equal:

$$(E_1\varepsilon)(n\,dz)dy = (E_2\varepsilon)dz\,dy$$

Therefore,

$$n = \frac{E_2}{E_1} \tag{8.12}$$

The ratio n is called the **modular ratio**.

This analysis shows that the actual cross section consisting of two materials can be transformed by use of the modular ratio into an equivalent cross section consisting of a single material. The actual cross section is transformed in the following manner: The area of Material 1 is unmodified, meaning that its original dimensions remain unchanged. The area of Material 2 is transformed into an equivalent area of Material 1 by multiplication of the actual *width* (i.e., the dimension that is parallel to the neutral axis) by the modular ratio n. The *height* of Material 2 (i.e., the dimension perpendicular to the neutral axis) is kept the same. This procedure produces a **transformed section**, made entirely of Material 1, that transmits the same force (for any given strain ε) as the actual cross section, which is composed of two materials, transmits.

Does the transformed section have the same neutral axis as the actual cross section? If the transformed cross section is equivalent to the actual cross section, then it must produce the same strain distribution. Therefore, it is essential that both cross sections have the same neutral axis location. For a homogeneous beam, the neutral axis was determined by summing forces in the x direction in Equation (8.4). Application of this same procedure to a beam made up of two materials gives

$$\Sigma F_x = \int_{A_1} \sigma_{x1}\,dA + \int_{A_2} \sigma_{x2}\,dA = 0$$

in which σ_{x1} is the stress in Material 1 and σ_{x2} is the stress in Material 2. In this equation, the first integral is evaluated over the cross-sectional area of Material 1 and the second integral is evaluated over the cross-sectional area of Material 2. From Equation (8.3), the normal stresses at y (measured from the neutral axis) for the two materials can be expressed in terms of the radius of curvature ρ as

$$\sigma_{x1} = -\frac{E_1}{\rho}y \quad \text{and} \quad \sigma_{x2} = -\frac{E_2}{\rho}y \tag{8.13}$$

Substituting these expressions for σ_{x1} and σ_{x2} gives

$$\Sigma F_x = -\int_{A_1} \frac{E_1}{\rho}y\,dA - \int_{A_2} \frac{E_2}{\rho}y\,dA = 0$$

The radius of curvature can be cancelled out so that this equation reduces to

$$E_1\int_{A_1} y\,dA + E_2\int_{A_2} y\,dA = 0$$

In this equation, the integrals represent the first moments of the two portions of the cross section with respect to the neutral axis. At this point, the modular ratio will be introduced so that the previous equation can be rewritten in terms of n:

$$E_1\int_{A_1} y\,dA + E_1\int_{A_2} yn\,dA = 0$$

This equation reduces to

$$\int_{A_1} y\,dA + \int_{A_2} yn\,dA = 0 \tag{8.14}$$

The area of the transformed cross section can be expressed as

$$\int_{A_1} dA + \int_{A_2} n\,dA = \int_{A_t} dA_t$$

so Equation (8.14) can be rewritten simply as

$$\int_{A_t} y\,dA_t = 0 \tag{8.15}$$

Therefore, the *neutral axis passes through the centroid of the transformed section,* just as it passes through the centroid of a homogeneous beam.

Does the transformed section have the same moment–curvature relationship as the actual cross section? From the relationships of Equation (8.13), the moment–curvature relationship for a beam of two materials is

$$\begin{aligned} M &= -\int_A y\sigma_x\,dA \\ &= -\int_{A_1} y\sigma_x\,dA - \int_{A_2} y\sigma_x\,dA \\ &= \frac{1}{\rho}\left[\int_{A_1} E_1 y^2\,dA + \int_{A_2} E_2 y^2\,dA\right] \end{aligned}$$

By the modular ratio, the elastic modulus of Material 2 can be expressed as $E_2 = nE_1$, reducing the preceding equation to

$$M = \frac{E_1}{\rho}\left[\int_{A_1} y^2\,dA + \int_{A_2} y^2 n\,dA\right]$$

The term in brackets is just the moment of inertia, I_t, of the transformed section about its neutral axis (which was previously shown to pass through the centroid). Therefore, the moment–curvature relationship can be written as

$$M = \frac{E_1 I_t}{\rho} \quad \text{where} \quad I_t = \int_{A_t} y^2\,dA_t \tag{8.16}$$

In other words, the moment–curvature relationship of the transformed cross section is equal to that of the actual cross section.

How are bending stresses calculated for each of the two materials, according to the transformed-section method? Equation (8.16) can be expressed as

$$\frac{1}{\rho} = \frac{M}{E_1 I_t}$$

and substituted into the stress relationships of Equation (8.13). Substituting into the first equation gives the bending stress at those locations corresponding to Material 1 in the actual cross section:

$$\sigma_{x1} = -\frac{E_1}{\rho}y = -\left(\frac{M}{E_1 I_t}\right)E_1 y = -\frac{My}{I_t} \tag{8.17}$$

Notice that the bending stress in Material 1 is calculated from the flexure formula. Recall that the actual area of Material 1 was not modified in developing the transformed section.

Substituting Equation (8.16) into the second equation of Equation (8.13) gives the bending stress at those locations corresponding to Material 2 in the actual cross section:

$$\sigma_{x2} = -\frac{E_2}{\rho}y = -\left(\frac{M}{E_1 I_t}\right)E_2 y = -\frac{E_2}{E_1}\frac{My}{I_t} = -n\frac{My}{I_t} \tag{8.18}$$

When the transformed-section method is used to calculate bending stresses at locations corresponding to Material 2 (i.e., the transformed material) in the actual cross section, the flexure formula must be multiplied by the modular ratio n.

For a cross section consisting of two materials (Figure 8.12*a*), the strains caused by a bending moment are distributed linearly over the depth of the cross section (Figure 8.12*b*), just as they are for a homogeneous beam. The corresponding normal stresses are also distributed linearly; however, there is a discontinuity at the intersection of the two materials (Figure 8.12*c*)—a consequence of the differing elastic moduli of the materials. To account for this discontinuity, in the transformed-section method the normal stresses for the material that was transformed (Material 2 in this instance) are calculated by multiplying the flexure formula by the modular ratio n.

To recapitulate, the procedure for calculating bending stresses by the transformed-section method depends upon whether or not the area of the material was transformed:

- If the area was not transformed, then simply calculate the associated bending stresses from the flexure formula.
- If the area was transformed, then multiply the flexure formula by n when calculating the associated bending stresses.

In this discussion, the actual cross section of the beam was transformed into an equivalent cross section consisting entirely of Material 1. It is also permissible to transform the cross section to one consisting entirely of Material 2. In that case, the modular ratio is defined as $n = E_1/E_2$. The bending stresses in Material 2 of the actual cross section will then be the same as the bending stresses in the corresponding portion of the transformed cross section. The bending stresses at those locations corresponding to Material 1 in the actual cross section will be obtained by multiplying the flexure formula by $n = E_1/E_2$.

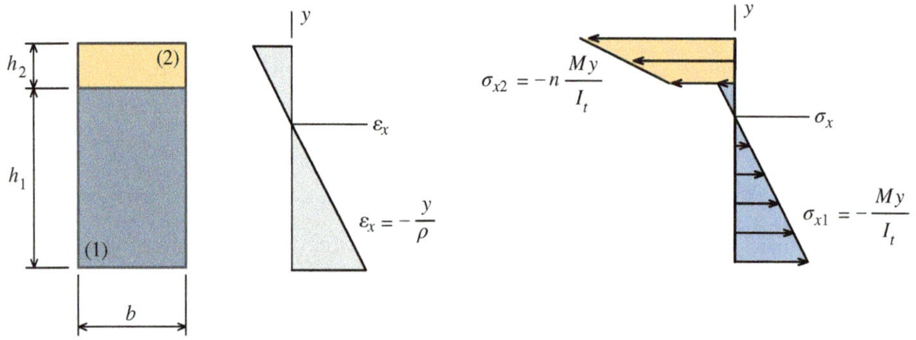

(*a*) Original cross section (*b*) Distribution of normal strains (*c*) Distribution of normal stresses

FIGURE 8.12 Beam with two materials: strain and stress distributions.

EXAMPLE 8.7

A cantilever beam 10 ft long carries a uniformly distributed load $w = 100$ lb/ft. The beam is constructed from 3 in. wide by 8 in. deep wood timber (1) that is reinforced on its upper surface by a 3 in. wide by 0.25 in. thick aluminum plate (2). The elastic modulus of the wood is $E = 1{,}700$ ksi, and the elastic modulus of the aluminum plate is $E = 10{,}200$ ksi. Determine the maximum bending stresses produced in the timber (1) and the aluminum plate (2).

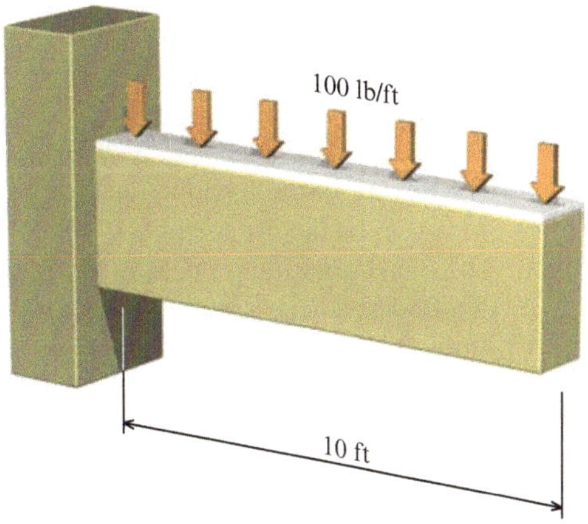

Cantilever beam with w = 100 lb/ft.

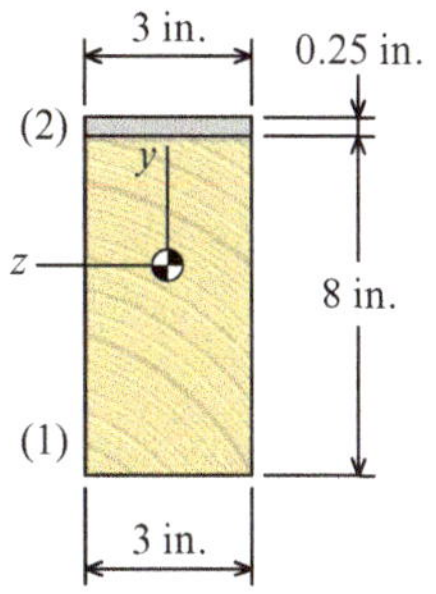

Cross-sectional dimensions.

Plan the Solution

The transformed-section method will be used to transform the cross section consisting of two materials into an equivalent cross section consisting of a single material. This transformed section will be used for calculation purposes. The centroid location and the moment of inertia of the transformed section about its centroid will be calculated. With these section properties, the flexure formula will be used to compute the bending stresses in both the wood and the aluminum for the maximum internal bending moment produced in the cantilever span.

SOLUTION

Modular Ratio

The transformation procedure is based on the ratio of the elastic moduli of the two materials, termed the *modular ratio* and denoted by n. The modular ratio is defined as the elastic modulus of the *transformed material* divided by the elastic modulus of the *reference material.* In this example, the stiffer material (i.e., the aluminum) will be transformed into an equivalent amount of the less stiff material (i.e., the wood); therefore, the wood will be used as the reference material. The modular ratio for this transformation is

$$n = \frac{E_{\text{trans}}}{E_{\text{ref}}} = \frac{E_2}{E_1}$$

$$= \frac{10{,}200 \text{ ksi}}{1{,}700 \text{ ksi}} = 6$$

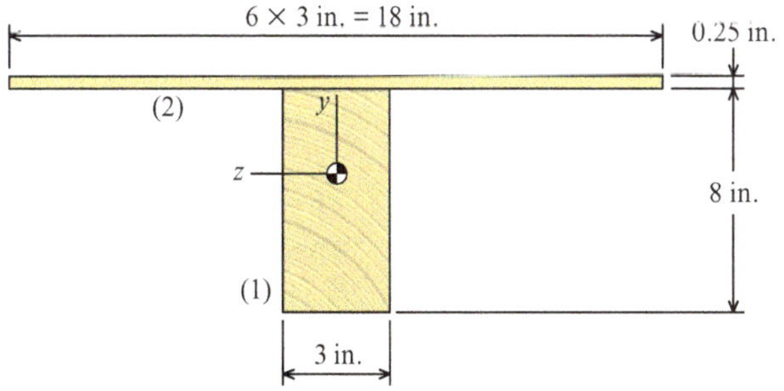

Transformed cross section.

The *width* of the aluminum portion of the cross section is multiplied by the modular ratio n. The resulting cross section, consisting solely of wood, is equivalent to the actual cross section, which consists of both wood and aluminum.

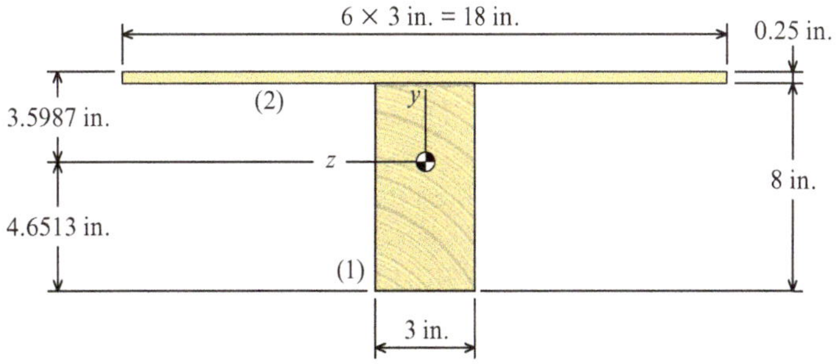

Section Properties

The centroid location for the transformed section is shown in the figure on the left. The moment of inertia of the transformed section about the z centroidal axis is $I_t = 192.5$ in.^{4}.

Maximum Bending Moment

The maximum bending moment for a 10 ft long cantilever beam with a uniformly distributed load $w = 100$ lb/ft is

$$M_{max} = -\frac{wL^2}{2} = -\frac{(100 \text{ lb/ft})(10 \text{ ft})^2}{2} = -5{,}000 \text{ lb}\cdot\text{ft} = -60{,}000 \text{ lb}\cdot\text{in.}$$

Flexure Formula

The flexure formula [Equation (8.7)] gives the bending stress at any coordinate location y; however, the flexure formula is valid only if the beam consists of a homogeneous material. The transformation process used to replace the aluminum plate with an equivalent amount of wood was necessary to obtain a homogeneous cross section that satisfies the limitations of the flexure formula.

The transformed section consisting entirely of wood is *equivalent* to the actual cross section. The transformed section is equivalent because the bending *strains* produced in the transformed section are identical to the strains produced in the actual cross section. The bending *stresses* in the transformed section, however, require an additional adjustment. The bending stresses computed for the original wood portion of the cross section [i.e., area (1)] are correctly computed from the flexure formula. The bending stresses computed for the aluminum plate must be multiplied by the modular ratio n to account for the difference in elastic moduli of the two materials.

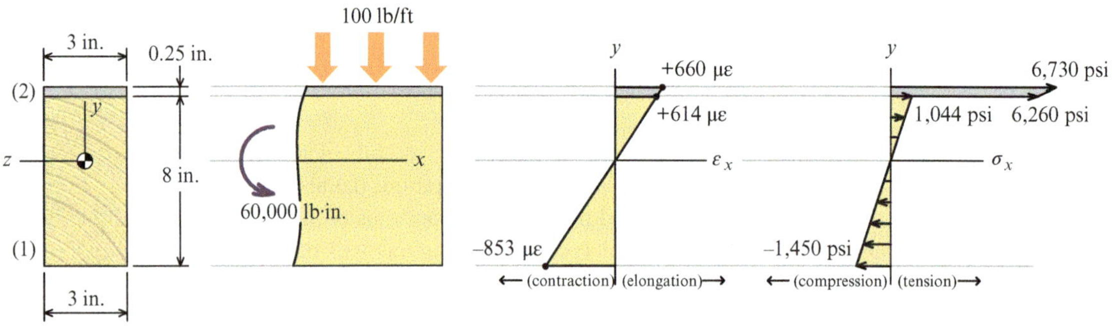

Beam cross section. **Profile view of beam.** **Bending strains.** **Bending stresses.**

Maximum Bending Stresses in the Wood

The maximum bending stress in the wood portion (1) of the cross section occurs at the lower surface of the beam. Since the wood was not transformed, Equation (8.17) is used to compute the maximum bending stress:

$$\sigma_{x1} = -\frac{My}{I_t} = -\frac{(-60{,}000 \text{ lb}\cdot\text{in.})(-4.6513 \text{ in.})}{192.5 \text{ in.}^4} = -1{,}450 \text{ psi} = 1{,}450 \text{ psi (C)} \quad \textbf{Ans.}$$

Maximum Bending Stresses in the Aluminum

The aluminum portion of the cross section was transformed in the analysis to an equivalent width of wood. While the bending *strains* for the transformed section are correct, the bending *stress* for the transformed material must be multiplied by the modular ratio n to account for the differing elastic moduli of the two materials. The maximum bending stress in the aluminum portion (2) of the cross section (a stress that occurs at the upper surface of the beam) is computed from Equation (8.18):

$$\sigma_{x2} = -n\frac{My}{I_t} = -6\frac{(-60{,}000 \text{ lb}\cdot\text{in.})(3.5987 \text{ in.})}{192.5 \text{ in.}^4} = 6{,}730 \text{ psi} = 6{,}730 \text{ psi (T)} \quad \textbf{Ans.}$$

Bending Stresses at the Intersection of the Two Materials

The joint between the timber (1) and the aluminum plate (2) occurs at $y = 3.3487$ in. At this location, the bending strain in both materials is identical: $\varepsilon_x = 614\ \mu\varepsilon$. Since the elastic modulus of the aluminum is six times greater than the elastic modulus of the wood, the bending stress in the aluminum, calculated as

$$\sigma_{x2} = -n\frac{My}{I_t} = -6\frac{(-60{,}000\text{ lb}\cdot\text{in.})(3.3487\text{ in.})}{192.5\text{ in.}^4} = 6{,}263\text{ psi} = 6{,}263\text{ psi (T)}$$

is six times greater than the bending stress in the wood:

$$\sigma_{x1} = -\frac{My}{I_t} = -\frac{(-60{,}000\text{ lb}\cdot\text{in.})(3.3487\text{ in.})}{192.5\text{ in.}^4} = 1{,}044\text{ psi} = 1{,}044\text{ psi (T)}$$

This result can also be seen by applying Hooke's law to each material. For a normal strain $\varepsilon_x = 614\ \mu\varepsilon$, the normal stress in wood timber (1), from Hooke's law, is

$$\sigma_{x1} = E_1\varepsilon_x = (1{,}700{,}000\text{ psi})(614\times10^{-6}\text{ in./in.}) = 1{,}044\text{ psi} = 1{,}044\text{ psi (T)}$$

and the normal stress in aluminum plate (2) is

$$\sigma_{x2} = E_2\varepsilon_x = (10{,}200{,}000\text{ psi})(614\times10^{-6}\text{ in./in.}) = 6{,}263\text{ psi} = 6{,}263\text{ psi (T)}$$

MecMovies

EXAMPLES

M8.16 Determine the bending stresses in a composite beam, using the transformed-section method.

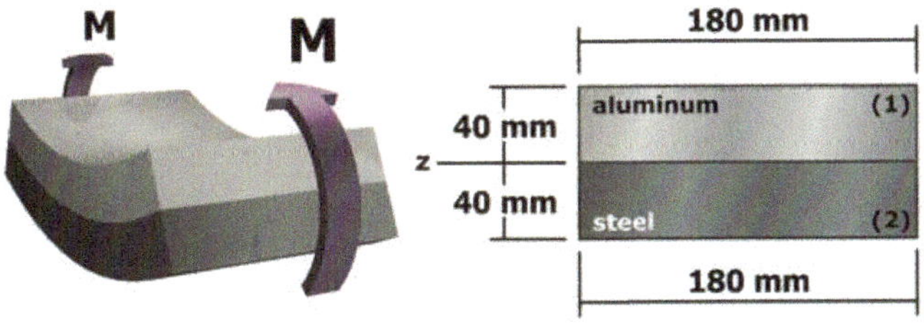

M8.17 Given allowable stresses for the aluminum and brass materials, determine the largest allowable moment that can be applied about the z axis to the beam cross section.

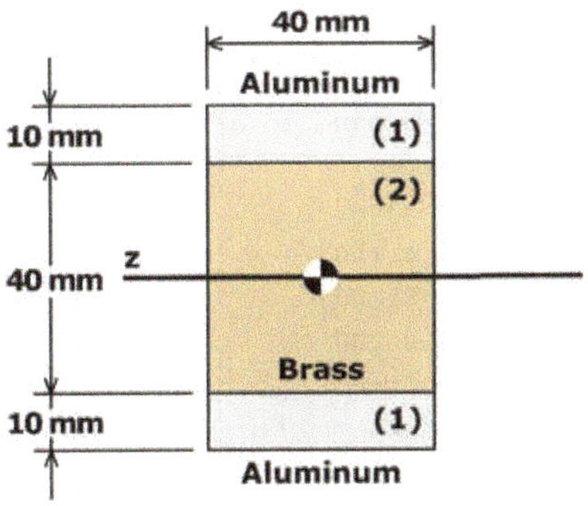

M8.18 Given allowable stresses for two materials, determine the largest allowable moment that can be applied about the horizontal axis of the beam cross section shown.

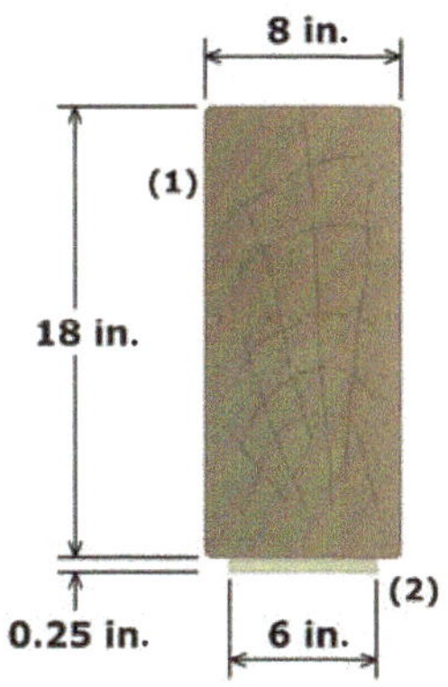

M8.19 Given allowable stresses for wood and steel materials, determine the largest allowable moment and, in turn, the maximum distributed load that can be applied to a simply supported beam.

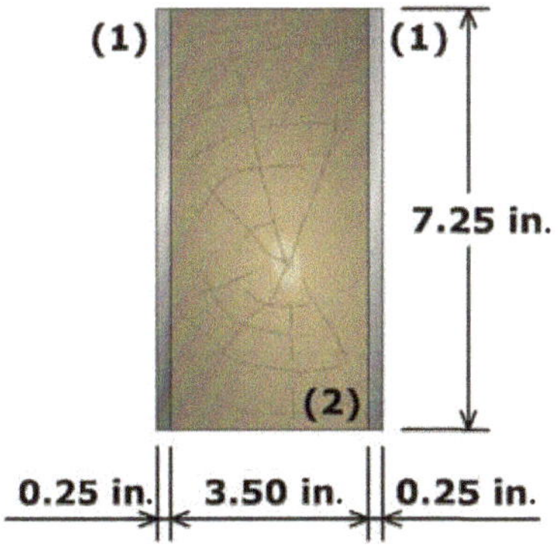

EXERCISES

M8.16 A composite-beam cross section consists of two rectangular bars securely bonded together. The beam is subjected to a specified bending moment M. Determine

(a) the vertical distance from K to the centroidal axis.
(b) the bending stress produced at H.
(c) the bending stress produced at K.

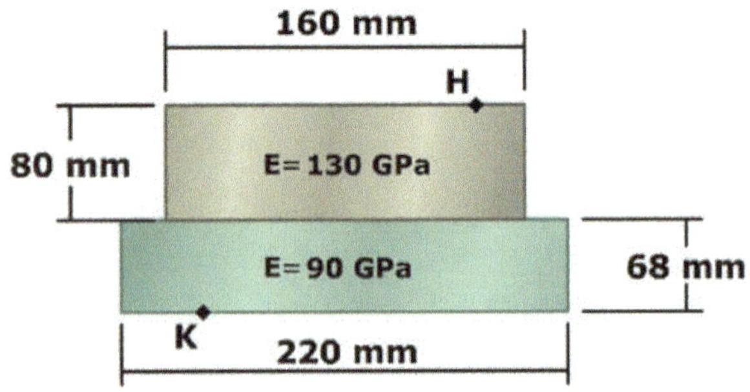

FIGURE M8.16

M8.17 A composite-beam cross section consists of two rectangular bars securely bonded together. From the indicated allowable stresses, determine

(a) the vertical distance from K to the centroidal axis.
(b) the maximum allowable bending moment M.
(c) the bending stress produced at H.
(d) the bending stress produced at K.

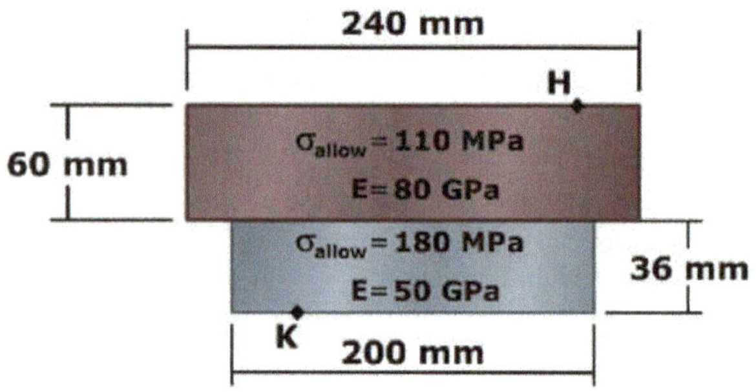

FIGURE M8.17

PROBLEMS

P8.37 A simply supported composite beam spans a distance $L = 24$ ft and carries a uniformly distributed load w, as shown in Figure P8.37*a*. The beam is fabricated by bolting two wooden boards ($b = 3$ in. and $d = 12$ in.) to the sides of a steel plate ($t = 0.5$ in. and $d = 12$ in.), as shown in Figure P8.37*b*. The moduli of elasticity of the wood and the steel are 1,800 ksi and 30,000 ksi, respectively. Neglect the weight of the beam in your calculations, and

(a) determine the maximum bending stresses produced in the wooden boards and the steel plate if $w = 400$ lb/ft.
(b) determine the largest acceptable magnitude for distributed load w. (Assume that the allowable bending stresses of the wood and the steel are 1,200 psi and 24,000 psi, respectively.)

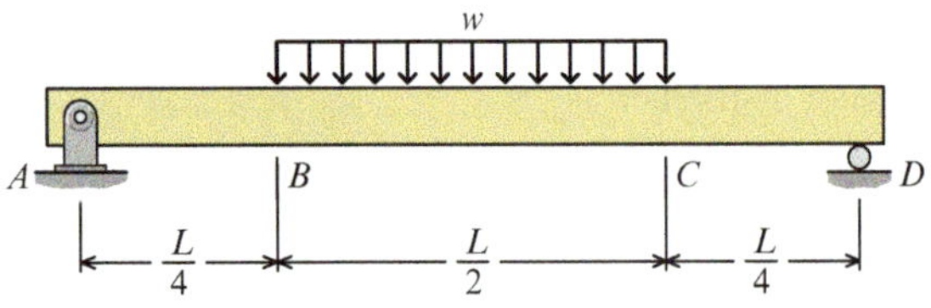

FIGURE P8.37*a*

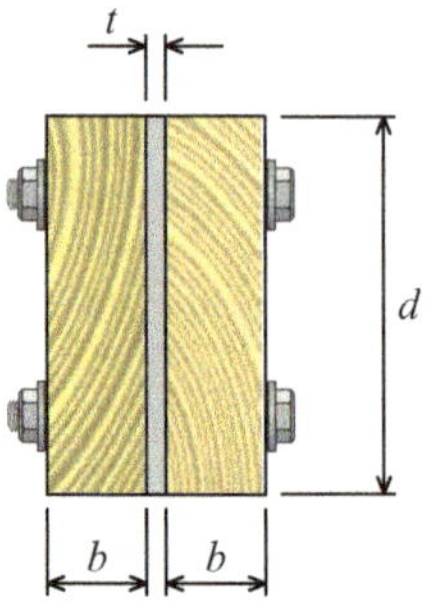

FIGURE P8.37*b*

P8.38 A flitch beam can be fabricated by sandwiching two steel plates (2) between three wooden boards (1), as shown in Figure P8.38. The dimensions of the cross section are $b = 1.5$ in., $d = 7.25$ in., and $t = 0.5$ in. The elastic moduli of the wood and steel are $E_1 = 1{,}800$ ksi and $E_2 = 30{,}000$ ksi, respectively. For a bending moment $M_z = +12$ kip · ft, determine

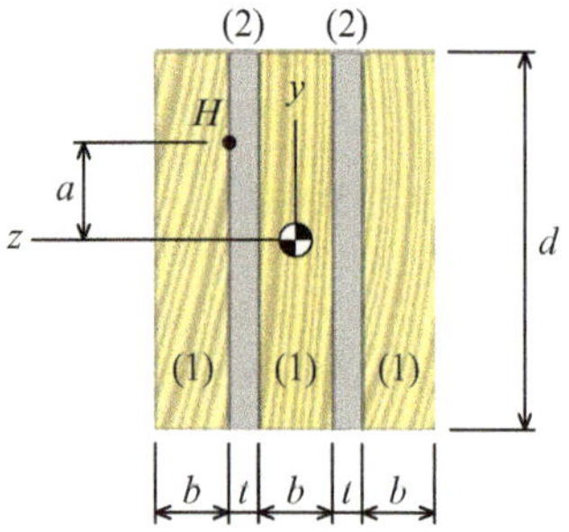

FIGURE P8.38

(a) the normal stress in each material at point H, which is located a distance $a = 1.75$ in. above the z centroidal axis.
(b) the normal strain in each material at point H.
(c) the maximum bending stress in each material.

P8.39 A composite beam consists of a bronze [$E = 105$ GPa] bar (2) attached rigidly to an aluminum alloy [$E = 70$ GPa] bar (1) as shown in Figure P8.39. The dimensions of the cross section are $b_1 = 60$ mm, $b_2 = 25$ mm, and $d = 40$ mm. The allowable stress of the aluminum alloy is 165 MPa, and the allowable stress of the bronze is 210 MPa. What is the magnitude of the allowable bending moment M_z that may be applied to the composite cross section?

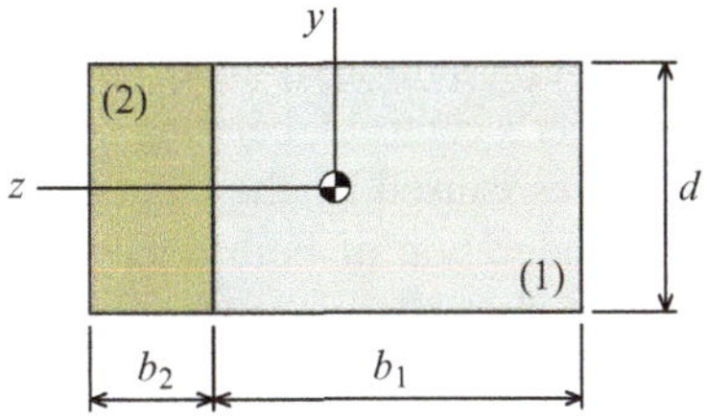

FIGURE P8.39

P8.40 Two aluminum alloy plates (2) are attached to the sides of a wooden beam (1) as shown in Figure P8.40. The dimensions of the composite cross section are $b_1 = 80$ mm, $d_1 = 240$ mm, $b_2 = 10$ mm, $d_2 = 120$ mm, and $a = 60$ mm. Determine the maximum bending stresses produced in both the wooden beam and the aluminum plates if a bending moment $M_z = +6{,}000$ N·m is applied about the z axis. Assume that $E_1 = 12.5$ GPa and $E_2 = 70$ GPa.

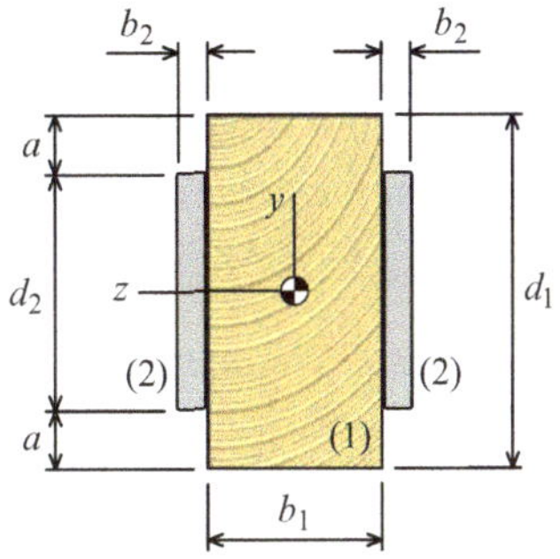

FIGURE P8.40

P8.41 A wooden beam (1) is reinforced with steel plates (2) rigidly attached to its top and bottom surfaces, as shown in Figure P8.40. The dimensions of the cross section are $b_1 = 6$ in., $d_1 = 12$ in., $b_2 = 4$ in., and $d_2 = 0.5$ in. The elastic moduli of the wood and steel are $E_1 = 1{,}250$ ksi and $E_2 = 30{,}000$ ksi, respectively. The allowable bending stresses of the wood and steel are 1,200 psi and 20,000 psi, respectively. Determine the largest concentrated vertical load that can be applied at midspan if a simply supported beam with this cross section spans 18 ft.

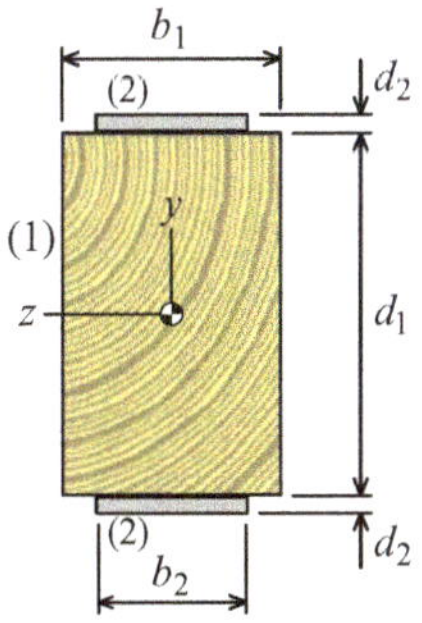

FIGURE P8.41

P8.42 A bronze [$E = 15{,}200$ ksi] bar is rigidly attached to a stainless steel [$E = 27{,}500$ ksi] bar to form a composite beam. The composite beam is subjected to a bending moment $M = 6{,}400$ lb·ft about the z axis (Figure P8.42*a*/43*a*). The dimensions of the beam cross section are $b = 4.5$ in., $d_B = 2.25$ in., and $d_S = 1.00$ in. (Figure P8.42*b*/8.43*b*). Determine

(a) the maximum bending stresses in the bronze and stainless steel bars.
(b) the normal stress in each bar at the surface where they contact each other.

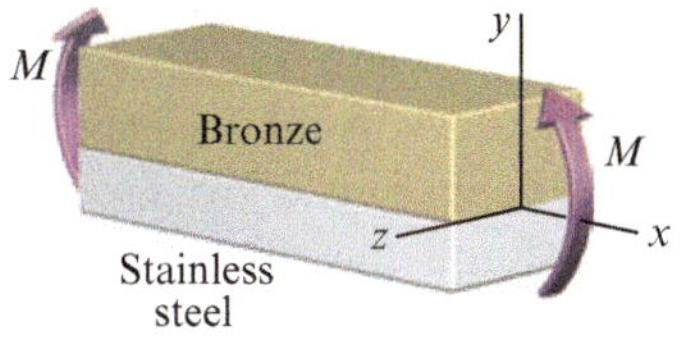

FIGURE P8.42*a*/43*a*

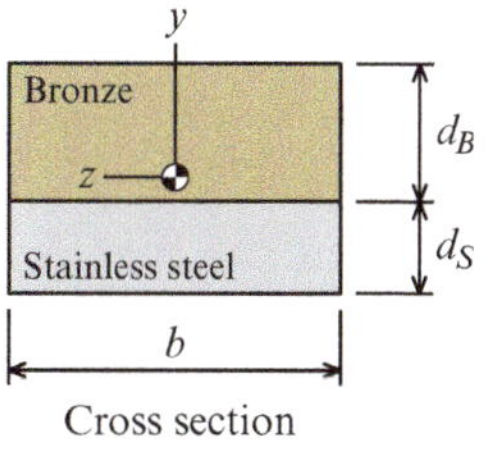

FIGURE P8.42*b*/43*b*

P8.43 A bronze [$E = 100$ GPa] bar is rigidly attached to a stainless steel [$E = 190$ GPa] bar to form a composite beam. The allowable bending stresses for the bronze and stainless steel bars are 180 MPa and 225 MPa, respectively. Determine the allowable bending moment M (Figure P8.42*a*/43*a*) that can be applied to the composite beam if the dimensions of the beam cross section are $b = 150$ mm, $d_B = 85$ mm, and $d_S = 40$ mm (Figure P8.42*b*/8.43*b*).

P8.44 A wooden beam (1) is reinforced on its lower surface by a steel plate (2) as shown in Figure P8.44. The dimensions of the cross section are $b_1 = 220$ mm, $d = 380$ mm, $b_2 = 160$ mm, and $t = 20$ mm. The elastic moduli of the wood and steel are $E_1 = 12.5$ GPa and $E_2 = 200$ GPa, respectively. The allowable bending stresses of the wood and steel are 7.5 MPa and 150 MPa, respectively. This cross section is used for a simply supported beam that spans 9 m. What is the largest uniformly distributed load that can be applied to the beam?

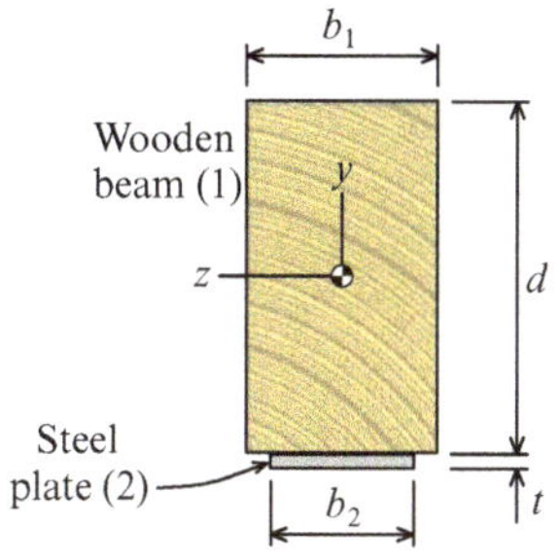

FIGURE P8.44

P8.45 A composite beam consists of a concrete slab (1) that is rigidly attached to the top flange of a W16 × 40 standard steel shape (2). The cross section of the beam is shown in Figure P8.45. The slab has a width $b = 48$ in., a thickness $t = 4$ in., and an elastic modulus of 3,500 ksi. The elastic modulus of the steel shape is 30,000 ksi, and the cross-sectional dimensions are presented in Appendix B. For a bending moment of $M_z = +150$ kip·ft, determine

(a) the location of the centroid for the transformed section, measured upward from point K.
(b) the normal stress at H in the concrete slab.
(c) the normal stress at K in the steel shape.

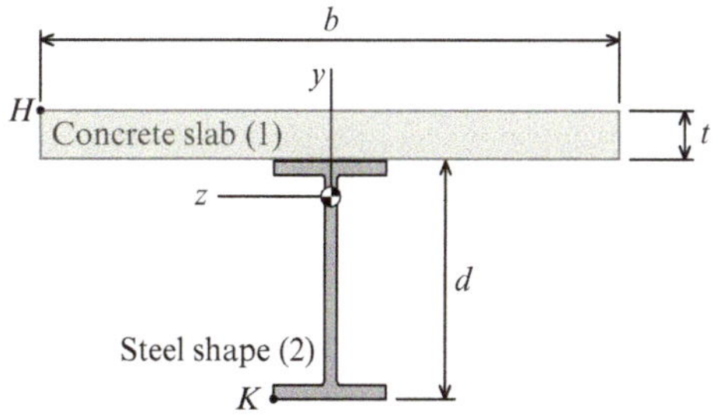

FIGURE P8.45

8.7 Bending Due to an Eccentric Axial Load

As discussed in Chapters 1, 4, and 5, an axial load whose line of action passes through the centroid of a cross section (termed a **centric axial load**) creates normal stress that is uniformly distributed across the cross-sectional area of a member. An **eccentric axial load** is a force whose line of action does not pass through the centroid of the cross section. When an axial force is offset from a member's centroid, bending stresses are created in the member in addition to the normal stresses caused by the axial force. Analysis of this type of bending, therefore, requires a consideration of both axial stresses and bending stresses. Many structures, including common objects such as signposts, clamps, and piers, are subjected to eccentric axial loads.

Suppose that the normal stresses acting on the section containing the C are to be determined for the object shown in Figure 8.13*a*. The analysis presented here assumes that the bending member has a plane of symmetry (see Figure 8.2*a*) and that all loads are applied in that plane.

The line of action of the axial load P does not pass through the centroid C; therefore, this object (between points H and K) is subjected to an eccentric axial load. The **eccentricity** between the line of action of P and the centroid C is denoted by the symbol e.

The internal forces acting on a cross section can be represented by an internal axial force F acting at the centroid of the cross section and an internal bending moment M acting in the plane of symmetry, as shown on the free-body diagram cut through C (Figure 8.13*b*).

Both the internal axial force F and the internal bending moment M produce normal stresses (Figure 8.14). These stresses must be combined in order to determine the complete stress distribution at the section of interest. The axial force F produces a normal stress $\sigma_x = F/A$ that is uniformly distributed over the entire cross section. The bending moment M produces a normal stress, given by the flexure formula $\sigma_x = -My/I_z$, that is linearly distributed over the depth of the cross section. The complete stress distribution is obtained by superposing the stresses produced by F and M as

$$\boxed{\sigma_x = \frac{F}{A} - \frac{My}{I_z}} \tag{8.19}$$

The sign conventions for F and M are the same as those presented in previous chapters. A positive internal axial force F produces tensile normal stresses. A positive internal bending moment produces compressive normal stresses for positive values of y.

An axial force whose line of action is separated from the centroid of the cross section by an eccentricity e produces an internal bending moment $M = P \times e$. Thus, for an eccentric axial force, Equation (8.19) can also be expressed as

$$\boxed{\sigma_x = \frac{F}{A} - \frac{(Pe)y}{I_z}} \tag{8.20}$$

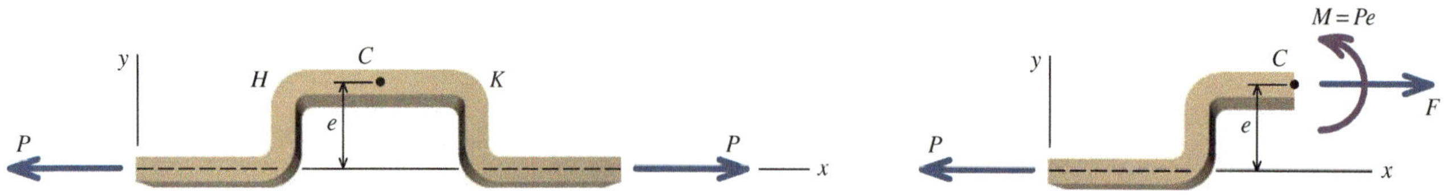

(*a*) Configuration of bending member

(*b*) Free-body diagram

FIGURE 8.13 Bending due to an eccentric axial load.

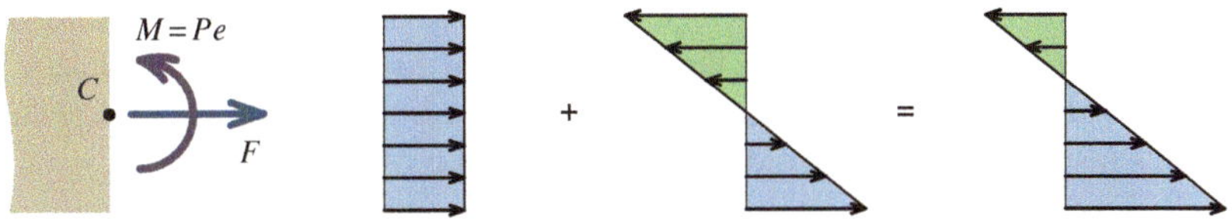

FIGURE 8.14 Normal stresses caused by an eccentric axial load.

Neutral-Axis Location

Whenever an internal axial force F acts simultaneously with an internal bending moment M, *the neutral axis is no longer located at the centroid of the cross section.* In fact, depending upon the magnitude of the force F, there may be no neutral axis at all. All normal stresses on the cross section may be either tension stresses or compression stresses. The location of the neutral axis can be determined by setting $\sigma_x = 0$ in Equation (8.19) and solving for the distance y measured from the centroid of the cross section.

Limitations

The stresses determined by this approach assume that the internal bending moment in the flexural member can be accurately calculated from the original undeformed dimensions. In other words, the deflections caused by the internal bending moment must be relatively small. If the flexural member is relatively long and slender, the lateral deflections caused by the eccentric axial load may significantly increase the eccentricity e, thereby amplifying the bending moment.

The use of Equations (8.19) and (8.20) should be consistent with Saint-Venant's principle. In practice, this condition means that stresses cannot be accurately calculated near points H and K in Figure 8.13a.

EXAMPLE 8.8

A structural member with a rectangular cross section 10 in. wide by 6 in. deep supports a 30 kip concentrated load as shown. Determine the distribution of normal stresses on section a–a of the member.

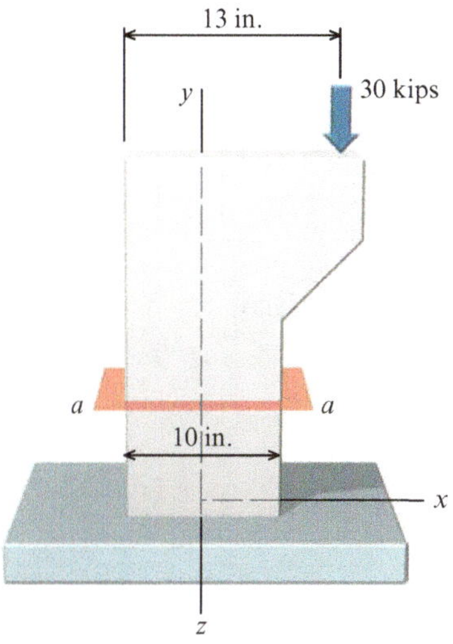

Plan the Solution

The internal forces acting on section a–a must be determined at the outset. The principle of *equivalent force systems* will be used to determine a force and a moment acting at the section of interest that together are equivalent to the single 30 kip concentrated load acting on the top of the structural member. Once the equivalent force and moment have been determined, the stresses produced at section a–a can be computed.

SOLUTION

Equivalent Force and Moment

The cross section of the structural member is rectangular; therefore, by symmetry, the centroid must be located 5 in. from the left side of the structural member. The 30 kip concentrated load is located 13 in. from the left side of

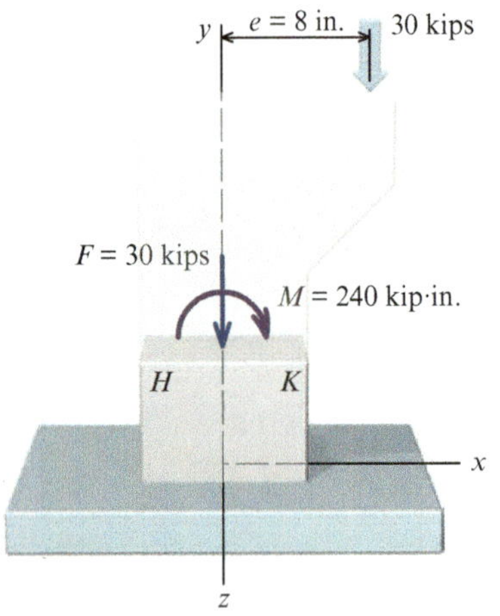

the structural member. Consequently, the concentrated load is located 8 in. to the right of the centroidal axis of the structural member. The distance between the line of action of the load and the centroidal axis of the member is commonly termed the *eccentricity e*. In this instance, the load is said to be located at an eccentricity $e = 8$ in.

Since its line of action does not coincide with the centroidal axis of the structural member, the 30 kip load produces bending in addition to axial compression. The equivalent force at section *a–a* is simply equal to the 30 kip load. The moment at section *a–a* that is required for equivalence is equal to the product of the load and the eccentricity *e*. Therefore, an internal force $F = 30$ kips and an internal bending moment $M = F \times e = (30 \text{ kips})(8 \text{ in.}) = 240 \text{ kip} \cdot \text{in.}$ acting at the centroid of section *a–a* are together equivalent to the 30 kip load applied to the top of the structural member.

Section Properties

The centroid location is known from symmetry. The area of the cross section is $A = (10 \text{ in.})(6 \text{ in.}) = 60 \text{ in.}^2$. The bending moment $M = 240 \text{ kip} \cdot \text{in.}$ acts about the z axis; therefore, the moment of inertia about the z axis must be determined in order to calculate the bending stresses:

$$I_z = \frac{(6 \text{ in.})(10 \text{ in.})^3}{12} = 500 \text{ in.}^4$$

Axial Stress

On section *a–a*, the internal force $F = 30$ kips (which acts along the y centroidal axis) produces a normal stress of

$$\sigma_{\text{axial}} = \frac{F}{A} = \frac{30 \text{ kips}}{60 \text{ in.}^2} = 0.5 \text{ ksi (C)}$$

which acts vertically (i.e., in the y direction). The axial stress is a compressive normal stress that is uniformly distributed over the entire section.

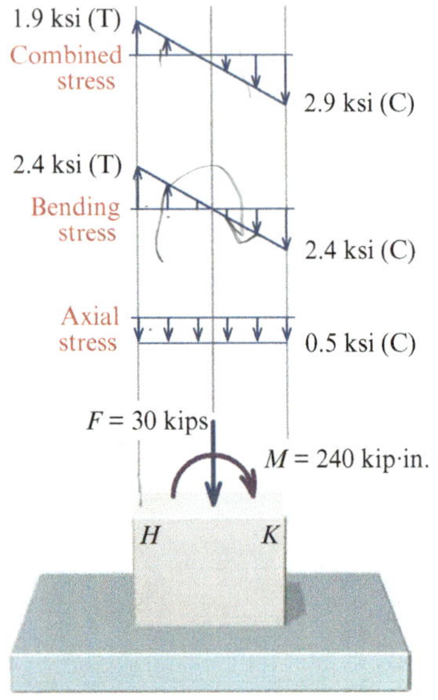

Bending Stress

The magnitude of the maximum bending stress on section *a–a* can be determined from the flexure formula:

$$\sigma_{\text{bend}} = \frac{Mc}{I_z} = \frac{(240 \text{ kip} \cdot \text{in.})(5 \text{ in.})}{500 \text{ in.}^4} = 2.4 \text{ ksi}$$

The bending stress acts in the vertical direction (i.e., in the y direction) and increases linearly with distance from the axis of bending. In the coordinate system defined for this problem, distance from the axis of bending is measured in the x direction from the z axis.

The sense of the bending stress (either tension or compression) can be readily determined by inspection, based on the direction of the internal bending moment M. In this instance, M causes compressive bending stresses on the K side of the structural member and tensile bending stresses on the H side.

Combined Normal Stresses

Since the axial and bending stresses are normal stresses that act in the same direction (i.e., the y direction), they can be directly added to give the combined stresses acting on section *a–a*. The combined normal stress on side H of the structural member is

$$\sigma_H = \sigma_{\text{axial}} + \sigma_{\text{bend}} = -0.5 \text{ ksi} + 2.4 \text{ ksi} = +1.9 \text{ ksi} = 1.9 \text{ ksi (T)} \qquad \textbf{Ans.}$$

and the combined normal stress on side K is

$$\sigma_K = \sigma_{\text{axial}} + \sigma_{\text{bend}} = -0.5 \text{ ksi} - 2.4 \text{ ksi} = -2.9 \text{ ksi} = 2.9 \text{ ksi (C)} \qquad \textbf{Ans.}$$

Neutral-Axis Location

For an eccentric axial load, the neutral axis (i.e., the location with zero stress) is not located at the centroid of the cross section. Although not requested in this example, the location of the axis of zero stress can be determined from the combined stress distribution. By the principle of similar triangles, the combined stress is zero at a distance of 3.958 in. from the left side of the structural member.

EXAMPLE 8.9

The C-clamp shown is made of an alloy that has a yield strength of 324 MPa in both tension and compression. Determine the allowable clamping force that the clamp can exert if a factor of safety of 3.0 is required.

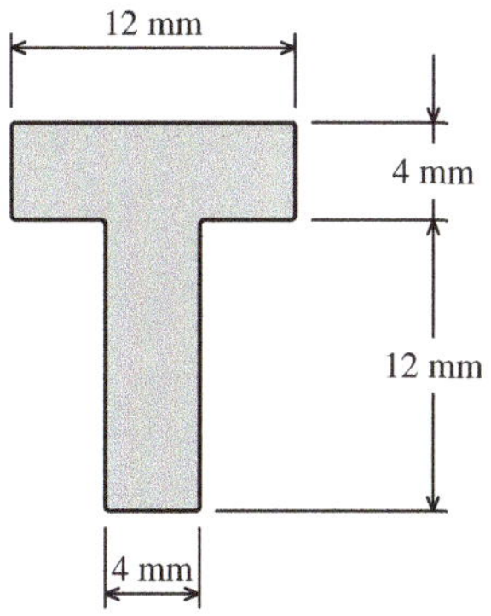

Plan the Solution

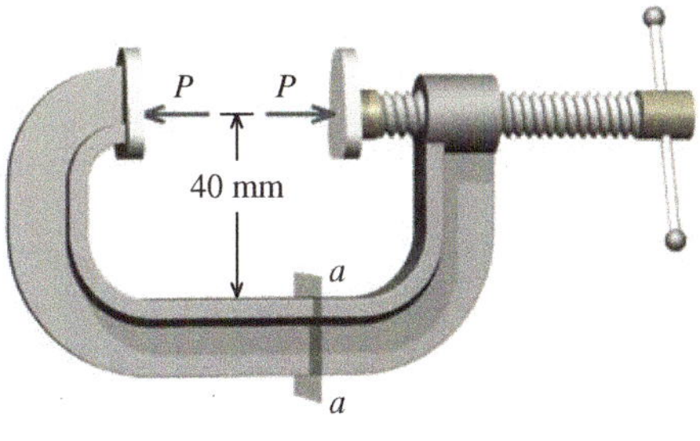

The location of the centroid for the tee-shaped cross section must be determined at the outset. Once the centroid has been located, the eccentricity e of the clamping force P can be determined and the equivalent force and moment acting on section a–a established. Expressions for the combined axial and bending stresses, written in terms of the unknown P, can be set equal to the allowable normal stress. From these expressions, the maximum allowable clamping force can be determined.

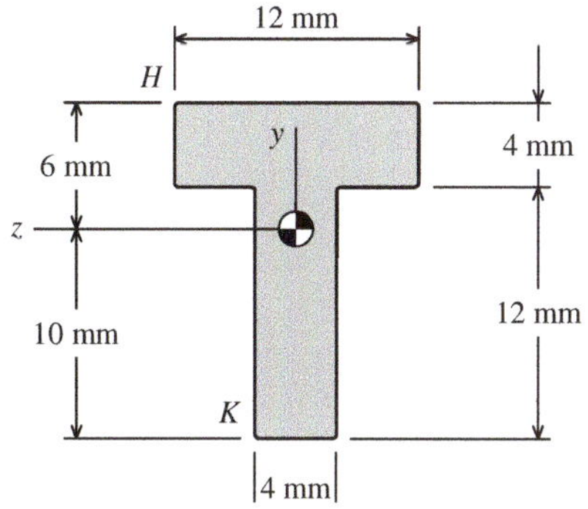

Section Properties

The centroid for the tee-shaped cross section is located as shown in the sketch on the left. The cross-sectional area is $A = 96 \text{ mm}^2$, and the moment of inertia about the z centroidal axis can be calculated as $I_z = 2{,}176 \text{ mm}^4$.

Allowable Normal Stress

The alloy used for the clamp has a yield strength of 324 MPa. Since a factor of safety of 3.0 is required, the allowable normal stress for this material is 108 MPa.

Internal Force and Moment

A free-body diagram cut through the clamp at section a–a is shown. The internal axial force F is equal to the clamping force P. The internal bending moment M is equal to the clamping force P times the eccentricity $e = 40 \text{ mm} + 6 \text{ mm} = 46 \text{ mm}$ between the centroid of section a–a and the line of action of P.

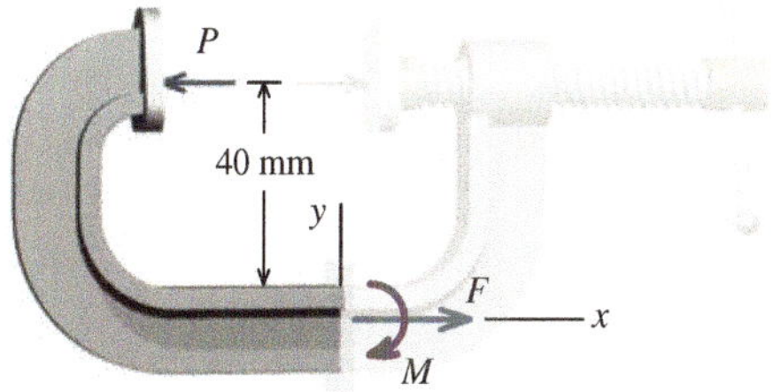

Axial Stress

On section *a–a*, the internal force *F* (which is equal to the clamping force *P*) produces a normal stress

$$\sigma_{\text{axial}} = \frac{F}{A} = \frac{P}{A} = \frac{P}{96\ \text{mm}^2}$$

This normal stress is uniformly distributed over the entire cross section. By inspection, the axial stress is a tensile stress.

Bending Stress

Since the tee shape is not symmetrical about its z axis, the bending stress on section *a–a* at the top of the flange (point *H*) will be different from the bending stress at the bottom of the stem (point *K*). At point *H*, the bending stress can be expressed in terms of the clamping force *P* as

$$\sigma_{\text{bend},H} = \frac{My}{I_z} = \frac{P(46\ \text{mm})(6\ \text{mm})}{2{,}176\ \text{mm}^4} = \frac{P}{7.88406\ \text{mm}^2}$$

By inspection, the bending stress at point *H* will be a tensile stress.

The bending stress at point *K* can be expressed as

$$\sigma_{\text{bend},K} = \frac{My}{I_z} = \frac{P(46\ \text{mm})(10\ \text{mm})}{2{,}176\ \text{mm}^4} = \frac{P}{4.73043\ \text{mm}^2}$$

By inspection, the bending stress at point *K* will be a compressive stress.

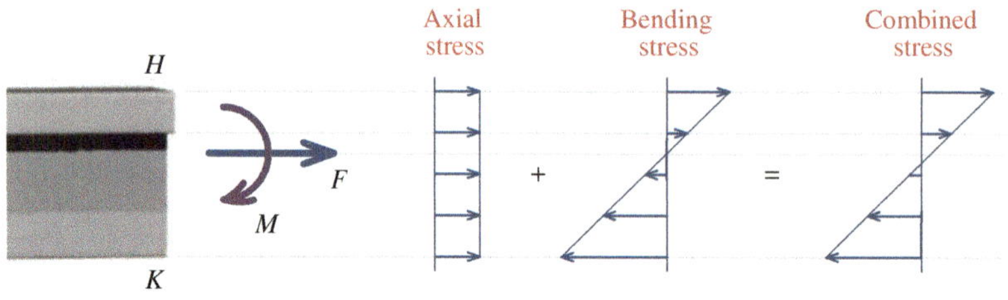

Combined Stress at *H*

The combined stress at point *H* can be expressed in terms of the unknown clamping force *P* as

$$\sigma_{\text{comb},H} = \frac{P}{96\ \text{mm}^2} + \frac{P}{7.88406\ \text{mm}^2} = P\left[\frac{1}{96\ \text{mm}^2} + \frac{1}{7.88406\ \text{mm}^2}\right] = \frac{P}{7.28572\ \text{mm}^2}$$

Note that the axial and bending stress expressions are added, since both are tensile stresses. The expression just obtained can be set equal to the allowable normal stress to obtain one possible value for *P*:

$$\frac{P}{7.28572\ \text{mm}^2} \leq 108\ \text{MPa} = 108\ \text{N/mm}^2 \qquad \therefore P \leq 787\ \text{N} \tag{a}$$

Combined Stress at K

The combined stress at point K is the sum of a tensile axial stress and a compressive bending stress:

$$\sigma_{\text{comb},K} = \frac{P}{96\text{ mm}^2} - \frac{P}{4.73043\text{ mm}^2} = P\left[\frac{1}{96\text{ mm}^2} - \frac{1}{4.73043\text{ mm}^2}\right] = -\frac{P}{4.97560\text{ mm}^2}$$

The negative sign indicates that the combined stress at K is a compressive normal stress. A second possible value for P can be derived from the expression

$$\frac{P}{4.97560\text{ mm}^2} \leq 108\text{ MPa} = 108\text{ N/mm}^2 \qquad \therefore P \leq 537\text{ N} \tag{b}$$

The negative signs can be omitted here because we are interested only in the magnitude of P.

Controlling Clamping Force

The allowable clamping force is the lesser of the two values obtained from Equations (a) and (b). For this clamp, the maximum allowable clamping force is $P = 537$ N. **Ans.**

MecMovies

EXAMPLES

M8.20 A C-clamp is expected to exert a maximum total clamping force of 400 N. The clamp cross section is 20 mm wide and 10 mm thick. Determine the maximum tension and compression stresses in the clamp.

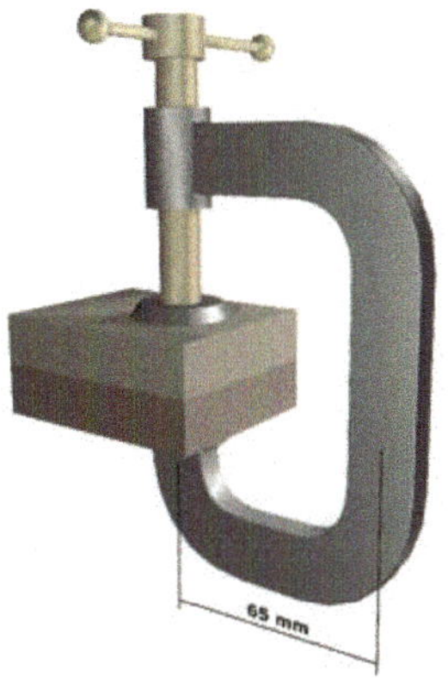

M8.21 A precast concrete beam is supported by a corbel on a concrete column. The reaction force at the end of the beam is 1,200 kN. This reaction force acts on the corbel at a distance of 240 mm from the column centerline. Determine the stresses at the base of the column at points a and b.

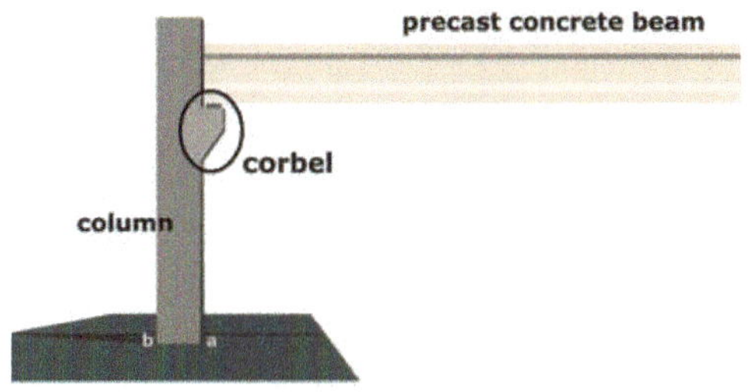

M8.22 A steel inverted tee shape is used as a boom for a wall bracket jib crane that can lift loads of up to 5 kN. The boom is pinned to the wall at *A*. At point *B*, the boom is supported by steel rod *BC*. The pin at *A* is located on the centroidal axis of the inverted tee, but at *B* the steel rod is connected to the tee 65 mm above the centroidal axis. Determine the normal stress at point *H*, located at the top-most edge of the inverted tee, 1.0 m from *A*, when the 5 kN crane load is in the position shown.

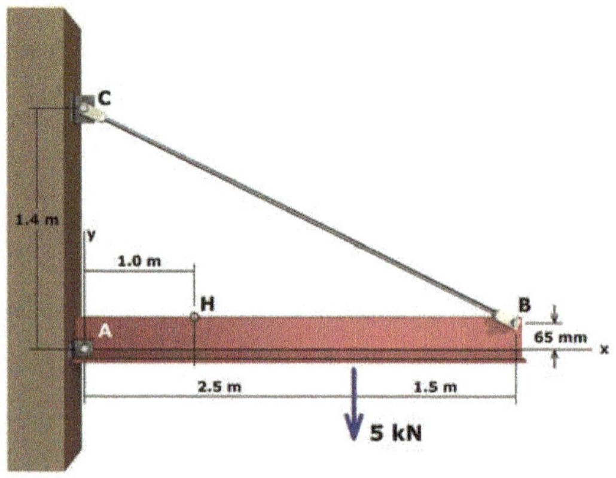

EXERCISES

M8.20 Determine the normal stresses at *A* and *B*.

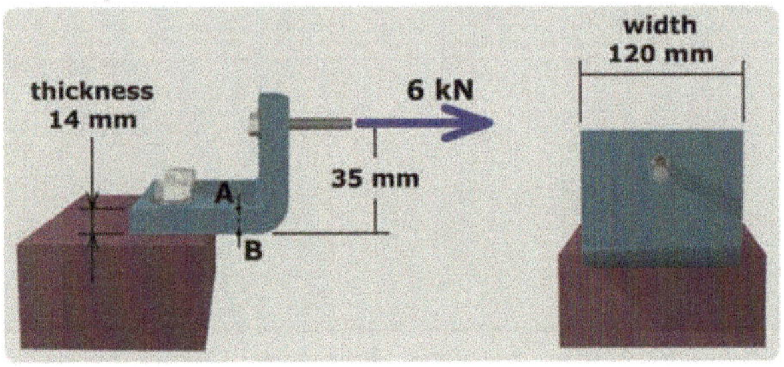

FIGURE M8.20

M8.21 Determine the normal stresses at *A* and *B*.

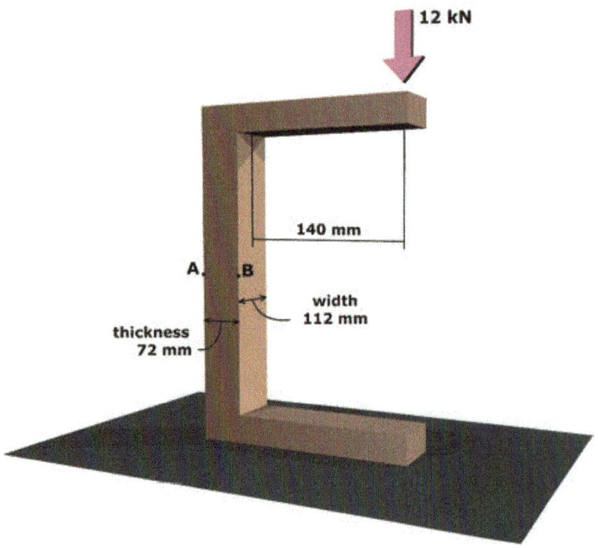

FIGURE M8.21

M8.22 Answer 10 questions concerning the structure shown subjected to various loads.

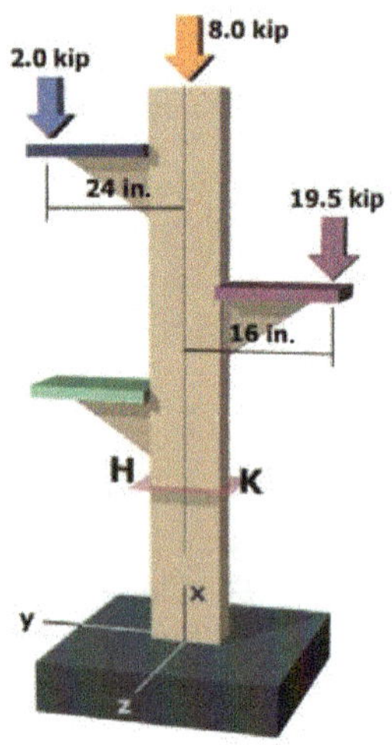

FIGURE M8.22

288

M8.23 Pipe *AB* (with outside diameter and wall thickness specified) supports a uniformly distributed load *w*. Determine the reaction forces at pin *A*, the axial force in member (1), and the normal stresses at points *H* and *K*, located a specified distance above pin *A*.

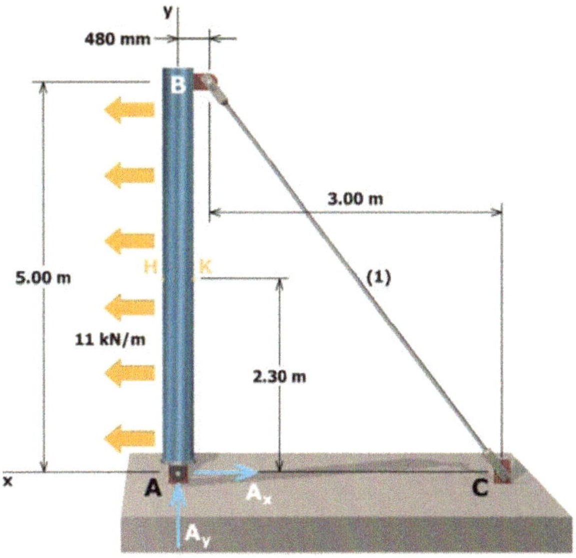

FIGURE M8.23

PROBLEMS

P8.46 A load $P = 1{,}200$ N acts on a curved steel rod as shown in Figure P8.46. The rod is solid and has a diameter $d = 15$ mm. Assume that $b = 35$ mm and $a = 175$ mm. Determine the normal stresses produced at points *H* and *K*.

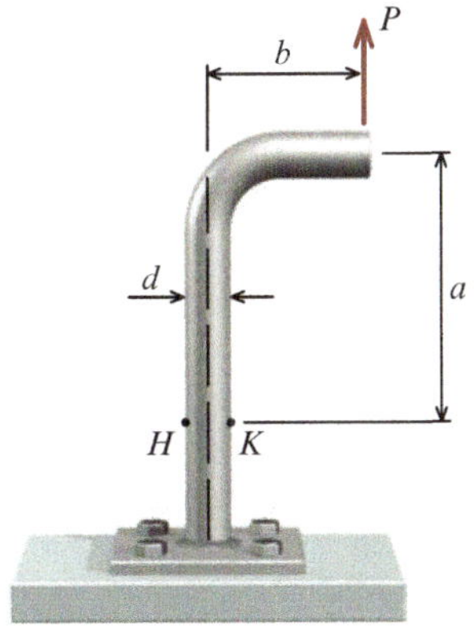

FIGURE P8.46

P8.47 A pulley is supported with the bar shown in Figure P8.47. The pulley has a diameter $D = 5$ in., and the tension in the pulley belt is $P = 45$ lb. Assume that $a = 3$ in. and $c = 6$ in. Determine the normal stresses produced at points *H* and *K*. The cross-sectional dimensions of the support bar at the section of interest are $d = 1.5$ in. by 0.25 in. thick.

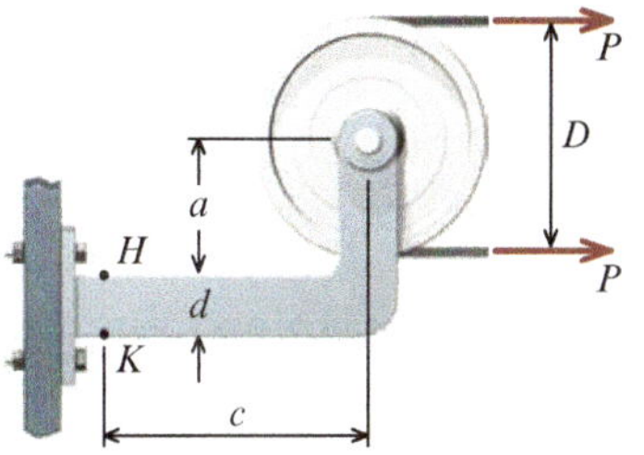

FIGURE P8.47

P8.48 A 30 mm diameter steel rod is formed into a machine part with the shape shown in Figure P8.48. A load $P = 2{,}500$ N is applied to the ends of the part. If the allowable normal stress is limited to 40 MPa, what is the maximum eccentricity *e* that may be used for the part?

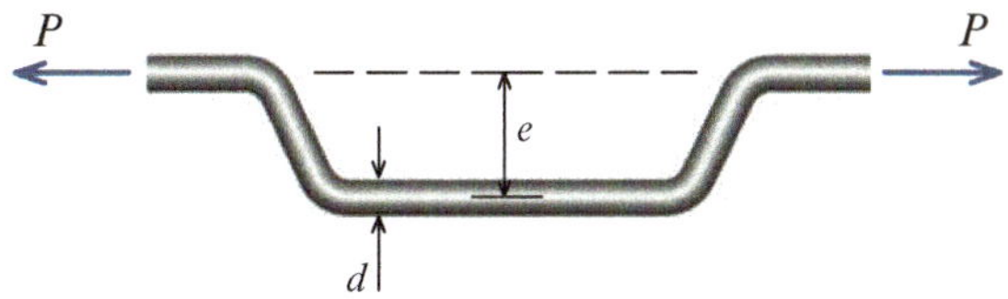

FIGURE P8.48

P8.49 The bent bar component shown in Figure P8.49 has a width $b = 70$ mm and a depth $d = 30$ mm. At the free end of the bar,

289

a load P is applied at an eccentricity $e = 100$ mm from the x centroidal axis. The yield strength of the bar is 310 MPa. If a factor of safety of 2.3 is required, what is the maximum load P that can be applied to the bar?

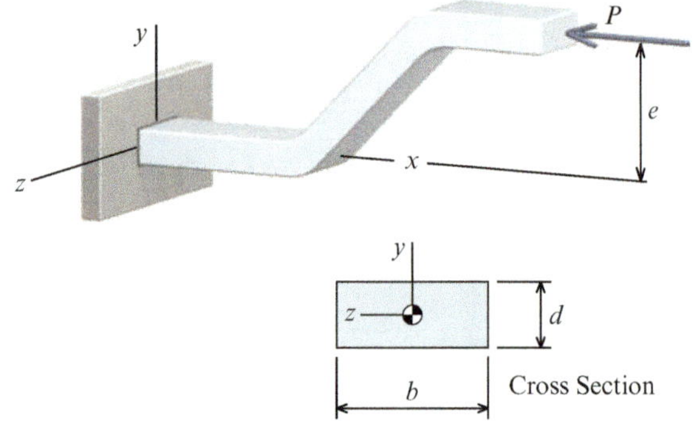

FIGURE P8.49

P8.50 A horizontal force $P = 1{,}200$ lb is applied to the rectangular bar shown in Figure P8.50. Calculate the normal stress at points H and K. Use the following values: $d = 1.50$ in., $b = 2.50$ in., and $e = 2.00$ in.

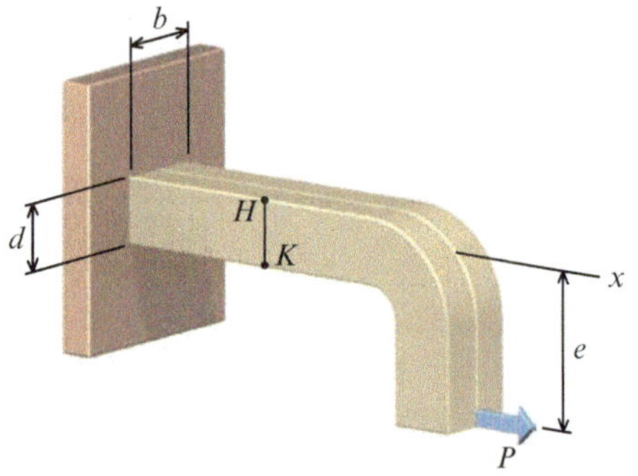

FIGURE P8.50

P8.51 The component shown in Figure P8.51 has a width $b = 35$ mm and a depth $d = 65$ mm. At the free end of the bar, a load P is applied at an eccentricity $e = 50$ mm from the y centroidal axis. The yield strength of the bar is 340 MPa. If a factor of safety of 2.3 is required, what is the maximum load P that can be applied to the bar?

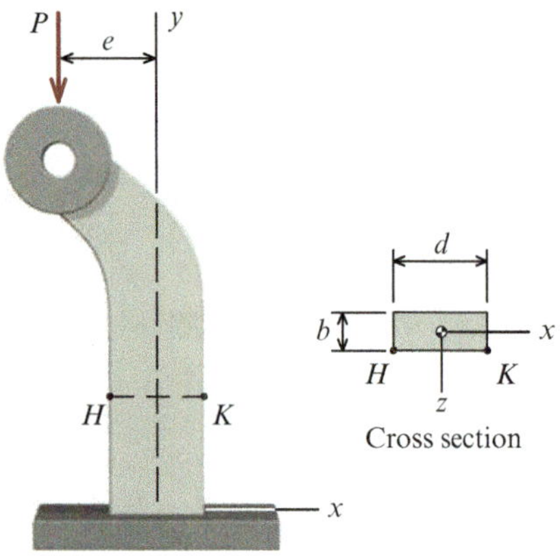

FIGURE P8.51

P8.52 A tubular steel column CD supports a horizontal cantilever arm ABC, as shown in Figure P8.52. Column CD has an outside diameter of 8.625 in. and a wall thickness of 0.365 in. The loads are $P_A = 400$ lb and $P_B = 600$ lb. The dimensions of the structure are $a = 6$ ft, $b = 8$ ft, and $c = 14$ ft. Determine the maximum compressive stress at the base of column CD.

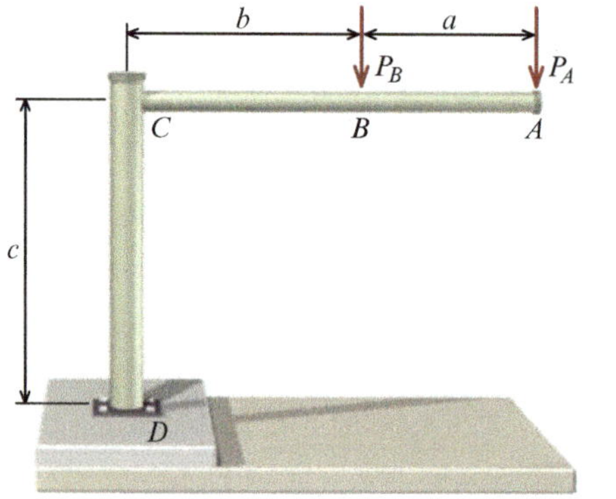

FIGURE P8.52

P8.53 A load $P = 32$ kN is applied parallel to the longitudinal axis of a rectangular structural tube as shown in Figure P8.53*a*. The cross-sectional dimensions of the structural tube (Figure P8.53*b*) are $b = 100$ mm, $d = 150$ mm, and $t = 5$ mm. Assume that $a = 500$ mm and $c = 40$ mm. Determine the normal stresses produced at points H and K.

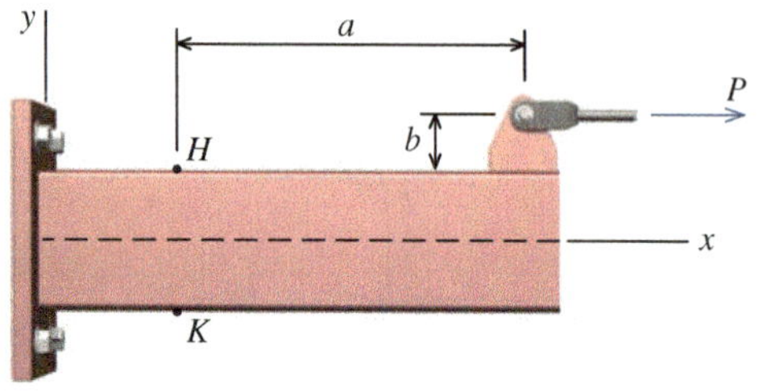

FIGURE P8.53*a*

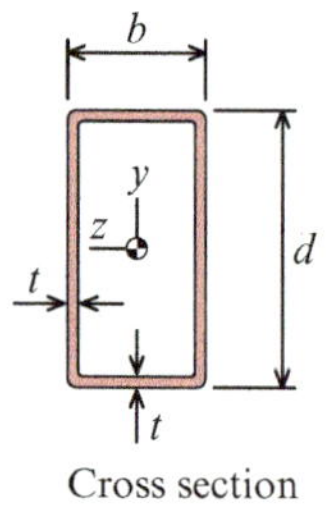

FIGURE P8.53*b*

P8.54 The bracket shown in Figure P8.54 is subjected to a load P = 1,300 lb. The bracket has a rectangular cross section with a width b = 3.00 in. and a thickness t = 0.375 in. If the tensile normal stress must be limited to 24,000 psi at section a–a, what is the maximum offset distance y that can be used?

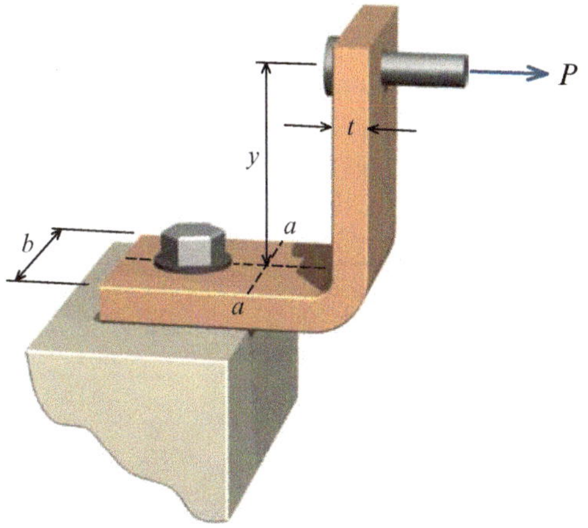

FIGURE P8.54

P8.55 A wall-mounted jib crane is shown in Figure P8.55*a*. The overall dimensions of the crane are a = 102 in., b = 52 in., c = 66 in., and e = 5.0 in. The cross-sectional dimensions of the crane rail (Figure P8.55*b*) are b_f = 4.0 in., t_f = 0.5 in., d = 6.0 in., and t_w = 0.3 in. Assume that a hoist load P = 2,000 lb is located as shown in Figure P8.55*a*. What is the maximum normal stress in the crane rail?

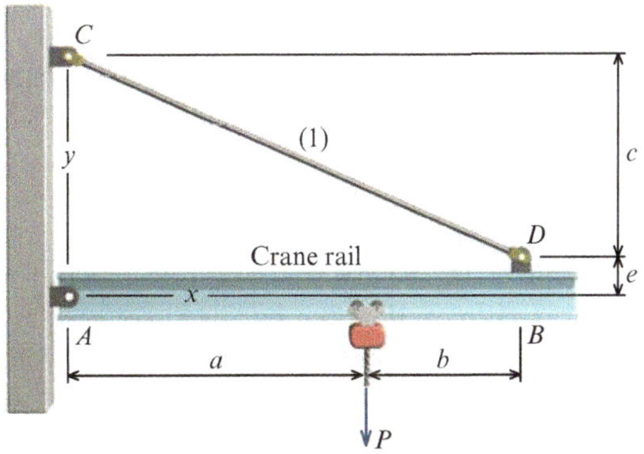

FIGURE P8.55*a*

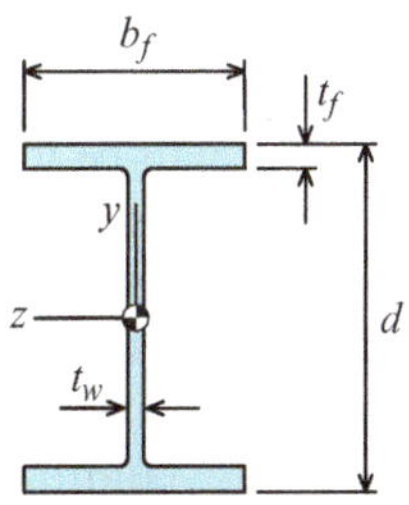

FIGURE P8.55*b*

P8.56 The U-shaped aluminum bar shown in Figure P8.56 is used as a dynamometer to determine the magnitude of the applied load P. The aluminum [E = 70 GPa] bar has a square cross section with dimensions a = 30 mm and b = 65 mm. The strain on the inner surface of the bar was measured and found to be 955 με. What is the magnitude of the load P?

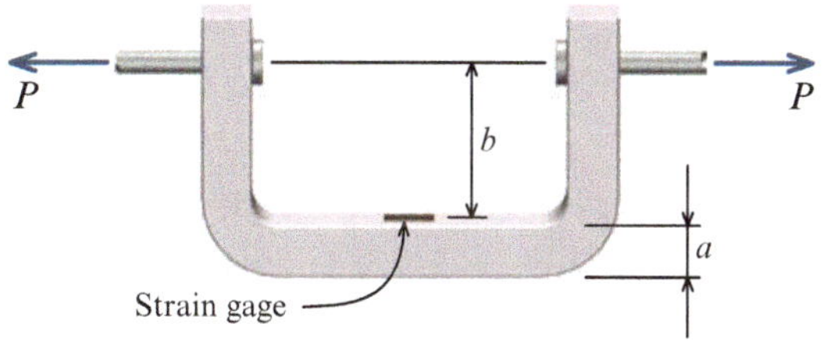

FIGURE P8.56

P8.57 The steel pipe shown in Figure P8.57 has an outside diameter of 195 mm, a wall thickness of 10 mm, an elastic modulus E = 200 GPa, and a coefficient of thermal expansion α = 11.7 × 10^{-6}/°C. Using a = 300 mm, b = 900 mm, and θ = 70°, calculate the normal strains at H and K after a load P = 40 kN has been applied and the temperature of the pipe has been increased by 25°C.

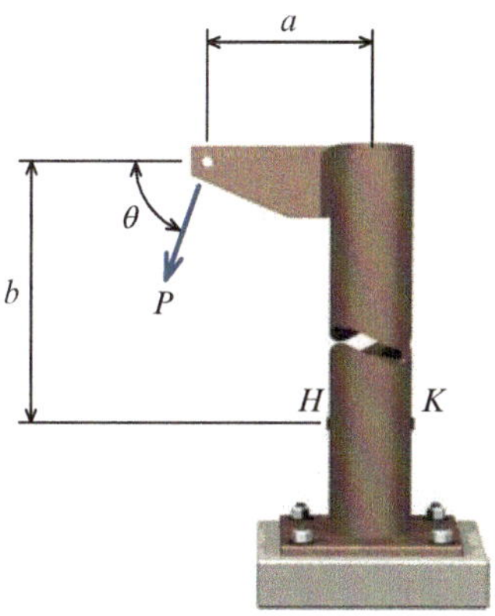

FIGURE P8.57

P8.58 A short length of a rolled-steel [E = 29 × 10^3 ksi] column supports a rigid plate on which two loads P and Q are applied as shown in Figure P8.58*a*/59*a*. The column cross section (Figure P8.58*b*/59*b*) has a depth d = 8.0 in., an area A = 5.40 in.^{2}, and a moment of inertia I_z = 57.5 in.^{4}. Normal strains are measured with strain gages H and K, which are attached on the centerline of the outer faces of the flanges. Load P is known to be 35 kips, and the strain in gage H is measured as $\varepsilon_H = 120 \times 10^{-6}$ in./in. Determine

(a) the magnitude of load Q.
(b) the expected strain reading in gage K.

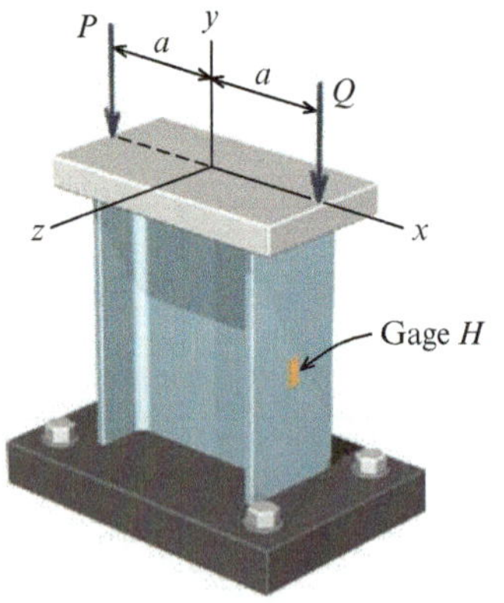

FIGURE P8.58*a*/59*a*

P8.59 A short length of a rolled-steel [$E = 29 \times 10^3$ ksi] column supports a rigid plate on which two loads P and Q are applied as shown in Figure P8.58*a*/59*a*. The column cross section (Figure P8.58*b*/59*b*) has a depth $d = 8.0$ in., an area $A = 5.40$ in.^{2}, and a moment of inertia $I_z = 57.5$ in.^{4}. Normal strains are measured with strain gages H and K, which are attached on the centerline of the outer faces of the flanges. The strains measured in the two gages are $\varepsilon_H = -530 \times 10^{-6}$ in./in. and $\varepsilon_K = -310 \times 10^{-6}$ in./in. Determine the magnitudes of loads P and Q.

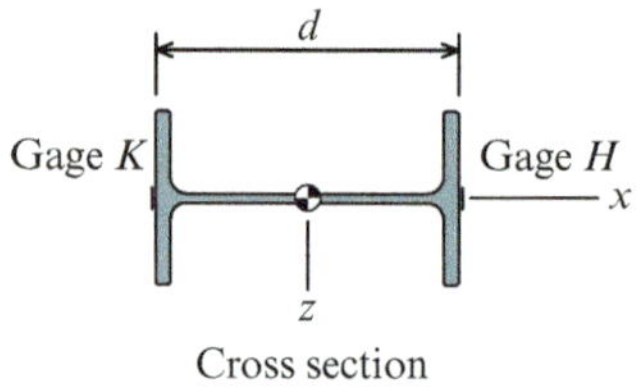

FIGURE P8.58*b*/59*b*

8.8 Unsymmetric Bending

In Sections 8.1 through 8.3, the theory of bending was developed for prismatic beams. In deriving this theory, beams were assumed to have a longitudinal plane of symmetry (Figure 8.2*a*), termed the **plane of bending**. Furthermore, loads acting on the beam, as well as the resulting curvatures and deflections, were assumed to act only in the plane of bending. If the beam cross section is unsymmetric or if the loads on the beam do not act in the plane of bending, then the theory of bending developed in Sections 8.1 through 8.3 is not valid.

To see why, consider the following thought experiment: Suppose that the unsymmetric flanged cross section shown in Figure 8.15*a* (termed a zee section) is subjected to equal-magnitude bending moments M_z that act as shown about the z axis. Suppose further that the beam bends only in the x–y plane in response to M_z and that the z axis is the neutral axis for bending. Then, if the latter supposition is correct, then the bending stresses shown in Figure 8.15*b* will be produced in the zee section. Compressive bending stresses will occur above the z axis, and tensile bending stresses will occur below the z axis.

Next, consider the stresses that act in the flanges of the zee section. Bending stresses will be uniformly distributed across the width of each flange. The internal resultant force of the compressive bending stresses acting in the top flange will be termed F_C (Figure 8.15*c*). Its line of action passes through the midpoint of the flange (in the horizontal direction) at a distance z_C from the y axis. Similarly, the internal resultant force of the tensile bending stresses in the bottom flange will be termed F_T, and its line of action is located a distance z_T from the y axis. Since these two forces are equal in magnitude, but act in opposite directions, they form an internal couple that creates a bending moment about the y axis. This internal moment about the y axis (i.e., acting in the x–z plane) is not counteracted by any external moment (since the applied moments M_z act about the z axis only); therefore, equilibrium is not satisfied. Consequently, bending of the unsymmetric beam cannot occur solely in the plane of the

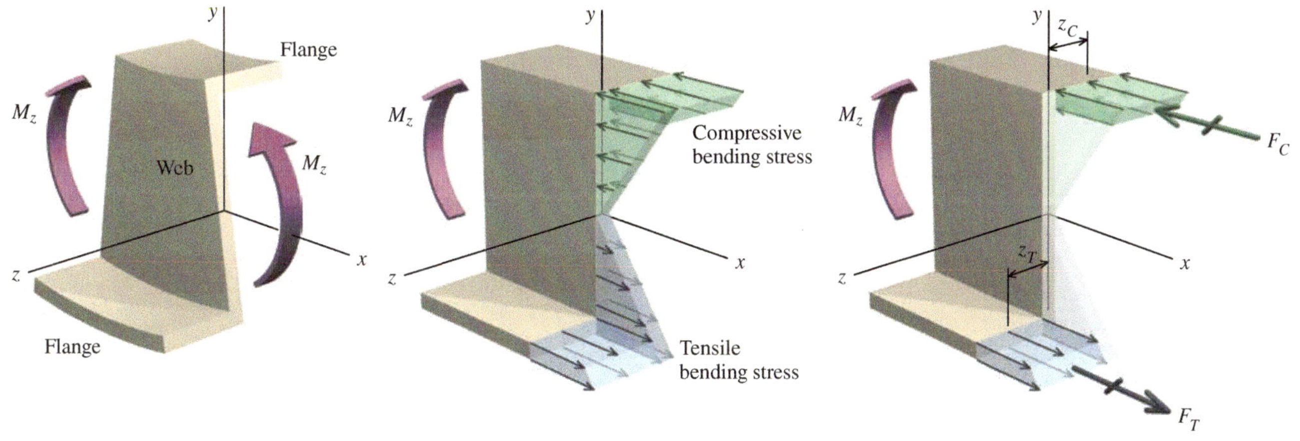

(*a*) Equal-magnitude bending moments M_z applied to the zee section

(*b*) Bending stresses produced in the zee section if bending were to occur in the x–y plane only

(*c*) Resultant forces produced by the bending stresses in the flanges

FIGURE 8.15 Unsymmetric-bending thought experiment.

applied loads (i.e., the x–y plane). This thought experiment shows that the unsymmetric beam must bend both in the plane of the applied moments M_z (i.e., the x–y plane) and in the out-of-plane direction (i.e., the x–z plane).

In this context, the term *arbitrary cross section* means "shapes that may not have axes of symmetry."

Prismatic Beams of Arbitrary Cross Section

A more general theory of bending is required for beams having an **arbitrary cross section**. We will assume that the beam is subjected to pure bending, that plane cross sections before bending remain plane after bending, and that bending stresses remain elastic. The cross section of such a beam is shown in Figure 8.16, and the longitudinal axis of the beam is defined as the x axis. In this derivation, the y and z axes will be assumed to be oriented vertically and horizontally, respectively. However, these axes may exist at any orientation, provided that they are orthogonal.

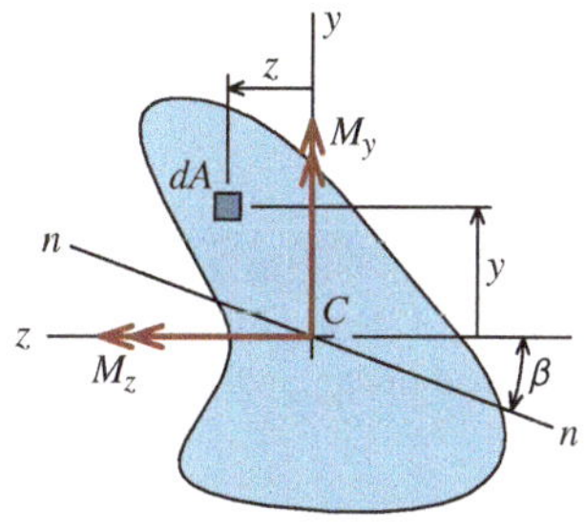

FIGURE 8.16 Bending of a beam with an arbitrary cross section.

Bending moments M_y and M_z will be assumed to act on the beam, causing it to curve in the x–z and x–y planes, respectively. The bending moments create normal stresses σ_x that are distributed linearly above and below the neutral axis n–n. As demonstrated in the preceding thought experiment, loads acting on an unsymmetrical beam may produce bending both within and perpendicular to the plane of loading.

Now, let $1/\rho_z$ denote the beam curvature in the x–y plane and $1/\rho_y$ denote the curvature in the x–z plane. Since cross sections that are planar before bending remain planar after bending, the normal strain in the longitudinal direction at any location (y, z) in the beam cross section can be expressed as

$$\varepsilon_x = -\frac{y}{\rho_z} - \frac{z}{\rho_y}$$

If the bending is elastic, then the bending stress σ_x is proportional to the bending strain, and the stress distribution over the cross section can be defined by

$$\sigma_x = E\varepsilon_x = -\frac{Ey}{\rho_z} - \frac{Ez}{\rho_y} \qquad \text{(a)}$$

To satisfy equilibrium, the resultant of all bending stresses must reduce to a zero net axial force:

$$\int_A \sigma_x \, dA = 0 \tag{b}$$

Also, the following moment equations must be satisfied:

$$\int_A z\sigma_x \, dA = M_y \tag{c}$$

$$\int_A y\sigma_x \, dA = -M_z \tag{d}$$

Next, substitute the expression for σ_x given by Equation (a) into Equation (b) to obtain

$$\int_A \left(-\frac{Ey}{\rho_z} - \frac{Ez}{\rho_y} \right) dA = \int_A \left(\frac{y}{\rho_z} + \frac{z}{\rho_y} \right) dA = \frac{1}{\rho_z} \int_A y \, dA + \frac{1}{\rho_y} \int_A z \, dA = 0 \tag{e}$$

This equation can be satisfied only if the neutral axis *n–n* passes through the centroid of the cross section.

Substitution of Equation (a) into Equation (c) then gives

$$\int_A z \left(-\frac{Ey}{\rho_z} - \frac{Ez}{\rho_y} \right) dA = -\frac{E}{\rho_z} \int_A yz \, dA - \frac{E}{\rho_y} \int_A z^2 \, dA = M_y \tag{f}$$

But the integral terms are simply the moment of inertia about the z axis and the product of inertia, respectively:

$$I_y = \int_A z^2 \, dA \qquad I_{yz} = \int_A yz \, dA$$

Moments of inertia and the product of inertia for areas are reviewed in Appendix A.

Therefore, Equation (f) can be rewritten as

$$-\frac{EI_{yz}}{\rho_z} - \frac{EI_y}{\rho_y} = M_y \tag{g}$$

Similarly, Equation (a) can be substituted into Equation (d) to give

$$-\frac{EI_z}{\rho_z} + \frac{EI_{yz}}{\rho_y} = M_z \tag{h}$$

where

$$I_z = \int_A y^2 \, dA$$

Equations (g) and (h) can be solved simultaneously to derive expressions for the curvatures in the x–y and x–z planes, respectively, due to bending moments M_y and M_z:

$$\frac{1}{\rho_z} = \frac{M_z I_y + M_y I_{yz}}{E(I_y I_z - I_{yz}^2)} \qquad \frac{1}{\rho_y} = -\frac{M_y I_z + M_z I_{yz}}{E(I_y I_z - I_{yz}^2)} \tag{i}$$

These curvature expressions can now be substituted into Equation (a) to give a general relationship for the bending stresses produced in a prismatic beam of arbitrary cross section subjected to bending moments M_y and M_z:

$$\sigma_x = -\frac{(M_z I_y + M_y I_{yz})y}{I_y I_z - I_{yz}^2} + \frac{(M_y I_z + M_z I_{yz})z}{I_y I_z - I_{yz}^2} \tag{8.21}$$

or

$$\sigma_x = \left(\frac{I_z z - I_{yz} y}{I_y I_z - I_{yz}^2}\right) M_y + \left(\frac{-I_y y + I_{yz} z}{I_y I_z - I_{yz}^2}\right) M_z \tag{8.22}$$

Neutral-Axis Orientation

The orientation of the neutral axis must be determined in order to locate points in the cross section where the normal stress has a maximum or minimum value. Since σ is zero on the neutral surface, the orientation of the neutral axis can be determined by setting Equation (8.21) equal to zero:

$$-(M_z I_y + M_y I_{yz})y + (M_y I_z + M_z I_{yz})z = 0$$

Solving for y then gives

$$y = \frac{M_y I_z + M_z I_{yz}}{M_z I_y + M_y I_{yz}} z$$

which is the equation of the neutral axis in the y–z plane. If the slope of the neutral axis is expressed as $dy/dz = \tan\beta$, then the orientation of the neutral axis is given by

$$\tan\beta = \frac{M_y I_z + M_z I_{yz}}{M_z I_y + M_y I_{yz}} \tag{8.23}$$

Beams with Symmetric Cross Sections

If a beam cross section has at least one axis of symmetry, then the product of inertia for the cross section is $I_{yz} = 0$. In this case, Equations (8.21) and (8.22) each reduce to

$$\sigma_x = \frac{M_y z}{I_y} - \frac{M_z y}{I_z} \tag{8.24}$$

and the neutral-axis orientation can be expressed by

$$\tan\beta = \frac{M_y I_z}{M_z I_y} \tag{8.25}$$

Notice that if the loading acts entirely in the x–y plane of the beam, then $M_y = 0$ and Equation (8.24) reduces to

$$\sigma_x = -\frac{M_z y}{I_z}$$

which is identical to the elastic flexure formula [Equation (8.7)] developed in Section 8.3.

Equation (8.24) is useful for the flexural analysis of many common cross-sectional shapes (e.g., a rectangle, W shape, C shape, WT shape) that are subjected to bending moments about two axes (e.g., M_y and M_z).

Principal Axes of Cross Sections

Since the principal axes are orthogonal, if either the y or z axis is a principal axis, then the other axis is necessarily a principal axis.

In the preceding derivation, the y and z axes were assumed to be oriented vertically and horizontally, respectively. However, any pair of orthogonal axes may be taken as y and z in using Equations (8.21) through (8.25). For any cross section, it can be shown that there are always two orthogonal centroidal axes for which the product of inertia $I_{yz} = 0$. These axes are called the **principal axes** of the cross section, and the corresponding planes of the beam are called the **principal planes of bending**. For bending moments applied in the principal planes, bending occurs only in those planes. If a beam is subjected to a bending moment that is not in a principal plane, then that bending moment can always be resolved into components that coincide with the two principal planes of the beam. Then, by superposition, the total bending stress at any (y, z) coordinate in the cross section can be obtained by algebraically adding the stresses produced by each moment component.

Limitations

The preceding discussion holds rigorously only for pure bending. During bending, shear stress and shear deformations will also occur in the cross section; however, these shear stresses do not greatly affect the bending action, and they can be neglected in the calculation of bending stresses by Equations (8.21) through (8.25).

EXAMPLE 8.10

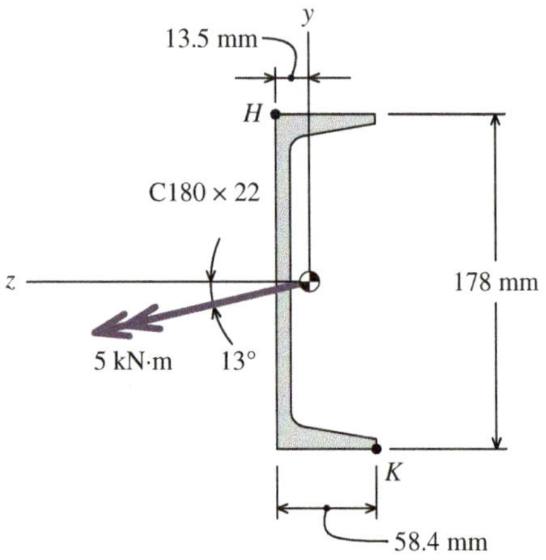

A standard steel C180 × 22 channel shape is subjected to a resultant bending moment $M = 5$ kN · m oriented at an angle of 13° with respect to the z axis, as shown. Calculate the bending stresses at points H and K, and determine the orientation of the neutral axis.

Plan the Solution

The section properties for the C180 × 22 channel shape can be obtained from Appendix B. Moment components in the y and z directions will be computed from the magnitude and orientation of the resultant bending moment. Since the channel shape has one axis of symmetry, the bending stresses at points H and K will be calculated from Equation (8.24) and the orientation of the neutral axis will be calculated from Equation (8.25).

SOLUTION

Section Properties

From Appendix B, the moments of inertia of the C180 × 22 shape are $I_y = 570{,}000$ mm^4 and $I_z = 11.3 \times 10^6$ mm^4. Since the shape has an axis of symmetry, the product of inertia is $I_{yz} = 0$. The depth and flange width of the C180 × 22 shape are $d = 178$ mm and $b_f = 58.4$ mm, respectively, and the distance from the back of the channel to its centroid is 13.5 mm. These dimensions are shown in the sketch.

Coordinates of Points *H* and *K*

The (*y*, *z*) coordinates of point *H* are

$$y_H = \frac{178 \text{ mm}}{2} = 89 \text{ mm} \qquad \text{and} \qquad z_H = 13.5 \text{ mm}$$

and the coordinates of point *K* are

$$y_K = -\frac{178 \text{ mm}}{2} = -89 \text{ mm} \qquad \text{and} \qquad z_K = 13.5 \text{ mm} - 58.4 \text{ mm} = -44.9 \text{ mm}$$

Moment Components

The bending moments about the *y* and *z* axes are as follows:

$$M_y = M \sin\theta = (5 \text{ kN}\cdot\text{m}) \sin(-13°) = -1.12576 \text{ kN}\cdot\text{m} = -1.12576 \times 10^6 \text{ N}\cdot\text{mm}$$

$$M_z = M \cos\theta = (5 \text{ kN}\cdot\text{m}) \cos(-13°) = 4.87185 \text{ kN}\cdot\text{m} = 4.87185 \times 10^6 \text{ N}\cdot\text{mm}$$

Bending Stresses at *H* and *K*

Since the C180 × 22 shape has an axis of symmetry, the bending stresses at points *H* and *K* can be computed from Equation (8.24). At point *H*, the bending stress is

$$\sigma_H = \frac{M_y z}{I_y} - \frac{M_z y}{I_z}$$

$$= \frac{(-1.12576 \times 10^6 \text{ N}\cdot\text{mm})(13.5 \text{ mm})}{570{,}000 \text{ mm}^4} - \frac{(4.87185 \times 10^6 \text{ N}\cdot\text{mm})(89 \text{ mm})}{11.3 \times 10^6 \text{ mm}^4}$$

$$= -65.0 \text{ MPa} = 65.0 \text{ MPa (C)}$$

Ans.

At point *K*, the bending stress is

$$\sigma_K = \frac{M_y z}{I_y} - \frac{M_z y}{I_z}$$

$$= \frac{(-1.12576 \times 10^6 \text{ N}\cdot\text{mm})(-44.9 \text{ mm})}{570{,}000 \text{ mm}^4} - \frac{(4.87185 \times 10^6 \text{ N}\cdot\text{mm})(-89 \text{ mm})}{11.3 \times 10^6 \text{ mm}^4}$$

$$= 127.0 \text{ MPa} = 127.0 \text{ MPa (T)}$$

Ans.

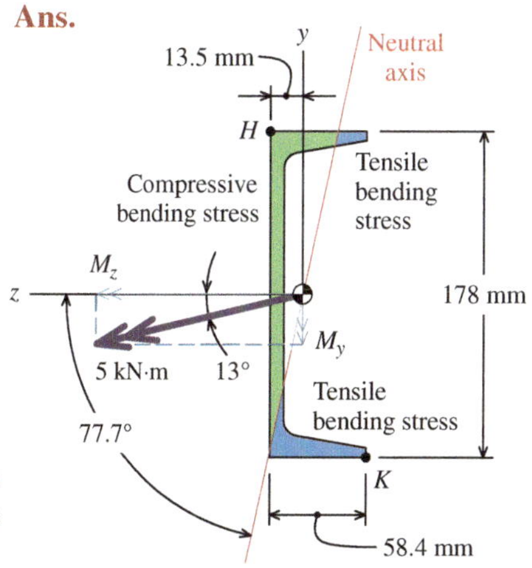

Orientation of the Neutral Axis

The orientation of the neutral axis can be calculated from Equation (8.25):

$$\tan\beta = \frac{M_y I_z}{M_z I_y} = \frac{(-1.12576 \text{ kN}\cdot\text{m})(11.3 \times 10^6 \text{ mm}^4)}{(4.87185 \text{ kN}\cdot\text{m})(570{,}000 \text{ mm}^4)} = -4.580949$$

$$\therefore \beta = -77.7°$$

Positive β angles are rotated clockwise from the *z* axis; therefore, the neutral axis is oriented as shown in the sketch, which has been shaded to indicate the tensile and compressive normal stress regions of the cross section.

297

EXAMPLE 8.11

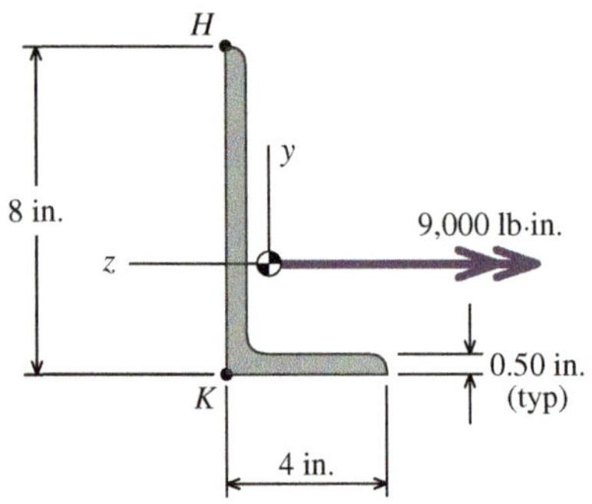

An unequal-leg angle shape is subjected to a bending moment $M = 9{,}000$ lb · in., oriented as shown. Calculate the bending stresses at points H and K, and determine the orientation of the neutral axis.

Plan the Solution

To begin the calculation, we must first locate the centroid of the angle shape. Then, the area moments of inertia I_y and I_z and the product of inertia I_{yz} must be computed with respect to the centroid location. The bending stresses at points H and K will be computed from Equation (8.21), and the orientation of the neutral axis will be computed from Equation (8.23).

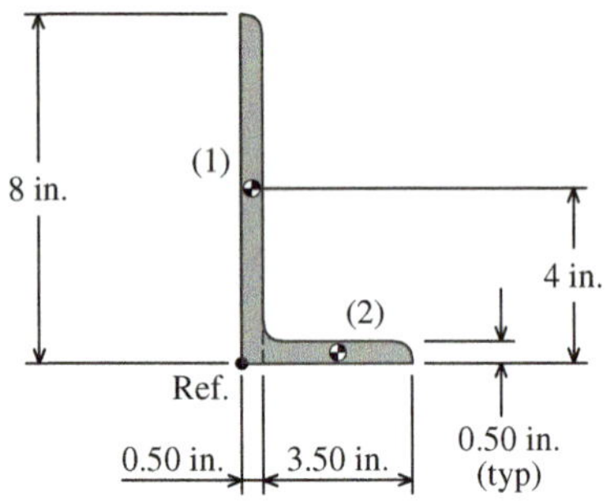

SOLUTION

Section Properties

The angle shape will be subdivided into two areas, (1) and (2), as shown. (**Note:** The fillets will be neglected in this calculation.) The corner of the angle will be used as the reference location (as indicated in the sketch) for calculations in both the horizontal and vertical directions. The location of the centroid in the vertical direction is calculated in the following manner:

	A_i (in.^{2})	y_i (in.)	y_iA_i (in.^{3})
(1)	4.00	4	16.00
(2)	1.75	0.25	0.4375
	5.75		16.4375

$$\bar{y} = \frac{\Sigma y_i A_i}{\Sigma A_i} = \frac{16.4375 \text{ in.}^3}{5.75 \text{ in.}^2} = 2.859 \text{ in.}$$

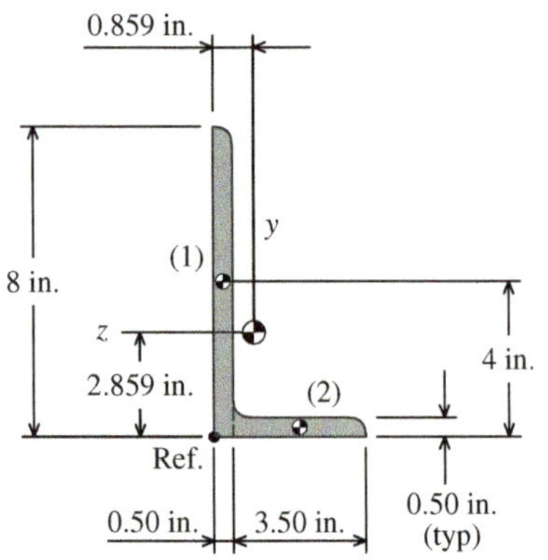

Similarly, the location of the centroid in the horizontal direction is calculated as follows:

	A_i (in.^{2})	z_i (in.)	z_iA_i (in.^{3})
(1)	4.00	−0.25	−1.00
(2)	1.75	−2.25	−3.9375
	5.75		−4.9375

$$\bar{z} = \frac{\Sigma z_i A_i}{\Sigma A_i} = \frac{-4.9375 \text{ in.}^3}{5.75 \text{ in.}^2} = -0.859 \text{ in.}$$

The location of the centroid for the angle shape is shown in the sketch. Next, the moment of inertia I_y is calculated for the angle shape about its y centroidal axis:

	A_i (in.^{2})	z_i (in.)	I_{yi} (in.^{4})	$\lvert d_i \rvert$ (in.)	$d_i^2A_i$ (in.^{4})	I_y (in.^{4})
(1)	4.00	−0.25	0.0833	0.609	1.4835	1.5668
(2)	1.75	−2.25	1.7865	1.391	3.3860	5.1725
						6.7393

Similarly, the moment of inertia I_z about the z centroidal axis is calculated as follows:

	A_i (in.^{2})	y_i (in.)	I_{zi} (in.^{4})	$\lvert d_i \rvert$ (in.)	$d_i^2 A_i$ (in.^{4})	I_z (in.^{4})
(1)	4.00	4	21.3333	1.1410	5.2075	26.5408
(2)	1.75	0.25	0.0365	2.6090	11.9120	11.9485
						38.4893

The product of inertia I_{yz} about the centroid is calculated thusly:

	A_i (in.^{2})	y_i (in.)	z_i (in.)	$\bar{y} - y_i$ (in.)	$\bar{z} - z_i$ (in.)	$I_{yz} = (\bar{y} - y_i)(\bar{z} - z_i)A_i$ (in.^{4})
(1)	4.00	4	−0.25	−1.1410	−0.6090	2.7795
(2)	1.75	0.25	−2.25	2.6090	1.3910	6.3510
						9.1304

Coordinates of Points *H* and *K*

The (y, z) coordinates of point H are

$$y_H = 8 \text{ in.} - 2.859 \text{ in.} = 5.141 \text{ in.} \qquad \text{and} \qquad z_H = 0.859 \text{ in.}$$

and the coordinates of point K are

$$y_K = -2.859 \text{ in.} \qquad \text{and} \qquad z_K = 0.859 \text{ in.}$$

Moment Components

The bending moment acts about the $-z$ axis; therefore,

$$M_z = -9{,}000 \text{ lb} \cdot \text{in.} \qquad \text{and} \qquad M_y = 0$$

Bending Stresses at *H* and *K*

Since the angle shape does not have an axis of symmetry, the bending stresses at points H and K must be computed from Equation (8.21) or Equation (8.22). Because $M_y = 0$, Equation (8.22) is the more convenient of the two equations in this instance. The bending stress at point H is therefore

$$\begin{aligned}
\sigma_H &= \left(\frac{I_z z - I_{yz} y}{I_y I_z - I_{yz}^2} \right) M_y + \left(\frac{-I_y y + I_{yz} z}{I_y I_z - I_{yz}^2} \right) M_z \\
&= 0 + \left[\frac{-(6.7393 \text{ in.}^4)(5.141 \text{ in.}) + (9.1304 \text{ in.}^4)(0.859 \text{ in.})}{(6.7393 \text{ in.}^4)(38.4893 \text{ in.}^4) - (9.1304 \text{ in.}^4)^2} \right] (-9{,}000 \text{ lb} \cdot \text{in.}) \\
&= 1{,}370 \text{ psi} = 1{,}370 \text{ psi (T)}
\end{aligned}$$

and the bending stress at point K is

$$\sigma_K = \left(\frac{I_z z - I_{yz} y}{I_y I_z - I_{yz}^2}\right) M_y + \left(\frac{-I_y y + I_{yz} z}{I_y I_z - I_{yz}^2}\right) M_z$$

$$= 0 + \left[\frac{-(6.7393\text{ in.}^4)(-2.859\text{ in.}) + (9.1304\text{ in.}^4)(0.859\text{ in.})}{(6.7393\text{ in.}^4)(38.4893\text{ in.}^4) - (9.1304\text{ in.}^4)^2}\right](-9{,}000\text{ lb}\cdot\text{in.})$$

$$= -1{,}386\text{ psi} = 1{,}386\text{ psi (C)}$$

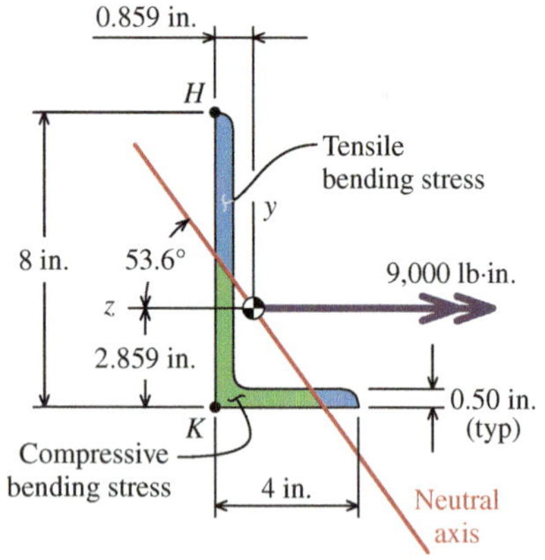

Orientation of the Neutral Axis

The orientation of the neutral axis can be calculated from Equation (8.23):

$$\tan\beta = \frac{M_y I_z + M_z I_{yz}}{M_z I_y + M_y I_{yz}} = \frac{0 + (-9{,}000\text{ lb}\cdot\text{in.})(9.1304\text{ in.}^4)}{(-9{,}000\text{ lb}\cdot\text{in.})(6.7393\text{ in.}^4) + 0} = 1.3548$$

$$\therefore \beta = 53.6^\circ$$

Positive β angles are rotated clockwise from the z axis; therefore, the neutral axis is oriented as shown in the sketch, which has been shaded to indicate the tensile and compressive normal stress regions of the cross section.

PROBLEMS

P8.60 A wooden beam with a rectangular cross section is subjected to bending moment magnitudes $M_z = 1{,}250$ lb · ft and $M_y = 460$ lb · ft, acting in the directions shown in Figure P8.60. The cross-sectional dimensions are $b = 4$ in. and $d = 7$ in. Determine

(a) the maximum magnitude of the bending stress in the beam.
(b) the angle β that the neutral axis makes with the $+z$ axis. Note that positive β angles rotate clockwise from the $+z$ axis.

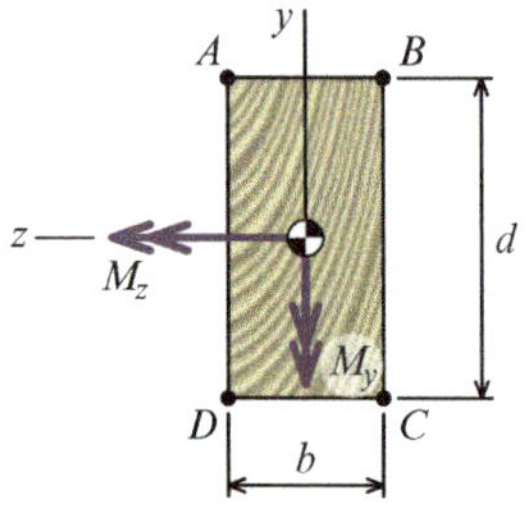

FIGURE P8.60

P8.61 A hollow-core concrete plank is subjected to bending moment magnitudes $M_z = 50$ kip · ft and $M_y = 20$ kip · ft, acting in the directions shown in Figure P8.61. The cross-sectional dimensions are $b = 24$ in., $h = 12$ in., $d = 7$ in., and $a = 5.5$ in. Determine

(a) the bending stress at B.
(b) the bending stress at C.
(c) the angle β that the neutral axis makes with the $+z$ axis. Note that positive β angles rotate clockwise from the $+z$ axis.

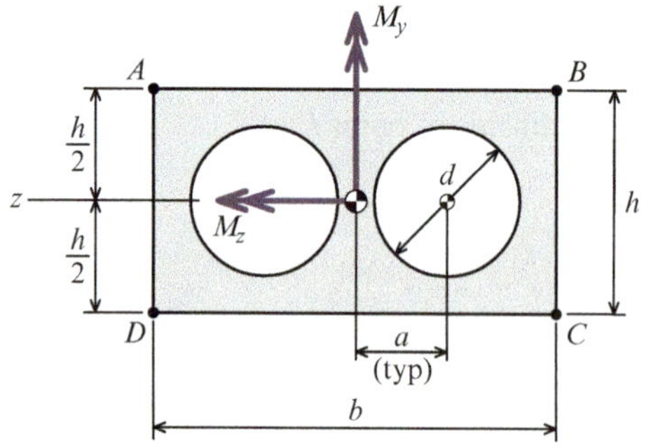

FIGURE P8.61

P8.62 The moment M acting on the cross section of a certain tee beam is oriented at an angle of $\theta = 55^\circ$ as shown in Figure P8.62. The dimensions of the cross section are $b_f = 180$ mm, $t_f = 16$ mm, $d = 200$ mm, and $t_w = 10$ mm. The allowable bending stress is 165 MPa. What is the largest bending moment M that can be applied as shown to this cross section?

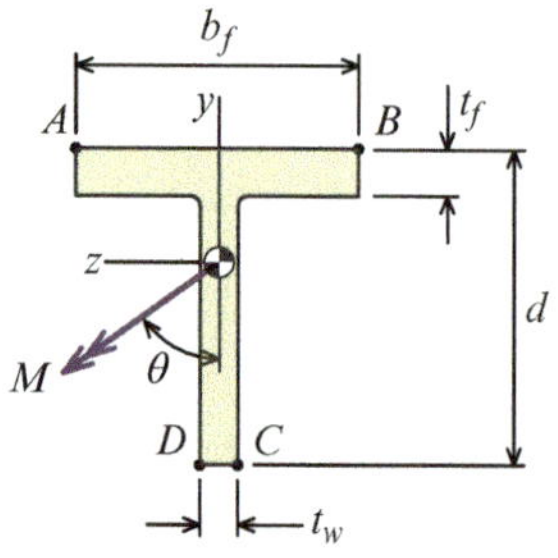

FIGURE P8.62

P8.63 A downward concentrated load P = 42 kN acts on wall BC of the box beam shown in Figure 8.63*a*. The load acts at a distance a = 35 mm below the z centroidal axis. The box has dimensions b = 80 mm, d = 120 mm, and t = 8 mm, as shown in Figure 8.63*b*. Determine the normal stress

(a) at corner A.
(b) at corner D.

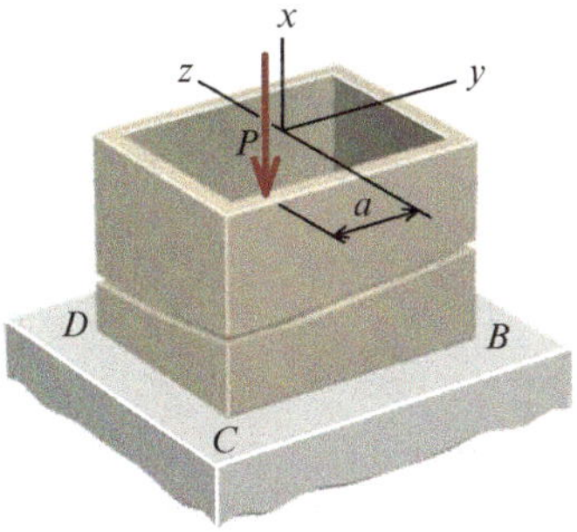

FIGURE P8.63*a*

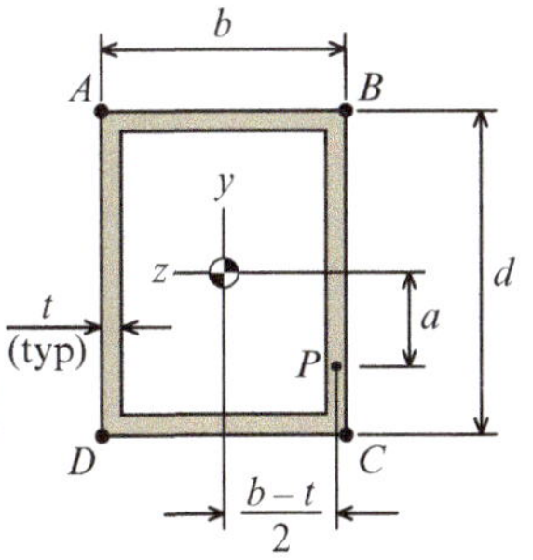

FIGURE P8.63*b*

P8.64 The cantilever beam shown in Figure P8.64*a* is subjected to load magnitudes P_z = 2.8 kips and P_y = 8.3 kips. The flanged cross section shown in Figure P8.64*b* has dimensions b_f = 8.0 in., t_f = 0.62 in., d = 10.0 in., and t_w = 0.35 in. Using L = 84 in., determine

(a) the bending stress at point A.
(b) the bending stress at point B.
(c) the angle β that the neutral axis makes with the +z axis. Note that positive β angles rotate clockwise from the +z axis.

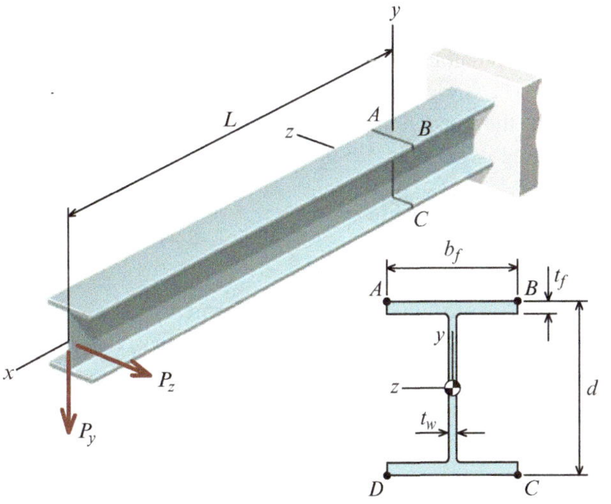

FIGURE P8.64*a*

FIGURE P8.64*b*

P8.65 The U-shaped cross section is oriented at an angle θ = 35° as shown in Figure P8.65. The dimensions of the cross section are b = 80 mm, d = 65 mm, and t = 5 mm. The allowable tensile bending stress is 150 MPa, and the allowable compressive bending stress is 80 MPa. What is the largest bending moment M_z that can be applied as shown to this cross section?

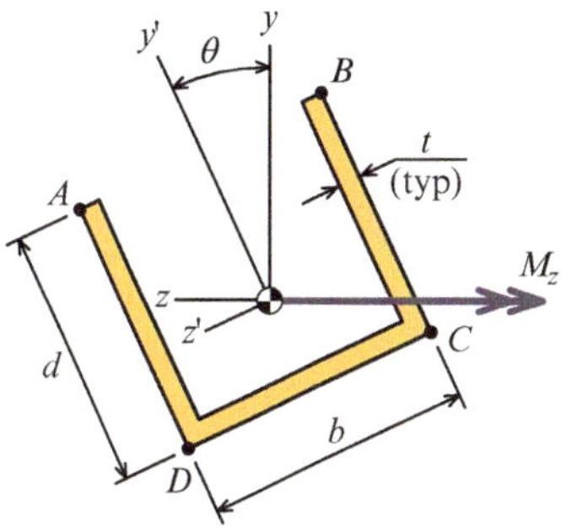

FIGURE P8.65

P8.66 The moment acting on the cross section of the zee shape shown in figure P8.66 has a magnitude of M_z = 5,600 lb·ft and is oriented at an angle θ = 24°. The dimensions of the cross section are b_f = 3.0 in., t_f = 0.35 in., d = 7.0 in., and t_w = 0.20 in. Determine

(a) the moments of inertia $I_{y'}$ and $I_{z'}$.
(b) the product of inertia $I_{y'z'}$.
(c) the principal moments of inertia and the angle from the +z' axis to the axis of the maximum moment of inertia.
(d) the bending stress at points A and C.
(e) the angle β that the neutral axis makes with the +z axis. Note that positive β angles rotate clockwise from the +z axis.

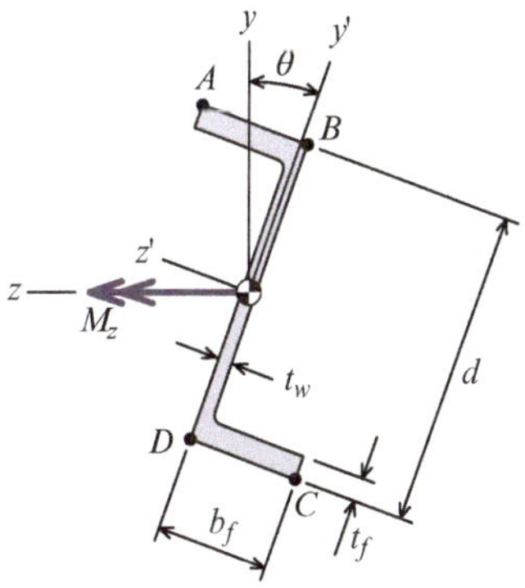

FIGURE P8.66

P8.67 A beam cross section is fabricated by rigidly attaching two lumber boards together as shown in Figure P8.67. The dimensions of the cross section are b_1 = 7.25 in., b_2 = 9.25 in., and t = 1.50 in. The moment acting on the cross section is M_z = 930 lb·ft. Determine

(a) the moments of inertia I_y and I_z.
(b) the product of inertia I_{yz}.

(c) the principal moments of inertia and the angle from the +z axis to the axis of the maximum moment of inertia.
(d) the maximum compressive bending stress in the cross section.

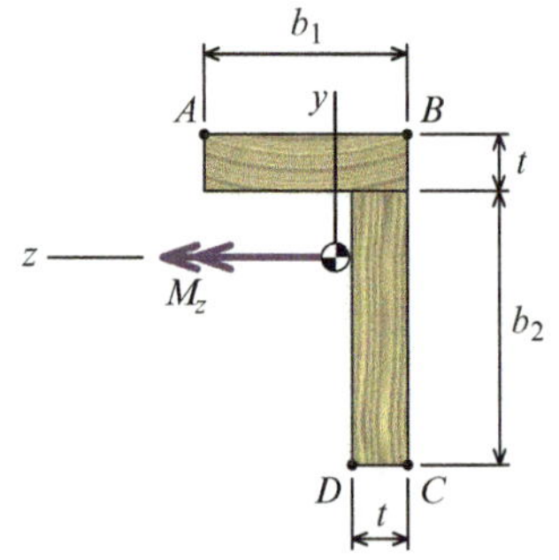

FIGURE P8.67

P8.68 The extruded beam cross section shown in Figure P8.68 is subjected to a bending moment of magnitude $M_z = 260$ N·m, acting in the indicated direction. The dimensions of the cross section are $b_1 = 40$ mm, $b_2 = 70$ mm, $d = 90$ mm, and $t = 10$ mm. Determine

(a) the moments of inertia I_y and I_z.
(b) the product of inertia I_{yz}.
(c) the principal moments of inertia and the angle from the +z axis to the axis of the maximum moment of inertia.
(d) the maximum tensile bending stress in the cross section.

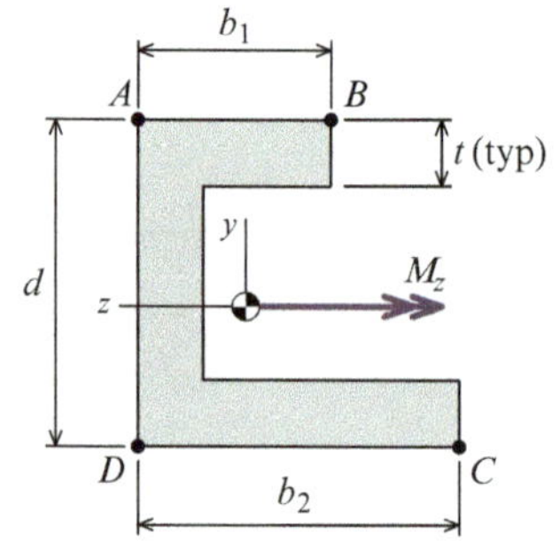

FIGURE P8.68

8.9 Stress Concentrations Under Flexural Loadings

In Section 5.7, it was shown that the introduction of a circular hole or other geometric discontinuity into an axially loaded member could cause a significant increase in the stress near the discontinuity. Similarly, increased stresses occur near any reduction in diameter of a circular shaft subjected to torsion. This phenomenon, termed **stress concentration**, occurs in flexural members as well.

In Section 8.3, it was shown that the normal stress magnitude in a beam of uniform cross section in a region of pure bending is given by Equation (8.10):

$$\sigma_{\text{max}} = \frac{Mc}{I_z}$$

The bending stress magnitude computed from Equation (8.10) is termed a **nominal stress** because it does not account for the stress-concentration phenomenon. Near notches, grooves, fillets, or any other abrupt change in cross section, the normal stress due to bending can be significantly greater. The relationship between the maximum bending stress at the discontinuity and the nominal stress computed from Equation (8.10) is expressed as the stress-concentration factor

For a rectangular beam, the nominal bending stress used in Equation (8.26) is the stress at its minimum depth. For a circular shaft, the nominal bending stress is computed for its minimum diameter.

$$K = \frac{\sigma_{\text{max}}}{\sigma_{\text{nom}}} \tag{8.26}$$

The nominal stress used in Equation (8.26) is the bending stress computed for the minimum depth or diameter of the flexural member at the location of the discontinuity. Since the factor K depends only upon the geometry of the member, curves can be developed that show the stress-concentration factor K as a function of the ratios of the parameters involved. Such curves for notches and fillets in rectangular cross sections subjected to pure bending are shown in Figures 8.17 and 8.18.[1] Similar curves for grooves and fillets in circular shafts subjected to pure bending are shown in Figures 8.19 and 8.20.[2]

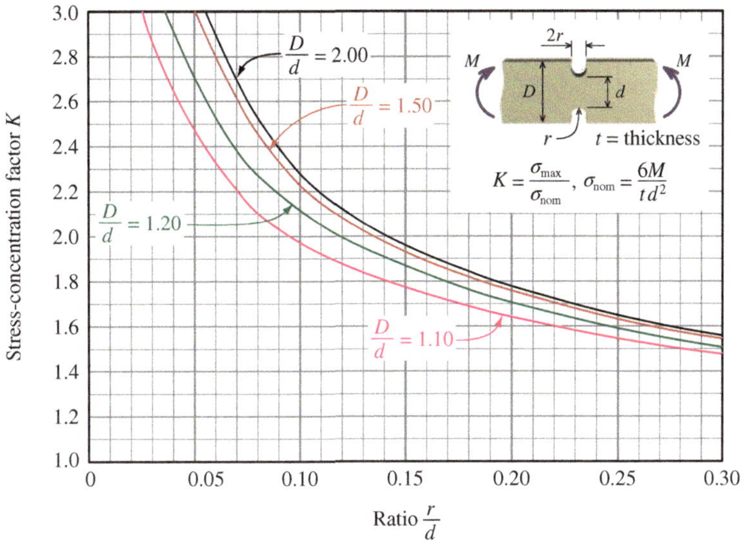

FIGURE 8.17 Stress-concentration factors K for bending of a flat bar with opposite U-shaped notches.

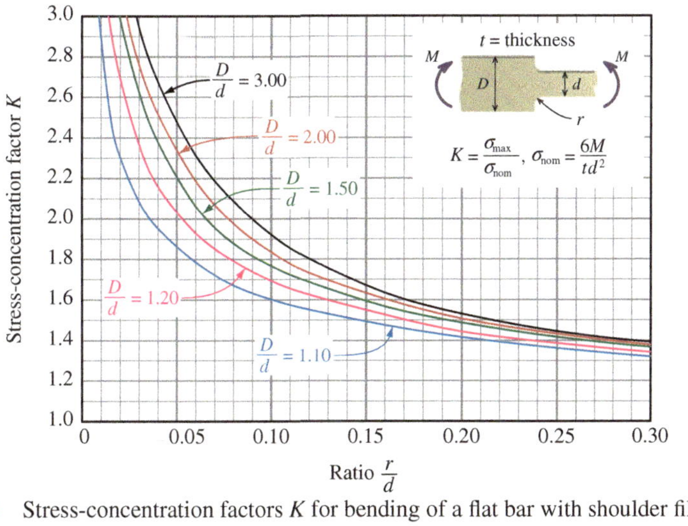

FIGURE 8.18 Stress-concentration factors K for bending of a flat bar with shoulder fillets.

[1] Adapted from Walter D. Pilkey, *Peterson's Stress Concentration Factors*, 2nd ed. (New York: John Wiley & Sons, Inc., 1997).

[2] *Ibid.*

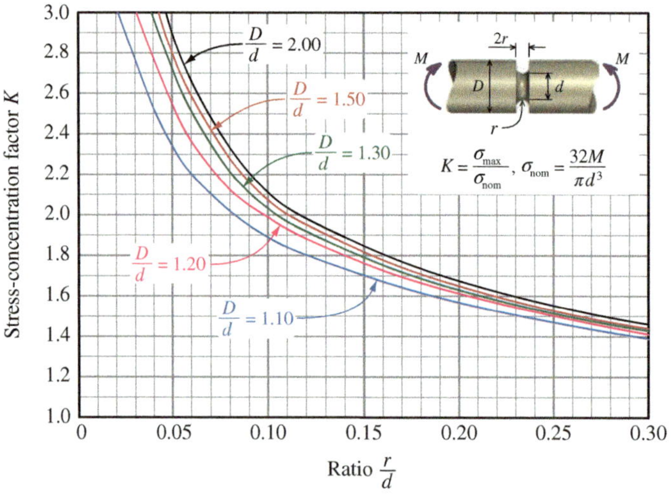

FIGURE 8.19 Stress-concentration factors K for bending of a circular shaft with a U-shaped groove.

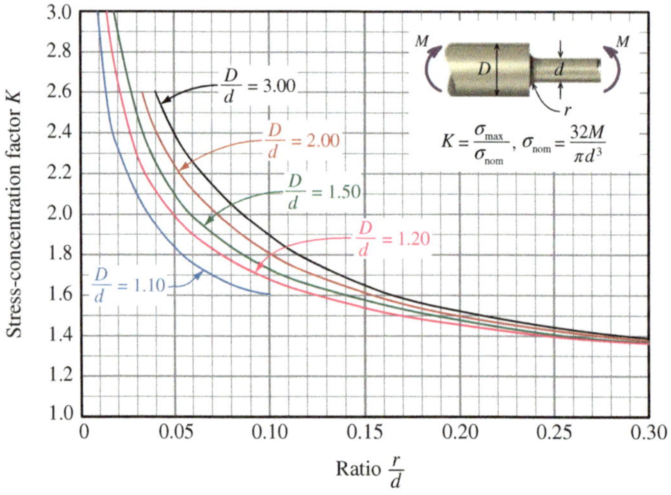

FIGURE 8.20 Stress-concentration factors K for bending of a stepped shaft with shoulder fillets.

EXAMPLE 8.12

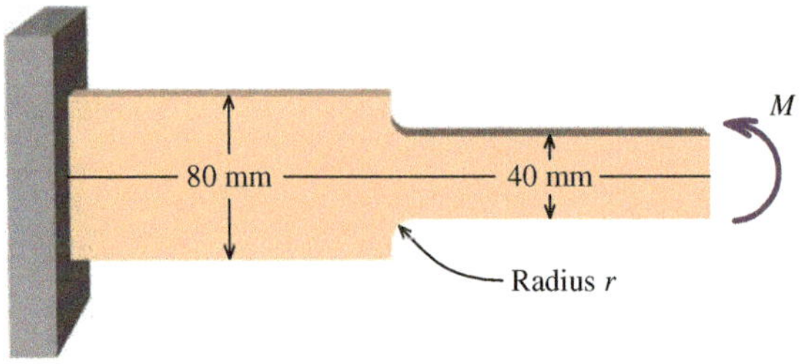

A cantilever spring made of SAE 4340 heat-treated steel is 50 mm thick. As shown in the figure, the depth of the rectangular cross section is reduced from 80 mm to 40 mm, with fillets at the transition. A factor of safety (FS) of 2.5 with respect to fracture is specified for the spring. Determine the maximum safe moment M for the spring if

(a) the fillet radius r is 4 mm.
(b) the fillet radius r is 12 mm.

SOLUTION

The ultimate strength σ_U for heat-treated SAE 4340 steel (see Appendix D for its properties) is 1,030 MPa. Thus, the allowable stress for the spring is

$$\sigma_{allow} = \frac{\sigma_U}{FS} = \frac{1{,}034 \text{ MPa}}{2.5} = 413.6 \text{ MPa}$$

The moment of inertia at the minimum spring depth is

$$I = \frac{(50 \text{ mm})(40 \text{ mm})^3}{12} = 266{,}667 \text{ mm}^4$$

An expression for the allowable bending-moment magnitude in terms of the stress-concentration factor K can be derived from Equation (8.26) by setting $\sigma_{max} = \sigma_{allow}$ and $\sigma_{nom} = M_{allow}c/I$:

$$K = \frac{\sigma_{max}}{\sigma_{nom}} = \frac{\sigma_{allow}}{\dfrac{M_{allow}c}{I}}$$

Solve this equation for M_{allow} to derive

$$M_{allow} = \frac{\sigma_{allow}I}{Kc} = \frac{(413.6 \text{ N/mm}^2)(266{,}667 \text{ mm}^4)}{K(20 \text{ mm})} = \frac{5{,}514{,}574 \text{ N}\cdot\text{mm}}{K} = \frac{5{,}515 \text{ N}\cdot\text{m}}{K}$$

With reference to the nomenclature used in Figure 8.18, the ratio of the maximum spring depth D to the reduced depth d is $D/d = 80/40 = 2.0$.

(a) Fillet Radius $r = 4$ mm

A stress-concentration factor $K = 1.84$ is obtained from Figure 8.18 with $D/d = 2.0$ and $r/d = 4/40 = 0.10$. The maximum allowable bending moment is thus

$$M = \frac{5{,}515 \text{ N}\cdot\text{m}}{K} = \frac{5{,}515 \text{ N}\cdot\text{m}}{1.84} = 2{,}997 \text{ N}\cdot\text{m} \qquad \textbf{Ans.}$$

(b) Fillet Radius $r = 12$ mm

For a 12 mm fillet, $r/d = 12/40 = 0.30$, and thus, the corresponding stress-concentration factor from Figure 8.18 is $K = 1.38$. Accordingly, the maximum allowable bending moment is

$$M = \frac{5{,}515 \text{ N}\cdot\text{m}}{K} = \frac{5{,}515 \text{ N}\cdot\text{m}}{1.38} = 3{,}996 \text{ N}\cdot\text{m} \qquad \textbf{Ans.}$$

PROBLEMS

P8.69 A stainless steel spring (shown in Figure P8.69/70) has a thickness of 3/4 in. and a change in depth at section B from $D = 1.50$ in. to $d = 1.25$ in. The radius of the fillet between the two sections is $r = 0.125$ in. If the bending moment applied to the spring is $M = 2{,}000$ lb·in., determine the maximum normal stress in the spring.

P8.70 A spring made of a steel alloy (shown in Figure P8.69/70) has a thickness of 25 mm and a change in depth at section B from $D = 75$ mm to $d = 50$ mm. If the radius of the fillet between the two sections is $r = 8$ mm, determine the maximum moment that the spring can resist if the maximum bending stress in the spring must not exceed 120 MPa.

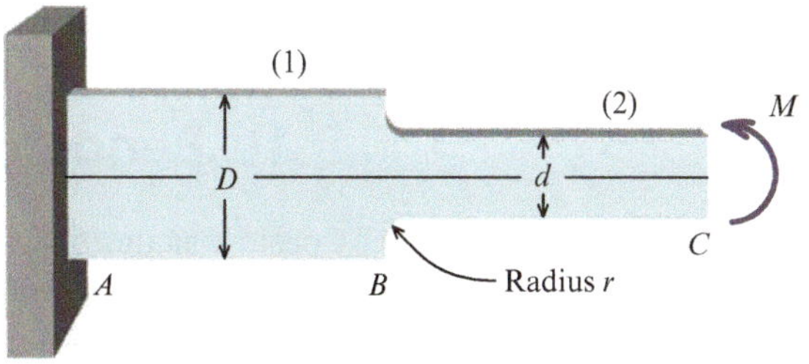

FIGURE P8.69/70

P8.71 The notched bar shown in Figure P8.71/72 is subjected to a bending moment $M = 300$ N · m. The major bar width is $D = 75$ mm, the minor bar width at the notches is $d = 50$ mm, and the radius of each notch is $r = 10$ mm. If the maximum bending stress in the bar must not exceed 90 MPa, determine the minimum required bar thickness b.

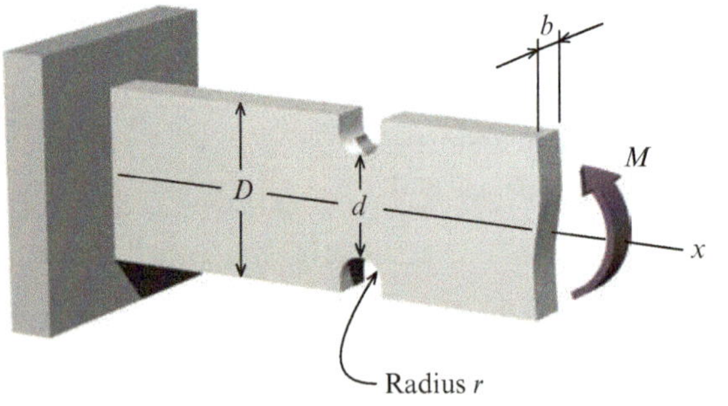

FIGURE P8.71/72

P8.72 The machine part shown in Figure P8.71/72 is made of cold-rolled 18-8 stainless steel. (See Appendix D for properties.) The major bar width is $D = 1.50$ in., the minor bar width at the notches is $d = 1.00$ in., the radius of each notch is $r = 0.125$ in., and the bar thickness is $b = 0.25$ in. Determine the maximum safe moment M that may be applied to the bar if a factor of safety of 2.5 with respect to failure by yield is specified.

P8.73 The C86100 bronze (see Appendix D for properties) shaft shown in Figure P8.73 is supported at each end by self-aligning bearings. The major shaft diameter is $D = 40$ mm, the minor shaft diameter is $d = 25$ mm, and the radius of the fillet between the major and minor diameter sections is $r = 5$ mm. The shaft length is $L = 500$ mm and the fillets are located at $x = 150$ mm. Determine the maximum load P that may be applied to the shaft if a factor of safety of 3.0 is specified.

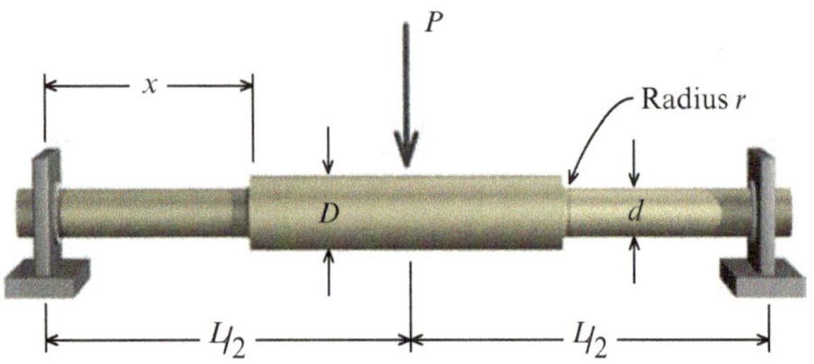

FIGURE P8.73

P8.74 The machine shaft shown in Figure P8.74/75 is made of 1020 cold-rolled steel. (See Appendix D for properties.) The major shaft diameter is $D = 1.000$ in., the minor shaft diameter is $d = 0.625$ in., and the radius of the fillet between the major and minor diameter sections is $r = 0.0625$ in. The fillets are located at $x = 4$ in. from C. If a load $P = 125$ lb is applied at C, determine the factor of safety in the fillet at B.

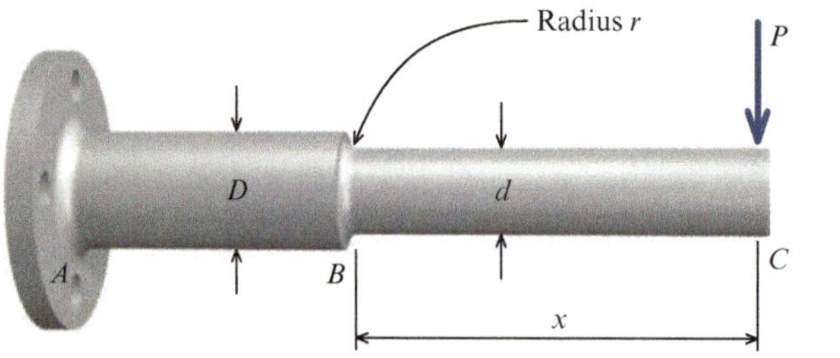

FIGURE P8.74/75

P8.75 The machine shaft shown in Figure P8.74/75 is made of 1020 cold-rolled steel. (See Appendix D for properties.) The major shaft diameter is $D = 30$ mm, the minor shaft diameter is $d = 20$ mm, and the radius of the fillet between the major and minor diameter sections is $r = 3$ mm. The fillets are located at $x = 90$ mm from C. Determine the maximum load P that can be applied to the shaft at C if a factor of safety of 1.5 is specified.

P8.76 The grooved shaft shown in Figure P8.76 is made of C86100 bronze. (See Appendix D for properties.) The major shaft diameter is $D = 50$ mm, the minor shaft diameter at the groove is $d = 34$ mm, and the radius of the groove is $r = 4$ mm. Determine the maximum allowable moment M that may be applied to the shaft if a factor of safety of 1.5 with respect to failure by yield is specified.

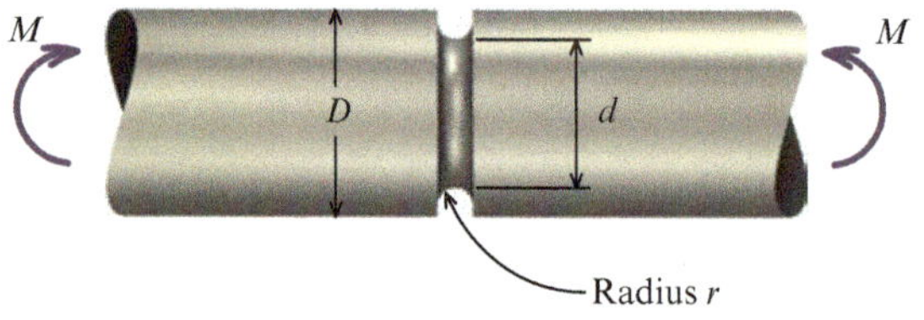

FIGURE P8.76

8.10 Bending of Curved Bars

Consider an unstressed curved bar (Figure 8.21*a*) of uniform cross section with a vertical axis of symmetry (Figure 8.21*b*). The outer and inner fibers of the beam are located at radial distances r_o and r_i from the center of curvature, O, respectively. The radius of curvature of the centroidal axis is denoted by r_c. We will focus our attention on a small portion of the bar located between cross sections a–c–e and b–d–f that are separated from each other by a small central angle θ.

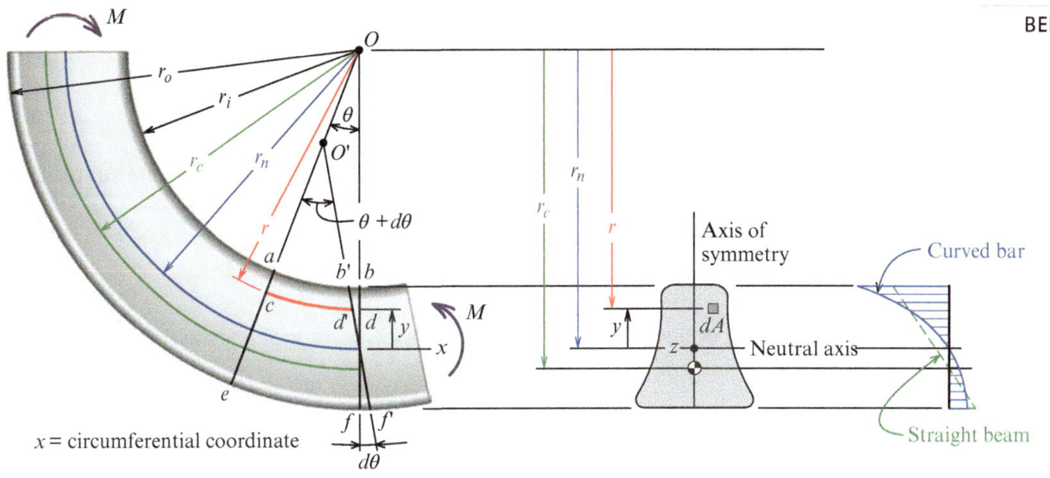

(a) Geometry of curved bar (b) Cross section (c) Circumferential strain and stress distribution

FIGURE 8.21 Curved bar in pure bending.

The ends of the bar are subjected to bending moments M that produce compressive normal stresses on the inner surface of the bar. After these moments are applied, the curvature of the bar changes and the center of curvature of the bar moves from its original location O to a new location O'. The movement of the center of curvature from O to O' is greatly exaggerated in Figure 8.21*a* to illustrate the changes in geometry more clearly.

We assume that plane cross sections (such as sections *a–c–e* and *b–d–f*) in the unstressed bar remain planar after the end moments M have been applied. On the basis of this assumption, the deformation of bar fibers must be linearly distributed with respect to a neutral surface (yet to be determined) that is located at a radius r_n from the center of curvature, O. Before the moments M are applied, the initial arc length of an arbitrary fiber cd of the bar can be expressed in terms of the radial distance r and the central angle θ as $cd = r\theta$. The initial arc length can also be expressed in terms of the distance from the neutral surface as $cd = (r_n - y)\theta$.

After the moments M are applied, the contraction of fiber cd can be expressed as $dd' = -y\,d\theta$, or, in terms of the radial distance to the neutral surface, $dd' = -(r_n - r)\,d\theta$. Therefore, the normal strain in the circumferential direction x for an arbitrary fiber of the bar is defined as

$$\varepsilon_x = \frac{dd'}{cd} = \frac{-y\,d\theta}{r\theta} = \frac{-(r_n - r)\,d\theta}{r\theta} \qquad \text{(a)}$$

Equation (a) shows that the circumferential normal strain ε_x does not vary linearly with the distance y from the neutral surface of the bar. Rather, the distribution of strain is *nonlinear*, as shown in Figure 8.21*c*. The physical reason for this distribution is that the initial lengths of circumferential fibers cd vary with y, being shorter toward the center of curvature, O. Thus, while *deformations* dd' are linear with respect to y, these elongations are divided by different initial lengths, so the strain ε_x is not directly proportional to y.

The circumferential normal stress acting on area dA can now be obtained from Hooke's law as

$$\sigma_x = -E\frac{(r_n - r)d\theta}{r\theta} \tag{b}$$

The location of the neutral axis (N.A.) follows from the condition that the summation of the forces acting perpendicular to the section must equal zero; that is,

$$\Sigma F_x = 0 \qquad \int_A \sigma_x dA = -\int_A E\frac{(r_n - r)d\theta}{r\theta}dA = 0$$

However, since r_n, E, θ, and $d\theta$ are constant at any one section of a stressed bar, they may be taken outside the integral to obtain

$$\int_A \sigma_x dA = -\frac{E\,d\theta}{\theta}\int_A \frac{r_n - r}{r}dA = -\frac{E\,d\theta}{\theta}\left(r_n\int_A \frac{dA}{r} - \int_A dA\right) = 0 \tag{c}$$

To satisfy equilibrium, the value of r_n must be

$$r_n = \frac{A}{\displaystyle\int_A \frac{dA}{r}} \tag{8.27}$$

where A is the cross-sectional area of the bar and r_n locates the neutral surface (and thus the neutral axis) of the curved bar relative to the center of curvature. (Note that the location of the neutral axis does not coincide with that of the centroidal axis.)

Formulas for the areas A and the radial distances r_n from the center of curvature, O, to the neutral axis are given in Table 8.2 for several typical cross sections. These formulas can be combined as necessary for a shape made up of several shapes. For example, a tee shape can be calculated as the sum of two rectangles, where Equation (8.27) is expanded by the addition of a second term. Shapes can also be treated as negative areas. For example, the radial distance to the neutral axis of a box-shaped cross section can be calculated by the combination of (a) an outer rectangle whose area A is treated as a positive quantity in Equation (8.27), and (b) an inner rectangle whose area is treated as a negative quantity. Clearly, the sum of the positive area of the outer rectangle and the negative area of the inner rectangle gives the area of the box shape.

After the location of the neutral axis is established, the equation for the circumferential normal stress distribution is obtained by equating the external moment M to the internal resisting moment developed by the stresses expressed in Equation (b). The summation of moments is made about the z axis, which is placed at the neutral-axis location found in Equation (8.27):

$$\Sigma M_z = 0 \qquad M + \int_A y\sigma_x\,dA = 0$$

The distance y can be expressed as $y = r_n - r$. The normal stress σ_x was defined in Equation (b). We substitute these expressions into the integral term to obtain

$$M = -\int_A \sigma_x(r_n - r)\,dA = \int_A E\frac{(r_n - r)^2\,d\theta}{r\theta}dA \tag{d}$$

Since the variables E, θ, and $d\theta$ are constant at any one section of a stressed bar, Equation (d) can be expressed as

$$M = \frac{E\,d\theta}{\theta}\int_A \frac{(r_n - r)^2}{r}dA \tag{e}$$

Table 8.2 Area and Radial Distance to Neutral Axis for Selected Cross Sections

Shape	Cross Section	Radius of Neutral Surface
A. Rectangle	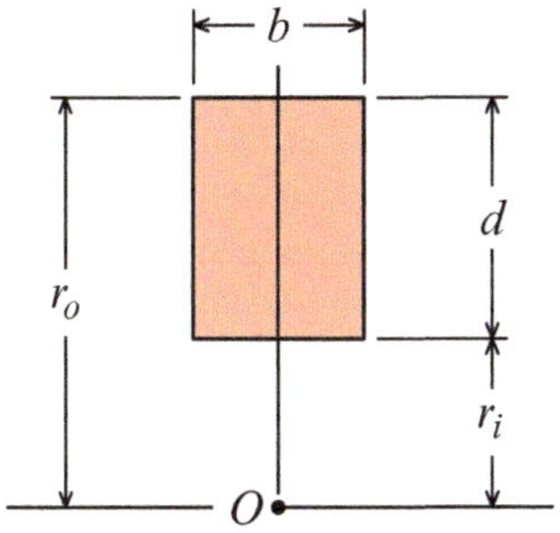	$r_n = \dfrac{A}{b \ln \dfrac{r_o}{r_i}}$ $A = bd$
B. Circle	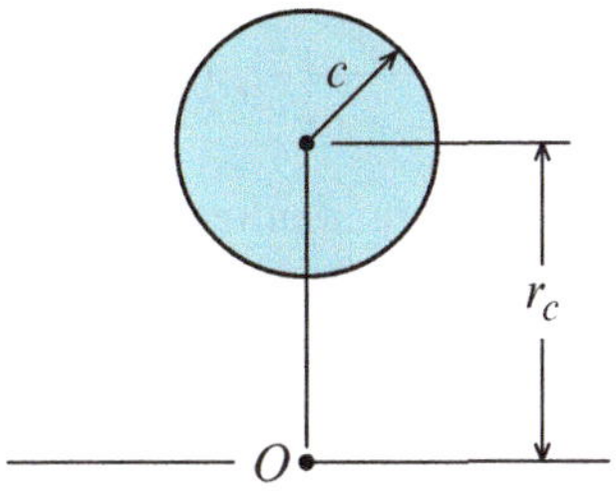	$r_n = \dfrac{A}{2\pi\left(r_c - \sqrt{r_c^2 - c^2}\right)}$ $A = \pi c^2$
C. Ellipse	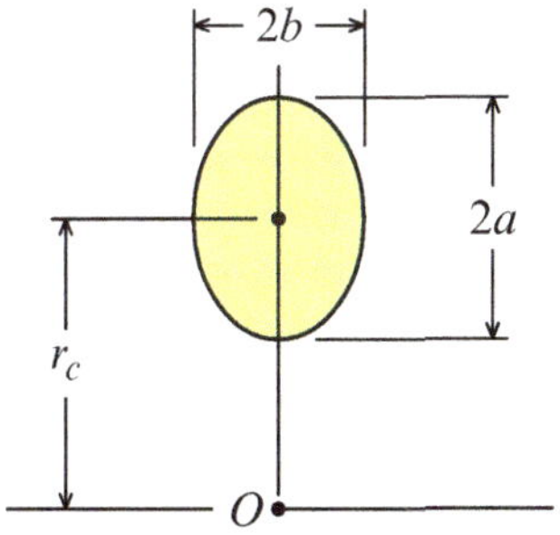	$r_n = \dfrac{A}{\dfrac{2\pi b}{a}\left(r_c - \sqrt{r_c^2 - a^2}\right)}$ $A = \pi ab$
D. Triangle	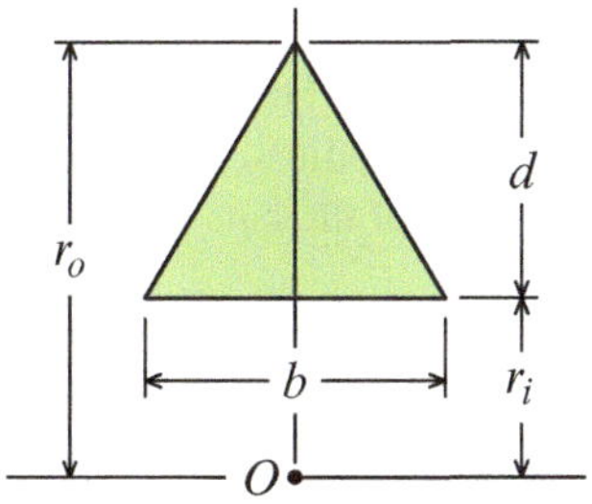	$r_n = \dfrac{A}{\dfrac{br_o}{d}\left(\ln \dfrac{r_o}{r_i}\right) - b}$ $A = \dfrac{1}{2}bd$
E. Trapezoid	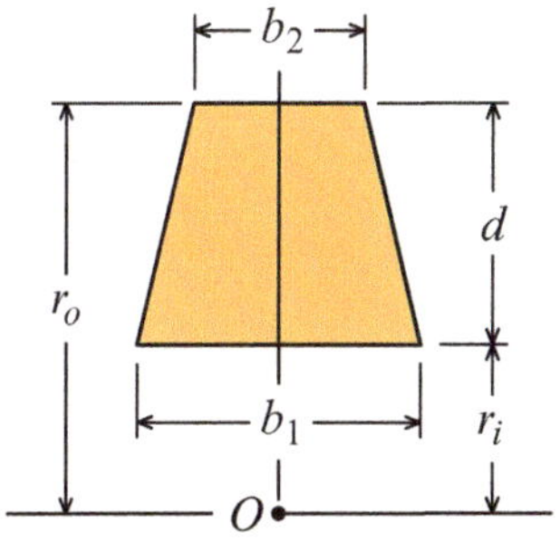	$r_n = \dfrac{A}{\dfrac{1}{d}\left[(b_1 r_o - b_2 r_i) \ln \dfrac{r_o}{r_i} - d(b_1 - b_2)\right]}$ $A = \dfrac{1}{2}(b_1 + b_2)d$

To remove θ from this expression, we once again use Equation (b), rewriting it, however, as

$$\frac{E\,d\theta}{\theta} = -\frac{\sigma_x r}{r_n - r} \tag{f}$$

Then we substitute Equation (f) into Equation (e) to obtain

$$M = -\frac{\sigma_x r}{r_n - r}\int_A \frac{(r_n - r)^2}{r}\,dA \tag{g}$$

Next, we expand Equation (g) as follows:

$$\begin{aligned} M &= -\frac{\sigma_x r}{r_n - r}\left(\int_A \frac{r_n^2}{r}\,dA - \int_A \frac{r_n r}{r}\,dA - \int_A \frac{r_n r}{r}\,dA + \int_A r\,dA\right) \\ &= -\frac{\sigma_x r r_n}{r_n - r}\left(r_n\int_A \frac{dA}{r} - \int_A dA\right) - \frac{\sigma_x r}{r_n - r}\left(-r_n\int_A dA + \int_A r\,dA\right) \end{aligned} \tag{h}$$

Notice that the first two integrals in Equation (h) are identical to the terms in parentheses in Equation (c). Accordingly, these two integrals vanish, leaving

$$M = -\frac{\sigma_x r}{r_n - r}\left(-r_n\int_A dA + \int_A r\,dA\right) \tag{i}$$

The first integral in Equation (i) is simply the area A. The second integral is the first moment of area about the center of curvature. From the definition of a centroid, this integral can be expressed as $r_c A$, where r_c is the radial distance from the center of curvature, O, to the centroid of the cross section. Thus,

$$M = -\frac{\sigma_x r}{r_n - r}(-r_n A + r_c A) = -\sigma_x r\,A\left(\frac{r_c - r_n}{r_n - r}\right)$$

Solving this equation for the circumferential normal stress created in a curved bar by a bending moment M gives

$$\sigma_x = -\frac{M(r_n - r)}{r\,A(r_c - r_n)} \tag{8.28}$$

A positive bending moment makes the curve tighter.

In this formula, the bending moment M is considered to be a positive value if it creates compressive normal stress on the inner surface (i.e., at radius r_i) of the curved bar. A positive bending moment decreases the radius of curvature of the bar.

EXAMPLE 8.13

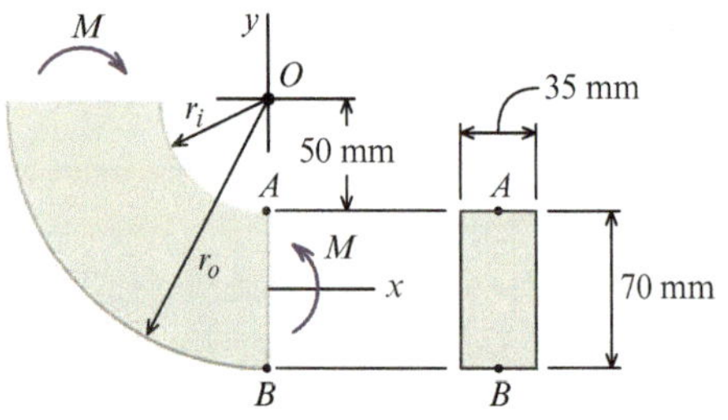

A curved bar with a rectangular cross section is subjected to a bending moment $M = 4{,}500$ N·m, acting in the direction shown. The bar has a width of 35 mm and a height of 70 mm. The inside radius of the curved bar is $r_i = 50$ mm.

(a) Determine the bending stresses in the curved bar at points A and B.

(b) Sketch the distribution of flexural stresses in the curved bar.

(c) Determine the percent error if the flexure formula for straight beams were used for part (a).

Plan the Solution

Begin by calculating the radial distances from the center of curvature, O, to the centroid of the cross section and to the neutral-axis location. Then, use Equation (8.28) to calculate the flexural stresses on the inner surface of the bar at A and on the outer surface at B. Repeat this calculation for radial values from $r = 50$ mm to $r = 120$ mm to obtain the flexural stress distribution. Next, determine the section modulus for the rectangle, and calculate the bending stresses at A and B using $\sigma_x = M/S$. Finally, calculate percent errors for the stresses at the inner and outer surfaces.

SOLUTION

The bar has a width of 35 mm and a depth of 70 mm; therefore, its cross-sectional area is $A = 2{,}450\text{ mm}^2$. The radial distance from the center of curvature, O, to the centroid of the cross section is

$$r_c = 50\text{ mm} + \frac{75\text{ mm}}{2} = 85\text{ mm}$$

Location of the Neutral Axis

Next, we will determine the location of the neutral axis r_n from Equation (8.27):

$$r_n = \frac{A}{\int_A \frac{dA}{r}}$$

The denominator can be integrated to yield

$$\int_A \frac{dA}{r} = \int_{50}^{120} \frac{(35\text{ mm})dr}{r}$$

$$= (35\text{ mm})\ln\left(\frac{120}{50}\right) = 30.641406\text{ mm}$$

Accordingly, the radial distance from the center of curvature, O, to the neutral axis of the curved bar is

$$r_n = \frac{A}{\int_A \frac{dA}{r}} = \frac{2{,}450\text{ mm}^2}{30.641406\text{ mm}} = 79.957167\text{ mm}$$

Note that it is important to carry additional significant digits with the value of r_n. In calculating bending stresses, the term $r_c - r_n$ will appear in the denominator of the bending stress equation. Since r_c and r_n are close in value to one another, the term $r_c - r_n$ will be relatively small and the extra significant digits for r_n are needed to avoid substantial rounding error.

(a) Bending Stresses

The curved bar is subjected to a bending moment $M = 4{,}500\text{ N}\cdot\text{m}$. The moment is positive, since it creates compressive normal stress on the inner surface (i.e., at radius r_i) of the curved bar. (Note that the moment reduces the radius of curvature of the bar, making the curve tighter.)

The bending stress is determined from Equation (8.28),

$$\sigma_x = -\frac{M(r_n - r)}{rA(r_c - r_n)}$$

where r is the distance from the center of curvature to the location where the stress is to be determined.

At point A, $r = 50$ mm. Thus, the bending stress at point A is

$$\sigma_x = -\frac{M(r_n - r)}{rA(r_c - r_n)} = -\frac{(4{,}500\text{ N}\cdot\text{m})(79.957167\text{ mm} - 50\text{ mm})(1{,}000\text{ mm/m})}{(50\text{ mm})(2{,}450\text{ mm}^2)(85\text{ mm} - 79.957167\text{ mm})}$$
$$= -218.224\text{ MPa} = 218\text{ MPa (C)} \qquad \textbf{Ans.}$$

At point B, $r = 120$ mm, making the bending stress at that point

$$\sigma_x = -\frac{M(r_n - r)}{rA(r_c - r_n)} = -\frac{(4{,}500\text{ N}\cdot\text{m})(79.957167\text{ mm} - 120\text{ mm})(1{,}000\text{ mm/m})}{(120\text{ mm})(2450\text{ mm}^2)(85\text{ mm} - 79.957167\text{ mm})}$$
$$= 121.539\text{ MPa} = 121.5\text{ MPa (T)} \qquad \textbf{Ans.}$$

(b) Flexural Stress Distribution

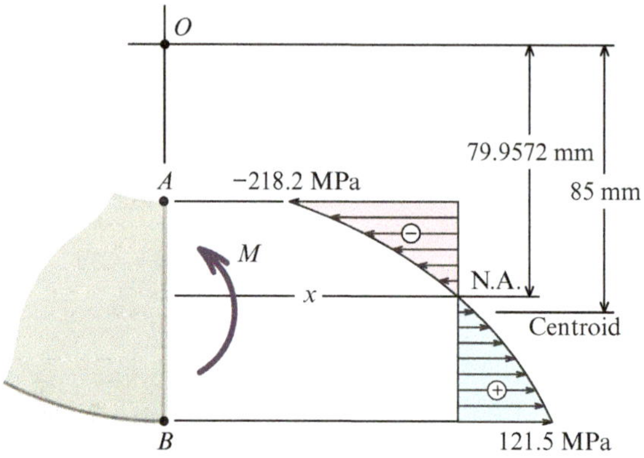

The bending stress created by $M = 4{,}500$ N·m in the curved rectangular bar is shown in the accompanying figure. Note that the distribution is nonlinear and that the neutral axis (N.A.) is not located at the centroid of the cross section.

(c) Comparison with Stresses from Straight-Beam Flexure Formula

For comparison, let's consider the bending stresses that we would have obtained from the flexure formula for initially straight beams. The section modulus of the rectangular shape is

$$S = \frac{bd^2}{6} = \frac{(35\text{ mm})(70\text{ mm})^2}{6} = 28{,}583.333\text{ mm}^3$$

If the beam were initially straight, the stresses at A and B for $M = 4{,}500$ N·m would be

$$\sigma_x = \pm\frac{M}{S} = \pm\frac{(4500\text{ N}\cdot\text{m})(1000\text{ mm/m})}{28{,}583.333\text{ mm}^3} = \pm 157.434\text{ MPa}$$

Therefore, the errors between the actual stress (determined from the curved-bar formula) and the stress obtained by using the flexure formula for straight beams is

$$\left|\frac{-157.434\text{ MPa} - (-218.224\text{ MPa})}{-218.224\text{ MPa}}\right|(100\%) = 27.9\%\text{ low at } A \qquad \textbf{Ans.}$$

$$\left|\frac{157.434\text{ MPa} - (121.539\text{ MPa})}{121.539\text{ MPa}}\right|(100\%) = 29.5\%\text{ high at } B \qquad \textbf{Ans.}$$

Superposition

Often, curved bars are loaded so that there is an axial force, as well as a moment, on the cross section, as is illustrated in the next example. The normal stress given by Equation (8.28) may then be algebraically added to the stress that is due to an internal axial force F. For this simple case of superposition, the total stress at a point located a distance r from the center of curvature, O, may be expressed as

$$\sigma_x = \frac{F}{A} - \frac{M(r_n - r)}{r\,A(r_c - r_n)} \tag{8.29}$$

A positive sign would be associated with a tensile internal force F, and a negative sign would be used for compressive internal forces. Note that the internal bending moment M produced by an external load P is calculated as the product of (the magnitude of) the force and the perpendicular distance between the line of action of P and the *centroid* of the cross section, not the neutral-axis location of the cross section.

EXAMPLE 8.14

The allowable stress for the clamp shown is 90 MPa. Use the following dimensions for the clamp:

$b_1 = 125$ mm	$d_1 = 25$ mm
$b_2 = 25$ mm	$d_2 = 75$ mm
$b_3 = 50$ mm	$d_3 = 25$ mm
$r_i = 150$ mm	$a = 20$ mm

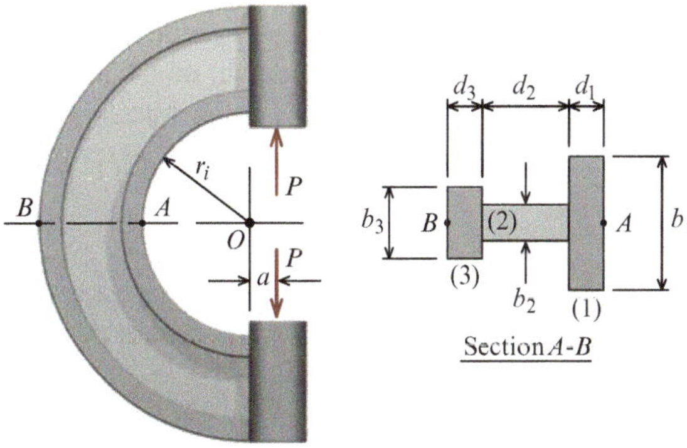

Determine the maximum permissible load P that the member can resist.

Plan the Solution

Begin by calculating the radial distances from the center of curvature, O, to the centroid of the cross section and to the neutral-axis location. Next, determine an expression for the bending moment in terms of the unknown load P and the perpendicular distance between the line of action of P and the centroid of the curved-bar cross section. The normal stresses at Section A–B will consist of both axial and bending stresses. Write an equation, similar to Equation (8.29), for the sum of the stresses in terms of the unknown load P, and use this equation to investigate the stresses at points A and B. At point A on the inner surface of the curved bar, both the axial and bending stresses will be tensile normal stresses. Set the sum of these two stresses equal to the specified allowable stresses, and solve for the allowable magnitude of load P. At point B, the axial stress will be a tensile normal stress while bending stresses will be compressive. Again, set the sum of the axial and flexural normal stresses equal to the allowable stress and solve for P. The smaller value obtained for the load P from these two calculations is the largest load that can be applied to the clamp.

SOLUTION

Centroid location and neutral-axis location for flanged shape

Shape	b (mm)	d (mm)	A_i (mm²)	r_i (mm)	r_o (mm)	r_{ci} (mm)	$r_{ci}A_i$ (mm³)	$b\ln\left(\frac{r_o}{r_i}\right)$ (mm)
Inner flange (1)	125	25	3,125	150	175	162.5	507,812.5	19.268835
Web (2)	25	75	1,875	175	250	212.5	398,437.5	8.916874
Outer flange (3)	50	25	1,250	250	275	262.5	328,125.0	4.765509
			6,250				1,234,375.0	32.951218

$$r_c = \frac{1{,}234{,}375.0\text{ mm}^3}{6{,}250\text{ mm}^2} = 197.500\text{ mm (from center of curvature to centroid)}$$

$$r_n = \frac{A}{\Sigma b\ln\left(\dfrac{r_o}{r_i}\right)} = \frac{6{,}250\text{ mm}^2}{32.951218\text{ mm}} = 189.674\text{ mm}$$

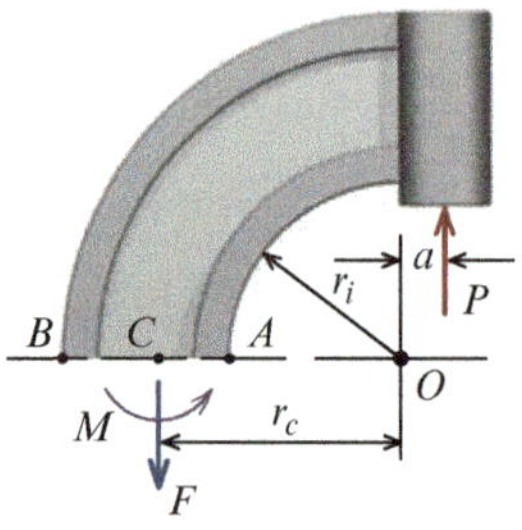

Internal force F and internal moment M at Section A–B

Draw a free-body diagram that cuts through the clamp at Section A–B. We will assume that a positive (i.e., tension) internal force F acts at the centroid of the cross section. We will also assume that a positive internal moment M exists at the section. Note that a positive internal moment creates compression on the inner surface of the curved bar.

We sum forces in the vertical direction to find that the internal force F equals the clamp force P:

$$+\uparrow\ \Sigma F = P - F = 0$$

$$\therefore\ F = P$$

Next, we sum moments about point C, which is located at the centroid of the cross section:

$$\Sigma M_C = P(r_c + a) + M = 0$$

$$\therefore\ M = -P(r_c + a) = -P(197.500\text{ mm} + 20\text{ mm}) = -P(217.5\text{ mm})$$

Note that moments in configurations such as this are always calculated as the product of a force and the perpendicular distance between the line of action of the force and the *centroid* of the cross section—not the neutral-axis location.

Axial and Bending Stresses

The internal axial force F will create a tensile normal stress $\sigma = F/A$ that is uniformly distributed on Section A–B. The bending stress is determined from Equation (8.28). Thus, the combined axial and bending stress on Section A–B is expressed as

$$\sigma = \frac{F}{A} - \frac{M(r_n - r)}{r\,A(r_c - r_n)} = \frac{P}{A} + \frac{P(217.5\text{ mm})(r_n - r)}{r\,A(r_c - r_n)} = \frac{P}{A}\left[1 + \frac{(217.5\text{ mm})(r_n - r)}{r(r_c - r_n)}\right]$$

where r is the distance from the center of curvature to the location where the stress is to be determined.

We now set the stress at point A ($r_A = 150$ mm) equal to the 90 MPa allowable stress and solve for the corresponding magnitude of load P:

$$90\text{ MPa} = \left|\frac{P}{6{,}250\text{ mm}^2}\left[1 + \frac{(217.5\text{ mm})(189.674\text{ mm} - 150\text{ mm})}{(150\text{ mm})(197.500\text{ mm} - 189.674\text{ mm})}\right]\right|$$

$$\therefore P \le 67{,}359\text{ N}$$

Finally, we repeat this process for point B ($r_B = 275$ mm) and solve for the corresponding magnitude of load P:

$$90\text{ MPa} = \left|\frac{P}{6{,}250\text{ mm}^2}\left[1 + \frac{(217.5\text{ mm})(189.674\text{ mm} - 275\text{ mm})}{(275\text{ mm})(197.500\text{ mm} - 189.674\text{ mm})}\right]\right|$$

$$\therefore P \le 73{,}788\text{ N}$$

This calculation demonstrates that the stress at point A, where tensile flexural stress is combined with the tensile axial stress, is the most highly stressed point on Section A–B. On the basis of this combination, the largest allowable load that can be applied to the clamp is $P_{allow} = 67.4$ kN. **Ans.**

PROBLEMS

P8.77 A curved bar with a rectangular cross section is subjected to a bending moment $M = 850$ N·m as shown in Figure P8.77. Use $b = 35$ mm, $d = 60$ mm, and $r_i = 100$ mm. Determine the normal stress at A and B.

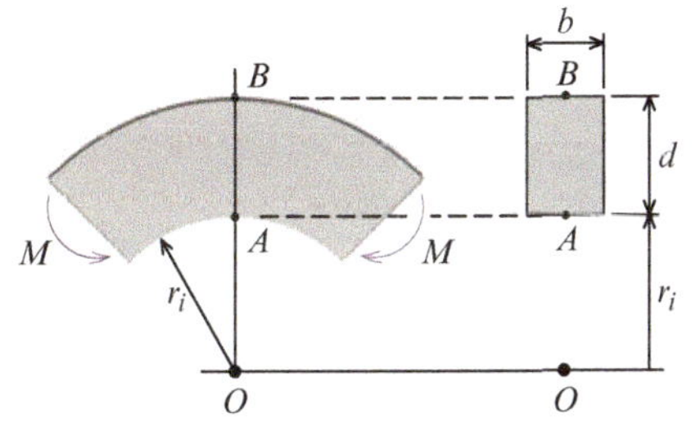

FIGURE P8.77

P8.78 The steel link shown in Figure P8.78 has a central angle $\beta = 120°$. Dimensions of the link are $b = 6$ mm, $d = 18$ mm, and $r_i = 30$ mm. If the allowable normal stress is 140 MPa, what is the largest load P that may be applied to the link?

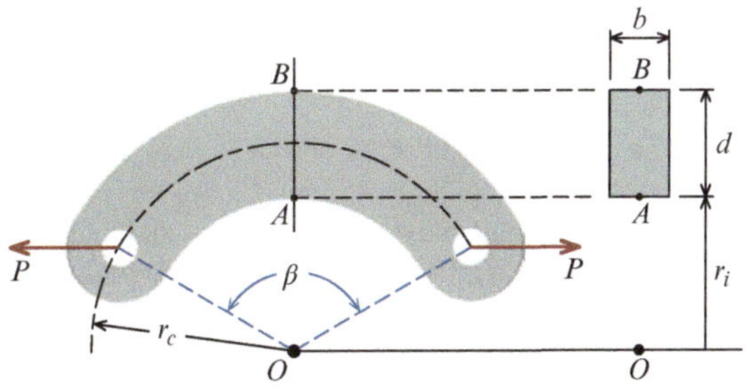

FIGURE P8.78

P8.79 The curved member shown in Figure P8.79 has a rectangular cross section with dimensions $b = 2$ in. and $d = 5$ in. The inside radius of the curved bar is $r_i = 4$ in. A load P is applied at a distance $a = 9$ in. from the center of curvature, O. If the tensile and compressive stresses in the member are not to exceed 24 ksi and 18 ksi, respectively, determine the value of the load P that may be safely supported by the member.

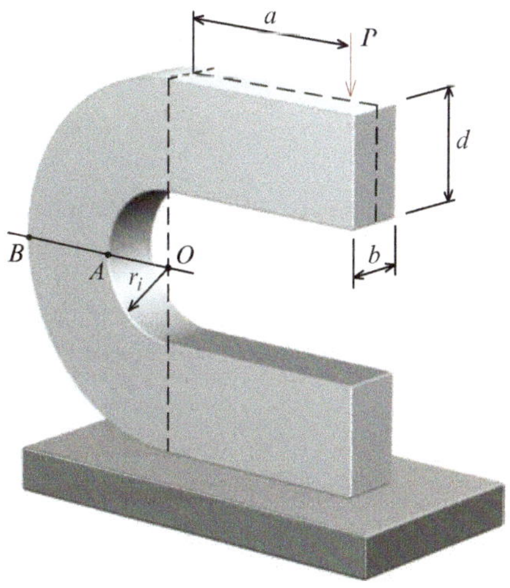

FIGURE P8.79

P8.80 The outside and inside diameters of the tube elbow shown in Figure P8.80 are $D = 1.75$ in. and $d = 1.50$ in., respectively, and the inner radius of the elbow is $r_i = 1.50$ in. Determine the stresses acting at points A and B when the tube is loaded by a moment $M = 2{,}500$ lb·in.

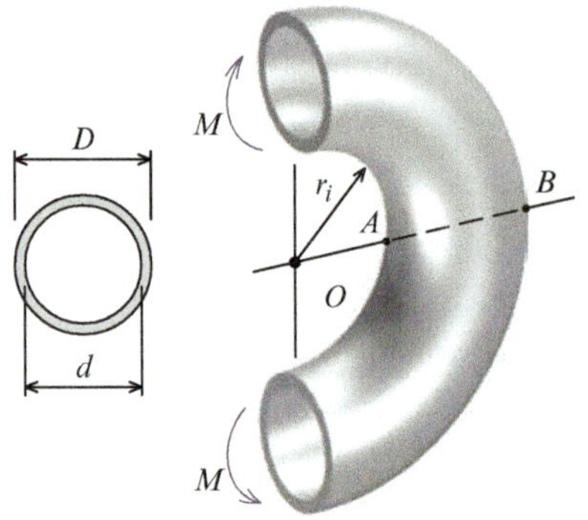

FIGURE P8.80

P8.81 Determine the stresses acting at points A and B for the curved solid rod shown in Figure P8.81. The diameter of the curved solid rod shown in Figure P8.81 $d = 0.875$ in. The inner radius of the bend in the rod is $r_i = 1.5$ in. A load $P = 90$ lb is applied at a distance $a = 5$ in. from the center of curvature, O. Determine the stresses acting at points A and B.

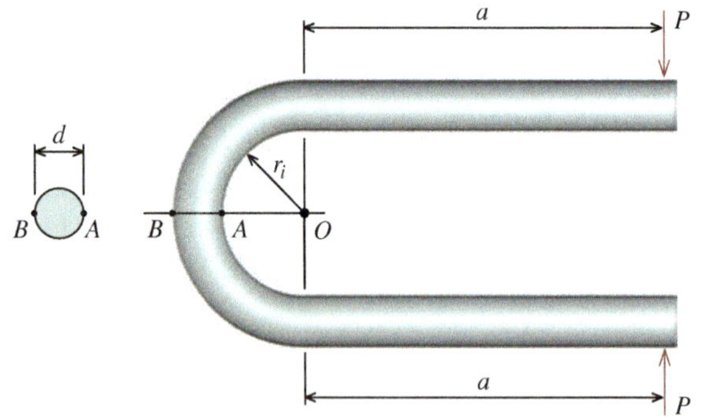

FIGURE P8.81

P8.82 A solid circular rod of diameter d is bent into a semicircle as shown in Figure P8.82. The radial distance to the centroid of the rod is to be $r_c = 3d$, and the curved rod is to support a load $P = 5{,}000$ N. If the allowable stress must be limited to 135 MPa, what is the smallest diameter d that may be used for the rod?

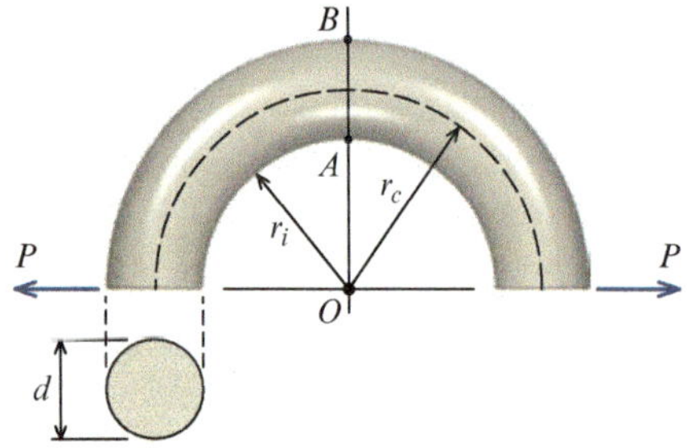

FIGURE P8.82

P8.83 The curved tee shape shown in Figure P8.83 is subjected to a bending moment $M = 2{,}700$ N · m. The dimensions of the cross section are $b_1 = 15$ mm, $d_1 = 70$ mm, $b_2 = 50$ mm, and $d_2 = 20$ mm. The radial distance from the center of curvature, O, to A is $r_i = 85$ mm. Determine

(a) the radial distance from O to the neutral axis.
(b) the stresses at points A and B.

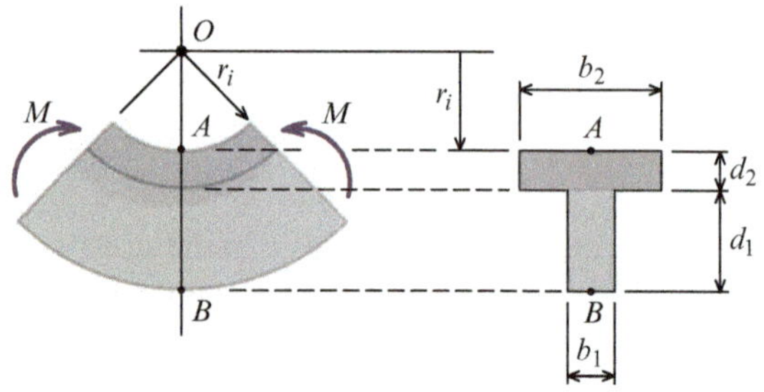

FIGURE P8.83

P8.84 The curved flanged shape shown in Figure P8.84/85 is subjected to a bending moment $M = 4{,}300$ N · m. Dimensions of the cross section are $b_1 = 75$ mm, $d_1 = 15$ mm, $b_2 = 15$ mm, $d_2 = 55$ mm, $b_3 = 35$ mm, and $d_3 = 15$ mm. The radial distance from the center of curvature, O, to A is $r_i = 175$ mm. Determine the stresses at points A and B.

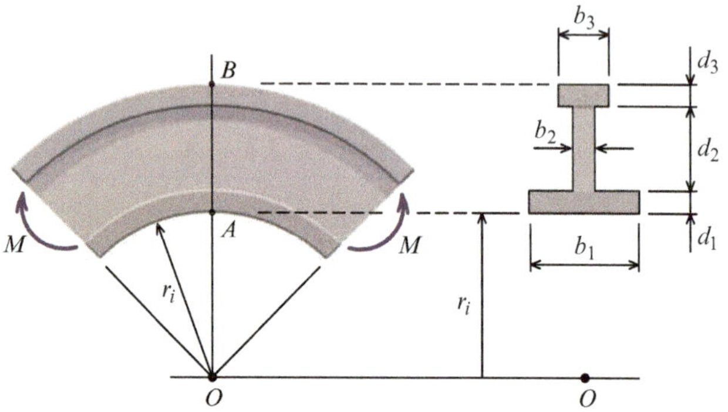

FIGURE P8.84/85

P8.85 The curved flanged shape shown in Figure P8.84/85 has cross-sectional dimensions $b_1 = 45$ mm, $d_1 = 9$ mm, $b_2 = 9$ mm, $d_2 = 36$ mm, $b_3 = 18$ mm, and $d_3 = 9$ mm. The radial distance from the center of curvature, O, to A is $r_i = 140$ mm. Determine

(a) the radial distance $r_c - r_n$ between the centroid and the neutral axis.
(b) the largest value of M if the allowable normal stress is 250 MPa.

P8.86 The curved tee shape shown in Figure P8.86 has cross-sectional dimensions $b_1 = 1.25$ in., $d_1 = 0.25$ in., $b_2 = 0.25$ in., and $d_2 = 1.00$ in. The radial distance from the center of curvature, O, to A is $r_i = 2.00$ in. Determine the stresses at A and B for an applied load $P = 560$ lb.

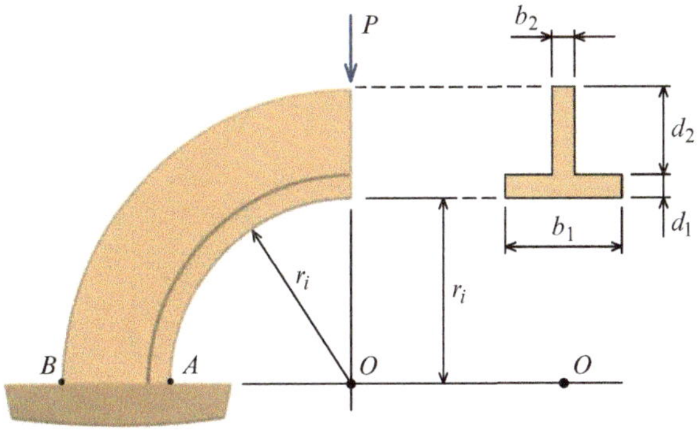

FIGURE P8.86

P8.87 The curved bar shown in Figure P8.87 has a trapezoidal cross section with dimensions $b_1 = 75$ mm, $b_2 = 30$ mm, and $d = 100$ mm. The radial distance from the center of curvature, O, to A is $r_i = 120$ mm. Determine

(a) the radial distance $r_c - r_n$ between the centroid and the neutral axis.
(b) the largest value of M if the allowable normal stress is 200 MPa.

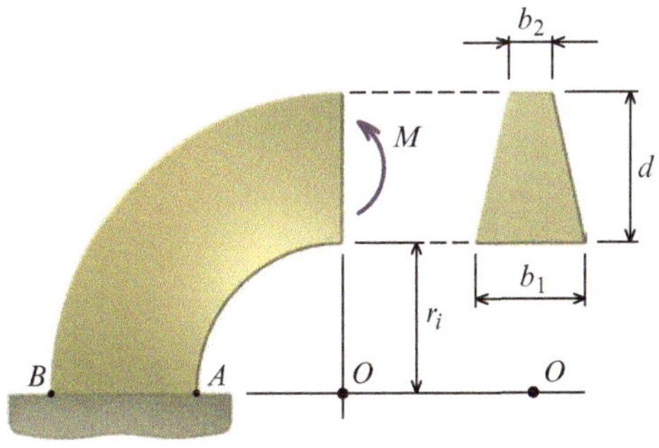

FIGURE P8.87

P8.88 The curved bar shown in Figure P8.88 has a trapezoidal cross section with dimensions $b_1 = 0.75$ in., $b_2 = 1.50$ in., and $d = 2.00$ in. The radial distance from O to A is $r_i = 1.50$ in. A load $P = 1{,}400$ lb acts on the bar. If the allowable stress is 18 ksi, what is the largest permissible distance a?

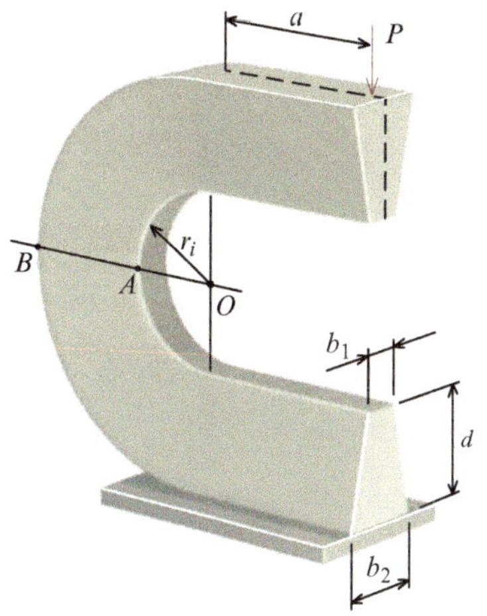

FIGURE P8.88

P8.89 The curved bar shown in Figure P8.89 has an elliptical cross section with dimensions $a = 0.9$ in. and $b = 0.4$ in. The inner radius of the curved bar is $r_i = 1.50$ in. For an applied moment $M = 1{,}700$ lb·in., determine

(a) the radial distance $r_c - r_n$ between the centroid and the neutral axis.
(b) the maximum compressive stress.

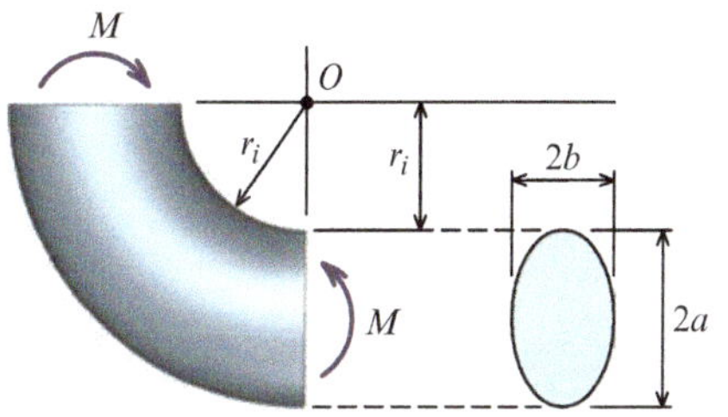

FIGURE P8.89

P8.90 The curved bar shown in Figure P8.90 has a triangular cross section with dimensions $b = 1.5$ in. and $d = 1.2$ in. The inner radius of the curved bar is $r_i = 3.8$ in. For an applied load $P = 70$ lb, determine the stresses at A and B.

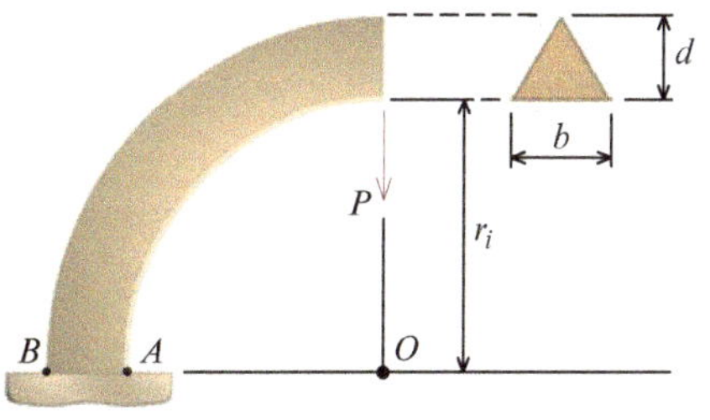

FIGURE P8.90

CHAPTER 9

Shear Stress In Beams

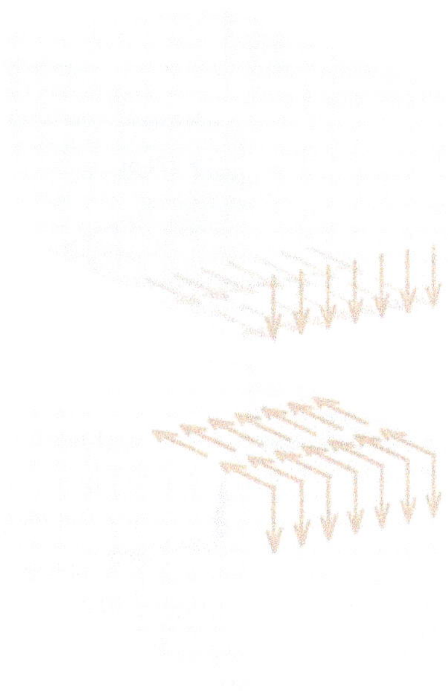

9.1 Introduction

For beams subjected to pure bending, only tensile and compressive normal stresses are developed in the flexural member. In most situations, however, loadings applied to a beam create nonuniform bending; that is, internal bending moments are accompanied by internal shear forces. As a consequence of nonuniform bending, shear stresses as well as normal stresses are produced in the beam. In this chapter, a method will be derived for determining the shear stresses produced by nonuniform bending. The method will also be adapted to consider beams fabricated from multiple pieces joined together by discrete fasteners.

9.2 Resultant Forces Produced by Bending Stresses

Before developing the equations that describe beam shear stresses, it is instructive to consider in more detail the resultant forces produced by bending stresses on portions of the beam cross section. Consider the simply supported beam shown in Figure 9.1, in which a concentrated load $P = 9{,}000$ N is applied at the middle of a 2 m long span. The shear-force and bending-moment diagrams for this span and loading are shown.

For this investigation, we will arbitrarily consider a 150 mm long segment BC of the beam. The segment located 300 mm from the left support, as shown in the figure. The beam is made up of two wooden boards, each having the same elastic modulus. The lower board will

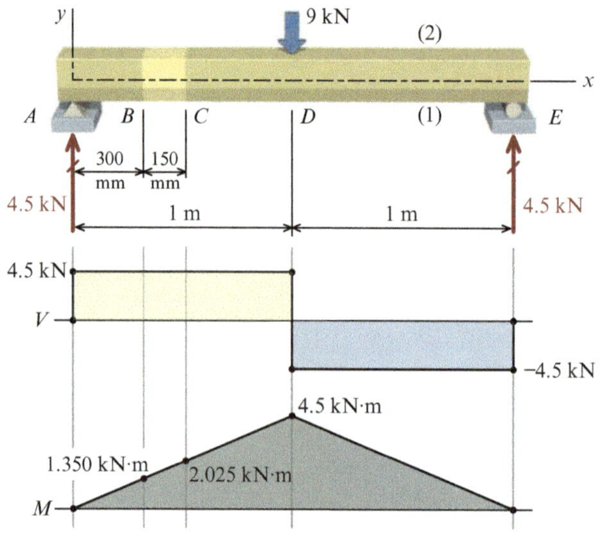

FIGURE 9.1 Simply supported beam with concentrated load applied at midspan.

be designated as member (1), and the upper board will be designated as member (2). The cross-sectional dimensions of the beam are shown in Figure 9.2.

The objective of this investigation is to determine the forces acting on member (1) at sections B and C.

From the bending-moment diagram, the internal bending moments at sections B and C are $M_B = 1.350$ kN · m and $M_C = 2.025$ kN · m, respectively. Both moments are positive; thus, beam segment BC will be deformed as shown in Figure 9.3a. Compressive normal stresses will be produced in the upper half of the beam cross section, and tensile normal stress will be produced in the lower half. The bending stress distribution over the depth of the cross section at these two locations can be determined from the flexure formula with the use of a moment of inertia $I_z = 33{,}750{,}000$ mm^4 about the z centroidal axis. The distribution of bending stresses is shown in Figure 9.3b.

To determine the forces acting on member (1), we will consider only those normal stresses acting between points b and c (on section B) and between points e and f (on section C). At B, the bending stress varies from 1.0 MPa (T) at b to 3.0 MPa (T) at c. At C, the bending stress varies from 1.5 MPa (T) at e to 4.5 MPa (T) at f.

From Figure 9.2, the cross-sectional area of member (1) is

$$A_1 = (50 \text{ mm})(120 \text{ mm}) = 6{,}000 \text{ mm}^2$$

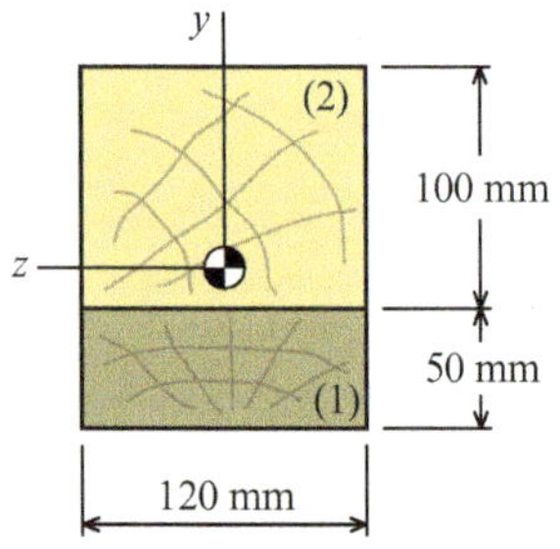

FIGURE 9.2 Cross-sectional dimensions of beam.

To determine the resultant force at section B that acts on this area, the stress distribution can be split into two components: a uniformly distributed portion having a magnitude of 1.0 MPa and a triangular portion having a maximum intensity of (3.0 MPa − 1.0 MPa) = 2.0 MPa. By this approach, the resultant force acting on member (1) at section B can be calculated as

$$\begin{aligned}\text{Resultant } F_B &= (1.0 \text{ N/mm}^2)(6{,}000 \text{ mm}^2) + \frac{1}{2}(2.0 \text{ N/mm}^2)(6{,}000 \text{ mm}^2)\\ &= 12{,}000 \text{ N} = 12 \text{ kN}\end{aligned}$$

Since the bending stresses are tensile, the resultant force acts in tension on section B.

Similarly, the stress distribution on section C can be split into two components: a uniformly distributed portion having a magnitude of 1.5 MPa and a triangular portion having a maximum intensity of (4.5 MPa − 1.5 MPa) = 3.0 MPa. The resultant force acting on member (1) at section C is found from

$$\begin{aligned}\text{Resultant } F_C &= (1.5\text{ N/mm}^2)(6{,}000\text{ mm}^2) + \frac{1}{2}(3.0\text{ N/mm}^2)(6{,}000\text{ mm}^2)\\ &= 18{,}000\text{ N} = 18\text{ kN}\end{aligned}$$

The resultant forces caused by the bending stresses on member (1) are shown in Figure 9.3*c*. Notice that the resultant forces are not equal in magnitude. *Why are these resultant forces unequal?* The resultant force on section C is larger than the resultant force on section B because the internal bending moment M_C is larger than M_B. The resultant forces F_B and F_C will be equal in magnitude only when the internal bending moments are the same on sections B and C. *Is member (1) of beam segment BC in equilibrium?* This portion of the beam is *not* in equilibrium, because $\Sigma F_x \neq 0$. *How much additional force is required to satisfy equilibrium?* An additional force of 6 kN in the horizontal direction is required to satisfy equilibrium for member (1). *Where must this additional force be located?* All normal stresses acting on the two vertical faces (*b–c* and *e–f*) have been considered in the calculations of F_B and F_C. The bottom horizontal face *c–f* is a free surface that has no stress acting on it. Therefore, the additional 6 kN horizontal force required to satisfy equilibrium must be located on horizontal surface *b–e*, as shown in Figure 9.4. This surface is the interface between member (1) and member (2). *What is the term given to a force that acts on a surface that is parallel to the line of action of the force?* The 6 kN horizontal force acting on surface *b–e* is termed a **shear force**. Notice that this force acts in the same direction as the resultants of the bending stress—that is, parallel to the x axis.

What lessons can be drawn from this simple investigation? In those beam spans in which the internal bending moment is not constant, the resultant forces acting on portions of the cross section will be unequal in magnitude. Equilibrium of these portions can be satisfied only by an additional shear force that is developed internally in the beam.

In the sections that follow, we will discover that this additional internal shear force required to satisfy equilibrium can be developed in two ways. The internal shear force can be the resultant of shear stresses developed in the beam, or it can be provided by individual fasteners such as bolts, nails, or screws.

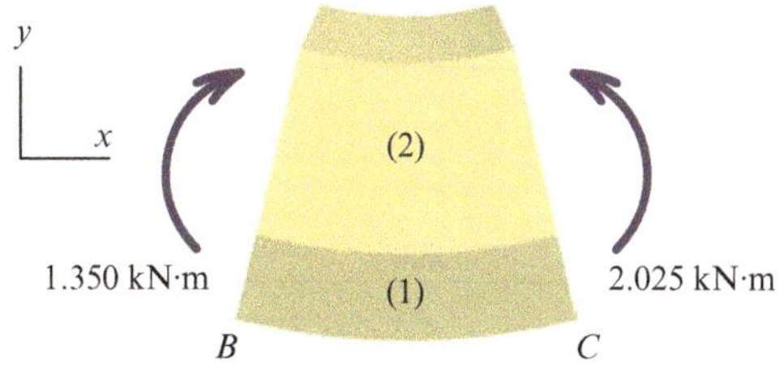

(*a*) Internal bending moments

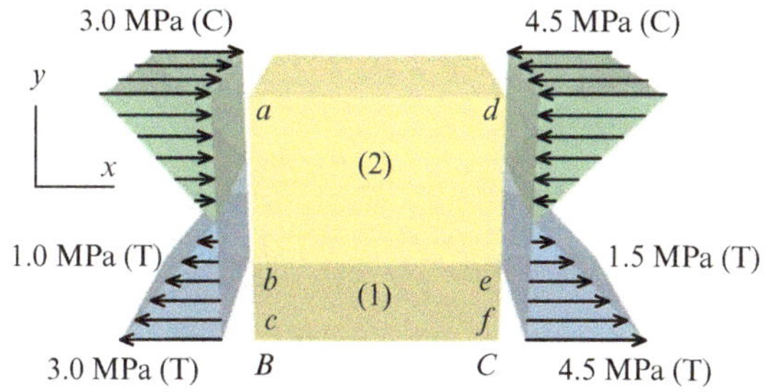

(*b*) Bending stresses

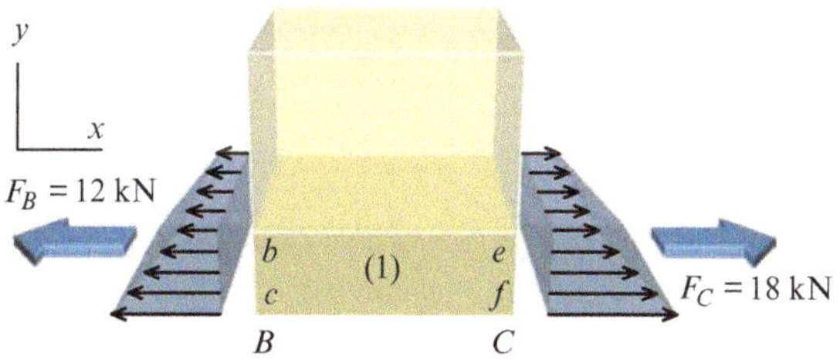

(*c*) Bending stress resultant forces

FIGURE 9.3 Moments, stresses, and forces acting on beam segment *BC*.

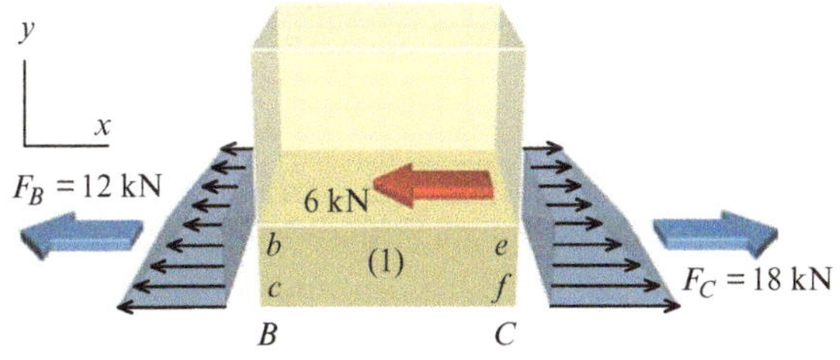

FIGURE 9.4 Free-body diagram of member (1).

EXAMPLE 9.1

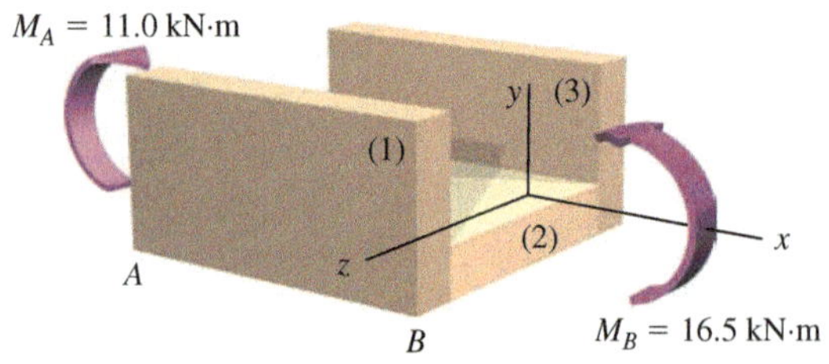

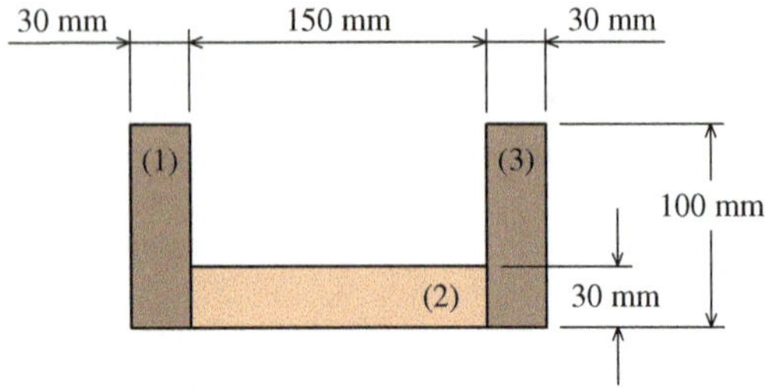

A beam segment is subjected to the internal bending moments shown. The cross-sectional dimensions of the beam are given.

(a) Sketch a side view of the beam segment, and plot the distribution of bending stresses acting at sections A and B. Indicate the magnitude of the key bending stresses in the sketch.

(b) Determine the resultant forces acting in the x direction on area (2) at sections A and B, and show these resultant forces in the sketch.

(c) *Is the specified area in equilibrium with respect to forces acting in the x direction?* If not, determine the horizontal force required to satisfy equilibrium for the specified area and show the location and direction of this force in the sketch.

Plan the Solution

After the section properties are computed, the normal stresses produced by the bending moment will be determined from the flexure formula. In particular, the bending stresses acting on area (2) will be calculated. From these stresses, the resultant forces acting in the horizontal direction at each end of the beam segment will be computed.

SOLUTION

(a) The centroid location in the z direction can be determined from symmetry. The centroid location in the y direction must be determined for the U-shaped cross section. The U shape is subdivided into rectangular shapes (1), (2), and (3), and the y centroid location is calculated from the following:

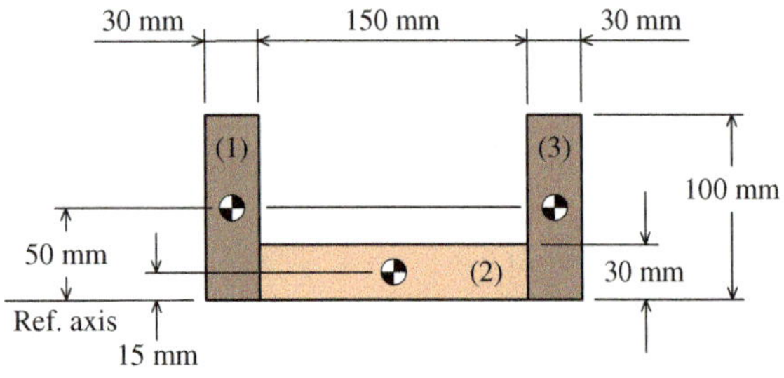

	A_i (mm^2)	y_i (mm)	$y_i A_i$ (mm^3)
(1)	3,000	50	150,000
(2)	4,500	15	67,500
(3)	3,000	50	150,000
	10,500		367,500

$$\bar{y} = \frac{\sum y_i A_i}{\sum A_i} = \frac{367{,}500 \text{ mm}^3}{10{,}500 \text{ mm}^2} = 35.0 \text{ mm}$$

Thus, the z centroidal axis is located 35.0 mm above the reference axis for the U-shaped cross section. Next, the moment of inertia about the z centroidal axis is calculated. The parallel-axis theorem is required, since the centroids of areas (1), (2), and (3) do not coincide with the z centroidal axis for the U shape. The complete calculation is summarized in the following table.

	I_{ci} (mm⁴)	$\lvert d_i \rvert$ (mm)	$d_i^2 A_i$ (mm⁴)	I_z (mm⁴)
(1)	2,500,000	15.0	675,000	3,175,000
(2)	337,500	20	1,800,000	2,137,500
(3)	2,500,000	15.0	675,000	3,175,000
				8,487,500

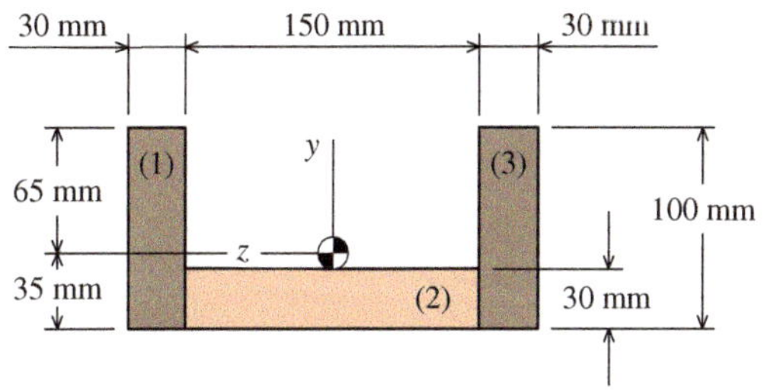

The moment of inertia of the U-shaped cross section about its z centroidal axis is $I_z = 8{,}487{,}500\ \text{mm}^4$.

For the positive bending moments M_A and M_B acting on the beam segment as shown, compressive normal stresses will be produced above the z centroidal axis and tensile normal stresses will occur below the z centroidal axis. The flexure formula [Equation (8.7)] is used to compute the bending stress at any coordinate location y. (Recall that the y coordinate axis has its origin at the centroid.) For example, the bending stress at the top of area (1) at section A is calculated with $y = 65$ mm:

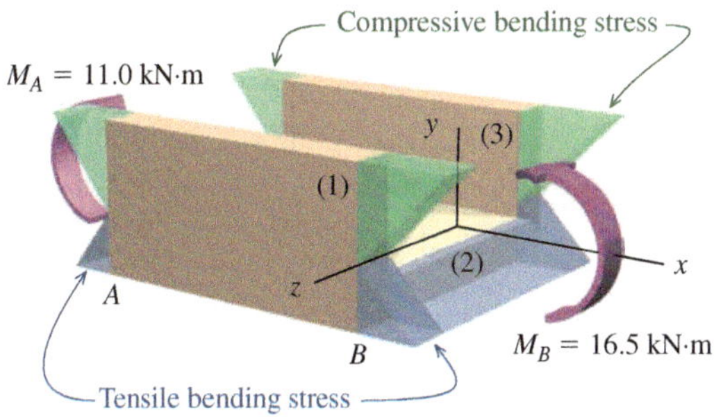

$$\sigma_x = -\frac{My}{I_z} = -\frac{(11\ \text{kN}\cdot\text{m})(65\ \text{mm})(1{,}000\ \text{N/kN})(1{,}000\ \text{mm/m})}{8{,}487{,}500\ \text{mm}^4}$$
$$= -84.2\ \text{MPa} = 84.2\ \text{MPa (C)}$$

The maximum tensile and compressive bending stresses at sections A and B are shown in the accompanying figure.

(b) Of particular interest in this example are the bending stresses acting on area (2) of the U-shaped cross section. The normal stresses acting on area (2) are shown in the following figure:

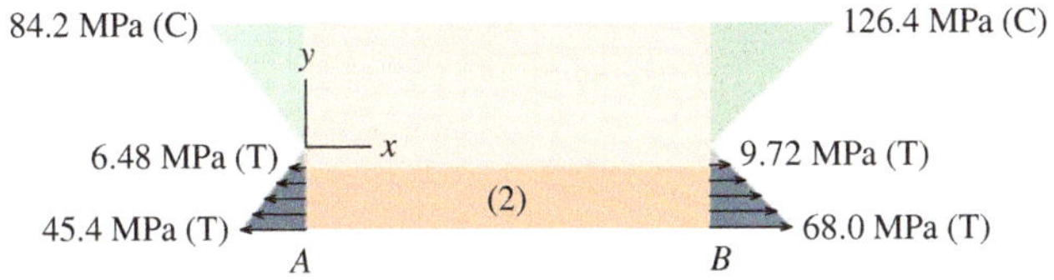

The resultant force of the bending stresses acting on area (2) must be determined at section A and at section B. The normal stresses acting on area (2) are all of the same sense (i.e., tension), and since these stresses are linearly distributed in the y direction, we need only determine the average stress intensity. The stress distribution is uniformly distributed across the z dimension of area (2). Therefore, the resultant force acting on area (2) can be determined from the product of the average normal stress and the area upon which it acts. Area (2) is 150 mm wide and 30 mm deep; therefore, $A_2 = 4{,}500\ \text{mm}^2$. On section A, the resultant force in the x direction is

$$F_A = \frac{1}{2}(6.48\ \text{MPa} + 45.4\ \text{MPa})(4{,}500\ \text{mm}^2) = 116{,}730\ \text{N} = 116.7\ \text{kN}$$

and on section B, the horizontal resultant force is

$$F_B = \frac{1}{2}(9.72\ \text{MPa} + 68.0\ \text{MPa})(4{,}500\ \text{mm}^2) = 174{,}870\ \text{N} = 174.9\ \text{kN}$$

as shown in the following figure:

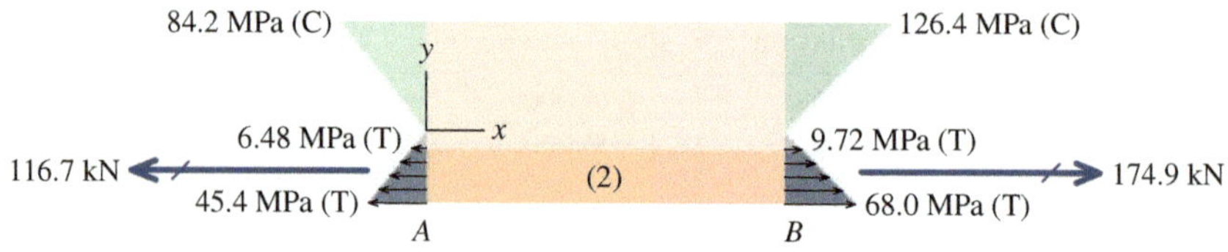

(c) Consider the equilibrium of area (2). In the x direction, the sum of the resultant forces is

$$\Sigma F_x = 174.9 \text{ kN} - 116.8 \text{ kN} = 58.2 \text{ kN} \neq 0$$

Area (2) is not in equilibrium. *What observations can be drawn from this situation?* Whenever a beam segment is subjected to nonuniform bending—that is, whenever the bending moments are changing along the span of the beam—portions of the beam cross section will require additional forces in order to satisfy equilibrium in the longitudinal direction. *Where can these additional forces be applied to area (2)?*

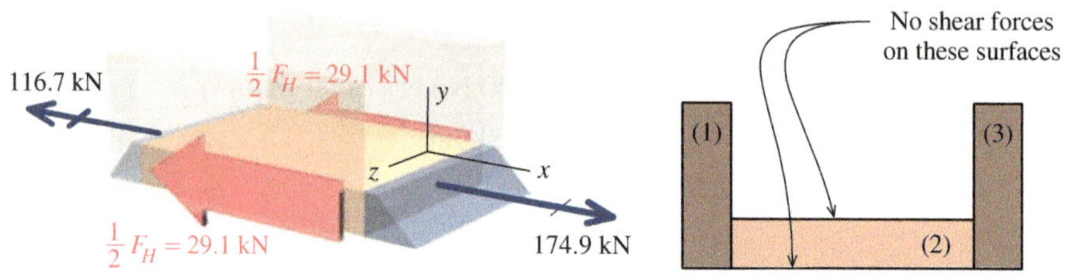

The additional force in the horizontal direction F_H required to satisfy equilibrium cannot emanate from the upper and lower surfaces of area (2), since these are free surfaces. Therefore, F_H must act at the boundary between areas (1) and (2), and at the boundary between areas (2) and (3). By symmetry, half of the horizontal force will act on each surface. Because F_H acts along the vertical sides of area (2), it is termed a shear force.

MecMovies

EXAMPLE

M9.1 Discussion of the horizontal shear force developed in a flexural member.

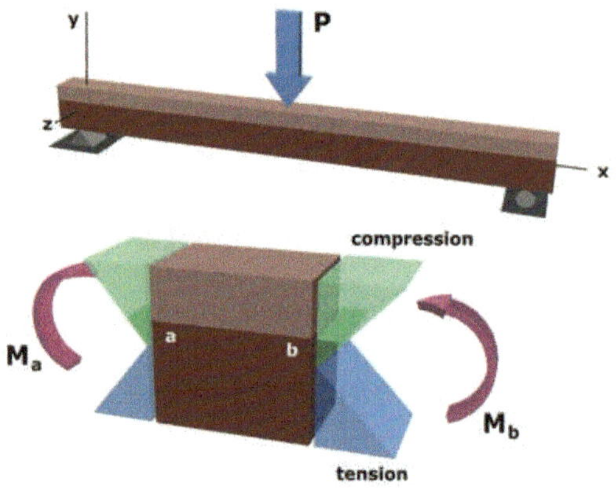

324

9.3 The Shear Stress Formula

In this section, a method for determining the shear stresses produced in a prismatic beam made of a homogeneous linear-elastic material will be developed. Consider the beam shown in Figure 9.5*a*. The beam is subjected to various loadings, and its cross section is shown in Figure 9.5*b*. In the development that follows, particular attention is focused on one portion of the cross section. Call that portion area A'.

Now consider a free-body diagram of the beam with cross section of length Δx and located some distance x from the origin (Figure 9.6*a*). The internal shear force and bending moment on the left side of the free-body diagram (section *a–b–c*) are designated as V and M, respectively. On the right side of the free-body diagram (section *d–e–f*), the internal shear force and bending moment are slightly different: $V + \Delta V$ and $M + \Delta M$. Equilibrium in the horizontal x direction will be considered here. The internal shear forces V and $V + \Delta V$ and the distributed load $w(x)$ act in the vertical direction; consequently, they will have no effect on equilibrium in the x direction, and they can be omitted in the subsequent analysis.

The normal stresses acting on this free-body diagram (Figure 9.6*b*) can be determined by the flexure formula. On the left side of the free-body diagram, the bending stresses due to the internal bending moment M are given by My/I_z, and on the right side, the internal bending moment $M + \Delta M$ creates bending stresses given by $(M + \Delta M)y/I_z$. The signs associated with the bending stresses will be determined by inspection. Above the neutral

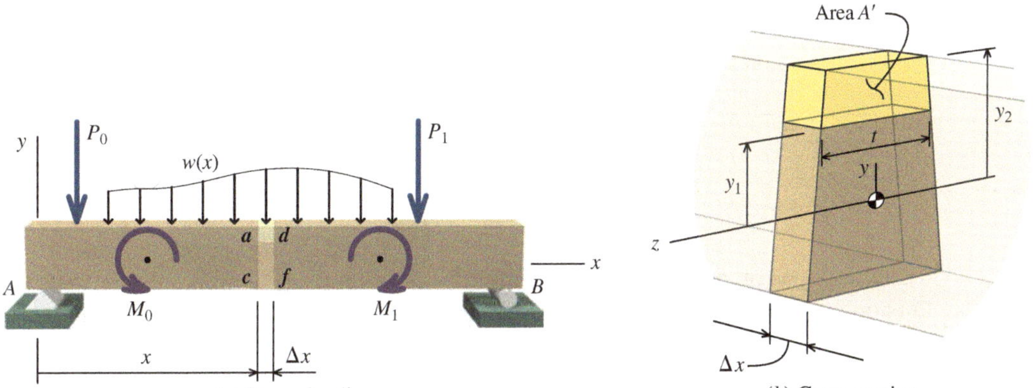

(*a*) Beam loading (*b*) Cross section

FIGURE 9.5 Prismatic beam subjected to nonuniform bending.

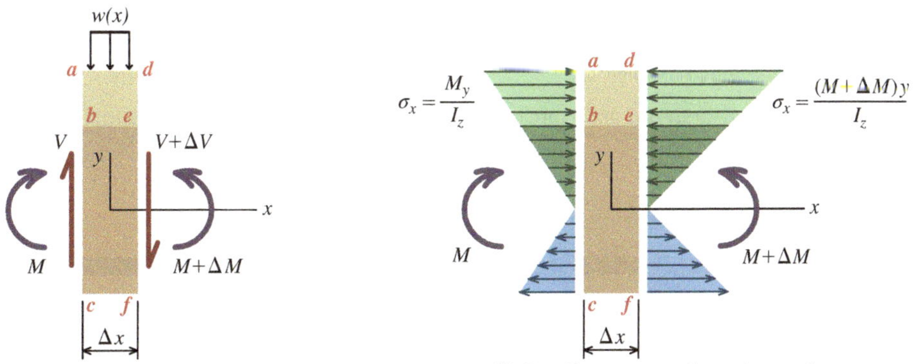

(*a*) Free-body diagram (*b*) Bending stresses due to internal bending moments

FIGURE 9.6 Free-body diagrams of beam segment.

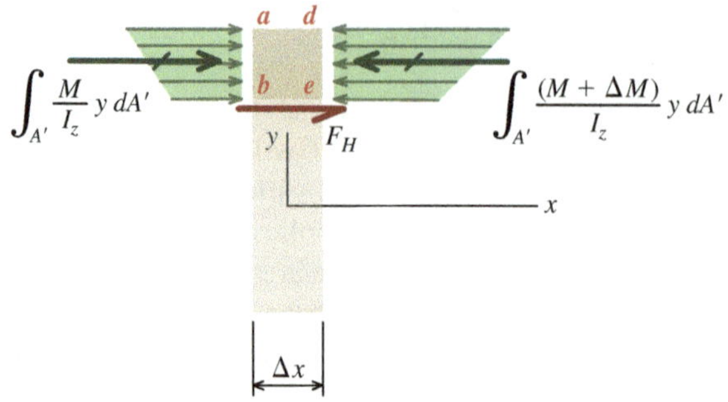

FIGURE 9.7 Free-body diagram of area A' (profile view).

axis, the internal bending moments produce compressive normal stresses, which act on the free-body diagram in the directions shown.

If a beam is in equilibrium, then any portion of the beam that we choose to consider must also be in equilibrium. We will consider a portion of the free-body diagram shown in Figure 9.6, starting at section *b–e* ($y = y_1$) in Figure 9.5*b* and *extending away from the neutral axis* (upward in this case) to the outermost boundary of the cross section ($y = y_2$) in Figure 9.5*b*. This is the portion of the cross section designated as A' in Figure 9.5*b*. A free-body diagram of area A' is shown in Figure 9.7.

The resultant force on sections *a–b* and *d–e* can be found by integrating the normal stresses acting on area A', which includes that portion of the cross-sectional area starting at $y = y_1$ and extending vertically to the top of the cross section at $y = y_2$. (See Figure 9.5*b*.) No force exists on section *a–d*; however, we shall assume that an internal horizontal force F_H could be present on section *b–e*. The equilibrium equation for the sum of forces acting on area A' in the x direction can be written as

$$\Sigma F_x = \int_{A'} \frac{M}{I_z} y\, dA' - \int_{A'} \frac{(M + \Delta M)}{I_z} y\, dA' + F_H = 0 \qquad \text{(a)}$$

where the signs of each term are determined by inspection of Figure 9.7. The integrals in Equation (a) can be expanded to give

$$\Sigma F_x = \int_{A'} \frac{M}{I_z} y\, dA' - \int_{A'} \frac{M}{I_z} y\, dA' - \int_{A'} \frac{\Delta M}{I_z} y\, dA' + F_H = 0 \qquad \text{(b)}$$

Canceling the leftmost two integrals and solving for F_H gives

$$F_H = \int_{A'} \frac{\Delta M}{I_z} y\, dA' \qquad \text{(c)}$$

With respect to area A', both ΔM and I_z are constant; therefore, Equation (c) can be simplified to

$$F_H = \frac{\Delta M}{I_z} \int_{A'} y\, dA' \qquad \text{(d)}$$

The moment-of-inertia term appearing in Equation (9.1) stems from the flexure formula, which was used to determine the bending stresses acting over the entire depth of the beam and over area A' in particular. For that reason, I_z is the moment of inertia *of the entire cross section* about the z axis.

The integral in Equation (d) is the *first moment of area A′ about the neutral axis of the cross section*. This quantity will be designated Q. More details concerning the calculation of Q will be presented in Section 9.4. By replacing the integral term with the designation Q, Equation (d) can be rewritten as

$$F_H = \frac{\Delta M Q}{I_z} \qquad (9.1)$$

What is the significance of Equation (9.1)? **If the internal bending moment in a beam is not constant** (i.e., $\Delta M \neq 0$), then an internal horizontal shear force F_H must exist at $y = y_1$ in order to satisfy equilibrium. Furthermore, note that the term Q pertains expressly to area A'. (See Figure 9.5*b*.) Since the value of Q changes with area A', so does F_H. In other words, at every value of y possible within a cross section, the internal shear force F_H required for equilibrium is unique.

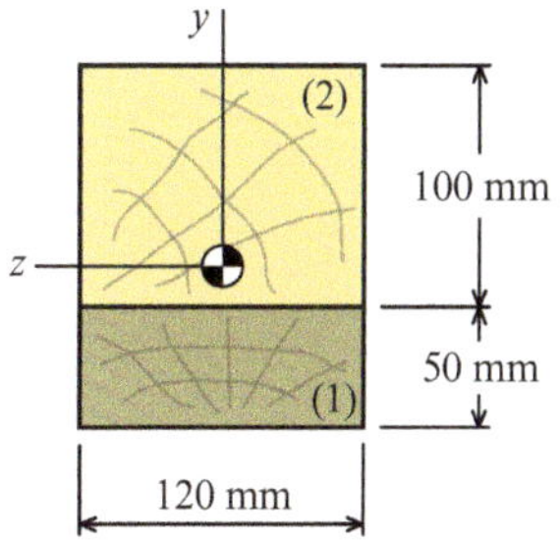

FIGURE 9.2 (repeated) Beam cross-sectional dimensions.

Before continuing, it may be helpful to apply Equation (9.1) to the problem discussed in Section 9.2. In that problem, the internal bending moments on the right and left sides of the beam segment (which had a length $\Delta x = 150$ mm) were $M_B = 1.350$ kN·m and $M_C = 2.025$ kN·m, respectively. From these two moments, $\Delta M = 2.025$ kN·m − 1.350 kN·m = 0.675 kN·m = 675 kN·mm. The moment of inertia was given as $I_z = 33{,}750{,}000$ mm⁴.

The area A' pertinent to this problem is simply the area of member (1), the 50 mm by 120 mm board at the bottom of the cross section (Figure 9.2, repeated here for convenience). The first moment of area, Q, is computed from $\int y\, dA'$. Let the width of member (1) be denoted by b. Since this width is constant, the differential area dA' can be conveniently expressed as $dA' = b\,dy$. In this instance, area A' starts at $y = -25$ mm and extends away from the neutral axis in a downward direction, to an outermost boundary at $y = -75$ mm. With $b = 120$ mm, we have

$$Q = \int_{-25}^{-75} by\, dy = \frac{b}{2}\left[y^2\right]_{-25}^{-75} = 300{,}000 \text{ mm}^3$$

and, from Equation (9.1), the horizontal shear force F_H required to keep member (1) in equilibrium is

$$F_H = \frac{\Delta M Q}{I_z} = \frac{(675 \text{ kN}\cdot\text{mm})(300{,}000 \text{ mm}^3)}{33{,}750{,}000 \text{ mm}^4} = 6 \text{ kN}$$

This result agrees with the horizontal force determined in Section 9.2.

Shear Stress in a Beam

Equation (9.1) can be extended to define the shear stress produced in a beam subjected to nonuniform bending. The surface upon which F_H acts has a length Δx. Depending upon the shape of the beam cross section, the width of area A' may vary, so the width of area A' at $y = y_1$ will be denoted by the variable t. (See Figure 9.5*b*.) Since stress is defined as force divided by area, the average horizontal shear stress acting on horizontal section *b–e* can be derived by dividing F_H in Equation (9.1) by the area of the surface upon which this force acts, which is $t\,\Delta x$:

$$\tau_{H,\text{avg}} = \frac{F_H}{t\,\Delta x} = \frac{\Delta M\, Q}{t\,\Delta x\, I_z} = \frac{\Delta M}{\Delta x}\frac{Q}{I_z t} \tag{e}$$

Implicit in equation (e) is the assumption that the shear stress is constant across the width of the cross section at any y position. That is, at any specific y position, the shear stress is constant for any z location. This derivation also assumes that the shear stresses τ are parallel to the vertical sides of the cross section (i.e., the y axis).

In the limit as $\Delta x \rightarrow 0$, $\Delta M/\Delta x$ can be expressed in terms of differentials as dM/dx, so Equation (e) can be enhanced to give the horizontal shear stress acting at any location x along the beam's span:

$$\tau_H = \frac{dM}{dx}\frac{Q}{I_z t} \tag{f}$$

Equation (f) defines the horizontal shear stress in a beam. *Note that shear stress will exist in a beam at those locations where the bending moment is not constant* (i.e., where $dM/dx \neq 0$). As discussed previously, the first moment of area, Q, varies in value for every possible y in the beam cross section. Depending upon the shape of the cross section, the width t may also vary with y. Consequently, the horizontal shear stress varies over the depth of the cross section at any location x along the beam span.

The simple investigation presented in Section 9.2 and the equations derived in this section have illustrated the concept that is essential to understanding shear stresses in beams:

> Horizontal shear forces and, consequently, horizontal shear stresses are caused in a flexural member at those locations where the internal bending moment is changing along the beam span. The imbalance in the resultant bending stress forces acting on a portion of the cross section requires an internal horizontal shear force for equilibrium.

Equation (f) gives an expression for the horizontal shear stress developed in a beam. Still, although the term dM/dx helps to clarify the source of shear stress in beams, it is awkward for calculation purposes. Then, *is there an equivalent expression for dM/dx?* Recall the relationships developed in Section 7.3 between the internal shear force and the internal bending moment. Equation (7.2) defined the following relationship:

$$\frac{dM}{dx} = V$$

In other words, wherever the bending moment is changing, there is an internal shear force *V*. The term dM/dx in Equation (f) can be replaced by the internal shear force V to give an expression for τ_H that is easier to use:

$$\tau_H = \frac{VQ}{I_z t} \tag{g}$$

The terms **horizontal shear stress** and **transverse shear stress** are both used in reference to beam shear stress. Since shear stresses on perpendicular planes must be equal in magnitude, these two terms are effectively synonyms in that both refer to the same numerical shear stress value.

Section 1.6 demonstrated that a shear stress never acts on just one surface. If there is a shear stress τ_H on a horizontal plane in the beam, then there is also a shear stress τ_V of the same magnitude on a vertical plane (Figure 9.8). Since the horizontal and vertical shear stresses are equal, we will let $\tau_H = \tau_V = \tau$; thus, Equation (g) can be simplified into a form commonly known as the **shear stress formula**:

$$\tau = \frac{VQ}{I_z t} \tag{9.2}$$

The moment of inertia I_z in Equation (9.2) is the moment of inertia *of the entire cross section* about the z axis.

Because Q varies with area A', the value of τ varies over the depth of the cross section. At the upper and lower boundaries of the cross section (e.g., points *a, c, d,* and *f* in Figure 9.8), the value of Q is zero, since area A' is zero. The maximum value of Q occurs at the neutral axis of the cross section. Accordingly, the largest shear stress τ is usually located at the neutral axis; however, this is not necessarily so. In Equation (9.2), the internal shear force V and the moment of inertia I_z are constant at any particular location x along the span. The value of Q is clearly dependent upon the particular y coordinate being considered. The term t (which is the width of the cross section in the z direction at any specific y location) in the denominator of Equation (9.2) can also vary over the depth of the cross section. Therefore, the maximum horizontal shear stress τ occurs at the y coordinate that has the largest value of Q/t—most often, but not necessarily, at the neutral axis.

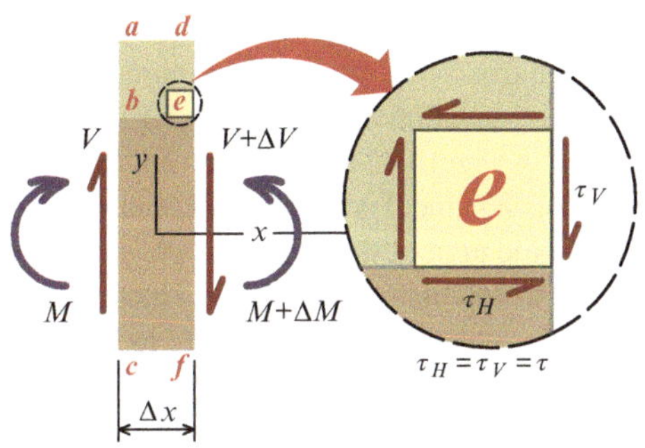

FIGURE 9.8 Shear stress at point *e*.

The direction of the shear stress acting on a transverse plane is the same as the direction of the internal shear force. As illustrated in Figure 9.8, the internal shear force acts downward on section *d–e–f*. The shear stress acts in the same direction on the vertical plane. Once the direction of the shear stress on one face has been determined, the shear stresses acting on the other planes are known.

Although the stress given by Equation (9.2) is associated with a particular point in a beam, it is averaged across the thickness t and hence is accurate only if t is not too great. For a rectangular section having a depth equal to twice its width, the maximum stress computed by methods that are more rigorous is about 3 percent greater than that given by Equation (9.2). If the cross section is square, the error is about 12 percent. If the width is four times the depth, the error is almost 100 percent! Furthermore, if the shear stress formula is applied to a cross section in which the sides of the beam are not parallel, such as a triangular section, the average stress is subject to additional error because the transverse variation of stress is greater when the sides are not parallel.

9.4 The First Moment of Area, Q

Calculation of the first moment of area, Q, for a specific y location in a beam cross section is initially one of the most confusing aspects associated with shear stress in flexural members. The reason for the confusion is that there is no unique value of Q for a particular cross section—there are many values of Q. For example, consider the box-shaped cross section shown in Figure 9.9*a*. In order to calculate the shear stress associated with the internal shear force V at points *a*, *b*, and *c*, three different values of Q must be determined.

What is Q? Q is a mathematical abstraction termed a first moment of area. Recall that a first-moment-of-area term appears as the numerator in the definition of a centroid:

$$\bar{y} = \frac{\int_A y\,dA}{\int_A dA} \tag{a}$$

Q is the first moment of area of only portion A' of the total cross-sectional area A. Equation (a) can be rewritten in terms of A' instead of the total area A and multiplied by the denominator of the right side (in terms of A') to give a useful formulation for Q:

$$Q = \int_{A'} y\,dA' = \bar{y}'\int_{A'} dA' = \bar{y}'A' \tag{9.3}$$

Here, $\bar{y}'$ is the distance *from the neutral axis* of the cross section to the centroid of area A'.

To determine Q at point *a* in Figure 9.9*a*, the cross-sectional area is subdivided at *a* by slicing parallel to the neutral axis (which is perpendicular to the direction of the internal shear force V). The area A' begins at this cut line and *extends away from the neutral axis* to the free surface of the beam. (Recall the free-body diagram in Figure 9.7 used to evaluate the horizontal equilibrium of area A' in the derivation of Equation 9.1.) The area A' to be used in calculating Q at point a is highlighted in Figure 9.9*b*. The centroid of the highlighted area relative to the neutral axis (the z axis in this instance) is determined, and Q is calculated from the product of this centroidal distance and the area of the shaded portion of the cross section.

A similar process is used to calculate Q at point b. The box shape is sectioned at b parallel to the neutral axis. (**Note:** V is always perpendicular to its corresponding neutral axis.) The area A' begins at this cut line and *extends away from the neutral axis* to the free surface, as shown in Figure 9.9*c*. The centroidal location $\bar{y}'$ of the highlighted area relative to the neutral axis is determined, and Q is calculated from $Q = \bar{y}'A'$.

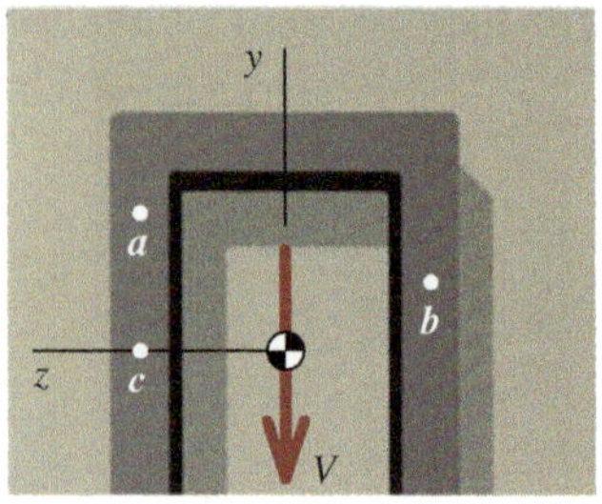

(*a*) Box shape

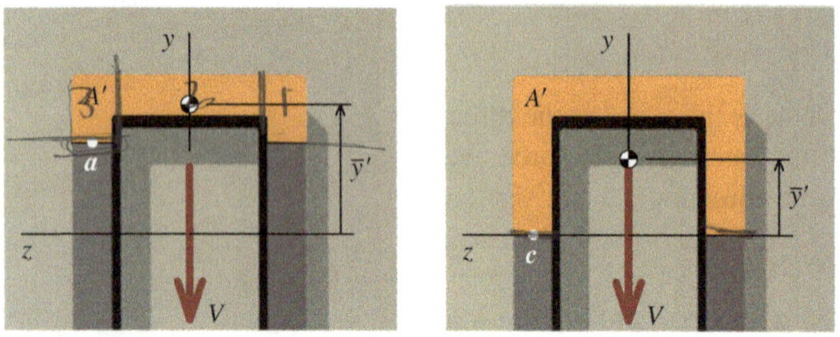

(*b*) Area A' for calculating Q at point a

(*d*) Area A' for calculating Q at point c

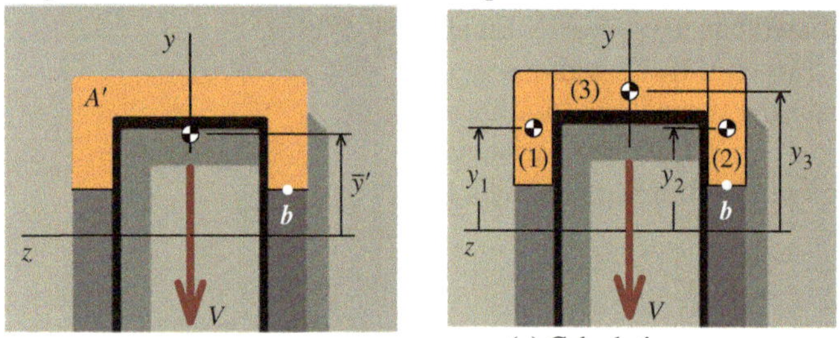

(*c*) Area A' for calculating Q at point b

(*e*) Calculation process

FIGURE 9.9 Calculating Q at different locations in a box-shaped cross section.

Generally, if the point of interest is above the neutral axis, it is convenient to consider an area A' that begins at the point and *extends upward*. If the point of interest is below the neutral axis, consider the area A' that begins at the point and *extends downward.*

Point c is located on the neutral axis for the box shape; thus, area A' begins at the neutral axis (Figure 9.9*d*). For points a and b, it was clear which direction was meant by the phrase "away from the neutral axis." However, in this instance c is actually on the neutral axis—a situation that raises the question, *Should area A' extend above the neutral axis or below the neutral axis?* The answer is, *Either direction will give the same Q at point c.* Although the area above the neutral axis is highlighted in Figure 9.9*d*, the area below the neutral axis would give the same result. The centroidal location $\bar{y}'$ of the highlighted area relative to the neutral axis is determined, and Q is calculated from $Q = \bar{y}'A'$.

The first moment of the total cross-sectional area A taken about the neutral axis must be zero (by definition of the neutral axis). While the illustrations given here have shown how Q can be calculated from an area A' *above* points a, b, and c, the first moment of the area *below* points a, b, and c is simply the negative. In other words, the value of Q calculated with the use of an area A' *below* points a, b, and c must have the same magnitude as Q calculated from an area A' *above* points a, b, and c. It is usually easier to calculate Q with the use of an area A' that extends away from the neutral axis, but there are exceptions.

Let us consider the calculation of Q at point b (Figure 9.9*c*) in more detail. The area A' can be divided into three rectangular areas (Figure 9.9*e*) so that $A' = A_1 + A_2 + A_3$. The centroid location of the highlighted area with respect to the neutral axis is

$$\bar{y}' = \frac{y_1A_1 + y_2A_2 + y_3A_3}{A_1 + A_2 + A_3}$$

The value of Q associated with point b is

$$Q = y'A' = \frac{y_1A_1 + y_2A_2 + y_3A_3}{A_1 + A_2 + A_3}(A_1 + A_2 + A_3) = y_1A_1 + y_2A_2 + y_3A_3$$

This result suggests a more direct calculation procedure that is often expedient. For cross sections that consist of i shapes,

$$Q = \sum_i y_iA_i \tag{9.4}$$

where y_i is the distance between the neutral axis and the centroid of shape i and A_i is the area of shape i.

9.5 Shear Stresses in Beams of Rectangular Cross Section

In this section, we consider beams of rectangular cross section, in order to develop some understanding of how shear stress is distributed over the depth of a beam. Consider a beam subjected to an internal shear force V. Keep in mind that a shear force exists only when the internal bending moment is not constant and that it is the variation in bending moment along the span that creates shear stress in a beam, as discussed in Section 9.3. The rectangular cross section shown in Figure 9.10*a* has width b and height h; therefore, the total cross-sectional area is $A = bh$. By symmetry, the centroid of the rectangle is located at midheight. The moment of inertia about the z centroidal axis (i.e., the neutral axis) is $I_z = bh^3/12$.

Shear stress in the beam will be determined from Equation (9.2). To investigate the distribution of τ over the cross section, the shear stress will be computed at an arbitrary height y from the neutral axis (Figure 9.10*b*). The first moment of area, Q, for the highlighted area A' can be expressed as

$$Q = \bar{y}'A' = \left[y + \frac{1}{2}\left(\frac{h}{2} - y\right)\right]\left(\frac{h}{2} - y\right)b = \frac{1}{2}\left(\frac{h^2}{4} - y^2\right)b \qquad \text{(a)}$$

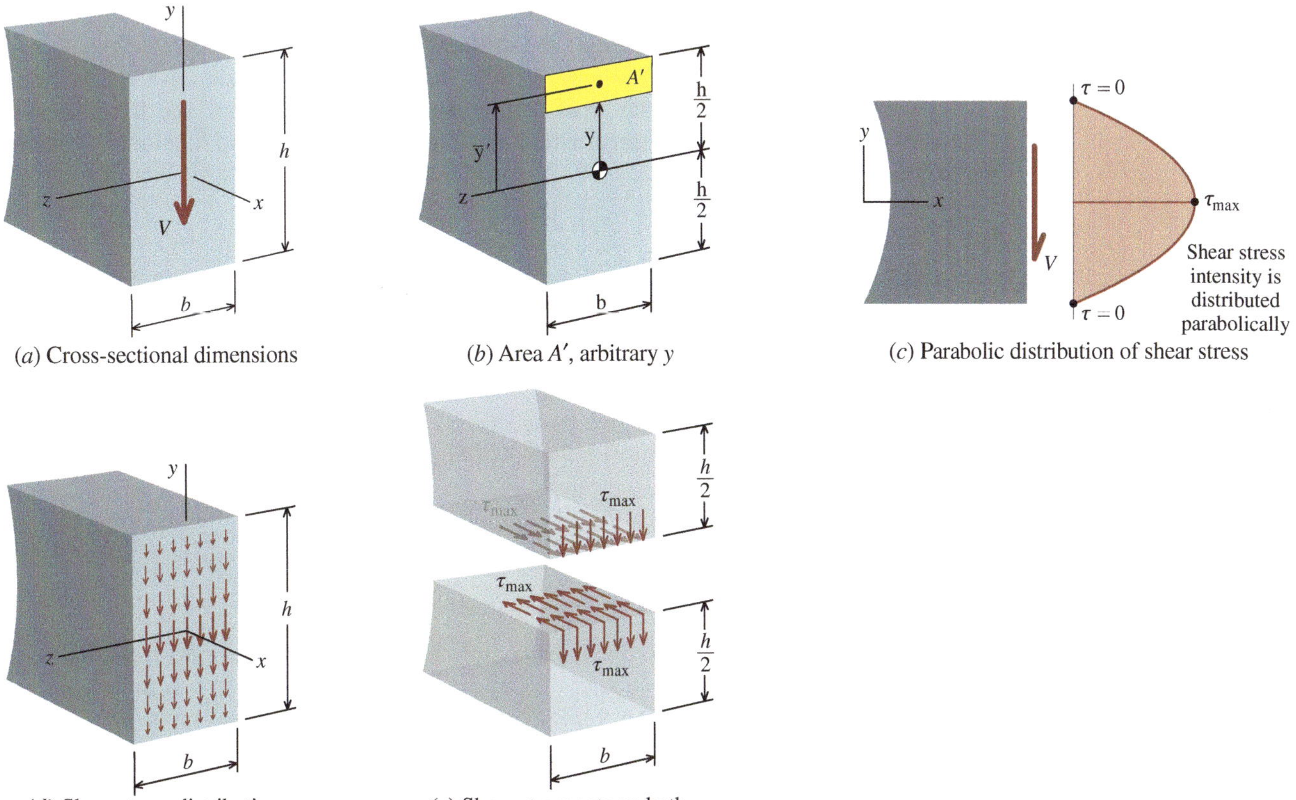

(*a*) Cross-sectional dimensions

(*b*) Area A', arbitrary y

(*c*) Parabolic distribution of shear stress

(*d*) Shear stress distribution

(*e*) Shear stress acts on both transverse and longitudinal planes

FIGURE 9.10 Shear distribution in a rectangular cross section.

The shear stress τ as a function of the vertical coordinate y can now be determined from the shear formula:

$$\tau = \frac{VQ}{I_z t} = \frac{V}{\left(\frac{1}{12}bh^3\right)b} \times \frac{1}{2}\left(\frac{h^2}{4} - y^2\right)b = \frac{6V}{bh^3}\left(\frac{h^2}{4} - y^2\right) \tag{9.5}$$

The accuracy of Equation (9.6) depends on the depth-to-width ratio of the cross section. For beams in which the depth is much greater than the width, Equation (9.6) can be considered exact. As the cross section approaches a square shape, the true maximum horizontal shear stress is somewhat greater than that given by Equation (9.6).

Equation (9.5) is a second-order equation, which indicates that τ is distributed parabolically with respect to y (Figure 9.10*c*). At $y = \pm h/2$, $\tau = 0$. The shear stress vanishes at the extreme fibers of the cross section, since $A' = 0$; consequently, $Q = 0$ at these locations. *There is no shear stress on a free surface of the beam.* The maximum horizontal shear stress occurs at $y = 0$, which is the neutral-axis location. At the neutral axis, the maximum horizontal shear stress in a rectangular cross section is given by

$$\tau_{max} = \frac{VQ}{I_z t} = \frac{V}{\left(\frac{1}{12}bh^3\right)b} \times \frac{1}{2}\left(\frac{h}{2}\right)\frac{bh}{2} = \frac{3V}{2bh} = \frac{3}{2}\frac{V}{A} \tag{9.6}$$

It is important to emphasize that Equation (9.6) gives the maximum horizontal shear stress *only for rectangular cross sections.* Note that the maximum horizontal shear stress at the neutral axis is 50 percent greater than the overall average shear stress given by $\tau = V/A$.

To summarize, the shear stress intensity associated with an internal shear force V in a rectangular cross section is distributed parabolically in the direction perpendicular to the neutral axis (i.e., in the y direction) and uniformly in the direction parallel to the neutral axis (i.e., in the z direction) (Figure 9.10*d*). The shear stress vanishes at the upper and lower edges of the rectangular cross section and peaks at the neutral-axis location. It is important to remember that shear stress acts on both transverse and longitudinal planes (Figure 9.10*e*).

The expression "maximum shear stress" in the context of beam shear stresses is problematic. In Chapter 12, the discussion of stress transformations will show that the state of stress existing at any point can be expressed by many different combinations of normal and shear stress, depending on the orientation of the plane surface upon which the stresses act. (This notion was introduced previously in Section 1.5, pertaining to axial members, and in Section 6.4, regarding torsion members.) Consequently, the expression "maximum shear stress," when applied to beams, could be interpreted to mean either

(a) the maximum value of $\tau = VQ/It$ for any coordinate y in the cross section, or
(b) the maximum shear stress at a particular point in the cross section when all possible plane surfaces that pass through the point are considered.

In this chapter, to preclude ambiguity, the expression "maximum horizontal shear stress" will be used in a context in which the maximum value of $\tau = VQ/It$ for any coordinate y in the cross section is to be determined. Since shear stresses on perpendicular planes must be equal in magnitude, it would be equally proper to use the expression "maximum transverse shear stress" for this purpose. In Chapter 12, the notion of stress transformations will be used to determine the maximum shear stress at a particular point. In Chapter 15, maximum normal and shear stresses at specific points in beams will be discussed in more detail.

MecMovies

EXAMPLE

M9.2 Derivation of the shear stress formula.

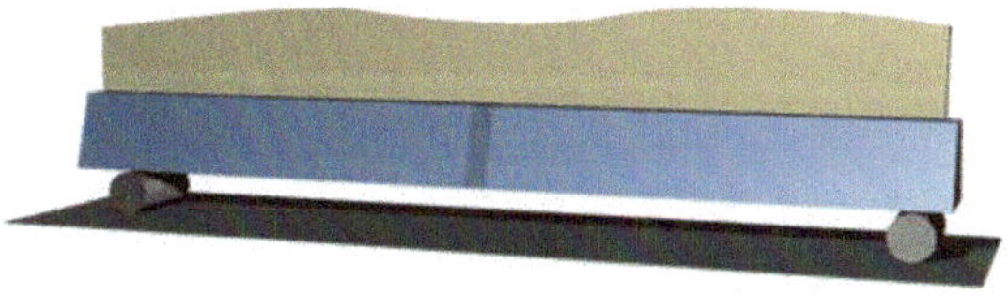

EXAMPLE 9.2

A 10 ft long simply supported laminated wooden beam consists of eight 1.5 in. by 6 in. planks glued together to form a section 6 in. wide by 12 in. deep, as shown. The beam carries a 9 kip concentrated load at midspan. Determine

(a) the average horizontal shear stress in the glue joints at b, c, and d, and
(b) the maximum horizontal shear stress in the cross section,

at section a–a, located 2.5 ft from pin support A.

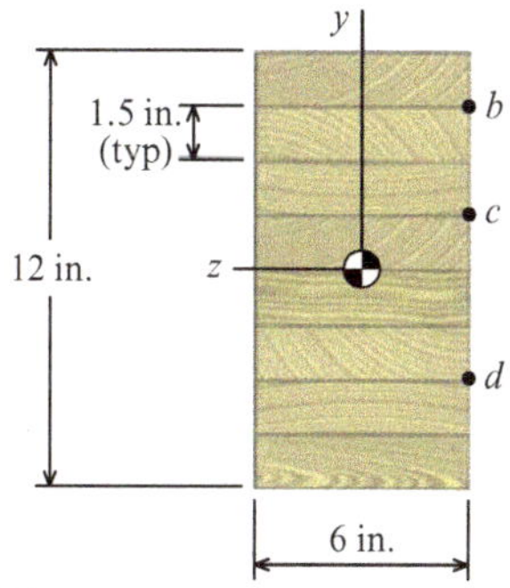

Plan the Solution

The transverse shear force V acting at section a–a can be determined from a shear-force diagram for the simply supported beam. To determine the horizontal shear stress in the indicated glue joints, the corresponding first moment of area, Q, must be calculated for each location. The average horizontal shear stress will then be determined by the shear stress formula given in Equation (9.2).

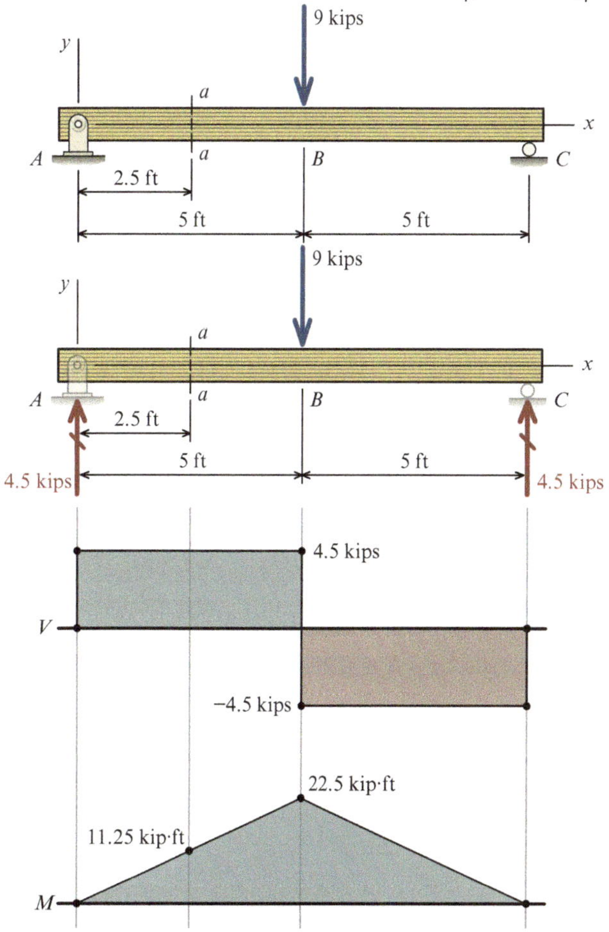

SOLUTION

Internal Shear Force at Section a–a

The shear-force and bending-moment diagrams can readily be constructed for the simply supported beam. From the shear-force diagram, the internal shear force V acting at section a–a is $V = 4.5$ kips.

Section Properties

The centroid location for the rectangular cross section can be determined from symmetry. The moment of inertia of the cross section about the z centroidal axis is equal to

$$I_z = \frac{bh^3}{12} = \frac{(6 \text{ in.})(12 \text{ in.})^3}{12} = 864 \text{ in.}^4$$

(a) Average Horizontal Shear Stress in Glue Joints

The shear stress formula is

$$\tau = \frac{VQ}{I_z t}$$

To determine the average horizontal shear stress in the glue joints at b, c, and d by the shear stress formula both, the first moment of area, Q, and the width of the stressed surface, t, must be determined for each location.

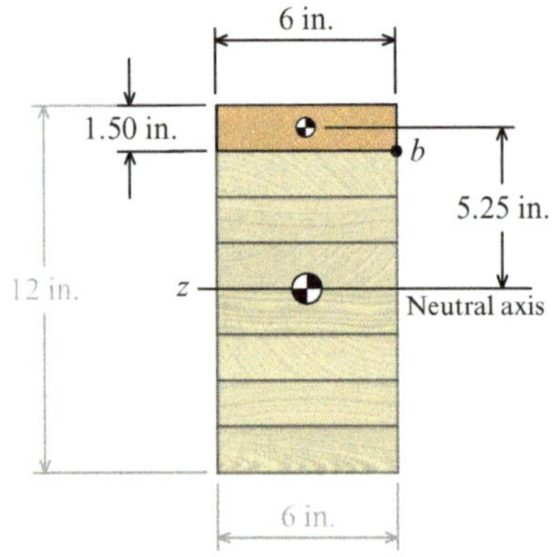

Shear stress in glue joint b: The portion of the cross section to be considered for the first moment of area, Q, about the neutral axis begins at point b and extends **away from the neutral axis**. The area to be considered for joint b is highlighted in the accompanying diagram. The area of the highlighted region is (1.50 in.)(6 in.) = 9 in.^{2}. The distance from the neutral axis to the centroid of the highlighted area is 5.25 in. The first moment of area corresponding to joint b is calculated as

$$Q_b = (1.50 \text{ in.})(6 \text{ in.})(5.25 \text{ in.}) = 47.25 \text{ in.}^3$$

The width of the glue joint is $t = 6$ in. From the shear stress formula, the average horizontal shear stress in glue joint b is computed as

$$\tau_b = \frac{VQ_b}{I_z t_b} = \frac{(4.5 \text{ kips})(47.25 \text{ in.}^3)}{(864 \text{ in.}^4)(6 \text{ in.})} = 0.0410 \text{ ksi} = 41.0 \text{ psi} \qquad \textbf{Ans.}$$

This shear stress acts in the x direction on the glue joint. (**Note:** The shear stress determined by the shear stress formula is an *average* shear stress because the shear stress actually varies somewhat in magnitude across the 6 in. width of the cross section. The variation is more pronounced for cross sections that are relatively short and wide.)

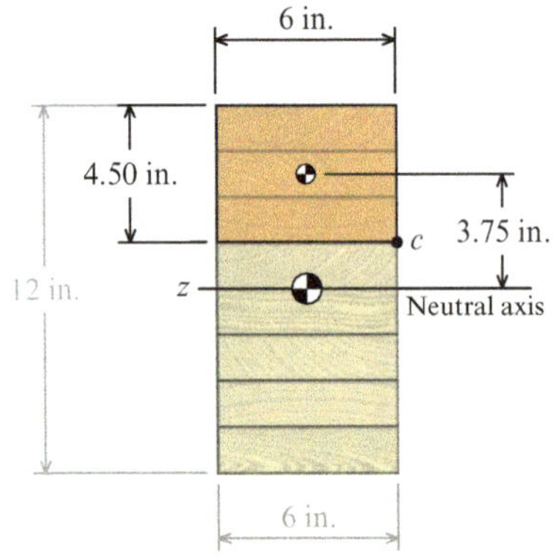

Shear stress in glue joint c: The area to be considered for joint c, highlighted in the figure to the left, begins at c and extends **away** from the neutral axis. The first moment of area corresponding to joint c is calculated as

$$Q_c = (4.50 \text{ in.})(6 \text{ in.})(3.75 \text{ in.}) = 101.25 \text{ in.}^3$$

The width of the glue joint is $t = 6$ in. From the shear stress formula, the average horizontal shear stress acting in the x direction in glue joint c is computed as

$$\tau_c = \frac{VQ_c}{I_z t_c} = \frac{(4.5 \text{ kips})(101.25 \text{ in.}^3)}{(864 \text{ in.}^4)(6 \text{ in.})} = 0.0879 \text{ ksi} = 87.9 \text{ psi} \qquad \textbf{Ans.}$$

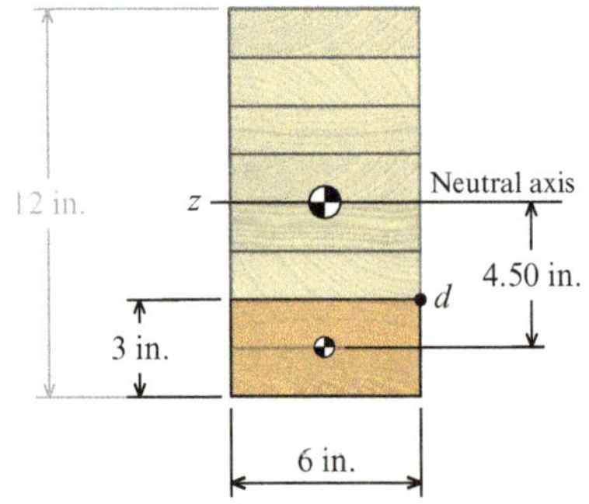

Shear stress in glue joint d: The area to be considered for joint d, highlighted in the accompanying figure, begins at d and extends **away** from the neutral axis. In this instance, however, the area extends downward from d, away from the z axis. The first moment of area corresponding to joint d is calculated as

$$Q_d = (3 \text{ in.})(6 \text{ in.})(4.50 \text{ in.}) = 81.0 \text{ in.}^3$$

The average horizontal shear stress acting in the x direction in glue joint d is computed as

$$\tau_d = \frac{VQ_d}{I_z t_d} = \frac{(4.5 \text{ kips})(81.0 \text{ in.}^3)}{(864 \text{ in.}^4)(6 \text{ in.})} = 0.0703 \text{ ksi} = 70.3 \text{ psi} \qquad \textbf{Ans.}$$

(b) Maximum Horizontal Shear Stress in Cross Section

The maximum horizontal shear stress in the rectangular cross section occurs at the neutral axis. To calculate Q, the area beginning at the z axis and extending upwards or extending downwards may be used, as shown in the following two figures:

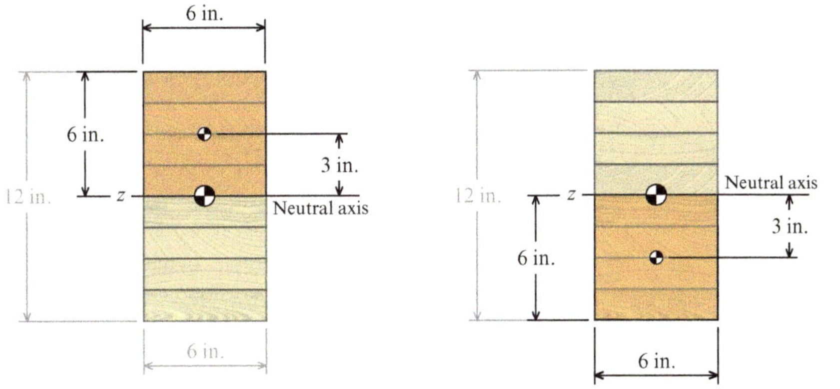

For either area, the first moment of area is calculated as

$$Q_{max} = (6 \text{ in.})(6 \text{ in.})(3 \text{ in.}) = 108 \text{ in.}^3$$

The maximum value of Q always occurs at the neutral-axis location. Also, the maximum horizontal shear stress *usually* occurs at the neutral axis. There are instances, however, in which the width t of the stressed surface varies over the depth of the cross section. In such instances, it is possible that the maximum horizontal shear stress will occur at a y location other than the neutral axis.

The maximum horizontal shear stress in the rectangular cross section is computed as

$$\tau_{max} = \frac{VQ_{max}}{I_z t} = \frac{(4.5 \text{ kips})(108 \text{ in.}^3)}{(864 \text{ in.}^4)(6 \text{ in.})} = 0.0938 \text{ ksi} = 93.8 \text{ psi} \qquad \textbf{Ans.}$$

PROBLEMS

For Problems P9.1 through P9.4, a beam segment subjected to internal bending moments at sections A and B is shown along with a sketch of the cross-sectional dimensions. For each problem,

(a) sketch a side view of the beam segment and plot the distribution of bending stresses acting at sections A and B. Indicate the magnitudes of key bending stresses in the sketch.
(b) determine the resultant forces acting in the x direction on the specified area at sections A and B, and show these resultant forces in the sketch.
(c) determine the horizontal force required to satisfy equilibrium for the specified area and show the location and direction of this force in the sketch.

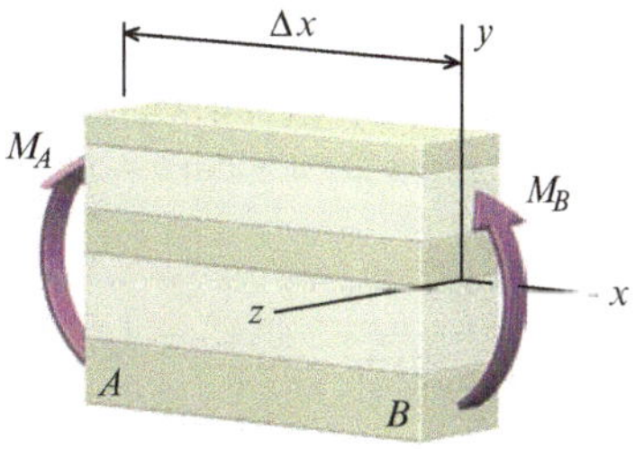

FIGURE P9.1*a*/2*a* Beam segment.

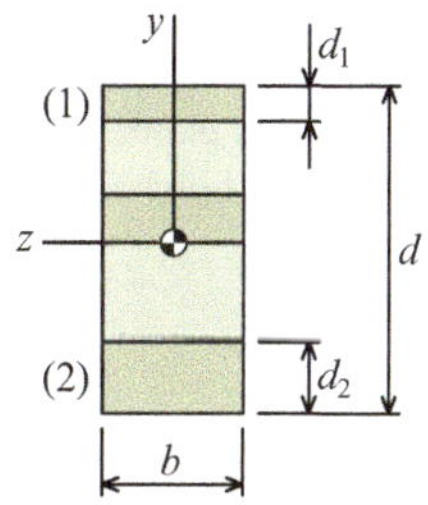

FIGURE P9.1*b*/2*b* Cross section.

P9.1 The beam segment shown in Figure P9.1*a*/2*a* is subjected to internal bending moments $M_A = 1{,}640$ lb·ft and $M_B = 760$ lb·ft. The segment length is $\Delta x = 7$ in. Consider area (1) shown in Figure P9.1*b*/2*b*, and use cross-sectional dimensions of $b = 3$ in., $d = 9$ in., and $d_1 = 1.5$ in.

P9.2 The beam segment shown in Figure P9.1*a*/2*a* is subjected to internal bending moments $M_A = 1{,}430$ N·m and $M_B = 3{,}140$ N·m. The segment length is $\Delta x = 180$ mm. Consider area (2) shown in Figure P9.1*b*/2*b*, and use cross-sectional dimensions of $b = 90$ mm, $d = 240$ mm, and $d_2 = 70$ mm.

P9.3 The beam segment shown in Figure P9.3*a*/4*a* is subjected to internal bending moments $M_A = -9.2$ kN·m and $M_B = -34.6$ kN·m. The segment length is $\Delta x = 300$ mm. Consider area (1) shown in Figure P9.3*b*/4*b*, and use cross-sectional dimensions of $b_1 = 250$ mm, $d_1 = 15$ mm, $b_2 = 10$ mm, $d_2 = 370$ mm, $b_3 = 150$ mm, and $d_3 = 15$ mm.

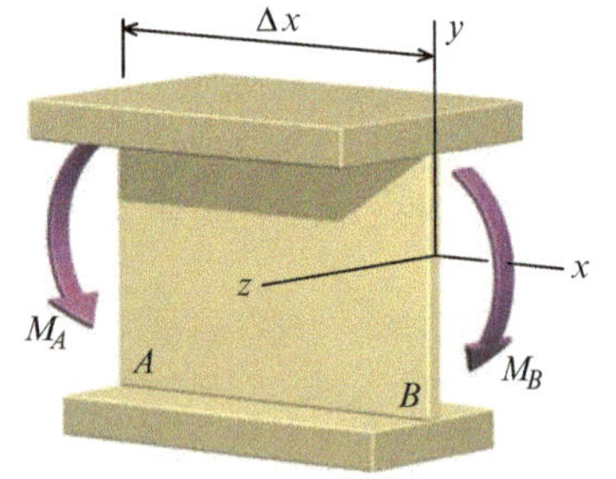

FIGURE P9.3*a*/4*a* Beam segment.

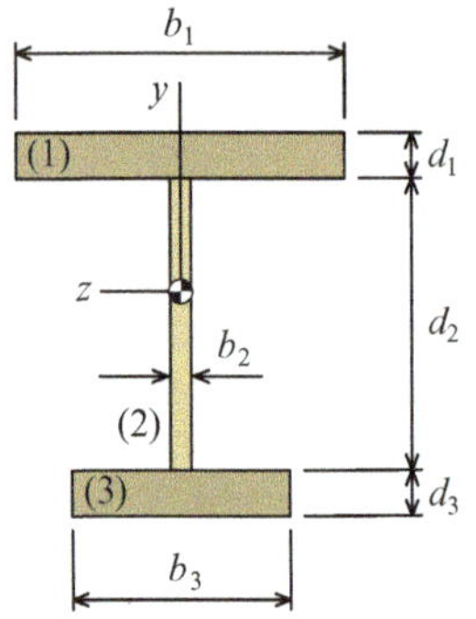

FIGURE P9.3*b*/4*b* Cross section.

P9.4 The beam segment shown in Figure P9.3*a*/4*a* is subjected to internal bending moments $M_A = -45.0$ kip·ft and $M_B = -26.0$ kip·ft. The segment length is $\Delta x = 9$ in. Consider area (3) shown in Figure P9.3*b*/4*b*, and use cross-sectional dimensions of $b_1 = 6.6$ in., $d_1 = 0.6$ in., $b_2 = 0.4$ in., $d_2 = 10.8$ in., $b_3 = 4.2$ in., and $d_3 = 0.6$ in.

P9.5 A cantilever wooden beam (Figure P9.5*a*) consists of eight 2 in. thick planks glued together to form a cross section that is 16 in. deep, as shown in Figure P9.5*b*. Each plank has a width $b = 5.5$ in.

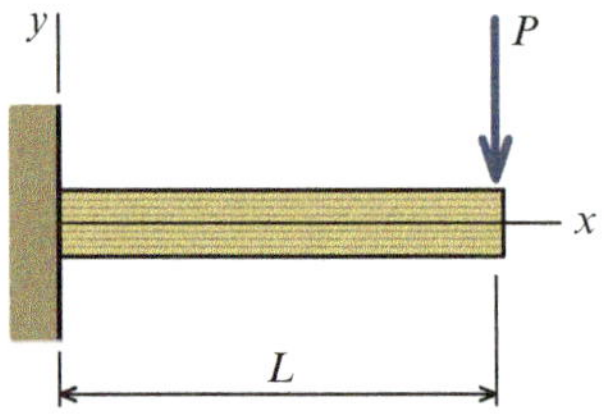

FIGURE P9.5*a* Cantilever beam.

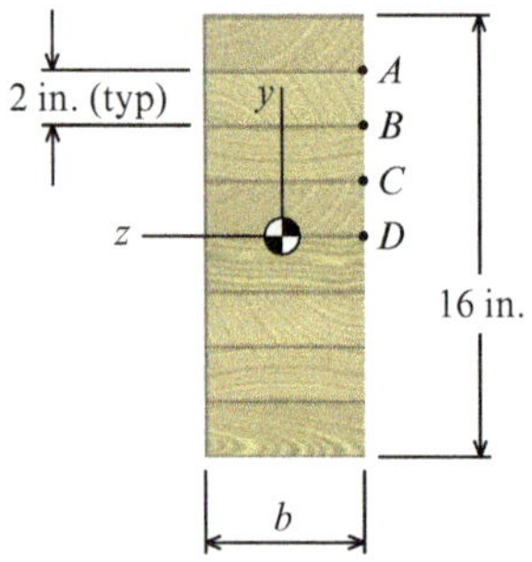

FIGURE P9.5*b* Cross-sectional dimensions.

The cantilever beam has a length $L = 6$ ft and supports a concentrated load $P = 3{,}800$ lb.

(a) Calculate the horizontal shear stress at points *A*, *B*, *C*, and *D*.
(b) From your results in (a), plot a graph showing the distribution of shear stresses from top to bottom of the beam.

P9.6 A 6 m long simply supported wooden beam carries a 36 kN concentrated load at *B*, as shown in Figure P9.6*a*. The cross-sectional dimensions of the beam, as shown in Figure P9.6*b*, are $b = 150$ mm, $d = 400$ mm, $a = 90$ mm, and $c = 30$ mm. Section *a–a* is located at $x = 0.8$ m from *B*. Determine

(a) the magnitude of the shear stress in the beam at point *H* in section *a–a*.
(b) the magnitude of the shear stress in the beam at point *K* in section *a–a*.
(c) the maximum horizontal shear stress that occurs in the beam at any location within its entire span.
(d) the maximum tensile bending stress that occurs in the beam at any location within its entire length.

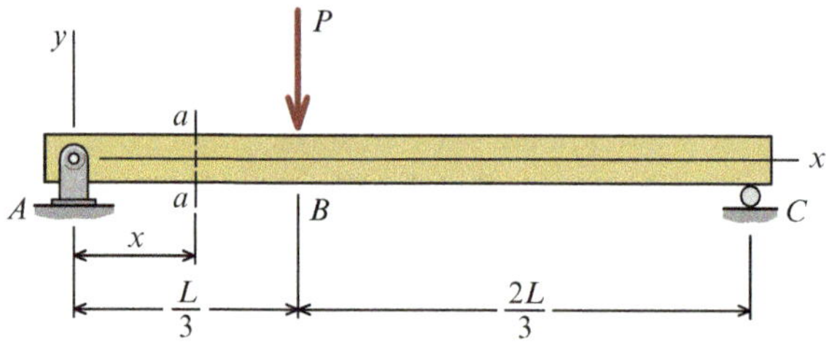

FIGURE P9.6*a* Simply supported wooden beam.

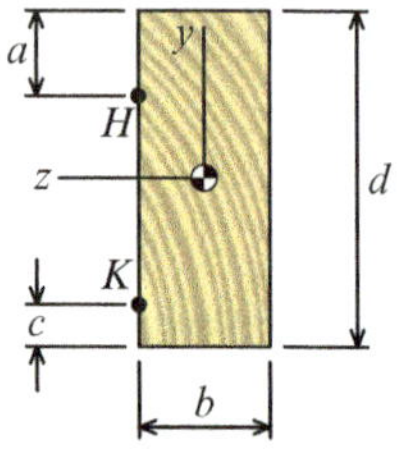

FIGURE P9.6*b* Cross-sectional dimensions.

P9.7 A 6 m long simply supported wooden beam carries a uniformly distributed load of 14 kN/m, as shown in Figure P9.7*a*. The cross-sectional dimensions of the beam, as shown in Figure P9.7*b*, are $b = 200$ mm, $d = 480$ mm, $y_H = 80$ mm, and $y_K = 160$ mm. Section *a–a* is located at $x = 1.2$ m from *B*.

(a) Determine the magnitude of the shear stress in the beam at point *H* in section *a–a*.
(b) Determine the magnitude of the shear stress in the beam at point *K* in section *a–a*.
(c) If the allowable shear stress for the wood is 950 kPa, what is the largest distributed load *w* that can be supported by the beam?

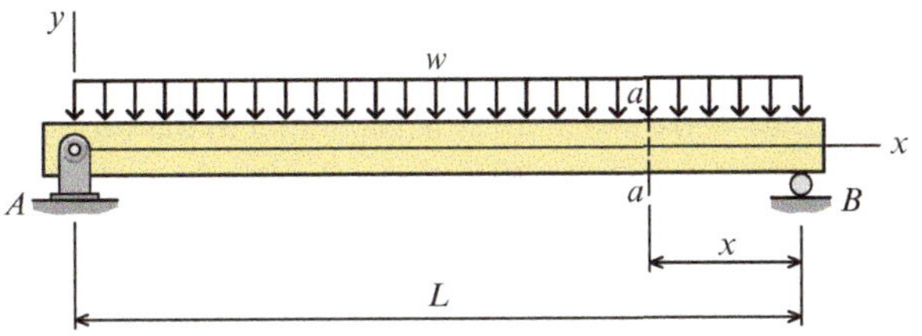

FIGURE P9.7*a* Simply supported wooden beam.

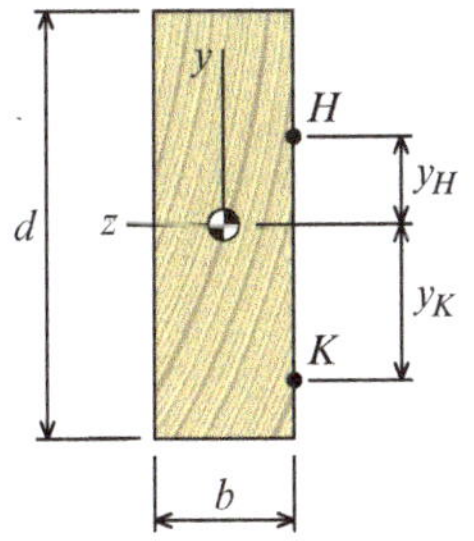

FIGURE P9.7*b* Cross-sectional dimensions.

P9.8 The simply supported wooden beam shown in Figure P9.8*a* carries a uniformly distributed load w = 780 lb/ft, as shown in Figure P9.8*a*, where L_{AB} = 8 ft and L_{BC} = 16 ft. The cross-sectional dimensions of the beam, as shown in Figure P9.8*b*, are b = 6 in., d = 20 in., y_H = 5 in., and y_K = 7 in. Section *a–a* is located at x = 4 ft from *A*.

(a) Determine the magnitude of the shear stress in the beam at point *H* in section *a–a*.
(b) Determine the magnitude of the shear stress in the beam at point *K* in section *a–a*.
(c) If the allowable shear stress for the wood is 150 psi, what is the largest distributed load w that can be supported by the beam?

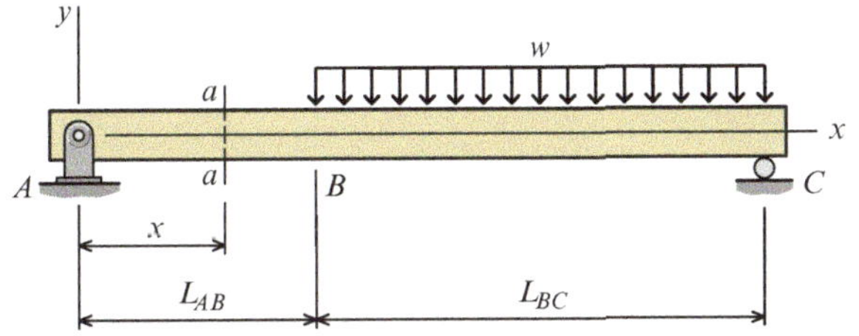

FIGURE P9.8*a* Simply supported wooden beam.

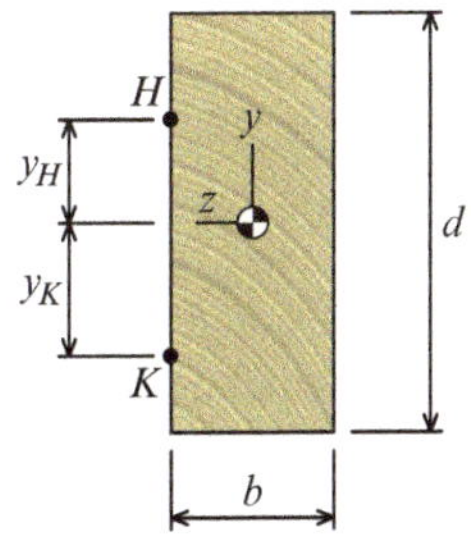

FIGURE P9.8*b* Cross-sectional dimensions.

P9.9 The wooden beam shown in Figure P9.9*a* supports loads w = 470 lb/ft and P = 1,000 lb on span lengths L_{AB} = 10 ft, L_{BC} = 5 ft, and L_{CD} = 5 ft. The beam cross section shown in Figure P9.9*b* has dimensions b_1 = 1.5 in., d_1 = 6.0 in., b_2 = 5.0 in., and d_2 = 16.0 in. Determine the magnitude and location of

(a) the maximum horizontal shear stress in the beam.
(b) the maximum tensile bending stress in the beam.

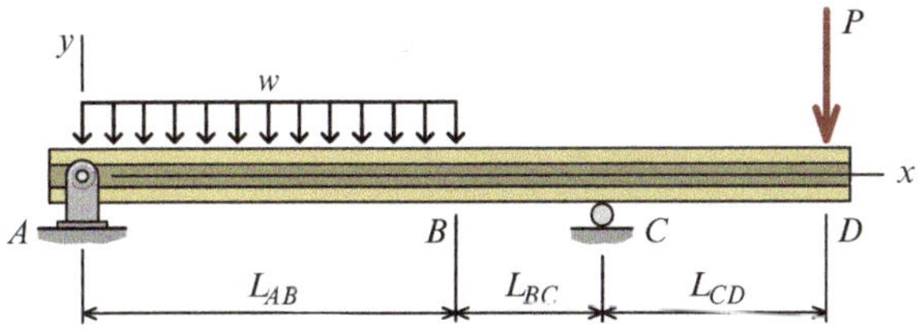

FIGURE P9.9*a* Simply supported wooden beam.

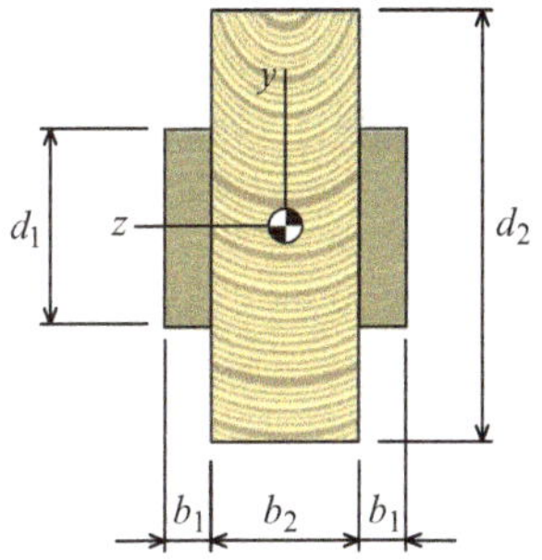

FIGURE P9.9*b* Cross-sectional dimensions.

P9.10 For the wooden beam shown in Figure P9.10*a*, assume that L = 1.2 m and P = 12 kN. The beam cross section shown in Figure P9.10*b* has dimensions b = 50 mm and d = 250 mm.

(a) Determine the maximum horizontal shear stress in the beam.
(b) If the allowable shear stress for the wood is 850 kPa, what is the minimum width b that is acceptable for this beam?

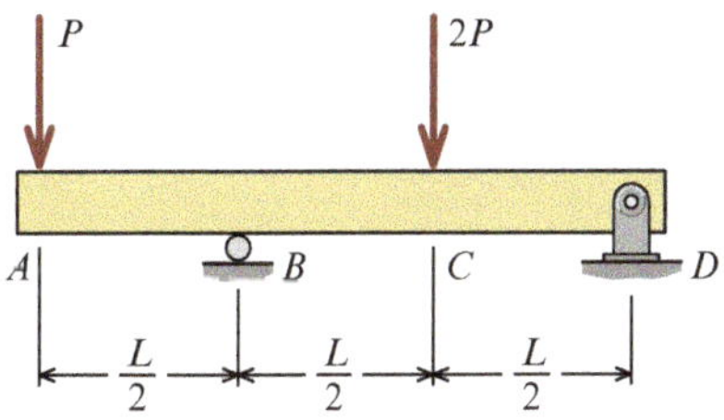

FIGURE P9.10*a* Simply supported wooden beam.

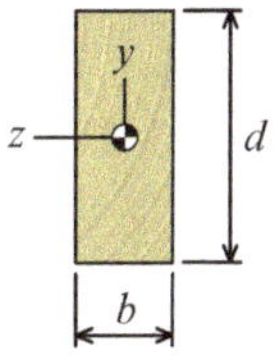

FIGURE P9.10*b* Cross-sectional dimensions.

9.6 Shear Stresses in Beams of Circular Cross Section

In beams with circular cross sections, transverse shear stress acts parallel to the y axis only over some, but not all, of the entire depth of the shape. Consequently, the shear stress formula is not applicable in general for a circular cross section. However, Equation (9.2) can be used to determine the shear stress acting at the neutral axis.

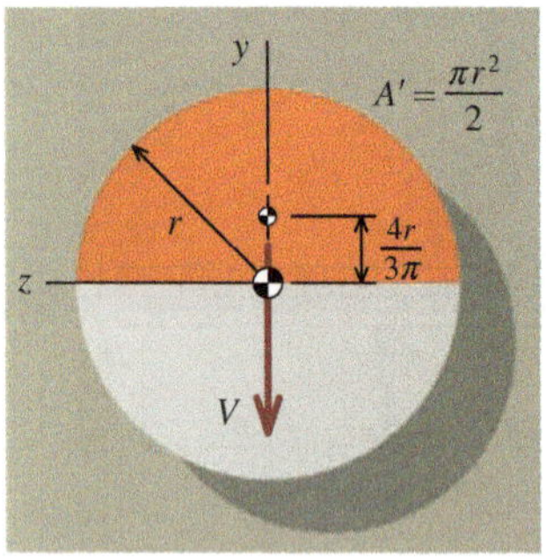

FIGURE 9.11 Solid circular cross section.

A **solid circular cross section** of radius r is shown in Figure 9.11. To use the shear stress formula, the value of Q for the highlighted semicircular area must be determined. The area of the semicircle is $A' = \pi r^2/2$. The distance from the neutral axis to the centroid of the semicircle is given by $\bar{y}' = 4r/3\pi$. Thus,

$$Q = \bar{y}'A' = \frac{4r}{3\pi}\frac{\pi r^2}{2} = \frac{2}{3}r^3 \tag{9.7}$$

or, in terms of the diameter $d = 2r$,

$$Q = \frac{1}{12}d^3 \tag{9.8}$$

The width of the circular cross section at the neutral axis is $t = 2r$, and the moment of inertia about the z axis is $I_z = \pi r^4/4 = \pi d^4/64$. Substituting these relationships into the shear stress formula gives the following expression for τ_{max} at the neutral axis of a solid circular cross section:

$$\tau_{max} = \frac{VQ}{I_z t} = \frac{V}{\pi r^4/4} \times \frac{2}{3}r^3 \times \frac{1}{2r} = \frac{4V}{3\pi r^2} = \frac{4V}{3A} \tag{9.9}$$

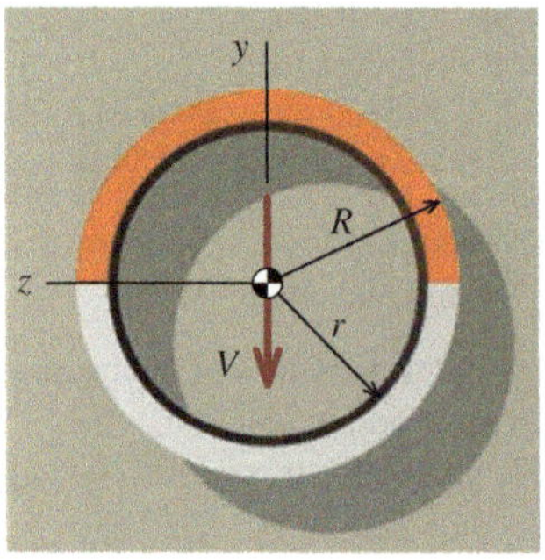

FIGURE 9.12 Hollow circular cross section.

A **hollow circular cross section** having outside radius R and inside radius r is shown in Figure 9.12. The results from Equations (9.7) and (9.8) can be used to determine Q for the highlighted area above the neutral axis:

$$Q = \frac{2}{3}[R^3 - r^3] = \frac{1}{12}[D^3 - d^3] \tag{9.10}$$

The width t of the hollow circular cross section at the neutral axis is two times the wall thickness, or $t = 2(R - r) = D - d$. The moment of inertia of the hollow circular shape about the z axis is

$$I_z = \frac{\pi}{4}[R^4 - r^4] = \frac{\pi}{64}[D^4 - d^4]$$

9.7 Shear Stresses in Webs of Flanged Beams

The elementary theory used to derive the shear stress formula is suitable for determining only the shear stress developed in the web of a flanged beam (if it is assumed that the beam is bent about its strong axis). A wide-flange beam shape is shown in Figure 9.13. To determine the shear stress at a point a located in the web of the cross section, the calculation for Q consists of finding the first moment of the two highlighted areas (1) and (2) about the neutral axis z (Figure 9.13b). A substantial portion of the total area of a flanged shape is concentrated in the flanges, so the first moment of area (1) about the z axis makes up a large percentage of Q. While Q increases as the value of y decreases, the change is not as pronounced in a flanged shape as it would be for a rectangular cross section. Consequently, the distribution of shear stress magnitudes over the depth of the web, while still parabolic, is relatively uniform (Figure 9.13*a*). The minimum horizontal shear stress occurs at the junction between the web and the flange, and the maximum horizontal shear stress occurs at the neutral axis. For wide-flange steel beams, the difference between the maximum and minimum web shear stresses is typically in the range of 10–60 percent.

In deriving the shear stress formula, it was assumed that the shear stress across the width of the beam (i.e., in the z direction) could be considered constant. This assumption, however, is not valid for the flanges of beams; therefore, shear stresses computed for the top and bottom flanges from Equation (9.2) and plotted in Figure 9.13*a* are fictitious. Shear stresses are developed in the flanges (1) of a wide-flange beam, but they act in the x and z directions, not the x and y directions. Shear stresses in thin-walled members, such as wide-flange shapes, will be discussed in more detail in Section 9.9.

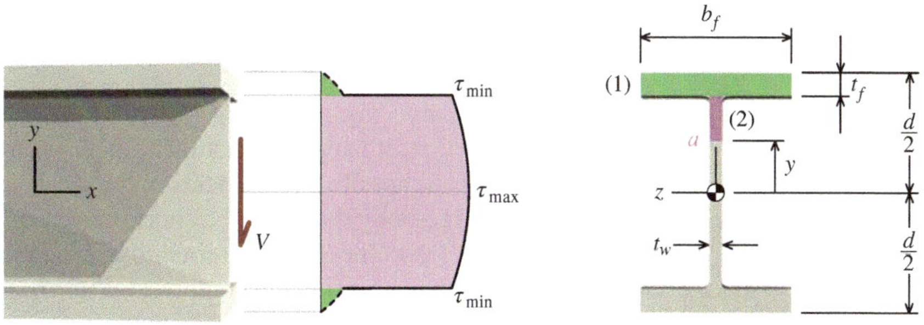

(*a*) Shear stress distribution (*b*) Cross section

FIGURE 9.13 Shear stress distribution in a wide-flange shape.

EXAMPLE 9.3

A concentrated load $P = 36$ kN is applied to the upper end of a pipe as shown. The outside diameter of the pipe is $D = 220$ mm, and the inside diameter is $d = 200$ mm. Determine the vertical shear stress on the y–z plane of the pipe wall.

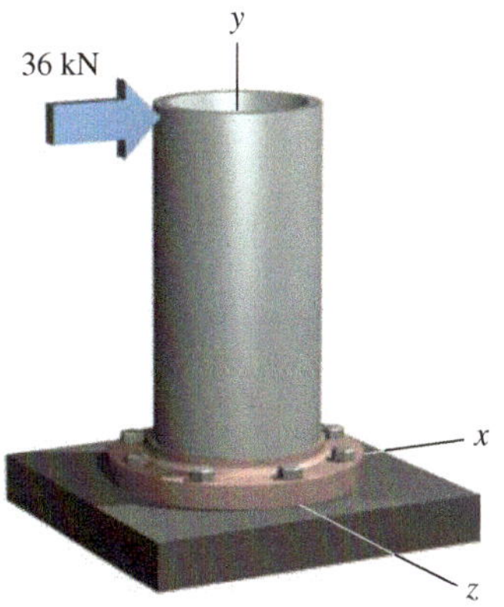

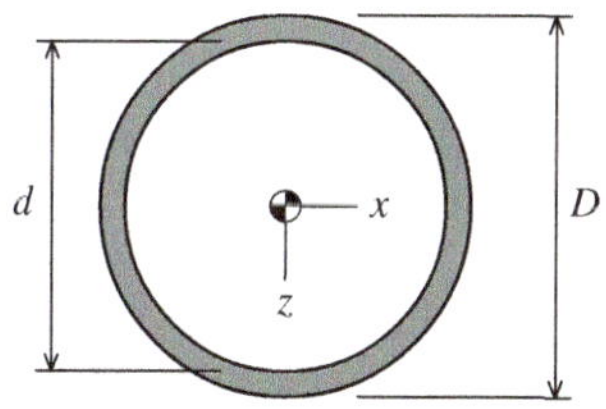

Pipe cross section.

Plan the Solution

The shear stress in a pipe shape can be determined from the shear stress formula [Equation (9.2)] with the use of the first moment of area, Q, calculated from Equation (9.10).

SOLUTION

Section Properties

The centroid location for the tubular cross section can be determined from symmetry. The moment of inertia of the cross section about the z centroidal axis is equal to

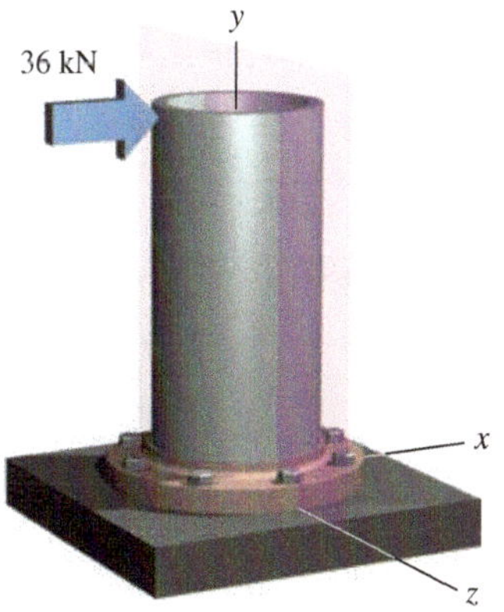

$$I_z = \frac{\pi}{64}[D^4 - d^4] = \frac{\pi}{64}[(220\text{ mm})^4 - (200\text{ mm})^4] = 36{,}450{,}329\text{ mm}^4$$

Equation (9.10) is used to compute the first moment of area for a pipe shape:

$$Q = \frac{1}{12}[D^3 - d^3] = \frac{1}{12}[(220\text{ mm})^3 - (200\text{ mm})^3] = 220{,}667\text{ mm}^3$$

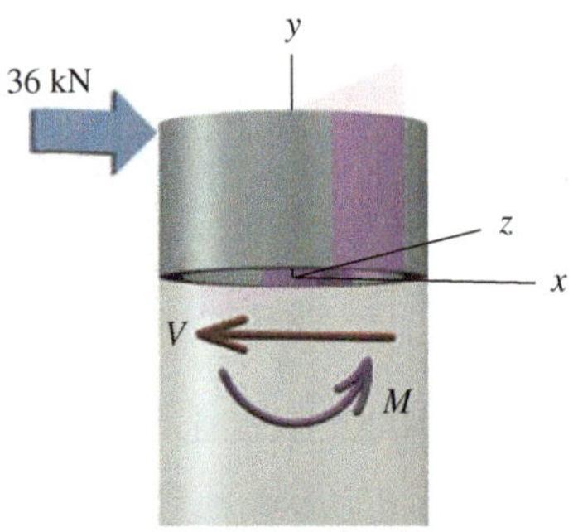

Pipe free-body diagram.

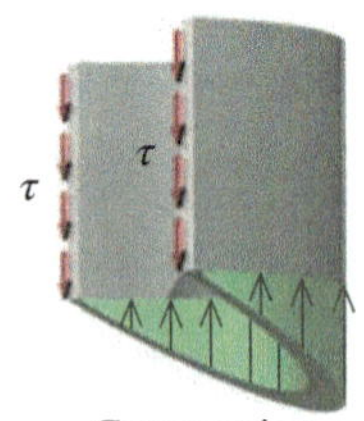

Stresses acting on the right half of the pipe.

Shear Stress Formula

The maximum vertical shear stress in this pipe occurs along the intersection of the y–z plane and the pipe wall. Note that the y–z plane is perpendicular to the direction of the shear force V, which acts in the x direction in this instance. The thickness t upon which the shear stress acts is equal to $t = D - d = 20$ mm. The maximum shear stress on this plane is computed from the shear stress formula:

$$\tau_{max} = \frac{VQ}{I_z t} = \frac{(36{,}000 \text{ N})(220{,}667 \text{ mm}^3)}{(36{,}450{,}329 \text{ mm}^4)(20 \text{ mm})} = 10.90 \text{ MPa} \qquad \textbf{Ans.}$$

Further Explanation

At first, it may be difficult for the student to visualize the shear stress acting in a pipe shape. To better understand the cause of shear stress in this situation, consider a free-body diagram of a short portion of the pipe near the point of application of the load. The 36 kN external load produces an internal bending moment M, which produces tensile and compressive normal stresses on the $-x$ and $+x$ portions of the pipe, respectively. We will investigate the equilibrium of half of the pipe.

Compressive normal stresses are created in the right half-pipe by the internal bending moment M. Equilibrium in the y direction requires a resultant force acting downward to resist the upward force created by the compressive normal stresses. This downward resultant force comes from shear stresses acting vertically in the wall of the pipe. For the example considered here, the shear stress has a magnitude of $\tau = 10.90$ MPa.

EXAMPLE 9.4

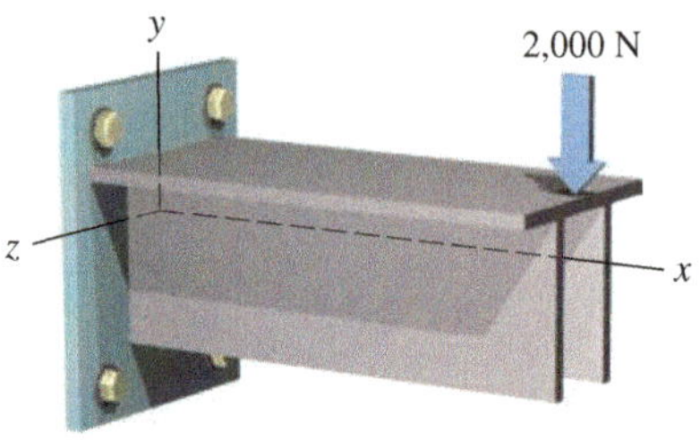

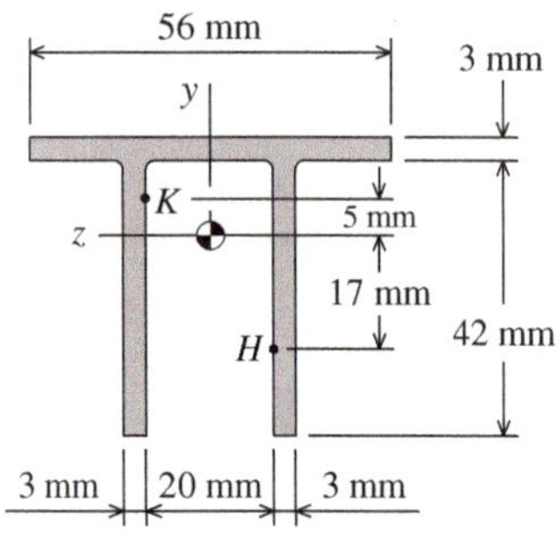

A cantilever beam is subjected to a concentrated load of 2,000 N. The cross-sectional dimensions of the double-tee shape are shown. Determine

(a) the shear stress at point H, which is located 17 mm below the centroid of the double-tee shape.
(b) the shear stress at point K, which is located 5 mm above the centroid of the double-tee shape.
(c) the maximum horizontal shear stress in the double-tee shape.

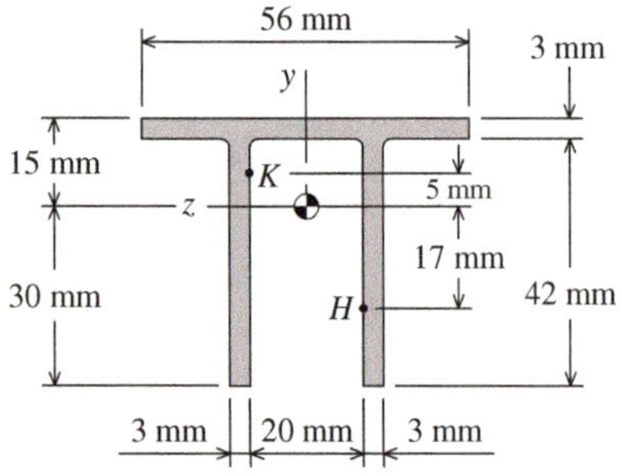

Plan the Solution

The shear stress in the double-tee shape can be determined from the shear stress formula [Equation (9.2)]. The challenge in this problem lies in determining the appropriate values of Q for each calculation.

SOLUTION

Section Properties

The centroid location for the double-tee cross section must be determined at the outset. The results are shown in the accompanying figure. The moment of inertia of the cross section about the z centroidal axis is $I_z = 88{,}200$ mm^4.

(a) Shear Stress at *H*

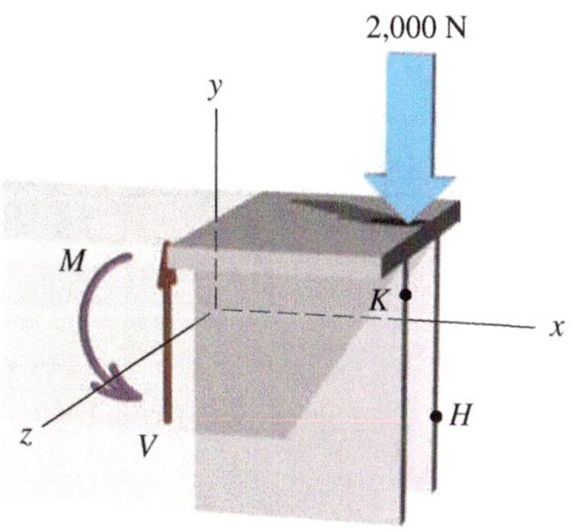

Before proceeding to the calculation of τ, it is helpful to visualize the source of the shear stresses produced in the flexural member. Consider a free-body diagram cut near the free end of the cantilever beam. The external 2,000 N concentrated load creates an internal shear force $V = 2{,}000$ N and an internal bending moment M, which varies over the cantilever span. To investigate the shear stresses produced in the double-tee cross section, the free body shown will be divided further in a manner similar to the analysis performed in the derivation presented in Section 9.3.

The shear stress acting at H is exposed by cutting the free body. The internal bending moment M produces compressive bending stresses that are linearly distributed over the stems of the double-tee shape. The resultant force from these compressive normal stresses tends to push the double-tee stems in the positive x direction. To satisfy equilibrium in the horizontal direction, shear stresses τ must act on the horizontal surfaces exposed at H. The magnitude of these shear stresses is found from the shear stress formula [Equation (9.2)].

In determining the proper value of the first moment of area, Q, for use in the shear stress formula, it is helpful to keep the free-body diagram in mind.

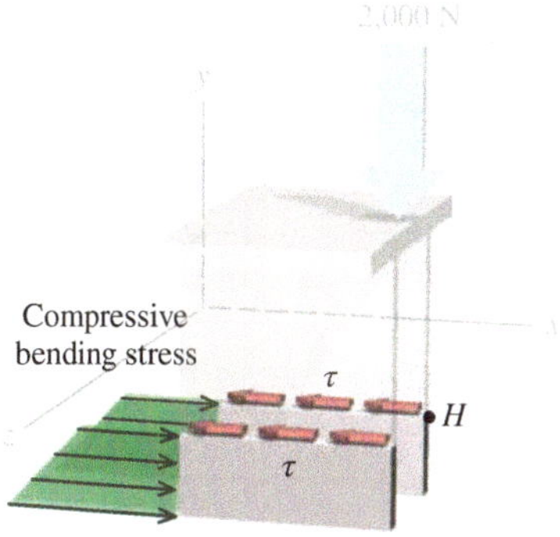

Calculating Q at point H: The double-tee cross section is shown in the next accompanying figure. Only a portion of the entire cross section is considered in the calculation of Q. To determine the proper area, slice through the cross section *parallel to the axis of bending* at point H and consider that portion of the cross section beginning at H and extending *away from the neutral axis.* Note that slicing through the section *parallel to the axis of bending* can also be described as slicing through the section *perpendicular to the direction of the internal shear force V.*

The area to be considered in the calculation of Q at point H is highlighted in the cross section. (This is the area denoted A' in the derivation of the shear stress formula in Section 9.3, particularly in Figures 9.5 and 9.7.)

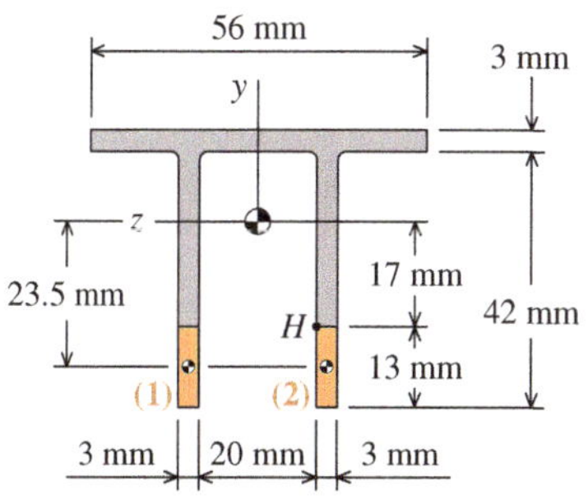

At point H, Q is the moment of areas (1) and (2) about the z centroidal axis (i.e., the neutral axis about which bending occurs). From the cross-sectional sketch,

$$Q_H = 2[(3 \text{ mm})(13 \text{ mm})(23.5 \text{ mm})] = 1{,}833 \text{ mm}^3$$

The shear stress acting at H can now be calculated from the shear stress formula:

$$\tau_H = \frac{VQ_H}{I_z t} = \frac{(2{,}000 \text{ N})(1{,}833 \text{ mm}^3)}{(88{,}200 \text{ mm}^4)(6 \text{ mm})} = 6.93 \text{ MPa} \qquad \textbf{Ans.}$$

Note that the term t in the shear stress formula is the width of the surface exposed in cutting the free-body diagram through point H. In slicing through the two stems of the double-tee shape, a surface 6 mm wide is exposed; therefore, $t = 6$ mm.

(b) Shear Stress at *K*

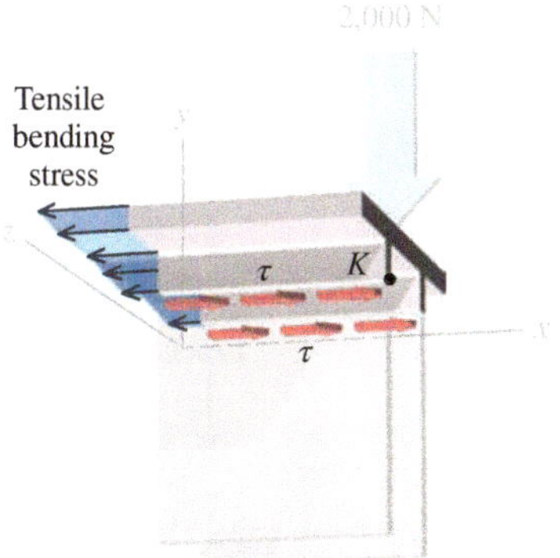

Consider again a free-body diagram cut near the free end of the cantilever beam. This free-body diagram will be further dissected by cutting a new free-body diagram, beginning at point K and extending *away from the neutral axis*, as shown in the next accompanying figure. The internal bending moment M produces tensile bending stresses that are linearly distributed over the stems and flange of the double-tee shape. The resultant force from these tensile normal stresses tends to pull this portion of the cross section in the $-x$ direction. Shear stresses τ must act on the horizontal surfaces exposed at K to satisfy equilibrium in the horizontal direction.

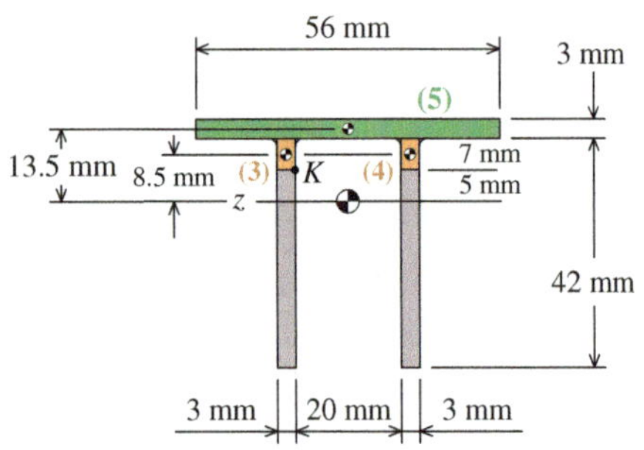

Calculating Q at point K: The area to be considered in the calculation of Q at point K is highlighted in the cross section. At point K, Q is the moment of areas (3), (4), and (5) about the z centroidal axis:

$$Q_K = 2[(3\text{ mm})(7\text{ mm})(8.5\text{ mm})] + (56\text{ mm})(3\text{ mm})(13.5\text{ mm}) = 2{,}625\text{ mm}^3$$

The shear stress acting at K is

$$\tau_K = \frac{VQ_K}{I_z t} = \frac{(2{,}000\text{ N})(2{,}625\text{ mm}^3)}{(88{,}200\text{ mm}^4)(6\text{ mm})} = 9.92\text{ MPa} \qquad \textbf{Ans.}$$

(c) Maximum Horizontal Shear Stress

The maximum value of Q corresponds to an area that begins at, and extends away from, the neutral axis. For this location, however, the words *extends away from the neutral axis* can mean either the area *above* or the area *below* the neutral axis. The value obtained for Q is the same in either case. For the double-tee cross section, the calculation of Q is somewhat simpler if we consider the highlighted area below the neutral axis. In that case,

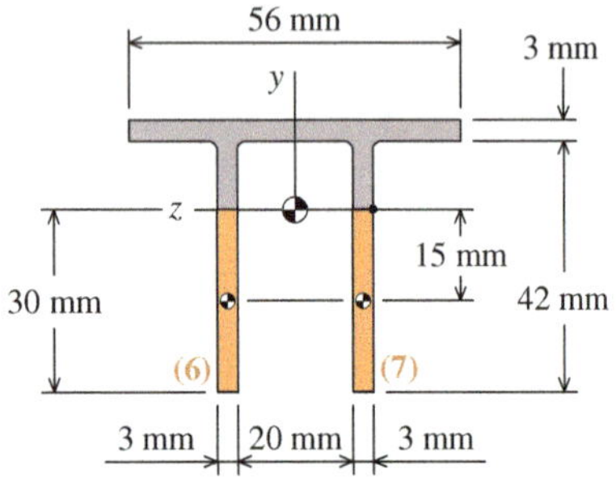

$$Q_{max} = 2[(3\text{ mm})(30\text{ mm})(15\text{ mm})] = 2{,}700\text{ mm}^3$$

The maximum horizontal shear stress in the double-tee shape is then

$$\tau_{max} = \frac{VQ_{max}}{I_z t} = \frac{(2{,}000\text{ N})(2{,}700\text{ mm}^3)}{(88{,}200\text{ mm}^4)(6\text{ mm})} = 10.20\text{ MPa} \qquad \textbf{Ans.}$$

MecMovies

EXAMPLES

M9.4 Determine the shear stress at points H and K for a simply supported beam that consists of the WT265 × 37 standard steel shape shown.

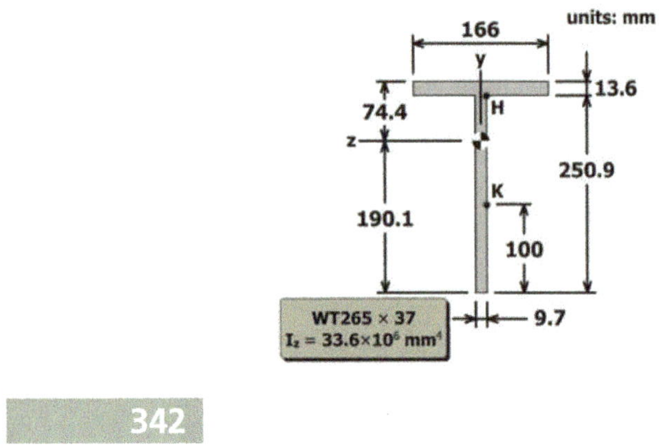

M9.5 Determine the distribution of shear stresses produced in a tee shape.

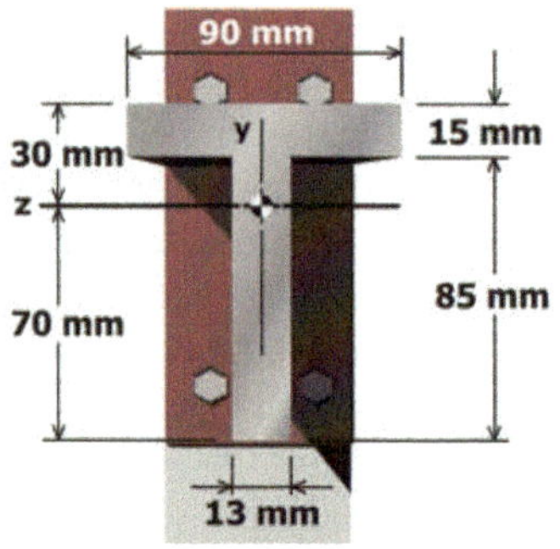

M9.6 Determine the maximum horizontal shear stress in a simply supported wide-flange beam.

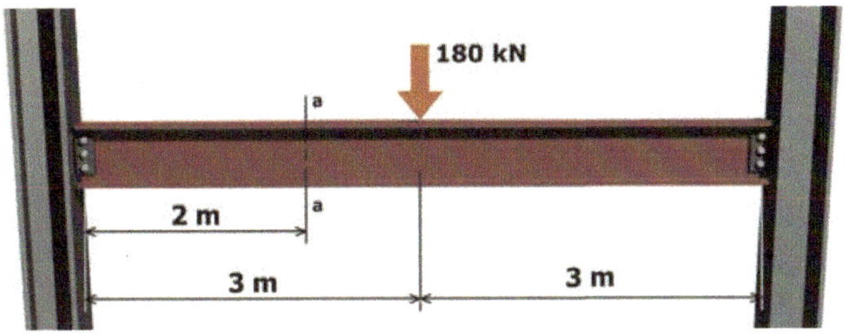

M9.7 Determine the shear stress at point H for a cantilever post that consists of a structural tube as shown.

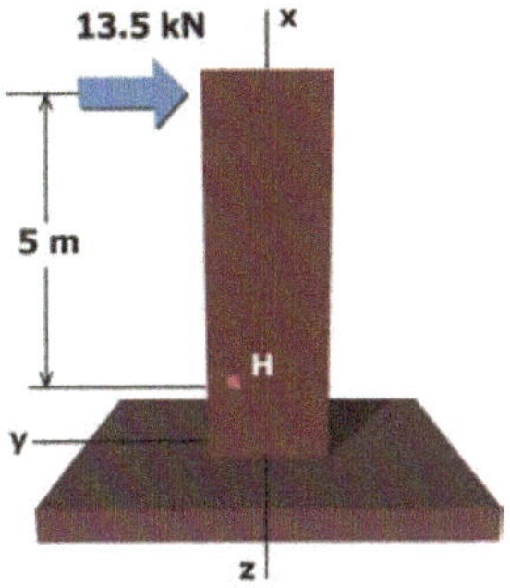

M9.8 Determine the normal and shear stresses at point H, which is located 3 in. above the centroidal axis of the wide-flange shape.

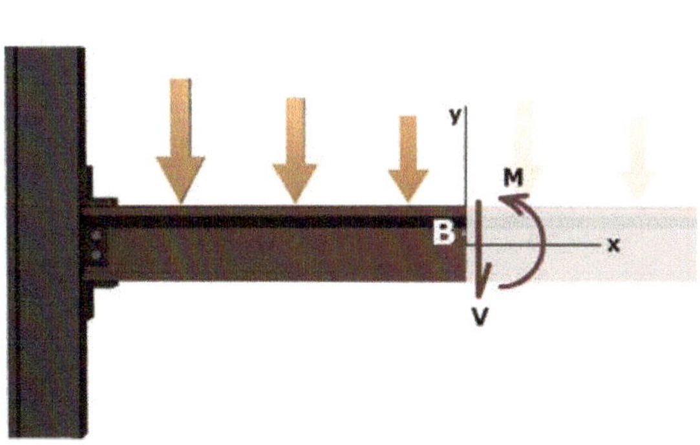

EXERCISES

M9.3 **Q-tile: The Q Section Property Game.** Score at least 90 percent on the Q-tile game.

FIGURE M9.3

M9.4 Determine the shear stresses acting at points H and K for a wide-flange shape subjected to an internal shear force V.

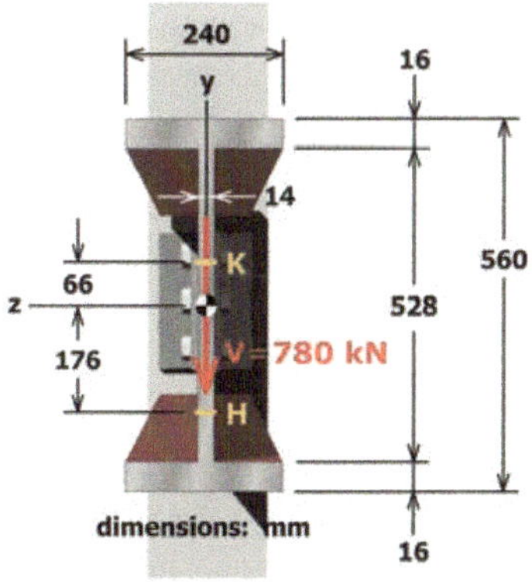

FIGURE M9.4

P9.11 A 0.375 in. diameter solid steel shaft supports two pulleys as shown in Figure P9.11. The shaft has a length $L = 9$ in., and the load applied to each pulley is $P = 60$ lb. The bearing at A can be idealized as a pin support, and the bearing at D can be idealized as a roller support. Determine the magnitude of the maximum horizontal shear stress in the shaft.

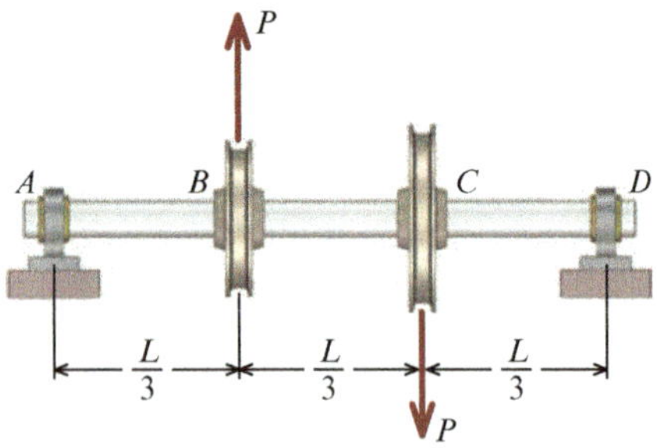

FIGURE P9.11

P9.12 A 25 mm diameter solid steel shaft supports loads $P_A = 1{,}000$ N, $P_C = 3{,}200$ N, and $P_E = 800$ N, as shown in Figure P9.12. Assume that $L_{AB} = 80$ mm, $L_{BC} = 200$ mm, $L_{CD} = 100$ mm, and $L_{DE} = 125$ mm. The bearing at B can be idealized as a roller support, and the bearing at D can be idealized as a pin support. Determine the magnitude and location of

(a) the maximum horizontal shear stress in the shaft.
(b) the maximum tensile bending stress in the shaft.

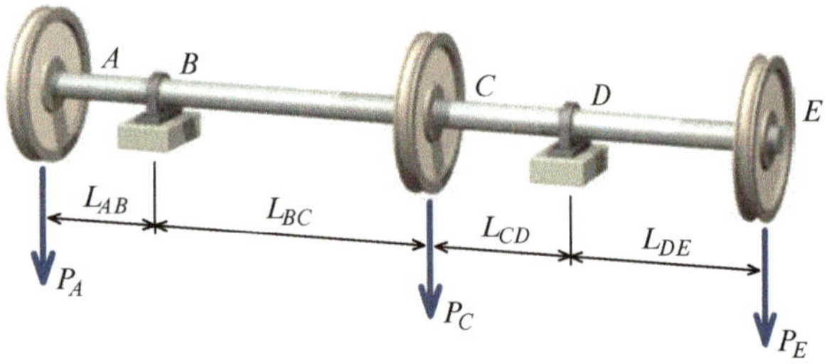

FIGURE P9.12

P9.13 An aluminum alloy pipe with an outside diameter of 4.0 in. and a wall thickness of 3/16 in. cantilevers upward a distance $L = 18$ ft, as shown in Figure P9.13. The pipe is subjected to a wind loading that increases from $w_A = 12$ lb/ft at its base to $w_B = 20$ lb/ft at its tip.

(a) Compute the value of Q for the pipe.
(b) What is the maximum vertical shear stress in the pipe?

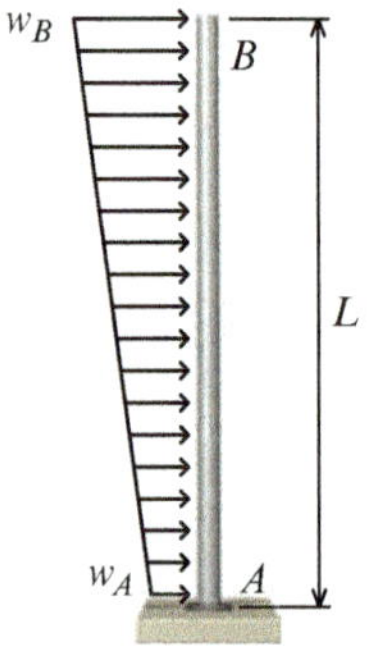

FIGURE P9.13

P9.14 The internal shear force at a certain section of a steel beam is $V = 215$ kN. The beam cross section shown in Figure P9.14 has dimensions $b_f = 300$ mm, $t_f = 19$ mm, $d = 390$ mm, and $t_w = 11$ mm. Determine

(a) the shear stress at point A, which is located at $y_A = 80$ mm below the centroid of the wide-flange shape.
(b) the maximum horizontal shear stress in the wide-flange shape.

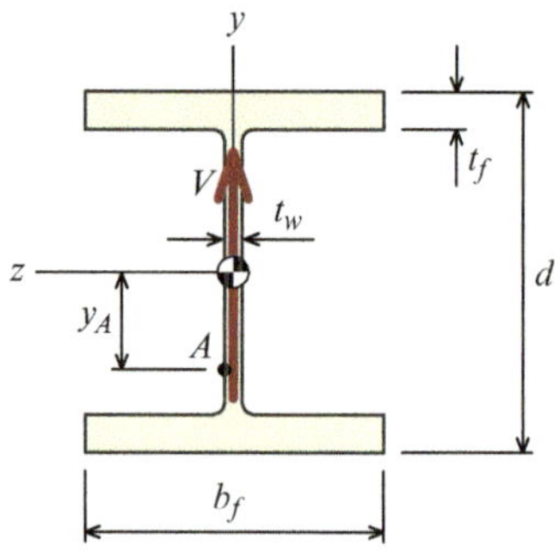

FIGURE P9.14

P9.15 The extruded cross section shown in Figure P9.15 is subjected to a shear force $V = 420$ N. The dimensions of the cross section are $b = 40$ mm, $d = 60$ mm, and $t = 4$ mm. Determine

(a) the value of Q associated with point H, located at $a = 13$ mm above the lower surface of the cross section.
(b) the horizontal shear stress at point H.
(c) the maximum horizontal shear stress in the cross section.

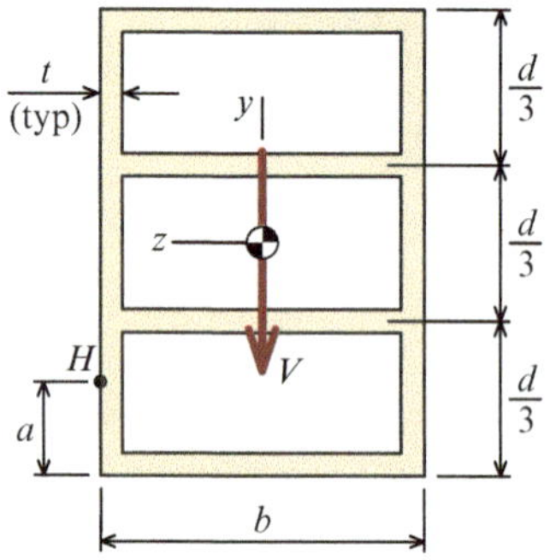

FIGURE P9.15

P9.16 The beam cross section shown in Figure P9.16 is subjected to a shear force $V = 215$ kN. The dimensions of the cross section are $b = 150$ mm, $d = 210$ mm, and $t = 6$ mm. Determine

(a) the horizontal shear stress at point H, located at $a = 60$ mm below the upper surface of the cross section.
(b) the maximum horizontal shear stress in the cross section.

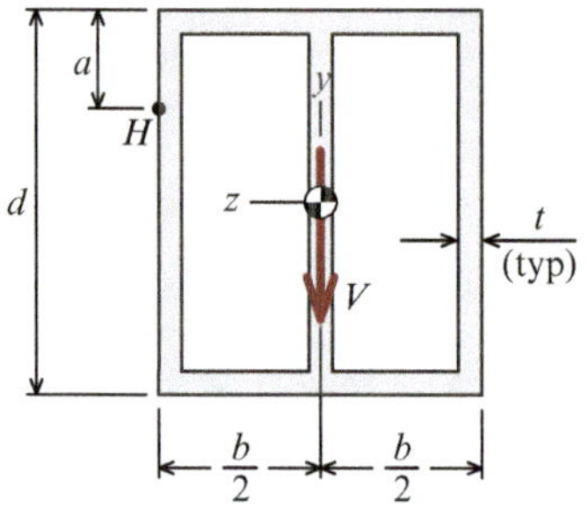

FIGURE P9.16

P9.17 The extruded plastic shape shown in Figure P9.17 will be subjected to an internal shear force $V = 300$ lb. The dimensions of the cross section are $b = 1.5$ in., $d = 3.0$ in., $c = 2.0$ in., and $t = 0.15$ in. Determine

(a) the horizontal shear stress at points A and B.
(b) the horizontal shear stress at point C, which is located at a distance $y_C = 1.0$ in. below the z centroidal axis.
(c) the maximum horizontal shear stress in the extrusion.

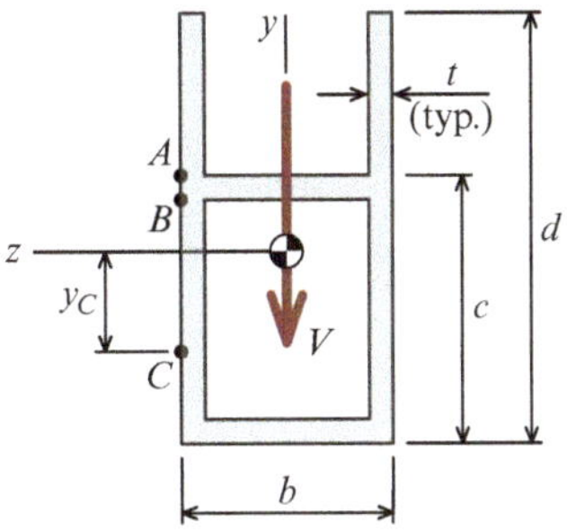

FIGURE P9.17

P9.18 The extruded plastic shape shown in Figure P9.18 has dimensions $D = 30$ mm, $d = 24$ mm, $a = 85$ mm, and $t = 7$ mm.

(a) Determine the horizontal shear stress at point A for a shear force $V = 320$ N.
(b) If the allowable shear stress for the plastic material is 1,100 kPa, what is the maximum shear force V that can be applied to the extrusion?

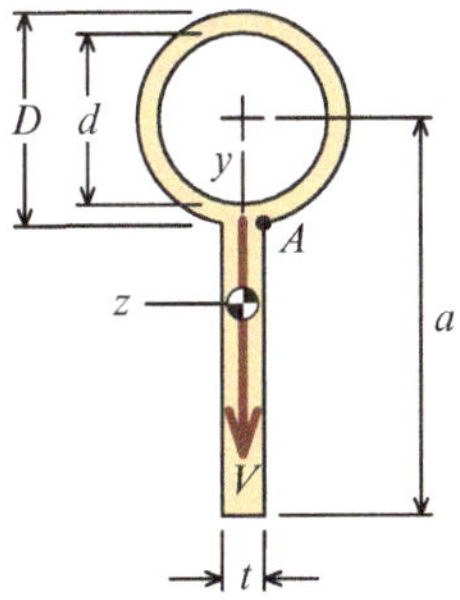

FIGURE P9.18

P9.19 The internal shear force at a certain section of a steel beam is $V = 9$ kN. The beam cross section, shown in Figure P9.19, has dimensions $b = 150$ mm, $c = 30$ mm, $d = 70$ mm, and $t = 6$ mm. Determine

(a) the shear stress at point H.
(b) the shear stress at point K.
(c) the maximum horizontal shear stress in the cross section.

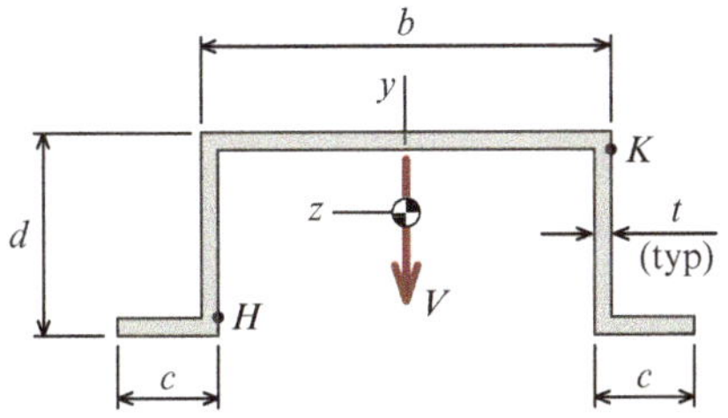

FIGURE P9.19

P9.20 The beam cross section shown in Figure P9.20a has been proposed for a short pedestrian bridge. The cross section will consist of two square tubes that are welded to a rectangular web plate. The dimensions of the cross section, as shown in Figure P9.20b, are $d = 410$ mm, $t_w = 12$ mm, $b = 120$ mm, and $t = 8.0$ mm. If the beam will be subjected to a shear force $V = 165$ kN, determine

(a) the shear stress at point H, located at $y_H = 160$ mm above the z centroidal axis.
(b) the shear stress at point K, located at $y_K = 70$ mm below the z centroidal axis.

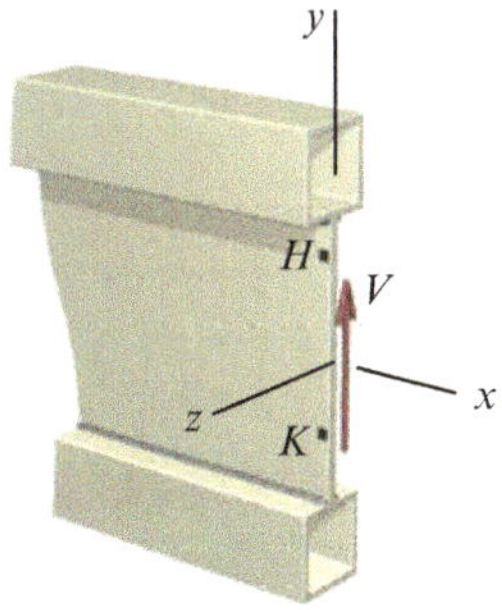

FIGURE P9.20a Beam segment.

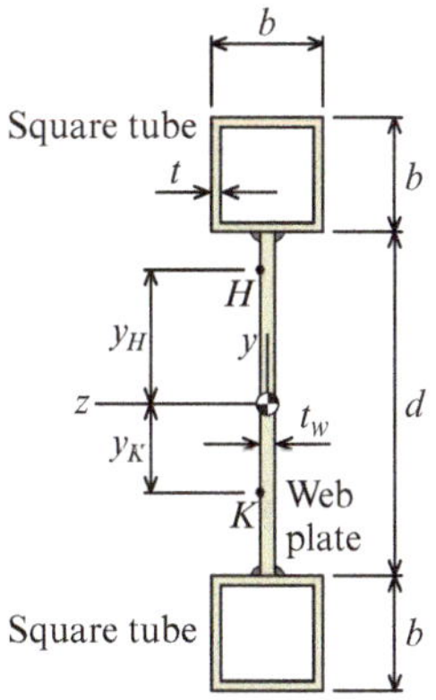

FIGURE P9.20b Cross-sectional dimensions.

P9.21 The cantilever beam shown in Figure P9.21a/22a is subjected to a concentrated load P. The cross-sectional dimensions of the rectangular tube shape shown in Figure P9.21b/22b are $b = 6$ in., $d = 10$ in., and $t = 0.375$ in.

(a) Compute the value of Q that is associated with point H, which is located to the right of the vertical centroidal axis at $z_H = 1.50$ in.
(b) If the allowable shear stress for the rectangular tube shape is 16 ksi, determine the maximum concentrated load P that can be applied as shown to the cantilever beam.

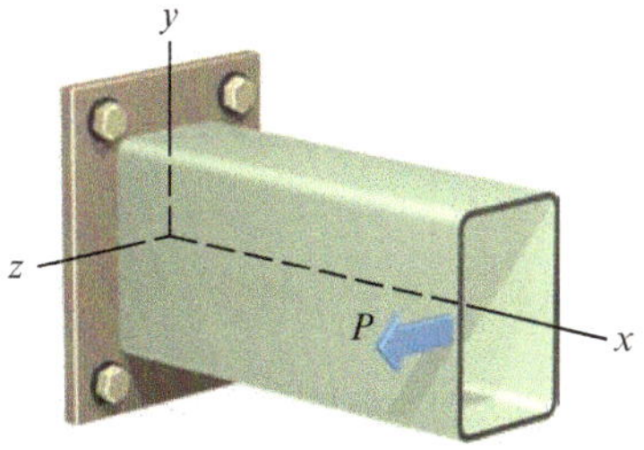

FIGURE P9.21a/22a Beam segment.

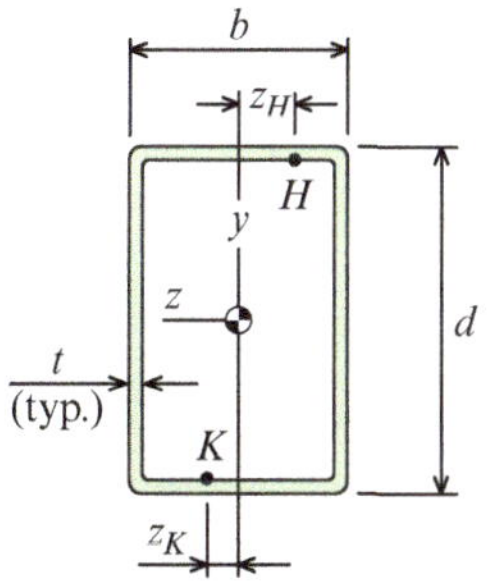

FIGURE P9.21b/22b Cross-sectional dimensions.

P9.22 The cantilever beam shown in Figure P9.21a/22a is subjected to a concentrated load $P = 25$ kips. The cross-sectional dimensions of the rectangular tube shape shown in Figure P9.21b/22b are $b = 5$ in., $d = 9$ in., and $t = 0.25$ in. Determine

(a) the shear stress at point K, which is located to the left of the vertical centroidal axis at $z_K = 1.00$ in.
(b) the maximum horizontal shear stress in the rectangular tube shape.

P9.23 A W14 × 34 standard steel section is used for a simple span of 6 ft 6 in. Determine the maximum uniform load that the section can support if the allowable bending stress is 30 ksi and the allowable shear stress is 18 ksi. Assume bending about the strong axis of the W shape.

P9.24 A simply supported beam with spans of L_{AB} = 9 ft and L_{BC} = 27 ft supports loads of w = 5 kips/ft and P = 30 kips, as shown in Figure P9.24*a*. The cross-sectional dimensions of the wide-flange shape shown in Figure P9.24*b* are b_f = 10 in., t_f = 0.64 in., d = 27.7 in., and t_w = 0.46 in. Determine the magnitude of

(a) the maximum horizontal shear stress in the beam.
(b) the maximum bending stress in the beam.

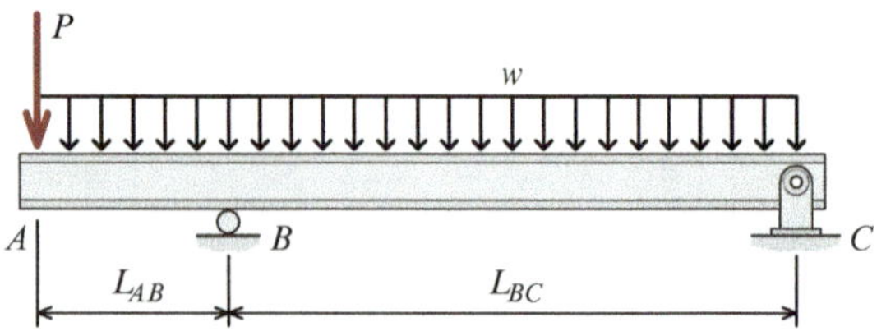

FIGURE P9.24*a*

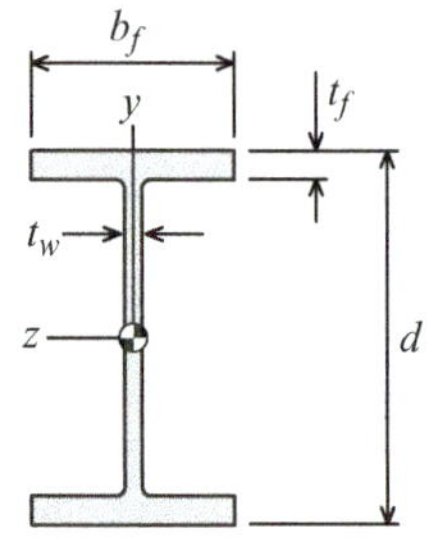

FIGURE P9.24*b*

P9.25 Vertical beam *AD* supports a horizontal load P = 120 kN as shown in Figure P9.25*a*. Dimensions of this structure are x_1 = 220 mm, x_2 = 5.78 m, y_1 = 0.85 m, y_2 = 3.15 m, and y_3 = 1.60 m. The beam is a rectangular tube (Figure P9.25*b*) with dimensions b = 100 mm, d = 180 mm, and t = 5 mm. Determine, at point *H*, which is located at a distance a = 40 mm to the right of the z centroidal axis,

(a) the vertical shear stress magnitude,
(b) the normal stress in the y direction produced by the bending moment. State whether it is a tensile or a compressive normal stress.
(c) the normal stress in the y direction produced by the axial force. State whether it is a tensile or a compressive normal stress.

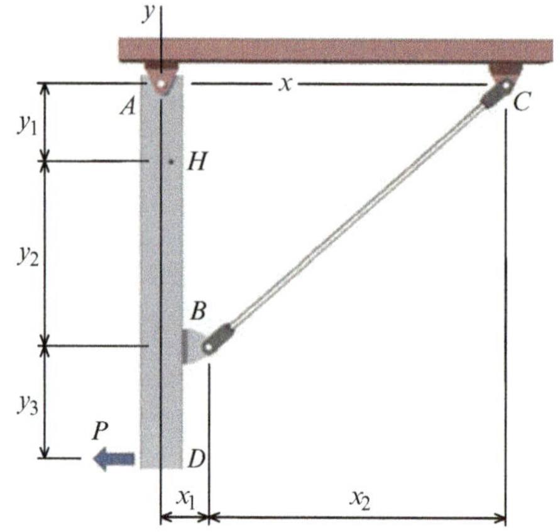

Figure P9.25*a*

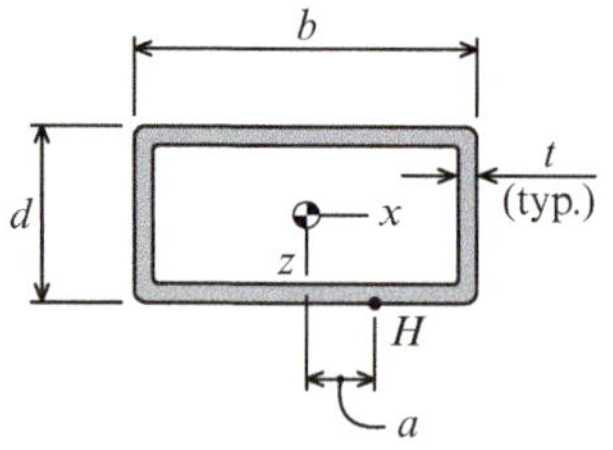

Cross section at point *H*. **Figure P9.25*b***

9.8 Shear Flow in Built-Up Members

While standard steel shapes and other specially formed cross sections are frequently used to construct beams, there are instances in which beams must be fabricated from components such as wooden boards or metal plates to suit a particular purpose. As has been shown in Section 9.2, nonuniform bending creates horizontal forces (i.e., forces parallel to the longitudinal axis of the beam) in each portion of the cross section. To satisfy equilibrium, additional horizontal forces must be developed internally between these parts. For a cross section made from disconnected components, fasteners such as nails, screws, bolts, or other individual connectors must be added so that the separate pieces act together as a unified flexural member.

The cross section of a built-up flexural member is shown in Figure 9.14*a*. Nails connect four wooden boards so that they act as a unified flexural member. As in Section 9.3, we will consider a length Δx of the beam, which is subjected to nonuniform bending (Figure 9.14*b*). Then, we will examine a portion A' of the cross section in order to assess

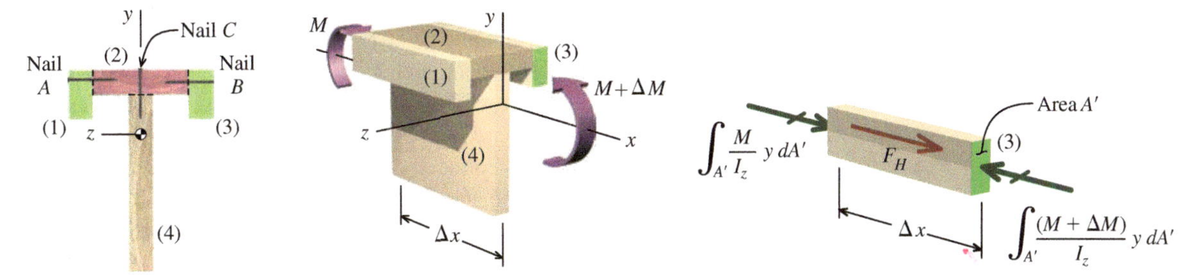

(*a*) Cross section made from four wooden boards

(*b*) Built-up beam subjected to nonuniform bending

(*c*) Free-body diagram of board (3)

FIGURE 9.14 Horizontal equilibrium of a built-up beam.

the forces that act in the longitudinal direction (i.e., the x direction). In this instance, we will consider board (3) as area A'. A free-body diagram of board (3) is shown in Figure 9.14*c*. Using an approach similar to that employed in the derivation presented in Section 9.3, Equation (9.1) relates the change in internal bending moment ΔM over a length Δx to the horizontal force required to satisfy equilibrium for area A':

$$F_H = \frac{\Delta M Q}{I_z}$$

The change in internal bending moment, ΔM, can be expressed as $\Delta M = (dM/dx)\Delta x = V\,\Delta x$, thereby allowing Equation (9.1) to be rewritten in terms of the internal shear force V:

The term I_z appearing in Equations (9.1), (9.11), and (9.12) is **always** the moment of inertia *of the entire cross section* about the z centroidal axis.

$$F_H = \frac{VQ}{I_z}\Delta x \tag{9.11}$$

Equation (9.11) relates the internal shear force V in a beam to the horizontal force F_H required to keep a specific portion of the cross section (area A') in equilibrium. The term Q is the first moment of area A' about the neutral axis, and I_z is the moment of inertia of the *entire cross section* about the neutral axis.

The force F_H required to keep board (3) (i.e., area A') in equilibrium must be supplied by nail B shown in Figure 9.14*a*, and it is the presence of individual fasteners (such as nails) that is unique to the design of built-up flexural members. In addition to using the flexure formula and the shear stress formula to consider bending stresses and shear stresses, the designer of a built-up flexural member must ensure that the fasteners which will connect the pieces together are adequate to transmit the horizontal forces required for equilibrium.

To facilitate this type of analysis, it is convenient to introduce a quantity known as **shear flow**. If both sides of Equation (9.11) are divided by Δx, then the **shear flow** q can be defined as

It is important to understand that shear flow stems from normal stresses created by internal bending moments that vary along the span of the beam. The term V appears in Equation (9.12) as a substitute for dM/dx. Shear flow acts parallel to the longitudinal axis of the beam—that is, in the same direction as the bending stresses.

$$\frac{F_H}{\Delta x} = q = \frac{VQ}{I_z} \tag{9.12}$$

The shear flow q is the *shear force per unit length of beam span* required to satisfy horizontal equilibrium for a specific portion of the cross section. Equation (9.12) is called the **shear flow formula**.

$\frac{F_H}{\Delta X} = q = \frac{VQ}{I_z}$ (9.12)

Analysis and Design of Fasteners

Built-up cross sections use individual fasteners such as nails, screws, or bolts to connect several components into a unified flexural member. One example of a built-up cross section is shown in Figure 9.14*a,* and several other examples are shown in Figure 9.15. Although these examples consist of wooden boards connected by nails, the principles are the same regardless of the beam material or type of fastener.

A consideration of fasteners usually involves one of the following objectives:

- Given the internal shear force V in the beam and the shear force capacity of a fastener, what is the proper spacing interval for fasteners along the beam span (i.e., in the longitudinal x direction)?
- Given the diameter and spacing interval s of the fasteners, what is the shear stress τ_f produced in each fastener for a given shear force V in the beam?
- Given the diameter, spacing interval s, and allowable shear stress of the fasteners, what is the maximum shear force V that is acceptable for the built-up member?

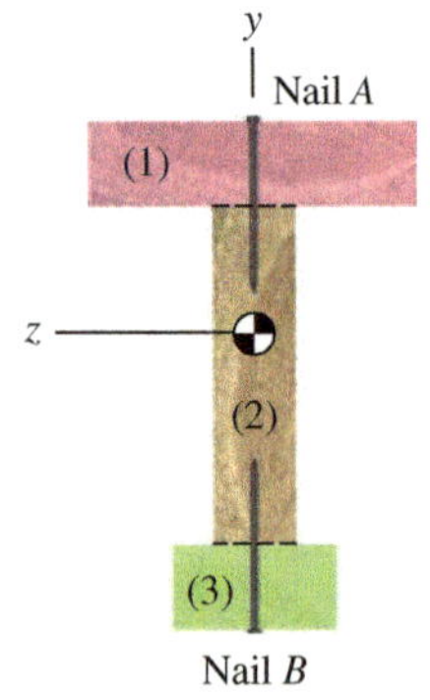

(*a*) I-shaped wooden beam cross section

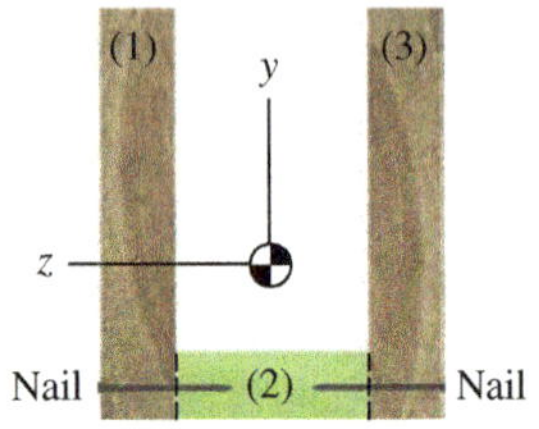

(*b*) U-shaped wooden beam cross section

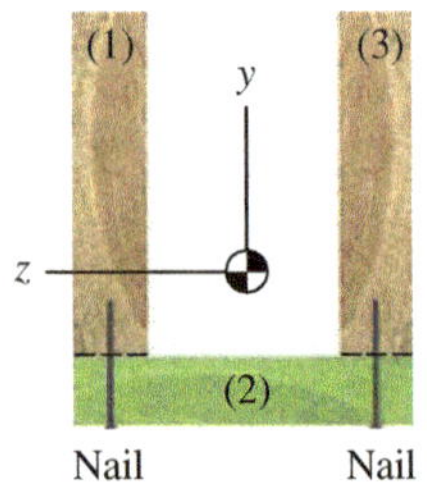

(*c*) Alternative U-shaped wooden beam cross section

FIGURE 9.15 Examples of built-up flexural members.

To address these objectives, an expression can be developed from Equation (9.12) that relates the resistance of the fastener to the horizontal force F_H required to keep an area A' in equilibrium. The length term Δx in Equation (9.12) will be set equal to the fastener spacing interval s along the x axis of the beam. In terms of the shear flow q, the total horizontal force that must be transmitted between connected parts over a beam interval s can be expressed as

$$F_H = qs \tag{a}$$

The internal horizontal force F_H must be transmitted between the boards or plates by the fasteners. (**Note:** The effect of friction between the connected parts is neglected.) The shear force that can be transmitted by a single fastener (e.g., a nail, screw, or bolt) will be denoted by V_f. Since more than one fastener could be used within the spacing interval s, the number of fasteners in the interval will be denoted by n_f. The resistance provided by n_f fasteners must be greater than or equal to the horizontal force F_H required to keep the connected part in equilibrium horizontally:

$$F_H \le n_f V_f \tag{b}$$

Combining Equation (a) with Inequality (b) gives a relationship between the shear flow q, the fastener spacing interval s, and the shear force that can be transmitted by a single fastener V_f. This relationship will be termed the **fastener force–spacing relationship** and is given by the inequality

$$qs \le n_f V_f \tag{9.13}$$

The average shear stress τ_f produced in a fastener can be expressed as

$$\tau_f = \frac{V_f}{A_f} \tag{c}$$

where the fastener is assumed to act in single shear and A_f is the cross-sectional area of the fastener. Using this relationship, we can rewrite Equation (9.13) in terms of the shear stress in the fastener:

$$qs \le n_f \tau_f A_f \tag{9.14}$$

This inequality will be termed the **fastener stress–spacing relationship**.

f subscript = individual fastener

Identifying the Proper Area for Q

In analyzing the shear flow q for a particular application, the most confusing decision often concerns which portion of the cross section to include in the calculation of Q, the first moment of area A'. The key to identifying the proper area A' is to determine which portion of the cross section is being held in place by the fastener.

Several built-up wooden beam cross sections are shown in Figure 9.14*a* and Figure 9.15. In each case, nails are used to connect the boards together into a unified flexural member. A vertical internal shear force V is assumed to act in the beam for each cross section.

For the tee shape shown in Figure 9.14*a*, board (1) is held in place by nail A. To analyze nail A, the designer must determine the shear flow q transmitted between board (1) and the remainder of the cross section. The proper Q for this purpose is the first moment of board (1)'s area about the z centroidal axis. Similarly, the shear flow associated with nail B requires Q for board (3) about the neutral axis. Nail C must transmit the shear flow arising from boards (1), (2), and (3) to the stem of the tee shape. Consequently, the proper Q for nail C involves boards (1), (2), and (3).

Figure 9.15*a* shows an I-shaped cross section that is fabricated by nailing flange boards (1) and (3) to web board (2). Nail A connects board (1) to the remainder of the cross section; therefore, the shear flow q associated with nail A is based on the first moment of area, Q, for board (1) about the z axis. Nail B connects board (3) to the remainder of the cross section. Since board (3) is smaller than board (1) and more distant from the z axis, a different value of Q will be calculated, resulting in a different value of q for board (3). Consequently, it is likely that the spacing interval s for nail B will be different from s for nail A. In both instances, I_z is the moment of inertia of the *entire cross section* about the z centroidal axis.

Figures 9.15*b* and 9.15*c* show alternative configurations for U-shaped cross sections in which board (2) is connected to the remainder of the cross section by two nails. The part held in place by the nails is board (2) in both configurations. Both alternatives have the same dimensions, the same cross-sectional area, and the same moment of inertia. However, the value of Q calculated for board (2) in Figure 9.15*b* will be smaller than that of Q for board (2) in Figure 9.15*c*. Consequently, the shear flow for the first configuration will be smaller than q for the alternative configuration.

EXAMPLE 9.5

A simply supported beam with an overhang supports a concentrated load of 500 lb at D. The beam is fabricated from two 2 in. by 8 in. wooden boards that are fastened together with lag screws spaced at 5 in. intervals along the length of the beam. The centroid location of the fabricated cross section is shown in the sketch, and the moment of inertia of the cross section about the z centroidal axis is $I_z = 290.667$ in.^{4}. Determine the shear force acting in the lag screws.

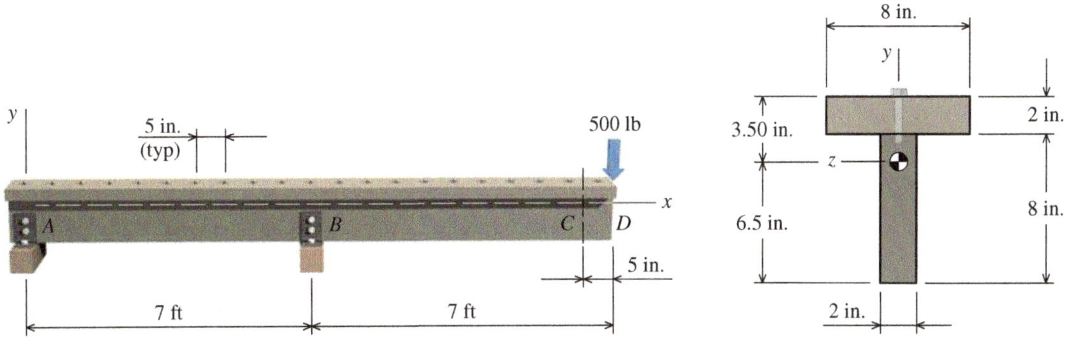

Plan the Solution

Whenever a cross section includes discrete fasteners (such as nails, screws, or bolts), the shear flow formula [Equation (9.12)] and the related fastener force–spacing relationship [Equation (9.13)] will be helpful in assessing the suitability of the fasteners for the intended purpose. To determine the shear force acting in the fasteners, we must first identify those portions of the cross section which are held in place by the fasteners. For the basic tee-shaped cross section considered here, it is evident that the top flange board is secured to the stem board by the lag screws. If the entire cross section is to be in equilibrium, the resultant force acting in the horizontal direction on the flange board must be transmitted to the stem board by shear forces in the fasteners. In the analysis that follows, a short length of the beam equal to the spacing interval of the lag screws will be considered to determine the shear force that must be supplied by each fastener to satisfy equilibrium.

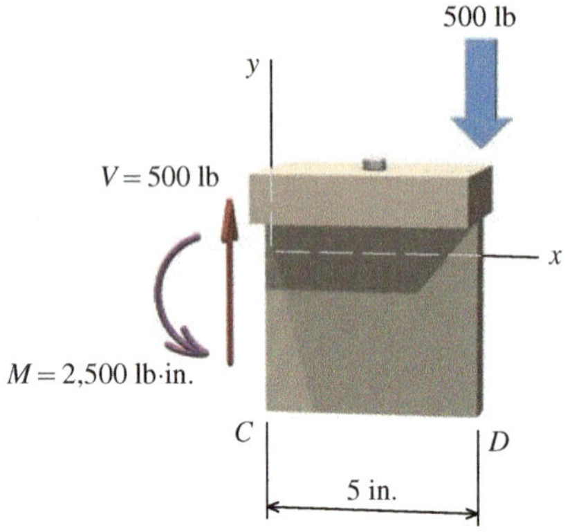

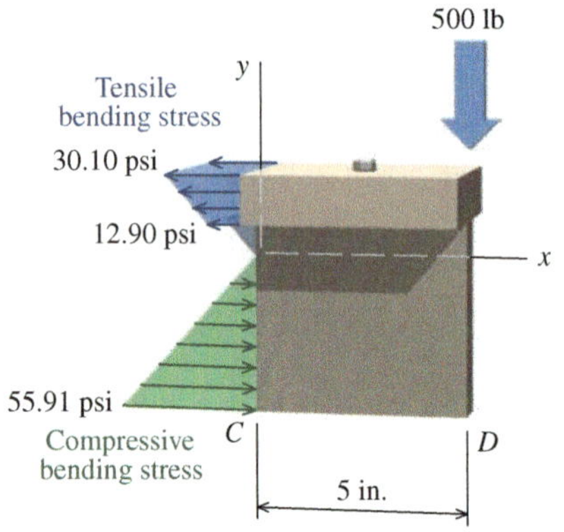

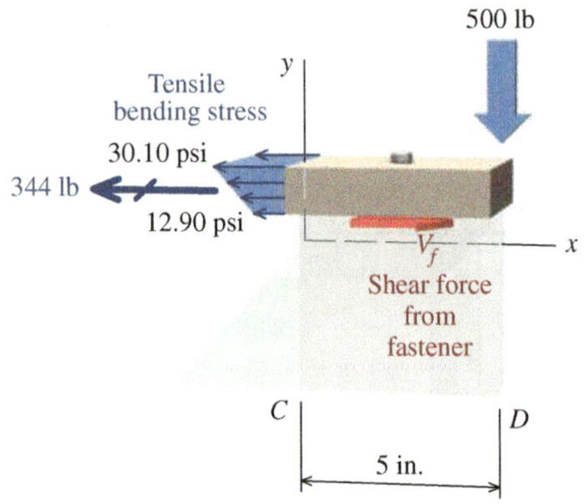

SOLUTION

Free-Body Diagram at *C*

To better understand the function of the fasteners, consider a free-body diagram (FBD) cut at section C, 5 in. from the end of the overhang. This FBD includes one lag screw fastener. The external 500 lb concentrated load creates an internal shear force $V = 500$ lb and an internal bending moment $M = 2{,}500$ lb · in. acting at C in the directions shown.

The internal bending moment $M = 2{,}500$ lb · in. creates tensile bending stresses above the neutral axis (i.e., the z centroidal axis) and compressive bending stresses below the neutral axis. These key normal stresses acting on the flange and the stem can be calculated from the flexure formula and are labeled in the figure.

The approach outlined in Section 9.2 can be used to compute the resultant horizontal force created by the tensile bending stresses acting on the flange. The resultant force has a magnitude of 344 lb, and it pulls the flange in the $-x$ direction. If the flange is to be in equilibrium, additional force acting in the $+x$ direction must be present. This added force is provided by the shear resistance of the lag screw. With that force denoted as V_f, equilibrium in the horizontal direction dictates that $V_f = 344$ lb.

In other words, equilibrium of the flange can be satisfied only if 344 lb of resistance from the stem *flows through the lag screw* into the flange. The magnitude of V_f determined here is applicable only to a 5 in. long segment of the beam. If a segment longer than 5 in. were considered, the internal bending moment M would be larger, which in turn would create larger bending stresses and a larger resultant force. Consequently, it is convenient to express the amount of force that must flow to the connected portion in terms of the horizontal resistance required per unit of beam span. The shear flow in this instance is

$$q = \frac{344 \text{ lb}}{5 \text{ in.}} = 68.8 \text{ lb/in.} \qquad \text{(a)}$$

The preceding discussion is intended to illuminate the behavior of a built-up beam. A basic understanding of the forces and stresses involved in this type of flexural member facilitates the proper use of the shear flow formula [Equation (9.12)] and the fastener force–spacing relationship [Equation (9.13)] to analyze and design fasteners in built-up flexural members.

Shear Flow Formula
The shear flow formula rewritten as

$$q = \frac{VQ}{I_z} \quad \text{(b)}$$

and the fastener force–spacing relationship

$$qs \leq n_f V_f \quad \text{(c)}$$

will be employed to determine the shear force V_f produced in the lag screws of the built-up beam. Appropriate values for the terms appearing in these equations will now be developed.

Beam internal shear force V: The shear-force and bending-moment diagrams for the simply supported beam are shown. The V diagram reveals that the internal shear force has a constant magnitude of $V = 500$ lb throughout the entire span of the beam.

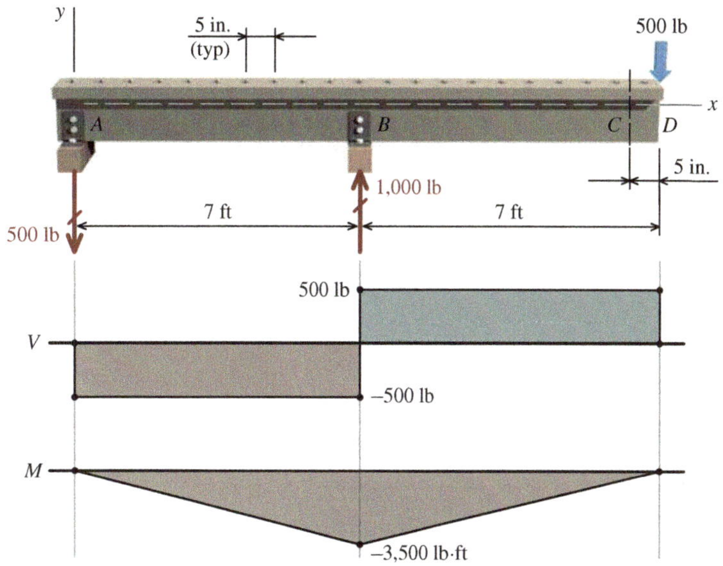

First moment of area, Q: Q is calculated for the portion of the cross section connected by the lag screw. Consequently, Q is calculated for the flange board in this situation:

$$Q = (8 \text{ in.})(2 \text{ in.})(2.5 \text{ in.}) = 40 \text{ in.}^3$$

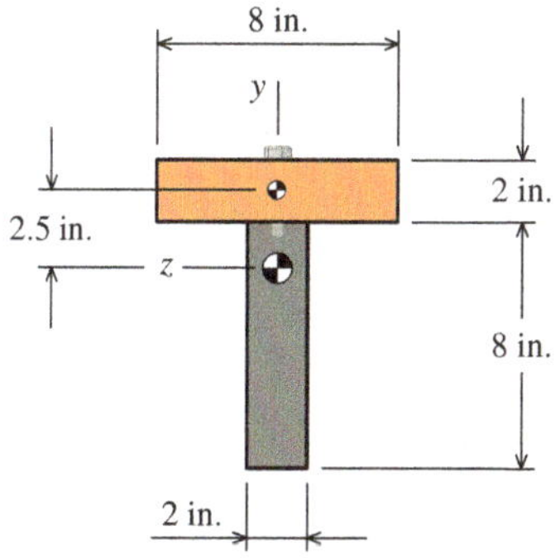

Fastener spacing interval s: The lag screws are installed at 5 in. intervals along the span; therefore, $s = 5$ in.

Shear flow q: The shear flow that must be transmitted from the stem to the flange through the fastener can be calculated from the shear flow formula:

$$q = \frac{VQ}{I_z} = \frac{(500 \text{ lb})(40 \text{ in.}^3)}{290.667 \text{ in.}^4} = 68.8 \text{ lb/in.} \quad \text{(d)}$$

Notice that the result obtained in Equation (d) from the shear flow formula is identical to the result obtained in Equation (a). While the shear flow formula provides a convenient format for calculation purposes, the underlying flexural behavior that it addresses may not be readily evident. The preceding investigation using an FBD of the beam at C may help to enhance one's understanding of this behavior.

Fastener shear force V_f: The shear force that must be provided by the fastener can be calculated from the fastener force–spacing relationship. The beam is fabricated with one lag screw installed in each 5 in. interval; it follows that $n_f = 1$, and we have

$$qs \leq n_f V_f$$

$$\therefore V_f = \frac{qs}{n_f} = \frac{(68.8 \text{ lb/in.})(5 \text{ in.})}{1 \text{ fastener}} = 344 \text{ lb per fastener} \qquad \textbf{Ans.}$$

EXAMPLE 9.6

An alternative cross section is proposed for the simply supported beam of Example 9.5. In the alternative design, the beam is fabricated from two 1 in. by 10 in. wooden boards nailed to a 2 in. by 6 in. flange board. The centroid location of the fabricated cross section is shown in the sketch, and the moment of inertia of the cross section about the z centroidal axis is I_z = 290.667 in.^{4}. If the allowable shear resistance of each nail is 80 lb, determine the maximum spacing interval s that is acceptable for the built-up beam.

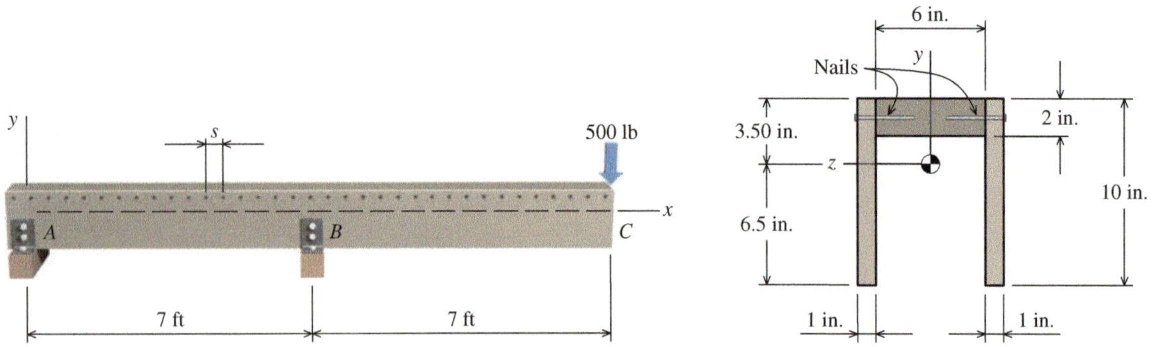

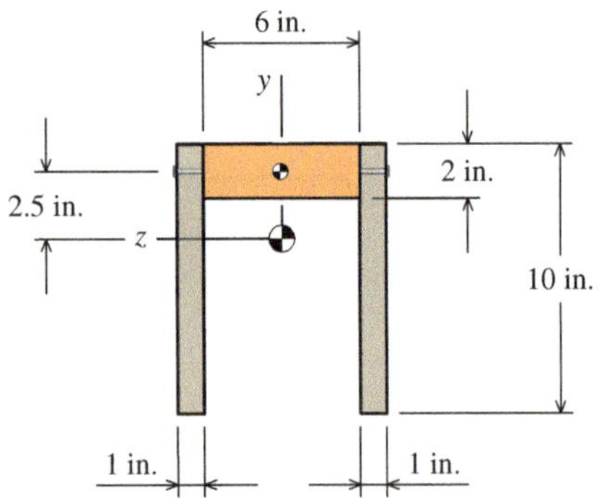

Plan the Solution

The shear flow formula [Equation (9.12)] and the fastener force–spacing relationship [Equation (9.13)] will be required to determine the maximum spacing interval s. Since the 2 in. by 6 in. flange board is held in place by the nails, the first moment of area, Q, as well as the shear flow q, will be based on this region of the cross section.

SOLUTION

Beam Internal Shear Force V

The shear-force and bending-moment diagrams for the simply supported beam are shown in Example 9.5. The shear force has a constant magnitude of V = 500 lb throughout the entire beam span.

First moment of area, Q: Q is calculated for the 2 in. by 6 in. flange board, which is the portion of the cross section held in place by the nails:

$$Q = (6 \text{ in.})(2 \text{ in.})(2.5 \text{ in.}) = 30 \text{ in.}^3$$

Shear flow q: The shear flow that must be transmitted through the pair of nails is

$$q = \frac{VQ}{I_z} = \frac{(500 \text{ lb})(30 \text{ in.}^3)}{290.667 \text{ in.}^4} = 51.6 \text{ lb/in.}$$

Maximum nail-spacing interval s: The maximum spacing interval for the nails can be calculated from the fastener force–spacing relationship [Equation (9.13)]. The beam is fabricated with two nails installed in each interval; consequently, $n_f = 2$, and we have

$$qs \le n_f V_f$$

$$\therefore s \le \frac{n_f V_f}{q} = \frac{(2 \text{ nails})(80 \text{ lb/nail})}{51.6 \text{ lb/in.}} = 3.10 \text{ in.} \qquad \textbf{Ans.}$$

Thus, pairs of nails must be installed at intervals less than or equal to 3.10 in. In practice, nails would be driven at 3 in. intervals.

MecMovies

EXAMPLES

M9.9 Determine the allowable shear force capacity of two wooden box beams fabricated with two different nail configurations.

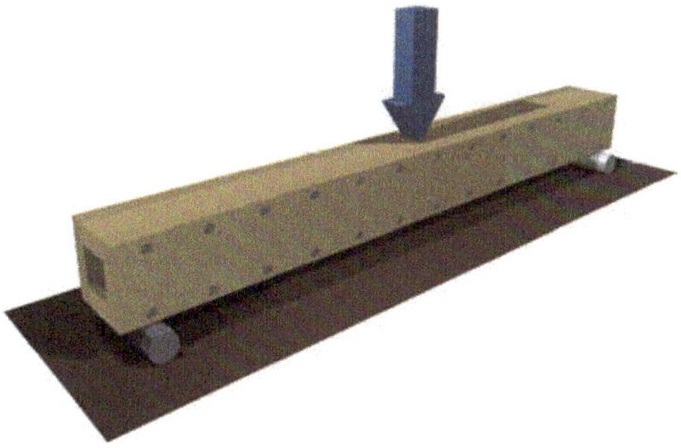

M9.10 Determine the maximum nail spacing that can be used to construct a U-shaped beam from wooden boards.

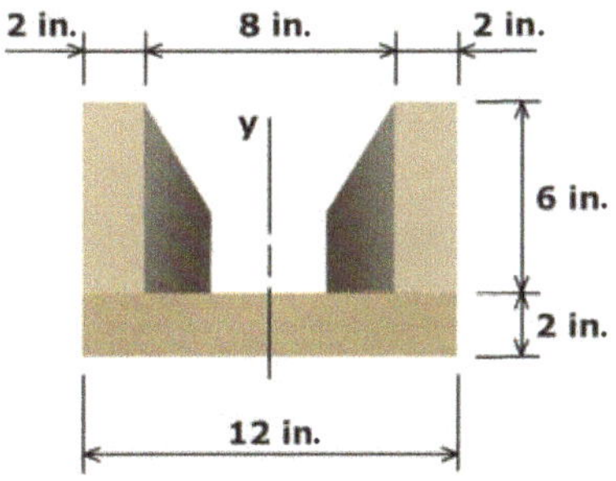

M9.11 Determine the maximum longitudinal bolt spacing required to support a 50 kip shear force.

M9.12 Determine the shear stress developed in the bolts used to connect two channel shapes back to back.

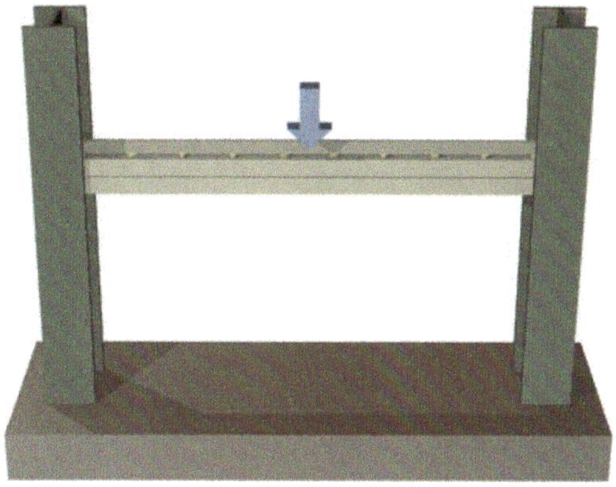

M9.13 Determine the shear stress in the bolts used to fabricate a box beam.

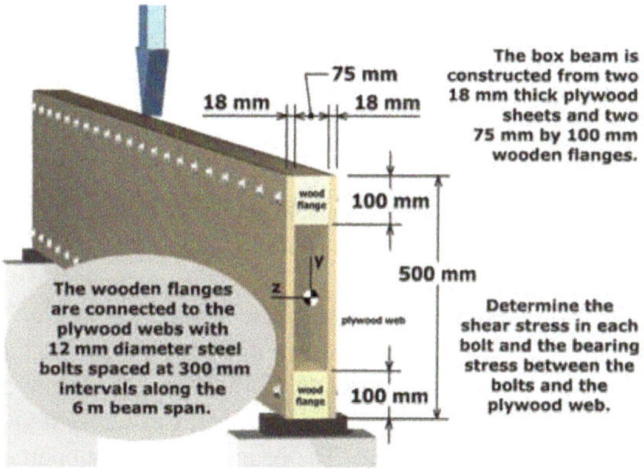

EXERCISES

M9.9 Five multiple-choice questions involving the calculation of Q for built-up beam cross sections.

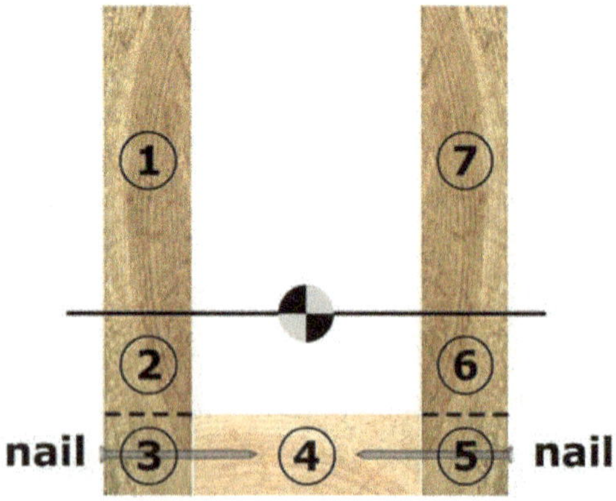

FIGURE M9.9

M9.10 Five multiple-choice questions pertaining to shear flow in built-up beam cross sections.

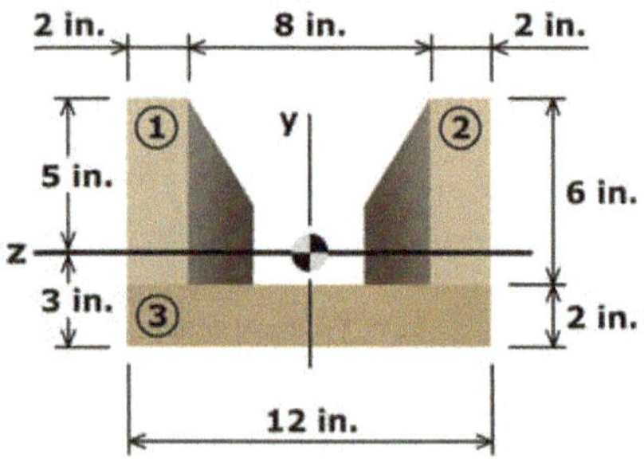

FIGURE M9.10

M9.11 Four multiple-choice questions pertaining to shear flow in built-up beam cross sections.

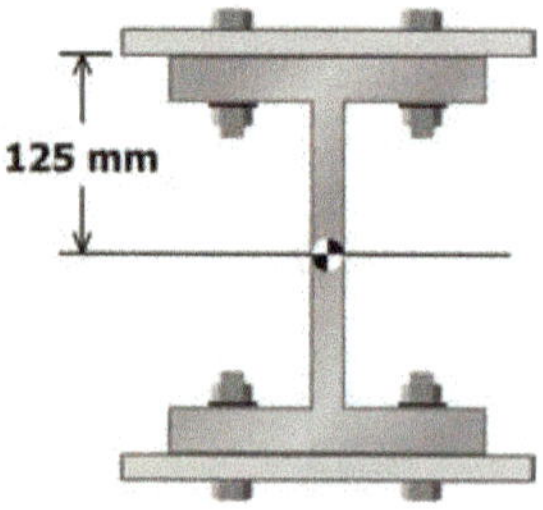

FIGURE M9.11

PROBLEMS

P9.26 A rectangular beam is fabricated by nailing together three pieces of dimension lumber as shown in Figure P9.26. Each board has a width $b = 4$ in. and a depth $d = 2$ in. Pairs of nails are driven along the beam at a spacing s, and each nail can safely transmit a force of 120 lb in direct shear. The beam is simply supported and carries a load of 800 lb at the center of a 9 ft span. Determine the maximum spacing s (along the length of the beam) permitted for each pair of nails.

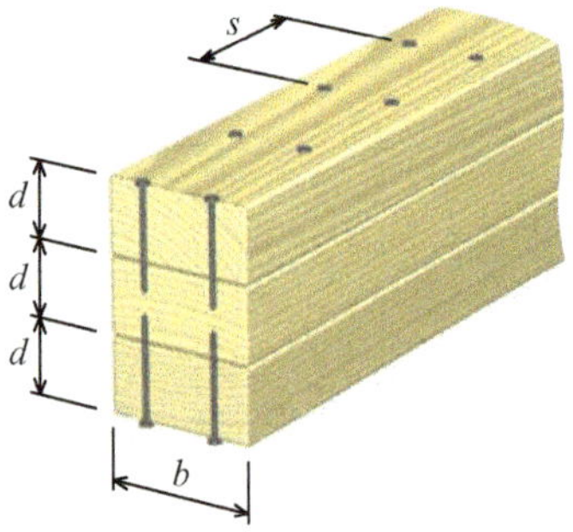

FIGURE P9.26

P9.27 A wooden I beam is fabricated from three pieces of dimension lumber as shown in Figure P9.27. The cross-sectional dimensions are $b_f = 90$ mm, $t_f = 40$ mm, $d_w = 240$ mm, and $t_w = 40$ mm. The beam will be used as a simply supported beam to carry a concentrated load P at the center of a 5 m span. The wood has an allowable bending stress of 8,300 kPa and an allowable shear stress of 620 kPa. The flanges of the beam are fastened to the web with screws that can safely transmit a force of 800 N in direct shear.

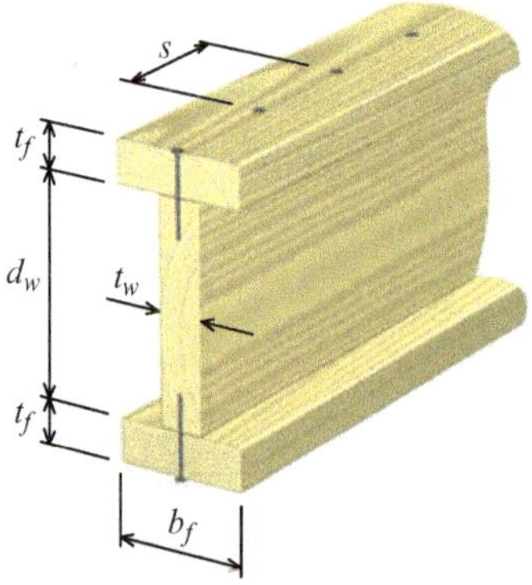

FIGURE P9.27

(a) If the screws are uniformly spaced at an interval $s = 200$ mm along the span, what is the maximum concentrated load P that can be supported by the beam, given the strength of the screw connections? Demonstrate that the maximum bending and shear stresses produced by P are acceptable.
(b) Determine the magnitude of the load P that produces the allowable bending stress in the span (i.e., $\sigma_b = 8{,}300$ kPa). What screw spacing s is required to support this load?

P9.28 The wooden box beam shown in Figure P9.28*a* is fabricated from four boards fastened together with nails (see Figure P9.28*b*) installed at a spacing s = 125 mm (see Figure P9.28*a*). Each nail can provide a resistance V_f = 500 N. In service, the beam will be installed so that bending occurs about the z axis. Determine the maximum shear force V that can be supported by the beam, given the shear capacity of the nailed connections.

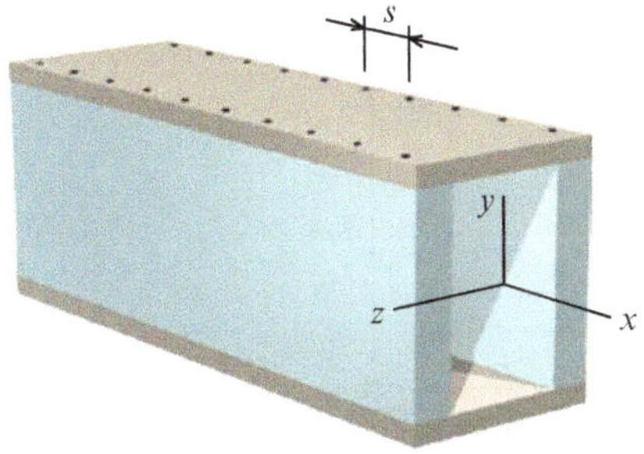

FIGURE P9.28*a*

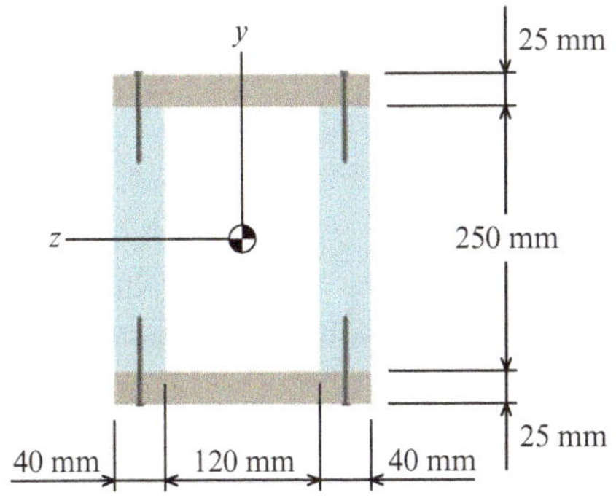

FIGURE P9.28*b*

P9.29 The wooden box beam shown in Figure P9.29*a* is fabricated from four boards fastened together with screws, as shown in Figure P9.29*b*. Each screw can provide a resistance of 800 N. In service, the beam will be installed so that bending occurs about the z axis and the maximum shear force in the beam will be 9 kN. Determine the maximum permissible spacing interval s (see Figure P9.29*a*) for the screws.

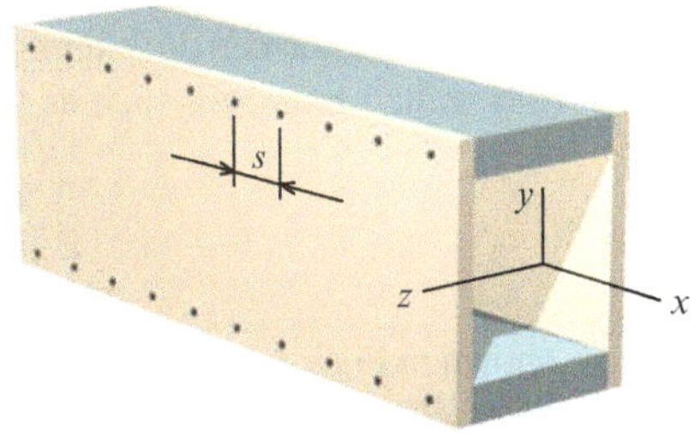

FIGURE P9.29*a*

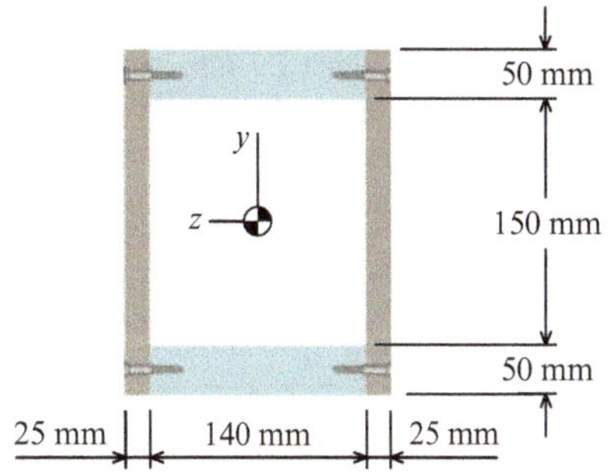

FIGURE P9.29*b*

P9.30 A wooden beam is fabricated by bolting together three members, as shown in Figure P9.30*a*/31*a*. The cross-sectional dimensions are shown in Figure P9.30*b*/31*b*. The 8 mm diameter bolts are spaced at intervals s = 200 mm along the x axis of the beam. If the internal shear force in the beam is V = 7 kN, determine the shear stress in each bolt.

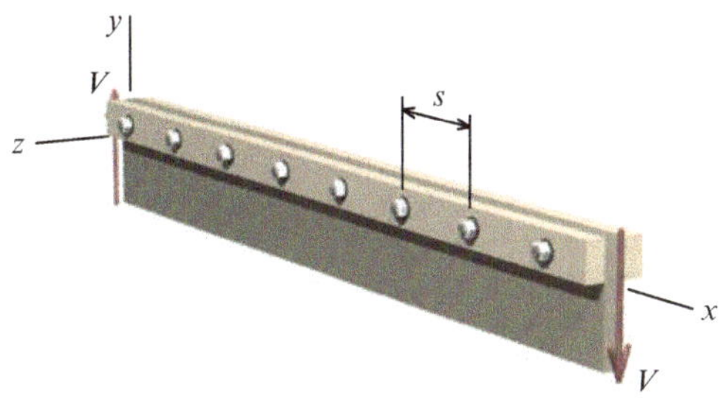

FIGURE P9.30*a*/31*a*

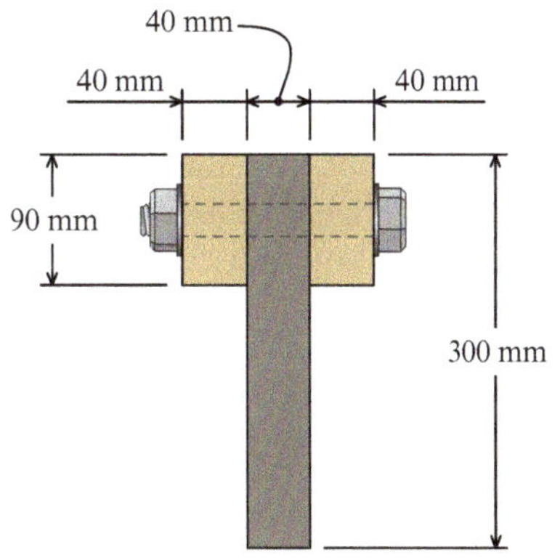

FIGURE P9.30*b*/31*b*

P9.31 A wooden beam is fabricated by bolting together three members, as shown in Figure P9.30*a*/31*a*. The cross-sectional dimensions are shown in Figure P9.30*b*/31*b*. The allowable shear stress of the wood is 850 kPa, and the allowable shear stress of the 10 mm diameter bolts is 40 MPa. Determine

(a) the maximum internal shear force V that the cross section can withstand, given the allowable shear stress in the wood.
(b) the maximum bolt spacing s required to develop the internal shear force computed in part (a).

P9.32 The cantilever beam shown in Figure P9.32*a* is fabricated by connecting two hat shapes with rivets that are spaced longitudinally at intervals of s along each side of the built-up shape. The cross section of the built-up shape is shown in Figure P9.32*b*, and the cross-sectional properties for a single hat shape are shown in Figure P9.32*c*. For an applied load $P = 6{,}800$ N, determine the maximum spacing interval s that can be used if each rivet has a shear strength of 1,700 N.

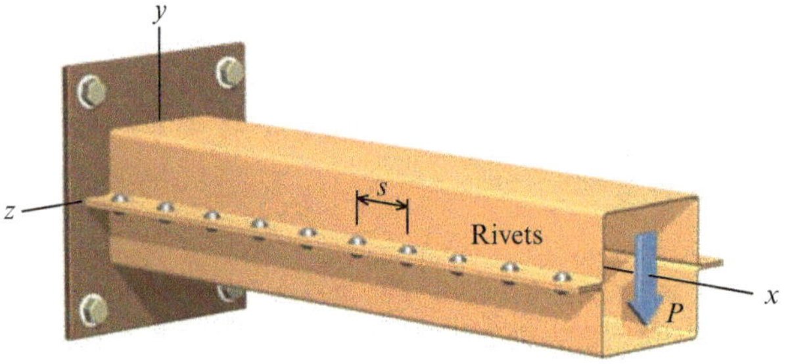

FIGURE P9.32*a* Riveted beam.

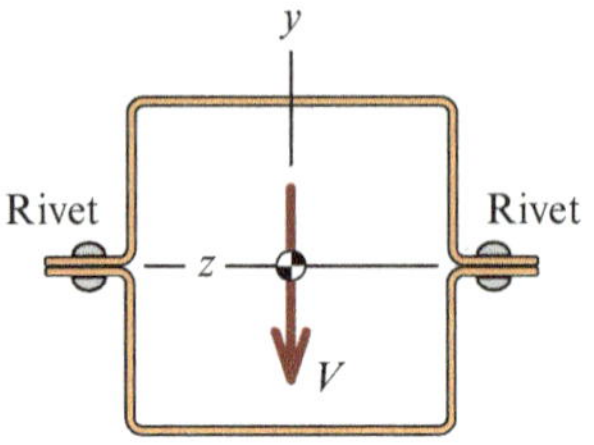

FIGURE P9.32*b* Cross section.

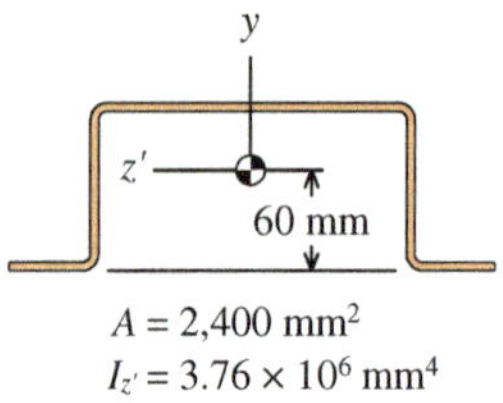

FIGURE P9.32*c* Cross-sectional properties for a single hat shape.

P9.33 A steel wide-flange beam in an existing structure is to be strengthened by adding a cover plate to its lower flange, as shown in Figure P9.33. The wide-flange shape has dimensions $b_f = 7.50$ in., $t_f = 0.57$ in., $d = 18.0$ in., and $t_w = 0.355$ in. The cover plate has a width $b = 15$ in. and $t = 1.0$ in. The cover plate is attached to the lower flange by pairs of 0.875 in. diameter bolts spaced at intervals s along the beam span. Bending occurs about the z centroidal axis.

(a) If the allowable bolt shear stress is 14 ksi, determine the maximum bolt spacing interval s required to support an internal shear force in the beam of $V = 55$ kips.

(b) If the allowable bending stress is 24 ksi, determine the allowable bending moment for the existing wide-flange shape, the allowable bending moment for the wide-flange shape with the added cover plate, and the percentage increase in moment capacity that is gained by adding the cover plate.

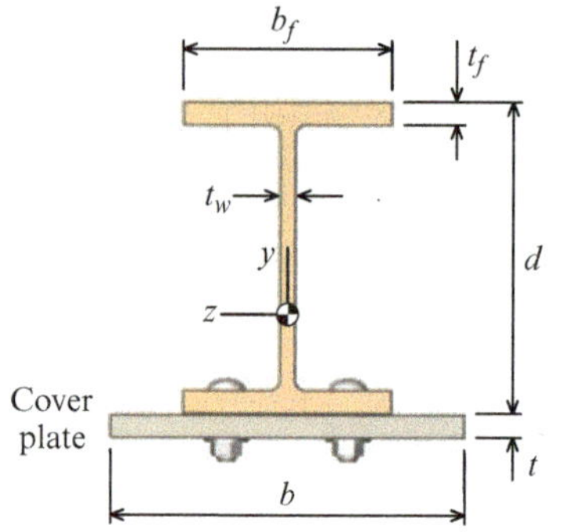

FIGURE P9.33

9.9 Shear Stress and Shear Flow in Thin-Walled Members

In the preceding discussion of built-up beams, the internal shear force F_H required for horizontal equilibrium of a specific portion and length of a flexural member was expressed by Equation (9.11):

$$F_H = \frac{VQ}{I_z}\Delta x$$

As shown in Figure 9.14*c*, the force F_H acts parallel to the bending stresses (i.e., in the x direction). The shear flow q was derived in Equation (9.12),

$$\frac{F_H}{\Delta x} = q = \frac{VQ}{I_z}$$

to express the shear force per unit length of beam span required to satisfy horizontal equilibrium for a specific portion of the cross section. In this section, these ideas will be applied to the analysis of average shear stress and shear flow in thin-walled members such as the flanges of wide-flange beam sections.

Shear Stress in Thin-Walled Sections

Consider the segment of length dx of the wide-flange beam shown in Figure 9.16*a*. The bending moments M and $M + dM$ produce compressive bending stresses in the upper flange of the

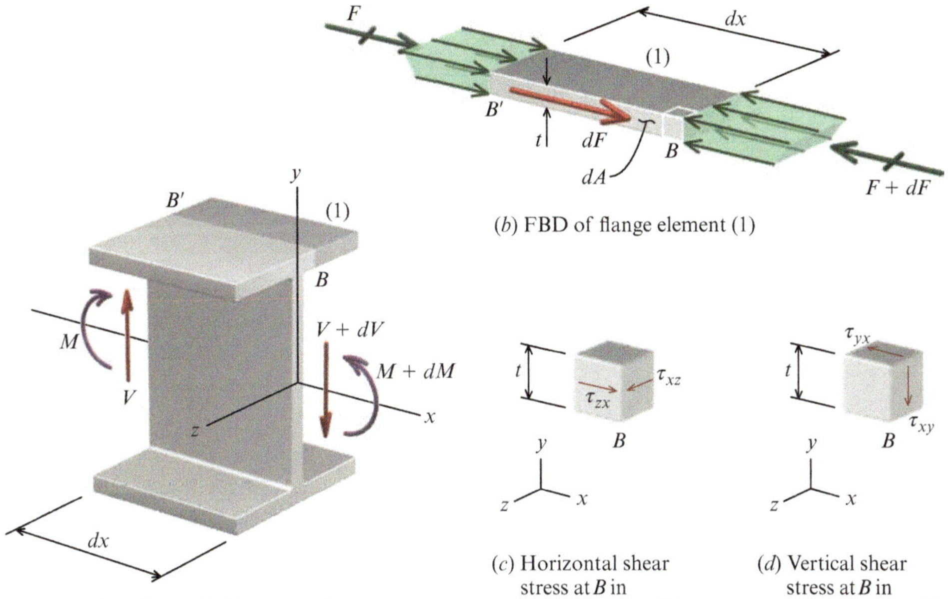

(*a*) Thin-walled beam section (*b*) FBD of flange element (1) (*c*) Horizontal shear stress at *B* in flange element (1) (*d*) Vertical shear stress at *B* in flange element (1)

FIGURE 9.16 Shear stresses in a thin-walled wide-flange beam.

member. Next, consider the FBD of a portion of the upper flange, element (1), shown in Figure 9.16*b*. On the back side of the beam segment, the bending moment M creates compressive normal stresses that act on the $-x$ face of flange element (1). The resultant of these normal stresses is the horizontal force F. Similarly, the bending moment $M + dM$ acting on the front side of the beam segment produces compressive normal stresses that act on the $+x$ face of flange element (1), and the resultant of these stresses is the horizontal force $F + dF$. Since the resultant force acting on the front side of element (1) is greater than the resultant force acting on the back side, an additional force dF must act on element (1) to satisfy equilibrium. This force can act only on the exposed surface BB' (because all other surfaces are free of stress). By a derivation similar to that used in obtaining Equation (9.11), the force dF can be expressed in terms of differentials as

$$dF = \frac{VQ}{I_z} dx \tag{9.15}$$

where Q is the first moment of the cross-sectional area of element (1) about the neutral axis of the beam section. The area of surface BB' is $dA = t\,dx$, and thus, the average shear stress acting on the longitudinal section BB' is

$$\tau = \frac{dF}{dA} = \frac{VQ}{I_z t} \tag{9.16}$$

Note that τ in this instance represents the average value of the shear stress acting on a z plane [i.e., the vertical surface BB' of element (1)] in the horizontal direction x—in other words, τ_{zx}. Since the flange is thin, the average shear stress τ_{zx} will not vary much over its thickness t. Consequently, τ_{zx} can be assumed to be constant. Because shear stresses acting on perpendicular planes must be equal (see Section 1.6), the shear stress τ_{xz} acting on an x face in the z direction must equal τ_{zx} at any point on the flange (Figure 9.16*c*). Accordingly, the horizontal shear stress τ_{xz} at any point on a transverse section of the flange can be obtained from Equation (9.16).

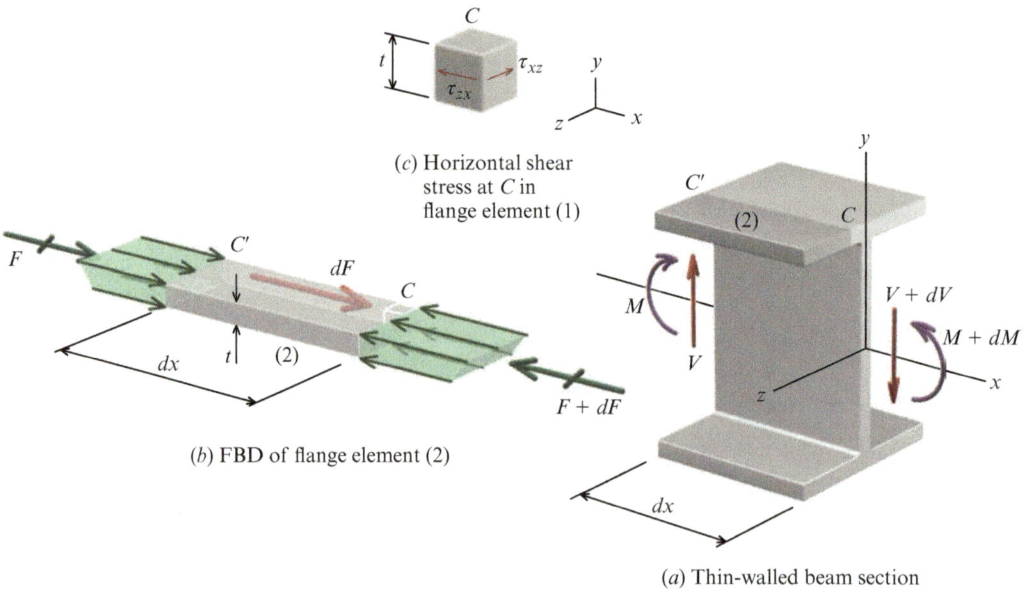

FIGURE 9.17 Thin-walled wide-flange beam.

The shear stress τ_{xy} acting on an x face in the vertical y direction at point B of the flange element is shown in Figure 9.16*d*. The top and bottom surfaces of the flange are free surfaces; thus, $\tau_{yx} = 0$. Since the flange is thin and the shear stresses on the top and bottom of the flange element are zero, the shear stress τ_{xy} through the thickness of the flange will be very small and thus can be neglected. Consequently, only the shear stresses (and shear flows) that act *parallel* to the free surfaces of the thin-walled section will be significant.

Next, consider point C on the upper flange of the beam segment shown in Figure 9.17*a*. An FBD of flange element (2) is shown in Figure 9.17*b*. With the same approach used for point B, it can be demonstrated that the shear stress τ_{xz} must act in the direction shown in Figure 9.17*c*. Similar analyses for points D and E on the lower flange of the cross section reveal that the shear stress τ_{xz} acts in the directions shown in Figure 9.18.

Equation (9.16) can be used to determine the shear stress in the flanges (Figure 9.19*a*) and the web (Figure 9.19*b*) of wide-flange shapes, in box beams (Figures 9.20*a* and 9.20*b*), in half-pipes (Figure 9.21), and in other thin-walled shapes, provided that the shear force V

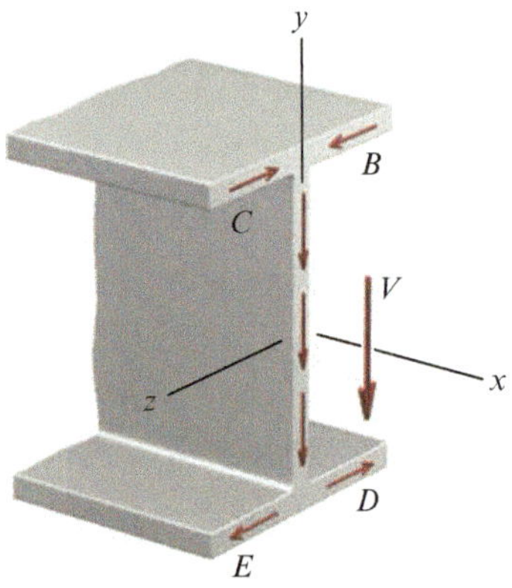

FIGURE 9.18 Shear stress directions at various locations in the cross section.

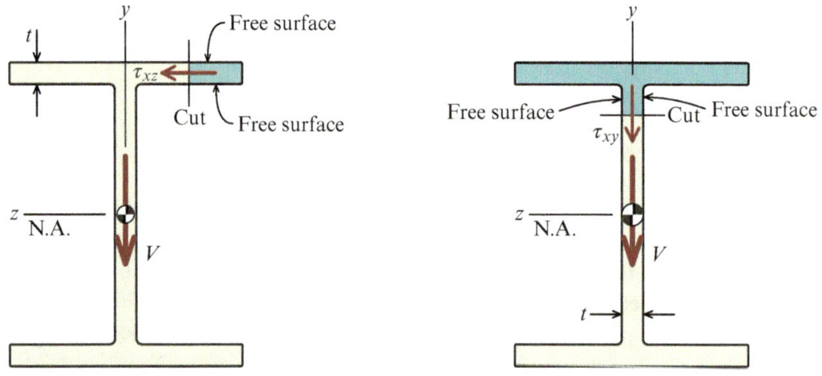

FIGURE 9.19 Shear stresses in a wide-flange shape.

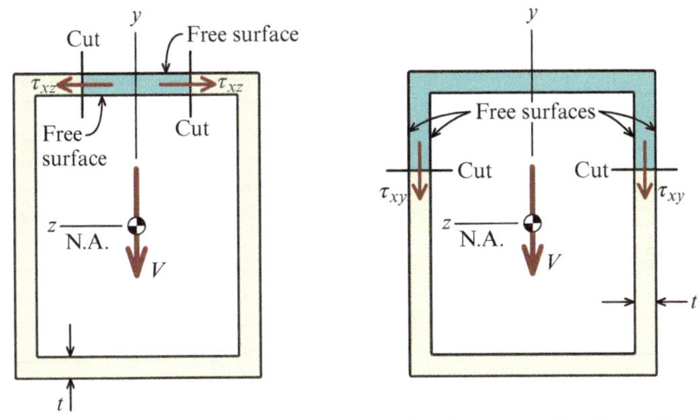

(*a*) Shear stress in upper wall (*b*) Shear stress in side walls

FIGURE 9.20 Shear stresses in a box-shaped cross section.

acts along an axis of symmetry of the cross section. For each shape, the cutting plane of the free-body diagram must be perpendicular to the free surface of the member. The shear stress acting *parallel* to the free surface can be calculated from Equation (9.16). (As discussed previously, the shear stress acting perpendicular to the free surface is negligible because of the thinness of the element and the proximity of the adjacent free surface.)

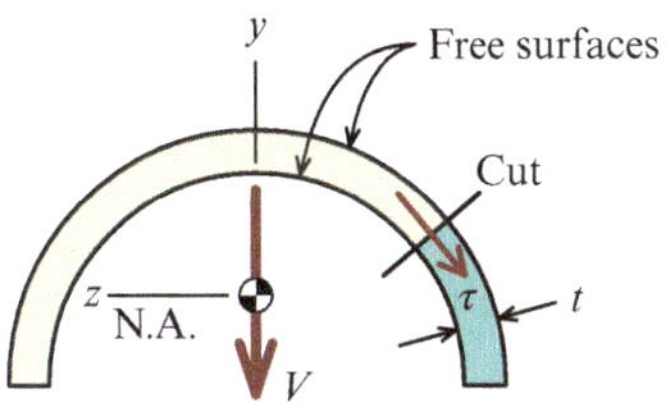

FIGURE 9.21 Shear stresses in a half-pipe cross section.

Shear Flow in Thin-Walled Sections

The shear flow along the top flange of the wide-flange shape shown in Figure 9.22*a* will be studied here. The product of the shear stress at any point in a thin-walled shape and the thickness t at that point is equal to the shear flow q:

$$\tau t = \left(\frac{VQ}{I_z t}\right) t = \frac{VQ}{I_z} = q \tag{9.17}$$

For a given cross section, the shear force V and the moment of inertia I_z in Equation (9.17) are constant. Thus, the shear flow at any location in the thin-walled shape depends only on the first moment of area, Q. Consider the shear flow acting on the shaded area, which is located a horizontal distance s from the tip of the flange. The shear flow acting at s can be calculated as

$$q = \frac{VQ}{I_z} = \frac{V}{I_z}\left(st\frac{d}{2}\right) = \frac{Vtd}{2I_z}s \tag{a}$$

Note that Q is the first moment of the shaded area about the neutral axis. From inspection of Equation (a), the distribution of shear flow along the top flange is a linear function of s. The maximum shear flow in the flange occurs at $s = b/2$:

$$(q_{\text{max}})_f = \frac{Vtd}{2I_z}\left(\frac{b}{2}\right) = \frac{Vbtd}{4I_z} \tag{b}$$

Note that $s = b/2$ is the centerline of the section. Since the cross section is assumed to be thin walled, centerline dimensions for the flange and web can be used in the calculation. This approximation procedure simplifies the calculations and is satisfactory for thin-walled cross sections. Owing to symmetry, similar analyses of the other three flange elements

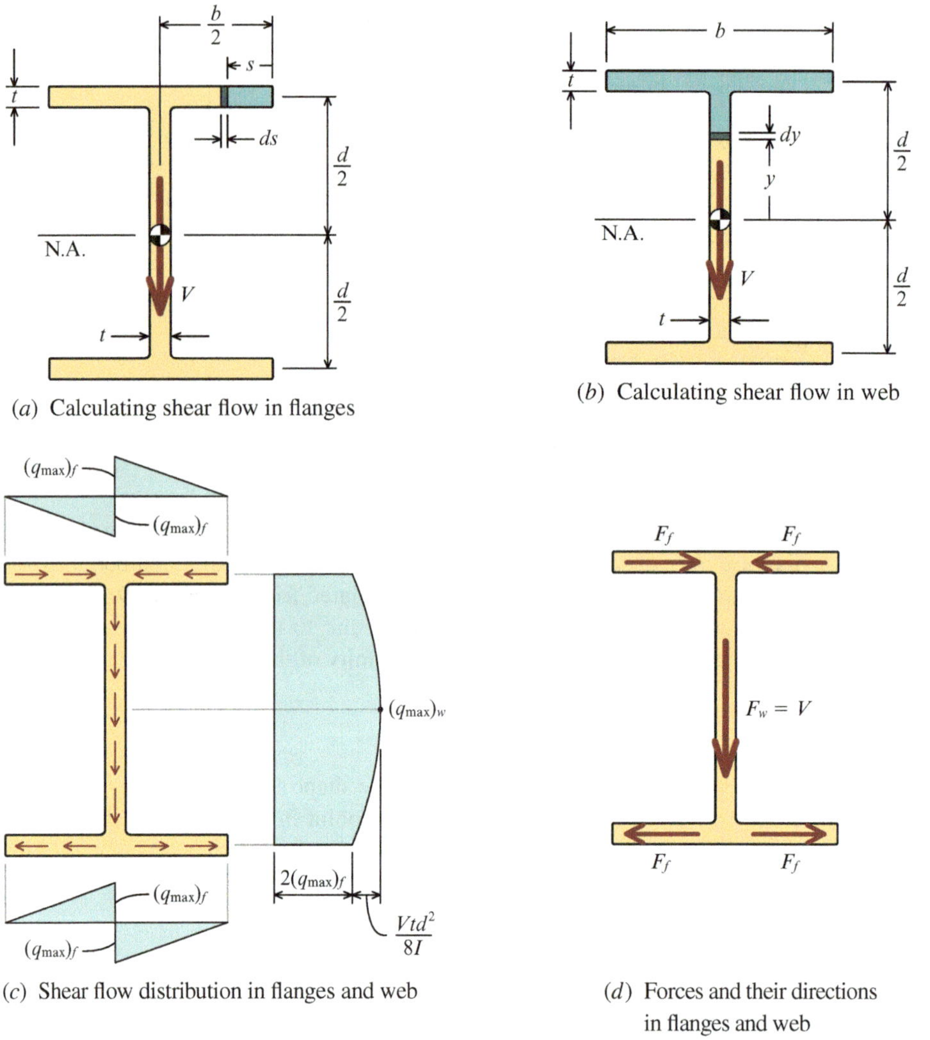

FIGURE 9.22 Wide-flange shape with equal flange and web thicknesses.

produce the same result for $(q_{max})_f$. The linear variation of shear flow in the flanges is shown in Figure 9.22*c*.

The total force developed in the upper left flange of Figure 9.22*a* can be determined by integration of Equation (a). The force on the differential element ds is $dF = q\,ds$. The total force acting on the upper left flange element is, therefore,

$$F_f = \int q\,ds = \int_0^{b/2} \frac{Vtd}{2I_z} s\,ds = \frac{Vb^2td}{16I_z}$$

This same result can also be obtained by calculating the area under the triangular distribution in Figure 9.22*c*, since q is a distribution of force per length:

$$F_f = \frac{1}{2}(q_{max})_f \frac{b}{2} = \frac{1}{2}\left(\frac{Vbtd}{4I_z}\right)\frac{b}{2} = \frac{Vb^2td}{16I_z}$$

Again on the basis of symmetry, the force F_f in each flange element will be the same. The flange forces are shown in their proper directions in Figure 9.22*d*. From the direction of these forces, it is evident that horizontal force equilibrium of the cross section is maintained.

Next, consider the web of the thin-walled cross section shown in Figure 9.22*b*. In the web, the shear flow is

$$\begin{aligned} q &= \frac{V}{I_z}\left[\frac{btd}{2} + \frac{1}{2}\left(\frac{d}{2} + y\right)\left(\frac{d}{2} - y\right)t\right] \\ &= \frac{Vbtd}{2I_z} + \frac{Vt}{2I_z}\left(\frac{d^2}{4} - y^2\right) \end{aligned} \qquad \text{(c)}$$

Using the expression for $(q_{max})_f$ derived in Equation (*b*), we can rewrite Equation (*c*) as the sum of the shear flows in the flange plus the change in shear flow over the depth of the web:

$$q = 2(q_{max})_f + \frac{Vt}{2I_z}\left(\frac{d^2}{4} - y^2\right)$$

The shear flow in the web increases parabolically from a minimum value of $(q_{min})_w = 2(q_{max})_f$ at $y = d/2$ to a maximum value of

$$(q_{max})_w = 2(q_{max})_f + \frac{Vtd^2}{8I_z}$$

at $y = 0$.

Again, it should be noted that the shear flow expression here has been based on the centerline dimensions of the cross section.

To determine the force in the web, Equation (c) must be integrated. Once more, with the centerline bounds of $y = \pm d/2$, the force in the web can be expressed as

$$\begin{aligned} F_w &= \int q\,dy = \int_{-d/2}^{d/2} \frac{Vt}{2I_z}\left[bd + \frac{d^2}{4} - y^2\right] dy \\ &= \frac{Vt}{2I_z}\left[bdy + \frac{d^2}{4}y - \frac{1}{3}y^3\right]_{-d/2}^{d/2} \\ &= \frac{Vt}{2I_z}\left[bd^2 + \frac{d^3}{6}\right] \end{aligned}$$

or

$$F_w = \frac{V}{I_z}\left[2bt\left(\frac{d}{2}\right)^2 + \frac{td^3}{12}\right] \qquad \text{(d)}$$

The moment of inertia for the thin-walled flanged shape can be expressed as

$$I_z = I_{flanges} + I_{web} = 2\left[\frac{bt^3}{12} + bt\left(\frac{d}{2}\right)^2\right] + \frac{td^3}{12}$$

Since t is small, the first term in the brackets can be neglected, so

$$I_z = 2bt\left(\frac{d}{2}\right)^2 + \frac{td^3}{12}$$

Substituting this expression into Equation (d) gives $F_w = V$, which is as expected. (See Figure 9.22*d*.)

It is useful to visualize shear flow in the same manner that one might visualize fluid flow in a network of pipes. In Figure 9.22*c*, the shear flows q in the two top flange elements are directed from the outermost edges toward the web. At the junction of the web and the flange, these shear flows turn the corner and flow down through the web. At the bottom flange, the flows split again and move outward toward the flange tips. Because the flow is always continuous in any structural section, it serves as a convenient method for determining the directions of shear stresses. For instance, if the shear force acts downward on the beam section of Figure 9.22*a*, then we can recognize immediately that the shear flow in the web must act downward as well. Also, since the shear flow must be continuous through the section, we can infer that (a) the shear flows in the upper flange must move toward the web and (b) the shear flows in the bottom flange must move away from the web. Using this simple technique to ascertain the directions of shear flows and shear stresses is easier than visualizing the directions of the forces acting on elements such as those in Figures 9.16*b* and 9.17*b*.

The preceding analysis demonstrates how shear stresses and shear flow in a thin-walled cross section can be calculated. The results offer a more complete understanding of how shear stresses are distributed throughout a beam that is subjected to shear forces. (Recall that, in Section 9.7, shear stresses in a wide-flange cross section were determined for the web only.) Three important conclusions should be drawn from this analysis:

1. The shear flow q is dependent on the value of Q, and Q will vary throughout the cross section. For beam cross-sectional elements that are perpendicular to the direction of the shear force V, q and hence τ will vary linearly in magnitude. Both q and τ will vary parabolically in cross-sectional elements that are parallel to or inclined toward the direction of V.
2. Shear flow will always act parallel to the free surfaces of the cross-sectional elements.
3. Shear flow is always continuous in any cross-sectional shape subjected to a shear force. Visualization of this flow pattern can be used to establish the direction of both q and τ in a shape. The flow is such that the shear flows in the various cross-sectional elements contribute to V while satisfying both horizontal and vertical equilibrium.

Closed Thin-Walled Sections

Flanged shapes such as wide-flange shapes (Figure 9.19) and tee shapes are classified as **open sections**, whereas box shapes (Figure 9.20) and circular pipe shapes are classified as **closed sections**. The distinction between open and closed sections is that closed shapes have a continuous periphery in which the shear flow is uninterrupted while open shapes do not. Now, consider beam cross sections that satisfy two conditions: (a) The cross section has at least one longitudinal plane of symmetry, and (b) the beam loads act in this plane of symmetry. For open sections, such as flanged shapes, satisfying these conditions, the shear flow and shear stress clearly must be zero at the tips of the flanges. For closed sections, such as box or pipe shapes, the locations at which the shear flow and the shear stress vanish are not so readily apparent.

A thin-walled box section subjected to a shear force V is shown in Figure 9.23*a*. The section is split vertically along its longitudinal plane of symmetry in Figure 9.23*b*. The shear flow in vertical walls of the box must flow parallel to the internal shear force V; thus, the shear flow in the top and bottom walls of the box must act in the directions shown. On the plane of symmetry, the shear stress at points B and B' must be equal; however, the shear flows act in opposite directions. Similarly, the shear stress at points C and C' must be equal, but they, too, act in opposing directions. Consequently, the only possible value of shear stress that can satisfy these constraints is $\tau = 0$. Since $q = \tau t$, the shear flow must also be zero at these points. From this analysis, we can conclude that the shear flow and the shear stress for a closed thin-walled beam section must be zero on a longitudinal plane of symmetry.

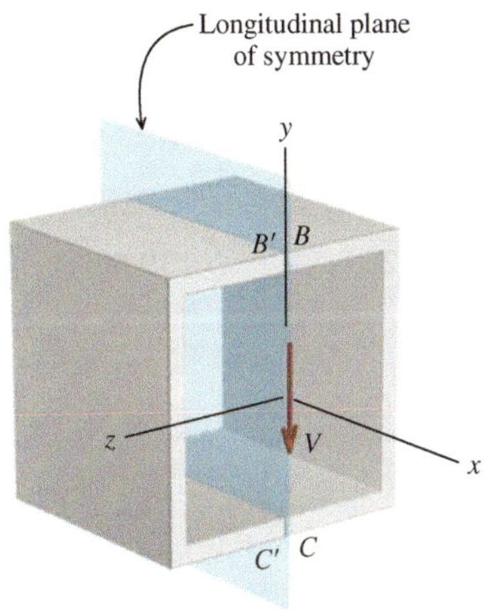

(a) Closed thin-walled section with a longitudinal plane of symmetry

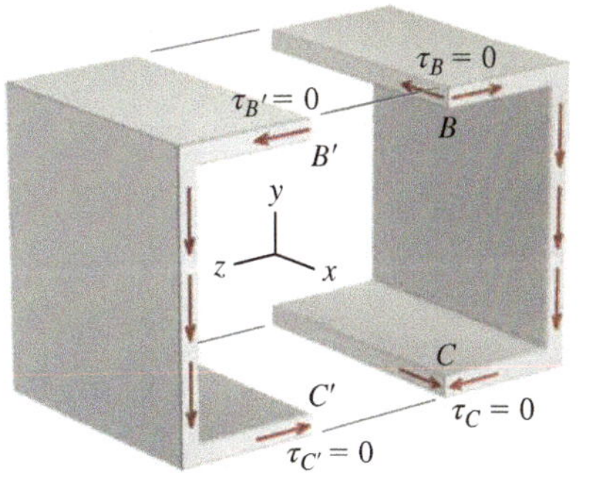

(b) Shear stresses at the plane of symmetry

FIGURE 9.23 Shear stress in a thin-walled box cross section.

EXAMPLE 9.7

A beam with the thin-walled inverted-tee-shaped cross section shown is subjected to a vertical shear force $V = 37$ kN. The location of the neutral axis is shown on the sketch, and the moment of inertia of the inverted-tee shape about the neutral axis is $I = 11{,}219{,}700$ mm^4. Determine the shear stresses in the tee stem at points a, b, c, and d, and in the tee flange at points e and f. Plot the distribution of shear stress in both the stem and flange.

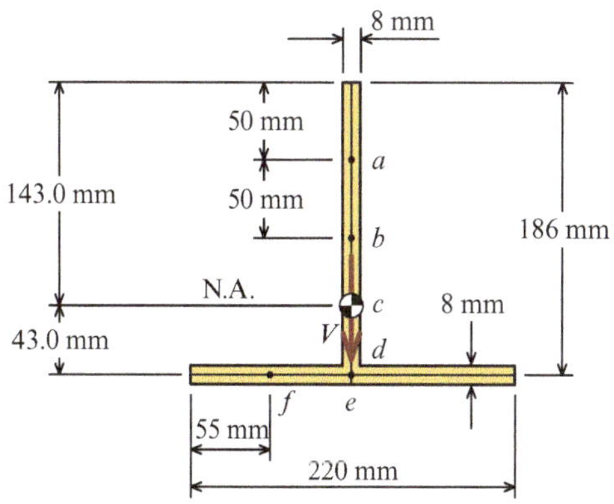

Plan the Solution

The location of the neutral axis and the moment of inertia of the inverted-tee shape about the neutral axis are given. The value of Q associated with each point will be determined from $Q = \bar{y}'A'$ for the applicable portion A' of the cross-sectional area. After Q is determined, the shear stress will be calculated from Equation (9.16).

SOLUTION

Points a, b, and c are located in the stem of the inverted-tee shape. A horizontal cutting plane that is perpendicular to the walls of the stem defines the boundary of area A'. For these locations, area A' begins at the cutting plane and reaches upward to the top of the stem. Point d is located at the junction of the stem and the flange. For this location, the area A' is simply the area of the flange. Point e is also at the junction of the stem and the flange; however, the shear stress in only the flange is to be determined at e. The area A' corresponding to point e extends from the left end of the flange to a vertical cutting plane located at the centerline of the stem. (Note that the centerline location for the cutting plane is acceptable because the shape is thin walled.) For point f in the flange, a vertical cutting plane defines the boundary of area A', which extends horizontally from the cutting plane to the outer edge of the flange. For all points, the first moment Q is the moment of the area A' about the neutral axis of the inverted-tee shape. The shear stress at each point is calculated from

$$\tau = \frac{VQ}{It}$$

where $V = 37$ kN and $I = 11{,}219{,}700$ mm^4. The thickness t is 8 mm for each location. The results of these analyses are summarized in the following table:

Point	Sketch	$\bar{y}'$ (mm)	A' (mm²)	Q (mm³)	τ (MPa)
a	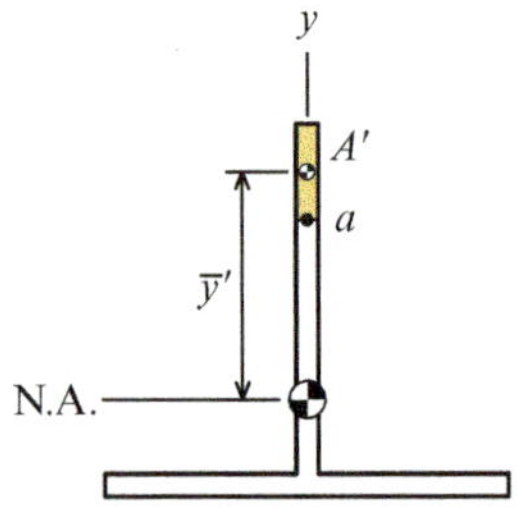	118.0	400	47,200	19.46
b	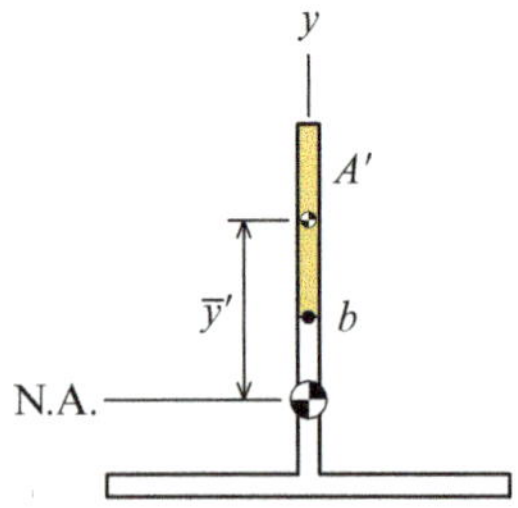	93.0	800	74,400	30.67
c	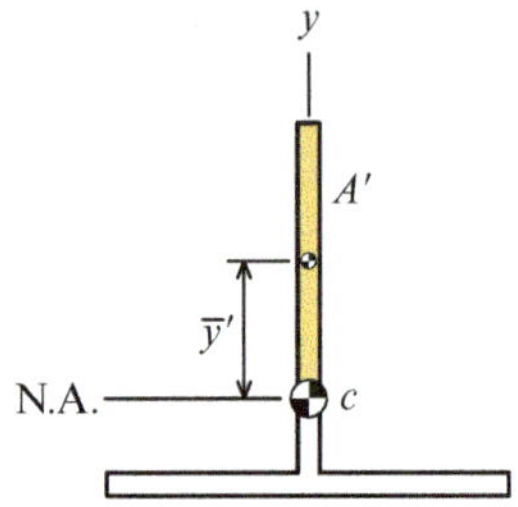	71.5	1,144	81,796	33.72
d	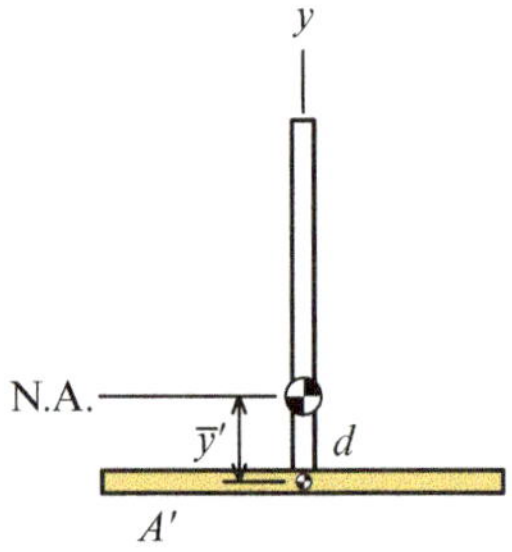	43.0	1,760	75,680	31.20
e	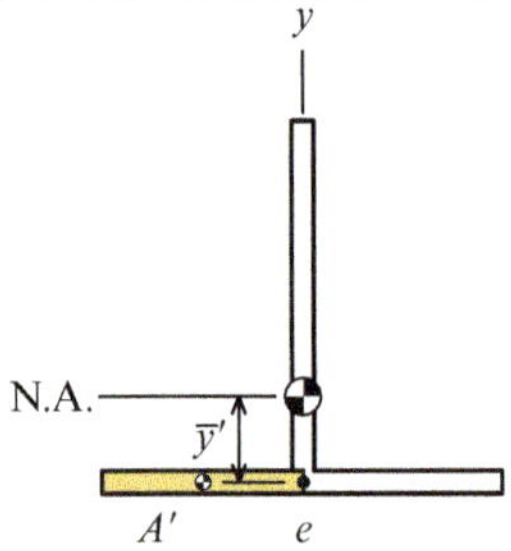	43.0	880	37,840	15.60
f	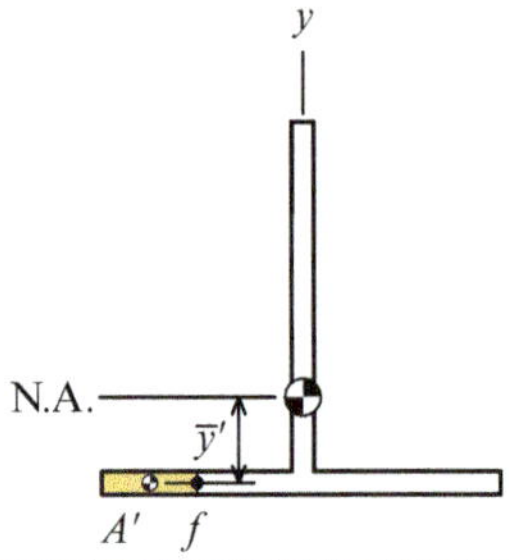	43.0	440	18,920	7.80

The directions and intensities of the shear stress in the inverted-tee shape are shown in the accompanying sketch. Note that the shear stress in the tee stem is distributed parabolically while the shear stress in the flange is distributed linearly. At the junction of the stem and the flange, the shear stress intensity is cut in half as shear flows outward in two opposing directions.

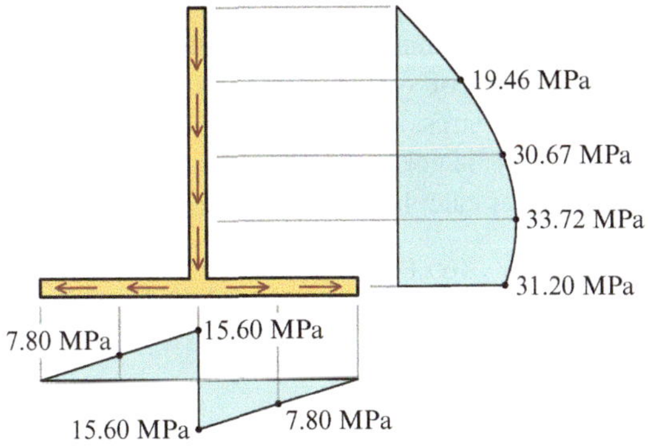

EXAMPLE 9.8

A beam with the cross section shown is subjected to a vertical shear force $V = 65$ kN. The location of the centroid appears on the sketch, and the moment of inertia of the cross section about the z centroidal axis is $I = 2{,}532{,}789$ mm^4. Determine the shear stress in the cross section at point a.

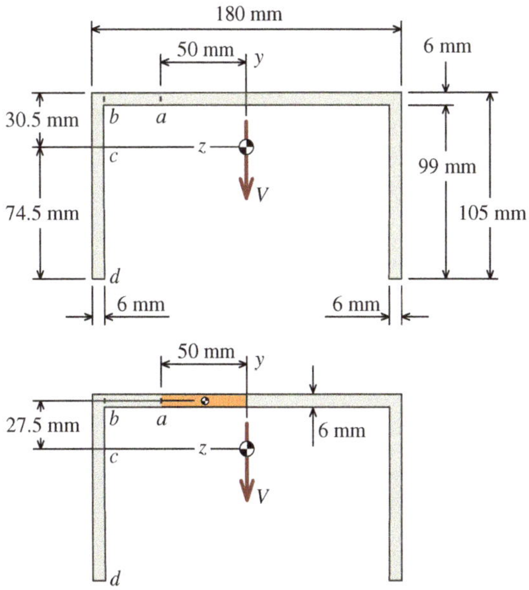

Plan the Solution

To calculate the shear stress at a, we must cut an FBD through the cross section so that the shear stress at a is exposed. We will consider three different FBDs that could be used to determine the quantity Q necessary to calculate the shear stress from the formula $\tau = VQ/It$.

SOLUTION

Approach 1

The first FBD that we will consider is obtained by making two vertical cuts through the cross section. One vertical cut will be made at point a, and a second vertical cut will be made on the y centroidal axis, which is an axis of symmetry for the cross section considered here.

The FBD exposed by these two cuts is shown in the accompanying figure. The pieces of the cross section have been separated in the figure to help illustrate the shear stresses on the interior surfaces. The directions of the shear flow and the shear stresses on the various surfaces are also indicated. Notice that shear stress on the vertical plane of symmetry is zero.

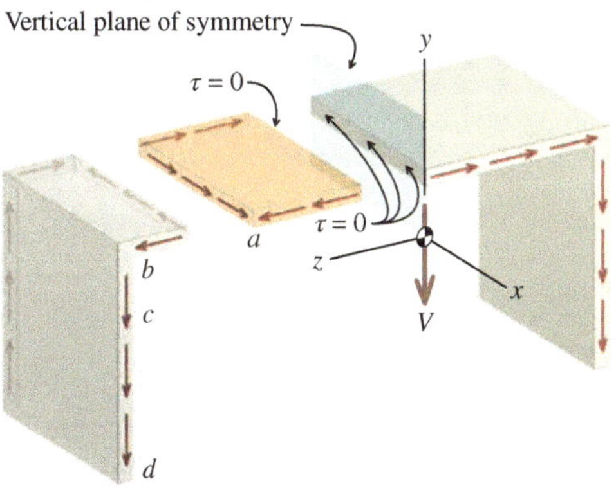

The shear stress at point a will be calculated from the formula $\tau = VQ/It$. **The shear force V and the moment of inertia I to be used in the formula are the same for any FBD that we might consider.**

For this FBD, the calculation of Q is straightforward:

$$Q = (27.5 \text{ mm})(50 \text{ mm} \times 6 \text{ mm}) = 8{,}250 \text{ mm}^3$$

Since we've made two vertical cuts through the cross section in defining our FBD, we might initially conclude that $t = 2 \times (6 \text{ mm}) = 12$ mm. However, a value of $t = 12$ mm is incorrect in this instance. To see why, note that one of our vertical cuts here was made on the y centroidal axis, which is a plane of symmetry for the shape. We know that the shear stress must be zero on a plane of symmetry that is parallel to the direction of the shear force. Accordingly, the shear stress on the cutting plane at the y axis must be zero, and a surface with zero shear stress is not included in the term t. The value of t needed for this FBD is $t = 6$ mm. The proper calculation for the shear stress at a is

$$\tau = \frac{VQ}{It} = \frac{(65{,}000 \text{ N})(8{,}250 \text{ mm}^3)}{(2{,}532{,}789 \text{ mm}^4)(6 \text{ mm})} = 35.29 \text{ MPa}$$

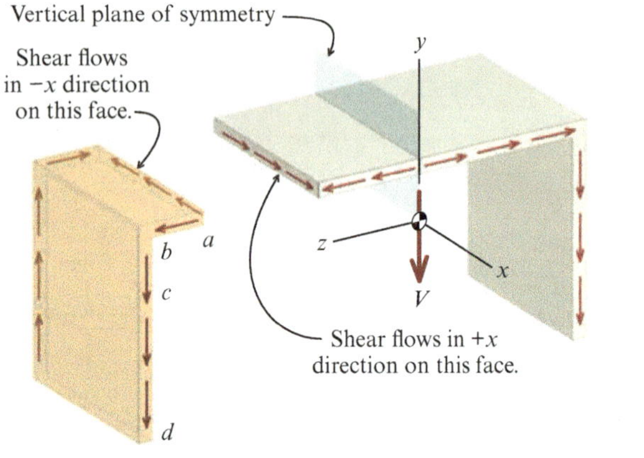

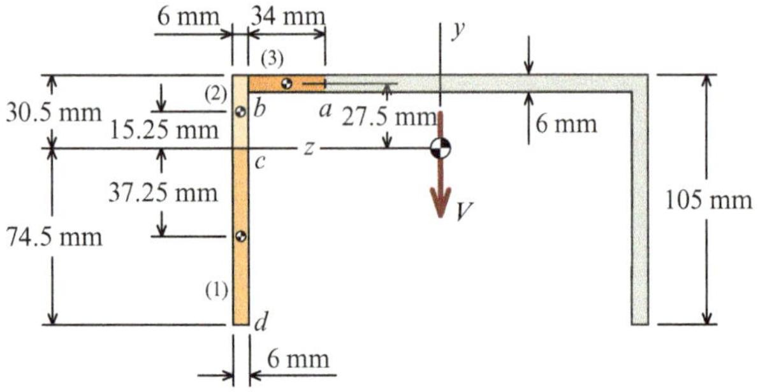

Approach 2

The second FBD that we will consider is obtained by making one vertical cut at point a. We will examine that portion of the cross section from point a to point b to point d.

The FBD exposed by this cut is shown in the accompanying figure. Again, the FBD has been separated to facilitate visualization of the shear stresses on the interior surfaces. The directions of the shear flow and the shear stresses on the various surfaces are indicated in the figure.

We begin the calculation at point d, where we know with certainty that the shear stress is zero. We will calculate the value of Q associated with point c by calculating the first moment of area (1) (i.e., the area between points c and d) about the neutral axis:

$$Q_c = (-37.25 \text{ mm})(6 \text{ mm} \times 74.5 \text{ mm}) = -16{,}650.75 \text{ mm}^3$$

Note that a negative value is used for the distance from the z centroidal axis to the centroid of area (1) because area (1) lies below the neutral axis.

Next, we consider area (2) in order to calculate the value of Q associated with point b. Since area (2) lies above the neutral axis, the distance from the z axis to the centroid of area (2) is a positive value, and we have

$$\begin{aligned} Q_b &= Q_c + (15.25 \text{ mm})(6 \text{ mm} \times 30.5 \text{ mm}) \\ &= -16{,}650.75 \text{ mm}^3 + 2{,}790.75 \text{ mm}^3 \\ &= -13{,}860 \text{ mm}^3 \end{aligned}$$

Finally, we consider area (3) in order to determine Q_a:

$$\begin{aligned} Q_a &= Q_b + (27.5 \text{ mm})(34 \text{ mm} \times 6 \text{ mm}) \\ &= -13{,}860 \text{ mm}^3 + 5{,}610 \text{ mm} \\ &= -8{,}250 \text{ mm}^3 \end{aligned}$$

Note that Q calculated from the FBD of Approach 2 has the same magnitude as Q determined with the FBD of Approach 1; however, a negative sign appears in the second calculation. To calculate the magnitude of the shear stress at a, we need only the magnitude of Q. However, what is the negative sign on Q for this FBD telling us? Consider the direction that the shear flows on the two faces exposed by the vertical cut. On the right-hand portion, shear stresses must flow in the positive x direction on the exposed z face to satisfy equilibrium. On the left-hand portion, shear stresses must flow in the negative x direction for equilibrium. Thus, the magnitudes of Q must be the same, but the directions that shear stresses flow on the two exposed z faces must be opposite to satisfy equilibrium.

For the second FBD, plainly only one vertical cut is made through the cross section and, consequently, $t = 6$ mm. The shear stress magnitude is calculated as

$$\tau = \frac{VQ}{It} = \frac{(65{,}000\ \text{N})(|-8{,}250\ \text{mm}^3|)}{(2{,}532{,}789\ \text{mm}^4)(6\ \text{mm})} = 35.29\ \text{MPa}$$

Approach 3

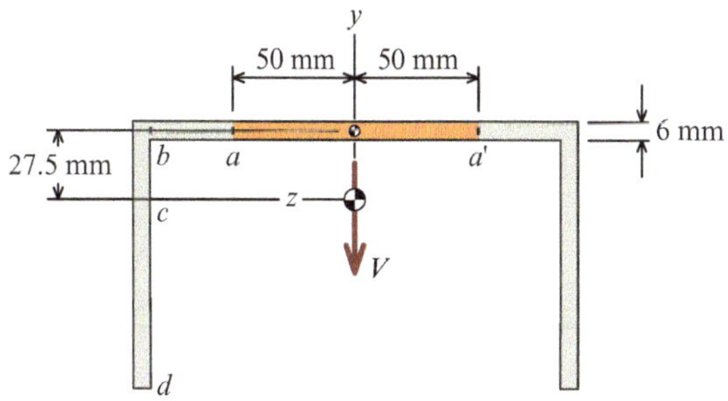

For the third approach, we will take advantage of the cross section's symmetry with respect to the y axis. We will make one vertical cut at point a and a second vertical cut at point a', which is located at the mirror image of point a on the opposite side of the plane of symmetry. We will consider the portion of the cross section that begins at point a and extends rightward to point a'. From symmetry, we know that the shear stress at points a and a' must be equal.

For this FBD,

$$\begin{aligned} Q &= 2 \times (27.5\ \text{mm})(50\ \text{mm} \times 6\ \text{mm}) \\ &= 16{,}500\ \text{mm}^3 \end{aligned}$$

We are making two cuts through the section; therefore, $t = 2 \times (6\ \text{mm}) = 12$ mm. The shear stress at a is thus

$$\tau = \frac{VQ}{It} = \frac{(65{,}000\ \text{N})(16{,}500\ \text{mm}^3)}{(2{,}532{,}789\ \text{mm}^4)(12\ \text{mm})} = 35.29\ \text{MPa}$$

Conclusions

What have we shown in this example? Clearly, there are several ways to calculate shear stress in a cross section.

In Approach 1, we made two vertical cuts through the cross section. The first vertical cut was at the point of interest (i.e., point a), and the second vertical cut was on the plane of symmetry. To use the first approach, we need to understand that shear stresses are zero on vertical planes of symmetry (i.e., planes of symmetry that are parallel to the direction of the shear force V). We must recognize that the term t should consider only thicknesses of surfaces upon which shear stresses act. If the shear stress on a surface is zero, then its thickness is not included in the value of t. All in all, this approach is not recommended for novices because of its potential for misinterpretation.

In Approach 2, we calculated Q from portions of the cross section both above and below the neutral axis. This approach gives a negative value for Q, which can be confusing at first glance. However, we need only remember to use the absolute value of Q in our stress calculations to use the second approach successfully.

With Approach 3, we take advantage of symmetry in choosing the portion of the cross section needed for our calculation of Q. The process of calculating Q is simplified, since the FBD is entirely on one side of the neutral axis and the sign of Q will be positive, thus avoiding the confusion that can occur in Approach 2. The thicknesses of both cut surfaces are included in the value of t, thereby avoiding the misunderstanding that could occur in Approach 1.

EXAMPLE 9.9

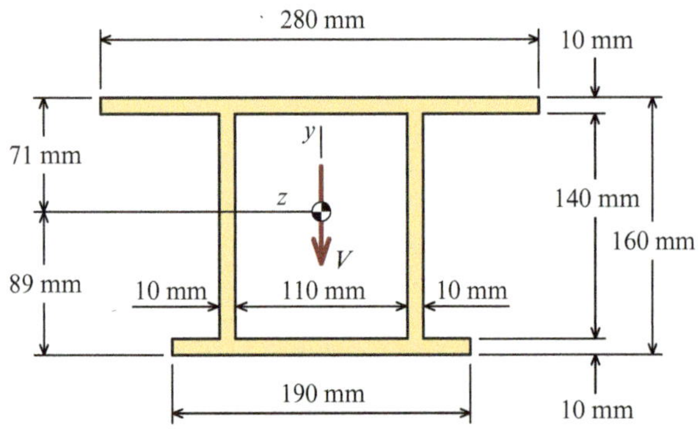

A beam with the cross section shown is subjected to a vertical shear force $V = 450$ kN. The location of the neutral axis appears on the sketch, and the moment of inertia of the cross section about the neutral axis is $I = 30{,}442{,}500$ mm^4.

(a) Determine the shear stresses in the cross section at points a through i in the accompanying sketch.

(b) Plot the distribution of shear stress in the cross section.

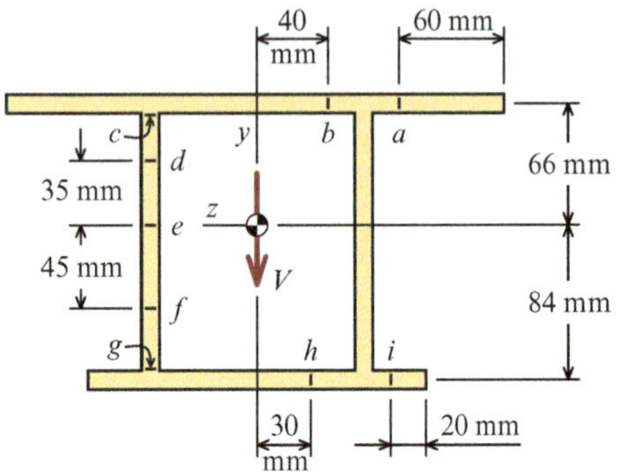

Plan the Solution

SOLUTION

(a) Shear stress at point *a*

Begin by making a vertical cut through the top flange of the cross section at point a. For our calculation of Q, we consider the flange area between a and the free surface at the rightmost edge of the flange. The vertical distance from the z axis to the centroid of this area is 71 mm – 10 mm/2 = 66 mm. Accordingly, for the area selected,

$$Q = (66 \text{ mm})(60 \text{ mm} \times 10 \text{ mm}) = 39{,}600 \text{ mm}^3$$

The shear stress at point a is

$$\tau_a = \frac{VQ}{It} = \frac{(450{,}000 \text{ N})(39{,}600 \text{ mm}^3)}{(30{,}442{,}500 \text{ mm}^4)(10 \text{ mm})} = 58.5 \text{ MPa} \quad \text{(a)}$$

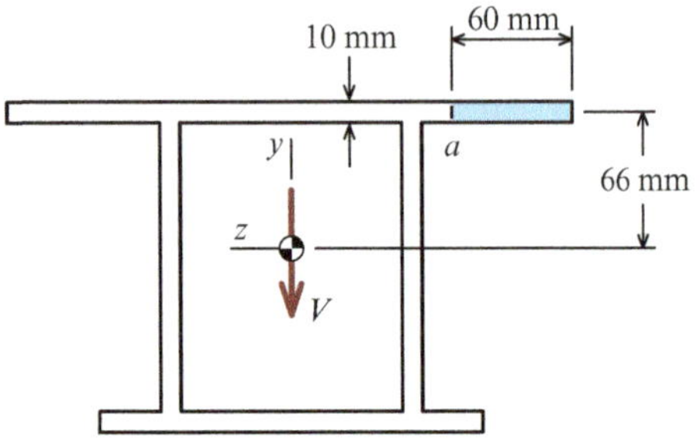

While this approach is simple and easy to understand for point a, other points in the cross section are not so directly isolated. For this cross section, approach 3 from Example 9.8 might be a more effective method for calculating shear stresses.

For any specific point at which we wish to calculate shear stress, we will also consider a mirror-image point located on the opposite side of the vertical axis of symmetry of the cross section (i.e., the y axis in this figure). Accordingly, in addition to point a, we will consider point a' when making our shear stress calculations. The specific areas are shown in the accompanying figure. Owing to symmetry, points a and a' must have the same shear stress magnitude.

For the highlighted area,

$$Q = 2 \times (66 \text{ mm})(60 \text{ mm} \times 10 \text{ mm}) = 79{,}200 \text{ mm}^3$$

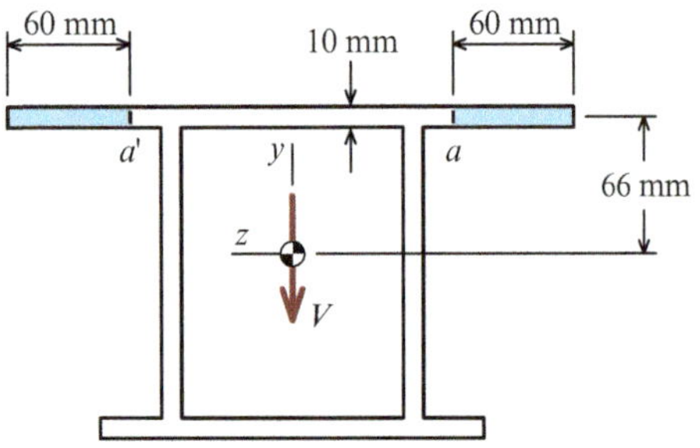

Two vertical cuts have been made through the upper flange to isolate these areas; therefore, $t = 2 \times (10 \text{ mm}) = 20$ mm. The shear stress at point a is

$$\tau_a = \frac{VQ}{It} = \frac{(450{,}000 \text{ N})(79{,}200 \text{ mm}^3)}{(30{,}442{,}500 \text{ mm}^4)(20 \text{ mm})} = 58.5 \text{ MPa} \quad \text{(b)}$$

Plainly, the results obtained in Equations (a) and (b) are the same. Note that values used in Equation (a) for the shear force V and the moment of inertia I are respectively identical to those used in Equation (b). **That is, the same values for the shear force V and the same values for the moment of inertia I are used in calculating shear stresses at each point in the cross section.**

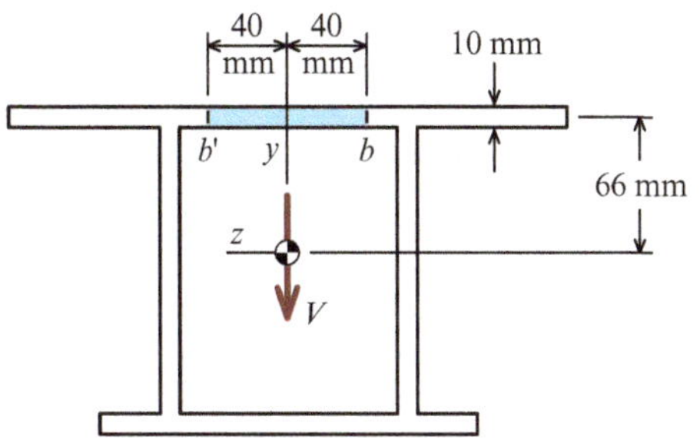

Shear stress at point b

We make one vertical cut through the top flange point b, located 40 mm to the right of the y axis, and a second vertical cut at the mirror point b', located 40 mm to the left of the y axis. From symmetry, we know that the shear stress magnitude must be the same at both locations. For our calculation of Q, we consider the flange area between b and b':

$$Q = 2 \times (66\text{ mm})(40\text{ mm} \times 10\text{ mm}) = 52{,}800\text{ mm}^3$$

We've made two vertical cuts through the flange to isolate this portion of the cross section; consequently, $t = 2 \times 10\text{ mm} = 20\text{ mm}$. The shear stress at point b (and also at point b') is

$$\tau_b = \frac{(450{,}000\text{ N})(52{,}800\text{ mm}^3)}{(30{,}442{,}500\text{ mm}^4)(2 \times 10\text{ mm})} = 39.0\text{ MPa}$$

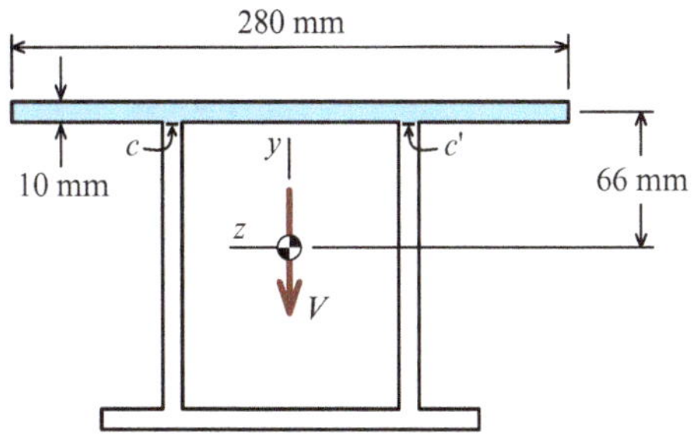

Shear stress at point c

Here, we cut horizontally through the webs at point c and mirror point c' to isolate the entire top flange. The value of Q for this location is

$$Q = (66\text{ mm})(280\text{ mm} \times 10\text{ mm}) = 184{,}800\text{ mm}^3$$

Also, because of the horizontal cuts through the webs, $t = 2 \times 10\text{ mm} = 20\text{ mm}$. The shear stress at point c (and also at point c') is

$$\tau_c = \frac{(450{,}000\text{ N})(184{,}800\text{ mm}^3)}{(30{,}442{,}500\text{ mm}^4)(2 \times 10\text{ mm})} = 136.6\text{ MPa}$$

Shear stress at points d–i

The shear stress at points d through i are calculated in the manner just demonstrated. For each point, a corresponding mirror point has been used to simplify and systematize the calculation process. The same values for the shear force V and the moment of inertia I are used in calculating shear stresses at each point in the cross section. Also, since the flanges and the webs are each 10 mm thick and since each area requires two cuts to isolate it, $t = 2 \times 10\text{ mm} = 20\text{ mm}$ at all locations. The results of the shear stress calculations for points d–i are summarized in the following table:

Point	Sketch	Calculation of $Q = \bar{y}'A'$	τ (MPa)
d	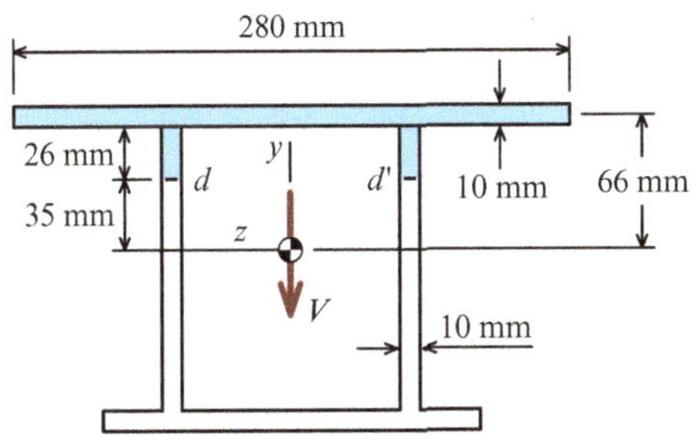	$(66\text{ mm})(280\text{ mm} \times 10\text{ mm})$ $+ 2 \times (48\text{ mm})(10\text{ mm} \times 26\text{ mm})$ $= 209{,}760\text{ mm}^3$	155.0

Point	Sketch	Calculation of $Q = \bar{y}'A'$	τ (MPa)
e	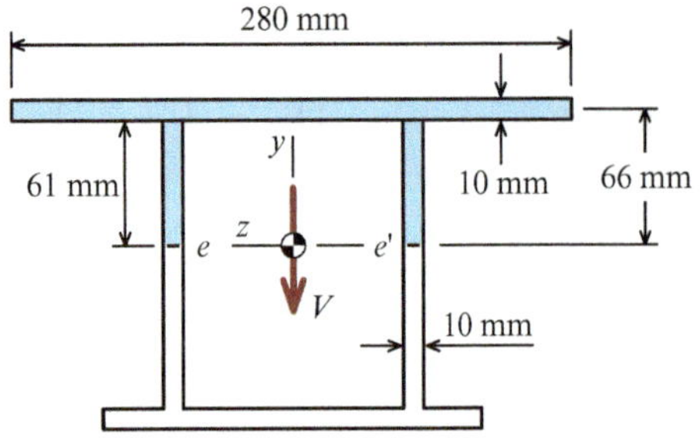	$(66\text{ mm})(280\text{ mm} \times 10\text{ mm}) + 2 \times (30.5\text{ mm})(10\text{ mm} \times 61\text{ mm}) = 222{,}010\text{ mm}^3$	164.1
f	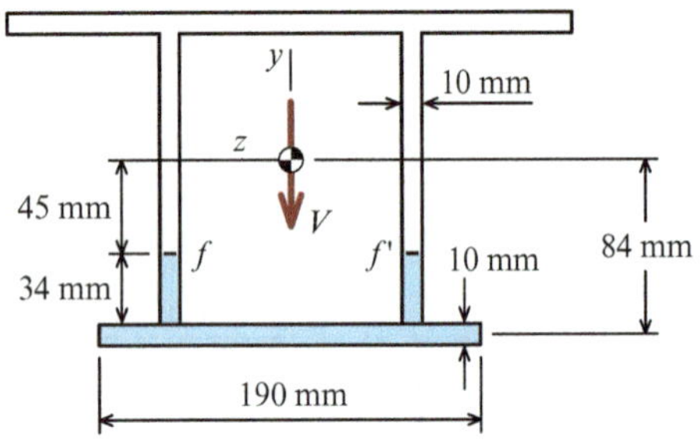	$(84\text{ mm})(190\text{ mm} \times 10\text{ mm}) + 2 \times (62\text{ mm})(10\text{ mm} \times 34\text{ mm}) = 201{,}760\text{ mm}^3$	149.1
g	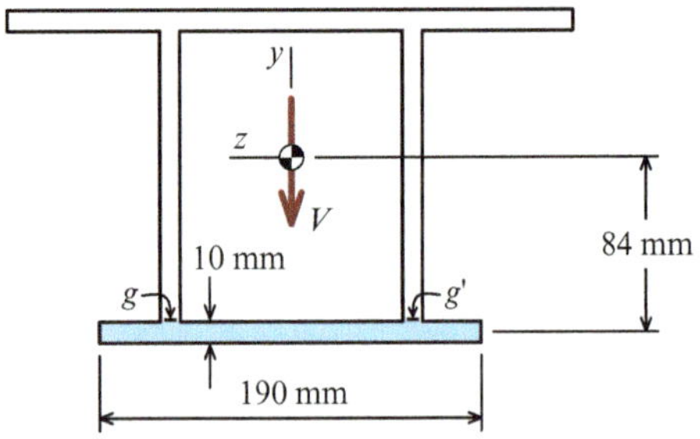	$(84\text{ mm})(190\text{ mm} \times 10\text{ mm}) = 159{,}600\text{ mm}^3$	118.0
h	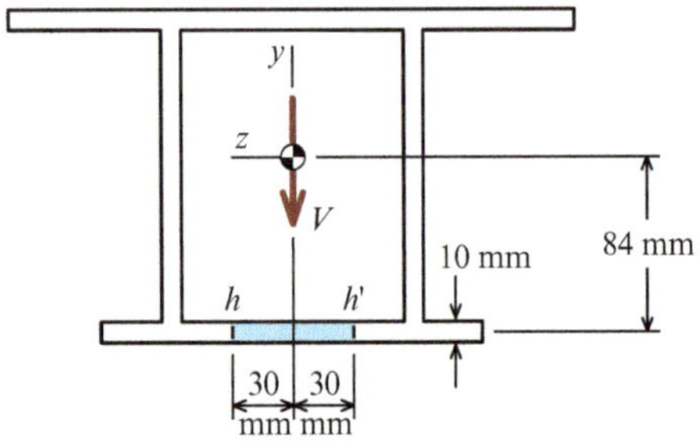	$2 \times (84\text{ mm})(30\text{ mm} \times 10\text{ mm}) = 50{,}400\text{ mm}^3$	37.3
i	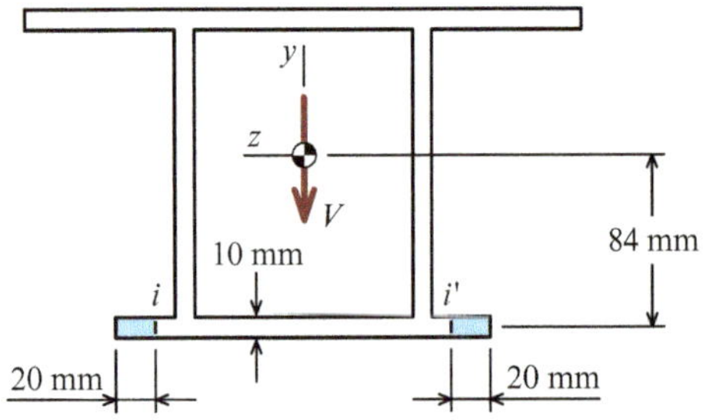	$2 \times (84\text{ mm})(20\text{ mm} \times 10\text{ mm}) = 33{,}600\text{ mm}^3$	24.8

(b) Distribution of shear stress
The distribution of shear stress and the direction of the shear flow is shown in the accompanying figure.

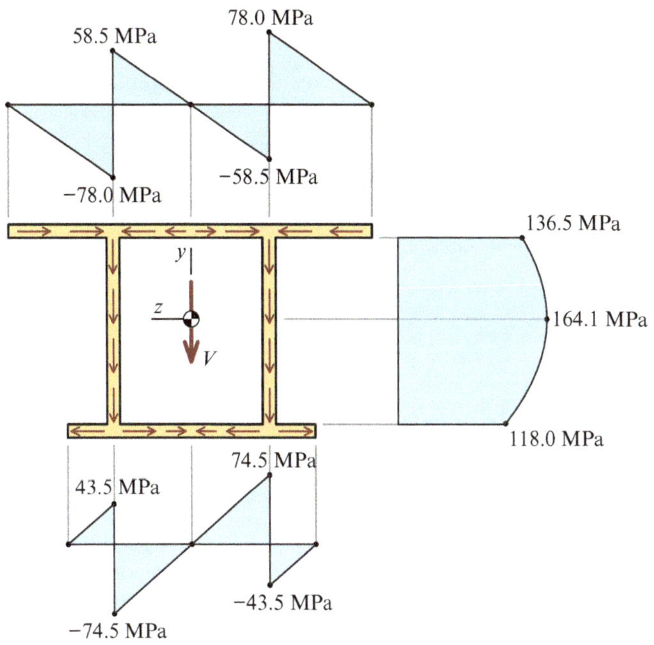

EXAMPLE 9.10

A 6061-T6 aluminum thin-walled tube is subjected to a vertical shear force $V = 21{,}000$ lb, as shown in the accompanying figure. The outside diameter of the tube is $D = 8.0$ in., and the inside diameter is $d = 7.5$ in. Plot the distribution of shear stress in the tube.

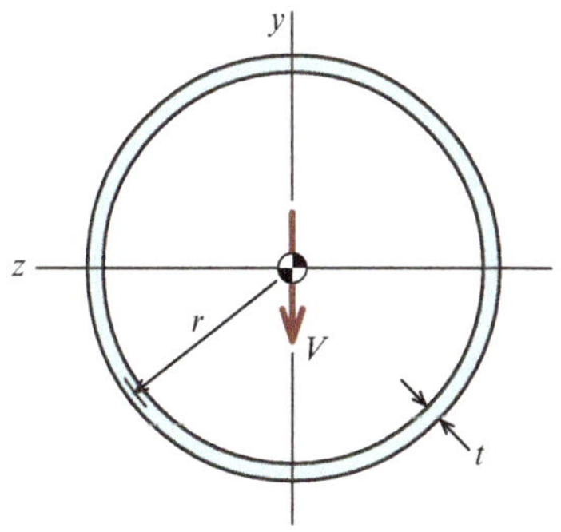

Plan the Solution

The shear stress distribution in the thin-walled tube will be calculated from the shear stress formula $\tau = VQ/It$. At the outset, an expression for the moment of inertia of a thin-walled tube will be derived. From the earlier discussion of shear stresses in closed thin-walled cross sections, the free-body diagram to be considered for the calculation of Q should be symmetric about the xy plane. On the basis of this free-body diagram, the first moment of area, Q, corresponding to an arbitrary location in the tube wall will be derived and the variation of shear stress will be determined.

SOLUTION

The shear stress in the tube will be determined from the shear stress formula $\tau = VQ/It$. The values for both I and Q can be determined by integration using polar coordinates. Since the tube is thin walled, the radius r of the tube is taken as the radius to the middle of the tube wall; therefore,

$$r = \frac{D + d}{4}$$

For a thin-walled tube, the radius r is much greater than the wall thickness t (i.e., $r \gg t$).

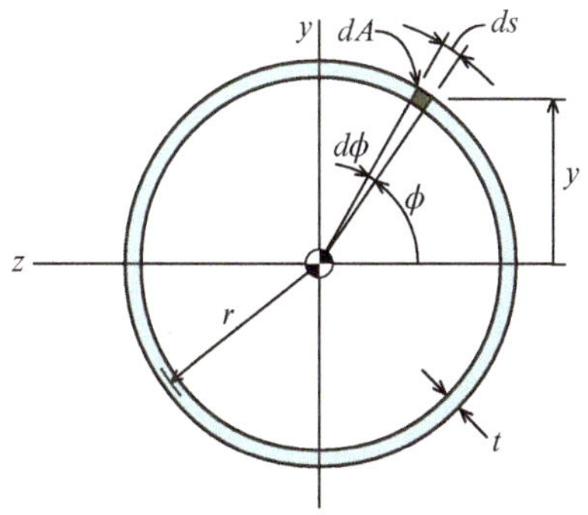

Moment of Inertia

From the sketch, observe that the distance y from the z axis to a differential area dA of the tube wall can be expressed as $y = r \sin \phi$. The differential area dA can be expressed as the product of the differential arclength ds and the tube thickness t; thus, $dA = t\,ds$. Furthermore, the differential arclength can be expressed as $ds = r\,d\phi$. As a result, the differential area can be expressed in polar coordinates r and ϕ as $dA = rt\,d\phi$. From these relationships for y and dA, the moment of inertia of the thin-walled tube can be derived as follows:

$$
\begin{aligned}
I_z &= \int y^2\, dA = \int_0^{2\pi} (r \sin\phi)^2 rt\, d\phi = r^3 t \int_0^{2\pi} \sin^2\phi\, d\phi \\
&= r^3 t\left[\frac{1}{2}\phi - \frac{1}{2}\sin\phi\,\cos\phi\right]_0^{2\pi} \\
&= \pi r^3 t
\end{aligned}
$$

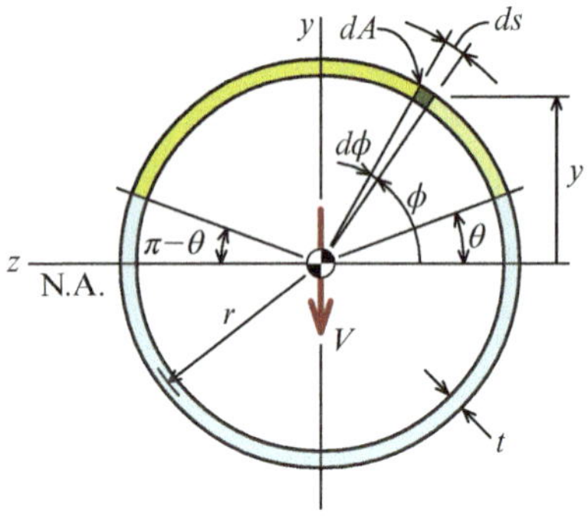

First Moment of Area, Q

The value of Q can also be determined by integration using polar coordinates. From the sketch on the left, the value of Q for the area of the cross section above the arbitrarily chosen sections defined by θ and $\pi - \theta$ will be determined. The free-body diagram to be considered for the calculation of Q should be symmetric about the xy plane.

From the definition of Q, the first moment of area dA about the neutral axis (N.A.) can be expressed as $dQ = y\,dA$. Substituting the previous expressions for y and dA into this definition gives the following expression for dQ in terms of r and ϕ:

$$dQ = y dA = (r \sin\phi) rt\, d\phi$$

The angle ϕ will vary between symmetric limits of θ and $\pi - \theta$. The following integration shows the derivation of a general expression for Q:

$$
\begin{aligned}
Q &= \int_\theta^{\pi-\theta} dQ = \int_\theta^{\pi-\theta} r^2 t \sin\phi\, d\phi \\
&= r^2 t[-\cos\phi]_\theta^{\pi-\theta} \\
&= 2r^2 t \cos\theta
\end{aligned}
$$

Shear Stress Expressions

The variation of the shear stress τ can now be expressed in terms of the angle θ:

$$\tau = \frac{VQ}{It} = \frac{V(2r^2 t\cos\theta)}{(\pi r^3 t)(2t)} = \frac{V}{\pi rt}\cos\theta$$

Note that the thickness term t in the shear stress equation is the total width of the surface exposed in cutting the free-body diagram. The free-body diagram considered between sections at θ and $\pi - \theta$ exposes a total width of two times the wall thickness; hence, the term $2t$ appears in the preceding shear stress equation.

For a thin-walled tube in which $r \gg t$, the cross-sectional area can be approximated by $A \cong 2\pi rt$. Thus, the shear stress can be expressed as

$$\tau = \frac{V}{A/2}\cos\theta = \frac{2V}{A}\cos\theta$$

and the maximum shear stress given by

$$\tau_{\max} = \frac{2V}{A}$$

at a value of $\theta = 0$.

Calculation of Shear Stress Distribution

The radius of the given aluminum tube is

$$r = \frac{D + d}{4} = \frac{8.00 \text{ in.} + 7.50 \text{ in.}}{4} = 3.875 \text{ in.}$$

Thus, the shear stress distribution is

$$\tau = \frac{V}{\pi r t}\cos\theta = \frac{21{,}000 \text{ lb}}{\pi(3.875 \text{ in.})(0.25 \text{ in.})}\cos\theta$$
$$= (6{,}900 \text{ psi})\cos\theta$$

The direction of the shear stress is shown in the next figure, along with a graph of the shear stress magnitude as a function of the angle θ:

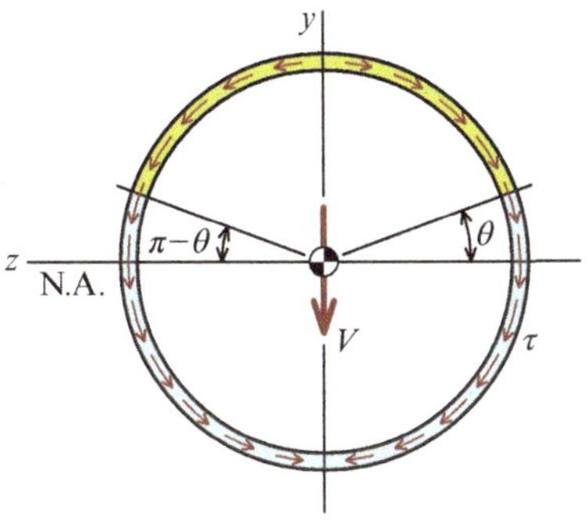

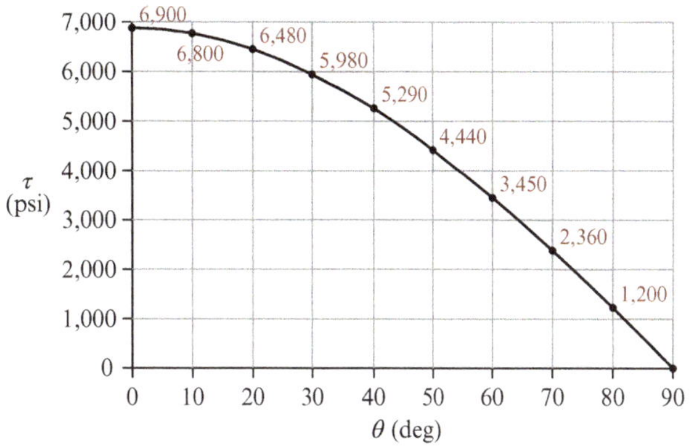

9.10 Shear Centers of Thin-Walled Open Sections

In Sections 8.1 through 8.3, the theory of bending was developed for prismatic beams. In deriving this theory, beams were assumed to have a longitudinal plane of symmetry (Figure 8.2*a*) and loads acting on the beam, as well as the resulting curvatures and deflections, were assumed to act only in the plane of bending. The only time that the requirement of symmetry was removed was in Section 8.8, where it was shown that the bending moment could be resolved into component moments about the principal axes of the cross section, provided that the loading was pure bending (i.e., no shear forces were present). However, unsymmetrical bending configurations in which shear forces were present were not considered.

If loads are applied in the plane of bending and the cross section is symmetric with respect to the plane of bending, twisting of the beam cannot occur. However, suppose that we consider bending of a beam that is (a) not symmetric with respect to the longitudinal plane of bending and (b) subjected to transverse shear forces in addition to bending moments. For beams such as this, the resultant of the shear stresses produced by the transverse loads will act in a plane that is parallel to, but offset from, the plane of loading. Whenever the resultant shear forces do not act in the plane of the applied loads, the beam will twist about its longitudinal axis in addition to bending about its neutral axis. Bending without twisting is possible, however, if the transverse loads pass through the **shear center**, which can be simply defined as the location (to the side of the longitudinal axis of the beam) where the transverse loads should be placed to avoid twisting of the cross section. In other words, transverse loads applied through the shear center cause no torsion of the beam.

Determination of the shear center location has important ramifications for beam design. Beam cross sections are generally configured to provide the greatest possible economy of material. As a result, they are frequently composed of thin plates arranged so that the resulting shape is strong in flexure. Wide-flange and channel shapes are designed with most of the material concentrated at the greatest practical distance from the neutral axis. This arrangement makes for an efficient flexural shape because most of the beam material is placed in the flanges, which are locations of high flexural stress. Less material is used in the web, which is near the neutral axis, where flexural stresses are low. The web serves primarily to carry shear force while also securing the flanges in position. An open cross section that is made up of thin plate elements may be strong in flexure, but it is extremely weak in torsion. If a beam twists as it bends, torsional shear stresses will be developed in the cross section, and generally, these shear stresses will be quite large in magnitude. For that reason, it is important for the beam designer to ensure that loads are applied in a manner that eliminates twisting of the beam. This aim can be accomplished when external loads act through the shear center of the cross section.

The exact location of the shear center for thick-walled unsymmetrical cross sections is difficult to obtain and is known only for a few cases.

The shear center of a cross section is always located on an axis of symmetry. The shear center for a beam cross section having two axes of symmetry coincides with the centroid of the section. For cross sections that are unsymmetrical about one axis or both axes, the shear center must be determined by computation or observation. The method of solution for thin-walled cross sections is conceptually simple: We will first assume that the beam cross section bends, but does not twist. On the basis of this assumption, the resultant internal shear forces in the thin-walled shape will be determined by a consideration of the shear flow produced in the shape. Equilibrium between the external load and the internal resultant forces must be maintained. From this requirement, the location of the external load necessary to satisfy equilibrium can then be computed.

Shear Center for a Channel Section

Consider the thin-walled channel shape used as a cantilever beam, as shown in Figure 9.24*a*. A vertical external load P that acts through the centroid of the cross section will

(*a*) Vertical load P acting through centroid

(*b*) Bending and twisting in response to the applied load

FIGURE 9.24 Bending and twisting of the cantilever beam.

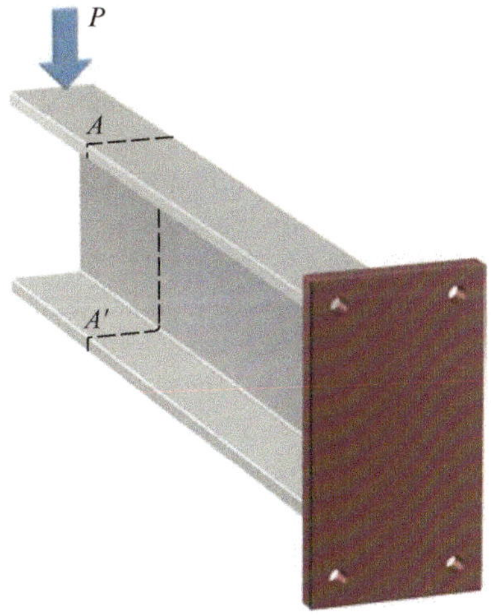

FIGURE 9.25 Rear view of cantilever beam.

cause the beam to both bend and twist, as depicted in Figure 9.24*b*. To better understand what causes the channel shape to twist, it is instructive to look at the internal shear flow produced in the beam in response to the applied load *P*.

The beam of Figure 9.24 is shown from the rear in Figure 9.25. The shear flow produced at Section *A–A′* in response to the external load *P* will be examined.

For the cantilever beam loaded as shown in Figure 9.26*a*, the upward internal shear force *V* must equal the downward external load *P*. The shear force *V* creates a shear flow *q* that acts in the web and the flanges in the directions shown in the figure.

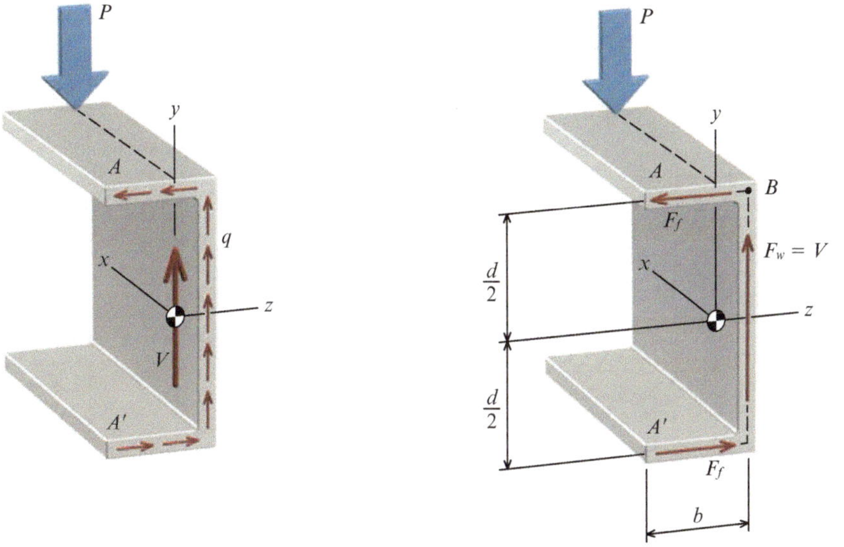

(*a*) Shear flow in the channel shape

(*b*) Resultant shear forces in the flanges and the web

FIGURE 9.26 Internal shear flow and resultant forces acting on Section *A–A′*.

The thickness of each flange is thin compared with the overall depth d of the channel shape; therefore, the vertical shearing force transmitted by each flange is small and can be neglected. (See Figure 9.16.) Consequently, the resultant shear force F_w determined by integrating the shear flow in the web must equal V. The resultant shear force F_f produced in each flange by the shear flow can be determined by integrating q over the width b of the channel flange. The directions of the resultant shear forces in the flanges and in the web are shown in Figure 9.26*b*. Since the forces F_f are equal in magnitude, but act in opposite directions, they form a couple that tends to twist the channel section about its longitudinal axis x. This couple, which arises from the resultant shear forces in the flanges, causes the channel to twist as it bends, as depicted in Figure 9.24*b*.

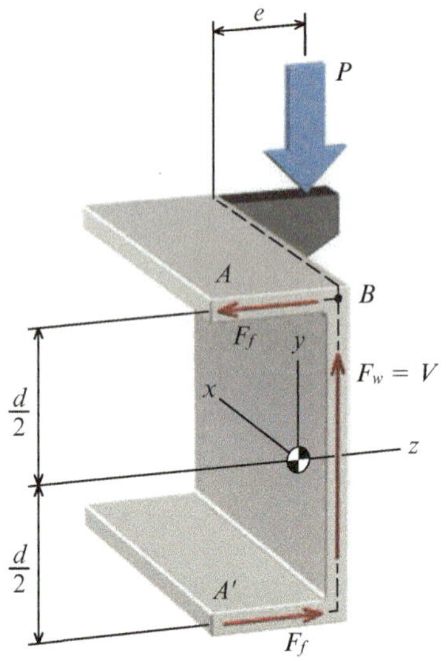

FIGURE 9.27 Shifting load P away from the centroid.

In Figure 9.27, the couple formed by the flange forces F_f causes the channel to twist in a counterclockwise direction. To counterbalance this twist, an equal clockwise torsional moment is required. A torsional moment can be produced by moving the external load P away from the centroid (i.e., to the right in Figure 9.27). Because there is moment equilibrium about point B (located at the top of the channel web), the beam will no longer have a tendency to twist when the clockwise moment Pe equals the counterclockwise moment $F_f\, d$. The distance e measured from the centerline of the channel web defines the location of the shear center O. Furthermore, the location of the shear center is solely a function of the cross-sectional geometry and dimensions, and does not depend upon the magnitude of the applied loading, as will be demonstrated in Example 9.11.

When the vertical external load P acts through the shear center O of the channel (Figure 9.28*a*), the cantilever beam bends without twisting (Figure 9.28*b*).

The shear center of a cross section is always located on an axis of symmetry. Thus, if the external load is applied in the horizontal direction through the centroid of the channel, as shown in Figure 9.29*a*, there is no tendency for the channel to twist as it bends (Figure 9.29*b*). The resultant shear forces in the flanges are equal in magnitude, and both act to oppose the applied load P. In the channel web, there

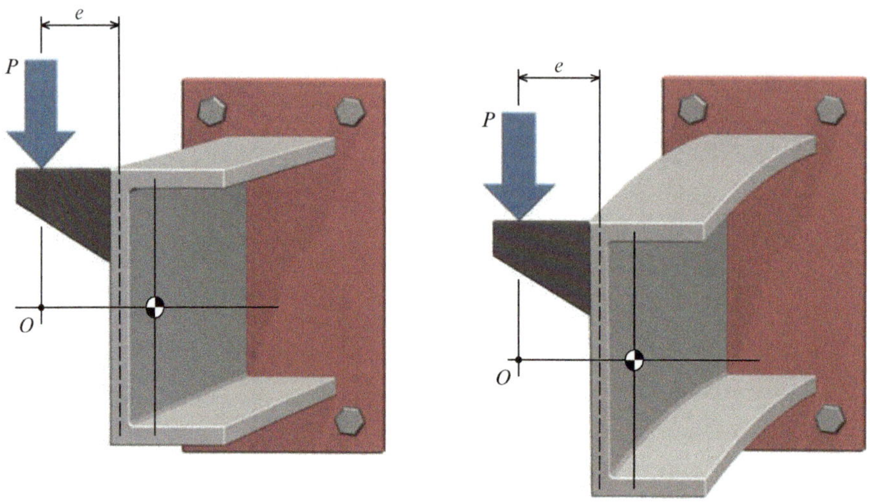

(*a*) External load P acting through the shear center O

(*b*) Bending without twisting in response to the applied load

FIGURE 9.28 Bending of the cantilever beam without twisting.

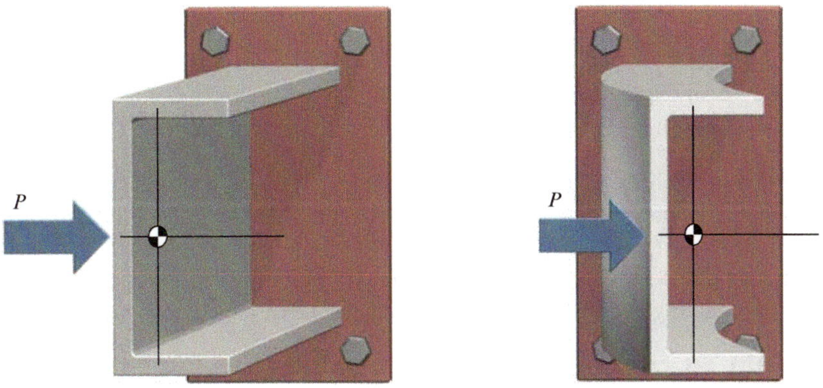

(*a*) External load *P* applied horizontally through the centroid (*b*) Cantilever beam bends without twisting

FIGURE 9.29 External load acting in a plane of symmetry.

are two equal resultant shear forces that act in opposite directions above and below the axis of symmetry.

In conclusion, as long as the external loads act through the shear center, the beam will bend without twisting. When this requirement is met, the stresses in the beam can be determined from the flexure formula.

Determination of the Shear Center Location

The location of the shear center for an unsymmetrical shape is computed as follows:

1. Determine how the shear "flows" in the various portions of the cross section.
2. Determine the distribution of shear flow q for each portion of the cross section from the shear flow equation $q = VQ/I$. Convert the shear flow into a force resultant by integrating q along the *length* of the cross-sectional element. The shear flow q will vary (*a*) linearly in elements that are perpendicular to the direction of the internal shear force V and (*b*) parabolically in elements that are parallel to or inclined toward the direction of V.
3. Alternatively, determine the distribution of shear stress τ from the shear stress equation $\tau = VQ/It$ and convert the shear stress into a force resultant by integrating τ over the *area* of the cross-sectional element.
4. Sketch the shear force resultants that act in each element of the cross section.
5. Determine the shear center location by summing moments about an arbitrary point (e.g., point B) on the cross section. Choose a convenient location for point B—one that eliminates as many force resultants from the moment equilibrium equation as possible.
6. Study the direction of rotation of the shear forces, and place the external force P at an eccentricity e from point B so that the direction of the moment Pe is opposite to that caused by the resultant shear forces.
7. Sum moments about point B, and solve for the eccentricity e.
8. If the cross section has an axis of symmetry, then the shear center lies at the point where this axis intersects the line of action of the external load. If the shape has no axes of symmetry, then rotate the cross section 90° and repeat the process to obtain another line of action for the external loads. The shear center lies at the intersection of these two lines.

EXAMPLE 9.11

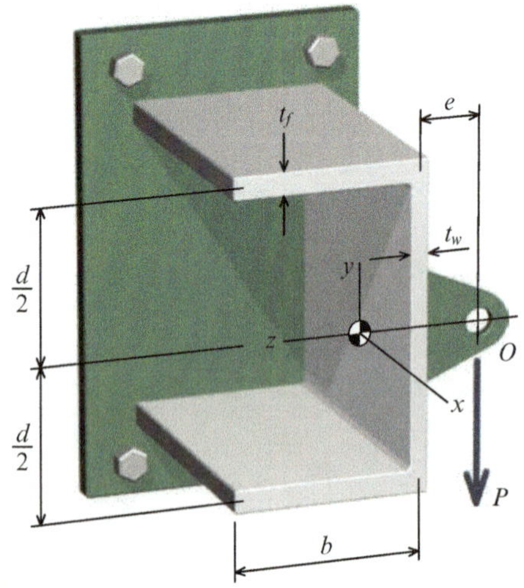

Derive an expression for the location of the shear center O for the channel shape shown.

Plan the Solution

From the concept of shear flow, the horizontal shear force produced in each channel flange will be determined. The twisting moment produced by these forces will be counteracted by the moment produced by the vertical external load P acting at a distance e from the centerline of the channel web.

SOLUTION

Since the applied load P is assumed to act at the shear center O, the channel shape will bend about the z axis (i.e., the neutral axis), but it will not twist about the x axis. To better understand both the forces that cause twisting in the thin-walled channel and the forces that counteract this twisting tendency, consider the rear face of the channel cross section.

The internal shear force V creates a shear flow q in the web and in the flanges. This shear flow is expressed by

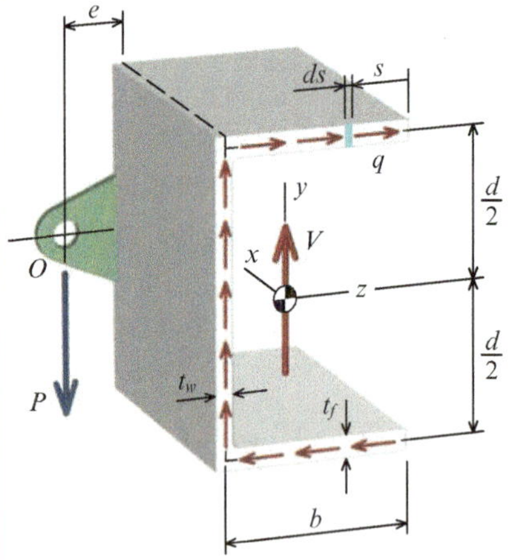

$$q = \frac{VQ}{I_z}$$

The shear force V and the shear flow q act in the directions shown in the figure to the left. The shear flow at any location in the thin-walled shape depends only on the first moment of area, Q.

Now, consider the shear flow in the upper flange. This shear flow acts in the shaded area, which is located a horizontal distance s from the tip of the flange. The shear flow acting at s can be calculated as

$$q = \frac{VQ}{I_z} = \frac{V}{I_z}\left(st_f\frac{d}{2}\right) = \frac{Vdt_f}{2I_z}s \tag{9.18}$$

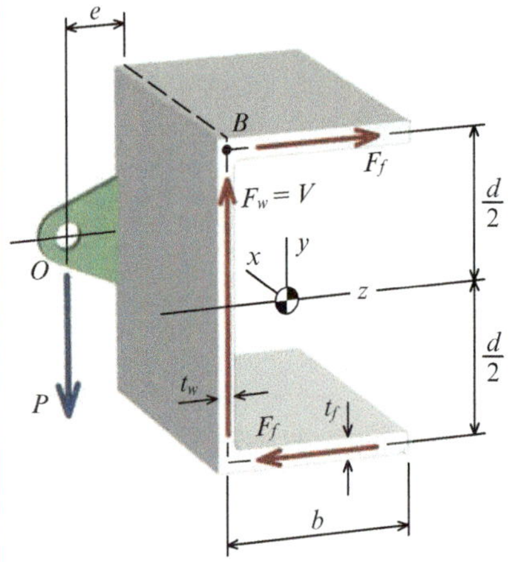

Notice that the magnitude of the shear flow varies linearly from the free surface at the flange tip, where $s = 0$, to a maximum value at the web, where $s = b$. The total horizontal force acting on the upper flange is determined by integrating the shear flow over the width of the top flange:

$$F_f = \int q\,ds = \int_0^b \frac{Vt_fd}{2I_z}s\,ds = \frac{Vb^2dt_f}{4I_z} \tag{a}$$

The force F_f in the lower flange will be the same magnitude as that in the upper flange; however, it will act in the opposite direction, thus maintaining equilibrium in the z direction. The couple created by the flange forces F_f tends to twist the channel shape in a clockwise direction, as shown in the figure in the accompanying figure.

The thickness t_f of each flange is thin compared with the overall depth d of the channel shape; therefore, the vertical shearing force transmitted by each flange is small and can be neglected. (See Figure 9.16.) Consequently, the resultant force F_w of the shear flow in the web must equal V. Moreover, the upward internal shear force V must equal the downward external load P to satisfy equilibrium in the y direction; hence, $P = V$.

The forces P and V, which are separated by a distance e, create a couple that tends to twist the channel shape in a counterclockwise direction. A moment equilibrium equation about point B can thus be written as

$$M_B = -F_f d + Pe = 0$$

In this equation, substitute $P = V$ and replace F_f with the expression derived in Equation (a) to get

$$Ve = \left(\frac{Vb^2 d t_f}{4I_z}\right) d$$

and then solve for e:

$$e = \frac{b^2 d^2 t_f}{4I_z} \qquad (9.19) \qquad \textbf{Ans.}$$

The distance e from the centerline of the channel web defines the location of the shear center O. Notice that the shear center location is dependent only on the dimensions and geometry of the cross section.

EXAMPLE 9.12

For the channel shape of Example 9.11, assume that $d = 8.00$ in., $b = 3.00$ in., $t_f = 0.125$ in., and $t_w = 0.125$ in. Determine the distribution of shear stress produced in the channel if a load $P = 900$ lb is applied at the shear center.

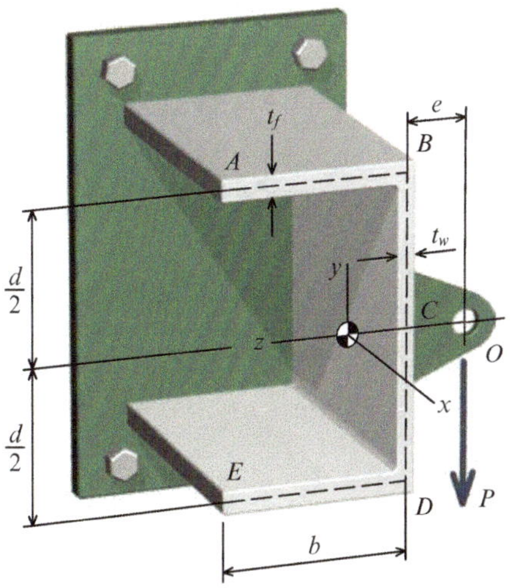

Plan the Solution

The moment of inertia of the thin-walled channel shape will be determined. The shear stress produced in each channel flange is linearly distributed; thus, only the maximum value, which occurs at points B and D, will need to be determined. The distribution of shear stress in the flange is parabolically distributed, with its minimum value occurring at points B and D, and its maximum value occurring at point C.

SOLUTION

Moment of Inertia

The moment of inertia for the channel shape can be expressed by the following equation:

$$I_z = \frac{t_w d^3}{12} + 2\left[\frac{b t_f^3}{12} + \left(\frac{d}{2}\right)^2 b t_f\right]$$

Note that, since the shape is thin walled, the centerline dimensions can be used in this calculation. Furthermore, the term containing t_f^3 can be neglected because it is very small; thus, the moment of inertia is calculated as

$$\begin{aligned} I_z &= \frac{t_w d^3}{12} + \frac{t_f b d^2}{2} \\ &= \frac{(0.125\text{ in.})(8.00\text{ in.})^3}{12} + \frac{(0.125\text{ in.})(3.00\text{ in.})(8.00\text{ in.})^2}{2} \\ &= 17.333\text{ in.}^4 \end{aligned}$$

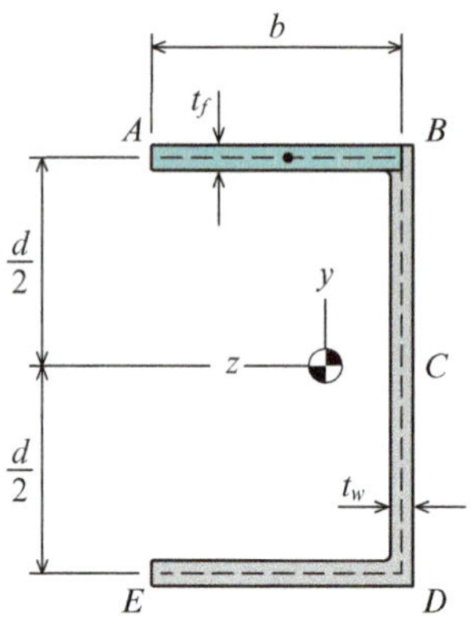

Shear Stress in the Flanges

The shear stress in the flanges will be distributed linearly, from zero at the flange tips (i.e., *A* and *E*) to a maximum value at the junction of the flange and the web (i.e., *B* and *D*). The first moment of area for point *B* can be calculated as

$$\begin{aligned} Q_B &= (bt_f)\frac{d}{2} \\ &= (3.00\text{ in.})(0.125\text{ in.})(4.00\text{ in.}) = 1.50\text{ in.}^3 \end{aligned}$$

and the shear stress at point *B* is thus

$$\begin{aligned} \tau_B &= \frac{VQ_B}{I_z t_f} \\ &= \frac{(900\text{ lb})(1.50\text{ in.}^3)}{(17.333\text{ in.}^4)(0.125\text{ in.})} = 623\text{ psi} \end{aligned}$$

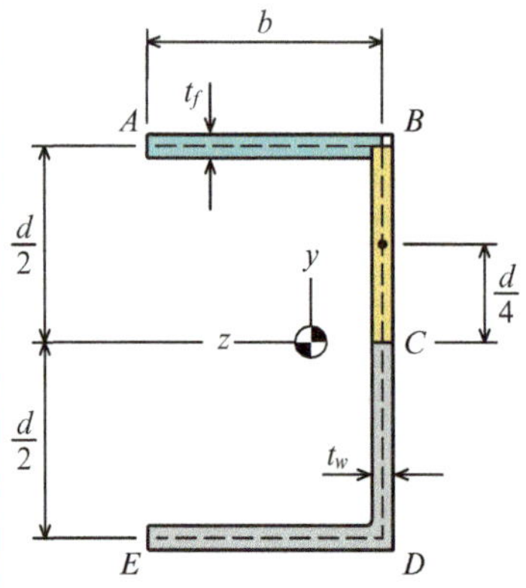

Shear Stress in the Web

The shear stress in the web will be distributed parabolically, from minimum values at points *B* and *D* to its maximum value at point *C*. The shear stress at point *B* in the web is

$$\begin{aligned} \tau_B &= \frac{VQ_B}{I_z t_w} \\ &= \frac{(900\text{ lb})(1.50\text{ in.}^3)}{(17.333\text{ in.}^4)(0.125\text{ in.})} = 623\text{ psi} \end{aligned}$$

The first moment of area for point *C* can be calculated as

$$\begin{aligned} Q_c &= (bt_f)\frac{d}{2} + \left(t_w \frac{d}{2}\right)\frac{d}{4} \\ &= 1.50\text{ in.}^3 + (0.125\text{ in.})(4.00\text{ in.})(2.00\text{ in.}) = 2.50\text{ in.}^3 \end{aligned}$$

and the shear stress at point *C* is

$$\begin{aligned} \tau_c &= \frac{VQ_c}{I_z t_w} \\ &= \frac{(900\text{ lb})(2.50\text{ in.}^3)}{(17.333\text{ in.}^4)(0.125\text{ in.})} = 1{,}038\text{ psi} \end{aligned}$$

Distribution of Shear Stress

The distribution of shear stress over the entire channel shape is plotted in the accompanying figure.

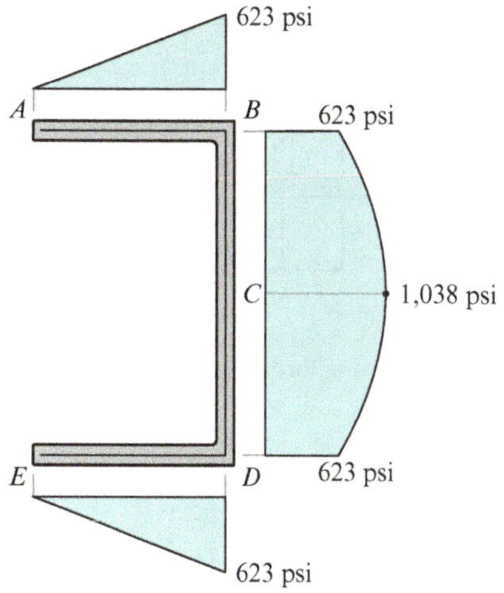

EXAMPLE 9.13

Consider the channel shape of Examples 9.11 and 9.12, shown again here. Neglecting stress concentrations, determine the maximum shear stress created in the shape if the load $P = 900$ lb is applied at the centroid of the section, which is located 0.643 in. to the left of the web centerline.

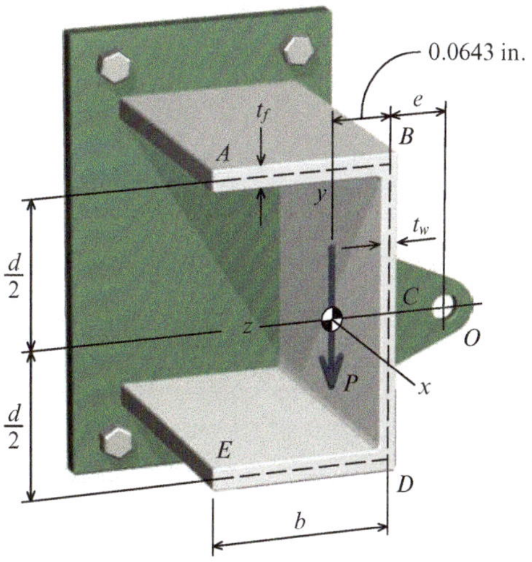

Plan the Solution

This example illustrates the considerable additional shear stress created in the channel when the external load does not act through the shear center. The distance from the channel centroid to the shear center O will be calculated and used to determine the magnitude of the torque that acts on the section. The shear stress created by this torque will be calculated from Equation (6.25). The total shear stress will be the sum of the shear stress due to bending, as determined in Equation 9.12, and the shear stress due to twisting.

SOLUTION

Shear Center

From Equation (9.19), the location of the shear center O for the channel is calculated as

$$e = \frac{b^2 d^2 t_f}{4 I_z} = \frac{(3.00 \text{ in.})^2 (8.00 \text{ in.})^2 (0.125 \text{ in.})}{4(17.333 \text{ in.}^4)} = 1.038 \text{ in.}$$

Equivalent Loading

We know that the channel will bend without twisting if load P is applied at the shear center O, and furthermore, we know how to determine the shear stresses in the channel shape for a load applied at the shear center. Therefore, it will be valuable to determine an equivalent loading that acts at the shear center. This equivalent loading will enable us to separate the loading into components that cause (*a*) bending and (*b*) torsion.

The actual load acts through the centroid, as shown in Figure (*a*). The equivalent load at the shear center consists of a force and a concentrated moment, as shown in Figure (*b*). The equivalent force at O is simply equal to the applied load P. The concentrated moment will be a torque of magnitude

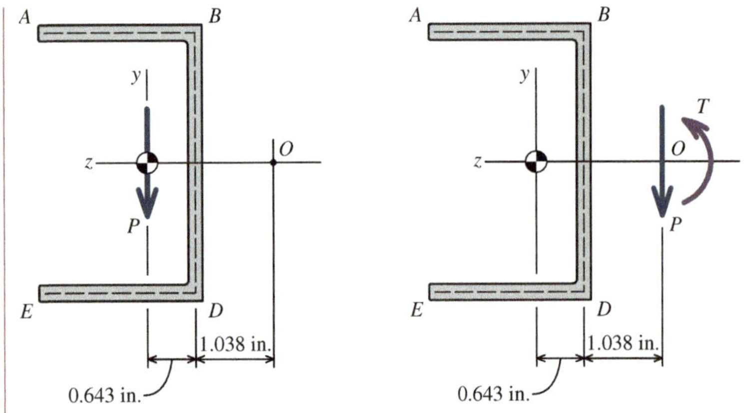

(*a*) Load acting through centroid

(*b*) Equivalent loading at shear center

$$T = (900 \text{ lb})(0.643 \text{ in.} + 1.038 \text{ in.}) = 1{,}513 \text{ lb}\cdot\text{in.}$$

Shear Stress due to Bending

The maximum shear stress due to bending caused by the 900 lb load was determined in Example 9.12. The flow of the shear stress is shown in Figure (*c*). Recall that the maximum shear stress due to this load occurred at the horizontal axis of symmetry and had a value

$$\tau_c = 1{,}038 \text{ psi}$$

Shear Stress due to Torsion

The torque T causes the member to twist, and the shear stress is greatest along the edges of the cross section. Recall that torsion of noncircular sections—particularly, narrow rectangular cross sections—was discussed in Section 6.11. That discussion revealed that the maximum shear stress and the shear stress distribution for a member of uniform thickness and arbitrary shape is equivalent to that of a rectangular bar with a large aspect ratio. (See Figure 6.20.) For the channel shape considered here, the shear stress can be calculated from Equation (6.25):

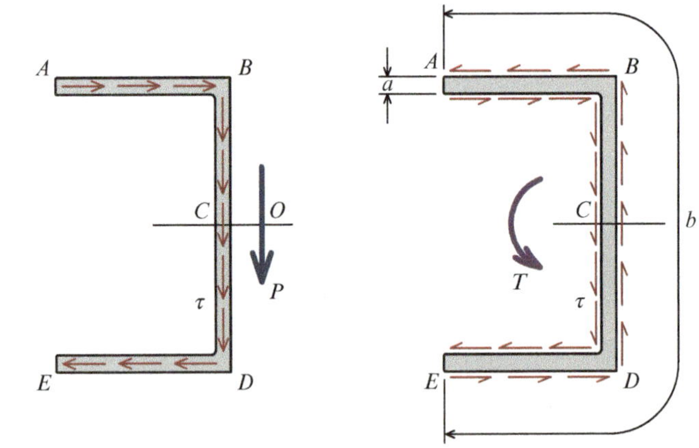

(*c*) Shear stress due to bending

(*d*) Shear stress due to torsion

$$a = 0.125 \text{ in.}$$

$$b = 3.00 \text{ in.} + 8.00 \text{ in.} + 3.00 \text{ in.} = 14.00 \text{ in.}$$

$$\tau_{max} = \frac{3T}{a^2 b} = \frac{3(1{,}513 \text{ lb}\cdot\text{in.})}{(0.125 \text{ in.})^2(14.00 \text{ in.})} = 20{,}750 \text{ psi}$$

Maximum Combined Shear Stress

The maximum stress due to the combined bending and twisting occurs at the neutral axis (i.e., point C) on the inside surface of the web. The value of this combined shear stress is

$$\tau_{max} = \tau_{bend} + \tau_{twist} = 1{,}038 \text{ psi} + 20{,}750 \text{ psi} = 21{,}788 \text{ psi} \qquad \textbf{Ans.}$$

EXAMPLE 9.14

Find the shear center O of the semicircular thin-walled cross section shown.

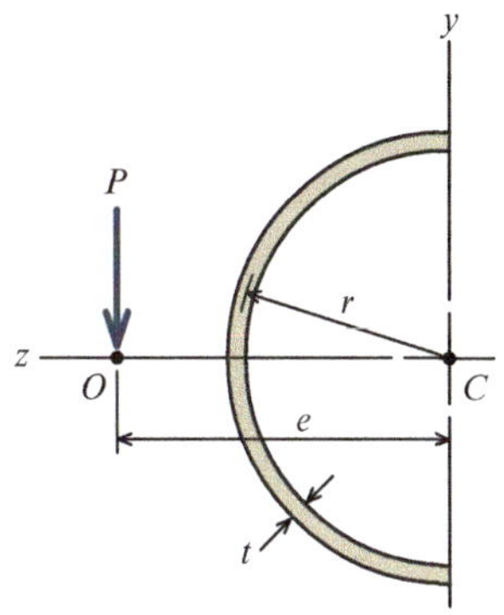

Plan the Solution

Shear stresses are created in the wall of the semicircular cross section in response to the applied load P. The moment produced by these shear stresses about the center C of the thin-walled cross section must equal the moment of the load P about center C if the section is to bend without twisting. We will develop an expression for the differential moment dM acting on an area dA of the wall. Then, we will integrate dM to determine the total twisting moment produced by the shear stresses and equate that expression to the moment created by the external load P acting at the shear center O. From this resulting equation, the location of the shear center O can be derived.

SOLUTION

Moment of Inertia

From the sketch, observe that the distance y from the z axis to a differential area dA of the wall can be expressed as $y = r\cos\phi$. The differential area dA can be expressed as the product of the differential arclength ds and the thickness t; thus, $dA = t\,ds$. Furthermore, the differential arclength can be expressed as $ds = r\,d\phi$. As a result, the differential area can be expressed in polar coordinates r and ϕ as $dA = r\,t\,d\phi$. From these relationships for y and dA, the moment of inertia of the semicircular thin-walled cross section can be derived as follows:

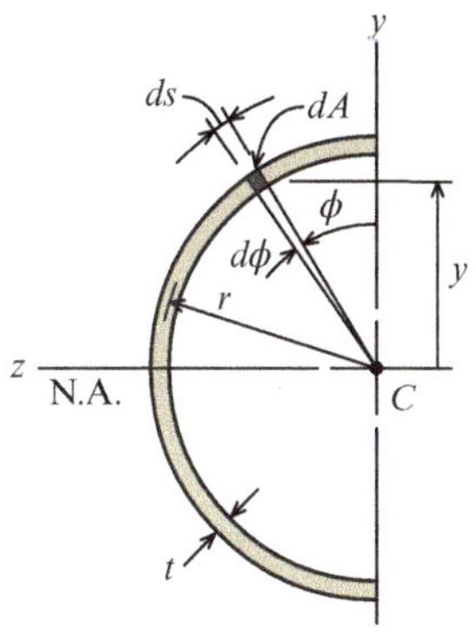

$$\begin{aligned} I_z &= \int y^2\,dA = \int_0^{\pi} (r\cos\phi)^2 rt\,d\phi = r^3 t\int_0^{\pi}\cos^2\phi\,d\phi \\ &= r^3 t\left[\frac{1}{2}\phi + \frac{1}{2}\sin\phi\cos\phi\right]_0^{\pi} \\ &= \frac{\pi r^3 t}{2} \end{aligned}$$

First Moment of Area, Q

The value of Q can also be determined by integration in polar coordinates. From the accompanying sketch, the value of Q for the area of the cross section above an arbitrarily chosen angle θ is to be determined.

From the definition of Q, the first moment of area dA about the neutral axis (N.A.) can be expressed as $dQ = y\,dA$. Substituting the previous expressions for y and dA into this equation gives the following equation for dQ in terms of r and ϕ:

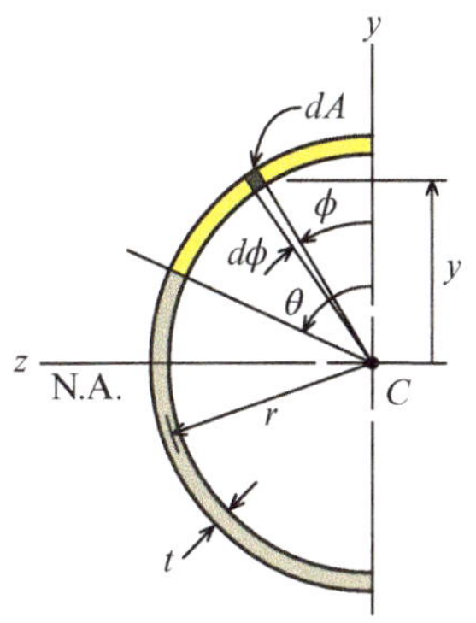

$$dQ = y\,dA = (r\cos\phi)rt\,d\phi$$

Integrating dQ between $\phi = 0$ and $\phi = \theta$ gives a general expression for Q:

$$\begin{aligned} Q &= \int_0^{\theta} dQ = \int_0^{\theta} r^2 t\cos\phi\,d\phi \\ &= r^2 t[\sin\phi]_0^{\theta} \\ &= r^2 t\sin\theta \end{aligned}$$

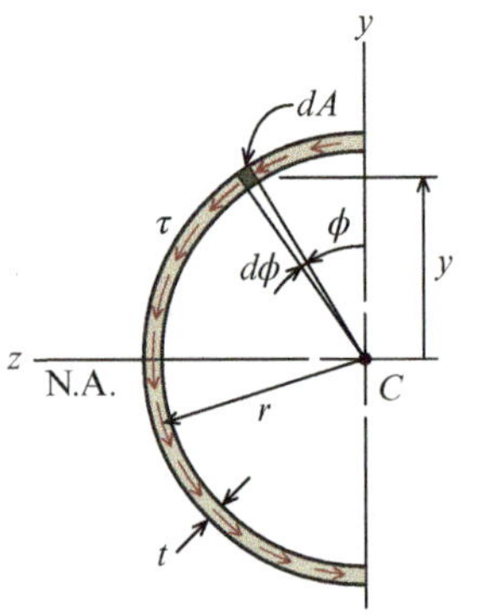

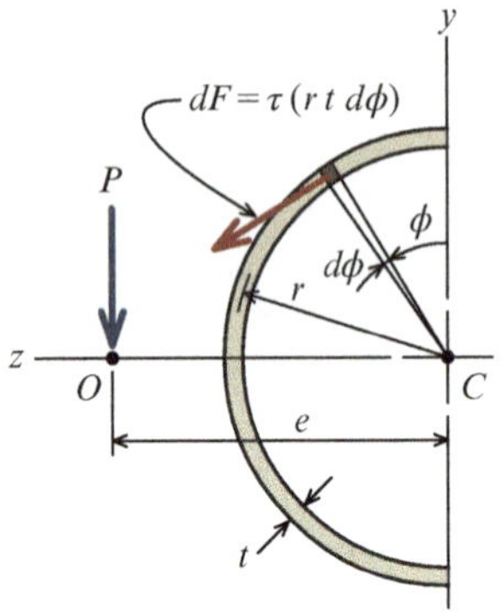

Shear Stress

The variation of the shear stress τ can now be expressed in terms of the angle ϕ:

$$\tau = \frac{VQ}{It} = \frac{V(r^2 t \sin\phi)}{\left(\frac{\pi r^3 t}{2}\right)t} = \frac{2V}{\pi r t}\sin\phi$$

Moments about C

The resultant force dF acting on the element of area dA is expressed as $dF = \tau dA = \tau\,(r\,t\,d\phi)$, or

$$dF = \frac{2rtV}{\pi r t}\sin\phi\, d\phi = \frac{2V}{\pi}\sin\phi\, d\phi$$

The moment of dF about point C is

$$dM_C = r\,dF = \frac{2rV}{\pi}\sin\phi\, d\phi$$

Integrating this expression between $\phi = 0$ and $\phi = \pi$ gives the moment produced by the shear stresses:

$$M_C = \int dM_C = \int_0^{\pi} \frac{2rV}{\pi}\sin\phi\, d\phi = \frac{4rV}{\pi}$$

To satisfy moment equilibrium, the moment M_C of the shear stress τ about the center C of the thin-walled cross section must equal the moment of the load P about that same point:

$$M_C = Pe$$

The resultant of the shear stress is the shear force V, which must equal the applied load P to satisfy vertical equilibrium. It follows that the distance e to the shear center is

$$e = \frac{M_C}{P} = \frac{M_C}{V} = \frac{4r}{\pi} \cong 1.27r \qquad \textbf{Ans.}$$

This result shows that the shear center O is located outside of the semicircular cross section.

Sections Consisting of Two Intersecting Thin Rectangles

Next, we will consider thin-walled open sections made up of two intersecting rectangles. Consider an equal-leg angle section, such as that shown in Figure 9.30. When a vertical

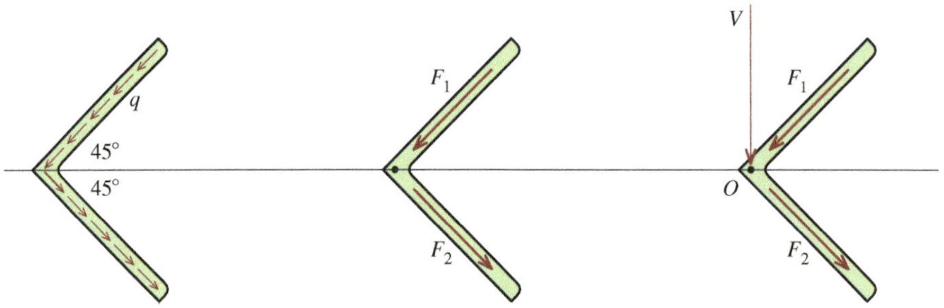

(*a*) Shear flow (*b*) Resultant shear forces (*c*) Shear center

FIGURE 9.30 Shear center of equal-leg angle shapes.

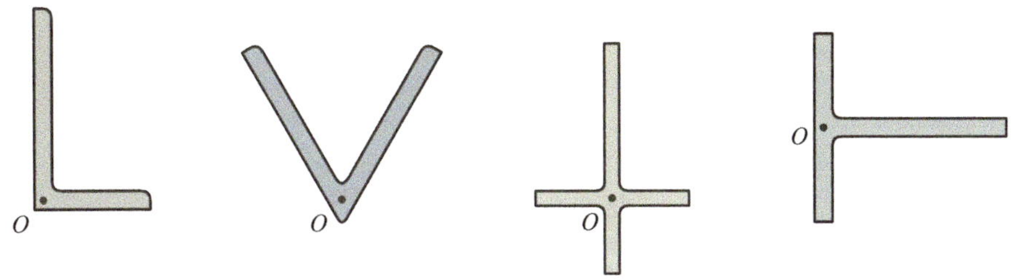

FIGURE 9.31 Various cross sections, each consisting of two thin rectangles.

shear force V is applied to the cross section, the shear flow q is directed along the centerline of each leg, parallel to the walls of the angle shape, as shown in Figure 9.30*a*. The resultant shear forces in the two legs are F_1 and F_2, as shown in Figure 9.30*b*. Horizontal equilibrium must be satisfied; therefore, the sum of the horizontal force components of F_1 and F_2 must be zero. Accordingly, forces F_1 and F_2 must be equal in magnitude. Moreover, the sum of the vertical force components of F_1 and F_2 must equal the vertical shear force acting in the beam.

Given that transverse loads applied through the shear center cause no torsion of the beam, where must a vertical load be placed so that the beam will not twist? The load must be placed at the point of intersection of forces F_1 and F_2. The intersection of the centerlines of the two legs must be the shear center, since the sum of the moments of forces F_1, F_2, and V about point O is zero.

A similar line of reasoning is applicable to all cross sections consisting of two intersecting thin rectangles, such as those shown in Figure 9.31. In each case, the resultant shear force must act along the centerline of the rectangle. Consequently, the point of intersection of the two centerlines defines the location of the shear center O.

PROBLEMS

P9.34 The tee shape shown in Figure P9.34 is constructed from two dimension lumber boards that are rigidly attached to each other. A vertical shear force V = 1,350 lb acts on the cross section. The dimensions of the tee shape are b_f = 7.25 in., t_f = 1.50 in., d = 10.75 in., and t_w = 1.50 in. Determine the shear flow magnitude and the shear stress magnitude that act in the tee shape at section 1–1.

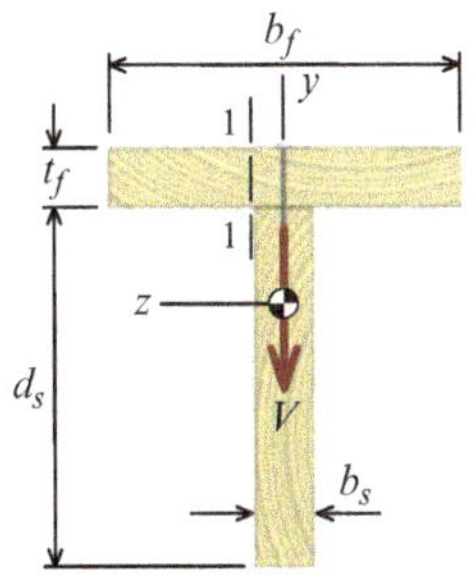

FIGURE P9.34

P9.35 The wide-flange shape shown in Figure P9.35 has dimensions b_f = 7.25 in., t_f = 0.8 in., d = 18.5 in., and t_w = 0.5 in. A vertical shear force V = 47,000 lb acts on the cross section. Determine the shear flow magnitude and the shear stress magnitude that act in the shape

(a) at section 1–1.
(b) at section 2–2.
(c) at section 3–3, where a = 1.75 in.

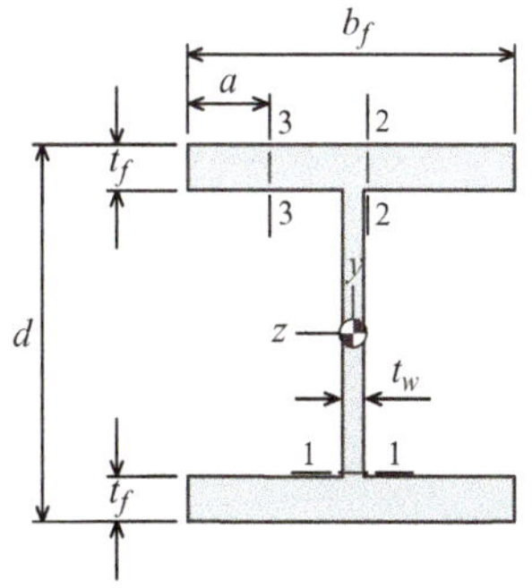

FIGURE P9.35

P9.36 A plastic extrusion has the shape shown in Figure P9.36, where $b_1 = 80$ mm, $b_2 = 50$ mm, $d = 65$ mm, and $t = 2$ mm. A vertical shear force $V = 600$ N acts on the cross section. Determine the shear flow magnitude and the shear stress magnitude that act in the shape

(a) at section 1–1.
(b) at section 2–2.
(c) at section 3–3 for $a = 15$ mm.

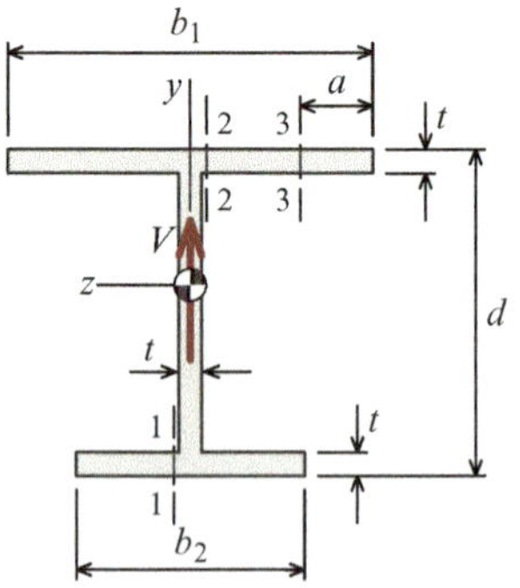

FIGURE P9.36

P9.37 Three plates are welded together to form the section shown in Figure P9.37. The dimensions of the cross section are $b = 4$ in., $d = 12$ in., $a = 3$ in., and $t = 0.5$ in. For a vertical shear force $V = 35$ kips, determine the shear flow through the welded surface at B.

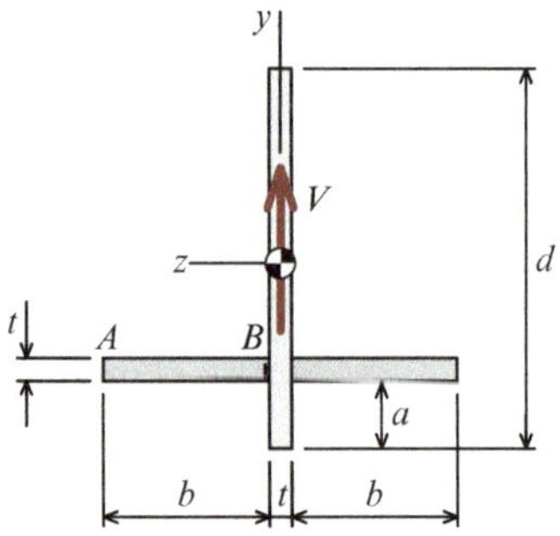

FIGURE P9.37

P9.38 An aluminum extrusion has the shape shown in Figure P9.38, where $b = 36$ mm, $d = 50$ mm, $a = 12$ mm, and $t = 3$ mm. A vertical shear force $V = 1{,}700$ N acts on the cross section. Determine the shear flow that acts in the shape at points A–D.

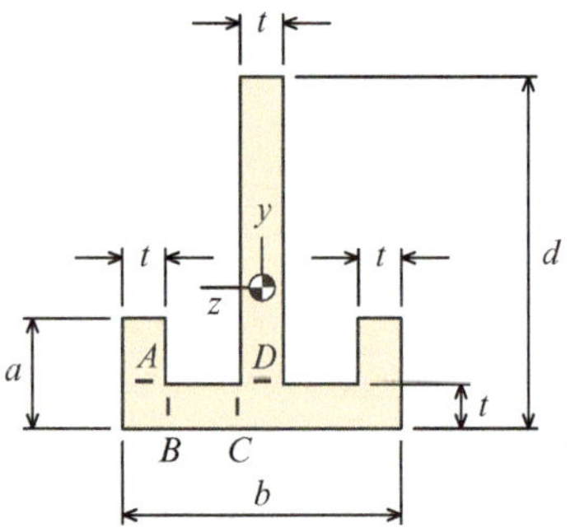

FIGURE P9.38

P9.39 An aluminum extrusion has the shape shown in Figure P9.39, where $b = 60$ mm, $d = 90$ mm, $a = 20$ mm, and $t = 3$ mm. A vertical shear force $V = 13$ kN acts on the cross section. Determine the shear stress in the shape

(a) at point A.
(b) at section 1–1.
(c) at section 2–2.
(d) at section 3–3.
(e) at section 4–4.
(f) at the neutral axis.

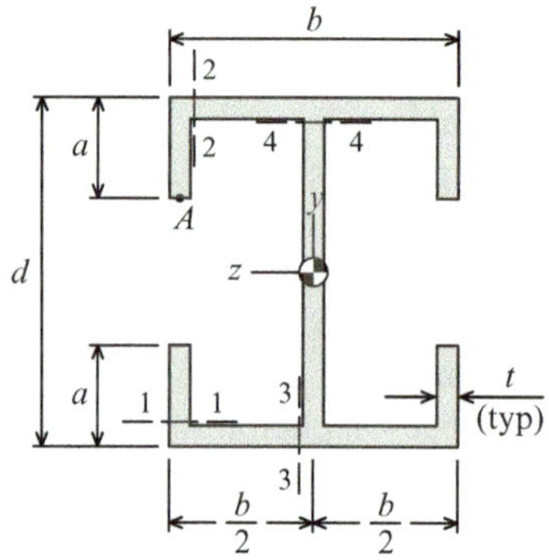

FIGURE P9.39

P9.40 The U shape shown in Figure P9.40 has dimensions $b = 250$ mm, $d = 120$ mm, $t_f = 19$ mm, and $t_w = 11$ mm. A vertical shear force $V = 130$ kN acts on the cross section. Determine the shear stress at points A–E. Assume that $a = 80$ mm.

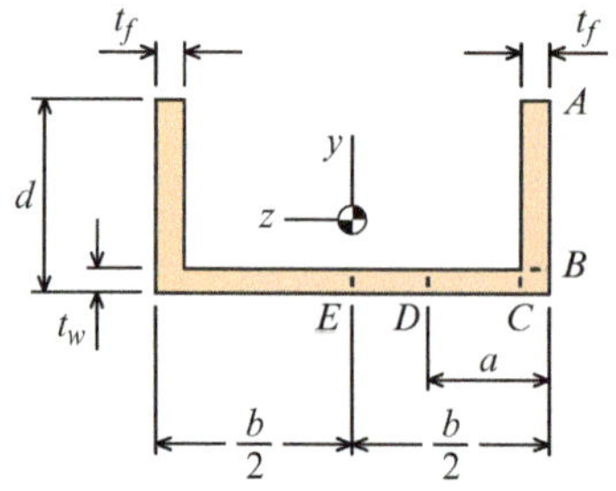

FIGURE P9.40

P9.41 The channel shape shown in Figure P9.41 has dimensions $b = 6.0$ in., $d = 3.0$ in., $a = 1.5$ in., $c = 2.0$ in., and $t = 0.25$ in. A vertical shear force $V = 47{,}000$ lb acts on the cross section. Determine

(a) the shear flow that acts in the shape at points A–G.
(b) the resultant horizontal force that acts between points A and B.

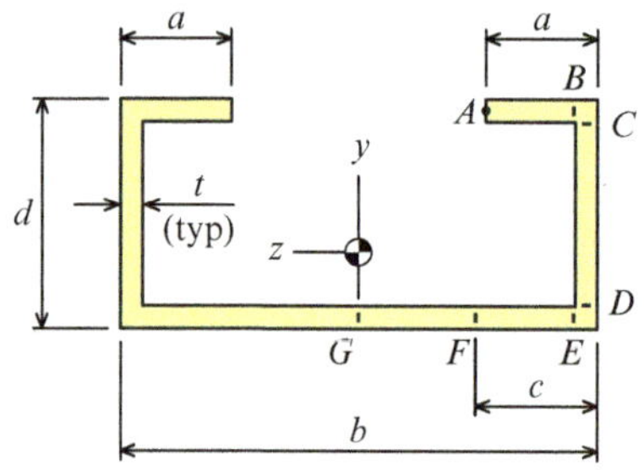

FIGURE P9.41

P9.42 The hat-shape extrusion shown in Figure P9.42 has dimensions $b = 50$ mm, $d = 40$ mm, $a = 30$ mm, and $t = 2$ mm. A vertical shear force $V = 600$ N acts on the cross section. Determine

(a) the shear stress at points A–D.
(b) the resultant horizontal force that acts between points A and B.

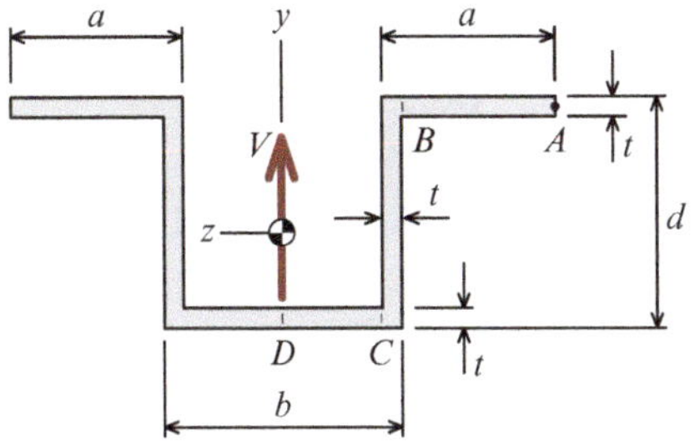

FIGURE P9.42

P9.43 Sheet metal is bent into the shape shown in Figure P9.43 and is then used as a beam. The shape has dimensions $b = 4.5$ in., $d = 7.0$ in., $a = 1.5$ in., and $t = 0.14$ in. A vertical shear force $V = 6{,}200$ lb acts on the cross section. Determine the shear stress that acts in the shape at points A–G.

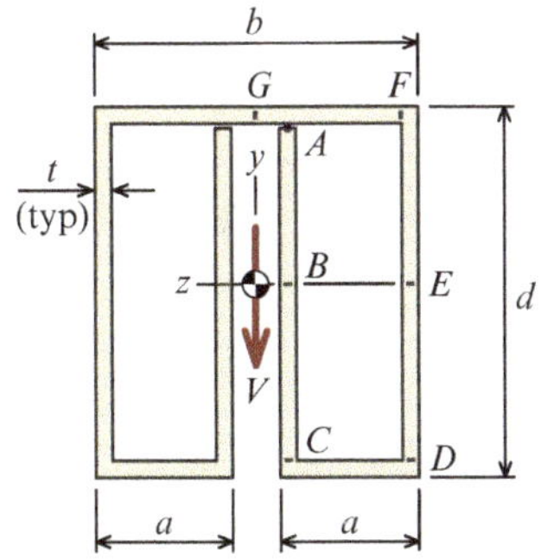

FIGURE P9.43

P9.44 Sheet metal is bent into the shape shown in Figure P9.44 and is then used as a beam. The shape has dimensions $b = 80$ mm, $d = 60$ mm, $a = 20$ mm, and $t = 3$ mm. A vertical shear force $V = 5{,}400$ N acts on the cross section. Determine the shear flow at points A–E.

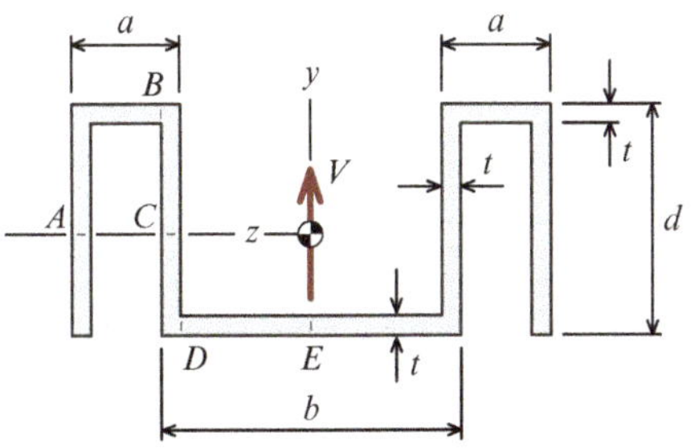

FIGURE P9.44

P9.45 Sheet metal is bent into the shape shown in Figure P9.45 and is then used as a beam. The shape has centerline dimensions $a = 120$ mm, $b = 80$ mm, and $t = 3$ mm. A vertical shear force $V = 6{,}400$ N acts on the cross section. Determine the shear stress magnitude that acts in the shape at points A–D.

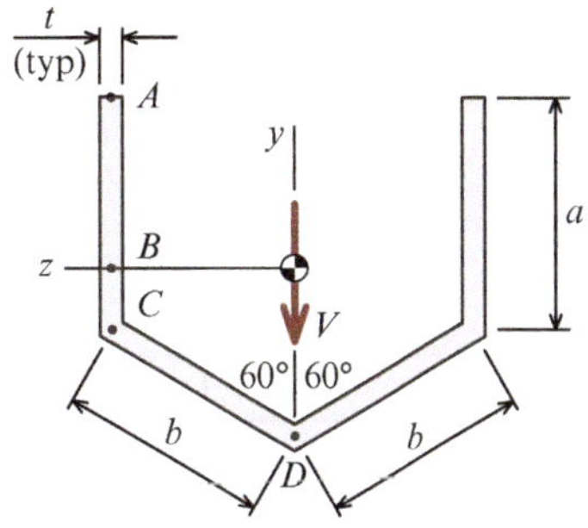

FIGURE P9.45

P9.46 Sheet metal is bent into the shape shown in Figure P9.46 and is then used as a beam. The shape has centerline dimensions $b = 3.0$ in., $d = 4.5$ in., $a = 1.8$ in., and $t = 0.12$ in. A vertical shear force $V = 480$ lb acts on the cross section. Determine

(a) the shear stress that acts in the shape at points A–D.
(b) the resultant horizontal force acting on the cross-sectional element between A and B.

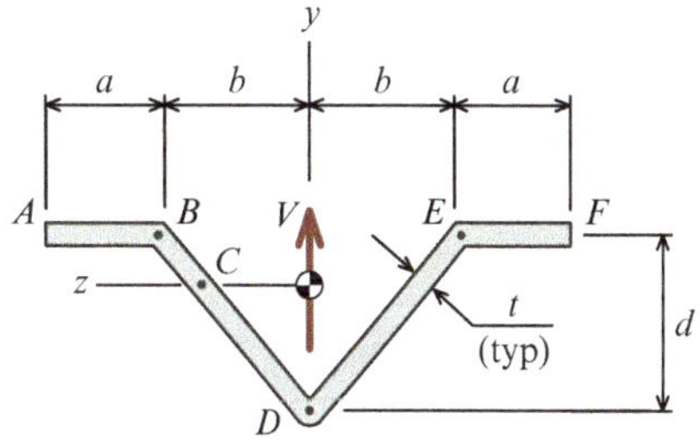

FIGURE P9.46

P9.47 The thin-walled sheet pile cross section shown in Figure P9.47 is subjected to a vertical shear force $V = 175$ kN. The shape has centerline dimensions $b = 160$ mm, $d = 350$ mm, $a = 120$ mm, and $t = 12$ mm. Determine

(a) the shear stress in the shape at points A–E.
(b) the resultant horizontal force acting on the cross-sectional element between A and B.

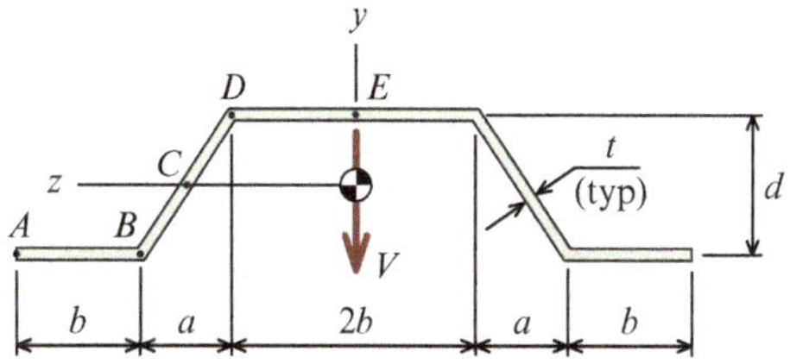

FIGURE P9.47

P9.48 The beam cross section shown in Figure P9.48 is subjected to a shear force $V = 12$ kips. The dimensions of the cross section are $b = 18$ in., $d = 10$ in., and $t = 0.4$ in. Using $a = 3$ in., calculate the shear stress magnitude at sections 1–1, 2–2, and 3–3.

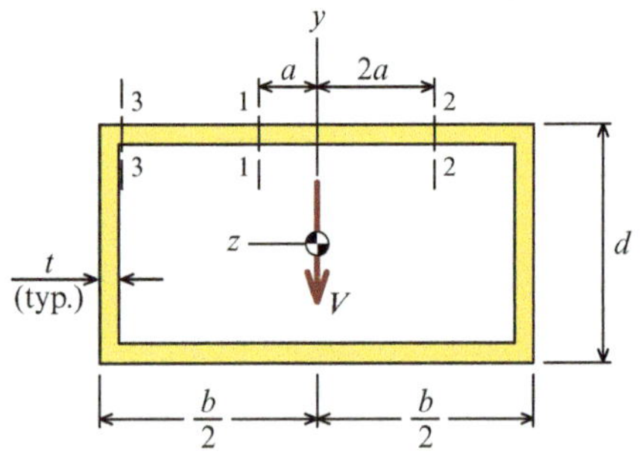

FIGURE P9.48

P9.49 The beam cross section shown in Figure P9.49 is subjected to a shear force $V = 12$ kips. The dimensions of the cross section are $b = 4$ in., $d = 7$ in., $a = 2$ in., and $t = 0.25$ in. Calculate the shear stress magnitude at sections 1–1, 2–2, and 3–3. Calculate the maximum shear stress.

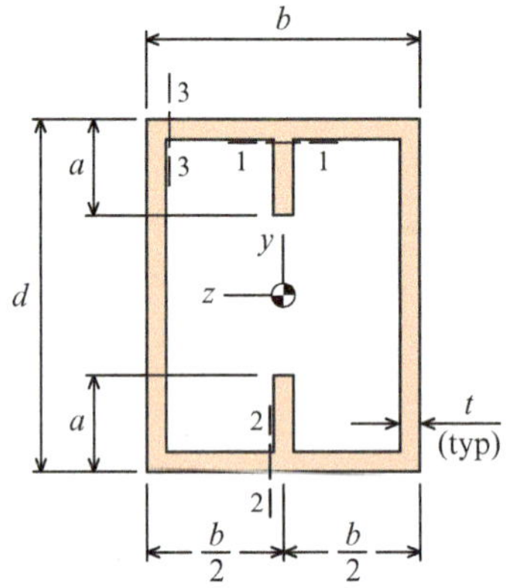

FIGURE P9.49

P9.50 An extruded plastic beam with the cross section shown in Figure P9.50 is subjected to a vertical shear force $V = 850$ N. The centerline dimensions of the cross section are $a = 20$ mm, $b = 40$ mm, $d = 50$ mm, and $t = 3$ mm. Determine the shear stress magnitude that acts in the shape at points A–E.

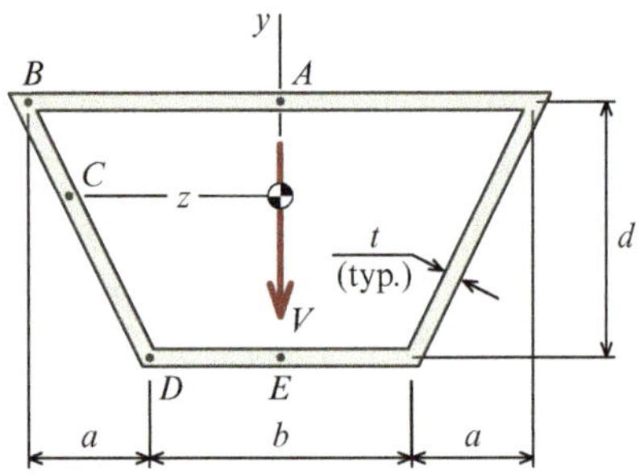

FIGURE P9.50

P9.51 The angle shown in Figure P9.51 is subjected to a vertical shear force $V = 3.5$ kips. Sketch the distribution of shear flow along the leg AB. Indicate the numerical value at all peaks.

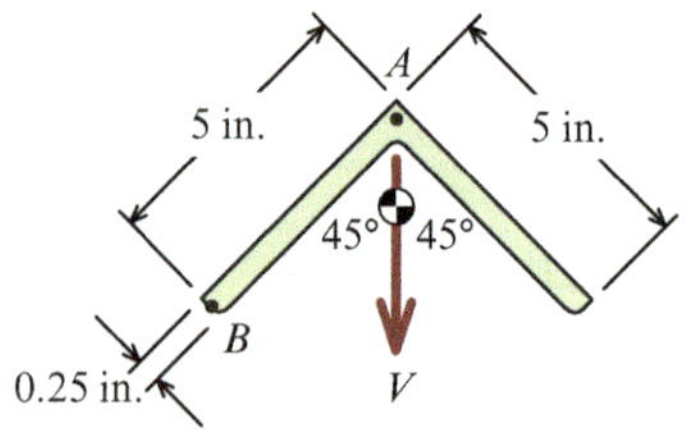

FIGURE P9.51

P9.52 The vertical shear force V acts on the thin-walled section shown in Figure P9.52. Sketch the shear flow diagram for the cross section. Assume that the wall thickness of the section is constant.

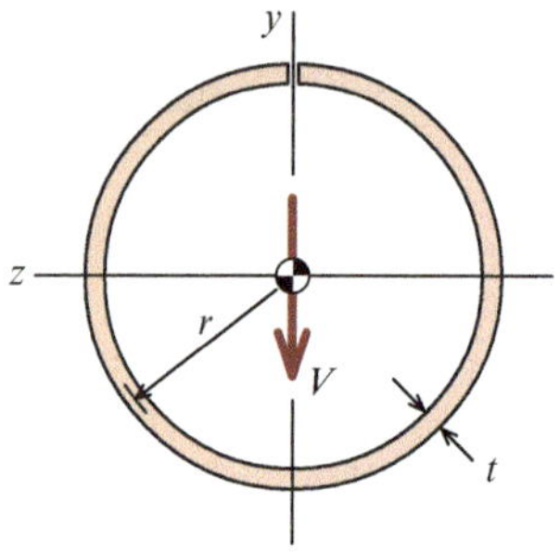

FIGURE P9.52

P9.53 Determine the location of the shear center O for the cross section shown in Figure P9.53.

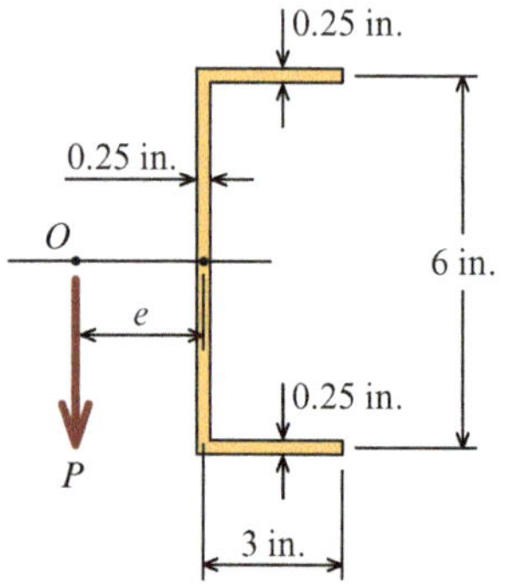

FIGURE P9.53

P9.54 Determine the location of the shear center O for the cross section shown in Figure P9.54. Assume a uniform thickness of $t = 4$ mm for all portions of the cross section. Use $a = 70$ mm, $b = 40$ mm, and $c = 90$ mm.

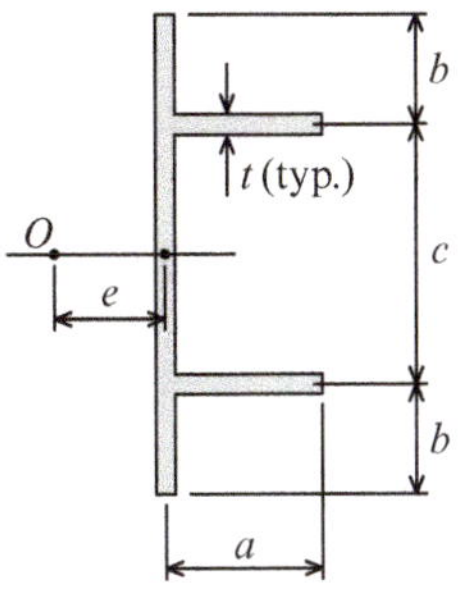

FIGURE P9.54

P9.55 An extruded beam has the cross section shown in Figure P9.55. Determine (a) the location of the shear center O and (b) the distribution of shear stress created by a shear force $V = 30$ kN.

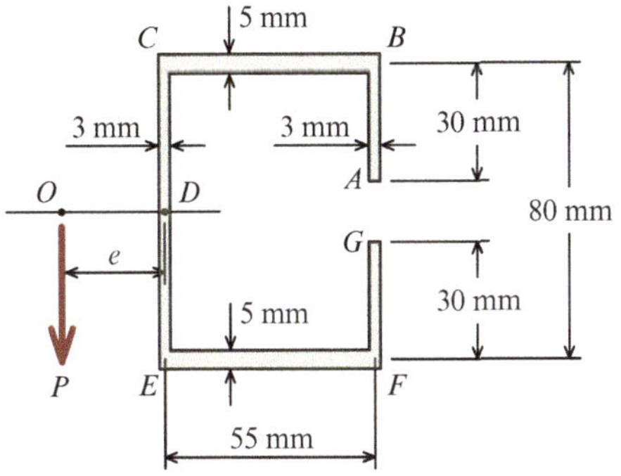

FIGURE P9.55

P9.56 An extruded beam has the cross section shown in Figure P9.56. Using dimensions of $b = 30$ mm, $h = 36$ mm, and $t = 5$ mm, calculate the location of the shear center O.

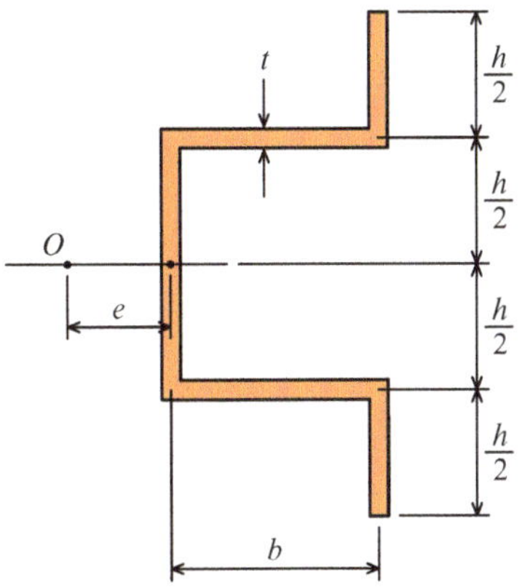

FIGURE P9.56

P9.57 An extruded beam has the cross section shown in Figure P9.57. For this shape, use dimensions of $b = 50$ mm, $h = 40$ mm, and $t = 3$ mm. What is the distance e to the shear center O?

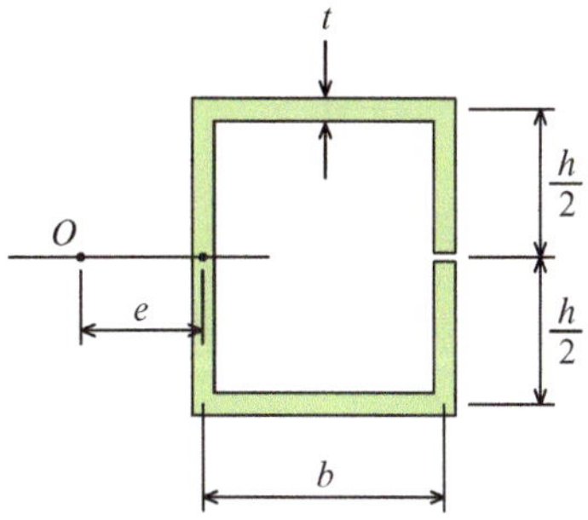

FIGURE P9.57

P9.58 An extruded beam has the cross section shown in Figure P9.58. The dimensions of this shape are $b = 45$ mm, $h = 75$ mm, and $t = 4$ mm. Assume that the thickness t is constant for all portions of the cross section. What is the distance e from the leftmost element to the shear center O?

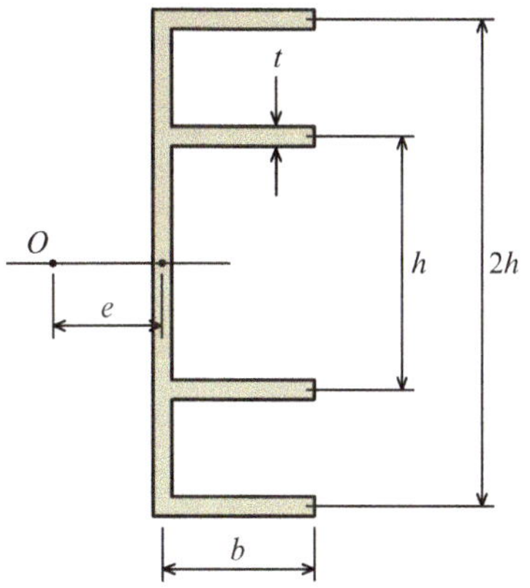

FIGURE P9.58

P9.59 Determine the location of the shear center of the cross section shown in Figure P9.59. Use dimensions of $a = 50$ mm, $b = 100$ mm, $h = 300$ mm, and $t = 5$ mm. Assume that the thickness t is constant for all portions of the cross section.

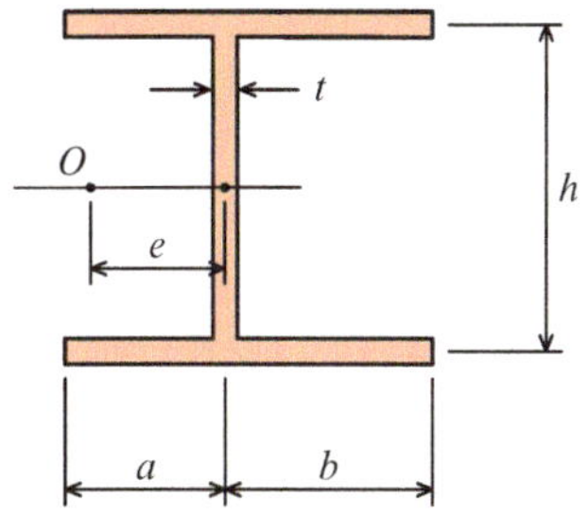

FIGURE P9.59

P9.60 Locate the shear center of the cross section shown in Figure P9.60. Assume that the web thickness is the same as the flange thickness.

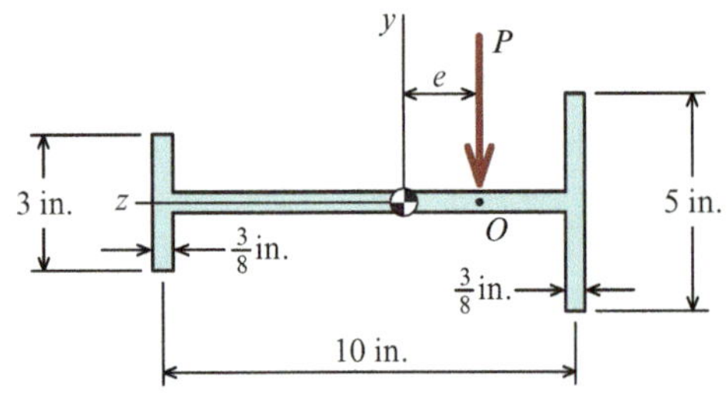

FIGURE P9.60

P9.61 Prove that the shear center of the Z section shown in Figure P9.61 is located at the centroid of the section.

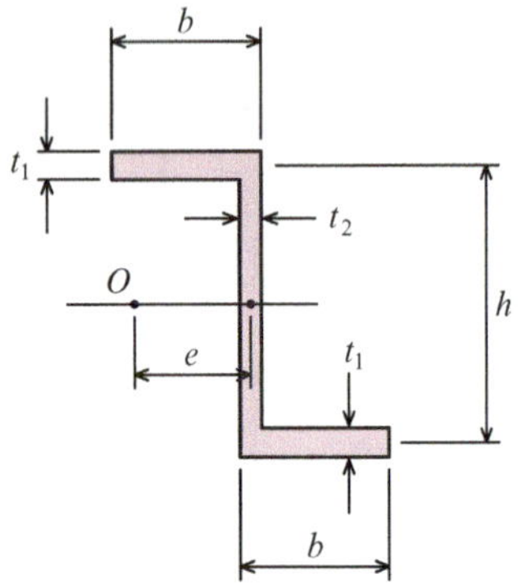

FIGURE P9.61

P9.62–P9.66 For each of Figures P9.62–P9.66, determine the location of the shear center O of a thin-walled beam of uniform thickness having the cross section shown.

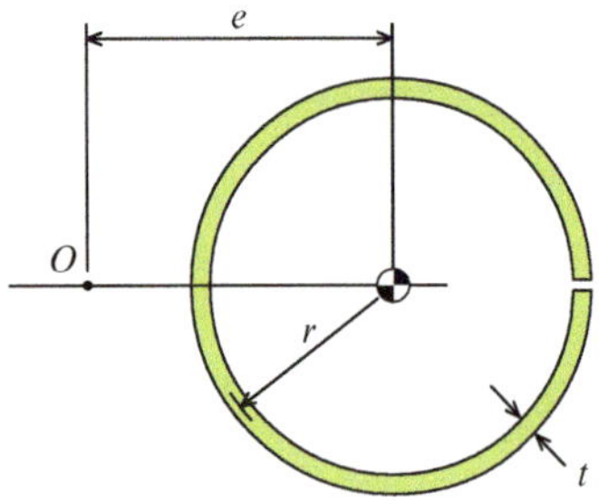

FIGURE P9.62

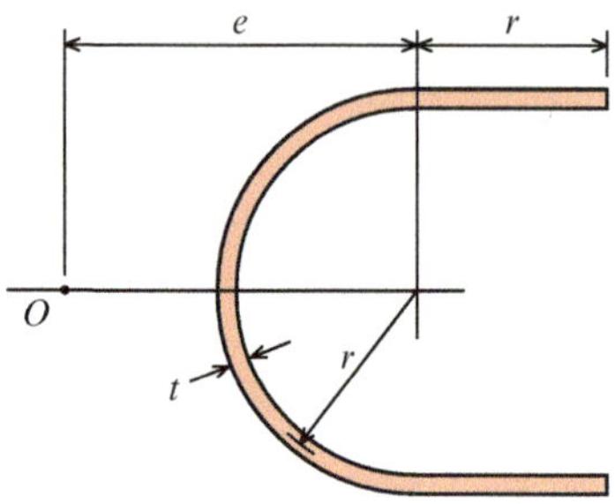

FIGURE P9.63

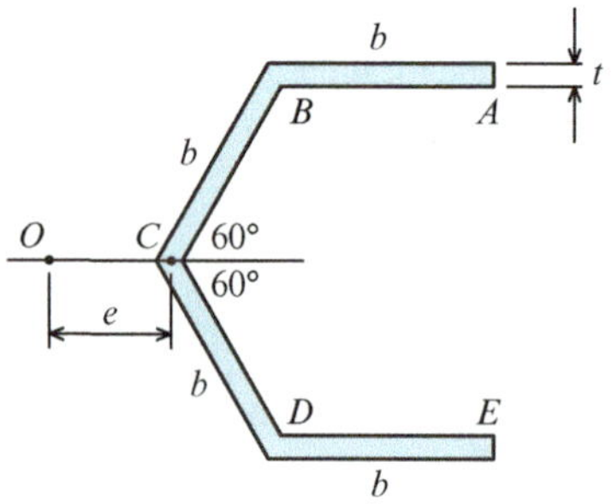

FIGURE P9.64

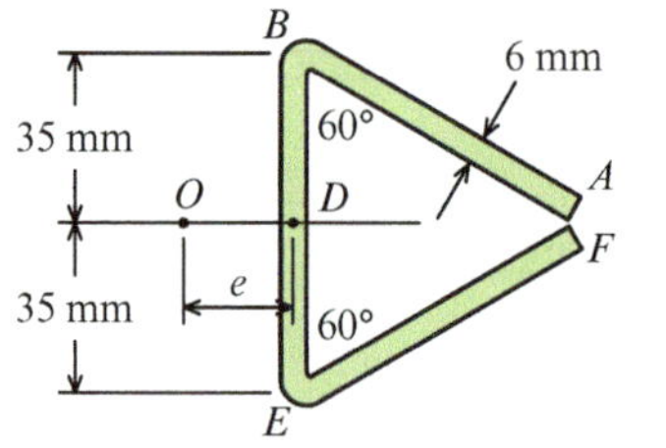

FIGURE P9.65

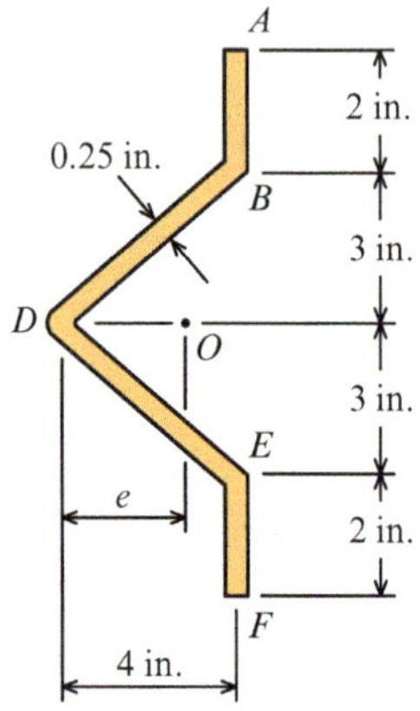

FIGURE P9.66

390

CHAPTER 10

Beam Deflections

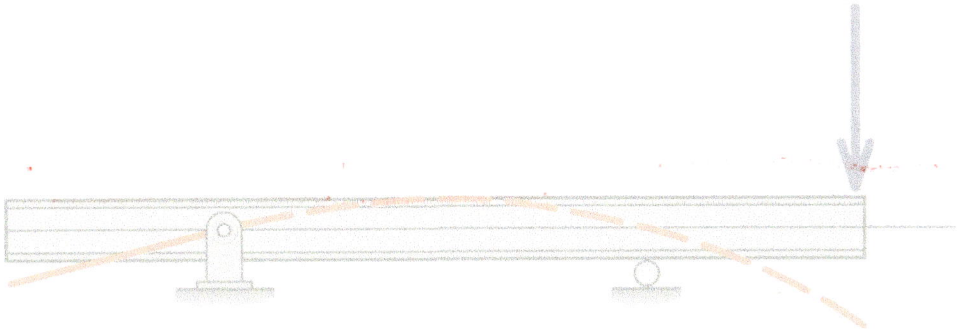

10.1 Introduction

Important relations between applied load and both normal and shear stresses developed in a beam were presented in Chapters 8 and 9. However, a design is normally not complete until the deflection of the beam has been determined for its particular load. While they generally do not create a safety risk in themselves, excessive beam deflections may impair the successful functioning of a structure in other ways. In building construction, excessive deflections can cause cracks in walls and ceilings. Doors and windows may not close properly. Floors may sag or vibrate noticeably as people walk on them. In many machines, beams and flexural components must deflect just the right amount for gears or other parts to make proper contact. In sum, the satisfactory design of a flexural component usually includes a specified maximum deflection in addition to a minimum load-carrying capacity.

The deflection of a beam depends on the stiffness of the material and the cross-sectional dimensions of the beam, as well as on the configuration of the applied loads and supports. Three common methods for calculating beam deflections are presented here: (1) the integration method, (2) the use of discontinuity functions, and (3) the superposition method.

In the discussion that follows, three coordinates will be used. As shown in Figure 10.1, the x axis (positive to the right) extends along the initially straight longitudinal axis of the beam. The x coordinate is used to locate a differential beam element, which has an undeformed width of dx. The v axis extends positive upward from the x axis. The v coordinate measures the displacement of the beam's neutral surface. The third coordinate is y, which is a localized coordinate with its origin at the neutral surface of the beam cross section. The y coordinate is measured positive upward, and it is used to describe specific locations within the beam cross section. The x and y coordinates are the same as those used in deriving the flexure formula in Chapter 8.

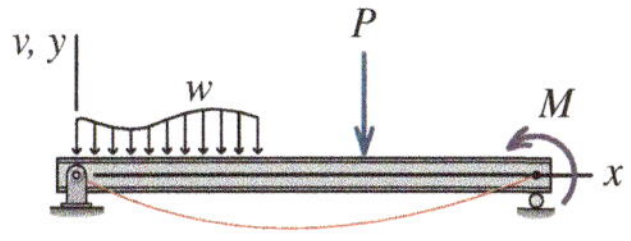

FIGURE 10.1 Coordinate system.

10.2 Moment–Curvature Relationship

When a straight beam is loaded and the action is elastic, the longitudinal centroidal axis of the beam becomes a curve, which is termed the **elastic curve**. The relationship between the internal bending moment and the curvature of the elastic curve was developed in Section 8.4. Equation 8.6 summarized the **moment–curvature** relationship:

$$\kappa = \frac{1}{\rho} = \frac{M}{EI_z}$$

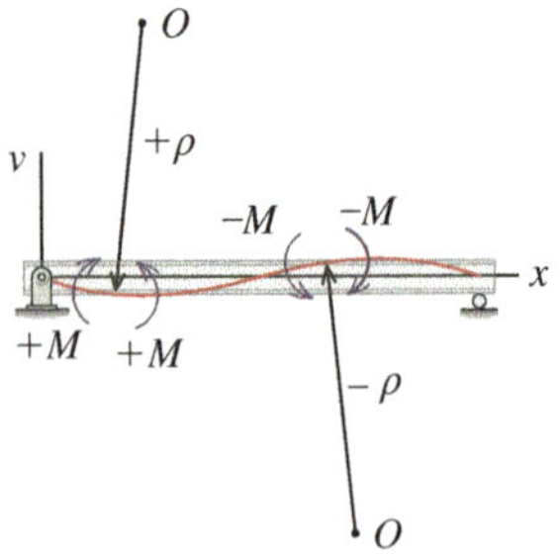

FIGURE 10.2 Radius of curvature ρ related to sign of M.

This equation relates the radius of curvature ρ of the neutral surface of the beam to the internal bending moment M (about the z axis), the elastic modulus of the material E, and the moment of inertia of the cross-sectional area, I_z. Since E and I_z are always positive, the sign of ρ is consistent with the sign of the bending moment. As shown in Figure 10.2, a positive bending moment M creates a radius of curvature ρ that extends above the beam—that is, in the positive v direction. When M is negative, ρ extends below the beam in a negative v direction.

10.3 The Differential Equation of the Elastic Curve

The relationship between the bending moment and the radius of curvature is applicable when the bending moment M is constant for a flexural component. For most beams, however, the bending moment varies along its span and a more general expression is required to express the deflection v as a function of the coordinate x.

From calculus, the curvature is defined as

$$\kappa = \frac{1}{\rho} = \frac{d^2v/dx^2}{[1 + (dv/dx)^2]^{3/2}}$$

For typical beams, the slope dv/dx is very small and its square can be neglected in comparison to unity. This approximation simplifies the curvature expression to

$$\kappa = \frac{1}{\rho} = \frac{d^2v}{dx^2}$$

and Equation (8.6) becomes

$$EI\frac{d^2v}{dx^2} = M(x) \tag{10.1}$$

This is the **differential equation of the elastic curve** for a beam. In general, the bending moment M will be a function of position x along the beam's span.

The differential equation of the elastic curve can also be obtained from the geometry of the deflected beam, as shown in Figure 10.3. The deflection v at point A on the elastic curve is shown in Figure 10.3a. Point A is located at a distance x from the origin. A second point, B, is located at a distance $x + dx$ from the origin, and it has deflection $v + dv$.

When the beam is bent, points along the beam both deflect and rotate. The **angle of rotation** θ of the elastic curve is the angle between the x axis and the tangent to the elastic curve, as shown for point A in the enlarged view of Figure 10.3*b*. Similarly, the angle of rotation at point B is $\theta + d\theta$, where $d\theta$ is the increase in rotation angle between points A and B.

The slope of the elastic curve is the first derivative dv/dx of the deflection v. From Figure 10.3*b*, the slope can also be defined as the vertical increment dv divided by the horizontal increment dx between points A and B. Since dv and dx are infinitesimally small, the first derivative dv/dx can be related to the rotation angle θ by the tangent function:

$$\frac{dv}{dx} = \tan\theta \tag{a}$$

Note that the slope dv/dx is positive when the tangent to the elastic curve slopes upward to the right.

In Figure 10.3*b*, the distance along the elastic curve between points A and B is denoted as ds, and from the definition of arclength, $ds = \rho\, d\theta$. If the angle of rotation θ is very small (as it would be for a beam with small deflections), then the distance ds along the elastic curve in Figure 10.3*b* is essentially the same as the increment dx along the x axis. Therefore, $dx = \rho\, d\theta$, or

$$\frac{1}{\rho} = \frac{d\theta}{dx} \tag{b}$$

Since $\tan\theta \approx \theta$ for small angles, Equation (a) can be approximated as

$$\frac{dv}{dx} \approx \theta \tag{c}$$

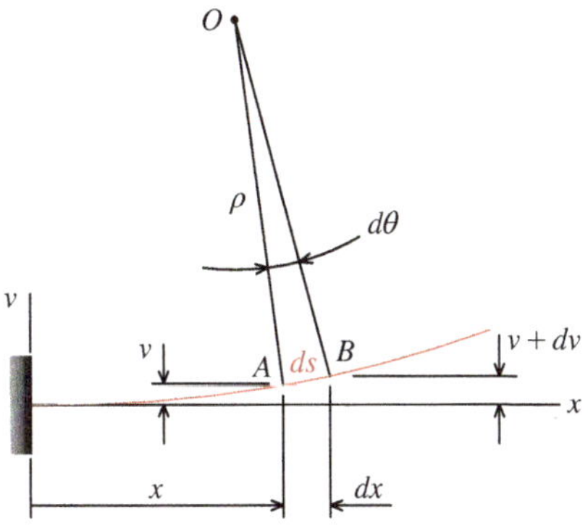

(*a*) Elastic curve

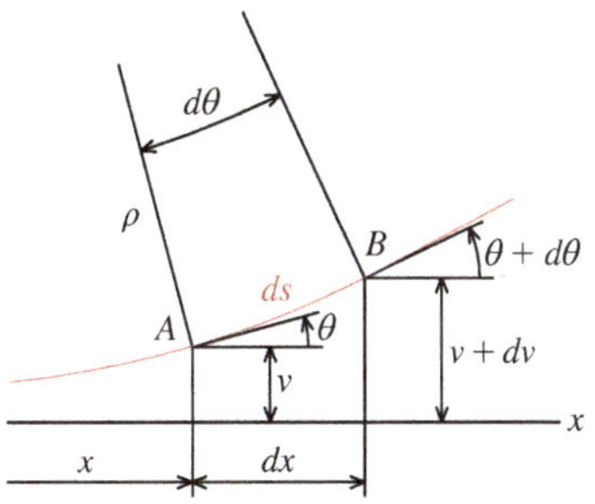

(*b*) Enlarged region around point A

FIGURE 10.3 Geometry of deflected beam.

Therefore, the beam angle of rotation θ (measured in radians) and the slope dv/dx are equal if beam deflections are small.

Taking the derivative of Equation (c) with respect to x gives

$$\frac{d^2v}{dx^2} = \frac{d\theta}{dx} \tag{d}$$

From Equation (b), $d\theta/dx = 1/\rho$. In addition, Equation (8.6) gives the relationship between M and ρ. Combining these equations gives

$$\frac{d^2v}{dx^2} = \frac{d\theta}{dx} = \frac{1}{\rho} = \frac{M}{EI} \tag{e}$$

or

$$EI\frac{d^2v}{dx^2} = M(x) \tag{10.1}$$

In general, the bending moment M will be a function of position x along the beam's span.

Sign Conventions

The sign convention for bending moments established in Section 7.2 (see Figure 10.4) will be used for Equation (10.1). Both E and I are always positive; therefore, the signs of the bending moment and the second derivative must be consistent. With the coordinate axes as shown in Figure 10.5, the beam slope changes from positive to negative in the segment from A to B; therefore, the second derivative is negative, which agrees with the sign convention of Section 7.2. For segment BC, both d^2v/dx^2 and M are seen to be positive.

Careful study of Figure 10.5 reveals that the signs of the bending moment and the second derivative are also consistent when the origin is selected at the right, with x positive to the left and v positive upward. However, the signs are inconsistent when v is positive downward. Consequently, v will always be chosen as positive upward for horizontal beams in this book.

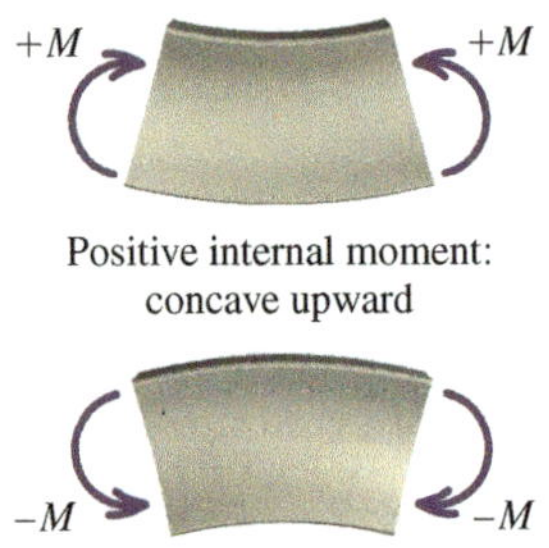

FIGURE 10.4 Bending-moment sign convention.

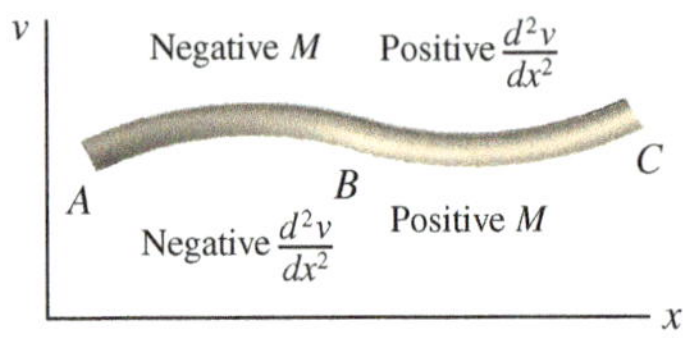

FIGURE 10.5 Relationship of d^2v/dx^2 to sign of M.

Relationship of Derivatives

Before proceeding with the solution of Equation (10.1), it is instructive to associate the successive derivatives of the elastic curve deflection v with the physical quantities that they represent in beam action:

$$\text{Deflection} = v$$

$$\text{Slope} = \frac{dv}{dx} = \theta$$

$$\text{Moment } M = EI\frac{d^2v}{dx^2} \quad \text{(from Equation 10.1)}$$

$$\text{Shear } V = \frac{dM}{dx} = EI\frac{d^3v}{dx^3} \quad \text{(for } EI \text{ constant)}$$

$$\text{Load } w = \frac{dV}{dx} = EI\frac{d^4v}{dx^4} \quad \text{(for } EI \text{ constant)}$$

The signs are as defined in Sections 7.2 and 7.3.

Starting from the load diagram, a method based on these differential relations was presented in Section 7.3 for constructing first the shear diagram V and then the moment diagram M. This method can be readily extended to the construction of the slope diagram θ and the beam deflection diagram v. From Equation (e), we obtain

$$\frac{d\theta}{dx} = \frac{M}{EI} \tag{f}$$

This equation can be integrated to give

$$\int_{\theta_A}^{\theta_B} d\theta = \int_{x_A}^{x_B} \frac{M}{EI}\,dx \qquad \therefore\ \theta_B - \theta_A = \int_{x_A}^{x_B} \frac{M}{EI}\,dx$$

The rightmost equation shows that the area under the moment diagram between any two points along the beam (with the added consideration of EI) gives the change in slope between the same two points. Likewise, the area under the slope diagram between two points along the beam gives the change in deflection between the points. These two equations have been used to construct the complete series of diagrams shown in Figure 10.6 for a simply

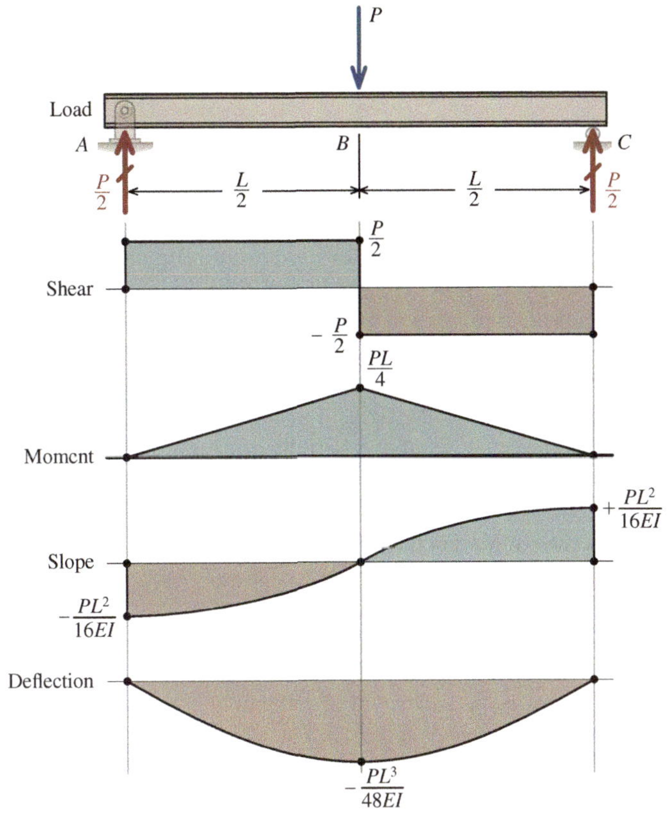

FIGURE 10.6 Relationship among beam diagrams.

supported beam with a concentrated load at midspan. The geometry of the beam was utilized to locate the points of zero slope and deflection, required as starting points for the construction. More commonly used methods for calculating beam deflections will be developed in succeeding sections.

Review of Assumptions

Before proceeding with specific methods for calculating beam deflections, it is helpful to keep in mind the assumptions used in developing the differential equation of the elastic curve. All of the limitations that apply to the flexure formula also apply to the calculation of deflections because the flexure formula was used in the derivation of Equation (10.1). It is further assumed that

1. The square of the slope of the beam is negligible compared with unity. This assumption means that beam deflections must be relatively small.
2. Plane cross sections of the beam remain planar as the beam deflects. This assumption means that beam deflections due to shear stresses are assumed negligible.
3. The values of E and I remain constant for any segment along the beam. If either E or I varies along the beam span, and if this variation can be expressed as a function of the distance x along the beam, a solution of Equation (10.1) that considers the variation may be possible.

10.4 Determining Deflections by Integration of a Moment Equation

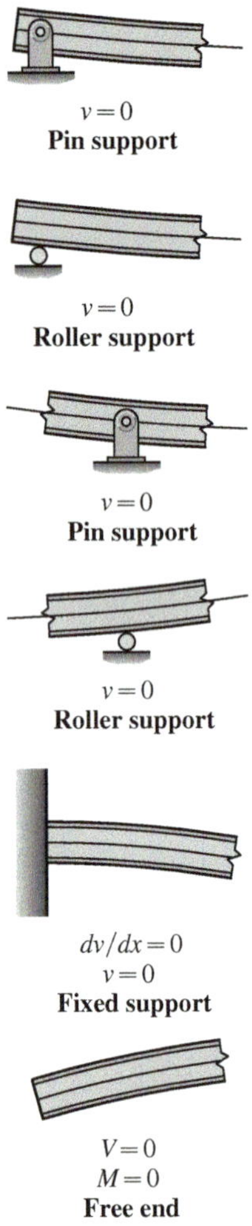

FIGURE 10.7 Boundary conditions.

Whenever the assumptions of the previous section are satisfied and the bending moment can be readily expressed as an integrable function of x, Equation (10.1) can be solved for the deflection v of the elastic curve at any location x along the beam's span. The procedure begins with the derivation of a bending-moment function $M(x)$ based on equilibrium considerations. A single function that is applicable to the entire span may be derived, or it may be necessary to derive several functions, each applicable only to a specific region of the beam span. The moment function is substituted into Equation (10.1) to define the differential equation. This type of differential equation can be solved by integration. Integration of Equation (10.1) produces an equation that defines the beam slope dv/dx. Integrating again produces an equation that defines the deflection v of the elastic curve. This approach to determining the elastic curve equation is called the **double-integration method**.

Each integration produces a constant of integration, and these constants must be evaluated from known conditions of slope and deflection. The types of conditions for which values of v and dv/dx are known can be grouped into three categories: boundary conditions, continuity conditions, and symmetry conditions.

Boundary Conditions

Boundary conditions are specific values of the deflection v or slope dv/dx that are known at particular locations along the beam span. As the term implies, boundary conditions are found at the lower and upper limits of the interval being considered. For example, a bending-moment equation $M(x)$ may be derived for a particular beam within a region $x_1 \leq x \leq x_2$. The boundary conditions, in this instance, would be found at $x = x_1$ and $x = x_2$.

> Boundary conditions are known slopes and deflections at the limits of the *bending-moment equation* $M(x)$. The term "boundary" refers to the bounds of $M(x)$, not necessarily the bounds of the beam. Although boundary conditions are found at beam supports, only those supports within the bounds of the bending-moment equation should be considered.

Figure 10.7 shows several supports and lists the boundary conditions associated with each. A pin or roller support represents a simple support at which the beam is restrained from deflecting transversely (either upward or downward for a horizontal beam); consequently, the beam deflection at either a pin or a roller must be $v = 0$. Neither a pin nor a roller, however, restrains a beam against rotation, and consequently, the beam slope at a simple support cannot be a boundary condition. At a fixed connection, the beam is restrained against both deflection and rotation; therefore, $v = 0$ and $dv/dx = 0$ at such a connection.

While boundary conditions involving a deflection v and a slope dv/dx are normally equal to zero at supports, there may be instances in which the engineer wishes to analyze the effects of support displacement on the beam. For instance, a common design concern is the possibility of **support settlement**, in which the compression of soil underneath a foundation causes the support to be displaced downward. To examine possibilities of this sort, nonzero boundary conditions may sometimes be specified.

> One boundary condition can be used to determine one and only one constant of integration.

Continuity Conditions

Many beams are subjected to abrupt changes in loading along the beam, such as concentrated loads, reactions, or even distinct changes in the intensity of a uniformly distributed load. The $M(x)$ equation for the region just to the left of an abrupt change will be different from the $M(x)$ equation for the region just to the right. As a result, it is not possible to derive a single equation for the bending moment (in terms of ordinary algebraic functions) that is valid for the entire beam length. This problem can be resolved by writing separate bending-moment equations for each segment of the beam. Although the segments are bounded by abrupt changes in load, the beam itself is continuous at such locations, and consequently, the deflection and the slope at the junction of two adjacent segments must match. This condition of matching is termed a **continuity condition**.

Symmetry Conditions

In some instances, beam supports and applied loads may be configured so that symmetry exists for the span. When symmetry exists, the value of the beam slope will be known at certain locations. For instance, a simply supported beam with a uniformly distributed load is symmetric. From symmetry, the slope of the beam at midspan must equal zero. Symmetry may also abbreviate the deflection analysis in that the elastic curve need be determined for only half of the span.

Each boundary, continuity, and symmetry condition produces an equation containing one or more of the constants of integration. In the double-integration method, two constants of integration are produced for each beam segment; therefore, two conditions are required to evaluate the constants.

Procedure for Double-Integration Method

Calculating the deflection of a beam by the double-integration method involves several steps, and the following sequence is strongly recommended:

1. **Sketch:** Sketch the beam, including supports, loads, and the x–v coordinate system. Sketch the approximate shape of the elastic curve. Pay particular attention to the slope and deflection of the beam at the supports.
2. **Support reactions:** For some beam configurations, it may be necessary to determine support reactions before proceeding to analyze specific beam segments. For these configurations, determine the beam reactions by considering the equilibrium of the entire beam. Show these reactions in their proper direction on the beam sketch.
3. **Equilibrium:** Select the segment or segments of the beam to be considered. For each segment, draw a free-body diagram (FBD) that cuts through the beam segment at some distance x from the origin. On the FBD, show all loads acting on the beam. If distributed loads act on the beam, then that portion of the distributed load which acts on the FBD must be shown at the outset. Include the internal bending moment M acting at the cut surface of the beam, and always show M acting in the positive direction. (See Figure 10.5.) The latter ensures that the bending-moment equation will have the correct sign. From the FBD, derive the bending-moment equation, taking care to note the interval to which it is applicable (e.g., $x_1 \le x \le x_2$).
4. **Integration:** For each segment, set the bending-moment equation equal to $EI\, d^2v/dx^2$. Integrate this differential equation twice, obtaining a slope equation dv/dx, a deflection equation v, and two constants of integration.

5. **Boundary and continuity conditions:** List the boundary conditions that are applicable to the bending-moment equation. If the analysis involves two or more beam segments, list the continuity conditions also. Remember that two conditions are required in order to evaluate the two constants of integration produced in each beam segment.
6. **Evaluate constants:** Use the boundary and continuity conditions to evaluate all constants of integration.
7. **Elastic curve and slope equations:** Replace the constants of integration arrived at in step 4 with the values obtained from the boundary and continuity conditions found in step 6. Check the resulting equations for dimensional homogeneity.
8. **Deflections and slopes at specific points:** Calculate the deflection at specific points when required.

The following examples illustrate the use of the double-integration method for calculating beam deflections:

EXAMPLE 10.1

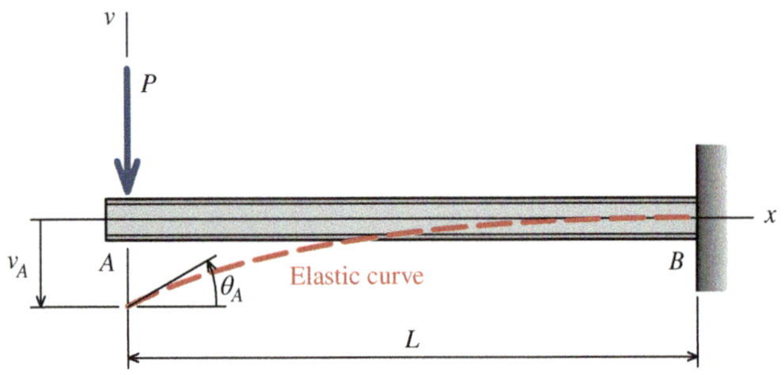

The cantilever beam shown is subjected to a concentrated load P at its free end. Determine the equation of the elastic curve, as well as the deflection and slope of the beam at A. Assume that EI is constant for the beam.

Plan the Solution

Consider a free-body diagram that cuts through the beam at a distance x from the free end of the cantilever. Write an equilibrium equation for the sum of moments, and from this equation, determine the equation for the bending moment M as it varies with x. Substitute M into Equation (10.1), and integrate twice. Use the boundary conditions known at the fixed end of the cantilever to evaluate the constants of integration.

SOLUTION

Equilibrium

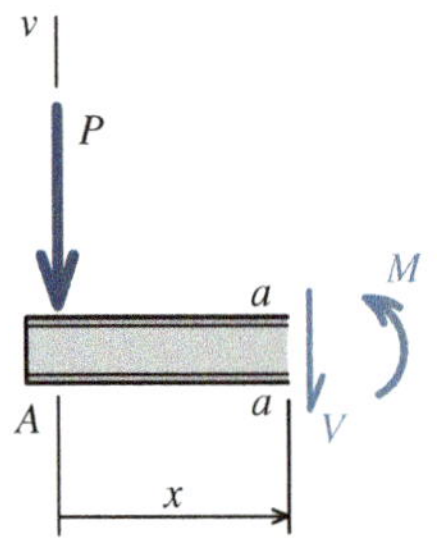

Cut through the beam at an arbitrary distance x from the origin, and draw a free-body diagram, taking care to show the internal moment M acting in the positive sense. The equilibrium equation for the sum of moments about section a–a is

$$\Sigma M_{a-a} = Px + M = 0$$

Therefore, the bending-moment equation for this beam is simply

$$M = -Px \qquad \text{(a)}$$

Notice that moment equation (a) is valid for all values of x for this particular beam. In other words, Equation (a) is valid in the interval $0 \leq x \leq L$. Substitute the expression for M into Equation (10.1) to obtain

$$EI\frac{d^2v}{dx^2} = -Px \qquad \text{(b)}$$

Integration

Equation (b) will be integrated twice. The first integration gives a general equation for the slope dv/dx of the beam:

$$EI\frac{dv}{dx} = -\frac{Px^2}{2} + C_1 \qquad \text{(c)}$$

Here, C_1 is a constant of integration. A second integration gives a general equation for the elastic curve v:

$$EIv = -\frac{Px^3}{6} + C_1x + C_2 \qquad \text{(d)}$$

In this equation, C_2 is a second constant of integration. The constants C_1 and C_2 must be evaluated before the slope and elastic curve equations are complete.

Boundary Conditions

Boundary conditions are values of the deflection v or slope dv/dx that are known at particular locations along the beam span. For this beam, the bending-moment equation M in Equation (a) is valid in the interval $0 \le x \le L$. The boundary conditions, therefore, are found at either $x = 0$ or $x = L$.

Consider the interval $0 \le x \le L$ for this beam and loading. At $x = 0$, the beam is unsupported. The beam will deflect downward, and as it deflects, the slope of the beam will no longer be zero. Consequently, neither the deflection v nor the slope dv/dx is known at $x = 0$. At $x = L$, the beam is supported by a fixed support. The fixed support at B prevents deflection and rotation; therefore, we know two bits of information with absolute certainty at $x = L$: $v = 0$ and $dv/dx = 0$. These are the two boundary conditions that will be used to evaluate the constants of integration C_1 and C_2.

Evaluate Constants

Substitute the boundary condition $dv/dx = 0$ at $x = L$ into Equation (c) to evaluate the constant C_1:

$$EI\frac{dv}{dx} = -\frac{Px^2}{2} + C_1 \quad \Rightarrow \quad EI(0) = -\frac{P(L)^2}{2} + C_1 \qquad \therefore\ C_1 = \frac{PL^2}{2}$$

Next, substitute the value of C_1 and the boundary condition $v = 0$ at $x = L$ into Equation (d), and solve for the second constant of integration C_2:

$$EIv = -\frac{Px^3}{6} + C_1x + C_2 \quad \Rightarrow \quad EI(0) = -\frac{P(L)^3}{6} + \frac{PL^2}{2}(L) + C_2 \qquad \therefore\ C_2 = -\frac{PL^3}{3}$$

Elastic Curve Equation

Substitute the expressions obtained for C_1 and C_2 into Equation (d) to complete the elastic curve equation:

$$EIv = -\frac{Px^3}{6} + \frac{PL^2}{2}x - \frac{PL^3}{3}, \quad \text{which simplifies to} \quad v = \frac{P}{6EI}[-x^3 + 3L^2x - 2L^3] \qquad \text{(e)}$$

Similarly, the beam slope equation from Equation (c) can be completed with the expression derived for C_1:

$$EI\frac{dv}{dx} = -\frac{Px^2}{2} + \frac{PL^2}{2}, \quad \text{which simplifies to} \quad \frac{dv}{dx} = \frac{P}{2EI}[L^2 - x^2] \qquad \text{(f)}$$

Beam Deflection and Slope at *A*

The deflection and slope of the beam at *A* are obtained by setting $x = 0$ in Equations (e) and (f). The beam deflection and slope at the free end of the cantilever are

$$v_A = -\frac{PL^3}{3EI} \quad \text{and} \quad \left(\frac{dv}{dx}\right)_A = \frac{PL^2}{2EI} \qquad \textbf{Ans.}$$

MecMovies

EXAMPLE

M10.2 Derive the equation for the elastic curve, and determine expressions for the slope and deflection of the beam at *B*. Use the double-integration method.

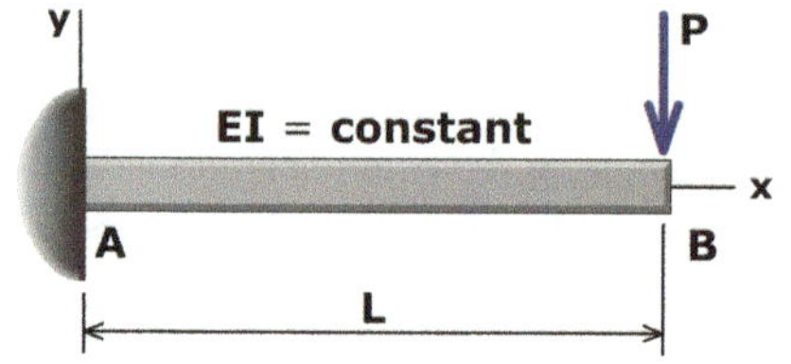

EXAMPLE 10.2

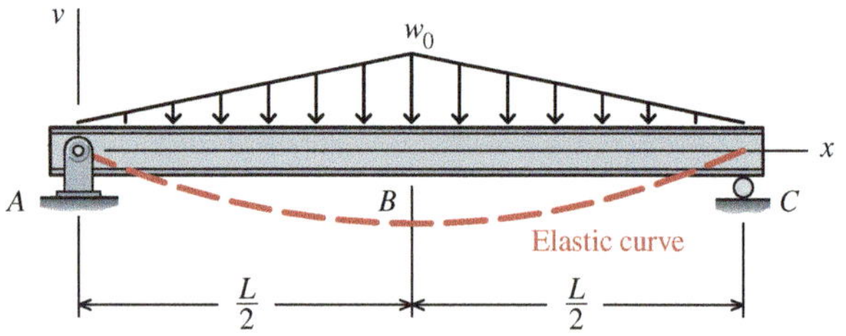

A simply supported beam is subjected to the linearly distributed load shown. Determine the equation of the elastic curve. Also, determine the deflection of the beam at midspan *B* and the slope at support *A*. Assume that *EI* is constant over the entire span of the beam.

Plan the Solution

Generally, two moment equations would be needed to define the complete variation of *M* over the entire span. However, in this case, the beam and loading are symmetrical. On the basis of symmetry, we need only solve for the elastic curve in the interval $0 \le x \le L/2$. The boundary conditions for this interval will be found at the pin support *A* and at midspan *B*.

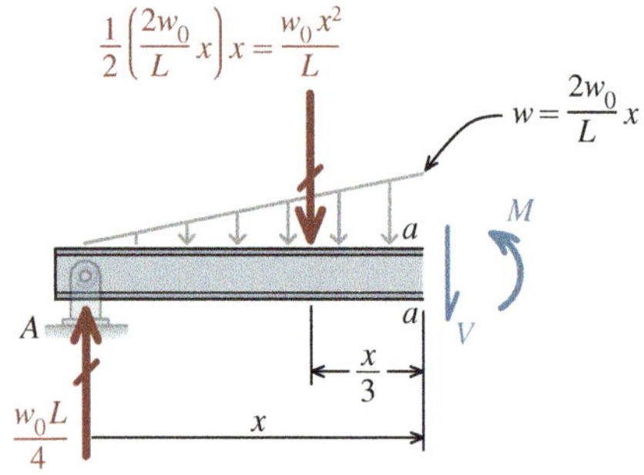

SOLUTION

Support Reactions

Since the beam is symmetrically supported and symmetrically loaded, the beam reactions at *A* and *C* are identical:

$$A_y = C_y = \frac{w_0 L}{4}$$

No loads act in the *x* direction; therefore, $A_x = 0$.

Equilibrium

Cut through the beam at an arbitrary distance *x* from the origin, and draw a free-body diagram, taking care to show the internal moment *M* acting in the positive direction. The equilibrium equation for the sum of moments about section *a–a* is

$$\Sigma M_{a-a} = \frac{1}{2}\left(\frac{2w_0 x}{L}\right)x\left(\frac{x}{3}\right) - \left(\frac{w_0 L}{4}\right)x + M = 0$$

Hence, the bending-moment equation for this beam is

$$M = \frac{w_0 Lx}{4} - \frac{w_0 x^3}{3L} \quad \text{(valid for } 0 \leq x \leq L/2\text{)} \tag{a}$$

Substitute this expression for M into Equation (10.1) to obtain

$$EI\frac{d^2v}{dx^2} = \frac{w_0 Lx}{4} - \frac{w_0 x^3}{3L} \tag{b}$$

Integration

To obtain the elastic curve equation, Equation (b) will be integrated twice. The first integration gives

$$EI\frac{dv}{dx} = \frac{w_0 Lx^2}{8} - \frac{w_0 x^4}{12L} + C_1 \tag{c}$$

where C_1 is a constant of integration. Integrating again gives

$$EIv = \frac{w_0 Lx^3}{24} - \frac{w_0 x^5}{60L} + C_1 x + C_2 \tag{d}$$

where C_2 is a second constant of integration.

Boundary Conditions

Moment equation (a) is valid only in the interval $0 \leq x \leq L/2$; therefore, the boundary conditions must be found in this same interval. At $x = 0$, the beam is supported by a pin connection; consequently, $v = 0$ at $x = 0$.

A common mistake in trying to solve this type of problem is to attempt to use the roller support at C as the second boundary condition. Although it is certainly true that the beam's deflection at C will be zero, we cannot use $v = 0$ at $x = L$ as a boundary condition for this problem. **Why?** We must choose a boundary condition that is within the bounds of the moment equation—that is, within the interval $0 \leq x \leq L/2$.

The second boundary condition required for evaluation of the constants of integration can be found from symmetry. The beam is symmetrically supported, and the loading is symmetrically placed on the span. Therefore, the slope of the beam at $x = L/2$ must be $dv/dx = 0$.

Evaluate Constants

Substitute the boundary condition $v = 0$ at $x = 0$ into Equation (d) to find that $C_2 = 0$.

Next, substitute the value of C_2 and the boundary condition $dv/dx = 0$ at $x = L/2$ into Equation (c), and solve for the constant of integration C_1:

$$EI\frac{dv}{dx} = \frac{w_0 Lx^2}{8} - \frac{w_0 x^4}{12L} + C_1 \quad \Rightarrow \quad EI(0) = \frac{w_0 L(L/2)^2}{8} - \frac{w_0 (L/2)^4}{12L} + C_1$$

$$\therefore C_1 = -\frac{5w_0 L^3}{192}$$

Elastic Curve Equation

Substitute the expressions obtained for C_1 and C_2 into Equation (d) to complete the elastic curve equation:

$$EIv = \frac{w_0 Lx^3}{24} - \frac{w_0 x^5}{60L} - \frac{5w_0 L^3}{192}x, \quad \text{which simplifies to} \quad v = \frac{w_0 x}{960EI}\left[40Lx^2 - \frac{16x^4}{L} - 25L^3\right] \tag{e}$$

Similarly, the beam slope equation from Equation (c) can be completed with the expression derived for C_1:

$$EI\frac{dv}{dx} = \frac{w_0 Lx^2}{8} - \frac{w_0 x^4}{12L} - \frac{5w_0 L^3}{193}, \quad \text{which simplifies to} \quad \frac{dv}{dx} = \frac{w_0 x}{192EI}\left[24Lx^2 - \frac{16x^4}{L} - 5L^3\right] \tag{f}$$

Beam Deflection at Midspan

The deflection of the beam at midspan B is obtained by setting $x = L/2$ in Equation (e):

$$EIv_B = \frac{w_0 L(L/2)^3}{24} - \frac{w_0 (L/2)^5}{60L} - \frac{5w_0 L^3}{192}(L/2)$$

$$\therefore v_B = -\frac{16w_0 L^4}{1{,}920EI} = -\frac{w_0 L^4}{120EI}$$ **Ans.**

Beam Slope at A

The slope of the beam at A is obtained by setting $x = 0$ in Equation (f):

$$EI\left(\frac{dv}{dx}\right)_A = \frac{w_0 L(0)^2}{8} - \frac{w_0 (0)^4}{12L} - \frac{5w_0 L^3}{192} \qquad \therefore \left(\frac{dv}{dx}\right)_A = -\frac{5w_0 L^3}{192}$$ **Ans.**

EXAMPLE 10.3

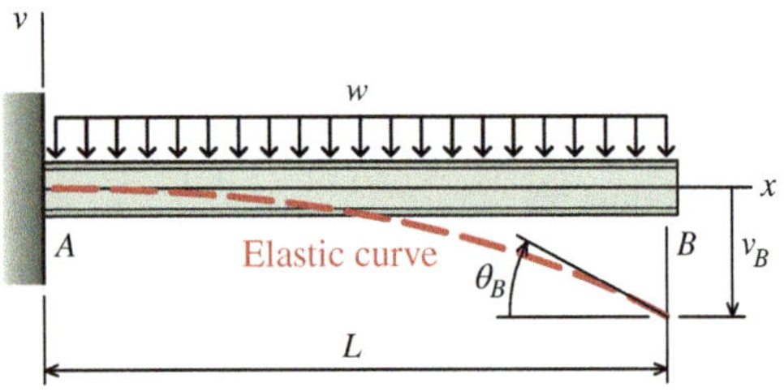

The cantilever beam shown is subjected to a uniformly distributed load w. Determine the equation of the elastic curve, as well as the deflection v_B and rotation angle θ_B of the beam at the free end of the cantilever. Assume that EI is constant for the beam.

Plan the Solution

In this example, we will consider a free-body diagram of the tip of the cantilever to illustrate how a simple coordinate transformation can simplify the analysis.

SOLUTION

Equilibrium

Before the elastic curve equation can be obtained, an equation describing the variation of the bending moment must be derived. Typically, one would begin this process by drawing a free-body diagram (FBD) of the left portion of the beam, such as the accompanying sketch. In order to complete this FBD, however, the vertical reaction force A_y and the moment reaction M_A must be determined. Perhaps it might be simpler to consider an FBD of the right portion of the cantilever, since the reactions at fixed support A do not appear on that FBD.

An FBD of the right portion of the cantilever beam is shown. A common mistake at this stage of the analysis is to define the beam length between section a–a and B as x. Note, however, that the origin of the x–v coordinate system is located at support A, with positive x extending to the right. Therefore, to be consistent with the defined coordinate system, the length of the beam segment must be denoted $L - x$. This simple coordinate transformation is the key to success with this type of problem.

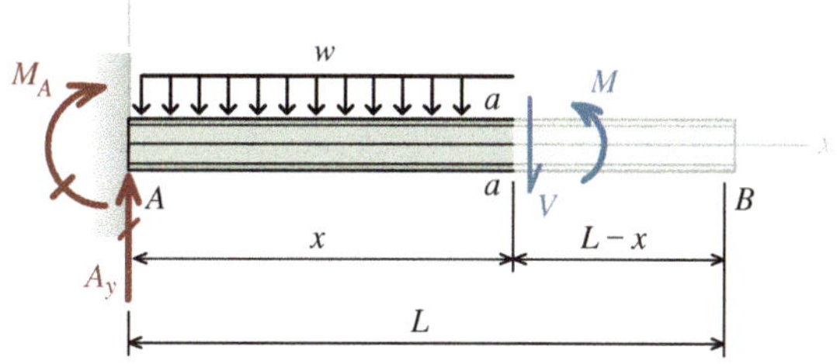

Free-body diagram of the left portion of the cantilever beam.

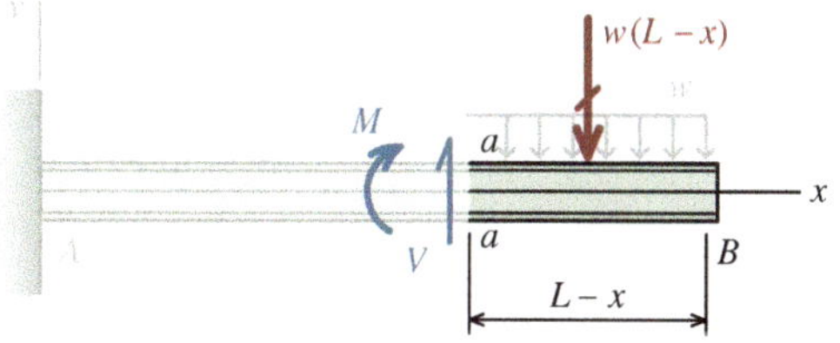

Free-body diagram of the right portion of the cantilever beam.

Accordingly, cut through the beam at section a–a, and consider the beam and its loads between a–a and the free end of the cantilever at B. Note that a clockwise internal moment M is shown acting on the beam segment at a–a. Clockwise is the positive direction for an internal moment acting on the left face of a bending element, and this direction is consistent with the sign convention shown in Figure 10.5.

The equilibrium equation for the sum of moments about a–a is

$$\Sigma M_{a-a} = -w(L - x)\left(\frac{L - x}{2}\right) - M = 0$$

Therefore, the bending-moment equation for this beam is

$$M = -\frac{w}{2}(L - x)^2 \qquad \text{(a)}$$

Notice that this equation is valid for the interval $0 \le x \le L$. Substitute the expression for M into Equation (10.1) to obtain

$$EI\frac{d^2v}{dx^2} = -\frac{w}{2}(L - x)^2 \qquad \text{(b)}$$

Integration

The first integration of Equation (b) gives

$$EI\frac{dv}{dx} = +\frac{w}{6}(L - x)^3 + C_1 \qquad \text{(c)}$$

where C_1 is a constant of integration. Note the sign change on the first term. Integrating again gives

$$EIv = -\frac{w}{24}(L - x)^4 + C_1x + C_2 \qquad \text{(d)}$$

where C_2 is a second constant of integration.

Boundary Conditions

Boundary conditions for the cantilever beam are

$$x = 0,\ v = 0 \qquad \text{and} \qquad x = 0,\ dv/dx = 0$$

Evaluate Constants

Substitute the boundary condition $dv/dx = 0$ at $x = 0$ into Equation (c) to evaluate the constant C_1:

$$EI\frac{dv}{dx} = \frac{w}{6}(L - x)^3 + C_1 \quad \Rightarrow \quad EI(0) = \frac{w}{6}(L - 0)^3 + C_1 \quad \therefore C_1 = -\frac{wL^3}{6}$$

Next, substitute the value of C_1 and the boundary condition $v = 0$ at $x = 0$ into Equation (d), and solve for the second constant of integration C_2:

$$EIv = -\frac{w}{24}(L - x)^4 + C_1x + C_2 \quad \Rightarrow \quad EI(0) = -\frac{w}{24}(L - 0)^4 - \frac{wL^3}{6}(0) + C_2$$

$$\therefore C_2 = \frac{wL^4}{24}$$

Elastic Curve Equation

Substitute the expressions obtained for C_1 and C_2 into Equation (d) to complete the elastic curve equation:

$$EIv = -\frac{w}{24}(L - x)^4 - \frac{wL^3}{6}x + \frac{wL^4}{24}, \quad \text{which simplifies to} \quad v = -\frac{wx^2}{24EI}(6L^2 - 4Lx + x^2) \qquad \text{(e)}$$

Similarly, the beam slope equation from Equation (c) can be completed with the expression derived for C_1:

$$EI\frac{dv}{dx} = \frac{w}{6}(L - x)^3 - \frac{wL^3}{6}, \quad \text{which simplifies to} \quad \frac{dv}{dx} = -\frac{wx}{6EI}(3L^2 - 3Lx + x^2) \qquad \text{(f)}$$

Beam Deflection at *B*

At the tip of the cantilever, $x = L$. Substituting this value into Equation (e) gives

$$EIv_B = -\frac{w}{24}[L - (L)]^4 - \frac{wL^3}{6}(L) + \frac{wL^4}{24} \qquad \therefore v_B = -\frac{wL^4}{8EI}$$ **Ans.**

Beam Rotation Angle at *B*

If the beam deflections are small, the rotation angle θ is equal to the slope dv/dx. Substituting $x = L$ into Equation (f) gives

$$EI\left(\frac{dv}{dx}\right)_B = \frac{w}{6}[L - (L)]^3 - \frac{wL^3}{6} \qquad \therefore \left(\frac{dv}{dx}\right)_B = -\frac{wL^3}{6EI} = \theta_B$$ **Ans.**

EXAMPLE 10.4

The simple beam shown supports a concentrated load P acting at distances a and b from the left and right supports, respectively. Determine the equations of the elastic curve. Also, determine the beam slopes at supports A and C. Assume that EI is constant for the beam.

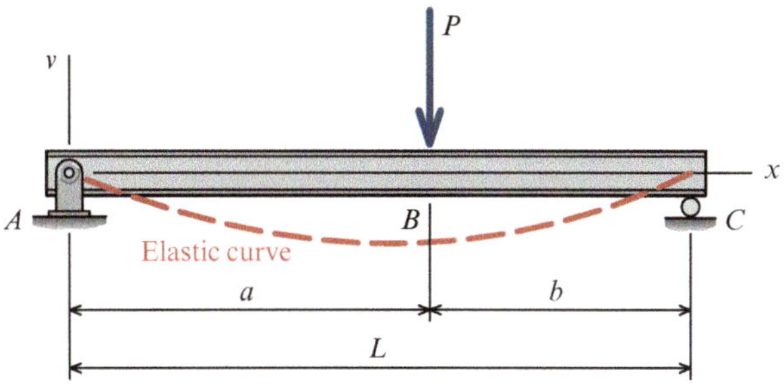

Plan the Solution

Two elastic curve equations will be required for this beam and loading: one curve that applies to the interval $0 \le x \le a$ and a second curve that applies to $a \le x \le L$. Altogether, four constants of integration will result from the double integration of two equations. Two of these constants can be evaluated from boundary conditions at the beam supports, where the beam deflections are known ($v = 0$ at $x = 0$ and $v = 0$ at $x = L$). The two remaining constants of integration will be found from *continuity conditions*. Since the beam is continuous, both sets of equations must produce the same beam slope and deflection at $x = a$, where the two elastic curves meet.

SOLUTION

Support Reactions

From equilibrium of the entire beam, the reactions at pin A and roller C are

$$A_x = 0, \qquad A_y = \frac{Pb}{L}, \qquad \text{and} \qquad C_y = \frac{Pa}{L}$$

Equilibrium

In this example, the bending moments are expressed by two equations, one for each segment of the beam. On the basis of the free-body diagrams shown here, the bending-moment equations for this beam are as follows:

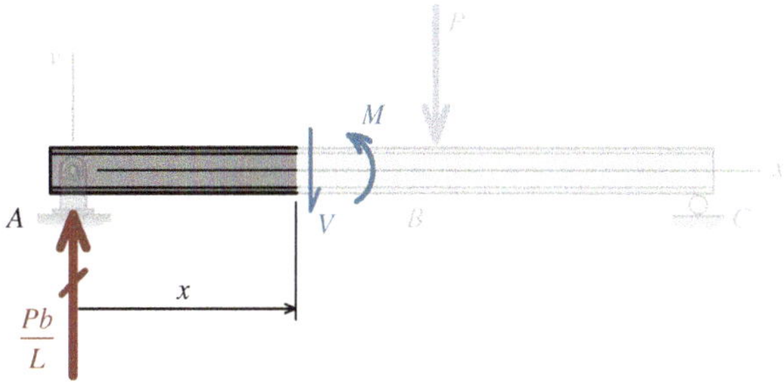

$$M = \frac{Pbx}{L} \qquad (0 \le x \le a) \tag{a}$$

$$M = \frac{Pbx}{L} - P(x - a) \qquad (a \le x \le L) \tag{b}$$

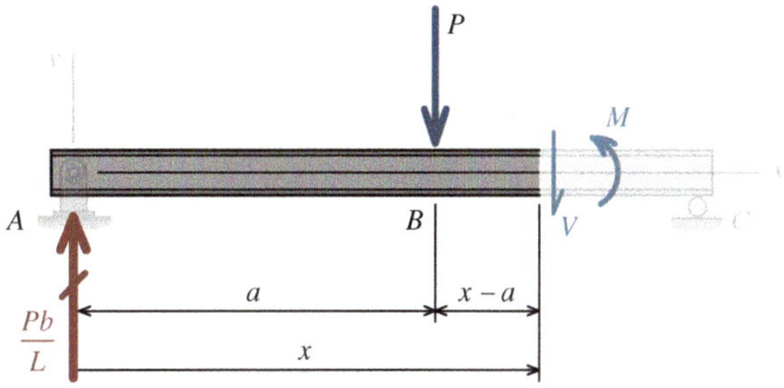

Integration over the Interval $0 \le x \le a$

Substitute Equation (a) into Equation (10.1) to obtain

$$EI\frac{d^2v}{dx^2} = \frac{Pbx}{L} \tag{c}$$

Integrate Equation (c) twice to obtain the following equations:

$$EI\frac{dv}{dx} = \frac{Pbx^2}{2L} + C_1 \tag{d}$$

$$EIv = \frac{Pbx^3}{6L} + C_1x + C_2 \tag{e}$$

Integration over the Interval $a \leq x \leq L$

Substitute Equation (b) into Equation (10.1) to obtain

$$EI\frac{d^2v}{dx^2} = \frac{Pbx}{L} - P(x-a) \tag{f}$$

Integration

Integrate Equation (f) twice to obtain the following equations:

$$EI\frac{dv}{dx} = \frac{Pbx^2}{2L} - \frac{P}{2}(x-a)^2 + C_3 \tag{g}$$

$$EIv = \frac{Pbx^3}{6L} - \frac{P}{6}(x-a)^3 + C_3x + C_4 \tag{h}$$

Equations (d), (e), (g), and (h) contain four constants of integration; therefore, four boundary and continuity conditions are required in order to evaluate the constants.

Continuity Conditions

The beam is a single, continuous member. Consequently, the two sets of equations found must produce the same slope and the same deflection at $x = a$. Consider slope equations (d) and (g). At $x = a$, these two equations must produce the same slope; therefore, set the two equations equal to each other, and substitute the value a for each variable x:

$$\frac{Pb(a)^2}{2L} + C_1 = \frac{Pb(a)^2}{2L} - \frac{P}{2}[(a)-a]^2 + C_3 \qquad \therefore C_1 = C_3 \tag{i}$$

Likewise, deflection equations (e) and (h) must give the same deflection v at $x = a$. Setting these equations equal to each other and substituting $x = a$ gives

$$\frac{Pb(a)^3}{6L} + C_1(a) + C_2 = \frac{Pb(a)^3}{6L} - \frac{P}{6}[(a)-a]^3 + C_3(a) + C_4 \qquad \therefore C_2 = C_4 \tag{j}$$

Boundary Conditions

At $x = 0$, the beam is supported by a pin connection; consequently, $v = 0$ at $x = 0$. Substitute this boundary condition into Equation (e) to find

$$EIv = \frac{Pbx^3}{6L} + C_1x + C_2 \quad \Rightarrow \quad EI(0) = \frac{Pb(0)^3}{6L} + C_1(0) + C_2 \quad \therefore C_2 = 0$$

Since $C_2 = C_4$ from Equation (j),

$$C_2 = C_4 = 0 \tag{k}$$

At $x = L$, the beam is supported by a roller connection; consequently, $v = 0$ at $x = L$. Substitute this boundary condition into Equation (h) to find

$$EIv = \frac{Pbx^3}{6L} - \frac{P}{6}(x-a)^3 + C_3x + C_4 \quad \Rightarrow \quad EI(0) = \frac{Pb(L)^3}{6L} - \frac{P}{6}(L-a)^3 + C_3(L) + C_4$$

Noting that $(L - a) = b$, simplify the latter equation to obtain

$$EI(0) = \frac{PbL^2}{6} - \frac{Pb^3}{6} + C_3L \qquad \therefore C_3 = -\frac{PbL^2}{6L} + \frac{Pb^3}{6L} = -\frac{Pb(L^2-b^2)}{6L}$$

Since $C_1 = C_3$,

$$C_1 = C_3 = -\frac{Pb(L^2 - b^2)}{6L} \tag{l}$$

Elastic Curve Equations

Substitute the expressions obtained for the constants of integration [i.e., Equations (k) and (l)] into Equations (e) and (h) to complete the elastic curve equations:

$$EIv = \frac{Pbx^3}{6L} - \frac{Pb(L^2 - b^2)}{6L}x, \quad \text{which simplifies to}$$
$$v = -\frac{Pbx}{6LEI}[L^2 - b^2 - x^2] \qquad (0 \le x \le a) \tag{m}$$

and

$$EIv = \frac{Pbx^3}{6L} - \frac{P}{6}(x-a)^3 - \frac{Pb(L^2 - b^2)}{6L}x, \quad \text{which simplifies to}$$
$$v = -\frac{Pbx}{6LEI}[L^2 - b^2 - x^2] - \frac{P(x-a)^3}{6EI} \qquad (a \le x \le L) \tag{n}$$

The slopes for the two portions of the beam can be determined by substituting the values for C_1 and C_3 into Equations (d) and (g), respectively, to obtain

$$EI\frac{dv}{dx} = \frac{Pbx^2}{2L} - \frac{Pb(L^2 - b^2)}{6L}$$
$$\therefore \frac{dv}{dx} = -\frac{Pb}{6LEI}(L^2 - b^2 - 3x^2) \qquad (0 \le x \le a) \tag{o}$$

Also,

$$EI\frac{dv}{dx} = \frac{Pbx^2}{2L} - \frac{P}{2}(x-a)^2 - \frac{Pb(L^2 - b^2)}{6L}$$
$$\therefore \frac{dv}{dx} = -\frac{Pb}{6LEI}(L^2 - b^2 - 3x^2) - \frac{P(x-a)^2}{2EI} \qquad (a \le x \le L) \tag{p}$$

The deflection v and slope dv/dx can be computed for any location x along the beam span from Equations (m), (n), (o), and (p).

Beam Slope at Supports

The slope of the beam can be determined at each support from Equations (o) and (p). At pin support A, the beam slope is found from Equation (o), with $x = 0$ and $a = L - b$:

$$\left(\frac{dv}{dx}\right)_A = -\frac{Pb}{6LEI}(L^2 - b^2) = -\frac{Pb}{6LEI}(L-b)(L+b) = -\frac{Pab(L+b)}{6LEI} \qquad \textbf{Ans.}$$

At roller support C, the beam slope is found from Equation (p), with $x = L$:

$$\left(\frac{dv}{dx}\right)_C = -\frac{Pb}{6LEI}(L^2 - b^2 - 3L^2) - \frac{P(L-a)^2}{2EI}$$
$$= \frac{Pb(2L^2 - 3bL + b^2)}{6LEI} = \frac{Pab(L+a)}{6LEI} \qquad \textbf{Ans.}$$

MecMovies

EXERCISE

M10.1 Beam Boundary Condition Game. Determine the appropriate boundary conditions needed to determine constants of integration for the double-integration method.

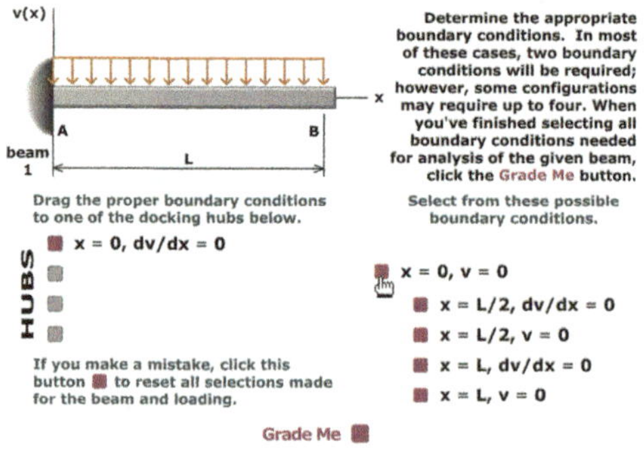

PROBLEMS

P10.1–P10.3 For the loading shown in Figures P10.1–P10.3, use the double-integration method to determine

(a) the equation of the elastic curve for the cantilever beam.
(b) the deflection at the free end.
(c) the slope at the free end.

Assume that EI is constant for each beam.

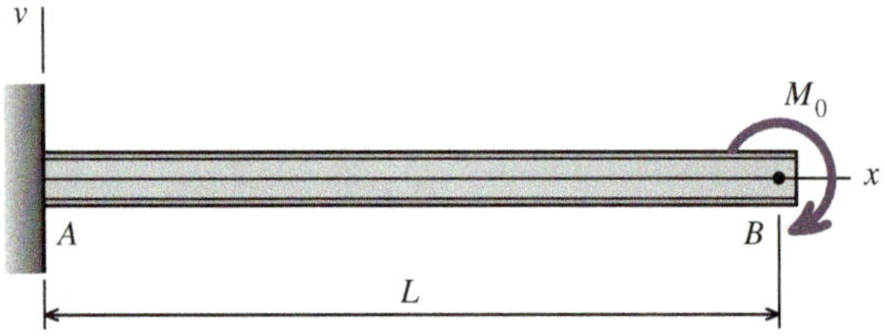

FIGURE P10.1

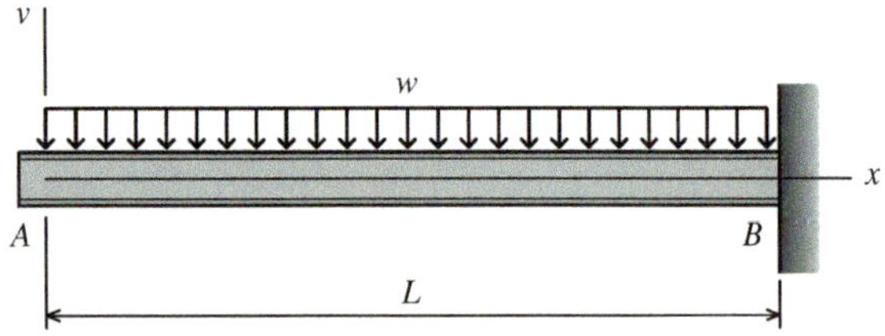

FIGURE P10.2

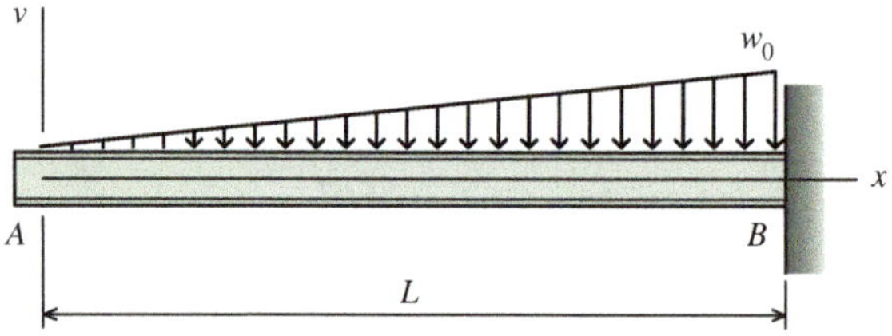

FIGURE P10.3

P10.4 For the beam and loading shown in Figure P10.4, use the double-integration method to determine

(a) the equation of the elastic curve for segment AB of the beam.
(b) the deflection at B.
(c) the slope at A.

Assume that EI is constant for the beam.

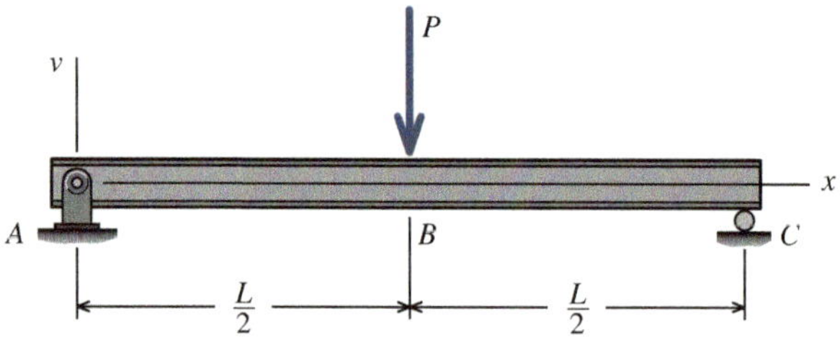

FIGURE P10.4

P10.5 For the beam and loading shown in Figure P10.5, use the double-integration method to determine

(a) the equation of the elastic curve for the beam.
(b) the slope at A.
(c) the slope at B.
(d) the deflection at midspan.

Assume that EI is constant for the beam.

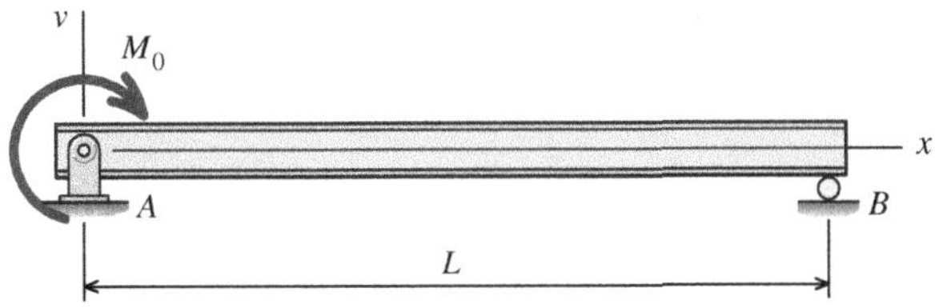

FIGURE P10.5

P10.6 For the beam and loading shown in Figure P10.6, use the double-integration method to determine

(a) the equation of the elastic curve for the beam.
(b) the maximum deflection.
(c) the slope at A.

Assume that EI is constant for the beam.

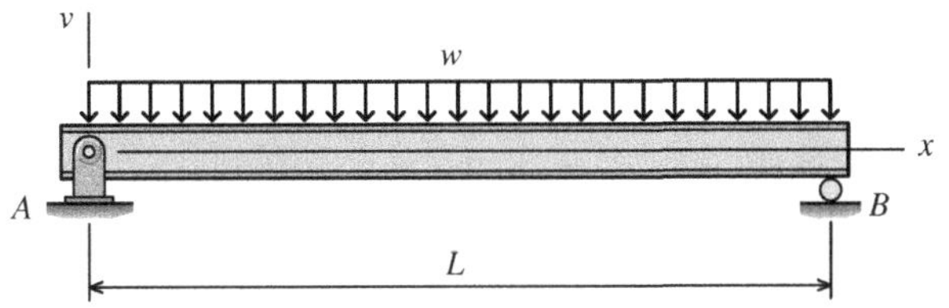

FIGURE P10.6

P10.7 For the simply supported steel beam [E = 200 GPa; I = 129 × 10^6 mm^4] shown in Figure P10.7, use the double-integration method to determine the deflection at B. Assume that L = 4 m, P = 60 kN, and w = 40 kN/m.

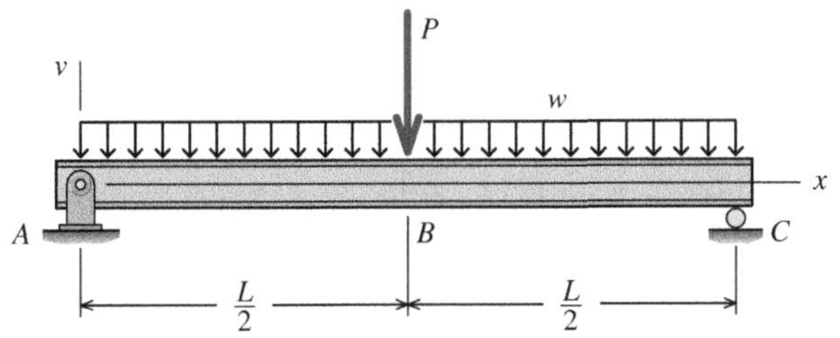

FIGURE P10.7

P10.8 For the cantilever steel beam [E = 200 GPa; I = 129 × 10^6 mm^4] shown in Figure P10.8, use the double-integration method to determine the deflection at A. Assume that L = 2.5 m, P = 50 kN, and w_0 = 90 kN/m.

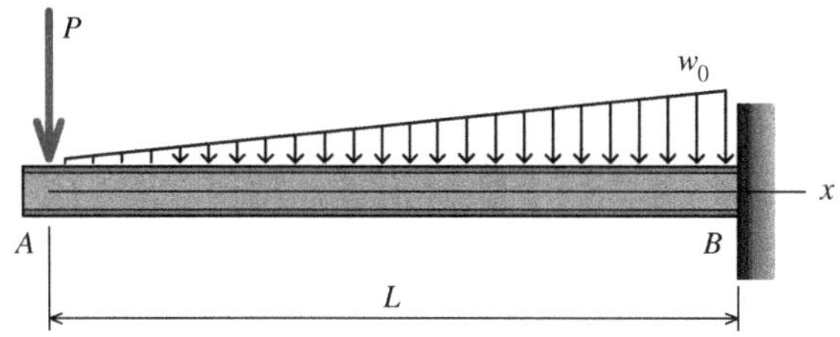

FIGURE P10.8

P10.9 For the beam and loading shown in Figure P10.9, use the double-integration method to determine

(a) the equation of the elastic curve for the cantilever beam.
(b) the deflection at the free end.
(c) the slope at the free end.

Assume that EI is constant for the beam.

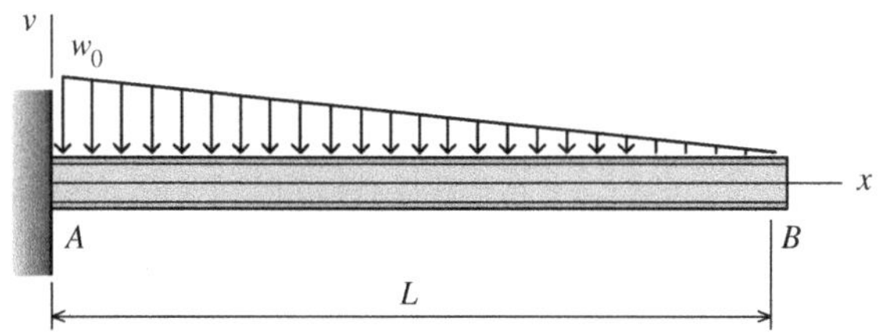

FIGURE P10.9

P10.10 For the beam and loading shown in Figure P10.10, use the double-integration method to determine

(a) the equation of the elastic curve for the cantilever beam.
(b) the deflection at B.
(c) the deflection at the free end.
(d) the slope at the free end.

Assume that EI is constant for the beam.

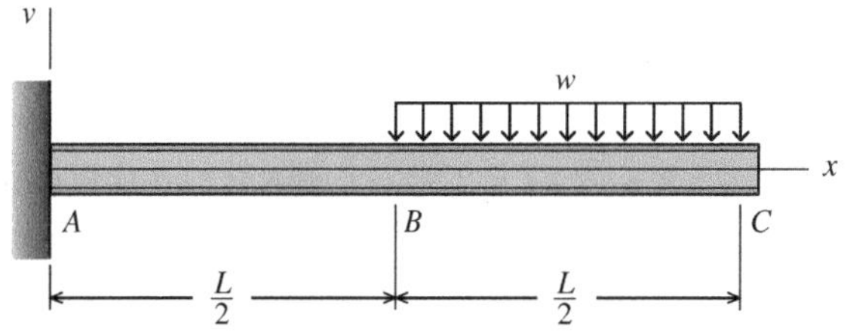

FIGURE P10.10

P10.11 For the beam and loading shown in Figure P10.11, use the double-integration method to determine

(a) the equation of the elastic curve for the beam.
(b) the deflection at B.

Assume that EI is constant for the beam.

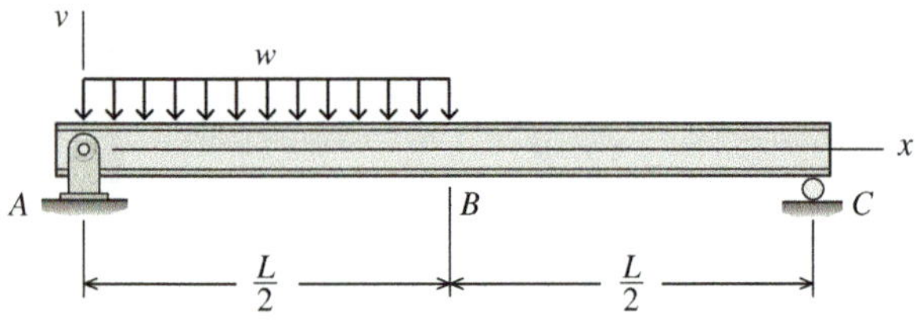

FIGURE P10.11

P10.12 For the beam and loading shown in Figure P10.12, use the double-integration method to determine

(a) the equation of the elastic curve for the beam.
(b) the location of the maximum deflection.
(c) the maximum beam deflection.

Assume that EI is constant for the beam.

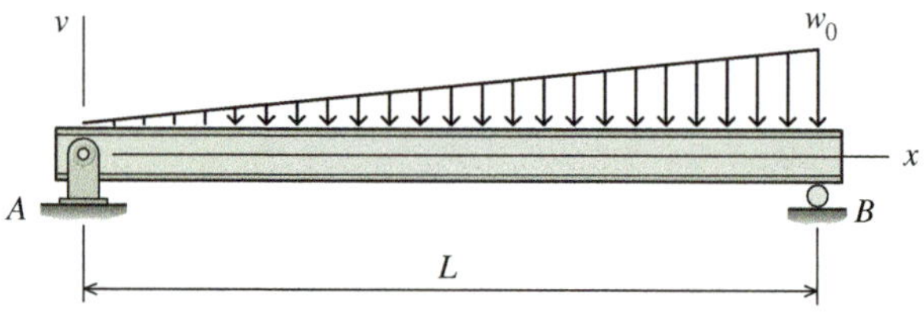

FIGURE P10.12

10.5 Determining Deflections by Integration of Shear-Force or Load Equations

In Section 10.3, the equation of the elastic curve was obtained by integrating the differential equation

$$EI\frac{d^2v}{dx^2} = M \tag{10.1}$$

and applying the appropriate bounding conditions to evaluate the two constants of integration. In a similar manner, the equation of the elastic curve can be obtained from shear-force or load equations. The differential equations that respectively relate the deflection v to the shear force V and load w are thus

$$EI\frac{d^3v}{dx^3} = V \tag{10.2}$$

and

$$EI\frac{d^4v}{dx^4} = w \tag{10.3}$$

where both V and w are functions of x. When Equation (10.2) or Equation (10.3) is used to obtain the equation of the elastic curve, either three or four integrations will be required instead of the two integrations required with Equation (10.1). These additional integrations will introduce additional constants of integration. The boundary conditions, however, now include conditions on the shear forces and bending moments, in addition to the conditions on slopes and deflections. The selection of a particular differential equation is usually based on mathematical convenience or personal preference. In those instances when the expression for the load is easier to write than the expression for the moment, Equation (10.3) would be preferred over Equation (10.1). The following example illustrates the use of Equation (10.3) for calculating beam deflections:

EXAMPLE 10.5

A beam is loaded and supported as shown. Assume that EI is constant for the beam. Determine

(a) the equation of the elastic curve in terms of w_0, L, x, E, and I.
(b) the deflection of the right end of the beam.
(c) the support reactions A_y and M_A at the left end of the beam.

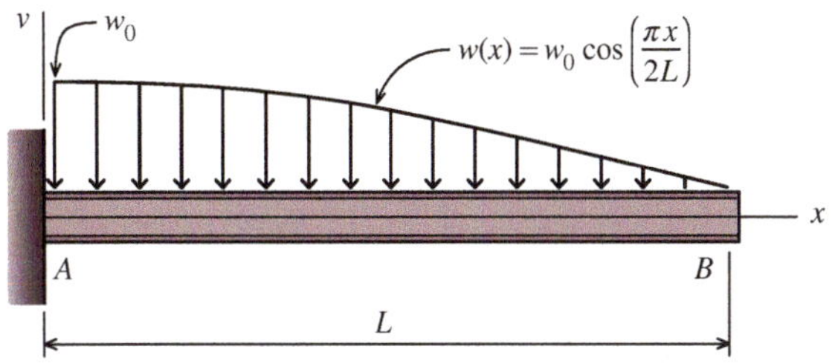

Plan the Solution

Since the equation for the load distribution is given and the moment equation is not easy to derive, Equation (10.3) will be used to determine the deflections.

SOLUTION

The upward direction is considered positive for a distributed load w; therefore, Equation (10.3) is written as

$$EI\frac{d^4v}{dx^4} = w(x) = -w_0\cos\left(\frac{\pi x}{2L}\right) \tag{a}$$

Integration

Equation (a) will be integrated four times to obtain the elastic curve equation:

$$EI\frac{d^3v}{dx^3} = V(x) = -\left(\frac{2w_0L}{\pi}\right)\sin\left(\frac{\pi x}{2L}\right) + C_1 \tag{b}$$

$$EI\frac{d^2v}{dx^2} = M(x) = \left(\frac{4w_0L^2}{\pi^2}\right)\cos\left(\frac{\pi x}{2L}\right) + C_1x + C_2 \tag{c}$$

$$EI\frac{dv}{dx} = EI\theta = \left(\frac{8w_0L^3}{\pi^3}\right)\sin\left(\frac{\pi x}{2L}\right) + C_1\frac{x^2}{2} + C_2x + C_3 \tag{d}$$

$$EIv = -\left(\frac{16w_0L^4}{\pi^4}\right)\cos\left(\frac{\pi x}{2L}\right) + C_1\frac{x^3}{6} + C_2\frac{x^2}{2} + C_3x + C_4 \tag{e}$$

Boundary Conditions and Constants

The four constants of integration are determined by applying the boundary conditions:

At $x = 0$, $v = 0$; therefore, $C_4 = \dfrac{16w_0L^4}{\pi^4}$

At $x = 0$, $\dfrac{dv}{dx} = 0$; therefore, $C_3 = 0$

At $x = L$, $V = 0$; therefore, $C_1 = \dfrac{2w_0L}{\pi}$

At $x = L$, $M = 0$; therefore, $C_2 = \dfrac{2w_0L^2}{\pi}$

411

Elastic Curve Equation

Substitute the expressions obtained for the constants of integration into Equation (e) to complete the elastic curve equation:

$$v = -\frac{w_0}{3\pi^4 EI}\left[48L^4 \cos\left(\frac{\pi x}{2L}\right) - \pi^3 Lx^3 + 3\pi^3 L^2x^2 - 48L^4\right] \qquad \textbf{Ans.}$$

Beam Deflection at Right End of Beam

The deflection of the beam at B is obtained by setting $x = L$ in the elastic curve equation:

$$v_B = -\frac{w_0}{3\pi^4 EI}[-\pi^3L^4 + 3\pi^3L^4 - 48L^4] = -\frac{(2\pi^3 - 48)w_0L^4}{3\pi^4 EI} = -0.04795\frac{w_0L^4}{EI} \qquad \textbf{Ans.}$$

Support Reactions at A

The shear force V and the bending moment M at any distance x from the support are given by the following equations derived from Equations (b) and (c):

$$V(x) = \frac{2w_0L}{\pi}\left[1 - \sin\left(\frac{\pi x}{2L}\right)\right]$$

$$M(x) = \frac{2w_0L}{\pi^2}\left[2L\cos\left(\frac{\pi x}{2L}\right) + \pi x - \pi L\right]$$

Thus, the support reactions at the left end of the beam (i.e., at $x = 0$) are

$$A_y = V_A = \frac{2w_0L}{\pi} \qquad \textbf{Ans.}$$

$$M_A = -\frac{2(\pi - 2)w_0L^2}{\pi^2} \qquad \textbf{Ans.}$$

PROBLEMS

P10.13 For the beam and loading shown in Figure P10.13, integrate the load distribution to determine

(a) the equation of the elastic curve for the beam.
(b) the maximum deflection of the beam.

Assume that EI is constant for the beam.

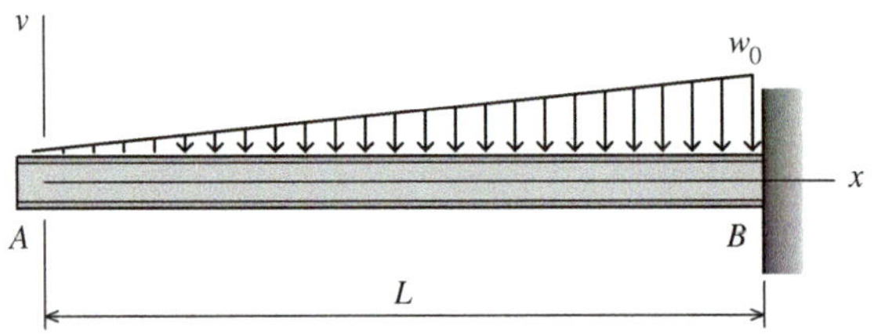

FIGURE P10.13

P10.14 For the beam and loading shown in Figure P10.14, integrate the load distribution to determine

(a) the equation of the elastic curve.
(b) the deflection at the left end of the beam.
(c) the support reactions B_y and M_B.

Assume that EI is constant for the beam.

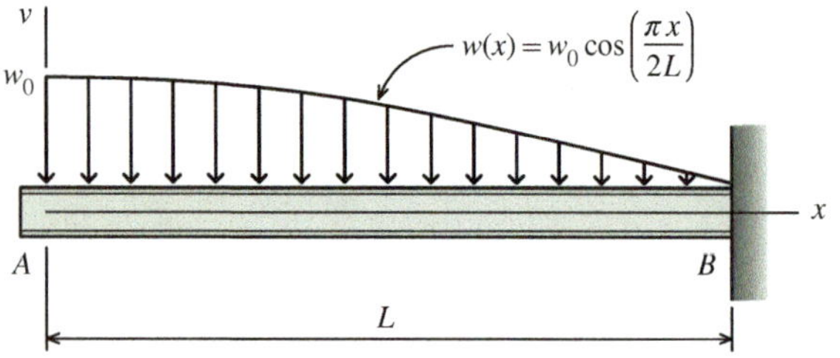

FIGURE P10.14

P10.15 For the beam and loading shown in Figure P10.15, integrate the load distribution to determine

(a) the equation of the elastic curve.
(b) the deflection midway between the supports.
(c) the slope at the left end of the beam.
(d) the support reactions A_y and B_y.

Assume that EI is constant for the beam.

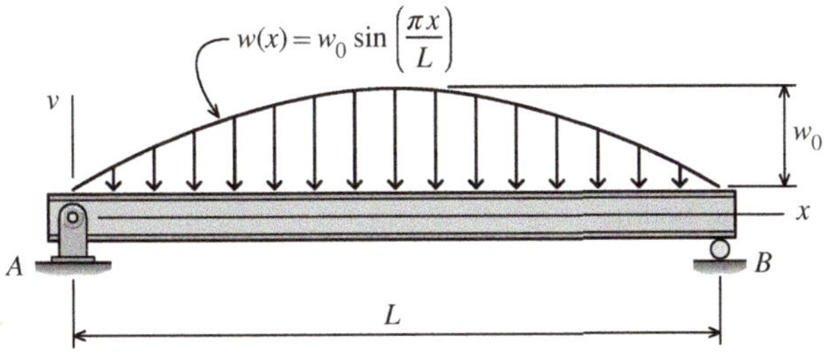

FIGURE P10.15

P10.16 For the beam and loading shown in Figure P10.16, integrate the load distribution to determine

(a) the equation of the elastic curve.
(b) the deflection midway between the supports.
(c) the slope at the left end of the beam.
(d) the support reactions A_y and B_y.

Assume that EI is constant for the beam.

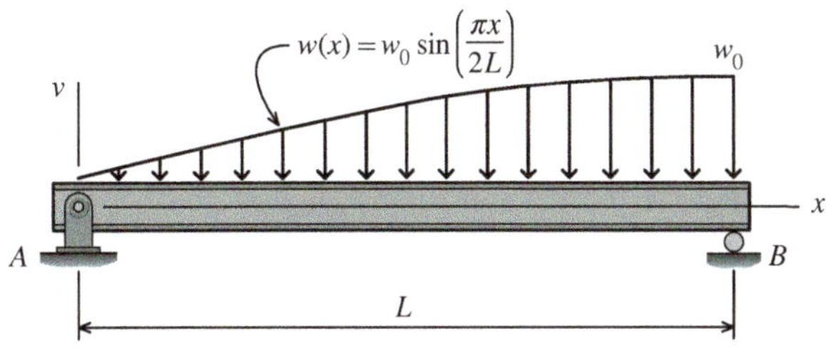

FIGURE P10.16

10.6 Determining Deflections by Using Discontinuity Functions

The integration procedures used to derive the elastic curve equations are relatively straightforward if the beam loading can be expressed as a single continuous function acting over the entire length of the beam. However, the procedures discussed in Sections 10.4 and 10.5 can become quite complicated and tedious for beams that carry multiple concentrated loads or segmented distributed loads. For example, the beam in Example 10.4 was loaded by a single concentrated load. In order to determine the elastic curve for this relatively uncomplicated beam and loading, moment equations had to be derived for two beam segments. Double integration of these two moment equations generated four constants of integration that had to be evaluated with the use of boundary conditions and continuity conditions. For beams that are more complicated, such as those with multiple concentrated loads or segmented distributed loads, it is evident that the computations required to derive all of the necessary equations and to solve for all of the constants of integration can become quite lengthy. The use of discontinuity functions, however, greatly simplifies the process. In this section, discontinuity functions will be used to determine the elastic curve for beams with several loads. These functions provide a versatile and efficient technique for the computation of deflections for both statically determinate and statically indeterminate beams with constant flexural rigidity EI. The use of discontinuity functions for statically indeterminate beams will be discussed in Section 11.4.

As discussed in Section 7.4, discontinuity functions allow all loads that act on the beam to be incorporated into a single load function $w(x)$ that is continuous for the entire length of the beam even though the loads may not be. Since $w(x)$ is a continuous function, the need for continuity conditions is eliminated, thus simplifying the calculation process. When the beam reaction forces and moments are included in $w(x)$, the constants of integration for both $V(x)$ or $M(x)$ are automatically determined without the need for explicit reference to boundary conditions. However, additional constants of integration arise in the double integration of $M(x)$ to obtain the elastic curve $v(x)$. Each integration produces one constant, and the two constants obtained must be evaluated by using the beam boundary conditions. Beginning with the moment–curvature relationship expressed in Equation (10.1), $M(x)$ is integrated to obtain $EIv'(x)$, producing a constant of integration that has the value $C_1 = EIv'(0)$. A second integration gives $EIv(x)$, and the resulting constant has the value $C_2 = EIv(0)$. For some beams, the slope or deflection or both may be known at $x = 0$, making it effortless to determine either C_1 or C_2. More typically, boundary conditions such as pin supports, roller supports, and fixed supports occur at locations other than $x = 0$. For such beams, it will be necessary to use two beam boundary conditions to develop equations containing the unknown constants C_1 and C_2. These equations are then solved simultaneously for C_1 and C_2.

The application of discontinuity functions to compute beam slopes and deflections is illustrated in Examples 10.6–10.8.

EXAMPLE 10.6

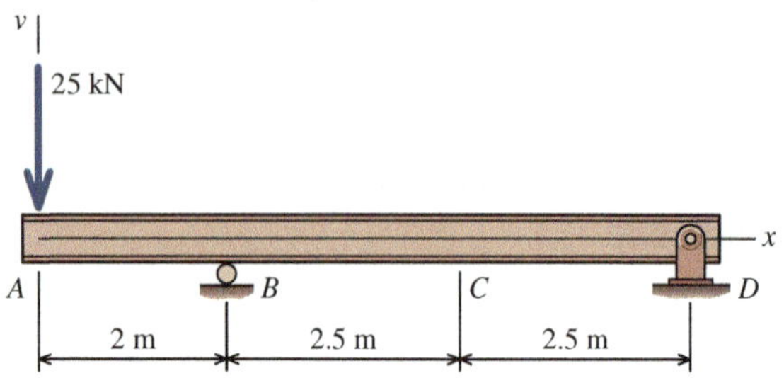

For the beam shown, use discontinuity functions to compute the deflection of the beam

(a) at A.
(b) at C.

Assume a constant value of $EI = 17 \times 10^3$ kN·m² for the beam.

Plan the Solution

Determine the reactions at the simple supports B and D. Using Table 7.2, write $w(x)$ expressions for the 25 kN concentrated load as well as the two support reactions. Integrate $w(x)$ four times to determine equations for the beam slope and deflection. Use the boundary conditions known at the simple supports to evaluate the constants of integration.

SOLUTION

Support Reactions

An FBD of the beam is shown. On the basis of this FBD, the beam reaction forces can be computed as follows:

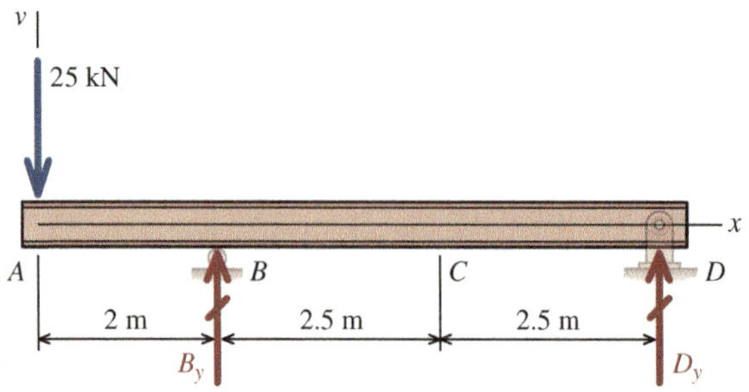

$$\Sigma M_B = (25\text{ kN})(2\text{ m}) + D_y(5\text{ m}) = 0$$
$$\therefore D_y = -10\text{ kN}$$
$$\Sigma F_y = B_y + D_y - 25\text{ kN} = 0$$
$$\therefore B_y = 35\text{ kN}$$

Discontinuity Expressions

25 kN concentrated load: Use case 2 of Table 7.2 to write the following expression for the 25 kN concentrated load:

$$w(x) = -25\text{ kN}\langle x - 0\text{ m}\rangle^{-1}$$

Reaction forces B_y and D_y: The upward reaction forces at B and D are also expressed by using case 2 of Table 7.2:

$$w(x) = 35\text{ kN}\langle x - 2\text{ m}\rangle^{-1} - 10\text{ kN}\langle x - 7\text{ m}\rangle^{-1}$$

Note that the term for the reaction force D_y will always have a value of zero in this example, since the beam is only 7 m long; therefore, this term may be omitted here.

Integrate the beam load expression: Integrate the load also expression,

$$w(x) = -25\text{ kN}\langle x - 0\text{ m}\rangle^{-1} + 35\text{ kN}\langle x - 2\text{ m}\rangle^{-1}$$

to obtain the shear-force function $V(x)$:

$$V(x) = \int w(x)\,dx = -25\text{ kN}\langle x - 0\text{ m}\rangle^0 + 35\text{ kN}\langle x - 2\text{ m}\rangle^0$$

Integrate again to obtain the bending-moment function $M(x)$:

$$M(x) = \int V(x)\,dx = -25\text{ kN}\langle x - 0\text{ m}\rangle^1 + 35\text{ kN}\langle x - 2\text{ m}\rangle^1$$

Note that, since $w(x)$ is written in terms of both the loads *and the reactions*, no constants of integration have been needed up to this point in the calculation. However, the next two integrations (which will produce functions for the beam slope and deflection) will require constants of integration that must be evaluated by using the beam boundary conditions.

From Equation (10.1), we can write

$$EI\frac{d^2v}{dx^2} = M(x) = -25\text{ kN}\langle x - 0\text{ m}\rangle^1 + 35\text{ kN}\langle x - 2\text{ m}\rangle^1$$

Integrate the moment function to obtain an expression for the beam slope:

$$EI\frac{dv}{dx} = -\frac{25\text{ kN}}{2}\langle x - 0\text{ m}\rangle^2 + \frac{35\text{ kN}}{2}\langle x - 2\text{ m}\rangle^2 + C_1 \quad \text{(a)}$$

Integrate again to obtain the beam deflection function:

$$EIv = -\frac{25\text{ kN}}{6}\langle x - 0\text{ m}\rangle^3 + \frac{35\text{ kN}}{6}\langle x - 2\text{ m}\rangle^3 + C_1x + C_2 \quad \text{(b)}$$

Evaluate the constants, using boundary conditions: Boundary conditions are specific values of the deflection v or slope dv/dx that are known at particular locations along the beam span. For this beam, the deflection v is known at the roller support ($x = 2$ m) and at the pin support ($x = 7$ m). Substitute the boundary condition $v = 0$ at $x = 2$ m into Equation (b) to obtain

$$-\frac{25\text{ kN}}{6}(2\text{ m})^3 + \frac{35\text{ kN}}{6}(0\text{ m})^3 + C_1(2\text{ m}) + C_2 = 0 \quad \text{(c)}$$

Next, substitute the boundary condition $v = 0$ at $x = 7$ m into Equation (b) to obtain

$$-\frac{25\text{ kN}}{6}(7\text{ m})^3 + \frac{35\text{ kN}}{6}(5\text{ m})^3 + C_1(7\text{ m}) + C_2 = 0 \quad \text{(d)}$$

Solve Equations (c) and (d) simultaneously for the two constants of integration C_1 and C_2:

$$C_1 = 133.3333\text{ kN}\cdot\text{m}^2 \qquad \text{and} \qquad C_2 = -233.3333\text{ kN}\cdot\text{m}^3$$

The beam slope and elastic curve equations are now complete:

$$EI\frac{dv}{dx} = -\frac{25\text{ kN}}{2}\langle x - 0\text{ m}\rangle^2 + \frac{35\text{ kN}}{2}\langle x - 2\text{ m}\rangle^2 + 133.3333\text{ kN}\cdot\text{m}^2$$

$$EIv = -\frac{25\text{ kN}}{6}\langle x - 0\text{ m}\rangle^3 + \frac{35\text{ kN}}{6}\langle x - 2\text{ m}\rangle^3 + (133.3333\text{ kN}\cdot\text{m}^2)\,x - 233.3333\text{ kN}\cdot\text{m}^3$$

(a) Beam Deflection at *A*

At the tip of the overhang, where $x = 0$ m, the beam deflection is

$$EIv_A = -\frac{25\text{ kN}}{6}\langle x - 0\text{ m}\rangle^3 + \frac{35\text{ kN}}{2}\langle x - 2\text{ m}\rangle^3 + (133.3333\text{ kN}\cdot\text{m}^2)x - 233.3333\text{ kN}\cdot\text{m}^3$$
$$= -233.3333\text{ kN}\cdot\text{m}^3$$

$$\therefore v_A = -\frac{233.3333\text{ kN}\cdot\text{m}^3}{17\times 10^3\text{ kN}\cdot\text{m}^2} = -0.013725\text{ m} = 13.73\text{ mm}\downarrow$$ **Ans.**

(b) Beam Deflection at *C*

At *C*, where $x = 4.5$ m, the beam deflection is

$$EIv_C = -\frac{25\text{ kN}}{6}(4.5\text{ m})^3 + \frac{35\text{ kN}}{6}(2.5\text{ m})^3 + (133.3333\text{ kN}\cdot\text{m}^2)(4.5\text{ m}) - 233.3333\text{ kN}\cdot\text{m}^3$$
$$= 78.1249\text{ kN}\cdot\text{m}^3$$

$$\therefore v_C = \frac{78.1249\text{ kN}\cdot\text{m}^3}{17\times 10^3\text{ kN}\cdot\text{m}^2} = 0.004596\text{ m} = 4.60\text{ mm}\uparrow$$ **Ans.**

EXAMPLE 10.7

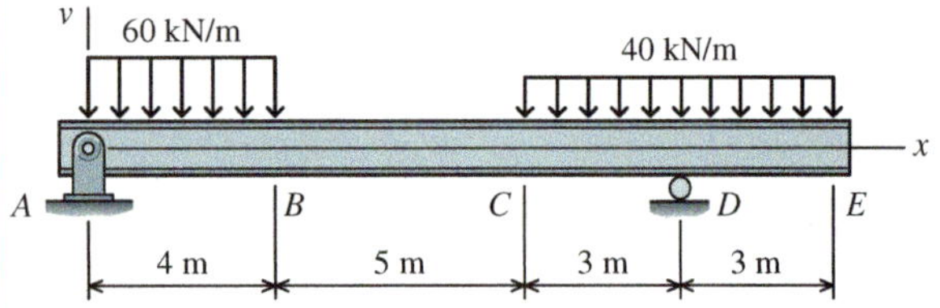

For the beam shown, use discontinuity functions to compute

(a) the slope of the beam at *A*.
(b) the deflection of the beam at *B*.

Assume a constant value of $EI = 125\times 10^3\text{ kN}\cdot\text{m}^2$ for the beam.

Plan the Solution

Determine the reactions at simple supports *A* and *D*. Using Table 7.2, write $w(x)$ expressions for the two uniformly distributed loads as well as the two support reactions. Integrate $w(x)$ four times to determine equations for the beam slope and deflection. Use the boundary conditions known at the simple supports to evaluate the constants of integration.

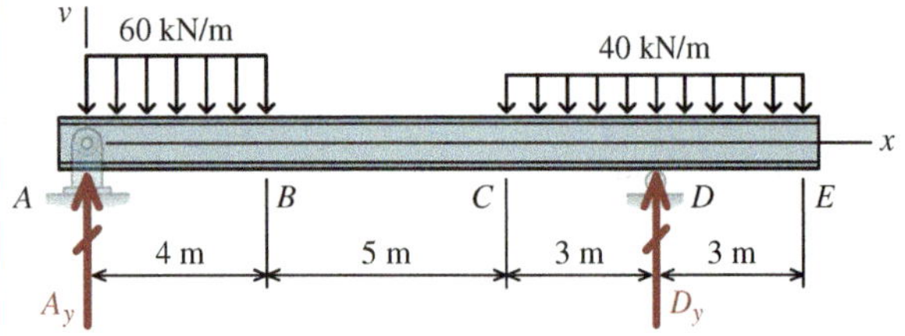

SOLUTION

Support Reactions

An FBD of the beam is shown. From this FBD, the beam reaction forces can be computed as follows:

$$\Sigma M_A = -(60\text{ kN/m})(4\text{ m})(2\text{ m}) - (40\text{ kN/m})(6\text{ m})(12\text{ m}) + D_y(12\text{ m}) = 0$$
$$\therefore D_y = 280\text{ kN}$$
$$\Sigma F_y = A_y + D_y - (60\text{ kN/m})(4\text{ m}) - (40\text{ kN/m})(6\text{ m}) = 0$$
$$\therefore A_y = 200\text{ kN}$$

Discontinuity Expressions

Distributed load between A and B: Use case 5 of Table 7.2 to write the following expression for the 60 kN/m distributed load:

$$w(x) = -60\text{ kN/m}\langle x - 0\text{ m}\rangle^0 + 60\text{ kN/m}\langle x - 4\text{ m}\rangle^0$$

Note that the second term in this expression is required in order to cancel out the first term for $x > 4$ m.

Distributed load between C and E: Again, use case 5 of Table 7.2 to write the following expression for the 40 kN/m distributed load:

$$w(x) = -40 \text{ kN/m}\langle x - 9 \text{ m}\rangle^0 + 40 \text{ kN/m}\langle x - 15 \text{ m}\rangle^0$$

The second term in this expression will have no effect, since the beam is only 15 m long; therefore, that term will be omitted from further consideration.

Reaction forces A_y and D_y: The upward reaction forces at A and D are expressed by using case 2 of Table 7.2:

$$w(x) = 200 \text{ kN}\langle x - 0 \text{ m}\rangle^{-1} + 280 \text{ kN}\langle x - 12 \text{ m}\rangle^{-1}$$

Integrate the beam load expression: The load expression $w(x)$ for the beam is thus

$$\begin{aligned} w(x) = {} & 200 \text{ kN}\langle x - 0 \text{ m}\rangle^{-1} - 60 \text{ kN/m}\langle x - 0 \text{ m}\rangle^0 + 60 \text{ kN/m}\langle x - 4 \text{ m}\rangle^0 \\ & -40 \text{ kN/m}\langle x - 9 \text{ m}\rangle^0 + 280 \text{ kN}\langle x - 12 \text{ m}\rangle^{-1} \end{aligned}$$

Integrate $w(x)$ to obtain the shear-force function $V(x)$:

$$\begin{aligned} V(x) = \int w(x)\,dx = {} & 200 \text{ kN}\langle x - 0 \text{ m}\rangle^0 - 60 \text{ kN/m}\langle x - 0 \text{ m}\rangle^1 + 60 \text{ kN/m}\langle x - 4 \text{ m}\rangle^1 \\ & -40 \text{ kN/m}\langle x - 9 \text{ m}\rangle^1 + 280 \text{ kN}\langle x - 12 \text{ m}\rangle^0 \end{aligned}$$

Then integrate again to obtain the bending-moment function $M(x)$:

$$\begin{aligned} M(x) = \int V(x)\,dx = {} & 200 \text{ kN}\langle x - 0 \text{ m}\rangle^1 - \frac{60 \text{ kN/m}}{2}\langle x - 0 \text{ m}\rangle^2 + \frac{60 \text{ kN/m}}{2}\langle x - 4 \text{ m}\rangle^2 \\ & - \frac{40 \text{ kN/m}}{2}\langle x - 9 \text{ m}\rangle^2 + 280 \text{ kN}\langle x - 12 \text{ m}\rangle^1 \end{aligned}$$

The inclusion of the reaction forces in the expression for $w(x)$ has automatically accounted for the constants of integration up to this point. However, the next two integrations (which will produce functions for the beam slope and deflection) will require constants of integration that must be evaluated from the beam boundary conditions.

From Equation (10.1), we can write

$$\begin{aligned} EI\frac{d^2v}{dx^2} = M(x) = {} & 200 \text{ kN}\langle x - 0 \text{ m}\rangle^1 - \frac{60 \text{ kN/m}}{2}\langle x - 0 \text{ m}\rangle^2 + \frac{60 \text{ kN/m}}{2}\langle x - 4 \text{ m}\rangle^2 \\ & - \frac{40 \text{ kN/m}}{2}\langle x - 9 \text{ m}\rangle^2 + 280 \text{ kN}\langle x - 12 \text{ m}\rangle^1 \end{aligned}$$

Integrate the moment function to obtain an expression for the beam slope:

$$\begin{aligned} EI\frac{dv}{dx} = {} & \frac{200 \text{ kN}}{2}\langle x - 0 \text{ m}\rangle^2 - \frac{60 \text{ kN/m}}{6}\langle x - 0 \text{ m}\rangle^3 + \frac{60 \text{ kN/m}}{6}\langle x - 4 \text{ m}\rangle^3 \\ & - \frac{40 \text{ kN/m}}{6}\langle x - 9 \text{ m}\rangle^3 + \frac{280 \text{ kN}}{2}\langle x - 12 \text{ m}\rangle^2 + C_1 \end{aligned} \qquad \text{(a)}$$

Integrate again to obtain the beam deflection function:

$$EIv = \frac{200\text{ kN}}{6}\langle x - 0\text{ m}\rangle^3 - \frac{60\text{ kN/m}}{24}\langle x - 0\text{ m}\rangle^4 + \frac{60\text{ kN/m}}{24}\langle x - 4\text{ m}\rangle^4 - \frac{40\text{ kN/m}}{24}\langle x - 9\text{ m}\rangle^4 + \frac{280\text{ kN}}{6}\langle x - 12\text{ m}\rangle^3 + C_1x + C_2 \quad \text{(b)}$$

Evaluate the constants, using boundary conditions: Boundary conditions are specific values of the deflection v or slope dv/dx that are known at particular locations along the beam span. For this beam, the deflection v is known at the pin support ($x = 0$ m) and at the roller support ($x = 12$ m). Substitute the boundary condition $v = 0$ at $x = 0$ m into Equation (b) to obtain

$$C_2 = 0$$

Next, substitute the boundary condition $v = 0$ at $x = 12$ m into Equation (b) to obtain the constant C_1:

$$\frac{200\text{ kN}}{6}(12\text{ m})^3 - \frac{60\text{ kN/m}}{24}(12\text{ m})^4 + \frac{60\text{ kN/m}}{24}(8\text{ m})^4 - \frac{40\text{ kN/m}}{24}(3\text{ m})^4 + C_1(12\text{ m}) = 0$$

$$\therefore C_1 = -1{,}322.0833\text{ kN}\cdot\text{m}^2$$

The beam slope and elastic curve equations are now complete:

$$EI\frac{dv}{dx} = \frac{200\text{ kN}}{2}\langle x - 0\text{ m}\rangle^2 - \frac{60\text{ kN/m}}{6}\langle x - 0\text{ m}\rangle^3 + \frac{60\text{ kN/m}}{6}\langle x - 4\text{ m}\rangle^3 - \frac{40\text{ kN/m}}{6}\langle x - 9\text{ m}\rangle^3 + \frac{280\text{ kN}}{2}\langle x - 12\text{ m}\rangle^2 - 1{,}322.0833\text{ kN}\cdot\text{m}^2$$

$$EIv = \frac{200\text{ kN}}{6}\langle x - 0\text{ m}\rangle^3 - \frac{60\text{ kN/m}}{24}\langle x - 0\text{ m}\rangle^4 + \frac{60\text{ kN/m}}{24}\langle x - 4\text{ m}\rangle^4 - \frac{40\text{ kN/m}}{24}\langle x - 9\text{ m}\rangle^4 + \frac{280\text{ kN}}{6}\langle x - 12\text{ m}\rangle^3 - (1{,}322.0833\text{ kN}\cdot\text{m}^2)x$$

(a) Beam Slope at *A*

The beam slope at A ($x = 0$ m) is

$$EI\left(\frac{dv}{dx}\right)_A = -1{,}322.0833\text{ kN}\cdot\text{m}^2$$

$$\therefore \left(\frac{dv}{dx}\right)_A = -\frac{1{,}322.0833\text{ kN}\cdot\text{m}^2}{125\times10^3\text{ kN}\cdot\text{m}^2} = -0.01058\text{ rad} \qquad \textbf{Ans.}$$

(b) Beam Deflection at *B*

The beam deflection at B ($x = 4$ m) is

$$EIv_B = \frac{200\text{ kN}}{6}(4\text{ m})^3 - \frac{60\text{ kN/m}}{24}(4\text{ m})^4 - (1{,}322.0833\text{ kN}\cdot\text{m}^2)(4\text{ m}) = -3{,}795\text{ kN}\cdot\text{m}^3$$

$$\therefore v_B = -\frac{3{,}795\text{ kN}\cdot\text{m}^3}{125\times10^3\text{ kN}\cdot\text{m}^2} = -0.030360\text{ m} = 30.4\text{ mm}\downarrow \qquad \textbf{Ans.}$$

EXAMPLE 10.8

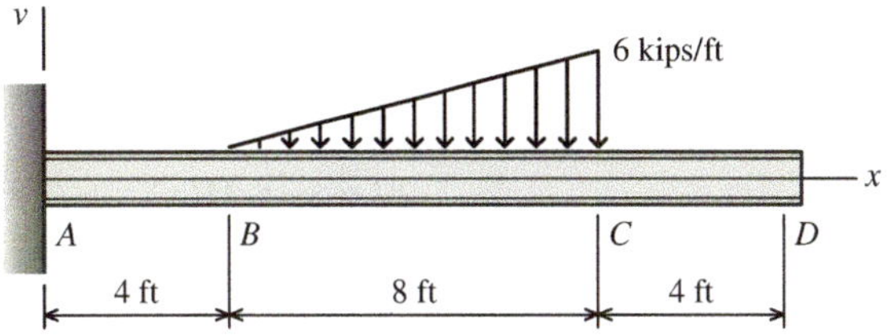

For the beam shown, use discontinuity functions to compute the deflection at D. Assume a constant value of EI = 192,000 kip · ft² for the beam.

Plan the Solution

Determine the reactions at the fixed support A. Using Table 7.2, write $w(x)$ expressions for the linearly distributed load as well as the two support reactions. Integrate $w(x)$ four times to determine equations for the beam slope and deflection. Use the boundary conditions known at the fixed support to evaluate the constants of integration.

SOLUTION

Support Reactions

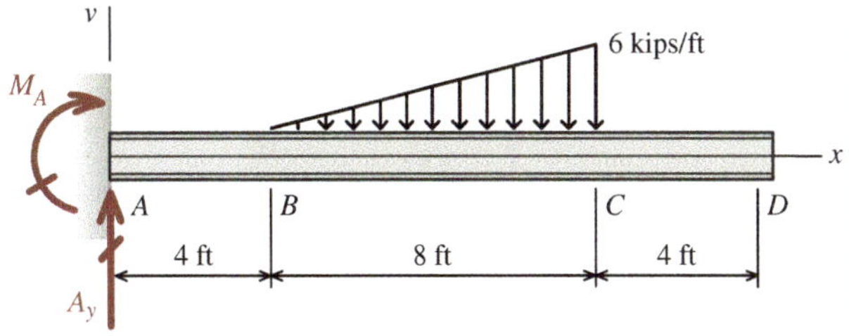

An FBD of the beam is shown. On the basis of this FBD, the beam reaction forces can be computed as follows:

$$\Sigma F_y = A_y - \frac{1}{2}(6\text{ kips/ft})(8\text{ ft}) = 0$$

$$\therefore A_y = 24\text{ kips}$$

$$\Sigma M_A = -M_A - \frac{1}{2}(6\text{ kips/ft})(8\text{ ft})\left[4\text{ ft} + \frac{2(8\text{ ft})}{3}\right] = 0$$

$$\therefore M_A = -224\text{ kip}\cdot\text{ft}$$

Discontinuity Expressions

Distributed load between B and C: Use case 6 of Table 7.2 to write the following expression for the distributed load:

$$w(x) = -\frac{6\text{ kips/ft}}{8\text{ ft}}\langle x - 4\text{ ft}\rangle^1 + \frac{6\text{ kips/ft}}{8\text{ ft}}\langle x - 12\text{ ft}\rangle^1 + 6\text{ kips/ft}\langle x - 12\text{ ft}\rangle^0$$

Reaction forces A_y and M_A: The reaction forces at A are expressed with the use of cases 1 and 2 of Table 7.2:

$$w(x) = -224\text{ kip}\cdot\text{ft}\langle x - 0\text{ ft}\rangle^{-2} + 24\text{ kips}\langle x - 0\text{ ft}\rangle^{-1}$$

Integrate the beam load expression: The load expression $w(x)$ for the beam is thus

$$w(x) = -224\text{ kip}\cdot\text{ft}\langle x - 0\text{ ft}\rangle^{-2} + 24\text{ kips}\langle x - 0\text{ ft}\rangle^{-1} - \frac{6\text{ kips/ft}}{8\text{ ft}}\langle x - 4\text{ ft}\rangle^1 + \frac{6\text{ kips/ft}}{8\text{ ft}}\langle x - 12\text{ ft}\rangle^1 + 6\text{ kips/ft}\langle x - 12\text{ ft}\rangle^0$$

Integrate $w(x)$ to obtain the shear-force function $V(x)$:

$$V(x) = \int w(x)\,dx = -224\text{ kip}\cdot\text{ft}\langle x - 0\text{ ft}\rangle^{-1} + 24\text{ kips}\langle x - 0\text{ ft}\rangle^0 - \frac{6\text{ kips/ft}}{2(8\text{ ft})}\langle x - 4\text{ ft}\rangle^2 + \frac{6\text{ kips/ft}}{2(8\text{ ft})}\langle x - 12\text{ ft}\rangle^2 + 6\text{ kips/ft}\langle x - 12\text{ ft}\rangle^1$$

Then integrate again to obtain the bending-moment function $M(x)$:

$$M(x) = \int V(x)\,dx = -224\text{ kip}\cdot\text{ft}\langle x - 0\text{ ft}\rangle^0 + 24\text{ kips}\langle x - 0\text{ ft}\rangle^1 - \frac{6\text{ kips/ft}}{6(8\text{ ft})}\langle x - 4\text{ ft}\rangle^3 + \frac{6\text{ kips/ft}}{6(8\text{ ft})}\langle x - 12\text{ ft}\rangle^3 + \frac{6\text{ kips/ft}}{2}\langle x - 12\text{ ft}\rangle^2$$

The inclusion of the reaction forces in the expression for $w(x)$ has automatically accounted for the constants of integration up to this point. However, the next two integrations (which will produce functions for the beam slope and deflection) will require constants of integration that must be evaluated from the beam boundary conditions.

From Equation (10.1), we can write

$$EI\frac{d^2v}{dx^2} = M(x) = -224\text{ kip}\cdot\text{ft}\langle x - 0\text{ ft}\rangle^0 + 24\text{ kips}\,\langle x - 0\text{ ft}\rangle^1$$
$$-\frac{6\text{ kips/ft}}{6(8\text{ ft})}\langle x - 4\text{ ft}\rangle^3 + \frac{6\text{ kips/ft}}{6(8\text{ ft})}\langle x - 12\text{ ft}\rangle^3 + \frac{6\text{ kips/ft}}{2}\langle x - 12\text{ ft}\rangle^2$$

Integrate the moment function to obtain an expression for the beam slope:

$$EI\frac{dv}{dx} = -224\text{ kip}\cdot\text{ft}\langle x - 0\text{ ft}\rangle^1 + \frac{24\text{ kips}}{2}\langle x - 0\text{ ft}\rangle^2$$
$$-\frac{6\text{ kips/ft}}{24(8\text{ ft})}\langle x - 4\text{ ft}\rangle^4 + \frac{6\text{ kips/ft}}{24(8\text{ ft})}\langle x - 12\text{ ft}\rangle^4 + \frac{6\text{ kips/ft}}{6}\langle x - 12\text{ ft}\rangle^3 + C_1 \qquad \text{(a)}$$

Integrate again to obtain the beam deflection function:

$$EIv = -\frac{224\text{ kip}\cdot\text{ft}}{2}\langle x - 0\text{ ft}\rangle^2 + \frac{24\text{ kips}}{6}\langle x - 0\text{ ft}\rangle^3$$
$$-\frac{6\text{ kips/ft}}{120(8\text{ ft})}\langle x - 4\text{ ft}\rangle^5 + \frac{6\text{ kips/ft}}{120(8\text{ ft})}\langle x - 12\text{ ft}\rangle^5 + \frac{6\text{ kips/ft}}{24}\langle x - 12\text{ ft}\rangle^4 + C_1x + C_2 \qquad \text{(b)}$$

Evaluate the constants, using boundary conditions: For this beam, the slope and the deflection are known at $x = 0$ ft. Substitute the boundary condition $dv/dx = 0$ at $x = 0$ ft into Equation (a) to obtain

$$C_1 = 0$$

Next, substitute the boundary condition $v = 0$ at $x = 0$ ft into Equation (b) to obtain the constant C_2:

$$C_2 = 0$$

The beam slope and elastic curve equations are now complete:

$$EI\frac{dv}{dx} = -224\text{ kip}\cdot\text{ft}\langle x - 0\text{ ft}\rangle^1 + \frac{24\text{ kips}}{2}\langle x - 0\text{ ft}\rangle^2$$
$$-\frac{6\text{ kips/ft}}{24(8\text{ ft})}\langle x - 4\text{ ft}\rangle^4 + \frac{6\text{ kips/ft}}{24(8\text{ ft})}\langle x - 12\text{ ft}\rangle^4 + \frac{6\text{ kips/ft}}{6}\langle x - 12\text{ ft}\rangle^3$$
$$EIv = -\frac{224\text{ kip}\cdot\text{ft}}{2}\langle x - 0\text{ ft}\rangle^2 + \frac{24\text{ kips}}{6}\langle x - 0\text{ ft}\rangle^3$$
$$-\frac{6\text{ kips/ft}}{120(8\text{ ft})}\langle x - 4\text{ ft}\rangle^5 + \frac{6\text{ kips/ft}}{120(8\text{ ft})}\langle x - 12\text{ ft}\rangle^5 + \frac{6\text{ kips/ft}}{24}\langle x - 12\text{ ft}\rangle^4$$

Beam deflection at D: The beam deflection at D (where $x = 16$ ft) is computed as follows:

$$EIv_D = -\frac{224\text{ kip}\cdot\text{ft}}{2}(16\text{ ft})^2 + \frac{24\text{ kips}}{6}(16\text{ ft})^3 - \frac{6\text{ kips/ft}}{120(8\text{ ft})}(12\text{ ft})^5 + \frac{6\text{ kips/ft}}{120(8\text{ ft})}(4\text{ ft})^5 + \frac{6\text{ kips/ft}}{24}(4\text{ ft})^4$$
$$= -13{,}772.8\text{ kip}\cdot\text{ft}^3$$

$$\therefore\ v_D = -\frac{13{,}772.8\text{ kip}\cdot\text{ft}^3}{192{,}000\text{ kip}\cdot\text{ft}^2} = -0.071733\text{ ft} = 0.861\text{ in.}\downarrow$$

Ans.

PROBLEMS

P10.17 For the beam and loading shown in Figure P10.17, use discontinuity functions to compute the deflection of the beam at D. Assume a constant value of $EI = 1{,}750 \text{ kip}\cdot\text{ft}^2$ for the beam.

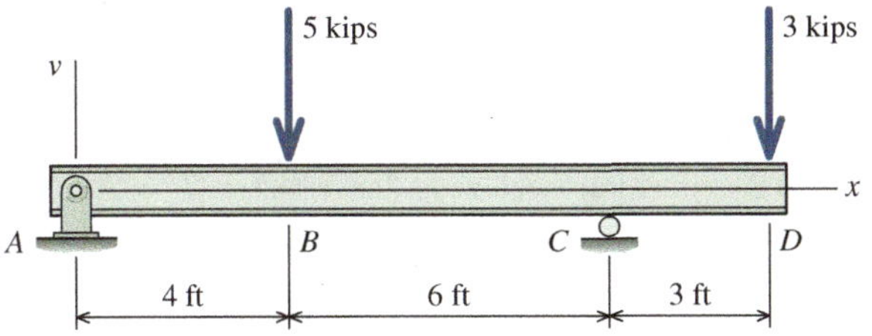

FIGURE P10.17

P10.18 The solid 30 mm diameter steel [E = 200 GPa] shaft shown in Figure P10.18 supports two pulleys. For the loading shown, use discontinuity functions to compute

(a) the deflection of the shaft at pulley B.
(b) the deflection of the shaft at pulley C.

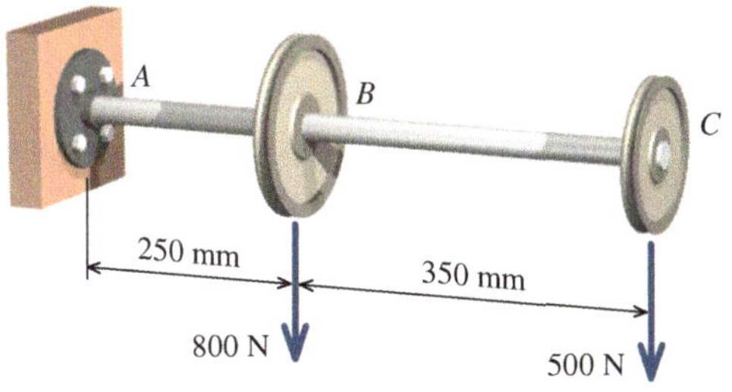

FIGURE P10.18

P10.19 For the beam and loading shown in Figure P10.19, use discontinuity functions to compute

(a) the slope of the beam at C.
(b) the deflection of the beam at C.

Assume a constant value of $EI = 560 \times 10^6 \text{ N}\cdot\text{mm}^2$ for the beam.

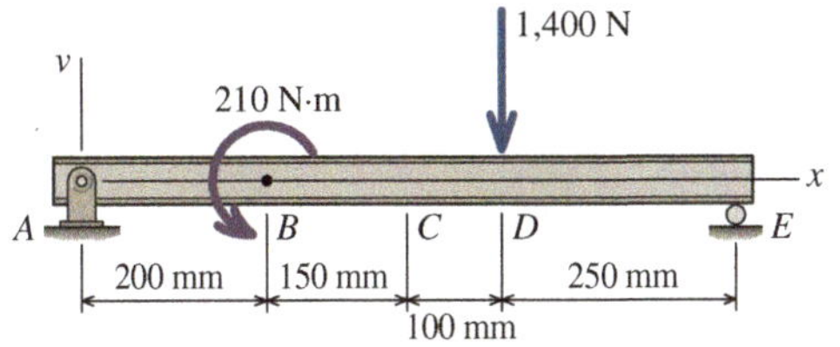

FIGURE P10.19

P10.20 The solid 30 mm diameter steel [E = 200 GPa] shaft shown in Figure P10.20 supports two belt pulleys. Assume that the bearing at A can be idealized as a pin support and that the bearing at E can be idealized as a roller support. For the loading shown, use discontinuity functions to compute

(a) the deflection of the shaft at pulley B.
(b) the deflection of the shaft at point C.

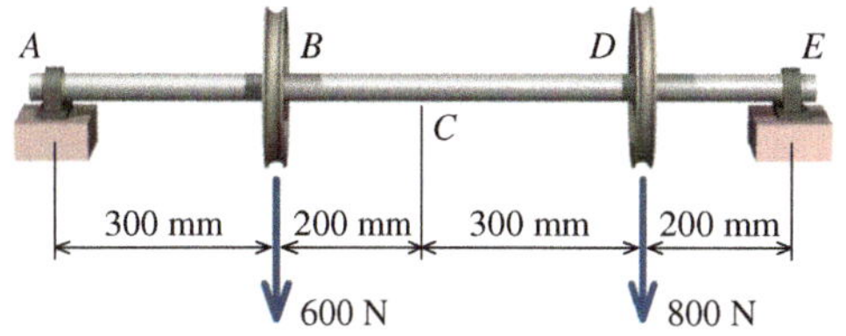

FIGURE P10.20

P10.21 The cantilever beam shown in Figure P10.21 consists of a W530 × 74 structural steel wide-flange shape [E = 200 GPa; $I = 410 \times 10^6 \text{ mm}^4$]. Use discontinuity functions to compute the deflection of the beam at C for the loading shown.

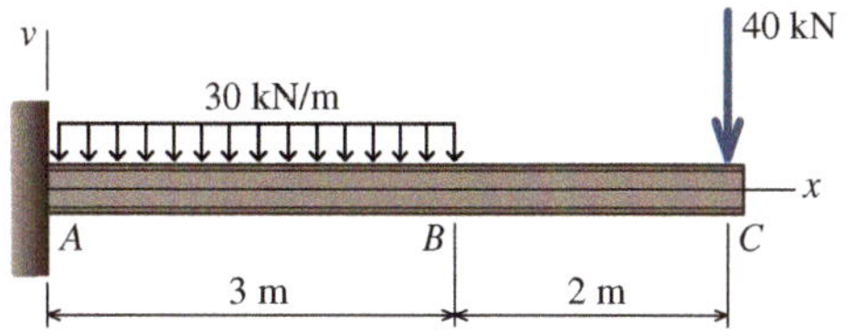

FIGURE P10.21

P10.22 The cantilever beam shown in Figure P10.22 consists of a W21 × 50 structural steel wide-flange shape [E = 29,000 ksi; $I = 984 \text{ in.}^4$]. Use discontinuity functions to compute the deflection of the beam at D for the loading shown.

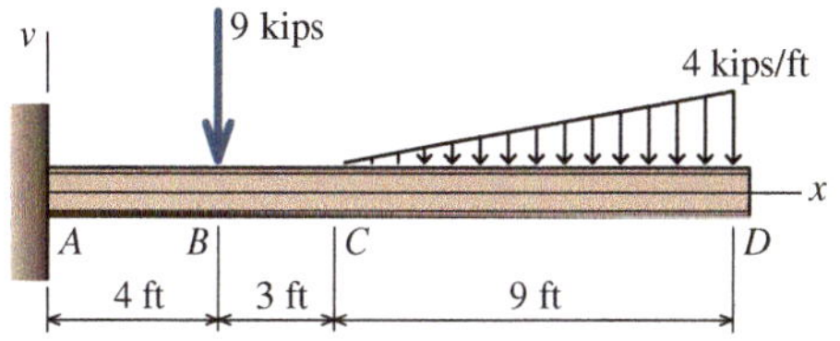

FIGURE P10.22

P10.23 The simply supported beam shown in Figure P10.23 consists of a W410 × 85 structural steel wide-flange shape [E = 200 GPa; $I = 316 \times 10^6 \text{ mm}^4$]. For the loading shown, use discontinuity functions to compute

(a) the slope of the beam at A.
(b) the deflection of the beam at midspan.

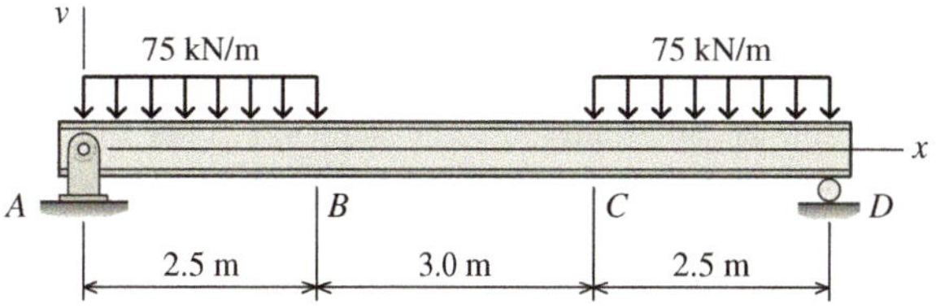

FIGURE P10.23

P10.24 The simply supported beam shown in Figure P10.24 consists of a W14 × 30 structural steel wide-flange shape [E = 29,000 ksi; I = 291 in.^{4}]. For the loading shown, use discontinuity functions to compute

(a) the slope of the beam at A.
(b) the deflection of the beam at midspan.

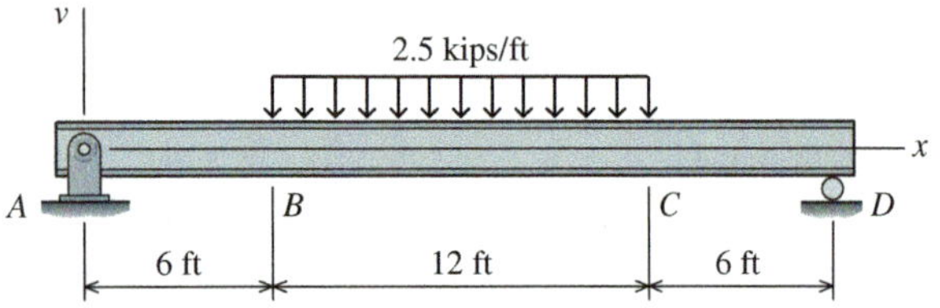

FIGURE P10.24

P10.25 The simply supported beam shown in Figure P10.25 consists of a W21 × 50 structural steel wide-flange shape [E = 29,000 ksi; I = 984 in.^{4}]. For the loading shown, use discontinuity functions to compute

(a) the slope of the beam at A.
(b) the deflection of the beam at B.

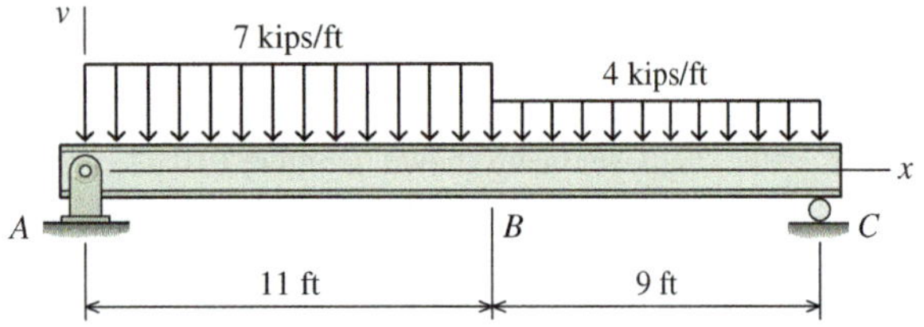

FIGURE P10.25

P10.26 The simply supported beam shown in Figure P10.26 consists of a W200 × 59 structural steel wide-flange shape [E = 200 GPa; I = 60.8 × 10^6 mm^4]. For the loading shown, use discontinuity functions to compute

(a) the deflection of the beam at C.
(b) the deflection of the beam at F.

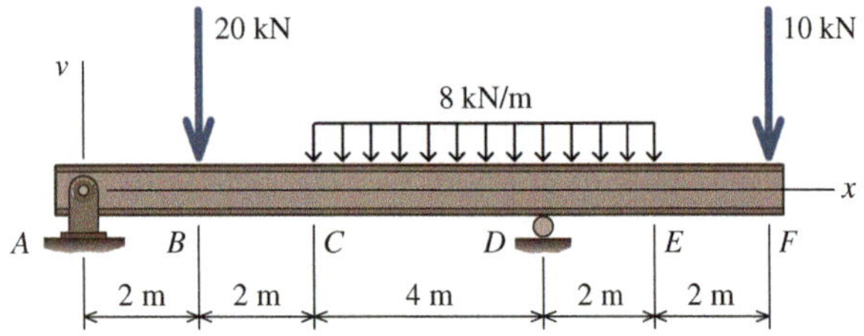

FIGURE P10.26

P10.27 The solid 0.50 in. diameter steel [E = 30,000 ksi] shaft shown in Figure P10.27 supports two belt pulleys. Assume that the bearing at B can be idealized as a pin support and that the bearing at D can be idealized as a roller support. For the loading shown, use discontinuity functions to compute

(a) the shaft deflection at pulley A.
(b) the shaft deflection at pulley C.

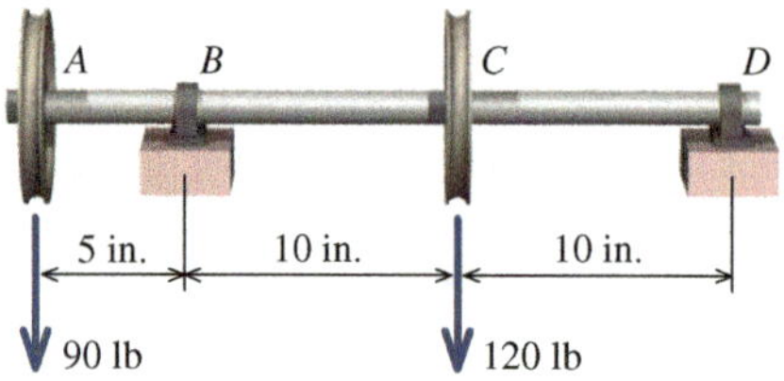

FIGURE P10.27

P10.28 The cantilever beam shown in Figure P10.28 consists of a W8 × 31 structural steel wide-flange shape [E = 29,000 ksi; I = 110 in.^{4}]. For the loading shown, use discontinuity functions to compute

(a) the slope of the beam at A.
(b) the deflection of the beam at A.

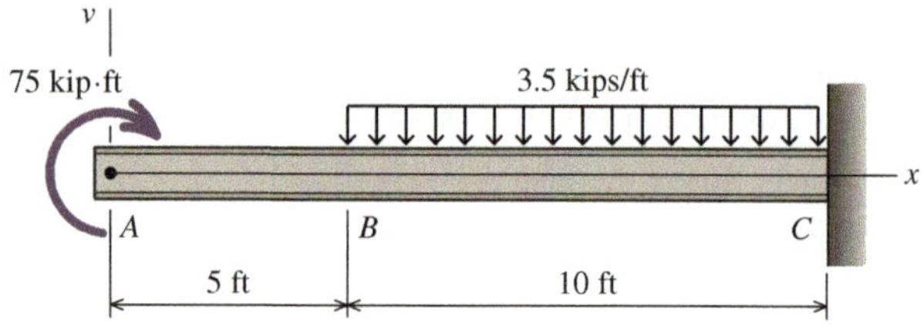

FIGURE P10.28

P10.29 The simply supported beam shown in Figure P10.29 consists of a W14 × 34 structural steel wide-flange shape [E = 29,000 ksi; I = 340 in.^{4}]. For the loading shown, use discontinuity functions to compute

(a) the slope of the beam at E.
(b) the deflection of the beam at C.

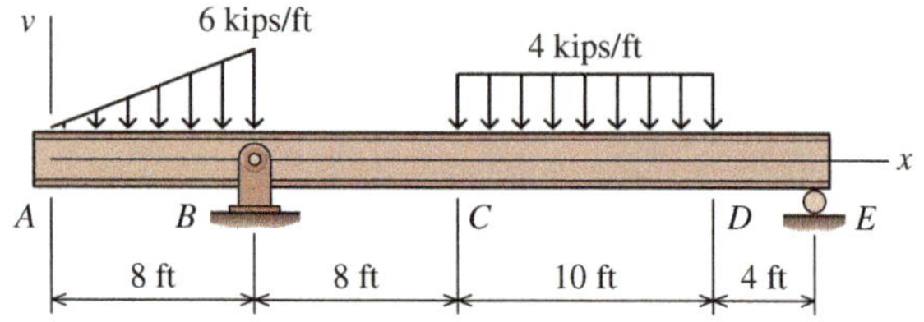

FIGURE P10.29

P10.30 For the beam and loading shown in Figure P10.30, use discontinuity functions to compute

(a) the deflection of the beam at A.
(b) the deflection of the beam at midspan (i.e., at x = 2.5 m).

Assume a constant value of EI = 1,500 kN · m^2 for the beam.

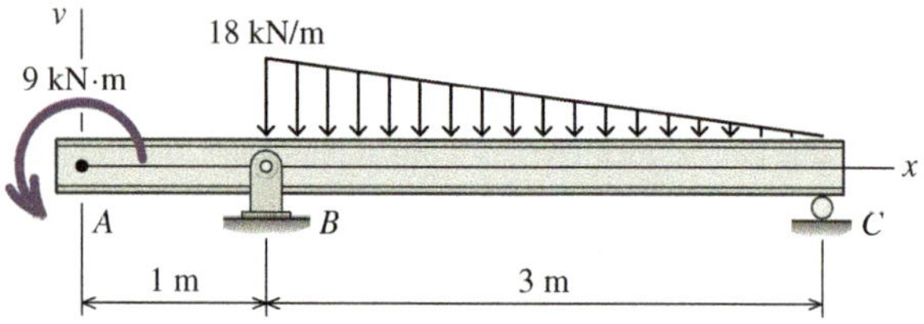

FIGURE P10.30

P10.31 For the beam and loading shown in Figure P10.31, use discontinuity functions to compute

(a) the slope of the beam at B.
(b) the deflection of the beam at A.

Assume a constant value of $EI = 133{,}000 \text{ kip} \cdot \text{ft}^2$ for the beam.

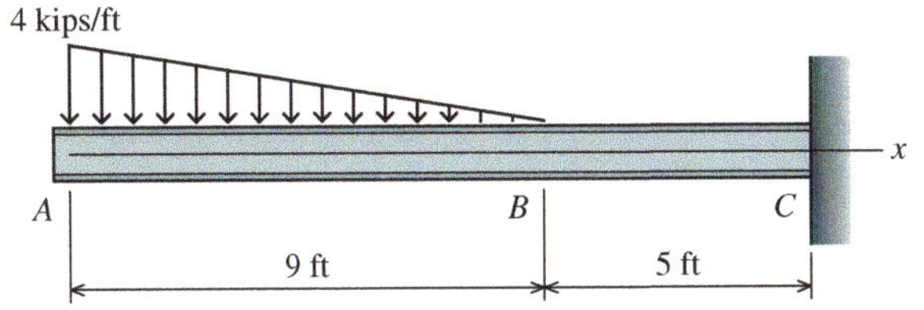

FIGURE P10.31

P10.32 For the beam and loading shown in Figure P10.32, use discontinuity functions to compute

(a) the slope of the beam at B.
(b) the deflection of the beam at C.

Assume a constant value of $EI = 34 \times 10^6 \text{ lb} \cdot \text{ft}^2$ for the beam.

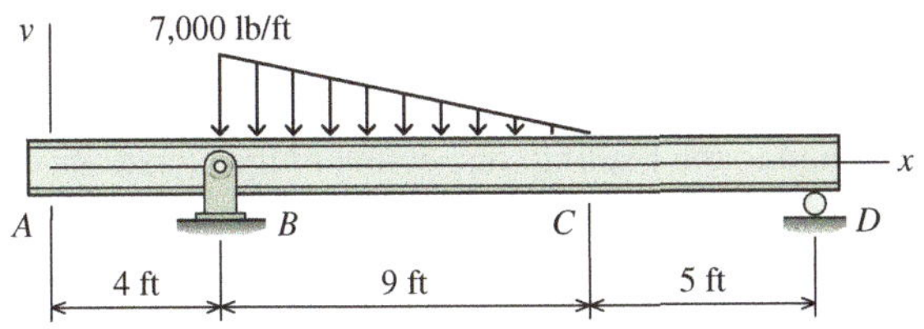

FIGURE P10.32

P10.33 For the beam and loading shown in Figure P10.33, use discontinuity functions to compute

(a) the slope of the beam at A.
(b) the deflection of the beam at B.

Assume a constant value of $EI = 370{,}000 \text{ kip} \cdot \text{ft}^2$ for the beam.

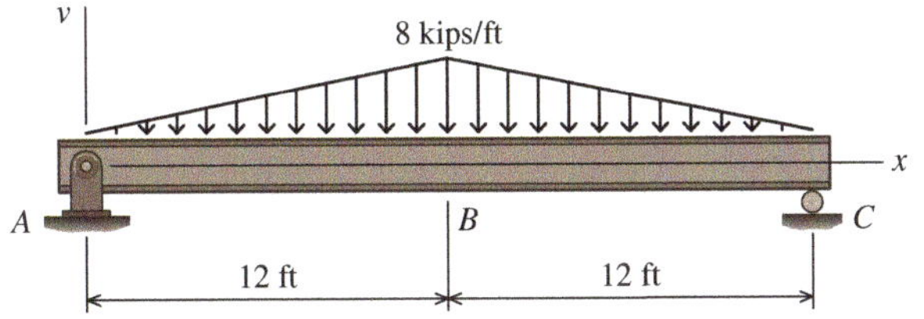

FIGURE P10.33

P10.34 For the beam and loading shown in Figure P10.34, use discontinuity functions to compute

(a) the slope of the beam at B.
(b) the deflection of the beam at B.

Assume a constant value of $EI = 110{,}000 \text{ kN} \cdot \text{m}^2$ for the beam.

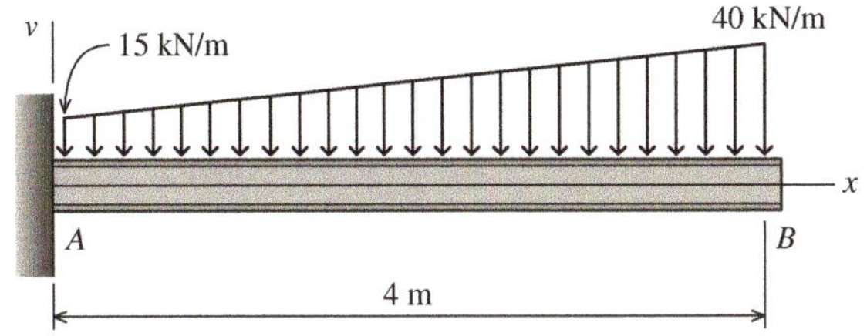

FIGURE P10.34

P10.35 For the beam and loading shown in Figure P10.35, use discontinuity functions to compute

(a) the deflection of the beam at A.
(b) the deflection of the beam at C.

Assume a constant value of $EI = 24{,}000 \text{ kN} \cdot \text{m}^2$ for the beam.

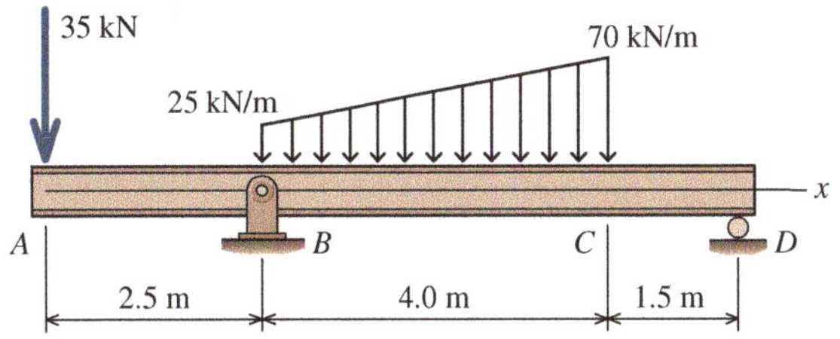

FIGURE P10.35

P10.36 For the beam and loading shown in Figure P10.36, use discontinuity functions to compute

(a) the slope of the beam at B.
(b) the deflection of the beam at A.

Assume a constant value of $EI = 54{,}000 \text{ kN} \cdot \text{m}^2$ for the beam.

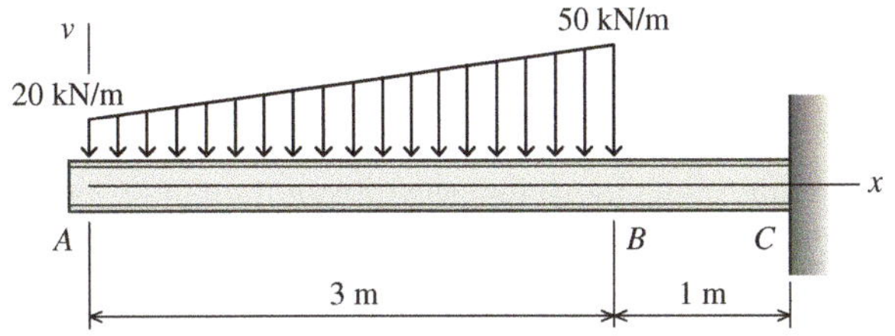

FIGURE P10.36

10.7 Determining Deflections by the Method of Superposition

The method of superposition is a practical and convenient method for obtaining beam deflections. The **principle of superposition** states that the combined effect of several loads acting simultaneously on an object can be computed from the sum of the effects produced by each load acting individually. How can this principle be used to compute beam deflections?

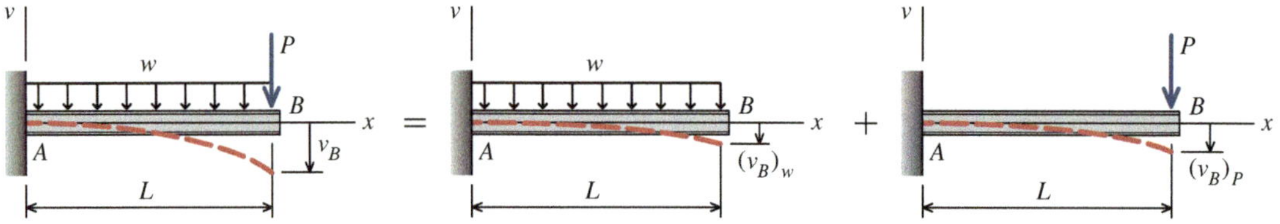

(*a*) Total load (*b*) Distributed load only (*c*) Concentrated load only

FIGURE 10.8 Superposition principle applied to beam deflections.

Consider a cantilever beam subjected to a uniformly distributed load and a concentrated load at its free end. To compute the deflection at B (Figure 10.8*a*), two separate deflection calculations can be performed. First, the cantilever beam deflection at B is calculated considering only the uniformly distributed load w (Figure 10.8*b*). Next, the deflection caused by the concentrated load P alone is computed (Figure 10.8*c*). The results of these two calculations are then added algebraically to give the deflection at B for the total load.

Beam deflection and slope equations for common support and load configurations are frequently tabulated in engineering handbooks and other reference materials. A table of equations for frequently used simply supported and cantilever beams is presented in Appendix C. (This table of common beam formulas is often referred to as a **beam table**.) Appropriate application of these equations enables the analyst to determine beam deflections for a wide variety of support and load configurations.

Several conditions must be satisfied if the principle of superposition is to be valid for beam deflections:

1. The deflection must be linearly related to the loading. Inspection of the equations found in Appendix C shows that all load variables (i.e., w, P, and M) are first-order variables.
2. Hooke's law must apply to the material, meaning that the relationship between stress and strain remains linear.
3. The loading must not significantly change the original geometry of the beam. This condition is satisfied if beam deflections are small.
4. Boundary conditions resulting from the sum of individual cases must be the same as the boundary conditions in the original beam configuration. In this context, boundary conditions are normally deflection or slope values at beam supports.

MecMovies

EXAMPLE

M10.7 Introduction to the superposition method with two elementary examples, one of a cantilever beam and the other of a simply supported beam.

Applying the Superposition Method

The superposition method can be a quick and powerful method for calculating beam deflections; however, its application may initially seem more like an art than an engineering calculation. Before proceeding, it may be helpful to consider various calculation skills that are often used in typical beam and loading configurations.

Skill 1—Using the slope to calculate the deflection: The beam slope at location *A* may be needed in order to calculate the beam deflection at location *B*.

Skill 2—Using both the deflection and slope values to calculate deflections: Both the beam slope and the beam deflection at location *A* may be needed to calculate the beam deflection at location *B*.

Skill 3—Using the elastic curve: Equations are given in beam tables for the beam slope and deflection at key locations, such as the free end of a cantilever beam and the midspan of a simply supported beam. There are many instances, however, when deflections must be computed at other locations. In these instances, deflections can be calculated from the elastic curve equation.

Skill 4—Using both cantilever and simply supported beam equations: For a simply supported beam with an overhang, both cantilever and simply supported beam equations are required to compute the deflection at the free end of the overhang.

Skill 5—Subtracting the load: For a beam with distributed loads on only a portion of a span, it may be expedient to consider first the distributed load over the entire span. Then, the load can be cancelled out on a portion of the span by adding the inverse of the load (i.e., a load equal in magnitude, but opposite in direction). This skill may also be useful for cases involving linearly distributed loadings (i.e., triangular loads).

Skill 6—Using known deflections at specific locations to compute unknown forces or moments: This skill is particularly useful in analyzing statically indeterminate beams.

Skill 7—Using known slopes at specific locations to compute unknown forces or moments: This skill is also useful in analyzing statically indeterminate beams.

Skill 8—A beam and loading configuration may often be subdivided in more than one manner: A given beam and loading may be subdivided and added in any manner that yields the same boundary conditions (i.e., deflection and/or slope at the supports) as those in the original beam configuration. Alternative approaches may require fewer calculations to produce the same results.

The skills in the preceding list are presented with examples and interactive problems in MecMovies M10.3 and M10.4 (8 Skills: Parts I and II) and in MecMovies M10.5 (Superposition Warm-Up).

MecMovies

EXAMPLES

M10.3 and 10.4 **8 Skills:** Parts I and II

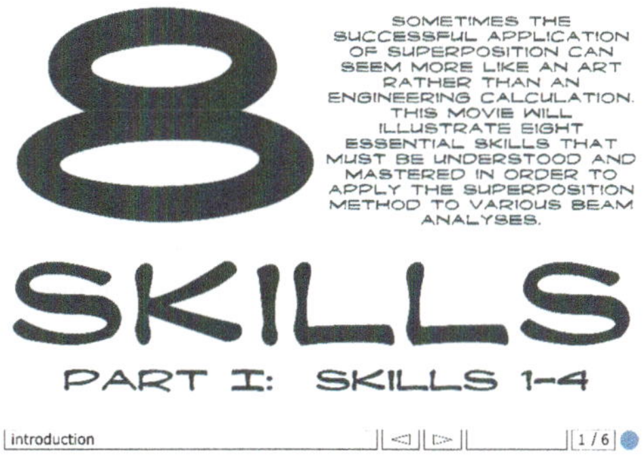

M10.5 Superposition Warm-Up. A series of examples and exercises that illustrate basic skills required for successful application of the superposition method to beam deflection problems.

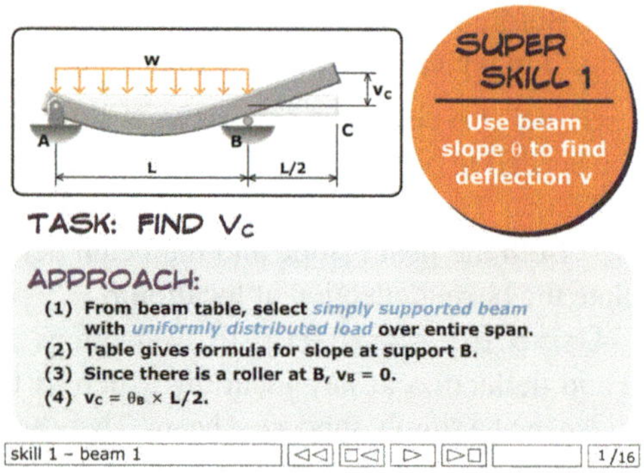

EXAMPLE 10.9

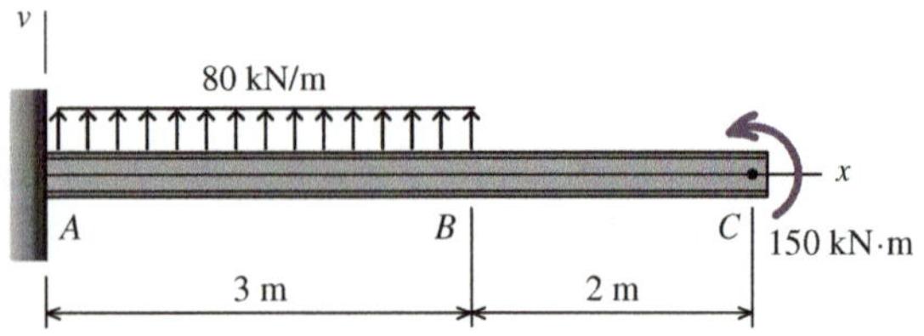

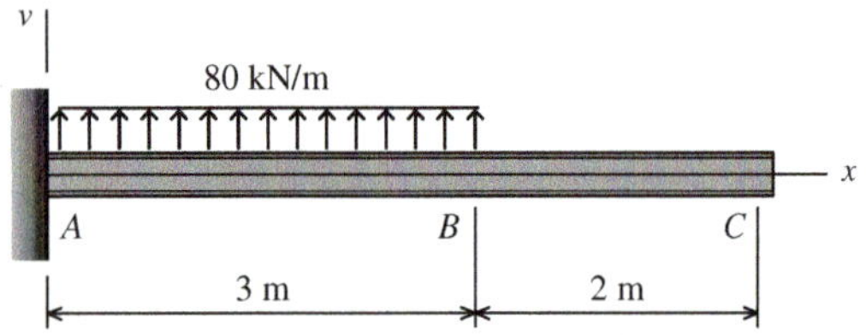

Case 1—Cantilever with uniform load.

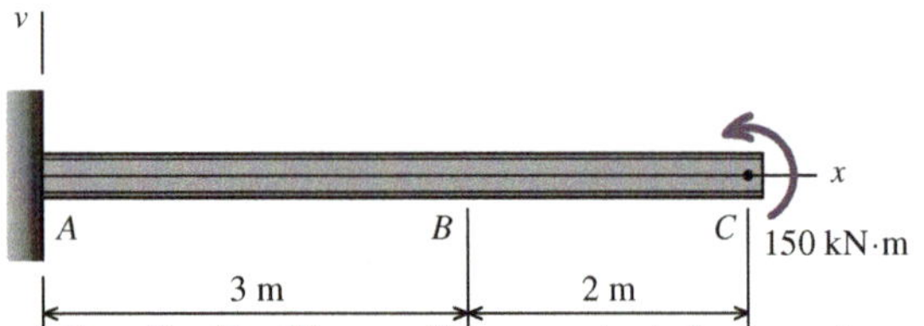

Case 2—Cantilever with concentrated moment.

The cantilever beam shown consists of a structural steel wide-flange shape [$E = 200$ GPa; $I = 650 \times 10^6$ mm^4]. For the loading shown, determine

(a) the beam deflection at point B.
(b) the beam deflection at point C.

Plan the Solution

To solve this problem, the given loading will be separated into two cases: (1) a cantilever beam with a uniformly distributed load and (2) a cantilever beam with a concentrated moment acting at the free end. Pertinent equations for these two cases are given in the beam table found in Appendix C. For case 1, we will use equations for the deflection and rotation angle at the free end of the cantilever to determine the beam deflections at B and C. For case 2, the elastic curve equation will be used to compute beam deflections at both locations.

SOLUTION

For this beam, the elastic modulus is $E = 200$ GPa and the moment of inertia is $I = 650 \times 10^6$ mm^4. Since the term EI will appear in all of the equations, it may be helpful to start by computing that value:

$$EI = (200 \text{ GPa})(650 \times 10^6 \text{ mm}^4) = 130 \times 10^{12} \text{ N}\cdot\text{mm}^2$$
$$= 130 \times 10^3 \text{ kN}\cdot\text{m}^2$$

As in all calculations, it is essential to use consistent units throughout the computations. This rule is particularly important in the superposition method. When we substitute numbers into the various equations obtained from the beam table, it is easy to lose track of the units. If this happens, you may find that you have calculated a beam deflection that seems absurd, such as a deflection of 1,000,000 mm for a beam that spans only 3 m. To avoid this situation, always be aware of the units associated with each variable and make sure that all units are consistent.

Case 1—Cantilever with Uniform Load

From the beam table in Appendix C, the deflection at the free end of a cantilever beam that is subjected to a uniformly distributed load over its entire span is given as

$$v_{max} = -\frac{wL^4}{8EI} \qquad \text{(a)}$$

The beam deflection at B can be calculated with equation (a); however, that equation alone will not be sufficient to calculate the deflection at C. For the beam considered here, the uniform load extends only between A and B. There are no loads acting on the beam between B and C, meaning that there will be no bending moment in the beam in that region. Since there is no moment, the beam will not be bent (i.e., curved), so its slope between B and C will be constant. Because the beam is continuous, its slope between B and C must equal the rotation angle of the beam at B caused by the uniformly distributed load. (**Note:** Since small deflections are assumed, the beam slope dv/dx is equal to the rotation angle θ and the terms "slope" and "rotation angle" will be used synonymously.)

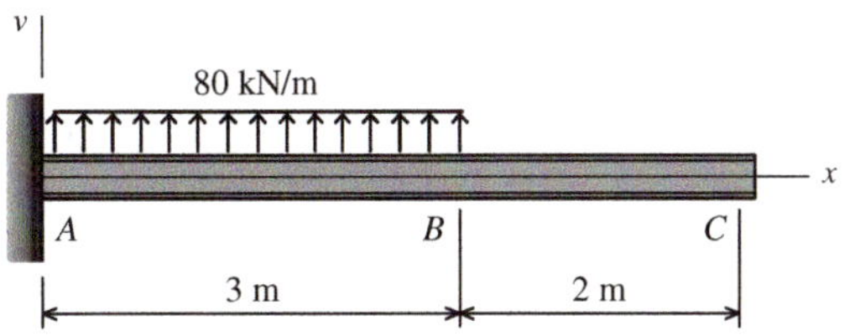

From the beam table in Appendix C, the slope at the free end of this cantilever beam is given as

$$\theta_{max} = -\frac{wL^3}{6EI} \qquad \text{(b)}$$

The beam deflection at C will be calculated from both Equation (a) and Equation (b).

Problem-Solving Tip: *Before beginning the calculation,* it is helpful to sketch the deflected shape of the beam. Next, make a list of the variables that appear in the standard equations along with the values applicable to the specific beam being analyzed. Make sure that the units are consistent at this point in the process. In this example, for instance, all force units will be expressed in terms of kilonewtons (kN) and all length units will be stated in terms of meters (m). Making a simple list of the variables appearing in the equations will greatly increase your likelihood of success, and it will **save you a lot of time** in checking your work.

Beam deflection at B: Equation (a) will be used to compute the beam deflection at B. For this beam, we have the following:

$$w = -80 \text{ kN/m}$$
$$L = 3 \text{ m}$$
$$EI = 130 \times 10^3 \text{ kN}\cdot\text{m}^2$$

Note: The distributed load w is negative in this instance because the distributed load on the beam acts opposite to the direction shown in the beam table. The cantilever span length L is taken as 3 m because that is the length of the uniformly distributed load.

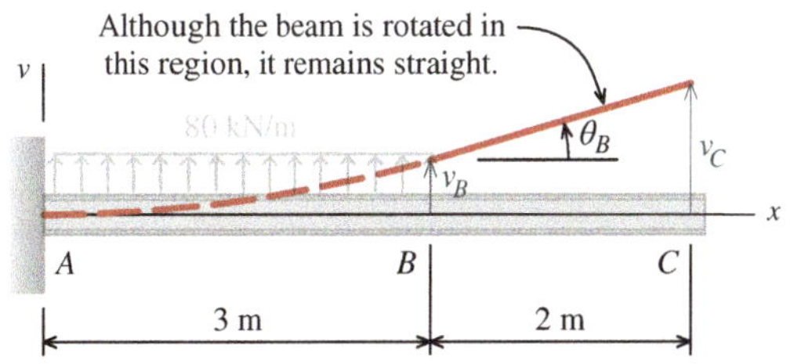

Substitute the preceding values into Equation (a) to find

$$v_B = -\frac{wL^4}{8EI} = -\frac{(-80 \text{ kN/m})(3 \text{ m})^4}{8(130 \times 10^3 \text{ kN}\cdot\text{m}^2)} = 6.231 \times 10^{-3} \text{ m} = 6.231 \text{ mm}$$

The positive value indicates an upward deflection, as expected.

Beam deflection at C: The beam deflection at C will be equal to the beam deflection at B plus an additional deflection caused by the slope of the beam between B and C. The rotation angle of the beam at B is given by Equation (b), using the same variables as before:

$$\theta_B = -\frac{wL^3}{6EI} = -\frac{(-80\text{ kN/m})(3\text{ m})^3}{6(130\times 10^3\text{ kN}\cdot\text{m}^2)} = 2.769\times 10^{-3}\text{ rad}$$

The deflection at C is computed from v_B, θ_B, and the length of the beam between B and C:

$$v_C = v_B + \theta_B(2\text{ m}) = (6.231\times 10^{-3}\text{ m}) + (2.769\times 10^{-3}\text{ rad})(2\text{ m})$$
$$= 11.769\times 10^{-3}\text{ m} = 11.769\text{ mm}$$

The positive value indicates an upward deflection.

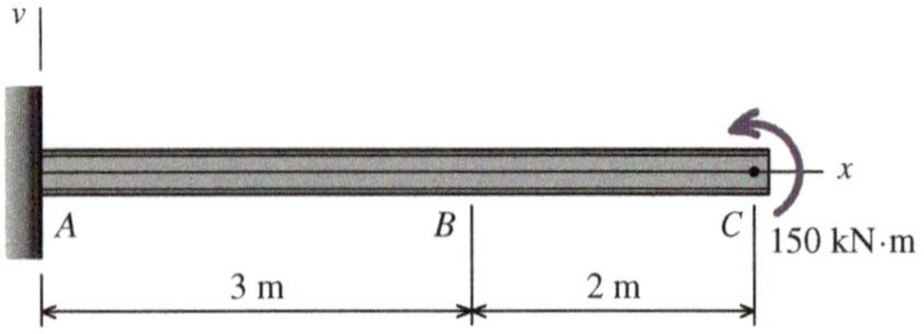

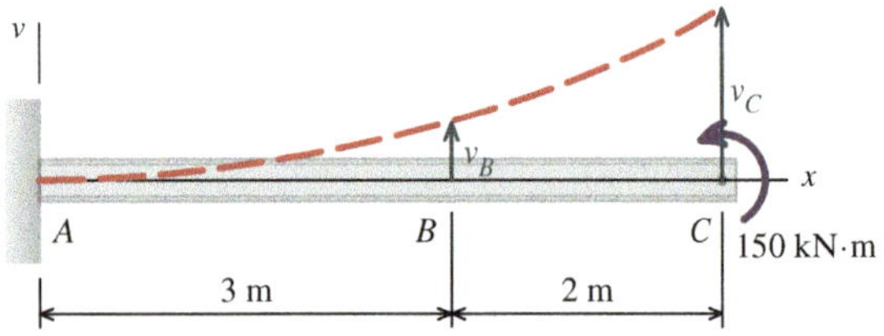

Case 2—Cantilever with Concentrated Moment

From the beam table in Appendix C, the elastic curve equation for a cantilever beam subjected to a concentrated moment applied at its free end is given as

$$v = -\frac{Mx^2}{2EI} \qquad \text{(c)}$$

Beam deflection at B: The elastic curve equation will be used to compute the beam deflections at both B and C for this case. For this beam,

$$M = -150\text{ kN}\cdot\text{m}$$
$$EI = 130\times 10^3\text{ kN}\cdot\text{m}^2$$

Note: The concentrated moment M is negative because it acts in the direction opposite that shown in the beam table.

Substitute the foregoing values into Equation (c), using $x = 3$ m to compute the beam deflection at B:

$$v_B = -\frac{Mx^2}{2EI} = -\frac{(-150\text{ kN}\cdot\text{m})(3\text{ m})^2}{2(130\times 10^3\text{ kN}\cdot\text{m}^2)} = 5.192\times 10^{-3}\text{ m} = 5.192\text{ m}$$

Beam deflection at C: Substitute the same values into Equation (c), using $x = 5$ m to compute the beam deflection at C:

$$v_C = -\frac{Mx^2}{2EI} = -\frac{(-150\text{ kN}\cdot\text{m})(5\text{ m})^2}{2(130\times 10^3\text{ kN}\cdot\text{m}^2)} = 14.423\times 10^{-3}\text{ m} = 14.423\text{ mm}$$

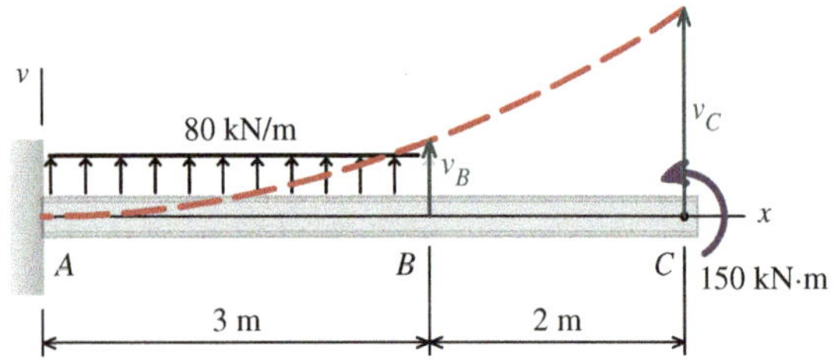

Combine the Two Cases

The deflections at B and C are found from the sum of the separate deflections in cases 1 and 2:

$$v_B = 6.231\text{ mm} + 5.192\text{ mm} = 11.42\text{ mm} \qquad \textbf{Ans.}$$

$$v_C = 11.769\text{ mm} + 14.423\text{ mm} = 26.2\text{ mm} \qquad \textbf{Ans.}$$

EXAMPLES

M10.8 Determine the maximum deflection of the cantilever beam. Assume that EI is constant for the beam.

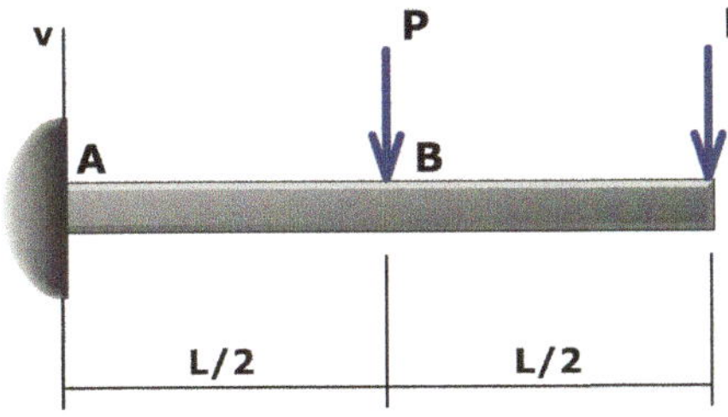

M10.9 Determine the deflection at point C on the beam shown. Assume that EI is constant for the beam.

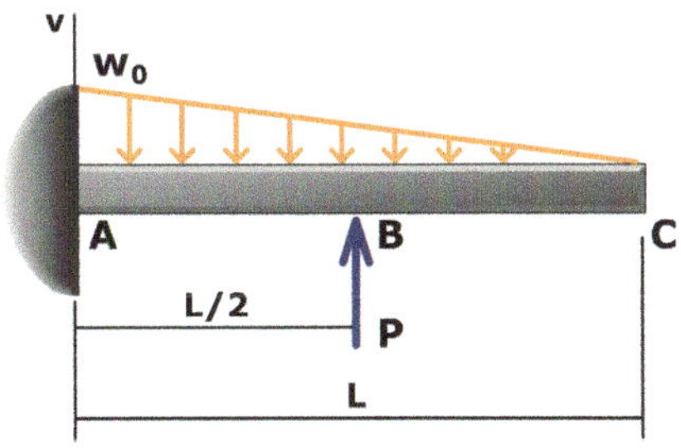

EXAMPLE 10.10

The simply supported beam shown consists of a W16 × 40 structural steel wide-flange shape [E = 29,000 ksi; I = 518 in.^{4}]. For the loading shown, determine the beam deflection at point C.

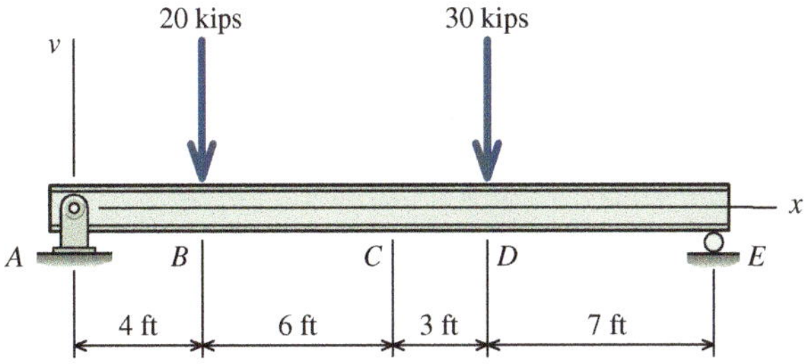

Plan the Solution

One of the standard configurations found in the beam tables is a simply supported beam with a concentrated load acting at a location other than the middle of the span. The elastic curve equation from this standard beam configuration will be used to compute the deflection for the beam considered here, which has two concentrated loads. However, the elastic curve equation must be applied differently for each load because it is applicable only for a portion of the total span.

SOLUTION

The solution of this beam deflection problem will be subdivided into two cases. In case 1, the 30 kip load acting on the simply supported beam will be considered. Case 2 will consider the 20 kip load. The elastic curve equation for a simply supported beam with a single concentrated load acting at a location other than the middle of the span is given in the beam table as

$$v = -\frac{Pbx}{6LEI}(L^2 - b^2 - x^2) \quad \text{for} \quad 0 \le x \le a \tag{a}$$

For this beam, the elastic modulus is E = 29,000 ksi and the moment of inertia is I = 518 in.^{4}. The term EI, which appears in all calculations, has the value

$$EI = (29{,}000 \text{ ksi})(518 \text{ in.}^4) = 15.022 \times 10^6 \text{ kip}\cdot\text{in.}^2$$

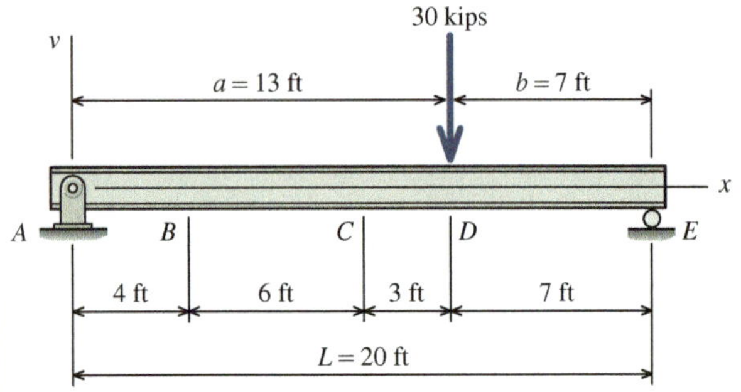

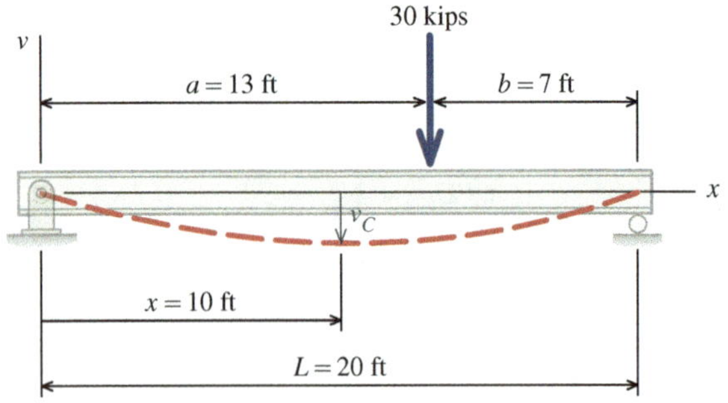

Case 1—30 kip Load on Simple Span

It is essential to note the interval upon which the elastic curve equation is applicable. Equation (a) gives the beam deflection at any distance x from the origin up to, but not past, the location of the concentrated load. That location is denoted by the term a in the equation. For this beam, $a = 13$ ft. Since point C is located at $x = 10$ ft, the elastic curve equation is applicable to this case.

The deflected shape of the beam is shown. List the variables that appear in the elastic curve equation along with their corresponding values:

$$P = 30 \text{ kips}$$
$$b = 7 \text{ ft} = 84 \text{ in.}$$
$$L = 20 \text{ ft} = 240 \text{ in.}$$
$$EI = 15.022 \times 10^6 \text{ kip·in.}^2$$

Beam deflection at C: At point C, $x = 10$ ft $= 120$ in. Therefore, the beam deflection at C for this case is

$$\begin{aligned} v_C &= -\frac{Pbx}{6LEI}(L^2 - b^2 - x^2) \\ &= -\frac{(30 \text{ kips})(84 \text{ in.})(120 \text{ in.})}{6(240 \text{ in.})(15.022 \times 10^6 \text{ kip·in.}^2)}[(240 \text{ in.})^2 - (84 \text{ in.})^2 - (120 \text{ in.})^2] \\ &= -0.5053 \text{ in.} \end{aligned}$$

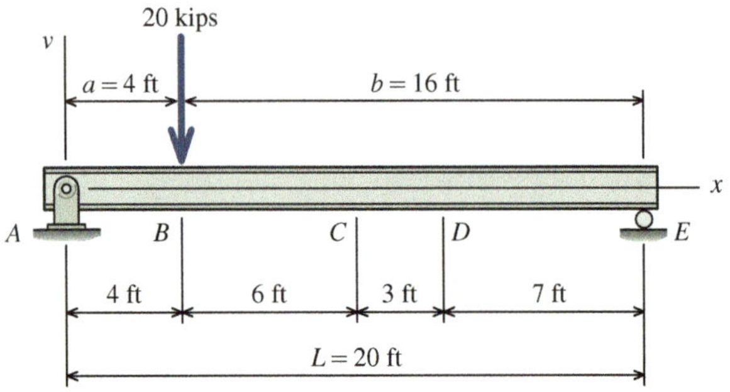

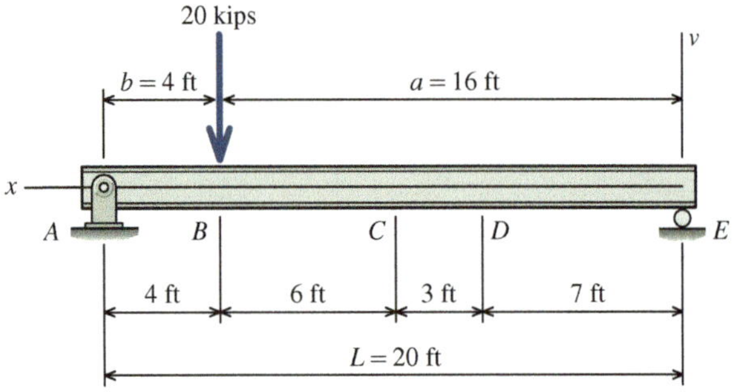

Case 2—20 kip Load on Simple Span

Next, consider the simply supported beam with only the 20 kip load. From this sketch, it is apparent that the distance a from the origin to the point of application of the load is $a = 4$ ft. Since C is located at $x = 10$ ft, the elastic curve equation is not applicable to this case, because $x > a$.

However, the elastic curve equation can be used for this case if we make a simple transformation. The origin of the x–v coordinate axes will be repositioned at the right end of the beam, and the positive x direction will be redefined as extending toward the pin support at the left end of the span. With this transformation, $x < a$ and the elastic curve equation can be used.

The variables that appear in the elastic curve equation and their corresponding values are as follows:

$$P = 20 \text{ kips}$$
$$b = 4 \text{ ft} = 48 \text{ in.}$$
$$L = 20 \text{ ft} = 240 \text{ in.}$$
$$EI = 15.022 \times 10^6 \text{ kip·in.}^2$$

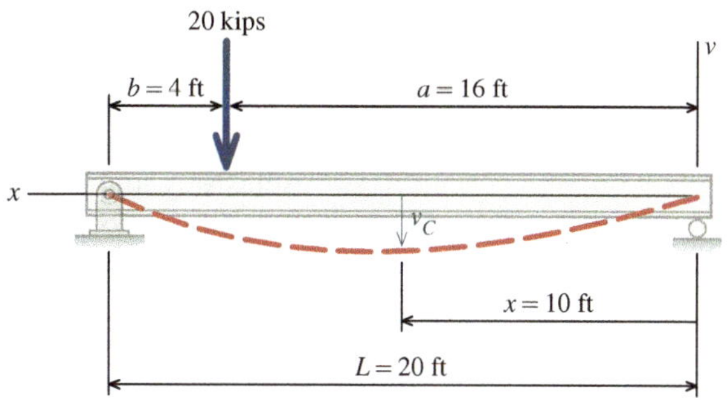

Beam deflection at C: At point C, $x = 10$ ft $= 120$ in. and the beam deflection there for this case is

$$
\begin{aligned}
v_C &= -\frac{Pbx}{6LEI}(L^2 - b^2 - x^2) \\
&= -\frac{(20 \text{ kips})(48 \text{ in.})(120 \text{ in.})}{6(240 \text{ in.})(15.022 \times 10^6 \text{ kip}\cdot\text{in.}^2)}[(240 \text{ in.})^2 - (48 \text{ in.})^2 - (120 \text{ in.})^2] \\
&= -0.2178 \text{ in.}
\end{aligned}
$$

Combine the Two Cases

The deflection at C is the sum of the separate deflections in cases 1 and 2.

$$v_C = -0.5053 \text{ in.} - 0.2178 \text{ in.} = -0.723 \text{ in.} \qquad \textbf{Ans.}$$

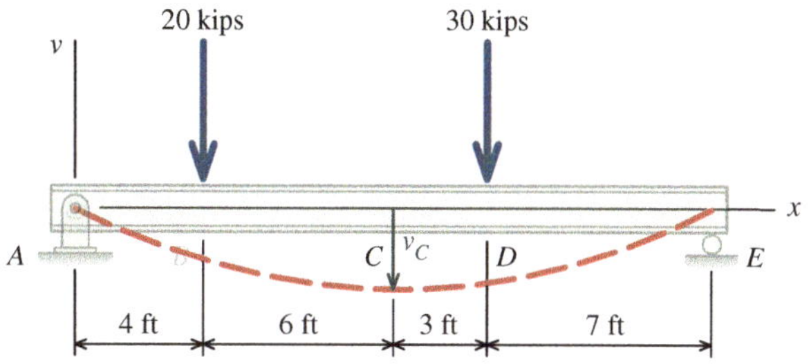

EXAMPLE 10.11

The simply supported beam shown consists of a W24 × 76 structural steel wide-flange shape [$E = 29{,}000$ ksi; $I = 2{,}100$ in.^{4}]. For the loading shown, determine

(a) the beam deflection at point C.
(b) the beam deflection at point A.
(c) the beam deflection at point E.

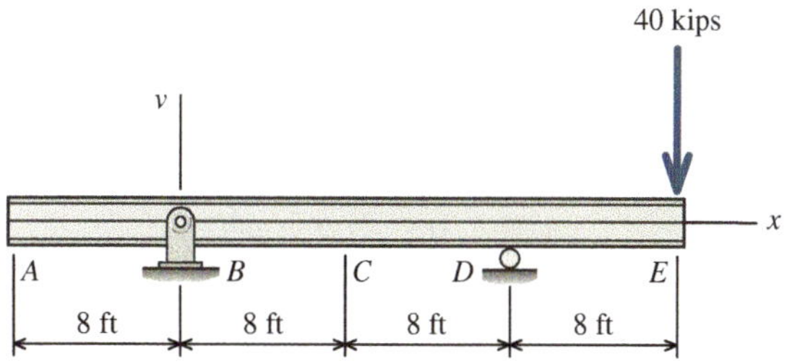

Plan the Solution

Before starting to solve this problem, sketch the deflected shape of the elastic curve. The 40 kip load causing the beam to bend downward at E, in turn causing the beam to bend upward between the simple supports. Since B is a pin support, the deflection of the beam at B will be zero, but the slope will not be zero.

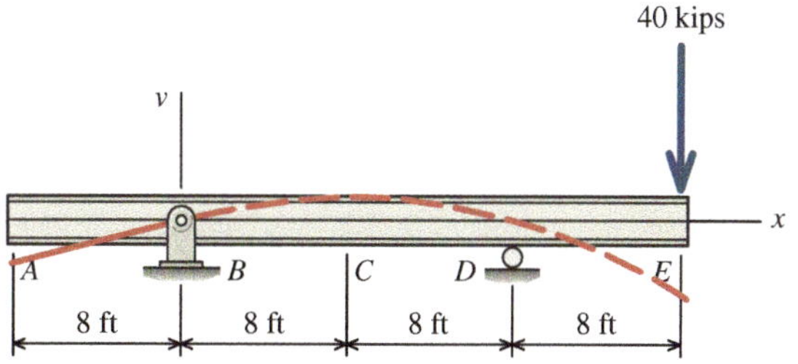

Let us consider the beam span between B and C in more detail. What is it exactly that causes the beam to bend upward in this region? Certainly, the 40 kip load is involved, but more precisely, the 40 kip load creates a bending moment, and it is this bending moment that causes the beam to bend upward. For that reason, the effect of a concentrated moment applied at one end of a simply supported span is the only consideration required in order to compute the beam deflection at C.

Next, consider the overhang span between A and B. No bending moments act in this portion of the beam; thus, the beam does not bend, but it does rotate because it is attached to the center span. The overhang portion AB rotates by an angle equal to the rotation angle θ_B, which occurs at the left end of the center span. The deflection of overhang AB is due exclusively to this rotation, and accordingly, the beam deflection at A can be calculated from the rotation angle θ_B of the center span.

Finally, consider the overhang span between D and E. The deflection at E is a combination of two effects. The more obvious effect is the deflection at the free end of a cantilever beam subjected to a concentrated load. This deflection, however, does not account for all of the deflection at E. The standard cantilever beam cases found in Appendix C assume that the beam does not rotate at the fixed support; in other words, the cantilever cases assume that the support is rigid. Overhang DE, however, is *not* connected to a rigid support. It is connected to center span BD, which is flexible. As the center span flexes, the overhang rotates downward, and this is the second effect that causes deflection at E. To calculate the beam deflection at E, we must consider both cases: the cantilever and the simply supported beam.

SOLUTION

For this beam, the elastic modulus is E = 29,000 ksi and the moment of inertia is I = 2,100 in.^{4}. The term EI, which appears in all the calculations, has the value

$$EI = (29{,}000 \text{ ksi})(2{,}100 \text{ in.}^4) = 60.9 \times 10^6 \text{ kip}\cdot\text{in.}^2$$

The bending moment produced at D by the 40 kip load is M = (40 kips)(8 ft) = 320 kip·ft = 3,840 kip·in.

Case 1—Upward Deflection of Center Span

The upward deflection of point C in the center span is computed from the elastic curve equation for a simply supported beam subjected to a concentrated moment at D:

$$v = -\frac{Mx}{6LEI}(x^2 - 3Lx + 2L^2) \qquad \text{(a)}$$

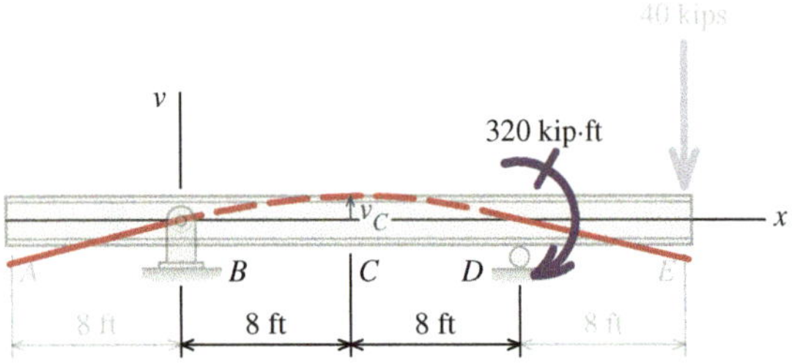

Beam deflection at C: Substitute the following values into Equation (a):

$$M = -320 \text{ kip}\cdot\text{ft} = -3{,}840 \text{ kip}\cdot\text{in.}$$
$$x = 8 \text{ ft} = 96 \text{ in.}$$
$$L = 16 \text{ ft} = 192 \text{ in.}$$
$$EI = 60.9 \times 10^6 \text{ kip}\cdot\text{in.}^2$$

Use these values to compute the beam deflection at C:

$$v_C = -\frac{Mx}{6LEI}(x^2 - 3Lx + 2L^2)$$
$$= -\frac{(-3{,}840 \text{ kip}\cdot\text{in.})(96 \text{ in.})}{6(192 \text{ in.})(60.9 \times 10^6 \text{ kip}\cdot\text{in.}^2)}[(96 \text{ in.})^2 - 3(192 \text{ in.})(96 \text{ in.}) + 2(192 \text{ in.})^2] = 0.1453 \text{ in.} \qquad \textbf{Ans.}$$

Case 2—Downward Deflection of Overhang *AB*

The downward deflection of point A on the overhang span is computed from the rotation angle produced at support B of the center span by the concentrated moment, which acts at D. In the beam table, the magnitude of the rotation angle at the end of the span *opposite* the concentrated moment is given by

$$\theta = \frac{ML}{6EI} \qquad \text{(b)}$$

By the values defined previously, the magnitude of the rotation angle at B is

$$\theta_B = \frac{ML}{6EI} = \frac{(3{,}840\ \text{kip}\cdot\text{in.})(192\ \text{in.})}{6(60.9 \times 10^6\ \text{kip}\cdot\text{in.}^2)} = 0.0020177\ \text{rad}$$

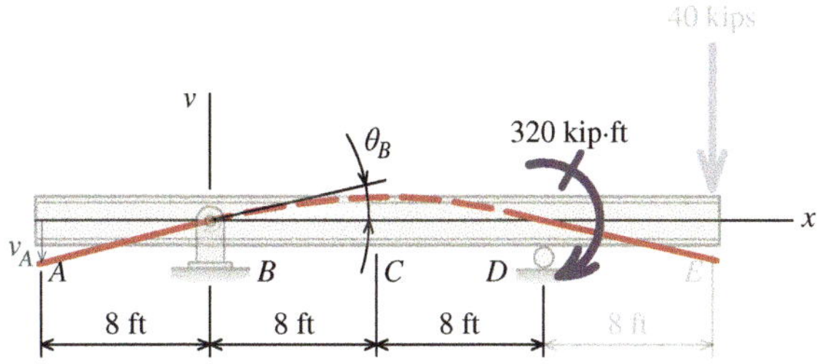

Beam deflection at A: By inspection, the rotation angle at B must be positive; that is, the beam slopes upward to the right at the pin support. Since there is no bending moment in overhang span AB, the beam will not bend between A and B. Its slope will be constant and equal to θ_B. The magnitude of the beam deflection at A is computed from the beam slope:

$$v_A = \theta_B L_{AB} = (0.0020177\ \text{rad})(96\ \text{in.}) = 0.1937\ \text{in.}$$

By inspection, the overhang will deflect downward at A; therefore, $v_A = -0.1937$ in. **Ans.**

Case 3—Downward Deflection of Overhang *DE*

The downward deflection of point E on the overhang span is computed from two considerations. First, consider a cantilever beam subjected to a concentrated load at its free end. The deflection at the tip of the cantilever is given by the equation

$$v_{max} = -\frac{PL^3}{3EI} \qquad \text{(c)}$$

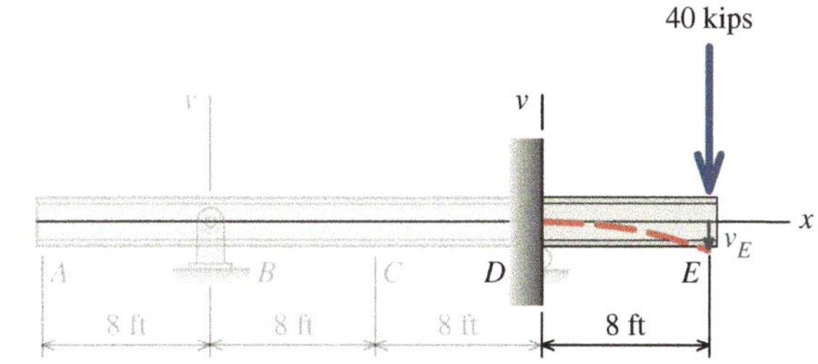

From the values

$$P = 40\ \text{kips}$$
$$L = 8\ \text{ft} = 96\ \text{in.}$$

and

$$EI = 60.9 \times 10^6\ \text{kip}\cdot\text{in.}^2$$

one component of the beam deflection at E can be computed as

$$v_E = -\frac{PL^3}{3EI} = -\frac{(40\ \text{kips})(96\ \text{in.})^3}{3(60.9 \times 10^6\ \text{kip}\cdot\text{in.}^2)} = -0.1937\ \text{in.} \qquad \text{(d)}$$

As discussed previously, this cantilever beam case does not account for all of the deflection at E. Specifically, Equation (c) assumes that the cantilever beam does not rotate at its support. However, since center span BD is flexible, overhang DE rotates downward as the center span bends. The magnitude of the rotation angle of the center span caused by the concentrated moment M can be computed from the following equation:

$$\theta = \frac{ML}{3EI} \qquad \text{(e)}$$

Note: Equation (e) gives the beam rotation angle *at the location of M* for a simply supported beam subjected to a concentrated moment applied at one end. With the values defined for case 2, the rotation angle of the center span at roller support D can be calculated as

$$\theta_D = \frac{ML}{3EI} = \frac{(3{,}840\ \text{kip}\cdot\text{in.})(192\ \text{in.})}{3(60.9 \times 10^6\ \text{kip}\cdot\text{in.}^2)} = 0.0040355\ \text{rad}$$

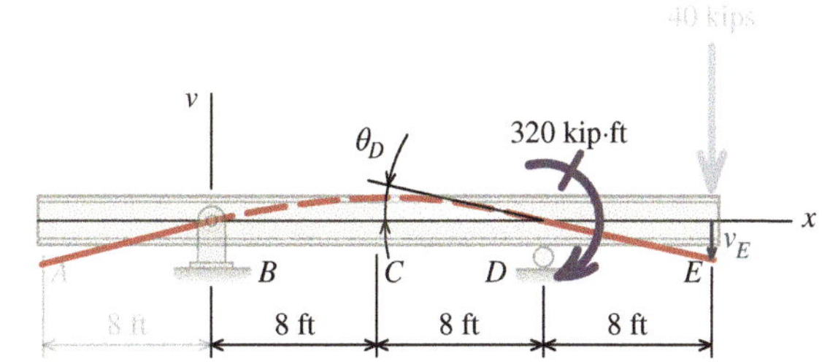

By inspection, the rotation angle at D must be negative; that is, the beam slopes downward to the right at the roller support. The magnitude of the beam deflection at E due to the center span rotation at D is computed from the beam slope and the length of overhang DE:

$$v_E = \theta_D L_{DE} = (0.0040355 \text{ rad})(96 \text{ in.}) = 0.3874 \text{ in.}$$

By inspection, the overhang will deflect downward at E; consequently, this deflection component is

$$\boxed{v_E = -0.3874 \text{ in.}} \qquad \text{(f)}$$

The total deflection at E is the sum of deflections (d) and (f):

$$v_E = -0.1937 \text{ in.} - 0.3874 \text{ in.} = -0.581 \text{ in.}$$ **Ans.**

MecMovies

EXAMPLE

M10.10 Determine expressions for the slope θ_C and the deflection v_C at end C of the beam shown. Assume that EI is constant for the beam.

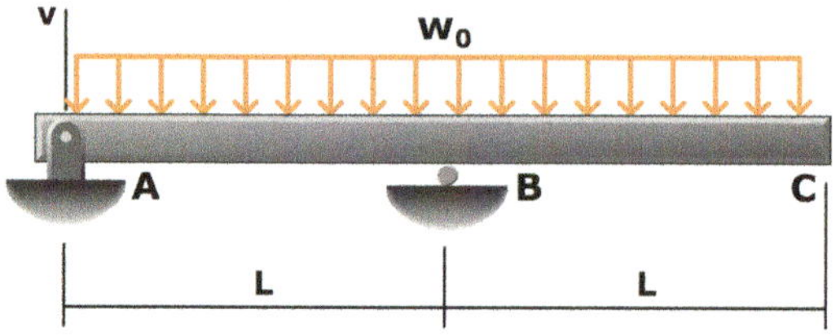

EXAMPLE 10.12

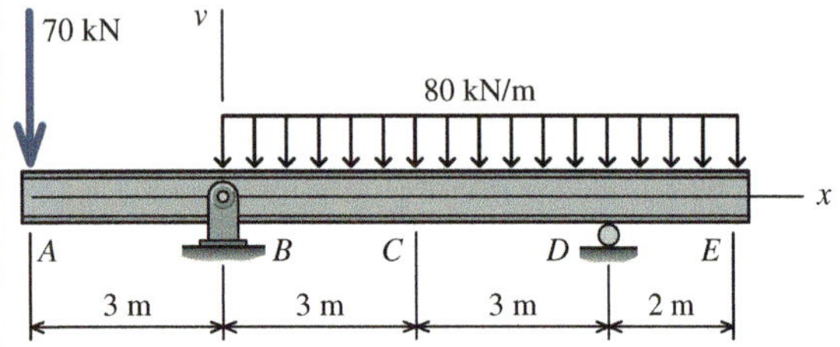

The simply supported beam shown consists of a W410 × 60 structural steel wide-flange shape [$E = 200$ GPa; $I = 216 \times 10^6$ mm^4]. For the loading shown, determine

(a) the beam deflection at point A.
(b) the beam deflection at point C.
(c) the beam deflection at point E.

Plan the Solution

Although the loading in this example is more complicated, the same general approach used to solve Example 10.8 will be used for this beam. The loading will be separated into three cases:

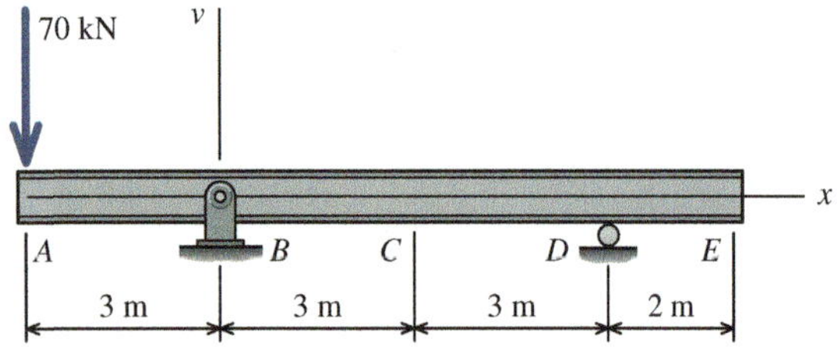

Case 1—Concentrated load on left overhang.

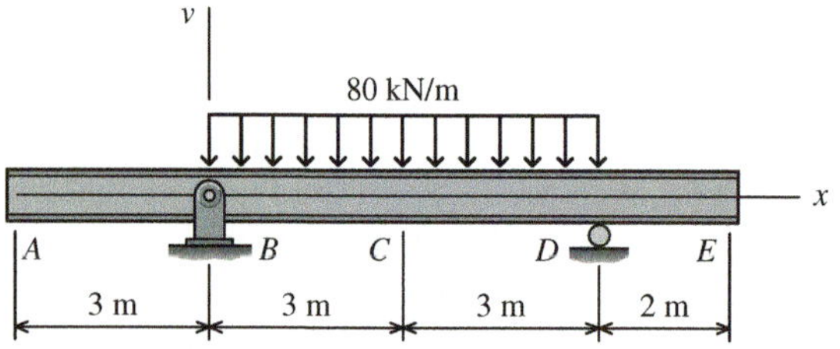

Case 2—Uniformly distributed load on center span.

The beam deflections at A, C, and E will be computed for each case with the use of standard equations from Appendix C for both the deflection and the slope. Cases 1 and 3 will require equations for both simply supported and cantilever beams, whereas case 2 will require equations only for simply supported beams. After completing the calculations for all three cases, the results will be added to give the final deflections at the three locations.

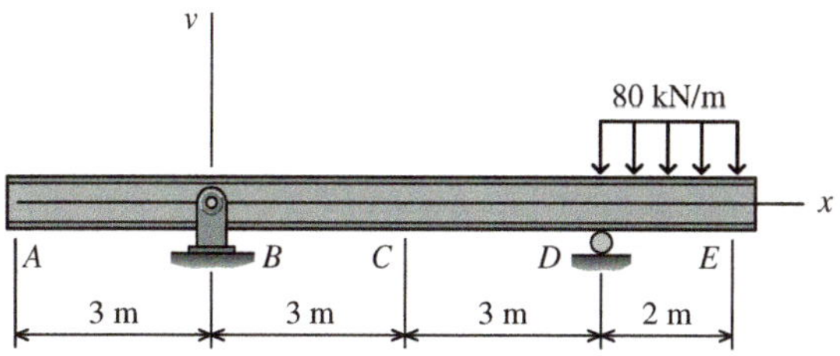

Case 3—Uniformly distributed load on right overhang.

SOLUTION

For this beam, the elastic modulus is $E = 200$ GPa and the moment of inertia is $I = 216 \times 10^6$ mm^4. Therefore,

$$EI = (200\,\text{GPa})(216 \times 10^6\,\text{mm}^4) = 43.2 + 10^{12}\,\text{N}\cdot\text{mm}^2 = 43.2 \times 10^3\,\text{kN}\cdot\text{m}^2$$

Case 1—Concentrated Load on Left Overhang

Both simply supported and cantilever beam equations will be required to compute deflections at A, but only simply supported beam equations will be necessary to compute the beam deflections at C and E.

Beam deflection at A: Consider the cantilever beam deflection at A of the 3 m long overhang. From Appendix C, the maximum deflection of a cantilever beam with a concentrated load applied at the tip is given as

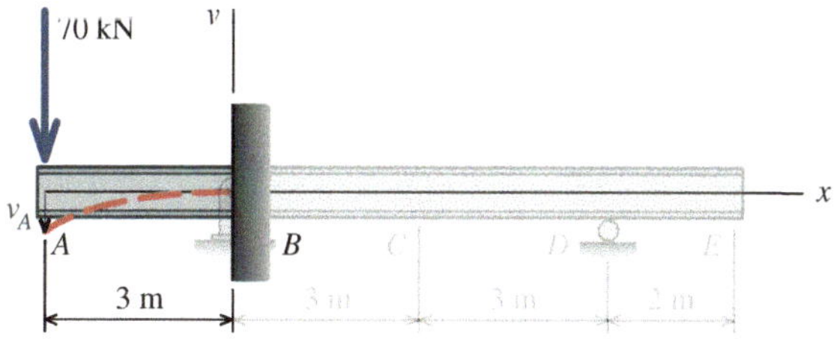

$$v_{max} = -\frac{PL^3}{3EI} \quad \text{(a)}$$

Equation (a) will be used to compute one portion of the beam deflection at A. We set

$$P = 70\text{ kN}$$
$$L = 3\text{ m}$$

and

$$EI = 43.2 \times 10^3\,\text{kN}\cdot\text{m}^2$$

The cantilever beam deflection at A will then be

$$v_A = -\frac{PL^3}{3EI} = -\frac{(70\text{ kN})(3\text{ m})^3}{3(43.2 \times 10^3\text{ kN}\cdot\text{m}^2)} = -14.583 \times 10^{-3}\text{ m} = -14.583\text{ mm}$$

This calculation implicitly assumes that the beam is fixed to a rigid support at B. However, the overhang is attached, not to a rigid support at B, but rather to a flexible beam that rotates in response to the moment produced by the 70 kN load. The rotation of the overhang at B must be accounted for in determining the deflection at A.

The moment at B due to the 70 kN load is $M = (70\text{ kN})(3\text{ m}) = 210\text{ kN}\cdot\text{m}$, and acts counterclockwise as shown. The rotation angles at the ends of the span of a simply supported beam subjected to a concentrated moment can be obtained from Appendix C:

$$\theta_1 = -\frac{ML}{3EI} \quad \text{(at the end where } M \text{ is applied)} \qquad \text{(b)}$$

$$\theta_2 = +\frac{ML}{6EI} \quad \text{(opposite the end where } M \text{ is applied)} \qquad \text{(c)}$$

The rotation angle at B is required in order to obtain the deflection at A. The rotation angle at D will be used later to calculate the deflection at E.

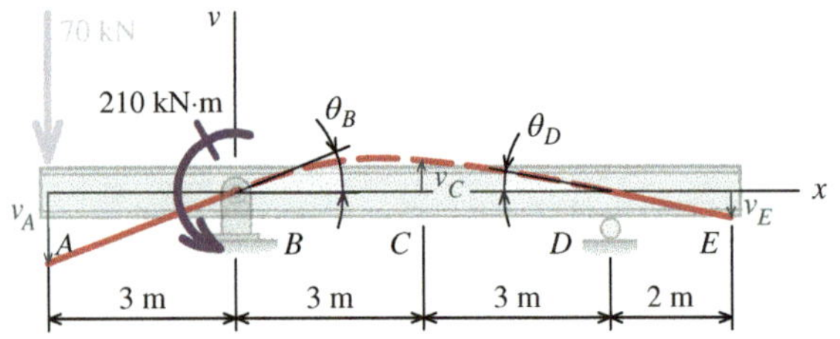

Using the variables and values

$$M = -210 \text{ kN}\cdot\text{m}$$
$$L = 6 \text{ m (i.e., the length of the center span)}$$

and

$$EI = 43.2 \times 10^3 \text{ kN}\cdot\text{m}^2$$

we calculate the rotation angle at B from Equation (b):

$$\theta_B = -\frac{ML}{3EI} = -\frac{(-210 \text{ kN}\cdot\text{m})(6 \text{ m})}{3(43.2 \times 10^3 \text{ kN}\cdot\text{m}^2)} = 9.722 \times 10^{-3} \text{ rad}$$

The beam deflection at A is computed from the rotation angle θ_B and the overhang length:

$$v_A = \theta_B x_{AB} = (9.722 \times 10^{-3} \text{ rad})(-3 \text{ m}) = -29.167 \times 10^{-3} \text{ m} = -29.167 \text{ mm}$$

Beam deflection at C: The beam deflection at C for this case is found from the elastic curve equation for a simply supported beam with a concentrated moment applied at one end. From Appendix C, the elastic curve equation is

$$v = -\frac{Mx}{6LEI}(x^2 - 3Lx + 2L^2) \qquad \text{(d)}$$

With the variables and values

$$M = -210 \text{ kN}\cdot\text{m}$$
$$x = 3 \text{ m}$$
$$L = 6 \text{ m (i.e., the length of the center span)}$$

and

$$EI = 43.2 \times 10^3 \text{ kN}\cdot\text{m}^2$$

the beam deflection at C is calculated from Equation (d):

$$\begin{aligned} v_C &= -\frac{Mx}{6LEI}(x^2 - 3Lx + 2L^2) \\ &= -\frac{(-210 \text{ kN}\cdot\text{m})(3 \text{ m})}{6(6 \text{ m})(43.2 \times 10^3 \text{ kN}\cdot\text{m}^2)}[(3 \text{ m})^2 - 3(6 \text{ m})(3 \text{ m}) + 2(6 \text{ m})^2] \\ &= 10.938 \times 10^{-3} \text{ m} = 10.938 \text{ mm} \end{aligned}$$

Beam deflection at E: For this case, the overhang at the right end of the span has no bending moment; therefore, it does not bend. The rotation angle at D given by Equation (c) and the overhang length are used to compute the deflection at E. With the variables and values

$$M = -210 \text{ kN}\cdot\text{m}$$
$$L = 6 \text{ m (i.e., the length of the simple span)}$$

and

$$EI = 43.2 \times 10^3 \text{ kN}\cdot\text{m}^2$$

the rotation angle at D is calculated from Equation (c):

$$\theta_D = +\frac{ML}{6EI} = \frac{(-210 \text{ kN}\cdot\text{m})(6 \text{ m})}{6(43.2 \times 10^3 \text{ kN}\cdot\text{m}^2)} = -4.861 \times 10^{-3} \text{ rad}$$

The beam deflection at E is computed from the rotation angle θ_D and the overhang length:

$$v_E = \theta_D x_{DE} = (-4.861 \times 10^{-3} \text{ rad})(2 \text{ m}) = -9.722 \times 10^{-3} \text{ m} = -9.722 \text{ mm}$$

Case 2—Uniformly Distributed Load on Center Span

For the uniformly distributed load acting on the center span, equations for the maximum deflection acting at midspan and the slopes at the ends of the span will be required.

Beam deflection at A: Since the uniformly distributed load acts only between the supports, there is no bending moment in the overhang spans. To compute the deflection at A, begin by computing the slope at the end of the simple span. From Appendix C, the rotation angles at the ends of the span are given by

$$\theta_1 = -\theta_2 = -\frac{wL^3}{24EI} \qquad \text{(e)}$$

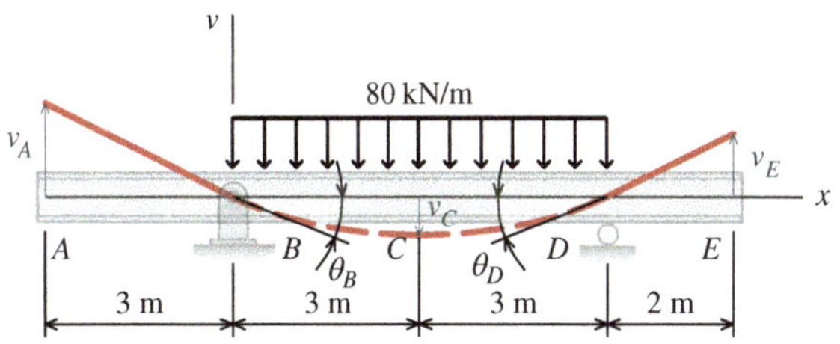

To compute the rotation angle at B, let

$$w = 80 \text{ kN/m}$$
$$L = 6 \text{ m}$$

and

$$EI = 43.2 \times 10^3 \text{ kN}\cdot\text{m}^2$$

and compute θ_B from Equation (e):

$$\theta_B = -\frac{wL^3}{24EI} = -\frac{(80 \text{ kN/m})(6 \text{ m})^3}{24(43.2 \times 10^3 \text{ kN}\cdot\text{m}^2)} = -16.667 \times 10^{-3} \text{ rad}$$

The beam deflection at A is computed from the rotation angle θ_B and the overhang length:

$$v_A = \theta_B x_{AB} = (-16.667 \times 10^{-3} \text{ rad})(-3 \text{ m}) = 50.001 \times 10^{-3} \text{ m} = 50.001 \text{ mm}$$

Beam deflection at C: The equation for the midspan deflection of a simply supported beam subjected to a uniformly distributed load can be obtained from Appendix C:

$$v_{max} = -\frac{5wL^4}{384EI} \qquad \text{(f)}$$

From Equation (f), the deflection at C for case 2 is

$$v_C = -\frac{5wL^4}{384EI} = -\frac{5(80 \text{ kN/m})(6 \text{ m})^4}{384(43.2 \times 10^3 \text{ kN}\cdot\text{m}^2)} = -31.250 \times 10^{-3} \text{ m} = -31.250 \text{ mm}$$

Beam deflection at E: The rotation angle at D is calculated from Equation (e):

$$\theta_D = \frac{wL^3}{24EI} = \frac{(80 \text{ kN/m})(6 \text{ m})^3}{24(43.2 \times 10^3 \text{ kN}\cdot\text{m}^2)} = 16.667 \times 10^{-3} \text{ rad}$$

The beam deflection at E is computed from the rotation angle θ_D and the overhang length:

$$v_E = \theta_D x_{DE} = (16.667 \times 10^{-3} \text{ rad})(2 \text{ m}) = 33.334 \times 10^{-3} \text{ m} = 33.334 \text{ mm}$$

Case 3—Uniformly Distributed Load on Right Overhang

Both simply supported and cantilever beam equations will be required to compute deflections at E; only simply supported beam equations will be necessary to compute the beam deflections at A and C.

Beam deflection at E: Consider the cantilever beam deflection at E of the 2 m long overhang. From Appendix C, the maximum deflection of a cantilever beam with a uniformly distributed load is given as

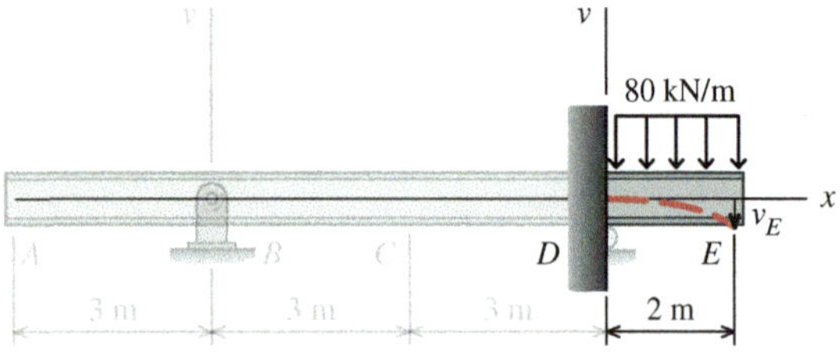

$$v_{max} = -\frac{wL^4}{8EI} \tag{g}$$

Let

$$w = -80\,\text{kN/m}$$

$$L = 2\,\text{m}$$

and

$$EI = 43.2 \times 10^3\ \text{kN}\cdot\text{m}^2$$

and use Equation (g) to compute one portion of the beam deflection at E:

$$v_E = -\frac{wL^4}{8EI} = -\frac{(80\ \text{kN/m})(2\ \text{m})^4}{8(43.2 \times 10^3\ \text{kN}\cdot\text{m}^2)} = -3.704 \times 10^{-3}\ \text{m} = -3.074\ \text{mm}$$

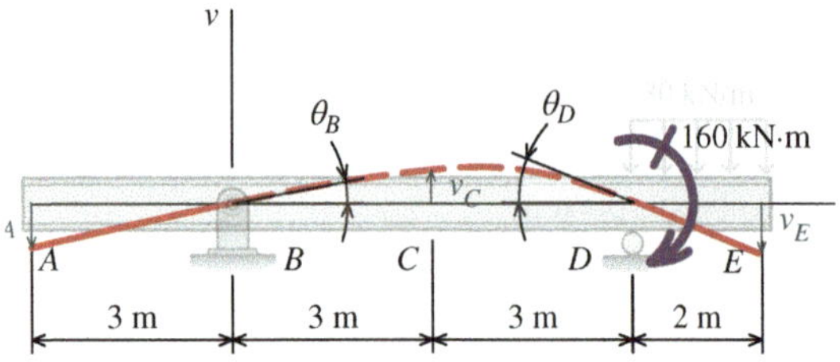

This calculation implicitly assumes that the beam is fixed to a rigid support at D. However, the overhang is not attached to a rigid support at D; rather, it is attached to a flexible beam that rotates in response to the moment produced by the 80 kN uniformly distributed load. The rotation of the overhang at D must be accounted for in determining the deflection at E.

The moment at D due to the 80 kN distributed load is $M = (0.5)(80\ \text{kN/m})(2\ \text{m})^2 = 160\ \text{kN}\cdot\text{m}$, which acts clockwise as shown. The rotation angles at the ends of the span of a simply supported beam subjected to a concentrated moment are given by Equations (b) and (c). Let

$$M = -160\ \text{kN}\cdot\text{m}$$

$$L = 6\ \text{m (i.e., the length of the center span)}$$

and

$$EI = 43.2 \times 10^3\ \text{kN}\cdot\text{m}^2$$

and use Equation (b) to compute the rotation angle at D:

$$\theta_D = \frac{ML}{3EI} = -\frac{(-160\ \text{kN}\cdot\text{m})(6\ \text{m})}{3(43.2 \times 10^3\ \text{kN}\cdot\text{m}^2)} = -7.407 \times 10^{-3}\ \text{rad}$$

The beam deflection at E is computed from the rotation angle θ_D and the overhang length:

$$v_E = \theta_D x_{DE} = (-7.407 \times 10^{-3}\ \text{rad})(2\ \text{m}) = -14.814 \times 10^{-3}\ \text{m} = -14.814\ \text{mm}$$

Beam deflection at C: The beam deflection at C for this case is found from the elastic curve equation [Equation (d)] for a simply supported beam with a concentrated moment applied at one end. With the variables and values

$$M = -160 \text{ kN}\cdot\text{m}$$
$$x = 3 \text{ m}$$
$$L = 6 \text{ m (i.e., the length of the center span)}$$

and

$$EI = 43.2 \times 10^3 \text{ kN}\cdot\text{m}^2$$

the beam deflection at C is calculated from Equation (d):

$$\begin{aligned} v_C &= -\frac{Mx}{6LEI}(x^2 - 3Lx + 2L^2) \\ &= -\frac{(-160 \text{ kN}\cdot\text{m})(3 \text{ m})}{6(6 \text{ m})(43.2 \times 10^3 \text{ kN}\cdot\text{m}^2)}[(3 \text{ m})^2 - 3(6 \text{ m})(3 \text{ m}) + 2(6 \text{ m})^2] \\ &= 8.333 \times 10^{-3} \text{ m} = 8.333 \text{ mm} \end{aligned}$$

Beam deflection at A: Use Equation (c) to compute the rotation angle at B:

$$\theta_B = -\frac{ML}{6EI} = -\frac{(-160 \text{ kN}\cdot\text{m})(6 \text{ m})}{6(43.2 \times 10^3 \text{ kN}\cdot\text{m}^2)} = 3.704 \times 10^{-3} \text{ rad}$$

The beam deflection at A is computed from the rotation angle θ_B and the length of the overhang:

$$v_A = \theta_B x_{AB} = (3.704 \times 10^{-3} \text{ rad})(-3 \text{ m}) = -11.112 \times 10^{-3} \text{ m} = -11.112 \text{ mm}$$

The following table summarizes the results of the method of superposition in this example:

Superposition Case	v_A **(mm)**	v_C **(mm)**	v_E **(mm)**
Case 1—Concentrated load on left overhang	−14.583 −29.167	10.938	−9.722
Case 2—Uniformly distributed load on center span	50.001	−31.250	33.334
Case 3—Uniformly distributed load on right overhang	−11.112	8.333	−3.704 −14.814
Total Beam Deflection	**−4.86**	**−11.98**	**5.09**

Ans.

MecMovies

EXAMPLES

M10.11 Determine an expression for the deflection of the beam at the midpoint of span BD. Assume that EI for the beam is constant throughout all spans.

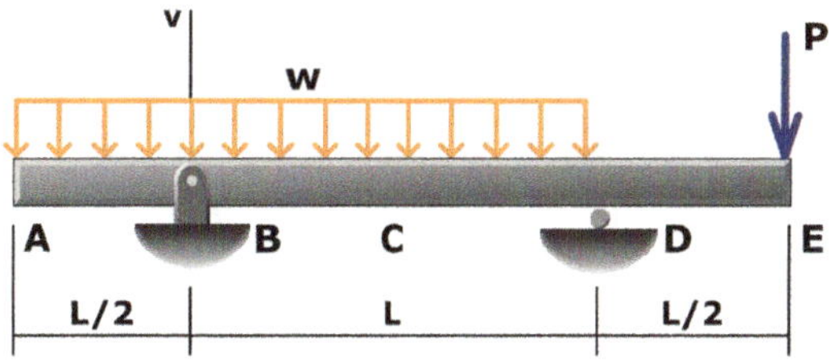

M10.12 Use the superposition method to determine the deflection of the beam at A. Assume that EI is constant across the beam.

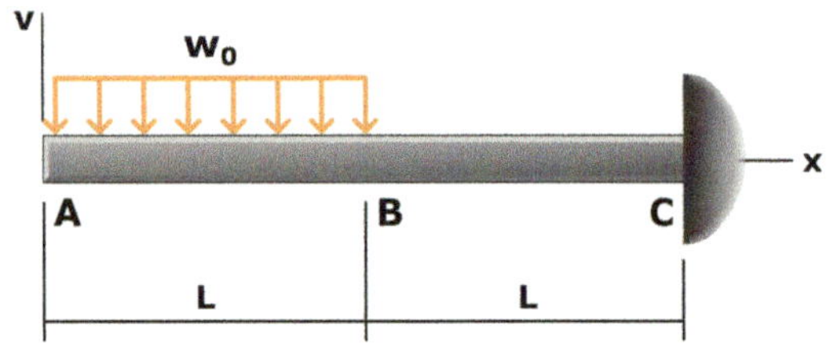

M10.13 Use the superposition method to determine the magnitude of the force P required to make the deflection of the beam equal to zero at B. Assume that EI is constant across the beam.

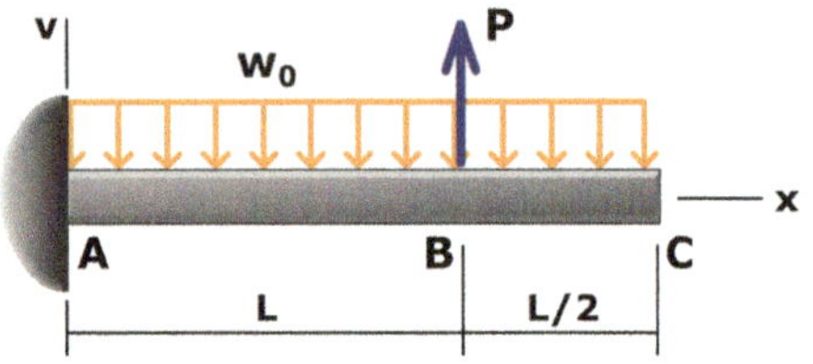

M10.14 Determine the magnitude of the moment M_0 such that the beam slope at A is zero. Assume that EI is constant for the beam.

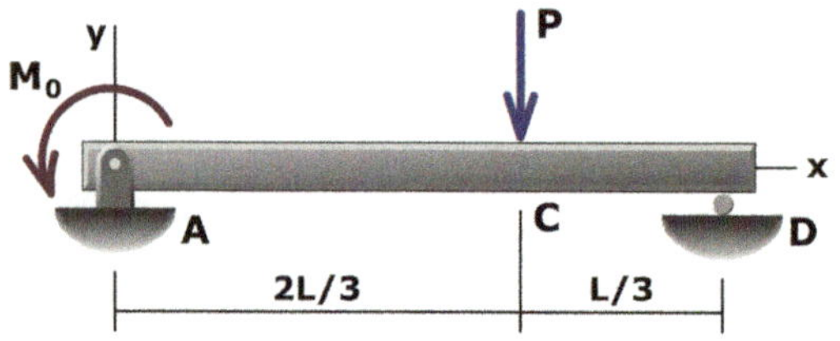

EXERCISES

M10.3 **8 Skills.** Part I: Skills 1–4. Series of skills necessary to solve beam deflection problems by the superposition method.

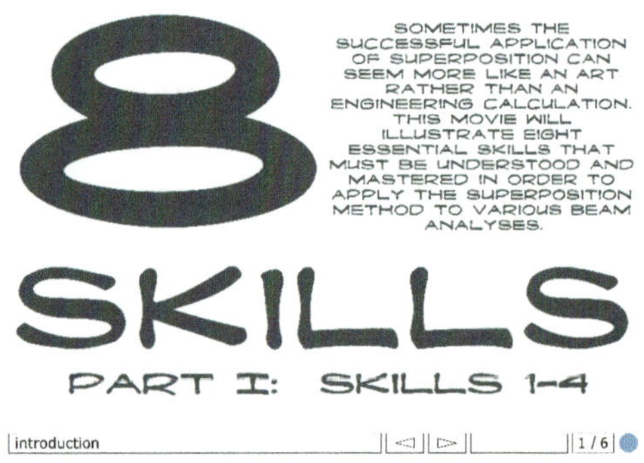

FIGURE M10.3

M10.4 **8 Skills.** Part II: Skills 5–8. Series of skills necessary to solve beam deflection problems by the superposition method.

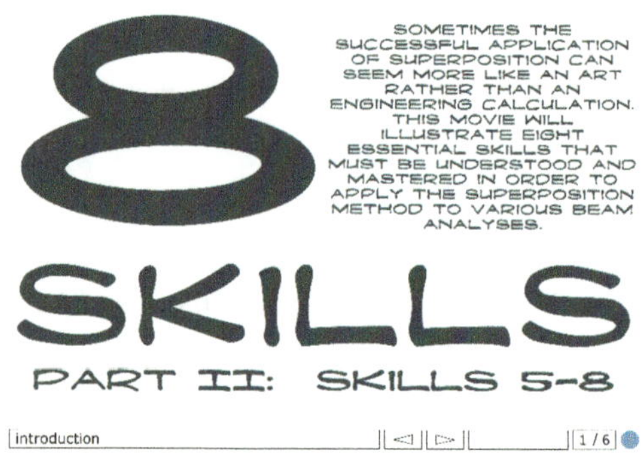

FIGURE M10.4

M10.5 Superposition Warm-Up. Examples and concept checkpoints pertaining to four basic superposition skills.

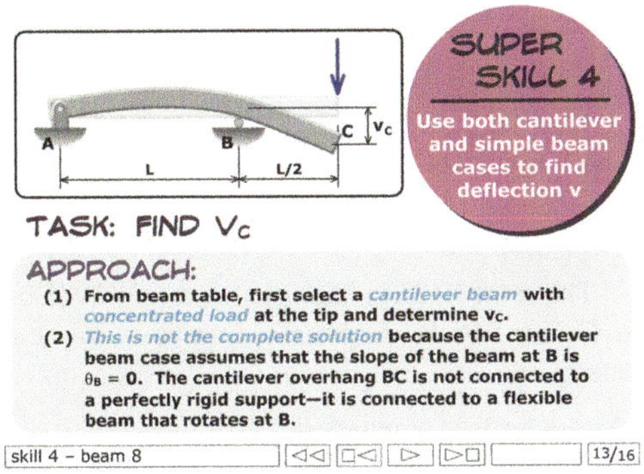

FIGURE M10.5

M10.6 One Simple Beam, One Load, Three Cases. Determine numeric values of beam deflections at various points in a simply supported beam with two overhangs. All deflections can be determined with superposition of no more than three basic deflection cases.

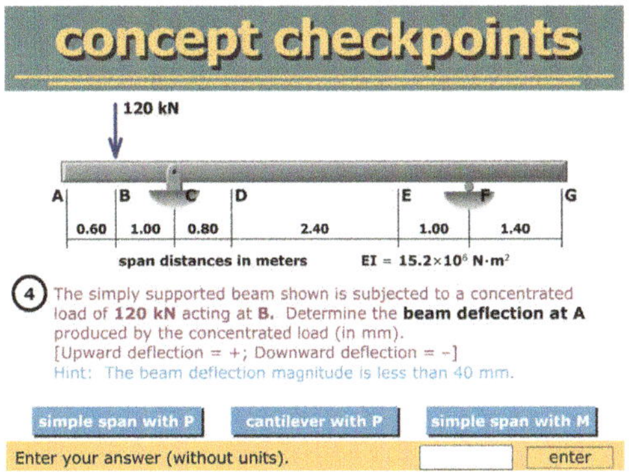

FIGURE M10.6

PROBLEMS

P10.37 For the beams and loadings shown in Figures P10.37*a–d*, determine the beam deflection at point *H*. Assume that $EI = 8 \times 10^4$ kN · m^2 is constant for each beam.

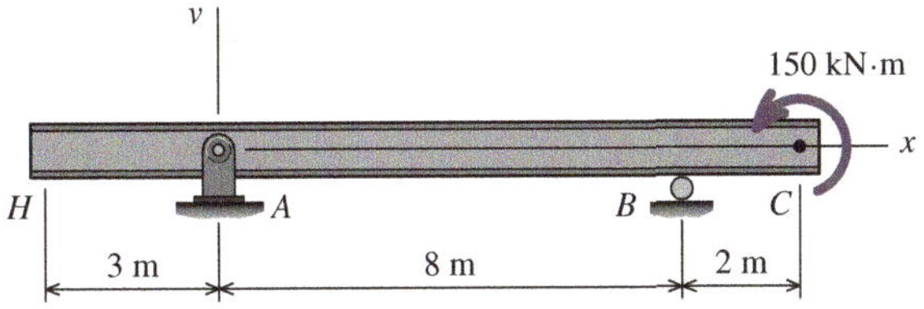

FIGURE P10.37*a*

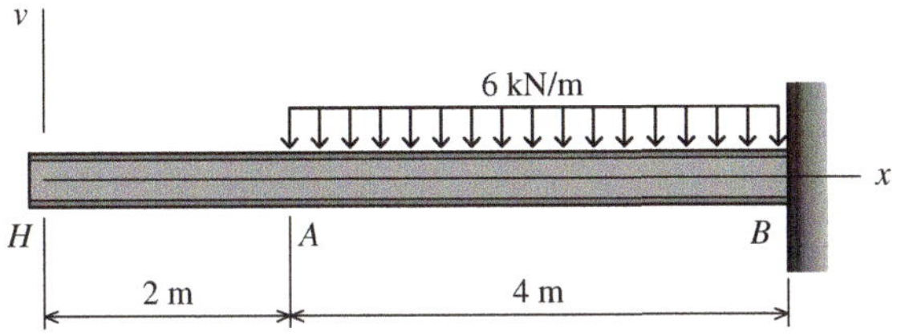

FIGURE P10.37*b*

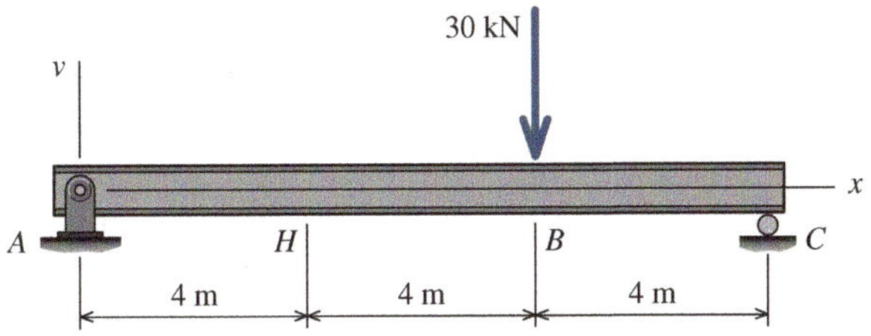

FIGURE P10.37*c*

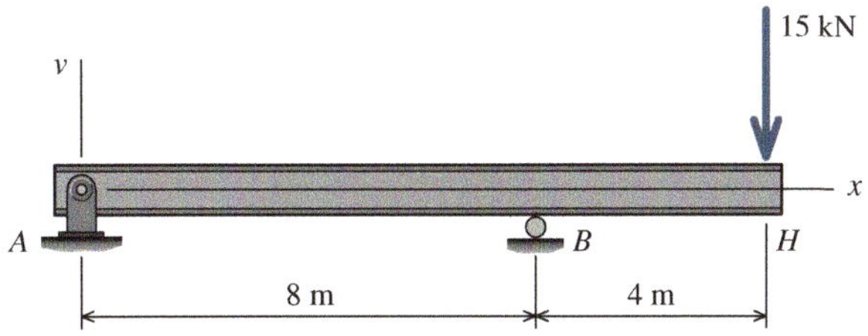

FIGURE P10.37*d*

P10.38 For the beams and loadings shown in Figures P10.38*a–d*, determine the beam deflection at point *H*. Assume that $EI = 6 \times 10^4$ kN · m^2 is constant for each beam.

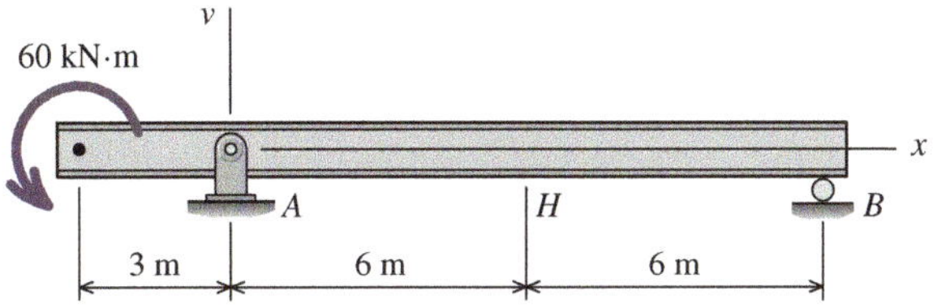

FIGURE P10.38*a*

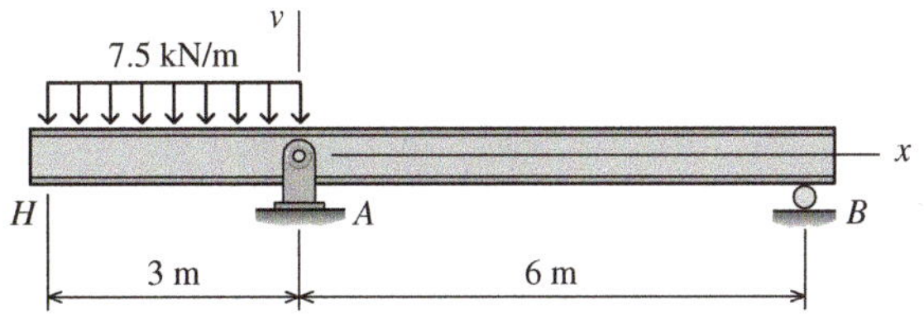

FIGURE P10.38*b*

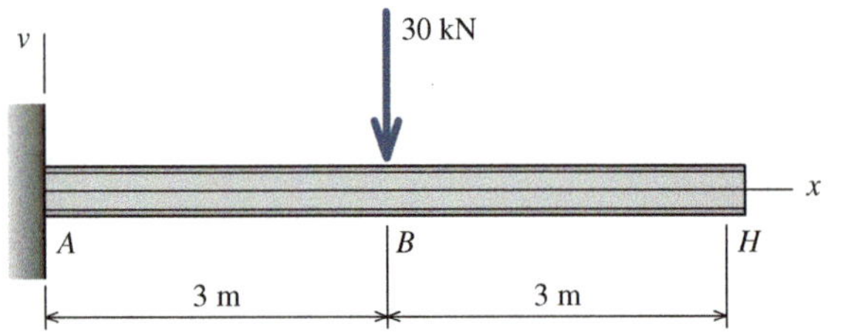

FIGURE P10.38*c*

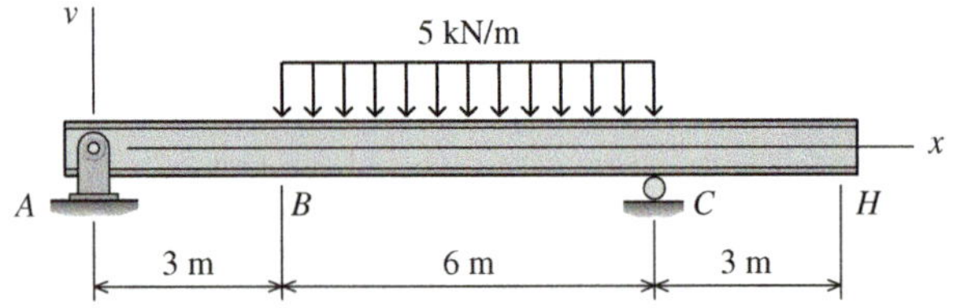

FIGURE P10.38*d*

P10.39 The simply supported beam shown in Figure P10.39 consists of a W24 × 94 structural steel wide-flange shape [$E = 29{,}000$ ksi; $I = 2{,}700$ in.^{4}]. For the loading shown, determine the beam deflection at point *C*.

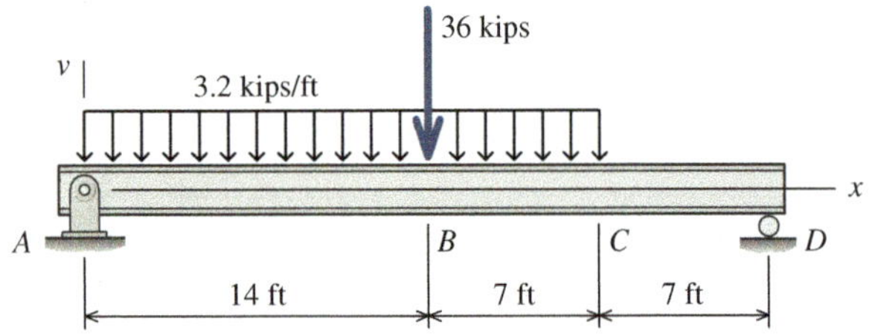

FIGURE P10.39

P10.40 The simply supported beam shown in Figure P10.40 consists of a W410 × 60 structural steel wide-flange shape [$E = 200$ GPa; $I = 216 \times 10^6$ mm^4]. For the loading shown, determine the beam deflection at point *B*.

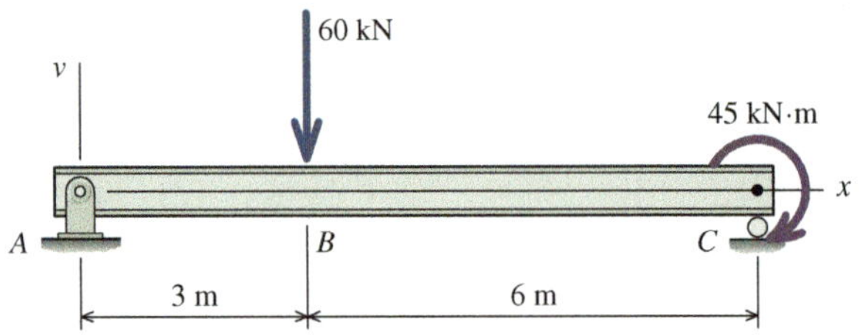

FIGURE P10.40

P10.41 The cantilever beam shown in Figure P10.41 consists of a rectangular structural steel tube shape [$E = 29{,}000$ ksi; $I = 476$ in.^{4}]. For the loading shown, determine

(a) the beam deflection at point *B*.
(b) the beam deflection at point *C*.

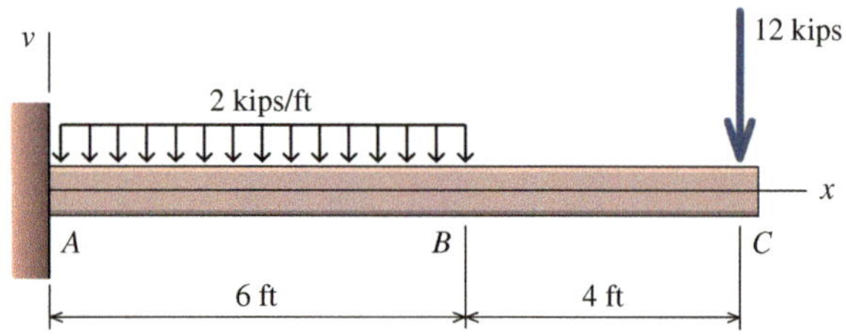

FIGURE P10.41

P10.42 The solid 1.125 in. diameter steel [$E = 29{,}000$ ksi] shaft shown in Figure P10.42 supports two pulleys. For the loading shown, determine

(a) the shaft deflection at pulley *B*.
(b) the shaft deflection at pulley *C*.

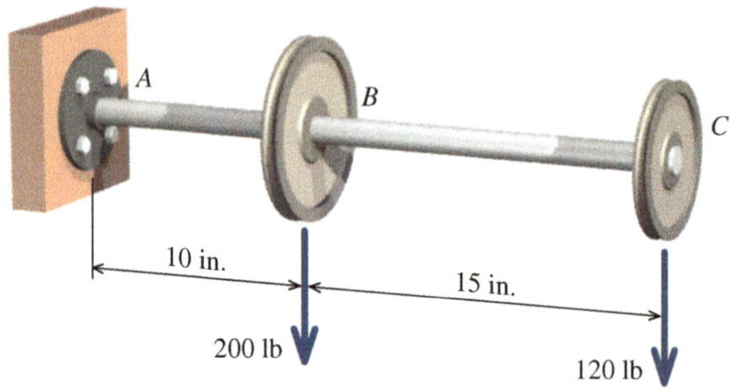

FIGURE P10.42

P10.43 The cantilever beam shown in Figure P10.43 consists of a rectangular structural steel tube shape [$E = 29{,}000$ ksi; $I = 1{,}710$ in.^{4}]. For the loading shown, determine

(a) the beam deflection at point *A*.
(b) the beam deflection at point *B*.

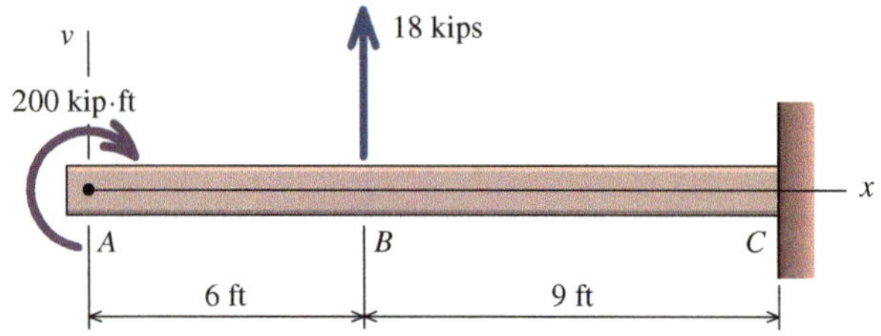

FIGURE P10.43

P10.44 The simply supported beam shown in Figure P10.44 consists of a W21 × 44 structural steel wide-flange shape [$E = 29{,}000$ ksi; $I = 843$ in.^{4}]. For the loading shown, determine

(a) the beam deflection at point *A*.
(b) the beam deflection at point *C*.

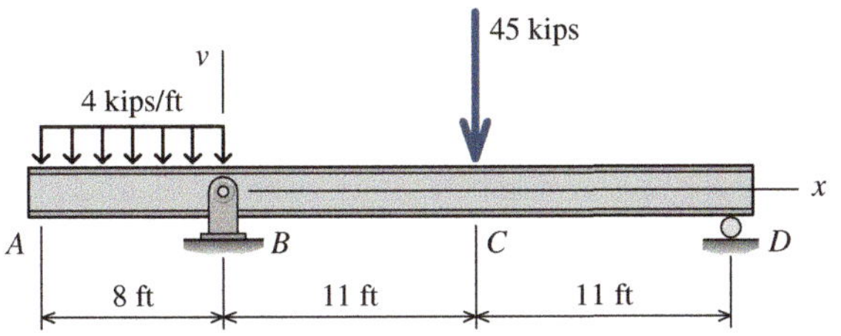

FIGURE P10.44

P10.45 The solid 30 mm diameter steel [E = 200 GPa] shaft shown in Figure P10.45 supports two belt pulleys. Assume that the bearing at B can be idealized as a roller support and that the bearing at D can be idealized as a pin support. For the loading shown, determine

(a) the shaft deflection at pulley A.
(b) the shaft deflection at pulley C.

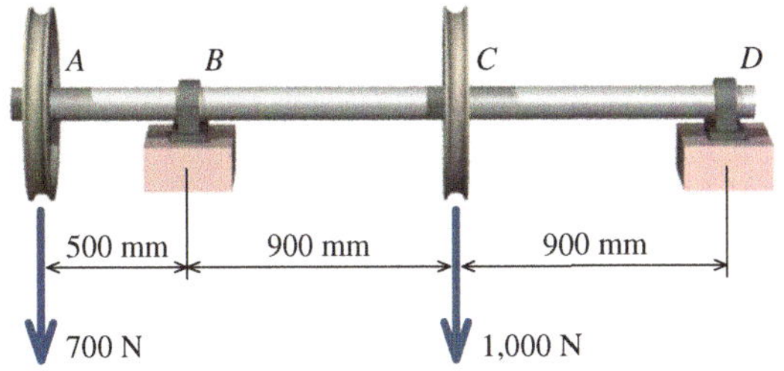

FIGURE P10.45

P10.46 The cantilever beam shown in Figure P10.46 consists of a W530 × 92 structural steel wide-flange shape [E = 200 GPa; $I = 552 \times 10^6$ mm^4]. For the loading shown, determine

(a) the beam deflection at point A.
(b) the beam deflection at point B.

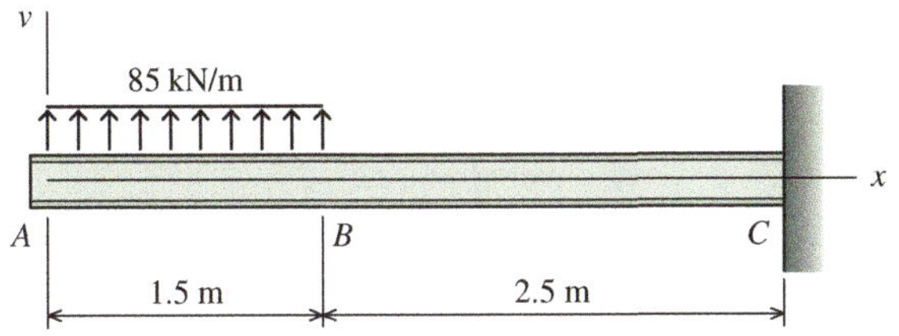

FIGURE P10.46

P10.47 The simply supported beam shown in Figure P10.47/48 consists of a W410 × 60 structural steel wide-flange shape [E = 200 GPa; $I = 216 \times 10^6$ mm^4]. For the loading shown, determine the beam deflection at point B.

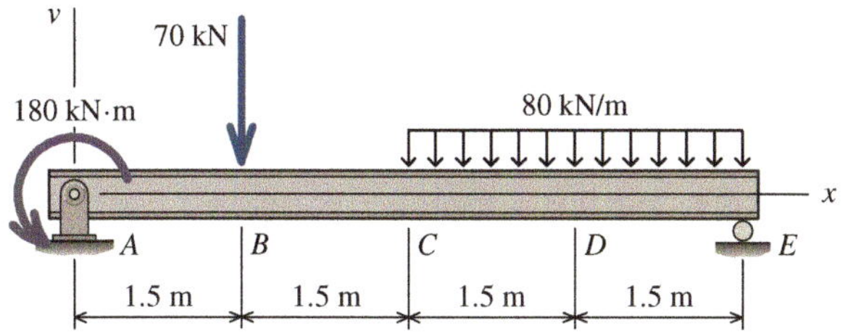

FIGURE P10.47/48

P10.48 The simply supported beam shown in Figure P10.47/48 consists of a W410 × 60 structural steel wide-flange shape [E = 200 GPa; $I = 216 \times 10^6$ mm^4]. For the loading shown, determine the beam deflection at point C.

P10.49 The simply supported beam shown in Figure P10.49/50 consists of a W530 × 66 structural steel wide-flange shape [E = 200 GPa; $I = 351 \times 10^6$ mm^4]. If w = 80 kN/m, determine

(a) the beam deflection at point A.
(b) the beam deflection at point C.

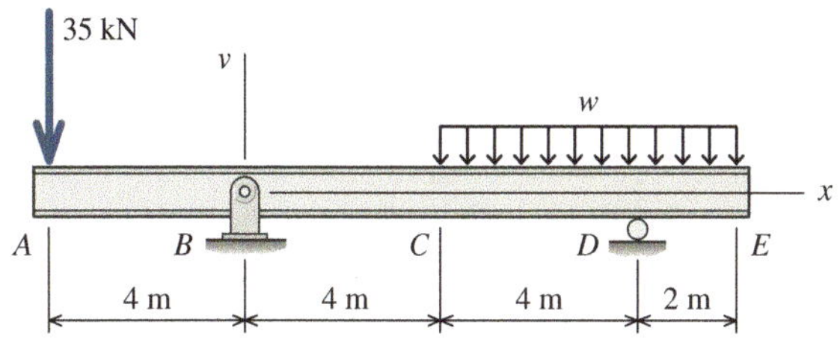

FIGURE P10.49/50

P10.50 The simply supported beam shown in Figure P10.49/50 consists of a W530 × 66 structural steel wide-flange shape [E = 200 GPa; $I = 351 \times 10^6$ mm^4]. If w = 90 kN/m, determine

(a) the beam deflection at point C.
(b) the beam deflection at point E.

P10.51 The cantilever beam shown in Figure P10.51/52 consists of a rectangular structural steel tube shape [E = 200 GPa; $I = 95 \times 10^6$ mm^4]. For the loading shown, determine the beam deflection at point B.

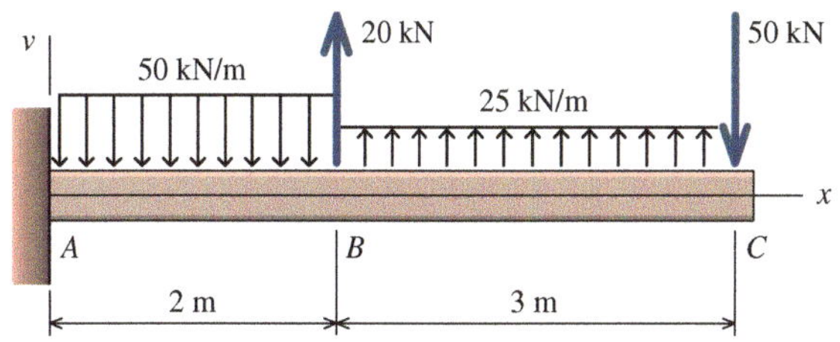

FIGURE P10.51/52

P10.52 The cantilever beam shown in Figure P10.51/52 consists of a rectangular structural steel tube shape [E = 200 GPa; $I = 95 \times 10^6$ mm^4]. For the loading shown, determine the beam deflection at point C.

P10.53 The simply supported beam shown in Figure P10.53/54 consists of a W10 × 30 structural steel wide-flange shape [E = 29,000 ksi; I = 170 in.^{4}]. If w = 5 kips/ft, determine

(a) the beam deflection at point A.
(b) the beam deflection at point C.

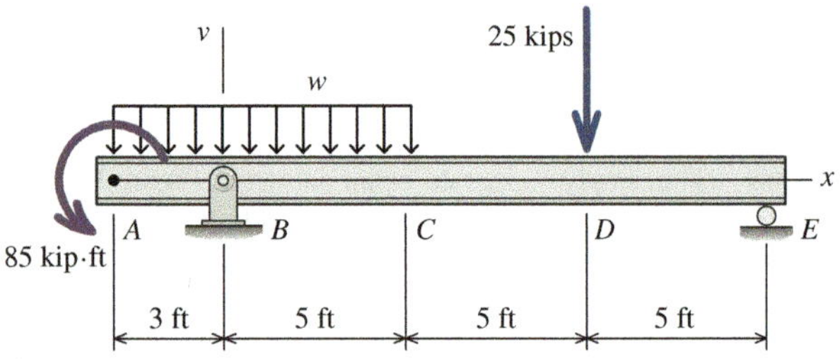

FIGURE P10.53/54

P10.54 The simply supported beam shown in Figure P10.53/54 consists of a W10 × 30 structural steel wide-flange shape [E = 29,000 ksi; I = 170 in.^{4}]. If w = 9 kips/ft, determine

(a) the beam deflection at point A.
(b) the beam deflection at point D.

P10.55 The simply supported beam shown in Figure P10.55 consists of a W21 × 44 structural steel wide-flange shape [E = 29,000 ksi; I = 843 in.^{4}]. For the loading shown, determine

(a) the beam deflection at point A.
(b) the beam deflection at point C.

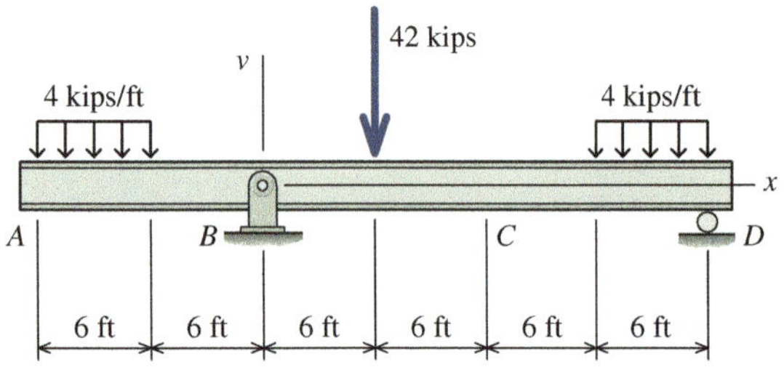

FIGURE P10.55

P10.56 The simply supported beam shown in Figure P10.56/57 consists of a W530 × 66 structural steel wide-flange shape [E = 200 GPa; I = 351 × 10^6 mm^4]. If w = 85 kN/m, determine the beam deflection at point B.

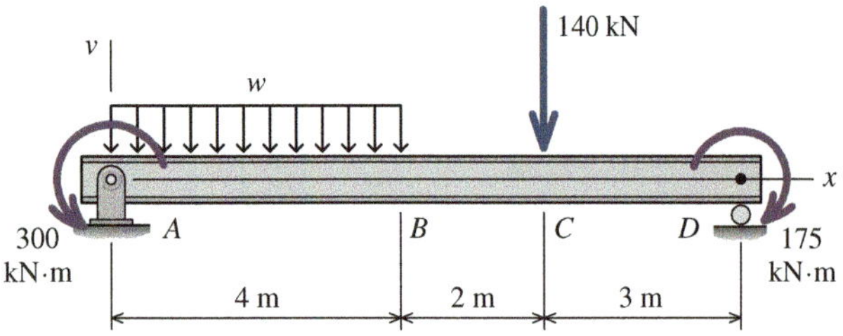

FIGURE P10.56/57

P10.57 The simply supported beam shown in Figure P10.56/57 consists of a W530 × 66 structural steel wide-flange shape [E = 200 GPa; I = 351 × 10^6 mm^4]. If w = 115 kN/m, determine the beam deflection at point C.

P10.58 A 25 ft long soldier beam is used as a key component of an earth retention system at an excavation site. The beam is subjected to a soil loading that is linearly distributed from 520 lb/ft to 260 lb/ft as shown in Figure P10.58. The beam can be idealized as a cantilever with a fixed support at A. Added support is supplied by a tieback anchor at B, that exerts a force of 5,000 lb on the beam. Determine the horizontal deflection of the beam at point C. Assume that EI = 5 × 10^8 lb·in.^{2}.

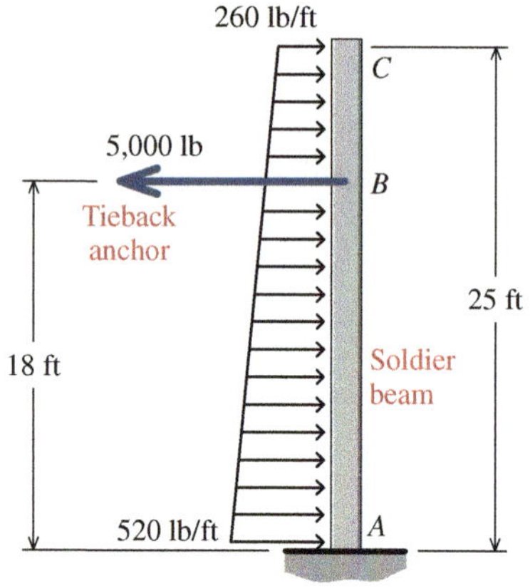

FIGURE P10.58

P10.59 A 25 ft long soldier beam is used as a key component of an earth retention system at an excavation site. The beam is subjected to a uniformly distributed soil loading of 260 lb/ft as shown in Figure P10.59. The beam can be idealized as a cantilever with a fixed support at A. Added support is supplied by a tieback anchor at B, that exerts a force of 4,000 lb on the beam. Determine the horizontal deflection of the beam at point C. Assume that EI = 5 × 10^8 lb·in.^{2}.

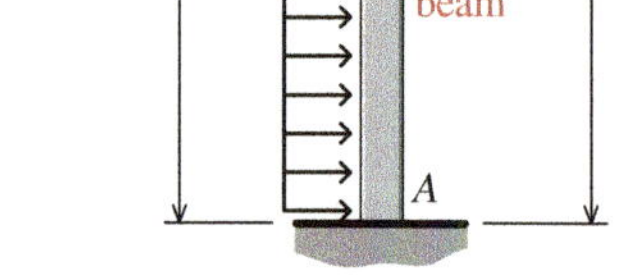

FIGURE P10.59

CHAPTER 11

Statically Indeterminate Beams

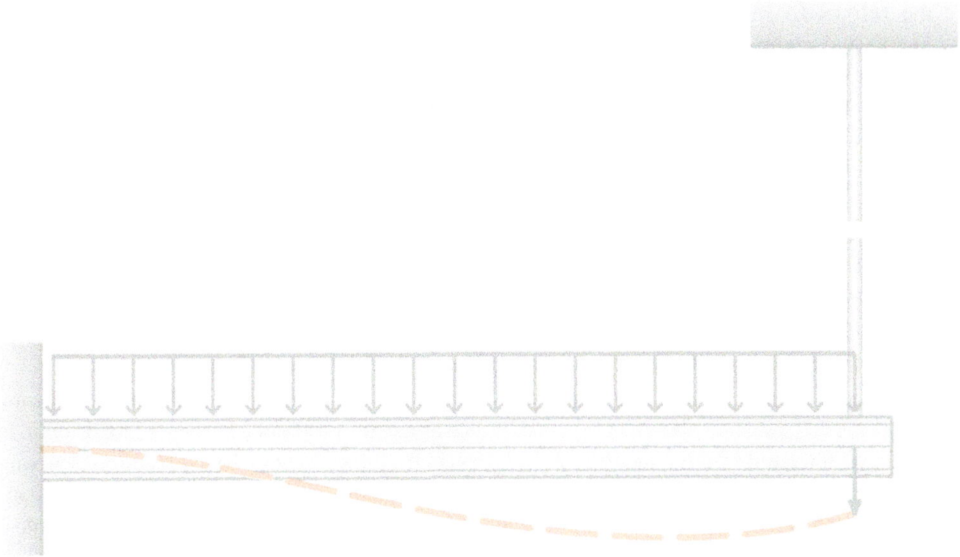

11.1 Introduction

A beam is classified as statically indeterminate if the number of unknown support reactions exceeds the available number of equilibrium equations. In such cases, the deformation of the loaded beam is used to derive additional relationships that are needed to evaluate the unknown reactions (or other unknown forces). The calculation methods presented in Chapter 10 will be employed, along with known beam slopes and deflections at supports (and other constraints), to develop compatibility equations. Together, the compatibility and equilibrium equations provide the basis needed to determine all beam reactions. Once all loads acting on the beam are known, the methods of Chapters 7 through 10 can be used to determine the required beam stresses and deflections.

11.2 Types of Statically Indeterminate Beams

A statically indeterminate beam is typically identified by the arrangement of its supports. Figure 11.1*a* shows a **propped cantilever** beam with loads and reactions illustrated. This type of beam has a fixed support at one end and a roller support at the opposite end. The fixed support provides three reactions: translation restraints A_x and A_y in the horizontal and vertical directions, respectively, and a restraint M_A against rotation. The roller support prevents translation in the vertical direction (B_y). Consequently, the propped cantilever has four unknown reactions. Three equilibrium equations can be developed for the beam ($\Sigma F_x = 0$, $\Sigma F_y = 0$, and $\Sigma M = 0$), but since there are more unknown reactions than there are equilibrium equations, the propped cantilever is classified as **statically indeterminate**. The number of reactions *in excess* of the number of equilibrium equations is termed the **degree of static indeterminacy**. Thus, the propped cantilever is said to be statically indeterminate to the first degree. The excess reactions are called **redundant reactions** or simply **redundants** because they are not essential to maintaining the equilibrium of the beam.

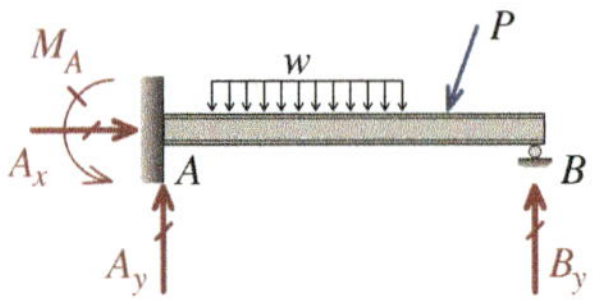

(*a*) Actual beam with loads and reactions

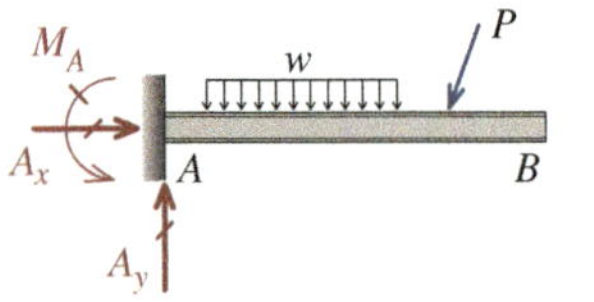

(*b*) Released beam if B_y is chosen as the redundant

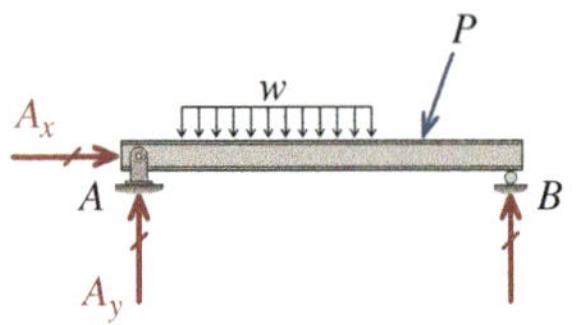

(*c*) Released beam if M_A is chosen as the redundant

FIGURE 11.1 Propped cantilever beam.

The general approach used to solve statically indeterminate beams involves selecting redundant reactions and developing an equation pertinent to each redundant on the basis of the deformed configuration of the loaded beam. To develop these geometric equations, redundant reactions are selected and removed from the beam. The beam that remains is called the **released beam**. The released beam must be **stable** (i.e., capable of supporting the loads) and statically determinate so that its reactions can be determined by equilibrium considerations. The effect of the redundant reactions is addressed separately, through knowledge about the deflections or rotations that must occur at the redundant support. For instance, we can know with certainty that the beam deflection at B must be zero, since the redundant support B_y prevents movement either up or down at this location.

As mentioned in the previous paragraph, the released beam must be stable and statically determinate. For example, the roller reaction B_y could be removed from the propped cantilever beam (Figure 11.1*b*), leaving a cantilever beam that is still capable of supporting the applied loads. In other words, the cantilever beam is stable. Alternatively, the moment reaction M_A could be removed from the propped cantilever (Figure 11.1*c*), leaving a simply supported beam with a pin support at A and a roller support at B. This released beam is also stable.

A special case arises if all of the loads act transverse to the longitudinal axis of the beam. The propped cantilever shown in Figure 11.2 is subjected to vertical (transverse) loads only. In this case, the equilibrium equation $\Sigma F_x = A_x = 0$ is trivial, so the horizontal reaction at A vanishes, leaving only three unknown reactions: A_y, B_y, and M_A. Even so, this beam is still statically indeterminate to the first degree, because only two equilibrium equations are available.

Another type of statically indeterminate beam is called a **fixed-end beam** or a **fixed-fixed beam** (Figure 11.3). The fixed connections at A and B each provide three reactions. Since there are only three equilibrium equations, this beam is statically indeterminate to the third degree. In the special case of transverse loads only (Figure 11.4), the fixed-end beam has four nonzero reactions but only two available equilibrium equations. Therefore, the fixed-end beam in Figure 11.4 is statically indeterminate to the second degree.

The beam shown in Figure 11.5*a* is called a **continuous beam** because it has more than one span and the beam is uninterrupted over the interior support. If only transverse loads act on the beam, it is statically indeterminate to the first degree. This beam could be released in two ways. In Figure 11.5*b*, the interior roller support at B is removed so that the released beam is simply supported at A and C, a stable configuration. In Figure 11.5*c*, the exterior support at C is removed. This released beam is also simply supported; however, it now has an overhang (from B to C). Nevertheless, the beam's configuration is stable.

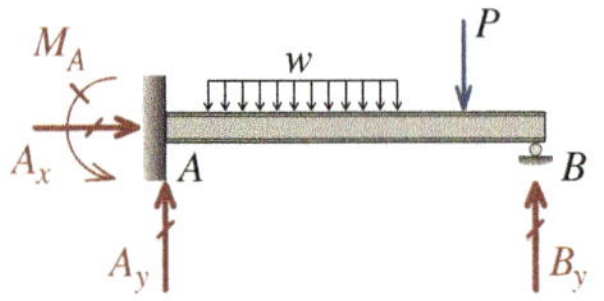

FIGURE 11.2 Propped cantilever subjected to transverse loads only.

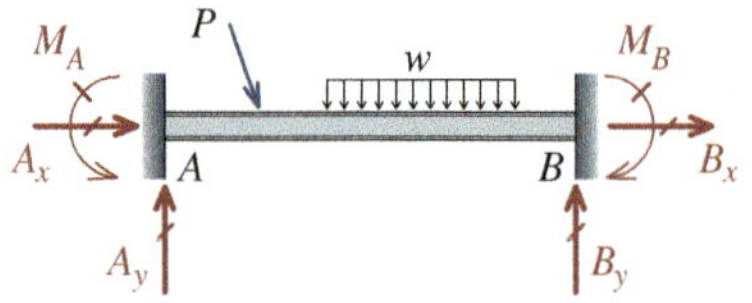

FIGURE 11.3 Fixed-end beam with load and reactions.

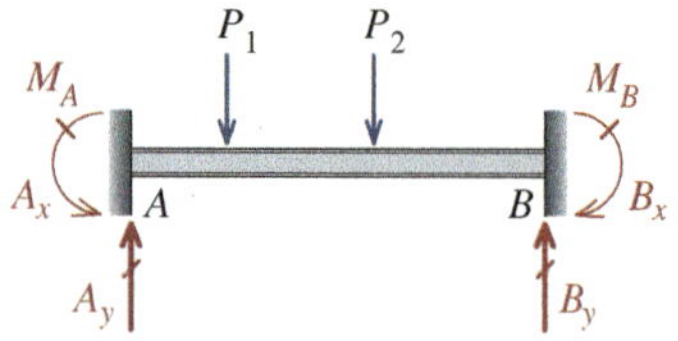

FIGURE 11.4 Fixed-end beam with transverse loads only.

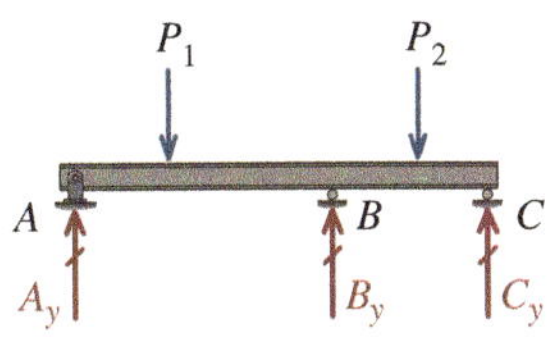

FIGURE 11.5*a* Continuous beam on three supports.

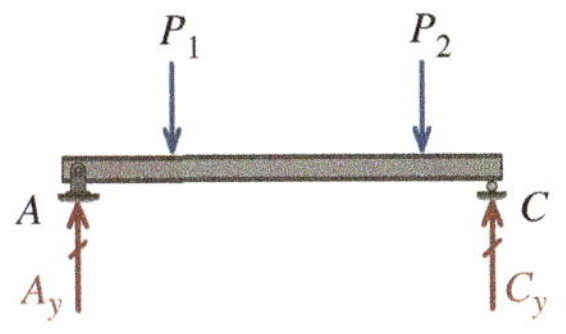

FIGURE 11.5*b* Released beam created by removing redundant B_y.

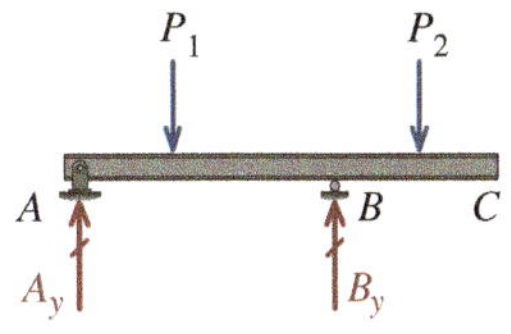

FIGURE 11.5*c* Released beam created by removing redundant C_y.

In the sections that follow, three methods for analyzing statically indeterminate beams will be discussed. In each case, the initial objective will be to determine the magnitude of the redundant reactions. After the redundants have been determined, the remaining reactions can be determined from equilibrium equations. After all of the reactions are known, the beam shear forces, bending moments, bending and shear stresses, and transverse deflections can be determined by the methods presented in Chapters 7 through 10.

11.3 The Integration Method

For statically determinate beams, known slopes and deflections were used to obtain boundary and continuity conditions, from which the constants of integration in the elastic curve equation could be evaluated. For statically indeterminate beams, the procedure is identical. However, the bending-moment equations derived at the outset of the procedure will contain reactions (or loads) that cannot be evaluated with the available equations of equilibrium. One additional boundary condition will be needed for the evaluation of each such unknown. For example, consider a transversely loaded beam with four unknown reactions that are to be solved simultaneously by the double-integration method. Two constants of integration will appear as the bending-moment equation is integrated twice; consequently, this statically indeterminate beam has six unknowns. Since a transversely loaded beam has only two nontrivial equilibrium equations, four additional equations must be derived from boundary (or continuity) conditions. Two boundary (or continuity) conditions will be required in order to solve for the constants of integration, and two extra boundary (or continuity) conditions will be needed in order to solve for two of the unknown reactions. Examples 11.1 and 11.2 illustrate the method.

EXAMPLE 11.1

A propped cantilever beam is loaded and supported as shown. Assume that EI is constant for the beam. Determine the reactions at supports A and B.

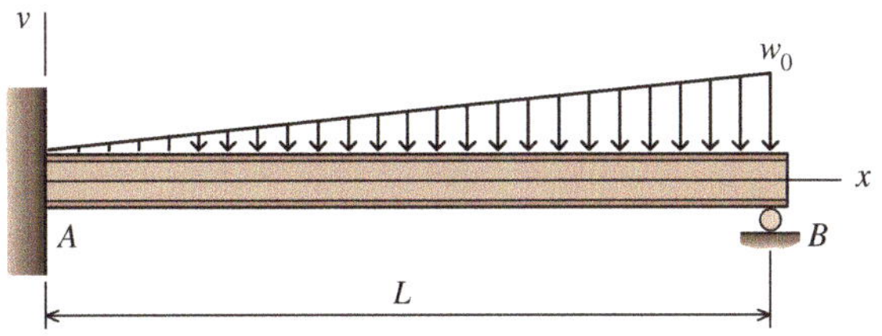

Plan the Solution

First, draw a free-body diagram (FBD) of the entire beam and write three equilibrium equations in terms of the four unknown reactions A_x, A_y, B_y, and M_A. Next, consider an FBD that cuts through the beam at a distance x from the origin. Write an equilibrium equation for the sum of moments, and from this equation, determine the equation for the bending moment M as it varies with x. Substitute M into Equation (10.1) and integrate twice, producing two constants of integration. At this point in the solution, there will be six unknowns, which will require that six equations be solved. In addition to the three equilibrium equations, three more equations will be obtained from three boundary conditions. These six equations will be solved to yield the constants of integration and the unknown beam reactions.

SOLUTION

Equilibrium

Consider an FBD of the entire beam. The equation for the sum of forces in the horizontal direction is trivial, since there are no loads in the x direction:

$$\Sigma F_x = A_x = 0$$

The sum of forces in the vertical direction yields

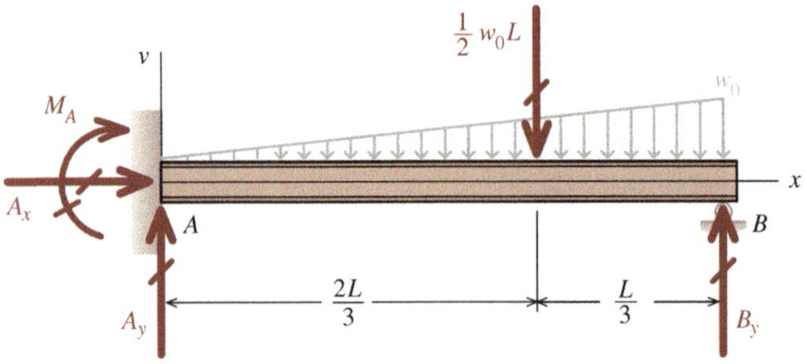

$$\Sigma F_y = A_y + B_y - \frac{1}{2}w_0 L = 0 \quad \text{(a)}$$

The sum of moments about roller support B gives

$$\Sigma M_B = \frac{1}{2}w_0 L\left(\frac{L}{3}\right) - A_y L - M_A = 0 \quad \text{(b)}$$

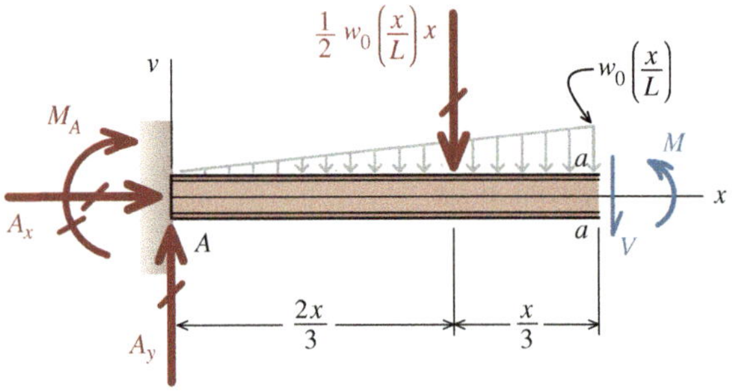

Next, cut a section through the beam at an arbitrary distance x from the origin and draw a free-body diagram, taking care to show the internal moment M acting in a positive direction on the exposed surface of the beam. The equilibrium equation for the sum of moments about section a–a is

$$\Sigma M = \frac{1}{2}w_0\left(\frac{x}{L}\right)x\left(\frac{x}{3}\right) - A_y x - M_A + M = 0$$

From this equation, the bending-moment equation can be expressed as

$$M = -\frac{w_0}{6L}x^3 + A_y x + M_A \quad (0 \le x \le L) \quad \text{(c)}$$

Substitute the expression for M into Equation (10.1) to obtain

$$EI\frac{d^2v}{dx^2} = -\frac{w_0}{6L}x^3 + A_y x + M_A \quad \text{(d)}$$

Integration

Equation (d) will be integrated twice to give

$$EI\frac{dv}{dx} = -\frac{w_0}{24L}x^4 + \frac{A_y}{2}x^2 + M_A x + C_1 \quad \text{(e)}$$

and

$$EIv = -\frac{w_0}{120L}x^5 + \frac{A_y}{6}x^3 + \frac{M_A}{2}x^2 + C_1 x + C_2 \quad \text{(f)}$$

Boundary Conditions

For this beam, Equation (c) is valid for the interval $0 \le x \le L$. The boundary conditions, therefore, are found at $x = 0$ and $x = L$. From the fixed support at A, the boundary conditions are $x = 0$, $dv/dx = 0$ and $x = 0$, $v = 0$. At roller support B, the boundary condition is $x = L$, $v = 0$.

Evaluate Constants

Substitute the boundary condition $x = 0$, $dv/dx = 0$ into Equation (e) to find $C_1 = 0$. Substitution of the value of C_1 and the boundary condition $x = 0$, $v = 0$ into Equation (f) gives $C_2 = 0$. Next, substitute the values of C_1 and C_2 and the boundary condition $x = L$, $v = 0$ into Equation (f) to obtain

$$EI(0) = -\frac{w_0}{120L}(L)^5 + \frac{A_y}{6}(L)^3 + \frac{M_A}{2}(L)^2$$

Solve this equation for M_A in terms of the reaction A_y:

$$\boxed{M_A = \frac{w_0 L^2}{60} - \frac{A_y L}{3}} \tag{g}$$

From equilibrium Equation (b), M_A can also be written as

$$\boxed{M_A = \frac{w_0 L^2}{6} - A_y L} \tag{h}$$

Solve for Reactions

Equate Equations (g) and (h):

$$\frac{w_0 L^2}{60} - \frac{A_y L}{3} = \frac{w_0 L^2}{6} - A_y L$$

Then solve for the vertical reaction force at A:

$$A_y = \frac{27}{120} w_0 L = \frac{9}{40} w_0 L \qquad \textbf{Ans.}$$

Substitute this result back into either Equation (g) or Equation (h) to determine the moment M_A:

$$M_A = -\frac{7}{120} w_0 L^2 \qquad \textbf{Ans.}$$

To determine the reaction force at roller B, substitute the result for A_y into Equation (a) and solve for B_y:

$$B_y = \frac{33}{120} w_0 L = \frac{11}{40} w_0 L \qquad \textbf{Ans.}$$

EXAMPLE 11.2

A beam is loaded and supported as shown. Assume that EI is constant for the beam. Determine the reactions at supports A and C.

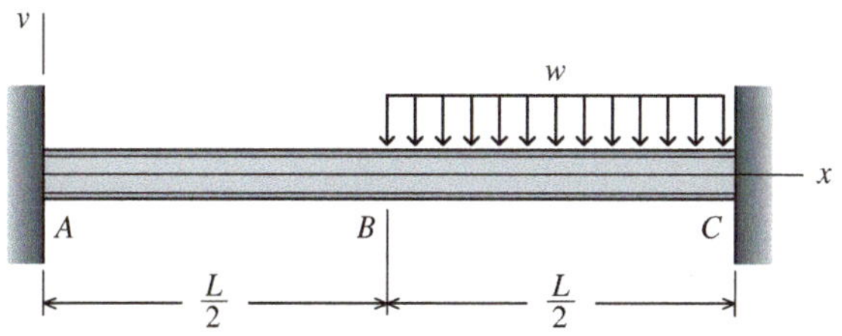

Plan the Solution

First, draw an FBD of the entire beam and develop two equilibrium equations in terms of the four unknown reactions A_y, C_y, M_A, and M_C. Two elastic curve equations must be derived for this beam and load. One curve applies to the interval $0 \leq x \leq L/2$, and the second curve applies to $L/2 \leq x \leq L$. Four constants of integration will result from the double integration of two equations; therefore, a total of eight unknowns must be determined. To solve for eight variables, eight equations are required. Four equations are obtained from the boundary conditions at the beam supports, where the beam deflection and slope are known. At the junction between the two intervals, two equations can be obtained from the continuity conditions at $x = L/2$. Finally, two nontrivial equations will be derived from equilibrium for the entire beam. These eight equations will be solved to yield the constants of integration and the unknown beam reactions.

SOLUTION

Equilibrium

Draw an FBD of the entire beam. Since no loads act in the horizontal direction, the reactions A_x and C_x will be omitted. Write two equilibrium equations:

$$\Sigma F_y = A_y + C_y - \frac{wL}{2} = 0 \tag{a}$$

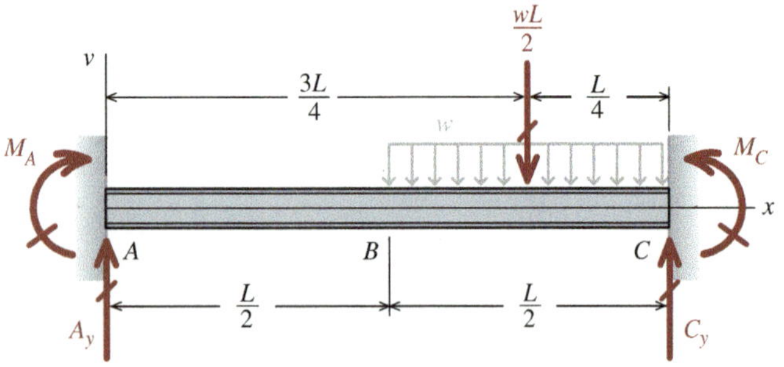

$$\Sigma M_C = \frac{wL}{2}\left(\frac{L}{4}\right) - A_yL - M_A + M_C = 0 \tag{b}$$

For this beam, two equations are required to describe the bending moments for the entire span. Draw two FBDs: one FBD that cuts through the beam between A and B, and the second FBD that cuts through the beam between B and C. From these two FBDs, derive the bending-moment equations and, in turn, the differential equations of the elastic curve.

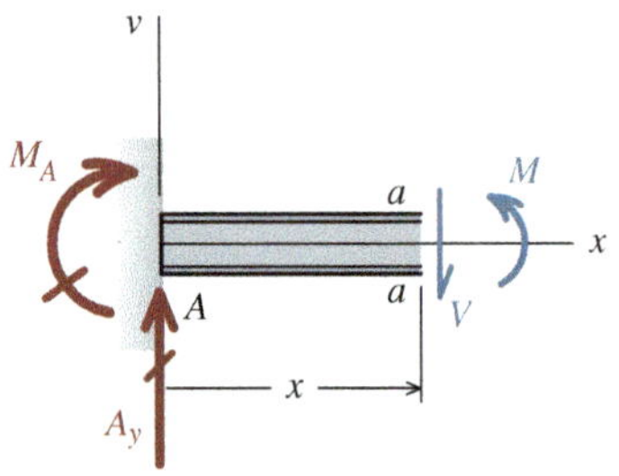

For the Interval $0 \le x \le L/2$ Between A and B,

$$M = A_yx + M_A$$

which gives the differential equation

$$EI\frac{d^2v}{dx^2} = A_yx + M_A \tag{c}$$

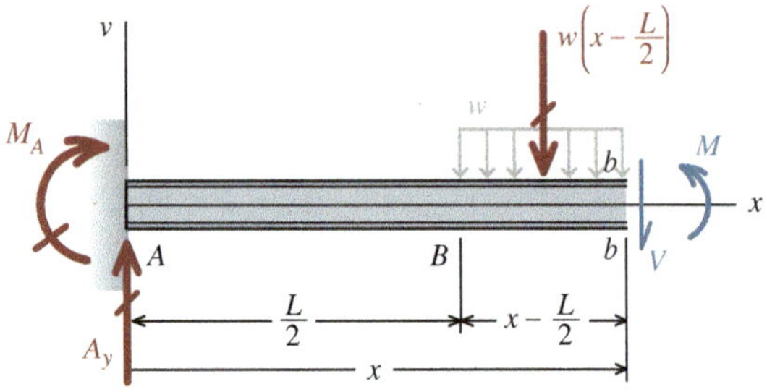

Integration

Integrate Equation (c) twice to obtain

$$EI\frac{dv}{dx} = \frac{A_y}{2}x^2 + M_Ax + C_1 \tag{d}$$

and

$$EIv = \frac{A_y}{6}x^3 + \frac{M_A}{2}x^2 + C_1x + C_2 \tag{e}$$

For the Interval $L/2 \le x \le L$ Between B and C,

$$M = -\frac{w}{2}\left(x - \frac{L}{2}\right)^2 + A_yx + M_A$$

which gives the differential equation

$$EI\frac{d^2v}{dx^2} = -\frac{w}{2}\left(x - \frac{L}{2}\right)^2 + A_yx + M_A \tag{f}$$

Integration

Integrate Equation (f) twice to obtain

$$EI\frac{dv}{dx} = -\frac{w}{6}\left(x - \frac{L}{2}\right)^3 + \frac{A_y}{2}x^2 + M_A x + C_3 \tag{g}$$

and

$$EIv = -\frac{w}{24}\left(x - \frac{L}{2}\right)^4 + \frac{A_y}{6}x^3 + \frac{M_A}{2}x^2 + C_3 x + C_4 \tag{h}$$

Boundary Conditions

There are four boundary conditions for this beam. Substituting $x = 0$, $dv/dx = 0$ into Equation (d) gives $C_1 = 0$, and substituting the value of C_1 and $x = 0$, $v = 0$ into Equation (e) gives $C_2 = 0$. Next, substitute the boundary condition $x = L$, $dv/dx = 0$ into Equation (g) to obtain the following expression for C_3:

$$C_3 = \frac{wL^3}{48} - \frac{A_y L^2}{2} - M_A L$$

Finally, substitute the boundary condition $x = L$, $v = 0$ and the expression obtained for C_3 into Equation (h) to obtain the following expression for C_4:

$$C_4 = -\frac{7wL^4}{384} + \frac{A_y L^3}{3} + \frac{M_A L^2}{2}$$

Continuity Conditions

The beam is a single, continuous member. Consequently, the two sets of equations must produce the same slope and the same deflection at $x = L/2$. Consider slope equations (d) and (g). At $x = L/2$, these two equations must produce the same slope; therefore, set the two equations equal to each other and substitute the value $L/2$ for each variable x:

$$\frac{A_y}{2}\left(\frac{L}{2}\right)^2 + M_A\left(\frac{L}{2}\right) = -\frac{w}{6}(0)^3 + \frac{A_y}{2}\left(\frac{L}{2}\right)^2 + M_A\left(\frac{L}{2}\right) + C_3$$

This equation reduces to

$$0 = C_3 = \frac{wL^3}{48} - \frac{A_y L^2}{2} - M_A L \qquad \therefore \frac{A_y L^2}{2} + M_A L = \frac{wL^3}{48} \tag{i}$$

Similarly, deflection equations (e) and (h) must produce the same deflection at $x = L/2$:

$$\frac{A_y}{6}\left(\frac{L}{2}\right)^3 + \frac{M_A}{2}\left(\frac{L}{2}\right)^2 = -\frac{w}{24}(0)^4 + \frac{A_y}{6}\left(\frac{L}{2}\right)^3 + \frac{M_A}{2}\left(\frac{L}{2}\right)^2 + C_3\left(\frac{L}{2}\right) + C_4$$

This equation reduces to

$$C_4 = -C_3\left(\frac{L}{2}\right) \qquad \therefore -\frac{7wL^4}{384} + \frac{A_y L^3}{3} + \frac{M_A L^2}{2} = -\left[\frac{wL^3}{48} - \frac{A_y L^2}{2} - M_A L\right]\left(\frac{L}{2}\right) \tag{j}$$

Solve for Reactions

Solve Equation (j) for the reaction force A_y:

$$A_y = \frac{36wL}{384} = \frac{3wL}{32} \qquad \textbf{Ans.}$$

Substitute the reaction force A_y into Equation (i) to solve for the moment at A:

$$M_A = \frac{wL^2}{48} - \frac{A_yL}{2} = \frac{wL^2}{48} - \frac{3wL^2}{64} = -\frac{10wL^2}{384} = -\frac{5wL^2}{192} \qquad \textbf{Ans.}$$

Substitute the reaction force A_y into Equation (a) to determine the reaction force C_y:

$$C_y = \frac{wL}{2} - A_y = \frac{wL}{2} - \frac{3wL}{32} = \frac{13wL}{32} \qquad \textbf{Ans.}$$

Finally, determine the reaction moment M_C from Equation (b):

$$M_C = M_A + A_yL - \frac{wL^2}{8} = -\frac{10wL^2}{384} + \frac{3wL^2}{32} - \frac{wL^2}{8} = -\frac{22wL^2}{384} = -\frac{11wL^2}{192} \qquad \textbf{Ans.}$$

PROBLEMS

P11.1 A beam is loaded and supported as shown in Figure P11.1. Use the double-integration method to determine the magnitude of the moment M_0 required to make the slope at the left end of the beam equal to zero.

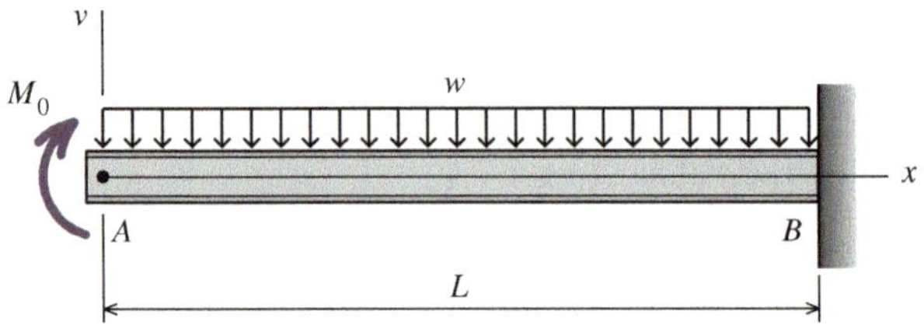

FIGURE P11.1

P11.2 When the load P is applied to the right end of the cantilever beam shown in Figure P11.2, the deflection at the right end of the beam is zero. Use the double-integration method to determine the magnitude of the load P.

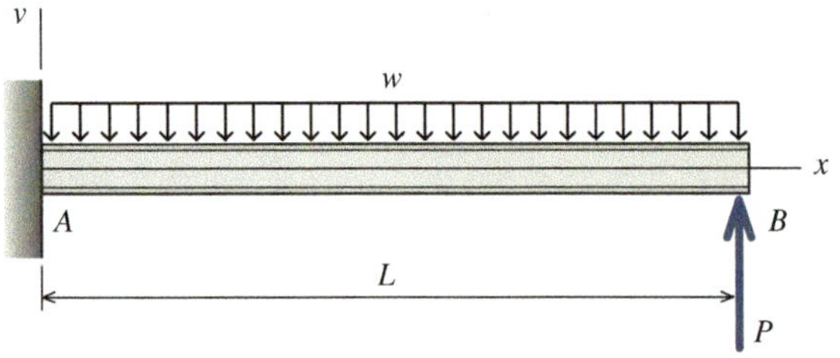

FIGURE P11.2

P11.3 A beam is loaded and supported as shown in Figure P11.3.

(a) Use the double-integration method to determine the reactions at supports A and B.

(b) Draw the shear-force and bending-moment diagrams for the beam.

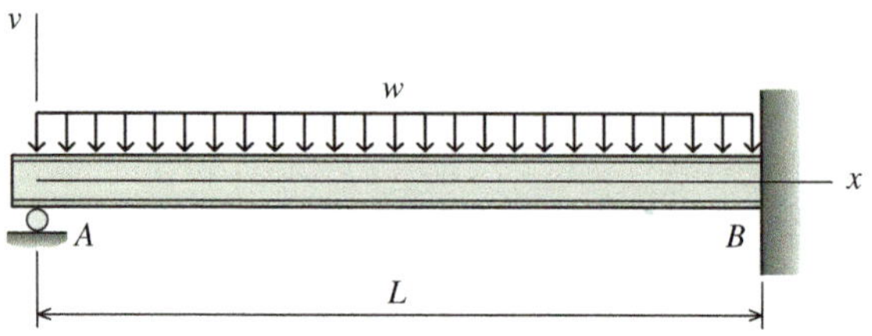

FIGURE P11.3

P11.4 A beam is loaded and supported as shown in Figure P11.4. Use the double-integration method to determine the reactions at supports A and B.

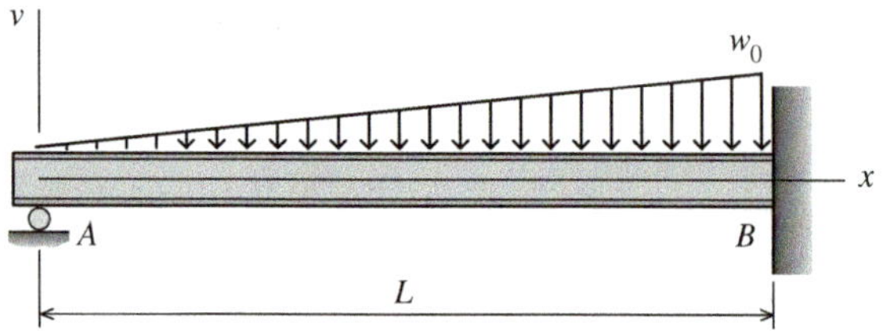

FIGURE P11.4

P11.5 A beam is loaded and supported as shown in Figure P11.5. Use the fourth-order integration method to determine the reaction at roller support *A*.

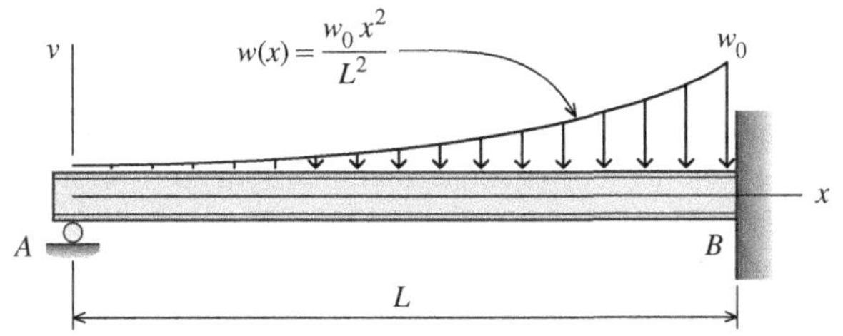

FIGURE P11.5

P11.6 A beam is loaded and supported as shown in Figure P11.6. Use the fourth-order integration method to determine the reactions at supports *A* and *B*.

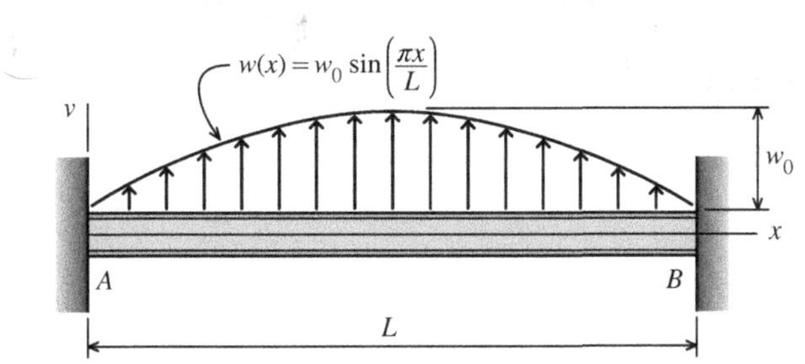

FIGURE P11.6

P11.7 A beam is loaded and supported as shown in Figure P11.7.

(a) Use the double-integration method to determine the reactions at supports *A* and *C*.
(b) Draw the shear-force and bending-moment diagrams for the beam.
(c) Determine the deflection in the middle of the span.

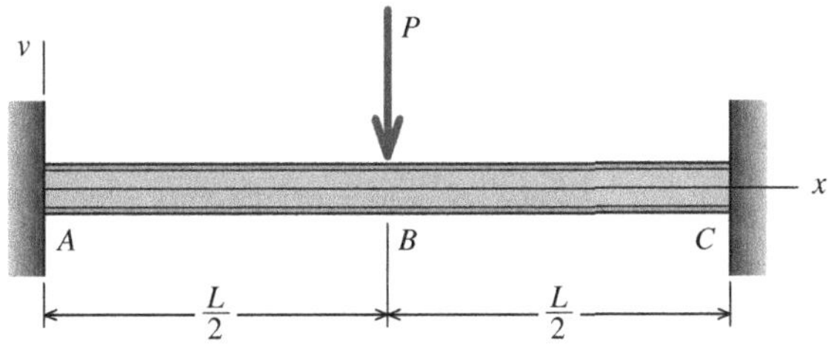

FIGURE P11.7

P11.8 A beam is loaded and supported as shown in Figure P11.8.

(a) Use the double-integration method to determine the reactions at supports *A* and *B*.
(b) Draw the shear-force and bending-moment diagrams for the beam.
(c) Determine the deflection in the middle of the span.

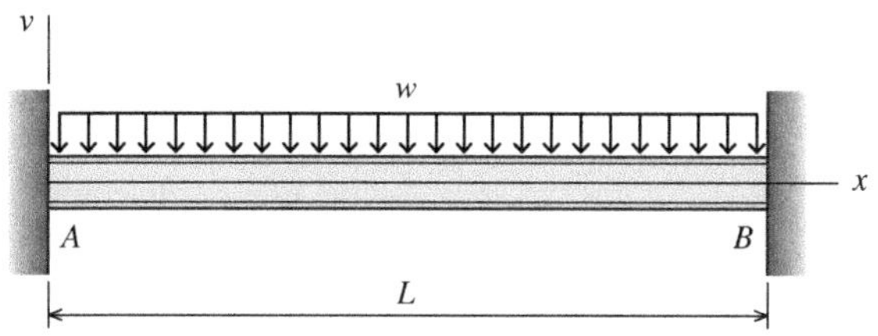

FIGURE P11.8

P11.9 A beam is loaded and supported as shown in Figure P11.9.

(a) Use the double-integration method to determine the reactions at supports *A* and *C*.
(b) Determine the deflection in the middle of the span.

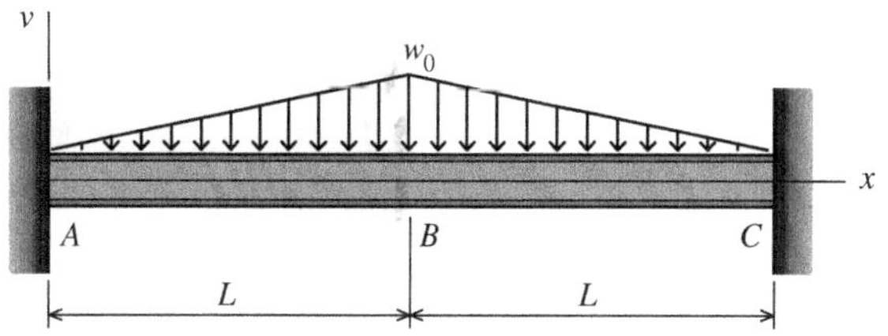

FIGURE P11.9

P11.10–P11.11 A beam is loaded and supported as shown in Figures P11.10 and P11.11.

(a) Use the double-integration method to determine the reactions at supports *A* and *C*.
(b) Draw the shear-force and bending-moment diagrams for the beam.

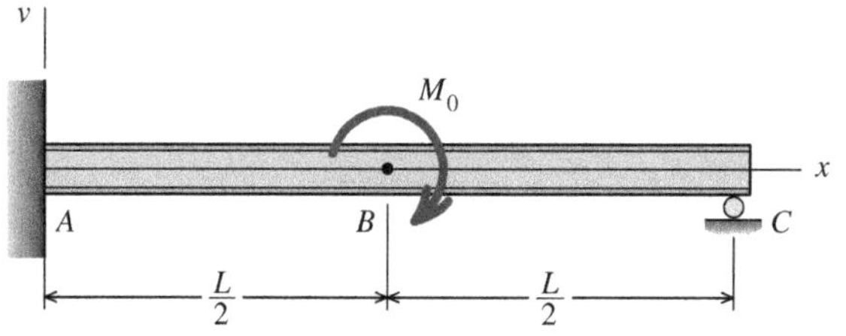

FIGURE P11.10

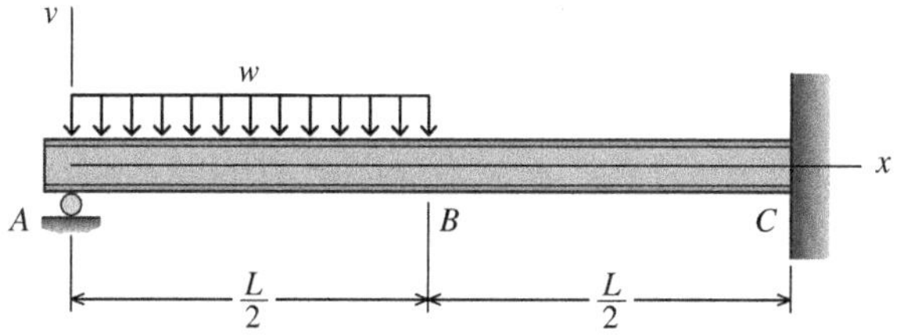

FIGURE P11.11

11.4 Use of Discontinuity Functions for Statically Indeterminate Beams

The use of discontinuity functions for statically determinate beam analysis has been discussed in Chapters 7 and 10. In Section 7.4, discontinuity functions were used to derive functions expressing the shear force and bending moment in beams. Beam deflections for statically determinate beams were computed with discontinuity functions in Section 10.6. In both sections, the reaction forces and moments were computed beforehand from equilibrium considerations, making it possible to incorporate known values into the load function $w(x)$ from the outset of the calculation process. The added difficulty posed by statically indeterminate beams is that the reactions cannot be determined from equilibrium considerations alone and, thus, known values for the reaction forces and moments cannot be included in $w(x)$.

For statically indeterminate beams, reaction forces and moments are initially expressed as unknown quantities in the load function $w(x)$. The integration process then proceeds in the manner described in Section 10.6, producing two constants of integration, C_1 and C_2. These integration constants, as well as the unknown beam reactions, must be computed in order to complete the elastic curve equation. Equations containing the constants C_1 and C_2 can be derived from the boundary conditions and, along with the beam equilibrium equations, are then solved simultaneously to evaluate C_1 and C_2 as well as to determine the beam reaction forces and moments. The solution process is demonstrated in Examples 11.3 and 11.4.

EXAMPLE 11.3

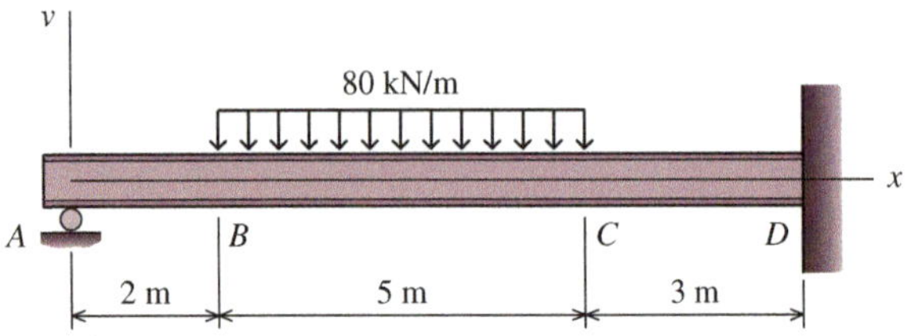

For the statically indeterminate beam shown, use discontinuity functions to determine

(a) the force and moment reactions at A and D.
(b) the deflection of the beam at C.

Assume a constant value of $EI = 120{,}000\ \text{kN}\cdot\text{m}^2$ for the beam.

Plan the Solution

The beam is statically indeterminate; therefore, the reaction forces at A and D cannot be determined solely from equilibrium considerations. From an FBD of the beam, two nontrivial equilibrium equations can be derived. However, since the beam is statically indeterminate, the reaction forces and moments can be stated only as unknowns. The distributed load on the beam, as well as the unknown reactions, will be expressed by discontinuity functions. The load function will be integrated twice to obtain the bending-moment function for the beam. In these first two integrations, constants of integration will not be necessary. The bending-moment function will then be integrated twice to obtain the elastic curve equation. Constants of integration must be considered in these two integrations. The three boundary conditions known at A and D, along with the two nontrivial equilibrium equations, will produce five equations that can be solved simultaneously to determine the three unknown reactions and the two constants of integration. After these quantities are found, the beam deflection at any location can be computed from the elastic curve equation.

SOLUTION

(a) Support Reactions

An FBD of the beam is shown. Since no forces act in the x direction, the ΣF_x equation will be omitted here. From the FBD, the beam reaction forces can be expressed by the following relationships:

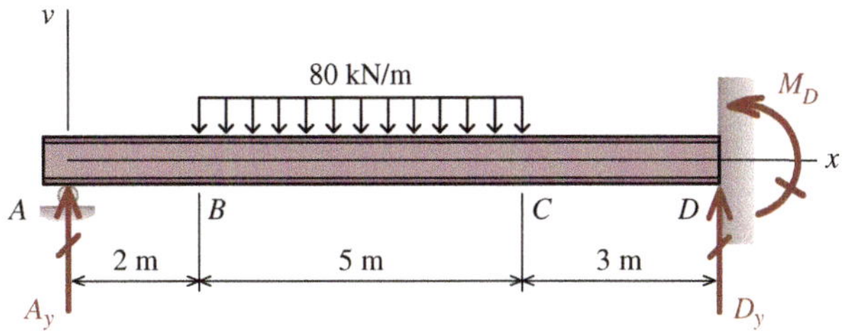

$$\Sigma F_y = A_y + D_y - (80 \text{ kN/m})(5 \text{ m}) = 0 \qquad \therefore A_y + D_y = 400 \text{ kN} \tag{a}$$

$$\Sigma M_D = -A_y(10 \text{ m}) + (80 \text{ kN/m})(5 \text{ m})(5.5 \text{ m}) + M_D = 0$$
$$\therefore M_D - A_y(10 \text{ m}) = -2{,}200 \text{ kN·m} \tag{b}$$

Discontinuity Expressions

Distributed load between B and C: Use case 5 of Table 7.2 to write the following expression for the distributed load:

$$w(x) = -80 \text{ kN/m}\langle x - 2 \text{ m}\rangle^0 + 80 \text{ kN/m}\langle x - 7 \text{ m}\rangle^0$$

Reaction forces A_y, D_y, and M_A: Since the beam is statically indeterminate, the reaction forces at A and D can be expressed only as unknown quantities at this time:

$$w(x) = A_y\langle x - 0 \text{ m}\rangle^{-1} + D_y\langle x - 10 \text{ m}\rangle^{-1} - M_D\langle x - 10 \text{ m}\rangle^{-2}$$

Note that the terms for D_y and M_D will always have the value zero in this example, since the beam is only 10 m long; therefore, these terms may be omitted here.

Integrate the beam load expression: The complete load expression $w(x)$ for the beam is thus

$$w(x) = A_y\langle x - 0 \text{ m}\rangle^{-1} - 80 \text{ kN/m}\langle x - 2 \text{ m}\rangle^0 + 80 \text{ kN/m}\langle x - 7 \text{ m}\rangle^0$$

The function $w(x)$ will be integrated to obtain the shear-force function $V(x)$:

$$V(x) = \int w(x)\,dx = A_y\langle x - 0 \text{ m}\rangle^0 - 80 \text{ kN/m}\langle x - 2 \text{ m}\rangle^1 + 80 \text{ kN/m}\langle x - 7 \text{ m}\rangle^1$$

Note that a constant of integration is not needed here, since the unknown reaction at A has been included in the function. The shear-force function is integrated to obtain the bending-moment function $M(x)$:

$$M(x) = \int V(x)\,dx = A_y\langle x - 0 \text{ m}\rangle^1 - \frac{80 \text{ kN/m}}{2}\langle x - 2 \text{ m}\rangle^2 + \frac{80 \text{ kN/m}}{2}\langle x - 7 \text{ m}\rangle^2$$

As before, a constant of integration is not needed for this result. However, the next two integrations (which will produce functions for the beam slope and deflection) will require constants of integration that must be evaluated from the beam boundary conditions.

From Equation (10.1), we can write

$$EI\frac{d^2v}{dx^2} = M(x) = A_y\langle x - 0 \text{ m}\rangle^1 - \frac{80 \text{ kN/m}}{2}\langle x - 2 \text{ m}\rangle^2 + \frac{80 \text{ kN/m}}{2}\langle x - 7 \text{ m}\rangle^2$$

Now integrate the moment function to obtain an expression for the beam slope:

$$EI\frac{dv}{dx} = \frac{A_y}{2}\langle x - 0 \text{ m}\rangle^2 - \frac{80 \text{ kN/m}}{6}\langle x - 2 \text{ m}\rangle^3 + \frac{80 \text{ kN/m}}{6}\langle x - 7 \text{ m}\rangle^3 + C_1 \tag{c}$$

Integrate again to obtain the beam deflection function:

$$EIv = \frac{A_y}{6}\langle x - 0\text{ m}\rangle^3 - \frac{80\text{ kN/m}}{24}\langle x - 2\text{ m}\rangle^4 + \frac{80\text{ kN/m}}{24}\langle x - 7\text{ m}\rangle^4 + C_1 x + C_2 \qquad \text{(d)}$$

Evaluate constants, using boundary conditions: For this beam, the deflection is known at $x = 0$ m. Substitute the boundary condition $v = 0$ at $x = 0$ m into Equation (d) to obtain constant C_2:

$$C_2 = 0 \qquad \text{(e)}$$

Next, substitute the boundary condition $v = 0$ at $x = 10$ m into Equation (d):

$$0 = \frac{A_y}{6}(10\text{ m})^3 - \frac{80\text{ kN/m}}{24}(8\text{ m})^4 + \frac{80\text{ kN/m}}{24}(3\text{ m})^4 + C_1(10\text{ m})$$
$$\therefore (166.6667\text{ m}^3)\, A_y + (10\text{ m})\, C_1 = 13{,}383.3333\text{ kN}\cdot\text{m}^3 \qquad \text{(f)}$$

Finally, substitute the boundary condition $dv/dx = 0$ at $x = 10$ m into Equation (c) to obtain

$$0 = \frac{A_y}{2}(10\text{ m})^2 - \frac{80\text{ kN/m}}{6}(8\text{ m})^3 + \frac{80\text{ kN/m}}{6}(3\text{ m})^3 + C_1$$
$$\therefore (50\text{ m}^2)\, A_y + C_1 = 6{,}466.6667\text{ kN}\cdot\text{m}^2 \qquad \text{(g)}$$

Equations (f) and (g) can be solved simultaneously to compute C_1 and A_y:

$$C_1 = -1{,}225.8333\text{ kN}\cdot\text{m}^3; \qquad A_y = 153.85\text{ kN}$$ **Ans.**

Now that A_y is known, the reactions D_y and M_D can be determined from Equations (a) and (b):

$$D_y = 400\text{ kN} - A_y = 400\text{ kN} - 153.85\text{ kN} = 246.15\text{ kN}$$ **Ans.**

$$M_D = A_y(10\text{ m}) - 2{,}200\text{ kN}\cdot\text{m} = (153.85\text{ kN})(10\text{ m}) - 2{,}200\text{ kN}\cdot\text{m}$$
$$= -661.50\text{ kN}\cdot\text{m}$$ **Ans.**

Equation (c) for the beam slope and Equation (d) for the elastic curve can now be completed:

$$EI\frac{dv}{dx} = \frac{153.85\text{ kN}}{2}\langle x - 0\text{ m}\rangle^2 - \frac{80\text{ kN/m}}{6}\langle x - 2\text{ m}\rangle^3 + \frac{80\text{ kN/m}}{6}\langle x - 7\text{ m}\rangle^3 - 1{,}225.8333\text{ kN}\cdot\text{m}^3 \qquad \text{(h)}$$

$$EIv = \frac{153.85\text{ kN}}{6}\langle x - 0\text{ m}\rangle^3 - \frac{80\text{ kN/m}}{24}\langle x - 2\text{ m}\rangle^4 + \frac{80\text{ kN/m}}{24}\langle x - 7\text{ m}\rangle^4 - (1{,}225.8333\text{ kN}\cdot\text{m}^3)x \qquad \text{(i)}$$

(b) Beam Deflection at C

From Equation (i), the beam deflection at C ($x = 7$ m) is computed as follows:

$$EIv_C = \frac{153.85\text{ kN}}{6}(7\text{ m})^3 - \frac{80\text{ kN/m}}{24}(5\text{ m})^4 - (1{,}225.8333\text{ kN}\cdot\text{m}^3)(7\text{ m})$$
$$= -1{,}869.075\text{ kN}\cdot\text{m}^3$$

$$\therefore v_C = -\frac{1{,}869.075\text{ kN}\cdot\text{m}^3}{120{,}000\text{ kN}\cdot\text{m}^2} = -0.015576\text{ m} = 15.58\text{ mm}\downarrow$$ **Ans.**

EXAMPLE 11.4

For the statically indeterminate beam shown, use discontinuity functions to determine

(a) the force reactions at B, D, and E.
(b) the deflection of the beam at A.
(c) the deflection of the beam at C.

Assume a constant value of $EI = 120{,}000\ \text{kN}\cdot\text{m}^2$ for the beam.

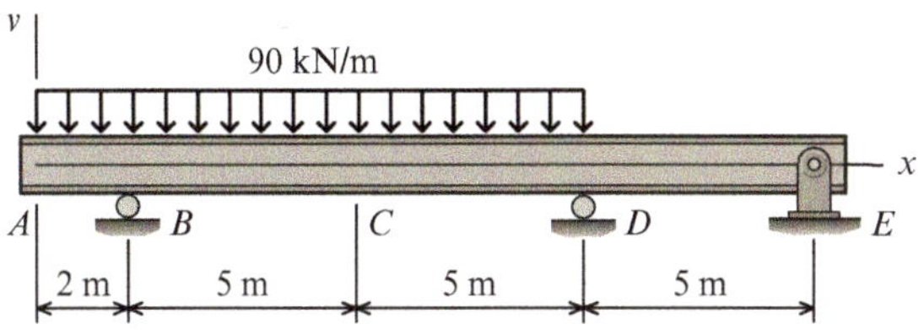

Plan the Solution

The beam is statically indeterminate; therefore, the reaction forces cannot be determined solely from equilibrium considerations. From an FBD of the beam, two nontrivial equilibrium equations can be derived. However, since the beam is statically indeterminate, the reaction forces can be stated only as unknowns. The distributed load on the beam, as well as the unknown reactions, will be expressed by discontinuity functions. The load function will be integrated twice to obtain the bending-moment function for the beam. In these first two integrations, constants of integration will not be necessary. The bending-moment function will then be integrated twice to obtain the elastic curve equation. Constants of integration must be considered in these two integrations. The three boundary conditions known at B, D, and E, along with the two nontrivial equilibrium equations, will produce five equations that can be solved simultaneously to determine the three unknown reactions and the two constants of integration. After these quantities are found, the beam deflection at any location can be computed from the elastic curve equation.

SOLUTION

(a) Support Reactions

An FBD of the beam is shown. Since no forces act in the x direction, the ΣF_x equation will be omitted here. From the FBD, the beam reaction forces can be expressed by the following relationships:

$$\Sigma F_y = B_y + D_y + E_y - (90\ \text{kN/m})(12\ \text{m}) = 0 \qquad \therefore B_y + D_y + E_y = 1{,}080\ \text{kN} \tag{a}$$

$$\Sigma M_E = -B_y(15\ \text{m}) - D_y(5\ \text{m}) + (90\ \text{kN/m})(12\ \text{m})(11\ \text{m}) = 0$$
$$\therefore B_y(15\ \text{m}) + D_y(5\ \text{m}) = 11{,}880\ \text{kN}\cdot\text{m} \tag{b}$$

Discontinuity Expressions

Distributed load between A and D: Use case 5 of Table 7.2 to write the following expression for the distributed load:

$$w(x) = -90\ \text{kN/m}\langle x - 0\ \text{m}\rangle^0 + 90\ \text{kN/m}\langle x - 12\ \text{m}\rangle^0$$

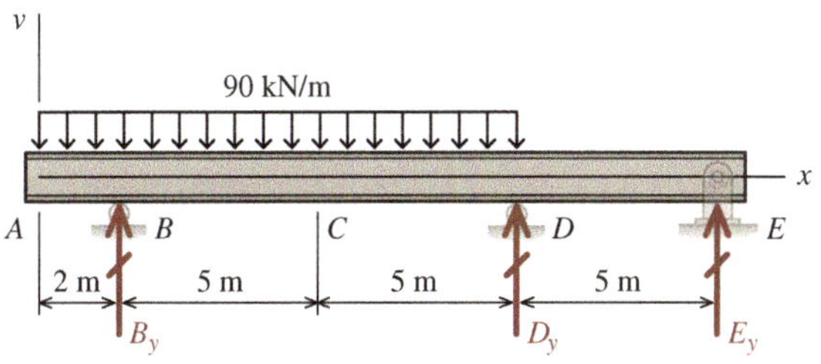

Reaction forces B_y, D_y, and E_y: Since the beam is statically indeterminate, the reaction forces at B, D, and E can be expressed only as unknown quantities at this time:

$$w(x) = B_y\langle x - 2\ \text{m}\rangle^{-1} + D_y\langle x - 12\ \text{m}\rangle^{-1} + E_y\langle x - 17\ \text{m}\rangle^{-1}$$

Note that the term for E_y will always have the value zero in this example, since the beam is only 17 m long; therefore, this term may be omitted here.

Integrate the beam load expression: The complete load expression $w(x)$ for the beam is thus

$$w(x) = -90 \text{ kN/m}\langle x - 0 \text{ m}\rangle^0 + B_y\langle x - 2 \text{ m}\rangle^{-1} + 90 \text{ kN/m}\langle x - 12 \text{ m}\rangle^0 + D_y\langle x - 12 \text{ m}\rangle^{-1}$$

The function $w(x)$ will be integrated to obtain the shear-force function $V(x)$:

$$V(x) = \int w(x)\,dx = -90 \text{ kN/m}\langle x - 0 \text{ m}\rangle^1 + B_y\langle x - 2 \text{ m}\rangle^0 + 90 \text{ kN/m}\langle x - 12 \text{ m}\rangle^1 + D_y\langle x - 12 \text{ m}\rangle^0$$

The shear-force function is integrated to obtain the bending-moment function $M(x)$:

$$M(x) = \int V(x)\,dx = -\frac{90 \text{ kN/m}}{2}\langle x - 0 \text{ m}\rangle^2 + B_y\langle x - 2 \text{ m}\rangle^1 + \frac{90 \text{ kN/m}}{2}\langle x - 12 \text{ m}\rangle^2 + D_y\langle x - 12 \text{ m}\rangle^1$$

Since the reactions have been included in these functions, constants of integration are not needed up to this point. However, the next two integrations (which will produce functions for the beam slope and deflection) will require constants of integration that must be evaluated from the beam boundary conditions.

From Equation (10.1), we can write

$$EI\frac{d^2v}{dx^2} = M(x) = -\frac{90 \text{ kN/m}}{2}\langle x - 0 \text{ m}\rangle^2 + B_y\langle x - 2 \text{ m}\rangle^1 + \frac{90 \text{ kN/m}}{2}\langle x - 12 \text{ m}\rangle^2 + D_y\langle x - 12 \text{ m}\rangle^1$$

Now integrate the moment function to obtain an expression for the beam slope:

$$EI\frac{dv}{dx} = -\frac{90 \text{ kN/m}}{6}\langle x - 0 \text{ m}\rangle^3 + \frac{B_y}{2}\langle x - 2 \text{ m}\rangle^2 + \frac{90 \text{ kN/m}}{6}\langle x - 12 \text{ m}\rangle^3 + \frac{D_y}{2}\langle x - 12 \text{ m}\rangle^2 + C_1 \quad \text{(c)}$$

Integrate again to obtain the beam deflection function:

$$EIv = -\frac{90 \text{ kN/m}}{24}\langle x - 0 \text{ m}\rangle^4 + \frac{B_y}{6}\langle x - 2 \text{ m}\rangle^3 + \frac{90 \text{ kN/m}}{24}\langle x - 12 \text{ m}\rangle^4 + \frac{D_y}{6}\langle x - 12 \text{ m}\rangle^3 + C_1x + C_2 \quad \text{(d)}$$

Evaluate constants, using boundary conditions: For this beam, substitute the boundary condition $v = 0$ at $x = 2$ m into Equation (d):

$$\begin{aligned} 0 &= -\frac{90 \text{ kN/m}}{24}(2 \text{ m})^4 + C_1(2 \text{ m}) + C_2 \\ \therefore C_1(2 \text{ m}) + C_2 &= 60 \text{ kN}\cdot\text{m}^3 \end{aligned} \quad \text{(e)}$$

Next, substitute the boundary condition $v = 0$ at $x = 12$ m into Equation (d):

$$\begin{aligned} 0 &= -\frac{90 \text{ kN/m}}{24}(12 \text{ m})^4 + \frac{B_y}{6}(10 \text{ m})^3 + C_1(12 \text{ m}) + C_2 \\ \therefore B_y(166.6667 \text{ m}^3) + C_1(12 \text{ m}) + C_2 &= 77{,}760 \text{ kN}\cdot\text{m}^3 \end{aligned} \quad \text{(f)}$$

Finally, substitute the boundary condition $v = 0$ at $x = 17$ m into Equation (d):

$$\begin{aligned} 0 &= -\frac{90 \text{ kN/m}}{24}(17 \text{ m})^4 + \frac{B_y}{6}(15 \text{ m})^3 + \frac{90 \text{ kN/m}}{24}(5 \text{ m})^4 + \frac{D_y}{6}(5 \text{ m})^3 + C_1(17 \text{ m}) + C_2 \\ \therefore B_y(562.5 \text{ m}^3) + D_y(20.8333 \text{ m}^3) + C_1(17 \text{ m}) + C_2 &= 310{,}860 \text{ kN}\cdot\text{m}^3 \end{aligned} \quad \text{(g)}$$

Five equations—Equations (a), (b), (e), (f), and (g)—must be solved simultaneously to determine the beam reaction forces at B, D, and E, as well as the two constants of integration C_1 and C_2:

$$C_1 = -1{,}880 \text{ kN}\cdot\text{m}^2 \quad \text{and} \quad C_2 = 3{,}820 \text{ kN}\cdot\text{m}^3$$

$$B_y = 579 \text{ kN} \qquad D_y = 639 \text{ kN} \qquad E_y = -138 \text{ kN}$$ **Ans.**

Equation (c) for the beam slope and Equation (d) for the elastic curve can now be completed:

$$EI\frac{dv}{dx} = -\frac{90 \text{ kN/m}}{6}\langle x - 0 \text{ m}\rangle^3 + \frac{579 \text{ kN}}{2}\langle x - 2 \text{ m}\rangle^2 + \frac{90 \text{ kN/m}}{6}\langle x - 12 \text{ m}\rangle^3 + \frac{639 \text{ kN}}{2}\langle x - 12 \text{ m}\rangle^2 - 1{,}880 \text{ kN}\cdot\text{m}^2 \quad \text{(h)}$$

$$EIv = -\frac{90 \text{ kN/m}}{24}\langle x - 0 \text{ m}\rangle^4 + \frac{579 \text{ kN}}{6}\langle x - 2 \text{ m}\rangle^3 + \frac{90 \text{ kN/m}}{24}\langle x - 12 \text{ m}\rangle^4 + \frac{639 \text{ kN}}{6}\langle x - 12 \text{ m}\rangle^3 - (1{,}880 \text{ kN}\cdot\text{m}^2)x + 3{,}820 \text{ kN}\cdot\text{m}^3 \quad \text{(i)}$$

(b) Beam Deflection at *A*

The beam deflection at *A* ($x = 0$ m) is computed from Equation (i):

$$EIv_A = 3{,}820 \text{ kN}\cdot\text{m}^3$$

$$\therefore v_A = \frac{3{,}820 \text{ kN}\cdot\text{m}^3}{120{,}000 \text{ kN}\cdot\text{m}^2} = 0.031833 \text{ m} = 31.8 \text{ mm} \uparrow \qquad \textbf{Ans.}$$

(c) Beam Deflection at *C*

From Equation (i), the beam deflection at *C* ($x = 7$ m) is computed as follows:

$$EIv_C = -\frac{90 \text{ kN/m}}{24}(7 \text{ m})^4 + \frac{579 \text{ kN}}{6}(5 \text{ m})^3 - (1{,}880 \text{ kN}\cdot\text{m}^2)(7 \text{ m}) + 3{,}820 \text{ kN}\cdot\text{m}^3$$
$$= -6{,}281.250 \text{ kN}\cdot\text{m}^3$$

$$\therefore v_C = -\frac{6{,}281.250 \text{ kN}\cdot\text{m}^3}{120{,}000 \text{ kN}\cdot\text{m}^2} = -0.052344 \text{ m} = 52.3 \text{ mm} \downarrow \qquad \textbf{Ans.}$$

PROBLEMS

P11.12 A propped cantilever beam is loaded as shown in Figure P11.12. Assume that $EI = 200{,}000 \text{ kN}\cdot\text{m}^2$, and use discontinuity functions to determine

(a) the reactions at *A* and *C*.
(b) the beam deflection at *B*.

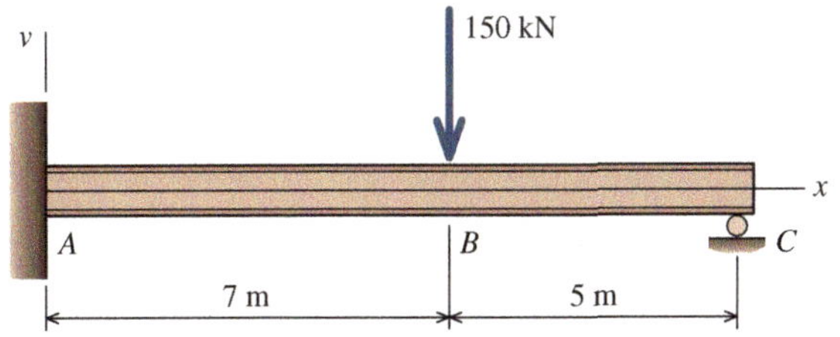

FIGURE P11.12

P11.13 A propped cantilever beam is loaded as shown in Figure P11.13. Assume that $EI = 200{,}000 \text{ kN}\cdot\text{m}^2$, and use discontinuity functions to determine

(a) the reactions at *A* and *B*.
(b) the beam deflection at *C*.

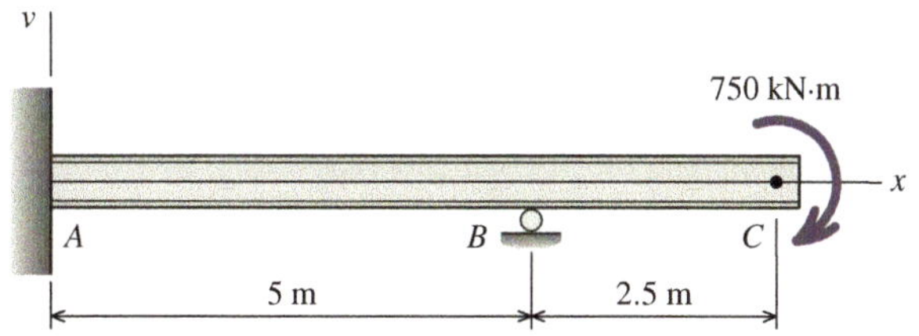

FIGURE P11.13

P11.14 A propped cantilever beam is loaded as shown in Figure P11.14. Assume that $EI = 100{,}000 \text{ kip}\cdot\text{ft}^2$, and use discontinuity functions to determine

(a) the reactions at *A* and *E*.
(b) the beam deflection at *C*.

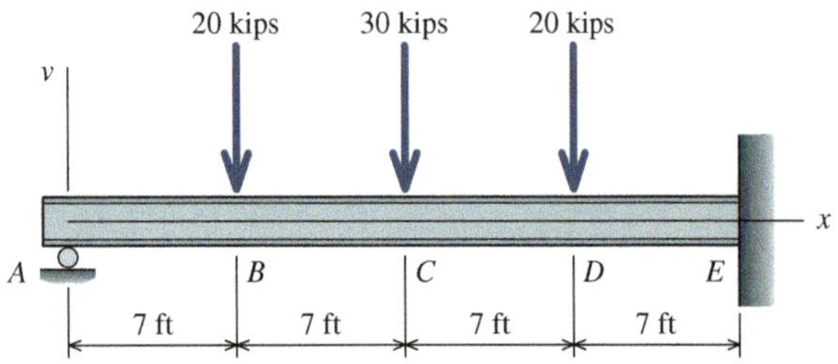

FIGURE P11.14

P11.15 A propped cantilever beam is loaded as shown in Figure P11.15. Assume that $EI = 100{,}000 \text{ kip}\cdot\text{ft}^2$, and use discontinuity functions to determine

(a) the reactions at A and B.
(b) the beam deflection at $x = 7$ ft.

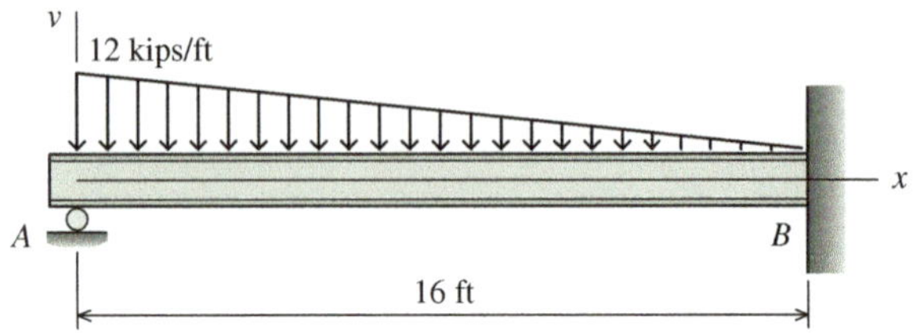

FIGURE P11.15

P11.16 A propped cantilever beam is loaded as shown in Figure P11.16. Assume that $EI = 200{,}000 \text{ kN}\cdot\text{m}^2$, and use discontinuity functions to determine

(a) the reactions at A and B.
(b) the beam deflection at C.

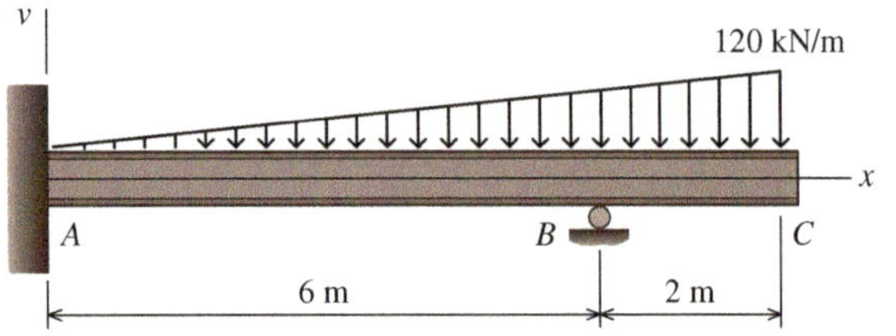

FIGURE P11.16

P11.17 For the beam shown in Figure P11.17, assume that $EI = 200{,}000 \text{ kN}\cdot\text{m}^2$ and use discontinuity functions to determine

(a) the reactions at A, C, and D.
(b) the beam deflection at B.

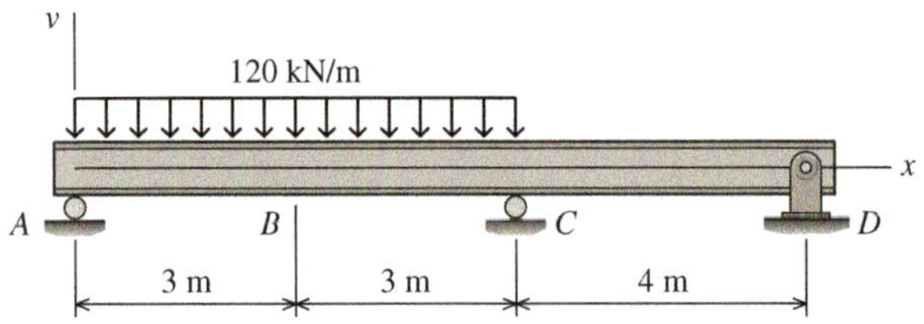

FIGURE P11.17

P11.18 For the beam shown in Figure P11.18, assume that $EI = 100{,}000 \text{ kip}\cdot\text{ft}^2$ and use discontinuity functions to determine

(a) the reactions at A, C, and D.
(b) the beam deflection at B.

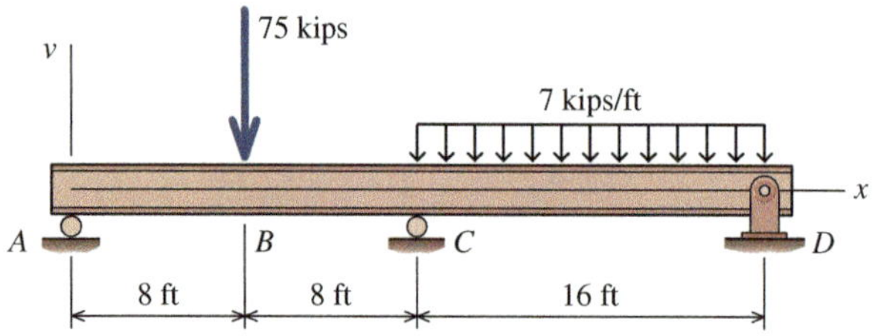

FIGURE P11.18

P11.19 For the propped cantilever beam shown in Figure P11.19, assume that $EI = 100{,}000 \text{ kip}\cdot\text{ft}^2$ and use discontinuity functions to determine

(a) the reactions at B and D.
(b) the beam deflection at C.

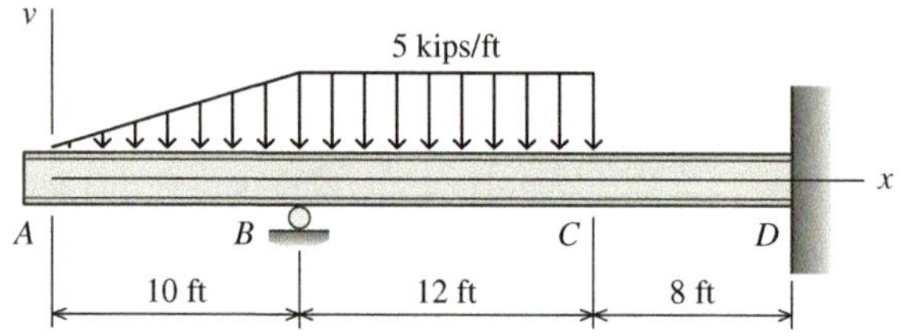

FIGURE P11.19

P11.20–P11.21 For the beams shown in Figures P11.20 and P11.21, assume that $EI = 200{,}000 \text{ kN}\cdot\text{m}^2$ and use discontinuity functions to determine

(a) the reactions at B, C, and D.
(b) the beam deflection at A.

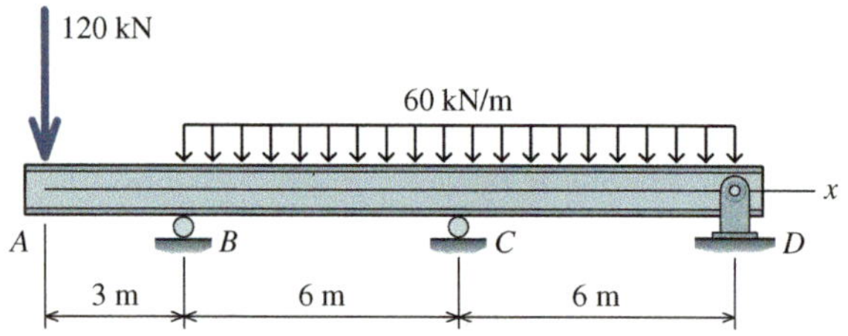

FIGURE P11.20

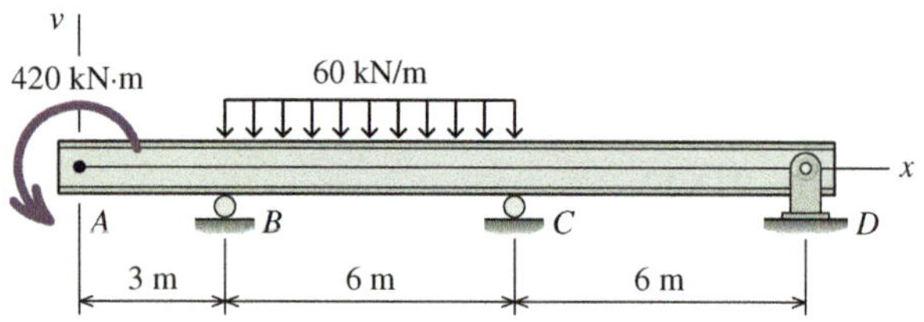

FIGURE P11.21

11.5 The Superposition Method

The concepts of **redundant reactions** and a **released beam** were introduced in Section 11.2. These notions can be combined with the principle of superposition to create a very powerful method for determining the support reactions of statically indeterminate beams. The general approach can be outlined as follows:

1. Redundant support reactions acting on the statically indeterminate beam are identified.
2. The selected redundant is removed from the structure, leaving a released beam that is stable and statically determinate.
3. The released beam subjected to the applied load is considered. The deflection or rotation (depending on the nature of the redundant) of the beam at the location of the redundant is determined.
4. Next, the released beam (without the applied load) is subjected to one of the redundant reactions and the deflection or the rotation of this beam-and-loading combination is determined at the location of the redundant. If more than one redundant exists, this step is repeated for each redundant.
5. By the principle of superposition, the actual loaded beam is equivalent to the sum of these individual cases.
6. To solve for the redundants, geometry-of-deformation equations are written for each of the locations where redundants act. The magnitude of the redundant can be obtained from the particular deformation equation.
7. Once the redundants are known, the other beam reactions can be determined from the equilibrium equations.

To clarify this approach, consider the propped cantilever beam shown in Figure 11.6*a*. The free-body diagram for this beam (Figure 11.6*b*) shows four unknown reactions. Three equilibrium equations can be written for the beam ($\Sigma F_x = 0$, $\Sigma F_y = 0$, and $\Sigma M = 0$); therefore, the beam is statically indeterminate to the first degree. Consequently, one additional equation must be developed in order to compute the reactions for the propped cantilever.

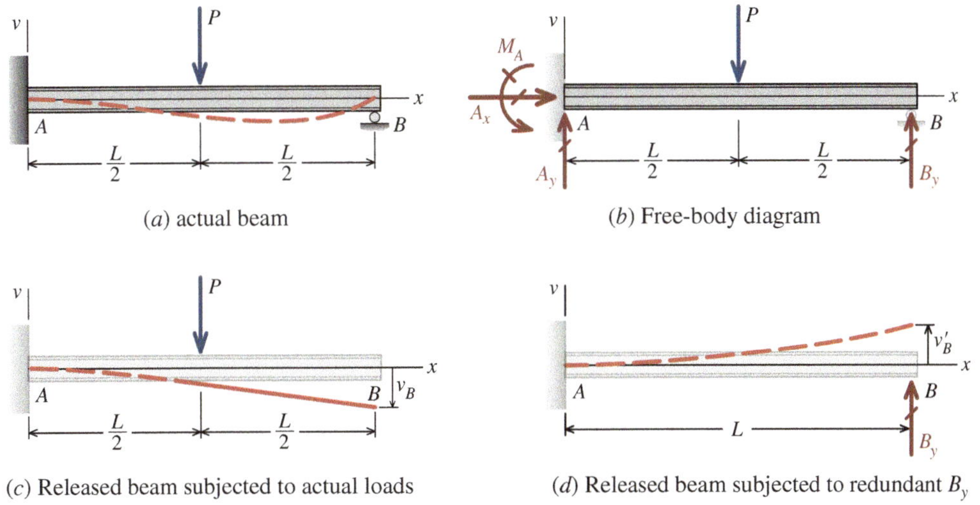

FIGURE 11.6 Superposition method applied to a propped cantilever beam.

The roller reaction B_y will be selected as the redundant. This reaction force is removed from the beam, leaving a cantilever as the released beam. Note that the released beam is stable and that it is statically determinate. Next, the deflection of the released beam at the location of the redundant is analyzed for two loading cases. The first case consists of the cantilever beam with applied load P, and the downward deflection v_B at the location of the redundant is determined (Figure 11.6*c*). The second case consists of the cantilever beam with only the redundant reaction force B_y, and the upward deflection v'_B caused by B_y is determined (Figure 11.6*d*).

By the principle of superposition, the sum of these two loading cases (Figures 11.6*c* and 11.6*d*) is equivalent to the actual beam (Figure 11.6*a*) if the sum of v_B and v'_B equals the actual beam deflection at B. The actual beam deflection at B is known beforehand: The deflection must be zero, since the beam is supported by a roller at B. From this fact, a geometry-of-deformation equation can be written for B in terms of the two loading cases:

$$v_B + v'_B = 0 \tag{a}$$

The deflections v_B and v'_B can be determined from equations given in the beam table found in Appendix C:

$$v_B = -\frac{5PL^3}{48EI} \quad \text{and} \quad v'_B = \frac{B_y L^3}{3EI} \tag{b}$$

These deflection expressions are substituted into Equation (a) to produce an equation based on the deflected geometry of the beam, but expressed in terms of the unknown reaction B_y. This **compatibility equation** can be solved for the value of the redundant:

$$-\frac{5PL^3}{48EI} + \frac{B_y L^3}{3EI} = 0 \qquad \therefore B_y = \frac{5}{16}P \tag{c}$$

Once the magnitude of B_y has been determined, the remaining reactions can be found from the equilibrium equations. The results are as follows:

$$A_x = 0 \qquad A_y = \frac{11}{16}P \qquad M_A = \frac{3}{16}PL \tag{d}$$

The choice of redundant is *arbitrary*, provided that the primary beam remains stable. Consider the previous propped cantilever beam (Figure 11.6*a*), which has four reactions (Figure 11.6*b*). (For convenience, the beam of Figure 11.6*a* is reproduced in Figure 11.7*a*.) Suppose that, instead of roller B, the moment reaction M_A is chosen as the redundant, leaving a simply supported span as the released beam. Removing M_A allows the beam to rotate freely at A; therefore, the rotation angle θ_A must be determined for the released beam subjected to the applied load P (Figure 11.7*b*). Next, the simple span is subjected to redundant M_A alone and the resulting rotation angle θ'_A is determined (Figure 11.7*c*).

Just as before, the sum of these two loading cases (Figures 11.7*b* and 11.7*c*) is equivalent to the actual beam (Figure 11.7*a*), provided that the rotations produced by the two separate loading cases add up to the actual beam rotation at A. Since the actual beam is fixed at A, the rotation angle must be zero, a condition that leads to the following geometry-of-deformation equation:

$$\theta_A + \theta'_A = 0 \tag{e}$$

Again from the beam table in Appendix C, the rotation angles for the two cases can be expressed as

$$\theta_A = -\frac{PL^2}{16EI} \quad \text{and} \quad \theta'_A = \frac{M_A L}{3EI} \tag{f}$$

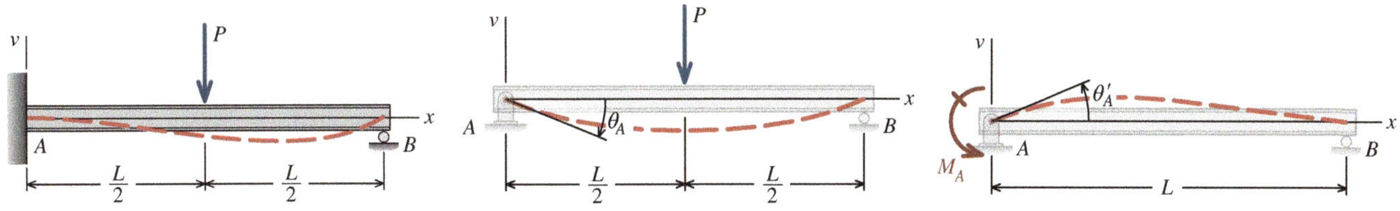

(a) Actual beam (b) Simply supported released beam (c) Redundant M_A applied to released beam

FIGURE 11.7 Superposition method for a propped cantilever beam, using a simply supported released beam.

Substituting these expressions into Equation (e) gives the following compatibility equation, which can be solved for the unknown redundant magnitude:

$$-\frac{PL^2}{16EI}+\frac{M_A L}{3EI}=0 \qquad \therefore M_A=\frac{3}{10}PL \tag{g}$$

The value for M_A is the same result computed previously. Once M_A has been determined, the remaining reactions can be computed from the equilibrium equations.

Examples 11.5–11.9 illustrate the application of the superposition method to determine support reactions for statically indeterminate beams.

MecMovies

EXAMPLE

M11.3 Use two different approaches of the superposition method to determine the roller reaction at A.

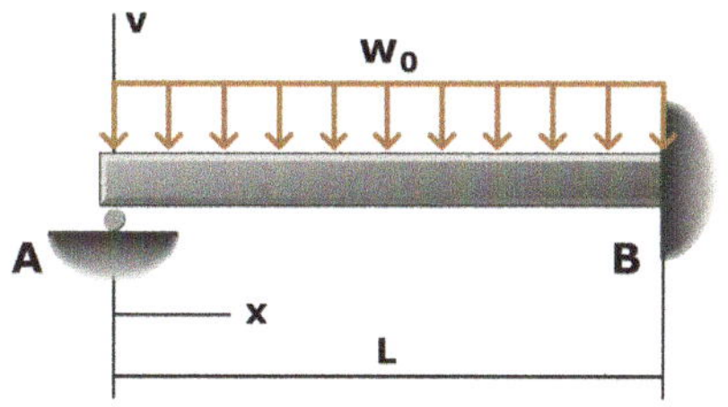

EXAMPLE 11.5

For the beam and loading shown, derive an expression for the reaction at support B. Assume that EI is constant for the beam.

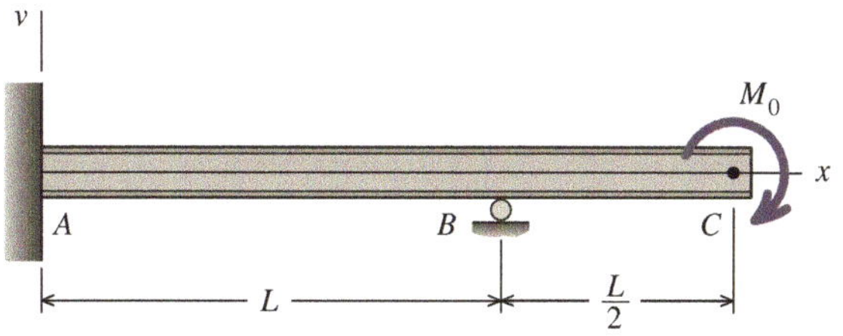

Plan the Solution

The propped cantilever has four unknown reaction forces (horizontal and vertical reaction forces at fixed reaction A, a moment reaction at A, and a vertical reaction force at roller B). Since only three equilibrium equations can be written for the beam, one additional equation must be developed in order to solve this problem. The additional equation will be developed by considering the deflected shape of the beam and, in particular, the known beam deflection at roller B. The roller support at B will be chosen as the redundant; therefore, the released beam will be a cantilever supported at A. The analysis will be subdivided into two cases. In the first case, the deflection at B produced by the concentrated moment M_0 will be determined. In the second case, the unknown roller reaction force will be applied to the cantilever beam at B and an expression for the corresponding beam deflection will be derived. These two deflection expressions will be added together in a compatibility equation to express the total beam deflection at B, which must equal zero, since B is a roller support. From this compatibility equation, the magnitude of the unknown roller reaction force at B can be determined.

SOLUTION

The beam will be analyzed as two cantilever beam cases. In both cases, the roller support at B will be removed, reducing the propped cantilever beam to a cantilever beam. In the first case, the concentrated moment M_0 acting at the tip of the cantilever will be considered. In the second case, the deflection caused by the roller reaction force at B will be considered.

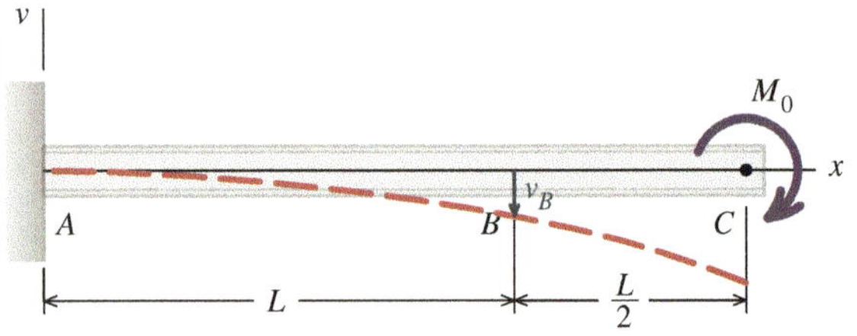

Case 1—Concentrated Moment at Tip of Cantilever

Remove the roller support at B and consider the cantilever beam ABC. From Appendix C, the elastic curve equation for a cantilever beam subjected to a concentrated moment acting at its free end is given by

$$v = -\frac{Mx^2}{2EI} \tag{a}$$

Use the elastic curve equation to compute the beam deflection at B. In Equation (a), let $M = M_0$ and $x = L$, and assume that EI is a constant for the beam. Substitute these values into Equation (a) to derive an expression for the beam deflection at B:

$$v_B = -\frac{M_0L^2}{2EI} \tag{b}$$

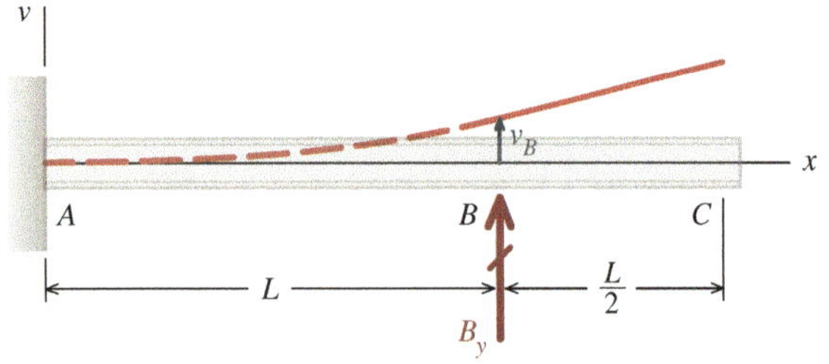

Case 2—Concentrated Force at Roller Support Location

By applying only redundant B_y to the cantilever beam, an expression for the resulting deflection at B is derived. From Appendix C, the maximum cantilever beam deflection produced by a concentrated force acting at the tip of the cantilever is given by the expression

$$v_{max} = -\frac{PL^3}{3EI} \tag{c}$$

In Equation (c), let $P = -B_y$ and $L = L$. Note that B_y is negative, since it acts upward, opposite to the direction assumed in the beam table. Substitute these values into Equation (c) to obtain an expression for the beam deflection at B in terms of the unknown roller reaction force B_y:

$$v_B = -\frac{(-B_y)L^3}{3EI} = \frac{B_yL^3}{3EI} \tag{d}$$

Compatibility Equation

Two expressions [Equations (b) and (d)] have been developed for the beam deflection at B. Add these two expressions, and set the result equal to the beam deflection at B, which is known to be zero at the roller support:

$$v_B = -\frac{M_0L^2}{2EI} + \frac{B_yL^3}{3EI} = 0 \tag{e}$$

Notice that EI appears in both terms; hence, it cancels out upon multiplying both sides of Equation (e) by EI. In other words, the specific value of EI has no effect on the magnitude of the roller force for this particular beam. The roller reaction B_y is the only unknown quantity in the compatibility equation, and thus, the roller reaction at B is

$$B_y = \frac{3M_0}{2L} \qquad \textbf{Ans.}$$

Once the reaction force at B is known, the beam is no longer statically indeterminate. The three remaining unknown reactions at fixed support A can be determined from the equilibrium equations.

EXAMPLE 11.6

For the beam and loading shown, derive an expression for the reaction at support B. Assume that EI is constant for the beam.

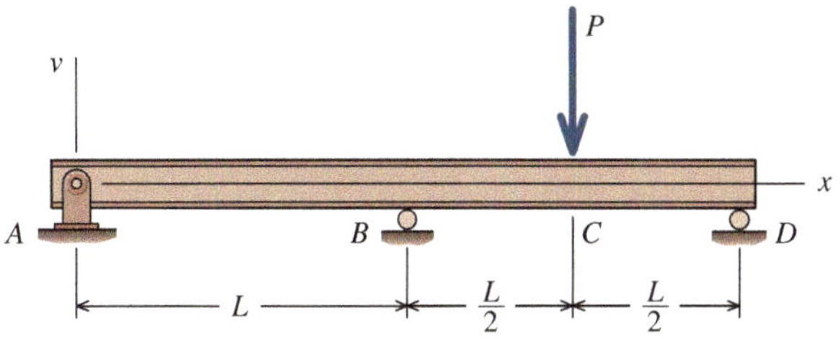

Plan the Solution

The beam considered here has four unknown reaction forces (horizontal and vertical reaction forces at pin A and vertical reaction forces at rollers B and D). Since there are only three equilibrium equations, a fourth equation must be developed. Although there are several approaches that could be used to develop this fourth equation, we will focus our attention on the roller at B. This roller will be chosen as the redundant reaction. Removing this redundant leaves a released beam that is simply supported at A and D. Two cases will then be analyzed. The first case consists of a simple beam AD subjected to a load P. The second case consists of a simple beam AD loaded at B with the unknown roller reaction. In both cases, equations for the beam deflection at B will be developed. These equations will be combined in a compatibility equation by using the fact that the beam deflection at B is known to be zero. From this compatibility equation, an equation for the unknown reaction force at B can be derived.

SOLUTION

Case 1—Simply Supported Beam with a Concentrated Load at C

Remove the roller support at B, and consider the simply supported beam AD with a concentrated load at C. The deflection of this beam at B must be determined. From Appendix C, the elastic curve equation for the beam is given as

$$v = -\frac{Pbx}{6LEI}(L^2 - b^2 - x^2) \qquad \text{(a)}$$

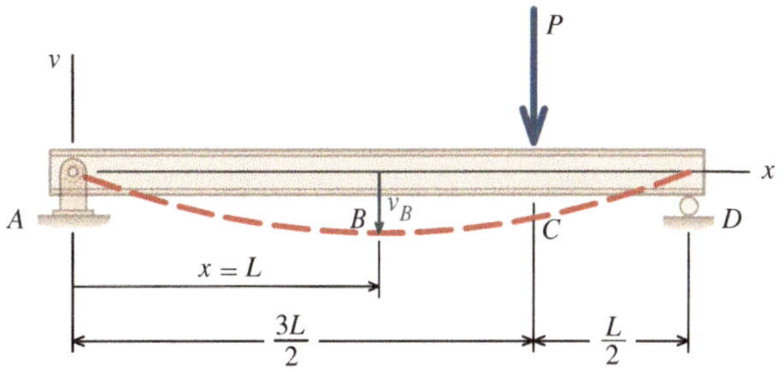

In this equation, the following values will be used:

$P = P$

$b = L/2$

$x = L$

$L = 2L$

$EI = \text{constant}$

465

Substitute these values into Equation (a), and derive the beam deflection at B:

$$v_B = -\frac{P(L/2)(L)}{6(2L)EI}[(2L)^2 - (L/2)^2 - (L)^2] = -\frac{PL}{24EI}\left[\frac{11}{4}L^2\right] = -\frac{11PL^3}{96EI} \qquad \text{(b)}$$

Case 2—Simply Supported Beam with Unknown Reaction Force at *B*

Consider the simply supported beam AD with the unknown roller reaction applied as a concentrated load at B. From Appendix C, the maximum deflection for a simply supported beam with a concentrated load at midspan is given as

$$v_{max} = -\frac{PL^3}{48EI} \qquad \text{(c)}$$

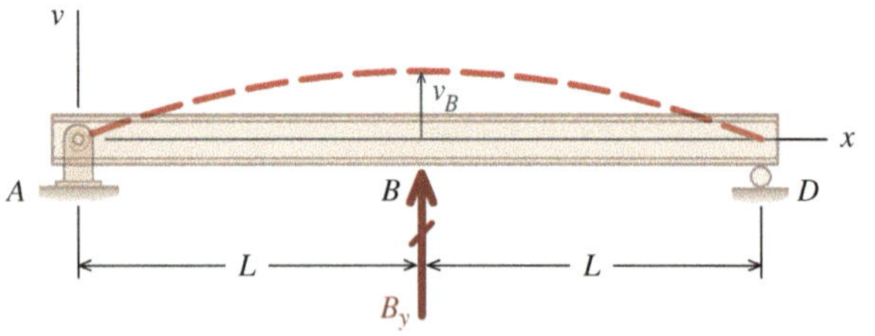

For this beam, let

$P = -B_y$ (negative, since B_y acts upward)

$L = 2L$

EI = constant

Substitute these values into Equation (c) to obtain the following expression for the beam deflection at B:

$$v_B = -\frac{(-B_y)(2L)^3}{48EI} = \frac{B_yL^3}{6EI} \qquad \text{(d)}$$

Compatibility Equation

Add Equations (b) and (d) to obtain an expression for the beam deflection at B. Since B is a roller support, the deflection at this location must be zero. Hence,

$$v_B = -\frac{11PL^3}{96EI} + \frac{B_yL^3}{6EI} = 0 \qquad \text{(e)}$$

As in the previous example, EI appears in both terms. Again, in other words, the specific value of EI has no effect on the magnitude of the reaction for the roller support at B. From the compatibility equation, the unknown roller reaction force can be expressed as

$$B_y = \frac{11}{16}P \qquad \textbf{Ans.}$$

EXAMPLE 11.7

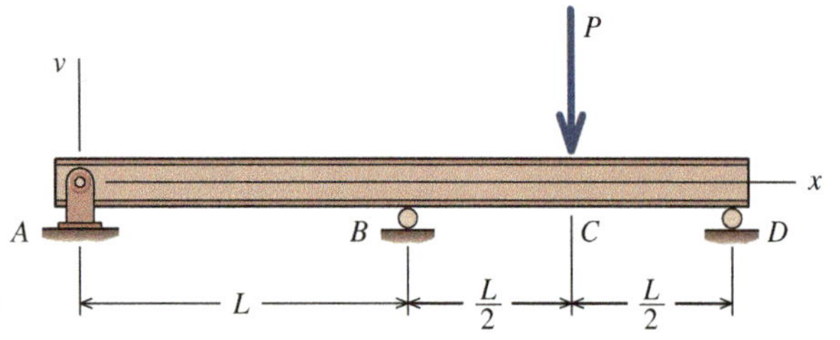

Consider the beam and loading from Example 11.6. The beam consists of a structural steel W530 × 66 wide-flange shape [E = 200 GPa; $I = 351 \times 10^6$ mm^4]. Assume that P = 240 kN and L = 5 m. Determine

(a) the reaction force at roller support B.
(b) the reaction force at B if the roller support settles 5 mm.

Plan the Solution

To answer part (a) of this problem, the equation developed for B_y in Example 11.6 will be used to calculate the reaction force. In part (b) of this example, the middle roller *settles* 5 mm, meaning that the roller support is displaced 5 mm downward. The compatibility equation developed in Example 11.6 assumed that the beam deflection at B was zero. In this instance, however, the compatibility equation must be revised to account for the 5 mm downward displacement.

SOLUTION

(a) From Example 11.6, the reaction force at B for this beam and loading configuration is given by

$$B_y = \frac{11}{16}P$$

Since $P = 240$ kN, the reaction force at B is $B_y = 165$ kN.

(b) The compatibility equation derived in Example 11.6 was

$$v_B = -\frac{11PL^3}{96EI} + \frac{B_y L^3}{6EI} = 0$$

This equation was based on the assumption that the beam deflection at roller support B would be zero. In part (b) of this example, however, the possibility that the support settles by 5 mm is being investigated. That possibility is a very practical consideration. All building structures rest on foundations. If these foundations are constructed on solid rock, there may be little or no settlement; however, foundations that rest on soil or sand will always settle to some extent. If all supports settle by the same amount, the structure will be displaced as a rigid body and there will be no effect on the internal forces and moments of the structure. However, if one support settles more that the others, then the reactions and internal forces in the structure will be affected. Part (b) of this example examines the change in reaction forces that would occur if the roller support at B were displaced downward 5 mm more than the displacements of supports A and C. This situation is termed *differential settlement*.

So, roller support B settles 5 mm. The beam is connected to this support; therefore, the beam deflection at B must be $v_B = -5$ mm. The compatibility equation from Example 11.6 will be revised to account for this nonzero beam deflection at B, giving

$$v_B = -\frac{11PL^3}{96EI} + \frac{B_y L^3}{6EI} = -5 \text{ mm}$$

and an expression for the reaction force at B can be derived:

$$\boxed{B_y = \frac{6EI}{L^3}\left[-5 \text{ mm} + \frac{11PL^3}{96EI}\right]} \qquad \text{(a)}$$

Unlike previous examples, EI does not cancel out of this equation. In other words, the magnitude of B_y will depend not only on the magnitude of the support settlement, but also on the flexural properties of the beam. In Equation (a), the following values will be used:

$$P = 240 \text{ kN} = 240{,}000 \text{ N}$$
$$L = 5 \text{ m} = 5{,}000 \text{ mm}$$
$$I = 351 \times 10^6 \text{ mm}^4$$
$$E = 200 \text{ GPa} = 200{,}000 \text{ MPa}$$

Substitute these values into Equation (a), and compute B_y. Pay particular attention to the units associated with each variable, and make sure that the calculation is dimensionally consistent. In this example, all force units will be converted to newtons and all length units will be expressed in millimeters. When all is said and done, we have

$$B_y = \frac{6(200{,}000\ \text{N/mm}^2)(351 \times 10^6\ \text{mm}^4)}{(5{,}000\ \text{mm})^3}\left[-5\ \text{mm} + \frac{11(240{,}000\ \text{N})(5{,}000\ \text{mm})^3}{96(200{,}000\ \text{N/mm}^2)(351 \times 10^6\ \text{mm}^4)}\right]$$

$$= (3{,}369.6\ \text{N/mm})[-5\ \text{mm} + 48.967\ \text{mm}]$$

$$= 148.152 \times 10^3\ \text{N} = 148.2\ \text{kN}$$ **Ans.**

The 5 mm settlement at support B decreases the reaction force B_y from 165 kN to 148.2 kN. The bending moments in the beam will also change because of the support settlement. If the roller support at B does not settle, the maximum positive bending moment in the beam is 243.75 kN·m and the maximum negative bending moment is −112.5 kN·m. A 5 mm settlement at roller B changes the maximum positive bending moment to 264.81 kN·m (an 8.6 percent increase) and the maximum negative bending moment to −70.38 kN·m (a 37 percent decrease). These values show that a relatively small differential settlement can produce significant changes in the bending moments produced in the beam. The engineer must be attentive to these potential variations.

EXAMPLE 11.8

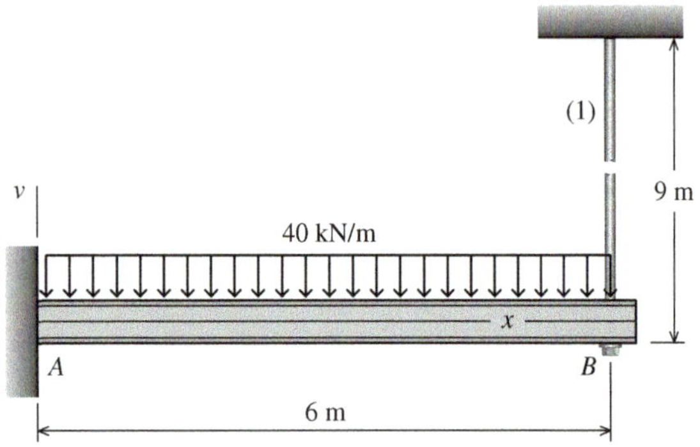

A structural steel [E = 200 GPa; I = 300 × 10⁶ mm⁴] beam supports a uniformly distributed load of 40 kN/m. The beam is fixed at the left end and supported by a 30 mm diameter, 9 m long solid aluminum [E_1 = 70 GPa] tie rod. Determine the tension in the tie rod and the deflection of the beam at B.

Plan the Solution

The cantilever beam is supported at B by a tie rod. Unlike a roller support, the tie rod is not rigid: It stretches in response to its internal tension force. Supports such as this are termed **elastic supports**. The beam deflection at B will not be zero in this instance; rather, the beam deflection will equal the elongation of the tie rod. To analyze this beam, select the reaction force provided by the tie rod as the redundant reaction. Removing this redundant leaves a cantilever as the released beam. Two cantilever beam cases will then be considered. In the first case, the downward deflection of the cantilever beam at B due to the distributed load will be calculated. The second case will consider the upward deflection at B produced by the internal force in the tie rod. The resulting two expressions will be added together in a compatibility equation, with the sum set equal to the downward deflection of the lower end of the tie rod—a deflection that is simply equal to the elongation of the rod. Since the elongation of the rod depends on the internal force in the rod, the compatibility equation will contain two terms that include the unknown tie rod force. Once the tie rod force has been computed from the compatibility equation, the deflection of the beam at B can be calculated.

SOLUTION

Case 1—Cantilever Beam with Uniformly Distributed Load

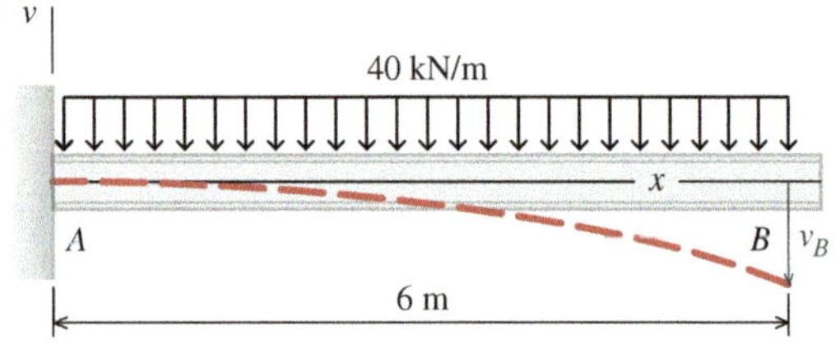

Remove the redundant tie rod support at B, and consider a cantilever beam subjected to a uniformly distributed load. The deflection of this beam at B must be determined. From Appendix C, the maximum beam deflection (which occurs at B) is given by

$$v_{max} = v_B = -\frac{wL^4}{8EI} \tag{a}$$

Case 2—Cantilever Beam with Concentrated Load

The tie rod provides the reaction force at B for the cantilever beam. Consider the cantilever beam subjected to this upward reaction force B_y. From Appendix C, the maximum beam deflection (which occurs at B) due to a concentrated load applied at the tip of the cantilever is given by

$$v_{max} = v_B = -\frac{PL^3}{3EI} = -\frac{(-B_y)L^3}{3EI} = \frac{B_y L^3}{3EI} \qquad \text{(b)}$$

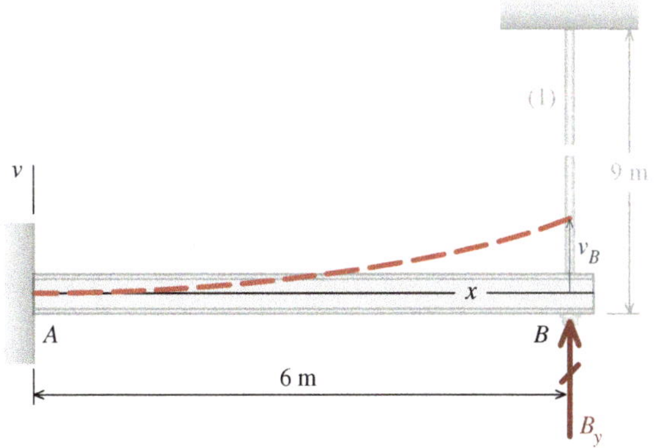

Compatibility Equation

The expressions developed for v_B from the two cases [Equations (a) and (b)] are combined in a compatibility equation:

$$v_B = -\frac{wL^4}{8EI} + \frac{B_y L^3}{3EI} \neq 0 \qquad \text{(c)}$$

In this instance, however, the beam deflection at B will not equal zero, as it would if there were a roller support at B. The beam is supported at B by an axial member that will stretch; consequently, we must determine how much the rod will stretch in this situation.

Consider a free-body diagram of the aluminum tie rod. In general, the elongation produced in rod (1) is given by

$$\delta_1 = \frac{F_1 L_1}{A_1 E_1}$$

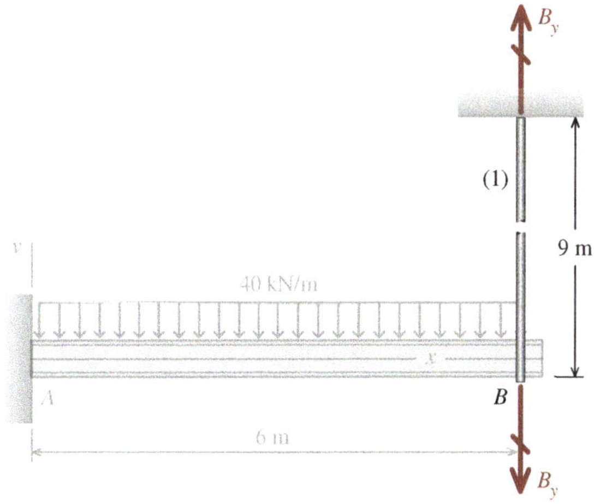

The tie rod exerts an upward force B_y on the cantilever beam. In turn, the beam exerts a force of equal magnitude, but opposite in direction, on the tie rod. Therefore, the deformation of rod (1) can be stated in terms of the unknown reaction force B_y as

$$\delta_1 = \frac{B_y L_1}{A_1 E_1}$$

As rod (1) elongates because of the force it carries, the lower end of the rod deflects downward. Since the beam is supported by the rod, the beam also deflects downward at this point. The compatibility equation [Equation (c)] must therefore be adjusted to account for the elongation of the tie rod:

$$v_B = -\frac{wL^4}{8EI} + \frac{B_y L^3}{3EI} = \frac{B_y L_1}{A_1 E_1} \qquad \text{(d)}$$

This equation is not quite correct. The error is a subtle, but important, one. **How is Equation (d) incorrect?**

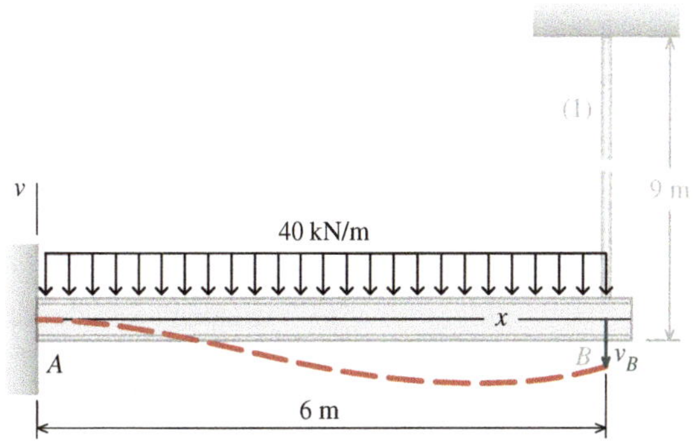

The upward direction has been defined as positive for beam deflections. When tie rod (1) elongates, point B (the lower end of the rod) moves downward. Since the compatibility equation pertains to deflections of the *beam*, the tie-rod term on the right-hand side of the equation should have a negative sign:

$$v_B = -\frac{wL^4}{8EI} + \frac{B_y L^3}{3EI} = -\frac{B_y L_1}{A_1 E_1} \qquad \text{(e)}$$

The only unknown term in this equation is the force in the tie rod—that is, B_y. Accordingly, solve the equation for B_y, to obtain

$$B_y\left[\frac{L^3}{3EI} + \frac{L_1}{A_1 E_1}\right] = \frac{wL^4}{8EI} \quad \text{(f)}$$

Before beginning the calculation, pay special attention to the terms L_1, A_1, and E_1. These are properties of the *tie rod*—not the beam. A common mistake in this type of problem is using the beam elastic modulus E for both the beam and the rod.

Now calculate the reaction force applied to the beam by the tie rod, using the following values:

Beam Properties	**Tie-Rod Properties**
$w = 40$ kN/m $= 40$ N/mm	$L_1 = 9$ m $= 9{,}000$ mm
$L = 6$ m $= 6{,}000$ mm	$d_1 = 30$ mm
$I = 300 \times 10^6$ mm^4	$A_1 = 706.858$ mm^2
$E = 200$ GPa $= 200{,}000$ N/mm^2	$E_1 = 70$ GPa $= 70{,}000$ N/mm^2

Substitute these values into Equation (f), and compute $B_y = 78{,}153.8$ N $= 78.2$ kN. Therefore, the internal axial force in the tie rod is 78.2 kN (T). **Ans.**

The deflection of the beam at B can be calculated from Equation (e) as

$$v_B = -\frac{B_y L_1}{A_1 E_1} = -\frac{(78{,}153.8\text{ N})(9{,}000\text{ mm})}{(706.858\text{ mm}^2)(70{,}000\text{ N/mm}^2)} = -14.22\text{ mm} = 14.22\text{ mm}\downarrow$$ **Ans.**

EXAMPLE 11.9

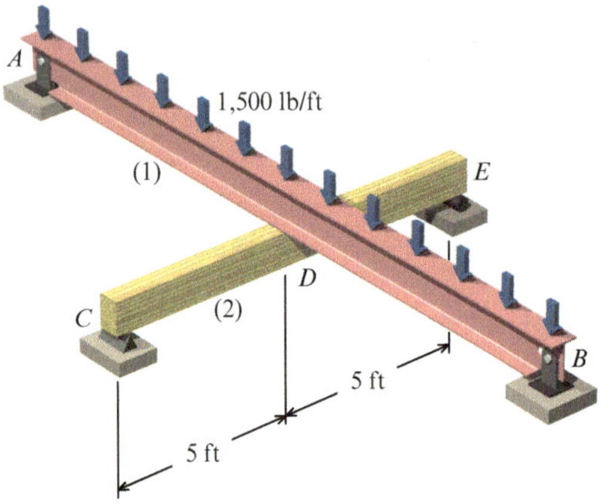

A 24 ft long W12 × 30 steel beam is supported at its ends by simple pin and roller supports and at midspan by a wooden beam, as shown in the accompanying figure. Steel [$E = 29 \times 10^6$ psi] beam (1) supports a uniformly distributed load of 1,500 lb/ft. Wooden [$E = 1.8 \times 10^6$ psi] beam (2) spans 10 ft between simple supports C and E. The steel beam rests on top of the wooden beam at the middle of the 10 ft span. The wooden beam has a cross section that is 6 in. wide and 10 in. deep. Determine

(a) the reaction force at point B applied by the wooden beam to the steel beam.

(b) the deflection of point D.

Plan the Solution

The wooden beam acts as an elastic support for the steel beam. Thus, the final deflection of the system will be determined by how much the wooden beam deflects downward in response to the force exerted on it by the steel beam. Begin by considering the steel beam. Remove the reaction force provided by the wooden beam, so that the released beam is a simply supported span with a uniformly distributed load. Determine an expression for the wooden beam's downward deflection. Next, consider the released beam with only the unknown upward reaction force at point D provided by the wooden beam. Determine an expression for the upward deflection of the simply supported steel beam due to a concentrated load acting at midspan. Next, consider the wooden beam. The upward reaction force exerted on the steel beam by the wooden beam causes the wooden beam to deflect downward. Determine an expression for the downward deflection of the wooden beam due to this unknown reaction force. Combine the three expressions for the deflection at D in a compatibility equation, and solve for the reaction force. Once the magnitude of the reaction force is known, the deflection at point D can be computed.

SOLUTION

Case 1—Simply Supported Steel Beam with Uniformly Distributed Load

Remove wooden beam (2), and consider simply supported steel beam (1) subjected to a uniformly distributed load of 1,500 lb/ft. The deflection of this beam at point D must be determined. From Appendix C, the deflection of beam (1) at midspan is given by

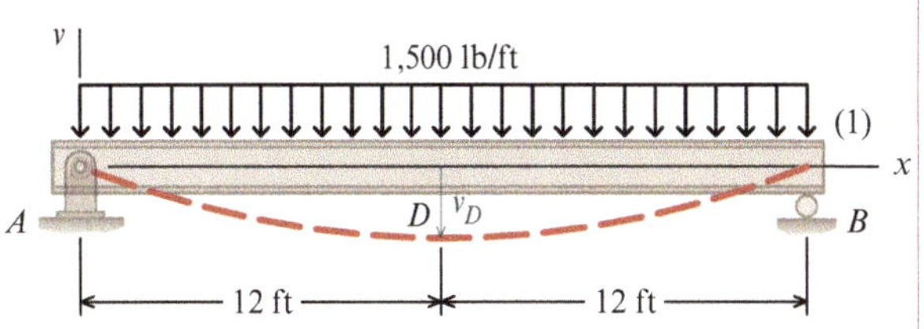

$$v_D = -\frac{5wL_1^4}{384E_1I_1} \tag{a}$$

Case 2—Simply Supported Steel Beam with Concentrated Load

Wooden beam (2) exerts an upward reaction force on the steel beam at D. Consider steel beam (1) subjected to this upward reaction force D_y. From Appendix C, the midspan deflection of a simply supported beam due to a concentrated load applied at midspan is given by

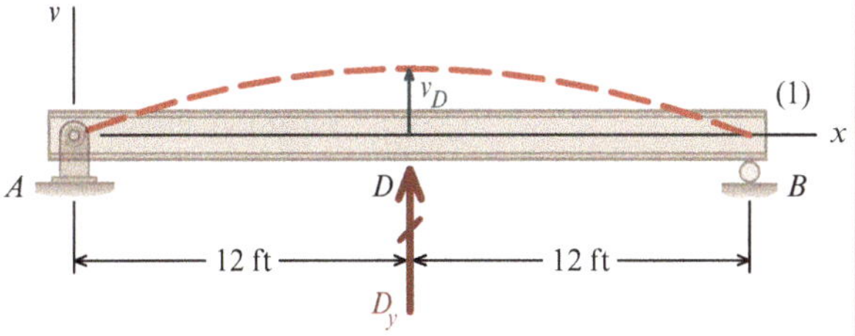

$$v_D = -\frac{PL_1^3}{48E_1I_1} = -\frac{(-D_y)L_1^3}{48E_1I_1} = \frac{D_yL_1^3}{48E_1I_1} \tag{b}$$

Case 3—Simply Supported Wooden Beam with Concentrated Load

Wooden beam (2) supplies an upward force to the steel beam at D. In reaction, steel beam (1) exerts a force of equal magnitude on the wooden beam, causing it to deflect downward. The downward deflection of beam (2) that is produced by reaction force D_y is given by

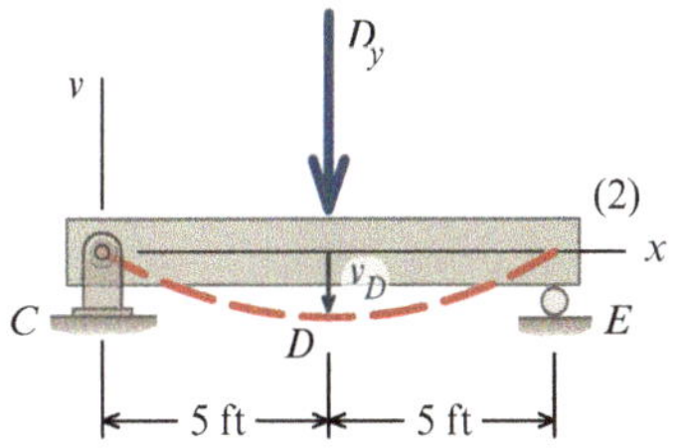

$$v_D = -\frac{D_yL_2^3}{48E_2I_2} \tag{c}$$

Compatibility Equation

The sum of the downward deflection of the steel beam due to the distributed load [Equation (a)] and the upward deflection produced by the reaction force supplied by the wooden beam [Equation (b)] must equal the downward deflection of the wooden beam [Equation (c)]. These three equations for the deflection at D are combined in a compatibility equation:

$$-\frac{5wL_1^4}{384E_1I_1} + \frac{D_yL_1^3}{48E_1I_1} = -\frac{D_yL_2^3}{48E_2I_2} \tag{d}$$

The only unknown term in this equation is the reaction force D_y. Consequently, solve the equation for D_y, to obtain

$$D_y\left[\frac{L_1^3}{48E_1I_1}+\frac{L_2^3}{48E_2I_2}\right]=\frac{5wL_1^4}{384E_1I_1} \tag{e}$$

Before beginning the calculation, pay special attention to the distinction between those properties which apply to the steel beam (i.e., L_1, I_1, and E_1) and those which apply to the wooden beam (i.e., L_2, I_2, and E_2). For instance, the flexural stiffness *EI* appears in each term, but *EI* for the wooden beam is much different than *EI* for the steel beam.

Calculate the reaction force exerted on steel beam (1), using the following values:

Steel Beam Properties	**Wood Beam Properties**
$w = 1{,}500$ lb/ft = 125 lb/in.	$L_2 = 10$ ft = 120 in.
$L_1 = 20$ ft = 240 in.	$I_2 = \dfrac{(6\text{ in.})(10\text{ in.})^3}{12} = 500\text{ in.}^4$
$I_1 = 238\text{ in.}^4$ (from Appendix B for W12 × 30)	$E_2 = 1.8 \times 10^6$ psi
$E_1 = 29 \times 10^6$ psi	

Substitute these values into Equation (e), and compute $D_y = 14{,}471.766$ lb = 14,470 lb. **Ans.**

The deflection of the system at *D* can now be calculated from Equation (c) as

$$v_D = -\frac{D_yL_2^3}{48E_2I_2} = -\frac{(14{,}471.766\text{ lb})(120\text{ in.})^3}{48(1.8\times10^6\text{ psi})(500\text{ in.}^4)} = -0.579\text{ in.} = 0.579\text{ in.}\downarrow$$ **Ans.**

MecMovies

EXAMPLE

M11.4 Determine the beam reactions for a simply supported beam with an elastic support at midspan.

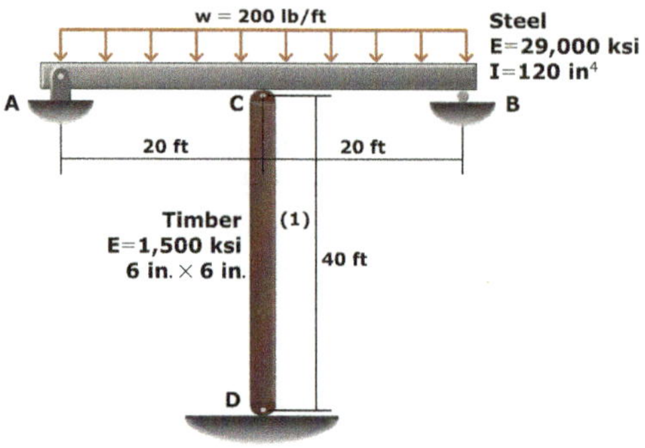

472

EXERCISES

M11.1 Propped Cantilevers. Determine the roller reaction for a propped cantilever. In each configuration, the roller reaction can be determined by the superposition of two cantilever cases: a cantilever with P and a cantilever with w.

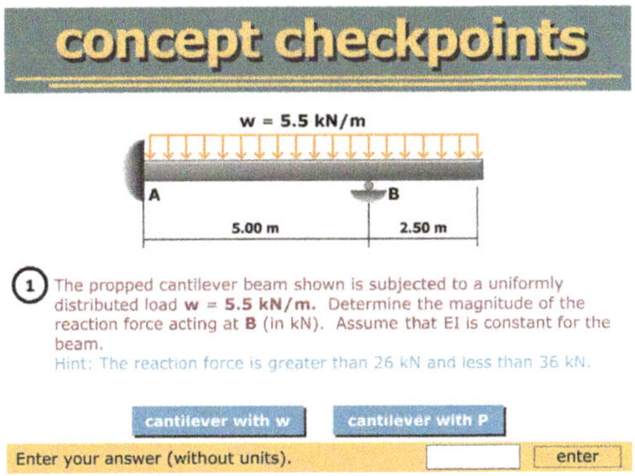

FIGURE M11.1

M11.2 Beam on Three Supports. Use superposition to determine one roller reaction for a simply supported beam on three supports.

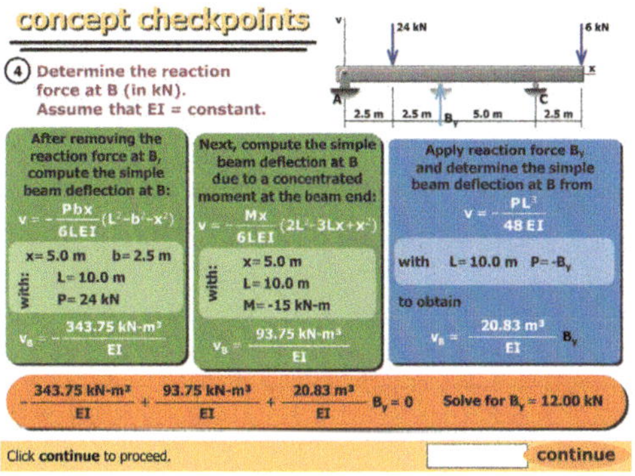

FIGURE M11.2

PROBLEMS

P11.22 For the beams and loadings shown, assume that $EI = 5.0 \times 10^4$ kN·m² is constant for each beam.

(a) For the beam in Figure P11.22*a*, determine the concentrated downward force P that will make the total beam deflection at B equal to zero (i.e., $v_B = 0$).

(b) For the beam in Figure P11.22*b*, determine the concentrated moment M that will make the total beam slope at A equal to zero (i.e., $\theta_A = 0$).

P11.23 For the beams and loadings shown, assume that $EI = 8.0 \times 10^6$ kip·in.² is constant for each beam.

(a) For the beam in Figure P11.23*a*, determine the concentrated downward force P that will make the total beam deflection at B equal to zero (i.e., $v_B = 0$).

(b) For the beam in Figure P11.23*b*, determine the concentrated moment M that will make the total beam slope at A equal to zero (i.e., $\theta_A = 0$).

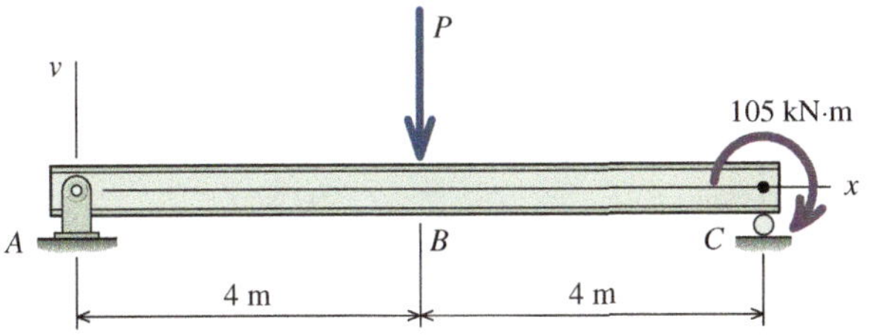

FIGURE P11.22*a*

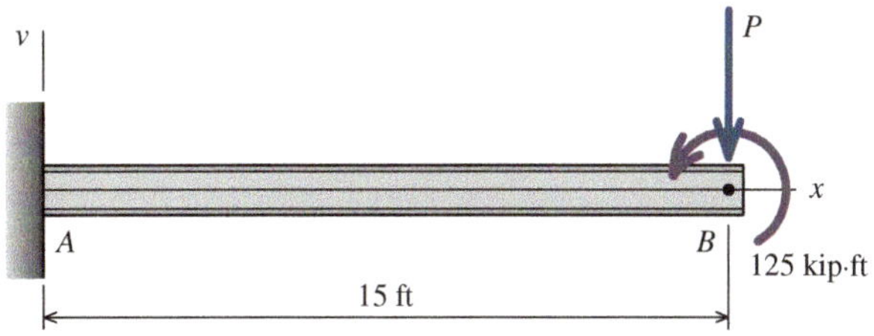

FIGURE P11.23*a*

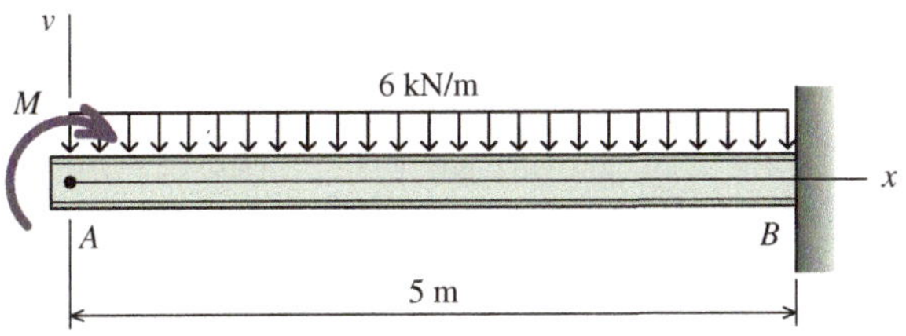

FIGURE P11.22*b*

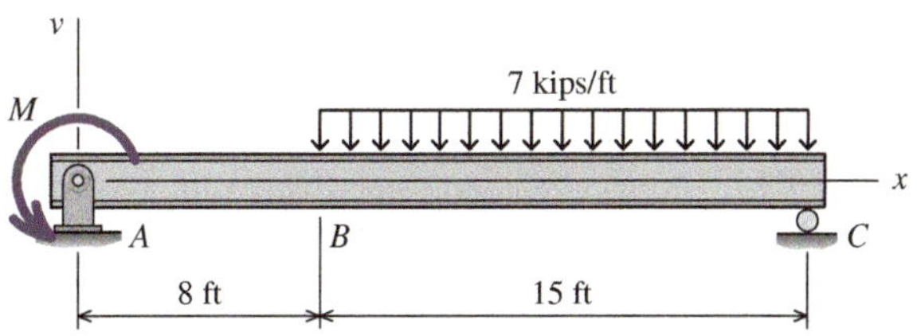

FIGURE P11.23*b*

P11.24–P11.26 For the beams and loadings shown in Figures P11.24–P11.26, derive an expression for the reactions at supports A and B. Assume that EI is constant for the beam.

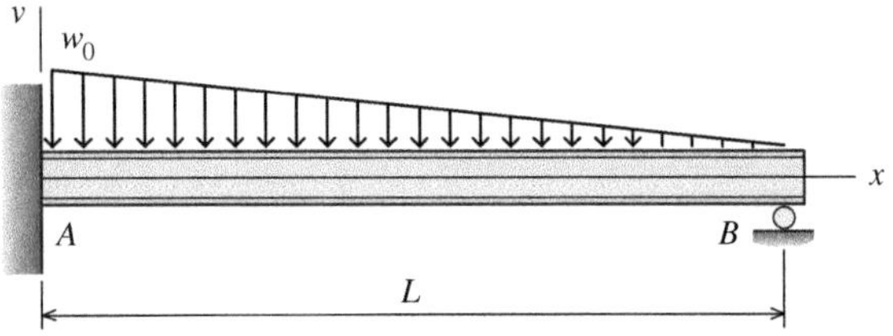

FIGURE P11.24

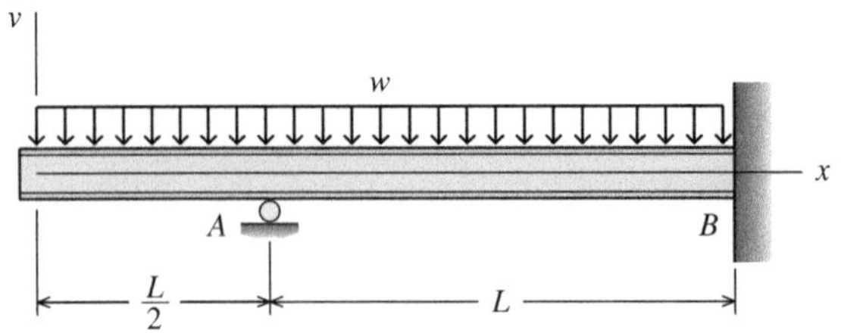

FIGURE P11.25

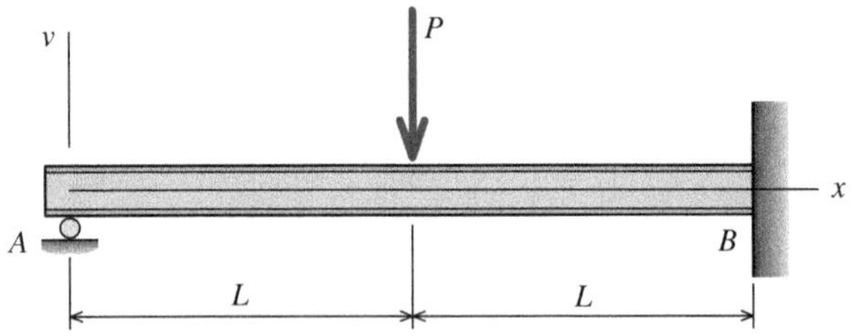

FIGURE P11.26

P11.27 For the beam and loading shown in Figure P11.27, derive an expression for the reactions at supports A and C. Assume that EI is constant for the beam.

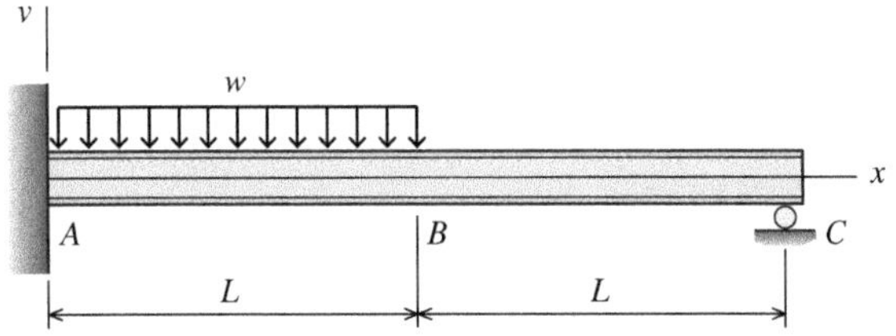

FIGURE P11.27

P11.28–P11.30 For the beams and loadings shown in Figures P11.28–P11.30, derive an expression for the reaction force at B. Assume that EI is constant for the beam. (*Reminder:* The roller symbol implies that both upward and downward displacements are restrained.)

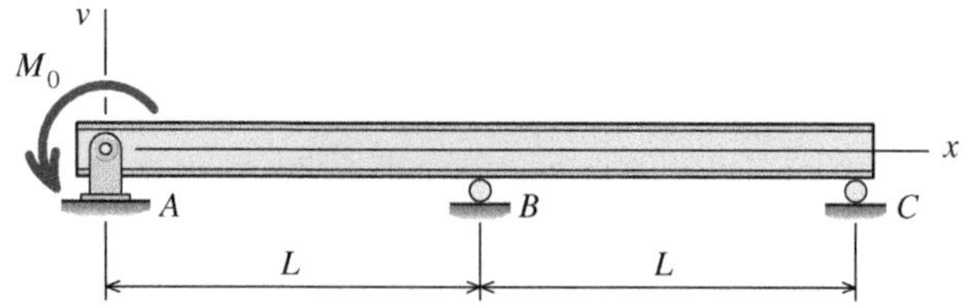

FIGURE P11.28

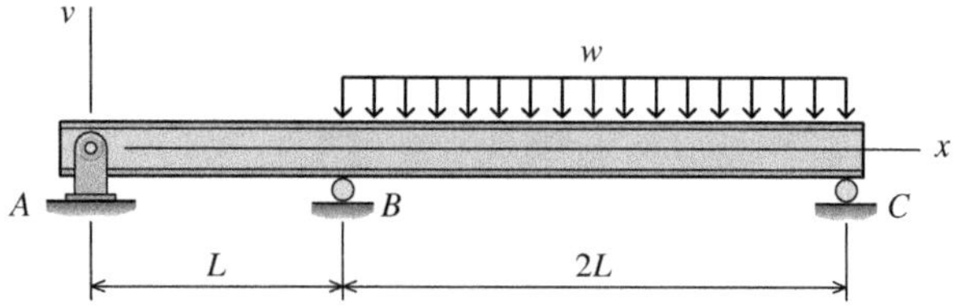

FIGURE P11.29

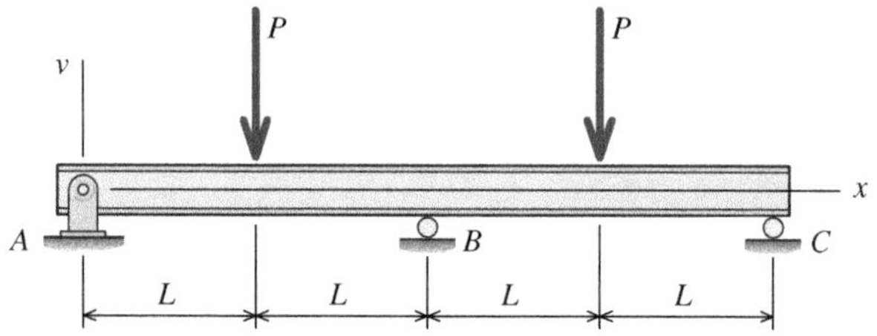

FIGURE P11.30

P11.31 The beam shown in Figure P11.31 consists of a W360 × 79 structural steel wide-flange shape [$E = 200$ GPa; $I = 225 \times 10^6$ mm^4]. For the loading shown, determine

(a) the reactions at A, B, and C.
(b) the magnitude of the maximum bending stress in the beam.

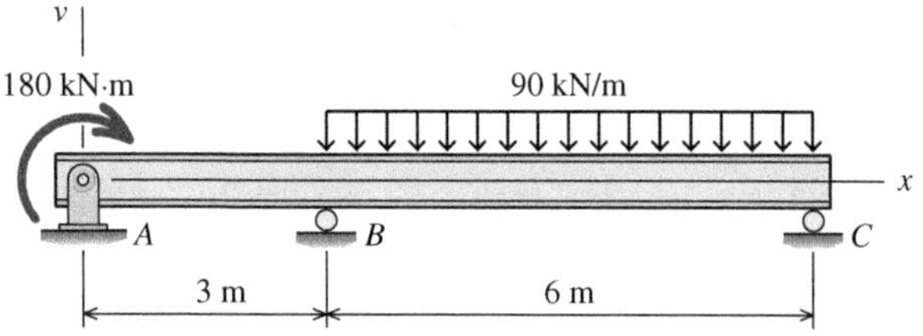

FIGURE P11.31

P11.32 The beam shown in Figure P11.32 consists of a W610 × 140 structural steel wide-flange shape [$E = 200$ GPa; $I = 1{,}120 \times 10^6$ mm^4]. For the loading shown, determine

(a) the reactions at A, B, and D.
(b) the magnitude of the maximum bending stress in the beam.

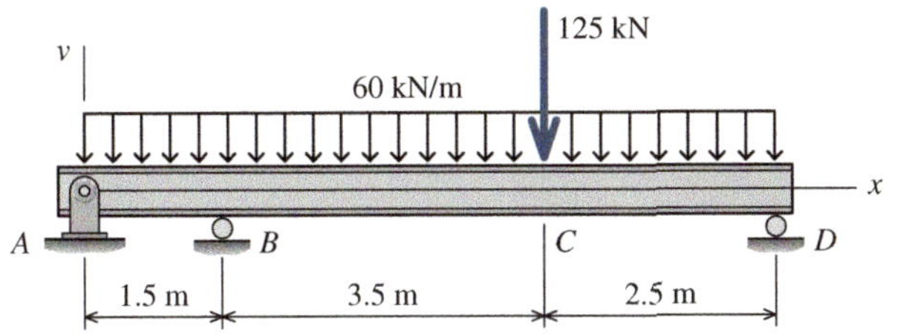

FIGURE P11.32

P11.33 A propped cantilever beam is loaded as shown in Figure P11.33. Assume that $EI = 24 \times 10^6$ kip · in.^{2}. Determine

(a) the reactions at B and C for the beam.
(b) the beam deflection at A.

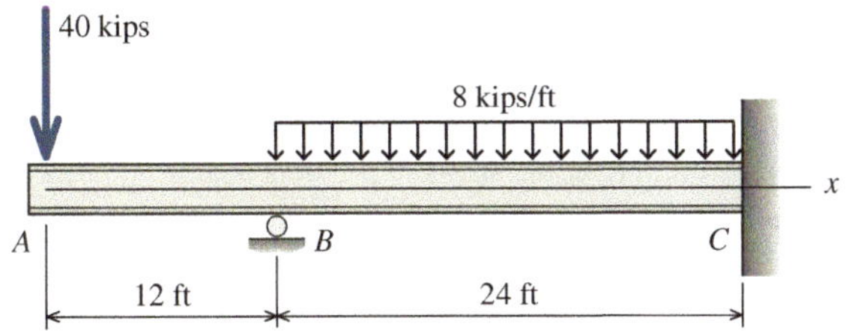

FIGURE P11.33

P11.34 The beam shown in Figure P11.34 consists of a W610 × 82 structural steel wide-flange shape [$E = 200$ GPa; $I = 562 \times 10^6$ mm^4]. For the loading shown, determine

(a) the reaction force at C.
(b) the beam deflection at A.

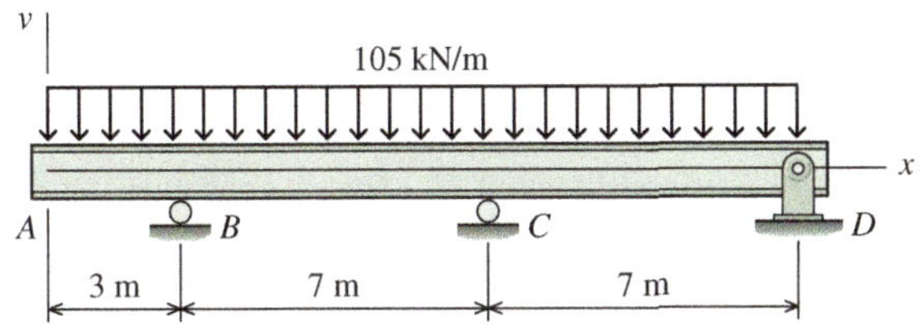

FIGURE P11.34

P11.35 The beam shown in Figure P11.35 consists of a W8 × 15 structural steel wide-flange shape [$E = 29{,}000$ ksi; $I = 48$ in.^{4}]. For the loading shown, determine

(a) the reactions at A and B.
(b) the magnitude of the maximum bending stress in the beam.

(*Reminder:* The roller symbol implies that both upward and downward displacements are restrained.)

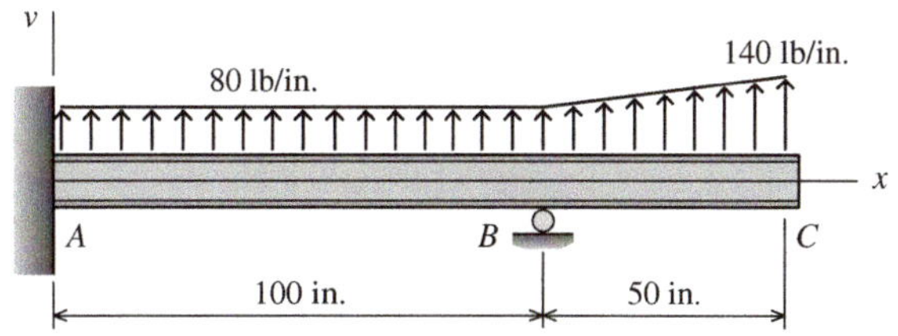

FIGURE P11.35

P11.36 The solid 1.00 in. diameter steel [$E = 29{,}000$ ksi] shaft shown in Figure P11.36 supports three belt pulleys. Assume that the bearing at A can be idealized as a pin support and that the bearings at C and E can be idealized as roller supports. For the loading shown, determine

(a) the reaction forces at bearings A, C, and E.
(b) the magnitude of the maximum bending stress in the shaft.

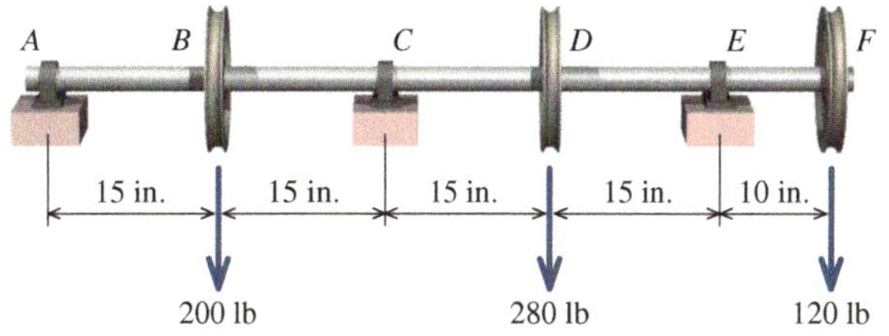

FIGURE P11.36

P11.37 The solid 1.00 in. diameter steel [$E = 29{,}000$ ksi] shaft shown in Figure P11.37 supports two belt pulleys. Assume that the bearing at E can be idealized as a pin support and that the bearings at B and C can be idealized as roller supports. For the loading shown, determine

(a) the reaction forces at bearings B, C, and E.
(b) the magnitude of the maximum bending stress in the shaft.

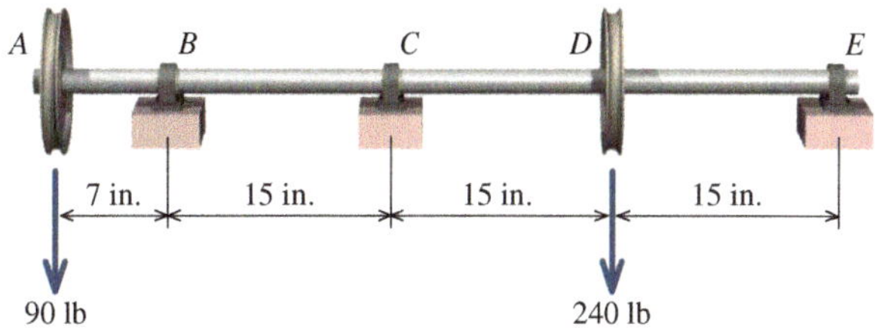

FIGURE P11.37

P11.38 The beam shown in Figure P11.38 consists of a W360 × 101 structural steel wide-flange shape [$E = 200$ GPa; $I = 301 \times 10^6$ mm^4]. For the loading shown, determine

(a) the reactions at A and B.
(b) the magnitude of the maximum bending stress in the beam.

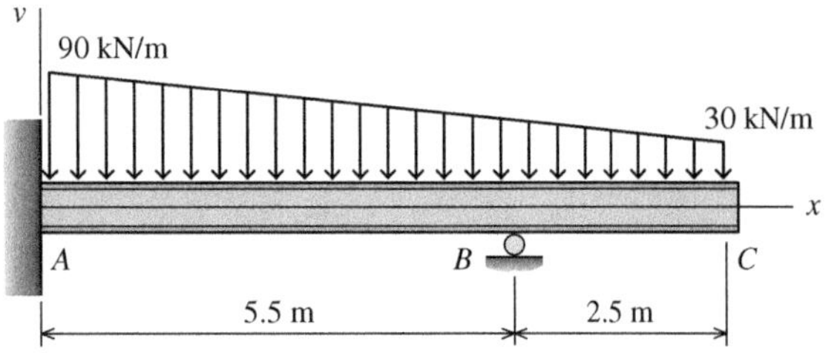

FIGURE P11.38

P11.39–P11.40 A W530 × 92 structural steel wide-flange shape [$E = 200$ GPa; $I = 554 \times 10^6$ mm^4] is loaded and supported as shown in Figures P11.39 and P11.40. Determine

(a) the force and moment reactions at supports *A* and *C*.
(b) the maximum bending stress in the beam.
(c) the deflection of the beam at *B*.

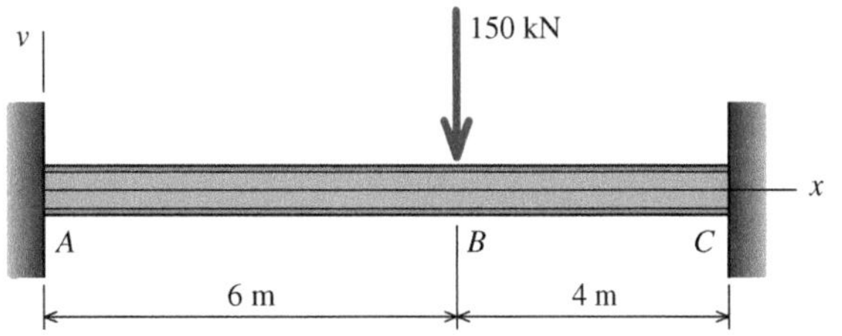

FIGURE P11.39

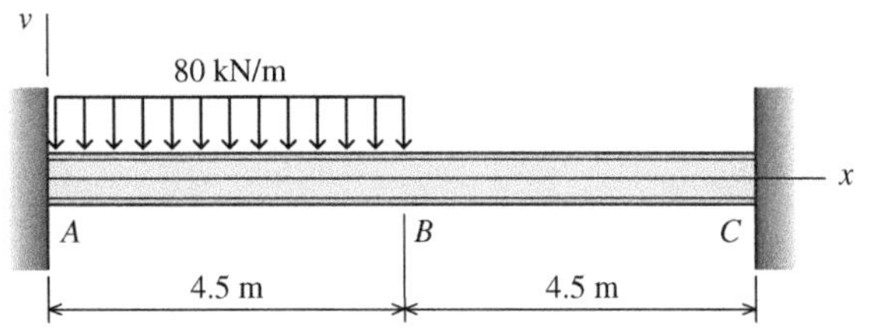

FIGURE P11.40

P11.41 A W360 × 72 structural steel [$E = 200$ GPa] wide-flange shape is loaded and supported as shown in Figure P11.41. The beam is supported at *B* by a 20 mm diameter solid aluminum [$E =$ 70 GPa] rod. A concentrated load of 40 kN is applied to the tip of the cantilever. Determine

(a) the force produced in the aluminum rod.
(b) the maximum bending stress in the beam.
(c) the deflection of the beam at *B*.

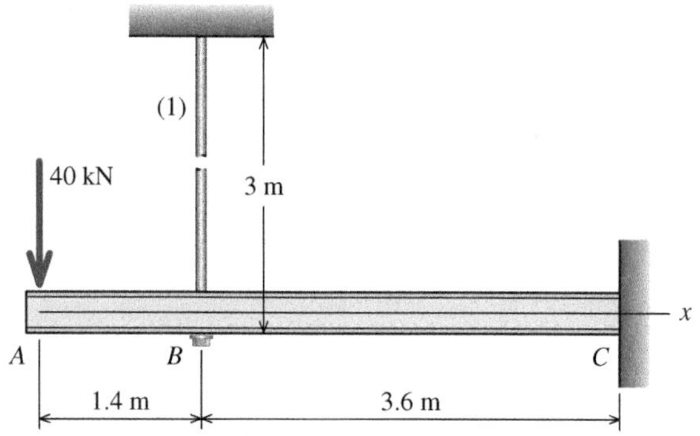

FIGURE P11.41

P11.42 A W18 × 55 structural steel [E = 29,000 ksi] wide-flange shape is loaded and supported as shown in Figure P11.42. The beam is supported at *C* by a 3/4 in. diameter aluminum [E = 10,000 ksi] rod that has no load before the distributed load is applied to the beam. A distributed load of 4 kips/ft is applied to the beam. Determine

(a) the force carried by the aluminum rod.
(b) the maximum bending stress in the steel beam.
(c) the deflection of the beam at *C*.

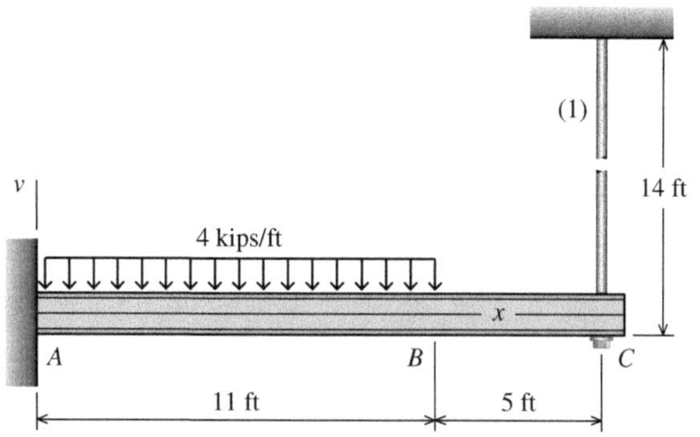

FIGURE P11.42

P11.43 A W250 × 32.7 structural steel [E = 200 GPa] wide-flange shape is loaded and supported as shown in Figure P11.43. A uniformly distributed load of 16 kN/m is applied to the beam, causing the roller support at *B* to settle downward (i.e., be displaced downward) by 15 mm. Determine

(a) the reactions at supports *A*, *B*, and *C*.
(b) the maximum bending stress in the beam.

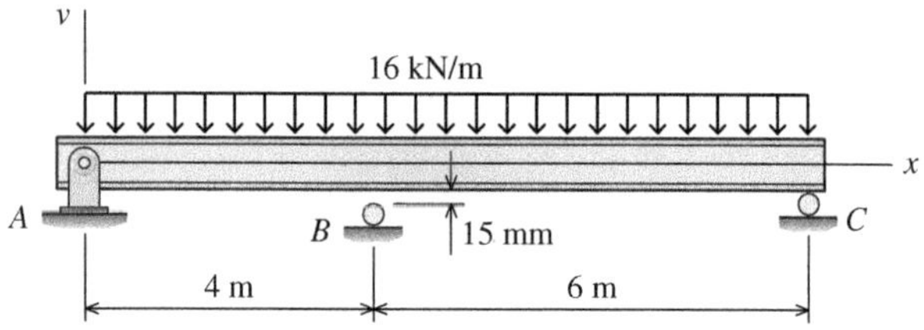

FIGURE P11.43

P11.44 A timber [$E = 12$ GPa] beam is loaded and supported as shown in Figure P11.44. The cross section of the timber beam is 100 mm wide and 300 mm deep. The beam is supported at B by a 12 mm diameter steel [$E = 200$ GPa] rod that has no load before the distributed load is applied to the beam. A distributed load of 7 kN/m is applied to the beam. Determine

(a) the force carried by the steel rod.
(b) the maximum bending stress in the timber beam.
(c) the deflection of the beam at B.

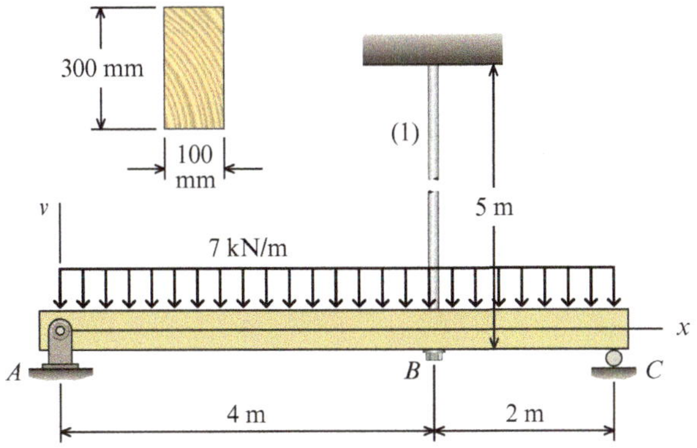

FIGURE P11.44

P11.45 A W360 × 72 structural steel [$E = 200$ GPa] wide-flange shape is loaded and supported as shown in Figure P11.45. The beam is supported at B by a timber [$E = 12$ GPa] post having a cross-sectional area of 20,000 mm². A uniformly distributed load of 50 kN/m is applied to the beam. Determine

(a) the reactions at supports A, B, and C.
(b) the maximum bending stress in the beam.
(c) the deflection of the beam at B.

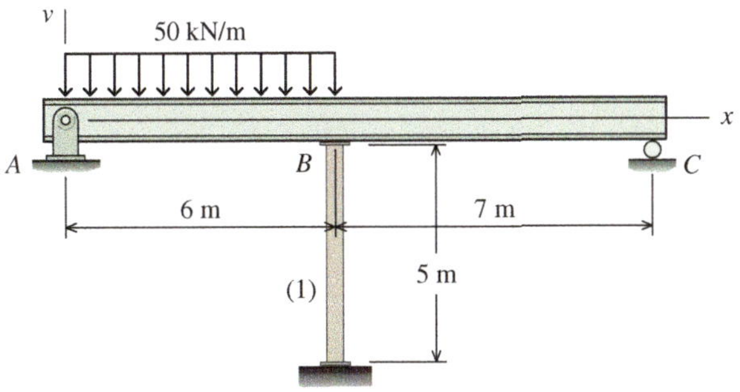

FIGURE P11.45

P11.46 A W530 × 66 structural steel [$E = 200$ GPa] wide-flange shape is loaded and supported as shown in Figure P11.46. A uniformly distributed load of 70 kN/m is applied to the beam, causing the roller support at B to settle downward (i.e., be displaced downward) by 10 mm. Determine

(a) the reactions at supports A and B.
(b) the maximum bending stress in the beam.

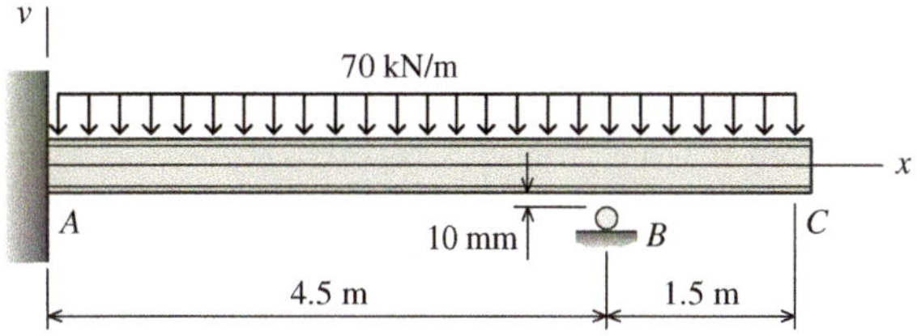

FIGURE P11.46

P11.47 Steel beam (1) carries a concentrated load $P = 13$ kips that is applied at midspan, as shown in Figure P11.47/48. The steel beam is supported at ends A and B by nondeflecting supports and at its middle by simply supported timber beam (2). In the unloaded condition, steel beam (1) touches, but exerts no force on, timber beam (2). The length of the steel beam is $L_1 = 30$ ft, and its flexural rigidity is $EI_1 = 7.2 \times 10^6$ kip·in.². The length and the flexural rigidity of the timber beam are $L_2 = 20$ ft and $EI_2 = 1.0 \times 10^6$ kip·in.², respectively. Determine the vertical reaction force that acts

(a) on the steel beam at A.
(b) on the timber beam at C.

P11.48 In Figure P11.47/48, a W10 × 45 steel beam (1) carries a concentrated load $P = 9$ kips that is applied at midspan. The steel beam is supported at ends A and B by nondeflecting supports and at its middle by simply supported timber beam (2) that is 8 in. wide and 12 in. deep. In the unloaded condition, steel beam (1) touches, but exerts no force on, timber beam (2). The length of the steel beam is $L_1 = 24$ ft, and its modulus of elasticity is $E_1 = 29 \times 10^3$ ksi. The length and the modulus of elasticity of the timber beam are $L_2 = 15$ ft and $E_2 = 1.8 \times 10^3$ ksi, respectively. Determine the maximum flexural stress

(a) in the steel beam.
(b) in the timber beam.
(c) in the steel beam if the timber beam is removed.

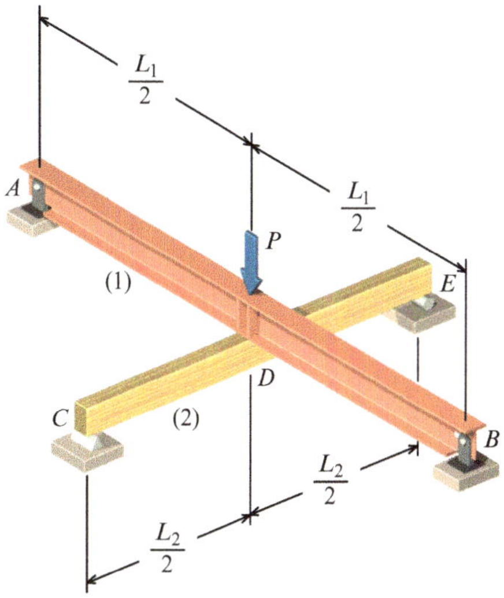

FIGURE P11.47/48

P11.49 Two steel beams support a concentrated load $P = 45$ kN, as shown in Figure P11.49/50. Beam (1) is supported by a fixed support at A and by a simply supported beam (2) at D. In the unloaded condition, beam (1) touches, but exerts no force on, beam (2). The beam lengths are $a = 4.0$ m, $b = 1.5$ m, and $L_2 = 6$ m. The flexural rigidities of the beams are $EI_1 = 40{,}000$ kN·m² and $EI_2 = 14{,}000$ kN·m². Determine the deflection of beam (1) (a) at D and (b) at B.

P11.50 Two steel beams support a concentrated load $P = 60$ kN, as shown in Figure P11.49/50. Beam (1) is supported by a fixed support at A and by a simply supported beam (2) at D. In the unloaded condition, beam (1) touches, but exerts no force on, beam (2). The beam lengths are $a = 5.0$ m, $b = 2.0$ m, and $L_2 = 8$ m. The flexural rigidities of the beams are $EI_1 = 40{,}000$ kN·m² and $EI_2 = 25{,}000$ kN·m². Determine

(a) the reactions on beam (1) at A.
(b) the reaction on beam (2) at C.

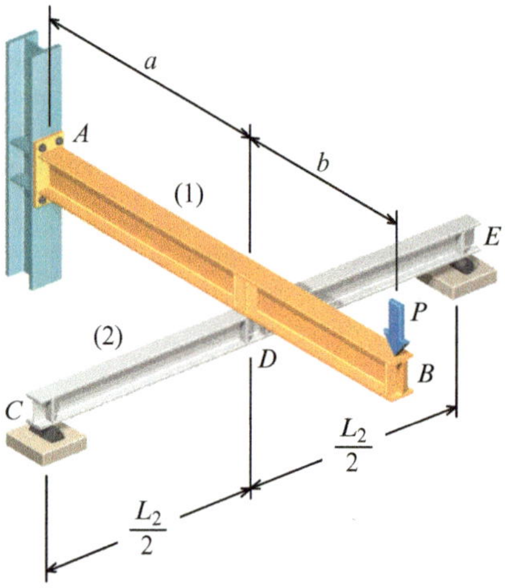

FIGURE P11.49/50

CHAPTER 12

Stress Transformations

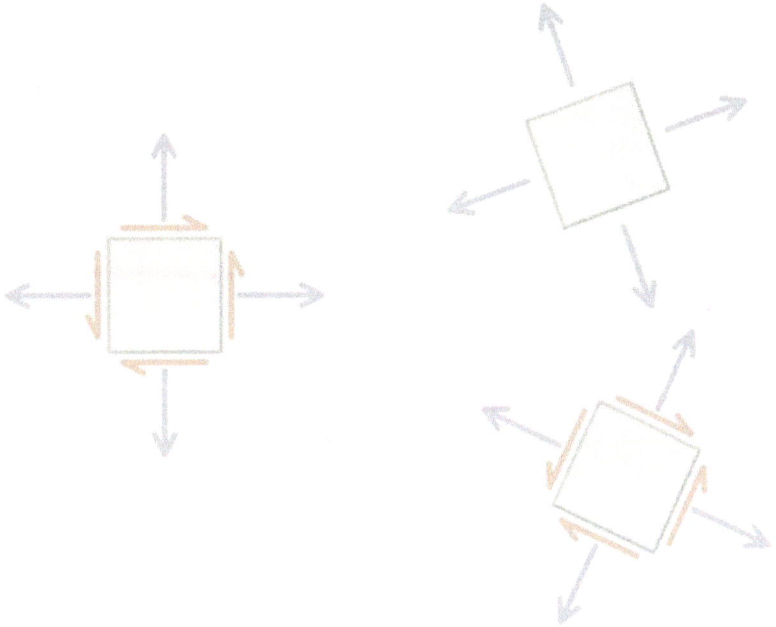

12.1 Introduction

In previous chapters, formulas were developed for normal and shear stresses that act on specific planes in axially loaded bars, circular shafts, and beams. For axially loaded bars, additional expressions were developed in Section 1.5 for the normal [Equation (1.8)] and shear [Equation (1.9)] stresses that act on inclined planes through the bar. The analysis presented in that section revealed that maximum normal stresses occur on transverse planes and that maximum shear stresses occur on planes inclined at 45° to the axis of the bar. (See Figure 1.4.) Similar expressions were developed for the case of pure torsion in a circular shaft. It was shown that maximum shear stresses [Equation (6.9)] occur on transverse planes of the torsion member, but that maximum tensile and compressive stresses [Equation (6.10)] occur on planes inclined at 45° to the axis of the member. (See Figure 6.9.) For both axial and torsion members, normal and shear stresses acting on specified planes were determined from a free-body diagram approach. Such an approach, while instructive, is not efficient for the determination of maximum normal and shear stresses, which are often required in a stress analysis. In this chapter, methods that are more powerful will be developed to determine

(a) normal and shear stresses acting on any specific plane passing through a point of interest, and
(b) maximum normal and shear stresses acting at any possible orientation at a point of interest.

12.2 Stress at a General Point in an Arbitrarily Loaded Body

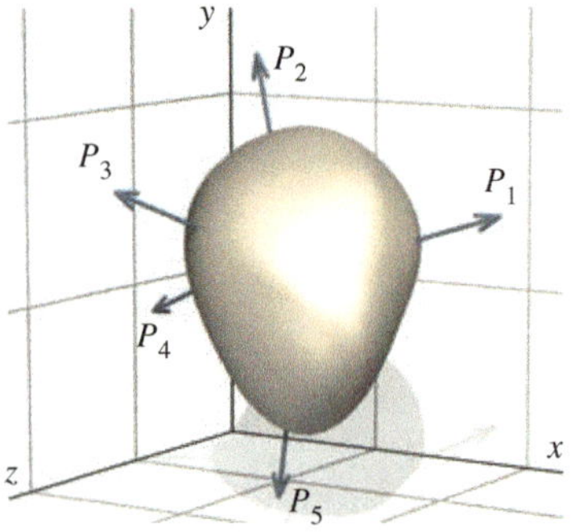

FIGURE 12.1 Solid body in equilibrium.

In Chapter 1, the concept of stress was introduced by considering the internal force distribution required to satisfy equilibrium in a portion of a bar under axial load. The nature of the force distribution led to uniformly distributed normal and shear stresses on transverse planes through the bar. (See Section 1.5.) In more complicated structural members or machine components, the stress distributions will not be uniform on arbitrary internal planes; therefore, a more general concept of the state of stress at a point is needed.

Consider a body of arbitrary shape that is in equilibrium under the action of a system of several applied loads P_1, P_2, and so on (Figure 12.1). The nature of the stresses created at an arbitrary interior point Q can be studied by cutting a section through the body at Q, where the cutting plane is parallel to the y–z plane, as shown in Figure 12.2*a*. This free body is subjected to some of the original loads (P_1, P_2, etc.), as well as to normal and shearing forces, distributed on the exposed plane surface. We will focus on a small portion ΔA of the exposed plane surface. The resultant force acting on ΔA can be resolved into components that act perpendicular and parallel, respectively, to the surface. The perpendicular component is a normal force ΔF_x, and the parallel component is a shear force ΔV_x. The subscript x is used to indicate that these forces act on a plane whose normal is in the x direction (termed the x plane).

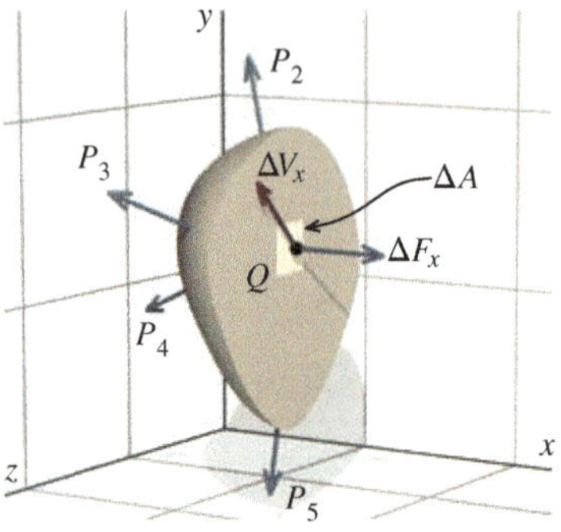

FIGURE 12.2*a* Resultant forces on area ΔA.

Although the direction of the normal force ΔF_x is well defined, the shear force ΔV_x could be oriented in any direction on the x plane. Therefore, ΔV_x will be resolved into two component forces, ΔV_{xy} and ΔV_{xz}, where the second subscript indicates that the shear forces on the x plane act in the y and z directions, respectively. The x, y, and z components of the normal and shear forces acting on ΔA are shown in Figure 12.2*b*.

If each force component is divided by the area ΔA, an average force per unit area is obtained. As ΔA is made smaller and smaller, three stress components are defined at point Q (Figure 12.3):

$$\sigma_x = \lim_{\Delta A \to 0} \frac{\Delta F_x}{\Delta A} \qquad \tau_{xy} = \lim_{\Delta A \to 0} \frac{\Delta V_{xy}}{\Delta A} \qquad \tau_{xz} = \lim_{\Delta A \to 0} \frac{\Delta V_{xz}}{\Delta A} \tag{12.1}$$

To reiterate, the first subscript on stresses σ_x, τ_{xy}, and τ_{xz} indicates that these stresses act on a plane whose normal is in the x direction. The second subscript on τ_{xy} and τ_{xz} indicates the direction in which the shear stress acts on the x plane.

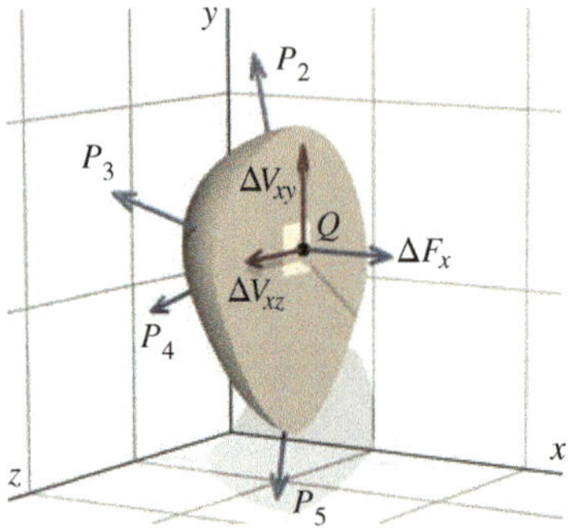

FIGURE 12.2*b* Resultant forces on area ΔA resolved into x, y, and z components.

$\sigma_x, \tau_{xy}, \tau_{xz}$
$\sigma_y, \tau_{yx}, \tau_{yz}$
$\sigma_z, \tau_{zx}, \tau_{zy}$

Next, suppose that a cutting plane parallel to the x–z plane is passed through the original body (from Figure 12.1). This cutting plane exposes a surface whose normal is in the y direction (Figure 12.4). According to the previous reasoning, three stresses are obtained on the y plane at Q: a normal stress σ_y acting in the y direction, a shear stress τ_{yx} acting on the y plane in the x direction, and a shear stress τ_{yz} acting on the y plane in the z direction.

Finally, a cutting plane parallel to the x–y plane is passed through the original body to expose a surface whose normal is in the z direction (Figure 12.5). Again, three stresses are obtained on the z plane at Q: a normal stress σ_z acting in the z direction, a shear stress τ_{zx} acting on the z plane in the x direction, and a shear stress τ_{zy} acting on the z plane in the y direction.

If a different set of coordinate axes (say, x'–y'–z') had been chosen in the previous discussion, then the stresses found at point Q would be different from those determined on the x, y, and z planes. Stresses in the x'–y'–z' coordinate system, however, are related to those in the x–y–z coordinate system, and through a mathematical process called **stress transformation**, stresses can be converted from one coordinate system to another. If the normal and shear stresses on the x, y, and z planes at point Q are known (Figures 12.3, 12.4, and 12.5), then the normal and shear stresses on any plane passing through point Q can be determined. For this reason, the stresses on these planes are called the **state of stress** at a point. The state of stress can be uniquely defined by three stress components acting on each of three mutually perpendicular planes.

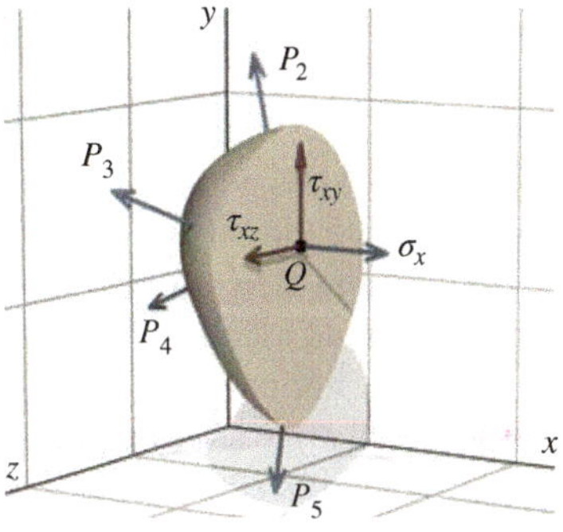

FIGURE 12.3 Stresses acting on an x plane at point Q in the body.

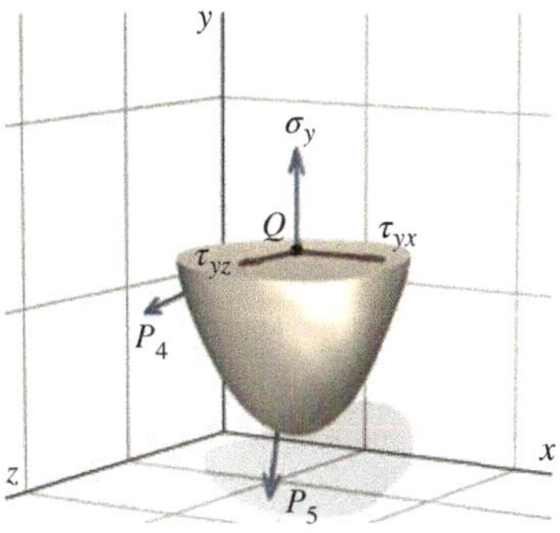

FIGURE 12.4 Stresses acting on a y plane at point Q in the body.

The state of stress at a point (such as point Q in the preceding figures) is conveniently represented by stress components acting on an *infinitesimally small* cubic element of material known as a **stress element** (Figure 12.6). The stress element is a *graphical symbol* that represents a point of interest in an object (such as a shaft or a beam). The six faces of the cubic element are each identified by the outward normal to the face. For example, the positive x face is the face whose outward normal is in the direction of the positive x axis. The coordinate axes x, y, and z are arranged as a right-hand coordinate system.

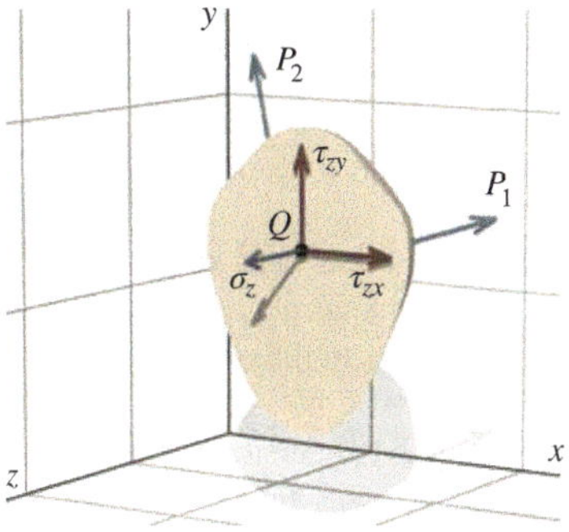

FIGURE 12.5 Stresses acting on a z plane at point Q in the body.

The stress components σ_x, σ_y, and σ_z are normal stresses that act on the faces that are perpendicular to the x, y, and z coordinate axes, respectively. There are six shear stress components acting on the cubic element: τ_{xy}, τ_{xz}, τ_{yx}, τ_{yz}, τ_{zx}, and τ_{zy}. However, only three of these shear stresses are independent, as will be demonstrated subsequently. *Specific values associated with stress components are dependent upon the orientation of the coordinate axes.* The state of stress shown in Figure 12.6 would be represented by a different set of stress components if the coordinate axes were rotated.

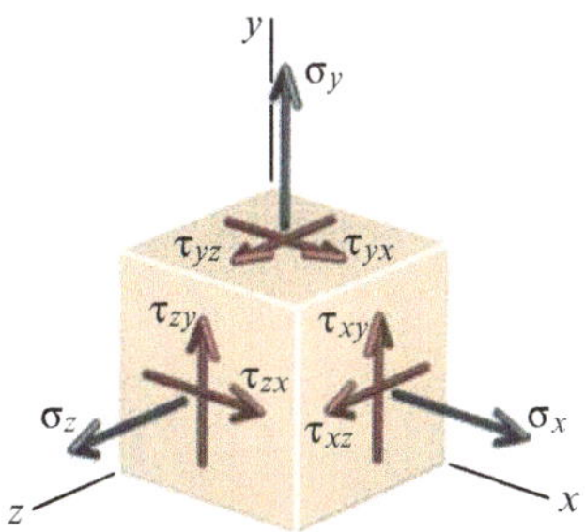

FIGURE 12.6 Stress element representing the state of stress at a point.

Stress Sign Conventions

+σ → tension
-σ → compression

Normal stresses are indicated by the symbol σ and a single subscript that indicates the plane on which the stress acts. The normal stress acting on a face of the stress element is positive if it points in the outward normal direction. In other words, normal stresses are positive if they cause tension in the material. Compressive normal stresses are negative.

Shear stresses are denoted by the symbol τ followed by two subscripts. The first subscript designates the plane on which the shear stress acts. The second subscript indicates the direction in which the stress acts. For example, τ_{xz} is shear stress on an x face acting in the z direction. The distinction between a positive and a negative shear stress depends on two considerations: (1) the face of the stress element upon which the shear stress acts and (2) the direction in which the stress acts.

A shear stress is positive if it

- acts in the *positive* coordinate direction on a *positive* face of the stress element, or
- acts in the *negative* coordinate direction on a *negative* face of the stress element.

For example, a shear stress on a positive x face that acts in a positive z direction is a positive shear stress. Similarly, a shear stress that acts in a negative x direction on a negative y face is also considered positive. The stresses shown on the stress element in Figure 12.6 are all positive.

MecMovies 12.5 presents an animated discussion of terminology used in stress transformations.

12.3 Equilibrium of the Stress Element

Figure 12.7*a* shows a two-dimensional projection of a stress element having width dx and height dy. The thickness of the stress element perpendicular to the x–y plane is dz. The stress element represents an infinitesimally small portion of a physical object. If an object is in equilibrium, then any portion of the object that one chooses to examine must also be in equilibrium, no matter how small that portion may be. Consequently, the stress element must be in equilibrium.

Keep in mind that the square shown in Figure 12.7 is simply a two-dimensional projection of the cube shown in Figure 12.6. In other words, we are seeing only one side of the infinitesimally small cube in Figure 12.7, but the stress element we are talking about is still a cube.

Equilibrium involves forces, not stresses. To consider equilibrium of the stress element in Figure 12.7*a*, the forces produced by the stresses that act on each face must be found by multiplying the stress acting on each face by the area of the face. These forces can then be considered on a free-body diagram of the element.

Since the stress element is infinitesimally small, we can assert that the normal stresses σ_x and σ_y acting on opposite faces of the stress element are equal in magnitude and aligned collinearly in pairs. Hence, the forces arising from normal stresses counteract each other, and equilibrium is assured with respect to both translation ($\Sigma F = 0$) and rotation ($\Sigma M = 0$).

Although the shear stress and shear force arrows in Figure 12.7 are shown slightly offset from the faces of the stress element, it should be understood that the shear stresses and the shear forces act directly on the face. The arrows are shown offset from the faces of the stress element for clarity.

Next, consider the shear stresses acting on the x and y faces of the stress element (Figure 12.7*b*). Suppose that a positive shear stress τ_{xy} acts on the positive x face of the stress element. Then, the shear force produced on the x face in the y direction by this stress is $V_{xy} = \tau_{xy}\,(dy\,dz)$ (where dz is the out-of-plane thickness of the element). To satisfy equilibrium in the y direction ($\Sigma F_y = 0$), the shear stress on the $-x$ face must act in the $-y$ direction. Similarly, a positive shear stress τ_{yx} acting on the positive y face of the stress element produces a shear force $V_{yx} = \tau_{yx}\,(dx\,dz)$ in the x direction. To satisfy equilibrium in the x direction ($\Sigma F_x = 0$), the shear stress on the $-y$ face must act in the $-x$ direction. Therefore, the shear stresses shown in Figure 12.7 satisfy equilibrium in the x and y directions.

The moments created by the shear stresses must also satisfy equilibrium. Consider the moments produced about point O, located at the lower left corner of the stress element. The lines of action of the shear forces acting on the $-x$ and $-y$ faces pass through point O; therefore, these forces do not produce moments. The shear force V_{yx} acting on the $+y$ face (a distance dy from point O) produces a clockwise moment $V_{yx}\,dy$. The shear force V_{xy} acting on the $+x$ face (a distance dx from point O) produces a counterclockwise moment equal to $V_{xy}\,dx$. Application of the equation $\Sigma M_O = 0$ yields

$$\Sigma M_O = V_{xy}dx - V_{yx}dy = \tau_{xy}(dy\,dz)dx - \tau_{yx}(dx\,dz)dy = 0$$

which reduces to

$$\boxed{\tau_{yx} = \tau_{xy}} \qquad (12.2)$$

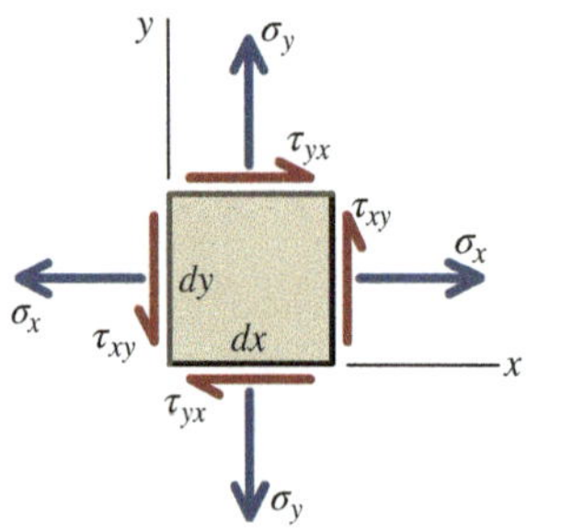

FIGURE 12.7*a*

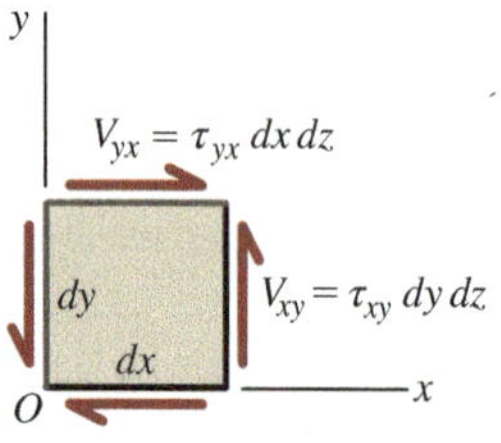

dz = out-of-plane thickness

FIGURE 12.7*b*

The result of this simple equilibrium analysis produces a significant conclusion:

If a shear stress exists on any plane, there must also be a shear stress of the same magnitude acting on an orthogonal plane (i.e., a plane perpendicular to the original plane).

From this conclusion, we can also assert that

$$\tau_{yx} = \tau_{xy} \qquad \tau_{yz} = \tau_{zy} \qquad \text{and} \qquad \tau_{xz} = \tau_{zx}$$

This analysis shows that the subscripts for shear stresses are *commutative*, meaning that the order of the subscripts may be interchanged. Consequently, only three of the six shear stress components acting on the cubic element in Figure 12.6 are independent.

12.4 Plane Stress

Significant insight into the nature of stress in a body can be gained from the study of a state known as two-dimensional stress or **plane stress**. For this case, two parallel faces of the stress element shown in Figure 12.6 are assumed to be free of stress. For purposes of analysis, assume that the faces perpendicular to the z axis (i.e., the $+z$ and $-z$ faces) are free of stress. Thus,

$$\sigma_z = \tau_{zx} = \tau_{zy} = 0$$

From Equation (12.2), however, the plane stress assumption also implies that

$$\tau_{xz} = \tau_{yz} = 0$$

since shear stresses acting on orthogonal planes must have the same magnitude. Therefore, only the σ_x, σ_y, and $\tau_{xy} = \tau_{yx}$ stress components appear in a plane stress analysis. For convenience, this state of stress is usually represented by the two-dimensional sketch shown in Figure 12.8. Keep in mind, however, that the sketch represents a three-dimensional block having thickness in the out-of-plane direction even though it is drawn as a two-dimensional square.

Many components commonly found in engineering design are subjected to plane stress. Thin plate elements such as beam webs and flanges are typically loaded in the plane of the element. Plane stress also describes the state of stress for *all free surfaces* of structural elements and machine components.

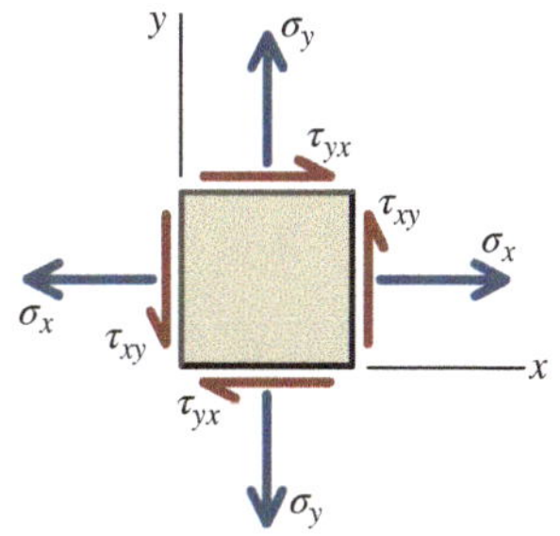

FIGURE 12.8

12.5 Generating the Stress Element

Sections 12.6 through 12.11 of this chapter discuss stress transformations, which are methods used to determine

(a) normal and shear stresses acting on any specific plane passing through a point of interest, and

(b) maximum normal and shear stresses acting at any possible orientation at a point of interest.

In discussing these methods, it is convenient to represent the state of stress at any particular point in a solid body by a stress element, such as that shown in Figure 12.8. Still, while the stress element is a convenient representation, it may be difficult at first for the student to connect the concept of the stress element to the topics presented in the previous chapters, such as the normal stresses produced by axial loads or bending moments, or the shear stresses produced by torsion or transverse shear in beams. Before proceeding to the methods used for stress transformations, it is helpful to consider how the analyst determines the stresses that appear on a stress element. This section focuses on solid components in which several internal loads or moments act simultaneously on a member's cross section. The method of superposition will be used to combine the various stresses acting at a particular point, and the results will be summarized on a stress element.

The analysis of the stresses produced by multiple internal loads or moments that act simultaneously on a member's cross section is usually referred to as **combined loadings**. In Chapter 15, combined loadings will be examined more completely. For instance, structures with multiple external loads will be considered, along with solid components that have three-dimensional geometry and loadings. Stress transformations will also be incorporated into the analysis in that discussion. The intent of this section is simply to introduce the reader to the process of evaluating the state of stress at a specific point. The process of putting together the stress element is demonstrated with the use of geometrically simple components and basic loadings.

EXAMPLE 12.1

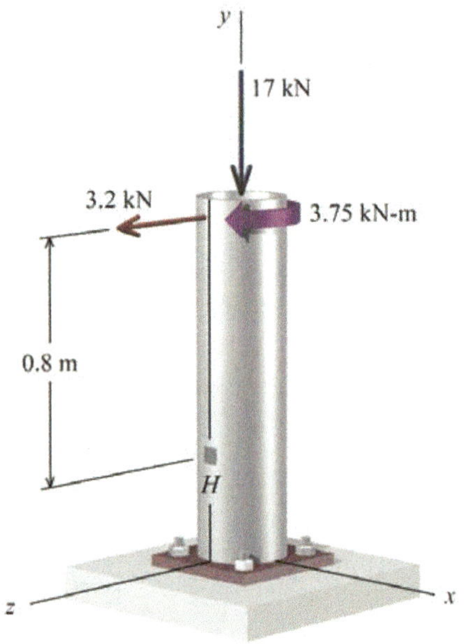

A vertical pipe column with an outside diameter $D = 114$ mm and an inside diameter $d = 102$ mm supports the loads shown. Determine the normal and shear stresses acting at point H, and show these stresses on a stress element.

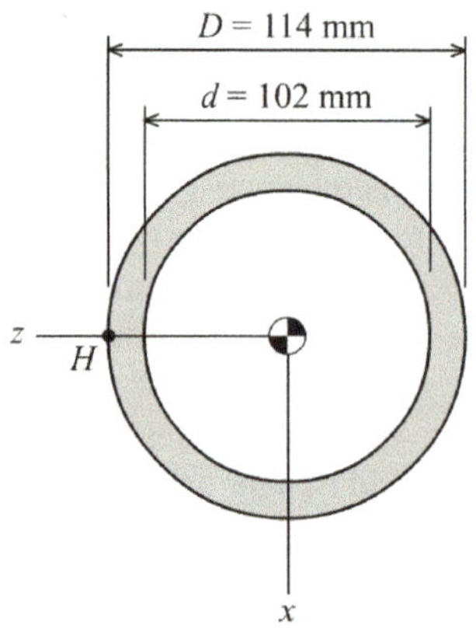

Column cross-sectional dimensions.

Plan the Solution

The cross-sectional properties will be computed for the pipe column. Each of the applied loads will be considered in turn. The normal and/or shear stresses created by each at point H will be computed. Both the magnitude and direction of the stress must be evaluated and shown on the proper face of a stress element. By the principle of superposition, the stresses will be combined appropriately so that the state of stress at point H is summarized succinctly by the stress element.

SOLUTION

Section Properties

The outside diameter of the pipe is $D = 114$ mm, and the inside diameter is $d = 102$ mm. The area, the moment of inertia, and the polar moment of inertia for the cross section are, respectively, as follows:

$$A = \frac{\pi}{4}[D^2 - d^2] = \frac{\pi}{4}[(114\text{ mm})^2 - (102\text{ mm})^2] = 2{,}035.752\text{ mm}^2$$

$$I = \frac{\pi}{64}[D^4 - d^4] = \frac{\pi}{64}[(114\text{ mm})^4 - (102\text{ mm})^4] = 2{,}977{,}287\text{ mm}^4$$

$$J = \frac{\pi}{32}[D^4 - d^4] = \frac{\pi}{32}[(114\text{ mm})^4 - (102\text{ mm})^4] = 5{,}954{,}575\text{ mm}^4$$

Stresses at *H*

The forces and moments acting at the section of interest will be evaluated sequentially to determine the type, magnitude, and direction of any stresses created at *H*.

The 17 kN axial force creates compressive normal stress, which acts in the *y* direction:

$$\sigma_y = \frac{F_y}{A} = \frac{17{,}000\ \text{N}}{2{,}035.752\ \text{mm}^2} = 8.351\ \text{MPa (C)}$$

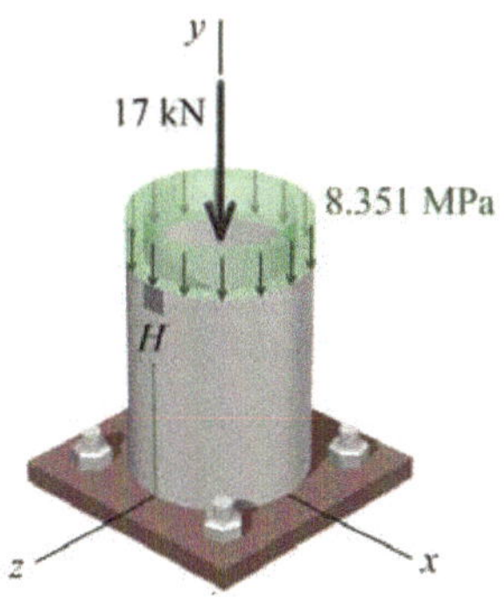

The 3.2 kN force acting in the positive *z* direction creates transverse shear stress (i.e., $\tau = VQ/It$) throughout the cross section of the pipe. However, the magnitude of the transverse shear stress is zero at point *H*.

The same 3.2 kN force acting in the positive *z* direction also creates a bending moment at the section where *H* is located. The magnitude of the bending moment is

$$M_x = (3.2\ \text{kN})(0.8\ \text{m}) = 2.56\ \text{kN}\cdot\text{m}$$

By inspection, we observe that this bending moment about the *x* axis creates compressive normal stress on the horizontal faces of the stress element at *H*:

$$\begin{aligned}\sigma_y &= \frac{M_x c}{I_x}\\ &= \frac{(2.56\ \text{kN}\cdot\text{m})(57\ \text{mm})(1{,}000\ \text{mm/m})(1{,}000\ \text{N/kN})}{2{,}977{,}287\ \text{mm}^4}\\ &= 49.011\ \text{MPa (C)}\end{aligned}$$

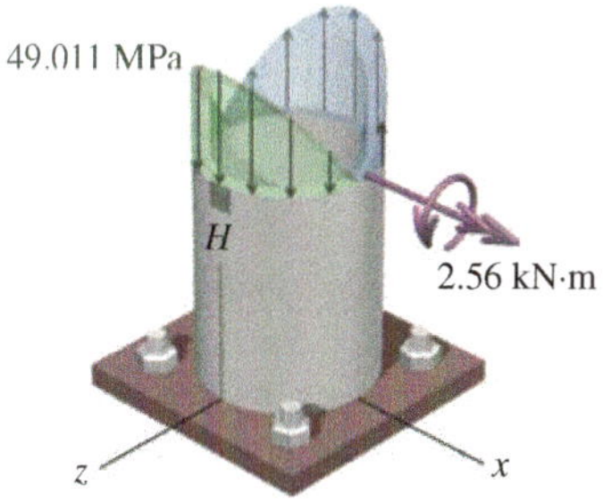

The 3.75 kN·m torque acting about the *y* axis creates shear stress at *H*. The magnitude of this shear stress can be calculated from the elastic torsion formula:

$$\begin{aligned}\tau &= \frac{Tc}{J}\\ &= \frac{(3.75\ \text{kN}\cdot\text{m})(57\ \text{mm})(1{,}000\ \text{mm/m})(1{,}000\ \text{N/kN})}{5{,}954{,}575\ \text{mm}^4}\\ &= 35.897\ \text{MPa}\end{aligned}$$

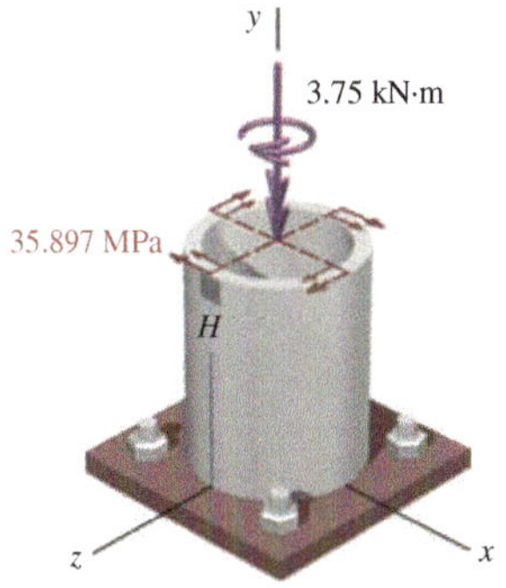

Combined Stresses at *H*

The normal and shear stresses acting at point *H* can be summarized on a stress element. Note that, at point *H*, the torsion shear stress acts in the −*x* direction on the +*y* face of the stress element. After the proper shear stress direction has been established on one face, the shear stress directions on the other three faces are known.

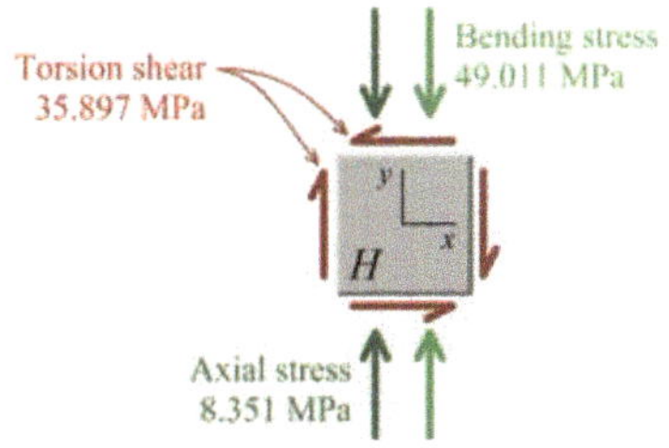

Multiple stresses acting at *H*.

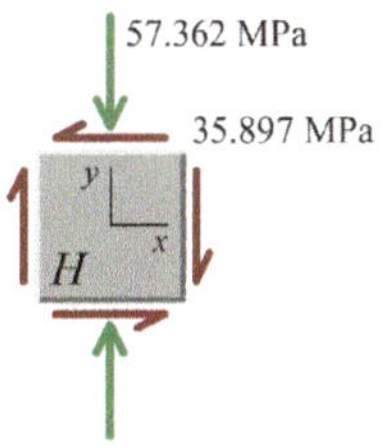

Summary of stresses acting at *H*.

PROBLEMS

P12.1 A compound shaft consists of segment (1), which has a diameter of 1.25 in., and segment (2), which has a diameter of 1.75 in. The shaft is subjected to a tensile axial load $P = 9$ kips and torques $T_A = 7$ kip·in. and $T_B = 28$ kip·in., which act in the directions shown in Figure P12.1. Determine the normal and shear stresses at (a) point H and (b) point K. For each point, show the stresses on a stress element.

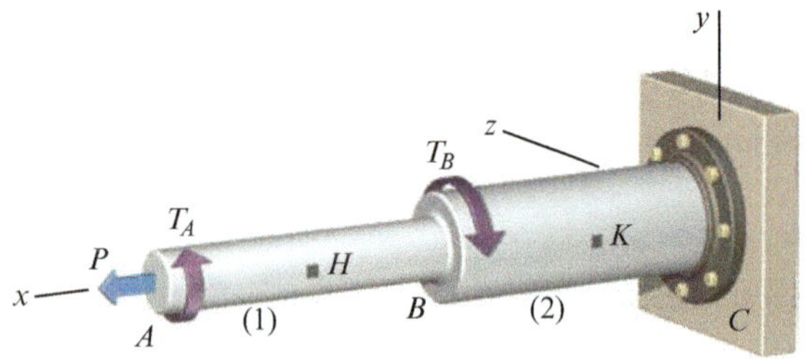

FIGURE P12.1

P12.2 The flanged member shown in Figure P12.2*a* is subjected to an internal axial force $P = 6{,}300$ lb, an internal shear force $V = 5{,}500$ lb, and an internal bending moment $M = 77{,}000$ lb·ft, acting in the directions shown. The dimensions of the cross section are $b_f = 10.00$ in., $t_f = 0.68$ in., $d = 12.00$ in., and $t_w = 0.32$ in. (Figure 12.2*b*). Determine the normal and shear stresses at point H, where $a = 2.50$ in.

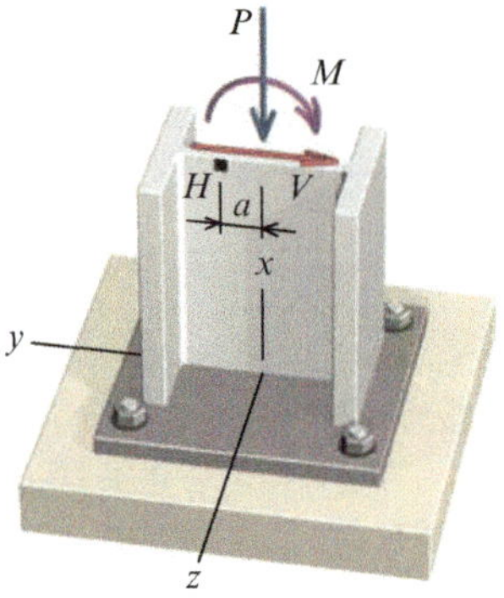

FIGURE P12.2*a*

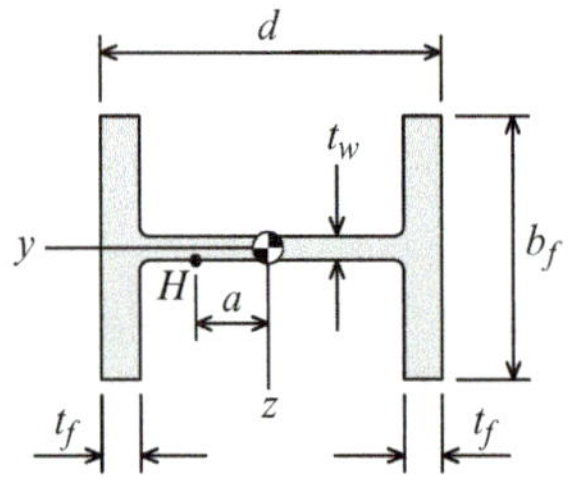

FIGURE P12.2*b*

P12.3 Figure 12.3*a* shows a doubly-symmetric beam whose cross section is shown in Figure P12.3*b*. The beam has been proposed for a short footbridge. The cross section will consist of two identical steel pipes that are securely welded to a steel web plate that has a thickness $t = 9.5$ mm. Each pipe has an outside diameter $d = 70$ mm and a wall thickness of 5 mm. The distance between the centers of the two pipes is $h = 370$ mm. Internal forces $P = 13$ kN, $V = 25$ kN, and $M = 9$ kN·m act in the directions shown in Figure P12.3*a*. Determine the stresses acting on horizontal and vertical planes

(a) at point H, which is located at a distance $y_H = 120$ mm above the z centroidal axis.
(b) at point K, which is located at a distance $y_K = 80$ mm below the z centroidal axis.

Show the stresses on a stress element for each point.

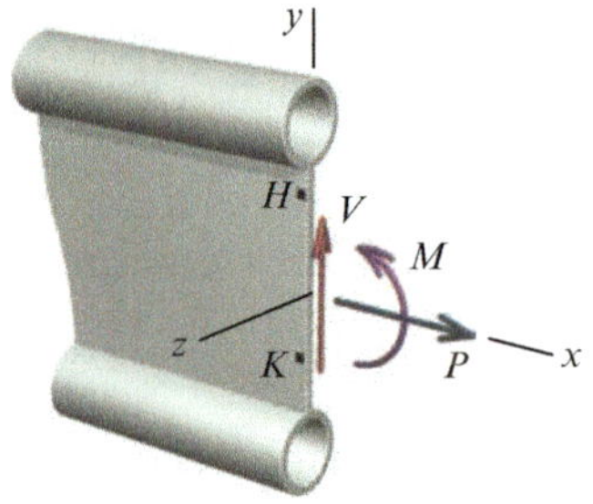

FIGURE P12.3*a*

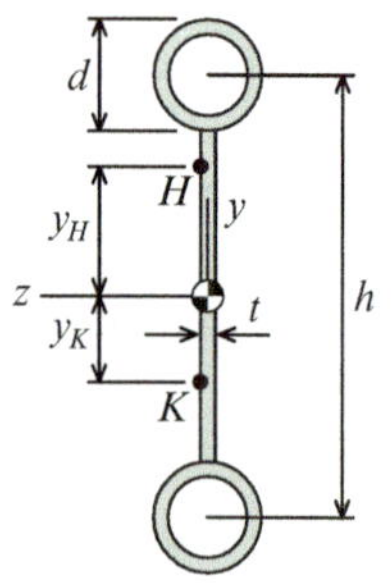

FIGURE P12.3*b*

P12.4 A hat-shaped flexural member is subjected to an internal axial force $P = 7{,}200$ N, an internal shear force $V = 6{,}000$ N, and an internal bending moment $M = 1{,}300$ N·m, acting as shown in Figure P12.4*a*. The dimensions of the cross section (Figure P12.4*b*) are $a = 20$ mm, $b = 100$ mm, $d = 55$ mm, and $t = 4$ mm. Determine the stresses acting on horizontal and vertical planes

(a) at point H, which is located at a distance $y_H = 20$ mm above the z centroidal axis.
(b) at point K, which is located at a distance $y_K = 12$ mm below the z centroidal axis.

Show the stresses on a stress element for each point.

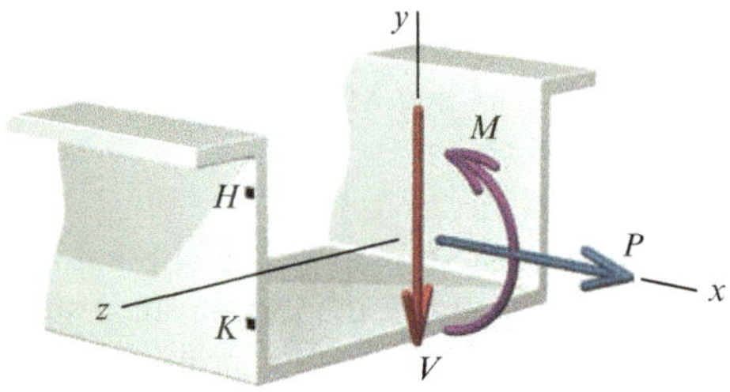

FIGURE P12.4*a*

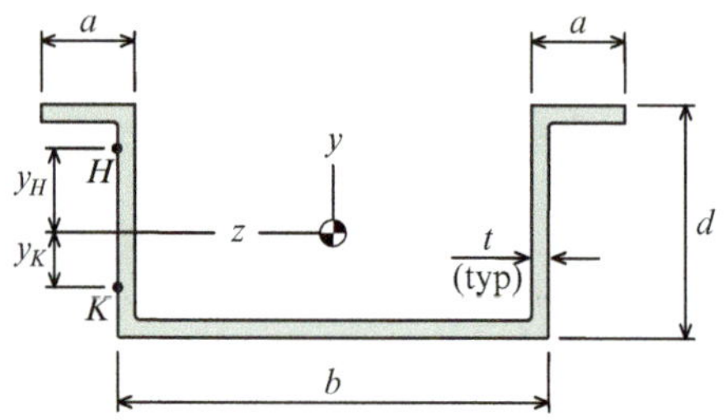

FIGURE P12.4*b*

P12.5 An extruded polymer flexural member is subjected to an internal axial force $P = 580$ lb, an internal shear force $V = 420$ lb, and an internal bending moment $M = 6{,}400$ lb·in., acting in the directions shown in Figure P12.5*a*. The cross-sectional dimensions (Figure P12.5*b*) of the extrusion are $b_1 = 2.0$ in., $t_1 = 0.6$ in., $b_2 = 4.0$ in., $t_2 = 0.4$ in., $d = 4.5$ in., and $t_w = 0.4$ in. Determine the normal and shear stresses acting on horizontal and vertical planes

(a) at point *H*, which is located at a distance $y_H = 0.8$ in. above the *z* centroidal axis.
(b) at point *K*, which is located at a distance $y_K = 1.1$ in. below the *z* centroidal axis.

Show the stresses on a stress element for each point.

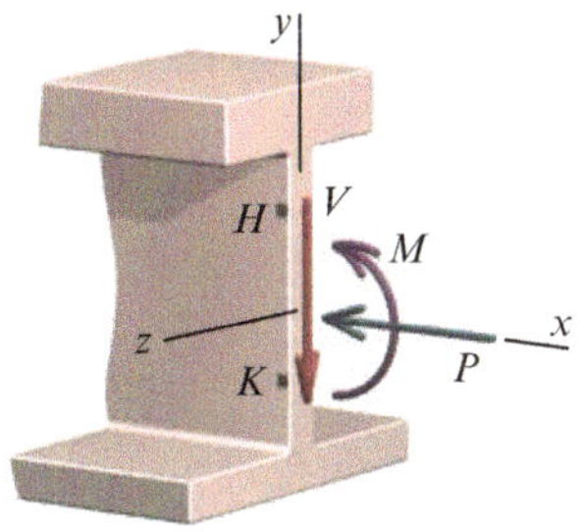

FIGURE P12.5*a*

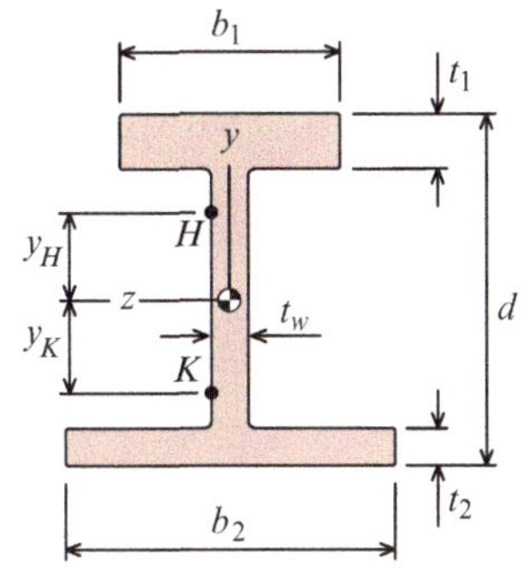

FIGURE P12.5*b*

P12.6 The machine component shown in Figure P12.6 is subjected to a load $P = 450$ lb that acts at an angle $\beta = 25°$. The dimensions of the component are $a = 4.0$ in., $b = 0.625$ in., $c = 2.0$ in., and $d = 2.5$ in. Determine the normal and shear stresses acting on horizontal and vertical planes

(a) at point *H*, which is located at a distance $y_H = 0.5$ in. above the *z* centroidal axis.
(b) at point *K*, which is located at a distance $y_K = 0.75$ in. below the *z* centroidal axis.

Show the stresses on a stress element for each point

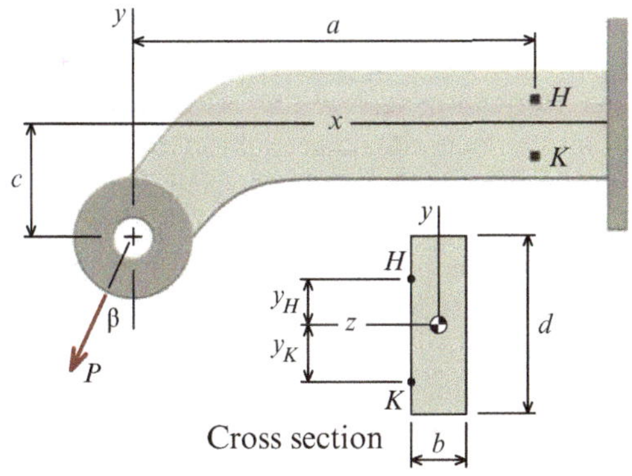

FIGURE P12.6

P12.7 A load $P = 900$ N exerted at an angle $\beta = 30°$ acts on the machine part shown in Figure P12.7*a*. The dimensions of the part are $a = 180$ mm, $b = 60$ mm, $c = 100$ mm, $d = 40$ mm, and $e = 10$ mm. The part has a uniform thickness of 8 mm (i.e., it is 8 mm thick in the *z* direction). Determine the normal and shear stresses acting on horizontal and vertical planes at points *H* and *K*, which are shown in detail in Figure P12.7*b*. For each point, show these stresses on a stress element.

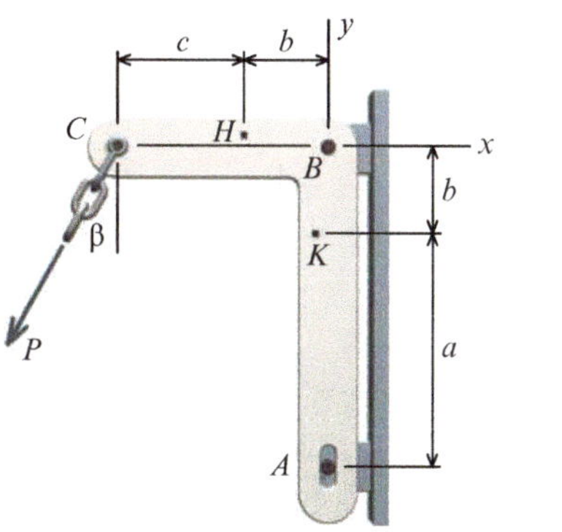

FIGURE P12.7*a*

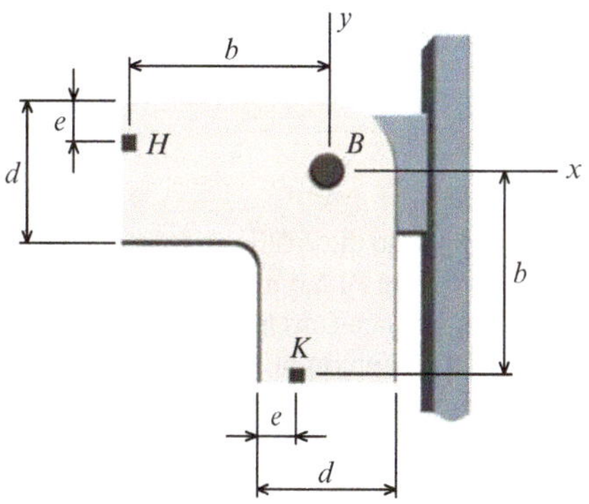

FIGURE P12.7*b* Detail at pin *B*.

P12.8 A hollow aluminum pipe is subjected to a vertical force $P_x = 1{,}500$ N, a horizontal force $P_y = 2{,}400$ N, and a concentrated torque $T = 80$ N·m, acting as shown in Figure P12.8. The outside and inside diameters of the pipe are 64 mm and 56 mm, respectively. Assume that $a = 90$ mm. Determine the normal and shear stresses on horizontal and vertical planes at (a) point *H* and (b) point *K*. For each point, show these stresses on a stress element.

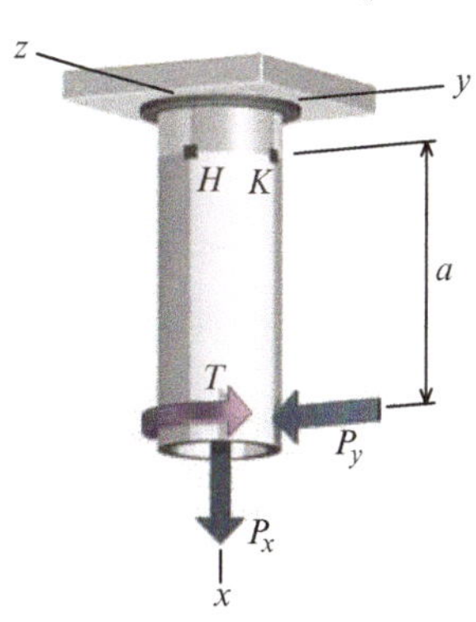

FIGURE P12.8

P12.9 A steel pipe with an outside diameter of 10.75 in. and an inside diameter of 9.50 in. is subjected to forces $P_x = 16$ kips, $P_z = 5$ kips, and $T = 23$ kip · ft, acting as shown in Figure P12.9. Assume that $a = 4$ ft. Determine the normal and shear stresses on horizontal and vertical planes at (a) point H and (b) point K. For each point, show these stresses on a stress element.

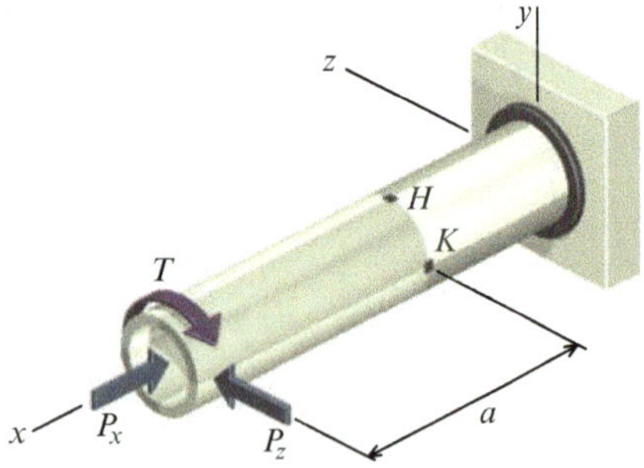

FIGURE P12.9

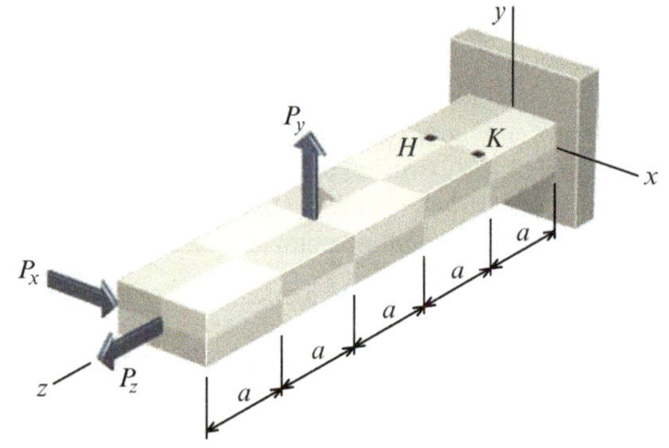

FIGURE P12.10*a*

P12.10 Concentrated loads $P_x = 3{,}300$ N, $P_y = 2{,}100$ N, and $P_z = 2{,}800$ N are applied to the cantilever beam in the locations and directions shown in Figure P12.10*a*. The beam cross section shown in Figure P12.10*b* has dimensions $b = 100$ mm and $d = 40$ mm. Using the value $a = 75$ mm, determine the normal and shear stresses at (a) point H and (b) point K. Show these stresses on a stress element.

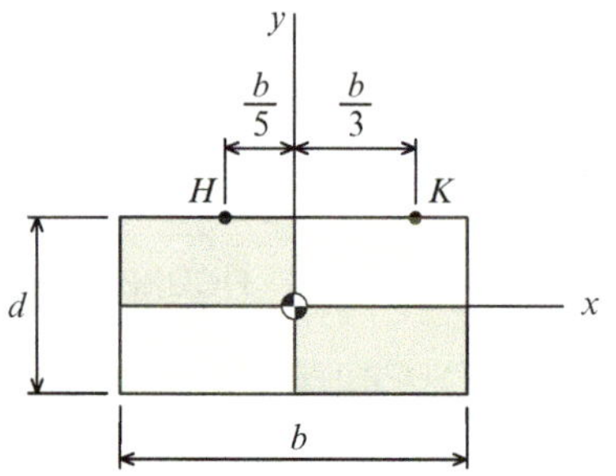

FIGURE P12.10*b*

12.6 Equilibrium Method for Plane Stress Transformations

As discussed in Sections 1.5 and 12.2, stress is not simply a vector quantity: Stress is dependent on the orientation of the plane surface upon which the stress acts. As shown in Section 12.2, the state of stress at a point in a material object subjected to plane stress is completely defined by three stress components—σ_x, σ_y, and τ_{xy}—acting on two orthogonal planes x and y defined with respect to x–y coordinate axes. The *same state of stress* at a point can be represented by different stress components—σ_n, σ_t, and τ_{nt}—acting on a different pair of orthogonal planes n and t, which are rotated with respect to the x and y planes. In other words, there is only one unique state of stress at a point, but the state of stress can have different representations, depending on the orientation of the axes used. The process of changing stresses from one set of coordinate axes to another is termed **stress transformation**.

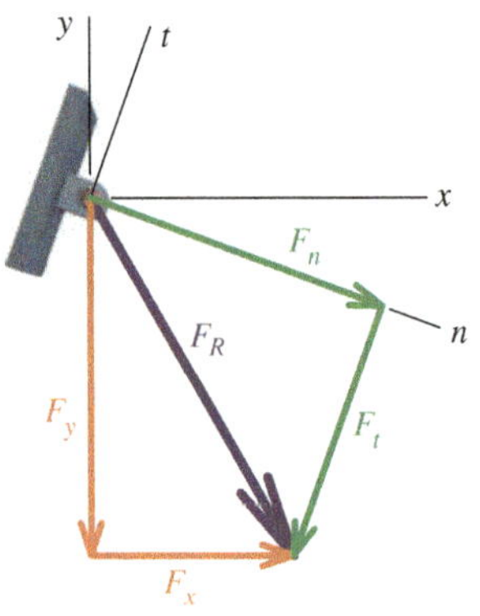

FIGURE 12.9

In some ways, the concept of stress transformation is analogous to vector addition. Suppose that there are two force components F_x and F_y, which are oriented parallel to the x and y axes, respectively (Figure 12.9). The sum of these two vectors is the resultant force F_R. Two different force components F_n and F_t, defined in an n–t coordinate system, could also be added together to produce the same resultant force F_R. In other words, the resultant force F_R could be expressed either as the sum of components F_x and F_y in an x–y coordinate system or as the sum of components F_n and F_t in an n–t coordinate system. The components are different in the two coordinate systems, but both sets of components represent the same resultant force.

In this vector addition illustration, the transformation of forces from one coordinate system (i.e., the x–y coordinate system) to a rotated n–t coordinate system must take into account the magnitude and direction of each force component. The transformation of stress components, however, is more complicated than vector addition. In considering stresses, the transformation must account for not only the magnitude and direction of each stress component, but also the orientation of the area upon which the stress component acts.

A more general approach to stress transformations will be developed in Section 12.7; however, at the outset, it is instructive to use equilibrium considerations to determine normal and shear stresses that act on an arbitrary plane. The solution method used here is similar to that developed in Section 1.5 for stresses on inclined sections of axial members. Example 12.2 illustrates this method for plane stress conditions.

EXAMPLE 12.2

At a given point in a machine component, the following stresses were determined: 150 MPa (T) on a vertical plane, 30 MPa (C) on a horizontal plane, and zero shear stress. Determine the stresses at this point on a plane having a slope of 3 vertical to 4 horizontal.

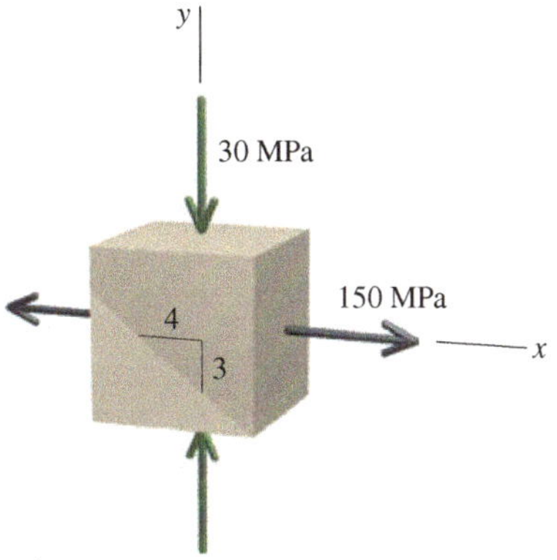

Plan the Solution

A free-body diagram of a wedge-shaped portion of the stress element will be investigated. Forces acting on vertical and horizontal planes will be derived from the given stresses and the areas of the wedge faces. Since the wedge-shaped portion of the stress element must satisfy equilibrium, the normal and shear stresses acting on the inclined surface can be determined.

SOLUTION

Sketch a free-body diagram of the wedge-shaped portion of the stress element. From the 3:4 slope of the inclined surface, the angle between the vertical face and the inclined surface is 53.13°. The area of the inclined surface will be designated dA. Accordingly, the area of the vertical face can be expressed as $dA \cos 53.13°$, and the area of the horizontal face can be expressed as $dA \sin 53.13°$. The *forces* acting on these areas are found from the product of the given stresses and the areas.

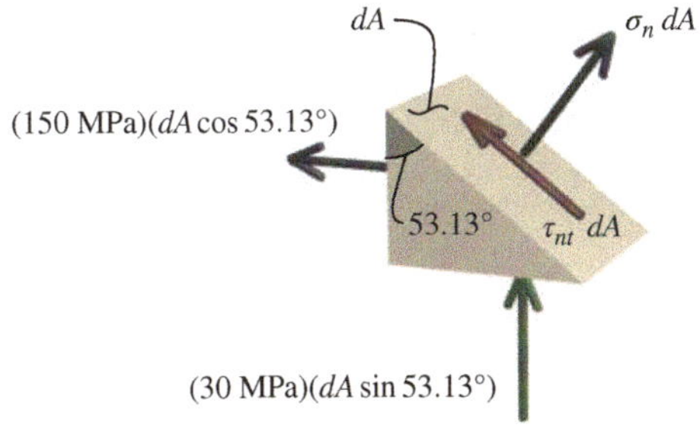

The forces acting on the vertical and horizontal faces of the wedge can be resolved into components acting in the n direction (i.e., the direction *normal* to the inclined plane) and the t direction (i.e., the direction parallel, or *tangential*, to the inclined plane).

From these force components, the sum of forces acting in the direction perpendicular to the inclined plane is

$$\Sigma F_n = \sigma_n \, dA + (30 \text{ MPa})(dA \sin 53.13°) \sin 53.13°$$
$$-(150 \text{ MPa})(dA \cos 53.13°) \cos 53.13° = 0$$

Notice that the area dA appears in each term; consequently, it will cancel out of the equation. From this equilibrium equation, the normal stress acting in the n direction is found to be

$$\sigma_n = 34.80 \text{ MPa (T)} \qquad \textbf{Ans.}$$

When forces are summed in the t direction, the equilibrium equation is

$$\Sigma F_t = \tau_{nt} \, dA + (30 \text{ MPa})(dA \sin 53.13°) \cos 53.13°$$
$$+ (150 \text{ MPa})(dA \cos 53.13°) \sin 53.13° = 0$$

Therefore, the shear stress on the n face of the wedge acting in the t direction is

$$\tau_{nt} = -86.4 \text{ MPa} \qquad \textbf{Ans.}$$

The negative sign indicates that the shear stress really acts in the negative t direction on the positive n face. Note that the normal stress should be designated as *tension* or *compression*. The presence of shear stresses on the horizontal and vertical planes, had there been any, would merely have required two more forces on the free-body diagram: one parallel to the vertical face and one parallel to the horizontal face. Note, however, that the magnitude of the shear stresses (not the forces) must be the same on any two orthogonal planes.

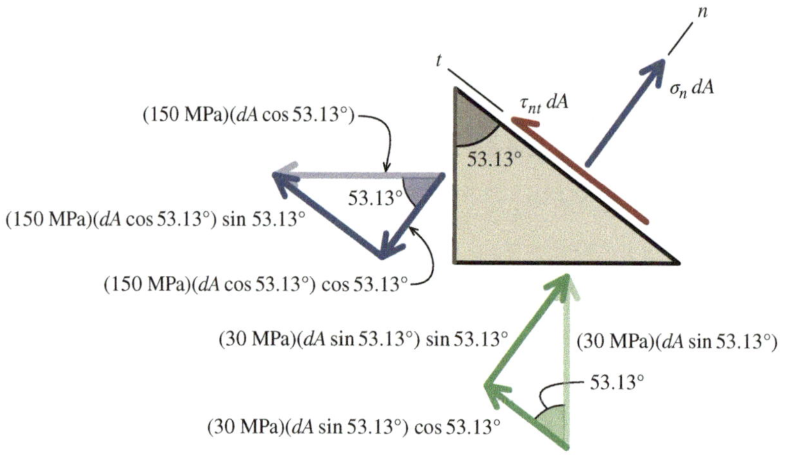

PROBLEMS

P12.11–P12.14 The stresses shown in Figures P12.11–P12.14 act at a point in a stressed body. Using the equilibrium equation approach, determine the normal and shear stresses at this point on the inclined plane shown.

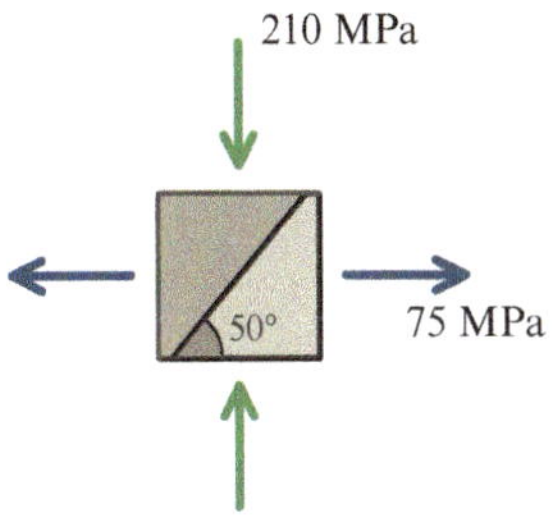

FIGURE P12.11

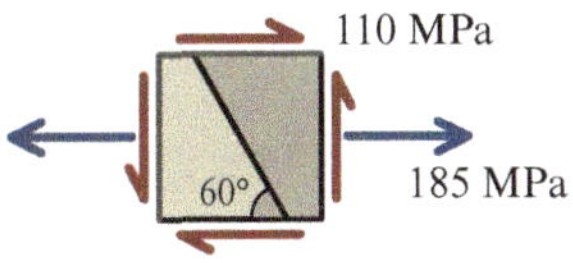

FIGURE P12.12

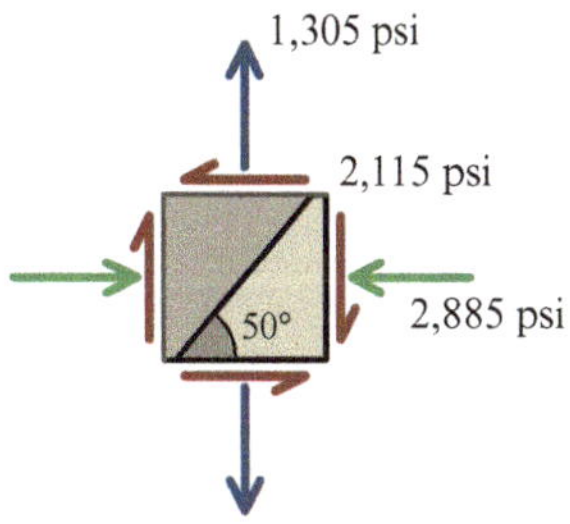

FIGURE P12.13

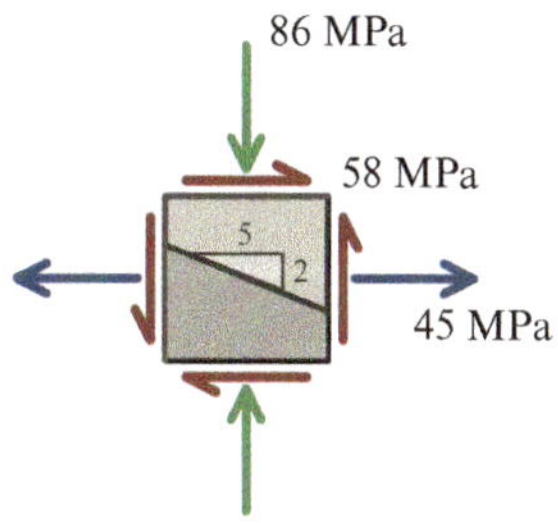

FIGURE P12.14

12.7 General Equations of Plane Stress Transformation

For a successful design, an engineer must be able to determine critical stresses at any point of interest in a material object. By the mechanics of materials theory developed for axial members, torsion members, and beams, normal and shear stresses at a point in a material object can be computed with reference to a particular coordinate system, such as an x–y coordinate system. Such a coordinate system, however, has no inherent significance with regard to the material used in a structural member. Failure of the material will occur in response to the largest stresses that are developed in the object, regardless of the orientation at which those critical stresses are acting. For instance, a designer has no assurance that a horizontal bending stress computed at a point in the web of a wide-flange beam will be the largest normal stress possible at the point. To find the critical stresses at a point in a material object, methods must be developed so that stresses acting at all possible orientations can be investigated.

MecMovies 12.1 presents an animated discovery example that illustrates the need for stress transformations.

Consider a state of stress represented by a plane stress element subjected to stresses σ_x, σ_y, and $\tau_{xy} = \tau_{yx}$, as shown in Figure 12.10*a*. *Keep in mind that the stress element is simply a convenient graphical symbol used to represent the state of stress at a specific point of interest in an object (such as a shaft or a beam).* To derive equations applicable to any orientation, we begin by defining a plane surface A–A oriented at some angle θ with respect to a reference axis x. The normal to surface A–A is termed the n axis. The axis parallel to surface A–A is termed the t axis. The z axis extends out of the plane of the stress element. Both the x–y–z and the n–t–z axes are arranged as right-hand coordinate systems. Given the σ_x, σ_y, and $\tau_{xy} = \tau_{yx}$ stresses acting on the x and y faces of the stress element, we will determine the normal and shear stress acting on surface A–A, known as the n face of the stress element. This process of changing stresses from one set of coordinate axes (i.e., x–y–z) to another set of axes (i.e., n–t–z) is termed **stress transformation**.

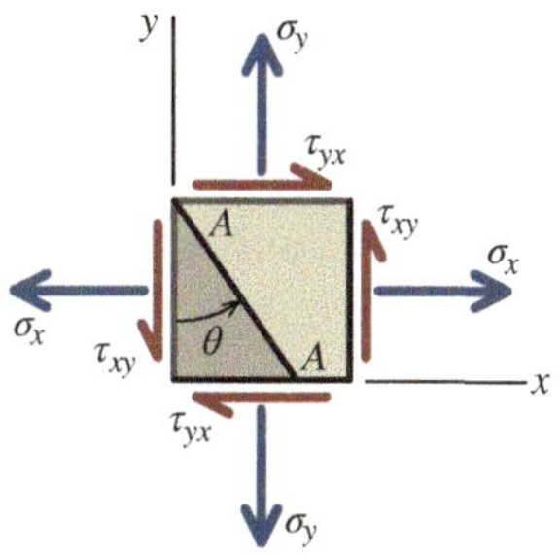

FIGURE 12.10*a*

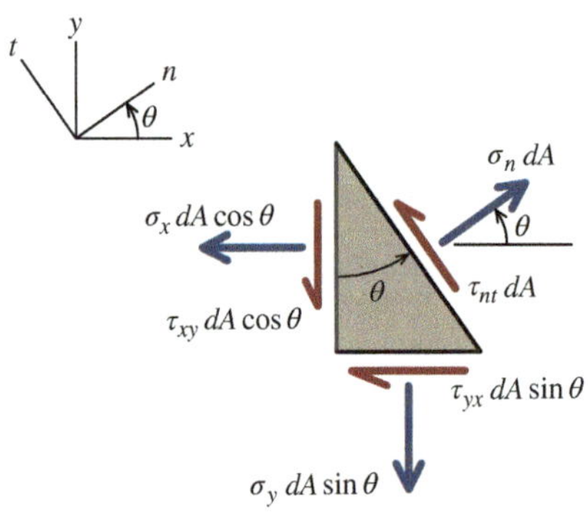

FIGURE 12.10*b*

Figure 12.10*b* is a free-body diagram of a wedge-shaped element in which the areas of the faces are dA for the inclined face (plane A–A), $dA\cos\theta$ for the vertical face (i.e., the x face), and $dA\sin\theta$ for the horizontal face (i.e., the y face). The equilibrium equation for the sum of forces in the n direction gives

$$\Sigma F_n = \sigma_n dA - \tau_{yx}(dA\,\sin\theta)\cos\theta - \tau_{xy}(dA\,\cos\theta)\sin\theta$$
$$- \sigma_x(dA\,\cos\theta)\cos\theta - \sigma_y(dA\,\sin\theta)\sin\theta = 0$$

Since $\tau_{yx} = \tau_{xy}$, this equation can be simplified to give the following expression for the normal stress acting on the n face of the wedge element:

$$\boxed{\sigma_n = \sigma_x\cos^2\theta + \sigma_y\sin^2\theta + 2\tau_{xy}\sin\theta\cos\theta} \tag{12.3}$$

From the free-body diagram in Figure 12.10*b*, the equilibrium equation for the sum of forces in the t direction gives

$$\Sigma F_t = \tau_{nt}\,dA - \tau_{xy}(dA\cos\theta)\cos\theta + \tau_{yx}(dA\sin\theta)\sin\theta$$
$$+ \sigma_x(dA\cos\theta)\sin\theta - \sigma_y(dA\sin\theta)\cos\theta = 0$$

Again from $\tau_{yx} = \tau_{xy}$, this equation can be simplified to give the following expression for the shear stress acting in the t direction on the n face of the wedge element:

$$\boxed{\tau_{nt} = -(\sigma_x - \sigma_y)\sin\theta\cos\theta + \tau_{xy}(\cos^2\theta - \sin^2\theta)} \tag{12.4}$$

The equations just derived for the normal stress and the shear stress can be written in an equivalent form by substituting the following double-angle identities from trigonometry:

$$\cos^2\theta = \frac{1}{2}(1 + \cos 2\theta)$$
$$\sin^2\theta = \frac{1}{2}(1 - \cos 2\theta)$$
$$2\sin\theta\cos\theta = \sin 2\theta$$

Using these double-angle identities, we can write Equation (12.3) as

$$\boxed{\sigma_n = \frac{\sigma_x + \sigma_y}{2} + \frac{\sigma_x - \sigma_y}{2}\cos 2\theta + \tau_{xy}\sin 2\theta} \tag{12.5}$$

MecMovies 12.6 presents an animated derivation of the plane stress transformation equations.

and Equation (12.4) as

$$\boxed{\tau_{nt} = -\frac{\sigma_x - \sigma_y}{2}\sin 2\theta + \tau_{xy}\cos 2\theta} \tag{12.6}$$

Equations (12.3), (12.4), (12.5), and (12.6) are called the **plane stress transformation equations**. They provide a means for determining normal and shear stresses on any plane whose outward normal is

(a) perpendicular to the z axis (i.e., the out-of-plane axis), and
(b) oriented at an angle θ with respect to the reference x axis.

Since the transformation equations were derived solely from equilibrium considerations, they are applicable to stresses in any kind of material, whether it is linear or nonlinear, elastic or inelastic.

Stress Invariance

The normal stress acting on the n face of the stress element shown in Figure 12.11 can be determined from Equation (12.5). The normal stress acting on the t face can also be obtained from Equation (12.5) by substituting $\theta + 90°$ in place of θ, giving the following equation:

$$\sigma_t = \frac{\sigma_x + \sigma_y}{2} - \frac{\sigma_x - \sigma_y}{2}\cos 2\theta - \tau_{xy}\sin 2\theta \tag{12.7}$$

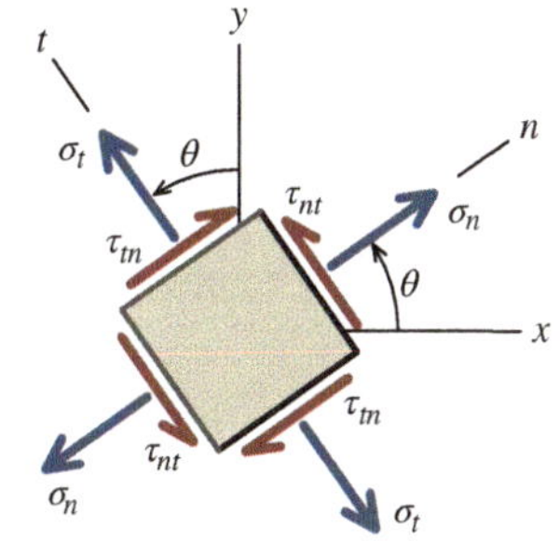

FIGURE 12.11

If the expressions for σ_n and σ_t [Equations (12.5) and (12.7)] are added, the following relationship is obtained:

$$\sigma_n + \sigma_t = \sigma_x + \sigma_y \tag{12.8}$$

Equation (12.8) shows that the sum of the normal stresses acting on any two orthogonal faces of a plane stress element is a constant value, independent of the angle θ. This mathematical characteristic of stress is termed **stress invariance**.

Stress is expressed with reference to specific coordinate systems. The stress transformation equations show that the n–t components of stress are different from the x–y components, even though both are representations of the same stress state. However, certain functions of stress components are not dependent on the orientation of the coordinate system. These functions, called **stress invariants**, have the same value regardless of which coordinate system is used. Two invariants, denoted I_1 and I_2, exist for plane stress:

$$\begin{aligned} I_1 &= \sigma_x + \sigma_y \qquad &(\text{or } I_1 = \sigma_n + \sigma_t) \\ I_2 &= \sigma_x\sigma_y - \tau_{xy}^2 \qquad &(\text{or } I_2 = \sigma_n\sigma_t - \tau_{nt}^2) \end{aligned} \tag{12.9}$$

MecMovies 12.5 presents an animated discussion of terminology used in stress transformations.

T → σ = +
C → σ = −

Sign Conventions

The sign conventions used in the development of the stress transformation equations must be rigorously followed. The sign conventions can be summarized as follows:

1. Tensile normal stresses are positive; compressive normal stresses are negative. All of the normal stresses shown in Figure 12.11 are positive.
2. A shear stress is positive if it
   - acts in the positive coordinate direction on a positive face of the stress element or
   - acts in the negative coordinate direction on a negative face of the stress element.

   All of the shear stresses shown in Figure 12.11 are positive. Shear stresses pointing in opposite directions are negative.

   An easy way to remember the shear stress sign convention is to use the directions associated with the two subscripts. The first subscript indicates the face of the stress element on which the shear stress acts. It will be either a positive face (plus) or a negative face (minus). The second subscript indicates the direction in which the stress acts, and it will be either a positive direction (plus) or a negative direction (minus).
   - A positive shear stress has subscripts that are either plus–plus or minus–minus.
   - A negative shear stress has subscripts that are either plus–minus or minus–plus.
3. Angles measured counterclockwise from the reference x axis are positive. Conversely, angles measured clockwise from the reference x axis are negative.
4. The n–t–z axes have the same order as the x–y–z axes. Both sets of axes form a right-hand coordinate system.

MecMovies 12.2 presents an interactive activity that focuses on the proper determination of the angle θ.

MecMovies 12.3 presents a game that tests understanding of the proper sign conventions and their use in the stress transformation equations.

EXAMPLE 12.3

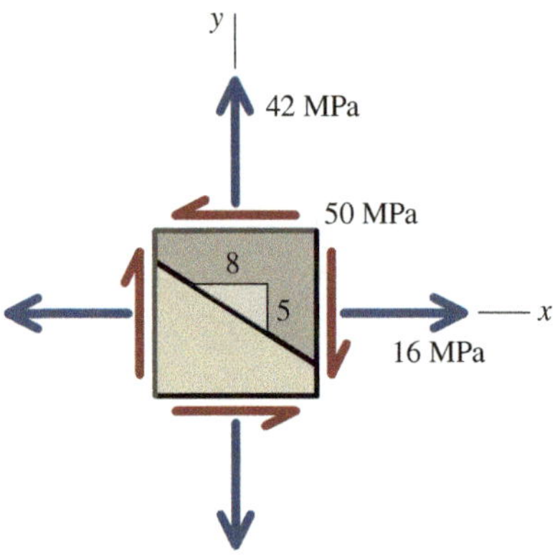

At a point on a structural member subjected to plane stress, normal and shear stresses exist on horizontal and vertical planes through the point as shown. Use the stress transformation equations to determine the normal and shear stress on the indicated plane surface.

Plan the Solution

Problems of this type are straightforward; however, the sign conventions used in deriving the stress transformation equations must be rigorously followed for a successful result. Particular attention should be given to identifying the proper value of θ, which is required to designate the inclination of the plane surface.

SOLUTION

The normal stress acting on the x face creates tension in the element; therefore, it is considered a positive normal stress ($\sigma_x = 16$ MPa) in the stress transformation equations. Likewise, the normal stress on the y face has a positive value of $\sigma_y =$ 42 MPa.

The 50 MPa shear stress on the positive x face acts in the negative y direction; therefore, this shear stress is considered negative when used in the stress transformation equations ($\tau_{xy} =$ −50 MPa). Note that the shear stress on the horizontal face is also negative. On the positive y face, the shear stress acts in the negative x direction; hence, $\tau_{yx} = -50 \text{ MPa} = \tau_{xy}$.

In this example, normal and shear stresses are to be calculated for a plane surface that has a slope of −5 (vertical) to 8 (horizontal). This slope information must be converted to the proper value of θ for use in the stress transformation equations.

A convenient way to determine θ is to find the angle between a vertical plane and the inclined surface. This angle will always be the same as the angle between the x axis and the n axis. For the surface specified here, the magnitude of the angle between a vertical plane and the inclined surface is

$$\tan\theta = \frac{8}{5} \qquad \therefore\ \theta = 58°$$

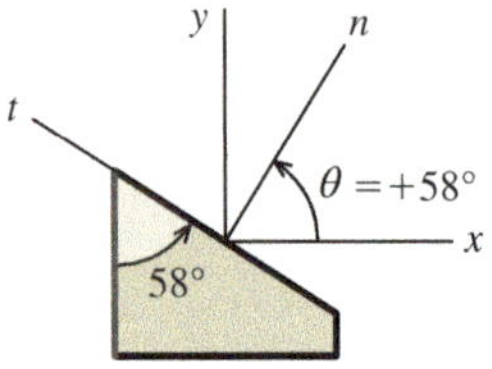

Notice that the preceding calculation determines only the magnitude of the angle. The proper sign for θ is determined by inspection. If the angle *from* the vertical plane *to* the inclined plane turns in a counterclockwise direction, the value of θ is positive. Therefore, $\theta = 58°$ for this example.

With the proper values for σ_x, σ_y, τ_{xy}, and θ now established, the normal and shear stresses acting on the inclined surface can be calculated. The normal stress in the n direction is found from Equation (12.3):

$$\begin{aligned}\sigma_n &= \sigma_x \cos^2\theta + \sigma_y \sin^2\theta + 2\tau_{xy} \sin\theta\cos\theta \\ &= (16\text{ MPa})\cos^2 58° + (42\text{ MPa})\sin^2 58° + 2(-50\text{ MPa})\sin 58°\cos 58° \\ &= -10.24\text{ MPa}\end{aligned}$$

Note that Equation (12.5) could also be used to obtain the same result:

$$\begin{aligned}\sigma_n &= \frac{\sigma_x + \sigma_y}{2} + \frac{\sigma_x - \sigma_y}{2}\cos 2\theta + \tau_{xy}\sin 2\theta \\ &= \frac{(16\text{ MPa}) + (42\text{ MPa})}{2} + \frac{(16\text{ MPa}) - (42\text{ MPa})}{2}\cos 2(58°) + (-50\text{ MPa})\sin 2(58°) \\ &= -10.24\text{ MPa}\end{aligned}$$

The choice of either Equation (12.3) or Equation (12.5) to calculate the normal stress acting on the inclined plane is a matter of personal preference.

The shear stress τ_{nt} acting on the n face in the t direction can be computed from Equation (12.4):

$$\begin{aligned}\tau_{nt} &= -(\sigma_x - \sigma_y)\sin\theta\cos\theta + \tau_{xy}\,(\cos^2\theta - \sin^2\theta)\\ &= -[(16\text{ MPa}) - (42\text{ MPa})]\sin 58^\circ\cos 58^\circ + (-50\text{ MPa})[\cos^2 58^\circ - \sin^2 58^\circ]\\ &= 33.6\text{ MPa}\end{aligned}$$

Alternatively, Equation (12.6) may be used:

$$\begin{aligned}\tau_{nt} &= -\frac{\sigma_x - \sigma_y}{2}\sin 2\theta + \tau_{xy}\cos 2\theta\\ &= -\frac{(16\text{ MPa}) - (42\text{ MPa})}{2}\sin 2(58^\circ) + (-50\text{ MPa})\cos 2(58^\circ)\\ &= 33.6\text{ MPa}\end{aligned}$$

To complete the problem, the stresses acting on the inclined plane are shown in a sketch. Since σ_n is negative, the normal stress acting in the n direction is shown as a compressive stress. The positive value of τ_{nt} indicates that the stress arrow points in the positive t direction on the positive n face. The arrows are labeled with the stress magnitude (i.e., absolute value). The signs associated with the stresses are indicated by the directions of the arrows.

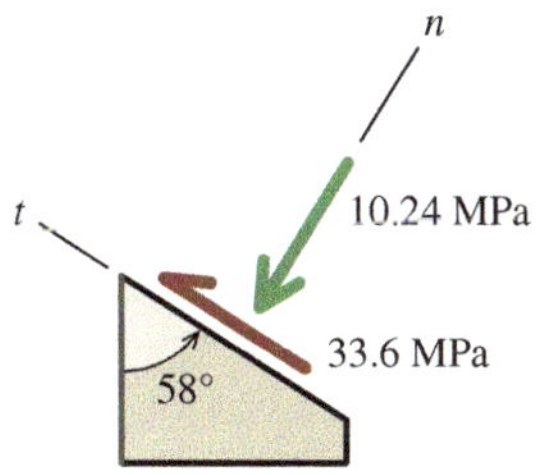

EXAMPLE 12.4

The stresses shown act at a point on the free surface of a machine component. Determine the normal stresses σ_x and σ_y and the shear stress τ_{xy} at the point.

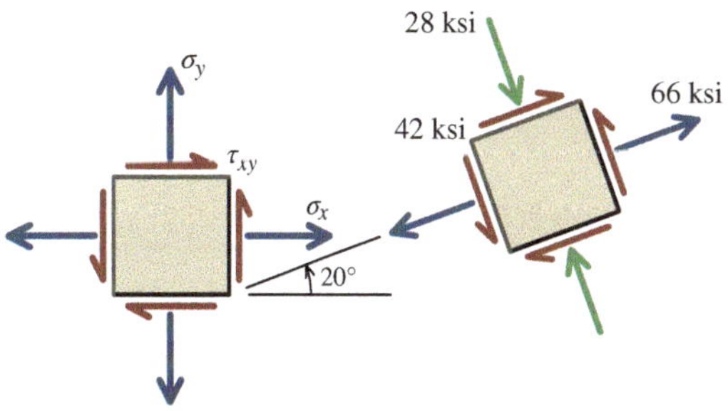

Plan the Solution

The stress transformation equations are written in terms of σ_x, σ_y, and τ_{xy}; however, the x and y directions do not necessarily have to be the horizontal and vertical directions, respectively. Any two orthogonal directions can be taken as x and y, as long as they define a right-hand coordinate system. To solve this problem, we will redefine the x and y axes, aligning them with the rotated element. The faces of the unrotated element will be redefined as the n and t faces.

SOLUTION

Redefine the x and y directions, aligning them with the rotated element. The axes of the unrotated element will be defined as the n and t directions.

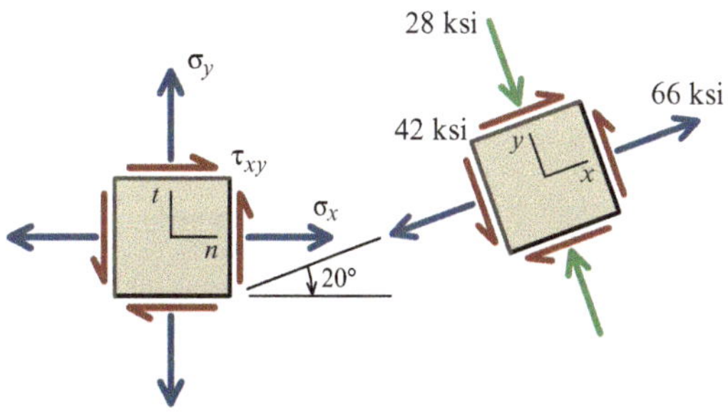

Accordingly, the stresses acting on the rotated element are now defined as

$$\begin{aligned}\sigma_x &= 66\text{ ksi}\\ \sigma_y &= -28\text{ ksi}\\ \tau_{xy} &= 42\text{ ksi}\end{aligned}$$

The angle θ from the redefined x axis to the n axis is 20° in a clockwise sense; therefore, $\theta = -20°$.

The normal stress on the vertical face of the unrotated element can be computed from Equation (12.3):

$$\begin{aligned}\sigma_n &= \sigma_x \cos^2\theta + \sigma_y \sin^2\theta + 2\tau_{xy}\sin\theta\cos\theta \\ &= (66\text{ ksi})\cos^2(-20°) + (-28\text{ ksi})\sin^2(-20°) + 2(42\text{ ksi})\sin(-20°)\cos(-20°) \\ &= 28.0\text{ ksi}\end{aligned}$$

The normal stress on the horizontal face of the unrotated element can be computed from Equation (12.3) if the angle θ is changed to $\theta = -20° + 90° = 70°$:

$$\begin{aligned}\sigma_t &= \sigma_x \cos^2\theta + \sigma_y \sin^2\theta + 2\tau_{xy}\sin\theta\cos\theta \\ &= (66\text{ ksi})\cos^2 70° + (-28\text{ ksi})\sin^2 70° + 2(42\text{ ksi})\sin 70°\cos 70° \\ &= 9.99\text{ ksi}\end{aligned}$$

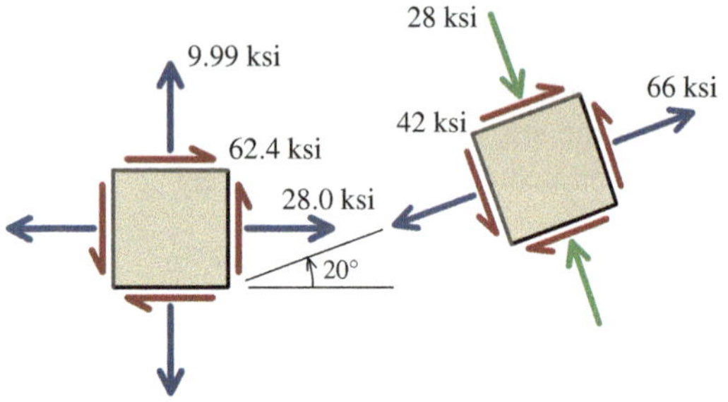

The shear stress on the unrotated element can be computed from Equation (12.4):

$$\begin{aligned}\tau_{nt} &= -(\sigma_x - \sigma_y)\sin\theta\cos\theta + \tau_{xy}(\cos^2\theta - \sin^2\theta) \\ &= -[(66\text{ ksi}) - (-28\text{ ksi})]\sin(-20°)\cos(-20°) \\ &\quad + (42\text{ ksi})[\cos^2(-20°) - \sin^2(-20°)] \\ &= 62.4\text{ ksi}\end{aligned}$$

The stresses acting on the horizontal and vertical planes are shown in the sketch.

MecMovies

EXAMPLES

M12.7 Determine the normal and shear stress acting on a specified plane surface.

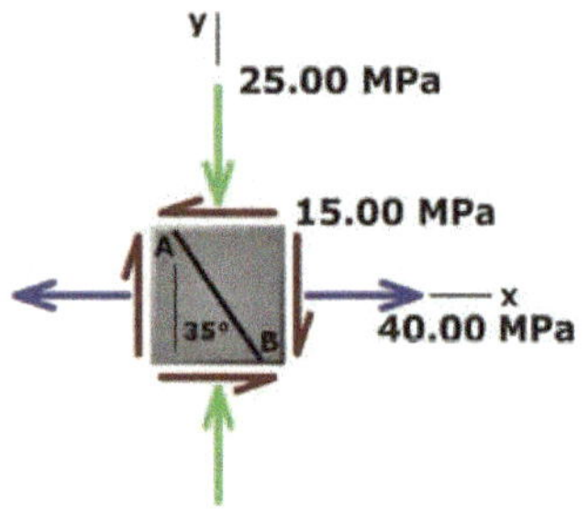

M12.8 Determine the normal and shear stress acting on a specified plane surface in a wooden object.

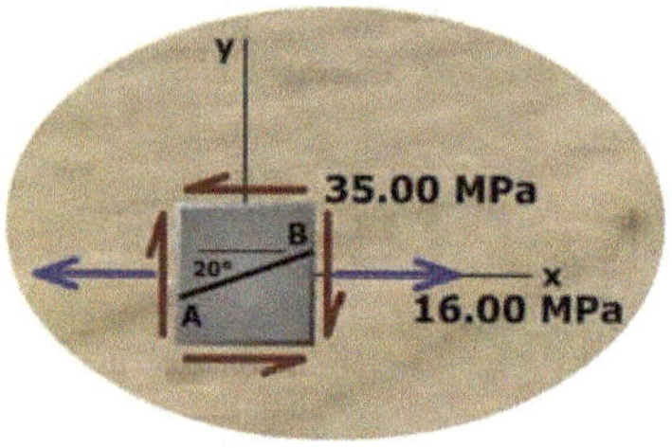

EXERCISES

M12.1 The Amazing Stress Camera. Interactive discovery activity that introduces the topic of stress transformations.

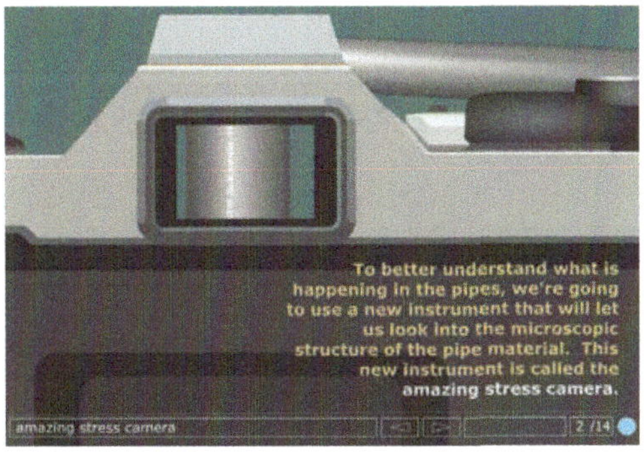

FIGURE M12.1

M12.2 Top–Drop–Sweep the Clock. Animated instruction teaching the proper method for determining θ. Eight easy multiple-choice questions.

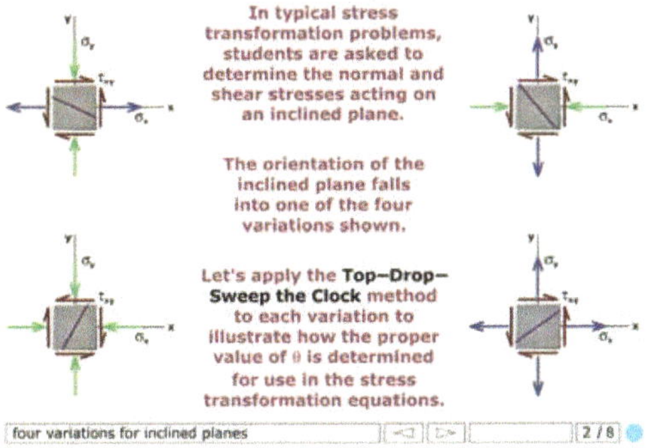

FIGURE M12.2

M12.3 Sign, Sign, Everywhere a Sign. A game that focuses on the correct sign conventions needed in the stress transformation equations. The game is won when two calculations for σ_n and τ_{nt} are correctly completed.

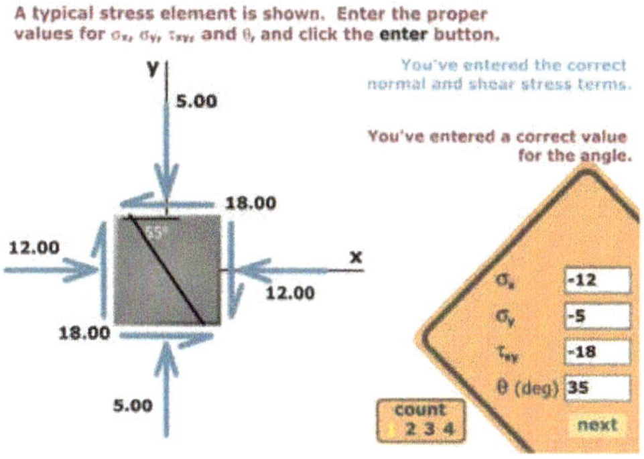

FIGURE M12.3

PROBLEMS

P12.15–P12.18 The stresses shown in Figures P12.15–P12.18 act at a point in a stressed body. Determine the normal and shear stresses at this point on the indicated inclined plane.

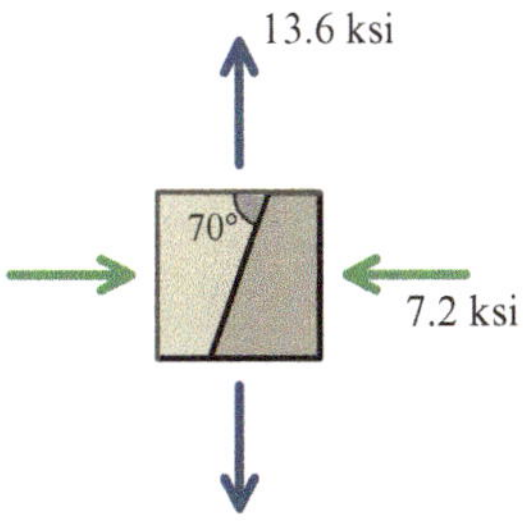

FIGURE P12.15

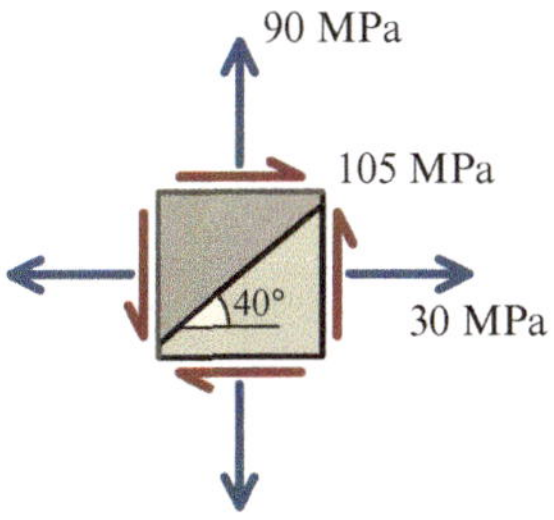

FIGURE P12.16

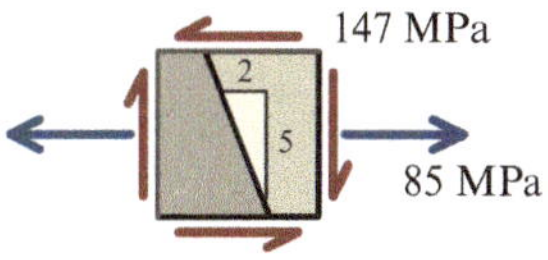

FIGURE P12.17

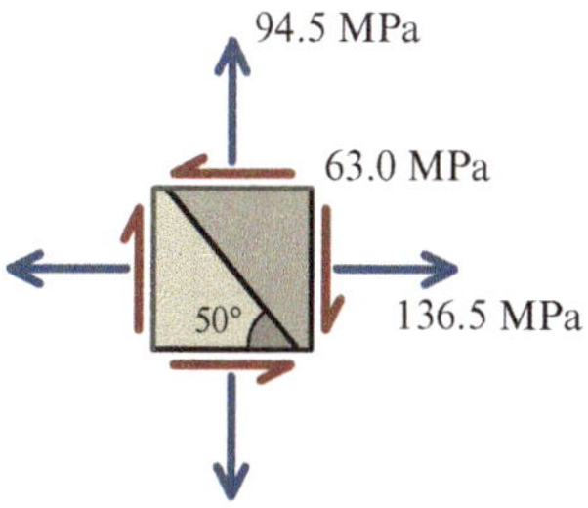

FIGURE P12.18

P12.19 Two steel plates of uniform cross section are welded together as shown in Figure P12.19. The plate dimensions are b = 5.5 in. and t = 0.375 in. An axial force P = 28.0 kips acts in the member. If a = 2.25 in., determine the magnitude of

(a) the normal stress that acts perpendicular to the weld seam.
(b) the shear stress that acts parallel to the weld seam.

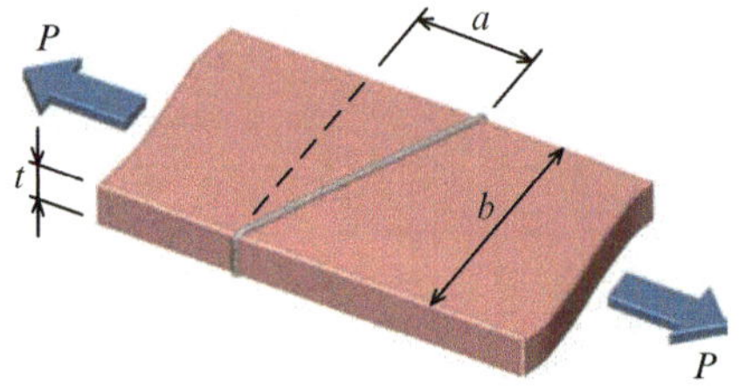

FIGURE P12.19

P12.20 The cylinder in Figure P12.20 consists of spirally wrapped steel plates that are welded at the seams. The cylinder has an outside diameter of 280 mm and a wall thickness of 7 mm. The cylinder is subjected to axial loads P = 90 kN. For a seam angle $\beta = 35°$, determine

(a) the normal stress perpendicular to the weld seams.
(b) the shear stress parallel to the weld seams.

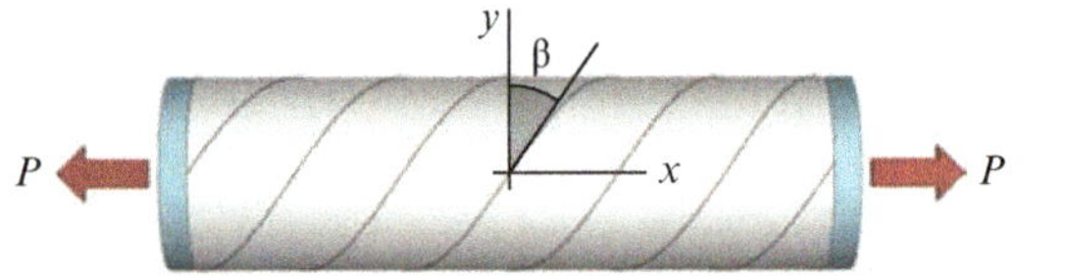

FIGURE P12.20

P12.21 A rectangular polymer plate of dimensions a = 150 mm and b = 360 mm is formed by gluing two triangular plates as shown in Figure P12.21. The plate is subjected to a tensile normal stress of 10 MPa in the long direction and a compressive normal stress of 6 MPa in the short direction. Determine

(a) the normal stress perpendicular to the glue seam.
(b) the shear stress parallel to the glue seam.

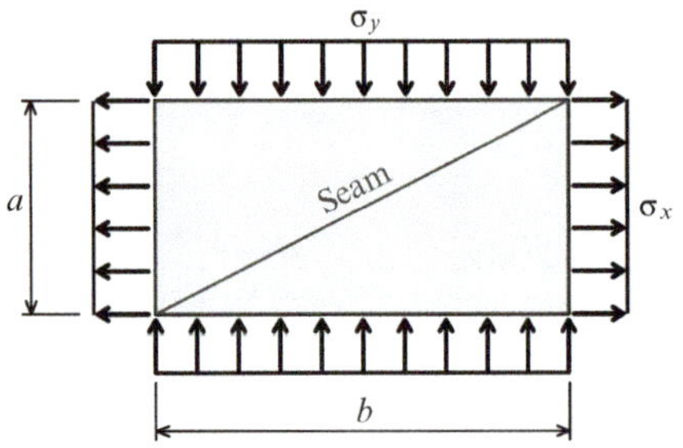

FIGURE P12.21

P12.22 The state of stress at a point in a solid component is shown in Figure P12.22. The stress magnitudes are σ_n = 80 MPa, σ_t = 40 MPa, and τ_{nt} = 32 MPa, acting in the directions indicated in the figure. Determine the normal and shear stress components acting on plane AB for a value of $\beta = 35°$.

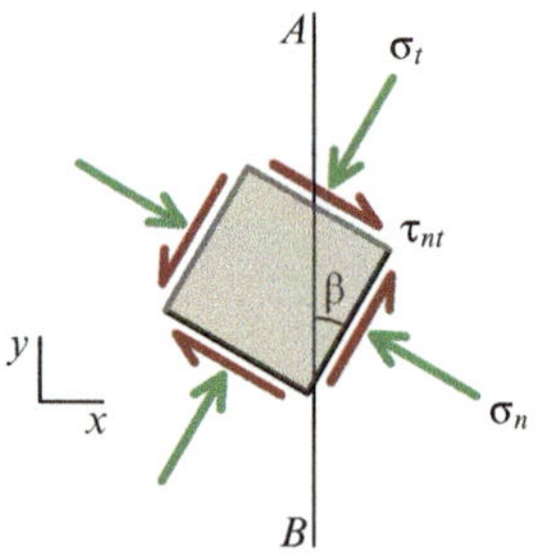

FIGURE P12.22

P12.23–P12.24 The stresses shown in Figures P12.23a and P12.24a act at a point on the free surface of a stressed body. Determine the normal stresses σ_n and σ_t and the shear stress τ_{nt} at this point if they act on the rotated stress element shown in Figures P12.23b and P12.24b.

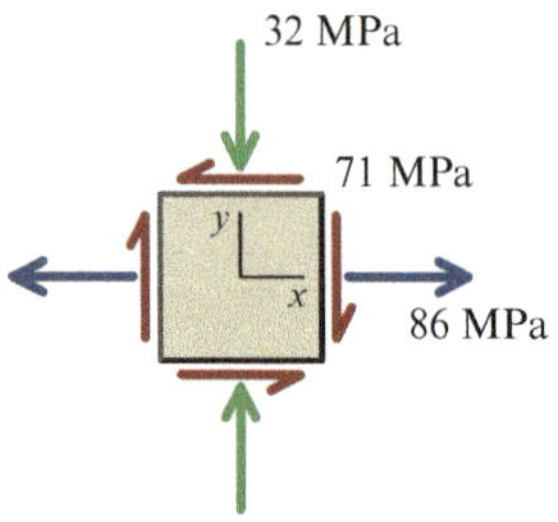

FIGURE P12.23a

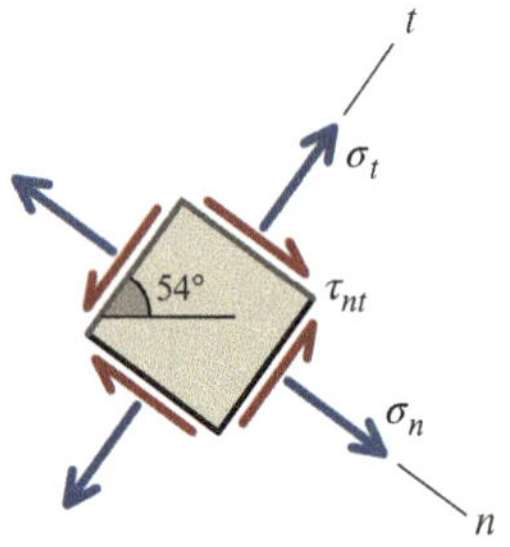

FIGURE P12.23b

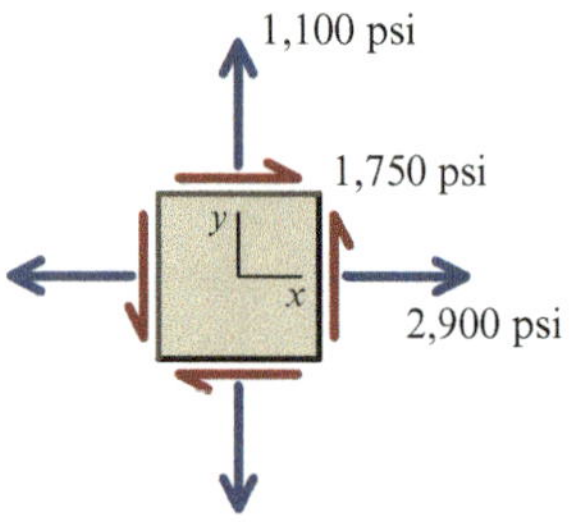

FIGURE P12.24a

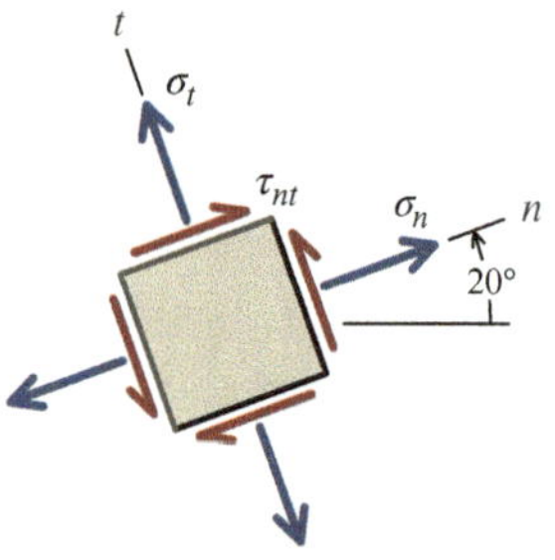

FIGURE P12.24b

P12.25–P12.26 The stresses shown in Figures P12.25 and P12.26 act at a point on the free surface of a machine component. Determine the normal stresses σ_x and σ_y and the shear stress τ_{xy} at the point.

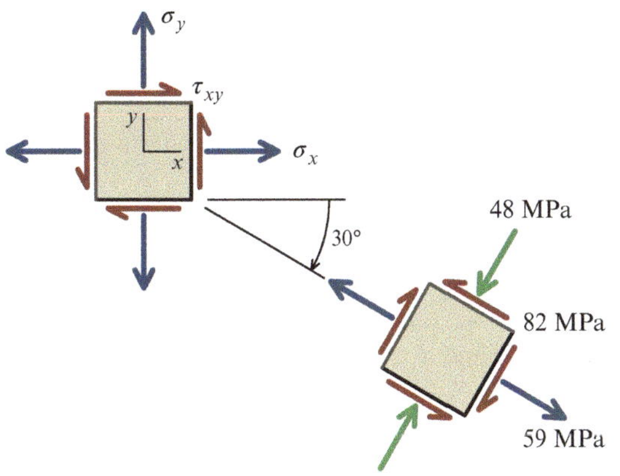

FIGURE P12.25

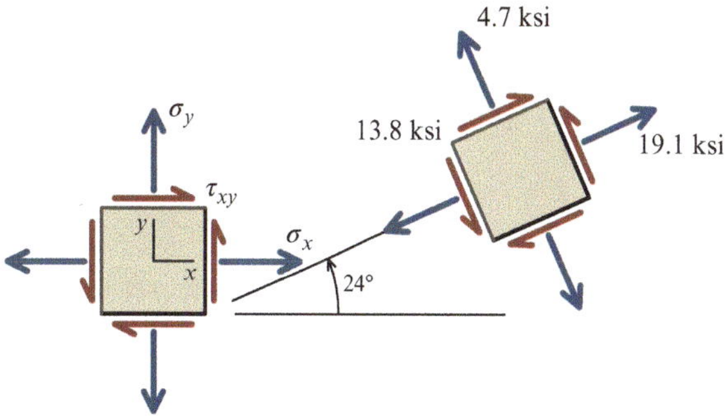

FIGURE P12.26

12.8 Principal Stresses and Maximum Shear Stress

The transformation equations for plane stress [Equations (12.3), (12.4), (12.5), and (12.6)] provide a means for determining the normal stress σ_n and the shear stress τ_{nt} acting on any plane through a point in a stressed body. For design purposes, the critical stresses at a point are often the maximum and minimum normal stresses and the maximum shear stress. The stress transformation equations can be used to develop additional relationships that indicate

(a) the orientations of planes where maximum and minimum normal stresses occur,
(b) the magnitudes of maximum and minimum normal stresses,
(c) the magnitudes of maximum shear stresses, and
(d) the orientations of planes where maximum shear stresses occur.

The transformation equations for plane stress were developed in Section 12.7. Equations (12.3) and (12.4), for normal stress and shear stress are, respectively, as follows:

$$\sigma_n = \sigma_x \cos^2\theta + \sigma_y \sin^2\theta + 2\tau_{xy}\sin\theta\cos\theta$$

$$\tau_{nt} = -(\sigma_x - \sigma_y)\sin\theta\cos\theta + \tau_{xy}(\cos^2\theta - \sin^2\theta)$$

These same equations can also be expressed in terms of double-angle trigonometric functions as Equations (12.5) and (12.6):

$$\sigma_n = \frac{\sigma_x + \sigma_y}{2} + \frac{\sigma_x - \sigma_y}{2}\cos 2\theta + \tau_{xy}\sin 2\theta$$

$$\tau_{nt} = -\frac{\sigma_x - \sigma_y}{2}\sin 2\theta + \tau_{xy}\cos 2\theta$$

Principal Planes

For a given state of plane stress, the stress components σ_x, σ_y, and τ_{xy} are constants. The dependent variables σ_n and τ_{nt} are actually functions of only one independent variable, θ. Therefore, the value of θ for which the normal stress σ_n is a maximum or a minimum can be determined by differentiating Equation (12.5) with respect to θ and setting the derivative equal to zero:

$$\frac{d\sigma_n}{d\theta} = -\frac{\sigma_x - \sigma_y}{2}(2\sin 2\theta) + 2\tau_{xy}\cos 2\theta = 0 \tag{12.10}$$

The solution of this equation gives the orientation $\theta = \theta_p$ of a plane where either a maximum or a minimum normal stress occurs:

$$\tan 2\theta_p = \frac{2\tau_{xy}}{\sigma_x - \sigma_y} \tag{12.11}$$

For a given set of stress components σ_x, σ_y, and τ_{xy}, Equation (12.11) can be satisfied by two values of $2\theta_p$, and these two values will be separated by 180°. Accordingly, the values of θ_p will differ by 90°. From this result, we can conclude that

(a) there will be only two planes where either a maximum or a minimum normal stress occurs, and
(b) these two planes will be 90° apart (i.e., orthogonal to each other).

Notice the similarity between the expressions for $d\sigma_n/d\theta$ in Equation (12.10) and τ_{nt} in Equation (12.6). Setting the derivative of σ_n equal to zero is equivalent to setting τ_{nt} equal to zero; therefore, the values of θ_p that are solutions of Equation (12.11) produce values of $\tau_{nt} = 0$ in Equation (12.6). This property leads us to another important conclusion:

Shear stress vanishes on planes where maximum and minimum normal stresses occur.

Planes free of shear stress are termed **principal planes**. The normal stresses acting on these planes—the maximum and minimum normal stresses—are called **principal stresses**.

The two values of θ_p that satisfy Equation (12.11) are called the **principal angles**. When $\tan 2\theta_p$ is positive, θ_p is positive and the principal plane defined by θ_p is rotated in a counterclockwise sense from the reference x axis. When $\tan 2\theta_p$ is negative, the rotation is clockwise. Observe that one value of θ_p will always be between positive and negative 45° (inclusive), and the second value will differ by 90°.

Magnitude of Principal Stresses

As mentioned, the normal stresses acting on the principal planes at a point in a stressed body are called *principal stresses*. The maximum normal stress (i.e., the most positive value algebraically) acting at a point is denoted as σ_{p1}, and the minimum normal stress (i.e., the most negative value algebraically) is denoted as σ_{p2}. There are two methods for computing the magnitudes of the normal stresses acting on the principal planes.

Method One. The first method is simply to substitute each of the θ_p values into either Equation (12.3) or Equation (12.5) and compute the corresponding normal stress. In addition to giving the value of the principal stress, this method has the advantage that it directly associates a principal stress magnitude with each of the principal angles.

Method Two. A general equation can be derived to give values for both σ_{p1} and σ_{p2}. To derive this general equation, values of $2\theta_p$ must be substituted into Equation (12.5). Equation (12.11) can be represented geometrically by the triangles shown in Figure 12.12.

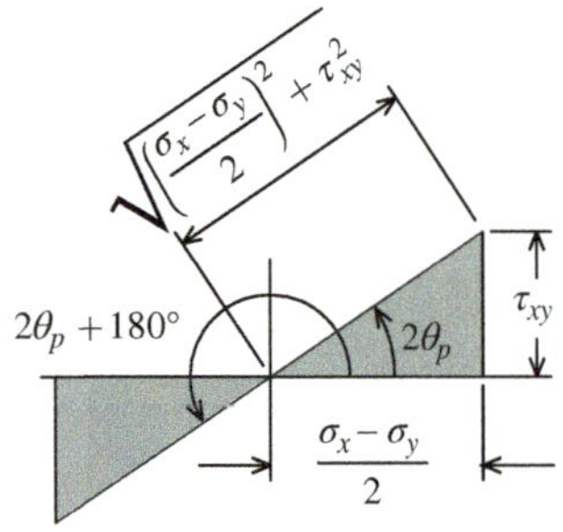

FIGURE 12.12

In this figure, we will assume that τ_{xy} and $(\sigma_x - \sigma_y)$ are both positive or both negative quantities. From trigonometry, expressions can be developed for $\sin 2\theta_p$ and $\cos 2\theta_p$, two terms that are needed for the solution of Equation (12.5):

$$\sin 2\theta_p = \frac{\tau_{xy}}{\sqrt{\left(\dfrac{\sigma_x - \sigma_y}{2}\right)^2 + \tau_{xy}^2}} \qquad \cos 2\theta_p = \frac{(\sigma_x - \sigma_y)/2}{\sqrt{\left(\dfrac{\sigma_x - \sigma_y}{2}\right)^2 + \tau_{xy}^2}}$$

When these functions of $2\theta_p$ are substituted into Equation (12.5) and simplified, one obtains

$$\sigma_{p1} = \frac{\sigma_x + \sigma_y}{2} + \sqrt{\left(\frac{\sigma_x - \sigma_y}{2}\right)^2 + \tau_{xy}^2}$$

A similar expression is obtained for σ_{p2} by repeating these steps with the principal angle $2\theta_p + 180°$:

$$\sigma_{p2} = \frac{\sigma_x + \sigma_y}{2} - \sqrt{\left(\frac{\sigma_x - \sigma_y}{2}\right)^2 + \tau_{xy}^2}$$

These two equations can then be combined into a single equation for the two in-plane principal stresses σ_{p1} and σ_{p2}:

$$\sigma_{p1,p2} = \frac{\sigma_x + \sigma_y}{2} \pm \sqrt{\left(\frac{\sigma_x - \sigma_y}{2}\right)^2 + \tau_{xy}^2} \tag{12.12}$$

Equation (12.12) does not directly indicate which principal stress, either σ_{p1} or σ_{p2}, is associated with each principal angle, and that is an important consideration. The solution of Equation (12.11) always gives a value of θ_p between +45° and −45° (inclusive). The principal stress associated with this value of θ_p can be determined from the following two-part rule:

- If the term $\sigma_x - \sigma_y$ is positive, then θ_p indicates the orientation of σ_{p1}.
- If the term $\sigma_x - \sigma_y$ is negative, then θ_p indicates the orientation of σ_{p2}.

The other principal stress is oriented perpendicular to θ_p.

The principal stresses determined from Equation (12.12) may both be positive, may both be negative, or may be of opposite signs. In naming the principal stresses, σ_{p1} is the more positive value algebraically (i.e., the algebraically larger value). If one or both of the principal stresses from Equation (12.12) are negative, σ_{p1} can have a smaller absolute value than σ_{p2}.

Shear Stresses on Principal Planes

As shown in the previous discussion, the values of θ_p that are solutions of Equation (12.11) will produce values of $\tau_{nt} = 0$ in Equation (12.6). Therefore, the shear stress on a principal plane must be zero. This is a very important conclusion.

This characteristic of principal planes can be restated in the following manner:

> *If a plane is a principal plane, then the shear stress acting on the plane must be zero.*

The converse of this statement is also true:

> *If the shear stress on a plane is zero, then that plane must be a principal plane.*

In many situations, a stress element (which represents the state of stress at a specific point) will have only normal stresses acting on its x and y faces. In these instances, one can conclude that the x and y faces must be principal planes because there is no shear stress acting on them.

Another important application of the preceding statement concerns the state of plane stress. As discussed in Section 12.4, a state of plane stress in the x–y plane means that there are no stresses acting on the z face of the stress element. Therefore,

$$\sigma_z = \tau_{zx} = \tau_{zy} = 0$$

If the shear stress on the z face is zero, one can conclude that the z face must be a principal plane. Consequently, the normal stress acting on the z face must be a principal stress—the third principal stress.

The Third Principal Stress

If the normal of a surface lies in the x–y plane, then the stresses that act on that surface are termed **in-plane stresses**.

In the previous discussion, the principal planes and principal stresses were determined for a state of plane stress. The two principal planes found from Equation (12.11) were oriented at angles of θ_p and $\theta_p \pm 90°$ with respect to the reference x axis, and they were oriented so that their outward normal was perpendicular to the z axis (i.e., the out-of-plane axis). The corresponding principal stresses determined from Equation (12.12) are called the **in-plane principal stresses**.

Although it is convenient to represent the stress element as a two-dimensional square, it is actually a three-dimensional cube with x, y, and z faces. For a state of plane stress, the stresses acting on the z face—σ_z, τ_{zx}, and τ_{zy}—are zero. So, since the shear stresses on the z face are zero, the normal stress acting on the z face must be a principal stress, even though its magnitude is zero. **A point subjected to plane stress has three principal stresses: the two in-plane principal stresses σ_{p1} and σ_{p2}, plus a third principal stress σ_{p3}, which acts in the out-of-plane direction and has a magnitude of zero.**

Orientation of Maximum In-Plane Shear Stress

To determine the planes where the maximum in-plane shear stress τ_{max} occurs, Equation (12.6) is differentiated with respect to θ and set equal to zero, yielding

$$\frac{d\tau_{nt}}{d\theta} = -(\sigma_x - \sigma_y)\cos 2\theta - 2\tau_{xy}\sin 2\theta = 0 \tag{12.13}$$

The solution of this equation gives the orientation $\theta = \theta_s$ of a plane where the shear stress is either a maximum or a minimum:

$$\tan 2\theta_s = -\frac{\sigma_x - \sigma_y}{2\tau_{xy}} \tag{12.14}$$

This equation defines two angles $2\theta_s$ that are 180° apart. Thus, the two values of θ_s are 90° apart. A comparison of Equations (12.14) and (12.11) reveals that the two tangent functions are negative reciprocals. For that reason, the values of $2\theta_p$ that satisfy Equation (12.11) are 90° away from the corresponding solutions $2\theta_s$ of Equation (12.14). Consequently, θ_p and θ_s are 45° apart. This property means that *the planes on which the maximum in-plane shear stresses occur are rotated 45° from the principal planes.*

Maximum In-Plane Shear Stress Magnitude

Similar to the way the principal stresses are computed, there are two methods for computing the magnitude of the maximum in-plane shear stress τ_{max}.

Method One. The first method is simply to substitute one of the θ_s values into either Equation (12.4) or Equation (12.6) and compute the corresponding shear stress. In addition to giving the value of the maximum in-plane shear stress, this method has the advantage that it directly associates a shear stress magnitude (including the proper sign) with the θ_s angle. Thus, given that shear stresses on orthogonal planes must be equal, the determination of the stress for only one θ_s angle is sufficient to define uniquely the shear stresses on both planes.

Since one is typically interested in finding both the principal stresses and the maximum in-plane shear stress, an efficient computational approach for finding both the magnitude and orientation of the maximum in-plane shear stress is as follows:

(a) From Equation (12.11), a specific value for θ_p will be known.
(b) Depending on the sign of θ_p and recognizing that θ_p and θ_s are always 45° apart, either add or subtract 45° to find an orientation of a maximum in-plane shear stress plane θ_s. To obtain an angle θ_s between +45° and –45° (inclusive), subtract 45° from a positive value of θ_p or add 45° to a negative value of θ_p.
(c) Substitute the resulting value of θ_s into either Equation (12.4) or Equation (12.6), and compute the corresponding shear stress. The result is τ_{max}, the maximum in-plane shear stress.
(d) The result obtained from either Equation (12.4) or Equation (12.6) for θ_s will furnish both the magnitude and the *sign* of the maximum in-plane shear stress τ_{max}. Obtaining the sign is particularly valuable in this method because Method Two offers no direct means for establishing the sign of τ_{max}.

Method Two. A general equation can be derived to give the magnitude of τ_{max} by substituting angle functions obtained from Equation (12.14) into Equation (12.6). The result is

$$\tau_{max} = -\frac{\sigma_x - \sigma_y}{2}\left[\frac{\pm(\sigma_x - \sigma_y)/2}{\sqrt{\left(\dfrac{\sigma_x - \sigma_y}{2}\right)^2 + \tau_{xy}^2}}\right] + \tau_{xy}\left[\frac{\mp\tau_{xy}}{\sqrt{\left(\dfrac{\sigma_x - \sigma_y}{2}\right)^2 + \tau_{xy}^2}}\right]$$

which reduces to

$$\tau_{max} = \pm\sqrt{\left(\frac{\sigma_x - \sigma_y}{2}\right)^2 + \tau_{xy}^2} \qquad (12.15)$$

Note that Equation (12.15) has the same magnitude as the second term of Equation (12.12).

From Equation (12.15), the sign of τ_{max} is ambiguous. The maximum shear stress differs from the minimum shear stress only in sign. Unlike normal stress, which can be either tension or compression, the sign of the maximum in-plane shear stress has no physical significance for the material behavior of a stressed body. The sign simply indicates the direction in which the shear stress acts on a particular plane surface.

A useful relation between the principal and the maximum in-plane shear stress is obtained from Equations (12.12) and (12.15) by subtracting the values for the two in-plane principal stresses and substituting the value of the radical from Equation (12.15). The result is

$$\tau_{max} = \frac{\sigma_{p1} - \sigma_{p2}}{2} \qquad (12.16)$$

In words, the maximum in-plane shear stress τ_{max} is equal in magnitude to one-half of the difference between the two in-plane principal stresses.

Normal Stresses on Maximum In-Plane Shear Stress Surfaces

Unlike principal planes, which are free of shear stress, planes subjected to τ_{max} usually have normal stresses. After substituting angle functions obtained from Equation (12.14) into Equation (12.5) and simplifying, we find that the normal stress acting on a plane of maximum in-plane shear stress is

$$\sigma_{avg} = \frac{\sigma_x + \sigma_y}{2} \tag{12.17}$$

The normal stress σ_{avg} is the same on both τ_{max} planes.

Absolute Maximum Shear Stress

In Equation (12.15), we derived an expression for the maximum shear stress magnitude acting in the plane of a body subjected to plane stress. We also found that the maximum in-plane shear stress τ_{max} is equal in magnitude to one-half the difference between the two in-plane principal stresses [Equation (12.16)]. Let us briefly consider a point in a stressed body in which stresses act in three directions. We ask the question, "What is the maximum shear stress for this more general state of stress?" We will denote the maximum shear stress magnitude on any plane that could be passed through the point as $\tau_{abs\ max}$ to differentiate it from the maximum in-plane shear stress τ_{max}. In the body at the point of interest, there will be three orthogonal planes with no shear stress: the principal planes. (See Section 12.11.) The normal stresses acting on these planes are termed principal stresses, and, in general, they each have unique algebraic values (i.e., $\sigma_{p1} \neq \sigma_{p2} \neq \sigma_{p3}$). Consequently, one principal stress will be the maximum algebraically (σ_{max}), one principal stress will be the minimum algebraically (σ_{min}), and the third principal stress will have a value in between these two extremes. The magnitude of the absolute maximum shear stress $\tau_{abs\ max}$ is equal to one-half of the difference between the maximum and minimum principal stresses:

$$\tau_{abs\ max} = \frac{\sigma_{max} - \sigma_{min}}{2} \tag{12.18}$$

Furthermore, $\tau_{abs\ max}$ acts on planes that bisect the angles between the maximum and minimum principal planes.

For example, if stresses act only in the x–y plane, then the z face of a stress element is a principal plane.

When a state of plane stress exists, normal and shear stresses on the out-of-plane face of a stress element are zero. Since no shear stresses act on it, the out-of-plane face is a principal plane and the principal stress acting on it is designated σ_{p3}. Therefore, two principal stresses σ_{p1} and σ_{p2} act in the plane of the stress, and the third principal stress, which acts in the out-of-plane direction, has a magnitude of $\sigma_{p3} = 0$. Thus, for plane stress, the magnitude of the absolute maximum shear stress can be determined from one of the following three conditions:

(a) If both σ_{p1} and σ_{p2} are positive, then

$$\tau_{abs\ max} = \frac{\sigma_{p1} - \sigma_{p3}}{2} = \frac{\sigma_{p1} - 0}{2} = \frac{\sigma_{p1}}{2}$$

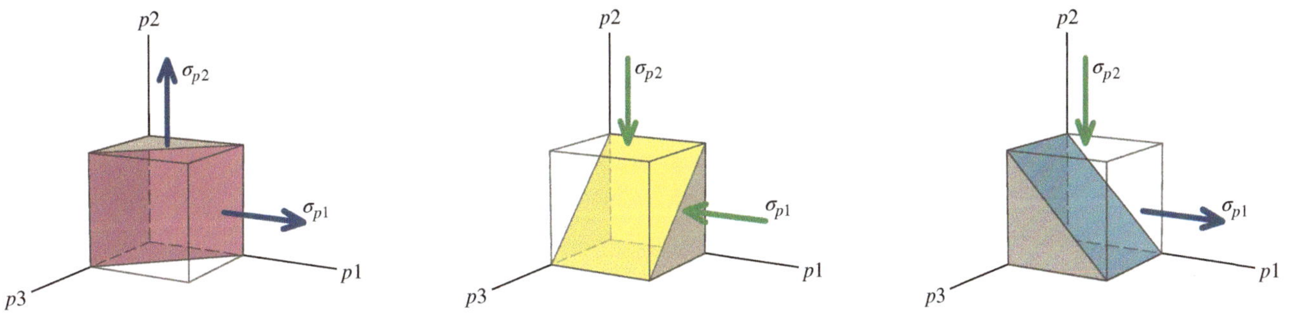

(*a*) If both σ_{p1} and σ_{p2} are positive (*b*) If both σ_{p1} and σ_{p2} are negative (*c*) If σ_{p1} is positive and σ_{p2} is negative

FIGURE 12.13 Planes of absolute maximum shear stress for plane stress.

(b) If both σ_{p1} and σ_{p2} are negative, then

$$\tau_{\text{abs max}} = \frac{\sigma_{p3} - \sigma_{p2}}{2} = \frac{0 - \sigma_{p2}}{2} = -\frac{\sigma_{p2}}{2}$$

(c) If σ_{p1} is positive and σ_{p2} is negative, then

$$\tau_{\text{abs max}} = \frac{\sigma_{p1} - \sigma_{p2}}{2}$$

These three possibilities are illustrated in Figure 12.13, in which one of the two orthogonal planes on which the maximum shear stress acts is highlighted in each example. Note that $\sigma_{p3} = 0$ in all three cases.

The direction of the absolute maximum shear stress can be determined by drawing a wedge-shaped block with two sides parallel to the planes having the maximum and minimum principal stresses, and with the third side at an angle of 45° with the other two sides. The direction of the maximum shear stress must oppose the larger of the two principal stresses.

Stress Invariance

A useful relationship between the principal stresses and the normal stresses on the orthogonal planes, shown in Figure 12.14, is obtained by adding the values for the two principal stresses as given in Equation (12.12). The result is

$$\boxed{\sigma_{p1} + \sigma_{p2} = \sigma_x + \sigma_y} \qquad (12.19)$$

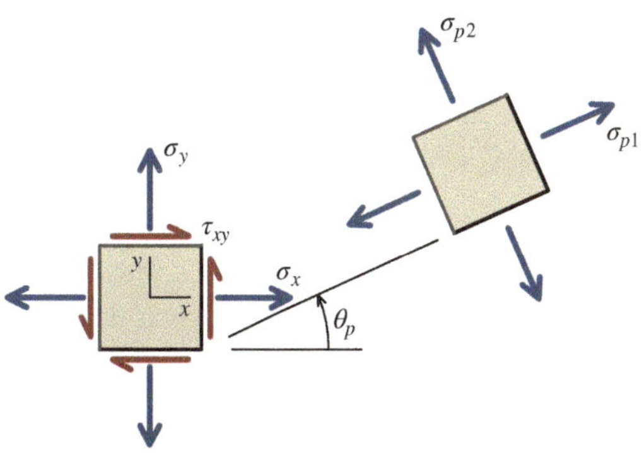

FIGURE 12.14

In words, *for plane stress, the sum of the normal stresses on any two orthogonal planes through a point in a body is constant and independent of the angle θ.*

12.9 Presentation of Stress Transformation Results

Principal stress and maximum in-plane shear stress results should be presented with a sketch that depicts the orientation of all stresses. Two sketch formats are generally used:

(a) two square stress elements or

(b) a single wedge-shaped element.

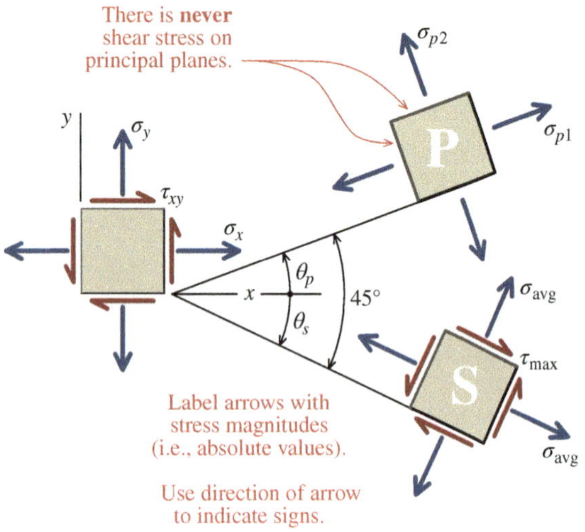

FIGURE 12.15

Two Square Stress Elements

Two square stress elements are sketched in Figure 12.15. One stress element shows the orientation and magnitude of the principal stresses, and a second element shows the orientation and magnitude of the maximum in-plane shear stress along with the associated normal stresses.

Principal Stress Element

- The principal stress element is shown rotated at the angle θ_p calculated from Equation (12.11):

$$\tan 2\theta_p = \frac{2\tau_{xy}}{\sigma_x - \sigma_y}$$

  This equation yields a value between +45° and –45° (inclusive).

- When θ_p is positive, the stress element is rotated in a counterclockwise sense from the reference x axis. When θ_p is negative, the rotation is clockwise.

- Note that the angle calculated from Equation (12.11) does not necessarily give the orientation of the σ_{p1} plane. Either σ_{p1} or σ_{p2} may act on the θ_p plane. The principal stress oriented at θ_p can be determined from the following rule:
  - If $\sigma_x - \sigma_y$ is positive, θ_p indicates the orientation of σ_{p1}.
  - If $\sigma_x - \sigma_y$ is negative, θ_p indicates the orientation of σ_{p2}.
- The other principal stress is shown on the perpendicular faces of the stress element.
- In the sketch, use the arrow direction to indicate whether the principal stress is tension or compression. Label the arrow with the absolute value of either σ_{p1} or σ_{p2}.
- *There is never a shear stress on the principal planes*; therefore, show no shear stress arrows on the principal stress element.

Maximum In-Plane Shear Stress Element

- Draw the maximum shear stress element rotated 45° from the principal stress element.
- If the principal stress element is rotated in a counterclockwise sense (i.e., positive θ_p) from the reference x axis, then the maximum shear stress element should be shown rotated 45° clockwise from the principal stress element. Therefore, the maximum shear stress element will be oriented an angle $\theta_s = \theta_p - 45°$ relative to the x axis.
- If the principal stress element is rotated in a clockwise sense (i.e., negative θ_p) from the reference x axis, then the maximum shear stress element should be shown rotated 45° counterclockwise from the principal stress element. Therefore, the maximum shear stress element will be oriented an angle $\theta_s = \theta_p + 45°$ relative to the x axis.

- Substitute the value of θ_s into either Equation (12.4) or (12.6), and compute τ_{max}.
- If τ_{max} is positive, draw the shear stress arrow on the θ_s face in the direction that tends to rotate the stress element counterclockwise. If τ_{max} is negative, the shear stress arrow on the θ_s face should tend to rotate the stress element clockwise. Label this arrow with the absolute value of τ_{max}.
- Once the shear stress arrow on the θ_s face has been established, draw appropriate shear stress arrows on the other three faces.
- Compute the average normal stress acting on the maximum in-plane shear stress planes from Equation (12.17).
- Show the average normal stress with arrows *acting on all four faces*. Use the direction of the arrow to indicate whether the average normal stress is tension or compression. Label a pair of the arrows with the magnitude of this stress.
- *In general, the maximum in-plane shear stress element will include both normal and shear stress arrows on all four faces.*

Wedge-Shaped Stress Element

A wedge-shaped stress element can be used to report both the principal stress and maximum in-plane shear stress results on a single element, as shown in Figure 12.16.

- The two orthogonal faces of the wedge element are used to report the orientation and magnitude of the principal stresses.
- Follow the procedures given previously for the *principal stress element* in order to specify the principal stresses acting on the two orthogonal faces of the wedge element. Since these two faces are principal planes, *there should not be a shear stress arrow on either face.*
- The sloped face of the wedge is oriented 45° away from the two orthogonal faces, and it is used to specify the maximum in-plane shear stress and the associated normal stress.
- Draw a shear stress arrow on the sloped face, and label the arrow with the magnitude of the maximum in-plane shear stress computed from Equation (12.15).
- There are several ways to determine the proper direction of the maximum in-plane shear stress arrow. One particularly easy way to construct a proper sketch is as follows: Begin the tail of the shear stress arrow at the σ_{p1} side of the wedge, and point the arrow toward the σ_{p2} side of the wedge.
- Compute the average normal stress acting on the maximum in-plane shear stress planes from Equation (12.17).
- Show the average normal stress on the sloped face of the wedge. Use the direction of the arrow to indicate whether the average normal stress is tension or compression. Label this arrow with the average normal stress magnitude.

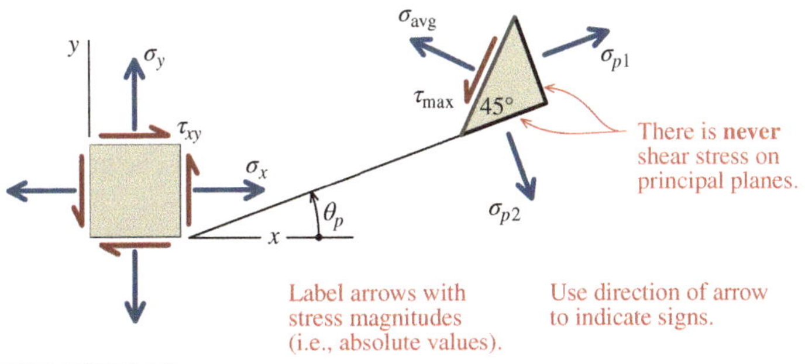

FIGURE 12.16

EXAMPLE 12.5

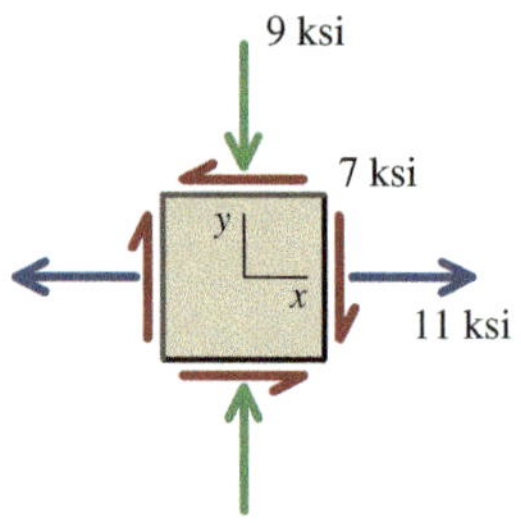

Consider a point in a structural member that is subjected to plane stress. Normal and shear stresses acting on horizontal and vertical planes at the point are shown.

(a) Determine the principal stresses and the maximum in-plane shear stress acting at the point.
(b) Show these stresses in an appropriate sketch.
(c) Determine the absolute maximum shear stress at the point.

Plan the Solution

The stress transformation equations derived in the preceding section will be used to compute the principal stresses and the maximum shear stress acting at the point.

SOLUTION

(a) From the given stresses, the values to be used in the stress transformation equations are $\sigma_x = 11$ ksi, $\sigma_y = -9$ ksi, and $\tau_{xy} = -7$ ksi. *The in-plane principal stress magnitudes* can be calculated from Equation (12.12):

$$\begin{aligned}\sigma_{p1,p2} &= \frac{\sigma_x + \sigma_y}{2} \pm \sqrt{\left(\frac{\sigma_x - \sigma_y}{2}\right)^2 + \tau_{xy}^2}\\ &= \frac{(11\text{ ksi}) + (-9\text{ ksi})}{2} \pm \sqrt{\left(\frac{(11\text{ ksi}) - (-9\text{ ksi})}{2}\right)^2 + (-7\text{ ksi})^2}\\ &= 13.21\text{ ksi},\quad -11.21\text{ ksi}\end{aligned}$$

Therefore, we have the following:

$$\sigma_{p1} = 13.21\text{ ksi} = 13.21\text{ ksi (T)}$$
$$\sigma_{p2} = -11.21\text{ ksi} = 11.21\text{ ksi (C)}$$

The *maximum in-plane shear stress* can be computed from Equation (12.15):

$$\begin{aligned}\tau_{\max} &= \pm\sqrt{\left(\frac{\sigma_x - \sigma_y}{2}\right)^2 + \tau_{xy}^2} = \pm\sqrt{\left(\frac{(11\text{ ksi}) - (-9\text{ ksi})}{2}\right)^2 + (-7\text{ ksi})^2}\\ &= \pm 12.21\text{ ksi}\end{aligned}$$

On the planes of maximum in-plane shear stress, the normal stress is simply the *average normal stress*, as given by Equation (12.17):

$$\sigma_{\text{avg}} = \frac{\sigma_x + \sigma_y}{2} = \frac{11\text{ ksi} + (-9\text{ ksi})}{2} = 1\text{ ksi} = 1\text{ ksi (T)}$$

(b) The principal stresses and the maximum in-plane shear stress must be shown in an appropriate sketch. The angle θ_p indicates the orientation of one principal plane relative to the reference x face. From Equation (12.11),

$$\tan 2\theta_p = \frac{2\tau_{xy}}{\sigma_x - \sigma_y} = \frac{2(-7\text{ ksi})}{11\text{ ksi} - (-9\text{ ksi})} = \frac{-14}{20}$$
$$\therefore\ \theta_p = -17.5°$$

Since θ_p is negative, the angle is turned clockwise. In other words, the *normal* of one principal plane is rotated 17.5° below the reference x axis. One of the in-plane principal stresses—either σ_{p1} or σ_{p2}—acts on this principal plane. To determine which principal stress acts at $\theta_p = -17.5°$, use the following two-part rule:

- If the term $\sigma_x - \sigma_y$ is positive, then θ_p indicates the orientation of σ_{p1}.
- If the term $\sigma_x - \sigma_y$ is negative, then θ_p indicates the orientation of σ_{p2}.

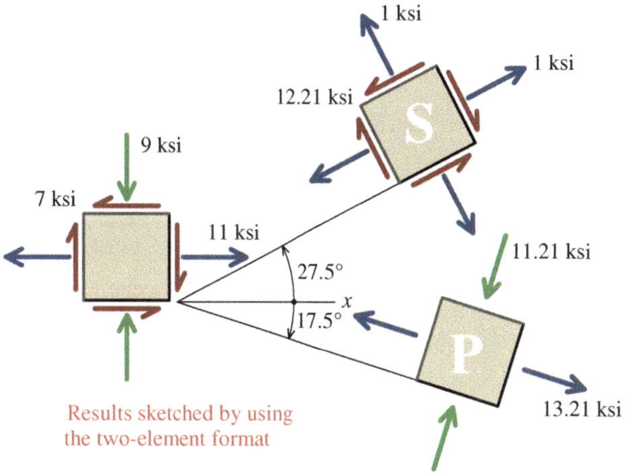

Since $\sigma_x - \sigma_y$ is positive, θ_p indicates the orientation of σ_{p1} = 13.21 ksi. The other principal stress, $\sigma_{p2} = -11.21$ ksi, acts on a perpendicular plane. The in-plane principal stresses are shown on the element labeled "P" in the figure. Note that there are never shear stresses acting on the principal planes.

The planes of maximum in-plane shear stress are always located 45° away from the principal planes; therefore, $\theta_s = 27.5°$. Although Equation (12.15) gives the magnitude of the maximum in-plane shear stress, it does not indicate the direction in which the shear stress acts on the plane defined by θ_s. To determine the direction of the shear stress, solve Equation (12.4) for τ_{nt}, using the values $\sigma_x = 11$ ksi, $\sigma_y = -9$ ksi, $\tau_{xy} = -7$ ksi, and $\theta = \theta_s = 27.5°$:

$$\begin{aligned}\tau_{nt} &= -(\sigma_x - \sigma_y)\sin\theta\cos\theta + \tau_{xy}(\cos^2\theta - \sin^2\theta) \\ &= -[(11\text{ ksi}) - (-9\text{ ksi})]\sin 27.5° \cos 27.5° + (-7\text{ ksi})[\cos^2 27.5° - \sin^2 27.5°] \\ &= -12.21\text{ ksi}\end{aligned}$$

Since τ_{nt} is negative, the shear stress acts in a negative t direction on a positive n face. Once the shear stress direction has been determined for one face, the shear stress direction is known for all four faces of the stress element. The maximum in-plane shear stress and the average normal stress are shown on the stress element labeled "S." Note that, unlike the principal stress element, normal stresses will usually be acting on the planes of maximum in-plane shear stress.

The principal stresses and the maximum in-plane shear stress can also be reported on a single wedge-shaped element, as shown in the accompanying sketch. This format can be somewhat easier to use than the two-element sketch format, particularly with regard to the direction of the maximum in-plane shear stress. The maximum in-plane shear stress and the associated average normal stress are shown on the sloped face of the wedge, which is rotated 45° from the principal planes. *The shear stress arrow on this face always starts on the σ_{p1} side of the wedge and points toward the σ_{p2} side of the wedge.* Once again, there is never a shear stress on the principal planes (i.e., the σ_{p1} and σ_{p2} sides of the wedge).

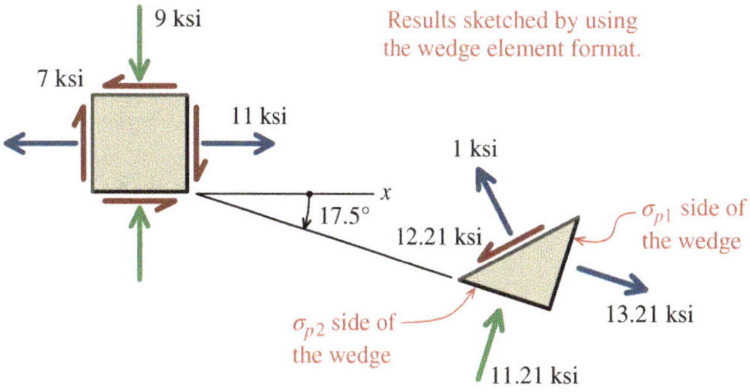

(c) For plane stress, such as the example presented here, the z face is free of stress. Therefore, $\tau_{zx} = 0$, $\tau_{zy} = 0$, and $\sigma_z = 0$. Since the shear stress on the z face is zero, the z face must be a principal plane with a principal stress $\sigma_{p3} = \sigma_z = 0$. The absolute maximum shear stress (considering all possible planes rather than simply those planes whose normal is perpendicular to the z axis) can be determined from the three principal stresses: $\sigma_{p1} = 13.21$ ksi, $\sigma_{p2} = -11.21$ ksi, and $\sigma_{p3} = 0$. The maximum principal stress (in an algebraic sense) is $\sigma_{max} = 13.21$ ksi, and the minimum principal stress is $\sigma_{min} = -11.21$ ksi. The absolute maximum shear stress can be computed from Equation (12.18):

$$\tau_{abs\ max} = \frac{\sigma_{max} - \sigma_{min}}{2} = \frac{13.21\text{ ksi} - (-11.21\text{ ksi})}{2} = 12.21\text{ ksi}$$

In this instance, the absolute maximum shear stress is equal to the maximum in-plane shear stress. That will always be the case whenever σ_{p1} is a positive value and σ_{p2} is a negative value. The absolute maximum shear stress will be greater than the maximum in-plane shear stress whenever σ_{p1} and σ_{p2} are either both positive or both negative.

EXAMPLE 12.6

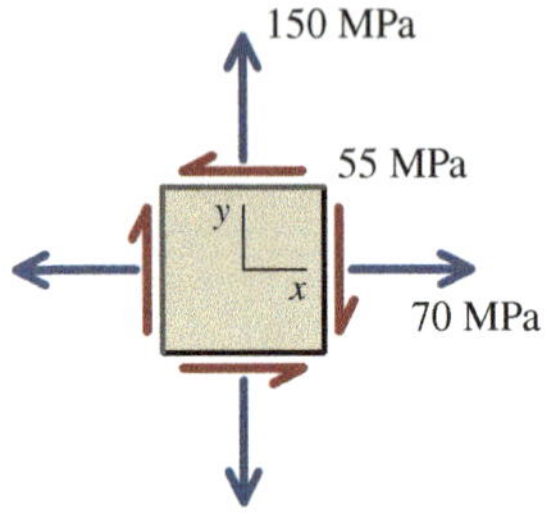

Consider a point in a structural member that is subjected to plane stress. Normal and shear stresses acting on horizontal and vertical planes at the point are shown.

(a) Determine the principal stresses and the maximum in-plane shear stress acting at the point.
(b) Show these stresses in an appropriate sketch.
(c) Determine the absolute maximum shear stress at the point.

Plan the Solution

The stress transformation equations derived in the preceding section will be used to compute the principal stresses and the maximum shear stress acting at the point.

SOLUTION

(a) From the given stresses, the values to be used in the stress transformation equations are $\sigma_x = 70$ MPa, $\sigma_y = 150$ MPa, and $\tau_{xy} = -55$ MPa. The *in-plane principal stresses* can be calculated from Equation (12.12):

$$\sigma_{p1,p2} = \frac{\sigma_x + \sigma_y}{2} \pm \sqrt{\left(\frac{\sigma_x - \sigma_y}{2}\right)^2 + \tau_{xy}^2}$$

$$= \frac{70\text{ MPa} + 150\text{ MPa}}{2} \pm \sqrt{\left(\frac{70\text{ MPa} - 150\text{ MPa}}{2}\right)^2 + (-55\text{ MPa})^2}$$

$$= 178.0\text{ MPa},\ 42.0\text{ MPa}$$

The *maximum in-plane shear stress* can be computed from Equation (12.15):

$$\tau_{max} = \pm\sqrt{\left(\frac{\sigma_x - \sigma_y}{2}\right)^2 + \tau_{xy}^2} = \pm\sqrt{\left(\frac{70\text{ MPa} - 150\text{ MPa}}{2}\right)^2 + (-55\text{ MPa})^2}$$

$$= \pm 68.0\text{ MPa}$$

On the planes of maximum in-plane shear stress, the normal stress is simply the *average normal stress*, as given by Equation (12.17):

$$\sigma_{avg} = \frac{\sigma_x + \sigma_y}{2} = \frac{70\text{ MPa} + 150\text{ MPa}}{2} = 110\text{ MPa} = 110\text{ MPa (T)}$$

(b) The principal stresses and the maximum in-plane shear stress must be shown in an appropriate sketch. The angle θ_p indicates the orientation of one principal plane relative to the reference x face. From Equation (12.11),

$$\tan 2\theta_p = \frac{2\tau_{xy}}{\sigma_x - \sigma_y} = \frac{2(-55\text{ MPa})}{70\text{ MPa} - 150\text{ MPa}} = \frac{-110}{-80}$$

$$\therefore \theta_p = 27.0°$$

The angle θ_p is positive; consequently, the angle is turned counterclockwise from the x axis. Since $\sigma_x - \sigma_y$ is negative, θ_p indicates the orientation of $\sigma_{p2} = 42.0$ MPa. The other principal stress, $\sigma_{p1} = 178.0$ MPa, acts on a perpendicular plane. The in-plane principal stresses are shown in the accompanying figure.

The maximum in-plane shear stress and the associated average normal stress are shown on the sloped face of the wedge, which is rotated 45° from the principal planes. Note that the arrow for τ_{max} starts on the σ_{p1} side of the wedge and points toward the σ_{p2} side.

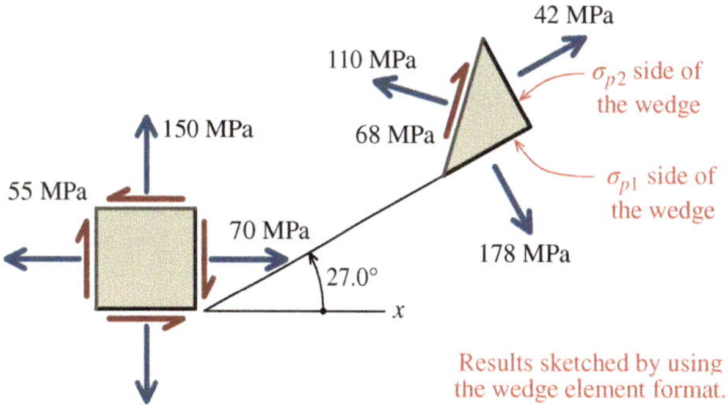

Results sketched by using the wedge element format.

(c) Since σ_{p1} and σ_{p2} are both positive values, the absolute maximum shear stress will be greater than the maximum in-plane shear stress. In this example, the three principal stresses are $\sigma_{p1} = 178$ MPa, $\sigma_{p2} = 42$ MPa, and $\sigma_{p3} = 0$. The maximum principal stress is $\sigma_{max} = 178$ MPa, and the minimum principal stress is $\sigma_{min} = 0$. The absolute maximum shear stress can be computed from Equation (12.18):

$$\tau_{abs\ max} = \frac{\sigma_{max} - \sigma_{min}}{2} = \frac{178\text{ MPa} - 0}{2} = 89.0\text{ MPa} \qquad \textbf{Ans.}$$

The absolute maximum shear stress acts on a plane whose normal does not lie in the x–y plane.

MecMovies

EXAMPLE

M12.9 Stress Transformation Learning Tool

Illustrates the correct usage of the stress transformation equations in determining stresses acting on a specified plane, principal stresses, and the maximum in-plane shear stress state for stress values specified by the user.

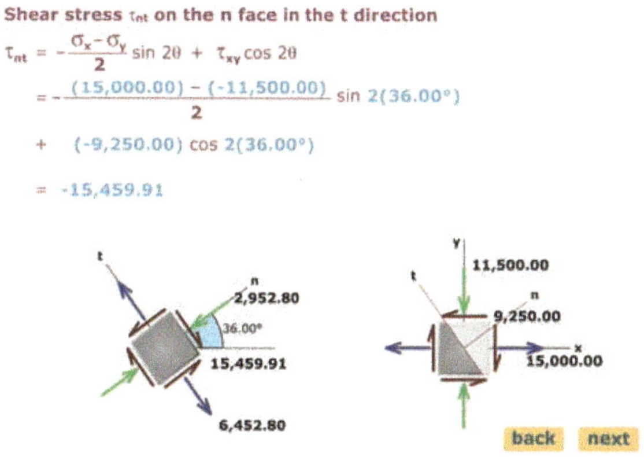

EXERCISE

M12.4 Sketching Stress Transformation Results. Score at least 100 points in this interactive activity. (Learning tool)

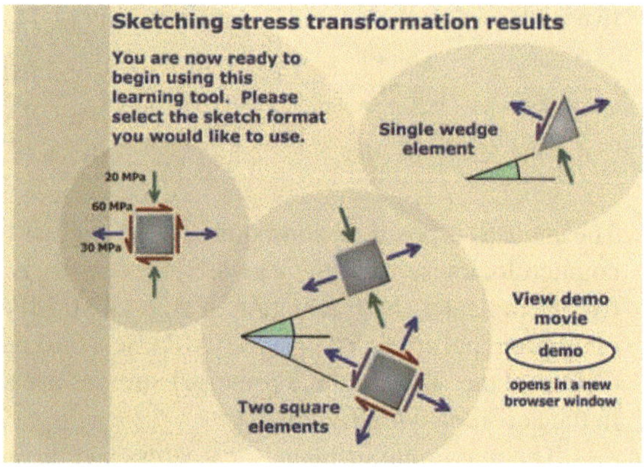

FIGURE M12.4

PROBLEMS

P12.27–P12.30 Consider a point in a structural member that is subjected to plane stress. Normal and shear stresses acting on horizontal and vertical planes at the point are shown in Figures P12.27–P12.30.

(a) Determine the principal stresses and the maximum in-plane shear stress acting at the point.
(b) Show these stresses on an appropriate sketch (e.g., see Figure 12.15 or Figure 12.16).
(c) Compute the absolute maximum shear stress at the point.

P12.31–P12.34 Consider a point in a structural member that is subjected to plane stress. Normal and shear stresses acting on horizontal and vertical planes at the point are shown in Figures P12.31–P12.34.

(a) Determine the principal stresses and the maximum in-plane shear stress acting at the point.
(b) Show these stresses on an appropriate sketch (e.g., see Figure 12.15 or Figure 12.16).
(c) Compute the absolute maximum shear stress at the point.

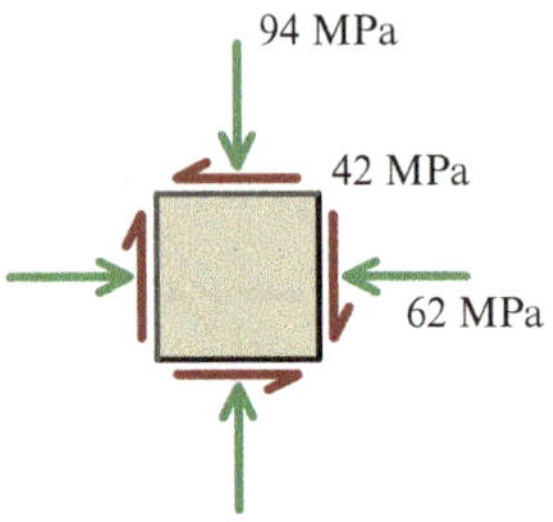

FIGURE P12.27

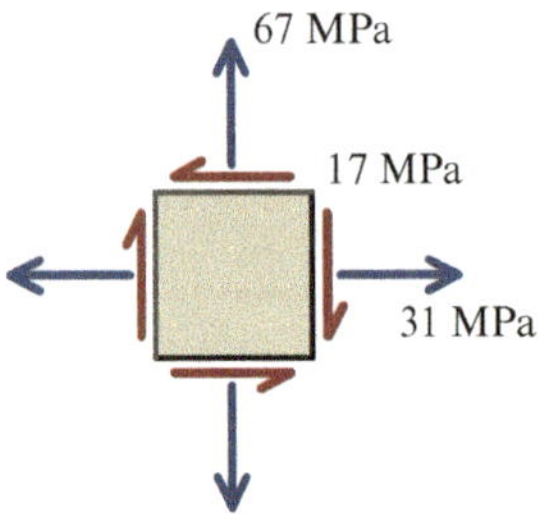

FIGURE P12.28

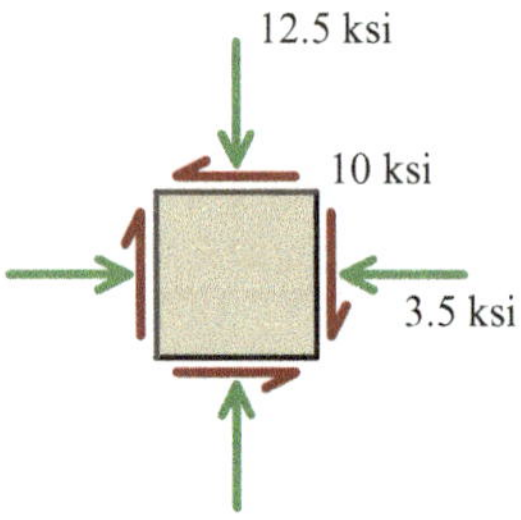

FIGURE P12.31

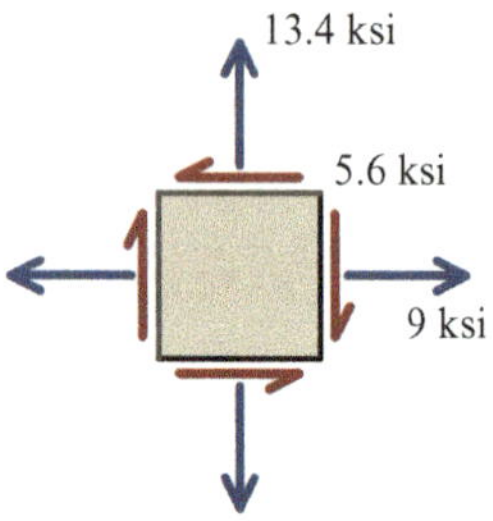

FIGURE P12.32

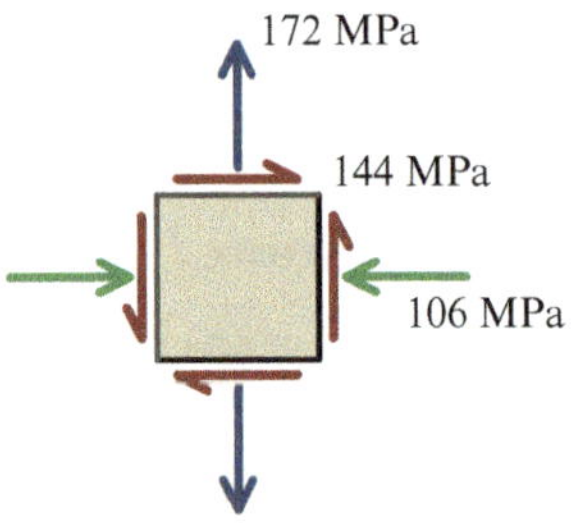

FIGURE P12.29

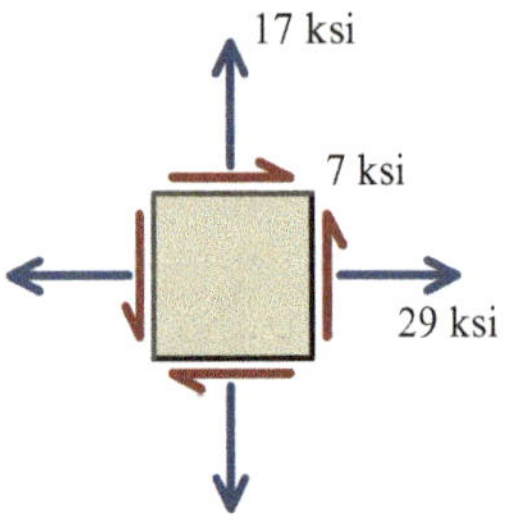

FIGURE P12.30

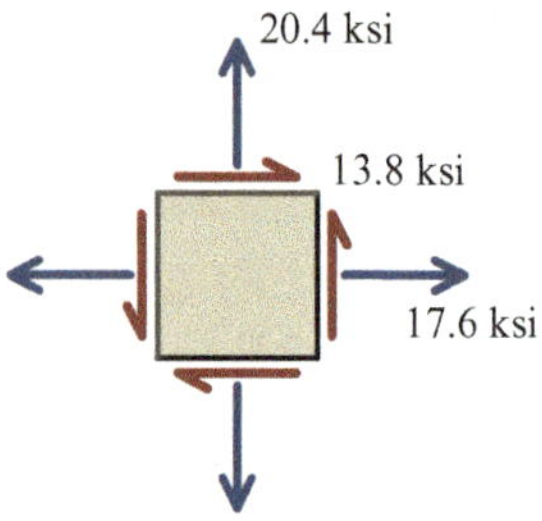

FIGURE P12.33

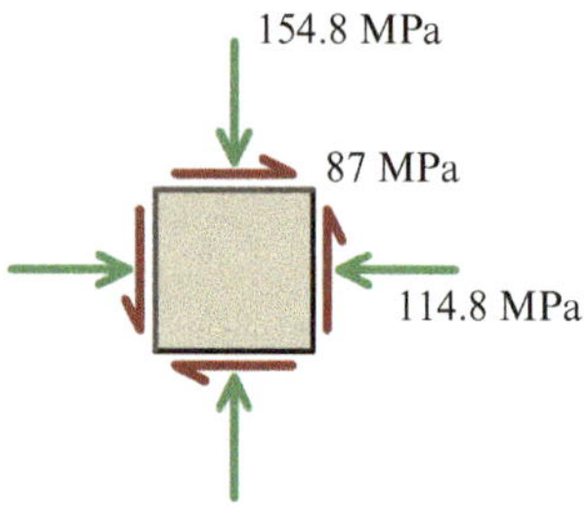

FIGURE P12.34

P12.35 A shear wall in a reinforced-concrete building is subjected to a vertical uniform load of intensity w and a horizontal force H, as shown in Figure P12.35*a*. As a consequence of these loads, the stresses at point A on the surface of the wall have the magnitudes $\sigma_y = 115.0$ MPa and $\tau_{xy} = 60.0$ MPa, acting in the directions shown on the stress element in Figure P12.35*b*.

(a) Determine the largest tension normal stress that acts at point A.
(b) What is the orientation of this stress with respect to the x axis?

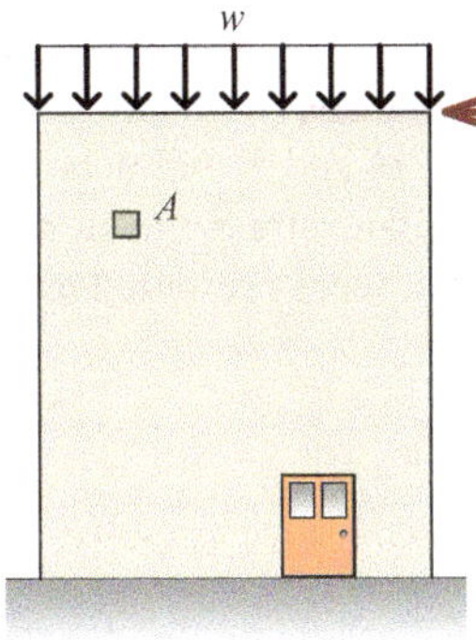

FIGURE P12.35*a*

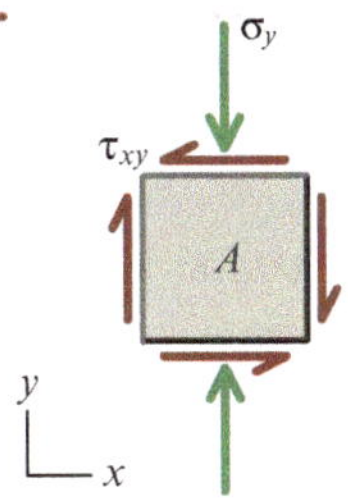

FIGURE P12.35*b*

P12.36 A 2 in. diameter shaft is supported by bearings at A and D, as shown in Figure P12.36. The bearings provide vertical reactions only. The shaft is subjected to an axial load $P = 3{,}600$ lb and a transverse load $Q = 250$ lb. The shaft length is $L = 48$ in. For point B, located on top of the shaft, determine

(a) the principal stresses.
(b) the maximum in-plane shear stress.

Show these results on a stress element.

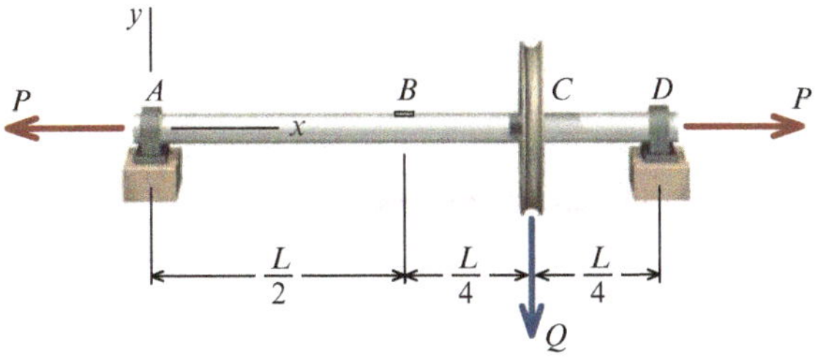

FIGURE P12.36

P12.37 The principal compressive stress on a vertical plane through a point in a wooden block is equal to three times the principal compression stress on a horizontal plane through the same point. The plane of the grain in the block is 25° clockwise from the vertical plane. If the normal and shear stresses must not exceed 400 psi (C) and 90 psi shear, respectively, determine the maximum allowable compressive stress on the horizontal plane.

P12.38 At a point on the free surface of a stressed body, a normal stress of 64 MPa (C) and an unknown positive shear stress exist on a horizontal plane. One principal stress at the point is 8 MPa (C). The absolute maximum shear stress at the point has a magnitude of 95 MPa. Determine the unknown stresses on the horizontal and vertical planes and the unknown principal stress at the point.

P12.39 At a point on the free surface of a stressed body, the normal stresses are 20 ksi (T) on a vertical plane and 30 ksi (C) on a horizontal plane. An unknown negative shear stress exists on the vertical plane. The absolute maximum shear stress at the point has a magnitude of 32 ksi. Determine the principal stresses and the shear stress on the vertical plane at the point.

P12.40 At a point on the free surface of a stressed body, a normal stress of 75 MPa (T) and an unknown negative shear stress exist on a horizontal plane. One principal stress at the point is 200 MPa (T). The maximum in-plane shear stress at the point has a magnitude of 85 MPa. Determine the unknown stresses on the vertical plane, the unknown principal stress, and the absolute maximum shear stress at the point.

P12.41 For the state of plane stress shown in Figure P12.41, determine (a) the largest value of τ_{xy} for which the maximum in-plane shear stress is equal to or less than 150 MPa and (b) the corresponding principal stresses.

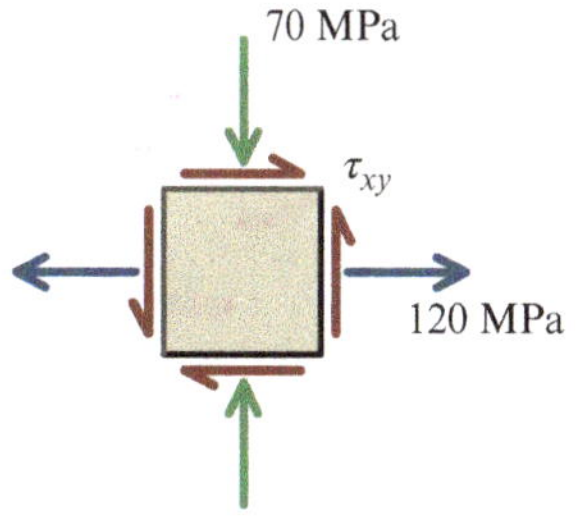

FIGURE P12.41

12.10 Mohr's Circle for Plane Stress

The process of changing stresses from one set of coordinate axes (i.e., x–y–z) to another set of axes (i.e., n–t–z) is termed stress transformation, and the general equations for plane stress transformation were developed in Section 12.7. The equations for computing the principal stresses and the maximum in-plane shear stress at a point in a stressed body were developed in Section 12.8. In the current section, a graphical procedure for plane stress transformations will be developed. In comparison with the various equations derived in Sections 12.7 and 12.8, this graphical procedure is much easier to remember and it provides a functional depiction of the relationships between stress components acting on different planes at a point.

MecMovies 12.15 presents an animated derivation of the Mohr's circle stress transformation equations.

The German civil engineer Otto Christian Mohr (1835–1918) developed a useful graphical interpretation of the stress transformation equation. This method is known as Mohr's circle. Although it will be used for plane stress transformations here, the Mohr's circle method is also valid for other transformations that are similar mathematically, such as area moments of inertia, mass moments of inertia, strain transformations, and three-dimensional stress transformations.

Derivation of the Circle Equation

Mohr's circle for plane stress is constructed with the normal stress σ plotted along the horizontal axis and the shear stress τ plotted along the vertical axis. The circle is constructed such that each point it represents is a combination of σ and τ that acts on one specific plane through a point in a stressed body. The general plane stress transformation equations, expressed with double-angle trigonometric functions, were presented as Equations (12.5) and (12.6), respectively, in Section 12.7:

$$\sigma_n = \frac{\sigma_x + \sigma_y}{2} + \frac{\sigma_x - \sigma_y}{2}\cos 2\theta + \tau_{xy}\sin 2\theta$$

$$\tau_{nt} = -\frac{\sigma_x - \sigma_y}{2}\sin 2\theta + \tau_{xy}\cos 2\theta$$

Equations (12.5) and (12.6) can be rewritten with terms involving 2θ on the right-hand side:

$$\sigma_n - \frac{\sigma_x + \sigma_y}{2} = \frac{\sigma_x - \sigma_y}{2}\cos 2\theta + \tau_{xy}\sin 2\theta$$

$$\tau_{nt} = -\frac{\sigma_x - \sigma_y}{2}\sin 2\theta + \tau_{xy}\cos 2\theta$$

Both equations can be squared, then added together, and simplified to give

$$\left(\sigma_n - \frac{\sigma_x + \sigma_y}{2}\right)^2 + \tau_{nt}^2 = \left(\frac{\sigma_x - \sigma_y}{2}\right)^2 + \tau_{xy}^2 \quad (12.20)$$

This is the equation of a circle in terms of the variables σ_n and τ_{nt}. The center of the circle is located on the σ axis (i.e., $\tau = 0$), at

$$C = \frac{\sigma_x + \sigma_y}{2} \quad (12.21)$$

The radius of the circle is given by the square root of the right-hand side of Equation (12.20):

$$R = \sqrt{\left(\frac{\sigma_x - \sigma_y}{2}\right)^2 + \tau_{xy}^2} \quad (12.22)$$

MecMovies 12.16 presents a step-by-step guide to constructing Mohr's circle for plane stress.

Equation (12.20) can be written in terms of C and R as

$$(\sigma_n - C)^2 + \tau_{nt}^2 = R^2 \quad (12.23)$$

which is the standard algebraic equation for a circle with radius R and center C.

Utility of Mohr's Circle

Mohr's circle can be used to determine stresses acting on any plane passing through a point. It is quite convenient for determining principal stresses and maximum shear stresses (both in-plane and absolute maximum shear stresses). If Mohr's circle is plotted to scale, measurements taken directly from the plot can be used to obtain stress values. However, Mohr's circle is probably most useful as a pictorial aid for the analyst who is performing analytical determinations of stresses and their directions at a point.

MecMovies 12.17 shows how principal stresses and principal planes are found with Mohr's circle.

Sign Conventions Used in Plotting Mohr's Circle

In constructing Mohr's circle, normal stresses are plotted as horizontal coordinates and shear stresses are plotted as vertical coordinates. Consequently, the horizontal axis is termed the σ axis and the vertical axis is termed the τ axis. To reiterate, Mohr's circle for plane stress is a circle plotted entirely in terms of the normal stress σ and shear stress τ.

MecMovies 12.18 illustrates how the maximum in-plane shear stress is found from Mohr's circle.

Normal Stresses. Tensile normal stresses are plotted on the right side of the τ axis, and compressive normal stresses are plotted on the left side of the τ axis. In other words, tensile normal stress is plotted as a positive value (algebraically) and compressive normal stress is plotted as a negative value.

Shear Stresses. A unique sign convention is required to determine whether a particular shear stress plots above or below the σ axis. The shear stress τ_{xy} acting on the x face must always equal the shear stress τ_{yx} acting on the y face. (See Section 12.3.) If a positive shear stress acts on the x face of the stress element, then a positive shear stress will also act on the y face, and vice versa. For shear stress, therefore, an ordinary sign convention (such as positive τ plots above the σ axis and negative τ plots below the σ axis) is not sufficient because

(a) the shear stresses on both the x and y faces will always have the same sign and
(b) the center of Mohr's circle must be located on the σ axis. [See Equation (12.20).]

To determine how a shear stress value should be plotted, one must consider both the face that the shear stress acts on and the direction in which the shear stress acts:

- If the shear stress acting on a face of the stress element tends to rotate the stress element in a clockwise direction, then the shear stress is plotted above the σ axis.
- If the shear stress tends to rotate the stress element in a counterclockwise direction, then the shear stress is plotted below the σ axis.

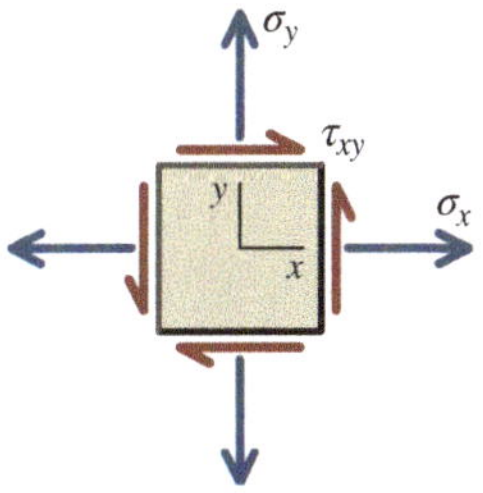

Basic Construction of Mohr's Circle

Mohr's circle can be constructed in several ways, depending on which stresses are known and which stresses are to be found. To illustrate the basic construction of Mohr's circle for plane stress, assume that stresses σ_x, σ_y, and τ_{xy} are known. Then, the following procedure can be used to construct the circle:

1. Identify the stresses acting on orthogonal planes at a point. These are usually the stresses σ_x, σ_y, and τ_{xy} acting on the x and y faces of the stress element. It is helpful to draw a stress element before beginning construction of Mohr's circle.
2. Draw a pair of coordinate axes. The σ axis is horizontal. The τ axis is vertical. It is not mandatory, but it is helpful, to construct Mohr's circle at least approximately to scale. Pick an appropriate stress interval for the data, and use the same interval for both σ and τ.

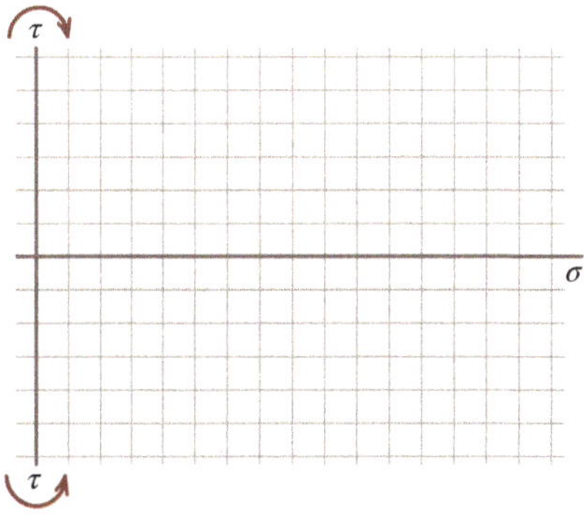

Label the upper half of the τ axis with a clockwise arrow. Label the lower half with a counterclockwise arrow. These symbols will help you remember the sign convention used in plotting shear stresses.

3. Plot the state of stress acting on the x face. If σ_x is positive (i.e., tension), then the point is plotted to the right of the τ axis. Conversely, a negative σ_x plots to the left of the τ axis.

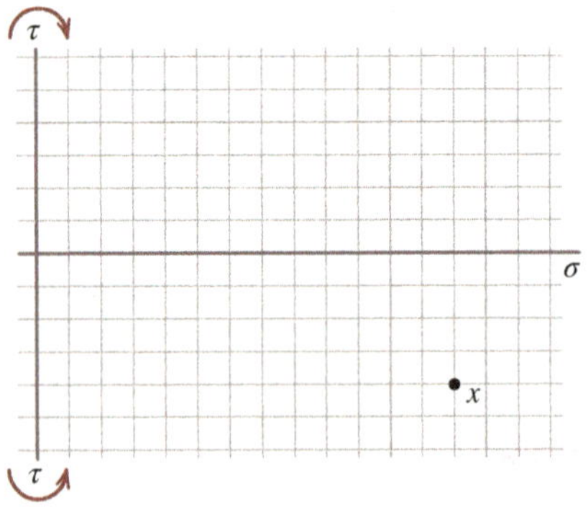

Correctly plotting the value of τ_{xy} is easier if you use the clockwise–counterclockwise sign convention. Look at the shear stress arrow on the x face. If this arrow tends to rotate the stress element clockwise, then plot the point above the σ axis. For the stress element shown here, the shear stress acting on the x face tends to rotate the element counterclockwise; therefore, the point should be plotted below the σ axis.

4. Label the point plotted in step 3 point x. This point represents the combination of normal and shear stress on a specific plane surface, specifically the x face of the stress element. Keep in mind that the coordinates used in plotting Mohr's circle are not spatial coordinates like x and y distances, which are more commonly used in other settings. Rather, the coordinates of Mohr's circle are σ and τ. To establish orientations of specific planes by using Mohr's circle, we must determine angles relative to some reference point, such as the point x, which represents the state of stress on the x face of the stress element. Consequently, it is very important to label the points as they are plotted.

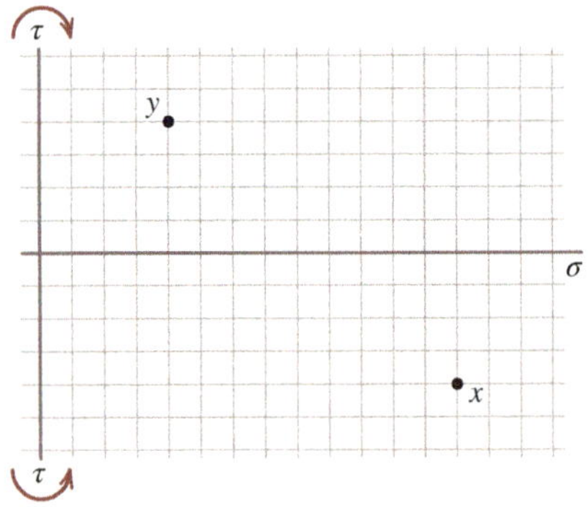

5. Plot the state of stress acting on the y face. Look at the shear stress arrow on the y face of the stress element shown in the previous figure. This arrow tends to rotate the element clockwise; therefore, the point is plotted above the σ axis. Label this point y, since it represents the combination of normal and shear stress acting on the y face of the stress element.

Notice that points x and y are both the same distance away from the σ axis—one point is above the σ axis, and the other point is below. This configuration will always be true because the shear stress acting on the x and y faces must always have the same magnitude. [See Section 12.3 and Equation (12.2).]

6. Draw a line connecting points x and y. The location where this line crosses the σ axis marks the center C of Mohr's circle.

The radius R of Mohr's circle is the distance from center C to point x or to point y. Moreover,

as shown by Equation (12.23), the center C of Mohr's circle will always lie on the σ axis.

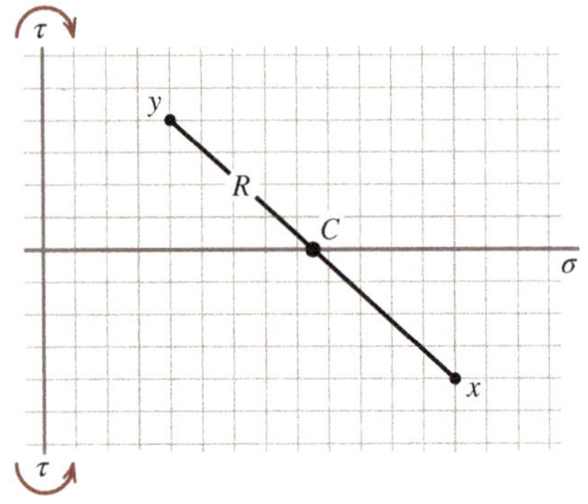

7. Draw a circle with center C and radius R. Every point on the circle represents a combination of σ and τ that exists at some orientation.

The equations used to derive Mohr's circle [Equations (12.5) and (12.6)] were expressed in terms of double-angle trigonometric functions. Consequently, all angular measures in Mohr's circle are double angles 2θ. Points x and y, which represent stresses on planes 90° apart in the x–y coordinate system, are 180° apart in the σ–τ coordinate system of Mohr's circle. Points at the ends of any diameter represent stresses on orthogonal planes in the x–y coordinate system.

8. Several points on Mohr's circle are of particular interest. The principal stresses are the extreme values of the normal stress that exist in the stressed body, given the specific set of stresses σ_x, σ_y, and τ_{xy} that act in the x and y directions. From Mohr's circle, the extreme values of σ are observed to occur at the two points where the circle crosses the σ axis. The more positive point (in an algebraic sense) is σ_{p1}, and the more negative point is σ_{p2}.

   Notice that the shear stress τ at both points is zero. As discussed previously, the shear stress τ is always zero on planes where the normal stress σ has a maximum or a minimum value.

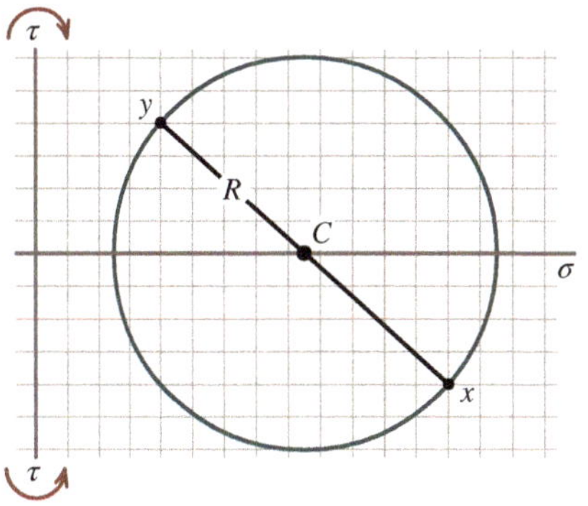

9. The geometry of Mohr's circle can be used to determine the orientation of the principal planes. From the geometry of the circle, the angle between point x and one of the principal stress points can be determined. The angle between point x and one of the principal stress points on the circle is $2\theta_p$. In addition to the magnitude of $2\theta_p$, the sense of the angle (either clockwise or counterclockwise) can be determined from the circle by inspection. The rotation of $2\theta_p$ *from* point x *to* the principal stress point should be determined.

   In the x–y coordinate system of the stress element, the angle between the x face of the stress element and a principal plane is θ_p, where θ_p rotates in the same sense (either clockwise or counterclockwise) in the x–y coordinate system as $2\theta_p$ does in Mohr's circle.

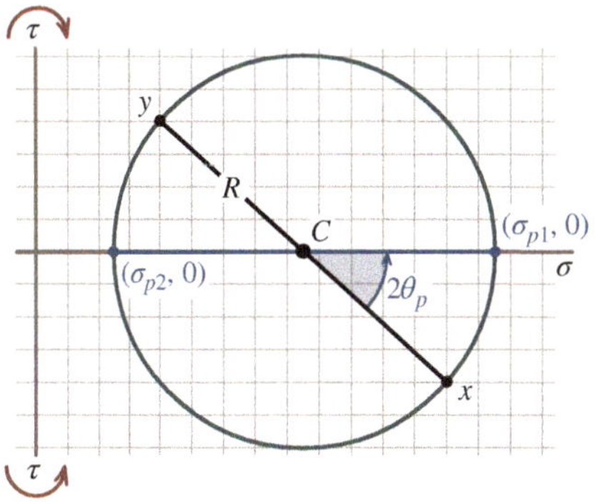

10. Two additional points of interest on Mohr's circle are the extreme shear stress values. The largest shear stress magnitudes will occur at points located at the top and at the bottom of the circle. Since the center C of the circle is always located on the σ axis, the largest possible value of τ is simply the radius R. Note that these two points occur directly above and directly below the center C. In contrast to the principal planes, which always have zero shear stress, the planes of maximum shear stress generally do have a normal stress. The magnitude of this normal stress is identical to the σ coordinate of the center C of the circle.

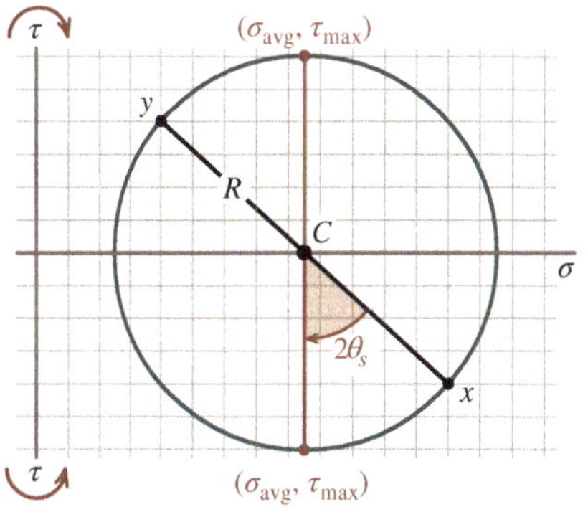

11. Notice that the angle between the principal stress points and the maximum shear stress points on Mohr's circle is 90°. Since angles in Mohr's circle are doubled, the actual angle between the principal planes and the maximum shear stress planes will always be 45°.

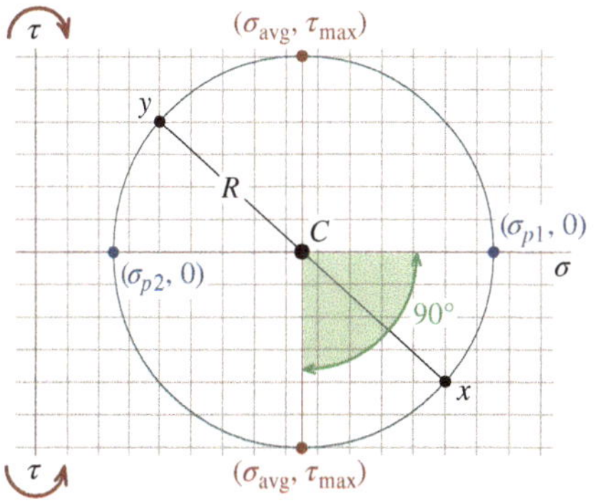

The stress transformation equations presented in Sections 12.7 and 12.8 and the Mohr's circle construction presented here are two methods for attaining the same result. The advantage offered by Mohr's circle is that it provides a concise visual summary of all stress combinations possible at any point in a stressed body. Since all stress calculations can be performed with the geometry of the circle and basic trigonometry, Mohr's circle provides an easy-to-remember tool for stress analysis. While developing mastery of stress analysis, the student may find it less confusing to avoid mixing the stress transformation equations presented in Sections 12.7 and 12.8 with Mohr's circle construction. Take advantage of Mohr's circle by using the geometry of the circle to compute all desired quantities rather than trying to merge the stress transformation equations into the Mohr's circle analysis.

MecMovies

EXAMPLE

M12.10 Coach Mohr's Circle of Stress

Learn to construct and use Mohr's circle to determine principal stresses, including the proper orientation of the principal stress planes.

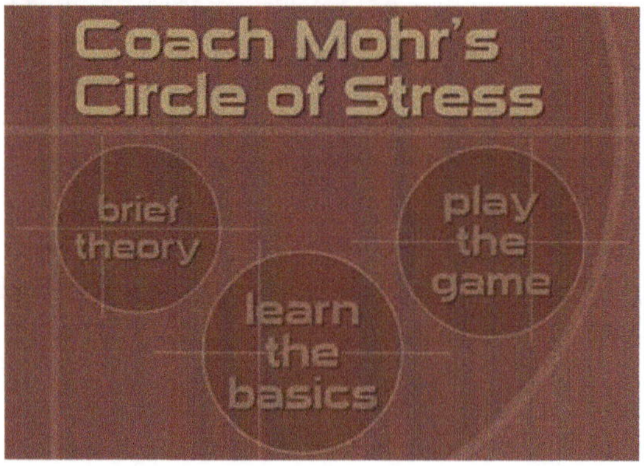

EXAMPLE 12.7

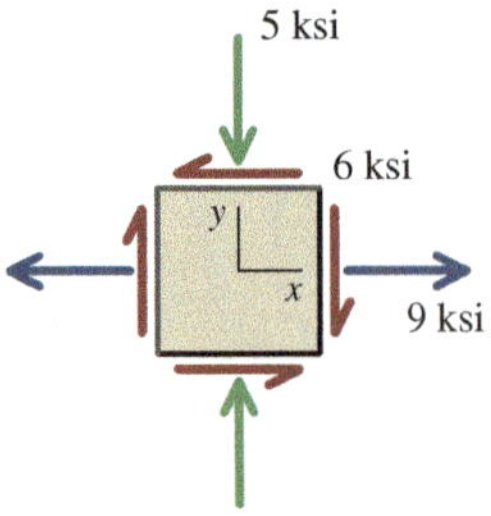

Principal and Maximum In-Plane Shear Stresses

Consider a point in a structural member that is subjected to plane stress. Normal and shear stresses acting on horizontal and vertical planes at the point are shown.

(a) Determine the principal stresses and the maximum in-plane shear stress acting at the point.

(b) Show these stresses in an appropriate sketch.

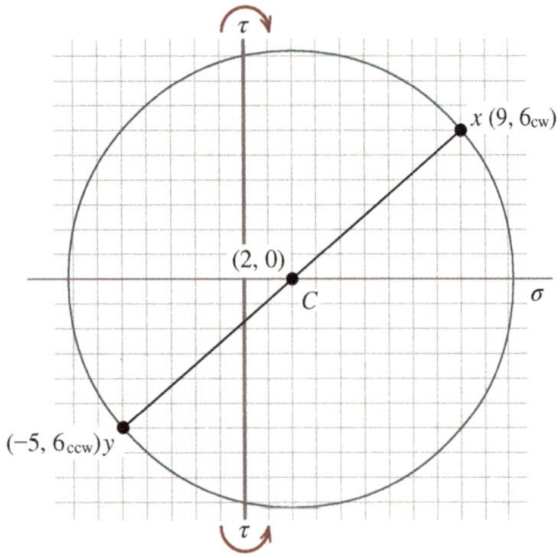

SOLUTION

Begin with the normal and shear stresses acting on the x face of the stress element. Since $\sigma_x = 9$ ksi is a tension stress, the point on Mohr's circle will be plotted to the right of the τ axis. The shear stress acting on the x face tends to rotate the stress element clockwise; therefore, point x on Mohr's circle is plotted above the σ axis.

On the y face, the normal stress $\sigma_y = -5$ ksi will be plotted to the left of the τ axis. The shear stress acting on the y face tends to rotate the stress element counterclockwise; therefore, point y on Mohr's circle is plotted below the σ axis.

Note: Attaching either a positive or a negative sign to τ values at points x and y does not add any useful information for the Mohr's circle stress analysis. Once the circle has been properly constructed, all computations are based on the geometry of the circle, irrespective of any signs. In this introductory example, the subscript "cw" has been added to the shear stress at point x simply to emphasize that the shear stress on the x face rotates the element clockwise. Similarly, the subscript "ccw" is meant to emphasize that the shear stress on the y face rotates the element counterclockwise.

Since points x and y are always the same distance above or below the σ axis, the center of Mohr's circle can be found by averaging the normal stresses acting on the x and y faces:

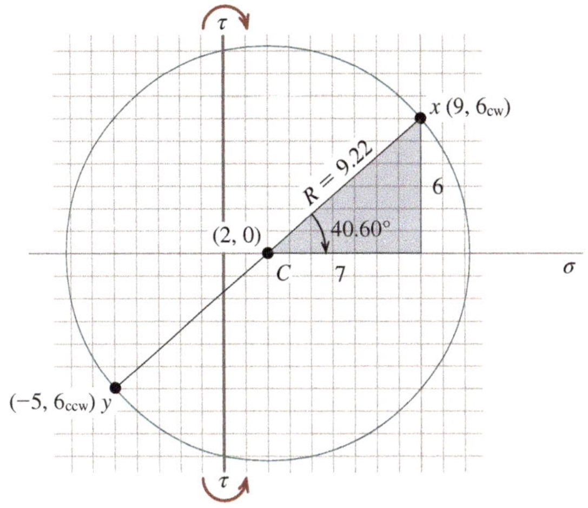

$$C = \frac{\sigma_x + \sigma_y}{2} = \frac{9 \text{ ksi} + (-5 \text{ ksi})}{2} = 2 \text{ ksi}$$

The center of Mohr's circle always lies on the σ axis.

The geometry of the circle is used to calculate the radius. The (σ, τ) coordinates of point x and center C are known. Use these coordinates with the Pythagorean theorem to calculate the hypotenuse of the shaded triangle:

$$R = \sqrt{(9 \text{ ksi} - 2 \text{ ksi})^2 + (6 \text{ ksi} - 0)^2}$$
$$= \sqrt{7^2 + 6^2} = 9.22 \text{ ksi}$$

The angle between the x–y diameter and the σ axis is $2\theta_p$, and it can be computed by means of the tangent function:

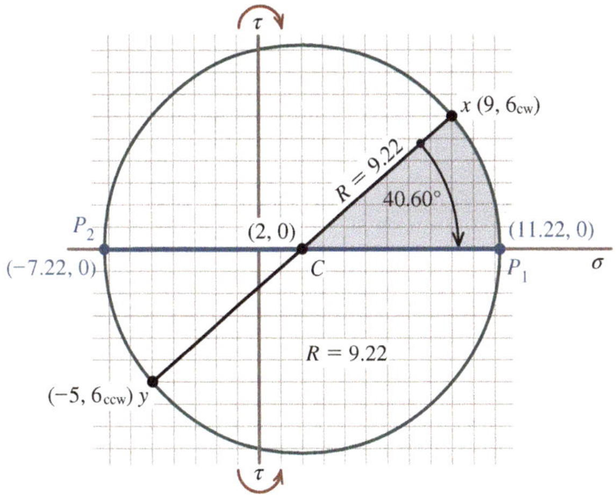

$$\tan 2\theta_p = \frac{6}{7} \qquad \therefore\ 2\theta_p = 40.60°$$

Note that this angle turns clockwise from point x to the σ axis.

The maximum value of σ (i.e., the most positive value algebraically) occurs at point P_1, where Mohr's circle crosses the σ axis. From the geometry of the circle,

$$\sigma_{p1} = C + R = 2 \text{ ksi} + 9.22 \text{ ksi} = 11.22 \text{ ksi}$$

The minimum value of σ (i.e., the most negative value algebraically) occurs at point P_2. Again, from the geometry of the circle,

$$\sigma_{p2} = C - R = 2 \text{ ksi} - 9.22 \text{ ksi} = -7.22 \text{ ksi}$$

The angle between point x and point P_1 was calculated as $2\theta_p = 40.60°$; however, angles in Mohr's circle are double angles, so, to determine the orientation of the principal planes in the x–y coordinate system, divide 40.60° by 2. Therefore, the principal stress σ_{p1} acts on a plane rotated 20.30° from the x face of the stress element. The 20.30° angle in the x–y coordinate system rotates in the same sense as $2\theta_p$ in Mohr's circle. In this example, the 20.30° angle rotates *clockwise* from the x axis.

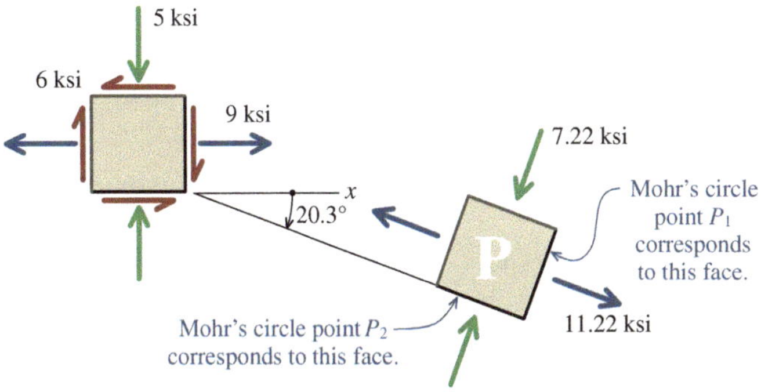

The principal stresses, as well as the orientation of the principal planes, are shown in the sketch.

The maximum values of τ occur at points S_1 and S_2, located at the bottom and at the top of Mohr's circle. The

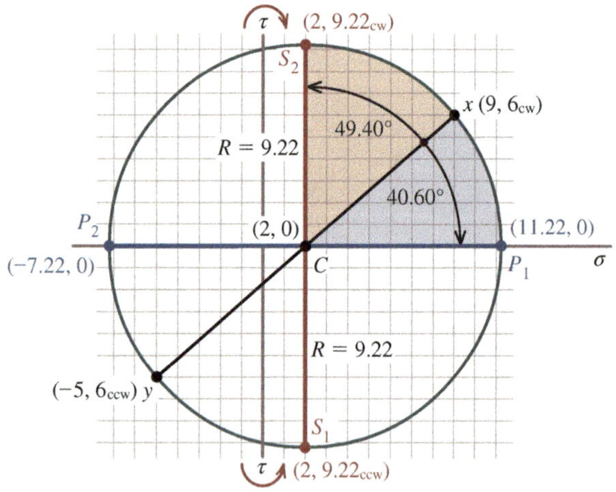

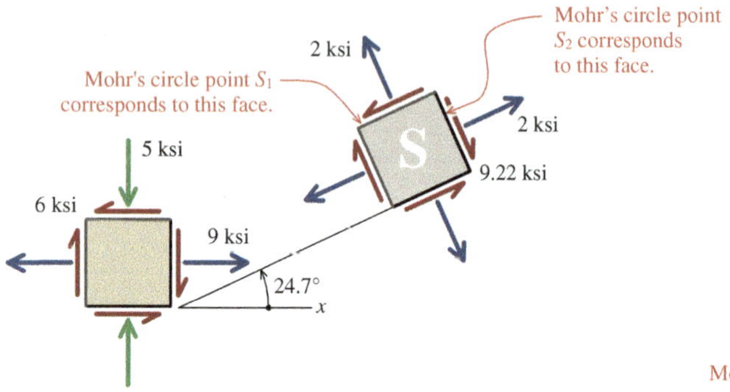

shear stress magnitude at these points is simply equal to the radius R of the circle. Notice that the normal stress at points S_1 and S_2 is not zero. Rather, the normal stress σ at these points is equal to the stress at the center C of the circle.

The angle between points P_1 and S_2 is 90°. Since the angle between point x and point P_1 was found to be 40.60°, the angle between point x and point S_2 must be 49.40°. This angle rotates in a counterclockwise direction.

One plane subjected to the maximum in-plane shear stress is oriented 24.70° counterclockwise from the x face. The magnitude of this shear stress is equal to the radius of the circle:

$$\tau_{max} = R = 9.22 \text{ ksi}$$

To determine the direction of the shear stress arrow acting on this face, note that point S_2 is on the upper half of the circle, above the σ axis. Consequently, the shear stress acting on this face rotates the stress element clockwise. Once the shear stress direction on one face has been determined, the shear stress directions on the other three faces are known.

A complete sketch showing the principal stresses, the maximum in-plane shear stress, and the orientations of the respective planes can now be prepared.

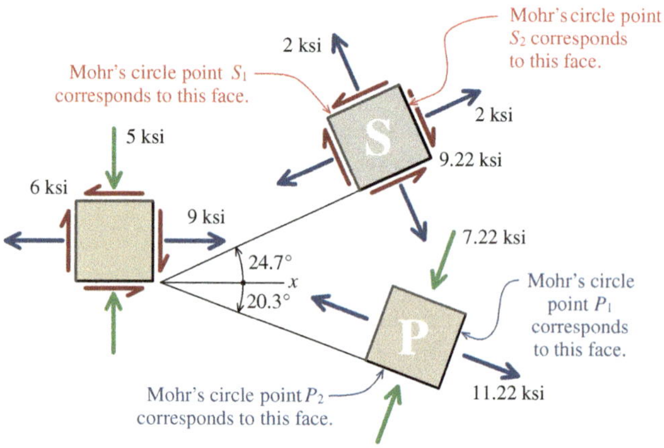

EXAMPLE 12.8

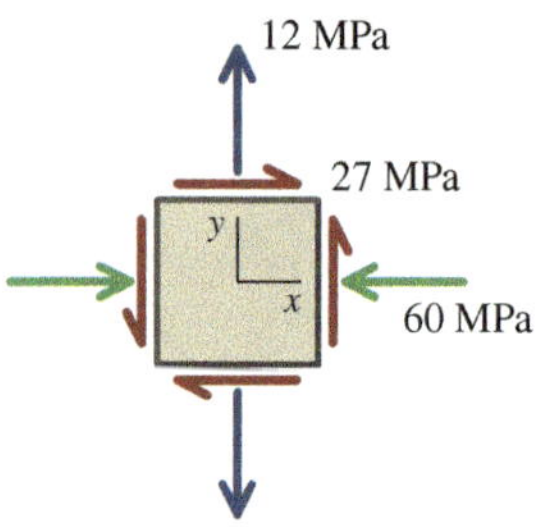

Principal and Maximum In-Plane Shear Stresses

Consider a point in a structural member that is subjected to plane stress. Normal and shear stresses acting on horizontal and vertical planes at the point are shown.

(a) Determine the principal stresses and the maximum in-plane shear stress acting at the point.
(b) Show these stresses in an appropriate sketch.

SOLUTION

Begin with the normal and shear stresses acting on the x face of the stress element. The normal stress is $\sigma_x = 60$ MPa (C), and the shear stress τ acting on the x face rotates the element counterclockwise; therefore, point x is located to the left of the τ axis and below the σ axis. On the y face, the normal stress is σ_y = 12 MPa (T), and the shear stress τ acting on the y face rotates the element clockwise; therefore, point y is located to the right of the τ axis and above the σ axis.

Note: In this introductory example, the subscript "ccw" has been added to the shear stress at point x simply to give further emphasis to the fact that the shear stress on the x face rotates the element counterclockwise. Similarly, the subscript "cw" added to the shear stress at point y is meant to call attention to the fact that the shear stress on the y face rotates the element clockwise.

The center of Mohr's circle can be found by averaging the normal stresses acting on the x and y faces:

$$C = \frac{\sigma_x + \sigma_y}{2} = \frac{(-60\text{ MPa}) + 12\text{ MPa}}{2} = -24\text{ MPa}$$

The radius R is found from the hypotenuse of the shaded triangle:

$$R = \sqrt{[(-60\text{ MPa}) - (-24\text{ MPa})]^2 + (27\text{ MPa} - 0)^2}$$
$$= \sqrt{36^2 + 27^2} = 45\text{ MPa}$$

The angle between the x–y diameter and the σ axis is $2\theta_p$, and it can be computed with the use of the tangent function:

$$\tan 2\theta_p = \frac{27}{36} \qquad \therefore\ 2\theta_p = 36.86°$$

Notice that this angle turns clockwise from point x to the σ axis.

The principal stresses are determined from the location of the center C and radius R of the circle:

$$\sigma_{p1} = C + R = -24\text{ MPa} + 45\text{ MPa} = 21\text{ MPa}$$
$$\sigma_{p2} = C - R = -24\text{ MPa} - 45\text{ MPa} = -69\text{ MPa}$$

The maximum values of τ occur at points S_1 and S_2, located at the bottom and at the top of Mohr's circle. The shear stress magnitude at these points is simply equal to the radius R of the circle, and the normal stress σ at these points is equal to the center C.

The angle between points P_2 and S_2 is 90°. Since the angle between point x and point P_2 is 36.86°, the angle between point x and point S_1 must be 53.14°. By inspection, this angle rotates in a counterclockwise direction.

The angle between point x and point P_2 was calculated as $2\theta_p$ = 36.86°. Thus, this principal plane

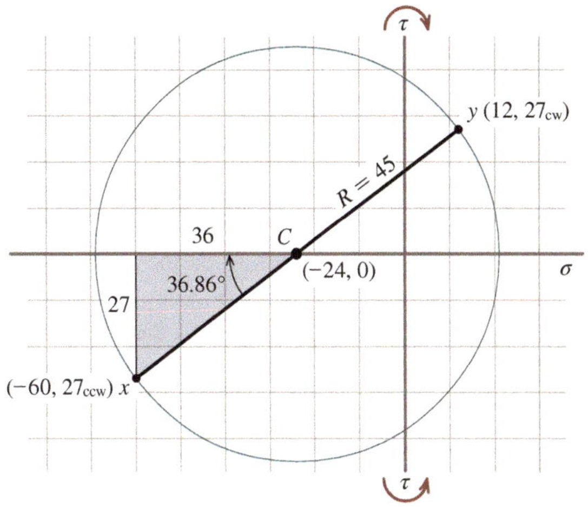

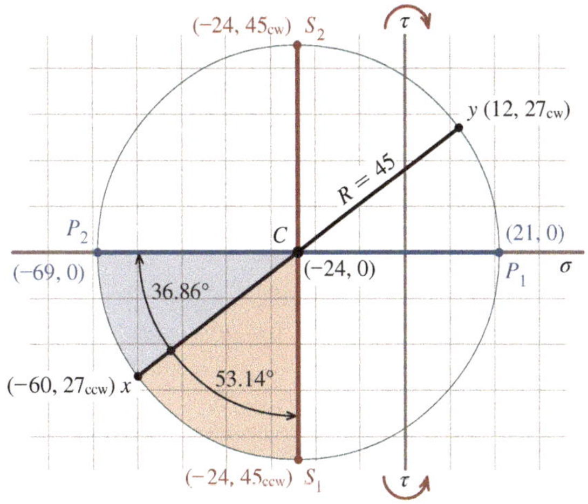

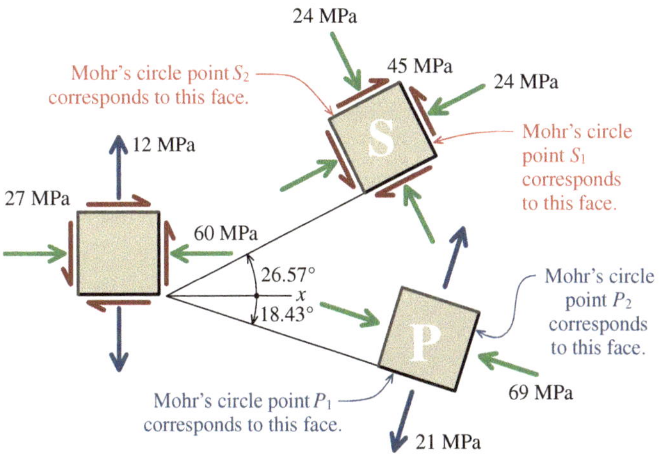

in the x–y coordinate system is rotated 18.43° clockwise from the x face of the stress element.

The angle between point x and point S_1 is 53.14°; therefore, this plane of maximum in-plane shear stress in the x–y coordinate system is rotated 26.57° counterclockwise from the x face of the stress element.

To determine the direction of the shear stress arrow acting on this face, note that point S_1 is on the lower half of the circle, below the σ axis. Consequently, the shear stress acting on the x face rotates the stress element counterclockwise.

A complete sketch showing the principal stresses, the maximum in-plane shear stress, and the orientations of the respective planes is given.

EXAMPLE 12.9

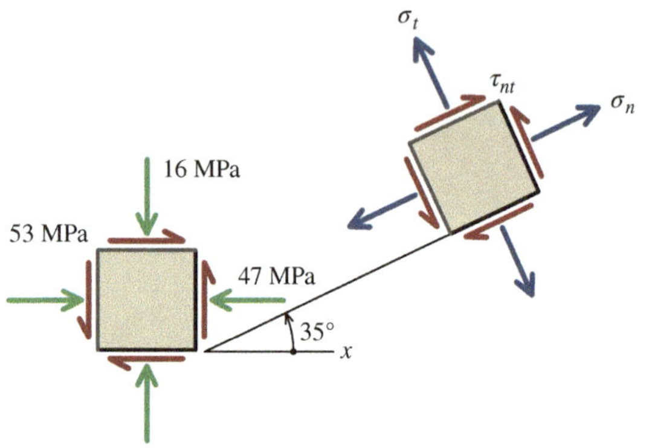

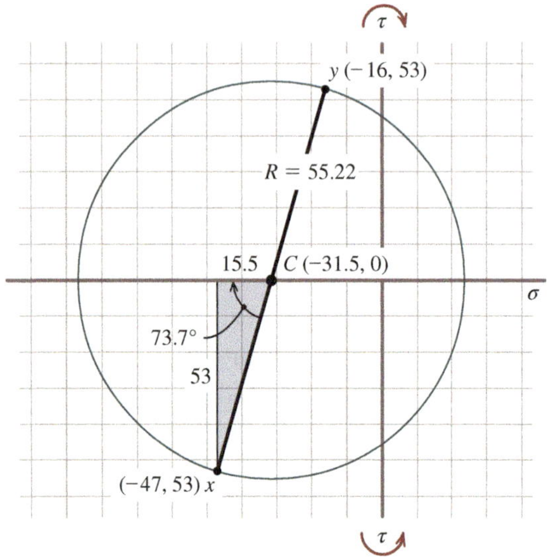

Stresses on an Inclined Plane

The stresses shown act at a point on the free surface of a stressed body.

(a) Determine the principal stresses and the maximum in-plane shear stress acting at the point.
(b) Show these stresses in an appropriate sketch.
(c) Determine the normal stresses σ_n and σ_t and the shear stress τ_{nt} that act on the rotated stress element.

SOLUTION

Construct Mohr's Circle

From the normal and shear stresses acting on the x and y faces of the stress element, Mohr's circle is constructed as shown.

The center of Mohr's circle is located at

$$C = \frac{-47 + (-16)}{2} = -31.5 \text{ MPa}$$

The radius R is found from the hypotenuse of the shaded triangle:

$$R = \sqrt{15.5^2 + 53^2} = 55.22 \text{ MPa}$$

The angle between the x–y diameter and the σ axis is $2\theta_p$, and it can be computed as follows:

$$\tan 2\theta_p = \frac{53}{15.5} \qquad \therefore\ 2\theta_p = 73.7° \text{ (cw)}$$

Principal and Maximum Shear Stress

The principal stresses (points P_1 and P_2) are determined from the location of the center C of the circle and the radius R:

$$\sigma_{p1} = C + R = -31.5 + 55.22 = 23.72 \text{ MPa}$$
$$\sigma_{p2} = C - R = -31.5 - 55.22 = -86.72 \text{ MPa}$$

The maximum in-plane shear stress corresponds to points S_1 and S_2 on Mohr's circle. The maximum in-plane shear stress magnitude is

$$\tau_{max} = R = 55.22 \text{ MPa}$$

and the normal stress acting on the planes of maximum shear stress is

$$\sigma_{avg} = C = -31.5 \text{ MPa}$$

A complete sketch showing the principal stresses, the maximum in-plane shear stress, and the orientations of the respective planes is shown.

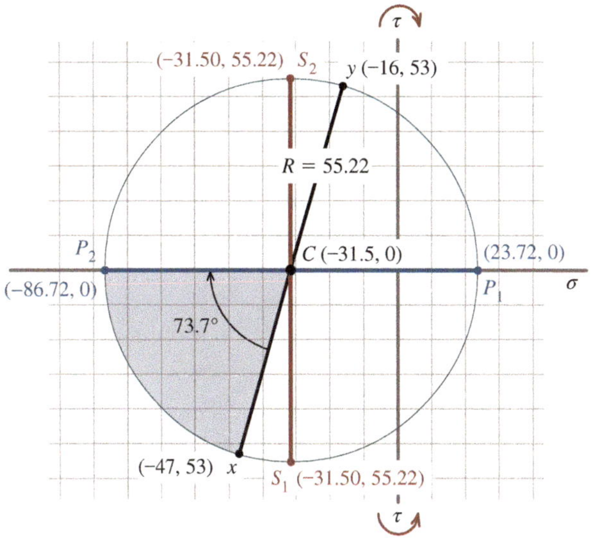

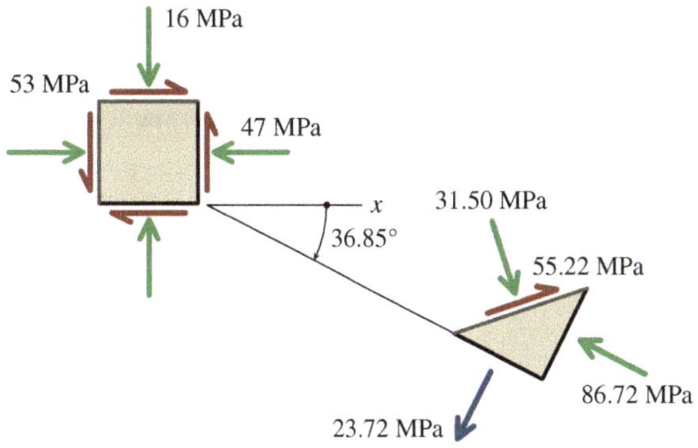

Determine σ_n, σ_t, and τ_{nt}

Next, the normal stresses σ_n and σ_t and the shear stress τ_{nt} acting on a stress element that is rotated 35° counterclockwise from the x direction, as shown in the accompanying sketch, must be determined.

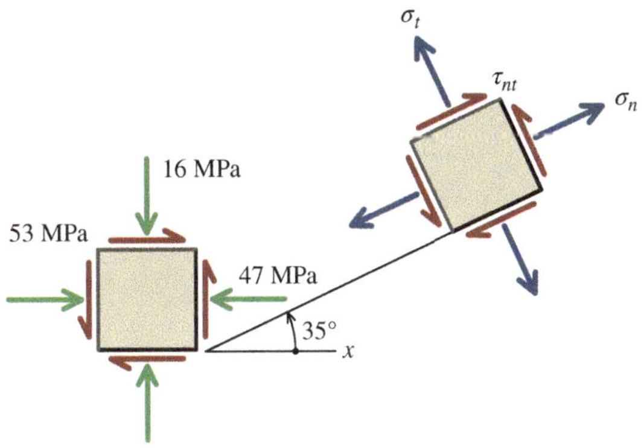

Begin at point x on Mohr's circle. This statement may seem obvious, but it is probably the most common mistake made in solving problems of this type.

In the x–y coordinate system, the 35° angle is rotated counterclockwise from the horizontal axis. As one transfers this angular measure to Mohr's circle, the natural tendency is to draw a diameter that is rotated 2(35°) = 70° counterclockwise from the horizontal axis. *This is incorrect!*

Remember that Mohr's circle is a plot in terms of the normal stress σ and shear stress τ. The horizontal axis in Mohr's circle does not necessarily correspond to the x face of the stress element. On Mohr's circle, the point labeled x is the one that corresponds to the x face. (This fact explains why it is so important to label the points as you construct Mohr's circle.)

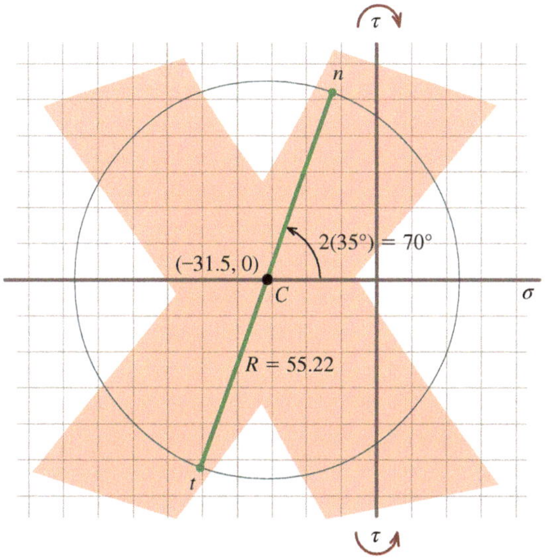

To determine stresses on the plane that is rotated 35° from the x face, a diameter that is rotated 2(35°) = 70° counterclockwise *from point x* is drawn on Mohr's circle. The point 70° away from point x should be labeled point n. The coordinates of this point are the normal and shear stresses acting on the n face of the rotated stress element. The other end of the diameter should be labeled point t, and its coordinates are σ and τ, acting on the t face of the rotated stress element.

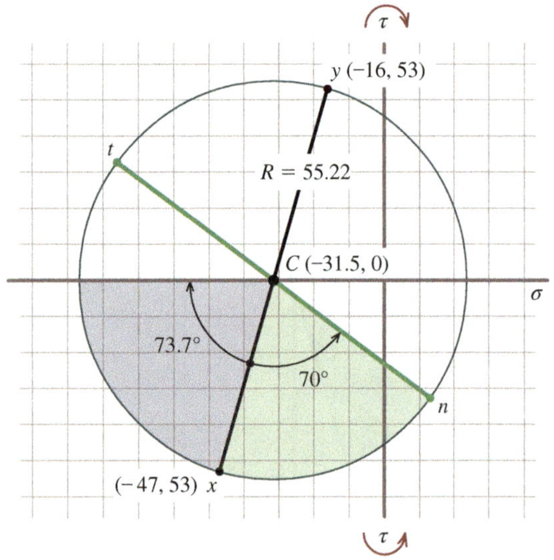

Begin at point x on Mohr's circle. Face n of the rotated stress element is oriented 35° counterclockwise from the x face. Since angles in Mohr's circle are doubled, point n is rotated 2(35°) = 70° counterclockwise from point x on the circle. The coordinates of point n are (σ_n, τ_{nt}). These coordinates will be determined from the geometry of the circle.

By inspection, the angle between the σ axis and point n is 180° − 73.7° − 70° = 36.3°. Keeping in mind that the coordinates of Mohr's circle are σ and τ, we find that the horizontal component of the line between the center C of the circle and point n is

$$\Delta\sigma = R\cos 36.3° = (55.22 \text{ MPa})\cos 36.3° = 44.50 \text{ MPa}$$

and the vertical component is

$$\Delta\tau = R\sin 36.3° = (55.22 \text{ MPa})\sin 36.3° = 32.69 \text{ MPa}$$

The normal stress on the n face of the rotated stress element can be computed by using the coordinates of the center C and $\Delta\sigma$:

$$\sigma_n = -31.5 \text{ MPa} + 44.50 \text{ MPa} = 13.0 \text{ MPa}$$

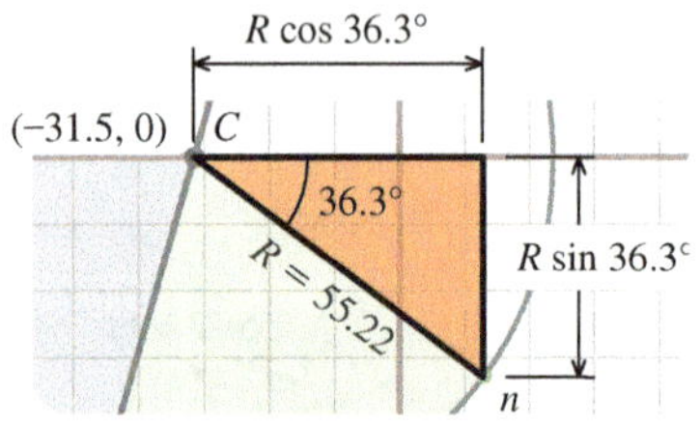

The shear stress is computed similarly:

$$\tau_{nt} = 0 + 32.69 \text{ MPa} = 32.69 \text{ MPa}$$

Since point n is located below the σ axis, the shear stress acting on the n face tends to rotate the stress element counterclockwise.

A similar procedure is used to determine the stresses at point t. The stress components relative to the center C of the circle are the same: $\Delta\sigma$ = 44.50 MPa and $\Delta\tau$ = 32.69 MPa. The normal stress on the t face of the rotated stress element is

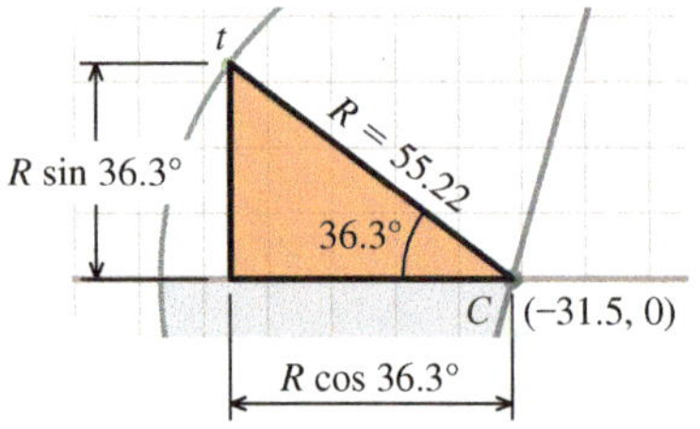

$$\sigma_t = -31.5 \text{ MPa} - 44.50 \text{ MPa} = -76.0 \text{ MPa}$$

Of course, the shear stress acting on the t face must be of the same magnitude as the shear stress acting on the n face. Since point t is located above the σ axis, the shear stress acting on the t face tends to rotate the stress element clockwise.

To determine the normal stress on the t face, we could also use the notion of *stress invariance*. Equation (12.8) shows that the sum of the normal stresses acting on any two orthogonal faces of a plane stress element is a constant value:

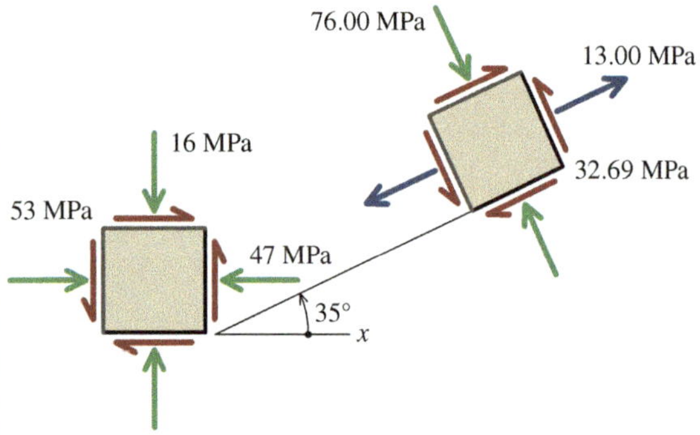

$$\sigma_n + \sigma_t = \sigma_x + \sigma_y$$

Therefore,

$$\begin{aligned}\sigma_t &= \sigma_x + \sigma_y - \sigma_n \\ &= -47 \text{ MPa} + (-16 \text{ MPa}) - 13 \text{ MPa} \\ &= -76 \text{ MPa}\end{aligned}$$

The normal and shear stresses acting on the rotated element are shown in the accompanying sketch.

EXAMPLE 12.10

Stresses on an Inclined Plane

The stresses shown act at a point on the free surface of a stressed body. Determine the normal stress σ_n and the shear stress τ_{nt} that act on the indicated plane surface.

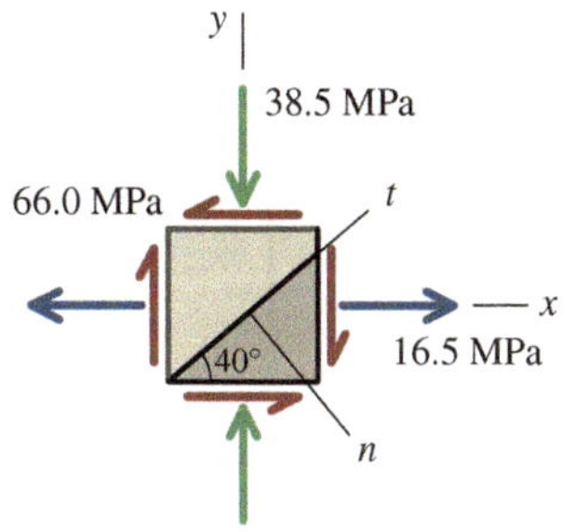

SOLUTION

From the normal and shear stresses acting on the x and y faces of the stress element, Mohr's circle is constructed as shown.

How Is the Orientation of the Inclined Plane Determined?

We must find the angle between the normal to the x face (i.e., the x axis) and the normal to the inclined plane (i.e., the n axis). The angle between the x and n axes is 50°; consequently, the inclined plane is oriented 50° clockwise from the x face.

On Mohr's circle, point n is located 100° clockwise from point x.

Using the coordinates of point x and the center C of the circle, we find the angle between point x and the σ axis to be 67.38°.

Consequently, the angle between point n and the σ axis must be 32.68°.

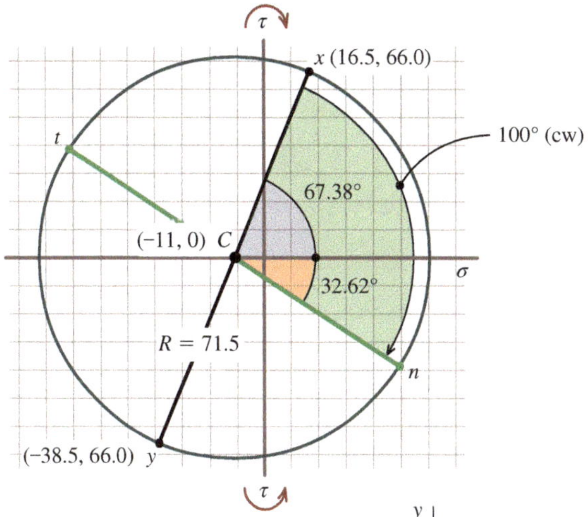

The horizontal component of the line between C and point n is

$$\Delta\sigma = R\cos 32.62° = (71.5 \text{ MPa})\cos 32.62° = 60.22 \text{ MPa}$$

and the vertical component is

$$\Delta\tau = R\sin 32.62° = (71.5 \text{ MPa})\sin 32.62° = 38.54 \text{ MPa}$$

The normal stress on the n face of the rotated stress element can be computed by using the coordinates of the center C and $\Delta\sigma$:

$$\sigma_n = -11.0 \text{ MPa} + 60.22 \text{ MPa} = 49.22 \text{ MPa}$$

The shear stress is computed similarly:

$$\tau_{nt} = 0 + 38.54 \text{ MPa} = 38.54 \text{ MPa}$$

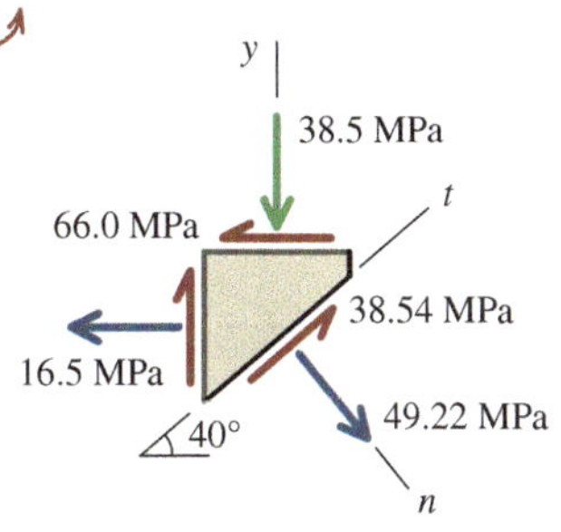

Since point n on Mohr's circle is located below the σ axis, the shear stress acting on the n face tends to rotate the stress element counterclockwise.

MecMovies

EXAMPLE

M12.19 Mohr's Circle Learning Tool

Illustrates the proper usage of Mohr's circle to determine stresses acting on a specified plane, principal stresses, and the maximum in-plane shear stress state for stress values specified by the user. Detailed "how-to" instructions.

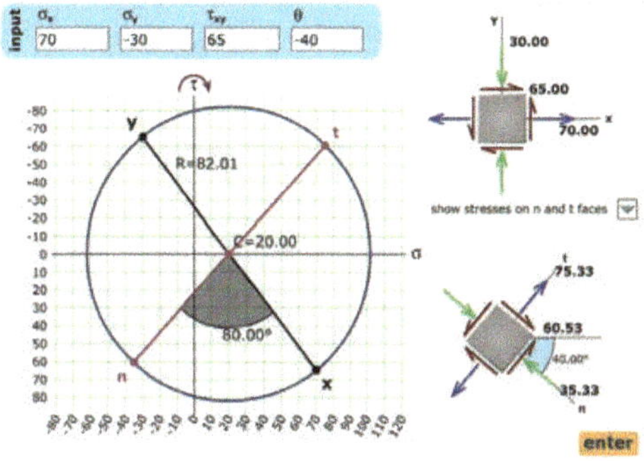

EXAMPLE 12.11

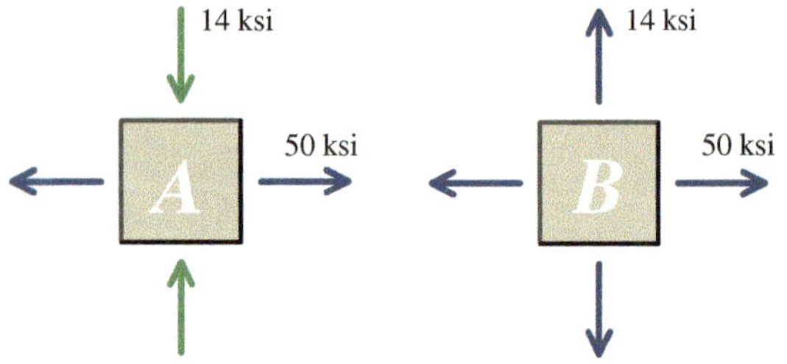

Absolute Maximum Shear Stress

Two elements subjected to plane stress are shown. Determine the absolute maximum shear stress for each element.

SOLUTION

In Section 12.8, it was shown that shear stress does not exist on planes where the maximum and minimum normal stresses occur. Furthermore, the following statement must also be true:

> *If the shear stress on a plane is zero, then that plane must be a principal plane.*

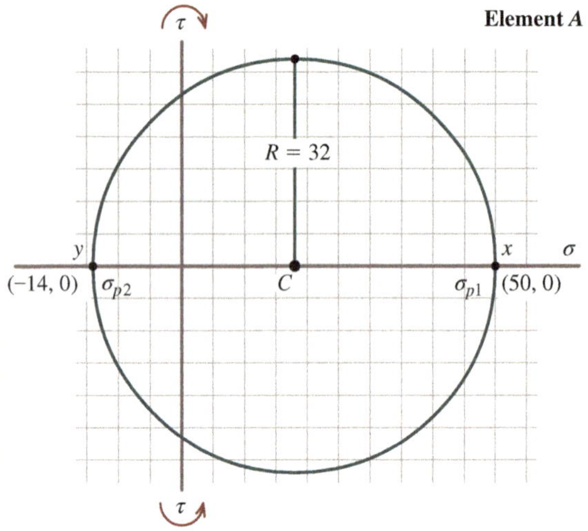

Since there is no shear stress acting on the x and y faces for both element A and element B, one can conclude that the stresses acting on these elements are principal stresses.

The Mohr's circle for element A is constructed as shown. Notice that point x is the principal stress σ_{p1} and point y is the principal stress σ_{p2}. This circle shows all possible combinations of σ and τ that occur in the x–y plane.

What is meant by the term *x–y plane*? This term refers to plane surfaces whose normals are *perpendicular to the z axis.*

The maximum in-plane shear stress for element A is simply equal to the radius of Mohr's circle; therefore, $\tau_{max} = 32$ ksi.

In the statement of the problem, we are told that element A is a point *subjected to plane stress*. From Section 12.4, we know that the term *plane stress* means that there are no stresses on the out-of-plane face of the stress element. In other words, there is no stress on the z face; hence, $\sigma_z = 0$, $\tau_{zx} = 0$, and $\tau_{zy} = 0$. We also know that a plane with no shear stress is, by definition, a principal plane. Therefore, the z face of the stress element is a principal plane, and the principal stress acting on this surface is the third principal stress: $\sigma_z = \sigma_{p3} = 0$.

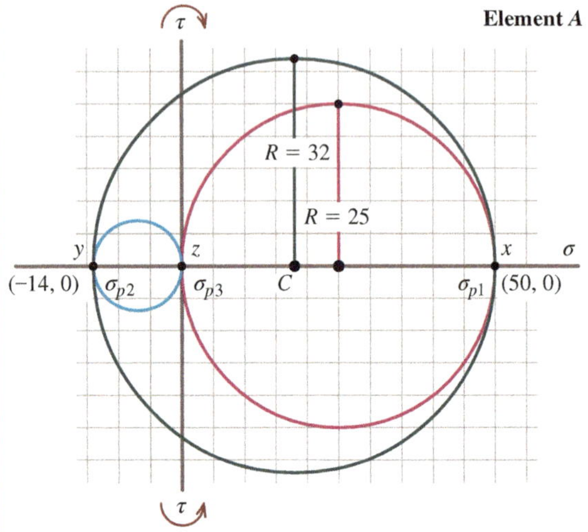

The state of stress on the z face can be plotted on Mohr's circle, and two additional circles can be constructed:

- The circle defined by σ_{p1} and σ_{p3} depicts all combinations of σ–τ that are possible on surfaces in the x–z plane (meaning plane surfaces whose normal is perpendicular to the y axis).
- The circle connecting σ_{p2} and σ_{p3} depicts all combinations of σ–τ that are possible on surfaces in the y–z plane (meaning plane surfaces whose normal is perpendicular to the x axis).

The maximum shear stress in the x–z plane is given by the radius of the Mohr's circle connecting points x and z, and the maximum shear stress in the y–z plane is given by the radius of the circle connecting points y and z. By inspection, both of these circles are smaller than the x–y circle. Consequently, the absolute maximum shear stress—that is, the largest shear stress that can occur on any possible plane—is equal to the maximum in-plane shear stress for element A.

For element A, the absolute maximum shear stress is $\tau_{abs\ max} = 32$ ksi.

The Mohr's circle for element B is constructed in the accompanying plot. This circle shows all possible combinations of σ and τ that occur in the x–y plane.

The maximum in-plane shear stress for element B is equal to the radius of Mohr's circle; therefore, $\tau_{max} = 18$ ksi.

As with element A, the z face of element B is also a principal plane, and therefore, $\sigma_z = \sigma_{p3} = 0$.

Two additional circles can be constructed. The maximum shear stress in the x–z plane is given by the radius of the Mohr's circle connecting points x and z, and the maximum shear stress in the y–z plane is given by the radius of the circle connecting points y and z.

By inspection, the larger of these two circles—the x–z circle—has a greater radius than the x–y circle. Consequently, the absolute maximum shear stress for element B is $\tau_{abs\ max} = 25$ ksi. For element B, the absolute maximum shear stress is greater than the maximum in-plane shear stress.

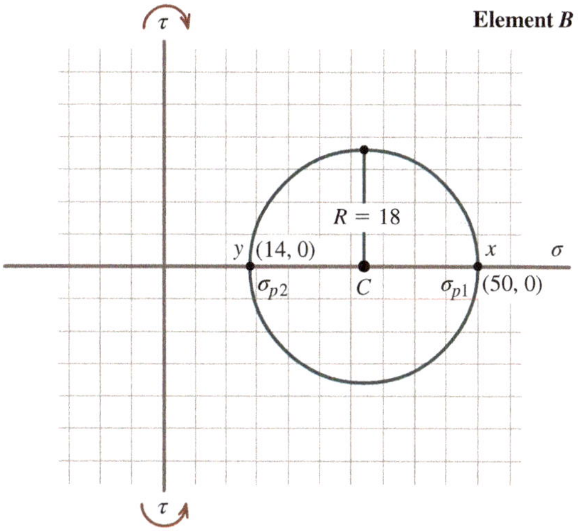

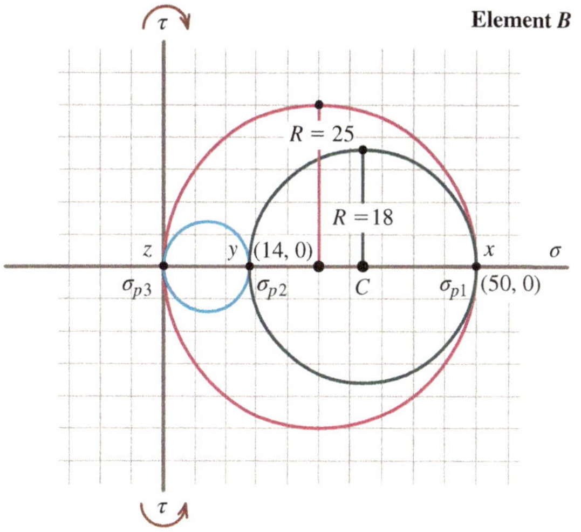

MecMovies

EXAMPLE

M12.13 Using Mohr's circle, interactively investigate a three-dimensional stress state at a point.

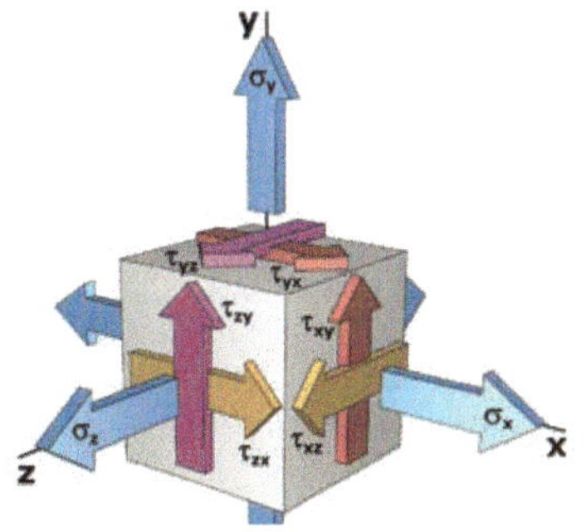

527

EXERCISES

M12.10 Coach Mohr's Circle of Stress. Learn to construct and use Mohr's circle to determine principal stresses, including the proper orientation of the principal stress planes.

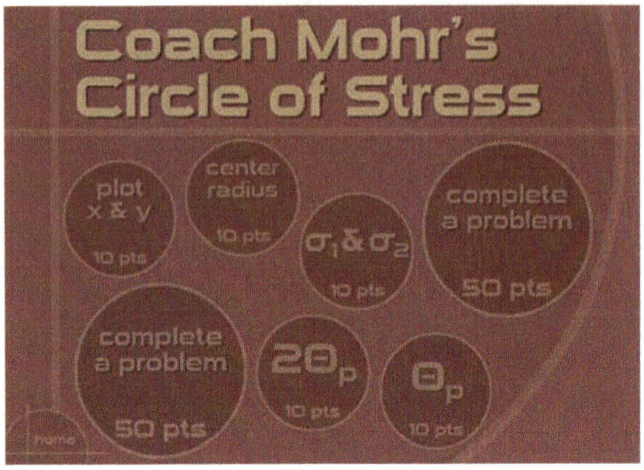

FIGURE M12.10

M12.11 Mohr's Circle Game. Score a minimum of 400 points (out of 450 points possible) in this game, which quizzes your ability to recognize correctly constructed Mohr's circles.

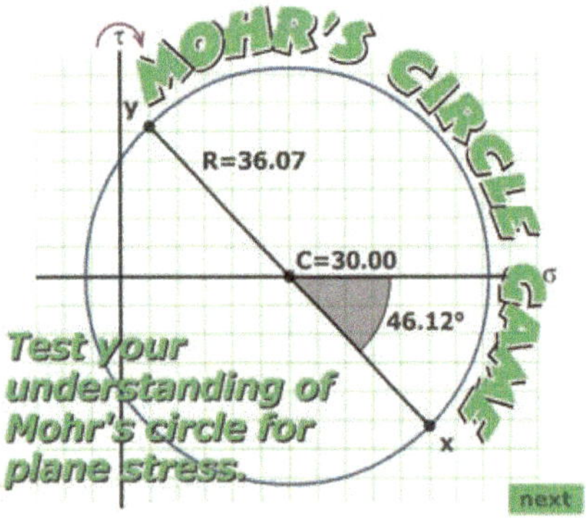

FIGURE M12.11

M12.12 Mohr's Circle Game. Score a minimum of 1,800 points (out of a possible 2,000 points) in this game, which quizzes your ability to recognize the principal stress element or maximum in-plane stress element that corresponds to a given Mohr's circle.

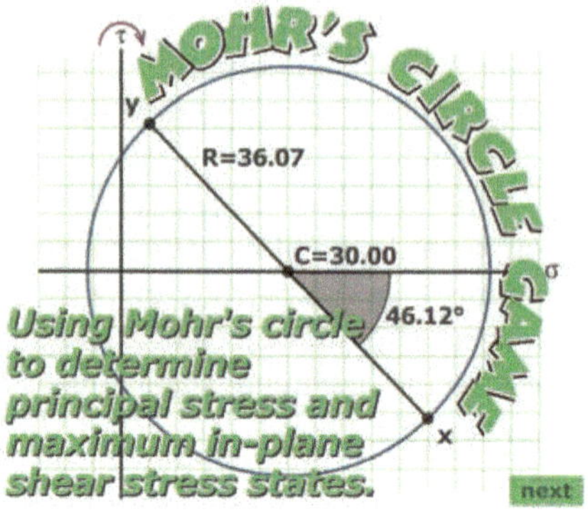

FIGURE M12.12

M12.13 Determine the principal stress magnitudes, the maximum in-plane shear stress magnitude, and the absolute maximum shear stress for a given state of stress.

M12.14 Sketching Stress Transformation Results. Score at least 100 points in this interactive activity.

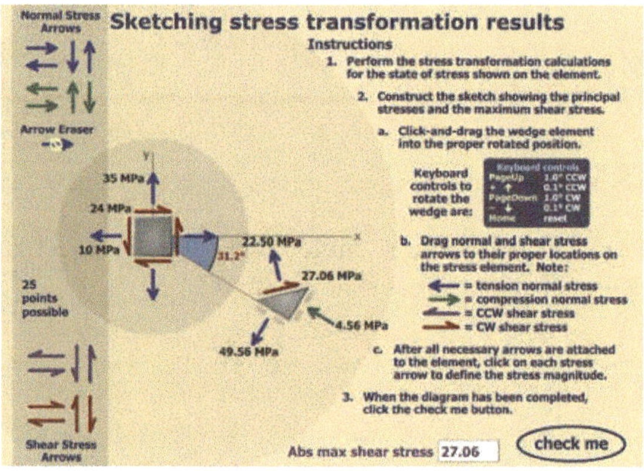

FIGURE M12.14

PROBLEMS

P12.42 Figure P12.42 shows Mohr's circle for a point in a physical object that is subjected to plane stress.

(a) Determine the stresses σ_x, σ_y, and τ_{xy}, and show them on a stress element.

(b) Determine the principal stresses and the maximum in-plane shear stress acting at the point, and show these stresses on an appropriate sketch (e.g., see Figure 12.15 or Figure 12.16).

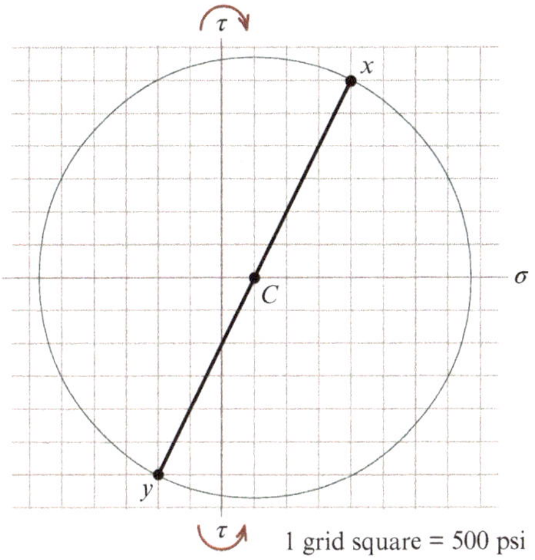

FIGURE P12.42

P12.43 Figure P12.43 shows Mohr's circle for a point in a physical object that is subjected to plane stress.

(a) Determine the stresses σ_x, σ_y, and τ_{xy}, and show them on a stress element.

(b) Determine the principal stresses and the maximum in-plane shear stress acting at the point, and show these stresses on an appropriate sketch (e.g., see Figure 12.15 or Figure 12.16).

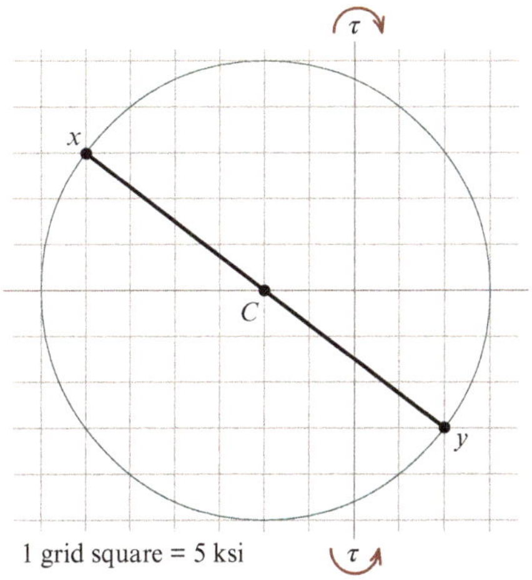

FIGURE P12.43

P12.44 Figure P12.44 shows Mohr's circle for a point in a physical object that is subjected to plane stress.

(a) Determine the stresses σ_x, σ_y, and τ_{xy}, and show them on a stress element.

(b) Determine the principal stresses and the maximum in-plane shear stress acting at the point, and show these stresses on an appropriate sketch (e.g., see Figure 12.15 or Figure 12.16).

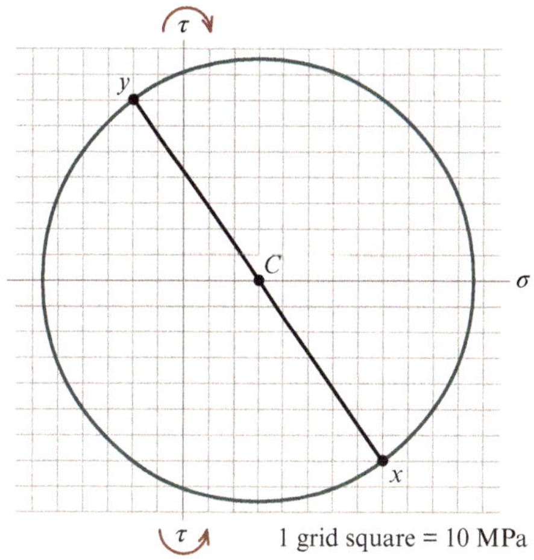

FIGURE P12.44

P12.45 Figure P12.45 shows Mohr's circle for a point in a physical object that is subjected to plane stress.

(a) Determine the stresses σ_x, σ_y, and τ_{xy}, and show them on a stress element.

(b) Determine the principal stresses and the maximum in-plane shear stress acting at the point, and show these stresses on an appropriate sketch (e.g., see Figure 12.15 or Figure 12.16).

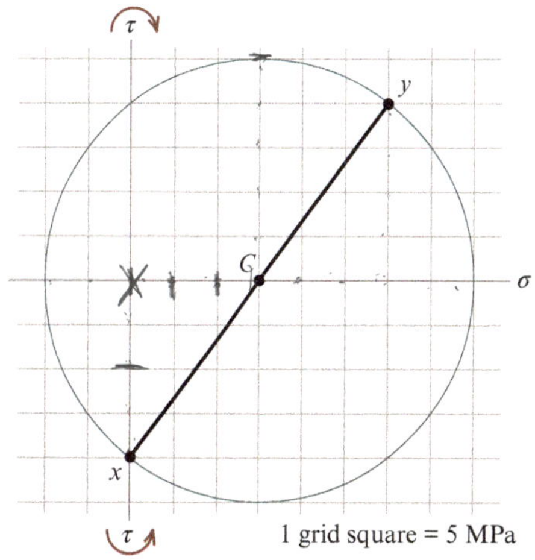

FIGURE P12.45

P12.46 Figure P12.46 shows Mohr's circle for a point in a physical object that is subjected to plane stress.

(a) Determine the stresses σ_x, σ_y, and τ_{xy}, and show them on a stress element.

(b) Determine the stresses σ_n, σ_t, and τ_{nt}, and show them on a stress element that is **properly rotated** with respect to the x–y element. The sketch must include the magnitude of the angle between the x and n axes and an indication of the direction of rotation (i.e., either clockwise or counterclockwise).

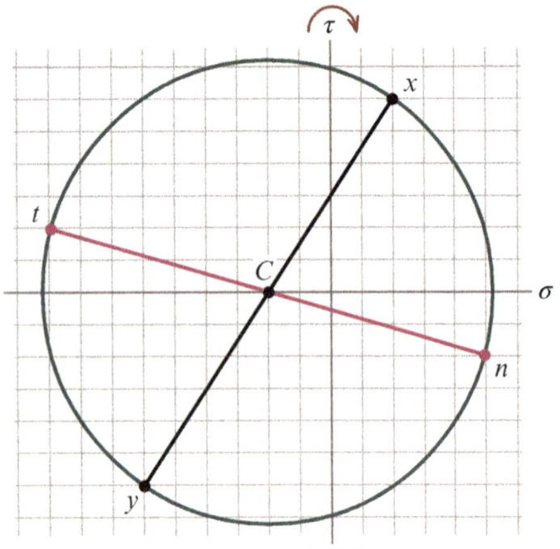

1 grid square = 10 MPa

FIGURE P12.46

P12.47 Figure P12.47 shows Mohr's circle for a point in a physical object that is subjected to plane stress.

(a) Determine the stresses σ_x, σ_y, and τ_{xy}, and show them on a stress element.

(b) Determine the stresses σ_n, σ_t, and τ_{nt}, and show them on a stress element that is **properly rotated** with respect to the x–y element. The sketch must include the magnitude of the angle between the x and n axes and an indication of the direction of rotation (i.e., either clockwise or counterclockwise).

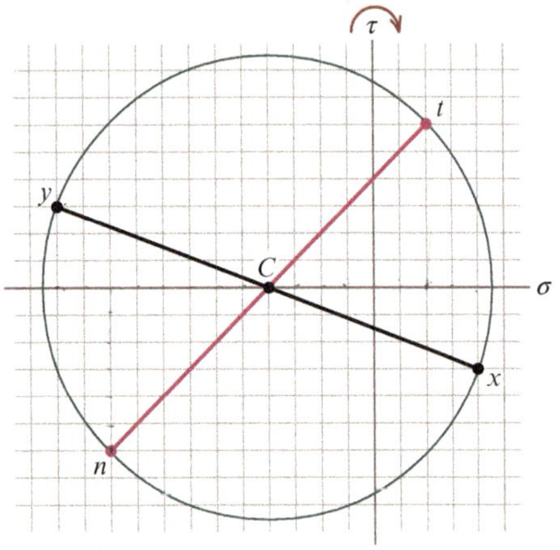

1 grid square = 3 MPa

FIGURE P12.47

P12.48–P12.51 Consider a point in a structural member that is subjected to plane stress. Normal and shear stresses acting on horizontal and vertical planes at the point are shown in Figures P12.48–P12.51.

(a) Draw Mohr's circle for this state of stress.

(b) Determine the principal stresses and the maximum in-plane shear stress acting at the point.

(c) Show these stresses in an appropriate sketch (e.g., see Figure 12.15 or Figure 12.16).

(d) Determine the absolute maximum shear stress at the point.

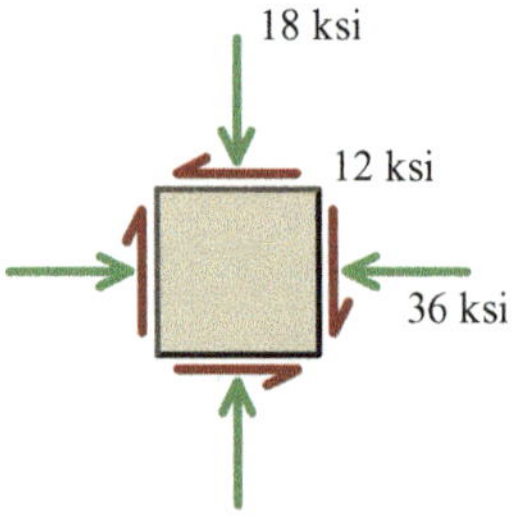

FIGURE P12.48

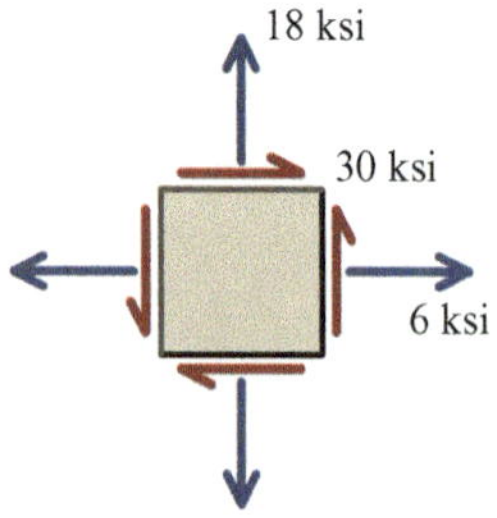

FIGURE P12.49

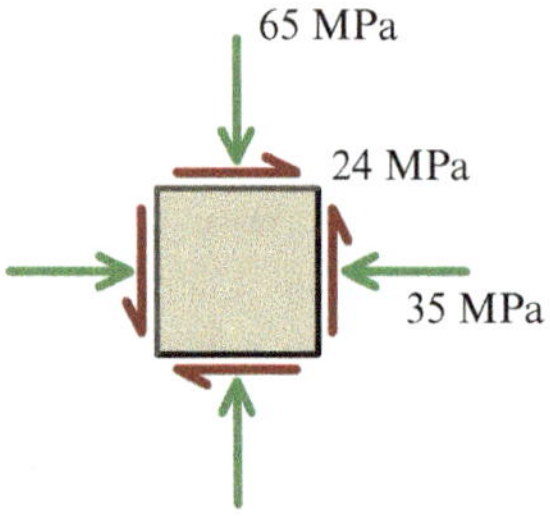

FIGURE P12.50

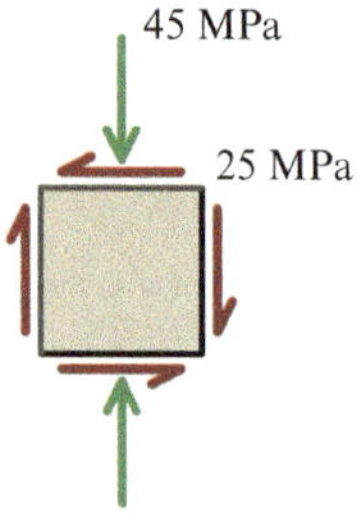

FIGURE P12.51

P12.52–P12.55 Consider a point in a structural member that is subjected to plane stress. Normal and shear stresses acting on horizontal and vertical planes at the point are shown in Figures P12.52–P12.55.

(a) Draw Mohr's circle for this state of stress.

(b) Determine the principal stresses and the maximum in-plane shear stress acting at the point.

(c) Show these stresses in an appropriate sketch (e.g., see Figure 12.15 or Figure 12.16).

(d) Determine the absolute maximum shear stress at the point.

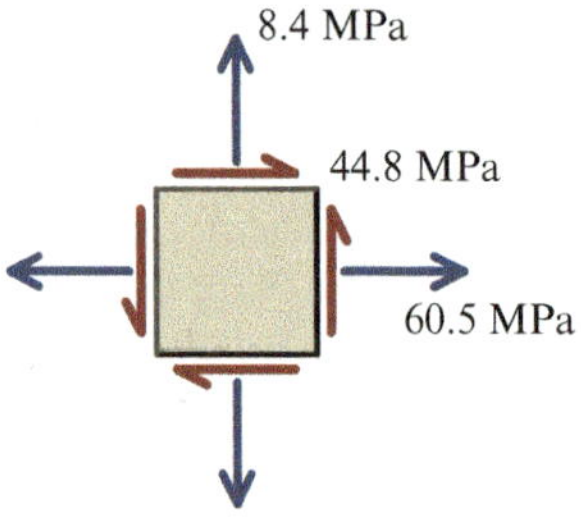

FIGURE P12.52

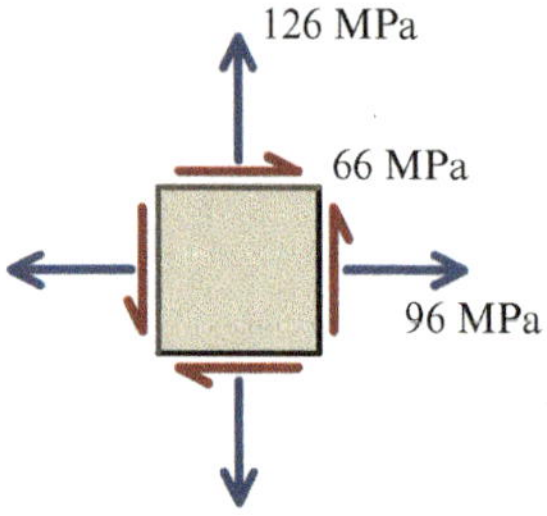

FIGURE P12.53

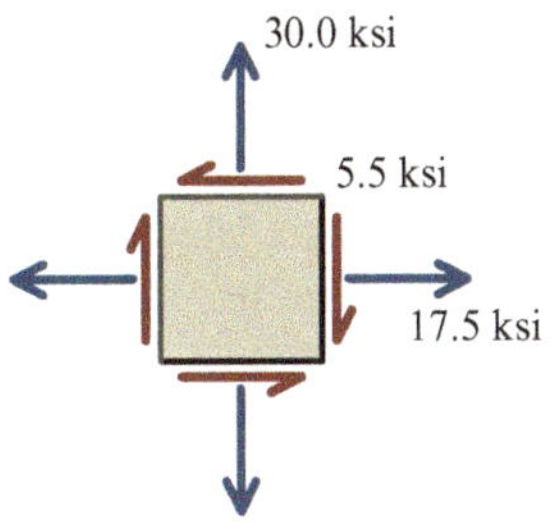

FIGURE P12.54

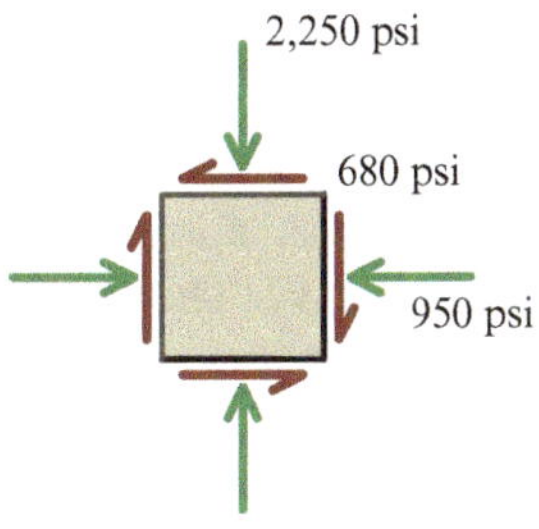

FIGURE P12.55

P12.56–P12.59 Consider a point in a structural member that is subjected to plane stress. Normal and shear stresses acting on horizontal and vertical planes at the point are shown in Figures P12.56–P12.59.

(a) Draw Mohr's circle for this state of stress.
(b) Determine the principal stresses and the maximum in-plane shear stress acting at the point, and show these stresses in an appropriate sketch (e.g., see Figure 12.15 or Figure 12.16).
(c) Determine the normal and shear stresses on the indicated plane, and show these stresses in an appropriate sketch.
(d) Determine the absolute maximum shear stress at the point.

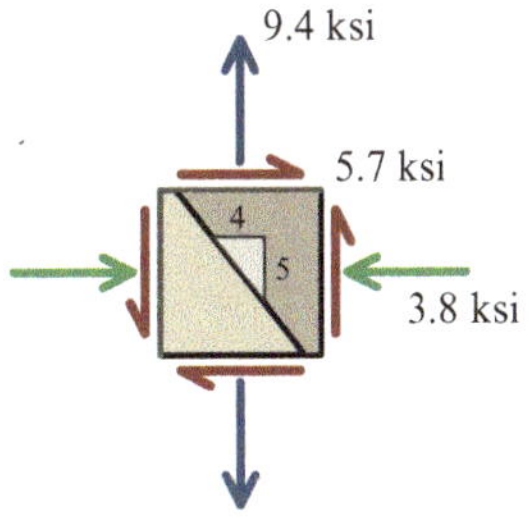

FIGURE P12.56

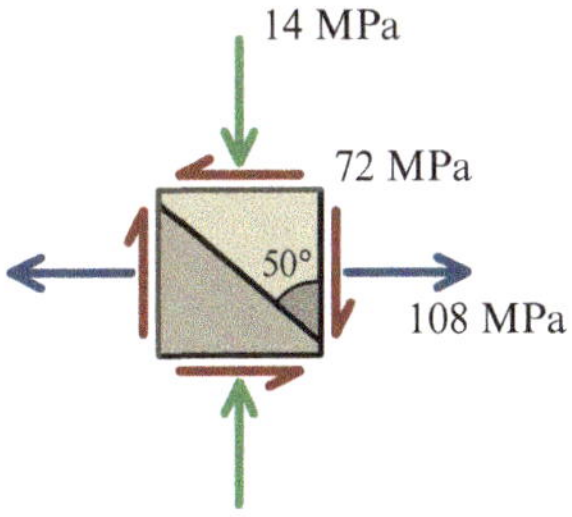

FIGURE P12.57

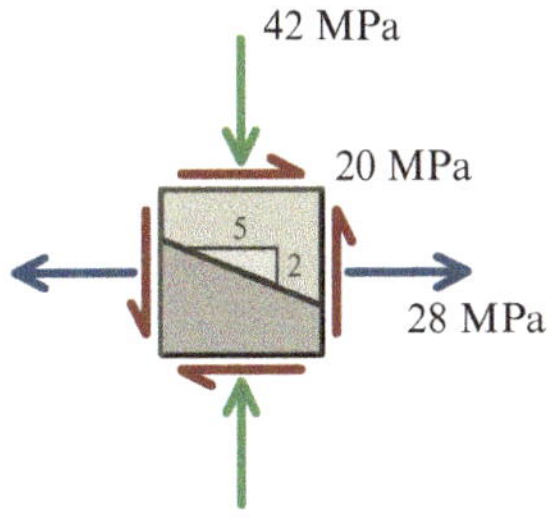

FIGURE P12.58

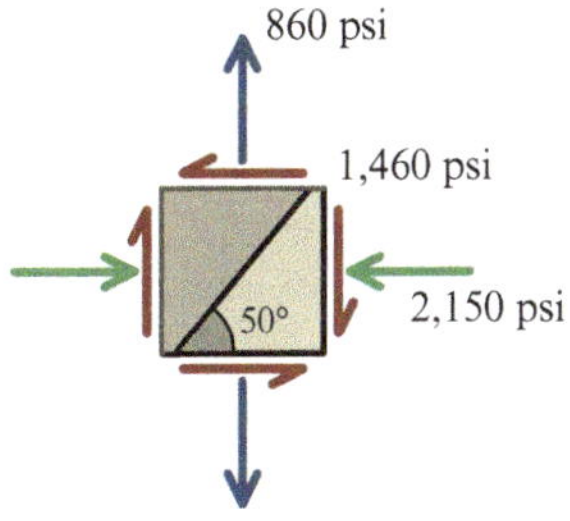

FIGURE P12.59

P12.60 At a point in a stressed body, the principal stresses are oriented as shown in Figure P12.60. Use Mohr's circle to determine

(a) the stresses on plane *a–a*.
(b) the stresses on the horizontal and vertical planes at the point.
(c) the absolute maximum shear stress at the point.

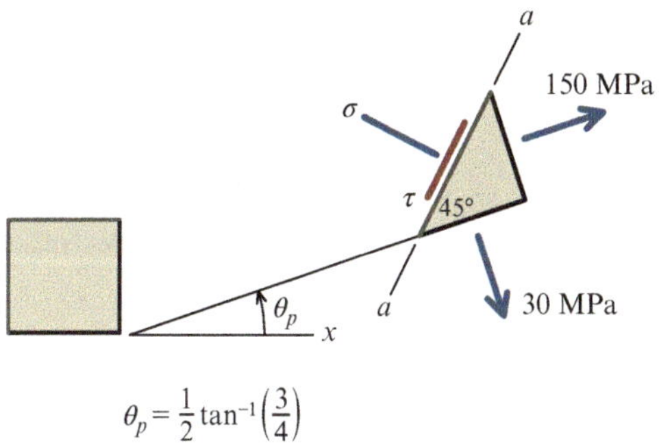

FIGURE P12.60

P12.61 At a point in a stressed body, the principal stresses are oriented as shown in Figure P12.61. Use Mohr's circle to determine

(a) the stresses on plane *a–a*.
(b) the stresses on the horizontal and vertical planes at the point.
(c) the absolute maximum shear stress at the point.

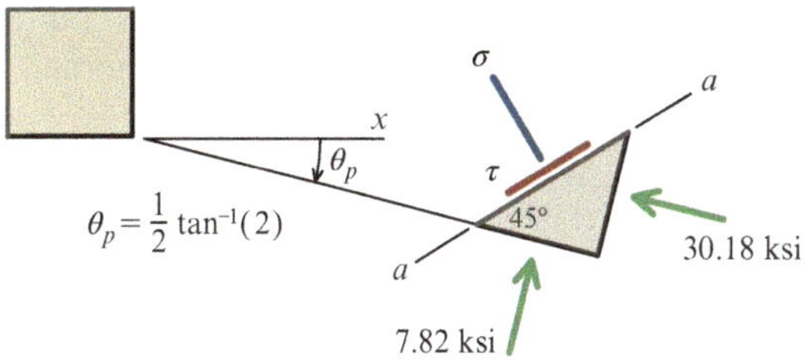

FIGURE P12.61

P12.62 A solid 1.50 in. diameter shaft is subjected to a torque T = 330 lb·ft and an axial load P acting as shown in Figure P12.62/63. If the largest tensile normal stress in the shaft must be limited to 12,000 psi, what is the largest load P that can be applied to the shaft?

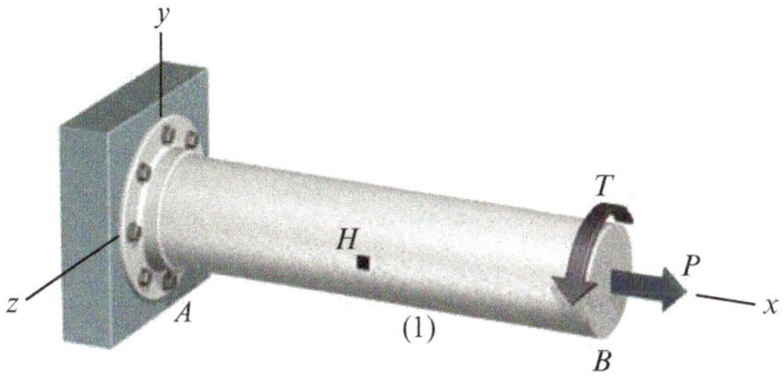

FIGURE P12.62/63

P12.63 A solid 20 mm diameter shaft is subjected to an axial load P = 30 kN and a torque T acting as shown in Figure P12.62/63. If the largest shear stress in the shaft must be limited to 90 MPa, what is the largest torque T that can be applied to the shaft?

12.11 General State of Stress at a Point

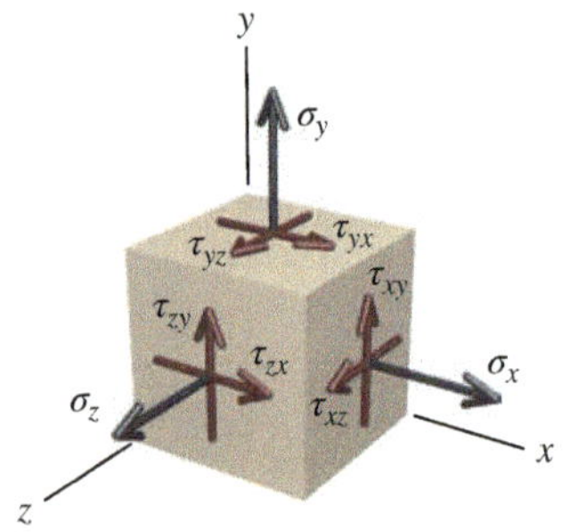

FIGURE 12.17

The general three-dimensional state of stress at a point was previously introduced in Section 12.2. This state of stress has three normal stress components and six shear stress components, as illustrated in Figure 12.17. The shear stress components are not all independent, however, since moment equilibrium requires that

$$\tau_{yx} = \tau_{xy} \qquad \tau_{yz} = \tau_{zy} \qquad \text{and} \qquad \tau_{xz} = \tau_{zx}$$

The stresses are all positive, according to the normal and shear stress sign conventions outlined in Section 12.2.

Normal and Shear Stresses

Expressions for the stresses on any oblique plane through a point, in terms of stresses on the reference x, y, and z planes, can be developed with the aid of the free-body diagram shown in Figure 12.18*a*. The n axis is normal to the oblique (shaded) face. The orientation of the n axis can be defined by three angles α, β, and γ as shown in Figure 12.18*b*. The area of the oblique face of the tetrahedral element is defined to be dA. Areas of the x, y, and z faces are thus $dA \cos \alpha$, $dA \cos \beta$, and $dA \cos \gamma$, respectively.[1] The resultant force F on the oblique face is $S\,dA$, where S is the resultant stress on the area. The resultant stress S is related to the normal and shear stress components on the oblique face by the expression

$$S = \sqrt{\sigma_n^2 + \tau_{nt}^2} \tag{12.24}$$

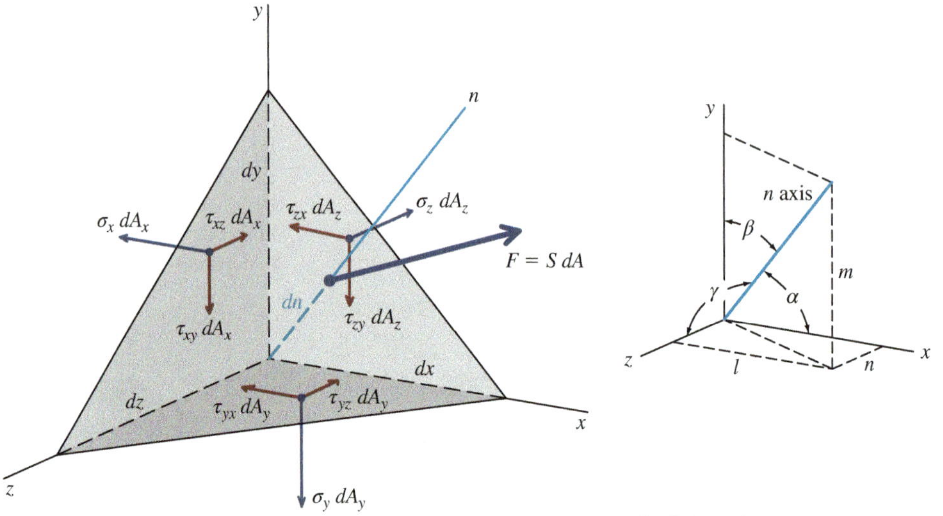

(a) Tetrahedron free-body diagram

(b) Orientation of n axis defined

FIGURE 12.18 Tetrahedron for deriving principal stresses on an oblique plane.

[1] These relationships can be established by considering the volume of the tetrahedron in Figure 12.18*a*. The volume of the tetrahedron can be expressed as $V = 1/3\ dn\ dA = 1/3\ dx\ dA_x = 1/3\ dy\ dA_y = 1/3\ dz\ dA_z$. However, the distance dn from the origin to the center of the oblique face can also be expressed as $dn = dx \cos \alpha = dy \cos \beta = dz \cos \gamma$. Thus, the areas of the tetrahedron faces can be expressed as $dA_x = dA \cos \alpha$, $dA_y = dA \cos \beta$, and $dA_z = dA \cos \gamma$.

The forces on the x, y, and z faces are shown as three components, each of whose magnitude is the product of the area by the appropriate stress. If we use l, m, and n to represent $\cos\alpha$, $\cos\beta$, and $\cos\gamma$, respectively, then the force equilibrium equations in the x, y, and z directions are as follows:

$$F_x = S_x\,dA = \sigma_x\,dA \cdot l + \tau_{yx}\,dA \cdot m + \tau_{zx}\,dA \cdot n$$
$$F_y = S_y\,dA = \sigma_y\,dA \cdot m + \tau_{zy}\,dA \cdot n + \tau_{xy}\,dA \cdot l$$
$$F_z = S_z\,dA = \sigma_z\,dA \cdot n + \tau_{xz}\,dA \cdot l + \tau_{yz}\,dA \cdot m$$

The terms:
$l = \cos\alpha$
$m = \cos\beta$
$n = \cos\gamma$
are called **direction cosines**.

From these equations, we obtain the three orthogonal components of the resultant stress:

$$\begin{aligned} S_x &= \sigma_x \cdot l + \tau_{yx} \cdot m + \tau_{zx} \cdot n \\ S_y &= \tau_{xy} \cdot l + \sigma_y \cdot m + \tau_{zy} \cdot n \\ S_z &= \tau_{xz} \cdot l + \tau_{yz} \cdot m + \sigma_z \cdot n \end{aligned} \tag{a}$$

The normal component σ_n of the resultant stress S equals $S_x \cdot l + S_y \cdot m + S_z \cdot n$; therefore, from Equation (a), it follows that the normal stress on any oblique plane through the point is

$$\sigma_n = \sigma_x l^2 + \sigma_y m^2 + \sigma_z n^2 + 2\tau_{xy} lm + 2\tau_{yz} mn + 2\tau_{zx}\, nl \tag{12.25}$$

The shear stress τ_{nt} on the oblique plane can be obtained from the relation $S^2 = \sigma_n^2 + \tau_{nt}^2$. Thus, for any given problem, the values of S and σ_n will be obtained from Equations (a) and (12.25).

Magnitude and Orientation of Principal Stresses

A principal plane was previously defined as a plane on which the shear stress τ_{nt} is zero. The normal stress σ_n on such a plane was defined as a principal stress σ_p. If the oblique plane of Figure 12.18 is a *principal plane*, then $S = \sigma_p$ and, in addition, $S_x = \sigma_p l$, $S_y = \sigma_p m$, and $S_z = \sigma_p n$. Substituting these components into Equation (a) then produces the following homogeneous linear equations in terms of the direction cosines l, m, and n:

$$\begin{aligned} (\sigma_x - \sigma_p)l + \tau_{yx} m + \tau_{zx} n &= 0 \\ (\sigma_y - \sigma_p)m + \tau_{zy} n + \tau_{xy} l &= 0 \\ (\sigma_z - \sigma_p)n + \tau_{xz} l + \tau_{yz} m &= 0 \end{aligned} \tag{b}$$

This set of equations has a nontrivial solution only if the determinant of the coefficients of l, m, and n is equal to zero. Thus,

$$\begin{vmatrix} (\sigma_x - \sigma_p) & \tau_{yx} & \tau_{zx} \\ \tau_{xy} & (\sigma_y - \sigma_p) & \tau_{zy} \\ \tau_{xz} & \tau_{yz} & (\sigma_z - \sigma_p) \end{vmatrix} = 0 \tag{12.26}$$

Expansion of the determinant yields the **stress cubic equation** for determining the principal stresses:

$$\sigma_p^3 - I_1\sigma_p^2 + I_2\sigma_p - I_3 = 0 \tag{12.27}$$

In Equation (12.27), the constants I_1, I_2, and I_3 are as follows:

$$I_1 = \sigma_x + \sigma_y + \sigma_z$$
$$I_2 = \sigma_x\sigma_y + \sigma_y\sigma_z + \sigma_z\sigma_x - \tau_{xy}^2 - \tau_{yz}^2 - \tau_{zx}^2 \qquad (12.28)$$
$$I_3 = \sigma_x\sigma_y\sigma_z + 2\tau_{xy}\tau_{yz}\tau_{zx} - (\sigma_x\tau_{yz}^2 + \sigma_y\tau_{zx}^2 + \sigma_z\tau_{xy}^2)$$

The roots of Equation (12.27) can be readily estimated by plotting a graph of the left-hand side of the equation as a function of σ.

I_1, I_2, and I_3 are stress invariants, because they are independent of how the coordinates x, y, and z are oriented in a given state of stress. Recall that stress invariants for plane stress were discussed in Section 12.7 and that the invariants I_1 and I_2 were given in Equation (12.9) for plane stress, where $\sigma_z = \tau_{yz} = \tau_{zx} = 0$. Equation (12.27) always has three real roots, which are the principal stresses at a given point.

The roots of Equation (12.27) can be found by a number of numerical methods. One simple method involves a procedure that begins by calculating the following constants:

$$Q = \frac{1}{3}I_1I_2 - I_3 - \frac{2}{27}I_1^3$$
$$R = \frac{1}{3}\sqrt{I_1^2 - 3I_2}$$
$$\phi = \cos^{-1}\left(-\frac{Q}{2R^3}\right)$$

Here, I_1, I_2, and I_3 are the three stress invariants and the angle ϕ is expressed in radians. The roots of Equation (12.27) are calculated as follows:

$$\sigma_a = 2R\left[\cos\left(\frac{\phi}{3}\right)\right] + \frac{I_1}{3}$$
$$\sigma_b = 2R\left[\cos\left(\frac{\phi}{3} + \frac{2\pi}{3}\right)\right] + \frac{I_1}{3}$$
$$\sigma_c = 2R\left[\cos\left(\frac{\phi}{3} + \frac{4\pi}{3}\right)\right] + \frac{I_1}{3}$$

The three roots (σ_a, σ_b, and σ_c) are then ranked algebraically. The algebraically largest root (i.e., the root with the most positive value) is σ_{p1}, and the algebraically smallest root (i.e., the root with the most negative value) is σ_{p3}. The remaining root is σ_{p2}.

Often, only the algebraic values of the principal stresses are needed. However, there are occasions in which the directions of the principal stresses are also desired. To determine the direction cosines that define each principal plane, we begin by rewriting Equation (b) in matrix form as

$$\begin{bmatrix} \sigma_x - \sigma_{pi} & \tau_{xy} & \tau_{xz} \\ \tau_{xy} & \sigma_y - \sigma_{pi} & \tau_{yz} \\ \tau_{xz} & \tau_{yz} & \sigma_z - \sigma_{pi} \end{bmatrix} \begin{Bmatrix} l_i \\ m_i \\ n_i \end{Bmatrix} = 0 \qquad \text{(c)}$$

Here, $i = 1$, 2, or 3, $\sigma_{pi} = \sigma_{p1}$, σ_{p2}, or σ_{p3}, and $\{l_i, m_i, n_i\}$ are the directions cosines corresponding to the respective principal stresses σ_{pi}. We next introduce a new term

$$k_i = \frac{1}{\sqrt{a_i^2 + b_i^2 + c_i^2}}$$

where a_i, b_i, and c_i are the cofactors of the elements on the first row of the matrix in Equation (c):

$$a_i = \begin{vmatrix} \sigma_y - \sigma_{pi} & \tau_{yz} \\ \tau_{yz} & \sigma_z - \sigma_{pi} \end{vmatrix}$$

$$b_i = -\begin{vmatrix} \tau_{xy} & \tau_{yz} \\ \tau_{xz} & \sigma_z - \sigma_{pi} \end{vmatrix}$$

$$c_i = \begin{vmatrix} \tau_{xy} & \sigma_y - \sigma_{pi} \\ \tau_{xz} & \tau_{yz} \end{vmatrix}$$

Thus, the following are the direction cosines giving the orientation of principal stress i:

$$l_i = a_i k_i$$
$$m_i = b_i k_i$$
$$n_i = c_i k_i$$

Each set of direction cosines defines the normal to a plane upon which a principal stress acts. Direction cosines must satisfy the relationship

$$l_i^2 + m_i^2 + n^2 = 1 \qquad \text{(d)}$$

It is clear that the direction cosines just derived will satisfy this requirement. Also, observe that the trivial solution ($l = m = n = 0$) is not possible, given that the direction cosines satisfy Equation (d). In addition, the following relations must be true if the planes are orthogonal:

$$l_1 l_2 + m_1 m_2 + n_1 n_2 = 0$$
$$l_2 l_3 + m_2 m_3 + n_2 n_3 = 0$$
$$l_3 l_1 + m_3 m_1 + n_3 n_1 = 0$$

In developing equations for maximum and minimum normal stresses, the special case will be considered in which $\tau_{xy} = \tau_{yz} = \tau_{zx} = 0$. No loss in generality is introduced by considering this special case, because it involves only a reorientation of the reference x, y, and z axes to coincide with the principal directions. Since the x, y, and z planes are now principal planes, the stresses σ_x, σ_y, and σ_z respectively become σ_{p1}, σ_{p2}, and σ_{p3}. Solving Equation (a) for the direction cosines thus yields

$$l = \frac{S_x}{\sigma_{p1}} \qquad m = \frac{S_y}{\sigma_{p2}} \qquad \text{and} \qquad n = \frac{S_z}{\sigma_{p3}}$$

Substituting these values into Equation (d), we obtain

$$\frac{S_x^2}{\sigma_{p1}^2} + \frac{S_y^2}{\sigma_{p2}^2} + \frac{S_z^2}{\sigma_{p3}^2} = 1 \qquad \text{(e)}$$

The plot of Equation (e) is the ellipsoid shown in Figure 12.19. Observe that the magnitude of σ_n is everywhere less than that of S (since $S^2 = \sigma_n^2 + \tau_{nt}^2$), except at the intercepts, where S is σ_{p1}, σ_{p2}, or σ_{p3}. Therefore, two of the principal stresses (σ_{p1} and σ_{p3} of Figure 12.19) are the maximum and minimum normal stresses at the point we are examining. The

S is the resultant stress acting on the oblique plane of Figure 12.19*a*. S_x, S_y, and S_z are the orthogonal components of the resultant stress S.

third principal stress is intermediate in value and has no particular significance. The preceding discussion demonstrates that the set of principal stresses includes the maximum and minimum normal stresses at the point.

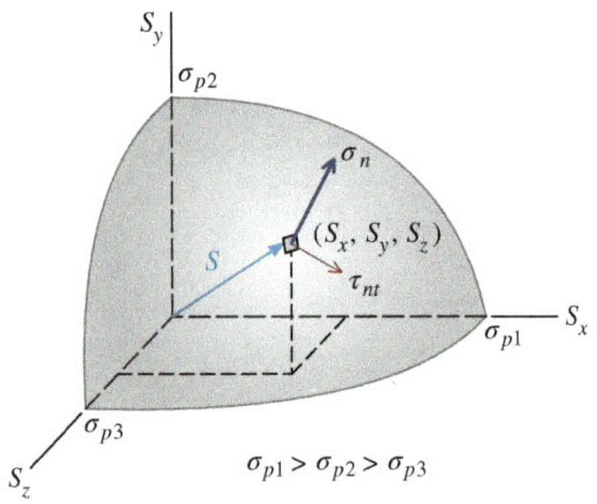

FIGURE 12.19

Magnitude and Orientation of Maximum Shear Stress

Continuing with the special case in which the given stresses σ_x, σ_y, and σ_z are principal stresses, we can develop equations for the maximum shear stress at the point under consideration. The resultant stress S on the oblique plane is given by the equation

$$S^2 = S_x^2 + S_y^2 + S_z^2$$

Substituting values for S_x, S_y, and S_z from Equation (a), with zero shear stresses, yields

$$S^2 = \sigma_x^2 l^2 + \sigma_y^2 m^2 + \sigma_z^2 n^2 \tag{f}$$

Also, from Equation (12.25),

$$\sigma_n^2 = (\sigma_x l^2 + \sigma_y m^2 + \sigma_z n^2)^2 \tag{g}$$

Since $S^2 = \sigma_n^2 + \tau_{nt}^2$, we obtain, from Equations (f) and (g), the following equation for the shear stress τ_{nt} on the oblique plane:

$$\tau_{nt} = \sqrt{\sigma_x^2 l^2 + \sigma_y^2 m^2 + \sigma_z^2 n^2 - (\sigma_x l^2 + \sigma_y m^2 + \sigma_z n^2)^2} \tag{12.29}$$

The planes on which maximum and minimum shear stresses occur can be found from Equation (12.29) by differentiating with respect to the direction cosines l, m, and n. One of the direction cosines in Equation (12.29) (e.g., n) can be eliminated by solving Equation (d) for n^2 and substituting into Equation (12.29). Thus,

$$\begin{aligned}\tau_{nt} = \{&(\sigma_x^2 - \sigma_z^2)l^2 + (\sigma_y^2 - \sigma_z^2)m^2 + \sigma_z^2 \\ &- [(\sigma_x - \sigma_z)l^2 + (\sigma_y - \sigma_z)m^2 + \sigma_z]^2\}^{1/2}\end{aligned} \tag{h}$$

Taking the partial derivatives of Equation (h), first with respect to l and then with respect to m and setting the partial derivatives equal to zero, we obtain the following equations for the direction cosines associated with planes having maximum and minimum shear stress:

$$l\left[\frac{1}{2}(\sigma_x - \sigma_z) - (\sigma_x - \sigma_z)l^2 - (\sigma_y - \sigma_z)m^2\right] = 0 \tag{i}$$

$$m\left[\frac{1}{2}(\sigma_y - \sigma_z) - (\sigma_x - \sigma_z)l^2 - (\sigma_y - \sigma_z)m^2\right] = 0 \tag{j}$$

One solution of these equations is, obviously, $l = m = 0$. Then, from Equation (d), $n = \pm 1$. Solutions different from zero are also possible for this set of equations. For instance, consider surfaces in which the direction cosine has the value $m = 0$. From Equation (i), $l = \pm\sqrt{1/2}$, and from Equation (d), $n = \pm\sqrt{1/2}$. Thus, the normal to this surface makes an angle of 45° with both the x and z axes, and is perpendicular to the y axis. This surface has

Table 12.1 Direction Cosines for Planes of Maximum and Minimum Shear Stress

	Minimum			Maximum		
	1	2	3	4	5	6
l	± 1	0	0	$\pm\sqrt{1/2}$	$\pm\sqrt{1/2}$	0
m	0	± 1	0	$\pm\sqrt{1/2}$	0	$\pm\sqrt{1/2}$
n	0	0	± 1	0	$\pm\sqrt{1/2}$	$\pm\sqrt{1/2}$
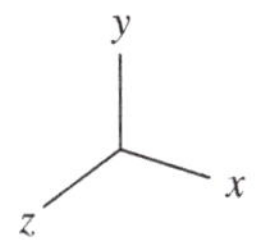	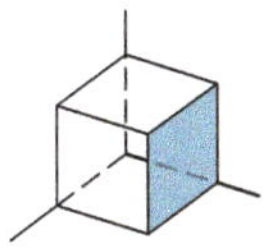	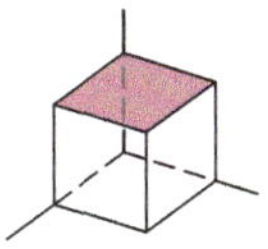	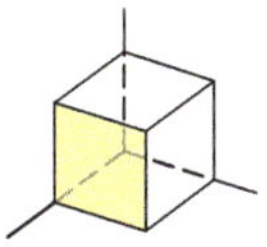	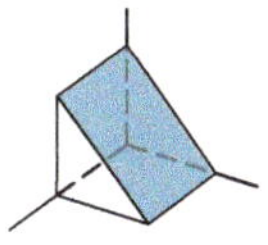	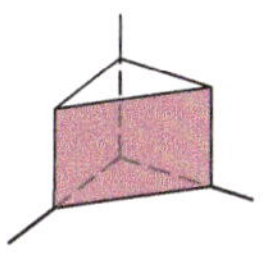	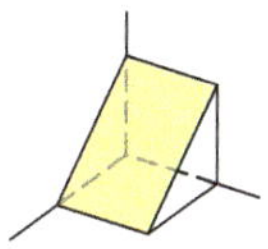

the largest shear stress of all surfaces whose normal is perpendicular to the y axis. Next, consider surfaces whose normal is perpendicular to the x axis; that is, the direction cosine has the value $l = 0$. From Equation (j), $m = \pm\sqrt{1/2}$, and from Equation (d), $n = \pm\sqrt{1/2}$. The normal to this surface makes an angle of 45° with both the y and z axes. This surface has the largest shear stress of all surfaces whose normal is perpendicular to the x axis. Repeating the preceding procedure by eliminating l and m in turn from Equation (h) yields other values for the direction cosines that make the shear stresses maximum or minimum. All of the possible combinations are listed in Table 12.1. In the last row of the table, the planes corresponding to the direction cosines in the column above are shown shaded. Note that in each case only one of the two possible planes is shown.

The first three columns of Table 12.1 give the direction cosines for planes of minimum shear stress. Since we are here considering the special case in which the given stresses σ_x, σ_y, and σ_z are principal stresses, columns 1, 2, and 3 are simply the principal planes for which the shear stress must be zero. Hence, the minimum shear stress is $\tau_{nt} = 0$.

To determine the magnitude of the maximum shear stress, values of direction cosines from Table 12.1 are substituted into Equation (12.29), replacing σ_x, σ_y, and σ_z with σ_{p1}, σ_{p2}, and σ_{p3}, respectively. Direction cosines from column 4 of Table 12.1 give the following expression for the maximum shear stress:

$$\tau_{\max} = \sqrt{\frac{1}{2}\sigma_{p1}^2 + \frac{1}{2}\sigma_{p2}^2 + 0 - \left(\frac{1}{2}\sigma_{p1} + \frac{1}{2}\sigma_{p2}\right)^2} = \frac{\sigma_{p1} - \sigma_{p2}}{2}$$

Similarly, direction cosines from columns 5 and 6 give

$$\tau_{\max} = \frac{\sigma_{p1} - \sigma_{p3}}{2} \quad \text{and} \quad \tau_{\max} = \frac{\sigma_{p2} - \sigma_{p3}}{2}$$

The shear stress of largest magnitude from these three possible results is $\tau_{\text{abs max}}$; hence, the absolute maximum shear stress can be expressed as

$$\tau_{\text{abs max}} = \frac{\sigma_{\max} - \sigma_{\min}}{2} \tag{12.30}$$

which confirms Equation (12.18). The maximum shear stress acts on the plane that bisects the angle between the maximum and minimum principal stresses.

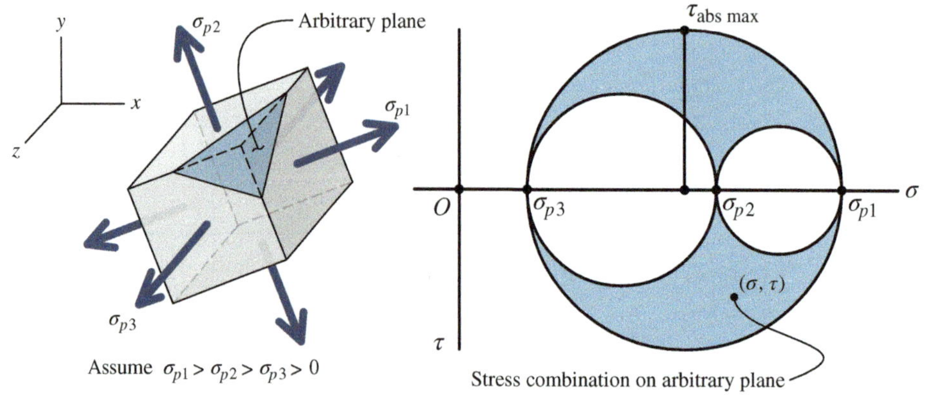

(*a*) Principal stress element (*b*) Mohr's circle

FIGURE 12.20

Application of Mohr's Circle to Three-Dimensional Stress Analysis

In Figure 12.20*a*, the principal stresses σ_{p1}, σ_{p2}, and σ_{p3} at a point are shown on a stress element. We will assume that the principal stresses have been ordered such that $\sigma_{p1} > \sigma_{p2} > \sigma_{p3} > 0$. Furthermore, observe that the principal planes represented by the stress element are rotated with respect to the *x*–*y*–*z* axes. From the three principal stresses, Mohr's circle can be plotted to visually represent the various stress combinations possible at the point (Figure 12.20*b*). Stress combinations for all possible planes plot either on one of the circles or in the shaded area. From Mohr's circle, the absolute maximum shear stress magnitude given by Equation (12.30) is evident.

PROBLEMS

P12.64 At a point in a solid body subjected to plane stress, $\sigma_x = 130$ MPa and $\sigma_y = 48$ MPa, acting as shown in Figure P12.64*a*. For plane *n* shown in Figure P12.64*b*, determine

(a) the resultant stress *S*.
(b) the normal stress σ_n and the shear stress τ_{nt}.

P12.65 At a point in a solid body subjected to plane stress, $\sigma_x = 9.2$ ksi, $\sigma_y = 35.8$ ksi, and $\tau_{xy} = 6.4$ ksi, acting as shown in Figure P12.65*a*. For plane *n* shown in Figure P12.65*b*, determine

(a) the resultant stress *S*.
(b) the normal stress σ_n and the shear stress τ_{nt}.

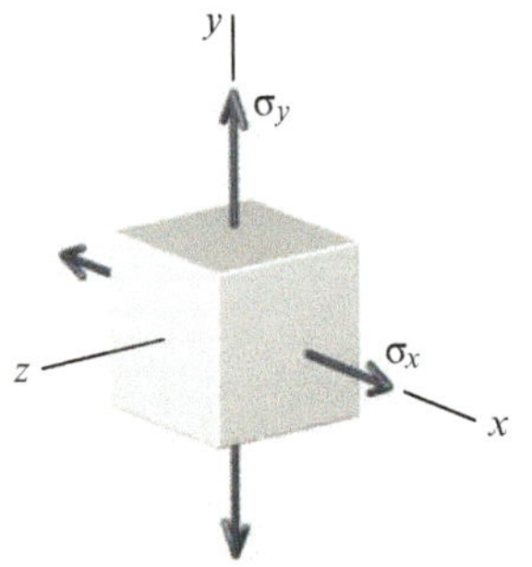

FIGURE P12.64*a*

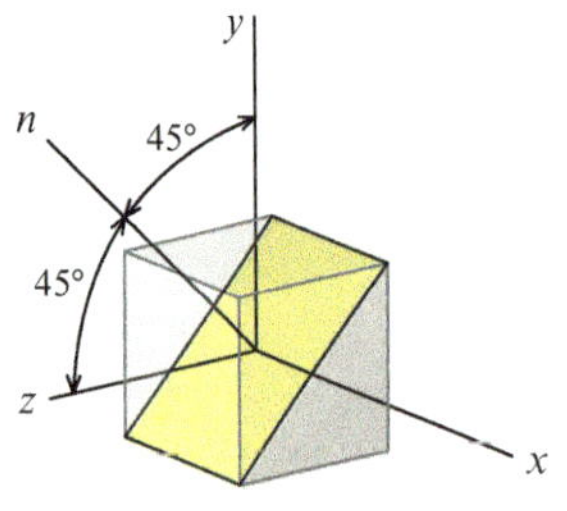

FIGURE P12.64*b*

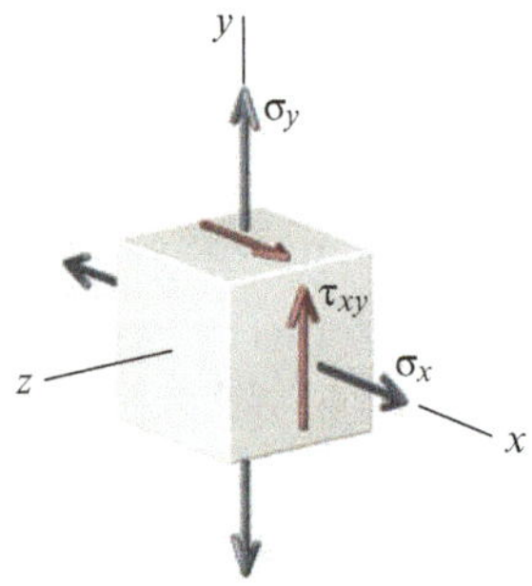

FIGURE P12.65*a*

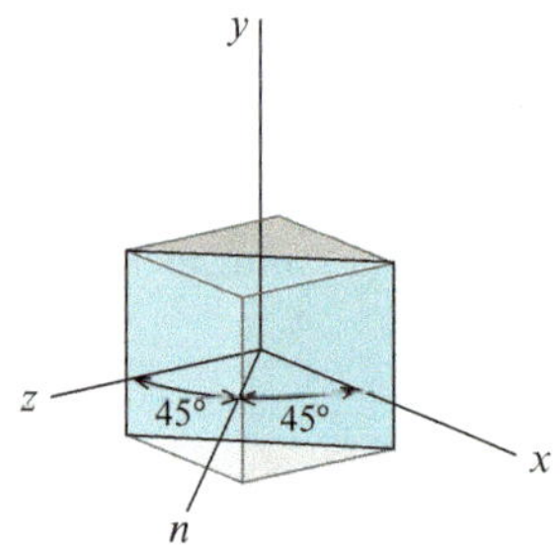

FIGURE P12.65*b*

P12.66 The stresses at point O in a solid body that is subjected to plane stress are $\sigma_x = 16.5$ ksi, $\sigma_y = -5.8$ ksi, and $\tau_{xy} = 8.7$ ksi. Consider Figure P12.66/67. The normal to the inclined surface is defined by vector OA, where point A has coordinates $a = 1.0$ in., $b = 1.4$ in., and $c = 2.5$ in. Determine

(a) the direction cosines of vector OA.
(b) the resultant stress S on the inclined plane.
(c) the normal and shear stresses on the inclined plane.

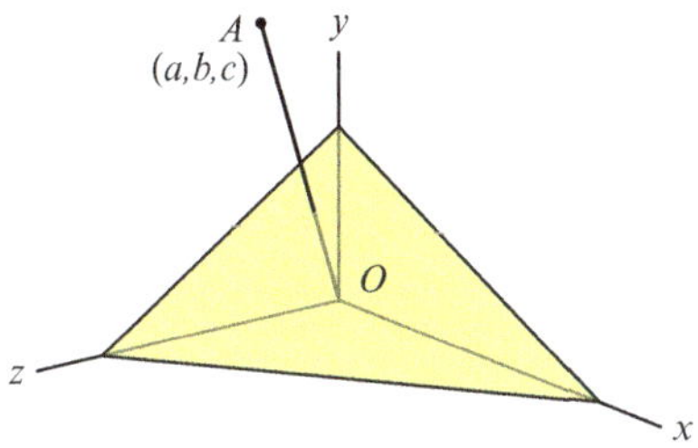

FIGURE P12.66/P12.67

P12.67 The stresses at point O in a solid body are $\sigma_x = -80$ MPa, $\sigma_y = -40$ MPa, $\sigma_z = -40$ MPa, $\tau_{xy} = 23$ MPa, $\tau_{xz} = -65$ MPa, and $\tau_{yz} = 49$ MPa. Consider Figure P12.66/67. The normal to the inclined surface is defined by vector OA, where point A has coordinates $a = 90$ mm, $b = 160$ mm, and $c = 220$ mm. Determine

(a) the direction cosines for vector OA.
(b) the resultant stress S on the inclined plane.
(c) the normal and shear stresses on the inclined plane.

P12.68 At a point in a stressed body, the known stresses are $\sigma_x = 10$ MPa, $\sigma_y = 0$, $\sigma_z = 0$, $\tau_{xy} = 24$ MPa, $\tau_{xz} = 38$ MPa, and $\tau_{yz} = 15$ MPa. Determine the normal and shear stresses on a plane whose outward normal makes equal angles with the x, y, and z axes.

P12.69 At a point in a stressed body, the known stresses are $\sigma_x = 9$ ksi (C), $\sigma_y = 13$ ksi (T), $\sigma_z = 22$ ksi (T), $\tau_{xy} = 4$ ksi, $\tau_{xz} = -19$ ksi, and $\tau_{yz} = 8$ ksi. Determine

(a) the stress invariants.
(b) the principal stresses and the absolute maximum shear stress at the point.

P12.70 At a point in a stressed body, the known stresses are $\sigma_x = 200$ MPa (C), $\sigma_y = 90$ MPa (C), $\sigma_z = 140$ MPa (C), $\tau_{xy} = -24$ MPa, $\tau_{xz} = 120$ MPa, and $\tau_{yz} = 56$ MPa. Determine

(a) the stress invariants.
(b) the principal stresses and the absolute maximum shear stress at the point.

P12.71 At a point in a stressed body, the known stresses are $\sigma_x = 75$ MPa (T), $\sigma_y = 30$ MPa (C), $\sigma_z = 60$ MPa (T), $\tau_{xy} = 35$ MPa, $\tau_{xz} = 15$ MPa, and $\tau_{yz} = -20$ MPa. Determine

(a) the stress invariants.
(b) the principal stresses and the absolute maximum shear stress at the point.
(c) the orientation of the plane on which the maximum tensile normal stress acts.

P12.72 At a point in a stressed body, the known stresses are $\sigma_x = 80$ MPa (C), $\sigma_y = 80$ MPa (C), $\sigma_z = 120$ MPa (C), $\tau_{xy} = 40$ MPa, $\tau_{xz} = -60$ MPa, and $\tau_{yz} = 50$ MPa. Determine

(a) the stress invariants.
(b) the principal stresses and the absolute maximum shear stress at the point.
(c) the orientation of the plane on which the maximum compressive normal stress acts.

CHAPTER 13

Strain Transformations

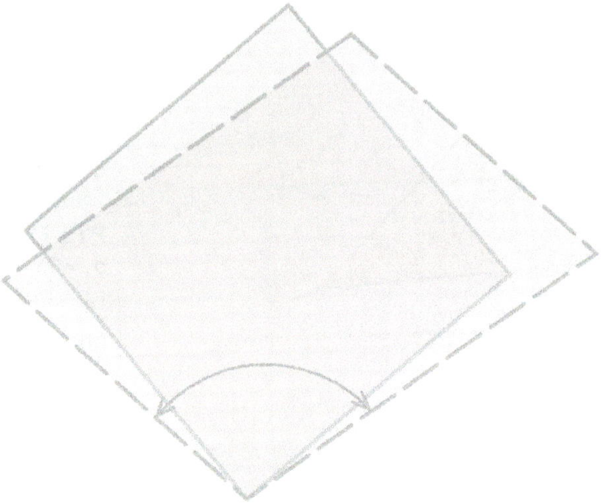

13.1 Introduction

The discussion of strain presented in Chapter 2 was useful in introducing the concept of strain as a measure of deformation. However, it was adequate only for one-directional loading. In many practical situations involving the design of structural or machine components, the configurations and loadings are such that strains occur in two or three directions simultaneously.

The complete state of strain at an arbitrary point in a body under load can be determined by considering the deformation associated with a small volume of material surrounding the point. For convenience, the volume, termed a **strain element**, is assumed to have the shape of a block. In the undeformed state, the faces of the strain element are oriented perpendicular to the x, y, and z reference axes, as shown in Figure 13.1*a*. Since the element is very small, deformations are assumed to be uniform. This assumption means that

(a) planes that initially are parallel to each other will remain parallel after deformation, and

(b) lines that are straight before deformation will remain straight after deformation, as shown in Figure 13.1*b*.

The final size of the deformed element is determined by the lengths of the three edges dx', dy', and dz'. The distorted shape of the element is determined by the angles θ'_{xy}, θ'_{xy}, and θ'_{zx} between faces.

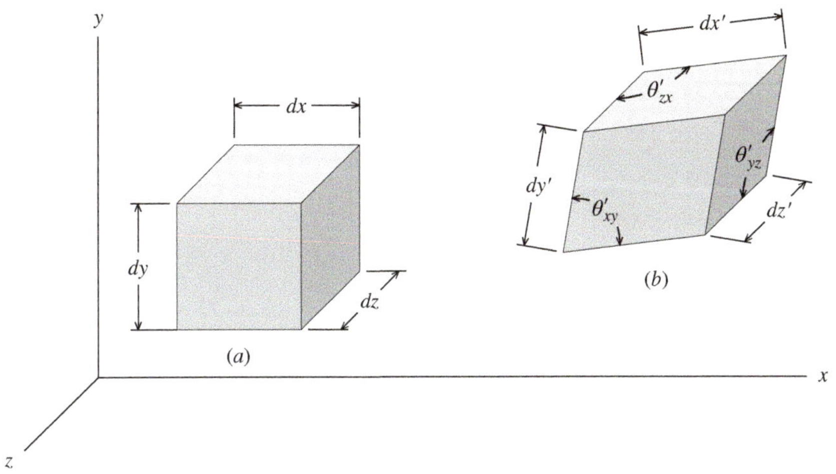

FIGURE 13.1

The Cartesian components of strain at the point can be expressed in terms of the deformations by using the definitions of normal and shear strain presented in Section 2.2. Thus, we have the following equations:

$$\begin{aligned} \varepsilon_x &= \frac{dx' - dx}{dx} & \gamma_{xy} &= \frac{\pi}{2} - \theta'_{xy} \\ \varepsilon_y &= \frac{dy' - dy}{dy} & \gamma_{yz} &= \frac{\pi}{2} - \theta'_{yz} \\ \varepsilon_z &= \frac{dz' - dz}{dz} & \gamma_{zx} &= \frac{\pi}{2} - \theta'_{zx} \end{aligned} \tag{13.1}$$

In a similar manner, the normal strain component associated with a line oriented in an arbitrary n direction and the shearing strain component associated with two *arbitrary* initially orthogonal lines oriented in the n and t directions in the undeformed element are respectively given by

$$\varepsilon_n = \frac{dn' - dn}{dn} \quad \text{and} \quad \gamma_{nt} = \frac{\pi}{2} - \theta'_{nt} \tag{13.2}$$

13.2 Plane Strain

Considerable insight into the nature of strain can be gained by considering a state of strain known as two-dimensional strain or **plane strain**. For this state, the x–y plane will be used as the reference plane. The length dz shown in Figure 13.1 does not change, and the angles θ'_{yz} and θ'_{zx} remain 90°. Thus, for the conditions of plane strain, $\varepsilon_z = \gamma_{xz} = \gamma_{yz} = 0$.

If the only deformations are those in the x–y plane, then three strain components may exist. Figure 13.2 shows an infinitesimal element of dimensions dx and dy that will be used to illustrate the strains existing at point O. In Figure 13.2a, the element subjected to a positive normal strain ε_x will elongate by the amount $\varepsilon_x\, dx$ in the horizontal direction. When subjected to a positive normal strain ε_y, the element will elongate by the amount $\varepsilon_y\, dy$ in the vertical direction (Figure 13.2b). Recall that positive normal strains create elongations, and negative normal strains create contractions, in the material.

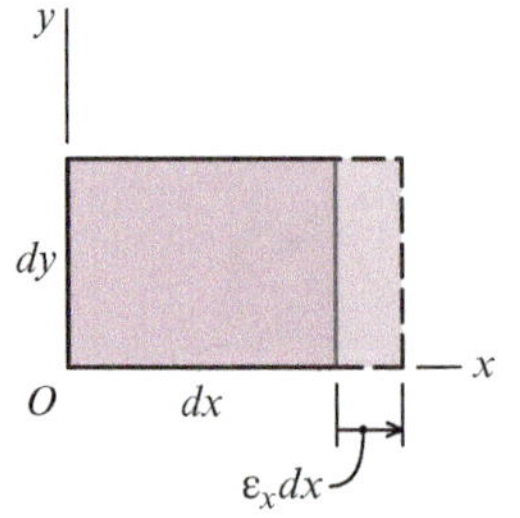

FIGURE 13.2*a*

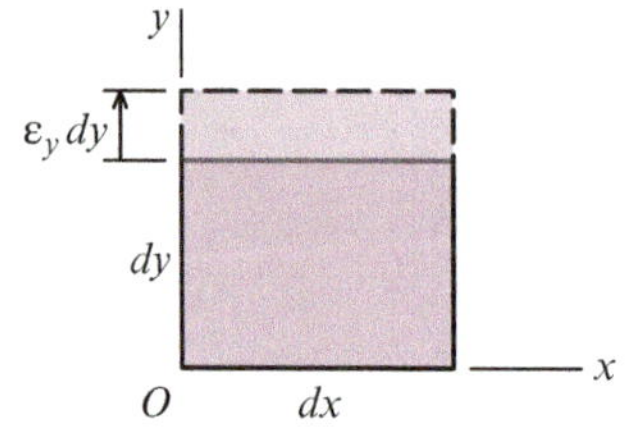

FIGURE 13.2*b*

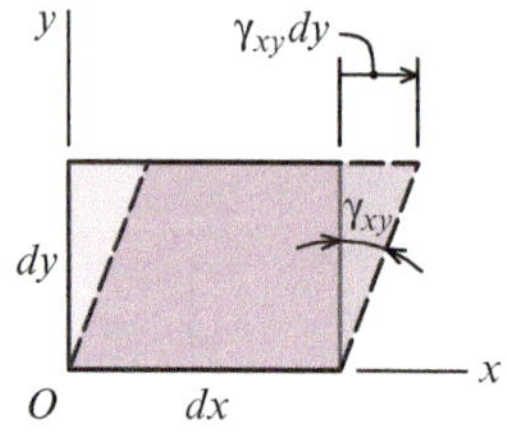

FIGURE 13.2*c*

The shear strain γ_{xy} shown in Figure 13.2*c* is a measure of the change in angle between the *x* and *y* axes, which are initially perpendicular to each other. Shear strains are considered positive when the angle between axes decreases and negative when the angle increases.

Note that the sign conventions for strain are consistent with the stress sign conventions. A positive normal stress (i.e., tensile normal stress) in the *x* direction causes a positive normal strain ε_x (i.e., elongation) (Figure 13.2*a*), a positive normal stress in the *y* direction creates a positive normal strain ε_y (Figure 13.2*b*), and a positive shear stress produces a positive shear strain γ_{xy} (Figure 13.2*c*).

13.3 Transformation Equations for Plane Strain

The state of plane strain at point *O* is defined by three strain components: ε_x, ε_y, and γ_{xy}. Transformation equations provide the means to determine normal and shear strains at point *O* for orthogonal axes rotated at any arbitrary angle θ.

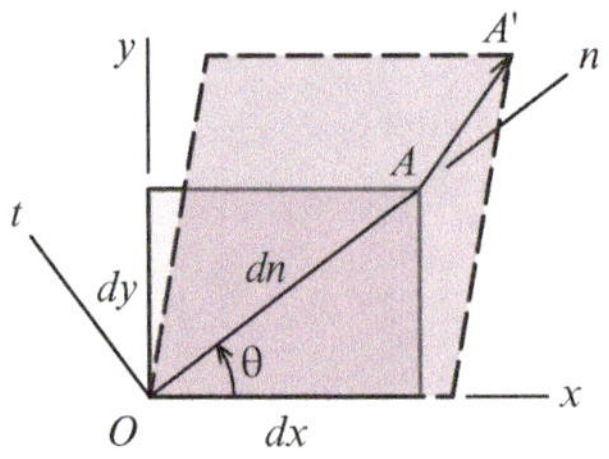

FIGURE 13.3

Equations that transform normal and shear strains from the *x*–*y* axes to any arbitrary orthogonal axes will be derived. To facilitate the derivation, the dimensions of the element are chosen such that the diagonal *OA* of the element coincides with the *n* axis (Figure 13.3). It is also convenient to assume that corner *O* is fixed and that the edge of the element along the *x* axis does not rotate.

When all three strain components (ε_x, ε_y, and ε_{xy}) occur simultaneously (Figure 13.3), corner *A* of the element is displaced to a new location denoted by *A*′. For clarity, the deformations are shown greatly exaggerated.

Transformation Equation for Normal Strain

The displacement vector from *A* to *A*′ (shown in Figure 13.3) is isolated and enlarged in Figure 13.4*a*. The horizontal component of vector **AA′** is composed of the deformations due to ε_x (see Figure 13.2*a*) and γ_{xy} (see Figure 13.2*c*). The vertical component of **AA′** is caused by ε_y (see Figure 13.2*b*).

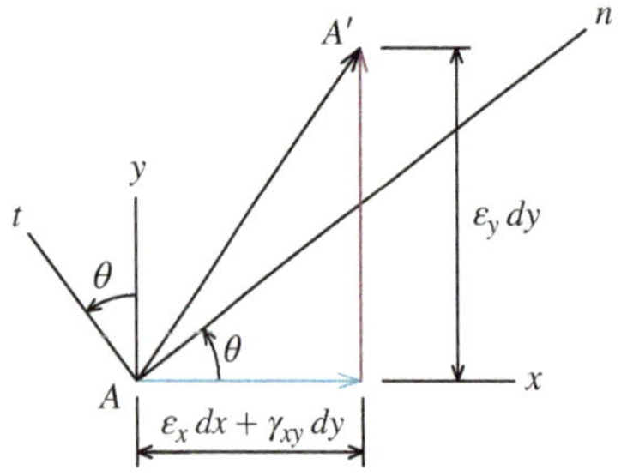

FIGURE 13.4*a*

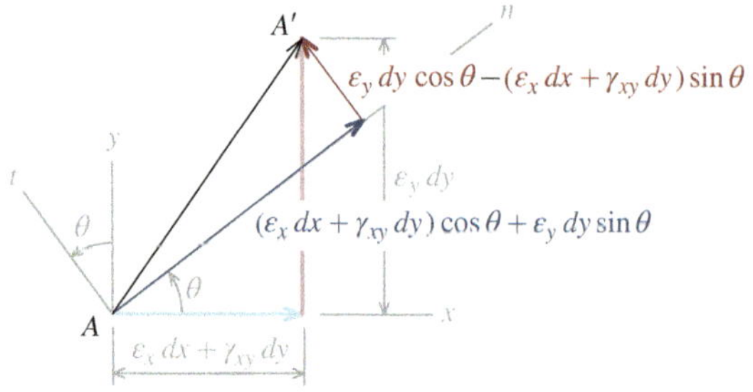

FIGURE 13.4*b*

Next, the displacement vector **AA′** will be resolved into components in the n and t directions. Unit vectors in the n and t directions are

$$\mathbf{n} = \cos\theta\,\mathbf{i} + \sin\theta\,\mathbf{j} \qquad \text{and} \qquad \mathbf{t} = -\sin\theta\,\mathbf{i} + \cos\theta\,\mathbf{j}$$

The displacement component in the n direction can be determined from the dot product:

$$\mathbf{AA'} \cdot \mathbf{n} = (\varepsilon_x\,dx + \gamma_{xy}\,dy)\cos\theta + \varepsilon_y\,dy\sin\theta \tag{a}$$

The displacement component in the t direction is

$$\mathbf{AA'} \cdot \mathbf{t} = \varepsilon_y\,dy\cos\theta - (\varepsilon_x\,dx + \gamma_{xy}\,dy)\sin\theta \tag{b}$$

The displacements in the n and t directions are shown in Figure 13.4*b*.

The displacement in the n direction represents the elongation of diagonal *OA* (see Figure 13.3) due to the normal and shear strains ε_x, ε_y, and γ_{xy}. The strain in the n direction can be found by dividing the elongation given in Equation (a) by the initial length dn of the diagonal:

$$\begin{aligned}\varepsilon_n &= \frac{(\varepsilon_x\,dx + \gamma_{xy}\,dy)\cos\theta + \varepsilon_y\,dy\sin\theta}{dn} \\ &= \left(\varepsilon_x\frac{dx}{dn} + \gamma_{xy}\frac{dy}{dn}\right)\cos\theta + \varepsilon_y\frac{dy}{dn}\sin\theta\end{aligned} \tag{c}$$

From Figure 13.3, $dx/dn = \cos\theta$ and $dy/dn = \sin\theta$. By substituting these relationships into Equation (c), the strain in the n direction can be expressed as

$$\varepsilon_n = \varepsilon_x\cos^2\theta + \varepsilon_y\sin^2\theta + \gamma_{xy}\sin\theta\cos\theta \tag{13.3}$$

Now recall the double-angle trigonometric identities:

$$\begin{aligned}\cos^2\theta &= \frac{1}{2}(1 + \cos 2\theta) \\ \sin^2\theta &= \frac{1}{2}(1 - \cos 2\theta) \\ 2\sin\theta\cos\theta &= \sin 2\theta\end{aligned}$$

From these identities, Equation (13.3) can also be expressed as

$$\varepsilon_n = \frac{\varepsilon_x + \varepsilon_y}{2} + \frac{\varepsilon_x - \varepsilon_y}{2}\cos 2\theta + \frac{\gamma_{xy}}{2}\sin 2\theta \tag{13.4}$$

Transformation Equation for Shear Strain

The component of the displacement vector **AA′** in the t direction [Equation (b)] represents an arc length through which the diagonal *OA* rotates. With this rotation angle denoted as α (Figure 13.5*a*), the arc length associated with radius dn can be expressed as

$$\alpha\,dn = \varepsilon_y\,dy\cos\theta - (\varepsilon_x\,dx + \gamma_{xy}\,dy)\sin\theta$$

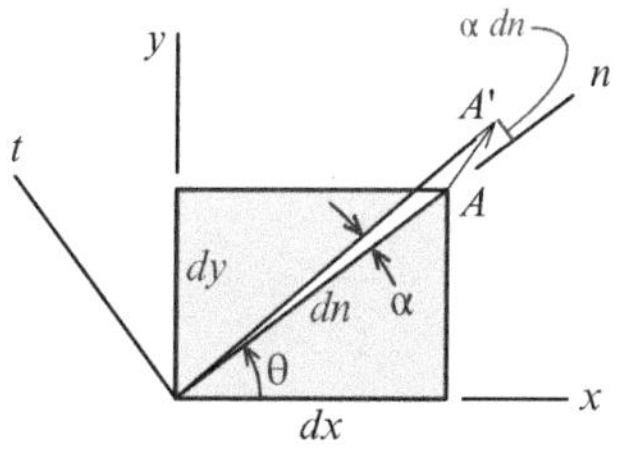

FIGURE 13.5*a*

Thus, diagonal OA rotates counterclockwise through an angle of

$$\begin{aligned}\alpha &= \varepsilon_y \frac{dy}{dn}\cos\theta - \left(\varepsilon_x \frac{dx}{dn} + \gamma_{xy}\frac{dy}{dn}\right)\sin\theta \\ &= \varepsilon_y \sin\theta\cos\theta - \varepsilon_x \sin\theta\cos\theta - \gamma_{xy}\sin^2\theta\end{aligned} \tag{d}$$

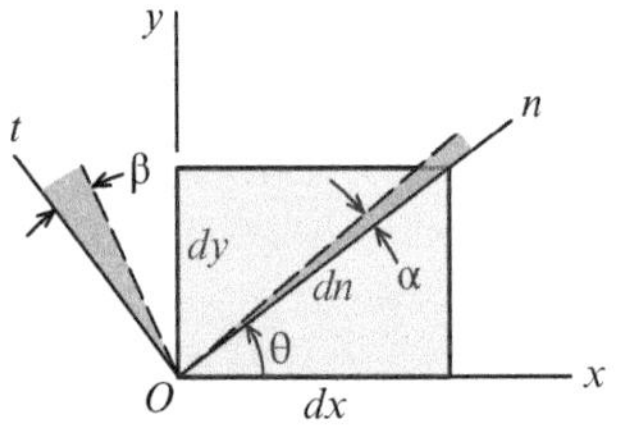

FIGURE 13.5*b*

The rotation angle β of a line element at right angles to OA (i.e., in the t direction as shown in Figure 13.5*b*) may be determined if the argument $\theta + 90°$ is substituted for θ in Equation (d):

$$\beta = -\varepsilon_y \sin\theta\cos\theta + \varepsilon_x \sin\theta\cos\theta - \gamma_{xy}\cos^2\theta \tag{e}$$

The rotation of β is clockwise. Since the positive direction for both α and β is counterclockwise, the shear strain γ_{nt}, which is the decrease in the right angle formed by the n and t axes, is the difference between α [Equation (d)] and β [Equation (e)]:

$$\gamma_{nt} = \alpha - \beta = 2\varepsilon_y \sin\theta\cos\theta - 2\varepsilon_x \sin\theta\cos\theta - \gamma_{xy}\sin^2\theta + \gamma_{xy}\cos^2\theta$$

Simplifying this equation gives

$$\gamma_{nt} = -2(\varepsilon_x - \varepsilon_y)\sin\theta\cos\theta + \gamma_{xy}(\cos^2\theta - \sin^2\theta) \tag{13.5}$$

or, in terms of the double-angle trigonometric functions,

$$\frac{\gamma_{nt}}{2} = -\frac{\varepsilon_x - \varepsilon_y}{2}\sin 2\theta + \frac{\gamma_{xy}}{2}\cos 2\theta \tag{13.6}$$

Comparison with Stress Transformation Equations

The strain transformation equations derived here are comparable to the stress transformation equations developed in Chapter 12. The corresponding variables in the two sets of transformation equations are listed in Table 13.1.

Table 13.1 Corresponding Variables in Stress and Strain Transformation Equations

Stresses	Strains
σ_x	ε_x
σ_y	ε_y
τ_{xy}	$\gamma_{xy}/2$
σ_n	ε_n
τ_{nt}	$\gamma_{nt}/2$

Strain Invariance

The normal strain in the t direction can be obtained from Equation (13.4) by substituting $\theta + 90°$ in place of θ, giving the following equation:

$$\varepsilon_t = \frac{\varepsilon_x + \varepsilon_y}{2} - \frac{\varepsilon_x - \varepsilon_y}{2}\cos 2\theta - \frac{\gamma_{xy}}{2}\sin 2\theta \tag{13.7}$$

If the expressions for ε_n and ε_t [Equations (13.4) and (13.7)] are added, the following relationship is obtained:

$$\varepsilon_n + \varepsilon_t = \varepsilon_x + \varepsilon_y \tag{13.8}$$

This equation shows that the sum of the normal strains acting in any two orthogonal directions is a constant value that is independent of the angle θ.

Sign Conventions

Equations (13.3) and (13.4) provide a means for determining the normal strain ε_n associated with a line oriented in an arbitrary n direction in the x–y plane. Equations (13.5) and (13.6)

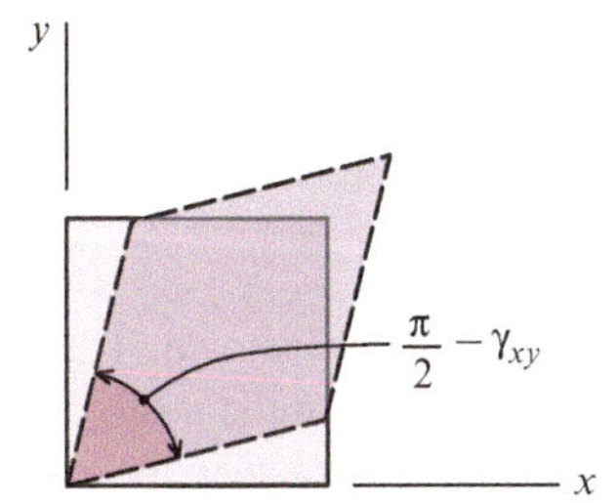

FIGURE 13.6*a* **Positive shear strain γ_{xy} at origin.**

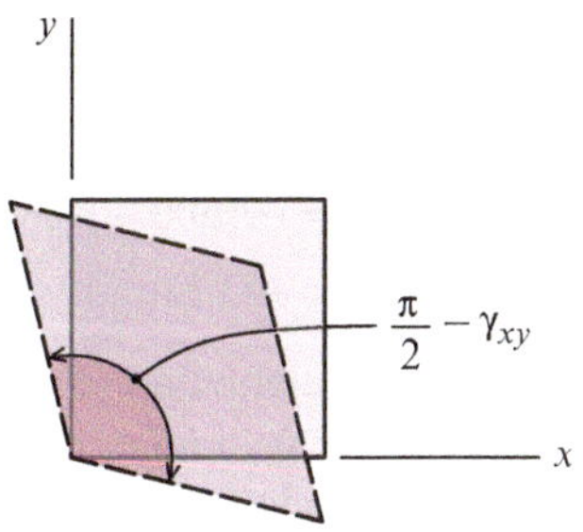

FIGURE 13.6*b* **Negative shear strain γ_{xy} at origin.**

allow the determination of the shear strain γ_{nt} associated with any two orthogonal lines oriented in the n and t directions in the x–y plane. With these equations, the sign conventions used in their development must be rigorously followed:

1. Normal strains that cause elongation are positive, and strains that cause contraction are negative.
2. A positive shear strain decreases the angle between the two lines at the origin of coordinates (Figure 13.6*a*). A negative shear strain increases the angle between the two lines (Figure 13.6*b*).
3. Angles measured counterclockwise from the reference x axis are positive. Conversely, angles measured clockwise from the reference x axis are negative.
4. The n–t–z axes have the same order as the x–y–z axes. Both sets of axes form a right-hand coordinate system.

EXAMPLE 13.1

An element of material at point O is subjected to a state of plane strain with the strains specified as $\varepsilon_x = 600$ με, $\varepsilon_y = -300$ με, and $\gamma_{xy} = 400$ μrad. The deflected shape of the element subjected to these strains is shown. Determine the strains acting at point O on an element that is rotated 40° counterclockwise from the original position.

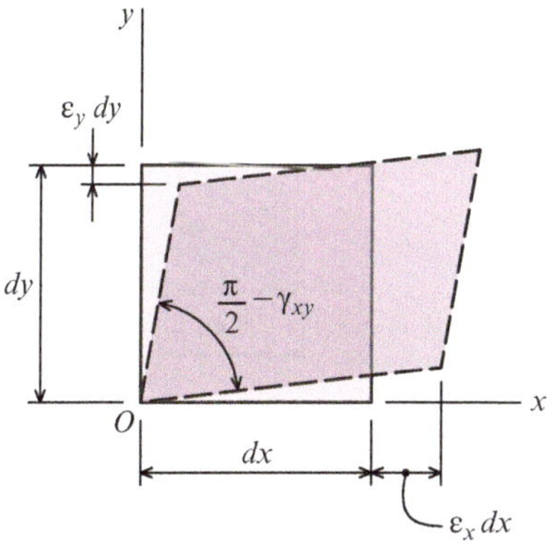

Plan the Solution

The strain transformation equations will be used to compute ε_n, ε_t, and γ_{nt}.

SOLUTION

The strain transformation equation

$$\varepsilon_n = \varepsilon_x \cos^2\theta + \varepsilon_y \sin^2\theta + \gamma_{xy} \sin\theta\cos\theta \qquad (13.3)$$

will be used to compute the normal strains ε_n and ε_t. Since counterclockwise angles are positive, the angle to be used in this instance is $\theta = 40°$. First,

$$\begin{aligned}\varepsilon_n &= (600)\cos^2(40°) + (-300)\sin^2(40°) + (400)\sin(40°)\cos(40°)\\ &= 425\ \mu\varepsilon\end{aligned}$$

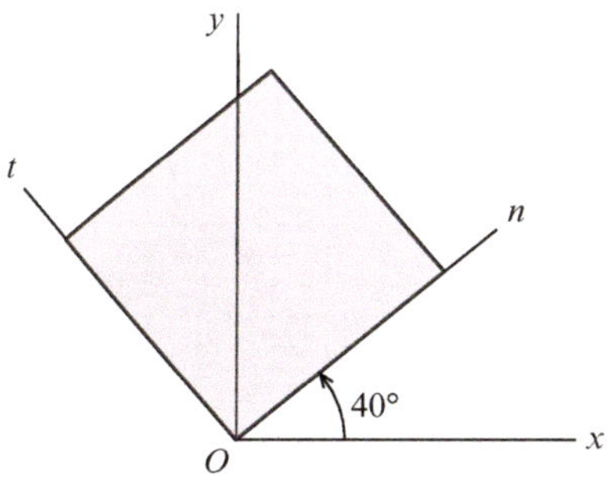

Next, to compute the normal strain ε_t, use an angle of $\theta = 40° + 90° = 130°$ in Equation (13.3):

$$\begin{aligned}\varepsilon_t &= (600)\cos^2(130°) + (-300)\sin^2(130°) + (400)\sin(130°)\cos(130°)\\ &= -125\ \mu\varepsilon\end{aligned}$$

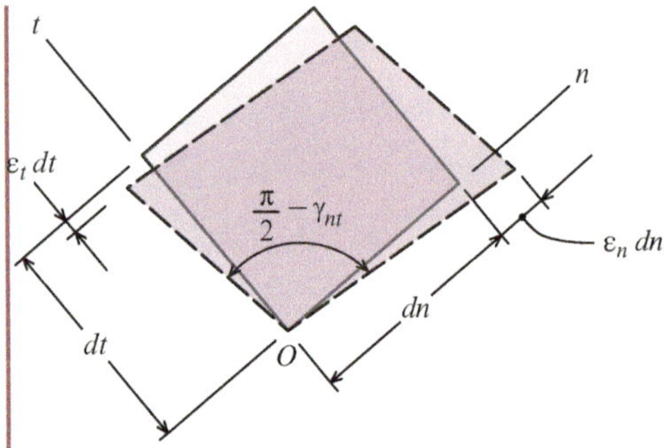

Finally, to compute the shear strain γ_{nt} from Equation (13.5), use an angle of $\theta = 40°$:

$$\begin{aligned}\gamma_{nt} &= -2[600 - (-300)]\sin(40°)\cos(40°) + (400)[\cos^2(40°) - \sin^2(40°)] \\ &= -817\ \mu\text{rad}\end{aligned}$$

The computed strains tend to distort the element as shown. The positive normal strain ε_n means that the element elongates in the n direction. The negative value for ε_t indicates that the element contracts in the t direction. Although it initially seems counterintuitive, note that the negative shear strain $\gamma_{nt} = -817\ \mu\text{rad}$ means that the angle between the n and t axes actually becomes greater than 90° at point O.

MecMovies

EXAMPLES

M13.1 The thin rectangular plate is uniformly deformed such that $\varepsilon_x = -700\ \mu\varepsilon$, $\varepsilon_y = -500\ \mu\varepsilon$, and $\gamma_{xy} = +900\ \mu\text{rad}$. Determine the normal strain

(a) along diagonal AC.
(b) along diagonal BD.

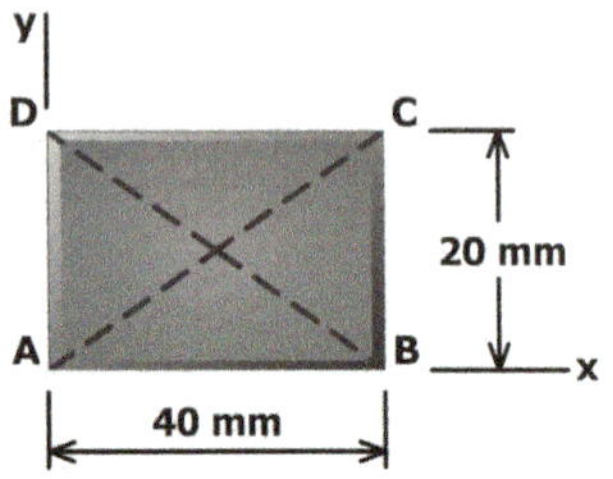

M13.2 The thin rectangular plate is subjected to strains $\varepsilon_x = +900\ \mu\varepsilon$, $\varepsilon_y = -600\ \mu\varepsilon$, and $\gamma_{xy} = -850\ \mu\text{rad}$. Determine the normal strains ε_n and ε_t, and the shear strain γ_{nt}, for $\theta = +50°$.

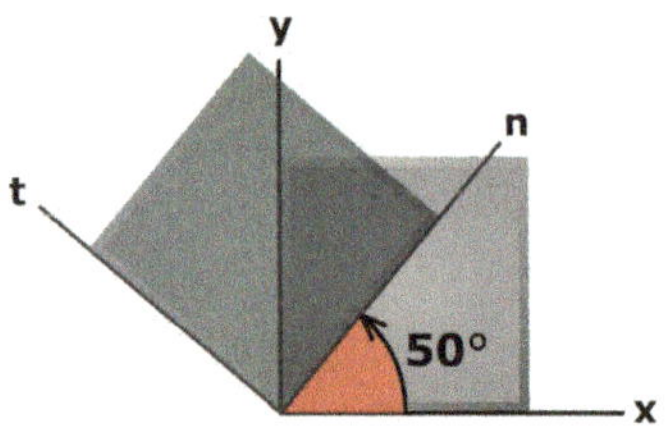

M13.3 A thin triangular plate is uniformly deformed such that, after deformation, the edges of the triangle are measured as AB = 300.30 mm, BC = 299.70 mm, and AC = 360.45 mm. Determine the strains ε_x, ε_y, and γ_{xy} in the plate.

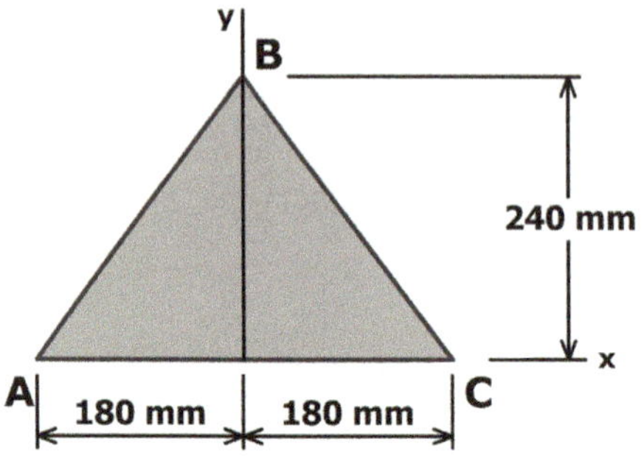

Before deformation

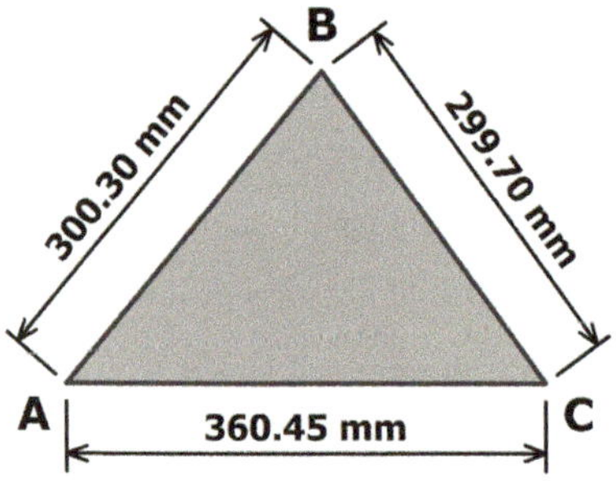

After deformation

13.4 Principal Strains and Maximum Shearing Strain

Given the similarity among Equations (13.3), (13.4), (13.5), and (13.6) for plane strain and Equations (12.5), (12.6), (12.7), and (12.8) for plane stress, it is not surprising that all of the relationships developed for plane stress can be applied to plane strain analysis, provided that the substitutions given in Table 13.1 are made. Expressions for the in-plane principal directions, the in-plane principal strains, and the maximum in-plane shear strain are as follows:

$$\tan 2\theta_p = \frac{\gamma_{xy}}{\varepsilon_x - \varepsilon_y} \tag{13.9}$$

$$\varepsilon_{p1,p2} = \frac{\varepsilon_x + \varepsilon_y}{2} \pm \sqrt{\left(\frac{\varepsilon_x - \varepsilon_y}{2}\right)^2 + \left(\frac{\gamma_{xy}}{2}\right)^2} \tag{13.10}$$

$$\frac{\gamma_{max}}{2} = \pm\sqrt{\left(\frac{\varepsilon_x - \varepsilon_y}{2}\right)^2 + \left(\frac{\gamma_{xy}}{2}\right)^2} \tag{13.11}$$

Equations (13.9), (13.10), and (13.11) are similar in form to Equations (12.11), (12.12), and (12.15), respectively. However, instances of τ_{xy} in the stress equations are replaced by $\gamma_{xy}/2$ in the strain equations. Be careful not to overlook these factors of 2 when switching between stress analysis and strain analysis.

In the preceding equations, normal strains that cause elongation (i.e., a stretching produced by a tensile stress) are positive. Positive shear strains decrease the angle between the element faces at the coordinate origin. (See Figure 13.3.)

As was true in plane stress transformations, Equation (13.10) does not indicate which principal strain, either ε_{p1} or ε_{p2}, is associated with the two principal angles. The solution of Equation (13.9) always gives a value of θ_p between –45° and +45° (inclusive). The principal strain associated with this value of θ_p can be determined from the following two-part rule:

- If the term $\varepsilon_x - \varepsilon_y$ is positive, then θ_p indicates the orientation of ε_{p1}.
- If the term $\varepsilon_x - \varepsilon_y$ is negative, then θ_p indicates the orientation of ε_{p2}.

The other principal strain is oriented perpendicular to θ_p.

The two principal strains determined from Equation (13.10) may be both positive, both negative, or positive and negative. In naming the principal strains, ε_{p1} is the more positive value algebraically (i.e., the algebraically larger value). If one or both of the principal strains from Equation (13.10) are negative, ε_{p1} can have a smaller absolute value than ε_{p2}.

Absolute Maximum Shear Strain

When a state of plane strain exists, ε_x, ε_y, and γ_{xy} may have nonzero values. However, strains in the z direction (i.e., the out-of-plane direction) are zero; thus, $\varepsilon_z = 0$ and $\gamma_{xz} = \gamma_{yz} = 0$. Equation (13.10) gives the two in-plane principal strains, and the third principal strain is $\varepsilon_{p3} = \varepsilon_z = 0$. An examination of Equations (13.10) and (13.11) reveals that the maximum in-plane shear strain is equal to the difference between the two in-plane principal strains:

$$\gamma_{max} = \varepsilon_{p1} - \varepsilon_{p2} \tag{13.12}$$

However, the magnitude of the absolute maximum shear strain for a plane strain element may be larger than the maximum in-plane shear strain, depending upon the relative magnitudes and signs of the principal strains. The absolute maximum shear strain can be determined from one of the three conditions shown in Table 13.2.

Table 13.2 Absolute Maximum Shear Strains

	Principal Strain Element	Absolute Maximum Shear Strain Element
(a) If both ε_{p1} and ε_{p2} are positive, then $\gamma_{\text{abs max}} = \varepsilon_{p1} - \varepsilon_{p3} = \varepsilon_{p1} - 0 = \varepsilon_{p1}$	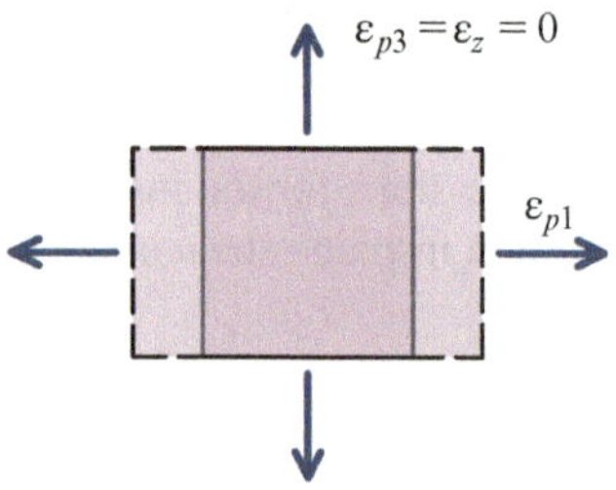	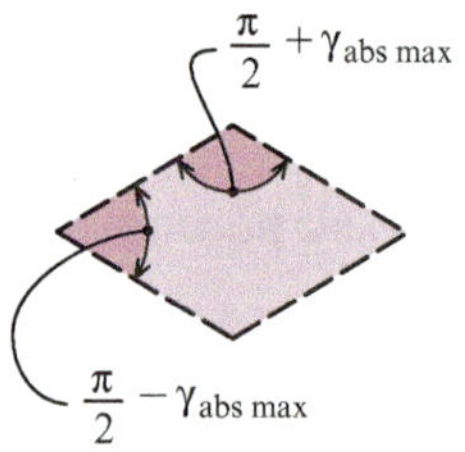
(b) If both ε_{p1} and ε_{p2} are negative, then $\gamma_{\text{abs max}} = \varepsilon_{p3} - \varepsilon_{p2} = 0 - \varepsilon_{p2} = -\varepsilon_{p2}$	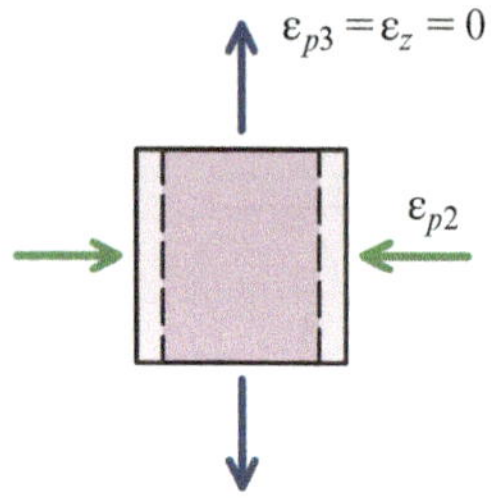	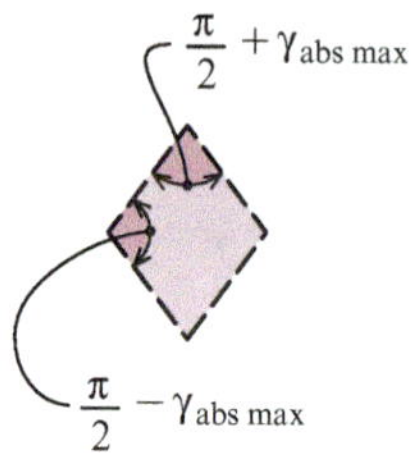
(c) If ε_{p1} is positive and ε_{p2} is negative, then $\gamma_{\text{abs max}} = \varepsilon_{p1} - \varepsilon_{p2}$	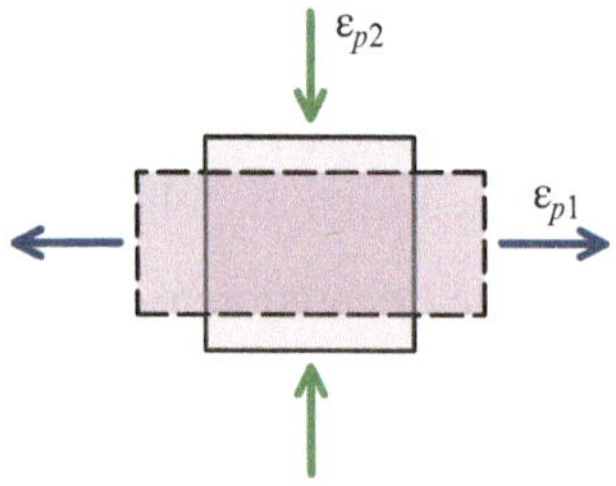	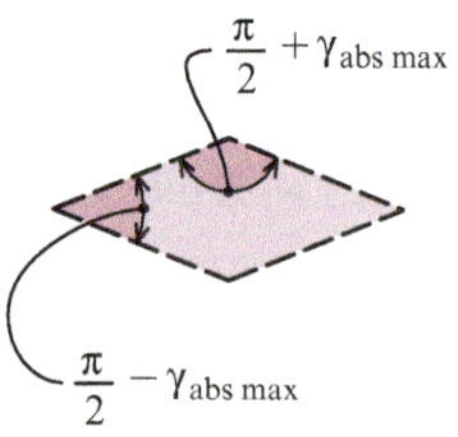

These conditions apply only to a state of *plane strain*. As will be shown in Sections 13.7 and 13.8, the third principal strain will not be zero for a state of *plane stress*.

13.5 Presentation of Strain Transformation Results

Principal strain and maximum in-plane shear strain results should be presented with a sketch that depicts the orientation of all strains. Strain results can be conveniently shown on a single element.

Draw an element rotated at the angle θ_p calculated from Equation (13.9), which will give a value between +45° and –45° (inclusive):

$$\tan 2\theta_p = \frac{\gamma_{xy}}{\varepsilon_x - \varepsilon_y}$$

- When θ_p is positive, the element is rotated in a counterclockwise sense from the reference x axis. When θ_p is negative, the rotation is clockwise.
- Note that the angle calculated from Equation (13.9) does not necessarily give the orientation of the ε_{p1} direction. Either ε_{p1} or ε_{p2} may act in the θ_p direction given by that equation. The principal strain oriented at θ_p can be determined from the following two-part rule:
    - If $\varepsilon_x - \varepsilon_y$ is positive, then θ_p indicates the orientation of ε_{p1}.
    - If $\varepsilon_x - \varepsilon_y$ is negative, then θ_p indicates the orientation of ε_{p2}.
- Elongate or contract the element into a rectangle according to the principal strains acting in the two orthogonal directions. If a principal strain is positive, then the element is elongated in that direction. The element is contracted if the principal strain is negative.
- Add either tension or compression arrows, labeled with the corresponding strain magnitudes, to each edge of the element.
- To show the distortion caused by the shear strain, draw a diamond shape inside of the rectangular principal strain element. The corners of the diamond should be located at the midpoint of each edge of the rectangle.
- The maximum in-plane shear strain calculated from Equation (13.11) or Equation (13.12) will be a positive value. Since a positive shear strain causes the angle between two axes to decrease, label one of the acute angles with the value $\pi/2 - \gamma_{max}$.

EXAMPLE 13.2

The strain components at a point in a body subjected to plane strain are $\varepsilon_x = -680\ \mu\varepsilon$, $\varepsilon_y = 320\ \mu\varepsilon$, and $\gamma_{xy} = -980$ m μrad. The deflected shape of an element that is subjected to these strains is shown. Determine the principal strains, the maximum in-plane shear strain, and the absolute maximum shear strain at point O. Show the principal strain deformations and the maximum in-plane shear strain distortion in a sketch.

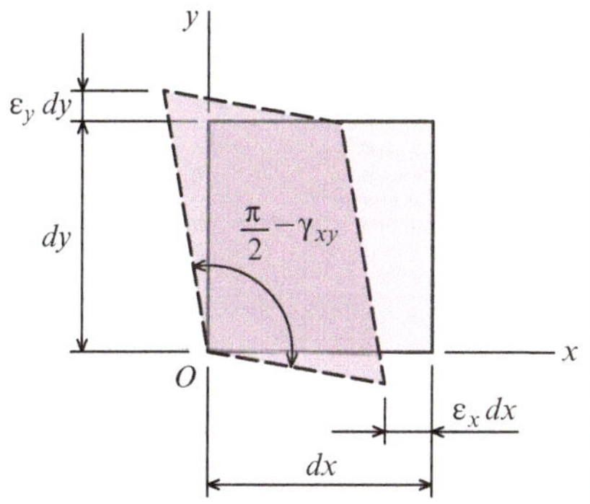

SOLUTION

From Equation (13.10), the in-plane principal strains are

$$\varepsilon_{p1,\,p2} = \frac{\varepsilon_x + \varepsilon_y}{2} \pm \sqrt{\left(\frac{\varepsilon_x - \varepsilon_y}{2}\right)^2 + \left(\frac{\gamma_{xy}}{2}\right)^2}$$

$$= \frac{-680 + 320}{2} \pm \sqrt{\left(\frac{-680 - 320}{2}\right)^2 + \left(\frac{-980}{2}\right)^2}$$

$$= -180 \pm 700$$

$$= 520\ \mu\varepsilon,\ -880\ \mu\varepsilon \qquad \textbf{Ans.}$$

and from Equation (13.11), the maximum in-plane shear strain is

$$\frac{\gamma_{max}}{2} = \pm\sqrt{\left(\frac{\varepsilon_x - \varepsilon_y}{2}\right)^2 + \left(\frac{\gamma_{xy}}{2}\right)^2}$$

$$= \pm\sqrt{\left(\frac{-680 - 320}{2}\right)^2 + \left(\frac{-980}{2}\right)^2}$$

$$= 700\ \mu\text{rad}$$

$$\therefore\ \gamma_{max} = 1{,}400\ \mu\text{rad} \qquad \textbf{Ans.}$$

The in-plane principal directions can be determined from Equation (13.9):

$$\tan 2\theta_p = \frac{\gamma_{xy}}{\varepsilon_x - \varepsilon_y} = \frac{-980}{-680 - 320} = \frac{-980}{-1000} \qquad \textbf{Note: } \varepsilon_x - \varepsilon_y < 0$$

$$\therefore\ 2\theta_p = 44.42° \qquad \text{and thus} \qquad \theta_p = 22.21°$$

Since $\varepsilon_x - \varepsilon_y < 0$, the angle θ_p is the angle between the x direction and the ε_{p2} direction.

The problem states that this is a *plane strain* condition. Therefore, the out-of-plane normal strain $\varepsilon_z = 0$ is the third principal strain ε_{p3}. Since ε_{p1} is positive and ε_{p2} is negative, the absolute maximum shear strain *is* the maximum in-plane shear strain. Therefore, the magnitude of the absolute maximum shear strain (see Table 13.2) is

$$\gamma_{\text{abs max}} = \varepsilon_{p1} - \varepsilon_{p2} = 1{,}400\ \mu\text{rad}$$

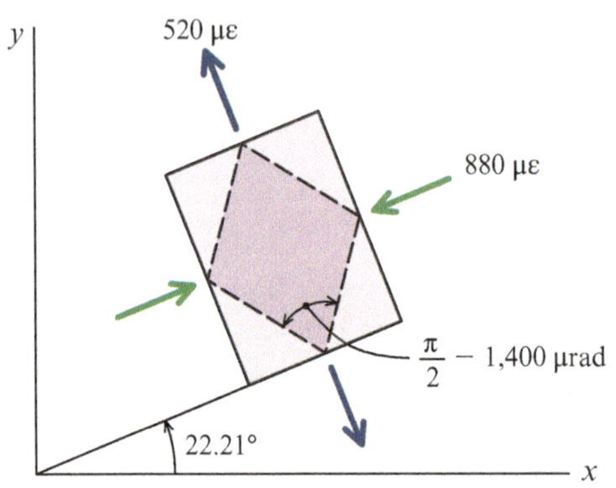

Sketch the Deformations and Distortions

The principal strains are oriented 22.21° counterclockwise from the x direction. The principal strain corresponding to this direction is $\varepsilon_{p2} = -880\ \mu\varepsilon$; therefore, the element contracts parallel to the 22.21° direction. In the perpendicular direction, the principal strain is $\varepsilon_{p1} = 520\ \mu\varepsilon$, which causes the element to elongate.

To show the distortion caused by the maximum in-plane shear strain, connect the midpoints of each of the rectangle's edges, to create a diamond. Two interior angles of this diamond will be acute angles (i.e., less than 90°), and two interior angles will be obtuse (i.e., greater than 90°). Use the positive value of $\gamma_{\max}$ obtained from Equation (13.11) to label one of the acute interior angles with $\pi/2 - \gamma_{\max}$. The obtuse interior angles will have a magnitude of $\pi/2 + \gamma_{\max}$. Note that the four interior angles of the diamond (or any quadrilateral) must total 2π radians (or 360°).

PROBLEMS

P13.1 The thin rectangular plate shown in Figure P13.1/2 is uniformly deformed such that $\varepsilon_x = 230\ \mu\varepsilon$, $\varepsilon_y = -480\ \mu\varepsilon$, and $\gamma_{xy} = -760\ \mu\text{rad}$. Using dimensions of $a = 20$ mm and $b = 25$ mm, determine the normal strain in the plate in the direction defined by

(a) points O and A.
(b) points O and C.

P13.2 The thin rectangular plate shown in Figure P13.1/2 is uniformly deformed such that $\varepsilon_x = -360\ \mu\varepsilon$, $\varepsilon_y = 770\ \mu\varepsilon$, and $\gamma_{xy} = 940\ \mu\text{rad}$. Using dimensions of $a = 25$ mm and $b = 40$ mm, determine the normal strain in the plate in the direction defined by

(a) points O and B.
(b) points O and D.

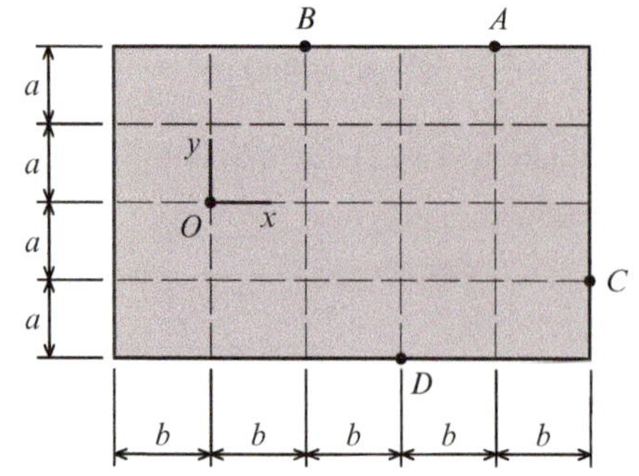

FIGURE P13.1/2

P13.3 The thin rectangular plate shown in Figure P13.3/4 is uniformly deformed such that ε_x = 120 με, ε_y = −860 με, and γ_{xy} = 1,100 μrad. If a = 25 mm, determine

(a) the normal strain ε_n in the plate.
(b) the normal strain ε_t in the plate.
(c) the shear strain γ_{nt} in the plate.

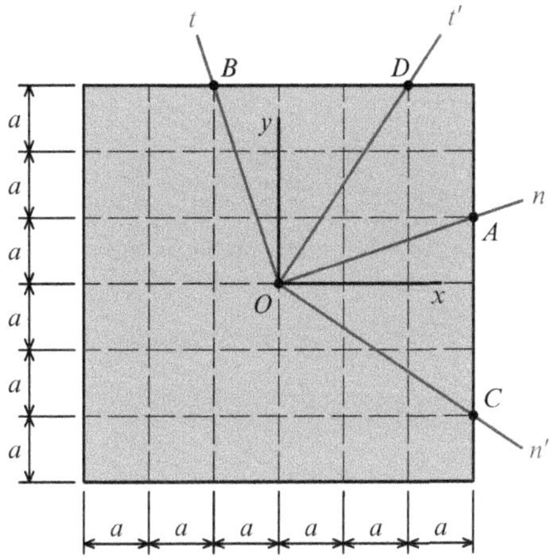

FIGURE P13.3/4

P13.4 The thin rectangular plate shown in Figure P13.3/4 is uniformly deformed such that ε_x = −890 με, ε_y = 440 με, and γ_{xy} = −310 μrad. If a = 50 mm, determine

(a) the normal strain $\varepsilon_{n'}$ in the plate.
(b) the normal strain $\varepsilon_{t'}$ in the plate.
(c) the shear strain $\gamma_{n't'}$ in the plate.

P13.5 The thin square plate shown in Figure P13.5/6 is uniformly deformed such that ε_n = 660 με, ε_t = 910 με, and γ_{nt} = 830 μrad. Determine

(a) the normal strain ε_x in the plate.
(b) the normal strain ε_y in the plate.
(c) the shear strain γ_{xy} in the plate.

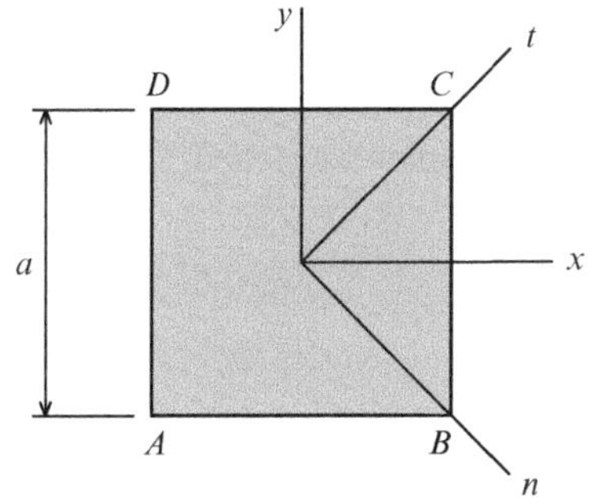

FIGURE P13.5/6

P13.6 The thin square plate shown in Figure P13.5/6 is uniformly deformed such that ε_x = 0 με, ε_y = 0 με, and γ_{xy} = −1,850 μrad. Using a = 650 mm, determine the deformed length of (a) diagonal AC and (b) diagonal BD.

P13.7–P13.10 The strain components ε_x, ε_y, and γ_{xy} are given for a point in a body subjected to **plane strain**. Determine the strain components ε_n, ε_t, and γ_{nt} at the point if the n–t axes are rotated with respect to the x–y axes by the amount, and in the direction, indicated by the angle θ shown in either Figure P13.7 or Figure P13.8. **Sketch the deformed shape of the element.**

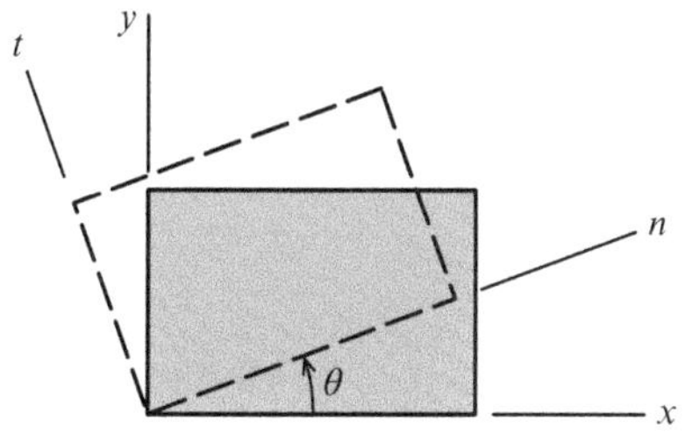

FIGURE P13.7

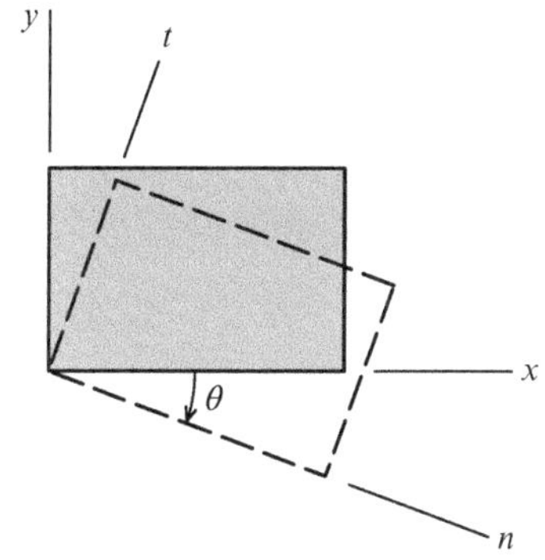

FIGURE P13.8

Problem	Figure	ε_x	ε_y	γ_{xy}	θ
P13.7	P13.7	−1,050 με	400 με	1,360 μrad	36°
P13.8	P13.8	−350 με	1,650 με	720 μrad	14°
P13.9	P13.7	−1,375 με	−1,825 με	650 μrad	15°
P13.10	P13.8	590 με	−1,670 με	−1,185 μrad	23°

P13.11–P13.15 The strain components ε_x, ε_y, and γ_{xy} are given for a point in a body subjected to **plane strain**. Determine the principal strains, the maximum in-plane shear strain, and the absolute maximum shear strain at the point. Show the angle θ_p, the principal strain deformations, and the maximum in-plane shear strain distortion on a sketch.

Problem	ε_x	ε_y	γ_{xy}
P13.11	−550 με	−285 με	940 μrad
P13.12	940 με	−360 με	830 μrad
P13.13	−270 με	510 με	1,150 μrad
P13.14	670 με	−280 με	−800 μrad
P13.15	960 με	650 με	350 μrad

13.6 Mohr's Circle for Plane Strain

The general strain transformation equations, expressed in terms of double-angle trigonometric functions, were presented in Section 13.3. Equation (13.4) is

$$\varepsilon_n = \frac{\varepsilon_x + \varepsilon_y}{2} + \frac{\varepsilon_x - \varepsilon_y}{2}\cos 2\theta + \frac{\gamma_{xy}}{2}\sin 2\theta$$

and Equation (13.6) is

$$\frac{\gamma_{nt}}{2} = -\frac{\varepsilon_x - \varepsilon_y}{2}\sin 2\theta + \frac{\gamma_{xy}}{2}\cos 2\theta$$

Equation (13.4) can be rewritten so that only terms involving 2θ appear on the right-hand side:

$$\varepsilon_n - \frac{\varepsilon_x + \varepsilon_y}{2} = \frac{\varepsilon_x - \varepsilon_y}{2}\cos 2\theta + \frac{\gamma_{xy}}{2}\sin 2\theta$$

$$\frac{\gamma_{nt}}{2} = -\frac{\varepsilon_x - \varepsilon_y}{2}\sin 2\theta + \frac{\gamma_{xy}}{2}\cos 2\theta$$

Both equations can be squared, then added together, and simplified to give

$$\left(\varepsilon_n - \frac{\varepsilon_x + \varepsilon_y}{2}\right)^2 + \left(\frac{\gamma_{nt}}{2}\right)^2 = \left(\frac{\varepsilon_x - \varepsilon_y}{2}\right)^2 + \left(\frac{\gamma_{xy}}{2}\right)^2 \qquad (13.13)$$

This is the equation of a circle in terms of the variables ε_n and $\gamma_{nt}/2$. It is similar in form to Equation (12.20), which was the basis of Mohr's circle for stress.

Mohr's circle for plane strain is constructed and used in much the same way as Mohr's circle for plane stress. The horizontal axis used in the construction is the ε axis, and the vertical axis is $\gamma/2$. The circle is centered at

$$C = \frac{\varepsilon_x + \varepsilon_y}{2}$$

on the ε axis, and it has a radius

$$R = \sqrt{\left(\frac{\varepsilon_x - \varepsilon_y}{2}\right)^2 + \left(\frac{\gamma_{xy}}{2}\right)^2}$$

There are two notable differences in constructing and using Mohr's circle for strain, compared with Mohr's circle for stress. First, note that the vertical axis for the strain circle is $\gamma/2$; hence, shear strain values must be divided by 2 before they are plotted. Second, the endpoints of the diameter have the coordinates $(\varepsilon_x, -\gamma_{xy}/2)$ and $(\varepsilon_y, \gamma_{xy}/2)$.

EXAMPLE 13.3

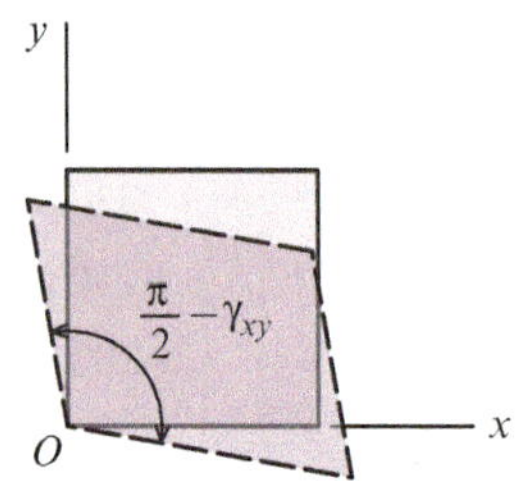

The strain components at a point in a body subjected to plane strain are $\varepsilon_x = 435\ \mu\varepsilon$, $\varepsilon_y = -135\ \mu\varepsilon$, and $\gamma_{xy} = -642\ \mu\text{rad}$. The deflected shape of an element that is subjected to these strains is shown. Determine the principal strains, the maximum in-plane shear strain, and the absolute maximum shear strain at point O. Show the principal strain deformations and the maximum in-plane shear strain distortion in a sketch.

SOLUTION

Plot point x as $(\varepsilon_x, -\gamma_{xy}/2) = (435\ \mu\varepsilon, 321\ \mu\text{rad})$, and plot point y as $(\varepsilon_y, \gamma_{xy}/2) = (-135\ \mu\varepsilon, -321\ \mu\text{rad})$. Connect points x and y to define the diameter of Mohr's circle for this strain state. (See accompanying diagram.)

Since points x and y are always the same distance above or below the ε axis, the center of Mohr's circle can be found by averaging the normal strains acting in the x and y directions:

$$C = \frac{\varepsilon_x + \varepsilon_y}{2} = \frac{435 + (-135)}{2} = 150\ \mu\varepsilon$$

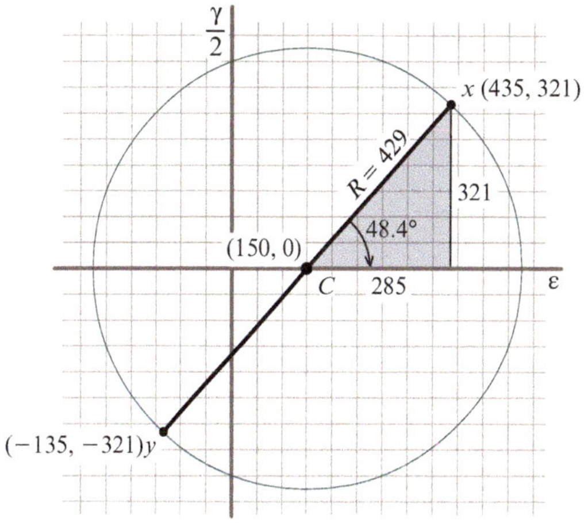

The center of Mohr's circle always lies on the ε axis.

The geometry of the circle is used to calculate the radius. The $(\varepsilon, \gamma/2)$ coordinates of both point x and center C are known. Use these coordinates with the Pythagorean theorem to calculate the hypotenuse of the shaded triangle:

$$R = \sqrt{(435 - 150)^2 + (321 - 0)^2}$$
$$= \sqrt{285^2 + 321^2} = 429\ \mu$$

Remember that the vertical coordinate used in plotting Mohr's circle is $\gamma/2$. The given shear strain is $\gamma_{xy} = -642\ \mu\text{rad}$; therefore, a vertical coordinate of 321 μrad is used in plotting Mohr's circle. The angle between the x–y diameter and the ε axis is $2\theta_p$, and its magnitude can be computed with the tangent function:

$$\tan 2\theta_p = \frac{321}{285} \qquad \therefore\ 2\theta_p = 48.4°$$

Note that this angle turns clockwise from point x to the ε axis.

The principal strains are determined from the location of the center C and the radius R of the circle:

$$\varepsilon_{p1} = C + R = 150\ \mu\varepsilon + 429\ \mu\varepsilon = 579\ \mu\varepsilon$$
$$\varepsilon_{p2} = C - R = 150\ \mu\varepsilon - 429\ \mu\varepsilon = -279\ \mu\varepsilon$$

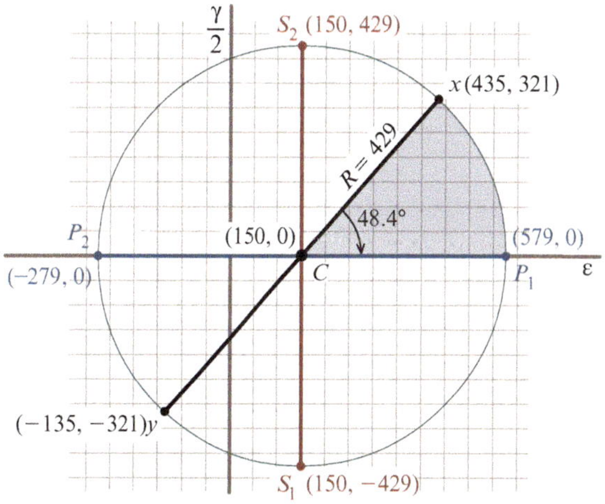

The maximum values of γ occur at points S_1 and S_2, located at the bottom and at the top of Mohr's circle. The shear strain magnitude at these points is equal to the *circle radius R times 2*; therefore, the maximum in-plane shear strain is

$$\gamma_{\max} = 2R = 2(429\ \mu) = 858\ \mu\text{rad}$$

The normal strain associated with the maximum in-plane shear strain is given by the center C of the circle:

$$\varepsilon_{\text{avg}} = C = 150\ \mu\varepsilon$$

The problem states that this is a *plane strain* condition. Therefore, the out-of-plane normal strain $\varepsilon_z = 0$ is the third principal strain ε_{p3}. Since ε_{p1} is positive and ε_{p2} is negative, the absolute maximum shear strain equals the maximum in-plane shear strain. Therefore, the magnitude of the absolute maximum shear strain (see Table 13.2) is

$$\gamma_{\text{abs max}} = \varepsilon_{p1} - \varepsilon_{p2} = 858\ \mu\text{rad} \qquad \textbf{Ans.}$$

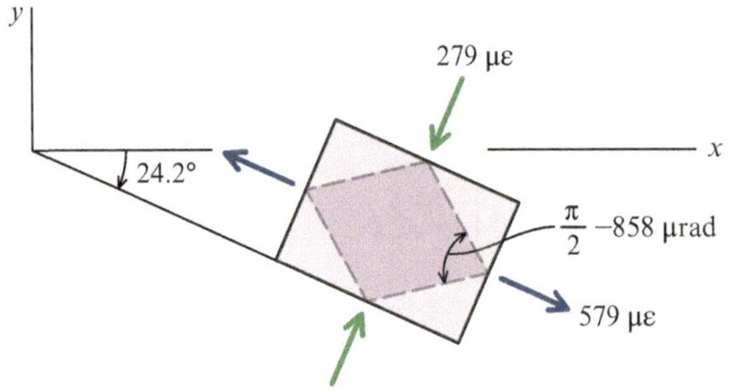

A complete sketch showing the principal strains, the maximum in-plane shear strain, and the orientations of the respective directions is given. The principal strains are shown by the solid rectangle, which has been elongated in the ε_{p1} direction (since $\varepsilon_{p1} = 579$ με) and contracted in the ε_{p2} direction (since $\varepsilon_{p2} = -279$ με).

The distortion caused by the maximum in-plane shear strain is shown by a diamond that connects the four midpoints of the principal strain element. Because the radius of Mohr's circle is $R = 429$ μ, the maximum in-plane shear strain is $\gamma_{max} = 2R = \pm 858$ μrad. A positive γ value causes the angle between adjacent edges of an element to decrease, forming an acute angle. Therefore, one of the acute angles in the distorted diamond shape is labeled with the positive value of γ_{max}: $\pi/2 - 858$ μrad.

MecMovies

EXAMPLE

M13.4 Coach Mohr's Circle of Strain

Learn to construct and use Mohr's circle to determine principal strains, including the proper orientations of the principal strain directions.

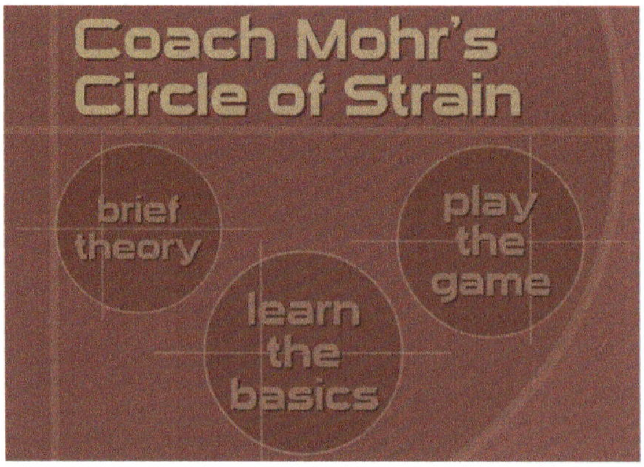

EXERCISE

M13.4 Coach Mohr's Circle of Strain. Learn to construct and use Mohr's circle to determine principal strains, including the proper orientations of the principal strain directions. (Game)

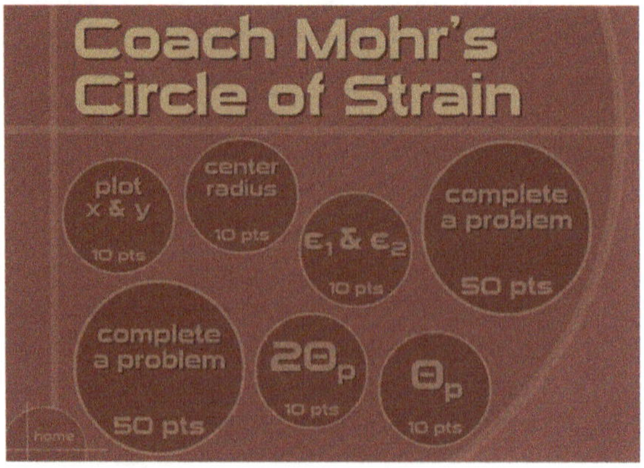

FIGURE M13.4

PROBLEMS

P13.16–P13.17 The principal strains are given for a point in a body subjected to **plane strain.** Construct Mohr's circle, and use it to

(a) determine the strains ε_x, ε_y, and γ_{xy}. (Assume that $\varepsilon_x > \varepsilon_y$.)
(b) determine the maximum in-plane shear strain and the absolute maximum shear strain.
(c) draw a sketch showing the angle θ_p, the principal strain deformations, and the maximum in-plane shear strain distortions.

Problem	ε_{p1}	ε_{p2}	θ_p
P13.16	780 με	590 με	35.66°
P13.17	−350 με	−890 με	−19.50°

P13.18–P13.24 The strain components ε_x, ε_y, and γ_{xy} are given for a point in a body subjected to **plane strain**. Using Mohr's circle, determine the principal strains, the maximum in-plane shear strain, and the absolute maximum shear strain at the point. Show the angle θ_p, the principal strain deformations, and the maximum in-plane shear strain distortion in a sketch.

Problem	ε_x	ε_y	γ_{xy}
P13.18	380 με	−770 με	−650 μrad
P13.19	760 με	590 με	−360 μrad
P13.20	−1,570 με	−430 με	−950 μrad
P13.21	475 με	685 με	−150 μrad
P13.22	670 με	455 με	−900 μrad
P13.23	0 με	320 με	260 μrad
P13.24	−180 με	−1,480 με	425 μrad

13.7 Strain Measurement and Strain Rosettes

Many engineered components are subjected to a combination of axial, torsion, and bending effects. Theories and procedures for calculating the stresses caused by each of these effects have been developed throughout this book. There are situations, however, in which the combination of effects is too complicated or uncertain to be confidently assessed with theoretical analysis alone. In these instances, an experimental analysis of component stresses is desired, either as an absolute determination of actual stresses or as validation for a numerical model that will be used for subsequent analyses. Stress cannot be measured. Strains, by contrast, can be measured directly through well-established experimental procedures. Once the strains in a component have been measured, the corresponding stresses can be calculated from stress–strain relationships, such as Hooke's law.

Strain Gages

Strains can be measured by using a simple component called a **strain gage**. The strain gage is a type of electrical resistor. Most commonly, strain gages are thin metal-foil grids that are bonded to the surface of a machine part or a structural element. When loads are applied, the object being tested elongates or contracts, creating normal strains. Since the strain gage is bonded to the object, it undergoes the same strain as the object. The electrical resistance of the metal-foil grid changes in proportion to its strain. Consequently, precise measurement of resistance change in the gage serves as an indirect measure of strain. The resistance change in a strain gage is very small—too small to be measured accurately with an ordinary ohmmeter; however, it can be measured accurately with a specific type of electrical circuit known as a **Wheatstone bridge**. For each type of gage, the relationship between strain and resistance change is determined through a calibration procedure performed by the manufacturer. Gage manufacturers report this property as a *gage factor* GF, which is defined as the ratio between the unit change in gage resistance R to the unit change in length L:

$$\text{GF} = \frac{\Delta R/R}{\Delta L/L} = \frac{\Delta R/R}{\varepsilon_{\text{avg}}}$$

In this equation, ΔR is the resistance change and ΔL is the change in length of the strain gage. The gage factor is constant for the small range of resistance change normally encountered, and most typical gages have a gage factor of about 2. Strain gages are very accurate, relatively inexpensive, and reasonably durable if they are properly protected from chemical attack, environmental conditions (such as temperature and humidity), and physical damage. Strain gages can measure normal static and dynamic strains as small as 1×10^{-6}.

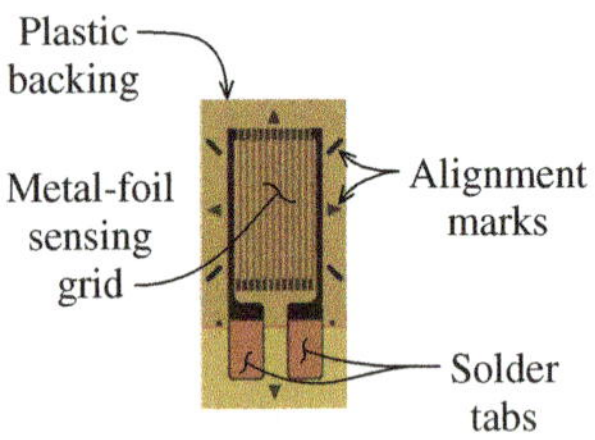

FIGURE 13.7

The photoetching process used to create the metal-foil grids is very versatile, enabling a wide variety of gage sizes and grid shapes to be produced. A typical single strain gage is shown in Figure 13.7. Since the foil itself is fragile and easily torn, the grid is bonded to a thin plastic backing film, which provides both strength and electrical insulation between the strain gage and the object being tested. For general-purpose strain gage applications, a polyimide plastic that is tough and flexible is used for the backing. Alignment markings are added to the backing to facilitate proper installation. Lead wires are attached to the solder tabs of the gage so that the change in resistance can be monitored with a suitable instrumentation system.

The objective of experimental stress analysis is to determine the state of stress at a specific point in the object being tested. In other words, the investigator ultimately wants to determine σ_x, σ_y, and τ_{xy} at a point. To accomplish this task, strain gages are used to determine ε_x, ε_y, and γ_{xy}, and then stress–strain relationships are used to compute the corresponding stresses. However, strain gages can measure normal strains in only one direction. Therefore, the question becomes "How can one determine three quantities (ε_x, ε_y, and γ_{xy}) by using a component that measures normal strain ε in only a single direction?"

The strain transformation equation for the normal strain ε_n at an arbitrary direction θ was derived as Equation (13.3) in Section 13.3:

$$\varepsilon_n = \varepsilon_n \cos^2\theta + \varepsilon_y \sin^2\theta + \gamma_{xy} \sin\theta\cos\theta$$

The rosette shown in Figure 13.8 is called a rectangular rosette because the angle between gages is 45°. The rectangular rosette is the most common rosette pattern.

Suppose that ε_n could be measured by a strain gage oriented at a known angle θ. Three unknown variables—ε_x, ε_y, and γ_{xy}—remain in Equation (13.3). To solve for these three unknowns, three equations in terms of ε_x, ε_y, and γ_{xy} are required. Those equations can be obtained by using three strain gages in combination, with each gage measuring the strain in a different direction. This combination of strain gages is called a **strain rosette**.

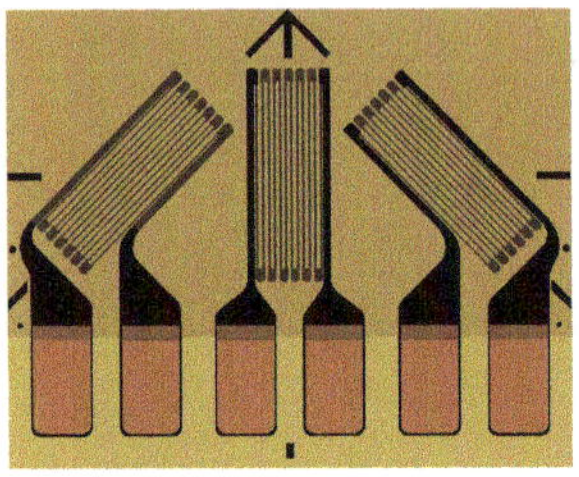

FIGURE 13.8 Typical strain rosette.

Strain Rosettes

A typical strain rosette is shown in Figure 13.8. The gage is configured so that the angles between each of the three gages are known. When the rosette is bonded to the object being tested, one of the three gages is aligned with a reference axis on the object—for example, along the longitudinal axis of a beam or a shaft. During the experimental test, strains are measured from each of the three gages. A strain transformation equation can be written for each of those gages in the notation indicated in Figure 13.9:

$$\begin{aligned}\varepsilon_a &= \varepsilon_x \cos^2\theta_a + \varepsilon_y \sin^2\theta_a + \gamma_{xy}\sin\theta_a\cos\theta_a\\ \varepsilon_b &= \varepsilon_x \cos^2\theta_b + \varepsilon_y \sin^2\theta_b + \gamma_{xy}\sin\theta_b\cos\theta_b\\ \varepsilon_c &= \varepsilon_x \cos^2\theta_c + \varepsilon_y \sin^2\theta_c + \gamma_{xy}\sin\theta_c\cos\theta_c\end{aligned} \tag{13.14}$$

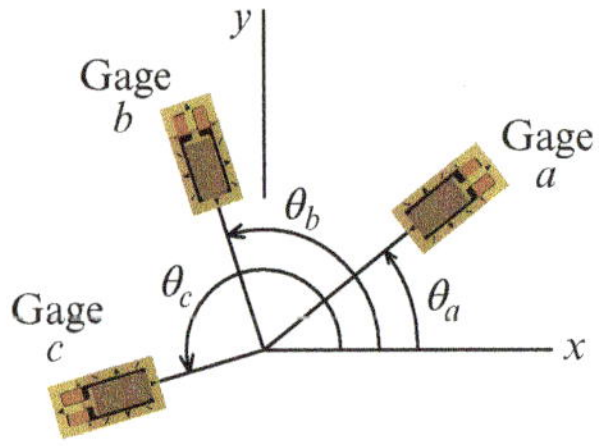

FIGURE 13.9

In this book, the angle used to identify the orientation of each rosette gage will always be measured counterclockwise from the reference x axis.

The three strain transformation equations in Equation (13.14) can be solved simultaneously to yield the values of ε_x, ε_y, and γ_{xy}. Once ε_x, ε_y, and γ_{xy} have been determined,

Equations (13.9), (13.10), and (13.11) or the corresponding Mohr's circle construction can be used to determine the in-plane principal strains, their orientations, and the maximum in-plane shear strain at a point.

Strains in the Out-of-Plane Direction

Rosettes are bonded to the surface of an object, and stresses in the out-of-plane direction on the free surface of an object are always zero. Consequently, a state of *plane stress* exists at the rosette. Whereas strains in the out-of-plane direction are zero for the plane strain condition, out-of-plane strains are not zero for plane stress.

The principal strain $\varepsilon_z = \varepsilon_{p3}$ can be determined from the measured in-plane data with the equation

$$\varepsilon_z = -\frac{\nu}{1-\nu}(\varepsilon_x + \varepsilon_y) \tag{13.15}$$

where ν = Poisson's ratio. The derivation of this equation will be presented in the next section, in the discussion of the generalized Hooke's law. The out-of-plane principal strain is important, since the absolute maximum shear strain at a point may be $(\varepsilon_{p1} - \varepsilon_{p2})$, $(\varepsilon_{p1} - \varepsilon_{p3})$, or $(\varepsilon_{p3} - \varepsilon_{p2})$, depending on the relative magnitudes and signs of the principal strains at the point. (See Section 13.4.)

EXAMPLE 13.4

A strain rosette consisting of three strain gages oriented as illustrated was mounted on the free surface of a steel machine component ($\nu = 0.30$). Under load, the following strains were measured:

$$\varepsilon_a = -600\ \mu\varepsilon \qquad \varepsilon_b = -900\ \mu\varepsilon \qquad \varepsilon_c = 700\ \mu\varepsilon$$

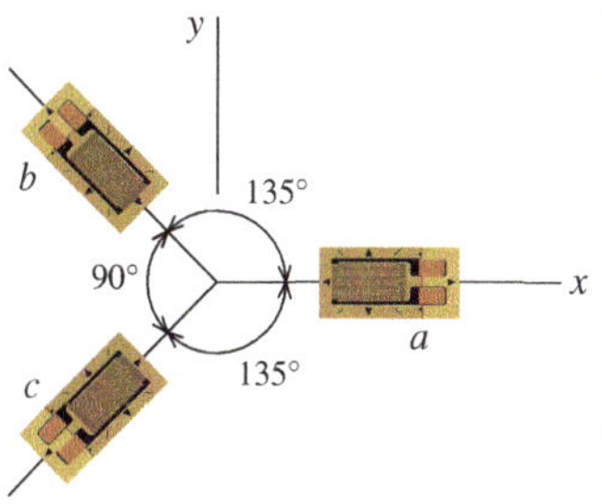

Determine the principal strains and the maximum shear strain at the point shown. Show the principal strain deformations and the maximum in-plane shear strain distortion in a sketch.

Plan the Solution

To compute the principal strains and the maximum in-plane shear strain, the strains ε_x, ε_y, and γ_{xy} must be determined. These normal and shear strains can be obtained from the rosette data by writing a strain transformation equation for each gage and then solving the three equations simultaneously. Since it is aligned with the x axis, gage a directly measures the normal strain ε_x, so the reduction of the strain gage data will actually involve solving only two equations simultaneously, for ε_y and γ_{xy}.

SOLUTION

The angles θ_a, θ_b, and θ_c must be determined for the three gages. Although it is not an absolute requirement, strain rosette problems such as this one are easier to solve if all angles θ are measured counterclockwise from the reference x axis. For the rosette configuration used in this problem, the three angles are $\theta_a = 0°$, $\theta_b = 135°$, and $\theta_c = 225°$. Using these angles, write a strain transformation equation for each gage, where ε_n is the experimentally measured strain value:

Equation for gage *a*:

$$-600 = \varepsilon_x \cos^2(0°) + \varepsilon_y \sin^2(0°) + \gamma_{xy} \sin(0°)\cos(0°) \tag{a}$$

Equation for gage *b*:

$$-900 = \varepsilon_x \cos^2(135°) + \varepsilon_y \sin^2(135°) + \gamma_{xy} \sin(135°)\cos(135°) \tag{b}$$

Equation for gage *c*:

$$700 = \varepsilon_x \cos^2(225°) + \varepsilon_y \sin^2(225°) + \gamma_{xy} \sin(225°)\cos(225°) \tag{c}$$

Since sin(0°) = 0, Equation (a) reduces to $\varepsilon_x = -600$ µε. Substitute this result into Equations (b) and (c), and collect constant terms on the left-hand side of the equations:

$$-600 = 0.5\varepsilon_y - 0.5\gamma_{xy}$$
$$1{,}000 = 0.5\varepsilon_y + 0.5\gamma_{xy}$$

Generally, the gage orientations used in common rosette patterns produce a pair of equations similar in form to these two equations, making them especially easy to solve simultaneously. To obtain ε_y, the two equations are added together to give $\varepsilon_y = 400$ µε. Subtracting the two equations results in $\gamma_{xy} = 1{,}600$ µrad. Therefore, the state of strain that exists at the point on the steel machine component can be summarized as $\varepsilon_x = -600$ µε, $\varepsilon_y = 400$ µε, and $\gamma_{xy} = 1{,}600$ µrad. These strains will be used to determine the principal strains and the maximum in-plane shear strain.

From Equation (13.10), the principal strains can be calculated as

$$\begin{aligned}\varepsilon_{p1,\,p2} &= \frac{\varepsilon_x + \varepsilon_y}{2} \pm \sqrt{\left(\frac{\varepsilon_x - \varepsilon_y}{2}\right)^2 + \left(\frac{\gamma_{xy}}{2}\right)^2}\\ &= \frac{-600 + 400}{2} \pm \sqrt{\left(\frac{-600 - 400}{2}\right)^2 + \left(\frac{1{,}600}{2}\right)^2}\\ &= -100 \pm 943\\ &= 843\ \mu\varepsilon,\ -1{,}043\ \mu\varepsilon\end{aligned}$$ **Ans.**

and from Equation (13.11), the maximum in-plane shear strain is

$$\begin{aligned}\frac{\gamma_{\max}}{2} &= \pm\sqrt{\left(\frac{\varepsilon_x - \varepsilon_y}{2}\right)^2 + \left(\frac{\gamma_{xy}}{2}\right)^2}\\ &= \pm\sqrt{\left(\frac{-600 - 400}{2}\right)^2 + \left(\frac{1{,}600}{2}\right)^2}\\ &= 943.4\ \mu\text{rad}\\ \therefore \gamma_{\max} &= 1{,}887\ \mu\text{rad}\end{aligned}$$ **Ans.**

The in-plane principal directions can be determined from Equation (13.9):

$$\tan 2\theta_p = \frac{\gamma_{xy}}{\varepsilon_x - \varepsilon_y} = \frac{1{,}600}{-600 - 400} = \frac{1{,}600}{-1{,}000}$$ **Note:** $\varepsilon_x - \varepsilon_y < 0$

$$\therefore 2\theta_p = -58.0° \quad \text{and thus} \quad \theta_p = -29.0°$$

Since $\varepsilon_x - \varepsilon_y < 0$, the angle θ_p is the angle between the *x* direction and the ε_{p2} direction.

The strain rosette is bonded to the *surface* of the steel machine component; therefore, the condition in this example is a *plane stress* condition. Accordingly, *the* *out-of-plane normal strain ε_z will not be zero.* The third principal strain ε_{p3} can be computed from Equation (13.15):

$$\varepsilon_{p3} = \varepsilon_z = -\frac{\nu}{1-\nu}(\varepsilon_x + \varepsilon_y) = -\frac{0.3}{1-0.3}(-600 + 400) = 85.7\ \mu\varepsilon$$

Because $\varepsilon_{p2} < \varepsilon_{p3} < \varepsilon_{p1}$ (see Table 13.2), the absolute maximum shear strain will equal the maximum in-plane shear strain:

$$\gamma_{\text{abs max}} = \varepsilon_{p1} - \varepsilon_{p2} = 843\ \mu\varepsilon - (-1{,}043\ \mu\varepsilon) = 1{,}887\ \mu\text{rad}$$

Sketch the Deformations and Distortions

The principal strains are oriented 29.0° clockwise from the x direction. Since $\varepsilon_x - \varepsilon_y < 0$, the principal strain corresponding to this direction is $\varepsilon_{p2} = -1{,}043\ \mu\varepsilon$. The element contracts in this direction. In the perpendicular direction, the principal strain is $\varepsilon_{p1} = 843\ \mu\varepsilon$, which means that the element elongates.

The distortion caused by the maximum in-plane shear strain is shown by the diamond that connects the midpoints of each of the rectangle's edges.

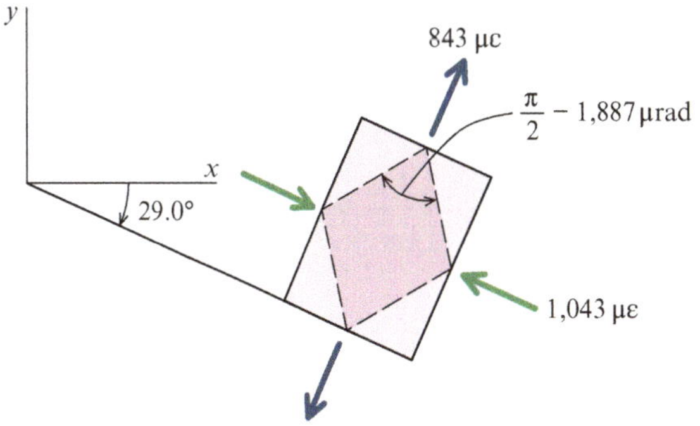

MecMovies

EXAMPLE

M13.5 The strain rosette shown was used to obtain normal strain data at a point on the free surface of a machine part. Determine

(a) the strain components ε_x, ε_y, and γ_{xy} at the point.

(b) the principal strains and the maximum shear strain at the point.

Example 1

$\varepsilon_a = -215\ \mu\varepsilon$

$\varepsilon_b = -130\ \mu\varepsilon$

$\varepsilon_c = 460\ \mu\varepsilon$

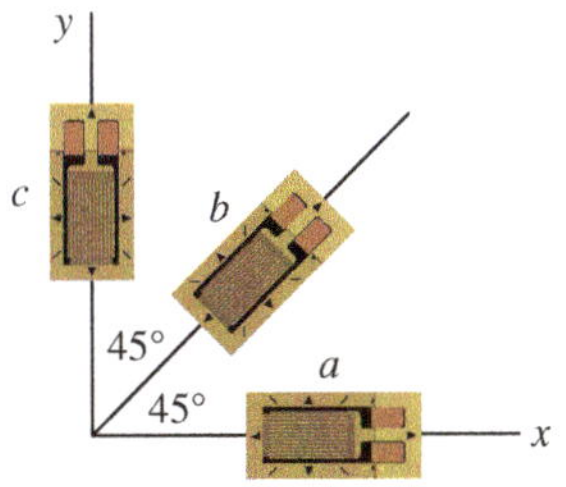

Example 2

$\varepsilon_a = 800\ \mu\varepsilon$

$\varepsilon_b = -200\ \mu\varepsilon$

$\varepsilon_c = 625\ \mu\varepsilon$

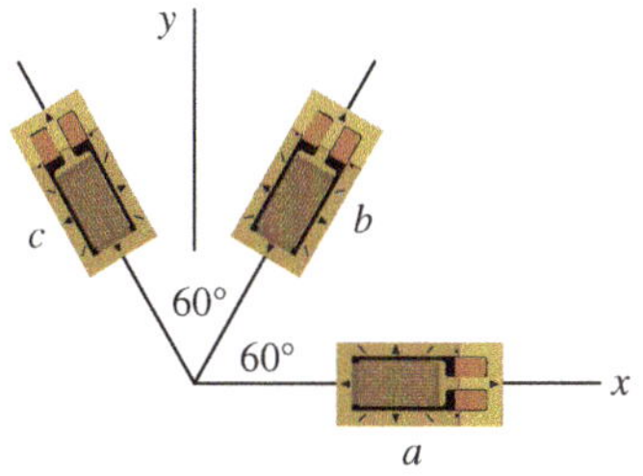

EXERCISE

M13.5 Strain Measurement with Rosettes. A strain rosette was used to obtain normal strain data at a point on the free surface of a machine part. Determine the normal strains, the shear strain, and the principal strains in the x–y plane.

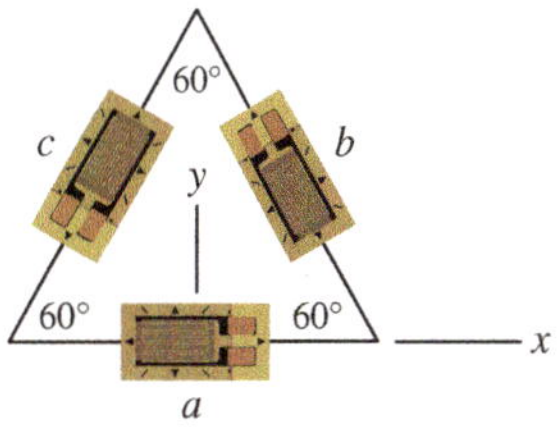

FIGURE M13.5

PROBLEMS

P13.25–P13.30 The strain rosette shown in Figures P13.25–P13.30 was used to obtain normal strain data at a point on the free surface of a machine part.

(a) Determine the strain components ε_x, ε_y, and γ_{xy} at the point.
(b) Determine the principal strains and the maximum in-plane shear strain at the point.
(c) Draw a sketch showing the angle θ_p, the principal strain deformations, and the maximum in-plane shear strain distortions.
(d) Determine the magnitude of the absolute maximum shear strain.

Problem	ε_a	ε_b	ε_c	ν
P13.25	410 με	−540 με	−330 με	0.30
P13.26	215 με	−710 με	−760 με	0.12
P13.27	510 με	415 με	430 με	0.33
P13.28	−960 με	−815 με	−505 με	0.33
P13.29	−360 με	−230 με	815 με	0.15
P13.30	775 με	−515 με	415 με	0.30

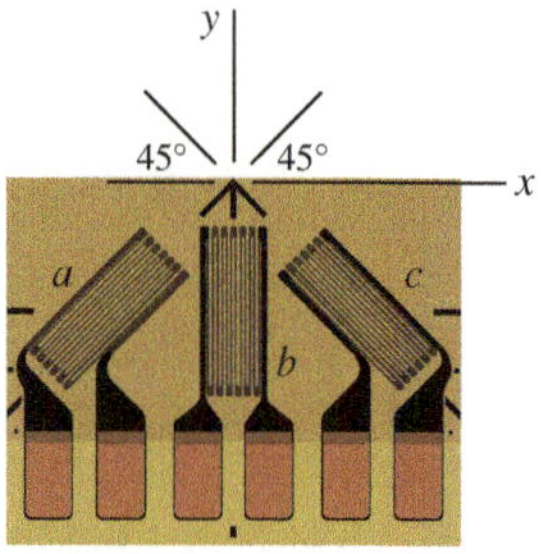

FIGURE P13.27

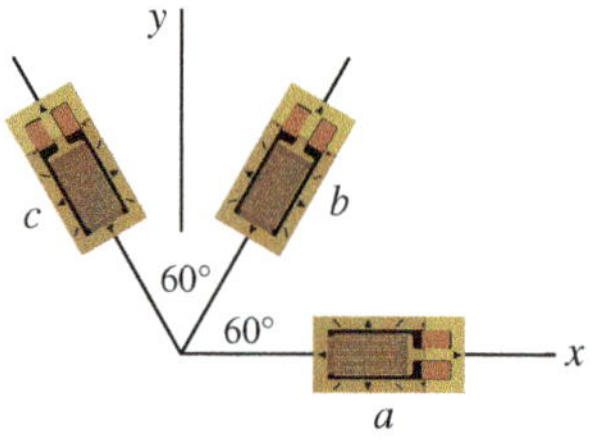

FIGURE P13.28

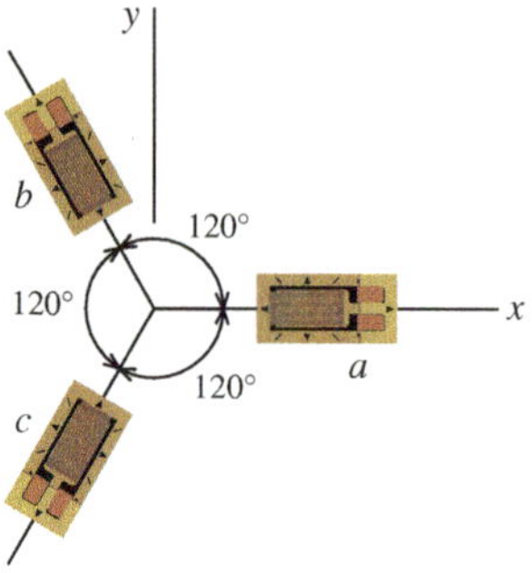

FIGURE P13.29

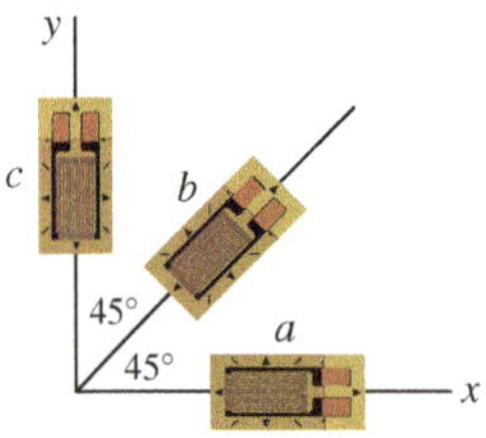

FIGURE P13.25

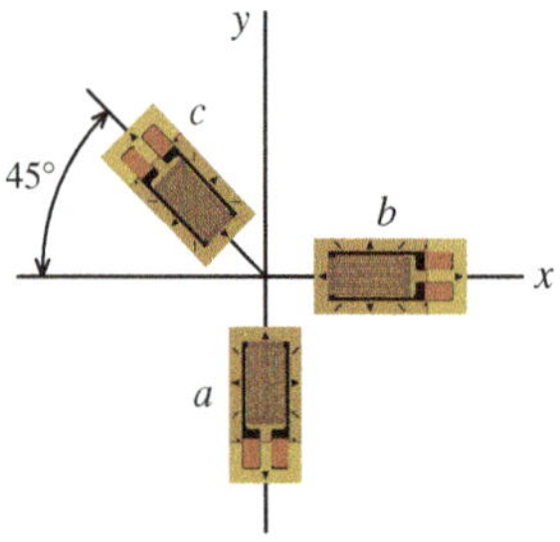

FIGURE P13.26

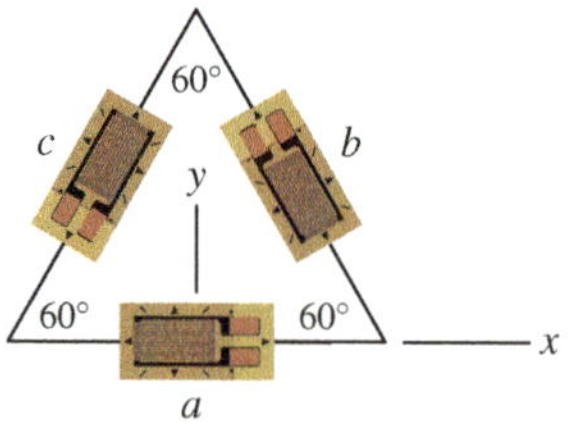

FIGURE P13.30

13.8 Generalized Hooke's Law for Isotropic Materials

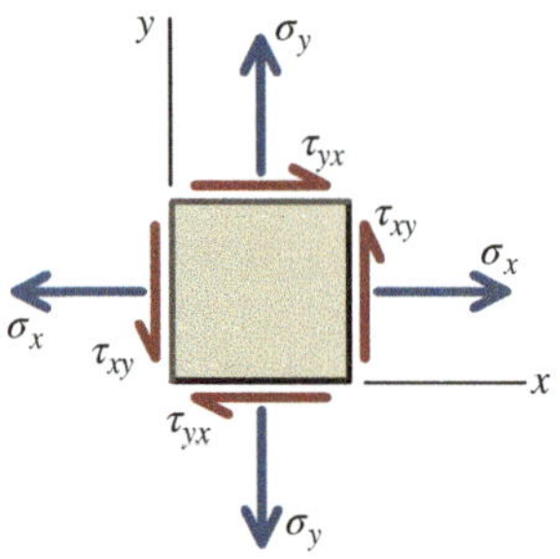

FIGURE 13.10

Hooke's law [see Equation (3.4)] can be extended to include the two-dimensional (Figure 13.10) and three-dimensional (Figure 13.11) states of stress often encountered in engineering practice. We will consider isotropic materials, which are materials with properties (such as the elastic modulus E and Poisson's ratio ν) that are independent of orientation. In other words, E and ν are the same in every direction for isotropic materials.

Figures 13.12*a–c* show a differential element of material subjected to three different normal stresses: σ_x, σ_y, and σ_z. In Figure 13.12*a*, a positive normal stress σ_x produces a positive normal strain (i.e., elongation) in the x direction:

$$\varepsilon_x = \frac{\sigma_x}{E}$$

Although stress is applied only in the x direction, normal strains are produced in the y and z directions because of the Poisson effect:

$$\varepsilon_y = -\nu\frac{\sigma_x}{E} \qquad \varepsilon_z = -\nu\frac{\sigma_x}{E}$$

Note that these strains in the transverse direction are negative (i.e., they denote contraction). If the element elongates in the x direction, then it contracts in the transverse directions, and vice versa.

Similarly, the normal stress σ_y produces strains not only in the y direction, but also in the transverse directions (Figure 13.12b):

$$\varepsilon_y = \frac{\sigma_y}{E} \qquad \varepsilon_x = -\nu\frac{\sigma_y}{E} \qquad \varepsilon_z = -\nu\frac{\sigma_y}{E}$$

Likewise, the normal stress σ_z produces the strains (Figure 13.12c)

$$\varepsilon_z = \frac{\sigma_z}{E} \qquad \varepsilon_x = -\nu\frac{\sigma_z}{E} \qquad \varepsilon_y = -\nu\frac{\sigma_z}{E}$$

If all three normal stresses σ_x, σ_y, and σ_z act on the element at the same time, the total deformation of the element can be determined by summing the deformations resulting from each normal stress. This procedure is based on the **principle of superposition**, which states that the effects of separate loadings can be added algebraically if two conditions are satisfied:

1. Each effect is linearly related to the load that produced it.
2. The effect of the first load does not significantly change the effect of the second load.

The first condition is satisfied if the stresses do not exceed the proportional limit for the material. The second condition is satisfied if the deformations are small, in which case the small changes in the areas of the faces of the element do not produce significant changes in the stresses.

Using the superposition principle, we can state the relationship between normal strains and normal stresses as follows:

$$\begin{aligned}\varepsilon_x &= \frac{1}{E}[\sigma_x - \nu(\sigma_y + \sigma_z)]\\ \varepsilon_y &= \frac{1}{E}[\sigma_y - \nu(\sigma_x + \sigma_z)]\\ \varepsilon_z &= \frac{1}{E}[\sigma_z - \nu(\sigma_x + \sigma_y)]\end{aligned} \tag{13.16}$$

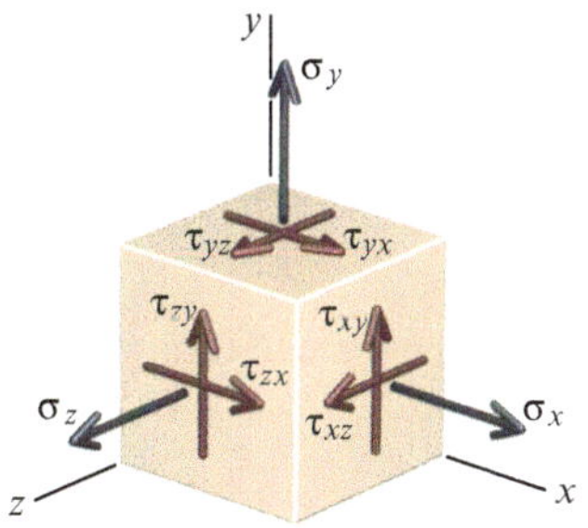

FIGURE 13.11

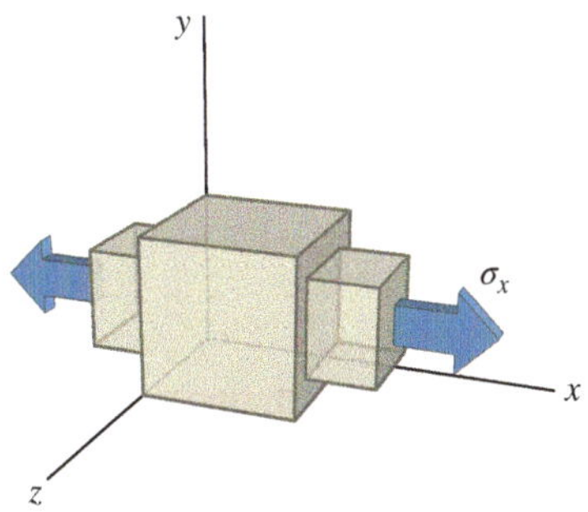

FIGURE 13.12a

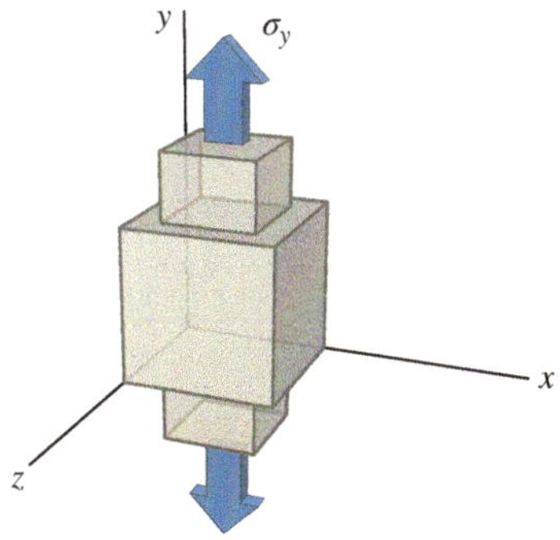

FIGURE 13.12b

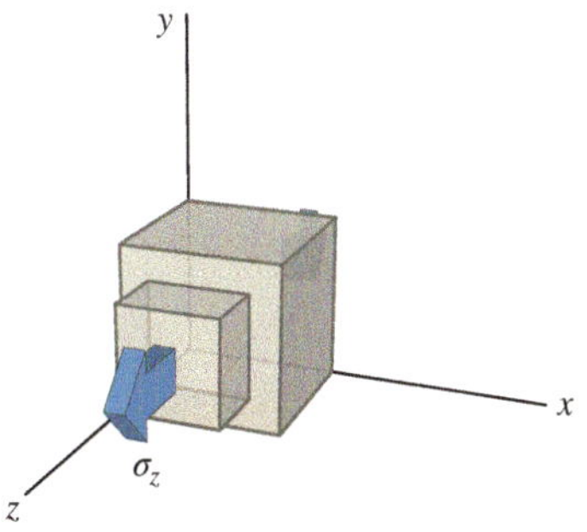

FIGURE 13.12c

561

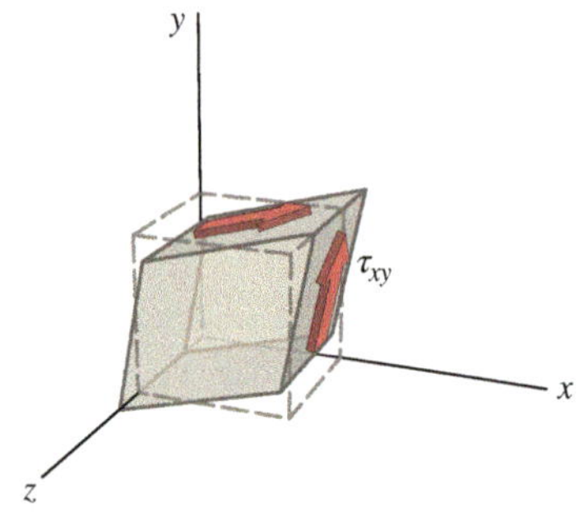

FIGURE 13.13*a*

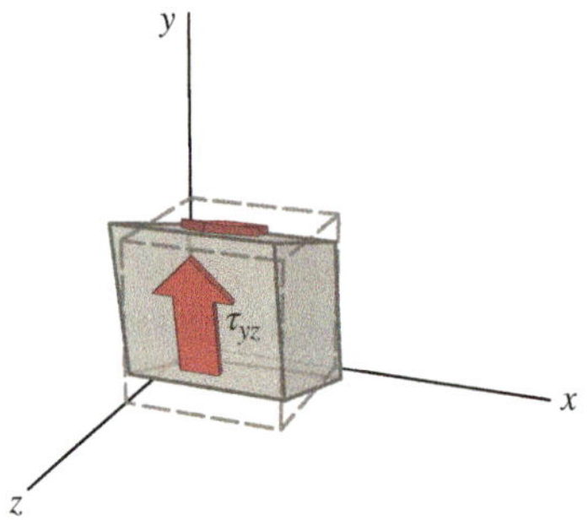

FIGURE 13.13*b*

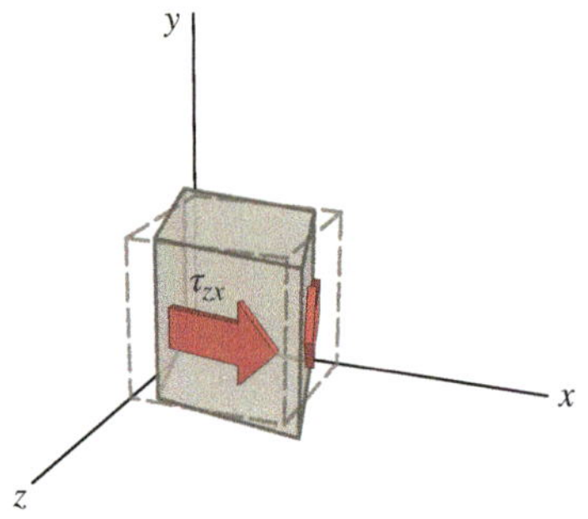

FIGURE 13.13*c*

The deformations produced in an element by the shear stresses τ_{xy}, τ_{yz}, and τ_{xz} are respectively shown in Figures 13.13*a–c*. There is no Poisson effect associated with shear strain; therefore, the relationship between shear strain and shear stress can be stated as

$$\gamma_{xy} = \frac{1}{G}\tau_{xy} \qquad \gamma_{yz} = \frac{1}{G}\tau_{yz} \qquad \gamma_{zx} = \frac{1}{G}\tau_{zx} \tag{13.17}$$

where G is the shear modulus, which is related to the elastic modulus E and Poisson's ratio ν by

$$G = \frac{E}{2(1+\nu)} \tag{13.18}$$

Equations (13.16) and (13.17) are known as the **generalized Hooke's law** for isotropic materials. Notice that the shear stresses do not affect the expressions for normal strain and that the normal stresses do not affect the expressions for shear strain; therefore, the normal and shear relationships are independent of each other. Furthermore, the shear strain expressions in Equation (13.17) are independent of each other, unlike the normal strain expressions in Equation (13.16), where all three normal stresses appear. For example, the shear strain γ_{xy} is affected solely by the shear stress τ_{xy}.

In addition, Equations (13.16) and (13.17) can be solved for the stresses in terms of the strains as follows:

$$\begin{aligned} \sigma_x &= \frac{E}{(1+\nu)(1-2\nu)}[(1-\nu)\varepsilon_x + \nu(\varepsilon_y + \varepsilon_z)] \\ \sigma_y &= \frac{E}{(1+\nu)(1-2\nu)}[(1-\nu)\varepsilon_y + \nu(\varepsilon_x + \varepsilon_z)] \\ \sigma_z &= \frac{E}{(1+\nu)(1-2\nu)}[(1-\nu)\varepsilon_z + \nu(\varepsilon_x + \varepsilon_y)] \end{aligned} \tag{13.19}$$

Similarly, Equation (13.17) can be solved for the stresses in terms of the strains as

$$\tau_{xy} = G\gamma_{xy} \qquad \tau_{yz} = G\gamma_{yz} \qquad \tau_{zx} = G\gamma_{zx} \tag{13.20}$$

Unit Volume Change

Consider an infinitesimal element with dimensions dx, dy, and dz (Figure 13.14). The initial volume of the cube is $V_i = dx\,dy\,dz$. After straining, the lengths of the three sides become $(1+\varepsilon_x)\,dx$, $(1+\varepsilon_y)\,dy$, and $(1+\varepsilon_z)\,dz$, respectively. The final volume of the cube after straining is thus

$$V_f = (1+\varepsilon_x)dx(1+\varepsilon_y)dy(1+\varepsilon_z)dz$$

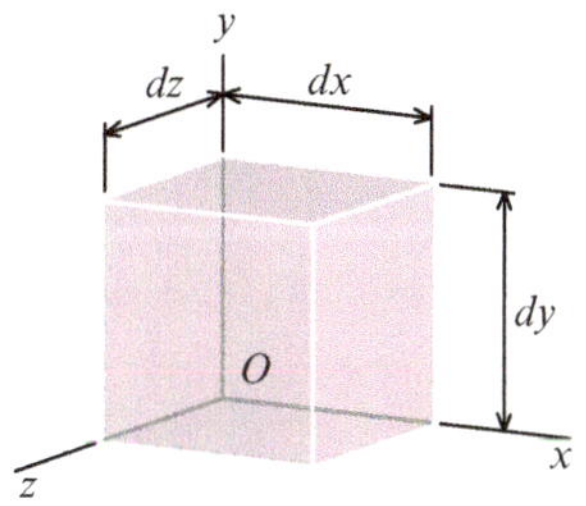

FIGURE 13.14 Infinitesimal element.

The change in volume of the cube is

$$\begin{aligned}\Delta V &= V_f - V_i \\ &= (1+\varepsilon_x)dx(1+\varepsilon_y)dy(1+\varepsilon_z)dz - dx\,dy\,dz \\ &= [1+\varepsilon_x+\varepsilon_y+\varepsilon_z+\varepsilon_x\varepsilon_y+\varepsilon_y\varepsilon_z+\varepsilon_z\varepsilon_x+\varepsilon_x\varepsilon_y\varepsilon_z - 1]\;dx\,dy\,dz \\ &= [\varepsilon_x+\varepsilon_y+\varepsilon_z+\varepsilon_x\varepsilon_y+\varepsilon_y\varepsilon_z+\varepsilon_z\varepsilon_x+\varepsilon_x\varepsilon_y\varepsilon_z]\;dx\,dy\,dz\end{aligned}$$

If the strains are small, we can neglect the higher order terms (i.e., $\varepsilon_x\varepsilon_y$, $\varepsilon_y\varepsilon_z$, $\varepsilon_z\varepsilon_x$, and $\varepsilon_x\varepsilon_y\varepsilon_z$) to find

$$\Delta V = \lfloor\varepsilon_x+\varepsilon_y+\varepsilon_z\rfloor\;dx\,dy\,dz$$

The change in volume per unit volume is called the *volumetric strain* or the *dilatation e*. It is defined as

$$e = \frac{\Delta V}{V} = \varepsilon_x+\varepsilon_y+\varepsilon_z \tag{13.21}$$

Note that the change in volume is not dependent on shear strains, which change the orthogonal shape of the infinitesimal element but not the volume.

Next, consider again the generalized Hooke's law equation (13.16):

$$\begin{aligned}\varepsilon_x &= \frac{1}{E}[\sigma_x - \nu(\sigma_y+\sigma_z)] \\ \varepsilon_y &= \frac{1}{E}[\sigma_y - \nu(\sigma_x+\sigma_z)] \\ \varepsilon_z &= \frac{1}{E}[\sigma_z - \nu(\sigma_x+\sigma_y)]\end{aligned}$$

If we add these equations together, we find that the dilatation can be expressed in terms of stresses and material constants as

$$e = \frac{1-2\nu}{E}(\sigma_x+\sigma_y+\sigma_z) \tag{13.22}$$

Since the dilatation e represents a change in volume, it is independent of the orientation of the element considered. It follows, then, that the quantities $\varepsilon_x+\varepsilon_y+\varepsilon_z$ and $\sigma_x+\sigma_y+\sigma_z$ are also independent of the orientation of the element.

An interesting fact about Poisson's ratio can be observed from the volumetric strain. Suppose we consider a material for which the volume doesn't change, regardless of the intensity of the stresses acting on it. In other words, the volumetric strain is zero (i.e., $e = 0$). In order for Equation (13.22) to be equal to zero for any possible combination of stresses, it must be true that

$$1-2\nu = 0 \qquad \therefore \nu = \frac{1}{2}$$

Hence, the largest possible value for Poisson's ratio is $\nu = 0.5$. A Poisson's ratio of $\nu = 0.5$ is found in materials such as rubber and soft biological tissues. For common engineering materials, Poisson's ratio is generally less than 0.5, since most materials demonstrate some change in volume when subjected to stress. A negative value of Poisson's ratio is theoretically possible, but materials with such a value are atypical. Several foams and certain crystals have negative Poisson's ratios.

Since $\nu < 0.5$ for most engineering materials, stretching a material in one direction will increase its volume. For example, consider an axially loaded rod in tension where the axial normal stress is $\sigma_x > 0$ and the transverse normal stresses are $\sigma_y = \sigma_z = 0$. From Equation (13.22), as long as $\nu < 0.5$, e must be greater than zero, and thus, the volume of the stretched rod will be greater than the volume of the unstretched rod.

Special Case of Hydrostatic Pressure. When a volume element of a material is subjected to the uniform pressure of a fluid, the pressure on the body is the same in all directions. Further, the fluid pressure always act perpendicular to any surface on which it acts. Shear stresses are not present, since the shear resistance of a fluid is zero.

When an elastic body is subjected to principal stresses such that $\sigma_x = \sigma_y = \sigma_z = -p$, this state of stress is referred to as the *hydrostatic stress state*. In that case, Equation (13.22) for the volumetric strain reduces to

$$e = -\frac{3(1-2\nu)}{E}p$$

The term *modulus* defines a ratio between a stress and a strain. Here, our stress is the hydrostatic compressive stress $-p$ and the strain is the volumetric strain. We can now define a new modulus, the *bulk modulus*, as

Notice that, for many metals, $\nu \approx \frac{1}{3}$. Thus, the bulk modulus K for many metals is approximately equal to the elastic modulus E.

$$K = \frac{-p}{e} = \frac{E}{3(1-2\nu)} \tag{13.23}$$

Note that a material under hydrostatic pressure can only decrease in volume; accordingly, the dilatation e must be negative and the bulk modulus is a positive constant.

EXAMPLE 13.5

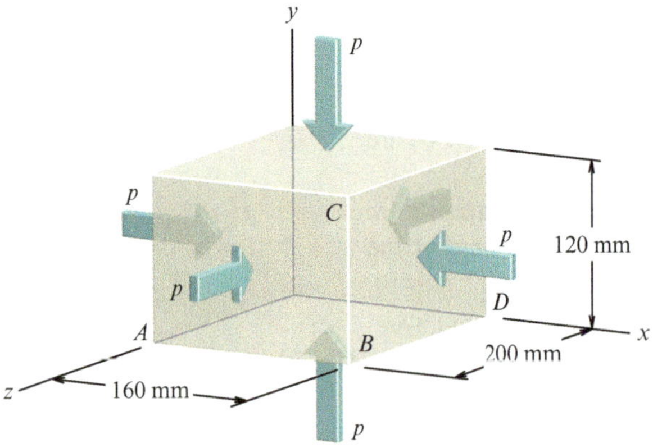

An aluminum alloy [$E = 73$ GPa; $\nu = 0.33$] block is subjected to a uniform pressure of $p = 35$ MPa as shown. Determine

(a) The change in length of sides AB, BC, and BD.
(b) The change in volume of the block.

Plan the Solution

This is a case of hydrostatic stress, so the stresses σ_x, σ_y, and σ_z are each equal to $-p$. Equations (13.16) will be used to calculate the strains ε_x, ε_y, and ε_z for the three directions, and from these strains and the initial dimensions of the block, the changes in length will be found. The change in volume will be found from the initial volume and the dilatation e, as defined in Equation (13.21).

SOLUTION

(a) Determine the change in length of sides *AB*, *BC*, and *BD*.

Normal stresses: The three normal stresses on the block (which are actually principal stresses) are equal:

$$\sigma_x = \sigma_y = \sigma_z = -p = -35 \text{ MPa}$$

Normal strains: From Equations (13.16), we find that ε_x for a hydrostatic stress state can be expressed as

$$\varepsilon_x = \frac{1}{E}[\sigma_x - \nu(\sigma_y + \sigma_z)] = \frac{1}{E}[-p - \nu(-p - p)]$$
$$= -\frac{p}{E}(1 - 2\nu)$$

Furthermore, we obtain this same expression for ε_y and ε_z. Consequently, we find that the normal strains are equal in each of the three orthogonal directions for a hydrostatic stress state. Thus, for a pressure $p = 35$ MPa, the strains in the aluminum alloy block are

$$\varepsilon_x = \varepsilon_y = \varepsilon_z = -\frac{35 \text{ MPa}}{73{,}000 \text{ MPa}}[1 - 2(0.33)] = -163.0 \times 10^{-6} \text{ mm/mm}$$

Deformations: In the *x*, *y*, and *z* directions, the deformations are, respectively, as follows:

$$\delta_{AB} = (160 \text{ mm})(-163.0 \times 10^{-6} \text{ mm/mm}) = -0.0261 \text{ mm}$$
$$\delta_{BC} = (120 \text{ mm})(-163.0 \times 10^{-6} \text{ mm/mm}) = -0.01956 \text{ mm} \qquad \textbf{Ans.}$$
$$\delta_{BD} = (200 \text{ mm})(-163.0 \times 10^{-6} \text{ mm/mm}) = -0.0326 \text{ mm}$$

(b) Determine the Change in Volume of the Block: From the normal strains just calculated, we can calculate the dilatation *e* (i.e., the volumetric strain) from Equation (13.21):

$$e = \varepsilon_x + \varepsilon_y + \varepsilon_z = 3(-163.0 \times 10^{-6}) = -489.0 \times 10^{-6}$$

The initial volume of the block was

$$V = (160 \text{ mm})(120 \text{ mm})(200 \text{ mm}) = 3.84 \times 10^6 \text{ mm}^3$$

The change in volume is calculated from the product of the dilatation and the initial volume:

$$\Delta V = eV = (-489.0 \times 10^{-6})(3.84 \times 10^6 \text{ mm}^3) = -1{,}878 \text{ mm}^3 \qquad \textbf{Ans.}$$

Notice that the volume of the block has decreased under the action of hydrostatic pressure.

EXAMPLE 13.6

A marble [$E = 55$ GPa; $\nu = 0.22$] cube has dimensions $a = 75$ mm. Compressive strains $\varepsilon_x = -650 \times 10^{-6}$ and $\varepsilon_y = \varepsilon_z = -370 \times 10^{-6}$ have been measured for the cube. Determine

(a) The normal stresses σ_x, σ_y, and σ_z acting on the *x*, *y*, and *z* faces of the cube.
(b) The maximum shear stress in the material.

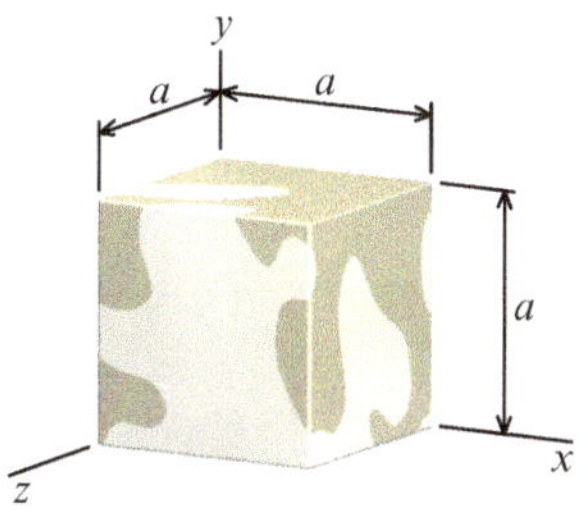

Plan the Solution

Equations (13.19) will be used to calculate the stresses σ_x, σ_y, and σ_z from the given strains ε_x, ε_y, and ε_z. Since there are no shear strains, we know that there are no shear stresses, and consequently, the stresses σ_x, σ_y, and σ_z must be principal stresses. From the three principal stresses, the maximum shear stress in the cube can be calculated.

SOLUTION

(a) Normal stresses: From Equations (13.19), calculate the normal stresses acting on the cube:

$$\begin{aligned}\sigma_x &= \frac{E}{(1+\nu)(1-2\nu)}[(1-\nu)\varepsilon_x + \nu(\varepsilon_y + \varepsilon_z)]\\ &= \frac{55{,}000\text{ MPa}}{(1+0.22)[1-2(0.22)]}[(1-0.22)(-650) + (0.22)(-370-370)](10^{-6})\\ &= -53.9\text{ MPa}\end{aligned}$$

Ans.

$$\begin{aligned}\sigma_y &= \frac{E}{(1+\nu)(1-2\nu)}[(1-\nu)\varepsilon_y + \nu(\varepsilon_x + \varepsilon_z)]\\ &= \frac{55{,}000\text{ MPa}}{(1+0.22)[1-2(0.22)]}[(1-0.22)(-370) + (0.22)(-650-370)](10^{-6})\\ &= -41.3\text{ MPa}\end{aligned}$$

Ans.

$$\begin{aligned}\sigma_z &= \frac{E}{(1+\nu)(1-2\nu)}[(1-\nu)\varepsilon_z + \nu(\varepsilon_x + \varepsilon_y)]\\ &= \frac{55{,}000\text{ MPa}}{(1+0.22)[1-2(0.22)]}[(1-0.22)(-370) + (0.22)(-650-370)](10^{-6})\\ &= -41.3\text{ MPa}\end{aligned}$$

Ans.

(b) Maximum shear stress in the material: There are no shear stresses acting on the x, y, or z faces of the marble cube; consequently, σ_x, σ_y, and σ_z must be principal stresses:

$$\begin{aligned}\sigma_{p1} &= \sigma_x = -53.9\text{ MPa}\\ \sigma_{p2} &= \sigma_y = -41.3\text{ MPa}\\ \sigma_{p3} &= \sigma_z = -41.3\text{ MPa}\end{aligned}$$

From Equation (12.18), we know that the absolute maximum shear stress is equal to one-half of the difference between the maximum and minimum principal stresses:

$$\tau_{\text{abs max}} = \frac{\sigma_{\text{max}} - \sigma_{\text{min}}}{2}$$

Thus, the absolute maximum shear stress in the marble cube is

$$\tau_{\text{abs max}} = \left|\frac{-53.9\text{ MPa} - (-41.3\text{ MPa})}{2}\right| = 6.30\text{ MPa}$$

Ans.

Special Case of Plane Stress

When stresses act only in the x–y plane (Figure 13.10), $\sigma_z = 0$ and $\tau_{yz} = \tau_{zx} = 0$. Consequently, Equation (13.16) reduces to the following:

$$\begin{aligned}\varepsilon_x &= \frac{1}{E}(\sigma_x - \nu\sigma_y)\\ \varepsilon_y &= \frac{1}{E}(\sigma_y - \nu\sigma_x)\\ \varepsilon_z &= -\frac{\nu}{E}(\sigma_x + \sigma_y)\end{aligned} \qquad (13.24)$$

MecMovies 13.7 presents an animated derivation of the generalized Hooke's Law equations for biaxial stress.

Also, Equation (13.17) becomes, simply,

$$\gamma_{xy} = \frac{1}{G}\tau_{xy} \tag{13.25}$$

Then, solving Equation (13.24) for the stresses in terms of the strains gives the following result:

$$\begin{aligned} \sigma_x &= \frac{E}{1-\nu^2}(\varepsilon_x + \nu\varepsilon_y) \\ \sigma_y &= \frac{E}{1-\nu^2}(\varepsilon_y + \nu\varepsilon_x) \end{aligned} \tag{13.26}$$

Equation (13.26) can be used to calculate normal stresses from measured or computed normal strains.

Note that the out-of-plane normal strain ε_z is generally not equal to zero for the plane stress condition. An expression for ε_z in terms of ε_x and ε_y was stated in Equation (13.15). This equation can be derived by substituting Equation (13.26) into the expression

$$\varepsilon_z = -\frac{\nu}{E}(\sigma_x + \sigma_y)$$

to give

$$\begin{aligned} \varepsilon_z &= -\frac{\nu}{E}(\sigma_x + \sigma_y) = -\frac{\nu}{E}\frac{E}{1-\nu^2}[(\varepsilon_x + \nu\varepsilon_y) + (\varepsilon_y + \nu\varepsilon_x)] \\ &= -\frac{\nu}{(1-\nu)(1+\nu)}[(1+\nu)\varepsilon_x + (1+\nu)\varepsilon_y] \\ &= -\frac{\nu}{1-\nu}(\varepsilon_x + \varepsilon_y) \end{aligned} \tag{13.27}$$

EXAMPLE 13.7

On the free surface of an aluminum [E = 10,000 ksi; ν = 0.33] component, three strain gages arranged as shown record the following strains: $\varepsilon_a = -420$ με, $\varepsilon_b = 380$ με, and $\varepsilon_c = 240$ με.

Determine the normal stress that acts along the axis of gage b (i.e., at an angle of $\theta = 45°$ with respect to the positive x axis).

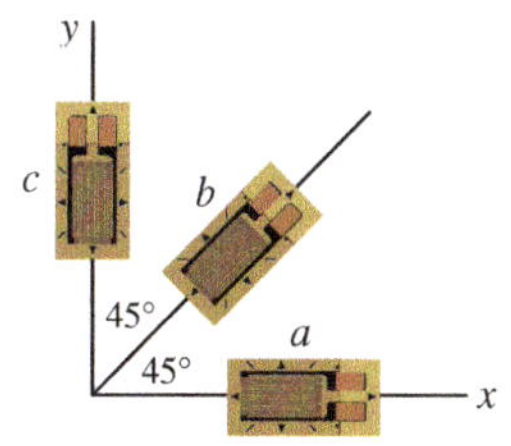

Plan the Solution

At first glance, one might be tempted to use the measured strain in gage b and the elastic modulus E to compute the normal strain acting in the specified direction. However, that approach is not correct because a state of uniaxial stress does not exist. In other words, the normal stress acting in the 45° direction is not the only stress acting in the material. To solve this problem, first reduce the strain rosette data to obtain ε_x, ε_y, and γ_{xy}. The stresses σ_x, σ_y, and τ_{xy} can then be calculated from Equations (13.26) and (13.25). Finally, the normal stress in the specified direction can be calculated from the stress transformation equation.

SOLUTION

From the geometry of the rosette, gage a measures the strain in the x direction and gage c measures the strain in the y direction. Therefore, $\varepsilon_x = -420$ με and $\varepsilon_y = 240$ με. To compute the shear strain γ_{xy}, write a strain transformation equation for gage b:

$$380 = \varepsilon_x \cos^2(45°) + \varepsilon_y \sin^2(45°) + \gamma_{xy} \sin(45°)\cos(45°)$$

Then solve for γ_{xy}:

$$380 = (-420)\cos^2(45°) + (240)\sin^2(45°) + \gamma_{xy}\sin(45°)\cos(45°)$$

$$\therefore \gamma_{xy} = \frac{380 + (420)(0.5) - (240)(0.5)}{0.5} = 940\ \mu\text{rad}$$

Since the strain rosette is bonded to the surface of the aluminum component, we have a plane stress condition. Use the generalized Hooke's law equation (13.26) and the material properties $E = 10{,}000$ ksi and $\nu = 0.33$ to compute the normal stresses σ_x and σ_y from the normal strains ε_x and ε_y:

$$\sigma_x = \frac{E}{1-\nu^2}(\varepsilon_x + \nu\varepsilon_y) = \frac{10{,}000\text{ ksi}}{1-(0.33)^2}[(-420\times 10^{-6}) + 0.33(240\times 10^{-6})] = -3.82\text{ ksi}$$

$$\sigma_y = \frac{E}{1-\nu^2}(\varepsilon_y + \nu\varepsilon_x) = \frac{10{,}000\text{ ksi}}{1-(0.33)^2}[(240\times 10^{-6}) + 0.33(-420\times 10^{-6})] = 1.138\text{ ksi}$$

Note: The strain measurements reported in microstrain units ($\mu\varepsilon$) must be converted to dimensionless quantities (i.e., in./in.) in making this calculation.

Before the shear stress τ_{xy} can be computed, the shear modulus G for the aluminum material must be calculated from Equation (13.18):

$$G = \frac{E}{2(1+\nu)} = \frac{10{,}000\text{ ksi}}{2(1+0.33)} = 3{,}760\text{ ksi}$$

Now we solve Equation (13.25) for the shear stress:

$$\tau_{xy} = G\gamma_{xy} = (3{,}760\text{ ksi})(940\times 10^{-6}) = 3.53\text{ ksi}$$

Finally, the normal stress acting in the direction of $\theta = 45°$ can be calculated with a stress transformation equation, such as Equation (12.5):

$$\begin{aligned}\sigma_n &= \sigma_x\cos^2\theta + \sigma_y\sin^2\theta + 2\tau_{xy}\sin\theta\cos\theta\\ &= (-3.82\text{ ksi})\cos^2 45° + (1.138\text{ ksi})\sin^2 45° + 2(3.53\text{ ksi})\sin 45°\cos 45°\\ &= 2.19\text{ ksi (T)}\end{aligned}$$

Ans.

EXAMPLE 13.8

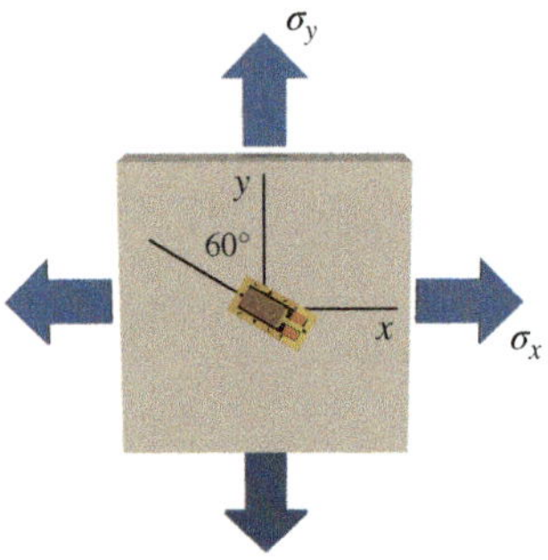

A thin steel [$E = 210$ GPa; $G = 80$ GPa] plate is subjected to biaxial stress. The normal stress in the x direction is known to be $\sigma_x = 70$ MPa. The strain gage measures a normal strain of 230 $\mu\varepsilon$ in the indicated direction on the free surface of the plate.

(a) Determine the magnitude of σ_y that acts on the plate.
(b) Determine the principal strains and the maximum in-plane shear strain in the plate. Show the principal strain deformations and the maximum in-plane shear strain distortion in a sketch.
(c) Determine the magnitude of the absolute maximum shear strain in the plate.

Plan the Solution

To begin this solution, we will write a strain transformation equation for the strain gage oriented as shown. The equation will express the strain ε_n measured by the gage in terms of the strains in the x and y directions. Since there is no shear stress τ_{xy} acting on the plate, the shear strain γ_{xy} will be zero and the strain transformation equation will be reduced to terms involving only ε_x and ε_y. Equations (13.24) from the generalized Hooke's law for

ε_x and ε_y in terms of σ_x and σ_y can be substituted into the strain transformation equation, producing an equation in which the only unknown will be σ_y. After solving for σ_y, we can again use Equation (13.24) to compute ε_x, ε_y, and ε_z. These values will then be used to determine the principal strains, the maximum in-plane shear strain, and the absolute maximum shear strain in the plate.

SOLUTION

(a) Normal Stress σ_y

The strain gage is oriented at an angle $\theta = 150°$. Using this angle, write a strain transformation equation for the gage, where the strain ε_n is the value measured by the gage:

$$230\ \mu\varepsilon = \varepsilon_x \cos^2(150°) + \varepsilon_y \sin^2(150°) + \gamma_{xy} \sin(150°)\cos(150°)$$

Note that the shear strain γ_{xy} is related to the shear stress τ_{xy} by Equation (13.25):

$$\gamma_{xy} = \frac{1}{G}\tau_{xy}$$

Since $\tau_{xy} = 0$, the shear strain γ_{xy} must also equal zero; thus, the strain transformation equation reduces to

$$230\ \mu\varepsilon = 230 \times 10^{-6}\ \text{mm/mm} = \varepsilon_x \cos^2(150°) + \varepsilon_y \sin^2(150°)$$

Equations (13.24) from the generalized Hooke's law define the following relationships between stresses and strains for a plane stress condition (which is observed to apply in this situation):

$$\varepsilon_x = \frac{1}{E}(\sigma_x - \nu\sigma_y) \qquad \text{and} \qquad \varepsilon_y = \frac{1}{E}(\sigma_y - \nu\sigma_x)$$

Substitute these expressions into the strain transformation equation, expand terms, and simplify:

$$\begin{aligned}
230 \times 10^{-6}\ \text{mm/mm} &= \varepsilon_x \cos^2(150°) + \varepsilon_y \sin^2(150°) \\
&= \frac{1}{E}(\sigma_x - \nu\sigma_y)\cos^2(150°) + \frac{1}{E}(\sigma_y - \nu\sigma_x)\sin^2(150°) \\
&= \frac{1}{E}[\sigma_x \cos^2(150°) - \nu\sigma_x \sin^2(150°)] + \frac{1}{E}[\sigma_y \sin^2(150°) - \nu\sigma_y \cos^2(150°)] \\
&= \frac{\sigma_x}{E}[\cos^2(150°) - \nu \sin^2(150°)] + \frac{\sigma_y}{E}[\sin^2(150°) - \nu \cos^2(150°)]
\end{aligned}$$

Solve for the unknown stress σ_y:

$$(230 \times 10^{-6}\ \text{mm/mm})E - \sigma_x[\cos^2(150°) - \nu \sin^2(150°)] = \sigma_y[\sin^2(150°) - \nu \cos^2(150°)]$$

$$\therefore \sigma_y = \frac{(230 \times 10^{-6}\ \text{mm/mm})E - \sigma_x[\cos^2(150°) - \nu \sin^2(150°)]}{\sin^2(150°) - \nu \cos^2(150°)}$$

Before computing the normal stress σ_y, the value of Poisson's ratio must be calculated from the elastic modulus E and the shear modulus G:

$$G = \frac{E}{2(1+\nu)} \qquad \therefore \nu = \frac{E}{2G} - 1 = \frac{210\ \text{GPa}}{2(80\ \text{GPa})} - 1 = 0.3125$$

The normal stress can now be computed:

$$\sigma_y = \frac{(230 \times 10^{-6}\ \text{mm/mm})(210{,}000\ \text{MPa}) - (70\ \text{MPa})[\cos^2(150°) - (0.3125)\sin^2(150°)]}{\sin^2(150°) - (0.3125)\cos^2(150°)} = 81.2\ \text{MPa}$$

Ans.

(b) Principal and Maximum In-Plane Shear Strains

The normal strains in the x, y, and z directions can be computed from Equations (13.24):

$$\varepsilon_x = \frac{1}{E}(\sigma_x - \nu\sigma_y) = \frac{1}{210{,}000 \text{ MPa}}[70 \text{ MPa} - (0.3125)(81.2 \text{ MPa})] = 212.5 \times 10^{-6} \text{ mm/mm}$$

$$\varepsilon_y = \frac{1}{E}(\sigma_y - \nu\sigma_x) = \frac{1}{210{,}000 \text{ MPa}}[81.2 \text{ MPa} - (0.3125)(70 \text{ MPa})] = 282.5 \times 10^{-6} \text{ mm/mm}$$

$$\varepsilon_z = -\frac{\nu}{E}(\sigma_x + \sigma_y) = -\frac{0.3125}{210{,}000 \text{ MPa}}[81.2 \text{ MPa} + 70 \text{ MPa}] = -225 \times 10^{-6} \text{ mm/mm}$$

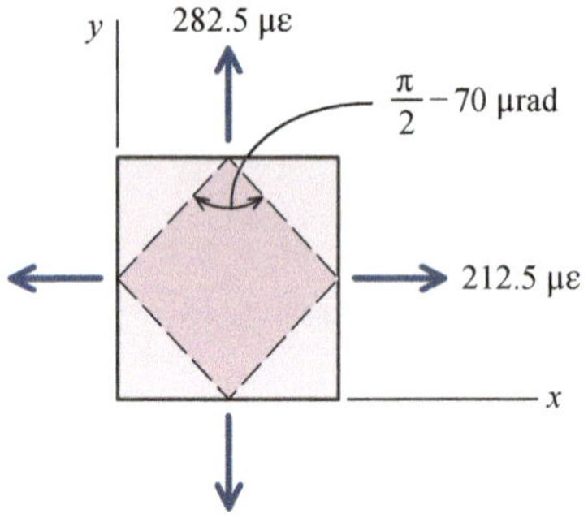

Since $\gamma_{xy} = 0$, the strains ε_x and ε_y are also the principal strains. *Why?* We know that there is never a shear strain associated with the principal strain directions. Conversely, we can also conclude that directions in which the shear strain is zero must also be principal strain directions. Therefore,

$$\varepsilon_{p1} = 282.5 \text{ }\mu\varepsilon \qquad \varepsilon_{p2} = 212.5 \text{ }\mu\varepsilon \qquad \varepsilon_{p3} = -225 \text{ }\mu\varepsilon \qquad \textbf{Ans.}$$

From Equation (13.12), the maximum in-plane shear strain can be determined from ε_{p1} and ε_{p2}:

$$\gamma_{max} = \varepsilon_{p1} - \varepsilon_{p2} = 282.5 - 212.5 = 70 \text{ }\mu\text{rad} \qquad \textbf{Ans.}$$

The in-plane principal strain deformations and the maximum in-plane shear strain distortion are shown in the sketch.

(c) Absolute Maximum Shear Strain

To determine the absolute maximum shear strain, three possibilities must be considered (see Table 13.2):

$$\gamma_{abs\ max} = \varepsilon_{p1} - \varepsilon_{p2} \qquad \text{(i)}$$

$$\gamma_{abs\ max} = \varepsilon_{p1} - \varepsilon_{p3} \qquad \text{(ii)}$$

$$\gamma_{abs\ max} = \varepsilon_{p2} - \varepsilon_{p3} \qquad \text{(iii)}$$

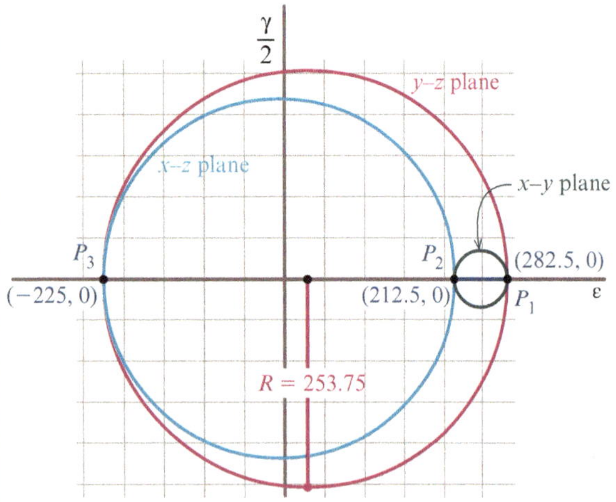

These possibilities can be readily visualized with Mohr's circle for strains. The combinations of ε and γ that are possible in the x–y plane are shown by the small circle between point P_1 (which corresponds to the y direction) and point P_2 (which represents the x direction). The radius of this circle is relatively small; therefore, the maximum shear strain in the x–y plane is small ($\gamma_{max} = 70$ μrad). The steel plate in this problem is subjected to *plane stress*, and consequently, the normal stress $\sigma_{p3} = \sigma_z = 0$. However, the normal strain in the z direction will not be zero. For this problem, $\varepsilon_{p3} = \varepsilon_z = -225$ με. When this principal strain is plotted on Mohr's circle (i.e., point P_3), it becomes evident that the out-of-plane shear strains will be much larger than the shear strain in the x–y plane.

The largest shear strain will thus occur in an out-of-plane direction—in this instance, a distortion in the y–z plane. Accordingly, the absolute maximum shear strain will be

$$\gamma_{abs\ max} = \varepsilon_{p1} - \varepsilon_{p3} = 282.5 - (-225) = 507.5 \text{ }\mu\text{rad} \qquad \textbf{Ans.}$$

EXAMPLE 13.9

On the free surface of a copper alloy [E = 115 GPa; ν = 0.307] component, three strain gages arranged as shown record the following strains at a point:

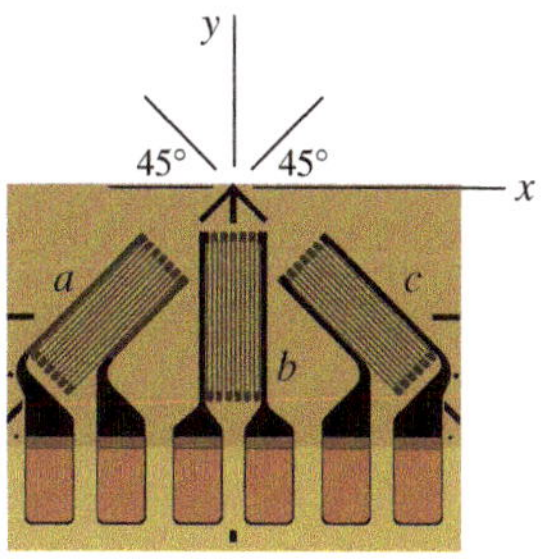

$$\varepsilon_a = 350\ \mu\varepsilon \qquad \varepsilon_b = 990\ \mu\varepsilon \qquad \varepsilon_c = 900\ \mu\varepsilon$$

(a) Determine the strain components ε_x, ε_y, and γ_{xy} at the point.
(b) Determine the principal strains and the maximum in-plane shear strain at the point.
(c) Using the results from part (b), determine the principal stresses and the maximum in-plane shear stress. Show these stresses in an appropriate sketch that indicates the orientation of the principal planes and the planes of maximum in-plane shear stress.
(d) Determine the magnitude of the absolute maximum shear stress at the point.

Plan the Solution

To solve this problem, first reduce the strain rosette data to obtain ε_x, ε_y, and γ_{xy}. Then, use Equations (13.9), (13.10), and (13.11) to determine the principal strains, the maximum in-plane shear strain, and the orientation of these strains. The principal stresses can be calculated from the principal strains with Equation (13.26), and the maximum in-plane shear stress can be computed from Equation (13.25).

SOLUTION

(a) Strain Components ε_x, ε_y, and γ_{xy}

To reduce the strain rosette data, the angles θ_a, θ_b, and θ_c must be determined for the three gages. For the rosette configuration used in this problem, the three angles are $\theta_a = 45°$, $\theta_b = 90°$, and $\theta_c = 135°$. (Alternatively, the angles $\theta_a = 225°$, $\theta_b = 270°$, and $\theta_c = 315°$ could be used.) Using these angles, write a strain transformation equation for each gage, where the strain ε_n is the experimentally measured value:

Equation for gage *a*:

$$350 = \varepsilon_x \cos^2(45°) + \varepsilon_y \sin^2(45°) + \gamma_{xy} \sin(45°)\cos(45°) \qquad \text{(a)}$$

Equation for gage *b*:

$$990 = \varepsilon_x \cos^2(90°) + \varepsilon_y \sin^2(90°) + \gamma_{xy} \sin(90°)\cos(90°) \qquad \text{(b)}$$

Equation for gage *c*:

$$900 = \varepsilon_x \cos^2(135°) + \varepsilon_y \sin^2(135°) + \gamma_{xy} \sin(135°)\cos(135°) \qquad \text{(c)}$$

Since $\cos(90°) = 0$, Equation (b) reduces to $\varepsilon_y = 990\ \mu\varepsilon$. Substitute this result into Equations (a) and (c), and collect the constant terms on the left-hand side of the equations:

$$-145 = 0.5\varepsilon_x + 0.5\gamma_{xy}$$
$$405 = 0.5\varepsilon_x - 0.5\gamma_{xy}$$

These two equations are added together to give $\varepsilon_x = 260\ \mu\varepsilon$. Subtracting the two equations gives $\gamma_{xy} = -550\ \mu\text{rad}$. Therefore, the state of strain that exists at the point on the copper alloy component can be summarized as $\varepsilon_x = 260\ \mu\varepsilon$, $\varepsilon_y = 990\ \mu\varepsilon$, and $\gamma_{xy} = -550\ \mu\text{rad}$. These strains will be used to determine the principal strains and the maximum in-plane shear strain. **Ans.**

(b) Principal and Maximum In-Plane Shear Strains

From Equation (13.10), the principal strains can be calculated as

$$\begin{aligned}\varepsilon_{p1,p2} &= \frac{\varepsilon_x + \varepsilon_y}{2} \pm \sqrt{\left(\frac{\varepsilon_x - \varepsilon_y}{2}\right)^2 + \left(\frac{\gamma_{xy}}{2}\right)^2} \\ &= \frac{260 + 990}{2} \pm \sqrt{\left(\frac{260 - 990}{2}\right)^2 + \left(\frac{-550}{2}\right)^2} \\ &= 625 \pm 457 \\ &= 1{,}082\ \mu\varepsilon,\ 168\ \mu\varepsilon\end{aligned}$$

Ans.

and from Equation (13.11), the maximum in-plane shear strain is

$$\begin{aligned}\frac{\gamma_{\max}}{2} &= \pm\sqrt{\left(\frac{\varepsilon_x - \varepsilon_y}{2}\right)^2 + \left(\frac{\gamma_{xy}}{2}\right)^2} \\ &= \pm\sqrt{\left(\frac{260 - 990}{2}\right)^2 + \left(\frac{-550}{2}\right)^2} \\ &= 457\ \mu\text{rad} \\ \therefore\ \gamma_{\max} &= 914\ \mu\text{rad}\end{aligned}$$

Ans.

The in-plane principal directions can be determined from Equation (13.9):

$$\tan 2\theta_p = \frac{\gamma_{xy}}{\varepsilon_x - \varepsilon_y} = \frac{-550}{260 - 990} = \frac{-550}{-730} \qquad \textbf{Note: } \varepsilon_x - \varepsilon_y < 0$$

$$\therefore\ 2\theta_p = 37.0° \qquad \text{and thus} \qquad \theta_p = 18.5°$$

Since $\varepsilon_x - \varepsilon_y < 0$, θ_p is the angle between the x direction and the ε_{p2} direction.

The strain rosette is bonded to the surface of the copper alloy component; therefore, we have a *plane stress* condition. Consequently, *the out-of-plane normal strain* ε_z *will not be zero*. The third principal strain ε_{p3} can be computed from Equation (13.15):

$$\varepsilon_{p3} = \varepsilon_z = -\frac{\nu}{1-\nu}(\varepsilon_x + \varepsilon_y) = -\frac{0.307}{1 - 0.307}(260 + 990) = -554\ \mu\varepsilon$$

Ans.

The absolute maximum shear strain will be the largest value obtained from three possibilities (see Table 13.2):

$$\gamma_{\text{abs max}} = \varepsilon_{p1} - \varepsilon_{p2} \qquad \text{or} \qquad \gamma_{\text{abs max}} = \varepsilon_{p1} - \varepsilon_{p3} \qquad \text{or} \qquad \gamma_{\text{abs max}} = \varepsilon_{p2} - \varepsilon_{p3}$$

In this instance, the absolute maximum shear strain will be

$$\gamma_{\text{abs max}} = \varepsilon_{p1} - \varepsilon_{p3} = 1{,}082 - (-554) = 1{,}636\ \mu\text{rad}$$

To better understand how $\gamma_{\text{abs max}}$ is determined in this instance, it is helpful to sketch Mohr's circle for strain. Strains in the x–y plane are represented by the solid circle with its center at $C = 625\ \mu\varepsilon$ and radius $R = 457\ \mu$. The principal strains in the x–y plane are $\varepsilon_{p1} = 1{,}082\ \mu\varepsilon$ and $\varepsilon_{p2} = 168\ \mu\varepsilon$.

Since the strain measurements were made on the free surface of the copper alloy component, we have a *plane stress* situation. In such a situation, the third principal stress σ_{p3} (which is the principal stress in the out-of-plane direction) will be zero; however, the third principal strain ε_{p3} (meaning the principal strain in the out-of-plane direction) will not be zero, because of the Poisson effect.

The third principal strain in this instance was $\varepsilon_{p3} = -554$ με. This point is plotted on the ε axis, and two additional Mohr's circles are constructed. As shown in the sketch, the circle defined by ε_{p3} and ε_{p1} is the largest circle. This result indicates that the absolute maximum shear strain $\gamma_{abs\ max}$ will not occur in the x–y plane.

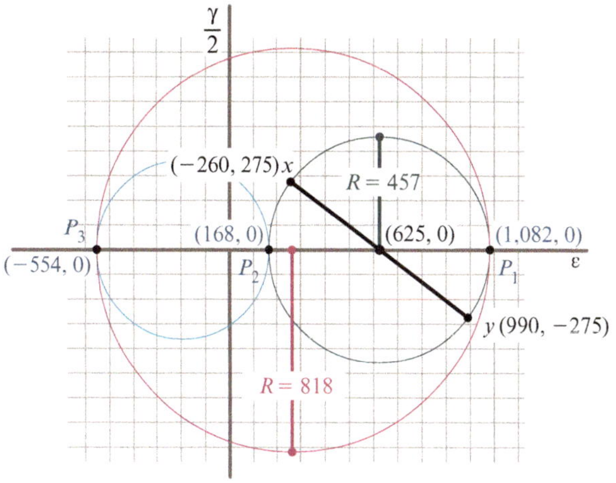

(c) Principal and Maximum In-Plane Shear Stress

The generalized Hooke's law equations are written in terms of the directions x and y in Equation (13.26); however, *these equations are applicable to any two orthogonal directions.* In this instance, the principal directions will be used. Given the material properties $E = 115$ GPa and $\nu = 0.307$, the principal stresses σ_{p1} and σ_{p2} can be computed from the principal strains ε_{p1} and ε_{p2}:

$$\sigma_{p1} = \frac{E}{1-\nu^2}(\varepsilon_{p1} + \nu\varepsilon_{p2}) = \frac{115{,}000 \text{ MPa}}{1-(0.307)^2}[(1{,}082 \times 10^{-6}) + 0.307(168 \times 10^{-6})] = 143.9 \text{ MPa} \quad \textbf{Ans.}$$

$$\sigma_{p2} = \frac{E}{1-\nu^2}(\varepsilon_{p2} + \nu\varepsilon_{p1}) = \frac{115{,}000 \text{ MPa}}{1-(0.307)^2}[(168 \times 10^{-6}) + 0.307(1{,}082 \times 10^{-6})] = 63.5 \text{ MPa} \quad \textbf{Ans.}$$

Note: The strain measurements reported in microstrain units (με) must be converted to dimensionless quantities (i.e., mm/mm) in making this calculation.

Before the maximum in-plane shear stress τ_{max} can be computed, the shear modulus G for the copper alloy material must be calculated from Equation (13.18):

$$G = \frac{E}{2(1+\nu)} = \frac{115{,}000 \text{ MPa}}{2(1+0.307)} = 44{,}000 \text{ MPa}$$

The maximum in-plane shear stress τ_{max} is calculated from Equation (13.25), which is solved for the stress:

$$\tau_{max} = G\gamma_{max} = (44{,}000 \text{ MPa})(914 \times 10^{-6}) = 40.2 \text{ MPa} \quad \textbf{Ans.}$$

Alternatively, the maximum in-plane shear stress τ_{max} can be calculated from the principal stresses:

$$\tau_{max} = \frac{\sigma_{p1} - \sigma_{p2}}{2} = \frac{143.9 - 63.5}{2} = 40.2 \text{ MPa} \quad \textbf{Ans.}$$

On the planes of maximum in-plane shear stress, the normal stress is

$$\sigma_{avg} = \frac{\sigma_{p1} + \sigma_{p2}}{2} = \frac{143.9 + 63.5}{2} = 103.7 \text{ MPa}$$

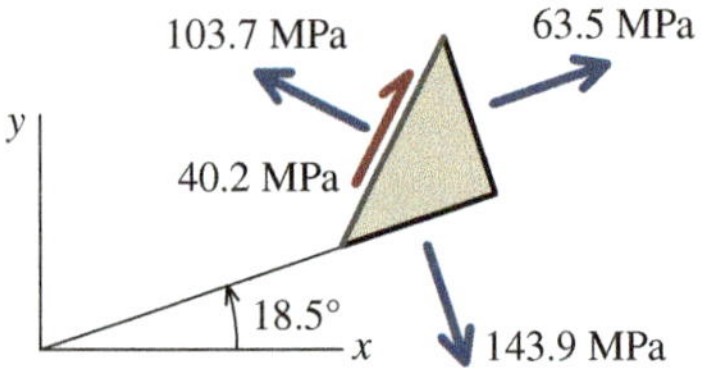

An appropriate sketch of the in-plane principal stresses, the maximum in-plane shear stress, and the orientation of these planes is shown.

(d) Absolute Maximum Shear Stress

The absolute maximum shear stress $\tau_{\text{abs max}}$ can be calculated from the absolute maximum shear strain:

$$\tau_{\text{abs max}} = G\gamma_{\text{abs max}} = (44{,}000 \text{ MPa})(1{,}636 \times 10^{-6}) = 72.0 \text{ MPa} \qquad \textbf{Ans.}$$

Alternatively, $\tau_{\text{abs max}}$ can be calculated from the principal stresses if we note that $\sigma_{p3} = \sigma_z = 0$ on the free surface of the copper alloy component:

$$\tau_{\text{abs max}} = \frac{\sigma_{p1} - \sigma_{p3}}{2} = \frac{143.9 - 0}{2} = 72.0 \text{ MPa} \qquad \textbf{Ans.}$$

Including Temperature Effects in the Generalized Hooke's Law

Temperature can be included in the three-dimensional stress–strain relations as follows:

$$\begin{aligned}\varepsilon_x &= \frac{1}{E}[\sigma_x - \nu(\sigma_y + \sigma_z)] + \alpha\Delta T \\ \varepsilon_y &= \frac{1}{E}[\sigma_y - \nu(\sigma_x + \sigma_z)] + \alpha\Delta T \\ \varepsilon_z &= \frac{1}{E}[\sigma_z - \nu(\sigma_x + \sigma_y)] + \alpha\Delta T\end{aligned} \tag{13.28}$$

Here, α is the coefficient of thermal expansion and ΔT is the change in temperature. Note that temperature change does not have any effect on shear stresses or shear strains.

In terms of strains, the three-dimensional stress–strain relations that include temperature effects can be written as follows:

$$\begin{aligned}\sigma_x &= \frac{E}{(1+\nu)(1-2\nu)}[(1-\nu)\varepsilon_x + \nu(\varepsilon_x + \varepsilon_z)] - \frac{E}{1-2\nu}\alpha\Delta T \\ \sigma_y &= \frac{E}{(1+\nu)(1-2\nu)}[(1-\nu)\varepsilon_y + \nu(\varepsilon_x + \varepsilon_z)] - \frac{E}{1-2\nu}\alpha\Delta T \\ \sigma_z &= \frac{E}{(1+\nu)(1-2\nu)}[(1-\nu)\varepsilon_z + \nu(\varepsilon_x + \varepsilon_y)] - \frac{E}{1-2\nu}\alpha\Delta T\end{aligned} \tag{13.29}$$

For the case of plane stress, the generalized Hooke's law relationships that include temperature can be expressed in terms of stresses as follows:

$$\begin{aligned}\varepsilon_x &= \frac{1}{E}(\sigma_x - \nu\sigma_y) + \alpha\Delta T \\ \varepsilon_y &= \frac{1}{E}(\sigma_y - \nu\sigma_x) + \alpha\Delta T \\ \varepsilon_z &= -\frac{\nu}{E}(\sigma_x + \sigma_y) + \alpha\Delta T\end{aligned} \tag{13.30}$$

In terms of strains, the Hooke's law relationships can be expressed as the following two equations:

$$\begin{aligned}\sigma_x &= \frac{E}{1-\nu^2}(\varepsilon_x + \nu\varepsilon_y) - \frac{E}{1-\nu}\alpha\Delta T \\ \sigma_y &= \frac{E}{1-\nu^2}(\varepsilon_y + \nu\varepsilon_x) - \frac{E}{1-\nu}\alpha\Delta T\end{aligned} \tag{13.31}$$

EXAMPLE 13.10

A 2014-T6 aluminum alloy [$E = 73$ GPa; $\nu = 0.33$; $\alpha = 23.0 \times 10^{-6}/°C$] block with dimensions of 250 mm by 250 mm by 90 mm is placed in the smooth rigid die fixture shown. Deformation of the block in the y and z directions is unrestricted; however, the deformation of the block is restricted to 0.20 mm in the x direction. The block is subjected to a uniformly distributed compressive stress of $\sigma_y = -360$ MPa, and the temperature is increased by 45°C. Determine

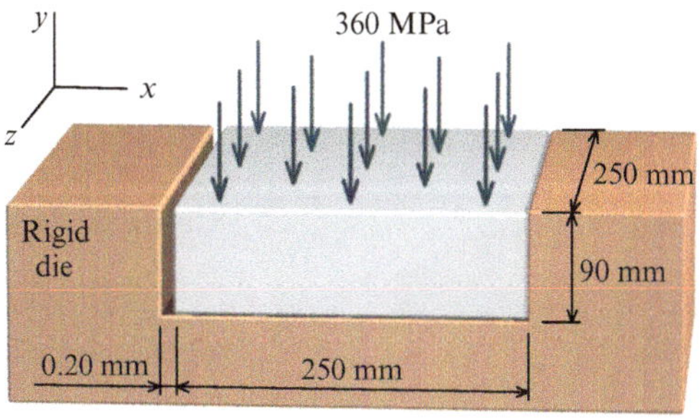

(a) The normal stress σ_x.
(b) The strains in the block.
(c) The change in volume of the block.

Plan the Solution

We know that the stress applied to the block in the y direction is $\sigma_y = -360$ MPa. Since the block is unrestrained in the z direction, we also know that the stress in the z direction is $\sigma_z = 0$, making this situation a plane stress problem. The first question to explore is whether the combination of the Poisson effect and the thermal expansion will cause the block to contact the rigid die in the x direction. If so, then we will know the value of the normal strain in the x direction.

SOLUTION

(a) Normal Stress σ_x

Let us assume that the combination of the Poisson effect and the thermal expansion will cause the block to expand so that it contacts the rigid die in the x direction. If so, then the normal strain in the x direction will be

$$\varepsilon_x = \frac{0.20 \text{ mm}}{250 \text{ mm}} = 800 \times 10^{-6} \text{ mm/mm}$$

We calculate σ_x from Equation (13.30), where ε_x is assumed to be 800×10^{-6} mm/mm:

$$\begin{aligned}
\varepsilon_x &= \frac{1}{E}(\sigma_x - \nu\sigma_y) + \alpha\Delta T \\
\therefore \sigma_x &= E(\varepsilon_x - \alpha\Delta T) + \nu\sigma_y \\
&= (73 \times 10^3 \text{ MPa})[800 \times 10^{-6} - (23 \times 10^{-6}/°\text{C})(45°\text{C})] + (0.33)(-360 \text{ MPa}) \\
&= -135.96 \text{ MPa} \qquad \textbf{Ans.}
\end{aligned}$$

The fact that σ_x is a compressive stress justifies our assumption that the block will contact the rigid die in the x direction.

(b) Strains in the Block

All three stresses are now known:

$$\sigma_x = -135.96 \text{ MPa} \qquad \sigma_y = -360 \text{ MPa} \qquad \sigma_z = 0 \text{ MPa}$$

We have determined that the normal strain in the x direction is $\varepsilon_x = 800 \times 10^{-6}$ mm/mm. Next, we will calculate ε_y and ε_z from Equations (13.28):

$$\begin{aligned}
\varepsilon_y &= \frac{1}{E}[\sigma_y - \nu(\sigma_x + \sigma_z)] + \alpha\Delta T \\
&= \frac{1}{73 \times 10^3 \text{ MPa}}[-360 \text{ MPa} - (0.33)(-135.96 \text{ MPa} + 0)] + (23 \times 10^{-6}/°\text{C})(45°\text{C}) \\
&= -3{,}282 \times 10^{-6} \text{ mm/mm} \qquad \textbf{Ans.}
\end{aligned}$$

$$\varepsilon_z = \frac{1}{E}[\sigma_z - \nu(\sigma_x + \sigma_y)] + \alpha \Delta T$$

$$= \frac{1}{73 \times 10^3 \text{ MPa}}[0 \text{ MPa} - (0.33)(-135.96 \text{ MPa} - 360 \text{ MPa})] + (23 \times 10^{-6}/°\text{C})(45°\text{C})$$

$$= 3{,}277 \times 10^{-6} \text{ mm/mm}$$ **Ans.**

(c) Change in Volume of the Block

Calculate the initial volume of the block as

$$V = (250 \text{ mm})(90 \text{ mm})(250 \text{ mm}) = 5.625 \times 10^6 \text{ mm}^3$$

Then, calculate the dilatation e from the three normal stresses and Equation (13.22):

$$e = \frac{1 - 2\nu}{E}(\sigma_x + \sigma_y + \sigma_z)$$

$$= \frac{1 - 2(0.33)}{73 \times 10^3 \text{ MPa}}(-135.96 \text{ MPa} - 360 \text{ MPa} + 0)$$

$$= 795.073 \times 10^{-6}$$

The change in volume of the block is thus

$$\Delta V = eV = (795.073 \times 10^{-6})(5.625 \times 10^6 \text{ mm}^3) = 4{,}472 \text{ mm}^3$$ **Ans.**

13.9 Generalized Hooke's Law for Orthotropic Materials

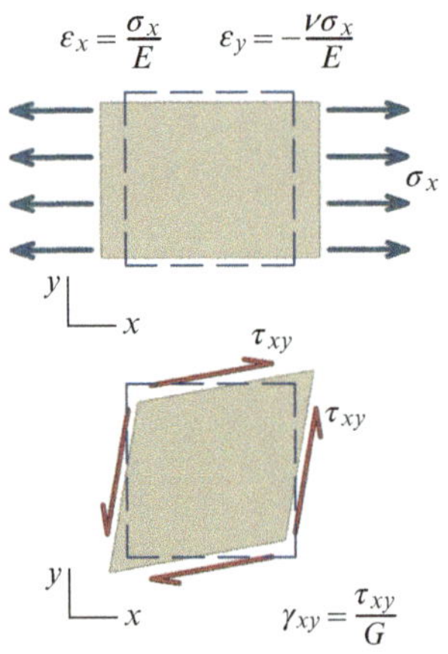

FIGURE 13.15 Isotropic material.

For a homogeneous, isotropic material, three material properties—the elastic modulus E, shear modulus G, and Poisson's ratio ν—are sufficient to describe the relationship of stress to strain. Further, of these three properties, only two are independent: For an isotropic material, a tensile normal stress causes an elongation in the direction of the stress and a contraction in the direction perpendicular to the stress (Figure 13.15), whereas shear stresses cause only shearing deformations. These types of deformations exist regardless of the direction of the stress.

For some materials, three material properties are not sufficient to describe the behavior of the material. For example, wood has three mutually perpendicular planes of material symmetry: (a) a plane parallel to the grain, (b) a plane tangential to the grain, and (c) a radial plane. Solids having material properties that are different in three mutually perpendicular directions at a point within their body are known as *orthotropic materials*. Such materials have three natural axes that are mutually perpendicular.

Figure 13.16 shows an orthotropic material subjected to a normal stress that acts in the direction of a natural axis of the material. The 1 and 2 directions in this figure represent natural axes of the material. Each of these directions has its own elastic modulus E_1 and E_2, and its own unique Poisson's ratio. For normal strain in the 2 direction resulting from loading in the 1 direction, Poisson's ratio is denoted ν_{12}. Similarly, Poisson's ratio ν_{21} gives the normal strain in the 1 direction that results from loading in the 2 direction. Just like an isotropic material, the orthotropic material elongates in the direction of the stress and contracts in the perpendicular direction. However, if the same magnitude of stress were applied perpendicular to the vertical natural axis of the material (i.e., in the 2 direction), the deformations in the horizontal and vertical directions would be much different.

Whereas elongations and contractions in an isotropic material are independent of the direction of the applied load, deformations of an orthotropic material depend very much on the directions of the natural axes, as well as on the direction of the applied load.

An applied stress that is not in the direction of one of the natural axes of an orthotropic material behaves as shown in Figure 13.17. A normal stress σ_x applied in the x direction produces an elongation ε_x in the x direction, a contraction ε_y in the y direction, and a shearing deformation γ_{xy}. Note that there will be a Poisson's ratio ν_{xy} relating the normal strain in the y direction to the normal strain in the x direction that occurs in response to loading in the x direction. The constant η_{xs} is termed a *shear coupling coefficient*. This material constant, the ratio of the shear strain γ_{xy} to the normal strain ε_x in the x direction, relates normal stress in the x direction to shear strain in the x–y plane.

A shear stress τ_{xy} produces an elongation ε_x, a contraction ε_y, and a shearing deformation γ_{xy}. The shear coupling coefficients η_{sx} and η_{sy} are the ratios of the normal strains ε_x and ε_y to the shear strain γ_{xy}, respectively, for the case of pure shear stress loading.

Thus, depending on the direction of the stress, there may exist a coupling among elongation, contraction, and shearing deformation. By contrast, no coupling exists for an isotropic material, regardless of the direction of stress.

Consider a thin piece of material (such as epoxy) reinforced by unidirectional fibers (such as graphite) that is subjected to plane stress. The material that holds the fibers in position is called the *matrix*. The combination of fibers embedded in a matrix creates a material that is orthotropic (Figure 13.18). Axes x, y, and z are in the principal material directions (i.e., the natural directions). Planes x–y, y–z, and x–z are planes of material symmetry. For plane stress, $\sigma_z = \tau_{xz} = \tau_{yz} = 0$.

To determine the stress–strain relations, assume that the behavior of the material is linear elastic and use the principal of superposition. First, suppose that the orthotropic specimen is subjected to a uniaxial loading of σ_x in the x direction, as shown in Figure 13.19*a*. Then the strains are

$$\varepsilon_x = \frac{\sigma_x}{E_x}$$

and

$$\varepsilon_y = -\nu_{xy}\varepsilon_x = -\nu_{xy}\frac{\sigma_x}{E_x}$$

where E_x is the elastic modulus for loading in the x direction and ν_{xy} is Poisson's ratio for loading in the x direction and strain in the y direction. For fibers directed as shown, ν_{xy} is often referred to as the *major Poisson's ratio*. Strains do occur in the z direction, but since the material is thin, ε_z is of no particular interest.

Next, consider a uniaxial loading of σ_y in the y direction as shown in Figure 13.19*b*. The strains are now

$$\varepsilon_y = \frac{\sigma_y}{E_y}$$

and

$$\varepsilon_x = -\nu_{yx}\varepsilon_y = -\nu_{yx}\frac{\sigma_y}{E_y}$$

where E_y is the elastic modulus for loading in the y direction and ν_{yx} is Poisson's ratio for loading in the y direction and strain in the x direction. For fibers directed as shown, ν_{yx} is often referred to as the *minor Poisson's ratio*.

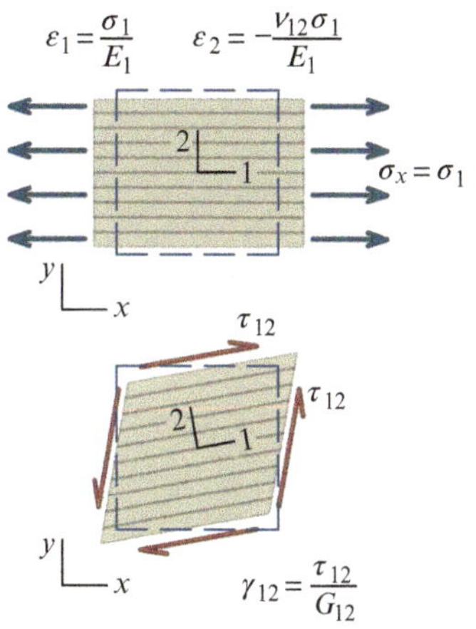

FIGURE 13.16 Orthotropic material loaded along principal material directions.

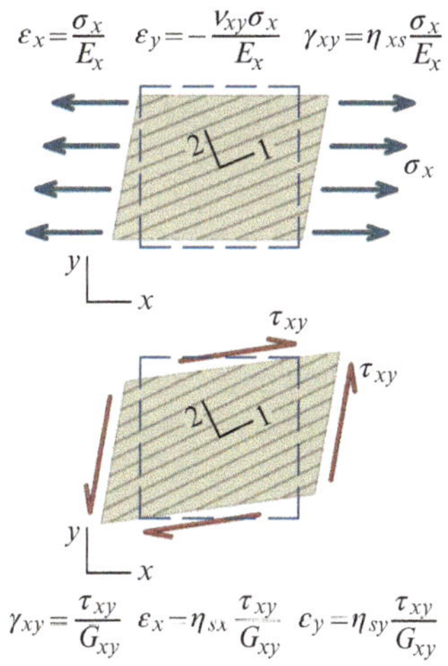

FIGURE 13.17 Orthotropic material loaded along nonprincipal material directions.

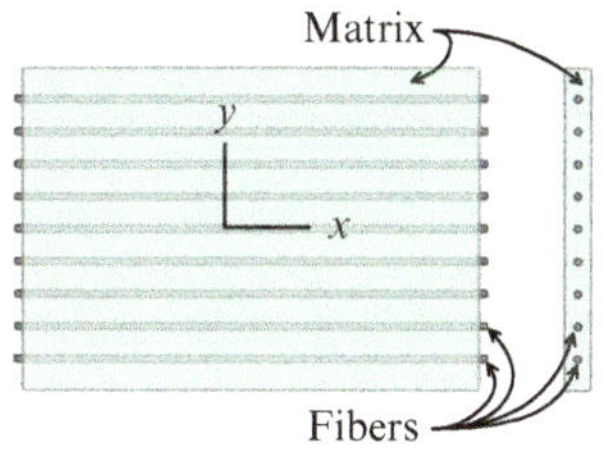

FIGURE 13.18 Matrix and fibers for a unidirectional layer.

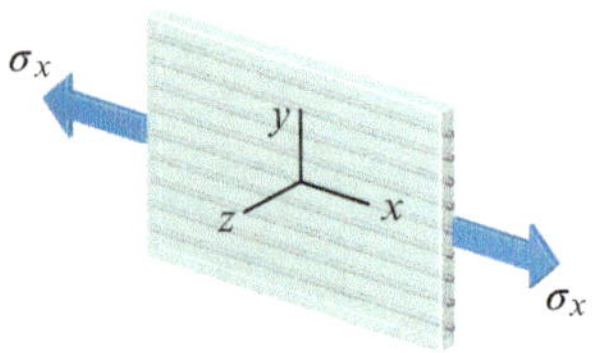

FIGURE 13.19*a* Loading in *x* direction.

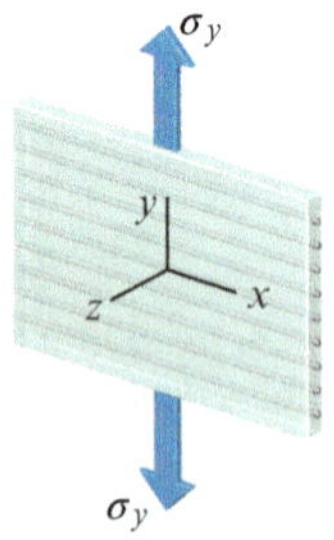

FIGURE 13.19*b* Loading in *y* direction.

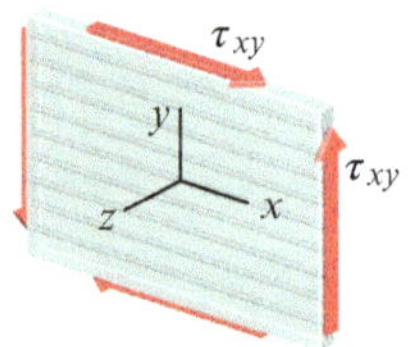

FIGURE 13.19*c* Shear stress loading in the *x*–*y* plane.

Finally, suppose the material is subjected to shear stress as shown in Figure 13.19*c*. Then the relation between shear stress and shear strain is

$$\gamma_{xy} = \frac{\tau_{xy}}{G_{xy}}$$

where G_{xy} is the shear modulus in the x–y plane.

Collecting these equations and using the principle of superposition, we can express the strains in terms of the stresses as follows:

$$\begin{aligned} \varepsilon_x &= \frac{\sigma_x}{E_x} - \nu_{yx}\frac{\sigma_y}{E_y} \\ \varepsilon_y &= \frac{\sigma_y}{E_y} - \nu_{xy}\frac{\sigma_x}{E_x} \\ \gamma_{xy} &= \frac{\tau_{xy}}{G_{xy}} \end{aligned} \tag{13.32}$$

Solving these equations for the stresses gives the following:

$$\begin{aligned} \sigma_x &= \frac{E_x}{1-\nu_{xy}\nu_{yx}}(\varepsilon_x + \nu_{yx}\varepsilon_y) \\ \sigma_y &= \frac{E_y}{1-\nu_{xy}\nu_{yx}}(\varepsilon_y + \nu_{xy}\varepsilon_x) \\ \tau_{xy} &= G_{xy}\gamma_{xy} \end{aligned} \tag{13.33}$$

Of the five material constants, only four are independent, because the major and minor Poisson's ratios are related reciprocally; that is,

$$\frac{\nu_{xy}}{E_x} = \frac{\nu_{yx}}{E_y}, \qquad \text{or} \qquad \nu_{yx} = \nu_{xy}\frac{E_y}{E_x} \tag{13.34}$$

Note that if the material is isotropic, then $E_x = E_y = E$, $\nu_{xy} = \nu_{xy} = \nu$, and $G_{xy} = G$. Thus, Equations (13.32) and (13.33) reduce to Equations (13.24), (13.25), and (13.26) for isotropic materials subjected to plane stress.

Some typical material properties for fiber-reinforced epoxy material with 60% unidirectional fibers by volume are shown in Table 13.3.

Table 13.3 Typical Material Properties for Fiber-Reinforced Epoxy Materials

Type	Material	E_x GPa (ksi)	E_y GPa (ksi)	G_{xy} GPa (ksi)	ν_{xy}
T-300	Graphite–Epoxy	132 (19,200)	10.3 (1,500)	6.5 (950)	0.25
GY-70	Graphite–Epoxy	320 (46,400)	5.5 (800)	4.1 (600)	0.25
E-Glass	Glass–Epoxy	45 (6,500)	12 (1,700)	4.4 (640)	0.25
Kevlar 49	Aramid–Epoxy	76 (11,000)	5.5 (800)	2.1 (300)	0.34

EXAMPLE 13.11

A unidirectional T-300 graphite–epoxy composite material is loaded with stresses $\sigma_x = 60$ ksi, $\sigma_y = 10$ ksi, and $\tau_{xy} = 5$ ksi in the principal material directions. The graphite fibers are aligned in the x direction. Determine the normal and shear strains in the principal material directions.

SOLUTION

The material properties from Table 3.1 are $E_x = 19{,}200$ ksi, $E_y = 1{,}500$ ksi, $G_{xy} = 950$ ksi, and $\nu_{xy} = 0.25$. The minor Poisson's ratio is

$$\nu_{yx} = \nu_{xy}\frac{E_y}{E_x} = (0.25)\frac{1{,}500 \text{ ksi}}{19{,}200 \text{ ksi}} = 0.0195$$

The normal strain in the direction of the fibers is

$$\varepsilon_x = \frac{\sigma_x}{E_x} - \nu_{yx}\frac{\sigma_y}{E_y} = \frac{60 \text{ ksi}}{19{,}200 \text{ ksi}} - (0.0195)\frac{10 \text{ ksi}}{1{,}500 \text{ ksi}}$$

$$= 2{,}995 \times 10^{-6} \text{ in./in.} \qquad \textbf{Ans.}$$

The normal strain perpendicular to the fibers is

$$\varepsilon_y = \frac{\sigma_y}{E_y} - \nu_{xy}\frac{\sigma_x}{E_x} = \frac{10 \text{ ksi}}{1{,}500 \text{ ksi}} - (0.25)\frac{60 \text{ ksi}}{19{,}200 \text{ ksi}}$$

$$= 5{,}886 \times 10^{-6} \text{ in./in.} \qquad \textbf{Ans.}$$

and the shear strain is calculated

$$\gamma_{xy} = \frac{\tau_{xy}}{G_{xy}} = \frac{5 \text{ ksi}}{950 \text{ ksi}} = 5{,}263 \times 10^{-6} \text{ rad} \qquad \textbf{Ans.}$$

MecMovies

EXAMPLE

M13.6 The strain rosette shown was used to obtain the following normal strain data at a point on the free surface of an aluminum [E = 70 GPa; ν = 0.33] plate: ε_a = 770 με, ε_b = 1,180 με, ε_c = −350 με.

(a) Determine the stress components σ_x, σ_y, and τ_{xy} at the point.
(b) Determine the principal stresses at the point.
(c) Show the principal stresses in an appropriate sketch.

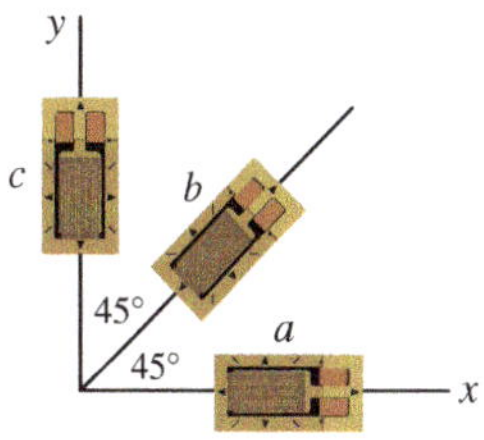

EXERCISE

M13.6 Principal Stresses from Rosette Data. A strain rosette was used to obtain normal strain data at a point on the free surface of a steel [E = 200 GPa; ν = 0.32] plate. Determine the normal strains, the shear strain, and the principal stresses in the x–y plane.

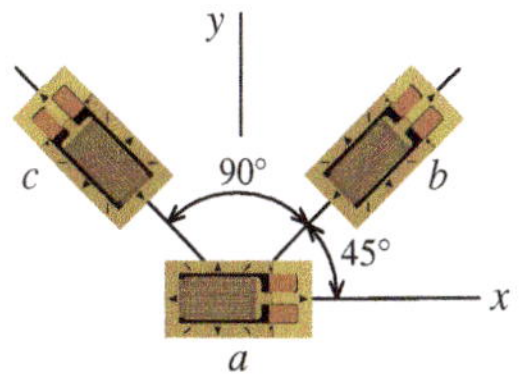

FIGURE M13.6

PROBLEMS

P13.31 An 8 mm thick brass [E = 83 GPa; ν = 0.33] plate is subjected to biaxial stress with σ_x = 180 MPa and σ_y = 65 MPa. The plate dimensions are b = 350 mm and h = 175 mm. (See Figure P13.31.) Determine

(a) the change in length of edges AB and AD.
(b) the change in length of diagonal AC.
(c) the change in thickness of the plate.

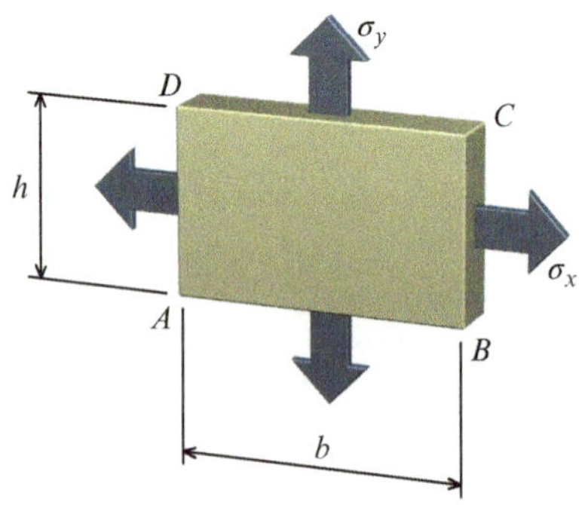

FIGURE P13.31

P13.32 A 0.75 in. thick polymer [E = 470,000 psi; ν = 0.37] casting is subjected to biaxial stresses σ_x = 2,500 psi and σ_y = 8,300 psi, acting in the directions shown in Figure P13.32. The dimensions of the casting are b = 12.0 in. and h = 8.0 in. Determine

(a) the change in length of edges AB and AD.
(b) the change in length of diagonal AC.
(c) the change in thickness of the plate.

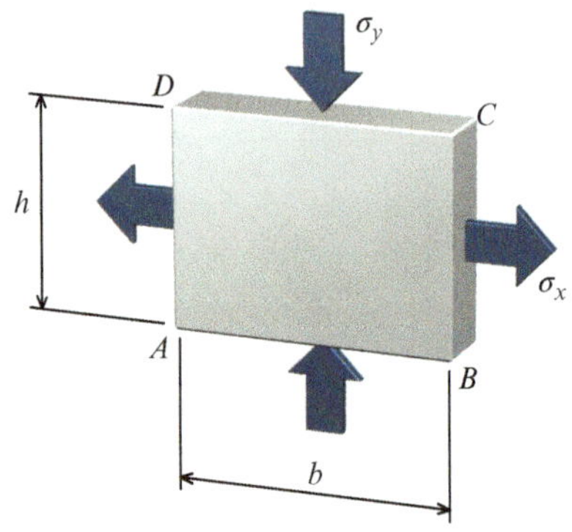

FIGURE P13.32

P13.33 A 100 mm by 100 mm square plate with a circle inscribed is subjected to normal stresses σ_x = 200 MPa and σ_y = 90 MPa, acting as shown in Figure P13.33. The plate material has an elastic modulus E = 70 GPa and a Poisson's ratio ν = 0.33. Assuming plane stress, determine the major and minor axes of the ellipse formed after deformation of the plate.

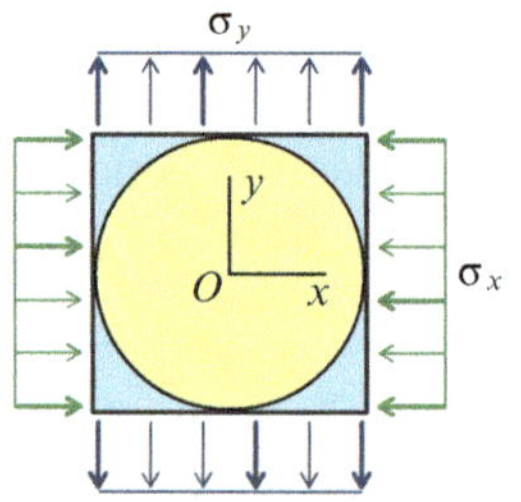

FIGURE P13.33

P13.34 A thin steel [E = 30,000 ksi; ν = 0.3] plate in a state of plane stress has dimensions of 8 in. in the x direction and 4 in. in the y direction. The plate increases in length in the x direction by 0.0015 in. and decreases in the y direction by 0.00028 in. Compute the normal stresses σ_x and σ_y that produced these deformations. Assume that $\tau_{xy} = 0$.

P13.35 A thin aluminum [E = 10,000 ksi; G = 3,800 ksi] plate is subjected to biaxial stress (Figure P13.35/36). The strains measured in the plate are ε_x = 810 με and ε_z = 1,350 με. Determine σ_x and σ_z.

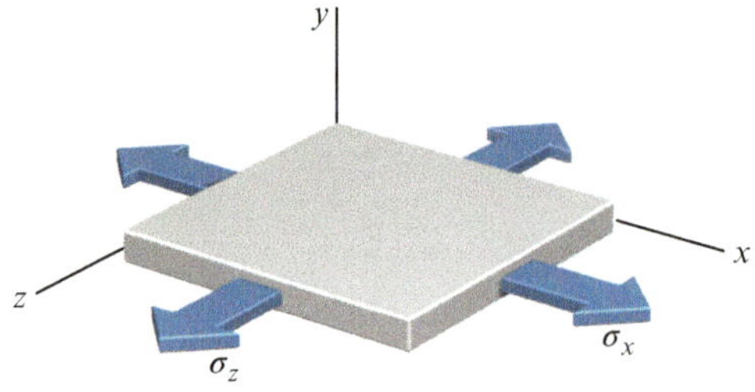

FIGURE P13.35/36

P13.36 A thin stainless steel plate [E = 190 GPa; G = 86 GPa] plate is subjected to biaxial stress (Figure P13.35/36). The strains measured in the plate are ε_x = 275 με and ε_z = 1,150 με. Determine σ_x and σ_z.

P13.37 The thin brass [E = 16,700 ksi; ν = 0.307] bar shown in Figure P13.37/38 is subjected to a normal stress σ_x = 19 ksi. A strain gage is mounted on the bar at an orientation of θ = 25° as shown in the figure. What normal strain reading would be expected from the strain gage at the specified stress?

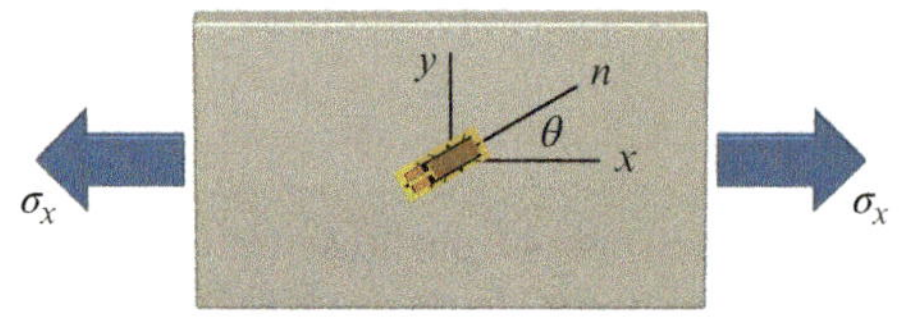

FIGURE P13.37/38

P13.38 A strain gage is mounted on a thin brass [E = 12,000 ksi; ν = 0.33] bar at an angle of θ = 35° as shown in Figure P13.37/38. If the strain gage records a normal strain ε_n = 470 με, what is the magnitude of the normal stress σ_x?

P13.39 A thin brass [E = 100 GPa; G = 39 GPa] plate is subjected to biaxial stress as shown in Figure P13.39/40. The normal stress in the y direction is known to be σ_y = 160 MPa. The strain gage measures a normal strain of 920 με at an orientation of θ = 35° in the indicated direction. What is the magnitude of σ_x that acts on the plate?

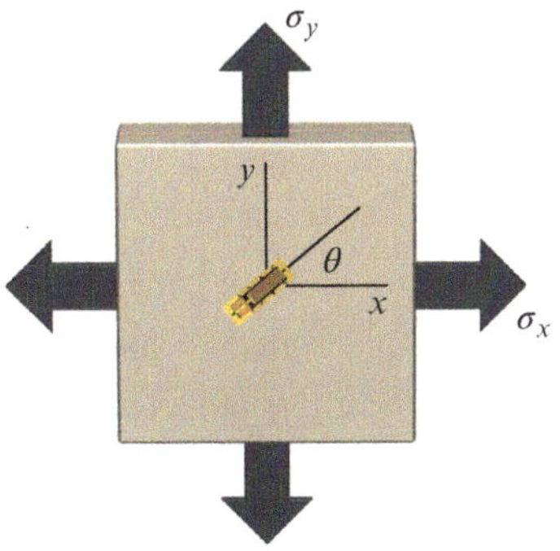

FIGURE P13.39/40

P13.40 A thin brass [E = 14,500 ksi; G = 5,500 ksi] plate is subjected to biaxial stress (Figure P13.39/40). The normal stress in the x direction is known to be twice as large as the normal stress in the y direction. The strain gage measures a normal strain of 775 με at an orientation of θ = 50° in the indicated direction. Determine the magnitudes of the normal stresses σ_x and σ_y acting on the plate.

P13.41 On the free surface of an aluminum [E = 10,000 ksi; ν = 0.33] component, the strain rosette shown in Figure P13.41 was used to obtain the following normal strain data: ε_a = 440 με, ε_b = 550 με, and ε_c = 870 με. Determine

(a) the normal stress σ_x.
(b) the normal stress σ_y.
(c) the shear stress τ_{xy}.

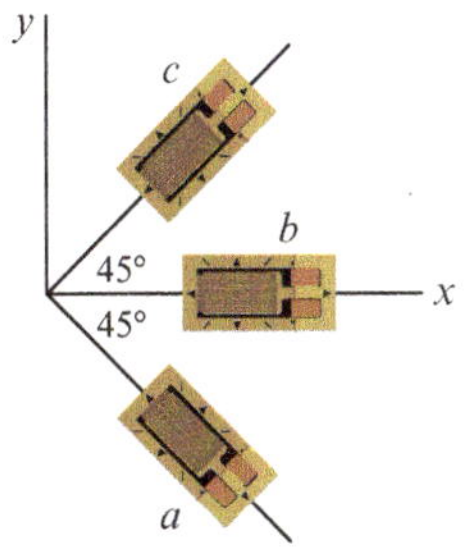

FIGURE P13.41

P13.42 On the free surface of an aluminum [E = 70 GPa; ν = 0.35] component, the strain rosette shown in Figure P13.42 was used to obtain the following normal strain data: ε_a = –300 με, ε_b = 735 με, and ε_c = 410 με. Determine

(a) the normal stress σ_x.
(b) the normal stress σ_y.
(c) the shear stress τ_{xy}.

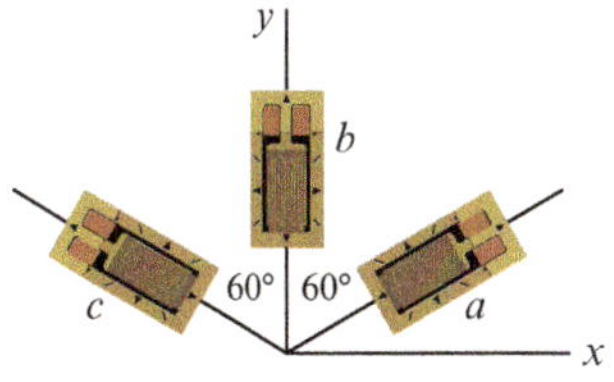

FIGURE P13.42

P13.43 On the free surface of a steel [E = 207 GPa; ν = 0.29] component, a strain rosette located at point A in Figure P13.43 was used to obtain the following normal strain data: ε_a = 133 με, ε_b = –92 με, and ε_c = –319 με. If θ = 50°, determine the stresses σ_n, σ_t, and τ_{nt} that act at point A.

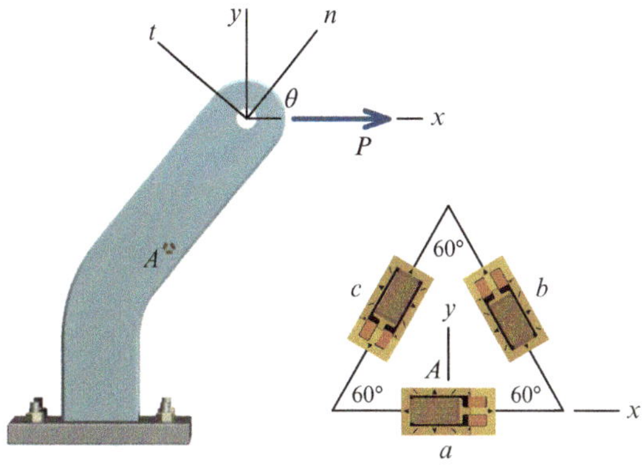

FIGURE P13.43

P13.44–P13.46 The strain components ε_x, ε_y, and γ_{xy} are given for a point on the free surface of a machine component. Determine the stresses σ_x, σ_y, and τ_{xy} at the point.

Problem	ε_x	ε_y	γ_{xy}	E	ν
P13.44	680 με	–320 με	–840 μrad	16,500 ksi	0.33
P13.45	–1,120 με	–890 με	1,300 μrad	207 GPa	0.30
P13.46	1,500 με	2,100 με	950 μrad	28,000 ksi	0.12

P13.47–P13.49 The strain rosettes shown in Figures P13.47–P13.49 were used to obtain normal strain data at a point on the free surface of a machine component. In each of Problems P13.47–P13.49, determine

(a) the stress components σ_x, σ_y, and τ_{xy} at the point.
(b) the principal stresses and the maximum in-plane shear stress at the point; show these stresses on an appropriate sketch that indicates the orientation of the principal planes and the planes of maximum in-plane shear stress.
(c) the magnitude of the absolute maximum shear stress at the point.

Problem	ε_a	ε_b	ε_c	E	ν
P13.47	−1,320 με	−870 με	340 με	100 GPa	0.28
P13.48	1,400 με	560 με	−1,270 με	210 GPa	0.31
P13.49	910 με	720 με	1,200 με	15,000 ksi	0.15

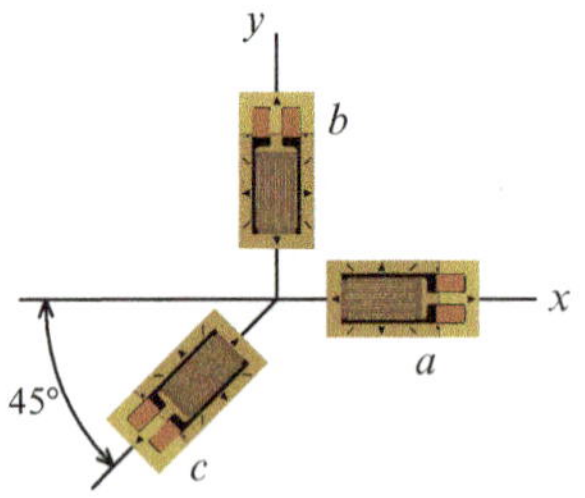

FIGURE P13.47

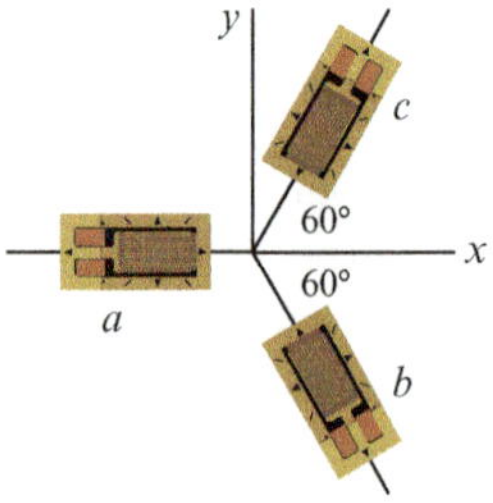

FIGURE P13.48

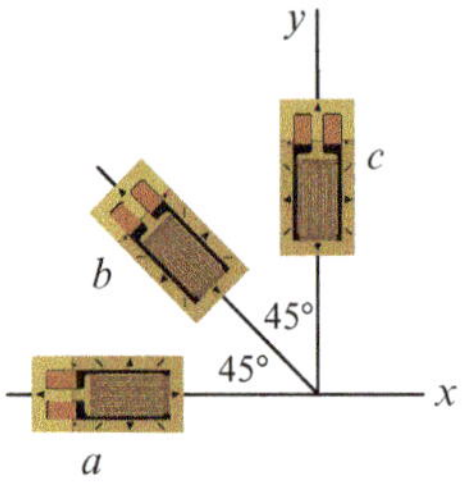

FIGURE P13.49

P13.50–P13.52 The strain rosettes shown in Figures P13.50–P13.52 were used to obtain normal strain data at a point on the free surface of a machine component. In each of Problems P13.50–P13.52,

(a) determine the strain components ε_x, ε_y, and γ_{xy} at the point.
(b) determine the principal strains and the maximum in-plane shear strain at the point.
(c) using the results from part (b), determine the principal stresses and the maximum in-plane shear stress. Show these stresses on an appropriate sketch that indicates the orientation of the principal planes and the planes of maximum in-plane shear stress.
(d) determine the magnitude of the absolute maximum shear stress at the point.

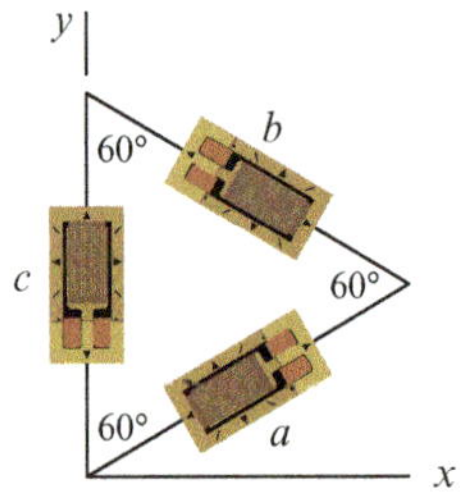

FIGURE P13.50

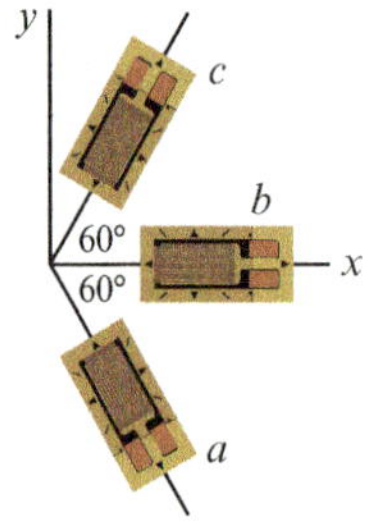

FIGURE P13.51

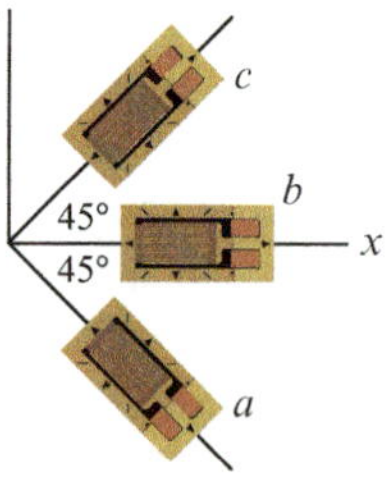

FIGURE P13.52

Problem	ε_a	ε_b	ε_c	E	ν
P13.50	−840 με	−1,775 με	665 με	9,000 ksi	0.24
P13.51	−680 με	220 με	−80 με	17,000 ksi	0.18
P13.52	55 με	−110 με	−35 με	212 GPa	0.30

P13.53 A block of 2014-T4 aluminum [E = 73 GPa; ν = 0.33] has a width a = 640 mm, a height b = 200 mm, and a thickness t = 160 mm. The block is constrained between two rigid, perfectly smooth surfaces as shown in Figure P13.53/54/55. The block is compressed by a normal stress σ_x = 210 MPa. Assuming plane stress, determine

(a) the average normal stress in the y direction.
(b) the change in the width a of the block.
(b) the change in the thickness t of the block.

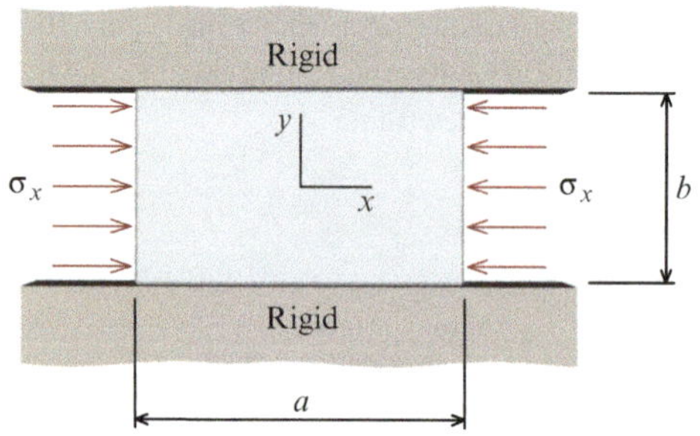

FIGURE P13.53/54/55

P13.54 A plate of stainless steel [E = 28,000 ksi; ν = 0.12] has a width a = 18 in. and a height b = 9 in. The plate is constrained between two rigid, perfectly smooth surfaces as shown in Figure P13.53/54/55. After a normal stress σ_x was applied, the width a of the plate was found to have decreased by 0.10 in. Assuming plane stress, determine the average normal stresses acting on the plate in the x and y directions.

P13.55 A plate of ductile cast iron [E = 168 GPa; ν = 0.32; α = 10.8×10^{-6}/°C] has a width a = 420 mm, a height b = 250 mm, and a thickness t = 30 mm. As shown in Figure P13.53/54/55, the plate is constrained between two rigid, perfectly smooth surfaces at an ambient temperature of 20°C. The plate is compressed by a constant normal stress σ_x = 300 MPa. Assume that plane stress conditions exist.

(a) Determine the normal strains ε_x and ε_z and the average normal stress σ_y at the ambient temperature.
(b) Determine the normal strains ε_x and ε_z and the average normal stress σ_y at a temperature of 150°C.
(c) At what temperature will the average normal stress in the y direction be reduced to zero?

P13.56 A thin aluminum alloy [E = 69 GPa; ν = 0.33; $\alpha = 23.6 \times 10^{-6}$/°C] plate with dimensions a = 1,700 mm and b = 1,000 mm is set in a rigid frictionless cavity as shown in Figure P13.56. At room temperature, there is a gap of c = 2 mm between the rigid cavity and three sides of the plate as shown in the figure. The plate is not constrained in the z direction. After the temperature of the plate has been raised by 140°C, what are the normal stresses σ_x and σ_y in the plate?

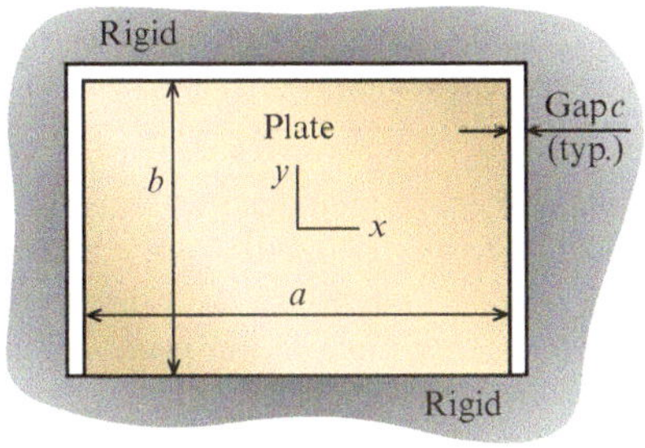

FIGURE P13.56

P13.57 At a point on the free surface of a ductile cast iron [E = 168 GPa; ν = 0.32; $\alpha = 10.8 \times 10^{-6}$/°C] machine part, the measured strains resulting from both a temperature decrease of 65°C and externally applied loads are $\varepsilon_x = -370$ με, $\varepsilon_y = -715$ με, and $\gamma_{xy} = 0$. Determine the stresses σ_x and σ_y at the point.

P13.58 A thin plate ($\sigma_z = 0$) of an aluminum alloy [E = 10,000 ksi; ν = 0.33; $\alpha = 13.1 \times 10^{-6}$/°F] is stretched until the strains in the x and y directions are 0.0010 in./in. and 0.0015 in./in., respectively. Then, the plate is rigidly held in the deformed position and heated until its temperature has increased by 80°F. Determine the final stresses in the plate.

P13.59 A thin plate ($\sigma_z = 0$) of an aluminum alloy [E = 10,000 ksi; ν = 0.33; $\alpha = 13.1 \times 10^{-6}$/°F] in the x–y plane is heated from an ambient temperature of 65°F to a final temperature of 350°F. Then, the plate is rigidly clamped in position so that it is restrained in both the x and y directions. Determine the absolute maximum shear stress in the plate when its temperature cools down to the ambient temperature.

P13.60 The principal stresses at a point are σ_x = 92 MPa, σ_y = 78 MPa, and σ_z = 66 MPa, acting as shown in Figure P13.60. The material is red brass, for which E = 115 GPa and ν = 0.307. Determine

(a) the principal strains.
(b) the dilatation.

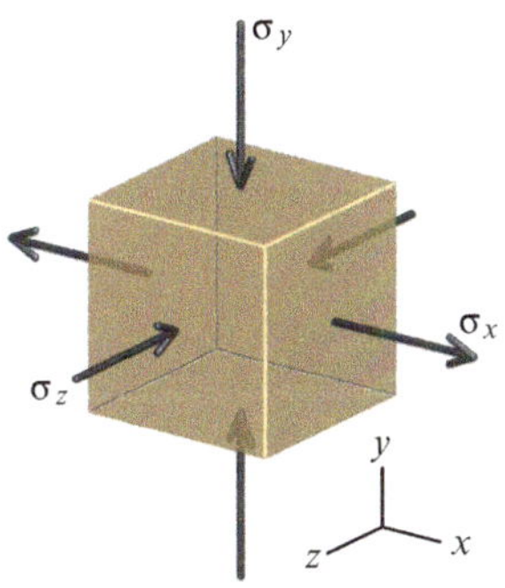

FIGURE P13.60

P13.61 The titanium [E = 16,500 ksi; ν = 0.33] block shown in Figure P13.61/62/63/64 has dimensions a = 6 in., b = 4 in., and c = 2 in. The block is subjected to triaxial stresses σ_x = −45 ksi, σ_y = −25 ksi, and σ_z = 15 ksi, acting on the x, y, and z faces, respectively. Determine (a) the changes Δa, Δb, and Δc in the dimensions of the block and (b) the change ΔV in the volume of the block.

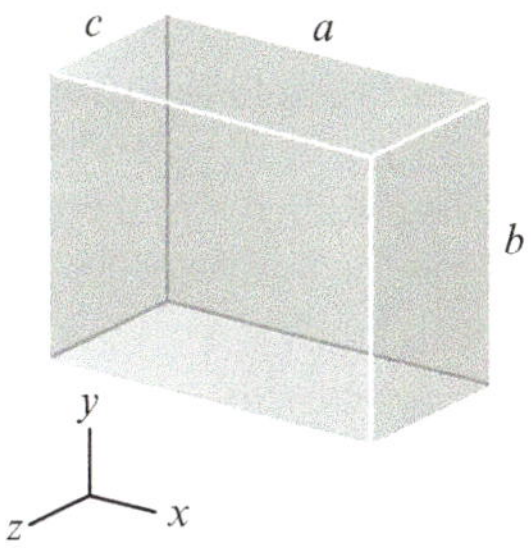

FIGURE P13.61/62/63/64

P13.62 The malleable cast iron [E = 26,000 ksi; ν = 0.27] block shown in Figure P13.61/62/63/64 has dimensions a = 9 in., b = 6 in., and c = 3 in. The block is subjected to normal stresses σ_y = −19 ksi and σ_z = −12 ksi; however, the block is constrained at its ends against displacement in the x direction. What normal stress develops in the x direction, and what are the strains in the y and z directions?

P13.63 The polymer [E = 3.7 GPa; ν = 0.33; $\alpha = 85 \times 10^{-6}$/°C] block shown in Figure P13.61/62/63/64 has dimensions a = 400 mm, b = 250 mm, and c = 50 mm. The block is subjected to a uniform temperature increase of 45°C. The block is completely free to expand in the y and z directions but is constrained at its ends against displacement in the x direction. Determine (a) the changes Δb and Δc in the dimensions of the block and (b) the normal stress in the x direction.

P13.64 The polymer [E = 3.7 GPa; ν = 0.33; $\alpha = 85 \times 10^{-6}$/°C] block shown in Figure P13.61/62/63/64 has dimensions a = 400 mm, b = 250 mm, and c = 50 mm. The block is subjected to a uniform temperature increase of 45°C. The block is completely free to expand in the z direction but is constrained at its ends against displacement in the x and y directions. Determine (a) the normal stresses in the x and y directions and (b) the change in the length of the block in the z direction.

P13.65 A solid plastic [$E = 45$ MPa; $\nu = 0.33$] rod with a diameter $d = 100$ mm is placed in a $D = 101$ mm diameter hole with rigid walls as shown in Figure P13.65. The rod has a length $L = 400$ mm. Determine the change in length of the rod after a load $P = 32$ kN is applied.

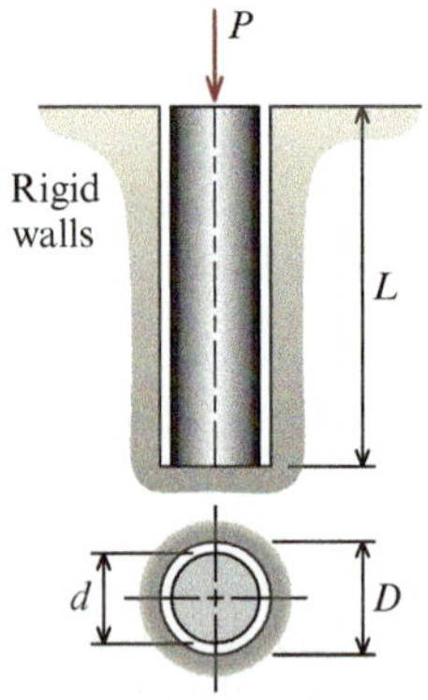

FIGURE P13.65

P13.66 A plate of epoxy reinforced with unidirectional Kevlar 49 fibers is loaded with stresses $\sigma_x = 350$ MPa, $\sigma_y = 17$ MPa, and $\tau_{xy} = 12$ MPa. The fibers are aligned in the x direction as shown in Figure P13.66/67/68. Determine the in-plane normal and shear strains for this loading.

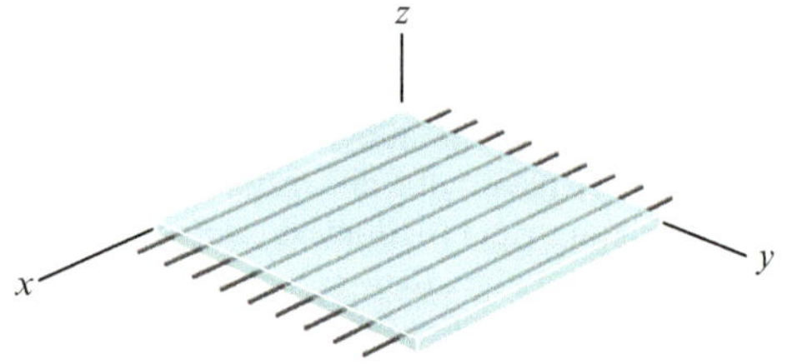

FIGURE P13.66/67/68

P13.67 A plate of epoxy reinforced with unidirectional T-300 fibers is loaded with stresses $\sigma_x = 40$ ksi, $\sigma_y = -10$ ksi, and $\tau_{xy} = 3$ ksi. The fibers are aligned in the x direction as shown in Figure P13.66/67/68. Determine the in-plane normal and shear strains for this loading.

P13.68 A plate of epoxy reinforced with unidirectional T-300 fibers is subjected to strains $\varepsilon_x = 1{,}380$ με, $\sigma_y = 5{,}250$ με, and $\tau_{xy} = 1{,}750$ μrad. The fibers are aligned in the x direction as shown in Figure P13.66/67/68. Determine the in-plane normal and shear stresses (in ksi) in the composite plate.

CHAPTER 14

Pressure Vessels

14.1 Introduction

Pressure vessels are used to hold fluids such as liquids or gases that must be stored at relatively high pressures. Pressure vessels may be found in settings such as chemical plants, airplanes, power plants, submersible vehicles, and manufacturing processes. Boilers, gas storage tanks, pulp digesters, aircraft fuselages, water distribution towers, inflatable boats, distillation towers, expansion tanks, and pipelines are examples of pressure vessels.

A pressure vessel can be described as *thin walled* when the ratio of the inside radius to the wall thickness is sufficiently large that the distribution of normal stress in the radial direction is essentially uniform across the vessel wall. Normal stress actually varies from a maximum value at the inside surface to a minimum value at the outside surface of the vessel wall. However, if the ratio of the inside radius to the wall thickness is greater than 10:1, it can be shown that the maximum normal stress is no more than 5 percent greater than the average normal stress. Therefore, a vessel can be classified as thin walled if the ratio of the inside radius to the wall thickness is greater than about 10:1 (i.e., $r/t > 10$).

Thin-walled pressure vessels are classified as **shell structures**. Shell structures derive a large measure of their strength from the shape of the structure itself. They can be defined as curved structures that support loads or pressures through stresses developed in two or more directions in the plane of the shell.

The wall comprising a pressure vessel is sometimes termed the *shell*.

Problems involving thin-walled vessels subject to fluid pressure p are readily solved with free-body diagrams of vessel sections *and the fluid contained therein*. Thin-walled spherical and cylindrical pressure vessels are considered in the sections that follow.

14.2 Thin-Walled Spherical Pressure Vessels

A typical thin-walled spherical pressure vessel is shown in Figure 14.1*a*. If the weights of the gas and vessel are negligible (a common situation), symmetry of loading and geometry requires that stresses be equal on sections that pass through the center of the sphere. Thus, on the small element shown in Figure 14.1*a*, $\sigma_x = \sigma_y = \sigma_t$. Furthermore, there are no shear stresses on any of these planes, since there are no loads to induce them. The normal stress component in a sphere is referred to as *tangential stress* and is commonly denoted σ_t.

The free-body diagram shown in Figure 14.1*b* can be used to evaluate the tangential stress σ_t in terms of the pressure p, the inside radius r, and the wall thickness t of the spherical vessel. The sphere is cut on a plane that passes through the center of the sphere to expose a hemisphere and the fluid contained within. The fluid pressure p acts horizontally against the plane circular area of the fluid contained in the hemisphere. The resultant force P from the internal pressure is the product of the fluid pressure p and the internal cross-sectional area of the sphere; that is,

$$P = p\pi r^2$$

where r is the *inside radius* of the sphere.

Because the fluid pressure and the sphere wall are symmetrical about the x axis, the tangential normal stress σ_t produced in the wall is uniform around the circumference. Since the vessel is thin walled, σ_t is assumed to be uniformly distributed across the wall thickness. For a thin-walled vessel, the exposed area of the sphere wall can be approximated by the product of the inner circumference $(2\pi r)$ and the wall thickness t of the sphere. The resultant force R from the internal stresses in the sphere wall can therefore be expressed as

$$R = \sigma_t(2\pi r t)$$

From a summation of forces in the x direction,

$$\Sigma F_x = R - P = \sigma_t(2\pi r t) - p\pi r^2 = 0$$

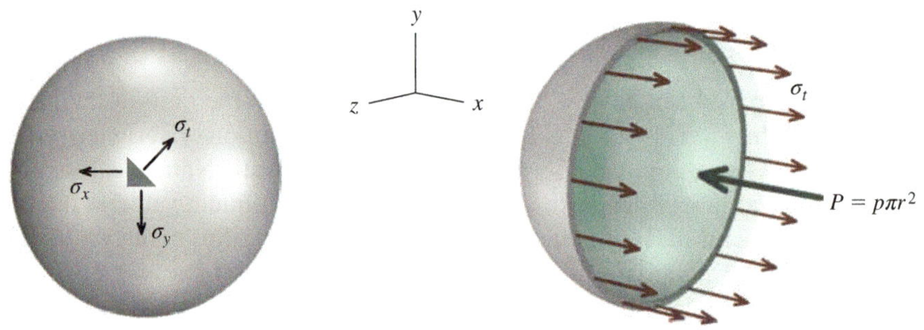

(*a*) Typical sphere (*b*) Free-body diagram exposing σ_t

FIGURE 14.1 Spherical pressure vessel.

From this equilibrium equation, an expression for the tangential normal stress in the sphere wall can be derived in terms of the inside radius r or the inside diameter d:

$$\sigma_t = \frac{pr}{2t} = \frac{pd}{4t} \tag{14.1}$$

Here, t is the wall thickness of the vessel.

By symmetry, a pressurized sphere is subjected to uniform tangential normal stresses σ_t in all directions.

Stresses on the Outer Surface

Commonly, pressures specified for a vessel are *gage* pressures, meaning that the pressure is measured with respect to atmospheric pressure. If a vessel at atmospheric pressure is subjected to a specified internal gage pressure, then the external pressure on the vessel is taken as zero while the internal pressure is equal to the gage pressure. Internal pressure in a spherical pressure vessel creates normal stress σ_t that acts in the circumferential direction of the shell. Since atmospheric pressure (i.e., zero gage pressure) exists on the outside of the sphere, the normal stress in the radial direction will be zero.

Pressure in the sphere creates no shear stress; therefore, the principal stresses are $\sigma_{p1} = \sigma_{p2} = \sigma_t$. Furthermore, no shear stress exists on free surfaces of the sphere, which means that any normal stress in the radial direction (perpendicular to the sphere wall) is also a principal stress. Since pressure outside the sphere is zero (assuming that the sphere is surrounded by atmospheric pressure), the normal stress in the radial direction due to external pressure is zero. Therefore, the third principal stress is $\sigma_{p3} = \sigma_{\text{radial}} = 0$. Consequently, the outer surface of the sphere (Figure 14.2) is in a condition of *plane stress*, which is also termed *biaxial stress* here.

Mohr's circle for the outer surface of a spherical pressure vessel (subjected to an internal gage pressure) is shown in Figure 14.3. Mohr's circle describing stresses in the plane of the sphere wall is a single point. Therefore, the maximum shear stress in the plane of the sphere wall is zero. The maximum *out-of-plane shear stress* is

$$\tau_{\text{abs max}} = \frac{1}{2}(\sigma_t - \sigma_{\text{radial}}) = \frac{1}{2}\left(\frac{pr}{2t} - 0\right) = \frac{pr}{4t} \tag{14.2}$$

Stresses on the Inner Surface

The stress σ_t on the inner surface of the spherical pressure vessel is the same as σ_t on the outer surface because the vessel is thin walled (Figure 14.2). Pressure exists inside the

b/c pressure outside = 0

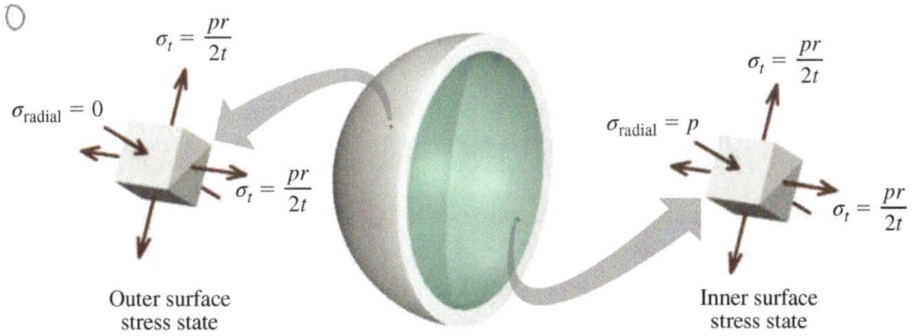

FIGURE 14.2 Stress elements on the outer and inner surfaces of a spherical pressure vessel.

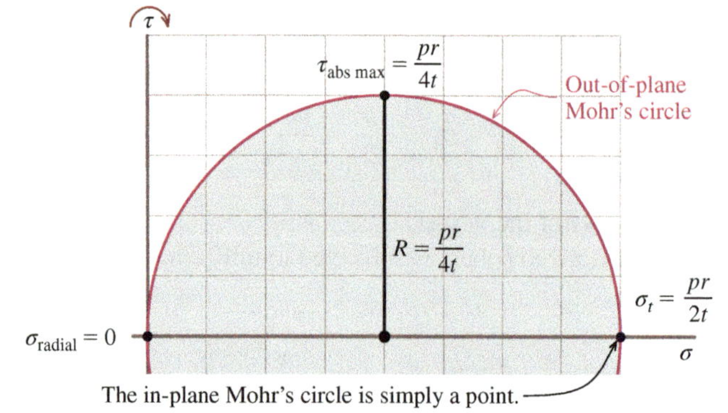

FIGURE 14.3 Mohr's circle for outer surface of a sphere.

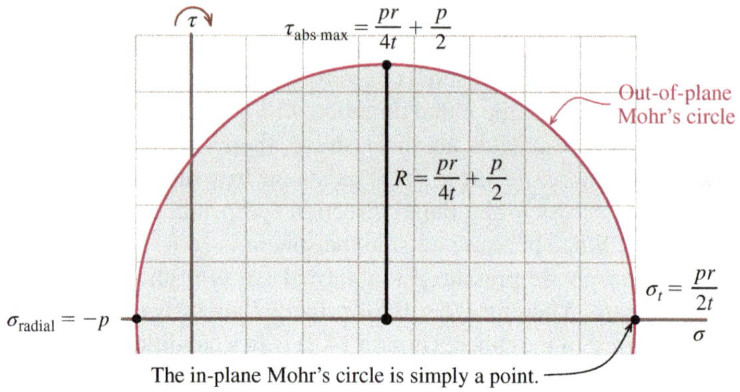

FIGURE 14.4 Mohr's circle for inner surface of a sphere.

vessel, and this pressure pushes on the sphere wall, creating a normal stress in the radial direction. The normal stress in the radial direction is equal and opposite to the pressure: $\sigma_{\text{radial}} = -p$. Thus, the inner surface is in a state of **triaxial stress**.

Mohr's circle for the inner surface of a spherical pressure vessel (subjected to an internal gage pressure) is shown in Figure 14.4. The maximum in-plane shear stress is zero. However, the maximum *out-of-plane shear stress* on the inner surface is increased because of the radial stress caused by the pressure:

$$\tau_{\text{abs max}} = \frac{1}{2}(\sigma_t - \sigma_{\text{radial}}) = \frac{1}{2}\left[\frac{pr}{2t} - (-p)\right] = \frac{pr}{4t} + \frac{p}{2} \tag{14.3}$$

14.3 Thin-Walled Cylindrical Pressure Vessels

A typical thin-walled cylindrical pressure vessel is shown in Figure 14.5*a*. The normal stress component on a transverse section is known as the **longitudinal stress**, which is denoted as σ_{long} or, simply, σ_l. The normal stress component on a longitudinal section is known as **hoop** or **circumferential stress** and is denoted as σ_{hoop} or, simply, σ_h. There are no shear stresses on transverse or longitudinal sections due to pressure alone.

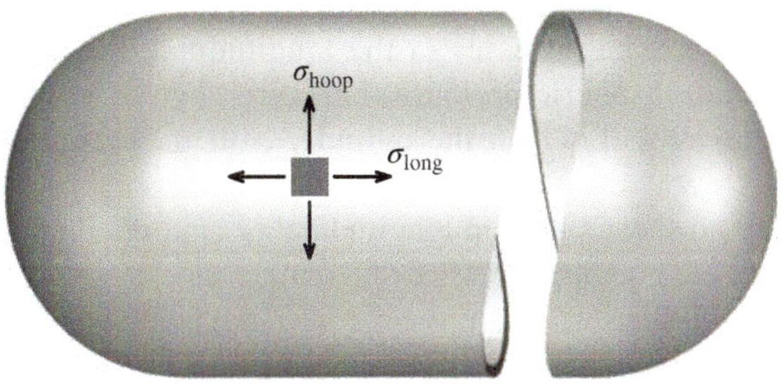

FIGURE 14.5*a* Cylindrical pressure vessel.

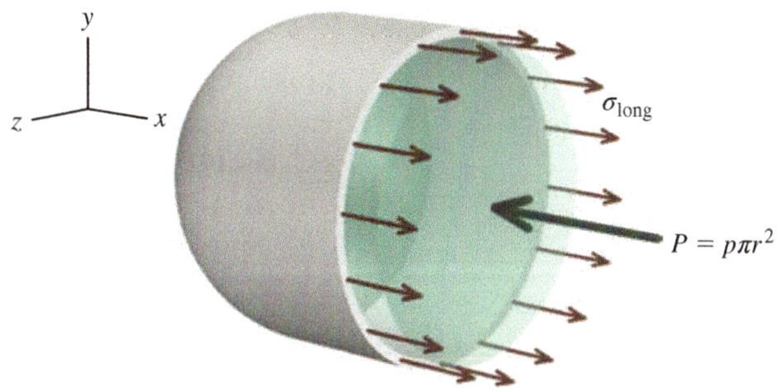

FIGURE 14.5*b* Free-body diagram exposing σ_{long}.

The free-body diagram used to determine the longitudinal stress (Figure 14.5*b*) is similar to the free-body diagram of Figure 14.1*b*, which was used for the sphere, and the results are the same. Specifically,

$$\boxed{\sigma_{long} = \frac{pr}{2t} = \frac{pd}{4t}} \tag{14.4}$$

To determine the stresses acting in the circumferential direction of the cylindrical pressure vessel, the free-body diagram shown in Figure 14.5*c* is considered. This free-body diagram exposes a longitudinal section of the cylinder wall.

There are two resultant forces P_x acting in the x direction, which are created by pressure acting on the semicircular ends of the free-body diagram. These forces are equal in magnitude, but opposite in direction; therefore, they cancel each other out.

In the lateral direction (i.e., the z direction), the resultant force P_z due to the pressure p acting on an internal area $2r\Delta x$ is

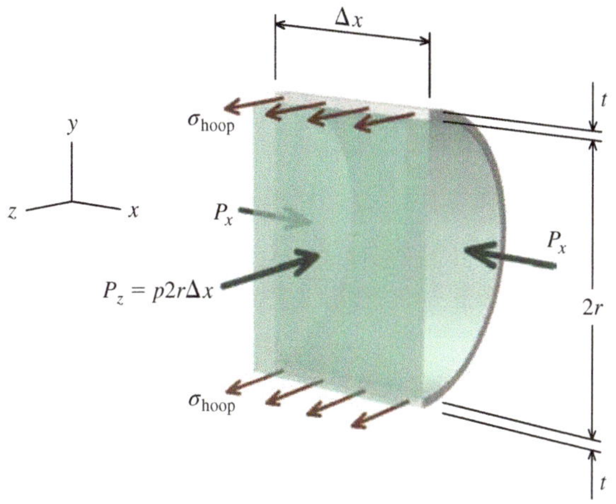

FIGURE 14.5*c* Free-body diagram exposing σ_{hoop}.

$$P_z = p2r\Delta x$$

where Δx is the length of the segment arbitrarily chosen for the free-body diagram.

The area of the cylinder wall exposed by the longitudinal section (i.e., the exposed z surfaces) is $2t\Delta x$. The internal pressure in the cylinder is resisted by normal stress that acts in the circumferential direction on these exposed surfaces. The total resultant force in the z direction from these circumferential stresses is

$$R_z = \sigma_{hoop}(2t\Delta x)$$

The summation of forces in the z direction gives

$$\Sigma F_z = R_z - P_z = \sigma_{hoop}(2t\Delta x) - p2r\Delta x = 0$$

From this equilibrium equation, an expression for the circumferential stress in the cylinder wall can be derived in terms of the inside radius r or the inside diameter d:

$$\boxed{\sigma_{hoop} = \frac{pr}{t} = \frac{pd}{2t}} \tag{14.5}$$

In a cylindrical pressure vessel, the hoop stress σ_{hoop} is twice as large as the longitudinal stress σ_{long}.

Stresses on the Outer Surface

Pressure in a cylindrical pressure vessel creates stresses in the longitudinal direction and in the circumferential direction. If atmospheric pressure (i.e., zero gage pressure) exists outside the cylinder, then no stress will act on the cylinder wall in the radial direction.

Since pressure in the vessel creates no shear stress on longitudinal or circumferential planes, the longitudinal and hoop stresses are principal stresses: $\sigma_{p1} = \sigma_{\text{hoop}}$ and $\sigma_{p2} = \sigma_{\text{long}}$. Furthermore, since no shear stress exists on free surfaces of the cylinder, any normal stress in the radial direction (perpendicular to the cylinder wall) is also a principal stress. Because pressure outside the cylinder is zero (assuming atmospheric pressure), the normal stress in the radial direction due to external pressure is zero. Therefore, the third principal stress is $\sigma_{p3} = \sigma_{\text{radial}} = 0$. The outer surface of the cylinder (Figure 14.6) is thus said to be in a state of *plane stress*, which can be termed *biaxial stress*.

Mohr's circle for the outer surface of a cylindrical pressure vessel (with internal pressure) is shown in Figure 14.7. The maximum *in-plane shear stresses* (i.e., stresses in the plane of the cylinder wall) occur on planes that are rotated at 45° with respect to the circumferential and longitudinal directions. From Mohr's circle, the magnitude of these shear stresses is

$$\tau_{\max} = \frac{1}{2}(\sigma_{\text{hoop}} - \sigma_{\text{long}}) = \frac{1}{2}\left(\frac{pr}{t} - \frac{pr}{2t}\right) = \frac{pr}{4t} \tag{14.6}$$

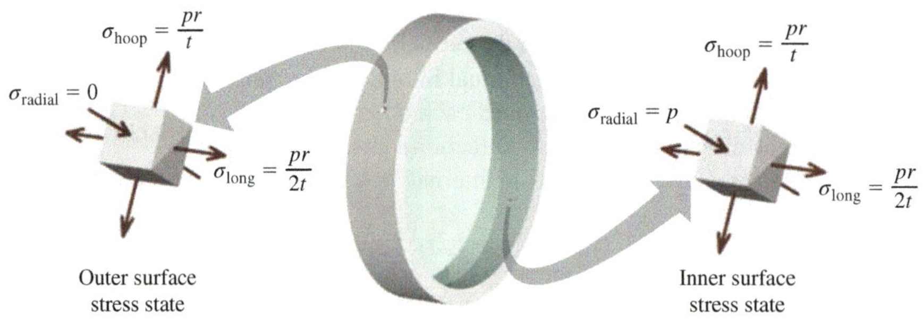

FIGURE 14.6 Stress elements on the outer and inner surfaces of a cylindrical pressure vessel.

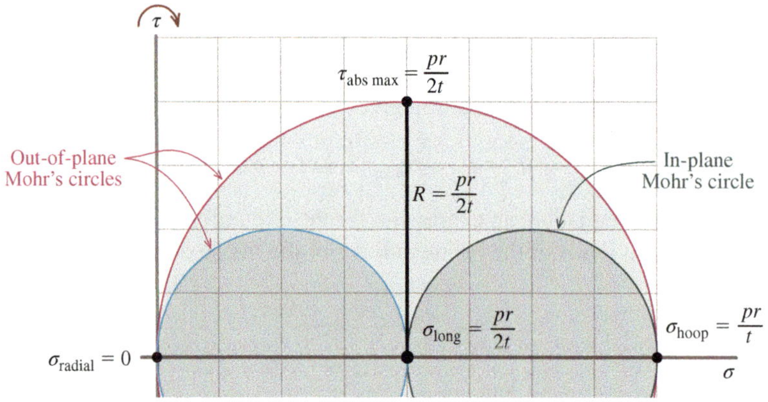

FIGURE 14.7 Mohr's circle for outer surface of a cylinder.

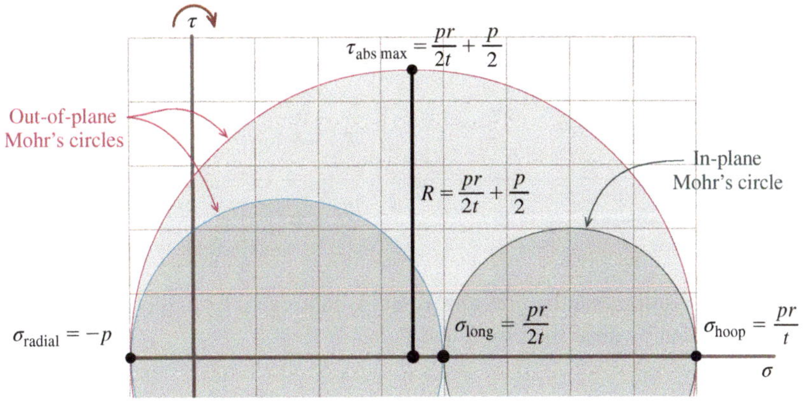

FIGURE 14.8 Mohr's circle for inner surface of a cylinder.

The maximum *out-of-plane shear stresses* occur on planes that are rotated at 45° with respect to the radial direction. The magnitude of these stresses is

$$\tau_{abs\,max} = \frac{1}{2}(\sigma_{hoop} - \sigma_{radial}) = \frac{1}{2}\left(\frac{pr}{t} - 0\right) = \frac{pr}{2t} \tag{14.7}$$

Stresses on the Inner Surface

The stresses σ_{long} and σ_{hoop} act on the inner surface of the cylindrical pressure vessel, and these stresses are the same as those on the outer surface because the vessel is assumed to be thin walled (Figure 14.6). Pressure inside the vessel pushes on the cylinder wall, creating a normal stress in the radial direction equal in magnitude to the internal pressure. Consequently, the inner surface is in a state of *triaxial stress* and the third principal stress is equal to $\sigma_{p3} = \sigma_{radial} = -p$.

Mohr's circle for the inner surface of a cylindrical pressure vessel (subjected to an internal gage pressure) is shown in Figure 14.8. The maximum *in-plane shear stresses* on the inner surface are the same as those on the outer surface. However, the maximum *out-of-plane shear stresses* on the inner surface are increased because of the radial stress caused by the pressure:

$$\tau_{abs\,max} = \frac{1}{2}(\sigma_{hoop} - \sigma_{radial}) = \frac{1}{2}\left[\frac{pr}{t} - (-p)\right] = \frac{pr}{2t} + \frac{p}{2} \tag{14.8}$$

14.4 Strains in Thin-Walled Pressure Vessels

Since pressure vessels are subjected to either biaxial stress (on outer surfaces) or triaxial stress (on inner surfaces), the generalized Hooke's law (Section 13.8) must be used to relate stress and strain. For the outer surface of a spherical pressure vessel, Equations (13.24) can be rewritten in terms of the tangential stress σ_t:

$$\varepsilon_t = \frac{1}{E}(\sigma_t - \upsilon\sigma_t) = \frac{1}{E}\left(\frac{pr}{2t} - \upsilon\frac{pr}{2t}\right) = \frac{pr}{2tE}(1 - \upsilon) \tag{14.9}$$

For the outer surface of a cylindrical pressure vessel, Equations (13.24) can be rewritten in terms of the longitudinal and hoop stresses:

$$\varepsilon_{\text{long}} = \frac{1}{E}(\sigma_{\text{long}} - \upsilon\sigma_{\text{hoop}}) = \frac{1}{E}\left(\frac{pr}{2t} - \upsilon\frac{pr}{t}\right) = \frac{pr}{2tE}(1 - 2\upsilon) \tag{14.10}$$

$$\varepsilon_{\text{hoop}} = \frac{1}{E}(\sigma_{\text{hoop}} - \upsilon\sigma_{\text{long}}) = \frac{1}{E}\left(\frac{pr}{t} - \upsilon\frac{pr}{2t}\right) = \frac{pr}{2tE}(2 - \upsilon) \tag{14.11}$$

These equations assume that the pressure vessel is fabricated from a homogeneous, isotropic material that can be described by E and ν.

MecMovies

EXAMPLES

M14.1 Derivation of equations for tangential normal stress due to pressure in a spherical pressure vessel.

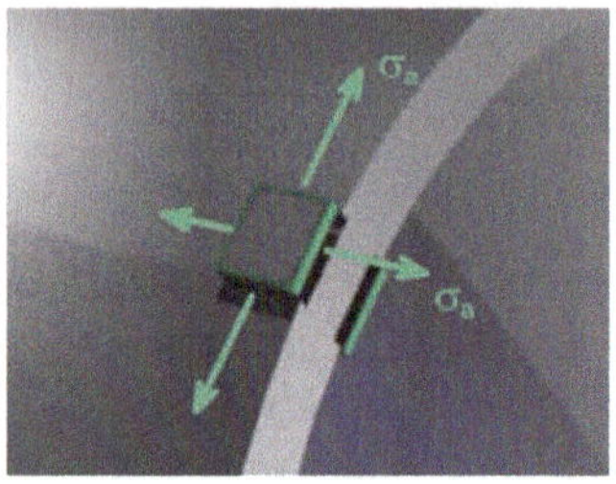

M14.2 Derivation of equations for longitudinal and circumferential stress due to pressure in a cylindrical pressure vessel.

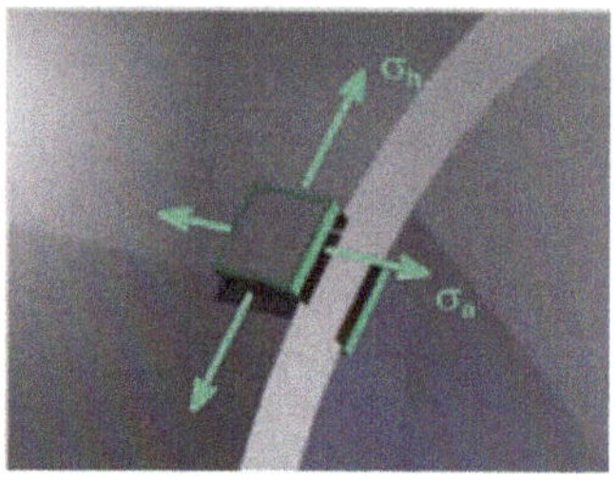

EXAMPLE 14.1

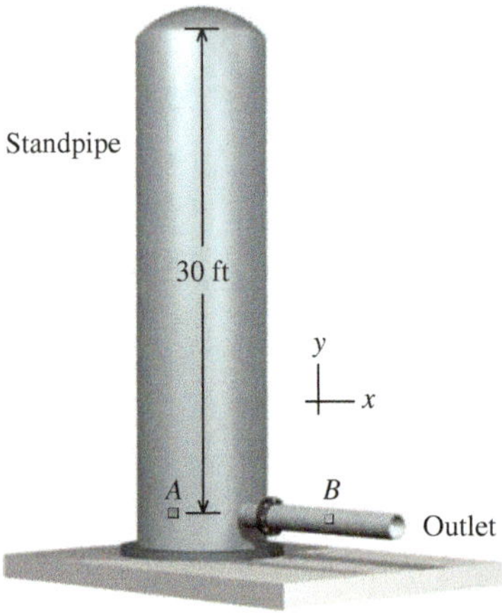

A standpipe with an inside diameter of 108 in. contains water, which has a weight density of 62.4 lb/ft^3. The column of water stands 30 ft above an outlet pipe, which has an outside diameter of 6.625 in. and an inside diameter of 6.065 in.

(a) Determine the longitudinal and hoop stresses in the outlet pipe at point B.
(b) If the maximum hoop stress in the standpipe at point A must be limited to 2,500 psi, determine the minimum wall thickness that can be used for the standpipe.

Plan the Solution

The fluid pressure at points A and B is found from the unit weight and the height of the fluid. Once the pressure is known, the equations for the longitudinal stress and the hoop stress will be used to determine the stresses in the outlet pipe and the minimum wall thickness required for the standpipe.

SOLUTION

Fluid Pressure

The fluid pressure is the product of the unit weight and the height of the fluid:

$$p = \gamma h = (62.4 \text{ lb/ft}^3)(30 \text{ ft}) = 1{,}872 \text{ lb/ft}^2 = 13 \text{ lb/in.}^2 = 13 \text{ psi}$$

Stresses in the Outlet Pipe

The longitudinal and circumferential stresses produced in a cylinder by fluid pressure are given by

$$\sigma_{\text{long}} = \frac{pd}{4t} \qquad \sigma_{\text{hoop}} = \frac{pd}{2t}$$

where d is the inside diameter of the cylinder and t is the wall thickness. For the outlet pipe, the wall thickness is $t = (6.625 \text{ in.} - 6.065 \text{ in.})/2 = 0.280$ in. The longitudinal stress is

$$\sigma_{\text{long}} = \frac{pd}{4t} = \frac{(13 \text{ psi})(6.065 \text{ in.})}{4(0.280 \text{ in.})} = 70.4 \text{ psi} \qquad \textbf{Ans.}$$

The hoop stress is twice as large:

$$\sigma_{\text{hoop}} = \frac{pd}{2t} = \frac{(13 \text{ psi})(6.065 \text{ in.})}{2(0.280 \text{ in.})} = 140.8 \text{ psi} \qquad \textbf{Ans.}$$

Stress Element at *B*

The longitudinal axis of the outlet pipe extends in the x direction; therefore, the longitudinal stress acts in the horizontal direction and the hoop stress acts in the vertical direction at point B.

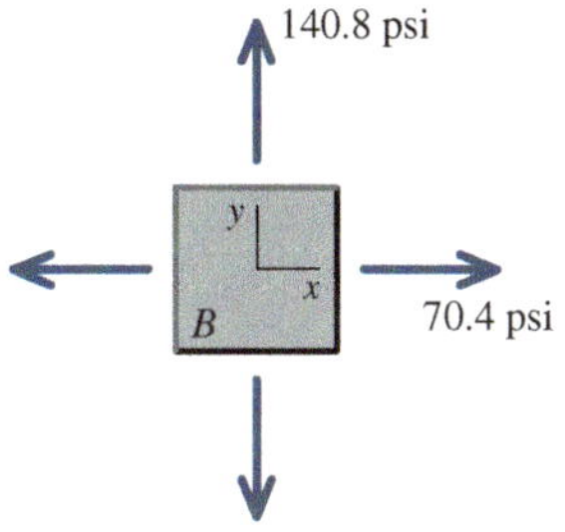

Minimum Wall Thickness for Standpipe

The maximum hoop stress in the standpipe must be limited to 2,500 psi:

$$\sigma_{\text{hoop}} = \frac{pd}{2t} \leq 2{,}500 \text{ psi}$$

This relationship is then solved for the minimum wall thickness t:

$$t \geq \frac{pd}{2\sigma_{\text{hoop}}} = \frac{(13 \text{ psi})(108 \text{ in.})}{2(2{,}500 \text{ psi})} = 0.281 \text{ in.} \qquad \textbf{Ans.}$$

EXAMPLE 14.2

A cylindrical pressure vessel with an outside diameter of 900 mm is constructed by spirally wrapping a 15 mm thick steel plate and butt-welding the mating edges of the plate. The butt-welded seams form an angle of 30° with a transverse plane through the cylinder. Determine the normal stress σ perpendicular to the weld and the shear stress τ parallel to the weld when the internal pressure in the vessel is 2.2 MPa.

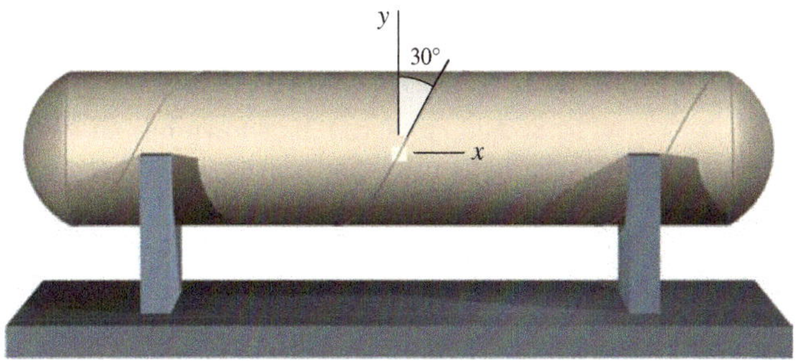

Plan the Solution

After computing the longitudinal and circumferential stresses in the cylinder wall, the stress transformation equations are used to determine the normal stress perpendicular to the weld and the shear stress parallel to the weld.

SOLUTION

The longitudinal and circumferential stresses produced in a cylinder by fluid pressure are given by

$$\sigma_{\text{long}} = \frac{pd}{4t} \qquad \sigma_{\text{hoop}} = \frac{pd}{2t}$$

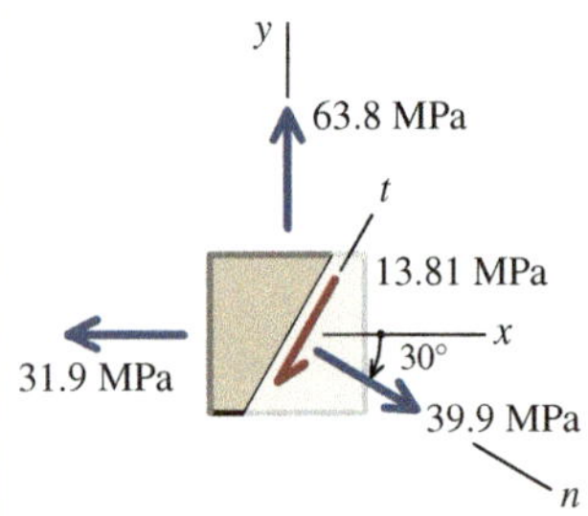

where d is the inside diameter of the cylinder and t is the wall thickness. The inside diameter of the cylinder is $d = 900 \text{ mm} - 2(15 \text{ mm}) = 870 \text{ mm}$. The longitudinal stress in the tank is

$$\sigma_{\text{long}} = \frac{pd}{4t} = \frac{(2.2 \text{ MPa})(870 \text{ mm})}{4(15 \text{ mm})} = 31.9 \text{ MPa}$$

The hoop stress is twice as large as the longitudinal stress:

$$\sigma_{\text{hoop}} = \frac{pd}{2t} = \frac{(2.2 \text{ MPa})(870 \text{ mm})}{2(15 \text{ mm})} = 63.8 \text{ MPa}$$

The weld seam is oriented at an angle of 30°, as shown. The normal stress perpendicular to the weld seam can be determined from Equation (12.3), with $\theta = -30°$:

$$\begin{aligned}\sigma_n &= \sigma_x \cos^2\theta + \sigma_y \sin^2\theta + 2\tau_{xy}\sin\theta\cos\theta \\ &= (31.9 \text{ MPa})\cos^2(-30°) + (63.8 \text{ MPa})\sin^2(-30°) \\ &= 39.9 \text{ MPa}\end{aligned}$$

Ans.

The shear stress parallel to the weld seam can be determined from Equation (12.4):

$$\begin{aligned}\tau_{nt} &= -(\sigma_x - \sigma_y)\sin\theta\cos\theta + \tau_{xy}(\cos^2\theta - \sin^2\theta) \\ &= -(31.9 \text{ MPa} - 63.8 \text{ MPa})\sin(-30°)\cos(-30°) \\ &= -13.81 \text{ MPa}\end{aligned}$$

Ans.

MecMovies

EXAMPLES

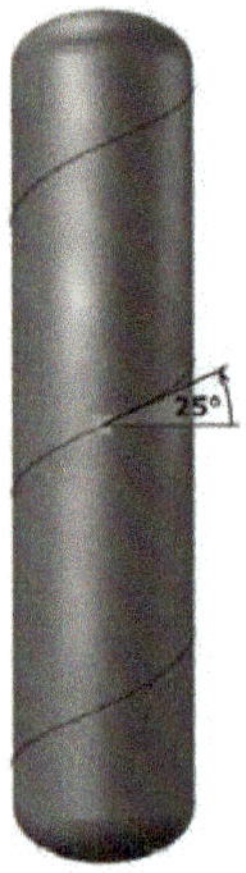

M14.3 The pressure tank shown has an outside diameter of 200 mm and a wall thickness of 5 mm. The tank has butt-welded seams forming an angle of $\beta = 25°$ with a transverse plane. For an internal gage pressure of $p = 1{,}500$ kPa, determine the normal stress perpendicular to the weld and the shear stress parallel to the weld.

M14.4 The strain gage shown is used to determine the gage pressure in the cylindrical steel tank. The tank has an outside diameter of 1,250 mm and a wall thickness of 15 mm, and is made of steel [$E = 200$ GPa; $\nu = 0.32$]. The gage is inclined at a 30° angle with respect to the longitudinal axis of the tank. Determine the pressure in the tank corresponding to a strain gage reading of 290 $\mu\varepsilon$.

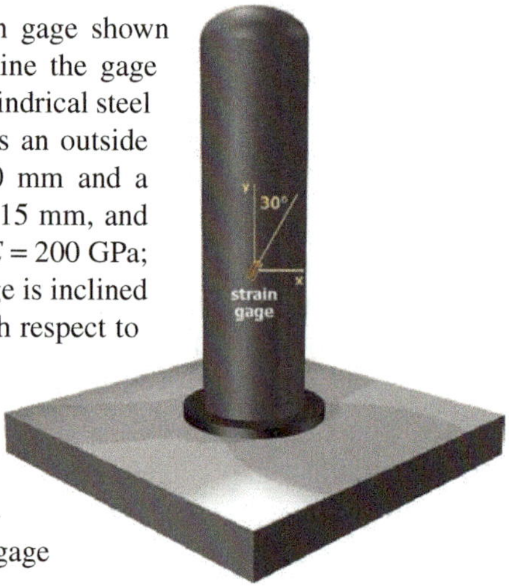

M14.5 A cylindrical steel [$E = 200$ GPa; $\nu = 0.3$] tank contains a fluid under pressure. The ultimate shear strength of the steel is 300 MPa, and a factor of safety of 4 is required. The fluid pressure must be carefully controlled to ensure that the shear stress in the cylinder does not exceed the allowable shear stress limit. To monitor the tank, a strain gage records the longitudinal strain in the tank. Determine the critical strain gage reading that must not be exceeded for safe operation of the tank.

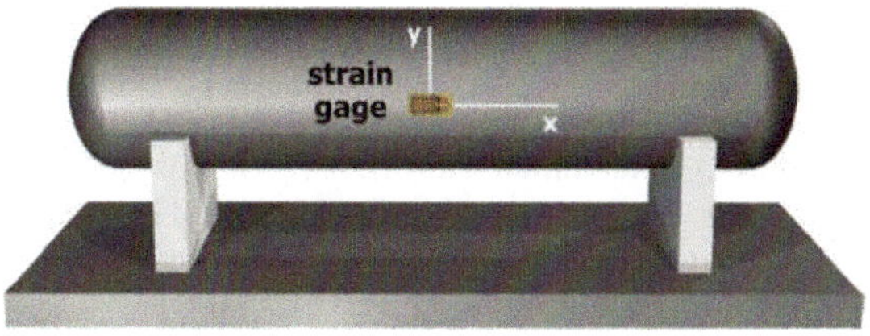

EXERCISES

M14.3 For an indicated internal gage pressure, determine the normal stress perpendicular to a weld and the shear stress parallel to a weld.

M14.4 The strain gage shown is used to determine the gage pressure in the cylindrical steel [$E = 200$ GPa; $\nu = 0.32$] tank. The tank has a specified outside diameter and wall thickness. Determine the strain gage reading for a specified internal tank pressure.

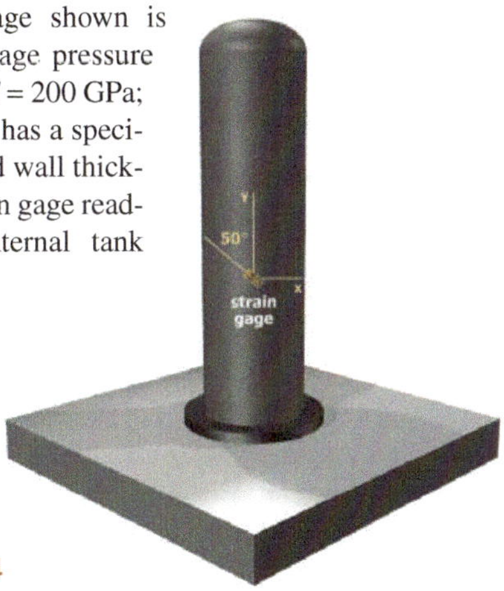

FIGURE M14.4

M14.5 A strain gage is used to monitor the strain in a spherical steel [$E = 210$ GPa; $\nu = 0.32$] tank, that contains a fluid under pressure. The ultimate shear strength of the steel is 560 MPa. Determine the factor of safety with respect to the ultimate shear strength if the strain gage reading is a specified value.

FIGURE M14.5

PROBLEMS

P14.1 Determine the normal stress in a ball (Figure P14.1) that has an outside diameter of 185 mm and a wall thickness of 3 mm when the ball is inflated to a gage pressure of 80 kPa.

FIGURE P14.1

P14.2 A spherical gas-storage tank with an inside diameter of 21 ft is being constructed to store gas under an internal pressure of 160 psi. The tank will be constructed from steel that has a yield strength of 50 ksi. If a factor of safety of 3.0 with respect to the yield strength is required, determine the minimum wall thickness required for the spherical tank.

P14.3 A spherical pressure vessel has an inside diameter of 3.25 m and a wall thickness of 8 mm. The tank will be constructed from structural steel that has a yield strength of 370 MPa. If a factor of safety of 3.0 with respect to the yield strength is required, determine the maximum allowable internal pressure.

P14.4 A spherical pressure vessel has an inside diameter of 6 m and a wall thickness of 15 mm. The vessel will be constructed from steel [$E = 200$ GPa; $\nu = 0.29$] that has a yield strength of 340 MPa. If the internal pressure in the vessel is 1,750 kPa, determine (a) the normal stress in the vessel wall, (b) the factor of safety with respect to the yield strength, (c) the normal strain in the sphere, and (d) the increase in the outside diameter of the vessel.

P14.5 The normal strain measured on the outside surface of a spherical pressure vessel is 670 με. The sphere has an outside diameter of 1.20 m and a wall thickness of 10 mm, and it will be fabricated from an aluminum alloy [E = 73 GPa; ν = 0.33]. Determine (a) the normal stress in the vessel wall and (b) the internal pressure in the vessel.

P14.6 A typical aluminum-alloy scuba-diving tank is shown in Figure P14.6. The outside diameter of the tank is 7.20 in. and the wall thickness is 0.55 in. If the air in the tank is pressurized to 3,500 psi, determine

(a) the longitudinal and hoop stresses in the wall of the tank.
(b) the maximum shear stress in the plane of the cylinder wall.
(c) the absolute maximum shear stress on the outer surface of the cylinder wall.

FIGURE P14.6

P14.7 A cylindrical boiler with an outside diameter of 2.75 m and a wall thickness of 32 mm is made of a steel alloy that has a yield stress of 340 MPa. Determine

(a) the maximum normal stress produced by an internal pressure of 2.3 MPa.
(b) the maximum allowable pressure if a factor of safety of 2.5 with respect to yield is required.

P14.8 When filled to capacity, the unpressurized storage tank shown in Figure P14.8 contains water to a height of h = 30 ft. The outside diameter of the tank is 12 ft and the wall thickness is 0.375 in. Determine the maximum normal stress and the absolute maximum shear stress on the outer surface of the tank at its base. (Weight density of water = 62.4 lb/ft^3.)

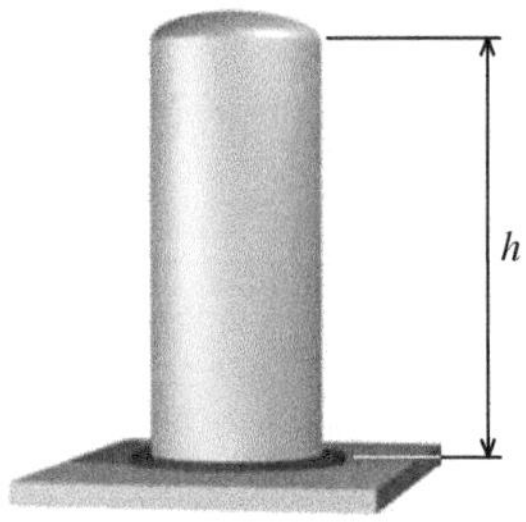

FIGURE P14.8

P14.9 A tall open-topped standpipe (Figure P14.9) has an inside diameter of 2,750 mm and a wall thickness of 6 mm. The standpipe contains water, which has a mass density of 1,000 kg/m^3.

(a) What height h of water will produce a circumferential stress of 16 MPa in the wall of the standpipe?
(b) What is the axial stress in the wall of the standpipe due to the water pressure?

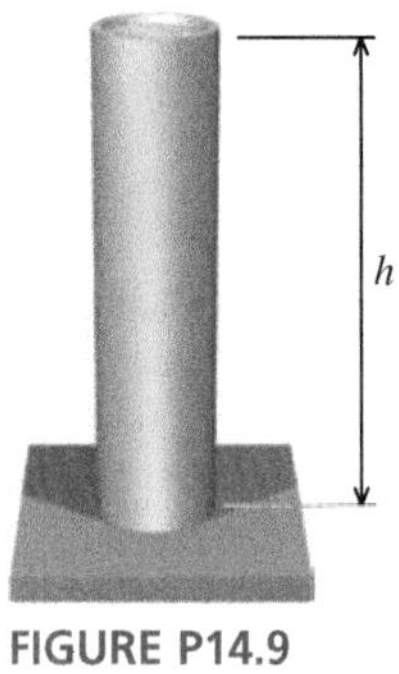

FIGURE P14.9

P14.10 The pressure tank in Figure P14.10/11 is fabricated from spirally wrapped metal plates that are welded at the seams in the orientation shown, where β = 50°. The tank has an inside diameter of 28.0 in. and a wall thickness of 0.375 in. Determine the largest gage pressure that can be used inside the tank if the allowable normal stress perpendicular to the weld is 18 ksi and the allowable shear stress parallel to the weld is 10 ksi.

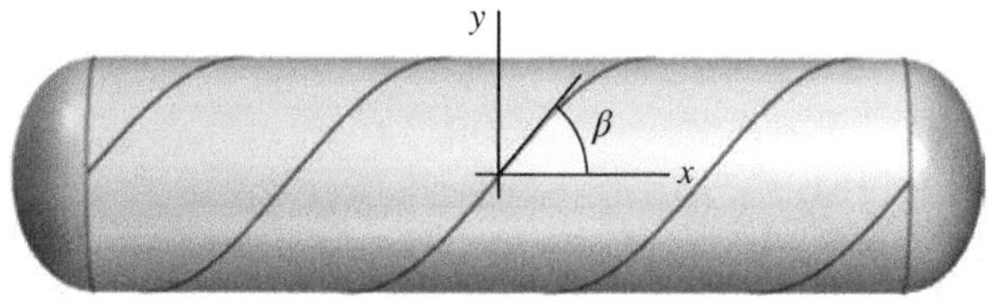

FIGURE P14.10/11

P14.11 The pressure tank in Figure P14.10/11 is fabricated from spirally wrapped metal plates that are welded at the seams in the orientation shown, where β = 40°. The tank has an inside diameter of 720 mm and a wall thickness of 8 mm. For a gage pressure of 2.15 MPa, determine (a) the normal stress perpendicular to the weld and (b) the shear stress parallel to the weld.

P14.12 The pressure tank in Figure P14.12/13 is fabricated from spirally wrapped metal plates that are welded at the seams in the orientation shown, where β = 35°. The tank has an outside diameter of 30.0 in. and a wall thickness of 7/16 in. For a gage pressure of 500 psi, determine (a) the normal stress perpendicular to the weld and (b) the shear stress parallel to the weld.

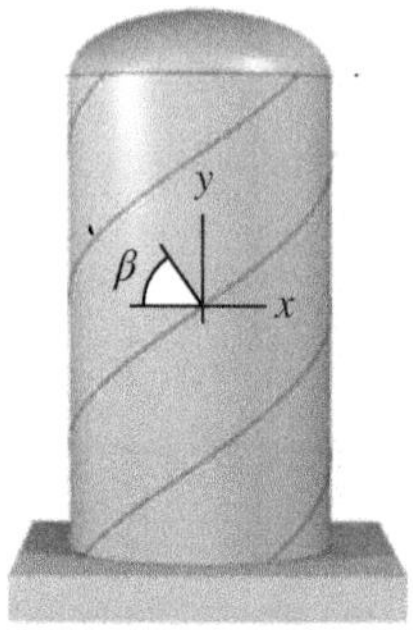

FIGURE P14.12/13

P14.13 The pressure tank in Figure P14.12/13 is fabricated from spirally wrapped metal plates that are welded at the seams in the orientation shown, where β = 55°. The tank has an inside diameter of 60 in. and a wall thickness of 0.25 in. Determine the largest allowable gage pressure if the allowable normal stress perpendicular to the weld is 12 ksi and the allowable shear stress parallel to the weld is 7 ksi.

P14.14 A strain gage is mounted to the outer surface of a thin-walled boiler as shown in Figure P14.14. The boiler has an inside diameter of 1,800 mm and a wall thickness of 20 mm, and it is made of stainless steel [E = 193 GPa; ν = 0.27]. Determine

(a) the internal pressure in the boiler when the strain gage reads 190 με.
(b) the maximum shear strain in the plane of the boiler wall.
(c) the absolute maximum shear strain on the outer surface of the boiler.

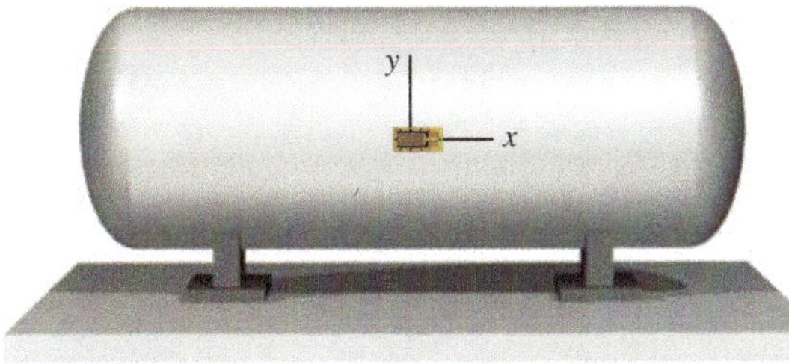

FIGURE P14.14

P14.15 A closed cylindrical tank containing a pressurized fluid has an inside diameter of 830 mm and a wall thickness of 10 mm. The stresses in the wall of the tank acting on a rotated element have the values shown in Figure P14.15. What is the fluid pressure in the tank?

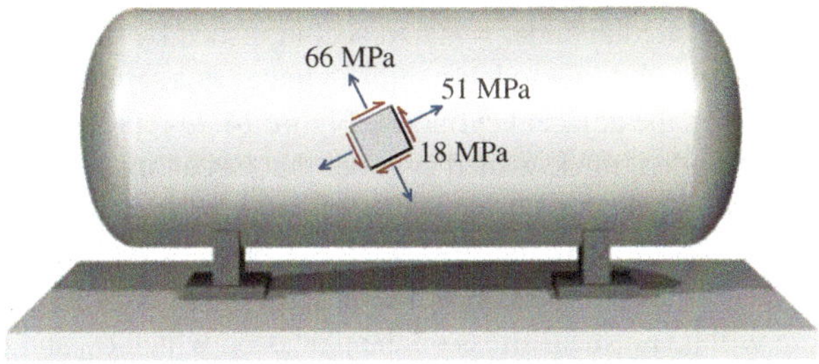

FIGURE P14.15

P14.16 A closed cylindrical vessel (Figure P14.16) contains a fluid at a pressure of 5.0 MPa. The cylinder, which has an outside diameter of 2,500 mm and a wall thickness of 20 mm, is fabricated from stainless steel [E = 193 GPa; ν = 0.27]. Determine the increase in both the diameter and the length of the cylinder.

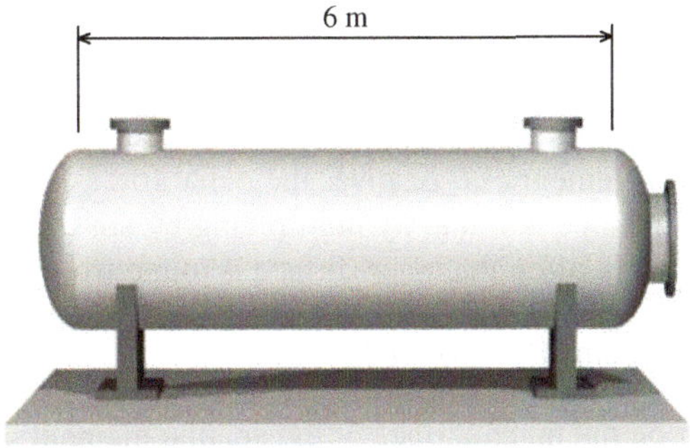

FIGURE P14.16

P14.17 A strain gage is mounted at an angle of θ = 25° with respect to the longitudinal axis of the cylindrical pressure vessel shown in Figure P14.17/18. The pressure vessel is fabricated from aluminum [E = 73 GPa; ν = 0.33], and it has an inside diameter of 1.6 m and a wall thickness of 8 mm. If the strain gage measures a normal strain of 885 με, determine

(a) the internal pressure in the cylinder.
(b) the absolute maximum shear stress on the outer surface of the cylinder.
(c) the absolute maximum shear stress on the inner surface of the cylinder.

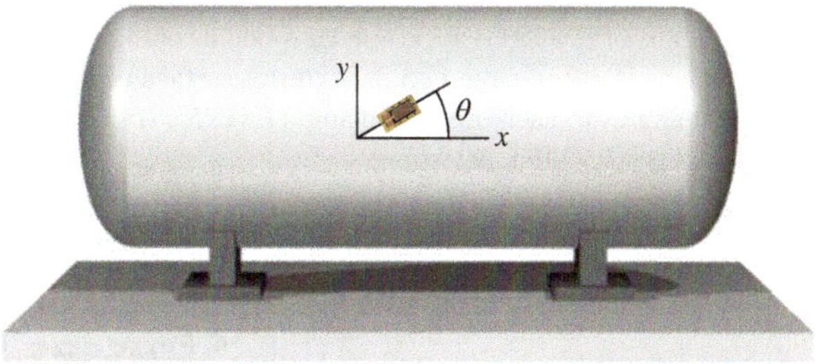

FIGURE P14.17/18

P14.18 A strain gage is mounted at an angle of θ = 30° with respect to the longitudinal axis of the cylindrical pressure vessel shown in Figure P14.17/18. The pressure vessel is fabricated from steel [E = 30,000 ksi; ν = 0.30], and it has an outside diameter of 37 in. and a wall thickness of 3/16 in. If the internal pressure in the cylinder is 200 psi, determine

(a) the expected strain gage reading (in με).
(b) the principal strains, the maximum shear strain, and the absolute maximum shear strain on the outer surface of the cylinder.

P14.19 The pressure vessel shown in Figure P14.19 consists of spirally wrapped steel plates that are welded at the seams in the orientation shown, where β = 35°. The cylinder has an inside diameter of 16 in. and a wall thickness of 0.375 in. The ends of the cylinder are capped by two rigid end plates. The gage pressure inside the cylinder is 260 psi, and tensile axial loads of P = 15 kips are applied to the rigid end caps. Determine

(a) the normal stress perpendicular to the weld seams.
(b) the shear stress parallel to the weld seams.
(c) the absolute maximum shear stress on the outer surface of the cylinder.

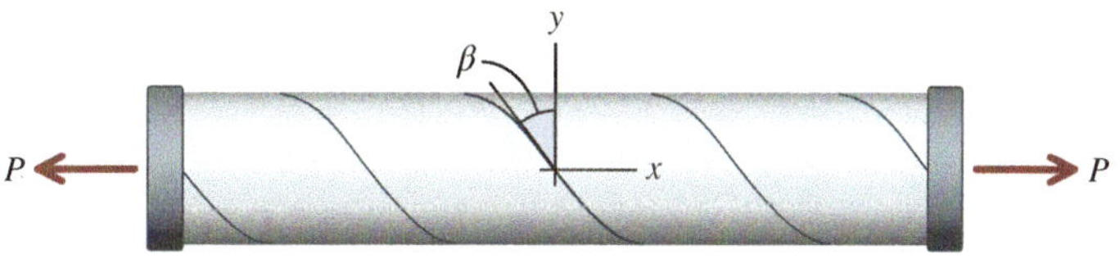

FIGURE P14.19

P14.20 The cylindrical pressure vessel shown in Figure P14.20/21 has an inside diameter of 610 mm and a wall thickness of 3 mm. The cylinder is made of an aluminum alloy that has an elastic modulus of $E = 70$ GPa and a shear modulus of $G = 26.3$ GPa. Two strain gages are mounted on the exterior surface of the cylinder at right angles to each other; however, the angle θ is not known. If the strains measured by the two gages are $\varepsilon_a = 360$ με and $\varepsilon_b = 975$ με, respectively, what is the pressure in the vessel? Notice that, when two orthogonal strains are measured, the angle θ is not needed to determine the normal stresses.

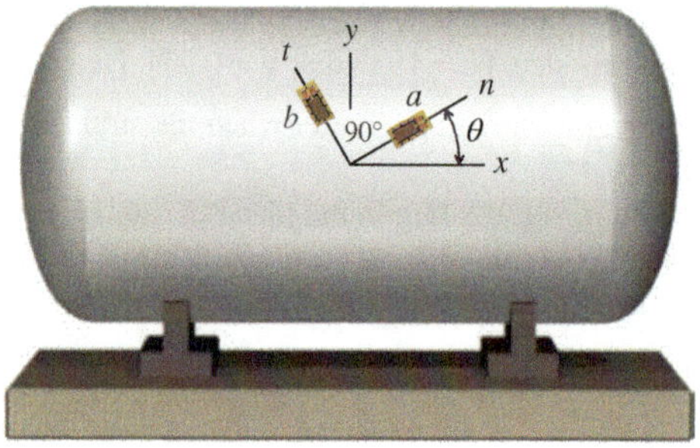

FIGURE P14.20/21

P14.21 The cylindrical pressure vessel shown in Figure P14.20/21 has an inside diameter of 900 mm and a wall thickness of 12 mm. The cylinder is made of an aluminum alloy that has an elastic modulus of $E = 70$ GPa and a shear modulus of $G = 26.3$ GPa. Two strain gages are mounted on the exterior surface of the cylinder at right angles to each other. The angle θ is 25°. If the pressure in the vessel is 1.75 MPa, determine

(a) the strains that act in the x and y directions.
(b) the strains expected in gages a and b.
(c) the normal stresses σ_n and σ_t.
(d) the shear stress τ_{nt}.

14.5 Stresses in Thick-Walled Cylinders

Equilibrium

The determination of stresses in a thick-walled cylinder was first accomplished by the French mathematician Gabriel Lamé in 1833. His solution models a thick cylinder as a series of thin, adjacent cylinders that each exert stress on the adjacent cylinders. Figure 14.9*a* represents the cross section of a relatively long, open-ended, thick-walled cylinder that has an inside radius a and an outside radius b. The cylinder is subjected to an inside pressure p_i and an outside pressure p_o. The circumferential stress σ_θ and the radial stress σ_r are to be determined at any radial distance r from the central axis O of the cylinder. Because of the symmetry of the cylinder about the central axis O, there will be no shear stress on the planes on which σ_θ and σ_r act. Consequently, the circumferential and radial stresses are principal stresses.

A free-body diagram that exposes both the circumferential and radial normal stresses is required. To expose the radial stresses, we will consider a thin cylinder of inner radius r, thickness dr, and arbitrary longitudinal length Δx. The circumferential stresses are exposed by cutting this cylinder in half with a vertical diametric plane. The free-body diagram shown in Figure 14.9*b* shows the forces acting on the annular element expressed in terms of σ_θ and σ_r. The condition of symmetry dictates that σ_θ and σ_r are independent of θ. Both σ_θ and σ_r are assumed to be tensile normal stresses.

In the lateral direction (i.e., the z direction), the resultant force due to the radial stress σ_r acting on the inner surface of the annular element is to be determined. Consider a differential element located at θ that has dimensions of dr in the radial direction, $r\,d\theta$ in the circumferential direction, and Δz in the longitudinal direction. The product of the radial stress σ_r and the area of the differential element $r\,d\theta\,\Delta z$ gives the radial force on the inner surface. The z component of this radial force is found by multiplying by $\sin\theta$. Thus, the resultant force in the z direction created by σ_r on the inner surface is found from

$$\int_0^\pi (r\Delta x)\sigma_r \sin\theta\, d\theta = \sigma_r 2r\Delta x$$

Similarly, the radial force acting on the outer surface of the annular element is found from the radial stress $\sigma_r + d\sigma_r$ acting on the area of the differential element $(r + dr)\,d\theta\,\Delta z$.

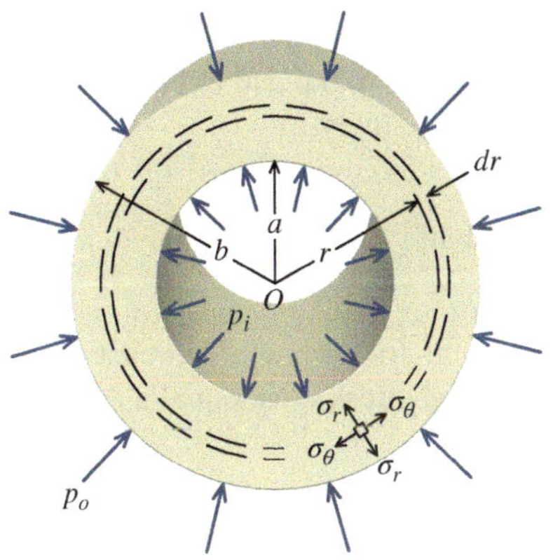

FIGURE 14.9*a* Cross section of a thick-walled cylinder.

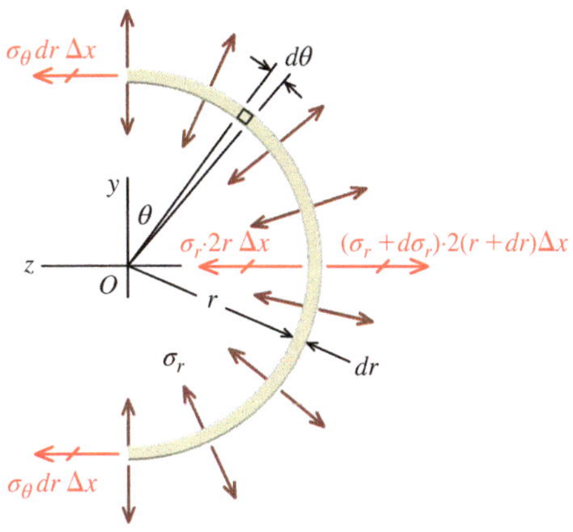

FIGURE 14.9*b* Free-body diagram exposing circumferential and radial normal stress.

The resultant force in the z direction created by $\sigma_r + d\sigma_r$ on the outer surface of the annular element can be expressed as

$$\int_0^{\pi} (r + dr)\Delta x(\sigma_r + d\sigma_r)\sin\theta\, d\theta = (\sigma_r + d\sigma_r)2(r + dr)\Delta x$$

Note that the thickness dr of the annular element and the incremental radial stress $d\sigma_r$ are independent of θ and are thus constants in this integration.

The circumferential normal stress σ_θ acts on the exposed z surfaces (i.e., on the y axis) at the top and bottom of the annular element. The area exposed at each of these sections is $dr\, \Delta x$. Since the thickness dr may be made infinitesimally small, σ_θ may be considered constant over the thickness of the element. Thus, the resultant force that is created by σ_θ acting in the z direction on each exposed surface is $\sigma_\theta\, dr\, \Delta x$.

The sum of all forces in the z direction can be expressed as

$$\sum F_z = 2\sigma_\theta dr\,\Delta x + \sigma_r \cdot 2r\,\Delta x - (\sigma_r + d\sigma_r)2(r + dr)\Delta x = 0$$

Since $2\Delta x$ appears in each term, it can be cancelled out. Further, the second-order term $d\sigma_r\, dr$ is negligibly small. Therefore, the relationship between the radial and circumferential stresses can be expressed as a function of the radius r as

$$\frac{d\sigma_r}{dr} + \frac{\sigma_r - \sigma_\theta}{r} = 0 \qquad (14.12)$$

Lamé Solution

It is reasonable to assume that strains in the longitudinal direction (i.e., the x direction in Figure 14.9*b*) of the thick-walled cylinder are equal. This means that plane transverse sections which are flat and parallel before pressures p_i and p_o are applied remain flat and parallel after the pressures are applied. This assumption will be true for at least a cylinder with open ends, and it will also be nearly true for a closed cylinder at sections well removed from the ends of the cylinder[1].

[1] Seely, Fred B., and James O. Smith. *Advanced Mechanics of Materials*, 2nd ed. (New York: John Wiley & Sons, Inc., 1952.)

Since the ends of the cylinder are open, the longitudinal stress $\sigma_{\text{long}} = \sigma_x = 0$. Therefore, a condition of plane stress exists in the cylinder wall. From the generalized Hooke's law, the relationship between the longitudinal strain $\varepsilon_{\text{long}}$ and the radial and circumferential stresses can be expressed as that for an open-ended cylinder, namely,

$$\varepsilon_{\text{long}} = -\frac{\nu}{E}(\sigma_r + \sigma_\theta) = \text{constant}$$

Poisson's ratio ν and the modulus of elasticity E are constants for the material constituting the cylinder. Hence, the sum of the radial and circumferential stress must equal a constant value. For convenience, we will denote this constant as 2α:

$$\sigma_r + \sigma_\theta = \text{constant} = 2\alpha$$

The circumferential stress can be expressed as

$$\sigma_\theta = 2\alpha - \sigma_r \tag{a}$$

The expression for σ_θ from Equation (a) can be substituted into Equation (14.12) to give

$$\frac{d\sigma_r}{dr} + 2\frac{\sigma_r - \alpha}{r} = 0 \tag{b}$$

Equation (b) may be written in the form

$$\frac{d\sigma_r}{\sigma_r - \alpha} = -2\frac{dr}{r} \tag{c}$$

Integration of Equation (c) yields

$$\ln(\sigma_r - \alpha) = -2\ln r + \ln\beta = \ln\frac{\beta}{r^2}$$

where β is a constant of integration. After exponentiation, we can express the radial stress as

$$\sigma_r = \alpha + \frac{\beta}{r^2} \tag{d}$$

The constants α and β are evaluated in terms of the pressures at the inner and outer surfaces, respectively, of the cylinder. On the inner surface at $r = a$, the radial normal stress is equal to the magnitude of the internal pressure; that is, $\sigma_r = -p_i$. The minus sign is used to convert a positive pressure to a compressive normal stress. On the outer surface at $r = b$, the radial normal stress is equal to the magnitude of the external pressure; that is, $\sigma_r = -p_o$. From these known boundary conditions, the values of α and β can be determined as

$$\begin{aligned} \alpha &= \frac{a^2 p_i - b^2 p_o}{b^2 - a^2} \\ \beta &= -\frac{a^2 b^2 (p_i - p_o)}{b^2 - a^2} \end{aligned} \tag{e}$$

The radial stress in the thick-walled cylinder is obtained by substituting the constants in Equation (e) into Equation (d):

$$\sigma_r = \frac{a^2 p_i - b^2 p_o}{b^2 - a^2} - \frac{a^2 b^2 (p_i - p_o)}{(b^2 - a^2) r^2} \tag{14.13}$$

Separating Equation (14.13) into portions pertaining to the internal pressure p_i and the external pressure p_o yields

$$\sigma_r = \frac{a^2 p_i}{b^2 - a^2}\left(1 - \frac{b^2}{r^2}\right) - \frac{b^2 p_o}{b^2 - a^2}\left(1 - \frac{a^2}{r^2}\right) \tag{14.14}$$

The circumferential stress in the thick-walled cylinder is obtained by substituting the constants in Equation (e) into Equation (a):

$$\sigma_\theta = \frac{a^2 p_i - b^2 p_o}{b^2 - a^2} + \frac{a^2 b^2 (p_i - p_o)}{(b^2 - a^2) r^2} \tag{14.15}$$

Separating Equation (14.15) as we did Equation (14.13) gives

$$\sigma_\theta = \frac{a^2 p_i}{b^2 - a^2}\left(1 + \frac{b^2}{r^2}\right) - \frac{b^2 p_o}{b^2 - a^2}\left(1 + \frac{a^2}{r^2}\right) \tag{14.16}$$

The stress equations (14.13) through (14.16) are together known as the Lamé solution.

Equations (14.13) through (14.16) can also be stated in terms of $(b/a)^2$ by dividing numerators and denominators by a^2. From the resulting equation, we observe that the radial and circumferential stresses depend on the ratio b/a of the outside radius b to the inside radius a of the cylinder, rather than on its absolute size.

Maximum Stresses when Internal and External Pressures Act Together

The maximum stresses occur either on the inner surface or the outer surface of a thick-walled cylinder.

If the internal pressure is greater than the external pressure, then the maximum radial stress occurs on the inner surface at $r = a$:

$$(\sigma_r)_{\max} = -p_i \tag{14.17}$$

So does the maximum circumferential stress:

$$(\sigma_\theta)_{\max} = \frac{(b^2 + a^2) p_i - 2b^2 p_o}{b^2 - a^2} \tag{14.18}$$

The maximum circumferential stress will be tensile.

If the external pressure is greater than the internal pressure, then the maximum radial stress occurs on the outer surface at $r = b$:

$$(\sigma_r)_{\max} = -p_o \tag{14.19}$$

The maximum circumferential stress still occurs on the inner surface at $r = a$; however, the circumferential stress will be compressive rather than tensile.

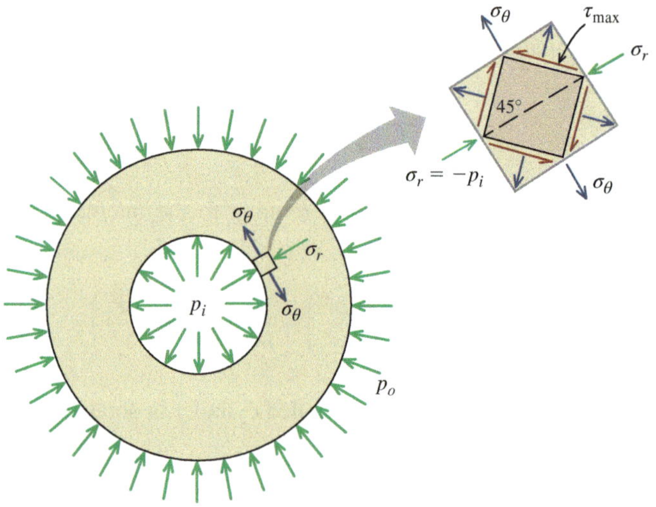

FIGURE 14.10 Maximum shear stress orientation in a thick-walled cylinder.

Maximum shear stress: Recall that the maximum shear stress at any point in a solid material equals one-half of the algebraic difference between the maximum and minimum principal stresses, which are the circumferential and radial stresses for the thick-walled cylinder. Thus, at any point in the cylinder, we find that the maximum shear stress is

$$\tau_{max} = \frac{1}{2}(\sigma_\theta - \sigma_r) = \frac{a^2 b^2 (p_i - p_o)}{(b^2 - a^2) r^2} \tag{14.20}$$

The largest value of τ_{max} is found at the inner surface where $r = a$. The effect of reducing the outside pressure p_o is clearly to increase τ_{max}. Consequently, the largest shear stress corresponds to $r = a$ and $p_o = 0$. Because σ_θ and σ_r are principal stresses, τ_{max} occurs on planes making an angle of 45° with the planes on which σ_θ and σ_r act. The orientation of planes of maximum shear stress is shown in Figure 14.10.

Relationships between inner surface and outer surface stresses: If we evaluate the circumferential stress σ_θ on the inside of a thick-walled cylinder at $r = a$, we find that

$$\sigma_{\theta i} = \frac{b^2 + a^2}{b^2 - a^2} p_i - \frac{2b^2}{b^2 - a^2} p_o$$

Next, we will evaluate σ_θ on the outside of the cylinder at $r = b$ to obtain

$$\sigma_{\theta o} = \frac{2a^2}{b^2 - a^2} p_i - \frac{b^2 + a^2}{b^2 - a^2} p_o$$

If $\sigma_{\theta o}$ is subtracted from $\sigma_{\theta i}$, we discover that the change in circumferential stresses from the inner surface to the outer surface always equals the difference between the internal and external pressures:

$$\sigma_{\theta i} - \sigma_{\theta o} = \frac{(b^2 - a^2) p_i - (b^2 - a^2) p_o}{b^2 - a^2} = p_i - p_o \tag{14.21}$$

Similarly, the change in radial stresses from the inner surface to the outer surface always equals the difference between the external and internal pressures:

$$\sigma_{ri} - \sigma_{ro} = \frac{-(b^2 - a^2)p_i + (b^2 - a^2)p_o}{b^2 - a^2} = p_o - p_i \tag{14.22}$$

Maximum Stresses for Internal Pressure Only

If the internal pressure is p_i and the external pressure is zero ($p_o = 0$), as in most hydraulic pipes, tanks, and other machinery, then Equations (14.14) and (14.16) respectively reduce to

$$\sigma_r = \frac{a^2 p_i}{b^2 - a^2}\left(1 - \frac{b^2}{r^2}\right) \tag{14.23}$$

and

$$\sigma_\theta = \frac{a^2 p_i}{b^2 - a^2}\left(1 + \frac{b^2}{r^2}\right) \tag{14.24}$$

The radial and circumferential normal stresses for this case are plotted in Figure 14.11*a*. The maximum radial stress occurs at $r = a$. Since $b^2/r^2 \geq 1$, the radial stress σ_r is negative (i.e., compressive) for all values of r except $r = b$, in which case the radial stress is zero. The circumferential stress σ_θ is positive (i.e., tensile) for all radii. Its maximum value also occurs at $r = a$.

Comparison of circumferential stresses in thick- and thin-walled cylinders subjected to internal pressure only: The average circumferential stress (i.e., hoop stress) in a thin-walled cylinder is given by Equation (14.5):

$$\sigma_{\text{hoop}} = \frac{pd}{2t}$$

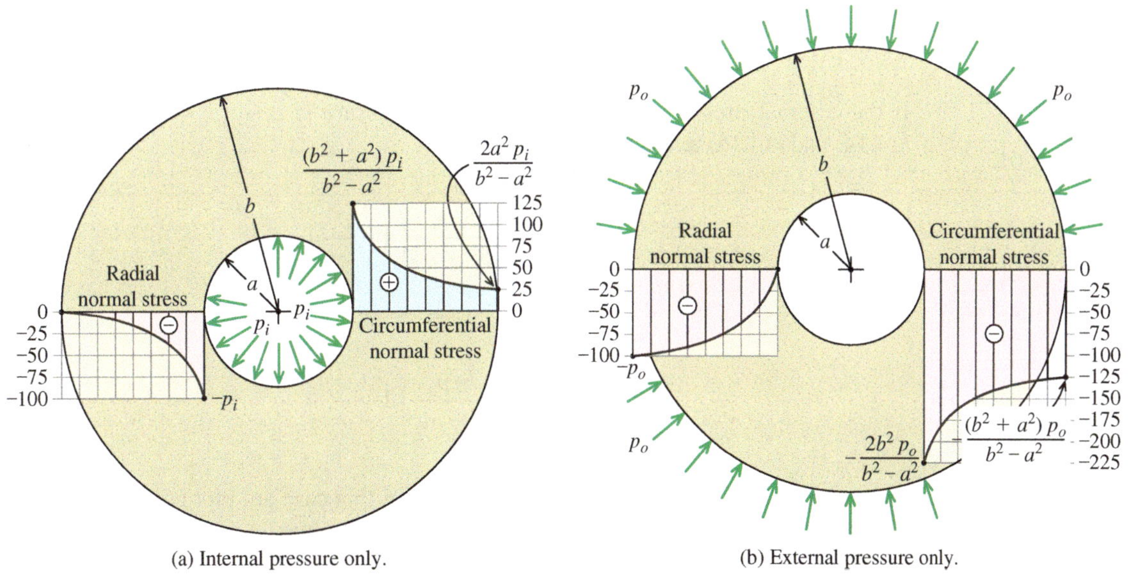

FIGURE 14.11 Radial and circumferential normal stresses in thick-walled cylinders subjected to internal and external pressure.

For convenience, let the ratio of the internal diameter to the wall thickness be denoted by the parameter k:

$$k = \frac{d}{t} = \frac{2a}{b - a} \tag{f}$$

Thus, the thin-walled circumferential stress can be expressed as

$$\sigma_{\text{hoop}} = \frac{p_i k}{2}$$

and the thick-walled circumferential stress can be expressed as

$$\sigma_\theta = \frac{p_i(k^2 + 2k + 2)}{2(k + 1)}$$

The percent error of the thin-walled average hoop stress relative to the thick-walled circumferential stress is therefore

$$\%\text{error} = \frac{\sigma_{\text{hoop}} - \sigma_\theta}{\sigma_\theta} \times 100\% = \frac{k + 2}{k^2 + 2k + 2} \times 100\% \tag{g}$$

Table 14.1 Comparison of Thin- and Thick-Walled Circumferential Stress

$\frac{d}{t}$	% error
1	60.0%
5	18.9%
10	9.8%
20	5.0%
50	2.0%
100	1.0%

Table 14.1 lists the percent error obtained from Equation (g) for a number of diameter-to-thickness ratios. This calculation shows that, when the internal diameter of a cylinder is at least 20 times the wall thickness, the average circumferential stress calculated by Equation (14.5) is approximately 5 percent less than the thick-walled stress value calculated from Equation (14.24). Thus, Equations (14.4) and (14.5) are typically appropriate for thin-walled pressure vessels when $d/t \geq 20$. The thick-walled stress equations should be used for cylinders when $d/t < 20$.

Maximum Stresses for External Pressure Only

If the external pressure is p_o and the internal pressure is zero ($p_i = 0$), then Equations (14.14) and (14.16) respectively reduce to

$$\sigma_r = -\frac{b^2 p_o}{b^2 - a^2}\left(1 - \frac{a^2}{r^2}\right) \tag{14.25}$$

and

$$\sigma_\theta = -\frac{b^2 p_o}{b^2 - a^2}\left(1 + \frac{a^2}{r^2}\right) \tag{14.26}$$

The radial and circumferential normal stresses for this case are plotted in Figure 14.11*b*. Note that, both σ_r and σ_θ are always compressive. The maximum radial stress occurs on the outer surface of the cylinder. However, the maximum circumferential stress occurs on the inner surface of the cylinder. The circumferential stress magnitude is always larger than the radial stress magnitude.

External Pressure on a Solid Circular Cylinder

Radial and circumferential stresses σ_r and σ_θ for this special case can be obtained from Equations (14.25) and (14.26) by letting the inside radius vanish ($a = 0$). Thus,

$$\sigma_r = -p_o$$

and

$$\sigma_\theta = -p_o$$

Both σ_r and σ_θ are always compressive. For this case, the stresses are independent of the radial position r and have a constant magnitude equal to the applied pressure.

Longitudinal Stress in Cylinder with Closed Ends

If the thick-walled cylinder has closed or capped ends and is subjected to an internal pressure p_i and an external pressure p_o, the longitudinal (i.e., acting in the x-direction in Figure 14.9b) stresses σ_{long} exist in addition to the radial and circumferential stresses. For a transverse section some distance from the ends, this stress may be assumed to be uniformly distributed over the wall thickness.

The magnitude of σ_{long} is determined from (a) the resultant force created by the internal pressure acting on the inside cross-sectional area, (b) the resultant force created by the external pressure acting on the outside cross-sectional area, and (c) the resultant x-directed longitudinal force in the cylinder wall. The external pressure resultant (b) and the resultant longitudinal force (c) act in opposition to the internal pressure resultant (a). For equilibrium as shown in Figure 14.12, the sum of forces in the longitudinal direction gives

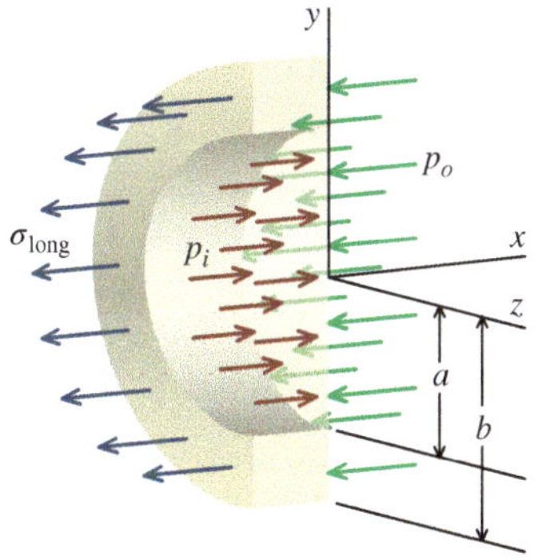

Note: Radial pressures omitted for clarity.

FIGURE 14.12 Longitudinal stress in a closed cylinder.

$$\Sigma F_x = \pi a^2 p_i - \pi b^2 p_o - \pi(b^2 - a^2)\sigma_{\text{long}} = 0$$

The resulting expression for the longitudinal stress is

$$\sigma_{\text{long}} = \frac{a^2 p_i - b^2 p_o}{b^2 - a^2} \tag{14.27}$$

End caps produce localized disturbances in stresses σ_r, σ_θ, and σ_{long}. The stress equations developed in Equations (14.13), (14.15), and (14.27) are accurate at a distance from the end caps roughly equal to the thickness $b - a$ when the cylinder wall is thick. Away from the ends, σ_{long} is not influenced by σ_r and σ_θ, nor does the presence of a constant σ_{long} alter the Lamé solution presented in Equations (14.13) and (14.15).[2] It is also assumed that the ends of the cylinder are not constrained; that is, $\varepsilon_x \neq 0$. Thus, in a closed cylinder, there are three principal stresses at any point in the cylinder wall, namely, σ_r, σ_θ, and σ_{long}.

In addition, notice that the expression given in Equation (14.27) is equal to one-half of the sum of Equations (14.13) and (14.15):

$$\sigma_{\text{long}} = \frac{1}{2}(\sigma_r + \sigma_\theta)$$

Interestingly, Equation (14.27) is equal to the Lamé constant α that was determined in Equation (e).

[2] Cook, Robert D., and Warren C. Young. *Advanced Mechanics of Materials*. (New York: Macmillan Publishing Co., 1985.)

14.6 Deformations in Thick-Walled Cylinders

Radial and circumferential deformations δ_r and δ_θ are important considerations in interference-fit connections such as the attachment of a gear to a shaft. When a thick-walled cylinder is subjected to internal pressure p_i or external pressure p_o, the circumference of the thin ring shown in Figure 14.9*a* will either elongate or contract. This deformation may be expressed in terms of the radial displacement δ_r of a point on the ring as

$$\delta_\theta = 2\pi\delta_r$$

The circumferential deformation δ_θ may also be expressed in terms of the circumferential strain ε_θ as

$$\delta_\theta = \varepsilon_\theta c$$

where $c = 2\pi r$ is the circumference of the thin ring. Thus,

$$\delta_r = \varepsilon_\theta r$$

For most interference-fit connections, the cylinder is open, which means that the longitudinal stress in the cylinder is $\sigma_{\text{long}} = 0$. As a result, a state of plane stress exists in the cylinder wall. Using the generalized Hooke's law for plane stress, we can express the circumferential strain in terms of the radial stress σ_r and the circumferential stress σ_θ:

$$\varepsilon_\theta = \frac{1}{E}(\sigma_\theta - \nu\sigma_r)$$

The radial displacement of a point in the cylinder wall is then obtained in terms of the radial and tangential stresses at that same point:

$$\delta_r = \frac{r}{E}(\sigma_\theta - \nu\sigma_r) \tag{14.28}$$

Radial Displacement for Internal Pressure Only

If the thick-walled cylinder is loaded only by an internal pressure, Equations (14.23) and (14.24) can be substituted into Equation (14.28) to obtain the following expression for the radial deformation:

$$\delta_r = \frac{a^2 p_i}{(b^2 - a^2)rE}[(1-\nu)r^2 + (1+\nu)b^2] \tag{14.29}$$

Radial Displacement for External Pressure Only

If only an external pressure acts on a thick-walled cylinder, Equations (14.25) and (14.26) can be substituted into Equation (14.28) to obtain the following expression for the radial deformation:

$$\delta_r = -\frac{b^2 p_o}{(b^2 - a^2)rE}[(1-\nu)r^2 + (1+\nu)a^2] \tag{14.30}$$

Radial Displacement for External Pressure on Solid Circular Cylinder

For a solid circular cylinder loaded by an external pressure, the radial and circumferential stresses equal the external pressure and both stresses are compressive. Thus, Equation (14.28) yields the following expression for the radial deformation:

$$\delta_r = -\frac{(1-\nu)p_o r}{E} \tag{14.31}$$

EXAMPLE 14.3

An open-ended high-pressure steel [$E = 200$ GPa, $\nu = 0.3$] tube for a boiler has an outside diameter of 273 mm and a wall thickness of 32 mm. The inside pressure in the boiler tube is $p_i = 45$ MPa.

(a) Determine the largest radial and circumferential stresses in the tube.
(b) Determine the maximum shear stress in the tube.
(c) Calculate the increase in the inside and outside diameters of the tube after the internal pressure has been applied.
(d) If the allowable normal stress in the tube material is 140 MPa, determine the minimum wall thickness required for the boiler tube with an outside diameter of 273 mm.

Plan the Solution

Since the cylinder is subjected to internal pressure only, Equations (14.23) and (14.24) will be used to compute the normal stresses. From Figure 14.11, we know that the stress with the largest magnitude will occur on the inner surface of the boiler tube. The radial stress will be compressive while the circumferential stress will be tensile. These two stresses are principal stresses. Since the tube is open ended, the longitudinal stress is zero; thus, the third principal stress is zero and a state of plane stress exists in the tube. Referring to Figure 14.12, we recognize (a) that the maximum shear stress can be determined from the two principal stresses and (b) that the planes upon which the maximum shear stress acts are oriented 45° away from the radial and circumferential planes.

SOLUTION

The outside diameter of the tube is $D = 273$ mm and the wall thickness is $t = 32$ mm. The inside diameter is thus 209 mm. Accordingly, the inner radius is $a = 104.5$ mm and the outer radius is $b = 136.5$ mm.

Radial and Circumferential Stresses

For internal pressure only, the radial and circumferential stresses will be largest on the inner surface of the tube at $r = 104.5$ mm. Therefore,

$$\sigma_r = \frac{a^2 p_i}{b^2 - a^2}\left(1 - \frac{b^2}{r^2}\right)$$

$$= \frac{(104.5\text{ mm})^2(45\text{ MPa})}{(136.5\text{ mm})^2 - (104.5\text{ mm})^2}\left[1 - \frac{(136.5\text{ mm})^2}{(104.5\text{ mm})^2}\right] = -45\text{ MPa} \qquad \textbf{Ans.}$$

$$\sigma_\theta = \frac{a^2 p_i}{b^2 - a^2}\left(1 + \frac{b^2}{r^2}\right)$$

$$= \frac{(104.5\text{ mm})^2(45\text{ MPa})}{(136.5\text{ mm})^2 - (104.5\text{ mm})^2}\left[1 + \frac{(136.5\text{ mm})^2}{(104.5\text{ mm})^2}\right] = 172.44\text{ MPa} \qquad \textbf{Ans.}$$

Maximum Shear Stress

The principal stresses are $\sigma_{p1} = \sigma_\theta = 172.44$ MPa and $\sigma_{p2} = \sigma_r = -45$ MPa. The maximum in-plane shear stress is thus

$$\tau_{max} = \frac{\sigma_{p1} - \sigma_{p2}}{2} = \frac{172.44 \text{ MPa} - (-45 \text{ MPa})}{2} = 108.72 \text{ MPa} \qquad \textbf{Ans.}$$

Since the tube is open ended, the longitudinal stress is zero. Therefore, $\sigma_{p3} = \sigma_{long} = 0$, and it follows that $\tau_{abs\,max} = \tau_{max}$.

Increase in Inside and Outside Diameters

The radial deformation for the case of internal pressure only is given by Equation (14.29).

The deformation of the inside radius is calculated with $r = a = 104.5$ mm:

$$\begin{aligned}\delta_r &= \frac{a^2 p_i}{(b^2 - a^2)rE}[(1-\nu)r^2 + (1+\nu)b^2] \\ &= \frac{(104.5 \text{ mm})^2(45 \text{ MPa})}{[(136.5 \text{ mm})^2 - (104.5 \text{ mm})^2](104.5 \text{ mm})(200{,}000 \text{ MPa})}[(1-0.3)(104.5 \text{ mm})^2 + (1+0.3)(136.5 \text{ mm})^2] \\ &= 0.0750 \text{ mm}\end{aligned}$$

The inside diameter increases by $2\delta_r = 0.150$ mm. **Ans.**

Use $r = b = 136.5$ mm to compute the deformation of the outside radius:

$$\begin{aligned}\delta_r &= \frac{a^2 p_i}{(b^2 - a^2)rE}[(1-\nu)r^2 + (1+\nu)b^2] \\ &= \frac{(104.5 \text{ mm})^2(45 \text{ MPa})}{[(136.5 \text{ mm})^2 - (104.5 \text{ mm})^2](136.5 \text{ mm})(200{,}000 \text{ MPa})}[(1-0.3)(136.5 \text{ mm})^2 + (1+0.3)(136.5 \text{ mm})^2] \\ &= 0.0870 \text{ mm}\end{aligned}$$

The outside diameter increases by $2\delta_r = 0.174$ mm. **Ans.**

Minimum Required Wall Thickness

The largest normal stress is the circumferential stress at the inner surface of the tube. An allowable normal stress of 140 MPa has been specified for the tube material, and the outside diameter of the tube is to remain as 273 mm (i.e., $b = 136.5$ mm). Begin with Equation (14.24). Substitute $r = a$ and simplify to obtain

$$\sigma_\theta = \frac{a^2 p_i}{b^2 - a^2}\left(1 + \frac{b^2}{r^2}\right) = \frac{b^2 + a^2}{b^2 - a^2} p_i$$

Therefore,

$$p_i(b^2 + a^2) = \sigma_\theta(b^2 - a^2)$$

First, solve for a^2:

$$a^2 = b^2\left(\frac{\sigma_\theta - p_i}{\sigma_\theta + p_i}\right)$$

Then, set σ_θ equal to the allowable normal stress and solve for a:

$$a = b\sqrt{\frac{\sigma_\theta - p_i}{\sigma_\theta + p_i}} = (136.5\text{ mm})\sqrt{\frac{140\text{ MPa} - 45\text{ MPa}}{140\text{ MPa} + 45\text{ MPa}}} = 97.8\text{ mm}$$

The wall thickness must have a minimum value of

$$t \geq b - a = 136.5\text{ mm} - 97.8\text{ mm} = 38.7\text{ mm}$$ **Ans.**

14.7 Interference Fits

In some practical applications, two or more concentric cylinders are arranged to fit one inside of the other. There are two main reasons for doing this. Firstly, it may be desirable to construct a thick-walled cylinder with a lining of a specific material that has particular physical or chemical properties, such as improved wear or corrosion resistance. In this case, the cylinders may be made to fit each other as closely as possible or an interference fit may be specified. An *interference fit* is a type of connection in which a critical radial dimension of the first part is slightly larger (or smaller) than the corresponding dimension of the second part. Force or temperature change is used to temporarily make the radial dimensions match so that the parts can be fit together, one inside of the other. The forces between the components joined in this manner are substantial, creating large friction forces that keep the pieces attached to each other so that the completed assembly acts as a single unit.

Secondly, two (or more) cylinders (often of the same material) may be assembled with interference fits between them. The compound cylinder is formed by "shrinking" an outer thick-walled cylinder (herein designated the jacket) over an inner cylinder (frequently termed the tube). This process creates a compound cylinder that can withstand higher internal pressures than a single cylinder with the same overall dimensions can. The process can be efficient when very high pressures, perhaps on the same order of magnitude as the allowable stress of the material, must be contained in the vessel.

A similar process, using either force or temperature, is frequently used to attach gears and pulleys to shafts in mechanical devices. When force is used, the connection is called a *press-fit* or *force-fit* connection. When temperature change is used, the process is termed *shrink fit*. In either case, the basic concept is the same. Parts that don't exactly match in one critical dimension (such as a radial dimension) are forced to fit together (with the use of either large external forces or temperature change). After the parts are combined, very large friction forces are created on the mating surfaces between the two components. These friction forces keep the components attached to each other, whether they consist of a jacket surrounding a tube or a gear attached to a shaft. For this discussion, we will refer to shrink-fit connections but understand that exactly the same concepts apply to press-fit or force-fit connections as well.

When a cylinder is shrink fitted onto another cylinder, the internal diameter of the larger cylinder (i.e., the jacket) is made slightly smaller than the external diameter of the smaller cylinder (i.e., the tube). Let the difference in preassembly dimensions between the inside radius of outer cylinder (i.e., the jacket) and the outside radius of the inner cylinder (i.e., the tube) be denoted as the *radial interference* δ. The geometry of the shrink-fit components, both before and after fabrication, is shown in Figure 14.13.

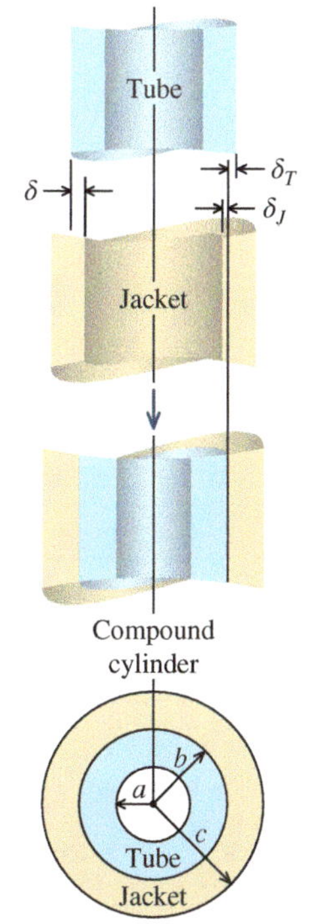

FIGURE 14.13 Interference between tube and jacket.

In the process of "shrinking" a jacket onto a tube, the internal diameter of the jacket is increased by heating the jacket enough so that it fits over the tube. (Alternatively, the tube could be cooled enough so that it fits inside of the jacket.) After the pieces are assembled and the temperature of both parts returns to the ambient temperature, the pressure between the two parts must be the same on the interface between the jacket and the tube. This pressure is termed the *contact pressure* p_c. The contact pressure creates stresses and deformations in both the jacket and the tube.

The determination of stresses in a compound cylinder requires the solution of a statically indeterminate problem that considers the radial interference δ and the radial deformations of the jacket δ_J (i.e., the inner cylinder) and the tube δ_T (i.e., the outer cylinder). The radial interference δ is related to radial deformations of the tube and jacket at the mating surface by the equation

$$\delta = |\delta_T| + |\delta_J| \quad \text{(a)}$$

After assembly, the inner cylinder is simply a thick-walled cylinder subjected to external pressure only, where the external pressure is the contact pressure p_c. For the tube, the inner radius will be denoted a and the outer radius will be denoted b. The magnitude of the radial deformation at the outer surface of the tube, $|d_T|$, is expressed by Equation (14.30), where we let $r = b$:

$$|\delta_T| = \frac{b\,p_c}{(b^2 - a^2)E}[(1 - \nu)b^2 + (1 + \nu)a^2] \quad \text{(b)}$$

After assembly, the outer cylinder is a thick-walled cylinder subjected to internal pressure only, where the internal pressure is the contact pressure p_c. The magnitude of the radial deformation at the inner surface of the jacket, $|d_J|$, is expressed by Equation (14.29). We will modify that equation to reflect the geometry shown in Figure 14.13, denoting the inner radius of the jacket as b and the outer radius as c, and setting $r = b$. Accordingly,

$$|\delta_J| = \frac{b\,p_c}{(c^2 - b^2)E}[(1 - \nu)b^2 + (1 + \nu)c^2] \quad \text{(c)}$$

Substituting Equations (b) and (c) into Equation (a) yields

$$\delta = \frac{2b^3\,p_c(c^2 - a^2)}{E(c^2 - b^2)(b^2 - a^2)} \quad (14.32)$$

It is convenient to solve Equation (14.32) for the contact pressure p_c:

$$p_c = \frac{E\delta(c^2 - b^2)(b^2 - a^2)}{2b^3(c^2 - a^2)} \quad (14.33)$$

In deriving Equations (14.32) and (14.33), we have assumed that both the tube and the jacket are made of the same material, and hence, they have the same values for E and ν. An expression for the contact pressure between cylinders of different materials will be presented subsequently.

Equation (14.32) provides a means for establishing the radial interference that can be tolerated without exceeding maximum allowable stress levels in the jacket or the tube. Equation (14.33) offers the means to determine the contact pressure between the tube and the jacket for a known interference. Once the contact pressure has been established, radial and circumferential stresses in both the tube and jacket can be determined.

Cylinder shrink fitted on a solid shaft: Shrink fitting or press fitting a gear, disk, or jacket to a solid shaft can be considered by setting $a = 0$ in Equation (14.33). With this revision, the contact pressure between a cylinder and a solid shaft can be expressed as

$$p_c = \frac{E\delta(c^2 - b^2)}{4bc^2} \tag{14.34}$$

Tube and jacket of different materials: If the tube and the jacket are of different materials, then the contact pressure between the two cylinders can be expressed as

$$p_c = \frac{\delta}{b\left[\dfrac{1}{E_T}\left(\dfrac{b^2 + a^2}{b^2 - a^2} - \nu_T\right) + \dfrac{1}{E_J}\left(\dfrac{c^2 + b^2}{c^2 - b^2} + \nu_J\right)\right]} \tag{14.35}$$

If the inner cylinder is a solid shaft (i.e., $a = 0$), then the contact pressure between a solid shaft of one material and an outer cylinder consisting of a second material is

$$p_c = \frac{\delta}{\dfrac{b}{E_T}(1 - \nu_T) + \dfrac{b}{E_J}\left(\dfrac{c^2 + b^2}{c^2 - b^2} + \nu_J\right)} \tag{14.36}$$

EXAMPLE 14.4

A steel [$E = 200$ GPa, $\nu = 0.3$] cylindrical jacket having an external diameter of 240 mm is shrink fitted over a steel tube having internal and external diameters of 80 mm and 160 mm, respectively. The radial interference is 0.050 mm. Calculate

(a) The pressure between the tube and the jacket.
(b) The maximum circumferential stress in the tube.
(c) The maximum circumferential stress in the jacket.

Plan the Solution

From the cylinder dimensions and the radial interference, calculate the contact pressure created by shrinking the jacket onto the tube. The contact pressure will be an external pressure for the tube and an internal pressure for the jacket. Once the contact pressure is established, the circumferential stresses in both the tube and the jacket can be calculated.

SOLUTION

The radii for the combined tube and jacket are $a = 40$ mm, $b = 80$ mm, and $c = 120$ mm.

Contact Pressure

The radial interference of $\delta = 0.050$ mm creates a contact pressure p_c on the interface between the two cylinders:

$$p_c = \frac{E\delta(c^2 - b^2)(b^2 - a^2)}{2b^3(c^2 - a^2)}$$

$$= \frac{(200{,}000\text{ MPa})(0.050\text{ mm})[(120\text{ mm})^2 - (80\text{ mm})^2][(80\text{ mm})^2 - (40\text{ mm})^2]}{2(80\text{ mm})^3[(120\text{ mm})^2 - (40\text{ mm})^2]}$$

$$= 29.30\text{ MPa}$$ **Ans.**

Maximum Circumferential Stress in the Tube

The smaller cylinder is subjected to external pressure only. Using an external pressure of $p_o = p_c = 29.30$ MPa in Equation (14.26) with values of $a = 40$ mm and $b = 80$ mm for the tube, we calculate the circumferential stress on the inner surface of the tube at $r = 40$ mm:

$$\sigma_\theta = -\frac{b^2 p_o}{b^2 - a^2}\left(1 + \frac{a^2}{r^2}\right)$$

$$= -\frac{(80\text{ mm})^2(29.30\text{ MPa})}{(80\text{ mm})^2 - (40\text{ mm})^2}\left[1 + \frac{(40\text{ mm})^2}{(40\text{ mm})^2}\right] = -78.125\text{ MPa} = -78.1\text{ MPa} \quad \textbf{Ans.}$$

Maximum Circumferential Stress in the Jacket

The larger cylinder is subjected to internal pressure only. Using an internal pressure of $p_i = p_c = 29.30$ MPa in Equation (14.24) with values of $a = 80$ mm and $b = 120$ mm for the jacket, we calculate the circumferential stress on the inner surface of the tube at $r = 80$ mm:

$$\sigma_\theta = \frac{a^2 p_i}{b^2 - a^2}\left(1 + \frac{b^2}{r^2}\right)$$

$$= \frac{(80\text{ mm})^2(29.30\text{ MPa})}{(120\text{ mm})^2 - (80\text{ mm})^2}\left[1 + \frac{(120\text{ mm})^2}{(80\text{ mm})^2}\right] = 76.172\text{ MPa} = 76.2\text{ MPa} \quad \textbf{Ans.}$$

EXAMPLE 14.5

Determine the maximum tensile stress in the compound cylinder of Example 14.4 after an internal pressure of 170 MPa is applied. Sketch a figure showing the variation of the circumferential stresses in the compound tube and jacket before and after the pressure is applied.

Plan the Solution

Stresses in the compound cylinder after the internal pressure is applied are due partly to the contact pressure (from the shrink-fit operation) and partly to the applied pressure. To determine the final stresses, the contact pressure results will be combined with the stresses found in the complete assembly subjected to internal pressure only.

SOLUTION

The maximum tensile stress will be a circumferential stress; however, it may occur either at the inside surface of the tube (i.e., the smaller cylinder) or at the inside surface of the jacket (i.e., the larger cylinder), depending on the magnitudes of the stresses associated with the two pressure loadings.

Maximum Circumferential Stress in the Tube

For the tube, the circumferential stress on its inner surface due to the shrink-fit pressure alone was calculated in Example 14.4 as $\sigma_\theta = -78.125$ MPa. Next, the circumferential stress created in the compound assembly by an internal pressure of $p_i = 170$ MPa is calculated from Equation (14.24) with $a = 40$ mm, $b = 120$ mm, and $r = 40$ mm:

$$\sigma_\theta = \frac{a^2 p_i}{b^2 - a^2}\left(1 + \frac{b^2}{r^2}\right)$$

$$= \frac{(40\text{ mm})^2(170\text{ MPa})}{(120\text{ mm})^2 - (40\text{ mm})^2}\left[1 + \frac{(120\text{ mm})^2}{(40\text{ mm})^2}\right] = 212.500\text{ MPa}$$

The circumferential stress on the inner surface of the tube is found by superimposing the stress due to the contact pressure and the stress caused by the internal pressure:

$$\begin{aligned}\sigma_\theta &= \sigma_\theta\,(\text{contact pressure}) + \sigma_\theta\,(\text{compound cylinder with } p_i)\\ &= -78.125 \text{ MPa} + 212.500 \text{ MPa}\\ &= 134.375 \text{ MPa}\end{aligned}$$

Maximum Circumferential Stress in the Jacket

For the jacket, the circumferential stress on its inner surface due to the shrink-fit contact pressure alone was calculated in Example 14.4 as $\sigma_\theta = 76.172$ MPa. Next, the circumferential stress created in the compound assembly by an internal pressure of $p_i = 170$ MPa is calculated from Equation (14.24) with $a = 40$ mm, $b = 120$ mm, and $r = 80$ mm:

$$\begin{aligned}\sigma_\theta &= \frac{a^2 p_i}{b^2 - a^2}\left(1 + \frac{b^2}{r^2}\right)\\ &= \frac{(40 \text{ mm})^2(170 \text{ MPa})}{(120 \text{ mm})^2 - (40 \text{ mm})^2}\left[1 + \frac{(120 \text{ mm})^2}{(80 \text{ mm})^2}\right] = 69.063 \text{ MPa}\end{aligned}$$

The circumferential stress on the inner surface of the jacket is found by superimposing the stress from the contact pressure and the stress from the internal pressure:

$$\begin{aligned}\sigma_\theta &= \sigma_\theta\,(\text{contact pressure}) + \sigma_\theta\,(\text{compound cylinder with } p_i)\\ &= 76.172 \text{ MPa} + 69.063 \text{ MPa}\\ &= 145.234 \text{ MPa}\end{aligned}$$

Maximum Tensile Stress in the Compound Cylinder

The maximum tensile stress occurs on the inner surface of the jacket:

$$(\sigma_\theta)_{\max} = 145.2 \text{ MPa}$$ **Ans.**

Variation of Circumferential Stresses in Compound Tube and Jacket

The variation of circumferential stresses in the compound cylinder is shown in Figure 14.14.

Take a closer look at the plot of the circumferential stress distribution due to the 170 MPa internal pressure in the figure. Notice that the material close to the inner surface of the compound cylinder is highly stressed while the stresses at the larger radii are relatively low. This effect becomes more pronounced as the ratio of the outside diameter to the inside diameter increases. From this observation, we can conclude that the use of a single huge cylinder to contain very high pressure is an inefficient use of material. By using a compound cylinder, stresses are more broadly distributed in the material. When the tube and jacket are first assembled, the shrink-fit contact pressure causes the inner cylinder to be in compression and the outer cylinder to be in tension. When the tensile circumferential stresses due to the 170-MPa internal pressure are superimposed on the stresses created by the contact pressure, the final distribution shows that circumferential stresses are spread to more of the cross section.

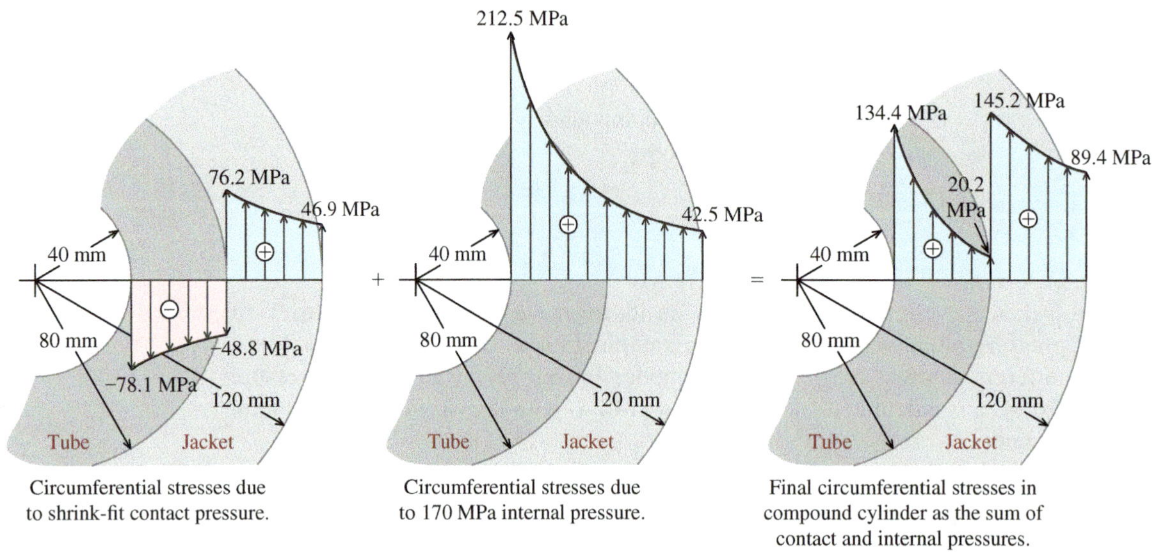

FIGURE 14.14 Variation of circumferential stresses in a compound cylinder.

PROBLEMS

P14.22 A thick-walled cylindrical tank with closed ends, an inside radius of $a = 250$ mm, and an outside radius of $b = 290$ mm is submerged to a depth of 40 m. Assume that the density of seawater is 1,100 kg/m^3. The tank also has an internal pressure of $p_i = 2$ MPa. Determine the maximum values of σ_r, σ_θ, and σ_{long}, as well as the maximum shear stress in the walls of the cylinder remote from its ends.

P14.23 An open-ended thick-walled cylindrical pressure vessel with an inside diameter of 250 mm and an outside diameter of 400 mm is subjected to an internal pressure of 75 MPa. Determine

(a) The circumferential stress at a point on the inside surface of the cylinder.
(b) The circumferential stress at a point on the outside surface of the cylinder.
(c) The maximum shear stress in the cylinder.

P14.24 A steel cylinder with an inside diameter of 8 in. and an outside diameter of 14 in. is subjected to an internal pressure of 25,000 psi. Determine

(a) The maximum shear stress in the cylinder.
(b) The radial and circumferential stresses at a point in the cylinder midway between the inside and outside surfaces.

P14.25 A hydraulic cylinder with an inside diameter of 200 mm and an outside diameter of 450 mm is made of steel [$E = 210$ GPa; $\nu = 0.30$]. For an internal pressure of 125 MPa, determine

(a) The maximum tensile stress in the cylinder.
(b) The change in internal diameter of the cylinder.

P14.26 A closed-ended cylindrical pressure vessel is to be designed to have an internal diameter of 125 mm and to be subjected to an internal gage pressure of 10 MPa. The maximum circumferential normal stress is not to exceed 100 MPa, and the maximum shear stress is not to exceed 60 MPa. Determine the minimum external diameter that may be used for the vessel.

P14.27 A cylindrical tank having an external diameter of 16 in. is to resist an internal pressure of 2,000 psi without the circumferential stress exceeding 20,000 psi. Determine the minimum wall thickness required for the tank.

P14.28 A cylindrical steel tank having an internal diameter of 4 ft is submerged in seawater (weight density = 64 lb/ft^3) to a depth of 10,000 ft. The internal gage pressure in the tank is 2,000 psi.

(a) If the allowable stress of the steel is 36 ksi, what is the minimum allowable outside diameter of the tank?
(b) What is the maximum tensile stress before the tank is immersed?

P14.29 A thick-walled cylinder with closed ends is subjected only to internal pressure p_i. Let $a = 36$ in. and $b = 40$ in., and determine the maximum permissible internal pressure p_i that may be contained in the cylinder if the allowable tensile stress is 22 ksi and the allowable shear stress is 12 ksi.

P14.30 A thick-walled cylindrical pressure vessel with an inside diameter of 200 mm and an outside diameter of 300 mm is made of a steel that has a yield strength of 430 MPa. Determine the maximum internal pressure that may be applied to the vessel if a factor of safety of 3.0 with respect to failure by yielding is required.

P14.31 An open-ended compound thick-walled cylinder is made by shrinkfitting an exterior cylindrical steel jacket onto an interior steel tube. The following data are provided: $E = 30{,}000$ ksi, $a = 20$ in., $b = 30$ in., and $c = 40$. The radial interference is 0.05 in. Determine the maximum values of σ_r and σ_θ produced in the compound cylinder by the shrink-fitting operation.

P14.32 A closed-ended compound cylinder is formed by shrinking a steel [$E = 200$ GPa; $\nu = 0.3$] tube that has an external diameter of 320 mm and an internal diameter of 250 mm onto a brass [$E = 100$ GPa; $\nu = 0.35$] tube that has an internal diameter of 200 mm. The difference between the diameters of the contacting surfaces before assembly is 0.10 mm. Determine the largest circumferential stress in the finished assembly after an internal pressure of $p_i = 50$ MPa is applied.

P14.33 A compound thick-walled steel [$E = 200$ GPa; $\nu = 0.30$] cylinder consists of a jacket with an inside diameter of 250 mm and an outside diameter of 300 mm shrunk onto a steel tube with an inside diameter of 150 mm. The initial shrinkage contact pressure is 28 MPa. Determine

(a) The initial radial interference required to produce the initial shrinkage pressure.
(b) The maximum tensile stress in the finished assembly after an internal pressure of $p_i = 200$ MPa is applied.

P14.34 A solid 40 mm diameter steel [$E = 200$ GPa] shaft is pressed into a steel gear hub that has an outside diameter of 120 mm.

(a) If a contact pressure of 180 MPa is required so that the gear will not slip on the shaft, what is the minimum radial interference needed to produce this shrinkage pressure?
(b) What is the maximum radial stress in the shaft?
(c) What is the largest circumferential stress in the gear hub?

CHAPTER 15

Combined Loads

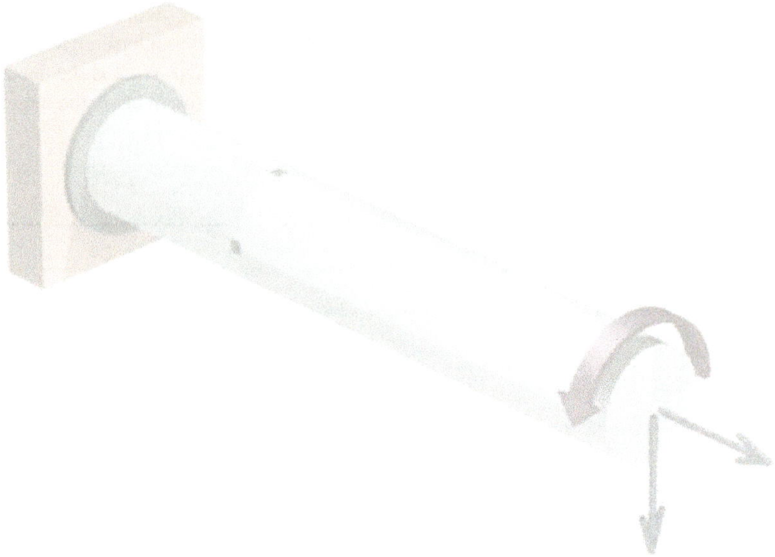

15.1 Introduction

The stresses and strains produced by three fundamental types of loads (axial, torsional, and flexural) have been analyzed in the preceding chapters. Many machine and structural components are subjected to a combination of these loads, and a procedure for calculating the ensuing stresses at a point on a specified section is required. One method is to replace the given force system with a statically equivalent system of forces and moments acting at the section of interest. The equivalent force system can be systematically evaluated to determine the type and magnitude of stresses produced at the point, and these stresses can be calculated by the methods developed in previous chapters. The combined effect can be obtained by the principle of superposition if the combined stresses do not exceed the proportional limit. Various combinations of loads that can be analyzed in this manner are discussed in the sections that follow.

15.2 Combined Axial and Torsional Loads

A shaft or some other machine component is subjected to both an axial and a torsional load in numerous situations. Examples include the drill rod for a well and the propeller shaft in a ship. Since radial and circumferential normal stresses are zero, the combination of axial and torsional loads creates plane stress conditions at any point in the body. Although axial normal stresses are identical at all points on the cross section, torsional shear stresses are greatest on the periphery of the shaft. For this reason, critical stresses are normally investigated on the outer surface of the shaft.

Example 15.1 illustrates the analysis of combined torsional and axial loads in a shaft.

EXAMPLE 15.1

A hollow circular shaft having an outside diameter of 4 in. and a wall thickness of 0.25 in. is loaded as shown. Determine the principal stresses and the maximum shear stress at points H and K.

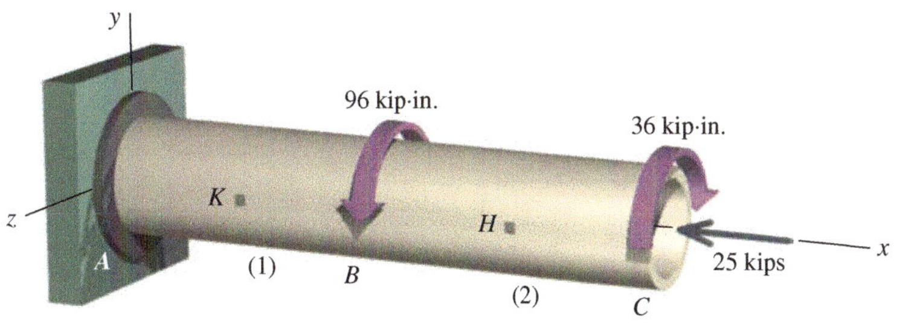

Plan the Solution

After computing the required section properties for the pipe shaft, we will determine the equivalent forces acting at point H. The normal and shear stresses created by the internal axial force and torque will be computed and shown in their proper directions on a stress element. Stress transformation calculations will be used to determine the principal stresses and maximum shear stress for the stress element at H. The process will be repeated for the stresses acting at K.

SOLUTION

Section Properties

The outside diameter D of the pipe is 4 in., and the wall thickness of the pipe is 0.25 in.; thus, the inside diameter is $d = 3.5$ in. The cross-sectional area of the pipe will be needed in order to calculate the normal stress caused by the axial force:

$$A = \frac{\pi}{4}[D^2 - d^2] = \frac{\pi}{4}[(4\text{ in.})^2 - (3.5\text{ in.})^2] = 2.9452\text{ in.}^2$$

The polar moment of inertia will be required in order to calculate the shear stress caused by the internal torques in the pipe:

$$J = \frac{\pi}{32}[D^4 - d^4] = \frac{\pi}{32}[(4\text{ in.})^4 - (3.5\text{ in.})^4] = 10.4004\text{ in.}^4$$

Equivalent Forces at H

The pipe will be sectioned just to the right of a stress element at H, and the equivalent forces and moments acting at the section of interest will be determined. This procedure is straightforward at H, where the equivalent force is simply the 25 kip axial force and the equivalent torque is equal to the 36 kip·in. torque applied at C.

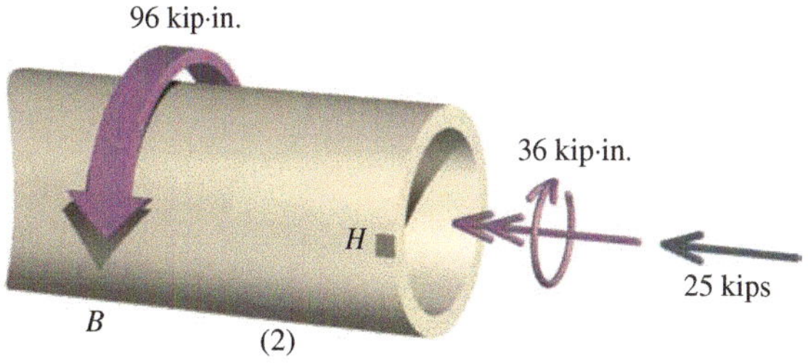

Normal and Shear Stresses at H

The normal and shear stresses at H can be calculated from the equivalent forces just shown. The 25 kip axial force creates a compressive normal stress

$$\sigma_{\text{axial}} = \frac{F}{A} = \frac{25\text{ kips}}{2.9452\text{ in.}^2} = 8.49\text{ ksi (C)}$$

The shear stress created by the 36 kip·in. torque is computed from the elastic torsion formula:

$$\tau = \frac{Tc}{J} = \frac{(36 \text{ kip·in.})(2 \text{ in.})}{10.4004 \text{ in.}^4} = 6.92 \text{ ksi}$$

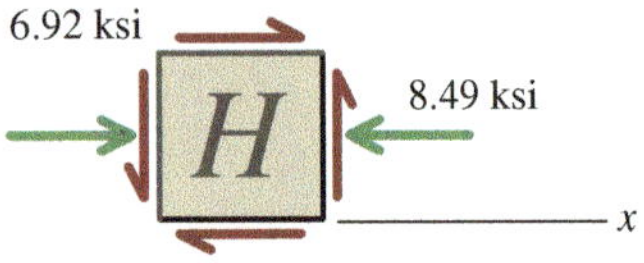

The normal and shear stresses acting at a point should be summarized on a stress element before one begins the stress transformation calculations. Often, it is more efficient to calculate the stress magnitudes from the appropriate formulae but to determine the proper direction of the stresses by inspection.

The axial stress acts in the same direction as the 25 kip force; therefore, the 8.49 ksi axial stress acts in compression in the x direction.

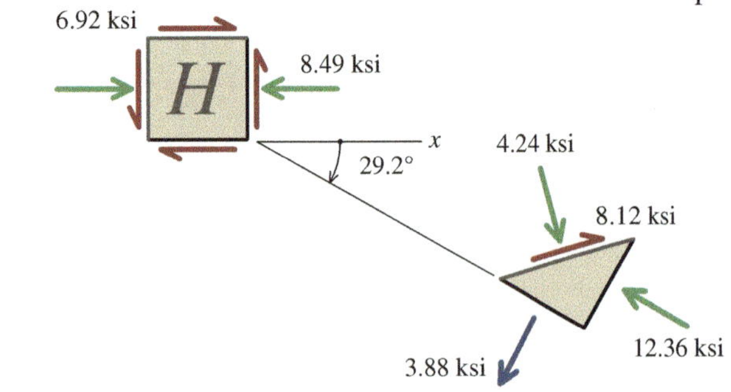

The direction of the torsion shear stress at the point of interest can be confusing to determine. Examine the illustration of the pipe, and note the direction of the equivalent torque acting at H. The shear stress arrow on the $+x$ face of the stress element acts in the same direction as the torque; therefore, the 6.92 ksi shear stress acts *upward* on the $+x$ face of the stress element. After the proper shear stress direction has been established on one face, the shear stress directions on the other three faces are known.

Stress Transformation Results at *H*

The principal stresses and the maximum shear stress at H can be determined from the stress transformation equations and procedures detailed in Chapter 12. The results of these calculations are shown in the accompanying figure.

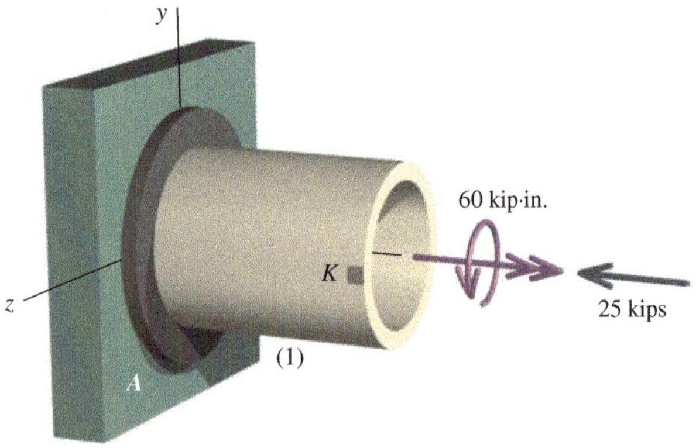

Equivalent Forces at *K*

The pipe will be sectioned just to the right of a stress element at K, and the equivalent forces and moments acting at the section of interest will be determined. While the equivalent force is simply the 25 kip axial force, the equivalent torque at K is the sum of the torques applied to the pipe shaft at B and C. The equivalent torque at the section of interest is 60 kip·in.

Normal and Shear Stresses at *K*

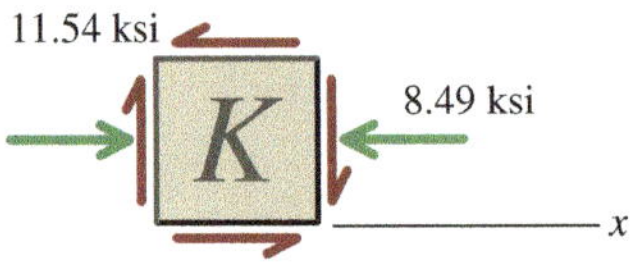

The normal and shear stresses at K can be calculated from the equivalent forces shown in the accompanying figure. The 25 kip axial force creates a compressive normal stress of 8.49 ksi. The 60 kip·in. equivalent torque creates a shear stress given by

$$\tau = \frac{Tc}{J} = \frac{(60 \text{ kip·in.})(2 \text{ in.})}{10.4004 \text{ in.}^4} = 11.54 \text{ ksi}$$

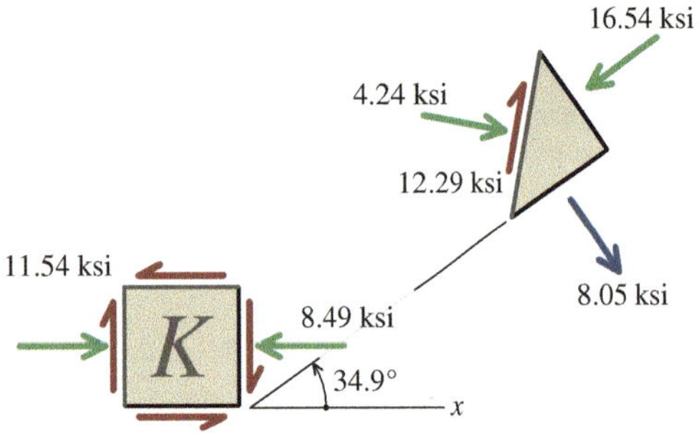

As at H, the 8.49 ksi axial stress acts in compression in the x direction. The equivalent torque at K creates a shear stress that acts *downward* on the $+x$ face of the stress element. The proper stress element for K is shown.

Stress Transformation Results at *K*

The principal stresses and the maximum shear stress at K are shown in the accompanying figure.

MecMovies

EXAMPLE

M15.1 A tubular shaft of outside diameter $D = 114$ mm and inside diameter $d = 102$ mm is subjected simultaneously to a torque $T = 5$ kN·m and an axial load $P = 40$ kN. Determine the principal stresses and the maximum shear stress at a typical point on the surface of the shaft.

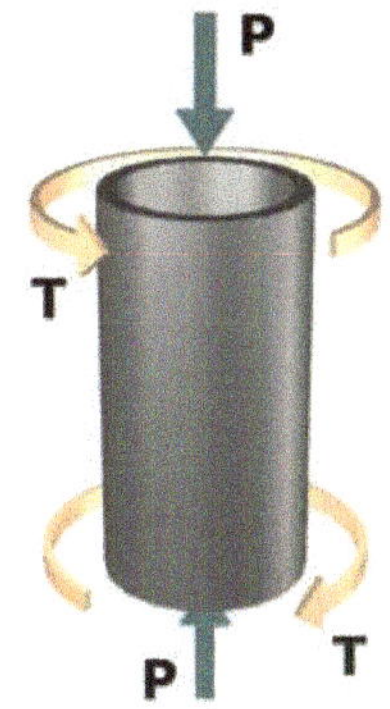

PROBLEMS

P15.1 A solid 1.50 in. diameter shaft is subjected to a torque T = 225 lb·ft and an axial load P = 5,500 lb, as shown in Figure P15.1/2.

(a) Determine the principal stresses and the maximum shear stress at point H on the surface of the shaft.
(b) Show the stresses of part (a) and their directions on an appropriate sketch.

P15.3 A hollow bronze [E = 15,200 ksi; ν = 0.34] shaft with an outside diameter of 2.50 in. and a wall thickness of 0.125 in. is subjected to a torque T = 720 lb·ft and an axial load P = 1,900 lb, as shown in Figure P15.3/4. Determine

(a) the strains ε_x, ε_y, and γ_{xy},
(b) the principal strains ε_{p1} and ε_{p2}, and
(c) the absolute maximum shear strain at point H on the outer surface of the shaft.

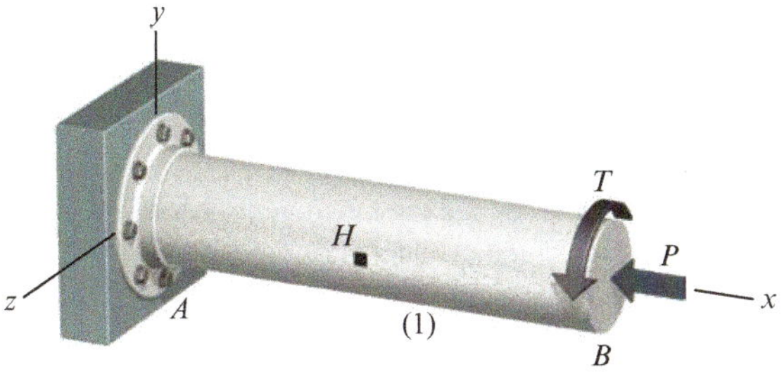

FIGURE P15.1/2

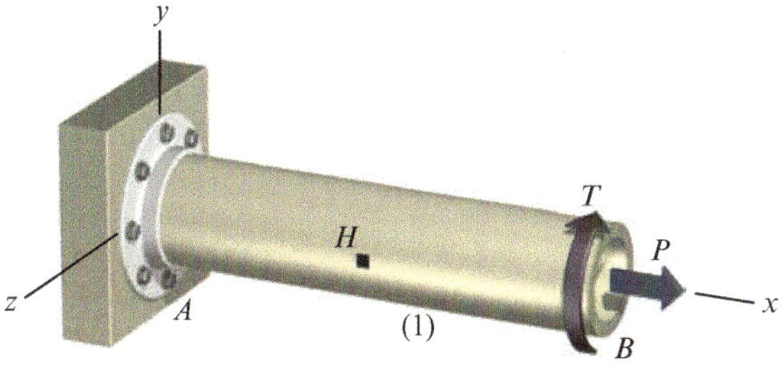

FIGURE P15.3/4

P15.2 A solid 19 mm diameter aluminum alloy [E = 70 GPa; ν = 0.33] shaft is subjected to a torque T = 60 N·m and an axial load P = 15 kN, as shown in Figure P15.1/2. Determine

(a) the strains ε_x, ε_y, and γ_{xy},
(b) the principal strains ε_{p1} and ε_{p2}, and
(c) the absolute maximum shear strain at point H on the outer surface of the shaft.

P15.4 A hollow bronze shaft with an outside diameter of 80 mm and a wall thickness of 5 mm is subjected to a torque of T = 620 N·m and an axial load of P = 9,500 N, as shown in Figure P15.3/4.

(a) Determine the principal stresses and the maximum shear stress at point H on the surface of the shaft.
(b) Show the stresses of part (a) and their directions on an appropriate sketch.

P15.5 A solid 1.50 in. diameter shaft is used in an aircraft engine to transmit 160 hp at 2,800 rpm to a propeller that develops a thrust of 1,800 lb. Determine the magnitudes of the principal stresses and the maximum shear stress at any point on the outside surface of the shaft.

P15.6 A short cast iron shaft 2 in. in diameter is subjected to a tensile force of 22,000 lb combined with a torque T. Find the torque T that the shaft can resist (a) if the allowable shear stress is 6,000 psi and (b) if the allowable tensile stress is 3,000 psi.

P15.7 A compound shaft consists of two tube segments. Segment (1) has an outside diameter of 40 mm and a wall thickness of 3.2 mm. Segment (2) has an outside diameter of 70 mm and a wall thickness of 4.0 mm. The shaft is subjected to a tensile load $P = 15$ kN and torques $T_A = 350$ N·m and $T_B = 900$ N·m, which act in the directions shown in Figure P15.7/8.

(a) Determine the principal stresses and the maximum shear stress at point K on the surface of the shaft.
(b) Show these stresses on an appropriate sketch.

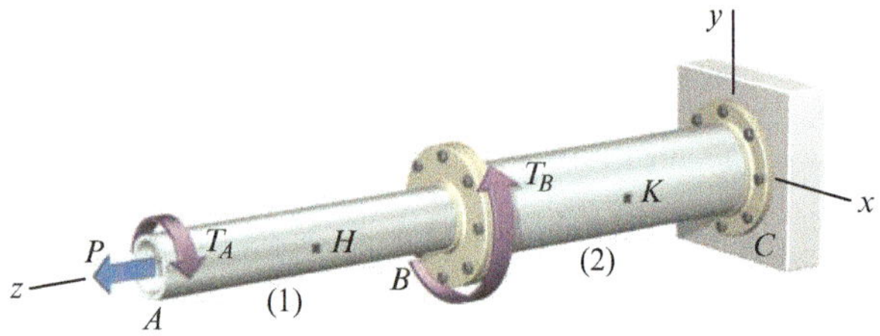

FIGURE P15.7/8

P15.8 A compound shaft consists of two stainless steel tube segments. Segment (1) has an outside diameter of 1.5 in. and a wall thickness of 0.120 in. Segment (2) has an outside diameter of 3.0 in. and a wall thickness of 0.120 in. The shaft is subjected to a tensile load $P = 6{,}200$ lb and torques $T_A = 420$ lb·ft and $T_B = 1{,}060$ lb·ft, which act in the directions shown in Figure P15.7/8. Determine the maximum compressive normal stress in the tube wall at (a) point H and (b) point K.

P15.9 A hollow shaft is subjected to an axial load P and a torque T, acting in the directions shown in Figure P15.9. The shaft is made of bronze [$E = 105$ GPa; $\nu = 0.34$], and it has an outside diameter of 55 mm and an inside diameter of 45 mm. A strain gage is mounted at an angle $\theta = 40°$ with respect to the longitudinal axis of the shaft, as shown in the figure.

(a) If $P = 13{,}000$ N and $T = 260$ N·m, what is the strain reading that would be expected from the gage?
(b) If the strain gage gives a reading of -195 με when the axial load has a magnitude of $P = 6{,}200$ N, what is the magnitude of the torque T applied to the shaft?

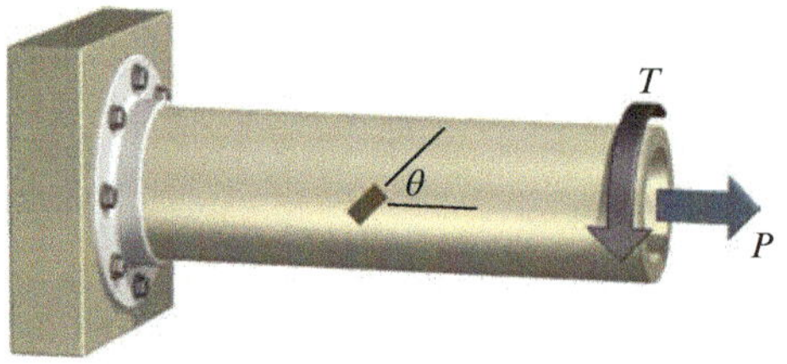

FIGURE P15.9

P15.10 A hollow shaft is subjected to an axial load P and a torque T, acting in the directions shown in Figure P15.10. The shaft is made of bronze [$E = 15{,}200$ ksi; $\nu = 0.34$], and it has an outside diameter of 2.50 in. and an inside diameter of 2.00 in. Strain gages a and b are mounted on the shaft at the orientations shown in the figure, where θ has a magnitude of 25°.

(a) If $P = 6$ kips and $T = 17$ kip·in., determine the strain readings that would be expected from the gages.
(b) If the strain gage readings are $\varepsilon_a = -1{,}100$ με and $\varepsilon_b = 720$ με, determine the axial force P and the torque T applied to the shaft.

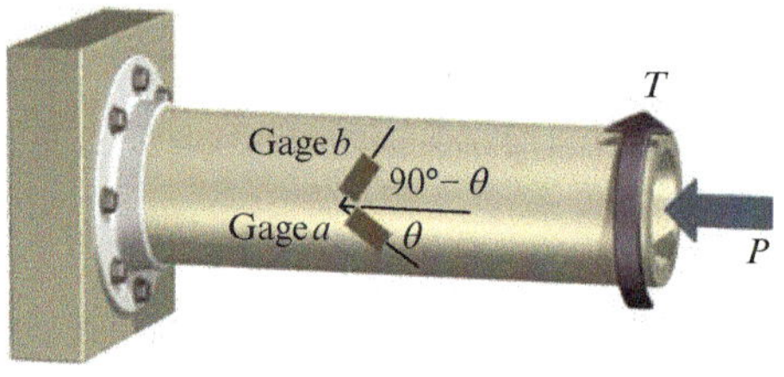

FIGURE P15.10

P15.11 The cylinder in Figure P15.11 is fabricated from spirally wrapped steel plates that are welded at the seams in the orientation shown. The cylinder has an inside diameter of 1,060 mm and a wall thickness of 10 mm. The cylinder is subjected to a compressive load $P = 2{,}200$ kN and a torque $T = 850$ kN·m, which acts in the direction shown. For a seam angle $\beta = 30°$, determine

(a) the normal stress perpendicular to the weld seams.
(b) the shear stress parallel to the weld seams.
(c) the principal stresses and the maximum shear stress on the outside surface of the cylinder.

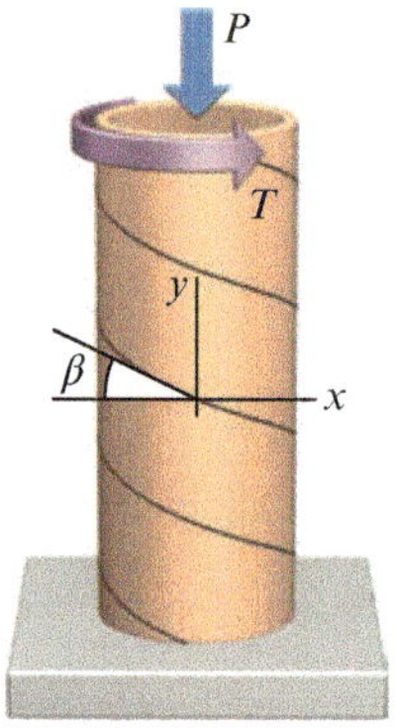

FIGURE P15.11

15.3 Principal Stresses in a Flexural Member

Procedures for locating the critical sections of a beam (i.e., the maximum internal shear force V and the maximum bending moment M) were presented in Chapter 7. Methods for calculating the bending stress at any point in a beam were presented in Sections 8.3 and 8.4. Methods for determining the horizontal and transverse shear stresses at any point in a beam were presented in Sections 9.5 through 9.7. However, the discussion of stresses in beams is incomplete without a consideration of the principal and maximum shear stresses that occur at the locations of maximum shear force and maximum bending moment.

The normal stress caused by flexure is largest on either the top or bottom surfaces of a beam, but the horizontal and transverse shear stress is zero at these locations. Consequently, the tensile and compressive normal stresses on the top and bottom surfaces of the beam are also principal stresses, and the corresponding maximum shear stress is equal to one-half of the bending stress [i.e., $\tau_{max} = (\sigma_p - 0)/2$]. On the neutral surface, the normal stress due to bending is zero; however, the largest horizontal and transverse shear stresses usually occur at the neutral surface. In this instance, the principal and maximum shear stresses are both equal to the horizontal shear stress. At points between these extremes, one might well wonder whether there are combinations of normal and shear stresses that create principal stresses larger than those at the extremes. Unfortunately, the magnitude of the principal stresses throughout a cross section cannot be expressed as a simple function of position for all sections; however, contemporary analytical software often provides insight into the distribution of principal stresses by means of color-coded stress contour plots.

Rectangular Cross Sections

For beams with a rectangular cross section, the largest principal stress is usually the maximum bending stress, which occurs on the top and bottom surfaces of the beam. The maximum shear stress usually occurs at the same location and has a magnitude equal to one-half of the bending stress. Although it may be of lesser intensity, the horizontal shear stress (calculated from $\tau = VQ/It$) at the neutral surface may also be a significant factor, particularly for materials having a horizontal plane of weakness, such as a typical timber beam.

Flanged Cross Sections

If the beam cross section is a flanged shape, then principal stresses at the junction between flange and web must also be investigated. When a beam with a flanged cross section is subjected to a combination of large V and large M, the bending and transverse shear stresses that occur at the junction of the flange and the web sometimes produce principal stresses that are greater than the maximum bending stress at the outermost surface of the flange. In general, at any point in a beam, a combination of large V, M, Q, and y, together with a small t, should suggest a check of the principal stresses at such a point. Otherwise, the maximum bending stress will very likely be the principal stress, and the maximum in-plane shear stress, will probably occur at the same point.

Stress Trajectories

Knowledge of the directions of the principal stresses also may aid in the prediction of the direction of cracks in a brittle material (e.g., concrete) and thus may also aid in the

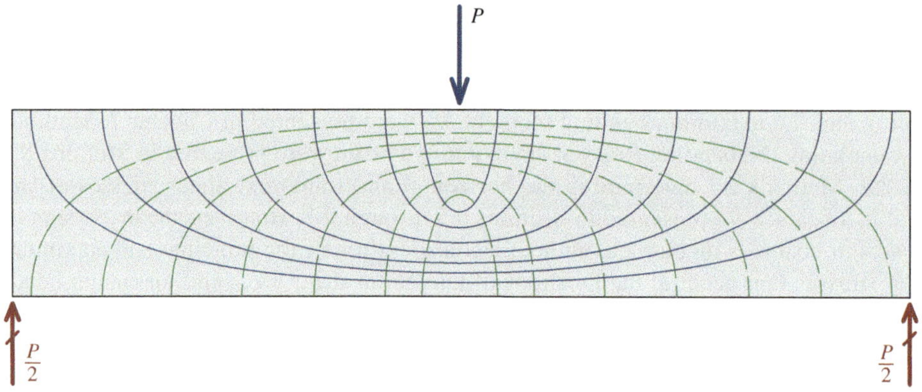

FIGURE 15.1 Stress trajectories for a simply supported beam subjected to a concentrated load at midspan.

design of reinforcement to carry the tensile stresses. Curves drawn with their tangents at each point in the directions of the principal stresses are called **stress trajectories**. Since there are generally two nonzero principal stresses at each point (in plane stress), there are two stress trajectories passing through each point. These curves will be perpendicular, since the principal stresses are orthogonal; one set of curves will represent the maximum stresses, whereas the other set of curves will represent the minimum stresses. The trajectories for a simply supported beam with a rectangular cross section where the beam is subjected to a concentrated load at midspan are shown in Figure 15.1. The dashed lines represent the directions of the compressive stresses, while the solid lines represent the tensile stress directions. Stress concentrations exist in the vicinities of the load and reactions, and consequently, the stress trajectories become much more complicated in those regions. Figure 15.1 omits the effect of stress concentrations.

General Calculation Procedures

To determine the principal stresses and the maximum shear stress at a particular point in a beam, the following procedures are useful:

1. Calculate the external beam reaction forces and moments (if any).
2. Determine the internal axial force (if applicable), shear force, and bending moment acting at the section of interest. To determine the internal forces, it may be expedient to construct the complete shear-force and bending-moment diagrams for the beam, or it may suffice to consider simply a free-body diagram that cuts through the beam at the section of interest.
3. Once the internal forces and moments are known, determine the magnitude of each normal stress and shear stress produced at the specific point of interest.
   a. Normal stresses are produced by an internal axial force F and by an internal bending moment M. The magnitude of the axial stress is given by $\sigma = F/A$, and the magnitude of the bending stress is given by the flexure formula $\sigma = -My/I$.
   b. Shear stress produced by nonuniform bending is calculated from $\tau = VQ/It$.
4. Summarize the stress calculation results on a stress element, taking care to identify the proper direction of each stress.

a. The normal stresses caused by F and M act in the longitudinal direction of the beam, either in tension or in compression.

b. The proper direction for the shear stress τ produced by nonuniform bending is sometimes challenging to establish. Determine the direction of the shear force V acting on a transverse plane at the point of interest (Figure 15.2). The transverse shear stress τ acts in the same direction on this plane. After the direction of the shear stress has been established on one face of the stress element, the shear stress directions on all four faces are known.

c. It is generally more reliable to use *inspection* to establish the directions of normal and shear stresses acting on the stress element. Consider the positive internal shear force V shown in Figure 15.2. (Recall that a positive V acts downward on the right-hand face of a beam segment and upward on the left-hand face.) Although the shear force V is *positive*, the corresponding shear stress τ_{xy} is considered *negative* according to the sign conventions used in the stress transformation equations.

5. Once all stresses on orthogonal planes through the point are known and summarized on a stress element, the methods of Chapter 12 can be used to calculate the principal stresses and the maximum shear stresses at the point.

Examples 15.2 and 15.3 illustrate the procedure.

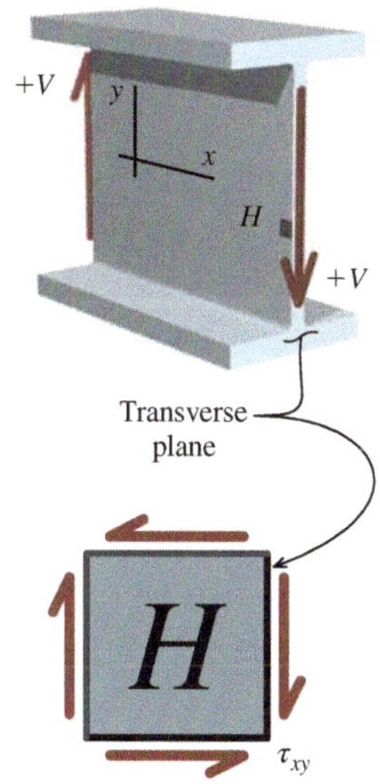

FIGURE 15.2 Correspondence between direction of V and direction of τ.

EXAMPLE 15.2

The simply supported wide-flange beam supports the loadings shown. Determine the principal stresses and maximum shear stress at points H and K. Show these stresses on a properly oriented stress element.

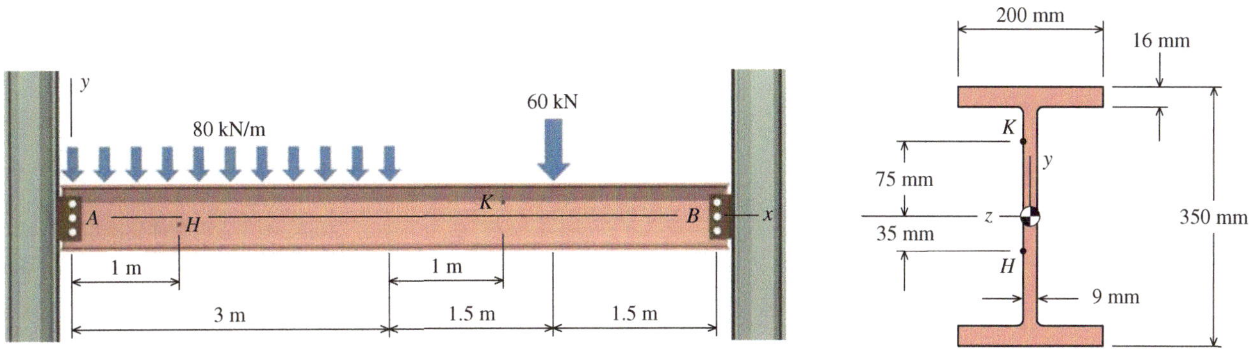

Plan the Solution

The moment of inertia of the wide-flange section will be computed from the cross-sectional dimensions. The shear-force and bending-moment diagrams will be constructed for the simply supported beam. From these diagrams, the internal shear force and the internal bending moment acting at the points of interest will be determined. The flexure formula and the shear stress formula will be used to compute the normal and shear stresses acting at each point. These stresses will be summarized on stress elements for each point, and then stress transformation calculations will be used to determine the principal stresses and maximum shear stress for the stress element at H. The process will be repeated for the stresses acting at K.

SOLUTION

Moment of Inertia

The moment of inertia for the wide-flange section is

$$I_z = \frac{(200\text{ mm})(350\text{ mm})^3}{12} - \frac{(191\text{ mm})(318\text{ mm})^3}{12} = 202.74 \times 10^6\text{ mm}^4$$

Shear-Force and Bending-Moment Diagrams

The shear-force and bending-moment diagrams for the simply supported beam are shown in the accompanying diagram.

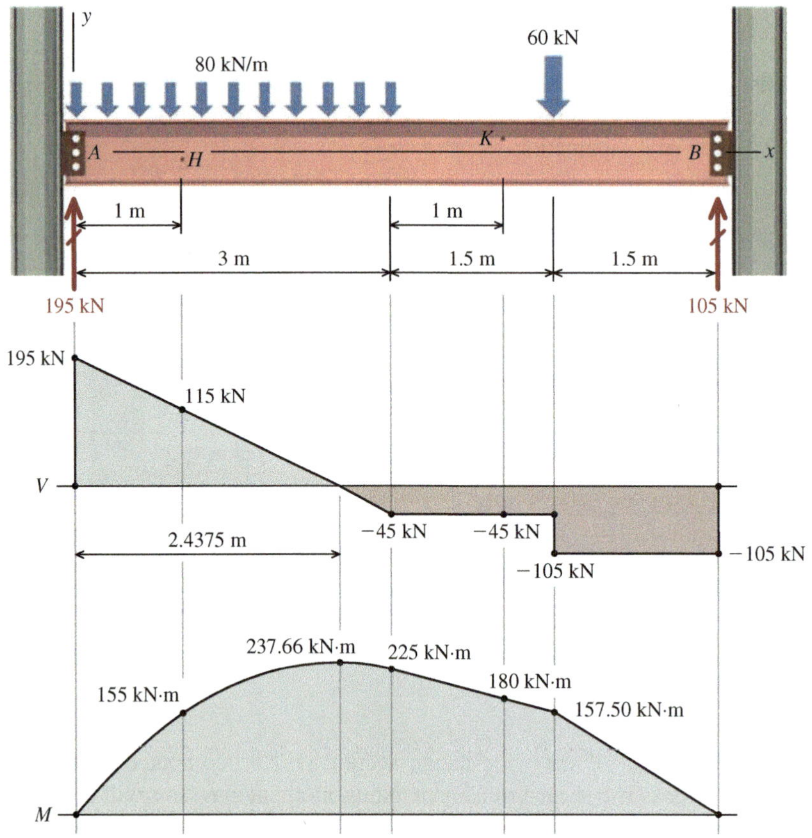

Shear Force and Bending Moment at H

At the location of point H, the internal shear force is $V = 115$ kN and the internal bending moment is $M = 155$ kN·m. These internal forces act in the directions shown.

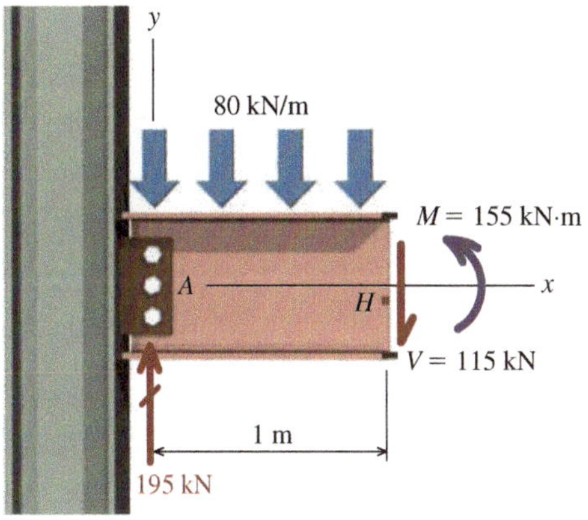

Normal and Shear Stresses at H

Point H is located 35 mm below the z centroidal axis; therefore, $y = -35$ mm. The bending stress at H can be calculated from the flexure formula:

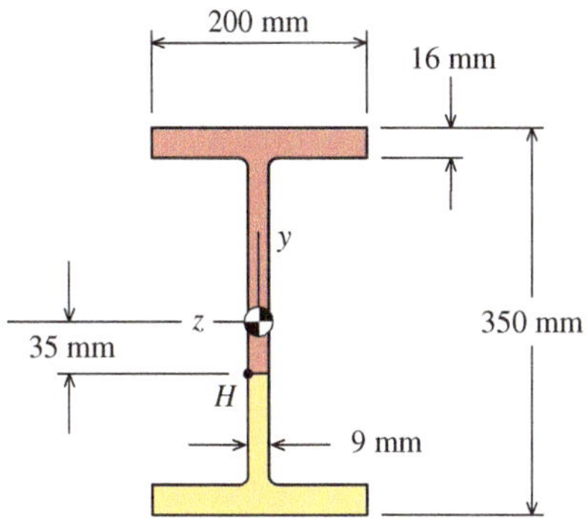

$$\sigma_x = -\frac{My}{I_z} = -\frac{(155 \text{ kN}\cdot\text{m})(-35 \text{ mm})(1{,}000 \text{ N/kN})(1{,}000 \text{ mm/m})}{202.74 \times 10^6 \text{ mm}^4}$$

$$= 26.76 \text{ MPa} = 26.76 \text{ MPa (T)}$$

Note that this tensile normal stress acts parallel to the longitudinal axis of the beam—that is, in the x direction.

Before the shear stress can be computed for point H, Q for the highlighted area must be calculated. The first moment of the highlighted area about the z centroidal axis is $Q = 642{,}652 \text{ mm}^3$. The shear stress at H due to beam flexure is then

$$\tau = \frac{VQ}{I_z t} = \frac{(115 \text{ kN})(642{,}652 \text{ mm}^3)(1{,}000 \text{ N/kN})}{(202.74 \times 10^6 \text{ mm}^4)(9 \text{ mm})} = 40.50 \text{ MPa}$$

This shear stress acts in the same direction as the internal shear force V. Therefore, on the *right face* of the stress element, the shear stress τ acts *downward*.

Stress Element for Point H

The tensile normal stress due to the bending moment acts on the x faces of the stress element. The shear stress acts *downward* on the $+x$ face of the stress element. After the proper shear stress direction has been established on one face, the shear stress directions on the other three faces are known.

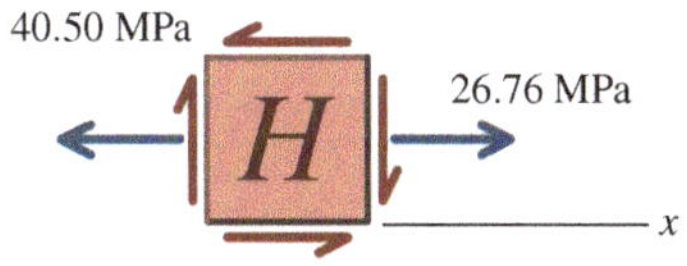

Stress Transformation Results at H

The principal stresses and the maximum shear stress at H can be determined from the stress transformation equations and procedures detailed in Chapter 12. The results of these calculations are shown in the accompanying figure.

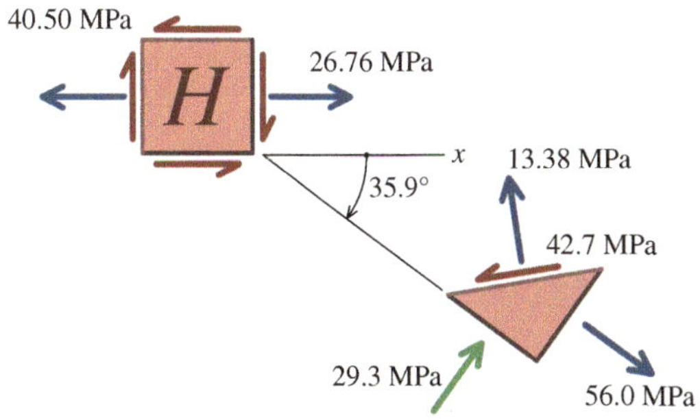

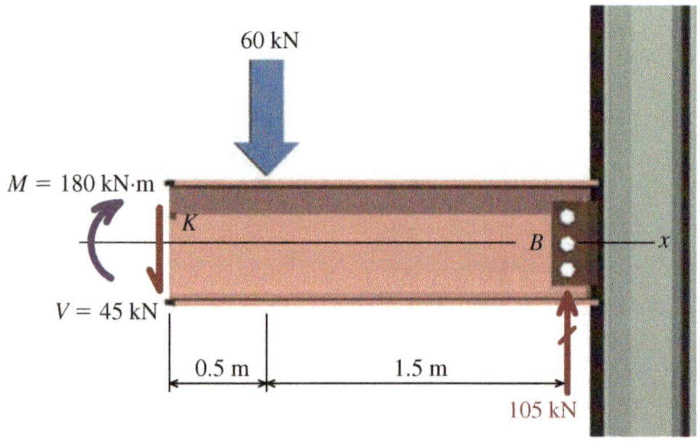

Shear Force and Bending Moment at *K*

At the location of point *K*, the internal shear force is $V = -45$ kN and the internal bending moment is $M = 180$ kN · m. These internal forces act in the directions shown.

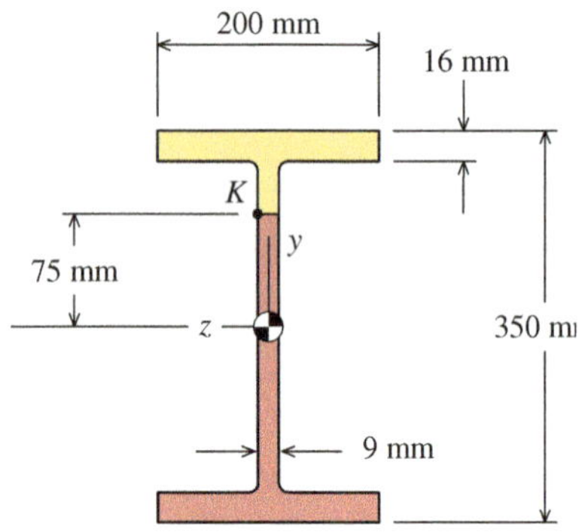

Normal and Shear Stresses at *K*

Point *K* is located 75 mm above the *z* centroidal axis; therefore, $y = 75$ mm. The bending stress at *K* can be calculated from the flexure formula:

$$\sigma_x = -\frac{My}{I_z} = -\frac{(180 \text{ kN}\cdot\text{m})(75 \text{ mm})(1{,}000 \text{ N/kN})(1{,}000 \text{ mm/m})}{202.74 \times 10^6 \text{ mm}^4}$$

$$= -66.6 \text{ MPa} = 66.6 \text{ MPa (C)}$$

Note that this compressive normal stress acts parallel to the longitudinal axis of the beam—that is, in the *x* direction.

To compute the shear stress at *K*, *Q* must be calculated for the highlighted area. The first moment of the highlighted area about the *z* centroidal axis is $Q = 622{,}852 \text{ mm}^3$. The shear stress at *K* due to beam flexure is then

$$\tau = \frac{VQ}{I_z t} = \frac{(45 \text{ kN})(622{,}852 \text{ mm}^3)(1{,}000 \text{ N/kN})}{(202.74 \times 10^6 \text{ mm}^4)(9 \text{ mm})} = 15.36 \text{ MPa}$$

Generally, the magnitude of *V* is used in this calculation and the direction of the shear stress is determined by inspection. The shear stress acts in the same direction as the internal shear force *V*. Therefore, on the *left face* of the stress element, the shear stress τ acts *downward.*

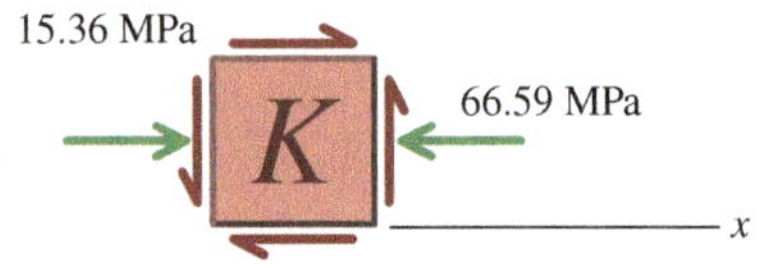

Stress Element for Point *K*

The compressive bending stress acts on the *x* faces of the stress element, and the shear stress acts *downward* on the $-x$ face of the stress element. After the proper shear stress direction has been established on one face, the shear stress directions on the other three faces are known.

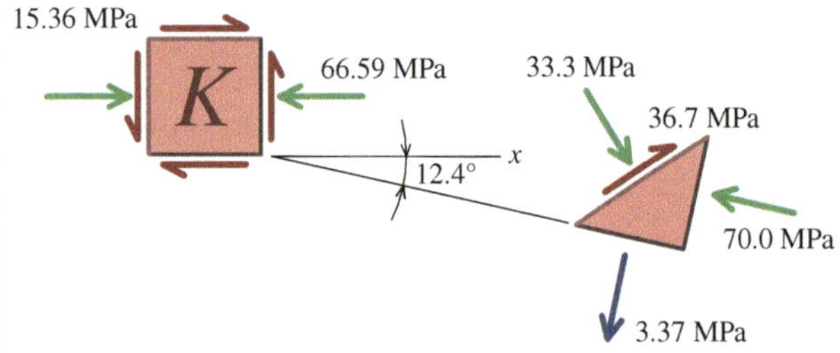

Stress Transformation Results at *K*

The principal stresses and the maximum shear stress at *K* are shown in the accompanying figure.

MecMovies

EXAMPLES

M15.2 A cantilever beam has a uniformly distributed load of 2 kips/ft. The beam cross section is a tee shape. Determine the principal stresses and the maximum shearing stress at point D, located 4 in. above the bottom of the tee stem and at a distance of 1 ft from the fixed support.

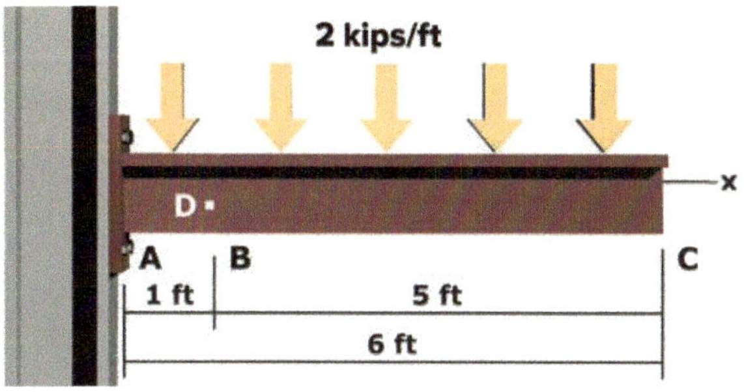

M15.3 A steel rectangular tube shape is used as a beam to support the loads shown. Determine the principal stresses and the maximum shear stress at point H, which is located 1 m to the right of pin support A.

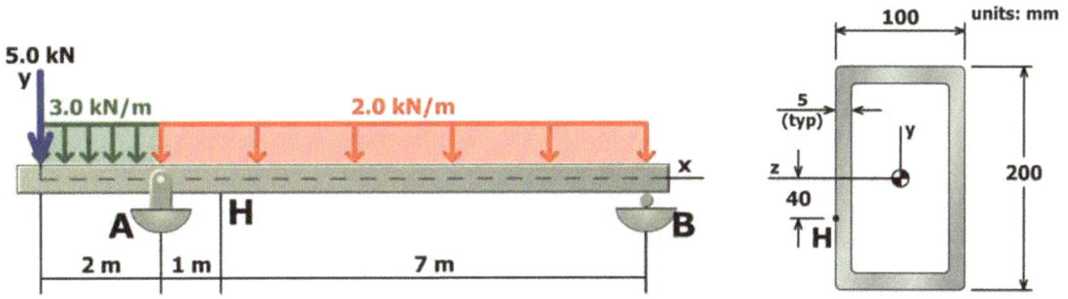

M15.4 A steel wide-flange beam carries loads that create an internal shear force of $V = 60$ kips and an internal bending moment of M = 150 kip·ft at a particular point along the span. Determine the normal and shear stresses that act at point B, located on the surface of the steel shape, 3 in. above the centroid.

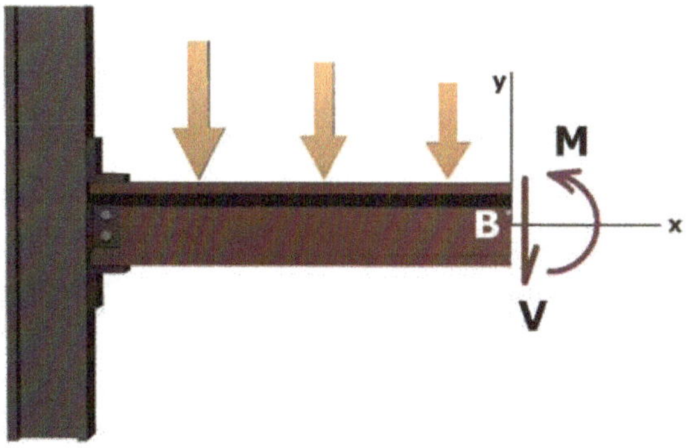

EXAMPLE 15.3

A steel hollow structural section (HSS) is supported by a pin connection at A and by an inclined steel rod (1) at B as shown. A concentrated load of 20 kips is applied to the beam at C. Determine the principal stresses and maximum shear stress acting at point H.

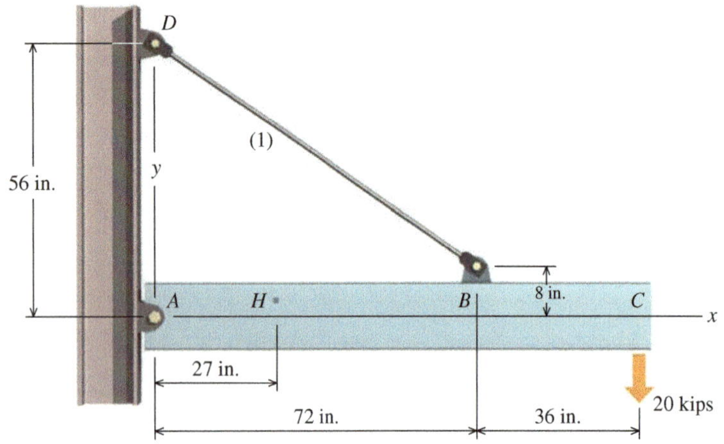

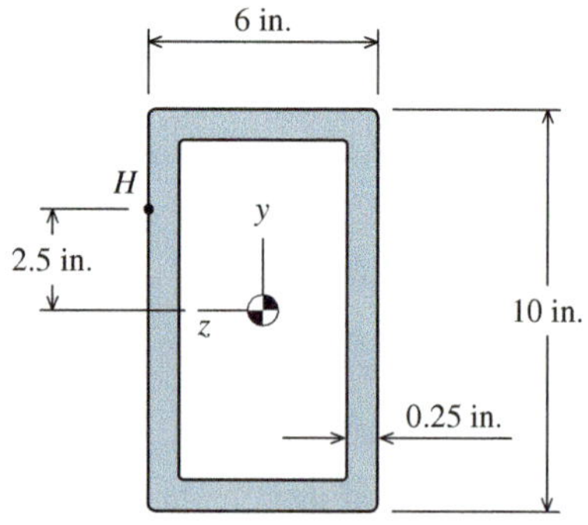

Cross-sectional dimensions.

Plan the Solution

Inclined rod (1), a two-force member, will provide a vertical reaction force to the beam at B. Since the rod is inclined, it creates an axial force that compresses the beam in the region between A and B. The rod is also connected 8 in. above the centerline of the HSS, and this eccentricity produces an additional bending moment in the beam. The analysis begins by calculating the beam reaction forces at A and B. Once these forces have been determined, a free-body diagram (FBD) that cuts through the beam at H will be drawn to establish the equivalent forces acting at the section of interest. The normal and shear stresses created by the equivalent forces will be calculated and shown on a stress element for point H. Stress transformations will be used to calculate the principal stresses and the maximum shear stresses at H.

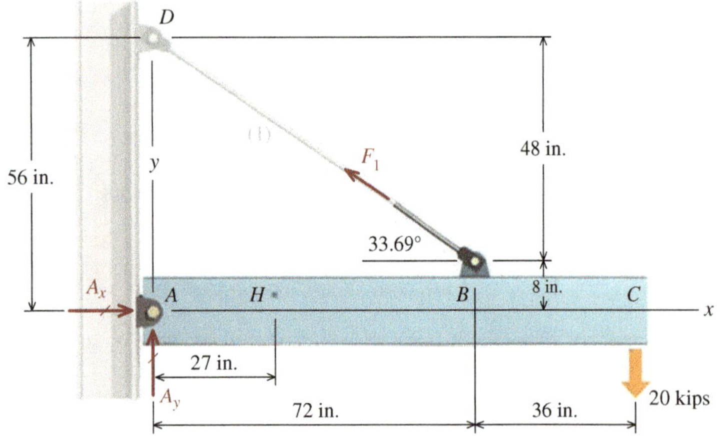

SOLUTION

Beam Reactions

An FBD of the beam is drawn, showing the horizontal and vertical reaction forces from the pin connection at A and the axial force in inclined rod (1), which is a two-force member. Note that the angle of rod (1) must take into account the 8 in. offset of the rod connection from the centerline of the HSS:

$$\tan\theta = \frac{56 \text{ in.} - 8 \text{ in.}}{72 \text{ in.}} = 0.66667$$

$$\therefore \theta = 33.69°$$

The following equilibrium equations can be developed from the FBD:

$$\Sigma F_x = A_x - F_1 \cos(33.69°) = 0 \quad \text{(a)}$$

$$\Sigma F_y = A_y + F_1 \sin(33.69°) - 20 \text{ kips} = 0 \quad \text{(b)}$$

$$\Sigma M_A = F_1 \sin(33.69°)(72 \text{ in.}) + F_1 \cos(33.69°)(8 \text{ in.}) - (20 \text{ kips})(72 \text{ in.} + 36 \text{ in.}) = 0 \quad \text{(c)}$$

From Equation (c), the internal axial force in rod (1) is found to be $F_1 = 46.4$ kips. This result can be substituted into Equations (a) and (b) to obtain the reactions at pin A: $A_x = 38.6$ kips and $A_y = -5.71$ kips. Since the value computed for A_y is negative, this reaction force actually acts opposite to the direction assumed initially.

FBD Exposing Internal Forces at *H*

An FBD that shows the external reaction forces at pin A is cut through the section containing point H. The internal forces acting at the section of interest can be calculated from this FBD.

The internal axial force is $F = 38.6$ kips, acting in compression. The internal shear force is $V = 5.71$ kips, acting upward on the exposed right face (i.e., the $+x$ face) shown in the FBD. The internal bending moment can be calculated by summing moments about the centerline of the HSS at the section containing point H:

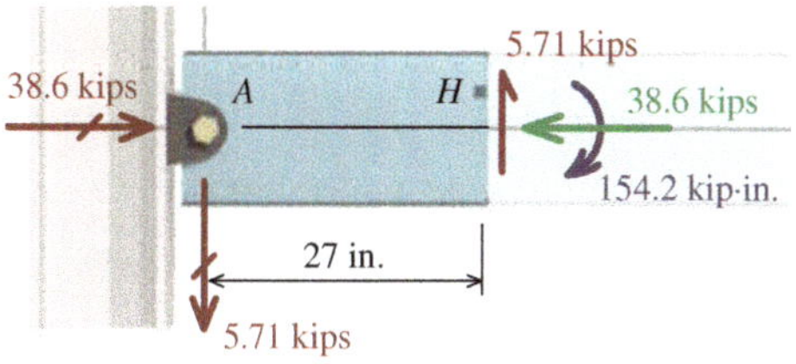

FBD at *H*.

$$\Sigma M_H = (5.71 \text{ kips})(27 \text{ in.}) - M = 0 \qquad \therefore M = 154.2 \text{ kip}\cdot\text{in.} \quad \text{(d)}$$

Section Properties

The cross-sectional area of the HSS is

$$A = (6 \text{ in.})(10 \text{ in.}) - (5.5 \text{ in.})(9.5 \text{ in.}) = 7.75 \text{ in.}^2$$

The moment of inertia of the cross-sectional area about the z centroidal axis is

$$I_z = \frac{(6 \text{ in.})(10 \text{ in.})^3}{12} - \frac{(5.5 \text{ in.})(9.5 \text{ in.})^3}{12} = 107.04 \text{ in.}^4$$

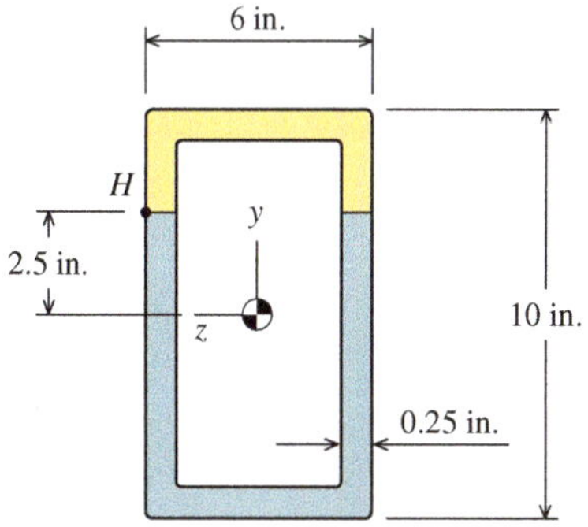

The first moment of area corresponding to point H is calculated for the highlighted area as

$$Q_H = 2(0.25 \text{ in.})(2.5 \text{ in.})(3.75 \text{ in.}) + (5.5 \text{ in.})(0.25 \text{ in.})(4.875 \text{ in.}) = 11.391 \text{ in.}^3$$

Stress Calculations

Axial stress due to F: The internal axial force $F = 38.6$ kips creates a uniformly distributed compressive normal stress that acts in the x direction. The stress magnitude is

$$\sigma_{\text{axial}} = \frac{F}{A} = \frac{38.6 \text{ kips}}{7.75 \text{ in.}^2} = 4.98 \text{ ksi (C)}$$

Bending stress due to M: The 154.2 kip · in. internal bending moment acting as shown creates tensile normal stresses above the z centroidal axis in the HSS. To compute the bending stress by the flexure formula, the bending moment has a value $M = -154.2$ kip · in. and the y coordinate for point H is $y = 2.5$ in. The bending stress is then

$$\sigma_{\text{bend}} = -\frac{My}{I_z} = -\frac{(-154.2 \text{ kip}\cdot\text{in.})(2.5 \text{ in.})}{107.04 \text{ in.}^4} = 3.60 \text{ ksi (T)}$$

Shear stress due to V: The shear stress at H associated with the 5.71 kip shear force can be calculated from the shear stress formula:

$$\tau_H = \frac{VQ}{I_z t} = \frac{(5.71 \text{ kips})(11.391 \text{ in.}^3)}{(107.04 \text{ in.}^4)(2 \times 0.25 \text{ in.})} = 1.215 \text{ ksi}$$

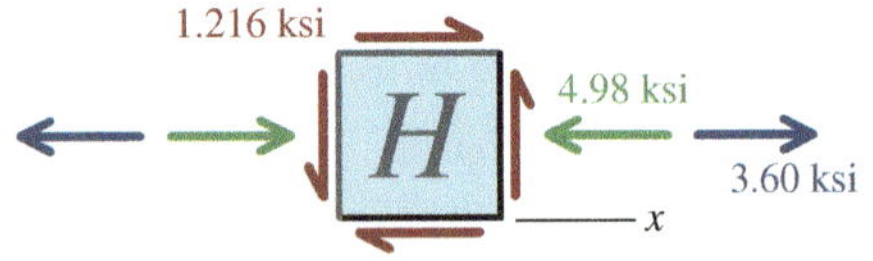

Stress element: The normal and shear stresses at H are shown on the stress element. The normal stresses due to both the axial force and the bending moment act in the x direction.

The direction of the shear stress on the stress element can be determined from the FBD at H. The internal shear force at H acts upward on the right face of the FBD. The shear stress due to $V = 5.71$ kips acts in the same direction—that is, upward on the right face of the stress element.

Stress Transformation Results at *H*

The principal stresses and the maximum shear stress at H can be determined from the stress transformation equations and procedures detailed in Chapter 12. The results of these calculations are shown in the accompanying figure.

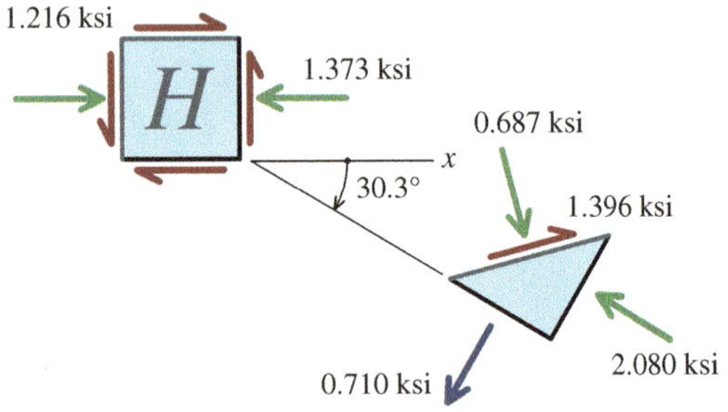

MecMovies

EXAMPLES

M15.2 The inverted tee shape is subjected to a transverse shear force V and a bending moment M, each acting in the direction shown. Determine the bending stress, the transverse shear stress magnitude, the principal stresses, and the maximum shear stress acting at location H.

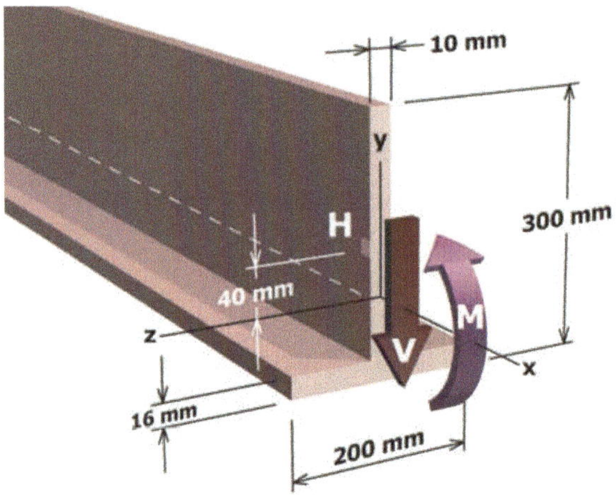

FIGURE M15.2

M15.3 The rectangular tube is subjected to a transverse shear force V and a bending moment M, each acting in the direction shown. Determine the bending stress, the transverse shear stress magnitude, the principal stresses, and the maximum shear stress acting at location H.

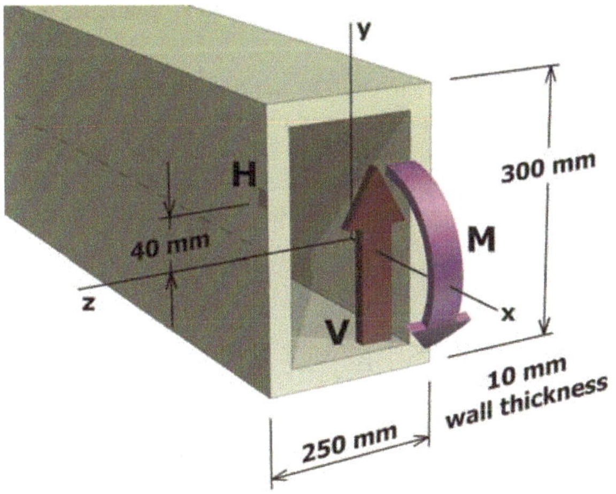

FIGURE M15.3

M15.4 The wide-flange shape is subjected to a transverse shear force V and a bending moment M, each acting in the direction shown. Determine the bending stress, the transverse shear stress magnitude, the principal stresses, and the maximum shear stress acting at location H.

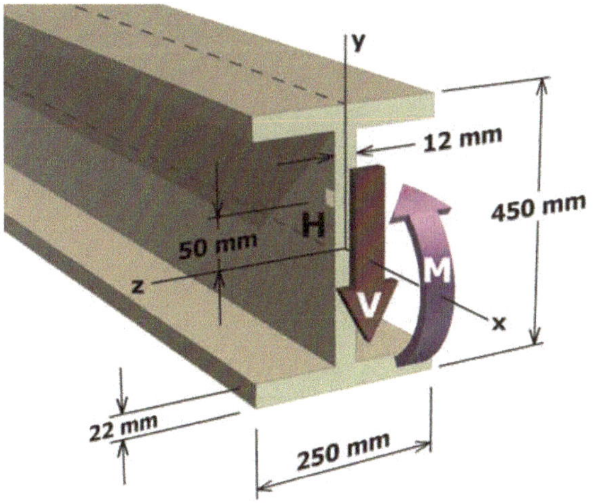

FIGURE M15.4

PROBLEMS

P15.12 The extruded flexural member shown in Figure P15.12*a* is subjected to a shear force $V = 3{,}600$ N and a bending moment $M = 550$ N·m. The cross-sectional dimensions of the shape are shown in Figure P15.12*b*, where $a = 15$ mm, $b = 8$ mm, $c = 30$ mm, and $d = 42$ mm. Determine the principal stresses and the maximum shear stress acting at point H, which is located at $h = 20$ mm below the upper surface of the member.

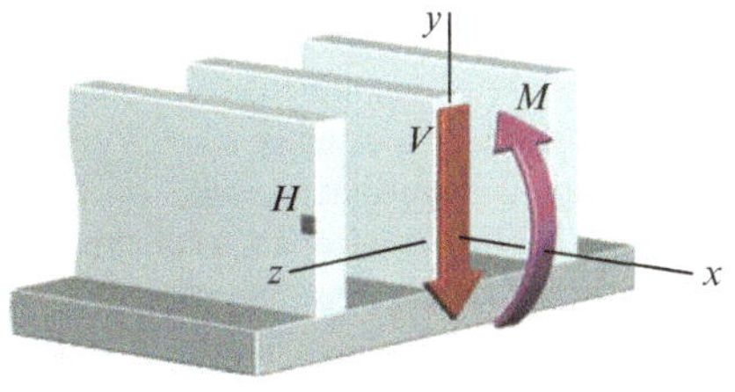

FIGURE P15.12*a*

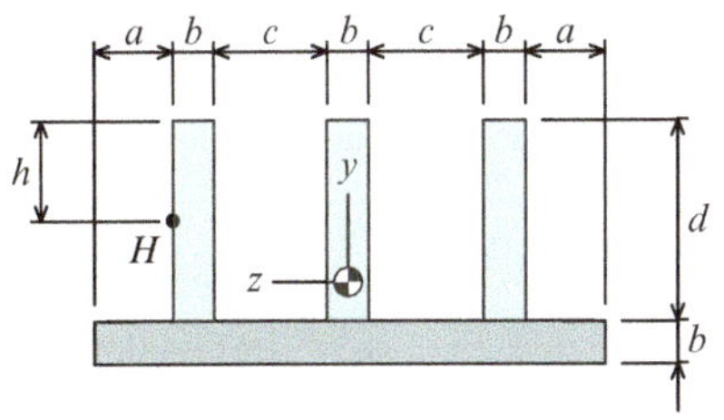

FIGURE P15.12*b*

P15.13 The beam cross section shown in Figure P15.13*a*/14*a* will consist of two square tubes that are welded to a rectangular web plate. The dimensions of the cross section as shown in Figure P15.13*b*/14*b* are $d = 240$ mm, $t_w = 12$ mm, $b = 80$ mm, and $t = 8.0$ mm. The beam is subjected to an axial force $P = 30$ kN, a shear force $V = 180$ kN, and a bending moment $M = 75$ kN·m. Determine the principal stresses and the maximum shear stress acting at point H, located at $y_H = 90$ mm above the z centroidal axis. Show these stresses on an appropriate sketch.

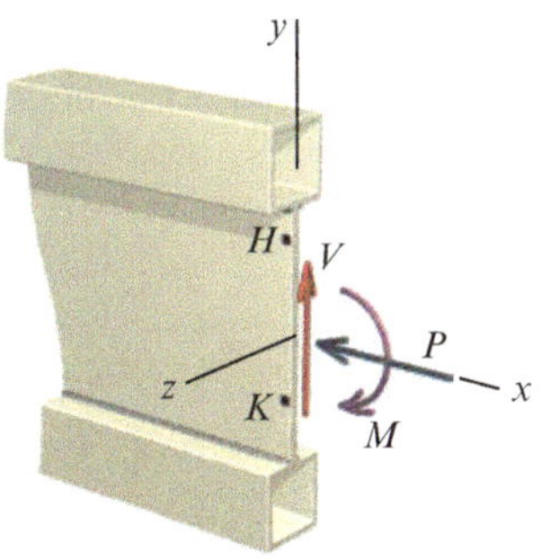

FIGURE P15.13*a*/14*a*

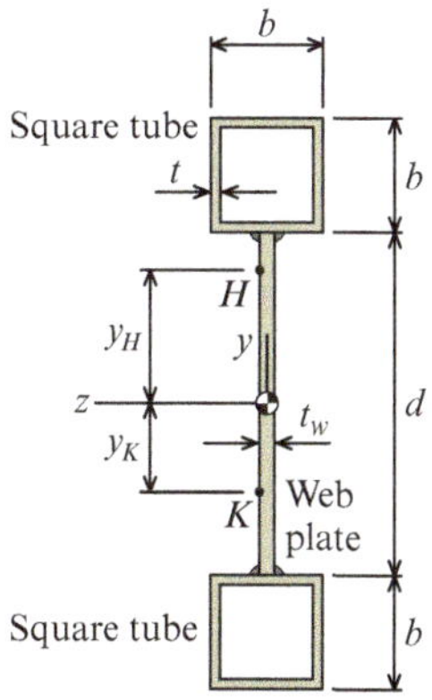

FIGURE P15.13*b*/14*b*

P15.14 The beam cross section shown in Figure P15.13*a*/14*a* will consist of two square tubes that are welded to a rectangular web plate. The dimensions of the cross section as shown in

Figure P15.13*b*/14*b* are $d = 240$ mm, $t_w = 12$ mm, $b = 80$ mm, and $t = 8.0$ mm. The beam is subjected to an axial force $P = 30$ kN, a shear force $V = 180$ kN, and a bending moment $M = 75$ kN·m. Determine the principal stresses and the maximum shear stress acting at point *K*, located at $y_K = 60$ mm below the *z* centroidal axis. Show these stresses on an appropriate sketch.

P15.15 The cantilever beam shown in Figure P15.15*a* is subjected to concentrated loads $P_x = 18$ kips and $P_y = 32$ kips. The cross-sectional dimensions of the rectangular tube shape shown in Figure P15.15*b* are $b = 8$ in., $d = 12$ in., and $t = 0.25$ in. Point *H* is located at a distance $a = 24$ in. to the left of the concentrated loads. Calculate the principal stresses and maximum in-plane shear stress at point *H* for the following values of *h*:

(a) $h = 4.0$ in.
(b) $h = 6.0$ in.
(c) $h = 8.0$ in.

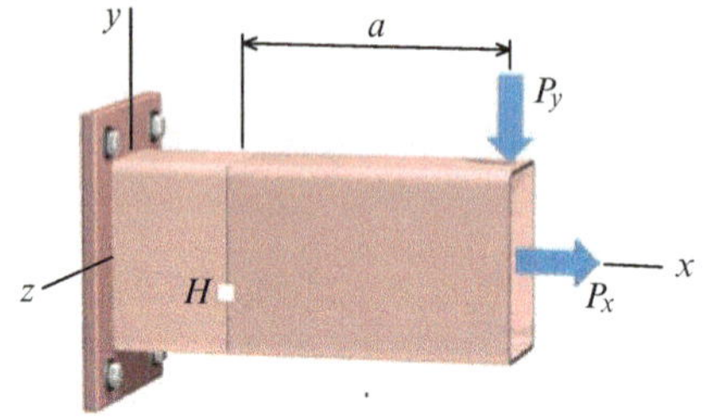

FIGURE P15.15*a*

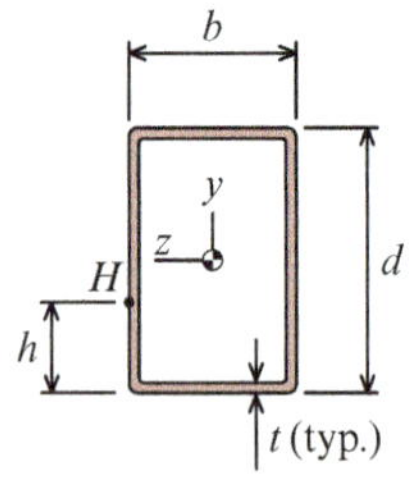

FIGURE P15.15*b*

P15.16 Loads $P_x = 13$ kips and $P_y = 20$ kips act on the titanium alloy [$E = 16{,}500$ ksi; $\nu = 0.33$] bar shown in Figure P15.16/17. The dimensions of the bar are $d = 4.0$ in., $t = 0.75$ in., and $L = 18$ in. Gages *a* and *b* are oriented at $\beta = 40°$ as shown in the figure. Point *H* is located a distance $h = 1.5$ in. above the bottom surface of the bar. Determine the normal strains expected in gages *a* and *b*.

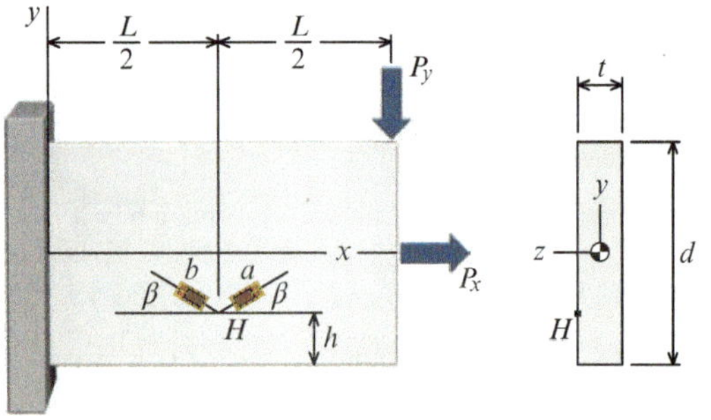

FIGURE P15.16/17

P15.17 Normal strain values $\varepsilon_a = 740$ με and $\varepsilon_b = -180$ με were recorded with the two strain gages mounted at point *H* on the aluminum alloy [$E = 10{,}000$ ksi; $\nu = 0.33$] bar shown in Figure P15.16/17. The dimensions of the bar are $d = 2.50$ in., $t = 0.375$ in., and $L = 12$ in. Gages *a* and *b* are oriented at $\beta = 35°$ as shown in the figure. Point *H* is located a distance $h = 1.0$ in. above the bottom surface of the bar. Determine the magnitudes of loads P_x and P_y.

P15.18 The beam shown in Figure P15.18/19 spans a distance $L = 30$ in., and its cross-sectional dimensions are $b = 1.0$ in. and $d = 4.0$ in. A load $P = 12{,}000$ lb is applied at midspan. Point *K* is located at a distance $a = 6$ in. from the roller support at *C*. Calculate the maximum compressive normal stress at point *K* for the following values of *k*:

(a) $k = 1.5$ in.
(b) $k = 2.0$ in.
(c) $k = 2.5$ in.

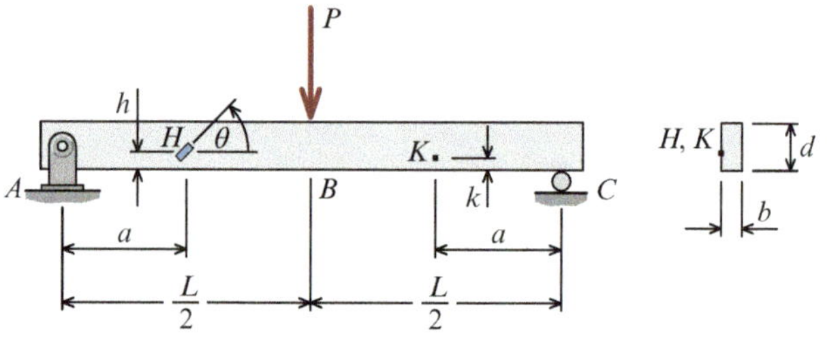

FIGURE P15.18/19

P15.19 A single strain gage is mounted on a simply supported beam at point *H* as shown in Figure P15.18/19. The beam is made of an aluminum alloy [$E = 70$ GPa; $\nu = 0.33$], and it spans a distance $L = 1.0$ m. The cross-sectional dimensions are $b = 20$ mm and $d = 60$ mm. Point *H* is located at a distance $a = 0.3$ m from the pin support at *A*, and the gage is aligned at an angle $\theta = 45°$ as shown. If a load $P = 9{,}000$ N is applied at midspan, what strains should be expected in the gage for the following values of *h*:

(a) $h = 20$ mm.
(b) $h = 30$ mm.
(c) $h = 40$ mm.

P15.20 The simply supported beam shown in Figure P15.20*a*/21*a* supports three concentrated loads. The loads at *B* and *C* each have a magnitude of $P = 25$ kips, and the load at *D* is $Q = 60$ kips. The beam span is $L = 32$ ft. The cross-sectional dimensions of the beam as shown in Figure P15.20*b*/21*b* are $b_f = 12.0$ in., $t_f = 0.85$ in., $d = 20.0$ in., $t_w = 0.50$ in., and $y_H = 3.5$ in. Determine the principal stresses and the maximum shear stress acting at point *H*, which is located at $x_H = 4$ ft from the left-hand support. Show these stresses on an appropriate sketch.

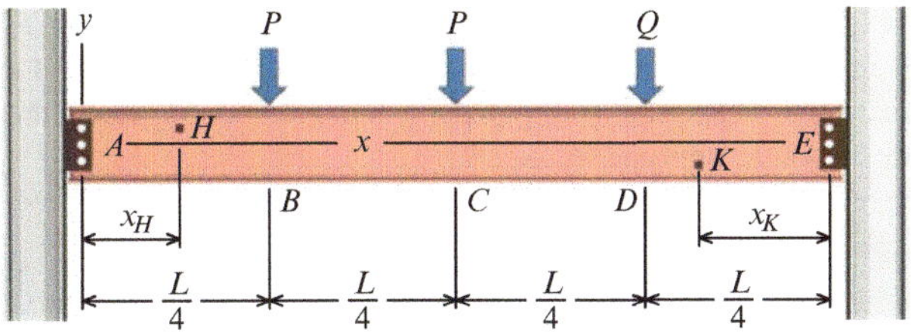

FIGURE P15.20*a*/21*a*

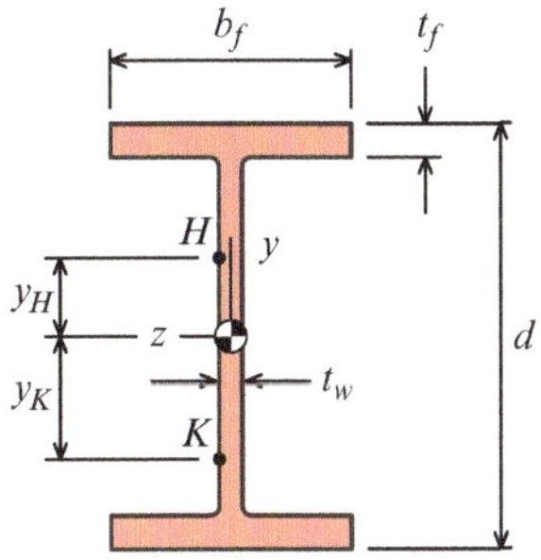

FIGURE P15.20*b*/21*b*

P15.21 The simply supported beam shown in Figure P15.20*a*/21*a* supports three concentrated loads. The loads at B and C each have a magnitude of $P = 70$ kN, and the load at D is $Q = 260$ kN. The beam span is $L = 10$ m. The cross-sectional dimensions of the beam as shown in Figure P15.20*b*/21*b* are $b_f = 300$ mm, $t_f = 20$ mm, $d = 500$ mm, $t_w = 12$ mm, and $y_K = 120$ mm. Determine the principal stresses and the maximum shear stress acting at point K, which is located at $x_K = 1.75$ m from the left-hand support. Show these stresses on an appropriate sketch.

P15.22 The beam shown in Figure P15.22*a* is supported by a tie bar at B and by a pin connection at C. The beam span is $L = 7$ m and the uniformly distributed load is $w = 22$ kN/m. The tie bar at B has an orientation of $\theta = 25°$. The cross-sectional dimensions of the beam shown in Figure P15.22*b* are $b_f = 130$ mm, $t_f = 12$ mm, $d = 360$ mm, $t_w = 6$ mm, and $y_H = 50$ mm. Determine the principal stresses and the maximum shear stress acting at point H. Show these stresses on an appropriate sketch.

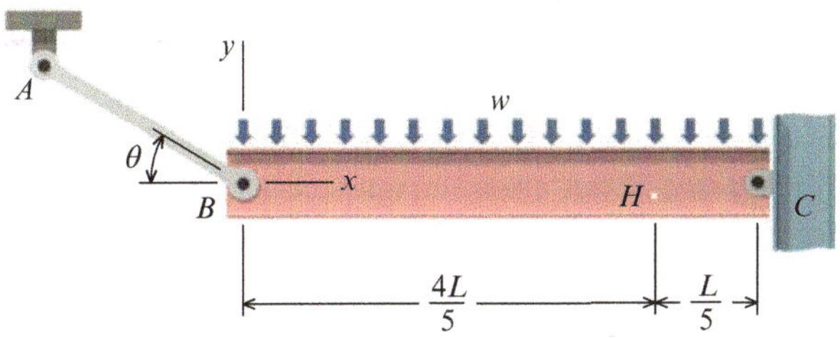

FIGURE P15.22*a*

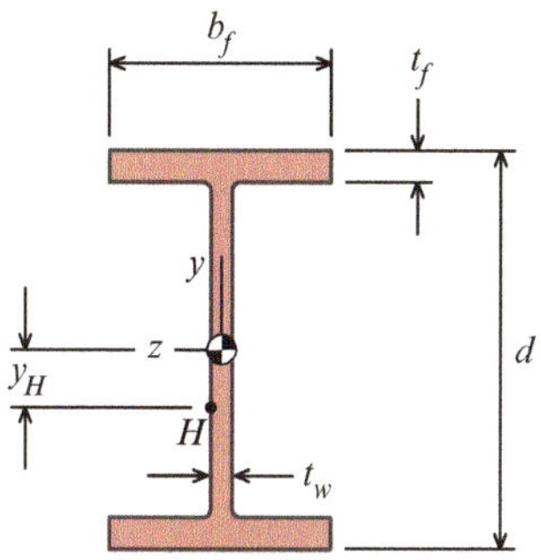

FIGURE P15.22*b*

P15.23 The bracket shown in Figure P15.23*a*/24*a* supports a pulley that has a diameter $D = 120$ mm. The tension in the pulley belt is $P = 200$ N, and the angle of the pulley belt is $\beta = 30°$ as shown. The bracket has dimensions $a = 90$ mm and $c = 150$ mm. The cross-sectional dimensions of the bracket are $b = 10$ mm and $d = 30$ mm as shown in Figure P15.23*b*/24*b*. Determine the principal stresses and the maximum shear stress acting at point H. Show these stresses on an appropriate sketch.

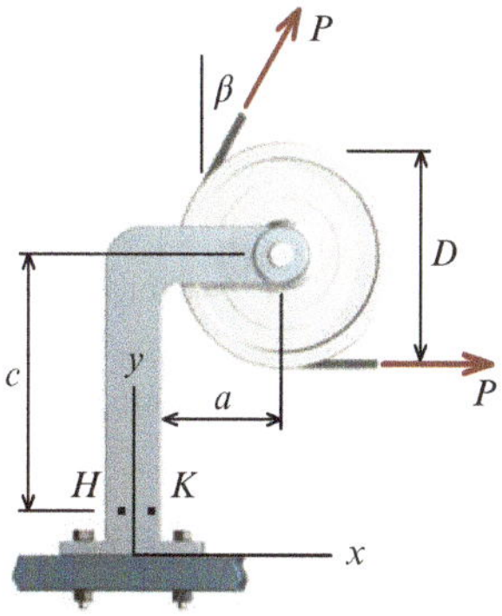

FIGURE P15.23*a*/24*a*

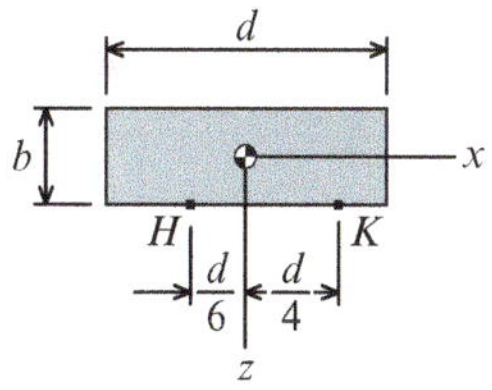

FIGURE P15.23*b*/24*b*

P15.24 The bracket shown in Figure P15.23*a*/24*a* supports a pulley that has a diameter $D = 6$ in. The tension in the pulley belt is $P = 75$ lb, and the angle of the pulley belt is $\beta = 30°$ as shown. The bracket has dimensions $a = 4.5$ in. and $c = 7.5$ in. The cross-sectional dimensions of the bracket are $b = 0.5$ in. and $d = 1.5$ in. as shown in Figure P15.23*b*/24*b*. Determine the principal stresses and the maximum shear stress acting at point K. Show these stresses on an appropriate sketch.

P15.25 A load $P = 75$ kN acting at an angle $\beta = 35°$ is supported by the structure shown in Figure P15.25*a*. The overall dimensions of the structure are $a = 2.4$ m, $b = 0.6$ m, $c = 1.5$ m, $e = 0.32$ m, and $x_1 = 2.2$ m. The cross-sectional dimensions of member *BC* (shown in Figure P15.25*b*) are $b_f = 160$ mm, $t_f = 15$ mm, $d = 300$ mm, $t_w = 10$ mm, and $h = 90$ mm. Determine the principal stresses and the maximum shear stress acting at point *H* in member *BC*. Show these stresses on an appropriate sketch.

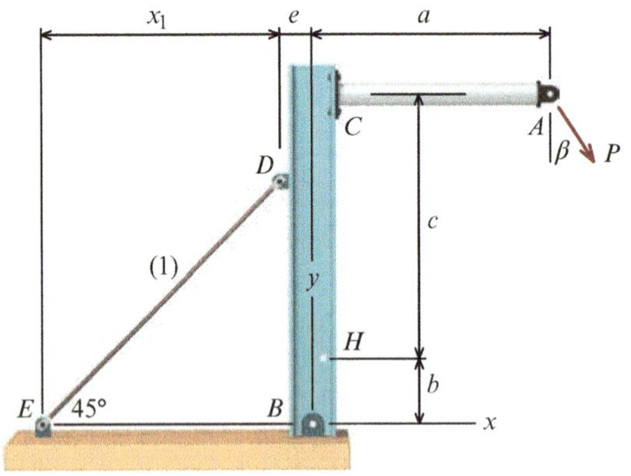

FIGURE P15.25*a*

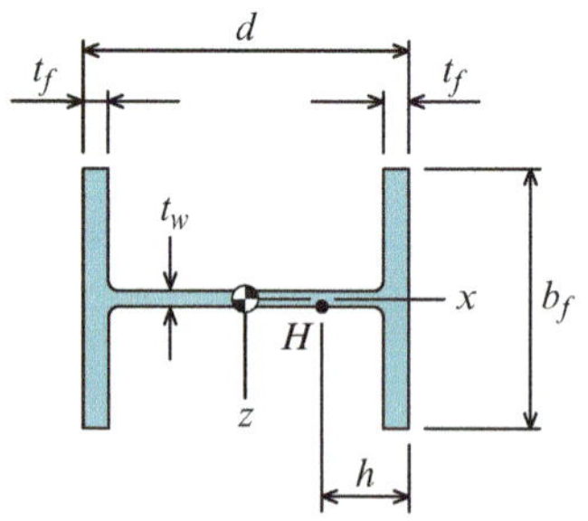

FIGURE P15.25*b*

15.4 General Combined Loadings

In numerous industrial situations, axial, torsional, and flexural loads act simultaneously on machine components and the combined effects of these loads must be analyzed to determine the critical stresses developed in the component. Although an experienced designer can usually predict one or more points where high stress is likely, the most severely stressed point on any particular cross section may not be obvious. As a result, it is almost always necessary to analyze the stresses at more than one point before the critical stresses in the component can be known.

Calculation Procedures

To determine the principal stresses and the maximum shear stress at a particular point in a component subjected to axial, torsion, bending, and pressure loads, the following procedures are useful:

1. Determine the statically equivalent forces and moments acting at the section of interest. In this step, a complicated three-dimensional component or structure subjected to multiple loads is reduced to a simple, prismatic member with no more than three forces and three moments acting at the section of interest.

    a. In finding the statically equivalent forces and moments, it is often convenient to consider the portion of the structure or component that extends from the section of interest to the free end of the structure. The statically equivalent forces at the section of interest are found by summing the loads that act on this portion of the structure (i.e., ΣF_x, ΣF_y, and ΣF_z). Note that these summations do not include the reaction forces.

    b. The statically equivalent moments can be more difficult to determine correctly than the statically equivalent forces, since both a load magnitude and a distance term make up each moment component. One approach is to consider each load on the structure, in turn. The magnitude of the moment, the axis about which the moment acts, and the sign of the moment must be assessed for each load. In addition, a single load on the structure may create unique moments about two axes. After all moment

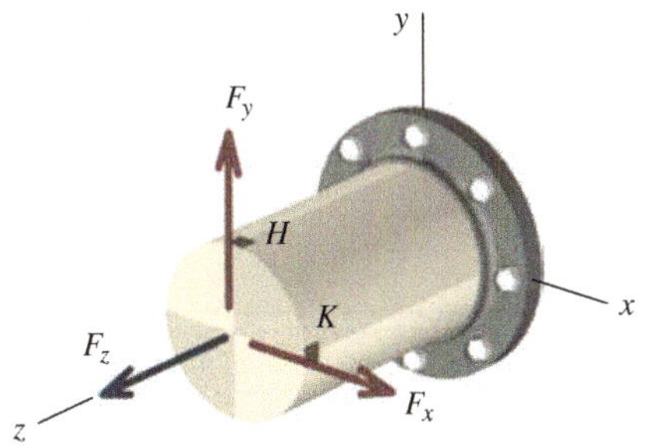

(*a*) Equivalent forces

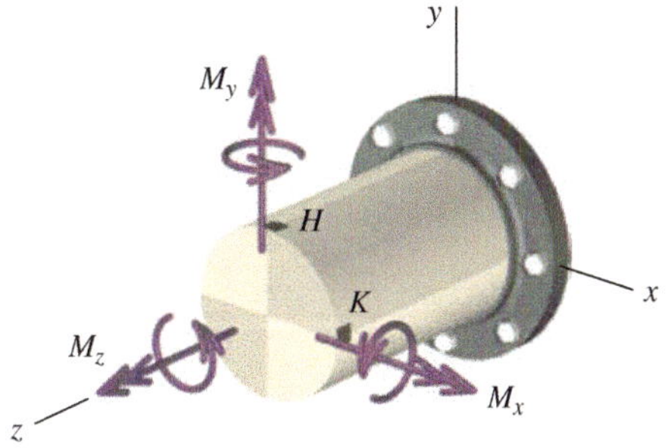

(*b*) Equivalent moments

FIGURE 15.3 Statically equivalent forces and moments at section of interest.

Note that the cross product is *not* commutative; therefore, the moment vector must be computed as $\mathbf{M} = \mathbf{r} \times \mathbf{F}$, not $\mathbf{M} = \mathbf{F} \times \mathbf{r}$.

components have been determined, the statically equivalent moments at the section of interest are found by summing the moment components in each direction (i.e., ΣM_x, ΣM_y, and ΣM_z).

c. As the geometry of the structure and of the loads becomes more complicated, it is often easier to use position vectors and force vectors to calculate equivalent moments. A position vector **r** from the section of interest to the specific point of application of the load is determined, along with a vector **F** describing the forces acting at that point. The moment vector **M** is computed from the cross product of the position and force vectors; that is, $\mathbf{M} = \mathbf{r} \times \mathbf{F}$. If loads are applied at more than one location on the structure, then multiple cross products must be computed.

2. After the statically equivalent forces and moments at the section of interest have been determined, *prepare two sketches* showing the magnitude and direction of all forces and moments acting at the section of interest. Typical sketches are shown in Figures 15.3 and 15.4. These sketches help organize and clarify the results before the stresses are computed.
3. Determine the stresses produced by each of the equivalent forces.
   a. An axial force (force F_z in Figure 15.3*a* and force F_y in Figure 15.4*a*) produces either tensile or compressive normal stress given by $\sigma = F/A$.
   b. Shear stresses computed with the equation $\tau = VQ/It$ are associated with shear forces (forces F_x and F_y in Figure 15.3*a* and forces F_x and F_z in Figure 15.4*a*). Use the direction of the shear force arrow on the section of interest to establish the direction of τ on the corresponding face of the stress element. Recall that τ associated with shear forces is parabolically distributed on a cross section (e.g., see Figure 9.10). For circular cross sections, Q is calculated from Equation (9.7) or Equation (9.8) for solid cross sections and Equation (9.10) for hollow cross sections.
4. Determine the stresses produced by the equivalent moments.
   a. Moments about the longitudinal axis of the component at the section of interest are termed *torques*. In Figure 15.3*b*, M_z is a torque; in Figure 15.4*b*, M_y is a torque. Torques produce shear stresses that are calculated from $\tau = Tc/J$, where J is the *polar moment of inertia*. Recall that the polar moment of inertia for a circular cross section is computed as follows:

$$J = \frac{\pi}{32} d^4 \qquad \text{(for solid circular sections)}$$

$$J = \frac{\pi}{32} [D^4 - d^4] \qquad \text{(for hollow circular section)}$$

Use the direction of the torque to determine the direction of τ on the transverse face of the stress element at the point of interest.

Note that the **area moment of inertia *I*** is used to calculate shear stresses associated with shear forces. Recall that these shear stresses arise from nonuniform bending in the flexural component.

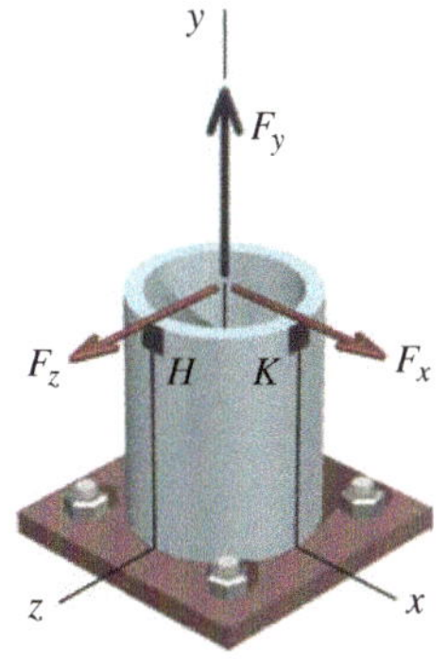

(*a*) Equivalent forces

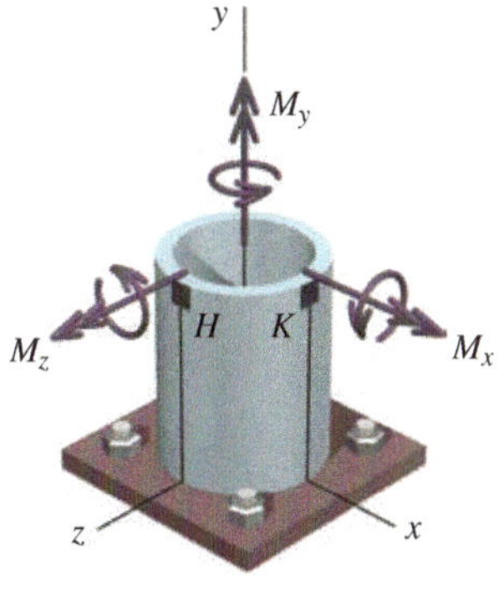

(*b*) Equivalent moments

FIGURE 15.4 Statically equivalent forces and moments at section of interest.

b. Bending moments produce normal stresses that are linearly distributed with respect to the axis of bending. In Figure 15.3*b*, M_x and M_y are bending moments; in Figure 15.4*b*, M_x and M_z are bending moments. Calculate the bending stress magnitude from $\sigma = My/I$, where I is the *area moment of inertia*. Recall that the area moment of inertia for a circular cross section is computed as follows:

$$I = \frac{\pi}{64} d^4 \qquad \text{(for solid circular sections)}$$

$$I = \frac{\pi}{64}[D^4 - d^4] \qquad \text{(for hollow circular sections)}$$

The sense of the stress (either tension or compression) can be determined by inspection. Recall that bending stresses act parallel to the longitudinal axis of the flexural member. Therefore, bending stresses in Figure 15.3*b* act in the z direction, while bending stresses in Figure 15.4*b* act in the y direction.

5. If the component is a hollow circular section that is subjected to internal pressure, then longitudinal and circumferential normal stresses are created. The longitudinal stress is calculated from $\sigma_{\text{long}} = pd/4t$, and the circumferential stress is given by $\sigma_{\text{hoop}} = pd/2t$, where d is the inside diameter of the component. Note that the term t in these two equations refers to the wall thickness of the pipe or tube. The term t appearing in the context of the shear stress equation $\tau = VQ/It$ has a different meaning. For a pipe, the term t in the equation $\tau = VQ/It$ is actually equal to the *wall thickness times 2!*
6. Using the principle of superposition, summarize the results on a stress element, taking care to identify the proper direction of each stress component. As stated previously, it is generally more reliable to use *inspection* to establish the direction of normal and shear stresses acting on the stress element.
7. Once the stresses on orthogonal planes through the point are known and summarized on a stress element, the methods of Chapter 12 can be used to calculate the principal stresses and the maximum shear stresses at the point.

For stresses below the proportional limit, the superposition principle allows similar types of stresses at a specific point to be added together. For instance, all normal stresses acting on the x face of a stress element can be added algebraically.

Examples 15.4–15.7 illustrate the procedure for the solution of elastic combined-load problems.

EXAMPLE 15.4

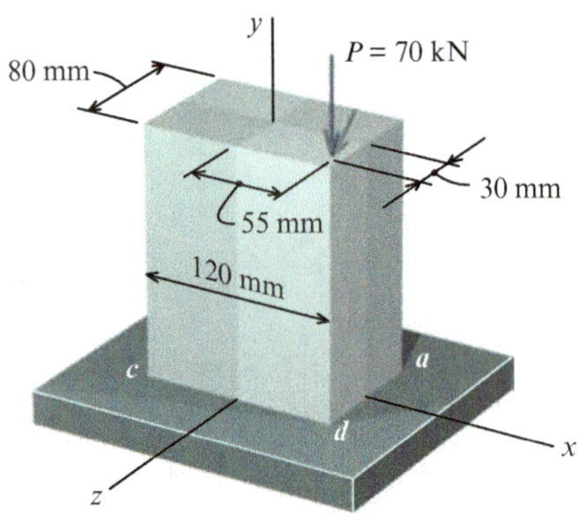

A short post supports a load P = 70 kN as shown. Determine the normal stresses at corners a, b, c, and d of the post.

Plan the Solution

The load P = 70 kN will create normal stresses at the corners of the post in three ways. The axial load P will create compressive normal stress that is distributed uniformly over the cross section. Since P is applied 30 mm away from the x centroidal axis and 55 mm away from the z centroidal axis, P will also create bending moments about these two axes. The moment about the x axis will create tensile and compressive normal stresses that will be linearly distributed across the 80 mm width of the post. The moment about the z axis will create tensile and compressive normal stresses that will be linearly distributed across the 120 mm depth of the cross section. The normal stresses created by the axial force and the bending moments will be determined at each of the four corners, and the results will be superimposed to give the normal stresses at a, b, c, and d.

SOLUTION

Section Properties

The cross-sectional area of the post is

$$A = (80\text{ mm})(120\text{ mm}) = 9{,}600\text{ mm}^2$$

The moment of inertia of the cross-sectional area about the x centroidal axis is

$$I_x = \frac{(120\text{ mm})(80\text{ mm})^3}{12} = 5.120 \times 10^6\text{ mm}^4$$

and the moment of inertia about the z centroidal axis is

$$I_z = \frac{(80\text{ mm})(120\text{ mm})^3}{12} = 11.52 \times 10^6\text{ mm}^4$$

Since $I_z > I_x$ for the coordinate axes shown, the x axis is termed the *weak axis* and the z axis is termed the *strong axis*.

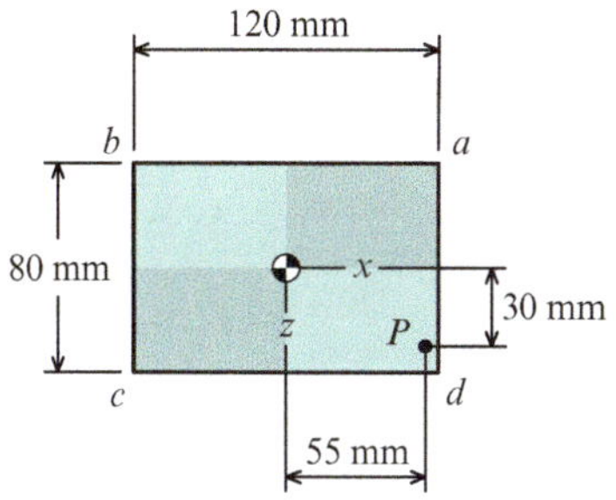

Cross-sectional dimensions and location of application of load.

Equivalent Forces in the Post

The vertical load $P = 70$ kN applied 30 mm from the x axis and 55 mm from the z axis is statically equivalent to an internal axial force $F =$ 70 kN, an internal bending moment $M_x = 2.10$ kN·m, and an internal bending moment $M_z = 3.85$ kN·m. The stresses created by each of these will be considered in turn.

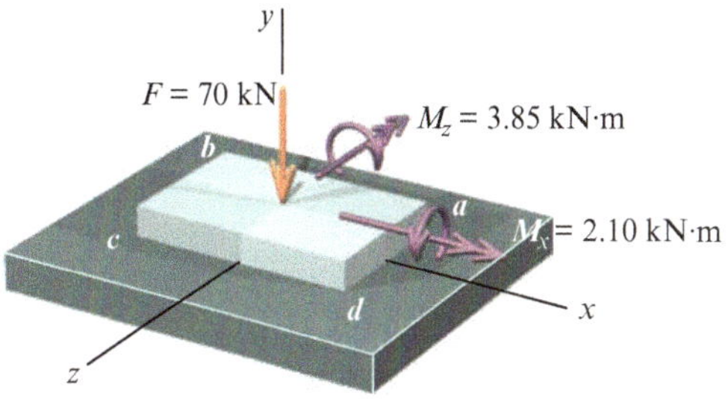

Axial Stress Due to F

The internal axial force $F = 70$ kN creates compressive normal stress that is uniformly distributed over the entire cross section. The magnitude of the stress is computed as

$$\sigma_{\text{axial}} = \frac{F}{A} = \frac{(70\text{ kN})(1{,}000\text{ N/kN})}{9{,}600\text{ mm}^2} = 7.29\text{ MPa (C)}$$

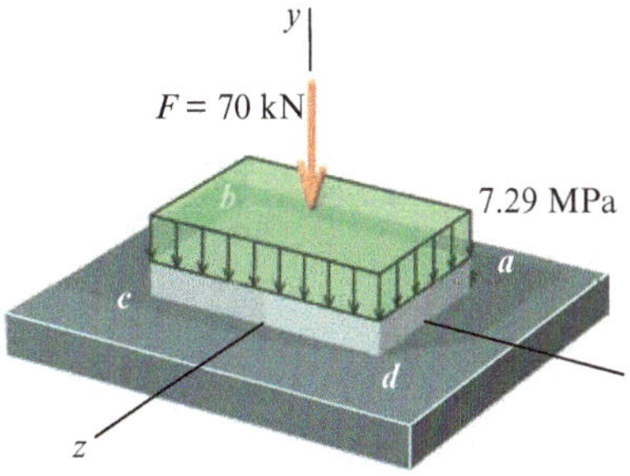

Bending Stress Due to M_x

The bending moment acting as shown about the x axis creates compressive normal stress on side cd, and tensile normal stress on side ab, of the post. The maximum bending stress occurs at a distance $z = \pm 40$ mm from the neutral axis (which is the x centroidal axis for M_x). The maximum bending stress magnitude is

$$\sigma_{\text{bend}} = \frac{M_x z}{I_x} = \frac{(2.10\text{ kN}\cdot\text{m})(\pm 40\text{ mm})(1{,}000\text{ N/kN})(1{,}000\text{ mm/m})}{5.120 \times 10^6\text{ mm}^4}$$
$$= \pm 16.41\text{ MPa}$$

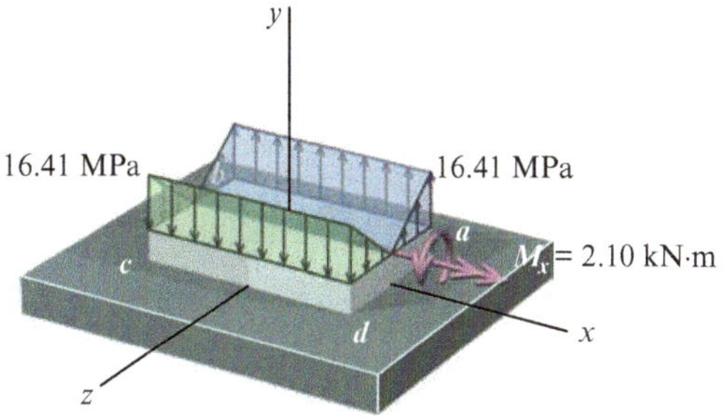

Bending Stress Due to M_z

The bending moment acting as shown about the z centroidal axis creates compressive normal stress on side ad, and tensile normal stress on side bc, of the post. The maximum bending stress occurs at a

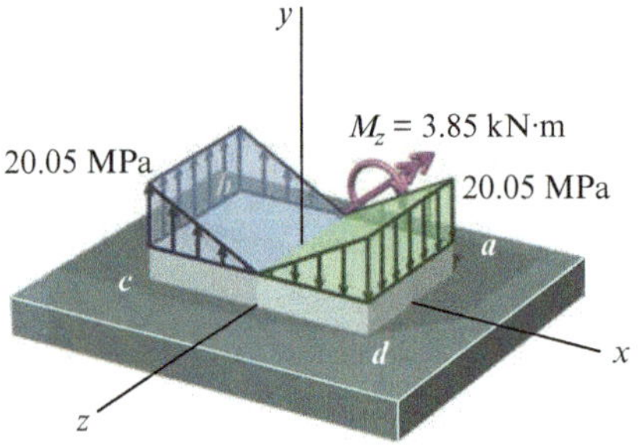

distance $x = \pm 60$ mm from the neutral axis (which is the z centroidal axis for M_z). The maximum bending stress magnitude is

$$\sigma_{\text{bend}} = \frac{M_z x}{I_z} = \frac{(3.85 \text{ kN}\cdot\text{m})(\pm 60 \text{ mm})(1{,}000 \text{ N/kN})(1{,}000 \text{ mm/m})}{11.52 \times 10^6 \text{ mm}^4}$$
$$= \pm 20.05 \text{ MPa}$$

Normal Stresses at Corners *a*, *b*, *c*, and *d*

The normal stresses acting at each of the four corners of the post can be determined by superimposing the preceding results. In all instances, the normal stresses act in the vertical direction—that is, the y direction. The sense of the stress, either tension or compression, can be determined by inspection.

Corner a:

$$\sigma_a = 7.29 \text{ MPa (C)} + 16.41 \text{ MPa (T)} + 20.05 \text{ MPa (C)}$$
$$= -7.29 \text{ MPa} + 16.41 \text{ MPa} - 20.05 \text{ MPa}$$
$$= -10.93 \text{ MPa} = 10.93 \text{ MPa (C)}$$ **Ans.**

Corner b:

$$\sigma_b = 7.29 \text{ MPa (C)} + 16.41 \text{ MPa (T)} + 20.05 \text{ MPa (T)}$$
$$= -7.29 \text{ MPa} + 16.41 \text{ MPa} + 20.05 \text{ MPa}$$
$$= 29.17 \text{ MPa} = 29.17 \text{ MPa (T)}$$ **Ans.**

Corner c:

$$\sigma_c = 7.29 \text{ MPa (C)} + 16.41 \text{ MPa (C)} + 20.05 \text{ MPa (T)}$$
$$= -7.29 \text{ MPa} - 16.41 \text{ MPa} + 20.05 \text{ MPa}$$
$$= -3.65 \text{ MPa} = 3.65 \text{ MPa (C)}$$ **Ans.**

Corner d:

$$\sigma_d = 7.29 \text{ MPa (C)} + 16.41 \text{ MPa (C)} + 20.05 \text{ MPa (C)}$$
$$= -7.29 \text{ MPa} - 16.41 \text{ MPa} - 20.05 \text{ MPa}$$
$$= -43.75 \text{ MPa} = 43.75 \text{ MPa (C)}$$ **Ans.**

MecMovies

EXAMPLES

M15.5 Two loads are applied as shown to the 80 mm by 45 mm cantilever beam. Determine the normal and shear stresses at point H.

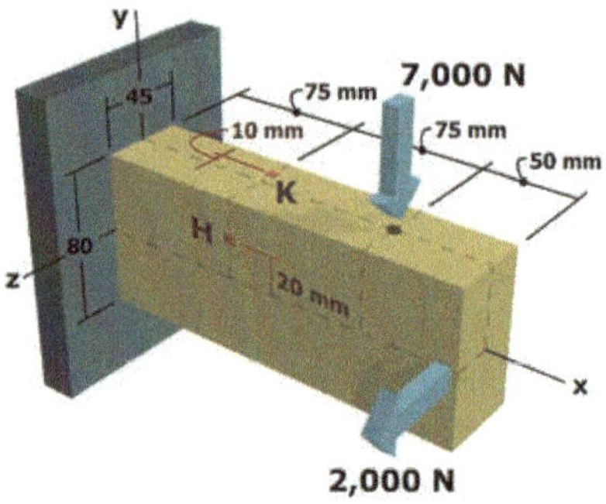

638

M15.7 A rectangular post has cross-sectional dimensions of 200 mm (height) by 80 mm (width). The post is subjected to a concentrated force of 10 kN acting in the x–y plane at an angle of 60° with the vertical direction. Determine the stresses that act in the x and y directions at point B, which is located on the front face of the post, 10 mm to the left of the longitudinal centerline.

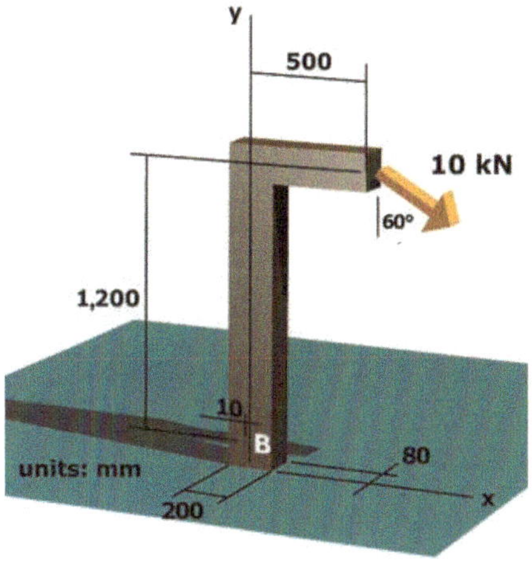

EXAMPLE 15.5

A 36 mm solid shaft supports a 640 N load as shown. Determine the principal stresses and the maximum shear stress at points H and K.

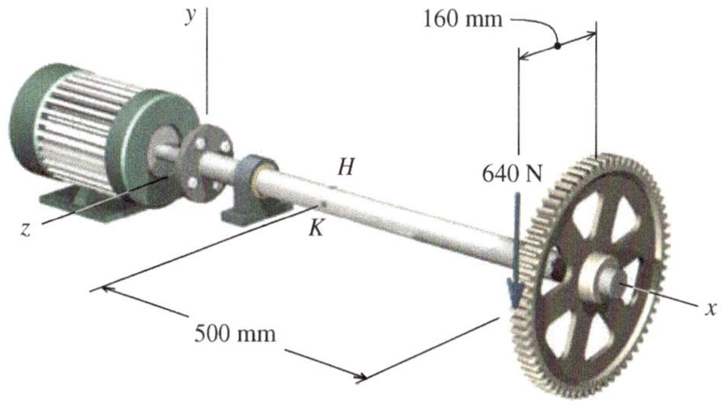

Plan the Solution

The 640 N load applied to the gear will create a vertical shear force, a torque, and a bending moment in the shaft at the section of interest. These internal forces will in turn create normal and shear stresses at points H and K, but because point H is located on the top of the shaft and point K is located on the side of the shaft, the states of stress will differ at the two points. We will begin the solution by determining a system of forces and moments that acts at the section of interest and that is statically equivalent to the 640 N load applied to the teeth of the gear. The normal and shear stresses created by this equivalent force system will be computed and shown in their proper directions on a stress element for both point H and point K. Stress transformation calculations will be used to determine the principal stresses and maximum shear stress for each stress element.

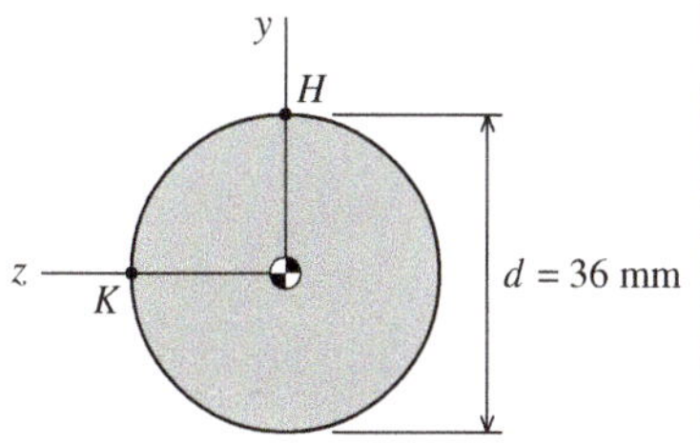

Shaft cross section.

SOLUTION

Equivalent Force System

A system of forces and moments that is statically equivalent to the 640 N load can be readily determined for the section of interest.

The equivalent force at the section is equal to the 640 N load on the gear. Since the line of action of the 640 N load does not pass through the section that includes points H and K, the moments produced by this load must be determined.

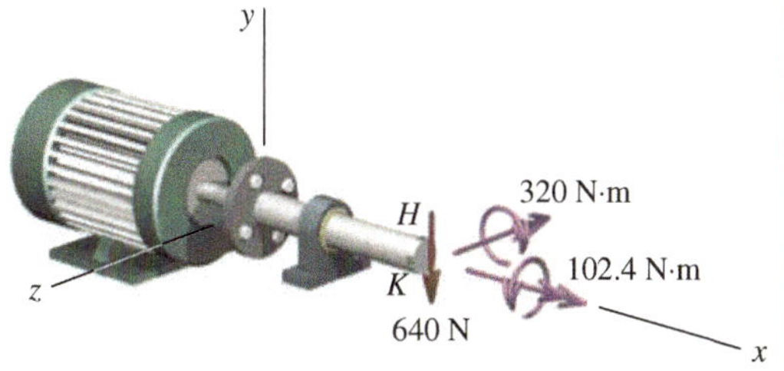

The moment about the x axis (i.e., a torque) is the product of the magnitude of the force and the distance in the z direction from the section of interest to the gear teeth: M_x = (640 N)(160 mm) =

102,400 N · mm = 102.4 N · m. Similarly, the moment about the z axis is the product of the magnitude of the force and the distance in the x direction from points H and K to the gear teeth: M_z = (640 N) (500 mm) = 320,000 N · mm = 320 N · m. By inspection, these moments act in the directions shown.

Alternative method: The geometry of this problem is relatively simple; therefore, the equivalent moments can be determined readily by inspection. For situations that are more complicated, it is sometimes easier to determine the equivalent moments from position vectors and force vectors.

The position vector **r** from the section of interest to the point of application of the load is **r** = 500 mm **i** + 160 mm **k**. The load acting on the gear teeth can be expressed as the force vector **F** = −640 N **j**. The equivalent moment vector **M** can be determined from the cross product **M** = **r** × **F**:

$$\mathbf{M} = \mathbf{r} \times \mathbf{F} = \begin{vmatrix} \mathbf{i} & \mathbf{j} & \mathbf{k} \\ 500 & 0 & 160 \\ 0 & -640 & 0 \end{vmatrix} = 102{,}400 \text{ N}\cdot\text{mm } \mathbf{i} - 320{,}000 \text{ N}\cdot\text{mm } \mathbf{k}$$

For the coordinate axes used here, the axis of the shaft extends in the x direction; therefore, the i component of the moment vector is recognized as a torque, while the k component is simply a bending moment.

Section Properties

The shaft diameter is 36 mm. The polar moment of inertia will be required in order to calculate the shear stress caused by the internal torque in the shaft:

$$J = \frac{\pi}{32} d^4 = \frac{\pi}{32}(36 \text{ mm})^4 = 164{,}896 \text{ mm}^4$$

The moment of inertia of the shaft about the z centroidal axis is

$$I_z = \frac{\pi}{64} d^4 = \frac{\pi}{64}(36 \text{ mm})^4 = 82{,}448 \text{ mm}^4$$

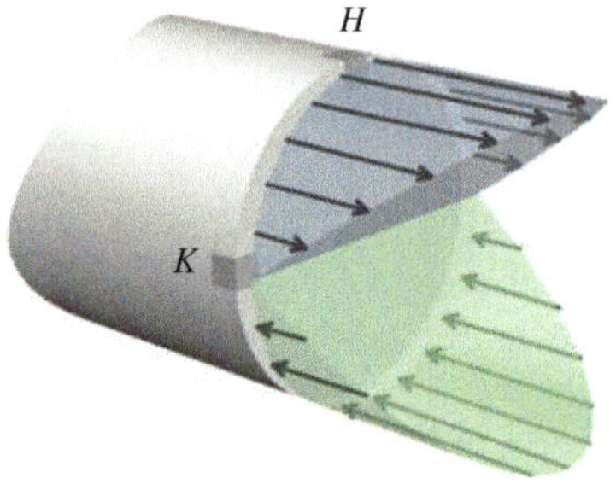

Normal Stresses at H

The 320 N · m bending moment acting about the z axis creates a normal stress that varies over the depth of the shaft. At point H, the bending stress can be computed from the flexure formula as

$$\sigma_x = \frac{Mc}{I_z} = \frac{(320{,}000 \text{ N}\cdot\text{mm})(18 \text{ mm})}{82{,}448 \text{ mm}^4} = 69.9 \text{ MPa (T)}$$

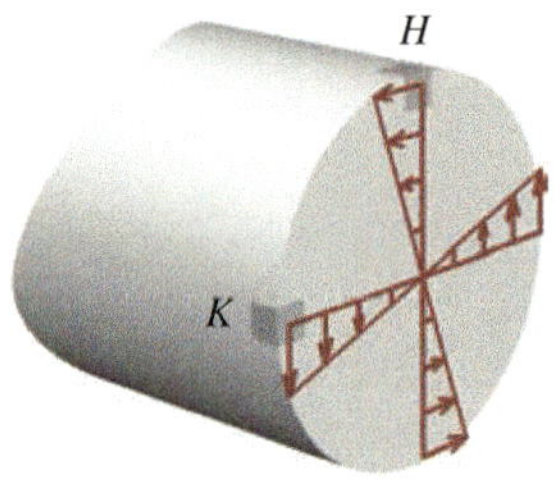

Shear Stress at H

The 102.4 N · m torque acting about the x axis creates shear stress at H. The magnitude of this shear stress can be calculated from the elastic torsion formula:

$$\tau = \frac{Tc}{J} = \frac{(102{,}400 \text{ N}\cdot\text{mm})(36 \text{ mm}/2)}{164{,}896 \text{ mm}^4} = 11.18 \text{ MPa}$$

The transverse shear stress associated with the 640 N shear force is zero at H.

640

Combined Stresses at H

The normal and shear stresses acting at point H can be summarized on a stress element.

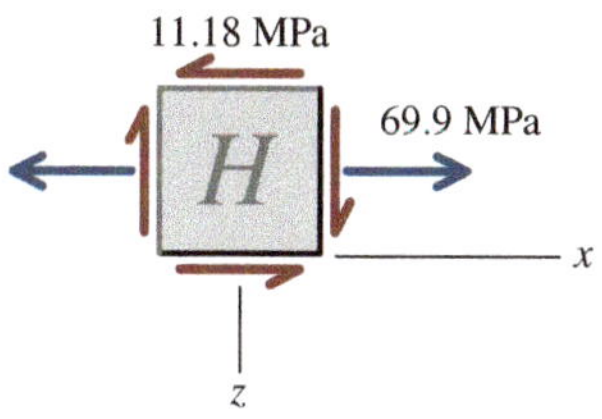

Stress Transformation Results at H

The principal stresses and the maximum shear stress at H are shown in the accompanying figure.

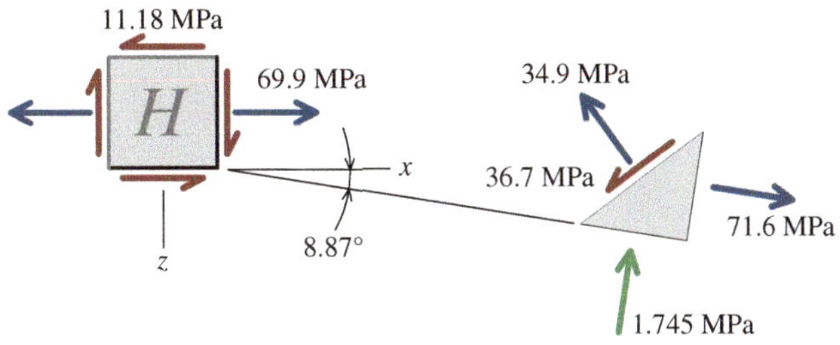

Normal Stresses at K

The 320 N·m bending moment acting about the z axis creates normal stress that varies over the depth of the shaft. Point K, however, is located on the z axis, which is the neutral axis for this bending moment. Consequently, the bending stress at K is zero.

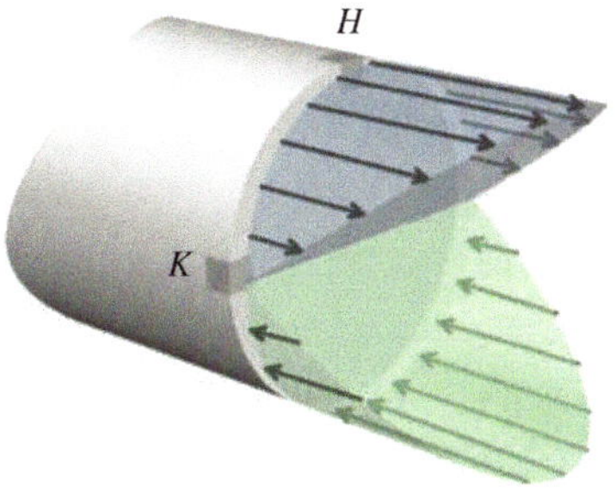

Shear Stresses at K

The 102.4 N·m torque acting about the x axis creates shear stress at K. The magnitude of this shear stress is the same as the stress magnitude at H: $\tau = 11.18$ MPa.

The 640 N shear force acting vertically at the section of interest is also associated with shear stress at point K. From Equation (9.8), the first moment of area for a solid circular cross section is

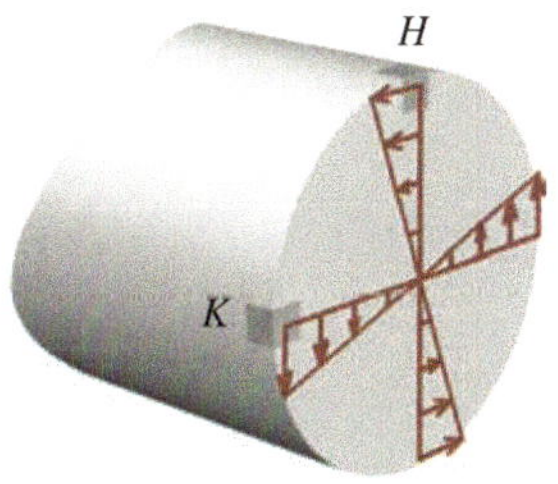

$$Q = \frac{d^3}{12} = \frac{(36\text{ mm})^3}{12} = 3{,}888\text{ mm}^3$$

The shear stress formula [Equation (9.2)] is used to calculate the shear stress:

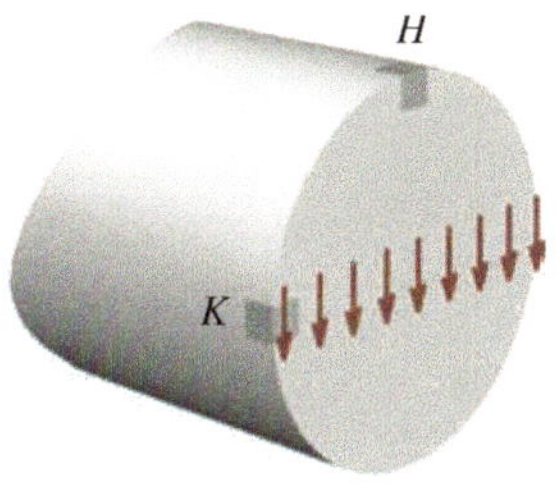

$$\tau = \frac{VQ}{I_z t} = \frac{(640\text{ N})(3{,}888\text{ mm}^3)}{(82{,}448\text{ mm}^4)(36\text{ mm})} = 0.838\text{ MPa}$$

Combined Stresses at K

The normal and shear stresses acting at point K can be summarized on a stress element. Note that, at point K, both shear stresses act *downward* on the $+x$ face of the stress element. After the proper shear stress direction has been established on one face, the shear stress directions on the other three faces are known.

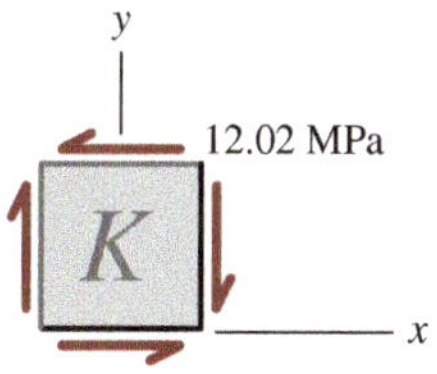

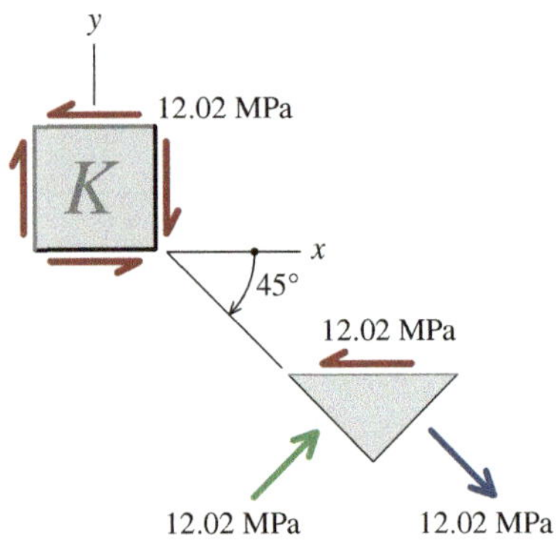

Stress Transformation Results at *K*

The principal stresses and the maximum shear stress at *K* can be determined from the stress transformation equations and procedures detailed in Chapter 12. The results of these calculations are shown in the accompanying figure.

EXAMPLE 15.6

A vertical pipe column with an outside diameter $D = 9.0$ in. and an inside diameter $d = 8.0$ in. supports the loads shown. Determine the principal stresses and the maximum shear stress at points *H* and *K*.

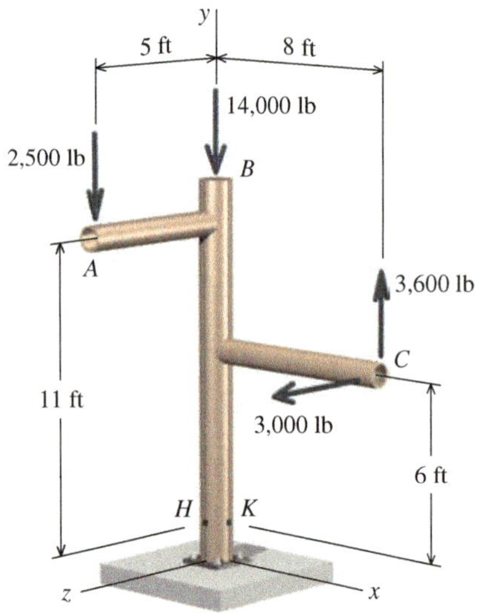

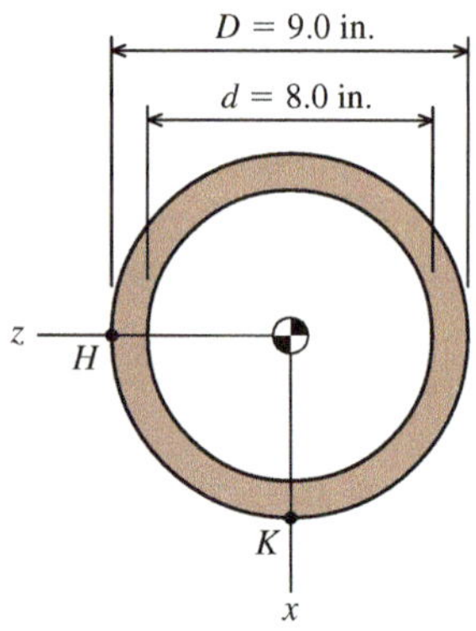

Column cross-sectional dimensions.

Plan the Solution

Several loads act on the structure, making the analysis seem complicated. However, it can be simplified considerably by first reducing the system of four loads to a statically determinate system of forces and moments acting at the section of interest. The normal and shear stresses created by this equivalent force system will be computed and shown in their proper directions on stress elements for points *H* and *K*. Stress transformation calculations will be used to determine the principal stresses and maximum shear stress for each stress element.

SOLUTION

Equivalent Force System

A system of forces and moments that is statically equivalent to the four loads applied at points A, B, and C can be readily determined for the section of interest.

The equivalent forces are simply equal to the applied loads. There is no force acting in the x direction. The sum of the forces in the y direction is

$$\Sigma F_y = -2{,}500 \text{ lb} - 14{,}000 \text{ lb} + 3{,}600 \text{ lb} = -12{,}900 \text{ lb}$$

In the z direction, the only force is the 3,000 lb load applied to point C. The equivalent forces acting at the section are shown in the accompanying figure.

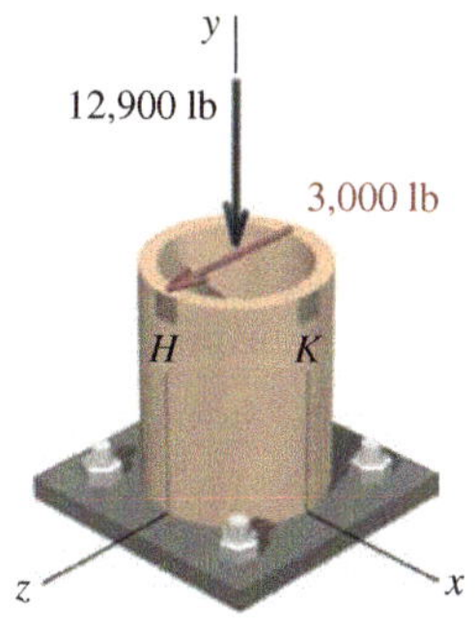

Equivalent forces at the section that contains points *H* and *K*.

The equivalent moments acting at the section of interest can be determined by considering each load in turn:

- The 2,500 lb load acting at A creates a moment of (2,500 lb)(5 ft) = 12,500 lb·ft, which acts about the $+x$ axis.
- The line of action of the 14,000 lb load passes through the section of interest; therefore, it creates no moments at H and K.
- The 3,600 lb load acting vertically at C creates a moment of (3,600 lb)(8 ft) = 28,800 lb·ft about the $+z$ axis.
- The 3,000 lb load acting horizontally at C creates two moment components.
  - One moment component has a magnitude of (3,000 lb)(8 ft) = 24,000 lb·ft and acts about the $-y$ axis.
  - A second moment component has a magnitude of (3,000 lb)(6 ft) = 18,000 lb·ft and acts about the $+x$ axis.
- The moments acting about the x axis can be summed to determine the equivalent moment:

$$M_x = 12{,}500 \text{ lb}\cdot\text{ft} + 18{,}000 \text{ lb}\cdot\text{ft} = 30{,}500 \text{ lb}\cdot\text{ft}$$

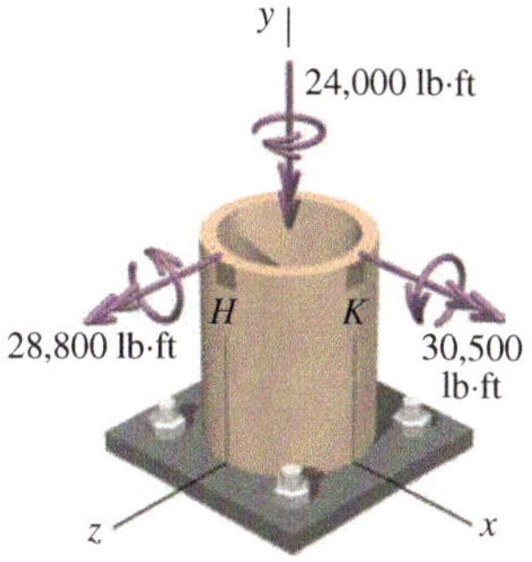

Equivalent moments at the section that contains points *H* and *K*.

For the coordinate system used here, the axis of the pipe column extends in the y direction. Therefore, the moment component acting about the y axis is recognized as a torque; the components about the x and z axes are simply bending moments.

Alternative method: The moments that are equivalent to the four-load system can be calculated systematically with the use of position and force vectors. The position vector $\mathbf{r}$ from the section of interest to point A is $\mathbf{r}_A = 11 \text{ ft}\,\mathbf{j} + 5 \text{ ft}\,\mathbf{k}$. The load at A can be expressed as the force vector $\mathbf{F}_A = -2{,}500 \text{ lb}\,\mathbf{j}$. The moment produced by the 2,500 lb load can be determined from the cross product $\mathbf{M}_A = \mathbf{r}_A \times \mathbf{F}_A$:

$$\mathbf{M}_A = \mathbf{r}_A \times \mathbf{F}_A = \begin{vmatrix} \mathbf{i} & \mathbf{j} & \mathbf{k} \\ 0 & 11 & 5 \\ 0 & -2{,}500 & 0 \end{vmatrix} = 12{,}500 \text{ lb}\cdot\text{ft}\,\mathbf{i}$$

The position vector from the section of interest to C is $\mathbf{r}_C = 8 \text{ ft}\,\mathbf{i} + 6 \text{ ft}\,\mathbf{j}$. The load at C can be expressed as $\mathbf{F}_C = 3{,}600 \text{ lb}\,\mathbf{j} + 3{,}000 \text{ lb}\,\mathbf{k}$. The moments can be determined from the cross product $\mathbf{M}_C = \mathbf{r}_C \times \mathbf{F}_C$:

$$\mathbf{M}_C = \mathbf{r}_C \times \mathbf{F}_C = \begin{vmatrix} \mathbf{i} & \mathbf{j} & \mathbf{k} \\ 8 & 6 & 0 \\ 0 & 3{,}600 & 3{,}000 \end{vmatrix} = 18{,}000 \text{ lb}\cdot\text{ft}\,\mathbf{i} - 24{,}000 \text{ lb}\cdot\text{ft}\,\mathbf{j} + 28{,}800 \text{ lb}\cdot\text{ft}\,\mathbf{k}$$

The equivalent moments at the section of interest are found from the sum of $\mathbf{M_A}$ and $\mathbf{M_C}$:

$$\mathbf{M} = \mathbf{M_A} + \mathbf{M_C} = 30{,}500 \text{ lb}\cdot\text{ft}\,\mathbf{i} - 24{,}000 \text{ lb}\cdot\text{ft}\,\mathbf{j} + 28{,}800 \text{ lb}\cdot\text{ft}\,\mathbf{k}$$

Section Properties

The outside diameter of the pipe column is $D = 9.0$ in., and the inside diameter is $d = 8.0$ in. The area, the moment of inertia, and the polar moment of inertia for the cross section are, respectively, as follows:

$$A = \frac{\pi}{4}[D^2 - d^2] = \frac{\pi}{4}[(9.0 \text{ in.})^2 - (8.0 \text{ in.})^2] = 13.352 \text{ in.}^2$$

$$I = \frac{\pi}{64}[D^4 - d^4] = \frac{\pi}{64}[(9.0 \text{ in.})^4 - (8.0 \text{ in.})^4] = 121.00 \text{ in.}^4$$

$$J = \frac{\pi}{32}[D^4 - d^4] = \frac{\pi}{32}[(9.0 \text{ in.})^4 - (8.0 \text{ in.})^4] = 242.00 \text{ in.}^4$$

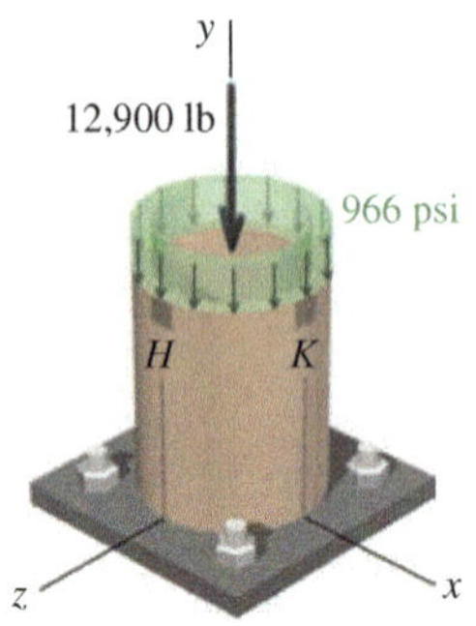

Stresses at H

The equivalent forces and moments acting at the section of interest will be evaluated sequentially to determine the type, magnitude, and direction of any stresses created at H.

The 12,900 lb axial force creates compressive normal stress, which acts in the y direction:

$$\sigma_y = \frac{F_y}{A} = \frac{12{,}900 \text{ lb}}{13.352 \text{ in.}^2} = 966 \text{ psi (C)}$$

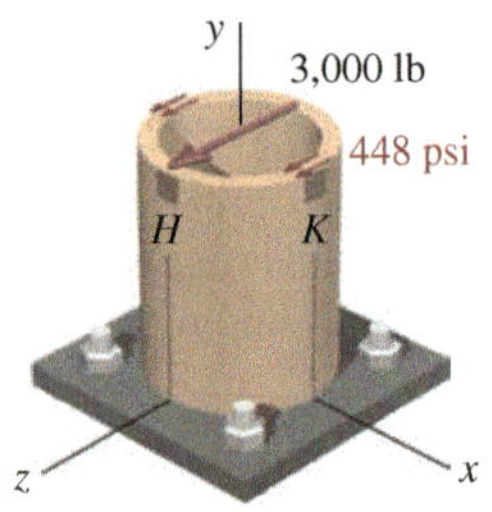

Although shear stresses are associated with the 3,000 lb shear force, the shear stress at point H is zero.

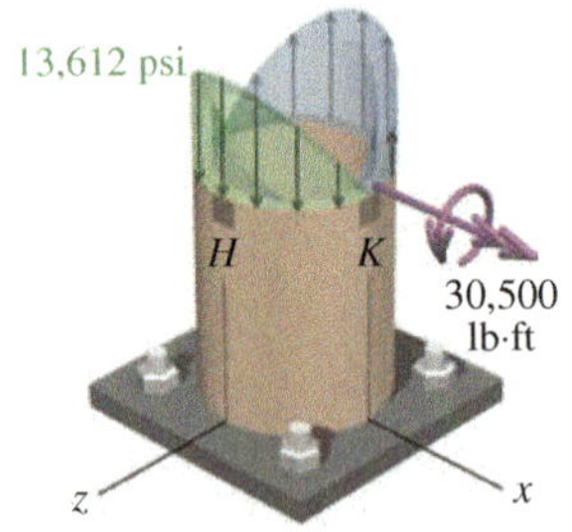

The 30,500 lb·ft bending moment about the x axis creates compressive normal stress at H:

$$\sigma_y = \frac{M_x c}{I_x} = \frac{(30{,}500 \text{ lb}\cdot\text{ft})(4.5 \text{ in.})(12 \text{ in./ft})}{121.0 \text{ in.}^4} = 13{,}612 \text{ psi (C)}$$

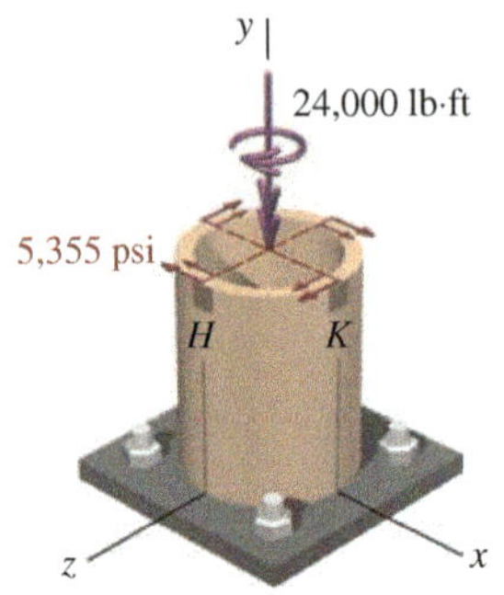

The 24,000 lb·ft torque acting about the y axis creates shear stress at H. The magnitude of this shear stress can be calculated from the elastic torsion formula:

$$\tau = \frac{Tc}{J} = \frac{(24{,}000 \text{ lb}\cdot\text{ft})(4.5 \text{ in.})(12 \text{ in./ft})}{242.0 \text{ in.}^4} = 5{,}355 \text{ psi}$$

The 28,800 lb · ft bending moment about the z axis creates bending stresses at the section of interest. Point H, however, is located on the neutral axis for this bending moment, and thus, the bending stress at H is zero.

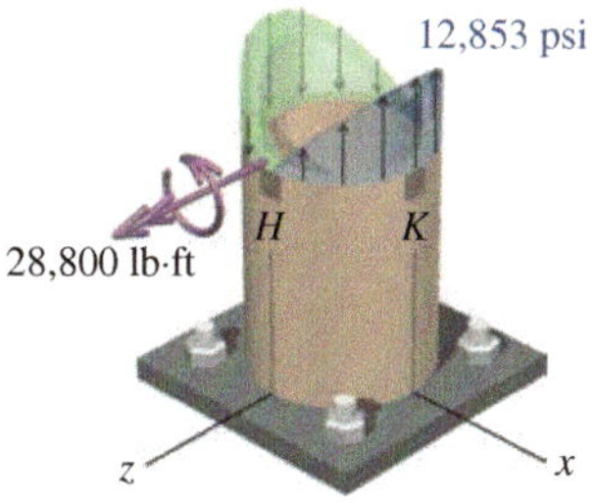

Combined Stresses at *H*

The normal and shear stresses acting at point H can be summarized on a stress element. Notice that the torsion shear stress acts in the $-x$ direction on the $+y$ face of the element. After the proper shear stress direction has been established on one face, the shear stress directions on the other three faces are known.

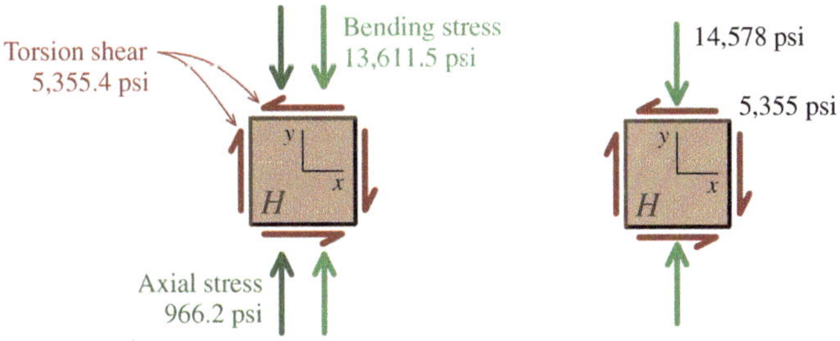

Stress Transformation Results at *H*

The principal stresses and the maximum shear stress at H can be determined from the stress transformation equations and procedures detailed in Chapter 12. The results of these calculations are shown in the accompanying figure.

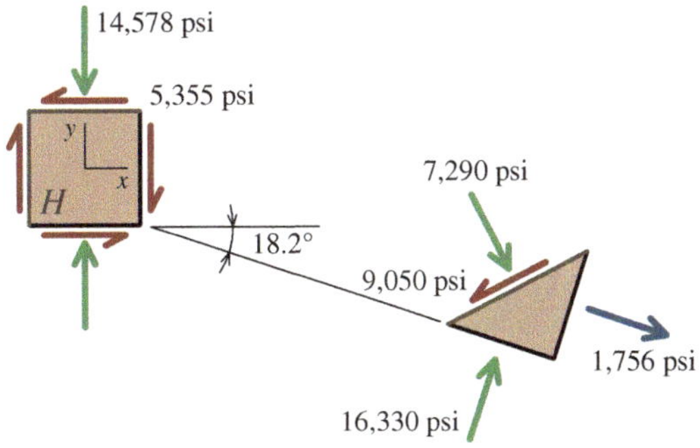

Stresses at *K*

The equivalent forces and moments acting at the section of interest will again be evaluated, this time to determine the type, magnitude, and direction of any stresses created at K.

The 12,900 lb axial force creates compressive normal stress, which acts in the y direction:

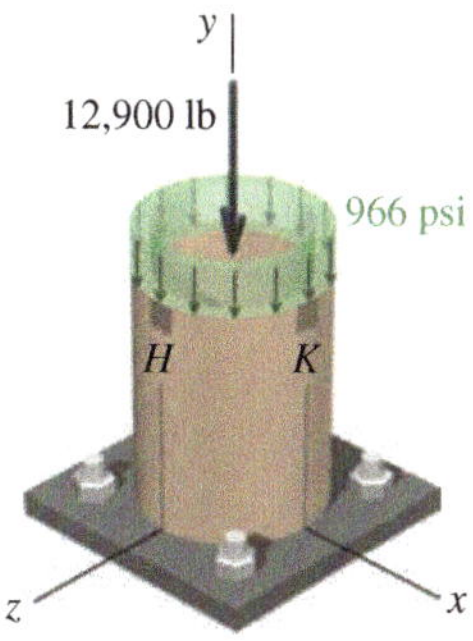

$$\sigma_y = \frac{F_y}{A} = \frac{12{,}900 \text{ lb}}{13.352 \text{ in.}^2} = 966 \text{ psi (C)}$$

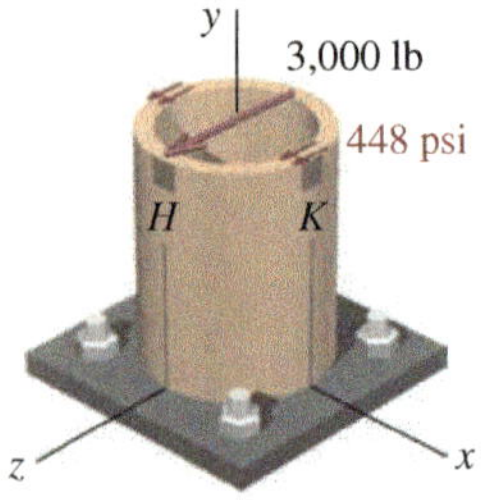

The 3,000 lb shear force acting horizontally at the section of interest is also associated with shear stress at point K. From Equation (9.10), the first moment of area for the hollow circular cross section is

$$Q = \frac{1}{12}[D^3 - d^3] = \frac{1}{12}[(9.0 \text{ in.})^3 - (8.0 \text{ in.})^3] = 18.083 \text{ in.}^3$$

The shear stress formula [Equation (9.2)] is used to calculate the shear stress:

$$\tau = \frac{VQ}{I_x t} = \frac{(3{,}000 \text{ lb})(18.083 \text{ in.}^3)}{(121.0 \text{ in.}^4)(9 \text{ in.} - 8 \text{ in.})} = 448 \text{ psi}$$

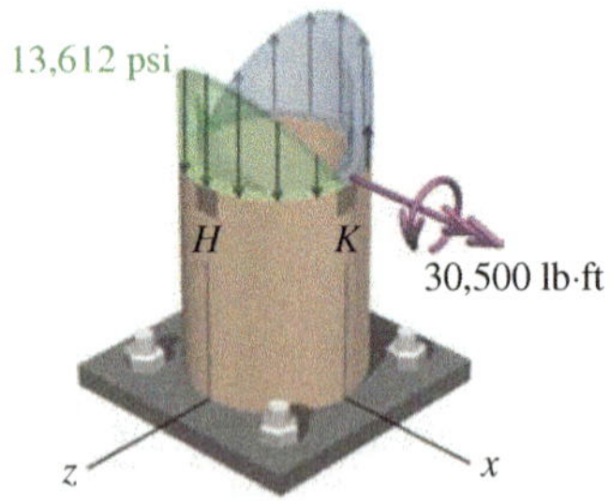

The 30,500 lb · ft bending moment about the x axis creates bending stresses at the section of interest. Point K, however, is located on the neutral axis for this bending moment, and consequently, the bending stress at K is zero.

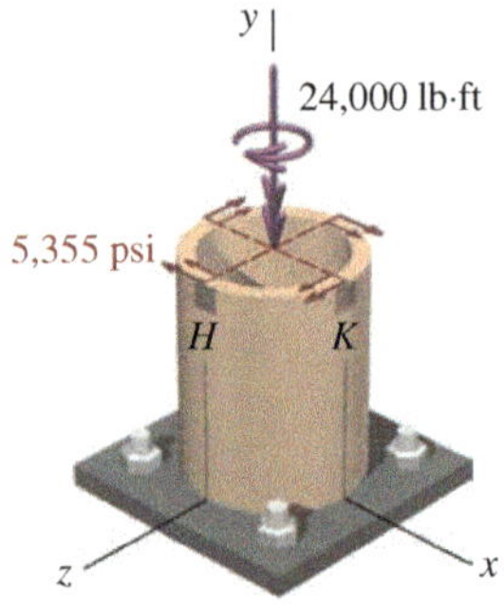

The 24,000 lb · ft torque acting about the y axis creates shear stress at K. The magnitude of this shear stress can be calculated from the elastic torsion formula:

$$\tau = \frac{Tc}{J} = \frac{(24{,}000 \text{ lb}\cdot\text{ft})(4.5 \text{ in.})(12 \text{ in./ft})}{242.0 \text{ in.}^4} = 5{,}355 \text{ psi}$$

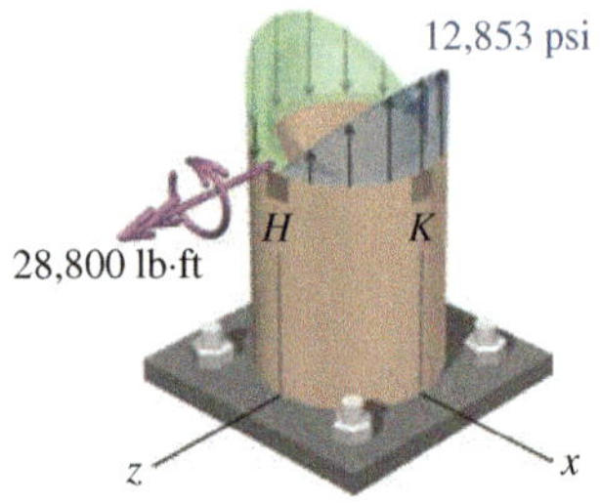

The 28,800 lb · ft bending moment about the z axis creates tensile normal stress at K:

$$\sigma_y = \frac{M_z c}{I_z} = \frac{(28{,}800 \text{ lb}\cdot\text{ft})(4.5 \text{ in.})(12 \text{ in./ft})}{121.0 \text{ in.}^4} = 12{,}853 \text{ psi (T)}$$

Combined Stresses at K

The normal and shear stresses acting at point K can be summarized on a stress element.

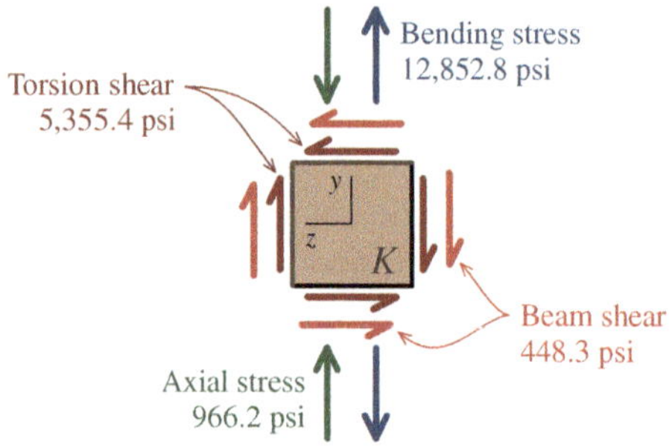

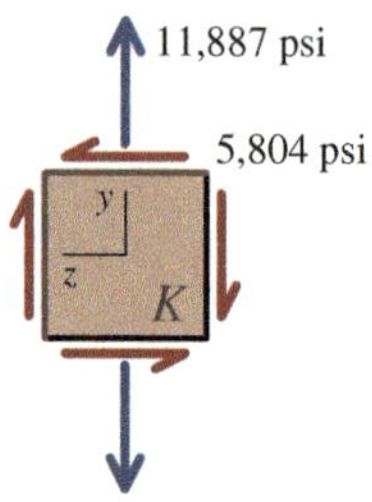

Stress Transformation Results at *K*

The principal stresses and the maximum shear stress at *K* are shown in the accompanying figure.

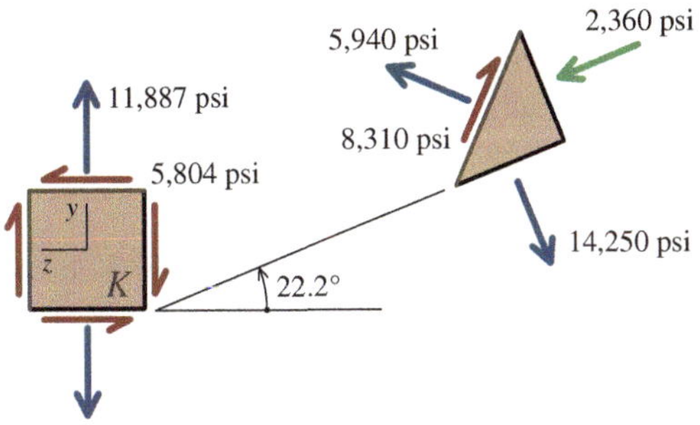

EXAMPLE 15.7

A piping system transports a fluid that has an internal pressure of 1,500 kPa. In addition to being subject to the fluid pressure, the piping supports a vertical load of 9 kN and a horizontal load of 13 kN (acting in the $+x$ direction) at flange *A*. The pipe has an outside diameter $D = 200$ mm and an inside diameter $d = 176$ mm. Determine the principal stresses, the maximum shear stress, and the absolute maximum shear stress at points *H* and *K*.

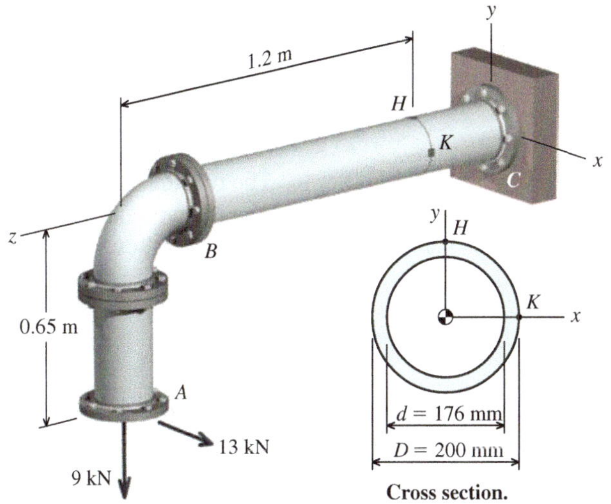

Cross section.

Plan the Solution

The analysis begins by determining the statically equivalent system of forces and moments acting internally at the section that contains points *H* and *K*. The normal and shear stresses created by this equivalent force system will be computed and shown in their proper directions on a stress element for both point *H* and point *K*. The internal pressure of the fluid also creates normal stresses, which act longitudinally and circumferentially in the pipe wall. These stresses will be computed and included on the stress elements for *H* and *K*. Stress transformation calculations will be used to determine the principal stresses, the maximum shear stress, and the absolute maximum shear stress for each stress element.

SOLUTION

Equivalent Force System

A system of forces and moments that is statically equivalent to the loads applied at flange *A* can be determined for the section of interest.

The equivalent forces are simply equal to the applied loads. A 13 kN force acts in the $+x$ direction, a 9 kN force acts in the $-y$ direction, and there is no force acting in the *z* direction.

The equivalent moments acting at the section of interest can be determined by considering each load in turn. The 9 kN load acting at *A* creates a moment of (9 kN)(1.2 m) = 10.8 kN·m, which acts about the $+x$ axis. The 13 kN load acting horizontally at *A* creates two moment components:

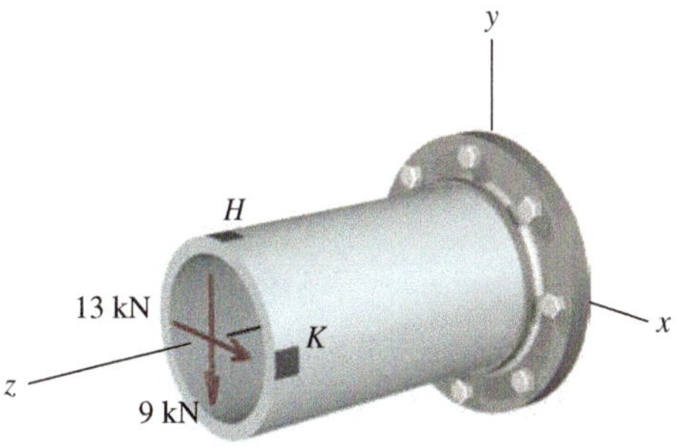

Equivalent forces at the section that contains points *H* and *K*.

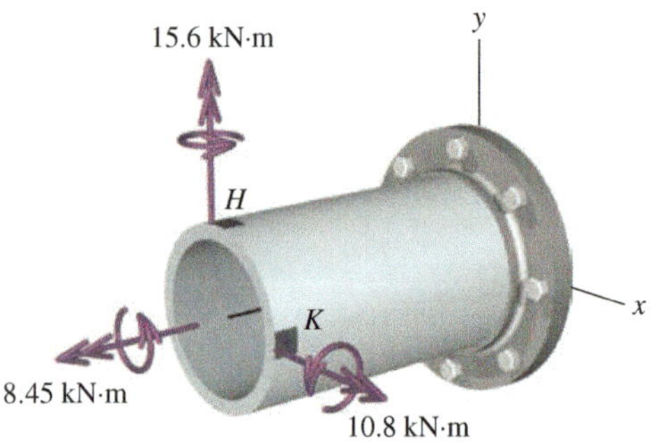

Equivalent moments at the section that contains points *H* and *K*.

- One moment component has a magnitude of (13 kN)(1.2 m) = 15.6 kN · m and acts about the +*y* axis.
- A second moment component has a magnitude of (13 kN)(0.65 m) = 8.45 kN · m and acts about the +*z* axis.

For the coordinate system used here, the longitudinal axis of the pipe extends in the *z* direction; therefore, the moment component acting about the *z* axis is recognized as a torque, while the components about the *x* and *y* axes are simply bending moments.

Alternative method: The moments equivalent to the two loads at *A* can be calculated systematically by the use of position and force vectors. The position vector **r** from the section of interest to point *A* is $\mathbf{r_A} = -0.65\text{ m}\,\mathbf{j} + 1.2\text{ m}\,\mathbf{k}$. The load at *A* can be expressed as the force vector $\mathbf{F_A} = 13\text{ kN}\,\mathbf{i} - 9\text{ kN}\,\mathbf{j}$. The moment produced by $\mathbf{F_A}$ can be determined from the cross product $\mathbf{M_A} = \mathbf{r_A} \times \mathbf{F_A}$:

$$\mathbf{M_A} = \mathbf{r_A} \times \mathbf{F_A} = \begin{vmatrix} \mathbf{i} & \mathbf{j} & \mathbf{k} \\ 0 & -0.65 & 1.2 \\ 13 & -9 & 0 \end{vmatrix}$$

$$= 10.8\text{ kN}\cdot\text{m}\,\mathbf{i} + 15.6\text{ kN}\cdot\text{m}\,\mathbf{j} + 8.45\text{ kN}\cdot\text{m}\,\mathbf{k}$$

Section Properties

The outside diameter of the pipe is $D = 200$ mm, and the inside diameter is $d = 176$ mm. The moment of inertia and the polar moment of inertia for the cross section are, respectively, as follows:

$$I = \frac{\pi}{64}[D^4 - d^4] = \frac{\pi}{64}[(200\text{ mm})^4 - (176\text{ mm})^4] = 31{,}439{,}853\text{ mm}^4$$

$$J = \frac{\pi}{32}[D^4 - d^4] = \frac{\pi}{32}[(200\text{ mm})^4 - (176\text{ mm})^4] = 62{,}879{,}706\text{ mm}^4$$

Stresses at *H*

The equivalent forces and moments acting at the section of interest will be sequentially evaluated to determine the type, magnitude, and direction of any stresses created at *H*. Transverse shear stress is associated with the 13 kN shear force acting in the $+x$ direction at the section of interest. From Equation (9.10), the first moment of area at the centroid for a hollow circular cross section is

$$Q = \frac{1}{12}[D^3 - d^3] = \frac{1}{12}[(200\text{ mm})^3 - (176\text{ mm})^3] = 212{,}352\text{ mm}^3$$

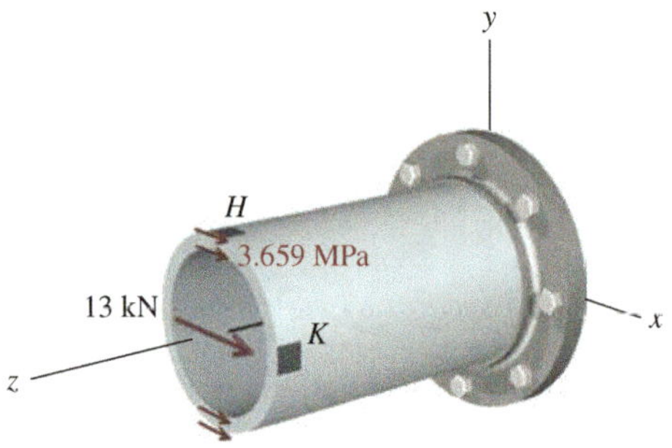

The shear stress formula [Equation (9.2)] is used to calculate the shear stress:

$$\tau = \frac{VQ}{I_y t} = \frac{(13\text{ kN})(212{,}352\text{ mm}^3)(1{,}000\text{ N/kN})}{(31{,}439{,}853\text{ mm}^4)(200\text{ mm} - 176\text{ mm})} = 3.659\text{ MPa}$$

Although shear stresses are associated with the 9 kN shear force that acts in the $-y$ direction, the shear stress at point *H* is zero.

The 10.8 kN·m (i.e., 10.8 × 10⁶ N·mm) bending moment about the x axis creates tensile normal stress at H:

$$\sigma_z = \frac{M_x c}{I_x} = \frac{(10.8 \times 10^6\ \text{N}\cdot\text{mm})(100\ \text{mm})}{31{,}439{,}853\ \text{mm}^4} = 34.351\ \text{MPa (T)}$$

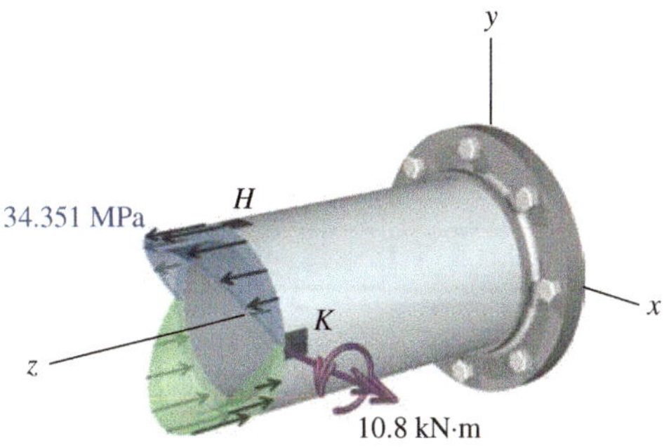

The 15.6 kN·m bending moment about the y axis creates bending stresses at the section of interest. Point H, however, is located on the neutral axis for this bending moment, and consequently, the bending stress at H is zero.

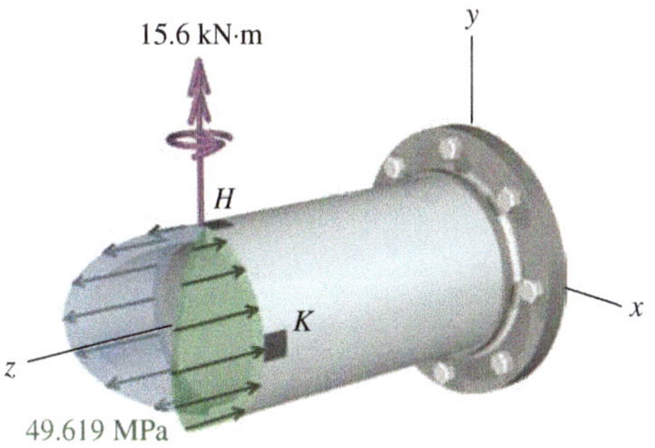

The 8.45 kN·m (i.e., 8.45 × 10⁶ N·mm) torque acting about the z axis creates shear stress at H. The magnitude of this shear stress can be calculated from the elastic torsion formula:

$$\tau = \frac{Tc}{J} = \frac{(8.45 \times 10^6\ \text{N}\cdot\text{mm})(100\ \text{mm})}{62{,}879{,}706\ \text{mm}^4} = 13.438\ \text{MPa}$$

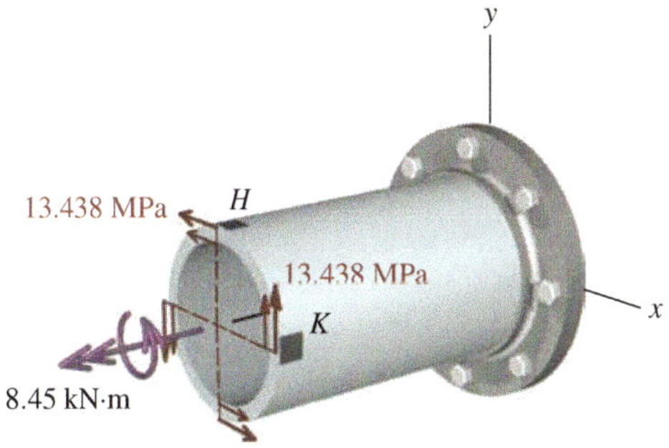

The 1,500 kPa internal fluid pressure creates tensile normal stresses in the 12 mm thick wall of the pipe. The longitudinal stress in the pipe wall is

$$\sigma_{\text{long}} = \frac{pd}{4t} = \frac{(1{,}500\ \text{kPa})(176\ \text{mm})}{4(12\ \text{mm})} = 5{,}500\ \text{kPa} = 5.500\ \text{MPa (T)}$$

and the circumferential stress is

$$\sigma_{\text{hoop}} = \frac{pd}{2t} = \frac{(1{,}500\ \text{kPa})(176\ \text{mm})}{2(12\ \text{mm})} = 11{,}000\ \text{kPa} = 11.000\ \text{MPa (T)}$$

Observe that the longitudinal stress acts in the z direction. At point H, the circumferential direction is the x direction.

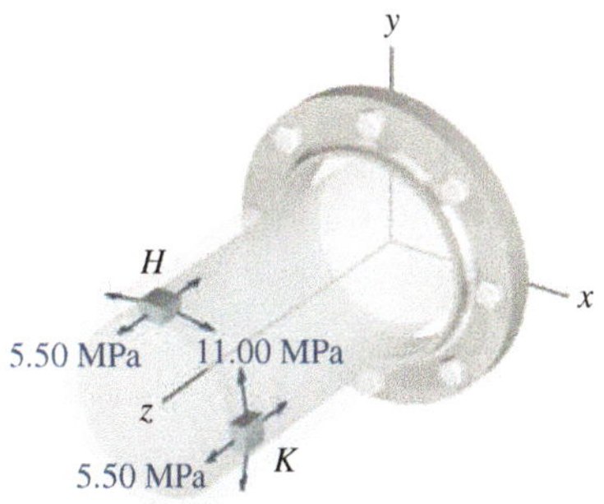

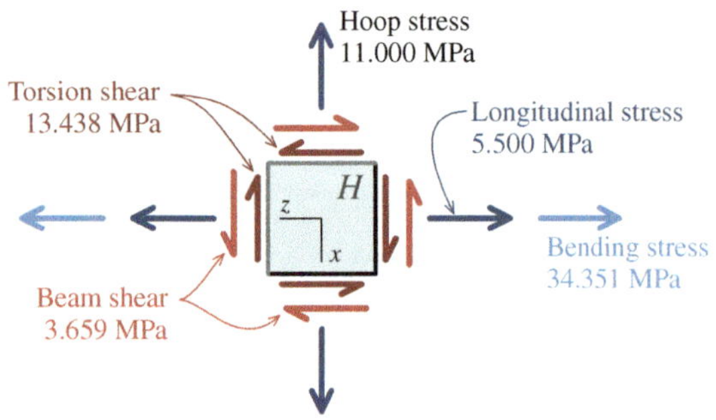

Combined Stresses at *H*

The normal and shear stresses acting at point *H* are summarized on a stress element. Note that, at point *H*, the torsion shear stress acts in the $-x$ direction on the $+z$ face of the stress element. The shear stress associated with the 13 kN shear force acts in the opposite direction.

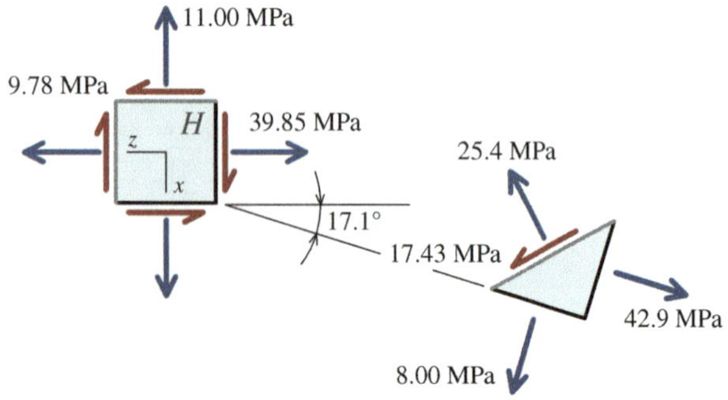

Stress Transformation Results at *H*

The principal stresses and the maximum shear stress at *H* can be determined from the stress transformation equations and procedures detailed in Chapter 12. The results of these calculations are shown in the accompanying figure.

The absolute maximum shear stress at *H* is 21.43 MPa.

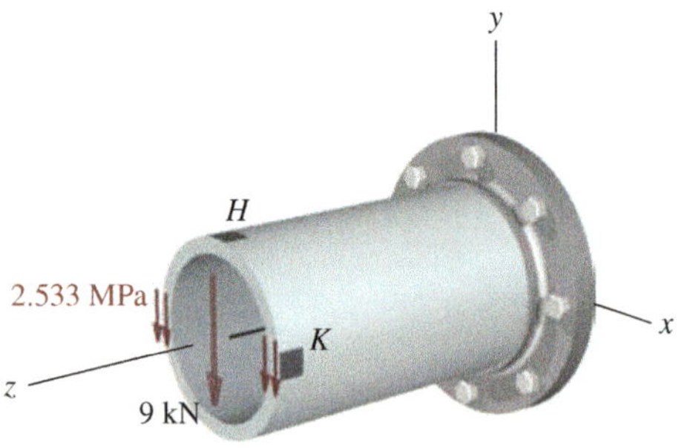

Stresses at *K*

Although shear stresses are associated with the 13 kN shear force that acts in the $-y$ direction, the shear stress at point *K* is zero.

Transverse shear stress is associated with the 9 kN shear force acting in the $-y$ direction at the section of interest. The shear stress formula [Equation (9.2)] is used to calculate the shear stress:

$$\tau = \frac{VQ}{I_y t} = \frac{(9\text{ kN})(212{,}352\text{ mm}^3)(1{,}000\text{ N/kN})}{(31{,}439{,}853\text{ mm}^4)(200\text{ mm} - 176\text{ mm})} = 2.533\text{ MPa}$$

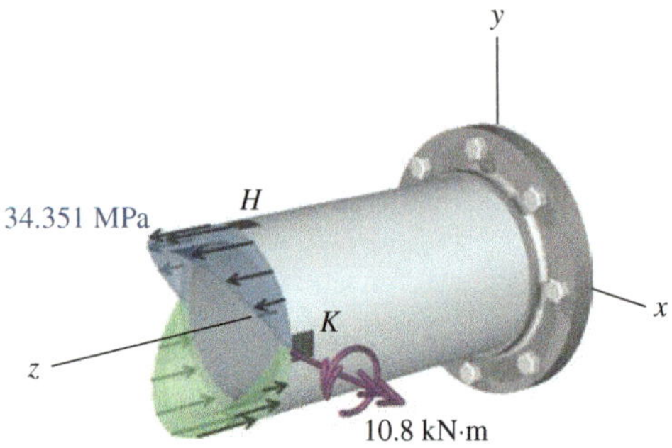

The 10.8 kN · m bending moment about the *x* axis creates bending stresses at the section of interest. Point *K*, however, is located on the neutral axis for this bending moment, and consequently, the bending stress at *K* is zero.

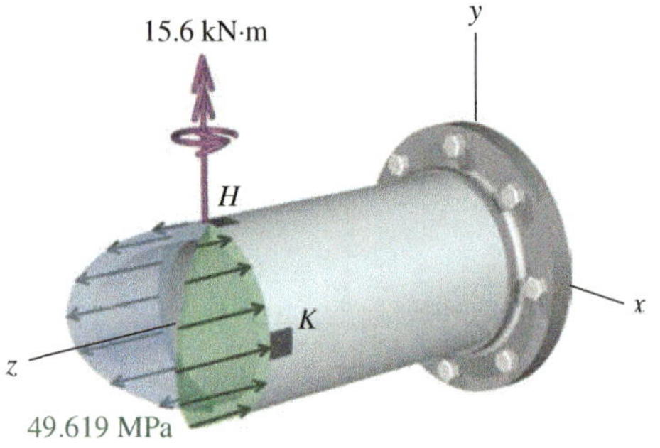

The 15.6 kN · m (i.e., 15.6×10^6 N · mm) bending moment about the *y* axis creates compressive normal stress at *K*:

$$\sigma_z = \frac{M_x c}{I_x} = \frac{(15.6 \times 10^6\text{ N}\cdot\text{mm})(100\text{ mm})}{31{,}439{,}853\text{ mm}^4} = 49.619\text{ MPa (C)}$$

The 8.45 kN·m (i.e., 8.45 × 10^6 N·mm) torque acting about the z axis creates shear stress at K. The magnitude of this shear stress can be calculated from the elastic torsion formula:

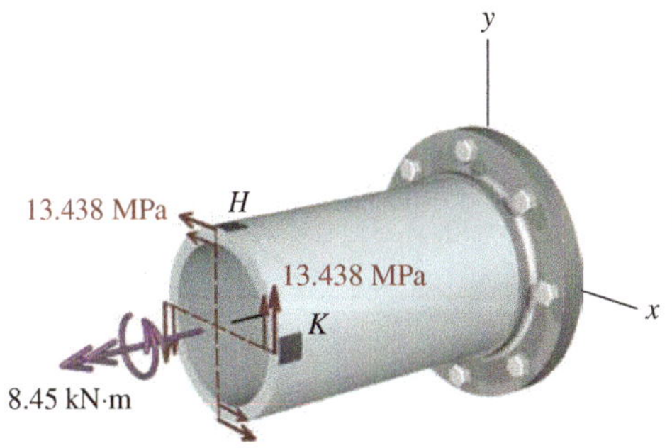

$$\tau = \frac{Tc}{J} = \frac{(8.45 \times 10^6 \text{ N·mm})(100 \text{ mm})}{62{,}879{,}706 \text{ mm}^4} = 13.438 \text{ MPa}$$

The 1,500 kPa internal fluid pressure creates tensile normal stresses in the 12 mm thick wall of the pipe. The longitudinal stress in the pipe wall is

$$\sigma_{\text{long}} = \frac{pd}{4t} = \frac{(1{,}500 \text{ kPa})(176 \text{ mm})}{4(12 \text{ mm})} = 5{,}500 \text{ kPa} = 5.500 \text{ MPa (T)}$$

and the circumferential stress is

$$\sigma_{\text{hoop}} = \frac{pd}{2t} = \frac{(1{,}500 \text{ kPa})(176 \text{ mm})}{2(12 \text{ mm})} = 11{,}000 \text{ kPa} = 11.000 \text{ MPa (T)}$$

Note that the longitudinal stress acts in the z direction. Furthermore, the circumferential direction at point K is the y direction.

Combined Stresses at *K*

The normal and shear stresses acting at point K are summarized on a stress element. Note that, at point K, the torsion shear stress acts in the $+y$ direction on the $+z$ face of the stress element. The transverse shear stress associated with the 9 kN shear force acts in the opposite direction.

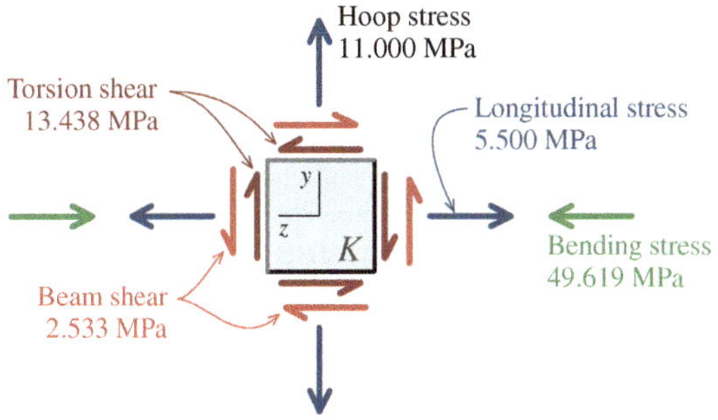

Stress Transformation Results at *K*

The principal stresses and the maximum shear stress at K can be determined from the stress transformation equations and procedures detailed in Chapter 12. The results of these calculations are shown in the accompanying figure.

The absolute maximum shear stress at K is 29.64 MPa.

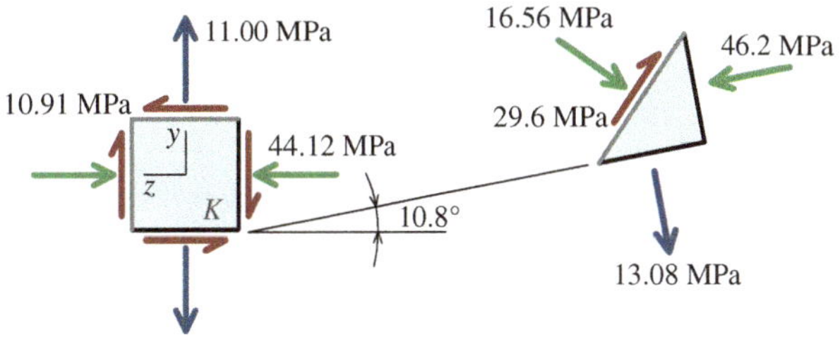

MecMovies

EXAMPLES

M15.6 A 12 kN force is applied to the component shown. Determine the internal forces acting at section a–a.

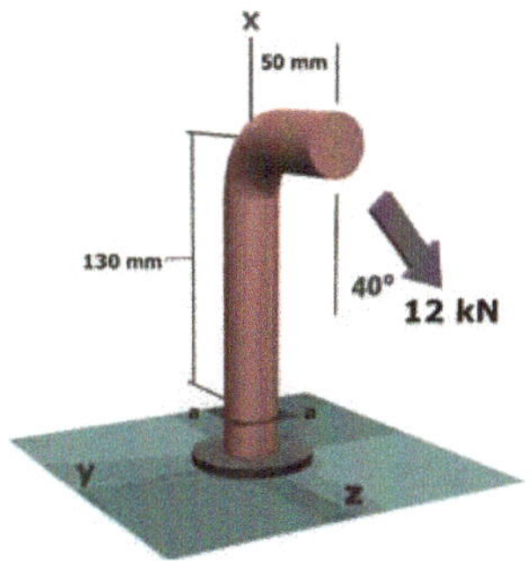

M15.5 Determine the stresses acting at point K in the beam.

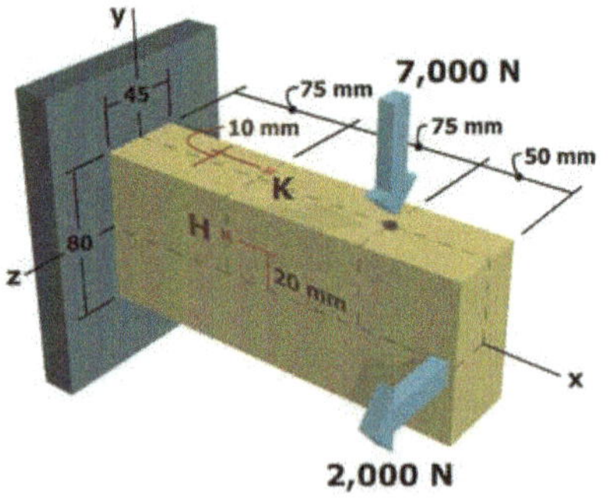

FIGURE M15.5

M15.6 Determine the internal forces (axial force, shear force, and torque) and the bending moments at a specific location in a member subjected to in-plane and out-of-plane forces.

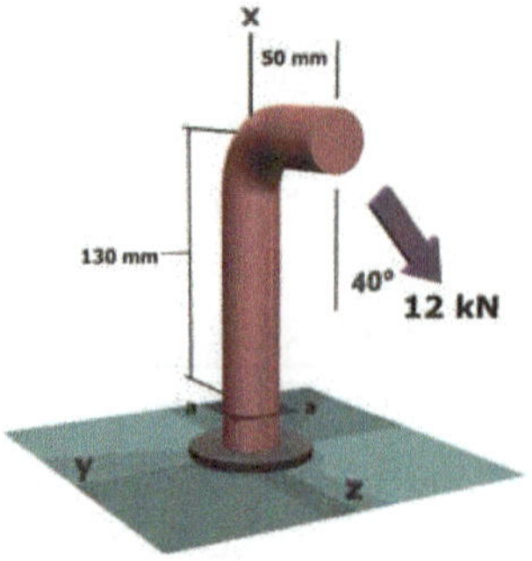

FIGURE M15.6

PROBLEMS

P15.26 A short rectangular post supports a compressive load $P = 35$ kN as shown in Figure P15.26*a*. A top view indicating the location where P is applied to the top of the post is shown in Figure P15.26*b*. The cross-sectional dimensions of the post are $b = 240$ mm and $d = 160$ mm. The load P is applied at distances $y_P = 60$ mm and $z_P = 50$ mm offset from the centroid of the post. Determine the normal stresses at corners A, B, C, and D of the post.

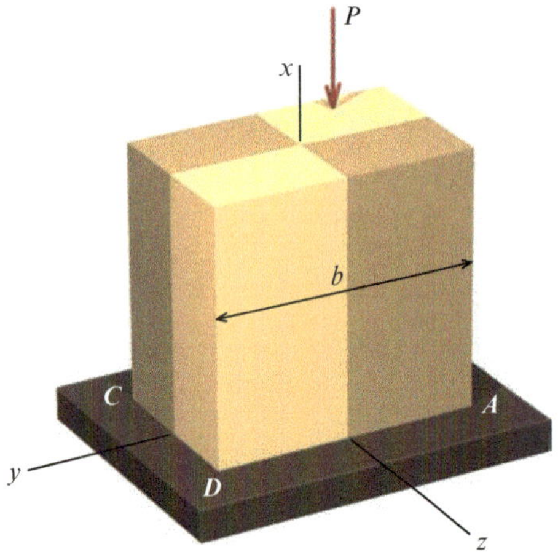

FIGURE P15.26*a*

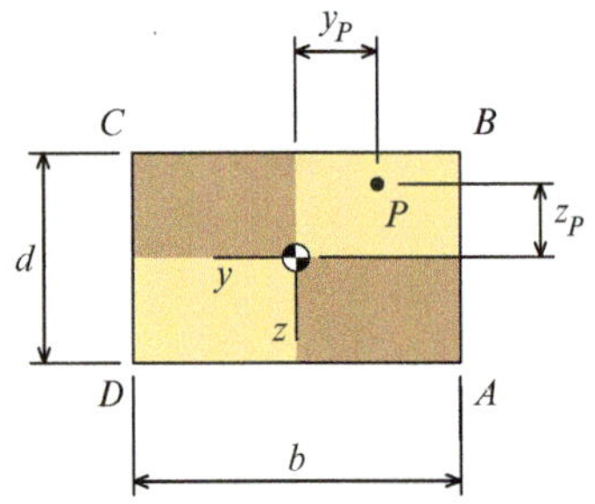

FIGURE P15.26*b* Top view of post.

P15.27 Three loads are applied to the rectangular tube shown in Figure P15.27*a*/28*a*. The load magnitudes are $P = 175$ kN, $Q = 60$ kN, and $R = 85$ kN. The dimensions of the tube, as shown in Figure P15.27*b*/28*b*, are $b = 150$ mm, $d = 200$ mm, $t = 10$ mm, and $z_K = 50$ mm. The location of load P is also shown in this figure. Use $a = 125$ mm, and determine the normal and shear stresses that act at point H. Show these stresses on a stress element.

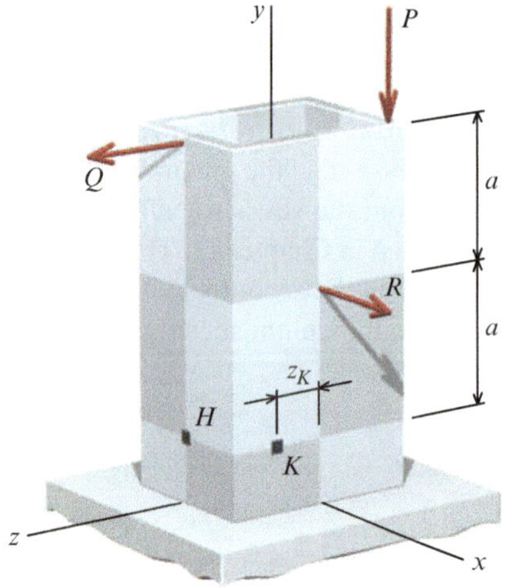

FIGURE P15.27*a*/28*a*

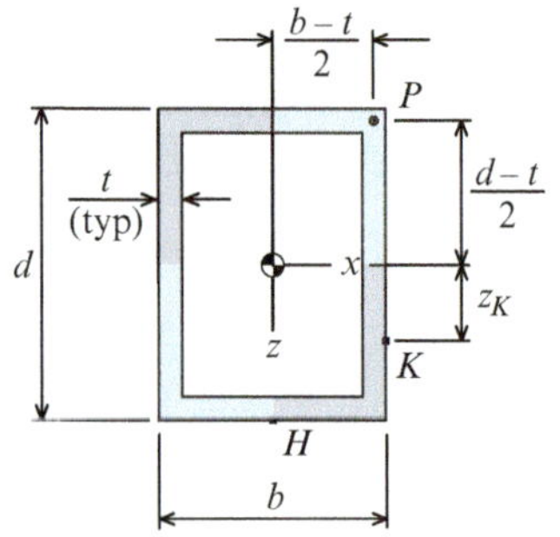

FIGURE P15.27*b*/28*b*

P15.28 Three loads are applied to the rectangular tube shown in Figure P15.27*a*/28*a*. The load magnitudes are $P = 175$ kN, $Q = 60$ kN, and $R = 85$ kN. The dimensions of the tube, as shown in Figure P15.27*b*/28*b*, are $b = 150$ mm, $d = 200$ mm, $t = 10$ mm, and $z_K = 50$ mm. The location of load *P* is also shown in this figure. Use $a = 125$ mm, and determine the normal and shear stresses that act at point *K*. Show these stresses on a stress element.

P15.29 Concentrated loads of $P_x = 33$ kips, $P_y = 29$ kips, and $P_z = 46$ kips are applied to the cantilever beam in the locations and directions shown in Figure P15.29*a*/30*a*. The beam cross section shown in Figure P15.29*b*/30*b* has dimensions $b = 9$ in. and $d = 4$ in. Using the value $a = 6.4$ in., determine the normal and shear stresses at point *H*. Show these stresses on a stress element.

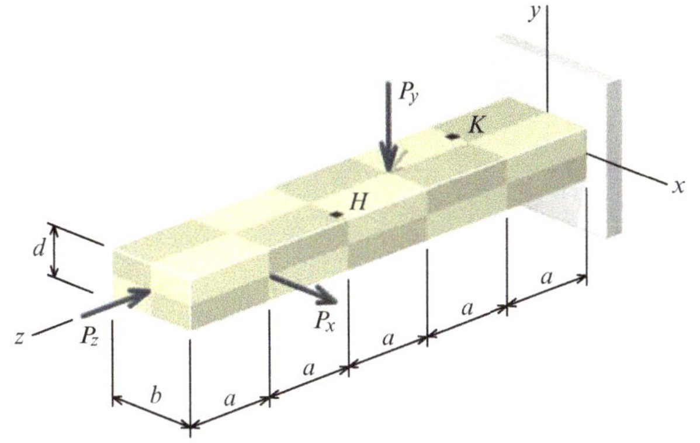

FIGURE P15.29*a*/30*a*

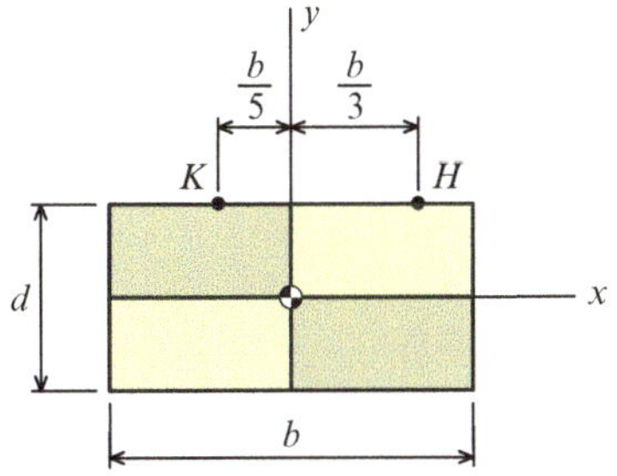

FIGURE P15.29*b*/30*b*

P15.30 Concentrated loads of $P_x = 33$ kips, $P_y = 29$ kips, and $P_z = 46$ kips are applied to the cantilever beam in the locations and directions shown in Figure P15.29*a*/30*a*. The beam cross section shown in Figure P15.29*b*/30*b* has dimensions $b = 9$ in. and $d = 4$ in. Using the value $a = 6.4$ in., determine the normal and shear stresses at point *K*. Show these stresses on a stress element.

P15.31 Consider the cantilever beam shown in Figure P15.31*a*/32*a*. The dimensions of the beam cross section and the location of point *K* are shown in Figure P15.31*b*/32*b*. Use the values $a = 2.15$ m, $b = 0.85$ m, $P_y = 13$ kN, and $P_z = 6$ kN, and determine the normal and shear stresses acting at point *K*.

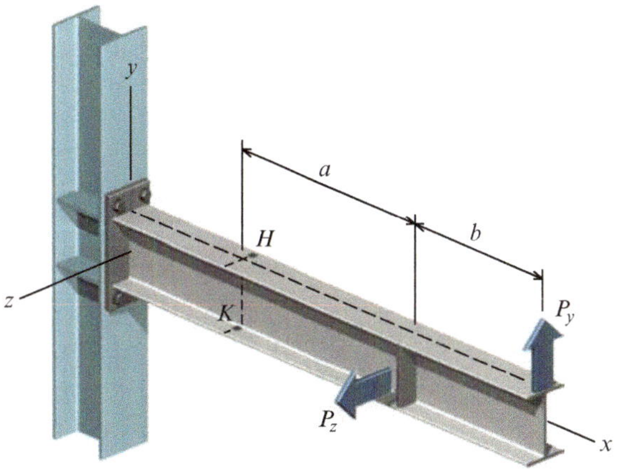

FIGURE P15.31*a*/32*a*

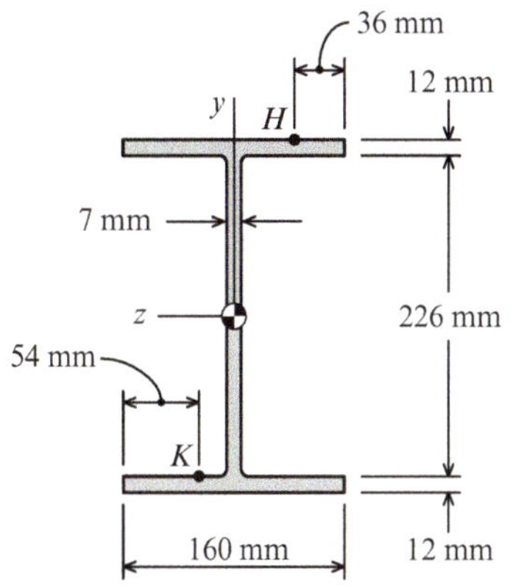

FIGURE P15.31*b*/32*b*

P15.32 Consider the cantilever beam shown in Figure P15.31*a*/32*a*. The dimensions of the beam cross section and the location of point *H* are shown in Figure P15.31*b*/32*b*. Use the values $a = 2.15$ m, $b = 0.85$ m, $P_y = 13$ kN, and $P_z - 6$ kN, and determine the normal and shear stresses acting at point *H*.

P15.33 A circular tube with an outside diameter of 4.0 in. and a wall thickness of 0.083 in. is subjected to loads $P_x = 1{,}360$ lb, $P_z = 950$ lb, and $T = 8{,}420$ lb·in. as shown in Figure P15.33. Using $a = 6$ in., determine the normal and shear stresses that exist at (a) point *H* and (b) point *K*.

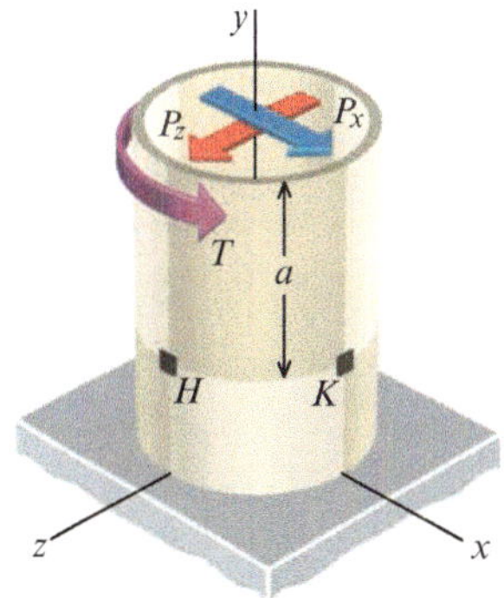

FIGURE P15.33

P15.34 Forces $P_x = 580$ lb and $P_y = 220$ lb act on the teeth of the gear shown in Figure P15.34. The gear has a radius $R = 6$ in., and the solid gear shaft has a diameter of 2.0 in. Using $a = 5$ in., determine the normal and shear stresses at (a) point H and (b) point K. Show the orientation of these stresses on an appropriate sketch.

FIGURE P15.34

P15.35 A 7 in. diameter pulley is attached to a solid 1.0 in. diameter steel shaft. The tension forces in the pulley belt are $T_2 = 160$ lb and $T_1 = 30$ lb, each acting in the x'–y' plane, as shown in Figure P15.35. Tension T_2 is oriented at $\beta = 30°$ with respect to the y' axis. Using $a = 7$ in., determine the normal and shear stresses at (a) point H and (b) point K. Show the orientation of these stresses on an appropriate sketch.

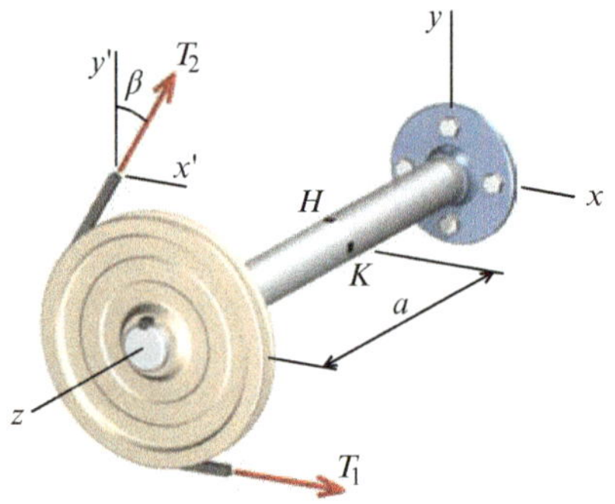

FIGURE P15.35

P15.36 A force $P_z = 780$ N is exerted on control arm BC, which is securely attached to the solid 30 mm diameter shaft shown in Figure P15.36. The dimensions of the assembly are $a = 150$ mm and $b = 120$ mm. Determine the absolute maximum shear stress in the shaft at (a) point H and (b) point K.

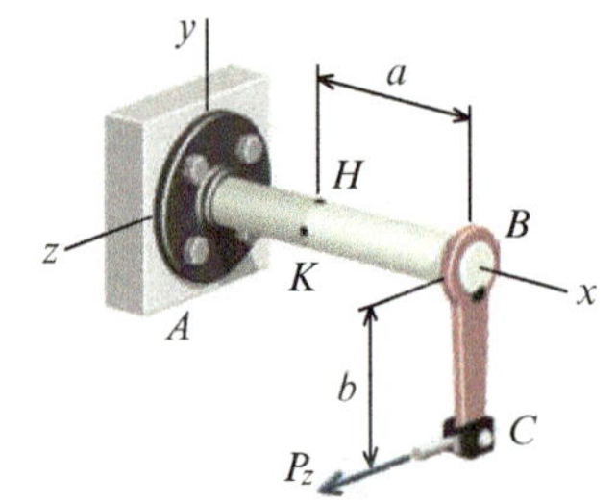

FIGURE P15.36

P15.37 Solid shaft AB of the pulley assembly shown in Figure P15.37/38 has a diameter of 35 mm. The length dimensions of the assembly are $a = 80$ mm, $b = 150$ mm, and $c = 115$ mm, and the diameter of the pulley is 130 mm. The pulley belt tension is $P = 950$ N. Determine the principal stresses and the maximum in-plane shear stress at point H on top of the shaft.

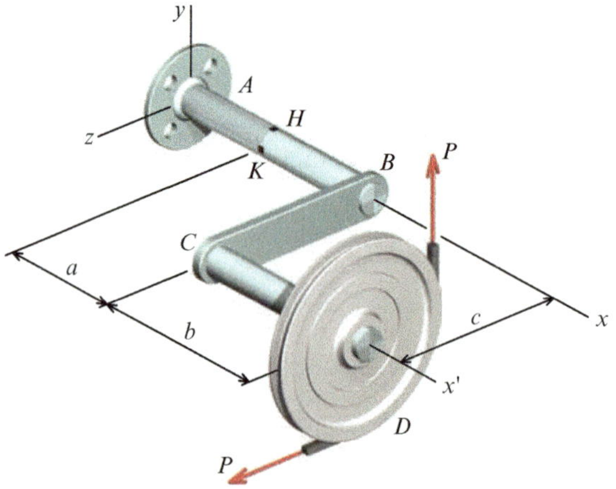

FIGURE P15.37/38

P15.38 Solid shaft AB of the pulley assembly shown in Figure P15.37/38 has a diameter of 1.75 in. The length dimensions of the assembly are $a = 3.0$ in., $b = 5.5$ in., and $c = 4.5$ in., and the diameter of the pulley is 5 in. The pulley belt tension is $P = 250$ lb. Determine the absolute maximum shear stress at point K, which is located on the side of the shaft.

P15.39 The tube shown in Figure P15.39/40 has an outside diameter of 3.50 in., a wall thickness of 0.12 in., and length dimensions of $a = 15$ in. and $b = 7$ in. The load magnitudes are $P_x = 850$ lb, $P_y = 375$ lb, and $P_z = 550$ lb. Determine the principal stresses on the side of the tube at point K, and show these results on a properly oriented stress element.

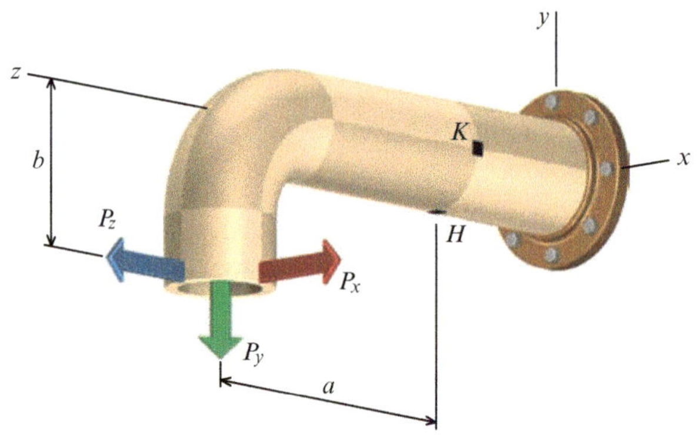

FIGURE P15.39/40

P15.40 The tube shown in Figure P15.39/40 has an outside diameter of 3.50 in., a wall thickness of 0.12 in., and length dimensions of $a = 15$ in. and $b = 7$ in. The load magnitudes are $P_x = 850$ lb,

$P_y = 375$ lb, and $P_z = 550$ lb. Determine the principal stresses on the underside of the tube at point *H*, and show these results on a properly oriented stress element.

P15.41 A solid steel shaft with an outside diameter of 25 mm is supported in flexible bearings at *A* and *D* as shown in Figure P15.41/42. Two pulleys are keyed to the shaft at *B* and *C*. Pulley *B* has a diameter $D_B = 350$ mm and belt tensions $T_1 = 842$ N and $T_2 = 234$ N. Pulley *C* has a diameter $D_C = 160$ mm and belt tensions $T_3 = 1{,}660$ N and $T_4 = 330$ N. The length dimensions of the shaft are $x_1 = 250$ mm, $x_2 = 120$ mm, $x_3 = 80$ mm, and $x_4 = 160$ mm. Determine the principal stresses and the absolute maximum shear stress at point *H* on the shaft.

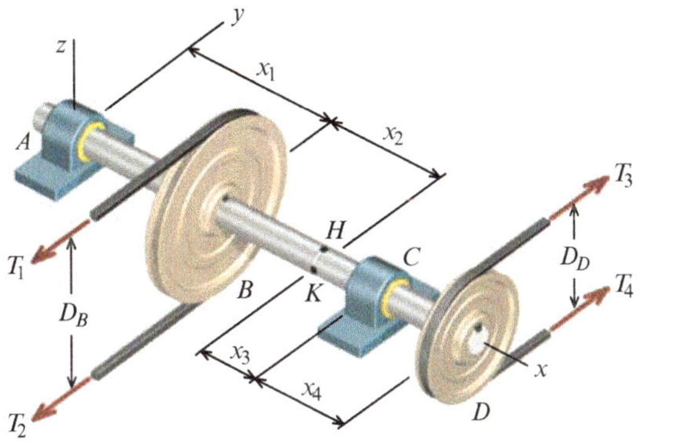

FIGURE P15.41/42

P15.42 A solid steel shaft with an outside diameter of 25 mm is supported in flexible bearings at *A* and *D* as shown in Figure P15.41/42. Two pulleys are keyed to the shaft at *B* and *C*. Pulley *B* has a diameter $D_B = 350$ mm and belt tensions $T_1 = 842$ N and $T_2 = 234$ N. Pulley *C* has a diameter $D_C = 160$ mm and belt tensions $T_3 = 1{,}660$ N and $T_4 = 330$ N. The length dimensions of the shaft are $x_1 = 250$ mm, $x_2 = 120$ mm, $x_3 = 80$ mm, and $x_4 = 160$ mm. Determine the principal stresses and the absolute maximum shear stress at point *K* on the shaft.

P15.43 A solid steel shaft with an outside diameter of 1.5 in. is supported in flexible bearings at *A* and *C* as shown in Figure P15.43/44. Two pulleys are keyed to the shaft at *B* and *D*. Pulley *B* has a diameter $D_B = 14$ in. and belt tensions $T_1 = 100$ lb and $T_2 = 220$ lb. Pulley *D* has a diameter $D_D = 7$ in. and belt tensions $T_3 = 60$ lb and $T_4 = 300$ lb. The length dimensions of the shaft are $x_1 = 8$ in., $x_2 = 4$ in., $x_3 = 3$ in., and $x_4 = 6$ in. Determine the normal and shear stresses at point *K* on the shaft. Show the orientation of these stresses on an appropriate sketch.

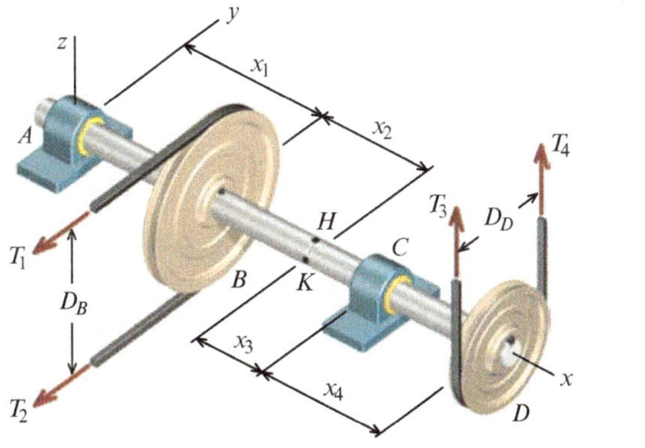

FIGURE P15.43/44

P15.44 A solid steel shaft with an outside diameter of 1.5 in. is supported in flexible bearings at *A* and *C* as shown in Figure P15.43/44. Two pulleys are keyed to the shaft at *B* and *D*. Pulley *B* has a diameter $D_B = 14$ in. and belt tensions $T_1 = 100$ lb and $T_2 = 220$ lb. Pulley *D* has a diameter $D_D = 7$ in. and belt tensions $T_3 = 60$ lb and $T_4 = 300$ lb. The length dimensions of the shaft are $x_1 = 8$ in., $x_2 = 4$ in., $x_3 = 3$ in., and $x_4 = 6$ in. Determine the normal and shear stresses at point *H* on the shaft. Show the orientation of these stresses on an appropriate sketch.

P15.45 A stainless steel pipe (Figure P15.45) with an outside diameter of 2.375 in. and a wall thickness of 0.109 in. is subjected to a bending moment $M = 50$ lb·ft and an internal pressure of 180 psi. Determine the absolute maximum shear stress on the outer surface of the pipe.

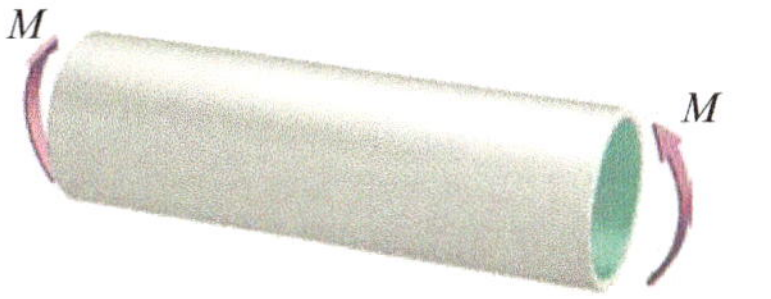

FIGURE P15.45

P15.46 The piping assembly shown in Figure P15.46/47/48/49 consists of stainless steel pipe that has an outside diameter of 6.675 in. and a wall thickness of 0.28 in. The assembly is subjected to concentrated loads $P_x = 320$ lb, $P_y = 410$ lb, and $P_z = 180$ lb, as well as an internal fluid pressure of 100 psi that acts in all of the pipes. The dimensions of the assembly are $y_1 = 2.0$ ft, $y_2 = 4.5$ ft, $z_1 = 3.5$ ft, and $z_2 = 3.0$ ft. Determine the normal and shear stresses on the outer surface of the pipe at (a) point *A* and (b) point *D*. Show these stresses on an appropriate sketch.

P15.47 The piping assembly shown in Figure P15.46/47/48/49 consists of stainless steel pipe that has an outside diameter of 6.675 in. and a wall thickness of 0.28 in. The assembly is subjected to concentrated loads $P_x = 320$ lb, $P_y = 410$ lb, and $P_z = 180$ lb, as well as an internal fluid pressure of 100 psi that acts in all of the pipes. The dimensions of the assembly are $y_1 = 2.0$ ft, $y_2 = 4.5$ ft, $z_1 = 3.5$ ft, and $z_2 = 3.0$ ft. Determine the normal and shear stresses on the outer surface of the pipe at (a) point *B* and (b) point *C*. Show these stresses on an appropriate sketch.

P15.48 The piping assembly shown in Figure P15.46/47/48/49 consists of stainless steel pipe that has an outside diameter of 200 mm and a wall thickness of 8 mm. The assembly is subjected to concentrated loads $P_x = 2{,}400$ N, $P_y = 0$, and $P_z = 1{,}100$ N, as well as an internal fluid pressure of 900 kPa that acts in all of the pipes. The dimensions of the assembly are $y_1 = 0.7$ m, $y_2 = 1.8$ m, $z_1 = 1.3$ m, and $z_2 = 1.1$ m. Determine the normal and shear stresses on the outer surface of the pipe at (a) point *B* and (b) point *C*. Show these stresses on an appropriate sketch.

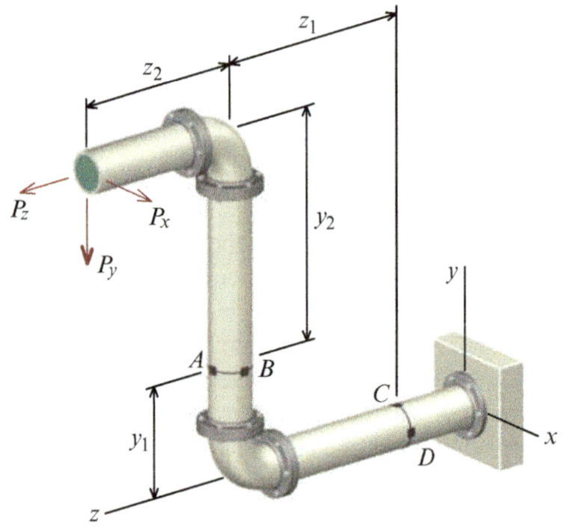

FIGURE P15.46/47/48/49

P15.49 The piping assembly shown in Figure P15.46/47/48/49 consists of stainless steel pipe that has an outside diameter of 200 mm and a wall thickness of 8 mm. The assembly is subjected to concentrated loads $P_x = 2{,}400$ N, $P_y = 0$, and $P_z = 1{,}100$ N, as well as an internal fluid pressure of 900 kPa that acts in all of the pipes. The dimensions of the assembly are $y_1 = 0.7$ m, $y_2 = 1.8$ m, $z_1 = 1.3$ m, and $z_2 = 1.1$ m. Determine the normal and shear stresses on the outer surface of the pipe at (a) point A and (b) point D. Show these stresses on an appropriate sketch.

P15.50 The piping assembly shown in Figure P15.50/51 consists of stainless steel pipe that has an outside diameter of 275 mm and a wall thickness of 9 mm. The assembly is subjected to concentrated loads $P_x = 3.2$ kN, $P_y = 5.4$ kN, and $P_z = 1.3$ kN, as well as an internal fluid pressure of 1,400 kPa that acts in all of the pipes. The dimensions of the assembly are $x_1 = 1.35$ m, $y_1 = 1.0$ m, $y_2 = 2.3$ m, and $z_1 = 1.6$ m. Determine the normal and shear stresses on the outer surface of the pipe at (a) point A and (b) point C. Show these stresses on an appropriate sketch.

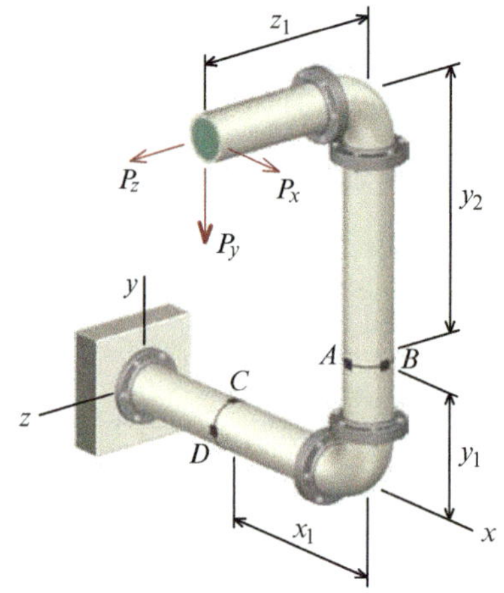

FIGURE P15.50/51

P15.51 The piping assembly shown in Figure P15.50/51 consists of stainless steel pipe that has an outside diameter of 275 mm and a wall thickness of 9 mm. The assembly is subjected to concentrated loads $P_x = 3.2$ kN, $P_y = 5.4$ kN, and $P_z = 1.3$ kN, as well as an internal fluid pressure of 1,400 kPa that acts in all of the pipes. The dimensions of the assembly are $x_1 = 1.35$ m, $y_1 = 1.0$ m, $y_2 = 2.3$ m, and $z_1 = 1.6$ m. Determine the normal and shear stresses on the outer surface of the pipe at (a) point B and (b) point D. Show these stresses on an appropriate sketch.

15.5 Theories of Failure

A tension test of an axially loaded member is easy to conduct, and the results, for many types of materials, are well known. When such a member fails, the failure occurs at a specific principal stress (i.e., the axial stress), a definite axial strain, a maximum shear stress equal to one-half of the axial stress, and a specific amount of strain energy per unit volume of stressed material. Since all of these limits are reached simultaneously for an axial load, it makes no difference which criterion (stress, strain, or energy) is used for predicting failure in another axially loaded member of the same material.

For an element subjected to biaxial or triaxial loading, however, the situation is more complicated because the limits of normal stress, normal strain, shear stress, and strain energy existing at failure are not reached simultaneously. In other words, the precise cause of failure, in general, is unknown. In such cases, it becomes important to determine the best criterion for predicting failure, because test results are difficult to obtain and the possible combinations of loads are endless. Several theories have been proposed for predicting the failure of various types of material subjected to many combinations of loads. Unfortunately,

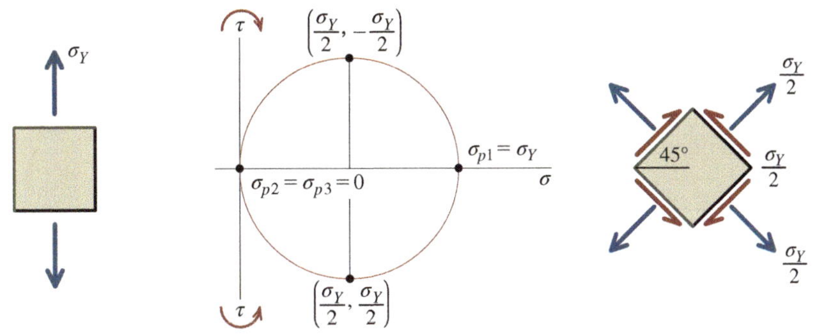

(a) Stress element at yield for test specimen (b) Mohr's circle for test specimen at yield (c) Maximum-shear-stress element

FIGURE 15.6 States of stress for a uniaxial tension test.

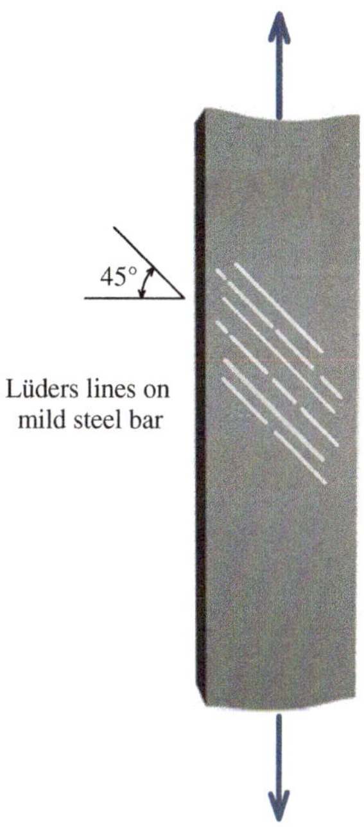

FIGURE 15.5 Lüders lines on a ductile tension test specimen.

no single theory agrees with test data for all types of materials and all combinations of loads. Several of the more common theories of failure are presented and briefly explained in the paragraphs that follow.

Ductile Materials

Maximum-Shear-Stress Theory.[1] When a flat bar of a ductile material such as mild steel is tested in uniaxial tension, yielding of the material is accompanied by lines that appear on the surface of the bar. These lines, known as **Lüders lines**, are caused by *slipping* (on a microscopic scale) that occurs along the planes of randomly ordered grains that make up the material. Lüders lines are oriented at 45° with respect to the longitudinal axis of the specimen (Figure 15.5). Therefore, if one assumes that slip is the failure mechanism associated with yielding of the material, then the stress that best characterizes this failure is the shear stress on the slip planes. In a uniaxial tension test, the state of stress at yield can be represented by the stress element shown in Figure 15.6*a*. The Mohr's circle corresponding to this state of stress is shown in Figure 15.6*b*. Mohr's circle reveals that the maximum shear stress in a uniaxial test specimen occurs at an orientation of 45° with respect to the load direction (Figure 15.6*c*), just as the Lüders lines do.

On the basis of these observations, the maximum-shear-stress theory predicts that failure will occur in a component that is subjected to any combination of loads when the maximum shear stress at any point in the object reaches the failure shear stress $\tau_f = \sigma_Y/2$, where σ_Y is determined by an axial tension or compression test of the same material. For ductile materials, the shearing elastic limit, as determined from a torsion test (for pure shear), is greater than one-half the tensile elastic limit. (An average value for τ_f is about $0.57\sigma_Y$.) Since the maximum-shear-stress theory is based on σ_Y obtained from an axial test, the theory errs on the conservative side.

The maximum-shear-stress theory is represented graphically in Figure 15.7 for an element subjected to biaxial principal stresses (i.e., plane stress). In the first and third quadrants, σ_{p1} and σ_{p2} have the same sign; therefore, the absolute maximum shear stress acts in an out-of-plane direction, and it has a magnitude that is equal to one-half of the numerically larger principal stress σ_{p1} or σ_{p2}, as explained in Section 12.8 [see Equation (12.18)]. In the second and fourth quadrants, where σ_{p1} and σ_{p2} are of opposite sign, the maximum shear

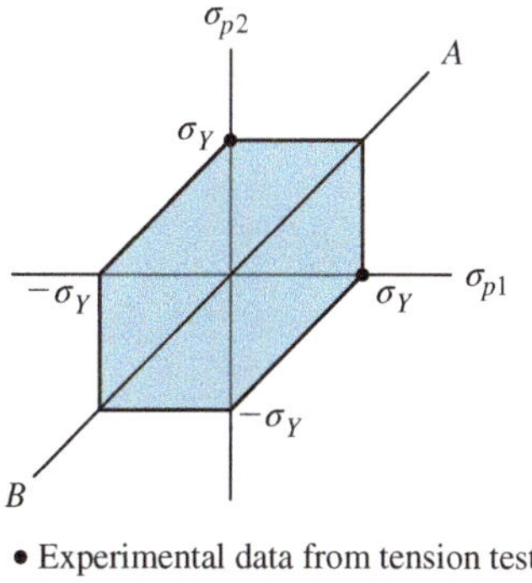

• Experimental data from tension test.

FIGURE 15.7 Failure diagram for maximum-shear-stress theory for an element subjected to plane stress.

[1]Sometimes called Coulomb's theory because it was originally stated by him in 1773. More frequently called the Tresca criterion or the Tresca–Guest yield surface because of the work of French elasticity theorist H. E. Tresca (1814–1885), which was advanced by the work of J. J. Guest in England in 1900.

stress is equal to one-half of the arithmetic sum of the two principal stresses (i.e., simply the radius of the in-plane Mohr's circle).

If the naming convention for principal stresses is followed (i.e., $\sigma_{p1} > \sigma_{p2}$), then all combinations of σ_{p1} and σ_{p2} will plot to the right of (i.e., below) line *AB* shown in Figure 15.7.

Therefore, the maximum-shear-stress theory applied to a *plane stress state* with in-plane principal stresses σ_{p1} and σ_{p2} predicts that yielding failure will occur under the following conditions:

- If σ_{p1} and σ_{p2} have the same sign, then failure will occur if $|\sigma_{p1}| \geq \sigma_Y$ or $|\sigma_{p2}| \geq \sigma_Y$.
- If σ_{p1} is positive and σ_{p2} is negative, then failure will occur if $\sigma_{p1} - \sigma_{p2} \geq \sigma_Y$.

Maximum-Distortion-Energy Theory.[2] The maximum-distortion-energy theory is founded on the concept of **strain energy**. The total strain energy per unit volume can be determined for a specimen subjected to any combination of loads. Further, the total strain energy can be broken down into two categories: strain energy that is associated with a change in volume of the specimen and strain energy that is associated with a change in shape, or *distortion*, of the specimen. This theory predicts that failure will occur when the strain energy causing distortion reaches the same intensity as the strain energy at failure found in axial tension or compression tests of the same material. Supporting evidence comes from experiments which reveal that homogeneous materials can withstand very high hydrostatic stresses (i.e., equal-intensity normal stresses in three orthogonal directions) without yielding. Based on this observation, the maximum-distortion-energy theory assumes that only the strain energy which produces a change of shape is responsible for the failure of the material. The strain energy of distortion is most readily computed by determining the total strain energy of the stressed material and subtracting the strain energy associated with the change in volume.

The load *P* must be applied slowly so that there is no kinetic energy associated with the application of the load. All work done by *P* is stored as potential energy in the strained bar.

The concept of strain energy is illustrated in Figure 15.8. A bar of uniform cross section subjected to a slowly applied axial load *P* is shown in Figure 15.8*a*. A load–deformation diagram for the bar is shown in Figure 15.8*b*. The work done in elongating the bar by an amount δ_2 is

$$W = \int_0^{\delta_2} P\,d\delta \tag{a}$$

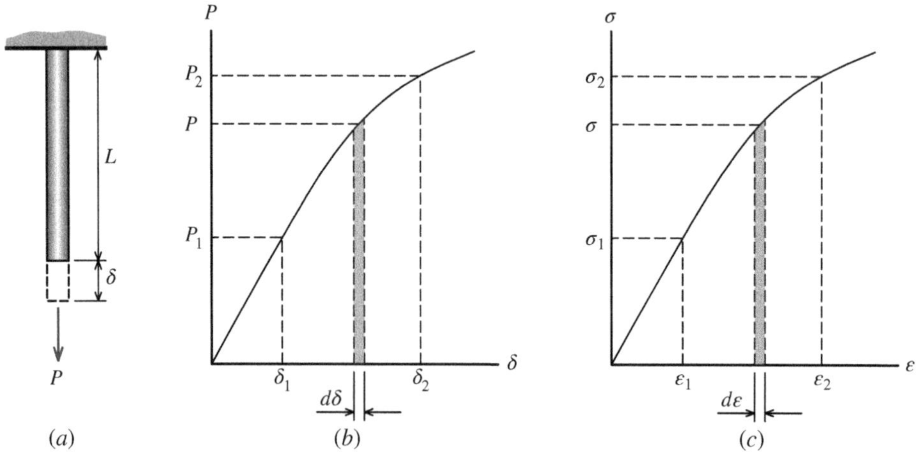

FIGURE 15.8 Concept of strain energy.

[2] Frequently called the Huber–von Mises–Hencky theory or the von Mises yield criterion because it was proposed by M. T. Huber of Poland in 1904 and, independently, by R. von Mises of Germany in 1913. The theory was further developed by H. Hencky and von Mises in Germany and the United States.

where P is some function of δ. The work done on the bar must equal the change in energy of the material,[3] and this energy change is termed the *strain energy* U because it involves the strained configuration of the material. If δ is expressed in terms of axial strain ($\delta = L\varepsilon$) and P is expressed in terms of axial stress ($P = A\sigma$), Equation (a) becomes

$$W = U = \int_0^{\varepsilon_2} (\sigma)(A)(L)\, d\varepsilon = AL\int_0^{\varepsilon_2} \sigma\, d\varepsilon \qquad \text{(b)}$$

where σ is a function of ε. (See Figure 15.8*c*.) If Hooke's law applies then,

$$\varepsilon = \sigma / E \qquad d\varepsilon = d\sigma / E$$

and Equation (b) becomes

$$U = \left(\frac{AL}{E}\right)\int_0^{\sigma_2} \sigma\, d\sigma$$

or

$$U = AL\left(\frac{\sigma_2^2}{2E}\right) \qquad \text{(c)}$$

Equation (c) gives the *elastic strain energy* (which is, in general, recoverable)[4] for axial loading of a material obeying Hooke's law. The quantity in parentheses, $\sigma_2^2/(2E)$, is the elastic the strain energy u in tension or compression per unit volume, or the *strain-energy density*, for a particular value of stress σ below the proportional limit of the material. Thus,

$$u = \frac{1}{2}\sigma\varepsilon = \frac{1}{2E}\sigma^2 = \frac{E}{2}\varepsilon^2 \qquad (15.1)$$

For shear loading, the expression would be identical except that σ would be replaced by τ, ε by γ, and E by G.

The concept of elastic strain energy can be extended to include biaxial and triaxial loadings by writing the expression for strain-energy density u as $1/2\sigma\varepsilon$ and adding the energies due to each of the stresses. Since energy is a positive scalar quantity, the addition is the arithmetic sum of the energies. For a system of triaxial principal stresses σ_{p1}, σ_{p2}, and σ_{p3}, the total elastic strain-energy density is

$$u = \frac{1}{2}\left[\sigma_{p1}\varepsilon_{p1} + \sigma_{p2}\varepsilon_{p2} + \sigma_{p3}\varepsilon_{p3}\right] \qquad \text{(d)}$$

When the generalized Hooke's law expressions for strains in terms of stresses from Equation (13.16) of Section 13.8 are substituted into Equation (d), the result is

$$u = \frac{1}{2E}\{\sigma_{p1}[\sigma_{p1} - v(\sigma_{p2} + \sigma_{p3})] + \sigma_{p2}[\sigma_{p2} - v(\sigma_{p3} + \sigma_{p1})] + \sigma_{p3}[\sigma_{p3} - v(\sigma_{p1} + \sigma_{p2})]\}$$

[3] Known as *Clapeyron's theorem*, after the French engineer B. P. E. Clapeyron (1799–1864).
[4] Elastic hysteresis is neglected here as an unnecessary complication.

from which it follows that

$$u = \frac{1}{2E}[\sigma_{p1}^2 + \sigma_{p2}^2 + \sigma_{p3}^2 - 2\nu(\sigma_{p1}\sigma_{p2} + \sigma_{p2}\sigma_{p3} + \sigma_{p3}\sigma_{p1})] \tag{15.2}$$

The total strain energy can be resolved into components associated with a volume change (u_v) and a distortion (u_d) by considering the principal stresses to be made up of two sets of stresses as indicated in Figures 15.9*a–c*. The state of stress depicted in Figure 15.9*c* will produce only distortion (no volume change) if the sum of the other three normal strains is zero—that is, if

$$\begin{aligned} E(\varepsilon_{p1} + \varepsilon_{p2} + \varepsilon_{p3})_d &= [(\sigma_{p1} - p) - \nu(\sigma_{p2} + \sigma_{p3} - 2p)] \\ &\quad + [(\sigma_{p2} - p) - \nu(\sigma_{p3} + \sigma_{p1} - 2p)] \\ &\quad + [(\sigma_{p3} - p) - \nu(\sigma_{p1} + \sigma_{p2} - 2p)] = 0 \end{aligned}$$

where p is the hydrostatic stress. This equation reduces to

$$(1 - 2\nu)(\sigma_{p1} + \sigma_{p2} + \sigma_{p3} - 3p) = 0$$

Therefore, the hydrostatic stress is

$$p = \frac{1}{3}(\sigma_{p1} + \sigma_{p2} + \sigma_{p3})$$

From Equation (13.16), the three normal strains due to the hydrostatic stress p are

$$\varepsilon_v = \frac{1}{E}(1 - 2\nu)p$$

and the energy resulting from the hydrostatic stress (i.e., the volume change) is

$$u_v = 3\left(\frac{p\varepsilon_v}{2}\right) = \frac{3}{2}\frac{1 - 2\nu}{E}p^2 = \frac{(1 - 2\nu)}{6E}(\sigma_{p1} + \sigma_{p2} + \sigma_{p3})^2$$

The energy resulting from the distortion (i.e., the change in shape) is

$$\begin{aligned} u_d &= u - u_v \\ &= \frac{1}{6E}\left[3(\sigma_{p1}^2 + \sigma_{p2}^2 + \sigma_{p3}^2) - 6\nu(\sigma_{p1}\sigma_{p2} + \sigma_{p2}\sigma_{p3} + \sigma_{p3}\sigma_{p1}) - (1 - 2\nu)(\sigma_{p1} + \sigma_{p2} + \sigma_{p3})^2\right] \end{aligned}$$

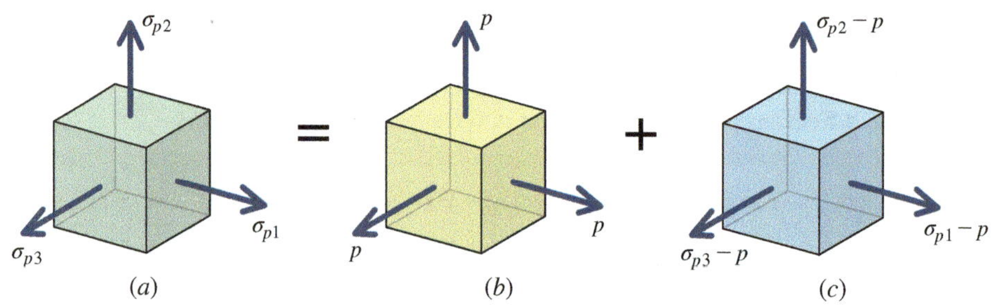

FIGURE 15.9 Expressing the state of stress in terms of volume-change components and distortion components.

When the third term in the brackets is expanded, this equation can be solved for u_d:

$$\begin{aligned} u_d &= \frac{1+v}{6E}\left[(\sigma_{p1}^2 - 2\sigma_{p1}\sigma_{p2} + \sigma_{p2}^2) + (\sigma_{p2}^2 - 2\sigma_{p2}\sigma_{p3} + \sigma_{p3}^2) + (\sigma_{p3}^2 - 2\sigma_{p3}\sigma_{p1} + \sigma_{p1}^2)\right] \\ &= \frac{1+v}{6E}\left[(\sigma_{p1} - \sigma_{p2})^2 + (\sigma_{p2} - \sigma_{p3})^2 + (\sigma_{p3} - \sigma_{p1})^2\right] \end{aligned} \quad \text{(e)}$$

The maximum-distortion-energy theory of failure assumes that inelastic action will occur whenever the energy given by Equation (e) exceeds the limiting value obtained from a tension test. In the tension test, only one of the principal stresses will be nonzero. If this stress is called σ_Y, then

$$(u_d)_Y = \frac{1+v}{3E}\sigma_Y^2$$

and when this value is substituted into Equation (e), the maximum-distortion-energy failure criterion is expressed as

$$\sigma_Y^2 = \frac{1}{2}\left[(\sigma_{p1} - \sigma_{p2})^2 + (\sigma_{p2} - \sigma_{p3})^2 + (\sigma_{p3} - \sigma_{p1})^2\right] \quad (15.3)$$

or

$$\sigma_Y^2 = \sigma_{p1}^2 + \sigma_{p2}^2 + \sigma_{p3}^2 - (\sigma_{p1}\sigma_{p2} + \sigma_{p2}\sigma_{p3} + \sigma_{p3}\sigma_{p1})$$

for failure by yielding. The maximum-distortion-energy failure criterion can be alternatively stated in terms of the normal stresses and shear stress on three arbitrary orthogonal planes:

$$\sigma_Y^2 = \frac{1}{2}\left[(\sigma_x - \sigma_y)^2 + (\sigma_y - \sigma_z)^2 + (\sigma_x - \sigma_z)^2 + 6(\tau_{xy}^2 + \tau_{yz}^2 + \tau_{xz}^2)\right] \quad (15.4)$$

When a state of plane stress exists (i.e., when $\sigma_{p3} = 0$), Equation (15.3) becomes

$$\sigma_Y^2 = \sigma_{p1}^2 - \sigma_{p1}\sigma_{p2} + \sigma_{p2}^2 \quad (15.5)$$

This last expression is the equation of an ellipse in the $\sigma_{p1} - \sigma_{p2}$ plane with its major axis along the line $\sigma_{p1} = \sigma_{p2}$, as shown in Figure 15.10. For comparison purposes, the failure hexagon for the maximum-shear-stress yield theory is also shown in dashed lines in Figure 15.10. While both theories predict failure at the six vertices of the hexagon, the maximum-shear-stress theory gives the more conservative estimate of the stresses required to produce yielding, since the hexagon falls inside the ellipse for all other combinations of stress.

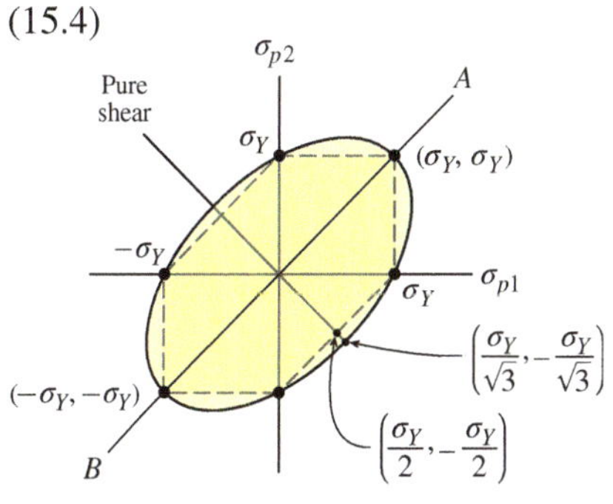

FIGURE 15.10 Failure diagram for maximum-distortion-energy theory for an element subjected to plane stress.

Mises Equivalent Stress. A convenient way to employ the maximum-distortion-energy theory is to establish an equivalent stress quantity σ_M that is defined as the square root of the right-hand side of Equation (15.3). This stress is called the **Mises equivalent stress** (or the **von Mises equivalent stress**)[5] and is given by

$$\sigma_M = \frac{\sqrt{2}}{2}\left[(\sigma_{p1} - \sigma_{p2})^2 + (\sigma_{p2} - \sigma_{p3})^2 + (\sigma_{p3} - \sigma_{p1})^2\right]^{1/2} \quad (15.6)$$

If the naming convention for principal stresses is followed (i.e., $\sigma_{p1} > \sigma_{p2}$), then all combinations of σ_{p1} and σ_{p2} will plot to the right of (i.e., below) line *AB* shown in Figure 15.10.

[5] After Richard Edler von Mises (1883–1953), Austrian mathematician and scientist who taught at the University of Istanbul and Harvard University.

Similarly, Equation (15.4) can also be used to compute the Mises equivalent stress:

$$\sigma_M = \frac{\sqrt{2}}{2}\left[(\sigma_x - \sigma_y)^2 + (\sigma_y - \sigma_z)^2 + (\sigma_x - \sigma_z)^2 + 6(\tau_{xy}^2 + \tau_{yz}^2 + \tau_{xz}^2)\right]^{1/2} \quad (15.7)$$

For the case of plane stress, the Mises equivalent stress can be expressed from Equation (15.5) as

$$\sigma_M = [\sigma_{p1}^2 - \sigma_{p1}\sigma_{p2} + \sigma_{p2}^2]^{1/2} \quad (15.8)$$

or it can be found from Equation (15.4) by setting $\sigma_z = \tau_{yz} = \tau_{xz} = 0$ to give

$$\sigma_M = [\sigma_x^2 - \sigma_x\sigma_y + \sigma_y^2 + 3\tau_{xy}^2]^{1/2} \quad (15.9)$$

To use the Mises equivalent stress, σ_M is calculated for the state of stress acting at any specific point in the component. The resulting value of σ_M is then compared with the tensile yield stress σ_Y, and if $\sigma_M > \sigma_Y$, then the material is predicted to fail according to the maximum-distortion-energy theory. The utility of the Mises equivalent stress has led to its widespread use in tabulated stress analysis results and in the form of color-coded stress contour plots common in finite element analysis results.

Brittle Materials

Unlike ductile materials, brittle materials tend to fail suddenly by fracture, with little evidence of yielding; therefore, the limiting stress appropriate for brittle materials is the fracture stress (or the ultimate strength) rather than the yield strength. Furthermore, the tensile strength of a brittle material is often different from its compressive strength.

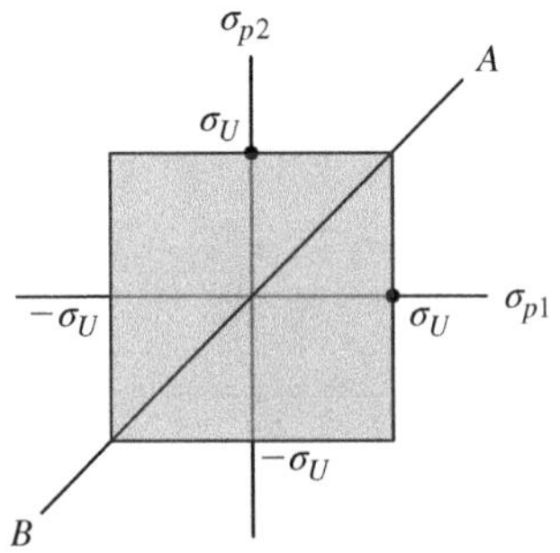

FIGURE 15.11 Failure diagram for maximum-normal-stress theory for an element subjected to plane stress.

Maximum-Normal-Stress Theory.[6] The maximum-normal-stress theory predicts that failure will occur in a specimen that is subjected to any combination of loads when the maximum normal stress at any point in the specimen reaches the axial failure stress determined from an axial tension or compression test of the same material.

The maximum-normal-stress theory is presented graphically in Figure 15.11 for an element subjected to biaxial principal stresses in the $p1$ and $p2$ directions. The limiting stress σ_U is the failure stress for the material the element is made of when it is loaded axially. According to this theory, any combination of biaxial principal stresses σ_{p1} and σ_{p2} represented by a point inside the square of Figure 15.11 is safe whereas any combination of stresses represented by a point outside of the square will cause failure of the element.

If the naming convention for principal stresses is followed (i.e., $\sigma_{p1} > \sigma_{p2}$), then all combinations of σ_{p1} and σ_{p2} will plot to the right of or below line AB shown in Figure 15.11.

Mohr's Failure Criterion. For many brittle materials, the ultimate tension and compression strengths are different, and in such cases, the maximum-normal-stress theory should not be used. An alternative failure theory, proposed by the German engineer Otto Mohr,[7] is called Mohr's failure criterion. To use this criterion, a uniaxial tension test and a uniaxial compression test are performed to establish the ultimate tensile strength σ_{UT} and ultimate

[6] Often called Rankine's theory after W. J. M. Rankine (1820–1872), an eminent engineering educator at Glasgow University in Scotland.
[7] The same Otto Mohr of Mohr circle fame.

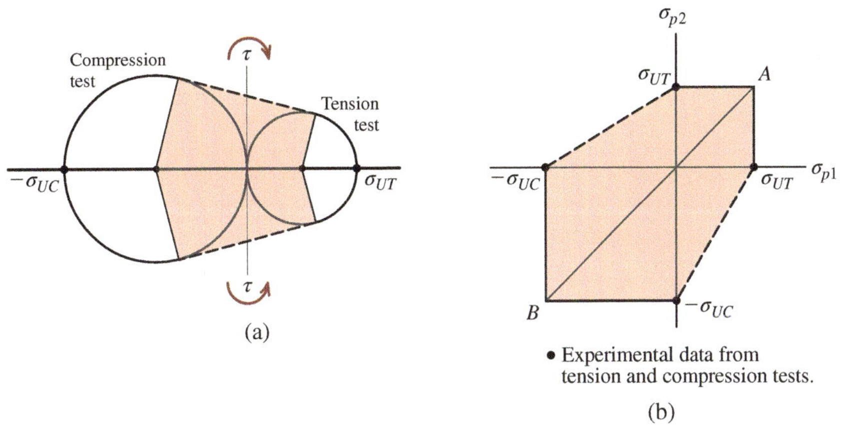

FIGURE 15.12 Mohr's failure criterion for an element subjected to plane stress.

compressive strength σ_{UC}, respectively, of the material. Mohr's circles for the tension and compression tests are shown in Figure 15.12*a*. Mohr's theory suggests that failure occurs in a material whenever Mohr's circle for the combination of stresses at a point in a body exceeds the "envelope" defined by the Mohr's circles for the tensile and compressive tests.

Mohr's failure criterion for a plane stress state may be represented on a graph of principal stresses in the $\sigma_{p1} - \sigma_{p2}$ plane (Figure 15.12*b*). The principal stresses for all Mohr's circles that have centers on the σ axis and are tangent to the dashed lines in Figure 15.12*a* will plot as points along the dashed lines in the $\sigma_{p1} - \sigma_{p2}$ plane of Figure 15.12*b*.

Mohr's failure criterion applied to a *plane stress state* with in-plane principal stresses σ_{p1} and σ_{p2} predicts that failure will occur under the following conditions:

- If σ_{p1} and σ_{p2} are both positive (i.e., tension), then failure will occur if $\sigma_{p1} \geq \sigma_{UT}$.
- If σ_{p1} and σ_{p2} are both negative (i.e., compression), then failure will occur if $\sigma_{p2} \leq -\sigma_{UC}$.

If the naming convention for principal stresses is followed (i.e., $\sigma_{p1} > \sigma_{p2}$), then all combinations of σ_{p1} and σ_{p2} will plot to the right of (i.e., below) line *AB* shown in Figure 15.12*b*. Stress states with $\sigma_{p1} > 0$ and $\sigma_{p2} < 0$ fall into the fourth quadrant of Figure 15.12*b*. In these cases, Mohr's failure criterion predicts that failure will occur for those combinations which plot on the dashed line—in other words, under the following condition:

- If σ_{p1} is positive and σ_{p2} is negative, then failure will occur if $\dfrac{\sigma_{p1}}{\sigma_{UT}} - \dfrac{\sigma_{p2}}{\sigma_{UC}} \geq 1$.

If torsion-test data are available, the dashed line in the fourth quadrant may be modified to incorporate these experimental data.

Examples 15.8 and 15.9 illustrate the application of the theories of failure in predicting the load-carrying capacity of a member:

EXAMPLE 15.8

The stresses on the free surface of a machine component are illustrated on the stress element shown. The component is made of 6061-T6 aluminum with a yield strength of $\sigma_Y = 270$ MPa.

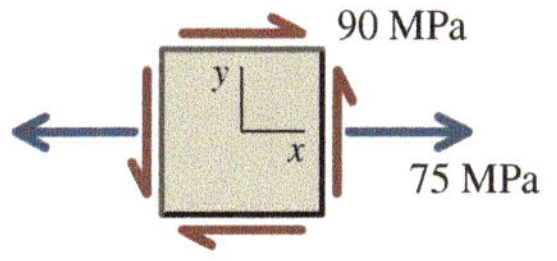

(a) What is the factor of safety predicted by the maximum-shear-stress theory of failure for the stress state shown? According to this theory, does the component fail?

(b) What is the value of the Mises equivalent stress for the given state of plane stress?

(c) What is the factor of safety predicted by the failure criterion of the maximum-distortion-energy theory of failure? According to this theory, does the component fail?

Plan the Solution

The principal stresses will be determined for the given state of stress. With these stresses, the maximum-shear-stress theory and the maximum-distortion-energy theory will be used to investigate the potential for failure.

SOLUTION

The principal stresses can be calculated from the stress transformation equations [Equation (12.12)] or from Mohr's circle, as discussed in Section 12.10. Equation 12.12 will be used here. From the stress element, the values to be used in the stress transformation equations are $\sigma_x = +75$ MPa, $\sigma_y = 0$ MPa, and $\tau_{xy} = +90$ MPa. The in-plane principal stresses are calculated as

$$\begin{aligned}\sigma_{p1,p2} &= \frac{\sigma_x + \sigma_y}{2} \pm \sqrt{\left(\frac{\sigma_x - \sigma_y}{2}\right)^2 + \tau_{xy}^2} \\ &= \frac{75\text{ MPa} + 0\text{ MPa}}{2} \pm \sqrt{\left(\frac{75\text{ MPa} - 0\text{ MPa}}{2}\right)^2 + (90\text{ MPa})^2} \\ &= 135.0\text{ MPa}, -60.0\text{ MPa}\end{aligned}$$

(a) Maximum-Shear-Stress Theory

Since σ_{p1} is positive and σ_{p2} is negative, failure will occur if $\sigma_{p1} - \sigma_{p2} \geq \sigma_Y$. For the principal stresses existing in the component,

$$\sigma_{p1} - \sigma_{p2} = 135.0\text{ MPa} - (-60.0\text{ MPa}) = 195.0\text{ MPa} < 270\text{ MPa}$$

Therefore, according to the maximum-shear-stress theory, the component does not fail. The factor of safety associated with this state of stress is

$$\text{FS} = \frac{270\text{ MPa}}{195.0\text{ MPa}} = 1.385$$ **Ans.**

(b) Mises Equivalent Stress

The Mises equivalent stress σ_M associated with the maximum-distortion-energy theory can be calculated from Equation (15.8) for the plane stress state considered here:

$$\begin{aligned}\sigma_M &= [\sigma_{p1}^2 - \sigma_{p1}\sigma_{p2} + \sigma_{p2}^2]^{1/2} \\ &= [(135.0\text{ MPa})^2 - (135.0\text{ MPa})(-60.0\text{ MPa}) + (-60.0\text{ MPa})^2]^{1/2} \\ &= 173.0\text{ MPa}\end{aligned}$$ **Ans.**

(c) Maximum-Distortion-Energy Theory Factor of Safety

The factor of safety for the maximum-distortion-energy theory can be calculated from the Mises equivalent stress:

$$\text{FS} = \frac{270\text{ MPa}}{173.0\text{ MPa}} = 1.561$$ **Ans.**

According to the maximum-distortion-energy theory, the component does not fail.

EXAMPLE 15.9

The stresses on the free surface of a machine component are illustrated on the stress element shown. The component is made of a brittle material with an ultimate tensile strength of 200 MPa and an ultimate compressive strength of 500 MPa. Use the Mohr failure criterion to determine whether this component is safe for the state of stress shown.

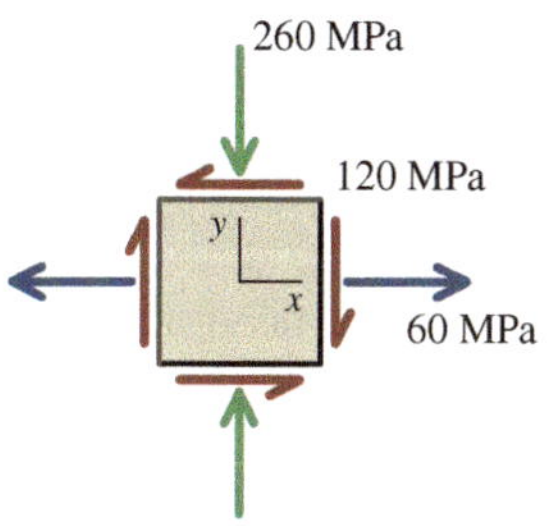

Plan the Solution

The principal stresses will be determined for the given state of stress. With these stresses, the Mohr failure criterion will be used to investigate the potential for failure.

SOLUTION

The principal stresses can be calculated from Equation 12.12:

$$\sigma_{p1,p2} = \frac{\sigma_x + \sigma_y}{2} \pm \sqrt{\left(\frac{\sigma_x - \sigma_y}{2}\right)^2 + \tau_{xy}^2}$$

$$= \frac{60\text{ MPa} + (-260\text{ MPa})}{2} \pm \sqrt{\left(\frac{60\text{ MPa} - (-260\text{ MPa})}{2}\right)^2 + (-120\text{ MPa})^2}$$

$$= 100\text{ MPa}, -300\text{ MPa}$$

Mohr Failure Criterion

Since σ_{p1} is positive and σ_{p2} is negative, failure will occur if

$$\frac{\sigma_{p1}}{\sigma_{UT}} - \frac{\sigma_{p2}}{\sigma_{UC}} \geq 1$$

For the principal stresses existing in the component,

$$\frac{\sigma_{p1}}{\sigma_{UT}} - \frac{\sigma_{p2}}{\sigma_{UC}} = \frac{100\text{ MPa}}{200\text{ MPa}} - \frac{(-300\text{ MPa})}{500\text{ MPa}} = 0.5 - (-0.6) = 1.1 > 1 \qquad \textbf{Ans.}$$

Therefore, according to the Mohr failure criterion, the component fails.

PROBLEMS

P15.52 The stresses on the surface of a structural steel beam are $\sigma_x = 38$ ksi, $\sigma_y = 0$, and $\tau_{xy} = 16.5$ ksi. The steel has a yield strength of $\sigma_Y = 50$ ksi.

(a) What is the factor of safety predicted by the maximum-shear-stress theory of failure for the stress state presented? Does the beam fail according to this theory?
(b) What is the value of the Mises equivalent stress for the given state of plane stress?
(c) What is the factor of safety predicted by the failure criterion of the maximum-distortion-energy theory of failure? According to this theory, does the beam fail?

P15.53 The stresses on the surface of a cast iron machine component are $\sigma_x = 85$ MPa, $\sigma_y = -230$ MPa, and $\tau_{xy} = 125$ MPa. The cast iron has a yield strength of $\sigma_Y = 350$ MPa.

(a) What is the factor of safety predicted by the maximum-shear-stress theory of failure for the stress state presented? According to this theory, does the component fail?
(b) What is the value of the Mises equivalent stress for the given state of plane stress?
(c) What is the factor of safety predicted by the failure criterion of the maximum-distortion-energy theory of failure? According to this theory, does the component fail?

P15.54 A thin-walled cylindrical pressure vessel of inside diameter $d = 36$ in. is fabricated from a material with a tensile yield strength of 40 ksi. The internal pressure in the cylinder is 1,600 psi. Assuming that the material obeys the von Mises criterion of yielding, and that there is to be a safety factor of 3.0 against yielding, determine the necessary wall thickness remote from the ends of the vessel.

P15.55 A thin-walled cylindrical pressure vessel of inside diameter d = 500 mm and wall thickness t = 5 mm is fabricated from a material with a tensile yield strength of 280 MPa. Determine the maximum internal pressure p that may be used in the cylinder according to

(a) the maximum-shear-stress theory.
(b) the maximum-distortion-energy theory.

P15.56 A solid 80 mm diameter circular shaft made of cold-rolled steel is subjected to the simultaneous action of a torque T = 19 kN · m, a bending moment M = 8.5 kN · m, and a compressive axial force P = 250 kN. The cold-rolled steel has a yield strength of σ_Y = 420 MPa, in both tension and compression. On the basis of

(a) the maximum-shear-stress theory and
(b) the maximum-distortion-energy theory,

is the shaft overstressed?

P15.57 A 75 mm diameter solid circular shaft rotating at 800 rpm transmits 400 kW and carries a tensile axial force P = 140 kN. The material that makes up the shaft has a tensile yield strength of σ_Y = 110 MPa. On the basis of

(a) the maximum-shear-stress theory and
(b) the maximum-distortion-energy theory,

is the shaft overstressed? State the factors of safety for each theory.

P15.58 A hollow circular shaft made of an aluminum alloy is subjected to a bending moment M = 1.2 kN · m and a torque T. The shaft has an outside diameter of 60 mm and an inside diameter of 40 mm. Assume that the aluminum alloy has a yield strength σ_Y = 275 MPa. Use

(a) the maximum-shear-stress theory and
(b) the maximum-distortion-energy theory

to determine the value of the torque T so that the shaft does not fail by yielding.

P15.59 A 3 in. diameter solid circular shaft made of stainless steel is subjected to the simultaneous action of a torque T = 6.0 kip · ft, a tensile axial force P = 55 kips, and a bending moment M. The stainless steel has a yield strength of σ_Y = 36 ksi, in both tension and compression. Use

(a) the maximum-shear-stress theory and
(b) the maximum-distortion-energy theory

to determine the value of the moment M so that the shaft does not fail by yielding.

P15.60 A solid circular shaft made of steel is subjected to a torque T = 780 lb · ft and a tensile axial force P = 13,500 lb. Assume that the steel has a yield strength σ_Y = 30,000 psi. Use

(a) the maximum-shear-stress theory and
(b) the maximum-distortion-energy theory

to determine the required diameter of the shaft so that the shaft does not fail by yielding.

P15.61 The stresses on the surface of a machine component will be in a state of plane stress with $\sigma_x = \sigma_y = -40$ ksi and τ_{xy} = 85 ksi. The ultimate failure strengths for the material that the component is made of are 100 ksi in tension and 190 ksi in compression. Use the Mohr failure criterion to determine whether this component is safe for the state of stress given. Support your answer with appropriate documentation.

P15.62 The stresses on the surface of a machine component will be in a state of plane stress with σ_x = 510 MPa, σ_y = −140 MPa, and τ_{xy} = 375 MPa. The ultimate failure strengths for the material that the component is made of are 700 MPa in tension and 420 MPa in compression. Use the Mohr failure criterion to determine whether this component is safe for the state of stress given. Support your answer with appropriate documentation.

P15.63 The state of stress at a point in a cast iron [σ_{UT} = 290 MPa; σ_{UC} = 650 MPa] component is σ_x = 0, σ_y = −180 MPa, and τ_{xy} = 200 MPa. Determine whether failure occurs at the point according to (a) the maximum-normal-stress theory and (b) the Mohr failure criterion.

CHAPTER 16

Columns

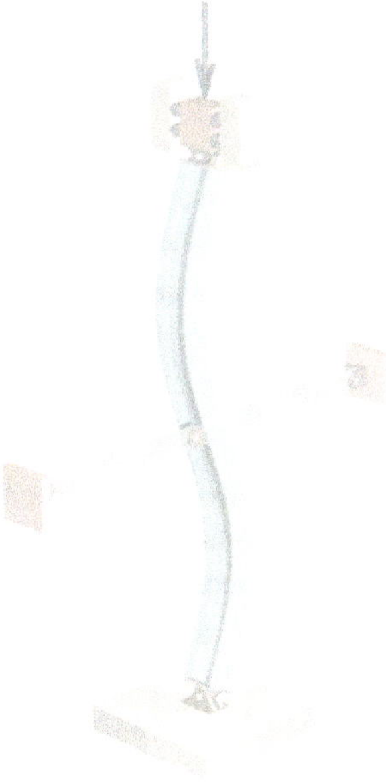

16.1 Introduction

In their simplest form, columns are long, straight, prismatic bars subjected to compressive, axial loads. As long as a column remains straight, it can be analyzed by the methods of Chapter 1; however, if a column begins to deform laterally, the deflection may become large and lead to catastrophic failure. This situation, called **buckling**, can be defined as the sudden large deformation of a structure due to a slight increase of an existing load under which the structure had exhibited little, if any, deformation before the load was increased.

A simple buckling "experiment" can be performed to illustrate this phenomenon, with a thin ruler, yardstick, or meterstick used to represent a column. A small compressive axial force applied to the ends of the column will cause no discernible effect. Gradually increase the magnitude of the compressive force applied to the ends of the column, however, and at some critical load, the column will suddenly bend laterally, or "bow out." The column has buckled. Once buckling occurs, a relatively small increase in compressive force will produce a relatively large lateral deflection, creating additional bending in the column. However, if the compressive force is removed, the column returns to its original straight shape. The buckling failure illustrated by this experiment is not a failure of the material. The fact that the column becomes straight again after the compressive force is removed demonstrates that the material remains elastic; that is, the stresses in the column have not exceeded the proportional limit of the material. Rather, the buckling failure is a **stability failure**: The column has transitioned from a stable equilibrium to an unstable one.

Stability of Equilibrium

The concept of stability of equilibrium with respect to columns can be investigated with the elementary column-buckling model shown in Figure 16.1*a*. In this figure, a column is modeled by two perfectly straight pin-connected rigid bars *AB* and *BC*. The column model is supported by a pin connection at *A* and by a slotted support at *C* that prevents lateral movement, but allows pin *C* to move freely in the vertical direction. In addition to the pin at *B*, a rotational spring with a spring constant *K* connects the bars, which are assumed to be perfectly aligned vertically before the axial load *P* is applied, making the column model initially straight.

Since the load *P* acts vertically and the column model is initially straight, there should be no tendency for pin *B* to move laterally as load *P* is applied. Furthermore, one might suppose that the magnitude of load *P* could be increased to any intensity without creating an effect in the rotational spring. However, common sense tells us that this supposition cannot be true: At some load *P*, the pin at *B* will move laterally. To investigate the issue further, we must examine the column model after pin *B* has been displaced laterally by a small amount.

In Figure 16.1*b*, the pin at *B* has been displaced slightly to the right so that each bar forms a small angle $\Delta\theta$ with the vertical. The rotational spring at *B* reacts to the angular rotation of $2\Delta\theta$ at *B*, tending to restore bars *AB* and *BC* to their initial vertical orientation. The question is whether the column model will return to its initial configuration from the displaced configuration after it is subjected to the axial load *P* or whether pin *B* will move farther to the right. If the column model returns to its initial configuration, the system is said to be *stable*. If pin *B* moves farther to the right, the system is said to be *unstable*.

To answer this question, consider the free-body diagram of bar *BC* shown in Figure 16.1*c*. In the displaced position, the forces *P* acting at pins *B* and *C* create a couple that tends to cause pin *B* to move farther away from its initial position. The moment created by this couple is called the destabilizing moment. In opposition, the rotational spring creates a *restoring moment M*, which tends to return the system to its initial vertical orientation. The moment produced by the rotational spring is equal to the product of the spring constant *K* and the angular rotation at *B*, which is $2\Delta\theta$. Therefore, the rotational spring produces a restoring moment $M = K(2\Delta\theta)$ at *B*. If the restoring moment is greater than the destabilizing

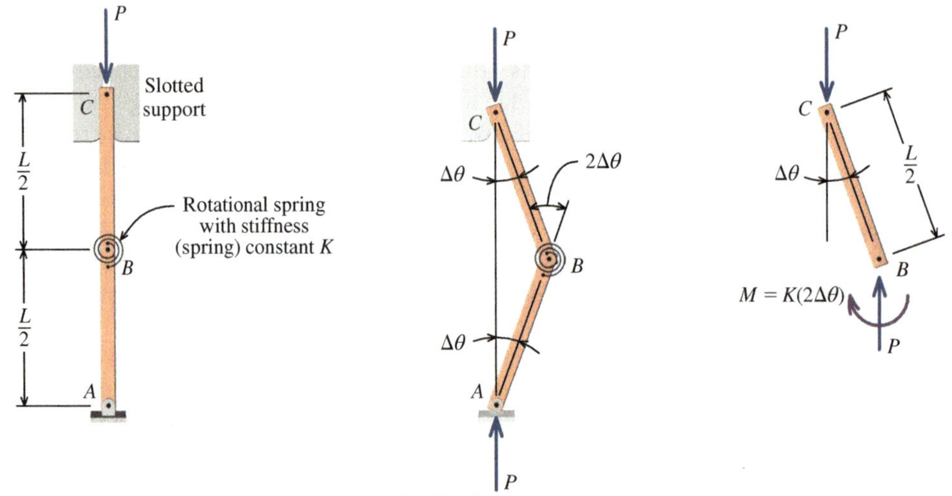

(*a*) Initial configuration (*b*) Displaced configuration (*c*) Free-body diagram

FIGURE 16.1 Elementary column-buckling model.

moment, then the system will tend to return to its initial configuration. However, if the upsetting moment is larger than the restoring moment, then the system will be unstable in the displaced configuration and pin B will move farther to the right until either equilibrium is attained or the model collapses. The magnitude of axial load P at which the restoring moment equals the destabilizing moment is called the **critical load** P_{cr}. To determine the critical load for the column model, consider the moment equilibrium of bar BC in Figure 16.1*c* for the load $P = P_{cr}$:

$$\Sigma M_B = P_{cr}(L/2)\sin\Delta\theta - K(2\Delta\theta) = 0 \tag{a}$$

Since the lateral displacement at B is assumed to be small, $\sin\Delta\theta \approx \Delta\theta$, and thus, Equation (a) can be simplified and solved for P_{cr}:

$$P_{cr}(L/2)\Delta\theta = K(2\Delta\theta)$$
$$\therefore\ P_{cr} = \frac{4K}{L} \tag{b}$$

If the load P applied to the column model is less than P_{cr}, then the restoring moment is greater than the destabilizing moment and the system is stable. However, if $P > P_{cr}$, then the system is unstable. At the point of transition, where $P = P_{cr}$, the system is neither stable nor unstable, but rather, is said to be in **neutral equilibrium**. The fact that $\Delta\theta$ does not appear in the second line of Equation (b) indicates that the critical load can be resisted at any value of $\Delta\theta$. In other words, pin B could be moved laterally to any position, and there would be no tendency for the column model either to return to the initial straight configuration or to move farther away from it.

Equation (b) also suggests that the stability of the elementary column-buckling model can be enhanced by *increasing the stiffness* K or by *decreasing the length* L. In the sections that follow, we will observe that these same relationships are applicable to the critical loads of actual columns.

The notions of stability and instability can be defined concisely in the following manner:

Stable—A small action produces a small effect.
Unstable—A small action produces a large effect.

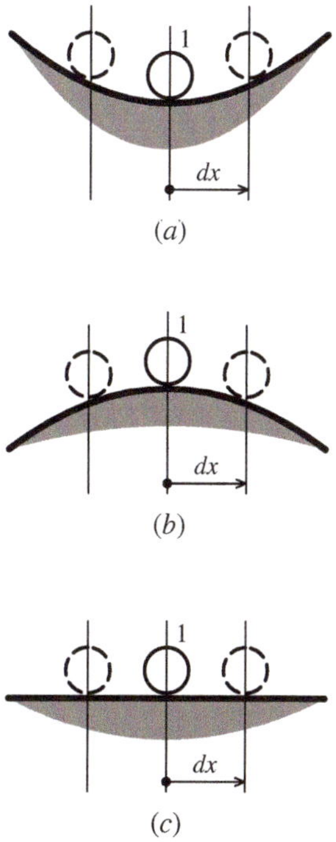

FIGURE 16.2 Concepts of (*a*) stable, (*b*) unstable, and (*c*) neutral equilibrium.

These notions and the concept of three equilibrium states can be illustrated by the equilibrium of a ball resting on three different surfaces, as shown in Figure 16.2. In all three cases, the ball is in equilibrium at position 1. To investigate the stability associated with each surface, the ball must be displaced an infinitesimally small distance dx to either side of position 1. In Figure 16.2*a*, a ball displaced laterally by dx and released would roll back to its initial position. In other words, a small action (i.e., displacing the ball by dx) produces a small effect (i.e., the ball rolls back a distance dx). Therefore, a ball at rest at position 1 on the concave upward surface of Figure 16.2*a* illustrates the notion of **stable equilibrium**. By contrast, the ball in Figure 16.2*b*, if displaced laterally by dx and released, would not return to position 1. Rather, the ball would roll farther away from position 1. In other words, a small action (i.e., displacing the ball by dx) produces a large effect (i.e., the ball rolls a large distance until it eventually reaches another equilibrium position). The ball at rest at position 1 on the concave downward surface of Figure 16.2*b* illustrates the notion of **unstable equilibrium**. The ball in Figure 16.2*c* is in a **neutral equilibrium** position on the horizontal plane because it will remain at any new position to which it is displaced, tending neither to return to nor to move farther from its original position.

Summary

Before a compressive load on a column is gradually increased from zero, the column is in a state of stable equilibrium. During this state, if the column is perturbed by small lateral deflections, it will return to its initial straight configuration when the load is removed. As the load is increased further, a critical value is reached at which the column is about to undergo large lateral deflections; that is, the column is at the transition between stable and unstable equilibrium. The maximum compressive load for which the column is in stable equilibrium is called the **critical buckling load**. The compressive load cannot be increased beyond this critical value unless the column is laterally restrained. For long, slender columns, the critical buckling load occurs at stress levels that are much lower than the proportional limit for the material, indicating that this type of buckling is an elastic phenomenon.

16.2 Buckling of Pin-Ended Columns

The stability of real columns will be investigated by analyzing a long, slender column with pinned ends (Figure 16.3*a*). The column is loaded by a compressive load P that passes through the centroid of the cross section at the ends. The pins at each end are frictionless, and the load is applied to the column by the pins. The column itself is perfectly straight and made of a linearly elastic material that is governed by Hooke's law. Since the column is assumed to have no imperfections, it is termed an **ideal column**. The ideal column in Figure 16.3*a* is assumed to be symmetric about the x–y plane, and any deflections occur in that plane.

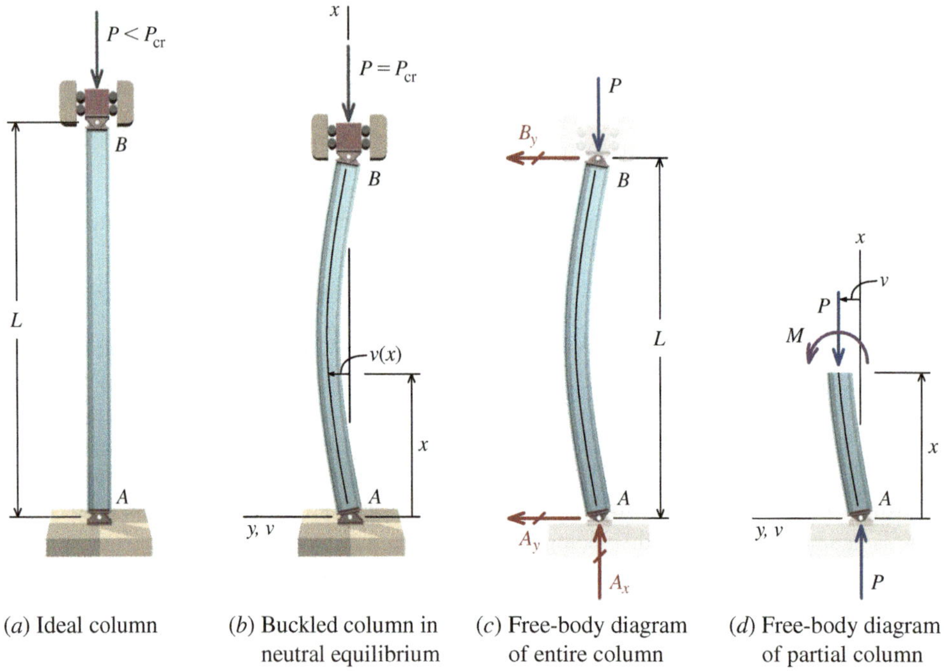

FIGURE 16.3 Buckling of an ideal column.

Buckled Configuration

If the compressive load P is less than the critical load P_{cr}, then the column will remain straight and will shorten in response to a uniform compressive axial stress $\sigma = P/A$. As long as $P < P_{cr}$, the column is in **stable equilibrium**. When the compressive load P is increased to the critical load P_{cr}, the column is at the transition point between stable and unstable equilibrium—a situation called **neutral equilibrium**. At $P = P_{cr}$, the deflected shape shown in Figure 16.3*b* also satisfies equilibrium. The value of the critical load P_{cr} and the shape of the buckled column will be determined by an analysis of this deflected shape.

Equilibrium of the Buckled Column

A free-body diagram of the entire buckled column is shown in Figure 16.3*c*. Summing forces in the vertical direction gives $A_x = P$, summing moments about A gives $B_y = 0$, and summing forces in the horizontal direction gives $A_y = 0$.

Next, consider a free-body diagram cut through the column at a distance x from the origin (Figure 16.3*d*). Since $A_y = 0$, any shear force V acting in the horizontal direction on the exposed surface of the column in this free-body diagram must also equal zero in order to satisfy equilibrium. Consequently, both the horizontal reaction A_y and a shear force V can be omitted from the free-body diagram in Figure 16.3*d*.

Differential Equation for Column Buckling

In the buckled column of Figure 16.3*d*, both the column deflection v and the internal bending moment M are shown in their positive directions. As defined in Section 10.2, the bending moment M creates positive curvature. From the free-body diagram in Figure 16.3*d*, the sum of moments about A is

$$\Sigma M_A = M + Pv = 0 \tag{a}$$

From Equation (10.1), the moment–curvature relationship (assuming small deflections) can be expressed as

$$M = EI\frac{d^2v}{dx^2} \tag{b}$$

Equation (b) can be substituted into Equation (a) to give

$$EI\frac{d^2v}{dx^2} + Pv = 0 \tag{16.1}$$

Equation (16.1) is the differential equation that dictates the deflected shape of an ideal column. This equation is a homogeneous second-order ordinary differential equation with constant coefficients that has boundary conditions $v(0) = 0$ and $v(L) = 0$.

Solution of the Differential Equation

Established methods are available for the solution of equations such as Equation (16.1). To use these methods, Equation (16.1) is first simplified by dividing by EI to obtain

$$\frac{d^2v}{dx^2} + \frac{P}{EI}v = 0$$

If we let

$$k^2 = \frac{P}{EI} \tag{16.2}$$

then Equation (16.1) can be rewritten as

$$\frac{d^2v}{dx^2} + k^2 v = 0$$

The general solution of this homogeneous equation is

$$v = C_1 \sin kx + C_2 \cos kx \tag{16.3}$$

where C_1 and C_2 are constants that must be evaluated with the use of the boundary conditions. From the boundary conditions $v(0) = 0$, we obtain

$$\begin{gathered} 0 = C_1 \sin(0) + C_2 \cos(0) = C_1(0) + C_2(1) \\ \therefore C_2 = 0 \end{gathered} \tag{c}$$

From the boundary conditions $v(L) = 0$, we obtain

$$0 = C_1 \sin(kL) \tag{d}$$

One solution of Equation (d) is $C_1 = 0$; however, this is a trivial solution, since it would imply that $v = 0$, and hence the column would remain perfectly straight. The other solution of Equation (d) is $\sin(kL) = 0$, so

$$kL = n\pi \qquad n = 1,2,3,\ldots \tag{e}$$

because the sine function equals zero for integer multiples of π.

Now, taking the square root of both sides of Equation (16.2) gives

$$k = \sqrt{\frac{P}{EI}}$$

This expression for k can be substituted into Equation (e) to give

$$\sqrt{\frac{P}{EI}}L = n\pi$$

which may be solved for the load P:

$$P = \frac{n^2\pi^2 EI}{L^2} \qquad n = 1,2,3,\ldots \tag{16.4}$$

Euler Buckling Load and Buckling Modes

The purpose of this analysis is to determine the minimum load P at which lateral deflections occur in the column; therefore, the smallest load P that causes buckling occurs for $n = 1$ in Equation (e), since that value of n gives the minimum value of P for a nontrivial solution. As in the elementary column-buckling model (see Figure 16.1), this load is called the critical buckling load; P_{cr} an ideal column, where

$$P_{cr} = \frac{\pi^2 EI}{L^2} \tag{16.5}$$

The critical load for an ideal column is known as the **Euler buckling load**, after the Swiss mathematician Leonhard Euler (1707–1783), who published the first solution of the equation for the buckling of long, slender columns in 1757. Equation (16.5) is also known as **Euler's formula**.

Equation (e) can be substituted into Equation (16.3) to describe the deflected shape of the buckled column:

$$v = C_1 \sin kx = C_1 \sin\left(\frac{n\pi}{L}x\right) \qquad n = 1,2,3,\ldots \tag{16.6}$$

An ideal pin-ended column subjected to a compressive axial load P is shown in Figure 16.4*a*. The deflected shape of the buckled column corresponding to the Euler buckling load given in Equation (16.5) is shown in Figure 16.4*b*. Note that the specific values for the constant C_1 cannot be obtained, since the exact deflected position of the buckled column is unknown. However, the deflections have been assumed to be small. The deflected shape is called the **mode shape**, and the buckled shape corresponding to $n = 1$ in Equation (16.6) is called the **first buckling mode**. By considering higher values of n in Equations (16.4) and (16.6), it is theoretically possible to obtain an infinite number of critical loads and corresponding mode shapes. The critical load and mode shape for the second buckling mode are illustrated in Figure 16.4*c*. The critical load for the second mode is four times greater than that of the first mode. However, buckled shapes for the higher modes are of no practical interest, since the column buckles upon reaching its lowest critical load value. Higher mode shapes can be attained only by providing lateral restraint to the column at intermediate locations to prevent the column from buckling in the first mode.

Euler Buckling Stress

The normal stress in the column at the critical load is

$$\sigma_{cr} = \frac{P_{cr}}{A} = \frac{\pi^2 EI}{AL^2} \tag{f}$$

The **radius of gyration** r is a section property defined as

$$r^2 = \frac{I}{A} \tag{16.7}$$

If the moment of inertia I is replaced by Ar^2, Equation (f) becomes

$$\sigma_{cr} = \frac{\pi^2 E(Ar^2)}{AL^2} = \frac{\pi^2 Er^2}{L^2} = \frac{\pi^2 E}{(L/r)^2} \tag{16.8}$$

The quantity L/r is termed the **slenderness ratio** and is determined for the axis about which bending tends to occur. For an ideal column with no intermediate bracing to restrain lateral deflection, buckling occurs about the axis of minimum moment of inertia (which corresponds to the minimum radius of gyration).

Note that Euler buckling is an *elastic phenomenon*. If the axial compressive load is removed from an ideal column that has buckled as described here, the column will return to its initial straight configuration. In Euler buckling, the critical stress σ_{cr} remains below the proportional limit for the material.

Graphs of Euler buckling stress [Equation (16.8)] are shown in Figure 16.5 for structural steel and for an aluminum alloy. Since Euler buckling is an elastic phenomenon,

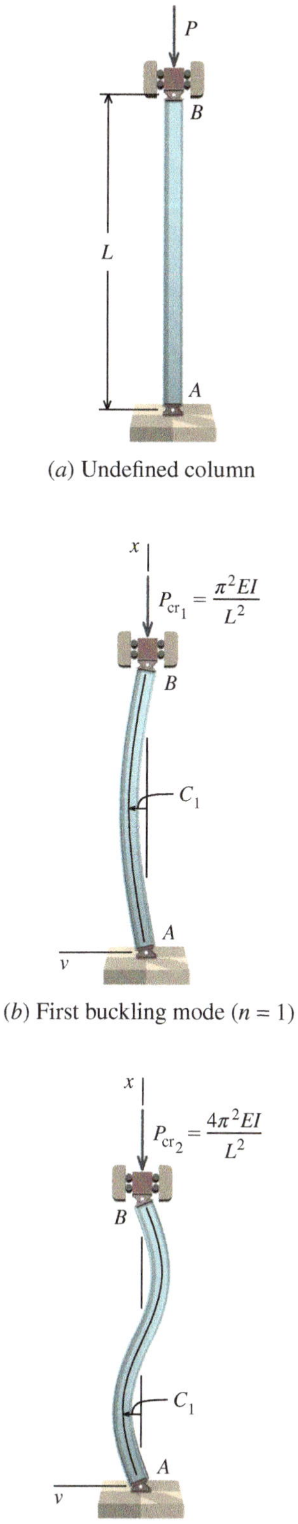

(*a*) Undefined column

(*b*) First buckling mode ($n = 1$)

(*c*) Second buckling mode ($n = 2$)

FIGURE 16.4 Two examples of buckling modes.

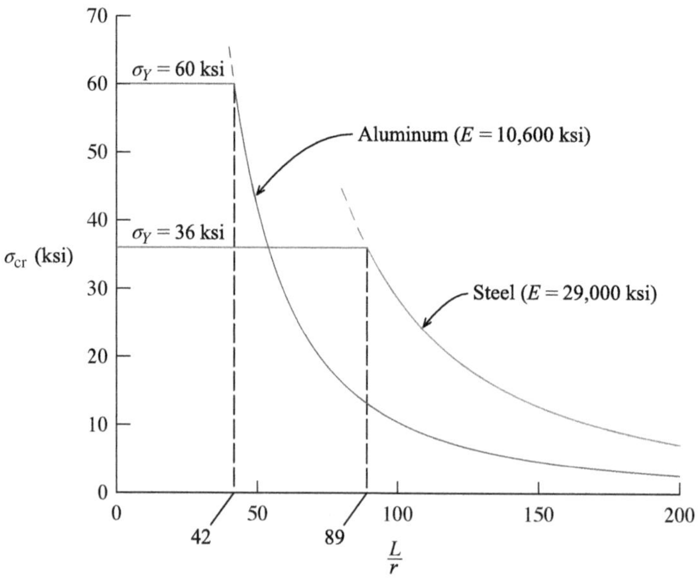

FIGURE 16.5 Graphs of Euler buckling stress for steel and an aluminum alloy.

Equation (16.8) is valid only when the critical stress is less than the proportional limit for the material, because the derivation of that equation is based on Hooke's law. Therefore, a horizontal line is drawn on the graph at the 36 ksi proportional limit stress for the structural steel and at the 60 ksi proportional limit for the aluminum alloy, and the respective Euler stress curves are truncated at these values.

Implications of Euler Buckling

An examination of Equations (16.5) and (16.8) reveals several implications for the buckling of an ideal column:

- The Euler buckling load is inversely related to the square of the column length. Therefore, the load that causes buckling decreases rapidly as the column length increases.
- The only material property that appears in Equations (16.5) and (16.8) is the elastic modulus E, which represents the *stiffness* of the material. One means of increasing the load-carrying capacity of a given column is to use a material with a higher value of E.
- Buckling occurs about the cross-sectional axis that corresponds to the *minimum moment of inertia* (which in turn corresponds to the minimum radius of gyration). Therefore, it is generally inefficient to select a member that has great disparity between the maximum and minimum moments of inertia for use as a column. This inefficiency can be mitigated if additional lateral bracing is provided to restrain lateral deflection about the weaker axis.
- Since the Euler buckling load is directly related to the moment of inertia I of the cross section, a column's load-carrying capacity can often be improved, without increasing its cross-sectional area, by employing thin-walled tubular shapes. Circular pipes and square hollow structural sections are particularly efficient in this regard. The radius of gyration r defined in Equation (16.7) provides a good measure of the relationship between moment of inertia and cross-sectional area. In choosing between two shapes of equal area for use as a column, it is helpful to keep in mind that the shape with the larger radius of gyration will be able to withstand more load before buckling.

- The Euler buckling load equation [Equation (16.5)] and the Euler buckling stress equation [Equation (16.8)] depend only on the column length L, the stiffness of the material (E), and the cross-sectional properties (I). The critical buckling load is independent of the strength of the material. For example, consider two round steel rods having the same diameter and length but differing strengths. Since E, I, and L are the same for both rods, the Euler buckling loads for the two rods will be identical. Consequently, there is no advantage in using the higher strength steel (which, presumably, is more expensive) instead of the lower strength steel in this instance.

The Euler buckling load as given by Equation (16.5) agrees well with experiment, but only for "long" columns for which the slenderness ratio L/r is large, typically in excess of 140 for steel columns. Whereas a "short" compression member can be treated as explained in Chapter 1, most practical columns are "intermediate" in length, and consequently, neither solution is applicable. These intermediate-length columns are analyzed by empirical formulas described in later sections. The slenderness ratio is the key parameter used to classify columns as long, intermediate, or short.

EXAMPLE 16.1

A 15 mm by 25 mm rectangular aluminum bar is used as a 650 mm long compression member. The ends of the compression member are pinned. Determine the slenderness ratio and the Euler buckling load for the compression member. Assume that $E = 70$ GPa.

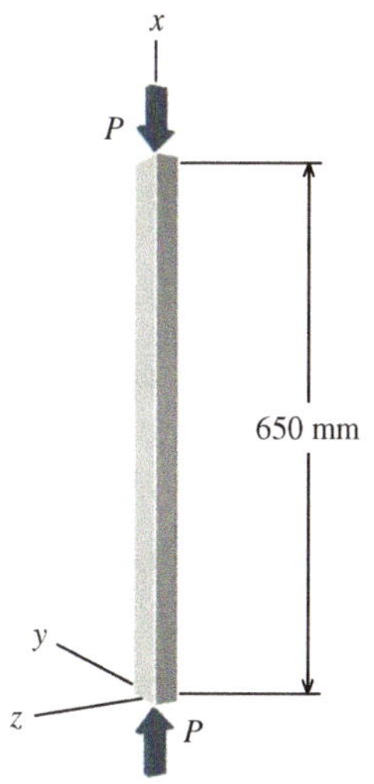

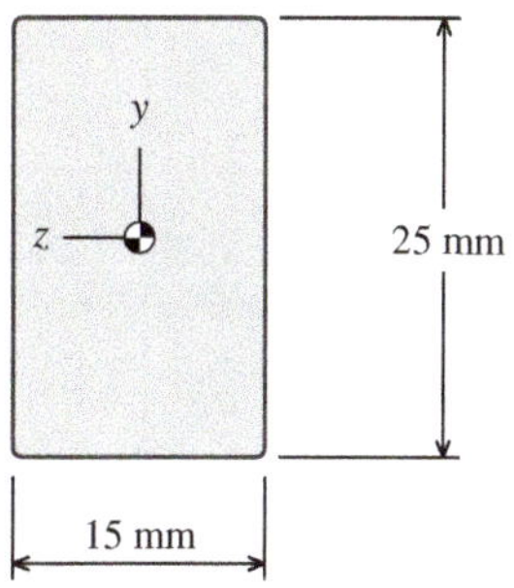

Cross section.

Plan the Solution

The aluminum bar will buckle about the weaker of the two principal axes for the cross-sectional shape of the compression member considered here. The smaller moment of inertia for the cross section occurs about the y axis; therefore, buckling will produce bending of the compression member in the x–z plane at the critical load P_{cr}.

SOLUTION

The cross-sectional area of the compression member is $A = (15 \text{ mm})(25 \text{ mm}) = 375 \text{ mm}^2$, and its moment of inertia about the y axis is

$$I_y = \frac{(25 \text{ mm})(15 \text{ mm})^3}{12} = 7{,}031.25 \text{ mm}^4$$

The slenderness ratio is equal to the length of the column divided by its radius of gyration. The radius of gyration for this cross section with respect to the y axis is

$$r_y = \sqrt{\frac{I_y}{A}} = \sqrt{\frac{7{,}031.25 \text{ mm}^4}{375 \text{ mm}^2}} = 4.330 \text{ mm}$$

and therefore, the slenderness ratio for buckling about the y axis is

$$\frac{L}{r_y} = \frac{650 \text{ mm}}{4.330 \text{ mm}} = 150.1 \qquad \textbf{Ans.}$$

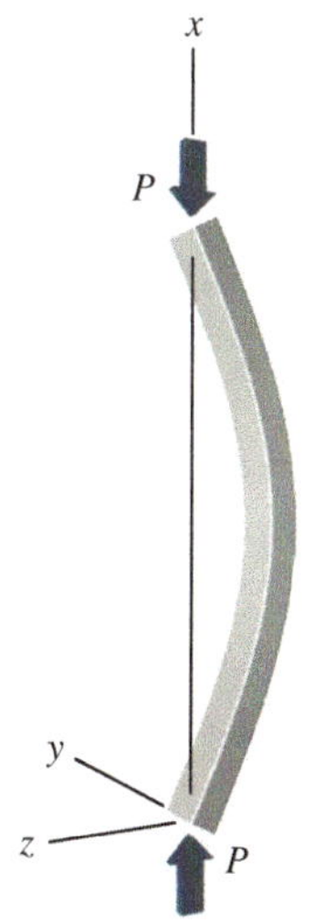

Note: The slenderness ratio is not necessary for determining the Euler buckling load in this instance; however, the slenderness ratio is an important parameter that is used in many empirical column formulas.

The Euler buckling load for this compression member can be calculated from Equation (16.5):

$$P_{cr} = \frac{\pi^2 EI}{L^2} = \frac{\pi^2 (70{,}000 \text{ N/mm}^2)(7{,}031.25 \text{ mm}^4)}{(650 \text{ mm})^2} = 11{,}498 \text{ N}$$
$$= 11.50 \text{ kN}$$ **Ans.**

When the compression member buckles, it bends in the x–z plane as shown.

EXAMPLE 16.2

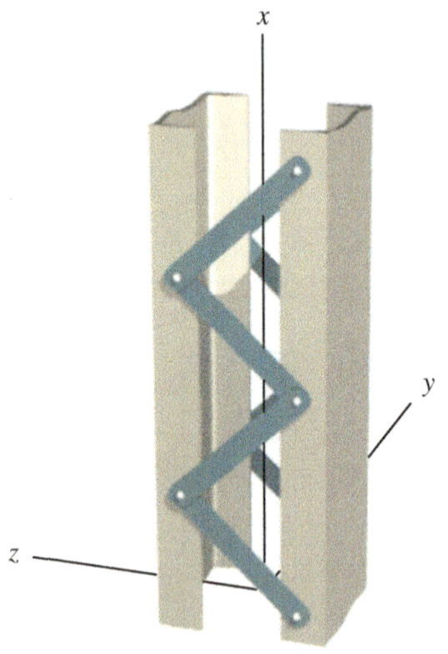

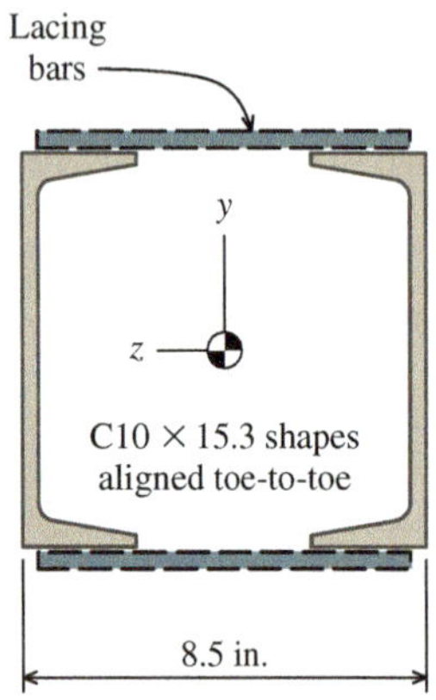

Cross section.

A 40 ft long column is fabricated by connecting two standard steel C10 × 15.3 channels (see Appendix B for cross-sectional properties) with lacing bars as shown. The ends of the column are pinned. Determine the Euler buckling load for the column. Assume that $E = 29{,}000$ ksi for the steel.

Plan the Solution

The column is built up from two standard steel channel shapes. The lacing bars serve only to connect the two shapes so that they act as a single structural unit. The bars do not add to the compressive strength of the column. Which principal axis of the cross section is the strong axis, and which is the weak axis? The answer is not evident by inspection; therefore, the moments of inertia about both axes must be calculated at the outset. Since both ends of the column are pinned, buckling will occur about the axis that corresponds to the smaller moment of inertia.

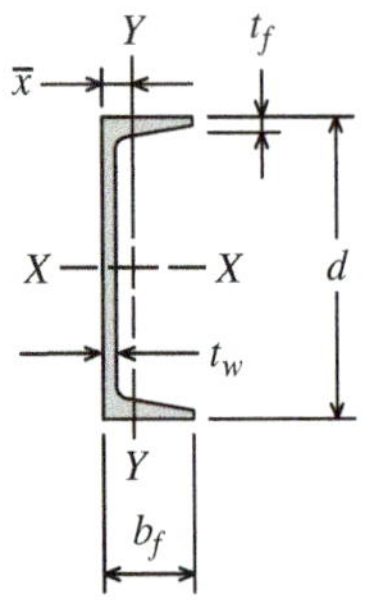

From Appendix B.

SOLUTION

The following section properties for a standard steel C10 × 15.3 channel are given in Appendix B:

$$A = 4.48 \text{ in.}^2$$
$$I_x = 67.3 \text{ in.}^4$$
$$I_y = 2.27 \text{ in.}^4$$
$$\bar{x} = 0.634 \text{ in.}$$

In Appendix B, the X–X axis is the strong axis for the channel and the Y–Y axis is the weak axis. For the coordinate system defined in this problem, the X–X axis will be denoted the z' axis and the Y–Y axis will be denoted the y' axis.

The cross-sectional area of the built-up column is equal to twice the area of a single channel shape:

$$A = 2(4.48 \text{ in.}^2) = 8.96 \text{ in.}^2$$

Similarly, the moment of inertia of the built-up column about the z axis is equal to twice that of a single channel shape about its strong axis (i.e., the z' axis):

$$I_z = 2(67.3 \text{ in.}^4) = 134.6 \text{ in.}^4$$

The horizontal distance from the y centroidal axis for the entire cross section to the back of one channel is 4.25 in. The distance from the back of the channel to its y' centroidal axis is given in Appendix B as 0.634 in. Therefore, the distance between the centroidal axis for the entire cross section and the centroidal axis for a single channel shape is equal to the difference in these two numbers: 4.25 in. − 0.634 in. = 3.616 in. This distance is shown in the figure to the right.

From the parallel-axis theorem, the moment of inertia of the built-up shape about its y centroidal axis is

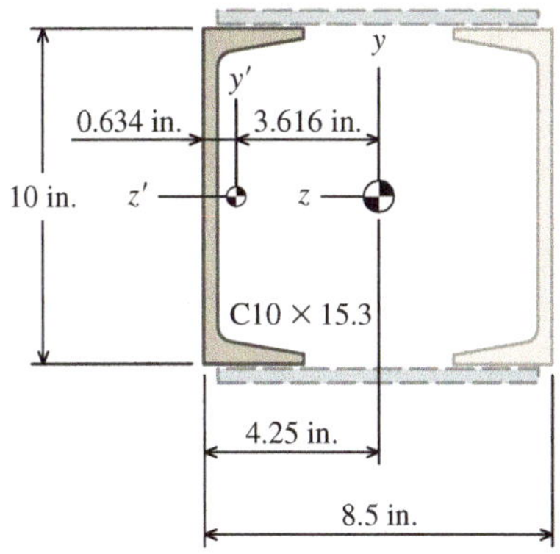

$$I_y = 2[2.27 \text{ in.}^4 + (3.616 \text{ in.})^2(4.48 \text{ in.})^2] = 121.6961 \text{ in.}^4$$

Since $I_y < I_z$, the built-up column will buckle about its y axis.

The Euler buckling load is calculated from Equation (16.5):

$$P_{cr} = \frac{\pi^2 EI}{L^2} = \frac{\pi^2(29{,}000 \text{ ksi})(121.6961 \text{ in.}^4)}{[(40 \text{ ft})(12 \text{ in./ft})]^2} = 151.2 \text{ kips} \qquad \textbf{Ans.}$$

PROBLEMS

P16.1 Determine the slenderness ratio and the Euler buckling load for round wooden dowels that are 1 m long and have a diameter of (a) 16 mm and (b) 25 mm. Assume that $E = 10$ GPa.

P16.2 An aluminum alloy tube with an outside diameter of 3.50 in. and a wall thickness of 0.30 in. is used as a 14 ft long column. Assume that $E = 10{,}000$ ksi and that pinned connections are used at each end of the column. Determine the slenderness ratio and the Euler buckling load for the column.

P16.3 A WT205 × 30 structural steel section (see Appendix B for cross-sectional properties) is used for a 6.5 m column. Assume pinned connections at each end of the column. Determine

(a) the slenderness ratio.
(b) the Euler buckling load. Use $E = 200$ GPa for the steel.
(c) the axial stress in the column when the Euler load is applied.

P16.4 Determine the maximum compressive load that an HSS6 × 4 × 1/4 structural steel column (see Appendix B for cross-sectional properties) can support if it is 24 ft long and a factor of safety of 1.92 is specified. Use $E = 29{,}000$ ksi for the steel.

P16.5 Two C12 × 25 structural steel channels (see Appendix B for cross-sectional properties) are used for a column that is 35 ft long. Assume pinned connections at each end of the column, and use $E = 29{,}000$ ksi for the steel. Determine the total compressive load required to buckle the two members if

(a) they act independently of each other.
(b) they are latticed back-to-back as shown in Figure P16.5.

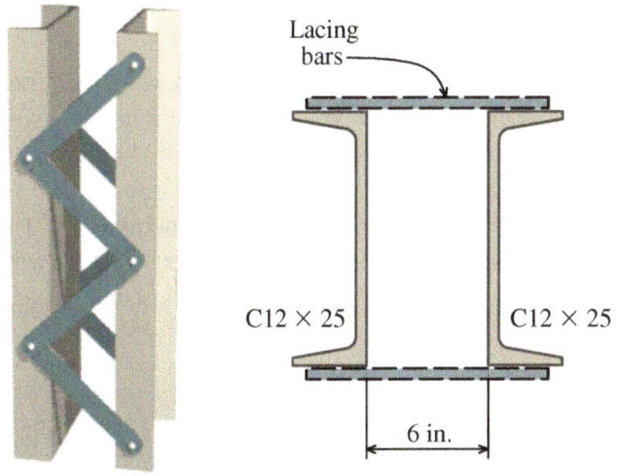

FIGURE P16.5

P16.6 Two L102 × 76 × 9.5 structural steel angles (see Appendix B for cross-sectional properties) are used as a compression member that is 4.5 m long. The angles are separated at intervals by spacer blocks and connected by bolts (as shown in Figure P16.6), which ensure that the double-angle shape acts as a unified structural member. Assume pinned connections at each end of the column, and use $E = 200$ GPa for the steel. Determine the Euler buckling load for the double-angle column if the spacer block thickness is

(a) 5 mm
(b) 20 mm.

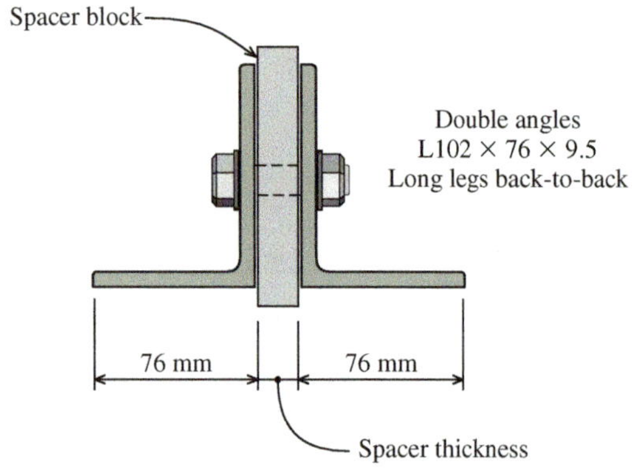

FIGURE P16.6

P16.7 A solid 0.5 in. diameter cold-rolled steel rod is pinned to fixed supports at A and B as shown in Figure P16.7. The length of the rod is $L = 24$ in., the rod's elastic modulus is $E = 30{,}000$ ksi, and its coefficient of thermal expansion is $\alpha = 6.6 \times 10^{-6}$ /°F. Determine the temperature increase ΔT that will cause the rod to buckle.

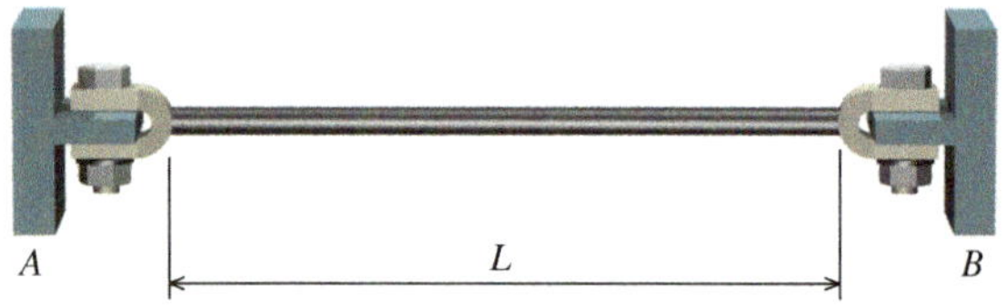

FIGURE P16.7

P16.8 Rigid beam ABC is supported by a pinned connection at A and by a timber post that is pin connected at B and D, as shown in Figure P16.8/9. A distributed load of $w = 2$ kips/ft acts on the 14 ft long beam, which has length dimensions $x_1 = 8$ ft and $x_2 = 6$ ft. The timber post has a length $L = 10$ ft, an elastic modulus $E = 1{,}800$ ksi, and a square cross section. If a factor of safety of 2.0 with respect to buckling is specified, determine the minimum width required for the square post.

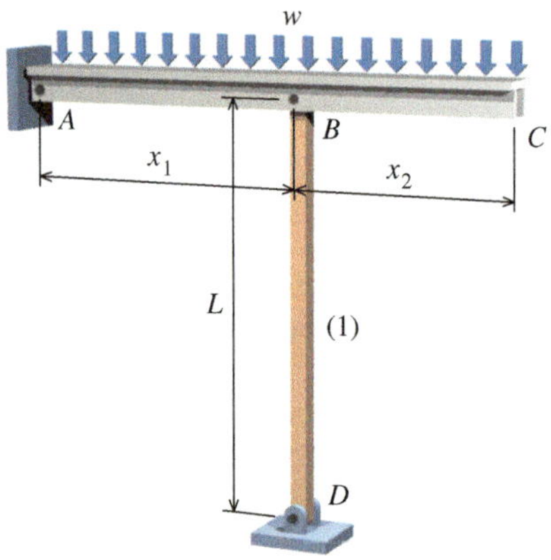

FIGURE P16.8/9

P16.9 Rigid beam ABC is supported by a pinned connection at A and by a 180 mm by 180 mm square timber post that is pin connected at B and D, as shown in Figure P16.8/9. The length dimensions of the beam are $x_1 = 3.6$ m and $x_2 = 2.8$ m. The timber post has a length $L = 4$ m and an elastic modulus $E = 12$ GPa. If a factor of safety of 2.0 with respect to buckling is specified, determine the magnitude of the maximum distributed load w that may be supported by the beam.

P16.10 A rigid beam is supported by a pinned connection at B and by an inclined strut that is pin connected at A and C as shown in Figure P16.10a. The dimensions of the structure are $a = 3.5$ m, $b = 2.5$ m, $c = 3.8$ m, and $e = 200$ mm. Loads on the structure are $P = 13$ kN and $w = 150$ kN/m. The strut is fabricated from two steel [$E = 200$ GPa] L127 × 76 × 12.7 angles, which are oriented with the long legs back-to-back, as shown in Figure P16.10b. The angles are separated and connected by spacer blocks, which are $s = 40$ mm thick. Determine

(a) the compressive force in the strut created by the loads acting on the beam.
(b) the slenderness ratios for the strut about the strong and weak axes of the double-angle shape.
(c) the minimum factor of safety in the strut with respect to buckling.

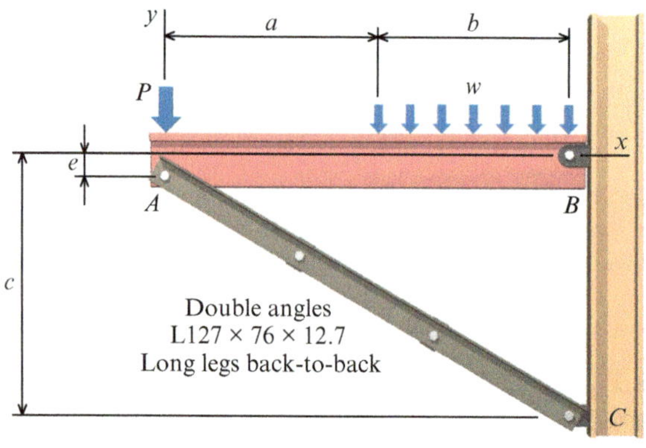

FIGURE P16.10a

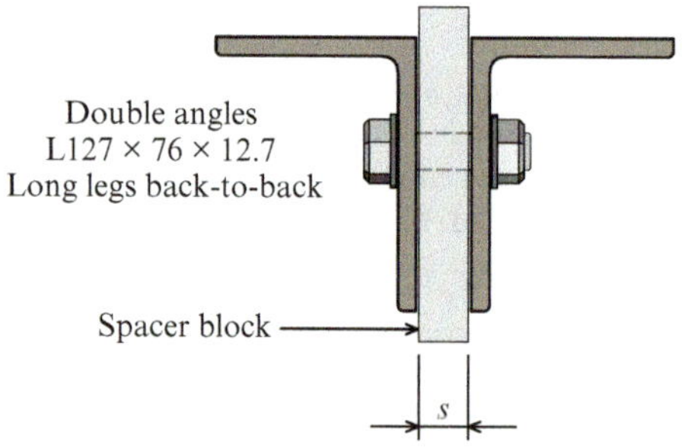

FIGURE P16.10b

P16.11 An assembly consisting of tie rod (1) and pipe strut (2) is used to support an 80 kip load, which is applied to joint *B*. Strut (2) is a pin-connected steel [$E = 29{,}000$ ksi] pipe with an outside diameter of 8.625 in. and a wall thickness of 0.322 in. For the loading shown in Figure P16.11, determine the factor of safety with respect to buckling for member (2).

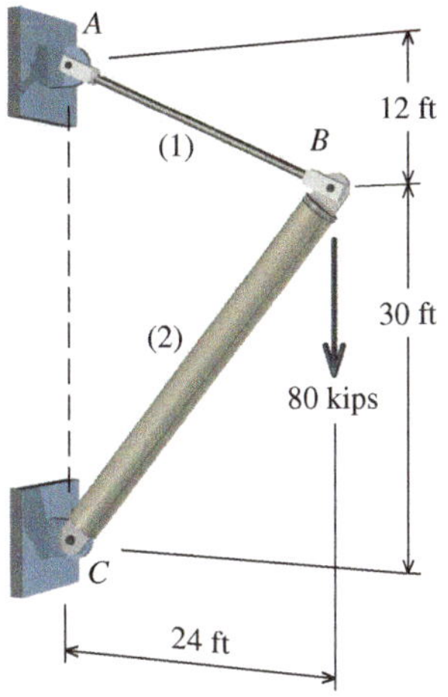

FIGURE P16.11

P16.12 A tie rod (1) and a structural steel WT shape (2) are used to support a load *P* as shown in Figure P16.12. Tie rod (1) is a solid 1.125 in. diameter steel rod, and member (2) is a WT8 × 20 structural shape oriented so that the tee stem points upward. Both the tie rod and the WT shape have an elastic modulus of 29,000 ksi and a yield strength of 36 ksi. Determine the maximum load *P* that can be applied to the structure if a factor of safety of 2.0 with respect to failure by yielding and a factor of safety of 3.0 with respect to failure by buckling are specified.

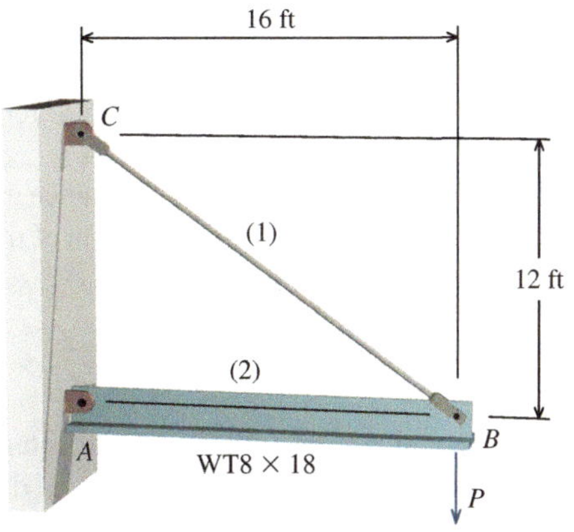

FIGURE P16.12

P16.13 A simple pin-connected truss is loaded and supported as shown in Figure P16.13. The dimensions of the truss are $a = 7$ ft and $b = 10$ ft. The loads on the truss are $P = 20$ kips, $Q = 12$ kips, and $R = 5$ kips. All members of the truss are aluminum [$E = 10{,}000$ ksi] pipes with an outside diameter of 4.00 in. and a wall thickness of 0.226 in. Consider all compression members, and determine the minimum factor of safety for the truss with respect to failure by buckling.

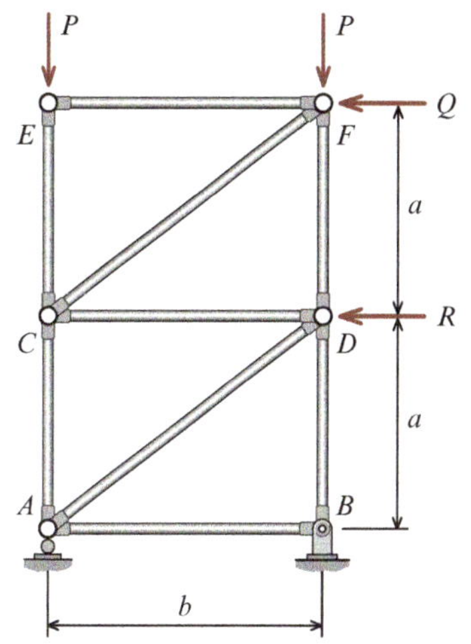

FIGURE P16.13

P16.14 A simple pin-connected wooden truss is loaded and supported as shown in Figure P16.14. The members of the truss are 150 mm by 150 mm square Douglas fir timbers that have an elastic modulus $E = 11$ GPa. Consider all compression members, and determine the minimum factor of safety for the truss with respect to failure by buckling.

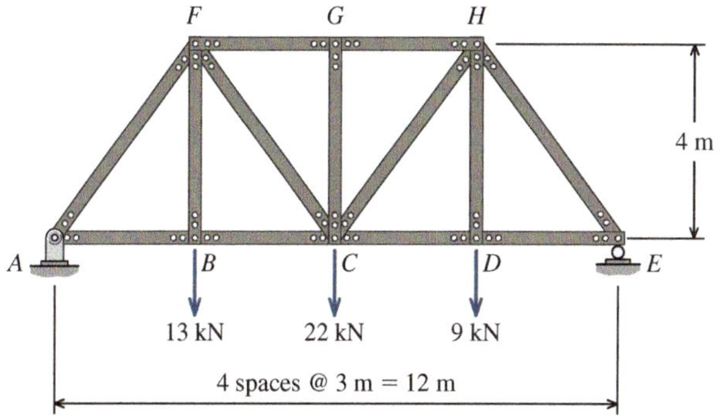

FIGURE P16.14

16.3 The Effect of End Conditions on Column Buckling

The Euler buckling formula expressed by either Equation (16.5) or Equation (16.8) was derived for an ideal column with pinned ends (i.e., ends with zero moment that are free to rotate, but are restrained against translation). Columns are commonly supported in other ways, as well and these different conditions at the ends of a column have a significant effect on the load at which buckling occurs. In this section, the effect of different *idealized end conditions* on the critical buckling load for a column will be investigated.

The critical buckling load for columns with various combinations of end conditions can be determined by the approach taken in Section 16.2 to analyze a column with pinned ends. In general, the column is assumed to be in a buckled condition and an expression for the internal bending moment in the buckled column is derived. From this equilibrium equation, a differential equation of the elastic curve can be expressed by means of the moment–curvature relationship [Equation (10.1)]. The differential equation can then be solved with the boundary conditions pertinent to the specific set of end conditions, and from the solution, the critical buckling load and the buckled shape of the column can be determined.

To illustrate this approach, the fixed–pinned column shown in Figure 16.6*a* will be analyzed to determine the critical buckling load and buckled shape of the column. Then, the *effective length concept* will be introduced. This concept provides a convenient way to determine the critical buckling load for columns with various end conditions.

Buckled Configuration

The fixed support at A prohibits both translation and rotation of the column at its lower end. The pinned support at B prohibits translation in the y direction, but allows the column to rotate at its upper end. When the column buckles, a moment reaction M_A must be developed, because rotation at A is prevented. On the basis of these constraints, the buckled shape of the column can be sketched as shown in Figure 16.6*b*. The value of the critical load P_{cr} and the shape of the buckled column will be determined from analysis of this deflected shape.

Equilibrium of the Buckled Column

A free-body diagram of the entire buckled column is shown in Figure 16.6*c*. Summing forces in the vertical direction gives $A_x = P$. Summing moments about A reveals that a horizontal reaction force B_y must exist at the upper end of the column as a consequence of the moment reaction M_A at the fixed support. The presence of B_y necessitates, in turn, a horizontal reaction force A_y at the base of the column, to satisfy equilibrium of forces in the horizontal direction.

Next, consider a free-body diagram cut through the column at a distance x from the origin (Figure 16.6*d*). We could consider either the lower or the upper portion of the column for further analysis, but here we will consider the upper portion.

Differential Equation for Column Buckling

In the buckled column of Figure 16.6*d*, both the column deflection v and the internal bending moment M are shown in their positive directions. From the free-body diagram in Figure 16.6*d*, the sum of moments about exposed surface O is

$$\Sigma M_O = -M - Pv + B_y(L - x) = 0 \quad \text{(a)}$$

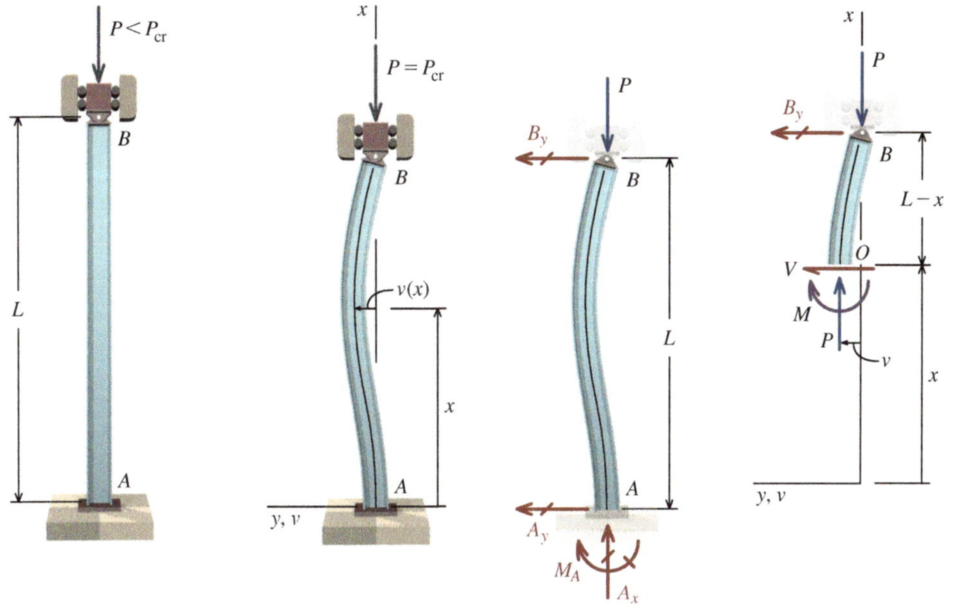

FIGURE 16.6 Buckling of a fixed–pinned column.

From Equation (10.1), the moment–curvature relationship (assuming small deflections) can be expressed as

$$M = EI\frac{d^2v}{dx^2} \tag{b}$$

which can be substituted into Equation (a) to give

$$EI\frac{d^2v}{dx^2} + Pv = B_y(L - x) \tag{16.9}$$

By dividing both sides of Equation (16.9) by EI and again substituting the term $k^2 = P/EI$, the differential equation for the fixed–pinned column can be expressed as

$$\frac{d^2v}{dx^2} + k^2v = \frac{B_y}{EI}(L - x) \tag{16.10}$$

Equation (16.10) is a nonhomogeneous second-order ordinary differential equation with constant coefficients that has boundary conditions $v(0) = 0$, $v'(0) = 0$, and $v(L) = 0$.

Solution of the Differential Equation

The general solution of Equation (16.10) is

$$v = C_1 \sin kx + C_2 \cos kx + \frac{B_y}{P}(L - x) \tag{16.11}$$

where the first two terms are the homogeneous solution (which is identical to the homogeneous solution for the pinned–pinned column) and the third term is the particular solution. The constants C_1 and C_2 must be evaluated with the use of the boundary conditions. From the boundary condition $v(0) = 0$, we obtain

$$0 = C_1 \sin(0) + C_2 \cos(0) + \frac{B_y}{P}(L) = C_2 + \frac{B_y L}{P} \tag{c}$$

From the boundary condition $v(L) = 0$, we obtain

$$0 = C_1 \sin(kL) + C_2 \cos(kL) + \frac{B_y}{P}(L - L)$$

which can be simplified to

$$0 = C_1 \tan(kL) + C_2 \tag{d}$$

The derivative of Equation (16.11) with respect to x is

$$\frac{dv}{dx} = C_1 k \cos kx - C_2 k \sin kx - \frac{B_y}{P}$$

From the boundary condition $v'(0) = 0$, the following expression is obtained:

$$0 = C_1 k \cos(0) - C_2 k \sin(0) - \frac{B_y}{P} = C_1 k - \frac{B_y}{P} \tag{e}$$

To obtain a nontrivial solution, B_y is eliminated from Equations (c) and (e), yielding an expression for C_2. From Equation (e), $B_y = C_1 kP$, and this expression can be substituted into Equation (c) to obtain

$$C_2 = -\frac{B_y L}{P} = -\frac{C_1 kPL}{P} = -C_1 kL \tag{f}$$

Upon substitution of this result into Equation (d), the following equation is obtained:

$$0 = C_1 \tan(kL) + C_2 = C_1 \tan(kL) - C_1 kL$$

This equation can be simplified to

$$\tan(kL) = kL \tag{16.12}$$

The solution of Equation (16.12) gives the critical buckling load for a fixed–pinned column. Since Equation (16.12) is a transcendental equation, it has no explicit solution. However, the following solution may be obtained by numerical methods:

$$kL = 4.4934 \tag{g}$$

Note that only the smallest value of kL that satisfies Equation (16.12) is of interest here. Since $k^2 = P/EI$, Equation (g) can then be expressed as

$$\sqrt{\frac{P}{EI}}L = 4.4934$$

and solved for the critical buckling load P_{cr}:

$$P_{cr} = \frac{20.1907\,EI}{L^2} = \frac{2.0457\pi^2 EI}{L^2} \tag{16.13}$$

The equation of the buckled column can be obtained by substituting $C_2 = -C_1 kL$ [Equation (f)] and $B_y/P = C_1 k$ [from Equation (e)] into Equation (16.11), yielding

$$\begin{aligned} v &= C_1 \sin kx - C_1 kL \cos kx + C_1 k(L - x) \\ &= C_1[\sin kx - kL \cos kx + k(L - x)] \\ &= C_1\left\{\sin\left(\frac{4.4934x}{L}\right) + 4.4934\left[1 - \frac{x}{L} - \cos\left(\frac{4.4934x}{L}\right)\right]\right\} \end{aligned} \tag{16.14}$$

The expression inside the braces is the shape of the first buckling mode for a fixed–pinned column. The constant C_1 cannot be evaluated; therefore, the amplitude of the curve is undefined, although deflections are assumed to be small.

Effective-Length Concept

From Equation (16.5), the Euler buckling load for a pinned–pinned column is

$$P_{cr} = \frac{\pi^2 EI}{L^2}$$

and from Equation (16.13), the critical buckling load for a fixed–pinned column is

$$P_{cr} = \frac{2.0457\pi^2 EI}{L^2}$$

A comparison of these two equations shows that the form of the critical load equation for a fixed–pinned column is nearly identical to the form of the Euler buckling load equation. Indeed, the two equations differ by only a constant. This similarity suggests that it is possible to relate the buckling loads of columns with various end conditions to the Euler buckling load.

The critical buckling load for a fixed–pinned column of length L is given by Equation (16.13), and this critical load is greater than the Euler buckling load for a pin-ended column of the same length L (assuming that EI is the same for both cases). *What would the length of an equivalent pin-ended column have to be in order for the equivalent pinned–pinned column to buckle at the same critical load as the actual fixed–pinned column?* Let L denote the length of the fixed–pinned column, and let L_e denote the length of the equivalent pin-ended column that buckles at the same critical load. Equating the two critical loads gives

$$\frac{2.0457\pi^2 EI}{L^2} = \frac{\pi^2 EI}{L_e^2}$$

or

$$L_e = 0.7L$$

Therefore, if the column length used in the Euler buckling load equation were modified to an *effective length* $L_e = 0.7L$, then the critical load calculated from Equation (16.5) would be identical to the critical load calculated from the actual column length in Equation (16.13). This idea of relating the critical buckling loads of columns with various end conditions to the Euler buckling loads is known as the **effective-length concept**.

The effective length L_e for any column is defined as the length of the equivalent pin-ended column. But what is meant by "equivalent" in this context? *An equivalent pin-ended column has the same critical buckling load and the same deflected shape as all or part of the actual column.*

Another way of expressing the idea of an effective column length is to consider points of zero internal bending moment. The pin-ended column, by definition, has zero internal bending moments at each end. The length L in the Euler buckling equation, therefore, is the distance between successive points of zero internal bending moment. All that is needed to adapt the Euler buckling equation for use with other end conditions is to replace L with L_e, where L_e is defined as the **effective length** of the column—that is, the distance between two successive points of zero internal bending moment. A point of zero internal bending moment is termed an **inflection point**.

The effective lengths of four common columns are shown in Figure 16.7. The pin-ended column is shown in Figure 16.7*a*, and by definition, the effective length L_e of this column is equal to its actual length L. The fixed–pinned column is shown in Figure 16.7*b*, and as was concluded in the preceding discussion, its effective length is $L_e = 0.7L$.

The ends of the column in Figure 16.7*c* are fixed. Since the deflection curve is symmetrical for this column, inflection points occur at distances of $L/4$ from each fixed end. The effective length is therefore represented by the middle half of the column length. Thus, the effective length L_e of a fixed–fixed column for use in the Euler buckling equation is equal to one-half of the actual length of the column, or $L_e = 0.5L$.

The column in Figure 16.7*d* is fixed at one end and free at the other end; consequently, the column has a zero internal bending moment only at the free end. If a mirror image of this column is visualized below the fixed end, however, the effective length between points of zero moment is seen to be twice the actual length of the column ($L_e = 2L$).

Effective-Length Factor

To simplify critical load calculations, many design codes employ a dimensionless coefficient K called the **effective-length factor**, which is defined as

$$L_e = KL \tag{16.15}$$

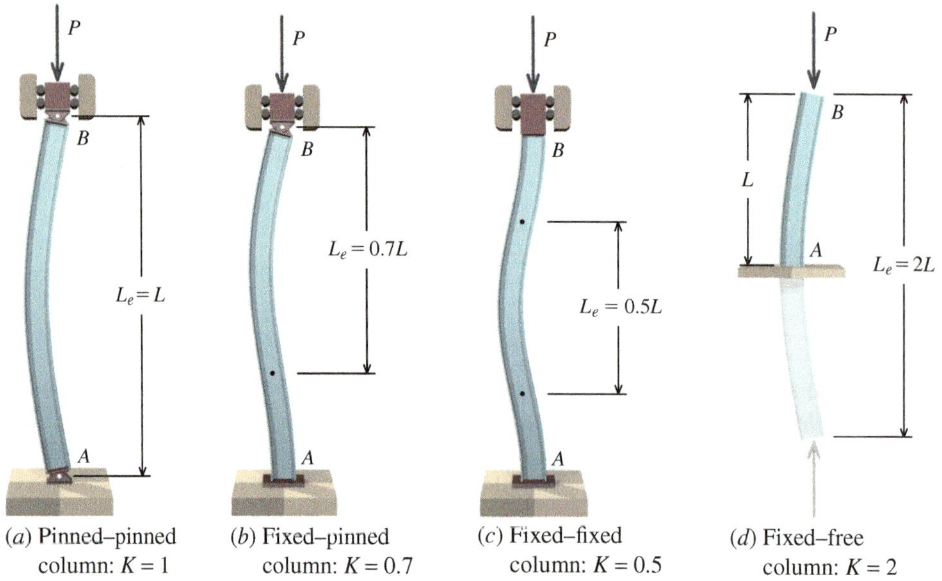

FIGURE 16.7 Effective lengths L_e and effective-length factors K for ideal columns with various end conditions.

where L is the actual length of the column. Effective-length factors are given in Figure 16.7 for four common types of column. With the effective-length factor known, the effect of end conditions on column capacity can readily be included in the critical buckling load equation:

$$P_{cr} = \frac{\pi^2 EI}{(KL)^2} \tag{16.16}$$

It can likewise be included in the critical buckling stress equation:

$$\sigma_{cr} = \frac{\pi^2 E}{(KL/r)^2} \tag{16.17}$$

In this equation, KL/r is the **effective-slenderness ratio**.

Practical Considerations

It is important to keep in mind that the column end conditions shown in Figure 16.7 are *idealizations*. A pin-ended column is usually loaded through a pin that, because of friction, is not completely free to rotate. Consequently, there will always be an indeterminate (though usually small) moment at the ends of a pin-ended column, and this moment will reduce the distance between the inflection points to a value less than L. Fixed-end connections theoretically provide perfect restraint against rotation. However, columns are typically connected to other structural members that have some measure of flexibility in themselves, so it is quite difficult to construct a real connection that prevents all rotation. Thus, a fixed–fixed column (Figure 16.7*c*) will have an effective length somewhat greater than $L/2$. Because of these practical considerations, the theoretical K factors given in Figure 16.7 are typically modified to account for the difference between the idealized and the realistic behavior of connections. Design codes that utilize effective-length factors therefore usually specify a recommended practical value for K factors in preference to the theoretical values.

EXAMPLE 16.3

A long, slender W8 × 24 structural steel shape (see Appendix B for cross-sectional properties) is used as a 35 ft long column. The column is supported in the vertical direction at base A and pinned at ends A and C against translation in the y and z directions. Lateral support is provided to the column so that deflection in the x–z plane is restrained at midheight B; however, the column is free to deflect in the x–y plane at B. Determine the maximum compressive load P that the column can support if a factor of safety of 2.5 is required. In your analysis, consider the possibility that buckling could occur about either the strong axis (i.e., the z axis) or the weak axis (i.e., the y axis) of the column. Assume that $E = 29{,}000$ ksi and $\sigma_Y = 36$ ksi.

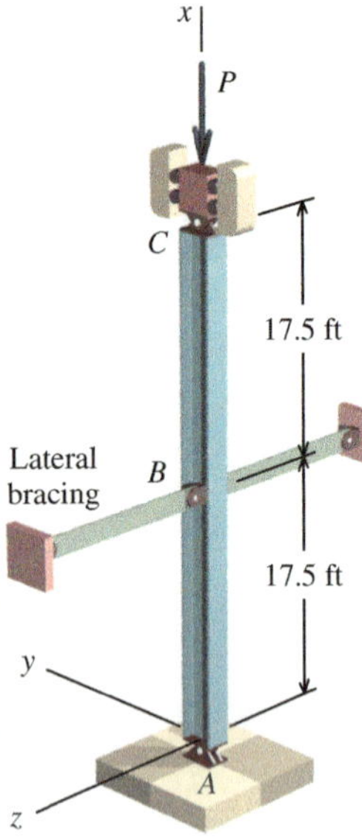

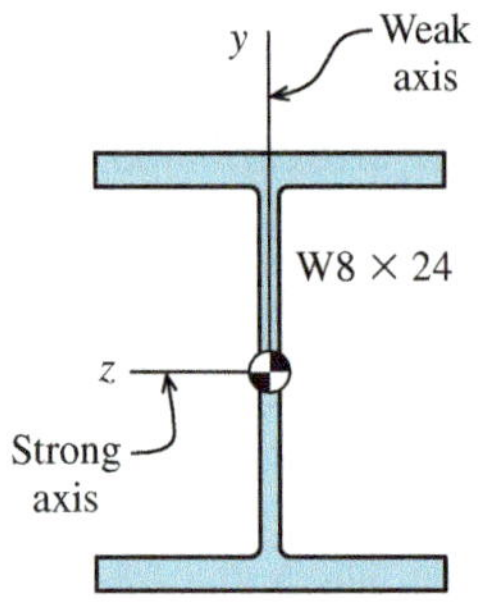

Plan the Solution

If the W8 × 24 column were supported only at its ends, then buckling about the weak axis of the cross section would be anticipated. However, additional lateral support is provided to this column, so the effective length with respect to buckling of the weak axis is reduced. For this reason, both the effective length and the radius of gyration with respect to both the strong and weak axes of the column must be considered. The critical buckling load will be dictated by the larger of the two effective-slenderness ratios.

SOLUTION

The following section properties can be obtained from Appendix B for the W8 × 24 structural steel shape:

$$I_z = 82.7 \text{ in.}^4 \qquad I_y = 18.3 \text{ in.}^4$$

$$r_z = 3.42 \text{ in.} \qquad r_y = 1.61 \text{ in.}$$

The subscripts for these properties have been modified to correspond to the axes shown on the cross section.

Buckling About the Strong Axis

The column could buckle about its strong axis, resulting in the buckled shape shown in which the column deflects in the x–y plane. For this manner of failure, the effective length of the column is $KL = 35$ ft. The critical buckling load is therefore

$$P_{cr} = \frac{\pi^2 EI_z}{(KL)_z^2} = \frac{\pi^2(29{,}000 \text{ ksi})(82.7 \text{ in.}^4)}{[(35 \text{ ft})(12 \text{ in./ft})]^2} = 134.2 \text{ kips}$$

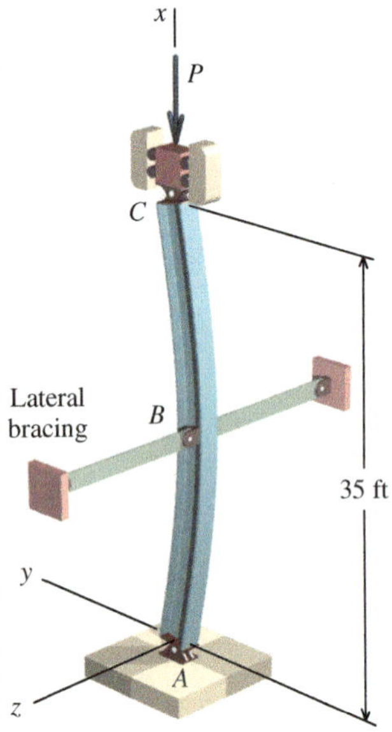

Buckling about the strong axis.

Although we are not required to determine P_{cr}, it is instructive to calculate the effective-slenderness ratio for buckling about the strong axis:

$$(KL/r)_z = \frac{(35 \text{ ft})(12 \text{ in./ft})}{3.42 \text{ in.}} = 122.8$$

Buckling About the Weak Axis

Alternatively, the column could buckle about its weak axis. In this case, the column deflection would occur in the x–z plane as shown in the accompanying figure. For this manner of failure, the effective length of the column is $KL = 17.5$ ft. The critical buckling load about the weak axis is therefore

$$P_{cr} = \frac{\pi^2 EI_y}{(KL)_y^2} = \frac{\pi^2(29{,}000 \text{ ksi})(18.3 \text{ in.}^4)}{[(17.5 \text{ ft})(12 \text{ in./ft})]^2} = 118.8 \text{ kips}$$

The effective-slenderness ratio for buckling about the weak axis is

$$(KL/r)_y = \frac{(17.5 \text{ ft})(12 \text{ in./ft})}{1.61 \text{ in.}} = 130.4$$

The critical load for the column is the smaller of the two load values:

$$P_{cr} = 118.8 \text{ kips}$$

Critical Stress

The critical load equation [Equation (16.16)] is valid only if the stresses in the column remain elastic; therefore, the critical buckling stress must be compared with the proportional limit of the material. For structural steel, the proportional limit is essentially equal to the yield stress.

The critical buckling stress will be computed with the use of the *larger* of the two effective-slenderness ratios:

$$\sigma_{cr} = \frac{\pi^2 E}{(KL/r)^2} = \frac{\pi^2(29{,}000 \text{ ksi})}{(130.4)^2} = 16.83 \text{ ksi} < 36 \text{ ksi} \qquad \text{O.K.}$$

Since the critical buckling stress of 16.83 ksi is less than the 36 ksi yield stress of the steel, the critical load calculations are valid.

Allowable Column Load

A factor of safety of 2.5 is required for this column. Therefore, the allowable axial load is

$$P_{allow} = \frac{118.8 \text{ kips}}{2.5} = 47.5 \text{ kips} \qquad \textbf{Ans.}$$

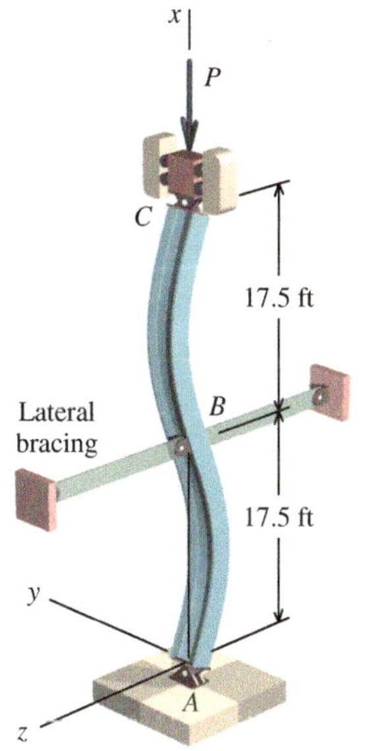

Buckling about the weak axis.

EXAMPLE 16.4

A W310 × 60 structural steel shape (see Appendix B for cross-sectional properties) is used as a column with an actual length $L = 9$ m. The column is fixed at base A. Lateral support is provided to the column, so deflection in the x–z plane is restrained at the upper end; however, the column is free to deflect in the x–y plane at B. Determine the critical buckling load P_{cr} of the column. Assume that $E = 200$ GPa and $\sigma_Y = 250$ MPa.

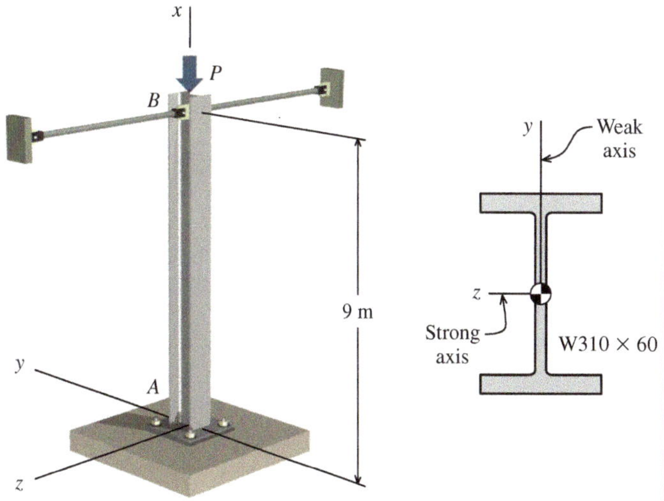

Plan the Solution

Although the actual length of the column is 9 m, the different end conditions with respect to the strong and weak axes of the cross section cause markedly different effective lengths for the two directions. Appropriate effective-length factors based on the column end conditions will be selected from Figure 16.7.

SOLUTION

Section Properties

The following section properties can be obtained from Appendix B for the W310 × 60 structural steel shape:

$$I_z = 128 \times 10^6 \text{ mm}^4 \qquad r_z = 130 \text{ mm} \qquad I_y = 18.4 \times 10^6 \text{ mm}^4 \qquad r_y = 49.3 \text{ mm}$$

The subscripts for these properties have been revised to correspond to the axes shown on the cross section.

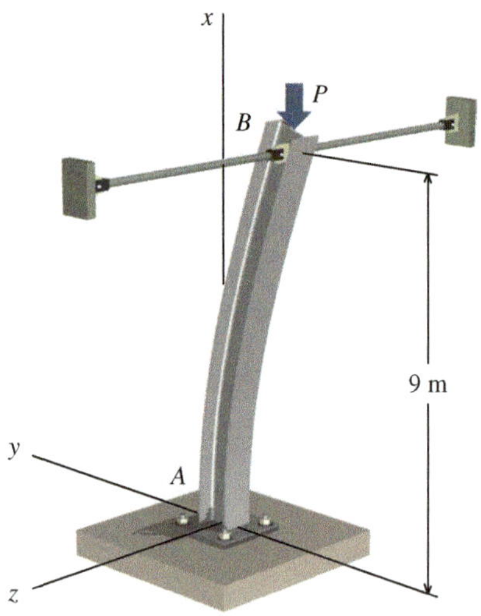

Buckling About the Strong Axis

The column could buckle about its strong axis, resulting in the buckled shape shown in which the column bends about its z axis and deflects in the x–y plane. For this manner of buckling, the column base is fixed and its upper end is free. From Figure 16.7, the appropriate effective-length factor is $K_z = 2.0$ and the effective length of the column is $(KL)_z = (2.0)(9 \text{ m}) = 18$ m. The critical buckling load is therefore

$$P_{cr} = \frac{\pi^2 EI_z}{(KL)_z^2} = \frac{\pi^2(200{,}000 \text{ N/mm}^2)(128 \times 10^6 \text{ mm}^4)}{[(2.0)(9 \text{ m})(1{,}000 \text{ mm/m})]^2}$$
$$= 779{,}821 \text{ N} = 780 \text{ kN}$$

The effective-slenderness ratio for buckling about the strong axis is

$$(KL/r)_z = \frac{(2.0)(9 \text{ m})(1{,}000 \text{ mm/m})}{130 \text{ mm}} = 138.5$$

Buckling About the Weak Axis

Alternatively, the column could buckle about its weak axis. In this case, the column would bend about its y axis and deflection would occur in the x–z plane as shown. For buckling about the weak axis, the column is fixed at A and pinned at B. From Figure 16.7, the appropriate effective-length factor is $K_y = 0.7$ and the effective length of the column is $(KL)_y = (0.7)(9 \text{ m}) = 6.3$ m. The critical buckling load about the weak axis is therefore

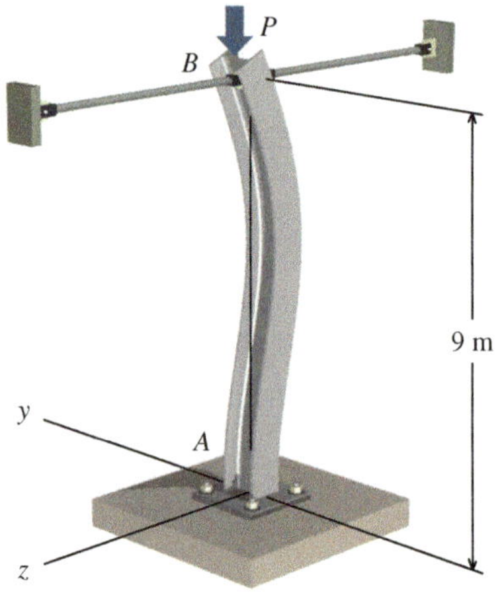

$$P_{cr} = \frac{\pi^2 EI_y}{(KL)_y^2} = \frac{\pi^2(200{,}000 \text{ N/mm}^2)(18.4 \times 10^6 \text{ mm}^4)}{[(0.7)(9 \text{ m})(1{,}000 \text{ mm/m})]^2}$$
$$= 915{,}096 \text{ N} = 915 \text{ kN}$$

The effective-slenderness ratio for buckling about the weak axis is

$$(KL/r)_y = \frac{(0.7)(9 \text{ m})(1{,}000 \text{ mm/m})}{49.3 \text{ mm}} = 127.8$$

The critical load for the column is the smaller of the two load values:

$$P_{cr} = 780 \text{ kN}$$ **Ans.**

Critical Stress

The critical load equation [Equation (16.16)] is valid only if the stresses in the column remain elastic; therefore, the critical buckling stress must be compared with the proportional limit of the material. For structural steel, the proportional limit is essentially equal to the yield stress.

The critical buckling stress will be computed with the use of the *larger* of the two effective-slenderness ratios:

$$\sigma_{cr} = \frac{\pi^2 E}{(KL/r)^2} = \frac{\pi^2(200{,}000 \text{ MPa})}{(138.5)^2} = 102.9 \text{ MPa} < 250 \text{ MPa} \quad \text{O.K.}$$

Since the critical buckling stress of 102.9 MPa is less than the 250 MPa yield stress of the steel, the critical load calculations are valid.

PROBLEMS

P16.15 An HSS152.4 × 101.6 × 6.4 structural steel [E = 200 GPa] section (see Appendix B for cross-sectional properties) is used as a column with an actual length of 6 m. The column is supported only at its ends and may buckle in any direction. If a factor of safety of 2 with respect to failure by buckling is specified, determine the maximum safe load for the column for the following end conditions:

(a) pinned–pinned
(b) fixed–free
(c) fixed–pinned
(d) fixed–fixed

P16.16 A W250 × 80 structural steel [E = 200 GPa] section (see Appendix B for cross-sectional properties) is used as a column with an actual length L = 12 m. The column is supported only at its ends and may buckle in any direction. The column is fixed at its base and pinned at its upper end, as shown in Figure P16.16. Determine the maximum load P that may be supported by the column if a factor of safety of 2.5 with respect to buckling is specified.

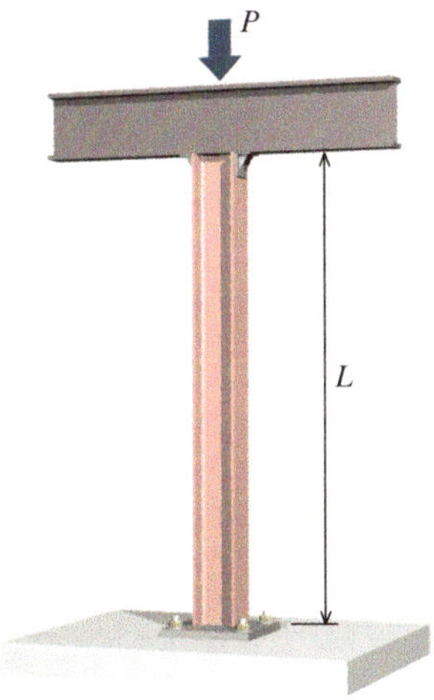

FIGURE P16.16

P16.17 A W14 × 53 structural steel [E = 29,000 ksi] section (see Appendix B for cross-sectional properties) is used as a column with an actual length L = 16 ft. The column is fixed at its base and unrestrained at its upper end, as shown in Figure P16.17. Determine the maximum load P that may be supported by the column if a factor of safety of 2.5 with respect to buckling is specified.

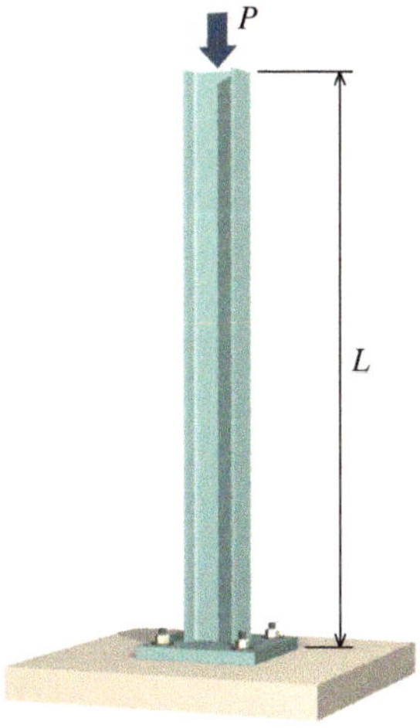

FIGURE P16.17

P16.18 A long, slender structural steel [E = 29,000 ksi] HSS8 × 4 × ¼ shape (see Appendix B for cross-sectional properties) is used as a 32 ft long column. The column is supported in the x direction at base A and pinned at ends A and C against translation in the y and z directions. Lateral support is provided to the column, so deflection in the x-z plane is restrained at midheight B; however, the column is free to deflect in the x–y plane at B (Figure P16.18). Determine the maximum compressive load the column can support if a factor of safety of 1.92 is required. In your analysis, consider the possibility that buckling could occur about either the strong axis (i.e., the z axis) or the weak axis (i.e., the y axis) of the steel column.

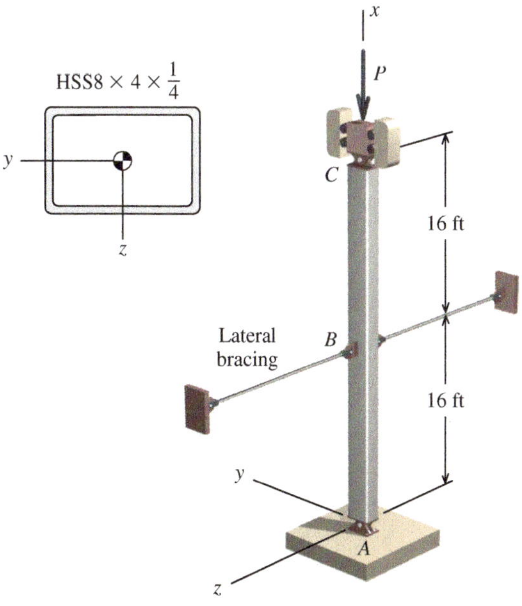

FIGURE P16.18

P16.19 The aluminum column shown in Figure P16.19 has a rectangular cross section and supports an axial load P. The base of the column is fixed. The support at the top allows rotation of the column in the x–y plane (i.e., bending about the strong axis) but prevents rotation in the x–z plane (i.e., bending about the weak axis).

(a) Determine the critical buckling load of the column for the following parameters: $L = 50$ in., $b = 0.50$ in., $h = 0.875$ in., and $E = 10{,}000$ ksi.

(b) Determine the ratio b/h for which the critical buckling load about the strong axis is the same as that about the weak axis.

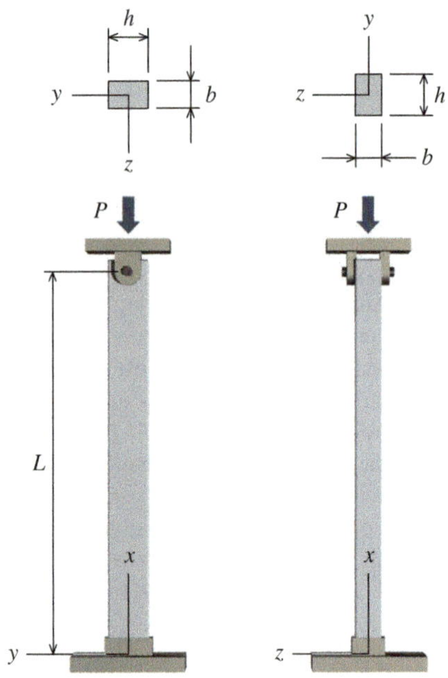

FIGURE P16.19

P16.20 The steel compression link shown in Figure P16.20 has a rectangular cross section and supports an axial load P. The supports allow rotation about the strong axis of the link cross section but prevent rotation about the weak axis. Determine the allowable compression load P if a factor of safety of 2.0 is specified. Use the following parameters: $L = 1{,}200$ mm, $b = 15$ mm, $h = 40$ mm, and $E = 200$ GPa.

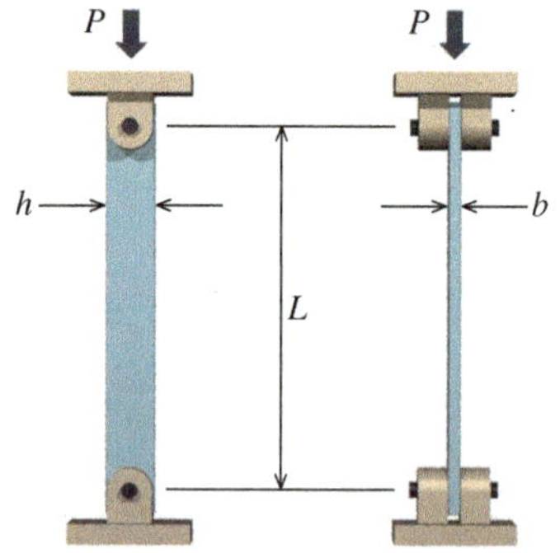

FIGURE P16.20

P16.21 A stainless steel pipe with an outside diameter of 100 mm and a wall thickness of 8 mm is rigidly attached to fixed supports at A and B as shown in Figure P16.21. The length of the pipe is $L = 8$ m, the elastic modulus of the pipe material is $E = 190$ GPa, and the coefficient of thermal expansion of the pipe is $\alpha = 17.3 \times 10^{-6}$ mm/mm/°C. Determine the temperature increase ΔT that will cause the pipe to buckle.

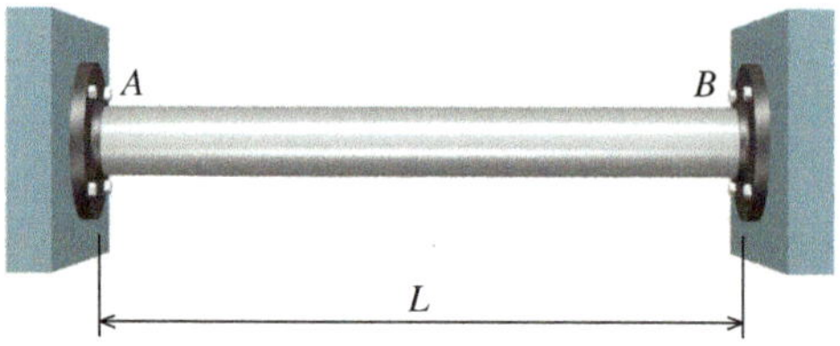

FIGURE P16.21

16.4 The Secant Formula

Many real columns do not behave as predicted by the Euler formula because of imperfections in the alignment of the loading. In this section, the effect of imperfect alignment is examined by considering an eccentric loading. We will consider a pinned–pinned column subjected to compressive forces acting at an eccentricity e from the centerline of the undeformed column, as shown in Figure 16.8a. (**Note:** The support symbols have been omitted from the figure for clarity.) When the eccentricity is nonzero, the free-body diagram for the column is as shown in Figure 16.8*b*. From this free-body diagram, the bending moment at any section can be expressed as

$$\Sigma M_A = M + Pv + Pe = 0$$
$$\therefore M = -Pv - Pe$$

If the stress does not exceed the proportional limit and deflections are small, the differential equation of the elastic curve becomes

$$EI\frac{d^2v}{dx^2} + Pv = -Pe$$

or

$$\frac{d^2v}{dx^2} + \frac{P}{EI}v = -\frac{P}{EI}e$$

As in the Euler derivation, the term P/EI will be denoted by k^2 [Equation (16.2)] so that the differential equation can be rewritten as

$$\frac{d^2v}{dx^2} + k^2v = -k^2e$$

The solution of this equation has the form

$$v = C_1 \sin kx + C_2 \cos kx - e \qquad \text{(a)}$$

Two boundary conditions exist for the column. At pin support A, the boundary condition $v(0) = 0$ gives

$$v(0) = 0 = C_1 \sin k(0) + C_2 \cos k(0) - e$$
$$\therefore C_2 = e$$

At pin support B, the boundary condition $v(L) = 0$ gives

$$v(L) = 0 = C_1 \sin kL + C_2 \cos kL - e = C_1 \sin kL - e(1 - \cos kL)$$
$$\therefore C_1 = e\left[\frac{1 - \cos kL}{\sin kL}\right]$$

Using the trigonometric identities

$$1 - \cos\theta = 2\sin^2\frac{\theta}{2} \qquad \text{and} \qquad \sin\theta = 2\sin\frac{\theta}{2}\cos\frac{\theta}{2}$$

allows Equation (a) to be rewritten as

$$C_1 = e\left[\frac{2\sin^2(kL/2)}{2\sin(kL/2)\cos(kL/2)}\right] = e\tan\frac{kL}{2}$$

With this expression for C_1, the solution of the differential equation preceding Equation (a) becomes

$$\begin{aligned} v &= e\tan\frac{kL}{2}\sin kx + e\cos kx - e \\ &= e\left[\tan\frac{kL}{2}\sin kx + \cos kx - 1\right] \end{aligned} \qquad (16.18)$$

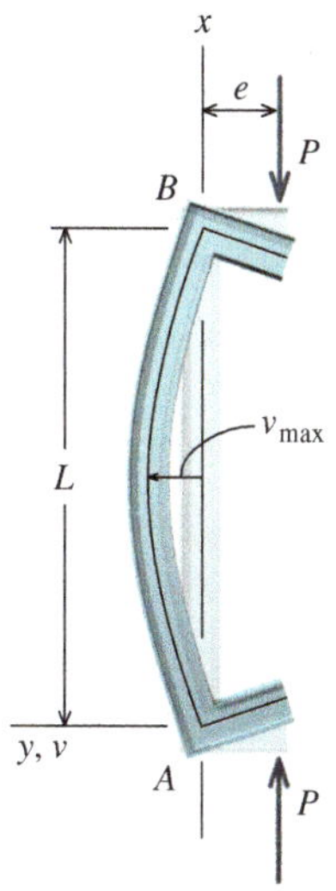

(*a*) Pinned–pinned column

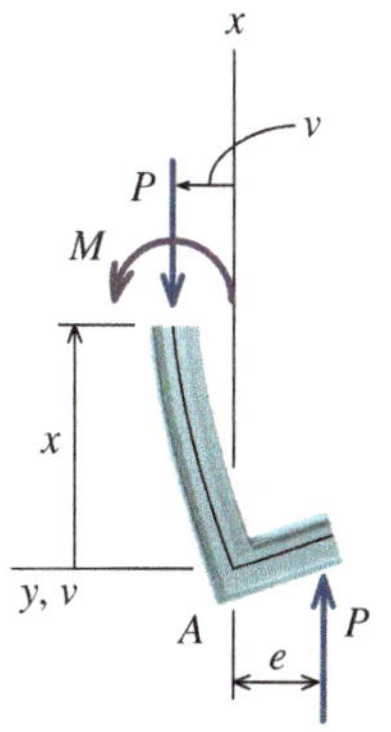

(*b*) Free-body diagram

FIGURE 16.8 Pinned–pinned column with eccentric load.

In this case, a relationship can be found between the pinned–pinned column's maximum deflection v_{max}, which occurs at $x = L/2$, and the load P. First,

$$\begin{aligned} v_{max} &= e\left[\tan\frac{kL}{2}\sin\frac{kL}{2} + \cos\frac{kL}{2} - 1\right] \\ &= e\left[\frac{\sin^2(kL/2)}{\cos(kL/2)} + \frac{\cos^2(kL/2)}{\cos(kL/2)} - 1\right] \\ &= e\left[\frac{1}{\cos(kL/2)} - 1\right] = e\left[\sec\frac{kL}{2} - 1\right] \end{aligned} \tag{b}$$

Then, since $k^2 = P/EI$, Equation (b) can be restated in terms of the load P and the flexural rigidity EI as

$$v_{max} = e\left[\sec\left(\frac{L}{2}\sqrt{\frac{P}{EI}}\right) - 1\right] \tag{c}$$

Equation (c) indicates that, for a given column in which E, I, and L are fixed and $e > 0$, the column exhibits lateral deflection for even small values of the load P. For any value of e, the quantity

$$\sec\left(\frac{L}{2}\sqrt{\frac{P}{EI}}\right) - 1$$

approaches positive or negative infinity as the argument of the function approaches $\pi/2$, $3\pi/2$, $5\pi/2, \ldots$, and the deflection v increases without bound, indicating that the critical load corresponds to one of these angles. If $\pi/2$ (the angle that yields the smallest load) is chosen, then

$$\frac{L}{2}\sqrt{\frac{P}{EI}} = \frac{\pi}{2}$$

or

$$\sqrt{\frac{P}{EI}} = \frac{\pi}{L}$$

from which it follows that

$$P_{cr} = \frac{\pi^2 EI}{L^2} \tag{16.19}$$

which is the Euler formula discussed in Section 16.2.

Unlike an Euler column, which deflects laterally only if P equals or exceeds the Euler buckling load, an eccentrically loaded column deflects laterally for any value of P. To illustrate this property, the quantities E, I, and L can be eliminated from Equation (b) by using Equation (16.19) to produce an expression for the maximum lateral column deflection in terms of P and P_{cr}. From Equation (16.19), let $EI = P_{cr}\, L^2/\pi^2$. Substituting this expression into Equation (b) then gives

$$v_{max} = e\left[\sec\left(\frac{L}{2}\sqrt{\frac{P}{EI}}\right) - 1\right] = e\left[\sec\left(\frac{L}{2}\sqrt{\frac{P}{P_{cr}}\frac{\pi^2}{L^2}}\right) - 1\right] = e\left[\sec\left(\frac{\pi}{2}\sqrt{\frac{P}{P_{cr}}}\right) - 1\right] \tag{d}$$

From this equation, it would appear that the maximum deflection becomes infinite as P approaches the Euler buckling load P_{cr}; however, under these conditions, the slope of the

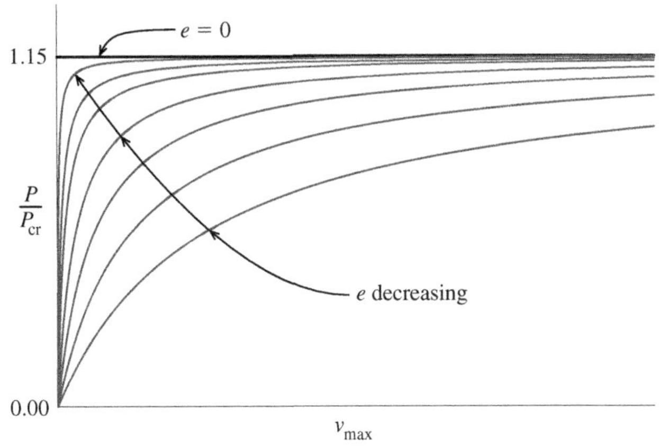

FIGURE 16.9 Load–deflection diagram for an eccentrically loaded column.

deflected column is no longer sufficiently small to be neglected in the expression for the curvature. As a result, accurate deflections can be obtained only by using the nonlinear form of the differential equation of the elastic curve.

Plots of Equation (d) for various values of eccentricity e are shown in Figure 16.9. These curves reveal that the maximum column deflection v_{max} is extremely small as e approaches zero until the load P approaches the Euler critical load P_{cr}. As P nears P_{cr}, v_{max} increases rapidly. In the limit as $e \to 0$, the curve degenerates into two lines that represent the straight unbuckled column ($P < P_{cr}$) and the buckled configuration ($P = P_{cr}$)—in other words, nothing more than Euler column buckling.

Secant Formula

In writing the elastic curve equation, it was assumed that stresses do not exceed the proportional limit. On the basis of this assumption, the maximum compressive stress can be obtained by superposition of the axial stress and the maximum bending stress. The maximum bending stress occurs on a section at the midspan of the column, where the bending moment attains its largest value, $M_{max} = P(e + v_{max})$. Thus, the magnitude of the maximum compressive stress in the column can be expressed as

$$\sigma_{max} = \frac{P}{A} + \frac{M_{max}c}{I} = \frac{P}{A} + \frac{P(e + v_{max})c}{Ar^2} \tag{e}$$

in which r is the radius of gyration of the column cross section about the axis of bending. If we now start with Equation (c),

$$v_{max} = e\left[\sec\left(\frac{L}{2}\sqrt{\frac{P}{EI}}\right) - 1\right]$$

add e to both sides, and perform some algebra on the right-hand side, we obtain the following expression for $e + v_{max}$:

$$e + v_{max} = e\sec\left(\frac{L}{2}\sqrt{\frac{P}{EI}}\right)$$

Using this expression allows Equation (e) to be written as

$$\sigma_{max} = \frac{P}{A}\left[1 + \frac{ec}{r^2}\sec\left(\frac{L}{2}\sqrt{\frac{P}{EI}}\right)\right]$$

which can be further simplified with the use of $I = Ar^2$ to give an expression for the maximum compressive stress in the deflected column:

$$\sigma_{\max} = \frac{P}{A}\left[1 + \frac{ec}{r^2}\sec\left(\frac{L}{2r}\sqrt{\frac{P}{EA}}\right)\right] \tag{16.20}$$

The maximum compressive stress $\sigma_{\max}$ occurs at midheight of the column on the inner (concave) side.

Equation (16.20) is known as the **secant formula**, and it relates the average force per unit area, P/A, that causes a specified maximum stress $\sigma_{\max}$ in a column to the dimensions of the column, the properties of the material the column is made of, and the eccentricity e from the centerline of the undeformed column. The term L/r is the same slenderness ratio found in the Euler buckling stress formula [Equation (16.8)]; thus, for columns with different end conditions (see Section 16.3), the secant formula can be restated as

$$\sigma_{\max} = \frac{P}{A}\left[1 + \frac{ec}{r^2}\sec\left(\frac{KL}{2r}\sqrt{\frac{P}{EA}}\right)\right] \tag{16.21}$$

The quantity ec/r^2 is called the **eccentricity ratio** and is seen to depend on the eccentricity of the load and the dimensions of the column. If the column is loaded precisely at its centroid then, $e = 0$ and $\sigma_{\max} = P/A$. It is virtually impossible, however, to eliminate all eccentricity that might result from various factors, such as initial crookedness of the column, minute flaws in the material, and a lack of uniformity of the cross section, as well as accidental eccentricity of the load.

To determine the maximum compressive load that can be applied at a given eccentricity to a particular column, the maximum compressive stress can be set equal to the yield stress in compression and Equation (16.20) can then be solved numerically for P/A. Figure 16.10 is a plot of the force per unit area, P/A, versus the slenderness ratio L/r for

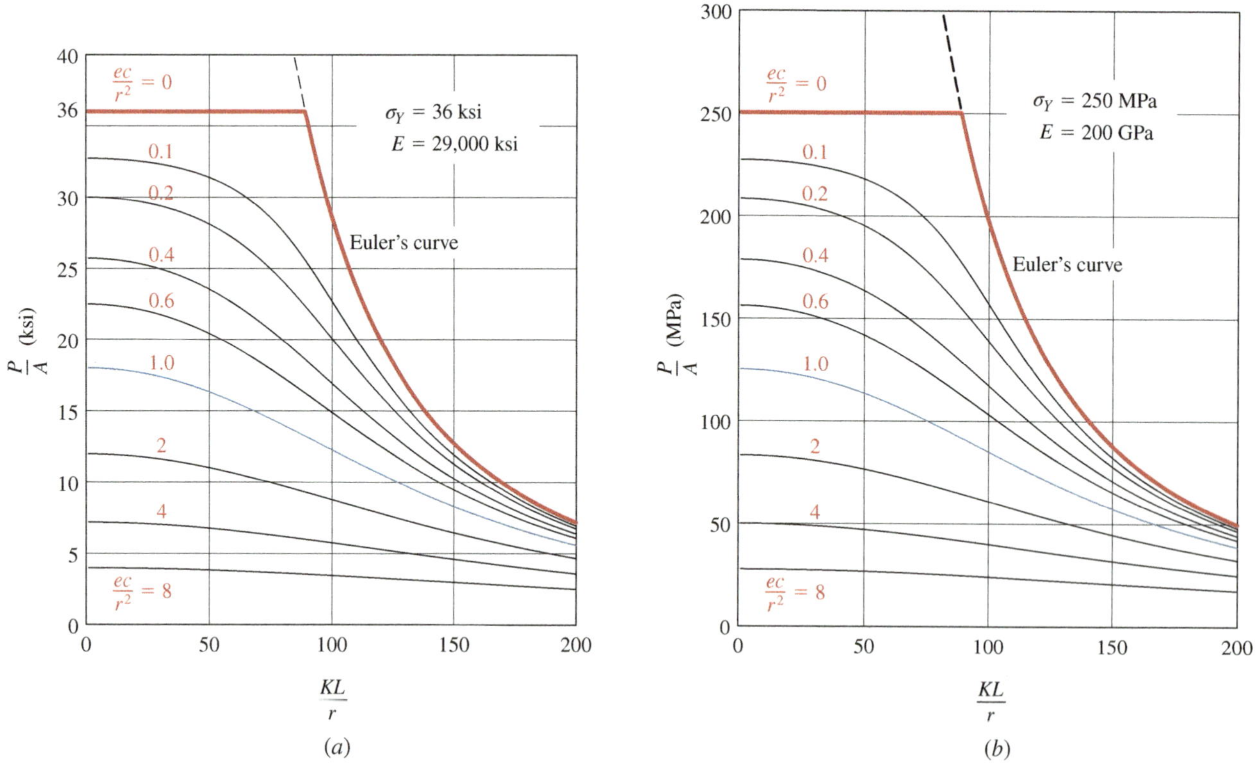

FIGURE 16.10 Average compression stress versus slenderness ratio, based on the secant formula.

several values of the eccentricity ratio ec/r^2. Figure 16.10*a* is plotted for structural steel having an elastic modulus $E = 29{,}000$ ksi and a compressive yield strength $\sigma_Y = 36$ ksi, and Figure 16.10*b* shows the corresponding curves in SI units.

The outer envelope of Figure 16.10, consisting of the horizontal line $P/A = 36$ ksi in Figure 16.10*a* or the horizontal line $P/A = 250$ MPa in Figure 16.10*b* and the Euler curve, corresponds to $e = 0$. The Euler curve is truncated at 36 ksi (250 MPa), since that value is the maximum allowable stress for the material. The curves presented in Figure 16.10 highlight the significance of the eccentricity of the load in reducing the maximum safe load in short and intermediate-length columns (i.e., columns with slenderness ratios less than about 126 for the steel assumed in Figure 16.10). For large slenderness ratios, the curves for the various eccentricity ratios tend to merge with the Euler curve. Consequently, the Euler formula can be used to analyze columns with large slenderness ratios. For a given problem, *the slenderness ratio must be computed* to determine whether or not the Euler equation is valid.

PROBLEMS

P16.22 An axial load P is applied to a solid 30 mm diameter steel rod AB as shown in Figure P16.22/23. For $L = 1.5$ m, $P = 18$ kN, and $e = 3.0$ mm, determine (a) the lateral deflection midway between A and B and (b) the maximum stress in the rod. Use $E = 200$ GPa.

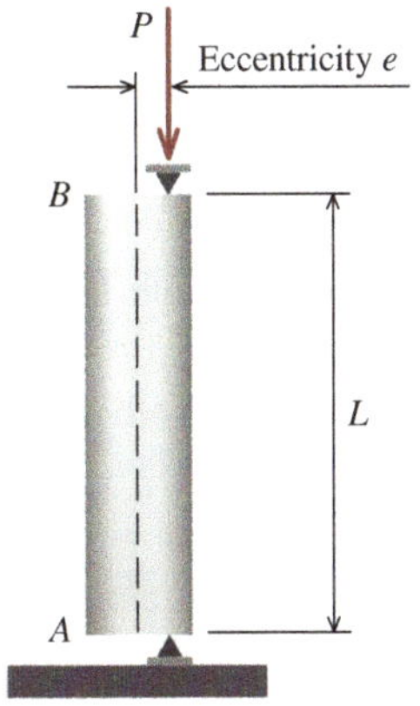

FIGURE P16.22/23

P16.23 A 4 ft long steel [$E = 29{,}000$ ksi; $\sigma_Y = 36$ ksi] tube supports an eccentrically applied axial load P, as shown in Figure P16.22/23. The tube has an outside diameter of 2.00 in. and a wall thickness of 0.15 in. For an eccentricity $e = 0.25$ in., determine (a) the maximum load P that can be applied without causing either buckling or yielding of the tube and (b) the corresponding maximum deflection midway between A and B.

P16.24 A square tube shape made of an aluminum alloy supports an eccentric compression load P that is applied at an eccentricity $e = 4.0$ in. from the centerline of the shape (Figure P16.24). The width of the tube is 3 in. and its wall thickness is 0.12 in. The column is fixed at its base and free at its upper end, and its length is $L = 8$ ft. For an applied load $P = 900$ lb, determine (a) the lateral deflection at the upper end of the column and (b) the maximum stress in the square tube. Use $E = 10 \times 10^6$ psi.

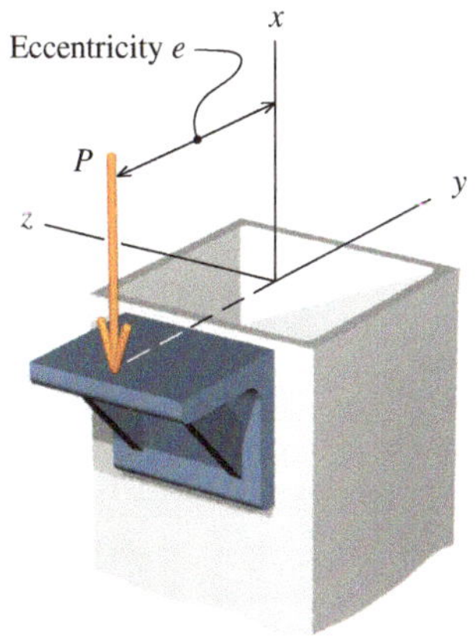

FIGURE P16.24

P16.25 A steel pipe (outside diameter = 130 mm; wall thickness = 12.5 mm) supports an axial load $P = 25$ kN, which is applied at an eccentricity $e = 175$ mm from the pipe centerline (Figure P16.25/26). The column is fixed at its base and free at its upper end, and its length is $L = 4.0$ m. Determine (a) the lateral deflection at the upper end of the column and (b) the maximum stress in the pipe. Use $E = 200$ GPa.

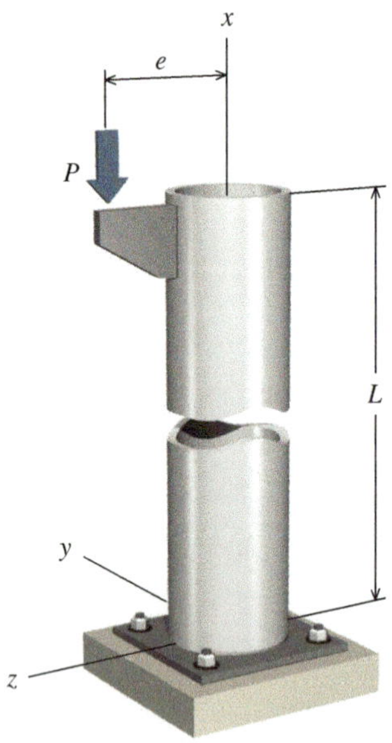

FIGURE P16.25/26

P16.26 A steel [$E = 200$ GPa] pipe with an outside diameter of 170 mm and a wall thickness of 7 mm supports an axial load P, which is applied at an eccentricity of $e = 150$ mm from the pipe centerline (Figure P16.25/26). The column is fixed at its base and free at its upper end, and its length is $L = 4.0$ m. The maximum compressive stress in the column must be limited to $\sigma_{max} = 80$ MPa.

(a) Use a trial-and-error approach or an iterative numerical solution to determine the allowable eccentric load P that can be applied.

(b) Determine the lateral deflection at the upper end of the column for the allowable load P.

16.5 Empirical Column Formulas—Centric Loading

The Euler buckling formulas for the critical buckling load [Equation (16.16)] and critical buckling stress [Equation (16.17)] were derived for ideal columns. In considering ideal columns, it was assumed that the column was perfectly straight, that the compression load was applied exactly at the centroid of the cross section, and that the column material remained below its proportional limit during buckling. Practical columns, however, rarely satisfy all of the conditions assumed for ideal columns. Although the Euler equations give reasonable predictions for the strength of long, slender columns, early researchers soon found that the strength of short and intermediate-length columns were not well predicted by these formulas. A representative graph of the results from numerous column load tests plotted as a function of slenderness ratio is shown in Figure 16.11. The

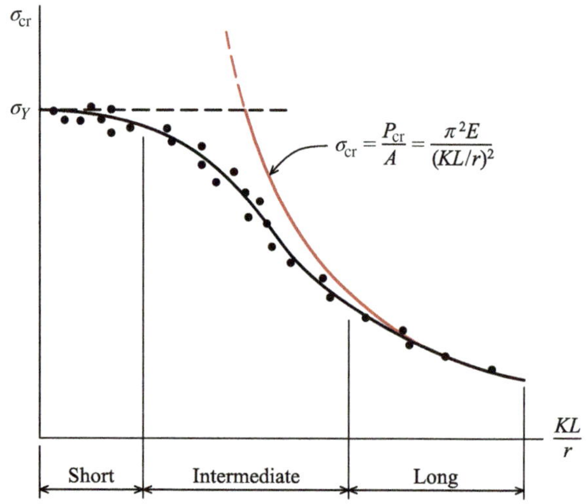

FIGURE 16.11 Representative column test data for a range of slenderness ratios.

graph shows a scattered range of values that transition from the yield stress for the very shortest columns to the Euler buckling stress for the very longest columns. In the broad range of slenderness ratios between these two extremes, neither the yield stress nor the Euler buckling stress is a good predictor of the strength of the column. Furthermore, most practical columns fall within this intermediate range of slenderness ratios. Consequently, practical column design is based primarily on empirical formulas that have been developed to represent the best fit of test results for a range of realistic full-size columns. These empirical formulas incorporate appropriate factors of safety, effective-length factors, and other modifying factors.

The strength of a column and the manner in which it fails are greatly dependent on its effective length. For example, consider the behavior of columns made of steel:

Short steel columns: A very short steel column may be loaded until the steel: reaches the yield stress; consequently, very short columns do not buckle. The strength of these members is the same in both compression and tension; however, the columns are so short that they have no practical value.

Intermediate-length steel columns: Most practical steel columns fall into this category. As the effective length (or slenderness ratio) increases, the cause of failure becomes more complicated. In steel columns—in particular, hot-rolled steel columns—the applied load may cause compression stresses that exceed the proportional limit in portions of the cross section; thus, the column will fail both by yielding and by buckling. These columns are said to buckle *inelastically*. The buckling strength of hot-rolled steel columns is particularly influenced by the presence of **residual stresses**—stresses that are "locked into" the steel shape during the manufacturing process because the steel flanges and webs cool faster than the fillet regions that connect them. Because of residual stress and other factors, the analysis and design of intermediate-length steel columns are based on empirical formulas developed from test results.

Long steel columns: Long, slender steel columns buckle *elastically*, since the Euler buckling stress is well below the proportional limit (even taking into account the presence of residual stress). Consequently, the Euler buckling equations are reliable predictors for long columns. Long, slender columns, however, are not very efficient, since the Euler buckling stress for these columns is much less than the proportional limit for the steel.

Several representative empirical design formulas for centrically loaded steel, aluminum, and wood columns will be presented to introduce basic aspects of column design.

Structural Steel Columns

Structural steel columns are designed in accordance with specifications published by the American Institute of Steel Construction (AISC). The AISC Allowable Stress Design[1] (ASD) procedure differentiates between short and intermediate-length columns and long columns. The transition point between these two categories is defined by an effective-slenderness ratio

$$\frac{KL}{r} = 4.71\sqrt{\frac{E}{\sigma_Y}}$$

This effective-slenderness ratio corresponds to an Euler buckling stress of $0.44\sigma_Y$.

[1] *Specification for Structural Steel Buildings*, ANSI/AISC 360-10, American Institute of Steel Construction, Chicago, 2010.

For short and intermediate-length columns, the AISC formula for the critical compression stress is

$$\sigma_{cr} = \left[0.658^{\frac{\sigma_Y}{\sigma_e}}\right]\sigma_Y \qquad \text{when } \frac{kL}{r} \le 4.71\sqrt{\frac{E}{\sigma_Y}} \tag{16.22}$$

where

$$\sigma_e = \frac{\pi^2 E}{\left(\frac{KL}{r}\right)^2} \tag{16.23}$$

is the elastic buckling stress (i.e., Euler stress).

For long columns with effective-slenderness ratios greater than $4.71\sqrt{E/\sigma_Y}$, the AISC formula simply multiplies the Euler buckling stress by a factor of 0.877 to account for initial column crookedness. This reduction accounts for the fact that no real column is perfectly straight. The AISC formula for the critical compression stress of long columns is

$$\sigma_{cr} = 0.877\sigma_e \qquad \text{when } \frac{KL}{r} > 4.71\sqrt{\frac{E}{\sigma_Y}} \tag{16.24}$$

The AISC recommends that the effective-slenderness ratios of columns not exceed 200.

The allowable compression stress for either short to intermediate-length or long columns is equal to the critical compression stress [given by either Equation (16.22) or (16.24)] divided by a factor of safety of 1.67:

$$\sigma_{allow} = \frac{\sigma_{cr}}{1.67} \tag{16.25}$$

Aluminum-Alloy Columns

The Aluminum Association publishes specifications for the design of aluminum-alloy structures. Euler's formula is the basis of the design equation for long columns, and straight lines are prescribed for short and intermediate-length columns. Design formulas are specified for each particular aluminum alloy and temper.[2] One of the most common alloys used in structural applications is 6061-T6, and the column design formulas for this alloy are given next. Each of these design formulas includes an appropriate safety factor.

For short columns with effective-slenderness ratios less than or equal to 9.5,

$$\begin{aligned}\sigma_{allow} &= 19 \text{ ksi}\\ &= 131 \text{ MPa} \qquad \text{where } \frac{KL}{r} \le 9.5\end{aligned} \tag{16.26}$$

[2] *Specifications for Aluminum Structures*, Aluminum Association, Inc., Washington, DC, 1986.

For intermediate-length columns with effective-slenderness ratios between 9.5 and 66,

$$\begin{aligned}\sigma_{\text{allow}} &= [20.2 - 0.125(KL/r)] \text{ ksi} \\ &= [139 - 0.868(KL/r)] \text{ MPa} \qquad \text{where } 9.5 < \frac{KL}{r} \leq 66\end{aligned} \tag{16.27}$$

For long columns with effective-slenderness ratios greater than 66,

$$\begin{aligned}\sigma_{\text{allow}} &= \frac{51{,}000}{(KL/r)^2} \text{ ksi} \\ &= \frac{351{,}000}{(KL/r)^2} \text{ MPa} \qquad \text{where } \frac{KL}{r} > 66\end{aligned} \tag{16.28}$$

Wood Columns

The design of wood structural members is governed by the document *National Design Specification for Wood Construction* with Supplement, published by the American Wood Council (AWC).[3] The *National Design Publication Specification* (NDS) provides a single formula for the design of rectangular wood columns. The format of this formula differs somewhat from that of the formulas for steel and aluminum in that the effective-slenderness ratio is expressed as KL/d, where d is the finished dimension of the rectangular cross section. The effective-slenderness ratio for wood columns must satisfy $KL/d \leq 50$. The NDS formula is

$$\sigma_{\text{allow}} = F_c \left\{ \frac{1 + (F_{cE}/F_c)}{2c} - \sqrt{\left[\frac{1 + (F_{cE}/F_c)}{2c}\right]^2 - \frac{F_{cE}/F_c}{c}} \right\} \tag{16.29}$$

where

F_c = allowable stress for compression parallel to grain,

$F_{cE} = \dfrac{0.822E'_{\text{min}}}{(KL/d)^2}$ = reduced Euler buckling stress,

E'_{min} = adjusted modulus of elasticity for stability calculations, and

$c = 0.8$ for sawn lumber.

The effective-slenderness ratio KL/d is taken as the larger of KL/d_1 or KL/d_2, where d_1 and d_2 are the two finished dimensions of the rectangular cross section.

Local Instability

All of the discussion so far has been concerned with the *overall stability* of the column, in which the entire column length deflects as a whole into a smooth curve. However, no discussion of compression loading is complete without mentioning *local instability*. Local instability occurs when *elements* of the cross section, such as a flange or a web, buckle because of the compressive load acting on them. Open sections such as angles, channels, and W sections are particularly sensitive to local instability, although it can be a concern with any thin plate or shell element. To address local instability, design specifications typically define limits on the acceptable width-to-thickness ratios for various types of cross-sectional elements.

[3] *National Design Specification for Wood* Construction, American Wood Council, Leesburg, VA, 2015.

EXAMPLE 16.5

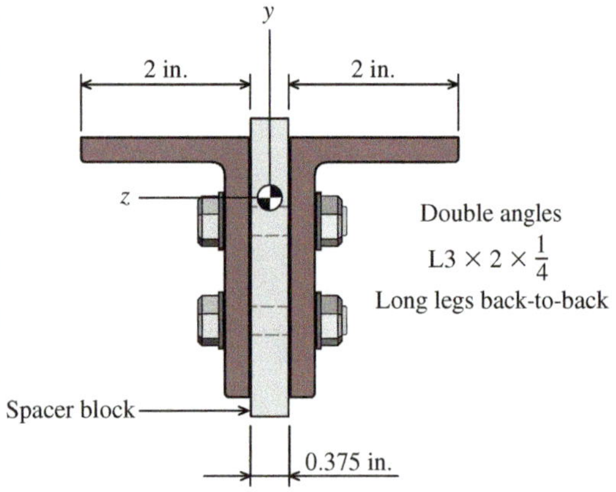

A compression chord of a small truss consists of two L3 × 2 × 1/4 steel angles arranged with long legs back-to-back as shown. The angles are separated at intervals by spacer blocks that are 0.375 in. thick. Determine the allowable axial load P_{allow} that may be supported by the compression chord if the effective length is

(a) $KL = 8$ ft.
(b) $KL = 12$ ft.

Use the AISC equations, and assume that $E = 29{,}000$ ksi and $\sigma_Y = 36$ ksi.

Plan the Solution

After computing the section properties for the built-up shape, we will use the AISC ASD formulas [i.e, Equations (16.22) through (16.25)] to determine the allowable axial loads.

SOLUTION

Section Properties

The following section properties can be obtained from Appendix B for the L3 × 2 × 1/4 structural steel shape:

$$A = 1.19 \text{ in.}^2 \qquad I_z = 1.09 \text{ in.}^4 \qquad r_z = 0.953 \text{ in.} \qquad I_y = 0.390 \text{ in.}^4$$

The subscripts for these properties have been adapted to correspond to the axes shown on the cross section. In addition, the distance from the back of the 3 in. leg to the centroid of the angle shape is given in Appendix B as $x = 0.487$ in. For the coordinate system defined here, this distance is measured in the z direction; therefore, we will denote the distance from the back of the 3 in. leg to the centroid of the angle shape as $z = 0.487$ in.

The double-angle shape is fabricated from two angles oriented back-to-back with a distance of 0.375 in. between them. The area of the double-angle shape is the sum of the areas of two angles; that is, $A = 2(1.19 \text{ in.}^2) = 2.38 \text{ in.}^2$. Additional section properties for this built-up shape must be determined.

Properties about the z axis for the double-angle shape: The z centroidal axis for the double-angle shape coincides with the centroidal axis of a single-angle shape. Therefore, the moment of inertia about the z centroidal axis for the double-angle shape is simply two times the single angle moment of inertia: $I_z = 2(1.09 \text{ in.}^4) = 2.18 \text{ in.}^4$. The radius of gyration about the z centroidal axis is the same as that for the single angle; therefore, $r_z =$ 0.953 in. for the double-angle shape.

Properties about the y axis for the double-angle shape: The y centroidal axis for the double-angle shape can be located by symmetry. Since the y centroids of the two individual angles do not coincide with the y centroidal axis for the double-angle shape, the moment of inertia about the vertical centroidal axis must be calculated with the parallel-axis theorem:

$$I_y = 2\left[0.390 \text{ in.}^4 + \left(\frac{0.375 \text{ in.}}{2} + 0.487 \text{ in.}\right)^2 (1.19 \text{ in.}^2)\right] = 1.8628 \text{ in.}^4$$

The radius of gyration about the y centroidal axis is computed from the double-angle moment of inertia I_y and area A:

$$r_y = \sqrt{\frac{I_y}{A}} = \sqrt{\frac{1.8628 \text{ in.}^4}{2.38 \text{ in.}^2}} = 0.885 \text{ in.}$$

Controlling slenderness ratio: Since $r_y < r_z$ for this double-angle shape, the effective-slenderness ratio for y-axis buckling will be larger than the effective-slenderness ratio for z-axis buckling. Therefore, buckling about the y centroidal axis will control for the compression chord member considered here.

AISC Allowable Stress Design Formulas

The AISC ASD formulas use an effective-slenderness ratio of

$$\frac{KL}{r} = 4.71\sqrt{\frac{E}{\sigma_Y}}$$

to differentiate between short and intermediate-length columns, on the one hand, and long columns, on the other. For $\sigma_Y = 36$ ksi, this parameter is calculated as

$$4.71\sqrt{\frac{E}{\sigma_Y}} = 4.71\sqrt{\frac{29{,}000\text{ ksi}}{36\text{ ksi}}} = 133.7$$

(a) Allowable axial load P_{allow} for $KL = 8$ ft: For an effective length $KL = 8$ ft, the controlling effective-slenderness ratio for the double-angle compression chord member is

$$\frac{KL}{r} = \frac{KL}{r_y} = (8\text{ ft})(12\text{ in./ft})/0.885\text{ in.} = 108.5$$

Since $KL/r_y \leq 133.7$, the column is considered an intermediate-length column, so the critical compression stress will be calculated with the use of Equation (16.22). The elastic buckling stress for a slenderness ratio of 108.5 is

$$\sigma_e = \frac{\pi^2 E}{\left(\dfrac{KL}{r}\right)^2} = \frac{\pi^2(29{,}000\text{ ksi})}{(108.5)^2} = 24.31\text{ ksi}$$

From Equation (16.22), the critical compression stress is

$$\sigma_{\text{cr}} = \left[0.658^{\frac{\sigma_Y}{\sigma_e}}\right]\sigma_Y = \left[0.658^{\left(\frac{36\text{ ksi}}{24.31\text{ ksi}}\right)}\right](36\text{ ksi}) = 19.37\text{ ksi}$$

The allowable compression stress is determined from Equation (16.25):

$$\sigma_{\text{allow}} = \frac{\sigma_{\text{cr}}}{1.67} = \frac{19.37\text{ ksi}}{1.67} = 11.60\text{ ksi}$$

From this allowable stress, the allowable axial load for an effective length $KL = 8$ ft is

$$P_{\text{allow}} = \sigma_{\text{allow}} A = (11.60\text{ ksi})(2.38\text{ in.}^2) = 27.6\text{ kips} \qquad \textbf{Ans.}$$

(b) Allowable axial load P_{allow} for $KL = 12$ ft: For an effective length $KL = 12$ ft, the controlling effective-slenderness ratio for the double-angle compression chord member is

$$\frac{KL}{r} = \frac{KL}{r_y} = (12\text{ ft})(12\text{ in./ft})/0.885\text{ in.} = 162.7$$

The elastic buckling stress for this slenderness ratio is

$$\sigma_e = \frac{\pi^2 E}{\left(\dfrac{KL}{r}\right)^2} = \frac{\pi^2(29{,}000\text{ ksi})}{(162.7)^2} = 10.81\text{ ksi}$$

Since $KL/ry > 133.7$, the column is classified as a long column, so Equation (16.24) is used to calculate the critical compression stress:

$$\sigma_{cr} = 0.877\ \sigma_e = 0.877(10.81\text{ ksi}) = 9.48\text{ ksi}$$

The allowable compression stress is determined from Equation (16.25):

$$\sigma_{allow} = \frac{\sigma_{cr}}{1.67} = \frac{9.48\text{ ksi}}{1.67} = 5.68\text{ ksi}$$

The allowable axial load for $KL = 12$ ft can be calculated from the allowable stress:

$$P_{allow} = \sigma_{allow} A = (5.68\text{ ksi})(2.38\text{ in.}^2) = 13.52\text{ kips}$$ **Ans.**

EXAMPLE 16.6

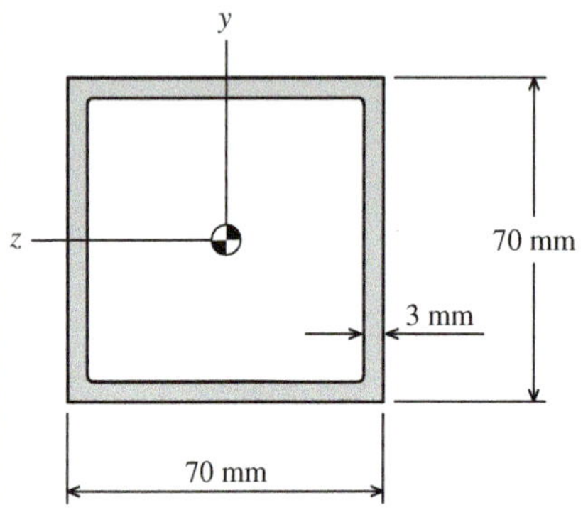

A square tube made of 6061-T6 aluminum alloy has the cross-sectional dimensions shown. Use the Aluminum Association column design formulas to determine the allowable axial load P_{allow} that may be supported by the tube if the effective length of the compression member is

(a) $KL = 1{,}500$ mm.
(b) $KL = 2{,}750$ mm.

Plan the Solution

After computing the section properties of the square tube, the Aluminum Association design formulas [Equations (16.26) through (16.28)] will be used to calculate the allowable loads for the specified effective lengths.

SOLUTION

Section Properties

The centroid of the square tube is found from symmetry. The cross-sectional area of the tube is

$$A = (70\text{ mm})^2 - (64\text{ mm})^2 = 804\text{ mm}^2$$

The moments of inertia about both the y and z centroidal axes are identical:

$$I_y = I_z = \frac{(70\text{ mm})^4}{12} - \frac{(64\text{ mm})^4}{12} = 602{,}732\text{ mm}^4$$

Similarly, the radii of gyration about both centroidal axes are the same:

$$r_y = r_z = \sqrt{\frac{602{,}732\text{ mm}^4}{804\text{ mm}^2}} = 27.38\text{ mm}$$

(a) Allowable axial load P_{allow} for $KL = 1{,}500$ mm: For an effective length $KL = 1{,}500$ mm, the effective-slenderness ratio for the square tube member is

$$\frac{KL}{r} = \frac{1{,}500\text{ mm}}{27.38\text{ mm}} = 54.8$$

Since this slenderness ratio is greater than 9.5 and less than 66, Equation (16.27) must be used to determine the allowable compression stress. The SI version of this equation can be used to give σ_{allow}:

$$\sigma_{allow} = [139 - 0.868(KL/r)]\text{ MPa} = [139 - 0.868(54.8)] = 91.43\text{ MPa}$$

From this allowable stress, the allowable axial load can be computed as

$$P_{allow} = \sigma_{allow} A = (91.43 \text{ N/mm}^2)(804 \text{ mm}^2) = 73{,}510 \text{ N} = 73.5 \text{ kN} \quad \textbf{Ans.}$$

(b) Allowable axial load P_{allow} for KL = 2,750 mm: For an effective length KL = 2,750 mm, the effective-slenderness ratio is

$$\frac{KL}{r} = \frac{2{,}750 \text{ mm}}{27.38 \text{ mm}} = 100.4$$

Since this slenderness ratio is greater than 66, the allowable compression stress is determined from Equation (16.28):

$$\sigma_{allow} = \frac{351{,}000}{(KL/r)^2} \text{ MPa} = \frac{351{,}000}{(100.4)^2} = 34.82 \text{ MPa}$$

The allowable axial load is therefore

$$P_{allow} = \sigma_{allow} A = (34.82 \text{ N/mm}^2)(804 \text{ mm}^2) = 27{,}995 \text{ N} = 28.0 \text{ kN} \quad \textbf{Ans.}$$

EXAMPLE 16.7

A sawn rectangular timber of visually graded No. 2 Spruce–Pine–Fir (SPF) wood has finished dimensions of 7.25 in. by 9.25 in. For this wood species and grade, the allowable compression stress parallel to the wood grain is F_c = 975 psi and the modulus of elasticity is E'_{min} = 400,000 psi. The timber column has a length L = 16 ft, and pinned connections are used at each end of the column. Use the AWC National Design Specification for Wood Construction (NDS) column design formula to determine the allowable axial load P_{allow} that may be supported by the column.

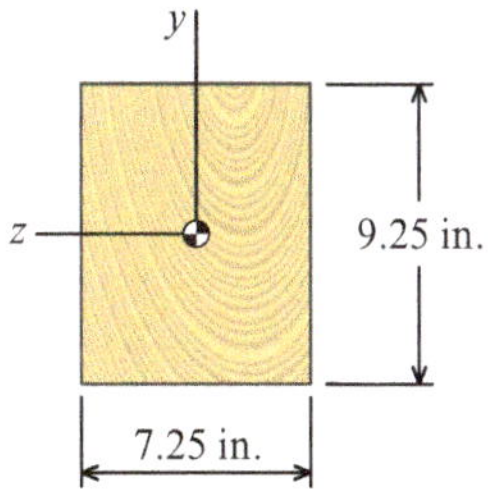

Plan the Solution

The AWC NDS column design formula given in Equation (16.29) will be used to compute the allowable axial load.

SOLUTION

The AWC NDS column design formula is

$$\sigma_{allow} = F_c \left\{ \frac{1 + (F_{cE}/F_c)}{2c} - \sqrt{\left[\frac{1 + (F_{cE}/F_c)}{2c}\right]^2 - \frac{(F_{cE}/F_c)}{c}} \right\}$$

where

F_c = allowable stress for compression parallel to grain,

$F_{cE} = \dfrac{0.822 E'_{min}}{(KL/d)^2}$ = reduced Euler buckling stress,

E'_{min} = adjusted modulus of elasticity for stability calculations, and

c = 0.8 for sawn lumber.

The finished dimensions of the timber column are 7.25 in. by 9.25 in. The smaller of these two dimensions is taken as d in the term KL/d. Since the column has pinned ends, the effective-length factor is K = 1.0; therefore,

$$\frac{KL}{d} = \frac{(1.0)(16 \text{ ft})(12 \text{ in./ft})}{7.25 \text{ in.}} = 26.48$$

The reduced Euler buckling stress term used in the AWC NDS formula has the value

$$F_{cE} = \frac{0.822E'_{\text{min}}}{(KL/d)^2} = \frac{0.822(400{,}000 \text{ psi})}{(26.48)^2} = 468.92 \text{ psi}$$

Also,

$$\frac{F_{cE}}{F_c} = \frac{468.92 \text{ psi}}{975 \text{ psi}} = 0.4809$$

This ratio, along with the values $F_c = 975$ psi and $c = 0.8$ (for sawn lumber), are used in the AWC NDS formula to calculate the allowable compression stress for the timber column:

$$\begin{aligned}
\sigma_{\text{allow}} &= F_c\left\{\frac{1+(F_{cE}/F_c)}{2c} - \sqrt{\left[\frac{1+(F_{cE}/F_c)}{2c}\right]^2 - \frac{F_{cE}/F_c}{c}}\right\} \\
&= (975 \text{ psi})\left\{\frac{1+(0.4809)}{2(0.8)} - \sqrt{\left[\frac{1+(0.4809)}{2(0.8)}\right]^2 - \frac{0.4809}{0.8}}\right\} \\
&= (975 \text{ psi})\left\{0.9256 - \sqrt{0.9256^2 - 0.6011}\right\} \\
&= 409.5 \text{ psi}
\end{aligned}$$

The allowable axial load that may be supported by the column is therefore

$$P_{\text{allow}} = \sigma_{\text{allow}}A = (409.5 \text{ psi})(7.25 \text{ in.})(9.25 \text{ in.}) = 27{,}462 \text{ lb} = 27{,}500 \text{ lb} \qquad \textbf{Ans.}$$

PROBLEMS

P16.27 Use the AISC equations to determine the allowable axial load P_{allow} that may be supported by a W8 × 48 wide-flange column for the following effective lengths: (a) $KL = 13$ ft and (b) $KL = 26$ ft. Assume that $E = 29{,}000$ ksi and $\sigma_Y = 50$ ksi.

P16.28 Use the AISC equations to determine the allowable axial load P_{allow} that may be supported by an HSS152.4 × 101.6 × 6.4 column for the following effective lengths: (a) $KL = 3.75$ m and (b) $KL = 7.5$ m. Assume that $E = 200$ GPa and $\sigma_Y = 320$ MPa.

P16.29 The 10 m long HSS304.8 × 203.2 × 9.5 (see Appendix B for cross-sectional properties) column shown in Figure P16.29/30 is fixed at base A with respect to bending about both the strong and weak axes of the HSS cross section. At upper end B, the column is restrained against rotation and translation in the x–z plane (i.e., bending about the weak axis) and against translation in the x–y plane (i.e., the column is free to rotate about the strong axis). Use the AISC equations to determine the allowable axial load P_{allow} that may be supported by the column, on the basis of (a) buckling in the x–y plane and (b) buckling in the x–z plane. Assume that $E = 200$ GPa and $\sigma_Y = 320$ MPa.

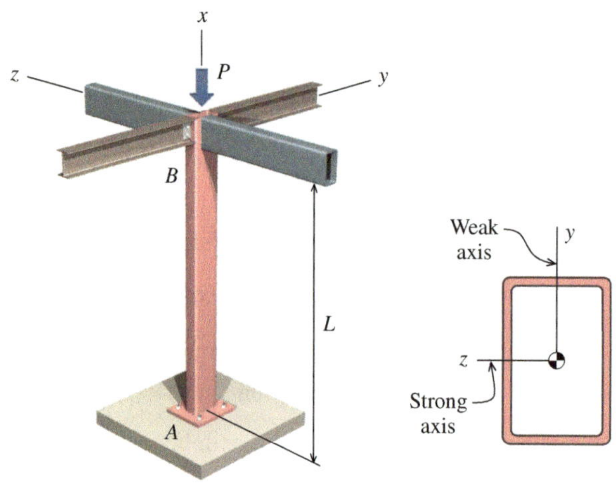

FIGURE P16.29/30

P16.30 The 25 ft long HSS6 × 4 × 1/8 (see Appendix B for cross-sectional properties) column shown in Figure P16.29/30 is fixed at base A with respect to bending about both the strong and weak axes of the HSS cross section. At upper end B, the column is restrained against rotation and translation in the x–z plane (i.e., bending about the weak axis) and against translation in the x–y plane (i.e., the column is free to rotate about the strong axis). Use the AISC equations to determine the allowable axial load P_{allow} that may be supported by the column, on the basis of (a) buckling in the x–y plane and (b) buckling in the x–z plane. Assume that $E = 29{,}000$ ksi and $\sigma_Y = 46$ ksi.

P16.31 A column with an effective length of 28 ft is fabricated by connecting two C15 × 40 steel channels (see Appendix B for cross-sectional properties) with lacing bars as shown in Figure P16.31/32. Use the AISC equations to determine the allowable axial load P_{allow} that may be supported by the column if $d = 10$ in. Assume that $E = 29{,}000$ ksi and $\sigma_Y = 36$ ksi.

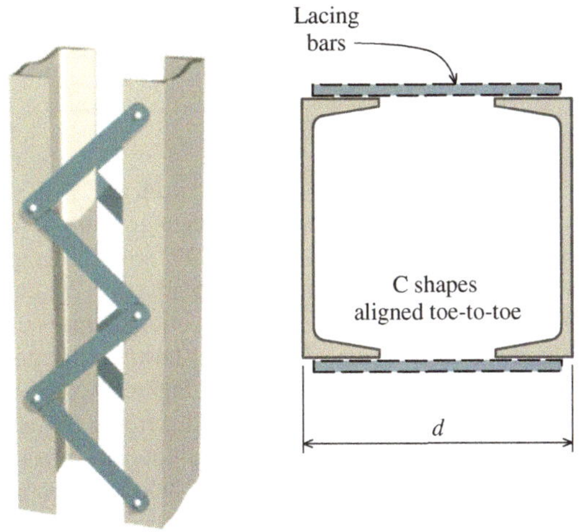

FIGURE P16.31/32

P16.32 A column is fabricated by connecting two C310 × 45 steel channels (see Appendix B for cross-sectional properties) with lacing bars as shown in Figure P16.31/32. (a) Determine the distance d required so that the moments of inertia for the section about the two principal axes are equal. (b) Using the value of d obtained in part (a), determine the allowable axial load P_{allow} that may be supported by a column with an effective length $KL = 9.5$ m. Use the AISC equations and assume that $E = 200$ GPa and $\sigma_Y = 340$ MPa.

P16.33 A compression chord of a small truss consists of two L127 × 76 × 12.7 steel angles arranged with long legs back-to-back as shown in Figure P16.33. The angles are separated at intervals by spacer blocks. (a) Determine the spacer thickness required so that the moments of inertia for the section about the two principal axes are equal. (b) Using the spacer thickness obtained in part (a), determine the allowable axial load P_{allow} that may be supported by a compression chord with an effective length $KL = 7$ m. Use the AISC equations and assume that $E = 200$ GPa and $\sigma_Y = 340$ MPa.

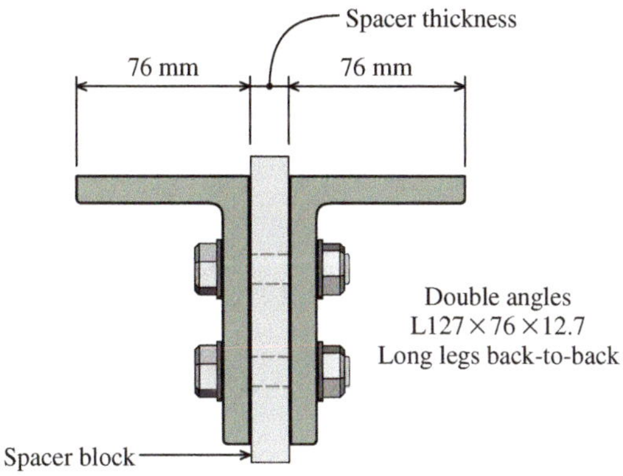

FIGURE P16.33

P16.34 Develop a list of three acceptable structural steel WT shapes (from those listed in Appendix B) that can be used as an 18 ft long pin-ended column to carry an axial compression load of 30 kips. Include the most economical WT8, WT9, and WT10.5 shapes on your list, and select the most economical shape from the available alternatives. Use the AISC equation for long columns [Eq. (P16.25)], and assume that $E = 29{,}000$ ksi and $\sigma_Y = 50$ ksi.

P16.35 A 6061-T6 aluminum-alloy pipe column with pinned ends has an outside diameter of 4.50 in. and a wall thickness of 0.237 in. Determine the allowable axial load P_{allow} that may be supported by the aluminum pipe column for the following effective lengths: (a) $KL = 7.5$ ft and (b) $KL = 15$ ft. Use the Aluminum Association column design formulas.

P16.36 A 6061-T6 aluminum-alloy rectangular tube shape has cross-sectional dimensions $b = 100$ mm, $d = 150$ mm, and $t = 5$ mm as shown in Figure P16.36. The tube is used as a compression member that is 7.5 m long. For buckling about the z axis, assume that both ends of the column are pinned. For buckling about the y axis, however, assume that both ends of the column are fixed. Determine the allowable axial load P_{allow} that may be supported by the rectangular tube. Use the Aluminum Association column design formulas.

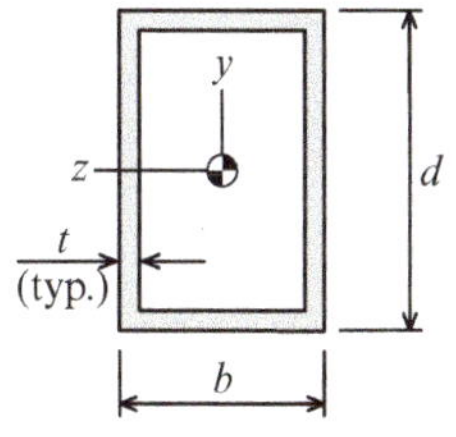

FIGURE P16.36

P16.37 The aluminum column shown in Figure P16.37 has a rectangular cross section and supports a compressive axial load P. The base of the column is fixed. The support at the top allows rotation of the column in the x–y plane (i.e., bending about the strong axis) but prevents rotation in the x–z plane (i.e., bending about the weak axis).

Determine the allowable axial load P_{allow} that may be applied to the column for the following parameters: $L = 60$ in., $b = 1.25$ in., and $h = 2.00$ in. Use the Aluminum Association column design formulas.

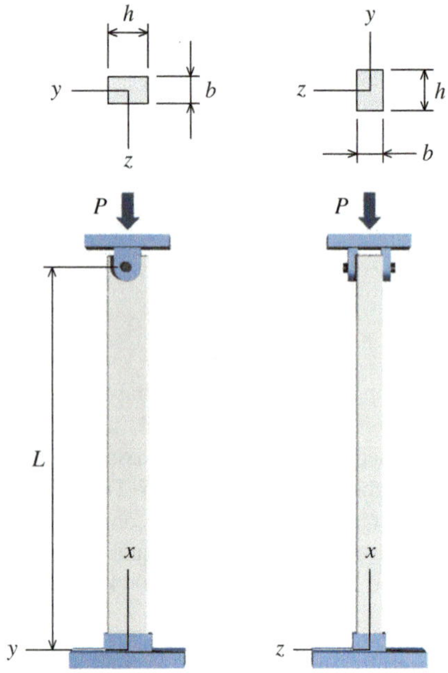

FIGURE P16.37

P16.38 A 6061-T6 aluminum-alloy wide-flange shape is used as a column of length $L = 5.5$ m. The column is fixed at base A. Pin-connected lateral bracing is present at B so that deflection in the x–z plane is restrained at the upper end of the column; however, the column is free to deflect in the x–y plane at B. (See Figure P16.38*a*.) The cross-sectional dimensions of the shape as shown in Figure P16.38*b* are $b_f = 130$ mm, $t_f = 9$ mm, $d = 200$ mm, and $t_w = 6$ mm. Use the Aluminum Association column design formulas to determine the allowable compressive load P_{allow} that the column can support. In your analysis, consider the possibility that buckling could occur about either the strong axis (i.e., the z axis) or the weak axis (i.e., the y axis) of the aluminum column.

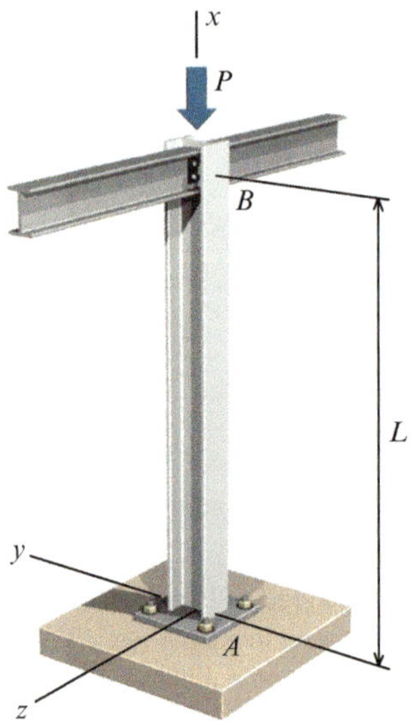

FIGURE P16.38*a*

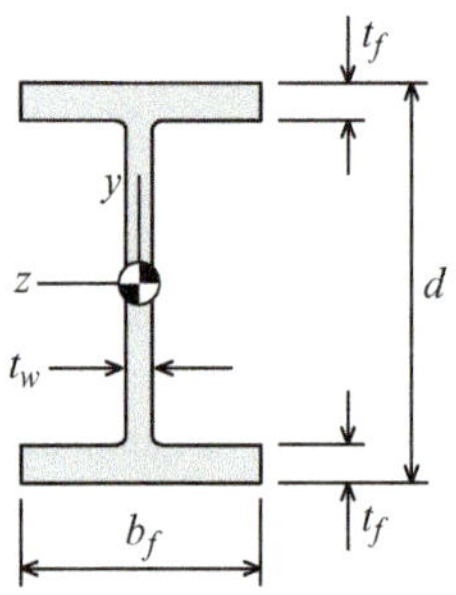

FIGURE P16.38*b*

P16.39 A wood post of rectangular cross section (Figure P16.39) consists of Select Structural grade Douglas fir lumber ($F_c = 1{,}150$ psi; $E'_{min} = 580{,}000$ psi). The finished dimensions of the post are $b = 5.5$ in. and $d = 7.5$ in. Assume pinned connections at each end of the post. Determine the allowable axial load P_{allow} that may be supported by the post for the following column lengths: (a) $L = 10$ ft, (b) $L = 16$ ft, and (c) $L = 24$ ft. Use the AWC NDS column design formula.

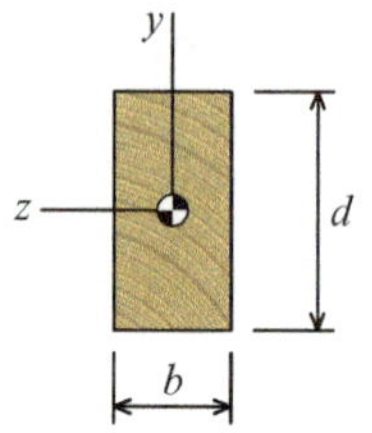

FIGURE P16.39

P16.40 A Select Structural grade Hem-Fir ($F_c = 10.3$ MPa; $E'_{min} = 4.0$ GPa) wood column of rectangular cross section has finished dimensions of $b = 100$ mm and $d = 235$ mm. The length of the column is $L = 4.75$ m. The column is fixed at base A. Pin-connected lateral bracing is present at B so that deflection in the x–z plane is restrained at the upper end of the column; however, the column is free to deflect in the x–y plane at B (see Figure P16.40). Use the AWC NDS column design formula to determine the allowable compressive load P_{allow} that the column can support. In your analysis, consider the possibility that buckling could occur about either the strong axis (i.e., the z axis) or the weak axis (i.e., the y axis) of the wood column.

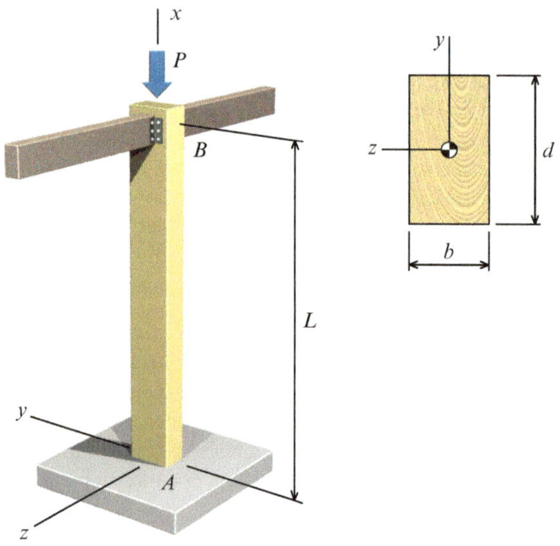

FIGURE P16.40

P16.41 A simple pin-connected wood truss is loaded and supported as shown in Figure P16.41. The members of the truss are square Douglas fir timbers (finished dimensions = 3.5 in. by 3.5 in.) with F_c = 1,350 psi and E'_{min} = 580,000 psi.

(a) For the loads shown, determine the axial forces produced in chord members *AF*, *FG*, *GH*, and *EH* and in web members *BG* and *DG*.
(b) Use the AWC NDS column design formula to determine the allowable compressive load P_{allow} for each of these members.
(c) Report the ratio $P_{\text{allow}}/P_{\text{actual}}$ for each of these members.

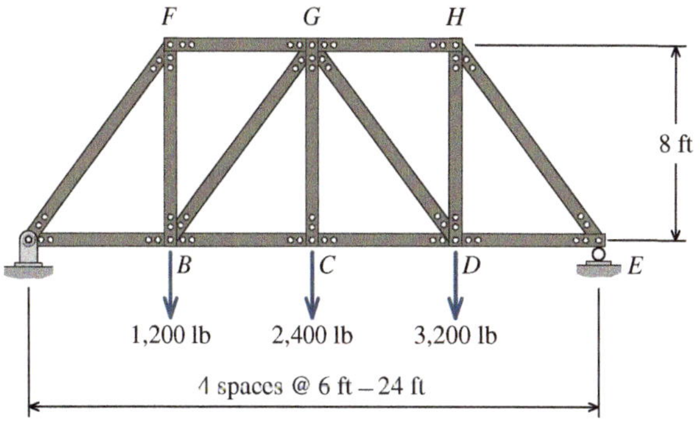

FIGURE P16.41

16.6 Eccentrically Loaded Columns

Although a given column will support its maximum load when the load is applied centrically, it is sometimes necessary to apply an eccentric load to a column. For example, a floor beam in a building may in turn be supported by an angle bolted or welded to the side of a column as shown in Figure 16.12. Since the reaction force from the beam acts at some eccentricity *e* from the centroid of the column, a bending moment is created in the column in addition to a compressive axial load. The bending moment applied to the column will increase the stress in the column and, in turn, decrease its load-carrying capacity. Three methods will be presented here for analyzing columns that are subjected to an eccentric axial load.

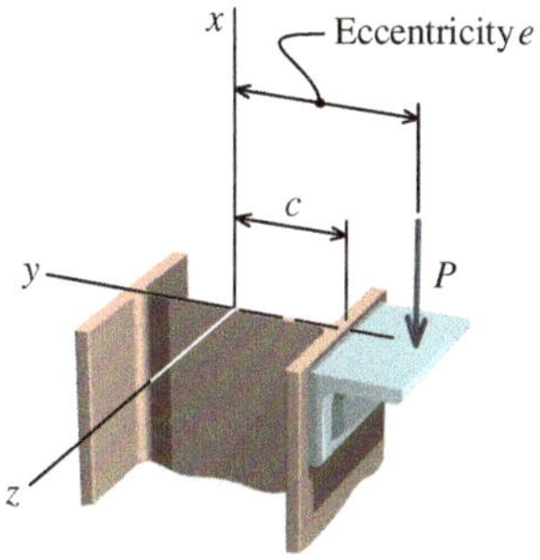

FIGURE 16.12 Column subjected to eccentric load *P*.

The Secant Formula

The secant formula [Equation (16.20)] was derived on the assumption that the applied load had an initial eccentricity *e*. If *e* is known, then its value can be substituted into the secant formula to determine the failure load (i.e., the load that causes incipient inelastic action). As mentioned previously, there is usually a small amount of unavoidable eccentricity that must be approximated when this formula is used for centric loads. The form of the secant formula makes it somewhat difficult to solve for the value of *P*/*A* that produces a specific maximum compression stress value; however, a number of equation-solving computer programs are available that can readily produce this sort of numerical solution.

Allowable-Stress Method

The topic of bending due to an eccentric axial load was discussed in Section 8.7. Figure 8.14 depicted the stress distributions caused by axial loads and by bending moments, and it illustrated the resulting stress distribution from the combined effects. Equation (8.19) was used to calculate the normal stress produced by the combination of an axial force and a bending moment. Buckling was not considered in Section 8.7; however, the approach taken in Equation (8.19) can be adapted for use in that context.

The allowable-stress method simply requires that the sum of the compressive axial stress and the compressive bending stress be less than the allowable compression stress prescribed by the pertinent column formula for centric loading. In this regard, Equation (8.19) can be restated as

$$\sigma_x = \frac{P}{A} + \frac{Mc}{I} \leq \sigma_{\text{allow}} \tag{16.30}$$

where the compressive stresses are treated as positive quantities. In Equation (16.30), σ_{allow} is the allowable stress calculated from one of the empirical design formulas presented in Section 16.5. The formula uses the largest value of the effective-slenderness ratio for the cross section, irrespective of the axis about which bending occurs. Values of c and I used in calculating the bending stress, however, do depend on the axis of bending. Thus, the allowable stress method and Equation (16.30) generally produce a conservative design.

Interaction Method

In an eccentrically loaded column, much of the total stress may be caused by the bending moment. However, the allowable bending stress is generally larger than the allowable compression stress. How, then, can some balance be attained between the two allowable stresses? Consider the axial stress $\sigma_a = P/A$. If the allowable axial stress for a member acting as a column is denoted by $(\sigma_{allow})_a$, then the area required for a given axial force P can be expressed as

$$A_a = \frac{P}{(\sigma_{allow})_a}$$

Next, consider the bending stress given by $\sigma_b = Mc/I$. The moment of inertia I can be expressed in terms of the area and the radius of gyration as $I = Ar^2$, where r is the radius of gyration in the plane of bending. Let the allowable bending stress be designated $(\sigma_{allow})_b$. Then the area required for a given bending moment M can be expressed as

$$A_b = \frac{Mc}{r^2(\sigma_{allow})_b}$$

Therefore, the total area required for a column subjected to an axial force and a bending moment can be expressed as the sum of the expressions for A_a and A_b:

$$A = A_a + A_b = \frac{P}{(\sigma_{allow})_a} + \frac{Mc}{r^2(\sigma_{allow})_b}$$

Dividing this equation by the total area A and letting $Ar^2 = I$ gives

$$\boxed{\frac{P/A}{(\sigma_{allow})_a} + \frac{Mc/I}{(\sigma_{allow})_b} = 1} \tag{16.31}$$

If the column has an axial load, but no bending moment (i.e., if it is a centrically loaded column), then Equation (16.31) indicates that the column is analyzed in accordance with the allowable axial stress. If the column has a bending moment, but no axial load (in other words, if it is truly a beam), then the normal stresses must satisfy the allowable bending stress. Between these two extremes, Equation (16.31) accounts for the relative importance of each normal stress component in relation to the combined effect. Equation (16.31) is known as an *interaction formula*, and the approach that it presents is a common method for considering the combined effect of an axial load and a bending moment in columns.

In Equation (16.31), $(\sigma_{allow})_a$ is the allowable axial stress given by one of the empirical column design formulas in Section 16.5 and $(\sigma_{allow})_b$ is the allowable bending stress. The AISC specifications use the general form of Equation (16.31) to analyze combined axial compression and bending; however, additional modification factors are added to that equation, depending on whether $(P/A)/(\sigma_{allow})_a$ is less than or greater than 0.2. Since the purpose of this discussion is to introduce the concept of interaction equations, rather than to teach the specific details of AISC steel column design, Equation (16.31) without additional factors will be used here to analyze columns subjected to both axial compression and bending moments.

EXAMPLE 16.8

The W12 × 58 structural steel column shown (see Appendix B for its cross-sectional properties) is fixed at its base and free at its upper end. At the top of the column, a load P is applied to a bracket at an eccentricity $e = 14$ in. from the centroidal axis of the wide-flange shape. Use the AISC ASD formulas given in Section 16.5, and assume that $E = 29{,}000$ ksi and $\sigma_Y = 36$ ksi.

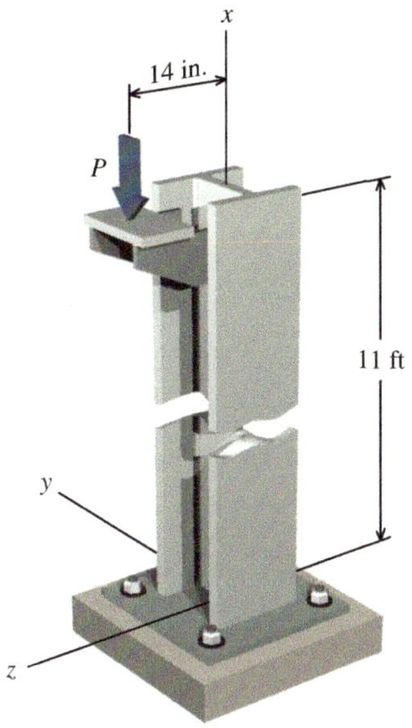

(a) Using the allowable-stress method, determine whether the column is safe for a load $P = 25$ kips. Report the results in the form of the stress ratio σ_x/σ_{allow}.
(b) Determine the magnitude of the largest eccentric load P that may be applied to the column according to the allowable stress method.
(c) Repeat the analysis, using the interaction method, and determine whether the column is safe for a load $P = 25$ kips. Assume that the allowable bending stress is $(\sigma_{allow})_b = 24$ ksi. Report the value of the interaction equation.
(d) Determine the magnitude of the largest eccentric load P that may be applied to the column according to the interaction method.

Plan the Solution

The section properties can be obtained from Appendix B for the W12 × 58 structural steel shape. From these properties, the compressive stresses due to the axial force and the bending moment can be determined for the specified 25 kip load, and the allowable compression stress can be determined from the AISC ASD formulas. These values, along with the specified allowable bending stress, can then be used in Equation (16.30) for the allowable-stress method, and in Equation (16.31) for the interaction method, to determine whether the column can safely carry $P = 25$ kips at the specified 14 in. eccentricity. To determine the largest acceptable eccentric load, the axial and bending stresses are specified in terms of P and the resulting equations are then solved for the maximum load magnitude.

SOLUTION

Section Properties

The following section properties can be obtained from Appendix B for the W12 × 58 structural steel shape:

$$A = 17.0 \text{ in.}^2 \qquad I_z = 475 \text{ in.}^4 \qquad r_z = 5.28 \text{ in.} \qquad I_y = 107 \text{ in.}^4 \qquad r_y = 2.51 \text{ in.}$$

The subscripts for these properties have been revised to correspond to the axes shown. In addition to the preceding values, the flange width of the W12 × 58 shape is $b_f = 10.0$ in.

Axial Stress Calculation

The 25 kip load will produce compressive normal stress in the column:

$$\sigma_{axial} = \frac{P}{A} = \frac{25 \text{ kips}}{17.0 \text{ in.}^2} = 1.47 \text{ ksi} \qquad \text{(a)}$$

Bending Stress Calculation

The eccentric axial load P applied at an eccentricity $e = 14$ in. will produce a bending moment $M_y = Pe$ about the y axis (i.e., the weak axis) of the wide-flange shape. The bending stress can be calculated from the flexure formula $\sigma_{bend} = M_y c/I_y$, where c is equal to half the flange width: $c = b_f/2 = 10.0 \text{ in.}/2 = 5.0$ in. For the specified axial load $P = 25$ kips, the maximum bending stress magnitude is

$$\sigma_{bend} = \frac{M_y c}{I_y} = \frac{Pec}{I_y} = \frac{(25 \text{ kips})(14 \text{ in.})(5.0 \text{ in.})}{107 \text{ in.}^4} = 16.36 \text{ ksi} \qquad \text{(b)}$$

Both tensile and compressive normal stresses will be produced by bending; however, the compressive normal stress is the focus of our interest here.

AISC Allowable-Stress Design Formulas

The AISC ASD formulas differentiate between short to intermediate-length columns, on the one hand, and long columns, on the other, according to an effective-slenderness ratio given by

$$4.71\sqrt{\frac{E}{\sigma_Y}} = 4.71\sqrt{\frac{29{,}000 \text{ ksi}}{36 \text{ ksi}}} = 133.681$$

From Figure 16.7, the appropriate effective-length factor for a column fixed at its base and free at its upper end is $K_y = K_z = 2.0$. The effective-slenderness ratios for buckling about the strong and weak axes, respectively, of the W12 × 58 are therefore

$$\frac{K_z L}{r_z} = \frac{(2.0)(11 \text{ ft})(12 \text{ in./ft})}{5.28 \text{ in.}} = 50.0 \quad \text{and} \quad \frac{K_y L}{r_y} = \frac{(2.0)(11 \text{ ft})(12 \text{ in./ft})}{2.51 \text{ in.}} = 105.2$$

The controlling effective-slenderness ratio for the column is 105.2. Since $KL/r_y \le 4.71\sqrt{E/\sigma_Y}$, the column is considered to be an intermediate-length column and the critical compression stress will be calculated from Equation (16.22). In this equation, the elastic buckling stress σ_e for the controlling effective-slenderness ratio is computed from Equation (16.23):

$$\sigma_e = \frac{\pi^2 E}{\left(\frac{KL}{r}\right)^2} = \frac{\pi^2 (29{,}000 \text{ ksi})}{(105.2)^2} = 25.86 \text{ ksi}$$

From Equation (16.22),

$$\sigma_{cr} = \left[0.658^{\left(\frac{\sigma_Y}{\sigma_e}\right)}\right]\sigma_Y = \left[0.658^{\left(\frac{36 \text{ ksi}}{25.86 \text{ ksi}}\right)}\right](36 \text{ ksi}) = 20.10 \text{ ksi}$$

Finally, the allowable compression stress is determined from Equation (16.25):

$$\sigma_{allow} = \frac{\sigma_{cr}}{1.67} = \frac{20.10 \text{ ksi}}{1.67} = 12.04 \text{ ksi} \qquad \text{(c)}$$

(a) Is the column safe for $P = 25$ kips, according to the allowable stress method? The allowable stress method simply requires that the sum of the compressive axial stress and the compressive bending stress be less than the allowable compression stress prescribed by the pertinent AISC ASD column formula for centric loading. The sum of the compressive axial stress and the compressive bending stress is

$$\sigma_x = 1.47 \text{ ksi} + 16.36 \text{ ksi} = 17.83 \text{ ksi (C)} \qquad \text{(d)}$$

Since σ_x is greater than the 12.04 ksi allowable compression stress, the column is **not safe** for $P = 25$ kips, according to the allowable stress method. The ratio between the allowable and actual stresses has the value

$$\frac{\sigma_x}{\sigma_{allow}} = \frac{17.83 \text{ ksi}}{12.04 \text{ ksi}} = 1.48 > 1 \quad \text{N.G.}$$

Ans.

(b) Magnitude of the largest eccentric load P: The axial and bending stresses in the allowable-stress-method equation can be expressed in terms of an unknown P:

$$\sigma_x = \frac{P}{A} + \frac{Pec}{I_y} = P\left[\frac{1}{A} + \frac{ec}{I_y}\right] \tag{e}$$

The largest load magnitude can be calculated by setting Equation (e) equal to the allowable compression stress from Equation (c) and solving for P:

$$\sigma_x = \sigma_{\text{allow}} = 12.04 \text{ ksi} = P\left[\frac{1}{A} + \frac{ec}{I_y}\right] = P\left[\frac{1}{17.0 \text{ in.}^2} + \frac{(14 \text{ in.})(5.0 \text{ in.})}{107 \text{ in.}^4}\right] = P[0.71303 \text{ in.}^{-2}]$$

$$\therefore P = 16.89 \text{ kips}$$ **Ans.**

(c) Is the column safe for P = 25 kips, according to the interaction method? In the interaction method, the axial stress is divided by the allowable compression stress, the bending stress is divided by the allowable bending stress, and the sum of the resulting two terms must not exceed 1:

$$\frac{P/A}{(\sigma_{\text{allow}})_a} + \frac{M_y c/I_y}{(\sigma_{\text{allow}})_b} = 1$$

The axial and bending stresses were computed in Equations (a) and (b). The allowable compression stress $(\sigma_{\text{allow}})_a$ was computed in Equation (c), and the allowable bending stress is specified as $(\sigma_{\text{allow}})_b = 24$ ksi. With these values, the interaction equation for the eccentrically loaded W12 × 58 column is

$$\frac{1.47 \text{ ksi}}{12.04 \text{ ksi}} + \frac{16.36 \text{ ksi}}{24 \text{ ksi}} = 0.1221 + 0.6817 = 0.8038 < 1 \quad \text{O.K.}$$ **Ans.**

Since the value of the interaction equation is less than 1, the column is safe for a load of $P = 25$ kips, according to the interaction method.

(d) Magnitude of the largest eccentric load P: The sum of the compressive axial and bending stresses for the eccentrically loaded W12 × 58 column can be expressed in terms of an unknown P:

$$\frac{P}{A(\sigma_{\text{allow}})_a} + \frac{Pec}{I_y(\sigma_{\text{allow}})_b} = P\left[\frac{1}{A(\sigma_{\text{allow}})_a} + \frac{ec}{I_y(\sigma_{\text{allow}})_b}\right] = 1 \tag{f}$$

Equation (f) can be solved for the largest load magnitude P:

$$P\left[\frac{1}{A(\sigma_{\text{allow}})_a} + \frac{ec}{I_y(\sigma_{\text{allow}})_b}\right] = P\left[\frac{1}{(17.0 \text{ in.}^2)(12.04 \text{ ksi})} + \frac{(14 \text{ in.})(5.0 \text{ in.})}{(107 \text{ in.}^4)(24 \text{ ksi})}\right] = P[0.032144 \text{ kip}^{-1}]$$

$$\therefore P = 31.1 \text{ kips}$$ **Ans.**

EXAMPLE 16.9

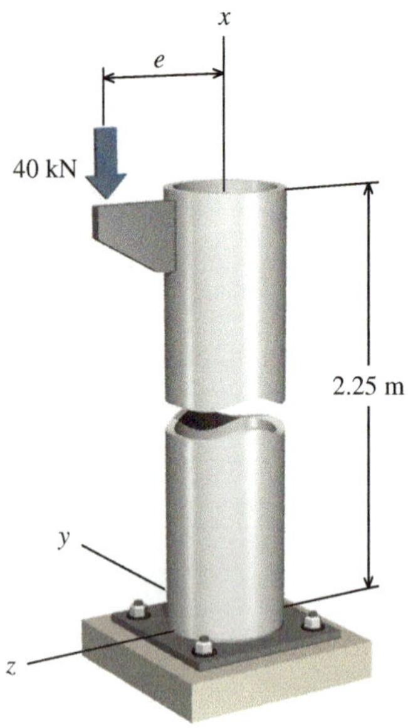

A 6061-T6 aluminum-alloy tube (outside diameter = 130 mm; wall thickness = 12.5 mm) supports an axial load $P = 40$ kN, which is applied at an eccentricity e from the centerline of the tube. The 2.25 m long tube is fixed at its base and free at its upper end. Apply the Aluminum Association equations given in Section 16.5, and assume that the allowable bending stress of the 6061-T6 alloy is 150 MPa. Determine the maximum value of eccentricity e that may be used

(a) according to the allowable stress method.
(b) according to the interaction method.

Plan the Solution

Compute the section properties of the tube, and then use the Aluminum Association equations to determine the allowable compression stress for the 2.25 m long fixed–free column. Express both the allowable-stress and interaction methods in terms of P and e, and solve for the allowable eccentricity e.

SOLUTION

Section Properties

The inside diameter of the tube is $d = 130\text{ mm} - 2(12.5\text{ mm}) = 105\text{ mm}$.

The cross-sectional area of the tube is

$$A = \frac{\pi}{4}[(130\text{ mm})^2 - (105\text{ mm})^2] = 4{,}614.2\text{ mm}^2$$

The moments of inertia about both the y and z centroidal axes are identical:

$$I_y = I_z = I = \frac{\pi}{64}[(130\text{ mm})^4 - (105\text{ mm})^4] = 8{,}053{,}246\text{ mm}^4$$

Similarly, the radii of gyration about both centroidal axes are the same:

$$r_y = r_z = r = \sqrt{\frac{8{,}053{,}246\text{ mm}^4}{4{,}614.2\text{ mm}^2}} = 41.78\text{ mm}$$

Allowable Compression Stress

From Figure 16.7, the effective-length factor for a fixed–free column is $K = 2.0$. Therefore, the effective-slenderness ratio for the 2.25 m long 6061-T6 tube is

$$\frac{KL}{r} = \frac{(2.0)(2{,}250\text{ mm})}{41.78\text{ mm}} = 107.7$$

Since this slenderness ratio is greater than 66, the allowable compression stress is determined from Equation (16.24):

$$\sigma_{\text{allow}} = \frac{351{,}000}{(KL/r)^2}\text{MPa} = \frac{351{,}000}{(107.7)^2} = 30.26\text{ MPa} \qquad \text{(a)}$$

(a) Maximum eccentricity based on the allowable-stress method: The axial and bending stresses in the allowable-stress-method equation can be expressed as

$$\sigma_x = \frac{P}{A} + \frac{Pec}{I} = P\left[\frac{1}{A} + \frac{ec}{I}\right] \qquad \text{(b)}$$

where c is the outside radius of the tube ($c = 130$ mm/2 = 65 mm). Now, set Equation (b) equal to the allowable compression stress determined in Equation (a), and solve for the maximum eccentricity e:

$$30.26 \text{ MPa} = (40{,}000 \text{ N})\left[\frac{1}{4{,}614.2 \text{ mm}^2} + \frac{(65 \text{ mm})e}{8{,}053{,}246 \text{ mm}^4}\right]$$

$$\frac{30.26 \text{ N/mm}^2}{40{,}000 \text{ N}} - \frac{1}{4{,}614.2 \text{ mm}^2} = \left[\frac{65 \text{ mm}}{8{,}053{,}246 \text{ mm}^4}\right]e$$

$$\therefore e_{max} = 66.9 \text{ mm}$$ **Ans.**

(b) Maximum eccentricity based on the interaction method: To determine the maximum eccentricity e, the interaction equation for axial and bending stresses is expressed as

$$\frac{P}{A(\sigma_{allow})_a} + \frac{Pec}{I(\sigma_{allow})_b} = P\left[\frac{1}{A(\sigma_{allow})_a} + \frac{ec}{I(\sigma_{allow})_b}\right] = 1 \qquad \text{(c)}$$

From Equation (a), the allowable compression stress was found to be 30.26 MPa therefore, $(\sigma_{allow})_a = 30.26$ MPa. The allowable bending stress was specified as $(\sigma_{allow})_b = 150$ MPa. The maximum allowable eccentricity e_{max} based on the interaction method can be computed with these values, along with $P = 40$ kN:

$$P\left[\frac{1}{A(\sigma_{allow})_a} + \frac{ec}{I(\sigma_{allow})_b}\right] = 1$$

$$(40{,}000 \text{ N})\left[\frac{1}{(4{,}614.2 \text{ mm}^2)(30.26 \text{ N/mm}^2)} + \frac{(65 \text{ mm})e}{(8{,}053{,}246 \text{ mm}^4)(150 \text{ N/mm}^2)}\right] = 1$$

$$\left[\frac{(65 \text{ mm})}{(8{,}053{,}246 \text{ mm}^4)(150 \text{ N/mm}^2)}\right]e = \frac{1}{40{,}000 \text{ N}} - \frac{1}{(4{,}614.2 \text{ mm}^2)(30.26 \text{ N/mm}^2)}$$

$$\therefore e_{max} = 332 \text{ mm}$$ **Ans.**

Since the effective-slenderness ratio of the tube is relatively large, the allowable compression stress computed in Equation (a) is relatively small. Because the allowable-stress method depends entirely on this allowable stress, the 66.9 mm maximum eccentricity that it arrived at is very conservative. In the interaction method, only the axial stress term (i.e., P/A) is directly affected by the small allowable compression stress. The bending stress component, which is a significant portion of the total stress, is divided by the 150 MPa allowable bending stress. Therefore, the maximum eccentricity determined from the interaction method is much larger than the eccentricity found from the allowable-stress method.

PROBLEMS

P16.42 The structural steel column shown in Figure P16.42/43 is fixed at its base and free at its upper end. At the top of the column, a load P is applied to the stiffened seat support at an eccentricity $e = 9$ in. from the centroidal axis of the wide-flange shape. Use the AISC equations given in Section 16.5, and assume that $E = 29{,}000$ ksi and $\sigma_Y = 36$ ksi. Employ the allowable-stress method to determine

(a) whether the column is safe for a load $P = 15$ kips. Report the results in the form of the stress ratio σ_x/σ_{allow}.
(b) the magnitude of the largest eccentric load P that may be applied to the column.

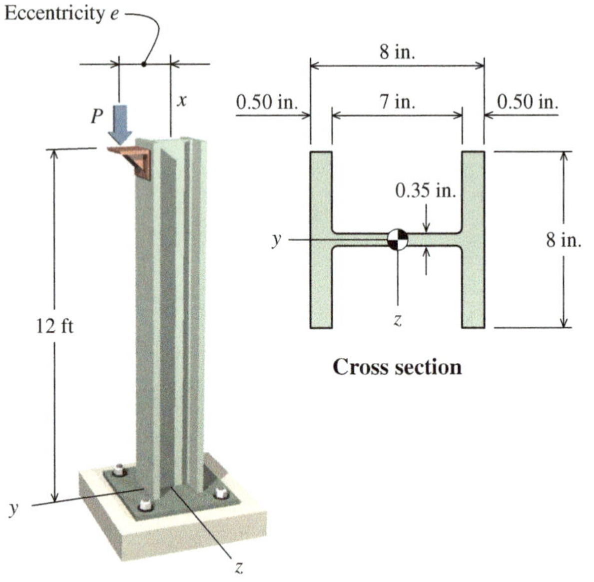

FIGURE P16.42/43

P16.43 The structural steel column shown in Figure P16.42/43 is fixed at its base and free at its upper end. At the top of the column, a load P is applied to the stiffened seat support at an eccentricity e from the centroidal axis of the wide-flange shape. Apply the AISC equations given in Section 16.5, and assume that $E = 29{,}000$ ksi and $\sigma_Y = 50$ ksi. Using the allowable-stress method, determine the maximum allowable eccentricity e if

(a) $P = 15$ kips.
(b) $P = 35$ kips.

P16.44 A W200 × 46.1 structural steel shape (see Appendix B for cross-sectional properties) is used as a column to support an eccentric axial load P. The column is 3.6 m long and is fixed at its base and free at its upper end. At the upper end of the column (see Figure P16.44), the load P is applied to a bracket at a distance $e = 170$ mm from the x axis, creating a bending moment about the strong axis (i.e., the z axis) of the W200 × 46.1 shape. Apply the AISC equations given in Section 16.5, and assume that $E = 200$ GPa and $\sigma_Y = 250$ MPa. On the basis of the allowable-stress method,

(a) determine whether the column is safe for a load $P = 125$ kN. Report the results in the form of the stress ratio σ_x/σ_{allow}.
(b) determine the magnitude of the largest eccentric load P that may be applied to the column.

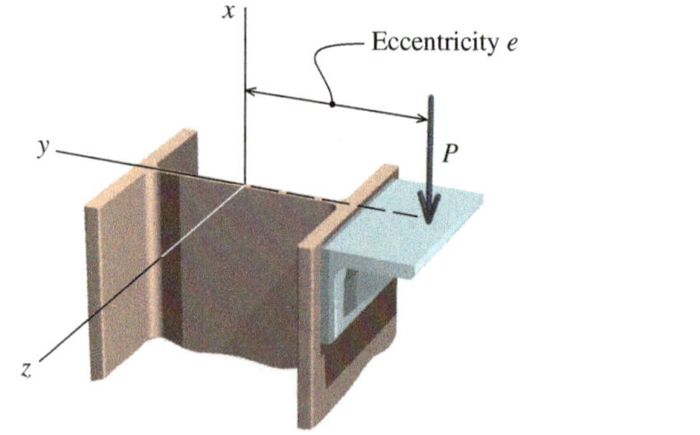

FIGURE P16.44

P16.45 An eccentric compression load $P = 32$ kN is applied at an eccentricity $e = 12$ mm from the centerline of a solid 45 mm diameter 6061-T6 aluminum-alloy rod. (See Figure P16.45/46.) Employing the interaction method and an allowable bending stress of 150 MPa, determine the longest effective length L that can be used.

P16.46 An eccentric compressive load $P = 13$ kips is applied at an eccentricity $e = 0.75$ in. from the centerline of a solid 6061-T6 aluminum-alloy rod. (See Figure P16.45/46.) The rod has an effective length of 45 in. Employing the interaction method and an allowable bending stress of 21 ksi, determine the smallest diameter that can be used.

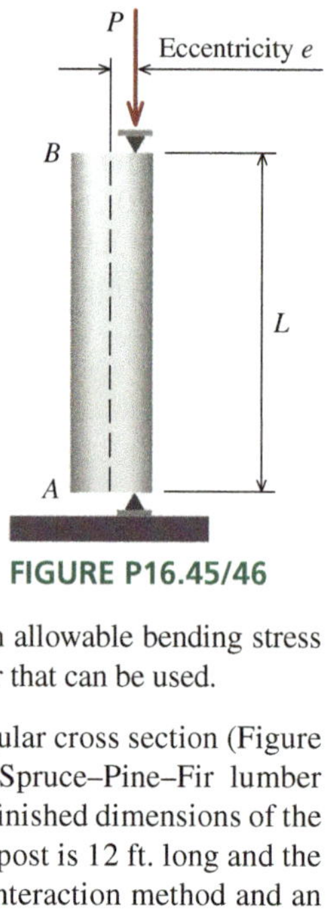

FIGURE P16.45/46

P16.47 A sawn wood post of rectangular cross section (Figure P16.47) consists of Select Structural Spruce–Pine–Fir lumber ($F_c = 700$ psi; $E'_{min} = 440{,}000$ psi). The finished dimensions of the post are $b = 5.5$ in. and $d = 7.25$ in. The post is 12 ft. long and the ends of the post are pinned. Using the interaction method and an allowable bending stress of 1,000 psi, determine the maximum allowable load that can be supported by the post if the load P acts at an eccentricity of $e = 6$ in. from the centerline of the post. Use the AWC NDS column design formula.

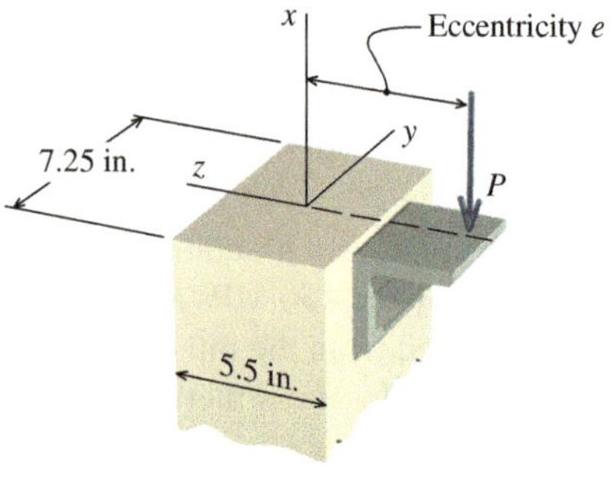

FIGURE P16.47

CHAPTER 17

Energy Methods

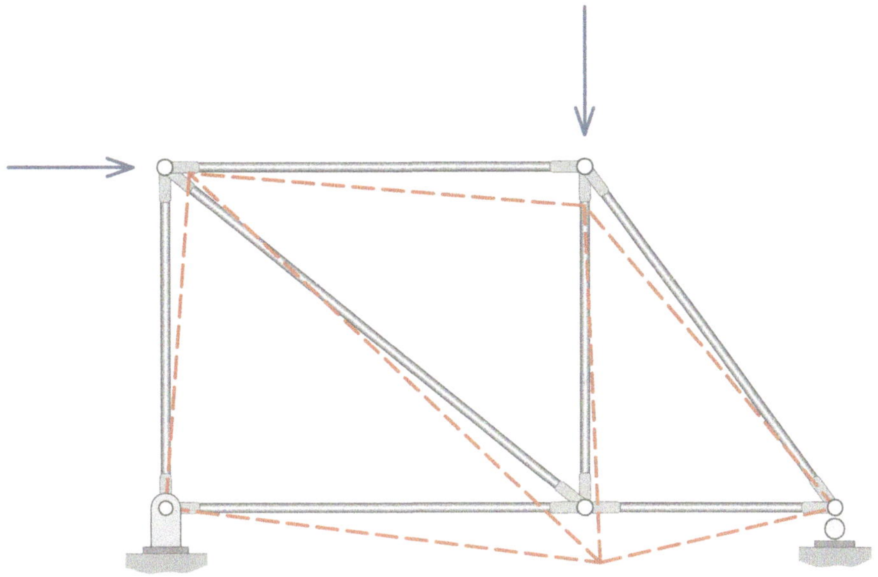

17.1 Introduction

When a solid body deforms as a consequence of applied loads, work is done on the body by these loads. Since the applied loads are external to the body, this work is called **external work**. As deformation occurs in the body, **internal work**, commonly referred to as **strain energy**, is stored within the body as potential energy. If the proportional limit of the material is not exceeded (i.e., if the material remains elastic), then no energy dissipation occurs and all strain energy is completely recoverable. For this situation, the principle of **conservation of energy** can be stated as follows: The work performed on an elastic body in static equilibrium by external forces is equal to the strain energy stored in the body.

From this principle, internal deformations in a body can be related to the external loads acting on the body. Energies related to axial, bending, torsional, and shear loadings will be considered next.

The load–deformation relationships that will be presented here are based on energy principles and will be limited to linearly elastic systems (although these energy principles are applicable to any conservative system).[1] These relationships make possible the application of powerful methods to the analysis of elastic bodies, particularly with regard to statically indeterminate structures, trusses, frames, and beams. Energy methods are also quite useful in investigating the effects of dynamic loads on solid bodies.

---

[1] A conservative system is a system in which the work done in moving a particle between two points is independent of the path taken and the final energy of the system is equal to the initial energy of the system.

The total energy of a system in static equilibrium and subjected to any combination of loads is the sum of the strain energies stored in the system as a result of each type of load. Consequently, energy methods make it possible to readily determine the total deformation of a solid body subjected to multiple loads—a situation frequently encountered in engineering applications.

17.2 Work and Strain Energy

Work of a Force

Work W is defined as the product of a force that acts on a particle (often, in a body) and the distance the particle (or body) moves in the direction of the force. For example, Figure 17.1 shows two forces acting on a body. As the body moves from initial position (a) to displaced position (b), a particle in it moves from location A to location A', a distance d_1, and another particle in it moves from location B to location B', a distance d_2.

Even though the particle acted on by force F_1 has moved a total distance d_1, the work done by this force is simply $W_1 = F_1 s_1$ because the work done by a force is defined as the product of the force and the distance that the particle it acts on has moved *in the direction that the force acts.* Similarly, the work done by force F_2 is $W_2 = F_2 s_2$. Work can be either a positive or a negative quantity. Positive work occurs when the particle moves in the same direction that the force acts. Negative work occurs when the particle moves in a direction opposite that in which the force acts. In Figure 17.1, the work done by forces F_1 and F_2 is positive if the body moves from position (a) to position (b). The work done by forces F_1 and F_2 is negative if the body moves from position (b) to position (a).

Next, consider a prismatic bar of length L that is subjected to a constant external load P as shown in Figure 17.2. The load will be applied to the bar very slowly—increasing from zero to its maximum value P—so that any dynamic or inertial effects due to motion are precluded. As the load is applied, the bar gradually elongates. The bar attains its maximum deformation δ when the full magnitude of P is reached. Thereafter, both the load and the deformation remain unchanged.

The work done by the load is the product of the magnitude of the force and the distance that the particle (or body) the force acts on moves; however, in this instance the force changes its magnitude from zero to its final value P. As a result, the work done by the load as the bar elongates is dependent on the manner in which the force and the corresponding

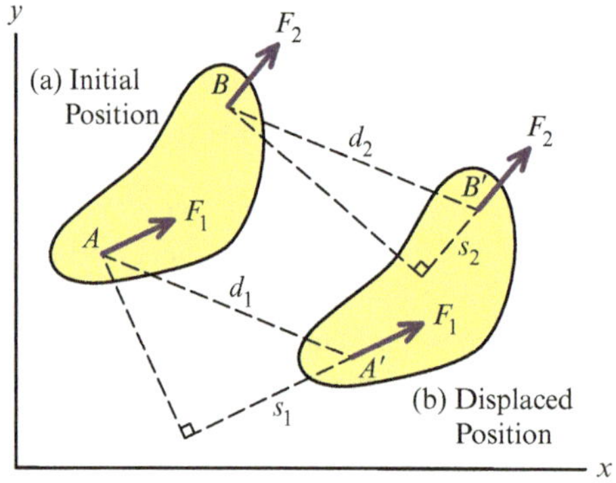

FIGURE 17.1 Forces acting on a body that changes position.

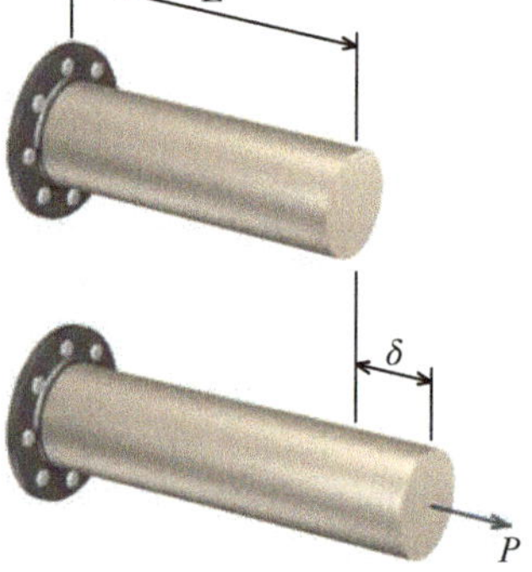

FIGURE 17.2 Prismatic bar with static load P.

deformation vary. This dependency is summarized in a **load–deformation diagram**, such as the one shown in Figure 17.3. The shape of the diagram depends upon the particular material being considered.

Now consider an arbitrary value of load P_1 between zero and the maximum value P. At the load P_1, the deformation of the bar is δ_1. From this state, an additional load increment dP will produce an increment of deformation $d\delta$. During this incremental deformation, the load P_1 will also move, and in so doing, it will perform work equal to $dW = P_1\, d\delta$. This work is shown in Figure 17.3 by the darkly shaded area beneath the load–deformation curve. The total work done by the load as it increases in magnitude from zero to P can be determined by summing together all such infinitely small increments:

$$W = \int_0^{\delta} P\, d\delta$$

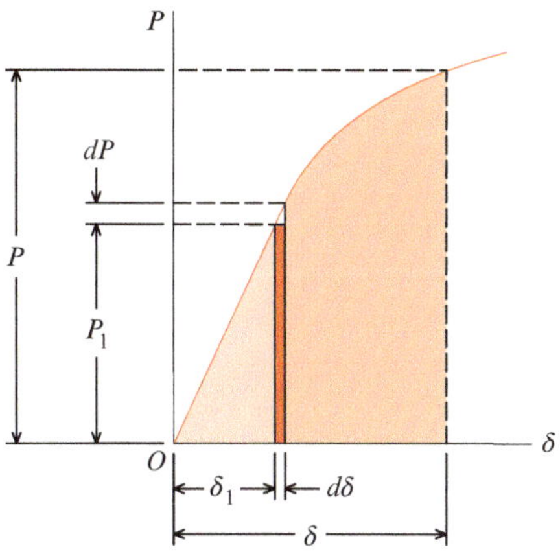

FIGURE 17.3 Load–deformation diagram.

The shaded area underneath the load–deformation curve represents the total work done. When the load–deformation diagram is linear (Figure 17.4), the work done by P is

$$W = \tfrac{1}{2} P\delta \qquad (17.1)$$

which is simply the area under the diagram.

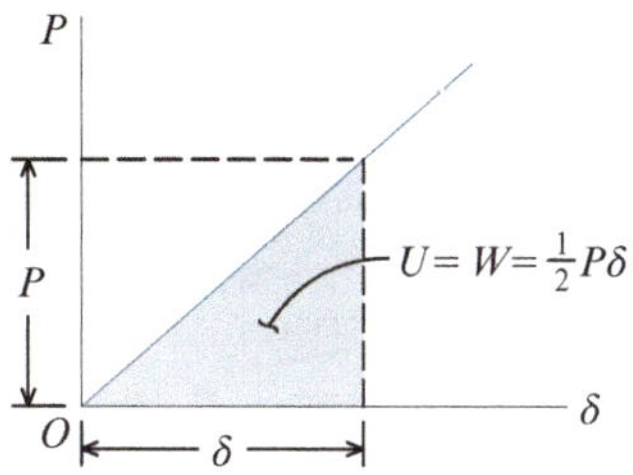

FIGURE 17.4 Linear load–deformation diagram.

Strain Energy

As external load P is applied to the bar, work is performed, and energy is expended. Since this work is performed by an external load, it is typically referred to as **external work**. The load causes the bar to deform, and in the process, it produces strains in the bar. The principle of conservation of energy asserts that energy in a closed system is never created or destroyed; rather, it is only transformed from one state to another. So, where does the energy expended by the work of the external load P go? The answer is that it is transformed into internal energy stored in the strains of the bar. The energy absorbed by the bar during the loading process is termed **strain energy**. In other words, strain energy is the energy that is stored in a material body as a consequence of the body's deformation. Provided that no energy is lost in the form of heat, the strain energy U is equal in magnitude to the external work W:

$$U = W = \int_0^{\delta_1} P\, d\delta \qquad (17.2)$$

Strain energy is sometimes referred to as **internal work** to distinguish it from the external work done by the load.

While external work may be either a positive or a negative quantity, strain energy is always a positive quantity.

An examination of Equation (17.2) reveals that work and energy are expressed in the same units—that is, the product of force and distance. In the SI, the unit of work and energy is the joule (J), which is equal to 1 N · m. In U.S. customary units, work and energy may be expressed in units of lb · ft., lb · in., kip · ft, or kip · in.

Because of its stored energy, the bar in Figure 17.2 is capable of doing work as it returns to its undeformed configuration after the load is removed. If the elastic limit is not exceeded, the bar will return to its original length. If the elastic limit is exceeded, as illustrated in Figure 17.5, a residual strain will remain after the load is removed. The total strain energy is always the area under the load–deformation curve (area *OABCDO*); however, only the elastic strain energy (the triangular area *BCD*) can be recovered. The other portion of the area under the curve (area *OABDO*) represents the strain energy that is spent in permanently deforming the material. This energy dissipates in the form of heat.

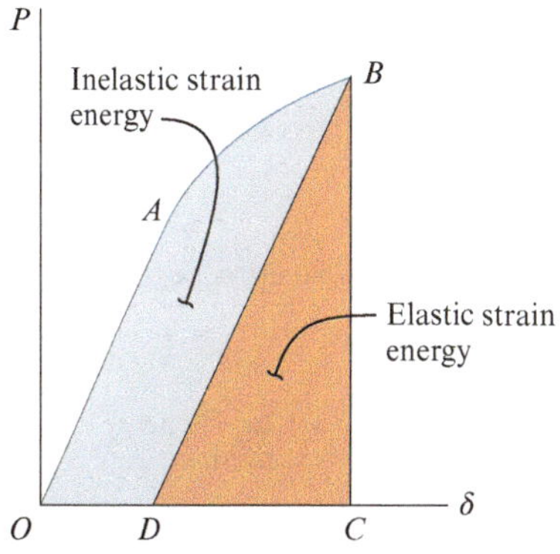

FIGURE 17.5 Elastic and inelastic strain energy.

Strain-Energy Density for Uniaxial Normal Stress

Axial deformation of an elastic bar affords a simple introduction to the concept of strain energy. For more complex situations, it is sometimes necessary to consider how stored strain energy is distributed throughout a deformed body. In such cases, a quantity called **strain-energy density** is convenient. Strain-energy density is defined as the strain energy per unit volume of material.

Consider a small volume element dV in a linearly elastic bar subjected to an axial load, as shown in Figure 17.6. The force acting on each x face of this element is $dF_x = \sigma_x\, dy\, dz$. If this force is applied gradually, like the force P shown in Figure 17.2, then the force on the element increases from zero to dF_x while the element elongates by the amount $d\delta_x = \varepsilon_x\, dx$. By Equation (17.1), the work done by dF_x can be expressed as

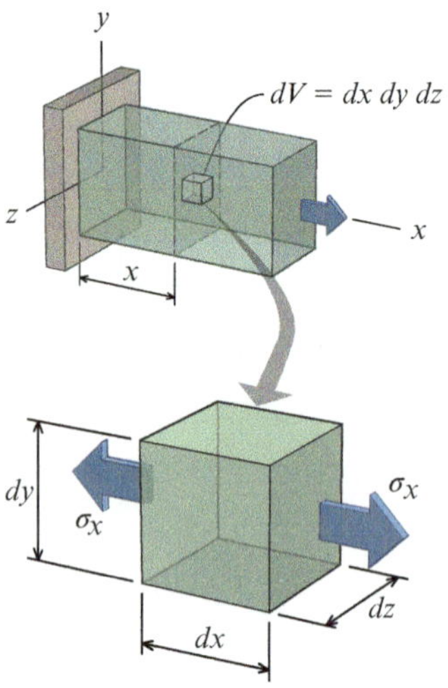

FIGURE 17.6 Volume element in uniaxial tension.

$$dW = \tfrac{1}{2}(\sigma_x\, dy\, dz)\varepsilon_x\, dx$$

Furthermore, by conservation of energy, the strain energy stored in the volume element must equal the external work:

$$dU = dW = \tfrac{1}{2}(\sigma_x\, dy\, dz)\varepsilon_x\, dx$$

The volume of the element is $dV = dx\, dy\, dz$. Thus, the strain energy in the volume element can be expressed as

$$dU = \tfrac{1}{2}\sigma_x \varepsilon_x\, dV$$

Note that strain energy must be a positive quantity because the normal stress and the normal strain always act in the same sense, either both positive (i.e., tension and elongation) or both negative (i.e., compression and contraction).

The strain-energy density u can be determined by dividing the strain energy dU by the volume dV of the element:

$$u = \frac{dU}{dV} = \tfrac{1}{2}\sigma_x \varepsilon_x \tag{17.3}$$

If the material is linearly elastic, then $\sigma_x = E\varepsilon_x$ and the strain-energy density can be expressed solely in terms of stress as

$$u = \frac{\sigma_x^2}{2E} \tag{17.4}$$

or strain as

$$u = \frac{E\varepsilon_x^2}{2} \tag{17.5}$$

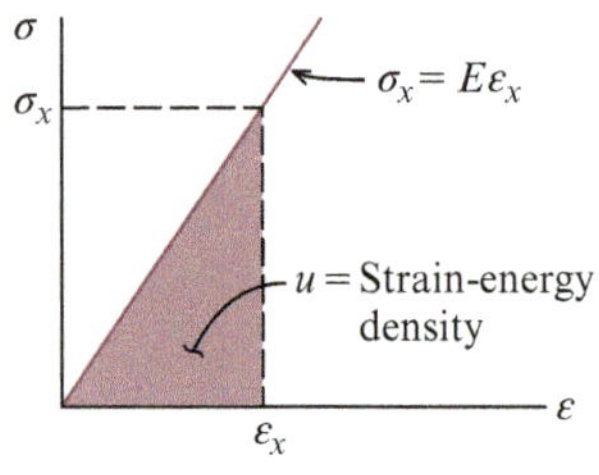

FIGURE 17.7 Strain-energy density for elastic materials.

Equations (17.4) and (17.5) have a straightforward geometric interpretation: Both are equal to the triangular area below the stress–strain curve for a linear elastic material (Figure 17.7). For materials that are not linearly elastic, the strain-energy density is still equal to the area under the stress–strain curve; however, the area under the curve must be evaluated by numerical or other methods.

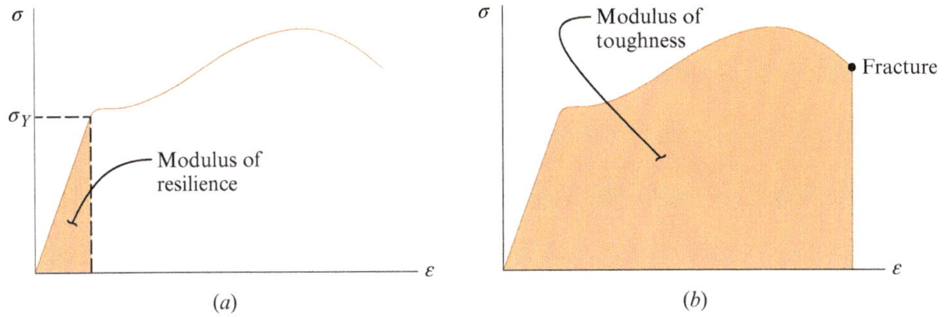

FIGURE 17.8 Geometric interpretation of (*a*) modulus of resilience and (*b*) modulus of toughness.

Strain-energy density has units of energy per unit volume. In the SI, an appropriate unit for strain-energy density is joules per cubic meter (J/m^3). In the U.S. customary system, units of $lb \cdot ft/ft^3$ or $lb \cdot in./in.^3$ are suitable. However, notice that all of these units reduce to stress units; therefore, strain-energy density can also be expressed in pascals (Pa) or pounds per square inch (psi).

The area under the straight-line portion of the stress–strain curve (Figure 17.8*a*), evaluated from zero to the proportional limit, represents a material property known as the **modulus of resilience**: the maximum strain-energy density that a material can store or absorb without exhibiting permanent deformations. In practice, the yield stress σ_Y, rather than the proportional limit, is generally used to determine the modulus of resilience.

The area under the entire stress–strain curve from zero to fracture (Figure 17.8*b*) gives a property known as the **modulus of toughness**. This modulus denotes the strain-energy density necessary to rupture the material. From the figure, it is evident that the modulus of toughness depends greatly on both the strength and the ductility of the material. A high modulus of toughness is particularly important when materials are subjected to dynamic or impact loads.

The total strain energy associated with uniaxial normal stress can be found by integrating the strain-energy density [Equation (17.4)] over the volume of the member:

$$U = \int_V \frac{\sigma_x^2}{2E}\, dV \tag{17.6}$$

Equation (17.6) can be used to determine strain energy for both axially loaded bars and beams in pure bending.

Strain-Energy Density for Shear Stress

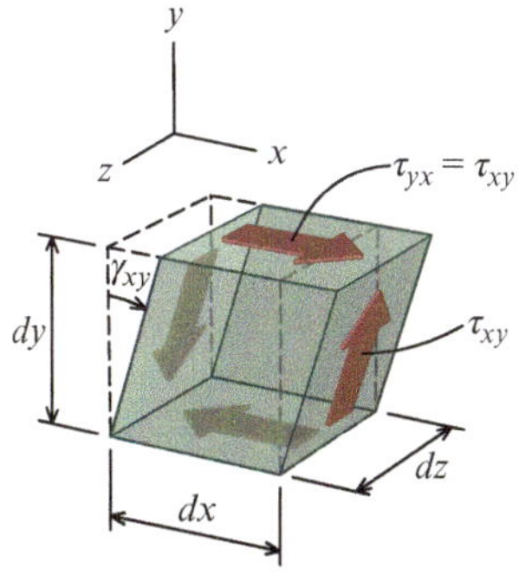

FIGURE 17.9 Volume element subjected to pure shear stress $\tau_{xy} = \tau_{yx}$.

Next, consider an elemental volume dV subjected to a shear stress $\tau_{xy} = \tau_{yx}$ (Figure 17.9). Notice that the shear stress on the upper face displaces the upper face of the element relative to the lower face. The vertical faces of the element do not displace relative to each other—they only rotate. Therefore, only the shear force acting on the upper face performs work as the element deforms. The shear force acting on the *y* face is $dF = \tau_{xy}\, dx\, dz$, and this force displaces through a horizontal distance $\gamma_{xy}\, dy$ relative to the bottom face. The work done by dF—and hence, the strain energy stored by the element—is

$$dU = \tfrac{1}{2}(\tau_{xy}\, dx\, dz)\gamma_{xy}\, dy$$

Since the volume of the element is $dV = dx\,dy\,dz$, the strain-energy density in pure shear is

$$u = \tfrac{1}{2}\tau_{xy}\gamma_{xy} \tag{17.7}$$

For a linearly elastic material, $\tau_{xy} = G\gamma_{xy}$ and the strain-energy density can then be expressed solely in terms of stress as

$$u = \frac{\tau_{xy}^2}{2G} \tag{17.8}$$

or strain as

$$u = \frac{G\gamma_{xy}^2}{2} \tag{17.9}$$

The total strain energy associated with shear stress can be found by integrating the strain-energy density [Equation (17.8)] over the volume of the member:

$$U = \int_V \frac{\tau_{xy}^2}{2G}\,dV \tag{17.10}$$

Equation (17.10) can be used to determine the strain energy for bars in torsion, as well as the strain energy associated with transverse shear stress in beams.

Although Equations (17.3) and (17.7) were derived for σ_x, ε_x, τ_{xy}, and γ_{xy}, additional strain-energy density expressions can be derived for the remaining stress components in a similar manner. The general expression for the strain-energy density of a linearly elastic body is

$$u = \frac{1}{2}[\sigma_x\varepsilon_x + \sigma_y\varepsilon_y + \sigma_z\varepsilon_z + \tau_{xy}\gamma_{xy} + \tau_{yz}\gamma_{yz} + \tau_{zx}\gamma_{zx}] \tag{17.11}$$

17.3 Elastic Strain Energy for Axial Deformation

The concept of strain energy was introduced in the previous section by considering the work done by a slowly applied axial load P in elongating a prismatic bar by an amount δ. If the load–deformation diagram is linear (Figure 17.4), then the external work W done in elongating the bar is

$$W = \tfrac{1}{2}P\delta$$

and since the strain energy stored in the bar must equal the external work, the strain energy U in the bar is given by

$$U = \tfrac{1}{2}P\delta$$

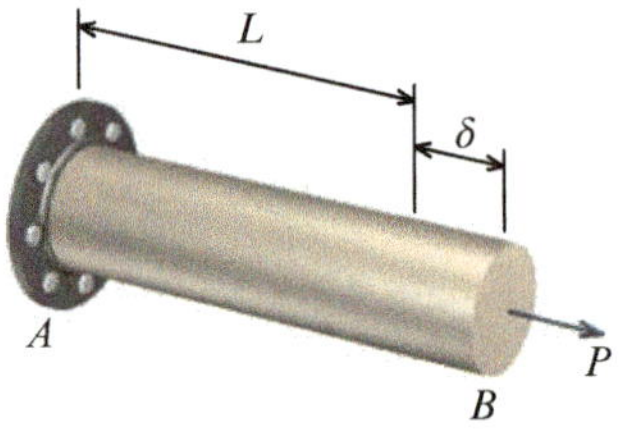

FIGURE 17.10 Prismatic bar with constant axial load P.

The prismatic bar shown in Figure 17.10 has a constant cross-sectional area A and modulus of elasticity E. When the load magnitude is such that the axial stress does not exceed the proportional limit for the material, the deformation of the bar is given by $\delta = PL/AE$. Consequently, the elastic strain energy of the bar can be expressed in terms of the force P as

$$U = \frac{P^2L}{2AE} \tag{17.12}$$

or in terms of the deformation δ as

$$U = \frac{AE\delta^2}{2L} \tag{17.13}$$

The total strain energy of a bar that consists of several segments (each having constant force, area, and elastic modulus) is equal to the sum of the strain energies in all the segments. For example, the strain energy in the multisegment bar shown in Figure 17.11 is equal to the sum of the strain energies in segment *AB* and segment *BC*. In general terms, the strain energy of a bar with *n* segments can be expressed as

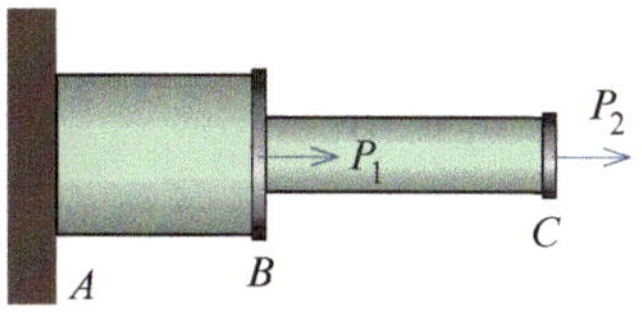

FIGURE 17.11 Bar with multiple prismatic segments.

$$U = \sum_{i=1}^{n} \frac{F_i^2 L_i}{2A_i E_i} \tag{17.14}$$

where F_i is the internal force in segment i and L_i, A_i, and E_i are, respectively, the length, area, and elastic modulus of segment i.

For a nonprismatic bar having a slightly tapered, variable cross section and a continuously varying axial force (Figure 17.12), then the total strain energy can be derived by integrating the strain energy in a differential element *dx* over the total length of the bar:

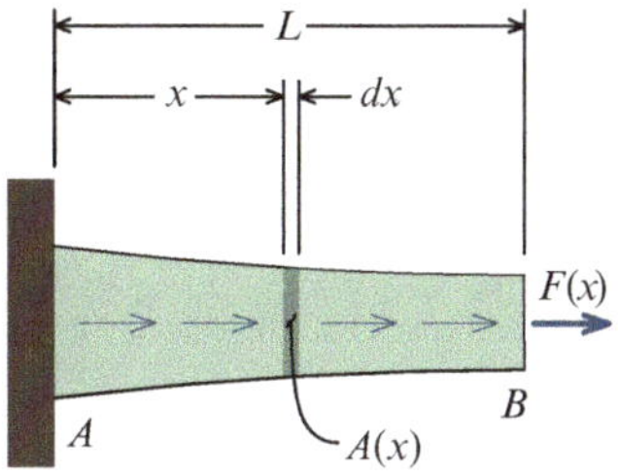

FIGURE 17.12 Nonprismatic bar subjected to varying axial loading.

$$U = \int_0^L \frac{[F(x)]^2}{2A(x)E}\,dx \tag{17.15}$$

Here, $F(x)$ is the internal force and $A(x)$ is the cross-sectional area at a distance x from the origin of the bar.

EXAMPLE 17.1

Segmented rod *ABC* is made of a brass that has a yield strength $\sigma_Y = 124$ MPa and a modulus of elasticity $E = 115$ GPa. The diameter of segment (1) is 25 mm, and the diameter of segment (2) is 15 mm. For the loading shown, determine the maximum strain energy that can be absorbed by the rod if no permanent deformation is caused.

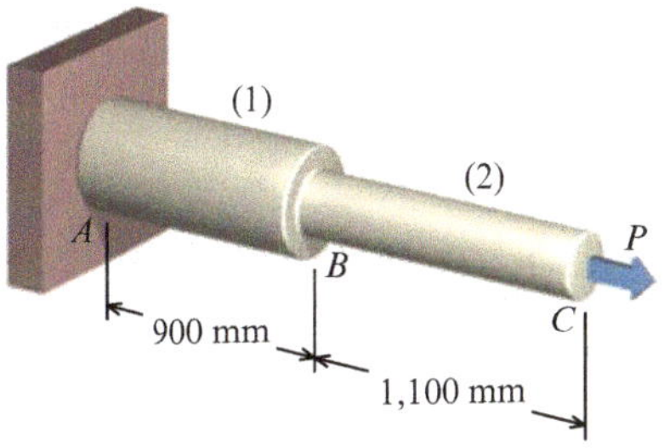

Plan the Solution

The maximum force that can be applied to the segmented rod will be dictated by the capacity of segment (2). From the yield strength and the cross-sectional area of segment (2), determine the maximum force *P*. The internal force in each segment will equal the external load. The strain energy in each segment can be calculated from the internal force along with the length, area, and elastic modulus of each segment. The total strain energy *U* is simply the sum of the strain energies in segments (1) and (2), as indicated by Equation (17.14).

SOLUTION

Compute the cross-sectional areas of segments (1) and (2):

$$A_1 = \frac{\pi}{4}d_1^2 = \frac{\pi}{4}(25\text{ mm})^2 = 490.874\text{ mm}^2$$

$$A_2 = \frac{\pi}{4}d_2^2 = \frac{\pi}{4}(15\text{ mm})^2 = 176.715\text{ mm}^2$$

The maximum force P that can be applied without causing any permanent deformation will be controlled by the smaller of these two areas. Therefore, P is calculated from the yield strength σ_Y and A_2:

$$P = \sigma_Y A_1 = (124 \text{ N/mm}^2)(176.715 \text{ mm})^2 = 21{,}912.61 \text{ N}$$

Now use Equation (17.14) to calculate the strain energy of each segment, as well as the total strain energy in the brass rod:

$$\begin{aligned} U &= \sum_{i=1}^{n} \frac{F_i^2 L_i}{2A_i E_i} = \frac{F_1^2 L_1}{2A_1 E_1} + \frac{F_2^2 L_2}{2A_2 E_2} \\ &= \frac{(21{,}912.61 \text{ N})^2 (900 \text{ mm})}{2(490.874 \text{ mm}^2)(115{,}000 \text{ N/mm}^2)} + \frac{(21{,}912.61 \text{ N})^2 (1{,}100 \text{ mm})}{2(176.715 \text{ mm}^2)(115{,}000 \text{ N/mm}^2)} \\ &= 3{,}827.7 \text{ N}\cdot\text{mm} + 12{,}995.1 \text{ N}\cdot\text{mm} \\ &= 16.82 \text{ N}\cdot\text{mm} = 16.82 \text{ J} \end{aligned}$$

Ans.

17.4 Elastic Strain Energy for Torsional Deformation

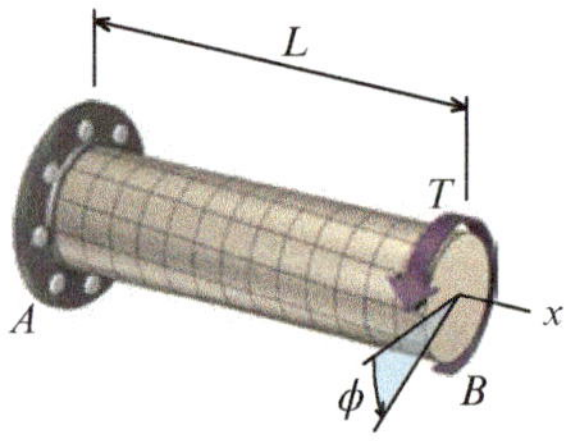

FIGURE 17.13 Prismatic shaft in pure torsion.

Consider a circular prismatic shaft of length L, and suppose the shaft is subjected to a torque T as shown in Figure 17.13. If the torque is applied gradually, the free end B of the shaft rotates through an angle ϕ. If the bar is linearly elastic, the relationship between the torque T and the rotation angle of the shaft will also be linear, as shown in the torque–rotation diagram of Figure 17.14 and as given by the torque–twist relationship $\phi = TL/JG$, where J is the polar moment of inertia of the cross-sectional area. The external work W done by the torque as it rotates through the angle ϕ is equal to the area of the shaded triangle. From the principle of conservation of energy, and with no dissipation of energy in the form of heat, the strain energy of the circular shaft is thus

$$U = W = \tfrac{1}{2}T\phi$$

From the torque–twist relationship $\phi = TL/JG$, the strain energy in the shaft can be expressed in terms of the torque T as

$$U = \frac{T^2 L}{2JG} \tag{17.16}$$

or in terms of the rotation angle ϕ as

$$U = \frac{JG\phi^2}{2L} \tag{17.17}$$

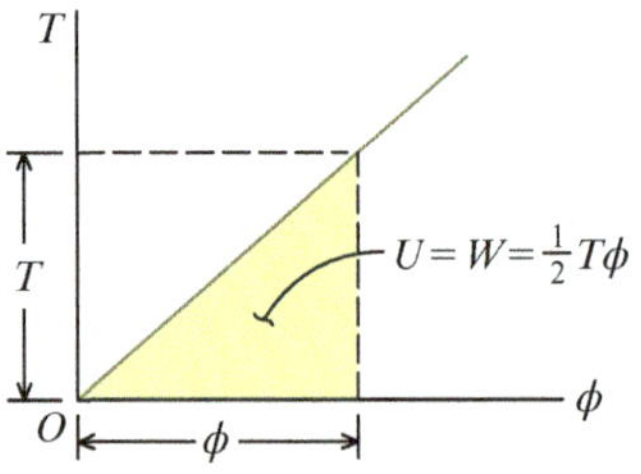

FIGURE 17.14 Torque–rotation diagram for a linearly elastic material.

Notice the parallels in form between Equations (17.12) and (17.13), on the one hand, which express the strain energy in a prismatic bar with a constant axial load, and Equations (17.16) and (17.17), on the other, which give the strain energy for a prismatic shaft with a constant torque.

The total strain energy of a shaft that consists of several segments (each having a constant torque, polar moment of inertia, and shear modulus) is equal to the sum of the strain energies in all the segments. The strain energy of a shaft with n segments can be expressed as

$$U = \sum_{i=1}^{n} \frac{T_i^2 L_i}{2J_i G_i} \tag{17.18}$$

where T_i is the internal force in segment i and L_i, J_i, and G_i are, respectively, the length, polar moment of inertia, and shear modulus of segment i.

For a nonprismatic shaft having a slightly tapered, variable cross section and a continuously varying internal torque, the strain energy can be derived by integrating the strain energy in a differential element dx over the total length of the shaft:

$$U = \int_0^L \frac{[T(x)]^2}{2J(x)G} dx \tag{17.19}$$

Here, $T(x)$ is the internal torque and $J(x)$ is the polar moment of inertia of the cross-sectional area at a distance x from the origin of the shaft.

EXAMPLE 17.2

Three identical shafts of identical torsional rigidity JG and length L are subjected to torques T as shown. What is the elastic strain energy stored in each shaft?

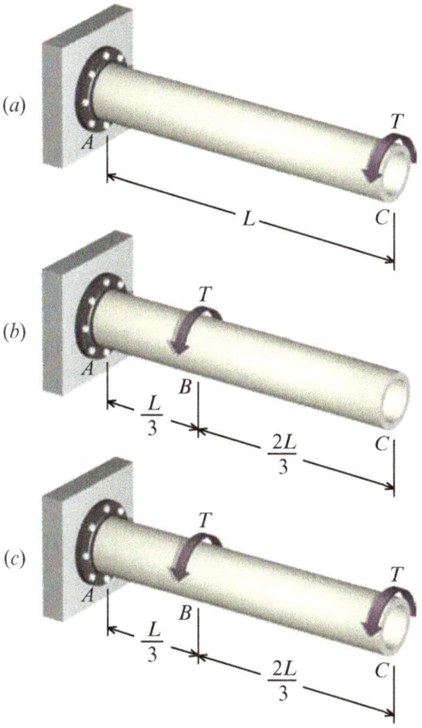

Plan the Solution

The elastic strain energy for cases (a) and (b) can be determined from Equation (17.16). The strain energy for case (c) can be found from Equation (17.18).

SOLUTION

From Equation (17.16), the strain energy for case (a) is

$$U_a = \frac{T^2 L}{2JG} \qquad \textbf{Ans.}$$

In case (b), strain energy is created in only one-third of the shaft, from A to B:

$$U_b = \frac{T^2(L/3)}{2JG} = \frac{T^2 L}{6JG} \qquad \textbf{Ans.}$$

In case (c), the internal torque in segment AB is $2T$ and the internal torque in segment BC is T. From Equation (17.18), the total strain energy in shaft ABC is then

$$U_c = \frac{(2T)^2(L/3)}{2JG} + \frac{T^2(2L/3)}{2JG} = \frac{2T^2 L}{3JG} + \frac{T^2 L}{3JG}$$

$$= \frac{T^2 L}{JG} \qquad \textbf{Ans.}$$

Notice that the sum of the strain energies for cases (a) and (b) does not equal the strain energy for case (c); that is, $U_c \neq U_a + U_b$. The torque term in Equations (17.16) and (17.18) is squared; thus, superposition is not valid for strain energies.

17.5 Elastic Strain Energy for Flexural Deformation

Consider an arbitrary axisymmetric prismatic beam such as the one depicted in Figure 17.15*a*. As the external load P acting on the beam is gradually intensified from zero to its maximum value, the internal bending moment M acting on a differential element dx increases steadily from zero to its final value. In response to the bending moment M, the sides of the differential element rotate by an angle $d\theta$ with respect to each other, as shown in Figure 17.15*b*. If the beam is linearly elastic, the relationship between the bending moment M and the rotation angle $d\theta$ of the beam will also be linear, as shown in the moment–rotation angle diagram of Figure 17.16. Therefore, the internal work and, hence, the strain energy stored in the differential element dx is

$$dU = \tfrac{1}{2} M d\theta$$

The following equation relates a beam's bending moment to its rotation angle:

$$\frac{d\theta}{dx} = \frac{M}{EI}$$

From this equation, the strain energy stored in the differential element dx can be stated as

$$dU = \frac{M^2}{2EI} dx$$

The strain energy in the entire beam is found by integrating over the length L of the beam:

$$U = \int_0^L \frac{M^2}{2EI} dx \qquad (17.20)$$

Note that the bending moment M may vary as a function of x.

When the quantity M/EI is not a continuous function of x over the entire length of the beam, the beam must be subdivided into segments in which M/EI is continuous. The integral on the right-hand side of Equation (17.20) is then evaluated as the sum of the integrals for all of the segments.

In the derivation of Equation (17.20), only the effects of bending moments were considered in evaluating the strain energy in a beam. Transverse shear forces, however, are also present in beams subjected to nonuniform bending, and these shear forces, too, will increase the strain energy stored in the beam. Fortunately, the strain energy associated with shear deformations is negligibly small compared with the strain energy of flexure for most ordinary beams, and consequently, it may be disregarded.

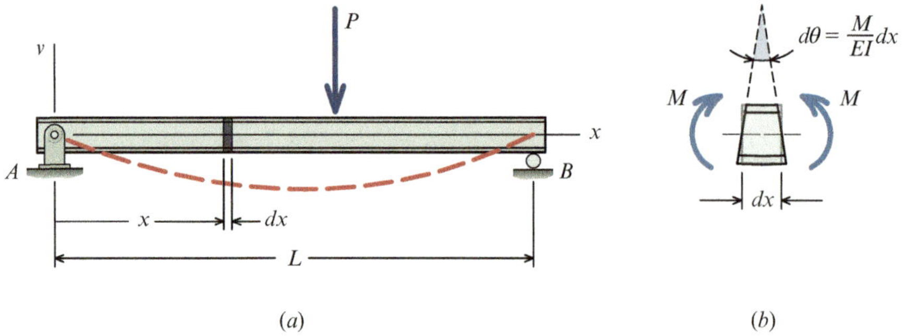

FIGURE 17.15 (*a*) Arbitrary prismatic beam. (*b*) Bending moments acting on a differential beam element.

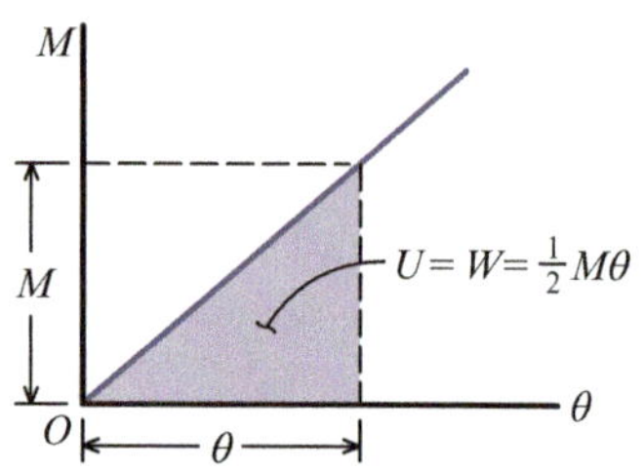

FIGURE 17.16 Moment–rotation angle diagram for a linearly elastic material.

EXAMPLE 17.3

A cantilever beam AB of length L and flexural rigidity EI supports the linearly distributed loading shown. Determine the elastic strain energy due to bending stored in this beam.

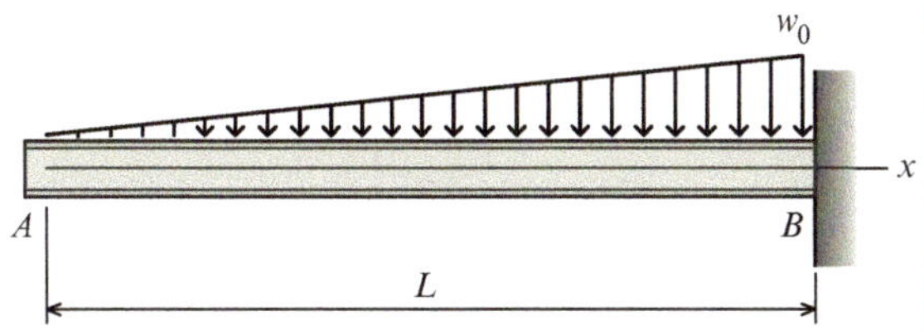

Plan the Solution

Consider a free-body diagram that cuts through the beam at a distance x from the free end of the cantilever. Derive the bending-moment equation $M(x)$, and then use it in Equation (17.20) to determine the elastic strain energy.

SOLUTION

Draw a free-body diagram that cuts through the beam at an arbitrary distance x from the origin. The equilibrium equation for the sum of moments about section a–a is

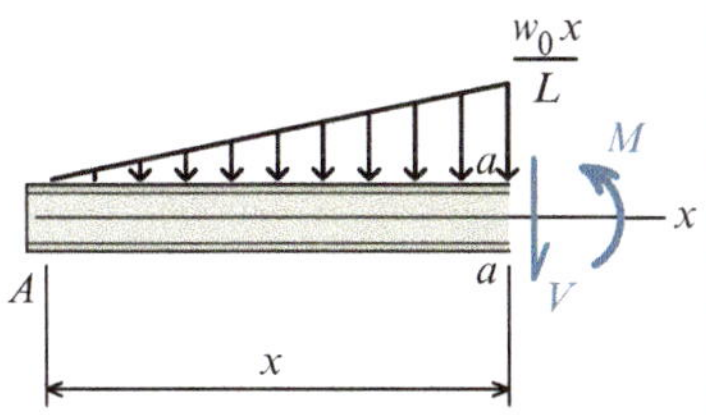

$$\sum M_{a-a} = \frac{w_0 x}{L}\left(\frac{x}{2}\right)\left(\frac{x}{3}\right) + M = 0$$

Therefore, the bending-moment equation for this beam is

$$M(x) = -\frac{w_0 x^3}{6L}$$

The elastic strain energy in a beam is given by Equation (17.20) as

$$U = \int_0^L \frac{M^2}{2EI}\,dx$$

For the cantilevered beam considered here,

$$U = \int_0^L \frac{1}{2EI}\left(-\frac{w_0 x^3}{6L}\right)^2 dx = \frac{w_0^2}{72EI\,L^2}\int_0^L x^6\,dx$$

or

$$U = \frac{w_0^2 L^7}{504EI\,L^2} = \frac{w_0^2 L^5}{504EI}$$ **Ans.**

EXAMPLE 17.4

A simply supported beam ABC of length L and flexural rigidity EI supports the concentrated load shown. What is the elastic strain energy due to bending that is stored in this beam?

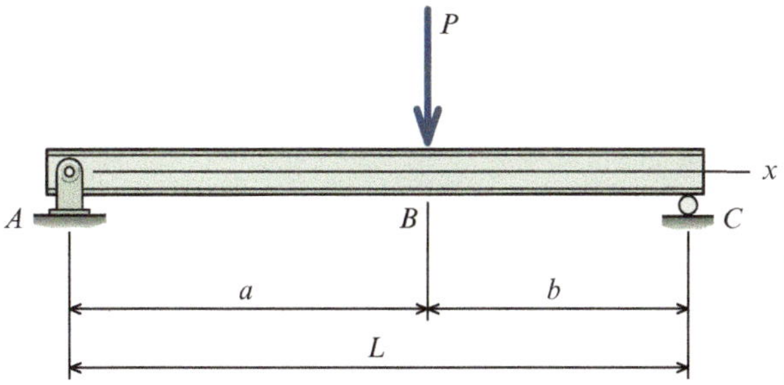

Plan the Solution

Determine the beam reactions from a free-body diagram of the entire beam. Then, consider two free-body diagrams that cut through the beam. The first free-body diagram cuts through the beam at a distance x from pin support A. From this diagram, derive the bending-moment equation $M(x)$ for segment AB of the beam. The second free-body diagram cuts through the beam at a distance x' from roller support C. From this diagram, derive the bending-moment equation $M(x')$ for segment BC of the beam. Substitute two moment expressions into Equation (17.20) to determine the elastic strain energy for the complete beam.

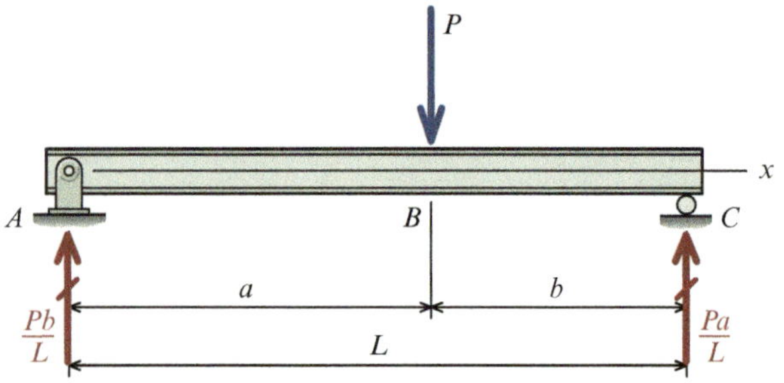

SOLUTION

From the free-body diagram for the entire beam, determine the vertical reaction force at A:

$$A_y = \frac{Pb}{L}$$

Also, determine the reaction force at C:

$$C_y = \frac{Pa}{L}$$

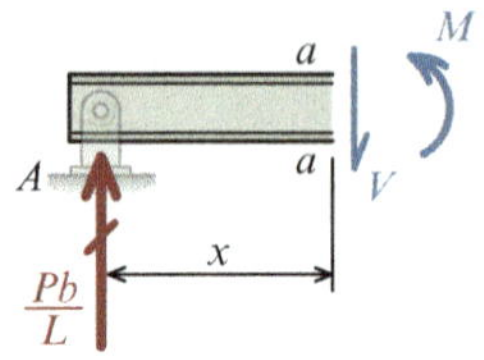

Note that the horizontal reaction force at A has been omitted, since $A_x = 0$.

Now draw a free-body diagram that cuts through the beam between A and B at a distance x from pin support A. Then, sum moments about section a–a to derive the bending-moment equation for segment AB of the beam:

$$\sum M_{a-a} = M - \frac{Pb}{L}x = 0$$

$$\therefore M = \frac{Pb}{L}x \qquad (0 \le x \le a)$$

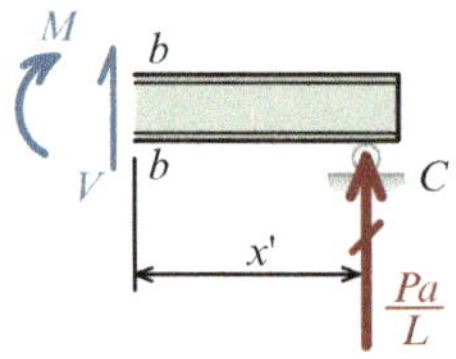

Similarly, draw a free-body diagram that cuts through the beam between B and C at a distance x' from roller support C. Then, sum moments about section b–b to derive the bending-moment equation for segment BC of the beam:

$$\sum M_{b-b} = -M + \frac{Pa}{L}x' = 0$$

$$\therefore M = \frac{Pa}{L}x' \qquad (0 \le x' \le b)$$

The total elastic strain energy in the beam is the sum of the elastic strain energies in segments AB and BC. By Equation (17.20),

$$\begin{aligned} U &= U_{AB} + U_{BC} \\ &= \frac{1}{2EI}\int_0^a \left(\frac{Pb}{L}x\right)^2 dx + \frac{1}{2EI}\int_0^b \left(\frac{Pa}{L}x'\right)^2 dx' \\ &= \frac{P^2b^2}{6L^2EI}a^3 + \frac{P^2a^2}{6L^2EI}b^3 \\ &= \frac{P^2a^2b^2}{6L^2EI}(a+b) \end{aligned}$$

or, since $a + b = L$,

$$U = \frac{P^2 a^2 b^2}{6LEI}$$ **Ans.**

This example demonstrates that the strain energy for a beam can be computed with *any* suitable x coordinate. For this beam, the bending moment equation for segment BC is much easier to derive and integrate if we consider a free-body diagram taken at the far end of the beam (around roller C).

PROBLEMS

P17.1 Determine the modulus of resilience for aluminum alloys with the following properties:

(a) 7075-T651 $E = 71.7$ GPa, $\sigma_Y = 503$ MPa
(b) 5082-H112 $E = 70.3$ GPa, $\sigma_Y = 190$ MPa
(c) 6262-T651 $E = 69.0$ GPa, $\sigma_Y = 241$ MPa

P17.2 Determine the modulus of resilience for each of the following metals:

(a) Red Brass UNS C23000 $E = 115$ GPa, $\sigma_Y = 125$ MPa
(b) Titanium Ti-6Al-4V (Grade 5) Annealed $E = 114$ GPa, $\sigma_Y = 830$ MPa
(c) 304 Stainless Steel $E = 193$ GPa, $\sigma_Y = 215$ MPa

P17.3 The compound solid steel rod shown in Figure P17.3/4 is subjected to a tensile force P. Assume that $E = 29{,}000$ ksi, $d_1 = 0.50$ in., $L_1 = 18$ in., $d_2 = 0.875$ in., $L_2 = 27$ in., and $P = 5.5$ kips. Determine

(a) the elastic strain energy in rod ABC.
(b) the strain-energy density in segments (1) and (2) of the rod.

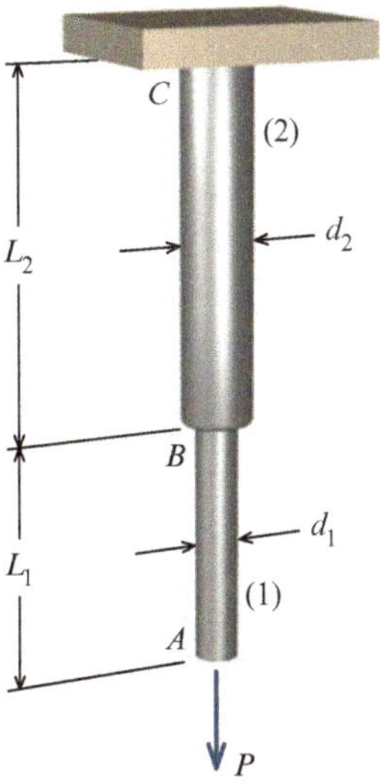

FIGURE P17.3/4

P17.4 The compound solid aluminum rod shown in Figure P17.3/4 is subjected to a tensile force P. Assume that $E = 69$ GPa, $d_1 = 16$ mm, $L_1 = 600$ mm, $d_2 = 25$ mm, $L_2 = 900$ mm, and $\sigma_Y = 276$ MPa. Calculate the largest amount of strain energy that can be stored in the rod without causing any yielding.

P17.5 A solid 2.5 m long stainless steel rod has a yield strength of 276 MPa and an elastic modulus of 193 GPa. A strain energy $U = 13$ N·m must be stored in the rod when a tensile load P is applied to rod. Calculate

(a) the maximum strain-energy density that can be stored in the solid rod if a factor of safety of 4.0 with respect to yielding is specified.
(b) the minimum diameter d required for the solid rod.

P17.6 A solid stepped shaft made of AISI 1020 cold-rolled steel ($G = 11{,}600$ ksi) is shown in Figure P17.6/7/8. The diameters of segments (1) and (2) are $d_1 = 2.25$ in. and $d_2 = 1.00$ in., respectively. The segment lengths are $L_1 = 36$ in. and $L_2 = 27$ in. Determine the elastic strain energy U stored in the shaft if the torque T_C produces a rotation angle of 4° at C.

P17.7 A solid stepped shaft made of AISI 1020 cold-rolled steel ($G = 80$ GPa) is shown in Figure P17.6/7/8. The diameters of segments (1) and (2) are $d_1 = 30$ mm and $d_2 = 15$ mm, respectively. The segment lengths are $L_1 = 320$ mm and $L_2 = 250$ mm. Determine the maximum torque T_C that can be applied to the shaft if the elastic strain energy must be limited to $U = 5.0$ J.

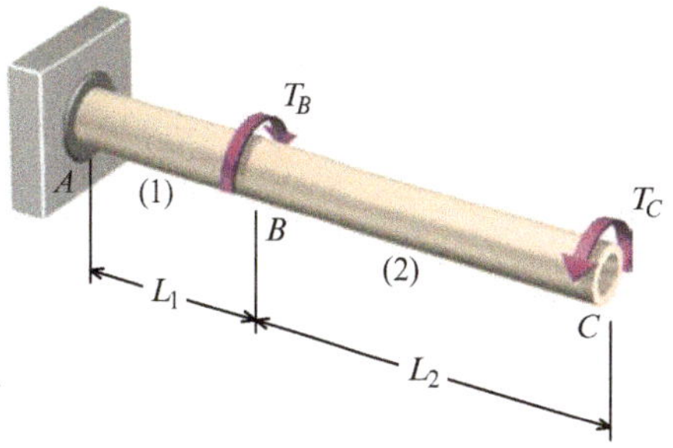

FIGURE P17.6/7/8

P17.8 A solid stepped shaft made of 2014-T4 aluminum ($G = 28$ GPa) is shown in Figure P17.6/7/8. The diameters of segments (1) and (2) are $d_1 = 20$ mm and $d_2 = 12$ mm, respectively. The segment lengths are $L_1 = 240$ mm and $L_2 = 180$ mm. Determine the elastic strain energy stored in the shaft when the maximum shear stress is 130 MPa.

P17.9 Determine the elastic strain energy of the prismatic beam *AB* shown in Figure P17.9 if $w = 6$ kN/m, $L = 5$ m, and $EI = 3 \times 10^7$ N·m^2.

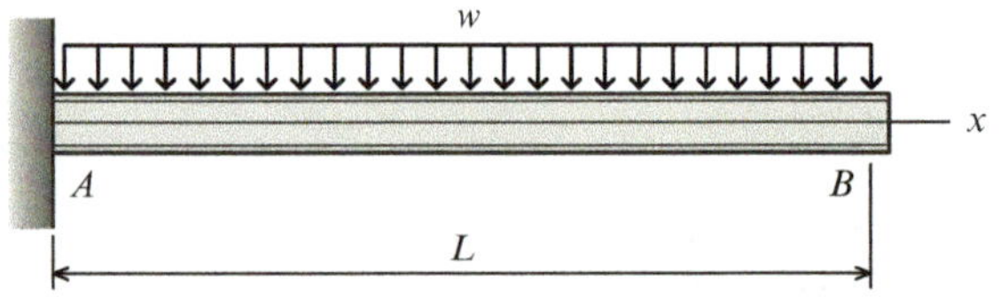

FIGURE P17.9

P17.10 Determine the elastic strain energy of the prismatic beam shown in Figure P17.10 if $w = 4{,}000$ lb/ft, $L = 18$ ft, and $EI = 1.33 \times 10^8$ lb·ft^2.

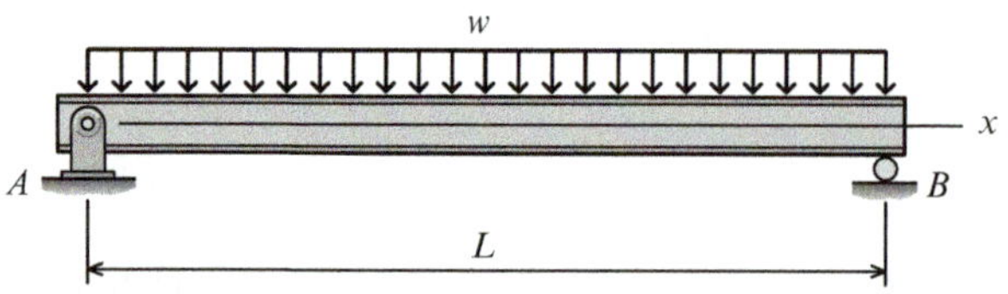

FIGURE P17.10

P17.11 Determine the elastic strain energy of the prismatic beam shown in Figure P17.11 if $P = 75$ kN, $L = 8$ m, and $EI = 5.10 \times 10^7$ N·m^2.

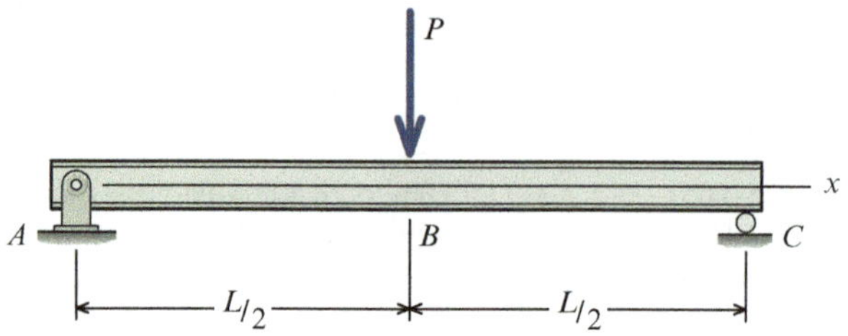

FIGURE P17.11

17.6 Impact Loading

When the motion of a body is changed (i.e., the body accelerates), the force necessary to produce the acceleration is called a **dynamic force** or a **dynamic load.** Following are some examples of a dynamic load:

- the force that an elbow in a pipeline exerts on the fluid in the pipe to change its direction of flow.
- the pressure on the wings of an airplane pulling out of a dive.
- the force of the wind on the exterior walls of a building.
- the weight of a vehicle as it rolls across a bridge.
- the force of a hammer striking the head of a nail.
- the collision of a ship and a bridge pier.
- the weight of a man jumping on a diving board.

A dynamic load may be expressed in terms of

- the mass of a body times the acceleration of its center of mass
- the rate of change of the momentum of a body, or
- the change in kinetic energy of a body.

A suddenly applied load is called an **impact load.** The last three examples of a dynamic load are considered impact loads. In each example, one object strikes another such that large forces are developed between the objects in a very short period. When subjected to impact loading, the loaded structure or system will vibrate until equilibrium is established if the material remains elastic.

In the loaded system, dynamic loading produces stresses and strains, the magnitude and distribution of which depend not only on the usual parameters (the dimensions of the member, the loading applied, the elastic modulus of the member, etc.), but also on the velocity of the strain waves that propagate through the solid material. This latter consideration, although very important when loads with extremely high velocities are applied, may often be neglected when the velocity of the impact load is relatively low. The loading may be considered a *low-velocity impact* when the loading time permits the material to act in the same manner as it does under static load—that is, when the relations between stress and strain and between load and deflection are essentially the same as those already developed for static loading. For low-velocity impacts, the time of application of the load is greater than several times the natural period of the loaded member. If the time of application of the load is short compared with the natural period of vibration of the member, the load is usually said to be a *high-velocity impact.*

Energy methods can be used to obtain solutions to many problems involving impacts in mechanics of materials, and such methods enable us to develop some insight into the significant differences between static and dynamic loading.

Investigation of Impact Loading with Simple Block-and-Spring Models

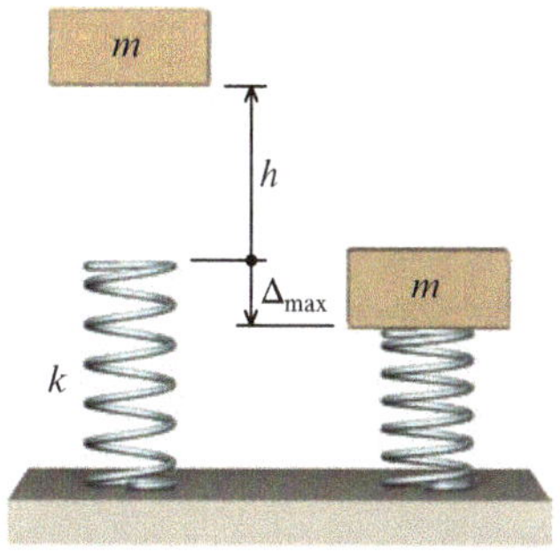

FIGURE 17.17 Freely falling weight and spring.

Freely Falling Weight: As an illustration of an elastic system subjected to an impact load, consider the simple block-and-spring system shown in Figure 17.17. A block having mass $m = W/g$ is initially positioned at a height h above a spring. Here, W represents the weight of the block and g the gravitational acceleration constant. At this initial position, the block's velocity is $v = 0$; hence, its kinetic energy is also zero. When the block is released from rest, it falls a distance h, where it first contacts the spring. The block continues to move downward, compressing the spring until its velocity is momentarily halted (i.e., $v = 0$). At this instant, the spring has been compressed an amount Δ_{max}, and the kinetic energy of the block is once again zero. If the mass of the spring is neglected and the spring responds elastically, then the principle of conservation of energy requires that the potential energy of the block at its initial position must be transformed into stored energy in the spring at its fully compressed position. In other words, the work done by gravity as the block moves downward a distance $h + \Delta_{max}$ is equal to the work required to compress the spring by Δ_{max}. The maximum force F_{max} developed in the spring is related to Δ_{max} by $F_{max} = k\,\Delta_{max}$, where k is the spring stiffness (expressed in units of force per unit of deflection). Therefore, from conservation of energy, and assuming no dissipation of energy at impact, the external work done by the weight W of the block as it moves downward must equal the internal work of the spring as it stores energy:

At the instant a moving body is stopped, its energy (both potential and kinetic) has been transformed into internal energy in the resisting system.

$$W(h + \Delta_{max}) = \tfrac{1}{2}(k\Delta_{max})\Delta_{max} = \tfrac{1}{2}k\Delta_{max}^2 \qquad \text{(a)}$$

Note that the factor one-half appears in Equation (a) because the force in the spring gradually increases from zero to its maximum value. Equation (a) can be rewritten as

$$\Delta_{max}^2 - \frac{2W}{k}\Delta_{max} - \frac{2W}{k}h = 0$$

This quadratic equation can be solved for Δ_{max}, and the positive root is

$$\Delta_{max} = \frac{W}{k} + \sqrt{\left(\frac{W}{k}\right)^2 + 2\left(\frac{W}{k}\right)h} \qquad \text{(b)}$$

Note that the negative root implies that the spring elongates when the block strikes it, which clearly makes no sense for this system. Therefore, the positive root is the only meaningful solution of equation (b).

If the block had been slowly and gradually placed on top of the spring, the deflection corresponding to the static force W acting on the spring would be $\Delta_{st} = W/k$. Therefore, Equation (b) can be restated in terms of the static spring deflection as

$$\Delta_{max} = \Delta_{st} + \sqrt{(\Delta_{st})^2 + 2\Delta_{st}h}$$

or

$$\Delta_{max} = \Delta_{st}\left[1 + \sqrt{1 + \frac{2h}{\Delta_{st}}}\right] \tag{17.21}$$

If Equation (17.21) is substituted into $F_{max} = k\,\Delta_{max}$, the dynamic force acting on the spring can be stated as

$$F_{max} = k\Delta_{st}\left[1 + \sqrt{1 + \frac{2h}{\Delta_{st}}}\right]$$

and since the weight of the block can be expressed in terms of the spring constant and the static deflection as $W = k\Delta_{st}$, it follows that

$$F_{max} = W\left[1 + \sqrt{1 + \frac{2h}{\Delta_{st}}}\right] \tag{17.22}$$

The force F_{max} is the dynamic force that acts on the spring. This force, along with its associated deflection Δ_{max}, occurs only for an instant. Unless the block rebounds off of the spring, the block will vibrate up and down until the motion dampens out and the block comes to equilibrium in the static deflected position Δ_{st}.

The expression in brackets in Equations (17.21) and (17.22) is termed an *impact factor*, which will be denoted here by the symbol n:

$$n = 1 + \sqrt{1 + \frac{2h}{\Delta_{st}}} \tag{17.23}$$

Thus, the maximum dynamic load F_{max} can be replaced by an **equivalent static load**, which is defined for the spring-and-block system as the product of the impact factor n and the actual static load W of the block:

$$F_{max} = nW$$

The notion of an impact factor is useful in that both the dynamic force F_{max} and the dynamic deflection

$$\Delta_{max} = n\Delta_{st}$$

can be readily expressed in terms of the static force and the static deflection for a particular impact factor. Expressed another way, the impact factor is simply the ratio of the dynamic effect to the static effect:

$$n = \frac{F_{max}}{W} = \frac{\Delta_{max}}{\Delta_{st}}$$

Special cases. Two extreme situations are of interest. First, if the drop height h for the block is much greater than the maximum spring deflection Δ_{max}, then the work term $W\Delta_{max}$ in Equation (a) can be neglected; thus,

$$Wh = \tfrac{1}{2}k\Delta_{max}^2$$

and the maximum spring deflection is

$$\Delta_{max} = \sqrt{\frac{2Wh}{k}} = \sqrt{2\Delta_{st}h}$$

For the other extreme, if the drop height h of the block is zero, then

$$\Delta_{max} = \Delta_{st}\left[1 + \sqrt{1 + \frac{2(0)}{\Delta_{st}}}\right] = 2\Delta_{st}$$

In other words, when the block is dropped from the top of the spring as a dynamic load, the spring deflection is twice as large as it would have been if the block were slowly and gradually placed on top of the spring. When a load is applied so gradually that the maximum deflection is the same as the static deflection, the impact factor is 1.0. However, if the load is applied suddenly, the effect produced in the elastic system is significantly amplified.

Impact from a Weight Moving Horizontally: By a procedure similar to that used for a freely falling weight, the impact load of a horizontally moving weight can be investigated. Suppose a block having mass $m = W/g$ slides on a smooth (i.e., frictionless) horizontal surface with a velocity v, as shown in Figure 17.18. The kinetic energy of the block before it contacts the spring is $\frac{1}{2}mv^2$. If the mass of the spring is neglected and the spring responds elastically, then the principle of conservation of energy requires that the kinetic energy of the block before it contacts the spring will be transformed into stored energy in the spring at its fully compressed position:

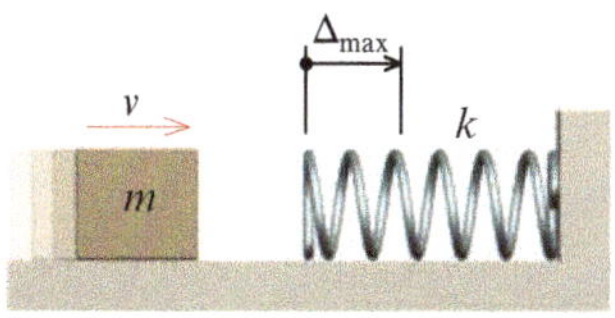

FIGURE 17.18 Horizontally moving weight and spring.

$$\frac{1}{2}\left(\frac{W}{g}\right)v^2 = \frac{1}{2}k\Delta_{max}^2$$

Thus,

$$\Delta_{max} = \sqrt{\frac{Wv^2}{gk}}$$

If we again define the static deflection of the spring caused by the weight of the block as $\Delta_{st} = W/k$ (note that this is *horizontal* deflection, which would occur in the spring from the application of a horizontal force equal in magnitude to the weight of the block), then the maximum spring deflection can be expressed as

$$\Delta_{max} = \Delta_{st}\sqrt{\frac{v^2}{g\Delta_{st}}} \tag{17.24}$$

and the impact force F_{max} of the block on the spring can be stated as

$$F_{max} = W\sqrt{\frac{v^2}{g\Delta_{st}}} \tag{17.25}$$

where the impact factor

$$n = \sqrt{\frac{v^2}{g\Delta_{st}}} \tag{17.26}$$

Significance: In the two cases just described, impact forces are imparted to an elastic spring by a freely falling weight and by a weight in motion. For these investigations, it has been assumed that the materials behave elastically and that no dissipation of energy (in the form of heat or sound or permanent deformation) takes place at the point of impact. The inertia of the resisting system has been neglected, and perfect rigidity of the block was implicit.

At first glance, the behavior of a spring system may not seem particularly useful or relevant. After all, mechanics of materials is the study of deformable solid materials such as axial members, shafts, and beams. However, the elastic behavior of these types of components is conceptually equivalent to the behavior of a spring. Therefore, the two models discussed previously can be seen as general cases that are widely applicable to common engineering components.

In this investigation, the deflection of a system is directly proportional to the magnitude of the applied force, regardless of whether that force is statically or dynamically applied. The block-and-spring models analyzed here show that the maximum dynamic response of a deformable solid can be determined from the product of its static response and an appropriate impact factor. Once the maximum deflection Δ_{max} due to impact has been determined, the maximum dynamic force can be found from $F_{max} = k\Delta_{max}$. The maximum dynamic load F_{max} can be considered an *equivalent static load* (a slowly applied load that will produce the same maximum deflection as the dynamic load). Provided that the assumption of identical material behavior under both static and dynamic loads is valid (a valid assumption for most mechanical-type loadings), the stress–strain diagram for any point in the loaded system does not change. Consequently, the stress and strain distributions produced by the equivalent static load will be the same as those produced by the dynamic load.

It is also significant that we have assumed no dissipation of energy during impact. There will always be some energy dissipation in the form of sound, heat, local deformations, and permanent distortions. Because of dissipation, less energy must be stored by the elastic system, and therefore, the actual maximum deflection due to impact loads is reduced. Given the assumptions delineated here, the actual impact factor will have a value somewhat less than that predicted by Equations (17.23) and (17.26); thus, the equivalent static load approach will be conservative. All in all, the equivalent static load approach provides the engineer a conservative, rational analysis of the stresses and strains produced by impact loading, using only the familiar equations found in mechanics of materials theory.

EXAMPLE 17.5

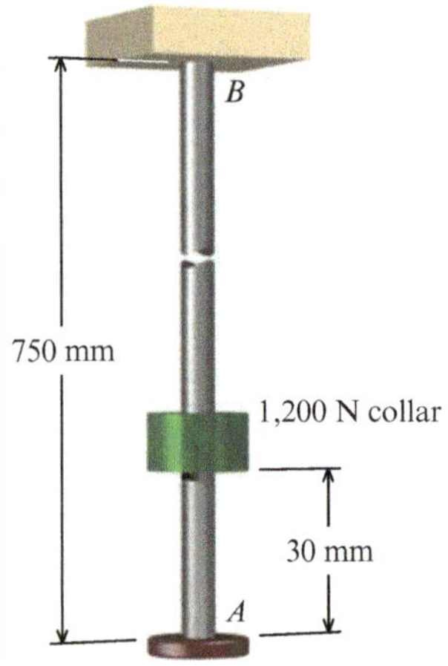

The 1,200 N collar shown is released from rest and slides without friction downward a distance of 30 mm, where it strikes a head fixed to the end of the rod. The AISI 1020 cold-rolled steel [E = 200 GPa] rod has a diameter of 15 mm and a length of 750 mm. Determine

(a) the axial deformation and the normal stress in the rod under static conditions; that is, the collar is gradually lowered until it contacts the rod head and comes to rest without impact.
(b) the maximum axial deformation of the rod if the collar is dropped from the height of 30 mm.
(c) the maximum dynamic force exerted on the rod by the collar.
(d) the maximum normal stress in the rod due to the dynamic force.
(e) the impact factor n.

Plan the Solution

The axial deformation and the normal stress in the rod under static conditions are determined with the weight of the collar applied to the lower end of the rod. Work and energy principles can be used to equate the work done by the 1,200 N weight as the collar falls 30 mm and elongates the rod to the strain energy stored in the rod at the instant of maximum deformation. From this energy balance, the maximum rod deformation can be calculated. The maximum deformation can then be used to determine the maximum dynamic force, the corresponding normal stress, and the impact factor n.

SOLUTION

(a) The static force applied to the rod is simply the weight W of the collar; thus,

$$F_{st} = W = 1{,}200 \text{ N}$$

The cross-sectional area of the rod is

$$A = \frac{\pi}{4}d^2 = \frac{\pi}{4}(15 \text{ mm})^2 = 176.7146 \text{ mm}^2$$

The axial deformation in the 15 mm diameter rod due to the 1,200 N weight of the collar is

$$\delta_{st} = \frac{F_{st}L}{AE} = \frac{(1{,}200 \text{ N})(750 \text{ mm})}{(176.7146 \text{ mm}^2)(200{,}000 \text{ N/mm}^2)} = 0.025465 \text{ mm} = 0.0255 \text{ mm} \quad \textbf{Ans.}$$

and the static normal stress is

$$\sigma_{st} = \frac{F_{st}}{A} = \frac{1{,}200 \text{ N}}{176.7146 \text{ mm}^2} = 6.79061 \text{ MPa} = 6.79 \text{ MPa} \quad \textbf{Ans.}$$

(b) The maximum rod deformation when the collar is dropped can be determined from work and energy principles. The external work done by the 1,200 N weight as the collar is dropped from height h must equal the strain energy stored by the rod at its maximum deformation. Recall from Section 17.3 that the strain energy stored in an axial member can be expressed in terms of the member deformation by Equation (17.13); thus,

$$\text{External work} = \text{Internal strain energy}$$

$$F_{st}(h + \delta_{max}) = \frac{AE\delta_{max}^2}{2L}$$

$$\frac{AE}{2L}\delta_{max}^2 - F_{st}(h + \delta_{max}) = 0$$

$$\delta_{max}^2 - 2\frac{F_{st}L}{AE}(h + \delta_{max}) = 0$$

Recognizing that the term $F_{st}L/AE$ is the static deformation δ_{st}, rewrite the last equation as

$$\delta_{max}^2 - 2\delta_{st}(h + \delta_{max}) = 0$$

Now multiply out the two factors of the middle term to get

$$\delta_{max}^2 - 2\delta_{st}\delta_{max} - 2\delta_{st}h = 0$$

and solve for δ_{max}, using the quadratic formula:

$$\delta_{max} = \frac{2\delta_{st} \pm \sqrt{(-2\delta_{st})^2 - 4(1)(-2\delta_{st}h)}}{2} = \delta_{st} \pm \sqrt{\delta_{st}^2 + 2\delta_{st}h}$$

From the positive root, the maximum rod deformation can now be expressed in terms of the static deformation and the drop height h as

$$\delta_{max} = \delta_{st} + \sqrt{\delta_{st}^2 + 2\delta_{st} h} \qquad \text{(a)}$$

The maximum axial deformation of the rod if the collar is dropped from the height of 30 mm can now be computed:

$$\begin{aligned}\delta_{max} &= 0.025465 \text{ mm} + \sqrt{(0.025465 \text{ mm})^2 + 2(0.025465 \text{ mm})(30 \text{ mm})}\\ &= 0.025465 \text{ mm} + 1.236345 \text{ mm}\\ &= 1.261810 \text{ mm} = 1.262 \text{ mm}\end{aligned}$$ **Ans.**

(c) The maximum dynamic force exerted on the rod is calculated from the maximum dynamic deformation. If it is assumed that the rod behaves elastically and that the stress–strain curve applicable to this dynamic load is the same as the stress–strain curve for a static load, the relationship of the force exerted on the rod and the deformation caused by the dynamic load is

$$\delta_{max} = \frac{F_{max} L}{AE}$$

Therefore, the maximum dynamic force is

$$\begin{aligned}F_{max} &= \delta_{max} \frac{AE}{L}\\ &= (1.261810 \text{ mm})\frac{(176.7146 \text{ mm}^2)(200{,}000 \text{ N/mm}^2)}{750 \text{ mm}}\\ &= 59{,}461.4 \text{ N} = 59{,}500 \text{ N}\end{aligned}$$ **Ans.**

The maximum dynamic normal stress in the rod is thus

$$\sigma_{max} = \frac{F_{max}}{A} = \frac{59{,}461.4 \text{ N}}{176.7146 \text{ mm}^2} = 336 \text{ MPa}$$ **Ans.**

(e) The impact factor n is simply the ratio of the dynamic effect to the static effect:

$$n = \frac{F_{max}}{F_{st}} = \frac{\delta_{max}}{\delta_{st}} = \frac{\sigma_{max}}{\sigma_{st}}$$

Hence,

$$n = \frac{1.261810 \text{ mm}}{0.025465 \text{ mm}} = 49.551$$ **Ans.**

SIMPLIFIED SOLUTION

An equation similar to Equation (17.23) can be derived for the impact factor n. Recall Equation (a):

$$\delta_{max} = \delta_{st} + \sqrt{\delta_{st}^2 + 2\delta_{st} h}$$

Multiplying the second term under the radical by δ_{st}/δ_{st} (i.e., 1), factoring and taking $\sqrt{\delta_{st}^2}\,(= \delta_{st})$ out of the radical, and factoring again yields

$$\delta_{max} = \delta_{st} + \sqrt{\delta_{st}^2 + \delta_{st}^2 \frac{2h}{\delta_{st}}} = \delta_{st} + \delta_{st}\sqrt{1 + \frac{2h}{\delta_{st}}} = \delta_{st}\left[1 + \sqrt{1 + \frac{2h}{\delta_{st}}}\right]$$

so that the impact factor can be written as

$$n = \frac{\delta_{max}}{\delta_{st}} = 1 + \sqrt{1 + \frac{2h}{\delta_{st}}}$$

Since the static deflection was calculated previously as $\delta_{st} = 0.025465$ mm, the impact factor for the 30 mm drop height is

$$n = 1 + \sqrt{1 + \frac{2(30 \text{ mm})}{0.025465 \text{ mm}}} = 49.551 \qquad \textbf{Ans.}$$

The static results can now be multiplied by the impact factor to give the dynamic deformation, force, and stress:

$$\delta_{max} = n\delta_{st} = 49.551(0.025465 \text{ mm}) = 1.262 \text{ mm} \qquad \textbf{Ans.}$$

$$F_{max} = nF_{st} = 49.551(1{,}200 \text{ N}) = 59{,}461 \text{ N} \qquad \textbf{Ans.}$$

$$\sigma_{max} = n\sigma_{st} = 49.551(6.79061 \text{ MPa}) = 336 \text{ MPa} \qquad \textbf{Ans.}$$

EXAMPLE 17.6

The simply supported beam shown in the figure spans 20 ft. The beam consists of a W14 × 34 shape of A992 steel [$E = 29 \times 10^6$ psi]. Calculate the deflection at B and the largest bending stress in the beam if the 3,000 lb load is

(a) applied statically.
(b) dropped from a height of $h = 5$ in.

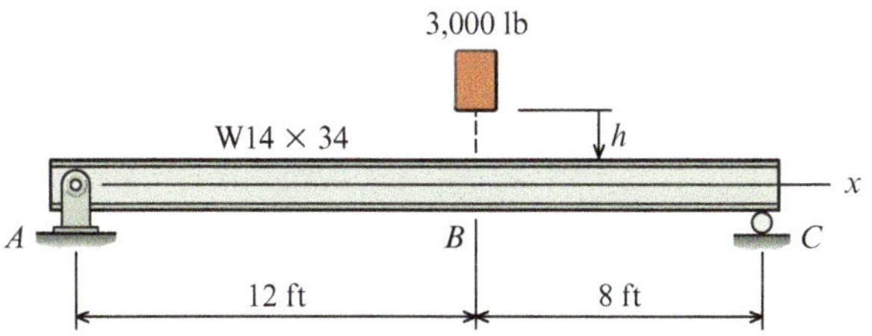

Plan the Solution

The beam deflection and the maximum normal stress at B due to a static loading are determined for the 3,000 lb load. Work and energy principles can be used to equate the work done by the load to the strain energy stored in the beam at the instant of maximum deflection. From this energy balance, the maximum beam deflection at B can be calculated. The maximum deformation can then be used to determine the maximum dynamic load, the corresponding bending stress, and the impact factor n.

SOLUTION

(a) Load Applied Statically. Applying the load statically means applying the 3,000 lb load slowly and gradually from a height of $h = 0$. The W14 × 34 shape used for the beam has the following section properties:

$$d = 14.0 \text{ in.}$$
$$I = 340 \text{ in.}^4$$

Deflection at B: The deflection formula found in Appendix C for a simply supported beam with a concentrated load applied anywhere will be used here. For this span, L = 20 ft = 240 in., a = 12 ft = 144 in., and b = 8 ft = 96 in. The 3,000 lb static load produces a downward deflection at B of

$$v_{st} = \frac{P_{st}a^2b^2}{3LEI} = \frac{(3{,}000 \text{ lb})(144 \text{ in.})^2(96 \text{ in.})^2}{3(240 \text{ in.})(29 \times 10^6 \text{ lb/in.}^2)(340 \text{ in.}^4)}$$
$$= 0.080757 \text{ in.} = 0.0808 \text{ in.} \qquad \textbf{Ans.}$$

Note: Throughout the previous chapters in this book, the symbol v has been used to denote deflection perpendicular to the longitudinal axis of a beam. In this chapter, velocity has been introduced as an additional consideration, and the symbol v is also used to denote velocity. In this example, beam deflections are being considered; therefore, the symbol v used in this context represents beam deflections.

Largest Bending Stress: From Example 17.4, the static beam reaction at A is

$$A_y = \frac{P_{st}b}{L}$$

Therefore, the maximum bending moment (which occurs at B) is

$$M_{st} = A_y a = \frac{P_{st}b}{L}a = \frac{(3{,}000 \text{ lb})(96 \text{ in.})}{240 \text{ in.}}(144 \text{ in.}) = 172{,}800 \text{ lb}\cdot\text{in.}$$

The largest bending stress in the beam is

$$\sigma_{st} = \frac{M_{st}c}{I} = \frac{(172{,}800 \text{ lb}\cdot\text{in.})(14 \text{ in.}/2)}{340 \text{ in.}^4} = 3{,}557.65 \text{ psi} = 3{,}560 \text{ psi}$$ **Ans.**

(b) Load Dropped from $h = 5$ in. The maximum beam deflection when the load is dropped can be determined from work and energy principles. The external work done by the 3,000 lb load dropped from height h must equal the strain energy stored by the beam at its maximum deflection. Recall from Section 17.5 that the strain energy stored in a flexural member can be expressed in terms of the member deformation by Equation (17.20):

$$U = \int_0^L \frac{M^2}{2EI}dx$$

The total elastic strain energy U for this type of beam and loading was derived in Example 17.4:

$$U = \frac{P^2a^2b^2}{6LEI}$$

Deflection at B: Equate the internal strain energy of the beam to the work done by gravity as the 3,000 lb load moves downward:

$$\text{External work} = \text{Internal strain enery}$$

$$P_{st}(h + v_{max}) = \frac{P_{max}^2 a^2b^2}{6LEI} = \frac{3LEI}{2a^2b^2}v_{max}^2$$

$$\frac{3LEI}{2a^2b^2}v_{max}^2 - P_{st}(h + v_{max}) = 0$$

$$v_{max}^2 - \frac{2P_{st}a^2b^2}{3LEI}(h + v_{max}) = 0$$

Since the static deflection at B is

$$v_{st} = \frac{P_{st}a^2b^2}{3LEI}$$

the quadratic equation in v_{max} can be rewritten as

$$v_{max}^2 - 2v_{st}(h + v_{max}) = 0$$

Multiplying the two factors of the second term on the left gives

$$v_{max}^2 - 2v_{st}v_{max} - 2v_{st}h = 0$$

Now solve for v_{max}, using the quadratic formula:

$$v_{max} = \frac{2v_{st} \pm \sqrt{(-2v_{st})^2 - 4(1)(-2v_{st}h)}}{2} = v_{st} \pm \sqrt{v_{st}^2 + 2v_{st}h} \qquad \text{(a)}$$

The maximum deflection of the beam at B if the load is dropped from the height of 5 in. can now be computed:

$$\begin{aligned} v_{max} &= 0.080757 \text{ in.} + \sqrt{(0.080757 \text{ in.})^2 + 2(0.080757 \text{ in.})(5 \text{ in.})} \\ &= 0.080757 \text{ in.} + 0.902270 \text{ in.} \\ &= 0.983027 \text{ in.} = 0.983 \text{ in.} \end{aligned}$$

Ans.

Largest Bending Stress: The maximum dynamic force P_{max} exerted on the beam is calculated from the maximum dynamic deflection v_{max}. If the beam behaves elastically and the stress–strain curve applicable to P_{max} is the same as the stress–strain curve for P, then the dynamic load is calculated as follows:

$$v_{max} = \frac{P_{max}a^2b^2}{3LEI}$$

$$\therefore P_{max} = \frac{3LEI}{a^2b^2}v_{max} = \frac{3(240 \text{ in.})(29 \times 10^6 \text{ lb/in.}^2)(340 \text{ in.}^4)}{(144 \text{ in.})^2(96 \text{ in.})^2}(0.983027 \text{ in.})$$

$$= 36{,}518.0 \text{ lb} = 36{,}500 \text{ lb}$$

The maximum dynamic bending moment in the beam is

$$M_{max} = \frac{P_{max}b}{L}a = \frac{(36{,}518.0 \text{ lb})(96 \text{ in.})}{240 \text{ in.}}(144 \text{ in.}) = 2.103437 \times 10^6 \text{ lb}\cdot\text{in.}$$

and the largest bending stress in the beam is

$$\sigma_{max} = \frac{M_{max}c}{I} = \frac{(2.103437 \times 10^6 \text{ lb}\cdot\text{in.})(14 \text{ in./2})}{340 \text{ in.}^4} = 43{,}306.1 \text{ psi} = 43{,}300 \text{ psi}$$

Ans.

Note that the impact factor is

$$n = \frac{v_{max}}{v_{st}} = \frac{0.983027 \text{ in.}}{0.080757 \text{ in.}} = 12.173$$

SIMPLIFIED SOLUTION

An equation similar to Equation (17.23) can be derived for the beam's impact factor n. Recall Equation (a):

$$v_{max} = v_{st} \pm \sqrt{v_{st}^2 + 2v_{st}h}$$

The right-hand side of this equation can be manipulated algebraically to give

$$v_{\max} = v_{\text{st}}\left[1 + \sqrt{1 + \frac{2h}{v_{\text{st}}}}\right]$$

so that the impact factor can be written as

$$n = \frac{v_{\max}}{v_{\text{st}}} = 1 + \sqrt{1 + \frac{2h}{v_{\text{st}}}}$$

The static deflection was calculated previously as $v_{\text{st}} = 0.080757$ in. On the basis of this static deflection, the impact factor for the 5 in. drop height is

$$n = 1 + \sqrt{1 + \frac{2(5 \text{ in.})}{0.080757 \text{ in.}}} = 12.173$$ **Ans.**

The static results can now be multiplied by the impact factor to give the dynamic deformation and bending stress:

$$v_{\max} = nv_{\text{st}} = 12.173(0.080757 \text{ in.}) = 0.983 \text{ in.}$$ **Ans.**

$$\sigma_{\max} = n\sigma_{\text{st}} = 12.173(3{,}557.65 \text{ psi}) = 43{,}300 \text{ psi}$$ **Ans.**

EXAMPLE 17.7

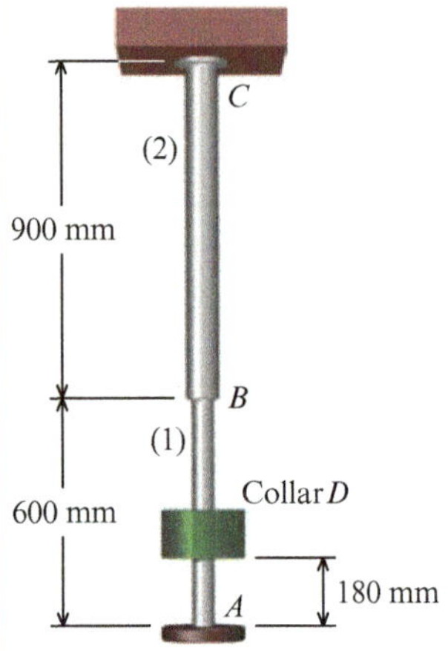

Collar D shown is released from rest and slides without friction downward a distance of 180 mm, where it strikes a head fixed to the end of compound rod ABC. The compound rod is made of aluminum [$E = 70$ GPa], and the diameters of rod segments (1) and (2) are 18 mm and 25 mm, respectively.

(a) Determine the largest mass of the collar for which the maximum normal stress in the rod is 240 MPa.

(b) If the diameter of rod segment (2) is reduced to 18 mm, what is the largest mass of the collar for which the maximum normal stress in the rod is 240 MPa?

Plan the Solution

From the maximum normal stress, calculate the maximum dynamic force allowed in the segment that has the smaller cross-sectional area, segment (1). Next, compute the total strain energy stored in the compound rod for this maximum load. Then, equate the total strain energy to the work performed by the maximum force on the rod to calculate the maximum deformation of the compound rod. Now use the dynamic deformation and the drop height to determine first the static deformation and then the static load. Then, determine the allowable mass from the static load. Finally, repeat the process for a rod that has a constant diameter of 18 mm, and compare the allowable masses for the two cases.

SOLUTION

(a) The areas of the two rod segments are as follows:

$$A_1 = \frac{\pi}{4}(18\text{ mm})^2 = 254.4690\text{ mm}^2$$

$$A_2 = \frac{\pi}{4}(25\text{ mm})^2 = 490.8739\text{ mm}^2$$

The maximum stress occurs in segment (1). The maximum dynamic load that can be applied to this segment without exceeding the 240 MPa limit is

$$P_{max} = \sigma_{max} A_1 = (240\text{ N/mm}^2)(254.4690\text{ mm}^2) = 61{,}072.6\text{ N}$$

For this dynamic load, the strain energy in compound rod *ABC* can be determined from Equation (17.14):

$$\begin{aligned} U_{total} &= \frac{F_1^2 L_1}{2A_1 E_1} + \frac{F_2^2 L_2}{2A_2 E_2} \\ &= \frac{(61{,}072.6\text{ N})^2}{2(70{,}000\text{ N/mm}^2)}\left[\frac{600\text{ mm}}{254.4690\text{ mm}^2} + \frac{900\text{ mm}}{490.8739\text{ mm}^2}\right] \\ &= 111{,}664.5\text{ N}\cdot\text{mm} \end{aligned}$$

Equate the strain energy stored in the compound rod to the work done by the falling collar to determine the maximum deformation of the entire rod due to the impact load:

$$\frac{1}{2}P_{max}\delta_{max} = 111{,}664.5\text{ N}\cdot\text{mm}$$

$$\delta_{max} = \frac{2(111{,}664.5\text{ N}\cdot\text{mm})}{61{,}072.6\text{ N}} = 3.6568\text{ mm}$$

The static deformation of the entire rod can be related to the dynamic deformation by

$$\delta_{max}^2 - 2\delta_{st}(h + \delta_{max}) = 0$$

which was derived in Example 17.5. The static deformation is thus

$$\delta_{st} = \frac{\delta_{max}^2}{2(h + \delta_{max})} = \frac{(3.6568\text{ mm})^2}{2(180\text{ mm} + 3.6568\text{ mm})} = 0.036405\text{ mm}$$

The static deformation of this compound rod can be expressed as

$$\delta_{st} = \frac{F_1 L_1}{A_1 E_1} + \frac{F_2 L_2}{A_2 E_2} = \frac{F_{st}}{E}\left[\frac{L_1}{A_1} + \frac{L_2}{A_2}\right]$$

$$\therefore F_{st} = \frac{(0.036405\text{ mm})(70{,}000\text{ N/mm}^2)}{\dfrac{600\text{ mm}}{254.4690\text{ mm}^2} + \dfrac{900\text{ mm}}{490.8739\text{ mm}^2}} = 608.00\text{ N}$$

Consequently, the largest mass that can be dropped is

$$m = \frac{F_{st}}{g} = \frac{608.00\text{ N}}{9.807\text{ m/s}^2} = 62.0\text{ kg}$$ **Ans.**

(b) If the rod has a constant diameter of 18 mm, the strain energy in the rod is

$$U_{\text{total}} = \frac{F^2 L}{2AE} = \frac{(61{,}072.6\text{ N})^2(1{,}500\text{ mm})}{2(70{,}000\text{ N/mm}^2)(254.4690\text{ mm}^2)}$$
$$= 157{,}043.9\text{ N}\cdot\text{mm}$$

Equate the strain energy stored in the prismatic rod to the work done by the falling collar to calculate the maximum deformation of the rod:

$$\frac{1}{2}P_{\text{max}}\delta_{\text{max}} = 157{,}043.9\text{ N}\cdot\text{mm}$$
$$\delta_{\text{max}} = \frac{2(157{,}043.9\text{ N}\cdot\text{mm})}{61{,}072.6\text{ N}} = 5.1429\text{ mm}$$

As before, compute the static deformation:

$$\delta_{\text{st}} = \frac{\delta_{\text{max}}^2}{2(h + \delta_{\text{max}})} = \frac{(5.1429\text{ mm})^2}{2(180\text{ mm} + 5.1429\text{ mm})} = 0.071430\text{ mm}$$

Now, from the relationship

$$\delta_{\text{st}} = \frac{F_{\text{st}} L}{AE}$$

compute the static load:

$$F_{\text{st}} = \frac{(0.071430\text{ mm})(254.4690\text{ mm}^2)(70{,}000\text{ N/mm}^2)}{1{,}500\text{ mm}} = 848.25\text{ N}$$

Consequently, the largest mass that can be dropped if the entire rod has a diameter of 18 mm is

$$m = \frac{F_{\text{st}}}{g} = \frac{848.25\text{ N}}{9.807\text{ m/s}^2} = 86.5\text{ kg}$$ **Ans.**

Note that the allowable mass for case (b) is about 40 percent larger than the allowable mass for case (a).

Comments: The results for cases (a) and (b) seem to be paradoxical because a larger mass can be dropped when some of the material in the rod is removed. This apparent discrepancy is probably best explained by considering strain-energy densities. The strain-energy density of the 18 mm diameter segment when it is subjected to the dynamic load is

$$u_1 = \frac{\sigma_1^2}{2E_1} = \frac{(240\text{ MPa})^2}{2(70{,}000\text{ MPa})} = 0.4114\text{ MPa}$$

This strain-energy density is represented by area *OCD* on the stress–strain diagram shown. The strain-energy density in the 25 mm diameter segment when it is subjected to the maximum dynamic load is

$$u_2 = \frac{\sigma_2^2}{2E_2} = \frac{(124.416\text{ MPa})^2}{2(70{,}000\text{ MPa})} = 0.1106\text{ MPa}$$

where

$$\sigma_2 = \frac{61{,}072.6\text{ N}}{490.8739\text{ mm}} = 124.416\text{ MPa}$$

This strain-energy density is represented by area *OAB* of the stress–strain diagram. The strain-energy density of the 25 mm diameter segment is roughly one-fourth the strain-energy density of the 18 mm diameter segment. When the 25 mm diameter is reduced to 18 mm, the volume of segment (2) is roughly halved. However, the strain energy absorbed by each unit volume of the remaining material is roughly quadrupled (i.e., area *OCD* compared with area *OAB*), resulting in a net gain in energy-absorbing capacity. For the rod considered here, this gain amounts to about a 40 percent increase in the allowable collar mass for the constant-diameter rod.

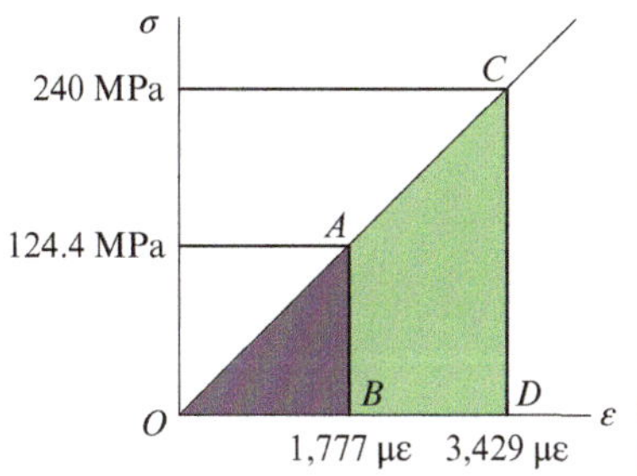

EXAMPLE 17.8

The cantilever post *AB* consists of a steel pipe that has an outside diameter of 33 mm and a wall thickness of 3 mm. A 30 kg block moving horizontally with a velocity v_0 strikes the post squarely at *B*. What is the maximum velocity v_0 for which the largest normal stress in the post does not exceed 190 MPa? Assume that $E = 200$ GPa for the steel pipe.

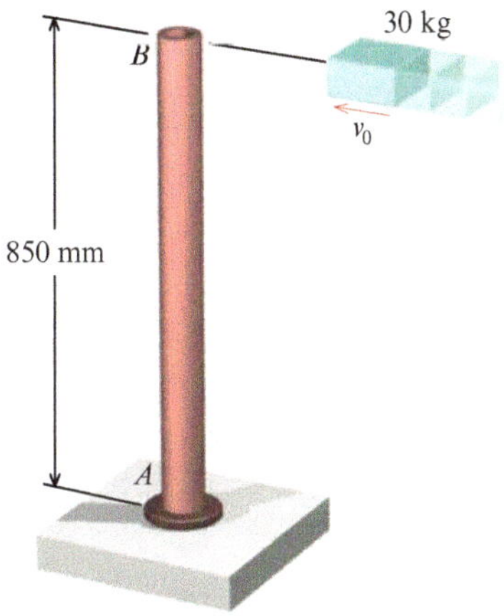

Plan the Solution

Calculate the maximum dynamic moment from the allowable normal stress and the section properties of the post. Then determine the maximum allowable dynamic load and the corresponding horizontal deflection of the post at *B*. Use conservation of energy to equate the work done on the post to the kinetic energy of the block, and solve for the maximum velocity v_0.

SOLUTION

The moment of inertia of the pipe is

$$I = \frac{\pi}{64}[(33\text{ mm})^4 - (27\text{ mm})^4] = 32{,}127.7\text{ mm}^4$$

The maximum dynamic moment that can be applied to the post at *A* without exceeding the 190 MPa limit is

$$M_{\text{max}} = \frac{\sigma_{\text{max}} I}{c} = \frac{(190\text{ N/mm}^2)(32{,}126.7\text{ mm}^4)}{33\text{ mm}/2} = 369{,}944\text{ N}\cdot\text{mm}$$

Since the cantilever post has a span of 850 mm, the maximum allowable dynamic load is

$$P_{\text{max}} = \frac{369{,}944\text{ N}\cdot\text{mm}}{850\text{ mm}} = 435.2\text{ N}$$

From Appendix C, the maximum horizontal deflection of the post at *B* can be calculated as

$$v_{\text{max}} = \frac{P_{\text{max}} L^3}{3EI} = \frac{(435.2\text{ N})(850\text{ mm})^3}{3(200{,}000\text{ N/mm}^2)(32{,}127.7\text{ mm}^4)} = 13.865\text{ mm}$$

Note: Throughout the previous chapters in this book, the symbol v has been used to denote deflection perpendicular to the longitudinal axis of a beam. In this chapter, velocity has been introduced as an additional consideration, and the symbol v is also used to denote velocity. In this example, both beam deflections and the velocity of a block are being considered. The context of the problem and the subscripts used with the symbol v clearly indicate whether a deflection or a velocity is intended; however, the reader is cautioned to examine the context in which the symbol v is used.

By the conservation of energy, the work that is performed on the post must equal the kinetic energy of the block:

$$\frac{1}{2}P_{\text{max}}v_{\text{max}} = \frac{1}{2}mv_0^2$$

Note: On the left-hand side of this equation, v_{max} is a displacement. On the right-hand side, v_0 is a velocity.

Therefore, the maximum velocity of the block must not exceed

$$v_0 = \sqrt{\frac{P_{\text{max}}v_{\text{max}}}{m}} = \sqrt{\frac{(435.2\text{ N})(0.013865\text{ m})}{30\text{ kg}}} = 0.448\text{ m/s}$$ **Ans.**

This problem can also be solved with the impact factor given in Equation (17.26). The weight of the block is (30 kg)(9.807 m/s^2) = 294.2 N. If this force were gradually applied horizontally to the post at B, the static deflection would be

$$v_{\text{st}} = \frac{P_{\text{st}}L^3}{3EI} = \frac{(294.2\text{ N})(850\text{ mm})^3}{3(200{,}000\text{ N/mm}^2)(32{,}127.7\text{ mm}^4)} = 9.373\text{ mm}$$

The impact factor can be calculated from the static and dynamic deflections at B:

$$n = \frac{v_{\text{max}}}{v_{\text{st}}} = \frac{13.866\text{ mm}}{9.373\text{ mm}} = 1.479$$

From Equation (17.26), the impact factor for a weight moving horizontally is

$$n = \sqrt{\frac{v_0^2}{g\,v_{\text{st}}}}$$

This equation can be solved for the maximum velocity v_0:

$$\begin{aligned} v_0 &= \sqrt{n^2 g v_{\text{st}}} \\ &= \sqrt{(1.479)^2(9.807\text{ m/s}^2)(0.009373\text{ m})} \\ &= 0.448\text{ m/s} \end{aligned}$$ **Ans.**

PROBLEMS

P17.12 A 19 mm diameter steel [$E = 200$ GPa] rod is required to absorb the energy of a 25 kg collar that falls $h = 75$ mm, as shown in Figure P17.12/13. Determine the minimum required rod length L so that the maximum stress in the rod does not exceed 210 MPa.

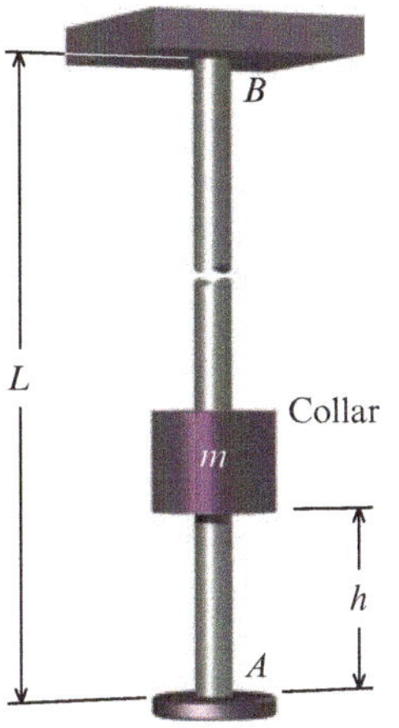

FIGURE P17.12/13

P17.13 In Figure P17.12/13, a 500 mm long steel [$E = 200$ GPa] rod is required to absorb the energy of a 16 kg mass that falls a distance h. If the diameter of the rod is 10 mm, what is the maximum drop height h so that the maximum stress in the rod does not exceed 210 MPa?

P17.14 A weight $W = 4{,}000$ lb falls from a height $h = 18$ in. onto the top of a 10 in. diameter wooden pole, as shown in Figure P17.14. The pole has a length $L = 24$ ft and a modulus of elasticity $E = 1.5 \times 10^6$ psi. For this problem, disregard any potential buckling effects. Calculate

(a) the impact factor n.
(b) the maximum shortening of the pole.
(c) the maximum compressive stress in the pole.

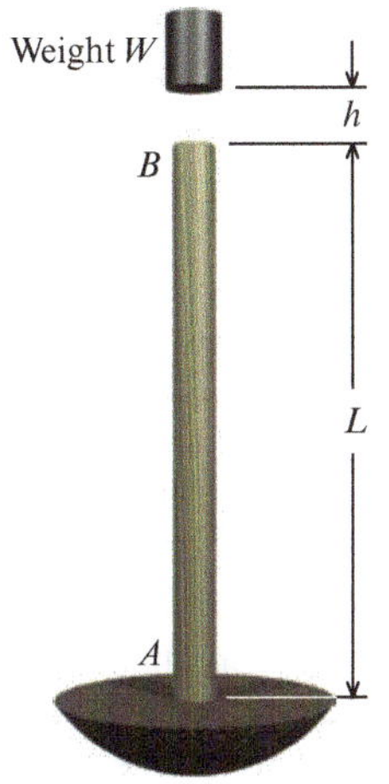

FIGURE P17.14

P17.15 As seen in Figure P17.15/16, collar D is released from rest and slides without friction downward a distance $h = 300$ mm, where it strikes a head fixed to the end of a compound rod ABC. Rod segment (1) is made of aluminum [$E_1 = 70$ GPa] and has a length $L_1 = 800$ mm and a diameter $d_1 = 12$ mm. Rod segment (2) is made of bronze [$E_2 = 105$ GPa] and has a length $L_2 = 1{,}300$ mm and a diameter $d_2 = 16$ mm. What is the allowable mass for collar D if the maximum normal stress in the aluminum rod segment must be limited to 200 MPa?

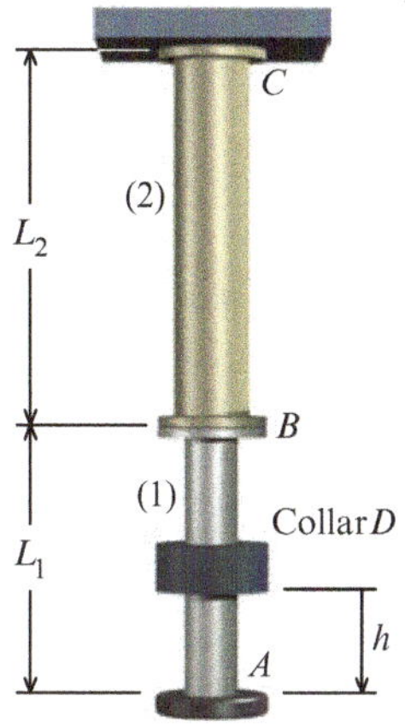

FIGURE P17.15/16

P17.16 Collar D shown in Figure P17.15/16 has a mass of 11 kg. When released from rest, the collar slides without friction downward a distance h, where it strikes a head fixed to the end of a compound rod ABC. Rod segment (1) is made of aluminum [$E_1 = 70$ GPa] and has a length $L_1 = 600$ mm and a diameter $d_1 = 12$ mm. Rod segment (2) is made of bronze [$E_2 = 105$ GPa] and has a length $L_2 = 1{,}000$ mm and a diameter $d_2 = 16$ mm. If the maximum normal stress in the aluminum rod segment must be limited to 250 MPa, determine the largest acceptable drop height h.

P17.17 In Figure P17.17, the 12 kg mass is falling at a velocity $v = 1.5$ m/s at the instant it is $h = 300$ mm above the spring-and-post

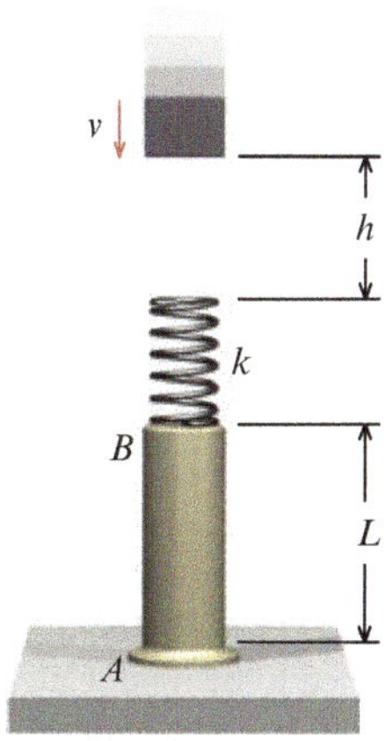

FIGURE P17.17

assembly. The solid bronze post has a length $L = 450$ mm, a diameter of 60 mm, and a modulus of elasticity $E = 105$ GPa. Compute the maximum stress in the bronze post and the impact factor

(a) if the spring has a stiffness $k = 5{,}000$ N/mm.
(b) if the spring has a stiffness $k = 500$ N/mm.

P17.18 The 32 mm diameter rod *AB* shown in Figure P17.18 has a length $L = 1.5$ m. The rod is made of bronze [$E = 105$ GPa] that has a yield stress $\sigma_Y = 330$ MPa. Collar *C* moves along the rod at a speed $v_0 = 3.5$ m/s until it strikes the rod end at *B*. If a factor of safety of 4 with respect to yield is required for the maximum normal stress in the rod, determine the maximum allowable mass for collar *C*.

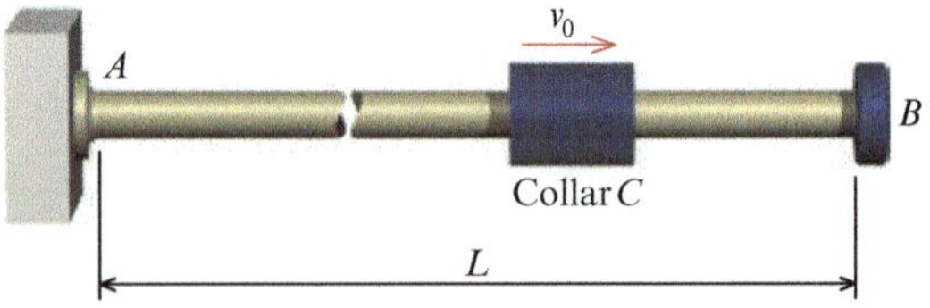

FIGURE P17.18

P17.19 The block *E* has a horizontal velocity $v_0 = 9$ ft/s when it squarely strikes the yoke *BD* that is connected to the 3/4 in. diameter rods *AB* and *CD*. (See Figure P17.19/20.) The rods are made of 6061-T6 aluminum that has a yield strength $\sigma_Y = 40$ ksi and an elastic modulus $E = 10{,}000$ ksi. Both rods have a length $L = 5$ ft. Yoke *BD* may be assumed to be rigid. What is the maximum allowable weight of block *E* if a factor of safety of 3 with respect to yield is required for the maximum normal stress in the rods?

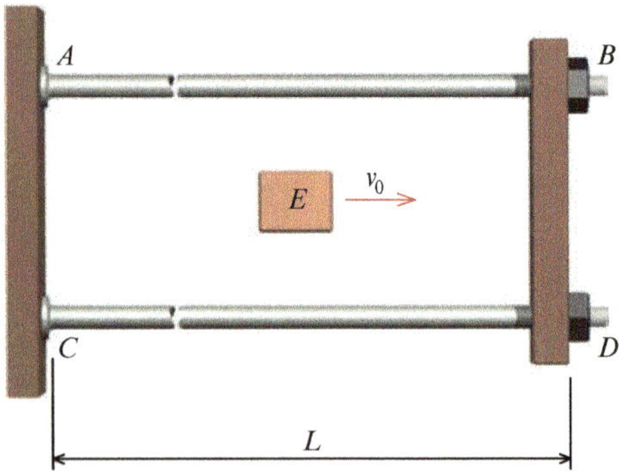

FIGURE P17.19/20

P17.20 In Figure P17.19/20, the 20 lb block *E* possesses a horizontal velocity v_0 when it squarely hits the yoke *BD* that is connected to the 1/4 in. diameter rods *AB* and *CD*. Both rods are made of 6061-T6 aluminum that has a yield strength $\sigma_Y = 40$ ksi and an elastic modulus $E = 10{,}000$ ksi, and both have a length $L = 30$ in. Yoke *BD* may be assumed to be rigid. Calculate the maximum allowable velocity v_0 of block *E* if a factor of safety of 3 with respect to yield is required for the maximum normal stress in the rods.

P17.21 The 120 kg block *C* shown in Figure P17.21 is dropped from a height *h* onto a wide-flange steel beam that spans $L = 6$ m. The steel beam has a moment of inertia $I = 125 \times 10^6$ mm^4, a depth $d = 300$ mm, a yield stress $\sigma_Y = 340$ MPa, and an elastic modulus $E = 200$ GPa. A factor of safety of 2.5 with respect to the yield stress is required for the maximum dynamic bending stress. If the falling block produces the maximum allowable dynamic bending stress, determine

(a) the equivalent static load.
(b) the maximum dynamic beam deflection at *A*.
(c) the maximum height *h* from which the 120 kg block *C* can be dropped.

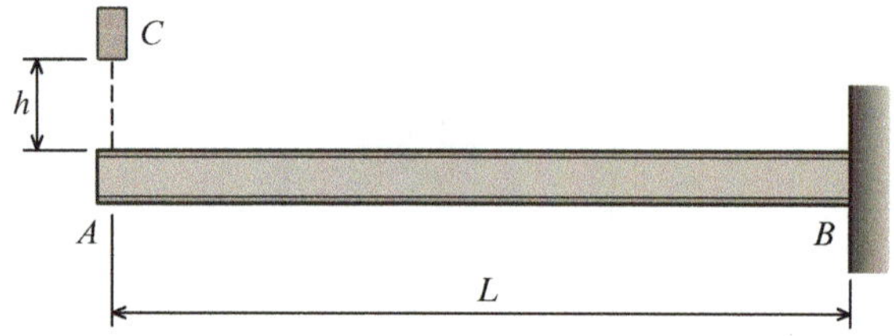

FIGURE P17.21

P17.22 The overhanging beam *ABC* shown in Figure P17.22/23 is made from an aluminum I shape, which has a moment of inertia $I = 25 \times 10^6$ mm^4, a depth $d = 200$ mm, and an elastic modulus $E = 70$ GPa. The beam spans are $a = 2.5$ m and $b = 1.5$ m. A block *D* with a mass of 90 kg is dropped from a height $h = 1.5$ m onto the free end of the overhang at *C*. Calculate

(a) the maximum bending stress in the beam.
(b) the maximum beam deflection at *C* due to the falling block.

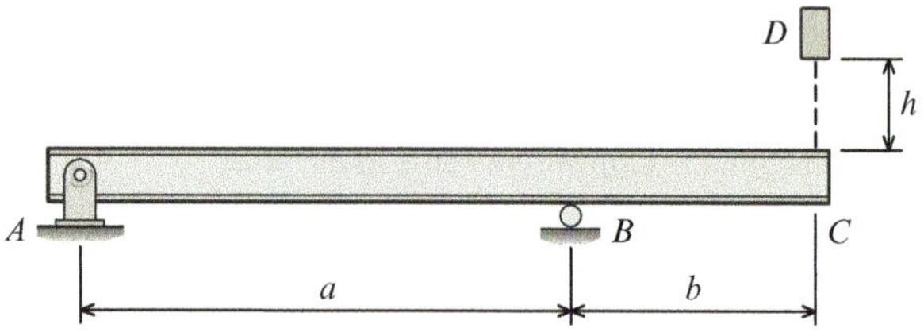

FIGURE P17.22/23

P17.23 In Figure P17.22/23, the overhanging beam *ABC*, made from an aluminum I shape, has a moment of inertia $I = 25 \times 10^6$ mm^4, a depth $d = 200$ mm, and an elastic modulus $E = 70$ GPa. The beam spans are $a = 3.5$ m and $b = 1.75$ m. A block *D* with a mass of 110 kg is dropped from a height *h* onto the free end of the overhang at *C*. If the maximum bending stress due to impact must not exceed 125 MPa, compute

(a) the maximum dynamic load allowed at *C*.
(b) the impact factor *n*.
(c) the maximum height *h* from which the 110 kg block *D* can be dropped.

P17.24 Figure P17.24 shows block D, weighing 200 lb, dropped from a height $h = 6$ ft onto a wide-flange steel beam that spans $L = 24$ ft with $a = 8$ ft and $b = 16$ ft. The steel beam has a moment of inertia $I = 300$ in.^{4}, a depth $d = 12$ in., and an elastic modulus $E = 29{,}000$ ksi. Determine

(a) the dynamic load applied to the beam.
(b) the maximum bending stress in the beam.
(c) the beam deflection at B due to the falling block.

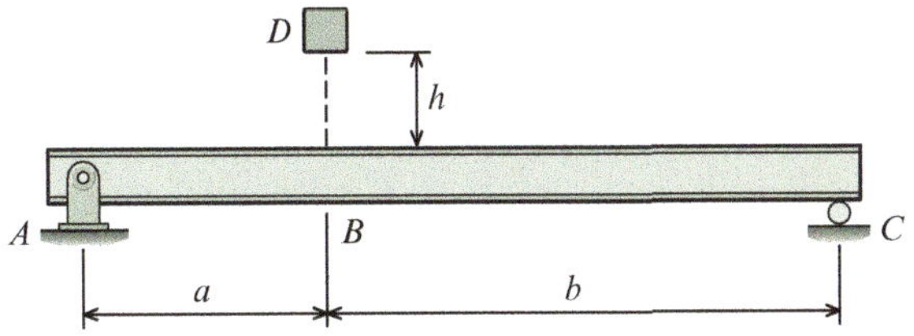

FIGURE P17.24

P17.25 A 75 lb block D at rest is dropped from a height $h = 2$ ft onto the top of the simply supported wooden beam. (See Figure P17.25.) The cross section of the beam is square—8 in. wide by 8 in. deep—and the modulus of elasticity of the wood is $E = 1{,}600$ ksi. The beam spans $L = 14$ ft and is supported at A and C by springs that each have a stiffness $k = 1{,}000$ lb/in. Assume that the springs at A and C do not restrain beam rotation. Compute

(a) the maximum beam deflection at B due to the falling block.
(b) the equivalent static load required to produce the same deflection as that in part (a).
(c) the maximum bending stress in the timber beam.

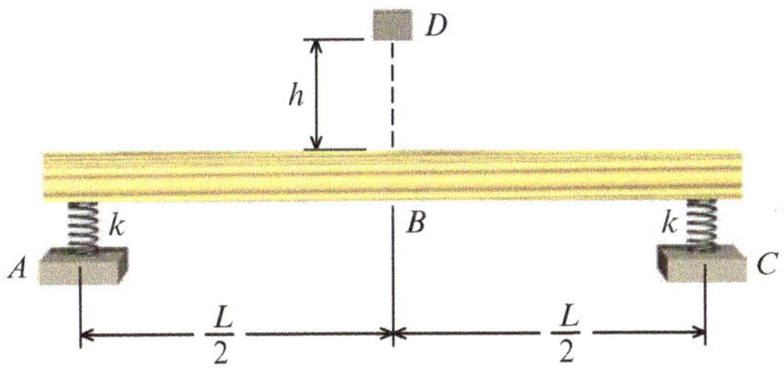

FIGURE P17.25

P17.26 The 120 kg block (Figure P17.26) is falling at 1.25 m/s when it is $h = 1{,}400$ mm above the spring that is located at midspan of the simply supported steel beam. The beam's moment of inertia is $I = 70 \times 10^6$ mm^4, its depth is $d = 250$ mm, and its elastic modulus is $E = 200$ GPa. $L = 5.5$ m is the beam span. The spring constant is $k = 100$ kN/m. Calculate

(a) the maximum beam deflection at B due to the falling block.
(b) the equivalent static load required to produce the same deflection as that in part (a).
(c) the maximum bending stress in the steel beam.

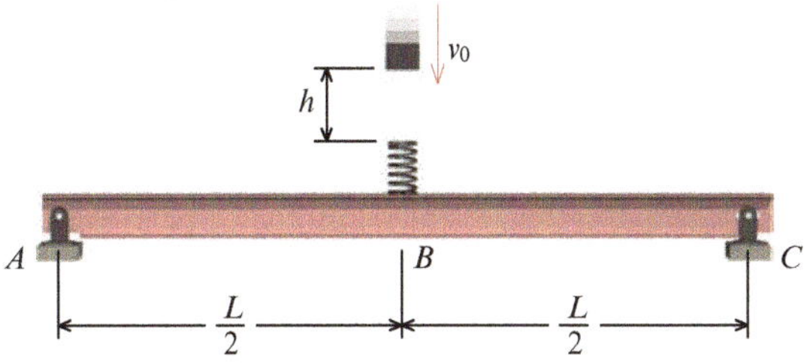

FIGURE P17.26

P17.27 The post AB shown in Figure P17.27/28 has a length $L = 2.25$ m. The post is made from a steel HSS that has a moment of inertia $I = 8.7 \times 10^6$ mm^4, a depth $d = 150$ mm, a yield strength $\sigma_Y = 315$ MPa, and an elastic modulus $E = 200$ GPa. A block with a mass $m = 25$ kg moves horizontally with a velocity v_0 and strikes the HSS post squarely at B. If a factor of safety of 1.5 is specified for the maximum bending stress, what is the largest acceptable velocity v_0 for the block?

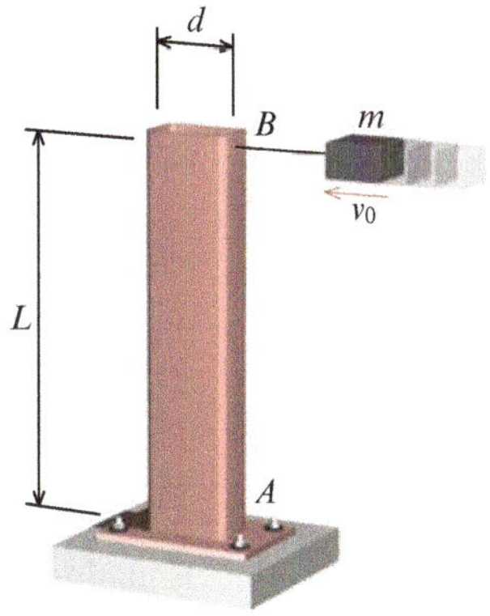

FIGURE P17.27/28

P17.28 In Figure P17.27/28, the post AB, with length $L = 4.2$ m, is made from a steel HSS with a moment of inertia $I = 24.4 \times 10^6$ mm^4, a depth $d = 200$ mm, a yield strength $\sigma_Y = 315$ MPa, and an elastic modulus $E = 200$ GPa. A block with a mass m moves horizontally with a velocity $v_0 = 4.5$ m/s and strikes the HSS post squarely at B. A factor of safety of 1.75 is specified for the maximum bending stress; determine the largest acceptable mass m for the block.

P17.29 The simply supported steel beam shown in Figure P17.29 is struck squarely at midspan by a 180 kg block moving horizontally with a velocity $v_0 = 2.5$ m/s. The beam's span is $L = 4$ m, its moment of inertia is $I = 15 \times 10^6$ mm^4, its depth is $d = 155$ mm, and its elastic modulus is $E = 200$ GPa. Compute

(a) the maximum dynamic load applied to the beam.
(b) the maximum bending stress in the steel beam.
(c) the maximum beam deflection at B due to the moving block.

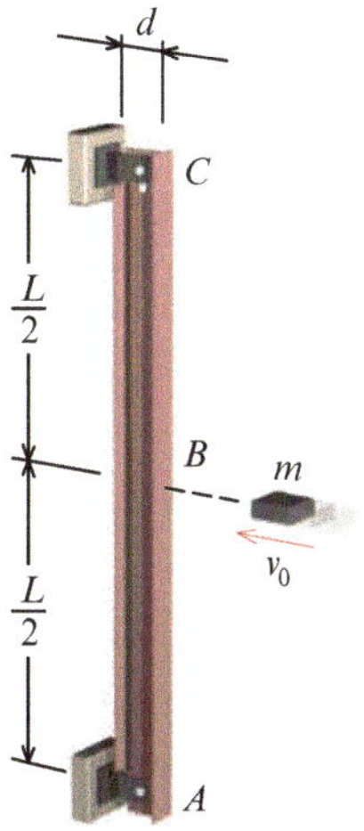

FIGURE P17.29

17.7 Work–Energy Method for Single Loads

As discussed in Section 17.2, the conservation-of-energy principle declares that energy in a closed system is never created or destroyed—it is only transformed from one state to another. For example, the work of an external load acting on a deformable body is transformed into internal strain energy, and provided that no energy is lost in the form of heat, the strain energy U is equal in magnitude to the external work W:

$$\boxed{W = U} \tag{17.27}$$

Equation (17.27) can be used to determine the deflection or slope of a member or structure under highly selective conditions—specifically, that the member or structure must be loaded by a single external concentrated force or concentrated moment. Corresponding displacements can be determined only at the location of the single load in the direction that the load acts. But why is this approach limited to a single external load or moment? The reason is that Equation (17.27) is the only equation available in that method. The strain energy U of the structure will be a single number. The work W performed by an external force acting on a deformable solid is one-half the product of the magnitude of the force and the displacement through which the solid moves in the direction of the force. (See Section 17.2.) Similarly, the work W performed by an external moment acting on a deformable solid is one-half the product of the magnitude of the moment and the angle through which the solid rotates. (See Section 17.5.) Consequently, if more than one external force or moment is applied, then W in Equation (17.27) will have more than one unknown deflection or rotation angle. Obviously, one equation cannot be solved for more than one unknown quantity.

Formulations for the strain energy were developed in Sections 17.3, 17.4, and 17.5 for axial deformation, torsional deformation, and flexural deformation, respectively. To recapitulate, the strain energy in prismatic axially loaded members can be determined from Equation (17.12):

$$U = \frac{P^2 L}{2AE}$$

For compound axial members and structures consisting of n prismatic axial members, the total strain energy in the member or structure can be computed with Equation (17.14):

$$U = \sum_{i=1}^{n} \frac{F_i^2 L_i}{2A_i E_i}$$

The strain energy in prismatic torsionally loaded members can be determined from Equation (17.16):

$$U = \frac{T^2 L}{2JG}$$

The total strain energy in compound torsional members can be computed from Equation (17.18):

$$U = \sum_{i=1}^{n} \frac{T_i^2 L_i}{2J_i G_i}$$

For a flexural member, the strain energy stored in the beam can be determined from Equation (17.20):

$$U = \int_0^L \frac{M^2}{2EI}\,dx$$

The external work done by a force acting on an axial member that deforms is

$$W = \tfrac{1}{2}P\delta$$

where δ is the distance (which equals the axial deformation) that the member moves *in the direction of the force.* The external work of a torque that acts on a shaft is

$$W = \tfrac{1}{2}T\phi$$

where ϕ is the rotation angle (in radians) through which the shaft rotates. For a beam subjected to a single external force, the work of the load is

$$W = \tfrac{1}{2}Pv$$

where v is the beam deflection at the location of the external force *in the direction of the load.* If the beam is subjected to a single external concentrated moment, the work of the external moment is

$$W = \tfrac{1}{2}M\theta$$

where θ is the beam slope (i.e., dv/dx) at the location of the external concentrated moment.

Another common use for the work–energy method involves the determination of deflections for simple trusses and for other assemblies of axial members. The work of a single external load acting on such a structure is

$$W = \tfrac{1}{2}P\Delta$$

where Δ is the deflection of the structure *in the direction that the force acts* at the location of the external load. To reiterate, the method described here and in the example that follows can be used only for structures subjected to a single external load, and only the deflection in the direction of the load can be determined.

While the work–energy method has limited application, it serves as a useful introduction to more powerful energy methods that will be developed in subsequent sections. These other energy methods can be used to perform a completely general deflection analysis on a member or structure.

EXAMPLE 17.9

A tie rod (1) and a pipe strut (2) are used to support a 50 kN load as shown. The cross-sectional areas are $A_1 = 650$ mm² for the tie rod and $A_2 = 925$ mm² for the pipe strut. Both members are made of structural steel that has an elastic modulus E = 200 GPa. Determine the vertical deflection of the two-member assembly at B.

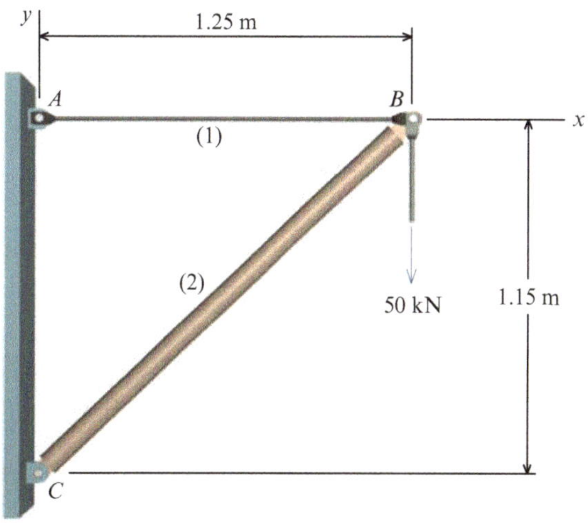

Plan the Solution

From a free-body diagram of joint B, the internal axial forces in members (1) and (2) can be calculated. From Equation (17.12), the strain energy of each member can be computed. The total strain energy in the assembly is found from the sum of the two strain energies. The total strain energy is then set equal to the work done by the 50 kN load as it deflects downward at B. From this conservation-of-energy equation, the unknown downward deflection of joint B can be determined.

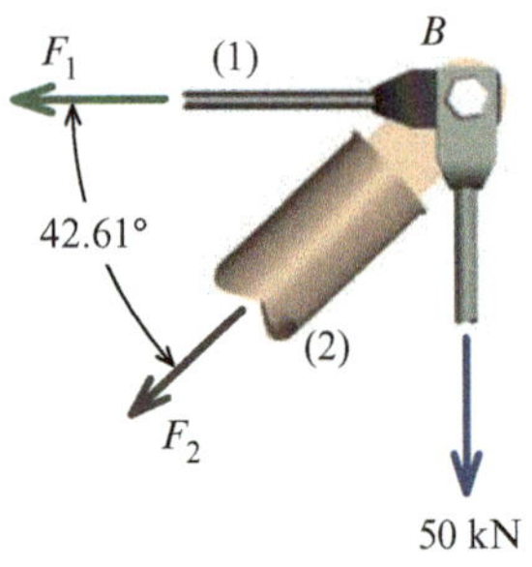

SOLUTION

The internal axial forces in members (1) and (2) can be calculated from equilibrium equations based on a free-body diagram of joint B. The sum of forces in the horizontal (x) direction can be written as

$$\Sigma F_x = -F_1 - F_2 \cos 42.61° = 0$$

and the sum of forces in the vertical (y) direction can be expressed as

$$\Sigma F_y = -F_2 \sin 42.61° - 50 \text{ kN} = 0$$
$$\therefore F_2 = -73.85 \text{ kN}$$

Substituting this result back into the preceding equation gives

$$F_1 = 54.36 \text{ kN}$$

The strain energy in tie rod (1) is

$$U_1 = \frac{F_1^2 L_1}{2A_1 E} = \frac{(54.36 \text{ kN})^2 (1.25 \text{ m})(1{,}000 \text{ N/kN})^2}{2(650 \text{ mm}^2)(200{,}000 \text{ N/mm}^2)} = 14.2068 \text{ N}\cdot\text{m}$$

The length of inclined pipe strut (2) is

$$L_2 = \sqrt{(1.25 \text{ m})^2 + (1.15 \text{ m})^2} = 1.70 \text{ m}$$

Thus, its strain energy is

$$U_2 = \frac{F_2^2 L_2}{2A_2 E} = \frac{(-73.85 \text{ kN})^2 (1.70 \text{ m})(1{,}000 \text{ N/kN})^2}{2(925 \text{ mm}^2)(200{,}000 \text{ N/mm}^2)} = 25.0581 \text{ N}\cdot\text{m}$$

The total strain energy of the two-bar assembly is, therefore,

$$U = U_1 + U_2 = 14.2068 \text{ N}\cdot\text{m} + 25.0581 \text{ N}\cdot\text{m} = 39.2649 \text{ N}\cdot\text{m}$$

The work of the 50 kN load can be expressed in terms of the downward deflection Δ of joint B as

$$W = \tfrac{1}{2}(50 \text{ kN})(1{,}000 \text{ N/kN})\Delta = (25{,}000 \text{ N})\Delta$$

From the conservation-of-energy principle, $W = U$; thus,

$$(25{,}000 \text{ N})\Delta = 39.2649 \text{ N}\cdot\text{m}$$
$$\therefore \Delta = 1.571 \times 10^{-3} \text{ m} = 1.571 \text{ mm}$$ **Ans.**

Compare this calculation method with the method demonstrated in Example 5.4, in which the same two-member assembly was considered. By the work–energy method, the downward deflection at B can be determined in a much simpler manner. However, the work–energy method cannot be used to determine the horizontal deflection of B.

PROBLEMS

P17.30 Determine the horizontal displacement of joint B of the two-bar assembly shown in Figure P17.30 if P = 80 kN. For this structure, x_1 = 3.0 m, y_1 = 3.5 m, and x_2 = 2.0 m. Assume that $A_1E_1 = 9.0 \times 10^4$ kN and $A_2E_2 = 38.0 \times 10^4$ kN.

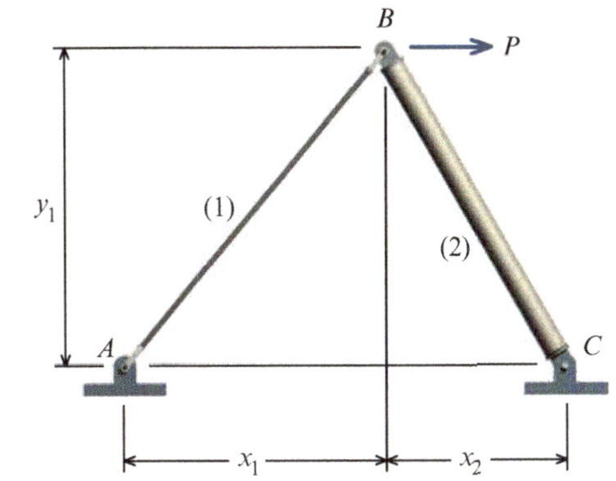

FIGURE P17.30

P17.31 Determine the vertical displacement of joint C of the truss shown in Figure P17.31 if P = 215 kN. For this structure, a = 3.5 m and b = 2.75 m. Assume that $AE = 8.50 \times 10^5$ kN for all members.

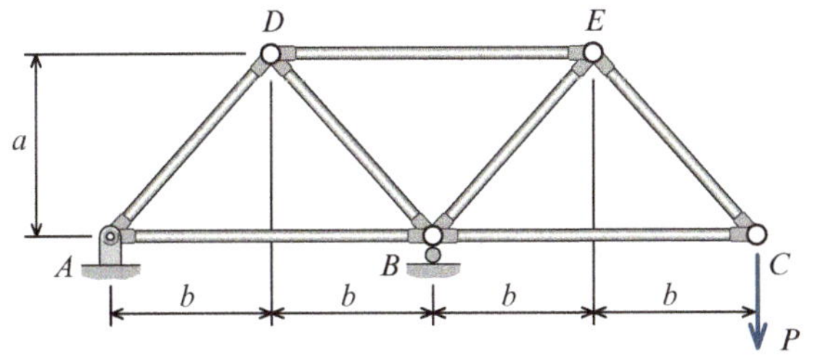

FIGURE P17.31

P17.32 Rigid bar BCD in Figure P17.32 is supported by a pin at C and by steel rod (1). A concentrated load P = 2.5 kips is applied to the lower end of aluminum rod (2), which is attached to the rigid bar at D. For this structure, a = 20 in. and b = 30 in. For steel rod (1), L_1 = 50 in., A_1 = 0.4 in.^{2}, and E_1 = 30,000 ksi. For aluminum rod (2), L_2 = 100 in., A_2 = 0.2 in.^{2}, and E_2 = 10,000 ksi. Determine the vertical displacement of point E.

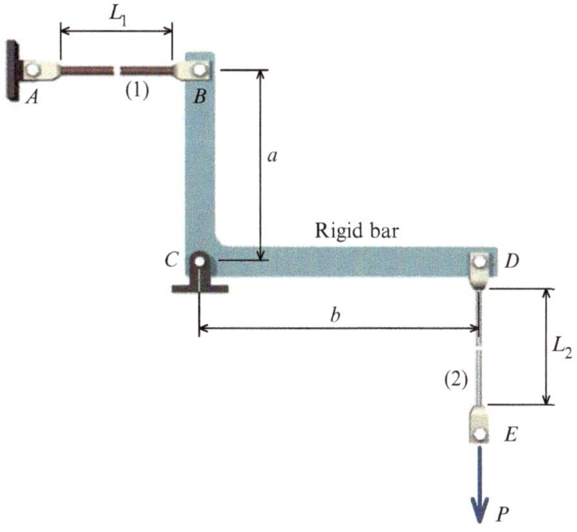

FIGURE P17.32

P17.33 Rigid bar ABC is supported by bronze rod (1) and aluminum rod (2) as shown in Figure P17.33. A concentrated load P = 90 kN is applied to the free end of aluminum rod (3). For this structure, a = 800 mm and b = 500 mm. For bronze rod (1), L_1 = 1.8 m, d_1 = 15 mm, and E_1 = 100 GPa. For aluminum rod (2), L_2 = 2.5 m, d_2 = 25 mm, and E_2 = 70 GPa. For aluminum rod (3), L_3 = 1.0 m, d_3 = 25 mm, and E_3 = 70 GPa. Determine the vertical displacement of point D.

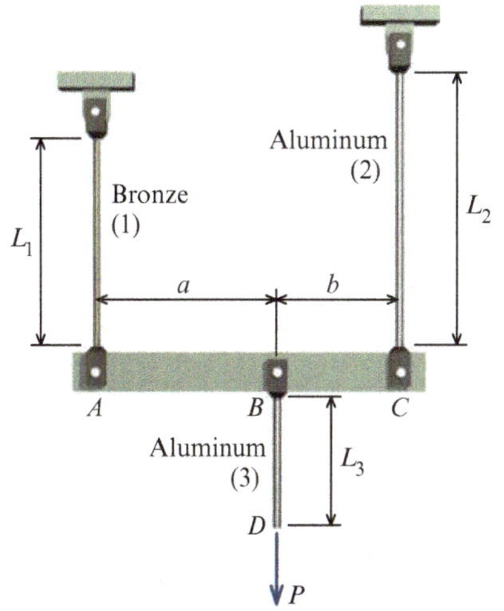

FIGURE P17.33

17.8 Method of Virtual Work

The method of virtual work is probably the most direct, versatile, and foolproof conservation-of-energy method for calculating deflections. The method may be used to determine deformations or deflections, at any location in a structure, that are caused by any type or combination of loads. The only limitation to the theory is that the principle of superposition must apply.

The principle of virtual work was first stated by Johann Bernoulli (1667–1748) in 1717. The term "virtual," as used in the context presented here, refers to an imaginary or hypothetical force or deformation, either finite or infinitesimal. Accordingly, the resulting work is only imaginary or hypothetical in nature. Before we discuss the virtual-work method in detail, some further discussion of work will be helpful.

Further Discussion of Work

As discussed in Section 17.2, work is defined as the product of a force that acts on a body and the distance that the body moves in the direction of the force. Work can be either a positive or a negative quantity. Positive work occurs when the body moves in the same direction as the force acts. Negative work occurs when the body moves opposite to the direction in which the force acts.

Consider the simple axial rod shown in Figure 17.19*a*. The rod is subjected to a load P_1. If applied gradually, the load increases in magnitude from zero to its final intensity P_1. The rod deforms in response to the increasing load, with each load increment dP producing an increment of deformation, $d\delta$. When the full magnitude of P_1 has been applied, the deformation of the rod is δ_1. The total work done by the load as it increases in magnitude from zero to P_1 can be determined from

$$W = \int_0^{\delta_1} P d\delta \tag{17.28}$$

As indicated by Equation (17.28), the work is equal to the area under the load–deformation diagram shown in Figure 17.19*b*. If the material behaves in a linear-elastic manner, the deformation varies linearly with the load as shown in Figure 17.19*c*. The work for linear-elastic behavior is given by the triangular area under the load–deformation diagram and is expressed as

$$W = \tfrac{1}{2} P_1 \delta_1$$

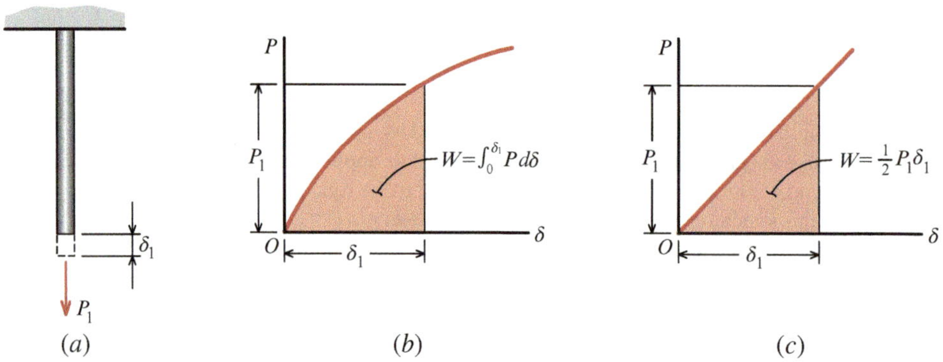

FIGURE 17.19 Work done by a single load on an axial rod.

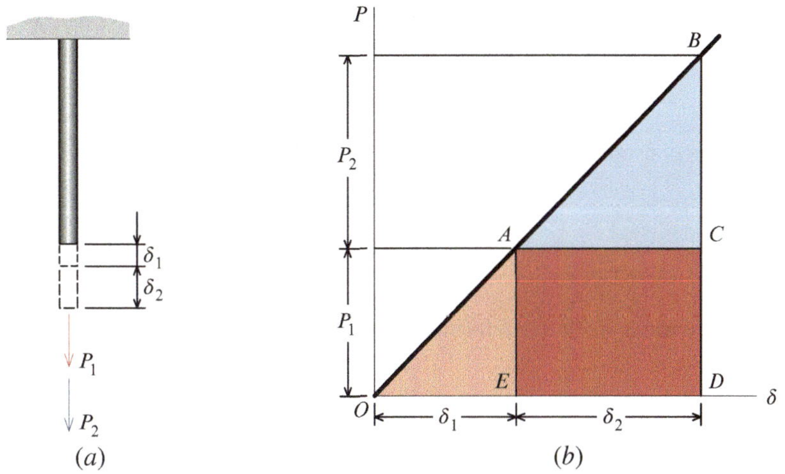

FIGURE 17.20 External work done by two loads on an axial rod.

Now, suppose that load P_1 has already been applied to the rod and a second load P_2 is gradually added, as shown in Figure 17.20*a*. The load P_2 causes the rod to elongate by an additional amount δ_2. The work done initially by the gradual application of the first load P_1 is

$$W = \tfrac{1}{2} P_1 \delta_1 \tag{a}$$

which corresponds to area *OAE* shown in Figure 17.20*b*. The work done by the gradual application of the second load P_2 is

$$W = \tfrac{1}{2} P_2 \delta_2 \tag{b}$$

which corresponds to area *ABC*. Area *ACDE*, the remaining area under the load–deformation diagram, represents the work performed by load P_1 as the rod deforms by the amount δ_2:

$$W = P_1 \delta_2 \tag{c}$$

Note that in this case load P_1 does not change its magnitude, because it was fully acting on the rod before load P_2 was applied.

To summarize, when a load is gradually applied, the expression for work contains the factor ½, as seen in Equations (a) and (b). Since the loads P_1 and P_2 increase from 0 to their maximum values, the terms $\tfrac{1}{2}P_1$ and $\tfrac{1}{2}P_2$ can be thought of as average loads. If a load is constant, however, the expression for work does not contain the factor ½, as seen in Equation (c). These two types of expressions—one with the factor ½ and the other without that factor—will be used to develop different methods for computing deflections.

The expressions for the work of concentrated moments are similar in form to those of concentrated forces. A concentrated moment does work when it rotates through an angle. The work dW that a concentrated moment M performs as it rotates through an incremental angle $d\theta$ is given by

$$dW = M d\theta$$

The total work of a gradually applied concentrated moment M through the rotation angle θ can be expressed by

$$W = \int_0^\theta M d\theta$$

If the material behaves linearly elastically, the work of a concentrated moment as it gradually increases in magnitude from 0 to its maximum value M can be expressed as

$$W = \frac{1}{2}M\theta$$

and if M remains constant during a rotation θ, the work is given by

$$W = M\theta$$

Principle of Virtual Work for Deformable Solids

The principle of virtual work for deformable solids can be stated as follows:

> If a deformable body is in equilibrium under a virtual-force system and remains in equilibrium while it is subjected to a set of small, compatible deformations, then the external virtual work done by the virtual external forces acting through the real external displacements (or rotations) is equal to the virtual internal work done by the virtual internal forces acting through the real internal displacements (or rotations).

There are three important provisions in the statement of this principle. First, the force system is *in equilibrium*, both *externally* and *internally*. Second, the set of deformations is *small*, implying that the deformations do not alter the geometry of the body significantly. Finally, the deformations of the structure are *compatible*, meaning that the elements of the structure must deform so that they do not break apart or become displaced away from the points of support. The parts of the body must stay connected after deformation and continue to satisfy the restraint conditions at the supports. These three conditions must always be satisfied in any application of the principle.

The principle of virtual work makes no distinction regarding the cause of the deformations: The principle holds whether the deformation is due to loads, temperature changes, misfits in lengths of members, or other causes. Further, the material may or may not follow Hooke's law.

To demonstrate the validity of the principle of virtual work, consider the statically determinate two-bar assembly shown in Figure 17.21*a*. The assembly is in equilibrium for an external virtual load P' that is applied at B. The free-body diagram of joint B is shown in Figure 17.21*b*. Since joint B is in equilibrium, the virtual external force P' and the virtual

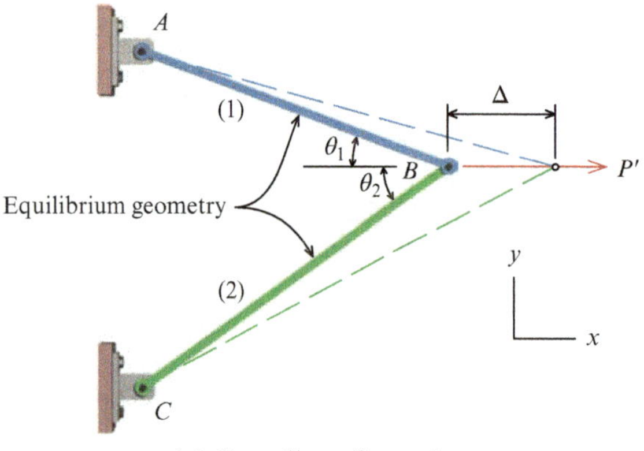

(*a*) Overall configuration

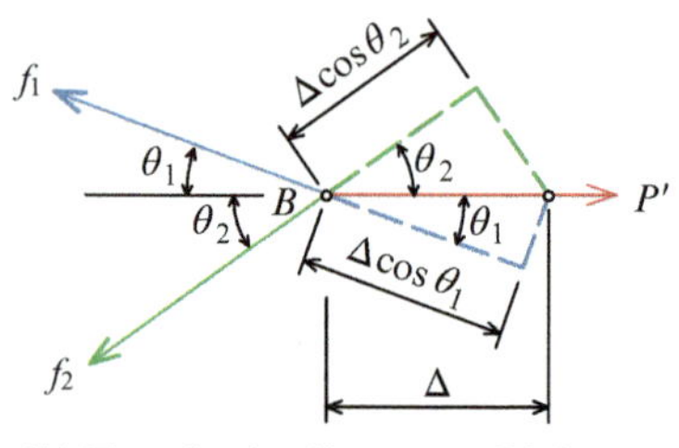

(*b*) Free-body diagram of joint B

FIGURE 17.21 Statically determinate two-bar assembly.

internal forces f_1 and f_2 acting in members (1) and (2), respectively, must satisfy the following two equilibrium equations:

$$\begin{aligned} \Sigma F_x &= P' - f_1 \cos\theta_1 - f_2 \cos\theta_2 = 0 \\ \Sigma F_y &= f_1 \sin\theta_1 - f_2 \sin\theta_2 = 0 \end{aligned} \qquad \text{(d)}$$

Next, we will assume that pin B is given a small real (as opposed to virtual) displacement Δ in the horizontal direction. Note that the displacement Δ is shown greatly exaggerated, for clarity, in Figures 17.21*a* and 17.21*b*. The real displacement Δ should be assumed to be small enough so that the displaced geometry of the two-bar assembly is essentially the same as the equilibrium geometry. Furthermore, the deformation of the two-bar assembly is compatible, meaning that bars (1) and (2) remain connected together at joint B and attached to their respective supports at A and C.

Since supports A and C do not move, the virtual forces f_1 and f_2 acting at these joints do not perform any work. The total virtual work for the two-bar assembly is thus equal to the algebraic sum of the separate bits of work performed by all the forces acting at joint B. The horizontal virtual external force P' moves the body it acts on through a real displacement Δ; thus, the work it performs is $P'\Delta$. Recalling that work is defined as the product of a force acting on a body and the distance that the body moves in the direction of the force, we observe that the virtual internal force f_1 in member (1) makes the body it acts on move through a distance $\Delta \cos \theta_1$ in a direction *opposite* to the direction of force f_1. Therefore, the virtual work done by the internal force in member (1) is *negative*, equal to $-f_1(\Delta \cos \theta_1)$. Similarly, the virtual internal force f_2 in member (2) makes the body *it* acts on move through a distance $\Delta \cos \theta_2$ in a direction *opposite* that of the force. Therefore, the virtual work done by the internal force in member (2) is $-f_2(\Delta \cos \theta_2)$. Consequently, the total work W_v done by the virtual forces acting at joint B is

$$W_v = P'\Delta - f_1(\Delta\cos\theta_1) - f_2(\Delta\cos\theta_2)$$

which can be restated as

$$W_v = (P' - f_1 \cos\theta_1 - f_2 \cos\theta_2)\Delta \qquad \text{(e)}$$

when Δ is factored out on the right-hand side.

The term in parentheses on the right-hand side of Equation (e) also appears in the equilibrium equation (d) for the sum of forces in the x direction; therefore, from Equation (d), we can conclude that the total virtual work for the two-bar assembly is $W_v = 0$. Then, from this observation, Equation (e) can be rewritten as

$$P'\Delta = f_1(\Delta\cos\theta_1) + f_2(\Delta\cos\theta_2) \qquad \text{(f)}$$

The term on the left-hand side of Equation (f) represents the virtual external work W_{ve} done by the virtual external load P' acting through the real external displacement Δ. On the right-hand side of Equation (f), the terms $\Delta \cos \theta_1$ and $\Delta \cos \theta_2$ are equal to the real internal deformations of bars (1) and (2), respectively. Consequently, the right-hand side of Equation (f) represents the virtual internal work W_{vi} of the virtual internal forces acting through the real internal displacements. As a result, Equation (f) can be restated as

$$W_{ve} = W_{vi} \qquad \text{(g)}$$

which is the mathematical statement of the principle of virtual work for deformable solids given in the box presented at the beginning of this section.

The general approach used to implement the principle of virtual work to determine deflections or deformations in a solid body can be described as follows:

1. Begin with the solid body to be analyzed. The solid body can be an axial member, a torsion member, a beam, a truss, a frame, or some other type of deformable solid. Initially, consider the solid body without external loads.
2. Apply an imaginary or hypothetical virtual external load to the solid body at the location where deflections or deformations are to be determined. Depending on the situation, this imaginary load may be a force, a torque, or a concentrated moment. For convenience, the imaginary load is assigned a "unit" magnitude, such as $P' = 1$.
3. The virtual load should be applied in the same direction as the desired deflection or deformation. For example, if the vertical deflection of a specific truss joint is desired, the virtual load should be applied in a vertical direction at that truss joint.
4. The virtual external load causes virtual internal forces throughout the body. These internal forces can be computed by the customary statics- or mechanics-of-materials techniques for any statically determinate system.
5. With the virtual load remaining on the body, apply the actual loads (i.e., the real loads) or introduce any specified deformations, such as those due to a change in temperature. These real external loads (or deformations) create real internal deformations, which can also be calculated by the customary mechanics-of-materials techniques for any statically determinate system.
6. As the solid body deflects or deforms in response to the real loads, the virtual external load and the virtual internal forces are displaced by some real amount. Consequently, the virtual external load and the virtual internal forces perform work. However, the virtual external load was present on the body, and the virtual internal forces were present in the body, before the real loads were applied. Accordingly, the work performed by them does not include the factor ½. [Refer to Equation (c) and Figure 17.20*b*.]
7. Conservation of energy, as shown in Equation (g), requires that the virtual external work equal the virtual internal work. From this relationship, the desired real external deflection or deformation can be determined.

Recalling that work is defined as the product of a force and a displacement, we can restate Equation (g) in words as

$$\begin{array}{c}\text{virtual}\\\text{external}\\\text{load}\end{array} \times \begin{array}{c}\text{real}\\\text{external}\\\text{displacement}\end{array} = \sum \left(\begin{array}{c}\text{virtual}\\\text{internal}\\\text{forces}\end{array} \times \begin{array}{c}\text{real}\\\text{internal}\\\text{displacements}\end{array} \right) \tag{17.29}$$

in which the terms *force* (or *load*) and *displacement* are used in a general sense and include moment and rotation, respectively. As Equation (17.29) indicates, the method of virtual work employs two independent systems: (a) a virtual-force system and (b) the real system of loads (or other effects) that create the deformations to be determined. To compute the deflection (or slope) at any location in a solid body or structure, a virtual-force system is chosen so that the desired deflection (or rotation) will be the only unknown in Equation (17.29).

Sections 17.9 and 17.10 illustrate the application of Equation (17.29) to trusses and beams, respectively.

17.9 Deflections of Trusses by the Virtual-Work Method

The method of virtual work is readily applied to structures such as trusses whose members are axially loaded. To develop the method, consider a truss that is subjected to two external loads P_1 and P_2 (Figure 17.22*a*). This truss consists of $j = 7$ axial members. The vertical deflection of the truss at joint *B* is to be determined.

Since the truss is statically determinate, the real internal force F_j created in each truss member by the application of real external loads P_1 and P_2 can be calculated by means of the method of joints. If F_j represents the real internal force in an arbitrary truss member *j* (e.g., member *CE* in Figure 17.22*a*), then the real internal deformation of the member is given by

$$\delta_j = \frac{F_j L_j}{A_j E_j}$$

in which *L*, *A*, and *E* denote the length, cross-sectional area, and elastic modulus, respectively, of member *j*. We will assume that each member has a constant cross-sectional area and that the load in each member is constant throughout the member's length.

Next, a virtual load system that is separate and independent from the real load system is carefully chosen so that the desired joint deflection can be determined. For this truss, the vertical deflection of joint *B* is desired. To obtain that deflection, first the real external loads P_1 and P_2 are removed from the truss and then a virtual external load having a magnitude of 1 is applied in a downward direction at joint *B*, as shown in Figure 17.22*b*. In response to this unit load, axial forces necessary to maintain equilibrium will be developed in each of the truss members.

To determine these forces, termed the *virtual internal forces* f_j, imagine that the truss is initially loaded only by the virtual external load (Figure 17.22*b*). Then, with the virtual load still in place, the real external loads P_1 and P_2 shown in Figure 17.22*a* are applied at joints *D* and *E*, respectively. Equation (17.29) can now be applied to express the virtual work done by the entire truss. The product of the virtual external load and the real external deflection Δ gives the virtual external work W_{ve}:

$$W_{ve} = 1 \cdot \Delta$$

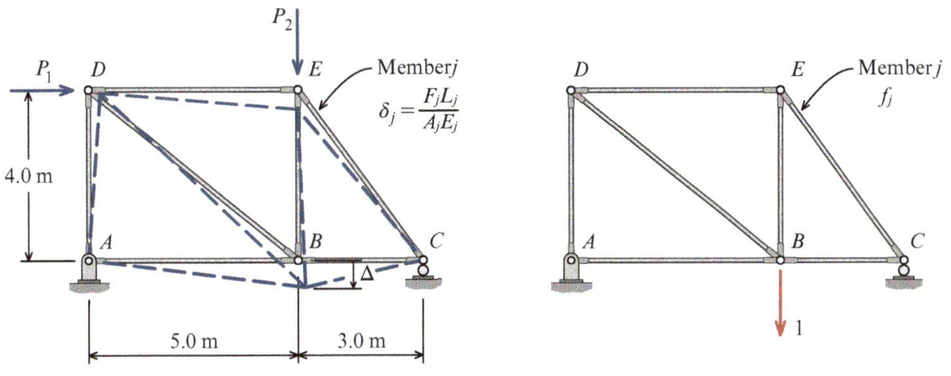

(*a*) Real system (*b*) Virtual system

FIGURE 17.22 Statically determinate truss.

The virtual internal work W_{vi} includes the work done by all truss members. For each member, the virtual internal work is the product of the virtual internal force f_j and the real internal deformation δ_j. Therefore, for the entire truss,

$$W_{vi} = \sum_j f_j \delta_j = \sum_j f_j \left(\frac{F_j L_j}{A_j E_j} \right)$$

Equating the external and internal virtual-work expressions gives

$$1 \cdot \Delta = \sum_j f_j \left(\frac{F_j L_j}{A_j E_j} \right) \tag{17.30}$$

in which

- 1 = virtual external unit load acting in the direction desired for Δ
- Δ = real joint displacement caused by the real loads that act on the truss
- f_j = virtual internal force created in truss member j when the truss is loaded with only the single virtual external unit load
- F_j = real internal force created in truss member j when the truss is loaded with all of the real loads
- L_j = length of truss member j
- A_j = cross-sectional area of truss member j
- E_j = elastic modulus of truss member j

The desired deflection Δ is the only unknown in Equation (17.30); consequently, we can determine its value by solving that equation.

Temperature Changes and Fabrication Errors

The length of an axial member changes in response to temperature changes. The axial deformation of truss member j due to a change in temperature ΔT_j can be expressed as

$$\delta_j = \alpha_j \Delta T_j L_j$$

where α_j is the coefficient of thermal expansion and L_j is the length of the member. Therefore, the displacement of a specific truss joint in response to temperature changes in some or all of the truss members can be determined from the virtual-work equation

$$1 \cdot \Delta = \sum_j f_j (\alpha_j \Delta T_j L_j) \tag{17.31}$$

Truss deflections due to fabrication errors can be determined by a simple substitution of changes in member lengths ΔL_j for δ_j in the virtual-work equation

$$1 \cdot \Delta = \sum_j f_j (\Delta L_j) \tag{17.32}$$

where ΔL_j is the difference in length of the member from its intended length when such difference is caused by a fabrication error.

The right-hand sides of Equations (17.30), (17.31), and (17.32) can be merged to consider trusses with combinations of external loads along with temperature changes or fabrication errors in some or all of their members:

$$1 \cdot \Delta = \sum_j f_j \left(\frac{F_j L_j}{A_j E_j} + \alpha_j \Delta T_j L_j + \Delta L_j \right) \tag{17.33}$$

Procedure for Analysis

The following procedure is recommended for calculating truss deflections by the virtual-work method:

1. **Real System:** If real external loads act on the truss, use the method of joints or the method of sections to determine the real internal forces in each truss member. Take care to be consistent in the signs associated with truss member forces and deformations. It is strongly recommended that tensile axial forces and elongation deformations be considered as positive quantities. In that case, a positive member force corresponds to an increase in member length. If this convention is followed, then increases in temperature and increases in member length due to fabrication errors should also be taken as positive quantities.
2. **Virtual System:** Begin by removing all real external loads that act on the truss. Then apply a single virtual unit load at the joint at which the deflection is desired. This unit load should act in the direction of the desired deflection. With the unit load in place and all real loads removed, analyze the truss to determine the member forces f_j produced in response to the virtual external load. The sign convention used for the member forces must be the same as that adopted in step 1.
3. **Virtual-Work Equation:** Apply the virtual-work equation, Equation (17.30), to determine the deflection at the desired joint due to real external loads. It is important to retain the algebraic sign for each of the f_j and F_j forces when these terms are substituted into the equation. If the right-hand side of Equation (17.30) turns out to be positive, then the displacement Δ is in the direction assumed for the virtual unit load. A negative result for the right-hand side of Equation (17.30) means that the displacement Δ actually acts opposite to the direction assumed for the virtual unit load.

   If the truss deflection is caused by temperature changes, then Equation (17.31) will be used. If the truss deflection is caused by fabrication errors, then Equation (17.32) is called for. Equation (17.33) can be used when a combination of real external loads, temperature changes, and fabrication misfits must be considered.

   The application of these virtual-work expressions can be facilitated by an arrangement of the real and virtual quantities into a tabular format, which will be demonstrated in subsequent examples.

EXAMPLE 17.10

Compute the vertical deflection at joint B for the truss shown in Figure 17.22a. Assume that $P_1 = 10$ kN and $P_2 = 40$ kN. For each member, the cross-sectional area is $A = 525$ mm^2 and the elastic modulus $E = 70$ GPa.

Plan the Solution

Calculate the length of each truss member. Determine the real internal forces F_j in all of the truss members, using an appropriate method, such as the method of joints. Remove both P_1 and P_2 from the truss, apply a unit load downward at joint B, and perform a second truss analysis to determine the member forces f_j created by the unit load. Construct a table of results from the two truss analyses, and then apply Equation (17.30) to determine the downward deflection Δ of joint B.

SOLUTION

A tabular format is a convenient way to organize the calculations. Compute the member lengths and record them in a column. Perform a truss analysis, using the real loads $P_1 =$ 10 kN and $P_2 = 40$ kN, and record the real internal forces F (i.e., the forces produced in the truss members by the real loads) in a second column. Note that tension member forces are assumed to be positive values here. These real internal forces will be used to calculate the real internal deformations. Accordingly, a positive force corresponds to elongation of the member.

Remove the real loads P_1 and P_2 from the truss. Since the downward deflection of the truss at joint B is to be determined, apply a downward virtual load of 1 kN at joint B, as shown in Figure P17.22*b*, and perform a second truss analysis. Again, use the sign convention that tension forces are positive. The member forces obtained from this second analysis are the virtual internal forces f. Record these results in a column.

Multiply the virtual internal force f by the real internal force F and the member length L for each truss member, and record the product in a final column. Sum all of the values for f that you have recorded, as well as all of the values in the final column, taking care to note the units that have been used.

Note that the cross-sectional area A and the elastic modulus E are the same for all members in this particular example. Therefore, they can be included after the two summations. If A and E differ for any truss members, additional columns will need to be added to the tabular format to account for the differences.

Following is the table produced by the preceding instructions for the truss shown in Figure 17.22a:

Member	L (m)	F (kN)	f (kN)	$f\,(FL)$ (kN$^2\cdot$m)
AB	5.0	10.0	0.0000	0.000
AD	4.0	−10.0	−0.3750	15.000
BC	3.0	22.5	0.4688	31.644
BD	6.403	16.008	0.6003	61.530
BE	4.0	−10.0	0.6250	−25.000
CE	5.0	−37.5	−0.7813	146.494
DE	5.0	−22.5	−0.4688	52.740
			$\sum f(FL) =$	282.408

Equation (17.30) can now be applied:

$$1 \cdot \Delta = \sum_j f_j \left(\frac{F_j L_j}{A_j E_j} \right) = \frac{1}{AE} \sum_j f_j \left(F_j L_j \right)$$

Recall that the left-hand side of this equation represents the external work performed by the virtual external load as the member it acts on moves through the real joint deflection at B. The right-hand side represents the internal work performed by the virtual internal forces f as the members they act on move through the real internal deformations that occur in them in response to the real external loads P_1 and P_2.

From the tabulated results,

$$(1\text{ kN})\cdot\Delta_B = \frac{(282.408\text{ kN}^2\cdot\text{m})(1{,}000\text{ N/kN})(1{,}000\text{ mm/m})}{(525\text{ mm}^2)(70{,}000\text{ N/mm}^2)}$$

$$\Delta_B = 7.68\text{ mm}$$ **Ans.**

Since the virtual load was applied in a downward direction at B, the positive value of the result confirms that joint B does become displaced downward.

EXAMPLE 17.11

For the truss shown, members BF, CF, CG, and DG each have cross-sectional areas of 750 mm². All other members have cross-sectional areas of 1,050 mm². The elastic modulus of all members is 70 GPa. Compute

(a) the horizontal deflection at joint G.
(b) the vertical deflection at joint G.

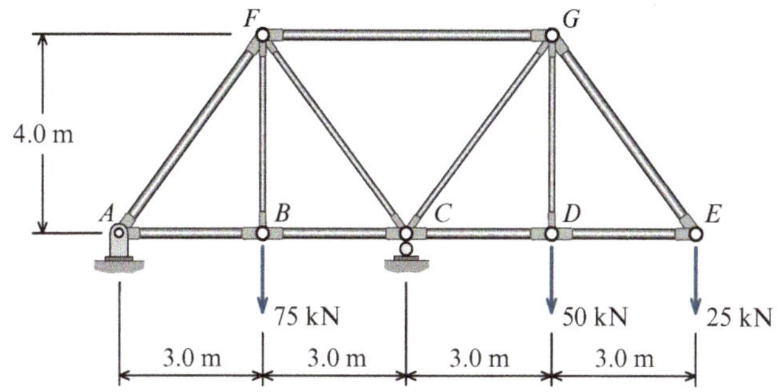

Plan the Solution

Calculate the length of each truss member. Determine the real internal forces F_j in all of the truss members, using an appropriate method, such as the method of joints or the method of sections. To compute the horizontal deflection at G, remove all loads from the truss, apply a unit load horizontally at joint G, and perform a second truss analysis to determine the member forces f_j created by the horizontal unit load. Construct a table of results from the two truss analyses, and then apply Equation (17.30) to determine the horizontal deflection Δ of joint G. To compute the vertical deflection at G, remove all loads from the truss, apply a unit load vertically at joint G, and perform a third truss analysis to determine the member forces f_j created by the vertical unit load. Construct a table of results from the two truss analyses, and then apply Equation (17.30) to calculate the vertical deflection Δ of joint G.

SOLUTION

(a) Horizontal Deflection of Joint G: Compute the member lengths and record them in a column. Note the member cross-sectional areas and record them in a second column. With the real external loads acting on the truss, perform a truss analysis to determine the real internal forces in each of the truss members. Record these real internal forces in a third column.

Remove the real loads from the truss. Since the horizontal deflection of the truss at joint G is to be determined, apply a horizontal virtual load of 1 kN at that joint. For this analysis, the virtual load will be directed to the right. Perform a second analysis of the truss in which the only load acting on the truss is the horizontal virtual load. This second analysis yields the virtual member forces f. Record these forces in a column.

Following is the table produced by the preceding instructions for the truss considered here:

Member	L (mm)	A (mm²)	F (kN)	f (kN)	$f\left(\frac{FL}{A}\right)$ (kN²/mm)
AB	3,000	1,050	−9.375	0.500	−13.393
AF	5,000	1,050	15.625	0.833	61.979
BC	3,000	1,050	−9.375	0.500	−13.393
BF	4,000	750	75.000	0	0
CD	3,000	1,050	−18.750	0	0
CF	5,000	750	−109.375	−0.833	607.396
CG	5,000	750	−93.750	0	0
DE	3,000	1,050	−18.750	0	0
DG	4,000	750	50.000	0	0
EG	5,000	1,050	31.250	0	0
FG	6,000	1,050	75.000	1.000	428.571
				$\sum f\left(\frac{FL}{A}\right) =$	1,071.161

Equation (17.30) can now be applied:

$$1 \cdot \Delta = \sum_j f_j \left(\frac{F_j L_j}{E_j} \right) = \frac{1}{E} \sum_j f_j \left(\frac{F_j L_j}{A_j} \right)$$

From the tabulated results,

$$(1 \text{ kN}) \cdot \Delta_G = \frac{(1{,}071.161 \text{ kN}^2/\text{mm})(1{,}000 \text{ N/kN})}{(70{,}000 \text{ N/mm}^2)}$$

$$\Delta_G = 15.30 \text{ mm} \rightarrow$$ **Ans.**

Since the virtual load was applied horizontally to the right at G, the positive value of the result confirms that joint G does become displaced to the right.

(b) Vertical Deflection of Joint G: Again, remove all loads from the truss. The vertical deflection of the truss at joint G is to be determined next; therefore, apply a vertical virtual load of 1 kN at joint G. For this analysis, the virtual load will be directed downward. Perform a third analysis of the truss in which the vertical virtual load is the only load acting on the truss. This analysis produces a different set of virtual internal forces f. Replace the previous values of f with these results, and recalculate the final column of the table:

Member	L (mm)	A (mm²)	F (kN)	f (kN)	$f\left(\frac{FL}{A}\right)$ (kN²/mm)
AB	3,000	1,050	−9.375	−0.375	10.045
AF	5,000	1,050	15.625	0.625	46.503
BC	3,000	1,050	−9.375	−0.375	10.045
BF	4,000	750	75.000	0	0.000
CD	3,000	1,050	−18.750	0	0.000
CF	5,000	750	−109.375	−0.625	455.729
CG	5,000	750	−93.750	−1.250	781.250
DE	3,000	1,050	−18.750	0	0.000
DG	4,000	750	50.000	0	0.000
EG	5,000	1,050	31.250	0	0.000
FG	6,000	1,050	75.000	0.750	321.429
				$\sum f\left(\frac{FL}{A}\right) =$	1,625.001

From the tabulated results,

$$(1\text{ kN})\cdot\Delta_G = \frac{(1{,}625.001\text{ kN}^2/\text{mm})(1{,}000\text{ N/kN})}{(70{,}000\text{ N/mm}^2)}$$

$$\Delta_G = 23.2\text{ mm}\downarrow$$ **Ans.**

EXAMPLE 17.12

For the truss shown, determine the vertical deflection of joint D if the temperature of the truss drops 90°F. Each member has a cross-sectional area of 1.25 in.², a coefficient of thermal expansion of 13.1×10^{-6}/°F, and an elastic modulus of 10,000 ksi.

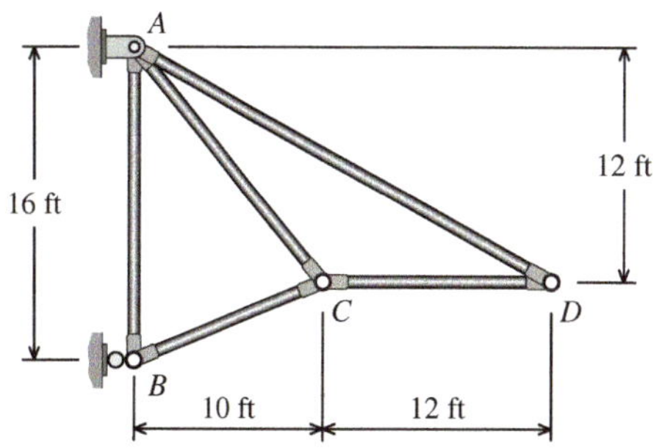

Plan the Solution

Calculate the length of each truss member. There are no external loads on the truss, but the temperature change will cause each member to contract in length. Determine the axial deformation that will occur in each member in response to a temperature change of $\Delta T = -90$°F. To compute the vertical deflection at D, apply a unit load vertically downward at joint D and perform truss analysis to determine the member forces f_j created by the vertical unit load. Construct a table of results that consists of the values for the member deformations and the virtual internal forces, and then apply Equation (17.31) to determine the vertical deflection Δ of joint D.

SOLUTION

Calculate the member lengths and record them in a column. The axial deformation of a truss member due to a change in temperature is given by $\delta = \alpha\,\Delta TL$. With this expression, calculate the real internal deformation produced in each member by the temperature change and record the values in a second column.

For this example, there are no real external loads that perform work on the truss. Since the vertical deflection of the truss at joint D is to be determined, apply a vertical virtual load of 1 kip in a downward direction at that joint. Perform a truss analysis and compute the corresponding virtual internal forces f. Record these results in a column.

Following is the table produced by the preceding instructions for the given truss:

Member	L (in.)	$\alpha\Delta TL$ (in.)	f (kips)	$f(\alpha\Delta TL)$ (kip·in.)
AB	192	−0.2264	0.550	−0.125
AC	187	−0.2210	−0.716	0.158
AD	301	−0.3545	2.088	−0.740
BC	129.244	−0.1524	−1.481	0.226
CD	144	−0.1698	−1.833	0.311
			$\sum f(\alpha\Delta TL) =$	−0.170

Equation (17.31) can now be applied:

$$1 \cdot \Delta = \sum_j f_j(\alpha_j \Delta T_j L_j)$$

The left-hand side of this equation represents the external work performed by the virtual external load as the member it acts on moves through the real joint deflection at D. The right-hand side of the equation represents the internal work performed by the virtual internal forces f as the members they act on move through the real internal deformations that occur in them in response to the temperature change.

From the tabulated results,

$$(1\text{ kip}) \cdot \Delta_D = -0.170\text{ kip} \cdot \text{in.}$$

$$\Delta_D = -0.170\text{ in.} = 0.170\text{ in.} \uparrow$$ **Ans.**

The virtual load at D was applied in a downward direction. The negative value obtained here means that joint D actually moves in the opposite direction—that is, upward.

17.10 Deflections of Beams by the Virtual-Work Method

The principle of virtual work can be used to determine the deflection of a beam. Consider a beam subjected to an arbitrary loading, as shown in Figure 17.23*a*. Assume that the vertical deflection of the beam at point B is desired. To determine this deflection, a virtual external unit load will first be applied to the beam at B in the direction of the desired deflection, as shown in Figure 17.23*b*. If the beam (as shown in Figure 17.23*b*) is then subjected to the

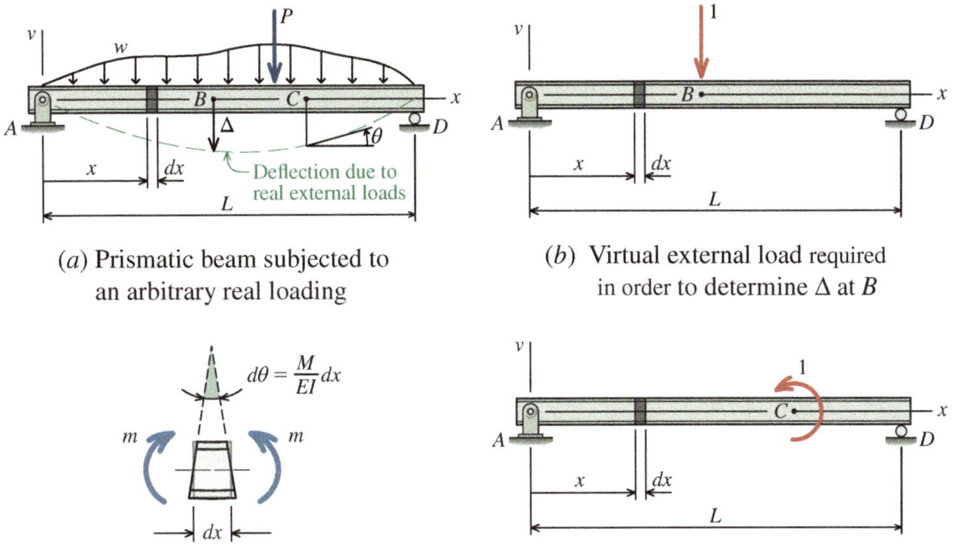

(*a*) Prismatic beam subjected to an arbitrary real loading

(*b*) Virtual external load required in order to determine Δ at *B*

(*c*) Internal work of virtual moment *m*

(*d*) Virtual external moment required in order to determine *θ* at *C*

FIGURE 17.23 Virtual-work method for beams.

deformations created by the real external loads (in Figure 17.23*a*), the virtual external work performed by the virtual external load as the beam moves downward through the real deflection Δ will be

$$W_{ve} = 1 \cdot \Delta \tag{a}$$

To obtain the virtual internal work, recall from Section 17.5 that the internal work of a beam is related to the moment and the rotation angle θ of the beam. Now consider a differential beam element dx located at a distance x from the left support, as shown in Figures 17.23*a* and 17.23*b*. When the real external loads are applied to the beam, bending moments M rotate the plane sections of the beam segment dx through an angle

$$d\theta = \frac{M}{EI}dx \tag{b}$$

When the beam with the virtual unit load (Figure 17.23*b*) is subjected to the real rotations caused by the external loading (Figure 17.23*a*), the virtual internal bending moment m acting on the element dx performs virtual work as the element undergoes the real rotation $d\theta$, as shown in Figure 17.23*c*. For beam element dx, the virtual internal work dW_{vi} performed by the virtual internal moment m as the element rotates through the real internal rotation angle $d\theta$ is

$$dW_{vi} = m\,d\theta \tag{c}$$

Note that the virtual moment m remains constant during the real rotation $d\theta$; therefore, Equation (c) does not contain the factor ½. (Compare Equation (c) with the expression for work in Figure 17.16.)

Now substitute the expression for $d\theta$ in Equation (b) into Equation (c) to obtain

$$dW_{vi} = m\left(\frac{M}{EI}\right)dx \tag{d}$$

Obtain the total virtual internal work done on the beam by integrating Equation (d) over the length of the beam:

$$W_{vi} = \int_0^L m\left(\frac{M}{EI}\right)dx \tag{17.34}$$

This expression represents the amount of virtual strain energy that is stored in the beam.

Finally, the virtual external work [Equation (a)] can be equated to the virtual internal work [Equation (17.34)], giving the virtual-work equation for beam deflections:

$$1 \cdot \Delta = \int_0^L m\left(\frac{M}{EI}\right)dx \tag{17.35}$$

The principle of virtual work can also be used to determine the angular rotation of a beam. Note that the slope of a beam can be expressed in terms of its angular rotation θ (measured in radians) as

$$\frac{dv}{dx} = \tan\theta$$

If the beam deflections are assumed to be small, as is typically the case, then $\tan\theta \cong \theta$ and the slope of the beam is equal to

$$\frac{dv}{dx} \cong \theta$$

The terms *angular rotation* and *slope* are thus effectively synonymous, provided that the beam deflections are small.

Now consider again a beam subjected to an arbitrary loading as shown in Figure 17.23*a*. Assume that the angular rotation θ of the beam at point C is desired. To determine θ, a virtual external unit moment will first be applied to the beam at C in the direction of the anticipated slope, as shown in Figure 17.23*d*. If this beam (as illustrated in Figure 17.23*d*) is then subjected to the deformations created by the real external loads (in Figure 17.23*a*), the virtual external work W_{ve} performed by the virtual external moment as the beam rotates counterclockwise through the real beam angular rotation θ is

$$W_{ve} = 1 \cdot \theta \tag{e}$$

The expression for the virtual internal work developed in Equation (17.34) remains the same as before, with the exception that m now represents the virtual internal moment created by the load of Figure 17.23*d*. Thus, the virtual-work equation for beam slopes is

$$1 \cdot \theta = \int_0^L m\left(\frac{M}{EI}\right)dx \tag{17.36}$$

In deriving Equation (17.34) for the virtual internal work performed in the beam, the internal work performed by virtual shear forces acting through real shear deformations has been neglected. Consequently, the virtual-work expressions in Equations (17.35) and (17.36) do not account for shear deformations in beams. However, shear deformations are very small for most common beams (with the exception of very deep beams), and they can be neglected in ordinary analyses.

In evaluating the integrals in Equations (17.35) and (17.36), a single integration over the entire length of the beam may not be possible. Concentrated forces or moments, or

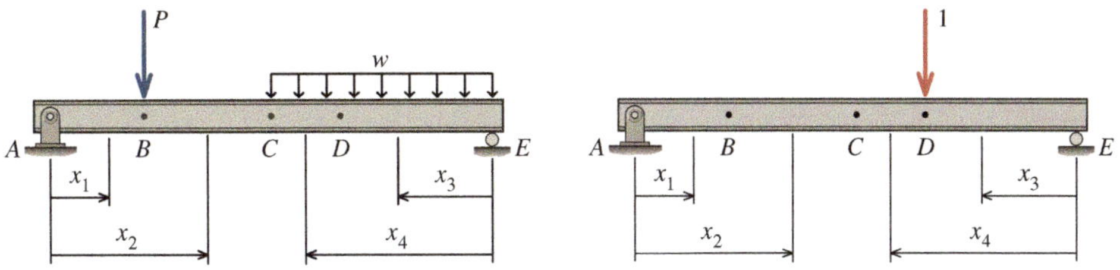

(a) Real loads (b) Virtual loads

FIGURE 17.24 Choice of x coordinates for integration of expressions in M and m.

distributed loadings spread across only a portion of the span, will cause discontinuities in the moment equation for a beam. For example, suppose that the deflection at point D is to be determined for the beam shown in Figure 17.24a. The real internal bending moments M could be expressed in equations written for segments AB, BC, and CE of the beam. From Figure 17.24b, however, we observe that the virtual internal moment m could be expressed with only two equations: one for segment AD and the other for segment DE. Note, though, that Equations (17.35) and (17.36) must be continuous functions throughout the length of the segment in order for a single integration to be carried out for each of them. Since, however, they are both discontinuous at D, segment CE of the beam must be further subdivided into segments CD and DE.

Typically, several x coordinates must be employed in order to express the moment equation for various regions of the beam span. To evaluate the integral in Equation (17.35), equations for the real internal bending moment M and the virtual internal bending moment m in each of segments AB, BC, CD, and DE of the beam must be derived. Separate x coordinates may be chosen to facilitate the formulation of moment equations for each of these segments. It is not necessary that these x coordinates all have the same origin; however, *it is necessary that the same x coordinate be used for both the real-moment and the virtual-moment equations* that are derived for any specific segment of the beam. For example, coordinate x_1, with origin at A, may be used with both Equation (17.35) and Equation (17.36) for segment AB of the beam. Then, a separate coordinate x_2, also with origin at A, may be used for the moment equations applicable to segment BC. Next, a third coordinate, x_3, with origin at E, may be used for segment DE of the beam, and a fourth coordinate, x_4, could be used to formulate the expressions for segment CD. In any case, each x coordinate should be chosen to facilitate the formulation of equations describing both the real internal moment M and the virtual internal moment m.

Procedure for Analysis

The following procedure is recommended for calculating beam deflections and slopes by the virtual-work method:

1. **Real System:** Draw a beam diagram showing all real loads.
2. **Virtual System:** Draw a diagram of the beam with all real loads removed. If a beam deflection is to be determined, apply a unit load at the location desired for the deflection. If a beam slope is to be determined, apply a unit moment at the desired location.
3. **Subdivide the Beam:** Examine both the real and virtual load systems. Also, consider any variations of the flexural rigidity EI that may exist in the beam. Divide the beam into segments so that the equations for the real and virtual loadings, as well as the flexural rigidity EI, are continuous in each segment.
4. **Derive Moment Equations:** For each segment of the beam, formulate an equation for the bending moment m produced by the virtual external load. Formulate a second equation expressing the variation in the bending moment M produced in the beam by

the real external loads. (Review Section 7.2 for a discussion on deriving bending-moment equations.) Note that the same x coordinate must be used in both equations. The origin for the x coordinate may be located anywhere on the beam and should be chosen so that the number of terms in the equation is minimized. Use the standard convention for bending-moment signs, illustrated in Figures 7.6 and 7.7, for both the virtual and real internal-moment equations.

5. **Virtual-Work Equation:** Determine the desired beam deflection by applying Equation (17.35), or compute the desired beam slope by applying Equation (17.36). If the beam has been divided into segments, then you can evaluate the integral on the right-hand side of Equation (17.35) or Equation (17.36) by algebraically adding the integrals for all segments of the beam. It is, of course, important to retain the algebraic sign of each integral calculated within a segment.

   If the algebraic sum of all of the integrals for the beam is positive, then Δ or θ is in the same direction as the virtual unit load or virtual unit moment. If a negative value is obtained, then the deflection or slope acts opposite to the direction of the virtual unit load or virtual unit moment.

   Examples 17.13–17.15 illustrate use of the virtual-work method to determine beam deflections and beam slopes.

EXAMPLE 17.13

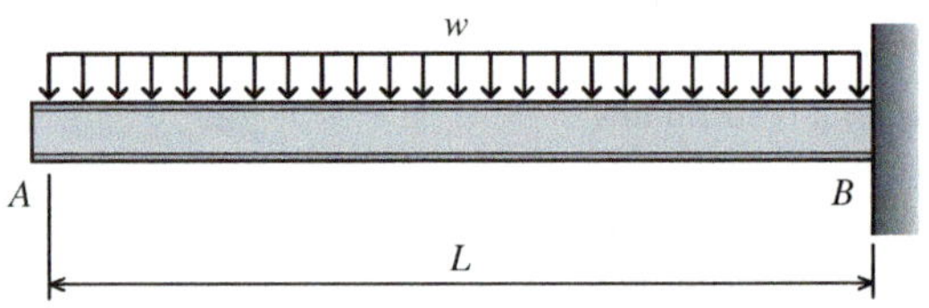

Calculate (a) the deflection and (b) the slope at end A of the cantilever beam shown. Assume that EI is constant.

Plan the Solution

The deflection at end A can be determined through the use of a virtual unit load acting downward at A. Consider the beam with the real load w removed and a virtual load applied at A. An equation for the variation of the virtual internal moment m can be derived, and this equation will be continuous over the entire length of the span. Next, consider the beam without the virtual load, but with the real load w reapplied. Derive an equation for the variation of the real internal moment M. This equation will also be continuous over the entire span. Therefore, the beam need not be subdivided in calculating its deflection and slope. Once equations for m and M are obtained, apply Equation (17.35) to compute the beam deflection Δ at A. To determine the slope of the beam at A, the virtual load will be a concentrated moment applied at that point. After deriving a new equation for m, use Equation (17.36) to calculate θ.

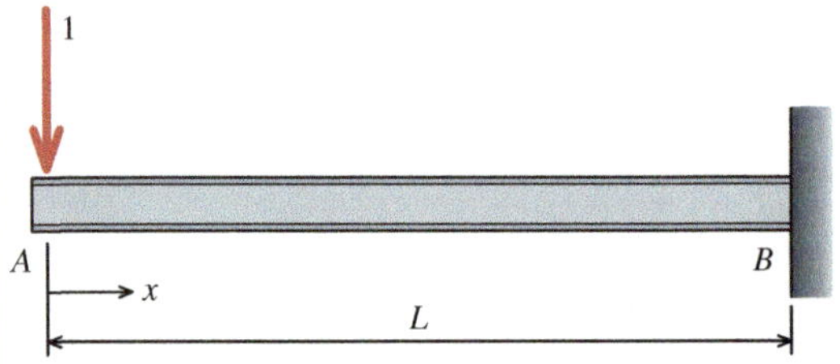

SOLUTION

(a) Virtual Moment m for Calculating the Beam Deflection: To determine the downward deflection of the cantilever beam, first remove the real load w from the beam and apply a virtual unit load downward at A.

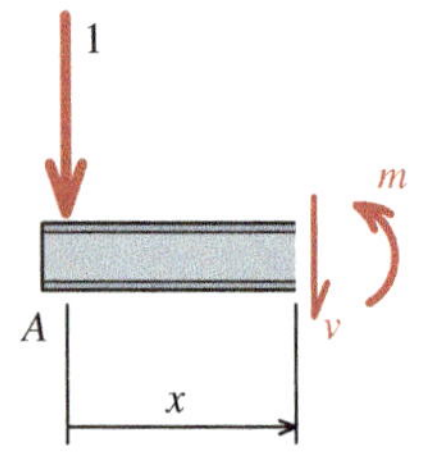

Next, draw a free-body diagram around end A of the beam. Place the origin of the x coordinate system at A. From the free-body diagram, derive the following equation for the virtual internal moment m:

$$m = -1x \qquad 0 \le x \le L$$

Real Moment M: Remove the virtual load and reapply the real load w.

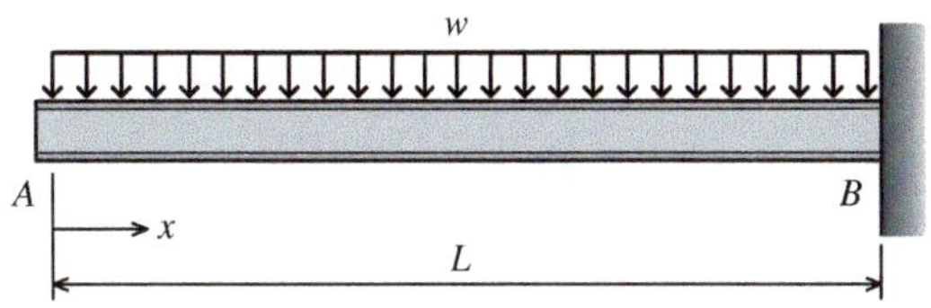

Again, draw a free-body diagram around end A of the beam. Note that the same x coordinate used to derive the virtual moment must be used to derive the real moment; therefore, the origin of the x coordinate system must be placed at A. From the free-body diagram, derive the following equation for the real internal moment M:

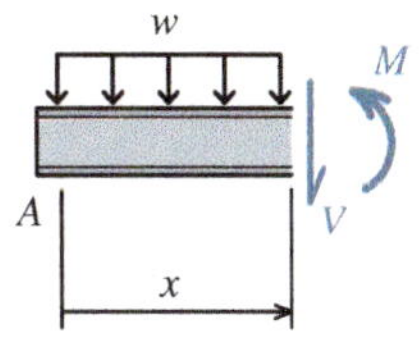

$$M = -\frac{wx^2}{2} \qquad 0 \le x \le L$$

Virtual-Work Equation for Beam Deflection: From Equation (17.35), the beam deflection at A can now be calculated:

$$1 \cdot \Delta_A = \int_0^L m\left(\frac{M}{EI}\right)dx = \int_0^L \frac{(-1x)(-wx^2/2)}{EI}\,dx = \frac{w}{2EI}\int_0^L x^3\,dx$$

$$\therefore\ \Delta_A = \frac{wL^4}{8EI} \downarrow \qquad \textbf{Ans.}$$

Since the result is a positive value, the deflection occurs in the same direction as was assumed for the unit load—that is, downward.

(b) Virtual Moment m for Calculating the Beam Slope: To compute the angular rotation of the cantilever beam at A, remove the real load w from the beam and apply a virtual unit moment at A. The unit moment will be applied counterclockwise in this instance because it is expected that the beam will slope upward from A.

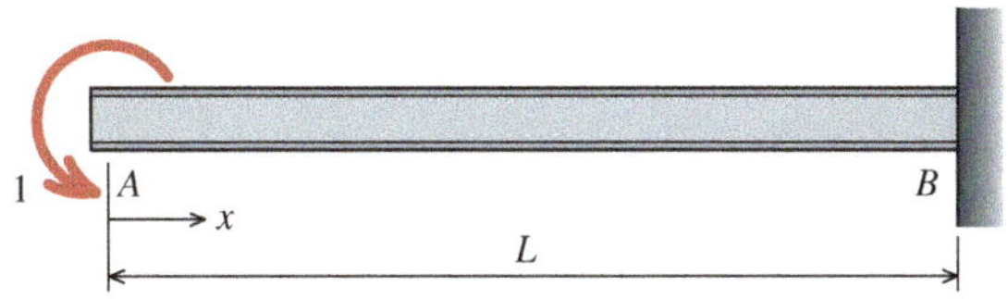

Again, draw a free-body diagram around end A of the beam, placing the origin of the x coordinate system at A. From the free-body diagram, derive the following equation for the virtual internal moment m:

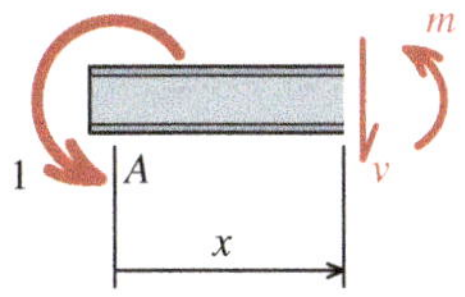

$$m = -1 \qquad 0 \le x \le L$$

Real Moment M: The real-moment equation M is the same as was derived previously.

Virtual-Work Equation for Beam Slope: From Equation (17.36), the beam slope at A can now be determined:

$$1 \cdot \theta_A = \int_0^L m\left(\frac{M}{EI}\right)dx = \int_0^L \frac{(-1)(-wx^2/2)}{EI}\,dx = \frac{w}{2EI}\int_0^L x^2\,dx$$

$$\therefore\ \theta_A = \frac{wL^3}{6EI}\text{(CCW)} \qquad \textbf{Ans.}$$

Since the result is a positive value, the angular rotation occurs in the same direction as was assumed for the unit moment—that is, counterclockwise.

EXAMPLE 17.14

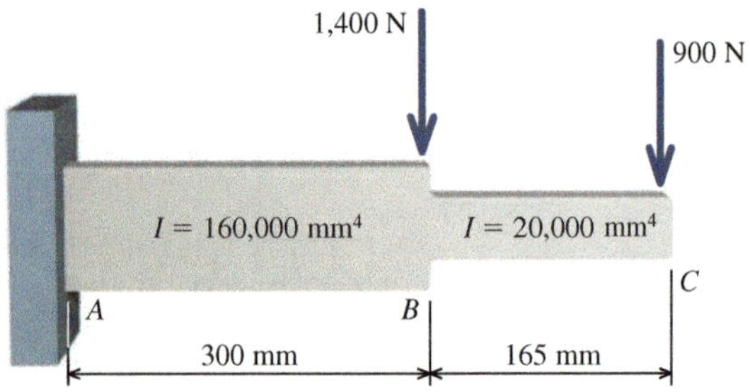

Calculate the deflection at end C of the cantilever beam shown. Assume that $E = 70$ GPa for the entire beam.

Plan the Solution

For the cantilever beam considered here, the virtual-moment equation will be continuous over the entire span, but the real-moment equation is discontinuous at B. Therefore, the beam must be considered in two segments: AB and BC. The moment equations will be simpler to derive if the origin of the x coordinate system is placed at the free end C.

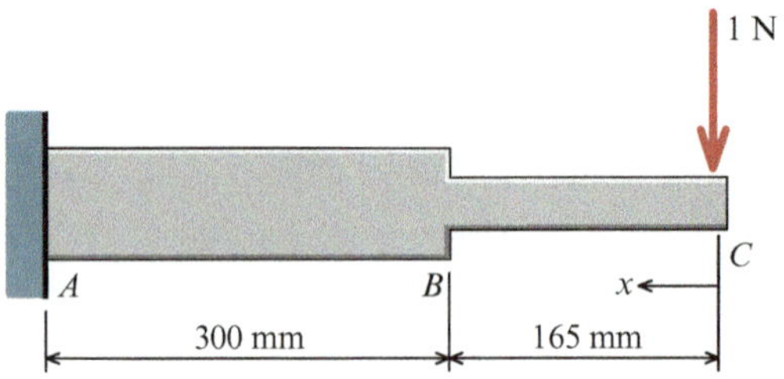

SOLUTION

Virtual Moment m: Remove the real loads from the beam and apply a virtual unit load downward at C, where the deflection is desired. At this stage of the calculation, it makes no difference that the beam depth changes along the span, since only the virtual moment m is needed here.

Now draw a free-body diagram around end C of the beam. Place the origin of the x coordinate system at C. From the free-body diagram, derive the following equation for the virtual internal moment m:

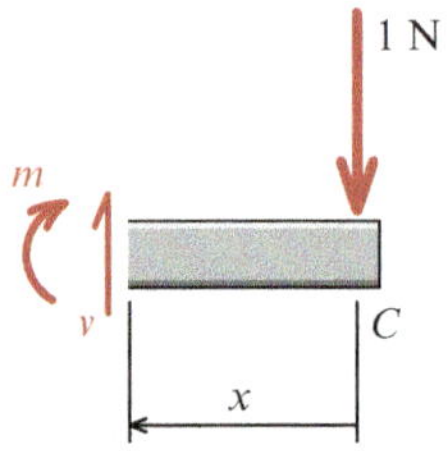

$$m = -(1\ \text{N})x \qquad 0 \leq x \leq 465\ \text{mm}$$

Real Moment M: Remove the virtual load, and reapply the real loads at B and C. Draw a free-body diagram around end C that cuts through segment BC of the beam. The same x coordinate used to derive the virtual moment must be used to derive the real moment; therefore, the origin of the x coordinate system must be placed at C. From the free-body diagram, derive the following equations for the real internal moment M:

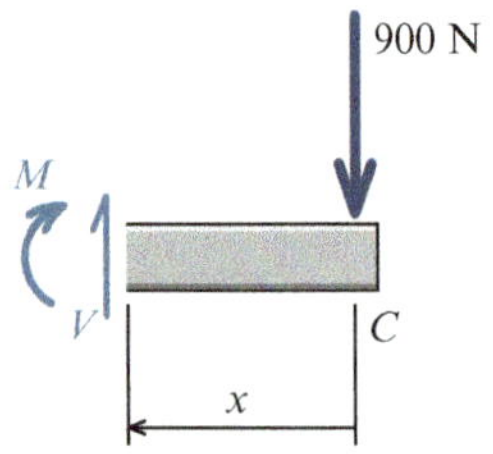

$$M = -(900\ \text{N})x \qquad 0 \leq x \leq 165\ \text{mm}$$

Now draw a second free-body diagram that cuts through segment AB of the beam and includes the free end of the cantilever. From the free-body diagram, derive the following equations for the real internal moment M:

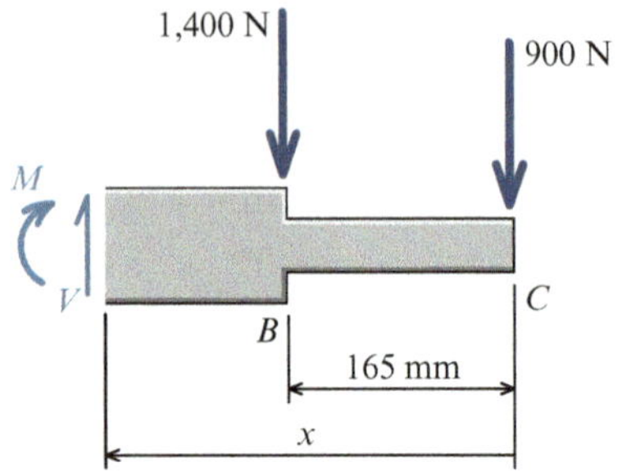

$$\begin{aligned} M &= -(1{,}400\ \text{N})(x - 165\ \text{mm}) - (900\ \text{N})x \\ &= -(2{,}300\ \text{N})x + (1{,}400\ \text{N})(165\ \text{mm}) \\ &\qquad 165\ \text{mm} < x \leq 465\ \text{mm} \end{aligned}$$

The moment of inertia differs for segments AB and BC. This difference will be incorporated into the calculation in the term M/EI. The equations developed so far, along with the limits of integration, are conveniently summarized in the following table:

Beam Segment	x Coordinate Origin	Limits (mm)	I	m (N·mm)	M (N·mm)	$\int m\left(\frac{M}{EI}\right)dx$
BC	C	0–165	20,000	$-1x$	$-900x$	$\frac{67{,}381.875\ \text{N}^2/\text{mm}}{E}$
AB	C	165–465	160,000	$-1x$	$-2{,}300x + 1{,}400(165)$	$\frac{323{,}817.188\ N^2/\text{mm}}{E}$
						$\frac{391{,}199.063\ \text{N}^2/\text{mm}}{E}$

Virtual-Work Equation: From Equation (17.35), the beam deflection at C can now be determined:

$$(1\ \text{N})\cdot\Delta_C = \frac{391{,}199.063\ \text{N}^2/\text{mm}}{E} = \frac{391{,}199.063\ \text{N}^2/\text{mm}}{70{,}000\ \text{N/mm}^2}$$

$$\therefore \Delta_C = 5.59\ \text{mm}\downarrow$$ **Ans.**

EXAMPLE 17.15

Compute the deflection at point C for the simply supported beam shown. Assume that $EI = 3.4 \times 10^5\ \text{kN}\cdot\text{m}^2$.

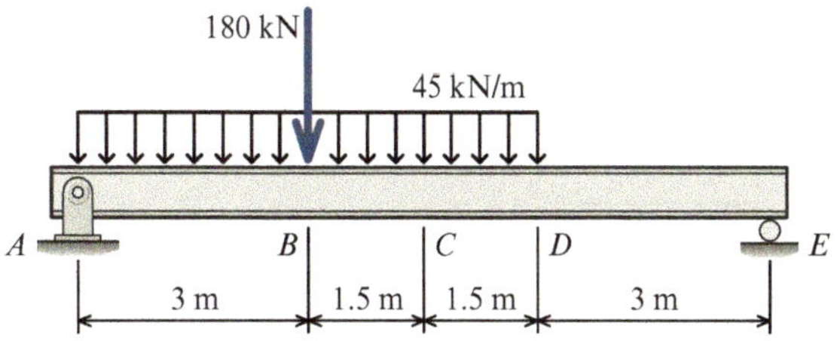

Plan the Solution

The real loadings are discontinuous at points B and D, while the virtual loading for the beam is discontinuous at C. Therefore, this beam must be considered in four segments: AB, BC, CD, and DE. To facilitate the derivation of moment equations, it will be convenient to locate the x coordinate origin at A for segments AB and BC and at E for segments CD and DE. To organize the calculation, it will also be convenient to summarize the pertinent equations in a tabular format.

SOLUTION

Virtual Moment m: Remove the real loads from the beam and apply a virtual unit load downward at C, where the deflection is desired. A free-body diagram of the beam is shown.

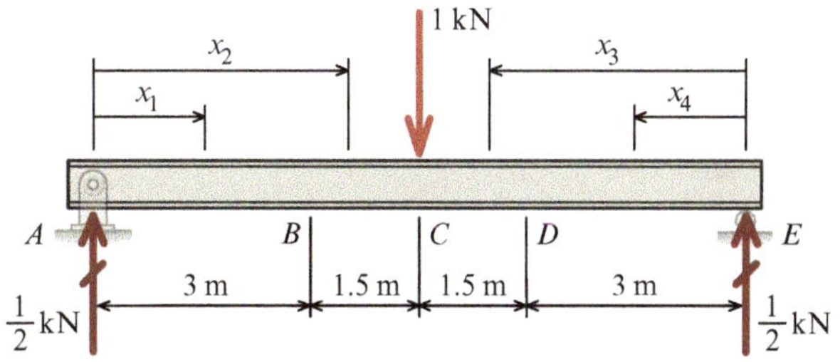

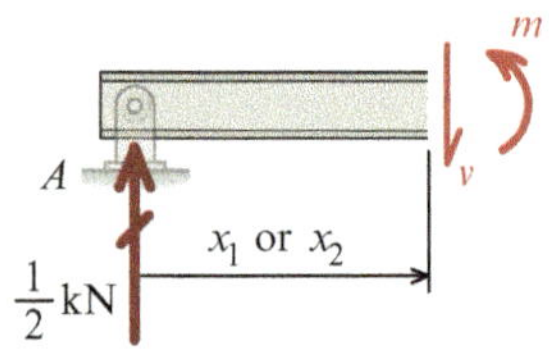

Draw a free-body diagram around end A of the beam. Place the origin of the x coordinate system at A when considering segments AB and BC. From the free-body diagram, derive the following equations for the virtual internal moment m:

$$m = \left(\frac{1}{2}\text{ kN}\right)x_1 \qquad 0\text{ m} \le x_1 \le 3\text{ m}$$

$$m = \left(\frac{1}{2}\text{ kN}\right)x_2 \qquad 3\text{ m} \le x_2 \le 4.5\text{ m}$$

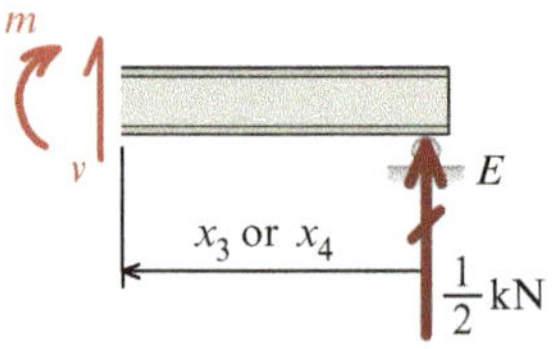

Now draw a free-body diagram around end E of the beam. Place the origin of the x coordinate system at E when considering segments CD and DE. Derive the following equations for the virtual internal moment m:

$$m = \left(\frac{1}{2}\text{ kN}\right)x_3 \qquad 3\text{ m} \le x_3 \le 4.5\text{ m}$$

$$m = \left(\frac{1}{2}\text{ kN}\right)x_4 \qquad 0\text{ m} \le x_4 \le 3\text{ m}$$

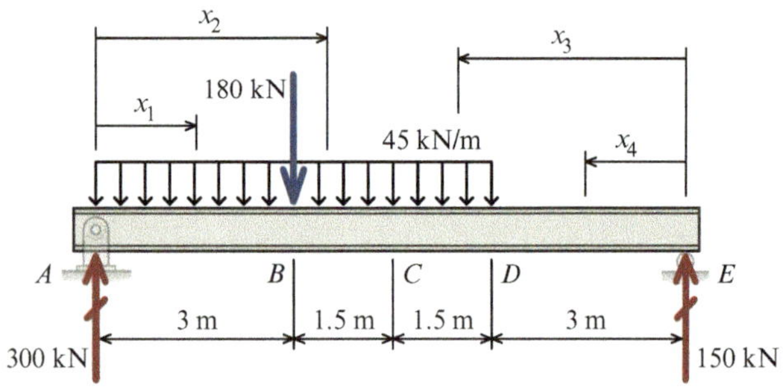

Real Moment M: Remove the virtual load and reapply the real loads. Determine the beam reactions, which are A_y = 300 kN and E_y = 150 kN, acting in the upward direction as shown.

Draw a free-body diagram that cuts through segment AB around support A of the beam. The same x coordinate used to derive the virtual moment must be used to derive the real moment; therefore, the origin of the x coordinate system must be placed at A. From the free-body diagram, derive the following equation for the real internal moment M in segment AB of the beam:

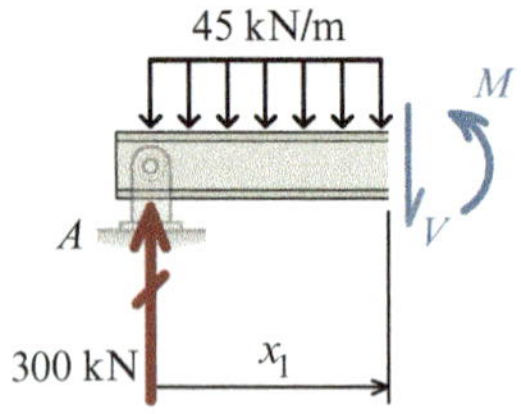

$$M = -\frac{45\text{ kN/m}}{2}x_1^2 + (300\text{ kN})x_1$$
$$0 \le x_1 \le 3\text{ m}$$

Repeat the process with a free-body diagram cut through segment BC of the beam. From the free-body diagram, derive the following equation for the real internal moment M in segment BC of the beam:

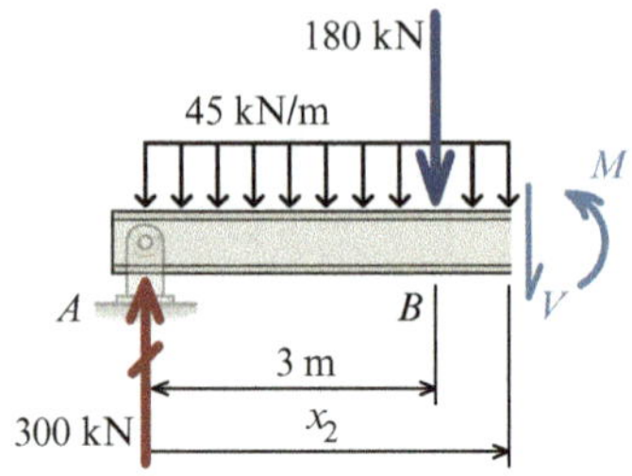

$$M = -\frac{45\text{ kN/m}}{2}x_2^2 - (180\text{ kN})(x_2 - 3\text{ m}) + (300\text{ kN})x_2$$
$$3\text{ m} \le x_2 \le 4.5\text{ m}$$

For segment CD, draw a free-body diagram that cuts through segment CD around support E of the beam. From the free-body diagram, derive the following equation for the real internal moment M in segment CD of the beam:

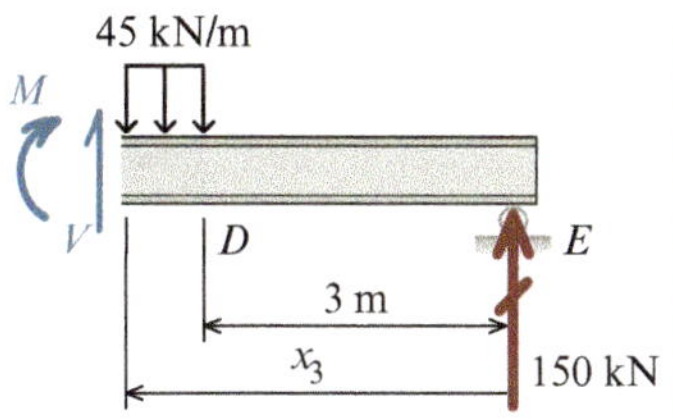

$$M = -\frac{45\text{ kN/m}}{2}(x_3 - 3\text{ m})^2 + (150\text{ kN})x_3$$
$$3\text{ m} \le x_3 \le 4.5\text{ m}$$

Finally, derive the following equation for the real internal moment M in segment DE of the beam:

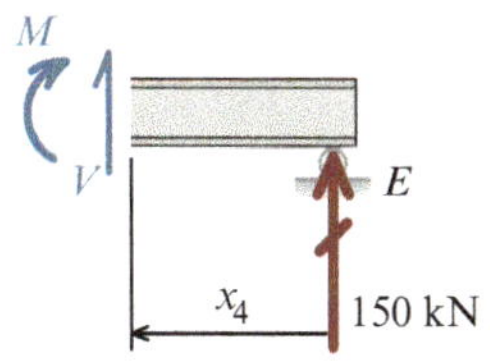

$$M = (150\text{ kN})x_4$$
$$0 \le x_4 \le 3\text{ m}$$

The equations for m and M, along with the appropriate limits of integration, are summarized in the table that follows. The results for the integration in each segment are also given.

	x Coordinate				
Beam Segment	**Origin**	**Limits (m)**	m **(kN·m)**	M **(kN·m)**	$\int m\left(\frac{M}{EI}\right)dx$
AB	A	0–3	$\frac{1}{2}x_1$	$-\frac{45}{2}x_1^2 + 300x_1$	$\frac{1{,}122.188\text{ kN}^2\text{-m}^3}{EI}$
BC	A	3–4.5	$\frac{1}{2}x_2$	$-\frac{45}{2}x_2^2 - 180(x_2 - 3) + 300x_2$	$\frac{1{,}875.762\text{ kN}^2\text{-m}^3}{EI}$
CD	E	3–4.5	$\frac{1}{2}x_3$	$-\frac{45}{2}(x_3 - 3)^2 + 150x_3$	$\frac{1{,}550.918\text{ kN}^2\text{-m}^3}{EI}$
DE	E	0–3	$\frac{1}{2}x_4$	$150x_4$	$\frac{675.0\text{ kN}^2\text{-m}^3}{EI}$
					$\frac{5{,}223.868\text{ kN}^2\text{-m}^3}{EI}$

PROBLEMS

P17.34 Determine the vertical displacement of joint B for the truss shown in Figure P17.34/35. Assume that each member has a cross-sectional area A = 1.25 in.^{2} and an elastic modulus E = 10,000 ksi. The loads acting on the truss arc P = 21 kips and Q = 7 kips.

P17.35 Determine the horizontal displacement of joint B for the truss shown in Figure P17.34/35. Assume that each member has a cross-sectional area A = 1.25 in.^{2} and an elastic modulus E = 10,000 ksi. The loads acting on the truss are P = 21 kips and Q = 7 kips.

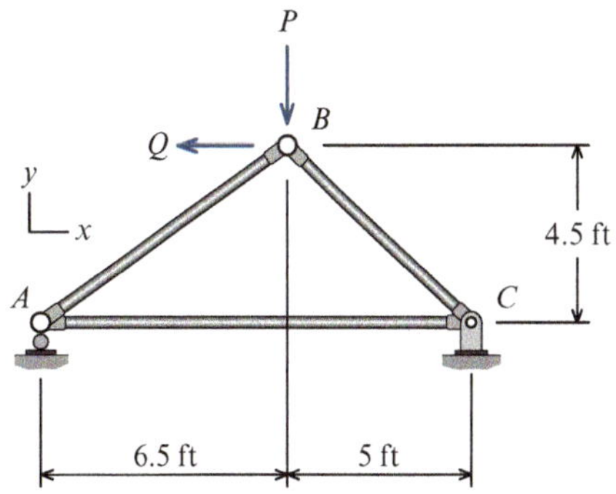

FIGURE P17.34/35

P17.36 Determine the vertical displacement of joint D for the truss shown in Figure P17.36/37. Assume that each member has a cross-sectional area A = 1.60 in.^{2} and an elastic modulus E = 29,000 ksi. The loads acting on the truss are P = 20 kips and Q = 30 kips.

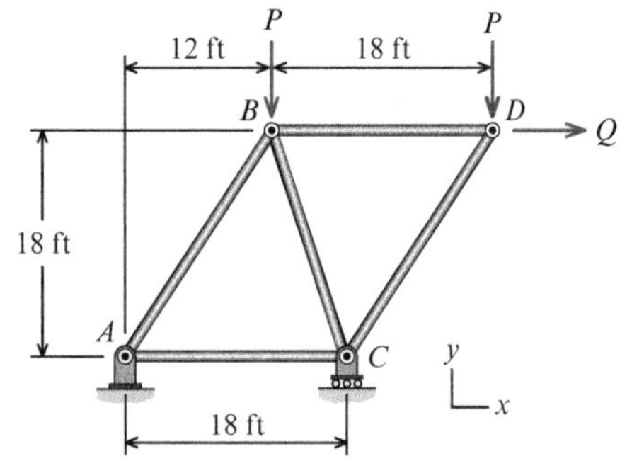

FIGURE P17.36/37

P17.37 Determine the horizontal displacement of joint D for the truss shown in Figure P17.36/37. Assume that each member has a cross-sectional area A = 1.60 in.^{2} and an elastic modulus E = 29,000 ksi. The loads acting on the truss are P = 20 kips and Q = 30 kips.

P17.38 Determine the horizontal displacement of joint A for the truss shown in Figure P17.38/39/40. Assume that each member has a cross-sectional area A = 750 mm^2 and an elastic modulus E = 70 GPa.

P17.39 Determine the vertical displacement of joint B for the truss shown in Figure P17.38/39/40. Assume that each member has a cross-sectional area A = 750 mm^2 and an elastic modulus E = 70 GPa.

P17.40 The truss shown in Figure P17.38/39/40 is constructed from aluminum [E = 70 GPa; α = 23.6 × 10^{-6}/°C] members that each have a cross-sectional area A = 750 mm^2. Determine the vertical displacement of joint A under two conditions:

(a) ΔT = 0°C.
(b) ΔT = +40°C.

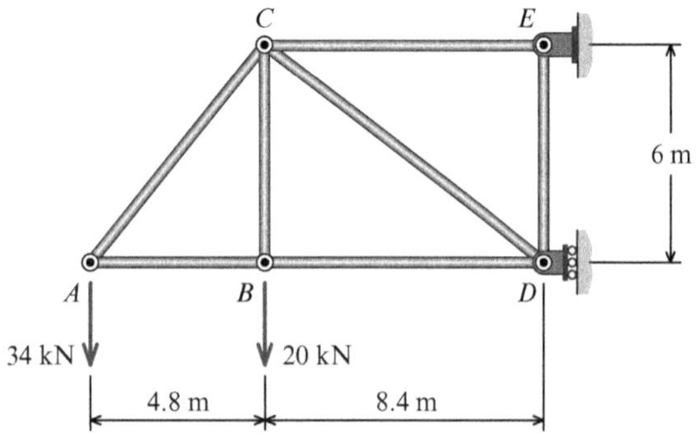

FIGURE P17.38/39/40

P17.41 The truss shown in Figure P17.41/42 is subjected to concentrated loads P = 320 kN and Q = 60 kN. Members AB, BC, DE, and EF each have a cross-sectional area A = 2,700 mm^2. All other members each have a cross-sectional area A = 1,060 mm^2. All members are made of steel [E = 200 GPa]. For the given loads, determine the horizontal displacement of (a) joint F and (b) joint B.

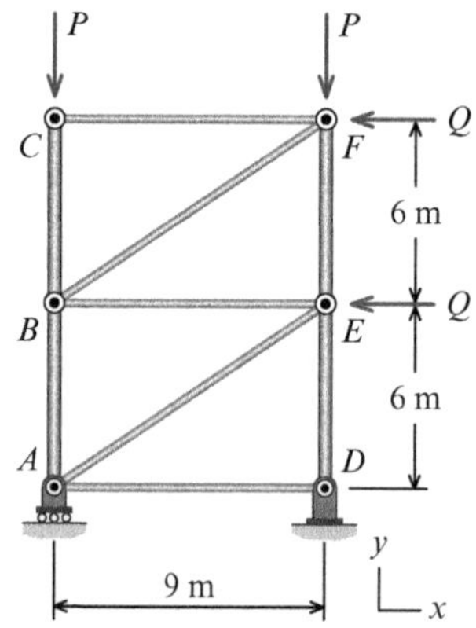

FIGURE P17.41/42

P17.42 The truss shown in Figure P17.41/42 is subjected to concentrated loads P = 200 kN and Q = 35 kN. Members AB, BC, DE, and EF each have a cross-sectional area A = 2,700 mm^2. All other members each have a cross-sectional area A = 1,060 mm^2. All members are made of steel [E = 200 GPa]. During construction, it was discovered that members AE and BF were fabricated 15 mm shorter than their intended length. For the given loads and the two member misfits, determine the horizontal displacement of (a) joint F and (b) joint B.

P17.43 The truss shown in Figure P17.43/44/45 is subjected to concentrated loads P = 160 kN and $2P$ = 320 kN. All members are made of steel [E = 200 GPa], and each member has a cross-sectional area A = 3,500 mm^2. Determine

(a) the horizontal displacement of joint A.
(b) the vertical displacement of joint A.

P17.44 The truss shown in Figure P17.43/44/45 is subjected to concentrated loads P = 160 kN and $2P$ = 320 kN. All members are made of steel [E = 200 GPa; α = 11.7 × 10^{-6}/°C], and each member has a cross-sectional area A = 3,500 mm^2. If the temperature of the truss increases by 30°C, determine

(a) the horizontal displacement of joint A.
(b) the vertical displacement of joint A.

P17.45 The truss shown in Figure P17.43/44/45 is subjected to concentrated loads P = 160 kN and $2P$ = 320 kN. All members are made of steel [E = 200 GPa; α = 11.7 × 10^{-6}/°C], and each member has a cross-sectional area A = 3,500 mm^2. Determine

(a) the vertical displacement of joint D.
(b) the vertical displacement of joint D if the temperature of the truss decreases by 40°C.

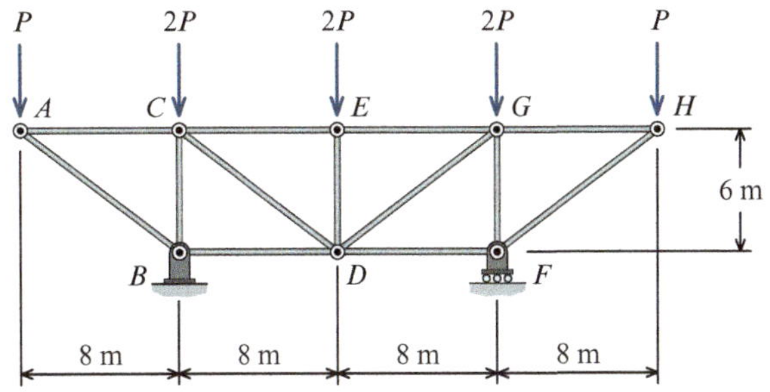

FIGURE P17.43/44/45

P17.46 Determine the deflection of the beam at B for the loading shown in Figure P17.46. Assume that EI is constant over the length of the beam.

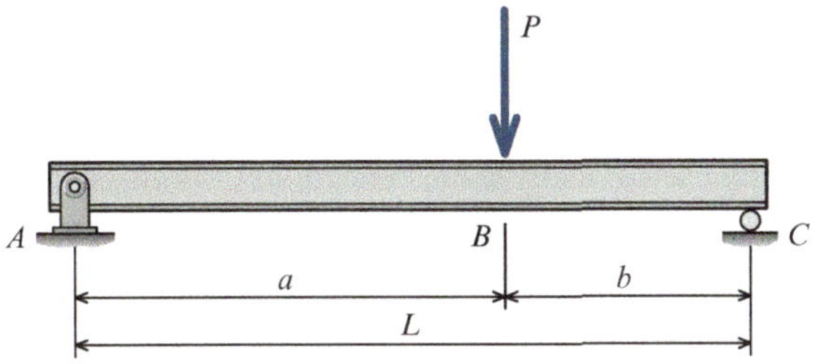

FIGURE P17.46

P17.47 Determine the slope of the beam at A for the loading shown in Figure P17.47. Assume that EI is constant over the length of the beam.

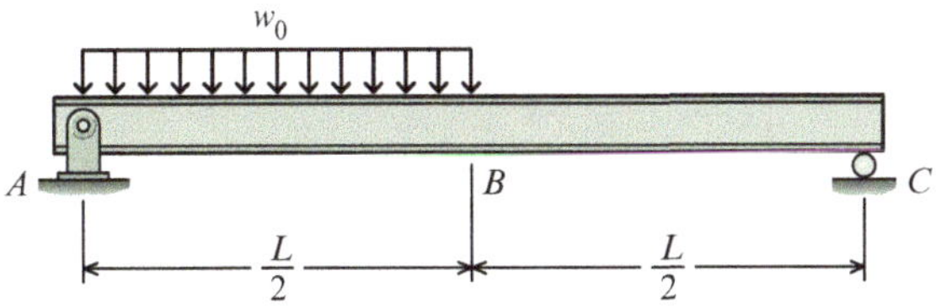

FIGURE P17.47

P17.48 Determine the slope and deflection of the beam at C for the loading shown in Figure P17.48. Assume that EI is constant over the length of the beam.

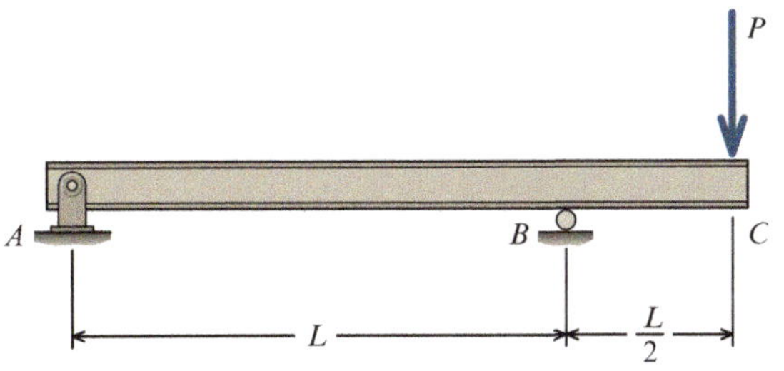

FIGURE P17.48

P17.49 Determine the deflection of the compound rod at C for the loading shown in Figure P17.49. Between A and B, the rod has a diameter of 30 mm. Between B and C, the rod diameter is 15 mm. Assume that $E = 200$ GPa for both segments of the compound rod.

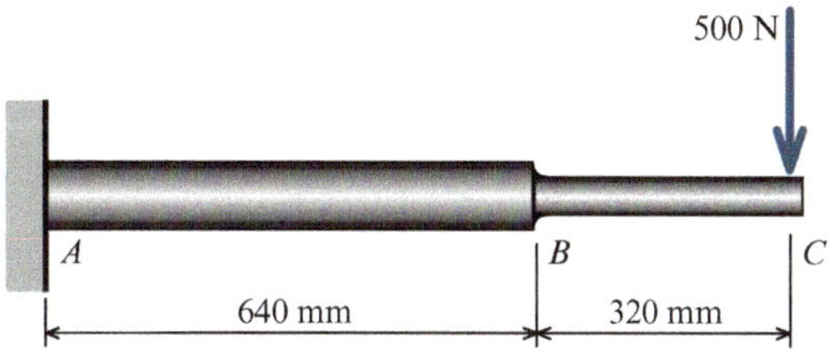

FIGURE P17.49

P17.50 The compound steel [$E = 200$ GPa] rod shown in Figure P17.50/51 has a diameter of 15 mm in each of segments AB and DE and a diameter of 30 mm in each of segments BC and CD. For the given loading, determine the slope of the compound rod at A.

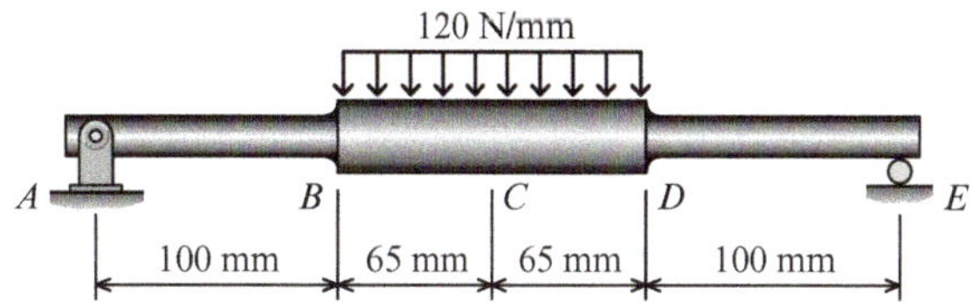

FIGURE P17.50/51

P17.51 The compound steel [$E = 200$ GPa] rod shown in Figure P17.50/51 has a diameter of 15 mm in each of segments AB and DE and a diameter of 30 mm in each of segments BC and CD. For the given loading, determine the deflection of the compound rod at C.

P17.52 A simply supported beam is shown in Figure P17.52. Assume that $EI = 15 \times 10^6$ kip·in.^{2} for the beam. Determine

(a) the deflection at A.
(b) the slope at C.

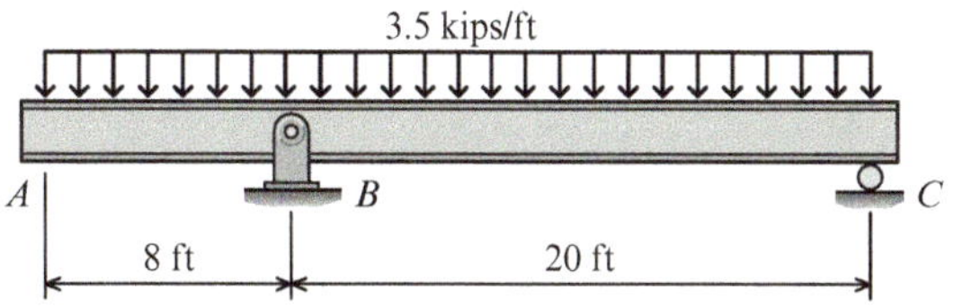

FIGURE P17.52

P17.53 A cantilever beam is loaded as shown in Figure P17.53. Assume that $EI = 74 \times 10^3$ kN·m^2 for the beam. Determine

(a) the slope at C.
(b) the deflection at C.

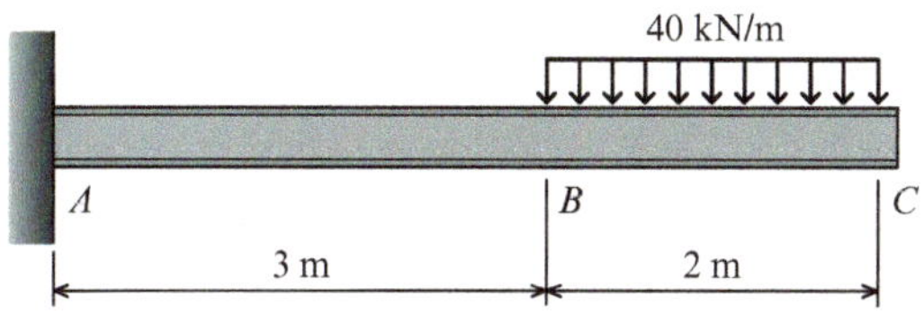

FIGURE P17.53

P17.54 Determine the minimum moment of inertia I required for the beam shown in Figure P17.54 if the maximum beam deflection must not exceed 35 mm. Assume that $E = 200$ GPa.

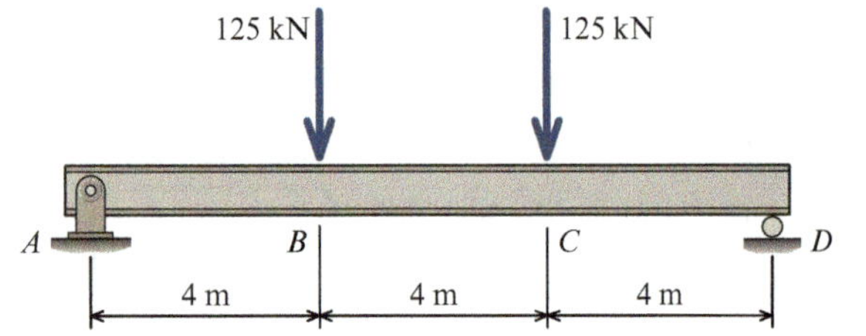

FIGURE P17.54

P17.55 Determine the minimum moment of inertia I required for the beam shown in Figure P17.55 if the maximum beam deflection must not exceed 0.5 in. Assume that $E = 29{,}000$ ksi.

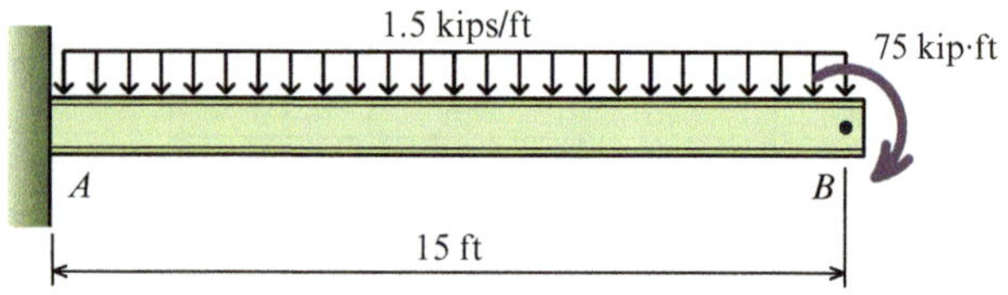

FIGURE P17.55

17.11 Castigliano's Second Theorem

Strain-energy techniques are frequently used to analyze the deflections of beams and structures. Of the many available methods, the application of Castigliano's second theorem, to be developed here, is one of the most widely used. The theorem was presented in 1873 by the Italian engineer Alberto Castigliano (1847–1884). Although we will derive it by considering the strain energy stored in beams, it is applicable to any structure for which the force–deformation relations are linear.[1] The method incorporates strain-energy principles developed earlier. Further, it is remarkably similar to the virtual-work method developed previously.

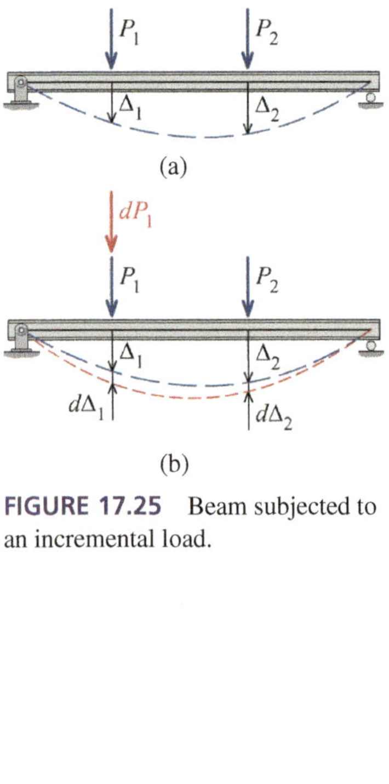

FIGURE 17.25 Beam subjected to an incremental load.

If the beam shown in Figure 17.25*a* is slowly and simultaneously loaded by two forces P_1 and P_2, with resulting deflections Δ_1 and Δ_2, the strain energy U of the beam is equal to the work done by the forces. Therefore,

$$U = \frac{1}{2}P_1\Delta_1 + \frac{1}{2}P_2\Delta_2$$

Recall that the factor ½ in each term is required because the loads build up from zero to their final magnitude. [See Equation (17.1).]

Let the force P_1 be increased by a small amount dP_1 while force P_2 remains constant, as shown in Figure 17.25*b*. The changes in deflection due to this incremental load will be denoted $d\Delta_1$ and $d\Delta_2$. The strain energy in the beam increases by the amount $\frac{1}{2}dP_1 d\Delta_1$ as the incremental force dP_1 deflects through the distance $d\Delta_1$. However, forces P_1 and P_2, which remain present on the beam, also perform work as the beam deflects. Altogether, the increase in the strain energy due to the application of dP_1 is

$$dU = P_1 d\Delta_1 + P_2 d\Delta_2 + \frac{1}{2}dP_1 d\Delta_1 \qquad \text{(a)}$$

[1]Castigliano's first theorem, which can be used to establish equations of equilibrium, will not be discussed in this text. That theorem is a powerful method for solving problems for statically indeterminate structures and has application in many computer-based analytical methods, such as finite-element analysis.

Therefore, the total strain energy in the beam is

$$U + dU = \frac{1}{2}P_1\Delta_1 + \frac{1}{2}P_2\Delta_2 + P_1 d\Delta_1 + P_2 d\Delta_2 + \frac{1}{2}dP_1 d\Delta_1 \qquad \text{(b)}$$

If the order of loading is reversed so that the incremental force dP_1 is applied first, followed by P_1 and P_2, the resulting strain energy will be

$$U + dU = \frac{1}{2}P_1\Delta_1 + \frac{1}{2}P_2\Delta_2 + dP_1\Delta_1 + \frac{1}{2}dP_1 d\Delta_1 \qquad \text{(c)}$$

Note that, since the beam is linearly elastic, the loads P_1 and P_2 cause the same deflections Δ_1 and Δ_2 regardless of whether or not any other load is acting on the beam. Because dP_1 remains constant at its point of application during the additional deflection Δ_1, the term $dP_1\Delta_1$ does not contain the factor ½.

Since elastic deformation is reversible and energy losses are neglected, the resulting strain energy must be independent of the order of loading. Hence, by equating Equations (b) and (c), we obtain

$$dP_1\Delta_1 = P_1 d\Delta_1 + P_2 d\Delta_2 \qquad \text{(d)}$$

Equations (a) and (d) can then be combined to give

$$dU = dP_1\Delta_1 + \frac{1}{2}dP_1 d\Delta_1 \qquad \text{(e)}$$

The term

$$\frac{1}{2}dP_1 d\Delta_1$$

is a second-order differential that may be neglected. Furthermore, the strain energy U is a function of both P_1 and P_2; therefore, the change in strain energy, dU, due to the incremental load dP_1 is expressed by the partial derivative of U with respect to P_1 as

$$dU = \frac{\partial U}{\partial P_1} dP_1$$

Equation (e) can then be written as

$$\frac{\partial U}{\partial P_1} dP_1 = dP_1\Delta_1$$

which can be simplified to

$$\frac{\partial U}{\partial P_1} = \Delta_1 \qquad \text{(f)}$$

For the general case in which there are n loads involved, Equation (f) is written as

$$\frac{\partial U}{\partial P_i} = \Delta_i (i = 1, \ldots, n) \tag{17.37}$$

Castigliano's second theorem applies to any elastic system at constant temperature and on unyielding supports and that obeys the law of superposition.

The following is a statement of Castigliano's second theorem:

If the strain energy of a linearly elastic structure is expressed in terms of the system of external loads, then the partial derivative of the strain energy with respect to a concentrated external load is the deflection of the structure at the point of application and in the direction of that load.

By a similar development, Castigliano's theorem can also be shown to be valid for applied moments and the resulting rotations (or changes in slope) of the structure. Thus, for n moments,

$$\frac{\partial U}{\partial M_i} = \theta_i (i = 1, \ldots, n) \tag{17.38}$$

If the deflection is required to be known either at a point where there is no unique point load or in a direction that is not aligned with the applied load, a dummy load acting in the proper direction is introduced at the desired point. We then obtain the deflection by first differentiating the strain energy with respect to the dummy load and then taking the limit as the dummy load approaches zero. Also, for the application of Equation (17.38), either a unique point moment or a dummy moment must be applied at point i. The moment will be in the direction of rotation at the point. Note that, if the loading consists of a number of point loads, all expressed in terms of a single parameter (e.g., P, $2P$, $3P$, wL, or $2wL$), and if the deflection at one of the applied loads is what is to be determined, then we must either write the moment equation with this load as a separate identifiable term or add a dummy load at the point so that the partial derivative can be taken with respect to that load only.

17.12 Calculating Deflections of Trusses by Castigliano's Theorem

The strain energy in an axial member was developed in Section 17.3. For compound axial members and structures consisting of n prismatic axial members, the total strain energy in the member or structure can be computed with Equation (17.14),

$$U = \sum_{i=1}^{n} \frac{F_i^2 L_i}{2A_i E_i}$$

To compute the deflection of a truss, the general expression for strain energy given by Equation (17.14) can be substituted into Equation (17.37) to obtain

$$\Delta = \frac{\partial}{\partial P} \sum \frac{F^2 L}{2AE}$$

where the subscripts i have been omitted. It is generally easier to differentiate before summing, in which case we express the preceding equation as

$$\Delta = \sum \frac{\partial F^2}{\partial P} \frac{L}{2AE}$$

The terms L, A, and E are constant for each particular member. Since the partial derivative $\partial F^2/\partial P = 2F(\partial F/\partial P)$, Castigliano's second theorem for trusses can be written as

$$\Delta = \sum \left(\frac{\partial F}{\partial P}\right)\frac{FL}{AE} \tag{17.39}$$

where

Δ = displacement of the truss joint
P = external force applied to the truss joint in the direction of Δ and *expressed as a variable*
F = internal axial force in a member caused by both the force P and the loads on the truss
L = length of the member
A = cross-sectional area of the member
E = elastic modulus of the member

To determine the partial derivative $\partial F/\partial P$, the external force P must be treated as a variable, not a specific numeric quantity. Consequently, each internal axial force F must be expressed as a function of P.

If the deflection is required at a joint at which there is no external load or if the deflection is required for a direction that is not aligned with the external load, then a dummy load must be added in the proper direction at the desired joint. We obtain the joint deflection by first differentiating the strain energy with respect to the dummy load and then taking the limit as the dummy load approaches zero.

Procedure for Analysis

The following procedure is recommended when Castigliano's second theorem is applied to calculate truss deflections:

1. **Load *P* Expressed as a Variable:** If an external load acts on the truss at the joint where deflections are to be calculated and in the direction of the desired deflection, then designate that load as the variable P. As a result, subsequent calculations will be performed in terms of the variable P rather than in terms of the actual numeric value known for this particular external load. Otherwise, apply a fictitious load (a dummy load) in the direction of the desired deflection at the particular joint. Designate this dummy load as P.
2. **Member Forces *F* in Terms of *P*:** Develop expressions for the internal axial force F created in each truss member by the actual external loads and the variable load P. It is likely that the expression for the internal force in a particular member will include both a numeric value and a function in terms of P. Assume that tensile forces are positive and compressive forces are negative.
3. **Partial Derivatives for Each Truss Member:** Differentiate the expressions for the truss-member forces F with respect to P to compute $\partial F/\partial P$.
4. **Substitute Numeric Value for *P*:** Substitute the actual numeric value for load P into the expressions for F and $\partial F/\partial P$ for each truss member. If a dummy load has been used for P, its numeric value is zero.
5. **Summation:** Perform the summation indicated by Equation (17.39) to calculate the desired joint deflection. A positive answer indicates that the deflection acts in the same direction as P, a negative answer in the opposite direction.

The use of Castigliano's theorem to compute truss joint deflections is illustrated in Examples 17.16 and 17.17.

EXAMPLE 17.16

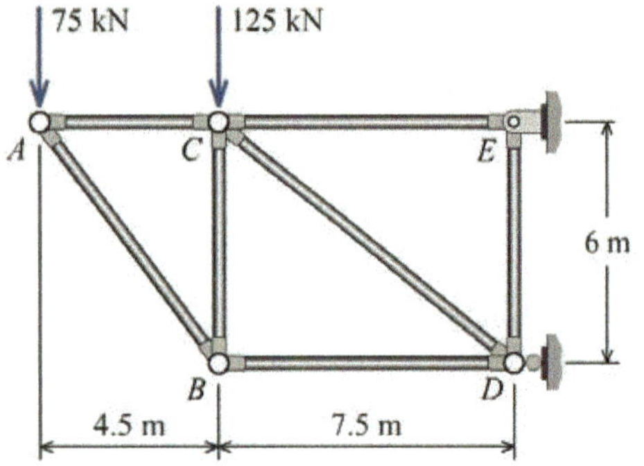

Determine the vertical deflection at joint A for the truss shown. For all members, the cross-sectional area is $A = 1{,}100\ \text{mm}^2$ and the elastic modulus is $E = 200$ GPa.

Plan the Solution

The vertical deflection is to be determined at joint A. Since a vertical load is already present at A on the truss, this load will be designated as P. Consequently, instead of performing a truss analysis with the 75 kN load at A, we will replace the 75 kN load with a variable load designated as P. With this variable load P, we will then perform a truss analysis at joint A, using an appropriate method, such as the method of joints. From the analysis, an expression in terms of P and any additional constant forces that arise from the 125 kN load applied at C will be obtained for each truss member.

Next, construct a table to organize the truss analysis results. For each truss member, record the complete member-force expressions F in one column. Four additional operations will then be performed:

(1) Differentiate the member-force expression F with respect to P, and record the partial derivatives.
(2) Substitute the actual 75 kN value for P in the member-force expression F and calculate the actual member force.
(3) Calculate the length of each truss member.
(4) Multiply the partial derivative $\partial F/\partial P$ by the actual member-force value F and the member length L.

The products $(\partial F/\partial P)FL$ for all truss members will then be summed. Since the area A and the elastic modulus E are the same for all members, these values will be introduced after the summation. Finally, Equation (17.39) will be applied to determine the downward deflection Δ of joint A.

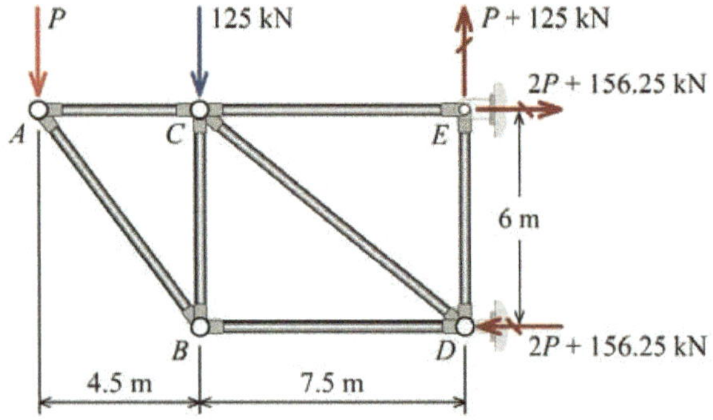

SOLUTION

We seek the vertical deflection of the truss at joint A. Because there is already a load applied at this joint in the desired direction, we will replace the 75 kN load with a variable load P. A free-body diagram of the truss with load P applied at joint A is shown.

We perform a truss analysis, using the loadings shown in the free-body diagram, to find the axial force in each truss member. With these loads, the member forces F will each be expressed as a unique function of P. Some member-force functions F may also contain constant terms arising from the 125 kN load that acts at joint C.

The tabular format that follows is a convenient way to organize the calculation of the truss deflection. The member name is shown in column (1), and the expression for the internal member force F in terms of the variable P is listed in column (2). Differentiate the function in column (2) with respect to P, and record the result in column (3). Next, substitute the actual value of $P = 75$ kN into the member-force functions in column (2), and record the result in column (4). These values are the actual member forces for the truss in response to the 75 kN and 125 kN loads, and they will be used for the term FL/AE found in Equation (17.39). Finally, calculate the member lengths and record them in column (5).

Castigliano's second theorem applied to trusses is expressed by Equation (17.39). For this particular truss, each member has a cross-sectional area $A = 1{,}100\ \text{mm}^2$ and an elastic modulus $E = 200$ GPa; therefore, the calculation process can be simplified by moving both A and E outside of the summation operation:

$$\Delta = \frac{1}{AE}\sum\left(\frac{\partial F}{\partial P}\right)FL$$

For each truss member, multiply the terms in column (3) by those in columns (4) and (5), and record the result in column (6). Then, sum these values for all of the members. Here is the table with all results shown:

(1)	(2)	(3)	(4)	(5)	(6)
Member	F (kN)	$\frac{\partial F}{\partial P}$	F (for $P = 75$ kN) (kN)	L (m)	$\left(\frac{\partial F}{\partial P}\right)FL$ (kN·m)
AB	$-1.25P$	−1.25	−93.75	7.5	878.91
AC	$0.75P$	0.75	56.25	4.5	189.84
BC	$1.00P$	1.00	75.00	6.0	450.00
BD	$-0.75P$	−0.75	−56.25	7.5	316.41
CD	$-1.60P - 200.10$	−1.60	−320.10	9.605	4,919.30
CE	$2.00P + 156.25$	2.00	306.25	7.5	4,593.75
DE	$1.00P + 125.00$	1.00	200.00	6.0	1,200.00
				$\sum\left(\frac{\partial F}{\partial P}\right)FL =$	12,548.20

Now apply Equation (17.39) to compute the deflection of joint A from the tabulated results:

$$\Delta_A = \frac{(12{,}548.20\ \text{kN}\cdot\text{m})(1{,}000\ \text{N/kN})(1{,}000\ \text{mm/m})}{(1{,}100\ \text{mm}^2)(200{,}000\ \text{N/mm}^2)}$$

$$= 57.0\ \text{mm}\downarrow$$ **Ans.**

Since the load P was applied in a downward direction at A, the positive value of the result confirms that joint A is in fact displaced downward.

EXAMPLE 17.17

Calculate the horizontal deflection at joint D for the truss shown. For all members, the cross-sectional area is $A = 3.7$ in.^{2} and the elastic modulus $E = 29{,}000$ ksi.

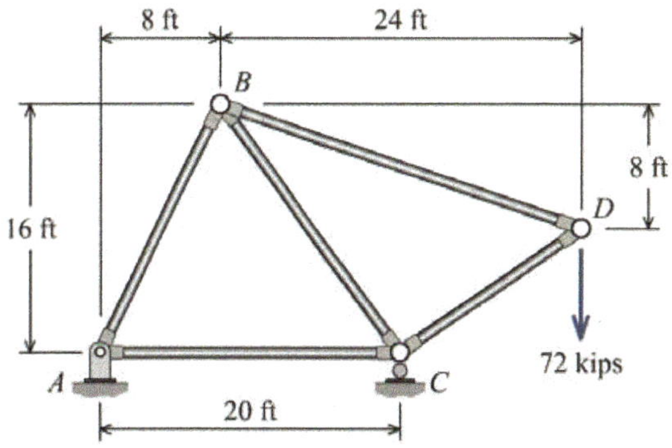

Plan the Solution

The horizontal deflection is to be determined at joint D. Since there is no external load in the horizontal direction at D, a dummy load P will be required. Apply such a load and include it in the truss analysis. Follow the procedure outlined in Example 17.16. However, when calculating the actual member force, substitute the value $P = 0$ kips in the member-force expressions. Finally, apply Equation (17.39) to determine the horizontal deflection Δ of joint D.

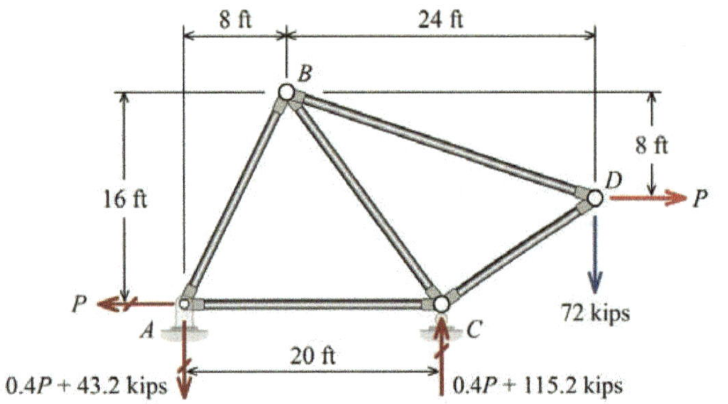

SOLUTION

We seek the horizontal deflection of the truss at joint D. Because no external load acts horizontally at joint D, a dummy load will be applied in the horizontal direction at that joint. A free-body diagram of the truss with load P applied at joint D is shown.

We perform a truss analysis, using the loadings shown in the free-body diagram, to find the axial force in each truss member. With these loads, the member forces F will each be expressed as a unique function of P.

In the table that follows, expressions for each member's internal force in terms of the variable P are listed in column (2). The partial derivative $\partial F/\partial P$ is listed in column (3). The actual force F in each member is calculated by substituting the value $P = 0$ kips into each force expression listed in column (2). The member forces are listed in column (4). Finally, the length of each member is shown in column (5).

Castigliano's second theorem applied to trusses is expressed by Equation (17.39). For this particular truss, each member has a cross-sectional area $A = 3.7$ in.^{2} and an elastic modulus $E = 29{,}000$ ksi; therefore, the calculation process can be simplified when both A and E are moved outside of the summation operation:

$$\Delta = \frac{1}{AE}\sum\left(\frac{\partial F}{\partial P}\right)FL$$

For each truss member, the terms in columns (3), (4), and (5) are multiplied together and recorded in column (6). Here is the table with all results shown:

(1)	(2)	(3)	(4)	(5)	(6)
Member	**F (kips)**	**$\frac{\partial F}{\partial P}$**	**F (for $P = 0$ kips) (kips)**	**L (ft)**	**$\left(\frac{\partial F}{\partial P}\right)FL$ (kip·ft)**
AB	$0.447P + 48.3$	0.447	48.3	17.9	386.46
AC	$0.8P - 21.6$	0.8	−21.6	20	−345.60
BC	$-0.778P - 84.0$	−0.778	−84.0	20	1,307.04
BD	$0.703P + 75.9$	0.703	75.9	25.3	1,349.95
CD	$0.401P - 86.5$	0.401	−86.5	14.4	−499.49
				$\sum\left(\frac{\partial F}{\partial P}\right)FL =$	2,198.36

Now apply Equation (17.39) to compute the horizontal deflection of joint D from the tabulated results:

$$\Delta_D = \frac{(2{,}198.36 \text{ kip}\cdot\text{ft})(12 \text{ in./ft})}{(3.7 \text{ in.}^2)(29{,}000 \text{ ksi})}$$

$$= 0.246 \text{ in.} \rightarrow$$

Ans.

Since the dummy load was applied rightward at D, the positive value of the result confirms that joint D is in fact displaced to the right.

17.13 Calculating Deflections of Beams by Castigliano's Theorem

The strain energy in a flexural member was developed in Section 17.5. The total strain energy in a beam of length L is given by Equation (17.20) as

$$U = \int_0^L \frac{M^2}{2EI} dx$$

To compute the deflection of a beam, the general expression for strain energy given by Equation (17.20) can be substituted into Equation (17.37) to obtain

$$\Delta = \frac{\partial}{\partial P} \int_0^L \frac{M^2}{2EI} dx$$

From the rules of calculus, the integral can be differentiated by differentiating inside the integral sign. If the elastic modulus E and the moment of inertia I are constant with respect to the applied load P, then

Differentiation inside the integral is permissible when P is not a function of x.

$$\frac{\partial}{\partial P} \int_0^L \frac{M^2}{2EI} dx = \int_0^L \left(\frac{\partial M^2}{\partial P} \right) \frac{1}{2EI} dx$$

Since the partial derivative $\partial M^2/\partial P = 2M(\partial M/\partial P)$, Castigliano's second theorem for beam deflections can be written as

$$\Delta = \int_0^L \left(\frac{\partial M}{\partial P} \right) \frac{M}{EI} dx \tag{17.40}$$

where

Δ = displacement of a point on the beam
P = external force applied to the beam in the direction of Δ and *expressed as a variable*
M = internal bending moment in the beam, expressed as a function of x and caused by both the force P and the loads on the beam
I = moment of inertia of the beam cross section about the neutral axis
E = elastic modulus of the beam
L = length of the beam

Similarly, Castigliano's second theorem can be used to compute the rotation angle (i.e., slope) of a beam from

$$\theta = \int_0^L \left(\frac{\partial M}{\partial M'} \right) \frac{M}{EI} dx \tag{17.41}$$

where θ is the rotation angle (or slope) of the beam at a point and M' is a concentrated moment applied to the beam in the direction of θ at the point of interest and *expressed as a variable*.

If the deflection is required at a point where there is no external load or if the deflection is required for a direction that is not aligned with the external load, then a dummy load must be added in the proper direction at the desired point. Likewise, if the slope is required at a point where there is no external concentrated moment, then a dummy moment must be added in the proper direction at the desired point.

Procedure for Analysis

The following procedure is recommended when Castigliano's second theorem is applied to calculate deflections for beams:

1. **Load P Expressed as a Variable:** If an external load acts on the beam at the point where deflections are to be calculated and in the direction of the desired deflection, then designate that load as the variable P. As a result, subsequent calculations will be performed in terms of the variable P rather than in terms of the actual numeric value known for this particular external load. Otherwise, apply a fictitious load (a dummy load) in the direction of the desired deflection at the particular point. Designate this dummy load as P.
2. **Beam Internal Moment M in Terms of P:** Establish appropriate x coordinates for regions of the beam where there is no discontinuity of force, distributed load, or concentrated moment. Develop expressions for the internal moment M in terms of the actual external loads and the variable load P. The expression M for the internal moment of a particular segment of the beam will likely include both a numeric value and a function in terms of P. Use the standard convention for bending-moment signs illustrated in Figures 7.6 and 7.7 for the M equations.
3. **Partial Derivatives:** Differentiate the expressions for the internal moment with respect to P in order to compute $\partial M/\partial P$.
4. **Substitute Numeric Value for P:** Substitute the actual numeric value for load P into the expressions for M and $\partial M/\partial P$. If a dummy load has been used for P, its numeric value is zero.
5. **Integration:** Perform the integration indicated in Equation (17.40) to calculate the desired deflection. A positive answer indicates that the deflection acts in the same direction as P, and a negative answer in the opposite direction.

The following procedure is recommended when Castigliano's second theorem is applied to calculate slopes for beams:

1. **Concentrated Moment M' Expressed as a Variable:** If an external concentrated moment acts on the beam at the point where slopes are to be calculated and in the direction of the desired beam rotation, then designate that concentrated moment as the variable M'. As a result, subsequent calculations will be performed in terms of the variable M' rather than in terms of the actual numeric value known for this particular external concentrated moment. Otherwise, apply a fictitious concentrated moment (a dummy moment) in the direction of the desired slope at the particular point. Designate this dummy moment as M'.
2. **Beam Internal Moment M in Terms of P:** Establish appropriate x coordinates for regions of the beam where there is no discontinuity of force, distributed load, or concentrated moment. Determine expressions for the internal moment M in terms of the actual external loads and the variable moment M'. Be aware that the expression M for the internal moment of a particular segment of the beam may include both a numeric value and a function in terms of M'. Use the standard convention for bending moment signs illustrated in Figures 7.6 and 7.7 for the M equations.
3. **Partial Derivatives:** Differentiate the expressions for the internal moment with respect to M' in order to compute $\partial M/\partial M'$.
4. **Substitute Numeric Value for M':** Substitute the actual numeric value for moment M' into the expressions for M and $\partial M/\partial M'$. If a dummy moment has been used for M', its numeric value is zero.

5. **Integration:** Perform the integration indicated in Equation (17.41) to calculate the desired slope. A positive answer indicates that the rotation acts in the same direction as M', a negative answer in the opposite direction.

The use of Castigliano's theorem to compute beam deflections and slopes is illustrated in Example 17.18.

EXAMPLE 17.18

Use Castigliano's second theorem to determine (a) the deflection and (b) the slope at end A of the cantilever beam shown. Assume that EI is constant.

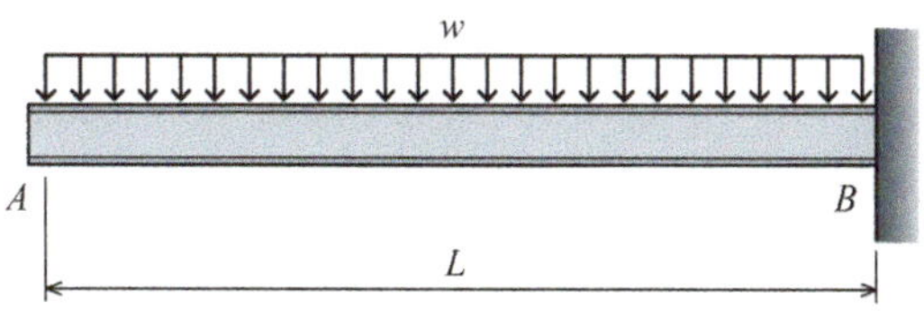

Plan the Solution

Since no external concentrated loads or concentrated moments act at A, dummy loads will be required for this problem. To determine the deflection at end A, a dummy load P acting downward will be applied at A. An expression for the internal moment M in the beam will be derived in terms of both the actual distributed load w and the dummy load P. The expression for M will then be differentiated with respect to P to obtain $\partial M/\partial P$. Next, the value $P = 0$ will be substituted in the expression for M, and then the latter will be multiplied by the partial derivative $\partial M/\partial P$. Finally, the resulting expression will be integrated over the beam length L to obtain the beam deflection at A. A similar procedure will then be used to determine the beam slope at A. The dummy load for this calculation will be a concentrated moment M' applied at A.

SOLUTION

(a) Calculation of Deflection: To determine the downward deflection of the cantilever beam, apply a dummy load P downward at A.

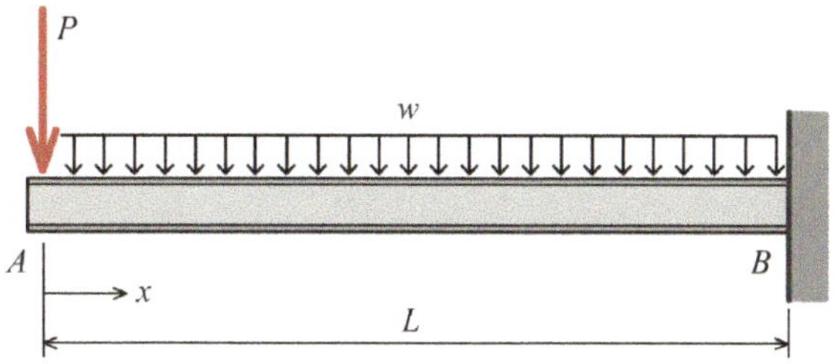

Draw a free-body diagram around end A of the beam. The origin of the x coordinate system will be placed at A. From the diagram, derive the following equation for the internal bending moment M:

$$M = -\frac{wx^2}{2} - Px \qquad 0 \le x \le L$$

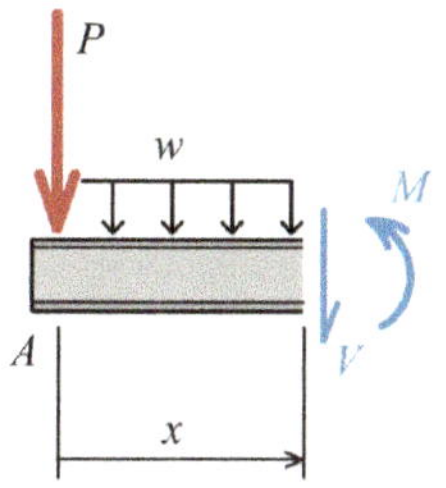

Next, differentiate this expression to obtain $\partial M/\partial P$:

$$\frac{\partial M}{\partial P} = -x$$

Substitute $P = 0$ into the bending-moment equation to obtain

$$M = -\frac{wx^2}{2}$$

Castigliano's second theorem applied to beam deflections is expressed by Equation (17.40). When the expressions just derived for $\partial M/\partial P$ and M are substituted, Equation (17.40) becomes

$$\Delta = \int_0^L \left(\frac{\partial M}{\partial P}\right)\frac{M}{EI}\,dx = \int_0^L -x\left(-\frac{wx^2}{2EI}\right)dx = \int_0^L \frac{wx^3}{2EI}\,dx$$

Now integrate this expression over the beam length L to determine the vertical beam deflection at A:

$$\Delta_A = \frac{wL^4}{8EI}\downarrow$$ **Ans.**

Since the result is a positive value, the deflection occurs in the direction assumed for the dummy load P—that is, downward.

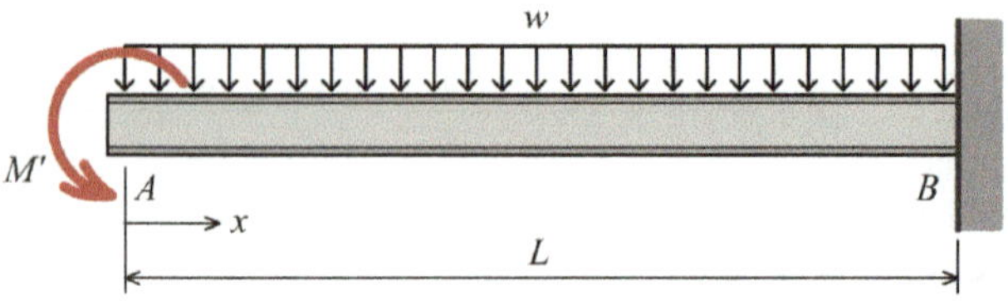

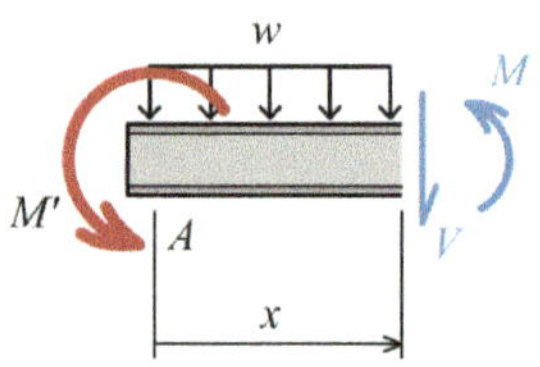

(b) Calculation of Slope: To determine the angular rotation of the cantilever beam at A, a dummy concentrated moment M' will be applied. Because the beam is expected to slope upward from A, the dummy moment will be applied counterclockwise in this instance.

Again, draw a free-body diagram around end A of the beam, placing the origin of the x coordinate system at A. From the diagram, derive the following equation for the internal bending moment M:

$$M = -\frac{wx^2}{2} - M' \qquad 0 \le x \le L$$

Next, differentiate this equation to obtain

$$\frac{\partial M}{\partial M'} = -1$$

Now substitute $M' = 0$ into the bending-moment equation to get

$$M = -\frac{wx^2}{2}$$

Castigliano's second theorem applied to beam slopes is expressed by Equation (17.41). When the expressions just derived for $\partial M/\partial M'$ and M are substituted, Equation (17.41) becomes

$$\theta = \int_0^L \left(\frac{\partial M}{\partial M'}\right)\frac{M}{EI}\,dx = \int_0^L -1\left(-\frac{wx^2}{2EI}\right)dx = \int_0^L \frac{wx^2}{2EI}\,dx$$

Integrate this expression over the beam length L to determine the beam slope at A:

$$\theta_A = \frac{wL^3}{6EI} \qquad \text{(ccw)}$$ **Ans.**

Since the result is a positive value, the angular rotation occurs in the same direction assumed for the dummy moment—that is, counterclockwise (ccw).

EXAMPLE 17.19

Compute the deflection at point C for the simply supported beam shown. Assume that $EI = 3.4 \times 10^5$ kN·m².

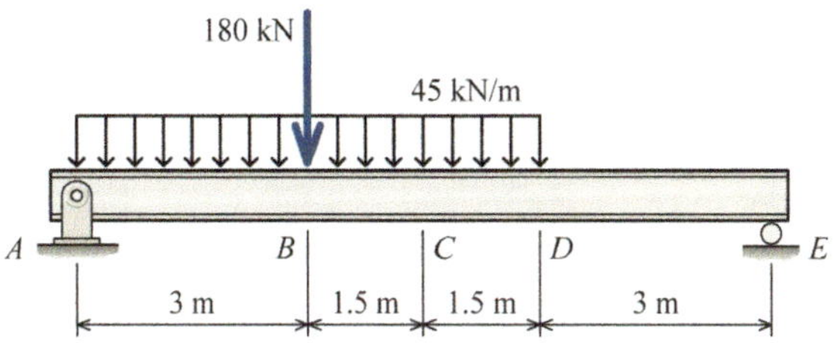

Plan the Solution

Since the deflection is desired at C and no external load acts at that location, a dummy load P will be required at C. With dummy load P placed at C, the bending-moment equation will be discontinuous at points B, C, and D. Therefore, this beam must be considered in four segments: AB, BC, CD, and DE. To facilitate the derivation of moment equations, it will be convenient to locate the origin of the x coordinate system at A for segments AB and BC, and at E for segments CD and DE. To organize the calculation, it will also be convenient to summarize the relevant equations in a tabular format.

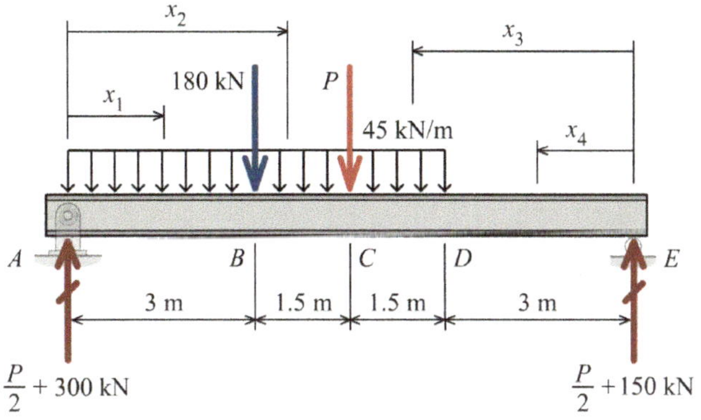

SOLUTION

Place a dummy load P at point C, which is located at the center of the 9 m beam span. Determine the beam reactions, taking care to include both the actual loads and the dummy load P. The reaction forces at A and E are shown on the free-body diagram of the beam.

Draw a free-body diagram around support A of the beam, cutting through segment AB. The origin of the x coordinate system will be placed at A. From the diagram, derive the following equation for the internal moment M in segment AB of the beam:

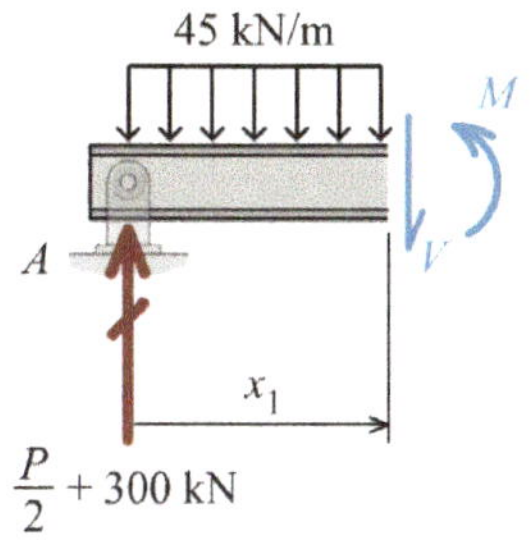

$$M = -\frac{45 \text{ kN/m}}{2} x_1^2 + \left(\frac{P}{2} + 300 \text{ kN}\right) x_1$$
$$0 \leq x_1 \leq 3 \text{ m}$$

Repeat the process with a free-body diagram cut through segment BC of the beam. From that diagram, derive the following equation for the internal moment M in segment BC of the beam:

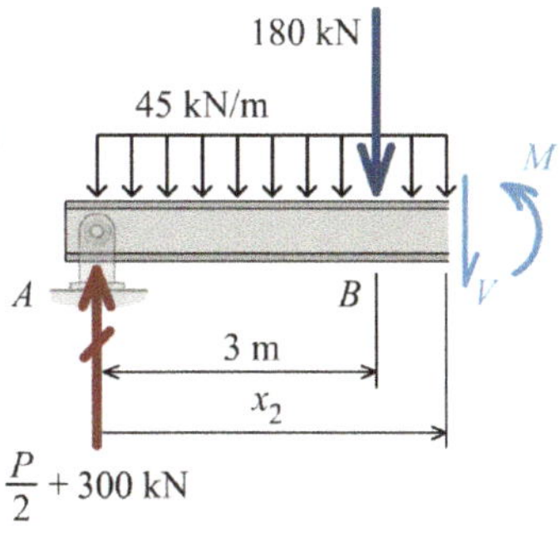

$$M = -\frac{45 \text{ kN/m}}{2} x_2^2 - (180 \text{ kN})(x_2 - 3 \text{ m}) + \left(\frac{P}{2} + 300 \text{ kN}\right) x_2$$
$$3 \text{ m} \leq x_2 \leq 4.5 \text{ m}$$

For segment CD, draw a free-body diagram around support E, cutting through segment CD of the beam. From the diagram, derive the following equation for the internal moment M in segment CD of the beam:

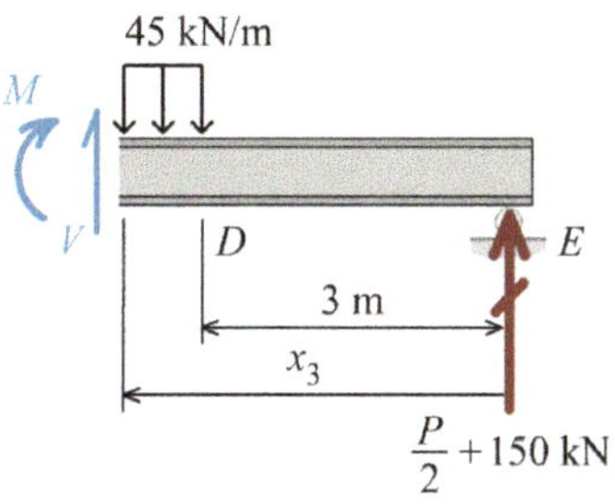

$$M = -\frac{45 \text{ kN/m}}{2}(x_3 - 3 \text{ m})^2 + \left(\frac{P}{2} + 150 \text{ kN}\right) x_3$$
$$3 \text{ m} \leq x_3 \leq 4.5 \text{ m}$$

Finally, derive the following equation for the real internal moment M in segment DE of the beam:

$$M = \left(\frac{P}{2} + 150 \text{ kN}\right)x_4$$
$$0 \leq x_4 \leq 3 \text{ m}$$

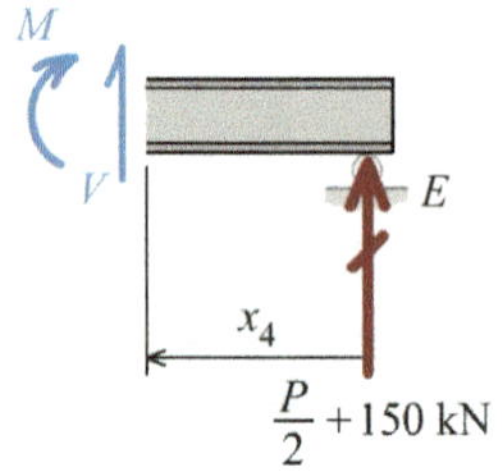

Now differentiate each equation in M with respect to P to obtain $\partial M/\partial P$. Then substitute $P = 0$ into each equation in M for the beam. The resulting expressions are summarized in the following table:

Beam Segment	M (kN·m)	$\frac{\partial M}{\partial P}$ (m)	M (for $P = 0$ kN) (kN·m)
AB	$-\frac{45}{2}x_1^2 + \left(\frac{P}{2} + 300\right)x_1$	$\frac{1}{2}x_1$	$-\frac{45}{2}x_1^2 + 300x_1$
BC	$-\frac{45}{2}x_2^2 - (180)(x_2 - 3) + \left(\frac{P}{2} + 300\right)x_2$	$\frac{1}{2}x_2$	$-\frac{45}{2}x_2^2 - 180(x_2 - 3) + 300x_2$
CD	$-\frac{45}{2}(x_3 - 3)^2 + \left(\frac{P}{2} + 150\right)x_3$	$\frac{1}{2}x_3$	$-\frac{45}{2}(x_3 - 3)^2 + 150x_3$
DE	$\left(\frac{P}{2} + 150\right)x_4$	$\frac{1}{2}x_4$	$150x_4$

Castigliano's second theorem applied to beam deflections is expressed by Equation (17.40). Substitute the expressions just derived for $\partial M/\partial P$ and M for each beam segment into Equation (17.40), taking care to note the appropriate limits of integration for each segment. These expressions, as well as the results of the integration, are summarized in the following table:

Beam Segment	*x* Coordinate		$\left(\frac{\partial M}{\partial P}\right)M$ (kN·m²)	$\int\left(\frac{\partial M}{\partial P}\right)\left(\frac{M}{EI}\right)dx$
	Origin	Limits (m)		
AB	*A*	0–3	$-11.25x_1^3 + 150x_1^2$	$\frac{1{,}122.188 \text{ kN}\cdot\text{m}^3}{EI}$
BC	*A*	3–4.5	$-11.25x_2^3 + 60x_2^2 + 270x_2$	$\frac{1{,}875.762 \text{ kN}\cdot\text{m}^3}{EI}$
CD	*E*	3–4.5	$-11.25x_3^3 + 142.5x_3^2 - 101.25x_3$	$\frac{1{,}550.918 \text{ kN}\cdot\text{m}^3}{EI}$
DE	*E*	0–3	$75x_4^2$	$\frac{675.0 \text{ kN}\cdot\text{m}^3}{EI}$
				$\frac{5{,}223.868 \text{ kN}\cdot\text{m}^3}{EI}$

From Equation (17.40), the beam deflection at C can now be determined:

$$\Delta_C = \frac{5{,}223.868 \text{ kN}\cdot\text{m}^3}{EI} = \frac{5{,}223.868 \text{ kN}\cdot\text{m}^3}{3.4 \times 10^5 \text{ kN}\cdot\text{m}^2}$$

$$\therefore \Delta_C = 15.3643 \times 10^{-3} \text{ m} = 15.36 \text{ mm} \downarrow \qquad \textbf{Ans.}$$

PROBLEMS

P17.56 Use Castigliano's second theorem to determine the vertical displacement of joint B for the truss shown in Figure P17.56/57. Assume that each member has a cross-sectional area $A = 0.85$ in.^{2} and an elastic modulus $E = 10{,}000$ ksi. The loads acting on the truss are $P = 17$ kips and $Q = 9$ kips.

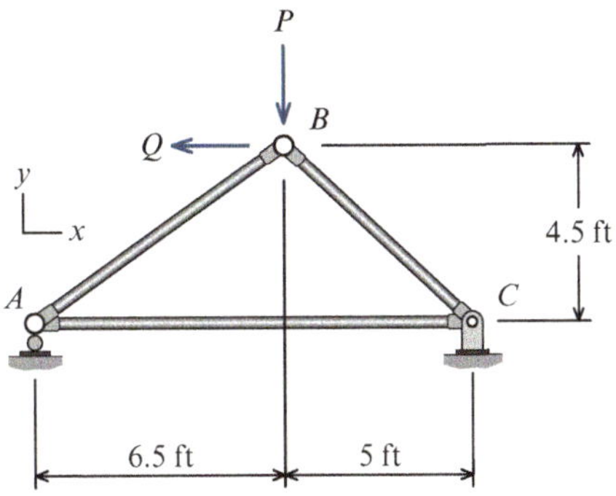

FIGURE P17.56/57

P17.57 Applying Castigliano's second theorem, find the horizontal displacement of joint B for the truss shown in Figure P17.56/57. Assume that each member has a cross-sectional area $A = 0.85$ in.^{2} and an elastic modulus $E = 10{,}000$ ksi, and that the loads acting on the truss are $P = 17$ kips and $Q = 9$ kips.

P17.58 Compute the vertical displacement of joint D for the truss shown in Figure P17.58/59. Assume that each member has a cross-sectional area $A = 2.25$ in.^{2} and an elastic modulus $E = 29{,}000$ ksi. The loads acting on the truss are $P = 13$ kips and $Q = 25$ kips. Employ Castigliano's second theorem.

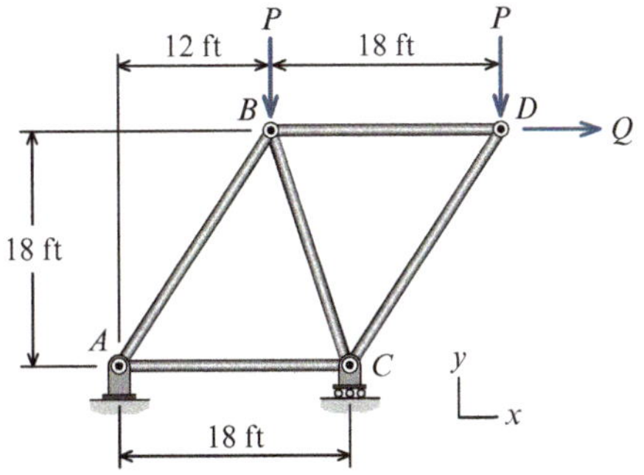

FIGURE P17.58/59

P17.59 In Figure P17.58/59, use Castigliano's second theorem to find the horizontal displacement of joint D for the truss. The assumptions are that each member has a cross-sectional area $A = 2.25$ in.^{2} and an elastic modulus $E = 29{,}000$ ksi and that the loads acting on the truss are $P = 13$ kips and $Q = 25$ kips.

P17.60 Employ Castigliano's second theorem to calculate the horizontal displacement of joint A for the truss shown in Figure P17.60/61. Make the assumption that each member has a cross-sectional area $A = 1{,}600$ mm^2 and an elastic modulus $E = 200$ GPa.

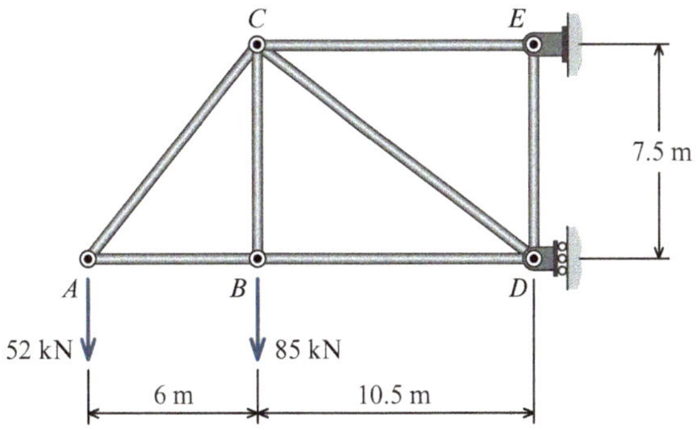

FIGURE P17.60/61

P17.61 In Figure P17.60/61, use Castigliano's second theorem to compute the vertical displacement of joint B for the truss. Each member is assumed to have a cross-sectional area $A = 1{,}600$ mm^2 and an elastic modulus $E = 200$ GPa.

P17.62 In Figure P17.62/63, the truss is subjected to concentrated loads $P = 200$ kN and $Q = 40$ kN. Members AB, BC, DE, and EF each have a cross-sectional area $A = 2{,}700$ mm^2, and all other members each have a cross-sectional area $A = 1{,}060$ mm^2. All members are made of steel [$E = 200$ GPa]. For the given loads, calculate the horizontal displacement of joint F by applying Castigliano's second theorem.

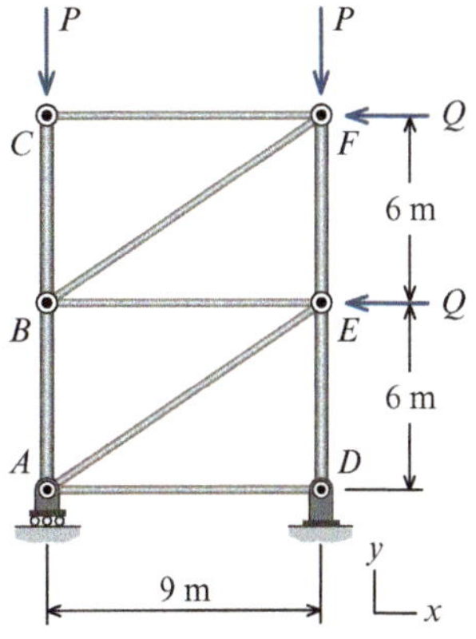

FIGURE P17.62/63

P17.63 The truss shown in Figure P17.62/63 is subjected to concentrated loads $P = 200$ kN and $Q = 40$ kN. Members AB, BC, DE, and EF each have a cross-sectional area $A = 2{,}700$ mm^2. All other members each have a cross-sectional area $A = 1{,}060$ mm^2. All members are made of steel [$E = 200$ GPa]. For the given loads, utilize Castigliano's second theorem to determine the horizontal displacement of joint B.

P17.64 The truss shown in Figure P17.64 is subjected to concentrated loads $P = 130$ kN and $2P = 260$ kN. All members are made of steel [$E = 200$ GPa], and each member has a cross-sectional area $A = 4{,}200$ mm^2. Use Castigliano's second theorem to determine

(a) the horizontal displacement of joint A.
(b) the vertical displacement of joint D.

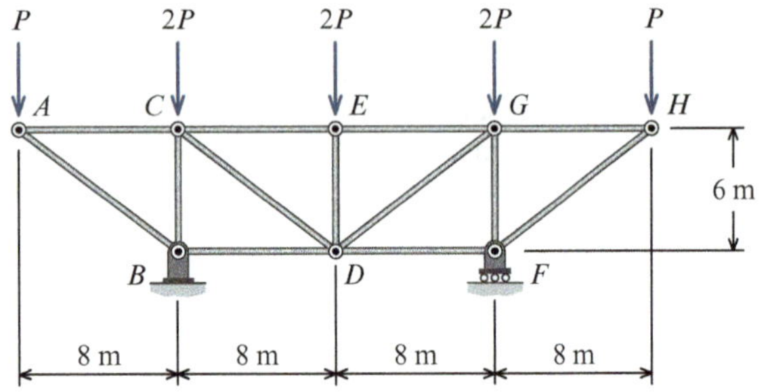

FIGURE P17.64

P17.65 Employing Castigliano's second theorem, calculate the slope of the beam at A for the loading shown in Figure P17.65. Assume that EI is constant over the length of the beam.

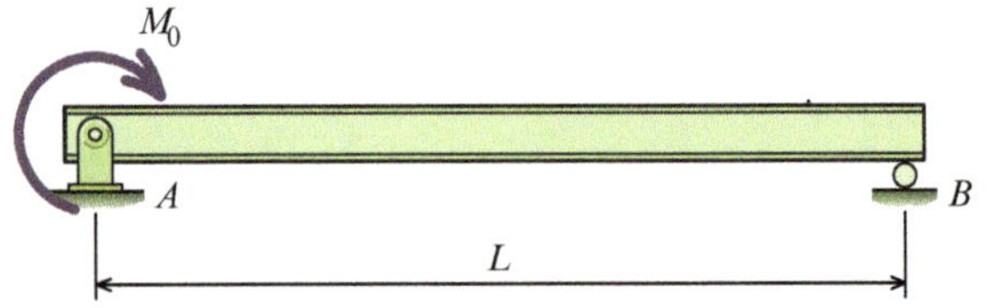

FIGURE P17.65

P17.66 Determine the deflection of the beam at A for the loading shown in Figure P17.66, utilizing Castigliano's second theorem. Assume that EI is constant over the length of the beam.

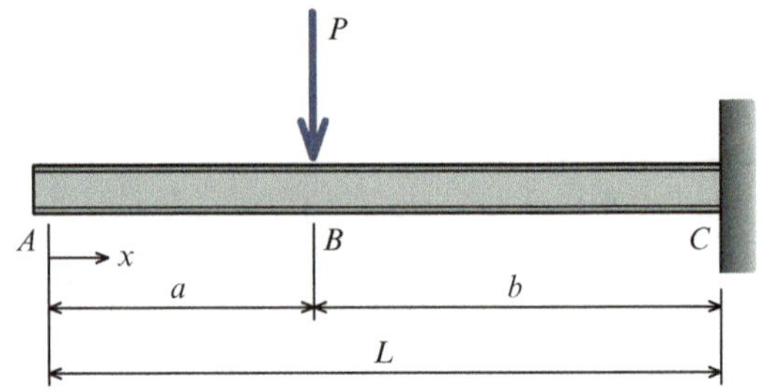

FIGURE P17.66

P17.67 Calculate the slope and the deflection of the beam at B for the loading shown in Figure P17.67. Use Castigliano's second theorem, and assume that EI is constant over the length of the beam.

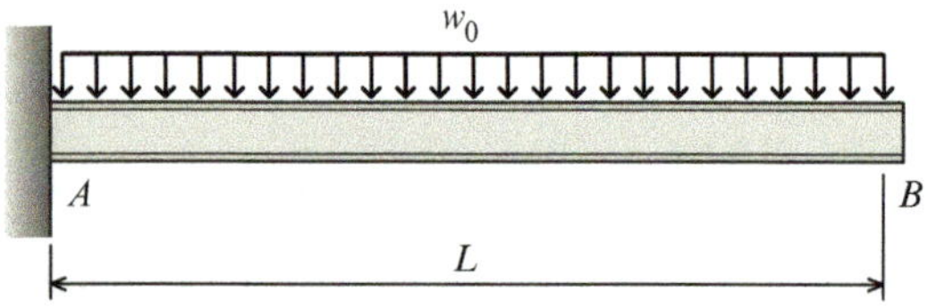

FIGURE P17.67

P17.68 Apply Castigliano's second theorem to compute the slope and deflection of the beam at C for the loading shown in Figure P17.68. Assume that EI is constant over the length of the beam.

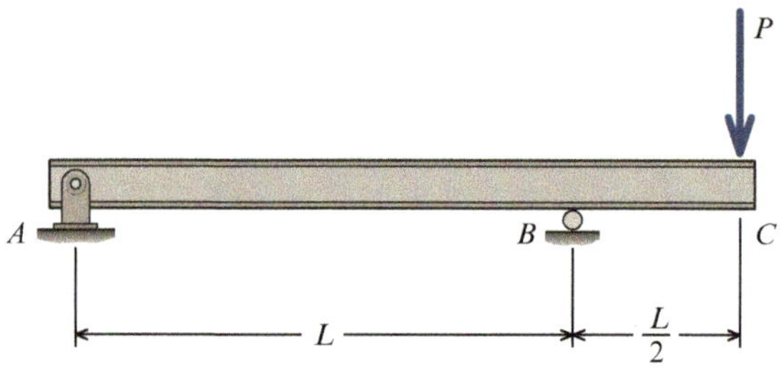

FIGURE P17.68

P17.69 What is the deflection of the compound rod at C for the loading shown in Figure P17.69? Between A and B, the rod's diameter is 35 mm, and between B and C, its diameter is 20 mm. Assume that $E = 200$ GPa for both segments of the compound rod, and use Castigliano's second theorem.

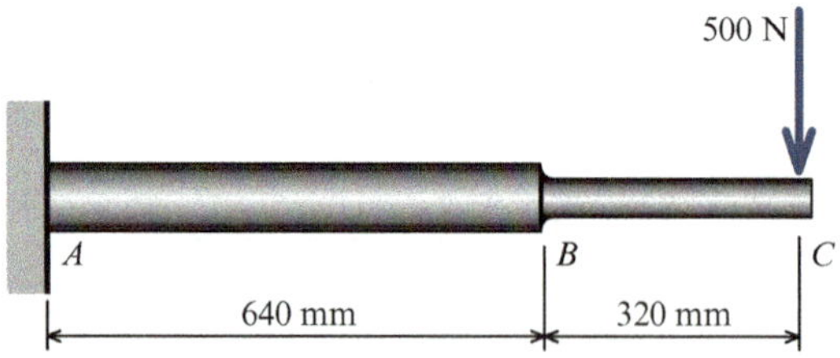

FIGURE P17.69

P17.70 The compound steel [$E = 200$ GPa] rod shown in Figure P17.70/71 has a diameter of 20 mm in each of segments AB and DE, and a diameter of 35 mm in each of segments BC and CD. For the given loading, employ Castigliano's second theorem to find the slope of the compound rod at A.

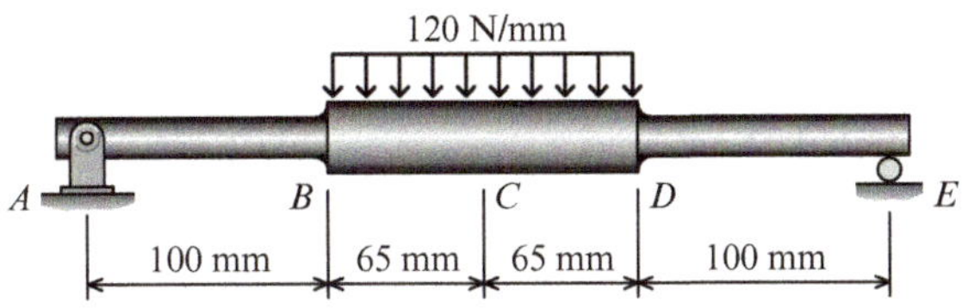

FIGURE P17.70/71

P17.71 Figure P17.70/71 shows a compound steel [$E = 200$ GPa] rod with a diameter of 20 mm in each of segments AB and DE, and a diameter of 35 mm in each of segments BC and CD. For the given loading, calculate the deflection of the compound rod at C, using Castigliano's second theorem.

P17.72 Figure P17.72 shows a simply supported beam. Assume that $EI = 15 \times 10^6$ kip·in.^{2} for the beam. Use Castigliano's second theorem to determine

(a) the deflection at A.
(b) the slope at C.

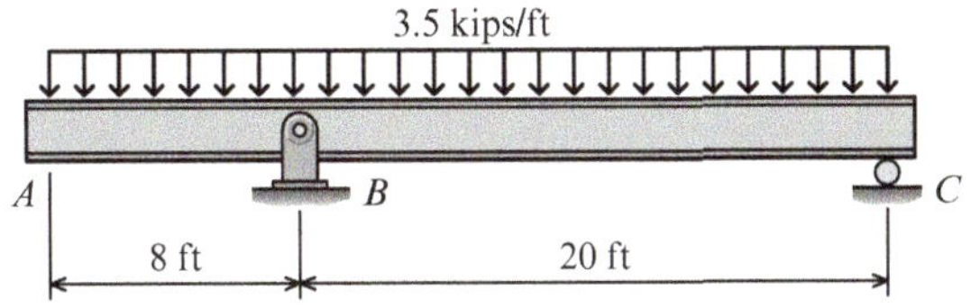

FIGURE P17.72

P17.73 A cantilever beam is loaded as shown in Figure P17.73. Assume that $EI = 74 \times 10^3$ kN · m^2 for the beam, and employ Castigliano's second theorem to find

(a) the slope at C.
(b) the deflection at C.

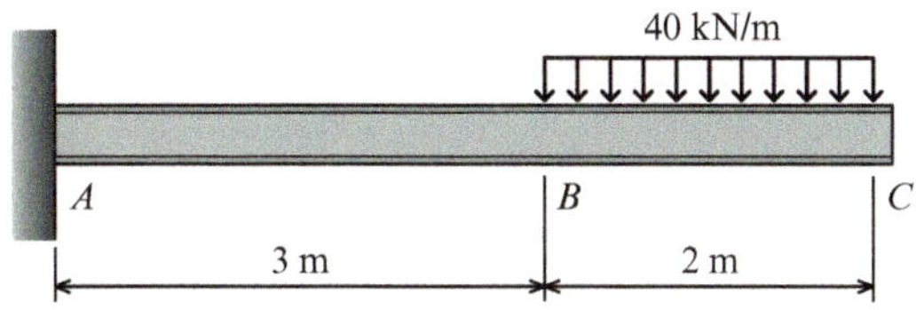

FIGURE P17.73

P17.74 Compute the minimum moment of inertia I required for the beam in Figure P17.74 if the maximum beam deflection must not exceed 35 mm. Assuming that $E = 200$ GPa, employ Castigliano's second theorem.

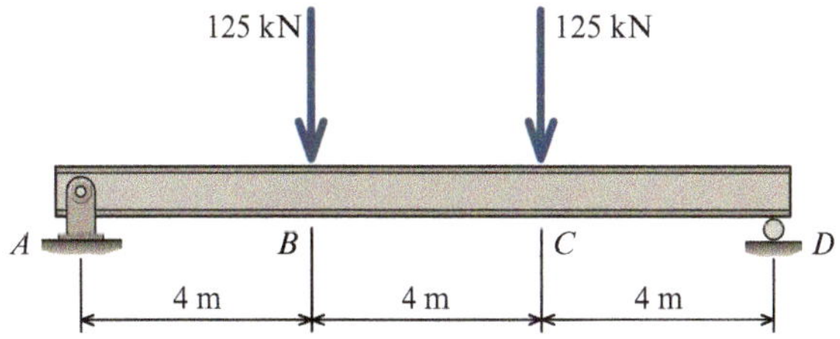

FIGURE P17.74

P17.75 In Figure P17.75, if the maximum beam deflection must not exceed 0.5 in., what is the minimum moment of inertia I required for the beam? Utilize Castigliano's second theorem, and assume that $E = 29{,}000$ ksi.

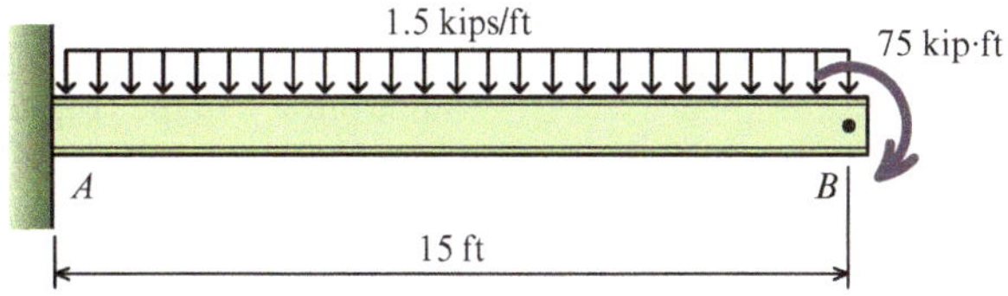

FIGURE P17.75

APPENDIX A

Geometric Properties of an Area

A.1 Centroid of an Area

The centroid of an area refers to the point that defines the geometric center of the area. For an arbitrary shape (Figure A.1*a*), the *x* and *y* coordinates of the centroid *c* are determined from the formulas:

The term *first moment* is used to describe $x\,dA$ since x is a term raised to the *first power*, as in $x^1 = x$. Another geometric property of an area, the moment of inertia, includes the term x^2, and hence, the area moment of inertia is sometimes referred to as the *second moment of area*.

$$\bar{x} = \frac{\int_A x\,dA}{\int_A dA} \qquad \bar{y} = \frac{\int_A y\,dA}{\int_A dA} \tag{A.1}$$

The expressions $x\,dA$ and $y\,dA$ are termed the *first moments of area dA* about the *y* and the *x* axis, respectively (Figure A.1*b*). The denominators in Equation (A.1) are expressions of the total area *A* of the shape.

The centroid will always lie on an axis of symmetry. In cases where an area has two axes of symmetry, the centroid will be found at the intersection of the two axes. Centroids for several common plane shapes are summarized in Table A.1.

Composite Areas

The cross-sectional area of many common mechanical and structural components can often be subdivided into a collection of simple shapes such as rectangles and circles. This subdivided area is termed a *composite area*. By virtue of the symmetry inherent in rectangles and circles, the centroid locations for these shapes are easily determined;

Table A.1 Properties of Plane Figures

1. Rectangle

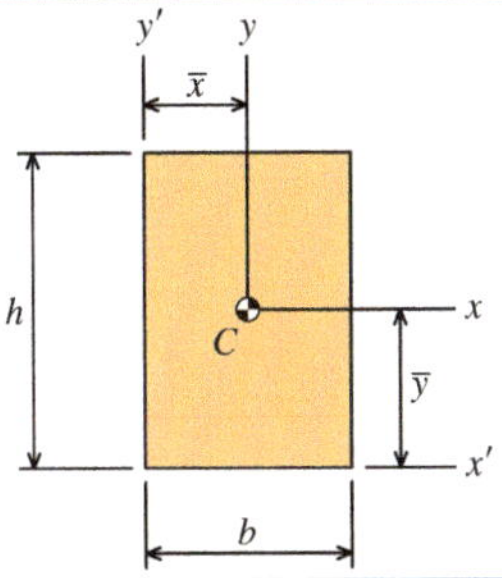

$$A = bh$$

$$\bar{y} = \frac{h}{2} \qquad I_x = \frac{bh^3}{12}$$

$$\bar{x} = \frac{b}{2} \qquad I_y = \frac{hb^3}{12}$$

$$I_{x'} = \frac{bh^3}{3} \qquad I_{y'} = \frac{hb^3}{3}$$

2. Right Triangle

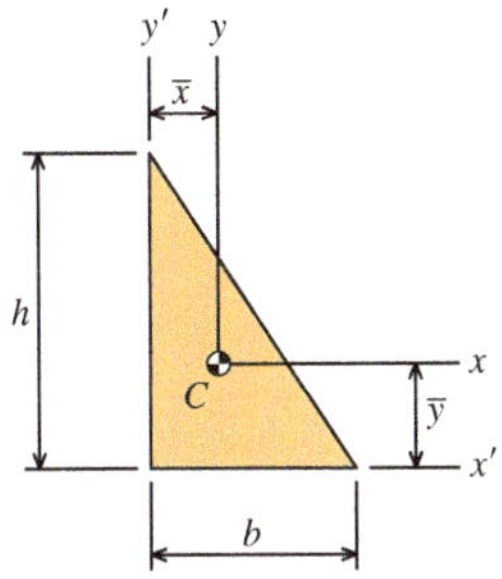

$$A = \frac{bh}{2}$$

$$\bar{y} = \frac{h}{3} \qquad I_x = \frac{bh^3}{36}$$

$$\bar{x} = \frac{b}{3} \qquad I_y = \frac{hb^3}{36}$$

$$I_{x'} = \frac{bh^3}{12} \qquad I_{y'} = \frac{hb^3}{12}$$

3. Triangle

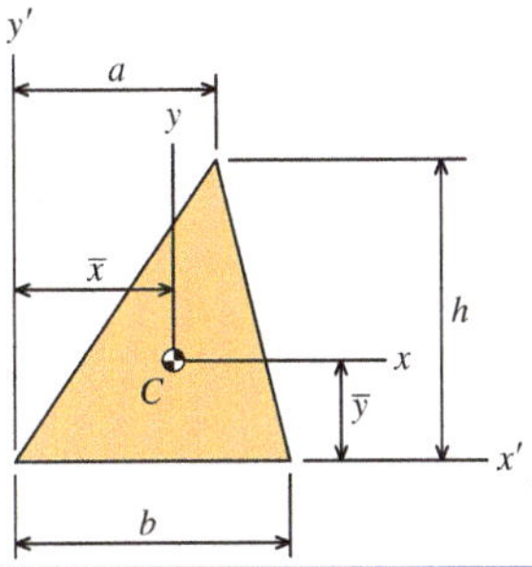

$$A = \frac{bh}{2}$$

$$\bar{y} = \frac{h}{3} \qquad I_x = \frac{bh^3}{36}$$

$$\bar{x} = \frac{(a+b)}{3} \qquad I_y = \frac{bh}{36}(a^2 - ab + b^2)$$

$$I_{x'} = \frac{bh^3}{12}$$

4. Trapezoid

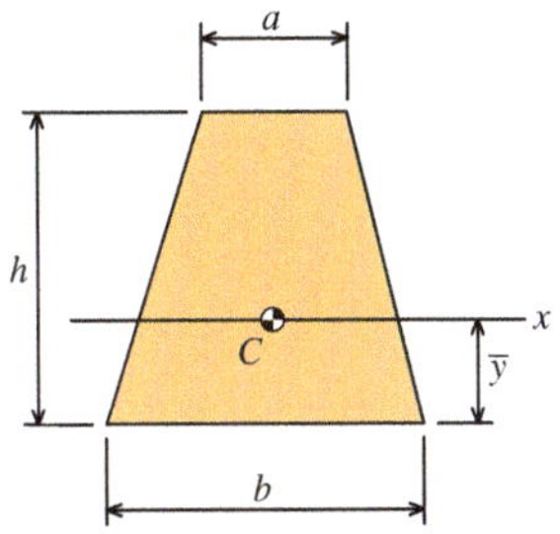

$$A = \frac{(a+b)h}{2}$$

$$\bar{y} = \frac{1}{3}\left(\frac{2a+b}{a+b}\right)h$$

$$I_x = \frac{h^3}{36(a+b)}(a^2 + 4ab + b^2)$$

5. Semicircle

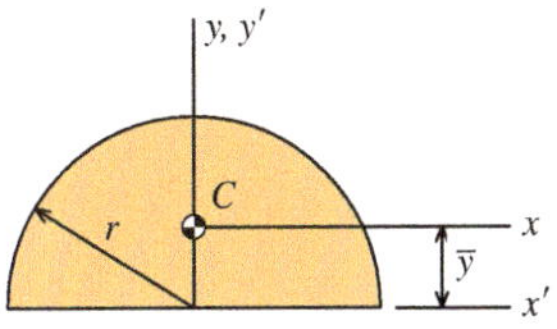

$$A = \frac{\pi r^2}{2}$$

$$\bar{y} = \frac{4r}{3\pi} \qquad I_x = \left(\frac{\pi}{8} - \frac{8}{9\pi}\right)r^4$$

$$I_{x'} = I_{y'} = \frac{\pi r^4}{8}$$

6. Circle

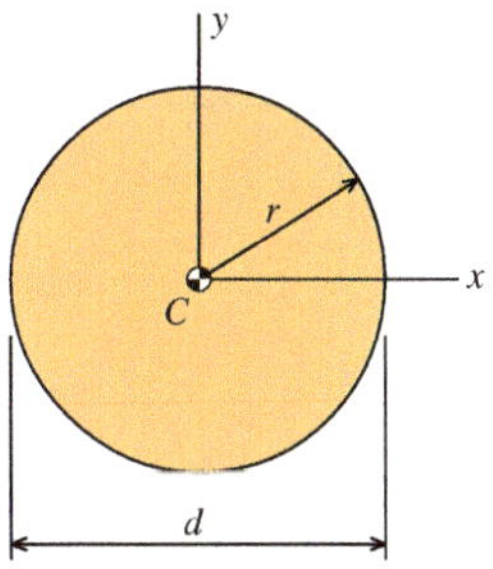

$$A = \pi r^2 = \frac{\pi d^2}{4}$$

$$I_x = I_y = \frac{\pi r^4}{4} = \frac{\pi d^4}{64}$$

7. Hollow Circle

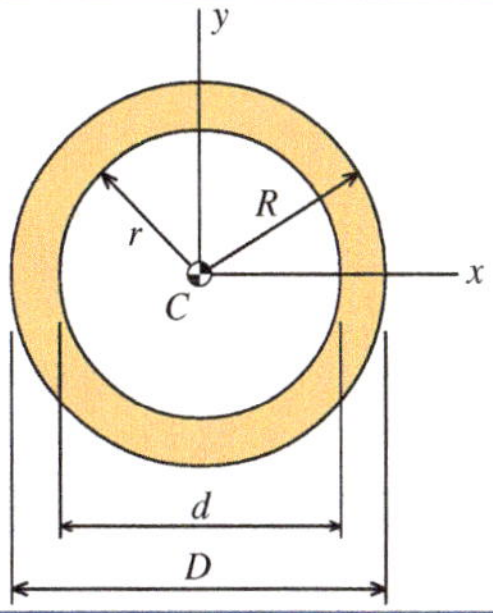

$$A = \pi(R^2 - r^2) = \frac{\pi}{4}(D^2 - d^2)$$

$$I_x = I_y = \frac{\pi}{4}(R^4 - r^4)$$

$$= \frac{\pi}{64}(D^4 - d^4)$$

8. Parabola

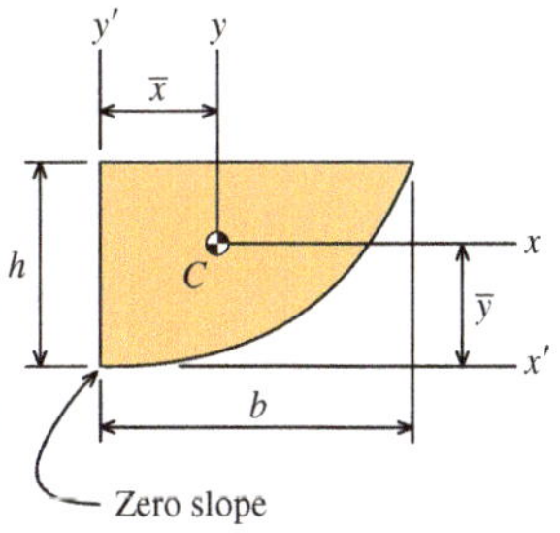

$$y' = \frac{h}{b^2}x'^2$$

$$A = \frac{2bh}{3}$$

$$\bar{x} = \frac{3b}{8} \qquad \bar{y} = \frac{3h}{5}$$

9. Parabolic Spandrel

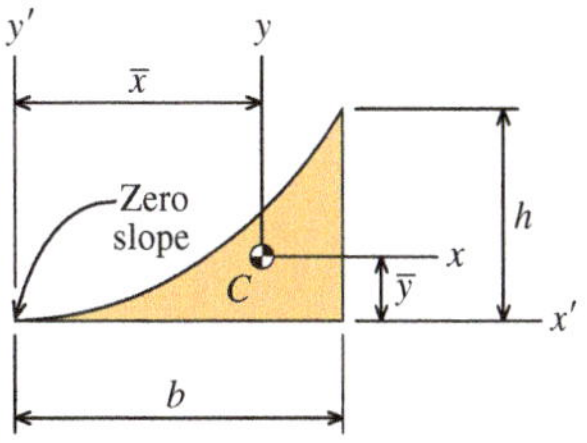

$$y' = \frac{h}{b^2}x'^2$$

$$A = \frac{bh}{3}$$

$$\bar{x} = \frac{3b}{4} \qquad \bar{y} = \frac{3h}{10}$$

10. General Spandrel

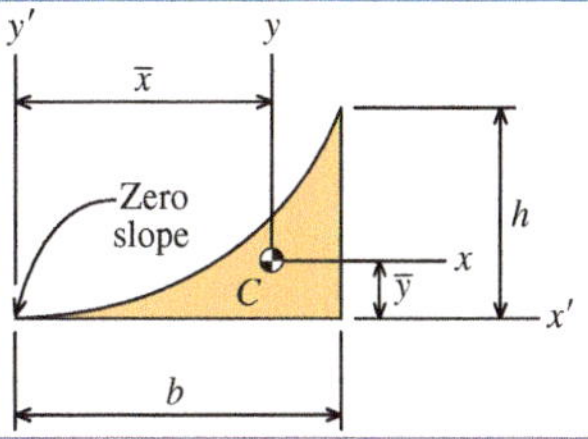

$$y' = \frac{h}{b^n}x'^n$$

$$A = \frac{bh}{n+1}$$

$$\bar{x} = \frac{n+1}{n+2}b \qquad \bar{y} = \frac{n+1}{4n+2}h$$

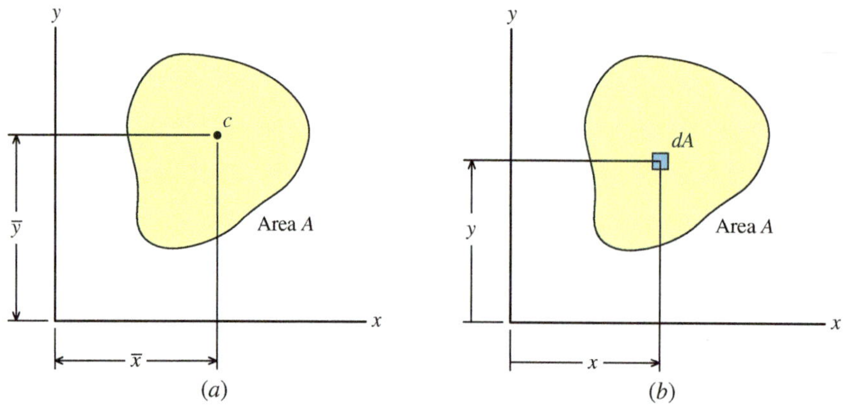

FIGURE A.1 Centroid of an area.

hence, the centroid calculation procedure for composite areas can be arranged so that integration is not necessary. Expressions analogous to Equation (A.1) in which the integral terms are replaced with summation terms can be used. For a composite area composed of i simple shapes, the centroid location can be computed with the following expressions:

$$\bar{x} = \frac{\Sigma x_i A_i}{\Sigma A_i} \qquad \bar{y} = \frac{\Sigma y_i A_i}{\Sigma A_i} \tag{A.2}$$

where x_i and y_i are the *algebraic distances* or coordinates measured from some defined reference axes to the centroids of each of the simple shapes comprising the composite area. The term ΣA_i represents the sum of the simple areas, which add up to the total area of the composite area. If a hole or region having no material lies within a composite area, then that hole is treated as a *negative area* in the calculation procedure.

MecMovies

EXAMPLE

A.1 The Centroids Game: Learning the Ropes

A game that helps to build proficiency in centroid calculations for composite areas made up of rectangles.

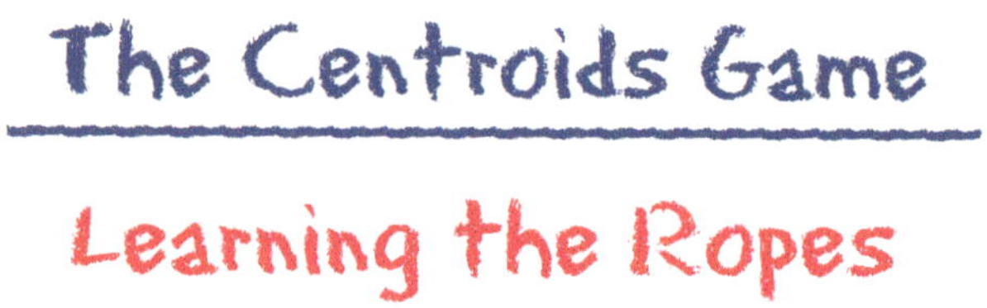

EXAMPLE A.1

Determine the location of the centroid for the flanged shape shown.

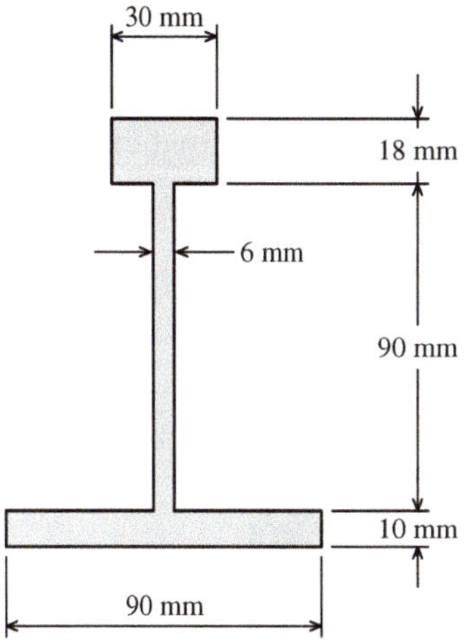

Plan the Solution

The centroid location in the horizontal direction can be determined from symmetry alone. To determine the vertical location of the centroid, the shape is subdivided into three rectangular areas. Using the lower edge of the shape as a reference axis, Equation (A.2) is then used to compute the centroid location.

SOLUTION

The centroid location in the horizontal direction can be determined from symmetry alone; however, the centroid location in the y direction must be calculated. The flanged shape is first subdivided into rectangular shapes (1), (2), and (3), and the area A_i for each of these shapes is computed. For calculation purposes, a reference axis must be established. In this example, the reference axis will be placed at the lower edge of the shape. The distance y_i in the vertical direction from the reference axis to the centroid of each rectangular area A_i is determined, and the product y_iA_i (termed the *first moment of area*) is computed. The centroid location $\bar{y}$ measured from the reference axis is computed as the sum of the first moments of area y_iA_i divided by the sum of the areas A_i. The calculation for the shape is summarized in the table below.

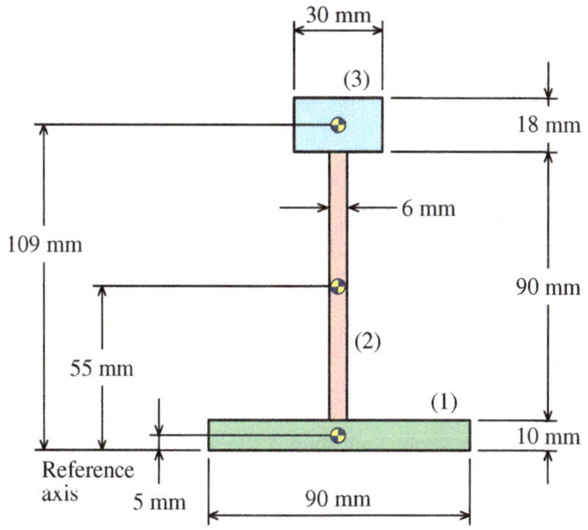

	A_i (mm²)	y_i (mm)	y_iA_i (mm³)
(1)	900	5	4,500
(2)	540	55	29,700
(3)	540	109	58,860
	1,980		93,060

$$\bar{y} = \frac{\Sigma y_i A_i}{\Sigma A_i} = \frac{93{,}060 \text{ mm}^3}{1{,}980 \text{ mm}^2} = 47.0 \text{ mm} \qquad \textbf{Ans.}$$

Therefore, the centroid is located 47.0 mm above the lower edge of the shape.

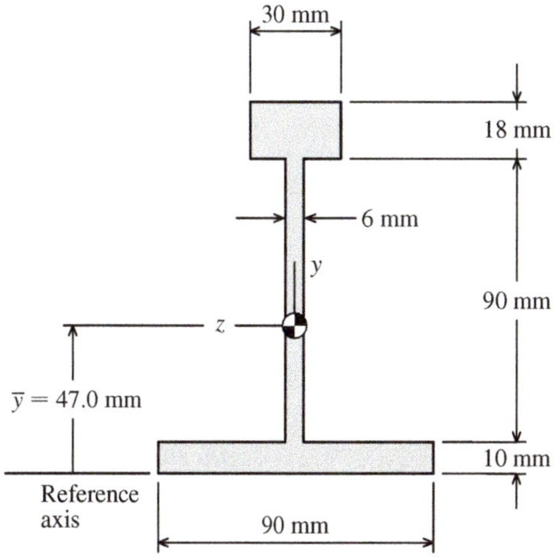

793

EXAMPLES

A.2 Animated example of the centroid calculation procedure for a tee shape.

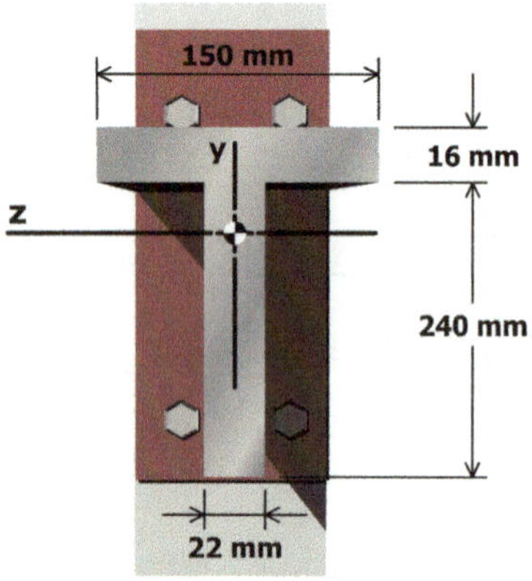

A.3 Animated example of the centroid calculation procedure for a U-shape.

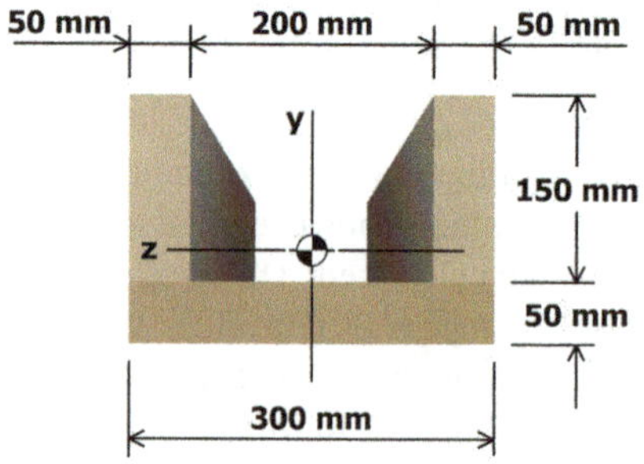

A.2 Moment of Inertia for an Area

The terms $\int x\,dA$ and $\int y\,dA$ appear in the definition of a centroid [Equation (A.1)], and these terms are called *first moments of area* about the y and x axes, respectively, because x and y are first-order terms. In mechanics of materials, several equations are derived that contain integrals of the form $\int x^2\,dA$ and $\int y^2\,dA$, and these terms are called *second moments of area* because x^2 and y^2 are second-order terms. However, the second moment of area is more commonly called the *moment of inertia* of an area.

The term *moment of inertia* is applied to the second moment of area because of its similarity to the moment of inertia of the mass of a body.

In Figure A.2, the moment of inertia for area A about the x axis is defined as

$$I_x = \int_A y^2\,dA \tag{A.3}$$

Similarly, the moment of inertia for area A about the y axis is defined as

$$I_y = \int_A x^2\,dA \tag{A.4}$$

A second moment expression can also be stated for a reference axis that is normal to the plane (such axes are called *poles*). In Figure A.2, the z axis that passes through origin O is perpendicular to the plane of the area A. An integral expression called the *polar moment of inertia* J can be written in terms of the distance r from the reference z axis to dA:

$$J = \int_A r^2\,dA \tag{A.5}$$

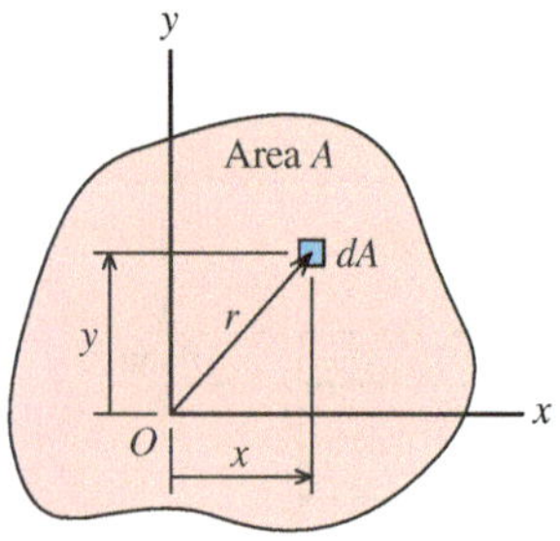

FIGURE A.2

Using the Pythagorean theorem, distance r is related to distances x and y by $r^2 = x^2 + y^2$. Accordingly, Equation (A.5) can be expressed as

$$J = \int_A r^2\,dA = \int_A (x^2 + y^2)\,dA = \int_A x^2\,dA + \int_A y^2\,dA$$

and thus,

$$J = I_y + I_x \tag{A.6}$$

Notice that the x and y axes can be any pair of mutually perpendicular axes intersecting at O.

From the definitions given in Equations (A.3), (A.4), and (A.5), moments of inertia are always positive terms that have dimensions of length raised to the fourth power (L^4). Common units are mm^4 and in.^{4}.

Area moments of inertia for several common plane shapes are summarized in Table A.1.

Parallel-Axis Theorem for an Area

When the moment of inertia of an area has been determined with respect to a given axis, the moment of inertia with respect to a parallel axis can be obtained by means of the **parallel-axis theorem**, provided that one of the axes passes through the centroid of the area.

The moment of inertia of the area in Figure A.3 about the b reference axis is

$$\begin{aligned} I_b &= \int_A (y + d)^2\,dA = \int_A y^2\,dA + 2d\int_A y\,dA + d^2\int_A dA \\ &= I_x + 2d\int_A y\,dA + d^2 A \end{aligned} \tag{a}$$

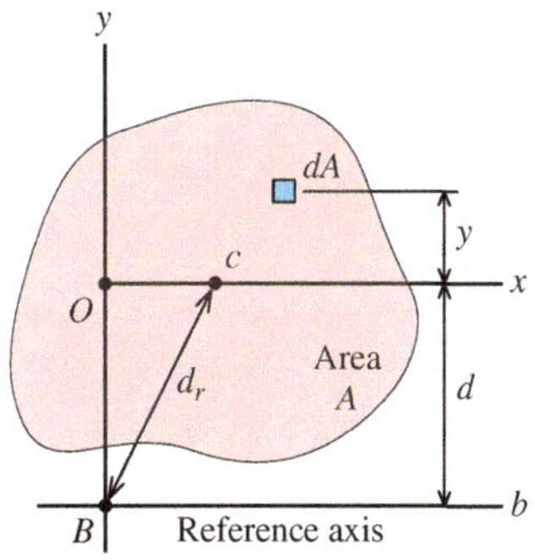

FIGURE A.3

The integral $\int_A y\,dA$ is simply the first moment of area A about the x axis. From Equation (A.1),

$$\int_A y\,dA = \bar{y}A$$

If the x axis passes through the centroid c of the area, then $\bar{y} = 0$ and Equation (a) reduces to

$$I_b = I_c + d^2 A \tag{A.7}$$

where I_c is the moment of inertia of area A about the centroidal axis that is parallel to the reference axis (i.e., the b axis in this instance), and d is the perpendicular distance between the two axes. In a similar manner it can be shown that the parallel-axis theorem is also applicable for polar moments of inertia:

$$J_b = J_c + d_r^2 A \tag{A.8}$$

The parallel-axis theorem states that the moment of inertia for an area about an axis is equal to the area's moment of inertia about a parallel axis passing through the centroid plus the product of the area and the square of the distance between the two axes.

Composite Areas

It is often necessary to calculate the moment of inertia for an irregularly shaped area. If such an area can be subdivided into a number of simple shapes such as rectangles, triangles, and circles, then the moment of inertia for the irregular area can be conveniently found by using the parallel-axis theorem. The moment of inertia for the composite area is equal to the sum of the moments of inertia for the constituent shapes:

$$I = \Sigma(I_c + d^2 A)$$

If an area such as a hole is removed from a larger area, then its moment of inertia must be subtracted in the summation above.

MecMovies

EXAMPLES

The Moment of Inertia Game: Starting from Square One
A.4 A game that helps to build proficiency in moment of inertia calculations for composite areas made up of rectangles.

A.5 Determine the centroid location and the moment of inertia about the centroidal axis for a tee shape.

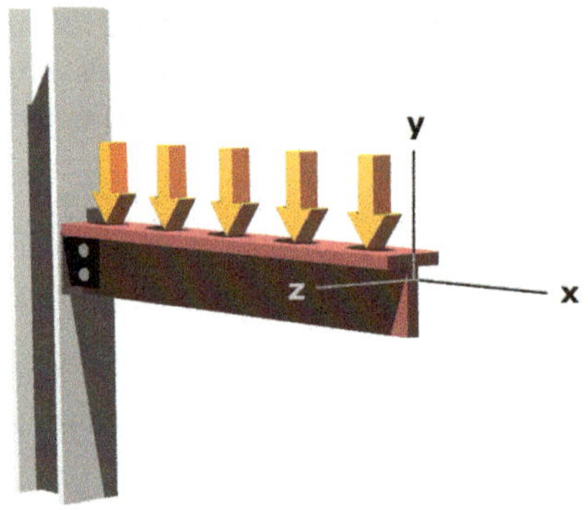

EXAMPLE A.2

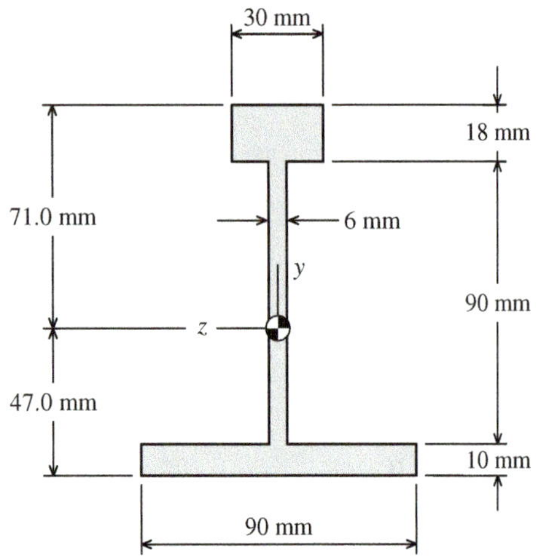

Determine the moment of inertia about the z and y axes for the flanged shape shown in Example A.2.

Plan the Solution

In Example A.1, the flanged shape was subdivided into three rectangles. The moment of inertia of a rectangle about its centroidal axis is given by $I_c = bh^3/12$. To compute I_z, this relationship will be used with the parallel-axis theorem to compute the moments of inertia for each of the three rectangles about the z centroidal axis of the flanged shape. These three terms will be added together to give I_z for the entire shape. The computation of I_y will be similar; however, the parallel-axis theorem will not be required since the centroids for all three rectangles lie on the y centroidal axis.

SOLUTION

(a) Moment of Inertia About the *z* Centroidal Axis

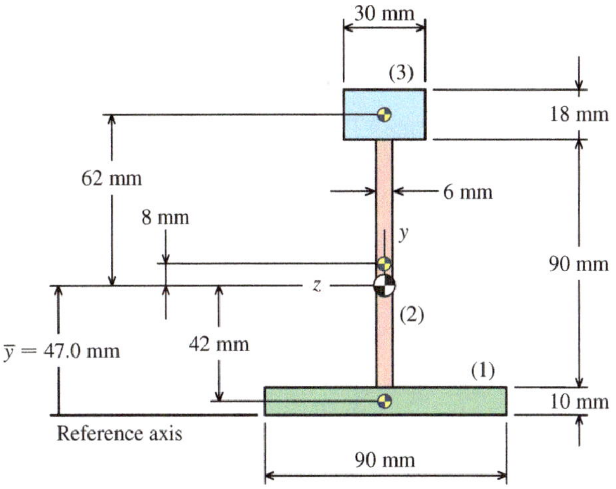

The moment of inertia I_{ci} of each rectangular shape about its own centroid must be computed to begin the calculation. The moment of inertia of a rectangle about its centroidal axis is given by the general equation $I_c = bh^3/12$, where b is the dimension parallel to the axis and h is the perpendicular dimension.

For example, the moment of inertia of area (1) about its horizontal centroidal axis is calculated as $I_{c1} = bh^3/12 = (90\text{ mm})(10\text{ mm})^3/12 = 7{,}500\text{ mm}^4$. Next, the perpendicular distance d_i between the z centroidal axis for the entire flanged shape and the z centroidal axis for area A_i must be determined. The term d_i is squared and multiplied by A_i and the result is added to I_{ci} to give the moment of inertia for each rectangular shape about the z centroidal axis of the entire flanged cross section. The results for all areas A_i are summed to determine the moment of inertia of the flanged cross section about its z centroidal axis. The complete calculation procedure is summarized in the table below.

	I_{ci} (mm^4)	$\|d_i\|$ (mm)	$d_i^2A_i$ (mm^4)	I_z (mm^4)
(1)	7,500	42.0	1,587,600	1,595,100
(2)	364,500	8.0	34,560	399,060
(3)	14,580	62.0	2,075,760	2,090,340
				4,084,500

Thus, the moment of inertia of the flanged shape about its z centroidal axis is $I_z = 4{,}080{,}000\text{ mm}^4$. **Ans.**

(b) Moment of Inertia About the *y* Centroidal Axis

As before, the moment of inertia I_{ci} of each rectangular shape about its own centroid must be computed to begin the calculation. However, it is the moment of inertia about the vertical centroidal axis that is required here. For example, the moment of inertia of area (1) about its vertical centroidal axis is calculated as $I_{c1} = bh^3/12 = (10\text{ mm})(90\text{ mm})^3/12 = 607{,}500\text{ mm}^4$. (Compared to the I_z calculation, notice that different values are associated with b and h in the standard formula $bh^3/12$.) The parallel-axis theorem is not needed for this calculation because the centroids of each rectangle lie on the y centroidal axis of the flanged shape. The complete calculation procedure is summarized in the table below.

	I_{ci} (mm^4)	$\|d_i\|$ (mm)	$d_i^2A_i$ (mm^4)	I_y (mm^4)
(1)	607,500	0	0	607,500
(2)	1,620	0	0	1,620
(3)	40,500	0	0	40,500
				649,620

The moment of inertia of the flanged shape about its y centroidal axis is thus $I_y = 650{,}000\text{ mm}^4$. **Ans.**

EXAMPLE A.3

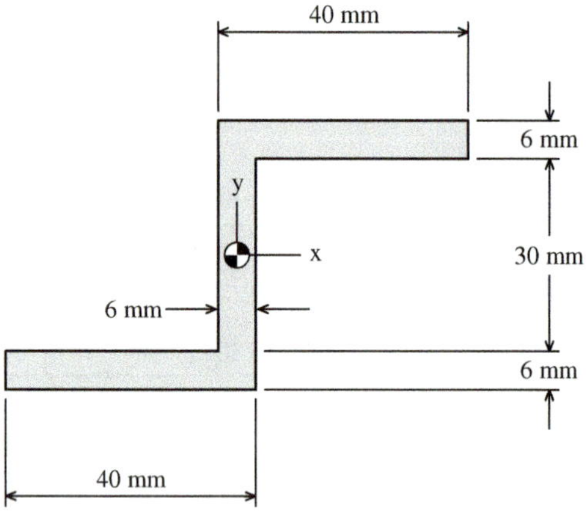

Determine the moment of inertia about the x and y centroidal axes for the zee shape shown.

Plan the Solution

After subdividing the zee shape into three rectangles, the moments of inertia I_x and I_y will be computed using $I_c = bh^3/12$ and the parallel-axis theorem.

SOLUTION

The centroid location for the zee shape is shown in the sketch. The complete calculation for I_x and I_y is summarized in the tables on the next page.

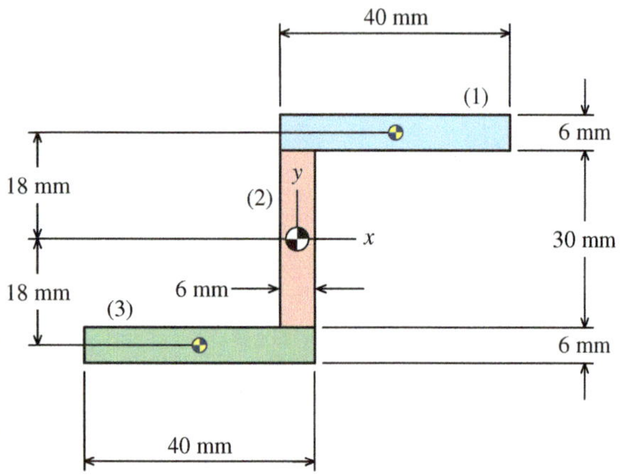

(a) Moment of Inertia About the x Centroidal Axis

	I_{ci} (mm^4)	$\lvert d_i \rvert$ (mm)	A_i (mm^2)	$d_i^2 A_i$ (mm^4)	I_z (mm^4)
(1)	720	18.0	240	77,760	78,480
(2)	13,500	0	180	0	13,500
(3)	720	18.0	240	77,760	78,480
					170,460

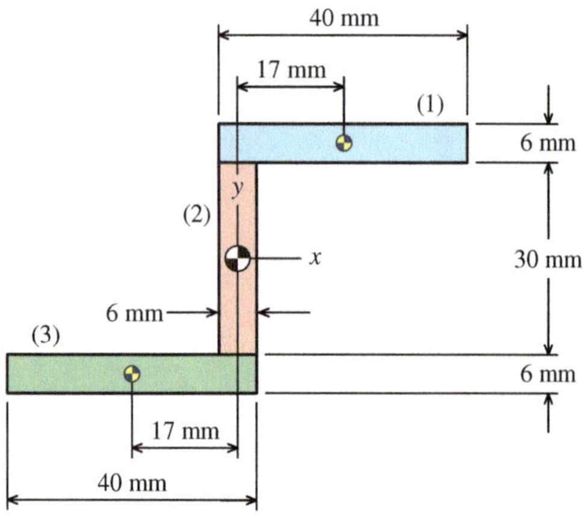

(b) Moment of Inertia About the y Centroidal Axis

	I_{ci} (mm^4)	$\lvert d_i \rvert$ (mm)	A_i (mm^2)	$d_i^2 A_i$ (mm^4)	I_z (mm^4)
(1)	32,000	17.0	240	69,360	101,360
(2)	540	0	180	0	540
(3)	32,000	17.0	240	69,360	101,360
					203,260

The moments of inertia for the zee shape are $I_x = 170{,}500$ mm^4 and $I_y = 203{,}000$ mm^4. **Ans.**

A.3 Product of Inertia for an Area

The product of inertia dI_{xy} of the area element dA in Figure A.4 with respect to the x and y axes is defined as the product of the two coordinates of the element multiplied by the area of the element. The product of inertia of the total area A is thus

$$I_{xy} = \int_A xy\,dA \tag{A.9}$$

The dimensions of the product of inertia are the same as those of the moment of inertia (i.e., length units raised to the fourth power). Whereas the moment of inertia is always positive, *the product of inertia can be positive, negative, or zero.*

The product of inertia for an area with respect to any two orthogonal axes is zero when either of the axes is an axis of symmetry. This statement can be demonstrated by Figure A.5 in which the area is symmetric with respect to the x axis. The products of inertia of the elements dA and dA' on opposite sides of the axis of symmetry will be equal in magnitude and opposite in sign; thus, they will cancel each other in the summation. The resulting product of inertia for the entire area will be zero.

The parallel-axis theorem for products of inertia can be derived from Figure A.6 in which the x' and y' axes pass through the centroid c and are parallel to the x and y axes. The product of inertia with respect to the x and y axes is

$$\begin{aligned} I_{xy} &= \int_A xy\,dA \\ &= \int_A (x_c + x')(y_c + y')\,dA \\ &= x_c y_c \int_A dA + x_c \int_A y'\,dA + y_c \int_A x'\,dA + \int_A x'y'\,dA \end{aligned}$$

The second and third integrals in the preceding equation are zero since x' and y' are centroidal axes. The last integral is the product of inertia of the area A with respect to its centroid. Consequently, the product of inertia is

$$I_{xy} = I_{x'y'} + x_c y_c A \tag{A.10}$$

The parallel-axis theorem for products of inertia can be stated as follows: *The product of inertia for an area with respect to any two orthogonal axes x and y is equal to the product of inertia of the area with respect to a pair of centroidal axes parallel to the x and y axes added to the product of the area and the two centroidal distances from the x and y axes.*

The product of inertia is used in determining principal axes of inertia, as discussed in the following section. The determination of the product of inertia is illustrated in the next two examples.

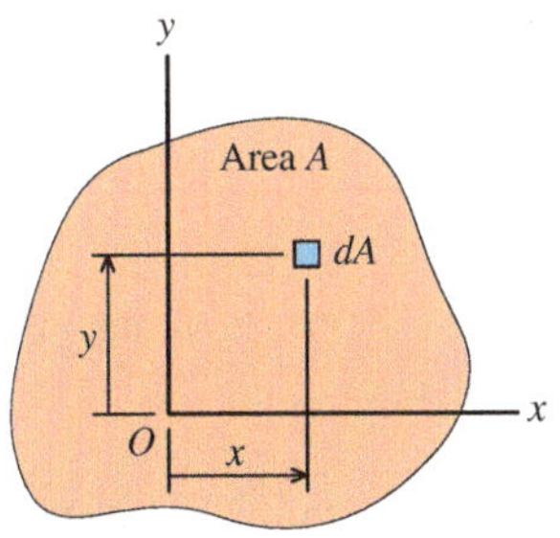

FIGURE A.4

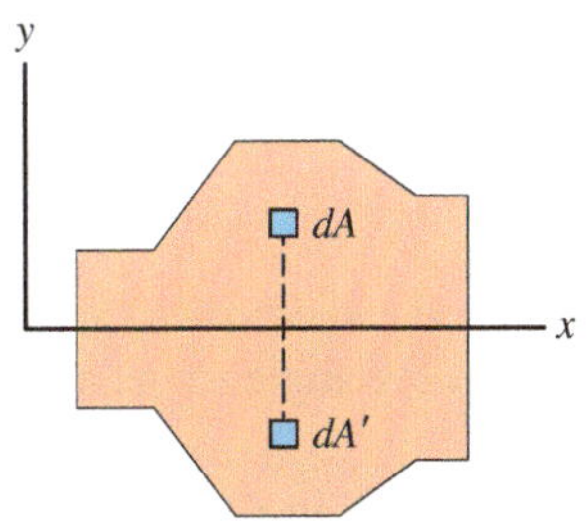

FIGURE A.5

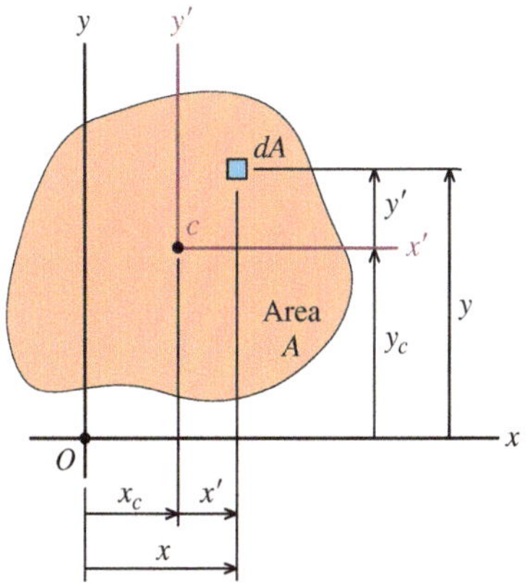

FIGURE A.6

EXAMPLE A.4

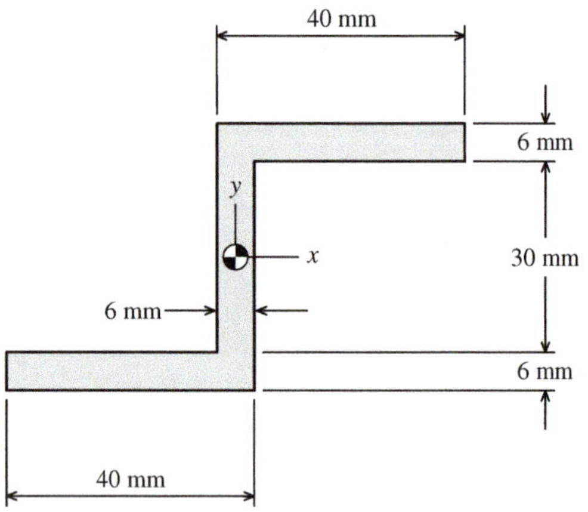

Determine the product of inertia for the zee shape shown in Example A.3.

Plan the Solution

The zee shape can be subdivided into three rectangles. Since rectangles are symmetric, their respective products of inertia about their own centroidal axes are zero. Consequently, the product of inertia for the entire zee shape will derive entirely from the parallel-axis theorem.

SOLUTION

The centroid location for the zee shape is shown in the sketch. In computing the product of inertia using the parallel-axis theorem [Equation (A.10)], it is essential that careful attention be paid to the signs of x_c and y_c. The terms x_c and y_c are measured *from* the centroid of the overall shape *to* the centroid of the individual area. The complete calculation for I_{xy} is summarized in the table on the next page.

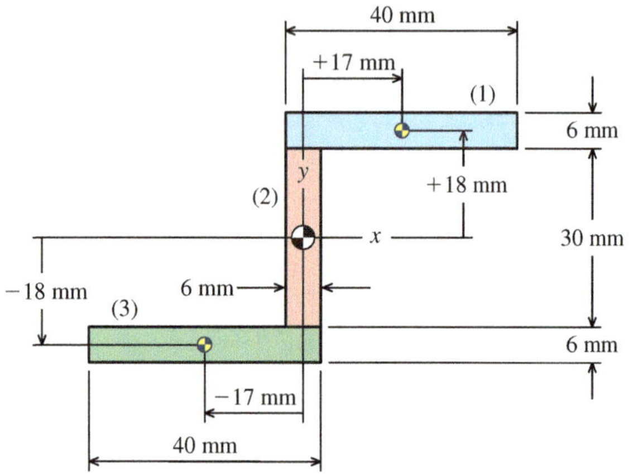

	$I_{x'y'}$ (mm⁴)	x_c (mm)	y_c (mm)	A_i (mm²)	$x_c y_c A_i$ (mm⁴)	I_{xy} (mm⁴)
(1)	0	17.0	18.0	240	73,440	73,440
(2)	0	0	0	180	0	0
(3)	0	−17.0	−18.0	240	73,440	73,440
						146,880

The product of inertia for the zee shape is thus

$I_{xy} = 146{,}900\ \text{mm}^4$. **Ans.**

EXAMPLE A.5

Determine the moments of inertia and the product of inertia for the unequal-leg angle shape shown with respect to the centroid of the area.

Plan the Solution

The unequal-leg angle is divided into two rectangles. The moments of inertia are computed about both the x and y axes. The product of inertia calculation is performed as demonstrated in Example A.4.

SOLUTION

The centroid location for the unequal-leg angle shape is shown in the sketch. The moment of inertia for the unequal-leg angle shape about the x centroidal axis is

$$I_x = \frac{(1\ \text{in.})(8\ \text{in.})^3}{12} + (1\ \text{in.})(8\ \text{in.})(1.346\ \text{in.})^2 + \frac{(5\ \text{in.})(1\ \text{in.})^3}{12} + (5\ \text{in.})(1\ \text{in.})(2.154\ \text{in.})^2$$

$$= 80.8\ \text{in.}^4$$ **Ans.**

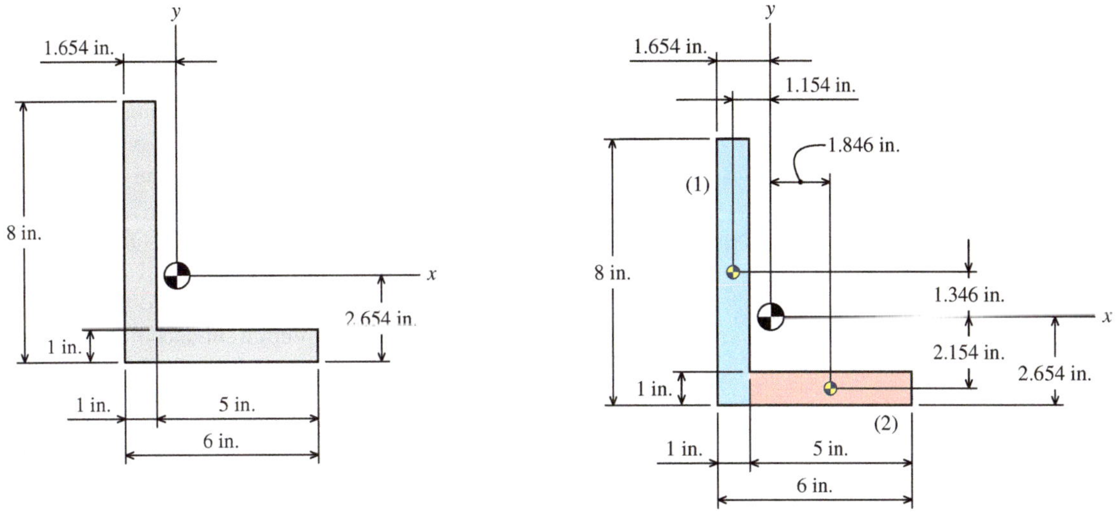

and the moment of inertia about the y centroidal axis is

$$I_y = \frac{(8 \text{ in.})(1 \text{ in.})^3}{12} + (8 \text{ in.})(1 \text{ in.})(1.154 \text{ in.})^2 + \frac{(1 \text{ in.})(5 \text{ in.})^3}{12} + (1 \text{ in.})(5 \text{ in.})(1.846 \text{ in.})^2$$

$$= 38.8 \text{ in.}^4$$ **Ans.**

In computing the product of inertia using the parallel-axis theorem [Equation (A.10)], it is essential that careful attention be paid to the signs of x_c and y_c. The terms x_c and y_c are measured *from* the centroid of the overall shape *to* the centroid of the individual area. The complete calculation for I_{xy} is summarized in the table below.

	$I_{x'y'}$ (in.^{4})	x_c (in.)	y_c (in.)	A_i (in.^{2})	$x_c y_c A_i$ (in.^{4})	I_{xy} (in.^{4})
(1)	0	−1.154	1.346	8.0	−12.426	−12.426
(2)	0	1.846	−2.154	5.0	−19.881	−19.881
						−32.307

The product of inertia for the unequal-leg angle shape is thus $I_{xy} = -32.3$ in.^{4}. **Ans.**

A.4 Principal Moments of Inertia

The moment of inertia of the area A in Figure A.7 with respect to the x' axis through O will, in general, vary with the angle θ. The x and y axes used to obtain Equation (A.6) were any pair of orthogonal axes in the plane of the area passing through O; therefore,

$$J = I_x + I_y = I_{x'} + I_{y'}$$

where x' and y' are any pair of orthogonal axes through O. Since the sum of $I_{x'}$ and $I_{y'}$ is a constant, $I_{x'}$ will be the maximum moment of inertia and the corresponding $I_{y'}$ will be the minimum moment of inertia for one particular value of θ.

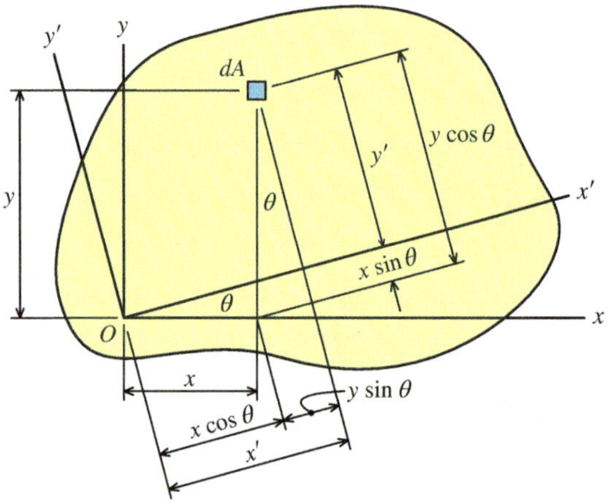

FIGURE A.7

The sets of axes for which the moments of inertia are maximum and minimum are called the *principal axes* of the area through point O and are designated as the $p1$ and $p2$ axes. The moments of inertia with respect to these axes are called the principal moments of inertia for the area and are designated I_{p1} and I_{p2}. There is only one set of principal axes for any area unless all axes have the same second moment, such as the diameters of a circle.

A convenient way to determine the principal moments of inertia for an area is to express $I_{x'}$ as a function of I_x, I_y, I_{xy}, and θ, and then set the derivative of $I_{x'}$ with respect to θ equal to zero to obtain the value of θ that gives the maximum and minimum moments of inertia. From Figure A.7,

$$dI_{x'} = y'^2 dA = (y\cos\theta - x\sin\theta)^2\, dA$$

and thus,

$$\begin{aligned} I_{x'} &= \cos^2\theta \int_A y^2\, dA - 2\sin\theta\cos\theta \int_A xy\, dA + \sin^2\theta \int_A x^2\, dA \\ &= I_x \cos^2\theta - 2I_{xy}\sin\theta\cos\theta + I_y \sin^2\theta \end{aligned}$$

which is commonly rearranged to the form

$$I_{x'} = I_x\cos^2\theta + I_y\sin^2\theta - 2I_{xy}\sin\theta\cos\theta \tag{A.11}$$

Equation (A.11) can be written in an equivalent form by substituting the following double-angle identities from trigonometry:

$$\begin{aligned} \cos^2\theta &= \frac{1}{2}(1 + \cos 2\theta) \\ \sin^2\theta &= \frac{1}{2}(1 - \cos 2\theta) \\ 2\sin\theta\cos\theta &= \sin 2\theta \end{aligned}$$

to give

$$I_{x'} = \frac{I_x + I_y}{2} + \frac{I_x - I_y}{2}\cos 2\theta - I_{xy}\sin 2\theta \tag{A.12}$$

The angle 2θ for which $I_{x'}$ is maximum can be obtained by setting the derivative of $I_{x'}$ with respect to θ equal to zero; thus,

$$\frac{dI_{x'}}{d\theta} = -(2)\frac{I_x - I_y}{2}\sin 2\theta - 2I_{xy}\cos 2\theta = 0$$

from which

$$\tan 2\theta_p = -\frac{2I_{xy}}{I_x - I_y} \tag{A.13}$$

where θ_p represents the two values of θ that locate the principal axes $p1$ and $p2$. *Positive values of θ indicate a counterclockwise rotation from the reference x axis.*

Notice that the two values of θ_p obtained from Equation (A.13) are 90° apart. The principal moments of inertia can be obtained by substituting these values of θ_p into Equation (A.12). From Equation (A.13),

$$\cos 2\theta_p = \mp \frac{(I_x - I_y)/2}{\sqrt{\left(\frac{(I_x - I_y)}{2}\right)^2 + I_{xy}^2}}$$

and

$$\sin 2\theta_p = \pm \frac{I_{xy}}{\sqrt{\left(\frac{(I_x - I_y)}{2}\right)^2 + I_{xy}^2}}$$

When these expressions are substituted in Equation (A.12), the principal moments of inertia reduce to

$$I_{p1,p2} = \frac{I_x + I_y}{2} \pm \sqrt{\left(\frac{I_x - I_y}{2}\right)^2 + I_{xy}^2} \qquad \text{(A.14)}$$

Equation (A.14) does not directly indicate which principal moment of inertia, either I_{p1} or I_{p2}, is associated with the two values of θ that locate the principal axes [Equation (A.13)]. The solution of Equation (A.13) always gives a value of θ_p between +45° and −45° (inclusive). The principal moment of inertia associated with this value of θ_p can be determined from the following two-part rule:

- If the term $I_x - I_y$ is positive, θ_p indicates the orientation of I_{p1}.
- If the term $I_x - I_y$ is negative, θ_p indicates the orientation of I_{p2}.

The principal moments of inertia determined from Equation (A.14) *will always be positive values*. In naming the principal moments of inertia, I_{p1} is the larger value algebraically.

The product of inertia of the element of area in Figure A.7 with respect to the x' and y' axes is

$$dI_{x'y'} = x'y'dA = (x\cos\theta + y\sin\theta)(y\cos\theta - x\sin\theta)\,dA$$

and the product of inertia for the area is

$$\begin{aligned} I_{x'y'} &= (\cos^2\theta - \sin^2\theta)\int_A xy\,dA + \sin\theta\cos\theta\int_A y^2\,dA - \sin\theta\cos\theta\int_A x^2\,dA \\ &= I_{xy}(\cos^2\theta - \sin^2\theta) + I_x\sin\theta\cos\theta - I_y\sin\theta\cos\theta \end{aligned}$$

which is commonly rearranged to the form

$$I_{x'y'} = (I_x - I_y)\sin\theta\cos\theta + I_{xy}(\cos^2\theta - \sin^2\theta) \qquad \text{(A.15)}$$

An equivalent form of Equation (A.15) is obtained with the substitution of double-angle trigonometric identities:

$$I_{x'y'} = \frac{I_x - I_y}{2}\sin 2\theta + I_{xy}\cos 2\theta \qquad \text{(A.16)}$$

The product of inertia $I_{x'y'}$ will be zero for values of θ given by

$$\tan 2\theta = -\frac{2I_{xy}}{I_x - I_y}$$

Notice that this expression is the same as Equation (A.13), which gives the orientation of the principal axes. Consequently, *the product of inertia is zero with respect to the principal axes.* Since the product of inertia is zero with respect to any axis of symmetry, it follows that **any axis of symmetry must also be a principal axis**.

EXAMPLE A.6

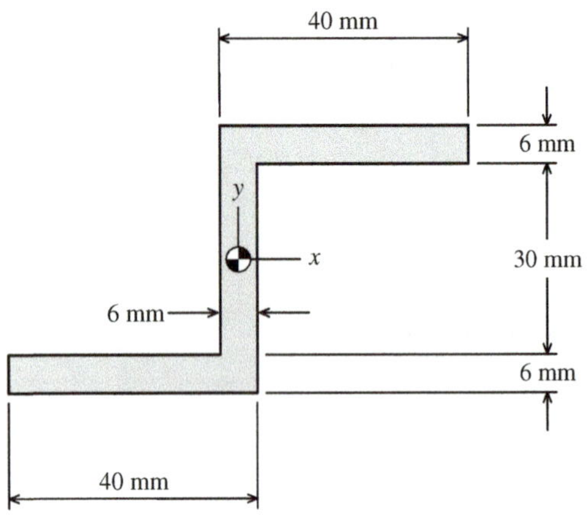

Determine the principal moments of inertia for the zee shape considered in Example A.4. Indicate the orientation of the principal axes.

Plan the Solution

Using the moments of inertia and the product of inertia determined in Examples A.3 and A.4, Equation (A.14) will give the magnitudes of I_{p1} and I_{p2}, and Equation (A.13) will define the orientation of the principal axes.

SOLUTION

From Examples A.3 and A.4, the moments of inertia and the product of inertia for the zee shape are

$$I_x = 170{,}460 \text{ mm}^4$$
$$I_y = 203{,}260 \text{ mm}^4$$
$$I_{xy} = 146{,}880 \text{ mm}^4$$

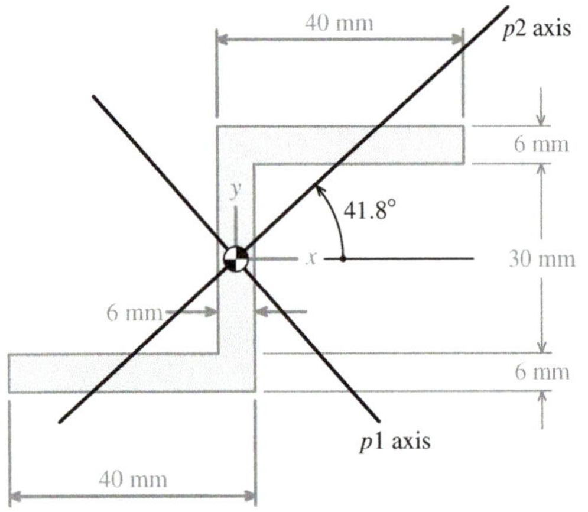

The principal moments of inertia can be calculated from Equation (A.14):

$$\begin{aligned} I_{p1,p2} &= \frac{I_x + I_y}{2} \pm \sqrt{\left(\frac{I_x - I_y}{2}\right)^2 + I_{xy}^2} \\ &= \frac{170{,}460 + 203{,}260}{2} \pm \sqrt{\left(\frac{170{,}460 - 203{,}260}{2}\right)^2 + (146{,}880)^2} \\ &= 186{,}860 \pm 147{,}793 \\ &= 335{,}000 \text{ mm}^4,\ 39{,}100 \text{ mm}^4 \end{aligned}$$

Ans.

The orientation of the principal axes is found from Equation (A.13):

$$\tan 2\theta_p = -\frac{2I_{xy}}{I_x - I_y} = -\frac{2(146{,}880)}{170{,}460 - 203{,}260} = 8.9561$$

$$\therefore\ 2\theta_p = 83.629°$$

Therefore, $\theta_p = 41.8°$. Since the denominator of this expression (i.e., $I_x - I_y$) is negative, the value obtained for θ_p gives the orientation of the $p2$ axis relative to the x axis. The positive value of θ_p indicates that the $p2$ axis is rotated 41.8° counterclockwise from the x axis.

The orientation of the principal axes is shown in the sketch.

EXAMPLE A.7

Determine the principal moments of inertia for the unequal-leg angle shape considered in Example A.5. Indicate the orientation of the principal axes.

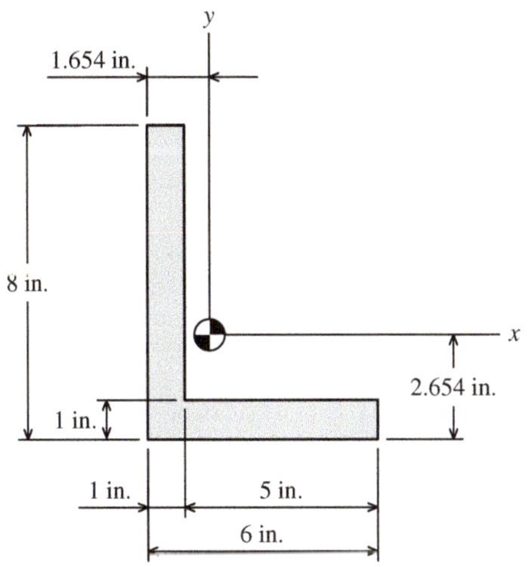

Plan the Solution

Using the moments of inertia and the product of inertia determined in Example A.5, Equation (A.14) will give the magnitudes of I_{p1} and I_{p2}, and Equation (A.13) will define the orientation of the principal axes.

SOLUTION

From Example A.5, the moments of inertia and the product of inertia for the unequal-leg angle shape are

$$I_x = 80.8 \text{ in.}^4$$
$$I_y = 38.8 \text{ in.}^4$$
$$I_{xy} = -32.3 \text{ in.}^4$$

The principal moments of inertia can be calculated from Equation (A.14):

$$I_{p1,p2} = \frac{I_x + I_y}{2} \pm \sqrt{\left(\frac{I_x - I_y}{2}\right)^2 + I_{xy}^2}$$

$$= \frac{80.8 + 38.8}{2} \pm \sqrt{\left(\frac{80.8 - 38.8}{2}\right)^2 + (-32.3)^2} \qquad \textbf{Ans.}$$

$$= 59.8 \pm 38.5$$

$$= 98.3 \text{ in.}^4,\ 21.3 \text{ in.}^4$$

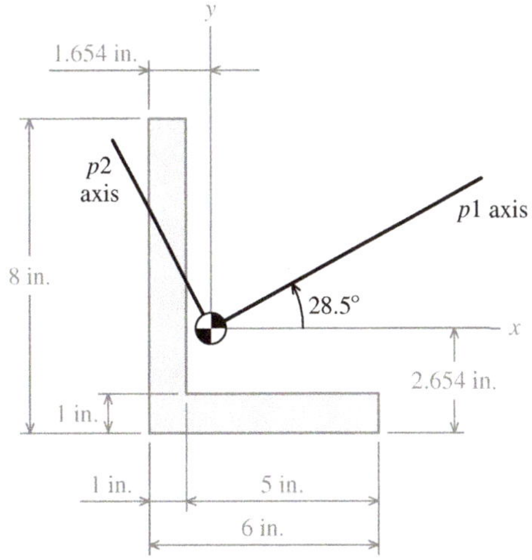

The orientation of the principal axes is found from Equation (A.13):

$$\tan 2\theta_p = -\frac{2I_{xy}}{I_x - I_y} = -\frac{2(-32.3)}{80.8 - 38.8} = 1.538095$$

$$\therefore 2\theta_p = 56.97°$$

Therefore, $\theta_p = 28.5°$. Since the denominator of this expression (i.e., $I_x - I_y$) is positive, the value obtained for θ_p gives the orientation of the $p1$ axis relative to the x axis. The positive value indicates that the $p1$ axis is rotated 28.5° counterclockwise from the x axis.

The orientation of the principal axes is shown in the sketch.

A.5 Mohr's Circle for Principal Moments of Inertia

The use of Mohr's circle for determining principal stresses was discussed in Section 12.9. A comparison of Equations (12.5) and (12.6) with Equations (A.12) and (A.16) suggests that a similar procedure can be used to obtain the principal moments of inertia for an area.

Figure A.8 illustrates the use of Mohr's circle for moments of inertia. Assume that I_x is greater than I_y and that I_{xy} is positive. Moments of inertia are plotted along the

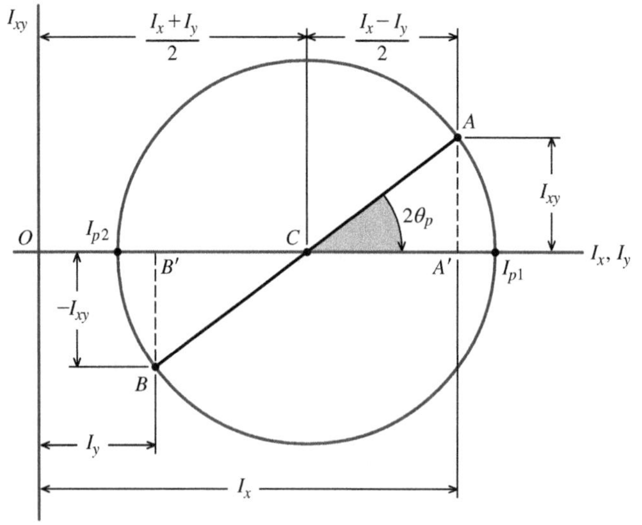

FIGURE A.8 Mohr's circle for moments of inertia.

horizontal axis, and products of inertia are plotted along the vertical axis. Moments of inertia are always positive and are plotted to the right of the origin. Products of inertia can be either positive or negative. Positive values are plotted above the horizontal axis. The horizontal distance OA' is equal to I_x, and the vertical distance $A'A$ is equal to I_{xy}. Similarly, horizontal distance OB' is equal to I_y and vertical distance $B'B$ is equal to $-I_{xy}$ (i.e., the algebraic negative of the product of inertia value, which can be either a positive or a negative number). The line AB intersects the horizontal axis at C, and line AB is the diameter of Mohr's circle. Each point on the circle represents $I_{x'}$ and $I_{x'y'}$ for one particular orientation of the x' and y' axes. As in Mohr's circle for stress analysis, angles in Mohr's circle are double angles 2θ. Thus, all angles on Mohr's circle are twice as large as the corresponding angles for the particular area.

Since the horizontal coordinate of each point on the circle represents a particular value of $I_{x'}$, the maximum and minimum moments of inertia are found where the circle intersects the horizontal axis. The maximum moment of inertia is I_{p1} and the minimum moment of inertia is I_{p2}. The center C of the circle is located at

$$C = \frac{I_x + I_y}{2}$$

and the circle radius is the length of CA, which can be found from the Pythagorean theorem:

$$CA = R = \sqrt{\left(\frac{I_x - I_y}{2}\right)^2 + I_{xy}^2}$$

The maximum moment of inertia I_{p1} is thus

$$I_{p1} = C + R = \frac{I_x + I_y}{2} + \sqrt{\left(\frac{I_x - I_y}{2}\right)^2 + I_{xy}^2}$$

and the minimum moment of inertia I_{p2} is

$$I_{p2} = C - R = \frac{I_x + I_y}{2} - \sqrt{\left(\frac{I_x - I_y}{2}\right)^2 + I_{xy}^2}$$

These expressions agree with Equation (A.14).

MecMovies

EXAMPLE

A.6 The theory and procedures for determining principal moments of inertia using Mohr's circle are presented in an interactive animation.

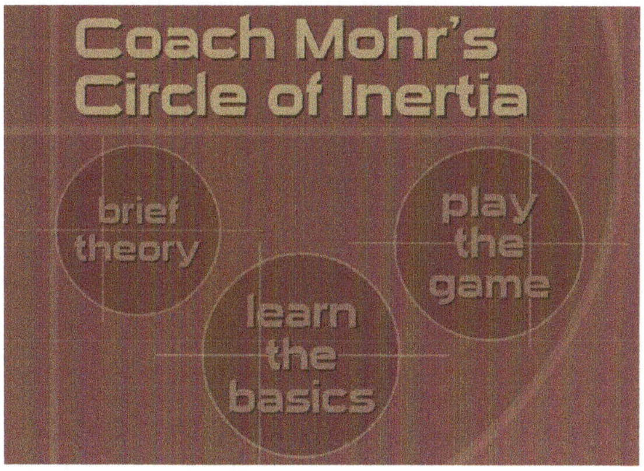

EXAMPLE A.8

Solve Example A.7 by means of Mohr's circle.

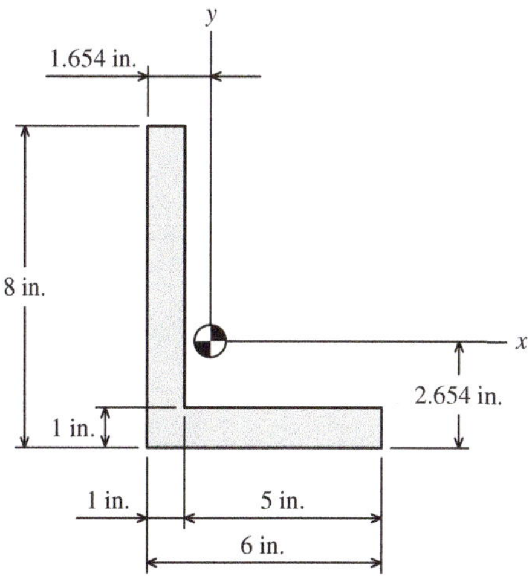

Plan the Solution

The moments of inertia and the product of inertia determined in Example A.5 will be used to construct Mohr's circle for moments of inertia.

SOLUTION

From Example A.5, the moments of inertia and the product of inertia for the unequal-leg angle shape are

$$I_x = 80.8 \text{ in.}^4$$
$$I_y = 38.8 \text{ in.}^4$$
$$I_{xy} = -32.3 \text{ in.}^4$$

Moments of inertia are plotted along the horizontal axis, and products of inertia are plotted on the vertical axis. Begin by plotting the point (I_x, I_{xy}) and labeling it x. Notice that since I_{xy} has a negative value, point x plots below the horizontal axis.

Next, plot the point $(I_y, -I_{xy})$ and label this point y. Since I_{xy} has a negative value, point y plots above the horizontal axis.

Draw the circle diameter that connects points x and y. The center of the circle is located where this diameter crosses the horizontal axis. Label the circle center as C. Using the circle center C, draw the circle that passes through points x and y. This is Mohr's circle for moments of inertia. Points on Mohr's circle represent possible combinations of moment of inertia and product of inertia.

The center of the circle is located midway between points x and y:

$$C = \frac{80.8 + 38.8}{2} = 59.8$$

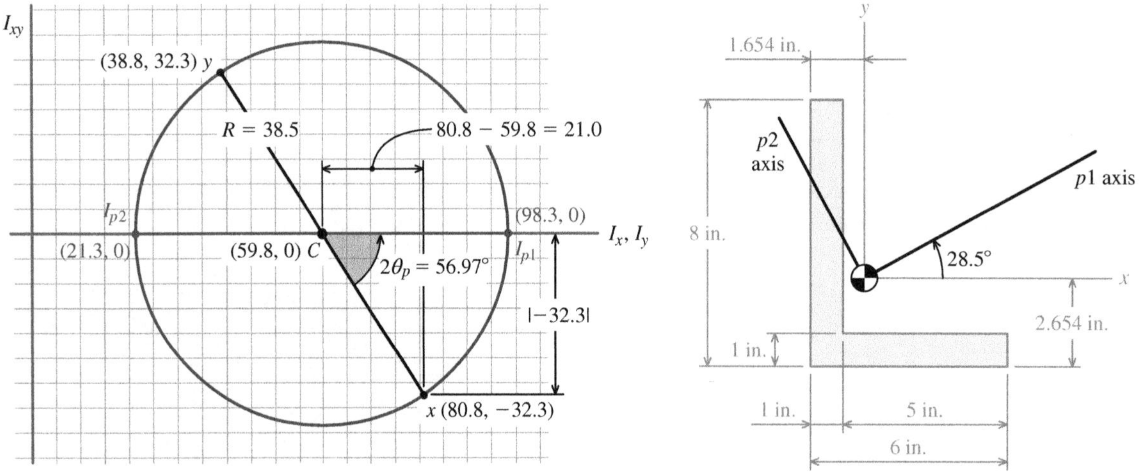

Using the coordinates of point x and center C, the radius R of the circle can be computed from the Pythagorean theorem:

$$R = \sqrt{\left(\frac{80.8 - 59.8}{2}\right)^2 + (-32.3)^2} = 38.5$$

The principal moments of inertia are given by

$$I_{p1} = C + R = 59.8 + 38.5 = 98.3 \qquad \text{and} \qquad I_{p2} = C - R = 59.8 - 38.5 = 21.3$$

The orientation of the principal axes is found from the angle between the radius to point x and the horizontal axis:

$$\tan 2\theta_p = \frac{|-32.3|}{80.8 - 59.8} = 1.538095$$

$$\therefore\ 2\theta_p = 56.97°$$

Note that the absolute value is used in the numerator because only the magnitude of $2\theta_p$ is needed here. From inspection of Mohr's circle, it is evident that the angle from point x to I_{p1} turns in a counterclockwise sense.

Finally, the results obtained from the Mohr's circle must be referred back to the actual unequal-leg angle shape. Since the angles found in Mohr's circle are doubled, the angle from the x axis to the axis of maximum moment of inertia is $\theta_p = 28.5°$, turned in a counterclockwise direction. The maximum moment of inertia for the unequal-leg angle shape occurs about the $p1$ axis. The axis of minimum moment of inertia I_{p2} is perpendicular to the $p1$ axis.

APPENDIX B

Geometric Properties of Structural Steel Shapes

Wide-Flange Sections or W Shapes—U.S. Customary Units

Designation	Area A	Depth d	Web thickness t_w	Flange width b_f	Flange thickness t_f	I_x	S_x	r_x	I_y	S_y	r_y
	in.^{2}	in.	in.	in.	in.	in.^{4}	in.^{3}	in.	in.^{4}	in.^{3}	in.
W24 × 94	27.7	24.3	0.515	9.07	0.875	2700	222	9.87	109	24.0	1.98
24 × 76	22.4	23.9	0.440	8.99	0.680	2100	176	9.69	82.5	18.4	1.92
24 × 68	20.1	23.7	0.415	8.97	0.585	1830	154	9.55	70.4	15.7	1.87
24 × 55	16.2	23.6	0.395	7.01	0.505	1350	114	9.11	29.1	8.30	1.34
W21 × 68	20.0	21.1	0.430	8.27	0.685	1480	140	8.60	64.7	15.7	1.80
21 × 62	18.3	21.0	0.400	8.24	0.615	1330	127	8.54	57.5	14.0	1.77
21 × 50	14.7	20.8	0.380	6.53	0.535	984	94.5	8.18	24.9	7.64	1.30
21 × 44	13.0	20.7	0.350	6.50	0.450	843	81.6	8.06	20.7	6.37	1.26
W18 × 55	16.2	18.1	0.390	7.53	0.630	890	98.3	7.41	44.9	11.9	1.67
18 × 50	14.7	18.0	0.355	7.50	0.570	800	88.9	7.38	40.1	10.7	1.65
18 × 40	11.8	17.9	0.315	6.02	0.525	612	68.4	7.21	19.1	6.35	1.27
18 × 35	10.3	17.7	0.300	6.00	0.425	510	57.6	7.04	15.3	5.12	1.22
W16 × 57	16.8	16.4	0.430	7.12	0.715	758	92.2	6.72	43.1	12.1	1.60
16 × 50	14.7	16.3	0.380	7.07	0.630	659	81.0	6.68	37.2	10.5	1.59
16 × 40	11.8	16.0	0.305	7.00	0.505	518	64.7	6.63	28.9	8.25	1.57
16 × 31	9.13	15.9	0.275	5.53	0.440	375	47.2	6.41	12.4	4.49	1.17
W14 × 68	20.0	14.0	0.415	10.0	0.720	722	103	6.01	121	24.2	2.46
14 × 53	15.6	13.9	0.370	8.06	0.660	541	77.8	5.89	57.7	14.3	1.92
14 × 48	14.1	13.8	0.340	8.03	0.595	484	70.2	5.85	51.4	12.8	1.91
14 × 34	10.0	14.0	0.285	6.75	0.455	340	48.6	5.83	23.3	6.91	1.53
14 × 30	8.85	13.8	0.270	6.73	0.385	291	42.0	5.73	19.6	5.82	1.49
14 × 26	7.69	13.9	0.255	5.03	0.420	245	35.3	5.65	8.91	3.55	1.08
14 × 22	6.49	13.7	0.230	5.00	0.335	199	29.0	5.54	7.00	2.80	1.04

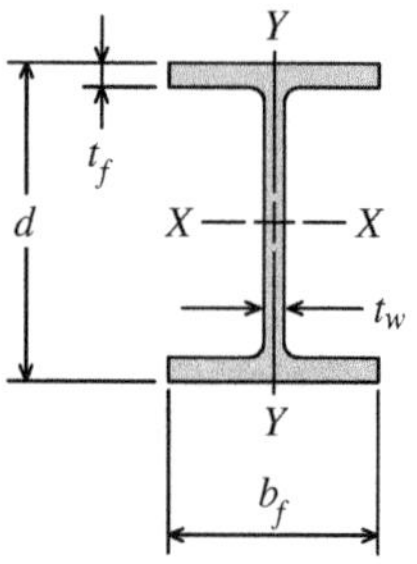

Designation	Area A	Depth d	Web thickness t_w	Flange width b_f	Flange thickness t_f	I_x	S_x	r_x	I_y	S_y	r_y
	in.^{2}	in.	in.	in.	in.	in.^{4}	in.^{3}	in.	in.^{4}	in.^{3}	in.
W12 × 58	17.0	12.2	0.360	10.0	0.640	475	78.0	5.28	107	21.4	2.51
12 × 50	14.6	12.2	0 370	8.08	0.640	391	64.2	5.18	56.3	13.9	1.96
12 × 40	11.7	11.9	0.295	8.01	0.515	307	51.5	5.13	44.1	11.0	1.94
12 × 30	8.79	12.3	0.260	6.52	0.440	238	38.6	5.21	20.3	6.24	1.52
12 × 26	7.65	12.2	0.230	6.49	0.380	204	33.4	5.17	17.3	5.34	1.51
12 × 22	6.48	12.3	0.260	4.03	0.425	156	25.4	4.91	4.66	2.31	0.848
12 × 14	4.16	11.9	0.200	3.97	0.225	86.6	14.9	4.62	2.36	1.19	0.753
W10 × 54	15.8	10.1	0.370	10.0	0.615	303	60.0	4.37	103	20.6	2.56
10 × 45	13.3	10.1	0.350	8.02	0.620	248	49.1	4.32	53.4	13.3	2.01
10 × 30	8.84	10.5	0.300	5.81	0.510	170	32.4	4.38	16.7	5.75	1.37
10 × 26	7.61	10.3	0.260	5.77	0.440	144	27.9	4.35	14.1	4.89	1.36
10 × 22	6.49	10.2	0.240	5.75	0.360	118	23.2	4.27	11.4	3.97	1.33
10 × 15	4.41	10.0	0.230	4.00	0.270	68.9	13.8	3.95	2.89	1.45	0.81
W8 × 48	14.1	8.50	0.400	8.11	0.685	184	43.2	3.61	60.9	15.0	2.08
8 × 40	11.7	8.25	0.360	8.07	0.560	146	35.5	3.53	49.1	12.2	2.04
8 × 31	9.12	8.00	0.285	8.00	0.435	110	27.5	3.47	37.1	9.27	2.02
8 × 24	7.08	7.93	0.245	6.50	0.400	82.7	20.9	3.42	18.3	5.63	1.61
8 × 15	4.44	8.11	0.245	4.01	0.315	48	11.8	3.29	3.41	1.70	0.876
W6 × 25	7.34	6.38	0.320	6.08	0.455	53.4	16.7	2.70	17.1	5.61	1.52
6 × 20	5.87	6.20	0.260	6.02	0.365	41.4	13.4	2.66	13.3	4.41	1.50
6 × 15	4.43	5.99	0.230	5.99	0.260	29.1	9.72	2.56	9.32	3.11	1.45
6 × 12	3.55	6.03	0.230	4.00	0.280	22.1	7.31	2.49	2.99	1.50	0.918

Wide-Flange Sections or W Shapes—SI Units

Designation	Area A	Depth d	Web thickness t_w	Flange width b_f	Flange thickness t_f	I_x	S_x	r_x	I_y	S_y	r_y
	mm^2	mm	mm	mm	mm	10^6 mm^4	10^3 mm^3	mm	10^6 mm^4	10^3 mm^3	mm
W610 × 140	17900	617	13.1	230	22.2	1120	3640	251	45.4	393	50.3
610 × 113	14500	607	11.2	228	17.3	874	2880	246	34.3	302	48.8
610 × 101	13000	602	10.5	228	14.9	762	2520	243	29.3	257	47.5
610 × 82	10500	599	10.0	178	12.8	562	1870	231	12.1	136	34.0
W530 × 101	12900	536	10.9	210	17.4	616	2290	218	26.9	257	45.7
530 × 92	11800	533	10.2	209	15.6	554	2080	217	23.9	229	45.0
530 × 74	9480	528	9.65	166	13.6	410	1550	208	10.4	125	33.0
530 × 66	8390	526	8.89	165	11.4	351	1340	205	8.62	104	32.0
W460 × 82	10500	460	9.91	191	16.0	370	1610	188	18.7	195	42.4
460 × 74	9480	457	9.02	191	14.5	333	1460	187	16.7	175	41.9
460 × 60	7610	455	8.00	153	13.3	255	1120	183	7.95	104	32.3
460 × 52	6650	450	7.62	152	10.8	212	944	179	6.37	83.9	31.0
W410 × 85	10800	417	10.9	181	18.2	316	1510	171	17.9	198	40.6
410 × 75	9480	414	9.65	180	16.0	274	1330	170	15.5	172	40.4
410 × 60	7610	406	7.75	178	12.8	216	1060	168	12.0	135	39.9
410 × 46.1	5890	404	6.99	140	11.2	156	773	163	5.16	73.6	29.7
W360 × 101	12900	356	10.5	254	18.3	301	1690	153	50.4	397	62.5
360 × 79	10100	353	9.40	205	16.8	225	1270	150	24.0	234	48.8
360 × 72	9100	351	8.64	204	15.1	201	1150	149	21.4	210	48.5
360 × 51	6450	356	7.24	171	11.6	142	796	148	9.70	113	38.9
360 × 44	5710	351	6.86	171	9.78	121	688	146	8.16	95.4	37.8
360 × 39	4960	353	6.48	128	10.7	102	578	144	3.71	58.2	27.4
360 × 32.9	4190	348	5.84	127	8.51	82.8	475	141	2.91	45.9	26.4

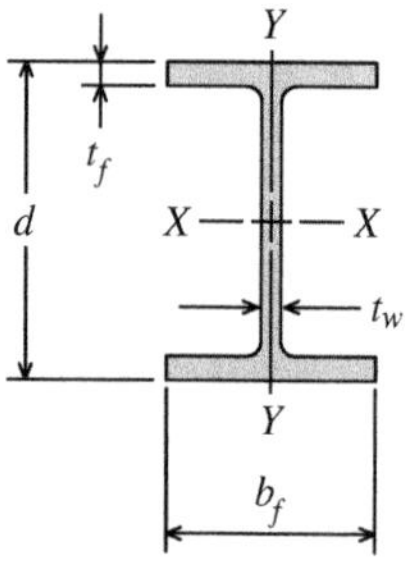

Designation	Area A	Depth d	Web thickness t_w	Flange width b_f	Flange thickness t_f	I_x	S_x	r_x	I_y	S_y	r_y
	mm^2	mm	mm	mm	mm	10^6 mm^4	10^3 mm^3	mm	10^6 mm^4	10^3 mm^3	mm
W310 × 86	11000	310	9.14	254	16.3	198	1280	134	44.5	351	63.8
310 × 74	9420	310	9.40	205	16.3	163	1050	132	23.4	228	49.8
310 × 60	7550	302	7.49	203	13.1	128	844	130	18.4	180	49.3
310 × 44.5	5670	312	6.60	166	11.2	99.1	633	132	8.45	102	38.6
310 × 38.7	4940	310	5.84	165	9.65	84.9	547	131	7.20	87.5	38.4
310 × 32.7	4180	312	6.60	102	10.8	64.9	416	125	1.94	37.9	21.5
310 × 21	2680	302	5.08	101	5.72	36.9	244	117	0.982	19.5	19.1
W250 × 80	10200	257	9.40	254	15.6	126	983	111	42.9	338	65.0
250 × 67	8580	257	8.89	204	15.7	103	805	110	22.2	218	51.1
250 × 44.8	5700	267	7.62	148	13.0	70.8	531	111	6.95	94.2	34.8
250 × 38.5	4910	262	6.60	147	11.2	59.9	457	110	5.87	80.1	34.5
250 × 32.7	4190	259	6.10	146	9.14	49.1	380	108	4.75	65.1	33.8
250 × 22.3	2850	254	5.84	102	6.86	28.7	226	100	1.20	23.8	20.6
W200 × 71	9100	216	10.2	206	17.4	76.6	708	91.7	25.3	246	52.8
200 × 59	7550	210	9.14	205	14.2	60.8	582	89.7	20.4	200	51.8
200 × 46.1	5880	203	7.24	203	11.0	45.8	451	88.1	15.4	152	51.3
200 × 35.9	4570	201	6.22	165	10.2	34.4	342	86.9	7.62	92.3	40.9
200 × 22.5	2860	206	6.22	102	8.00	20	193	83.6	1.42	27.9	22.3
W150 × 37.1	4740	162	8.13	154	11.6	22.2	274	68.6	7.12	91.9	38.6
150 × 29.8	3790	157	6.60	153	9.27	17.2	220	67.6	5.54	72.3	38.1
150 × 22.5	2860	152	5.84	152	6.60	12.1	159	65.0	3.88	51.0	36.8
150 × 18	2290	153	5.84	102	7.11	9.2	120	63.2	1.24	24.6	23.3

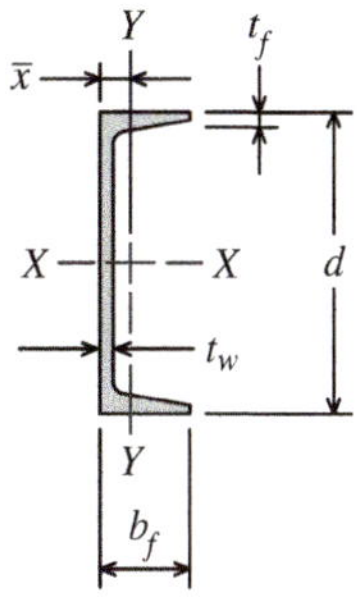

American Standard Channels or C Shapes—U.S. Customary Units

Designation	Area A	Depth d	Web thickness t_w	Flange width b_f	Flange thickness t_f	Centroid $\bar{x}$	I_x	S_x	r_x	I_y	S_y	r_y
	in.^{2}	in.	in.	in.	in.	in.	in.^{4}	in.^{3}	in.	in.^{4}	in.^{3}	in.
C15 × 50	14.7	15	0.716	3.72	0.650	0.799	404	53.8	5.24	11.0	3.77	0.865
15 × 40	11.8	15	0.520	3.52	0.650	0.778	348	46.5	5.45	9.17	3.34	0.883
15 × 33.9	10.0	15	0.400	3.40	0.650	0.788	315	42.0	5.62	8.07	3.09	0.901
C12 × 30	8.81	12	0.510	3.17	0.501	0.674	162	27.0	4.29	5.12	2.05	0.762
12 × 25	7.34	12	0.387	3.05	0.501	0.674	144	24.0	4.43	4.45	1.87	0.779
12 × 20.7	6.08	12	0.282	2.94	0.501	0.698	129	21.5	4.61	3.86	1.72	0.797
C10 × 30	8.81	10	0.673	3.03	0.436	0.649	103	20.7	3.42	3.93	1.65	0.668
10 × 25	7.34	10	0.526	2.89	0.436	0.617	91.1	18.2	3.52	3.34	1.47	0.675
10 × 20	5.87	10	0.379	2.74	0.436	0.606	78.9	15.8	3.66	2.80	1.31	0.690
10 × 15.3	4.48	10	0.240	2.60	0.436	0.634	67.3	13.5	3.87	2.27	1.15	0.711
C9 × 20	5.87	9	0.448	2.65	0.413	0.583	60.9	13.5	3.22	2.41	1.17	0.640
9 × 15	4.41	9	0.285	2.49	0.413	0.586	51.0	11.3	3.40	1.91	1.01	0.659
9 × 13.4	3.94	9	0.233	2.43	0.413	0.601	47.8	10.6	3.49	1.75	0.954	0.666
C8 × 18.7	5.51	8	0.487	2.53	0.390	0.565	43.9	11.0	2.82	1.97	1.01	0.598
8 × 13.7	4.04	8	0.303	2.34	0.390	0.554	36.1	9.02	2.99	1.52	0.848	0.613
8 × 11.5	3.37	8	0.220	2.26	0.390	0.572	32.5	8.14	3.11	1.31	0.775	0.623
C7 × 14.7	4.33	7	0.419	2.30	0.366	0.532	27.2	7.78	2.51	1.37	0.772	0.561
7 × 12.2	3.6	7	0.314	2.19	0.366	0.525	24.2	6.92	2.60	1.16	0.696	0.568
7 × 9.8	2.87	7	0.210	2.09	0.366	0.541	21.2	6.07	2.72	0.957	0.617	0.578
C6 × 13	3.81	6	0.437	2.16	0.343	0.514	17.3	5.78	2.13	1.05	0.638	0.524
6 × 10.5	3.08	6	0.314	2.03	0.343	0.500	15.1	5.04	2.22	0.86	0.561	0.529

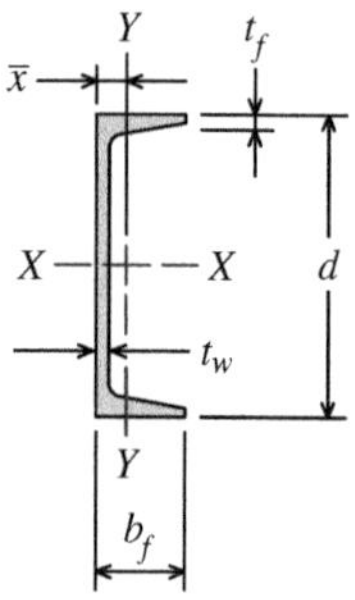

American Standard Channels or C Shapes—SI Units

Designation	Area A	Depth d	Web thickness t_w	Flange width b_f	Flange thickness t_f	Centroid $\bar{x}$	I_x	S_x	r_x	I_y	S_y	r_y
	mm^2	mm	mm	mm	mm	mm	10^6 mm^4	10^3 mm^3	mm	10^6 mm^4	10^3 mm^3	mm
C380 × 74	9480	381	18.2	94.5	16.5	20.3	168	882	133	4.58	61.8	22.0
380 × 60	7610	381	13.2	89.4	16.5	19.8	145	762	138	3.82	54.7	22.4
380 × 50.4	6450	381	10.2	86.4	16.5	20.0	131	688	143	3.36	50.6	22.9
C310 × 45	5680	305	13.0	80.5	12.7	17.1	67.4	442	109	2.13	33.6	19.4
310 × 37	4740	305	9.83	77.5	12.7	17.1	59.9	393	113	1.85	30.6	19.8
310 × 30.8	3920	305	7.16	74.7	12.7	17.7	53.7	352	117	1.61	28.2	20.2
C250 × 45	5680	254	17.1	77.0	11.1	16.5	42.9	339	86.9	1.64	27.0	17.0
250 × 37	4740	254	13.4	73.4	11.1	15.7	37.9	298	89.4	1.39	24.1	17.1
250 × 30	3790	254	9.63	69.6	11.1	15.4	32.8	259	93.0	1.17	21.5	17.5
250 × 22.8	2890	254	6.10	66.0	11.1	16.1	28.0	221	98.3	0.945	18.8	18.1
C230 × 30	3790	229	11.4	67.3	10.5	14.8	25.3	221	81.8	1.00	19.2	16.3
230 × 22	2850	229	7.24	63.2	10.5	14.9	21.2	185	86.4	0.795	16.6	16.7
230 × 19.9	2540	229	5.92	61.7	10.5	15.3	19.9	174	88.6	0.728	15.6	16.9
C200 × 27.9	3550	203	12.4	64.3	9.91	14.4	18.3	180	71.6	0.820	16.6	15.2
200 × 20.5	2610	203	7.70	59.4	9.91	14.1	15.0	148	75.9	0.633	13.9	15.6
200 × 17.1	2170	203	5.59	57.4	9.91	14.5	13.5	133	79.0	0.545	12.7	15.8
C180 × 22	2790	178	10.6	58.4	9.30	13.5	11.3	127	63.8	0.570	12.7	14.2
180 × 18.2	2320	178	7.98	55.6	9.30	13.3	10.1	113	66.0	0.483	11.4	14.4
180 × 14.6	1850	178	5.33	53.1	9.30	13.7	8.82	100	69.1	0.398	10.1	14.7
C150 × 19.3	2460	152	11.1	54.9	8.71	13.1	7.20	94.7	54.1	0.437	10.5	13.3
150 × 15.6	1990	152	7.98	51.6	8.71	12.7	6.29	82.6	56.4	0.358	9.19	13.4

Shapes Cut from Wide-Flange Sections or WT Shapes

Designation	Area A	Depth d	Web thickness t_w	Flange width b_f	Flange thickness t_f	Centroid $\bar{y}$	I_x	S_x	r_x	I_y	S_y	r_y
	in.^{2}	in.	in.	in.	in.	in.	in.^{4}	in.^{3}	in.	in.^{4}	in.^{3}	in.
WT12 × 47	13.8	12.2	0.515	9.07	0.875	2.99	186	20.3	3.67	54.5	12.0	1.98
12 × 38	11.2	12.0	0.440	8.99	0.680	3.00	151	16.9	3.68	41.3	9.18	1.92
12 × 34	10.0	11.9	0.415	8.97	0.585	3.06	137	15.6	3.70	35.2	7.85	1.87
12 × 27.5	8.10	11.8	0.395	7.01	0.505	3.50	117	14.1	3.80	14.5	4.15	1.34
WT10.5 × 34	10.0	10.6	0.430	8.27	0.685	2.59	103	12.9	3.20	32.4	7.83	1.80
10.5 × 31	9.13	10.5	0.400	8.24	0.615	2.58	93.8	11.9	3.21	28.7	6.97	1.77
10.5 × 25	7.36	10.4	0.380	6.53	0.535	2.93	80.3	10.7	3.30	12.5	3.82	1.30
10.5 × 22	6.49	10.3	0.350	6.50	0.450	2.98	71.1	9.68	3.31	10.3	3.18	1.26
WT9 × 27.5	8.10	9.06	0.390	7.53	0.630	2.16	59.5	8.63	2.71	22.5	5.97	1.67
9 × 25	7.33	9.00	0.355	7.50	0.570	2.12	53.5	7.79	2.70	20.0	5.35	1.65
9 × 20	5.88	8.95	0.315	6.02	0.525	2.29	44.8	6.73	2.76	9.55	3.17	1.27
9 × 17.5	5.15	8.85	0.300	6.00	0.425	2.39	40.1	6.21	2.79	7.67	2.56	1.22
WT8 × 28.5	8.39	8.22	0.430	7.12	0.715	1.94	48.7	7.77	2.41	21.6	6.06	1.60
8 × 25	7.37	8.13	0.380	7.07	0.630	1.89	42.3	6.78	2.40	18.6	5.26	1.59
8 × 20	5.89	8.01	0.305	7.00	0.505	1.81	33.1	5.35	2.37	14.4	4.12	1.56
8 × 15.5	4.56	7.94	0.275	5.53	0.440	2.02	27.5	4.64	2.45	6.2	2.24	1.17

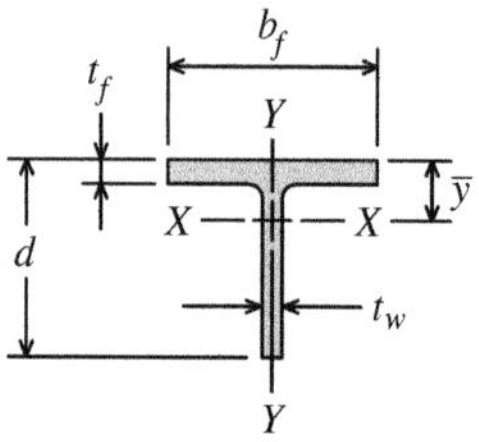

Designation	Area A	Depth d	Web thickness t_w	Flange width b_f	Flange thickness t_f	Centroid $\bar{y}$	I_x	S_x	r_x	I_y	S_y	r_y
	mm^2	mm	mm	mm	mm	mm	10^6 mm^4	10^3 mm^3	mm	10^6 mm^4	10^3 mm^3	mm
WT305 × 70	8900	310	13.1	230	22.2	75.9	77.4	333	93.2	22.7	197	50.3
305 × 56.5	7230	305	11.2	228	17.3	76.2	62.9	277	93.5	17.2	150	48.8
305 × 50.5	6450	302	10.5	228	14.9	77.7	57.0	256	94.0	14.7	129	47.5
305 × 41	5230	300	10.0	178	12.8	88.9	48.7	231	96.5	6.04	68.0	34.0
WT265 × 50.5	6450	269	10.9	210	17.4	65.8	42.9	211	81.3	13.5	128	45.7
265 × 46	5890	267	10.2	209	15.6	65.5	39.0	195	81.5	11.9	114	45.0
265 × 37	4750	264	9.65	166	13.6	74.4	33.4	175	83.8	5.20	62.6	33.0
265 × 33	4190	262	8.89	165	11.4	75.7	29.6	159	84.1	4.29	52.1	32.0
WT230 × 41	5230	230	9.91	191	16.0	54.9	24.8	141	68.8	9.37	97.8	42.4
230 × 37	4730	229	9.02	191	14.5	53.8	22.3	128	68.6	8.32	87.7	41.9
230 × 30	3790	227	8.00	153	13.3	58.2	18.6	110	70.1	3.98	51.9	32.3
230 × 26	3320	225	7.62	152	10.8	60.7	16.7	102	70.9	3.19	42.0	31.0
WT205 × 42.5	5410	209	10.9	181	18.2	49.3	20.3	127	61.2	8.99	99.3	40.6
205 × 37.5	4750	207	9.65	180	16.0	48.0	17.6	111	61.0	7.74	86.2	40.4
205 × 30	3800	203	7.75	178	12.8	46.0	13.8	87.7	60.2	5.99	67.5	39.6
205 × 23.05	2940	202	6.99	140	11.2	51.3	11.4	76.0	62.2	2.58	36.7	29.7

Hollow Structural Sections or HSS Shapes

Designation	Depth d	Width b	Wall thickness (nom.) t	Weight per foot	Area A	I_x	S_x	r_x	I_y	S_y	r_y
	in.	in.	in.	lb/ft	in.^{2}	in.^{4}	in.^{3}	in.	in.^{4}	in.^{3}	in.
HSS12 × 8 × 1/2	12	8	0.5	62.3	17.2	333	55.6	4.41	178	44.4	3.21
× 8 × 3/8	12	8	0.375	47.8	13.2	262	43.7	4.47	140	35.1	3.27
× 6 × 1/2	12	6	0.5	55.5	15.3	271	45.2	4.21	91.1	30.4	2.44
× 6 × 3/8	12	6	0.375	42.7	11.8	215	35.9	4.28	72.9	24.3	2.49
HSS10 × 6 × 1/2	10	6	0.5	48.7	13.5	171	34.3	3.57	76.8	25.6	2.39
× 6 × 3/8	10	6	0.375	37.6	10.4	137	27.4	3.63	61.8	20.6	2.44
× 4 × 1/2	10	4	0.5	41.9	11.6	129	25.8	3.34	29.5	14.7	1.59
× 4 × 3/8	10	4	0.375	32.5	8.97	104	20.8	3.41	24.3	12.1	1.64
HSS8 × 4 × 1/2	8	4	0.5	35.1	9.74	71.8	17.9	2.71	23.6	11.8	1.56
× 4 × 3/8	8	4	0.375	27.4	7.58	58.7	14.7	2.78	19.6	9.80	1.61
× 4 × 1/4	8	4	0.25	19.0	5.24	42.5	10.6	2.85	14.4	7.21	1.66
× 4 × 1/8	8	4	0.125	9.85	2.70	22.9	5.73	2.92	7.90	3.95	1.71
HSS6 × 4 × 3/8	6	4	0.375	22.3	6.18	28.3	9.43	2.14	14.9	7.47	1.55
× 4 × 1/4	6	4	0.25	15.6	4.30	20.9	6.96	2.20	11.1	5.56	1.61
× 4 × 1/8	6	4	0.125	8.15	2.23	11.4	3.81	2.26	6.15	3.08	1.66
× 3 × 3/8	6	3	0.375	19.7	5.48	22.7	7.57	2.04	7.48	4.99	1.17
× 3 × 1/4	6	3	0.25	13.9	3.84	17.0	5.66	2.10	5.70	3.80	1.22
× 3 × 1/8	6	3	0.125	7.30	2.00	9.43	3.14	2.17	3.23	2.15	1.27

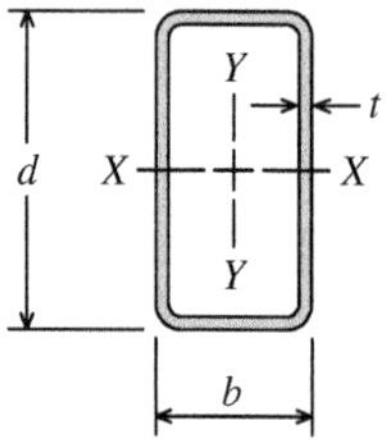

Designation	Depth d	Width b	Wall thickness (nom.) t	Mass per meter	Area A	I_x	S_x	r_x	I_y	S_y	r_y
	mm	mm	mm	kg/m	mm^2	10^6 mm^4	10^3 mm^3	mm	10^6 mm^4	10^3 mm^3	mm
HSS304.8 × 203.2 × 12.7	304.8	203.2	12.7	137	11100	139	911	112	74.1	728	81.5
× 203.2 × 9.5	304.8	203.2	9.53	105	8520	109	716	114	58.3	575	83.1
× 152.4 × 12.7	304.8	152.4	12.7	122	9870	113	741	107	37.9	498	62.0
× 152.4 × 9.5	304.8	152.4	9.53	94.2	7610	89.5	588	109	30.3	398	63.2
HSS254 × 152.4 × 12.7	254	152.4	12.7	107	8710	71.2	562	90.7	32.0	420	60.7
× 152.4 × 9.5	254	152.4	9.53	82.9	6710	57.0	449	92.2	25.7	338	62.0
× 101.6 × 12.7	254	101.6	12.7	92.4	7480	53.7	423	84.8	12.3	241	40.4
× 101.6 × 9.5	254	101.6	9.53	71.7	5790	43.3	341	86.6	10.1	198	41.7
HSS203.2 × 101.6 × 12.7	203.2	101.6	12.7	77.4	6280	29.9	293	68.8	9.82	193	39.6
× 101.6 × 9.5	203.2	101.6	9.53	60.4	4890	24.4	241	70.6	8.16	161	40.9
× 101.6 × 6.4	203.2	101.6	6.35	41.9	3380	17.7	174	72.4	5.99	118	42.2
× 101.6 × 3.2	203.2	101.6	3.18	21.7	1740	9.53	93.9	74.2	3.29	64.7	43.4
HSS152.4 × 101.6 × 9.5	152.4	101.6	9.53	49.2	3990	11.8	155	54.4	6.20	122	39.4
× 101.6 × 6.4	152.4	101.6	6.35	34.4	2770	8.70	114	55.9	4.62	91.1	40.9
× 101.6 × 3.2	152.4	101.6	3.18	18.0	1440	4.75	62.4	57.4	2.56	50.5	42.2
× 76.2 × 9.5	152.4	76.2	9.53	43.5	3540	9.45	124	51.8	3.11	81.8	29.7
× 76.2 × 6.4	152.4	76.2	6.35	30.6	2480	7.08	92.8	53.3	2.37	62.3	31.0
× 76.2 × 3.2	152.4	76.2	3.18	16.1	1290	3.93	51.5	55.1	1.34	35.2	32.3

Angle Shapes or L Shapes

Designation	Weight per foot	Area A	I_x	S_x	r_x	y	I_y	S_y	r_y	x	r_z	tan α
	lb/ft	in.^{2}	in.^{4}	in.^{3}	in.	in.	in.^{4}	in.^{3}	in.	in.	in.	
L5 × 5 × 3/4	23.6	6.94	15.7	4.52	1.50	1.52	15.7	4.52	1.50	1.52	0.972	1.00
× 5 × 1/2	16.2	4.75	11.3	3.15	1.53	1.42	11.3	3.15	1.53	1.42	0.980	1.00
× 5 × 3/8	12.3	3.61	8.76	2.41	1.55	1.37	8.76	2.41	1.55	1.37	0.986	1.00
L5 × 3 × 1/2	12.8	3.75	9.43	2.89	1.58	1.74	2.55	1.13	0.824	0.746	0.642	0.357
× 3 × 3/8	9.80	2.86	7.35	2.22	1.60	1.69	2.01	0.874	0.838	0.698	0.646	0.364
× 3 × 1/4	6.60	1.94	5.09	1.51	1.62	1.64	1.41	0.600	0.853	0.648	0.652	0.371
L4 × 4 × 1/2	12.8	3.75	5.52	1.96	1.21	1.18	5.52	1.96	1.21	1.18	0.776	1.00
× 4 × 3/8	9.80	2.86	4.32	1.50	1.23	1.13	4.32	1.50	1.23	1.13	0.779	1.00
× 4 × 1/4	6.60	1.94	3.00	1.03	1.25	1.08	3.00	1.03	1.25	1.08	0.783	1.00
L4 × 3 × 5/8	13.6	3.89	6.01	2.28	1.23	1.37	2.85	1.34	0.845	0.867	0.631	0.534
× 3 × 3/8	8.50	2.48	3.94	1.44	1.26	1.27	1.89	0.851	0.873	0.775	0.636	0.551
× 3 × 1/4	5.80	1.69	2.75	0.988	1.27	1.22	1.33	0.585	0.887	0.725	0.639	0.558
L3 × 3 × 1/2	9.40	2.75	2.20	1.06	0.895	0.929	2.20	1.06	0.895	0.929	0.580	1.00
× 3 × 3/8	7.20	2.11	1.75	0.825	0.910	0.884	1.75	0.825	0.910	0.884	0.581	1.00
× 3 × 1/4	4.90	1.44	1.23	0.569	0.926	0.836	1.23	0.569	0.926	0.836	0.585	1.00
L3 × 2 × 1/2	7.70	2.25	1.92	1.00	0.922	1.08	0.667	0.470	0.543	0.580	0.425	0.413
× 2 × 3/8	5.90	1.73	1.54	0.779	0.937	1.03	0.539	0.368	0.555	0.535	0.426	0.426
× 2 × 1/4	4.10	1.19	1.09	0.541	0.953	0.980	0.390	0.258	0.569	0.487	0.431	0.437

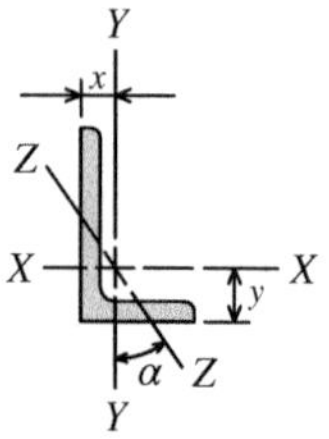

Designation	Mass per meter	Area A	I_x	S_x	r_x	y	I_y	S_y	r_y	x	r_z	tan α
	kg/m	mm^2	10^6 mm^4	10^3 mm^3	mm	mm	10^6 mm^4	10^3 mm^3	mm	mm	mm	
L127 × 127 × 19	35.1	4480	6.53	74.1	38.1	38.6	6.53	74.1	38.1	38.6	24.7	1.00
× 127 × 12.7	24.1	3060	4.70	51.6	38.9	36.1	4.70	51.6	38.9	36.1	24.9	1.00
× 127 × 9.5	18.3	2330	3.65	39.5	39.4	34.8	3.65	39.5	39.4	34.8	25.0	1.00
L127 × 76 × 12.7	19.0	2420	3.93	47.4	40.1	44.2	1.06	18.5	20.9	18.9	16.3	0.357
× 76 × 9.5	14.5	1850	3.06	36.4	40.6	42.9	0.837	14.3	21.3	17.7	16.4	0.364
× 76 × 6.4	9.80	1250	2.12	24.7	41.1	41.7	0.587	9.83	21.7	16.5	16.6	0.371
L102 × 102 × 12.7	19.0	2420	2.30	32.1	30.7	30.0	2.30	32.1	30.7	30.0	19.7	1.00
× 102 × 9.5	14.6	1850	1.80	24.6	31.2	28.7	1.80	24.6	31.2	28.7	19.8	1.00
× 102 × 6.4	9.80	1250	1.25	16.9	31.8	27.4	1.25	16.9	31.8	27.4	19.9	1.00
L102 × 76 × 15.9	20.2	2510	2.50	37.4	31.2	34.8	1.19	22.0	21.5	22.0	16.0	0.534
× 76 × 9.5	12.6	1600	1.64	23.6	32.0	32.3	0.787	13.9	22.2	19.7	16.2	0.551
× 76 × 6.4	8.60	1090	1.14	16.2	32.3	31.0	0.554	9.59	22.5	18.4	16.2	0.558
L76 × 76 × 12.7	14.0	1770	0.916	17.4	22.7	23.6	0.916	17.4	22.7	23.6	14.7	1.00
× 76 × 9.5	10.7	1360	0.728	13.5	23.1	22.5	0.728	13.5	23.1	22.5	14.8	1.00
× 76 × 6.4	7.30	929	0.512	9.32	23.5	21.2	0.512	9.32	23.5	21.2	14.9	1.00
L76 × 51 × 12.7	11.5	1450	0.799	16.4	23.4	27.4	0.278	7.70	13.8	14.7	10.8	0.413
× 51 × 9.5	8.80	1120	0.641	12.8	23.8	26.2	0.224	6.03	14.1	13.6	10.8	0.426
× 51 × 6.4	6.10	768	0.454	8.87	24.2	24.9	0.162	4.23	14.5	12.4	10.9	0.437

APPENDIX C

Table of Beam Slopes and Deflections

	Beam	Slope	Deflection	Elastic Curve
1	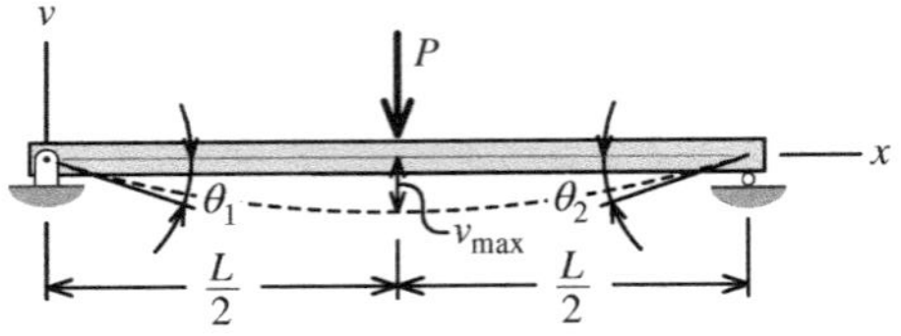	$\theta_1 = -\theta_2 = -\dfrac{PL^2}{16EI}$	$v_{max} = -\dfrac{PL^3}{48EI}$	$v = -\dfrac{Px}{48EI}(3L^2 - 4x^2)$ for $0 \le x \le \dfrac{L}{2}$
2	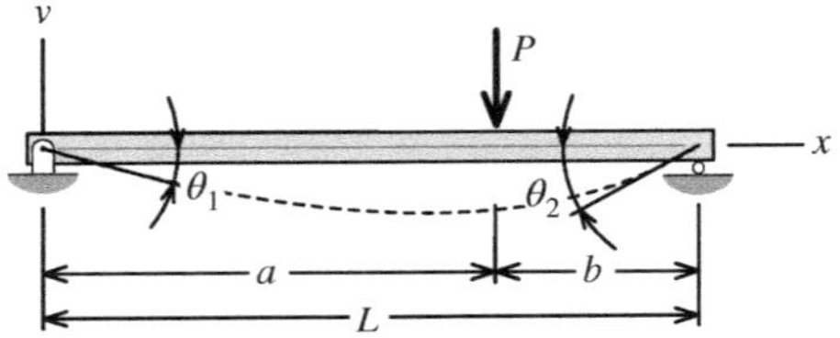	$\theta_1 = -\dfrac{Pb(L^2 - b^2)}{6LEI}$ $\theta_2 = +\dfrac{Pa(L^2 - a^2)}{6LEI}$	$v = -\dfrac{Pa^2b^2}{3LEI}$ at $x = a$	$v = -\dfrac{Pbx}{6LEI}(L^2 - b^2 - x^2)$ for $0 \le x \le a$
3	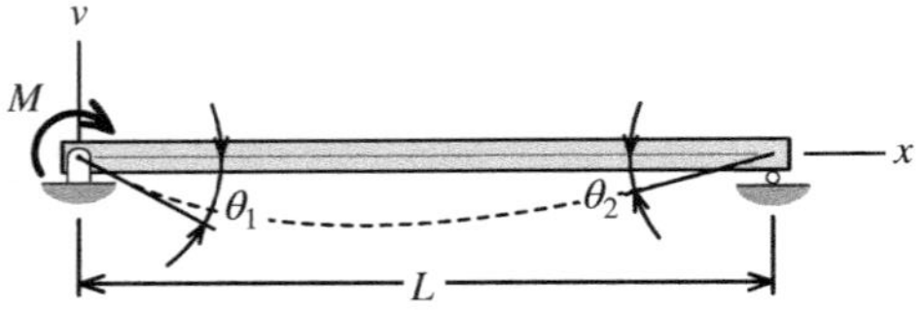	$\theta_1 = -\dfrac{ML}{3EI}$ $\theta_2 = +\dfrac{ML}{6EI}$	$v_{max} = -\dfrac{ML^2}{9\sqrt{3}\,EI}$ at $x = L\left(1 - \dfrac{\sqrt{3}}{3}\right)$	$v = -\dfrac{Mx}{6LEI}(2L^2 - 3Lx + x^2)$
4	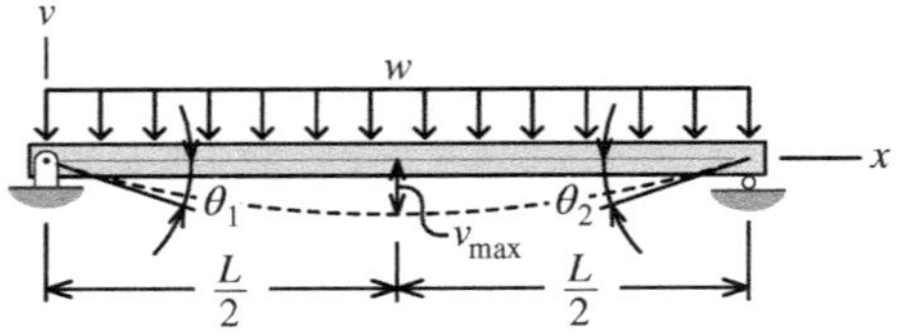	$\theta_1 = -\theta_2 = -\dfrac{wL^3}{24EI}$	$v_{max} = -\dfrac{5wL^4}{384EI}$	$v = -\dfrac{wx}{24EI}(L^3 - 2Lx^2 + x^3)$
5	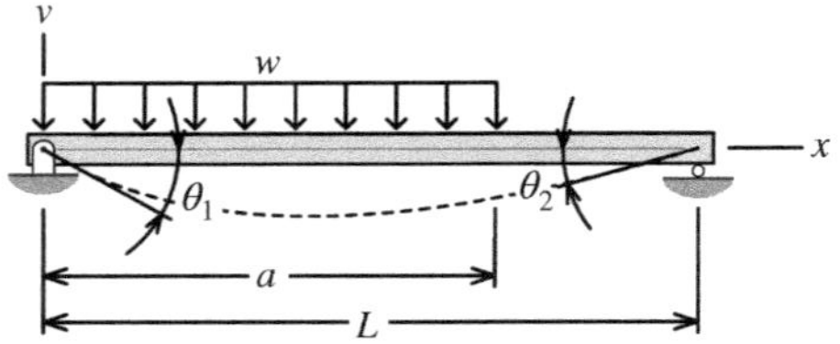	$\theta_1 = -\dfrac{wa^2}{24LEI}(2L - a)^2$ $\theta_2 = +\dfrac{wa^2}{24LEI}(2L^2 - a^2)$	$v = -\dfrac{wa^3}{24LEI}(4L^2 - 7aL + 3a^2)$ at $x = a$	$v = -\dfrac{wx}{24LEI}(Lx^3 - 4aLx^2 + 2a^2x^2 + 4a^2L^2 - 4a^3L + a^4)$ for $0 \le x \le a$ $v = -\dfrac{wa^2}{24LEI}(2x^3 - 6Lx^2 + a^2x + 4L^2x - a^2L)$ for $a \le x \le L$
6	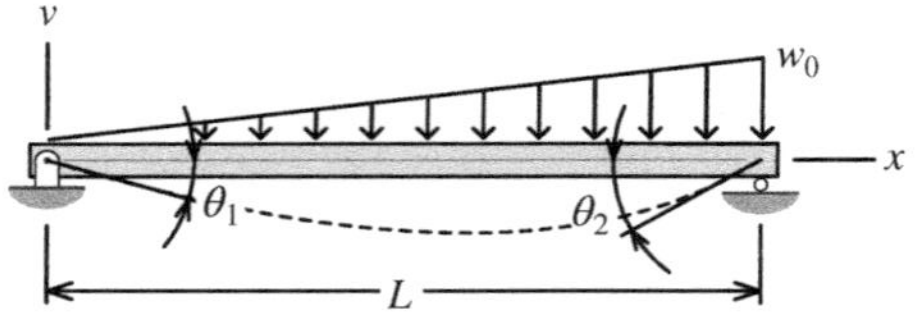	$\theta_1 = -\dfrac{7w_0L^3}{360EI}$ $\theta_2 = +\dfrac{w_0L^3}{45EI}$	$v_{max} = -0.00652\dfrac{w_0L^4}{EI}$ at $x = 0.5193L$	$v = -\dfrac{w_0x}{360LEI}(7L^4 - 10L^2x^2 + 3x^4)$

Cantilever Beams

	Beam	Slope	Deflection	Elastic Curve
7	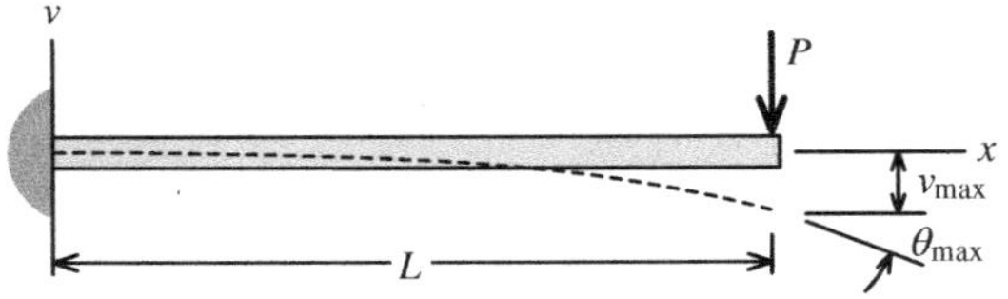	$\theta_{max} = -\dfrac{PL^2}{2EI}$	$v_{max} = -\dfrac{PL^3}{3EI}$	$v = -\dfrac{Px^2}{6EI}(3L - x)$
8	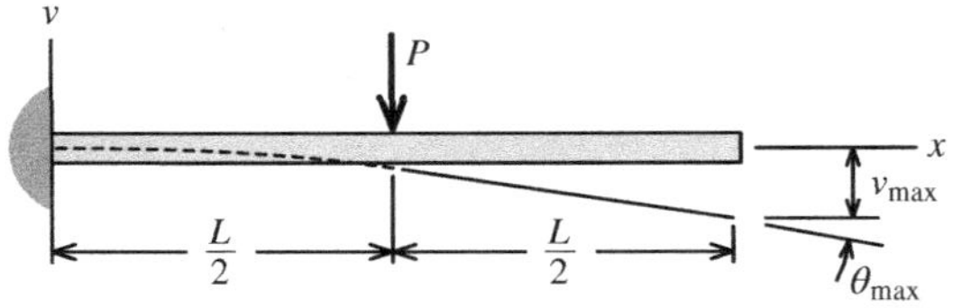	$\theta_{max} = -\dfrac{PL^2}{8EI}$	$v_{max} = -\dfrac{5PL^3}{48EI}$	$v = -\dfrac{Px^2}{12EI}(3L - 2x)$ for $0 \le x \le \dfrac{L}{2}$ $v = -\dfrac{PL^2}{48EI}(6x - L)$ for $\dfrac{L}{2} \le x \le L$
9	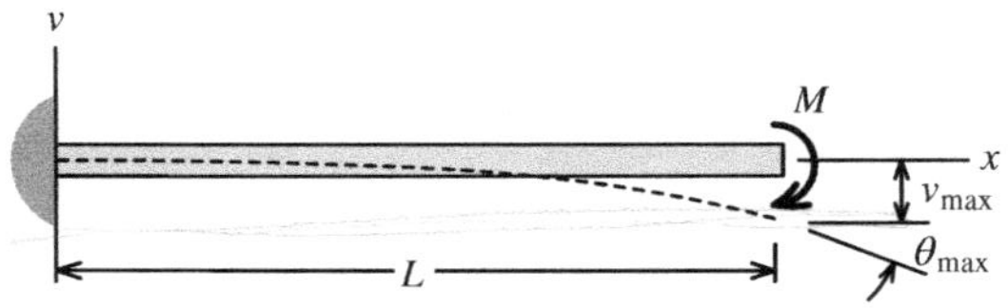	$\theta_{max} = -\dfrac{ML}{EI}$	$v_{max} = -\dfrac{ML^2}{2EI}$	$v = -\dfrac{Mx^2}{2EI}$
10	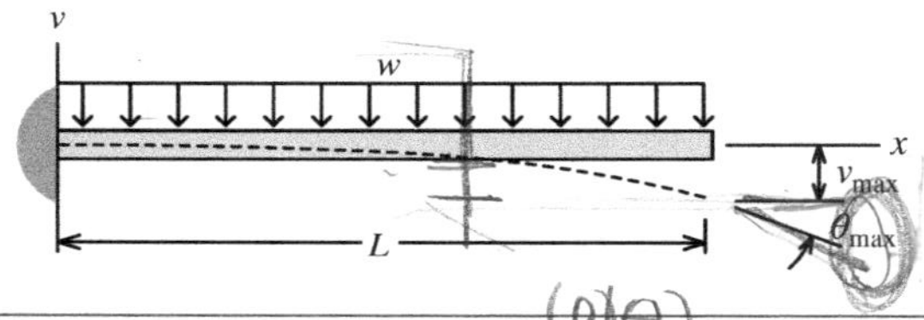	$\theta_{max} = -\dfrac{wL^3}{6EI}$	$v_{max} = -\dfrac{wL^4}{8EI}$	$v = -\dfrac{wx^2}{24EI}(6L^2 - 4Lx + x^2)$
11	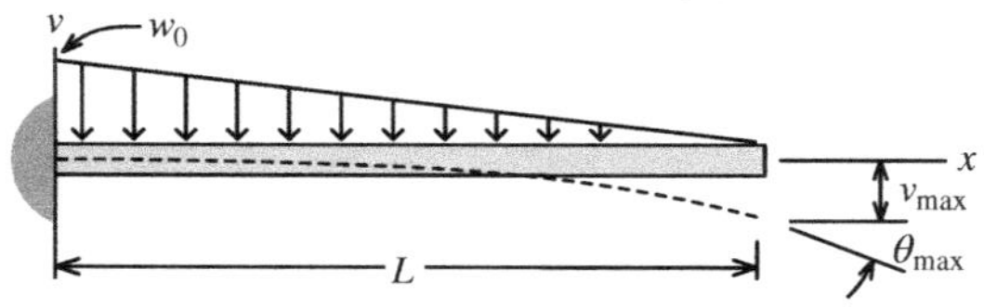	$\theta_{max} = -\dfrac{w_0L^3}{24EI}$	$v_{max} = -\dfrac{w_0L^4}{30EI}$	$v = -\dfrac{w_0x^2}{120LEI}(10L^3 - 10L^2x + 5Lx^2 - x^3)$

APPENDIX D

Average Properties of Selected Materials

Mechanical properties of metalic engineering materials vary significantly as a result of mechanical working, heat treatment, chemical content, and various other factors. The values presented in Table D.1*a* and D.1*b* should be considered *representative values* that are intended for educational purposes only. Commercial design applications should be based on appropriate values for specific materials and specific usages rather than the average values given here.

Table D.1*a* Average Properties of Selected Materials (U.S. Customary Units)

Materials	Specific weight (lb/ft³)	Yield strength (ksi)[a b]	Ultimate strength (ksi)[a]	Modulus of elasticity (1,000 ksi)	Shear modulus (1,000 ksi)	Poisson's ratio	Percent elongation over 2-in. gage length	Coefficient of thermal expansion (10^{-6}/°F)
Aluminum Alloys								
Alloy 2014-T4 (A92014)	175	42	62	10.6	4	0.33	20	12.8
Alloy 2014-T6 (A92014)	175	60	70	10.6	4	0.33	13	12.8
Alloy 6061-T6 (A96061)	170	40	45	10	3.8	0.33	17	13.1
Brass								
Red Brass C23000	550	18	44	16.7	6.4	0.307	45	10.4
Red Brass C83600	550	17	37	12	4.5	0.33	30	10.0
Bronze								
Bronze C86100	490	48	95	15.2	6.5	0.34	20	12.2
Bronze C95400 TQ50	465	45	90	16	6	0.316	8	9.0
Cast Iron								
Gray, ASTM A48 Grade 20	450		20	12.2	5	0.22	<1	5.0
Ductile, ASTM A536 80-55-06	450	55	80	24.4	9.3	0.32	6	6.0
Malleable, ASTM A220 45008	450	45	65	26	10.2	0.27	8	6.7
Steel								
Structural, ASTM-A36	490	36	58	29	11.2	0.3	21	6.5
Structural, ASTM-A992	490	50	65	29	11.2	0.3	21	6.5
AISI 1020, Cold-rolled	490	62	90	30	11.6	0.29	15	6.5
AISI 1040, Hot-rolled	490	60	90	30	11.5	0.3	25	6.3
AISI 1040, Cold-rolled	490	82	97	30	11.5	0.3	16	6.3
AISI 1040, WQT 900	490	90	118	30	11.5	0.3	22	6.3
AISI 4140, OQT 1100	490	131	147	30	11.5	0.3	16	6.2
AISI 5160, OQT 700	490	238	263	30	11.5	0.3	9	6.2
SAE 4340, Heat-treated	490	132	150	31	12	0.29	20	6.0
Stainless (18-8) annealed	490	36	85	28	12.5	0.12	55	9.6
Stainless (18-8) cold-rolled	490	165	190	28	12.5	0.12	8	9.6
Titanium								
Alloy (6% Al, 4%V)	280	120	130	16.5	6.2	0.33	10	5.3
Plastics								
ABS	66	6	5.5	0.3	–	–	36	48.8
Nylon 6/6	69	9	–	0.2	–	–	–	65.6
Polycarbonate	90	16	17	1.1	–	–	–	14.5
Polyethylene, Low-density	58	1.4	1.7	0.029	–	–	–	100
Polyethylene, High-density	60	3.3	4.3	0.128	–	–	721	88
Polypropylene	71	11	12	0.9	–	–	4	22.6
Polystyrene	73	7.5	7.5	0.54	0.2	0.33	39	47.2
Vinyl, rigid PVC	81	6.7	5.5	0.41	0.145	0.42	100	35

[a]For ductile metals, it is customary to assume that the properties in compression have the same values as those in tension.
[b]For most metals, this is the 0.2% offset value.

Table D.1*b* Average Properties of Selected Materials (SI Units)

Materials	Specific weight (kN/m³)	Yield strength (MPa)[a b]	Ultimate strength (MPa)[a]	Modulus of elasticity (GPa)	Shear modulus (GPa)	Poisson's ratio	Percent elongation over 50-mm gage length	Coefficient of thermal expansion ($10^{-6}/°C$)
Aluminum Alloys								
Alloy 2014-T4 (A92014)	27	290	427	73	28	0.33	20	23.0
Alloy 2014-T6 (A92014)	27	414	483	73	28	0.33	13	23.0
Alloy 6061-T6 (A96061)	27	276	310	69	26	0.33	17	23.6
Brass								
Red Brass C23000	86	124	303	115	44	0.307	45	18.7
Red Brass C83600	86	117	255	83	31	0.33	30	18.0
Bronze								
Bronze C86100	77	331	655	105	45	0.34	20	22.0
Bronze C95400 TQ50	73	310	621	110	41	0.316	8	16.2
Cast Iron								
Gray, ASTM A48 Grade 20	71		138	84	34	0.22	<1	9.0
Ductile, ASTM A536 80-55-06	71	379	552	168	64	0.32	6	10.8
Malleable, ASTM A220 45008	71	310	448	179	70	0.27	8	12.1
Steel								
Structural, ASTM-A36	77	250	400	200	77.2	0.3	21	11.7
Structural, ASTM-A992	77	345	450	200	77.2	0.3	21	11.7
AISI 1020, Cold-rolled	77	427	621	207	80	0.29	15	11.7
AISI 1040, Hot-rolled	77	414	621	207	80	0.3	25	11.3
AISI 1040, Cold-rolled	77	565	669	207	80	0.3	16	11.3
AISI 1040, WQT 900	77	621	814	207	80	0.3	22	11.3
AISI 4140, OQT 1100	77	903	1,014	207	80	0.3	16	11.2
AISI 5160, OQT 700	77	1,641	1,813	207	80	0.3	9	11.2
SAE 4340, Heat-treated	77	910	1,034	214	83	0.29	20	10.8
Stainless (18-8) annealed	77	248	586	193	86	0.12	55	17.3
Stainless (18-8) cold-rolled	77	1,138	1,310	193	86	0.12	8	17.3
Titanium								
Alloy (6% Al, 4%V)	44	827	896	114	43	0.33	10	9.5
Plastics								
ABS	1,060	41	38	2.1	–	–	36	88
Nylon 6/6	1,105	62	–	1.4	–	–	–	118
Polycarbonate	1,440	110	117	7.6	–	–	–	26
Polyethylene, Low-density	930	9.7	11.7	0.2	–	–	–	180
Polyethylene, High-density	960	22.8	29.6	0.9	–	–	721	158
Polypropylene	1,140	75.8	82.7	6.2	–	–	4	40.7
Polystyrene	1,170	52	52	3.7	1.4	0.33	39	85
Vinyl, rigid PVC	1,300	46	38	2.8	1.0	0.42	100	63

[a]For ductile metals, it is customary to assume that the properties in compression have the same values as those in tension.
[b]For most metals, this is the 0.2% offset value.

Table D.2 Typical Properties of Selected Wood Construction Materials

	Allowable Stresses								Modulus of Elasticity			
	Bending		Tension parallel to grain		Horizontal shear		Compression parallel to grain		E		E_{min}	
Type and grade	psi	MPa	psi	MPa	psi	MPa	psi	MPa	ksi	GPa	ksi	GPa
Framing Lumber: 2 in. to 4 in. thick by 2 in. and wider												
Douglas Fir-Larch												
Select Structural		10.3	1,000	6.9	180	1.24	1,900	13.1	1,900	13.1	690	4.76
No. 2	900	6.2	575	4.0	180	1.24	1,350	9.3	1,600	11.0	580	4.00
Hem-Fir												
Select Structural	1,400	9.7	925	6.4	150	1.03	1,500	10.3	1,600	11.0	580	4.00
No. 2	850	5.9	525	3.6	150	1.03	1,300	9.0	1,300	9.0	470	3.24
Spruce-Pine-Fir (South)												
Select Structural	1,300	9.0	575	4.0	135	0.93	1,200	8.3	1,300	9.0	470	3.24
No. 2	775	5.3	350	2.4	135	0.93	1,000	6.9	1,100	7.6	400	2.76
Western Cedars												
Select Structural	1,000	6.9	600	4.1	155	1.07	1,000	6.9	1,100	7.6	400	2.76
No. 2	700	4.8	425	2.9	155	1.07	650	4.5	1,000	6.9	370	2.55
Beams: 5 in. and thicker, width more than 2 in. greater than thickness												
Douglas Fir-Larch												
Select Structural	1,600	11.0	950	6.6	170	1.17	1,100	7.6	1,600	11.0	580	4.00
No. 2	875	6.0	425	2.9	170	1.17	600	4.1	1,300	9.0	470	3.24
Hem-Fir												
Select Structural	1,300	9.0	750	5.2	140	0.97	925	6.4	1,300	9.0	470	3.24
No. 2	675	4.7	350	2.4	140	0.97	500	3.4	1,100	7.6	400	2.76
Spruce-Pine-Fir (South)												
Select Structural	1,050	7.2	625	4.3	125	0.86	675	4.7	1,200	8.3	440	3.03
No. 2	575	4.0	300	2.1	125	0.86	375	2.6	1,000	6.9	370	2.55
Western Cedars												
Select Structural	1,150	7.9	675	4.7	140	0.97	875	6.0	1,000	6.9	370	2.55
No. 2	625	4.3	325	2.2	140	0.97	475	3.3	800	5.5	290	2.00
Posts: 5 in. by 5 in. and larger, width not more than 2 in. greater than thickness												
Douglas Fir-Larch												
Select Structural	1,500	10.3	1,000	6.9	170	1.17	1,150	7.9	1,600	11.0	580	4.00
No. 2	750	5.2	475	3.3	170	1.17	700	4.8	1,300	9.0	470	3.24
Hem-Fir												
Select Structural	1,200	8.3	800	5.5	140	0.97	975	6.7	1,300	9.0	470	3.24
No. 2	575	4.0	375	2.6	140	0.97	575	4.0	1,100	7.6	400	2.76
Spruce-Pine-Fir (South)												
Select Structural	1,000	6.9	675	4.7	125	0.86	700	4.8	1,200	8.3	440	3.03
No. 2	475	3.3	325	2.2	125	0.86	425	2.9	1,000	6.9	370	2.55
Western Cedars												
Select Structural	1,100	7.6	725	5.0	140	0.97	925	6.4	1,000	6.9	370	2.55
No. 2	550	3.8	350	2.4	140	0.97	550	3.8	800	5.5	290	2.00

APPENDIX E

Fundamental Mechanics of Materials Equations

Common Greek letters

α	Alpha	μ	Mu
β	Beta	ν	Nu
γ	Gamma	π	Pi
Δ, δ	Delta	ρ	Rho
ε	Epsilon	Σ, σ	Sigma
θ	Theta	τ	Tau
κ	Kappa	ϕ	Phi
λ	Lambda	ω	Omega

Basic definitions

Average normal stress in an axial member

$$\sigma_{\text{avg}} = \frac{F}{A}$$

Average direct shear stress

$$\tau_{\text{avg}} = \frac{V}{A_V}$$

Average bearing stress

$$\sigma_b = \frac{F}{A_b}$$

Average normal strain in an axial member

$$\varepsilon_{\text{long}} = \frac{\Delta L}{L} = \frac{\delta}{L}$$

$$\varepsilon_{\text{lat}} = \frac{\Delta d}{d} \quad \text{or} \quad \frac{\Delta t}{t} \quad \text{or} \quad \frac{\Delta h}{h}$$

Average normal strain caused by temperature change

$$\varepsilon_T = \alpha \, \Delta T$$

Average shear strain

$$\gamma = \text{change in angle from } \frac{\pi}{2} \text{ rad}$$

Hooke's law (one-dimensional)

$$\sigma = E\varepsilon \quad \text{and} \quad \tau = G\gamma$$

Poisson's ratio

$$\nu = -\frac{\varepsilon_{\text{lat}}}{\varepsilon_{\text{long}}}$$

Relationship between E, G, and ν

$$G = \frac{E}{2(1+\nu)}$$

Definition of allowable stress

$$\sigma_{\text{allow}} = \frac{\sigma_{\text{failure}}}{\text{FS}} \quad \text{or} \quad \tau_{\text{allow}} = \frac{\tau_{\text{failure}}}{\text{FS}}$$

Factor of safety

$$\text{FS} = \frac{\sigma_{\text{failure}}}{\sigma_{\text{actual}}} \quad \text{or} \quad \text{FS} = \frac{\tau_{\text{failure}}}{\tau_{\text{actual}}}$$

Axial deformation

Deformation in axial members

$$\delta = \frac{FL}{AE} \quad \text{or} \quad \delta = \sum_i \frac{F_i L_i}{A_i E_i}$$

Force-temperature-deformation relationship

$$\delta = \frac{FL}{AE} + \alpha \Delta T L$$

Torsion

Maximum torsion shear stress in a circular shaft

$$\tau_{\text{max}} = \frac{Tc}{J}$$

where the polar moment of inertia J is defined as:

$$J = \frac{\pi}{2}[R^4 - r^4] = \frac{\pi}{32}[D^4 - d^4]$$

Angle of twist in a circular shaft

$$\phi = \frac{TL}{JG} \quad \text{or} \quad \phi = \sum_i \frac{T_i L_i}{J_i G_i}$$

Power transmission in a shaft

$$P = T\omega$$

Power units and conversion factors

$$1 \text{ W} = \frac{1 \text{ N}\cdot\text{m}}{\text{s}} \qquad 1 \text{ hp} = \frac{550 \text{ lb}\cdot\text{ft}}{\text{s}} = \frac{6{,}600 \text{ lb}\cdot\text{in.}}{\text{s}}$$

$$1 \text{ Hz} = \frac{1 \text{ rev}}{\text{s}} \qquad 1 \text{ rev} = 2\pi \text{ rad}$$

$$1 \text{ rpm} = \frac{2\pi \text{ rad}}{60 \text{ s}}$$

Gear relationships between gears A and B

$$\frac{T_A}{R_A} = \frac{T_B}{R_B} \qquad R_A\phi_A = -R_B\phi_B \qquad R_A\omega_A = R_B\omega_B$$

$$\text{Gear ratio} = \frac{R_A}{R_B} = \frac{D_A}{D_B} = \frac{N_A}{N_B}$$

Six rules for constructing shear-force and bending-moment diagrams

Rule 1: $\Delta V = P_0$

Rule 2: $\Delta V = V_2 - V_1 = \int_{x_1}^{x_2} w(x)\,dx$

Rule 3: $\dfrac{dV}{dx} = w(x)$

Rule 4: $\Delta M = M_2 - M_1 = \int_{x_1}^{x_2} V\,dx$

Rule 5: $\dfrac{dM}{dx} = V$

Rule 6: $\Delta M = -M_0$

Flexure

Flexural strain and stress

$$\varepsilon_x = -\frac{1}{\rho}y \qquad \sigma_x = -\frac{E}{\rho}y$$

Flexure Formula

$$\sigma_x = -\frac{My}{I_z} \quad \text{or} \quad \sigma_{\max} = \frac{Mc}{I} = \frac{M}{S} \quad \text{where } S = \frac{I}{c}$$

Transformed-section method for beams of two materials [where material (2) is transformed into an equivalent amount of material (1)]

$$n = \frac{E_2}{E_1} \qquad \sigma_{x1} = -\frac{My}{I_{\text{transformed}}} \qquad \sigma_{x2} = -n\frac{My}{I_{\text{transformed}}}$$

Bending due to eccentric axial load

$$\sigma_x = \frac{F}{A} - \frac{My}{I_z}$$

Unsymmetric bending of arbitrary cross sections

$$\sigma_x = \left[\frac{I_z z - I_{yz} y}{I_y I_z - I_{yz}^2}\right] M_y + \left[\frac{-I_y y + I_{yz} z}{I_y I_z - I_{yz}^2}\right] M_z$$

or

$$\sigma_x = -\frac{(M_z I_y + M_y I_{yz})y}{I_y I_z - I_{yz}^2} + \frac{(M_y I_z + M_z I_{yz})z}{I_y I_z - I_{yz}^2}$$

$$\tan\beta = \frac{M_y I_z + M_z I_{yz}}{M_z I_y + M_y I_{yz}}$$

Unsymmetric bending of symmetric cross sections

$$\sigma_x = \frac{M_y z}{I_y} - \frac{M_z y}{I_z} \qquad \tan\beta = \frac{M_y I_z}{M_z I_y}$$

Bending of curved bars

$$\sigma_x = -\frac{M(r_n - r)}{rA(r_c - r_n)} \qquad \text{where } r_n = \frac{A}{\int_A \frac{dA}{r}}$$

Horizontal shear stress associated with bending

$$\tau_H = \frac{VQ}{It} \qquad \text{where } Q = \Sigma\,\bar{y}_i A_i$$

Shear flow formula

$$q = \frac{VQ}{I}$$

Shear flow, fastener spacing, and fastener shear relationship

$$qs \le n_f V_f = n_f \tau_f A_f$$

For circular cross sections,

$$Q = \frac{2}{3}r^3 = \frac{1}{12}d^3 \qquad \text{(solid sections)}$$

$$Q = \frac{2}{3}[R^3 - r^3] = \frac{1}{12}[D^3 - d^3] \qquad \text{(hollow sections)}$$

Beam deflections

Elastic curve relations between w, V, M, θ, and v for constant EI

Deflection $= v$

Slope $= \dfrac{dv}{dx} = \theta$ (for small deflections)

Moment $M = EI\dfrac{d^2v}{dx^2}$

Shear $V = \dfrac{dM}{dx} = EI\dfrac{d^3v}{dx^3}$

Load $w = \dfrac{dV}{dx} = EI\dfrac{d^4v}{dx^4}$

Plane stress transformations

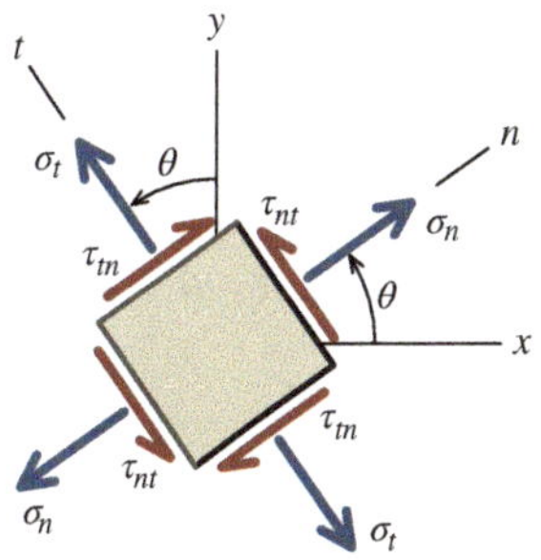

Stresses on an arbitrary plane

$$\sigma_n = \sigma_x\cos^2\theta + \sigma_y\sin^2\theta + 2\tau_{xy}\sin\theta\cos\theta$$

$$\sigma_t = \sigma_x\sin^2\theta + \sigma_y\cos^2\theta - 2\tau_{xy}\sin\theta\cos\theta$$

$$\tau_{nt} = -(\sigma_x - \sigma_y)\sin\theta\cos\theta + \tau_{xy}(\cos^2\theta - \sin^2\theta)$$

or

$$\sigma_n = \frac{\sigma_x + \sigma_y}{2} + \frac{\sigma_x - \sigma_y}{2}\cos 2\theta + \tau_{xy}\sin 2\theta$$

$$\sigma_t = \frac{\sigma_x + \sigma_y}{2} - \frac{\sigma_x - \sigma_y}{2}\cos 2\theta - \tau_{xy}\sin 2\theta$$

$$\tau_{nt} = -\frac{\sigma_x - \sigma_y}{2}\sin 2\theta + \tau_{xy}\cos 2\theta$$

Principal stress magnitudes

$$\sigma_{p1,p2} = \frac{\sigma_x + \sigma_y}{2} \pm \sqrt{\left(\frac{\sigma_x - \sigma_y}{2}\right)^2 + \tau_{xy}^2}$$

Orientation of principal planes

$$\tan 2\theta_p = \frac{2\tau_{xy}}{\sigma_x - \sigma_y}$$

Maximum in-plane shear stress magnitude

$$\tau_{max} = \pm\sqrt{\left(\frac{\sigma_x - \sigma_y}{2}\right)^2 + \tau_{xy}^2} \quad \text{or} \quad \tau_{max} = \frac{\sigma_{p1} - \sigma_{p2}}{2}$$

$$\sigma_{avg} = \frac{\sigma_x + \sigma_y}{2}$$

$$\tan 2\theta_s = -\frac{\sigma_x - \sigma_y}{2\tau_{xy}} \qquad \text{note: } \theta_s = \theta_p \pm 45°$$

Absolute maximum shear stress magnitude

$$\tau_{abs\,max} = \frac{\sigma_{max} - \sigma_{min}}{2}$$

Normal stress invariance

$$\sigma_x + \sigma_y = \sigma_n + \sigma_t = \sigma_{p1} + \sigma_{p2}$$

Plane strain transformations

Strain in arbitrary directions

$$\varepsilon_n = \varepsilon_x \cos^2\theta + \varepsilon_y \sin^2\theta + \gamma_{xy}\sin\theta\cos\theta$$

$$\varepsilon_t = \varepsilon_x \sin^2\theta + \varepsilon_y \cos^2\theta - \gamma_{xy}\sin\theta\cos\theta$$

$$\gamma_{nt} = -2(\varepsilon_x - \varepsilon_y)\sin\theta\cos\theta + \gamma_{xy}(\cos^2\theta - \sin^2\theta)$$

or

$$\varepsilon_n = \frac{\varepsilon_x + \varepsilon_y}{2} + \frac{\varepsilon_x - \varepsilon_y}{2}\cos 2\theta + \frac{\gamma_{xy}}{2}\sin 2\theta$$

$$\varepsilon_t = \frac{\varepsilon_x + \varepsilon_y}{2} - \frac{\varepsilon_x - \varepsilon_y}{2}\cos 2\theta - \frac{\gamma_{xy}}{2}\sin 2\theta$$

$$\gamma_{nt} = -(\varepsilon_x - \varepsilon_y)\sin 2\theta + \gamma_{xy}\cos 2\theta$$

Principal strain magnitudes

$$\varepsilon_{p1,p2} = \frac{\varepsilon_x + \varepsilon_y}{2} \pm \sqrt{\left(\frac{\varepsilon_x - \varepsilon_y}{2}\right)^2 + \left(\frac{\gamma_{xy}}{2}\right)^2}$$

Orientation of principal strains

$$\tan 2\theta_p = \frac{\gamma_{xy}}{\varepsilon_x - \varepsilon_y}$$

Maximum in-plane shear strain

$$\gamma_{max} = \pm 2\sqrt{\left(\frac{\varepsilon_x - \varepsilon_y}{2}\right)^2 + \left(\frac{\gamma_{xy}}{2}\right)^2} \quad \text{or} \quad \gamma_{max} = \varepsilon_{p1} - \varepsilon_{p2}$$

$$\varepsilon_{avg} = \frac{\varepsilon_x + \varepsilon_y}{2}$$

Normal strain invariance

$$\varepsilon_x + \varepsilon_y = \varepsilon_n + \varepsilon_t = \varepsilon_{p1} + \varepsilon_{p2}$$

Generalized Hooke's law

Normal stress/normal strain relationships

$$\varepsilon_x = \frac{1}{E}[\sigma_x - \nu(\sigma_y + \sigma_z)]$$

$$\varepsilon_y = \frac{1}{E}[\sigma_y - \nu(\sigma_x + \sigma_z)]$$

$$\varepsilon_z = \frac{1}{E}[\sigma_z - \nu(\sigma_x + \sigma_y)]$$

$$\sigma_x = \frac{E}{(1+\nu)(1-2\nu)}[(1-\nu)\varepsilon_x + \nu(\varepsilon_y + \varepsilon_z)]$$

$$\sigma_y = \frac{E}{(1+\nu)(1-2\nu)}[(1-\nu)\varepsilon_y + \nu(\varepsilon_x + \varepsilon_z)]$$

$$\sigma_z = \frac{E}{(1+\nu)(1-2\nu)}[(1-\nu)\varepsilon_z + \nu(\varepsilon_x + \varepsilon_y)]$$

Shear stress/shear strain relationships

$$\gamma_{xy} = \frac{1}{G}\tau_{xy}; \qquad \gamma_{yz} = \frac{1}{G}\tau_{yz}; \qquad \gamma_{zx} = \frac{1}{G}\tau_{zx}$$

where

$$G = \frac{E}{2(1+\nu)}$$

Volumetric strain or Dilatation

$$e = \frac{\Delta V}{V} = \varepsilon_x + \varepsilon_y + \varepsilon_z = \frac{1-2\nu}{E}(\sigma_x + \sigma_y + \sigma_z)$$

Bulk modulus

$$K = \frac{E}{3(1-2\nu)}$$

Normal stress/normal strain relationships for plane stress

$$\varepsilon_x = \frac{1}{E}(\sigma_x - \nu\sigma_y)$$

$$\varepsilon_y = \frac{1}{E}(\sigma_y - \nu\sigma_x)$$

$$\varepsilon_z = -\frac{\nu}{E}(\sigma_x + \sigma_y)$$

$$\varepsilon_z = -\frac{\nu}{1-\nu}(\varepsilon_x + \varepsilon_y)$$

or

$$\sigma_x = \frac{E}{1-\nu^2}(\varepsilon_x + \nu\varepsilon_y)$$

$$\sigma_y = \frac{E}{1-\nu^2}(\varepsilon_y + \nu\varepsilon_x)$$

Shear stress/shear strain relationships for plane stress

$$\gamma_{xy} = \frac{1}{G}\tau_{xy} \quad \text{or} \quad \tau_{xy} = G\gamma_{xy}$$

Thin-walled pressure vessels

Tangential stress and strain in spherical pressure vessel

$$\sigma_t = \frac{pr}{2t} = \frac{pd}{4t} \qquad \varepsilon_t = \frac{pr}{2tE}(1 - \nu)$$

Longitudinal and circumferential stresses in cylindrical pressure vessels

$$\sigma_{\text{long}} = \frac{pr}{2t} = \frac{pd}{4t} \qquad \varepsilon_{\text{long}} = \frac{pr}{2tE}(1 - 2\nu)$$

$$\sigma_{\text{hoop}} = \frac{pr}{t} = \frac{pd}{2t} \qquad \varepsilon_{\text{hoop}} = \frac{pr}{2tE}(2 - \nu)$$

Thick-walled pressure vessels

Radial stress in thick-walled cylinder

$$\sigma_r = \frac{a^2 p_i - b^2 p_o}{b^2 - a^2} - \frac{a^2 b^2 (p_i - p_o)}{(b^2 - a^2) r^2}$$

or

$$\sigma_r = \frac{a^2 p_i}{b^2 - a^2}\left(1 - \frac{b^2}{r^2}\right) - \frac{b^2 p_o}{b^2 - a^2}\left(1 - \frac{a^2}{r^2}\right)$$

Circumferential stress in thick-walled cylinder

$$\sigma_\theta = \frac{a^2 p_i - b^2 p_o}{b^2 - a^2} + \frac{a^2 b^2 (p_i - p_o)}{(b^2 - a^2) r^2}$$

or

$$\sigma_\theta = \frac{a^2 p_i}{b^2 - a^2}\left(1 + \frac{b^2}{r^2}\right) - \frac{b^2 p_o}{b^2 - a^2}\left(1 + \frac{a^2}{r^2}\right)$$

Maximum shear stress

$$\tau_{\max} = \frac{1}{2}(\sigma_\theta - \sigma_r) = \frac{a^2 b^2 (p_i - p_o)}{(b^2 - a^2) r^2}$$

Longitudinal normal stress in closed cylinder

$$\sigma_{\text{long}} = \frac{a^2 p_i - b^2 p_o}{b^2 - a^2}$$

Radial displacement for internal pressure only

$$\delta_r = \frac{a^2 p_i}{(b^2 - a^2) r E}\left[(1 - \nu) r^2 + (1 + \nu) b^2\right]$$

Radial displacement for external pressure only

$$\delta_r = -\frac{b^2 p_o}{(b^2 - a^2) r E}\left[(1 - \nu) r^2 + (1 + \nu) a^2\right]$$

Radial displacement for external pressure on solid cylinder

$$\delta_r = -\frac{(1 - \nu) p_o r}{E}$$

Contact pressure for interference fit connection of thick cylinder onto a thick cylinder

$$p_c = \frac{E\delta (c^2 - b^2)(b^2 - a^2)}{2b^3 (c^2 - a^2)}$$

Contact pressure for interference fit connection of thick cylinder onto a solid cylinder

$$p_c = \frac{E\delta (c^2 - b^2)}{2bc^2}$$

Failure theories

Mises equivalent stress for plane stress

$$\sigma_M = \left[\sigma_{p1}^2 - \sigma_{p1}\sigma_{p2} + \sigma_{p2}^2\right]^{1/2} = \left[\sigma_x^2 - \sigma_x\sigma_y + \sigma_y^2 + 3\tau_{xy}^2\right]^{1/2}$$

Column buckling

Euler buckling load

$$P_{cr} = \frac{\pi^2 EI}{(KL)^2}$$

Euler buckling stress

$$\sigma_{cr} = \frac{\pi^2 E}{(KL/r)^2}$$

Radius of gyration

$$r^2 = \frac{I}{A}$$

Secant formula

$$\sigma_{\max} = \frac{P}{A}\left[1 + \frac{ec}{r^2}\sec\left(\frac{KL}{2r}\sqrt{\frac{P}{EA}}\right)\right]$$

Answers to Odd Numbered Problems

Chapter 1

P1.1 $\sigma = 122.2$ MPa

P1.3 $t_{\text{min}} = 16.09$ mm

P1.5 $b_2 = 27.1$ mm

P1.7 $d_1 = 1.382$ in., $d_2 = 0.853$ in., $d_3 = 1.596$ in.

P1.9 $\sigma_1 = 93.2$ MPa (T)

P1.11 $w = 13.01$ kN/m

P1.13 $P_{\text{max}} = 55.2$ kips

P1.15 (a) $\sigma = 4{,}380$ psi (T)
(b) $\sigma = 1{,}730$ psi (T)

P1.17 $\tau_{\text{avg}} = 101.9$ psi

P1.19 $\tau_{\text{avg}} = 17{,}740$ psi

P1.21 $d_{\text{min}} = 0.824$ in.

P1.23 $a \geq 324$ mm

P1.25 5 in. × 5 in. plate at A, 6 in. × 6 in. plate at B

P1.27 $\sigma_b = 20.6$ MPa

P1.29 (a) $\tau_{\text{avg}} = 18{,}050$ psi
(b) $\sigma_b = 28{,}400$ psi

P1.31 (a) $\tau_B = 11.58$ MPa
(b) $\sigma_b = 18.19$ MPa

P1.33 (a) $d_{\text{min}} = 14.42$ mm
(b) $d_{\text{min}} = 16.33$ mm
(c) $d_{\text{min}} = 6.60$ mm

P1.35 $d_{\text{min}} = 19.54$ mm

P1.37 $\sigma_n = 37.6$ psi, $\tau_{nt} = 53.7$ psi

P1.39 $P_{\text{max}} = 152.3$ kips

P1.41 $\tau_{nt} = 11.75$ MPa

Chapter 2

P2.1 $\varepsilon_1 = 825$ µε

P2.3 $\varepsilon_{AB} = 8{,}300$ µε

P2.5 $\varepsilon_1 = -1{,}103$ µε

P2.7 (a) $\varepsilon_1 = 900$ µε
(b) $\varepsilon_1 = 1{,}127$ µε
(c) $\varepsilon_1 = 275$ µε

P2.9 (a) $\delta = \dfrac{\gamma L^2}{6E}$
(b) $\varepsilon_{\text{avg}} = \dfrac{\gamma L}{6E}$
(c) $\varepsilon_{\text{max}} = \dfrac{\gamma L}{3E}$

P2.11 $\gamma_{P'} = -9{,}170$ µrad

P2.13 $\gamma_Q = -3{,}750$ µrad

P2.15 (a) $\varepsilon_{QS} = -4{,}410$ µε
(b) $\gamma_{P'} = 5{,}430$ µrad

P2.17 (a) $\Delta w = 1.896$ mm
(b) $\Delta d = 1.138$ mm

P2.19 96.1°F

P2.21 $u_D = 0.522$ mm ←

Chapter 3

P3.1 (a) $\sigma_{PL} = 43.0$ ksi
(b) $E = 15{,}860$ ksi
(c) $\nu = 0.343$

P3.3 (a) $E = 2.92$ GPa
(b) $\nu = 0.410$
(c) $\Delta b = 0.1900$ mm

P3.5 $E = 3.19$ GPa

P3.7 $P = 35.3$ kips (c)

P3.9 (a) permanent set = 0.0035 mm/mm
(b) bar length unloaded = 351.225 mm
(c) $\sigma_{PL} = 444$ MPa

P3.11 $P = 42{,}000$ lb

P3.13 (a) $E = 30{,}000$ ksi
(b) $\sigma_{PL} = 60$ ksi
(c) $\sigma_U = 159$ ksi
(d) $\sigma_Y = 80$ ksi
(e) $\sigma_{\text{fracture}} = 135$ ksi
(f) true $\sigma_{\text{fracture}} = 270$ ksi

P3.15 (a) $E = 100.5$ GPa
(b) $\sigma_{PL} = 247$ MPa
(c) $\sigma_Y = 417$ MPa
(d) $\sigma_U = 557$ MPa
(e) $\sigma_{\text{fracture}} = 547$ MPa
(f) true $\sigma_{\text{fracture}} = 909$ MPa

P3.17 (a) $P = 756$ kN
(b) $\Delta d = 0.1923$ mm

P3.19 $P = 16.99$ kips

Chapter 4

P4.1 $d_{\text{min}} = 29.5$ mm

P4.3 $P_{\text{allow}} = 53.6$ lb

P4.5 $P_{\text{max}} = 42.0$ kips
bar (1): FS = 1.670,
bar (2): FS = 2.22

P4.7 (a) FS = 3.64
(b) FS = 2.53
(c) FS = 4.05

P4.9 (a) $A_{\text{min}} = 104.4$ mm^2
(b) $d_{\text{min}} = 13.68$ mm

P4.11 $w_{\text{max}} = 44.1$ kN/m

P4.13 (a) $b_{\text{min}} = 184.4$ mm
(b) $b_{\text{min}} = 176.4$ mm

Chapter 5

P5.1 $d_{\text{min}} = 53.5$ mm

P5.3 $P = 19.12$ kips

P5.5 $u_B = 0.0867$ in.

P5.7 (a) $v_B = -8.43$ mm↓
(b) $v_C = -12.88$ mm↓

P5.9 $u_A = 0.1207$ in. →

P5.11 $P = 89.3$ lb

P5.13 $\delta = 0.0721$ in.

P5.15 $\delta = 90.6 \times 10^{-6}$ in. ↓

P5.17 (a) $v_A = 40.1$ mm ↑
(b) $v_D = 10.49$ mm ↓

P5.19 (a) $u_D = 6.00$ mm
(b) $\delta_1 = 5.26$ mm, $\delta_2 =$ 6.00 mm, $\delta_3 = 5.69$ mm
(c) $P = 363$ kN

P5.21 (a) $d_{\text{min}} = 1.309$ in.

P5.23 $\sigma_1 = 66.5$ MPa, $\sigma_2 = 137.4$ MPa

P5.25 (a) $P = 103.0$ kips
(b) $u_B = 0.1725$ in. ←

P5.27 (a) $\sigma_1 = 120.1$ MPa (T), $\sigma_2 = 55.0$ MPa (T)
(b) $v_D = 0.991$ mm ↓

P5.29 (a) $F_1 = 6.67$ kips, $F_2 =$ 5.33 kips, $F_3 = 4.00$ kips
(b) $v_B = 0.1956$ in. ↓

P5.31 (a) $\sigma_1 = 23.2$ ksi (T), $\sigma_2 = 16.02$ ksi (C)
(b) $u_D = 0.0586$ in. →

P5.33 (a) $\sigma_1 = 176.8$ MPa (T), $\sigma_2 = 140.2$ MPa (T)
(b) $u_A = 4.27$ mm →

P5.35 (a) $F_1 = 23.1$ kN (T), $F_2 = 38.8$ kN (T)
(b) $v_{ABC} = 4.22$ mm ↓

P5.37 (a) $\sigma_1 = 11.79$ ksi (T), $\sigma_2 = 22.8$ ksi (T)
(b) $v_A = 0.0471$ in. ↓

P5.39 (a) $\delta = -13.78$ mm
(b) $\Delta d = -0.0962$ mm

P5.41 $T = 202$°C

P5.43 $v_D = 0.550$ in. ↓

P5.45 $\Delta T = -41.2$°C

P5.47 $\tau_{\text{pin}} = 29.4$ ksi

P5.49 (a) $d_2 = 35.1$ mm
(b) $\Delta T = -53.4$°C

P5.51 (a) $\sigma_1 = 14.27$ ksi (T), $\sigma_2 = 13.23$ ksi (T)
(b) $v_D = 0.497$ in. ↓

P5.53 $\varepsilon_1 = 777$ με, $\varepsilon_2 = -994$ με

P5.55 $P_{\text{allow}} = 30.0$ kN

P5.57 $r_{\text{min}} = 9$ mm

Chapter 6

P6.1 $\tau_{\text{max}} = 56.6$ MPa

P6.3 $d_{\text{min}} = 5.35$ in.

P6.5 (b) $d_{\text{min}} = 1.848$ in.

P6.7 (a) $\tau_{\text{max}} = 1{,}880$ psi
(b) $\phi = 0.0253$ rad

P6.9 $T_A = 5{,}980$ N·m

P6.11 $d_{\text{min}} = 3.44$ in.

P6.13 $\phi_{C/A} = 0.0290$ rad

P6.15 $T_B = 591$ lb·ft

P6.17 (a) $\tau_A = 77.4$ MPa
(b) $\tau_B = 51.6$ MPa

P6.19 $\tau_1 = 10{,}430$ psi, $\tau_2 = 6{,}520$ psi

P6.21 $d_1 = 2.39$ in., $d_2 = 1.904$ in.

P6.23 (a) $T_1 = -10{,}430$ N·m, $T_2 = -6{,}000$ N·m
(b) $\phi_1 = -0.0810$ rad, $\phi_2 = -0.0349$ rad
(c) $\phi_B = -0.0810$ rad, $\phi_C = 0.1409$ rad
(d) $\phi_D = 0.1758$ rad

P6.25 $P = 29.5$ hp, $\tau_{\text{max}} = 4{,}680$ psi

P6.27 33.6 Hz

P6.29 (a) $d_{\text{min}} = 33.8$ mm
(b) $d_{\text{max}} = 31.7$ mm
(c) 48.1%

P6.31 (a) $d_{\text{min}} = 29.5$ mm
(b) $d_{\text{min}} = 77.4$ mm

P6.33 (a) $\tau_1 = 48.9$ MPa, $\tau_2 = 15.44$ MPa
(b) $P = 63.7$ kW, 576 rpm
(c) $T_A = 1{,}056$ N·m

P6.35 (a) $\tau_1 = 7{,}370$ psi, $\tau_2 = 4{,}420$ psi, $\tau_3 = 1{,}475$ psi
(b) $\phi_{E/B} = 0.349$ rad

P6.37 (a) $d_{\text{min}} = 41.8$ mm
(b) $d_{\text{min}} = 30.9$ mm

(c) $\phi_{E/C} = 0.1408$ rad
(d) $T_A = 232$ N·m, 34.3 Hz

P6.39 (a) $T_{max} = 84.7$ kip·in.
(b) $T_1 = 45.4$ kip·in., $T_2 = 39.3$ kip·in.
(c) $\phi = 0.0769$ rad

P6.41 (a) $\tau_1 = 4{,}230$ psi
(b) $\tau_2 = 5{,}090$ psi
(c) $\tau_3 = 9{,}780$ psi
(d) $\phi_B = 0.0446$ rad
(e) $\phi_C = -0.01043$ rad

P6.43 (a) $T_{allow} = 73.8$ lb·ft
(b) $T_1 = 44.7$ lb·ft, $T_2 = 29.1$ lb·ft
(c) $\phi_B = 0.1103$ rad

P6.45 (a) $T_{0,allow} = 14.02$ kip·in.
(b) $\tau_1 = 7.27$ ksi
(c) $\tau_2 = 1.082$ ksi

P6.47 (a) $\tau_1 = 36.0$ MPa
(b) $\tau_3 = 23.3$ MPa
(c) $\phi_E = -0.0208$ rad
(d) $\phi_C = 0.0776$ rad

P6.49 $\tau = 40.4$ MPa

P6.51 $r_{min} = 0.25$ in.

P6.53 $T_{max} = 310$ N·m

P6.55 (a) $T_{max} = 11.23$ kip·in.
(b) $\phi = 0.0988$ rad

P6.57 (a) $\tau_1 = 103.9$ MPa, $\tau_2 = 68.0$ MPa, $\tau_3 = 85.8$ MPa
(b) $\phi_1 = 0.223$ rad, $\phi_2 = 0.0907$ rad, $\phi_3 = 0.1399$ rad

P6.59 $t_{min} = 0.0726$ in.

P6.61 $\tau_A = 41.8$ MPa, $\tau_B = 83.7$ MPa, $\tau_C = 50.2$ MPa

Chapter 7

P7.1 (a) $V = w_0(L - x)$;

$$M = -\frac{w_0}{2}(L^2 + x^2) + w_0 L x$$

P7.3 (a) $0 \le x < a$: $V = -w_a x$, $M = -\dfrac{w_a}{2}x^2$

$a \le x < a + b$: $V = -(w_a - w_b)a - w_b x$,

$$M = -\frac{w_b}{2}x^2 - (w_a - w_b)ax + \frac{(w_a - w_b)a^2}{2}$$

P7.5 (a) $0 \le x < a$: $V = -P$, $M = -Px$
$a \le x < a + b$: $V = -P$, $M = M_B - Px$

P7.7 (a) $A_y = \dfrac{w_0}{6(a+b)}[a^2 + 3ab + 3b^2]$

$0 \le x < a$:

$$V = A_y - \frac{w_0 x^2}{2a},\ M = A_y x - \frac{w_0 x^3}{6a}$$

$a \le x < a + b$:

$$V = A_y - \frac{w_0 a}{2} - w_0(x - a),$$

$$M = A_y x - \frac{w_0}{2}(x - a)^2 - \frac{w_0 a}{2}\left(x - \frac{2a}{3}\right)$$

P7.9

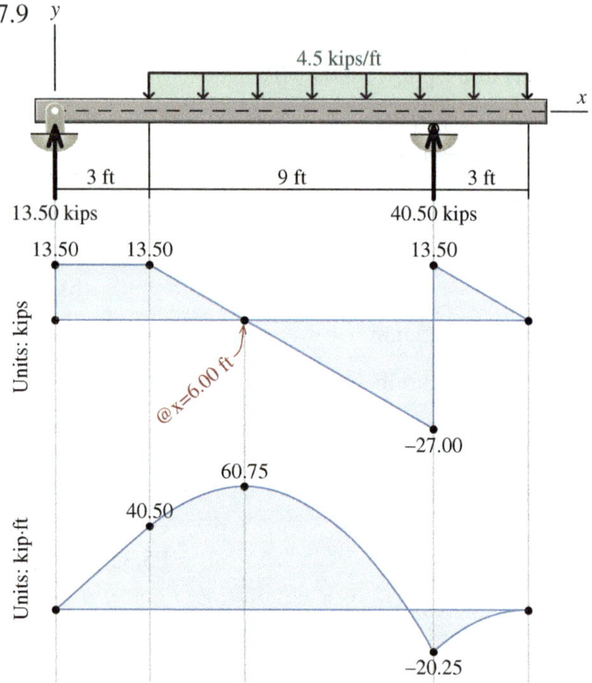

P7.11

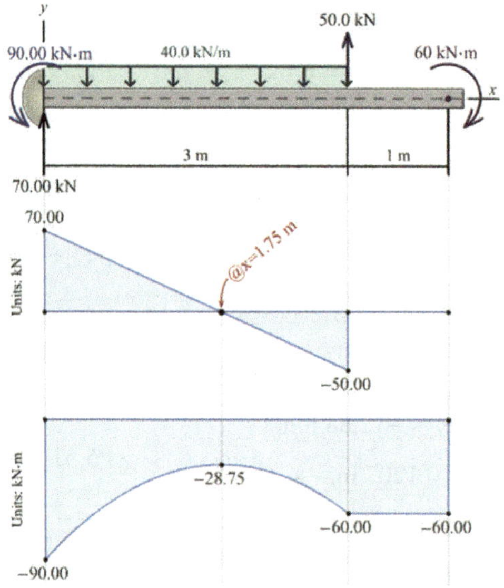

P7.13

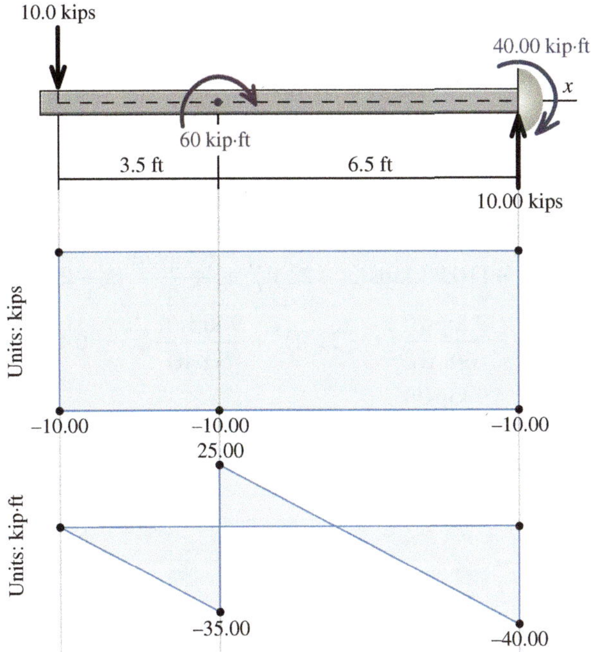

P7.15

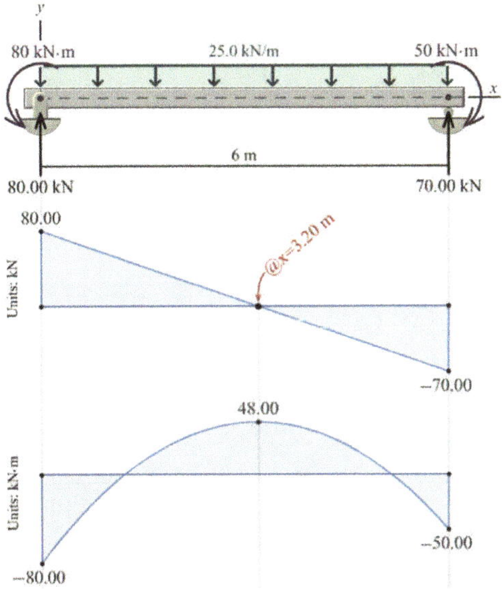

P7.17 (a) $V_{max} = 16.50$ kips
(b) $M_{max} = 33.0$ kip · ft

P7.19 (a) $V = 85.5$ kN, $M = 30.4$ kN · m
(b) $V = -74.5$ kN, $M = 52.4$ kN · m

P7.21 (a) $V = 93.7$ kN, $M = 23.9$ kN · m
(b) $V = -125.1$ kN, $M = 75.7$ kN · m

P7.23 $V_{max} = 20.0$ kips, $M_{max} = 119.3$ kip · ft

P7.25 $V_{max} = 55.0$ kN, $M_{max} = -50.0$ kN · m

P7.27 $V_{max} = -245$ kN, $M_{max} = -208$ kN · m

P7.29

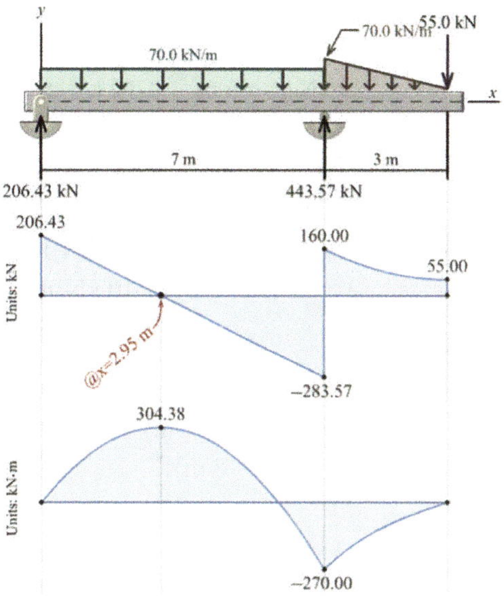

P7.31

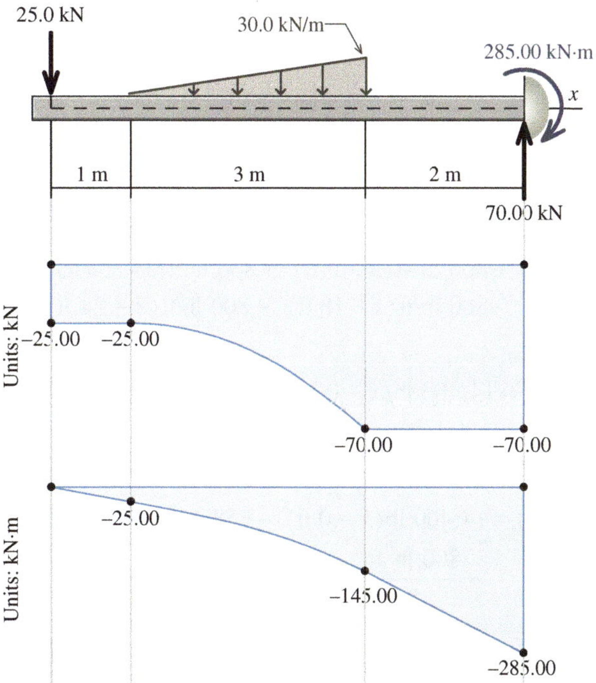

P7.33 (a)

$$w(x) = -10\text{ kN}\langle x - 0\text{ m}\rangle^{-1} + 29\text{ kN}\langle x - 2.5\text{ m}\rangle^{-1} - 35\text{ kN}\langle x - 5.5\text{ m}\rangle^{-1} + 16\text{ kN}\langle x - 7.5\text{ m}\rangle^{-1}$$

(b)

$$V(x) = -10\text{ kN}\langle x - 0\text{ m}\rangle^{0} + 29\text{ kN}\langle x - 2.5\text{ m}\rangle^{0} - 35\text{ kN}\langle x - 5.5\text{ m}\rangle^{0} + 16\text{ kN}\langle x - 7.5\text{ m}\rangle^{0}$$

$$M(x) = -10\text{ kN}\langle x - 0\text{ m}\rangle^{1} + 29\text{ kN}\langle x - 2.5\text{ m}\rangle^{1} - 35\text{ kN}\langle x - 5.5\text{ m}\rangle^{1} + 16\text{ kN}\langle x - 7.5\text{ m}\rangle^{1}$$

P7.35 (a)

$$w(x) = -5\text{ kN}\langle x-0\text{ m}\rangle^{-1} + 20\text{ kN}\cdot\text{m}\langle x-3\text{ m}\rangle^{-2} + 5\text{ kN}\langle x-6\text{ m}\rangle^{-1} + 10\text{ kN}\cdot\text{m}\langle x-6\text{ m}\rangle^{-2}$$

(b)

$$V(x) = -5\text{ kN}\langle x-0\text{ m}\rangle^{0} + 20\text{ kN}\cdot\text{m}\langle x-3\text{ m}\rangle^{-1} + 5\text{ kN}\langle x-6\text{ m}\rangle^{0} + 10\text{ kN}\cdot\text{m}\langle x-6\text{ m}\rangle^{-1}$$

$$M(x) = -5\text{ kN}\langle x-0\text{ m}\rangle^{1} + 20\text{ kN}\cdot\text{m}\langle x-3\text{ m}\rangle^{0} + 5\text{ kN}\langle x-6\text{ m}\rangle^{1} + 10\text{ kN}\cdot\text{m}\langle x-6\text{ m}\rangle^{0}$$

P7.37 (a)

$$w(x) = 83\text{ kN}\langle x-0\text{ m}\rangle^{-1} - 25\text{ kN/m}\langle x-0\text{ m}\rangle^{0} + 25\text{ kN/m}\langle x-4\text{ m}\rangle^{0} - 32\text{ kN}\langle x-6\text{ m}\rangle^{-1} + 49\text{ kN}\langle x-8\text{ m}\rangle^{-1}$$

(b)

$$V(x) = 83\text{ kN}\langle x-0\text{ m}\rangle^{0} - 25\text{ kN/m}\langle x-0\text{ m}\rangle^{1} + 25\text{ kN/m}\langle x-4\text{ m}\rangle^{1} - 32\text{ kN}\langle x-6\text{ m}\rangle^{0} + 49\text{ kN}\langle x-8\text{ m}\rangle^{0}$$

$$M(x) = 83\text{ kN}\langle x-0\text{ m}\rangle^{1} - \frac{25\text{ kN/m}}{2}\langle x-0\text{ m}\rangle^{2} + \frac{25\text{ kN/m}}{2}\langle x-4\text{ m}\rangle^{2} - 32\text{ kN}\langle x-6\text{ m}\rangle^{1} + 49\text{ kN}\langle x-8\text{ m}\rangle^{1}$$

P7.39 (a)

$$w(x) = 14{,}400\text{ lb}\langle x-0\text{ ft}\rangle^{-1} - 158{,}400\text{ lb}\cdot\text{ft}\langle x-0\text{ ft}\rangle^{-2} - 800\text{ lb}\cdot\text{ft}\langle x-0\text{ ft}\rangle^{0} + 800\text{ lb/ft}\langle x-12\text{ ft}\rangle^{0} - 800\text{ lb/ft}\langle x-18\text{ ft}\rangle^{0} + 800\text{ lb/ft}\langle x-24\text{ ft}\rangle^{0}$$

(b)

$$V(x) = 14{,}400\text{ lb}\langle x-0\text{ ft}\rangle^{0} - 158{,}400\text{ lb}\cdot\text{ft}\langle x-0\text{ ft}\rangle^{-1} - 800\text{ lb}\cdot\text{ft}\langle x-0\text{ ft}\rangle^{1} + 800\text{ lb/ft}\langle x-12\text{ ft}\rangle^{1} - 800\text{ lb/ft}\langle x-18\text{ ft}\rangle^{1} + 800\text{ lb/ft}\langle x-24\text{ ft}\rangle^{1}$$

$$M(x) = 14{,}400\text{ lb}\langle x-0\text{ ft}\rangle^{1} - 158{,}400\text{ lb}\cdot\text{ft}\langle x-0\text{ ft}\rangle^{0} - \frac{800\text{ lb}\cdot\text{ft}}{2}\langle x-0\text{ ft}\rangle^{2} + \frac{800\text{ lb/ft}}{2}\langle x-12\text{ ft}\rangle^{2} - \frac{800\text{ lb/ft}}{2}\langle x-18\text{ ft}\rangle^{2} + \frac{800\text{ lb/ft}}{2}\langle x-24\text{ ft}\rangle^{2}$$

P7.41 (a)

$$w(x) = 57.27\text{ kips}\langle x-0\text{ ft}\rangle^{-1} - 6\text{ kips/ft}\langle x-0\text{ ft}\rangle^{0} + 110.73\text{ kips}\langle x-22\text{ ft}\rangle^{-1} + 6\text{ kips/ft}\langle x-22\text{ ft}\rangle^{0} - \frac{9\text{ kips/ft}}{8\text{ ft}}\langle x-22\text{ ft}\rangle^{1} + \frac{9\text{ kips/ft}}{8\text{ ft}}\langle x-30\text{ ft}\rangle^{1} + 9\text{ kips/ft}\langle x-30\text{ ft}\rangle^{0}$$

(b)

$$V(x) = 57.27\text{ kips}\langle x-0\text{ ft}\rangle^{0} - 6\text{ kips/ft}\langle x-0\text{ ft}\rangle^{1} + 110.73\text{ kips}\langle x-22\text{ ft}\rangle^{0} + 6\text{ kips/ft}\langle x-22\text{ ft}\rangle^{1} - \frac{9\text{ kips/ft}}{2(8\text{ ft})}\langle x-22\text{ ft}\rangle^{2} + \frac{9\text{ kips/ft}}{2(8\text{ ft})}\langle x-30\text{ ft}\rangle^{2} + 9\text{ kips/ft}\langle x-30\text{ ft}\rangle^{1}$$

$$M(x) = 57.27\text{ kips}\langle x-0\text{ ft}\rangle^{1} - \frac{6\text{ kips/ft}}{2}\langle x-0\text{ ft}\rangle^{2} + 110.73\text{ kips}\langle x-22\text{ ft}\rangle^{1} + \frac{6\text{ kips/ft}}{2}\langle x-22\text{ ft}\rangle^{2} - \frac{9\text{ kips/ft}}{6(8\text{ ft})}\langle x-22\text{ ft}\rangle^{3} + \frac{9\text{ kips/ft}}{6(8\text{ ft})}\langle x-30\text{ ft}\rangle^{3} + \frac{9\text{ kips/ft}}{2}\langle x-30\text{ ft}\rangle^{2}$$

(c)

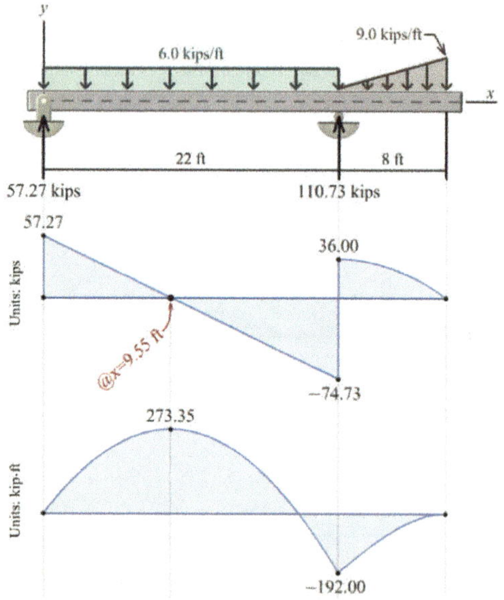

P7.43 (a)

$$w(x) = -9\text{ kN}\cdot\text{m}\langle x-0\text{ m}\rangle^{-2} + 21\text{ kN}\langle x-1\text{ m}\rangle^{-1} - 18\text{ kN/m}\langle x-1\text{ m}\rangle^{0} + \frac{18\text{ kN/m}}{3\text{ m}}\langle x-1\text{ m}\rangle^{1} - \frac{18\text{ kN/m}}{3\text{ m}}\langle x-4\text{ m}\rangle^{1} + 6\text{ kN}\langle x-4\text{ m}\rangle^{-1}$$

(b)

$$V(x) = -9\text{ kN}\cdot\text{m}\langle x-0\text{ m}\rangle^{-1} + 21\text{ kN}\langle x-1\text{ m}\rangle^{0} - 18\text{ kN/m}\langle x-1\text{ m}\rangle^{1} + \frac{18\text{ kN/m}}{2(3\text{ m})}\langle x-1\text{ m}\rangle^{2} - \frac{18\text{ kN/m}}{2(3\text{ m})}\langle x-4\text{ m}\rangle^{2} + 6\text{ kN}\langle x-4\text{ m}\rangle^{0}$$

$$M(x) = -9\text{ kN}\cdot\text{m}\langle x-0\text{ m}\rangle^{0} + 21\text{ kN}\langle x-1\text{ m}\rangle^{1} - \frac{18\text{ kN/m}}{2}\langle x-1\text{ m}\rangle^{2} + \frac{18\text{ kN/m}}{6(3\text{ m})}\langle x-1\text{ m}\rangle^{3} - \frac{18\text{ kN/m}}{6(3\text{ m})}\langle x-4\text{ m}\rangle^{3} + 6\text{ kN}\langle x-4\text{ m}\rangle^{1}$$

(c)

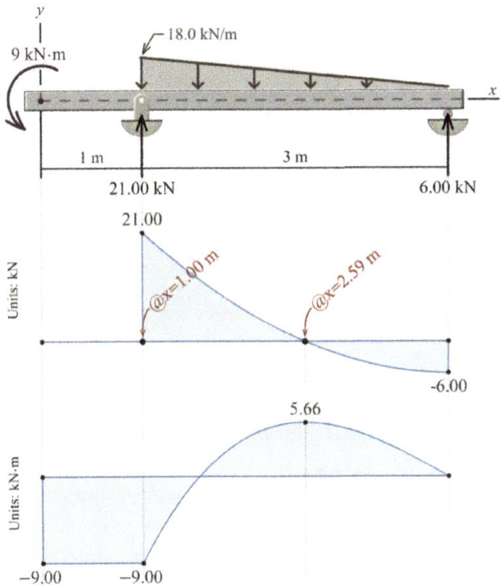

(c)

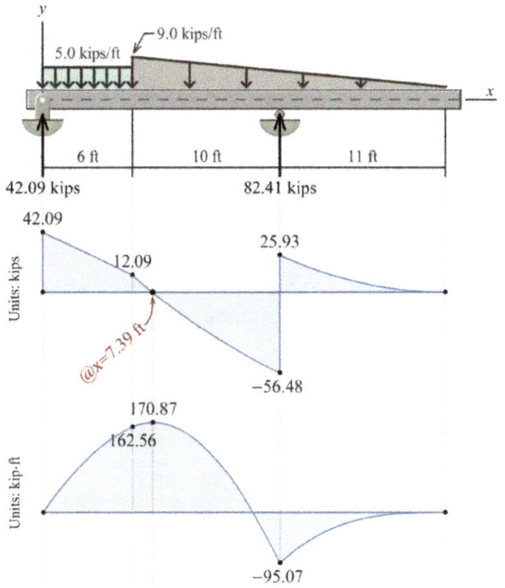

P7.45

(a) $$w(x) = 42.09 \text{ kips}\langle x - 0 \text{ ft}\rangle^{-1} - 5 \text{ kips/ft}\langle x - 0 \text{ ft}\rangle^{0} + 5 \text{ kips/ft}\langle x - 6 \text{ ft}\rangle^{0} - 9 \text{ kips/ft}\langle x - 6 \text{ ft}\rangle^{0} + \frac{9 \text{ kips/ft}}{21 \text{ ft}}\langle x - 6 \text{ ft}\rangle^{1} + 82.41 \text{ kips}\langle x - 16 \text{ ft}\rangle^{-1} - \frac{9 \text{ kips/ft}}{21 \text{ ft}}\langle x - 27 \text{ ft}\rangle^{1}$$

(b) $$V(x) = 42.09 \text{ kips}\langle x - 0 \text{ ft}\rangle^{0} - 5 \text{ kips/ft}\langle x - 0 \text{ ft}\rangle^{1} + 5 \text{ kips/ft}\langle x - 6 \text{ ft}\rangle^{1} - 9 \text{ kips/ft}\langle x - 6 \text{ ft}\rangle^{1} + \frac{9 \text{ kips/ft}}{2(21 \text{ ft})}\langle x - 6 \text{ ft}\rangle^{2} + 82.41 \text{ kips}\langle x - 16 \text{ ft}\rangle^{0} - \frac{9 \text{ kips/ft}}{2(21 \text{ ft})}\langle x - 27 \text{ ft}\rangle^{2}$$

$$M(x) = 42.09 \text{ kips}\langle x - 0 \text{ ft}\rangle^{1} - \frac{5 \text{ kips/ft}}{2}\langle x - 0 \text{ ft}\rangle^{2} + \frac{5 \text{ kips/ft}}{2}\langle x - 6 \text{ ft}\rangle^{2} - \frac{9 \text{ kips/ft}}{2}\langle x - 6 \text{ ft}\rangle^{2} + \frac{9 \text{ kips/ft}}{6(21 \text{ ft})}\langle x - 6 \text{ ft}\rangle^{3} + 82.41 \text{ kips}\langle x - 16 \text{ ft}\rangle^{1} - \frac{9 \text{ kips/ft}}{6(21 \text{ ft})}\langle x - 27 \text{ ft}\rangle^{3}$$

P7.47 (a)

$$w(x) = -30 \text{ kN/m}\langle x - 0 \text{ m}\rangle^{0} - 40 \text{ kN/m}\langle x - 0 \text{ m}\rangle^{0} + \frac{40 \text{ kN/m}}{7.0 \text{ m}}\langle x - 0 \text{ m}\rangle^{1} + 234.24 \text{ kN}\langle x - 1.5 \text{ m}\rangle^{-1} + 30 \text{ kN/m}\langle x - 7 \text{ m}\rangle^{0} - \frac{40 \text{ kN/m}}{7.0 \text{ m}}\langle x - 7 \text{ m}\rangle^{1} + 215.76 \text{ kN}\langle x - 7 \text{ m}\rangle^{-1} - 50 \text{ kN/m}\langle x - 7.0 \text{ m}\rangle^{0} + 50 \text{ kN/m}\langle x - 9.0 \text{ m}\rangle^{0}$$

(b)

$$V(x) = -30 \text{ kN/m}\langle x - 0 \text{ m}\rangle^{1} - 40 \text{ kN/m}\langle x - 0 \text{ m}\rangle^{1} + \frac{40 \text{ kN/m}}{2(7.0 \text{ m})}\langle x - 0 \text{ m}\rangle^{2} + 234.24 \text{ kN}\langle x - 1.5 \text{ m}\rangle^{0} + 30 \text{ kN/m}\langle x - 7 \text{ m}\rangle^{1} - \frac{40 \text{ kN/m}}{2(7.0 \text{ m})}\langle x - 7 \text{ m}\rangle^{2} + 215.76 \text{ kN}\langle x - 7 \text{ m}\rangle^{0} - 50 \text{ kN/m}\langle x - 7.0 \text{ m}\rangle^{1} + 50 \text{ kN/m}\langle x - 9.0 \text{ m}\rangle^{1}$$

$$M(x) = -\frac{30 \text{ kN/m}}{2}\langle x - 0 \text{ m}\rangle^{2} - \frac{40 \text{ kN/m}}{2}\langle x - 0 \text{ m}\rangle^{2} + \frac{40 \text{ kN/m}}{6(7.0 \text{ m})}\langle x - 0 \text{ m}\rangle^{3} + 234.24 \text{ kN}\langle x - 1.5 \text{ m}\rangle^{1} + \frac{30 \text{ kN/m}}{2}\langle x - 7 \text{ m}\rangle^{2} - \frac{40 \text{ kN/m}}{6(7.0 \text{ m})}\langle x - 7 \text{ m}\rangle^{3} + 215.76 \text{ kN}\langle x - 7 \text{ m}\rangle^{1} - \frac{50 \text{ kN/m}}{2}\langle x - 7.0 \text{ m}\rangle^{2} + \frac{50 \text{ kN/m}}{2}\langle x - 9.0 \text{ m}\rangle^{2}$$

(c)

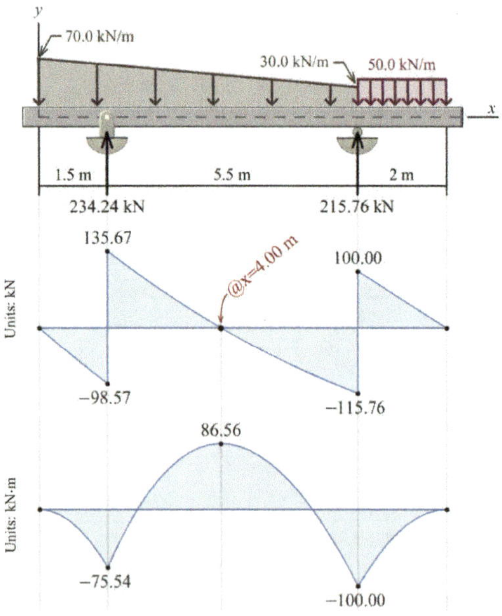

Chapter 8

P8.1 $\sigma = 1.979$ ksi

P8.3 $\sigma = 443$ MPa

P8.5 (a) $\bar{y} = 5.84$ in. above the bottom surface, $I_z = 256.60$ in.^{4}, $S_z = 43.97$ in.^{3}
(b) at H, $\sigma_x = 8.07$ ksi (T)
(c) at K, $\sigma_x = 4.55$ ksi (C)
(d) $\sigma_x = 12.28$ ksi (T)

P8.7 $\sigma_x = 27.6$ MPa

P8.9 $\sigma_x = 5.06$ ksi (T)

P8.11 At H: $\sigma_x = 12.54$ MPa (C),
At K: $\sigma_x = 16.96$ MPa (T)

P8.13 $\sigma_x = 2{,}300$ psi (T)
$\sigma_x = 2{,}900$ psi (C)

P8.15 $\sigma_x = 6{,}630$ psi at C

P8.17 $\sigma_x = 141.0$ MPa at A

P8.19 $\sigma_x = 26.9$ ksi at C

P8.21 $\sigma_x = 142.7$ MPa

P8.23 (a) $\sigma_x = 1{,}006$ psi (T)
(b) $\sigma_x = 798$ psi (C)

P8.25 $w_0 = 70.2$ kN/m

P8.27 $d_{min} = 24.0$ mm

P8.29 $b_{min} = 5.55$ in.

P8.31 $w_0 = 638$ lb/ft

P8.33 (a) answer not given
(b) W21 × 50

P8.35 (a) answer not given
(b) W360 × 44

P8.37 (a) $\sigma_{wood} = 768$ psi,
$\sigma_{steel} = 12{,}380$ psi
(b) $w_{max} = 625$ lb/ft

P8.39 $M_z = 3.64$ kN·m

P8.41 $P_{max} = 10{,}960$ lb

P8.43 $M_{allow} = 73.1$ kN·m

P8.45 (a) $\bar{y} = 14.55$ in.
(b) $\sigma_H = 924$ psi (C)
(c) $\sigma_K = 21{,}200$ psi (T)

P8.47 $\sigma_H = 3{,}840$ psi (T), $\sigma_K = 3{,}360$ psi (C)

P8.49 $P_{max} = 13.48$ kN

P8.51 $P_{max} = 59.9$ kN

P8.53 $\sigma_H = 49.9$ MPa (T), $\sigma_K = 23.2$ MPa (C)

P8.55 $\sigma_{max} = 7{,}550$ psi (C)

P8.57 $\varepsilon_H = -201\ \mu\varepsilon$, $\varepsilon_K = 721\ \mu\varepsilon$

P8.59 $P = 29.1$ kips, $Q = 36.7$ kips

P8.61 (a) $\sigma_B = 972$ psi(C)
(b) $\sigma_C = 972$ psi (T)
(c) $\beta = 6.87°$

P8.63 (a) $\sigma_A = 22.9$ MPa (T)
(b) $\sigma_D = 9.01$ MPa (C)

P8.65 $M_z = 1{,}333$ N·m

P8.67 (a) $I_y = 100.6$ in.^{4}, $I_z = 277$ in.^{4}
(b) $I_{yz} = 94.2$ in.^{4}
(c) $I_{p1} = 318$ in.^{4}, $I_{p2} = 59.8$ in.^{4}, $\theta_p = -23.4°$
(d) $\sigma = 334$ psi (C)

P8.69 $\sigma_{max} = 17.31$ ksi

P8.71 $b_{min} = 14.08$ mm

P8.73 $P_{max} = 1{,}525$ N

P8.75 $P_{max} = 1{,}572$ N

P8.77 $\sigma_A = 47.8$ MPa (C), $\sigma_B = 34.9$ MPa (T)

P8.79 $P_{max} = 6.89$ kips

P8.81 $\sigma_A = 11{,}560$ psi (C), $\sigma_B = 7{,}930$ psi (T)

P8.83 (a) $r_n = 112.5$ mm
(b) $\sigma_A = 76.2$ MPa (C), $\sigma_B = 84.3$ MPa (T)

P8.85 (a) 1.9382 mm
(b) $M_{max} = 2{,}390$ N·m

P8.87 (a) 4.684 mm
(b) $M_{max} = 15.46$ kN·m

P8.89 (a) 0.08757 in.
(b) $\sigma = 9.30$ ksi (C)

Chapter 9

P9.1 (b) $F_{1A} = 1{,}822$ lb (C), $F_{1B} = 844$ lb (C)
(c) $F_H = 978$ lb directed from B to A required for equilibrium of area (1).

P9.3 (b) $F_{1A} = 21.9$ (T), $F_{1B} = 82.5$ kN (T)
(c) $F_H = 60.5$ kN directed from B to A required for equilibrium of area (1).

P9.5 $\tau_A = 28.3$ psi, $\tau_B =$ not given, $\tau_C =$ not given, $\tau_D = 64.8$ psi

P9.7 (a) $\tau_H = 350$ kPa
(b) $\tau_K = 219$ kPa
(c) $w_{max} = 20.3$ kN/m

P9.9 (a) $\tau_{max} = 34.5$ psi
(c) $\sigma_x = 455$ psi (T)

P9.11 $\tau_{max} = 483$ psi

P9.13 (a) $Q = 1.3638$ in.^{3}
(b) $\tau_{max} = 256$ psi

P9.15 (a) $Q_H = 6{,}028$ mm^3
(b) $\tau_H = 123.7$ psi
(c) $\tau_{max} = 173.7$ psi

P9.17 (a) $\tau_A = 327$ psi, $\tau_B = 449$ psi
(b) $\tau_C = 341$ psi
(c) $\tau_{max} = 484$ psi

P9.19 (a) $\tau_H = 8.25$ MPa
(b) $\tau_K = 11.70$ MPa
(c) $\tau_{max} = 13.07$ MPa

P9.21 (a) $Q_H = 12.29$ in.^{3}
(b) $P_{max} = 62.6$ kips

P9.23 $w_{max} = 19.66$ kips/ft

P9.25 (a) $\tau_{max} = 25.6$ MPa
(b) $\sigma = 121.0$ MPa (T)
(c) $\sigma = 41.5$ MPa (C)

P9.27 (a) $P_{max} = 2{,}990$ N
(b) $P_{max} = 7.81$ kN, $s_{max} = 76.5$ mm

P9.29 $s_{max} = 52.8$ mm

P9.31 (a) $V_{max} = 6.85$ kN
(b) $s_{max} = 281$ mm

P9.33 (a) $s_{max} = 6.37$ in.
(b) Wide-flange shape alone: $M_{allow} = 175.9$ kip·ft, Wide-flange shape with cover plate: $M_{allow} = 211$ kip·ft, percentage increase = 20.0%

P9.35 (a) $q = 2{,}170$ lb/in., $\tau = 4{,}350$ psi
(b) $q = 1{,}012$ lb/in., $\tau = 1{,}264$ psi
(c) $q = 524$ lb/in., $\tau = 656$ psi

P9.37 $q = 1{,}280$ lb/in.

P9.39 (a) $\tau_A = 0$ MPa
(b) $\tau = 6.96$ MPa
(c) $\tau = 8.56$ MPa
(d) $\tau = 22.1$ MPa
(e) $\tau = 45.8$ MPa
(f) $\tau = 56.6$ MPa

P9.41 (a) $q_A = 0$ lb/in., $q_B = 5{,}320$ lb/in., $q_C = 6{,}380$ lb/in., $q_D = 8{,}260$ lb/in., $q_E = 7{,}570$ lb/in., $q_F = 2{,}750$ lb/in., $q_G = 0$ lb/in.
(b) 3,320 lb

P9.43 $\tau_A = 0$ psi, $\tau_B = 1{,}210$ psi, $\tau_C = 236$ psi, $\tau_D = 1{,}497$ psi, $\tau_E = 2{,}940$ psi, $\tau_F = 1{,}625$ psi, $\tau_G = 0$ psi

P9.45 $\tau_A = 0$ MPa, $\tau_B = 9.77$ MPa, $\tau_C = 8.87$ MPa, $\tau_D = 0$ MPa

P9.47 (a) $\tau_A = 0$ MPa, $\tau_B = 15.03$ MPa, $\tau_C = 23.7$ MPa, $\tau_D = 15.03$ MPa, $\tau_E = 0$ MPa
(b) 14.43 kN

P9.49 $\tau_{1\text{-}1} = 1.266$ ksi, $\tau_{2\text{-}2} = 0.762$ ksi, $\tau_{3\text{-}3} = 2.43$ ksi, $\tau_{max} = 4.30$ ksi

P9.51 $q_{max} = 742$ lb/in.

P9.53 $e = 1.125$ in.

P9.55 (a) $e = 32.7$ mm
(b) stiffener: $\tau_A = 0$ MPa, $\tau_B = 19.82$ MPa
flange: $\tau_B = 11.89$ MPa, $\tau_C = 70.0$ MPa
web: $\tau_C = 116.7$ MPa, $\tau_D = 137.9$ MPa

P9.57 $e = 30.3$ mm

P9.59 $e = 18.75$ mm

P9.61 $e = 0$

P9.63 $e = \left(\dfrac{\pi + 3}{\pi + 4}\right)2r$

P9.65 $e = 20.2$ mm

Chapter 10

P10.1 (a) $v = -\dfrac{M_0 x^2}{2EI}$

(b) $v_B = -\dfrac{M_0 L^2}{2EI}$

(c) $\theta_B = -\dfrac{M_0 L}{EI}$

P10.3 (a)
$$v = -\frac{w_0}{120LEI}(x^5 - 5L^4 x + 4L^5)$$

(b) $v_A = -\dfrac{w_0 L^4}{30EI}$

(c) $\theta_A = \dfrac{w_0 L^3}{24EI}$

P10.5 (a)
$$v = -\frac{M_0 x}{6LEI}(x^2 - 3Lx + 2L^2)$$

(b) $\theta_A = -\dfrac{M_0 L}{3EI}$

(c) $\theta_B = \dfrac{M_0 L}{6EI}$

(d) $v_{x=L/2} = -\dfrac{M_0 L^2}{16EI}$

P10.7 $v_B = -8.27$ mm

P10.9 (a)

$$v = -\frac{w_0 x^2}{120 L EI}[-x^3 + 5Lx^2 - 10L^2 x + 10L^3]$$

(b) $v_B = -\dfrac{w_0 L^4}{30EI}$

(c) $\theta_B = -\dfrac{w_0 L^3}{24EI}$

P10.11 (a)

$$v = -\frac{wx}{384EI}[16x^3 - 24Lx^2 + 9L^3]$$

$(0 \le x \le L/2)$

$$v = -\frac{wL}{384EI}[8x^3 - 24Lx^2 + 17L^2 x - L^3]$$

$(L/2 \le x \le L)$

(b) $v_B = -\dfrac{5wL^4}{768EI}$

P10.13 (a)

$$v = -\frac{w_0}{120LEI}(x^5 - 5L^4 x + 4L^5)$$

(b) $v_{max} = -\dfrac{w_0 L^4}{30EI}$

P10.15 (a) $v = -\dfrac{w_0 L^4}{\pi^4 EI}\sin\dfrac{\pi x}{L}$

(b) $v_{L/2} = -\dfrac{w_0 L^4}{\pi^4 EI}$

(c) $\theta_A = -\dfrac{w_0 L^3}{\pi^3 EI}$

(d) $A_y = B_y = \dfrac{w_0 L}{\pi} \uparrow$

P10.17 $v_D = 0.226$ in. ↓

P10.19 (a) $\theta_C = -0.00915$ rad
(b) $v_C = 8.15$ mm ↓

P10.21 $v_C = 27.3$ mm ↓

P10.23 (a) $\theta_A = -0.01174$ rad
(b) $v_{midspan} = 27.7$ mm ↓

P10.25 (a) $\theta_A = -0.00994$ rad
(b) $v_{midspan} = 0.712$ in. ↓

P10.27 (a) $v_A = 0.0407$ in. ↓
(b) $v_C = 0.0951$ in. ↓

P10.29 (a) $\theta_E = 0.01326$ rad
(b) $v_C = 0.858$ in. ↓

P10.31 (a) $\theta_B = 0.00575$ rad
(b) $v_A = 1.028$ in. ↓

P10.33 (a) $\theta_A = -0.00778$ rad
(b) $v_B = 0.717$ in. ↓

P10.35 (a) $v_A = 6.77$ mm ↑
(b) $v_C = 11.30$ mm ↓

P10.37 (a) $v_H = 7.50$ mm ↑
(b) $v_H = 4.00$ mm ↓
(c) $v_H = 9.33$ mm ↓
(d) $v_H = 12.00$ mm ↓

P10.39 $v_C = 0.584$ in. ↓

P10.41 (a) $v_B = 0.257$ in. ↓
(b) $v_C = 0.577$ in. ↓

P10.43 (a) $v_A = 1.089$ in. ↑
(b) $v_C = 0.435$ in. ↑

P10.45 (a) $v_A = 4.14$ mm ↓
(b) $v_C = 6.37$ mm ↓

P10.47 $v_B = 6.06$ mm ↓

P10.49 (a) $v_A = 1.520$ mm ↓
(b) $v_C = 13.30$ mm ↓

P10.51 $v_B = 2.89$ mm ↓

P10.53 (a) $v_A = 0.1967$ in. ↓
(b) $v_C = 0.1765$ in. ↓

P10.55 (a) $v_A = 0.733$ in. ↓
(b) $v_C = 0.214$ in. ↓

P10.57 $v_C = 41.0$ mm ↓

P10.59 $v_C = 1.325$ in. →

Chapter 11

P11.1 $M_0 = \dfrac{wL^2}{6}$ (CW)

P11.3 (a) $A_y = \dfrac{3wL}{8} \uparrow$, $B_y = \dfrac{5wL}{8} \uparrow$, $M_B = \dfrac{wL^2}{8}$(CW)

P11.5 $A_y = \dfrac{w_0 L}{24} \uparrow$

P11.7 (a) $A_y = C_y = \dfrac{P}{2}$, $M_A = \dfrac{PL}{8}$ (CCW), $M_C = \dfrac{PL}{8}$(CW)

(c) $v_B = -\dfrac{PL^3}{192EI}$

P11.9 (a) $A_y = C_y = \dfrac{w_0 L}{2}$

$M_A = \dfrac{5w_0 L^2}{24}$ (CCW), $M_C = \dfrac{5w_0 L^2}{24}$(CW)

(b) $v_B = -\dfrac{7w_0 L^4}{240EI}$

P11.11 (a) $A_y = \dfrac{41wL}{128} \uparrow$, $C_y = \dfrac{23wL}{128} \uparrow$, $M_C = \dfrac{7wL^2}{128}$(CW)

P11.13 (a) $A_y = 225$ kN ↓,
$M_A = 375$ kN·m (CW),
$B_y = 225$ kN ↑
(b) $v_C = 23.4$ mm ↓

P11.15 (a) $A_y = 52.8$ kips ↑,
$B_y = 43.2$ kips ↑,
$M_B = 179.2$ kip·ft (CW)
(b) $v = 0.285$ in. ↓

P11.17 (a) $A_y = 306$ kN ↑,
$C_y = 495$ kN ↑,
$D_y = 81.0$ kN ↓
(b) $v_B = 6.48$ mm ↓

P11.19 (a) $B_y = 65.9$ kips ↑,
$D_y = 19.13$ kips ↑,
$M_D = 105.9$ kip·ft (CW)
(b) $v_C = 0.211$ in. ↓

P11.21 (a) $B_y = 245$ kN ↑,
$C_y = 120.0$ kN ↑,
$D_y = 5.00$ kN ↓
(b) $v_A = 14.40$ mm ↓

P11.23 (a) $P = 12.50$ kips
(b) $M = 310$ kip·ft

P11.25 $A_y = 17wL/16 \uparrow$, $B_y = 7wL/16 \uparrow$, $M_B = wL^2/16$ (CW)

P11.27 $A_y = \dfrac{57wL}{64} \uparrow$, $C_y = \dfrac{7wL}{64} \uparrow$

$M_A = \dfrac{9wL^2}{32}$ (CCW)

P11.29 $B_y = 3wL/2 \uparrow$

P11.31 (a) $A_y = 160.0$ kN $\downarrow$, $B_y = 480$ kN $\uparrow$, $C_y = 220$ kN $\uparrow$
(b) $\sigma_{max} = 235$ MPa

P11.33 (a) $B_y = 142.0$ kips $\uparrow$, $C_y = 90.0$ kips $\uparrow$,
$M_C = 336$ kip·ft (CW)
(b) $v_B = 2.16$ in. $\downarrow$

P11.35 (a) $A_y = 2{,}750$ lb $\downarrow$, $B_y = 10{,}750$ lb $\downarrow$,
$M_A = 25{,}000$ lb·in. (CW)
(b) $\sigma_{max} = 12{,}670$ psi

P11.37 (a) $B_y = 79.0$ lb $\uparrow$, $C_y = 157.5$ lb $\uparrow$, $E_y = 93.5$ lb $\uparrow$
(b) $\sigma_{max} = 14{,}290$ psi

P11.39 (a) $A_y = 52.8$ kN $\uparrow$, $C_y = 97.2$ kN $\uparrow$,
$M_A = 144$ kN·m (CCW), $M_C = 216$ kN·m (CW)
(b) $\sigma_{max} = 103.9$ MPa (at C)
(c) $v_B = 6.24$ mm $\downarrow$

P11.41 (a) $F_1 = 46.8$ kN (T)
(b) $\sigma_{max} = 48.9$ MPa
(c) $v_B = 6.39$ mm $\downarrow$

P11.43 (a) $A_y = 22.6$ kN $\uparrow$, $B_y = 95.7$ kN $\uparrow$, $C_y = 41.7$ kN $\uparrow$
(b) $\sigma_{max} = 143.6$ MPa

P11.45 (a) $A_y = 134.5$ kN $\uparrow$, $B_y = 178.8$ kN $\uparrow$,
$C_y = 13.31$ kN $\downarrow$
(b) $\sigma_{max} = 157.9$ MPa
(c) $v_B = 3.73$ mm $\downarrow$

P11.47 (a) $A_y = 4.43$ kips $\uparrow$
(b) $C_y = 2.07$ kips $\uparrow$

P11.49 (a) $v_D = 14.10$ mm $\downarrow$
(b) $v_B = 25.8$ mm $\downarrow$

Chapter 12

P12.1 (a) $\sigma_x = 3.74$ ksi, $\tau_{xy} = 6.65$ ksi
(b) $\sigma_x = 1.273$ ksi, $\tau_{xy} = -3.96$ ksi

P12.3 (a) $\sigma_x = -9.04$ MPa, $\tau_{xy} = 6.48$ MPa
(b) $\sigma_x = 10.45$ MPa, $\tau_{xy} = 7.56$ MPa

P12.5 (a) $\sigma_x = -547$ psi, $\tau_{xy} = -273$ psi
(b) $\sigma_x = 425$ psi, $\tau_{xy} = -263$ psi

P12.7 Point H: $\sigma_x = 19.67$ MPa, $\sigma_y = 0$ MPa, $\tau_{xy} = 2.74$ MPa.
Point K: $\sigma_x = 0$ MPa, $\sigma_y = -21.9$ MPa,
$\tau_{xy} = 1.827$ MPa

P12.9 (a) $\sigma_x = -805$ psi, $\tau_{xy} = 2{,}400$ psi
(b) $\sigma_x = 4{,}240$ psi, $\tau_{xy} = 2{,}900$ psi

P12.11 $\sigma_n = -42.8$ MPa, $\tau_{nt} = 140.3$ MPa

P12.13 $\sigma_n = 929$ psi, $\tau_{nt} = -2{,}430$ psi

P12.15 $\sigma_n = -4.77$ ksi, $\tau_{nt} = -6.69$ ksi

P12.17 $\sigma_n = -28.1$ MPa, $\tau_{nt} = -135.8$ MPa

P12.19 $\sigma_n = 11.63$ ksi, $\tau_{nt} = 4.76$ ksi

P12.21 $\sigma_n = -3.63$ MPa, $\tau_{nt} = 5.68$ MPa

P12.23 $\sigma_n = 112.8$ MPa, $\sigma_t = -58.8$ MPa, $\tau_{nt} = 34.2$ MPa

P12.25 $\sigma_x = -38.8$ MPa, $\sigma_y = 49.8$ MPa, $\tau_{xy} = -87.3$ MPa

P12.27

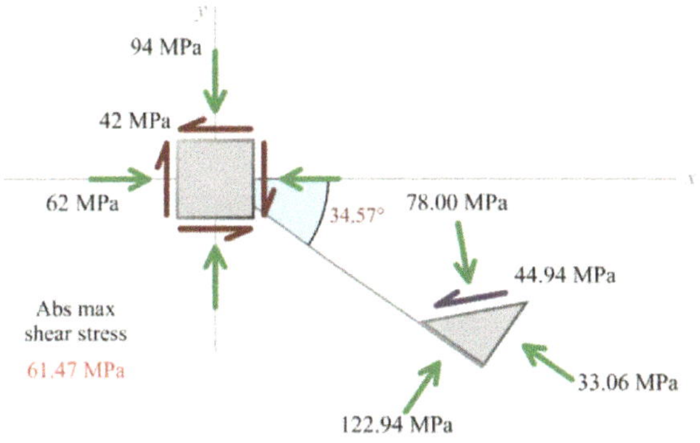

P12.29

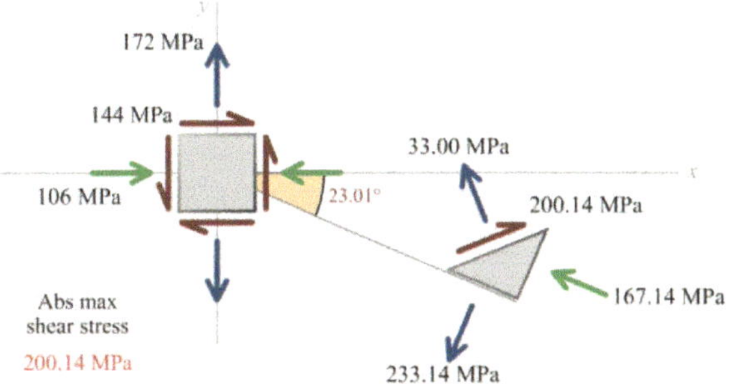

P12.31 (a) $\sigma_{p1} = 2.97$ ksi, $\sigma_{p2} = -18.97$ ksi,
$\theta_p = -32.9°$, $\tau_{max} = 10.97$ ksi
(c) $\tau_{abs\ max} = 10.97$ ksi

P12.33 (a) $\sigma_{p1} = 32.9$ ksi, $\sigma_{p2} = 5.13$ ksi,
$\theta_p = -42.1°$, $\tau_{max} = 13.87$ ksi
(c) $\tau_{abs\ max} = 16.44$ ksi

P12.35 (a) 25.6 MPa
(b) 23.1° clockwise from x axis

P12.37 $\sigma_y = -117.5$ psi

P12.39 $\tau_{xy} = -19.98$ ksi, $\sigma_{p1} = 27.0$ ksi, $\sigma_{p2} = -37.0$ ksi, $\sigma_{p3} = 0$

P12.41 (a) max. $\tau_{xy} = 116.1$ MPa
(b) $\sigma_{p1} = 175.0$ MPa, $\sigma_{p2} = -125.0$ MPa

P12.43

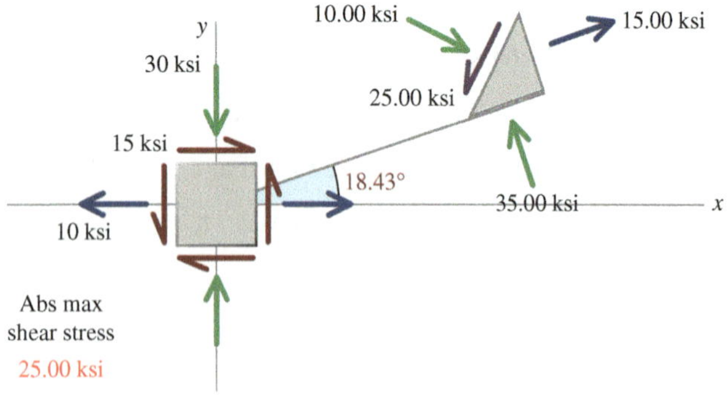

P12.45

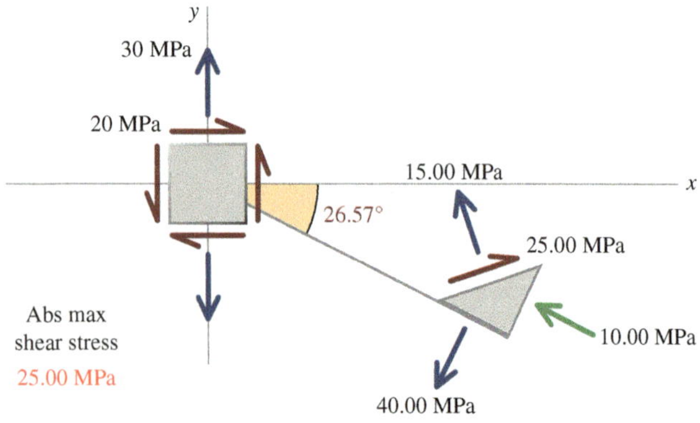

P12.47

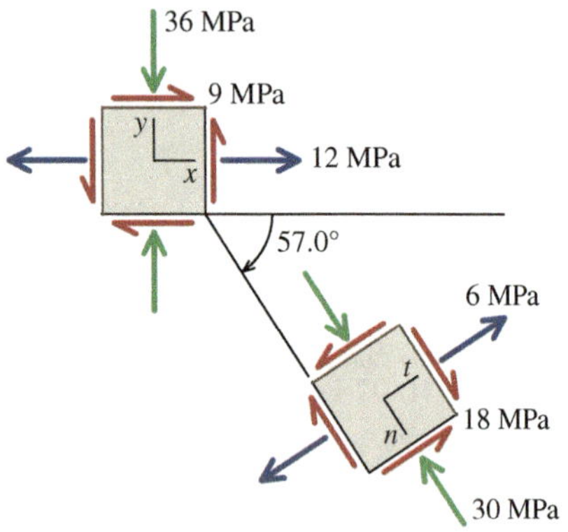

P12.49

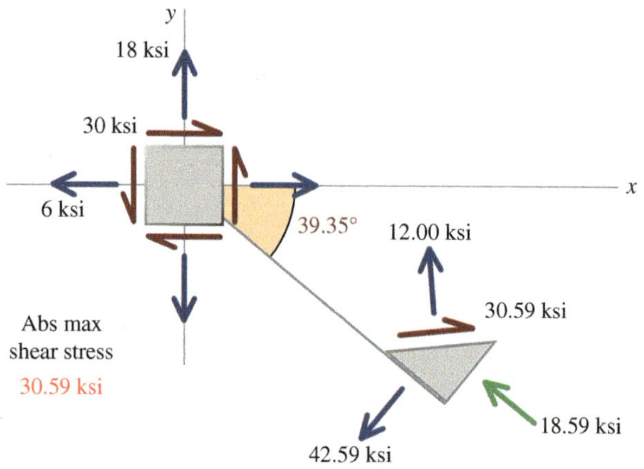

P12.51

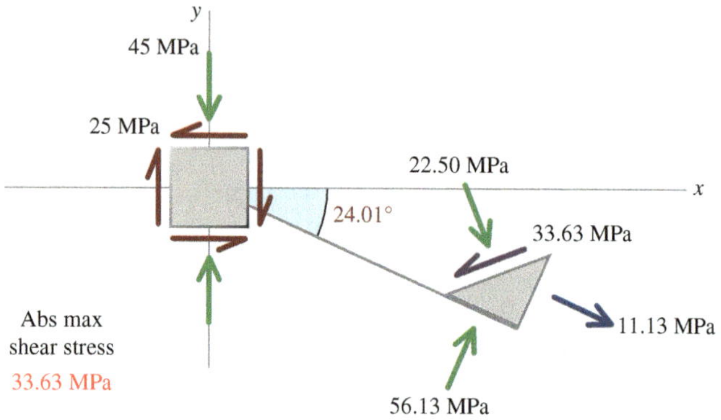

P12.53 (b) $\sigma_{p1} = 178.7$ MPa, $\sigma_{p2} = 43.3$ MPa, $\tau_{max} = 67.7$ MPa, $\theta_p = -38.6°$ (to σ_{p2})
(d) $\tau_{abs\ max} = 89.3$ MPa

P12.55 (b) $\sigma_{p1} = -659$ psi, $\sigma_{p2} = -2{,}540$ psi, $\tau_{max} = 941$ psi, $\theta_p = -23.1°$ (to σ_{p1})
(d) $\tau_{abs\ max} = 1{,}270$ psi

P12.57 (b) $\sigma_{p1} = 141.4$ MPa, $\sigma_{p2} = -47.4$ MPa, $\tau_{max} = 94.4$ MPa, $\theta_p = -24.9°$ (to σ_{p1})
(c) $\sigma_n = -34.5$ MPa, $\tau_{nt} = -47.6$ MPa
(d) $\tau_{abs\ max} = 94.4$ MPa

P12.59 (b) $\sigma_{p1} = 4{,}830$ psi, $\sigma_{p2} = -6{,}860$ psi, $\tau_{max} = 5{,}850$ psi, $\theta_p = 14.87°$ (to σ_{p1})
(c) $\sigma_n = -2{,}690$ psi, $\tau_{nt} = -5{,}600$ psi
(d) $\tau_{abs\ max} = 5{,}850$ psi

P12.61 (a) $\sigma = -19.00$ ksi, $\tau = 11.18$ ksi (arrow points to the right)
(b) $\sigma_x = -24.0$ ksi, $\sigma_y = -14.00$ ksi, $\tau_{xy} = 10.00$ ksi
(c) $\tau_{abs\ max} = 15.09$ ksi

P12.63 $T_{max} = 119.8$ N·m

P12.65 (a) $S = 7.92$ ksi
(b) $\sigma_n = 4.60$ ksi, $\tau_{nt} = 6.45$ ksi

P12.67 (a) $l = 0.3141$, $m = 0.5584$, $n = 0.7678$
(b) $S = 70.3$ MPa
(c) $\sigma_n = -25.2$ MPa, $\tau_{nt} = 65.6$ MPa

P12.69 (a) $I_1 = 26$, $I_2 = -470$, $I_3 = -8{,}259$
(b) $\sigma_{p1} = 32.6$ MPa, $\sigma_{p2} = 12.93$ MPa, $\sigma_{p3} = -19.57$ MPa, $\tau_{abs\ max} = 26.1$ MPa

P12.71 (a) $I_1 = 105$, $I_2 = -1{,}400$, $I_3 = -252{,}750$
(b) $\sigma_{p1} = 88.6$ MPa, $\sigma_{p2} = 62.2$ MPa, $\sigma_{p3} = -45.8$ MPa, $\tau_{abs\ max} = 67.2$ MPa
(c) $l_1 = 0.9187$, $m_1 = 0.2152$, $n_1 = 0.3312$

Chapter 13

P13.1 (a) $\varepsilon_{OA} = -243$ με
(b) $\varepsilon_{OC} = 349$ με

P13.3 (a) $\varepsilon_n = 352$ με
(b) $\varepsilon_t = -1{,}092$ με
(c) $\gamma_{nt} = 292$ μrad

P13.5 (a) $\varepsilon_n = 1{,}200$ με
(b) $\varepsilon_t = 370$ με
(c) $\gamma_{nt} = 250$ μrad

P13.7 $\varepsilon_n = 97.7$ με, $\varepsilon_t = -748$ με, $\gamma_{nt} = 1{,}799$ μrad

P13.9 $\varepsilon_n = -1{,}243$ με, $\varepsilon_t = -1{,}957$ με, $\gamma_{nt} = 338$ μrad

P13.11 $\varepsilon_{p1} = 70.8$ με, $\varepsilon_{p2} = -906$ με, $\gamma_{max} = 977$ μrad, $\theta_p = -37.1°$, $\gamma_{abs\ max} = 977$ μrad

P13.13 $\varepsilon_{p1} = 815$ με, $\varepsilon_{p2} = -575$ με, $\gamma_{max} = 1{,}390$ μrad, $\theta_p = -27.9°$, $\gamma_{abs\ max} = 1{,}390$ μrad

P13.15 $\varepsilon_{p1} = 1{,}039$ με, $\varepsilon_{p2} = 571$ με, $\gamma_{max} = 468$ μrad, $\theta_p = 24.2°$, $\gamma_{abs\ max} = 1{,}039$ μrad

P13.17 (a) $\varepsilon_x = -410$ με, $\varepsilon_y = -830$ με, $\gamma_{xy} = -340$ μrad
(b) $\gamma_{max} = 540$ μrad, $\gamma_{abs\ max} = 890$ μrad

P13.19 $\varepsilon_{p1} = 874$ με, $\varepsilon_{p2} = 476$ με, $\gamma_{max} = 398$ μrad, $\theta_p = -32.4°$, $\gamma_{abs\ max} = 874$ μrad

P13.21 $\varepsilon_{p1} = 709$ με, $\varepsilon_{p2} = 451$ με, $\gamma_{max} = 258$ μrad, $\theta_p = 17.77°$, $\gamma_{abs\ max} = 709$ μrad

P13.23 $\varepsilon_{p1} = 366$ με, $\varepsilon_{p2} = -46.2$ με, $\gamma_{max} = 412$ μrad, $\theta_p = -19.55°$, $\gamma_{abs\ max} = 412$ μrad

P13.25 (a) $\varepsilon_x = 410$ με, $\varepsilon_y = -330$ με, $\gamma_{xy} = -1{,}160$ μrad
(b) $\varepsilon_{p1} = 728$ με, $\varepsilon_{p2} = -648$ με, $\varepsilon_{p3} = -34.3$ με, $\gamma_{max} = 1{,}376$ μrad, $\theta_p = -28.7°$ (to ε_{p1})
(d) $\gamma_{abs\ max} = 1{,}376$ μrad

P13.27 (a) $\varepsilon_x = 525$ με, $\varepsilon_y = 415$ με, $\gamma_{xy} = 80$ μrad
(b) $\varepsilon_{p1} = 538$ με, $\varepsilon_{p2} = 402$ με, $\varepsilon_{p3} = -463$ με, $\gamma_{max} = 136.0$ μrad, $\theta_p = 18.01°$ (to ε_{p1})
(d) $\gamma_{abs\ max} = 1{,}001$ μrad

P13.29 (a) $\varepsilon_x = -360$ με, $\varepsilon_y = 510$ με, $\gamma_{xy} = 1{,}207$ μrad
(b) $\varepsilon_{p1} = 819$ με, $\varepsilon_{p2} = -669$ με, $\varepsilon_{p3} = -26.5$ με, $\gamma_{max} = 1{,}488$ μrad, $\theta_p = -27.1°$ (to ε_{p2})
(d) $\gamma_{abs\ max} = 1{,}488$ μrad

P13.31 (a) $\delta_{AB} = 0.669$ mm, $\delta_{AD} = 0.01181$ mm
(b) $\delta_{AC} = 0.603$ mm
(c) $\delta_{thick} = -0.00779$ mm

P13.33 Major axis = 100.223 mm, Minor axis = 99.672 mm

P13.35 $\sigma_x = 13.73$ ksi, $\sigma_z = 17.84$ ksi

P13.37 $\varepsilon_n = 872$ με

P13.39 $\sigma_x = 120.4$ MPa

P13.41 (a) $\sigma_x = 8.99$ ksi
(b) $\sigma_y = 10.57$ ksi
(c) $\tau_{xy} = 1.617$ ksi

P13.43 $\sigma_n = -54.0$ MPa, $\sigma_t = -0.0180$ MPa, $\tau_{nt} = -32.0$ MPa

P13.45 $\sigma_x = -316$ MPa, $\sigma_y = -279$ MPa, $\tau_{xy} = 103.5$ MPa

P13.47 (a) $\sigma_x = -169.7$ MPa, $\sigma_y = -134.5$ MPa, $\tau_{xy} = 112.1$ MPa
(b) $\sigma_{p1} = -38.6$ MPa, $\sigma_{p2} = -266$ MPa, $\tau_{max} = 113.5$ MPa, $\theta_p = -40.54°$ (to σ_{p2})
(c) $\tau_{abs\ max} = 132.8$ MPa

P13.49 (a) $\sigma_x = 16.73$ ksi, $\sigma_y = 20.5$ ksi, $\tau_{xy} = 4.37$ ksi
(b) $\sigma_{p1} = 23.4$ ksi, $\sigma_{p2} = 13.86$ ksi, $\tau_{max} = 4.76$ ksi, $\theta_p = -33.3°$ (to σ_{p2})
(c) $\tau_{abs\ max} = 11.69$ ksi

P13.51 (a) $\varepsilon_x = 220$ με, $\varepsilon_y = -580$ με, $\gamma_{xy} = 693$ μrad
(b) $\varepsilon_{p1} = 349$ με, $\varepsilon_{p2} = -709$ με, $\varepsilon_{p3} = 79.0$ με, $\gamma_{max} = 1{,}058$ μrad
(c) $\sigma_{p1} = 3.89$ ksi, $\sigma_{p2} = -11.36$ ksi, $\tau_{max} = 7.62$ ksi, $\theta_p = 20.5°$ (to σ_{p1})
(d) $\tau_{abs\ max} = 7.62$ ksi

P13.53 (a) $\sigma_y = -69.3$ MPa
(b) $\Delta a = -1.641$ mm
(c) $\Delta t = 0.202$ mm

P13.55 (a) $\varepsilon_x = -1{,}603$ με, $\varepsilon_z = 754$ με, $\sigma_y = -96.0$ MPa
(b) $\varepsilon_x = 250$ με, $\varepsilon_z = 2{,}610$ με, $\sigma_y = -332$ MPa
(c) −32.9°C

P13.57 $\sigma_x = 61.4$ MPa, $\sigma_y = 17.45$ MPa

P13.59 $\tau_{abs\ max} = 27.9$ ksi

P13.61 (a) $\Delta a = -0.01516$ in., $\Delta b = -0.00366$ in., $\Delta c = 0.00462$ in.
(b) $\Delta V = -0.0544$ in.^{3}

P13.63 (a) $\Delta b = 1.272$ mm, $\Delta c = 0.254$ mm
(b) $\sigma_x = -14.15$ MPa

P13.65 $\Delta L = -28.4$ mm

P13.67 $\varepsilon_x = 2{,}210\ \mu\varepsilon$, $\varepsilon_y = -7{,}190\ \mu\varepsilon$, $\gamma_{xy} = 3{,}160\ \mu\text{rad}$

Chapter 14

P14.1 $\sigma_t = 1.193$ MPa

P14.3 $p_{max} \leq 1.214$ MPa

P14.5 (a) $\sigma_t = 73.0$ MPa
(b) $p = 2.47$ MPa

P14.7 (a) $\sigma_{hoop} = 96.5$ MPa
(b) $p \leq 3.24$ MPa

P14.9 (a) $h = 7.12$ m
(b) $\sigma_{long} = 0$

P14.11 (a) $\sigma_n = 76.8$ MPa (T)
(b) $\tau_{nt} = -23.8$ MPa

P14.13 $p_{allow} = 150.5$ psi

P14.15 $p = 1.880$ MPa

P14.17 (a) $p = 2.24$ MPa
(b) $\tau_{abs\ max} = 111.9$ MPa
(c) $\tau_{abs\ max} = 113.0$ MPa

P14.19 (a) $\sigma_n = 4{,}210$ psi (T)
(b) $\tau_{nt} = 938$ psi
(c) $\tau_{abs\ max} = 2{,}900$ psi

P14.21 (a) $\varepsilon_x = 158.6\ \mu\varepsilon$, $\varepsilon_y = 782\ \mu\varepsilon$, $\gamma_{xy} = 0\ \mu\text{rad}$
(b) $\varepsilon_a = 270\ \mu\varepsilon$, $\varepsilon_b = 671\ \mu\varepsilon$
(c) $\sigma_n = 38.7$ MPa, $\sigma_t = 59.8$ MPa
(d) $\tau_{nt} = 12.57$ MPa

P14.23 (a) $\sigma_\theta = 171.2$ MPa
(b) $\sigma_\theta = 96.2$ MPa
(c) $\tau_{max} = 123.1$ MPa

P14.25 (a) $\sigma_\theta = 186.5$ MPa
(b) $\Delta d = 0.213$ mm

P14.27 $t_{min} \geq 0.764$ in.

P14.29 $p_{max} = 2{,}280$ psi

P14.31 Tube: $\sigma_r = -8{,}100$ psi, $\sigma_\theta = -29{,}200$ psi
Jacket: $\sigma_r = -8{,}100$ psi, $\sigma_\theta = 28{,}900$ psi

P14.33 (a) $\delta = 0.1342$ mm
(b) $\sigma_\theta = 338$ MPa

Chapter 15

P15.1 (a) $\sigma_{p1} = 2{,}810$ psi, $\sigma_{p2} = -5{,}920$ psi, $\tau_{max} = 4{,}360$ psi, $\theta_p = -55.4°$ (to σ_{p1}) or $\theta_p = 34.6°$ (to σ_{p2})

P15.3 (a) $\varepsilon_x = 134.0\ \mu\varepsilon$, $\varepsilon_y = -45.6\ \mu\varepsilon$, $\gamma_{xy} = 1{,}444\ \mu\text{rad}$
(b) $\varepsilon_{p1} = 772\ \mu\varepsilon$, $\varepsilon_{p2} = -683\ \mu\varepsilon$
(c) $\gamma_{abs\ max} = 1{,}455\ \mu\text{rad}$

P15.5 $\sigma_{p1} = 5{,}970$ psi, $\sigma_{p2} = -4{,}950$ psi, $\tau_{max} = 5{,}460$ psi

P15.7 (a) $\sigma_{p1} = 32.1$ MPa, $\sigma_{p2} = -14.04$ MPa, $\tau_{max} = 23.1$ MPa, $\theta_p = 33.5°$ (to σ_{p1})

P15.9 (a) $\varepsilon = -110.9\ \mu\varepsilon$
(b) $T = 328$ N·m

P15.11 (a) $\sigma_n = -7.11$ MPa
(b) $\tau_{nt} = -53.7$ MPa
(c) $\sigma_{p1} = 26.4$ MPa, $\sigma_{p2} = -93.1$ MPa, $\tau_{max} = 59.7$ MPa, $\theta_p = 28.0°$ (to σ_{p1})

P15.13 $\sigma_{p1} = 73.2$ MPa, $\sigma_{p2} = -27.5$ MPa, $\tau_{max} = 50.4$ MPa, $\theta_p = 31.5°$ (to σ_{p1})

P15.15 (a) $\sigma_{p1} = 3.81$ ksi, $\sigma_{p2} = -9.59$ ksi, $\tau_{max} = 6.70$ ksi
(b) $\sigma_{p1} = 7.35$ ksi, $\sigma_{p2} = -5.50$ ksi, $\tau_{max} = 6.43$ ksi
(c) $\sigma_{p1} = 12.41$ ksi, $\sigma_{p2} = -2.94$ ksi, $\tau_{max} = 7.68$ ksi

P15.17 $P_x = 11.57$ kips, $P_y = 2.40$ kips

P15.19 (a) $\varepsilon = 84.5\ \mu\varepsilon$
(b) $\varepsilon = -106.9\ \mu\varepsilon$
(c) $\varepsilon = -274\ \mu\varepsilon$

P15.21 $\sigma_{p1} = 87.7$ MPa, $\sigma_{p2} = -21.8$ MPa, $\tau_{max} = 54.7$ MPa, $\theta_p = 26.5°$ (to σ_{p1})

P15.23 $\sigma_{p1} = 6.80$ MPa, $\sigma_{p2} = -0.262$ MPa, $\tau_{max} = 3.53$ MPa, $\theta_p = -11.10°$ (to σ_{p2})

P15.25 $\sigma_{p1} = 7.81$ MPa, $\sigma_{p2} = -44.8$ MPa, $\tau_{max} = 26.3$ MPa, $\theta_p = -22.7°$ (to σ_{p1})

P15.27 $\sigma_y = -22.1$ MPa, $\tau_{xy} = 33.3$ MPa

P15.29 $\sigma_x = 2{,}280$ psi, $\tau_{xz} = 1{,}295$ psi

P15.31 $\sigma_x = 31.1$ MPa, $\tau_{xz} = 2.10$ MPa

P15.33 (a) $\sigma_y = -5{,}820$ psi, $\tau_{xy} = 6{,}960$ psi
(b) $\sigma_y = -8{,}330$ psi, $\tau_{yz} = -2{,}440$ psi

P15.35 (a) $\sigma_z = -9{,}880$ psi, $\tau_{xz} = 186.7$ psi
(b) $\sigma_z = -7{,}840$ psi, $\tau_{yz} = 235$ psi

P15.37 $\sigma_{p1} = 2.50$ MPa, $\sigma_{p2} = -54.4$ MPa, $\tau_{max} = 28.5$ MPa

P15.39 $\sigma_{p1} = 421$ psi, $\sigma_{p2} = -12{,}240$ psi,

P15.41 $\sigma_{p1} = 37.3$ MPa, $\sigma_{p2} = -37.3$ MPa, $\tau_{abs\ max} = 37.3$ MPa

P15.43 $\sigma_x = 1{,}545$ psi, $\sigma_z = 0$ psi, $\tau_{xz} = -1{,}376$ psi

P15.45 $\tau_{abs\ max} = 1{,}159$ psi

P15.47 (a) $\sigma_y = -1{,}529$ psi, $\sigma_z = 1{,}092$ psi, $\tau_{yz} = -603$ psi
(b) $\sigma_x = 1{,}092$ psi, $\sigma_z = 5{,}910$ psi, $\tau_{xz} = -1{,}559$ psi

P15.49 (a) $\sigma_x = 10.35$ MPa, $\sigma_y = -3.71$ MPa, $\tau_{xy} = 6.92$ MPa
(b) $\sigma_z = -21.5$ MPa, $\sigma_y = 10.35$ MPa, $\tau_{yz} = -13.47$ MPa

P15.51 (a) $\sigma_z = 19.99$ MPa, $\sigma_y = -5.92$ MPa, $\tau_{yz} = -4.94$ MPa
(b) $\sigma_x = 17.37$ MPa, $\sigma_y = 19.99$ MPa, $\tau_{xy} = -14.78$ MPa

P15.53 (a) FS = 0.870; the component fails.
(b) $\sigma_M = 356$ MPa
(c) FS = 0.984; the component fails.

P15.55 (a) 5.60 MPa
(b) 6.47 MPa

P15.57 (a) Overstressed. FS = 0.920
(b) Acceptable. FS = 1.050

P15.59 (a) 42.0 kip·in.
(b) 51.6 kip·in.

P15.61 not safe; interaction equation = 1.108

P15.63 (a) safe; $\sigma_{p1} = 129.3$ MPa, $\sigma_{p2} = -309$ MPa
(b) safe; interaction equation = 0.922

Chapter 16

P16.1 (a) $L/r = 250$, $P_{cr} = 318$ N
(b) $L/r = 160$, $P_{cr} = 1{,}892$ N

P16.3 (a) $L/r = 164.1$
(b) $P_{cr} = 280$ kN
(c) $\sigma_{cr} = 73.6$ MPa

P16.5 (a) $P_{cr} = 14.44$ kips
(b) $P_{cr} = 336$ kips

P16.7 $\Delta T = 40.6$°F

P16.9 $w = 56.9$ kN/m

P16.11 FS = 1.334

P16.13 $FS_{min} = 1.061$

P16.15 (a) $P_{allow} = 126.7$ kN
(b) $P_{allow} = 31.7$ kN
(c) $P_{allow} = 258$ kN
(d) $P_{allow} = 507$ kN

P16.17 $P_{allow} = 44.8$ kips

P16.19 (a) $P_{cr} = 1{,}439$ lb
(b) $b/h = 0.714$

P16.21 $\Delta T = 38.0$°C

P16.23 (a) $P_{cr} = 16.02$ kips
(b) $v_{max} = 0.1630$ in.

P16.25 (a) $v_{max} = 0.1630$ in.
(b) $\sigma_{max} = 45.6$ MPa

P16.27 (a) $P_{allow} = 280$ kips
(b) $P_{allow} = 94.2$ kips

P16.29 (a) $P_{allow} = 1{,}264$ kN
(b) $P_{allow} = 1{,}277$ kN

P16.31 $P_{allow} = 370$ kips

P16.33 (a) 31.1 mm
(b) $P_{allow} = 166.3$ kN

P16.35 (a) $P_{allow} = 40.5$ kips
(b) $P_{allow} = 11.39$ kips

P16.37 $P_{allow} = 18.45$ kips

P16.39 (a) $P_{allow} = 30{,}400$ lb
(b) $P_{allow} = 14{,}800$ lb
(c) $P_{allow} = 6{,}930$ lb

P16.41 (a) not given
(b) not given
(c) ratio of P_{allow}/P_{actual}: chord members $AF = 1.272$, $FG = 4.76$, $GH = 3.54$, $EH = 0.946$; web members $BG = 2.17$, $DG = 5.27$

P16.43 (a) $e_{max} = 12.20$ in.
(b) $e_{max} = 3.55$ in.

P16.45 $L_{max} = 1{,}255$ mm

P16.47 $P_{max} = 4{,}440$ lb

Chapter 17

P17.1 (a) $u_r = 1{,}764$ kJ/m^3
(b) $u_r = 257$ kJ/m^3
(c) $u_r = 421$ kJ/m^3

P17.3 (a) $U = 71.2$ lb·in.
(b) $u_1 = 13.53$ lb·in./in.^{3}, $u_2 = 1.442$ lb·in./in.^{3}

P17.5 (a) $u = 12.33$ kJ/m^3
(b) $d_{min} = 23.2$ mm

P17.7 $T_C = 121.4$ N·m

P17.9 $U = 93.8$ J

P17.11 $U = 588$ J

P17.13 $h_{max} = 27.1$ mm

P17.15 $m_{max} = 13.97$ kg

P17.17 (a) $\sigma_{max} = 7.83$ MPa
(b) $\sigma_{max} = 2.51$ MPa

P17.19 $W_{max} = 31.2$ lb

P17.21 (a) $P_{equiv} = 18.89$ kN
(b) $v_{max} = 54.4$ mm
(c) $h_{max} = 382$ mm

P17.23 (a) $P_{max} = 17.86$ kN
(b) $n = 16.55$
(c) $h_{max} = 398$ mm

P17.25 (a) $v_{max} = 0.430$ in.
(b) $P_{equiv} = 2{,}380$ lb
(c) $\sigma_{max} = 1{,}169$ psi

P17.27 $v_{0,max} = 3.20$ m/s

P17.29 (a) $P_{max} = 50.3$ kN
(b) $\sigma_{max} = 260$ MPa
(c) $v_{max} = 22.4$ mm

P17.31 $\Delta_C = 12.44$ mm ↓

P17.33 $\Delta_D = 6.46$ mm ↓

P17.35 $\Delta_B = 0.1642$ in. ←

P17.37 $\Delta_D = 0.816$ in. →

P17.39 $\Delta_B = 62.9$ mm ↓

P17.41 (a) $\Delta_F = 18.54$ mm ← (b) $\Delta_B = 15.00$ mm ←

P17.43 (a) $\Delta_A = 4.88$ mm ← (b) $\Delta_A = 12.85$ mm ↓

P17.45 (a) $\Delta_D = 10.46$ mm ↓ (b) $\Delta_D = 10.46$ mm ↓

P17.47 $\theta_A = \dfrac{3w_0L^3}{128EI}$ (CW)

P17.49 $\Delta_C = 28.8$ mm ↓

P17.51 $\Delta_C = 6.27$ mm ↓

P17.53 (a) $\theta_C = 0.00883$ rad (CW) (b) $\Delta_C = 31.9$ mm ↓

P17.55 $I_{min} = 2{,}140$ in.^{4}

P17.57 $\Delta_B = 0.241$ in. ←

P17.59 $\Delta_D = 0.450$ in. →

P17.61 $\Delta_B = 32.3$ mm ↓

P17.63 $\Delta_B = 10.10$ mm ←

P17.65 $\theta_A = \dfrac{M_0L}{3EI}$ (CW)

P17.67 $\theta_B = \dfrac{w_0L^3}{6EI}$ (CW),

$\Delta_B = \dfrac{w_0L^4}{8EI}$ ↓

P17.69 $\Delta_C = 13.12$ mm ↓

P17.71 $\Delta_C = 2.22$ mm ↓

P17.73 (a) $\theta_C = 0.00883$ rad (CW) (b) $\Delta_C = 31.9$ mm ↓

P17.75 $I_{min} = 2{,}140$ in.^{4}

Index

C

D

E

F

G

H

I

J

K

L

M

N

O

P

Q

R

S

T

U

V

W

Y